COMPLETE SOLUTIONS MAN
VOLUME ONE, FOR STEWAR

CALCULUS

THIRD EDITION

JAMES STEWART
McMaster University

DANIEL ANDERSON
University of Iowa

DANIEL DRUCKER
Wayne State University

with the assistance of
Andy Bulman-Fleming

Brooks/Cole Publishing Company

 An International Thomson Publishing Company

Pacific Grove • Albany • Bonn • Boston • Cincinnati • Detroit • London • Madrid • Melbourne
Mexico City • New York • Paris • San Francisco • Singapore • Tokyo • Toronto • Washington

For more information, contact:

BROOKS/COLE PUBLISHING COMPANY
511 Forest Lodge Road
Pacific Grove, CA 93950
USA

International Thomson Editores
Campos Eliseos 385, Piso 7
Col. Polanco
11560 México D. F. México

International Thomson Publishing Europe
Berkshire House 168-173
High Holborn
London WC1V 7AA
England

International Thomson Publishing GmbH
Königswinterer Strasse 418
3227 Bonn
Germany

Thomas Nelson Australia
102 Dodds Street
South Melbourne, 3205
Victoria, Australia

International Thomson Publishing Asia
221 Henderson Road
#05-10 Henderson Building
Singapore 0315

Nelson Canada
1120 Birchmount Road
Scarborough, Ontario
Canada M1K 5G4

International Thomson Publishing Japan
Hirakawacho Kyowa Building, 3F
2-2-1 Hirakawacho
Chiyoda-ku, Tokyo 102
Japan

Printed in the United States of America

5 4 3 2 1

ISBN 0-534-21799-0

Preface

I have edited this solutions manual by comparing the solutions provided by Daniel Anderson and Daniel Drucker with my own solutions and those of McGill University students Andy Bulman-Fleming and Alex Taler. Andy also produced this book using EXP Version 3.0 for Windows. I thank him and the staff of TECHarts for producing the diagrams.

JAMES STEWART

CONTENTS

REVIEW AND PREVIEW

EXERCISES 1

1. $f(x) = 2x^2 + 3x - 4$, so $f(0) = 2(0)^2 + 3(0) - 4 = -4$, $f(2) = 2(2)^2 + 3(2) - 4 = 10$,
$f(\sqrt{2}) = 2(\sqrt{2})^2 + 3(\sqrt{2}) - 4 = 3\sqrt{2}$,
$f(1+\sqrt{2}) = 2(1+\sqrt{2})^2 + 3(1+\sqrt{2}) - 4 = 2(1+2+2\sqrt{2}) + 3 + 3\sqrt{2} - 4 = 5 + 7\sqrt{2}$,
$f(-x) = 2(-x)^2 + 3(-x) - 4 = 2x^2 - 3x - 4$,
$f(x+1) = 2(x+1)^2 + 3(x+1) - 4 = 2(x^2 + 2x + 1) + 3x + 3 - 4 = 2x^2 + 7x + 1$,
$2f(x) = 2(2x^2 + 3x - 4) = 4x^2 + 6x - 8$, and
$f(2x) = 2(2x)^2 + 3(2x) - 4 = 2(4x^2) + 6x - 4 = 8x^2 + 6x - 4$.

2. $g(x) = x^3 + 2x^2 - 3$, so $g(0) = 0^3 + 2(0)^2 - 3 = -3$, $g(3) = 3^3 + 2(3)^2 - 3 = 42$,
$g(-x) = (-x)^3 + 2(-x)^2 - 3 = -x^3 + 2x^2 - 3$, and $g(1+h) = (1+h)^3 + 2(1+h)^2 - 3 = h^3 + 5h^2 + 7h$.

3. $f(x) = x - x^2$, so $f(2+h) = 2 + h - (2+h)^2 = 2 + h - 4 - 4h - h^2 = -(h^2 + 3h + 2)$,
$f(x+h) = x + h - (x+h)^2 = x + h - x^2 - 2xh - h^2$, and
$$\frac{f(x+h) - f(x)}{h} = \frac{x + h - x^2 - 2xh - h^2 - x + x^2}{h} = \frac{h - 2xh - h^2}{h} = 1 - 2x - h.$$

4. $f(x) = \dfrac{x}{x+1}$, so $f(2+h) = \dfrac{2+h}{2+h+1} = \dfrac{2+h}{3+h}$, $f(x+h) = \dfrac{x+h}{x+h+1}$, and
$$\frac{f(x+h) - f(x)}{h} = \frac{\dfrac{x+h}{x+h+1} - \dfrac{x}{x+1}}{h} = \frac{(x+h)(x+1) - x(x+h+1)}{h(x+h+1)(x+1)} = \frac{1}{(x+h+1)(x+1)}.$$

5. $f(x) = \sqrt{x}$, $0 \le x \le 4$

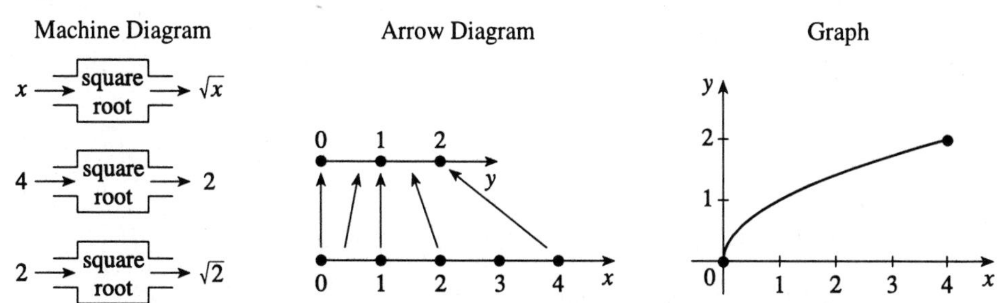

6. $f(x) = 2/x, 1 \le x \le 4$

Machine Diagram

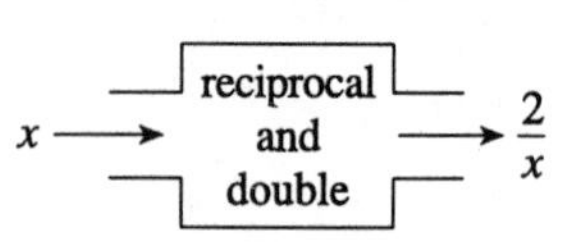

Arrow Diagram

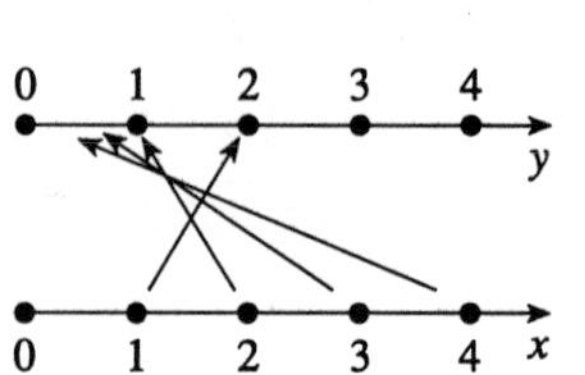

Graph

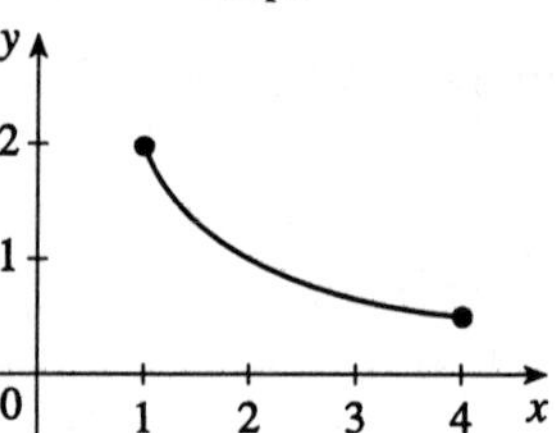

7. The range of f is the set of values of f, $\{0, 1, 2, 4\}$.

Arrow Diagram

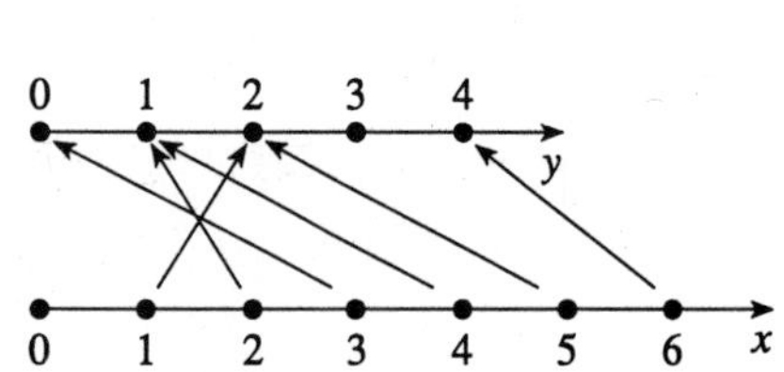

Graph

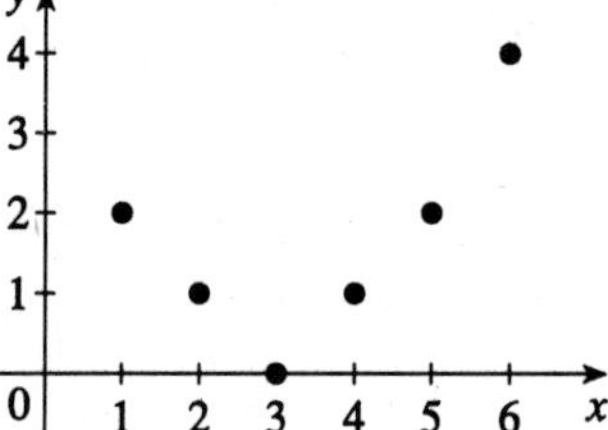

8. $f(x) = 2x + 7, -1 \le x \le 6$. The domain is $[-1, 6]$. If $-1 \le x \le 6$, then $5 = 2(-1) + 7 \le 2x + 7 \le 2(6) + 7 = 19$, so the range is $[5, 19]$.

9. $f(x) = 6 - 4x, -2 \le x \le 3$. The domain is $[-2, 3]$. If $-2 \le x \le 3$, then $14 = 6 - 4(-2) \ge 6 - 4x \ge 6 - 4(3) = -6$, so the range is $[-6, 14]$.

10. $g(x) = \dfrac{2}{3x - 5}$. This is defined when $3x - 5 \ne 0$, so the domain of f is $\left\{x \mid x \ne \frac{5}{3}\right\} = \left(-\infty, \frac{5}{3}\right) \cup \left(\frac{5}{3}, \infty\right)$, and the range is $\{y \mid y \ne 0\} = (-\infty, 0) \cup (0, \infty)$.

11. $h(x) = \sqrt{2x - 5}$ is defined when $2x - 5 \ge 0$ or $x \ge \frac{5}{2}$, so the domain is $\left[\frac{5}{2}, \infty\right)$ and the range is $[0, \infty)$.

12. $h(x) = \sqrt[4]{7 - 3x}$. This is defined when $7 - 3x \ge 0$ or $x \le \frac{7}{3}$, so the domain is $\left(-\infty, \frac{7}{3}\right]$. The range is $[0, \infty)$ since every non-negative real number is the fourth root of another.

13. $F(x) = \sqrt{1 - x^2}$ is defined when $1 - x^2 \ge 0 \quad \Leftrightarrow \quad x^2 \le 1 \quad \Leftrightarrow \quad |x| \le 1 \quad \Leftrightarrow \quad -1 \le x \le 1$, so the domain is $[-1, 1]$ and the range is $[0, 1]$.

14. $F(x) = 1 - \sqrt{x}$ is defined when $x \ge 0$, so the domain is $[0, \infty)$. Now $\sqrt{x}$ takes on all values ≥ 0, so $-\sqrt{x}$ takes on all values ≤ 0 and $1 - \sqrt{x}$ takes on all values ≤ 1, so the range is $(-\infty, 1]$.

15. $f(x) = \dfrac{x + 2}{x^2 - 1}$ is defined for all x except when $x^2 - 1 = 0 \quad \Leftrightarrow \quad x = 1$ or $x = -1$, so the domain is $\{x \mid x \ne \pm 1\}$.

16. $f(x) = x^4/(x^2 + x - 6)$ is defined for all x except when $0 = x^2 + x - 6 = (x+3)(x-2) \Leftrightarrow x = -3$ or 2, so the domain is $\{x \in \mathbb{R} \mid x \neq -3, 2\}$.

17. $g(x) = \sqrt[4]{x^2 - 6x}$ is defined when $0 \leq x^2 - 6x = x(x-6) \Leftrightarrow x \geq 6$ or $x \leq 0$, so the domain is $(-\infty, 0] \cup [6, \infty)$.

18. $g(x) = \sqrt{x^2 - 2x - 8}$ is defined when $0 \leq x^2 - 2x - 8 = (x-4)(x+2) \Leftrightarrow x \geq 4$ or $x \leq -2$, so the domain is $(-\infty, -2] \cup [4, \infty)$.

19. $\phi(x) = \sqrt{\dfrac{x}{\pi - x}}$ is defined when $\dfrac{x}{\pi - x} \geq 0$. So either $x \leq 0$ and $\pi - x < 0$ ($\Leftrightarrow x > \pi$), which is impossible, or $x \geq 0$ and $\pi - x > 0$ ($\Leftrightarrow x < \pi$), and so the domain is $[0, \pi)$.

20. $\phi(x) = \sqrt{\dfrac{x^2 - 2x}{x - 1}}$ is defined when $0 \leq \dfrac{x^2 - 2x}{x - 1} = \dfrac{x(x-2)}{x-1}$. Constructing a table:

Interval	x	$x-1$	$x-2$	$x(x-2)/(x-1)$
$x < 0$	$-$	$-$	$-$	$-$
$0 < x < 1$	$+$	$-$	$-$	$+$
$1 < x < 2$	$+$	$+$	$-$	$-$
$x > 2$	$+$	$+$	$+$	$+$

So the domain is $[0, 1) \cup [2, \infty)$.

21. $f(t) = \sqrt[3]{t - 1}$ is defined for every t, since every real number has a cube root. The domain is the set of all real numbers.

22. $f(t) = \sqrt{t^2 + 1}$ is defined for every t, since $t^2 + 1$ is always positive. The domain is the set of all real numbers.

23. $f(x) = 3 - 2x$. Domain is $\mathbb{R}$.

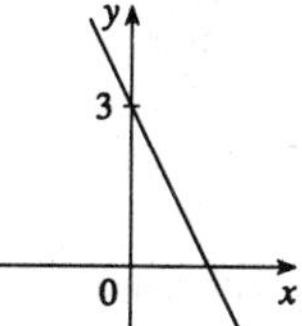

24. $f(x) = \dfrac{x+3}{2}$, $-2 \leq x \leq 2$. Domain is $[-2, 2]$.

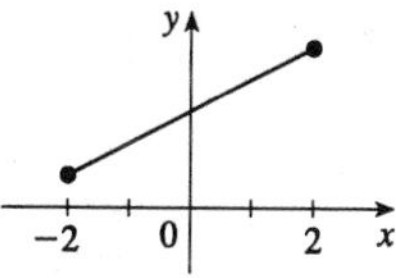

25. $f(x) = x^2 + 2x - 1 = (x^2 + 2x + 1) - 2 = (x+1)^2 - 2$, so the graph is a parabola with vertex at $(-1, -2)$. The domain is $\mathbb{R}$.

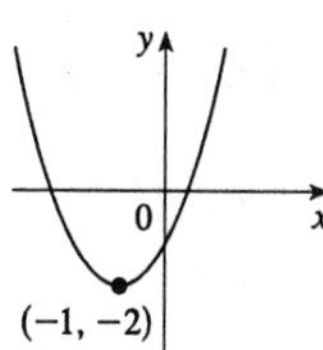

26. $f(x) = -x^2 + 6x - 7 = -(x^2 - 6x + 9) + 2 = -(x-3)^2 + 2$, so the graph is a parabola with vertex at $(3, 2)$. The domain is $\mathbb{R}$.

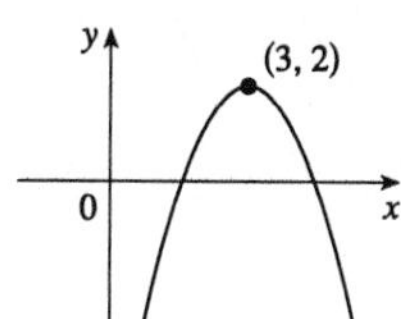

27. $g(x) = \sqrt{-x}$. The domain is $\{x \mid -x \geq 0\} = (-\infty, 0]$.

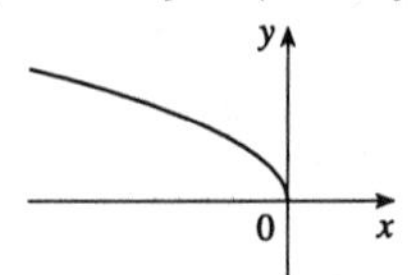

28. $g(x) = \sqrt{6 - 2x}$. The domain is $\{x \mid 6 - 2x \geq 0\} = (-\infty, 3]$.

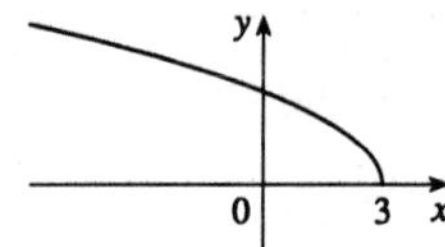

29. $h(x) = \sqrt{4 - x^2}$. Now $y = \sqrt{4 - x^2} \;\Rightarrow\; y^2 = 4 - x^2 \;\Leftrightarrow\; x^2 + y^2 = 4$, so the graph is the top half of a circle of radius 2. The domain is $\{x \mid 4 - x^2 \geq 0\} = [-2, 2]$.

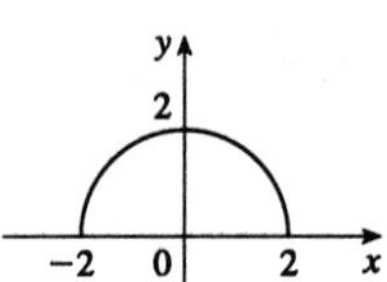

30. $h(x) = \sqrt{x^2 - 4}$. Now $y = \sqrt{x^2 - 4} \;\Rightarrow\; y^2 = x^2 - 4 \;\Leftrightarrow\; x^2 - y^2 = 4$, so the graph is the top half of a hyperbola. The domain is $\{x \mid x^2 - 4 \geq 0\} = (-\infty, -2] \cup [2, \infty)$.

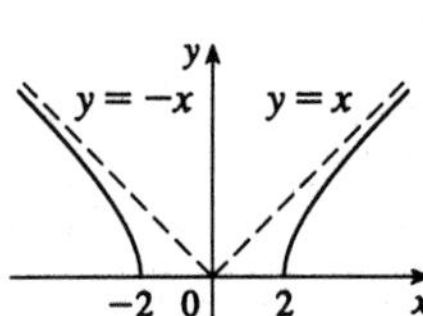

31. $F(x) = \dfrac{1}{x}$. The domain is $\{x \mid x \neq 0\}$.

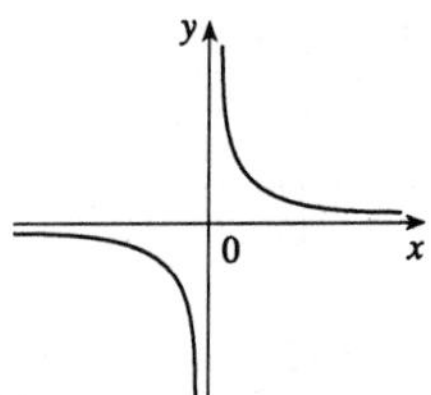

32. $F(x) = \dfrac{2}{x + 4}$. The domain is $\{x \mid x \neq -4\}$.

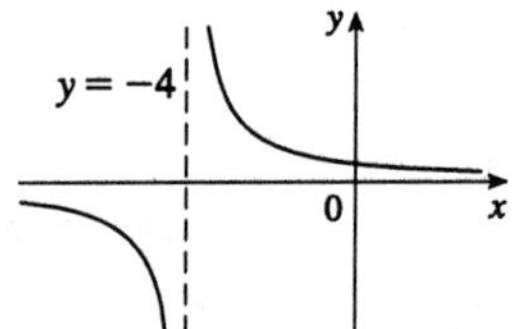

33. $G(x) = |x| + x = \begin{cases} 2x & \text{if } x \geq 0 \\ 0 & \text{if } x < 0 \end{cases}$

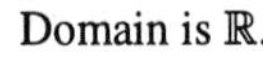
Domain is $\mathbb{R}$.

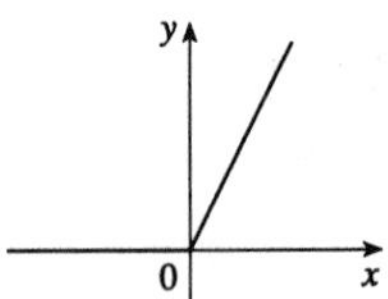

34. $G(x) = |x| - x = \begin{cases} 0 & \text{if } x \geq 0 \\ -2x & \text{if } x < 0 \end{cases}$

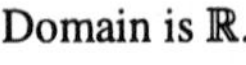
Domain is $\mathbb{R}$.

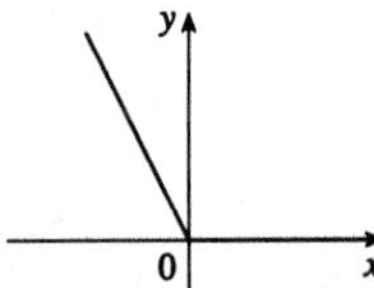

35. $H(x) = |2x| = \begin{cases} 2x & \text{if } x \geq 0 \\ -2x & \text{if } x < 0 \end{cases}$

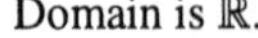
Domain is $\mathbb{R}$.

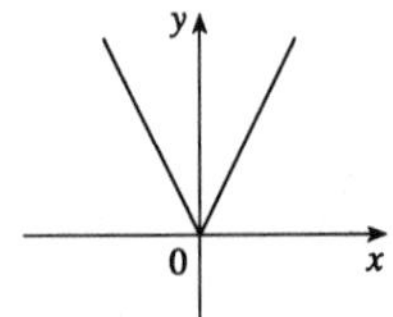

36. $H(x) = |2x - 3| = \begin{cases} 2x - 3 & \text{if } x \geq \frac{3}{2} \\ 3 - 2x & \text{if } x < \frac{3}{2} \end{cases}$

Domain is $\mathbb{R}$.

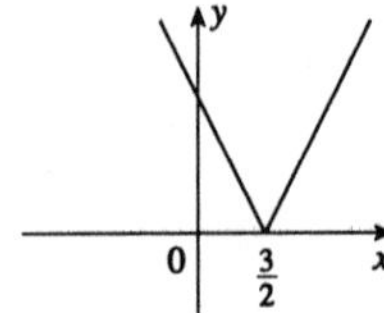

37. $f(x) = \dfrac{x}{|x|} = \begin{cases} 1 & \text{if } x > 0 \\ -1 & \text{if } x < 0 \end{cases}$

Domain is $\{x \mid x \neq 0\}$.

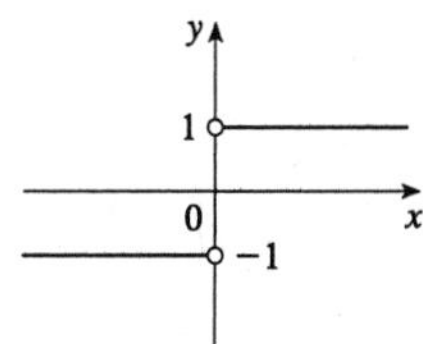

38. $f(x) = |x^2 - 1| = \begin{cases} x^2 - 1 & \text{if } x > 1 \text{ or } x < -1 \\ -x^2 + 1 & \text{if } -1 \leq x \leq 1 \end{cases}$

Domain is $\mathbb{R}$.

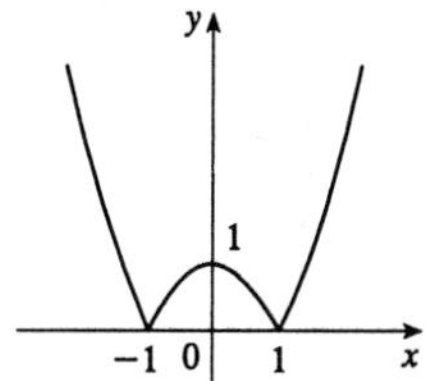

39. $f(x) = \dfrac{x^2 - 1}{x - 1} = \dfrac{(x+1)(x-1)}{x-1}$, so for $x \neq 1$, $f(x) = x + 1$. Domain is $\{x \mid x \neq 1\}$.

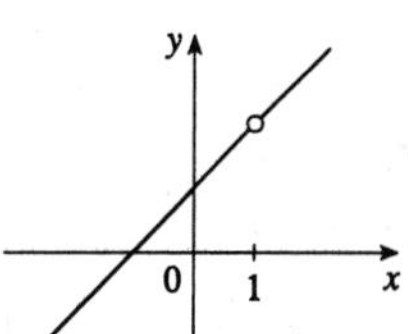

40. $f(x) = \dfrac{x^2 + 5x + 6}{x + 2} = \dfrac{(x+3)(x+2)}{x+2}$, so for $x \neq -2$, $f(x) = x + 3$. Domain is $\{x \mid x \neq -2\}$.

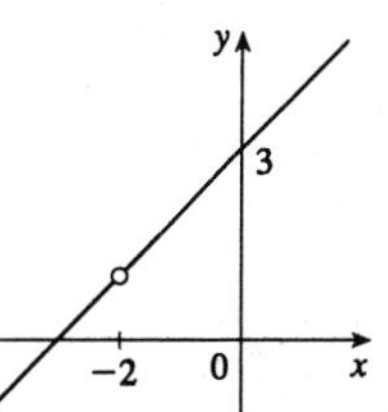

41. $f(x) = \begin{cases} 0 & \text{if } x < 2 \\ 1 & \text{if } x \geq 2 \end{cases}$ Domain is $\mathbb{R}$.

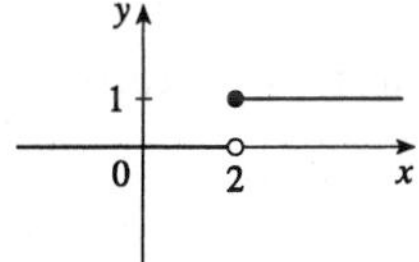

42. $f(x) = \begin{cases} 1 & \text{if } -1 \leq x \leq 1 \\ -1 & \text{if } x > 1 \text{ or } x < -1 \end{cases}$ Domain is $\mathbb{R}$.

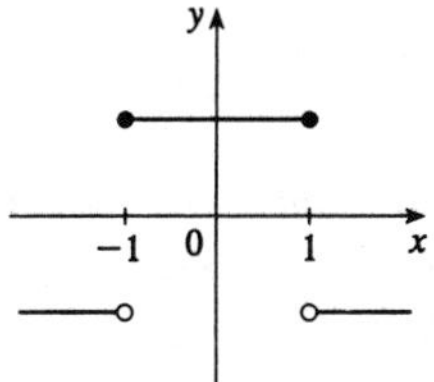

43. $f(x) = \begin{cases} x & \text{if } x \leq 0 \\ x + 1 & \text{if } x > 0 \end{cases}$ Domain is $\mathbb{R}$.

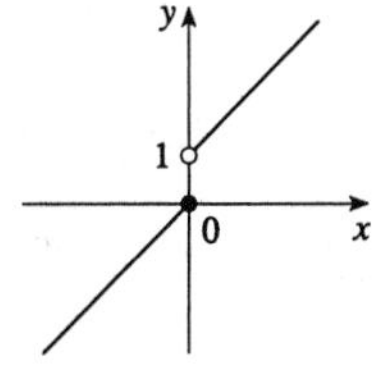

44. $f(x) = \begin{cases} 2x + 3 & \text{if } x < -1 \\ 3 - x & \text{if } x \geq -1 \end{cases}$ Domain is $\mathbb{R}$.

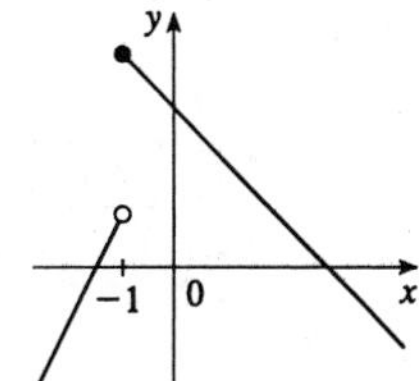

45. $f(x) = \begin{cases} -1 & \text{if } x < -1 \\ x & \text{if } -1 \le x \le 1 \\ 1 & \text{if } x > 1 \end{cases}$ Domain is $\mathbb{R}$.

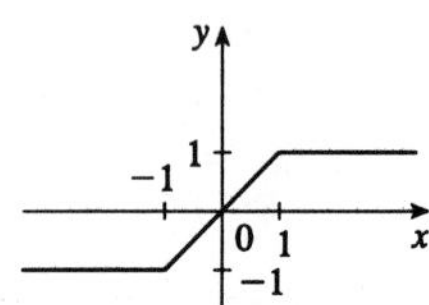

46. 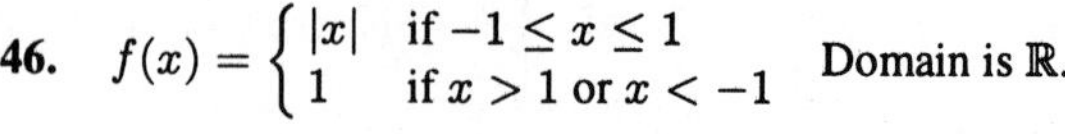

$f(x) = \begin{cases} |x| & \text{if } -1 \le x \le 1 \\ 1 & \text{if } x > 1 \text{ or } x < -1 \end{cases}$ Domain is $\mathbb{R}$.

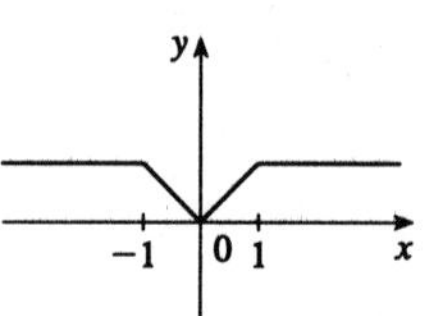

47. $f(x) = \begin{cases} x+2 & \text{if } x \le -1 \\ x^2 & \text{if } x > -1 \end{cases}$ Domain is $\mathbb{R}$.

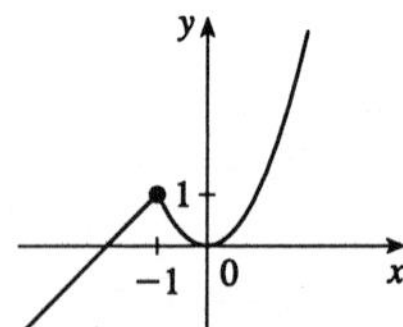

48. 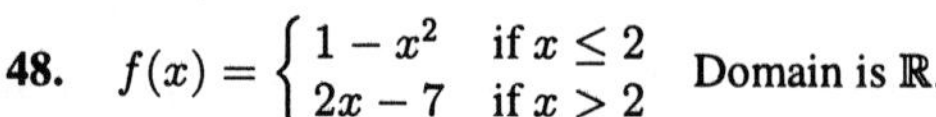

$f(x) = \begin{cases} 1 - x^2 & \text{if } x \le 2 \\ 2x - 7 & \text{if } x > 2 \end{cases}$ Domain is $\mathbb{R}$.

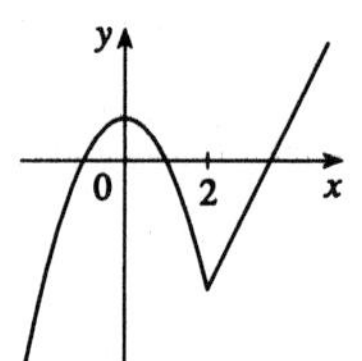

49. $f(x) = \begin{cases} -1 & \text{if } x \le -1 \\ 3x+2 & \text{if } -1 < x < 1 \\ 7 - 2x & \text{if } x \ge 1 \end{cases}$ Domain is $\mathbb{R}$.

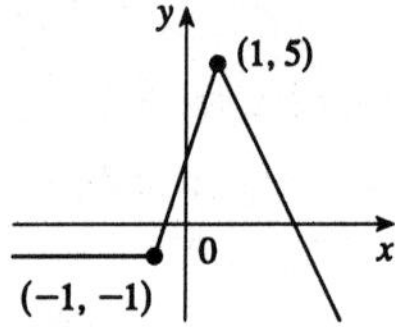

50. 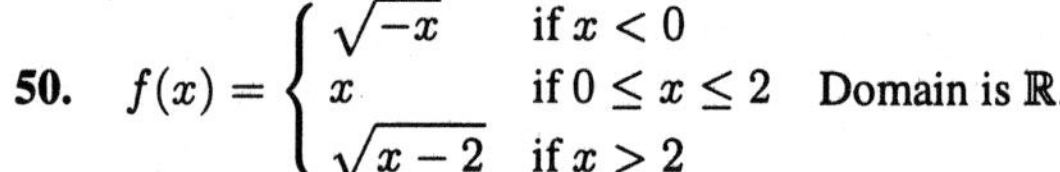

$f(x) = \begin{cases} \sqrt{-x} & \text{if } x < 0 \\ x & \text{if } 0 \le x \le 2 \\ \sqrt{x-2} & \text{if } x > 2 \end{cases}$ Domain is $\mathbb{R}$.

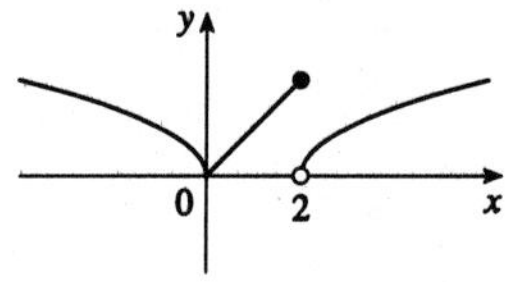

51. Yes, the curve is the graph of a function. The domain is $[-3, 2]$ and the range is $[-2, 2]$.

52. No, this is not the graph of a function by the Vertical Line Test.

53. No, this is not the graph of a function since for $x = -1$ there are infinitely many points on the curve.

54. Yes, this is the graph of a function with domain $[-3, 2]$ and range $\{-2\} \cup (0, 3]$.

55. The slope of this line segment is $\dfrac{-6-1}{4-(-2)} = -\dfrac{7}{6}$, so its equation is $y - 1 = -\frac{7}{6}(x+2)$. The function is $f(x) = -\frac{7}{6}x - \frac{4}{3}$, $-2 \le x \le 4$.

56. The slope of this line segment is $\dfrac{3-(-2)}{6-(-3)} = \dfrac{5}{9}$, so its equation is $y + 2 = \frac{5}{9}(x+3)$. The function is $f(x) = \frac{5}{9}x - \frac{1}{3}$, $-3 \le x \le 6$.

57. $x + (y-1)^2 = 0 \quad \Leftrightarrow \quad y - 1 = \pm\sqrt{-x}$. The bottom half is given by the function $f(x) = 1 - \sqrt{-x}$, $x \le 0$.

58. $(x-1)^2 + y^2 = 1 \quad \Rightarrow \quad y = \pm\sqrt{1-(x-1)^2} = \pm\sqrt{2x - x^2}$. The top half is given by the function $f(x) = \sqrt{2x - x^2}$, $0 \le x \le 2$.

59. For $-1 \le x \le 2$, the graph is the line with slope 1 and y-intercept 1, that is, the line $y = x + 1$. For $2 < x \le 4$, the graph is the line with slope $-\frac{3}{2}$ and x-intercept 4, so $y = -\frac{3}{2}(x-4) = -\frac{3}{2}x + 6$. So the function is
$$f(x) = \begin{cases} x+1 & \text{if } -1 \le x \le 2 \\ -\frac{3}{2}x + 6 & \text{if } 2 < x \le 4 \end{cases}$$

60. For $x \le 0$, the graph is the line $y = 2$. For $0 < x \le 1$, the graph is the line with slope -2 and y-intercept 2, that is, the line $y = -2x + 2$. For $x > 1$, the graph is the line with slope 1 and x-intercept 1, that is, the line $y = 1(x-1) = x - 1$. So the function is $f(x) = \begin{cases} 2 & \text{if } x \le 0 \\ -2x+2 & \text{if } 0 < x \le 1 \\ x - 1 & \text{if } 1 < x. \end{cases}$

61. Let the length and width of the rectangle be L and W respectively. Then the perimeter is $2L + 2W = 20$, and the area is $A = LW$. Solving the first equation for W in terms of L gives $W = \dfrac{20 - 2L}{2} = 10 - L$. Thus $A(L) = L(10 - L) = 10L - L^2$. Since lengths are positive, the domain of A is $0 < L < 10$.

62. Let the length and width of the rectangle be L and W respectively. Then the area is $LW = 16$, so that $W = 16/L$. The perimeter is $P = 2L + 2W$, so $P(L) = 2L + 2(16/L) = 2L + 32/L$, and the domain of P is $L > 0$, since lengths must be positive quantities.

63. Let the length of a side of the equilateral triangle be x. Then by the Pythagorean Theorem, the height y of the triangle satisfies $y^2 + \left(\frac{1}{2}x\right)^2 = x^2$, so that $y = \frac{\sqrt{3}}{2}x$. Thus the area of the triangle is $A = \frac{1}{2}xy$ and so $A(x) = \frac{1}{2}\left(\frac{\sqrt{3}}{2}x\right)x = \frac{\sqrt{3}}{4}x^2$, with domain $x > 0$.

64. Let the volume of the cube be V and the length of an edge be L. Then $V = L^3$ so $L = \sqrt[3]{V}$, and the surface area is $S(V) = 6\left(\sqrt[3]{V}\right)^2 = 6V^{2/3}$, with domain $V > 0$.

65. Let each side of the base of the box have length x, and let the height of the box be h. Since the volume is 2, we know that $2 = hx^2$, so that $h = 2/x^2$, and the surface area is $S = x^2 + 4xh$. Thus $S(x) = x^2 + 4x(2/x^2) = x^2 + 8/x$, with domain $x > 0$.

66. The area of the window is $A = xh + \dfrac{\pi x^2}{8}$, where h is the height of the rectangular portion of the window. The perimeter is $P = 2h + x + \frac{1}{2}\pi x = 30 \quad \Leftrightarrow \quad 2h = 30 - x - \frac{1}{2}\pi x \quad \Leftrightarrow \quad h = \frac{1}{4}(60 - 2x - \pi x)$. Thus $A(x) = x\dfrac{60 - 2x - \pi x}{4} + \dfrac{\pi x^2}{8} = 15x - x^2\left(\dfrac{\pi + 4}{8}\right)$ with domain $0 < x < \dfrac{60}{\pi + 2}$, since lengths must be positive quantities, and in the limiting case where $h \to 0$, $x + \dfrac{\pi x}{2} < 30 \quad \Rightarrow \quad x < \dfrac{60}{\pi + 2}$.

67. The height of the box is x and the length and width are $L = 20 - 2x$, $W = 12 - 2x$. Then $V = LWx$ and so $V(x) = (20 - 2x)(12 - 2x)(x) = 4(10 - x)(6 - x)(x) = 4x\left(60 - 16x + x^2\right) = 4x^3 - 64x^2 + 240x$, with domain $0 < x < 6$.

68.

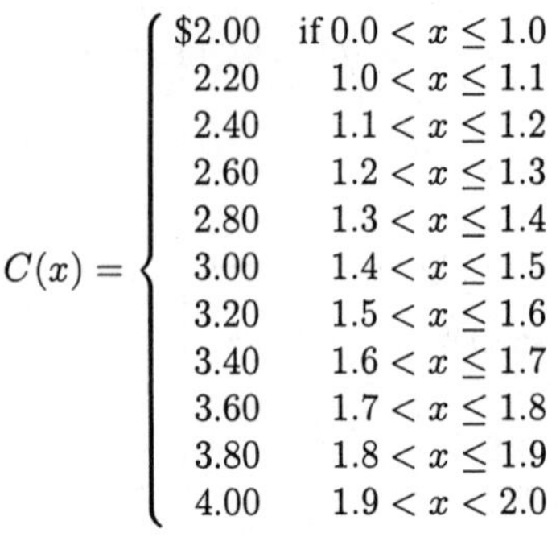

$$C(x) = \begin{cases} \$2.00 & \text{if } 0.0 < x \le 1.0 \\ 2.20 & 1.0 < x \le 1.1 \\ 2.40 & 1.1 < x \le 1.2 \\ 2.60 & 1.2 < x \le 1.3 \\ 2.80 & 1.3 < x \le 1.4 \\ 3.00 & 1.4 < x \le 1.5 \\ 3.20 & 1.5 < x \le 1.6 \\ 3.40 & 1.6 < x \le 1.7 \\ 3.60 & 1.7 < x \le 1.8 \\ 3.80 & 1.8 < x \le 1.9 \\ 4.00 & 1.9 < x < 2.0 \end{cases}$$

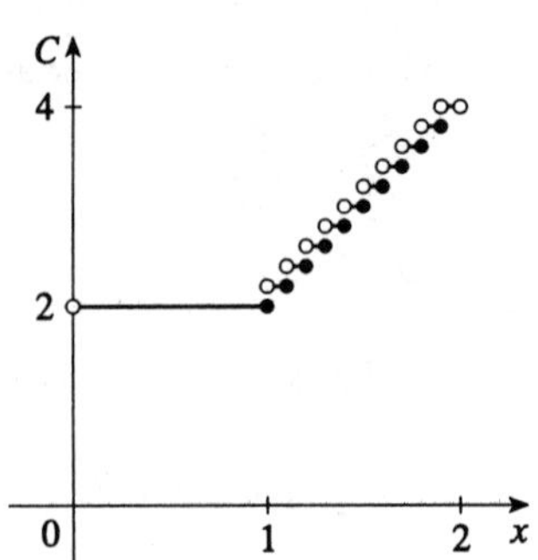

69. **(a)** T is a linear function of h, so $T = mh + b$ with m and b constants. We know two points on the graph of T as a function of h: $(h, T) = (0{,}20)$ and $(h{,}T) = (1{,}10)$. The slope of the line passing through these two points is $(10 - 20)/(1 - 0) = -10$ and the line's T-intercept is 20, so the slope-intercept form of the equation of the line is $T = -10h + 20$.

(b) The slope is $m = -10°\,\mathrm{C/km}$, and it represents the rate of change of temperature with respect to height.

(c)

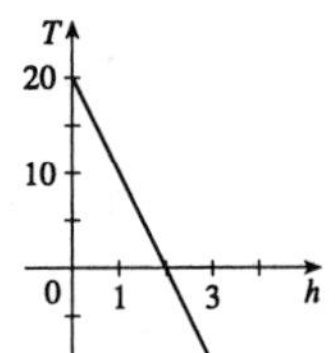

At a height of $h = 2.5$ km, the temperature is

$$\begin{aligned} T &= -10(2.5) + 20 \\ &= -25 + 20 \\ &= -5°\,\mathrm{C}. \end{aligned}$$

70. **(a)** We have two points $(480, 380)$ and $(800, 460)$ $\Rightarrow$ $m = \dfrac{460 - 380}{800 - 480} = \dfrac{80}{320} = \dfrac{1}{4}$. Thus the equation is $C - 380 = \frac{1}{4}(d - 480)$ $\Leftrightarrow$ $4C - 1520 = d - 480$ $\Leftrightarrow$ $4C = d + 1040$ or $C = \frac{1}{4}d + 260$.

(b) For $d = 1500$, $C = \frac{1}{4}(1500) + 260 = 635$. The predicted cost is \$635.

(c)

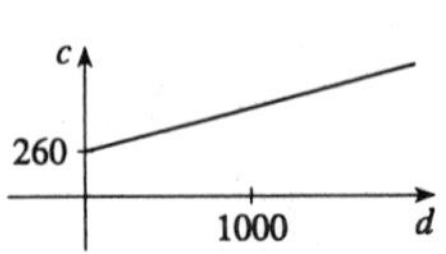

The slope of the line represents the cost per mile driven (for gas, oil, tires, ...).

(d) The y-intercept represents fixed costs associated with owning an automobile, including license, insurance, and depreciation.

(e) A linear model is suitable because the total cost is fixed expenses plus expenses per mile.

71. The water will cool down almost to freezing as the ice melts. Then, when the ice has melted, the water will slowly warm up to room temperature.

72. The summer solstice (the longest day of the year) is around June 21, and the winter solstice (the shortest day) is around December 22.

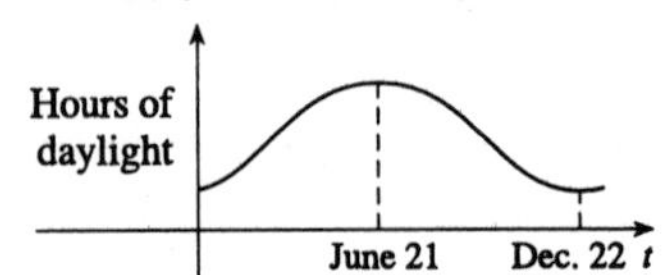

73. Of course, this graph depends strongly on the geographical location!

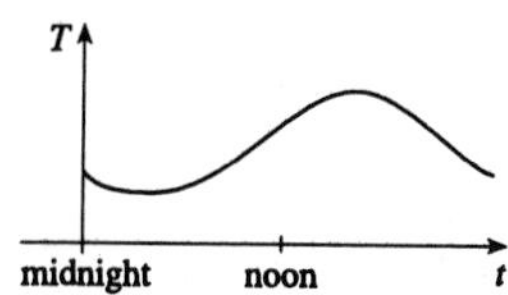

74.

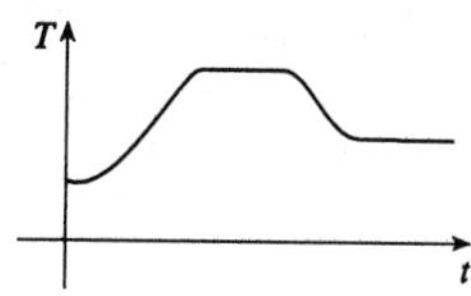

75. **(a)**

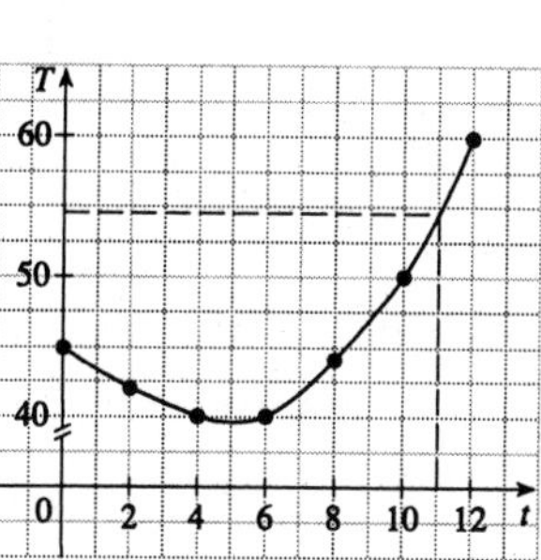

(b) $T(11) \approx 54°\text{F}$

76. **(a)**

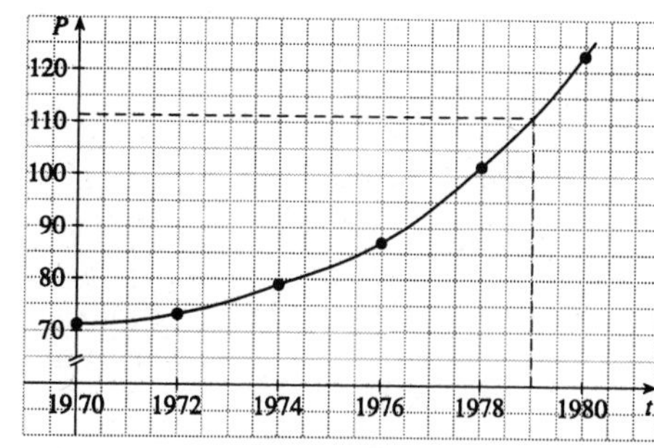

(b) $P(1979) \approx 112{,}000$ people.

77. $f(-x) = \dfrac{1}{(-x)^2} = \dfrac{1}{x^2} = f(x)$, so f is an even function.

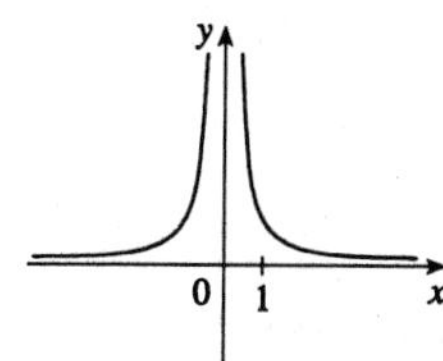

78. $f(-x) = (-x)^{-3} = -(x^{-3}) = -f(x)$, so f is odd.

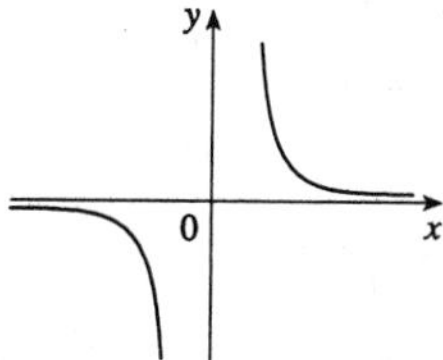

79. $f(-x) = (-x)^2 + (-x) = x^2 - x$. Since this is neither $f(x)$ nor $-f(x)$, the function f is neither even nor odd.

80. $f(-x) = (-x)^4 - 4(-x)^2 = x^4 - 4x^2$
$= f(x)$, so f is even.

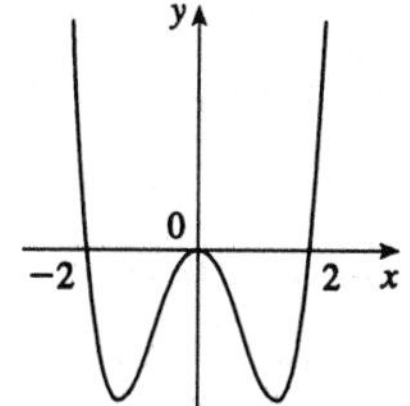

81. $f(-x) = (-x)^3 - (-x) = -x^3 + x = -(x^3 - x)$
$= -f(x)$, so f is odd.

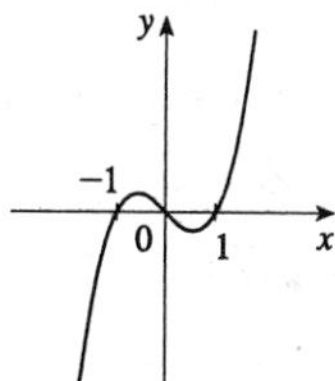

82. $f(-x) = 3(-x)^3 + 2(-x)^2 + 1 = -3x^3 + 2x^2 + 1$. Since this is neither $f(x)$ nor $-f(x)$, the function f is neither even nor odd.

NOTE: For the rest of this section, "$D =$ " stands for "The domain of the function is".

83. $f(x) = x^3 + 2x^2$; $g(x) = 3x^2 - 1$. $D = \mathbb{R}$ for both f and g.

$(f+g)(x) = x^3 + 2x^2 + 3x^2 - 1 = x^3 + 5x^2 - 1$, $D = \mathbb{R}$.

$(f-g)(x) = x^3 + 2x^2 - (3x^2 - 1) = x^3 - x^2 + 1$, $D = \mathbb{R}$.

$(fg)(x) = (x^3 + 2x^2)(3x^2 - 1) = 3x^5 + 6x^4 - x^3 - 2x^2$, $D = \mathbb{R}$.

$(f/g)(x) = (x^3 + 2x^2)/(3x^2 - 1)$, $D = \left\{x \mid x \neq \pm\frac{1}{\sqrt{3}}\right\}$.

84. $f(x) = \sqrt{1+x}$, $D = [-1, \infty)$. $g(x) = \sqrt{1-x}$, $D = (-\infty, 1]$.

$(f+g)(x) = \sqrt{1+x} + \sqrt{1-x}$, $D = (-\infty, 1] \cap [-1, \infty) = [-1, 1]$.

$(f-g)(x) = \sqrt{1+x} - \sqrt{1-x}$, $D = [-1, 1]$.

$(fg)(x) = \sqrt{1+x} \cdot \sqrt{1-x} = \sqrt{1-x^2}$, $D = [-1, 1]$.

$(f/g)(x) = \sqrt{1+x}/\sqrt{1-x}$, $D = [-1, 1)$.

85. $f(x) = x$, $g(x) = 1/x$

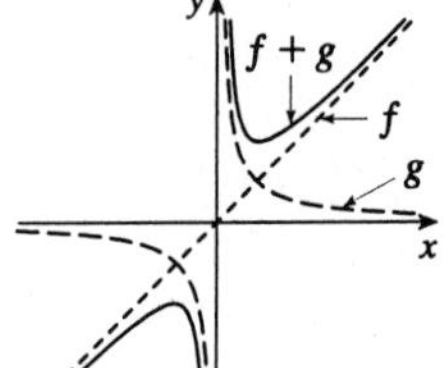

86. $f(x) = x^3$, $g(x) = -x^2$

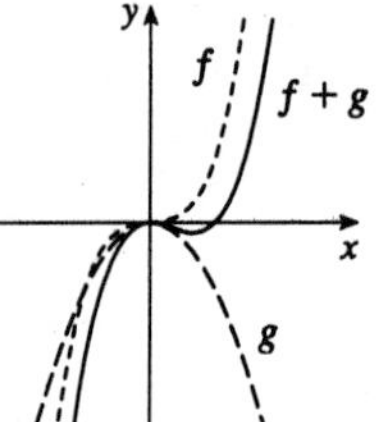

87. $f(x) = 2x^2 - x$, $g(x) = 3x + 2$. $D = \mathbb{R}$ for both f and g, and hence for their composites.

$(f \circ g)(x) = f(g(x)) = f(3x+2) = 2(3x+2)^2 - (3x+2) = 18x^2 + 21x + 6$.

$(g \circ f)(x) = g(f(x)) = g(2x^2 - x) = 3(2x^2 - x) + 2 = 6x^2 - 3x + 2$.

$(f \circ f)(x) = f(f(x)) = f(2x^2 - x) = 2(2x^2 - x)^2 - (2x^2 - x) = 8x^4 - 8x^3 + x$.

$(g \circ g)(x) = g(g(x)) = g(3x+2) = 3(3x+2) + 2 = 9x + 8$.

88. $f(x) = \sqrt{x-1}$. $D = [1, \infty)$; $g(x) = x^2$, $D = \mathbb{R}$.

$(f \circ g)(x) = f(g(x)) = f(x^2) = \sqrt{x^2 - 1}$, $D = \{x \in \mathbb{R} \mid g(x) \in [1, \infty)\} = (-\infty, -1] \cup [1, \infty)$.

$(g \circ f)(x) = g(f(x)) = g\left(\sqrt{x-1}\right) = \left(\sqrt{x-1}\right)^2 = x - 1$, $D = [1, \infty)$.

$(f \circ f)(x) = f(f(x)) = f\left(\sqrt{x-1}\right) = \sqrt{\sqrt{x-1} - 1}$, $D = \left\{x \in [1, \infty) \mid \sqrt{x-1} \geq 1\right\} = [2, \infty)$.

$(g \circ g)(x) = g(g(x)) = g(x^2) = (x^2)^2 = x^4$, $D = \mathbb{R}$.

89. $f(x) = 1/x$, $D = \{x \mid x \neq 0\}$; $g(x) = x^3 + 2x$, $D = \mathbb{R}$.

$(f \circ g)(x) = f(g(x)) = f(x^3 + 2x) = 1/(x^3 + 2x)$, $D = \{x \mid x^3 + 2x \neq 0\} = \{x \mid x \neq 0\}$.

$(g \circ f)(x) = g(f(x)) = g(1/x) = 1/x^3 + 2/x$, $D = \{x \mid x \neq 0\}$.

$(f\circ f)(x)=f(f(x))=f(1/x)=\dfrac{1}{1/x}=x$, $D=\{x\mid x\neq 0\}$.

$(g\circ g)(x)=g(g(x))=g(x^3+2x)=(x^3+2x)^3+2(x^3+2x)=x^9+6x^7+12x^5+10x^3+4x$, $D=\mathbb{R}$.

90. $f(x)=\dfrac{1}{x-1}$, $D=\{x\mid x\neq 1\}$; $g(x)=\dfrac{x-1}{x+1}$, $D=\{x\mid x\neq -1\}$.

$(f\circ g)(x)=f\left(\dfrac{x-1}{x+1}\right)=\left(\dfrac{x-1}{x+1}-1\right)^{-1}=\left(\dfrac{-2}{x+1}\right)^{-1}=\dfrac{-x-1}{2}$, $D=\{x\mid x\neq -1\}$.

$(g\circ f)(x)=g\left(\dfrac{1}{x-1}\right)=\dfrac{1/(x-1)-1}{1/(x-1)+1}=\dfrac{2-x}{x}$, $D=\{x\mid x\neq 0,1\}$.

$(f\circ f)(x)=f\left(\dfrac{1}{x-1}\right)=\dfrac{1}{1/(x-1)-1}=\dfrac{x-1}{2-x}$, $D=\{x\mid x\neq 1,2\}$.

$(g\circ g)(x)=g\left(\dfrac{x-1}{x+1}\right)=\dfrac{(x-1)/(x+1)-1}{(x-1)/(x+1)+1}=-\dfrac{1}{x}$, $D=\{x\mid x\neq 0,-1\}$.

91. $f(x)=\sqrt[3]{x}$, $D=\mathbb{R}$; $g(x)=1-\sqrt{x}$, $D=[0,\infty)$.

$(f\circ g)(x)=f(g(x))=f(1-\sqrt{x})=\sqrt[3]{1-\sqrt{x}}$, $D=[0,\infty)$.

$(g\circ f)(x)=g(f(x))=g(\sqrt[3]{x})=1-x^{1/6}$, $D=[0,\infty)$.

$(f\circ f)(x)=f(f(x))=f(\sqrt[3]{x})=x^{1/9}$, $D=\mathbb{R}$.

$(g\circ g)(x)=g(g(x))=g(1-\sqrt{x})=1-\sqrt{1-\sqrt{x}}$, $D=\{x\geq 0\mid 1-\sqrt{x}\geq 0\}=[0,1]$.

92. $f(x)=\sqrt{x^2-1}$, $D=(-\infty,-1]\cup[1,\infty)$; $g(x)=\sqrt{1-x}$, $D=(-\infty,1]$.

$(f\circ g)(x)=f(g(x))=f\left(\sqrt{1-x}\right)=\sqrt{\left(\sqrt{1-x}\right)^2-1}=\sqrt{-x}$,

$D=\left\{x\leq 1\mid \sqrt{1-x}\in(-\infty,-1]\cup[1,\infty)\right\}=(-\infty,0]$.

$(g\circ f)(x)=g(f(x))=g\left(\sqrt{x^2-1}\right)=\sqrt{1-\sqrt{x^2-1}}$,

$D=\left\{x\in(-\infty,-1]\cup[1,\infty)\mid \sqrt{x^2-1}\in(-\infty,1]\right\}=\left[-\sqrt{2},-1\right]\cup\left[1,\sqrt{2}\right]$.

$(f\circ f)(x)=f(f(x))=f\left(\sqrt{x^2-1}\right)=\sqrt{x^2-2}$, $D=\left(-\infty,-\sqrt{2}\right]\cup\left[\sqrt{2},\infty\right)$

$(g\circ g)(x)=g(g(x))=g\left(\sqrt{1-x}\right)=\sqrt{1-\sqrt{1-x}}$, $D=[0,1]$

93. $f(x)=\dfrac{x+2}{2x+1}$, $D=\left\{x\mid x\neq -\frac{1}{2}\right\}$; $g(x)=\dfrac{x}{x-2}$, $D=\{x\mid x\neq 2\}$.

$(f\circ g)(x)=f(g(x))=f\left(\dfrac{x}{x-2}\right)=\dfrac{x/(x-2)+2}{2x/(x-2)+1}=\dfrac{3x-4}{3x-2}$, $D=\left\{x\mid x\neq 2,\frac{2}{3}\right\}$.

$(g\circ f)(x)=g(f(x))=g\left(\dfrac{x+2}{2x+1}\right)=\dfrac{(x+2)/(2x+1)}{(x+2)/(2x+1)-2}=\dfrac{-x-2}{3x}$, $D=\left\{x\mid x\neq 0,-\frac{1}{2}\right\}$.

$(f\circ f)(x)=f(f(x))=f\left(\dfrac{x+2}{2x+1}\right)=\dfrac{(x+2)/(2x+1)+2}{2(x+2)/(2x+1)+1}=\dfrac{5x+4}{4x+5}$, $D=\left\{x\mid x\neq -\frac{1}{2},-\frac{5}{4}\right\}$.

$(g\circ g)(x)=g(g(x))=g\left(\dfrac{x}{x-2}\right)=\dfrac{x/(x-2)}{x/(x-2)-2}=\dfrac{x}{4-x}$, $D=\{x\mid x\neq 2,4\}$.

94. $f(x) = 1/\sqrt{x}$, $D = (0, \infty)$; $g(x) = x^2 - 4x$, $D = \mathbb{R}$.

$(f \circ g)(x) = f(g(x)) = f(x^2 - 4x) = 1/\sqrt{x^2 - 4x}$, $D = \{x \mid x^2 - 4x > 0\} = (-\infty, 0) \cup (4, \infty)$.

$(g \circ f)(x) = g(f(x)) = g\left(\frac{1}{\sqrt{x}}\right) = \frac{1}{x} - \frac{4}{\sqrt{x}}$, $D = (0, \infty)$.

$(f \circ f)(x) = f(f(x)) = f\left(\frac{1}{\sqrt{x}}\right) = \frac{1}{\sqrt{1/\sqrt{x}}} = x^{1/4}$, $D = (0, \infty)$.

$(g \circ g)(x) = g(g(x)) = g(x^2 - 4x) = (x^2 - 4x)^2 - 4(x^2 - 4x) = x^4 - 8x^3 + 12x^2 + 16x$, $D = \mathbb{R}$.

95. $(f \circ g \circ h)(x) = f(g(h(x))) = f(g(x - 1)) = f\left(\sqrt{x-1}\right) = \sqrt{x-1} - 1$

96. $(f \circ g \circ h)(x) = f(g(h(x))) = f(g(x^2 + 2)) = f\left((x^2+2)^3\right) = \dfrac{1}{(x^2+2)^3}$

97. $(f \circ g \circ h)(x) = f(g(h(x))) = f(g(\sqrt{x})) = f(\sqrt{x} - 5) = (\sqrt{x} - 5)^4 + 1$

98. $(f \circ g \circ h)(x) = f(g(h(x))) = f(g(\sqrt[3]{x})) = f\left(\dfrac{\sqrt[3]{x}}{\sqrt[3]{x} - 1}\right) = \sqrt{\dfrac{\sqrt[3]{x}}{\sqrt[3]{x} - 1}}$

99. Let $g(x) = x - 9$ and $f(x) = x^5$. Then $(f \circ g)(x) = (x - 9)^5 = F(x)$.

100. Let $g(x) = \sqrt{x}$ and $f(x) = x + 1$. Then $(f \circ g)(x) = \sqrt{x} + 1 = F(x)$.

101. Let $g(x) = x^2$ and $f(x) = \dfrac{x}{x+4}$. Then $(f \circ g)(x) = \dfrac{x^2}{x^2+4} = G(x)$.

102. Let $g(x) = x + 3$ and $f(x) = \dfrac{1}{x}$. Then $(f \circ g)(x) = \dfrac{1}{x+3} = G(x)$.

103. Let $h(x) = x^2$, $g(x) = x + 1$ and $f(x) = \dfrac{1}{x}$. Then $(f \circ g \circ h)(x) = \dfrac{1}{x^2+1} = H(x)$.

104. Let $h(x) = \sqrt{x}$, $g(x) = x - 1$ and $f(x) = \sqrt[3]{x}$. Then $(f \circ g \circ h)(x) = \sqrt[3]{\sqrt{x} - 1} = H(x)$.

105. Let r be the radius of the ripple in cm. The area of the ripple is $A = \pi r^2$ but, as a function of time, $r = 60t$. Thus, $A = \pi(60t)^2 = 3600\pi t^2$.

106. Let r be the radius of the spherical balloon in cm. Then the volume is $V = \frac{4}{3}\pi r^3$ and, as a function of time, $r = t$. Thus $V = \frac{4}{3}\pi t^3$.

107. We need a function g so that $f(g(x)) = 3(g(x)) + 5 = h(x) = 3x^2 + 3x + 2 = 3(x^2 + x) + 2$ $= 3(x^2 + x - 1) + 5$. So we see that $g(x) = x^2 + x - 1$.

108. We need a function g so that $g(f(x)) = g(x + 4) = h(x) = 4x - 1 = 4(x + 4) - 17$. So we see that the function g must be $g(x) = 4x - 17$.

109. The function $g(x) = x$ has domain $(-\infty, \infty)$. However, the function $f \circ f$, where $f(x) = 1/x$, has for its domain $(-\infty, 0) \cup (0, \infty)$ even though the rule is the same: $(f \circ f)(x) = f(1/x) = x$.

EXERCISES 2

1. **(a)** $f(x) = \sqrt[5]{x}$ is a root function.

(b) $g(x) = \sqrt{1 - x^2}$ is an algebraic function because it is a root of a polynomial.

(c) $h(x) = x^9 + x^4$ is a polynomial of degree 9.

(d) $r(x) = \dfrac{x^2 + 1}{x^3 + x}$ is a rational function because it is a ratio of polynomials

(e) $s(x) = \tan 2x$ is a trigonometric function.

(f) $t(x) = \log_{10} x$ is a logarithmic function.

2. **(a)** $y = (x - 6)/(x + 6)$ is a rational function because it is a ratio of polynomials.

(b) $y = x + x^2/\sqrt{x - 1}$ is an algebraic function because it involves polynomials and roots of polynomials.

(c) $y = 10^x$ is an exponential function (notice that x is the *exponent*).

(d) $y = x^{10}$ is a power function (notice that x is the *base*).

(e) $y = 2t^6 + t^4 - \pi$ is a polynomial of degree 6.

(f) $y = \cos\theta + \sin\theta$ is a trigonometric function.

3. **(a)** To graph $y = f(2x)$ we compress the graph of f horizontally by a factor of 2.

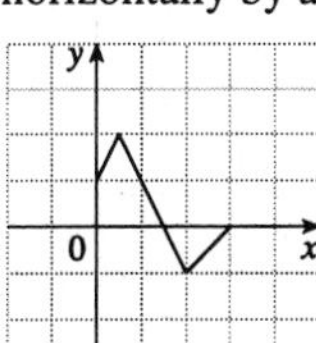

(b) To graph $y = f\left(\frac{1}{2}x\right)$ we expand the graph of f horizontally by a factor of 2.

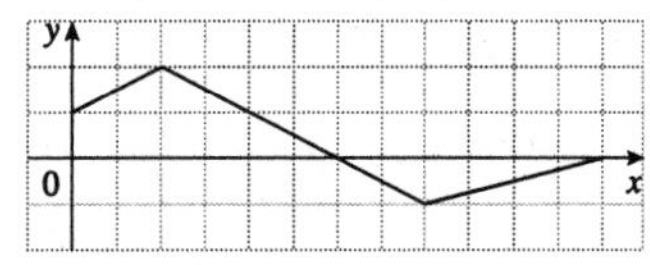

(c) To graph $y = f(-x)$ we reflect the graph of f about the y-axis.

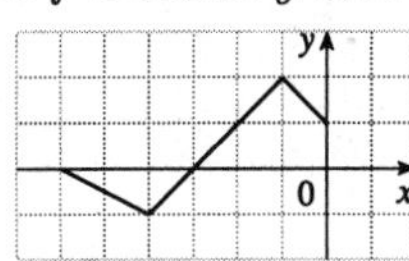

(d) To graph $y = -f(-x)$ we reflect the graph of f about the y-axis, then about the x-axis.

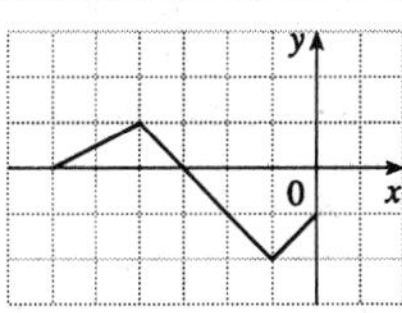

4. **(a)** To graph $y = f(x + 4)$ we shift the graph of f 4 units to the left.

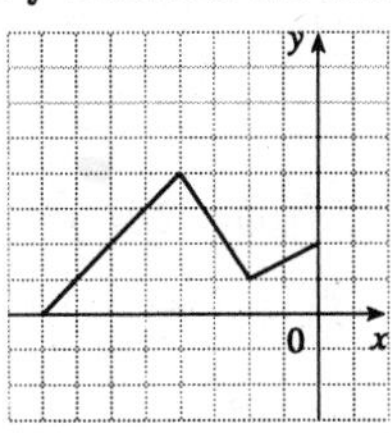

(b) To graph $y = f(x) + 4$ we shift the graph of f 4 units upward.

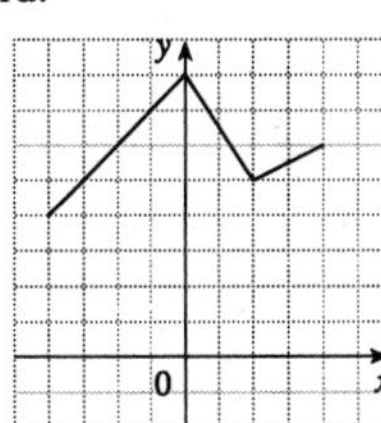

(c) To graph $y = 2f(x)$ we stretch the graph of f vertically by a factor of 2.

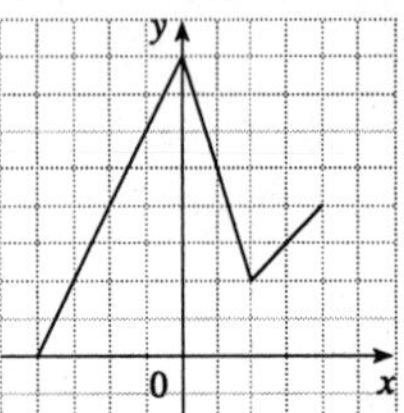

(d) To graph $y = -\frac{1}{2}f(x) + 3$ we compress the graph of f vertically by a factor of 2, then reflect the resulting graph about the x-axis, then shift the resulting graph 3 units upward.

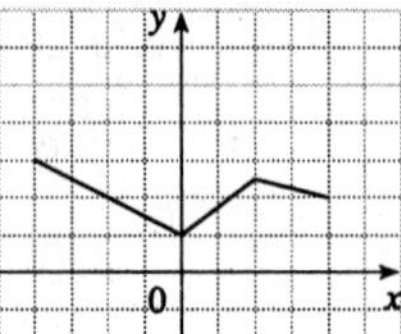

5. $y = -1/x$

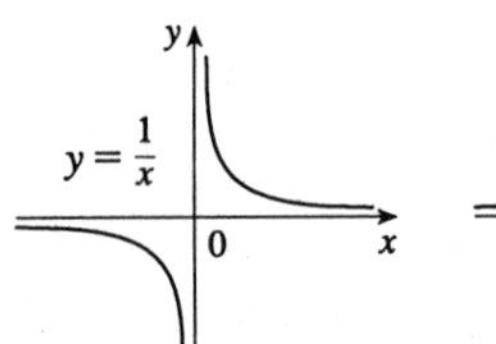

6. $y = -x^3$

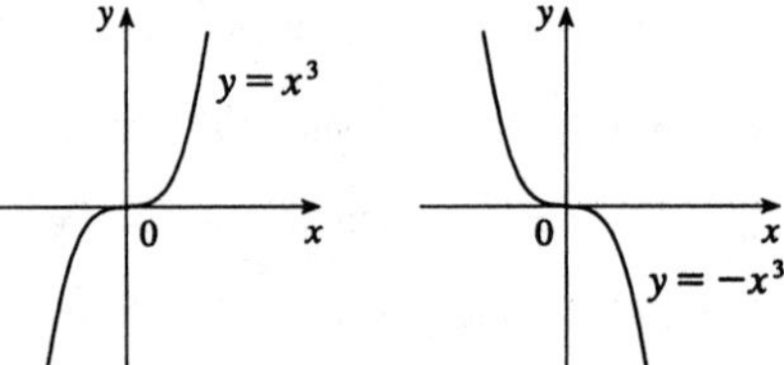

7. $y = 2\sin x$

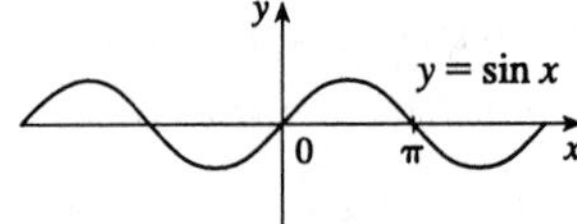

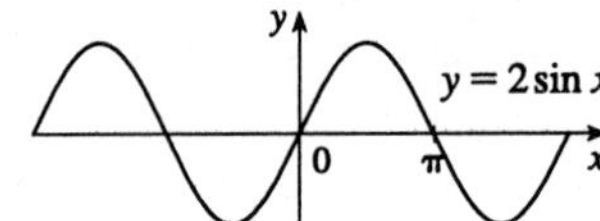

8. $y = 1 + \sqrt{x}$

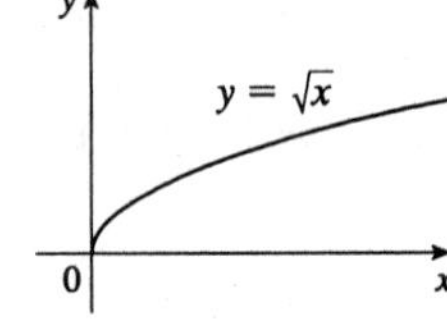

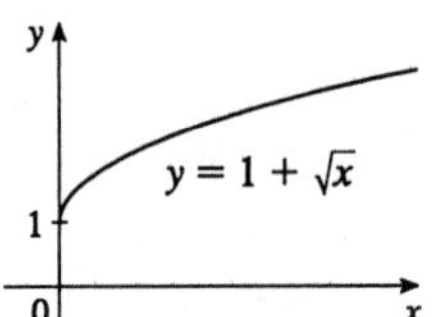

9. $y = (x-1)^3 + 2$

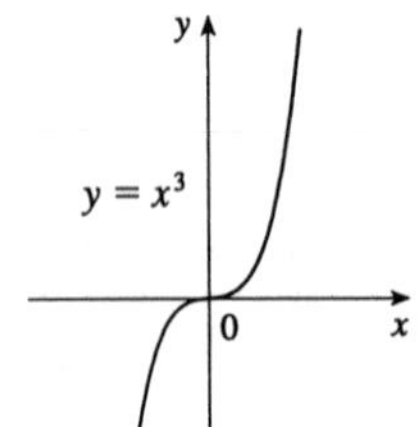

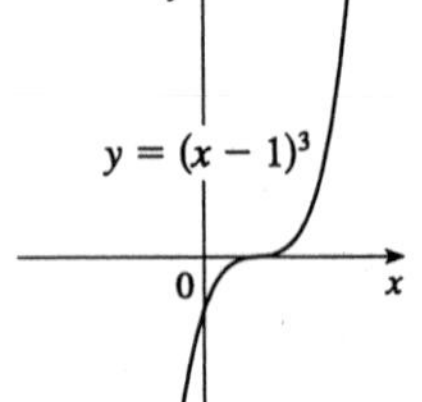

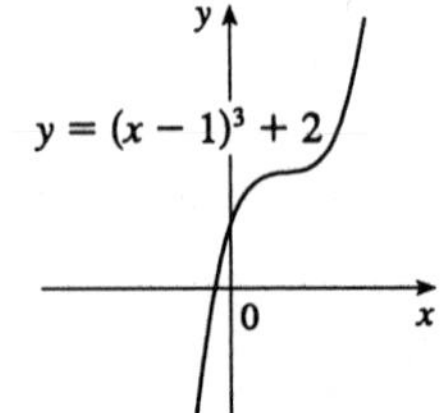

10. $y = 2 - \cos x$

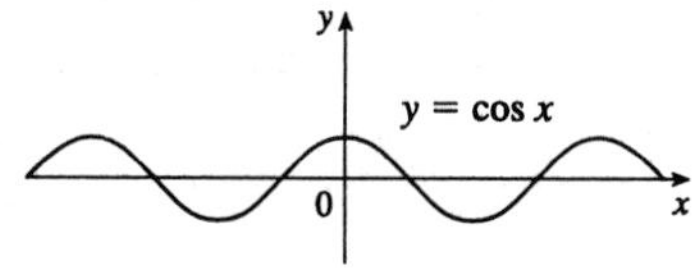

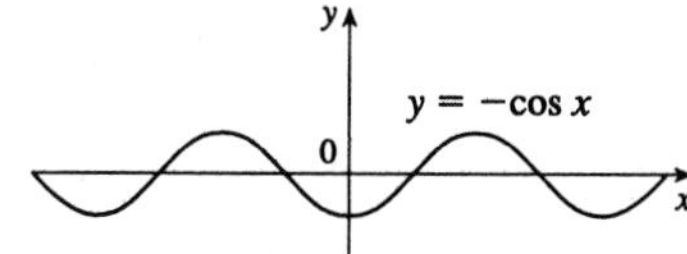

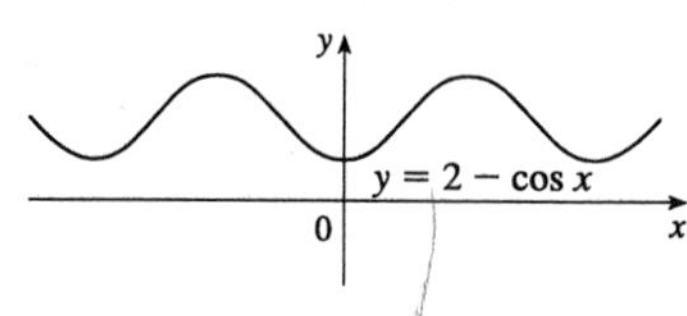

11. $y = \tan 2x$

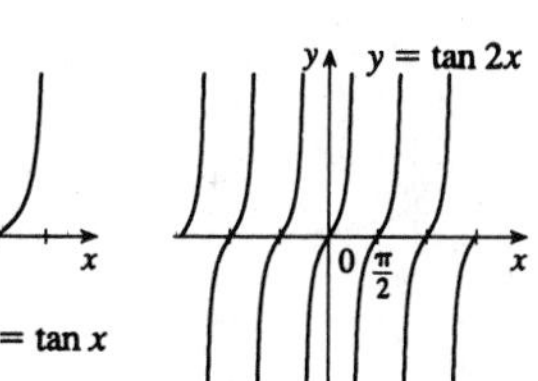

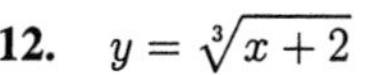

12. $y = \sqrt[3]{x+2}$

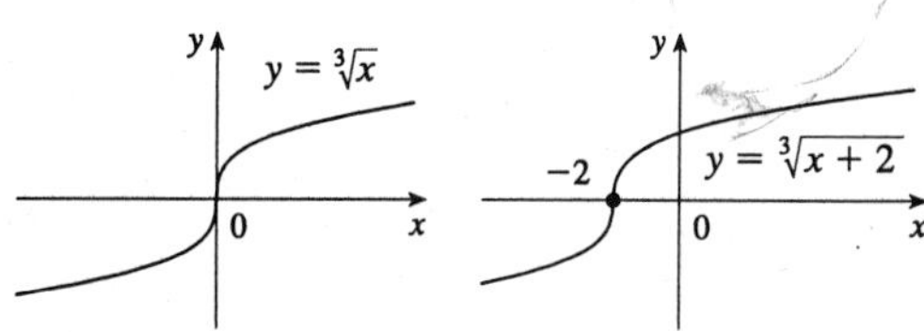

13. $y = \cos(x/2)$

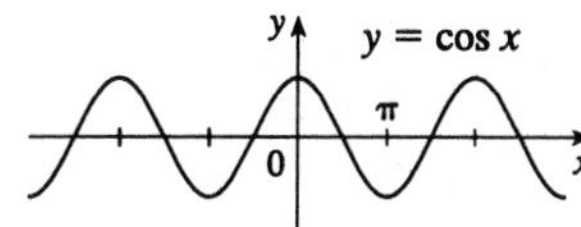

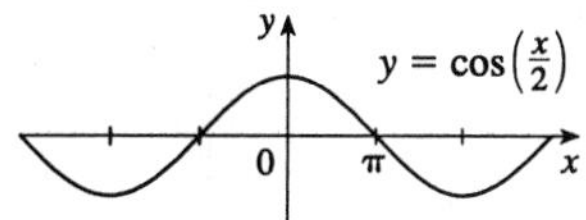

14. $y = x^2 + x + 1 = \left(x + \frac{1}{2}\right)^2 + \frac{3}{4}$

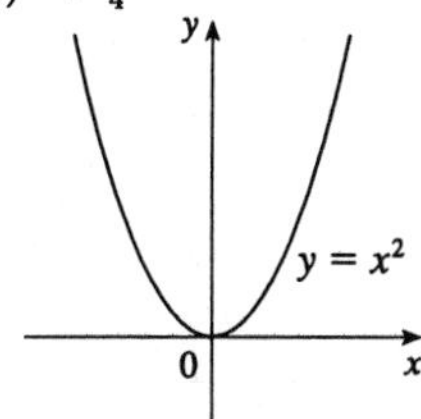

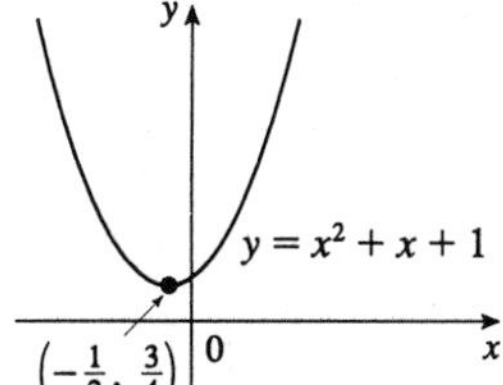

15. $y = \dfrac{1}{x-3}$

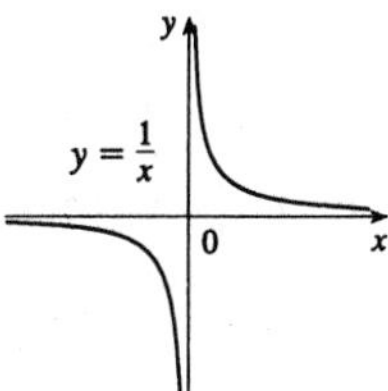

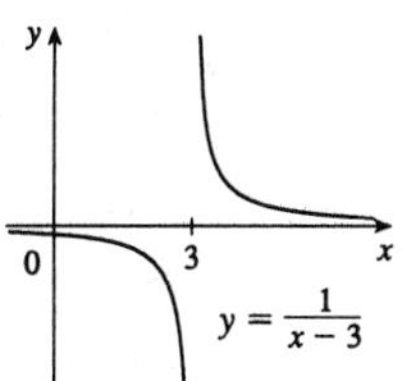

16. $y = -2\sin \pi x$

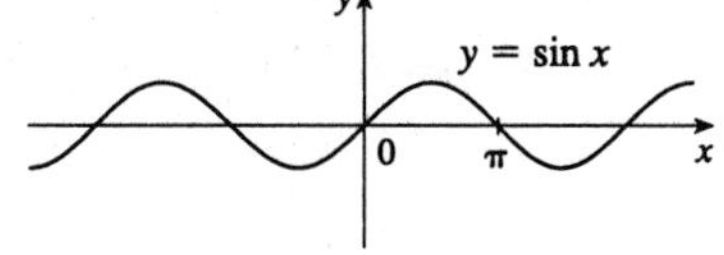

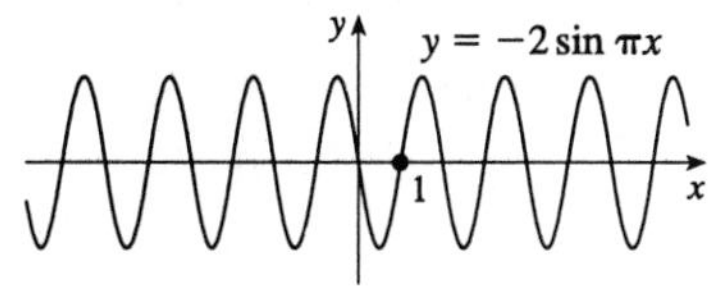

17. $y = \frac{1}{3}\sin\left(x - \frac{\pi}{6}\right)$

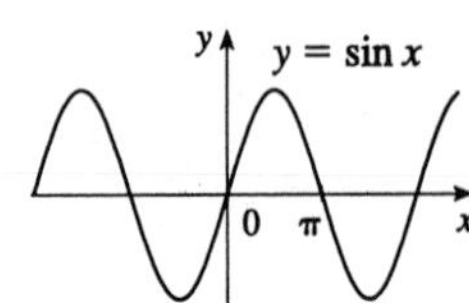

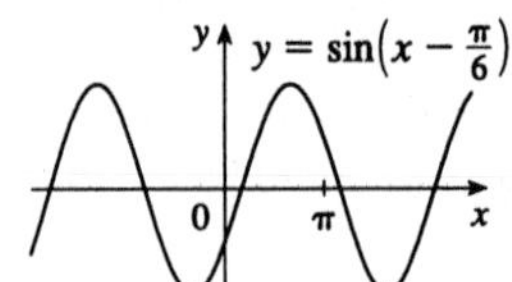

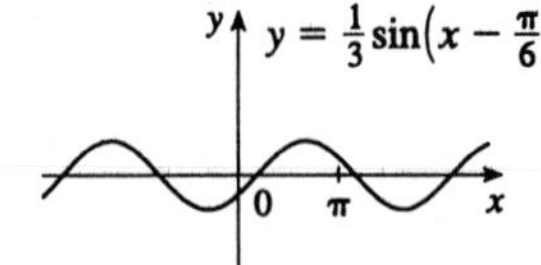

18. $y = 2 + \dfrac{1}{x+1}$

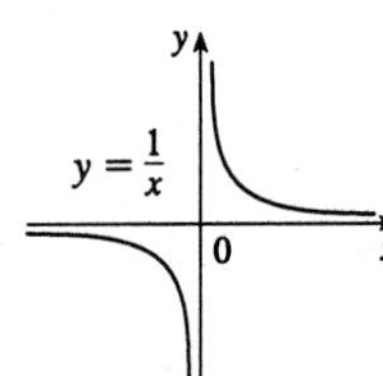

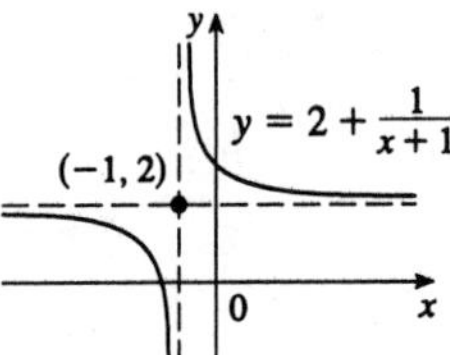

19. $y = 1 + 2x - x^2 = -(x-1)^2 + 2$

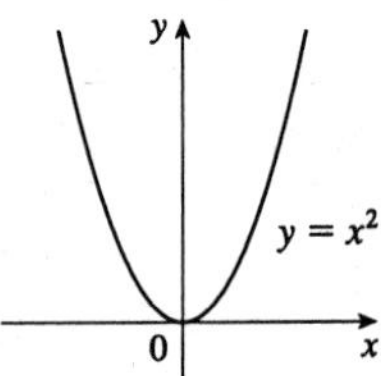

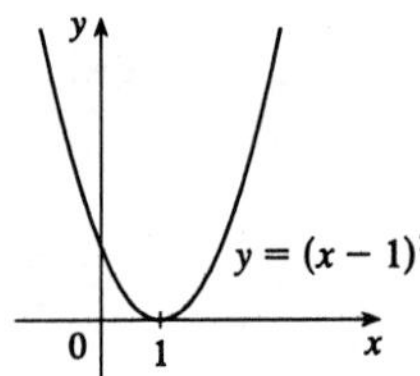

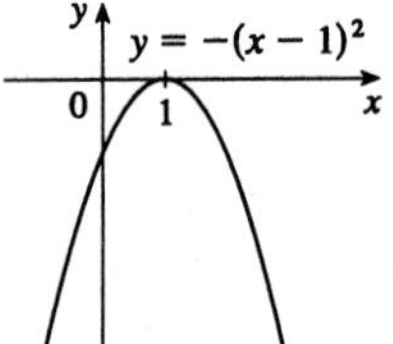

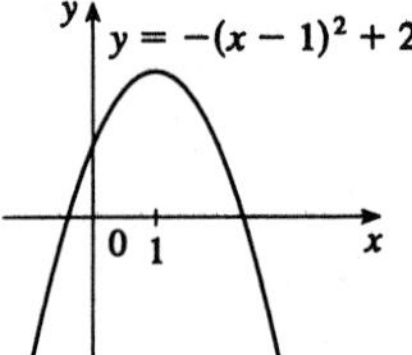

20. $y = \frac{1}{2}\sqrt{x+4} - 3$

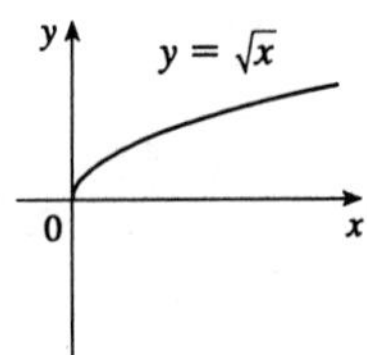

21. $y = 2 - \sqrt{x+1}$

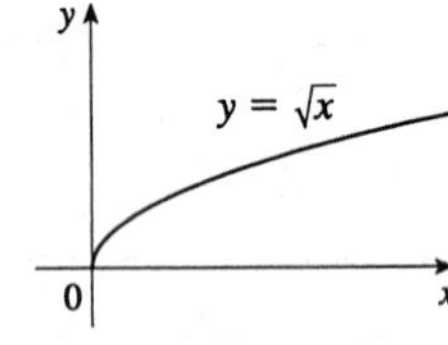

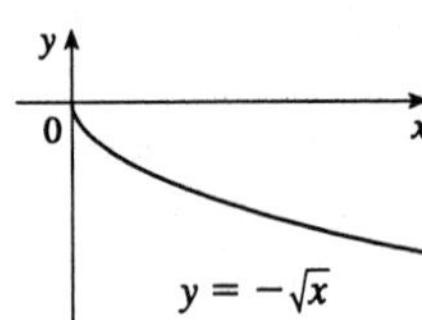

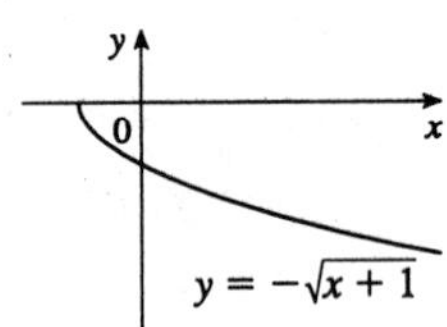

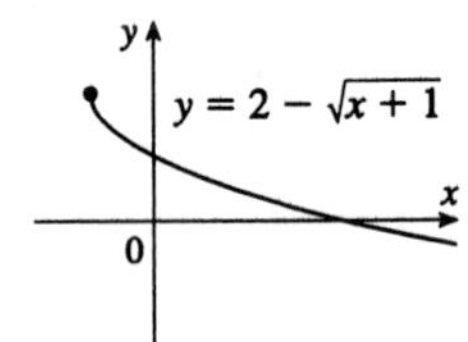

22. $y = 1 - (x-8)^6$

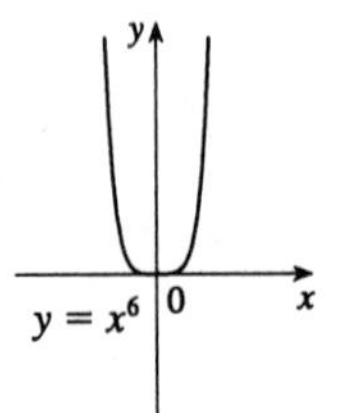

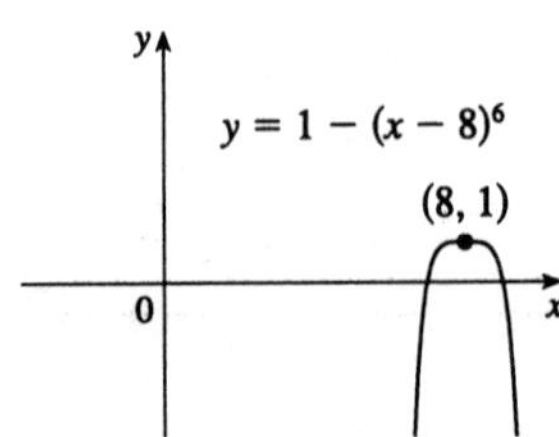

23.

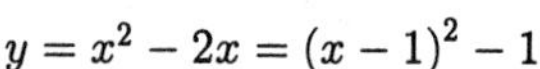
$y = x^2 - 2x = (x-1)^2 - 1$

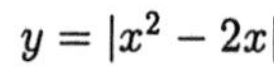
$y = |x^2 - 2x|$

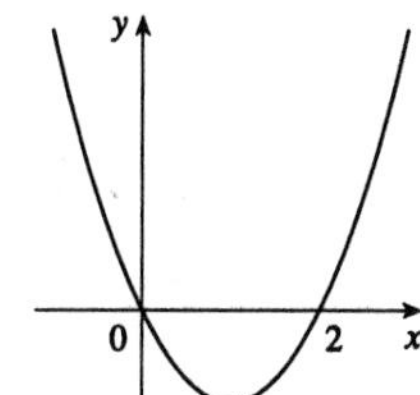

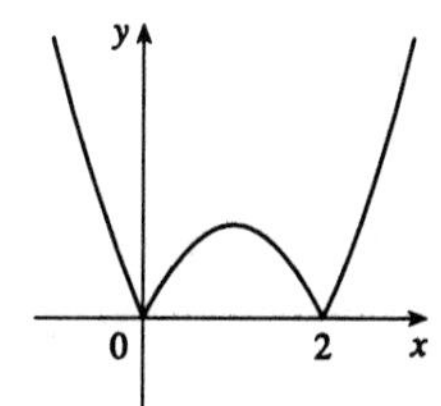

24. $y = |\cos x|$

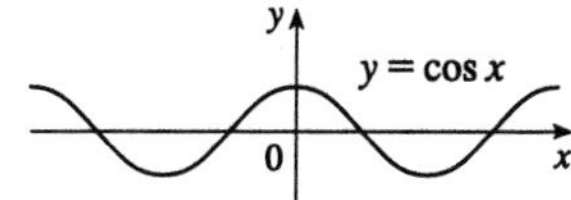

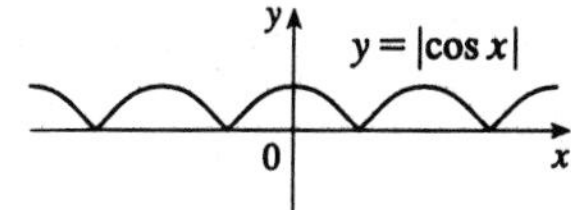

25. $y = ||x| - 1|$

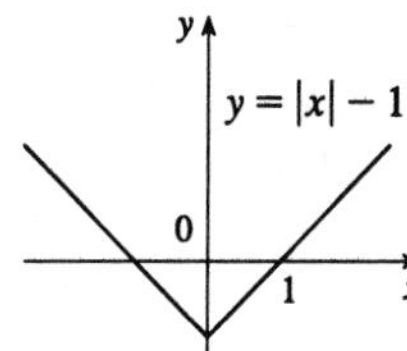

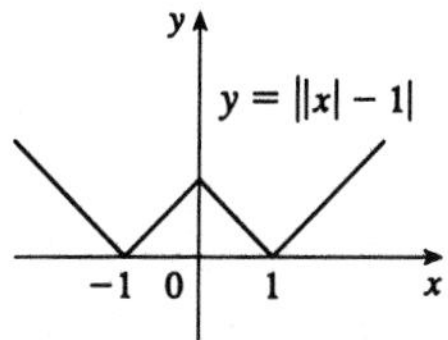

26. $y = |||x| - 2| - 1|$

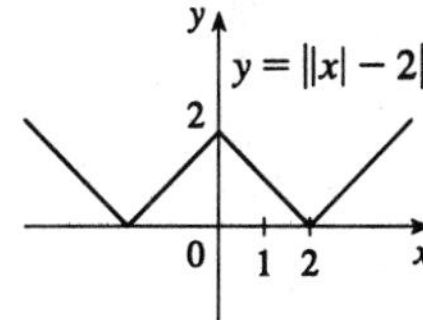

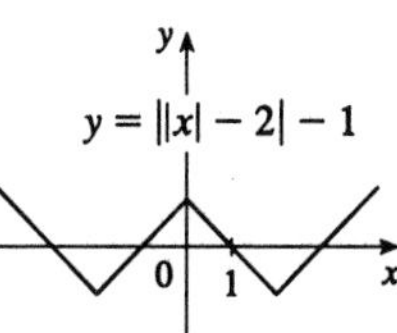

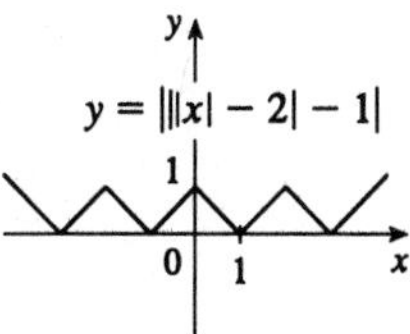

27. **(a)** To obtain $y = f(|x|)$, the portion of $y = f(x)$ right of the y-axis is reflected in the y-axis.

(b) $y = \sin|x|$

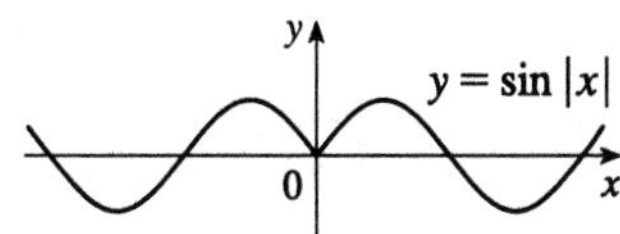

28. 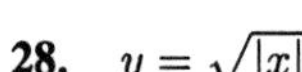 $y = \sqrt{|x|}$

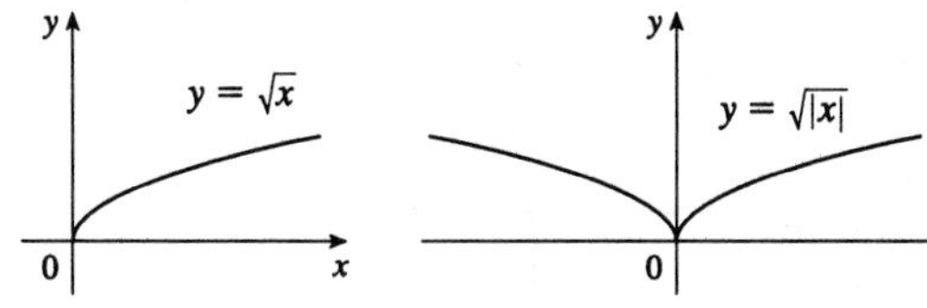

29. 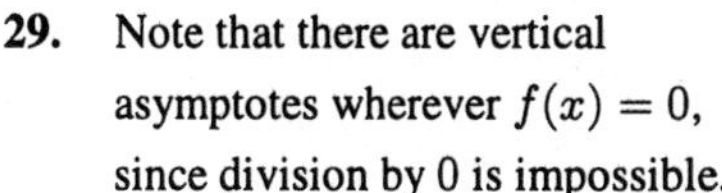 Note that there are vertical asymptotes wherever $f(x) = 0$, since division by 0 is impossible.

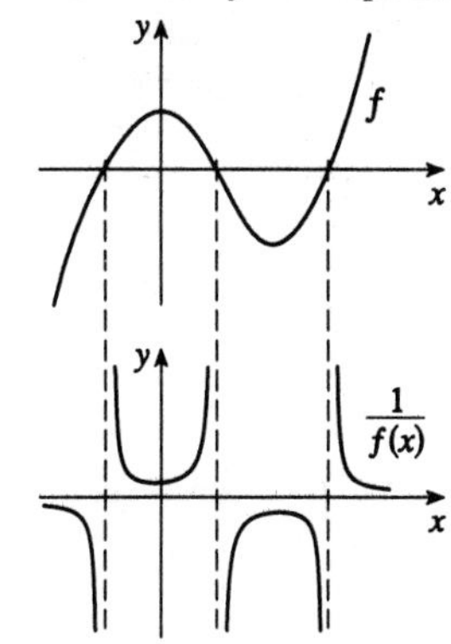

EXERCISES 3

1. $f(x) = x^4 + 2$

(a) $[-2, 2]$ by $[-2, 2]$ **(b)** $[0, 4]$ by $[0, 4]$ **(c)** $[-4, 4]$ by $[-4, 4]$

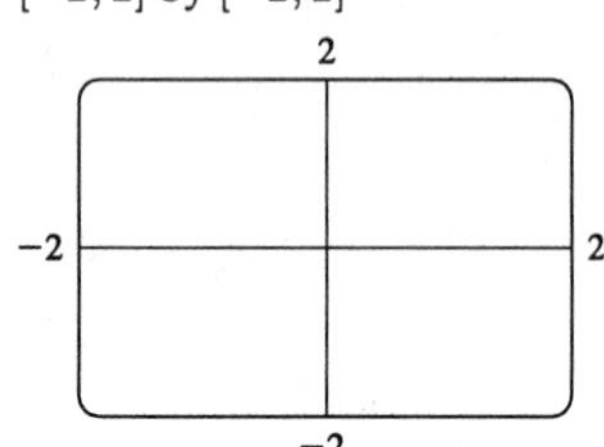

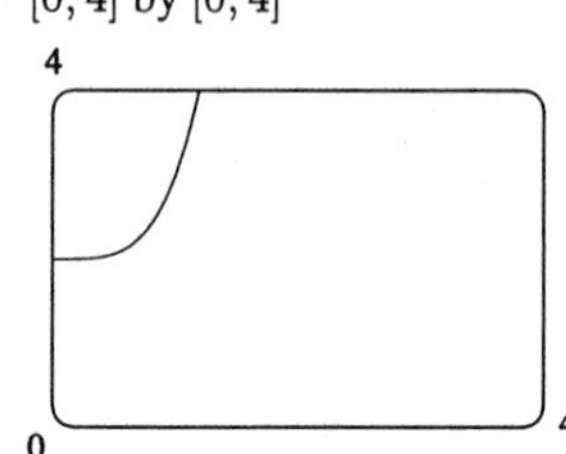

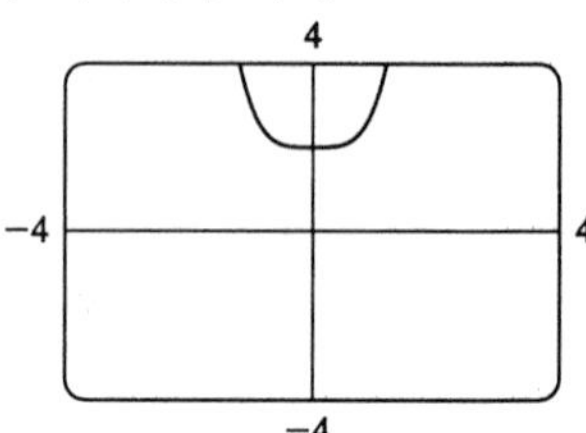

(d) $[-8, 8]$ by $[-4, 40]$ **(e)** $[-40, 40]$ by $[-80, 800]$

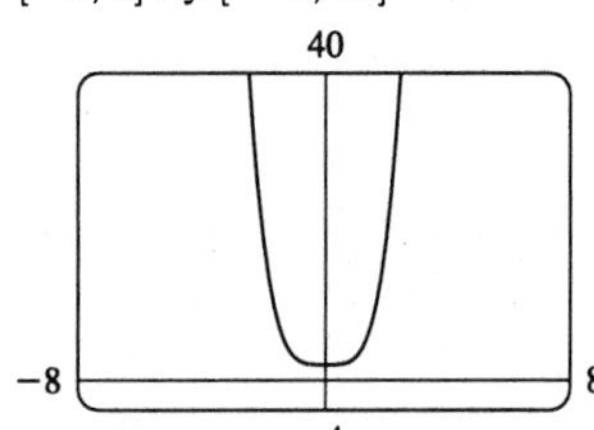

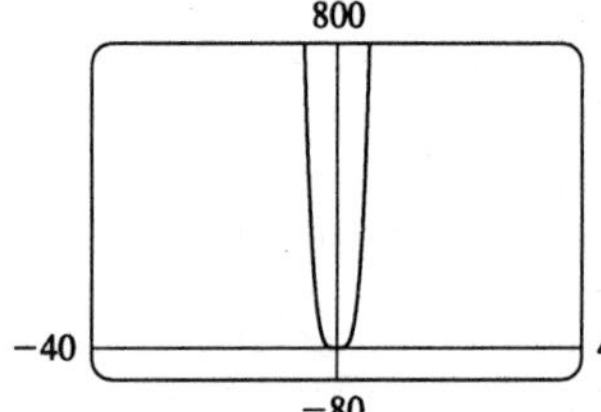

The most appropriate graph is produced in viewing rectangle (d).

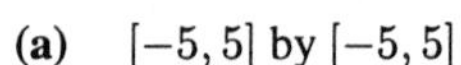
2. $f(x) = x^2 + 7x + 6$

(a) $[-5, 5]$ by $[-5, 5]$ **(b)** $[0, 10]$ by $[-20, 100]$

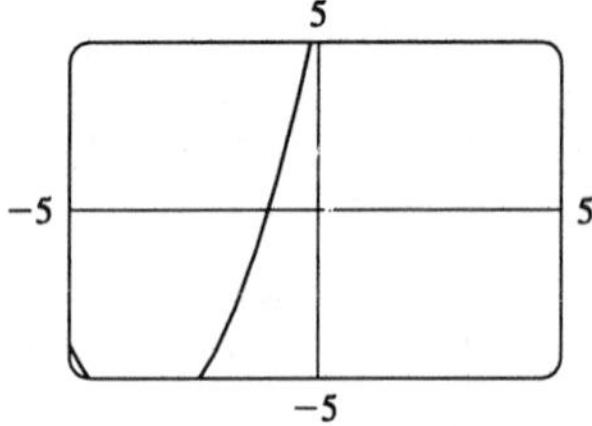

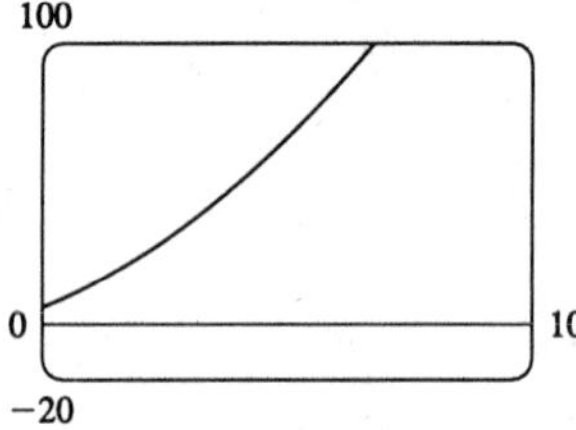

(c) $[-15, 8]$ by $[-20, 200]$ **(d)** $[-10, 3]$ by $[-100, 20]$

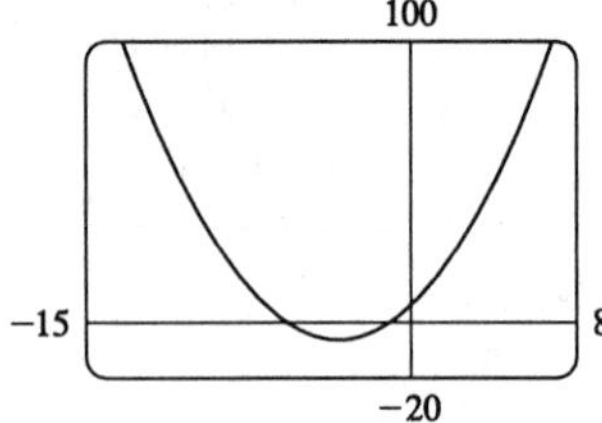

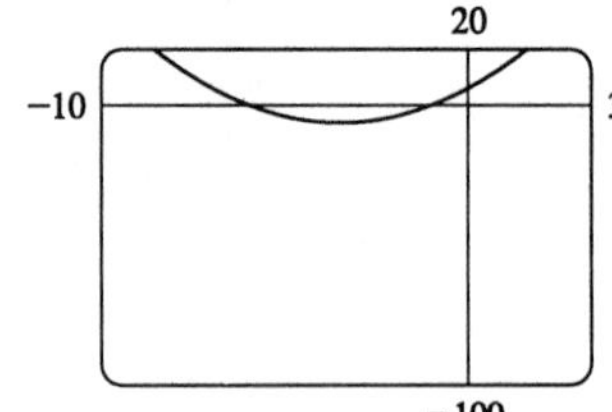

The most appropriate graph is produced in viewing rectangle (c).

3. $f(x) = 10 + 25x - x^3$

(a) $[-4, 4]$ by $[-4, 4]$

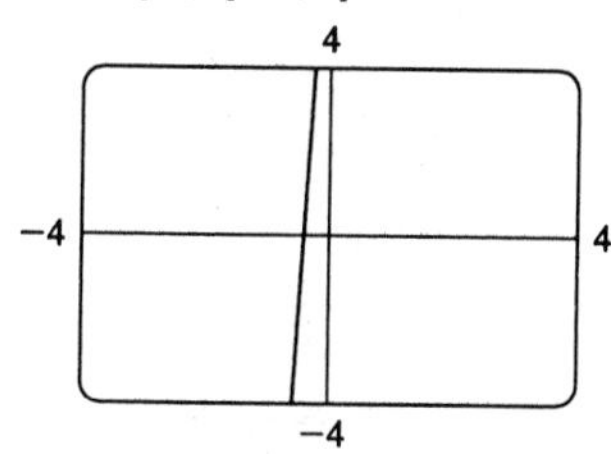

(b) $[-10, 10]$ by $[-10, 10]$

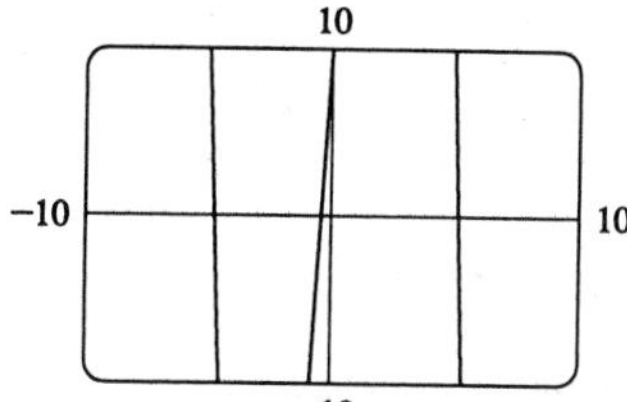

(c) $[-20, 20]$ by $[-100, 100]$

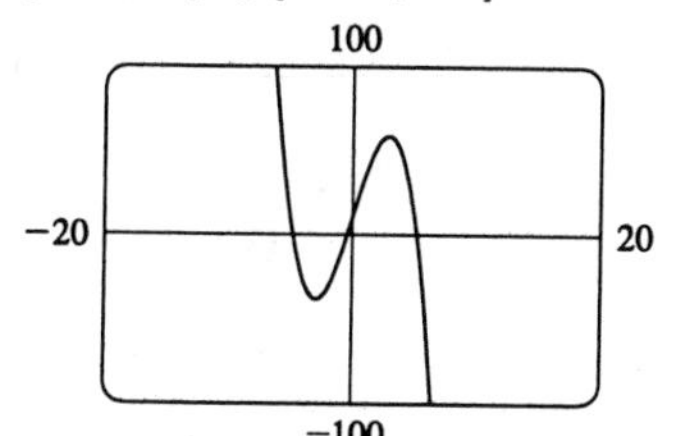

(d) $[-100, 100]$ by $[-200, 200]$

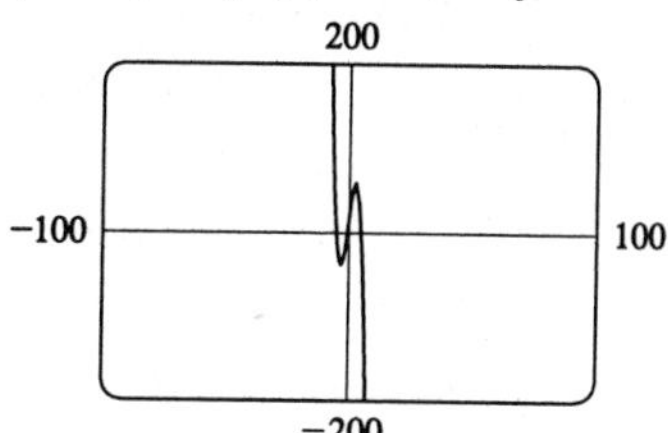

The most appropriate graph is produced in viewing rectangle (c).

4. $f(x) = \sqrt{8x - x^2}$

(a) $[-4, 4]$ by $[-4, 4]$

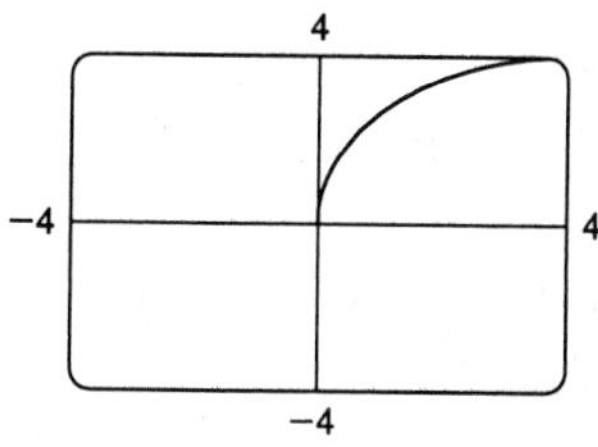

(b) $[-5, 5]$ by $[0, 100]$

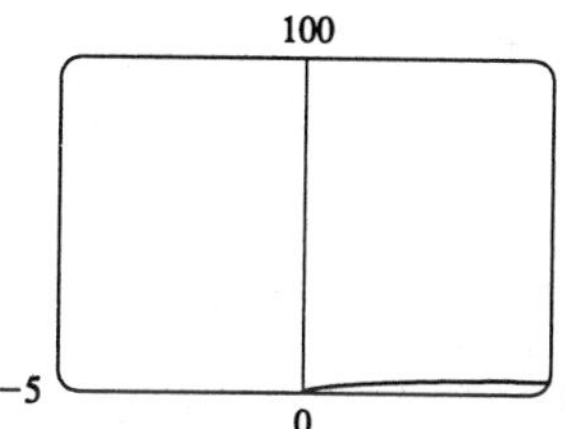

(c) $[-10, 10]$ by $[-10, 40]$

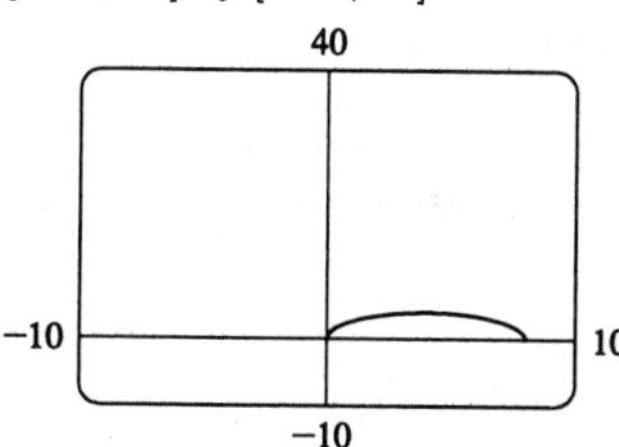

(d) $[-2, 10]$ by $[-2, 6]$

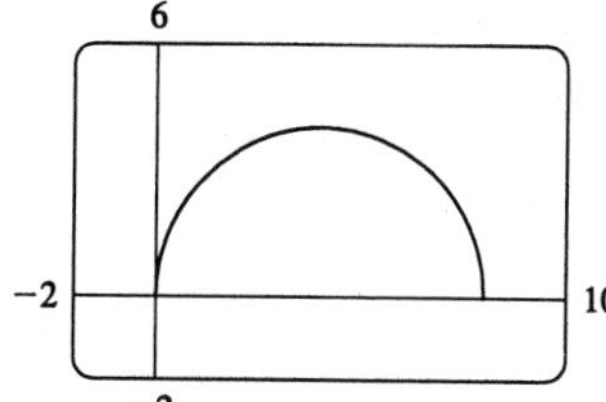

The most appropriate graph is produced in viewing rectangle (d).

5. $f(x) = 4 + 6x - x^2$

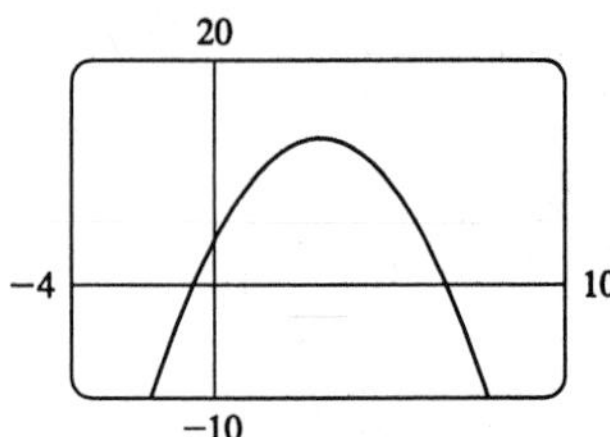

Note that many similar rectangles give equally good views of the function.

6. $f(x) = 0.3x^2 + 1.7x - 3$

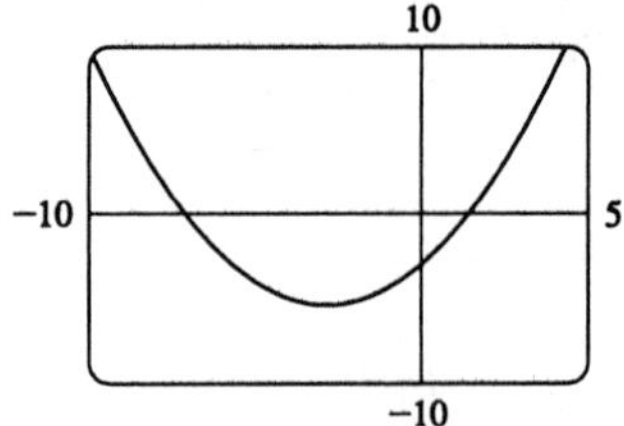

Note that many similar rectangles give equally good views of the function.

7. $f(x) = \sqrt[4]{256 - x^2}$ To find an appropriate viewing rectangle, we calculate f's domain and range: $256 - x^2 \geq 0 \quad \Leftrightarrow \quad x^2 \leq 256 \quad \Leftrightarrow$ $|x| \leq 16 \quad \Leftrightarrow \quad -16 \leq x \leq 16$, so the domain is $[-16, 16]$. Also, $0 \leq \sqrt[4]{256 - x^2} \leq \sqrt[4]{256} = 4$, so the range is $[0, 4]$. Thus we choose the viewing rectangle to be $[-20, 20]$ by $[-2, 6]$.

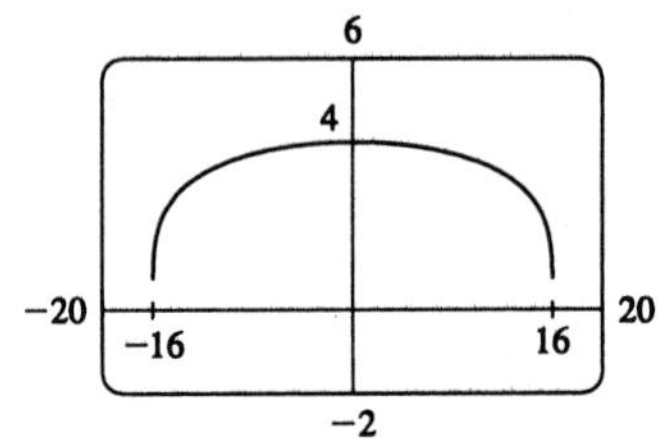
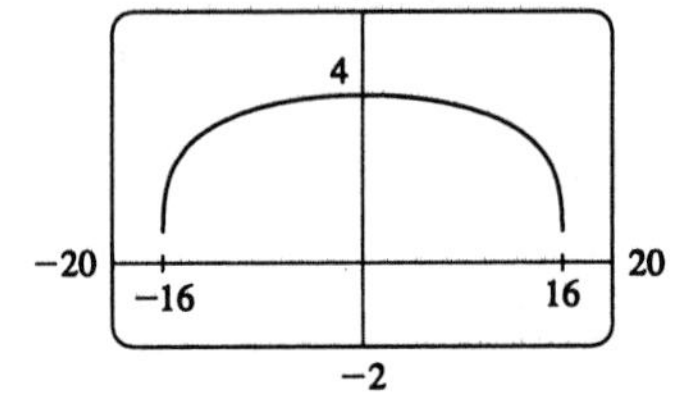

8. $f(x) = \sqrt{12x - 17}$

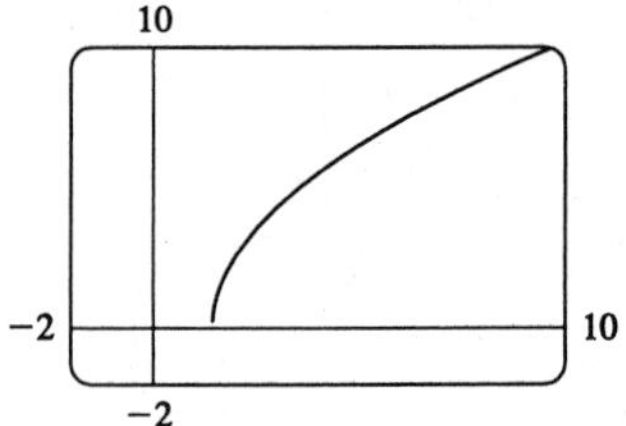

9. $f(x) = 0.01x^3 - x^2 + 5$

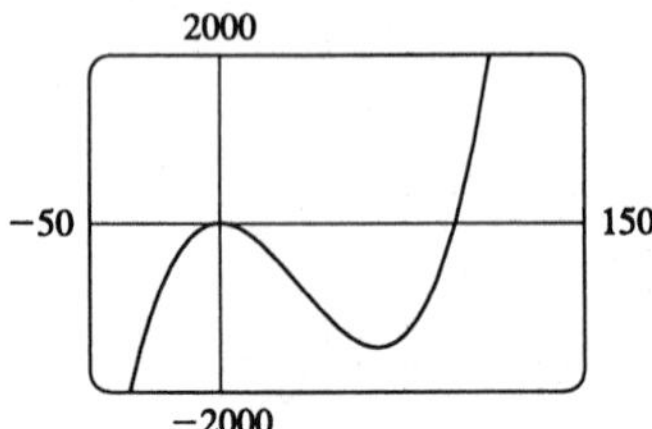

10. $f(x) = x(x + 6)(x - 9)$

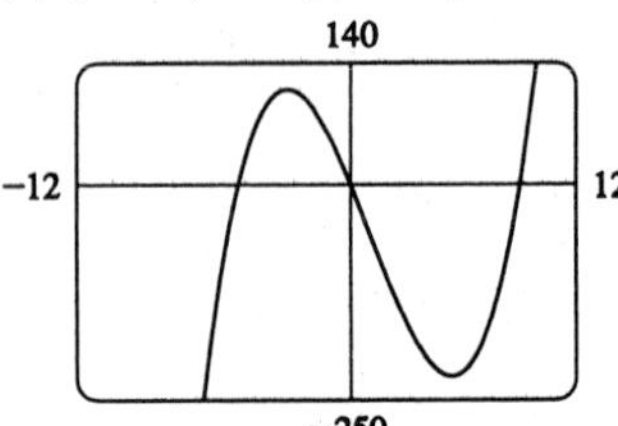

11. $y = \dfrac{1}{x^2 + 25}$

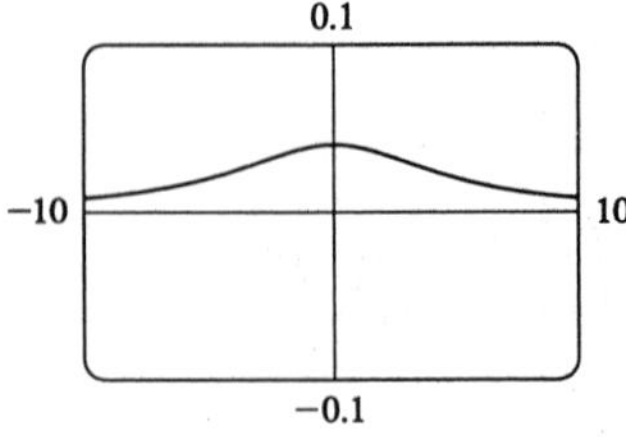

12. $y = \dfrac{x}{x^2 + 25}$

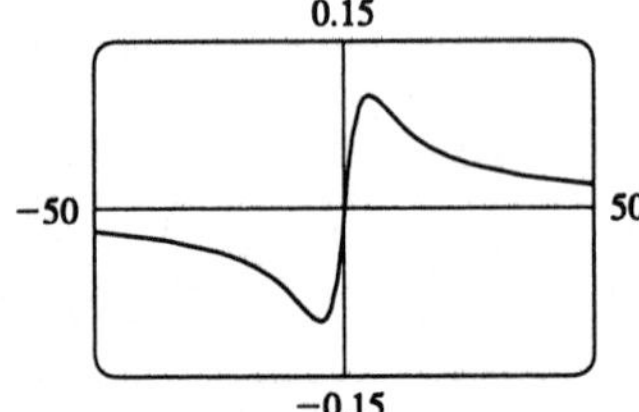

13. $y = x^4 - 4x^3$

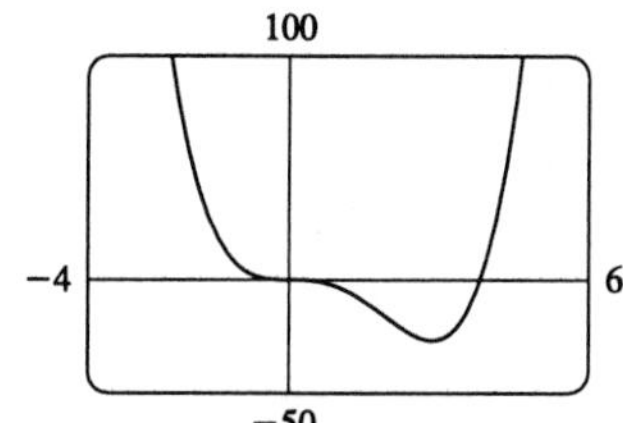

14. $y = x^3 + 1/x$

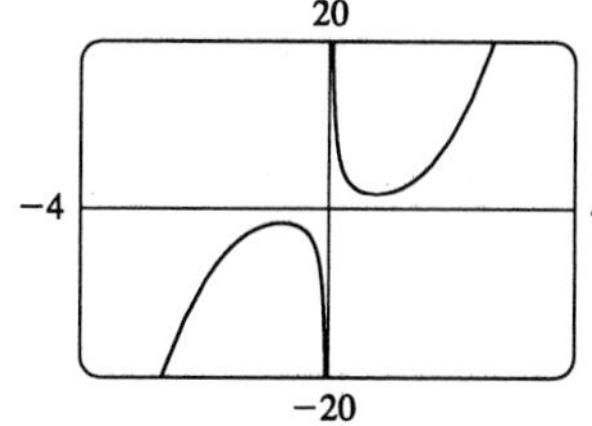

15. $y = \dfrac{2x - 1}{x + 3}$

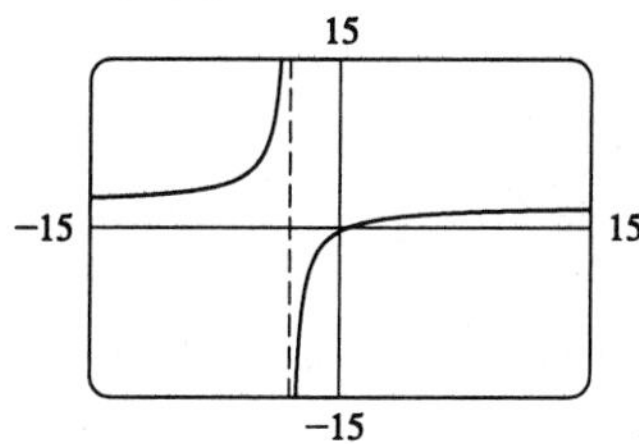

16. $y = 2x - |x^2 - 5|$

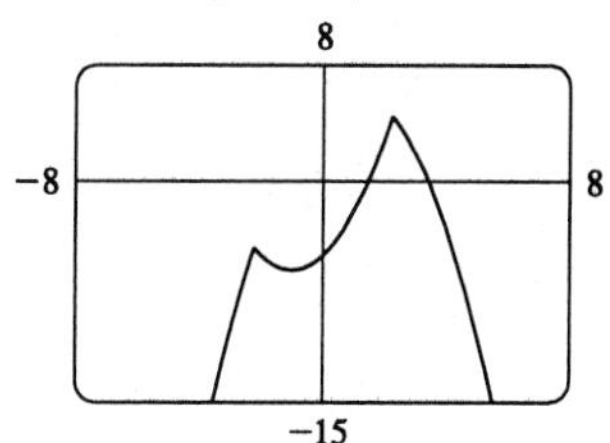

17. $f(x) = \cos(100x)$

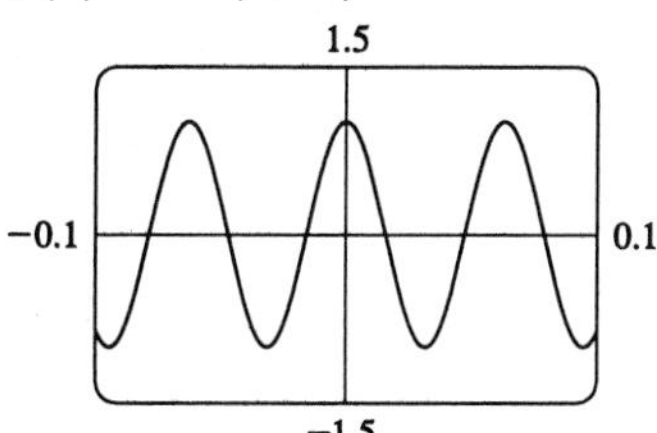

18. $f(x) = 3\sin(120x)$

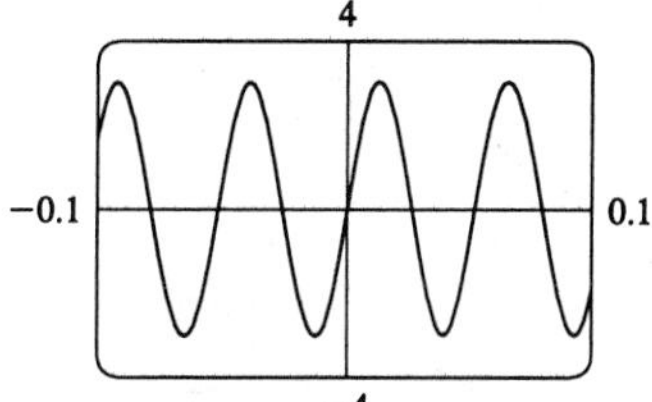

19. $f(x) = \sin(x/40)$

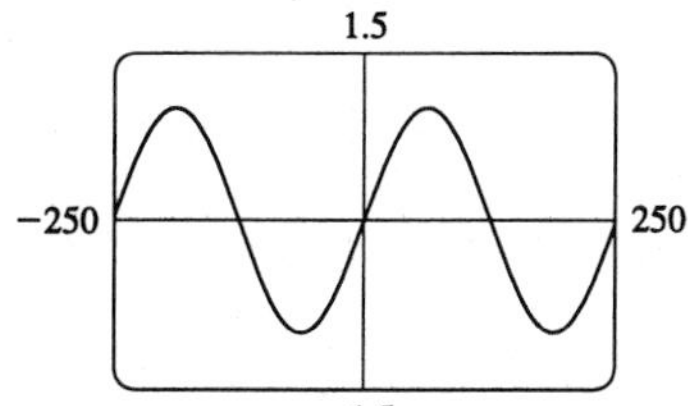

20. $f(x) = \tan(25x)$

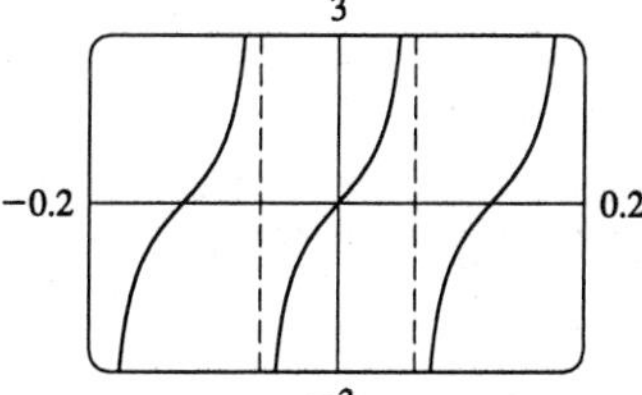

21. $y = 3^{\cos(x^2)}$

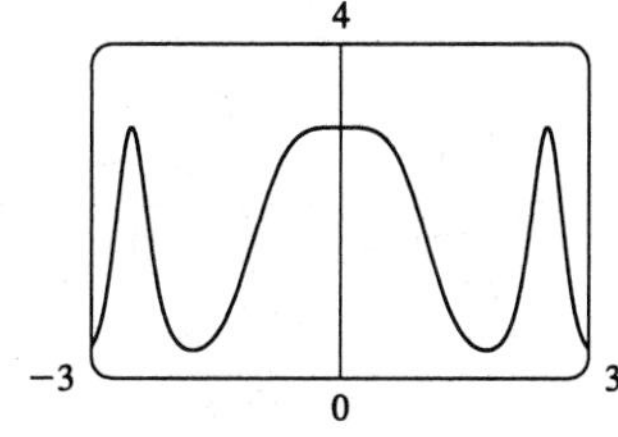

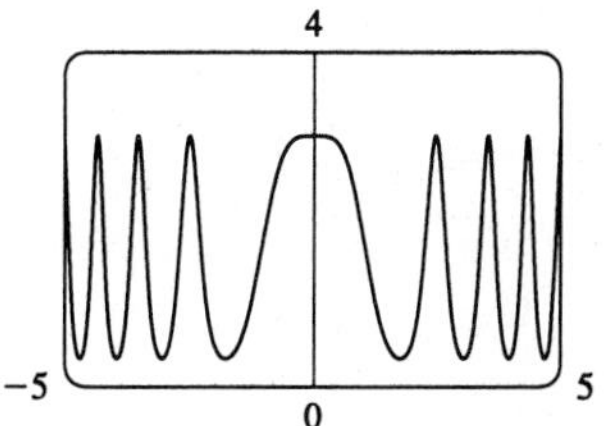

22. $y = x^2 + 0.02\sin(50x)$

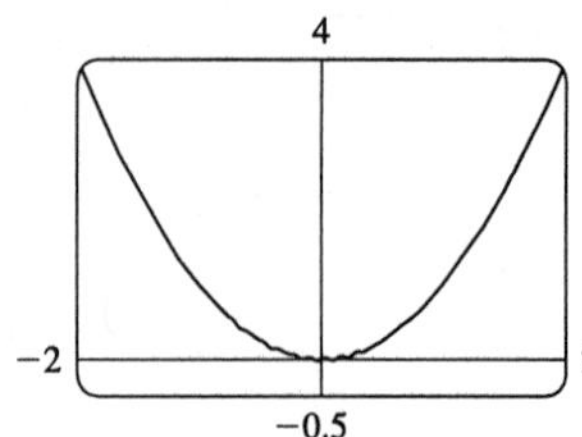

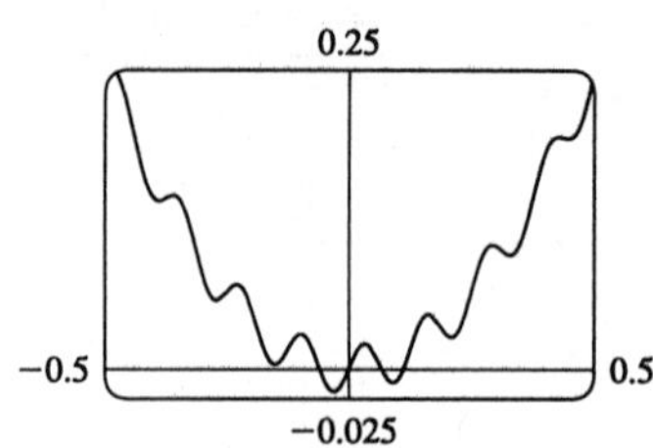

23. $y = \pm\sqrt{\dfrac{1-4x^2}{2}}$

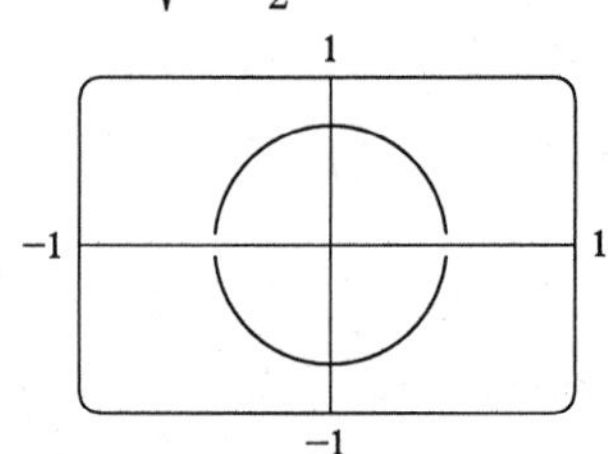

24. $y = \pm\sqrt{1+9x^2}$

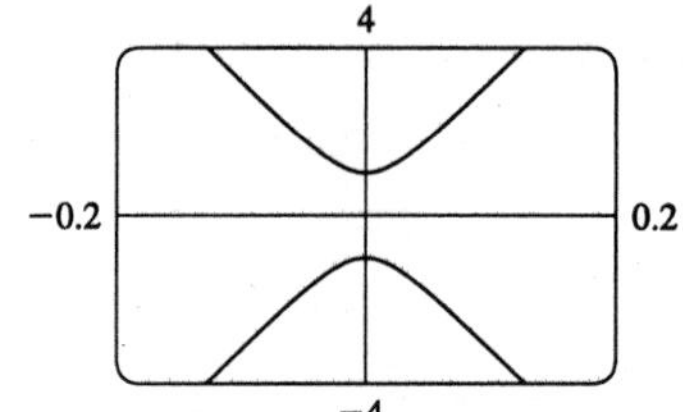

25. In Maple, we can use the procedure

```
f:=proc(x)
    if x<=1 then x^3-2*x+1
    else (x-1)^(1/3) fi
  end;
```

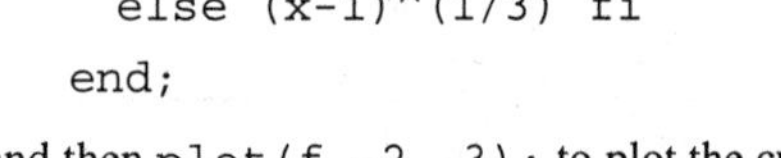

and then `plot(f,-2..3);` to plot the curve.
To define f in Mathematica, we can use
`f[x_]:=If[x<=1,x^3-2*x+1,(x-1)^(1/3)].`

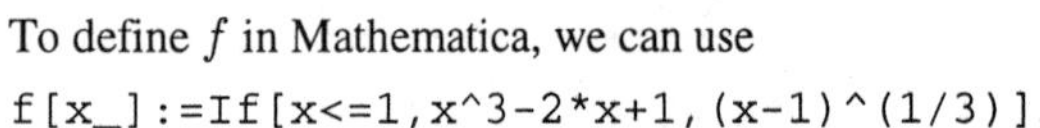

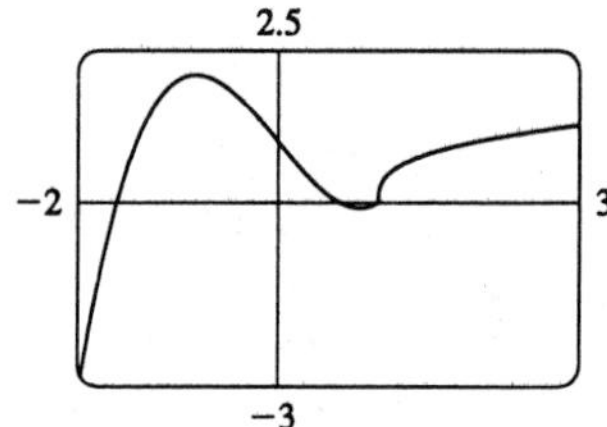

26. In Maple, we can use the procedure

```
f:=proc(x)
    if x<0 then sin(x)
    elif 0<x and x<2 then (2*x-x^2)^(1/3)
    else (x-2) fi
  end;
```

and then `plot(f,-2..4);` to plot the curve.

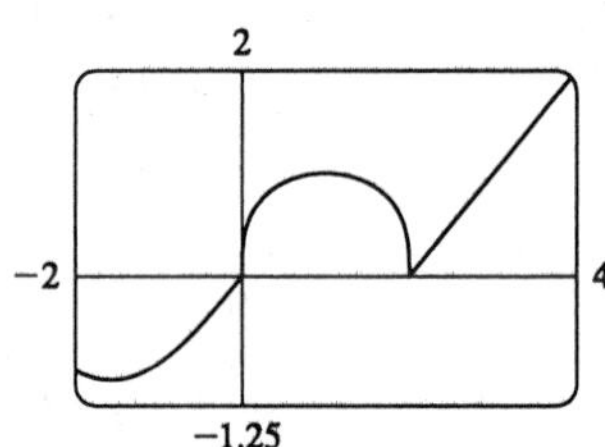

In Mathematica, we can define f with the command
`f[x_]:=Which[x<0,Sin[x],x>2,x-1,True,(2*x-x^2)^(1/3)].` If we use a TI calculator, we use the equation `y=(x<0)(sin(x))+(x>0 and x<2)((2*x-x^2)^(1/3)+(x>2)(x-2).`

27. We first graph $f(x) = 3x^3 + x^2 + x - 2$ in the viewing rectangle $[-2, 2] \times [-30, 30]$ to find the approximate value of the root. The only root appears to be between $x = 0.5$ and $x = 1$, so we graph f again in the rectangle $[0.5, 1] \times [-1, 1]$ (or use the cursor on a graphing calculator). From the second graph, it appears that the only solution to the equation $f(x) = 0$ is 0.67, to 2 decimal places.

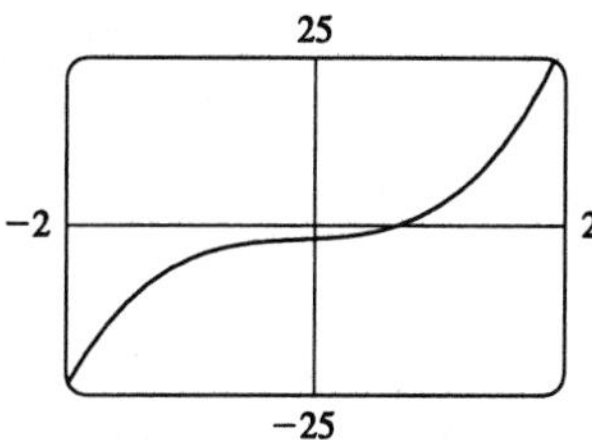

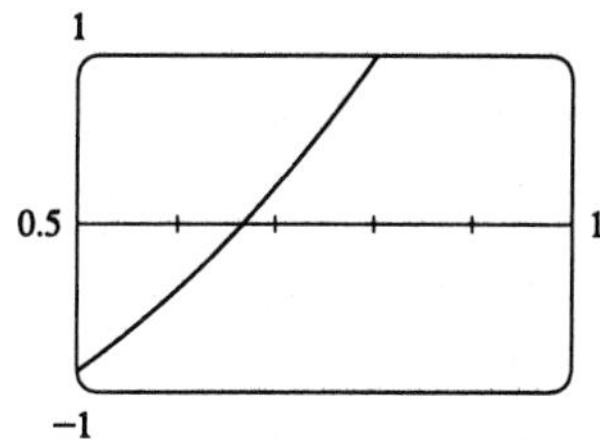

28. We graph both $f(x) = x^4 + 8x + 16$ and $g(x) = 2x^3 + 8x^2$, and zoom in on the roots, or use the cursor. From the graphs, it appears that to two decimal places, the solutions are -2, -1.24, 2 and 3.24.

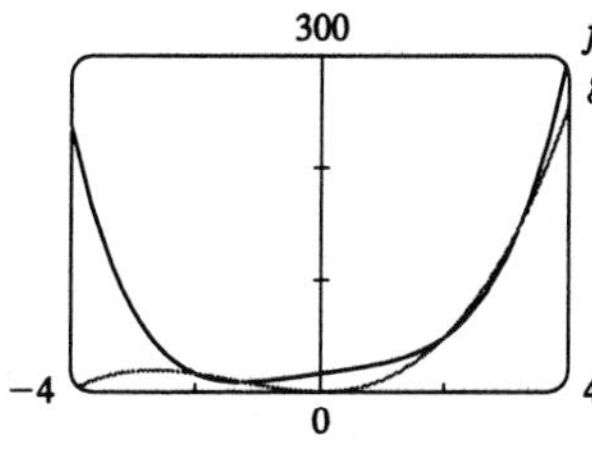

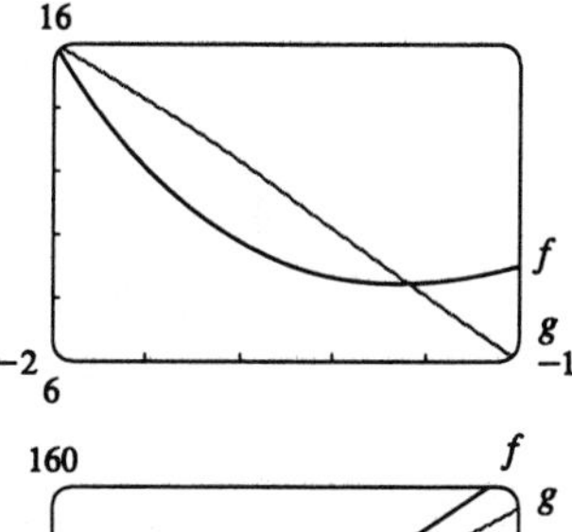

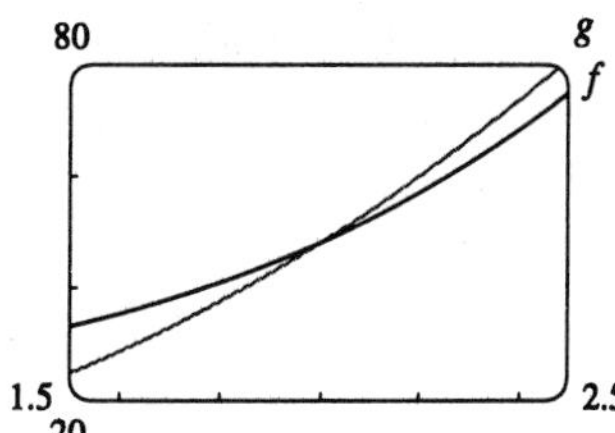

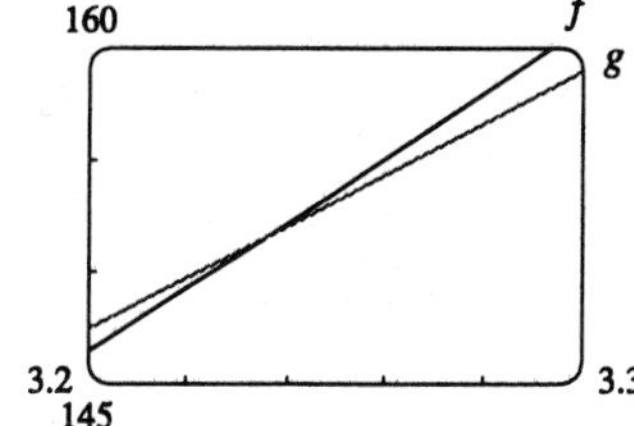

29. From the first graph, it appears that the roots lie near ± 2. We zoom in (or use the cursor) and find that the solutions are -1.90, 0 and 1.90, to two decimal places.

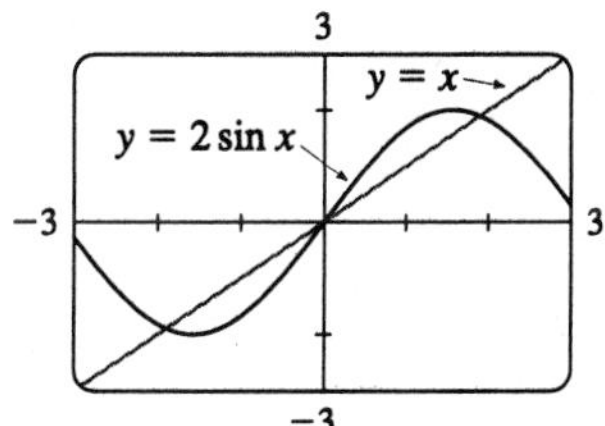

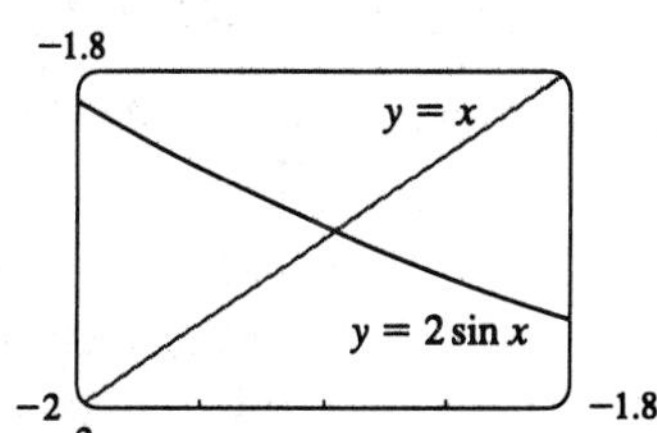

30. It appears that to 2 decimal places, the solutions are 0, 1.11 and 3.70.

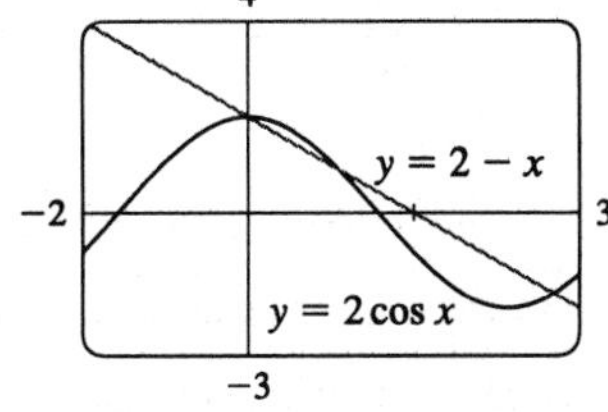

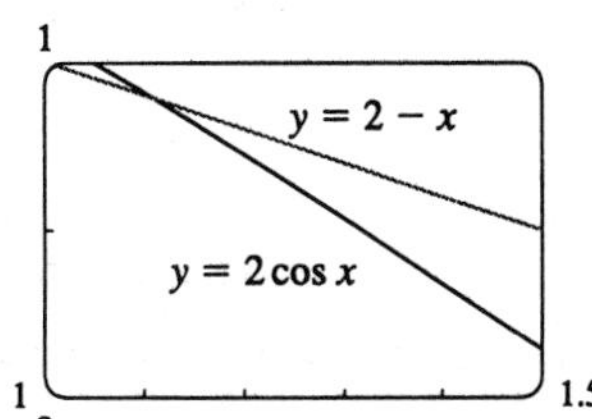

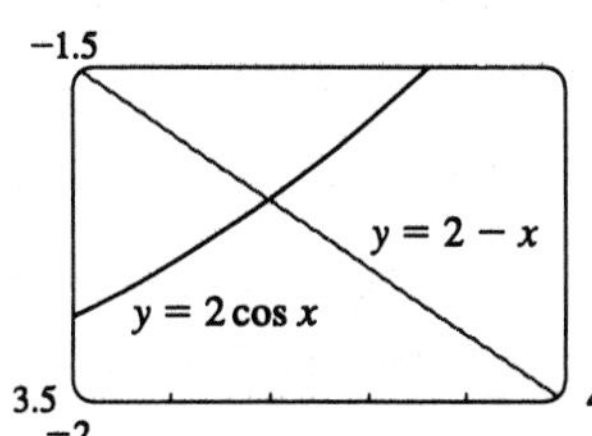

31. $g(x) = x^3/10$ is eventually larger than $f(x) = 10x^2$.

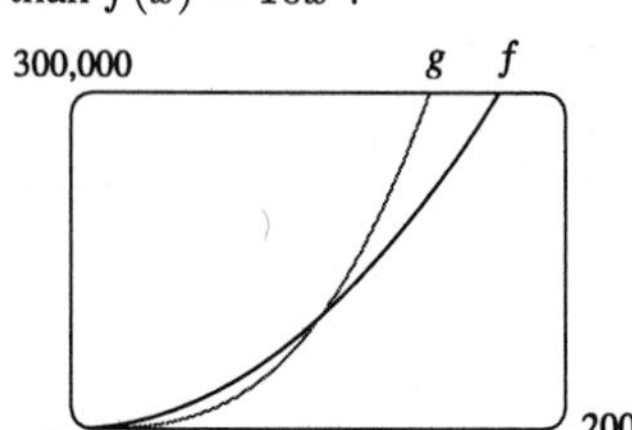

32. $f(x) = x^4 - 100x^3$ is eventually larger than $g(x) = x^3$.

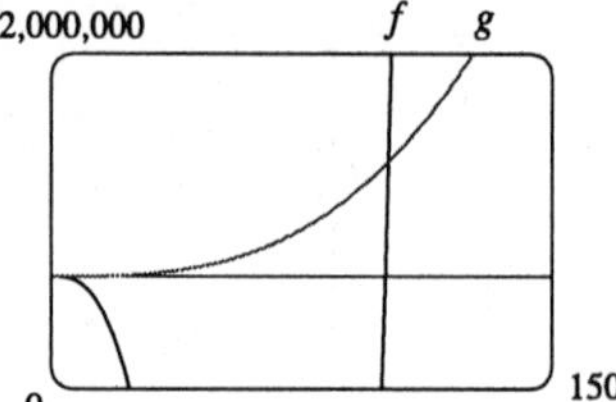

33. **(a)** **(i)** $[0, 5]$ by $[0, 20]$ **(ii)** $[0, 25]$ by $[0, 10^7]$ **(iii)** $[0, 50]$ by $[0, 10^8]$

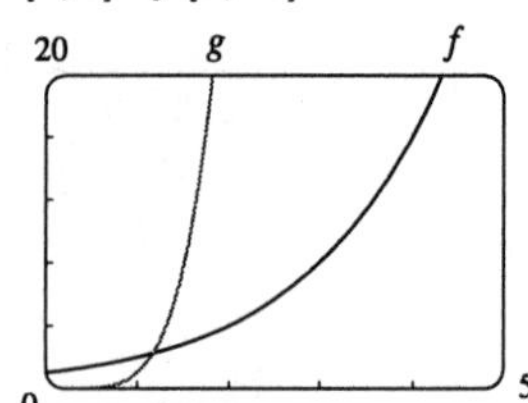

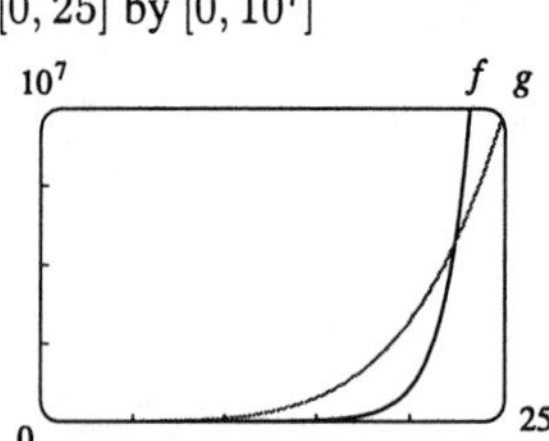

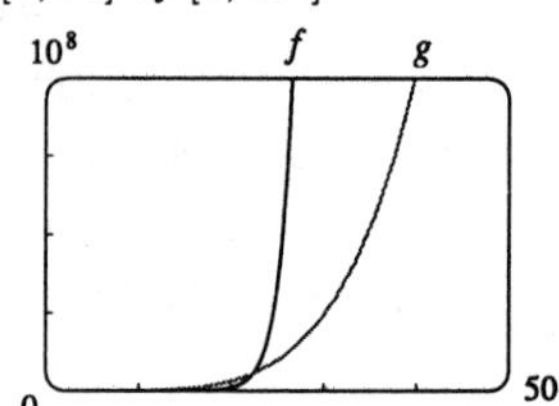

As x gets large, f grows much more quickly than g.

(b) From the graphs in part (a), it appears that the two solutions are $x \approx 1.2$ and 22.4.

34. **(a)** **(i)** $[-4, 4]$ by $[0, 20]$ **(ii)** $[0, 10]$ by $[0, 5000]$ **(iii)** $[0, 20]$ by $[0, 10^5]$

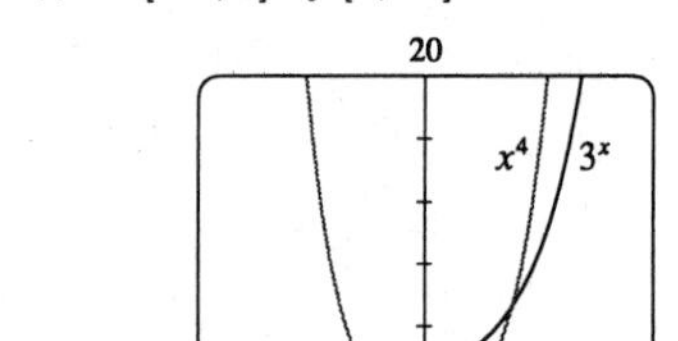

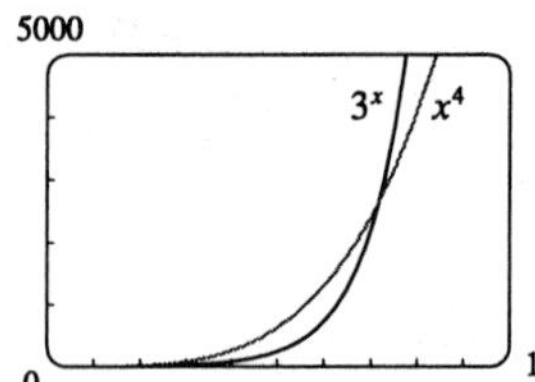

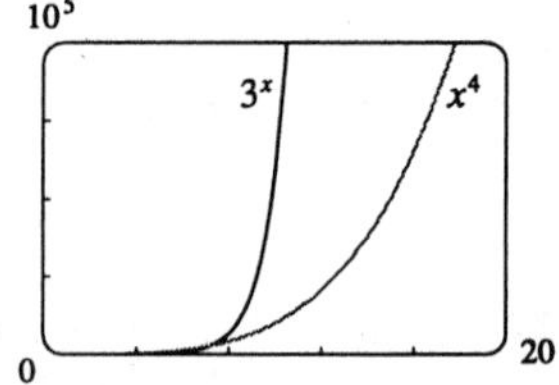

(b) From the graphs in (a), it appears that the three solutions are $x \approx -0.80$, 1.52 and 7.17.

35. We see from the graph of $y = |\sin x - x|$ that there are two solutions to the equation $|\sin x - x| = 0.1$: $x \approx -0.85$ and $x \approx 0.85$. The condition $|\sin x - x| < 0.1$ holds for any x lying between these two values.

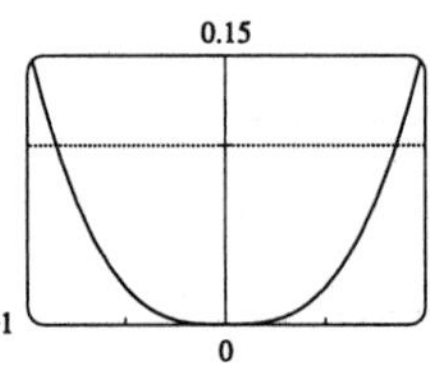

36.

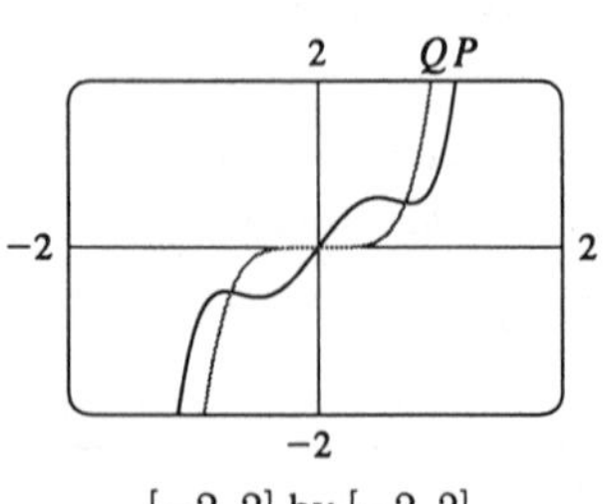

$[-2, 2]$ by $[-2, 2]$

10,000
−10
10
−10,000

$[-10, 10]$ by $[-10{,}000, 10{,}000]$

$P(x) = 3x^5 - 5x^3 + 2x$, $Q(x) = 3x^5$

These graphs are significantly different only in the region close to the origin. The larger a viewing rectangle one chooses, the more similar the two graphs look.

37. **(a)** The root functions

$y = \sqrt{x}$, $y = \sqrt[4]{x}$ and $y = \sqrt[6]{x}$

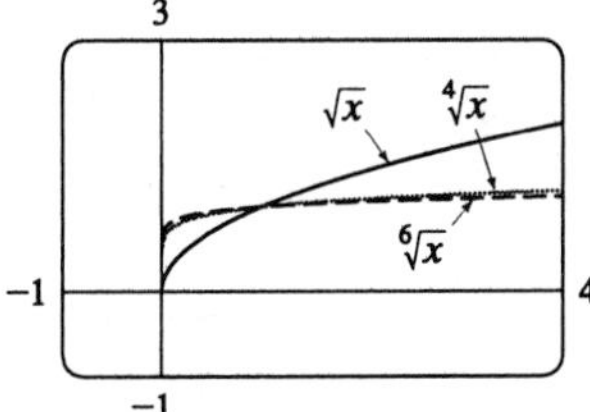

(b) The root functions

$y = x$, $y = \sqrt[3]{x}$ and $y = \sqrt[5]{x}$

(c) The root functions

$y = \sqrt{x}$, $y = \sqrt[3]{x}$, $y = \sqrt[4]{x}$ and $y = \sqrt[5]{x}$

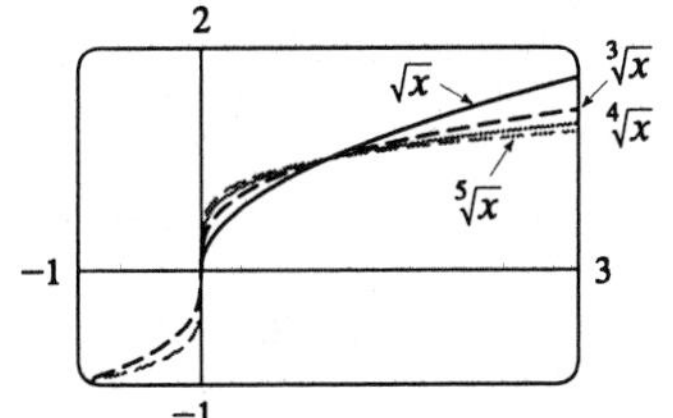

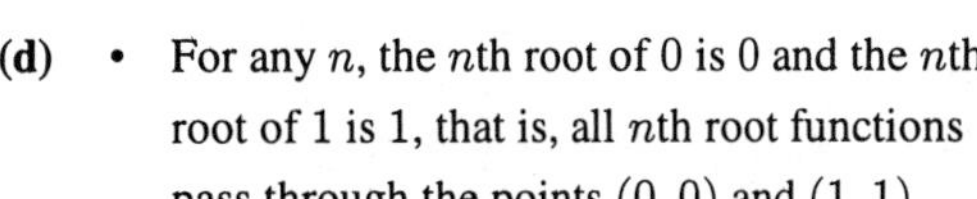

(d)
- For any n, the nth root of 0 is 0 and the nth root of 1 is 1, that is, all nth root functions pass through the points $(0, 0)$ and $(1, 1)$.
- For odd n, the domain of the nth root function is $\mathbb{R}$, while for even n, it is $\{x \in \mathbb{R} \mid x \geq 0\}$.
- Graphs of even root functions look similar to that of $\sqrt{x}$, while those of odd root functions resemble that of $\sqrt[3]{x}$.
- As n increases, the graph of $\sqrt[n]{x}$ becomes steeper near 0 and flatter for $x > 1$.

38. **(a)** The functions $y = 1/x$ and $y = 1/x^3$

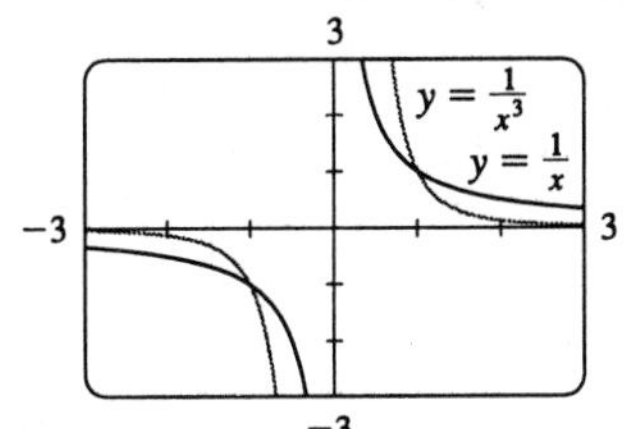

(b) The functions $y = 1/x^2$ and $y = 1/x^4$

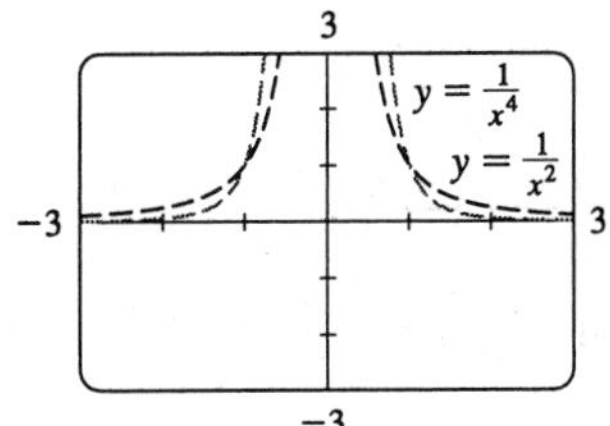

(c) The functions $y = 1/x$, $y = 1/x^2$, $y = 1/x^3$ and $y = 1/x^4$

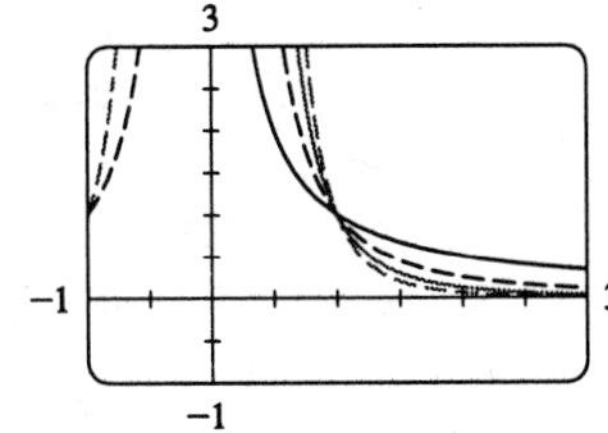

(d)
- The graphs of all functions of the form $y = 1/x^n$ pass through the point $(1, 1)$.
- If n is even, the graph of the function is entirely above the x-axis. The graphs of $1/x^n$ for n even are similar to one another.
- If n is odd, the function is positive for positive x and negative for negative x. The graphs of $1/x^n$ for n odd are similar to one another.
- As n increases, the graphs of $1/x^n$ approach 0 faster as $x \to \infty$.

39. $f(x) = x^4 + cx^2 + x$
If $c < 0$, there are three humps: two minimum points and a maximum point. These humps get flatter as c increases, until at $c = 0$ two of the humps disappear and there is only one minimum point. This single hump then moves to the right and approaches the origin as c increases.

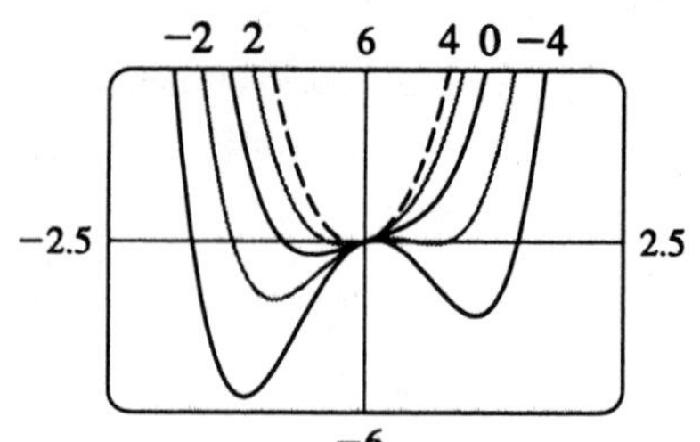

40. $f(x) = \sqrt{1 + cx^2}$
If $c < 0$, the function is only defined on $\left[-1/\sqrt{-c}, 1/\sqrt{-c}\right]$, and its graph is the top half of an ellipse. If $c = 0$, the graph is the line $y = 1$. If $c > 0$, the graph is the top half of a hyperbola. As c approaches 0, these curves become flatter and approach the line $y = 1$.

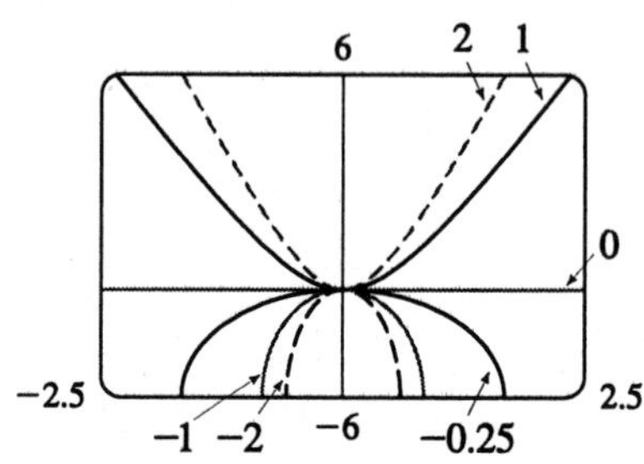

41. $y = x^n 2^{-x}$
As n increases, the maximum of the function moves further from the origin, and gets larger. Note, however, that regardless of n, the function approaches 0 as $x \to \infty$.

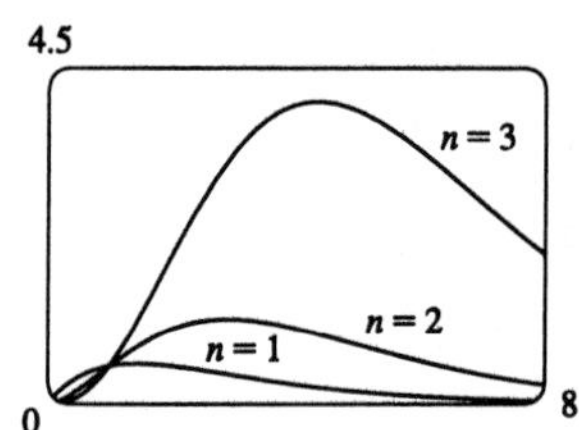

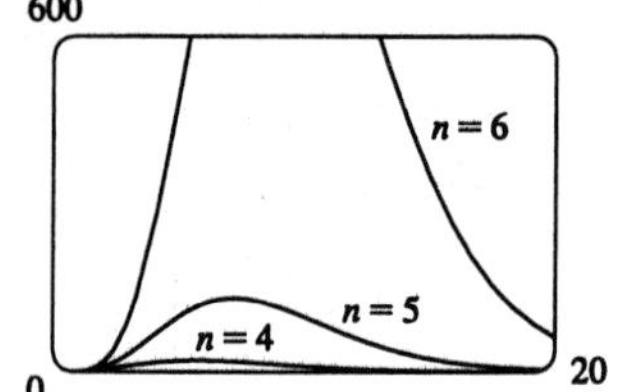

42. $y = \dfrac{|x|}{\sqrt{c - x^2}}$. The "bullet" becomes broader as c increases.

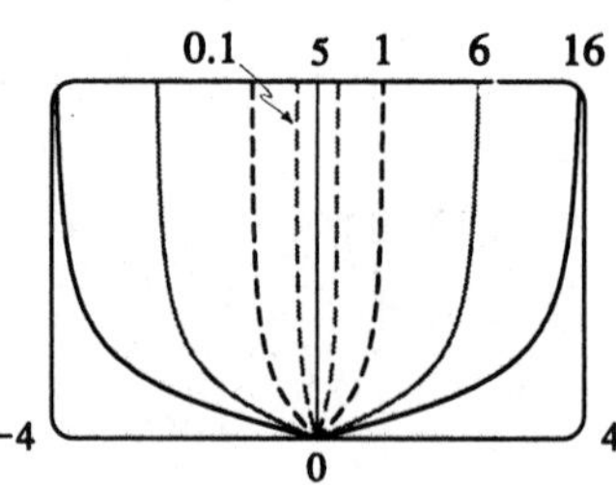

43. $y^2 = cx^3 + x^2$
If $c < 0$, the loop is to the right of the origin, and if c is positive, it is to the left. In both cases, the closer c is to 0, the larger the loop is. (In the limiting case, $c = 0$, the loop is "infinite," that is, it doesn't close.) Also, the larger $|c|$ is, the steeper the slope is on the loopless side of the origin.

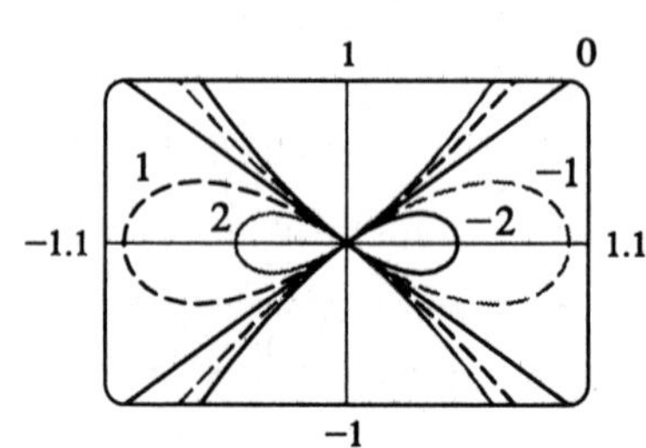

44. **(a)** $y = \sin(\sqrt{x})$

This function is not periodic; it oscillates less frequently as x increases.

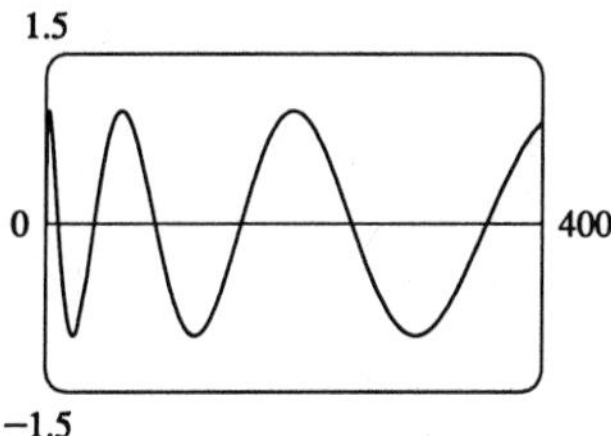

(b) $y = \sin(x^2)$

This function oscillates more frequently as $|x|$ increases. Note also that this function is even, whereas $\sin x$ is odd.

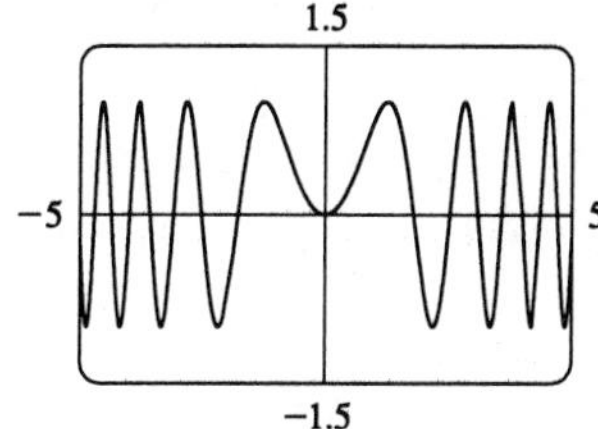

EXERCISES 4

1. As in Example 2, $|x+1| = \begin{cases} -x-1 & \text{if } x < -1 \\ x+1 & \text{if } x \geq 1 \end{cases}$ and $|x+4| = \begin{cases} -x-4 & \text{if } x < -4 \\ x+4 & \text{if } x \geq -4 \end{cases}$

Therefore we consider three cases: $x < -4$, $-4 \leq x < -1$, and $x \geq -1$.

If $x < -4$, we must have $-x-1-x-4 \leq 5 \quad \Leftrightarrow \quad x \geq -5$.

If $-4 \leq x < -1$, we must have $-x-1+x+4 \leq 5 \quad \Leftrightarrow \quad 3 \leq 5$.

If $x \geq -1$, we must have $x+1+x+4 \leq 5 \quad \Leftrightarrow \quad x \leq 0$.

These conditions together imply that $-5 \leq x \leq 0$.

2. $|x-1| = \begin{cases} 1-x & \text{if } x < 1 \\ x-1 & \text{if } x \geq 1 \end{cases}$ and $|x-3| = \begin{cases} 3-x & \text{if } x < 3 \\ x-3 & \text{if } x \geq 3 \end{cases}$

Therefore we consider the three cases $x < 1$, $1 \leq x < 3$, and $x \geq 3$.

If $x < 1$, we must have $1-x-(3-x) \geq 5 \quad \Leftrightarrow \quad 0 \geq 7$, which is false.

If $1 \leq x < 3$, we must have $x-1-(3-x) \geq 5 \quad \Leftrightarrow \quad x \geq \frac{9}{2}$, which is false because $x < 3$.

If $x \geq 3$, we must have $x-1-(x-3) \geq 5 \quad \Leftrightarrow \quad 2 \geq 5$, which is false.

All three cases lead to falsehoods, so the inequality has no solution.

3. $|2x-1| = \begin{cases} 1-2x & \text{if } x < \frac{1}{2} \\ 2x-1 & \text{if } x \geq \frac{1}{2} \end{cases}$ and $|x+5| = \begin{cases} -x-5 & \text{if } x < -5 \\ x+5 & \text{if } x \geq -5 \end{cases}$

Therefore we consider the cases $x < -5$, $-5 \leq x < \frac{1}{2}$, and $x \geq \frac{1}{2}$.

If $x < -5$, we must have $1-2x-(-x-5) = 3 \quad \Leftrightarrow \quad x = 3$, which is false, since we are considering $x < -5$.

If $-5 \leq x < \frac{1}{2}$, we must have $1-2x-(x+5) = 3 \quad \Leftrightarrow \quad x = -\frac{7}{3}$.

If $x \geq \frac{1}{2}$, we must have $2x-1-(x+5) = 3 \quad \Leftrightarrow \quad x = 9$.

So the two solutions of the equation are $x = -\frac{7}{3}$ and $x = 9$.

4. *Case (i):* $|2x+1|+5=10 \Leftrightarrow |2x+1|=5 \Leftrightarrow 2x+1=\pm 5 \Leftrightarrow 2x=\pm 5-1 \Leftrightarrow x=2$ or $x=-3$.

Case (ii): $|2x+1|+5=-10 \Leftrightarrow |2x+1|=-15$, which has no solution since $|a|\geq 0$ for all a.

Substituting into the original equation gives

$x=-3$: $||2(-3)+1|+5| = ||-5|+5| = |5+5| = 10$ and $x=2$: $||2\cdot 2+1|+5| = |5+5| = 10$.

Therefore the solutions are $x=-3$ and $x=2$.

5. The final digit in 947^{362} is determined by 7^{362}, since $947^{362}=(900+40+7)^{362}$, and every term in the expansion of this expression is the product of powers of either 40 or 900 or both, the last digits of which are all zero, except the term 7^{362}. Looking at the first few powers of 7 we see: $7^1=7$, $7^2=49$, $7^3=343$, $7^4=2401$, $7^5=16{,}807$, $7^6=117{,}649$ and it appears that the final digit follows a cyclical pattern, namely $7\to 9\to 3\to 1\to 7\to 9$, of length 4. Since $362\div 4=90$ with remainder 2, the final digit is the second in the cycle, that is, 9.

6. $8^{15}\cdot 5^{37}=(2^3)^{15}\cdot 5^{37}=2^{45}\cdot 5^{37}=2^8\cdot 2^{37}\cdot 5^{37}=2^8\cdot 10^{37}=256\times 10^{37}$. Therefore, the number has $3+37=40$ digits.

7. $f_0(x)=x^2$ and $f_{n+1}(x)=f_0(f_n(x))$ for $n=0,1,2,\ldots$.

$f_1(x)=f_0(f_0(x))=f_0(x^2)=(x^2)^2=x^4$, $f_2(x)=f_0(f_1(x))=f_0(x^4)=(x^4)^2=x^8$,

$f_3(x)=f_0(f_2(x))=f_0(x^8)=(x^8)^2=x^{16},\ldots$. Thus a general formula is $f_n(x)=x^{2^{n+1}}$.

8. $f_0(x)=\dfrac{1}{2-x}$ and $f_{n+1}=f_0\circ f_n$ for $n=0,1,2,\ldots$.

$$f_1(x)=f_0\left(\frac{1}{2-x}\right)=\frac{1}{2-\dfrac{1}{2-x}}=\frac{2-x}{2(2-x)-1}=\frac{2-x}{3-2x},$$

$$f_2(x)=f_0\left(\frac{2-x}{3-2x}\right)=\frac{1}{2-\dfrac{2-x}{3-2x}}=\frac{3-2x}{2(3-2x)-(2-x)}=\frac{3-2x}{4-3x},$$

$$f_3(x)=f_0\left(\frac{3-2x}{4-3x}\right)=\frac{1}{2-\dfrac{3-2x}{4-3x}}=\frac{4-3x}{2(4-3x)-(3-2x)}=\frac{4-3x}{5-4x},\ldots$$

Thus the general formula is $f_n(x)=\dfrac{n+1-nx}{n+2-(n+1)x}$.

So $f_{100}(x)=\dfrac{100+1-100x}{100+2-101x}=\dfrac{101-100x}{102-101x}$ and $f_{100}(3)=\dfrac{101-100\cdot 3}{102-101\cdot 3}=\dfrac{-199}{-201}=\dfrac{199}{201}$.

9. $f(x)=|x^2-4|x|+3|$. If $x\geq 0$, then $f(x)=|x^2-4x+3|=|(x-1)(x-3)|$.

Case (i): If $0<x\leq 3$, then $f(x)=x^2-4x+3$.

Case (ii): If $1<x\leq 3$, then $f(x)=-(x^2+4x+3)=-x^2-4x-3$.

Case (iii): If $x>3$ then $f(x)=x^2-4x+3$.

This enables us to sketch the graph for $x \geq 0$. Then we use the fact that f is an even function to reflect this part of the graph about the y-axis to obtain the entire graph. Or, we could consider also the cases $x < -3$, $-3 \leq x < -1$, and $-1 \leq x < 0$.

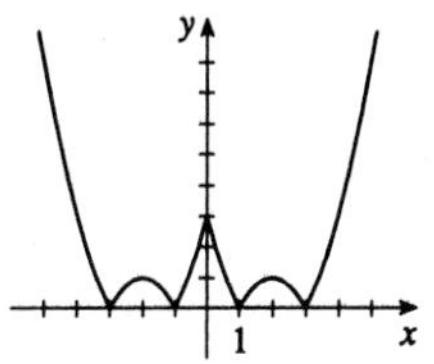

10. $|x^2 - 1| = \begin{cases} 1 - x^2 & \text{if } |x| < 1 \\ x^2 - 1 & \text{if } |x| \geq 1 \end{cases}$ and

$|x^2 - 4| = \begin{cases} 4 - x^2 & \text{if } |x| < 2 \\ x^2 - 4 & \text{if } |x| \geq 2 \end{cases}$

So for $0 \leq |x| < 1, g(x) = 1 - x^2 - (4 - x^2) = -3$,
for $1 \leq |x| < 2, g(x) = x^2 - 1 - (4 - x^2) = 2\,x^2 - 5$,
and for $|x| \geq 2, g(x) = x^2 - 1 - (x^2 - 4) = 3$.

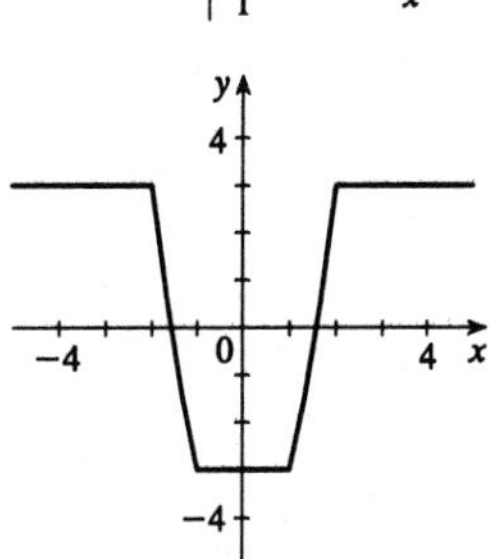

11. $\left[\sqrt{3 + 2\sqrt{2}} - \sqrt{3 - 2\sqrt{2}}\right]^2 = \left(3 + 2\sqrt{2}\right) + \left(3 - 2\sqrt{2}\right) - 2\sqrt{\left(3 + 2\sqrt{2}\right)\left(3 - 2\sqrt{2}\right)}$

$= 6 - 2\sqrt{9 + 6\sqrt{2} - 6\sqrt{2} - 4\sqrt{2}\sqrt{2}} = 6 - 2 = 4$. So the given expression is $\sqrt{4} = 2$.

12. $\dfrac{\sqrt{2} + \sqrt{6}}{\sqrt{2 + \sqrt{3}}} = \sqrt{\dfrac{\left(\sqrt{2} + \sqrt{6}\right)^2}{2 + \sqrt{3}}} = \sqrt{\dfrac{2 + 2\sqrt{12} + 6}{2 + \sqrt{3}}} = \sqrt{\dfrac{8 + 4\sqrt{3}}{2 + \sqrt{3}}} = 2.$

13. 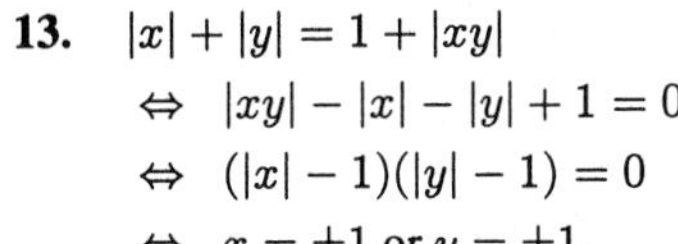
$|x| + |y| = 1 + |xy|$
$\Leftrightarrow \; |xy| - |x| - |y| + 1 = 0$
$\Leftrightarrow \; (|x| - 1)(|y| - 1) = 0$
$\Leftrightarrow \; x = \pm 1$ or $y = \pm 1$.

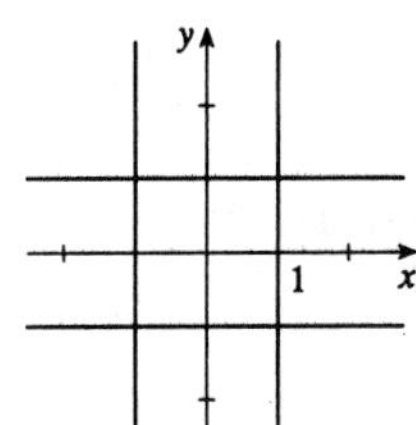

14. 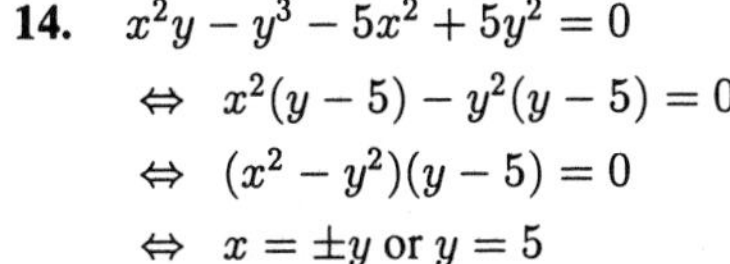
$x^2 y - y^3 - 5x^2 + 5y^2 = 0$
$\Leftrightarrow \; x^2(y - 5) - y^2(y - 5) = 0$
$\Leftrightarrow \; (x^2 - y^2)(y - 5) = 0$
$\Leftrightarrow \; x = \pm y$ or $y = 5$

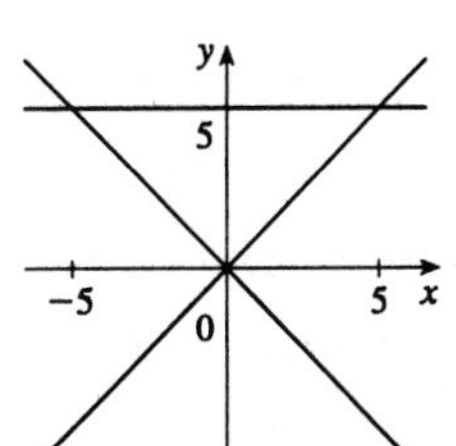

15. $|x| + |y| \leq 1$. The boundary of the region has equation $|x| + |y| = 1$. In quadrants I, II, III, IV, this becomes the lines $x + y = 1$, $-x + y = 1$, $-x - y = 1$, and $x - y = 1$ respectively.

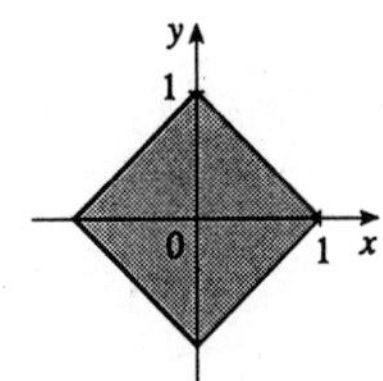

16. $|x-y|+|x|-|y| \le 2$

Case (i):	$x > y > 0$	$\Leftrightarrow$	$x-y+x-y \le 2$	$\Leftrightarrow$	$x-y \le 1$	$\Leftrightarrow$	$y \ge x-1$
Case (ii):	$y > x > 0$	$\Leftrightarrow$	$y-x+x-y \le 2$	$\Leftrightarrow$	$y > x > 0$	$\Leftrightarrow$	$0 \le 2$
Case (iii):	$x > 0$ and $y < 0$	$\Leftrightarrow$	$x-y+x+y \le 2$	$\Leftrightarrow$	$2x \le 2$	$\Leftrightarrow$	$x \le 1$
Case (iv):	$x < 0$ and $y > 0$	$\Leftrightarrow$	$y-x-x-y \le 2$	$\Leftrightarrow$	$-2x \le 2$	$\Leftrightarrow$	$x \ge -1$
Case (v):	$y < x < 0$	$\Leftrightarrow$	$x-y-x+y \le 2$	$\Leftrightarrow$	$0 \le 2$	$\Leftrightarrow$	$y < x < 0$
Case (vi):	$x < y < 0$	$\Leftrightarrow$	$y-x-x+y \le 2$	$\Leftrightarrow$	$y-x \le 1$	$\Leftrightarrow$	$y \le x+1$

Note: Instead of considering cases (iv), (v), and (vi), we could have noted that the region is unchanged if x and y are replaced by $-x$ and $-y$, so the region is symmetric about the origin. Therefore we need only draw cases (i), (ii), and (iii), and rotate through 180° about the origin.

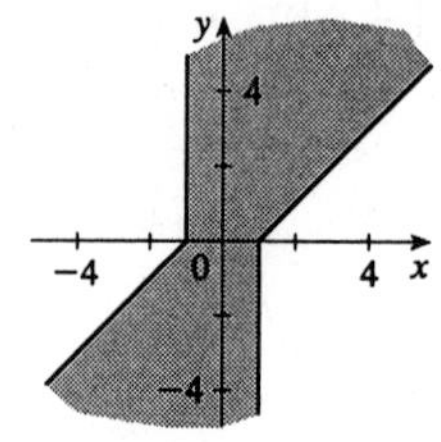

17. **(a)** The amount of ribbon needed is equal to the circumference of the earth, which is about $2\pi r = 2\pi(3960) \approx 24{,}880$ mi.

(b) The additional ribbon needed is $2\pi(r+1) - 2\pi(r) = 2\pi(1) \approx 6.3$ ft. (!)

18. Each second, George runs $\frac{1}{50}$th of a lap and Sue runs $\frac{1}{30}$th. We want to find the time at which Sue has run exactly one lap further than George. Let t be the elapsed time in seconds. Then, using the formula distance = velocity × time (where distance is measured in laps), and setting Sue's distance equal to George's distance plus one lap, we get $\frac{1}{30}t = \frac{1}{50}t + 1 \quad \Leftrightarrow \quad t = 75$ s.

19. Let x represent the length of the base in cm and h the length of the altitude in cm. By the Pythagorean Theorem, $3^2 + x^2 = 5^2 \quad \Leftrightarrow \quad x^2 = 16 \quad \Leftrightarrow \quad x = 4$ and so the area of the triangle is $\frac{1}{2} \cdot 4 \cdot 3 = 6$. But the area of the triangle is also $\frac{1}{2} \cdot 5 \cdot h = 6 \quad \Leftrightarrow \quad h = 2.4$ and hence the length of the altitude is 2.4 cm.

20. Let the lengths of the three sides in cm be a, b, and c. The area of the triangle is $\frac{1}{2}ab$ and is also $\frac{1}{2}c \cdot 12$, so $\frac{1}{2}ab = 6c \quad \Leftrightarrow \quad ab = 12c$. Since the perimeter of the triangle is 60, $a+b+c = 60 \quad \Leftrightarrow$ $(a+b)^2 = (60-c)^2$, and also from the Pythagorean Theorem, $a^2+b^2 = c^2 \quad \Leftrightarrow$ $(a+b)^2 = c^2 + 2ab = c^2 + 24c$. Since the left-hand sides are equivalent, we can equate the right-hand sides and get $(60-c)^2 = c^2 + 24c \quad \Leftrightarrow \quad 3600 - 120c + c^2 = c^2 + 24c \quad \Leftrightarrow \quad 144c = 3600 \quad \Leftrightarrow \quad c = 25$. Substituting back into our perimeter equation gives $a + b + 25 = 60 \quad \Leftrightarrow \quad a + b = 35 \quad \Leftrightarrow \quad a = 35 - b$, and from the area of the triangle, $ab = (35-b)b = 12 \cdot 25 = 300 \quad \Leftrightarrow \quad b^2 - 35b + 300 = 0 \quad \Leftrightarrow$ $b = 15$ or $b = 20$, and hence $a = 35 - b = 20$ or 15.

Therefore the lengths of the 3 sides of the triangle are 15, 20, and 25 cm.

21. We use a proof by contradiction. Assume that $\sqrt{3}$ is rational. So $\sqrt{3} = p/q$ for some integers p and q. This fraction is assumed to be in lowest form (that is, p and q have no common factors). Then $3q^2 = p^2$. We see that

3 must be one of the prime factors of p^2, and thus of p itself. So $p = 3k$ for some integer k. Substituting this into the previous equation, we get $3q^2 = (3k)^2 = 9k^2 \quad \Leftrightarrow \quad q^2 = 3k^2$. So 3 is one of the prime factors of q^2, and thus of q itself. So 3 divides both p and q. But this contradicts our assumption that the fraction p/q was in lowest form So $\sqrt{3}$ is not expressible as a ratio of integers, that is, $\sqrt{3}$ is irrational.

22. Let x and y be the two numbers. Then $x + y = 4$ and $xy = 1$.

$$x^3 + y^3 = (x+y)(x^2 - xy + y^2) = (x+y)(x^2 + y^2 - xy) = (x+y)\left[(x+y)^2 - 2xy - xy\right]$$
$$= (x+y)\left[(x+y)^2 - 3xy\right] = 4(4^2 - 3 \cdot 1) = 4 \cdot 13 = 52$$

23. The statement is false. Here is one particular counterexample:

	First Half	Second Half	Whole Season
Player A	1/99	1/1	2/100 = .020
Player B	0/1	98/99	98/100 = .980

24. Proof by contradiction: Suppose the assertion is false that is, everybody knows a different number of people. Then we can order the people so that P_1 knows 1 person (self), P_2 knows 2 people, …, and P_n knows n people. But then P_n knows everyone. So P_n knows P_1 and P_1 knows P_n, and therefore P_1 knows 2 people. This is a contradiction, so the given assertion must be true.

25. The odometer reading is proportional to the number of tire revolutions. Let r_1 represent the number of revolutions made by the tire on the 400 mi trip and r_2 represent the number of revolutions made by the tire on the 390 mi trip. Then $r_1 = 400k$ and $r_2 = 390k$ for some constant k. Let d be the actual distance traveled, R_1 be the radius of normal tires, and R_2 be the radius of snow tires. Then $d = 2\pi R_1 r_1 = 2\pi R_2 r_2 \Leftrightarrow R_1 r_1 = R_2 r_2$ $\Leftrightarrow \quad R_2 = \dfrac{R_1 r_1}{r_2} = \dfrac{15 \cdot 400k}{390k} = \dfrac{15 \cdot 40}{39} \approx 15.38$. So the radius of the snow tires is about 15.4 in.

26. Let s represent the volume of a spoonful of liquid and c the volume of a cup of liquid. After moving a spoonful of cream into the cup of coffee there is a volume $s + c$ of mixture having proportions $\dfrac{s}{s+c}$ of cream and $\dfrac{c}{s+c}$ of coffee. Then removing a spoonful of mixture removes $s \cdot \dfrac{s}{s+c}$ of cream and $s \cdot \dfrac{c}{s+c}$ of coffee, putting $\dfrac{sc}{s+c}$ coffee in the cream. Of the original volume (s) of cream placed in the coffee, there is now $s - \dfrac{s \cdot s}{s+c} = \dfrac{s(s+c) - s \cdot s}{s+c} = \dfrac{sc}{s+c}$ left in the coffee. Therefore the amount of coffee in the cream is the same as the amount of cream in the coffee.

Alternately, look at the drawing of the spoonful of coffee and cream mixture being returned to the cup of cream. Suppose it is possible to separate the cream from the coffee as shown. Then you can see that the coffee going into the cream occupies the same space as the cream that was left in the coffee.

CHAPTER ONE

EXERCISES 1.1

1. **(a)** Slopes of the secant lines:

x	m_{PQ}
0	$\dfrac{2.6-1.3}{0-3} \approx -0.43$
1	$\dfrac{2.0-1.3}{1-3} = -0.35$
2	$\dfrac{1.1-1.3}{2-3} = 0.2$
4	$\dfrac{2.1-1.3}{4-3} = 0.8$
5	$\dfrac{3.5-1.3}{5-3} = 1.1$

(b) The slope of the tangent line at P is about $\dfrac{2.5-0}{5-0.6} \approx 0.57$.

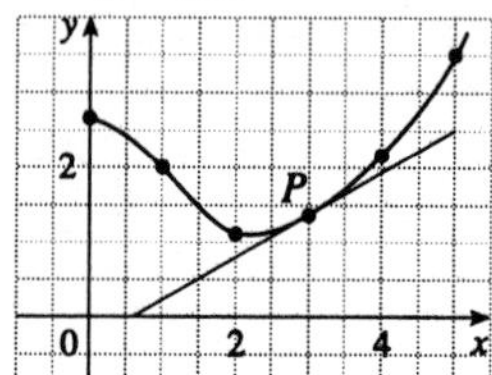

2. For the curve $f(x) = 1 + x + x^2$ and the point $P(1, 3)$:

(a)

	x	Q	m_{PQ}
(i)	2	$(2, 7)$	4
(ii)	1.5	$(1.5, 4.75)$	3.5
(iii)	1.1	$(1.1, 3.31)$	3.1
(iv)	1.01	$(1.01, 3.0301)$	3.01
(v)	1.001	$(1.001, 3.003001)$	3.001

	x	Q	m_{PQ}
(vi)	0	$(0, 1)$	2
(vii)	0.5	$(0.5, 1.75)$	2.5
(viii)	0.9	$(0.9, 2.71)$	2.9
(ix)	0.99	$(0.99, 2.9701)$	2.99
(x)	0.999	$(0.999, 2.997001)$	2.999

(b) The slope appears to be 3.

(c) $y - 3 = 3(x - 1)$ or $y = 3x$

3. For the curve $y = \sqrt{x}$ and the point $P(4, 2)$:

(a)

	x	Q	m_{PQ}
(i)	5	$(5, 2.236068)$	0.236068
(ii)	4.5	$(4.5, 2.121320)$	0.242641
(iii)	4.1	$(4.1, 2.024846)$	0.248457
(iv)	4.01	$(4.01, 2.002498)$	0.249844
(v)	4.001	$(4.001, 2.000250)$	0.249984

	x	Q	m_{PQ}
(vi)	3	$(3, 1.732051)$	0.267949
(vii)	3.5	$(3.5, 1.870829)$	0.258343
(viii)	3.9	$(3.9, 1.974842)$	0.251582
(ix)	3.99	$(3.99, 1.997498)$	0.250156
(x)	3.999	$(3.999, 1.999750)$	0.250016

(b) The slope appears to be $\frac{1}{4}$.

(c) $y - 2 = \frac{1}{4}(x - 4)$ or $x - 4y + 4 = 0$

4. For the curve $y = 1/x$ and the point $P(0.5, 2)$:

(a)

	x	Q	m_{PQ}
(i)	2	$(2, 0.5)$	-1
(ii)	1	$(1, 1)$	-2
(iii)	0.9	$(0.9, 1.111111)$	-2.222222
(iv)	0.8	$(0.8, 1.25)$	-2.5
(v)	0.7	$(0.7, 1.428571)$	-2.857143

	x	Q	m_{PQ}
(vi)	0.6	$(0.6, 1.666667)$	-3.333333
(vii)	0.55	$(0.55, 1.818182)$	-3.636364
(viii)	0.51	$(0.51, 1.960784)$	-3.921569
(ix)	0.45	$(0.45, 2.222222)$	-4.444444
(x)	0.49	$(0.49, 2.040816)$	-4.081633

(b) The slope appears to be -4.

(c) $y - 2 = -4(x - 0.5)$ or $4x + y = 4$

(d)

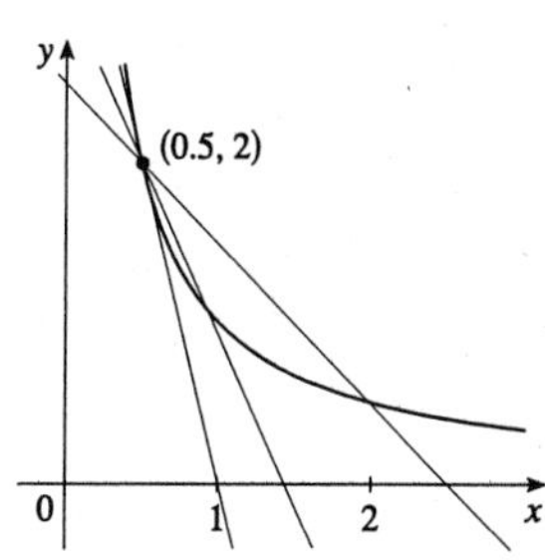

5. **(a)** At $t = 2$, $y = 40(2) - 16(2)^2 = 16$. The average velocity between times 2 and $2 + h$ is

$$\frac{40(2+h) - 16(2+h)^2 - 16}{h} = \frac{-24h - 16h^2}{h} = -24 - 16h, \text{ if } h \neq 0.$$

(i) $h = 0.5$, -32 ft/s **(ii)** $h = 0.1$, -25.6 ft/s

(iii) $h = 0.05$, -24.8 ft/s **(iv)** $h = 0.01$, -24.16 ft/s

(b) The instantaneous velocity when $t = 2$ is -24 ft/s.

6. The average velocity between t and $t + h$ seconds is

$$\frac{58(t+h) - 0.83(t+h)^2 - (58t - 0.83t^2)}{h} = \frac{58h - 1.66th - 0.83h^2}{h} = 58 - 1.66t - 0.83h, \text{ if } h \neq 0.$$

(a) Here $t = 1$, so the average velocity is $58 - 1.66 - 0.83h = 56.34 - 0.83h$.

$[1, 2]: h = 1$ so 55.51 m/s. $[1, 1.5]: h = 0.5$ so 55.925 m/s.

$[1, 1.1]: h = 0.1$ so 56.257 m/s. $[1, 1.01]: h = 0.01$ so 56.3317 m/s.

$[1, 1.001]: h = 0.001$ so 56.33917 m/s.

(b) The instantaneous velocity after 1 second is 56.34 m/s.

7. Average velocity between times 1 and $1 + h$ is

$$\frac{s(1+h) - s(1)}{h} = \frac{(1+h)^3/6 - 1/6}{h} = \frac{h^3 + 3h^2 + 3h}{6h} = \frac{h^2 + 3h + 3}{6} \text{ if } h \neq 0.$$

(a) **(i)** $v_{av} = \dfrac{2^2 + 3(2) + 3}{6} = \dfrac{13}{6}$ ft/s **(ii)** $v_{av} = \dfrac{1^2 + 3(1) + 3}{6} = \dfrac{7}{6}$ ft/s

(iii) $v_{av} = \dfrac{(0.5)^2 + 3(0.5) + 3}{6} = \dfrac{19}{24}$ ft/s **(iv)** $v_{av} = \dfrac{(0.1)^2 + 3(0.1) + 3}{6} = \dfrac{331}{600}$ ft/s

(b) As h approaches 0, the velocity approaches $\frac{1}{2}$ ft/s.

(c) & (d)

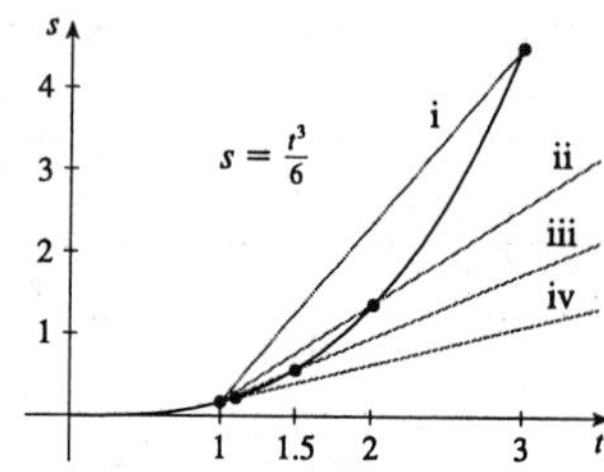

8. Average velocity between times $t = 2$ and $t = 2 + h$ is given by $\dfrac{s(2+h) - s(2)}{h}$.

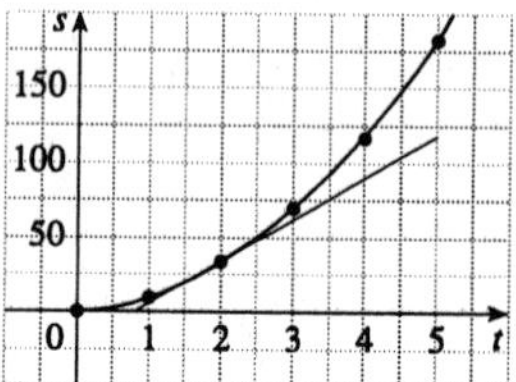

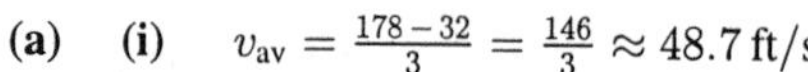

(a) **(i)** $v_{\text{av}} = \frac{178-32}{3} = \frac{146}{3} \approx 48.7 \text{ ft/s}$

(ii) $v_{\text{av}} = \frac{119-32}{2} = \frac{87}{2} = 43.5 \text{ ft/s}$

(iii) $v_{\text{av}} = \frac{70-32}{1} = 38 \text{ ft/s}$

(b) The instantaneous velocity at $t = 2$ is about $\frac{118-0}{5-0.8} \approx 28$ ft/s.

EXERCISES 1.2

1. **(a)** $\lim_{x\to 1} f(x) = 3$ **(b)** $\lim_{x\to 3^-} f(x) = 2$ **(c)** $\lim_{x\to 3^+} f(x) = -2$
(d) $\lim_{x\to 3} f(x)$ doesn't exist **(e)** $f(3) = 1$ **(f)** $\lim_{x\to -2^-} f(x) = -1$
(g) $\lim_{x\to -2^+} f(x) = -1$ **(h)** $\lim_{x\to -2} f(x) = -1$ **(i)** $f(-2) = -3$

2. **(a)** $\lim_{x\to -2^-} g(x) = -1$ **(b)** $\lim_{x\to -2^+} g(x) = 1$ **(c)** $\lim_{x\to -2} g(x)$ doesn't exist
(d) $g(-2) = 1$ **(e)** $\lim_{x\to 2^-} g(x) = 1$ **(f)** $\lim_{x\to 2^+} g(x) = 2$
(g) $\lim_{x\to 2} g(x)$ doesn't exist **(h)** $g(2) = 2$ **(i)** $\lim_{x\to 4^+} g(x)$ doesn't exist
(j) $\lim_{x\to 4^-} g(x) = 2$ **(k)** $g(0)$ doesn't exist **(l)** $\lim_{x\to 0} g(x) = 0$

3. **(a)** $\lim_{x\to 3} f(x) = 2$ **(b)** $\lim_{x\to 1} f(x) = -1$ **(c)** $\lim_{x\to -3} f(x) = 1$
(d) $\lim_{x\to 2^-} f(x) = 1$ **(e)** $\lim_{x\to 2^+} f(x) = 2$ **(f)** $\lim_{x\to 2} f(x)$ doesn't exist

4. **(a)** $\lim_{x\to 1} g(x) = 0$ **(b)** $\lim_{x\to 0} g(x)$ doesn't exist **(c)** $\lim_{x\to 2} g(x) = 1$
(d) $\lim_{x\to -2} g(x) = 0$ **(e)** $\lim_{x\to -1^-} g(x) = 1$ **(f)** $\lim_{x\to -1} g(x)$ doesn't exist

5. **(a)** $\lim_{x\to 3} f(x) = \infty$ **(b)** $\lim_{x\to 7} f(x) = -\infty$ **(c)** $\lim_{x\to -4} f(x) = -\infty$
(d) $\lim_{x\to -9^-} f(x) = \infty$ **(e)** $\lim_{x\to -9^+} f(x) = -\infty$

(f) The equations of the vertical asymptotes: $x = -9$, $x = -4$, $x = 3$, $x = 7$

6. $\lim_{t \to 12^-} f(t) = 150$ mg and $\lim_{t \to 12^+} f(t) = 300$ mg. These limits show that there is an abrupt change in the amount of drug in the patient's bloodstream at $t = 12$ h. The left-hand limit represents the amount of the drug just before the fourth injection. The right-hand limit represents the amount of the drug just after the fourth injection.

7. **(a)**

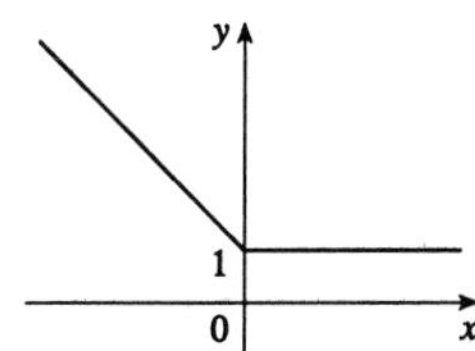

(b) **(i)** $\lim_{x \to 0^-} f(x) = 1$

(ii) $\lim_{x \to 0^+} f(x) = 1$

(iii) $\lim_{x \to 0} f(x) = 1$

8. **(a)**

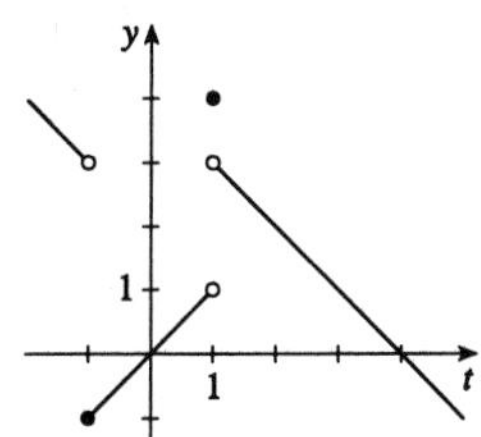

(b) **(i)** $\lim_{x \to -1^-} g(x) = 3$

(ii) $\lim_{x \to -1^+} g(x) = -1$

(iii) $\lim_{x \to -1} g(x)$ doesn't exist

(iv) $\lim_{x \to 1^-} g(x) = 1$

(v) $\lim_{x \to 1^+} g(x) = 3$

(vi) $\lim_{x \to 1} g(x)$ doesn't exist

9. For $g(x) = \dfrac{x-1}{x^3-1}$:

x	$g(x)$
0.2	0.806452
0.4	0.641026
0.6	0.510204
0.8	0.409836
0.9	0.369004
0.99	0.336689

x	$g(x)$
1.8	0.165563
1.6	0.193798
1.4	0.229358
1.2	0.274725
1.1	0.302115
1.01	0.330022

It appears that $\lim_{x \to 1} \dfrac{x-1}{x^3-1} = 0.\overline{3} = \dfrac{1}{3}$.

10. For $g(x) = \dfrac{1-x^2}{x^2+3x-10}$:

x	$g(x)$
3	-1
2.1	-4.8028
2.01	-43.368
2.001	-429.08
2.0001	-4286.2
2.00001	-42858

It appears that $\lim_{x \to 2^+} \dfrac{1-x^2}{x^2+3x-10} = -\infty$.

11. For $F(x) = \dfrac{(1/\sqrt{x}) - \frac{1}{5}}{x - 25}$:

x	$F(x)$
26	−0.003884
25.5	−0.003941
25.1	−0.003988
25.05	−0.003994
25.01	−0.003999

x	$F(x)$
24	−0.004124
24.5	−0.004061
24.9	−0.004012
24.95	−0.004006
24.99	−0.004001

It appears that $\lim\limits_{x \to 25} F(x) = -0.004$.

12. For $f(t) = \dfrac{\sqrt[3]{t} - 1}{\sqrt{t} - 1}$:

t	$f(t)$
2	0.627505
1.5	0.643905
1.1	0.661358
1.01	0.666114
1.001	0.666611

It appears that $\lim\limits_{t \to 1} \dfrac{\sqrt[3]{t} - 1}{\sqrt{t} - 1} = 0.\overline{6} = \dfrac{2}{3}$.

13. For $f(x) = \dfrac{1 - \cos x}{x^2}$:

x	$f(x)$
1	0.459698
0.5	0.489670
0.4	0.493369
0.3	0.496261
0.2	0.498336
0.1	0.499583
0.05	0.499896
0.01	0.499996

It appears that $\lim\limits_{x \to 0} \dfrac{1 - \cos x}{x^2} = 0.5$.

14. For $g(x) = \dfrac{\cos x - 1}{\sin x}$:

x	$g(x)$
1	−0.546302
0.5	−0.255342
0.4	−0.202710
0.3	−0.151135
0.2	−0.100335
0.1	−0.050042
0.05	−0.025005
0.01	−0.005000

It appears that $\lim\limits_{x \to 0} \dfrac{\cos x - 1}{\sin x} = 0$.

15. $\lim\limits_{x \to 5^+} \dfrac{6}{x - 5} = \infty$ since $(x - 5) \to 0$ as $x \to 5^+$ and $\dfrac{6}{x - 5} > 0$ for $x > 5$.

16. $\lim\limits_{x \to 5^-} \dfrac{6}{x - 5} = -\infty$ since $(x - 5) \to 0$ as $x \to 5^-$ and $\dfrac{6}{x - 5} < 0$ for $x < 5$.

17. $\lim\limits_{x \to 3} \dfrac{1}{(x - 3)^8} = \infty$ since $(x - 3) \to 0$ as $x \to 3$ and $\dfrac{1}{(x - 3)^8} > 0$.

18. $\lim\limits_{x \to 0} \dfrac{x - 1}{x^2(x + 2)} = -\infty$ since $x^2 \to 0$ as $x \to 0$ and $\dfrac{x - 1}{x^2(x + 2)} < 0$ for $0 < x < 1$ and for $-2 < x < 0$.

19. $\lim\limits_{x \to -2^+} \dfrac{x - 1}{x^2(x + 2)} = -\infty$ since $(x + 2) \to 0$ as $x \to 2^+$ and $\dfrac{x - 1}{x^2(x + 2)} < 0$ for $-2 < x < 0$.

20. $\lim\limits_{x \to \pi^-} \csc x = \lim\limits_{x \to \pi^-} (1/\sin x) = \infty$ since $\sin x \to 0$ as $x \to \pi^-$ and $\csc x > 0$ for $0 < x < \pi$.

21. **(a)**

x	$f(x)$
0.5	−1.14
0.9	−3.69
0.99	−33.7
0.999	−333.7
0.9999	−3333.7
0.99999	−33,333.7

x	$f(x)$
1.5	0.42
1.1	3.02
1.01	33.0
1.001	333.0
1.0001	3333.0
1.00001	33,333.3

From these calculations, it seems that $\lim_{x \to 1^-} f(x) = -\infty$ and $\lim_{x \to 1^+} f(x) = \infty$.

(b) If x is slightly smaller than 1, then $x^3 - 1$ will be a negative number close to 0, and the reciprocal of $x^3 - 1$, that is, $f(x)$, will be a negative number with large absolute value. So $\lim_{x \to 1^-} f(x) = -\infty$.

If x is slightly larger than 1, then $x^3 - 1$ will be a small positive number, and its reciprocal, $f(x)$, will be a large positive number. So $\lim_{x \to 1^+} f(x) = \infty$.

(c) It appears from the graph of f that $\lim_{x \to 1^-} f(x) = -\infty$ and $\lim_{x \to 1^+} f(x) = \infty$.

22. **(a)** $y = \dfrac{x}{x^2 - x - 2} = \dfrac{x}{(x-2)(x+1)}$. Therefore, as $x \to -1^+$ or $x \to 2^+$, the denominator approaches 0, and $y > 0$ for $x < -1$ and for $x > 2$, so $\lim_{x \to -1^+} y = \lim_{x \to 2^+} y = \infty$. Also, as $x \to -1^-$ or $x \to 2^-$, the denominator approaches 0 and $y < 0$ for $-1 < x < 2$, so $\lim_{x \to -1^-} y = \lim_{x \to 2^-} y = -\infty$.

(b)

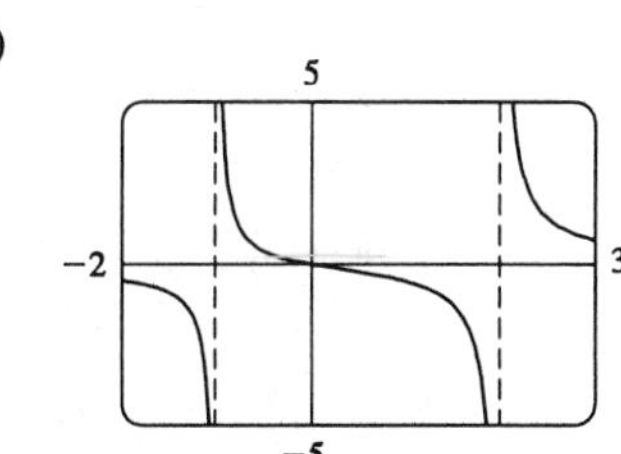

23. Let $h(x) = (1 + x)^{1/x}$.

x	$h(x)$
1.0	2.0
0.1	2.593742
0.01	2.704814
0.001	2.716924
0.0001	2.718146
0.00001	2.718268
0.000001	2.718280
0.0000001	2.718282
0.00000001	2.718282
0.000000001	2.718282

It appears that $\lim_{x \to 0} (1 + x)^{1/x} \approx 2.71828$.

24. For the curve $y = 2^x$ and the points $P(0, 1)$ and $Q(x, 2^x)$:

x	Q	m_{PQ}
0.5	$(0.5, 1.4142136)$	0.82843
0.1	$(0.1, 1.0717735)$	0.71773
0.05	$(0.05, 1.0352649)$	0.70530
0.01	$(0.01, 1.0069556)$	0.69556
0.005	$(0.005, 1.0034718)$	0.69435
0.001	$(0.001, 1.0006934)$	0.69339
0.0005	$(0.0005, 1.0003466)$	0.69327
0.0001	$(0.0001, 1.0000693)$	0.69317

The slope appears to be about 0.693.

25. For $f(x) = x^2 - (2^x/1000)$:

(a)

x	$f(x)$
1	0.998000
0.8	0.638259
0.6	0.358484
0.4	0.158680
0.2	0.038851
0.1	0.008928
0.05	0.001465

It appears that $\lim_{x \to 0} f(x) = 0$.

(b)

x	$f(x)$
0.04	0.000572
0.02	−0.000614
0.01	−0.000907
0.005	−0.000978
0.003	−0.000993
0.001	−0.001000

It appears that $\lim_{x \to 0} f(x) = -0.001$.

26. $h(x) = \dfrac{\tan x - x}{x^3}$.

(a)

x	$h(x)$
1.0	0.55740773
0.5	0.37041992
0.1	0.33467210
0.05	0.33366704
0.01	0.33335000
0.005	0.33333600

(b) It seems that $\lim_{x \to 0} h(x) = \frac{1}{3}$.

(c)

x	$h(x)$
0.001	0.33300000
0.0005	0.33360000
0.0004	0.33281250
0.0003	0.33333333
0.0002	0.33750000
0.0001	0.30000000
0.00005	0.32000000
0.00001	0.00000000
0.000001	0.00000000
0.0000001	0.00000000

Here the values will vary from one calculator to another. Every calculator will eventually give *false values*. (See Appendix G.)

27. From the following graphs, it seems that $\lim_{x \to 0} \dfrac{\tan(4x)}{x} = 4$.

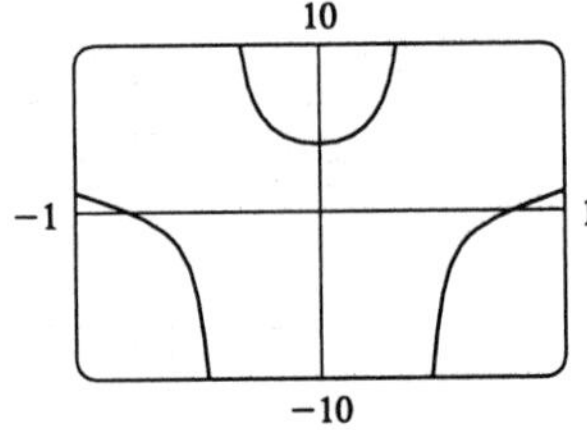

5

−0.2

0.2

3.5

28. From the following graphs, it seems that $\lim_{x \to 0} \dfrac{6^x - 2^x}{x} \approx 1.10$.

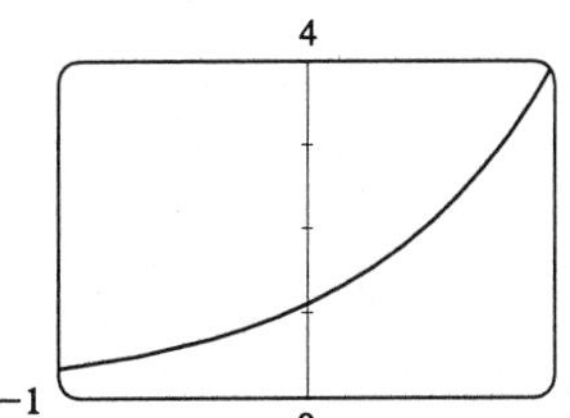

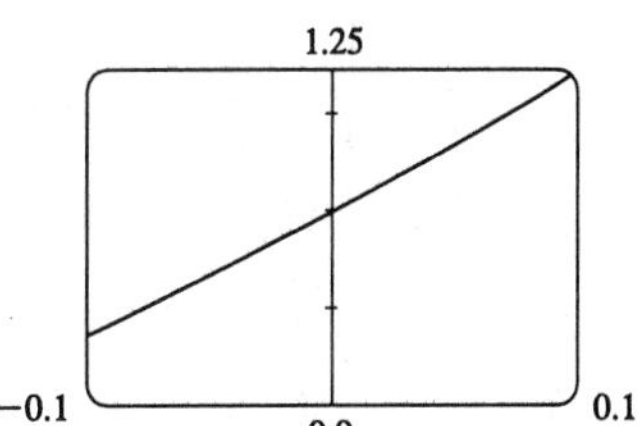

29. No matter how many times we zoom in towards the origin, the graphs appear to consist of almost-vertical lines. This indicates more and more frequent oscillations as $x \to 0$.

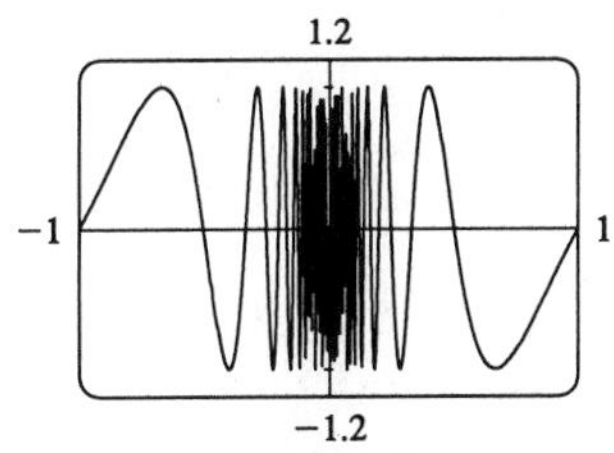

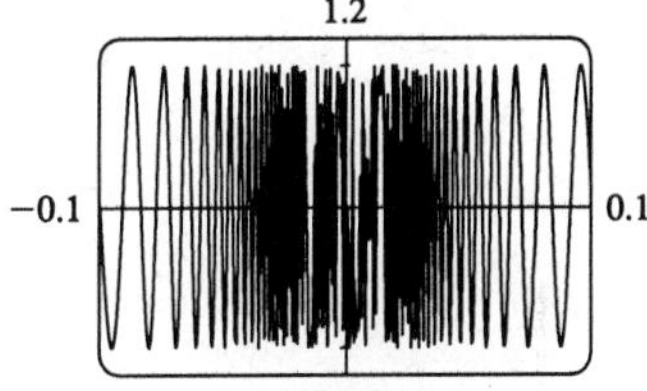

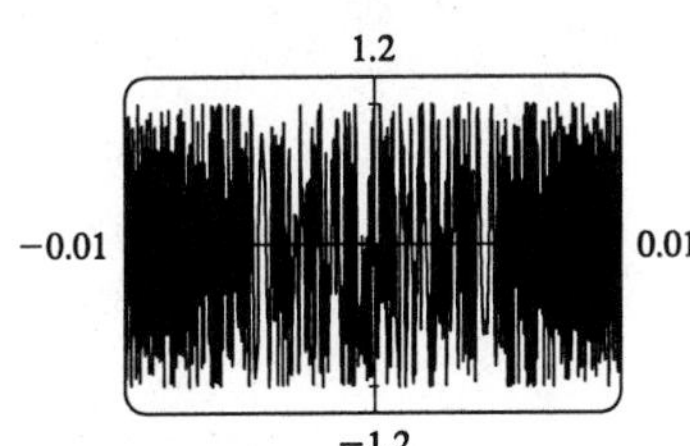

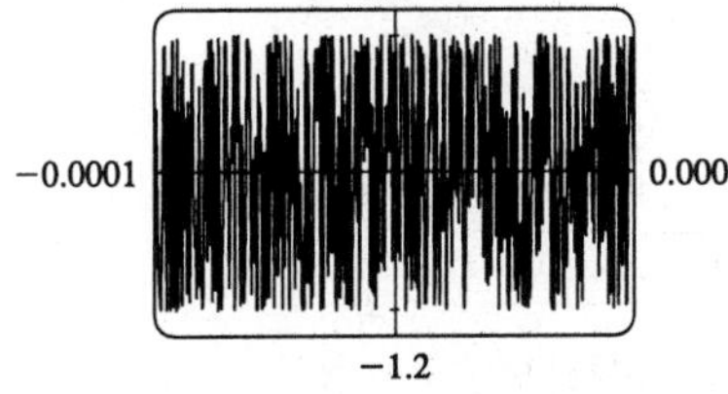

30. Just as in Exercise 26, when we take a small enough viewing rectangle we get incorrect output.

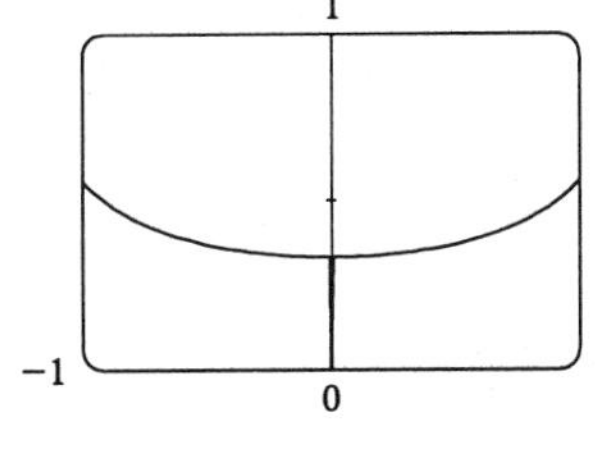

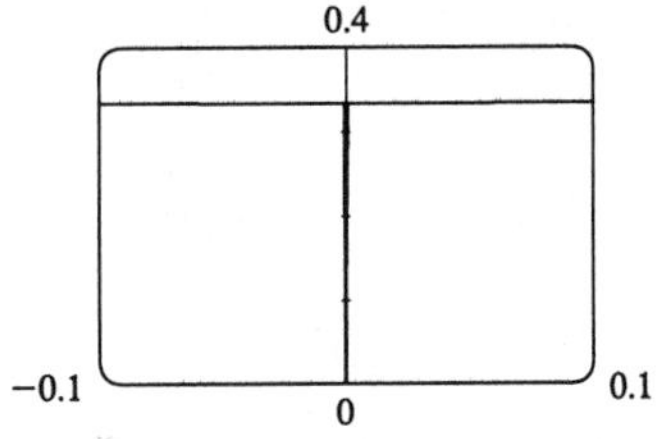

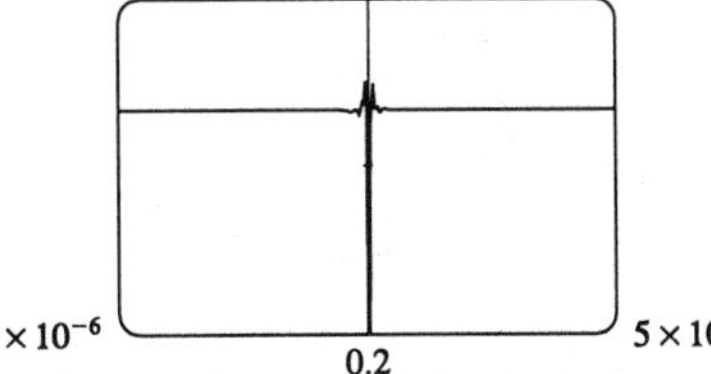

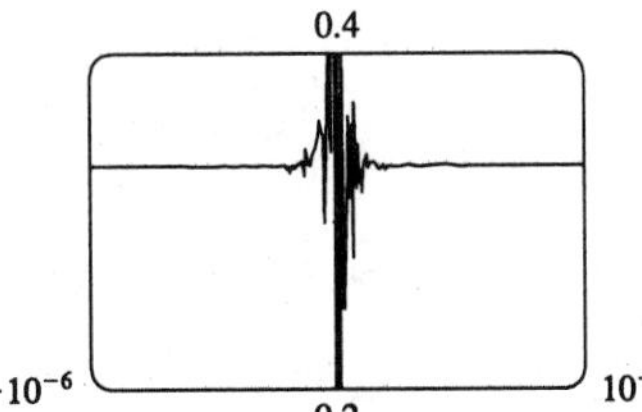

EXERCISES 1.3

1. $\lim_{x\to 4}(5x^2-2x+3) = \lim_{x\to 4}5x^2 - \lim_{x\to 4}2x + \lim_{x\to 4}3$ (Limit Laws 2 & 1)

$= 5\lim_{x\to 4}x^2 - 2\lim_{x\to 4}x + 3$ (3 & 7)

$= 5(4)^2 - 2(4) + 3 = 75$ (9 & 8)

2. $\lim_{x\to -3}(x^3+2x^2+6) = \lim_{x\to -3}x^3 + \lim_{x\to -3}2x^2 + \lim_{x\to -3}6$ (1)

$= (-3)^3 + 2\lim_{x\to -3}x^2 + 6$ (9, 3 & 7)

$= -27 + 2(-3)^2 + 6 = -3$ (9)

3. $\lim_{x\to 2}(x^2+1)(x^2+4x) = \lim_{x\to 2}(x^2+1)\lim_{x\to 2}(x^2+4x)$ (4)

$= \left(\lim_{x\to 2}x^2 + \lim_{x\to 2}1\right)\left(\lim_{x\to 2}x^2 + 4\lim_{x\to 2}x\right)$ (1 & 3)

$= \left[(2)^2+1\right]\left[(2)^2+4(2)\right] = 60$ (9, 7 & 8)

4. $\lim_{x\to -2}(x^2+x+1)^5 = \left[\lim_{x\to -2}(x^2+x+1)\right]^5$ (6)

$= \left(\lim_{x\to -2}x^2 + \lim_{x\to -2}x + \lim_{x\to -2}1\right)^5$ (1)

$= \left[(-2)^2+(-2)+1\right]^5 = 243$ (9, 8 & 7)

5. $\lim_{x\to -1}\dfrac{x-2}{x^2+4x-3} = \dfrac{\lim_{x\to -1}(x-2)}{\lim_{x\to -1}(x^2+4x-3)}$ (5)

$= \dfrac{\lim_{x\to -1}x - \lim_{x\to -1}2}{\lim_{x\to -1}x^2 + 4\lim_{x\to -1}x - \lim_{x\to -1}3}$ (2, 1 & 3)

$= \dfrac{(-1)-2}{(-1)^2+4(-1)-3} = \dfrac{1}{2}$ (8, 7 & 9)

6. $\lim_{t\to -2}\dfrac{t^3-t^2-t+10}{t^2+3t+2} = \lim_{t\to -2}\dfrac{(t+2)(t^2-3t+5)}{(t+2)(t+1)} = \lim_{t\to -2}\dfrac{t^2-3t+5}{t+1}$

$= \dfrac{\lim_{t\to -2}(t^2-3t+5)}{\lim_{t\to -2}(t+1)}$ (5)

$= \dfrac{\lim_{t\to -2}t^2 - 3\lim_{t\to -2}t + \lim_{t\to -2}5}{\lim_{t\to -2}t + \lim_{t\to -2}1}$ (1, 2 & 3)

$= \dfrac{(-2)^2-3(-2)+5}{(-2)+1} = -15$ (9, 8 & 7)

7. $$\lim_{x\to -1}\sqrt{x^3+2x+7}=\sqrt{\lim_{x\to -1}(x^3+2x+7)} \qquad (11)$$

$$=\sqrt{\lim_{x\to -1}x^3+2\lim_{x\to -1}x+\lim_{x\to -1}7} \qquad (1\ \&\ 3)$$

$$=\sqrt{(-1)^3+2(-1)+7}=2 \qquad (9, 8\ \&\ 6)$$

8. $$\lim_{x\to 64}\left(\sqrt[3]{x}+3\sqrt{x}\right)=\lim_{x\to 64}\sqrt[3]{x}+3\lim_{x\to 64}\sqrt{x} \qquad (1\ \&\ 3)$$

$$=\sqrt[3]{64}+3\sqrt{64}=28 \qquad (10\ \&\ 8)$$

9. $$\lim_{t\to -2}(t+1)^9(t^2-1)=\lim_{t\to -2}(t+1)^9\lim_{t\to -2}(t^2-1) \qquad (4)$$

$$=\left[\lim_{t\to -2}(t+1)\right]^9\lim_{t\to -2}(t^2-1) \qquad (6)$$

$$=\left[\lim_{t\to -2}t+\lim_{t\to -2}1\right]^9\left[\lim_{t\to -2}t^2-\lim_{t\to -2}1\right] \qquad (1\ \&\ 2)$$

$$=[(-2)+1]^9[(-2)^2-1]=-3 \qquad (8, 7\ \&\ 9)$$

10. $$\lim_{r\to 3}(r^4-7r+4)^{2/3}=\left[\lim_{r\to 3}(r^4-7r+4)\right]^{2/3} \qquad (6\ \&\ 11)$$

$$=\left(\lim_{r\to 3}r^4-7\lim_{r\to 3}r+\lim_{r\to 3}4\right)^{2/3} \qquad (2, 1\ \&\ 3)$$

$$=\left[(3)^4-7(3)+4\right]^{2/3}=16 \qquad (9, 8\ \&\ 7)$$

11. $$\lim_{w\to -2}\sqrt[3]{\frac{4w+3w^3}{3w+10}}=\sqrt[3]{\lim_{w\to -2}\frac{4w+3w^3}{3w+10}} \qquad (11)$$

$$=\sqrt[3]{\frac{\lim_{w\to -2}(4w+3w^3)}{\lim_{w\to -2}(3w+10)}} \qquad (5)$$

$$=\sqrt[3]{\frac{4\lim_{w\to -2}w+3\lim_{w\to -2}w^3}{3\lim_{w\to -2}w+\lim_{w\to -2}10}} \qquad (1\ \&\ 3)$$

$$=\sqrt[3]{\frac{4(-2)+3(-2)^3}{3(-2)+10}}=-2 \qquad (8, 9\ \&\ 7)$$

12. $$\lim_{y\to 3}\frac{3(8y^2-1)}{2y^2(y-1)^4}=\frac{\lim_{y\to 3}3(8y^2-1)}{\lim_{y\to 3}2y^2(y-1)^4} \qquad (5)$$

$$=\frac{3\left(8\lim_{y\to 3}y^2-\lim_{y\to 3}1\right)}{2\lim_{y\to 3}y^2\cdot\lim_{y\to 3}(y-1)^4} \qquad (2, 3\ \&\ 4)$$

$$=\frac{3[8(3)^2-1]}{2(3)^2(3-1)^4}=\frac{71}{96} \qquad (9, 7\ \&\ 6)$$

13. **(a)** $\lim_{x\to a}[f(x)+h(x)] = \lim_{x\to a} f(x) + \lim_{x\to a} h(x) = -3+8=5$

(b) $\lim_{x\to a}[f(x)]^2 = \left[\lim_{x\to a} f(x)\right]^2 = (-3)^2 = 9$

(c) $\lim_{x\to a}\sqrt[3]{h(x)} = \sqrt[3]{\lim_{x\to a} h(x)} = \sqrt[3]{8} = 2$

(d) $\lim_{x\to a}\frac{1}{f(x)} = \frac{1}{\lim_{x\to a} f(x)} = \frac{1}{-3} = -\frac{1}{3}$

(e) $\lim_{x\to a}\frac{f(x)}{h(x)} = \frac{\lim_{x\to a} f(x)}{\lim_{x\to a} h(x)} = \frac{-3}{8} = -\frac{3}{8}$

(f) $\lim_{x\to a}\frac{g(x)}{f(x)} = \frac{\lim_{x\to a} g(x)}{\lim_{x\to a} f(x)} = \frac{0}{-3} = 0$

(g) The limit does not exist, since $\lim_{x\to a} g(x) = 0$ but $\lim_{x\to a} f(x) \neq 0$.

(h) $\lim_{x\to a}\frac{2f(x)}{h(x)-f(x)} = \frac{2\lim_{x\to a} f(x)}{\lim_{x\to a} h(x) - \lim_{x\to a} f(x)} = \frac{2(-3)}{8-(-3)} = -\frac{6}{11}$

14. **(a)** The left-hand side of the equation is not defined for $x=2$, but the right-hand side is.

(b) Since the equation holds for all $x \neq 2$, it follows that both sides of the equation approach the same limit as $x \to 2$, just as in Example 4. Remember that in finding $\lim_{x\to a} f(x)$, we never consider $x = a$.

15. $\lim_{x\to -3}\frac{x^2-x+12}{x+3}$ does not exist since $x+3 \to 0$ but $x^2-x+12 \to 24$ as $x \to -3$.

16. $\lim_{x\to -3}\frac{x^2-x-12}{x+3} = \lim_{x\to -3}\frac{(x+3)(x-4)}{x+3} = \lim_{x\to -3}(x-4) = -3-4 = -7$

17. $\lim_{x\to -1}\frac{x^2-x-2}{x+1} = \lim_{x\to -1}\frac{(x+1)(x-2)}{x+1} = \lim_{x\to -1}(x-2) = -3$

18. $\lim_{x\to 1}\frac{x^2-x-2}{x+1} = \frac{1^2-1-2}{1+1} = -1$

19. $\lim_{t\to 1}\frac{t^3-t}{t^2-1} = \lim_{t\to 1}\frac{t(t^2-1)}{t^2-1} = \lim_{t\to 1} t = 1$

20. $\lim_{x\to -1}\frac{x^2-x-3}{x+1}$ does not exist since as $x \to -1$, numerator $\to -1$ and denominator $\to 0$.

21. $\lim_{h\to 0}\frac{(h-5)^2-25}{h} = \lim_{h\to 0}\frac{(h^2-10h+25)-25}{h} = \lim_{h\to 0}\frac{h^2-10h}{h} = \lim_{h\to 0}(h-10) = -10$

22. $\lim_{x\to 1}\frac{x^3-1}{x^2-1} = \lim_{x\to 1}\frac{(x-1)(x^2+x+1)}{(x-1)(x+1)} = \lim_{x\to 1}\frac{x^2+x+1}{x+1} = \frac{1^2+1+1}{1+1} = \frac{3}{2}$

23. $\lim_{h\to 0}\frac{(1+h)^4-1}{h} = \lim_{h\to 0}\frac{\left(1+4h+6h^2+4h^3+h^4\right)-1}{h} = \lim_{h\to 0}\frac{4h+6h^2+4h^3+h^4}{h}$

$= \lim_{h\to 0}\left(4+6h+4h^2+h^3\right) = 4$

24. $\lim_{h\to 0}\frac{(2+h)^3-8}{h} = \lim_{h\to 0}\frac{(8+12h+6h^2+h^3)-8}{h} = \lim_{h\to 0}\frac{12h+6h^2+h^3}{h} = \lim_{h\to 0}\left(12+6h+h^2\right) = 12$

25. $\lim_{x\to -2}\frac{x+2}{x^2-x-6} = \lim_{x\to -2}\frac{x+2}{(x-3)(x+2)} = \lim_{x\to -2}\frac{1}{x-3} = -\frac{1}{5}$

26. $\lim\limits_{x\to1}\dfrac{x^2+x-2}{x^2-3x+2}=\lim\limits_{x\to1}\dfrac{(x+2)(x-1)}{(x-2)(x-1)}=\lim\limits_{x\to1}\dfrac{x+2}{x-2}=\dfrac{1+2}{1-2}=-3$

27. $\lim\limits_{t\to9}\dfrac{9-t}{3-\sqrt{t}}=\lim\limits_{t\to9}\dfrac{\left(3+\sqrt{t}\right)\left(3-\sqrt{t}\right)}{3-\sqrt{t}}=\lim\limits_{t\to9}\left(3+\sqrt{t}\right)=3+\sqrt{9}=6$

28. $\lim\limits_{t\to2}\dfrac{t^2+t-6}{t^2-4}=\lim\limits_{t\to2}\dfrac{(t+3)(t-2)}{(t+2)(t-2)}=\lim\limits_{t\to2}\dfrac{t+3}{t+2}=\dfrac{5}{4}$

29. $\lim\limits_{t\to0}\dfrac{\sqrt{2-t}-\sqrt{2}}{t}=\lim\limits_{t\to0}\dfrac{\sqrt{2-t}-\sqrt{2}}{t}\cdot\dfrac{\sqrt{2-t}+\sqrt{2}}{\sqrt{2-t}+\sqrt{2}}=\lim\limits_{t\to0}\dfrac{-t}{t\left(\sqrt{2-t}+\sqrt{2}\right)}=\lim\limits_{t\to0}\dfrac{-1}{\sqrt{2-t}+\sqrt{2}}$

$$=-\frac{1}{2\sqrt{2}}=-\frac{\sqrt{2}}{4}$$

30. $\lim\limits_{x\to2}\dfrac{x^4-16}{x-2}=\lim\limits_{x\to2}\dfrac{(x+2)(x-2)(x^2+4)}{x-2}=\lim\limits_{x\to2}(x+2)(x^2+4)$

$$=\lim_{x\to2}(x+2)\lim_{x\to2}(x^2+4)=(2+2)(2^2+4)=32$$

31. $\lim\limits_{x\to9}\dfrac{x^2-81}{\sqrt{x}-3}=\lim\limits_{x\to9}\dfrac{(x-9)(x+9)}{\sqrt{x}-3}=\lim\limits_{x\to9}\dfrac{\left(\sqrt{x}-3\right)\left(\sqrt{x}+3\right)(x+9)}{\sqrt{x}-3}$

$$=\lim_{x\to9}\left(\sqrt{x}+3\right)(x+9)=\lim_{x\to9}\left(\sqrt{x}+3\right)\lim_{x\to9}(x+9)=\left(\sqrt{9}+3\right)(9+9)=108$$

32. $\lim\limits_{x\to1}\left(\dfrac{1}{x-1}-\dfrac{2}{x^2-1}\right)=\lim\limits_{x\to1}\dfrac{(x+1)-2}{(x-1)(x+1)}=\lim\limits_{x\to1}\dfrac{x-1}{(x-1)(x+1)}=\lim\limits_{x\to1}\dfrac{1}{x+1}=\dfrac{1}{2}$

33. $\lim\limits_{t\to0}\left[\dfrac{1}{t\sqrt{1+t}}-\dfrac{1}{t}\right]=\lim\limits_{t\to0}\dfrac{1-\sqrt{1+t}}{t\sqrt{1+t}}=\lim\limits_{t\to0}\dfrac{\left(1-\sqrt{1+t}\right)\left(1+\sqrt{1+t}\right)}{t\sqrt{t+1}\left(1+\sqrt{1+t}\right)}=\lim\limits_{t\to0}\dfrac{-t}{t\sqrt{1+t}\left(1+\sqrt{1+t}\right)}$

$$=\lim_{t\to0}\frac{-1}{\sqrt{1+t}\left(1+\sqrt{1+t}\right)}=\frac{-1}{\sqrt{1+0}\left(1+\sqrt{1+0}\right)}=-\frac{1}{2}$$

34. $\lim\limits_{h\to0}\dfrac{(3+h)^{-1}-3^{-1}}{h}=\lim\limits_{h\to0}\dfrac{\dfrac{1}{3+h}-\dfrac{1}{3}}{h}=\lim\limits_{h\to0}\dfrac{3-(3+h)}{h(3+h)\,3}=\lim\limits_{h\to0}\dfrac{-h}{h(3+h)\,3}$

$$=\lim_{h\to0}\left[-\frac{1}{3(3+h)}\right]=-\frac{1}{\lim\limits_{h\to0}[3(3+h)]}=-\frac{1}{3(3+0)}=-\frac{1}{9}$$

35. $\lim\limits_{x\to0}\dfrac{x}{\sqrt{1+3x}-1}=\lim\limits_{x\to0}\dfrac{x\left(\sqrt{1+3x}+1\right)}{\left(\sqrt{1+3x}-1\right)\left(\sqrt{1+3x}+1\right)}=\lim\limits_{x\to0}\dfrac{x\left(\sqrt{1+3x}+1\right)}{3x}$

$$=\lim_{x\to0}\frac{\sqrt{1+3x}+1}{3}=\frac{\sqrt{1}+1}{3}=\frac{2}{3}$$

36. $\lim\limits_{x\to2}\dfrac{1/x-\frac{1}{2}}{x-2}=\lim\limits_{x\to2}\dfrac{2-x}{2x(x-2)}=\lim\limits_{x\to2}\dfrac{-1}{2x}=-\dfrac{1}{4}$

37. $\lim\limits_{x\to2}\dfrac{x-\sqrt{3x-2}}{x^2-4}=\lim\limits_{x\to2}\dfrac{\left(x-\sqrt{3x-2}\right)\left(x-\sqrt{3x-2}\right)}{(x^2-4)\left(x-\sqrt{3x-2}\right)}=\lim\limits_{x\to2}\dfrac{x^2-3x+2}{(x^2-4)\left(x+\sqrt{3x-2}\right)}$

$$=\lim_{x\to2}\frac{(x-2)(x-1)}{(x-2)(x+2)\left(x+\sqrt{3x-2}\right)}=\lim_{x\to2}\frac{(x-1)}{(x+2)\left(x+\sqrt{3x-2}\right)}=\frac{1}{4\left(2+\sqrt{4}\right)}=\frac{1}{16}$$

38. $\lim\limits_{x\to1}\dfrac{\sqrt{x}-x^2}{1-\sqrt{x}}=\lim\limits_{x\to1}\dfrac{\sqrt{x}(1-x^{3/2})}{1-\sqrt{x}}=\lim\limits_{x\to1}\dfrac{\sqrt{x}(1-\sqrt{x})(1+\sqrt{x}+x)}{1-\sqrt{x}}$ (difference of cubes)

$=\lim\limits_{x\to1}[\sqrt{x}(1+\sqrt{x}+x)]=\lim\limits_{x\to1}[1(1+1+1)]=3$

Another Method: We "add and subtract" 1 in the numerator, and then split up the fraction:

$$\lim_{x\to1}\frac{\sqrt{x}-x^2}{1-\sqrt{x}}=\lim_{x\to1}\frac{(\sqrt{x}-1)+(1-x^2)}{1-\sqrt{x}}=\lim_{x\to1}\left[-1+\frac{(1-x)(1+x)}{1-\sqrt{x}}\right]$$

$$=\lim_{x\to1}\left[-1+\frac{(1-\sqrt{x})(1+\sqrt{x})(1+x)}{1-\sqrt{x}}\right]=-1+(1+\sqrt{1})(1+1)=3$$

39. Let $f(x)=-x^2$, $g(x)=x^2\cos 20\pi x$ and $h(x)=x^2$. Then $-1\le\cos 20\pi x\le1 \quad\Rightarrow\quad f(x)\le g(x)\le h(x)$. So since $\lim\limits_{x\to0}f(x)=\lim\limits_{x\to0}h(x)=0$, by the Squeeze Theorem we have $\lim\limits_{x\to0}(x^2\cos 20\pi x)=\lim\limits_{x\to0}g(x)=0$.

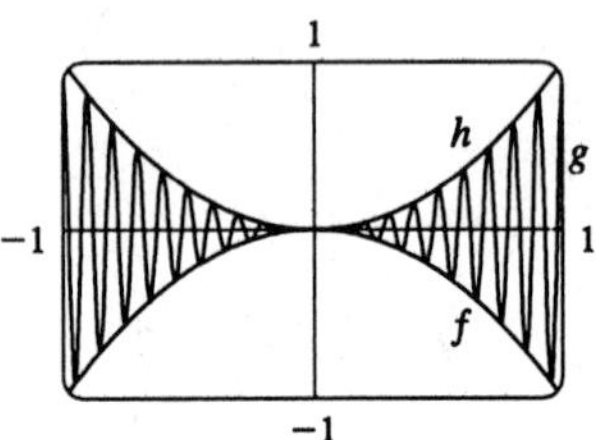

40. Let $f(x)=-\sqrt{x^3+x^2}$, $g(x)=\sqrt{x^3+x^2}\sin(\pi/x)$, and $h(x)=\sqrt{x^3+x^2}$. Then $-1\le\sin(\pi/x)\le1 \quad\Rightarrow\quad f(x)\le g(x)\le h(x)$. So since $\lim\limits_{x\to0}f(x)=\lim\limits_{x\to0}h(x)=0$, by the Squeeze Theorem we have $\lim\limits_{x\to0}g(x)=0$.

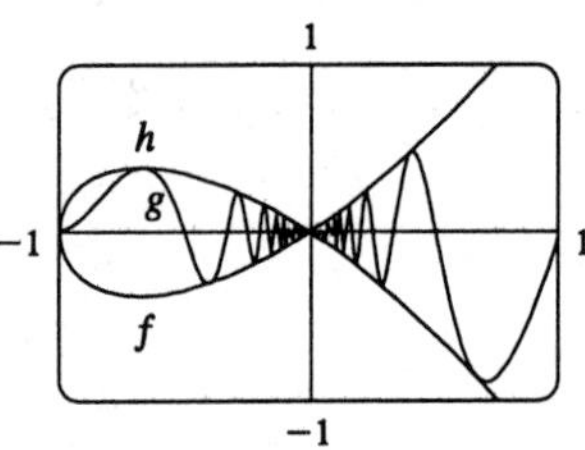

41. $1\le f(x)\le x^2+2x+2$ for all x. But $\lim\limits_{x\to-1}1=1$ and $\lim\limits_{x\to-1}(x^2+2x+2)=\lim\limits_{x\to-1}x^2+2\lim\limits_{x\to-1}x+\lim\limits_{x\to-1}2$ $=(-1)^2+2(-1)+2=1$. Therefore, by the Squeeze Theorem, $\lim\limits_{x\to-1}f(x)=1$.

42. $3x\le f(x)\le x^3+2$ for $0\le x\le2$. Also $\lim\limits_{x\to1}3x=3$ and $\lim\limits_{x\to1}(x^3+2)=\lim\limits_{x\to1}x^3+\lim\limits_{x\to1}2=1^3+2=3$. Therefore, by the Squeeze Theorem, $\lim\limits_{x\to1}f(x)=3$.

43. $-1\le\sin(1/x)\le1 \quad\Rightarrow\quad -x^2\le x^2\sin(1/x)\le x^2$. Since $\lim\limits_{x\to0}(-x^2)=0$ and $\lim\limits_{x\to0}x^2=0$, we have $\lim\limits_{x\to0}x^2\sin(1/x)=0$ by the Squeeze Theorem.

44. $-1\le\cos x\le1 \quad\Rightarrow\quad 0\le\cos^4x\le1 \quad\Rightarrow\quad 0\le\sqrt{x}\cos^4x\le\sqrt{x}$

But $\lim\limits_{x\to0^+}0=\lim\limits_{x\to0^+}\sqrt{x}=0$. So by the Squeeze Theorem, $\lim\limits_{x\to0^+}\sqrt{x}\cos^4x=0$.

45. $\lim\limits_{x\to4^-}\sqrt{16-x^2}=\sqrt{\lim\limits_{x\to4^-}16-\lim\limits_{x\to4^-}x^2}=\sqrt{16-4^2}=0$

46. $\lim\limits_{x\to-1.5^+}(\sqrt{3+2x}+x)=\sqrt{\lim\limits_{x\to-1.5^+}3+2\lim\limits_{x\to-1.5^+}x}+\lim\limits_{x\to-1.5^+}x=\sqrt{3+2(-1.5)}-1.5=-1.5$

47. If $x > -4$, then $|x+4| = x+4$, so $\lim\limits_{x\to -4^+} |x+4| = \lim\limits_{x\to -4^+} (x+4) = -4+4 = 0.$
If $x < -4$, then $|x+4| = -(x+4)$, so $\lim\limits_{x\to -4^-} |x+4| = \lim\limits_{x\to -4^-} -(x+4) = 4-4 = 0.$
Therefore $\lim\limits_{x\to -4} |x+4| = 0.$

48. If $x < -4$, then $|x+4| = -(x+4)$, so $\lim\limits_{x\to -4^-} \dfrac{|x+4|}{x+4} = \lim\limits_{x\to -4^-} \dfrac{-(x+4)}{x+4} = \lim\limits_{x\to -4^-} (-1) = -1.$

49. If $x > 2$, then $|x-2| = x-2$, so $\lim\limits_{x\to 2^+} \dfrac{|x-2|}{x-2} = \lim\limits_{x\to 2^+} \dfrac{x-2}{x-2} = \lim\limits_{x\to 2^+} 1 = 1.$

If $x < 2$, then $|x-2| = -(x-2)$ so $\lim\limits_{x\to 2^-} \dfrac{|x-2|}{x-2} = \lim\limits_{x\to 2^-} \dfrac{-(x-2)}{x-2} = \lim\limits_{x\to 2^-} -1 = -1.$

The right and left limits are different, so $\lim\limits_{x\to 2} \dfrac{|x-2|}{x-2}$ does not exist.

50. If $x > \frac{3}{2}$, then $|2x-3| = 2x-3$, so

$$\lim_{x\to 1.5^+} \frac{2x^2-3x}{|2x-3|} = \lim_{x\to 1.5^+} \frac{2x^2-3x}{2x-3} = \lim_{x\to 1.5^+} \frac{x(2x-3)}{2x-3} = \lim_{x\to 1.5^+} x = 1.5.$$

If $x < \frac{3}{2}$, then $|2x-3| = 3-2x$, so

$$\lim_{x\to 1.5^-} \frac{2x^2-3x}{|2x-3|} = \lim_{x\to 1.5^-} \frac{2x^2-3x}{-(2x-3)} = \lim_{x\to 1.5^-} \frac{x(2x-3)}{-(2x-3)} = \lim_{x\to 1.5^-} -x = -1.5.$$

The right and left limits are different, so $\lim\limits_{x\to 1.5} \dfrac{2x^2-3x}{|2x-3|}$ does not exist.

51. $[\![x]\!] = -2$ for $-2 \le x < -1$, so $\lim\limits_{x\to -2^+} [\![x]\!] = \lim\limits_{x\to -2^+} (-2) = -2$

52. $[\![x]\!] = -2$ for $-2 \le x < -1$, so $\lim\limits_{x\to -2^+} [\![x]\!] = \lim\limits_{x\to -2^+} (-2) = -2$. Also $[\![x]\!] = -3$ for $-3 \le x < -2$, so $\lim\limits_{x\to -2^-} [\![x]\!] = \lim\limits_{x\to -2^-} (-3) = -3$. The right and left limits are different, so $\lim\limits_{x\to -2} [\![x]\!]$ does not exist.

53. $[\![x]\!] = -3$ for $-3 \le x < -2$, so $\lim\limits_{x\to -2.4} [\![x]\!] = \lim\limits_{x\to -2.4} (-3) = -3.$

54. $\lim\limits_{x\to 8^+} \left(\sqrt{x-8} + [\![x+1]\!]\right) = \lim\limits_{x\to 8^+} \sqrt{x-8} + \lim\limits_{x\to 8^+} [\![x+1]\!] = 0+9 = 9$ because $[\![x+1]\!] = 9$ for $8 \le x < 9$.

55. $\lim\limits_{x\to 1^+} \sqrt{x^2+x-2} = \sqrt{\lim\limits_{x\to 1^+} x^2 + \lim\limits_{x\to 1^+} x - \lim\limits_{x\to 1^+} 2} = \sqrt{1^2+1-2} = 0$

Notice that the domain of $\sqrt{x^2+x-2}$ is $(-\infty, -2] \cup [1, \infty)$.

56. $\lim\limits_{x\to -2^-} \sqrt{x^2+x-2} = \sqrt{\lim\limits_{x\to -2^-} x^2 + \lim\limits_{x\to -2^-} x - \lim\limits_{x\to -2^-} 2} = \sqrt{(-2)^2-2-2} = 0$

Notice that the domain of $\sqrt{x^2+x-2}$ is $(-\infty, -2] \cup [1, \infty)$.

57. Since $|x| = -x$ for $x < 0$, we have $\lim\limits_{x\to 0^-} \left(\dfrac{1}{x} - \dfrac{1}{|x|}\right) = \lim\limits_{x\to 0^-} \left(\dfrac{1}{x} - \dfrac{1}{-x}\right) = \lim\limits_{x\to 0^-} \dfrac{2}{x}$, which does not exist since the denominator $\to 0$ and the numerator does not.

58. Since $|x| = x$ for $x > 0$, we have $\lim\limits_{x\to 0^+} \left(\dfrac{1}{x} - \dfrac{1}{|x|}\right) = \lim\limits_{x\to 0^+} \left(\dfrac{1}{x} - \dfrac{1}{x}\right) = \lim\limits_{x\to 0^+} 0 = 0.$

59. (a)

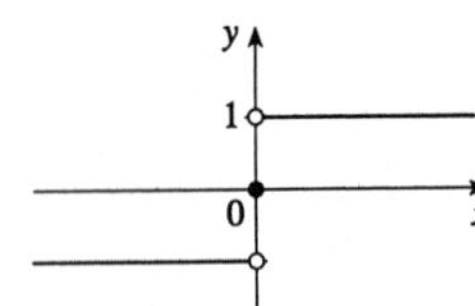

(b) (i) Since $\operatorname{sgn} x = 1$ for $x > 0$, $\lim\limits_{x\to 0^+} \operatorname{sgn} x = \lim\limits_{x\to 0^+} 1 = 1$.

(ii) Since $\operatorname{sgn} x = -1$ for $x < 0$, $\lim\limits_{x\to 0^-} \operatorname{sgn} x = \lim\limits_{x\to 0^-} -1 = -1$.

(iii) Since $\lim\limits_{x\to 0^-} \operatorname{sgn} x \neq \lim\limits_{x\to 0^+} \operatorname{sgn} x$, $\lim\limits_{x\to 0} \operatorname{sgn} x$ does not exist.

(iv) Since $|\operatorname{sgn} x| = 1$ for $x \neq 0$, $\lim\limits_{x\to 0} |\operatorname{sgn} x| = \lim\limits_{x\to 0} 1 = 1$.

60. (a) $\lim\limits_{x\to 1^-} f(x) = \lim\limits_{x\to 1^-} (x^2 - 2x + 2) = \lim\limits_{x\to 1^-} x^2 - 2\lim\limits_{x\to 1^-} x + \lim\limits_{x\to 1^-} 2$

$= 1^2 - 2 + 2 = 1$

$\lim\limits_{x\to 1^+} f(x) = \lim\limits_{x\to 1^+} (3 - x) = \lim\limits_{x\to 1^+} 3 - \lim\limits_{x\to 1^+} x = 3 - 1 = 2$

(b) $\lim\limits_{x\to 1} f(x)$ does not exist because $\lim\limits_{x\to 1^-} f(x) \neq \lim\limits_{x\to 1^+} f(x)$.

(c)

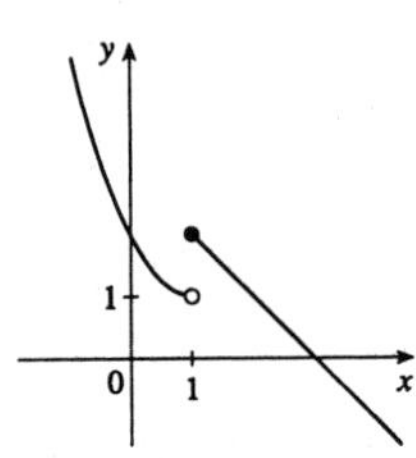

61. (a) $\lim\limits_{x\to -1^-} g(x) = \lim\limits_{x\to -1^-} (-x^3) = -(-1)^3 = 1$,

$\lim\limits_{x\to -1^+} g(x) = \lim\limits_{x\to -1^+} (x+2)^2 = (-1+2)^2 = 1$

(b) By part (a), $\lim\limits_{x\to -1} g(x) = 1$.

(c)

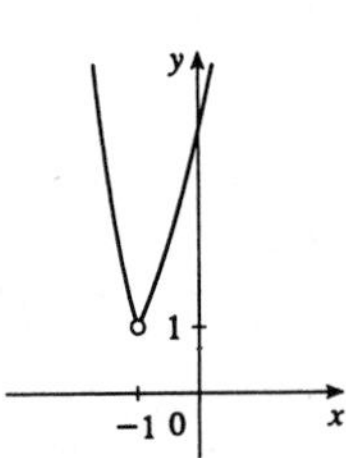

62. (a) (i) $\lim\limits_{x\to 0^+} h(x) = \lim\limits_{x\to 0^+} x^2 = 0^2 = 0$

(ii) $\lim\limits_{x\to 0^-} h(x) = \lim\limits_{x\to 0^-} x = 0$ So $\lim\limits_{x\to 0} h(x) = 0$.

(iii) $\lim\limits_{x\to 1} h(x) = \lim\limits_{x\to 1} x^2 = 1^2 = 1$

(iv) $\lim\limits_{x\to 2^-} h(x) = \lim\limits_{x\to 2^-} x^2 = 2^2 = 4$

(v) $\lim\limits_{x\to 2^+} h(x) = \lim\limits_{x\to 2^+} (8 - x) = 8 - 2 = 6$

(vi) Since $\lim\limits_{x\to 2^-} h(x) \neq \lim\limits_{x\to 2^+} h(x)$, $\lim\limits_{x\to 2} h(x)$ does not exist.

(b)

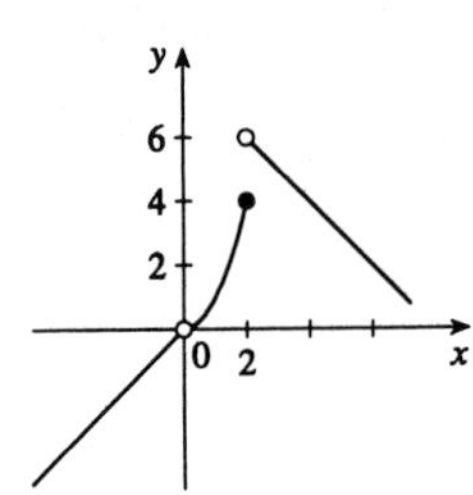

63. (a) (i) $[\![x]\!] = n - 1$ for $n - 1 \le x < n$, so $\lim\limits_{x\to n^-} [\![x]\!] = \lim\limits_{x\to n^-} (n-1) = n - 1$.

(ii) $[\![x]\!] = n$ for $n \le x < n + 1$, so $\lim\limits_{x\to n^+} [\![x]\!] = \lim\limits_{x\to n^+} n = n$.

(b) $\lim\limits_{x\to a} [\![x]\!]$ exists $\Leftrightarrow$ a is not an integer.

64. (a)

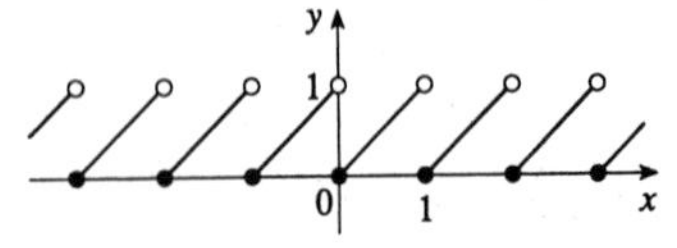

(b) (i) $\lim\limits_{x\to n^-} f(x) = \lim\limits_{x\to n^-} (x - [\![x]\!]) = \lim\limits_{x\to n^-} [x - (n-1)]$

$= n - (n - 1) = 1$

(ii) $\lim\limits_{x\to n^+} f(x) = \lim\limits_{x\to n^+} (x - [\![x]\!]) = \lim\limits_{x\to n^+} (x - n)$

$= n - n = 0$

(c) $\lim\limits_{x\to a} f(x)$ exists $\Leftrightarrow$ a is not an integer.

(d) $\text{FRAC}(x) = f(x)$ if $x \geq 0$ or if x is a negative integer, but for other negative x we have $\text{FRAC}(x) = f(x) - 1$.

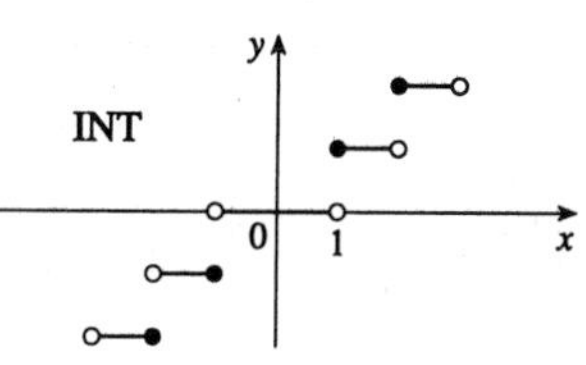

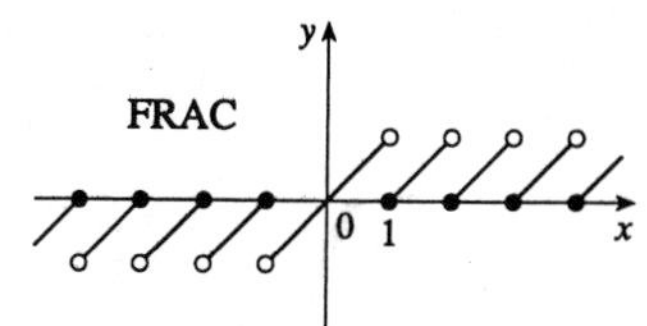

65. **(a)** **(i)** $\lim_{x\to1^+}\dfrac{x^2-1}{|x-1|} = \lim_{x\to1^+}\dfrac{x^2-1}{x-1} = \lim_{x\to1^+}(x+1) = 2$

(ii) $\lim_{x\to1^-}\dfrac{x^2-1}{|x-1|} = \lim_{x\to1^-}\dfrac{x^2-1}{-(x-1)} = \lim_{x\to1^-}-(x+1) = -2$

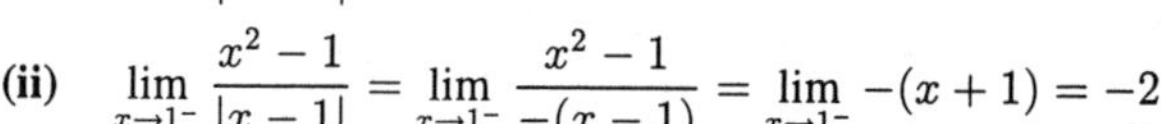

(b) No, $\lim_{x\to1}F(x)$ does not exist since $\lim_{x\to1^+}F(x) \neq \lim_{x\to1^-}F(x)$.

(c)

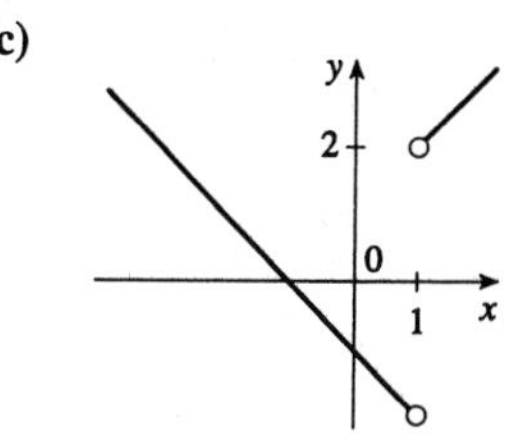

66. **(a)**

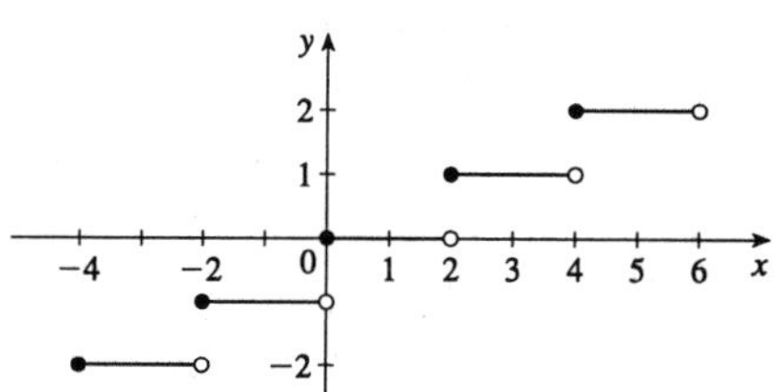

(b) **(i)** $\lim_{x\to1^+}g(x) = 0$ since $[\![x/2]\!] = 0$ for $0 \leq x < 2$.

(ii) $\lim_{x\to1^-}g(x) = 0$ since $[\![x/2]\!] = 0$ for $0 \leq x < 2$.

(iii) $\lim_{x\to1}g(x) = 0$ since $[\![x/2]\!] = 0$ for $0 \leq x < 2$.

(iv) $\lim_{x\to2^+}g(x) = 1$ since $[\![x/2]\!] = 1$ for $2 \leq x < 4$.

(v) $\lim_{x\to2^-}g(x) = 0$ since $[\![x/2]\!] = 0$ for $0 \leq x < 2$.

(vi) $\lim_{x\to2}g(x)$ does not exist because $\lim_{x\to2^+}g(x) \neq \lim_{x\to2^-}g(x)$.

(c) $\lim_{x\to a}g(x)$ exists except when a is an even integer.

67. Since $p(x)$ is a polynomial, $p(x) = a_0 + a_1x + a_2x^2 + \cdots + a_nx^n$. Thus, by the Limit Laws,

$$\lim_{x\to a}p(x) = \lim_{x\to a}\left(a_0 + a_1x + a_2x^2 + \cdots + a_nx^n\right) = a_0 + a_1\lim_{x\to a}x + a_2\lim_{x\to a}x^2 + \cdots + a_n\lim_{x\to a}x^n$$

$$= a_0 + a_1a + a_2a^2 + \cdots + a_na^n = p(a).$$ Thus, for any polynomial p, $\lim_{x\to a}p(x) = p(a)$.

68. Let $r(x) = \dfrac{p(x)}{q(x)}$ where $p(x)$ and $q(x)$ are any polynomials, and suppose that $q(a) \neq 0$. Thus

$$\lim_{x\to a}r(x) = \lim_{x\to a}\frac{p(x)}{q(x)} = \frac{\lim_{x\to a}p(x)}{\lim_{x\to a}q(x)} \text{ (Limit Law 5) } = \frac{p(a)}{q(a)} \text{ (Exercise 67) } = r(a).$$

69. Observe that $0 \leq f(x) \leq x^2$ for all x, and $\lim_{x\to0}0 = 0 = \lim_{x\to0}x^2$. So, by the Squeeze Theorem, $\lim_{x\to0}f(x) = 0$.

70. Let $f(x) = [\![x]\!]$ and $g(x) = -[\![x]\!]$. Then $\lim_{x\to3}f(x)$ and $\lim_{x\to3}g(x)$ do not exist (Example 11) but

$\lim_{x\to3}[f(x) + g(x)] = \lim_{x\to3}([\![x]\!] - [\![x]\!]) = \lim_{x\to3}0 = 0$.

71. Let $f(x) = H(x)$ and $g(x) = 1 - H(x)$, where H is the Heaviside function defined in Example 1.2.7. Then $\lim_{x\to0}f(x)$ and $\lim_{x\to0}g(x)$ do not exist but $\lim_{x\to0}[f(x)g(x)] = \lim_{x\to0}0 = 0$.

72. $\lim_{x\to 2}\dfrac{\sqrt{6-x}-2}{\sqrt{3-x}-1}=\lim_{x\to 2}\left(\dfrac{\sqrt{6-x}-2}{\sqrt{3-x}-1}\cdot\dfrac{\sqrt{6-x}+2}{\sqrt{6-x}+2}\cdot\dfrac{\sqrt{3-x}+1}{\sqrt{3-x}+1}\right)$

$$=\lim_{x\to 2}\frac{(2-x)\left(\sqrt{3-x}+1\right)}{(2-x)\left(\sqrt{6-x}+2\right)}=\lim_{x\to 2}\frac{\sqrt{3-x}+1}{\sqrt{6-x}+2}=\frac{1}{2}$$

73. Let $t=\sqrt[3]{1+cx}$. Then $t\to 1$ as $x\to 0$ and $t^3=1+cx \quad\Rightarrow\quad x=(t^3-1)/c$. (If $c=0$, then the limit is obviously 0.) Therefore

$$\lim_{x\to 0}\frac{\sqrt[3]{1+cx}-1}{x}=\lim_{t\to 1}\frac{t-1}{(t^3-1)/c}=\lim_{t\to 1}\frac{c(t-1)}{(t-1)(t^2+t+1)}=\lim_{t\to 1}\frac{c}{t^2+t+1}=\frac{c}{1^2+1+1}=\frac{c}{3}.$$

Another Method: Multiply numerator and denominator by $(1+cx)^{2/3}+(1+cx)^{1/3}+1$.

74. Let $t=\sqrt[6]{x}$. Then $t\to 1$ as $x\to y$, so

$$\lim_{x\to 1}\frac{\sqrt[3]{x}-1}{\sqrt{x}-1}=\lim_{t\to 1}\frac{t^2-1}{t^3-1}=\lim_{t\to 1}\frac{(t-1)(t+1)}{(t-1)(t^2+t+1)}=\lim_{t\to 1}\frac{t+1}{t^2+t+1}=\frac{1+1}{1^2+1+1}=\frac{2}{3}.$$

Another Method: Multiply both numerator and denominator by $\sqrt{x}+1$.

75. Since the denominator approaches 0 as $x\to -2$, the limit will exist only if the numerator also approaches 0 as $x\to -2$. In order for this to happen, we need $\lim_{x\to -2}(3x^2+ax+a+3)=0 \quad\Leftrightarrow$

$3(-2)^2+a(-2)+a+3=0 \quad\Leftrightarrow\quad 12-2a+a+3=0 \quad\Leftrightarrow\quad a=15$. With $a=15$, the limit becomes

$$\lim_{x\to -2}\frac{3x^2+15x+18}{x^2+x-2}=\lim_{x\to -2}\frac{3(x+2)(x+3)}{(x-1)(x+2)}=\frac{3(-2+3)}{-2-1}=-1.$$

76. First rationalize the numerator: $\lim_{x\to 0}\dfrac{\sqrt{ax+b}-2}{x}=\lim_{x\to 0}\dfrac{ax+b-4}{x\left(\sqrt{ax+b}+2\right)}$. Now since the denominator approaches 0 as $x\to 0$, the limit will exist only if the numerator also approaches 0 as $x\to 0$. So we require that $a(0)+b-4=0 \quad\Rightarrow\quad b=4$. So the equation becomes $\lim_{x\to 0}\dfrac{a}{\sqrt{ax+4}+2}=1 \quad\Rightarrow\quad \dfrac{a}{\sqrt{4}+2}=1 \quad\Rightarrow$ $a=4$. Therefore, the solution is $a=b=4$.

77. $y-1<[\![y]\!]\leq y$, so $x^2\left(\dfrac{1}{4x^2}-1\right)<x^2\left[\!\!\left[\dfrac{1}{4x^2}\right]\!\!\right]\leq x^2\left(\dfrac{1}{4x^2}\right)=\dfrac{1}{4} \quad (x\neq 0)$. But

$\lim_{x\to 0}x^2\left(\dfrac{1}{4x^2}-1\right)=\lim_{x\to 0}\left(\dfrac{1}{4}-x^2\right)=\dfrac{1}{4}$ and $\lim_{x\to 0}\dfrac{1}{4}=\dfrac{1}{4}$. So by the Squeeze Theorem, $\lim_{x\to 0}x^2\left[\!\!\left[\dfrac{1}{4x^2}\right]\!\!\right]=\dfrac{1}{4}$.

78. *Solution 1:* First, we find the coordinates of P and Q as functions of r. Then we can find the equation of the line determined by these two points, and thus find the x-intercept (the point R), and take the limit as $r\to 0$.

The coordinates of P are $(0,r)$. The point Q is the point of intersection of the two circles $x^2+y^2=r^2$ and $(x-1)^2+y^2=1$. Eliminating y from these equations, we get $r^2-x^2=1-(x-1)^2 \quad\Leftrightarrow\quad r^2=1+2x-1 \quad\Leftrightarrow\quad x=\frac{1}{2}r^2$. Substituting back into the equation of the shrinking circle to find the y-coordinate, we get $\left(\frac{1}{2}r^2\right)^2+y^2=r^2 \quad\Leftrightarrow\quad y^2=r^2\left(1-\frac{1}{4}r^2\right) \quad\Leftrightarrow\quad y=r\sqrt{1-\frac{1}{4}r^2}$ (the positive y-value). So the coordinates of Q are $\left(\frac{1}{2}r^2, r\sqrt{1-\frac{1}{4}r^2}\right)$. The equation of the line joining P and Q is thus $y - r = \dfrac{r\sqrt{1-\frac{1}{4}r^2} - r}{\frac{1}{2}r^2 - 0}(x - 0)$.

We set $y = 0$ in order to find the x-intercept, and get

$x = -r\dfrac{\frac{1}{2}r^2}{r\left(\sqrt{1-\frac{1}{4}r^2} - 1\right)} = \dfrac{-\frac{1}{2}r^2\left(\sqrt{1-\frac{1}{4}r^2} + 1\right)}{1 - \frac{1}{4}r^2 - 1} = 2\left(\sqrt{1-\frac{1}{4}r^2} + 1\right)$. Now we take the limit as $r \to 0^+$:

$\lim\limits_{r\to 0^+} x = \lim\limits_{r\to 0^+} 2\left(\sqrt{1-\frac{1}{4}r^2} + 1\right) = \lim\limits_{r\to 0^+} 2\left(\sqrt{1} + 1\right) = 4$. So the limiting position of R is the point $(4, 0)$.

Solution 2: We add a few lines to the diagram, as shown. Note that $\angle PQS = 90°$ (subtended by diameter PS). So $\angle SQR = 90° = \angle OQT$ (subtended by diameter OT). It follows that $\angle OQS = \angle TQR$. Also $\angle PSQ = 90° - \angle SPQ = \angle ORP$. Since ΔQOS is isosceles, so is ΔQTR, implying that $QT = TR$. As the circle shrinks, the point Q plainly approaches the origin, so the point R must approach a point twice as far from the origin as T, that is, the point $(4, 0)$, as above.

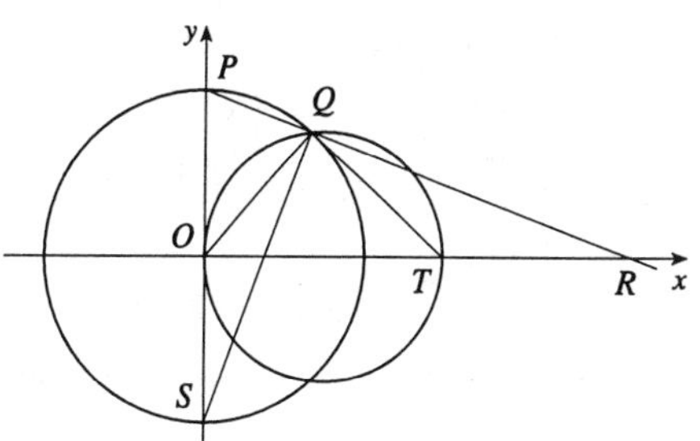

EXERCISES 1.4

1. **(a)** $|(6x+1) - 19| < 0.1 \quad\Leftrightarrow\quad |6x - 18| < 0.1 \quad\Leftrightarrow\quad 6|x - 3| < 0.1 \quad\Leftrightarrow\quad |x - 3| < (0.1)/6 = \frac{1}{60}$

(b) $|(6x+1) - 19| < 0.01 \quad\Leftrightarrow\quad |x - 3| < (0.01)/6 = \frac{1}{600}$

2. **(a)** $|(8x-5) - 11| < 0.01 \Leftrightarrow |8x - 16| < 0.01 \Leftrightarrow 8|x - 2| < 0.01 \Leftrightarrow |x - 2| < (0.01)/8 = \frac{1}{800}$

(b) $|(8x-5) - 11| < 0.001 \quad\Leftrightarrow\quad |x - 2| < (0.001)/8 = \frac{1}{8000}$

(c) $|(8x-5) - 11| < 0.0001 \quad\Leftrightarrow\quad |x - 2| < (0.0001)/8 = \frac{1}{80{,}000}$

3. On the left side, we need $|x - 2| < \left|\frac{10}{7} - 2\right| = \frac{4}{7}$. On the right side, we need $|x - 2| < \left|\frac{10}{3} - 2\right| = \frac{4}{3}$. For both of these conditions to be satisfied at once, we need the more restrictive of the two to hold, that is, $|x - 2| < \frac{4}{7}$. So we can choose $\delta = \frac{4}{7}$, or any smaller positive number.

4. The left-hand question mark is the positive solution of $x^2 = \frac{1}{2}$, that is, $x = \frac{1}{\sqrt{2}}$, and the right-hand question mark is the positive solution of $x^2 = \frac{3}{2}$, that is, $x = \sqrt{\frac{3}{2}}$. On the left side, we need $|x - 1| < \left|\frac{1}{\sqrt{2}} - 1\right| \approx 0.293$. On the right side, we need $|x - 1| < \left|\sqrt{\frac{3}{2}} - 1\right| \approx 0.224$ (rounding down to be safe). The more restrictive of these two conditions must apply, so we choose $\delta = 0.224$ (or any smaller positive number).

5. $\left|\sqrt{4x+1} - 3\right| < 0.5 \quad \Leftrightarrow \quad 2.5 < \sqrt{4x+1} < 3.5.$
We plot the three parts of this inequality on the same screen and identify the x-coordinates of the points of intersection using the cursor. It appears that the inequality holds for $1.32 \le x \le 2.81$. Since $|2 - 1.32| = 0.68$ and $|2 - 2.81| = 0.81$, we choose $0 < \delta \le \min\{0.68, 0.81\} = 0.68$.

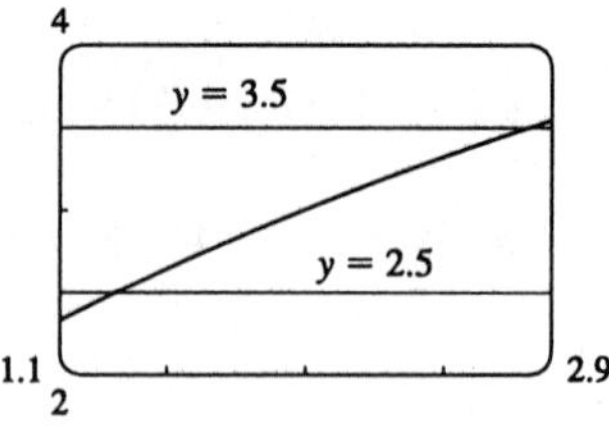

6. $\left|\sin x - \frac{1}{2}\right| < 0.1 \quad \Leftrightarrow \quad 0.4 < \sin x < 0.6.$
From the graph, we see that for this inequality to hold, we need $0.42 \le x \le 0.64$. So since $|0.5 - 0.42| = 0.08$ and $|0.5 - 0.64| = 0.14$, we choose $0 < \delta \le \min\{0.08, 0.14\} = 0.08$.

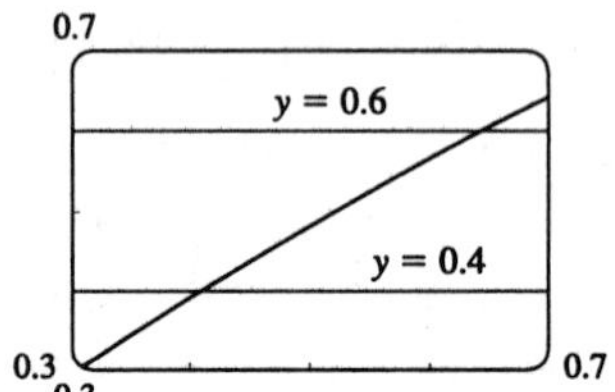

7. For $\epsilon = 1$, the definition of a limit requires that we find δ such that $|(4 + x - 3x^3) - 2| < 1 \quad \Leftrightarrow$ $1 < 4 + x - 3x^3 < 3$ whenever $|x - 1| < \delta$. If we plot the graphs of $y = 1$, $y = 4 + x - 3x^3$ and $y = 3$ on the same screen, we see that we need $0.86 \le x \le 1.11$. So since $|1 - 0.86| = 0.14$ and $|1 - 1.11| = 0.11$, we choose $\delta = 0.11$ (or any smaller positive number). For $\epsilon = 0.1$, we must find δ such that $|(4 + x - 3x^3) - 2| < 0.1 \quad \Leftrightarrow \quad 1.9 < 4 + x - 3x^3 < 2.1$ whenever $|x - 1| < \delta$. From the graph, we see that we need $0.988 \le x \le 1.012$. So since $|1 - 0.988| = 0.012$ and $|1 - 1.012| = 0.012$, we must choose $\delta = 0.012$ (or any smaller positive number) for the inequality to hold.

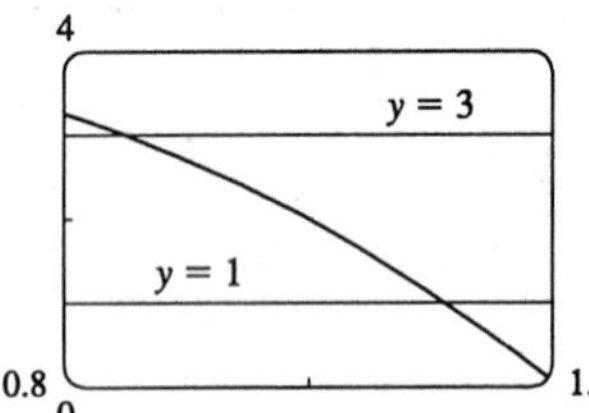

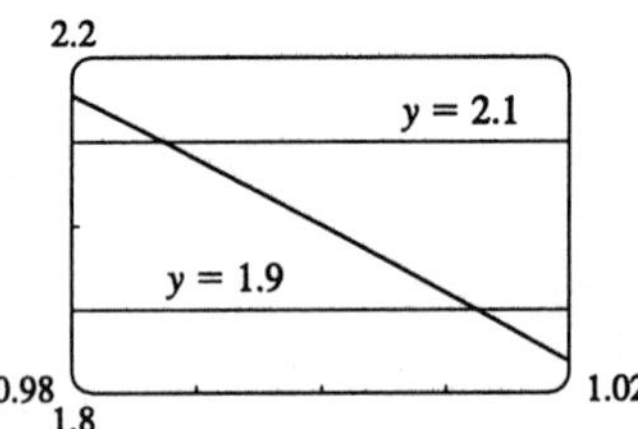

8. For $\epsilon = 0.5$, we need $1.91 \le x \le 2.125$. So since $|2 - 1.91| = 0.09$ and $|2 - 2.125| = 0.125$, we can take $0 < \delta \le 0.09$. For $\epsilon = 0.1$, we need $1.980 \le 2.021$. So since $|2 - 1.980| = 0.02$ and $|2 - 2.021| = 0.021$, we can take $\delta = 0.02$ (or any smaller positive number).

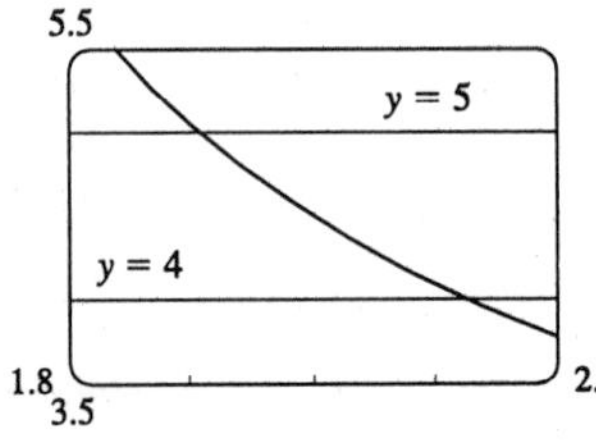

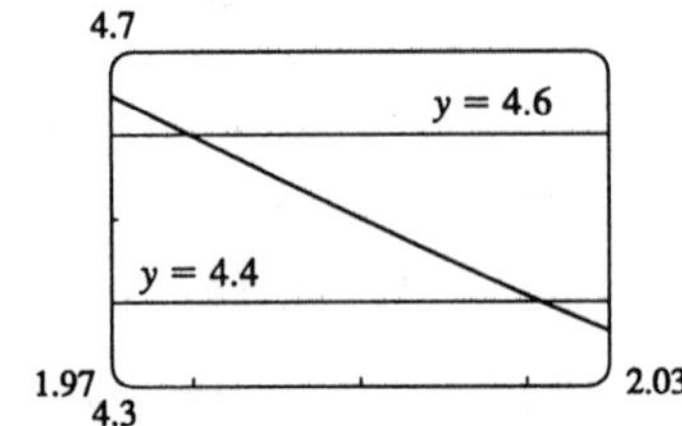

9. From the graph, we see that $\dfrac{x}{(x^2+1)(x-1)^2} > 100$ whenever $0.93 \le x \le 1.07$. So since $|1-0.93| = 0.7$ and $|1-1.07| = 0.7$, we can take $\delta = 0.07$ (or any smaller positive number).

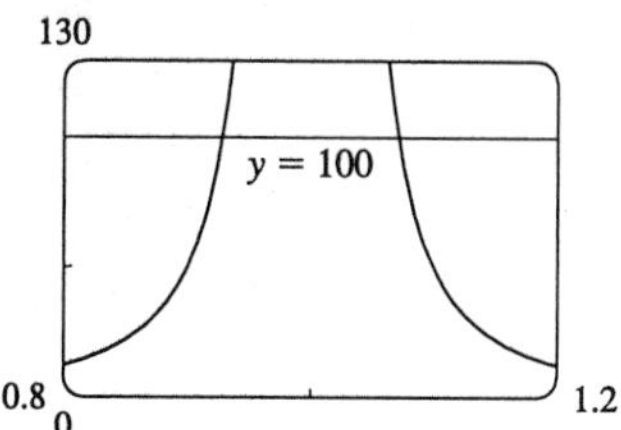

10. For $M = 100$, we need $1.48 \le x \le \frac{\pi}{2} \approx 1.5708$, so since $\left|\frac{\pi}{2} - 1.48\right| \approx 0.09$ we choose $0 < \delta \le 0.09$. For $M = 1000$, we need $1.54 \le x \le \frac{\pi}{2}$, so since $\left|\frac{\pi}{2} - 1.54\right| \approx 0.03$, we choose $\delta \le 0.03$.

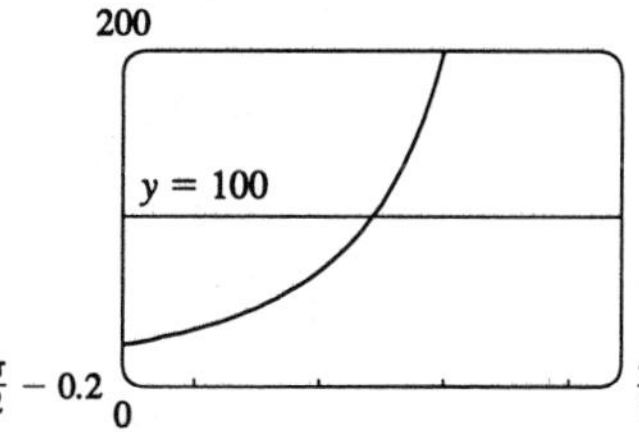

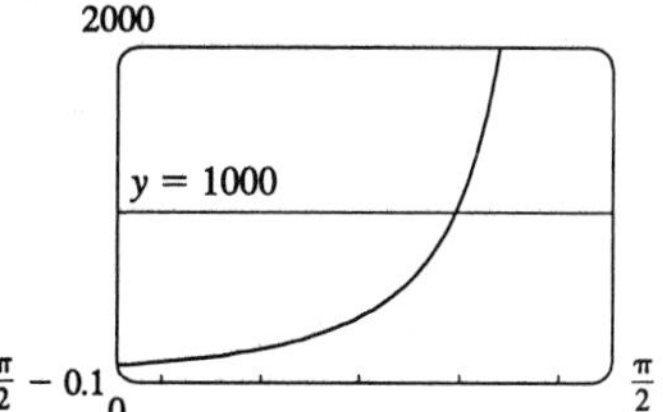

11. Given $\epsilon > 0$, we need $\delta > 0$ such that if $|x-2| < \delta$, then $|(3x-2)-4| < \epsilon \Leftrightarrow |3x-6| < \epsilon \Leftrightarrow 3|x-2| < \epsilon \Leftrightarrow |x-2| < \epsilon/3$. So if we choose $\delta = \epsilon/3$, then $|x-2| < \delta \Rightarrow |(3x-2)-4| < \epsilon$. Thus $\lim_{x\to 2}(3x-2) = 4$ by the definition of a limit.

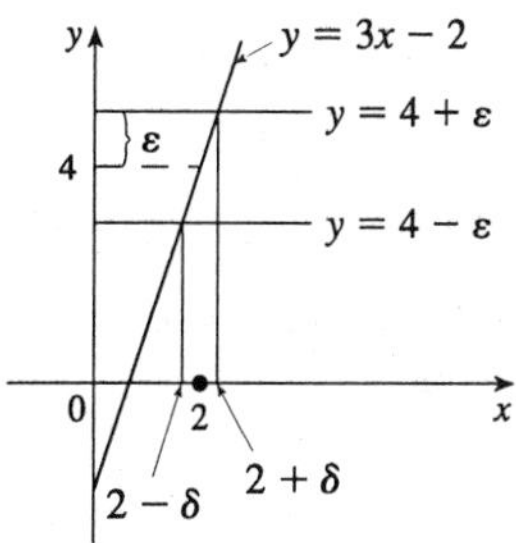

12. Given $\epsilon > 0$, we need $\delta > 0$ such that if $|x-4| < \delta$, then $|(5-2x)-(-3)| < \epsilon \Leftrightarrow |-2x+8| < \epsilon \Leftrightarrow 2|x-4| < \epsilon \Leftrightarrow |x-4| < \epsilon/2$. So choose $\delta = \epsilon/2$. Then $|x-4| < \delta \Rightarrow |(5-2x)-(-3)| < \epsilon$. Thus $\lim_{x\to 4}(5-2x) = -3$ by the definition of a limit.

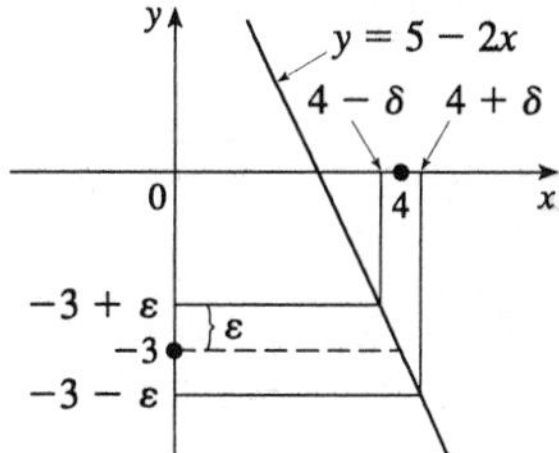

13. Given $\epsilon > 0$, we need $\delta > 0$ such that if $|x-(-1)| < \delta$, then $|(5x+8)-3| < \epsilon \Leftrightarrow |5x+5| < \epsilon \Leftrightarrow 5|x+1| < \epsilon \Leftrightarrow |x-(-1)| < \epsilon/5$. So if we choose $\delta = \epsilon/5$, then $|x-(-1)| < \delta \Rightarrow |(5x+8)-3| < \epsilon$. Thus $\lim_{x\to -1}(5x+8) = 3$ by the definition of a limit.

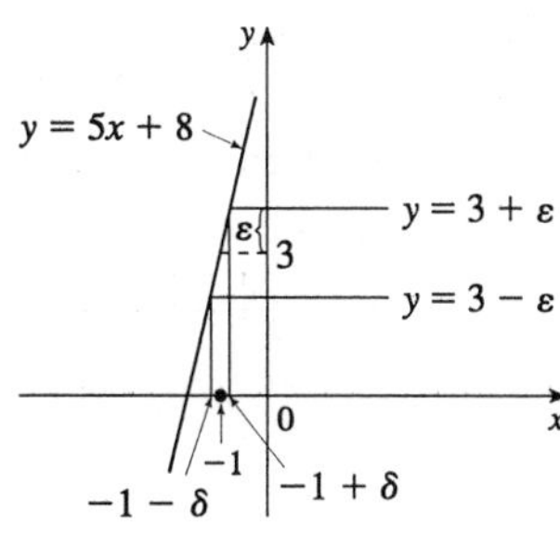

14. Given $\epsilon > 0$, we need $\delta > 0$ such that if $|x-(-1)| < \delta$, then $|(3-4x)-7| < \epsilon \Leftrightarrow |-4x-4| < \epsilon \Leftrightarrow 4|x+1| < \epsilon \Leftrightarrow |x-(-1)| < \epsilon/4$. So choose $\delta = \epsilon/4$. Then $|x-(-1)| < \delta \Rightarrow |(3-4x)-7| < \epsilon$. Thus $\lim_{x\to -1}(3-4x) = 7$ by the definition of a limit.

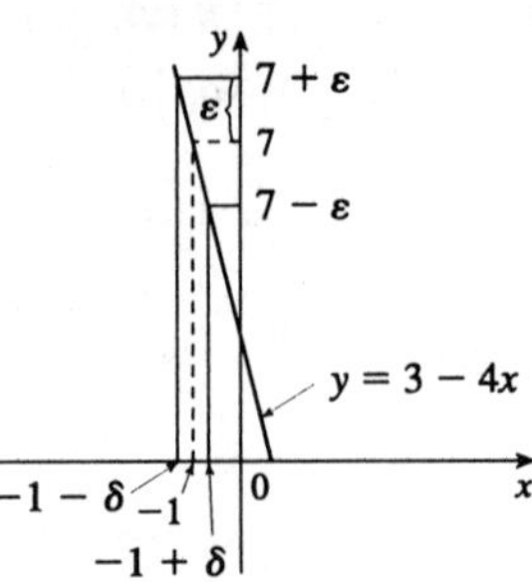

15. Given $\epsilon > 0$, we need $\delta > 0$ such that if $|x-2| < \delta$ then $\left|\frac{x}{7}-\frac{2}{7}\right| < \epsilon \Leftrightarrow \frac{1}{7}|x-2| < \epsilon \Leftrightarrow |x-2| < 7\epsilon$. So take $\delta = 7\epsilon$. Then $|x-2| < \delta \Rightarrow \left|\frac{x}{7}-\frac{2}{7}\right| < \epsilon$. Thus $\lim_{x\to 2}\frac{x}{7} = \frac{2}{7}$ by the definition of a limit.

16. Given $\epsilon > 0$, we need $\delta > 0$ such that if $|x-4| < \delta$ then $\left|\left(\frac{x}{3}+1\right)-\frac{7}{3}\right| < \epsilon \Leftrightarrow \frac{1}{3}|x-4| < \epsilon \Leftrightarrow |x-4| < 3\epsilon$. So take $\delta = 3\epsilon$. Then $|x-4| < \delta \Rightarrow \left|\left(\frac{x}{3}+1\right)-\frac{7}{3}\right| < \epsilon$. Thus $\lim_{x\to 4}\left(\frac{x}{3}+1\right) = \frac{7}{3}$ by the definition of a limit.

17. Given $\epsilon > 0$, we need $\delta > 0$ such that if $|x-(-5)| < \delta$ then $\left|\left(4-\frac{3}{5}x\right)-7\right| < \epsilon \Leftrightarrow \frac{3}{5}|x+5| < \epsilon \Leftrightarrow |x-(-5)| < \frac{5}{3}\epsilon$. So take $\delta = \frac{5}{3}\epsilon$. Then $|x-(-5)| < \delta \Rightarrow \left|\left(4-\frac{3}{5}x\right)-7\right| < \epsilon$. Thus $\lim_{x\to -5}\left(4-\frac{3}{5}x\right) = 7$ by the definition of a limit.

18. Given $\epsilon > 0$, we need $\delta > 0$ such that if $|x-2| < \delta$ then $\left|\frac{x^2+x-6}{x-2}-5\right| < \epsilon \Leftrightarrow \left|\frac{(x-2)(x+3)}{x-2}-5\right| < \epsilon \Leftrightarrow |x+3-5| < \epsilon$ (for $x \neq 2$) $\Leftrightarrow |x-2| < \epsilon$. So take $\delta = \epsilon$, and certainly $|x-2| < \delta \Rightarrow |x-2| < \epsilon$. Thus $\lim_{x\to 2}\frac{x^2+x-6}{x-2} = 5$, by the definition of a limit.

19. Given $\epsilon > 0$, we need $\delta > 0$ such that if $|x-a| < \delta$ then $|x-a| < \epsilon$. So $\delta = \epsilon$ will work.

20. Given $\epsilon > 0$, we need $\delta > 0$ such that if $|x-a| < \delta$ then $|c-c| < \epsilon$. But $|c-c| = 0$, so this will be true no matter what δ we pick.

21. Given $\epsilon > 0$, we need $\delta > 0$ such that if $|x| < \delta$ then $|x^2-0| < \epsilon \Leftrightarrow x^2 < \epsilon \Leftrightarrow |x| < \sqrt{\epsilon}$. Take $\delta = \sqrt{\epsilon}$. Then $|x-0| < \delta \Rightarrow |x^2-0| < \epsilon$. Thus $\lim_{x\to 0}x^2 = 0$ by the definition of a limit.

22. Given $\epsilon > 0$, we need $\delta > 0$ such that if $|x| < \delta$ then $|x^3 - 0| < \epsilon \quad\Leftrightarrow\quad |x|^3 < \epsilon \quad\Leftrightarrow\quad |x| < \sqrt[3]{\epsilon}$. Take $\delta = \sqrt[3]{\epsilon}$. Then $|x - 0| < \delta \quad\Rightarrow\quad |x^3 - 0| < \delta^3 = \epsilon$. Thus $\lim_{x\to 0} x^3 = 0$ by the definition of a limit.

23. Given $\epsilon > 0$, we need $\delta > 0$ such that if $|x - 0| < \delta$ then $||x| - 0| < \epsilon$. But $||x|| = |x|$. So this is true if we pick $\delta = \epsilon$.

24. Given $\epsilon > 0$, we need $\delta > 0$ such that if $9 - \delta \le x < 9$, then $\left|\sqrt[4]{9-x} - 0\right| < \epsilon \quad\Leftrightarrow\quad \sqrt[4]{9-x} < \epsilon \quad\Leftrightarrow\quad 9 - x < \epsilon^4 \quad\Leftrightarrow\quad 9 - \epsilon^4 < x < 9$. So take $\delta = \epsilon^4$. Then $9 - \delta \le x < 9 \quad\Rightarrow\quad \left|\sqrt[4]{9-x} - 0\right| < \epsilon$. Thus $\lim_{x\to 9^-} \sqrt[4]{9-x} = 0$ by the definition of a limit.

25. Given $\epsilon > 0$, we need $\delta > 0$ such that if $|x - 2| < \delta$, then $|(x^2 - 4x + 5) - 1| < \epsilon \quad\Leftrightarrow\quad |x^2 - 4x + 4| < \epsilon \quad\Leftrightarrow\quad |(x-2)^2| < \epsilon$. So take $\delta = \sqrt{\epsilon}$. Then $|x - 2| < \delta \quad\Leftrightarrow\quad |x - 2| < \sqrt{\epsilon} \quad\Leftrightarrow\quad |(x-2)^2| < \epsilon$. So $\lim_{x\to 2}(x^2 - 4x + 5) = 1$ by the definition of a limit.

26. Given $\epsilon > 0$, we need $\delta > 0$ such that if $|x - 3| < \delta$, then $|(x^2 + x - 4) - 8| < \epsilon \quad\Leftrightarrow\quad |x^2 + x - 12| < \epsilon \quad\Leftrightarrow\quad |(x-3)(x+4)| < \epsilon$. Notice that if $|x - 3| < 1$, then $-1 < x - 3 < 1 \quad\Rightarrow\quad 6 < x + 4 < 8 \Rightarrow |x + 4| < 8$. So take $\delta = \min\{1, \epsilon/8\}$. Then $|x - 3| < \delta \quad\Leftrightarrow\quad |(x-3)(x+4)| \le |8(x-3)| = 8 \cdot |x - 3| < 8\delta \le \epsilon$. So $\lim_{x\to 3}(x^2 + x - 4) = 8$ by the definition of a limit.

27. Given $\epsilon > 0$, we need $\delta > 0$ such that if $|x - (-2)| < \delta$ then $|(x^2 - 1) - 3| < \epsilon$ or upon simplifying we need $|x^2 - 4| < \epsilon$ whenever $|x + 2| < \delta$. Notice that if $|x + 2| < 1$, then $-1 < x + 2 < 1 \quad\Rightarrow\quad -5 < x - 2 < -3 \quad\Rightarrow\quad |x - 2| < 5$. So take $\delta = \min\{\epsilon/5, 1\}$. Then $|x - 2| < 5$ and $|x + 2| < \epsilon/5$, so
$|(x^2 - 1) - 3| = |(x+2)(x-2)| = |x+2||x-2| < (\epsilon/5)(5) = \epsilon$.
Therefore, by the definition of a limit, $\lim_{x\to -2}(x^2 - 1) = 3$.

28. Given $\epsilon > 0$, we need $\delta > 0$ such that if $|x - 2| < \delta$, then $|x^3 - 8| < \epsilon$. Now $|x^3 - 8| = |(x-2)(x^2 + 2x + 4)|$. If $|x - 2| < 1$, that is, $1 < x < 3$, then $x^2 + 2x + 4 < 3^2 + 2(3) + 4 = 19$ and so $|x^3 - 8| = |x - 2|(x^2 + 2x + 4) < 19|x - 2|$. So if we take $\delta = \min\left\{1, \frac{\epsilon}{19}\right\}$, then $|x - 2| < \delta \quad\Rightarrow\quad |x^3 - 8| = |x - 2|(x^2 + 2x + 4) < \frac{\epsilon}{19} \cdot 19 = \epsilon$. So by the definition of a limit, $\lim_{x\to 2} x^3 = 8$.

29. Given $\epsilon > 0$, we let $\delta = \min\left\{2, \frac{\epsilon}{8}\right\}$. If $0 < |x - 3| < \delta$ then $|x - 3| < 2 \quad\Rightarrow\quad 1 < x < 5 \quad\Rightarrow\quad |x + 3| < 8$. Also $|x - 3| < \frac{\epsilon}{8}$, so $|x^2 - 9| = |x+3||x-3| < 8 \cdot \frac{\epsilon}{8} = \epsilon$. Thus $\lim_{x\to 3} x^2 = 9$.

30. *1. Guessing a value for δ* Let $\epsilon > 0$ be given. We have to find a number $\delta > 0$ such that $\left|\frac{1}{x} - \frac{1}{2}\right| < \epsilon$ whenever $0 < |x-2| < \delta$. But $\left|\frac{1}{x} - \frac{1}{2}\right| = \left|\frac{2-x}{2x}\right| = \frac{|x-2|}{|2x|} < \epsilon$. We find a positive constant C such that $\frac{1}{|2x|} < C \quad \Rightarrow \quad \frac{|x-2|}{|2x|} < C|x-2|$ and we can make $C|x-2| < \epsilon$ by taking $|x-2| < \frac{\epsilon}{C} = \delta$. We restrict x to lie in the interval $|x-2| < 1 \quad \Rightarrow \quad 1 < x < 3$ so $1 > \frac{1}{x} > \frac{1}{3} \quad \Rightarrow \quad \frac{1}{6} < \frac{1}{2x} < \frac{1}{2} \quad \Rightarrow \quad \frac{1}{|2x|} < \frac{1}{2}$. So $C = \frac{1}{2}$ is suitable. Thus we should choose $\delta = \min\{1, 2\epsilon\}$.

2. *Showing that δ works* Given $\epsilon > 0$ we let $\delta = \min\{1, 2\epsilon\}$. If $0 < |x-2| < \delta$, then $|x-2| < 1 \quad \Rightarrow$ $1 < x < 3 \quad \Rightarrow \quad \frac{1}{|2x|} < \frac{1}{2}$ (as in part 1). Also $|x-2| < 2\epsilon$, so $\left|\frac{1}{x} - \frac{1}{2}\right| = \frac{|x-2|}{|2x|} < \frac{1}{2} \cdot 2\epsilon = \epsilon$. This shows that $\lim_{x\to 2} (1/x) = \frac{1}{2}$.

31. *1. Guessing a value for δ* Given $\epsilon > 0$, we must find $\delta > 0$ such that $\left|\sqrt{x} - \sqrt{a}\right| < \epsilon$ whenever $0 < |x-a| < \delta$. But $\left|\sqrt{x} - \sqrt{a}\right| = \frac{|x-a|}{\sqrt{x}+\sqrt{a}} < \epsilon$ (from the hint). Now if we can find a positive constant C such that $\sqrt{x} + \sqrt{a} > C$ then $\frac{|x-a|}{\sqrt{x}+\sqrt{a}} < \frac{|x-a|}{C} < \epsilon$, and we take $|x-a| < C\epsilon$. We can find this number by restricting x to lie in some interval centered at a. If $|x-a| < \frac{1}{2}a$, then $\frac{1}{2}a < x < \frac{3}{2}a \quad \Rightarrow$ $\sqrt{x} + \sqrt{a} > \sqrt{\frac{1}{2}a} + \sqrt{a}$, and so $C = \sqrt{\frac{1}{2}a} + \sqrt{a}$ is a suitable choice for the constant. So $|x-a| < \left(\sqrt{\frac{1}{2}a} + \sqrt{a}\right)\epsilon$. This suggests that we let $\delta = \min\left\{\frac{1}{2}a, \left(\sqrt{\frac{1}{2}a} + \sqrt{a}\right)\epsilon\right\}$.

2. *Showing that δ works* Given $\epsilon > 0$, we let $\delta = \min\left\{\frac{1}{2}a, \left(\sqrt{\frac{1}{2}a} + \sqrt{a}\right)\epsilon\right\}$. If $0 < |x-a| < \delta$, then $|x-a| < \frac{1}{2}a \quad \Rightarrow \quad \sqrt{x} + \sqrt{a} > \sqrt{\frac{1}{2}a} + \sqrt{a}$ (as in part 1). Also $|x-a| < \left(\sqrt{\frac{1}{2}a} + \sqrt{a}\right)\epsilon$, so $\left|\sqrt{x} + \sqrt{a}\right| = \frac{|x-a|}{\sqrt{x}+\sqrt{a}} < \frac{\left(\sqrt{a/2} + \sqrt{a}\right)\epsilon}{\left(\sqrt{a/2} + \sqrt{a}\right)} = \epsilon$. Therefore $\lim_{x\to a} \sqrt{x} = \sqrt{a}$ by the definition of a limit.

32. Suppose that $\lim_{t\to 0} H(t) = L$. Given $\epsilon = \frac{1}{2}$, there exists $\delta > 0$ such that $0 < |t| < \delta \quad \Rightarrow \quad |H(t) - L| < \frac{1}{2}$ $\Leftrightarrow \quad L - \frac{1}{2} < H(t) < L + \frac{1}{2}$. For $0 < t < \delta$, $H(t) = 1$, so $1 < L + \frac{1}{2} \quad \Rightarrow \quad L > \frac{1}{2}$. For $-\delta < t < 0$, $H(t) = 0$, so $L - \frac{1}{2} < 0 \quad \Rightarrow \quad L < \frac{1}{2}$. This contradicts $L > \frac{1}{2}$. Therefore $\lim_{t\to 0} H(t)$ does not exist.

33. Suppose that $\lim_{x\to 0} f(x) = L$. Given $\epsilon = \frac{1}{2}$, there exists $\delta > 0$ such that $0 < |x| < \delta \quad \Rightarrow \quad |f(x) - L| < \frac{1}{2}$. Take any rational number r with $0 < |r| < \delta$. Then $f(r) = 0$, so $|0 - L| < \frac{1}{2}$, so $L \le |L| < \frac{1}{2}$. Now take any irrational number s with $0 < |s| < \delta$. Then $f(s) = 1$, so $|1 - L| < \frac{1}{2}$. Hence $1 - L < \frac{1}{2}$, so $L > \frac{1}{2}$. This contradicts $L < \frac{1}{2}$, so $\lim_{x\to 0} f(x)$ does not exist.

34. First suppose that $\lim_{x\to a} f(x) = L$. Then, given $\epsilon > 0$ there exists $\delta > 0$ so that $0 < |x-a| < \delta \quad \Rightarrow$ $|f(x) - L| < \epsilon$. Then $a - \delta < x < a \quad \Rightarrow \quad 0 < |x-a| < \delta$ so $|f(x) - L| < \epsilon$. Thus $\lim_{x\to a^-} f(x) = L$. Also $a < x < a + \delta \quad \Rightarrow \quad 0 < |x-a| < \delta$ so $|f(x) - L| < \epsilon$. Hence $\lim_{x\to a^+} f(x) = L$.

Now suppose $\lim_{x\to a^-} f(x) = L = \lim_{x\to a^+} f(x)$. Let $\epsilon > 0$ be given. Since $\lim_{x\to a^-} f(x) = L$, there exists $\delta_1 > 0$ so that $a - \delta_1 < x < a \quad\Rightarrow\quad |f(x) - L| < \epsilon$. Since $\lim_{x\to a^+} f(x) = L$, there exists $\delta_2 > 0$ so that $a < x < a + \delta_2 \quad\Rightarrow$ $|f(x) - L| < \epsilon$. Let δ be the smaller of δ_1 and δ_2. Then $0 < |x - a| < \delta \quad\Rightarrow\quad a - \delta_1 < x < a$ or $a < x < a + \delta_2$ so $|f(x) - L| < \epsilon$. Hence $\lim_{x\to a} f(x) = L$.

So we have proved that $\lim_{x\to a} f(x) = L \quad\Leftrightarrow\quad \lim_{x\to a^-} f(x) = L = \lim_{x\to a^+} f(x)$.

35. $\dfrac{1}{(x+3)^4} > 10{,}000 \quad\Leftrightarrow\quad (x+3)^4 < \dfrac{1}{10{,}000} \quad\Leftrightarrow\quad |x - (-3)| = |x+3| < \dfrac{1}{10}$

36. Given $M > 0$, we need $\delta > 0$ such that $|x+3| < \delta \quad\Rightarrow\quad 1/(x+3)^4 > M$. Now $\dfrac{1}{(x+3)^4} > M \quad\Leftrightarrow$ $(x+3)^4 < \dfrac{1}{M} \quad\Leftrightarrow\quad |x+3| < \dfrac{1}{\sqrt[4]{M}}$. So take $\delta = \dfrac{1}{\sqrt[4]{M}}$. Then $0 < |x+3| < \delta = \dfrac{1}{\sqrt[4]{M}} \quad\Rightarrow$ $\dfrac{1}{(x+3)^4} > M$, so $\lim_{x\to -3} \dfrac{1}{(x+3)^4} = \infty$.

37. Let $N < 0$ be given. Then, for $x < -1$, we have $\dfrac{5}{(x+1)^3} < N \quad\Leftrightarrow\quad \dfrac{5}{N} < (x+1)^3 \quad\Leftrightarrow$ $\sqrt[3]{\dfrac{5}{N}} < x + 1$. Let $\delta = -\sqrt[3]{\dfrac{5}{N}}$. Then $-1 - \delta < x < -1 \quad\Rightarrow\quad \sqrt[3]{\dfrac{5}{N}} < x + 1 < 0 \quad\Rightarrow\quad \dfrac{5}{(x+1)^3} < N$, so $\lim_{x\to -1^-} \dfrac{5}{(x+1)^3} = -\infty$.

38. **(a)** Let M be given. Since $\lim_{x\to a} f(x) = \infty$, there exists $\delta_1 > 0$ such that $0 < |x - a| < \delta_1 \quad\Rightarrow$ $f(x) > M + 1 - c$. Since $\lim_{x\to a} g(x) = c$, there exists $\delta_2 > 0$ such that $0 < |x - a| < \delta_2 \quad\Rightarrow$ $|g(x) - c| < 1 \quad\Rightarrow\quad g(x) > c - 1$. Let δ be the smaller of δ_1 and δ_2. Then $0 < |x - a| < \delta \quad\Rightarrow$ $f(x) + g(x) > (M + 1 - c) + (c - 1) = M$. Thus $\lim_{x\to a} [f(x) + g(x)] = \infty$.

(b) Let $M > 0$ be given. Since $\lim_{x\to a} g(x) = c > 0$, there exists $\delta_1 > 0$ such that $0 < |x - a| < \delta_1 \quad\Rightarrow$ $|g(x) - c| < c/2 \;\Rightarrow\; g(x) > c/2$. Since $\lim_{x\to a} f(x) = \infty$, there exists $\delta_2 > 0$ such that $0 < |x - a| < \delta_2$ $\Rightarrow\; f(x) > 2M/c$. Let $\delta = \min\{\delta_1, \delta_2\}$. Then $0 < |x - a| < \delta \;\Rightarrow\; f(x)\, g(x) > \dfrac{2M}{c}\,\dfrac{c}{2} = M$, so $\lim_{x\to a} f(x)\, g(x) = \infty$.

(c) Let $N < 0$ be given. Since $\lim_{x\to a} g(x) = c < 0$, there exists $\delta_1 > 0$ such that $0 < |x - a| < \delta_1 \quad\Rightarrow$ $|g(x) - c| < -c/2 \quad\Rightarrow\quad g(x) < c/2$. Since $\lim_{x\to a} f(x) = \infty$, there exists $\delta_2 > 0$ such that $0 < |x - a| < \delta_2 \quad\Rightarrow\quad f(x) > 2N/c$. (Note that $c < 0$ and $N < 0 \quad\Rightarrow\quad 2N/c > 0$.) Let $\delta = \min\{\delta_1, \delta_2\}$. Then $0 < |x - a| < \delta \quad\Rightarrow\quad f(x) > 2N/c \quad\Rightarrow\quad f(x)\, g(x) < \dfrac{2N}{c} \cdot \dfrac{c}{2} = N$, so $\lim_{x\to a} f(x)\, g(x) = -\infty$.

EXERCISES 1.5

1. **(a)** f is discontinuous at $-5, -3, -1, 3, 5, 8$ and 10.

(b) f is continuous from the left at -5 and -3, and continuous from the right at 8.

It is continuous on neither side at -1, 3, 5, and 10.

2. g is continuous on $[-6, -5]$, $(-5, -3)$, $(-3, -2]$, $(-2, 1)$, $(1, 3)$, $[3, 5]$, $(5, 7]$, $(7, 8)$, and $(8, 9]$.

3. $\lim\limits_{x\to 3}(x^4 - 5x^3 + 6) = \lim\limits_{x\to 3} x^4 - 5\lim\limits_{x\to 3} x^3 + \lim\limits_{x\to 3} 6 = 3^4 - 5(3^3) + 6 = -48 = f(3)$. Thus f is continuous at 3.

4. $\lim\limits_{x\to 2} f(x) = \lim\limits_{x\to 2}\left[x^2 + (x-1)^9\right] = \lim\limits_{x\to 2} x^2 + \left(\lim\limits_{x\to 2} x - \lim\limits_{x\to 2} 1\right)^9 = 2^2 + (2-1)^9 = 5 = f(2)$. Thus f is continuous at 2.

5. $\lim\limits_{x\to 5} f(x) = \lim\limits_{x\to 5}\left(1 + \sqrt{x^2 - 9}\right) = \lim\limits_{x\to 5} 1 + \sqrt{\lim\limits_{x\to 5} x^2 - \lim\limits_{x\to 5} 9} = 1 + \sqrt{5^2 - 9} = 5 = f(5)$. Thus f is continuous at 5.

6. $\lim\limits_{x\to 4} g(x) = \lim\limits_{x\to 4}\dfrac{x+1}{2x^2-1} = \dfrac{\lim\limits_{x\to 4} x + \lim\limits_{x\to 4} 1}{2\lim\limits_{x\to 4} x^2 - \lim\limits_{x\to 4} 1} = \dfrac{4+1}{2(4)^2 - 1} = \dfrac{5}{31} = g(4)$. So g is continuous at 4.

7. $\lim\limits_{t\to -8} g(t) = \lim\limits_{t\to -8}\dfrac{\sqrt[3]{t}}{(t+1)^4} = \dfrac{\sqrt[3]{\lim\limits_{t\to -8} t}}{\left(\lim\limits_{t\to -8} t + 1\right)^4} = \dfrac{\sqrt[3]{-8}}{(-8+1)^4} = -\dfrac{2}{2401} = g(-8)$. Thus g is continuous at -8.

8. For $a > 1$ we have $\lim\limits_{x\to a} f(x) = \lim\limits_{x\to a}\left(x + \sqrt{x-1}\right) = \lim\limits_{x\to a} x + \sqrt{\lim\limits_{x\to a} x - \lim\limits_{x\to a} 1} = a + \sqrt{a-1} = f(a)$, so f is continuous on $(1, \infty)$. A similar calculation shows that $\lim\limits_{x\to 1^+} f(x) = 1 = f(1)$, so f is continuous from the right at 1. Thus f is continuous on $[1, \infty)$.

9. For $-4 < a < 4$ we have $\lim\limits_{x\to a} f(x) = \lim\limits_{x\to a} x\sqrt{16 - x^2} = \lim\limits_{x\to a} x\sqrt{\lim\limits_{x\to a} 16 - \lim\limits_{x\to a} x^2} = a\sqrt{16 - a^2} = f(a)$, so f is continuous on $(-4, 4)$. Similarly, we get $\lim\limits_{x\to 4^-} f(x) = 0 = f(4)$ and $\lim\limits_{x\to -4^+} f(x) = 0 = f(-4)$, so f is continuous from the left at 4 and from the right at -4. Thus f is continuous on $[-4, 4]$.

10. For $a < 3$, $\lim\limits_{x\to a} F(x) = \lim\limits_{x\to a}\dfrac{x+1}{x-3} = \dfrac{\lim\limits_{x\to a} x + \lim\limits_{x\to a} 1}{\lim\limits_{x\to a} x - \lim\limits_{x\to a} 3} = \dfrac{a+1}{a-3} = F(a)$, so F is continuous on $(-\infty, 3)$.

11. For any $a \in \mathbb{R}$ we have $\lim\limits_{x\to a} f(x) = \lim\limits_{x\to a}(x^2-1)^8 = \left(\lim\limits_{x\to a} x^2 - \lim\limits_{x\to a} 1\right)^8 = (a^2-1)^8 = f(a)$. Thus f is continuous on $(-\infty, \infty)$.

12. $f(x) = \dfrac{x^2 - 1}{x + 1}$ is discontinuous at -1 because $f(-1)$ is not defined.

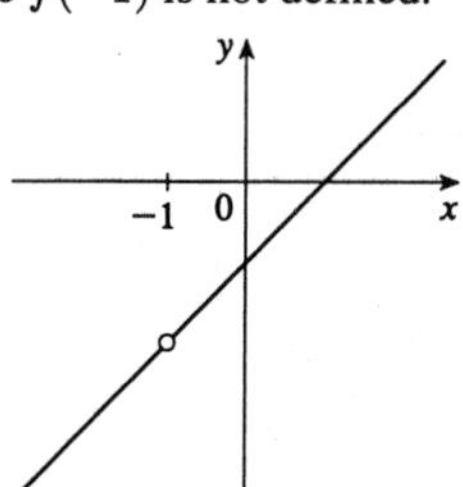

13. $f(x) = -\dfrac{1}{(x-1)^2}$ is discontinuous at 1 since $f(1)$ is not defined.

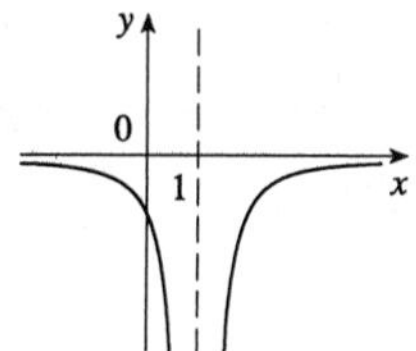

14. Since $f(x) = \dfrac{x^2 - 1}{x + 1}$ for $x \neq -1$, we have $\lim\limits_{x \to -1} f(x) = \lim\limits_{x \to -1} \dfrac{x^2 - 1}{x + 1} = \lim\limits_{x \to -1} (x - 1)$ $= -2$. But $f(-1) = 6$, so $\lim\limits_{x \to -1} f(x) \neq f(-1)$. Therefore f is discontinuous at -1.

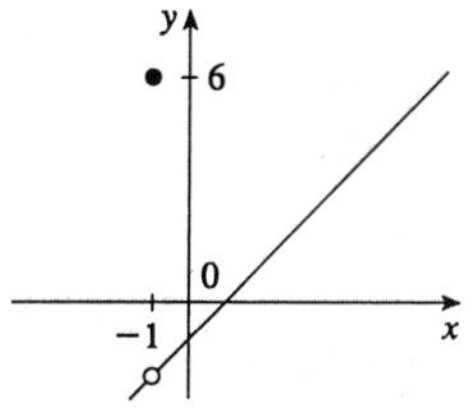

15. $\lim\limits_{x \to 1} f(x) = \lim\limits_{x \to 1} \left[-\dfrac{1}{(x-1)^2}\right]$ does not exist. Therefore f is discontinuous at 1.

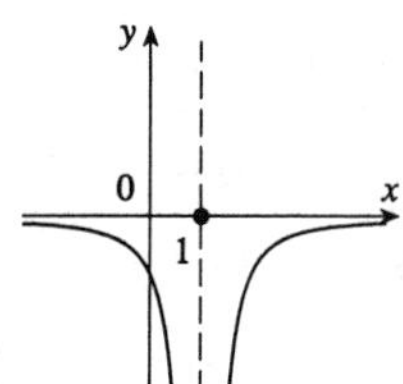

16. Since $f(x) = \dfrac{x^2 - 2x - 8}{x - 4}$ if $x \neq 4$,

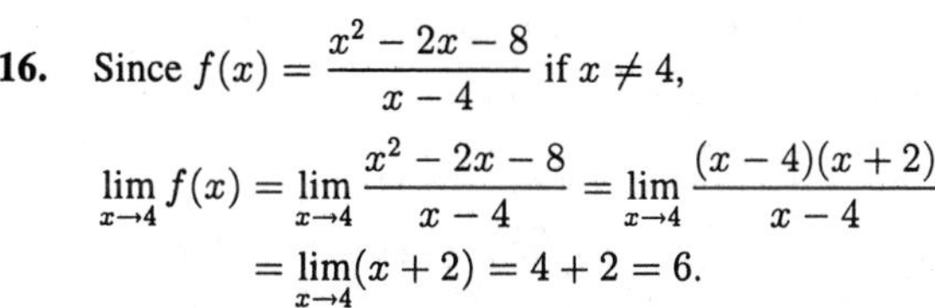

$$\lim_{x \to 4} f(x) = \lim_{x \to 4} \frac{x^2 - 2x - 8}{x - 4} = \lim_{x \to 4} \frac{(x-4)(x+2)}{x - 4}$$
$$= \lim_{x \to 4} (x + 2) = 4 + 2 = 6.$$

But $f(4) = 3$, so $\lim\limits_{x \to 4} f(x) \neq f(4)$. Therefore f is discontinuous at 4.

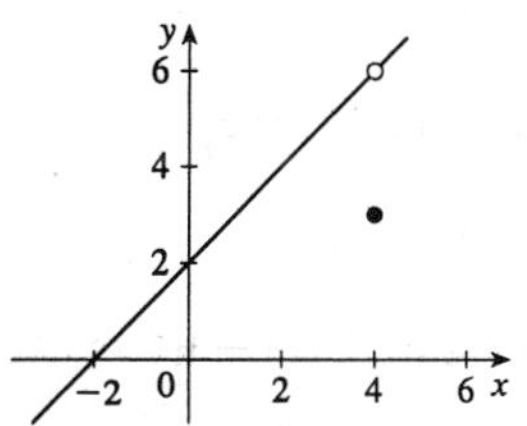

17. Since $f(x) = x^2 - 2$ for $x \neq -3$,

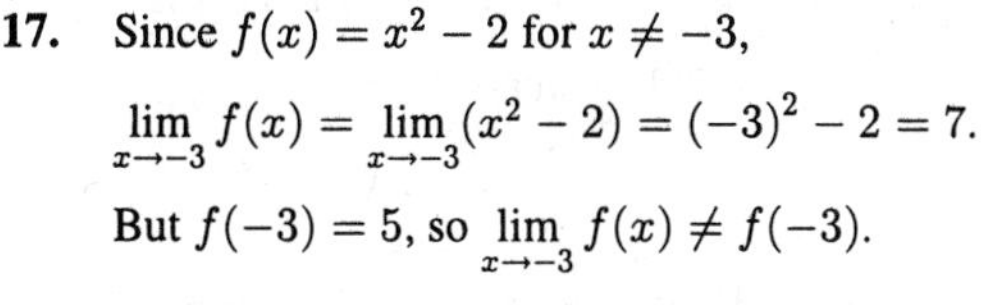

$\lim\limits_{x \to -3} f(x) = \lim\limits_{x \to -3} (x^2 - 2) = (-3)^2 - 2 = 7.$

But $f(-3) = 5$, so $\lim\limits_{x \to -3} f(x) \neq f(-3)$.

Therefore f is discontinuous at -3.

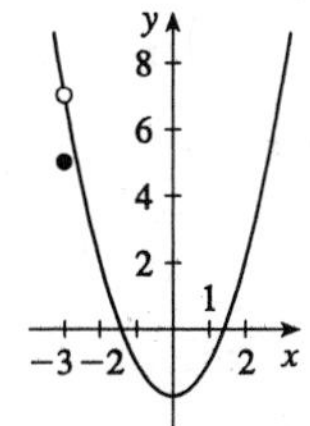

18. $\lim_{x\to 2^-} f(x) = \lim_{x\to 2^-}(1-x) = 1-2 = -1$ and $\lim_{x\to 2^+} f(x) = \lim_{x\to 2^+}(x^2-2x) = (2)^2 - 2(2) = 0$. Since $\lim_{x\to 2^-} f(x) \neq \lim_{x\to 2^+} f(x)$, $\lim_{x\to 2} f(x)$ does not exist and therefore f is discontinuous at 2 [by Note 2 after Definition 1].

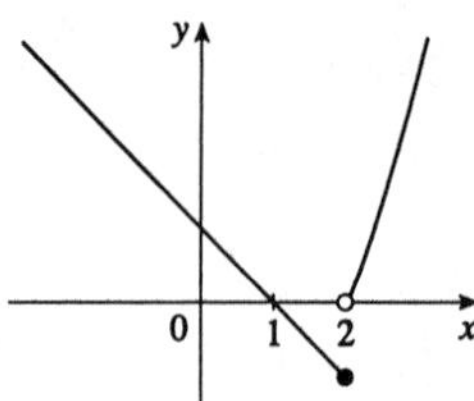

19. $f(x) = (x+1)(x^3+8x+9)$ is a polynomial, so by Theorem 5 it is continuous on $\mathbb{R}$.

20. $G(x) = \dfrac{x^4+17}{6x^2+x-1}$ is a rational function, so by Theorem 5 it is continuous on its domain, which is $\{x \mid (3x-1)(2x+1) \neq 0\} = \left\{x \mid x \neq -\frac{1}{2}, \frac{1}{3}\right\}$.

21. $g(x) = x+1$, a polynomial, is continuous (by Theorem 5) and $f(x) = \sqrt{x}$ is continuous on $[0,\infty)$ by Theorem 6, so $f(g(x)) = \sqrt{x+1}$ is continuous on $[-1,\infty)$ by Theorem 8. By Theorem 4 #5, $H(x) = 1/\sqrt{x+1}$ is continuous on $(-1,\infty)$.

22. $G(t) = 25 - t^2$ is a polynomial, so it is continuous (Theorem 5). $F(x) = \sqrt{x}$ is continuous by Theorem 6. So, by Theorem 8, $F(G(t)) = \sqrt{25-t^2}$ is continuous on its domain, which is $\{t \mid 25 - t^2 \geq 0\} = \{t \mid |t| \leq 5\} = [-5,5]$. Also, $2t$ is continuous on $\mathbb{R}$, so by Theorem 4 #1, $f(t) = 2t + \sqrt{25-t^2}$ is continuous on its domain, which is $[-5,5]$.

23. $g(x) = x-1$ and $G(x) = x^2 - 2$ are both polynomials, so by Theorem 5 they are continuous. Also $f(x) = \sqrt[5]{x}$ is continuous by Theorem 6, so $f(g(x)) = \sqrt[5]{x-1}$ is continuous on $\mathbb{R}$ by Theorem 8. Thus the product $h(x) = \sqrt[5]{x-1}(x^2-2)$ is continuous on $\mathbb{R}$ by Theorem 4 #4.

24. $G(t) = t^2 - 4$ is continuous since it is a polynomial (Theorem 5). $F(x) = \sqrt{x}$ is continuous by Theorem 6. So, by Theorem 8, $F(G(t)) = \sqrt{t^2-4}$ is continuous on its domain, which is $D = \{t \mid t^2 - 4 \geq 0\} = \{t \mid |t| \geq 2\}$. Also t is continuous so $t + \sqrt{t^2-4}$ is continuous on D by Theorem 4 #1. Thus by Theorem 4 #5, $g(t) = 1/\left(t+\sqrt{t^2-4}\right)$ is continuous on its domain, which is $\left\{t \in D \mid t + \sqrt{t^2-4} \neq 0\right\}$. But if $t + \sqrt{t^2-4} = 0$, then $\sqrt{t^2-4} = -t \;\Rightarrow\; t^2 - 4 = t^2 \;\Rightarrow\; -4 = 0$ which is false. So the domain of g is $\{t \in D \mid |t| \geq 2\} = (-\infty, -2] \cup [2, \infty)$.

25. Since the discriminant of t^2+t+1 is negative, t^2+t+1 is always positive. So the domain of $F(t)$ is $\mathbb{R}$. By Theorem 5 the polynomial $(t^2+t+1)^3$ is continuous. By Theorems 6 and 8 the composition $F(t) = \sqrt{(t^2+t+1)^3}$ is continuous on $\mathbb{R}$.

26. $H(x) = \sqrt{(x-2)/(5+x)}$. The domain is $\{x \mid (x-2)/(5+x) > 0\} = (-\infty, -5) \cup [2,\infty)$ by the methods of Appendix A. By Theorem 5 the rational function $(x-2)/(5+x)$ is continuous. Since the square root function is continuous (Theorem 6), the composition $H(x) = \sqrt{(x-2)/(5+x)}$ is continuous on its domain by Theorem 8.

27. $g(x) = x^3 - x$ is continuous on $\mathbb{R}$ since it is a polynomial [Theorem 5(a)], and $f(x) = |x|$ is continuous on $\mathbb{R}$ by Example 9(a). So $L(x) = |x^3 - x|$ is continuous on $\mathbb{R}$ by Theorem 8.

28. **(a)** $x^2 - x$ and $(x^3 + x)^9$ are continuous on $(-\infty, \infty)$ since they are polynomials (Theorem 5). By Theorems 6 and 8, $\sqrt[5]{x^2 - x}$ is continuous on $(-\infty, \infty)$. Finally, by Theorem 4 #1, $\sqrt[5]{x^2 - x} + (x^3 + x)^9$ is continuous on $(-\infty, \infty)$.

(b) Since the function is continuous, $\lim\limits_{x \to 1} f(x) = f(1)$, so

$$\lim_{x \to 1} \left[\sqrt[5]{x^2 - x} + (x^3 + x)^9\right] = \sqrt[5]{1^2 - 1} + (1^3 + 1)^9 = 2^9 = 512.$$

29. f is continuous on $(-\infty, 3)$ and $(3, \infty)$ since on each of these intervals it is a polynomial.

Also $\lim\limits_{x \to 3^+} f(x) = \lim\limits_{x \to 3^+} (5 - x) = 2$ and $\lim\limits_{x \to 3^-} f(x) = \lim\limits_{x \to 3^-} (x - 1) = 2$, so $\lim\limits_{x \to 3} f(x) = 2$.

Since $f(3) = 5 - 3 = 2$, f is also continuous at 3. Thus f is continuous on $(-\infty, \infty)$.

30. g is continuous on $(-\infty, 0)$, $(0, 1)$ and $(1, \infty)$ since on each of these intervals it is a polynomial. Now $\lim\limits_{x \to 0^-} g(x) = \lim\limits_{x \to 0^-} x = 0 = \lim\limits_{x \to 0^+} x^2 = \lim\limits_{x \to 0^+} g(x)$, so $\lim\limits_{x \to 0} g(x) = 0 = 0^2 = g(0)$. Also $\lim\limits_{x \to 1^-} g(x) = \lim\limits_{x \to 1^-} x^2 = 1 = \lim\limits_{x \to 1^+} x^3 = \lim\limits_{x \to 1^+} g(x)$, so $\lim\limits_{x \to 1} g(x) = 1 = 1^2 = g(1)$. So g is also continuous at 0 and 1. Thus g is continuous on $(-\infty, \infty)$.

31. f is continuous on $(-\infty, 0)$ and $(0, \infty)$ since on each of these intervals it is a polynomial. Now $\lim\limits_{x \to 0^-} f(x) = \lim\limits_{x \to 0^-} (x - 1)^3 = -1$ and $\lim\limits_{x \to 0^+} f(x) = \lim\limits_{x \to 0^+} (x + 1)^3 = 1$. Thus $\lim\limits_{x \to 0} f(x)$ does not exist, so f is discontinuous at 0. Since $f(0) = 1$, f is continuous from the right at 0.

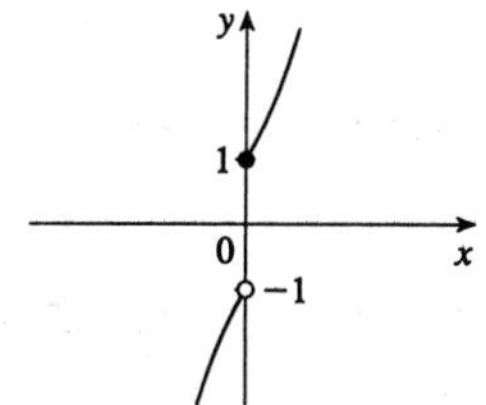

32. f is continuous on $(-\infty, -1)$, $(-1, 1)$ and $(1, \infty)$ since on each of these intervals it is a polynomial. Now $\lim\limits_{x \to -1^-} f(x) = \lim\limits_{x \to -1^-} (2x + 1) = -1$ and $\lim\limits_{x \to -1^+} f(x) = \lim\limits_{x \to -1^+} 3x = -3$, so f is discontinuous at -1. Since $f(-1) = -1$, f is continuous from the left at -1. Also $\lim\limits_{x \to 1^-} f(x) = \lim\limits_{x \to 1^-} 3x = 3$ and $\lim\limits_{x \to 1^+} f(x) = \lim\limits_{x \to 1^+} (2x - 1) = 1$, so f is discontinuous at 1. Since $f(1) = 1$, f is continuous from the right at 1.

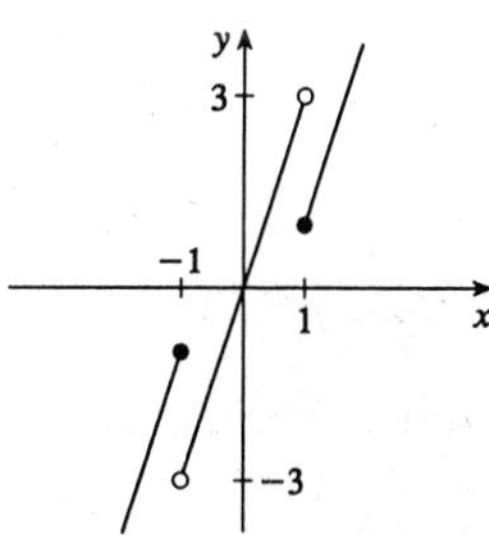

33. f is continuous on $(-\infty,-1)$, $(-1,1)$ and $(1,\infty)$. Now

$\lim_{x\to-1^-} f(x) = \lim_{x\to-1^-} \frac{1}{x} = -1$ and $\lim_{x\to-1^+} f(x) = \lim_{x\to-1^+} x = -1$,

so $\lim_{x\to-1} f(x) = -1 = f(-1)$ and f is continuous at -1.

Also $\lim_{x\to1^-} f(x) = \lim_{x\to1^-} x = 1$ and $\lim_{x\to1^+} f(x) = \lim_{x\to1^+} \frac{1}{x^2} = 1$,

so $\lim_{x\to1} f(x) = 1 = f(1)$ and f is continuous at 1.

Thus f has no discontinuities.

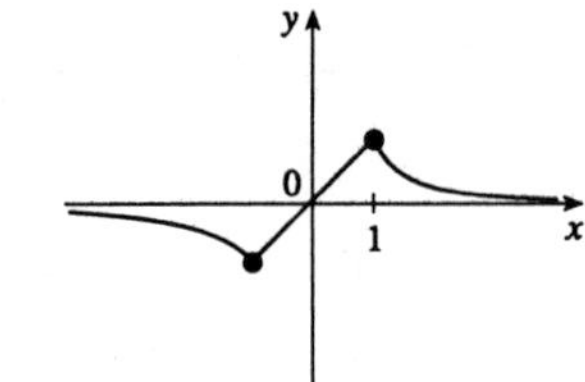

34. f is continuous on $(-\infty,0)$, $(0,1)$ and $(1,\infty)$. Since f is not defined at $x=0$, f is continuous neither from the right nor the left at 0. Also $\lim_{x\to1^-} f(x) = \lim_{x\to1^-} 1 = 1$ and

$\lim_{x\to1^+} f(x) = \lim_{x\to1^+} \sqrt{x} = 1$, so $\lim_{x\to1} f(x) = 1 = f(1)$ and

f is continuous at 1.

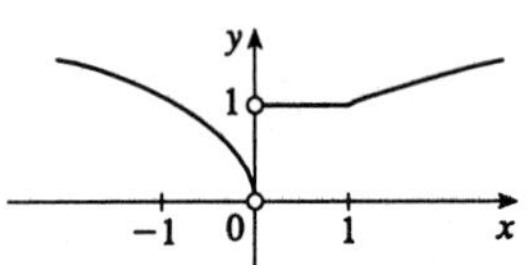

35. $f(x) = [\![2x]\!]$ is continuous except when $2x = n \quad\Leftrightarrow$

$x = n/2$, n an integer. In fact, $\lim_{x\to n/2^-} [\![2x]\!] = n-1$

and $\lim_{x\to n/2^+} [\![2x]\!] = n = f(n)$, so f is continuous only

from the right at $n/2$.

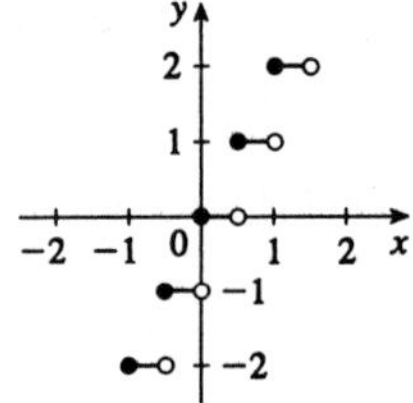

36. The salary function has discontinuities at $t = 6$, 12, 18, and 24, but is continuous from the right at 6, 12, and 18.

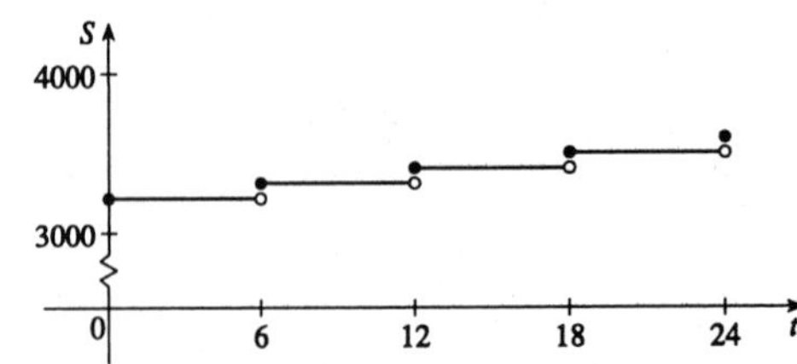

37. f is continuous on $(-\infty,3)$ and $(3,\infty)$. Now $\lim_{x\to3^-} f(x) = \lim_{x\to3^-} (cx+1) = 3c+1$ and

$\lim_{x\to3^+} f(x) = \lim_{x\to3^+} (cx^2-1) = 9c-1$. So f is continuous $\quad\Leftrightarrow\quad 3c+1 = 9c-1 \quad\Leftrightarrow\quad 6c = 2 \quad\Leftrightarrow$

$c = \frac{1}{3}$. Thus for f to be continuous on $(-\infty,\infty)$, $c = \frac{1}{3}$.

38. The functions x^2-c^2 and $cx+20$, considered on the intervals $(-\infty,4)$ and $[4,\infty)$ respectively, are continuous for any value of c. So the only possible discontinuity is at $x=4$. For the function to be continuous at $x=4$, the left-hand and right-hand limits must be the same. Now $\lim_{x\to4^-} g(x) = \lim_{x\to4^-} (x^2-c^2) = 16-c^2$ and

$\lim_{x\to4^-} (x^2-c^2) = g(4) = 4c+20$. Thus $4^2-c^2 = 4c+20 \quad\Leftrightarrow\quad c^2+4c+4=0 \quad\Leftrightarrow\quad c=-2$.

39. The functions $2x$, $cx^2 + d$ and $4x$ are continuous on their own domains, so the only possible problems occur at $x = 1$ and $x = 2$. The left- and right-hand limits at these points must be the same in order for $\lim_{x \to 1} h(x)$ and $\lim_{x \to 2} h(x)$ to exist. So we must have $2 \cdot 1 = c(1)^2 + d$ and $c(2)^2 + d = 4 \cdot 2$. From the first of these equations we get $d = 2 - c$. Substituting this into the second, we get $4c + (2 - c) = 8 \quad \Leftrightarrow \quad c = 2$. Back-substituting into the first to get d, we find that $d = 0$.

40. **(a)** $f(x) = \dfrac{x^2 - 2x - 8}{x + 2} = \dfrac{(x - 4)(x + 2)}{x + 2}$ has a removable discontinuity at -2 because $g(x) = x - 4$ is continuous on $\mathbb{R}$ and $f(x) = g(x)$ for $x \neq -2$. [The discontinuity is removed by defining $f(-2) = -6$.]

(b) $f(x) = \dfrac{x - 7}{|x - 7|} \quad \Rightarrow \quad \lim_{x \to 7^-} f(x) = -1$ and $\lim_{x \to 7^+} f(x) = 1$. Thus $\lim_{x \to 7} f(x)$ does not exist, so the discontinuity is not removable. (It is a jump discontinuity.)

(c) $f(x) = \dfrac{x^3 + 64}{x + 4} = \dfrac{(x + 4)(x^2 - 4x + 16)}{x + 4}$ has a removable discontinuity at -4 because $g(x) = x^2 - 4x + 16$ is continuous on $\mathbb{R}$ and $f(x) = g(x)$ for $x \neq -4$. [The discontinuity is removed by defining $f(-4) = 48$.]

(d) $f(x) = \dfrac{3 - \sqrt{x}}{9 - x} = \dfrac{3 - \sqrt{x}}{(3 - \sqrt{x})(3 + \sqrt{x})}$ has a removable discontinuity at 9 because $g(x) = 1/(3 + \sqrt{x})$ is continuous on $\mathbb{R}$ and $f(x) = g(x)$ for $x \neq 9$. [The discontinuity is removed by defining $f(9) = \frac{1}{6}$.]

41. **(a)** $\lim_{x \to 1^-} f(x) = \lim_{x \to 1^-} (1 - x^2) = 0$ and $\lim_{x \to 1^+} f(x) = \lim_{x \to 1^+} (1 + x/2) = \frac{3}{2}$. Thus $\lim_{x \to 1} f(x)$ does not exist, so f is not continuous at 1.

(b) $f(0) = 1$ and $f(2) = 2$. For $0 \leq x \leq 1$, f takes the values in $[0, 1]$. For $1 < x \leq 2$, f takes the values in $(1.5, 2]$. Thus f does not take on the value 1.5 $\big($or any other value in $(1, 1.5]\big)$.

42. $f(x) = x^2$ is continuous on the interval $[1, 2]$ and $f(1) = 1$ and $f(2) = 4$. Since $1 < 2 < 4$, there is a number c in $(1, 2)$ such that $f(c) = c^2 = 2$ by the Intermediate Value Theorem.

43. $f(x) = x^3 - x^2 + x$ is continuous on $[2, 3]$ and $f(2) = 6$, $f(3) = 21$. Since $6 < 10 < 21$, there is a number c in $(2, 3)$ such that $f(c) = 10$ by the Intermediate Value Theorem.

44. $g(x) = x^5 - 2x^3 + x^2 + 2$ is continuous on $[-2, -1]$ and $g(-2) = -10$, $g(-1) = 4$. Since $-10 < -1 < 4$, there is a number c in $(-2, -1)$ such that $g(c) = -1$ by the Intermediate Value Theorem.

45. $f(x) = x^3 - 3x + 1$ is continuous on $[0, 1]$ and $f(0) = 1$, $f(1) = -1$. Since $-1 < 0 < 1$, there is a number c in $(0, 1)$ such that $f(c) = 0$ by the Intermediate Value Theorem. Thus there is a root of the equation $x^3 - 3x + 1 = 0$ in the interval $(0, 1)$.

46. $f(x) = x^5 - 2x^4 - x - 3$ is continuous on $[2, 3]$ and $f(2) = -5$, $f(3) = 75$. Since $-5 < 0 < 75$, there is a number c in $(2, 3)$ such that $f(c) = 0$ by the Intermediate Value Theorem. Thus there is a root of the equation $x^5 - 2x^4 - x - 3 = 0$ in the interval $(2, 3)$.

47. $f(x) = x^3 + 2x - (x^2 + 1) = x^3 + 2x - x^2 - 1$ is continuous on $[0, 1]$ and $f(0) = -1$, $f(1) = 1$. Since $-1 < 0 < 1$, there is a number c in $(0, 1)$ such that $f(c) = 0$ by the Intermediate Value Theorem. Thus there is a root of the equation $x^3 + 2x - x^2 - 1 = 0$, or equivalently, $x^3 + 2x = x^2 + 1$, in the interval $(0, 1)$.

48. $f(x) = x^2 - \sqrt{x+1}$ is continuous on $[1, 2]$ and $f(1) = 1 - \sqrt{2}$, $f(2) = 4 - \sqrt{3}$. Since $1 - \sqrt{2} < 0 < 4 - \sqrt{3}$, there is a number c in $(1, 2)$ such that $f(c) = 0$ by the Intermediate Value Theorem. Thus there is a root of the equation $x^2 - \sqrt{x+1} = 0$, or $x^2 = \sqrt{x+1}$, in the interval $(1, 2)$.

49. **(a)** $f(x) = x^3 - x + 1$ is continuous on $[-2, -1]$ and $f(-2) = -5$, $f(-1) = 1$. Since $-5 < 0 < 1$, there is a number c in $(-2, 1)$ such that $f(c) = 0$ by the Intermediate Value Theorem. Thus there is a root of the equation $x^3 - x + 1 = 0$ in the interval $(-2, -1)$.

(b) $f(-1.33) \approx -0.0226$ and $f(-1.32) \approx 0.0200$, so there is a root between -1.33 and -1.32.

50. **(a)** $f(x) = x^5 - x^2 + 2x + 3$ is continuous on $[-1, 0]$ and $f(-1) = -1$, $f(0) = 3$. Since $-1 < 0 < 3$, there is a number c in $(-1, 0)$ such that $f(c) = 0$ by the Intermediate Value Theorem. Thus there is a root of the equation $x^5 - x^2 + 2x + 3 = 0$ in the interval $(-1, 0)$.

(b) $f(-0.88) \approx -0.0062$ and $f(-0.87) \approx 0.0047$, so there is a root between -0.88 and -0.87.

51. **(a)** Let $f(x) = x^5 - x^2 - 4$. Then $f(1) = 1^5 - 1^2 - 4 = -4 < 0$ and $f(2) = 2^5 - 2^2 - 4 = 24 > 0$. So by the Intermediate Value Theorem, there is a number c in $(1, 2)$ such that $c^5 - c^2 - 4 = 0$.

(b) We can see from the graphs that, correct to three decimal places, the root is $x \approx 1.434$.

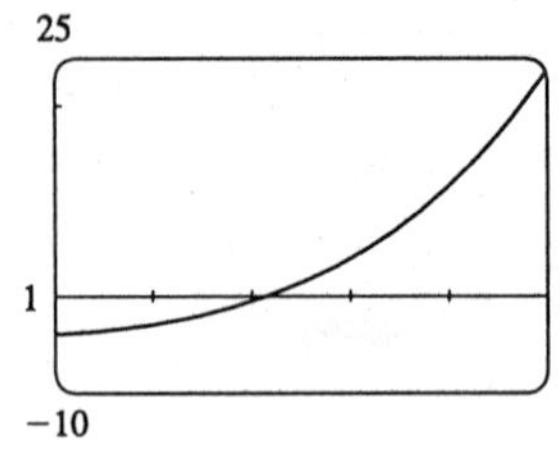

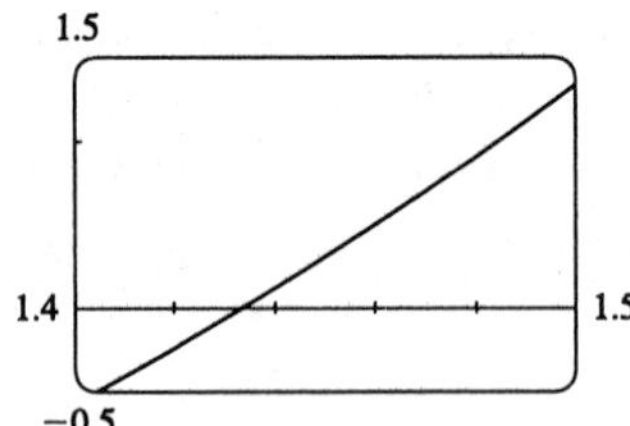

52. **(a)** Let $f(x) = \sqrt{x-5} - \dfrac{1}{x+3}$. Then $f(5) = \sqrt{0} - \dfrac{1}{5+3} = -\dfrac{1}{8} < 0$ and $f(6) = \sqrt{6-5} - \dfrac{1}{6+3} = \dfrac{8}{9} > 0$. So by the Intermediate Value Theorem, there is a number c in $(5, 6)$ such that $f(c) = 0$. This implies that $\dfrac{1}{c-3} = \sqrt{c-5}$.

(b) We can see from the graphs that, correct to three decimal places, one root is $x = 5.016$.
Note that instead of graphing the left- and right-hand-sides separately, we could have graphed the function f from part (a).

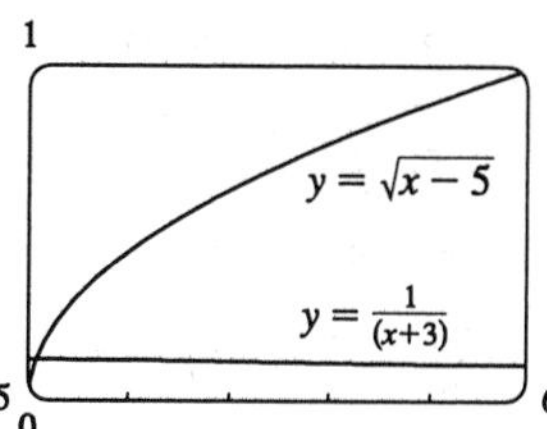

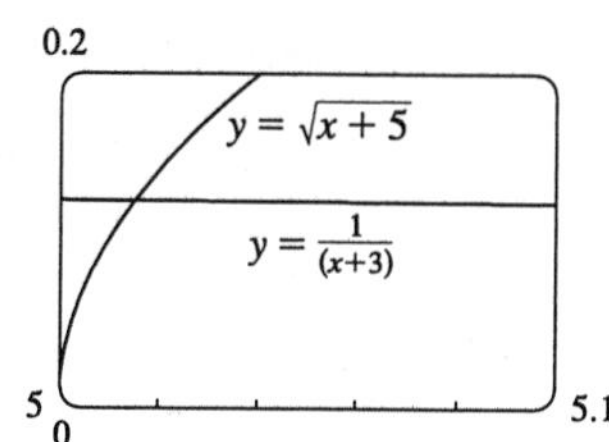

53. $(\Rightarrow)$ If f is continuous at a, then by Theorem 7 with $g(h) = a + h$, we have

$$\lim_{h\to 0} f(a+h) = f\left(\lim_{h\to 0}(a+h)\right) = f(a).$$

$(\Leftarrow)$ Let $\epsilon > 0$. Since $\lim_{h\to 0} f(a+h) = f(a)$, there exists $\delta > 0$ such that $|h| < \delta \;\Rightarrow$

$|f(a+h) - f(a)| < \epsilon$. So if $|x - a| < \delta$, then $|f(x) - f(a)| = |f(a + (x-a)) - f(a)| < \epsilon$. Thus $\lim_{x\to a} f(x) = f(a)$ and so f is continuous at a.

54. **(a)** Since f is continuous at a, $\lim_{x\to a} f(x) = f(a)$. Thus, using the Constant Multiple Law of Limits, we have $\lim_{x\to a}(cf)(x) = \lim_{x\to a} cf(x) = c\lim_{x\to a} f(x) = cf(a) = (cf)(a)$. Therefore cf is continuous at a.

(b) Since f and g are continuous at a, $\lim_{x\to a} f(x) = f(a)$ and $\lim_{x\to a} g(x) = g(a)$. Since $g(a) \neq 0$, we can use the Quotient Law of limits: $\lim_{x\to a}\left(\dfrac{f}{g}\right)(x) = \lim_{x\to a}\dfrac{f(x)}{g(x)} = \dfrac{\lim_{x\to a} f(x)}{\lim_{x\to a} g(x)} = \dfrac{f(a)}{g(a)} = \left(\dfrac{f}{g}\right)(a)$. Thus $\dfrac{f}{g}$ is continuous at a.

55. $f(x) = \begin{cases} 0 & \text{if } x \text{ is rational} \\ 1 & \text{if } x \text{ is irrational} \end{cases}$ is continuous nowhere. For, given any number a and any $\delta > 0$, the interval $(a-\delta, a+\delta)$ contains both infinitely many rational and infinitely many irrational numbers. Since $f(a) = 0$ or 1, there are infinitely many numbers x with $|x-a| < \delta$ and $|f(x) - f(a)| = 1$. Thus $\lim_{x\to a} f(x) \neq f(a)$. [In fact, $\lim_{x\to a} f(x)$ does not even exist.]

56. $g(x) = \begin{cases} 0 & \text{if } x \text{ is rational} \\ x & \text{if } x \text{ is irrational} \end{cases}$ is continuous at 0. To see why, note that $-|x| \le g(x) \le |x|$, so by the Squeeze Theorem $\lim_{x\to 0} g(x) = 0 = g(0)$. But g is continuous nowhere else. For if $a \neq 0$ and $\delta > 0$, the interval $(a-\delta, a+\delta)$ contains both infinitely many rational and infinitely many irrational numbers. Since $g(a) = 0$ or a, there are infinitely many numbers x with $|x-a| < \delta$ and $|g(x) - g(a)| > |a|/2$. Thus $\lim_{x\to a} g(x) \neq g(a)$.

57. If there is such a number, it satisfies the equation $x^3 + 1 = x \;\Leftrightarrow\; x^3 - x + 1 = 0$.
Let the LHS of this equation be called $f(x)$. Now $f(-2) = (-2)^3 - (-2) + 1 = -5 < 0$, and $f(-1) = (-1)^3 - (-1) + 1 = 1 > 0$. Note also that $f(x)$ is a polynomial, and thus continuous. So by the Intermediate Value Theorem, there is a number c between -2 and -1 such that $f(c) = 0$, so that $c = c^3 + 1$.

58. **(a)** Assume that f is continuous on the interval I. Then for $a \in I$, $\lim_{x \to a} |f(x)| = \left|\lim_{x \to a} f(x)\right| = |f(a)|$ by Theorem 7. (If a is an endpoint of I, use the appropriate one-sided limit.) So $|f|$ is continuous on I.

(b) No, the converse is false. For example, the function $f(x) = \begin{cases} 1 & \text{if } x \geq 0 \\ -1 & \text{if } x < 0 \end{cases}$ is not continuous at $x = 0$, but $|f(x)| = 1$ is continuous on $\mathbb{R}$.

59. Define $u(t)$ to be the monk's distance from the monastery, as a function of time, on the first day, and define $d(t)$ to be his distance from the monastery, as a function of time, on the second day. Let D be the distance from the monastery to the top of the mountain. From the given information we know that $u(0) = 0$, $d(12) = D$, $u(0) = D$ and $d(12) = 0$. Now consider the function $u - d$, which is clearly continuous (assuming that the monk does not use his mental powers to instantaneously transport himself). We calculate that $(u - d)(0) = -D$ and $(u - d)(12) = D$. So by the Intermediate Value Theorem there must be some time t_0 between 0 and 12 such that $(u - d)(t_0) = 0 \quad \Leftrightarrow \quad u(t_0) = d(t_0)$. So at time t_0 after 7:00 A.M., the monk will be at the same place on both days.

60. **(a)** Here are a few possibilities:

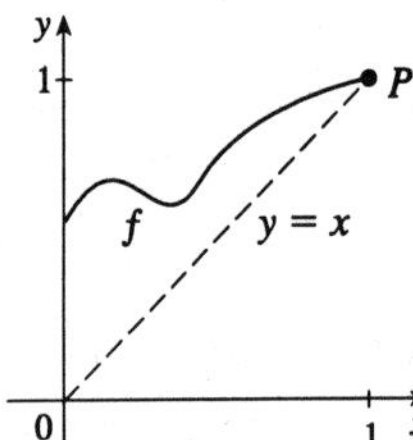

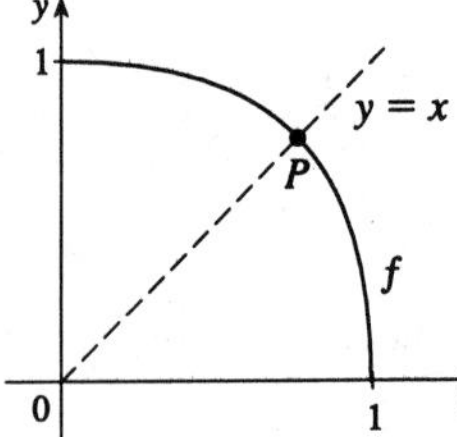

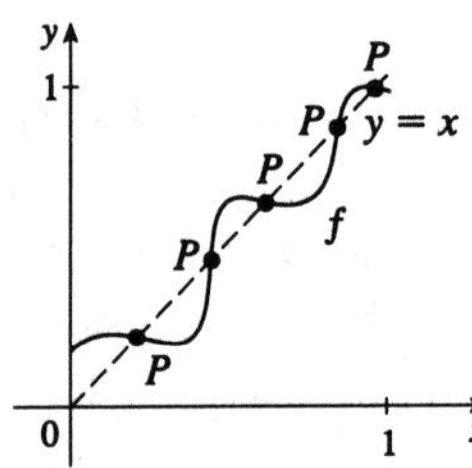

(b) The "obstacle" is the line $x = y$ (see diagram). Any intersection of the graph of f with the line $y = x$ constitutes a fixed point, and if the graph of the function does not cross the line somewhere in $(0, 1)$, then it must either start at $(0, 0)$ (in which case 0 is a fixed point) or finish at $(1, 1)$ (in which case 1 is a fixed point).

(c) Consider the function $F(x) = f(x) - x$, where f is any continuous function with domain $[0, 1]$ and range in $[0, 1]$. We shall prove that f has a fixed point.

Now if $f(0) = 0$ then we are done: f has a fixed point (the number 0), which is what we are trying to prove. So assume $f(0) \neq 0$. For the same reason we can assume that $f(1) \neq 1$. Then $F(0) > 0$ and $F(1) < 0$ (since $0 \leq f(x) \leq 1$ for all x in $[0, 1]$). So by the Intermediate Value Theorem, there exists some number c in the interval $(0, 1)$ such that $F(c) = 0$. So $f(c) = c$, and therefore f has a fixed point.

EXERCISES 1.6

1. **(a)** **(i)** $m = \lim_{x \to -3} \dfrac{x^2 + 2x - 3}{x - (-3)} = \lim_{x \to -3} \dfrac{(x+3)(x-1)}{x+3} = \lim_{x \to -3}(x-1) = -4$

(ii) $m = \lim_{h \to 0} \dfrac{(-3+h)^2 + 2(-3+h) - 3}{h} = \lim_{h \to 0} \dfrac{9 - 6h + h^2 - 6 + 2h - 3}{h}$

$= \lim_{h \to 0} \dfrac{h(h-4)}{h} = \lim_{h \to 0}(h - 4) = -4$

(b) The equation of the tangent line is $y - 3 = -4(x+3)$ or $y = -4x - 9$.

(c)

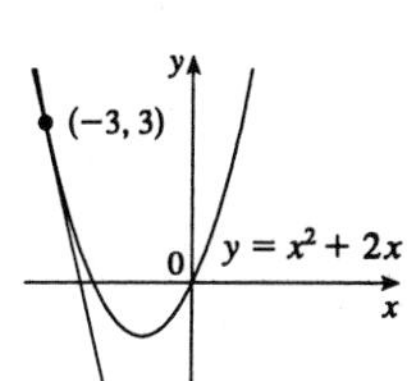

2. **(a)** **(i)** $m = \lim_{x \to -1} \dfrac{f(x) - f(-1)}{x - (-1)} = \lim_{x \to -1} \dfrac{x^3 + 1}{x + 1} = \lim_{x \to -1} \dfrac{(x+1)(x^2 - x + 1)}{x + 1}$

$= \lim_{x \to -1}(x^2 - x + 1) = 3$

(ii) $m = \lim_{h \to 0} \dfrac{f(-1+h) - f(-1)}{h} = \lim_{h \to 0} \dfrac{(-1+h)^3 + 1}{h}$

$= \lim_{h \to 0} \dfrac{h^3 - 3h^2 + 3h - 1 + 1}{h} = \lim_{h \to 0}(h^2 - 3h + 3) = 3$

(b) $y - (-1) = 3[x - (-1)]$

$\Rightarrow \quad y = 3x + 2$

(c)

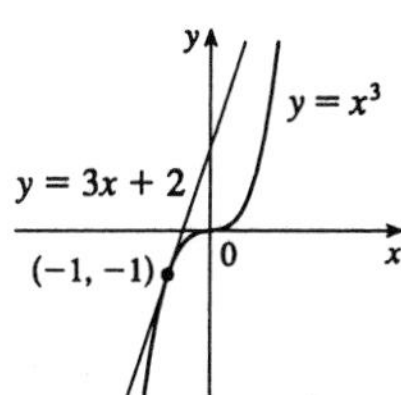

3. Using (1), $m = \lim_{x \to -2} \dfrac{1 - 2x - 3x^2 + 7}{x + 2} = \lim_{x \to -2} \dfrac{-3x^2 - 2x + 8}{x + 2} = \lim_{x \to -2} \dfrac{(-3x + 4)(x + 2)}{x + 2}$

$= \lim_{x \to -2}(-3x + 4) = 10$. Thus the equation of the tangent is $y + 7 = 10(x + 2)$ or $y = 10x + 13$.

4. Using (1), $m = \lim_{x \to 1} \dfrac{1/\sqrt{x} - 1}{x - 1} = \lim_{x \to 1} \dfrac{-(\sqrt{x} - 1)}{\sqrt{x}(\sqrt{x} - 1)(\sqrt{x} + 1)} = \lim_{x \to 1} \dfrac{-1}{\sqrt{x}(\sqrt{x} + 1)} = -\dfrac{1}{2}$.

Thus the equation of the tangent line is $y - 1 = -\frac{1}{2}(x - 1)$ or $x + 2y = 3$.

5. Using (1), $m = \lim_{x \to -2} \dfrac{1/x^2 - \frac{1}{4}}{x + 2} = \lim_{x \to -2} \dfrac{4 - x^2}{4x^2(x + 2)} = \lim_{x \to -2} \dfrac{(2 - x)(2 + x)}{4x^2(x + 2)} = \lim_{x \to -2} \dfrac{2 - x}{4x^2} = \dfrac{1}{4}$. Thus the equation of the tangent is $y - \frac{1}{4} = \frac{1}{4}(x + 2)$ or $x - 4y + 3 = 0$.

6. Using (1), $m = \lim_{x \to 0} \dfrac{x/(1 - x) - 0}{x - 0} = \lim_{x \to 0} \dfrac{x}{x(1 - x)} = \lim_{x \to 0} \dfrac{1}{1 - x} = 1$. Thus the equation of the tangent line is $y - 0 = 1(x - 0)$ or $y = x$.

7. **(a)** $m = \lim_{x \to a} \dfrac{\dfrac{2}{x+3} - \dfrac{2}{a+3}}{x-a} = \lim_{x \to a} \dfrac{2(a-x)}{(x-a)(x+3)(a+3)} = \lim_{x \to a} \dfrac{-2}{(x+3)(a+3)} = \dfrac{-2}{(a+3)^2}$

(b) **(i)** $a = -1 \Rightarrow m = \dfrac{-2}{(-1+3)^2} = -\dfrac{1}{2}$ **(ii)** $a = 0 \Rightarrow m = \dfrac{-2}{(0+3)^2} = -\dfrac{2}{9}$

(iii) $a = 1 \Rightarrow m = \dfrac{-2}{(1+3)^2} = -\dfrac{1}{8}$

8. **(a)** Using (1), $m = \lim_{x \to a} \dfrac{1 + x + x^2 - (1 + a + a^2)}{x - a} = \lim_{x \to a} \dfrac{x + x^2 - a - a^2}{x - a}$

$= \lim_{x \to a} \dfrac{x - a + (x-a)(x+a)}{x-a} = \lim_{x \to a} \dfrac{(x-a)(1+x+a)}{x-a} = \lim_{x \to a}(1 + x + a) = 1 + 2a.$

(b) **(i)** $x = -1 \Rightarrow m = 1 + 2(-1) = -1$

(ii) $x = -\frac{1}{2} \Rightarrow m = 1 + 2\left(-\frac{1}{2}\right) = 0$

(iii) $x = 1 \Rightarrow m = 1 + 2(1) = 3$

(c)

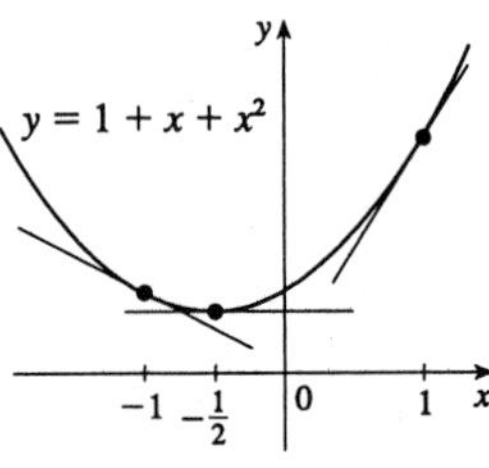

9. **(a)** Using (1), $m = \lim_{x \to a} \dfrac{(x^3 - 4x + 1) - (a^3 - 4a + 1)}{x - a} = \lim_{x \to a} \dfrac{(x^3 - a^3) - 4(x - a)}{x - a}$

$= \lim_{x \to a} \dfrac{(x-a)(x^2 + ax + a^2) - 4(x-a)}{x - a} = \lim_{x \to a}(x^2 + ax + a^2 - 4) = 3a^2 - 4.$

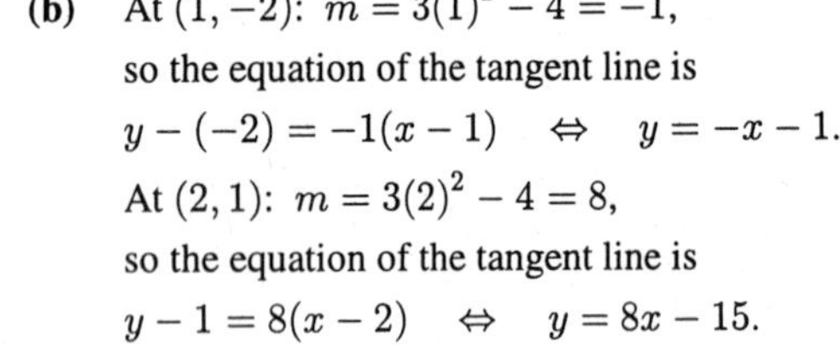

(b) At $(1, -2)$: $m = 3(1)^2 - 4 = -1$,
so the equation of the tangent line is
$y - (-2) = -1(x - 1) \Leftrightarrow y = -x - 1.$
At $(2, 1)$: $m = 3(2)^2 - 4 = 8$,
so the equation of the tangent line is
$y - 1 = 8(x - 2) \Leftrightarrow y = 8x - 15.$

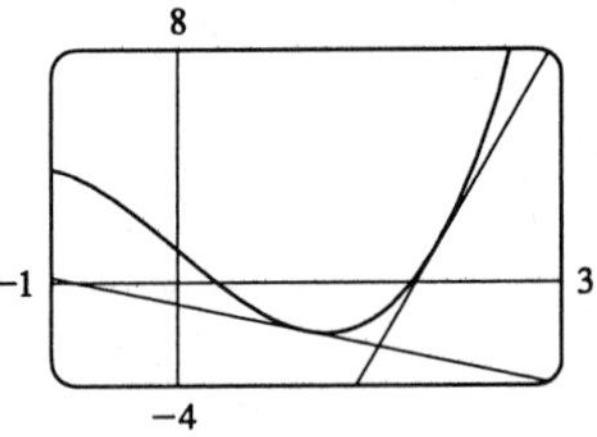

10. **(a)** Using (1), $m = \lim_{x \to a} \dfrac{\dfrac{1}{\sqrt{5-2x}} - \dfrac{1}{\sqrt{5-2a}}}{x - a} = \lim_{x \to a} \dfrac{\sqrt{5-2a} - \sqrt{5-2x}}{(x-a)\sqrt{5-2x}\sqrt{5-2a}}$

$= \lim_{x \to a} \dfrac{2(x-a)}{(x-a)\sqrt{(5-2x)(5-2a)}\left[\sqrt{5-2x} + \sqrt{5-2a}\right]}$

$= \lim_{x \to a} \dfrac{2}{(5-2x)\sqrt{5-2a} + (5-2a)\sqrt{5-2x}} = \dfrac{1}{(5-2a)^{3/2}} = (5-2a)^{-3/2}.$

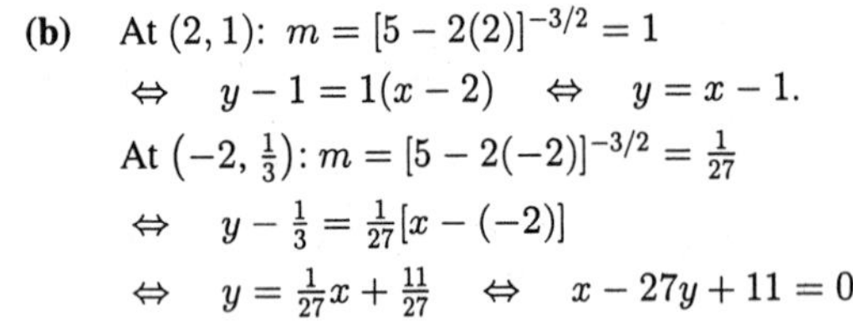

(b) At $(2, 1)$: $m = [5 - 2(2)]^{-3/2} = 1$
$\Leftrightarrow y - 1 = 1(x - 2) \Leftrightarrow y = x - 1.$
At $\left(-2, \frac{1}{3}\right)$: $m = [5 - 2(-2)]^{-3/2} = \frac{1}{27}$
$\Leftrightarrow y - \frac{1}{3} = \frac{1}{27}[x - (-2)]$
$\Leftrightarrow y = \frac{1}{27}x + \frac{11}{27} \Leftrightarrow x - 27y + 11 = 0.$

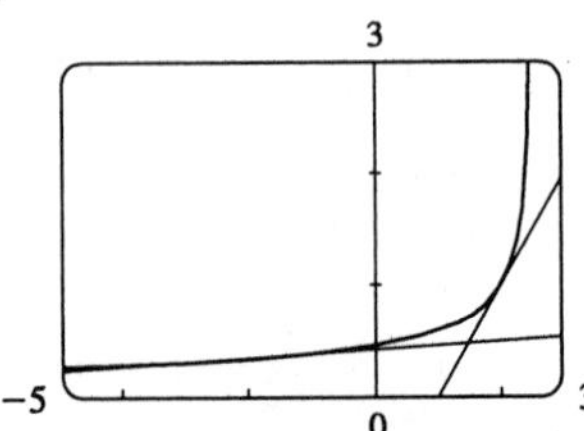

11. Let $s(t) = 40t - 16t^2$. $v(2) = \lim_{t\to 2} \dfrac{s(t) - s(2)}{t - 2} = \lim_{t\to 2} \dfrac{40t - 16t^2 - 16}{t - 2} = \lim_{t\to 2} \dfrac{8(t-2)(-2t+1)}{t-2}$
$= \lim_{t\to 2} 8(-2t+1) = -24$. Thus, the instantaneous velocity when $t = 2$ is -24 ft/s.

12. **(a)** $v(1) = \lim_{h\to 0} \dfrac{h(1+h) - h(1)}{h} = \lim_{h\to 0} \dfrac{58 + 58h - 0.83 - 1.66h - 0.83h^2 - 57.17}{h} = \lim_{h\to 0}(56.34 - 0.83h)$
$= 56.34$ m/s

(b) $v(a) = \lim_{h\to 0} \dfrac{h(a+h) - h(a)}{h} = \lim_{h\to 0} \dfrac{58a + 58h - 0.83a^2 - 1.66ah - 0.83h^2 - 58a + 0.83a^2}{h}$
$= \lim_{h\to 0}(58 - 1.66a - 0.83h) = 58 - 1.66a$ m/s.

(c) The arrow strikes the moon when the height is 0, that is, $58t - 0.83t^2 = 0 \quad \Leftrightarrow \quad t(58 - 0.83t) = 0$
$\Leftrightarrow \quad t = \frac{58}{0.83} \approx 69.9$ s (since t can't be 0).

(d) $58 - 1.66(69.88) \approx -58$ m/s. Thus, the arrow will have a velocity of -58 m/s.

13. $v(a) = \lim_{h\to 0} \dfrac{s(a+h) - s(a)}{h} = \lim_{h\to 0} \dfrac{4(a+h)^3 + 6(a+h) + 2 - (4a^3 + 6a + 2)}{h}$
$= \lim_{h\to 0} \dfrac{4a^3 + 12a^2h + 12ah^2 + 4h^3 + 6a + 6h + 2 - 4a^3 - 6a - 2}{h}$
$= \lim_{h\to 0} \dfrac{12a^2h + 12ah^2 + 4h^3 + 6h}{h} = \lim_{h\to 0}\left(12a^2 + 12ah + 4h^2 + 6\right) = 12a^2 + 6$

So $v(1) = 12(1)^2 + 6 = 18$ m/s, $v(2) = 12(2)^2 + 6 = 54$ m/s, and $v(3) = 12(3)^2 + 6 = 114$ m/s.

14. **(a)** The average velocity between times t and $t + h$ is

$$\frac{(t+h)^2 - 8(t+h) + 18 - (t^2 - 8t + 18)}{h} = \frac{t^2 + 2th + h^2 - 8t - 8h + 18 - t^2 + 8t - 18}{h}$$
$$= \frac{2th + h^2 - 8h}{h} = 2t + h - 8.$$

(i) $[3, 4]$: $t = 3$ and $h = 1$ so the average velocity is $2(3) + 1 - 8 = -1$ m/s.

(ii) $[3.5, 4]$: $t = 3.5$, $h = 0.5$ so the average velocity is $2(3.5) + 0.5 - 8 = -0.5$ m/s.

(iii) $[4, 5]$: $t = 4$, $h = 1$ so the average velocity is $2(4) + 1 - 8 = 1$ m/s.

(iv) $[4, 4.5]$: $t = 4$, $h = 0.5$ so the average velocity is $2(4) + 0.5 - 8 = 0.5$ m/s.

(b) $v(t) = \lim_{h\to 0} \dfrac{s(t+h) - s(t)}{h}$
$= \lim_{h\to 0}(2t + h - 8) = 2t - 8$,
so $v(4) = 0$.

(c)

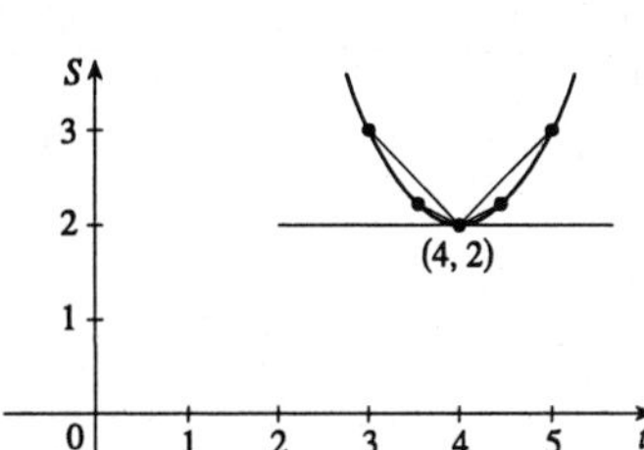

15. **(a)** Since the slope of the tangent at $s = 0$ is 0, the car's initial velocity was 0.

(b) The slope of the tangent is greater at C than at B, so the car was going faster at C.

(c) Near A, the tangent lines are becoming steeper as x increases, so the velocity was increasing, so the car was speeding up. Near B, the tangent lines are becoming less steep, so the car was slowing down. The steepest tangent near C is the one right at C, so at C the car had just finished speeding up, and was about to start slowing down.

(d) Between D and E, the slope of the tangent is 0, so the car did not move during that time.

16. The slope of the tangent (that is, the rate of change of temperature with respect to time) at $t = 1$ h seems to be about $\dfrac{75 - 168}{132 - 0} \approx -0.7°\,\mathrm{F/min}$.

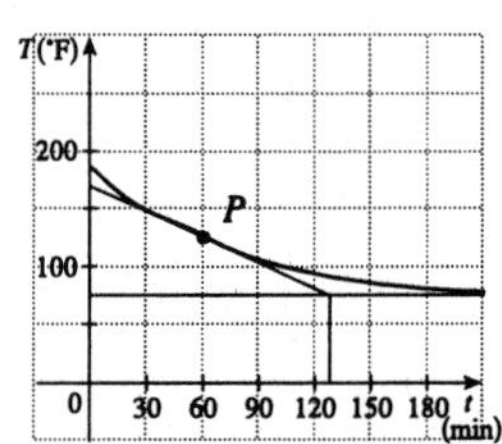

17. **(a)** **(i)** $[8, 11]$: $\dfrac{7.9 - 11.5}{3} = -1.2°/\mathrm{h}$ **(ii)** $[8, 10]$: $\dfrac{9.0 - 11.5}{2} = -1.25°/\mathrm{h}$

(iii) $[8, 9]$: $\dfrac{10.2 - 11.5}{1} = -1.3°/\mathrm{h}$

(b) The instantaneous rate of change is approximately $-1.6°/\mathrm{h}$ at 8 P.M.

18. **(a)** **(i)** $\dfrac{164 - 117}{4} = 11.75$

(ii) $\dfrac{150 - 117}{3} = 11$

(iii) $\dfrac{137 - 117}{2} = 10$

(iv) $\dfrac{126 - 117}{1} = 9$

(b)

The tangent line at 1992 has a slope of approximately 7 (thousand people per year).

19. **(a)** **(i)** $\dfrac{\Delta C}{\Delta x} = \dfrac{C(105) - C(100)}{5} = \dfrac{6601.25 - 6500}{5} = \$20.25/\mathrm{unit}.$

(ii) $\dfrac{\Delta C}{\Delta x} = \dfrac{C(101) - C(100)}{1} = \dfrac{6520.05 - 6500}{1} = \$20.05/\mathrm{unit}.$

(b) $\dfrac{C(100 + h) - C(100)}{h} = \dfrac{5000 + 10(100 + h) + 0.05(100 + h)^2 - 6500}{h} = 20 + 0.05h,\ h \neq 0.$ So as h approaches 0, the rate of change of C approaches \$20/unit.

20. Rate of change between 20 and $20 + h$ minutes is

$$\frac{V(20 + h) - V(20)}{h} = \frac{100{,}000\left[1 - \frac{1}{60}(20 + h)\right]^2 - 100{,}000\left(1 - \frac{20}{60}\right)^2}{h} = -\frac{20{,}000}{9} + \frac{2500h}{9}.$$

As h approaches 0, the rate of flow approaches $-\frac{20{,}000}{9} \approx -2222.2\ \mathrm{gal/min}$.

REVIEW EXERCISES FOR CHAPTER 1

1. False, since $\lim\limits_{x\to 4}\dfrac{2x}{x-4}$ and $\lim\limits_{x\to 4}\dfrac{8}{x-4}$ do not exist.

2. False, since $\lim\limits_{x\to 1}(x^2+5x-6)=0$ (Limit Law #5).

3. True by Limit Law #5, since $\lim\limits_{x\to 1}(x^2+2x-4)=-1\neq 0$.

4. True, for if $\lim\limits_{x\to 5}\dfrac{f(x)}{g(x)}=L$ were to exist, then we would have

$2=\lim\limits_{x\to 5}f(x)=\lim\limits_{x\to 5}\dfrac{f(x)}{g(x)}g(x)=\lim\limits_{x\to 5}\dfrac{f(x)}{g(x)}\lim\limits_{x\to 5}g(x)=L\cdot 0=0$, which is a contradiction.

5. False. For example, let $f(x)=\begin{cases}x^2+1 & \text{if } x\neq 0\\ 2 & \text{if } x=0\end{cases}$

Then $f(x)>1$ for all x, but $\lim\limits_{x\to 0}f(x)=\lim\limits_{x\to 0}(x^2+1)=1$.

6. False. It is possible that there is a discontinuity at $x=4$, for example $f(x)=\begin{cases}7 & \text{if } x\neq 4\\ 23 & \text{if } x=4\end{cases}$

In this case, $\lim\limits_{x\to 4}f(x)=7$, but $f(4)=23$.

7. True, by the definition of a limit with $\epsilon=1$.

8. False. For example, let $f(x)=1$ for $x\neq 6$ and $f(6)=2$, and let $g(x)=f(x)$. Then $\lim\limits_{x\to 6}f(x)g(x)=\lim\limits_{x\to 6}f(x)\cdot\lim\limits_{x\to 6}g(x)=1\cdot 1=1$, but $f(6)g(6)=2\cdot 2=4$.

9. True. See Exercise 1.3.67.

10. False. The function f must be *continuous* in order to use the Intermediate Value Theorem. For example, let $f(x)=\begin{cases}1 & \text{if } 0\leq x<3\\ -1 & \text{if } x=3\end{cases}$ There is no number $c\in[0,3]$ with $f(c)=0$.

11. True by Theorem 1.5.7 with $a=2$, $b=5$, and $g(x)=4x^2-11$.

12. True by the Intermediate Value Theorem with $a=-1$, $b=1$, $N=\pi$, since $4>\pi>3$.

13. $\lim\limits_{x\to 4}\sqrt{x+\sqrt{x}}=\sqrt{4+\sqrt{4}}=\sqrt{6}$ since the function is continuous.

14. $\lim\limits_{x\to 0^-}\sqrt{-x}=\sqrt{-0}=0$ since $\sqrt{-x}$ is continuous on $(-\infty,0]$.

15. $\lim\limits_{t\to -1}\dfrac{t+1}{t^3-t}=\lim\limits_{t\to -1}\dfrac{t+1}{t(t+1)(t-1)}=\lim\limits_{t\to -1}\dfrac{1}{t(t-1)}=\dfrac{1}{(-1)(-2)}=\dfrac{1}{2}$

16. $\lim\limits_{t\to 4}\dfrac{t-4}{t^2-3t-4}=\lim\limits_{t\to 4}\dfrac{t-4}{(t-4)(t+1)}=\lim\limits_{t\to 4}\dfrac{1}{t+1}=\dfrac{1}{4+1}=\dfrac{1}{5}$

17. $\lim\limits_{h\to 0}\dfrac{(1+h)^2-1}{h}=\lim\limits_{h\to 0}\dfrac{1+2h+h^2-1}{h}=\lim\limits_{h\to 0}\dfrac{2h+h^2}{h}=\lim\limits_{h\to 0}(2+h)=2$

18. $\lim\limits_{h\to 0}\dfrac{(1+h)^{-2}-1}{h}=\lim\limits_{h\to 0}\dfrac{1-(1+h)^2}{h(1+h)^2}=\lim\limits_{h\to 0}\dfrac{-2h-h^2}{h(1+h)^2}=\lim\limits_{h\to 0}\dfrac{-2-h}{(1+h)^2}=\dfrac{-2-0}{(1+0)^2}=-2$

19. $\lim_{x\to -1}\dfrac{x^2-x-2}{x^2+3x-2}=\dfrac{\lim_{x\to -1}(x^2-x-2)}{\lim_{x\to -1}(x^2+3x-2)}=\dfrac{(-1)^2-(-1)-2}{(-1)^2+3(-1)-2}=\dfrac{0}{-4}=0$

20. $\lim_{x\to 1}\dfrac{x^2-x-2}{x^2+3x+2}=\lim_{x\to -1}\dfrac{(x+1)(x-2)}{(x+1)(x+2)}=\lim_{x\to -1}\dfrac{x-2}{x+2}=\dfrac{-1-2}{-1+2}=-3$

21. $\lim_{t\to 6}\dfrac{17}{(t-6)^2}=\infty$ since $(t-6)^2\to 0$ and $\dfrac{17}{(t-6)^2}>0$.

22. $\lim_{x\to -6^+}\dfrac{x}{x+6}=-\infty$ since $x+6\to 0$ as $x\to -6^+$ and $\dfrac{x}{x+6}<0$ for $-6<x<0$.

23. $\lim_{s\to 16}\dfrac{4-\sqrt{s}}{s-16}=\lim_{s\to 16}\dfrac{4-\sqrt{s}}{(\sqrt{s}+4)(\sqrt{s}-4)}=\lim_{s\to 16}\dfrac{-1}{\sqrt{s}+4}=\dfrac{-1}{\sqrt{16}+4}=-\dfrac{1}{8}$

24. $\lim_{v\to 2}\dfrac{v^2+2v-8}{v^4-16}=\lim_{v\to 2}\dfrac{(v+4)(v-2)}{(v+2)(v-2)(v^2+4)}=\lim_{v\to 2}\dfrac{v+4}{(v+2)(v^2+4)}=\dfrac{2+4}{(2+2)(2^2+4)}=\dfrac{3}{16}$

25. $\lim_{x\to 8^-}\dfrac{|x-8|}{x-8}=\lim_{x\to 8^-}\dfrac{-(x-8)}{x-8}=\lim_{x\to 8^-}(-1)=-1$

26. $\lim_{x\to 9^+}\left(\sqrt{x-9}+[\![x+1]\!]\right)=\lim_{x\to 9^+}\sqrt{x-9}+\lim_{x\to 9^+}[\![x+1]\!]=\sqrt{9-9}+10=10$

27. $\lim_{x\to 0}\dfrac{1-\sqrt{1-x^2}}{x}\cdot\dfrac{1+\sqrt{1-x^2}}{1+\sqrt{1-x^2}}=\lim_{x\to 0}\dfrac{1-(1-x^2)}{x\left(1+\sqrt{1-x^2}\right)}=\lim_{x\to 0}\dfrac{x^2}{x\left(1+\sqrt{1-x^2}\right)}=\lim_{x\to 0}\dfrac{x}{1+\sqrt{1-x^2}}=0$

28. $\lim_{x\to 2}\dfrac{\sqrt{x+2}-\sqrt{2x}}{x(x-2)}\cdot\dfrac{\sqrt{x+2}+\sqrt{2x}}{\sqrt{x+2}+\sqrt{2x}}=\lim_{x\to 2}\dfrac{-(x-2)}{x(x-2)\left(\sqrt{x+2}+\sqrt{2x}\right)}=\lim_{x\to 2}\dfrac{-1}{x\left(\sqrt{x+2}+\sqrt{2x}\right)}=-\dfrac{1}{8}$

29. Given $\epsilon>0$, we need $\delta>0$ so that if $|x-5|<\delta$ then $|(7x-27)-8|<\epsilon$ $\Leftrightarrow$ $|7x-35|<\epsilon$ $\Leftrightarrow$ $|x-5|<\epsilon/7$. So take $\delta=\epsilon/7$. Then $|x-5|<\delta$ $\Rightarrow$ $|(7x-27)-8|<\epsilon$. Thus $\lim_{x\to 5}(7x-27)=8$ by the definition of a limit.

30. Given $\epsilon>0$ we must find $\delta>0$ so that if $|x-0|<\delta$, then $|\sqrt[3]{x}-0|<\epsilon$. Now $|\sqrt[3]{x}-0|=|\sqrt[3]{x}|<\epsilon$ $\Rightarrow$ $|x|=|\sqrt[3]{x}|^3<\epsilon^3$. So take $\delta=\epsilon^3$, then $|x-0|=|x|<\epsilon^3$ $\Rightarrow$ $|\sqrt[3]{x}-0|=|\sqrt[3]{x}|=\sqrt[3]{|x|}<\sqrt[3]{\epsilon^3}=\epsilon$. Therefore, by the definition of a limit, $\lim_{x\to 0}\sqrt[3]{x}=0$.

31. Given $\epsilon>0$, we need $\delta>0$ so that if $|x-2|<\delta$ then $|x^2-3x-(-2)|<\epsilon$. First, note that if $|x-2|<1$, then $-1<x-2<1$, so $0<x-1<2$ $\Rightarrow$ $|x-1|<2$. Now let $\delta=\min\{\epsilon/2,1\}$. Then $|x-2|<\delta$ $\Rightarrow$ $|x^2-3x-(-2)|=|(x-2)(x-1)|=|x-2||x-1|<(\epsilon/2)(2)=\epsilon$.

Thus $\lim_{x\to 2}(x^2-3x)=-2$ by the definition of a limit.

32. Given $M>0$, we need $\delta>0$ such that if $0<x-4<\delta$ then $\dfrac{2}{\sqrt{x-4}}>M$. This is true $\Leftrightarrow$ $\sqrt{x-4}<2/M$ $\Leftrightarrow$ $x-4<4/M^2$. So if we choose $\delta=4/M^2$, then $0<x-4<\delta$ $\Rightarrow$ $\dfrac{2}{\sqrt{x-4}}>M$. So by the definition of a limit, $\lim_{x\to 4^+}\dfrac{2}{\sqrt{x-4}}=\infty$.

33. Since $2x - 1 \le f(x) \le x^2$ for $0 < x < 3$ and $\lim_{x\to 1}(2x-1) = 1 = \lim_{x\to 1} x^2$, we have $\lim_{x\to 1} f(x) = 1$ by the Squeeze Theorem.

34. We use the Squeeze Theorem. Let $f(x) = -x^2$, $g(x) = x^2 \cos \frac{1}{x^2}$ and $h(x) = x^2$. Then since $\left|\cos \frac{1}{x^2}\right| \le 1$ for $x \ne 0$, we have $f(x) \le g(x) \le h(x)$ for $x \ne 0$, and so $\lim_{x\to 0} f(x) = \lim_{x\to 0} h(x) = 0 \quad \Rightarrow \quad \lim_{x\to 0} g(x) = 0$ by the Squeeze Theorem.

35. **(a)** $f(x) = \sqrt{-x}$ if $x < 0$, $f(x) = 3 - x$ if $0 \le x < 3$, $f(x) = (x-3)^2$ if $x > 3$. So

(i) $\lim_{x\to 0^+} f(x) = \lim_{x\to 0^+} (3 - x) = 3$

(ii) $\lim_{x\to 0^-} f(x) = \lim_{x\to 0^-} \sqrt{-x} = 0$

(iii) Because of (i) and (ii), $\lim_{x\to 0} f(x)$ does not exist.

(iv) $\lim_{x\to 3^-} f(x) = \lim_{x\to 3^-} (3 - x) = 0$

(v) $\lim_{x\to 3^+} f(x) = \lim_{x\to 3^+} (x-3)^2 = 0$

(vi) Because of (iv) and (v), $\lim_{x\to 3} f(x) = 0$.

(b) f is discontinuous at 0 since $\lim_{x\to 0} f(x)$ does not exist. f is discontinuous at 3 since $f(3)$ does not exist.

(c)

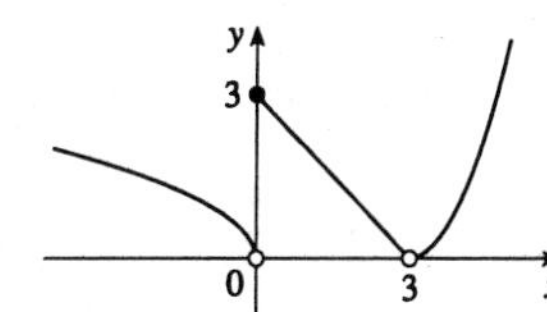

36. **(a)** $g(x) = 2x - x^2$ if $0 \le x \le 2$,
$g(x) = 2 - x$ if $2 < x \le 3$,
$g(x) = x - 4$ if $3 < x < 4$,
$g(x) = \pi$ if $x \ge 4$. Therefore
$\lim_{x\to 2^-} g(x) = \lim_{x\to 2^-} (2x - x^2) = 0$ and
$\lim_{x\to 2^+} g(x) = \lim_{x\to 2^+} (2 - x) = 0$.
Thus $\lim_{x\to 2} g(x) = 0 = g(2)$, so g is continuous at 2. $\lim_{x\to 3^-} g(x) = \lim_{x\to 3^-} (2 - x) = -1$ and $\lim_{x\to 3^+} g(x) = \lim_{x\to 3^+} (x - 4) = -1$. Thus $\lim_{x\to 3} g(x) = -1 = g(3)$, so g is continuous at 3. $\lim_{x\to 4^-} g(x) = \lim_{x\to 4^-} (x - 4) = 0$ and $\lim_{x\to 4^+} g(x) = \lim_{x\to 4^+} \pi = \pi$. Thus $\lim_{x\to 4} g(x)$ does not exist, so g is discontinuous at 4. But $\lim_{x\to 4^+} g(x) = \pi = g(4)$, so g is continuous from the right at 4.

(b)

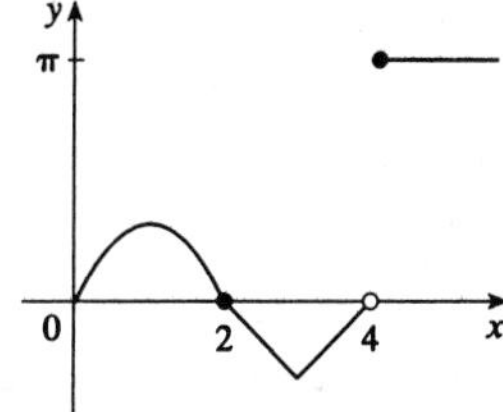

37. $f(x) = \dfrac{x+1}{x^2+x+1}$ is rational so it is continuous on its domain which is $\mathbb{R}$.
(Note that $x^2 + x + 1 = 0$ has no real roots.)

38. $x^2 - 9$ is continuous on $\mathbb{R}$ since it is a polynomial and $\sqrt{x}$ is continuous on $[0, \infty)$, so the composition $\sqrt{x^2 - 9}$ is continuous on $\{x \mid x^2 - 9 \ge 0\} = (-\infty, -3] \cup [3, \infty)$. Note that $x^2 - 2 \ne 0$ on this set and so the quotient function $g(x) = \dfrac{\sqrt{x^2-9}}{x^2-2}$ is continuous on $(-\infty, -3] \cup [3, \infty)$.

39. $f(x) = 2x^3 + x^2 + 2$ is a polynomial, so it is continuous on $[-2, -1]$ and $f(-2) = -10 < 0 < 1 = f(-1)$. So by the Intermediate Value Theorem there is a number c in $(-2, -1)$ such that $f(c) = 0$, that is, the equation $2x^3 + x^2 + 2 = 0$ has a root in $(-2, -1)$.

40. $f(x) = x^4 + 1 - 1/x$ is continuous on $[0.5, 1]$ and $f(0.5) = -\frac{15}{16} < 0$, $f(1) = 1 > 0$. So by the Intermediate Value Theorem there is a number c in $(0.5, 1)$ such that $f(c) = 0$, that is, the equation $x^4 + 1 = 1/x$ has a root in $(0.5, 1)$.

41. **(a)** The slope of the tangent line at $(2, 1)$ is $\lim\limits_{x\to 2} \dfrac{f(x) - f(2)}{x - 2} = \lim\limits_{x\to 2} \dfrac{9 - 2x^2 - 1}{x - 2}$

$$= \lim_{x\to 2} \frac{8 - 2x^2}{x - 2} = \lim_{x\to 2} \frac{-2(x^2 - 4)}{x - 2} = \lim_{x\to 2} \frac{-2(x - 2)(x + 2)}{x - 2} = \lim_{x\to 2} -2(x + 2) = -8.$$

(b) The equation of this tangent line is $y - 1 = -8(x - 2)$ or $8x + y = 17$.

42. For a general point with x-coordinate a, we have $m = \lim\limits_{x\to a} \dfrac{\dfrac{2}{1 - 3x} - \dfrac{2}{1 - 3a}}{x - a} = \lim\limits_{x\to a} \dfrac{2(1 - 3a) - 2(1 - 3x)}{(1 - 3a)(1 - 3x)(x - a)}$

$= \lim\limits_{x\to a} \dfrac{6(x - a)}{(1 - 3a)(1 - 3x)(x - a)} = \lim\limits_{x\to a} \dfrac{6}{(1 - 3a)(1 - 3x)} = \dfrac{6}{(1 - 3a)^2}$. For $a = 0$, $m = 6$ and $f(0) = 2$, so the equation of the tangent line is $y - 2 = 6(x - 0)$ or $y = 6x + 2$. For $a = -1$, $m = \frac{3}{8}$ and $f(-1) = \frac{1}{2}$, so the equation of the tangent line is $y - \frac{1}{2} = \frac{3}{8}(x + 1)$ or $3x - 8y + 7 = 0$.

43. **(a)** $s = 1 + 2t + t^2/4$. The average velocity over the time interval $[1, 1 + h]$ is

$\dfrac{s(1 + h) - s(1)}{h} = \dfrac{1 + 2(1 + h) + (1 + h)^2/4 - 13/4}{h} = \dfrac{10h + h^2}{4h} = \dfrac{10 + h}{4}$. So for the following intervals the average velocities are:

(i) $[1, 3]$: $(10 + 2)/4 = 3\text{ m/s}$

(ii) $[1, 2]$: $(10 + 1)/4 = 2.75\text{ m/s}$

(iii) $[1, 1.5]$: $(10 + 0.5)/4 = 2.625\text{ m/s}$

(iv) $[1, 1.1]$: $(10 + 0.1)/4 = 2.525\text{ m/s}$

(b) When $t = 1$ the velocity is $\lim\limits_{h\to 0} \dfrac{s(1 + h) - s(1)}{h} = \lim\limits_{h\to 0} \dfrac{10 + h}{4} = 2.5\text{ m/s}$.

44. **(a)** When V increases from 200 in^3 to 250 in^3, we have $\Delta V = 250 - 200 = 50\text{ in}^3$, and since $P = \dfrac{800}{V}$,

$\Delta P = P(250) - P(200) = \frac{800}{250} - \frac{800}{200} = 3.2 - 4 = -0.8\text{ lb/in}^2$.

So the average rate of change is $\dfrac{\Delta P}{\Delta V} = \dfrac{-0.8}{50} = -0.016\,\dfrac{\text{lb/in}^2}{\text{in}^3}$.

(b) Since $V = 800/P$, the instantaneous rate of change of V with respect to P is

$$\lim_{h\to 0} \frac{\Delta V}{\Delta P} = \lim_{h\to 0} \frac{V(P + h) - V(P)}{h} = \lim_{h\to 0} \frac{\dfrac{800}{P + h} - \dfrac{800}{P}}{h} = \lim_{h\to 0} \frac{800[P - (P + h)]}{h(P + h)P} = \lim_{h\to 0} \frac{-800}{(P + h)P}$$

$= -\dfrac{800}{P^2}$, which is proportional to the inverse square of P.

45. The inequality $\left|\dfrac{x+1}{x-1}-3\right|<0.2$ is equivalent to the double inequality $2.8<\dfrac{x+1}{x-1}<3.2$. Graphing the functions $y=2.8$, $y=|(x+1)/(x-1)|$ and $y=3.2$ on the interval $[1.9,2.15]$, we see that the inequality holds whenever $1.91<x<2.11$ (approximately). So since $|2-1.91|=0.09$ and $|2-2.15|=0.15$, any positive $\delta\leq 0.09$ will do.

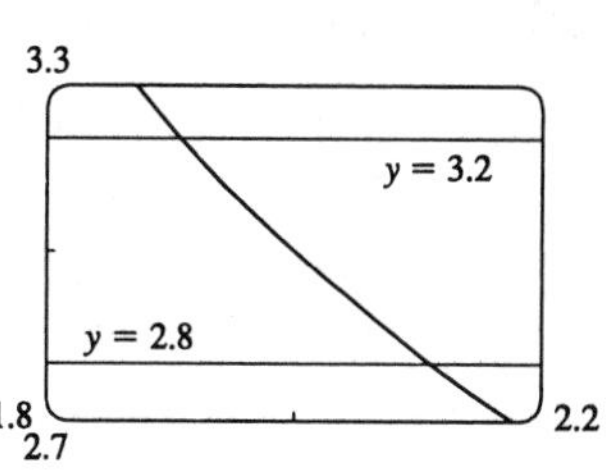

46. The slope of the tangent to $y=\dfrac{x+1}{x-1}$ is

$$\lim_{h\to 0}\frac{\dfrac{(x+h)+1}{(x+h)-1}-\dfrac{x+1}{x-1}}{h}=\lim_{h\to 0}\frac{(x-1)(x+h+1)-(x+1)(x+h-1)}{h(x-1)(x+h-1)}$$

$$=\lim_{h\to 0}\frac{-2h}{h(x-1)(x+h-1)}=-\frac{2}{(x-1)^2}.$$

So at $(2,3)$, $m=-\dfrac{2}{(2-1)^2}=-2\quad\Rightarrow$

$y-3=-2(x-2)\quad\Rightarrow\quad y=-2x+7.$

At $(-1,0)$, $m=-\dfrac{2}{(-1-1)^2}=-\dfrac{1}{2}$

$\Rightarrow\quad y=-\frac{1}{2}(x+1)\quad\Rightarrow\quad y=-\frac{1}{2}x-\frac{1}{2}.$

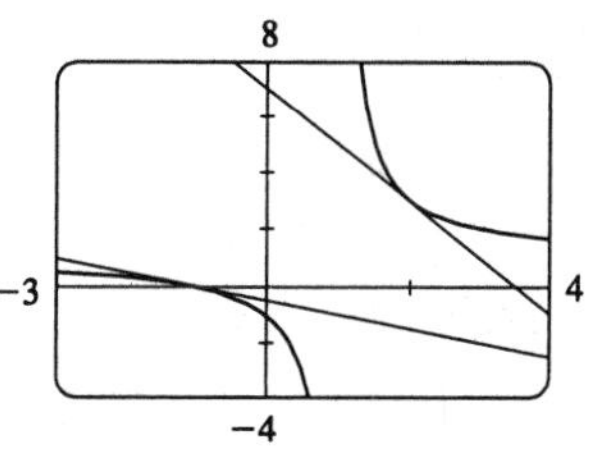

47. $|f(x)|\leq g(x)\quad\Leftrightarrow\quad -g(x)\leq f(x)\leq g(x)$ and $\lim\limits_{x\to a}g(x)=0=\lim\limits_{x\to a}-g(x)$.

Thus, by the Squeeze Theorem, $\lim\limits_{x\to a}f(x)=0$.

48. **(a)** Note that f is an even function since $f(x)=f(-x)$. Now for any integer n, $[\![n]\!]+[\![-n]\!]=n-n=0$, and for any real number k which is not an integer, $[\![k]\!]+[\![-k]\!]=[\![k]\!]+(-[\![k]\!]-1)=-1$. So $\lim\limits_{x\to a}f(x)$ exists (and is equal to -1) for all values of a.

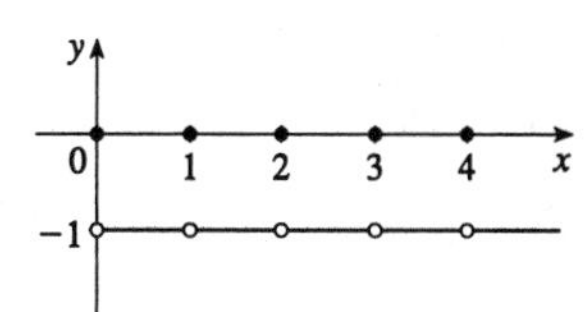

(b) f is discontinuous at all integers.

49. $\lim\limits_{x\to a}f(x)=\lim\limits_{x\to a}\left(\frac{1}{2}[f(x)+g(x)]+\frac{1}{2}[f(x)-g(x)]\right)=\frac{1}{2}\lim\limits_{x\to a}[f(x)+g(x)]+\frac{1}{2}\lim\limits_{x\to a}[f(x)-g(x)]=\frac{1}{2}\cdot 2+\frac{1}{2}\cdot 1$
$=\frac{3}{2}$, and $\lim\limits_{x\to a}g(x)=\lim\limits_{x\to a}([f(x)+g(x)]-f(x))=\lim\limits_{x\to a}[f(x)+g(x)]-\lim\limits_{x\to a}f(x)=2-\frac{3}{2}=\frac{1}{2}$.

So $\lim\limits_{x\to a}[f(x)g(x)]=\left[\lim\limits_{x\to a}f(x)\right]\left[\lim\limits_{x\to a}g(x)\right]=\frac{3}{2}\cdot\frac{1}{2}=\frac{3}{4}$.

Alternate Solution: Since $\lim\limits_{x\to a}[f(x)+g(x)]$ and $\lim\limits_{x\to a}[f(x)-g(x)]$ exist, we must have

$\lim\limits_{x\to a}[f(x)+g(x)]^2=\left(\lim\limits_{x\to a}[f(x)+g(x)]\right)^2$ and $\lim\limits_{x\to a}[f(x)-g(x)]^2=\left(\lim\limits_{x\to a}[f(x)-g(x)]\right)^2$, so

$\lim\limits_{x\to a}[f(x)\,g(x)]=\lim\limits_{x\to a}\frac{1}{4}\left([f(x)+g(x)]^2-[f(x)-g(x)]^2\right)$ (since all of the f^2 and g^2 cancel)

$=\frac{1}{4}\left(\lim\limits_{x\to a}[f(x)+g(x)]^2-\lim\limits_{x\to a}[f(x)-g(x)]^2\right)=\frac{1}{4}(2^2-1^2)=\frac{3}{4}.$

CHAPTER TWO

EXERCISES 2.1

1. $f'(2) = \lim_{h\to 0} \frac{f(2+h) - f(2)}{h} = \lim_{h\to 0} \frac{3(2+h)^2 - 5(2+h) - [3(2)^2 - 5(2)]}{h}$

$= \lim_{h\to 0} \frac{12 + 12h + 3h^2 - 10 - 5h - 12 + 10}{h} = \lim_{h\to 0} \frac{3h^2 + 7h}{h} = \lim_{h\to 0} (3h + 7) = 7$

So the equation of the tangent line at $(2, 2)$ is $y - 2 = 7(x - 2)$ or $7x - y = 12$.

2. $g'(0) = \lim_{h\to 0} \frac{g(0+h) - g(0)}{h} = \lim_{h\to 0} \frac{1 - (0+h)^3 - (1 - 0^3)}{h} = \lim_{h\to 0} \frac{1 - h^3 - 1}{h} = \lim_{h\to 0}(-h^2) = 0.$

So the equation of the tangent line is $y - 1 = 0(x - 0)$ or $y = 1$.

3. **(a)** $F'(1) = \lim_{x\to 1} \frac{F(x) - F(1)}{x - 1} = \lim_{x\to 1} \frac{x^3 - 5x + 1 - (-3)}{x - 1} = \lim_{x\to 1} \frac{x^3 - 5x + 4}{x - 1} = \lim_{x\to 1} \frac{(x-1)(x^2 + x - 4)}{x - 1}$

$= \lim_{x\to 1} (x^2 + x - 4) = -2.$

So the equation of the tangent line at $(1, -3)$ is $y - (-3) = -2(x - 1)$ $\Leftrightarrow$ $y = -2x - 1$.

Note: Instead of using Equation 3 to compute $F'(1)$, we could have used Equation 1.

(b)

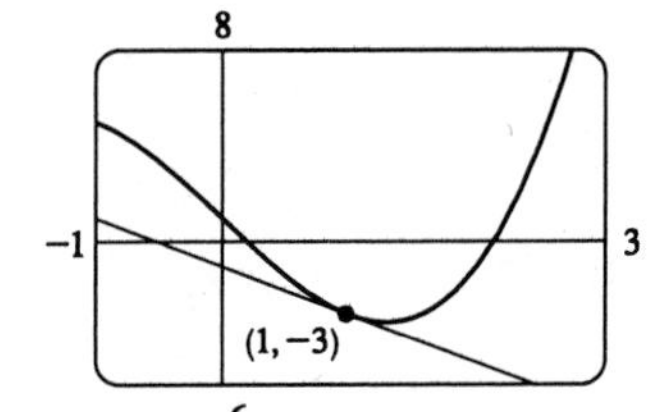

4. **(a)** $G'(a) = \lim_{h\to 0} \frac{G(a+h) - G(a)}{h} = \lim_{h\to 0} \frac{\frac{a+h}{1 + 2(a+h)} - \frac{a}{1 + 2a}}{h}$

$= \lim_{h\to 0} \frac{a + 2a^2 + h + 2ah - a - 2a^2 - 2ah}{h(1 + 2a + 2h)(1 + 2a)} = \lim_{h\to 0} \frac{1}{(1 + 2a + 2h)(1 + 2a)} = (1 + 2a)^{-2}.$

So the slope of the tangent at the point $\left(-\frac{1}{4}, -\frac{1}{2}\right)$ is $m = \left[1 + 2\left(-\frac{1}{4}\right)\right]^{-2} = 4$, and thus its equation is $y + \frac{1}{2} = 4\left(x + \frac{1}{4}\right)$ or $y = 4x + \frac{1}{2}$.

(b)

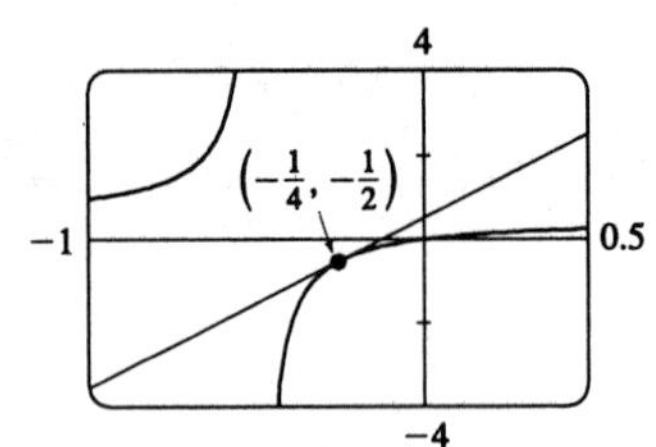

5. $v(2) = f'(2) = \lim_{h\to 0} \frac{f(2+h) - f(2)}{h} = \lim_{h\to 0} \frac{(2+h)^2 - 6(2+h) - 5 - (2^2 - 6(2) - 5)}{h}$

$= \lim_{h\to 0} \frac{4 + 4h + h^2 - 12 - 6h - 5 - 4 + 12 + 5}{h} = \lim_{h\to 0} \frac{h^2 - 2h}{h} = \lim_{h\to 0} (h - 2) = -2\text{ m/s}$

6. $$v(2) = f'(2) = \lim_{h\to 0} \frac{f(2+h) - f(2)}{h} = \lim_{h\to 0} \frac{2(2+h)^3 - (2+h) + 1 - \left(2(2)^3 - 2 + 1\right)}{h}$$
$$= \lim_{h\to 0} \frac{2h^3 + 12h^2 + 24h + 16 - 2 - h + 1 - 15}{h} = \lim_{h\to 0} \frac{2h^3 + 12h^2 + 23h}{h}$$
$$= \lim_{h\to 0} \left(2h^2 + 12h + 23\right) = 23 \text{ m/s}$$

7. $$f'(a) = \lim_{h\to 0} \frac{f(a+h) - f(a)}{h} = \lim_{h\to 0} \frac{1 + (a+h) - 2(a+h)^2 - (1 + a - 2a^2)}{h}$$
$$= \lim_{h\to 0} \frac{h - 4ah - 2h^2}{h} = \lim_{h\to 0} (1 - 4a - 2h) = 1 - 4a$$

8. $$f'(a) = \lim_{h\to 0} \frac{f(a+h) - f(a)}{h} = \lim_{h\to 0} \frac{(a+h)^3 + 3(a+h) - (a^3 + 3a)}{h}$$
$$= \lim_{h\to 0} \frac{3a^2h + 3ah^2 + h^3 + 3h}{h} = \lim_{h\to 0} \left(3a^2 + 3ah + h^2 + 3\right) = 3a^2 + 3$$

9. $$f'(a) = \lim_{h\to 0} \frac{f(a+h) - f(a)}{h} = \lim_{h\to 0} \frac{\dfrac{a+h}{2(a+h) - 1} - \dfrac{a}{2a - 1}}{h}$$
$$= \lim_{h\to 0} \frac{(a+h)(2a-1) - a(2a + 2h - 1)}{h(2a + 2h - 1)(2a - 1)} = \lim_{h\to 0} \frac{-h}{h(2a + 2h - 1)(2a - 1)}$$
$$= \lim_{h\to 0} \frac{-1}{(2a + 2h - 1)(2a - 1)} = -\frac{1}{(2a-1)^2}$$

10. $$f'(a) = \lim_{h\to 0} \frac{f(a+h) - f(a)}{h} = \lim_{h\to 0} \frac{\dfrac{a+h}{(a+h)^2 - 1} - \dfrac{a}{a^2 - 1}}{h}$$
$$= \lim_{h\to 0} \frac{(a+h)(a^2 - 1) - a(a^2 + 2ah + h^2 - 1)}{h(a^2 - 1)(a^2 + 2ah + h^2 - 1)} = \lim_{h\to 0} \frac{h(-a^2 - 1 - ah)}{h(a^2 - 1)(a^2 + 2ah + h^2 - 1)}$$
$$= \lim_{h\to 0} \frac{-a^2 - 1 - ah}{(a^2 - 1)(a^2 + 2ah + h^2 - 1)} = \frac{-a^2 - 1}{(a^2 - 1)(a^2 - 1)} = -\frac{a^2 + 1}{(a^2 - 1)^2}$$

11. $$f'(a) = \lim_{h\to 0} \frac{f(a+h) - f(a)}{h} = \lim_{h\to 0} \frac{\dfrac{2}{\sqrt{3 - (a+h)}} - \dfrac{2}{\sqrt{3-a}}}{h} = \lim_{h\to 0} \frac{2\left(\sqrt{3-a} - \sqrt{3-a-h}\right)}{h\sqrt{3-a-h}\,\sqrt{3-a}}$$
$$= \lim_{h\to 0} \frac{2\left(\sqrt{3-a} - \sqrt{3-a-h}\right)}{h\sqrt{3-a-h}\,\sqrt{3-a}} \cdot \frac{\sqrt{3-a} + \sqrt{3-a-h}}{\sqrt{3-a} + \sqrt{3-a-h}}$$
$$= \lim_{h\to 0} \frac{2[3 - a - (3 - a - h)]}{h\sqrt{3-a-h}\,\sqrt{3-a}\left(\sqrt{3-a} + \sqrt{3-a-h}\right)}$$
$$= \lim_{h\to 0} \frac{2}{\sqrt{3-a-h}\,\sqrt{3-a}\left(\sqrt{3-a} + \sqrt{3-a-h}\right)}$$
$$= \frac{2}{\sqrt{3-a}\,\sqrt{3-a}\left(2\sqrt{3-a}\right)} = \frac{1}{(3-a)^{3/2}}$$

12. $f'(a) = \lim_{h\to 0}\dfrac{f(a+h)-f(a)}{h} = \lim\dfrac{\sqrt{a+h-1}-\sqrt{a-1}}{h}$

$= \lim_{h\to 0}\dfrac{\sqrt{a+h-1}-\sqrt{a-1}}{h}\cdot\dfrac{\sqrt{a+h-1}+\sqrt{a-1}}{\sqrt{a+h-1}+\sqrt{a-1}} = \lim_{h\to 0}\dfrac{(a+h-1)-(a-1)}{h\left(\sqrt{a+h-1}+\sqrt{a-1}\right)}$

$= \lim_{h\to 0}\dfrac{1}{\sqrt{a+h-1}+\sqrt{a-1}} = \dfrac{1}{\sqrt{a-1}+\sqrt{a-1}} = \dfrac{1}{2\sqrt{a-1}}$

13. By Equation 1, $\lim_{h\to 0}\dfrac{\sqrt{1+h}-1}{h} = f'(1)$ where $f(x)=\sqrt{x}$. [Or $f'(0)$ where $f(x)=\sqrt{1+x}$; the answers to Exercises 13-18 are not unique.]

14. $\lim_{h\to 0}\dfrac{(2+h)^3-8}{h} = f'(2)$ where $f(x)=x^3$.

15. $\lim_{x\to 1}\dfrac{x^9-1}{x-1} = f'(1)$ where $f(x)=x^9$. (See Equation 3.)

16. $\lim_{x\to 3\pi}\dfrac{\cos x+1}{x-3\pi} = f'(3\pi)$ where $f(x)=\cos x$. (See Equation 3.)

17. $\lim_{t\to 0}\dfrac{\sin\left(\frac{\pi}{2}+t\right)-1}{t} = f'\left(\frac{\pi}{2}\right)$ where $f(x)=\sin x$.

18. $\lim_{x\to 0}\dfrac{3^x-1}{x} = f'(0)$ where $f(x)=3^x$.

19. $f'(x) = \lim_{h\to 0}\dfrac{f(x+h)-f(x)}{h} = \lim_{h\to 0}\dfrac{5(x+h)+3-(5x+3)}{h} = \lim_{h\to 0}\dfrac{5h}{h} = \lim_{h\to 0} 5 = 5.$

Domain of f = domain of $f' = \mathbb{R}$.

20. $f'(x) = \lim_{h\to 0}\dfrac{f(x+h)-f(x)}{h} = \lim_{h\to 0}\dfrac{18-18}{h} = \lim_{h\to 0} 0 = 0.$ Domain of f = domain of $f' = \mathbb{R}$.

21. $f'(x) = \lim_{h\to 0}\dfrac{f(x+h)-f(x)}{h} = \lim_{h\to 0}\dfrac{(x+h)^3-(x+h)^2+2(x+h)-(x^3-x^2+2x)}{h}$

$= \lim_{h\to 0}\dfrac{3x^2h+3xh^2+h^3-2xh-h^2+2h}{h} = \lim_{h\to 0}\left(3x^2+3xh+h^2-2x-h+2\right) = 3x^2-2x+2$

Domain of f = domain of $f' = \mathbb{R}$.

22. $f'(x) = \lim_{h\to 0}\dfrac{\sqrt{6-(x+h)}-\sqrt{6-x}}{h} = \lim_{h\to 0}\dfrac{\sqrt{6-x-h}-\sqrt{6-x}}{h}\cdot\dfrac{\sqrt{6-x-h}+\sqrt{6-x}}{\sqrt{6-x-h}+\sqrt{6-x}}$

$= \lim_{h\to 0}\dfrac{-h}{h\left(\sqrt{6-x-h}+\sqrt{6-x}\right)} = -\dfrac{1}{2\sqrt{6-x}}$

Domain of $f = \{x \mid x \le 6\} = (-\infty, 6]$ and domain of $f' = \{x \mid x < 6\} = (-\infty, 6)$.

23. $g'(x) = \lim_{h\to 0}\dfrac{g(x+h)-g(x)}{h} = \lim_{h\to 0}\dfrac{\sqrt{1+2(x+h)}-\sqrt{1+2x}}{h}\left[\dfrac{\sqrt{1+2(x+h)}+\sqrt{1+2x}}{\sqrt{1+2(x+h)}+\sqrt{1+2x}}\right]$

$= \lim_{h\to 0}\dfrac{1+2x+2h-(1+2x)}{h\left[\sqrt{1+2(x+h)}+\sqrt{1+2x}\right]} = \lim_{h\to 0}\dfrac{2}{\sqrt{1+2(x+h)}+\sqrt{1+2x}} = \dfrac{1}{\sqrt{1+2x}}$

Domain of $g = \left[-\frac{1}{2},\infty\right)$, domain of $g' = \left(-\frac{1}{2},\infty\right)$.

24. $f'(x) = \lim_{h\to 0} \frac{f(x+h)-f(x)}{h} = \lim_{h\to 0} \frac{\frac{x+h+1}{x+h-1} - \frac{x+1}{x-1}}{h}$

$= \lim_{h\to 0} \frac{(x+h+1)(x-1)-(x+1)(x+h-1)}{h(x+h-1)(x-1)} = \lim_{h\to 0} \frac{-2h}{h(x+h-1)(x-1)}$

$= \lim_{h\to 0} \frac{-2}{(x+h-1)(x-1)} = \frac{-2}{(x-1)^2}$

Domain of f = domain of $f' = \{x \mid x \neq 1\}$.

25. $G'(x) = \lim_{h\to 0} \frac{G(x+h)-G(x)}{h} = \lim_{h\to 0} \frac{\frac{4-3(x+h)}{2+(x+h)} - \frac{4-3x}{2+x}}{h}$

$= \lim_{h\to 0} \frac{(4-3x-3h)(2+x)-(4-3x)(2+x+h)}{h(2+x+h)(2+x)} = \lim_{h\to 0} \frac{-10h}{h(2+x+h)(2+x)}$

$= \lim_{h\to 0} \frac{-10}{(2+x+h)(2+x)} = \frac{-10}{(2+x)^2}$

Domain of G = domain of $G' = \{x \mid x \neq -2\}$.

26. $g'(x) = \lim_{h\to 0} \frac{g(x+h)-g(x)}{h} = \lim_{h\to 0} \frac{\frac{1}{(x+h)^2} - \frac{1}{x^2}}{h} = \lim_{h\to 0} \frac{x^2-(x+h)^2}{h(x+h)^2 x^2} = \lim_{h\to 0} \frac{-2xh-h^2}{h(x+h)^2x^2}$

$= \lim_{h\to 0} \frac{-2x-h}{(x+h)^2x^2} = \frac{-2x}{x^4} = -2x^{-3}$

Domain of g = domain of $g' = \{x \mid x \neq 0\}$.

27. $f'(x) = \lim_{h\to 0} \frac{f(x+h)-f(x)}{h} = \lim_{h\to 0} \frac{(x+h)^4-x^4}{h} = \lim_{h\to 0} \frac{4x^3h+6x^2h^2+4xh^3+h^4}{h}$

$= \lim_{h\to 0} (4x^3+6x^2h+4xh^2+h^3) = 4x^3$

Domain of f = domain of $f' = \mathbb{R}$.

28. $F'(x) = \lim_{h\to 0} \frac{F(x+h)-F(x)}{h} = \lim_{h\to 0} \frac{\frac{1}{\sqrt{x+h-1}} - \frac{1}{\sqrt{x-1}}}{h}$

$= \lim_{h\to 0} \frac{\sqrt{x-1}-\sqrt{x+h-1}}{h\sqrt{x+h-1}\sqrt{x-1}} \cdot \frac{\sqrt{x-1}+\sqrt{x+h-1}}{\sqrt{x-1}+\sqrt{x+h-1}}$

$= \lim_{h\to 0} \frac{-h}{h\sqrt{x+h-1}\sqrt{x-1}\left(\sqrt{x-1}+\sqrt{x+h-1}\right)} = \frac{-1}{\sqrt{x-1}\sqrt{x-1}\left(2\sqrt{x-1}\right)}$

$= -\frac{1}{2(x-1)^{3/2}}$

Domain of F = domain of $F' = (1, \infty)$.

29. $f(x) = x \quad \Rightarrow \quad f'(x) = \lim\limits_{h \to 0} \dfrac{x+h-x}{h} = \lim\limits_{h \to 0} 1 = 1$

$f(x) = x^2 \quad \Rightarrow \quad f'(x) = \lim\limits_{h \to 0} \dfrac{(x+h)^2 - x^2}{h} = \lim\limits_{h \to 0} \dfrac{2xh + h^2}{h} = \lim\limits_{h \to 0}(2x + h) = 2x$

$f(x) = x^3 \quad \Rightarrow \quad f'(x) = \lim\limits_{h \to 0} \dfrac{(x+h)^3 - x^3}{h} = \lim\limits_{h \to 0} \dfrac{3x^2h + 3xh^2 + h^3}{h} = \lim\limits_{h \to 0}(3x^2 + 3xh + h^2) = 3x^2$

$f(x) = x^4 \quad \Rightarrow \quad f'(x) = 4x^3$ from Exercise 31.

Guess: The derivative of $f(x) = x^n$ is $f'(x) = nx^{n-1}$. Test for $n = 5$: $f(x) = x^5 \quad \Rightarrow$

$$f'(x) = \lim_{h \to 0} \frac{(x+h)^5 - x^5}{h} = \lim_{h \to 0} \frac{5x^4h + 10x^3h^2 + 10x^2h^3 + 5xh^4 + h^5}{h}$$
$$= \lim_{h \to 0}\left(5x^4 + 10x^3h + 10x^2h^2 + 5xh^3 + h^4\right) = 5x^4$$

30. **(a)** $$f'(t) = \lim_{h \to 0} \frac{f(t+h) - f(t)}{h} = \lim_{h \to 0} \frac{\dfrac{6}{1+(t+h)^2} - \dfrac{6}{1+t^2}}{h} = \lim_{h \to 0} \frac{6 + 6t^2 - 6 - 6(t+h)^2}{h\left[1+(t+h)^2\right](1+t^2)}$$

$$= \lim_{h \to 0} \frac{-12th - 6h^2}{h\left[1+(t+h)^2\right](1+t^2)}$$
$$= \lim_{h \to 0} \frac{-12t - 6h}{\left[1+(t+h)^2\right](1+t^2)}$$
$$= \frac{-12t}{(1+t^2)^2}$$

Notice that f has a horizontal tangent when $t = 0$. This corresponds to $f'(0) = 0$.

(b)

6, −2, 2, −4, f, f′

31. **(a)** $$f'(x) = \lim_{h \to 0} \frac{f(x+h) - f(x)}{h} = \lim_{h \to 0} \frac{\left[x + h - \left(\dfrac{2}{x+h}\right)\right] - \left[x - \left(\dfrac{2}{x}\right)\right]}{h}$$

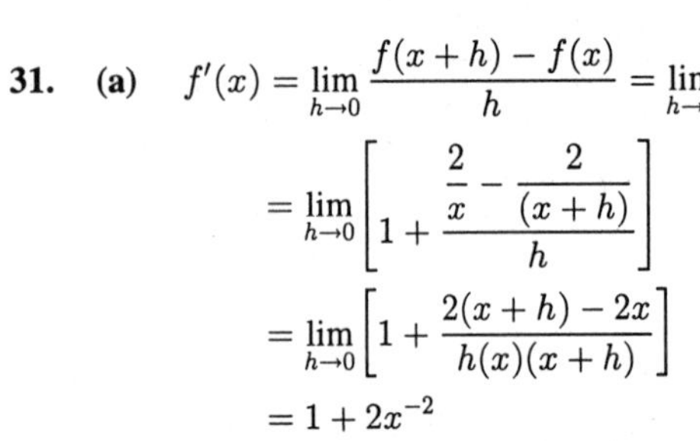

$$= \lim_{h \to 0}\left[1 + \frac{\dfrac{2}{x} - \dfrac{2}{(x+h)}}{h}\right]$$
$$= \lim_{h \to 0}\left[1 + \frac{2(x+h) - 2x}{h(x)(x+h)}\right]$$
$$= 1 + 2x^{-2}$$

Notice that when f has steep tangent lines, $f'(x)$ is very large. When f is flatter, $f'(x)$ is smaller.

(b)

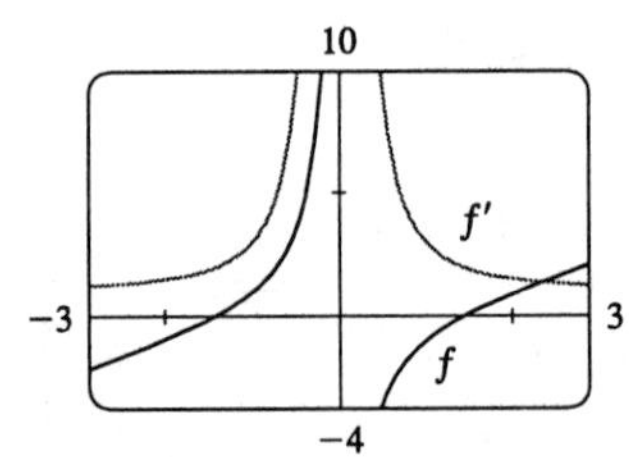

32. From the graph of f, it appears that

(a) $f'(0) \approx 2$ **(b)** $f'(1) \approx 0.7$

(c) $f'(2) \approx 0$ **(d)** $f'(3) \approx -1$

(e) $f'(4) \approx -1.3$ **(f)** $f'(5) \approx 0$

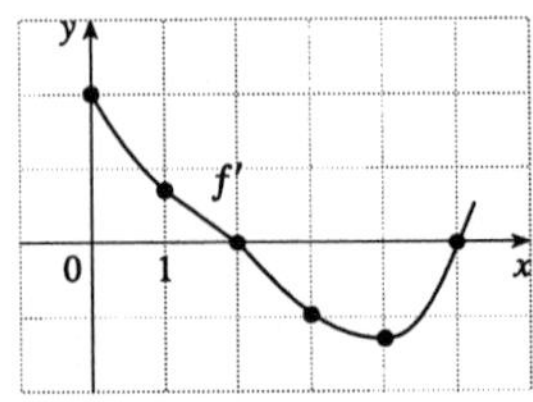

33. From the graph of f, it appears that

(a) $f'(1) \approx -2$ **(b)** $f'(2) \approx 0.8$

(c) $f'(3) \approx -1$ **(d)** $f'(4) \approx -0.5$

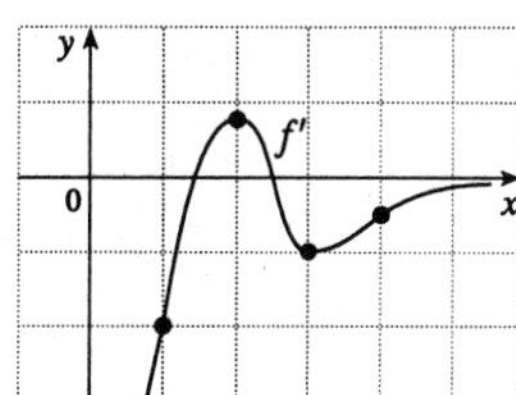

34. (a)$'$ = (ii), since from left to right, the slopes of the tangents to graph (a) start out negative, become 0, then positive, then 0, then negative again. The actual function values in graph (ii) follow the same pattern.

(b)$'$ = (iv), since from left to right, the slopes of the tangents to graph (b) start out at a fixed positive quantity, then suddenly become negative, then positive again. The discontinuities in graph (iv) indicate sudden changes in the slopes of the tangents.

(c)$'$ = (i), since the slopes of the tangents to graph (c) are negative for $x < 0$ and positive for $x > 0$, as are the function values of graph (i).

(d)$'$ = (iii), since from left to right, the slopes of the tangents to graph (d) are positive, then 0, then negative, then 0, then positive, then 0, then negative again, and the function values in graph (iii) follow the same pattern.

35.

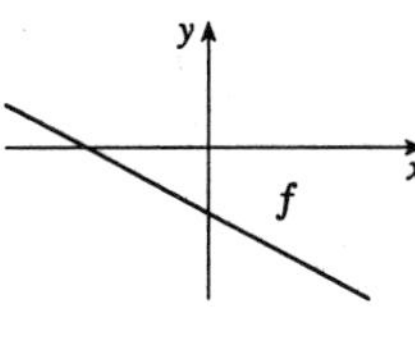

36.

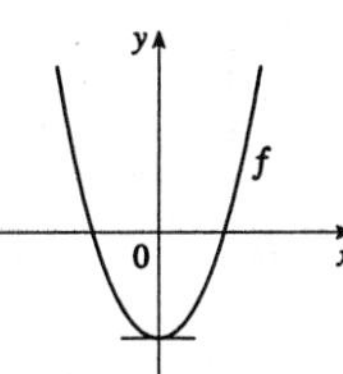

37.

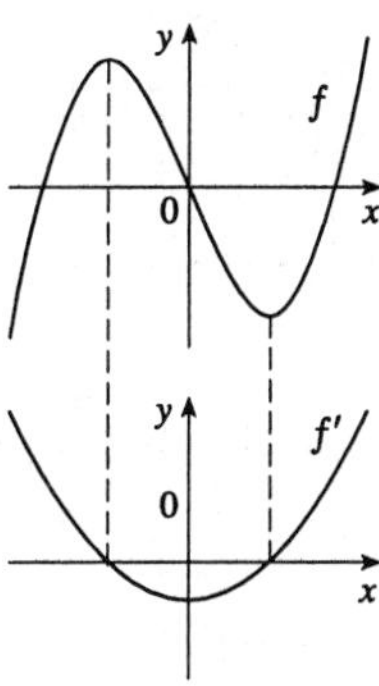

38.

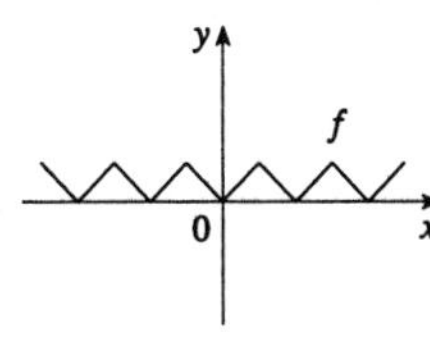

39.

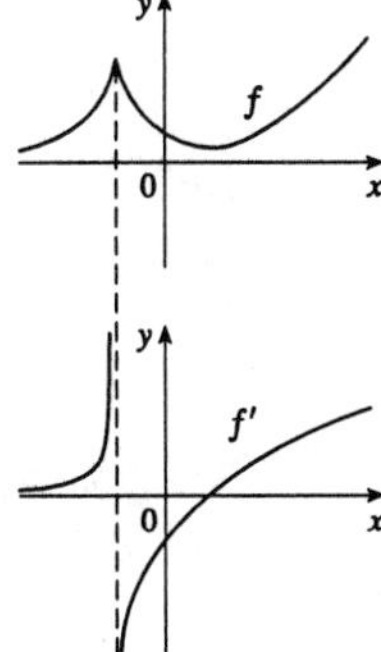

40.

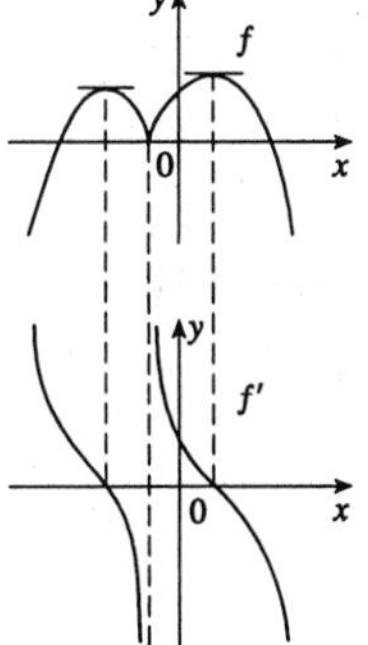

41.

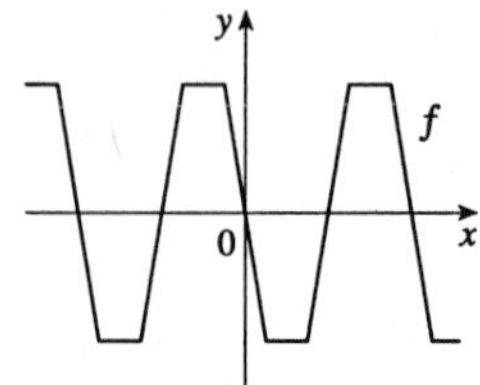

42.

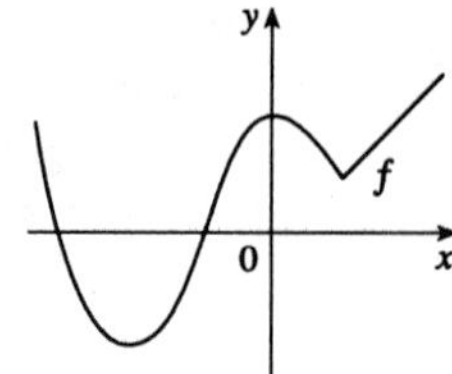

43.

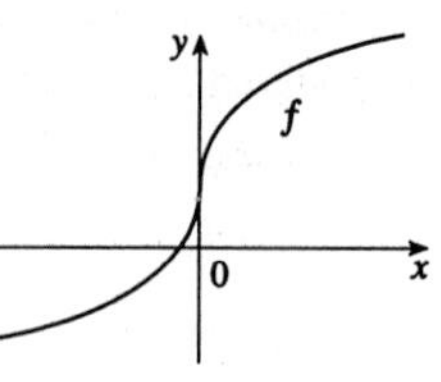

44. See Figure 3 in Section 2.4.

45. $f'(1) = \lim\limits_{h\to 0} \dfrac{f(1+h) - f(1)}{h} = \lim\limits_{h\to 0} \dfrac{3^{1+h} - 3^1}{h}$ So let $F(h) = \dfrac{3^{1+h} - 3}{h}$. We calculate:

h	$F(h)$
0.1	3.48
0.01	3.314
0.001	3.297
0.0001	3.296
−0.1	3.12
−0.01	3.278
−0.001	3.294
−0.0001	3.296

We estimate that $f'(1) \approx 3.296$.

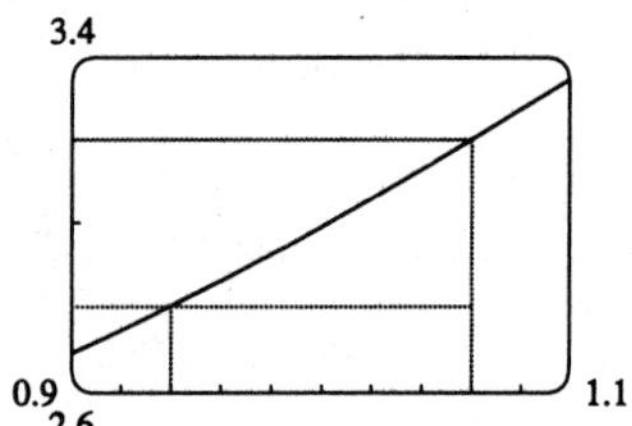

From the graph we estimate that the slope of the tangent is about $\dfrac{3.2 - 2.8}{1.06 - 0.94} = \dfrac{0.4}{0.12} \approx 3.3$.

46. $g'\left(\frac{\pi}{4}\right) = \lim\limits_{h\to 0} \dfrac{g\left(\frac{\pi}{4} + h\right) - g\left(\frac{\pi}{4}\right)}{h} = \lim\limits_{h\to 0} \dfrac{\tan\left(\frac{\pi}{4} + h\right) - \tan\left(\frac{\pi}{4}\right)}{h}$. So let $G(h) = \dfrac{\tan\left(\frac{\pi}{4} + h\right) - 1}{h}$. We calculate:

h	$G(h)$
0.1	2.23
0.01	2.020
0.001	2.002
0.0001	2.0002
−0.1	1.82
−0.01	1.980
−0.001	1.998
−0.0001	1.9998

We estimate that $g'\left(\frac{\pi}{4}\right) = 2$.

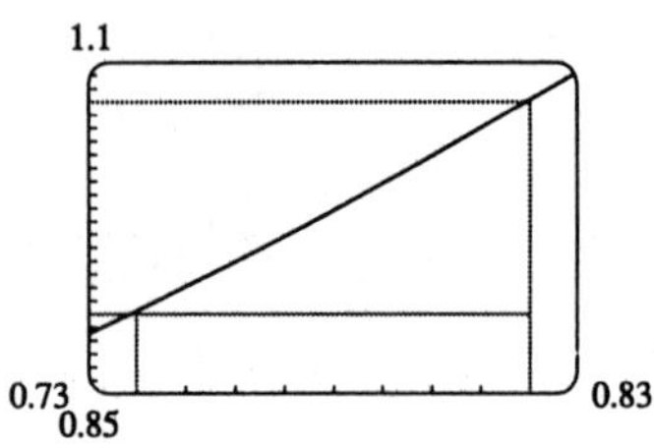

From the graph we estimate that the slope of the tangent is about $\dfrac{1.07 - 0.91}{0.82 - 0.74} = \dfrac{0.16}{0.08} = 2$.

47. We plot the points given by the data in the table, then sketch the rough shape of the curve. To estimate the derivative $f'(x)$, we draw the tangent line to the curve at x. It appears that $f'(0.1) \approx -5$, $f'(0.2) \approx 4$, $f'(0.3) \approx 8$, $f'(0.4) \approx 9$, $f'(0.5) \approx 5$, $f'(0.6) \approx -0.5$, and $f'(0.7) \approx -8$.

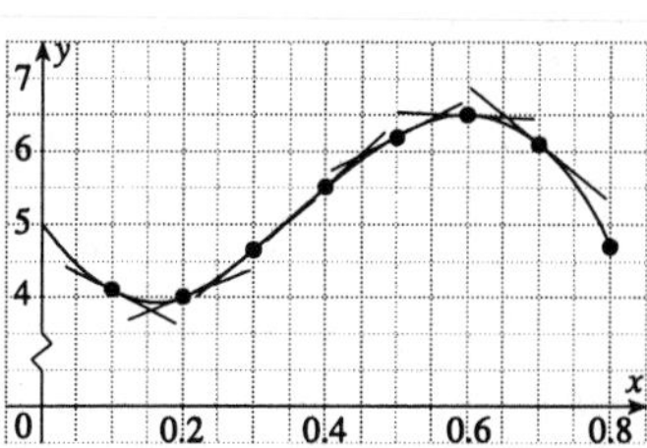

48. We plot the points given by the data in the table, then sketch the rough shape of the curve. To estimate the derivative $f'(x)$, we draw the tangent line to the curve at x. It appears that $g'(2) \approx 1.1$, $g'(4) \approx 0.6$, $g'(6) \approx -0.5$, $g'(8) \approx -1.1$, $g'(10) \approx -0.4$, $g'(12) \approx -0.2$, and $g'(14) \approx -0.1$.

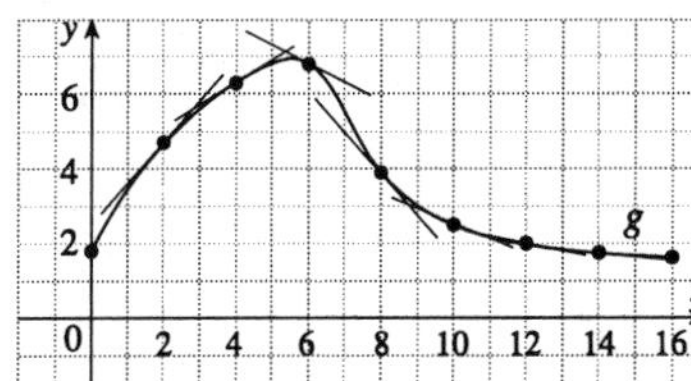

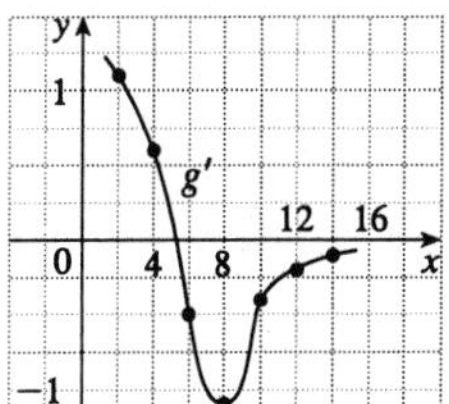

49. **(a)** $f'(a) = \lim\limits_{x \to a} \dfrac{f(x) - f(a)}{x - a} = \lim\limits_{x \to a} \dfrac{x^{1/3} - a^{1/3}}{x - a} = \lim\limits_{x \to a} \dfrac{x^{1/3} - a^{1/3}}{\left(x^{1/3} - a^{1/3}\right)\left(x^{2/3} + a^{2/3} + x^{1/3}a^{1/3}\right)}$

$$= \lim_{x \to a} \frac{1}{x^{2/3} + a^{2/3} + x^{1/3}a^{1/3}} = \lim_{x \to a} \frac{1}{3x^{2/3}} = \frac{1}{3a^{2/3}}$$

(b) $f'(0) = \lim\limits_{h \to 0} \dfrac{f(0+h) - f(0)}{h} = \lim\limits_{h \to 0} \dfrac{\sqrt[3]{h} - 0}{h} = \lim\limits_{h \to 0} \dfrac{1}{h^{2/3}}$. This limit does not exist, and therefore $f'(0)$ does not exist.

(c) $\lim\limits_{x \to 0} |f'(x)| = \lim\limits_{x \to 0} \dfrac{1}{3x^{2/3}} = \infty$. Also f is continuous at $x = 0$ (root function), so f has a vertical tangent at $x = 0$.

50. **(a)** Using Equation 3, $g'(0) = \lim\limits_{x \to 0} \dfrac{g(x) - g(0)}{x - 0} = \lim\limits_{x \to 0} \dfrac{x^{2/3}}{x} = \lim\limits_{x \to 0} \dfrac{1}{x^{-1/3}}$, which does not exist.

(b) Using Equation 3, $g'(a) = \lim\limits_{x \to a} \dfrac{g(x) - g(a)}{x - a} = \lim\limits_{x \to a} \dfrac{x^{2/3} - a^{2/3}}{x - a}$

$$= \lim_{x \to a} \frac{\left(x^{1/3} - a^{1/3}\right)\left(x^{1/3} + a^{1/3}\right)}{\left(x^{1/3} - a^{1/3}\right)\left(x^{2/3} + a^{2/3} + x^{1/3}a^{1/3}\right)}$$

$$= \lim_{x \to a} \frac{x^{1/3} + a^{1/3}}{x^{2/3} + a^{2/3} + x^{1/3}a^{1/3}} = \lim_{x \to a} \frac{2x^{1/3}}{3x^{2/3}} = \frac{2}{3a^{1/3}}.$$

(c) $g(x) = x^{2/3}$ is continuous at $x = 0$ and $\lim\limits_{x \to 0} |g'(x)| = \lim\limits_{x \to 0} \dfrac{2}{3|x|^{1/3}} = \infty$. This shows that g has a vertical tangent line at $x = 0$.

(d)

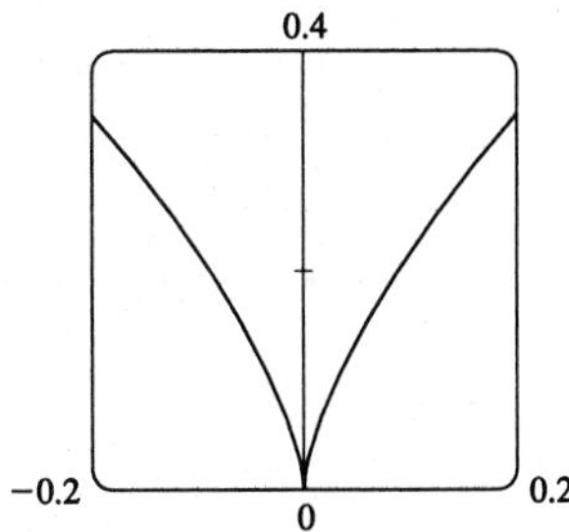

51. f is not differentiable at $x = -1$ or at $x = 11$ because the graph has vertical tangents at those points; at $x = 4$, because there is a discontinuity there; and at $x = 8$, because the graph has a corner there.

52. **(a)** g is discontinuous at $x = -2$ (a removable discontinuity), at $x = 0$ (g is not defined there), and at $x = 5$ (a jump discontinuity).

(b) g is not differentiable at the above points (by Theorem 8), and also at $x = -1$ (corner), at $x = 2$ (vertical tangent), and at $x = 4$ (vertical tangent.)

53. $f(x) = |x-6| = \begin{cases} 6-x & \text{if } x < 6 \\ x-6 & \text{if } x \geq 6. \end{cases}$ $\lim\limits_{x\to 6^+} \dfrac{f(x)-f(6)}{x-6} = \lim\limits_{x\to 6^+} \dfrac{|x-6|-0}{x-6} = \lim\limits_{x\to 6^+} \dfrac{x-6}{x-6} = \lim\limits_{x\to 6^+} 1 = 1.$

But $\lim\limits_{x\to 6^-} \dfrac{f(x)-f(6)}{x-6} = \lim\limits_{x\to 6^-} \dfrac{|x-6|-0}{x-6}$

$= \lim\limits_{x\to 6^-} \dfrac{6-x}{x-6} = \lim\limits_{x\to 6^-} (-1) = -1.$

So $f'(6) = \lim\limits_{x\to 6} \dfrac{f(x)-f(6)}{x-6}$ does not exist.

However $f'(x) = \begin{cases} -1 & \text{if } x < 6 \\ 1 & \text{if } x > 6. \end{cases}$

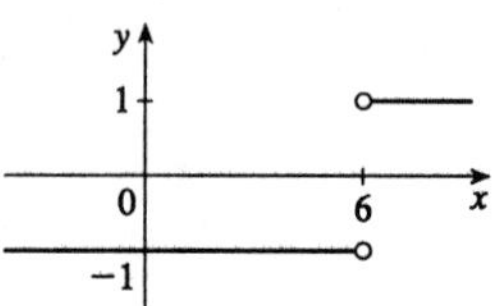

54. $f(x) = [\![x]\!]$ is not continuous at any integer n (see Example 11 and Exercise 63 in Section 1.3) so f is not differentiable at n by Theorem 8. If a is not an integer, then f is constant on an open interval containing a, so $f'(a) = 0$. Thus $f'(x) = 0$, x not an integer.

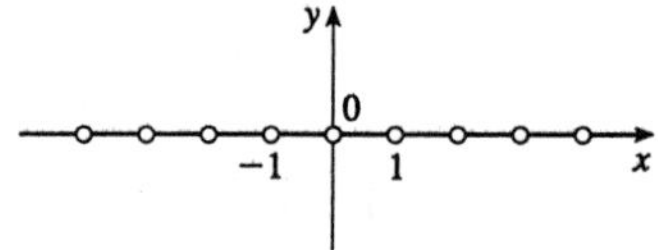

55. **(a)** $f(x) = x|x| = \begin{cases} x^2 & \text{if } x \geq 0 \\ -x^2 & \text{if } x < 0 \end{cases}$

(b) Since $f(x) = x^2$ for $x \geq 0$, we have $f'(x) = 2x$ for $x > 0$. Since $f(x) = -x^2$ for $x < 0$, we have $f'(x) = -2x$ for $x < 0$. At $x = 0$, we have

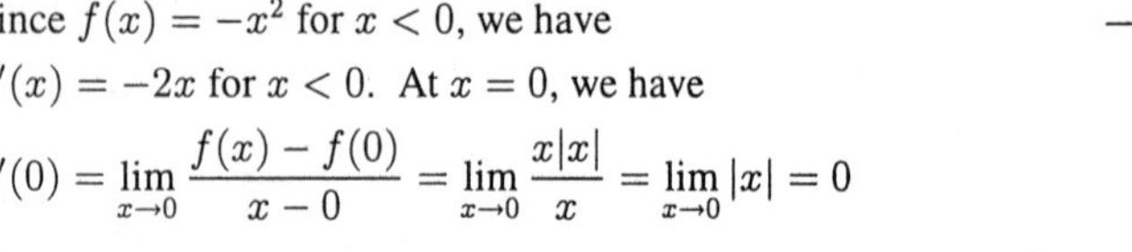

$f'(0) = \lim\limits_{x\to 0} \dfrac{f(x)-f(0)}{x-0} = \lim\limits_{x\to 0} \dfrac{x|x|}{x} = \lim\limits_{x\to 0} |x| = 0$

(by Example 1.3.8). So f is differentiable at 0. Thus f is differentiable for all x.

(c) From part (b) we have $f'(x) = \left\{ \begin{array}{ll} 2x & x \geq 0 \\ -2x & x < 0 \end{array} \right\} = 2|x|.$

56. $g(x) = \begin{cases} (x^3-x)/(x^2+x) & \text{if } x < 1, x \neq 0, x \neq -1 \\ 0 & \text{if } x = 0 \\ 1-x & \text{if } x \geq 1. \end{cases}$ These expressions show that f is differentiable on the intervals $(-\infty, -1)$, $(-1, 0)$ and $(0, \infty)$. Note that if $x < 1$ and $x \neq 0$, then $g(x) = \dfrac{x(x+1)(x-1)}{x(x+1)} = x - 1.$

g is discontinuous (and so not differentiable) at $x = -1$ since it is not defined there. Also

$\lim_{x\to 0} g(x) = \lim_{x\to 0}(x-1) = -1$, whereas $g(0) = 0$, so g is discontinuous (and therefore not differentiable) at $x = 0$. Also g is not differentiable at 1 because

$$\lim_{h\to 0^+} \frac{g(1+h)-g(1)}{h} = \lim_{h\to 0^+} \frac{-h-0}{h} = \lim_{h\to 0^+}(-1) = -1,$$

but $\lim_{h\to 0^-} \dfrac{g(1+h)-g(1)}{h} = \lim_{h\to 0^-} \dfrac{h-0}{h} = \lim_{h\to 0^-} 1 = 1$,

so $g'(1)$ does not exist.

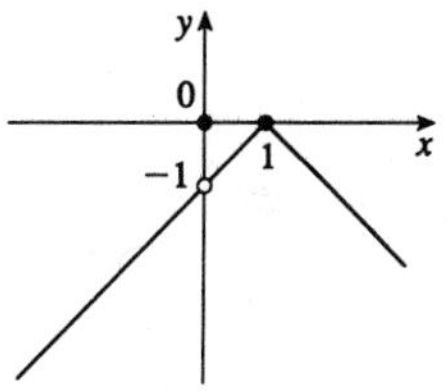

57. **(a)**
$$f'_-(0.6) = \lim_{h\to 0^-} \frac{f(0.6+h)-f(0.6)}{h} = \lim_{h\to 0^-} \frac{|5(0.6+h)-3|-|3-3|}{h}$$
$$= \lim_{h\to 0^-} \frac{|3+5h-3|}{h} = \lim_{h\to 0^-} \frac{-5h}{h} = -5$$
$$f'_+(0.6) = \lim_{h\to 0^+} \frac{f(0.6+h)-f(0.6)}{h} = \lim_{h\to 0^+} \frac{|5(0.6+h)-3|-|3-3|}{h}$$
$$= \lim_{h\to 0^+} \frac{5h}{h} = 5$$

(b) Since $f'_-(0.6) \neq f'_+(0.6)$, $f'(0.6)$ does not exist.

58. **(a)** $f'_-(4) = \lim_{h\to 0^-} \dfrac{f(4+h)-f(4)}{h} = \lim_{h\to 0^-} \dfrac{5-(4+h)-1}{h} = \lim_{h\to 0^-} \dfrac{-h}{h} = -1$ and

$$f'_+(4) = \lim_{h\to 0^+} \frac{f(4+h)-f(4)}{h} = \lim_{h\to 0^+} \frac{\dfrac{1}{5-(4+h)}-1}{h} = \lim_{h\to 0^+} \frac{1-(1-h)}{h(1-h)} = \lim_{h\to 0^+} \frac{1}{1-h} = 1$$

(b)

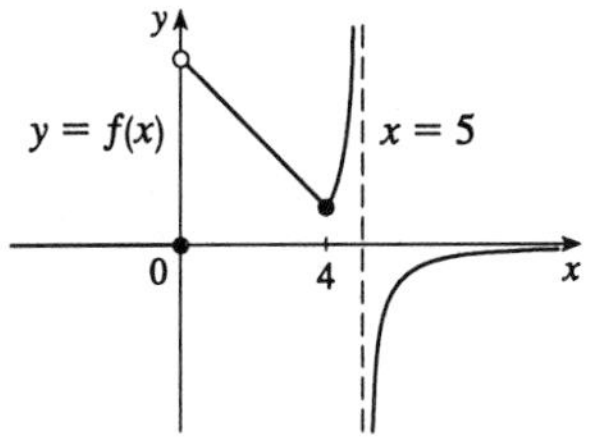

y
y = f(x)
x = 5
0
4
x

(c) $f(x) = \begin{cases} 0 & \text{if } x \le 0 \\ 5-x & \text{if } 0 < x < 4 \\ 1/(5-x) & \text{if } x \ge 4. \end{cases}$ These expressions show that f is continuous on the intervals $(-\infty, 0)$, $(0, 4)$, $(4, 5)$ and $(5, \infty)$. Since $\lim_{x\to 0^+} f(x) = \lim_{x\to 0^+}(5-x) = 5 \neq 0 = \lim_{x\to 0^-} f(x)$, $\lim_{x\to 0} f(x)$ does not exist, so f is discontinuous (and therefore not differentiable) at 0. At 4 we have $\lim_{x\to 4^-} f(x) = \lim_{x\to 4^-}(5-x) = 1$ and $\lim_{x\to 4^+} f(x) = \lim_{x\to 4^+} \dfrac{1}{5-x} = 1$, so $\lim_{x\to 4} f(x) = 1 = f(4)$ and f is continuous at 4. Since $f(5)$ is not defined, f is discontinuous at 5.

(d) From (a), f is not differentiable at 4 since $f'_-(4) \neq f'_+(4)$, and from (c), f is not differentiable at 0 or 5.

59. Since $f(x) = x\sin(1/x)$ when $x \neq 0$ and $f(0) = 0$, we have

$f'(0) = \lim_{h\to 0} \dfrac{f(0+h)-f(0)}{h} = \lim_{h\to 0} \dfrac{h\sin(1/h)-0}{h} = \lim_{h\to 0} \sin(1/h)$. This limit does not exist since $\sin(1/h)$ takes the values -1 and 1 on any interval containing 0. (Compare with Example 1.2.4.)

60. Since $f(x) = x^2\sin(1/x)$ when $x \neq 0$ and $f(0) = 0$, we have

$f'(0) = \lim_{h\to 0} \dfrac{f(0+h)-f(0)}{h} = \lim_{h\to 0} \dfrac{h^2\sin(1/h)-0}{h} = \lim_{h\to 0} h\sin(1/h) = 0$. (See Example 1.3.12.)

61. **(a)** If f is even, then $f'(-x) = \lim\limits_{h\to 0} \dfrac{f(-x+h) - f(-x)}{h} = \lim\limits_{h\to 0} \dfrac{f(x-h) - f(x)}{h}$

$= -\lim\limits_{h\to 0} \dfrac{f(x-h) - f(x)}{-h}$ [let $\Delta x = -h$] $= -\lim\limits_{\Delta x\to 0} \dfrac{f(x+\Delta x) - f(x)}{\Delta x} = -f'(x)$. Therefore f' is odd.

(b) If f is odd, then $f'(-x) = \lim\limits_{h\to 0} \dfrac{f(-x+h) - f(-x)}{h} = \lim\limits_{h\to 0} \dfrac{-f(x-h) + f(x)}{h}$

$= \lim\limits_{h\to 0} \dfrac{f(x-h) - f(x)}{-h}$ [let $\Delta x = -h$] $= \lim\limits_{\Delta x\to 0} \dfrac{f(x+\Delta x) - f(x)}{\Delta x} = f'(x)$. Therefore f' is even.

62. **(a)**

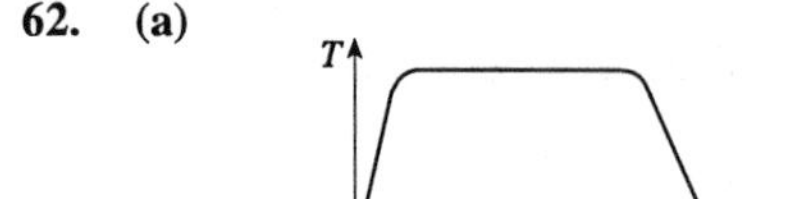

(c)

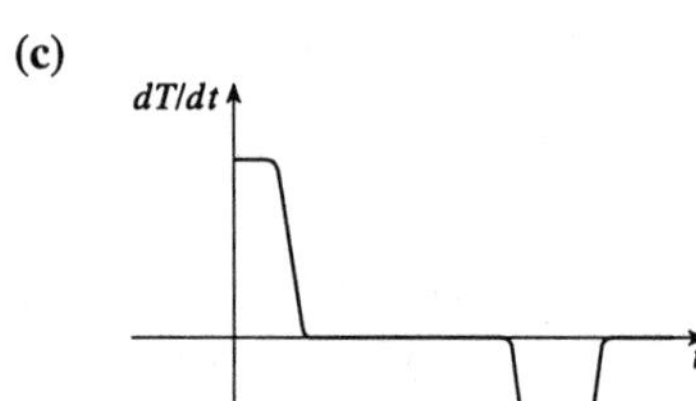

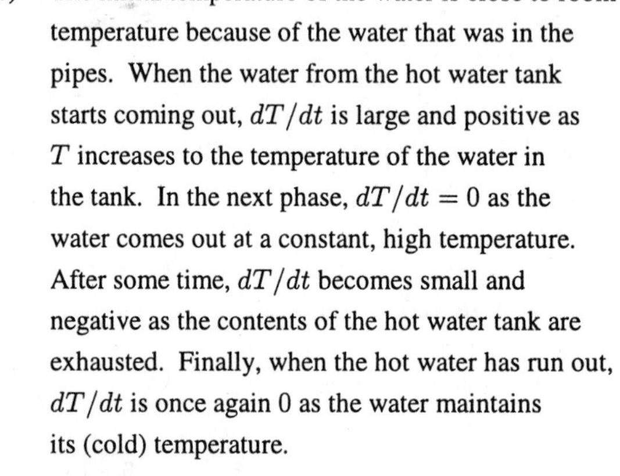

(b) The initial temperature of the water is close to room temperature because of the water that was in the pipes. When the water from the hot water tank starts coming out, dT/dt is large and positive as T increases to the temperature of the water in the tank. In the next phase, $dT/dt = 0$ as the water comes out at a constant, high temperature. After some time, dT/dt becomes small and negative as the contents of the hot water tank are exhausted. Finally, when the hot water has run out, dT/dt is once again 0 as the water maintains its (cold) temperature.

63. From the diagram, we see that the slope of the tangent is equal to $\tan\phi$, and also that $0 < \phi < \frac{\pi}{2}$. We know (see Exercise 29) that the derivative of $f(x) = x^2$ is $f'(x) = 2x$. So the slope of the tangent to the curve at the point $(1, 1)$ is 2. So ϕ is the angle between 0 and $\frac{\pi}{2}$ whose tangent is 2, that is, $\phi = \tan^{-1} 2 \approx 63°$.

64. **(a)** Put $x = 0$ and $y = 0$ in the equation: $f(0) = f(0+0) = f(0) + f(0) + 0^2 \cdot 0 + 0 \cdot 0^2 = 2f(0)$. Subtracting $f(0)$ from each side of this equation gives $f(0) = 0$.

(b) $f'(0) = \lim\limits_{h\to 0} \dfrac{f(0+h) - f(0)}{h} = \lim\limits_{h\to 0} \left[\dfrac{f(0) + f(h) + 0 + 0 - f(0)}{h}\right]$

$= \lim\limits_{h\to 0} \dfrac{f(h)}{h} = \lim\limits_{x\to 0} \dfrac{f(x)}{x} = 1.$

(c) $f'(x) = \lim\limits_{h\to 0} \dfrac{f(x+h) - f(h)}{h} = \lim\limits_{h\to 0} \dfrac{[f(x) + f(h) + x^2h + xh^2] - f(x)}{h}$

$= \lim\limits_{h\to 0} \dfrac{f(h) + x^2h + xh^2}{h} = \lim\limits_{h\to 0} \left[\dfrac{f(h)}{h} + x^2 + xh\right] = 1 + x^2.$

EXERCISES 2.2

1. $f(x) = x^2 - 10x + 100 \quad \Rightarrow \quad f'(x) = 2x - 10$

2. $g(x) = x^{100} + 50x + 1 \quad \Rightarrow \quad g'(x) = 100x^{99} + 50$

3. $V(r) = \frac{4}{3}\pi r^3 \quad \Rightarrow \quad V'(r) = \frac{4}{3}\pi(3r^2) = 4\pi r^2$

4. $s(t) = t^8 + 6t^7 - 18t^2 + 2t \quad \Rightarrow \quad s'(t) = 8t^7 + 6(7t^6) - 18(2t) + 2 = 8t^7 + 42t^6 - 36t + 2$

5. $F(x) = (16x)^3 = 4{,}096x^3 \quad \Rightarrow \quad F'(x) = 4{,}096(3x^2) = 12{,}288x^2$

6. $G(y) = (y^2 + 1)(2y - 7) \quad \Rightarrow$
$G'(y) = (y^2 + 1)D(2y - 7) + (2y - 7)D(y^2 + 1) = (y^2 + 1)(2) + (2y - 7)(2y) = 6y^2 - 14y + 2$

7. $Y(t) = 6t^{-9} \quad \Rightarrow \quad Y'(t) = 6(-9)t^{-10} = -54t^{-10}$

8. $R(x) = \dfrac{\sqrt{10}}{x^7} = \sqrt{10}x^{-7} \quad \Rightarrow \quad R'(x) = -7\sqrt{10}x^{-8} = -\dfrac{7\sqrt{10}}{x^8}$

9. $g(x) = x^2 + \dfrac{1}{x^2} = x^2 + x^{-2} \quad \Rightarrow \quad g'(x) = 2x + (-2)x^{-3} = 2x - \dfrac{2}{x^3}$

10. $f(t) = \sqrt{t} - \dfrac{1}{\sqrt{t}} = t^{1/2} - t^{-1/2} \quad \Rightarrow \quad f'(t) = \frac{1}{2}t^{-1/2} - \left(-\frac{1}{2}t^{-3/2}\right) = \dfrac{1}{2\sqrt{t}} + \dfrac{1}{2t\sqrt{t}}$

11. $h(x) = \dfrac{x+2}{x-1} \quad \Rightarrow \quad h'(x) = \dfrac{(x-1)D(x+2) - (x+2)D(x-1)}{(x-1)^2} = \dfrac{x - 1 - (x+2)}{(x-1)^2} = \dfrac{-3}{(x-1)^2}$

12. $f(u) = \dfrac{1-u^2}{1+u^2} \quad \Rightarrow$
$f'(u) = \dfrac{(1+u^2)D(1-u^2) - (1-u^2)D(1+u^2)}{(1+u^2)^2} = \dfrac{(1+u^2)(-2u) - (1-u^2)(2u)}{(1+u^2)^2} = \dfrac{-4u}{(1+u^2)^2}$

13. $G(s) = (s^2 + s + 1)(s^2 + 2) \quad \Rightarrow \quad G'(s) = (2s+1)(s^2+2) + (s^2+s+1)(2s) = 4s^3 + 3s^2 + 6s + 2$

14. $H(t) = \sqrt[3]{t}(t+2) = t^{4/3} + 2t^{1/3} \quad \Rightarrow \quad H'(t) = \frac{4}{3}t^{1/3} + \frac{2}{3}t^{-2/3}$. *Another Method:* Use the Product Rule.

15. $y = \dfrac{x^2 + 4x + 3}{\sqrt{x}} = x^{3/2} + 4x^{1/2} + 3x^{-1/2} \quad \Rightarrow$
$y' = \frac{3}{2}x^{1/2} + 4\left(\frac{1}{2}\right)x^{-1/2} + 3\left(-\frac{1}{2}\right)x^{-3/2} = \frac{3}{2}\sqrt{x} + \dfrac{2}{\sqrt{x}} - \dfrac{3}{2x\sqrt{x}}$. *Another Method:* Use the Quotient Rule.

16. $y = \dfrac{\sqrt{x} - 1}{\sqrt{x} + 1} \quad \Rightarrow \quad y' = \dfrac{(\sqrt{x}+1)\left[1/(2\sqrt{x})\right] - (\sqrt{x}-1)\left[1/(2\sqrt{x})\right]}{(\sqrt{x}+1)^2} = \dfrac{1}{\sqrt{x}(\sqrt{x}+1)^2}$

17. $y = \sqrt{5x} = \sqrt{5}x^{1/2} \quad \Rightarrow \quad y' = \sqrt{5}\left(\frac{1}{2}\right)x^{-1/2} = \dfrac{\sqrt{5}}{2\sqrt{x}}$

18. $y = x^{4/3} - x^{2/3} \quad \Rightarrow \quad y' = \frac{4}{3}x^{1/3} - \frac{2}{3}x^{-1/3}$

19. $y = \dfrac{1}{x^4 + x^2 + 1} \quad\Rightarrow\quad y' = \dfrac{(x^4 + x^2 + 1)(0) - 1(4x^3 + 2x)}{(x^4 + x^2 + 1)^2} = -\dfrac{4x^3 + 2x}{(x^4 + x^2 + 1)^2}$

20. $y = x^2 + x + x^{-1} + x^{-2} \quad\Rightarrow\quad y' = 2x + 1 - x^{-2} - 2x^{-3}$

21. $y = ax^2 + bx + c \quad\Rightarrow\quad y' = 2ax + b$

22. $y = A + \dfrac{B}{x} + \dfrac{C}{x^2} = A + Bx^{-1} + Cx^{-2} \quad\Rightarrow\quad y' = -Bx^{-2} - 2Cx^{-3} = -\dfrac{B}{x^2} - 2\dfrac{C}{x^3}$

23. $y = \dfrac{3t - 7}{t^2 + 5t - 4} \quad\Rightarrow\quad y' = \dfrac{(t^2 + 5t - 4)(3) - (3t - 7)(2t + 5)}{(t^2 + 5t - 4)^2} = \dfrac{-3t^2 + 14t + 23}{(t^2 + 5t - 4)^2}$

24. $y = \dfrac{4t + 5}{2 - 3t} \quad\Rightarrow\quad y' = \dfrac{(2 - 3t)(4) - (4t + 5)(-3)}{(2 - 3t)^2} = \dfrac{23}{(2 - 3t)^2}$

25. $y = x + \sqrt[5]{x^2} = x + x^{2/5} \quad\Rightarrow\quad y' = 1 + \frac{2}{5}x^{-3/5} = 1 + \dfrac{2}{5\sqrt[5]{x^3}}$

26. $y = x^4 - \sqrt[4]{x} = x^4 - x^{1/4} \quad\Rightarrow\quad y' = 4x^3 - \frac{1}{4}x^{-3/4}$

27. $u = x^{\sqrt{2}} \quad\Rightarrow\quad u' = \sqrt{2}\,x^{\sqrt{2}-1}$

28. $u = \sqrt[3]{t^2} + 2\sqrt{t^3} = t^{2/3} + 2t^{3/2} \quad\Rightarrow\quad u' = \frac{2}{3}t^{-1/3} + 2\left(\frac{3}{2}\right)t^{1/2} = \dfrac{2}{3\sqrt[3]{t}} + 3\sqrt{t}$

29. $v = x\sqrt{x} + \dfrac{1}{x^2\sqrt{x}} = x^{3/2} + x^{-5/2} \quad\Rightarrow\quad v' = \frac{3}{2}x^{1/2} - \frac{5}{2}x^{-7/2} = \frac{3}{2}\sqrt{x} - \dfrac{5}{2x^3\sqrt{x}}$

30. $v = \dfrac{6}{\sqrt[3]{t^5}} = 6t^{-5/3} \quad\Rightarrow\quad v' = 6\left(-\frac{5}{3}\right)t^{-8/3} = -\dfrac{10}{\sqrt[3]{t^8}}$

31. $f(x) = \dfrac{x}{x + c/x} \quad\Rightarrow\quad f'(x) = \dfrac{(x + c/x)(1) - x(1 - c/x^2)}{(x + c/x)^2} = \dfrac{2cx}{(x^2 + c)^2}$

32. $f(x) = \dfrac{ax + b}{cx + d} \quad\Rightarrow\quad f'(x) = \dfrac{(cx + d)(a) - (ax + b)(c)}{(cx + d)^2} = \dfrac{ad - bc}{(cx + d)^2}$

33. $f(x) = \dfrac{x^5}{x^3 - 2} \quad\Rightarrow\quad f'(x) = \dfrac{(x^3 - 2)(5x^4) - x^5(3x^2)}{(x^3 - 2)^2} = \dfrac{2x^4(x^3 - 5)}{(x^3 - 2)^2}$

34. $s = \sqrt{t}(t^3 - \sqrt{t} + 1) = t^{7/2} - t + t^{1/2} \Rightarrow s' = \frac{7}{2}t^{5/2} - 1 + \dfrac{1}{2\sqrt{t}}$. *Another Method:* Use the Product Rule.

35. $P(x) = a_n x^n + a_{n-1}x^{n-1} + \cdots + a_2x^2 + a_1x + a_0 \quad\Rightarrow$
$P'(x) = na_nx^{n-1} + (n-1)a_{n-1}x^{n-2} + \cdots + 2a_2x + a_1$

36. $y = f(x) = \dfrac{x}{x - 3} \quad\Rightarrow\quad f'(x) = \dfrac{(x - 3)\,1 - x(1)}{(x - 3)^2} = \dfrac{-3}{(x - 3)^2}$. So the slope of the tangent line at $(6, 2)$ is $f'(6) = -\frac{1}{3}$ and its equation is $y - 2 = -\frac{1}{3}(x - 6)$ or $x + 3y = 12$.

37. $y = f(x) = x + \dfrac{4}{x} \quad\Rightarrow\quad f'(x) = 1 - \dfrac{4}{x^2}$. So the slope of the tangent line at $(2, 4)$ is $f'(2) = 0$ and its equation is $y - 4 = 0$ or $y = 4$.

38. $y = f(x) = x^{5/2} \quad \Rightarrow \quad f'(x) = \frac{5}{2}x^{3/2}$. So the slope of the tangent line at $(4, 32)$ is $f'(4) = 20$ and its equation is $y - 32 = 20(x - 4)$ or $y = 20x - 48$.

39. $y = f(x) = x + \sqrt{x} \quad \Rightarrow \quad f'(x) = 1 + \frac{1}{2}x^{-1/2}$. So the slope of the tangent line at $(1, 2)$ is $f'(1) = 1 + \frac{1}{2}(1) = \frac{3}{2}$ and its equation is $y - 2 = \frac{3}{2}(x - 1)$ or $y = \frac{3}{2}x + \frac{1}{2}$ or $3x - 2y + 1 = 0$.

40. **(a)** $y = f(x) = \dfrac{x}{1 + x^2} \quad \Rightarrow$

$$f'(x) = \frac{(1 + x^2)1 - x(2x)}{(1 + x^2)^2} = \frac{1 - x^2}{(1 + x^2)^2}$$

So the slope of the tangent line at the point $(3, 0.3)$ is

$$\frac{1 - 3^2}{(1 + 3^2)^2} = -\frac{2}{25} = -0.08 \text{ and its equation is}$$

$y - 0.3 = -0.08(x - 3)$, or $y = -0.08x + 0.54$.

(b)

0.75

(3, 0.3)

−2 5

−0.5

41. **(a)** $y = f(x) = \dfrac{1}{1 + x^2} \quad \Rightarrow \quad f'(x) = \dfrac{-2x}{(1 + x^2)^2}$.

So the slope of the tangent line at the point $\left(-1, \frac{1}{2}\right)$ is

$$f'(-1) = \frac{-2(-1)}{\left[1 + (-1)^2\right]^2} = \frac{1}{2} \text{ and its equation is}$$

$y - \frac{1}{2} = \frac{1}{2}(x + 1)$ or $y = \frac{1}{2}x + 1$ or $x - 2y + 2 = 0$.

(b)

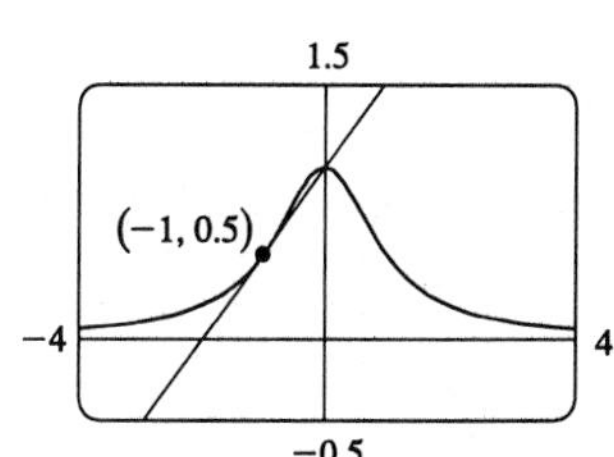

42. **(a)** $f(x) = \dfrac{x}{x^2 - 1}$

$$f'(x) = \frac{(x^2 - 1)1 - x(2x)}{(x^2 - 1)^2} = \frac{-x^2 - 1}{(x^2 - 1)^2}$$

(b)

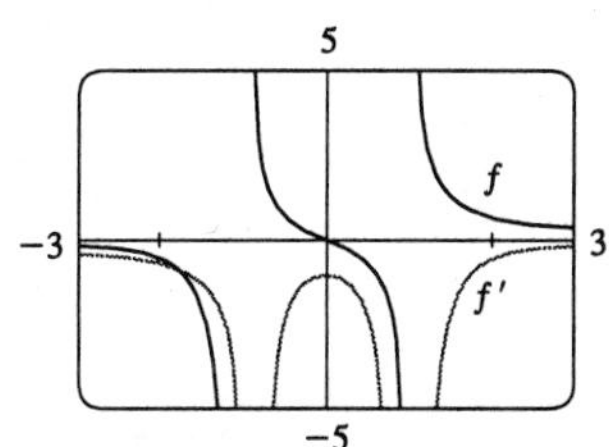

Notice that the slopes of all tangents to f are negative and $f'(x) < 0$ always.

43. **(a)** $f(x) = 3x^{15} - 5x^3 + 3 \quad \Rightarrow$

$$f'(x) = 3 \cdot 15x^{14} - 5 \cdot 3x^2 = 45x^{14} - 15x^2$$

(b)

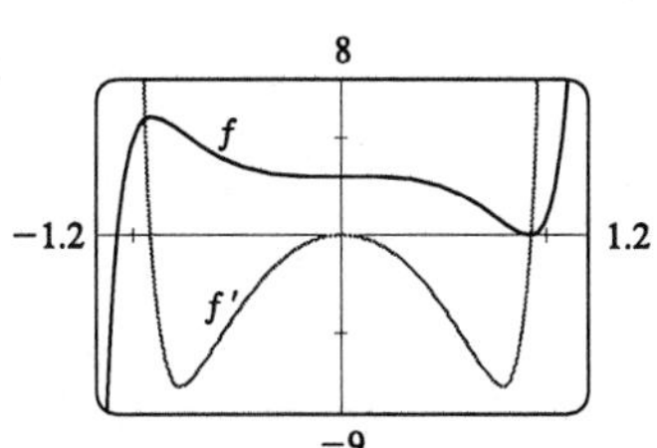

Notice that $f'(x) = 0$ when f has a horizontal tangent.

44. $y = \dfrac{x-1}{x+1} \quad \Rightarrow \quad y' = \dfrac{(x+1)(1)-(x-1)(1)}{(x+1)^2} = \dfrac{2}{(x+1)^2}$. If the tangent intersects the curve when $x = a$, then its slope is $2/(a+1)^2$. But if the tangent is parallel to $x - 2y = 1$, its slope is $\dfrac{1}{2}$. Thus $\dfrac{2}{(a+1)^2} = \dfrac{1}{2} \Rightarrow$ $(a+1)^2 = 4 \Rightarrow a+1 = \pm 2 \Rightarrow a = 1$ or -3.

When $a = 1$, $y = 0$ and the equation of the tangent is $y = \frac{1}{2}(x-1)$ or $x - 2y - 1 = 0$.

When $a = -3$, $y = 2$ and the equation of the tangent is $y - 2 = \frac{1}{2}(x+3)$ or $x - 2y + 7 = 0$.

45. $y = x\sqrt{x} = x^{3/2} \quad \Rightarrow \quad y' = \frac{3}{2}\sqrt{x}$ so the tangent line is parallel to $3x - y + 6 = 0$ when $\frac{3}{2}\sqrt{x} = 3 \quad \Leftrightarrow$ $\sqrt{x} = 2 \quad \Leftrightarrow \quad x = 4$. So the point is $(4, 8)$.

46. $f(x) = 2x^3 - 3x^2 - 6x + 87$ has a horizontal tangent when $f'(x) = 6x^2 - 6x - 6 = 0 \quad \Leftrightarrow \quad x^2 - x - 1 = 0$ $\Leftrightarrow \quad x = \frac{1 \pm \sqrt{5}}{2}$.

47. $y = x^3 - x^2 - x + 1$ has a horizontal tangent when $y' = 3x^2 - 2x - 1 = 0 \quad (3x+1)(x-1) = 0 \Leftrightarrow x = 1$ or $-\frac{1}{3}$. Therefore the points are $(1, 0)$ and $\left(-\frac{1}{3}, \frac{32}{27}\right)$.

48.

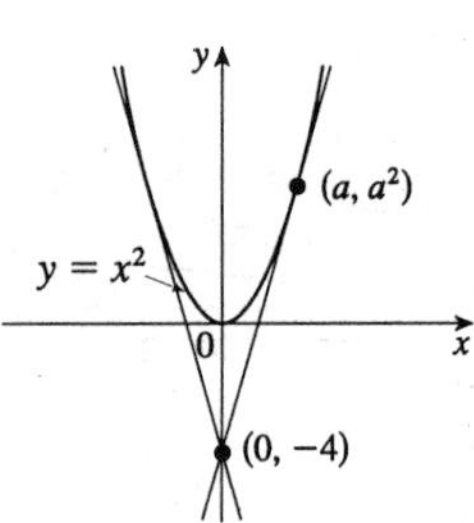

Let (a, a^2) be a point of intersection. The tangent line has slope $2a$ and equation $y - (-4) = 2a(x - 0) \quad \Rightarrow \quad y = 2ax - 4$. Since (a, a^2) also lies on the line, $a^2 = 2a(a) - 4$, or $a^2 = 4$. So $a = \pm 2$ and the points are $(2, 4)$ and $(-2, 4)$.

49. If $y = f(x) = \dfrac{x}{x+1}$ then $f'(x) = \dfrac{(x+1)(1) - x(1)}{(x+1)^2} = \dfrac{1}{(x+1)^2}$. When $x = a$, the equation of the tangent line is $y - \dfrac{a}{a+1} = \dfrac{1}{(a+1)^2}(x - a)$. This line passes through $(1, 2)$ when $2 - \dfrac{a}{a+1} = \dfrac{1}{(a+1)^2}(1 - a) \quad \Leftrightarrow$ $2(a+1)^2 = a(a+1) + (1-a) = a^2 + 1 \quad \Leftrightarrow \quad a^2 + 4a + 1 = 0$. The quadratic formula gives the roots of this equation as $-2 \pm \sqrt{3}$, so there are two such tangent lines, which touch the curve at $\left(-2+\sqrt{3}, \frac{1-\sqrt{3}}{2}\right)$ and $\left(-2-\sqrt{3}, \frac{1+\sqrt{3}}{2}\right)$.

50. If $y = x^2 + x$, then $y' = 2x + 1$. If the point at which a tangent meets the parabola is $(a, a^2 + a)$, then the slope of the tangent is $2a + 1$. But since it passes through $(2, -3)$, the slope must also be $\dfrac{a^2 + a + 3}{a - 2}$. Therefore $2a + 1 = \dfrac{a^2 + a + 3}{a - 2}$. Solving this equation for a we get $a^2 + a + 3 = 2a^2 - 3a - 2 \quad \Leftrightarrow$ $a^2 - 4a - 5 = (a-5)(a+1) = 0 \quad \Leftrightarrow \quad a = 5$ or -1. If $a = -1$, the point is $(-1, 0)$ and the slope is -1, so the equation is $y - 0 = (-1)(x+1)$ or $x + y + 1 = 0$. If $a = 5$, the point is $(5, 30)$ and the slope is 11, so the equation is $y - 30 = 11(x - 5)$ or $11x - y = 25$.

51. $y = 6x^3 + 5x - 3 \quad \Rightarrow \quad m = y' = 18x^2 + 5$, but $x^2 \geq 0$ for all x so $m \geq 5$ for all x.

52. The sides of the groove must be tangent to the parabola $y = 16x^2$. $y' = 32x = 1.75$ when $x = \frac{1.75}{32} = \frac{7}{128}$, which implies that $y = 16\left(\frac{7}{128}\right)^2 = \frac{49}{1024}$. Therefore the points of contact are $\left(\pm\frac{7}{128}, \frac{49}{1024}\right)$.

53. $y = f(x) = 1 - x^2 \quad \Rightarrow \quad f'(x) = -2x$, so the tangent line at $(2, -3)$ has slope $f'(2) = -4$. The normal line has slope $-1/(-4) = \frac{1}{4}$ and equation $y + 3 = \frac{1}{4}(x - 2)$ or $x - 4y = 14$.

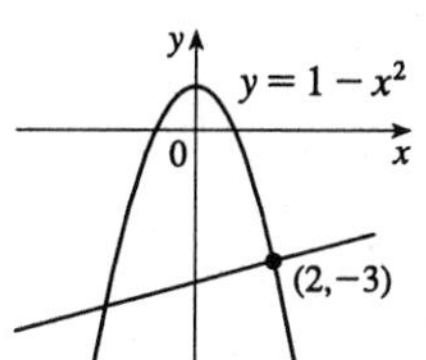

54. $y = f(x) = \dfrac{1}{x - 1} \quad \Rightarrow$

$$f'(x) = \frac{(x - 1)(0) - 1(1)}{(x - 1)^2} = \frac{-1}{(x - 1)^2},$$

so the tangent line at $(2, 1)$ has slope $f'(2) = -1$. The normal line has slope $-1/(-1) = 1$ and equation $y - 1 = 1(x - 2)$ or $y = x - 1$.

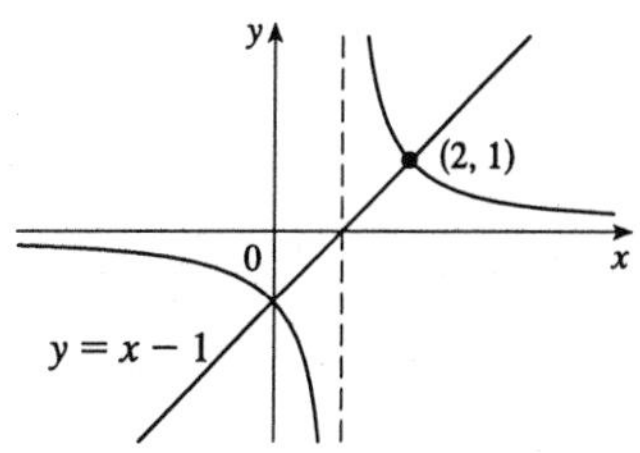

55. $y = f(x) = \sqrt[3]{x} = x^{1/3} \quad \Rightarrow \quad f'(x) = \frac{1}{3}x^{-2/3}$, so the tangent line at $(-8, -2)$ has slope $f'(-8) = \frac{1}{12}$. The normal line has slope $-1/\left(\frac{1}{12}\right) = -12$ and equation $y + 2 = -12(x + 8) \quad \Leftrightarrow \quad 12x + y + 98 = 0$.

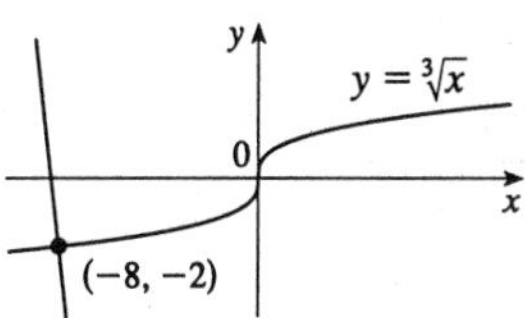

56. The slope of the tangent line to the curve $y = f(x)$ at $(a, f(a))$ is $f'(a)$, so the slope of the normal line is $-1/f'(a)$ [if $f'(a) \neq 0$] and its equation is $y - f(a) = -[1/f'(a)](x - a)$ or $x + f'(a)y = a + f(a)f'(a)$. [If $f'(a) = 0$, the normal is $x = a$.]

57. If the normal line has slope 16, then the tangent has slope $-\frac{1}{16}$, so $y' = 4x^3 = -\frac{1}{16} \quad \Rightarrow \quad x^3 = -\frac{1}{64} \quad \Rightarrow$ $x = -\frac{1}{4}$. The point is $\left(-\frac{1}{4}, \frac{1}{256}\right)$.

58.

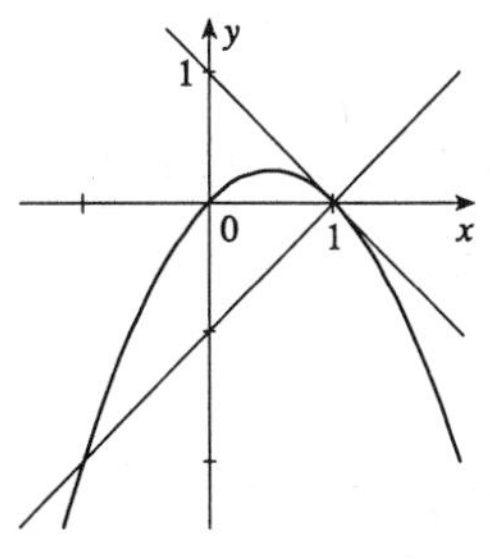

$y = f(x) = x - x^2 \quad \Rightarrow \quad f'(x) = 1 - 2x$. So $y'(1) = 1 - 2(1) = -1$, and the slope of the normal line is the negative reciprocal of that of the tangent line, that is, $-1/(-1) = 1$. So the equation of the normal line at $(1, 0)$ is $y - 0 = 1(x - 1) \quad \Leftrightarrow \quad y = x - 1$. Substituting this into the equation of the parabola, we obtain $x - 1 = x - x^2 \quad \Leftrightarrow \quad x = \pm 1$. The solution $x = -1$ is the one we require. Substituting $x = -1$ into the equation of the parabola to find the y-coordinate, we have $y = x - x^2 = -1 - (-1)^2 = -2$. So the point of intersection is $(-1, -2)$, as shown in the sketch.

59. **(a)** $(fg)'(5) = f'(5)g(5) + f(5)g'(5) = 6(-3) + 1(2) = -16$

(b) $\left(\dfrac{f}{g}\right)'(5) = \dfrac{f'(5)g(5) - f(5)g'(5)}{(g(5))^2} = \dfrac{6(-3) - 1(2)}{(-3)^2} = -\dfrac{20}{9}$

(c) $\left(\dfrac{g}{f}\right)'(5) = \dfrac{g'(5)f(5) - g(5)f'(5)}{(f(5))^2} = \dfrac{2(1) - (-3)(6)}{1^2} = 20$

60. **(a)** $(f+g)'(3) = f'(3) + g'(3) = -6 + 5 = -1$

(b) $(fg)'(3) = f'(3)g(3) + f(3)g'(3) = -6 \cdot 2 + 4 \cdot 5 = -12 + 20 = 8$

(c) $\left(\dfrac{f}{g}\right)'(3) = \dfrac{f'(3)\,g(3) - f(3)\,g'(3)}{[g(3)]^2} = \dfrac{-6 \cdot 2 - 4 \cdot 5}{2^2} = \dfrac{-12 - 20}{4} = -8$

(d) $\left(\dfrac{f}{f-g}\right)'(3) = \dfrac{f'(3)[f(3) - g(3)] - f(3)[f'(3) - g'(3)]}{[f(3) - g(3)]^2} = \dfrac{-6(4-2) - 4(-6-5)}{(4-2)^2} = 8$

61. **(a)** $u(x) = f(x)g(x)$, so $u'(1) = f(1)g'(1) + g(1)f'(1) = 2 \cdot (-1) + 1 \cdot 2 = 0$

(b) $v(x) = f(x)/g(x)$, so $v'(5) = \dfrac{g(5)f'(5) - f(5)g'(5)}{[g(5)]^2} = \dfrac{2(-\frac{1}{3}) - 3 \cdot \frac{2}{3}}{2^2} = -\dfrac{2}{3}$

62. **(a)** $y = x^2 f(x) \quad \Rightarrow \quad y' = 2x\,f(x) + x^2 f'(x)$

(b) $y = \dfrac{f(x)}{x^2} \quad \Rightarrow \quad y' = \dfrac{f'(x)\,x^2 - 2x\,f(x)}{x^4}$

(c) $y = \dfrac{x^2}{f(x)} \quad \Rightarrow \quad y' = \dfrac{2x\,f(x) - x^2 f'(x)}{[f(x)]^2}$

(d) $y = \dfrac{1 + x\,f(x)}{\sqrt{x}} \quad \Rightarrow \quad y' = \dfrac{[f(x) + x\,f'(x)]\sqrt{x} - \dfrac{1}{2\sqrt{x}}[1 + xf(x)]}{x}$

$= \dfrac{2x[f(x) + xf'(x)] - [1 + xf(x)]}{2x^{3/2}} = \dfrac{xf(x) + 2x^2\,f'(x) - 1}{2x^{3/2}}$

63. **(a)** $(fgh)' = [(fg)h]' = (fg)'h + (fg)h' = (f'g + fg')h + (fg)h' = f'gh + fg'h + fgh'$

(b) Putting $f = g = h$ in part (a), we have

$\dfrac{d}{dx}[f(x)]^3 = (fff)' = f'ff + ff'f + fff' = 3fff' = 3[f(x)]^2 f'(x)$.

64. $y = (x+5)(x^2+7)(x-3)$. Using Exercise 63(a), we have

$y' = (x^2+7)(x-3)(x+5)' + (x+5)(x-3)(x^2+7)' + (x+5)(x^2+7)(x-3)'$

$= (x^2+7)(x-3) + (x+5)(x-3)(2x) + (x+5)(x^2+7) = 4x^3 + 6x^2 - 16x + 14$.

65. $y = \sqrt{x}(x^4 + x + 1)(2x - 3)$. Using Exercise 63(a), we have

$y' = \dfrac{1}{2\sqrt{x}}(x^4 + x + 1)(2x - 3) + \sqrt{x}(4x^3 + 1)(2x - 3) + \sqrt{x}(x^4 + x + 1)(2)$

$= (x^4 + x + 1)\dfrac{2x - 3}{2\sqrt{x}} + \sqrt{x}\left[(4x^3 + 1)(2x - 3) + 2(x^4 + x + 1)\right]$.

66. $y = (x^4 + 3x^3 + 17x + 82)^3$. Using Exercise 63(b), we have
$y' = 3(x^4 + 3x^3 + 17x + 82)^2 D(x^4 + 3x^3 + 17x + 82) = 3(x^4 + 3x^3 + 17x + 82)^2(4x^3 + 9x^2 + 17)$.

67. $f(x) = 2 - x$ if $x \le 1$ and $f(x) = x^2 - 2x + 2$ if $x > 1$. Now we compute the right- and left-hand derivatives defined in Exercise 2.1.57:

$$f'_-(1) = \lim_{h \to 0^-} \frac{f(1+h) - f(1)}{h} = \lim_{h \to 0^-} \frac{2 - (1+h) - 1}{h} = \lim_{h \to 0^-} \frac{-h}{h} = \lim_{h \to 0^-} -1 = -1 \text{ and}$$

$$f'_+(1) = \lim_{h \to 0^+} \frac{f(1+h) - f(1)}{h} = \lim_{h \to 0^+} \frac{(1+h)^2 - 2(1+h) + 2 - 1}{h} = \lim_{h \to 0^+} \frac{h^2}{h} = \lim_{h \to 0^+} h = 0.$$

Thus $f'(1)$ does not exist since $f'_-(1) \ne f'_+(1)$, so f is not differentiable at 1. But $f'(x) = -1$ for $x < 1$ and $f'(x) = 2x - 2$ if $x > 1$.

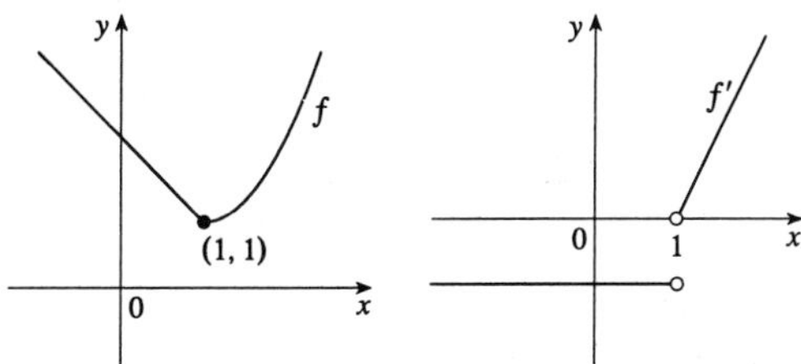

68. $g(x) = \begin{cases} 1 - 2x & \text{if } x < -1 \\ x^2 & \text{if } -1 \le x \le 1 \\ x & \text{if } x > 1 \end{cases}$

$$\lim_{h \to 0^-} \frac{g(-1+h) - g(-1)}{h} = \lim_{h \to 0^-} \frac{[-1 - 2(-1+h)] - 1}{h} = \lim_{h \to 0^-} \frac{-2h}{h} = \lim_{h \to 0^-} (-2) = -2 \text{ and}$$

$$\lim_{h \to 0^+} \frac{g(-1+h) - g(-1)}{h} = \lim_{h \to 0^+} \frac{(-1+h)^2 - 1}{h} = \lim_{h \to 0^+} \frac{-2h + h^2}{h} = \lim_{h \to 0^+} (-2 + h) = -2,$$

so g is differentiable at -1 and $g'(-1) = -2$.

$$\lim_{h \to 0^-} \frac{g(1+h) - g(1)}{h} = \lim_{h \to 0^-} \frac{(1+h)^2 - 1}{h} = \lim_{h \to 0^-} \frac{2h + h^2}{h} = \lim_{h \to 0^-} (2 + h) = 2 \text{ and}$$

$$\lim_{h \to 0^+} \frac{g(1+h) - g(1)}{h} = \lim_{h \to 0^+} \frac{(1+h) - 1}{h} = \lim_{h \to 0^+} \frac{h}{h} = \lim_{h \to 0^+} 1 = 1, \text{ so } g'(1) \text{ does not exist.}$$

Thus g is differentiable except when $x = 1$, and $g'(x) = \begin{cases} -2 & \text{if } x \le -1 \\ 2x & \text{if } -1 < x < 1 \\ 1 & \text{if } x > 1. \end{cases}$

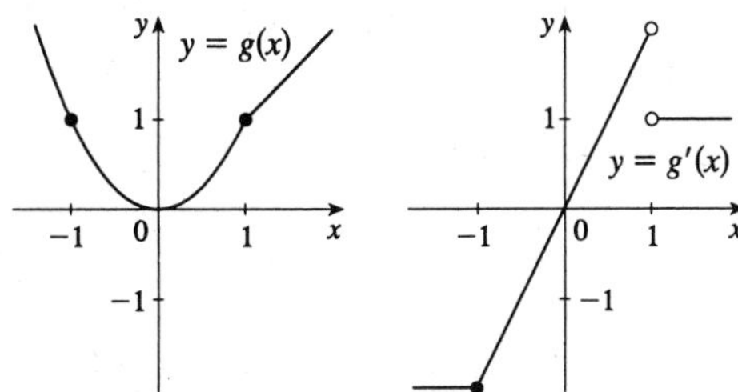

69. **(a)** Note that $x^2 - 9 < 0$ for $x^2 < 9 \quad \Leftrightarrow \quad |x| < 3 \quad \Leftrightarrow \quad -3 < x < 3$. So

$$f(x) = \begin{cases} x^2 - 9 & \text{if } x \le -3 \\ -x^2 + 9 & \text{if } -3 < x < 3 \\ x^2 - 9 & \text{if } x \ge 3 \end{cases} \quad \Rightarrow \quad f'(x) = \begin{cases} 2x & \text{if } x < -3 \\ -2x & \text{if } -3 < x < 3 \\ 2x & \text{if } x > 3. \end{cases}$$

To show that $f'(3)$ does not exist we investigate $\lim\limits_{h \to 0} \dfrac{f(3+h) - f(3)}{h}$ by computing the left- and right-hand derivatives defined in Exercise 2.1.57.

$$f'_-(3) = \lim_{h \to 0^-} \frac{f(3+h) - f(3)}{h} = \lim_{h \to 0^-} \frac{\left(-(3+h)^2 + 9\right) - 0}{h} = \lim_{h \to 0^-} (-6 + h) = -6 \text{ and}$$

$$f'_+(3) = \lim_{h \to 0^+} \frac{f(3+h) - f(3)}{h} = \lim_{h \to 0^+} \frac{\left[(3+h)^2 + 9\right] - 0}{h} = \lim_{h \to 0^+} \frac{6h + h^2}{h} = \lim_{h \to 0^+} (6 + h) = 6.$$

Since the left and right limits are different, $\lim\limits_{h \to 0} \dfrac{f(3+h) - f(3)}{h}$ does not exist, that is, $f'(3)$ does not exist. Similarly, $f'(-3)$ does not exist. Therefore f is not differentiable at 3 or at -3.

(b)

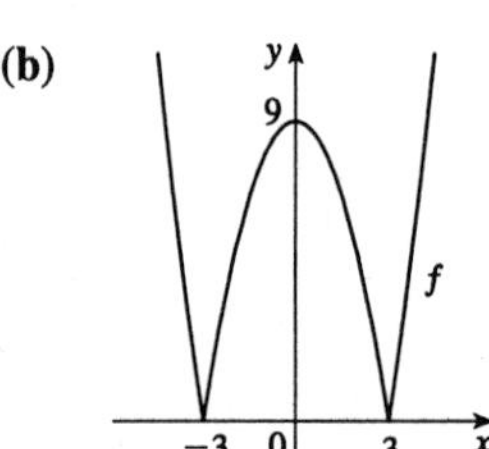

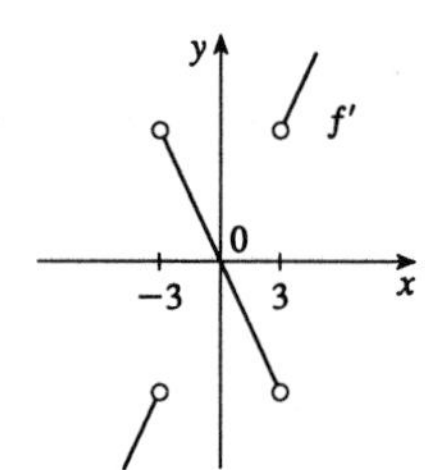

70. If $x \ge 1$, then $h(x) = |x - 1| + |x + 2| = x - 1 + x + 2 = 2x + 1$.

If $-2 < x < 1$, then $h(x) = -(x - 1) + x + 2 = 3$.

If $x \le -2$, then $h(x) = -(x - 1) - (x + 2) = -2x - 1$. Therefore

$$h(x) = \begin{cases} -2x - 1 & \text{if } x \le -2 \\ 3 & \text{if } -2 < x < 1 \\ 2x + 1 & \text{if } x \ge 1 \end{cases} \quad \Rightarrow \quad h'(x) = \begin{cases} -2 & \text{if } x < -2 \\ 0 & \text{if } -2 < x < 1 \\ 2 & \text{if } x > 1 \end{cases}$$

To see that $h'(1) = \lim\limits_{x \to 1} \dfrac{h(x) - h(1)}{x - 1}$ does not exist, observe that $\lim\limits_{x \to 1^-} \dfrac{h(x) - h(1)}{x - 1} = \lim\limits_{x \to 1^-} \dfrac{3 - 3}{3 - 1} = 0$ but $\lim\limits_{x \to 1^+} \dfrac{h(x) - h(1)}{x - 1} = \lim\limits_{x \to 1^+} \dfrac{2x - 2}{x - 1} = 2$. Similarly, $h'(-2)$ does not exist.

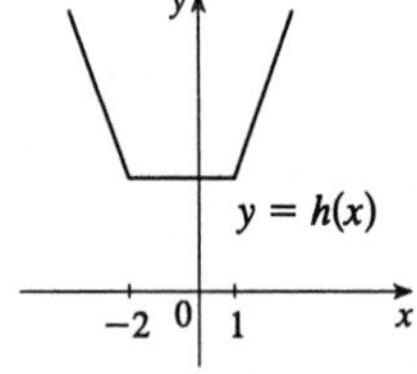

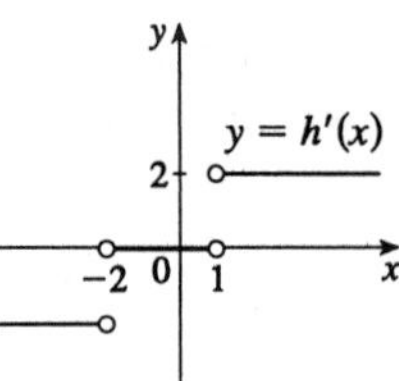

71. $y = f(x) = ax^2 \quad \Rightarrow \quad f'(x) = 2ax$. So the slope of the tangent to the parabola at $x = 2$ is $m = 2a(2) = 4a$. The slope of the given line is seen to be -2, so we must have $4a = -2 \quad \Leftrightarrow \quad a = -\frac{1}{2}$. So the point in question has y-coordinate $-\frac{1}{2} \cdot 2^2 = -2$. Now we simply require that the given line, whose equation is $2x + y = b$, pass through the point $(2, -2)$: $2(2) + (-2) = b \quad \Leftrightarrow \quad b = 2$. So we must have $a = -\frac{1}{2}$ and $b = 2$.

72. f is clearly differentiable for $x < 2$ and for $x > 2$. For $x < 2$, $f'(x) = 2x$, so $f'_-(2) = 4$. For $x > 2$, $f'(x) = m$, so $f'_+(2) = m$. For f to be differentiable at $x = 2$, we need $4 = f'_-(2) = f'_+(2) = m$. So $f(x) = 4x + b$. We must also have continuity at $x = 2$, so $4 = f(2) = \lim_{x \to 2^+} f(x) = \lim_{x \to 2^+} (4x + b) = 8 + b$.

Hence $b = -4$.

73. $F = f/g \quad \Rightarrow \quad f = Fg \quad \Rightarrow \quad f' = F'g + Fg' \quad \Rightarrow \quad F' = \dfrac{f' - Fg'}{g} = \dfrac{f' - (f/g)g'}{g} = \dfrac{f'g - fg'}{g^2}$

74. **(a)** Here $y = \dfrac{c}{x}$. Let $P = \left(a, \dfrac{c}{a}\right)$. The slope of the tangent line at $x = a$ is $f'(a) = -\dfrac{c}{a^2}$. Its equation is $y - \dfrac{c}{a} = -\dfrac{c}{a^2}(x - a)$ or $y = -\dfrac{c}{a^2}x + \dfrac{2c}{a}$. So the y-intercept is $\dfrac{2c}{a}$. Setting $y = 0$ gives $x = 2a$, so the x-intercept is $2a$. The midpoint of the line segment joining $\left(0, \dfrac{2c}{a}\right)$ to $(2a, 0)$ is $\left(a, \dfrac{c}{a}\right) = P$.

(b) We know the x- and y-intercepts of the tangent line from part (a), so the area of the triangle bounded by the axes and the tangent is $\frac{1}{2}xy = \frac{1}{2} \cdot (2a)(2c/a) = 2c$, a constant.

75. *Solution 1:* Let $f(x) = x^{1000}$. Then, by the definition of the derivative,
$f'(1) = \lim_{x \to 1} \dfrac{f(x) - f(1)}{x - 1} = \lim_{x \to 1} \dfrac{x^{1000} - 1}{x - 1}$. But this is just the limit we want to find, and we know (from the Power Rule) that $f'(x) = 1000x^{999}$, so $f'(1) = 1000(1)^{999} = 1000$. So $\lim_{x \to 1} \dfrac{x^{1000} - 1}{x - 1} = 1000$.

Solution 2: Note that $(x^{1000} - 1) = (x - 1)\left(x^{999} + x^{998} + x^{997} + \cdots + x^2 + x + 1\right)$. So

$$\begin{aligned}\lim_{x \to 1} \frac{x^{1000} - 1}{x - 1} &= \lim_{x \to 1} \frac{(x - 1)\left(x^{999} + x^{998} + x^{997} + \cdots + x^2 + x + 1\right)}{x - 1} \\ &= \lim_{x \to 1} \left(x^{999} + x^{998} + x^{997} + \cdots + x^2 + x + 1\right) = 1 + 1 + 1 + \cdots + 1 + 1 + 1 = 1000 \text{ as above.}\end{aligned}$$

76.

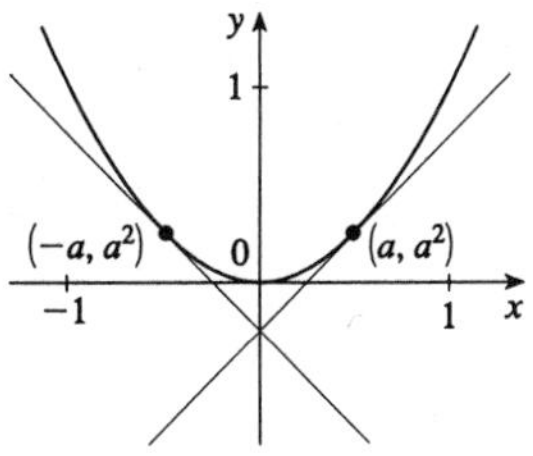

In order for the two tangents to intersect on the y-axis, the points of tangency must be at equal distances from the y-axis, since the parabola $y = x^2$ is symmetric about the y-axis. Say the points of tangency are (a, a^2) and $(-a, a^2)$, for some $a > 0$. Then since the derivative of $y = x^2$ is $dy/dx = 2x$, the left-hand tangent has slope $-2a$ and equation $y - a^2 = -2a(x + a)$, or $y = -2ax - a^2$, and similarly the right-hand tangent line has equation $y = 2ax - a^2$. So the two lines intersect at $(0, -a^2)$.

Now if the lines are perpendicular, then the product of their slopes is -1, so $(-2a)(2a) = -1 \quad \Leftrightarrow \quad a^2 = \frac{1}{4} \quad \Leftrightarrow \quad a = \frac{1}{2}$. So the lines intersect at $\left(0, -\frac{1}{4}\right)$.

EXERCISES 2.3

1. **(a)** $v(t) = f'(t) = 2t - 6$ **(b)** $v(2) = 2(2) - 6 = -2$ ft/s

(c) It is at rest when $v(t) = 2t - 6 = 0 \quad \Leftrightarrow \quad t = 3$.

(d) It moves in the positive direction when $2t - 6 > 0 \quad \Leftrightarrow \quad t > 3$.

(e) Distance in positive direction $= |f(4) - f(3)| = |1 - 0| = 1$ ft

Distance in negative direction $= |f(3) - f(0)| = |0 - 9| = 9$ ft

Total distance traveled $= 1 + 9 = 10$ ft

(f)

2. **(a)** $v(t) = f'(t) = 12t^2 - 18t + 6$ **(b)** $v(2) = 12(2)^2 - 18(2) + 6 = 18$ ft/s

(c) It is at rest when $v(t) = 12t^2 - 18t + 6 = 6(2t - 1)(t - 1) = 0 \quad \Rightarrow \quad t = \frac{1}{2}$ or 1.

(d) It moves in the positive direction when $v(t) = 6(2t - 1)(t - 1) > 0 \quad \Leftrightarrow \quad 0 \le t < \frac{1}{2}$ or $t > 1$.

(e) Distance in positive direction $= \left|f\left(\frac{1}{2}\right) - f(0)\right| + |f(4) - f(1)| = |3.25 - 2| + |138 - 3| = 136.25$ ft

Distance in negative direction $= \left|f(1) - f\left(\frac{1}{2}\right)\right| = |3 - 3.25| = 0.25$ ft

Total distance traveled $= 136.25 + 0.25 = 136.5$ ft

(f)

3. **(a)** $v(t) = f'(t) = 6t^2 - 18t + 12$ **(b)** $v(2) = 6(2)^2 - 18(2) + 12 = 0$ ft/s

(c) It is at rest when $v(t) = 6t^2 - 18t + 12 = 6(t - 1)(t - 2) = 0 \quad \Leftrightarrow \quad t = 1$ or 2.

(d) It moves in the positive direction when $6(t - 1)(t - 2) > 0 \quad \Leftrightarrow \quad 0 \le t < 1$ or $t > 2$.

(e) Distance in positive direction $= |f(4) - f(2)| + |f(1) - f(0)| = |33 - 5| + |6 - 1| = 33$ ft

Distance in negative direction $= |f(2) - f(1)| = |5 - 6| = 1$ ft

Total distance traveled $= 33 + 1 = 34$ ft

(f)

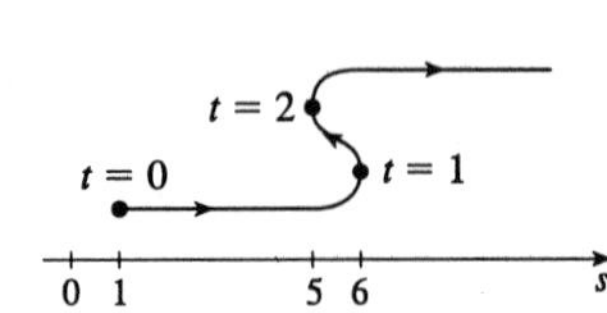

4. **(a)** $v(t) = f'(t) = 4t^3 - 4$ **(b)** $v(2) = 4(2)^3 - 4 = 28$ ft/s

(c) It is at rest when $v(t) = 4(t^3 - 1) = 4(t-1)(t^2+t+1) = 0 \quad \Leftrightarrow \quad t = 1$.

(d) It moves in the positive direction when $4(t^3 - 1) > 0 \quad \Leftrightarrow \quad t > 1$.

(e) Distance in positive direction $= |f(4) - f(1)| = |241 - (-2)| = 243$ ft

Distance in negative direction $= |f(1) - f(0)| = |-2 - 1| = 3$ ft

Total distance traveled $= 243 + 3 = 246$ ft

(f)

5. **(a)** $v(t) = s'(t) = \dfrac{(t^2+1)(1) - t(2t)}{(t^2+1)^2} = \dfrac{1-t^2}{(t^2+1)^2}$ **(b)** $v(2) = \dfrac{1-(2)^2}{(2^2+1)^2} = -\dfrac{3}{25}$ ft/s

(c) It is at rest when $v = 0 \quad \Leftrightarrow \quad 1 - t^2 = 0 \quad \Leftrightarrow \quad t = 1$.

(d) It moves in the positive direction when $v > 0 \quad \Leftrightarrow \quad 1 - t^2 > 0 \quad \Leftrightarrow \quad t^2 < 1 \quad \Leftrightarrow \quad 0 \le t < 1$.

(e) Distance in positive direction $= |s(1) - s(0)| = \left|\frac{1}{2} - 0\right| = \frac{1}{2}$ ft

Distance in negative direction $= |s(4) - s(1)| = \left|\frac{4}{17} - \frac{1}{2}\right| = \frac{9}{34}$ ft

Total distance traveled $= \frac{1}{2} + \frac{9}{34} = \frac{13}{17}$ ft

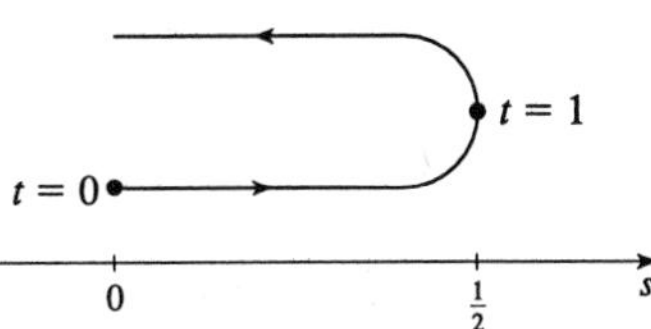

6. $s = f(t) = \sqrt{t}(5 - 5t + 2t^2) = 5t^{1/2} - 5t^{3/2} + 2t^{5/2}$

(a) $v(t) = f'(t) = \frac{5}{2}t^{-1/2} - \frac{15}{2}t^{1/2} + 5t^{3/2} = \frac{5}{2}t^{-1/2}(1 - 3t + 2t^2)$.

(b) $v(2) = \frac{5}{2}(2)^{-1/2}\left[1 - 3(2) + 2(2)^2\right] = 15/(2\sqrt{2}) = 15\sqrt{2}/4$ ft/s.

(c) It is at rest when $v = 0 \quad \Leftrightarrow \quad 1 - 3t + 2t^2 = (2t-1)(t-1) = 0 \quad \Leftrightarrow \quad t = \frac{1}{2}$ or 1.

(d) It moves in the positive direction when $v > 0 \;\Leftrightarrow\; (2t-1)(t-1) > 0 \;\Leftrightarrow\; 0 \le t < \frac{1}{2}$ or $t > 1$.

(e) Distance in positive direction $= \left|f\left(\frac{1}{2}\right) - f(0)\right| + |f(4) - f(1)| = \left|\frac{3}{\sqrt{2}}\right| + |34 - 2| = 32 + \frac{3}{2}\sqrt{2}$

Distance in negative direction $= \left|f(1) - f\left(\frac{1}{2}\right)\right| = \left|2 - \frac{3}{2}\sqrt{2}\right| = \frac{3}{2}\sqrt{2} - 2$

Total distance traveled $= 32 + 3\sqrt{2} - 2 \approx 34.24$ ft

(f)

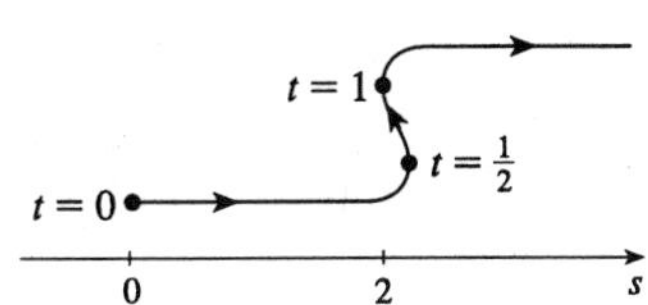

7. $s(t) = t^3 - 4.5t^2 - 7t \quad \Rightarrow \quad v(t) = s'(t) = 3t^2 - 9t - 7 = 5 \quad \Leftrightarrow \quad 3t^2 - 9t - 12 = 0 \quad \Leftrightarrow$
$3(t-4)(t+1) = 0 \quad \Leftrightarrow \quad t = 4$ or -1. Since $t \geq 0$, the particle reaches a velocity of 5 m/s at $t = 4$ s.

8. **(a)** At maximum height the velocity of the ball is 0 ft/s. $v(t) = s'(t) = 80 - 32t = 0 \;\Leftrightarrow\; 32t = 80 \;\Leftrightarrow$
$t = \frac{5}{2}$. So the maximum height is $s\left(\frac{5}{2}\right) = 80\left(\frac{5}{2}\right) - 16\left(\frac{5}{2}\right)^2 = 200 - 100 = 100$ ft.

(b) $s(t) = 80t - 16t^2 = 96 \quad \Leftrightarrow \quad 16t^2 - 80t + 96 = 0 \quad \Leftrightarrow \quad 16(t-3)(t-2) = 0$. So the ball has a height of 96 ft on the way up at $t = 2$ and on the way down at $t = 3$. At these times the velocities are $v(2) = 80 - 32(2) = 16$ ft/s and $v(3) = 80 - 32(3) = -16$ ft/s respectively.

9. **(a)** $V(x) = x^3$, so the average rate of change is:

(i) $\dfrac{V(6) - V(5)}{6 - 5} = 6^3 - 5^3 = 216 - 125 = 91$ **(ii)** $\dfrac{V(5.1) - V(5)}{5.1 - 5} = \dfrac{(5.1)^3 - 5^3}{0.1} = 76.51$

(iii) $\dfrac{V(5.01) - V(5)}{5.01 - 5} = \dfrac{(5.01)^3 - 5^3}{0.01} = 75.1501$

(b) $V'(x) = 3x^2$, $V'(5) = 75$

(c) The surface area is $S(x) = 6x^2$, so $V'(x) = 3x^2 = \frac{1}{2}(6x^2) = \frac{1}{2}S(x)$.

10. **(a)** $A(r) = \pi r^2$, so the average rate of change is:

(i) $\dfrac{A(3) - A(2)}{3 - 2} = \dfrac{9\pi - 4\pi}{1} = 5\pi$ **(ii)** $\dfrac{A(2.5) - A(2)}{2.5 - 2} = \dfrac{6.25\pi - 4\pi}{0.5} = 4.5\pi$

(iii) $\dfrac{A(2.1) - A(2)}{2.1 - 2} = \dfrac{4.41\pi - 4\pi}{0.1} = 4.1\pi$

(b) $A'(r) = 2\pi r$, so $A'(2) = 4\pi$.

(c) The circumference is $C(r) = 2\pi r = A'(r)$.

11. After t seconds the radius is $r = 60t$, so the area is $A(t) = \pi(60t)^2 = 3600\pi t^2 \quad \Rightarrow \quad A'(t) = 7200\pi t \quad \Rightarrow$

(a) $A'(1) = 7200\pi$ cm²/s **(b)** $A'(3) = 21{,}600\pi$ cm²/s **(c)** $A'(5) = 36{,}000\pi$ cm²/s

12. **(a)** $V(r) = \frac{4}{3}\pi r^3 \quad \Rightarrow$ the average rate of change is:

(i) $\dfrac{V(8) - V(5)}{8 - 5} = \dfrac{\frac{4}{3}\pi(512) - \frac{4}{3}\pi(125)}{3} = 172\pi\ \mu\text{m}^3/\mu\text{m}$

(ii) $\dfrac{V(6) - V(5)}{6 - 5} = \dfrac{\frac{4}{3}\pi(216) - \frac{4}{3}\pi(125)}{1} = 121.\overline{3}\pi\ \mu\text{m}^3/\mu\text{m}$

(iii) $\dfrac{V(5.1) - V(5)}{5.1 - 5} = \dfrac{\frac{4}{3}\pi(5.1)^3 - \frac{4}{3}\pi(5)^3}{0.1} = 102.01\overline{3}\pi\ \mu\text{m}^3/\mu\text{m}$

(b) $V'(r) = 4\pi r^2$, so $V'(5) = 100\pi\ \mu\text{m}^3/\mu\text{m}$.

13. $S(r) = 4\pi r^2 \quad \Rightarrow \quad S'(r) = 8\pi r$

(a) $S'(1) = 8\pi$ ft²/ft **(b)** $S'(2) = 16\pi$ ft²/ft **(c)** $S'(3) = 24\pi$ ft²/ft

14. $V(r) = \frac{4}{3}\pi r^3 \quad \Rightarrow \quad V'(r) = 4\pi r^2 = S(r)$

15. $f(x) = 3x^2$, so the linear density at x is $\rho(x) = f'(x) = 6x$.

(a) $\rho(1) = 6\,\text{kg/m}$

(b) $\rho(2) = 12\,\text{kg/m}$

(c) $\rho(3) = 18\,\text{kg/m}$

16. $V(t) = 5000(1 - t/40)^2 \quad\Rightarrow\quad V'(t) = 5000(2)\left(1 - \frac{1}{40}t\right)\left(-\frac{1}{40}\right) = -250\left(1 - \frac{1}{40}t\right)$

(a) $V'(5) = -250\left(1 - \frac{5}{40}\right) = -218.75\,\text{gal/min}$

(b) $V'(10) = -250\left(1 - \frac{10}{40}\right) = -187.5\,\text{gal/min}$

(c) $V'(20) = -250\left(1 - \frac{20}{40}\right) = -125\,\text{gal/min}$

17. $Q(t) = t^3 - 2t^2 + 6t + 2$, so the current is $Q'(t) = 3t^2 - 4t + 6$.

(a) $Q'(0.5) = 3(0.5)^2 - 4(0.5) + 6 = 4.75\,\text{A}$

(b) $Q'(1) = 3(1)^2 - 4(1) + 6 = 5\,\text{A}$

18. $\dfrac{dF}{dr} = \dfrac{-2GmM}{r^3}$

19. **(a)** $PV = C \quad\Rightarrow\quad V = \dfrac{C}{P} \quad\Rightarrow\quad \dfrac{dV}{dP} = -\dfrac{C}{P^2}$

(b) $\beta = -\dfrac{1}{V}\dfrac{dV}{dP} = -\dfrac{1}{V}\left(-\dfrac{C}{P^2}\right) = \dfrac{C}{(PV)P} = \dfrac{C}{CP} = \dfrac{1}{P}$

20. **(a)** **(i)** $\dfrac{C(6) - C(2)}{6 - 2} = \dfrac{0.0295 - 0.0570}{4} = -0.006875\,\text{moles/L/min}$

(ii) $\dfrac{C(4) - C(2)}{4 - 2} = \dfrac{0.0408 - 0.0570}{2} = -0.0081\,\text{moles/L/min}$

(iii) $\dfrac{C(2) - C(0)}{2 - 0} = \dfrac{0.0570 - 0.0800}{2} = -0.0115\,\text{moles/L/min}$

(b) Slope $\approx -\dfrac{0.077}{7.8} \approx -0.01\,\text{moles/L/min}$

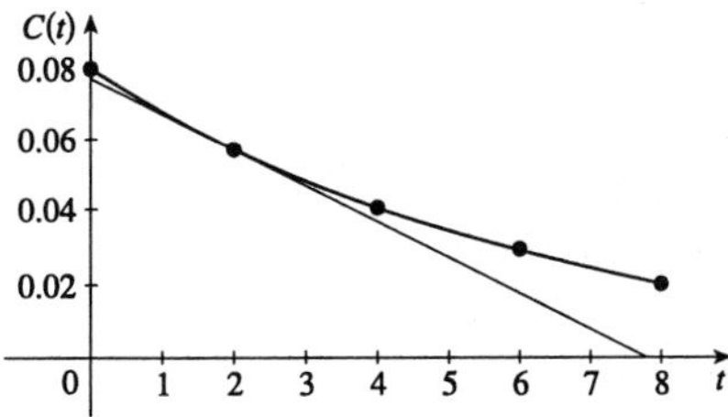

21. **(a)** rate of reaction $= \dfrac{d[C]}{dt} = \dfrac{a^2k(akt+1) - (a^2kt)(ak)}{(akt+1)^2} = \dfrac{a^2k(akt + 1 - akt)}{(akt+1)^2} = \dfrac{a^2k}{(akt+1)^2}$

(b) $a - x = a - \dfrac{a^2kt}{akt+1} = \dfrac{a^2kt + a - a^2kt}{akt+1} = \dfrac{a}{akt+1}$.

So $k(a-x)^2 = k\left(\dfrac{a}{akt+1}\right)^2 = \dfrac{a^2k}{(akt+1)^2} = \dfrac{dx}{dt}$.

22. $\frac{1}{p} = \frac{1}{f} - \frac{1}{q} \quad \Leftrightarrow \quad \frac{1}{p} = \frac{q-f}{fq} \quad \Leftrightarrow \quad p = \frac{fq}{q-f}$. So $\frac{dp}{dq} = \frac{f(q-f) - fq}{(q-f)^2} = -\frac{f^2}{(q-f)^2}$.

23. $m(t) = 5 - 0.02t^2 \quad \Rightarrow \quad m'(t) = -0.04t \quad \Rightarrow \quad m'(1) = -0.04$

24. $n(t) = 100 + 24t + 2t^2 \quad \Rightarrow \quad n'(t) = 24 + 4t \quad \Rightarrow \quad n'(2) = 32$ bacteria/h

25. $v(r) = \frac{P}{4\eta\ell}\left(R^2 - r^2\right) \quad \Rightarrow \quad v'(r) = \frac{P}{4\eta\ell}(-2r) = -\frac{Pr}{2\eta\ell}$.

When $\ell = 3$, $P = 3000$ and $\eta = 0.027$, we have $v'(0.005) = -\frac{3000(0.005)}{2(0.027)(3)} \approx -92.6\,\frac{\text{cm/s}}{\text{cm}}$.

26. **(a)**
$$S = \frac{dR}{dx} = \frac{\left(1 + 4x^{0.4}\right) D\left(40 + 24x^{0.4}\right) - \left(40 + 24x^{0.4}\right) D\left(1 + 4x^{0.4}\right)}{\left(1 + 4x^{0.4}\right)^2}$$
$$= \frac{\left(1 + 4x^{0.4}\right)\left(9.6x^{-0.6}\right) - \left(40 + 24x^{0.4}\right)\left(1.6x^{-0.6}\right)}{\left(1 + 4x^{0.4}\right)^2}$$
$$= -\frac{54.4x^{-0.6}}{\left(1 + 4x^{0.4}\right)^2}$$

(b)

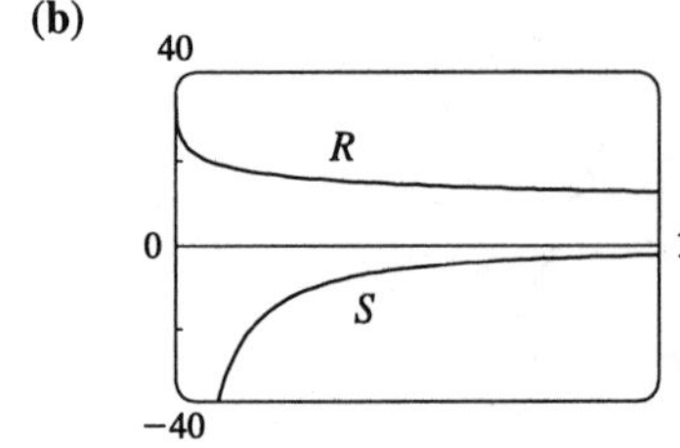

At low levels of brightness, R is quite large [$R(0) = 40$] and is quickly decreasing, that is, S is negative with large absolute value. This is to be expected: at low levels of brightness, the eye is more sensitive to slight changes than it is at higher levels of brightness.

27. $C(x) = 420 + 1.5x + 0.002x^2 \quad \Rightarrow \quad C'(x) = 1.5 + 0.004x \quad \Rightarrow$

$C'(100) = 1.5 + (0.004)(100) = \$1.90/\text{item}$

$C(101) - C(100) = (420 + 151.5 + 20.402) - (420 + 150 + 20) = \$1.902/\text{item}$

28. $C(x) = 1200 + \frac{x}{10} + \frac{x^2}{10{,}000} \quad \Rightarrow \quad C'(x) = \frac{1}{10} + \frac{x}{5000} \quad \Rightarrow \quad C'(100) = \frac{1}{10} + \frac{100}{5000} = \$0.12/\text{item}$

$C(101) - C(100) = (1200 + 10.1 + 1.0201) - (1200 + 10 + 1) = \$0.1201$

29. $C(x) = 2000 + 3x + 0.01x^2 + 0.0002x^3 \quad \Rightarrow \quad C'(x) = 3 + 0.02x + 0.0006x^2 \quad \Rightarrow$

$C'(100) = 3 + 0.02(100) + 0.0006(10{,}000) = 3 + 2 + 6 = \$11/\text{item}$

$C(101) - C(100) = (2000 + 303 + 102.1 + 206.0602) - (2000 + 300 + 100 + 200)$

$= 11.0702 \approx \$11.07/\text{item}$

30. $C(x) = 2500 + 2\sqrt{x} \quad \Rightarrow \quad C'(x) = \frac{1}{\sqrt{x}} \quad \Rightarrow \quad C'(100) = \frac{1}{\sqrt{100}} = \$0.10/\text{item}$.

$C(101) - C(100) = \left(2500 + 2\sqrt{101}\right) - (2500 + 20) \approx \$0.0998$

EXERCISES 2.4

1. $\lim_{x\to 0}(x^2+\cos x) = \lim_{x\to 0} x^2 + \lim_{x\to 0}\cos x = 0^2 + \cos 0 = 0 + 1 = 1$

2. $\lim_{x\to 0}\cos(\sin x) = \cos\left(\lim_{x\to 0}\sin x\right) = \cos 0 = 1$

3. $\lim_{x\to \pi/3}(\sin x - \cos x) = \sin\frac{\pi}{3} - \cos\frac{\pi}{3} = \frac{\sqrt{3}}{2} - \frac{1}{2}$

4. $\lim_{x\to\pi} x^2\sec x = \left(\lim_{x\to\pi} x^2\right)\left(\lim_{x\to\pi}\sec x\right) = \pi^2\sec\pi = \pi^2(-1) = -\pi^2$

5. $\lim_{x\to\pi/4}\frac{\sin x}{3x} = \frac{\sin(\pi/4)}{3\pi/4} = \frac{1/\sqrt{2}}{3\pi/4} = \frac{2\sqrt{2}}{3\pi}$

6. $\lim_{x\to 0}\frac{\sin x}{3x} = \frac{1}{3}\lim_{x\to 0}\frac{\sin x}{x} = \frac{1}{3}\cdot 1 = \frac{1}{3}$

7. $\lim_{t\to 0}\frac{\sin 5t}{t} = \lim_{t\to 0}\frac{5\sin 5t}{5t} = 5\lim_{t\to 0}\frac{\sin 5t}{5t} = 5\cdot 1 = 5$

8. $\lim_{t\to 0}\frac{\sin 8t}{\sin 9t} = \lim_{t\to 0}\frac{8\left(\frac{\sin 8t}{8t}\right)}{9\left(\frac{\sin 9t}{9t}\right)} = \frac{8\lim_{t\to 0}\frac{\sin 8t}{8t}}{9\lim_{t\to 0}\frac{\sin 9t}{9t}} = \frac{8\cdot 1}{9\cdot 1} = \frac{8}{9}$

9. $\lim_{\theta\to 0}\frac{\sin(\cos\theta)}{\sec\theta} = \frac{\sin\left(\lim_{\theta\to 0}\cos\theta\right)}{\lim_{\theta\to 0}\sec\theta} = \frac{\sin 1}{1} = \sin 1$

10. $\lim_{\theta\to 0}\frac{\cos\theta - 1}{\sin\theta} = \lim_{\theta\to 0}\frac{\frac{\cos\theta-1}{\theta}}{\frac{\sin\theta}{\theta}} = \frac{\lim_{\theta\to 0}\frac{\cos\theta-1}{\theta}}{\lim_{\theta\to 0}\frac{\sin\theta}{\theta}} = \frac{0}{1} = 0$

11. $\lim_{x\to\pi/4}\frac{\tan x}{4x} = \frac{\tan(\pi/4)}{4(\pi/4)} = \frac{1}{\pi}$

12. $\lim_{x\to 0}\frac{\tan x}{4x} = \lim_{x\to 0}\frac{1}{4}\frac{\sin x}{x}\cdot\frac{1}{\cos x} = \frac{1}{4}\lim_{x\to 0}\frac{\sin x}{x}\lim_{x\to 0}\frac{1}{\cos x} = \frac{1}{4}\cdot 1\cdot 1 = \frac{1}{4}$

13. $\lim_{\theta\to 0}\frac{\sin^2\theta}{\theta} = \lim_{\theta\to 0}\left(\frac{\sin\theta}{\theta}\right)\sin\theta = \lim_{\theta\to 0}\frac{\sin\theta}{\theta}\lim_{\theta\to 0}\sin\theta = 1\cdot 0 = 0$

14. $\lim_{h\to 0}\frac{\sin 5h}{\tan 3h} = \lim_{h\to 0}\frac{5\frac{\sin 5h}{5h}}{3\frac{\tan 3h}{3h}} = \frac{5}{3}\frac{\lim_{h\to 0}\frac{\sin 5h}{5h}}{\lim_{h\to 0}\frac{\sin 3h}{3h}\cdot\lim_{h\to 0}\frac{1}{\cos 3h}} = \frac{5}{3}\cdot\frac{1}{1\cdot 1} = \frac{5}{3}$

15. $\lim_{x\to 0}\frac{\tan 3x}{3\tan 2x} = \lim_{x\to 0}\frac{\frac{\tan 3x}{3x}}{2\frac{\tan 2x}{2x}} = \frac{1}{2}\frac{\lim_{x\to 0}\frac{\sin 3x}{3x}\cdot\frac{1}{\cos 3x}}{\lim_{x\to 0}\frac{\sin 2x}{2x}\cdot\lim_{x\to 0}\frac{1}{\cos 2x}} = \frac{1}{2}\frac{1\cdot 1}{1\cdot 1} = \frac{1}{2}$

16. $\lim_{t\to 0}\frac{\sin^2 3t}{t^2} = \lim_{t\to 0}9\left(\frac{\sin 3t}{3t}\right)^2 = 9\left(\lim_{t\to 0}\frac{\sin 3t}{3t}\right)^2 = 9(1)^2 = 9$

17. $\dfrac{d}{dx}(\csc x) = \dfrac{d}{dx}\left(\dfrac{1}{\sin x}\right) = \dfrac{(\sin x)(0) - 1(\cos x)}{\sin^2 x} = \dfrac{-\cos x}{\sin^2 x} = -\dfrac{1}{\sin x} \cdot \dfrac{\cos x}{\sin x} = -\csc x \cot x$

18. $\dfrac{d}{dx}(\sec x) = \dfrac{d}{dx}\left(\dfrac{1}{\cos x}\right) = \dfrac{(\cos x)(0) - 1(-\sin x)}{\cos^2 x} = \dfrac{\sin x}{\cos^2 x} = \dfrac{1}{\cos x} \cdot \dfrac{\sin x}{\cos x} = \sec x \tan x$

19. $\dfrac{d}{dx}(\cot x) = \dfrac{d}{dx}\left(\dfrac{\cos x}{\sin x}\right) = \dfrac{(\sin x)(-\sin x) - (\cos x)(\cos x)}{\sin^2 x} = -\dfrac{\sin^2 x + \cos^2 x}{\sin^2 x} = -\dfrac{1}{\sin^2 x} = -\csc^2 x$

20. $y = \cos x - 2\tan x \quad \Rightarrow \quad dy/dx = -\sin x - 2\sec^2 x$

21. $y = \sin x + \cos x \quad \Rightarrow \quad dy/dx = \cos x - \sin x$

22. $y = x \csc x \quad \Rightarrow \quad dy/dx = \csc x - x \csc x \cot x = \csc x\,(1 - x \cot x)$

23. $y = \csc x \cot x \quad \Rightarrow \quad dy/dx = (-\csc x \cot x)\cot x + \csc x\,(-\csc^2 x) = -\csc x\,(\cot^2 x + \csc^2 x)$

24. $y = \dfrac{\sin x}{1 + \cos x} \quad \Rightarrow$

$$\frac{dy}{dx} = \frac{(1 + \cos x)\cos x - \sin x\,(-\sin x)}{(1 + \cos x)^2} = \frac{\cos x + \cos^2 x + \sin^2 x}{(1 + \cos x)^2} = \frac{\cos x + 1}{(1 + \cos x)^2} = \frac{1}{1 + \cos x}$$

25. $y = \dfrac{\tan x}{x} \quad \Rightarrow \quad \dfrac{dy}{dx} = \dfrac{x \sec^2 x - \tan x}{x^2}$

26. $y = \dfrac{\tan x - 1}{\sec x} \quad \Rightarrow \quad \dfrac{dy}{dx} = \dfrac{\sec x \sec^2 x - (\tan x - 1)\sec x \tan x}{\sec^2 x} = \dfrac{\sec x\,(\sec^2 x - \tan^2 x + \tan x)}{\sec^2 x} = \dfrac{1 + \tan x}{\sec x}$

Another Method: Write $y = \sin x - \cos x$. Then $y' = \cos x + \sin x$.

27. $y = \dfrac{x}{\sin x + \cos x} \quad \Rightarrow$

$$\frac{dy}{dx} = \frac{(\sin x + \cos x) - x(\cos x - \sin x)}{(\sin x + \cos x)^2} = \frac{(1 + x)\sin x + (1 - x)\cos x}{\sin^2 x + \cos^2 x + 2\sin x \cos x} = \frac{(1 + x)\sin x + (1 - x)\cos x}{1 + \sin 2x}$$

28. $y = 2x\left(\sqrt{x} - \cot x\right) \quad \Rightarrow \quad \dfrac{dy}{dx} = 2\left(\sqrt{x} - \cot x\right) + 2x\left(\dfrac{1}{2\sqrt{x}} + \csc^2 x\right) = 3\sqrt{x} - 2\cot x + 2x\csc^2 x$

29. $y = x^{-3} \sin x \tan x \quad \Rightarrow$

$$\frac{dy}{dx} = -3x^{-4} \sin x \tan x + x^{-3} \cos x \tan x + x^{-3} \sin x \sec^2 x = x^{-4} \sin x\,(-3\tan x + x + x\sec^2 x)$$

30. $y = x \sin x \cos x \Rightarrow \dfrac{dy}{dx} = \sin x \cos x + x \cos x \cos x + x \sin x\,(-\sin x) = \sin x \cos x + x\cos^2 x - x\sin^2 x$

31. $y = \dfrac{x^2 \tan x}{\sec x} \quad \Rightarrow$

$$\frac{dy}{dx} = \frac{\sec x\,(2x \tan x + x^2 \sec^2 x) - x^2 \tan x \sec x \tan x}{\sec^2 x} = \frac{2x \tan x + x^2(\sec^2 x - \tan^2 x)}{\sec x} = \frac{2x\tan x + x^2}{\sec x}.$$

Another Method: Write $y = x^2 \sin x$. Then $y' = 2x \sin x + x^2 \cos x$.

32. $y = 2\sin x \quad \Rightarrow \quad y' = 2\cos x \quad \Rightarrow$ The slope of the tangent line at $\left(\frac{\pi}{6}, 1\right)$ is $2\cos\frac{\pi}{6} = 2 \cdot \frac{\sqrt{3}}{2} = \sqrt{3}$ and the equation is $y - 1 = \sqrt{3}\left(x - \frac{\pi}{6}\right)$ or $y = \sqrt{3}x + 1 - \frac{\sqrt{3}\pi}{6}$.

33. $y = \tan x \quad \Rightarrow \quad y' = \sec^2 x \quad \Rightarrow$ The slope of the tangent line at $\left(\frac{\pi}{4}, 1\right)$ is $\sec^2 \frac{\pi}{4} = 2$ and the equation is $y - 1 = 2\left(x - \frac{\pi}{4}\right)$ or $4x - 2y = \pi - 2$.

34. $y = \sec x - 2\cos x \quad \Rightarrow \quad y' = \sec x \tan x + 2\sin x \quad \Rightarrow$ The slope of the tangent line at $\left(\frac{\pi}{3}, 1\right)$ is $\sec \frac{\pi}{3} \tan \frac{\pi}{3} + 2\sin \frac{\pi}{3} = 2\sqrt{3} + 2 \cdot \frac{\sqrt{3}}{2} = 3\sqrt{3}$ and the equation is $y - 1 = 3\sqrt{3}\left(x - \frac{\pi}{3}\right)$ or $y = 3\sqrt{3}x + 1 - \pi\sqrt{3}$.

35. **(a)** $y = x\cos x \quad \Rightarrow$
$y' = x(-\sin x) + \cos x\,(1) = \cos x - x\sin x$
So the slope of the tangent at the point $(\pi, -\pi)$ is $\cos\pi - \pi\sin\pi = -1 - \pi(0) = -1$, and its equation is $y + \pi = -(x - \pi)$ $\Leftrightarrow \quad y = -x$.

(b)

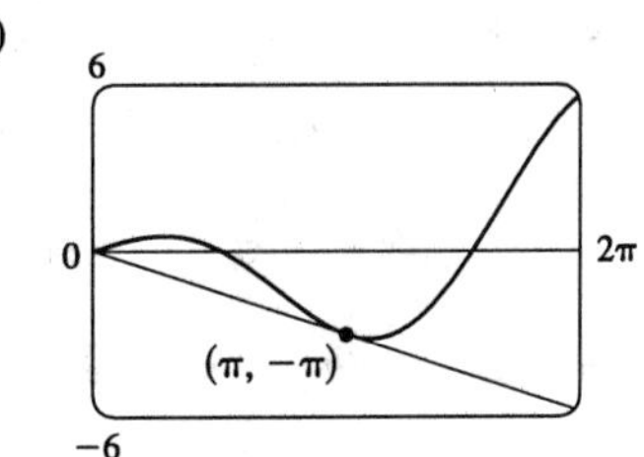

36. **(a)** $f(x) = 2x + \cot x \quad \Rightarrow \quad f'(x) = 2 - \csc^2 x$.

(b) Notice that $f'(x) = 0$ when f has a horizontal tangent. Also $f'(x)$ is large negative when the graph of f is steep.

37. $y = x + 2\sin x$ has a horizontal tangent when $y' = 1 + 2\cos x = 0 \quad \Leftrightarrow \quad \cos x = -\frac{1}{2} \quad \Leftrightarrow$ $x = (2n+1)\pi \pm \frac{\pi}{3}$, n an integer.

38. $y = \dfrac{\cos x}{2 + \sin x} \quad \Rightarrow \quad y' = \dfrac{-\sin x\,(2 + \sin x) - \cos x \cos x}{(2 + \sin x)^2} = \dfrac{-2\sin x - \sin^2 x - \cos^2 x}{(2 + \sin x)^2} = \dfrac{-2\sin x - 1}{(2 + \sin x)^2} = 0$ when $-2\sin x - 1 = 0 \quad \Leftrightarrow \quad \sin x = -\frac{1}{2} \quad \Leftrightarrow \quad x = \frac{11\pi}{6} + 2n\pi$ or $x = \frac{7\pi}{6} + 2n\pi$, n an integer. So $y = \frac{1}{\sqrt{3}}$ or $y = -\frac{1}{\sqrt{3}}$. So the points on the curve with horizontal tangents are: $\left(\frac{11\pi}{6} + 2n\pi, \frac{1}{\sqrt{3}}\right)$, $\left(\frac{7\pi}{6} + 2n\pi, -\frac{1}{\sqrt{3}}\right)$, n an integer.

39. From the diagram we can see that $\sin\theta = 10/x$ $\Leftrightarrow \quad x = 10\sin\theta$. But we want to find the rate of change of x with respect to θ, that is, $dx/d\theta$. Taking the derivative of the above expression, $dx/d\theta = 10(\cos\theta)$. So when $\theta = \frac{\pi}{3}$, $dx/d\theta = 10\cos\frac{\pi}{3} = 10\left(\frac{1}{2}\right) = 5$ ft/rad.

40. (a) $\dfrac{dF}{d\theta} = \dfrac{(\mu\sin\theta + \cos\theta)(0) - \mu W(\mu\cos\theta - \sin\theta)}{(\mu\sin\theta + \cos\theta)^2} = \dfrac{\mu W(\sin\theta - \mu\cos\theta)}{(\mu\sin\theta + \cos\theta)^2}$

(b) $\dfrac{dF}{d\theta} = 0 \quad\Rightarrow\quad \mu W(\sin\theta - \mu\cos\theta) = 0 \quad\Rightarrow\quad \sin\theta = \mu\cos\theta \quad\Rightarrow\quad \tan\theta = \mu \quad\Rightarrow\quad \theta = \tan^{-1}\mu$

(c)

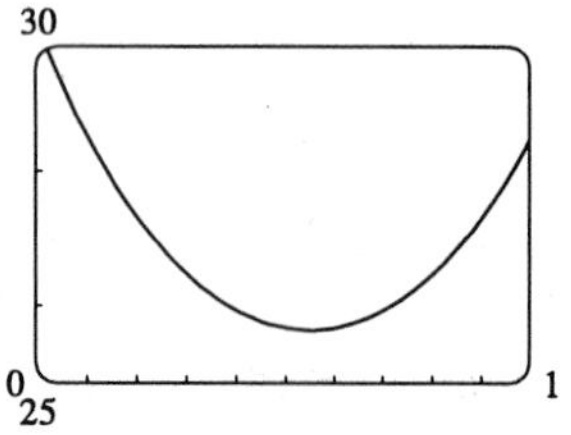

From the graph, we see that $\dfrac{dF}{d\theta} = 0 \quad\Rightarrow\quad \theta \approx 0.54.$ Checking this with part (b), we calculate $\theta = \tan^{-1}0.6 \approx 0.54$. So the graph is consistent with part (b).

41. $\displaystyle\lim_{\theta\to 0}\frac{\cos\theta - 1}{\theta} = \lim_{\theta\to 0}\frac{1 - 2\sin^2(\theta/2) - 1}{\theta} = \lim_{\theta\to 0}\frac{-\sin^2(\theta/2)}{\theta/2} = -\lim_{\theta\to 0}\frac{\sin(\theta/2)}{\theta/2}\lim_{\theta\to 0}\sin(\theta/2) = -1\cdot 0 = 0$

42. $\displaystyle f(x) = \cos x \quad\Rightarrow\quad f'(x) = \lim_{h\to 0}\frac{f(x+h) - f(x)}{h} = \lim_{h\to 0}\frac{\cos(x+h) - \cos x}{h}$

$$= \lim_{h\to 0}\frac{\cos x\cos h - \sin x\sin h - \cos x}{h} = \lim_{h\to 0}\left(\cos x\,\frac{\cos h - 1}{h} - \sin x\,\frac{\sin h}{h}\right)$$

$$= \cos x\lim_{h\to 0}\frac{\cos h - 1}{h} - \sin x\lim_{h\to 0}\frac{\sin h}{h} = (\cos x)(0) - (\sin x)(1) = -\sin x$$

43. $\displaystyle\lim_{x\to 0}\frac{\cot 2x}{\csc x} = \lim_{x\to 0}\frac{\cos 2x\sin x}{\sin 2x} = \lim_{x\to 0}\cos 2x\left[\frac{(\sin x)/x}{(\sin 2x)/x}\right] = \lim_{x\to 0}\cos 2x\left[\frac{\lim\limits_{x\to 0}[(\sin x)/x]}{2\lim\limits_{x\to 0}[(\sin 2x)/2x]}\right] = 1\cdot\frac{1}{2\cdot 1} = \frac{1}{2}$

44. Using the identity $\sin^2\theta = \frac{1}{2}(1 - \cos 2\theta)$, or $1 - \cos x = 2\sin^2(x/2)$, we have

$$\lim_{x\to 0}\frac{1 - \cos x}{2x^2} = \lim_{x\to 0}\frac{2\sin^2(x/2)}{2x^2} = \frac{1}{4}\left[\lim_{x\to 0}\frac{\sin(x/2)}{x/2}\right]^2 = \tfrac{1}{4}(1)^2 = \tfrac{1}{4}.$$

Another Method: Multiply numerator and denominator by $1 + \cos x$.

45. $\displaystyle\lim_{x\to\pi}\frac{\tan x}{\sin 2x} = \lim_{x\to\pi}\frac{\sin x}{\cos x\,(2\sin x\cos x)} = \lim_{x\to\pi}\frac{1}{2\cos^2 x} = \frac{1}{2(-1)^2} = \frac{1}{2}$

46. $\displaystyle\lim_{x\to\pi/4}\frac{\sin x - \cos x}{\cos 2x} = \lim_{x\to\pi/4}\frac{\sin x - \cos x}{\cos^2 x - \sin^2 x} = \lim_{x\to\pi/4}\frac{\sin x - \cos x}{(\cos x + \sin x)(\cos x - \sin x)} = \lim_{x\to\pi/4}\frac{-1}{\cos x + \sin x}$

$$= \frac{-1}{\cos\frac{\pi}{4} + \sin\frac{\pi}{4}} = \frac{-1}{\sqrt{2}}$$

47. Divide numerator and denominator by θ. ($\sin\theta$ also works.)

$$\lim_{\theta\to 0}\frac{\sin\theta}{\theta + \tan\theta} = \lim_{\theta\to 0}\frac{\dfrac{\sin\theta}{\theta}}{1 + \dfrac{\sin\theta}{\theta}\cdot\dfrac{1}{\cos\theta}} = \frac{\lim\limits_{\theta\to 0}\dfrac{\sin\theta}{\theta}}{1 + \lim\limits_{\theta\to 0}\dfrac{\sin\theta}{\theta}\lim\limits_{\theta\to 0}\dfrac{1}{\cos\theta}} = \frac{1}{1 + 1\cdot 1} = \frac{1}{2}$$

48. $\displaystyle\lim_{x\to 1}\frac{\sin(x-1)}{x^2 + x - 2} = \lim_{x\to 1}\frac{\sin(x-1)}{(x+2)(x-1)} = \lim_{x\to 1}\frac{1}{x+2}\lim_{x\to 1}\frac{\sin(x-1)}{x-1} = \frac{1}{3}\cdot 1 = \frac{1}{3}$

49. $\displaystyle \lim_{x\to 0}\frac{\cos x\sin x-\tan x}{x^2\sin x}=\lim_{x\to 0}\frac{\cos x\sin x-\dfrac{\sin x}{\cos x}}{x^2\sin x}=\lim_{x\to 0}\frac{\cos^2x\sin x-\sin x}{x^2\sin x\cos x}=\lim_{x\to 0}\frac{\cos^2x-1}{x^2\cos x}$

$\displaystyle =\lim_{x\to 0}\left(\frac{-\sin^2x}{x^2}\right)\frac{1}{\cos x}=-\left[\lim_{x\to 0}\frac{\sin x}{x}\right]^2\left[\lim_{x\to 0}\frac{1}{\cos x}\right]=-1$

50. $\displaystyle \lim_{y\to 0}\left(\lim_{x\to 0}\frac{\cos x\sin y}{x-y}\right)=\lim_{y\to 0}\left(\frac{1\cdot\sin y}{0-y}\right)=\lim_{y\to 0}\left(-\frac{\sin y}{y}\right)=-1$

Note that $\displaystyle \lim_{y\to 0}\left(\lim_{x\to 0}\frac{\cos x\sin y}{x-y}\right)\neq\lim_{x\to 0}\left(\lim_{y\to 0}\frac{\cos x\sin y}{x-y}\right)=\lim_{x\to 0}\left(\frac{0}{x}\right)=0.$

51. $\displaystyle \lim_{x\to 0}\frac{\sin(\sin x)}{\sin x}=\lim_{\sin x\to 0}\frac{\sin(\sin x)}{\sin x}$ since as $x\to 0$, $\sin x\to 0$. So we make the substitution $y=\sin x$, and see that $\displaystyle \lim_{x\to 0}\frac{\sin(\sin x)}{\sin x}=\lim_{y\to 0}\frac{\sin y}{y}=1.$

52. $\displaystyle \lim_{x\to 0}\frac{\sin(\sin x)}{x}=\lim_{x\to 0}\frac{\sin(\sin x)}{\sin x}\cdot\frac{\sin x}{x}=\left[\lim_{x\to 0}\frac{\sin(\sin x)}{\sin x}\right]\left[\lim_{x\to 0}\frac{\sin x}{x}\right]=1\cdot 1=1$

See Exercise 51 for a proof that $\displaystyle \lim_{x\to 0}\frac{\sin(\sin x)}{\sin x}=1.$

53. **(a)** $\displaystyle \frac{d}{dx}\tan x=\frac{d}{dx}\frac{\sin x}{\cos x}\;\Rightarrow\;\sec^2 x=\frac{\cos x\cos x-\sin x\,(-\sin x)}{\cos^2x}=\frac{\cos^2x+\sin^2x}{\cos^2x}.$ So $\displaystyle \sec^2 x=\frac{1}{\cos^2x}.$

(b) $\displaystyle \frac{d}{dx}\sec x=\frac{d}{dx}\frac{1}{\cos x}\;\Rightarrow\;\sec x\tan x=\frac{(\cos x)(0)-1(-\sin x)}{\cos^2x}.$ So $\displaystyle \sec x\tan x=\frac{\sin x}{\cos^2x}.$

(c) $\displaystyle \frac{d}{dx}(\sin x+\cos x)=\frac{d}{dx}\frac{1+\cot x}{\csc x}\;\Rightarrow$

$\displaystyle \cos x-\sin x=\frac{\csc x\,(-\csc^2 x)-(1+\cot x)(-\csc x\cot x)}{\csc^2 x}=\frac{-\csc^2 x+\cot^2 x+\cot x}{\csc x}$

So $\displaystyle \cos x-\sin x=\frac{\cot x-1}{\csc x}.$

54. Let $|PR|=x$. Then we get the following formulas for r and h in terms of θ and x:

$\displaystyle \sin\frac{\theta}{2}=\frac{r}{x}\;\Rightarrow\; r=x\sin\frac{\theta}{2}$ and $\displaystyle \cos\frac{\theta}{2}=\frac{h}{x}\;\Rightarrow\; h=x\cos\frac{\theta}{2}.$

Now $A(\theta)=\frac{1}{2}\pi r^2$ and $B(\theta)=\frac{1}{2}(2r)h=rh$. So

$$\lim_{\theta\to 0^+}\frac{A(\theta)}{B(\theta)}=\lim_{\theta\to 0^+}\frac{\frac{1}{2}\pi r^2}{rh}=\lim_{\theta\to 0^+}\frac{\frac{1}{2}\pi x\sin(\theta/2)}{x\cos(\theta/2)}=\lim_{\theta\to 0^+}\tfrac{\pi}{2}\tan(\theta/2)=0.$$

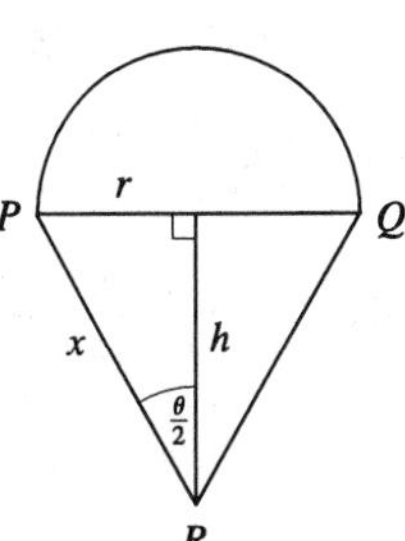

55. By the definition of radian measure, $s=r\theta$, where r is the radius of the circle. By drawing the bisector of the angle θ, we can see that $\displaystyle \sin\frac{\theta}{2}=\frac{d/2}{r}\;\Rightarrow\; d=2r\sin\frac{\theta}{2}.$ So

$$\lim_{\theta\to 0^+}\frac{s}{d}=\lim_{\theta\to 0^+}\frac{r\theta}{2r\sin(\theta/2)}=\lim_{\theta\to 0^+}\frac{2\cdot(\theta/2)}{2\sin(\theta/2)}=\lim_{\theta\to 0}\frac{\theta/2}{\sin(\theta/2)}=1.$$

[This is just the reciprocal of the limit $\displaystyle \lim_{x\to 0}\frac{\sin x}{x}=1$ combined with the fact that as $\theta\to 0$, $\dfrac{\theta}{2}\to 0$ also.]

EXERCISES 2.5

1. $y = u^2$, $u = x^2 + 2x + 3$

(a) $\dfrac{dy}{dx} = \dfrac{dy}{du}\dfrac{du}{dx} = 2u(2x+2) = 4u(x+1)$. When $x = 1$, $u = 1^2 + 2(1) + 3 = 6$, so
$\left.\dfrac{dy}{dx}\right|_{x=1} = 4(6)(1+1) = 48$.

(b) $y = u^2 = (x^2+2x+3)^2 = x^4 + 4x^2 + 9 + 4x^3 + 6x^2 + 12x = x^4 + 4x^3 + 10x^2 + 12x + 9$, so
$\dfrac{dy}{dx} = 4x^3 + 12x^2 + 20x + 12$ and $\left.\dfrac{dy}{dx}\right|_{x=1} = 4(1)^3 + 12(1)^2 + 20(1) + 12 = 48$.

2. $y = u^2 - 2u + 3$, $u = 5 - 6x$

(a) $\dfrac{dy}{dx} = \dfrac{dy}{du}\dfrac{du}{dx} = (2u-2)(-6) = 12(1-u)$. When $x = 1$, $u = -1$, so $\left.\dfrac{dy}{dx}\right|_{x=1} = 12[1-(-1)] = 24$.

(b) $y = (5-6x)^2 - 2(5-6x) + 3 = 36x^2 - 48x + 18$, so $\dfrac{dy}{dx} = 72x - 48$ and $\left.\dfrac{dy}{dx}\right|_{x=1} = 72(1) - 48 = 24$.

3. $y = u^3$, $u = x + 1/x$

(a) $\dfrac{dy}{dx} = \dfrac{dy}{du}\dfrac{du}{dx} = 3u^2\left(1 - \dfrac{1}{x^2}\right)$. When $x = 1$, $u = 1 + \dfrac{1}{1} = 2$, so $\left.\dfrac{dy}{dx}\right|_{x=1} = 3(2)^2\left(1 - \dfrac{1}{1^2}\right) = 0$.

(b) $y = u^3 = \left(x + \dfrac{1}{x}\right)^3 = x^3 + 3x^2\left(\dfrac{1}{x}\right) + 3x\left(\dfrac{1}{x}\right)^2 + \left(\dfrac{1}{x}\right)^3 = x^3 + 3x + 3x^{-1} + x^{-3}$, so
$\dfrac{dy}{dx} = 3x^2 + 3 - 3x^{-2} - 3x^{-4}$ and $\left.\dfrac{dy}{dx}\right|_{x=1} = 3(1)^2 + 3 - 3(1)^{-2} - 3(1)^{-4} = 0$.

4. $y = u - u^2$, $u = \sqrt{x} + \sqrt[3]{x}$

(a) $\dfrac{dy}{dx} = \dfrac{dy}{du}\dfrac{du}{dx} = (1-2u)\left(\frac{1}{2}x^{-1/2} + \frac{1}{3}x^{-2/3}\right)$. When $x = 1$, $u = 2$, so
$\left.\dfrac{dy}{dx}\right|_{x=1} = (1 - 2\cdot 2)\left[\frac{1}{2}(1)^{-1/2} + \frac{1}{3}(1)^{-2/3}\right] = -\dfrac{5}{2}$.

(b) $y = u - u^2 = \left(\sqrt{x} + \sqrt[3]{x}\right) - \left(\sqrt{x} + \sqrt[3]{x}\right)^2 = x^{1/2} + x^{1/3} - x - 2x^{5/6} - x^{2/3}$, so
$\dfrac{dy}{dx} = \frac{1}{2}x^{-1/2} + \frac{1}{3}x^{-2/3} - 1 - \frac{5}{3}x^{-1/6} - \frac{2}{3}x^{-1/3}$ and $\left.\dfrac{dy}{dx}\right|_{x=1} = \dfrac{1}{2} + \dfrac{1}{3} - 1 - \dfrac{5}{3} - \dfrac{2}{3} = -\dfrac{5}{2}$.

5. $F(x) = \left(x^2 + 4x + 6\right)^5 \quad \Rightarrow$
$F'(x) = 5\left(x^2 + 4x + 6\right)^4 \dfrac{d}{dx}\left(x^2 + 4x + 6\right) = 5\left(x^2 + 4x + 6\right)^4(2x+4) = 10\left(x^2 + 4x + 6\right)^4(x+2)$

6. $F(x) = \left(x^3 - 5x\right)^4 \quad \Rightarrow \quad F'(x) = 4\left(x^3 - 5x\right)^3 \dfrac{d}{dx}\left(x^3 - 5x\right) = 4\left(x^3 - 5x\right)^3\left(3x^2 - 5\right)$

7. $G(x) = (3x-2)^{10}(5x^2-x+1)^{12} \Rightarrow$

$$G'(x) = 10(3x-2)^9(3)(5x^2-x+1)^{12} + (3x-2)^{10}(12)(5x^2-x+1)^{11}(10x-1)$$
$$= 30(3x-2)^9(5x^2-x+1)^{12} + 12(3x-2)^{10}(5x^2-x+1)^{11}(10x-1)$$

[This can be simplified to $6(3x-2)^9(5x^2-x+1)^{11}(85x^2-51x+9)$.]

8. $g(t) = (6t^2+5)^3(t^3-7)^4 \Rightarrow g'(t) = 3(6t^2+5)^2(12t)(t^3-7)^4 + (6t^2+5)^3(4)(t^3-7)^3(3t^2)$
$$= 36t(6t^2+5)^2(t^3-7)^4 + 12t^2(6t^2+5)^3(t^3-7)^3$$

9. $f(t) = (2t^2-6t+1)^{-8} \Rightarrow f'(t) = -8(2t^2-6t+1)^{-9}(4t-6) = -16(2t^2-6t+1)^{-9}(2t-3)$

10. $f(t) = \dfrac{1}{(t^2-2t-5)^4} = (t^2-2t-5)^{-4} \Rightarrow f'(t) = -4(t^2-2t-5)^{-5}(2t-2) = \dfrac{8(1-t)}{(t^2-2t-5)^5}$

11. $g(x) = \sqrt{x^2-7x} = (x^2-7x)^{1/2} \Rightarrow g'(x) = \frac{1}{2}(x^2-7x)^{-1/2}(2x-7) = \dfrac{2x-7}{2\sqrt{x^2-7x}}$

12. $k(x) = \sqrt[3]{1+\sqrt{x}} = (1+x^{1/2})^{1/3} \Rightarrow k'(x) = \frac{1}{3}(1+x^{1/2})^{-2/3}(\frac{1}{2})x^{-1/2} = \dfrac{1}{6\sqrt{x}\sqrt[3]{(1+\sqrt{x})^2}}$

13. $h(t) = (t-1/t)^{3/2} \Rightarrow h'(t) = \frac{3}{2}(t-1/t)^{1/2}(1+1/t^2)$

14. $F(s) = \sqrt{s^3+1}(s^2+1)^4 = (s^3+1)^{1/2}(s^2+1)^4 \Rightarrow$

$$F'(s) = \tfrac{1}{2}(s^3+1)^{-1/2}(3s^2)(s^2+1)^4 + (s^3+1)^{1/2}(4)(s^2+1)^3(2s) = \frac{3s^2(s^2+1)^4}{2\sqrt{s^3+1}} + 8s(s^2+1)^3\sqrt{s^3+1}$$

15. $F(y) = \left(\dfrac{y-6}{y+7}\right)^3 \Rightarrow F'(y) = 3\left(\dfrac{y-6}{y+7}\right)^2 \dfrac{(y+7)(1)-(y-6)(1)}{(y+7)^2} = 3\left(\dfrac{y-6}{y+7}\right)^2 \dfrac{13}{(y+7)^2} = \dfrac{39(y-6)^2}{(y+7)^4}$

16. $s(t) = \left(\dfrac{t^3+1}{t^3-1}\right)^{1/4} \Rightarrow s'(t) = \dfrac{1}{4}\left(\dfrac{t^3+1}{t^3-1}\right)^{-3/4} \dfrac{3t^2(t^3-1)-(t^3+1)(3t^2)}{(t^3-1)^2} = \dfrac{1}{4}\left(\dfrac{t^3+1}{t^3-1}\right)^{-3/4} \dfrac{-6t^2}{(t^3-1)^2}$

17. $f(z) = (2z-1)^{-1/5} \Rightarrow f'(z) = -\frac{1}{5}(2z-1)^{-6/5}(2) = -\frac{2}{5}(2z-1)^{-6/5}$

18. $f(x) = \dfrac{x}{\sqrt{7-3x}} \Rightarrow$

$$f'(x) = \frac{\sqrt{7-3x} - x(\frac{1}{2})(7-3x)^{-1/2}(-3)}{7-3x} = \frac{1}{\sqrt{7-3x}} + \frac{3x}{2(7-3x)^{3/2}} \quad \left[\text{or } \frac{14-3x}{2(7-3x)^{3/2}}\right]$$

19. $y = (2x-5)^4(8x^2-5)^{-3} \Rightarrow y' = 4(2x-5)^3(2)(8x^2-5)^{-3} + (2x-5)^4(-3)(8x^2-5)^{-4}(16x)$
$$= 8(2x-5)^3(8x^2-5)^{-3} - 48x(2x-5)^4(8x^2-5)^{-4}$$

[This simplifies to $8(2x-5)^3(8x^2-5)^{-4}(-4x^2+30x-5)$.]

20. $y = (x^2+1)(x^2+2)^{1/3} \Rightarrow$

$$y' = 2x(x^2+2)^{1/3} + (x^2+1)(\tfrac{1}{3})(x^2+2)^{-2/3}(2x) = 2x(x^2+2)^{1/3}\left[1 + \frac{x^2+1}{3(x^2+2)}\right]$$

21. $y = \tan 3x \quad \Rightarrow \quad y' = \sec^2 3x \, \frac{d}{dx}(3x) = 3\sec^2 3x$

22. $y = 4\sec 5x \quad \Rightarrow \quad y' = 4\sec 5x \tan 5x\,(5) = 20\sec 5x \tan 5x$

23. $y = \cos(x^3) \quad \Rightarrow \quad y' = -\sin(x^3)(3x^2) = -3x^2 \sin(x^3)$

24. $y = \cos^3 x = (\cos x)^3 \quad \Rightarrow \quad y' = 3(\cos x)^2(-\sin x) = -3\cos^2 x \sin x$

25. $y = (1 + \cos^2 x)^6 \quad \Rightarrow \quad y' = 6(1+\cos^2 x)^5 2\cos x\,(-\sin x) = -12\cos x \sin x\,(1+\cos^2 x)^5$

26. $y = \tan(x^2) + \tan^2 x \quad \Rightarrow \quad y' = \sec^2(x^2)(2x) + 2\tan x \sec^2 x$

27. $y = \cos(\tan x) \quad \Rightarrow \quad y' = -\sin(\tan x)\sec^2 x$

28. $y = \sin(\sin x) \quad \Rightarrow \quad y' = \cos(\sin x)\cos x$

29. $y = \sec^2 2x - \tan^2 2x \quad \Rightarrow \quad y' = 2\sec 2x\,(\sec 2x \tan 2x)(2) - 2\tan 2x \sec^2(2x)(2) = 0$

Easier method: $y = \sec^2 2x - \tan^2 2x = 1 \quad \Rightarrow \quad y' = 0$

30. $y = \sqrt{1 + 2\tan x} \quad \Rightarrow \quad y' = \frac{1}{2}(1+2\tan x)^{-1/2}\, 2\sec^2 x = \dfrac{\sec^2 x}{\sqrt{1+2\tan x}}$

31. $y = \csc\dfrac{x}{3} \quad \Rightarrow \quad y' = -\dfrac{1}{3}\csc\dfrac{x}{3}\cot\dfrac{x}{3}$

32. $y = \cot\sqrt[3]{1+x^2} \quad \Rightarrow \quad y' = -\csc^2\left(\sqrt[3]{1+x^2}\right)\left(\frac{1}{3}\right)(1+x^2)^{-2/3}(2x) = -\dfrac{2x\csc^2\left(\sqrt[3]{1+x^2}\right)}{3(1+x^2)^{2/3}}$

33. $y = \sin^3 x + \cos^3 x \quad \Rightarrow \quad y' = 3\sin^2 x \cos x + 3\cos^2 x(-\sin x) = 3\sin x \cos x\,(\sin x - \cos x)$

34. $y = \sin^2(\cos 4x) \quad \Rightarrow \quad y' = 2\sin(\cos 4x)\cos(\cos 4x)(-\sin 4x)(4) = -4\sin 4x \sin(2\cos 4x)$

35. $y = \sin\dfrac{1}{x} \quad \Rightarrow \quad y' = \cos\dfrac{1}{x}\left(-\dfrac{1}{x^2}\right) = -\dfrac{1}{x^2}\cos\dfrac{1}{x}$

36. $y = \dfrac{\sin^2 x}{\cos x} \quad \Rightarrow$

$$y' = \frac{\cos x\,(2\sin x \cos x) - \sin^2 x\,(-\sin x)}{\cos^2 x} = \frac{\sin x\,(2\cos^2 x + \sin^2 x)}{\cos^2 x} = \frac{\sin x\,(1+\cos^2 x)}{\cos^2 x} = \sin x\,(1 + \sec^2 x)$$

Another Method: $y = \tan x \sin x \quad \Rightarrow \quad y' = \sec^2 x \sin x + \tan x \cos x = \sec^2 x \sin x + \sin x$

37. $y = \dfrac{1+\sin 2x}{1-\sin 2x} \quad \Rightarrow \quad y' = \dfrac{(1-\sin 2x)(2\cos 2x) - (1+\sin 2x)(-2\cos 2x)}{(1-\sin 2x)^2} = \dfrac{4\cos 2x}{(1-\sin 2x)^2}$

38. $y = x\sin\dfrac{1}{x} \quad \Rightarrow \quad y' = \sin\dfrac{1}{x} + x\cos\dfrac{1}{x}\left(-\dfrac{1}{x^2}\right) = \sin\dfrac{1}{x} - \dfrac{1}{x}\cos\dfrac{1}{x}$

39. $y = \tan^2(x^3) \quad \Rightarrow \quad y' = 2\tan(x^3)\sec^2(x^3)(3x^2) = 6x^2\tan(x^3)\sec^2(x^3)$

40. $y = \left(\sin\sqrt{x^2+1}\right)^{\sqrt{2}} \quad \Rightarrow$

$$y' = \sqrt{2}\left(\sin\sqrt{x^2+1}\right)^{\sqrt{2}-1}\left(\cos\sqrt{x^2+1}\right)\left(\tfrac{1}{2}\right)(x^2+1)^{-1/2}(2x) = \sqrt{2}x\left(\sin\sqrt{x^2+1}\right)^{\sqrt{2}-1}\frac{\cos\sqrt{x^2+1}}{\sqrt{x^2+1}}$$

41. $y = \cos^2(\cos x) + \sin^2(\cos x) = 1 \quad \Rightarrow \quad y' = 0.$

42. $y = \sin(\sin(\sin x)) \quad \Rightarrow \quad y' = \cos(\sin(\sin x))\dfrac{d}{dx}(\sin(\sin x)) = \cos(\sin(\sin x))\cos(\sin x)\cos x$

43. $y = \sqrt{x + \sqrt{x}} \quad \Rightarrow \quad y' = \frac{1}{2}\left(x + \sqrt{x}\right)^{-1/2}\left(1 + \frac{1}{2}x^{-1/2}\right) = \dfrac{1}{2\sqrt{x + \sqrt{x}}}\left(1 + \dfrac{1}{2\sqrt{x}}\right)$

44. $y = \sqrt{x + \sqrt{x + \sqrt{x}}} \quad \Rightarrow \quad y' = \frac{1}{2}\left(x + \sqrt{x + \sqrt{x}}\right)^{-1/2}\left[1 + \frac{1}{2}\left(x + \sqrt{x}\right)^{-1/2}\left(1 + \frac{1}{2}x^{-1/2}\right)\right]$

45. $f(x) = \left[x^3 + (2x - 1)^3\right]^3 \quad \Rightarrow$
$f'(x) = 3\left[x^3 + (2x - 1)^3\right]^2\left[3x^2 + 3(2x - 1)^2(2)\right] = 9\left[x^3 + (2x - 1)^3\right]^2\left[9x^2 - 8x + 2\right]$

46. $g(t) = \sqrt[4]{(1 - 3t)^4 + t^4} \quad \Rightarrow$
$g'(t) = \frac{1}{4}\left((1 - 3t)^4 + t^4\right)^{-3/4}\left[4(1 - 3t)^3(-3) + 4t^3\right] = \left((1 - 3t)^4 + t^4\right)^{-3/4}\left[t^3 - 3(1 - 3t^3)\right]$

47. $y = \sin\left(\tan\sqrt{\sin x}\right) \quad \Rightarrow \quad y' = \cos\left(\tan\sqrt{\sin x}\right)\left(\sec^2\sqrt{\sin x}\right)\left(\dfrac{1}{2\sqrt{\sin x}}\right)(\cos x)$

48. $y = \sqrt{\cos(\sin^2 x)} \quad \Rightarrow \quad y' = \frac{1}{2}\left(\cos(\sin^2 x)\right)^{-1/2}\left[-\sin(\sin^2 x)\right](2\sin x\cos x) = -\dfrac{\sin(\sin^2 x)\sin x\cos x}{\sqrt{\cos(\sin^2 x)}}$

49. $y = f(x) = \left(x^3 - x^2 + x - 1\right)^{10} \quad \Rightarrow \quad f'(x) = 10\left(x^3 - x^2 + x - 1\right)^9\left(3x^2 - 2x + 1\right)$. The slope of the tangent at $(1, 0)$ is $f'(1) = 0$ and its equation is $y - 0 = 0(x - 1)$ or $y = 0$.

50. $y = f(x) = \sqrt{x + 1/x} \quad \Rightarrow \quad f'(x) = \dfrac{1}{2}\left(x + \dfrac{1}{x}\right)^{-1/2}\left(1 - \dfrac{1}{x^2}\right)$. The slope of the tangent at $\left(1, \sqrt{2}\right)$ is $f'(1) = 0$ and its equation is $y - \sqrt{2} = 0(x - 1)$ or $y = \sqrt{2}$.

51. $y = f(x) = \dfrac{8}{\sqrt{4 + 3x}} \quad \Rightarrow \quad f'(x) = 8\left(-\frac{1}{2}\right)(4 + 3x)^{-3/2}(3) = -12(4 + 3x)^{-3/2}$. The slope of the tangent at $(4, 2)$ is $f'(4) = -\frac{3}{16}$ and its equation is $y - 2 = -\frac{3}{16}(x - 4)$ or $3x + 16y = 44$.

52. $y = f(x) = \sin x + \cos 2x \quad \Rightarrow \quad f'(x) = \cos x - 2\sin 2x$. The slope of the tangent at $\left(\frac{\pi}{6}, 1\right)$ is $f'\left(\frac{\pi}{6}\right) = \frac{\sqrt{3}}{2} - 2\left(\frac{\sqrt{3}}{2}\right) = -\frac{\sqrt{3}}{2}$ and its equation is $y - 1 = -\frac{\sqrt{3}}{2}\left(x - \frac{\pi}{6}\right)$ or $\sqrt{3}x + 2y = 2 + \frac{\sqrt{3}}{6}\pi$.

53. **(a)** 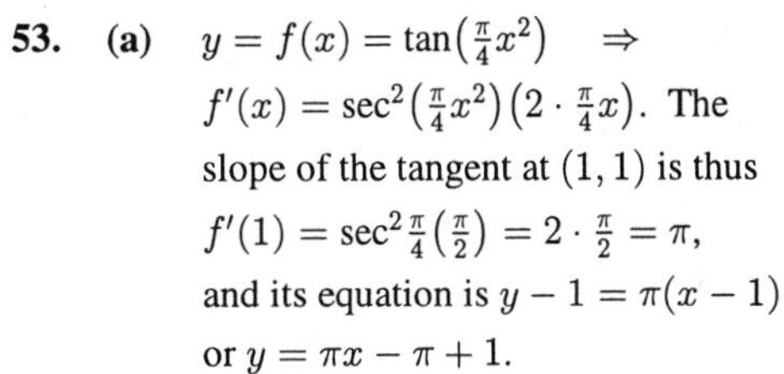
$y = f(x) = \tan\left(\frac{\pi}{4}x^2\right) \quad \Rightarrow$
$f'(x) = \sec^2\left(\frac{\pi}{4}x^2\right)\left(2 \cdot \frac{\pi}{4}x\right)$. The slope of the tangent at $(1, 1)$ is thus $f'(1) = \sec^2\frac{\pi}{4}\left(\frac{\pi}{2}\right) = 2 \cdot \frac{\pi}{2} = \pi$, and its equation is $y - 1 = \pi(x - 1)$ or $y = \pi x - \pi + 1$.

(b)

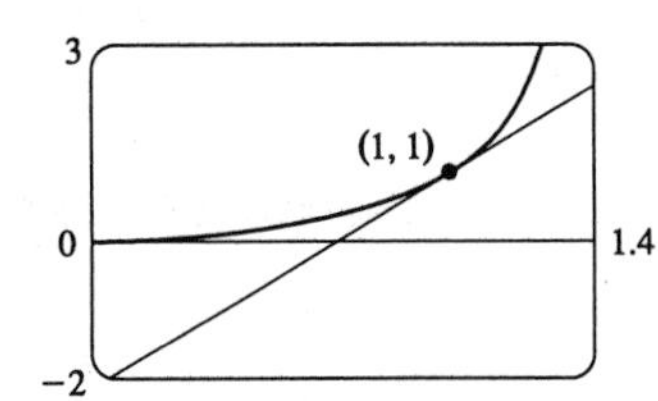

54. **(a)** For $x > 0$, $y = f(x) = \dfrac{x}{\sqrt{2 - x^2}} \quad \Rightarrow$

$$f'(x) = \frac{\sqrt{2 - x^2}(1) - x\left(\frac{1}{2}\right)(2 - x^2)^{-1/2}(-2x)}{2 - x^2}$$

$= \dfrac{2}{(2 - x^2)^{3/2}}$. So at $(1, 1)$, the slope of the tangent is $f'(1) = \dfrac{2}{(2 - 1)^{3/2}} = 2$ and its equation is $y - 1 = 2(x - 1) \quad \Leftrightarrow \quad y = 2x - 1$.

(b)

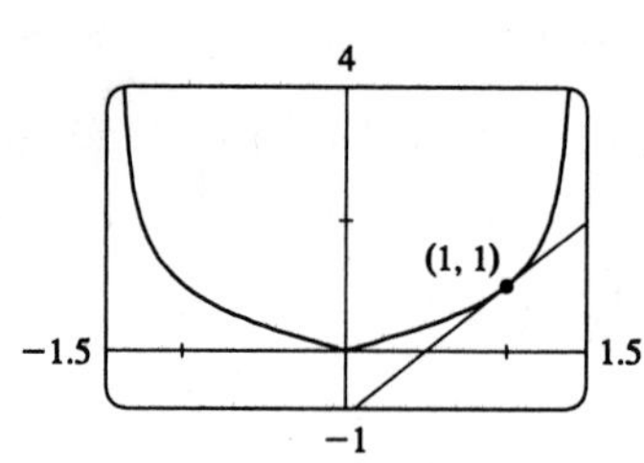

55. **(a)** $f(x) = \dfrac{\sqrt{1 - x^2}}{x} \quad \Rightarrow$

$$f'(x) = \frac{x \cdot \frac{1}{2}(1 - x^2)^{-1/2}(-2x) - \sqrt{1 - x^2}}{x^2}$$

$$= \frac{-1}{\sqrt{1 - x^2}} - \frac{\sqrt{1 - x^2}}{x^2}$$

$$= \frac{-x^2 - \sqrt{1 - x^2}\sqrt{1 - x^2}}{x^2\sqrt{1 - x^2}} = \frac{-1}{x^2\sqrt{1 - x^2}}$$

(b)

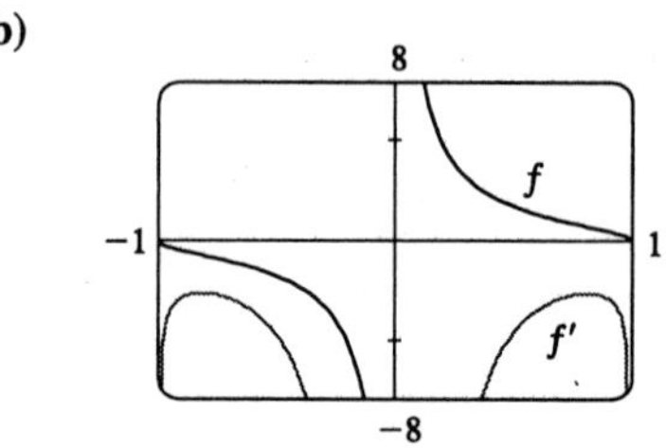

Notice that all tangents to the graph of f have negative slopes and $f'(x) < 0$ always.

56. **(a)** $f(x) = \dfrac{1}{\cos^2 \pi x + 9 \sin^2 \pi x} = \dfrac{1}{1 + 8 \sin^2 \pi x} \quad \Rightarrow$

$$f'(x) = -\left(1 + 8 \sin^2 \pi x\right)^{-2}(16 \sin \pi x)(\cos \pi x)\pi$$

$$= \frac{-16\pi \sin \pi x \cos \pi x}{\left(1 + 8 \sin^2 \pi x\right)^2}$$

(b)

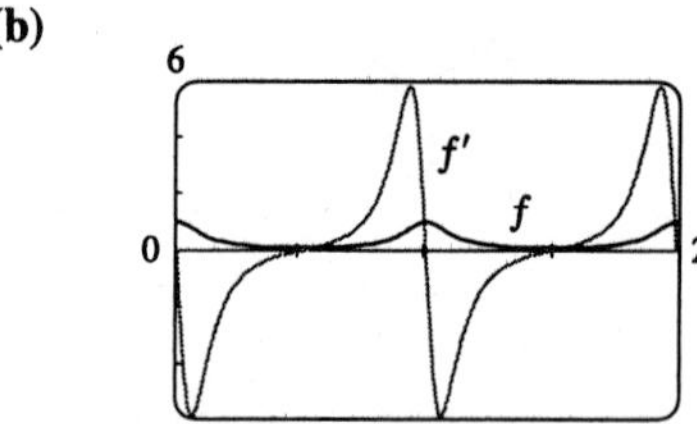

Notice that $f'(x) = 0$ when f has horizontal tangents.

57. For the tangent line to be horizontal $f'(x) = 0$. $f(x) = 2 \sin x + \sin^2 x \quad \Rightarrow$

$f'(x) = 2 \cos x + 2 \sin x \cos x = 0 \quad \Leftrightarrow \quad 2 \cos x\,(1 + \sin x) = 0 \quad \Leftrightarrow \quad \cos x = 0$

or $\sin x = -1$, so $x = \left(n + \frac{1}{2}\right)\pi$ or $\left(2n + \frac{3}{2}\right)\pi$ where n is any integer. So the points on the curve with a horizontal tangent are $\left(\left(2n + \frac{1}{2}\right)\pi, 3\right)$ and $\left(\left(2n + \frac{3}{2}\right)\pi, -1\right)$ where n is any integer.

58. $f(x) = \sin 2x - 2 \sin x \quad \Rightarrow \quad f'(x) = 2 \cos 2x - 2 \cos x = 4 \cos^2 x - 2 \cos x - 2$, and

$4 \cos^2 x - 2 \cos x - 2 = 0 \quad \Leftrightarrow \quad (\cos x - 1)(4 \cos x + 2) = 0 \quad \Leftrightarrow$

$\cos x = 1$ or $\cos x = -\frac{1}{2}$. So $x = 2n\pi$ or $(2n + 1)\pi \pm \frac{\pi}{3}$, n any integer.

59. $F(x) = f(g(x)) \quad \Rightarrow \quad F'(x) = f'(g(x))\,g'(x)$, so $F'(3) = f'(g(3))g'(3) = f'(6)g'(3) = 7 \cdot 4 = 28$.

60. $w = u \circ v \quad \Rightarrow \quad w'(x) = u'(v(x))v'(x)$, so $w'(0) = u'(v(0))v'(0) = u'(2)v'(0) = 4 \cdot 5 = 20$.

61. $s(t) = 10 + \frac{1}{4}\sin(10\pi t) \quad \Rightarrow \quad$ the velocity after t seconds is
$v(t) = s'(t) = \frac{1}{4}\cos(10\pi t)(10\pi) = \frac{5\pi}{2}\cos(10\pi t)$ cm/s.

62. **(a)** $s = A\cos(\omega t + \delta) \quad \Rightarrow \quad$ velocity $= s' = -\omega A \sin(\omega t + \delta)$.

(b) If $A \neq 0$ and $\omega \neq 0$, then $s' = 0 \;\Leftrightarrow\; \sin(\omega t + \delta) = 0 \;\Leftrightarrow\; \omega t + \delta = n\pi \;\Leftrightarrow\; t = \dfrac{n\pi - \delta}{\omega}$, n an integer.

63. **(a)** $\dfrac{dB}{dt} = \left(0.35\cos\dfrac{2\pi t}{5.4}\right)\left(\dfrac{2\pi}{5.4}\right) = \dfrac{7\pi}{54}\cos\dfrac{2\pi t}{5.4}$ **(b)** At $t = 1$, $\dfrac{dB}{dt} = \dfrac{7\pi}{54}\cos\dfrac{2\pi}{5.4} \approx 0.16$.

64. $f = \dfrac{1}{2L}\sqrt{\dfrac{T}{\rho}}$

(a) $\dfrac{df}{dL} = -\dfrac{1}{2L^2}\sqrt{\dfrac{T}{\rho}}$ **(b)** $\dfrac{df}{dT} = \dfrac{1}{2L}\dfrac{1}{2\sqrt{T}}\dfrac{1}{\sqrt{\rho}} = \dfrac{1}{4L\sqrt{T\rho}}$

(c) $\dfrac{df}{d\rho} = \dfrac{1}{2L}\sqrt{T}\left(-\frac{1}{2}\rho^{-3/2}\right) = -\dfrac{\sqrt{T}}{4L\rho^{3/2}}$

65. **(a)** Since h is differentiable on $[0, \infty)$ and $\sqrt{x}$ is differentiable on $(0, \infty)$, it follows that $G(x) = h(\sqrt{x})$ is differentiable on $(0, \infty)$.

(b) By the Chain Rule, $G'(x) = h'(\sqrt{x})\dfrac{d}{dx}\sqrt{x} = \dfrac{h'(\sqrt{x})}{2\sqrt{x}}$.

66. **(a)** $F(x) = f(x^\alpha) \quad \Rightarrow \quad F'(x) = f'(x^\alpha)\dfrac{d}{dx}(x^\alpha) = f'(x^\alpha)\alpha x^{\alpha - 1}$

(b) $G(x) = [f(x)]^\alpha \quad \Rightarrow \quad G'(x) = \alpha[f(x)]^{\alpha-1}f'(x)$

67. **(a)** $F(x) = f(\cos x) \quad \Rightarrow \quad F'(x) = f'(\cos x)\dfrac{d}{dx}(\cos x) = -\sin x\, f'(\cos x)$

(b) $G(x) = \cos(f(x)) \quad \Rightarrow \quad G'(x) = -\sin(f(x))f'(x)$

68. $g(t) = (f(\sin t))^2 \quad \Rightarrow \quad g'(t) = 2f(\sin t)f'(\sin t)\cos t$

69. $g(x) = f(b + mx) + f(b - mx) \quad \Rightarrow$
$g'(x) = f'(b + mx)D(b + mx) + f'(b - mx)D(b - mx) = mf'(b + mx) - mf'(b - mx)$
So $g'(0) = mf'(b) - mf'(b) = 0$.

70. $80\dfrac{dy}{dx} = \dfrac{d}{dx}y^5 = 5y^4\dfrac{dy}{dx} \;\Leftrightarrow\; 80 = 5y^4$ $\left(\text{Note that } \dfrac{dy}{dx} \neq 0 \text{ since the curve never has a horizontal tangent}\right)$

$\Leftrightarrow \quad y^4 = 16 \quad \Leftrightarrow \quad y = 2 \quad$ (since $y > 0$ for all x)

71. **(a)** If f is even, then $f(x) = f(-x)$. Using the Chain Rule to differentiate this equation, we get $f'(x) = f'(-x)\dfrac{d}{dx}(-x) = -f'(-x)$. Thus $f'(-x) = -f'(x)$, so f' is odd.

(b) If f is odd, then $f(x) = -f(-x)$. Differentiating this equation, we get $f'(x) = -f'(-x)(-1) = f'(-x)$, so f' is even.

72. $\left[\dfrac{f(x)}{g(x)}\right]' = \left[f(x)[g(x)]^{-1}\right]' = f'(x)[g(x)]^{-1} + (-1)[g(x)]^{-2}g'(x)f(x)$

$$= \frac{f'(x)}{g(x)} - \frac{f(x)\,g'(x)}{[g(x)]^2} = \frac{f'(x)\,g(x) - f(x)\,g'(x)}{[g(x)]^2}$$

73. $\dfrac{d}{dx}(\sin^n x \cos nx) = n\sin^{n-1}x\cos x\cos nx + \sin^n x\,(-n\sin nx)$

$$= n\sin^{n-1}x\,(\cos nx\cos x - \sin nx\sin x) = n\sin^{n-1}x\cos[(n+1)x]$$

74. $\dfrac{d}{dx}(\cos^n x\cos nx) = n\cos^{n-1}x(-\sin x)\cos nx + \cos^n x(-n\sin nx)$

$$= -n\cos^{n-1}x\,(\cos nx\sin x + \sin nx\cos x) = -n\cos^{n-1}x\sin[(n+1)x]$$

75. $f(x) = |x| = \sqrt{x^2} \quad\Rightarrow\quad f'(x) = \frac{1}{2}(x^2)^{-1/2}(2x) = x/\sqrt{x^2} = x/|x|.$

76. Using Exercise 75, we have $f(x) = \dfrac{|x|}{x} \quad\Rightarrow$

$$f'(x) = \frac{x(x/|x|) - |x|(1)}{x^2} = \frac{x(|x|/x) - |x|}{x^2} = \frac{|x| - |x|}{x^2} = 0. \quad \left(\text{Note that } \frac{x}{|x|} = \frac{|x|}{x}.\right)$$

Another Method: Use the fact that $\dfrac{|x|}{x} = \begin{cases} 1 & \text{if } x > 0 \\ -1 & \text{if } x < 0. \end{cases}$

77. Using Exercise 75, we have $h(x) = x|2x-1| \quad\Rightarrow$

$$h'(x) = |2x-1| + x\frac{2x-1}{|2x-1|}(2) = |2x-1| + \frac{2x(2x-1)}{|2x-1|}.$$

78. **(a)**

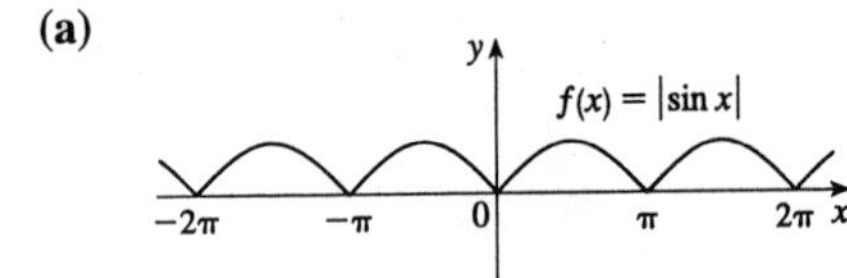

(b) f is not differentiable when $x = n\pi$, n an integer.

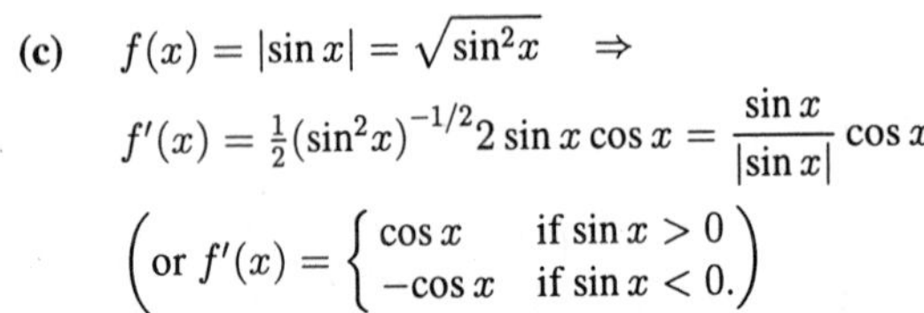

(c) $f(x) = |\sin x| = \sqrt{\sin^2 x} \quad\Rightarrow$

$$f'(x) = \tfrac{1}{2}(\sin^2 x)^{-1/2}\,2\sin x\cos x = \frac{\sin x}{|\sin x|}\cos x$$

$$\left(\text{or } f'(x) = \begin{cases} \cos x & \text{if } \sin x > 0 \\ -\cos x & \text{if } \sin x < 0. \end{cases}\right)$$

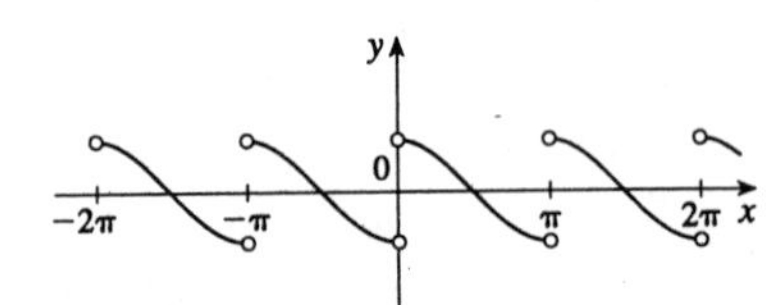

(d) g is not differentiable at 0.

$$g(x) = \sin|x| = \sin\sqrt{x^2} \quad\Rightarrow$$

$$g'(x) = \cos|x|\cdot\frac{x}{|x|} = \frac{x}{|x|}\cos x$$

$$\left(\text{or } g'(x) = \begin{cases} \cos x & \text{if } x > 0 \\ -\cos x & \text{if } x < 0 \end{cases}\right)$$

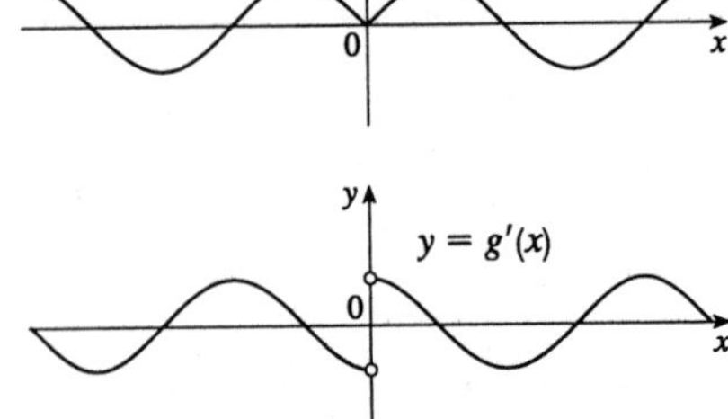

79. Since $\theta^\circ = \left(\frac{\pi}{180}\right)\theta$ rad, we have $\dfrac{d}{d\theta}(\sin\theta^\circ) = \dfrac{d}{d\theta}\left(\sin\frac{\pi}{180}\theta\right) = \frac{\pi}{180}\cos\frac{\pi}{180}\theta = \frac{\pi}{180}\cos\theta^\circ$.

80. **(a)** $u(x) = f(g(x)) \Rightarrow u'(x) = f'(g(x))g'(x)$. So $u'(1) = f'(g(1))g'(1) = f'(3)\,g'(1) = \left(-\frac{1}{4}\right)(-3) = \frac{3}{4}$.

(b) $v(x) = g(f(x)) \Rightarrow v'(x) = g'(f(x))f'(x)$. So $v'(1) = g'(f(1))f'(1) = g'(2)\,f'(1)$, which does not exist since $g'(2)$ does not exist.

(c) $w(x) = g(g(x)) \Rightarrow w'(x) = g'(g(x))g'(x)$. So $w'(1) = g'(g(1))g'(1) = g'(3)g'(1) = \left(\frac{2}{3}\right)(-3) = -2$.

EXERCISES 2.6

1. **(a)** $x^2 + 3x + xy = 5 \Rightarrow 2x + 3 + y + xy' = 0 \Rightarrow y' = -\dfrac{2x+y+3}{x}$

(b) $x^2 + 3x + xy = 5 \Rightarrow y = \dfrac{5 - x^2 - 3x}{x} = \dfrac{5}{x} - x - 3 \Rightarrow y' = -\dfrac{5}{x^2} - 1$

(c) $y' = -\dfrac{2x+y+3}{x} = \dfrac{-2x - 3 - (-3 - x + 5/x)}{x} = -1 - \dfrac{5}{x^2}$

2. **(a)** $\dfrac{x^2}{2} + \dfrac{y^2}{4} = 1 \Rightarrow x + \dfrac{y}{2}y' = 0 \Rightarrow y' = -\dfrac{2x}{y}$

(b) $\dfrac{y^2}{4} = 1 - \dfrac{x^2}{2} \Rightarrow y^2 = 4 - 2x^2 \Rightarrow y = \pm\sqrt{4 - 2x^2} \Rightarrow y' = \pm\dfrac{1}{2\sqrt{4-2x^2}}(-4x) = \mp\dfrac{2x}{\sqrt{4-2x^2}}$

(c) $y' = \dfrac{-2x}{y} = \dfrac{-2x}{\pm\sqrt{4-2x^2}} = \mp\dfrac{2x}{\sqrt{4-2x^2}}$

3. **(a)** $2y^2 + xy = x^2 + 3 \Rightarrow 4yy' + y + xy' = 2x \Rightarrow y' = \dfrac{2x - y}{x + 4y}$

(b) Use the quadratic formula: $2y^2 + xy - (x^2 + 3) = 0 \Rightarrow$

$$y = \frac{-x \pm \sqrt{x^2 + 8(x^2+3)}}{4} = \frac{-x \pm \sqrt{9x^2+24}}{4} \Rightarrow y' = \frac{1}{4}\left(-1 \pm \frac{9x}{\sqrt{9x^2+24}}\right)$$

(c) $y' = \dfrac{2x - y}{x + 4y} = \dfrac{2x - \frac{1}{4}\left(-x \pm \sqrt{9x^2+24}\right)}{x + \left(-x \pm \sqrt{9x^2+24}\right)} = \dfrac{1}{4}\left(-1 \pm \dfrac{9x}{\sqrt{9x^2+24}}\right)$

4. **(a)** $\sqrt{x} + \sqrt{y} = 4 \Rightarrow \dfrac{1}{2\sqrt{x}} + \dfrac{1}{2\sqrt{y}}y' = 0 \Rightarrow y' = -\dfrac{\sqrt{y}}{\sqrt{x}}$

(b) $\sqrt{y} = 4 - \sqrt{x} \Rightarrow y = \left(4 - \sqrt{x}\right)^2 = 16 - 8\sqrt{x} + x \Rightarrow y' = 1 - \dfrac{4}{\sqrt{x}}$

(c) $y' = -\dfrac{\sqrt{y}}{\sqrt{x}} = -\dfrac{4 - \sqrt{x}}{\sqrt{x}} = -\dfrac{4}{\sqrt{x}} + 1$

5. $x^2 - xy + y^3 = 8 \Rightarrow 2x - y - xy' + 3y^2y' = 0 \Rightarrow y' = \dfrac{y - 2x}{3y^2 - x}$

6. $\sqrt{xy}-2x=\sqrt{y} \quad\Rightarrow\quad \dfrac{y+xy'}{2\sqrt{xy}}-2=\dfrac{y'}{2\sqrt{y}} \quad\Rightarrow\quad \dfrac{y-4\sqrt{xy}}{2\sqrt{xy}}=\dfrac{\sqrt{x}-x}{2\sqrt{xy}}\,y' \quad\Rightarrow\quad y'=\dfrac{y-4\sqrt{xy}}{\sqrt{x}-x}$

7. $2y^2+\sqrt[3]{xy}=3x^2+17 \quad\Rightarrow\quad 4yy'+\frac{1}{3}x^{-2/3}y^{1/3}+\frac{1}{3}\,x^{1/3}y^{-2/3}y'=6x \quad\Rightarrow$

$y'=\dfrac{6x-\frac{1}{3}\,x^{-2/3}y^{1/3}}{4y+\frac{1}{3}x^{1/3}y^{-2/3}}=\dfrac{18x-x^{-2/3}y^{1/3}}{12y+x^{1/3}y^{-2/3}}$

8. $y^5+3x^2y^2+5x^4=12 \quad\Rightarrow\quad 5y^4y'+6xy^2+6x^2yy'+20x^3=0 \quad\Rightarrow\quad y'=-\dfrac{20x^3+6xy^2}{5y^4+6x^2y}$

9. $x^4+y^4=16 \quad\Rightarrow\quad 4x^3+4y^3y'=0 \quad\Rightarrow\quad y'=-\dfrac{x^3}{y^3}$

10. $\sqrt{x+y}+\sqrt{xy}=6 \quad\Rightarrow\quad \frac{1}{2}(x+y)^{-1/2}(1+y')+\frac{1}{2}(xy)^{-1/2}(y+xy')=0 \quad\Rightarrow$

$y'=-\dfrac{(x+y)^{-1/2}+(xy)^{-1/2}y}{(x+y)^{-1/2}+(xy)^{-1/2}x}=-\dfrac{\sqrt{xy}+y\sqrt{x+y}}{\sqrt{xy}+x\sqrt{x+y}}$

11. $\dfrac{y}{x-y}=x^2+1 \quad\Rightarrow\quad 2x=\dfrac{(x-y)y'-y(1-y')}{(x-y)^2}=\dfrac{xy'-y}{(x-y)^2} \quad\Rightarrow\quad y'=\dfrac{y}{x}+2(x-y)^2$

Another Method: Write the equation as $y=(x-y)(x^2+1)=x^3+x-yx^2-y$. Then $y'=\dfrac{3x^2+1-2xy}{x^2+2}$.

12. $x\sqrt{1+y}+y\sqrt{1+2x}=2x \quad\Rightarrow\quad \sqrt{1+y}+x\dfrac{1}{2\sqrt{1+y}}y'+y'\sqrt{1+2x}+y\dfrac{2}{2\sqrt{1+2x}}=2 \quad\Rightarrow$

$y'=\dfrac{2-\sqrt{1+y}-y/\sqrt{1+2x}}{\sqrt{1+2x}+x/(2\sqrt{1+y})}$

13. $\cos(x-y)=y\sin x \quad\Rightarrow\quad -\sin(x-y)(1-y')=y'\sin x+y\cos x \quad\Rightarrow\quad y'=\dfrac{\sin(x-y)+y\cos x}{\sin(x-y)-\sin x}$

14. $x\sin y+\cos 2y=\cos y \quad\Rightarrow\quad \sin y+(x\cos y)\,y'-(2\sin 2y)\,y'=(-\sin y)\,y' \quad\Rightarrow$

$y'=\dfrac{\sin y}{2\sin 2y-x\cos y-\sin y}$

15. $xy=\cot(xy) \quad\Rightarrow\quad y+xy'=-\csc^2(xy)(y+xy') \quad\Rightarrow\quad (y+xy')[1+\csc^2(xy)]=0 \quad\Rightarrow\quad y+xy'=0$

$\Rightarrow\quad y'=-y/x$

16. $x\cos y+y\cos x=1 \;\Rightarrow\; \cos y+x(-\sin y)\,y'+y'\cos x-y\sin x=0 \;\Rightarrow\; y'=\dfrac{y\sin x-\cos y}{\cos x-x\sin y}$

17. $y^4+x^2y^2+yx^4=y+1 \;\Rightarrow\; 4y^3+2x\dfrac{dx}{dy}y^2+2x^2y+x^4+4yx^3\dfrac{dx}{dy}=1 \;\Rightarrow\; \dfrac{dx}{dy}=\dfrac{1-4y^3-2x^2y-x^4}{2xy^2+4yx^3}$

18. $(x^2+y^2)^2=ax^2y \quad\Rightarrow\quad 2(x^2+y^2)\left(2x\dfrac{dx}{dy}+2y\right)=2ayx\dfrac{dx}{dy}+ax^2 \quad\Rightarrow\quad \dfrac{dx}{dy}=\dfrac{ax^2-4y(x^2+y^2)}{4x(x^2+y^2)-2axy}$

19. $x[f(x)]^3+xf(x)=6 \quad\Rightarrow\quad [f(x)]^3+3x[f(x)]^2f'(x)+f(x)+xf'(x)=0 \quad\Rightarrow$

$f'(x)=-\dfrac{[f(x)]^3+f(x)}{3x[f(x)]^2+x} \quad\Rightarrow\quad f'(3)=-\dfrac{(1)^3+1}{3(3)(1)^2+3}=-\dfrac{1}{6}$

20. $[g(x)]^2 + 12x = x^2 g(x) \quad \Rightarrow \quad 2g(x)g'(x) + 12 = 2xg(x) + x^2 g'(x) \quad \Rightarrow \quad g'(x) = \dfrac{2xg(x) - 12}{2g(x) - x^2} \quad \Rightarrow$
$g'(4) = \dfrac{2(4)(12) - 12}{2(12) - (4)^2} = \dfrac{21}{2}$

21. $\dfrac{x^2}{16} - \dfrac{y^2}{9} = 1 \quad \Rightarrow \quad \dfrac{x}{8} - \dfrac{2yy'}{9} = 0 \quad \Rightarrow \quad y' = \dfrac{9x}{16y}$. When $x = -5$ and $y = \frac{9}{4}$ we have $y' = \dfrac{9(-5)}{16(9/4)} = -\dfrac{5}{4}$
so the equation of the tangent is $y - \frac{9}{4} = -\frac{5}{4}(x+5)$ or $5x + 4y + 16 = 0$.

22. $\dfrac{x^2}{9} + \dfrac{y^2}{36} = 1 \quad \Rightarrow \quad \dfrac{2x}{9} + \dfrac{yy'}{18} = 0 \quad \Rightarrow \quad y' = -\dfrac{4x}{y}$. When $x = -1$ and $y = 4\sqrt{2}$ we have
$y' = -\dfrac{4(-1)}{4\sqrt{2}} = \dfrac{1}{\sqrt{2}}$ so the equation of the tangent line is $y - 4\sqrt{2} = \frac{1}{\sqrt{2}}(x+1)$ or $x - \sqrt{2}y + 9 = 0$.

23. $y^2 = x^3(2-x) = 2x^3 - x^4 \quad \Rightarrow \quad 2yy' = 6x^2 - 4x^3 \quad \Rightarrow \quad y' = \dfrac{3x^2 - 2x^3}{y}$. When $x = y = 1$,
$y' = \dfrac{3(1)^2 - 2(1)^3}{1} = 1$, so the equation of the tangent line is $y - 1 = 1(x-1)$ or $y = x$.

24. $x^{2/3} + y^{2/3} = 4 \quad \Rightarrow \quad \frac{2}{3}x^{-1/3} + \frac{2}{3}y^{-1/3}y' = 0 \quad \Rightarrow \quad y' = -\dfrac{\sqrt[3]{y}}{\sqrt[3]{x}}$. When $x = -3\sqrt{3}$ and $y = 1$, we have
$y' = -\dfrac{1}{\left(-3\sqrt{3}\right)^{1/3}} = \dfrac{1}{\sqrt{3}}$, so the equation of the tangent is $y - 1 = \frac{1}{\sqrt{3}}\left(x + 3\sqrt{3}\right)$ or $y = \frac{1}{\sqrt{3}}x + 4$.

25. $2(x^2+y^2)^2 = 25(x^2 - y^2) \quad \Rightarrow \quad 4(x^2+y^2)(2x + 2yy') = 25(2x - 2yy') \quad \Rightarrow \quad y' = \dfrac{25x - 4x(x^2+y^2)}{25y + 4y(x^2+y^2)}$.
When $x = 3$ and $y = 1$, $y' = \dfrac{75 - 120}{25 + 40} = -\dfrac{9}{13}$ so the equation of the tangent is $y - 1 = -\frac{9}{13}(x - 3)$ or
$9x + 13y = 40$.

26. $x^2y^2 = (y+1)^2(4 - y^2) \quad \Rightarrow \quad 2xy^2 + 2x^2yy' = 2(y+1)y'(4-y^2) + (y+1)^2(-2yy') \quad \Rightarrow$
$y' = \dfrac{xy^2}{(y+1)(4-y^2) - y(y+1)^2 - x^2y} = 0$ when $x = 0$. So the equation of the tangent line at $(0, -2)$ is
$y + 2 = 0(x - 0)$ or $y = -2$.

27. **(a)** $y^2 = 5x^4 - x^2 \quad \Rightarrow \quad 2yy' = 5(4x^3) - 2x \quad \Rightarrow$
$y' = \dfrac{10x^3 - x}{y}$. So at the point $(1, 2)$ we have
$y' = \dfrac{10(1)^3 - 1}{2} = \dfrac{9}{2}$, and the equation of
the tangent line is $y - 2 = \frac{9}{2}(x-1) \quad \Leftrightarrow \quad y = \frac{9}{2}x - \frac{5}{2}$.

(b)

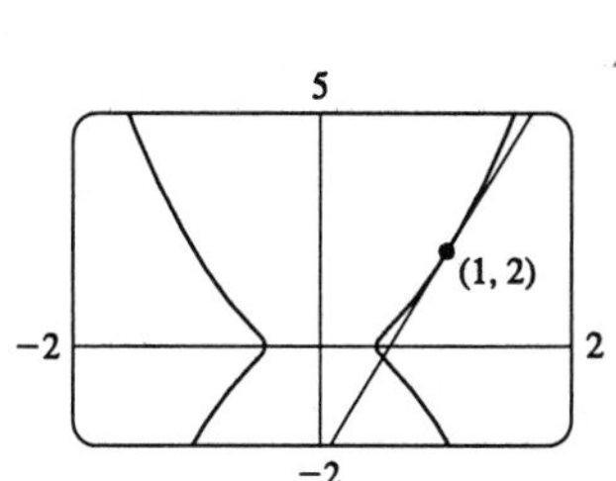

28. (a) $y^2 = x^3 + 3x^2 \Rightarrow 2yy' = 3x^2 + 3(2x) \Rightarrow y' = \dfrac{3x^2 + 6x}{2y}$. So at the point $(1, -2)$ we have $y' = \dfrac{3(1)^2 + 6(1)}{2(-2)} = -\dfrac{9}{4}$, and the equation of the tangent is $y - (-2) = -\frac{9}{4}(x - 1) \Leftrightarrow y = -\frac{9}{4}x + \frac{1}{4}$.

(b) The curve has a horizontal tangent where $y' = 0 \Leftrightarrow 3x^2 + 6x = 0 \Leftrightarrow 3x(x + 2) = 0 \Leftrightarrow x = 0$ or $x = -2$. But note that at $x = 0$, $y = 0$ also, so the derivative does not exist. At $x = -2$, $y^2 = (-2)^3 + 3(-2)^2 = -8 + 12 = 4$, so $y = \pm 2$. So the two points at which the curve has a horizontal tangent are $(-2, -2)$ and $(-2, 2)$.

(c)

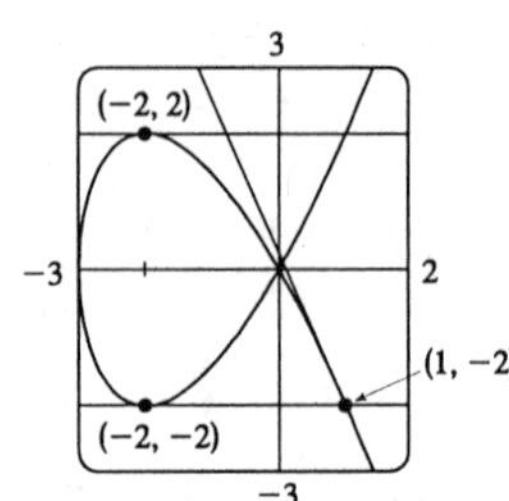

29. From Exercise 25, a tangent to the lemniscate will be horizontal $\Rightarrow y' = 0 \Rightarrow 25x - 4x(x^2 + y^2) = 0 \Rightarrow x^2 + y^2 = \frac{25}{4}$. (Note that $x = 0 \Rightarrow y = 0$ and there is no horizontal tangent at the origin.) Putting this in the equation of the lemniscate, we get $x^2 - y^2 = \frac{25}{8}$. Solving these two equations we have $x^2 = \frac{75}{16}$ and $y^2 = \frac{25}{16}$, so the four points are $\left(\pm\frac{5\sqrt{3}}{4}, \pm\frac{5}{4}\right)$.

30. $\dfrac{x^2}{a^2} + \dfrac{y^2}{b^2} = 1 \Rightarrow \dfrac{2x}{a^2} + \dfrac{2yy'}{b^2} = 0 \Rightarrow y' = -\dfrac{b^2x}{a^2y} \Rightarrow$ the equation of the tangent at (x_0, y_0) is $y - y_0 = \dfrac{-b^2x_0}{a^2y_0}(x - x_0)$. Multiplying both sides by $\dfrac{y_0}{b^2}$ gives $\dfrac{y_0y}{b^2} - \dfrac{y_0^2}{b^2} = -\dfrac{x_0x}{a^2} + \dfrac{x_0^2}{a^2}$. Since (x_0, y_0) lies on the ellipse we have $\dfrac{x_0x}{a^2} + \dfrac{y_0y}{b^2} = \dfrac{x_0^2}{a^2} + \dfrac{y_0^2}{b^2} = 1$.

31. $\dfrac{x^2}{a^2} - \dfrac{y^2}{b^2} = 1 \Rightarrow \dfrac{2x}{a^2} - \dfrac{2yy'}{b^2} = 0 \Rightarrow y' = \dfrac{b^2x}{a^2y} \Rightarrow$ the equation of the tangent at (x_0, y_0) is $y - y_0 = \dfrac{b^2x_0}{a^2y_0}(x - x_0)$. Multiplying both sides by $\dfrac{y_0}{b^2}$ gives $\dfrac{y_0y}{b^2} - \dfrac{y_0^2}{b^2} = \dfrac{x_0x}{a^2} - \dfrac{x_0^2}{a^2}$. Since (x_0, y_0) lies on the hyperbola, we have $\dfrac{x_0x}{a^2} - \dfrac{y_0y}{b^2} = \dfrac{x_0^2}{a^2} - \dfrac{y_0^2}{b^2} = 1$.

32. $\sqrt{x} + \sqrt{y} = \sqrt{c} \Rightarrow \dfrac{1}{2\sqrt{x}} + \dfrac{y'}{2\sqrt{y}} = 0 \Rightarrow y' = -\dfrac{\sqrt{y}}{\sqrt{x}} \Rightarrow$ the equation of the tangent line at (x_0, y_0) is $y - y_0 = -\dfrac{\sqrt{y_0}}{\sqrt{x_0}}(x - x_0)$. Now $x = 0 \Rightarrow y = y_0 - \dfrac{\sqrt{y_0}}{\sqrt{x_0}}(-x_0) = y_0 + \sqrt{x_0}\sqrt{y_0}$, so the y-intercept is $y_0 + \sqrt{x_0}\sqrt{y_0}$. Also $y = 0 \Rightarrow -y_0 = -\dfrac{\sqrt{y_0}}{\sqrt{x_0}}(x - x_0)$, so the x-intercept is $x_0 + \sqrt{x_0}\sqrt{y_0}$. The sum of the intercepts is $\left(y_0 + \sqrt{x_0}\sqrt{y_0}\right) + \left(x_0 + \sqrt{x_0}\sqrt{y_0}\right) = x_0 + 2\sqrt{x_0}\sqrt{y_0} + y_0 = \left(\sqrt{x_0} + \sqrt{y_0}\right)^2 = \left(\sqrt{c}\right)^2 = c$.

33. If the circle has radius r, its equation is $x^2 + y^2 = r^2 \Rightarrow 2x + 2yy' = 0 \Rightarrow y' = -\dfrac{x}{y}$, so the slope of the tangent line at $P(x_0, y_0)$ is $-\dfrac{x_0}{y_0}$. The slope of OP is $\dfrac{y_0}{x_0} = \dfrac{-1}{-x_0/y_0}$, so the tangent is perpendicular to OP.

34. $y^q = x^p \Rightarrow qy^{q-1}y' = px^{p-1} \Rightarrow y' = \dfrac{px^{p-1}}{qy^{q-1}} = \dfrac{px^{p-1}y}{qy^q} = \dfrac{px^{p-1}x^{p/q}}{qx^p} = \dfrac{p}{q}x^{(p/q)-1}$

35. $2x^2 + y^2 = 3$ and $x = y^2$ intersect when $2x^2 + x - 3 = (2x+3)(x-1) = 0 \quad \Leftrightarrow \quad x = -\frac{3}{2}$ or 1, but $-\frac{3}{2}$ is extraneous. $2x^2 + y^2 = 3 \quad \Rightarrow \quad 4x + 2yy' = 0 \quad \Rightarrow \quad y' = -2x/y$ and $x = y^2 \quad \Rightarrow \quad 1 = 2yy' \quad \Rightarrow$ $y' = 1/(2y)$. At $(1,1)$ the slopes are $m_1 = -2$ and $m_2 = \frac{1}{2}$, so the curves are orthogonal there. By symmetry they are also orthogonal at $(1,-1)$.

36. $x^2 - y^2 = 5$ and $4x^2 + 9y^2 = 72$ intersect when $4x^2 + 9(x^2 - 5) = 72 \quad \Leftrightarrow \quad 13x^2 = 117 \quad \Leftrightarrow \quad x = \pm 3$, so there are four points of intersection: $(\pm 3, \pm 2)$. $x^2 - y^2 = 5 \quad \Rightarrow \quad 2x - 2yy' = 0 \quad \Rightarrow \quad y' = x/y$ and $4x^2 + 9y^2 = 72 \quad \Rightarrow \quad 8x + 18yy' = 0 \quad \Rightarrow \quad y' = -4x/9y$. At $(3,2)$ the slopes are $m_1 = \frac{3}{2}$ and $m_2 = -\frac{2}{3}$, so the curves are orthogonal there. By symmetry, they are also orthogonal at $(3,-2)$, $(-3,2)$ and $(-3,-2)$.

37. $x^2 + y^2 = r^2$ is a circle with center O and $ax + by = 0$ is a line through O. By Exercise 35, the curves are orthogonal.

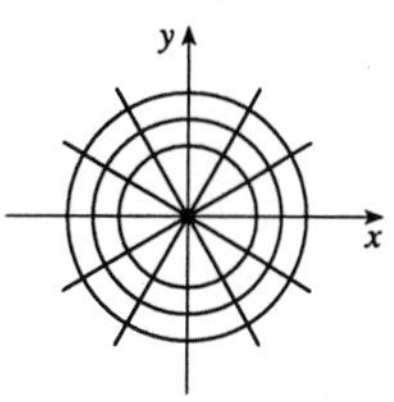

38. The circles $x^2 + y^2 = ax$ and $x^2 + y^2 = by$ intersect at the origin where the tangents are vertical and horizontal. If (x_0, y_0) is the other point of intersection, then $ax_0 = x_0^2 + y_0^2 = by_0$ (★).

Now $x^2 + y^2 = ax \Rightarrow 2x + 2yy' = a \Rightarrow y' = \dfrac{a - 2x}{2y}$ and $x^2 + y^2 = by$

$\Rightarrow \quad 2x + 2yy' = by' \quad \Rightarrow \quad y' = \dfrac{2x}{b - 2y}$. Thus the curves are orthogonal

at $(x_0, y_0) \quad \Leftrightarrow \quad \dfrac{a - 2x_0}{2y_0} = -\dfrac{b - 2y_0}{2x_0} \quad \Leftrightarrow \quad 2ax_0 - 4x_0^2 = 4y_0^2 - 2by_0$

$\Leftrightarrow \quad 2ax_0 + 2by_0 = 4(x_0^2 + y_0^2)$, which is true by (★).

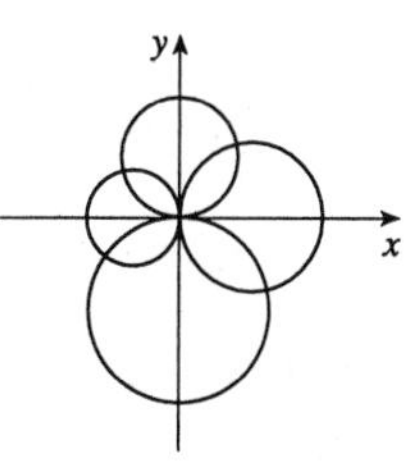

39. $y = cx^2 \quad \Rightarrow \quad y' = 2cx$ and $x^2 + 2y^2 = k$ $\Rightarrow \ 2x + 4yy' = 0 \ \Rightarrow \ y' = -\dfrac{x}{2y} = -\dfrac{x}{2cx^2}$ $= -1/(2cx)$, so the curves are orthogonal.

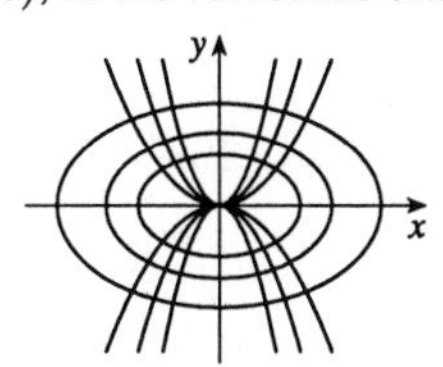

40. $y = ax^3 \ \Rightarrow \ y' = 3ax^2$ and $x^2 + 3y^2 = b \ \Rightarrow$

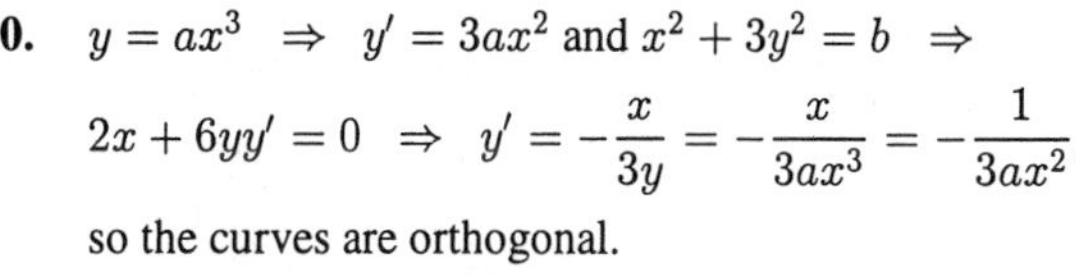

$2x + 6yy' = 0 \ \Rightarrow \ y' = -\dfrac{x}{3y} = -\dfrac{x}{3ax^3} = -\dfrac{1}{3ax^2}$,

so the curves are orthogonal.

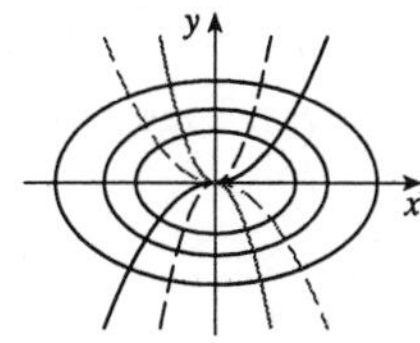

41. $y = 0 \quad \Rightarrow \quad x^2 + x(0) + 0^2 = 3 \quad \Leftrightarrow \quad x = \pm\sqrt{3}$. So the graph of the ellipse crosses the x-axis at the points $(\pm\sqrt{3}, 0)$. Using implicit differentiation to find y', we get $2x - xy' - y + 2yy' = 0 \ \Rightarrow \ y'(2y - x) = y - 2x$ $\Rightarrow \quad y' = \dfrac{y - 2x}{2y - x}$. So $y'(\sqrt{3}, 0) = \dfrac{0 - 2\sqrt{3}}{2(0) - \sqrt{3}} = 2$, and $y'(-\sqrt{3}, 0) = \dfrac{0 + 2\sqrt{3}}{2(0) + \sqrt{3}} = 2 = y'(\sqrt{3}, 0)$. So the tangent lines at these points are parallel.

42. **(a)** We use implicit differentiation to find $y' = \dfrac{y-2x}{2y-x}$ as in Exercise 41. The slope of the tangent line at $(-1,1)$ is $m = \dfrac{1-2(-1)}{2(1)-(-1)} = 1$, so the slope of the normal line is $-1/m = -1$, and its equation is $y - 1 = -(x+1) \quad \Leftrightarrow \quad y = -x$. Substituting this into the equation of the ellipse, we get $x^2 - x(-x) + (-x)^2 = 3 \quad \Rightarrow$ $3x^2 = 3 \quad \Leftrightarrow \quad x = \pm 1$. So the normal line must intersect the ellipse again at $x = 1$, and since the equation of the line is $y = -x$, the other point of intersection must be $(1, -1)$.

(c)

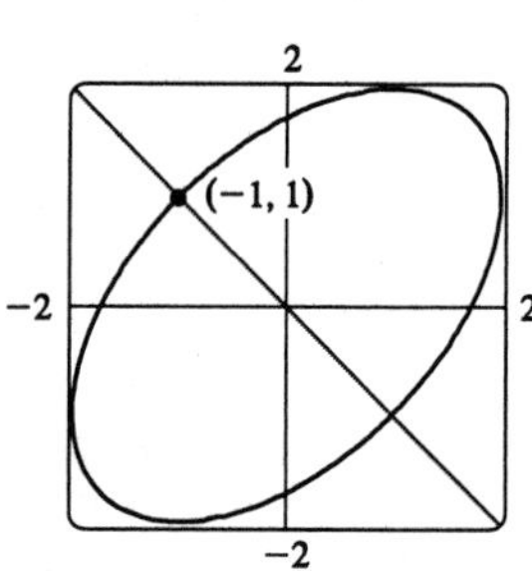

43. $x^2y^2 + xy = 2 \quad \Rightarrow \quad 2xy^2 + 2x^2yy' + y + xy' = 0 \quad \Leftrightarrow \quad y'(2x^2y + x) = -2xy^2 - y \quad \Leftrightarrow$ $y' = -\dfrac{2xy^2 + y}{2x^2y + x}$. So $-\dfrac{2xy^2 + y}{2x^2y + x} = -1 \quad \Leftrightarrow \quad 2xy^2 + y = 2x^2y + x \quad \Leftrightarrow \quad y(2xy+1) = x(2xy+1) \quad \Leftrightarrow$ $(2xy+1)(y-x) = 0 \quad \Leftrightarrow \quad y = x$ or $xy = -\frac{1}{2}$. But $xy = -\frac{1}{2} \quad \Rightarrow \quad x^2y^2 + xy = \frac{1}{4} - \frac{1}{2} \neq 2$ so we must have $x = y$. Then $x^2y^2 + xy = 2 \quad \Rightarrow \quad x^4 + x^2 = 2 \quad \Leftrightarrow \quad x^4 + x^2 - 2 = 0 \quad \Leftrightarrow \quad (x^2+2)(x^2-1) = 0$. So $x^2 = -2$, which is impossible, or $x^2 = 1 \quad \Leftrightarrow \quad x = \pm 1$. So the points on the curve where the tangent line has a slope of -1 are $(-1,-1)$ and $(1,1)$.

44. Using implicit differentiation, $2x + 8yy' = 0$ so $y' = -\dfrac{x}{4y}$. Let (a,b) be a point on $x^2 + 4y^2 = 36$ whose tangent line passes through $(12,3)$. The tangent line is then $y - 3 = -\dfrac{a}{b}(x - 12)$, or $b - 3 = -\dfrac{a}{4b}(a - 12)$. Multiplying both sides by $4b$ gives $4b^2 - 12b = -a^2 + 12a$ so $4b^2 + a^2 = 12(a+b)$. But $4b^2 + a^2 = 36$, so $36 = 12(a+b)$, so $a + b = 3$ so $b = 3 - a$. Substituting into $x^2 + 4y^2 = 36$ gives $a^2 + 4(3-a)^2 = 36$, so $a^2 + 36 - 24a + 4a^2 = 36$. Hence $0 = 5a^2 - 24a = a(5a - 24)$, so $a = 0$ or $a = \frac{24}{5}$. Thus if $a = 0$, $b = 3 - 0 = 3$ while if $a = \frac{24}{5}$, $b = 3 - \frac{24}{5} = -\frac{9}{5}$. So the two points are $(0,3)$ and $\left(\frac{24}{5}, -\frac{9}{5}\right)$. A check shows that both points satisfy the necessary hypothesis.

45. We use implicit differentiation to find y': $2x + 4(2yy') = 0 \quad \Rightarrow \quad y' = -\dfrac{x}{4y}$. Now let h be the height of the lamp, and let (a,b) be the point of tangency of the line passing through the points $(3,h)$ and $(-5,0)$. This line has slope $(h-0)/[3-(-5)] = \frac{1}{8}h$. But the slope of the tangent line through the point (a,b) can be expressed as $y' = -\dfrac{a}{4b}$, or as $\dfrac{b-0}{a-(-5)} = \dfrac{b}{a+5}$ [since the line passes through $(-5,0)$ and (a,b)], so $-\dfrac{a}{4b} = \dfrac{b}{a+5} \quad \Leftrightarrow$ $4b^2 = -a^2 - 5a \quad \Leftrightarrow \quad a^2 + 4b^2 = -5a$. But $a^2 + 4b^2 = 5$, since (a,b) is on the ellipse, so $5 = -5a \quad \Leftrightarrow$ $a = -1$. Then $4b^2 = -1 - 5(-1) = 4 \quad \Rightarrow \quad b = 1$, since the point is on the top half of the ellipse. So $\dfrac{h}{8} = \dfrac{b}{a+5} = \dfrac{1}{-1+5} = \dfrac{1}{4} \quad \Rightarrow \quad h = 2$. So the lamp is located 2 units above the x-axis.

EXERCISES 2.7

1. $a = f, b = f', c = f''$. We can see this because where a has a horizontal tangent, $b = 0$, and where b has a horizontal tangent, $c = 0$. We can immediately see that c can be neither f nor f', since at the points where c has a horizontal tangent, neither a nor b is equal to 0.

2. We can immediately see that a is the graph of the acceleration function, since at the points where a has a horizontal tangent, neither c nor b is equal to 0. Next, we note that $a = 0$ at the point where b has a horizontal tangent, so b must be the graph of the velocity function, so that $b' = a$. We conclude that c is the graph of the position function. This is in accordance with the fact that near the right side of the graph, $b = 0$ and c has a horizontal tangent.

3. $f(x) = x^4 - 3x^3 + 16x \quad \Leftrightarrow \quad f'(x) = 4x^3 - 9x^2 + 16 \quad \Rightarrow \quad f''(x) = 12x^2 - 18x$

4. $f(t) = t^{10} - 2t^7 + t^4 - 6t + 8 \quad \Rightarrow \quad f'(t) = 10t^9 - 14t^6 + 4t^3 - 6 \quad \Rightarrow \quad f''(t) = 90t^8 - 84t^5 + 12t^2$

5. $h(x) = \sqrt{x^2+1} \quad \Rightarrow \quad h'(x) = \frac{1}{2}\left(x^2+1\right)^{-1/2}(2x) = \dfrac{x}{\sqrt{x^2+1}} \quad \Rightarrow$

$$h''(x) = \frac{\sqrt{x^2+1} - x\left(x/\sqrt{x^2+1}\right)}{x^2+1} = \frac{x^2+1-x^2}{\left(x^2+1\right)^{3/2}} = \frac{1}{\left(x^2+1\right)^{3/2}}$$

6. $G(r) = \sqrt{r} + \sqrt[3]{r} \quad \Rightarrow \quad G'(r) = \frac{1}{2}r^{-1/2} + \frac{1}{3}r^{-2/3} \quad \Rightarrow \quad G''(r) = -\frac{1}{4}r^{-3/2} - \frac{2}{9}r^{-5/3}$

7. $F(s) = (3s+5)^8 \quad \Rightarrow \quad F'(s) = 8(3s+5)^7(3) = 24(3s+5)^7 \quad \Rightarrow \quad F''(s) = 168(3s+5)^6(3) = 504(3s+5)^6$

8. $g(u) = 1/\sqrt{1-u} = (1-u)^{-1/2} \quad \Rightarrow \quad g'(u) = -\frac{1}{2}(1-u)^{-3/2}(-1) = \frac{1}{2}(1-u)^{-3/2} \quad \Rightarrow$
$g''(u) = -\frac{3}{4}(1-u)^{-5/2}(-1) = \frac{3}{4}(1-u)^{-5/2}$.

9. $y = \dfrac{x}{1-x} \quad \Rightarrow \quad y' = \dfrac{1(1-x) - x(-1)}{(1-x)^2} = \dfrac{1}{(1-x)^2} \quad \Rightarrow \quad y'' = -2(1-x)^{-3}(-1) = \dfrac{2}{(1-x)^3}$

10. $y = x^\pi \quad \Rightarrow \quad y' = \pi x^{\pi-1} \quad \Rightarrow \quad y'' = \pi(\pi-1)x^{\pi-2}$

11. $y = \left(1-x^2\right)^{3/4} \quad \Rightarrow \quad y' = \frac{3}{4}\left(1-x^2\right)^{-1/4}(-2x) = -\frac{3}{2}x\left(1-x^2\right)^{-1/4} \quad \Rightarrow$
$y'' = -\frac{3}{2}\left(1-x^2\right)^{-1/4} - \frac{3}{2}x\left(-\frac{1}{4}\right)\left(1-x^2\right)^{-5/4}(-2x) = -\frac{3}{2}\left(1-x^2\right)^{-1/4} - \frac{3}{4}x^2\left(1-x^2\right)^{-5/4}$
$= \frac{3}{4}\left(1-x^2\right)^{-5/4}\left(x^2-2\right)$

12. $y = \dfrac{x^2}{x+1} \quad \Rightarrow \quad y' = \dfrac{(x+1)2x - x^2}{(x+1)^2} = \dfrac{x^2+2x}{(x+1)^2} \quad \Rightarrow$

$$y'' = \frac{(x+1)^2(2x+2) - \left(x^2+2x\right)(2)(x+1)}{(x+1)^4} = \frac{2(x+1)\left[(x+1)^2 - \left(x^2+2x\right)\right]}{(x+1)^4} = \frac{2}{(x+1)^3}$$

13. $H(t) = \tan^3(2t-1) \quad \Rightarrow \quad H'(t) = 3\tan^2(2t-1)\sec^2(2t-1)(2) = 6\tan^2(2t-1)\sec^2(2t-1) \quad \Rightarrow$
$H''(t) = 12\tan(2t-1)\sec^2(2t-1)(2)\sec^2(2t-1) + 6\tan^2(2t-1)2\sec(2t-1)\sec(2t-1)\tan(2t-1)(2)$
$= 24\tan(2t-1)\sec^4(2t-1) + 24\tan^3(2t-1)\sec^2(2t-1)$

14. $g(s) = s^2 \cos s \quad \Rightarrow \quad g'(s) = 2s \cos s - s^2 \sin s \quad \Rightarrow$

$g''(s) = 2 \cos s - 2s \sin s - 2s \sin s - s^2 \cos s = (2 - s^2) \cos s - 4s \sin s$

15. **(a)** $f(x) = 2 \cos x + \sin^2 x \quad \Rightarrow \quad f'(x) = 2(-\sin x) + 2 \sin x (\cos x) = \sin 2x - 2 \sin x \quad \Rightarrow$

$f''(x) = 2 \cos 2x - 2 \cos x = 2(\cos 2x - \cos x)$

(b)

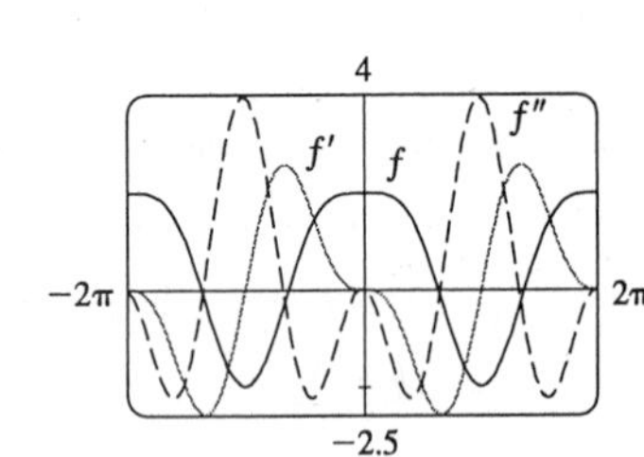

We can see that our answers are plausible, since f has horizontal tangents where $f'(x) = 0$, and f' has horizontal tangents where $f''(x) = 0$.

16. **(a)** $f(x) = \dfrac{x}{x^2 + 1} \quad \Rightarrow \quad f'(x) = \dfrac{(x^2 + 1) - x(2x)}{(x^2 + 1)^2} = \dfrac{1 - x^2}{(x^2 + 1)^2} \quad \Rightarrow$

$$f''(x) = \frac{(x^2 + 1)^2(-2x) - (1 - x^2)(2)(x^2 + 1)(2x)}{(x^2 + 1)^4} = \frac{2x(2x^2 - 2 - x^2 - 1)}{(x^2 + 1)^3} = \frac{2x(x^2 - 3)}{(x^2 + 1)^3}$$

(b)

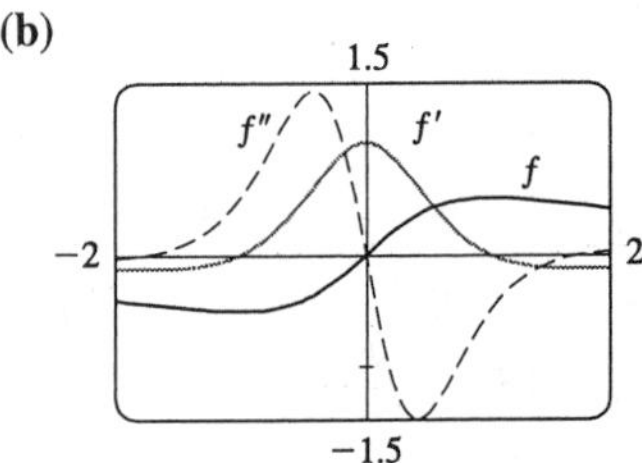

We can see that our answers are plausible, since f has horizontal tangents where $f'(x) = 0$, and f' has horizontal tangents where $f''(x) = 0$.

17. $y = \sqrt{5t - 1} \quad \Rightarrow \quad y' = \frac{1}{2}(5t - 1)^{-1/2}(5) = \frac{5}{2}(5t - 1)^{-1/2} \quad \Rightarrow$

$y'' = -\frac{5}{4}(5t - 1)^{-3/2}(5) = -\frac{25}{4}(5t - 1)^{-3/2} \quad \Rightarrow \quad y''' = \frac{75}{8}(5t - 1)^{-5/2}(5) = \frac{375}{8}(5t - 1)^{-5/2}$

18. $y = \dfrac{1 - x}{1 + x} \quad \Rightarrow \quad y' = \dfrac{(1 + x)(-1) - (1 - x)}{(1 + x)^2} = \dfrac{-2}{(1 + x)^2} = -2(1 + x)^{-2} \quad \Rightarrow \quad y'' = 4(1 + x)^{-3} \quad \Rightarrow$

$y''' = -12(1 + x)^{-4}$

19. $f(x) = (2 - 3x)^{-1/2} \quad \Rightarrow \quad f(0) = 2^{-1/2} = \frac{1}{\sqrt{2}}$

$f'(x) = -\frac{1}{2}(2 - 3x)^{-3/2}(-3) = \frac{3}{2}(2 - 3x)^{-3/2} \quad \Rightarrow \quad f'(0) = \frac{3}{2}(2)^{-3/2} = \frac{3}{4\sqrt{2}}$

$f''(x) = -\frac{9}{4}(2 - 3x)^{-5/2}(-3) = \frac{27}{4}(2 - 3x)^{-5/2} \quad \Rightarrow \quad f''(0) = \frac{27}{4}(2)^{-5/2} = \frac{27}{16\sqrt{2}}$

$f'''(x) = \frac{405}{8}(2 - 3x)^{-7/2} \quad \Rightarrow \quad f'''(0) = \frac{405}{8}(2)^{-7/2} = \frac{405}{64\sqrt{2}}$

20. $g(t) = (2 - t^2)^6 \quad \Rightarrow \quad g(0) = 2^6 = 64$

$g'(t) = 6(2 - t^2)^5(-2t) = -12t(2 - t^2)^5 \quad \Rightarrow \quad g'(0) = 0$

$g''(t) = -12(2 - t^2)^5 + 120t^2(2 - t^2)^4 \quad \Rightarrow \quad g''(0) = -12(2)^5 = -384$

$g'''(t) = 360t(2 - t^2)^4 - 960t^3(2 - t^2)^3 \quad \Rightarrow \quad g'''(0) = 0$

21. $f(\theta) = \cot\theta \quad\Rightarrow\quad f'(\theta) = -\csc^2\theta \quad\Rightarrow\quad f''(\theta) = -2\csc\theta(-\csc\theta\cot\theta) = 2\csc^2\theta\cot\theta \quad\Rightarrow$

$f'''(\theta) = 2(-2\csc^2\theta\cot\theta)\cot\theta + 2\csc^2\theta(-\csc^2\theta) = -2\csc^2\theta(2\cot^2\theta + \csc^2\theta) \quad\Rightarrow$

$f'''\left(\frac{\pi}{6}\right) = -2(2)^2\left[2\left(\sqrt{3}\right)^2 + (2)^2\right] = -80$

22. $g(x) = \sec x \quad\Rightarrow\quad g'(x) = \sec x\tan x \quad\Rightarrow$

$g''(x) = (\sec x\tan x)\tan x + \sec x\sec^2 x = \sec x\,(\sec^2 x - 1 + \sec^2 x) = 2\sec^3 x - \sec x \quad\Rightarrow$

$g'''(x) = 6\sec^2 x(\sec x\tan x) - \sec x\tan x = (6\sec^2 x - 1)\sec x\tan x \quad\Rightarrow$

$g^{(4)}(x) = (12\sec x\sec x\tan x)(\sec x\tan x) + (6\sec^2 x - 1)(2\sec^3 x - \sec x) \quad\Rightarrow$

$g^{(4)}\left(\frac{\pi}{4}\right) = 12\left(\sqrt{2}\right)^3(1)^2 + \left[6\left(\sqrt{2}\right)^2 - 1\right]\left[2\left(\sqrt{2}\right)^3 - \sqrt{2}\right] = 57\sqrt{2}$

23. $x^3 + y^3 = 1 \quad\Rightarrow\quad 3x^2 + 3y^2y' = 0 \quad\Rightarrow\quad y' = -\dfrac{x^2}{y^2} \quad\Rightarrow$

$y'' = -\dfrac{2xy^2 - 2x^2yy'}{y^4} = -\dfrac{2xy^2 - 2x^2y(-x^2/y^2)}{y^4} = -\dfrac{2xy^3 + 2x^4}{y^5} = -\dfrac{2x(y^3 + x^3)}{y^5} = -\dfrac{2x}{y^5}$, since x and y must satisfy the original equation, $x^3 + y^3 = 1$.

24. $\sqrt{x} + \sqrt{y} = 1 \quad\Rightarrow\quad \dfrac{1}{2\sqrt{x}} + \dfrac{y'}{2\sqrt{y}} = 0 \quad\Rightarrow\quad y' = -\sqrt{y}/\sqrt{x} \quad\Rightarrow$

$y'' = -\dfrac{\sqrt{x}\left[1/(2\sqrt{y})\right]y' - \sqrt{y}\left[1/(2\sqrt{x})\right]}{x} = -\dfrac{\sqrt{x}(1/\sqrt{y})(-\sqrt{y}/\sqrt{x}) - \sqrt{y}(1/\sqrt{x})}{2x} = \dfrac{1 + \sqrt{y}/\sqrt{x}}{2x}$

$= \dfrac{\sqrt{x} + \sqrt{y}}{2x\sqrt{x}} = \dfrac{1}{2x\sqrt{x}}$ since x and y must satisfy the original equation, $\sqrt{x} + \sqrt{y} = 1$.

25. $x^2 + 6xy + y^2 = 8 \quad\Rightarrow\quad 2x + 6y + 6xy' + 2yy' = 0 \quad\Rightarrow\quad y' = -\dfrac{x + 3y}{3x + y} \quad\Rightarrow$

$y'' = -\dfrac{(1 + 3y')(3x + y) - (x + 3y)(3 + y')}{(3x + y)^2} = \dfrac{8(y - xy')}{(3x + y)^2} = \dfrac{8\left[y - x(-x - 3y)/(3x + y)\right]}{(3x + y)^2}$

$= \dfrac{8[y(3x + y) + x(x + 3y)]}{(3x + y)^3} = \dfrac{8(x^2 + 6xy + y^2)}{(3x + y)^3} = \dfrac{64}{(3x + y)^3}$, since x and y must satisfy the original equation, $x^2 + 6xy + y^2 = 8$.

26. $\dfrac{x^2}{a^2} - \dfrac{y^2}{b^2} = 1 \quad\Rightarrow\quad \dfrac{2x}{a^2} - \dfrac{2yy'}{b^2} = 0 \quad\Rightarrow\quad y' = \dfrac{b^2x}{a^2y} \quad\Rightarrow\quad y'' = \dfrac{b^2}{a^2}\dfrac{y - xy'}{y^2} = \dfrac{b^2}{a^2}\dfrac{y - x\dfrac{b^2x}{a^2y}}{y^2}$

$= \dfrac{b^2}{a^2}\dfrac{a^2y^2 - b^2x^2}{a^2y^3} = \dfrac{b^4}{a^2y^3}\left(\dfrac{y^2}{b^2} - \dfrac{x^2}{a^2}\right) = -\dfrac{b^4}{a^2y^3}$ since x and y must satisfy the original equation.

27. $f(x) = x - x^2 + x^3 - x^4 + x^5 - x^6 \quad\Rightarrow\quad f'(x) = 1 - 2x + 3x^2 - 4x^3 + 5x^4 - 6x^5 \quad\Rightarrow$

$f''(x) = -2 + 6x - 12x^2 + 20x^3 - 30x^4 \quad\Rightarrow\quad f'''(x) = 6 - 24x + 60x^2 - 120x^3 \quad\Rightarrow$

$f^{(4)}(x) = -24 + 120x - 360x^2 \quad\Rightarrow\quad f^{(5)}(x) = 120 - 720x \quad\Rightarrow\quad f^{(6)}(x) = -720 \quad\Rightarrow$

$f^{(n)}(x) = 0$ for $7 \le n \le 73$.

28. $f(x) = \sqrt{x} = x^{1/2} \quad\Rightarrow\quad f'(x) = \frac{1}{2}x^{-1/2} \quad\Rightarrow\quad f''(x) = \frac{1}{2}\left(-\frac{1}{2}\right)x^{-3/2} \quad\Rightarrow$

$f'''(x) = \frac{1}{2}\left(-\frac{1}{2}\right)\left(-\frac{3}{2}\right)x^{-5/2} \quad\Rightarrow\quad f^{(4)}(x) = \frac{1}{2}\left(-\frac{1}{2}\right)\left(-\frac{3}{2}\right)\left(-\frac{5}{2}\right)x^{-7/2} = -\frac{1\cdot 3\cdot 5}{2^4}x^{-7/2} \quad\Rightarrow$

$f^{(5)}(x) = \frac{1}{2}\left(-\frac{1}{2}\right)\left(-\frac{3}{2}\right)\left(-\frac{5}{2}\right)\left(-\frac{7}{2}\right)x^{-9/2} = \frac{1\cdot 3\cdot 5\cdot 7}{2^5}x^{-9/2} \quad\Rightarrow\quad \cdots \quad\Rightarrow$

$f^{(n)}(x) = \frac{1}{2}\left(-\frac{1}{2}\right)\left(-\frac{3}{2}\right)\cdots\left(\frac{1}{2} - n + 1\right)x^{-(2n-1)/2} = (-1)^{n-1}\frac{1\cdot 3\cdot 5\cdot \cdots \cdot (2n-3)}{2^n}x^{-(2n-1)/2}$

29. $f(x) = x^n \quad\Rightarrow\quad f'(x) = nx^{n-1} \quad\Rightarrow\quad f''(x) = n(n-1)\,x^{n-2} \quad\Rightarrow\quad \cdots \quad\Rightarrow$

$f^{(n)}(x) = n(n-1)(n-2)\cdots 2\cdot 1\,x^{n-n} = n!$

30. $f(x) = (1-x)^{-2} \quad\Rightarrow\quad f'(x) = -2(1-x)^{-3}(-1) = 2(1-x)^{-3} \quad\Rightarrow$

$f''(x) = 2(-3)(1-x)^{-4}(-1) = 2\cdot 3(1-x)^{-4} \quad\Rightarrow\quad f'''(x) = 2\cdot 3(-4)(1-x)^{-5}(-1) = 2\cdot 3\cdot 4(1-x)^{-5}$

$\Rightarrow\quad \cdots \quad\Rightarrow\quad f^{(n)}(x) = 2\cdot 3\cdot 4\cdot \cdots \cdot n(n+1)(1-x)^{-(n+2)} = \dfrac{(n+1)!}{(1-x)^{n+2}}$

31. $f(x) = 1/(3x^3) = \frac{1}{3}x^{-3} \quad\Rightarrow\quad f'(x) = \frac{1}{3}(-3)x^{-4} \quad\Rightarrow\quad f''(x) = \frac{1}{3}(-3)(-4)x^{-5} \quad\Rightarrow$

$f'''(x) = \frac{1}{3}(-3)(-4)(-5)x^{-6} \quad\Rightarrow\quad \cdots \quad\Rightarrow$

$f^{(n)}(x) = \frac{1}{3}(-3)(-4)\cdots[-(n+2)]x^{-(n+3)} = \dfrac{(-1)^n\cdot 3\cdot 4\cdot 5\cdot \cdots \cdot (n+2)}{3x^{n+3}} = \dfrac{(-1)^n(n+2)!}{6x^{n+3}}$

32. $D\sin x = \cos x \quad\Rightarrow\quad D^2\sin x = -\sin x \quad\Rightarrow\quad D^3\sin x = -\cos x \quad\Rightarrow\quad D^4\sin x = \sin x$

The derivatives of $\sin x$ occur in a cycle of four. Since $99 = 4(24) + 3$, we have $D^{99}\sin x = D^3\sin x = -\cos x$.

33. In general, $Df(2x) = 2f'(2x)$, $D^2f(2x) = 4f''(2x)$, $\cdots$, $D^nf(2x) = 2^nf^{(n)}(2x)$. Since $f(x) = \cos x$ and $50 = 4(12) + 2$, we have $f^{(50)}(x) = f^{(2)}(x) = -\cos x$, so $D^{50}\cos 2x = -2^{50}\cos 2x$.

34. Let $f(x) = x\sin x$ and $h(x) = \sin x$, so $f(x) = xh(x)$. Then $f'(x) = h(x) + xh'(x)$,

$f''(x) = h'(x) + h'(x) + xh''(x) = 2h'(x) + xh''(x)$,

$f'''(x) = 2h''(x) + h''(x) + xh'''(x) = 3h''(x) + xh'''(x), \cdots, f^{(n)}(x) = nh^{(n-1)}(x) + xh^{(n)}(x)$. Since $34 = 4(8) + 2$, we have $h^{(34)}(x) = h^{(2)}(x) = D^2\sin x = -\sin x$ and $h^{(35)}(x) = -\cos x$. Thus

$D^{(35)}\,x\sin x = 35h^{(34)}(x) + xh^{(35)}(x) = -35\sin x - x\cos x$.

35. **(a)** $s = t^3 - 3t \quad\Rightarrow\quad v(t) = s'(t) = 3t^2 - 3 \quad\Rightarrow\quad a(t) = v'(t) = 6t$

(b) $a(1) = 6(1) = 6\text{ m/s}^2$

(c) $v(t) = 3t^2 - 3 = 0$ when $t^2 = 1$, that is, $t = 1$ and $a(1) = 6\text{ m/s}^2$.

36. **(a)** $s = t^2 - t + 1 \quad\Rightarrow\quad v(t) = s'(t) = 2t - 1 \quad\Rightarrow\quad a(t) = v'(t) = 2$

(b) $a(1) = 2\text{ m/s}^2$

(c) $v(t) = 2t - 1 = 0$ when $t = \frac{1}{2}$ and $a\left(\frac{1}{2}\right) = 2\text{ m/s}^2$.

37. **(a)** $s = At^2 + Bt + C \quad \Rightarrow \quad v(t) = s'(t) = 2At + B \quad \Rightarrow \quad a(t) = v'(t) = 2A$

(b) $a(1) = 2A\,\text{m/s}^2$

(c) The acceleration at these instants is $2A\,\text{m/s}^2$, since $a(t)$ is constant.

38. **(a)** $s = 2t^3 - 7t^2 + 4t + 1 \quad \Rightarrow \quad v(t) = s'(t) = 6t^2 - 14t + 4 \quad \Rightarrow \quad a(t) = v'(t) = 12t - 14$

(b) $a(1) = 12 - 14 = -2\,\text{m/s}^2$

(c) $v(t) = 2(3t^2 - 7t + 2) = 2(3t - 1)(t - 2) = 0$ when $t = \frac{1}{3}$ or 2 and $a\left(\frac{1}{3}\right) = 12\left(\frac{1}{3}\right) - 14 = -10\,\text{m/s}^2$, $a(2) = 12(2) - 14 = 10\,\text{m/s}^2$.

39. **(a)** $s(t) = t^4 - 4t^3 + 2 \quad \Rightarrow \quad v(t) = s'(t) = 4t^3 - 12t^2 \quad \Rightarrow$
$a(t) = v'(t) = 12t^2 - 24t = 12t(t - 2) = 0$ when $t = 0$ or 2.

(b) $s(0) = 2\,\text{m}$, $v(0) = 0\,\text{m/s}$, $s(2) = -14\,\text{m}$, $v(2) = -16\,\text{m/s}$

40. **(a)** $s(t) = 2t^3 - 9t^2 \quad \Rightarrow \quad v(t) = s'(t) = 6t^2 - 18t \quad \Rightarrow \quad a(t) = v'(t) = 12t - 18 = 0$ when $t = 1.5$.

(b) $s(1.5) = -13.5\,\text{m}$, $v(1.5) = -13.5\,\text{m/s}$

41. **(a)** $y(t) = A\sin\omega t \quad \Rightarrow \quad v(t) = y'(t) = A\omega\cos\omega t \quad \Rightarrow \quad a(t) = v'(t) = -A\omega^2\sin\omega t$

(b) $a(t) = -A\omega^2\sin\omega t = -\omega^2 y(t)$

(c) $|v(t)| = A\omega|\cos\omega t|$ is a maximum when $\cos\omega t = \pm 1 \;\Leftrightarrow\; \sin\omega t = 0 \;\Leftrightarrow\; a(t) = -A\omega^2\sin^2\omega t = 0$.

42. $a(t) = v'(t) = \frac{1}{2}[2gs(t) + c]^{-1/2}[2gs'(t)] = \dfrac{gv(t)}{\sqrt{2gs(t) + c}}$ but $\sqrt{2gs(t) + c} = v(t)$, so $a(t) = \dfrac{gv(t)}{v(t)} = g$, a constant.

43. Let $P(x) = ax^2 + bx + c$. Then $P'(x) = 2ax + b$ and $P''(x) = 2a$.
$P''(2) = 2 \quad \Rightarrow \quad 2a = 2 \quad \Rightarrow \quad a = 1$. $P'(2) = 3 \quad \Rightarrow \quad 4a + b = 4 + b = 3 \quad \Rightarrow \quad b = -1$.
$P(2) = 5 \quad \Rightarrow \quad 2^2 - 2 + c = 5 \quad \Rightarrow \quad c = 3$. So $P(x) = x^2 - x + 3$.

44. Let $Q(x) = ax^3 + bx^2 + cx + d$. Then $Q'(x) = 3ax^2 + 2bx + c$, $Q''(x) = 6ax + 2b$ and $Q'''(x) = 6a$. Thus $Q(1) = a + b + c + d = 1$, $Q'(1) = 3a + 2b + c = 3$, $Q''(1) = 6a + 2b = 6$ and $Q'''(1) = 6a = 12$. Solving these four equations in four unknowns a, b, c and d we get $a = 2$, $b = -3$, $c = 3$ and $d = -1$, so $Q(x) = 2x^3 - 3x^2 + 3x - 1$.

45. $P(x) = c_n x^n + c_{n-1}x^{n-1} + \cdots + c_1 x + c_0 \quad \Rightarrow \quad P'(x) = nc_n x^{n-1} + (n-1)c_{n-1}x^{n-2} + \cdots \quad \Rightarrow$
$P''(x) = n(n-1)c_n x^{n-2} + \cdots \quad \Rightarrow \quad P^{(n)}(x) = n(n-1)(n-2)\cdots(1)c_n x^{n-n} = n!\,c_n$ which is a constant.
Therefore $P^{(m)}(x) = 0$ for $m > n$.

46. $f(x) = |x^2 - x| = \begin{cases} x^2 - x & \text{if } x \le 0 \text{ or } x \ge 1 \\ x - x^2 & \text{if } 0 < x < 1 \end{cases}$ Domain $= \mathbb{R}$

$f'(x) = \begin{cases} 2x - 1 & \text{if } x < 0 \text{ or } x > 1 \\ 1 - 2x & \text{if } 0 < x < 1 \end{cases}$ Domain $= \{x \mid x \ne 0, 1\}$

[Or use Exercise 2.5.75 and the Chain Rule: $f'(x) = \dfrac{x^2 - x}{|x^2 - x|}(2x - 1)$.]

$f''(x) = \begin{cases} 2 & \text{if } x < 0 \text{ or } x > 1 \\ -2 & \text{if } 0 < x < 1 \end{cases}$ Domain $= \{x \mid x \ne 0, 1\}$

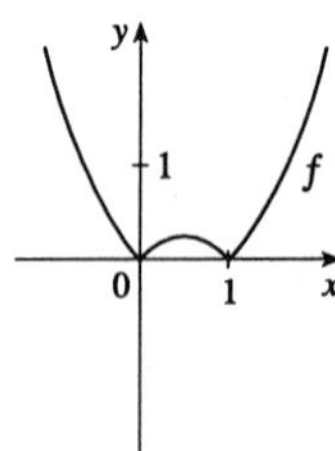

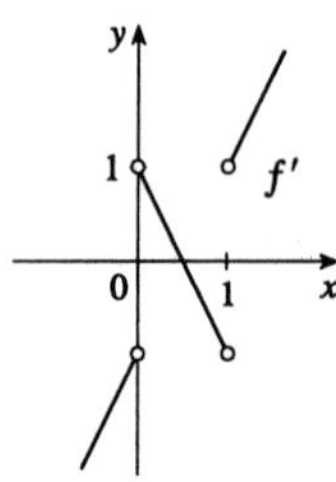

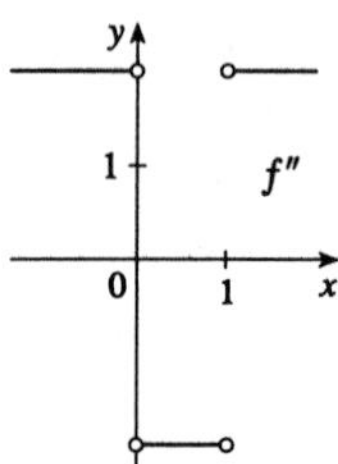

47. $f(x) = xg(x^2) \quad \Rightarrow \quad f'(x) = g(x^2) + xg'(x^2)2x = g(x^2) + 2x^2g'(x^2) \quad \Rightarrow$

$f''(x) = 2xg'(x^2) + 4xg'(x^2) + 4x^3g''(x^2) = 6xg'(x^2) + 4x^3g''(x^2)$

48. $f(x) = \dfrac{g(x)}{x} \quad \Rightarrow \quad f'(x) = \dfrac{xg'(x) - g(x)}{x^2} \quad \Rightarrow$

$$f''(x) = \frac{x^2[g'(x) + xg''(x) - g'(x)] - 2x[xg'(x) - g(x)]}{x^4} = \frac{x^2g''(x) - 2xg'(x) + 2g(x)}{x^3}$$

49. $f(x) = g(\sqrt{x}) \;\Rightarrow\; f'(x) = \dfrac{g'(\sqrt{x})}{2\sqrt{x}} \;\Rightarrow\; f''(x) = \dfrac{\dfrac{g''(\sqrt{x})}{2\sqrt{x}} \cdot 2\sqrt{x} - \dfrac{g'(\sqrt{x})}{\sqrt{x}}}{4x} = \dfrac{\sqrt{x}\,g''(\sqrt{x}) - g'(\sqrt{x})}{4x\sqrt{x}}$

50.

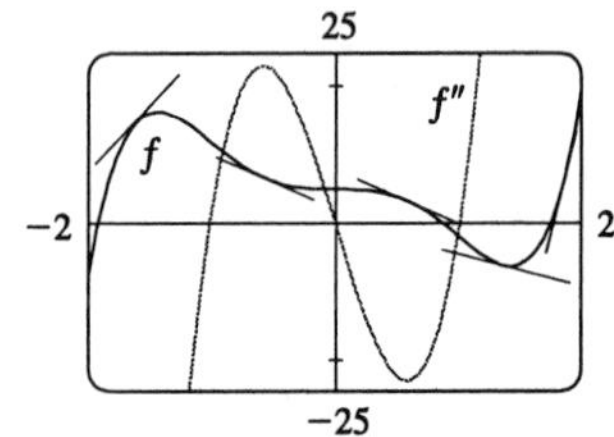

$f(x) = 3x^5 - 10x^3 + 5 \quad \Rightarrow \quad f'(x) = 15x^4 - 30x^2 \quad \Rightarrow$
$f''(x) = 60x^3 - 60x = 60x(x^2 - 1) = 60x(x + 1)(x - 1)$
So $f''(x) > 0$ when $-1 < x < 0$ or $x > 1$, and on these intervals the graph of f lies above its tangent lines; and $f''(x) < 0$ when $x < -1$ or $0 < x < 1$, and on these intervals the graph of f lies below its tangent lines.

51. (a) $f(x) = \dfrac{1}{x^2 + x} \quad \Rightarrow \quad f'(x) = \dfrac{-(2x + 1)}{(x^2 + x)^2} \quad \Rightarrow$

$$f''(x) = \frac{(x^2 + x)^2(-2) + (2x + 1)(2)(x^2 + x)(2x + 1)}{(x^2 + x)^4} = \frac{2(3x^2 + 3x + 1)}{(x^2 + x)^3} \quad \Rightarrow$$

$$f'''(x) = \frac{(x^2 + x)^3(2)(6x + 3) - 2(3x^2 + 3x + 1)(3)(x^2 + x)^2(2x + 1)}{(x^2 + x)^6}$$

$$= \frac{-6(4x^3 + 6x^2 + 4x + 1)}{(x^2 + x)^4} \quad \Rightarrow$$

$$f^{(4)}(x) = \frac{(x^2+x)^4(-6)(12x^2+12x+4)+6(4x^3+6x^2+4x+1)(4)(x^2+x)^3(2x+1)}{(x^2+x)^8}$$
$$= \frac{24(5x^4+10x^3+10x^2+5x+1)}{(x^2+x)^5}$$

$f^{(5)}(x) = ?$

(b) $f(x) = \dfrac{1}{x(x+1)} = \dfrac{1}{x} - \dfrac{1}{x+1} \Rightarrow f'(x) = -x^{-2} + (x+1)^{-2} \Rightarrow f''(x) = 2x^{-3} - 2(x+1)^{-3}$

$\Rightarrow f'''(x) = (-3)(2)x^{-4} + (3)(2)(x+1)^{-4} \Rightarrow \cdots \Rightarrow f^{(n)}(x) = (-1)^n n!\left[x^{-(n+1)} - (x+1)^{-(n+1)}\right]$

52. **(a)** Use the Product Rule repeatedly: $F = fg \Rightarrow F' = f'g + fg' \Rightarrow$

$F'' = (f''g + f'g') + (f'g' + fg'') = f''g + 2f'g' + fg''$.

(b) $F''' = f'''g + f''g' + 2(f''g' + f'g'') + f'g'' + fg''' = f'''g + 3f''g' + 3f'g'' + fg''' \Rightarrow$

$$F^{(4)} = f^{(4)}g + f'''g' + 3(f'''g' + f''g'') + 3(f''g'' + f'g''') + f'g''' + fg^{(4)}$$
$$= f^{(4)}g + 4f'''g' + 6f''g'' + 4f'g''' + fg^{(4)}.$$

(c) By analogy with the Binomial Theorem, we make the guess:

$F^{(n)} = f^{(n)}g + nf^{(n-1)}g' + \binom{n}{2}f^{(n-2)}g'' + \cdots + \binom{n}{k}f^{(n-k)}g^{(k)} + \cdots + nf'g^{(n-1)} + fg^{(n)}$, where

$\binom{n}{k} = \dfrac{n!}{k!(n-k)!} = \dfrac{n(n-1)(n-2)\cdots(n-k+1)}{k!}$.

53. The Chain Rule says that $\dfrac{dy}{dx} = \dfrac{dy}{du}\dfrac{du}{dx}$, so

$$\frac{d^2y}{dx^2} = \frac{d}{dx}\left(\frac{dy}{dx}\right) = \frac{d}{dx}\left(\frac{dy}{du}\frac{du}{dx}\right) = \left[\frac{d}{dx}\left(\frac{dy}{du}\right)\right]\frac{du}{dx} + \frac{dy}{du}\frac{d}{dx}\left(\frac{du}{dx}\right) \quad \text{(Product Rule)}$$
$$= \left[\frac{d}{du}\left(\frac{dy}{du}\right)\frac{du}{dx}\right]\frac{du}{dx} + \frac{dy}{du}\frac{d^2u}{dx^2} = \frac{d^2y}{du^2}\left(\frac{du}{dx}\right)^2 + \frac{dy}{du}\frac{d^2u}{dx^2}.$$

54. From Exercise 53, $\dfrac{d^2y}{dx^2} = \dfrac{d^2y}{du^2}\left(\dfrac{du}{dx}\right)^2 + \dfrac{dy}{du}\dfrac{d^2u}{dx^2} \Rightarrow$

$$\frac{d^3y}{dx^3} = \frac{d}{dx}\frac{d^2y}{dx^2} = \frac{d}{dx}\left[\frac{d^2y}{du^2}\left(\frac{du}{dx}\right)^2\right] + \frac{d}{dx}\left[\frac{dy}{du}\frac{d^2u}{dx^2}\right]$$
$$= \left[\frac{d}{dx}\left(\frac{d^2y}{du^2}\right)\right]\left(\frac{du}{dx}\right)^2 + \left[\frac{d}{dx}\left(\frac{du}{dx}\right)^2\right]\frac{d^2y}{du^2} + \left[\frac{d}{dx}\left(\frac{dy}{du}\right)\right]\frac{d^2u}{dx^2} + \left[\frac{d}{dx}\left(\frac{d^2u}{dx^2}\right)\right]\frac{dy}{du}$$
$$= \left[\frac{d}{du}\left(\frac{d^2y}{du^2}\right)\frac{du}{dx}\right]\left(\frac{du}{dx}\right)^2 + 2\frac{du}{dx}\frac{d^2u}{dx^2}\frac{d^2y}{du^2} + \left[\frac{d}{du}\left(\frac{dy}{du}\right)\frac{du}{dx}\right]\left(\frac{d^2u}{dx^2}\right) + \frac{d^3u}{dx^3}\frac{dy}{du}$$
$$= \frac{d^3y}{du^3}\left(\frac{du}{dx}\right)^3 + 3\frac{du}{dx}\frac{d^2u}{dx^2}\frac{d^2y}{du^2} + \frac{dy}{du}\frac{d^3u}{dx^3}.$$

EXERCISES 2.8

1. $V = x^3 \quad \Rightarrow \quad \dfrac{dV}{dt} = 3x^2\dfrac{dx}{dt}$

2. $A = \pi r^2 \quad \Rightarrow \quad \dfrac{dA}{dt} = 2\pi r\dfrac{dr}{dt}$

3. $xy = 1 \quad \Rightarrow \quad x\dfrac{dy}{dt} + y\dfrac{dx}{dt} = 0$. If $\dfrac{dx}{dt} = 4$ and $x = 2$, then $y = \dfrac{1}{2}$, so $\dfrac{dy}{dt} = -\dfrac{y}{x}\dfrac{dx}{dt} = -\dfrac{1/2}{2}(4) = -1$.

4. $x^2 + 3xy + y^2 = 1 \quad \Rightarrow \quad 2x\dfrac{dx}{dt} + 3y\dfrac{dx}{dt} + 3x\dfrac{dy}{dt} + 2y\dfrac{dy}{dt} = 0 \quad \Rightarrow \quad \dfrac{dx}{dt} = -\dfrac{3x+2y}{2x+3y}\dfrac{dy}{dt}$. When $y = 1$, we have $x^2 + 3x = 0 \quad \Rightarrow \quad x = 0$ or -3. If $\dfrac{dy}{dt} = 2$ and $x = 0$ and $y = 1$, then $\dfrac{dx}{dt} = -\dfrac{3(0)+2(1)}{2(0)+3(1)}(2) = -\dfrac{4}{3}$. If $x = -3$, then $\dfrac{dx}{dt} = -\dfrac{3(-3)+2(1)}{2(-3)+3(1)}(2) = -\dfrac{14}{3}$.

5. If the radius is r and the diameter x, then $V = \frac{4}{3}\pi r^3 = \frac{\pi}{6}x^3 \quad \Rightarrow \quad -1 = \dfrac{dV}{dt} = \dfrac{\pi}{2}x^2\dfrac{dx}{dt} \quad \Rightarrow \quad \dfrac{dx}{dt} = -\dfrac{2}{\pi x^2}$.
When $x = 10$, $\dfrac{dx}{dt} = -\dfrac{2}{\pi(100)} = -\dfrac{1}{50\pi}$. So the rate of decrease is $\dfrac{1}{50\pi}\dfrac{\text{cm}}{\text{min}}$.

6. If the radius is r and the diameter x, then $S = 4\pi r^2 = \pi x^2 \quad \Rightarrow \quad -1 = \dfrac{dS}{dt} = 2\pi x\dfrac{dx}{dt} \quad \Rightarrow \quad \dfrac{dx}{dt} = -\dfrac{1}{2\pi x}$.
When $x = 10$, $\dfrac{dx}{dt} = -\dfrac{1}{20\pi}$. So the rate of decrease is $\dfrac{1}{20\pi}$ cm/min.

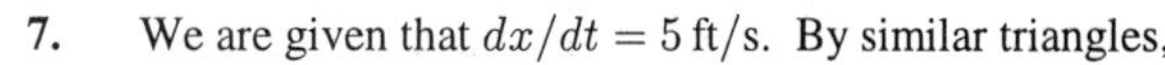

7. We are given that $dx/dt = 5$ ft/s. By similar triangles,
$\dfrac{15}{6} = \dfrac{x+y}{y} \quad \Rightarrow \quad y = \frac{2}{3}x$.

(a) The shadow moves at a rate of
$\dfrac{d}{dt}(x+y) = \dfrac{d}{dt}\left(x + \frac{2}{3}x\right) = \dfrac{5}{3}\dfrac{dx}{dt} = \frac{5}{3}(5) = \frac{25}{3}$ ft/s.

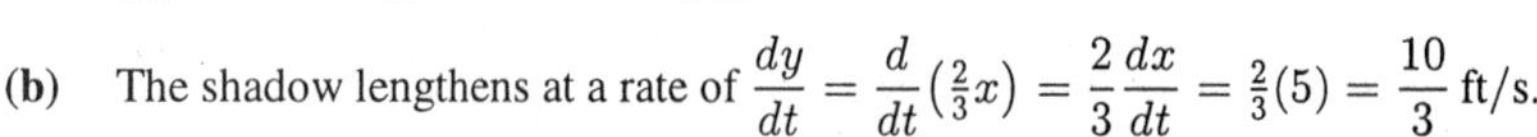

(b) The shadow lengthens at a rate of $\dfrac{dy}{dt} = \dfrac{d}{dt}\left(\frac{2}{3}x\right) = \dfrac{2}{3}\dfrac{dx}{dt} = \frac{2}{3}(5) = \dfrac{10}{3}$ ft/s.

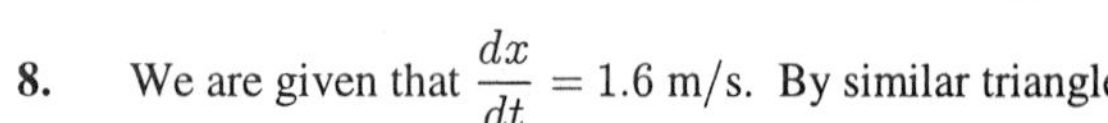

8. We are given that $\dfrac{dx}{dt} = 1.6$ m/s. By similar triangles,
$\dfrac{y}{12} = \dfrac{2}{x} \quad \Rightarrow \quad y = \dfrac{24}{x} \quad \Rightarrow \quad \dfrac{dy}{dt} = -\dfrac{24}{x^2}\dfrac{dx}{dt} = -\dfrac{24}{x^2}(1.6)$.
When $x = 8$, $\dfrac{dy}{dt} = -\dfrac{24(1.6)}{64} = -0.6$ m/s, so the shadow is decreasing at a rate of 0.6 m/s.

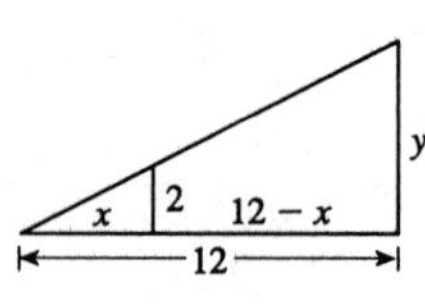

9. We are given that $dx/dt = 500$ mi/h. By the Pythagorean Theorem,
$y^2 = x^2 + 1$, so $2y\dfrac{dy}{dt} = 2x\dfrac{dx}{dt} \quad \Rightarrow \quad \dfrac{dy}{dt} = \dfrac{x}{y}\dfrac{dx}{dt} = 500\dfrac{x}{y}$.
When $y = 2$, $x = \sqrt{3}$, so $\dfrac{dy}{dt} = 500\left(\frac{\sqrt{3}}{2}\right) = 250\sqrt{3}$ mi/h.

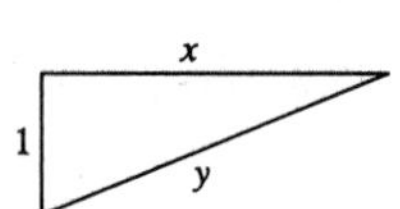

10. We are given that $\dfrac{dx}{dt} = 24\,\text{ft/s}$.

(a) $y^2 = (90 - x)^2 + 90^2 \;\Rightarrow\; 2y\,\dfrac{dy}{dt} = 2(90 - x)\left(-\dfrac{dx}{dt}\right)$.

When $x = 45$, $y = 45\sqrt{5}$, so $\dfrac{dy}{dt} = \dfrac{45}{45\sqrt{5}}(-24) = -\dfrac{24}{\sqrt{5}}$,

so the distance from second base is decreasing at a rate of $\frac{24}{\sqrt{5}} \approx 10.7\,\text{ft/s}$.

(b) $z^2 = x^2 + 90^2 \;\Rightarrow\; 2z\dfrac{dz}{dt} = 2x\dfrac{dx}{dt}$. When $x = 45$, $z = 45\sqrt{5}$, so $\dfrac{dz}{dt} = \dfrac{45}{45\sqrt{5}}(24) = \dfrac{24}{\sqrt{5}} \approx 10.7\,\text{ft/s}$.

11. We are given that $\dfrac{dx}{dt} = 60\text{ mi/h}$ and $\dfrac{dy}{dt} = 25\text{ mi/h}$.

$z^2 = x^2 + y^2 \quad\Rightarrow\quad 2z\dfrac{dz}{dt} = 2x\dfrac{dx}{dt} + 2y\dfrac{dy}{dt}$. After 2 hours, $x = 120$ and $y = 50 \quad\Rightarrow\quad z = 130$, so

$$\frac{dz}{dt} = \frac{1}{z}\left(x\frac{dx}{dt} + y\frac{dy}{dt}\right) = \frac{120(60) + 50(25)}{130} = 65\,\text{mi/h}.$$

12. We are given that $\dfrac{dx}{dt} = 35\text{ km/h}$ and $\dfrac{dy}{dt} = 25\text{ km/h}$.

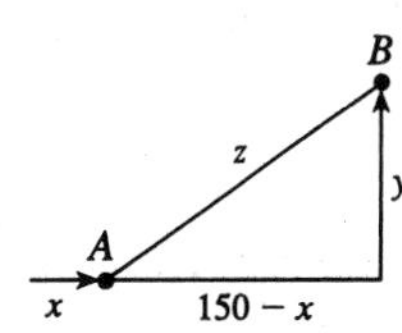

$z^2 = (150 - x)^2 + y^2 \quad\Rightarrow\quad 2z\dfrac{dz}{dt} = 2(150 - x)\left(-\dfrac{dx}{dt}\right) + 2y\dfrac{dy}{dt}$.

At 4:00 P.M., $x = 140$ and $y = 100 \quad\Rightarrow\quad z = \sqrt{10{,}100}$. So

$$\frac{dz}{dt} = \frac{1}{z}\left[(x - 150)\frac{dx}{dt} + y\frac{dy}{dt}\right] = \frac{-10(35) + 100(25)}{\sqrt{10{,}100}} = \frac{215}{\sqrt{101}} \approx 21.4\,\text{km/h}.$$

13. We are given that $\dfrac{dx}{dt} = 35\text{ km/h}$ and $\dfrac{dy}{dt} = 25\text{ km/h}$.

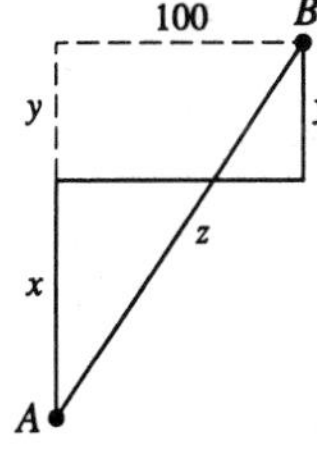

$z^2 = (x + y)^2 + 100^2 \;\Rightarrow\; 2z\dfrac{dz}{dt} = 2(x + y)\left(\dfrac{dx}{dt} + \dfrac{dy}{dt}\right)$.

At 4:00 P.M., $x = 140$ and $y = 100 \quad\Rightarrow\quad z = 260$, so

$$\frac{dz}{dt} = \frac{x + y}{z}\left(\frac{dx}{dt} + \frac{dy}{dt}\right) = \frac{140 + 100}{260}(35 + 25) = \frac{720}{13} \approx 55.4\text{ km/h}.$$

14. We are given that $\dfrac{dx}{dt} = 4\,\text{ft/s}$ and $\dfrac{dy}{dt} = 5\,\text{ft/s}$.

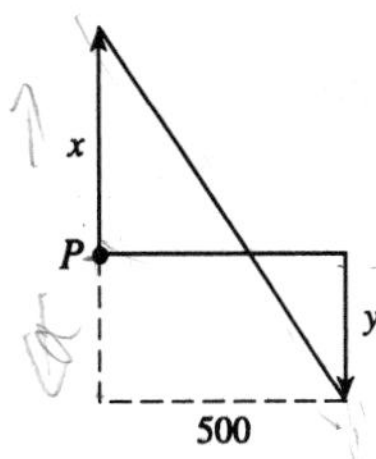

$z^2 = (x + y)^2 + 500^2 \;\Rightarrow\; 2z\dfrac{dz}{dt} = 2(x + y)\left(\dfrac{dx}{dt} + \dfrac{dy}{dt}\right)$.

15 minutes after the woman starts, we have $x = 4 \cdot 20 \cdot 60 = 4800$,

$y = 5 \cdot 15 \cdot 60 = 4500 \quad\Rightarrow\quad z = \sqrt{9300^2 + 500^2}$, so

$$\frac{dz}{dt} = \frac{x + y}{z}\left(\frac{dx}{dt} + \frac{dy}{dt}\right) = \frac{9300}{\sqrt{86{,}740{,}000}}(5 + 4) = \frac{837}{\sqrt{8674}} \approx 8.99\,\text{ft/s}.$$

15. $A = \dfrac{bh}{2}$, where b is the base and h is the altitude. We are given that $\dfrac{dh}{dt} = 1$ and $\dfrac{dA}{dt} = 2$. So $2 = \dfrac{dA}{dt} = \dfrac{b}{2}\dfrac{dh}{dt} + \dfrac{h}{2}\dfrac{db}{dt} = \dfrac{b}{2} + \dfrac{h}{2}\dfrac{db}{dt} \quad\Rightarrow\quad \dfrac{db}{dt} = \dfrac{4-b}{h}$. When $h = 10$ and $A = 100$, we have $b = 20$, so $\dfrac{db}{dt} = \dfrac{4-20}{10} = -1.6$ cm/min.

16. Given $\dfrac{dy}{dt} = -1$, find $\dfrac{dx}{dt}$ when $x = 8$. $y^2 = x^2 + 1 \;\Rightarrow\; 2y\dfrac{dy}{dt} = 2x\dfrac{dx}{dt}$ $\Rightarrow\; \dfrac{dx}{dt} = \dfrac{y}{x}\dfrac{dy}{dt} = -\dfrac{y}{x}$. When $x = 8$, $y = \sqrt{65}$, so $\dfrac{dx}{dt} = -\dfrac{\sqrt{65}}{8}$. Thus the boat approaches the dock at $\frac{\sqrt{65}}{8}$ m/s.

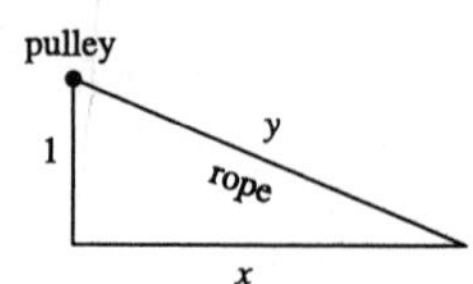

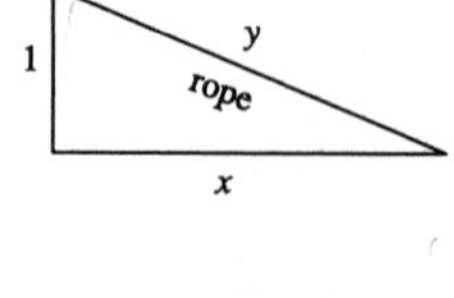

17. If $C =$ the rate at which water is pumped in, then $\dfrac{dV}{dt} = C - 10{,}000$, where $V = \frac{1}{3}\pi r^2 h$ is the volume at time t. By similar triangles, $\dfrac{r}{2} = \dfrac{h}{6}$ $\Rightarrow\; r = \frac{1}{3}h \;\Rightarrow\; V = \frac{1}{3}\pi\left(\frac{1}{3}h\right)^2 h = \frac{\pi}{27}h^3 \;\Rightarrow\; \dfrac{dV}{dt} = \frac{\pi}{9}h^2\dfrac{dh}{dt}$. When $h = 200$, $\dfrac{dh}{dt} = 20$, so $C - 10{,}000 = \frac{\pi}{9}(200)^2(20) \;\Rightarrow$ $C = 10{,}000 + \frac{800{,}000}{9}\pi \approx 2.89 \times 10^5\ \text{cm}^3/\text{min}$.

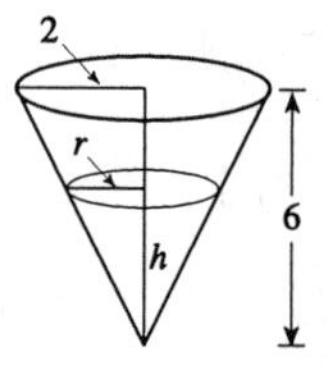

18. By similar triangles, $\dfrac{3}{1} = \dfrac{b}{h}$, so $b = 3h$. The trough has volume $V = \frac{1}{2}bh(10) = 5(3h)h = 15h^2 \;\Rightarrow\; 12 = \dfrac{dV}{dt} = 30h\dfrac{dh}{dt}$ $\Rightarrow\; \dfrac{dh}{dt} = \dfrac{2}{5h}$. When $h = \dfrac{1}{2}$, $\dfrac{dh}{dt} = \dfrac{2}{5\cdot\frac{1}{2}} = \dfrac{4}{5}$ ft/min.

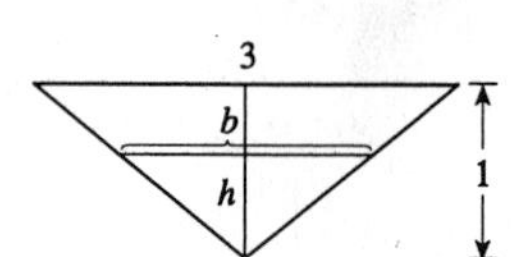

19. $V = \frac{1}{2}[0.3 + (0.3 + 2a)]h(10)$, where $\dfrac{a}{h} = \dfrac{0.25}{0.5} = \dfrac{1}{2}$ so $2a = h \;\Rightarrow\; V = 5(0.6 + h)h = 3h + 5h^2 \;\Rightarrow$ $0.2 = \dfrac{dV}{dt} = (3 + 10h)\dfrac{dh}{dt} \;\Rightarrow\; \dfrac{dh}{dt} = \dfrac{0.2}{3 + 10h}$. When $h = 0.3$, $\dfrac{dh}{dt} = \dfrac{0.2}{3 + 10(0.3)} = \dfrac{0.2}{6}$ m/min $= \dfrac{10}{3}$ cm/min.

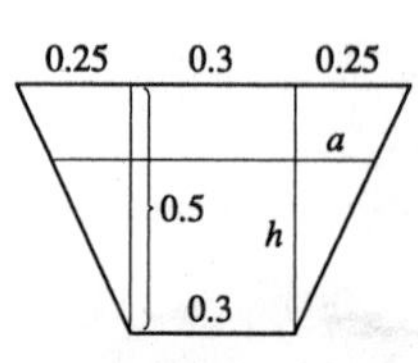

20. $V = \frac{1}{2}(20)(b + 12)\,h = 10(b + 12)h$ and, from similar triangles, $x = h$ and $\dfrac{y}{h} = \dfrac{16}{6} = \dfrac{8}{3}$, so $b = h + 12 + \dfrac{8h}{3} = 12 + \dfrac{11h}{3}$. Thus $V = 10\left(24 + \dfrac{11h}{3}\right)h = 240h + \dfrac{110h^2}{3}$ and so $0.8 = \dfrac{dV}{dt} = \left(240 + \dfrac{220}{3}h\right)\dfrac{dh}{dt}$. When $h = 5$, $\dfrac{dh}{dt} = \dfrac{0.8}{240 + 5(220/3)} = \dfrac{3}{2275} \approx 0.00132$ ft/min.

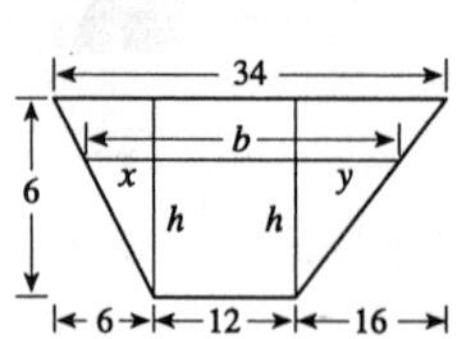

21. We are given that $\frac{dV}{dt} = 30\ \text{ft}^3/\text{min}$. $V = \frac{1}{3}\pi\left(\frac{h}{2}\right)^2 h = \frac{h^3\pi}{12}$

$\Rightarrow \quad 30 = \frac{dV}{dt} = \frac{h^2\pi}{4}\frac{dh}{dt} \quad \Rightarrow \quad \frac{dh}{dt} = \frac{120}{\pi h^2}$.

When $h = 10$ ft, $\frac{dh}{dt} = \frac{120}{10^2\pi} = \frac{6}{5\pi} \approx 0.38$ ft/min.

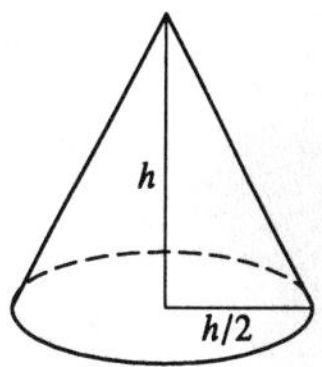

22. We are given $dx/dt = 8$ ft/s. $x = 100\cot\theta \quad \Rightarrow$

$\frac{dx}{dt} = -100\csc^2\theta\,\frac{d\theta}{dt} \quad \Rightarrow \quad \frac{d\theta}{dt} = -\frac{\sin^2\theta}{100}\cdot 8$. When $y = 200$,

$\sin\theta = \frac{100}{200} = \frac{1}{2} \quad \Rightarrow \quad \frac{d\theta}{dt} = -\frac{(1/2)^2}{100}\cdot 8 = -\frac{1}{50}$ rad/s.

The angle is decreasing at a rate of $\frac{1}{50}$ rad/s.

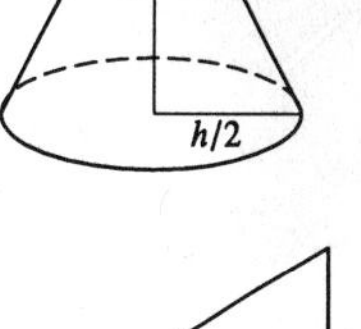

23. $A = \frac{1}{2}bh$, but $b = 5$ m and $h = 4\sin\theta$ so $A = 10\sin\theta$.

We are given $\frac{d\theta}{dt} = 0.06$ rad/s. $\frac{dA}{dt} = 10\cos\theta\,\frac{d\theta}{dt} = 0.6\cos\theta$.

When $\theta = \frac{\pi}{3}$, $\frac{dA}{dt} = 10(0.06)\left(\cos\frac{\pi}{3}\right) = (0.6)\left(\frac{1}{2}\right) = 0.3\ \text{m}^2/\text{s}$.

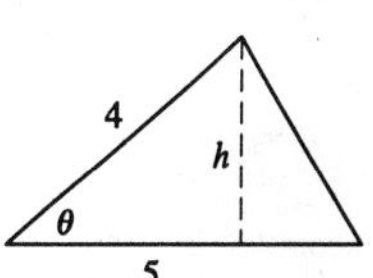

24. We are given $d\theta/dt = 2°/\text{min} = \frac{\pi}{90}$ rad/min. By the Law of Cosines,

$x^2 = 12^2 + 15^2 - 2(12)(15)\cos\theta = 369 - 360\cos\theta \quad \Rightarrow$

$2x\,\frac{dx}{dt} = 360\sin\theta\,\frac{d\theta}{dt}$. When $\theta = 60°$, $x = \sqrt{369 - 360\cos 60°} = \sqrt{189}$,

so $\frac{dx}{dt} = \frac{360\sin 60°}{2\left(3\sqrt{21}\right)}\frac{\pi}{90} = \frac{\pi\sqrt{3}}{3\sqrt{21}} = \frac{\sqrt{7}\pi}{21}$ m/min.

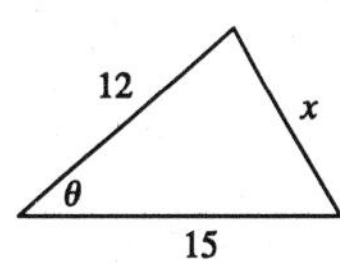

25. $PV = C \quad \Rightarrow \quad P\frac{dV}{dt} + V\frac{dP}{dt} = 0 \quad \Rightarrow \quad \frac{dV}{dt} = -\frac{V}{P}\frac{dP}{dt}$. When $V = 600$, $P = 150$ and $\frac{dP}{dt} = 20$, we have $\frac{dV}{dt} = -\frac{600}{150}(20) = -80$, so the volume is decreasing at a rate of $80\ \text{cm}^3/\text{min}$.

26. $PV^{1.4} = C \quad \Rightarrow \quad V^{1.4}\frac{dP}{dt} + 1.4PV^{0.4}\frac{dV}{dt} = 0 \quad \Rightarrow \quad \frac{dV}{dt} = -\frac{V}{1.4P}\frac{dP}{dt}$. When $V = 400$, $P = 80$ and $\frac{dP}{dt} = -10$, we have $\frac{dV}{dt} = -\frac{400}{1.4(80)}(-10) = \frac{250}{7}$ so the volume is increasing at a rate of $\frac{250}{7} \approx 36\ \text{cm}^3/\text{min}$.

27. **(a)** By the Pythagorean Theorem, $4000^2 + y^2 = \ell^2$. Differentiating with respect to t, we obtain $2y\frac{dy}{dt} = 2\ell\frac{d\ell}{dt}$. We know that $\frac{dy}{dt} = 600$, so when $y = 3000$ and $\ell = 5000$,

$\frac{d\ell}{dt} = \frac{y(dy/dt)}{\ell} = \frac{3000(600)}{5000} = \frac{1800}{5} = 360$ ft/s.

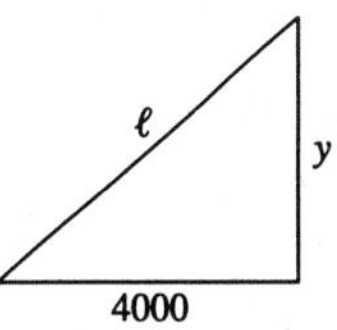

(b) Here $\tan\theta = y/4000$, so $\sec^2\theta\,\frac{d\theta}{dt} = \frac{d}{dt}\tan\theta = \frac{1}{4000}\frac{dy}{dt} \quad \Rightarrow \quad \frac{d\theta}{dt} = \frac{\cos^2\theta}{4000}\frac{dy}{dt}$. When $y = 3000$, $\frac{dy}{dt} = 600$, $z = 5000$ and $\cos\theta = \frac{4000}{z} = \frac{4000}{5000} = \frac{4}{5}$, so $\frac{d\theta}{dt} = \frac{(4/5)^2}{4000}(600) = 0.096$ rad/s.

28. Using the Pythagorean Theorem twice, we have $\sqrt{x^2+12^2}+\sqrt{y^2+12^2}=39$, the total length of the rope. Differentiating with respect to t, we get $\dfrac{x}{\sqrt{x^2+12^2}}\dfrac{dx}{dt}+\dfrac{y}{\sqrt{y^2+12^2}}\dfrac{dy}{dt}=0$, so $\dfrac{dy}{dt}=-\dfrac{x\sqrt{y^2+12^2}}{y\sqrt{x^2+12^2}}\dfrac{dx}{dt}$. Now when $x=5$,

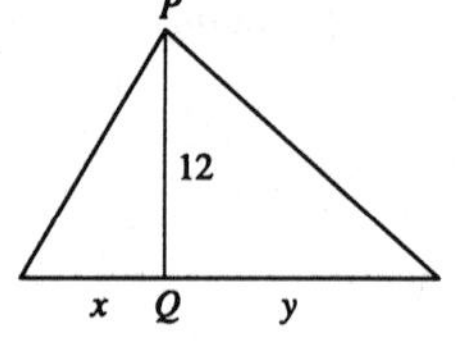

$39=\sqrt{5^2+12^2}+\sqrt{y^2+12^2}=13+\sqrt{y^2+12^2} \quad\Leftrightarrow\quad \sqrt{y^2+12^2}=26$, and

$y=\sqrt{26^2-12^2}=\sqrt{532}=2\sqrt{133}$. So when $x=5$, $\dfrac{dy}{dt}=-\dfrac{5(26)}{2\sqrt{133}(13)}=-\dfrac{10\sqrt{133}}{133}\approx -0.87$ ft/s.

So cart B is moving towards Q at about 0.87 ft/s.

29. We are given that $\dfrac{dx}{dt}=2$ ft/s. $x=10\sin\theta \quad\Rightarrow\quad \dfrac{dx}{dt}=10\cos\theta\,\dfrac{d\theta}{dt}$.

When $\theta=\frac{\pi}{4}$, $\dfrac{d\theta}{dt}=\dfrac{2}{10\left(1/\sqrt{2}\right)}=\dfrac{\sqrt{2}}{5}$ rad/s.

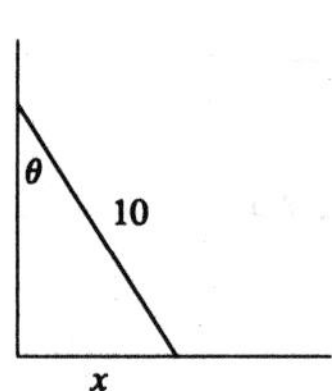

30. We are given that $\dfrac{d\theta}{dt}=4(2\pi)=8\pi$ rad/s.

$x=3\tan\theta \quad\Rightarrow\quad \dfrac{dx}{dt}=3\sec^2\theta\,\dfrac{d\theta}{dt}$.

When $x=1$, $\tan\theta=3$, so $\sec^2\theta=1+\left(\dfrac{1}{3}\right)^2=\dfrac{10}{9}$

and $\dfrac{dx}{dt}=3\left(\dfrac{10}{9}\right)(8\pi)=\dfrac{80\pi}{3}$ km/min (or 1600π km/h).

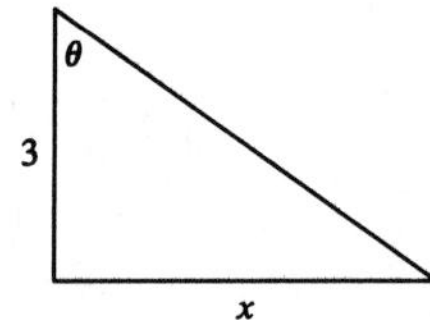

31. We are given that $\dfrac{dx}{dt}=30$ km/h. By the Law of Cosines,

$y^2=x^2+1-2x\cos 120^\circ=x^2+1-2x\left(-\frac{1}{2}\right)=x^2+x+1$,

so $2y\dfrac{dy}{dt}=2x\dfrac{dx}{dt}+\dfrac{dx}{dt} \quad\Rightarrow\quad \dfrac{dy}{dt}=\dfrac{2x+1}{2y}\dfrac{dx}{dt}$. After 1 minute,

$x=\frac{300}{60}=5 \;\Rightarrow\; y=\sqrt{31} \;\Rightarrow$

$\dfrac{dy}{dt}=\dfrac{2(5)+1}{2\sqrt{31}}(300)=\dfrac{1650}{\sqrt{31}}\approx 296$ km/h.

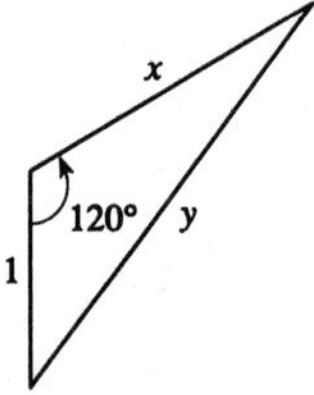

32. We are given that $\dfrac{dx}{dt}=3$ mi/h and $\dfrac{dy}{dt}=2$ mi/h.

By the Law of Cosines, $z^2=x^2+y^2-2xy\cos 45^\circ=x^2+y^2-\sqrt{2}xy$

$\Rightarrow\quad 2z\dfrac{dz}{dt}=2x\dfrac{dx}{dt}+2y\dfrac{dy}{dt}-\sqrt{2}x\dfrac{dy}{dt}-\sqrt{2}y\dfrac{dx}{dt}$.

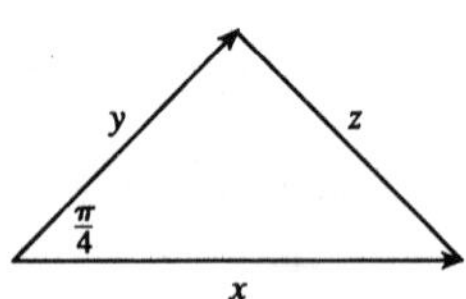

After 15 minutes, we have $x = \frac{3}{4}$ and $y = \frac{1}{2}$ $\Rightarrow$ $z = \dfrac{\sqrt{13 - 6\sqrt{2}}}{4}$ and

$$\frac{dz}{dt} = \frac{2}{\sqrt{13 - 6\sqrt{2}}}\left[2\left(\tfrac{3}{4}\right)3 + 2\left(\tfrac{1}{2}\right)2 - \sqrt{2}\left(\tfrac{3}{4}\right)2 - \sqrt{2}\left(\tfrac{1}{2}\right)\right] = \sqrt{13 - 6\sqrt{2}} \approx 2.125 \text{ mi/h}.$$

33. Let the distance between the runner and the friend be ℓ.
Then by the Law of Cosines,
$\ell^2 = 200^2 + 100^2 - 2 \cdot 200 \cdot 100 \cdot \cos\theta = 50{,}000 - 40{,}000 \cos\theta$ (★).
Differentiating implicitly with respect to t, we obtain
$2\ell \dfrac{d\ell}{dt} = -40{,}000(-\sin\theta)\dfrac{d\theta}{dt}$. Now if D is the distance run when the angle is θ radians, then $D = 100\theta$, so $\theta = \frac{1}{100}D$ $\Rightarrow$
$\dfrac{d\theta}{dt} = \dfrac{1}{100}\dfrac{dD}{dt} = \dfrac{7}{100}$. To substitute into the expression for $\dfrac{d\ell}{dt}$, we must

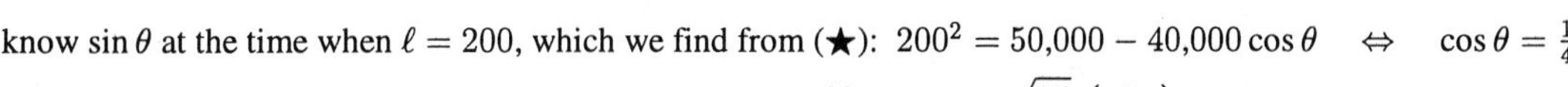
know $\sin\theta$ at the time when $\ell = 200$, which we find from (★): $200^2 = 50{,}000 - 40{,}000 \cos\theta$ $\Leftrightarrow$ $\cos\theta = \frac{1}{4}$

$\Rightarrow$ $\sin\theta = \sqrt{1 - \left(\frac{1}{4}\right)^2} = \frac{\sqrt{15}}{4}$. Substituting, we get $2\ell\dfrac{d\ell}{dt} = 40{,}000\dfrac{\sqrt{15}}{4}\left(\dfrac{7}{100}\right)$ $\Rightarrow$

$\dfrac{d\ell}{dt} = \dfrac{700\sqrt{15}}{2 \cdot 200} = \dfrac{7\sqrt{15}}{4} \approx 6.78$ m/s. Whether the distance between them is increasing or decreasing depends on the direction in which the runner is running.

34. The hour hand of a clock goes around once every 12 hours or, in radians per hour, $\frac{2\pi}{12} = \frac{\pi}{6}$ rad/h.
The minute hand goes around once an hour, or at the rate of 2π rad/h.
So the angle θ between them (measuring clockwise from the minute hand to the hour hand) is changing at the rate of $\dfrac{d\theta}{dt} = \dfrac{\pi}{6} - 2\pi = -\dfrac{11\pi}{6}$ rad/h.

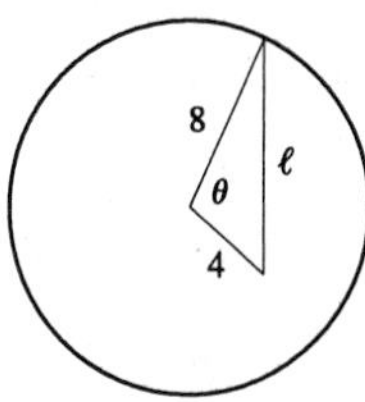

Now, to relate θ to ℓ, we use the Law of Cosines: $\ell^2 = 4^2 + 8^2 - 2 \cdot 4 \cdot 8 \cdot \cos\theta = 80 - 64\cos\theta$ (★).
Differentiating implicitly with respect to t, we get $2\ell\dfrac{d\ell}{dt} = -64(-\sin\theta)\dfrac{d\theta}{dt}$. At 1:00, the angle between the two hands is one-twelfth of the circle, that is, $\frac{2\pi}{12} = \frac{\pi}{6}$ radians. We use (★) to find ℓ at 1:00:
$\ell = \sqrt{80 - 64\cos\frac{\pi}{6}} = \sqrt{80 - 32\sqrt{3}}$. Substituting, we get $2\ell\dfrac{d\ell}{dt} = 64\sin\frac{\pi}{6}\left(-\frac{11\pi}{6}\right)$ $\Rightarrow$

$$\frac{d\ell}{dt} \approx -\frac{64\left(\frac{1}{2}\right)\left(-\frac{11\pi}{6}\right)}{2\sqrt{80 - 32\sqrt{3}}} = -\frac{88\pi}{3\sqrt{80 - 32\sqrt{3}}} \approx -18.6.$$

So the distance between the tips of the hands is decreasing at a rate of 18.6 mm/h at 1:00.

EXERCISES 2.9

1. $y = x^5 \Rightarrow dy = 5x^4 dx$

2. $y = \sqrt[4]{x} = x^{1/4} \Rightarrow dy = \frac{1}{4}x^{-3/4}dx$

3. $y = \sqrt{x^4 + x^2 + 1} \Rightarrow dy = \frac{1}{2}(x^4 + x^2 + 1)^{-1/2}(4x^3 + 2x)dx = \dfrac{2x^3 + x}{\sqrt{x^4 + x^2 + 1}}dx$

4. $y = \dfrac{x-2}{2x+3} \Rightarrow dy = \dfrac{(2x+3) - (x-2)(2)}{(2x+3)^2}dx = \dfrac{7}{(2x+3)^2}dx$

5. $y = \sin 2x \Rightarrow dy = 2\cos 2x\, dx$

6. $y = x\tan x \Rightarrow dy = (\tan x + x\sec^2 x)dx$

7. **(a)** $y = 1 - x^2 \Rightarrow dy = -2x\, dx$

(b) When $x = 5$ and $dx = \frac{1}{2}$, $dy = -2(5)\left(\frac{1}{2}\right) = -5$.

8. **(a)** $y = x^4 - 3x^3 + x - 1 \Rightarrow dy = (4x^3 - 9x^2 + 1)dx$

(b) When $x = 2$ and $dx = 0.1$, $dy = \left[4(2)^3 - 9(2)^2 + 1\right](0.1) = -0.3$.

9. **(a)** $y = (x^2 + 5)^3 \Rightarrow dy = 3(x^2 + 5)^2\, 2x\, dx = 6x(x^2 + 5)^2\, dx$

(b) When $x = 1$ and $dx = 0.05$, $dy = 6(1)(1^2 + 5)^2(0.05) = 10.8$.

10. **(a)** $y = \sqrt{1 - x} \Rightarrow dy = \frac{1}{2}(1 - x)^{-1/2}(-1)dx = -\dfrac{1}{2\sqrt{1-x}}\,dx$

(b) When $x = 0$ and $dx = 0.02$, $dy = -\frac{1}{2}(0.02) = -0.01$.

11. **(a)** $y = \cos x \Rightarrow dy = -\sin x\, dx$

(b) When $x = \frac{\pi}{6}$ and $dx = 0.05$, $dy = -\frac{1}{2}(0.05) = -0.025$.

12. **(a)** $y = \sin x \Rightarrow dy = \cos x\, dx$

(b) When $x = \frac{\pi}{6}$ and $dx = -0.1$, $dy = \frac{\sqrt{3}}{2}(-0.1) = -\frac{\sqrt{3}}{20}$.

13. $y = x^2, x = 1, \Delta x = 0.5 \Rightarrow$

$\Delta y = (1.5)^2 - 1^2 = 1.25.$

$dy = 2x\, dx = 2(1)(0.5) = 1$

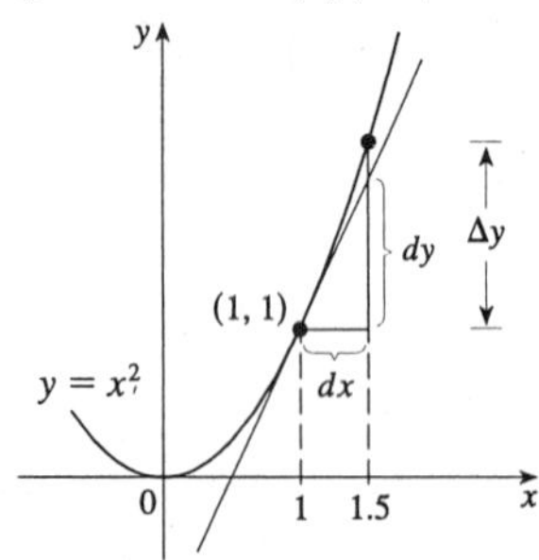

14. $y = \sqrt{x}, x = 1, \Delta x = 1 \Rightarrow$

$\Delta y = \sqrt{2} - \sqrt{1} = \sqrt{2} - 1$
≈ 0.414

$dy = \dfrac{1}{2\sqrt{x}}dx = \frac{1}{2}(1) = 0.5$

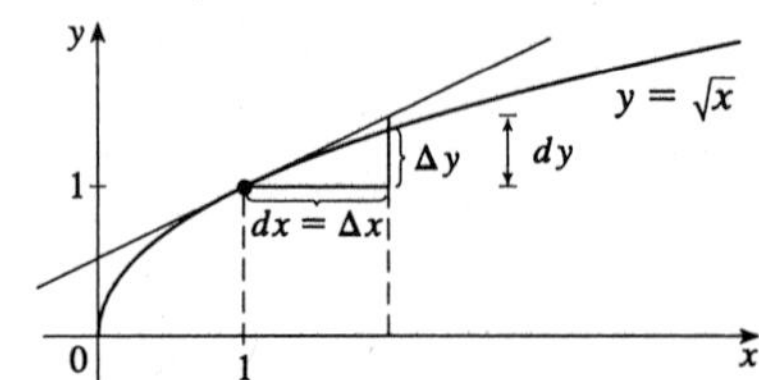

15. $y = 6 - x^2,\ x = -2,\ \Delta x = 0.4 \quad \Rightarrow$
$\Delta y = \left(6 - (-1.6)^2\right) - \left(6 - (-2)^2\right) = 1.44$
$dy = -2x\,dx = -2(-2)(0.4) = 1.6$

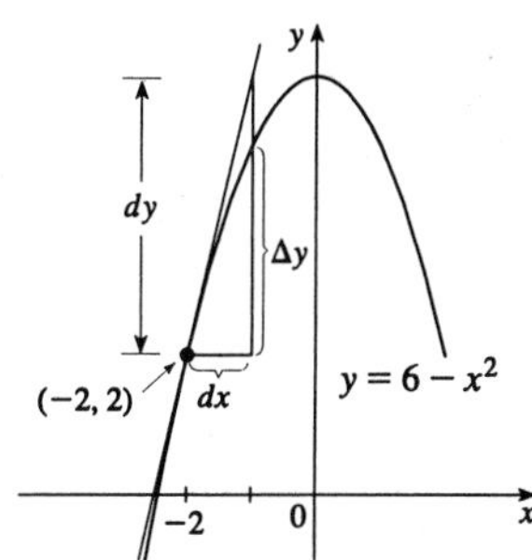

16. $y = \dfrac{16}{x},\ x = 4,\ \Delta x = -1 \quad \Rightarrow$
$\Delta y = \dfrac{16}{3} - \dfrac{16}{4} = \dfrac{4}{3}.$
$dy = -\left(\dfrac{16}{x^2}\right)dx = -\left(\dfrac{16}{4^2}\right)(-1) = 1$

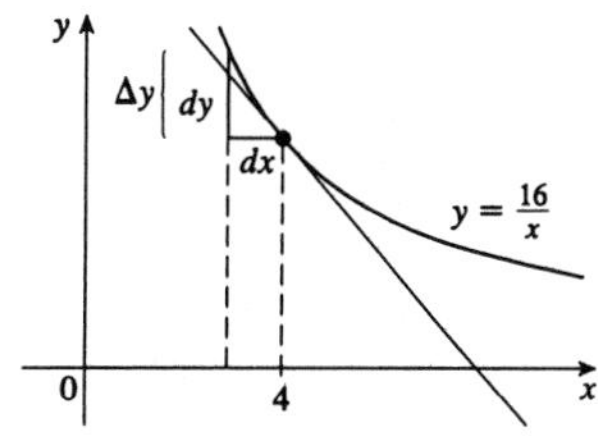

17. $y = f(x) = 2x^3 + 3x - 4,\ x = 3 \quad \Rightarrow \quad dy = (6x^2 + 3)dx = 57\,dx$

$\Delta x = 1 \quad \Rightarrow \quad \Delta y = f(4) - f(3) = 136 - 59 = 77,\ dy = 57(1) = 57,\ \Delta y - dy = 77 - 57 = 20$

$\Delta x = 0.5 \quad \Rightarrow \quad \Delta y = f(3.5) - f(3) = 92.25 - 59 = 33.25,\ dy = 57(0.5) = 28.5,$

$\Delta y - dy = 33.25 - 28.5 = 4.75$

$\Delta x = 0.1 \quad \Rightarrow \quad \Delta y = f(3.1) - f(3) = 64.882 - 59 = 5.882,\ dy = 57(0.1) = 5.7,$

$\Delta y - dy = 5.882 - 5.7 = 0.182$

$\Delta x = 0.01 \quad \Rightarrow \quad \Delta y = f(3.01) - f(3) = 59.571802 - 59 = 0.571802,\ dy = 57(0.01) = 0.57,$

$\Delta y - dy = 0.571802 - 0.57 = 0.001802$

18. $y = f(x) = x^4 + x^2 + 1,\ x = 1 \quad \Rightarrow \quad dy = (4x^3 + 2x)dx = 6\,dx$

$\Delta x = 1 \quad \Rightarrow \quad \Delta y = f(2) - f(1) = 21 - 3 = 18,\ dy = 6(1) = 6,\ \Delta y - dy = 18 - 6 = 12$

$\Delta x = 0.5 \quad \Rightarrow \quad \Delta y = f(1.5) - f(1) = 8.3125 - 3 = 5.3125,\ dy = 6(0.5) = 3,$

$\Delta y - dy = 5.3125 - 3 = 2.3125$

$\Delta x = 0.1 \quad \Rightarrow \quad \Delta y = f(1.1) - f(1) = 3.6741 - 3 = 0.6741,\ dy = 6(0.1) = 0.6,$

$\Delta y - dy = 0.6741 - 0.6 = 0.0741$

$\Delta x = 0.01 \quad \Rightarrow \quad \Delta y = f(1.01) - f(1) = 3.06070401 - 3 = 0.06070401,\ dy = 6(0.01) = 0.06,$

$\Delta y - dy = 0.06070401 - 0.06 = 0.00070401$

19. $y = f(x) = \sqrt{x} \quad \Rightarrow \quad dy = \dfrac{1}{2\sqrt{x}}\,dx$. When $x = 36$ and $dx = 0.1$, $dy = \frac{1}{2\sqrt{36}}(0.1) = \frac{1}{120}$, so

$\sqrt{36.1} = f(36.1) \approx f(36) + dy = \sqrt{36} + \frac{1}{120} \approx 6.0083.$

20. $y = f(x) = \sqrt[3]{x} + \sqrt[4]{x} \quad \Rightarrow \quad dy = \left(\frac{1}{3}x^{-2/3} + \frac{1}{4}x^{-3/4}\right)dx$. If $x = 1$ and $dx = 0.02$, then
$dy = \left(\frac{1}{3} + \frac{1}{4}\right)(0.02) = \frac{7}{12}(0.02)$. Thus $\sqrt[3]{1.02} + \sqrt[4]{1.02} = f(1.02) \approx f(1) + dy = 2 + \frac{7}{12}(0.02) \approx 2.0117.$

21. $y = f(x) = 1/x \quad \Rightarrow \quad dy = (-1/x^2)dx$. When $x = 10$ and $dx = 0.1$, $dy = \left(-\frac{1}{100}\right)(0.1) = -0.001$, so $\frac{1}{10.1} = f(10.1) \approx f(10) + dy = 0.1 - 0.001 = 0.099$.

22. $y = f(x) = x^6 \quad \Rightarrow \quad dy = 6x^5\,dx$. When $x = 2$ and $dx = -0.03$, $dy = 6(2)^5(-0.03) = -5.76$, so $(1.97)^6 = f(1.97) \approx f(2) + dy = 64 - 5.76 = 58.24$.

23. $y = f(x) = \sin x \quad \Rightarrow \quad dy = \cos x\,dx$. When $x = \frac{\pi}{3}$ and $dx = -\frac{\pi}{180}$, $dy = \cos\frac{\pi}{3}\left(-\frac{\pi}{180}\right) = -\frac{\pi}{360}$, so $\sin 59° = f\left(\frac{59}{180}\pi\right) \approx f\left(\frac{\pi}{3}\right) + dy = \frac{\sqrt{3}}{2} - \frac{\pi}{360} \approx 0.857$.

24. $y = f(x) = \cos x \quad \Rightarrow \quad dy = -\sin x\,dx$. When $x = \frac{\pi}{6}$ and $dx = \frac{1.5}{180}\pi$, $dy = -\sin\frac{\pi}{6}\left(\frac{1.5}{180}\pi\right) = -\frac{1}{2}\left(\frac{\pi}{120}\right) = -\frac{\pi}{240}$, so $\cos 31.5° = f\left(\frac{31.5}{180}\pi\right) \approx f\left(\frac{\pi}{6}\right) + dy = \frac{\sqrt{3}}{2} - \frac{\pi}{240} \approx 0.853$.

25. **(a)** If x is the edge length, then $V = x^3 \quad \Rightarrow \quad dV = 3x^2\,dx$. When $x = 30$ and $dx = 0.1$, $dV = 3(30)^2(0.1) = 270$, so the maximum error is about 270 cm^3.

(b) $S = 6x^2 \quad \Rightarrow \quad dS = 12x\,dx$. When $x = 30$ and $dx = 0.1$, $dS = 12(30)(0.1) = 36$, so the maximum error is about 36 cm^2.

26. **(a)** $A = \pi r^2 \quad \Rightarrow \quad dA = 2\pi r\,dr$. When $r = 24$ and $dr = 0.2$, $dA = 2\pi(24)(0.2) = 9.6\pi$, so the maximum error is about $9.6\pi \approx 30\ \text{cm}^2$.

(b) Relative error $= \dfrac{\Delta A}{A} \approx \dfrac{dA}{A} = \dfrac{9.6\pi}{\pi(24)^2} = \dfrac{1}{60} \approx 0.0167$

27. **(a)** For a sphere of radius r, the circumference is $C = 2\pi r$ and the surface area is $S = 4\pi r^2$, so $r = C/(2\pi)$ $\Rightarrow \quad S = 4\pi(C/2\pi)^2 = C^2/\pi \quad \Rightarrow \quad dS = (2/\pi)C\,dC$. When $C = 84$ and $dC = 0.5$, $dS = \frac{2}{\pi}(84)(0.5) = \frac{84}{\pi}$, so the maximum error is about $\frac{84}{\pi} \approx 27\ \text{cm}^2$.

(b) Relative error $\approx \dfrac{dS}{S} = \dfrac{84/\pi}{84^2/\pi} = \dfrac{1}{84} \approx 0.012$

28. **(a)** $V = \frac{4}{3}\pi r^3 = \frac{4}{3}\pi\left(\dfrac{C}{2\pi}\right)^3 = \dfrac{C^3}{6\pi^2} \quad \Rightarrow \quad dV = \dfrac{1}{2\pi^2}C^2\,dC$. When $C = 84$ and $dC = 0.5$, $dV = \left[1/(2\pi^2)\right](84)^2(0.5) = 1764/\pi^2$, so the maximum error is about $1764/\pi^2 \approx 179\ \text{cm}^3$.

(b) Relative error $\approx \dfrac{dV}{V} = \dfrac{1764/\pi^2}{(84)^3/(6\pi^2)} = \dfrac{1}{56} \approx 0.018$

29. **(a)** $V = \pi r^2 h \quad \Rightarrow \quad \Delta V \approx dV = 2\pi rh\,dr = 2\pi rh\,\Delta r$

(b) $\Delta V = \pi(r + \Delta r)^2 h - \pi r^2 h$, so the error is $\Delta V - dv = \pi(r + \Delta r)^2 h - \pi r^2 h - 2\pi rh\,\Delta r = \pi(\Delta r)^2 h$

30. $V = \frac{2}{3}\pi r^3 \quad \Rightarrow \quad dV = 2\pi r^2\,dr$. When $r = 25$ and $dr = 0.0005$, $dV = 2\pi(25)^2(0.0005) = \frac{5\pi}{8}$, so the amount of paint is about $\frac{5\pi}{8} \approx 2\ \text{m}^3$.

31. $L(x) = f(1) + f'(1)(x - 1)$. $f(x) = x^3 \quad \Rightarrow \quad f'(x) = 3x^2$ so $f(1) = 1$ and $f'(1) = 3$. So $L(x) = 1 + 3(x - 1) = 3x - 2$.

32. $f(x) = 1/\sqrt{2+x} = (2+x)^{-1/2} \quad \Rightarrow \quad f'(x) = -\frac{1}{2}(2+x)^{-3/2}$ so $f(0) = \frac{1}{\sqrt{2}}$ and $f'(0) = -1/(4\sqrt{2})$. So $L(x) = f(0) + f'(0)(x-0) = \frac{1}{\sqrt{2}} - \frac{1}{4\sqrt{2}}(x-0) = \frac{1}{\sqrt{2}}\left(1 - \frac{1}{4}x\right)$.

33. $f(x) = 1/x \quad \Rightarrow \quad f'(x) = -1/x^2$. So $f(4) = \frac{1}{4}$ and $f'(4) = -\frac{1}{16}$.
So $L(x) = f(4) + f'(4)(x-4) = \frac{1}{4} + \left(-\frac{1}{16}\right)(x-4) = \frac{1}{2} - \frac{1}{16}x$.

34. $f(x) = \sqrt[3]{x} = x^{1/3} \quad \Rightarrow \quad f'(x) = \frac{1}{3}x^{-2/3}$ so $f(-8) = -2$ and $f'(-8) = \frac{1}{12}$.
So $L(x) = -2 + \frac{1}{12}(x+8) = \frac{1}{12}x - \frac{4}{3}$.

35. $f(x) = \sqrt{1+x} \quad \Rightarrow \quad f'(x) = \dfrac{1}{2\sqrt{1+x}}$ so $f(0) = 1$ and $f'(0) = \frac{1}{2}$.

So $f(x) \approx f(0) + f'(0)(x-0) = 1 + \frac{1}{2}(x-0) = 1 + \frac{1}{2}x$.

36. $f(x) = \sin x \quad \Rightarrow \quad f'(x) = \cos x$ so $f(0) = 0$ and $f'(0) = 1$.
So $f(x) \approx f(0) + f'(0)(x-0) = 0 + 1(x-0) = x$.

37. $f(x) = \dfrac{1}{(1+2x)^4} \quad \Rightarrow \quad f'(x) = \dfrac{-8}{(1+2x)^5}$ so $f(0) = 1$ and $f'(0) = -8$.

So $f(x) \approx f(0) + f'(0)(x-0) = 1 + (-8)(x-0) = 1 - 8x$.

38. $f(x) = \dfrac{1}{\sqrt{4-x}} \quad \Rightarrow \quad f'(x) = \dfrac{1}{2(4-x)^{3/2}}$ so $f(0) = \frac{1}{2}$ and $f'(0) = \frac{1}{16}$.

So $f(x) \approx \frac{1}{2} + \frac{1}{16}(x-0) = \frac{1}{2} + \frac{1}{16}x$.

39. $f(x) = \sqrt{1-x} \quad \Rightarrow \quad f'(x) = \dfrac{-1}{2\sqrt{1-x}}$ so $f(0) = 1$ and $f'(0) = -\frac{1}{2}$. Therefore

$\sqrt{1-x} = f(x) \approx f(0) + f'(0)(x-0) = 1 + \left(-\frac{1}{2}\right)(x-0) = 1 - \frac{1}{2}x$.
So $\sqrt{0.9} = \sqrt{1-0.1} \approx 1 - \frac{1}{2}(0.1) = 0.95$ and $\sqrt{0.99} = \sqrt{1-0.01} \approx 1 - \frac{1}{2}(0.01) = 0.995$.

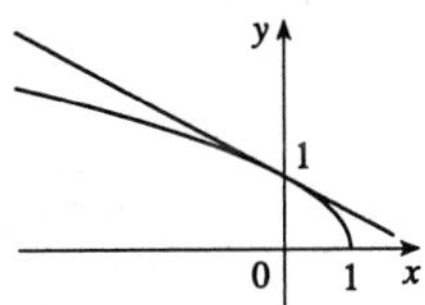

40. $g(x) = \sqrt[3]{1+x} = (1+x)^{1/3} \quad \Rightarrow \quad g'(x) = \frac{1}{3}(1+x)^{-2/3}$ so $g(0) = 1$ and $g'(0) = \frac{1}{3}$.
Therefore $\sqrt[3]{1+x} \approx 1 + \frac{1}{3}x$, $\sqrt[3]{0.95} = \sqrt[3]{1+(-0.05)} \approx 1 + \frac{1}{3}(-0.05) = \frac{59}{60} \approx 0.9833$, and
$\sqrt[3]{1.1} = \sqrt[3]{1+0.1} \approx 1 + \frac{1}{3}(0.1) = \frac{31}{30} \approx 1.0333$.

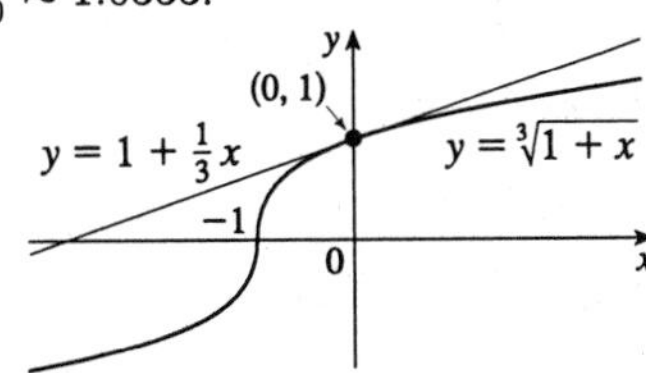

41. We need $\sqrt{1+x} - 0.1 < 1 + \frac{1}{2}x < \sqrt{1+x} + 0.1$. By zooming in or using a cursor, we see that this is true when $-0.69 < x < 1.09$.

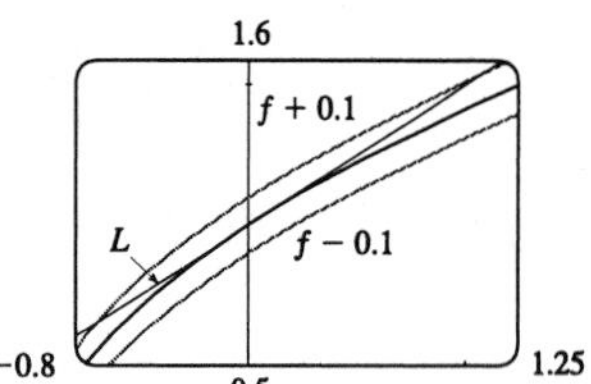

42. We need $\sin x - 0.1 < x < \sin x + 0.1$, which is true when $-0.85 < x < 0.85$.

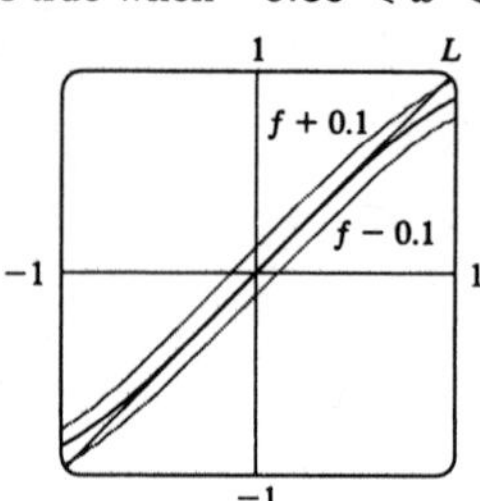

43. We need $1/(1+2x)^4 - 0.1 < 1 - 8x$ and $1 - 8x < 1/(1+2x)^4 + 0.1$, which both hold when $-0.045 < x < 0.055$.

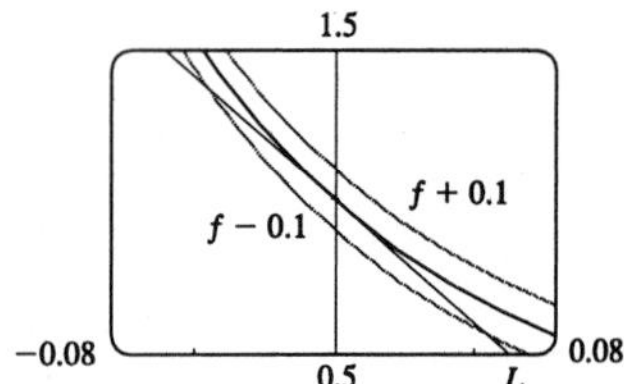

44. We need $1/\sqrt{4-x} - 0.1 < \frac{1}{2} + \frac{1}{16}x$ and $\frac{1}{2} + \frac{1}{16}x < 1/\sqrt{4-x} + 0.1$, which both hold when $-3.9 < x < 2.1$.

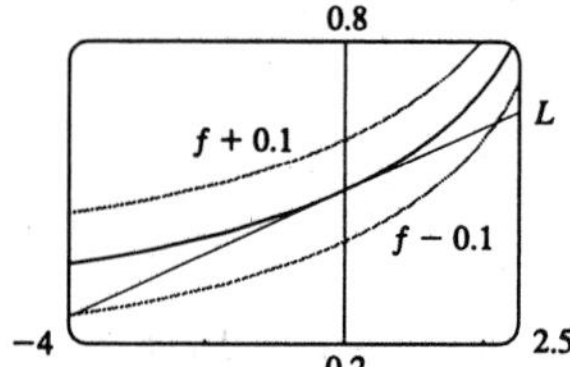

45. Using (10) with $f(x) = 1/x$, $f'(x) = -1/x^2$, and $f''(x) = 2/x^3$,

$$1/x \approx f(4) + f'(4)(x-4) + \tfrac{1}{2} f''(4)(x-4)^2 = \tfrac{1}{4} + (-1)4^{-2}(x-4) + \tfrac{1}{2}(2)4^{-3}(x-4)^2$$
$$= \tfrac{1}{4} - \tfrac{1}{16}(x-4) + \tfrac{1}{64}(x-4)^2$$

46. Using (10) with $f(x) = \sqrt[3]{x}$, $f'(x) = \frac{1}{3}x^{-2/3}$, and $f''(x) = -\frac{2}{9}x^{-5/3}$,

$$\sqrt[3]{x} \approx f(-8) + f'(-8)(x+8) + \tfrac{1}{2}f''(-8)(x+8)^2 = -2 + \left(\tfrac{1}{3}\right)(-8)^{-2/3}(x+8) + \tfrac{1}{2}\left(-\tfrac{2}{9}\right)(-8)^{-5/3}(x+8)^2$$
$$= -2 + \tfrac{1}{12}(x+8) + \tfrac{1}{288}(x+8)^2.$$

47. $f(x) = \sec x$, $f'(x) = \sec x \tan x$, and $f''(x) = \sec x \tan^2 x + \sec^3 x$, so

$$\sec x \approx f(0) + f'(0)(x) + \tfrac{1}{2} f''(x)(x)^2 = \sec 0 + \sec 0 \tan 0(x) + \tfrac{1}{2}[\sec 0(\sec^2 0) + \tan 0(\sec 0 \tan 0)]x^2$$
$$= \tfrac{1}{2}x^2 + 1.$$

48. $f(x) = \sin x$, $f'(x) = \cos x$, and $f''(x) = -\sin x$, so

$$\sin x \approx f\left(\tfrac{\pi}{6}\right) + f'\left(\tfrac{\pi}{6}\right)\left(x - \tfrac{\pi}{6}\right) + \tfrac{1}{2}f''\left(\tfrac{\pi}{6}\right)\left(x - \tfrac{\pi}{6}\right)^2 = \sin\tfrac{\pi}{6} + \cos\tfrac{\pi}{6}\left(x - \tfrac{\pi}{6}\right) + \tfrac{1}{2}\left(-\sin\tfrac{\pi}{6}\right)\left(x - \tfrac{\pi}{6}\right)^2$$
$$= \tfrac{1}{2} + \tfrac{\sqrt{3}}{2}\left(x - \tfrac{\pi}{6}\right) - \tfrac{1}{4}\left(x - \tfrac{\pi}{6}\right)^2.$$

49. $f(x) = \sqrt{x}$, $f'(x) = \dfrac{1}{2\sqrt{x}}$, and $f''(x) = \dfrac{1}{-4x\sqrt{x}}$,

so the linear approximation is

$$\sqrt{x} \approx f(1) + f'(1)(x-1) = \sqrt{1} + \tfrac{1}{2\sqrt{1}}(x-1) = 1 + \tfrac{1}{2}(x-1),$$

and the quadratic approximation is

$$\sqrt{x} \approx f(1) + f'(1)(x-1) + \tfrac{1}{2} f''(1)(x-1)^2 = 1 + \tfrac{1}{2}(x-1) - \tfrac{1}{8}(x-1)^2.$$

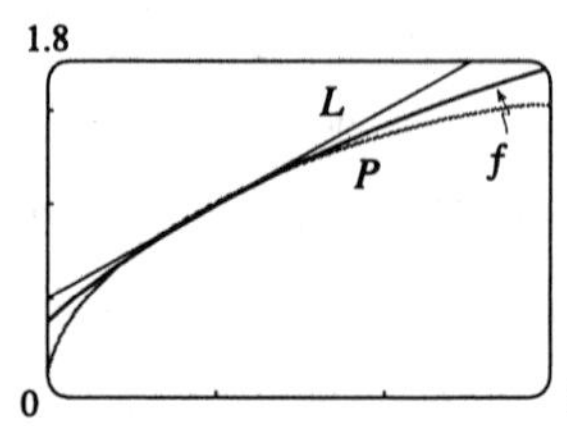

50. $f(x) = \tan x$, $f'(x) = \sec^2 x$, and $f''(x) = 2\sec^2 x \tan x$, so the linear approximation is

$$\tan x \approx f\left(\tfrac{\pi}{4}\right) + f'\left(\tfrac{\pi}{4}\right)\left(x - \tfrac{\pi}{4}\right) = \tan\tfrac{\pi}{4} + \sec^2\tfrac{\pi}{4}\left(x - \tfrac{\pi}{4}\right)$$
$$= 1 + 2\left(x - \tfrac{\pi}{4}\right),$$

and the quadratic approximation is

2.5 P f L 0 1.2

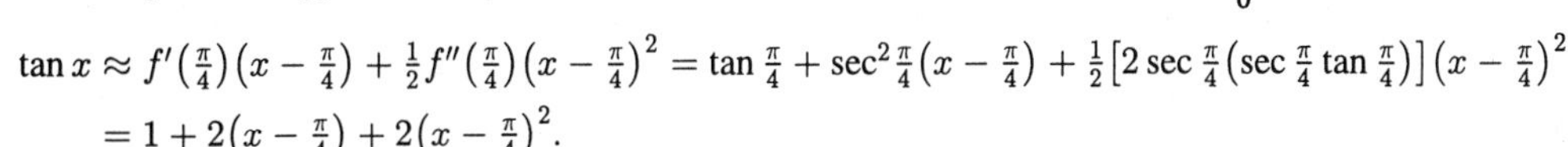

$$\tan x \approx f'\left(\tfrac{\pi}{4}\right)\left(x - \tfrac{\pi}{4}\right) + \tfrac{1}{2}f''\left(\tfrac{\pi}{4}\right)\left(x - \tfrac{\pi}{4}\right)^2 = \tan\tfrac{\pi}{4} + \sec^2\tfrac{\pi}{4}\left(x - \tfrac{\pi}{4}\right) + \tfrac{1}{2}\left[2\sec\tfrac{\pi}{4}\left(\sec\tfrac{\pi}{4}\tan\tfrac{\pi}{4}\right)\right]\left(x - \tfrac{\pi}{4}\right)^2$$
$$= 1 + 2\left(x - \tfrac{\pi}{4}\right) + 2\left(x - \tfrac{\pi}{4}\right)^2.$$

51. **(a)** $f(x) = \cos x \;\Rightarrow\; f'(x) = -\sin x \;\Rightarrow\; f''(x) = -\cos x.$

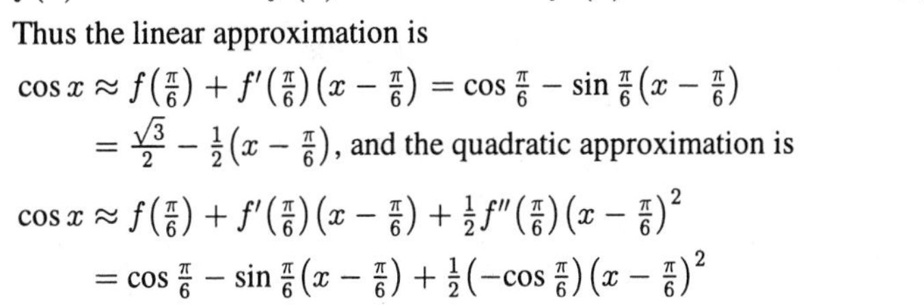

Thus the linear approximation is

$$\cos x \approx f\left(\tfrac{\pi}{6}\right) + f'\left(\tfrac{\pi}{6}\right)\left(x - \tfrac{\pi}{6}\right) = \cos\tfrac{\pi}{6} - \sin\tfrac{\pi}{6}\left(x - \tfrac{\pi}{6}\right)$$
$$= \tfrac{\sqrt{3}}{2} - \tfrac{1}{2}\left(x - \tfrac{\pi}{6}\right),$$ and the quadratic approximation is

$$\cos x \approx f\left(\tfrac{\pi}{6}\right) + f'\left(\tfrac{\pi}{6}\right)\left(x - \tfrac{\pi}{6}\right) + \tfrac{1}{2}f''\left(\tfrac{\pi}{6}\right)\left(x - \tfrac{\pi}{6}\right)^2$$
$$= \cos\tfrac{\pi}{6} - \sin\tfrac{\pi}{6}\left(x - \tfrac{\pi}{6}\right) + \tfrac{1}{2}\left(-\cos\tfrac{\pi}{6}\right)\left(x - \tfrac{\pi}{6}\right)^2$$
$$= \tfrac{\sqrt{3}}{2} - \tfrac{1}{2}\left(x - \tfrac{\pi}{6}\right) - \tfrac{\sqrt{3}}{4}\left(x - \tfrac{\pi}{6}\right)^2.$$

(b)

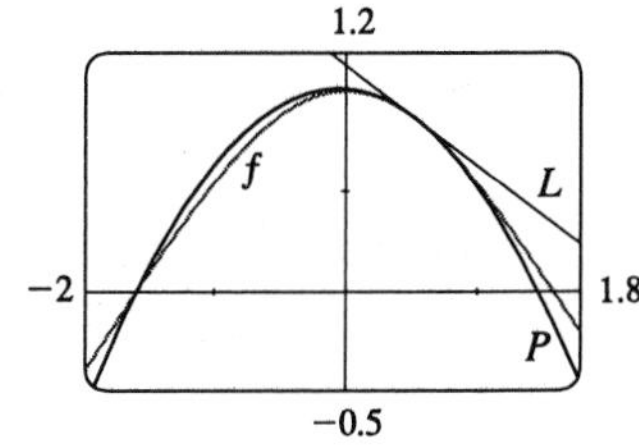

(c) We need $\cos x - 0.1 < \tfrac{\sqrt{3}}{2} - \tfrac{1}{2}\left(x - \tfrac{\pi}{6}\right) < \cos x + 0.1$. From the graph, it appears that the linear approximation has the required accuracy when $0.06 < x < 1.03$.

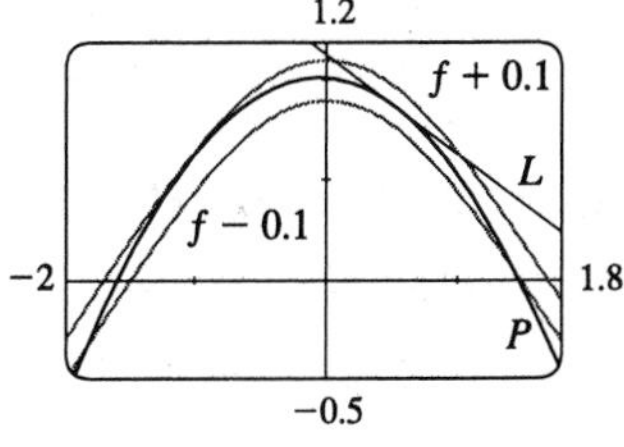

(d) We need $\cos x - 0.1 < \tfrac{\sqrt{3}}{2} - \tfrac{1}{2}\left(x - \tfrac{\pi}{6}\right) - \tfrac{\sqrt{3}}{4}\left(x - \tfrac{\pi}{6}\right)^2 < \cos x + 0.1$. From the graph, it appears that this is true when $-1.82 < x < 1.48$.

52. **(a)** $f(x) = 1/(1 + x^2) \;\Rightarrow\; f'(x) = -2x/(1 + x^2)^2 \;\Rightarrow$

$$f''(x) = \frac{-2(1 + x^2)^2 + 2x(2)(1 + x^2)(2x)}{(1 + x^2)^4} = \frac{6x^2 - 2}{(1 + x^2)^3}.$$

Thus the linear approximation is

$$1/(1 + x^2) \approx f(1) + f'(1)(x - 1)$$
$$= \frac{1}{1 + 1^2} + \frac{-2(1)}{(1 + 1^2)^2}(x - 1) = \tfrac{1}{2} - \tfrac{1}{2}(x - 1).$$

(b)

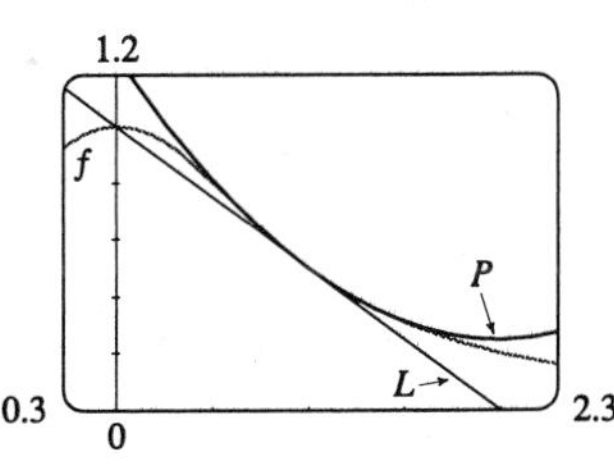

The quadratic approximation is

$$1/(1 + x^2) \approx f(1) + f'(1)(x - 1) + \tfrac{1}{2}f''(1)(x - 1) = \tfrac{1}{2} - \tfrac{1}{2}(x - 1) + \tfrac{1}{4}(x - 1)^2.$$

(c) From the graph, it appears that $\dfrac{1}{1 + x^2} - 0.1 < \tfrac{1}{2} - \tfrac{1}{2}(x - 1)$ and $\tfrac{1}{2} - \tfrac{1}{2}(x - 1) < \dfrac{1}{1 + x^2} + 0.1$ when $-0.15 < x < 1.67$.

(d) From the graph, it appears that

$$\frac{1}{1 + x^2} - 0.1 < \tfrac{1}{2} - \tfrac{1}{2}(x - 1) + \tfrac{1}{4}(x - 1)^2 < \frac{1}{1 + x^2} + 0.1$$

when $0.20 < x < 2.24$.

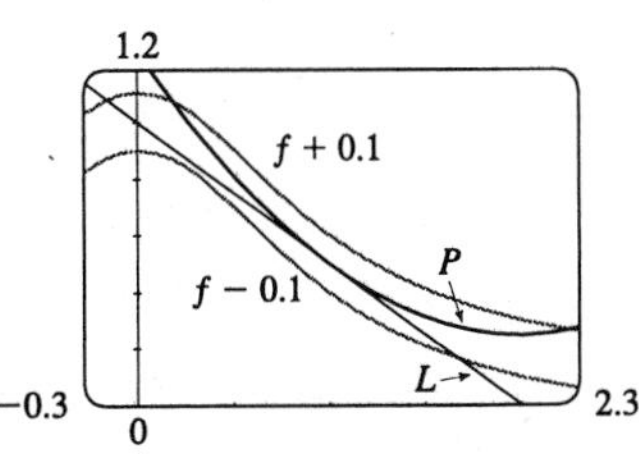

(e)

x	$f(x)$	linear	quadratic
0.9	0.55249	0.5500	0.5525
1.1	0.45249	0.4500	0.4525
1.2	0.40984	0.4000	0.4100
1.3	0.37175	0.3500	0.3725

53. **(a)** $dc = \frac{dc}{dx}\,dx = 0\,dx = 0$

(b) $d(cu) = \frac{d}{dx}(cu)dx = c\frac{du}{dx}\,dx = c\,du$

(c) $d(u+v) = \frac{d}{dx}(u+v)dx = \left(\frac{du}{dx} + \frac{dv}{dx}\right)dx = \frac{du}{dx}\,dx + \frac{dv}{dx}\,dx = du + dv$

(d) $d(uv) = \frac{d}{dx}(uv)dx = \left(u\frac{dv}{dx} + v\frac{du}{dx}\right)dx = u\frac{dv}{dx}\,dx + v\frac{du}{dx}\,dx = u\,dv + v\,du$

(e) $d\left(\frac{u}{v}\right) = \frac{d}{dx}\left(\frac{u}{v}\right)dx = \frac{v\frac{du}{dx} - u\frac{dv}{dx}}{v^2}\,dx = \frac{v\frac{du}{dx}\,dx - u\frac{dv}{dx}\,dx}{v^2} = \frac{v\,du - u\,dv}{v^2}$

(f) $d(x^n) = \frac{d}{dx}(x^n)dx = nx^{n-1}\,dx$

54. **(a)** The linear approximation is $f(x) \approx f(1) + f'(1)(x-1) = 2 + \sqrt{1^3+1}(x-1) = 2 + \sqrt{2}(x-1)$. So at $x = 1.1$, $f(x) \approx 2 + 0.1\sqrt{2} \approx 2.1414$.

(b) The true value of $f(1.1)$ is greater than the linear estimate, since the derivative of the function is getting larger while the derivative of the approximation is constant.

(c) $f''(x) = [f'(x)]' = \frac{3x^2}{2\sqrt{x^3+1}}$. So the quadratic approximation at $x = 1.1$ is

$f(1.1) \approx 2 + \sqrt{2}(0.1) + \frac{1}{2}\left(\frac{3}{2\sqrt{2}}\right)(0.1)^2 \approx 2.1467.$

55. $P(x) = a_0 + a_1x + a_2x^2 + a_3x^3 + \cdots + a_nx^n \quad \Rightarrow$

$P'(x) = a_1 + 2a_2x + 3a_3x^2 + \cdots \quad \Rightarrow$

$P''(x) = 2a_2 + 2\cdot 3a_3x + 3\cdot 4a_4x + \cdots \quad \Rightarrow$

$P'''(x) = 2\cdot 3a_3 + 2\cdot 3\cdot 4a_4x + \cdots \quad \Rightarrow$

$P^{(k)}(x) = 2\cdot 3\cdot 4\cdot\cdots\cdot ka_k + 2\cdot 3\cdot 4\cdot\cdots\cdot k\cdot(k+1)a_{k+1}x + \cdots \quad \Rightarrow$

$P^{(n)}(x) = n!\,a_n$. Therefore $P^{(k)}(0) = f^{(k)}(0) = k!\,a_k$, and so $a_k = \frac{f^{(k)}(0)}{k!}$ for $k = 1, 2, \ldots, n$.

Now let $f(x) = \sin x$. Then $f'(x) = \cos x$, $f''(x) = -\sin x$ and $f'''(x) = -\cos x$. So the Taylor polynomial of degree 3 for $\sin x$ is $P(x) = \sin 0 + \frac{\cos 0}{1!}x + \frac{-\sin 0}{2!}x^2 + \frac{-\cos 0}{3!}x^3 = x - \frac{x^3}{6}$.

EXERCISES 2.10

1.

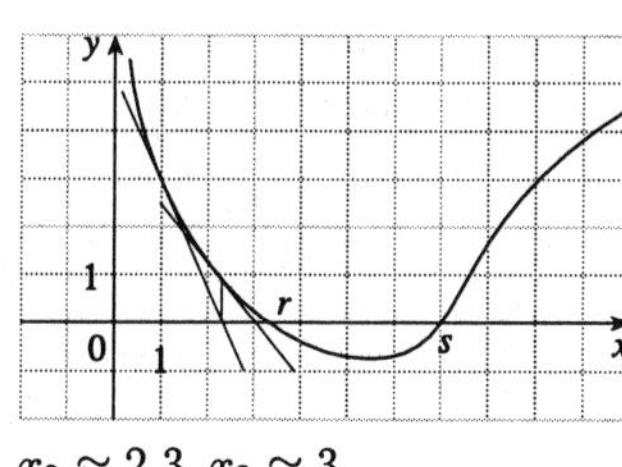

$x_2 \approx 2.3$, $x_3 \approx 3$

2.

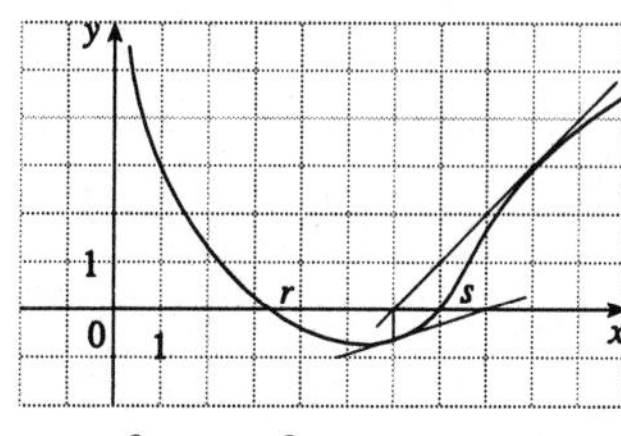

$x_2 \approx 6$, $x_3 \approx 8$

3. $f(x) = x^3 + x + 1 \quad \Rightarrow \quad f'(x) = 3x^2 + 1$, so $x_{n+1} = x_n - \dfrac{x_n^3 + x_n + 1}{3x_n^2 + 1}$. $x_1 = -1 \quad \Rightarrow$

$$x_2 = -1 - \frac{-1 - 1 + 1}{3 \cdot 1 + 1} = -0.75 \quad \Rightarrow \quad x_3 = -0.75 - \frac{(-0.75)^3 - 0.75 + 1}{3(-0.75)^2 + 1} \approx -0.6860$$

4. $f(x) = x^3 + x^2 + 2 = 0 \quad \Rightarrow \quad f'(x) = 3x^2 + 2x$, so $x_{n+1} = x_n - \dfrac{x_n^3 + x_n^2 + 2}{3x_n^2 + 2x_n}$. $x_1 = -2 \quad \Rightarrow$

$$x_2 = -2 - \frac{-2}{8} = -1.75 \quad \Rightarrow \quad x_3 = -1.75 - \frac{f(-1.75)}{f'(-1.75)} = -1.6978$$

5. $f(x) = x^5 - 10 \quad \Rightarrow \quad f'(x) = 5x^4$, so $x_{n+1} = x_n - \dfrac{x_n^5 - 10}{5x_n^4}$. $x_1 = 1.5 \quad \Rightarrow$

$$x_2 = 1.5 - \frac{(1.5)^5 - 10}{5(1.5)^4} \approx 1.5951 \quad \Rightarrow \quad x_3 = 1.5951 - \frac{f(1.5951)}{f'(1.5951)} \approx 1.5850$$

6. $f(x) = x^7 - 100 \quad \Rightarrow \quad f'(x) = 7x^6$, so $x_{n+1} = x_n - \dfrac{x_n^7 - 100}{7x_n^6}$. $x_1 = 2 \quad \Rightarrow$

$$x_2 = 1 - \frac{128 - 100}{7 \cdot 64} = 1.9375 \quad \Rightarrow \quad x_3 = 1 - \frac{(1.9375)^7 - 100}{7(1.9375)^6} \approx 1.9308$$

7. Finding $\sqrt[4]{22}$ is equivalent to finding the positive root of $x^4 - 22 = 0$ so we take $f(x) = x^4 - 22 \quad \Rightarrow$ $f'(x) = 4x^3$ and $x_{n+1} = x_n - \dfrac{x_n^4 - 22}{4x_n^3}$. Taking $x_1 = 2$, we get $x_2 \approx 2.1875$, $x_3 \approx 2.166059$, $x_4 \approx 2.165737$ and $x_5 \approx 2.165737$. Thus $\sqrt[4]{22} \approx 2.165737$ to six decimal places.

8. Finding $\sqrt[10]{100}$ is equivalent to finding the positive root of $x^{10} - 100 = 0$, so we take $f(x) = x^{10} - 100 \quad \Rightarrow$ $f'(x) = 10x^9$ and $x_{n+1} = x_n - \dfrac{x_n^{10} - 100}{10x^9}$. Taking $x_1 = 1.5$, we get $x_2 \approx 1.610123$, $x_3 \approx 1.586600$, $x_4 \approx 1.584901$, $x_5 \approx 1.584893$, and $x_6 \approx 1.584893$. Thus $\sqrt[10]{100} \approx 1.584893$ to six decimal places.

9. $f(x) = x^3 - 2x - 1 \quad \Rightarrow \quad f'(x) = 3x^2 - 2$, so $x_{n+1} = x_n - \dfrac{x_n^3 - 2x_n - 1}{3x_n^2 - 2}$. Taking $x_1 = 1.5$, we get $x_2 \approx 1.631579$, $x_3 \approx 1.618184$, $x_4 \approx 1.618034$ and $x_5 \approx 1.618034$. So the root is 1.618034 to six decimal places.

10. $f(x) = x^4 + x^3 - 22x^2 - 2x + 41 \quad \Rightarrow \quad f'(x) = 4x^3 + 3x^2 - 44x - 2$, so

$x_{n+1} = x_n - \dfrac{x_n^4 + x_n^3 - 22x_n^2 - 2x_n + 41}{4x_n^3 + 3x_n^2 - 44x_n - 2}$. Taking $x_1 = 1.5$, we get $x_2 \approx 1.435864$, $x_3 \approx 1.435476$,

$x_4 \approx 1.435467$. So the root in the interval $[1, 2]$ is 1.435476 to six decimal places.

11. From the graph it appears that there is a root near 2, so we take $x_1 = 2$. Write the equation as $f(x) = 2\sin x - x = 0$. Then $f'(x) = 2\cos x - 1$, so $x_{n+1} = x_n - \dfrac{2\sin x_n - x_n}{2\cos x_n - 1} \quad \Rightarrow \quad x_1 = 2$, $x_2 \approx 1.900996$, $x_3 \approx 1.895512$, $x_4 \approx 1.895494$, and $x_5 \approx 1.895494$. So the root is 1.895494 to 6 decimal places.

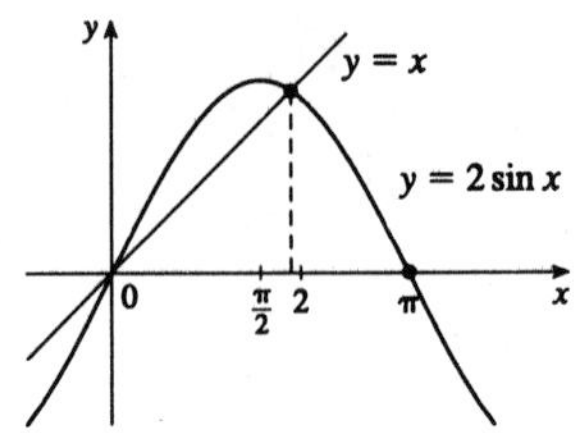

12. From the graph, it appears there is a root near 4.5. So we take $x_1 = 4.5$. Write the equation as $f(x) = \tan x - x = 0$. Then $f'(x) = \sec^2 x - 1$, so $x_{n+1} = x_n - \dfrac{\tan x_n - x_n}{\sec^2 x_n - 1}$. $x_1 = 4.5$, $x_2 \approx 4.493614$, $x_3 \approx 4.493410$, $x_4 \approx 4.493409$ and $x_5 \approx 4.493409$. To six decimal places, the root is 4.493409.

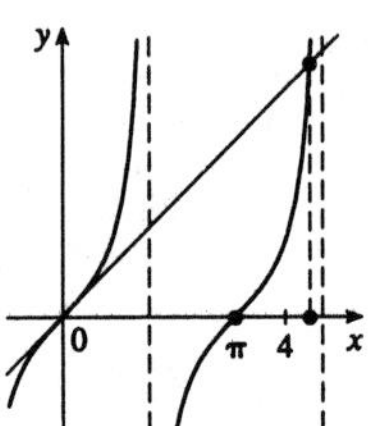

13. $f(x) = x^3 - 4x + 1 \quad \Rightarrow \quad f'(x) = 3x^2 - 4$, so $x_{n+1} = x_n - \dfrac{x_n^3 - 4x_n + 1}{3x_n^2 - 4}$. Observe that $f(-3) = -14$,

$f(-2) = 1$, $f(0) = 1$, $f(1) = -2$ and $f(2) = 1$ so there are roots in $[-3, -2]$, $[0, 1]$ and $[1, 2]$.

$[-3, -2]$
$x_1 = -2$
$x_2 = -2.125$
$x_3 \approx -2.114975$
$x_4 \approx -2.114908$
$x_5 \approx -2.114908$

$[0, 1]$
$x_1 = 0$
$x_2 = 0.25$
$x_3 \approx 0.254098$
$x_4 \approx 0.254102$
$x_5 \approx 0.254102$

$[1, 2]$
$x_1 = 2$
$x_2 = 1.875$
$x_3 \approx 1.860979$
$x_4 \approx 1.860806$
$x_5 \approx 1.860806$

To six decimal places, the roots are -2.114908, 0.254102 and 1.860806.

14. $f(x) = x^5 - 5x + 2 \quad \Rightarrow \quad f'(x) = 5x^4 - 5$, so $x_{n+1} = x_n - \dfrac{x_n^5 - 5x_n + 2}{5x_n^4 - 5}$. Observe that $f(-2) = -20$,

$f(-1) = 6$, $f(0) = 2$, $f(1) = -2$ and $f(2) = 24$ so there are roots in $[-2, -1]$, $[0, 1]$ and $[1, 2]$. A sketch shows that these are the only intervals with roots.

$[-2, -1]$
$x_1 = -1.5$
$x_2 \approx -1.593846$
$x_3 \approx -1.582241$
$x_4 \approx -1.582036$
$x_5 \approx -1.582036$

$[0, 1]$
$x_1 = 0.5$
$x_2 = 0.4$
$x_3 \approx 0.402102$
$x_4 \approx 0.402102$

$[1, 2]$
$x_1 = 1.5$
$x_3 \approx 1.396923$
$x_3 \approx 1.373078$
$x_4 \approx 1.371885$
$x_5 \approx 1.371882$
$x_6 \approx 1.371882$

To six decimal places, the roots are -1.582036, 0.402102 and 1.371882.

15. $f(x) = x^4 + x^2 - x - 1 \quad \Rightarrow \quad f'(x) = 4x^3 + 2x - 1$, so $x_{n+1} = x_n - \dfrac{x_n^4 + x_n^2 - x_n - 1}{4x_n^3 + 2x_n - 1}$. Note that $f(1) = 0$, so $x = 1$ is a root. Also $f(-1) = 2$ and $f(0) = -1$, so there is a root in $[-1, 0]$. A sketch shows that these are the only roots. Taking $x_1 = -0.5$, we have $x_2 = -0.575$, $x_3 \approx -0.569867$, $x_4 \approx -0.569840$ and $x_5 \approx -0.569840$. The roots are 1 and -0.569840, to six decimal places.

16. $f(x) = (x-2)^4 - \frac{1}{2}x \quad \Rightarrow \quad f'(x) = 4(x-2)^3 - \frac{1}{2}$, so $x_{n+1} = x_n - \dfrac{(x_n - 2)^4 - \frac{1}{2}x_n}{4(x_n - 2)^3 - \frac{1}{2}}$. Observe that $f(1) = \frac{1}{2}$, $f(2) = -1$, $f(3) = -\frac{1}{2}$ and $f(4) = 14$ so there are roots in $[1, 2]$ and $[3, 4]$ and a sketch shows that these are the only roots. Taking $x_1 = 1$, we get $x_2 \approx 1.111111$, $x_3 \approx 1.131883$, $x_4 \approx 1.132529$ and $x_5 \approx 1.132529$. Taking $x_1 = 3$, we get $x_2 \approx 3.142857$, $x_3 \approx 3.118267$, $x_4 \approx 3.117350$, $x_5 \approx 3.117349$ and $x_6 \approx 3.117349$. To six decimal places, the roots are 1.132529 and 3.117349.

17.

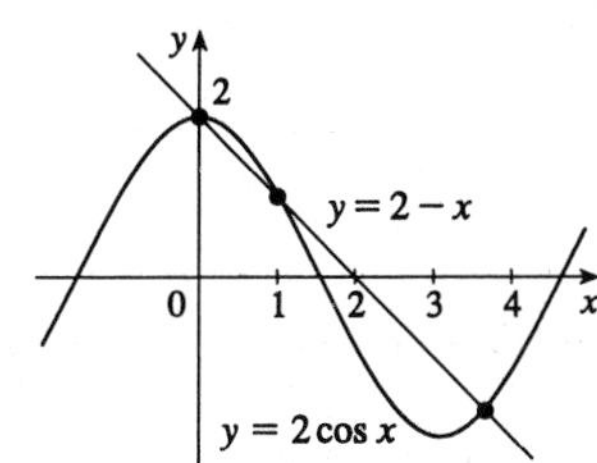

Clearly $x = 0$ is a root. From the sketch, there appear to be roots near 1 and 3.5. Write the equation as $f(x) = 2\cos x + x - 2 = 0$. Then $f'(x) = -2\sin x + 1$, so $x_{n+1} = x_n - \dfrac{2\cos x_n + x_n - 2}{1 - 2\sin x_n}$. Taking $x_1 = 1$, we get $x_2 \approx 1.118026$, $x_3 \approx 1.109188$, $x_4 \approx 1.109144$ and $x_5 \approx 1.109144$. Taking $x_1 = 3.5$, we get $x_2 \approx 3.719159$, $x_3 \approx 3.698331$, $x_4 \approx 3.698154$ and $x_5 \approx 3.698154$. To six decimal places the roots are 0, 1.109144 and 3.698154.

18.

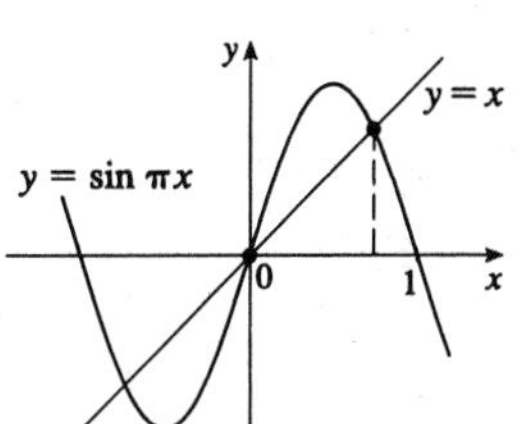

Clearly $x = 0$ is a root. From the sketch, there appear to be roots near -0.75 and 0.75. Write the equation as $f(x) = \sin \pi x - x = 0$. Then $f'(x) = \pi \cos \pi x - 1$, so $x_{n+1} = x_n - \dfrac{\sin \pi x_n - x_n}{\pi \cos \pi x_n - 1}$. Taking $x_1 = 0.75$ we get $x_2 \approx 0.736685$, $x_3 \approx 0.736484$, $x_4 \approx 0.736484$. To six decimal places, the roots are 0, 0.736484 and -0.736484.

19.

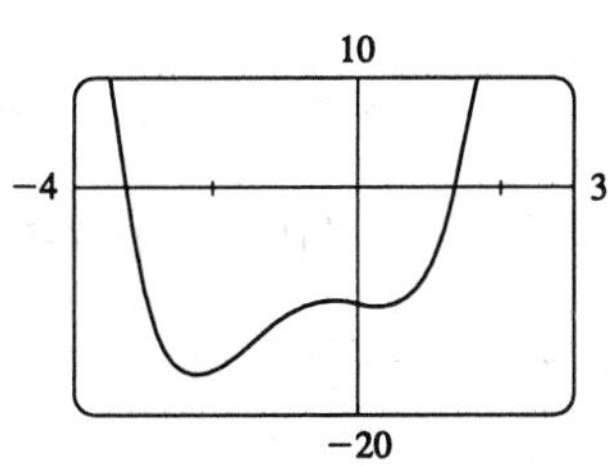

From the graph, there appear to be roots near -3.2 and 1.4. Let $f(x) = x^4 + 3x^3 - x - 10 \quad \Rightarrow \quad f'(x) = 4x^3 + 9x^2 - 1$, so $x_{n+1} = x_n - \dfrac{x_n^4 + 3x_n^3 - x_n - 10}{4x_n^3 + 9x_n^2 - 1}$. Taking $x_1 = -3.2$, we get $x_2 \approx -3.20617358$, $x_3 \approx -3.20614267 \approx x_4$. Taking $x_1 = 1.4$, we get $x_2 \approx 1.37560834$, $x_3 \approx 1.37506496$, $x_4 \approx 1.37506470 \approx x_5$. To eight decimal places, the roots are -3.20614267 and 1.37506470.

20.

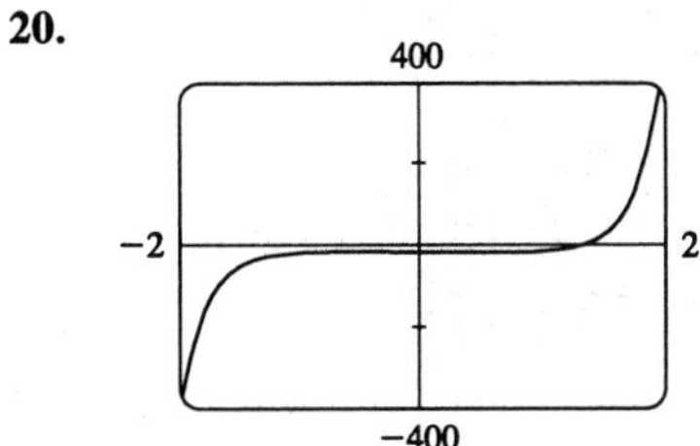

From the graph, we see that the only root appears to be near 1.25. Let $f(x) = x^9 - x^6 + 2x^4 + 5x - 14$. Then $f'(x) = 9x^8 - 6x^5 + 8x^3 + 5$, so $x_{n+1} = x_n - \dfrac{x_n^9 - x_n^6 + 2x_n^4 + 5x_n - 14}{9x_n^8 - 6x_n^5 + 8x_n^3 + 5}$. Taking $x_1 = 1.25$, we get $x_2 \approx 1.23626314$, $x_3 \approx 1.23571823$, $x_4 \approx 1.23571742 \approx x_5$. To eight decimal places, the root of the equation is 1.23571742.

21. 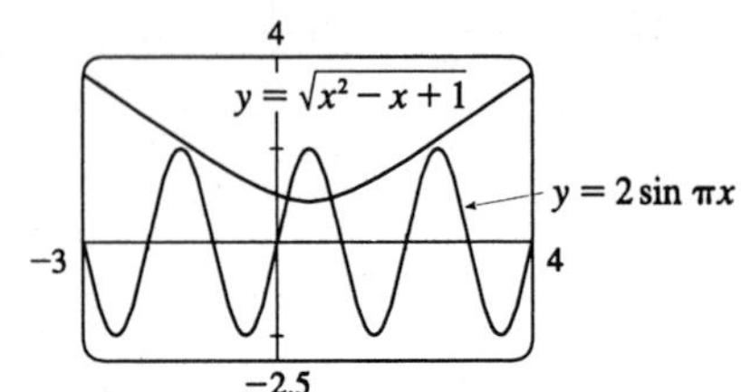

From the graph, we see that there are roots of this equation near 0.2 and 0.8. Let $f(x) = \sqrt{x^2 - x + 1} - 2\sin \pi x \quad \Rightarrow$

$$f'(x) = \frac{2x - 1}{2\sqrt{x^2 - x + 1}} - 2\pi \cos \pi x, \text{ so}$$

$$x_{n+1} = x_n - \frac{\sqrt{x_n^2 - x_n + 1} - 2\sin \pi x_n}{\dfrac{2x_n - 1}{2\sqrt{x_n^2 - x_n + 1}} - 2\pi \cos \pi x_n}$$

Taking $x_1 = 0.2$, we get $x_2 \approx 0.152120155$, $x_3 \approx 0.154380674$, $x_4 \approx 0.154385001 \approx x_5$. Taking $x_1 = 0.8$, we get $x_2 \approx 0.847879845$, $x_3 \approx 0.845619326$, $x_4 \approx 0.845614998 \approx x_5$. So, to eight decimal places, the roots of the equation are 0.15438500 and 0.84561500.

22. 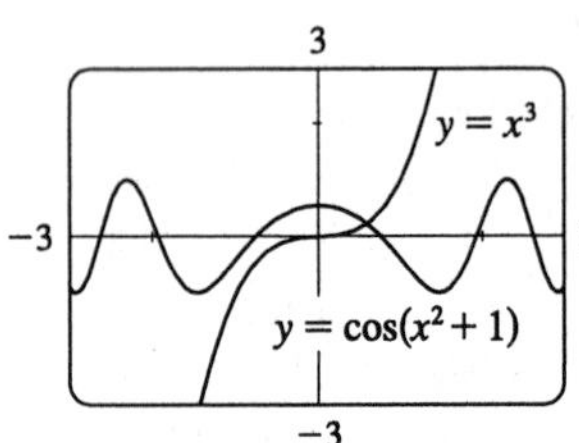

From the graph, we see that the only root of this equation is near 0.6. Let $f(x) = \cos(x^2 + 1) - x^3 \quad \Rightarrow$

$f'(x) = -2x\sin(x^2 + 1) - 3x^2$, so

$x_{n+1} = x_n + \dfrac{\cos(x_n^2 + 1) - x_n^3}{2x_n \sin(x_n^2 + 1) + 3x_n^2}$. Taking $x_1 = 0.6$, we get $x_2 \approx 0.596999547$, $x_3 \approx 0.596987767 \approx x_3$. To eight decimal places, the root of the equation is 0.596987767.

23. **(a)** $f(x) = x^2 - a \quad \Rightarrow \quad f'(x) = 2x$, so Newton's Method gives

$$x_{n+1} = x_n - \frac{x_n^2 - a}{2x_n} = x_n - \tfrac{1}{2}x_n + \frac{a}{2x_n} = \frac{1}{2}\left(x_n + \frac{a}{x_n}\right).$$

(b) Using (a) with $x_1 = 30$, we get $x_2 \approx 31.666667$, $x_3 \approx 31.622807$, $x_4 \approx 31.622777$ and $x_5 \approx 31.622777$. So $\sqrt{1000} \approx 31.622777$.

24. **(a)** $f(x) = \dfrac{1}{x} - a \quad \Rightarrow \quad f'(x) = -\dfrac{1}{x^2}$, so $x_{n+1} = x_n - \dfrac{1/x_n - a}{-1/x_n^2} = x_n + x_n - ax_n^2 = 2x_n - ax_n^2$.

(b) Using (a) with $a = 1.6894$ and $x_1 = 0.6$, we get $x_2 \approx 0.588576$, $x_3 \approx 0.588789$ and $x_4 \approx 0.588789$. So $1/1.6984 \approx 0.588789$.

25. If we attempt to compute x_2 we get $x_2 = x_1 - \dfrac{f(x_1)}{f'(x_1)}$, but $f(x) = x^3 - 3x + 6 \quad \Rightarrow$ $f'(x_1) = 3x_1^2 - 3 = 3(1)^2 - 3 = 0$. For Newton's Method to work $f'(x_n) \neq 0$ (no horizontal tangents).

26. $x^3 - x = 1 \quad \Rightarrow \quad x^3 - x - 1 = 0$. So $f(x) = x^3 - x - 1 \quad \Rightarrow \quad f'(x) = 3x^2 - 1$ so

$$x_{n+1} = x_n - \frac{x_n^3 - x_n - 1}{3x_n^2 - 1}$$

(a) $x_1 = 1$, $x_2 = 1.5$, $x_3 \approx 1.347826$, $x_4 \approx 1.325200$, $x_5 \approx 1.324718 \approx x_6$

(b) $x_1 = 0.6$, $x_2 = 17.9$, $x_3 \approx 11.946802$, $x_4 \approx 7.985520$, $x_5 \approx 5.356909$, $x_6 \approx 3.624996$, $x_7 \approx 2.505589$, $x_8 \approx 1.820129$, $x_9 \approx 1.461044$, $x_{10} \approx 1.339323$, $x_{11} \approx 1.324913$, $x_{12} \approx 1.324718 \approx x_{13}$

(c) $x_1 = 0.57$, $x_2 \approx -54.165455$, $x_3 \approx -36.114293$, $x_4 \approx -24.082094$, $x_5 \approx -16.063387$, $x_6 \approx -10.721483$, $x_7 \approx -7.165534$, $x_8 \approx -4.801704$, $x_9 \approx -3.233425$, $x_{10} \approx -2.193674$, $x_{11} \approx -1.496867$, $x_{12} \approx -0.997546$, $x_{13} \approx -0.496305$, $x_{14} \approx -2.894162$, $x_{15} \approx -1.967962$, $x_{16} \approx -1.341355$, $x_{17} \approx -0.870187$, $x_{18} \approx -0.249949$, $x_{19} \approx -1.192219$, $x_{20} \approx -0.731952$, $x_{21} \approx 0.355213$, $x_{22} \approx -1.753322$, $x_{23} \approx -1.189420$, $x_{24} \approx -0.729123$, $x_{25} \approx 0.377844$, $x_{26} \approx -1.937872$, $x_{27} \approx -1.320350$, $x_{28} \approx -0.851919$, $x_{29} \approx -0.200959$, $x_{30} \approx -1.119386$, $x_{31} \approx -0.654291$, $x_{32} \approx 1.547009$, $x_{33} \approx 1.360050$, $x_{34} \approx 1.325828$, $x_{35} \approx 1.324719$, $x_{36} \approx 1.324718 \approx x_{37}$.

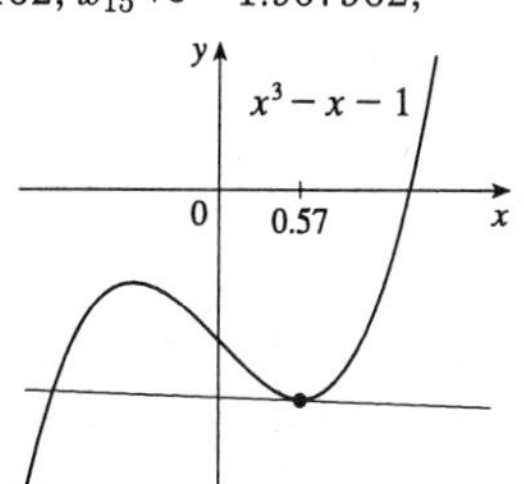

27. For $f(x) = x^{1/3}$, $f'(x) = \frac{1}{3}x^{-2/3}$ and $x_{n+1} = x_n - \dfrac{f(x_n)}{f'(x_n)} = x_n - \dfrac{x_n^{1/3}}{\frac{1}{3}x_n^{-2/3}} = x_n - 3x_n = -2x_n$. Therefore each successive approximation becomes twice as large as the previous one in absolute value, so the sequence of approximations fails to converge to the root, which is 0.

28. According to Newton's Method, for $x_n > 0$,

$$x_{n+1} = x_n - \frac{\sqrt{x_n}}{1/(2\sqrt{x_n})} = x_n - 2x_n = -x_n \text{ and for}$$

$$x_n < 0,\ x_{n+1} = x_n - \frac{-\sqrt{-x_n}}{1/(2\sqrt{-x_n})} = x_n - [-2(-x_n)] = -x_n.$$

So we can see that after choosing any value x_1 the subsequent values will alternate between $-x_1$ and x_1 and never approach the root.

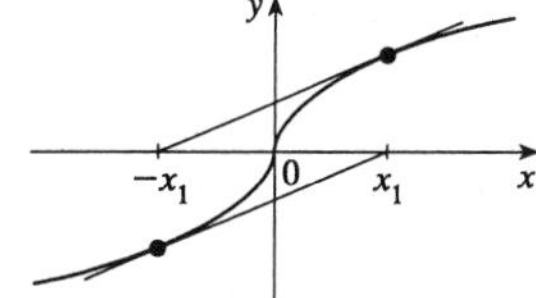

29. The volume of the silo, in terms of its radius, is $V(r) = \pi r^2(30) + \frac{1}{2}\left(\frac{4}{3}\pi r^3\right) = 30\pi r^2 + \frac{2}{3}\pi r^3$. From a graph of V, we see that $V(r) = 15{,}000$ at $r \approx 11$ ft. Now we use Newton's Method to solve the equation $V(r) - 15{,}000 = 0$. First we must calculate

$$\frac{dV}{dr} = 60\pi r + 2\pi r^2, \text{ so } r_{n+1} = r_n - \frac{30\pi r_n^2 + \frac{2}{3}\pi r_n^3 - 15{,}000}{60\pi r_n + 2\pi r_n^2}.$$

Taking $r_1 = 11$, we get $r_2 = 11.2853$, $r_3 = 11.2807 \approx r_4$. So in order for the silo to hold 15,000 ft³ of grain, its radius must be about 11.2807 ft.

56,000
$y = 15{,}000$
0 20

30. Let the radius of the circle be r. Then (by Equation 3 in Appendix D) $5 = r\theta$ and so $r = 5/\theta$. From the Law of Cosines we get $4^2 = r^2 + r^2 - 2 \cdot r \cdot r \cdot \cos\theta \quad\Rightarrow\quad 16 = 2r^2(1 - \cos\theta) = 2(5/\theta)^2(1 - \cos\theta)$. Multiplying by θ^2 gives $16\theta^2 = 50(1 - \cos\theta)$, so we take $f(\theta) = 16\theta^2 + 50\cos\theta - 50 \quad\Rightarrow\quad f'(\theta) = 32\theta - 50\sin\theta$. The formula for Newton's Method is $\theta_{n+1} = \theta_n - \dfrac{16\theta^2 + 50\cos\theta - 50}{32\theta - 50\sin\theta}$, and we take $\theta_1 = 2$, giving $\theta_2 = 2.2565$, $\theta_3 = 2.2610$, $\theta_4 = 2.2620$, $\theta_5 = 2.2622 \approx \theta_6$. So correct to four decimal places, the angle is 2.2622 radians $\approx 130°$.

31. In this case, $A = 18{,}000$, $R = 375$, and $n = 60$. So the formula becomes $18{,}000 = \dfrac{375}{x}\left[1 - (1 + x)^{-60}\right] \Leftrightarrow$ $48x = 1 - (1 + x)^{-60} \Leftrightarrow 48x(1 + x)^{60} - (1 + x)^{60} + 1 = 0$. Let the LHS be called $f(x)$, so that $f'(x) = 48x(60)(1 + x)^{59} + 48(1 + x)^{60} - 60(1 + x)^{59} = 12(1 + x)^{59}(244x - 1)$. So we use Newton's Method with $x_{n+1} = x_n - \dfrac{48x(1 + x)^{60} - (1 + x)^{60} + 1}{12(1 + x)^{59}(244x - 1)}$ and $x_1 = 1\% = 0.01$.

We get $x_2 = 0.0082202$, $x_3 \approx 0.0076802$, $x_4 \approx 0.0076290$, $x_5 \approx 0.0076286 \approx x_6$. So the dealer is charging a monthly interest rate of 0.76286%.

32. **(a)** $p(x) = x^5 - (2 + r)x^4 + (1 + 2r)x^3 - (1 - r)x^2 + 2(1 - r)x + r - 1 \Rightarrow$ $p'(x) = 5x^4 - 4(2 + r)x^3 + 3(1 + 2r)x^2 - 2(1 - r)x + 2(1 - r)$. So we use $x_{n+1} = x_n - \dfrac{x_n^5 - (2 + r)x_n^4 + (1 + 2r)x_n^3 - (1 - r)x_n^2 + 2(1 - r)x_n + r - 1}{5x_n^4 - 4(2 + r)x_n^3 + 3(1 + 2r)x_n^2 - 2(1 - r)x_n + 2(1 - r)}$. We substitute in the value $r \approx 3.04042 \times 10^{-6}$ in order to evaluate the approximations numerically. The libration point L_1 is slightly less than 1 AU from the sun, so we take $x_1 = 0.95$ as our first approximation, and get $x_2 \approx 0.96682$, $x_3 \approx 0.97770$, $x_4 \approx 0.98451$, $x_5 \approx 0.98820$, $x_6 \approx 0.98976$, $x_7 \approx 0.98998$, $x_8 \approx 0.98999 \approx x_9$. So, to five decimal places, L_1 is located 0.98999 AU from the sun (or 0.01001 AU from Earth).

(b) In this case we use Newton's Method with the function $p(x) - 2rx^2 = x^5 - (2 + r)x^4 + (1 + 2r)x^3 - (1 + r)x^2 + 2(1 - r)x + r - 1 \Rightarrow$ $[p(x) - 2rx^2]' = 5x^4 - 4(2 + r)x^3 + 3(1 + 2r)x^2 - 2(1 + r)x + 2(1 - r)$. So $x_{n+1} = x_n - \dfrac{x_n^5 - (2 + r)x_n^4 + (1 + 2r)x_n^3 - (1 + r)x_n^2 + 2(1 - r)x_n + r - 1}{5x_n^4 - 4(2 + r)x_n^3 + 3(1 + 2r)x_n^2 - 2(1 + r)x_n + 2(1 - r)}$. Again, we substitute $r \approx 3.04042 \times 10^{-6}$. L_2 is slightly more than 1 AU from the sun and, judging from the result of part (a), probably less than 0.02 AU from Earth. So we take $x_1 = 1.02$ and get $x_2 \approx 1.01422$, $x_3 \approx 1.01118$, $x_4 \approx 1.01018$, $x_5 \approx 1.01008 \approx x_6$. So, to five decimal places, L_2 is located 1.01008 AU from the sun (or 0.01008 AU from Earth).

REVIEW EXERCISES FOR CHAPTER 2

1. False; see the warning after Theorem 2.1.8.

2. True. This is the Sum Rule.

3. False. See the discussion before the Product Rule.

4. True. This is the Chain Rule.

5. True, by the Chain Rule.

6. False, $\frac{d}{dx}f(\sqrt{x}) = \frac{f'(\sqrt{x})}{2\sqrt{x}}$ by the Chain Rule.

7. False. $f(x) = |x^2 + x| = x^2 + x$ for $x \geq 0$ or $x \leq -1$ and $|x^2 + x| = -(x^2 + x)$ for $-1 < x < 0$. So $f'(x) = 2x + 1$ for $x > 0$ or $x < -1$ and $f'(x) = -(2x + 1)$ for $-1 < x < 0$. But $|2x + 1| = 2x + 1$ for $x \geq -\frac{1}{2}$ and $|2x + 1| = -2x - 1$ for $x < -\frac{1}{2}$.

8. True. $f'(r)$ exists $\Rightarrow$ f is differentiable at r $\Rightarrow$ f is continuous at r $\Rightarrow$ $\lim\limits_{x\to r} f(x) = f(r)$.

9. True. $g(x) = x^5 \quad \Rightarrow \quad g'(x) = 5x^4 \quad \Rightarrow \quad g'(2) = 5(2)^4 = 80$, and by the definition of the derivative, $\lim\limits_{x\to 2} \frac{g(x) - g(2)}{x - 2} = g'(2) = 80$.

10. False. $\frac{d^2y}{dx^2}$ is the second derivative while $\left(\frac{dy}{dx}\right)^2$ is the first derivative squared. For example, if $y = x$, then $\frac{d^2y}{dx^2} = 0$, but $\left(\frac{dy}{dx}\right)^2 = 1$.

11. False. A tangent to the parabola has slope $\frac{dy}{dx} = 2x$, so at $(-2, 4)$ the slope of the tangent is $2(-2) = -4$ and the equation is $y - 4 = -4(x + 2)$. [The equation $y - 4 = 2x(x + 2)$ is not even linear!]

12. True. $D(\tan^2 x) = 2\tan x\,(\sec^2 x)$, and $D(\sec^2 x) = 2\sec x\,(\sec x \tan x) = 2\tan x\,(\sec^2 x)$. We can also show this by differentiating the identity $\tan^2 x + 1 = \sec^2 x$: we get $\frac{d}{dx}(\tan^2 x + 1) = \frac{d}{dx}\tan^2 x = \frac{d}{dx}\sec^2 x$.

13. $f(x) = x^3 + 5x + 4 \quad \Rightarrow$

$$f'(x) = \lim_{h\to 0} \frac{f(x+h) - f(x)}{h} = \lim_{h\to 0} \frac{(x+h)^3 + 5(x+h) + 4 - (x^3 + 5x + 4)}{h}$$

$$= \lim_{h\to 0} \frac{3x^2h + 3xh^2 + h^3 + 5h}{h} = \lim_{h\to 0}(3x^2 + 3xh + h^2 + 5) = 3x^2 + 5$$

14. $f(x) = \frac{4 - x}{3 + x} \quad \Rightarrow$

$$f'(x) = \lim_{h\to 0} \frac{f(x+h) - f(x)}{h} = \lim_{h\to 0} \frac{\frac{4 - (x+h)}{3 + (x+h)} - \frac{4 - x}{3 + x}}{h}$$

$$= \lim_{h\to 0} \frac{(4 - x - h)(3 + x) - (4 - x)(3 + x + h)}{h(3 + x + h)(3 + x)} = \lim_{h\to 0} \frac{-7h}{h(3 + x + h)(3 + x)}$$

$$= \lim_{h\to 0} \frac{-7}{(3 + x + h)(3 + x)} = -\frac{7}{(3 + x)^2}$$

15. $f(x) = \sqrt{3-5x} \quad \Rightarrow$

$$f'(x) = \lim_{h\to 0} \frac{f(x+h)-f(x)}{h} = \lim_{h\to 0} \frac{\sqrt{3-5(x+h)} - \sqrt{3-5x}}{h}$$

$$= \lim_{h\to 0} \frac{\sqrt{3-5x-5h} - \sqrt{3-5x}}{h} \left(\frac{\sqrt{3-5x-5h} + \sqrt{3-5x}}{\sqrt{3-5x-5h} + \sqrt{3-5x}} \right)$$

$$= \lim_{h\to 0} \frac{-5h}{h\left(\sqrt{3-5x-5h} + \sqrt{3-5x}\right)} = \lim_{h\to 0} \frac{-5}{\sqrt{3-5x-5h} + \sqrt{3-5x}} = \frac{-5}{2\sqrt{3-5x}}$$

16. $f(x) = x\sin x \quad \Rightarrow$

$$f'(x) = \lim_{h\to 0} \frac{f(x+h)-f(x)}{h} = \lim_{h\to 0} \frac{(x+h)\sin(x+h) - x\sin x}{h}$$

$$= \lim_{h\to 0} \frac{(x+h)(\sin x\cos h + \cos x \sin h) - x\sin x}{h}$$

$$= \lim_{h\to 0} \frac{x\sin x\,(\cos h - 1) + x\cos x\sin h + h(\sin x\cos h + \sin h\cos x)}{h}$$

$$= x\sin x \lim_{h\to 0} \frac{\cos h - 1}{h} + x\cos x \lim_{h\to 0} \frac{\sin h}{h} + \sin x \lim_{h\to 0} \cos h + \cos x \lim_{h\to 0} \sin h$$

$$= x\sin x\,(0) + x\cos x\,(1) + \sin x\,(1) + \cos x\,(0) = x\cos x + \sin x$$

17. $y = (x+2)^8(x+3)^6 \quad \Rightarrow \quad y' = 6(x+3)^5(x+2)^8 + 8(x+2)^7(x+3)^6 = 2(7x+18)(x+2)^7(x+3)^5$

18. $y = \sqrt[3]{x} + 1/\sqrt[3]{x} = x^{1/3} + x^{-1/3} \quad \Rightarrow \quad y' = \frac{1}{3}x^{-2/3} - \frac{1}{3}x^{-4/3}$

19. $y = \dfrac{x}{\sqrt{9-4x}} \quad \Rightarrow \quad y' = \dfrac{\sqrt{9-4x} - x\left[-4/\left(2\sqrt{9-4x}\right)\right]}{9-4x} = \dfrac{9-4x+2x}{(9-4x)^{3/2}} = \dfrac{9-2x}{(9-4x)^{3/2}}$

20. $y = \left(x + 1/x^2\right)^{\sqrt{7}} \quad \Rightarrow \quad y' = \sqrt{7}\left(x+1/x^2\right)^{\sqrt{7}-1}\left(1 - 2/x^3\right)$

21. $x^2y^3 + 3y^2 = x - 4y \quad \Rightarrow \quad 2xy^3 + 3x^2y^2y' + 6yy' = 1 - 4y' \quad \Rightarrow \quad y' = \dfrac{1-2xy^3}{3x^2y^2 + 6y + 4}$

22. $y = \left(1 - x^{-1}\right)^{-1} \quad \Rightarrow \quad y' = -\left(1-x^{-1}\right)^{-2}x^{-2} = -(x-1)^{-2}$

23. $y = \sqrt{x\sqrt{x\sqrt{x}}} = \left[x\left(x^{3/2}\right)^{1/2}\right]^{1/2} = \left[x\left(x^{3/4}\right)\right]^{1/2} = x^{7/8} \quad \Rightarrow \quad y' = \frac{7}{8}x^{-1/8}$

24. $y = -2/\sqrt[4]{x^3} = -2x^{-3/4} \quad \Rightarrow \quad y' = (-2)\left(-\frac{3}{4}\right)x^{-7/4} = \frac{3}{2}x^{-7/4}$

25. $y = \dfrac{x}{8-3x} \quad \Rightarrow \quad y' = \dfrac{(8-3x) - x(-3)}{(8-3x)^2} = \dfrac{8}{(8-3x)^2}$

26. $y\sqrt{x-1} + x\sqrt{y-1} = xy \quad \Rightarrow \quad y'\sqrt{x-1} + y\dfrac{1}{2\sqrt{x-1}} + \sqrt{y-1} + x\dfrac{1}{2\sqrt{x-1}}y' = y + xy' \quad \Rightarrow$

$$y' = \frac{y - \sqrt{y-1} - y/\left(2\sqrt{x-1}\right)}{\sqrt{x-1} - x + x/\left(2\sqrt{y-1}\right)}$$

27. $y = (x\tan x)^{1/5} \quad \Rightarrow \quad y' = \frac{1}{5}(x\tan x)^{-4/5}\left(\tan x + x\sec^2 x\right)$

28. $y = \sin(\cos x) \quad \Rightarrow \quad y' = \cos(\cos x)(-\sin x) = -\sin x \cos(\cos x)$

29. $x^2 = y(y+1) = y^2 + y \quad \Rightarrow \quad 2x = 2yy' + y' \quad \Rightarrow \quad y' = 2x/(2y+1)$

30. $y = \left(x + \sqrt{x}\right)^{-1/3} \quad \Rightarrow \quad y' = -\frac{1}{3}\left(x + \sqrt{x}\right)^{-4/3}\left(1 + \dfrac{1}{2\sqrt{x}}\right)$

31. $y = \dfrac{(x-1)(x-4)}{(x-2)(x-3)} = \dfrac{x^2 - 5x + 4}{x^2 - 5x + 6} \quad \Rightarrow$

$$y' = \frac{(x^2 - 5x + 6)(2x - 5) - (x^2 - 5x + 4)(2x - 5)}{(x^2 - 5x + 6)^2} = \frac{2(2x - 5)}{(x - 2)^2(x - 3)^2}$$

32. $y = \sqrt{\sin\sqrt{x}} \quad \Rightarrow \quad y' = \frac{1}{2}\left(\sin\sqrt{x}\right)^{-1/2}\left(\cos\sqrt{x}\right)\left(\dfrac{1}{2\sqrt{x}}\right) = \dfrac{\cos\sqrt{x}}{4\sqrt{x \sin\sqrt{x}}}$

33. $y = \tan\sqrt{1-x} \quad \Rightarrow \quad y' = \left(\sec^2\sqrt{1-x}\right)\left(\dfrac{1}{2\sqrt{1-x}}\right)(-1) = -\dfrac{\sec^2\sqrt{1-x}}{2\sqrt{1-x}}$

34. $y = \dfrac{1}{\sin(x - \sin x)} \quad \Rightarrow \quad y' = -\dfrac{\cos(x - \sin x)(1 - \cos x)}{\sin^2(x - \sin x)}$

35. $y = \sin\left(\tan\sqrt{1+x^3}\right) \quad \Rightarrow \quad y' = \cos\left(\tan\sqrt{1+x^3}\right)\left(\sec^2\sqrt{1+x^3}\right)\left[3x^2/\left(2\sqrt{1+x^3}\right)\right]$

36. $y = \dfrac{(x+\lambda)^4}{x^4 + \lambda^4} \quad \Rightarrow \quad y' = \dfrac{(x^4 + \lambda^4)(4)(x+\lambda)^3 - (x+\lambda)^4(4x^3)}{(x^4 + \lambda^4)^2} = \dfrac{4(x+\lambda)^3(\lambda^4 - \lambda x^3)}{(x^4 + \lambda^4)^2}$

37. $y = \cot(3x^2 + 5) \quad \Rightarrow \quad y' = -\csc^2(3x^2 + 5)(6x) = -6x\csc^2(3x^2 + 5)$

38. $y = (\sin mx)/x \quad \Rightarrow \quad y' = (mx\cos mx - \sin mx)/x^2$

39. $y = \cos^2(\tan x) \quad \Rightarrow \quad y' = 2\cos(\tan x)[-\sin(\tan x)]\sec^2 x = -\sin(2\tan x)\sec^2 x$

40. $x\tan y = y - 1 \quad \Rightarrow \quad \tan y + \left(x\sec^2 y\right) y' = y' \quad \Rightarrow \quad y' = \dfrac{\tan y}{1 - x\sec^2 y}$

41. $f(x) = (2x-1)^{-5} \quad \Rightarrow \quad f'(x) = -5(2x-1)^{-6}(2) = -10(2x-1)^{-6} \quad \Rightarrow$

$f''(x) = 60(2x-1)^{-7}(2) = 120(2x-1)^{-7} \quad \Rightarrow \quad f''(0) = 120(-1)^{-7} = -120$

42. $g(t) = \csc 2t \quad \Rightarrow \quad g'(t) = -2\csc 2t\cot 2t \quad \Rightarrow$

$g''(t) = -2(-2\csc 2t\cot 2t)\cot 2t - 2\csc 2t(-2\csc^2 2t) = 4\csc 2t(\cot^2 2t + \csc^2 2t) = 8\csc^3 2t - 4\csc 2t \quad \Rightarrow$

$g'''(t) = 24\csc^2 2t(-2\csc 2t\cot 2t) - 4(-2\csc 2t\cot 2t) = -48\csc^3 2t\cot 2t + 8\csc 2t\cot 2t \quad \Rightarrow$

$g'''\left(-\frac{\pi}{8}\right) = -48\left(-\sqrt{2}\right)^3(-1) + 8\left(-\sqrt{2}\right)(-1) = -88\sqrt{2}$

43. $x^6 + y^6 = 1 \quad \Rightarrow \quad 6x^5 + 6y^5y' = 0 \quad \Rightarrow \quad y' = -\dfrac{x^5}{y^5} \quad \Rightarrow$

$$y'' = -\frac{5x^4y^5 - x^5(5y^4y')}{y^{10}} = -\frac{5x^4y^5 - 5x^5y^4(-x^5/y^5)}{y^{10}} = -\frac{5x^4y^6 + 5x^{10}}{y^{11}} = -\frac{5x^4(y^6 + x^6)}{y^{11}} = -\frac{5x^4}{y^{11}}$$

44. $f(x) = (2-x)^{-1} \quad \Rightarrow \quad f'(x) = (2-x)^{-2} \quad \Rightarrow \quad f''(x) = 2(2-x)^{-3} \quad \Rightarrow \quad f'''(x) = 2 \cdot 3(2-x)^{-4}$ $\Rightarrow \quad f^{(4)}(x) = 2 \cdot 3 \cdot 4(2-x)^{-5}$. In general, $f^{(n)}(x) = 2 \cdot 3 \cdot 4 \cdots n(2-x)^{-(n+1)} = n!/(2-x)^{(n+1)}$.

45. $\displaystyle \lim_{x \to 0} \frac{\sec x}{1 - \sin x} = \frac{\sec 0}{1 - \sin 0} = \frac{1}{1 - 0} = 1$

46. $\displaystyle \lim_{t \to 0} \frac{t^3}{\tan^3 2t} = \lim_{t \to 0} \frac{t^3 \cos^3 2t}{\sin^3 2t} = \lim_{t \to 0} \cos^3 2t \cdot \frac{1}{8 \dfrac{\sin^3 2t}{(2t)^3}} = \lim_{t \to 0} \frac{\cos^3 2t}{8 \left(\lim\limits_{t \to 0} \dfrac{\sin 2t}{2t} \right)^3} = \frac{1}{8 \cdot 1^3} = \frac{1}{8}$

47. $y = \dfrac{x}{x^2 - 2} \quad \Rightarrow \quad y' = \dfrac{(x^2-2) - x(2x)}{(x^2-2)^2} = \dfrac{-x^2-2}{(x^2-2)^2}$. When $x = 2$, $y' = \dfrac{-2^2-2}{(2^2-2)^2} = -\dfrac{3}{2}$, so the equation of the tangent at $(2,1)$ is $y - 1 = -\frac{3}{2}(x-2)$ or $3x + 2y - 8 = 0$.

48. $\sqrt{x} + \sqrt{y} = 3 \quad \Rightarrow \quad \dfrac{1}{2\sqrt{x}} + \dfrac{1}{2\sqrt{y}} y' = 0 \quad \Rightarrow \quad y' = -\dfrac{\sqrt{y}}{\sqrt{x}}$. At $(4,1)$, $y' = -\frac{1}{2}$, so the tangent is $y - 1 = -\frac{1}{2}(x-4)$ or $x + 2y = 6$.

49. $y = \tan x \quad \Rightarrow \quad y' = \sec^2 x$. When $x = \frac{\pi}{3}$, $y' = 2^2 = 4$, so the equation of the tangent line at $\left(\frac{\pi}{3}, \sqrt{3}\right)$ is $y - \sqrt{3} = 4\left(x - \frac{\pi}{3}\right)$ or $y = 4x + \sqrt{3} - \frac{4}{3}\pi$.

50. $y = x\sqrt{1+x^2} \quad \Rightarrow \quad y' = \sqrt{1+x^2} + x^2/\sqrt{1+x^2}$. When $x = 1$, $y = \sqrt{2} + \frac{1}{\sqrt{2}} = \frac{3\sqrt{2}}{2}$, so the equation of the tangent line at $\left(1, \sqrt{2}\right)$ is $y - \sqrt{2} = \frac{3\sqrt{2}}{2}(x-1)$ or $3\sqrt{2}x - 2y = \sqrt{2}$.

51. $y = \sin x + \cos x \quad \Rightarrow \quad y' = \cos x - \sin x = 0 \quad \Leftrightarrow \quad \cos x = \sin x$ and $0 \le x \le 2\pi \quad \Leftrightarrow \quad x = \frac{\pi}{4}$ or $\frac{5\pi}{4}$, so the points are $\left(\frac{\pi}{4}, \sqrt{2}\right)$ and $\left(\frac{5\pi}{4}, -\sqrt{2}\right)$.

52. $x^2 + 2y^2 = 1 \;\Rightarrow\; 2x + 4yy' = 0 \;\Rightarrow\; y' = -x/(2y) = 1 \;\Leftrightarrow\; x = -2y$. Since the points lie on the ellipse, we have $(-2y)^2 + 2y^2 = 1 \quad \Rightarrow \quad 6y^2 = 1 \quad \Rightarrow \quad y = \pm\frac{1}{\sqrt{6}}$. The points are $\left(-\frac{2}{\sqrt{6}}, \frac{1}{\sqrt{6}}\right)$ and $\left(\frac{2}{\sqrt{6}}, -\frac{1}{\sqrt{6}}\right)$.

53. $f(x) = (x-a)(x-b)(x-c) \quad \Rightarrow \quad f'(x) = (x-b)(x-c) + (x-a)(x-c) + (x-a)(x-b)$. So $\dfrac{f'(x)}{f(x)} = \dfrac{(x-b)(x-c) + (x-a)(x-c) + (x-a)(x-b)}{(x-a)(x-b)(x-c)} = \dfrac{1}{x-a} + \dfrac{1}{x-b} + \dfrac{1}{x-c}$.

54. **(a)** $\cos 2x = \cos^2 x - \sin^2 x \quad \Rightarrow \quad -2\sin 2x = -2\cos x \sin x - 2\sin x \cos x \quad \Leftrightarrow \quad \sin 2x = 2\sin x \cos x$

(b) $\sin(x+a) = \sin x \cos a + \cos x \sin a \quad \Rightarrow \quad \cos(x+a) = \cos x \cos a - \sin x \sin a$.

55. **(a)** $h'(x) = f'(x)g(x) + f(x)g'(x) \quad \Rightarrow \quad h'(2) = f'(2)g(2) + f(2)g'(2) = (-2)(5) + (3)(4) = 2$

(b) $F'(x) = f'(g(x))g'(x) \quad \Rightarrow \quad F'(2) = f'(g(2))g'(2) = f'(5)(4) = 11 \cdot 4 = 44$

56. **(a)**

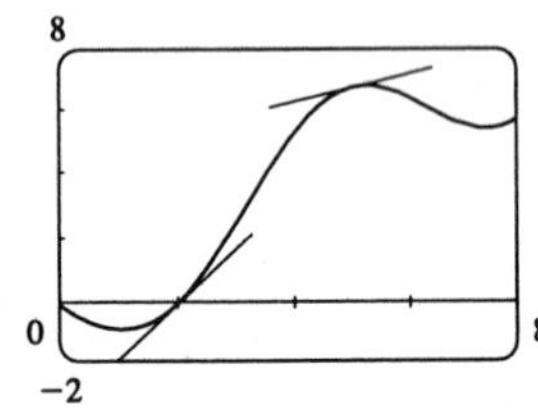

(b) The average rate of change is larger on $[2, 3]$.

(c) The instantaneous rate of change (the slope of the tangent) is larger at $x = 2$.

(d) $f'(x) = 1 - 2\cos x$, so $f'(2) = 1 - \cos 2 \approx 1.8323$ and $f'(5) = 1 - \cos 5 \approx 0.4327$. So $f'(2) > f'(5)$, as predicted in part (c).

57. The graph of a has tangent lines with positive slope for $x < 0$ and negative slope for $x > 0$, and the values of c fit this pattern, so c must be the graph of the derivative of the function for a. The graph of c has horizontal tangent lines to the left and right of the x-axis and b has zeros at these points. Hence b is the graph of the derivative of the function for c. Therefore a is the graph of f, c is the graph of f', and b is the graph of f''.

58.

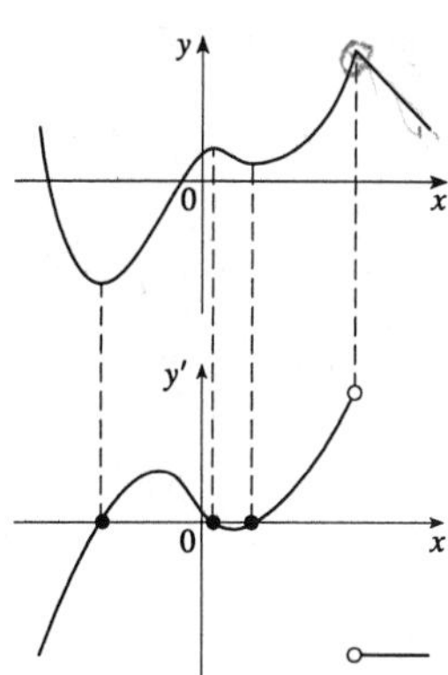

59. **(a)** $f(x) = x\sqrt{5-x} \quad\Rightarrow\quad f'(x) = \dfrac{-x}{2\sqrt{5-x}} + \sqrt{5-x} = \dfrac{10-3x}{2\sqrt{5-x}}$

(b) At $(1,2)$: $f'(1) = 1\left(\dfrac{-1}{2\sqrt{5-1}}\right) + \sqrt{4} = \dfrac{7}{4}$. So the equation of the tangent is

$y - 2 = \frac{7}{4}(x-1) \quad\Leftrightarrow\quad y = \frac{7}{4}x + \frac{1}{4}$.

At $(4,4)$: $f'(4) = 4\left(\frac{-1}{2\sqrt{1}}\right) + \sqrt{1} = -1$.

So the equation of the tangent is

$y - 4 = -(x-4) \quad\Leftrightarrow\quad y = -x + 8$.

(c)

(d)

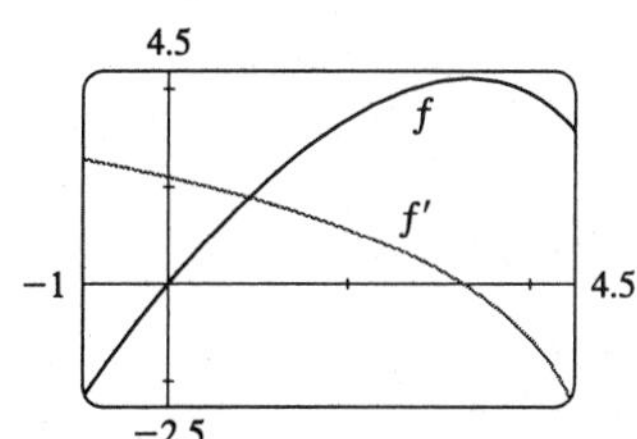

The graphs look reasonable, since f' is positive where f has tangents with positive slope, and f' is negative where f has tangents with negative slope.

60. **(a)** $f(x) = 4x - \tan x \quad\Rightarrow\quad f'(x) = 4 - \sec^2 x \quad\Rightarrow$

$f''(x) = -2\sec x(\sec x \tan x) = -2\sec^2 x \tan x$.

(b) We can see that our answers are reasonable, since the graph of f' is 0 where that of f has a horizontal tangent, and the graph of f' is positive where the graph of f has tangents with positive slope and negative where the graph of f has tangents with negative slope. The same correspondence holds between the graphs of f' and f''.

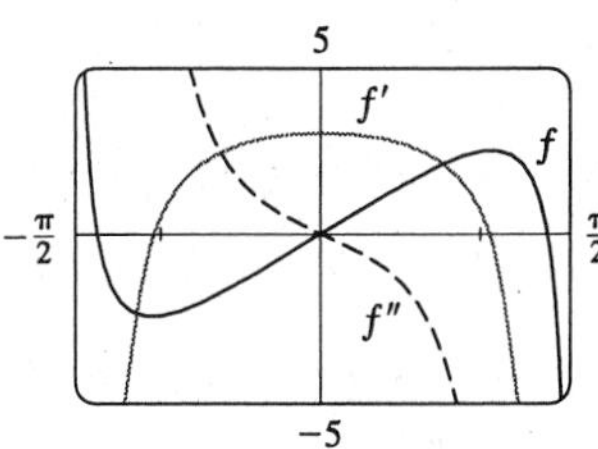

61. $f(x) = x^2 g(x) \quad\Rightarrow\quad f'(x) = 2xg(x) + x^2 g'(x)$

62. $f(x) = g(x^2) \quad\Rightarrow\quad f'(x) = g'(x^2)(2x)$

63. $f(x) = (g(x))^2 \quad\Rightarrow\quad f'(x) = 2g(x)\,g'(x)$

64. $f(x) = x^a g(x^b) \quad\Rightarrow\quad f'(x) = ax^{a-1}g(x^b) + x^a g'(x^b)\left(bx^{b-1}\right) = ax^{a-1}g(x^b) + bx^{a+b-1}g'(x^b)$

65. $f(x) = g(g(x)) \quad \Rightarrow \quad f'(x) = g'(g(x))g'(x)$

66. $f(x) = g(\tan\sqrt{x}) \quad \Rightarrow \quad f'(x) = g'(\tan\sqrt{x}) \cdot \dfrac{d}{dx}(\tan\sqrt{x})$

$= g'(\tan\sqrt{x}) \cdot \sec^2\sqrt{x} \cdot \dfrac{d}{dx}(\sqrt{x}) = \dfrac{g'(\tan\sqrt{x})\sec^2\sqrt{x}}{2\sqrt{x}}$

67. $h(x) = \dfrac{f(x)g(x)}{f(x) + g(x)} \quad \Rightarrow$

$h'(x) = \dfrac{[f'(x)g(x) + f(x)g'(x)][f(x) + g(x)] - f(x)g(x)[f'(x) + g'(x)]}{[f(x) + g(x)]^2} = \dfrac{f'(x)[g(x)]^2 + g'(x)[f(x)]^2}{[f(x) + g(x)]^2}$

68. $h(x) = \sqrt{\dfrac{f(x)}{g(x)}} \quad \Rightarrow \quad h'(x) = \dfrac{f'(x)g(x) - f(x)g'(x)}{2\sqrt{f(x)/g(x)}\,[g(x)]^2} = \dfrac{f'(x)g(x) - f(x)g'(x)}{2[g(x)]^{3/2}\sqrt{f(x)}}$

69. Using the Chain Rule repeatedly, $h(x) = f(g(\sin 4x)) \quad \Rightarrow$

$h'(x) = f'(g(\sin 4x)) \cdot \dfrac{d}{dx}(g(\sin 4x)) = f'(g(\sin 4x)) \cdot g'(\sin 4x) \cdot \dfrac{d}{dx}(\sin 4x)$

$= f'(g(\sin 4x))g'(\sin 4x)(\cos 4x)(4)$

70. **(a)** $x = \sqrt{b^2 + c^2t^2} \quad \Rightarrow \quad v(t) = x' = \left[1/\left(2\sqrt{b^2 + c^2t^2}\right)\right]2c^2t = c^2t/\sqrt{b^2 + c^2t^2} \quad \Rightarrow$

$a(t) = v'(t) = \dfrac{c^2\sqrt{b^2 + c^2t^2} - c^2t\left(c^2t/\sqrt{b^2 + c^2t^2}\right)}{b^2 + c^2t^2} = \dfrac{b^2c^2}{(b^2 + c^2t^2)^{3/2}}$

(b) $v(t) > 0$ for $t > 0$, so the particle always moves in the positive direction.

71. **(a)** $y = t^3 - 12t + 3 \quad \Rightarrow \quad v(t) = y' = 3t^2 - 12 \quad \Rightarrow \quad a(t) = v'(t) = 6t$

(b) $v(t) = 3(t^2 - 4) > 0$ when $t > 2$, so it moves upward when $t > 2$ and downward when $0 \le t < 2$.

(c) Distance upward $= y(3) - y(2) = -6 - (-13) = 7$,

Distance downward $= y(0) - y(2) = 3 - (-13) = 16$. Total distance $= 7 + 16 = 23$

72. **(a)** $V = \frac{1}{3}\pi r^2 h \quad \Rightarrow \quad dV/dh = \frac{1}{3}\pi r^2$ **(b)** $dV/dr = \frac{2}{3}\pi rh$

73. $\rho = x(1 + \sqrt{x}) = x + x^{3/2} \quad \Rightarrow \quad d\rho/dx = 1 + \frac{3}{2}\sqrt{x}$, so the density when $x = 4$ is $1 + \frac{3}{2}\sqrt{4} = 4$ kg/m.

74. $C(x) = 950 + 12x + 0.01x^2 \quad \Rightarrow \quad C'(x) = 12 + 0.02x$, so the marginal cost when $x = 200$ is $C'(200) = 12 + 0.02(200) = 12 + 4 = \$16$/unit. The cost to produce the 201st unit is $C(201) - C(200) = 3766.01 - 3750 = \$16.01$.

75. If x = edge length, then $V = x^3 \quad \Rightarrow \quad dV/dt = 3x^2\,dx/dt = 10 \quad \Rightarrow \quad dx/dt = 10/(3x^2)$ and $S = 6x^2$ $\Rightarrow \quad dS/dt = (12x)dx/dt = 12x\,[10/(3x^2)] = 40/x$. When $x = 30$, $dS/dt = \frac{40}{30} = \frac{4}{3}$ cm^2/min.

76. Given $dV/dt = 2$, find dh/dt when $h = 5$. $V = \frac{1}{3}\pi r^2 h$ and, from similar triangles, $\dfrac{r}{h} = \dfrac{3}{10}$ $\Rightarrow$ $V = \dfrac{\pi}{3}\left(\dfrac{3h}{10}\right)^2 h = \dfrac{3\pi}{100}h^3$, so

$$2 = \frac{dV}{dt} = \frac{9\pi}{100}h^2\,\frac{dh}{dt} \quad\Rightarrow\quad \frac{dh}{dt} = \frac{200}{9\pi h^2} = \frac{200}{9\pi(5)^2} = \frac{8}{9\pi}\text{ cm/s}$$

when $h = 5$.

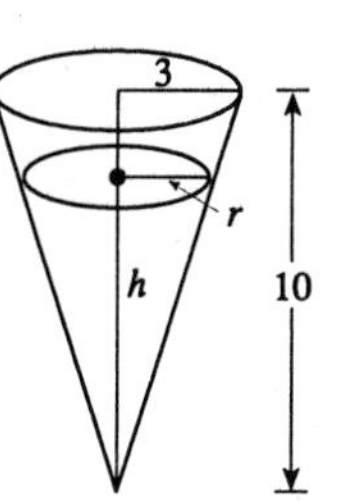

77. Given $dh/dt = 5$ and $dx/dt = 15$, find dz/dt. $z^2 = x^2 + h^2$ $\Rightarrow$

$$2z\,\frac{dz}{dt} = 2x\,\frac{dx}{dt} + 2h\,\frac{dh}{dt} \quad\Rightarrow\quad \frac{dz}{dt} = \frac{1}{z}(15x + 5h).$$

When $t = 3$, $h = 45 + 3(5) = 60$ and $x = 15(3) = 45$ $\Rightarrow$ $z = 75$, so $\dfrac{dz}{dt} = \frac{1}{75}[15(45) + 5(60)] = 13$ ft/s.

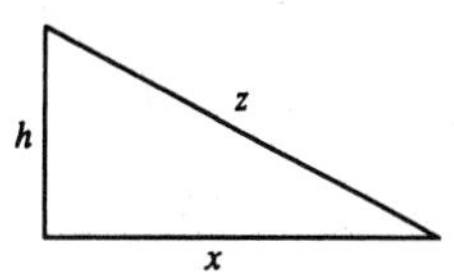

78. We are given $dz/dt = 30$ ft/s. By similar triangles, $\dfrac{y}{z} = \dfrac{4}{\sqrt{241}}$ $\Rightarrow$ $y = \dfrac{4}{\sqrt{241}}z$, so $\dfrac{dy}{dt} = \dfrac{4}{\sqrt{241}}\dfrac{dz}{dt} = \dfrac{120}{\sqrt{241}} \approx 7.7$ ft/s.

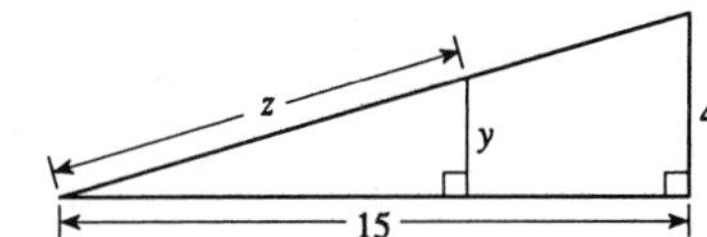

79. We are given $d\theta/dt = -0.25$ rad/h. $x = 400\cot\theta$ $\Rightarrow$ $\dfrac{dx}{dt} = -400\csc^2\theta\,\dfrac{d\theta}{dt}$. When $\theta = \dfrac{\pi}{6}$, $\dfrac{dx}{dt} = -400(2)^2(-0.25) = 400$ ft/h.

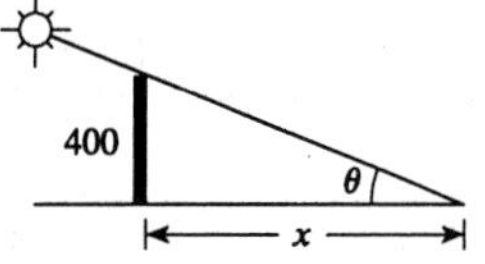

80. $y = (4 - x^2)^{3/2}$ $\Rightarrow$ $dy = \frac{3}{2}(4 - x^2)^{1/2}(-2x)dx = -3x(t4 - x^2)^{1/2}dx$.

81. $y = x^3 - 2x^2 + 1$ $\Rightarrow$ $dy = (3x^2 - 4x)dx$. When $x = 2$ and $dx = 0.2$, $dy = \left[3(2)^2 - 4(2)\right](0.2) = 0.8$.

82. Let $y = f(x) = 8 + \sqrt{x}$. Then $dy = \left[1/(2\sqrt{x})\right]dx$. When $x = 144$ and $dx = \Delta x = -0.4$, $dy = \frac{1}{24}(-0.4) = -\frac{1}{60}$. Thus $8 + \sqrt{143.6} = f(143.6) \approx f(144) + dy = 20 - \frac{1}{60} \approx 19.983$.

83. $f(x) = \sqrt[3]{1 + 3x} = (1 + 3x)^{1/3}$ $\Rightarrow$ $f'(x) = (1 + 3x)^{-2/3}$ so $L(x) = f(0) + f'(0)(x - 0) = 1^{1/3} + 1^{-2/3}x = 1 + x$. Thus $\sqrt[3]{1 + 3x} \approx 1 + x$ $\Rightarrow$ $\sqrt[3]{1.03} = \sqrt[3]{1 + 3(0.01)} \approx 1 + (0.01) = 1.01$.

84. $A = x^2 + \frac{1}{2}\pi\left(\frac{1}{2}x\right)^2 = \left(1 + \frac{\pi}{8}\right)x^2$ $\Rightarrow$ $dA = \left(2 + \frac{\pi}{4}\right)x\,dx$. When $x = 60$ and $dx = 0.1$, $dA = \left(2 + \frac{\pi}{4}\right)60(0.1) = 12 + \frac{3\pi}{2}$ so the maximum error $\approx 12 + \frac{3\pi}{2} \approx 16.7$ cm^2.

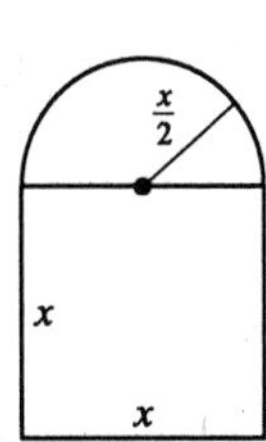

85. The linear approximation is $\sqrt[3]{1+3x} \approx 1+x$, so for the required accuracy we want
$\sqrt[3]{1+3x} - 0.1 < 1+x < \sqrt[3]{1+3x} + 0.1$.
From the graph, it appears that this is true when $-0.23 < x < 0.40$.

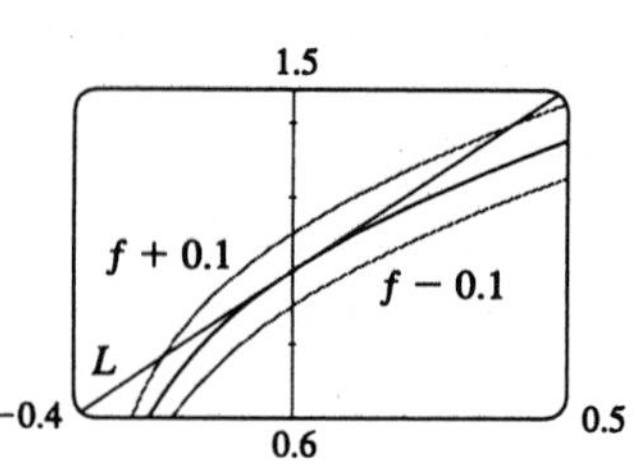

86. (a) $f(x) = \sqrt{25-x^2} \quad \Rightarrow \quad f'(x) = \dfrac{-2x}{2\sqrt{25-x^2}} = -x\left(25-x^2\right)^{-1/2} \quad \Rightarrow$

$f''(x) = -x\left(-\frac{1}{2}\right)\left(25-x^2\right)^{-3/2}(-2x) - \left(25-x^2\right)^{-1/2} = \dfrac{-25}{\left(25-x^2\right)^{3/2}}$. So the linear approximation to $f(x)$ near 3 is

$f(x) \approx f(3) + f'(3)(x-3) = 4 - \frac{3}{4}(x-3)$,

and the quadratic approximation is

$$f(x) \approx f(3) + f'(3)(x-3) + \tfrac{1}{2}f''(3)(x-3)^2$$
$$= 4 - \tfrac{3}{4}(x-3) - \tfrac{25}{128}(x-3)^2.$$

(b)

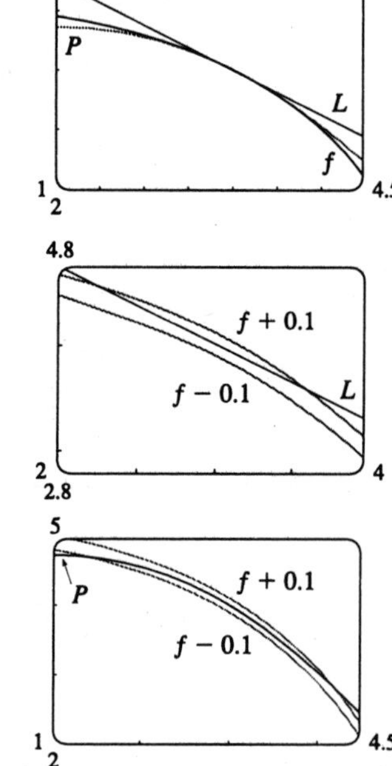

(c) For the required accuracy, we want
$\sqrt{25-x^2} - 0.1 < 4 - \frac{3}{4}(x-3)$
and $4 - \frac{3}{4}(x-3) < \sqrt{25-x^2} + 0.1$.
From the graph, it appears that these both hold for $2.24 < x < 3.66$.

(d) We want
$\sqrt{25-x^2} - 0.1 < 4 - \frac{3}{4}(x-3) - \frac{25}{128}(x-3)^2$
and $4 - \frac{3}{4}(x-3) - \frac{25}{128}(x-3)^2 < \sqrt{25-x^2} + 0.1$.
From the graph it appears that these both hold for $1.41 < x < 4.18$

87. $f(x) = x^4 + x - 1 \quad \Rightarrow \quad f'(x) = 4x^3 + 1 \quad \Rightarrow \quad x_{n+1} = x_n - \dfrac{x_n^4 + x_n - 1}{4x_n^3 + 1}$. If $x_1 = 0.5$ then $x_2 \approx 0.791667$, $x_3 \approx 0.729862$, $x_4 \approx 0.724528$, $x_5 \approx 0.724492$ and $x_6 \approx 0.724492$, so, to six decimal places, the root is 0.724492.

88. $f(x) = x - 6\cos x \quad \Rightarrow \quad f'(x) = 1 + 6\sin x \quad \Rightarrow \quad x_{n+1} = x_n - \dfrac{x_n - 6\cos x_n}{1 + 6\sin x_n}$. From the graphs of $y = \cos x$ and $y = x/6$, it appears that there are roots near 1, -2 and -4. If $x_1 = 1$, then $x_2 \approx 1.370620$, $x_3 \approx 1.344812$, $x_4 \approx 1.344751$ and $x_5 \approx 1.344751$. If $x_1 = -2$, then $x_2 \approx -1.888486$, $x_3 \approx -1.891518$, $x_4 \approx -1.891520$ and $x_5 \approx -1.891520$. If $x_1 = -4$, then $x_2 \approx -3.985898$, $x_3 \approx -3.985826$ and $x_4 \approx -3.985826$. So, to six decimal places, the roots are 1.344751, -1.891520 and -3.985826.

89. $y = x^6 + 2x^2 - 8x + 3$ has a horizontal tangent when $y' = 6x^5 + 4x - 8 = 0$. Let $f(x) = 6x^5 + 4x - 8$. Then $f'(x) = 30x^4 + 4$, so $x_{n+1} = x_n - \dfrac{6x_n^5 + 4x_n - 8}{30\,x_n^4 + 4}$. A sketch shows that the root is near 1, so we take $x_1 = 1$. Then $x_2 \approx 0.9412$, $x_3 \approx 0.9341$, $x_4 \approx 0.9340$ and $x_5 \approx 0.9340$. Thus, to four decimal places, the point is $(0.9340, -2.0634)$.

90. $\displaystyle\lim_{x\to 1} \frac{x^{17} - 1}{x - 1} = \frac{d}{dx} x^{17}\bigg|_{x=1} = 17(1)^{16} = 17$

91. $\displaystyle\lim_{h\to 0} \frac{(2+h)^6 - 64}{h} = \frac{d}{dx} x^6\bigg|_{x=2} = 6(2)^5 = 192$

92. $\displaystyle\lim_{\theta\to\pi/3} \frac{\cos\theta - 0.5}{\theta - \pi/3} = \frac{d}{d\theta}\cos\theta\bigg|_{\theta=\pi/3} = -\sin\tfrac{\pi}{3} = -\tfrac{\sqrt{3}}{2}$

93. Differentiating the expression for $g(x)$ and using the Chain Rule repeatedly, we obtain

$$\begin{aligned} g(x) &= f(x^3 + f(x^2 + f(x))) \quad \Rightarrow \\ g'(x) &= f'\big(x^3 + f(x^2 + f(x))\big)\big(x^3 + f(x^2 + f(x))\big)' \\ &= f'\big(x^3 + f(x^2 + f(x))\big)\Big(3x^2 + f'\big(x^2 + f(x)\big)\big[x^2 + f(x)\big]'\Big) \\ &= f'\big(x^3 + f(x^2 + f(x))\big)\big(3x^2 + f'\big(x^2 + f(x)\big)[2x + f'(x)]\big). \text{ So} \\ g'(1) &= f'\big(1^3 + f(1^2 + f(1))\big)\big(3\cdot 1^2 + f'\big(1^2 + f(1)\big)[2\cdot 1 + f'(1)]\big) \\ &= f'(1 + f(1+1))(3 + f'(1+1)[2 + f'(1)]) = f'(1+2)[3 + f'(2)(2+1)] = 3(3 + 2\cdot 3) = 27. \end{aligned}$$

94. Differentiating the first given equation implicitly with respect to x and using the Chain Rule, we obtain $f(g(x)) = x \quad\Rightarrow\quad f'(g(x))g'(x) = 1 \quad\Rightarrow\quad g'(x) = \dfrac{1}{f'(g(x))}$. Using the second given equation to expand the denominator of this expression gives $g'(x) = \dfrac{1}{1 + [f(g(x))]^2}$. But the first given equation states that $f(g(x)) = x$, so $g'(x) = \dfrac{1}{1 + x^2}$.

95.

$$\begin{aligned} \lim_{x\to 0}\frac{\sqrt{1+\tan x} - \sqrt{1+\sin x}}{x^3} &= \lim_{x\to 0}\frac{\big(\sqrt{1+\tan x} - \sqrt{1+\sin x}\big)\big(\sqrt{1+\tan x} + \sqrt{1+\sin x}\big)}{x^3\big(\sqrt{1+\tan x} + \sqrt{1+\sin x}\big)} \\ &= \lim_{x\to 0}\frac{(1+\tan x) - (1+\sin x)}{x^3\big(\sqrt{1+\tan x} + \sqrt{1+\sin x}\big)} = \lim_{x\to 0}\frac{\sin x(1/\cos x - 1)\cos x}{x^3\big(\sqrt{1+\tan x} + \sqrt{1+\sin x}\big)\cos x} \\ &= \lim_{x\to 0}\frac{\sin x\,(1-\cos x)(1+\cos x)}{x^3\big(\sqrt{1+\tan x} + \sqrt{1+\sin x}\big)\cos x\,(1+\cos x)} \\ &= \lim_{x\to 0}\frac{\sin x\cdot\sin^2 x}{x^3\big(\sqrt{1+\tan x} + \sqrt{1+\sin x}\big)\cos x\,(1+\cos x)} \\ &= \left(\lim_{x\to 0}\frac{\sin x}{x}\right)^3 \lim_{x\to 0}\frac{1}{\big(\sqrt{1+\tan x} + \sqrt{1+\sin x}\big)\cos x\,(1+\cos x)} \\ &= 1^3\cdot\frac{1}{\big(\sqrt{1} + \sqrt{1}\big)\cdot 1\cdot(1+1)} = \frac{1}{4} \end{aligned}$$

96. 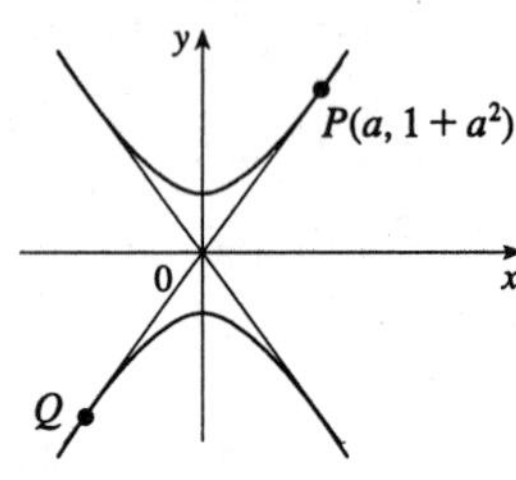

Let P have coordinates $(a, 1+a^2)$. By symmetry, Q has coordinates $(-a, -(1+a^2))$. The tangent line to $y = 1 + x^2$ through P has slope $2a$ and the tangent line through Q to $y = -1 - x^2$ has slope $-2(-a) = 2a$. Therefore

$$m_{PQ} = \frac{1 + a^2 - (-1 - a^2)}{a - (-a)} = \frac{1 + a^2}{a} = 2a \quad \Rightarrow$$

$2a^2 = 1 + a^2 \quad \Rightarrow \quad a^2 = 1 \quad \Rightarrow \quad a = \pm 1$. Therefore the points are $(1, 2)$, $(-1, -2)$ and, by symmetry, $(1, -2)$ and $(-1, 2)$.

97. We are given that $|f(x)| \le x^2$ for all x. In particular, $|f(0)| \le 0$, but $|a| \ge 0$ for all a. The only conclusion is that $f(0) = 0$. Now $\left|\dfrac{f(x) - f(0)}{x - 0}\right| = \left|\dfrac{f(x)}{x}\right| = \dfrac{|f(x)|}{|x|} \le \dfrac{x^2}{|x|} = \dfrac{|x^2|}{|x|} \quad \Rightarrow \quad -|x| \le \dfrac{f(x) - f(0)}{x - 0} \le |x|$. But $\lim\limits_{x \to 0} -|x| = 0 = \lim\limits_{x \to 0} |x|$, so by the Squeeze Theorem, $\lim\limits_{x \to 0} \dfrac{f(x) - f(0)}{x - 0} = 0$. So by the definition of the derivative, f is differentiable at 0 and, furthermore, $f'(0) = 0$.

98. Let (b, c) be on the curve, that is, $b^{2/3} + c^{2/3} = a^{2/3}$. Now $x^{2/3} + y^{2/3} = a^{2/3} \quad \Rightarrow \quad \frac{2}{3}x^{-1/3} + \frac{2}{3}y^{-1/3}\dfrac{dy}{dx} = 0$, so $\dfrac{dy}{dx} = -\dfrac{y^{1/3}}{x^{1/3}} = -\left(\dfrac{y}{x}\right)^{1/3}$, so at (b, c) the slope of the tangent line is $-\left(\dfrac{c}{b}\right)^{1/3}$ and the equation of the tangent line is $y - c = -\left(\dfrac{c}{b}\right)^{1/3}(x - b)$ or $y = -\left(\dfrac{c}{b}\right)^{1/3} x + \left(c + b^{2/3}c^{1/3}\right)$. Setting $y = 0$, we find that the x-intercept is $b^{1/3}c^{2/3} + b$ and setting $x = 0$ we find that the y-intercept is $c + b^{2/3}c^{1/3}$. So the length of the tangent line between these two points is $\sqrt{\left[b^{1/3}(c^{2/3} + b^{2/3})\right]^2 + \left[c^{1/3}(c^{2/3} + b^{2/3})\right]^2} = \sqrt{b^{2/3}(a^{2/3})^2 + c^{2/3}(a^{2/3})^2}$
$= \sqrt{(b^{2/3} + c^{2/3})a^{4/3}} = \sqrt{a^{2/3}a^{4/3}} = \sqrt{a^2} = a = \text{constant}$.

PROBLEMS PLUS (page 178)

1. Let a be the x-coordinate of Q. Then $y = 1 - x^2 \Rightarrow$ the slope at $Q = y'(a) = -2a$. But since the triangle is equilateral, $\angle ACB = 60°$, so that the slope at Q is $\tan 120° = -\sqrt{3}$. Therefore we must have that $-2a = -\sqrt{3} \Rightarrow a = \frac{\sqrt{3}}{2}$. Therefore the point Q has coordinates $\left(\frac{\sqrt{3}}{2}, 1 - \left(\frac{\sqrt{3}}{2}\right)^2\right) = \left(\frac{\sqrt{3}}{2}, \frac{1}{4}\right)$ and by symmetry P has coordinates $\left(-\frac{\sqrt{3}}{2}, \frac{1}{4}\right)$.

2. **(a)** This is a finite geometric sum with $a = 1$, $r = x$, and it has $n + 1$ terms. So $1 + x + \cdots + x^n = \dfrac{1 - x^{n+1}}{1 - x}$.

(b) We differentiate the sum in part (a):

$$1 + 2x + 3x^2 + \cdots + nx^{n-1} = \frac{d}{dx}(1 + x + \cdots + x^n) = \frac{d}{dx}\left(\frac{1 - x^{n+1}}{1 - x}\right)$$

$$= \frac{(1-x)[-(n+1)x^n] - (1 - x^{n+1})(-1)}{(1-x)^2} = \frac{1 - (n+1)x^n + nx^{n+1}}{(1-x)^2}$$

3. $1 + x + x^2 + \cdots + x^{100} = \dfrac{1 - x^{101}}{1 - x}$ $(x \neq 1)$. If $x = 1$, then the sum is clearly equal to $101 > 0$. If $x \geq 0$, then we have a sum of positive terms which is clearly positive. And if $x < 0$ then $x^{101} < 0 \Rightarrow 1 - x > 0$ and $1 - x^{101} > 0 \Rightarrow \dfrac{1 - x^{101}}{1 - x} > 0$. Therefore $1 + x + x^2 + \cdots + x^{100} = \dfrac{1 - x^{101}}{1 - x} \geq 0$ for all x.

4. **(a)** We consider three cases.

Case (i): If $x \geq 2$, then $|x + 1| = x + 1$ and $|x - 2| = x - 2$. So $|x+1| + |x-2| < 7 \Leftrightarrow x + 1 + x - 2 < 7 \Leftrightarrow 2x < 8 \Leftrightarrow x < 4$. So $x \in [2, 4)$.

Case (ii): If $-1 \leq x < 2$, then $|x + 1| = x + 1$, but $|x - 2| = -(x - 2)$. So $|x+1| + |x-2| < 7 \Leftrightarrow x + 1 - (x - 2) < 7 \Leftrightarrow 3 < 7$. Since this is always true, this case leads to $x \in [-1, 2)$.

Case (iii): If $x < -1$, then $|x + 1| = -(x + 1)$ and $|x - 2| = -(x - 2)$. So $|x+1| + |x-2| < 7 \Leftrightarrow -(x+1) - (x-2) < 7 \Leftrightarrow x > -3$.

Combining these three cases, we obtain the solution: $x \in (-3, 4)$.

(b) Consider the same three cases as in part (a) to get

$$f(x) = \begin{cases} -2x + 1 & \text{if } x < -1 \\ 3 & \text{if } -1 \leq x < 2 \\ 2x - 1 & \text{if } x \geq 2. \end{cases}$$

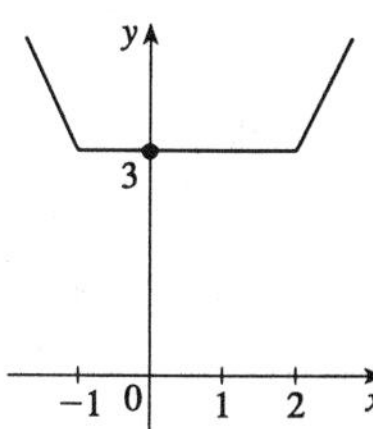

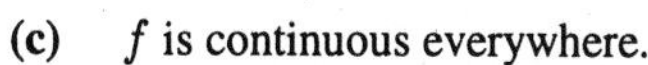

(c) f is continuous everywhere.

(d) From the graph in part (b) it appears that f is differentiable at all x except -1 and 2. This can be confirmed by computing the left- and right-hand derivatives at -1 and 2: $f'_-(-1) = -2$ but $f'_+(-1) = 0$, and similarly $f'_-(2) \neq f'_+(2)$.

(e) $g(x) = |x| + |x+1| + |x-1|$.
For $x \geq 1$, $g(x) = x + (x+1) + (x-1) = 3x$.
For $0 \leq x < 1$, $g(x) = x + (x+1) - (x-1) = x + 2$.
For $-1 \leq x < 0$, $g(x) = -x + (x+1) - (x-1) = -x + 2$.
For $x < -1$, $g(x) = -x - (x+1) - (x-1) = -3x$.

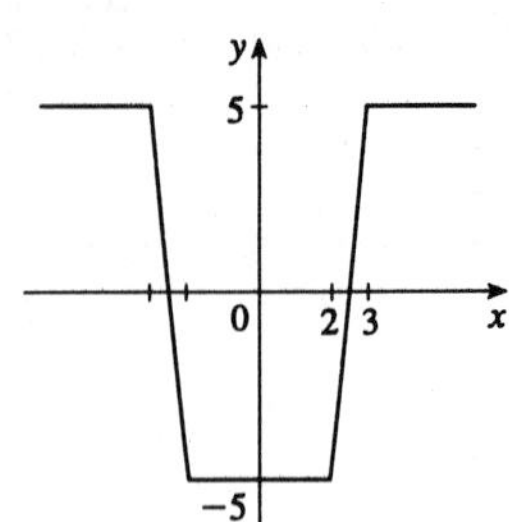

(f) As in part (d) we compute the left- and right-hand derivatives at -1, 0, and 1 to find that g is differentiable everywhere except at $x = 0, \pm 1$. This can also be seen from the graph.

5. For $-\frac{1}{2} < x < \frac{1}{2}$ we have $2x - 1 < 0$, so $|2x-1| = -(2x-1)$ and $2x + 1 > 0 \quad \Rightarrow \quad |2x+1| = 2x+1$.
Therefore, $\lim\limits_{x \to 0} \dfrac{|2x-1| - |2x+1|}{x} = \lim\limits_{x \to 0} \dfrac{-(2x-1) - (2x+1)}{x} = \lim\limits_{x \to 0} \dfrac{-4x}{x} = \lim\limits_{x \to 0}(-4) = -4$.

6. We find the equation of the parabola by substituting the point $(-100, 100)$, at which the car is situated, into the general equation $y = ax^2$: $100 = a(-100)^2 \quad \Rightarrow \quad a = \frac{1}{100}$. Now we find the equation of a tangent to the parabola at the point (x_0, y_0): $dy/dx = \frac{1}{100}(2x) = \frac{1}{50}x$, so the equation of the tangent is $y - y_0 = \frac{1}{50}x_0(x - x_0)$. Since the point (x_0, y_0) is on the parabola, we must have $y_0 = \frac{1}{100}x_0^2$, so the equation of the tangent can be simplified to y $= \frac{1}{100}x_0^2 + \frac{1}{50}x_0(x - x_0)$. We want the statue to be located on the tangent line, so we substitute its coordinates $(100, 50)$ into this equation: $50 = \frac{1}{100}x_0^2 + \frac{1}{50}x_0(100 - x_0) \quad \Rightarrow \quad x_0^2 - 200x_0 + 5000 = 0$ $\Rightarrow \quad x_0 = \frac{1}{2}\left[200 \pm \sqrt{200^2 - 4(5000)}\right] \quad \Rightarrow \quad x_0 = 100 \pm 50\sqrt{2}$. But $x_0 < 100$, so the car's headlights illuminate the statue when it is located at the point $\left(100 - 50\sqrt{2}, 150 - 100\sqrt{2}\right) \approx (29.3, 8.6)$, that is, about 29.3 m east and 8.6 m north of the origin.

7. $V = \frac{4}{3}\pi r^3 \quad \Rightarrow \quad \dfrac{dV}{dt} = 4\pi r^2 \dfrac{dr}{dt}$. But $\dfrac{dV}{dt}$ is proportional to the surface area, so $\dfrac{dV}{dt} = k \cdot 4\pi r^2$ for some constant k. Therefore $4\pi r^2 \dfrac{dr}{dt} = k \cdot 4\pi r^2 \quad \Rightarrow \quad \dfrac{dr}{dt} = k = \text{constant} \quad \Rightarrow \quad r = kt + r_0$. To find k we use the fact that when $t = 3$, $r = 3k + r_0$ and $V = \frac{1}{2}V_0 \quad \Rightarrow \quad \frac{4}{3}\pi(3k + r_0)^3 = \frac{1}{2} \cdot \frac{4}{3}\pi r_0^3 \quad \Rightarrow \quad (3k + r_0)^3 = \frac{1}{2}r_0^3$ $\Rightarrow \quad (3k + r_0) = \frac{1}{\sqrt[3]{2}}r_0 \quad \Rightarrow \quad k = \frac{1}{3}r_0\left(\frac{1}{\sqrt[3]{2}} - 1\right)$. Therefore $r = \frac{1}{3}r_0\left(\frac{1}{\sqrt[3]{2}} - 1\right)t + r_0$. When the snowball has melted completely we have $r = 0 \quad \Rightarrow \quad \frac{1}{3}r_0\left(\frac{1}{\sqrt[3]{2}} - 1\right)t + r_0 = 0$ which gives $t = \dfrac{3\sqrt[3]{2}}{\sqrt[3]{2} - 1}$. Therefore it takes $\dfrac{3\sqrt[3]{2}}{\sqrt[3]{2} - 1} - 3 = \dfrac{3}{\sqrt[3]{2} - 1} \approx 11$ hours and 33 minutes longer.

8.
$$\lim_{x \to a} \frac{f(x) - f(a)}{\sqrt{x} - \sqrt{a}} = \lim_{x \to a}\left(\frac{f(x) - f(a)}{\sqrt{x} - \sqrt{a}} \cdot \frac{\sqrt{x} + \sqrt{a}}{\sqrt{x} + \sqrt{a}}\right) = \lim_{x \to a}\left(\frac{f(x) - f(a)}{x - a} \cdot \left(\sqrt{x} + \sqrt{a}\right)\right)$$
$$= \lim_{x \to a} \frac{f(x) - f(a)}{x - a} \cdot \lim_{x \to a}\left(\sqrt{x} + \sqrt{a}\right) = f'(a) \cdot \left(\sqrt{a} + \sqrt{a}\right) = 2\sqrt{a}\, f'(a)$$

9. We use mathematical induction. Let S_n be the statement that $\dfrac{d^n}{dx^n}(\sin^4 x + \cos^4 x) = 4^{n-1}\cos(4x + n\pi/2)$. S_1 is true because $\dfrac{d}{dx}(\sin^4 x + \cos^4 x) = 4\sin^3 x\cos x - 4\cos^3 x\sin x = 4\sin x\cos x(\sin^2 x - \cos^2 x)$

$$= -4\sin x\cos x\cos 2x = -2\sin 2x\cos 2x = -\sin 4x$$
$$= \cos\left[\tfrac{\pi}{2} - (-4x)\right] = \cos\left(\tfrac{\pi}{2} + 4x\right)4^{n-1}\cos\left(4x + n\tfrac{\pi}{2}\right) \text{ when } n = 1.$$

Now assume S_k is true, that is, $\dfrac{d^k}{dx^k}(\sin^4 x + \cos^4 x) = 4^{k-1}\cos\left(4x + k\frac{\pi}{2}\right)$. Then

$$\frac{d^{k+1}}{dx^{k+1}}(\sin^4 x + \cos^4 x) = \frac{d}{dx}\left[\frac{d^k}{dx^k}(\sin^4 x + \cos^4 x)\right] = \frac{d}{dx}\left[4^{k-1}\cos\left(4x + k\tfrac{\pi}{2}\right)\right]$$
$$= -4^{k-1}\sin\left(4x + k\tfrac{\pi}{2}\right)\cdot\frac{d}{dx}\left(4x + k\tfrac{\pi}{2}\right) = -4^k\sin\left(4x + k\tfrac{\pi}{2}\right)$$
$$= 4^k\sin\left(-4x - k\tfrac{\pi}{2}\right) = 4^k\cos\left[\tfrac{\pi}{2} - \left(-4x - k\tfrac{\pi}{2}\right)\right]$$
$$= 4^k\cos\left[4x + (k+1)\tfrac{\pi}{2}\right] \text{ which shows that } S_{k+1} \text{ is true.}$$

Therefore $\dfrac{d_n}{dx_n}(\sin^4 x + \cos^4 x) = 4^{n-1}\cos\left(4x + n\frac{\pi}{2}\right)$ for every positive integer n by mathematical induction.

Another Proof: First write

$\sin^4 x + \cos^4 x = (\sin^2 x + \cos^2 x)^2 - 2\sin^2 x\cos^2 x = 1 - \frac{1}{2}\sin^2 2x = 1 - \frac{1}{4}(1 - \cos 4x) = \frac{3}{4} + \frac{1}{4}\cos 4x.$

Then we have $\dfrac{d^n}{dx^n}(\sin^4 x + \cos^4 x) = \dfrac{d^n}{dx^n}\left(\frac{3}{4} + \frac{1}{4}\cos 4x\right) = \frac{1}{4}\cdot 4^n\cos\left(4x + n\frac{\pi}{2}\right) = 4^{n-1}\cos\left(4x + n\frac{\pi}{2}\right).$

10. If we divide $1 - x$ into x^n by long division, we find that

$f(x) = \dfrac{x^n}{1 - x} = -x^{n-1} - x^{n-2} - \cdots - x - 1 + \dfrac{1}{1 - x}$. This can also be seen by multiplying the last expression by $1 - x$ and canceling terms on the right-hand side. So we let $g(x) = 1 + x + x^2 + \cdots + x^{n-1}$, so that $f(x) = \dfrac{1}{1 - x} - g(x) \quad\Rightarrow\quad f^{(n)}(x) = \left(\dfrac{1}{1 - x}\right)^{(n)} - g^{(n)}(x)$. But g is a polynomial of degree $(n - 1)$, so its nth derivative will be 0, and therefore $f^{(n)}(x) = \left(\dfrac{1}{1 - x}\right)^{(n)}$. Now

$\dfrac{d}{dx}(1 - x)^{-1} = (-1)(1 - x)^{-2}(-1) = (1 - x)^{-2}$, $\dfrac{d^2}{dx^2}(1 - x)^{-1} = (-2)(1 - x)^{-3}(-1) = 2(1 - x)^{-3}$,

$\dfrac{d^3}{dx^3}(1 - x)^{-1} = (-3)\cdot 2(1 - x)^{-4}(-1) = 3\cdot 2(1 - x)^{-4}$, $\dfrac{d^4}{dx^4}(1 - x)^{-1} = 4\cdot 3\cdot 2(1 - x)^{-5}$, and so on. So after n differentiations, we will have $f^{(n)}(x) = \left(\dfrac{1}{1 - x}\right)^{(n)} = \dfrac{n!}{(1 - x)^{n+1}}$.

11. It seems from the figure that as P approaches the point $(0, 2)$ from the right, $x_T \to \infty$ and $y_T \to 2^+$. As P approaches the point $(3, 0)$ from the left, it appears that $x_T \to 3^+$ and $y_T \to \infty$. So we guess that $x_T \in (3, \infty)$ and $y_T \in (2, \infty)$. It is more difficult to estimate the range of values for x_N and y_N. We might perhaps guess that $x_N \in (0, 3)$, and $y_N \in (-\infty, 0)$ or $(-2, 0)$.

In order to actually solve the problem, we implicitly differentiate the equation of the ellipse to find the equation of the tangent line: $\dfrac{x^2}{9} + \dfrac{y^2}{4} = 1 \quad\Rightarrow\quad \dfrac{2x}{9} + \dfrac{2y}{4}y' = 0$, so $y' = -\dfrac{4}{9}\dfrac{x}{y}$. So at the point (x_0, y_0) on the ellipse, the equation of the tangent line is $y - y_0 = -\dfrac{4\,x_0}{9\,y_0}(x - x_0)$ or $4x_0x + 9y_0y = 4x_0^2 + 9y_0^2$. This can be written as $\dfrac{x_0x}{9} + \dfrac{y_0y}{4} = \dfrac{x_0^2}{9} + \dfrac{y_0^2}{4} = 1$, because (x_0, y_0) lies on the ellipse. So an equation of the tangent line is $\dfrac{x_0x}{9} + \dfrac{y_0y}{4} = 1$.

Therefore the x-intercept x_T for the tangent line is given by $\dfrac{x_0x_T}{9} = 1 \quad \Leftrightarrow \quad x_T = \dfrac{9}{x_0}$, and the y-intercept y_T is given by $\dfrac{y_0y_T}{4} = 1 \quad \Leftrightarrow \quad y_T = \dfrac{4}{y_0}$.

So as x_0 takes on all values in $(0, 3)$, x_T takes on all values in $(3, \infty)$, and as y_0 takes on all values in $(0, 2)$, y_T takes on all values in $(2, \infty)$.

At the point (x_0, y_0) on the ellipse, the slope of the normal line is $-\dfrac{1}{y'(x_0, y_0)} = \dfrac{9\,y_0}{4\,x_0}$, and its equation is $y - y_0 = \dfrac{9\,y_0}{4\,x_0}(x - x_0)$. So the x-intercept x_N for the normal line is given by $0 - y_0 = \dfrac{9\,y_0}{4\,x_0}(x_N - x_0) \quad \Rightarrow$ $x_N = -\dfrac{4x_0}{9} + x_0 = \dfrac{5x_0}{9}$, and the y-intercept y_N is given by $y_N - y_0 = \dfrac{9\,y_0}{4\,x_0}(0 - x_0) \quad \Rightarrow$ $y_N = -\dfrac{9y_0}{4} + y_0 = -\dfrac{5}{4}$.

So as x_0 takes on all values in $(0, 3)$, x_N takes on all values in $\left(0, \frac{5}{3}\right)$, and as y_0 takes on all values in $(0, 2)$, y_N takes on all values in $\left(-\frac{5}{2}, 0\right)$.

12. As in Problem 11, we differentiate implicitly, and get $y' = -\dfrac{b^2x}{a^2y}$. So for a point (x_0, y_0) on the ellipse in the first quadrant, the equation of the normal line is $y - y_0 = \dfrac{a^2y_0}{b^2x_0}(x - x_0)$.

Setting $y = 0$ gives $x_N = -y_0\dfrac{b^2x_0}{a^2y_0} + x_0 = \left(1 - \dfrac{b^2}{a^2}\right)x_0 = \left(\dfrac{a^2 - b^2}{a^2}\right)x_0 = \dfrac{c^2}{a^2}x_0$. Since $0 < x_0 < a$, we have $0 < x_N < \dfrac{c^2}{a^2}a = \dfrac{c^2}{a} = ec$, that is, $0 < x_n < ec$. But $e < 1$, so this says that x_N has to lie to the left of the focus $(c, 0)$ of the ellipse.

Setting $x = 0$ gives $y_N = -\dfrac{a^2}{b^2}y_0 + y_0 = \left(\dfrac{b^2 - a^2}{b^2}\right)y_0 = -\left(\dfrac{c}{b}\right)^2 y_0 = -\dfrac{a^2}{b^2}e^2y_0$.

So as y takes on all values between 0 and b, y_N takes on all values between 0 and $-\dfrac{c^2}{b} = -\dfrac{a^2e^2}{b}$. [In particular, $y_N > -b$, so $(0, y_N)$ has to lie within the ellipse.]

13. (a)
$$\begin{aligned} D &= \left\{x \mid 3 - x \ge 0, 2 - \sqrt{3 - x} \ge 0, 1 - \sqrt{2 - \sqrt{3 - x}} \ge 0\right\} \\ &= \left\{x \mid 3 \ge x, 2 \ge \sqrt{3 - x}, 1 \ge \sqrt{2 - \sqrt{3 - x}}\right\} = \left\{x \mid 3 \ge x, 4 \ge 3 - x, 1 \ge 2 - \sqrt{3 - x}\right\} \\ &= \left\{x \mid x \le 3, x \ge -1, 1 \le \sqrt{3 - x}\right\} = \{x \mid x \le 3, x \ge -1, 1 \le 3 - x\} \\ &= \{x \mid x \le 3, x \ge -1, x \le 2\} = \{x \mid -1 \le x \le 2\} = [-1, 2] \end{aligned}$$

(b) $f(x)=\sqrt{1-\sqrt{2-\sqrt{3-x}}}\quad\Rightarrow$

$$f'(x)=\frac{1}{2\sqrt{1-\sqrt{2-\sqrt{3-x}}}}\frac{d}{dx}\left(1-\sqrt{2-\sqrt{3-x}}\right)$$

$$=\frac{1}{2\sqrt{1-\sqrt{2-\sqrt{3-x}}}}\cdot\frac{-1}{2\sqrt{2-\sqrt{3-x}}}\frac{d}{dx}\left(2-\sqrt{3-x}\right)$$

$$=-\frac{1}{8\sqrt{1-\sqrt{2-\sqrt{3-x}}}\sqrt{2-\sqrt{3-x}}\sqrt{3-x}}$$

14. **(a)** $|x^2-4|=\begin{cases}-x^2+4 & \text{if } -2<x<2\\ x^2-4 & \text{if } x\le -2 \text{ or } x\ge 2\end{cases}$ and $|x^2-9|=\begin{cases}-x^2+9 & \text{if } -3<x<3\\ x^2-9 & \text{if } x\le -3 \text{ or } x\ge 3.\end{cases}$

So $g(x)=\begin{cases}x^2-4-(x^2-9) & \text{if } x\le -3 \text{ or } x\ge 3\\ x^2-4-(-x^2+9) & \text{if } -3<x\le -2 \\ & \text{or } 2\le x<3\\ -x^2+4-(-x^2+9) & \text{if } -2<x<2\end{cases}$

$=\begin{cases}5 & \text{if } x\le -3 \text{ or } x\ge 3\\ 2x^2-13 & \text{if } -3<x\le -2 \text{ or } 2\le x<3\\ -5 & \text{if } -2<x<2\end{cases}$

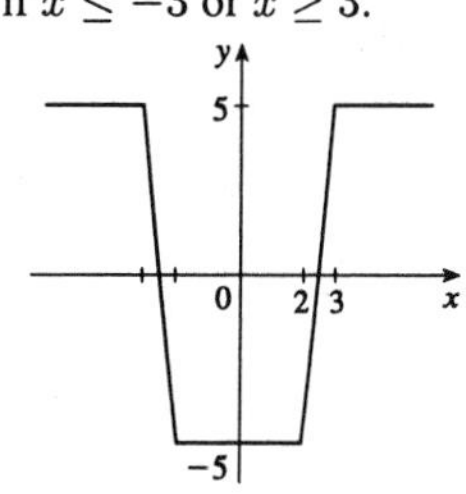

(b) $g'(x)=\begin{cases}0 & \text{if } x<-3 \text{ or } x>3\\ 4x & \text{if } -3<x<-2 \text{ or } 2<x<3\\ 0 & \text{if } -2<x<2\end{cases}$

15. **(a)** For $x\ge 0$,

$h(x)=|x^2-6|x|+8|=|x^2-6x+8|=|(x-2)(x-4)|=\begin{cases}x^2-6x+8 & \text{if } 0\le x\le 2\\ -x^2+6x-8 & \text{if } 2<x<4\\ x^2-6x+8 & \text{if } x\ge 4\end{cases}$

and for $x<0$,

$h(x)=\left|x^2-6|x|+8\right|=|x^2+6x+8|$

$=|(x+2)(x+4)|=\begin{cases}x^2+6x+8 & \text{if } x\le -4\\ -x^2-6x-8 & \text{if } -4<x<-2\\ x^2+6x+8 & \text{if } -2\le x<0\end{cases}$

Or: Use the fact that h is an even function and reflect the part of the graph for $x\ge 0$ about the y-axis.

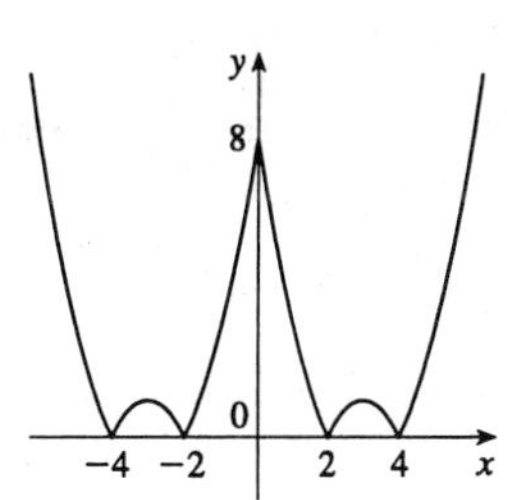

(b) To find where h is differentiable we check the points 0, 2 and 4 by computing the left- and right-hand derivatives:

$$h'_-(0)=\lim_{t\to 0^-}\frac{h(0+t)-h(0)}{t}=\lim_{t\to 0^-}\frac{(0+t)^2+6(0+t)+8-8}{t}=\lim_{t\to 0^-}(t+6)=6$$

$$h'_+(0)=\lim_{t\to 0^+}\frac{h(0+t)-h(0)}{t}=\lim_{t\to 0^+}\frac{(0+t)^2-6(0+t)+8-8}{t}=\lim_{t\to 0^+}(t-6)=-6$$

$\neq h'_-(0)$, so h is not differentiable at 0. Similarly h is not differentiable at ± 2 or ± 4. This can also be seen from the graph.

16. **(a)** $[\![x]\!]^2 + [\![y]\!]^2 = 1$. Since $[\![x]\!]^2$ and $[\![y]\!]^2$ are positive integers or 0, there are only 4 cases:

Case (i): $[\![x]\!] = 1, [\![y]\!] = 0 \quad \Rightarrow \quad 1 \le x < 2$ and $0 \le y < 1$

Case (ii): $[\![x]\!] = -1, [\![y]\!] = 0 \quad \Rightarrow \quad -1 \le x < 0$ and $0 \le y < 1$

Case (iii): $[\![x]\!] = 0, [\![y]\!] = 1 \quad \Rightarrow \quad 0 \le x < 1$ and $1 \le y < 2$

Case (iv): $[\![x]\!] = 0, [\![y]\!] = -1 \quad \Rightarrow \quad 0 \le x < 1$ and $-1 \le y < 0$

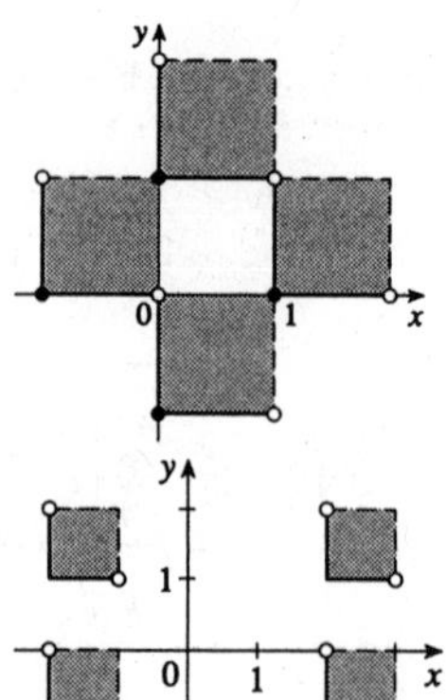

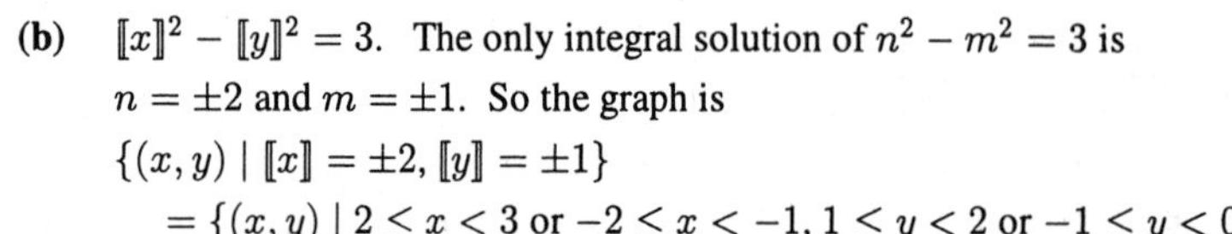

(b) $[\![x]\!]^2 - [\![y]\!]^2 = 3$. The only integral solution of $n^2 - m^2 = 3$ is $n = \pm 2$ and $m = \pm 1$. So the graph is

$\{(x, y) \mid [\![x]\!] = \pm 2, [\![y]\!] = \pm 1\}$

$= \{(x, y) \mid 2 \le x < 3 \text{ or } -2 \le x < -1, 1 \le y < 2 \text{ or } -1 \le y < 0\}$.

(c) $[\![x + y]\!]^2 = 1 \quad \Rightarrow$

$[\![x + y]\!] = \pm 1 \quad \Rightarrow$

$1 \le x + y < 2$ or $-1 \le x + y < 0$

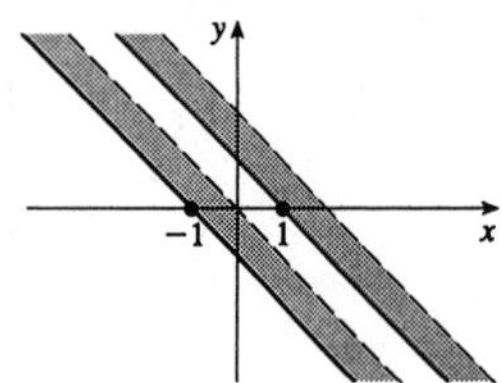

(d) For $n \le x < n + 1$, $[\![x]\!] = n$. Then $[\![x]\!] + [\![y]\!] = 1$

$\Rightarrow \quad [\![y]\!] = 1 - n \quad \Rightarrow \quad 1 - n \le y < 2 - n$.

Choosing integer values for n produces the graph.

17.

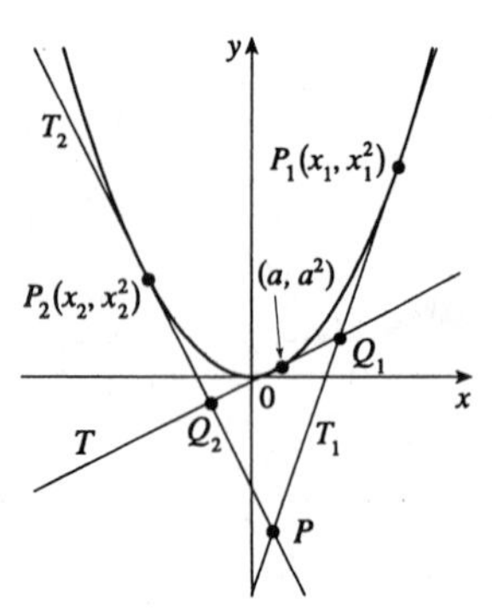

The equation of T_1 is $y - x_1^2 = 2x_1(x - x_1) = 2x_1x - 2x_1^2$ or $y = 2x_1x - x_1^2$. The equation of T_2 is $y = 2x_2x - x_2^2$. Solving for the point of intersection, we get $2x(x_1 - x_2) = x_1^2 - x_2^2 \quad \Rightarrow x = \frac{1}{2}(x_1 + x_2)$. Therefore the coordinates of P are $\left(\frac{1}{2}(x_1 + x_2), x_1x_2\right)$. So if the point of contact of T is (a, a^2), then Q_1 is $\left(\frac{1}{2}(a + x_1), ax_1\right)$ and Q_2 is $\left(\frac{1}{2}(a + x_2), ax_2\right)$. Therefore

$|PQ_1|^2 = \frac{1}{4}(a - x_2)^2 + x_1^2(a - x_2)^2 = (a - x_2)^2\left(\frac{1}{4} + x_1^2\right)$ and

$|PP_1|^2 = \frac{1}{4}(x_1 - x_2^2) + x_1^2(x_1 - x_2)^2 = (x_1 - x_2)^2\left(\frac{1}{4} + x_1^2\right)$.

So $\dfrac{|PQ_1|^2}{|PP_1|^2} = \dfrac{(a - x_2)^2}{(x_1 - x_2)^2}$, and similarly $\dfrac{|PQ_2|^2}{|PP_2|^2} = \dfrac{(x_1 - a)^2}{(x_1 - x_2)^2}$.

Finally, $\dfrac{|PQ_1|}{|PP_1|} + \dfrac{|PQ_2|}{|PP_2|} = \dfrac{a - x_2}{x_1 - x_2} + \dfrac{x_1 - a}{x_1 - x_2} = 1$.

18. $\displaystyle\lim_{x \to 0} \frac{\sin(3 + x)^2 - \sin 9}{x} = f'(3)$ where $f(x) = \sin x^2$. Now $f'(x) = \left(\cos x^2\right)(2x)$, so $f'(3) = 6\cos 9$.

19. **(a)** Since f is differentiable at 0, f is continuous at 0 so $f(0) = \displaystyle\lim_{x \to 0} f(x) = \lim_{x \to 0} \frac{f(x)}{x} \cdot x = \lim_{x \to 0} \frac{f(x)}{x} \cdot \lim_{x \to 0} x = 4 \cdot 0 = 0$.

(b) $f'(0) = \displaystyle\lim_{x \to 0} \frac{f(x) - f(0)}{x - 0} = \lim_{x \to 0} \frac{f(x)}{x} = 4$ [since $f(0) = 0$ from (a)]

(c) $\displaystyle\lim_{x \to 0} \frac{g(x)}{f(x)} = \lim_{x \to 0} \frac{g(x)/x}{f(x)/x} = \frac{\lim_{x \to 0}[g(x)/x]}{\lim_{x \to 0}[f(x)/x]} = \frac{2}{4} = \frac{1}{2}$

20. **(a)** *Solution 1:* We introduce a coordinate system and drop a perpendicular from P, as shown. We see from $\angle NCP$ that $\tan 2\theta = \dfrac{y}{1-x}$, and from $\angle NBP$ that $\tan\theta = y/x$. Using the double-angle formula for tangents, we get $\dfrac{y}{1-x} = \tan 2\theta = \dfrac{2\tan\theta}{1-\tan^2\theta} = \dfrac{2(y/x)}{1-(y/x)^2}$. After a bit of simplification, this becomes $\dfrac{1}{1-x} = \dfrac{2x}{x^2-y^2} \Leftrightarrow$

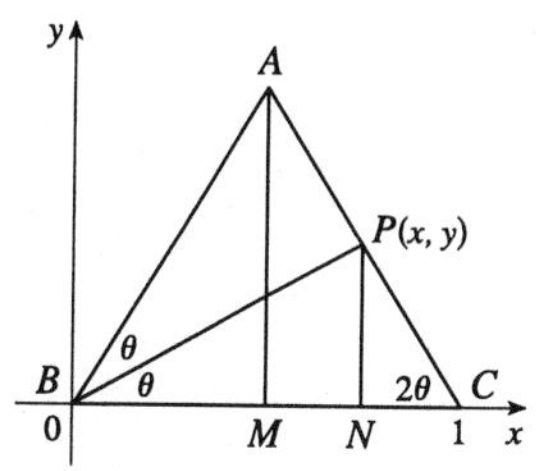

$y^2 = x(3x-2)$. As the altitude AM decreases in length, the point P will approach the x-axis, that is, $y \to 0$, so the limiting location of P must be one of the roots of the equation $x(3x-2)=0$. Obviously it is not $x=0$ (the point P can never be to the left of the altitude AM, which it would have to be in order to approach 0) so it must be $3x-2=0$, that is, $x=\frac{2}{3}$.

Solution 2: We add a few lines to the original diagram, as shown. Now note that $\angle BPQ = \angle PBC$ (alternate angles; $QP \parallel BC$ by symmetry) and similarly $\angle CQP = \angle QCB$. So the triangles ΔBPQ and ΔCQP are isosceles, and the line segments BQ, QP and PC are all of equal length. As $|AM| \to 0$, P and Q approach points on the base, and the point P is seen to approach a position two-thirds of the way between B and C, as above.

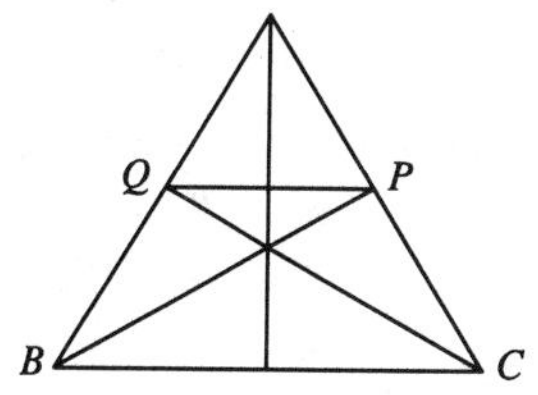

(b) The equation $y^2 = x(3x-2)$ calculated in part (a) is the equation of the curve traced out by P. Note that P only traces out the part of the curve with $0 \le y < 1$, since as $|AM| \to \infty$, $\theta \to \pi/2$.

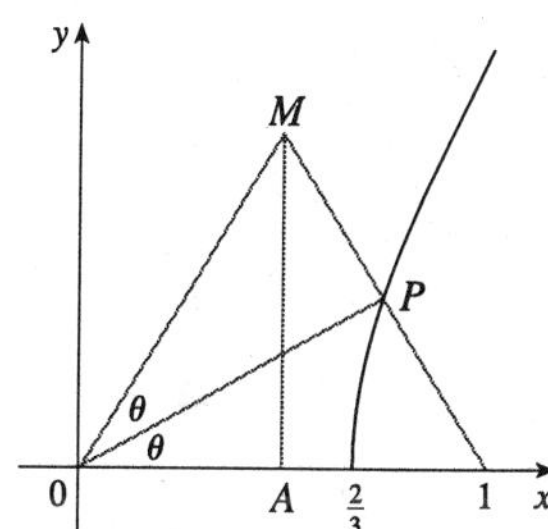

21. $\displaystyle \lim_{x\to 0} \frac{\sin(a+2x) - 2\sin(a+x) + \sin a}{x^2} = \lim_{x\to 0} \frac{\sin a\cos 2x + \cos a \sin 2x - 2\sin a\cos x - 2\cos a\sin x + \sin a}{x^2}$

$$= \lim_{x\to 0} \frac{\sin a(\cos 2x - 2\cos x + 1) + \cos a(\sin 2x - 2\sin x)}{x^2}$$

$$= \lim_{x\to 0} \frac{\sin a(2\cos^2 x - 1 - 2\cos x + 1) + \cos a(2\sin x\cos x - 2\sin x)}{x^2}$$

$$= \lim_{x\to 0} \frac{\sin a(2\cos x)(\cos x - 1) + \cos a(2\sin x)(\cos x - 1)}{x^2}$$

$$= \lim_{x\to 0} \frac{2(\cos x - 1)[\sin a\cos x + \cos a\sin x](\cos x + 1)}{x^2(\cos x + 1)}$$

$$= \lim_{x\to 0} \frac{-2\sin^2 x\,[\sin(a+x)]}{x^2(\cos x + 1)} = -2\lim_{x\to 0}\left(\frac{\sin x}{x}\right)^2 \cdot \frac{\sin(a+x)}{\cos x + 1}$$

$$= -2(1)^2 \frac{\sin(a+0)}{\cos 0 + 1} = -\sin a$$

22. **(a)** $\dfrac{f(x+h)-f(x-h)}{2h} = m_{PQ}$

(b)
$$\begin{aligned}
&\lim_{h\to 0}\frac{f(x+h)-f(x-h)}{2h} \\
&= \lim_{h\to 0}\frac{[f(x+h)-f(x)]+[f(x)-f(x-h)]}{2h} \\
&= \frac{1}{2}\left[\lim_{h\to 0}\frac{f(x+h)-f(x)}{h} + \lim_{h\to 0}\frac{f(x)-f(x-h)}{h}\right] \\
&= \frac{1}{2}\left[\lim_{h\to 0}\frac{f(x+h)-f(x)}{h} + \lim_{h\to 0}\frac{f(x-h)-f(x)}{-h}\right] \\
&= \frac{1}{2}\left[\lim_{h\to 0}\frac{f(x+h)-f(x)}{h} + \lim_{k\to 0}\frac{f(x+k)-f(x)}{k}\right] \quad \left(\begin{array}{c}\text{where}\\ k=-h\end{array}\right) \\
&= \tfrac{1}{2}[f'(x)+f'(x)] = f'(x).
\end{aligned}$$

(c) Take $f(x)=|x|$. Then by Example 8 in Section 2.1, $f'(0)$ does not exist.
However, $\displaystyle\lim_{h\to 0}\frac{f(0+h)-f(0-h)}{2h} = \lim_{h\to 0}\frac{|h|-|-h|}{2h} = \lim_{h\to 0}\frac{|h|-|h|}{2h} = \lim_{h\to 0} 0 = 0.$

23. **(a)** If the two lines L_1 and L_2 have slopes m_1 and m_2 and angles of inclination ϕ_1 and ϕ_2, then $m_1=\tan\phi_1$ and $m_2=\tan\phi_2$. The figure shows that $\phi_2=\phi_1+\alpha$ and so $\alpha=\phi_2-\phi_1$. Therefore, using the identity for $\tan(x-y)$, we have

$\tan\alpha = \tan(\phi_2-\phi_1) = \dfrac{\tan\phi_2-\tan\phi_1}{1+\tan\phi_2\tan\phi_1}$ and so $\tan\alpha = \dfrac{m_2-m_1}{1+m_1m_2}$.

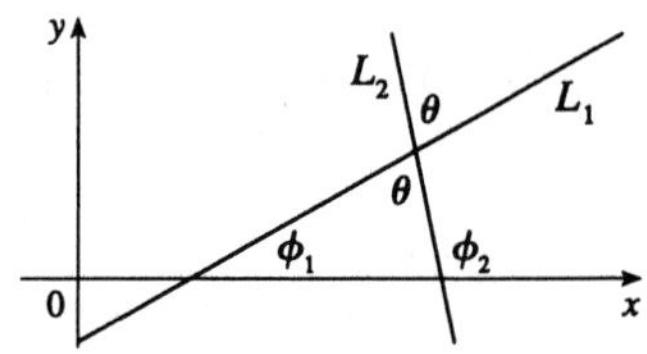

(b) **(i)** The parabolas intersect when $x^2=(x-2)^2 \;\Rightarrow\; x=1$. If $y=x^2$, then $y'=2x$, so the slope of the tangent to $y=x^2$ at $(1,1)$ is $m_1=2(1)=2$. If $y=(x-2)^2$, then $y'=2(x-2)$, so the slope of the tangent to $y=(x-2)^2$ at $(1,1)$ is $m_2=2(1-2)=-2$. Therefore
$\tan\alpha = \dfrac{m_2-m_1}{1+m_1m_2} = \dfrac{-2-2}{1+2(-2)} = \dfrac{4}{3}$ and so $\alpha=\tan^{-1}\frac{4}{3}\approx 53°$.

(ii) $x^2-y^2=3$ and $x^2-4x+y^2+3=0$ intersect when $x^2-4x+x^2=0 \;\Leftrightarrow\; 2x(x-2)=0$ $\Rightarrow\; x=0$ or 2, but 0 is extraneous. If $x^2-y^2=3$ then $2x-2yy'=0 \;\Rightarrow\; y'=x/y$ and $x-4x+y^2+3=0 \;\Rightarrow\; 2x-4+2yy'=0 \;\Rightarrow\; y'=\dfrac{2-x}{y}$. At $(2,1)$ the slopes are $m_1=2$ and $m_2=0$, so $\tan\alpha=\dfrac{0-2}{1+2\cdot 0}=-2 \;\Rightarrow\; \alpha\approx 117°$. At $(2,-1)$ the slopes are $m_1=-2$ and $m_2=0$, so $\tan\alpha=\dfrac{0-(-2)}{1+(-2)(0)}=2 \;\Rightarrow\; \alpha\approx 63°$.

24. $2yy'=4p \;\Rightarrow\; y'=\dfrac{2p}{y} \;\Rightarrow$ slope of tangent at P is $\dfrac{2p}{y_1}$. Slope of FP is $\dfrac{y_1}{x_1-p}$, so by the formula from 23(a),
$$\tan\alpha = \frac{-2p/y_1 + y_1/(x_1-p)}{1+(2p/y_1)\left[y_1/(x_1-p)\right]} = \frac{-2p(x_1-p)+y_1^2}{y_1(x_1-p)+2py_1} = \frac{-2px_1+2p^2+4px_1}{y_1(p+x_1)} = \frac{2p(p+x_1)}{y_1(p+x_1)}$$
$= \dfrac{2p}{y_1}$ = slope of tangent at $P=\tan\beta$. Since $0\le\alpha,\beta\le\frac{\pi}{2}$, this proves that $\alpha=\beta$.

25. Since $\angle ROQ = \angle OQP = \theta$, the triangle QOR is isosceles, so $|QR| = |RO| = x$. By the Law of Cosines, $x^2 = x^2 + r^2 - 2rx\cos\theta$. Hence $2rx\cos\theta = r^2$, so $x = \dfrac{r^2}{2r\cos\theta} = \dfrac{r}{2\cos\theta}$. Note that as $y \to 0^+$, $\theta \to 0^+$ (since $\sin\theta = y/r$), and hence $x \to \dfrac{r}{2\cos 0} = \dfrac{r}{2}$. Thus as P is taken closer and closer to the x-axis, the point R approaches the midpoint of the radius AO.

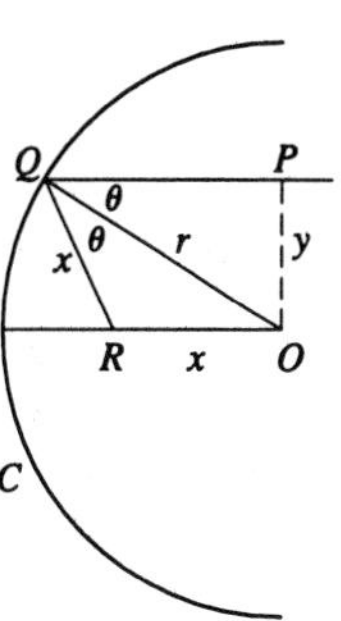

26. Suppose that $y = mx + c$ is a tangent line to the ellipse. Then it intersects the ellipse at only one point, so the discriminant of the equation $\dfrac{x^2}{a^2} + \dfrac{(mx+c)^2}{b^2} = 1 \Leftrightarrow (b^2 + a^2m^2)x^2 + 2mca^2x + a^2c^2 - a^2b^2 = 0$ must be 0, that is, $0 = (2mca^2)^2 - 4(b^2 + a^2m^2)(a^2c^2 - a^2b^2) = 4a^4c^2m^2 - 4a^2b^2c^2 + 4a^2b^4 + 4a^4c^2m^2 - 4a^2b^4$
$= 4a^2b^2(a^2m^2 + b^2 - c^2)$. Therefore, $a^2m^2 + b^2 - c^2 = 0$.

Now if a point (α, β) lies on the line $y = mx + c$, then $c = \beta - m\alpha$, so from above,

$$0 = a^2m^2 + b^2 - (\beta - m\alpha)^2 = (a^2 - \alpha^2)m^2 + 2\alpha\beta m + b^2 - \beta^2 \quad\Leftrightarrow\quad m^2 + \frac{2\alpha\beta}{a^2 - \alpha^2}m + \frac{b^2 - \beta^2}{a^2 - \alpha^2} = 0.$$

(a) Suppose that the two tangent lines from the point (α, β) to the ellipse have slopes m and $\dfrac{1}{m}$. Then m and $\dfrac{1}{m}$ are roots of the equation $z^2 + \dfrac{2\alpha\beta}{a^2 - \alpha^2}z + \dfrac{b^2 - \beta^2}{a^2 - \alpha^2} = 0$. This implies that $(z - m)\left(z - \dfrac{1}{m}\right) = 0$ $\Leftrightarrow \quad z^2 - \left(m + \dfrac{1}{m}\right)z + m\left(\dfrac{1}{m}\right) = 0$, so equating the constant terms in the two quadratic equations, we get $\dfrac{b^2 - \beta^2}{a^2 - \alpha^2} = m\left(\dfrac{1}{m}\right) = 1$, and hence $b^2 - \beta^2 = a^2 - \alpha^2$. So (α, β) lies on the hyperbola $x^2 - y^2 = a^2 - b^2$.

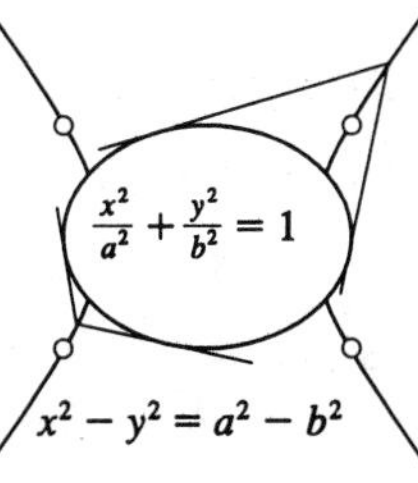

(b) If the two tangent lines from the point (α, β) to the ellipse have slopes m and $-1/m$, then m and $-1/m$ are roots of the quadratic equation, and so $(z - m)(z + 1/m) = 0$, and equating the constant terms as in part (a), we get $\dfrac{b^2 - \beta^2}{a^2 - \alpha^2} = -1$, and hence $b^2 - \beta^2 = \alpha^2 - a^2$. So the point (α, β) lies on the circle $x^2 + y^2 = a^2 + b^2$.

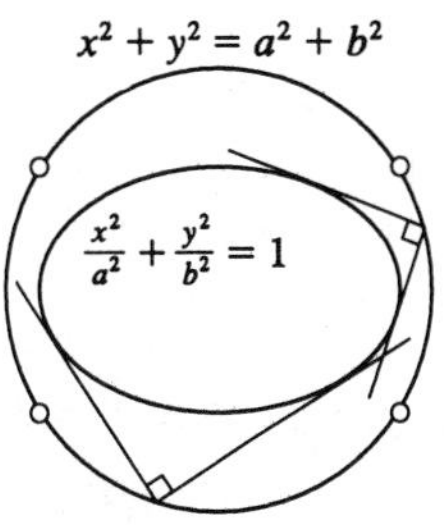

27. $y = x^4 - 2x^2 - x \quad\Rightarrow\quad y' = 4x^3 - 4x - 1$. The equation of the tangent line at $x = a$ is $y - (a^4 - 2a^2 - a) = (4a^3 - 4a - 1)(x - a)$ or $y = (4a^3 - 4a - 1)x + (-3a^4 + 2a^2)$ and similarly for $x = b$. So if at $x = a$ and $x = b$ we have the same tangent line, then $4a^3 - 4a - 1 = 4b^3 - 4b - 1$ and $-3a^4 + 2a^2 = -3b^4 + 2b^2$. The first equation gives $a^3 - b^3 = a - b \quad\Rightarrow\quad (a - b)(a^2 + ab + b^2) = (a - b)$. Assuming $a \neq b$, we have $1 = a^2 + ab + b^2$. The second equation gives $3(a^4 - b^4) = 2(a^2 - b^2) \quad\Rightarrow$

$3(a^2 - b^2)(a^2 + b^2) = 2(a^2 - b^2)$ which is true if $a = -b$. Substituting into $1 = a^2 + ab + b^2$ gives $1 = a^2 - a^2 + a^2 \quad \Rightarrow \quad a = \pm 1$ so that $a = 1$ and $b = -1$ or vice versa. It is easily verified that the points $(1, -2)$ and $(-1, 0)$ have a common tangent line.

As long as there are only two such points, we are done. So we show that these are in fact the only two such points. Suppose that $a^2 - b^2 \neq 0$. Then $3(a^2 - b^2)(a^2 + b^2) = 2(a^2 - b^2)$ gives $3(a^2 + b^2) = 2$ or $a^2 + b^2 = \frac{2}{3}$. Thus $ab = (a^2 + ab + b^2) - (a^2 + b^2) = 1 - \frac{2}{3} = \frac{1}{3}$, so $b = \dfrac{1}{3a}$. Hence $a^2 + \dfrac{1}{9a^2} = \dfrac{2}{3}$, so $9a^4 + 1 = 6a^2 \quad \Rightarrow \quad 0 = 9a^4 - 6a^2 + 1 = (3a^2 - 1)^2$. So $3a^2 - 1 = 0$, so $a^2 = \dfrac{1}{3} \Rightarrow b^2 = \dfrac{1}{9a^2} = \dfrac{1}{3} = a^2$, contradicting our assumption that $a^2 \neq b^2$.

28. Suppose that the normal lines at the three points (a_1, a_1^2), (a_2, a_2^2), and (a_3, a_3^2) intersect at a common point. Now if one of the $\{a_i\}$ is 0 (suppose $a_1 = 0$) then by symmetry $a_2 = -a_3$, so $a_1 + a_2 + a_3 = 0$. So we can assume that none of the $\{a_i\}$ is 0.

The slope of the tangent line at (a_i, a_i^2) is $2a_i$, so the slope of the normal line is $-1/(2a_i)$ and its equation is $y - a_i^2 = -\dfrac{1}{2a_i}(x - a_i)$. We solve for the x-coordinate of the intersection of the normal lines from (a_1, a_1^2) and (a_2, a_2^2): $y = a_1^2 - \dfrac{1}{2a_1}(x - a_1) = a_2^2 - \dfrac{1}{2a_2}(x - a_2) \quad \Rightarrow \quad x\left(\dfrac{1}{2a_2} - \dfrac{1}{2a_1}\right) = a_2^2 - a_1^2 \quad \Rightarrow$ $x\left(\dfrac{a_1 - a_2}{2a_1 a_2}\right) = -(a_1 - a_2)(a_1 + a_2) \quad \Leftrightarrow \quad x = -2a_1 a_2(a_1 + a_2)$ ($\bigstar$). Similarly, solving for the x-coordinate of the intersections of the normal lines from (a_1, a_1^2) and (a_3, a_3^2)gives $x = -2a_1 a_3(a_1 + a_3)$ ($\dagger$). Solving ($\bigstar$) and ($\dagger$) gives $x = a_2(a_1 + a_2) = a_3(a_1 + a_3) \quad \Leftrightarrow$ $a_1(a_2 + a_3) = a_3^2 - a_2^2 = -(a_2 + a_3)(a_2 - a_3) \quad \Leftrightarrow \quad a_1 = -(a_2 + a_3) \quad \Leftrightarrow \quad a_1 + a_2 + a_3 = 0$.

29. Because of the periodic nature of the lattice points, it suffices to consider the points in the 5×2 grid shown. We can see that the minimum value of r occurs when there is a line with slope $\frac{2}{5}$ which touches the circle centered at $(3, 1)$ and the circles centered at $(0, 0)$ and $(5, 2)$. To find P, the point at which the line is tangent to the circle at $(0, 0)$, we simultaneously solve $x^2 + y^2 = r^2$ and $y = -\frac{5}{2}x \quad \Rightarrow \quad x^2 + \frac{25}{4}x^2 = r^2 \quad \Rightarrow$ $x^2 = \frac{4}{29}r^2 \quad \Rightarrow \quad x = \frac{2}{\sqrt{29}}r$, $y = -\frac{5}{\sqrt{29}}r$. To find Q, we either use symmetry or solve $(x-3)^2 + (y-1)^2 = r^2$ and $y - 1 = -\frac{5}{2}(x - 3)$. As above, we get $x = 3 - \frac{2}{\sqrt{29}}r$, $y = 1 + \frac{5}{\sqrt{29}}r$. Now the slope of the line PQ is $\frac{2}{5}$, so

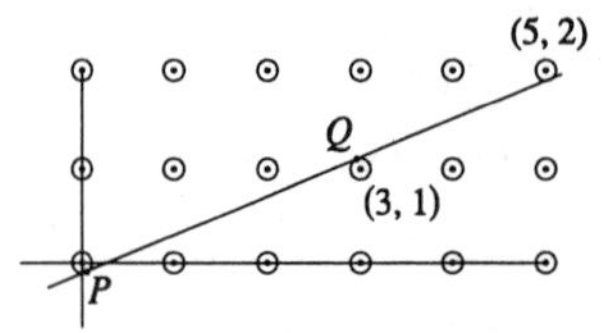

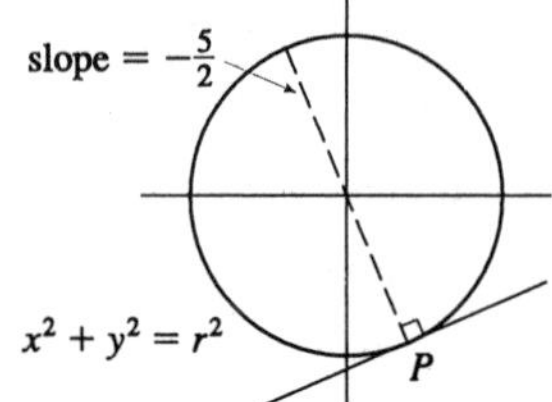

$$m_{PQ} = \frac{1 + \frac{5}{\sqrt{29}}r - \left(-\frac{5}{\sqrt{29}}r\right)}{3 - \frac{2}{\sqrt{29}}r - \frac{2}{\sqrt{29}}r} = \frac{1 + \frac{10}{\sqrt{29}}r}{3 - \frac{4}{\sqrt{29}}r} = \frac{\sqrt{29} + 10r}{3\sqrt{29} - 4r} = \frac{2}{5} \quad \Rightarrow \quad 5\sqrt{29} + 50r = 6\sqrt{29} - 8r \quad \Leftrightarrow$$

$58r = \sqrt{29} \quad \Leftrightarrow \quad r = \frac{\sqrt{29}}{58}$. So the minimum value of r for which any line with slope $\frac{2}{5}$ intersects circles with radius r centered at the lattice points on the plane is $r = \frac{\sqrt{29}}{58}$.

CHAPTER THREE

EXERCISES 3.1

1. Absolute maximum at e; absolute minimum at d; local maxima at b, e; local minima at d, s.

2. Absolute maximum at e; absolute minimum at t; local maxima at c, e, s; local minima at b, c, d, r.

3. Absolute maximum value is $f(4) = 4$; absolute minimum value is $f(7) = 0$; local maximum values are $f(4) = 4$ and $f(6) = 3$; local minimum values are $f(2) = 1$ and $f(5) = 2$.

4. Absolute maximum value is $f(7) = 5$; absolute minimum value is $f(1) = 0$; local maximum values are $f(0) = 2$, $f(3) = 4$, and $f(5) = 3$; local minimum values are $f(1) = 0$, $f(4) = 2$, and $f(6) = 1$.

5. $f(x) = 1 + 2x$, $x \geq -1$.
Absolute minimum $f(-1) = -1$; no local minimum. No local or absolute maximum.

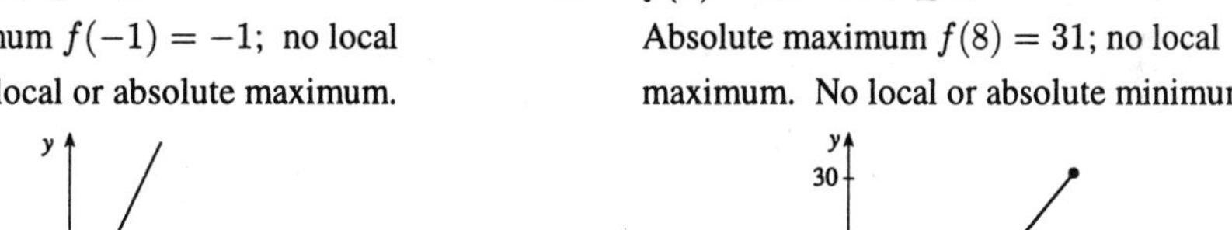

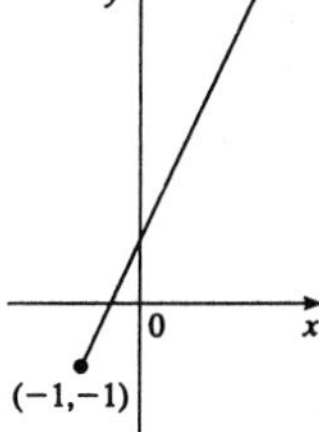

6. $f(x) = 4x - 1$, $x \leq 8$.
Absolute maximum $f(8) = 31$; no local maximum. No local or absolute minimum.

7. $f(x) = 1 - x^2$, $0 < x < 1$. No extremum.

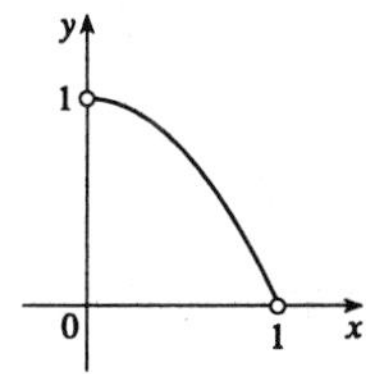

8. $f(x) = 1 - x^2$, $0 < x \leq 1$.
Absolute minimum $f(1) = 0$; no local minimum. No absolute or local maximum.

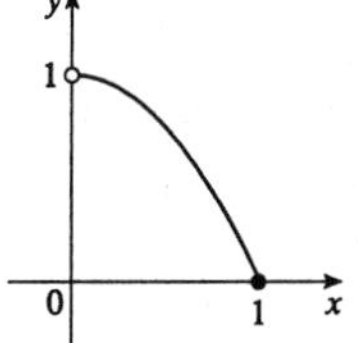

9. $f(x) = 1 - x^2$, $0 \leq x < 1$.
Absolute maximum $f(0) = 1$; no local maximum. No absolute or local minimum.

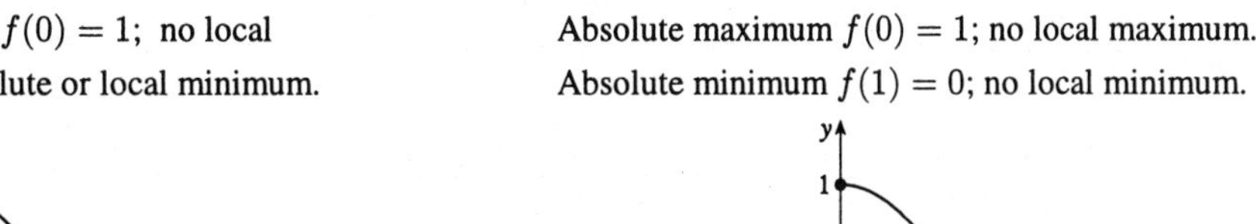

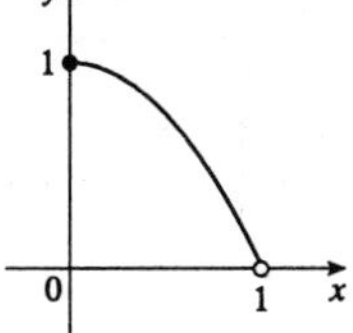

10. $f(x) = 1 - x^2$, $0 \leq x \leq 1$.
Absolute maximum $f(0) = 1$; no local maximum.
Absolute minimum $f(1) = 0$; no local minimum.

11. $f(x) = 1 - x^2$, $-2 \le x \le 1$.
Absolute and local maximum $f(0) = 1$.
Absolute minimum $f(-2) = -3$;
no local minimum.

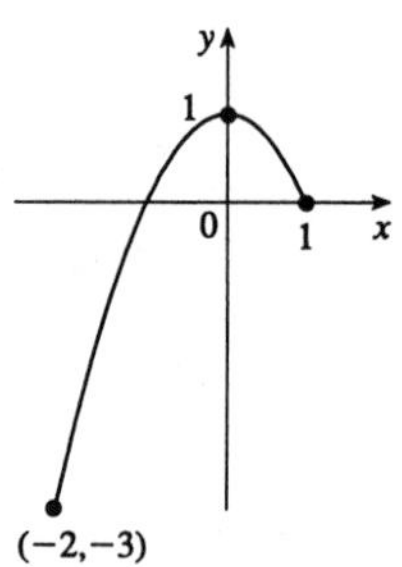

12. $f(x) = 1 + (x+1)^2$, $-2 \le x < 5$.
No absolute or local maximum.
Absolute and local minimum $f(-1) = 1$.

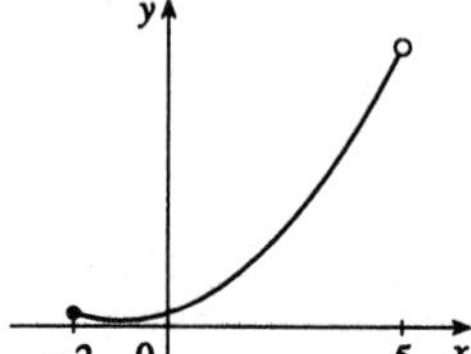

13. $f(t) = 1/t$, $0 < t < 1$. No extremum.

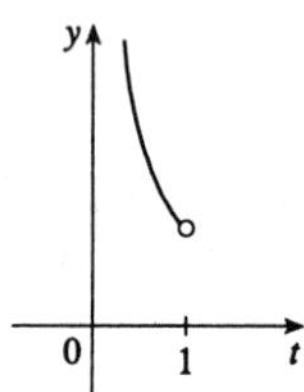

14. $f(t) = 1/t$, $0 < t \le 1$.
Absolute minimum $f(1) = 1$; no local minimum. No local or absolute maximum.

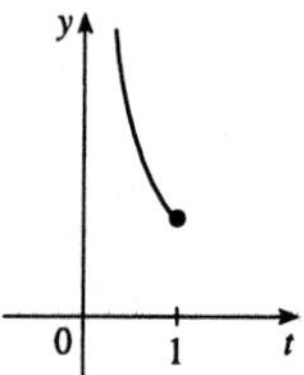

15. $f(\theta) = \sin\theta$, $-2\pi \le \theta \le 2\pi$. Absolute and local maxima $f\left(-\frac{3\pi}{2}\right) = f\left(\frac{\pi}{2}\right) = 1$. Absolute and local minima $f\left(-\frac{\pi}{2}\right) = f\left(\frac{3\pi}{2}\right) = -1$.

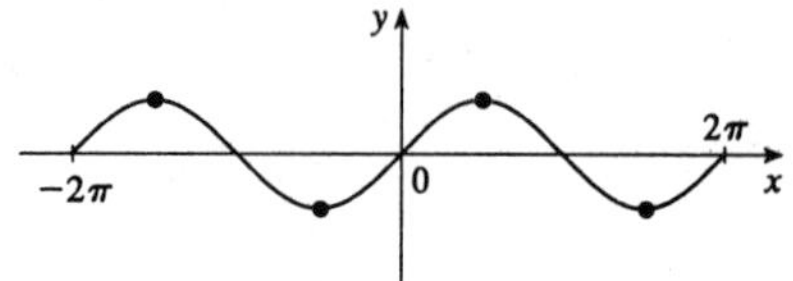

16. $f(\theta) = \tan\theta$, $-\frac{\pi}{4} \le \theta < \frac{\pi}{2}$.
Absolute minimum $f\left(-\frac{\pi}{4}\right) = -1$; no local minimum. No absolute or local maximum.

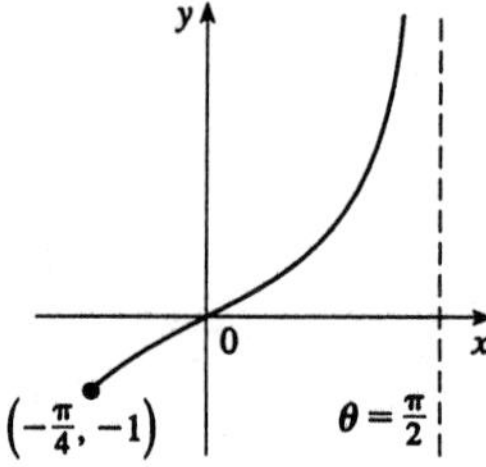

17. $f(x) = x^5$. No extremum.

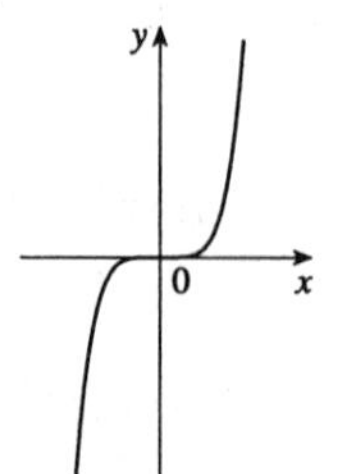

18. $f(x) = 2 - x^4$. Local and absolute maximum $f(0) = 2$. No local or absolute minimum.

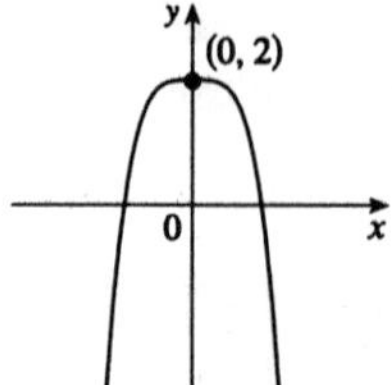

19. $f(x) = \begin{cases} 2x & \text{if } 0 \le x < 1 \\ 2 - x & \text{if } 1 \le x \le 2 \end{cases}$

Absolute minima $f(0) = f(2) = 0$; no local minimum. No absolute or local maximum.

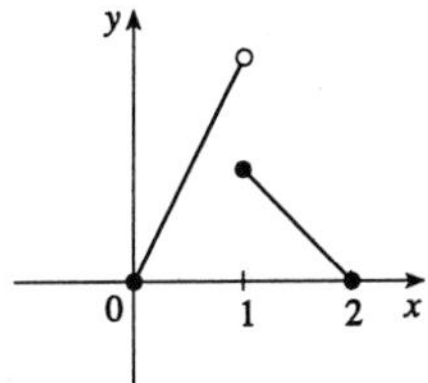

20. $f(x) = \begin{cases} x^2 & \text{if } -1 \le x < 0 \\ 2 - x^2 & \text{if } 0 \le x \le 1 \end{cases}$

Absolute and local maximum $f(0) = 2$. No absolute or local minimum.

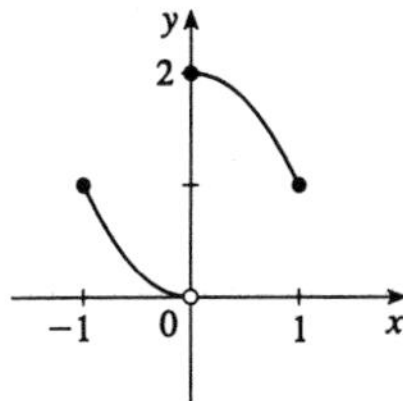

21. $f(x) = 2x - 3x^2 \quad \Rightarrow \quad f'(x) = 2 - 6x = 0 \quad \Leftrightarrow \quad x = \frac{1}{3}$. So the critical number is $\frac{1}{3}$.

22. $f(x) = 5 + 8x \quad \Rightarrow \quad f'(x) = 8 \neq 0$. No critical number.

23. $f(x) = x^3 - 3x + 1 \quad \Rightarrow \quad f'(x) = 3x^2 - 3 = 3(x^2 - 1) = 3(x + 1)(x - 1)$. So the critical numbers are ± 1.

24. $f(x) = 4x^3 - 9x^2 - 12x + 3 \quad \Rightarrow \quad f'(x) = 12x^2 - 18x - 12 = 6(2x^2 - 3x - 2) = 6(2x + 1)(x - 2)$. So the critical numbers are $x = -\frac{1}{2}, 2$.

25. $f(t) = 2t^3 + 3t^2 + 6t + 4 \quad \Rightarrow \quad f'(t) = 6t^2 + 6t + 6$. But $t^2 + t + 1 = 0$ has no real solutions since $b^2 - 4ac = 1 - 4(1)(1) = -3 < 0$. No critical number.

26. $f(t) = t^3 + 6t^2 + 3t - 1 \quad \Rightarrow \quad f'(t) = 3t^2 + 12t + 3 = 3(t^2 + 4t + 1)$. By the quadratic formula, solutions are $t = \left(-4 \pm \sqrt{12}\right)/2 = -2 \pm \sqrt{3}$. Critical numbers are $t = -2 \pm \sqrt{3}$.

27. $s(t) = 2t^3 + 3t^2 - 6t + 4 \quad \Rightarrow \quad s'(t) = 6t^2 + 6t - 6 = 6(t^2 + t - 1)$. By the quadratic formula, the critical numbers are $t = \left(-1 \pm \sqrt{5}\right)/2$.

28. $s(t) = t^4 + 4t^3 + 2t^2 \quad \Rightarrow \quad s'(t) = 4t^3 + 12t^2 + 4t = 4t(t^2 + 3t + 1) = 0$ when $t = 0$ or $t^2 + 3t + 1 = 0$. By the quadratic formula, the critical numbers are $t = 0, \left(-3 \pm \sqrt{5}\right)/2$.

29. $g(x) = \sqrt[9]{x} = x^{1/9} \quad \Rightarrow \quad g'(x) = \frac{1}{9}x^{-8/9} = 1\Big/\left(9\sqrt[9]{x^8}\right) \neq 0$, but $g'(0)$ does not exist, so $x = 0$ is a critical number.

30. $g(x) = |x + 1| \quad \Rightarrow \quad g'(x) = 1$ if $x > -1$, $g'(x) = -1$ if $x < -1$, but $g'(-1)$ does not exist, so $x = -1$ is a critical number.

31. $g(t) = 5t^{2/3} + t^{5/3} \quad \Rightarrow \quad g'(t) = \frac{10}{3}t^{-1/3} + \frac{5}{3}t^{2/3}$. $g'(0)$ does not exist, so $t = 0$ is a critical number. $g'(t) = \frac{5}{3}t^{-1/3}(2 + t) = 0 \quad \Leftrightarrow \quad t = -2$, so $t = -2$ is also a critical number.

32. $g(t) = \sqrt{t}(1 - t) = t^{1/2} - t^{3/2} \quad \Rightarrow \quad g'(t) = \dfrac{1}{2\sqrt{t}} - \frac{3}{2}\sqrt{t}$. $g'(0)$ does not exist, so $t = 0$ is a critical number. $0 = g'(t) = \dfrac{1 - 3t}{2\sqrt{t}} \quad \Rightarrow \quad t = \frac{1}{3}$, so $t = \frac{1}{3}$ is also a critical number.

33. $f(r) = \dfrac{r}{r^2+1} \quad\Rightarrow\quad f'(r) = \dfrac{1(r^2+1) - r(2r)}{(r^2+1)^2} = \dfrac{-r^2+1}{(r^2+1)^2} = 0 \quad\Leftrightarrow\quad r^2 = 1 \quad\Leftrightarrow\quad r = \pm 1$, so these are the critical numbers.

34. $f(z) = \dfrac{z+1}{z^2+z+1} \quad\Rightarrow\quad f'(z) = \dfrac{1(z^2+z+1) - (z+1)(2z+1)}{(z^2+z+1)^2} = \dfrac{-z^2-2z}{(z^2+z+1)^2} = 0 \quad\Leftrightarrow$ $z(z+2) = 0 \quad\Rightarrow\quad z = 0, -2$ are the critical numbers. (Note that $z^2 + z + 1 \neq 0$ since the discriminant < 0.)

35. $F(x) = x^{4/5}(x-4)^2 \quad\Rightarrow\quad F'(x) = \frac{4}{5}x^{-1/5}(x-4)^2 + 2x^{4/5}(x-4) = \dfrac{(x-4)(7x-8)}{5x^{1/5}} = 0$ when $x = 4, \frac{8}{7}$ and $F'(0)$ does not exist. Critical numbers are $0, \frac{8}{7}, 4$.

36. $G(x) = \sqrt[3]{x^2-x} \quad\Rightarrow\quad G'(x) = \frac{1}{3}(x^2-x)^{-2/3}(2x-1)$. $G'(x)$ does not exist when $x^2 - x = 0$ or $x = 0, 1$. $G'(x) = 0 \quad\Leftrightarrow\quad 2x - 1 = 0 \quad\Leftrightarrow\quad x = \frac{1}{2}$. So the critical numbers are $x = 0, \frac{1}{2}, 1$.

37. $f(\theta) = \sin^2(2\theta) \quad\Rightarrow\quad f'(\theta) = 2\sin(2\theta)\cos(2\theta)(2) = 2\sin 4\theta = 0 \quad\Leftrightarrow\quad \sin 4\theta = 0 \quad\Leftrightarrow\quad 4\theta = n\pi$, n an integer. So $\theta = n\pi/4$ are the critical numbers.

38. $g(\theta) = \theta + \sin\theta \quad\Rightarrow\quad g'(\theta) = 1 + \cos\theta = 0 \quad\Leftrightarrow\quad \cos\theta = -1$. The critical numbers are $\theta = (2n+1)\pi$, n an integer.

39. $f(x) = x^2 - 2x + 2$, $[0, 3]$. $f'(x) = 2x - 2 = 0 \quad\Leftrightarrow\quad x = 1$. $f(0) = 2$, $f(1) = 1$, $f(3) = 5$. So $f(3) = 5$ is the absolute maximum and $f(1) = 1$ is the absolute minimum.

40. $f(x) = 1 - 2x - x^2$, $[-4, 1]$. $f'(x) = -2 - 2x = 0 \quad\Leftrightarrow\quad x = -1$. $f(-4) = -7$, $f(-1) = 2$, $f(1) = -2$. So $f(-4) = -7$ is the absolute minimum, $f(-1) = 2$ is the absolute maximum.

41. $f(x) = x^3 - 12x + 1$, $[-3, 5]$. $f'(x) = 3x^2 - 12 = 3\left(x^2 - 4\right) = 3(x+2)(x-2) = 0 \quad\Leftrightarrow\quad x = \pm 2$. $f(-3) = 10$, $f(-2) = 17$, $f(2) = -15$, $f(5) = 66$. So $f(2) = -15$ is the absolute minimum and $f(5) = 66$ is the absolute maximum.

42. $f(x) = 4x^3 - 15x^2 + 12x + 7$, $[0, 3]$. $f'(x) = 12x^2 - 30x + 12 = 6(2x-1)(x-2) = 0 \quad\Leftrightarrow\quad x = \frac{1}{2}, 2$. $f(0) = 7$, $f\left(\frac{1}{2}\right) = \frac{39}{4}$, $f(2) = 3$, $f(3) = 16$. So $f(3) = 16$ is the absolute maximum and $f(2) = 3$ the absolute minimum.

43. $f(x) = 2x^3 + 3x^2 + 4$, $[-2, 1]$. $f'(x) = 6x^2 + 6x = 6x(x+1) = 0 \quad\Leftrightarrow\quad x = -1, 0$. $f(-2) = 0$, $f(-1) = 5$, $f(0) = 4$, $f(1) = 9$. So $f(1) = 9$ is the absolute maximum and $f(-2) = 0$ is the absolute minimum.

44. $f(x) = 18x + 15x^2 - 4x^3$, $[-3, 4]$. $f'(x) = 18 + 30x - 12x^2 = 6(3-x)(1+2x) = 0 \quad\Leftrightarrow\quad x = 3, -\frac{1}{2}$. $f(-3) = 189$, $f\left(-\frac{1}{2}\right) = -\frac{19}{4}$, $f(3) = 81$, $f(4) = 56$. So $f(-3) = 189$ is the absolute maximum and $f\left(-\frac{1}{2}\right) = -\frac{19}{4}$ is the absolute minimum.

45. $f(x) = x^4 - 4x^2 + 2$, $[-3, 2]$. $f'(x) = 4x^3 - 8x = 4x(x^2 - 2) = 0 \quad \Leftrightarrow \quad x = 0, \pm\sqrt{2}$. $f(-3) = 47$, $f\left(-\sqrt{2}\right) = -2$, $f(0) = 2$, $f\left(\sqrt{2}\right) = -2$, $f(2) = 2$, so $f\left(\pm\sqrt{2}\right) = -2$ is the absolute minimum and $f(-3) = 47$ is the absolute maximum.

46. $f(x) = 3x^5 - 5x^3 - 1$, $[-2, 2]$. $f'(x) = 15x^4 - 15x^2 = 15x^2(x+1)(x-1) = 0 \quad \Leftrightarrow \quad x = -1, 0, 1$. $f(-2) = -57$, $f(-1) = 1$, $f(0) = -1$, $f(1) = -3$, $f(2) = 55$. So $f(-2) = -57$ is the absolute minimum and $f(2) = 55$ is the absolute maximum.

47. $f(x) = x^2 + \dfrac{2}{x}$, $\left[\frac{1}{2}, 2\right]$. $f'(x) = 2x - \dfrac{2}{x^2} = 2\dfrac{x^3 - 1}{x^2} = 0 \quad \Leftrightarrow \quad x = 1$. $f\left(\frac{1}{2}\right) = \frac{17}{4}$, $f(1) = 3$, $f(2) = 5$. So $f(1) = 3$ is the absolute minimum and $f(2) = 5$ is the absolute maximum.

48. $f(x) = \sqrt{9 - x^2}$, $[-1, 2]$. $f'(x) = -x/\sqrt{9 - x^2} = 0 \quad \Leftrightarrow \quad x = 0$. $f(-1) = 2\sqrt{2}$, $f(0) = 3$, $f(2) = \sqrt{5}$. So $f(2) = \sqrt{5}$ is the absolute minimum and $f(0) = 3$ is the absolute maximum.

49. $f(x) = x^{4/5}$, $[-32, 1]$. $f'(x) = \frac{4}{5}x^{-1/5} \quad \Rightarrow \quad f'(x) \neq 0$ but $f'(0)$ does not exist, so 0 is the only critical number. $f(-32) = 16$, $f(0) = 0$, $f(1) = 1$. So $f(0) = 0$ is the absolute minimum and $f(-32) = 16$ is the absolute maximum.

50. $f(x) = \dfrac{x}{x+1}$, $[1, 2]$. $f'(x) = \dfrac{(x+1) - x}{(x+1)^2} = \dfrac{1}{(x+1)^2} \neq 0 \quad \Rightarrow \quad$ no critical numbers. $f(1) = \frac{1}{2}$ and $f(2) = \frac{2}{3}$, so $f(1) = \frac{1}{2}$ is the absolute minimum and $f(2) = \frac{2}{3}$ the absolute maximum.

51. $f(x) = \sin x + \cos x$, $\left[0, \frac{\pi}{3}\right]$. $f'(x) = \cos x - \sin x = 0 \quad \Leftrightarrow \quad x = \frac{\pi}{4}$. $f(0) = 1$, $f\left(\frac{\pi}{4}\right) = \sqrt{2}$, $f\left(\frac{\pi}{3}\right) = \frac{\sqrt{3}+1}{2}$. So $f(0) = 1$ is the absolute minimum and $f\left(\frac{\pi}{4}\right) = \sqrt{2}$ the absolute maximum.

52. $f(x) = x - 2\cos x$, $[-\pi, \pi]$. $f'(x) = 1 + 2\sin x = 0 \quad \Leftrightarrow \quad \sin x = -\frac{1}{2} \quad \Leftrightarrow \quad x = -\frac{5\pi}{6}, -\frac{\pi}{6}$. $f(-\pi) = 2 - \pi \approx -1.14$, $f\left(-\frac{5\pi}{6}\right) = \sqrt{3} - \frac{5\pi}{6} \approx -0.886$, $f\left(-\frac{\pi}{6}\right) = -\frac{\pi}{6} - \sqrt{3} \approx -2.26$, $f(\pi) = \pi + 2 \approx 5.14$. So $f\left(-\frac{\pi}{6}\right) = -\frac{\pi}{6} - \sqrt{3}$ is the absolute minimum and $f(\pi) = \pi + 2$ the absolute maximum.

53.

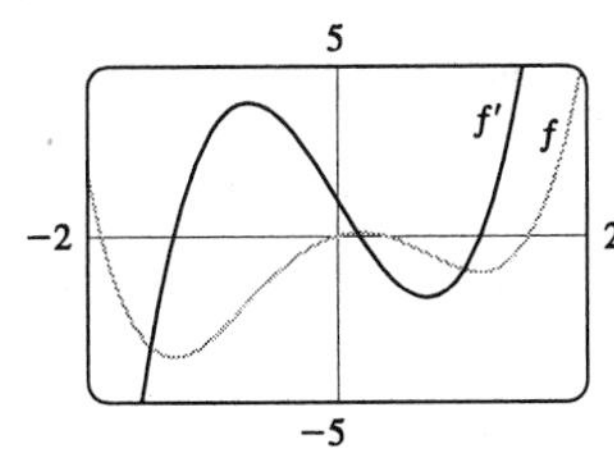

We see that $f'(x) = 0$ at about $x = -1.3, 0.2$, and 1.1. Since f' exists everywhere, these are the only critical numbers

54.

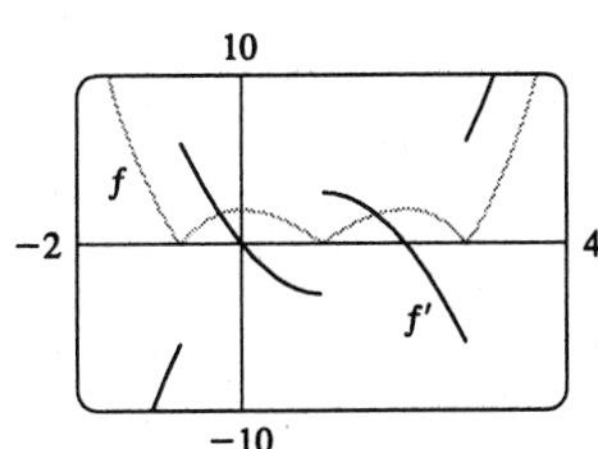

We see that $f'(x) = 0$ at about $x = 0.0$ and 2.0, and that $f'(x)$ does not exist at about $x = -0.7$, 1.0, and 2.7, so the critical numbers of f are about $-0.7, 0.0, 1.0, 2.0$, and 2.7.

55. **(a)**

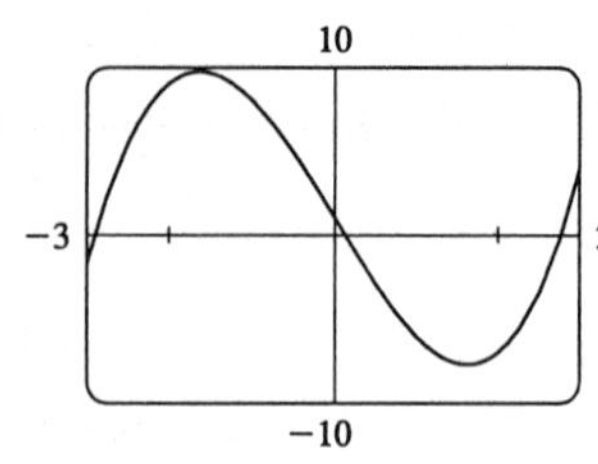

From the graph, it appears that the absolute maximum value is about $f(-1.63) = 9.71$, and the absolute minimum value is about $f(1.63) = -7.71$.

(b) $f(x) = x^3 - 8x + 1 \quad \Rightarrow \quad f'(x) = 3x^2 - 8$.
So $f'(x) = 0 \quad \Rightarrow \quad x = \pm\frac{2\sqrt{6}}{3} \quad \Rightarrow$
$f(x) = \left(\pm\frac{2\sqrt{6}}{3}\right)^3 - 8\left(\pm\frac{2\sqrt{6}}{3}\right) + 1 = \pm\frac{16\sqrt{6}}{9} \mp \frac{16\sqrt{6}}{3} + 1$
$= 1 + \frac{32\sqrt{6}}{9}$ (maximum) or $1 - \frac{32\sqrt{6}}{9}$ (minimum).

(From the graph, we see that the extreme values do not occur at the endpoints.)

56. **(a)**

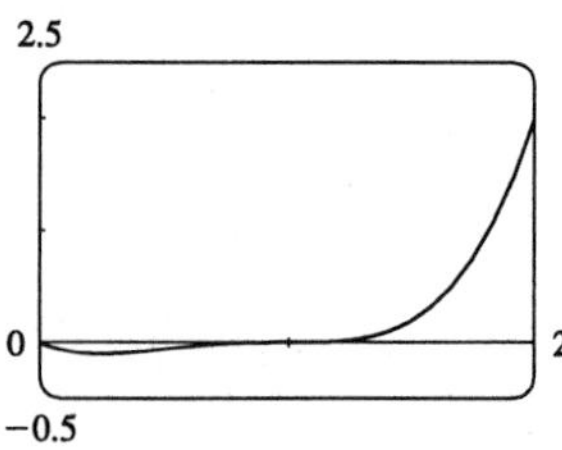

From the graph, it appears that the absolute maximum value is $f(2) = 2$, and that the absolute minimum value is about $f(0.25) = -0.11$.

(b) $f(x) = x^4 - 3x^3 + 3x^2 - x \quad \Rightarrow$
$f'(x) = 4x^3 - 9x^2 + 6x - 1 = (4x - 1)(x - 1)^2$.
So $f'(x) = 0 \quad \Rightarrow \quad x = \frac{1}{4}$ or $x = 1$. Now

$f(1) = 1^4 - 3 \cdot 1^3 + 3 \cdot 1^2 - 1 = 0$ (not an extremum) and $f\left(\frac{1}{4}\right) = \left(\frac{1}{4}\right)^4 - 3\left(\frac{1}{4}\right)^3 + 3\left(\frac{1}{4}\right)^2 - \frac{1}{4} = -\frac{27}{256}$ (minimum). At the right endpoint we have $f(2) = 2^4 - 3 \cdot 2^3 + 3 \cdot 2^2 - 2 = 2$ (maximum).

57. **(a)**

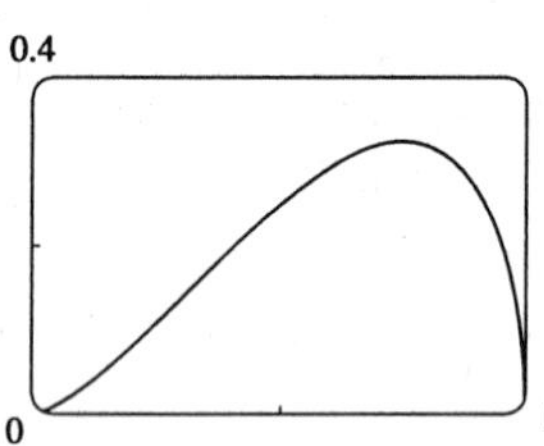

From the graph, it seems that the absolute maximum value is about $f(0.75) = 0.32$, and the absolute minimum value is $f(0) = f(1) = 0$.

(b) $f(x) = x\sqrt{x - x^2} \quad \Rightarrow \quad f'(x) = \dfrac{x - 2x^2}{2\sqrt{x - x^2}} + \sqrt{x - x^2}$.

So $f'(x) = 0 \quad \Rightarrow \quad \sqrt{x - x^2}\left[\dfrac{x - 2x^2}{2(x - x^2)} + 1\right] = 0$.

So either $\sqrt{x - x^2} = 0 \quad \Rightarrow \quad x = 0$ or 1, giving $f(0) = 0$ (minimum), or $\dfrac{x - 2x^2}{2(x - x^2)} + 1 = 0 \quad \Rightarrow$

$3 - 4x = 0 \quad \Rightarrow \quad x = \frac{3}{4}$, and $f\left(\frac{3}{4}\right) = \frac{3}{4}\sqrt{\frac{3}{4} - \left(\frac{3}{4}\right)^2} = \frac{3\sqrt{3}}{16}$ (maximum).

58. **(a)**

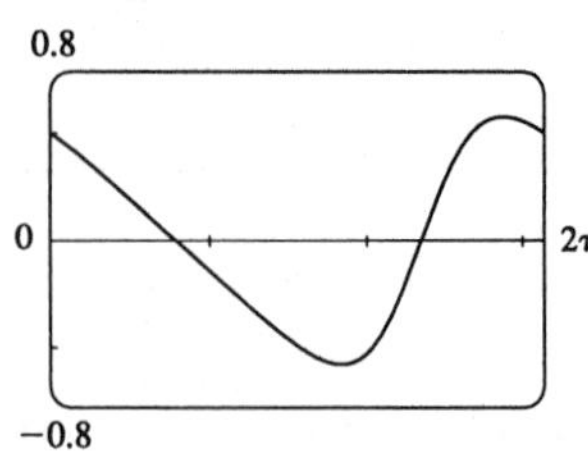

From the graph, it appears that the absolute maximum value is about $f(5.76) = 0.58$, and the absolute minimum value is about $f(3.67) = -0.58$.

(b) $f(x) = \dfrac{\cos x}{2 + \sin x} \quad \Rightarrow$

$f'(x) = \dfrac{(2 + \sin x)(-\sin x) - (\cos x)(\cos x)}{(2 + \sin x)^2} = \dfrac{-1 - 2\sin x}{(2 + \sin x)^2}$.

So $f'(x) = 0 \quad \Rightarrow \quad 1 = -2\sin x \quad \Rightarrow \quad x = \frac{7\pi}{6}$ or $\frac{11\pi}{6}$. Now $f\left(\frac{7\pi}{6}\right) = \frac{-\sqrt{3}/2}{3/2} = -\frac{1}{\sqrt{3}}$ (minimum), and $f\left(\frac{11\pi}{6}\right) = \frac{\sqrt{3}/2}{3/2} = \frac{1}{\sqrt{3}}$ (maximum).

59. $f(x) = [\![x]\!]$ is discontinuous at every integer n (See Example 11 and Exercise 63 in Section 1.3), so that $f'(n)$ does not exist. For all other real numbers a, $[\![x]\!]$ is constant on an open interval containing a, and so $f'(a) = 0$. Therefore every real number is a critical number of $f(x) = [\![x]\!]$.

60. **(a)** $f(x) = 3x^4 - 28x^3 + 6x^2 + 24x \quad \Rightarrow \quad f'(x) = 12x^3 - 84x^2 + 12x + 24 \quad \Rightarrow$ $f''(x) = 36x^2 - 168x + 12$. Now to solve $f'(x) = 0$, try $x_1 = \frac{1}{2} \quad \Rightarrow \quad x_2 = x_1 - \dfrac{f'(x_1)}{f''(x_1)} = \dfrac{2}{3} \quad \Rightarrow$ $x_3 \approx 0.6455 \quad \Rightarrow \quad x_4 \approx 0.6452 \quad \Rightarrow \quad x_5 \approx 0.6452$. Now try $x_1 = 6 \quad \Rightarrow \quad x_2 = 7.12 \quad \Rightarrow$ $x_3 \approx 6.8353 \quad \Rightarrow \quad x_4 \approx 6.8102 \quad \Rightarrow \quad x_5 \approx 6.8100$. Finally try $x_1 = -0.5 \quad \Rightarrow \quad x_2 \approx -0.4571$ $\Rightarrow \quad x_3 \approx -0.4552 \quad \Rightarrow \quad x_4 \approx -0.4552$. Therefore $x = -0.455$, 6.810 and 0.645 are all critical numbers correct to three decimal places.

(b) $f(-1) = 13$, $f(7) = -1939$, $f(6.810) \approx -1949.07$, $f(-0.455) \approx -6.912$, $f(0.645) \approx 10.982$. Therefore $f(6.810) \approx -1949.07$ is the absolute minimum correct to two decimal places.

61. The density is defined as $\rho = \dfrac{\text{mass}}{\text{volume}} = \dfrac{1000}{V(T)}$ (in g/cm^3). But a critical point of ρ will also be a critical point of V $\left[\text{since } \dfrac{d\rho}{dT} = -1000V^{-2}\dfrac{dV}{dT} \text{ and } V \text{ is never } 0\right]$, and V is easier to differentiate than ρ.

$V(T) = 999.87 - 0.06426T + 0.0085043T^2 - 0.0000679T^3 \quad \Rightarrow$

$V'(T) = -0.06426 + 0.0170086T - 0.0002037T^2$. Setting this equal to 0 and using the quadratic formula to find T, we get $T = \dfrac{-0.0170086 \pm \sqrt{0.0170086^2 - 4 \cdot 0.0003037 \cdot 0.06426}}{2(-0.0002037)} \approx 3.9665°$ or $79.5318°$. Since we are only interested in the region $0° \le T \le 30°$, we check the density ρ at the endpoints and at $3.9665°$: $\rho(0) \approx \dfrac{1000}{999.87} \approx 1.00013$; $\rho(30) \approx \dfrac{1000}{1003.7641} \approx 0.99625$; $\rho(3.9665) \approx \dfrac{1000}{999.7447} \approx 1.000255$. So water has its maximum density at about $3.9665°$ C.

62. $F = \dfrac{\mu W}{\mu \sin\theta + \cos\theta} \quad \Rightarrow \quad \dfrac{dF}{d\theta} = \dfrac{(\mu\sin\theta + \cos\theta)(0) - \mu W(\mu\cos\theta - \sin\theta)}{(\mu\sin\theta + \cos\theta)^2} = \dfrac{-\mu W(\mu\cos\theta - \sin\theta)}{(\mu\sin\theta + \cos\theta)^2}$. So $\dfrac{dF}{d\theta} = 0 \quad \Rightarrow \quad \mu\cos\theta - \sin\theta = 0 \quad \Rightarrow \quad \mu = \tan\theta$. To evaluate F at this point, we calculate $F(\theta) = \dfrac{\mu W}{\mu\sin\theta + \cos\theta} = \dfrac{\mu W}{\cos\theta(\mu\tan\theta + 1)} = \dfrac{\sec\theta}{\mu\tan\theta + 1}\mu W$, so when $\tan\theta = \mu$, $\sec^2\theta = \tan^2\theta + 1 = \mu^2 + 1 \Rightarrow \sec\theta = \sqrt{\mu^2 + 1}$, and hence $\dfrac{\sec\theta}{\mu\tan\theta + 1}\mu W = \dfrac{\sqrt{\mu^2+1}}{\mu^2 + 1}\mu W = \dfrac{\mu}{\sqrt{\mu^2+1}}W$.

We compare this with the value of F at the endpoints: $F(0) = \mu W$ and $F\left(\frac{\pi}{2}\right) = W$. Now because $\dfrac{\mu}{\sqrt{\mu^2+1}} \le 1$ and $\dfrac{\mu}{\sqrt{\mu^2+1}} \le \mu$, we have that $\dfrac{\mu}{\sqrt{\mu^2+1}} \le$ each of $F(0)$ and $F\left(\frac{\pi}{2}\right)$. Hence $\dfrac{\mu}{\sqrt{\mu^2+1}}W$ is the absolute minimum value of $F(\theta)$, and it occurs when $\tan\theta = \mu$.

63. $f(x) = x^5$. $f'(x) = 5x^4 \quad \Rightarrow \quad f'(0) = 0$ so 0 is a critical number. But $f(0) = 0$ and f takes both positive and negative values in any open interval containing 0, so f does not have a local extremum at 0.

64. $g(x) = 2 + (x-5)^3 \Rightarrow g'(x) = 3(x-5)^2 \Rightarrow f'(5) = 0$, so 5 is a critical number. But $g(5) = 2$ and g takes on values > 2 and values < 2 in any open interval containing 5, so g does not have a local extremum at 5.

65. $f(x) = x^{101} + x^{51} + x + 1 \Rightarrow f'(x) = 101x^{100} + 51x^{50} + 1 \geq 1$ for all x, so $f'(x) = 0$ has no solution. Thus $f(x)$ has no critical number, so $f(x)$ can have no local extremum.

66. Suppose that f has a minimum value at c, so $f(x) \geq f(c)$ for all x near c. Then $g(x) = -f(x) \leq -f(c)$ for all x near c, so $g(x)$ has a maximum value at c.

67. If f has a local minimum at c, then $g(x) = -f(x)$ has a local maximum at c, so $g'(c) = 0$ by the case of Fermat's Theorem proved in the text. Thus $f'(c) = -g'(c) = 0$.

68. **(a)** $f(x) = ax^3 + bx^2 + cx + d$, $a \neq 0$. So $f'(x) = 3ax^2 + 2bx + c$ is a quadratic and hence has either 2, 1, or 0 real roots, so $f(x)$ has either 2, 1 or 0 critical numbers.

Case (i) (2 critical numbers): $f(x) = x^3 - 3x$
$\Rightarrow f'(x) = 3x^2 - 3$, so $x = -1, 1$ are critical numbers.

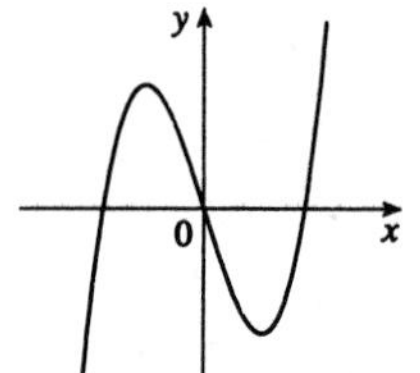

Case (ii) (1 critical number): $f(x) = x^3$,
$f'(x) = 3x^2$, $x = 0$ is the only critical number.

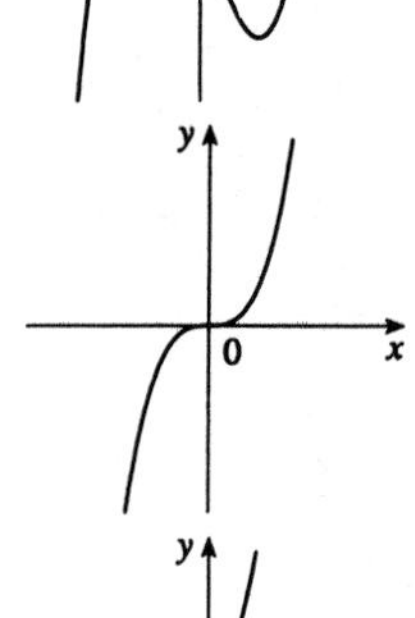

Case (iii) (no critical numbers): $f(x) = x^3 + 3x$,
$f'(x) = 3x^2 + 3$, no real roots.

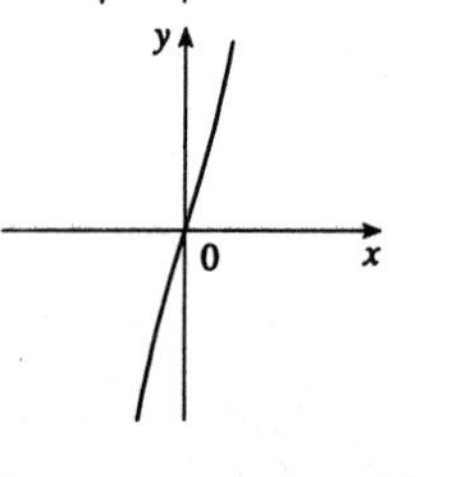

(b) Since there are at most two critical numbers, it can have at most two local extreme values and by (i) this can occur. By (iii) it can have no local extremum. However, if there is only one critical number, then there is no local extremum.

69.

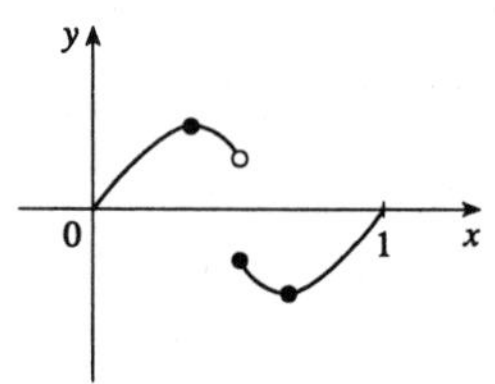

EXERCISES 3.2

1. $f(x) = x^3 - x$, $[-1, 1]$. f, being a polynomial, is continuous on $[-1, 1]$ and differentiable on $(-1, 1)$. Also $f(-1) = 0 = f(1)$. $f'(c) = 3c^2 - 1 = 0 \quad \Rightarrow \quad c = \pm\frac{1}{\sqrt{3}}$.

2. $f(x) = x^3 + x^2 - 2x + 1$, $[-2, 0]$. f, being a polynomial, is continuous on $[-2, 0]$ and differentiable on $(-2, 0)$. Also $f(-2) = 1 = f(0)$. $f'(c) = 3c^2 + 2c - 2 = 0 \quad \Rightarrow \quad c = \frac{-1 \pm \sqrt{7}}{3}$, but only $\frac{-1-\sqrt{7}}{3}$ lies in the interval $(-2, 0)$.

3. $f(x) = \cos 2x$, $[0, \pi]$. f is continuous on $[0, \pi]$ and differentiable on $(0, \pi)$. Also $f(0) = 1 = f(\pi)$. $f'(c) = -2\sin 2c = 0 \quad \Rightarrow \quad \sin 2c = 0 \quad \Rightarrow \quad 2c = \pi \quad \Rightarrow \quad c = \frac{\pi}{2}$ [since $c \in (0, \pi)$].

4. $f(x) = \sin x + \cos x$, $[0, 2\pi]$. Since $\sin x$ and $\cos x$ are continuous on $[0, 2\pi]$ and differentiable on $(0, 2\pi)$ so is their sum $f(x)$. $f(0) = 1 = f(2\pi)$. $f'(c) = \cos c - \sin c = 0 \quad \Leftrightarrow \quad \cos c = \sin c \quad \Leftrightarrow \quad c = \frac{\pi}{4}$ or $\frac{5\pi}{4}$.

5. $f(x) = 1 - x^{2/3}$. $f(-1) = 1 - (-1)^{2/3} = 1 - 1 = 0 = f(1)$. $f'(x) = -\frac{2}{3}x^{-1/3}$, so $f'(c) = 0$ has no solution. This does not contradict Rolle's Theorem, since $f'(0)$ does not exist, and so f is not differentiable on $[-1, 1]$.

6. $f(x) = (x-1)^{-2}$. $f(0) = (0-1)^{-2} = 1 = (2-1)^{-2} = f(2)$. $f'(x) = -2(x-1)^{-3} \quad \Rightarrow \quad f'(x)$ is never 0. This does not contradict Rolle's Theorem since $f'(1)$ does not exist.

7. $\dfrac{f(8) - f(0)}{8 - 0} = \dfrac{6 - 4}{8} = \dfrac{1}{4}$. The values of c which satisfy $f'(c) = \frac{1}{4}$ seem to be about $c = 0.8, 3.2, 4.4$, and 6.1.

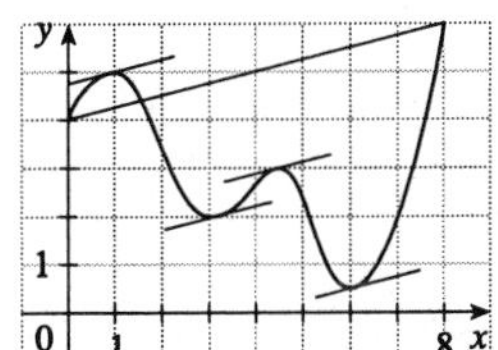

8. $\dfrac{f(7) - f(0)}{7 - 0} = \dfrac{2 - 4}{7} = -\dfrac{2}{7}$. The values of c which satisfy $f'(c) = -\frac{2}{7}$ seem to be about $c = 1.2, 2.8, 4.7$, and 5.8.

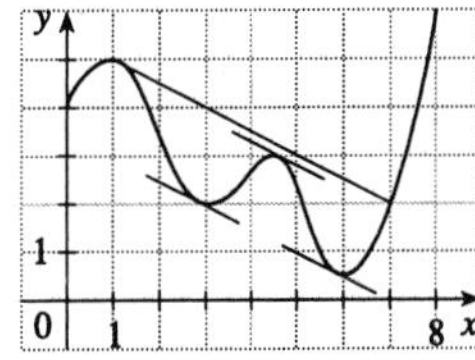

9. (a), (b) The equation of the secant line is

$$y - 5 = \frac{8.5 - 5}{8 - 1}(x - 1) \quad \Leftrightarrow \quad y = \tfrac{1}{2}x + \tfrac{9}{2}.$$

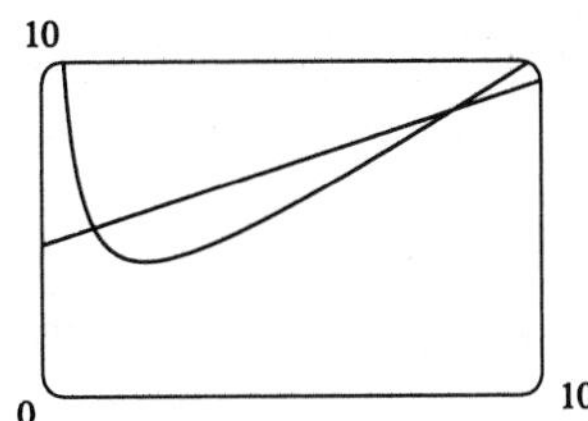

(c) $f(x) = x + 4/x \quad \Rightarrow \quad f'(x) = 1 - 4/x^2$. So $f'(c) = \frac{1}{2} \quad \Rightarrow \quad c^2 = 8 \Rightarrow c = 2\sqrt{2}$, and $f(c) = 2\sqrt{2} + \frac{4}{2\sqrt{2}} = 3\sqrt{2}$. Thus the equation of the tangent is $y - 3\sqrt{2} = \frac{1}{2}\left(x - 2\sqrt{2}\right)$ $\Leftrightarrow \quad y = \frac{1}{2}x + 2\sqrt{2}$.

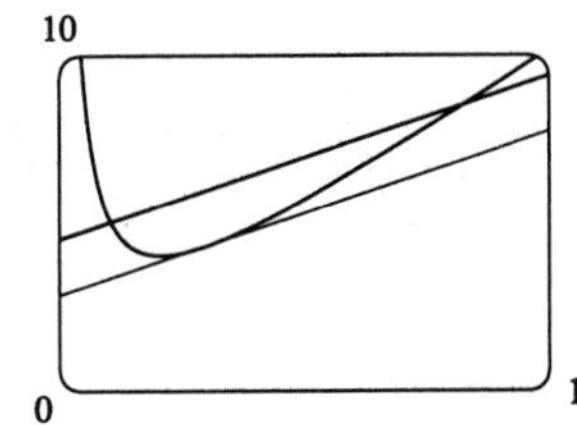

10. (a)

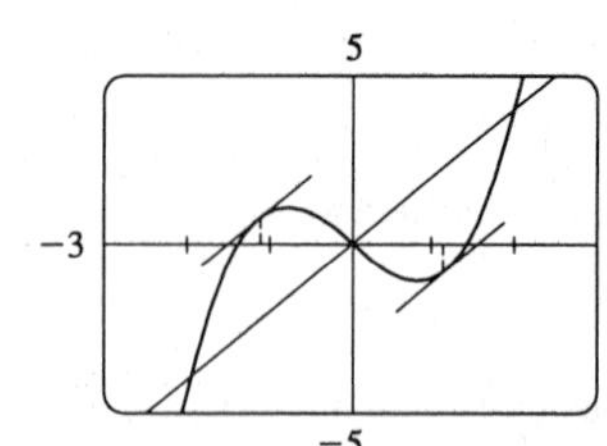

It seems that the tangent lines are parallel to the secant at $x \approx \pm 1.2$.

(b) The slope of the secant line is 2, and its equation is $y = 2x$. $f(x) = x^3 - 2x \Rightarrow f'(x) = 3x^2 - 2$, so we solve $f'(c) = 2 \Rightarrow 3c^2 = 4 \Rightarrow c = \pm\frac{2\sqrt{3}}{3} \approx 1.155$. Our estimates were off by about 0.045 in each case.

11. $f(x) = 1 - x^2$, $[0, 3]$. f, being a polynomial, is continuous on $[0, 3]$ and differentiable on $(0, 3)$.
$\dfrac{f(3) - f(0)}{3 - 0} = \dfrac{-8 - 1}{3} = -3$ and $-3 = f'(c) = -2c \Rightarrow c = \frac{3}{2}$.

12. $f(x) = 2x^3 + x^2 - x - 1$, $[0, 2]$. f, being a polynomial, is continuous on $[0, 2]$ and differentiable on $(0, 2)$.
$\dfrac{f(2) - f(0)}{2 - 0} = \dfrac{17 - (-1)}{2} = 9$ and $9 = f'(c) = 6c^2 + 2c - 1 \Rightarrow 0 = 6c^2 + 2c - 10 \Rightarrow$
$c = \dfrac{-2 \pm \sqrt{244}}{2} = \dfrac{-1 \pm \sqrt{61}}{6}$, but only $\dfrac{-1 + \sqrt{61}}{6}$ lies in $(0, 2)$.

13. $f(x) = 1/x$, $[1, 2]$. f, being a rational function, is continuous on $[1, 2]$ and differentiable on $(1, 2)$.
$\dfrac{f(2) - f(1)}{2 - 1} = \dfrac{\frac{1}{2} - 1}{1} = -\dfrac{1}{2}$ and $-\dfrac{1}{2} = f'(c) = -\dfrac{1}{c^2} \Rightarrow c^2 = 2 \Rightarrow c = \sqrt{2}$ (since c must lie in $[1, 2]$).

14. $f(x) = \sqrt{x}$, $[1, 4]$. $f(x)$ is continuous on $[1, 4]$ and differentiable on $(1, 4)$. $\dfrac{f(4) - f(1)}{4 - 1} = \dfrac{2 - 1}{3} = \dfrac{1}{3}$ and
$\dfrac{1}{3} = f'(c) = \dfrac{1}{2\sqrt{c}} \Rightarrow \sqrt{c} = \dfrac{3}{2} \Rightarrow c = \left(\dfrac{3}{2}\right)^2 = \dfrac{9}{4}$.

15. 1 and $x - 1$ are continuous on $\mathbb{R}$ by Theorem 1.5.5, $\sqrt[3]{x}$ is continuous on $\mathbb{R}$ by Theorem 1.5.6; therefore $f(x) = 1 + \sqrt[3]{x - 1}$ is continuous on $\mathbb{R}$ by Theorems 1.5.8 and 1.5.4(1), and hence continuous on $[2, 9]$. $f'(x) = \frac{1}{3}(x - 1)^{-2/3}$, so that f is differentiable for all $x \neq 1$ and so f is differentiable on $(2, 9)$. By the Mean Value Theorem, there exists a number c such that $f'(c) = \frac{1}{3}(c - 1)^{-2/3} = \dfrac{f(9) - f(2)}{9 - 2} = \dfrac{3 - 2}{7} = \dfrac{1}{7} \Rightarrow$
$\frac{1}{3}(c - 1)^{-2/3} = \frac{1}{7} \Rightarrow (c - 1)^2 = \left(\frac{7}{3}\right)^3 \Rightarrow c = \pm\left(\frac{7}{3}\right)^{3/2} + 1 \Rightarrow c = \left(\frac{7}{3}\right)^{3/2} + 1 \approx 4.564$ since $c \in [2, 9]$.

16. $f(x) = x^4 - 6x^3 + 4x - 1$ is continuous and differentiable on $\mathbb{R}$ since it is a polynomial. So by the Mean Value Theorem there exists a number c such that $f'(c) = 4c^3 - 18c^2 + 4 = \dfrac{f(1) - f(0)}{1 - 0} = \dfrac{-2 - (-1)}{1} = -1 \Rightarrow$
$4c^3 - 18c^2 + 5 = 0$. Now let $g(x) = 4x^3 - 18x^2 + 5 \Rightarrow g'(x) = 12x^2 - 36x$, and try $x_1 = 4 \Rightarrow$
$x_2 = x_1 - \dfrac{g(x_1)}{g'(x_1)} = 4.5625 \Rightarrow x_3 \approx 4.4432 \Rightarrow x_4 \approx 4.4365 \Rightarrow x_5 \approx 4.4365$. Now try $x_1 = 0.5$
$\Rightarrow x_2 \approx 0.5667 \Rightarrow x_3 \approx 0.5635 \Rightarrow x_4 \approx 0.5635$. Finally try $x_1 = -0.5$ to find that it is a root. These are the only numbers since a polynomial of degree three has at most three real roots. Therefore the numbers, correct to two decimal places, are $c = 4.44$, 0.56, and -0.50.

17. $f(x) = |x - 1|$. $f(3) - f(0) = |3 - 1| - |0 - 1| = 1$. Since $f'(c) = -1$ if $c < 1$ and $f'(c) = 1$ if $c > 1$, $f'(c)(3 - 0) = \pm 3$ and so is never equal to 1. This does not contradict the Mean Value Theorem since $f'(1)$ does not exist.

18. $f(x) = \dfrac{x+1}{x-1}$. $f(2) - f(0) = 3 - (-1) = 4$. $f'(x) = \dfrac{1(x-1) - 1(x+1)}{(x+1)^2} = \dfrac{-2}{(x+1)^2}$. Since $f'(x) < 0$ for all x (except $x = -1$), $f'(c)(2-0)$ is always < 0 and hence cannot equal 4. This does not contradict the Mean Value Theorem since f is not continuous at $x = 1$.

19. $f(x) = x^5 + 10x + 3 = 0$. Since f is continuous and $f(-1) = -8$ and $f(0) = 3$, the equation has at least one root in $(-1, 0)$ by the Intermediate Value Theorem. Suppose that the equation has more than one root; say a and b are both roots with $a < b$. Then $f(a) = 0 = f(b)$ so by Rolle's Theorem $f'(x) = 5x^4 + 10 = 0$ has a root in (a, b). But this is impossible since clearly $f'(x) \geq 10 > 0$ for all real x.

20. $f(x) = 3x - 2 + \cos\left(\frac{\pi}{2}x\right) = 0$. Since f is continuous and $f(0) = -1$ and $f(1) = 1$, the equation has at least one root in $(0, 1)$ by the Intermediate Value Theorem. Suppose it has more than one root; say $a < b$ are both roots. Then $f(a) = 0 = f(b)$, so by Rolle's Theorem, $f'(x) = 3 - \frac{\pi}{2}\sin\left(\frac{\pi}{2}x\right) = 0$ has a root in (a, b). But this is impossible since $-\sin x \geq -1 \quad \Rightarrow \quad f'(x) \geq 3 - \frac{\pi}{2} > 0$ for all real x.

21. $f(x) = x^5 - 6x + c = 0$. Suppose that $f(x)$ has two roots a and b with $-1 \leq a < b \leq 1$. Then $f(a) = 0 = f(b)$, so by Rolle's Theorem there is a number d in (a, b) with $f'(d) = 0$. Now $0 = f'(d) = 5d^4 - 6 \quad \Rightarrow \quad d = \pm\sqrt[4]{\frac{6}{5}}$, which are both outside $[-1, 1]$ and hence outside (a, b). Thus $f(x)$ can have at most one root in $[-1, 1]$.

22. Suppose that $f(x) = x^4 + 4x + c = 0$ has three distinct real roots a, b, d where $a < b < d$. Then $f(a) = f(b) = f(d) = 0$. By Rolle's Theorem there are numbers c_1 and c_2 with $a < c_1 < b$ and $b < c_2 < d$ and $0 = f'(c_1) = f'(c_2)$, so $f'(x) = 0$ must have at least two real solutions. However $0 = f'(x) = 4x^3 + 4 = 4(x^3 + 1) = 4(x+1)(x^2 - x + 1)$ has as its only real solution $x = -1$. Thus $f(x)$ can have at most two real roots.

23. **(a)** Suppose that a cubic polynomial $P(x)$ has roots $a_1 < a_2 < a_3 < a_4$, so $P(a_1) = P(a_2) = P(a_3) = P(a_4)$. By Rolle's Theorem there are numbers c_1, c_2, c_3 with $a_1 < c_1 < a_2$, $a_2 < c_2 < a_3$ and $a_3 < c_3 < a_4$ and $P'(c_1) = P'(c_2) = P'(c_3) = 0$. Thus the second-degree polynomial $P'(x)$ has 3 distinct real roots, which is impossible.

(b) We prove by induction that a polynomial of degree n has at most n real roots. This is certainly true for $n = 1$. Suppose that the result is true for all polynomials of degree n and let $P(x)$ be a polynomial of degree $n+1$. Suppose that $P(x)$ has more than $n+1$ real roots, say $a_1 < a_2 < a_3 < \cdots < a_{n+1} < a_{n+2}$. Then $P(a_1) = P(a_2) = \cdots = P(a_{n+2}) = 0$. By Rolle's Theorem there are real numbers $c_1, \ldots, c_{n+1}$ with $a_1 < c_1 < a_2, \ldots, \; a_{n+1} < c_{n+1} < a_{n+2}$ and $P'(c_1) = \cdots = P'(c_{n+1}) = 0$. Thus the nth degree polynomial $P'(x)$ has at least $n+1$ roots. This contradiction shows that $P(x)$ has at most $n+1$ real roots.

24. **(a)** Suppose that $f(a) = f(b) = 0$ where $a < b$. By Rolle's Theorem applied to f on $[a, b]$ there is a number c such that $a < c < b$ and $f'(c) = 0$.

(b) Suppose that $f(a) = f(b) = f(c) = 0$ where $a < b < c$. By Rolle's Theorem applied to $f(x)$ on $[a, b]$ and $[b, c]$ there are numbers $a < d < b$ and $b < e < c$ with $f'(d) = 0$ and $f'(e) = 0$. By Rolle's Theorem applied to $f'(x)$ on $[d, e]$ there is a number g with $d < g < e$ such that $f''(g) = 0$.

(c) Suppose that f is n times differentiable on $\mathbb{R}$ and has $n + 1$ distinct real roots. Then $f^{(n)}$ has at least one real root.

25. By the Mean Value Theorem, $f(4) - f(1) = f'(c)(4 - 1)$ for some $c \in (1, 4)$. But for every $c \in (1, 4)$ we have $f'(c) \geq 2$. Putting $f'(c) \geq 2$ into the above equation and substituting $f(1) = 10$, we get $f(4) = f(1) + f'(c)(4 - 1) = 10 + 3f'(c) \geq 10 + 3 \cdot 2 = 16$. So the smallest possible value of $f(4)$ is 16.

26. By the Mean Value Theorem, $\dfrac{f(5) - f(2)}{5 - 2} = f'(c)$ for some $c \in (2, 5)$. Since $1 \leq f'(x) \leq 4$, we have $1 \leq \dfrac{f(5) - f(2)}{5 - 2} \leq 4$ or $1 \leq \dfrac{f(5) - f(2)}{3} \leq 4$ or $3 \leq f(5) - f(2) \leq 12$.

27. Suppose that such a function f exists. By the Mean Value Theorem there is a number $0 < c < 2$ with $f'(c) = \dfrac{f(2) - f(0)}{2 - 0} = \dfrac{5}{2}$. But this is impossible since $f'(x) \leq 2 < \frac{5}{2}$ for all x, so no such function can exist.

28. Let $h = f - g$. Then since f and g are continuous on $[a, b]$ and differentiable on (a, b), so is h, and thus h satisfies the assumptions of the Mean Value Theorem. Therefore there is a number c with $a < c < b$ such that $h(b) = h(b) - h(a) = h'(c)(b - a)$. Since $h'(c) < 0$, $h'(c)(b - a) < 0$, so $f(b) - g(b) = h(b) < 0$ and hence $f(b) < g(b)$.

29. We use Exercise 28 with $f(x) = \sqrt{1 + x}$, $g(x) = 1 + \frac{1}{2}x$, and $a = 0$. Notice that $f(0) = 1 = g(0)$ and $f'(x) = \dfrac{1}{2\sqrt{1 + x}} < \dfrac{1}{2} = g'(x)$ for $x > 0$. So by Exercise 28, $f(b) < g(b) \Rightarrow \sqrt{1 + b} < 1 + \frac{1}{2}b$ for $b > 0$.

Another Method: Apply the Mean Value Theorem directly to either $f(x) = 1 + \frac{1}{2}x - \sqrt{1 + x}$ or $g(x) = \sqrt{1 + x}$ on $[0, b]$.

30. f satisfies the conditions for the Mean Value Theorem, so we use this theorem on the interval $[-b, b]$: $\dfrac{f(b) - f(-b)}{b - (-b)} = f'(c)$ for some $c \in (-b, b)$. But since f is odd, $f(-b) = -f(b)$. Substituting this into the above equation, we get $\dfrac{f(b) + f(b)}{2b} = f'(c) \Rightarrow \dfrac{f(b)}{b} = f'(c)$.

31. Let $f(x) = \sin x$ and let $b < a$. Then $f(x)$ is continuous on $[b, a]$ and differentiable on (b, a). By the Mean Value Theorem, there is a number $c \in (b, a)$ with $\sin a - \sin b = f(a) - f(b) = f'(c)(a - b) = (\cos c)(a - b)$. Thus $|\sin a - \sin b| \leq |\cos c||b - a| \leq |a - b|$. If $a < b$, then $|\sin a - \sin b| = |\sin b - \sin a| \leq |b - a| = |a - b|$. If $a = b$, both sides of the inequality are 0.

32. Suppose that $f'(x) = c$. Let $g(x) = cx$, so $g'(x) = c$. Then, by Corollary 7, $f(x) = g(x) + d$, where d is a constant, so $f(x) = cx + d$.

33. For $x > 0$, $f(x) = g(x)$, so $f'(x) = g'(x)$. For $x < 0$, $f'(x) = (1/x)' = -1/x^2$ and $g'(x) = (1 + 1/x)' = -1/x^2$, so again $f'(x) = g'(x)$. However, the domain of $g(x)$ is not an interval $\left[\text{it is } (-\infty, 0) \cup (0, \infty)\right]$ so we cannot conclude that $f - g$ is constant (in fact it is not).

34. Let $v(t)$ be the velocity of the car t hours after 2:00 P.M. Then $\dfrac{v(1/6) - v(0)}{1/6 - 0} = \dfrac{50 - 30}{1/6} = 120$. By the Mean Value Theorem there is a number $0 < c < \frac{1}{6}$ with $v'(c) = 120$. Since $v'(t)$ is the acceleration at time t, the acceleration c hours after 2:00 P.M. is exactly $120 \text{ mi}/\text{h}^2$.

35. Let $g(t)$ and $h(t)$ be the position functions of the two runners and let $f(t) = g(t) - h(t)$. By hypothesis $f(0) = g(0) - h(0) = 0$ and $f(b) = g(b) - h(b) = 0$ where b is the finishing time. Then by Rolle's Theorem, there is a time $0 < c < b$ with $0 = f'(c) = g'(c) - h'(c)$. Hence $g'(c) = h'(c)$, so at time c, both runners have the same velocity $g'(c) = h'(c)$.

EXERCISES 3.3

1. **(a)** $f'(x) > 0$ for $x < 0$ and $x > 3$, so f is increasing on $(-\infty, 0]$ and $[3, \infty)$. $f'(x) < 0$ for $0 < x < 3$, so f is decreasing on $[0, 3]$.

(b) f has a local maximum where f' changes from positive to negative, at $x = 0$, and a local minimum where f' changes from negative to positive, at $x = 3$.

2. **(a)** $f'(x) > 0$ for $-1 < x < 3$ and $x > 4$, so f is increasing on $[-1, 3]$ and $[4, \infty)$. $f'(x) < 0$ for $x < -1$ and $3 < x < 4$, so f is decreasing on $(-\infty, -1]$ and $[3, 4]$.

(b) f has a local maximum where f' changes from positive to negative, at $x = 3$, and local minima where f' changes from negative to positive, at $x = -1$ and $x = 4$.

3. $f(x) = 20 - x - x^2$, $f'(x) = -1 - 2x = 0 \quad \Rightarrow$ $x = -\frac{1}{2}$ (the only critical number)

(a) $f'(x) > 0 \quad \Leftrightarrow \quad -1 - 2x > 0 \quad \Leftrightarrow \quad x < -\frac{1}{2}$, $f'(x) < 0 \quad \Leftrightarrow \quad x > -\frac{1}{2}$, so f is increasing on $\left(-\infty, -\frac{1}{2}\right]$ and decreasing on $\left[-\frac{1}{2}, \infty\right)$.

(b) By the First Derivative Test, $f\left(-\frac{1}{2}\right) = 20.25$ is a local maximum.

(c)

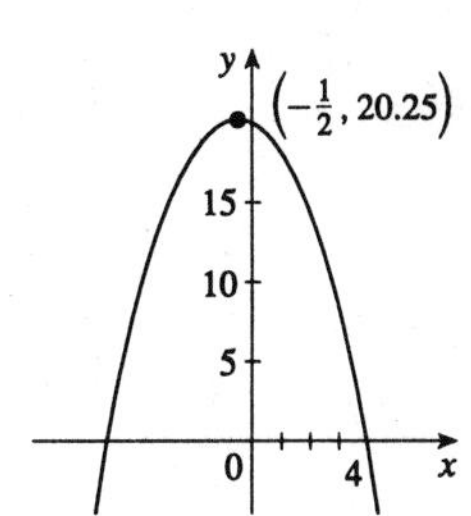

4. $f(x) = x^3 - x + 1$. $f'(x) = 3x^2 - 1 = 0 \quad \Rightarrow \quad x = \pm\frac{1}{\sqrt{3}}$ (the only critical numbers)

(a) $f'(x) > 0 \Leftrightarrow 3x^2 > 1 \Leftrightarrow |x| > \frac{1}{3} \Leftrightarrow x < -\frac{1}{\sqrt{3}}$ or $x > \frac{1}{\sqrt{3}}$ and $f'(x) < 0 \quad \Leftrightarrow -\frac{1}{\sqrt{3}} < x < \frac{1}{\sqrt{3}}$. So f is increasing on $\left(-\infty, -\frac{1}{\sqrt{3}}\right]$ and $\left[\frac{1}{\sqrt{3}}, \infty\right)$ and decreasing on $\left[-\frac{1}{\sqrt{3}}, \frac{1}{\sqrt{3}}\right]$.

(b) By the First Derivative Test, $f\left(-\frac{1}{\sqrt{3}}\right) = 1 + \frac{2}{3\sqrt{3}}$ is a local maximum and $f\left(\frac{1}{\sqrt{3}}\right) = 1 - \frac{2}{3\sqrt{3}}$ is a local minimum.

(c)

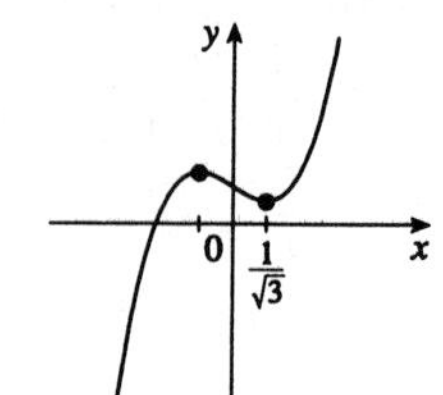

5. $f(x) = x^3 + x + 1 \quad \Rightarrow \quad f'(x) = 3x^2 + 1 > 0$ for all $x \in \mathbb{R}$.

(a) f is increasing on $\mathbb{R}$.

(b) f has no local extremum.

(c)

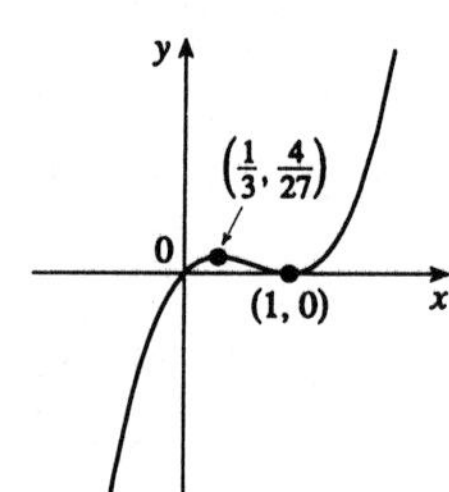

6. $f(x) = x^3 - 2x^2 + x$. $f'(x) = 3x^2 - 4x + 1 = (3x-1)(x-1)$. So the critical numbers are $x = \frac{1}{3}, 1$.

(a) $f'(x) > 0 \quad \Leftrightarrow \quad (3x-1)(x-1) > 0 \quad \Leftrightarrow$
$x < \frac{1}{3}$ or $x > 1$ and $f'(x) < 0 \quad \Leftrightarrow \quad \frac{1}{3} < x < 1$.
So f is increasing on $\left(-\infty, \frac{1}{3}\right]$ and $[1, \infty)$ and
f is decreasing on $\left[\frac{1}{3}, 1\right]$.

(b) The local maximum is $f\left(\frac{1}{3}\right) = \frac{4}{27}$ and the local minimum is $f(1) = 0$.

(c)

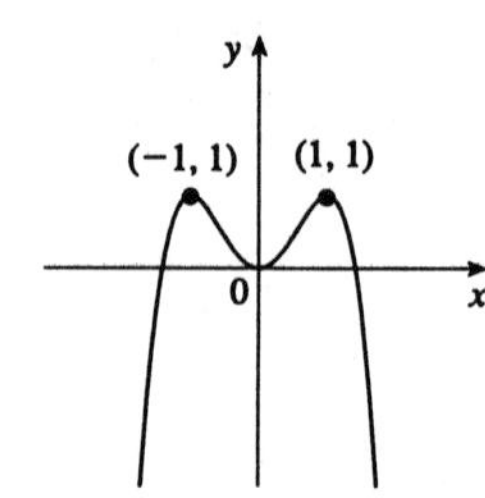

7. $f(x) = 2x^2 - x^4$. $f'(x) = 4x - 4x^3 = 4x(1-x^2) = 4x(1+x)(1-x)$, so the critical numbers are $x = 0, \pm1$.

(a)

Interval	$4x$	$1+x$	$1-x$	$f'(x)$	f
$x < -1$	$-$	$-$	$+$	$+$	increasing on $(-\infty, -1]$
$-1 < x < 0$	$-$	$+$	$+$	$-$	decreasing on $[-1, 0]$
$0 < x < 1$	$+$	$+$	$+$	$+$	increasing on $[0, 1]$
$x > 1$	$+$	$+$	$-$	$-$	decreasing on $[1, \infty)$

(b) Local maximum $f(-1) = 1$, local minimum $f(0) = 0$, local maximum $f(1) = 1$.

(c)

(−1, 1) (1, 1)
y
0
x

8. $f(x) = x^2(1-x)^2$. $0 = f'(x) = 2x(1-x)^2 + x^2[2(1-x)(-1)] = 2x(1-x)(1-2x)$.
So the critical numbers are $x = 0, \frac{1}{2}, 1$.

(a)

Interval	$2x$	$1-x$	$1-2x$	$f'(x)$	f
$x < 0$	$-$	$+$	$+$	$-$	decreasing on $(-\infty, 0]$
$0 < x < \frac{1}{2}$	$+$	$+$	$+$	$+$	increasing on $\left[0, \frac{1}{2}\right]$
$\frac{1}{2} < x < 1$	$+$	$+$	$-$	$-$	decreasing on $\left[\frac{1}{2}, 1\right]$
$x > 1$	$+$	$-$	$-$	$+$	increasing on $[1, \infty)$

(c)

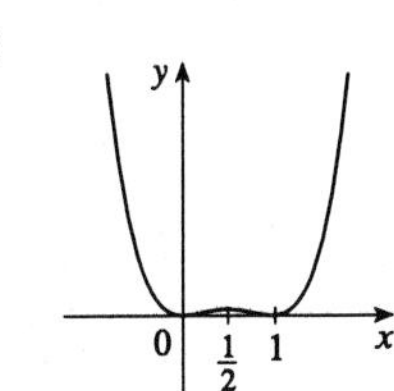

(b) Local minimum $f(0) = 0$, local maximum $f\left(\frac{1}{2}\right) = \frac{1}{16}$,
local minimum $f(1) = 0$.

9. $f(x) = x^3(x-4)^4$. $f'(x) = 3x^2(x-4)^4 + x^3\left[4(x-4)^3\right] = x^2(x-4)^3(7x-12)$.
The critical numbers are $x = 0, 4, \frac{12}{7}$.

(a) $x^2(x-4)^2 \geq 0$ so $f'(x) \geq 0 \quad \Leftrightarrow \quad (x-4)(7x-12) \geq 0$
$\Leftrightarrow \quad x \leq \frac{12}{7}$ or $x \geq 4$. $f'(x) \leq 0 \quad \Leftrightarrow \quad \frac{12}{7} \leq x \leq 4$.
So f is increasing on $\left(-\infty, \frac{12}{7}\right]$ and $[4, \infty)$ and decreasing on $\left[\frac{12}{7}, 4\right]$.

(b) Local maximum $f\left(\frac{12}{7}\right) = 12^3 \cdot \dfrac{16^4}{7^7} \approx 137.5$, local minimum $f(4) = 0$.

(c)

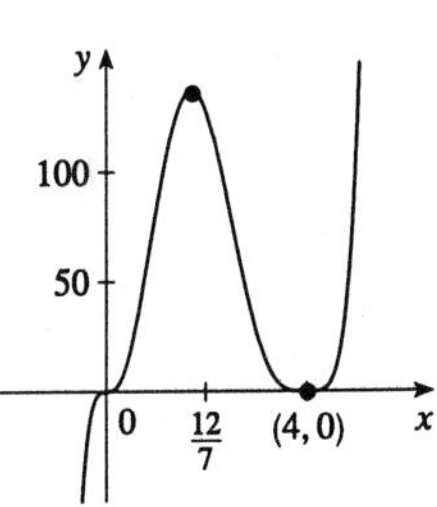

10. $f(x) = 3x^5 - 25x^3 + 60x$.
$f'(x) = 15x^4 - 75x^2 + 60 = 15\left(x^4 - 5x^2 + 4\right) = 15(x^2-4)(x^2-1) = 15(x-2)(x+2)(x+1)(x-1)$.

So the critical numbers are $x = \pm 2, \pm 1$.

(a)

Interval	$x+2$	$x-2$	$x+1$	$x-1$	$f'(x)$	f
$x < -2$	$-$	$-$	$-$	$-$	$+$	increasing on $(-\infty, -2]$
$-2 < x < -1$	$+$	$-$	$-$	$-$	$-$	decreasing on $[-2, -1]$
$-1 < x < 1$	$+$	$-$	$+$	$-$	$+$	increasing on $[-1, 2]$
$1 < x < 2$	$+$	$-$	$+$	$+$	$-$	decreasing on $[1, 2]$
$x > 2$	$+$	$+$	$+$	$+$	$+$	increasing on $[2, \infty)$

(b) Local maximum $f(-2) = -16$, local minimum $f(-1) = -38$,
local maximum $f(1) = 38$, local minimum $f(2) = 16$.

(c)

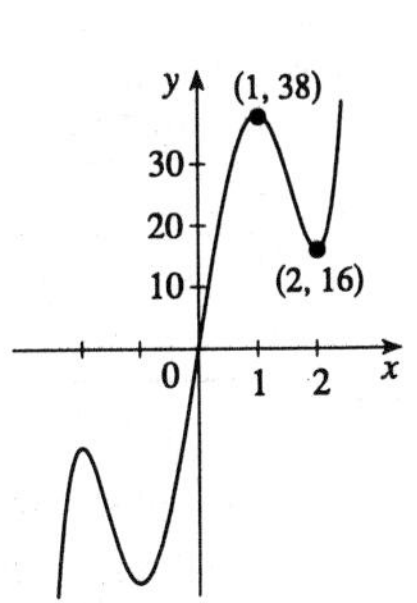

11. $f(x) = x\sqrt{6-x}$. $f'(x) = \sqrt{6-x} + x\left(-\dfrac{1}{2\sqrt{6-x}}\right) = \dfrac{3(4-x)}{2\sqrt{6-x}}$. Critical numbers are $x = 4, 6$.

(a) $f'(x) > 0 \iff 4 - x > 0$ (and $x < 6$) $\iff x < 4$ and
$f'(x) < 0 \iff 4 - x < 0$ (and $x < 6$) $\iff 4 < x < 6$.
So f is increasing on $(-\infty, 4]$ and decreasing on $[4, 6]$.

(b) Local maximum $f(4) = 4\sqrt{2}$

(c)

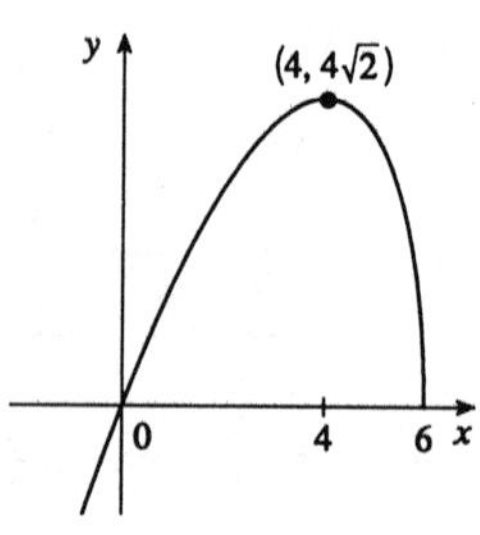

12. $f(x) = x\sqrt{1-x^2}$. $f'(x) = \sqrt{1-x^2} - \dfrac{x^2}{\sqrt{1-x^2}} = \dfrac{1-2x^2}{\sqrt{1-x^2}}$. Critical numbers are $\pm\frac{1}{\sqrt{2}}$ and ± 1.

(a) $f'(x) > 0 \iff 1 - 2x^2 > 0 \iff x^2 < \frac{1}{2} \iff |x| < \frac{1}{\sqrt{2}}$
$\iff -\frac{1}{\sqrt{2}} < x < \frac{1}{\sqrt{2}}$. $f'(x) < 0 \iff -1 < x < -\frac{1}{\sqrt{2}}$ or
$\frac{1}{\sqrt{2}} < x < 1$. So f is increasing on $\left[-\frac{1}{\sqrt{2}}, \frac{1}{\sqrt{2}}\right]$ and
decreasing on $\left[-1, -\frac{1}{\sqrt{2}}\right]$ and $\left[\frac{1}{\sqrt{2}}, 1\right]$.

(b) Local minimum $f\left(-\frac{1}{\sqrt{2}}\right) = -\frac{1}{2}$, local maximum $f\left(\frac{1}{\sqrt{2}}\right) = \frac{1}{2}$.

(c)

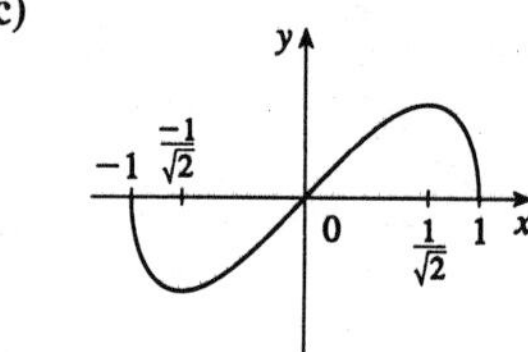

13. $f(x) = x^{1/5}(x+1)$. $f'(x) = \frac{1}{5}x^{-4/5}(x+1) + x^{1/5} = \frac{1}{5}x^{-4/5}(6x+1)$. The critical numbers are $x = 0, -\frac{1}{6}$.

(a) $f'(x) > 0 \iff 6x + 1 > 0$ $(x \neq 0)$ $\iff x > -\frac{1}{6}$ $(x \neq 0)$
and $f'(x) < 0 \iff x < -\frac{1}{6}$. So f is increasing
on $\left[-\frac{1}{6}, \infty\right)$ and decreasing on $\left(-\infty, -\frac{1}{6}\right]$.

(b) Local minimum $f\left(-\frac{1}{6}\right) = -\dfrac{5}{6^{6/5}} \approx -0.58$

(c)

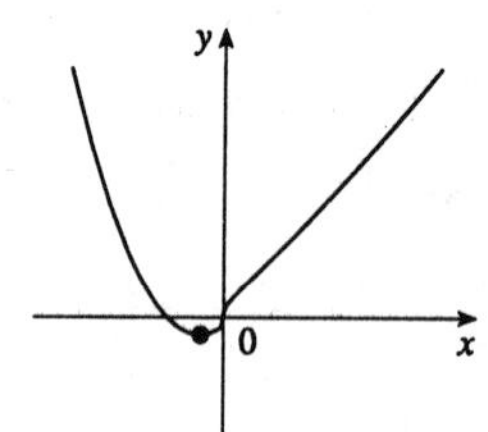

14. $f(x) = x^{2/3}(x-2)^2$. Domain is $\mathbb{R}$. $f'(x) = \frac{2}{3}x^{-1/3}(x-2)^2 + x^{2/3}[2(x-2)] = \frac{2}{3}x^{-1/3}(x-2)(4x-2)$.
Critical numbers are $x = 0, \frac{1}{2}, 2$.

(a)

Interval	$x^{-1/3}$	$x-2$	$2x-1$	$f'(x)$	f
$x < 0$	$-$	$-$	$-$	$-$	decreasing on $(-\infty, 0]$
$0 < x < \frac{1}{2}$	$+$	$-$	$-$	$+$	increasing on $\left[0, \frac{1}{2}\right]$
$\frac{1}{2} < x < 2$	$+$	$-$	$+$	$-$	decreasing on $\left[\frac{1}{2}, 2\right]$
$x > 2$	$+$	$+$	$+$	$+$	increasing on $[2, \infty)$

(b) Local minimum $f(0) = 0$, local maximum $f\left(\frac{1}{2}\right) = \left(\frac{9}{4}\right)^{4/3} \approx 1.42$,
local minimum $f(2) = 0$.

(c)

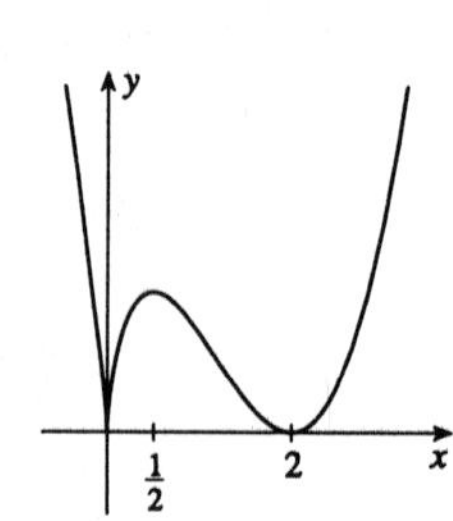

15. $f(x) = x\sqrt{x - x^2}$. The domain of f is $\{x \mid x(1-x) \geq 0\} = [0, 1]$.

$f'(x) = \sqrt{x - x^2} + x\dfrac{1 - 2x}{2\sqrt{x - x^2}} = \dfrac{x(3 - 4x)}{2\sqrt{x - x^2}}$. So the critical numbers are $x = 0, \frac{3}{4}, 1$.

(a) $f'(x) > 0 \Leftrightarrow 3 - 4x > 0 \Leftrightarrow$ $0 < x < \frac{3}{4}$. $f'(x) < 0 \Leftrightarrow \frac{3}{4} < x < 1$. So f is increasing on $\left[0, \frac{3}{4}\right]$ and decreasing on $\left[\frac{3}{4}, 1\right]$.

(b) Local maximum $f\left(\frac{3}{4}\right) = \frac{3\sqrt{3}}{16}$

(c)

16. $f(x) = \sqrt[3]{x} - \sqrt[3]{x^2} = x^{1/3} - x^{2/3}$. $f'(x) = \frac{1}{3}x^{-2/3} - \frac{2}{3}x^{-1/3} = \frac{1}{3}x^{-2/3}\left(1 - 2x^{1/3}\right)$.
So the critical numbers are $x = 0, \frac{1}{8}$.

(a) $f'(x) > 0 \Leftrightarrow 1 - 2x^{1/3} > 0 \Leftrightarrow$ $\frac{1}{2} > x^{1/3} \Leftrightarrow x < \frac{1}{8}$ $(x \neq 0)$ $f'(x) < 0 \Leftrightarrow x > \frac{1}{8}$. So f is increasing on $\left(-\infty, \frac{1}{8}\right]$ and decreasing on $\left[\frac{1}{8}, \infty\right)$.

(b) Local maximum $f\left(\frac{1}{8}\right) = \frac{1}{4}$

(c)

17. $f(x) = x - 2\sin x, 0 \leq x \leq 2\pi$. $f'(x) = 1 - 2\cos x$. So $f'(x) = 0 \Leftrightarrow \cos x = \frac{1}{2} \Leftrightarrow x = \frac{\pi}{3}$ or $\frac{5\pi}{3}$.

(a) $f'(x) > 0 \Leftrightarrow 1 - 2\cos x > 0 \Leftrightarrow \frac{1}{2} > \cos x \Leftrightarrow$ $\frac{\pi}{3} < x < \frac{5\pi}{3}$. $f'(x) < 0 \Leftrightarrow 0 \leq x < \frac{\pi}{3}$ or $\frac{5\pi}{3} < x \leq 2\pi$. So f is increasing on $\left[\frac{\pi}{3}, \frac{5\pi}{3}\right]$ and decreasing on $\left[0, \frac{\pi}{3}\right]$ and $\left[\frac{5\pi}{3}, 2\pi\right]$.

(b) Local minimum $f\left(\frac{\pi}{3}\right) = \frac{\pi}{3} - \sqrt{3} \approx -0.68$,
local maximum $f\left(\frac{5\pi}{3}\right) = \sqrt{3} + \frac{5\pi}{3} \approx 6.97$

(c)

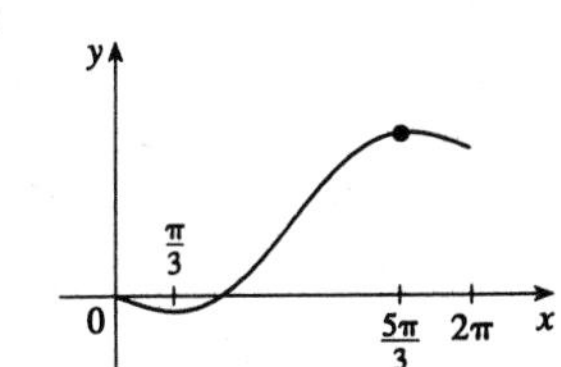

18. $f(x) = x + \cos x, 0 \leq x \leq 2\pi \Rightarrow f'(x) = 1 - \sin x$.
The only critical number is $x = \frac{\pi}{2}$.

(a) $f'(x) > 0 \Leftrightarrow 1 > \sin x \Leftrightarrow$ $x \neq \frac{\pi}{2}$. So f is increasing on $[0, 2\pi]$.

(b) No local maximum or minimum

(c)

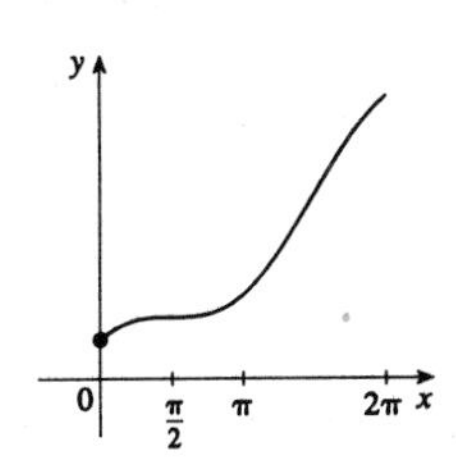

19. $f(x) = \sin^4 x + \cos^4 x, 0 \leq x \leq 2\pi$.

$f'(x) = 4\sin^3 x \cos x - 4\cos^3 x \sin x = -4\sin x \cos x(\cos^2 x - \sin^2 x) = -2\sin 2x \cos 2x = -\sin 4x$.

$f'(x) = 0 \Leftrightarrow \sin 4x = 0 \Leftrightarrow 4x = n\pi \Leftrightarrow x = n\frac{\pi}{4}$. So the critical numbers are $0, \frac{\pi}{4}, \frac{\pi}{2}, \frac{3\pi}{4}, \pi, \frac{5\pi}{4}, \frac{3\pi}{2}, \frac{7\pi}{4}, 2\pi$.

(a) $f'(x) > 0 \Leftrightarrow \sin 4x < 0 \Leftrightarrow \frac{\pi}{4} < x < \frac{\pi}{2}$ or $\frac{3\pi}{4} < x < \pi$ or $\frac{5\pi}{4} < x < \frac{3\pi}{2}$ or $\frac{7\pi}{4} < x < 2\pi$. f is increasing on these intervals. f is decreasing on $\left[0, \frac{\pi}{4}\right], \left[\frac{\pi}{2}, \frac{3\pi}{4}\right], \left[\pi, \frac{5\pi}{4}\right], \left[\frac{3\pi}{2}, \frac{7\pi}{4}\right]$.

(b) Local maxima $f\left(\frac{\pi}{2}\right) = f(\pi) = f\left(\frac{3\pi}{2}\right) = 1$,
local minima $f\left(\frac{\pi}{4}\right) = f\left(\frac{3\pi}{4}\right) = f\left(\frac{5\pi}{4}\right) = f\left(\frac{7\pi}{4}\right) = \frac{1}{2}$.

(c)

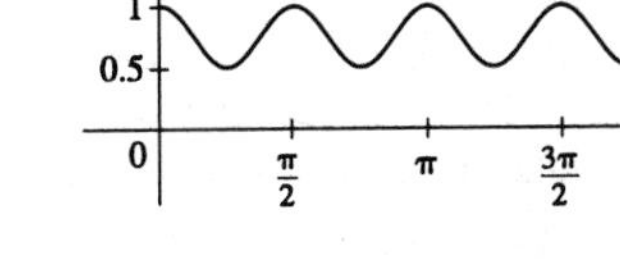

20. $f(x) = x\sin x + \cos x$, $-\pi \le x \le \pi$.

$f'(x) = \sin x + x\cos x - \sin x = x\cos x$, $f'(x) = 0 \Leftrightarrow x = -\frac{\pi}{2}, 0, \frac{\pi}{2}$.

(a) $f'(x) > 0 \Leftrightarrow x\cos x > 0 \Leftrightarrow -\pi \le x < -\frac{\pi}{2}$ or $0 < x < \frac{\pi}{2}$.
So f is increasing on $\left[-\pi, -\frac{\pi}{2}\right]$ and $\left[0, \frac{\pi}{2}\right]$ and decreasing on $\left[-\frac{\pi}{2}, 0\right]$ and $\left[\frac{\pi}{2}, \pi\right]$.

(b) Local maxima $f\left(-\frac{\pi}{2}\right) = f\left(\frac{\pi}{2}\right) = \frac{\pi}{2}$, local minimum $f(0) = 1$,

(c)

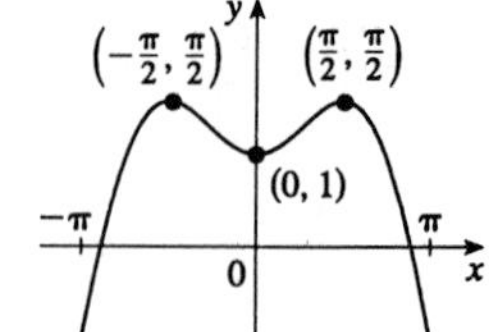

21. $f(x) = x^3 + 2x^2 - x + 1$. $f'(x) = 3x^2 + 4x - 1 = 0 \Rightarrow x = \frac{-4 \pm \sqrt{28}}{6} = \frac{-2 \pm \sqrt{7}}{3}$. Now $f'(x) > 0$ for $x < \frac{-2-\sqrt{7}}{3}$ or $x > \frac{-2+\sqrt{7}}{3}$ and $f'(x) < 0$ for $\frac{-2-\sqrt{7}}{3} < x < \frac{-2+\sqrt{7}}{3}$. f is increasing on $\left(-\infty, \frac{-2-\sqrt{7}}{3}\right]$ and $\left[\frac{-2+\sqrt{7}}{3}, \infty\right)$ and decreasing on $\left[\frac{-2-\sqrt{7}}{3}, \frac{-2+\sqrt{7}}{3}\right]$.

22. $f(x) = x^5 + 4x^3 - 6$. $f'(x) = 5x^4 + 12x^2 > 0$ for all $x \ne 0$. So f is increasing on $\mathbb{R}$.

23. $f(x) = x^6 + 192x + 17$. $f'(x) = 6x^5 + 192 = 6(x^5 + 32)$. So $f'(x) > 0 \Leftrightarrow x^5 > -32 \Leftrightarrow x > -2$ and $f'(x) < 0 \Leftrightarrow x < -2$. So f is increasing on $[-2, \infty)$ and decreasing on $(-\infty, -2]$.

24. $f(x) = 2\tan x - \tan^2 x$. $f'(x) = 2\sec^2 x - 2\tan x\sec^2 x = 2\sec^2 x(1 - \tan x)$. So $f'(x) > 0 \Leftrightarrow 1 - \tan x > 0 \Leftrightarrow \tan x < 1 \Leftrightarrow x \in \left(n\pi - \frac{\pi}{2}, n\pi + \frac{\pi}{4}\right)$, n an integer. So f is increasing on $\left(n\pi - \frac{\pi}{2}, n\pi + \frac{\pi}{4}\right]$, n an integer, and decreasing on $\left[n\pi + \frac{\pi}{4}, n\pi + \frac{\pi}{2}\right)$, n an integer.

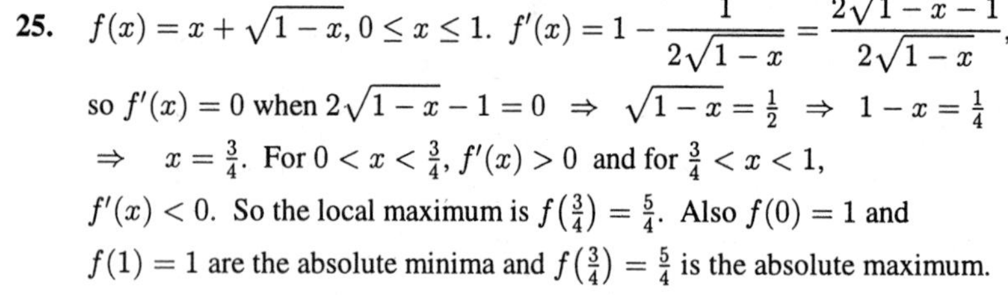

25. $f(x) = x + \sqrt{1-x}$, $0 \le x \le 1$. $f'(x) = 1 - \dfrac{1}{2\sqrt{1-x}} = \dfrac{2\sqrt{1-x} - 1}{2\sqrt{1-x}}$, so $f'(x) = 0$ when $2\sqrt{1-x} - 1 = 0 \Rightarrow \sqrt{1-x} = \frac{1}{2} \Rightarrow 1 - x = \frac{1}{4} \Rightarrow x = \frac{3}{4}$. For $0 < x < \frac{3}{4}$, $f'(x) > 0$ and for $\frac{3}{4} < x < 1$, $f'(x) < 0$. So the local maximum is $f\left(\frac{3}{4}\right) = \frac{5}{4}$. Also $f(0) = 1$ and $f(1) = 1$ are the absolute minima and $f\left(\frac{3}{4}\right) = \frac{5}{4}$ is the absolute maximum.

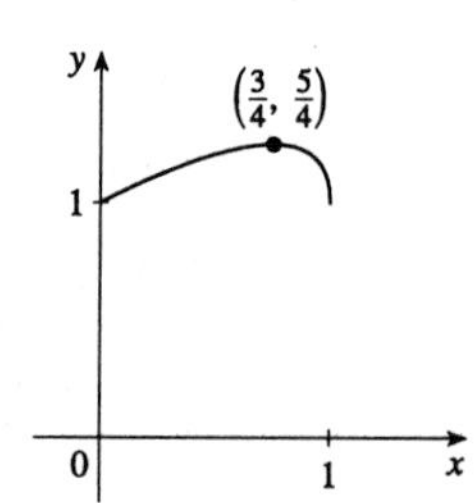

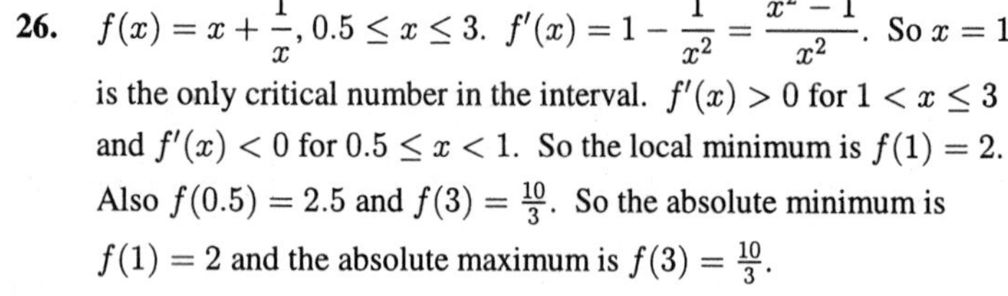

26. $f(x) = x + \dfrac{1}{x}$, $0.5 \le x \le 3$. $f'(x) = 1 - \dfrac{1}{x^2} = \dfrac{x^2 - 1}{x^2}$. So $x = 1$ is the only critical number in the interval. $f'(x) > 0$ for $1 < x \le 3$ and $f'(x) < 0$ for $0.5 \le x < 1$. So the local minimum is $f(1) = 2$. Also $f(0.5) = 2.5$ and $f(3) = \frac{10}{3}$. So the absolute minimum is $f(1) = 2$ and the absolute maximum is $f(3) = \frac{10}{3}$.

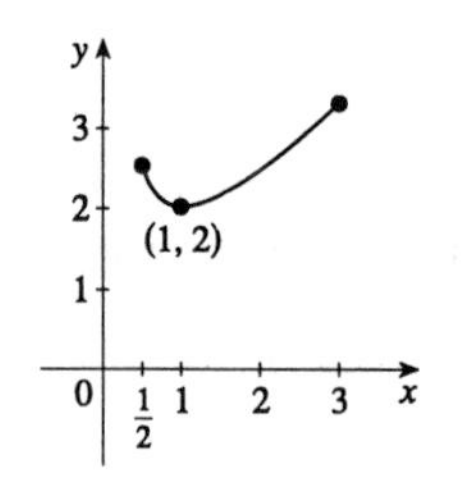

27. $g(x) = \dfrac{x}{x^2 + 1}$, $-5 \le x \le 5$. $g'(x) = \dfrac{(x^2 + 1) - x(2x)}{(x^2 + 1)^2} = \dfrac{1 - x^2}{(x^2 + 1)^2}$.
The critical numbers are $x = \pm 1$. $g'(x) > 0 \Leftrightarrow x^2 < 1 \Leftrightarrow -1 < x < 1$ and $g'(x) < 0 \Leftrightarrow x < -1$ or $x > 1$.
So $g(-1) = -\frac{1}{2}$ is a local minimum and $g(1) = \frac{1}{2}$ is a local maximum.
Also $g(-5) = -\frac{5}{26}$ and $g(5) = \frac{5}{26}$. So $g(-1) = -\frac{1}{2}$ is the absolute minimum and $g(1) = \frac{1}{2}$ is the absolute maximum.

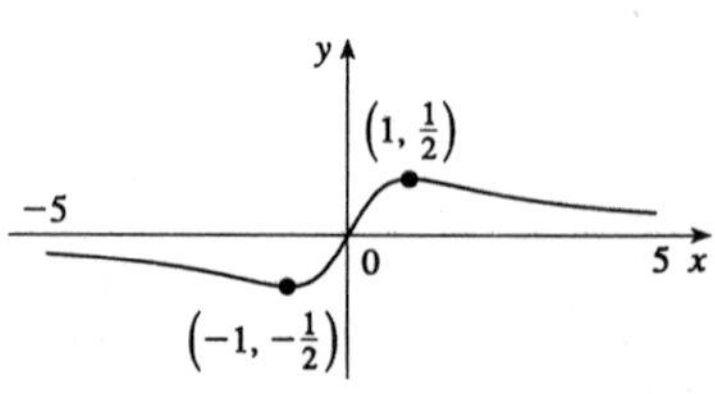

28. $g(x) = \sin x - \cos x$, $-\frac{\pi}{2} \le x \le \frac{\pi}{2}$. Here $g'(x) = \cos x + \sin x = 0$ when $x = -\frac{\pi}{4}$. $g'(x) > 0$ when $x > -\frac{\pi}{4}$ and $g'(x) < 0$ when $x < -\frac{\pi}{4}$ $(-\frac{\pi}{2} \le x \le \frac{\pi}{2})$. So $g\left(-\frac{\pi}{4}\right) = -\sqrt{2}$ is a local minimum. Now $g\left(-\frac{\pi}{2}\right) = -1$ and $g\left(\frac{\pi}{2}\right) = 1$. So $g\left(-\frac{\pi}{4}\right) = -\sqrt{2}$ is the absolute minimum and $g\left(\frac{\pi}{2}\right) = 1$ is the absolute maximum.

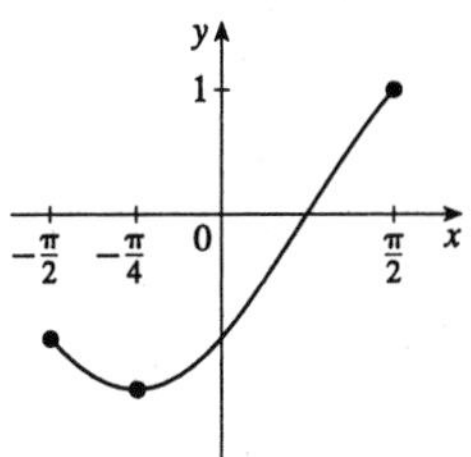

29. (a) It seems from the graph that f is increasing on $(-\infty, -0.67]$ and $[0.67, \infty)$ and decreasing on $[-0.67, 0.67]$, that the local maximum at $x \approx -0.67$ is $f(-0.67) \approx 2.53$, and that the local minimum at $x \approx 0.67$ is $f(0.67) \approx 1.47$.

(b)

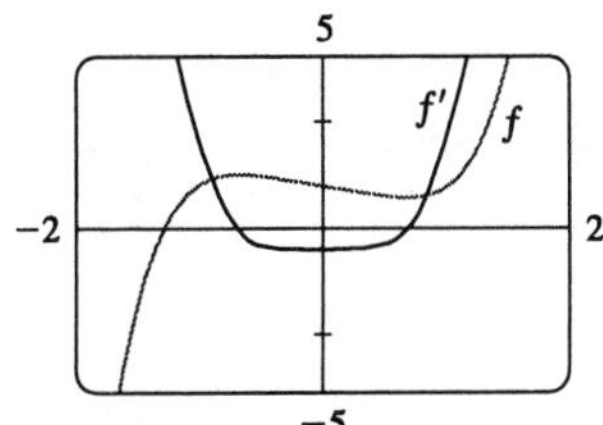

The graph of f' confirms our estimates in part (a), by the First Derivative Test.

(c) $f(x) = x^5 - x + 2 \quad \Rightarrow \quad f'(x) = 5x^4 - 1$. So there are critical points where $f'(x) = 0 \quad \Rightarrow \quad 5x^4 = 1 \quad \Rightarrow \quad x = \pm 5^{-1/4}$. Now f' changes from positive to negative at $x = -5^{-1/4}$, and then back to positive at $x = 5^{-1/4}$, so f is increasing on $\left(-\infty, -5^{-1/4}\right]$ and $\left[5^{-1/4}, \infty\right)$ and decreasing on $\left[-5^{-1/4}, 5^{-1/4}\right]$. The local maximum is $f\left(-5^{-1/4}\right) = 4 \cdot 5^{-5/4} + 2 \approx 2.535$, and the local minimum is $f\left(5^{-1/4}\right) = -4 \cdot 5^{-5/4} + 2 \approx 1.465$.

30. (a) It seems from the graph that f is increasing on $[0, 0.52]$ and $[2.6, 2\pi]$, and decreasing on $[0.52, 2.6]$, that the local maximum is $f(0.52) \approx 2.3$, and that the local minimum is $f(2.6) \approx 0.9$.

(b)

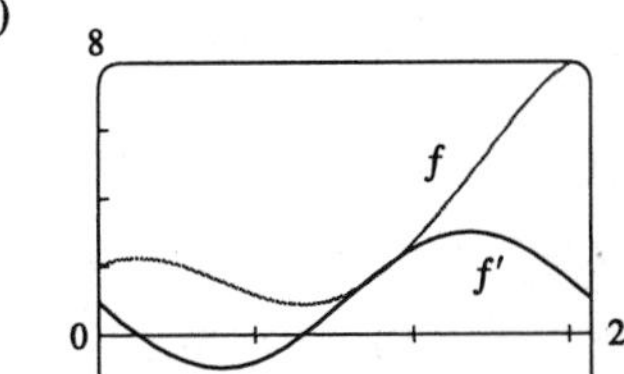

The graph of f' confirms our estimates in part (a), by the First Derivative Test.

(c) $f(x) = x + 2\cos x \quad \Rightarrow \quad f'(x) = 1 - 2\sin x$. So the critical points occur where $f'(x) = 0 \quad \Rightarrow \quad \sin x = \frac{1}{2} \quad \Rightarrow \quad x = \frac{\pi}{6}$ or $\frac{5\pi}{6}$. f' changes from positive to negative at $\frac{\pi}{6}$, and then back to positive at $\frac{5\pi}{6}$, so f is increasing on $\left[0, \frac{\pi}{6}\right]$ and $\left[\frac{5\pi}{6}, 2\pi\right]$ and decreasing on $\left[\frac{\pi}{6}, \frac{5\pi}{6}\right]$. The extreme values are $f\left(\frac{\pi}{6}\right) = \frac{\pi}{6} + 2\cos\frac{\pi}{6} = \frac{\pi}{6} + \sqrt{3}$ (local maximum) and $f\left(\frac{5\pi}{6}\right) = \frac{5\pi}{6} + 2\cos\frac{5\pi}{6} = \frac{5\pi}{6} - \sqrt{3}$ (local minimum).

31. Let $f(x) = x + \dfrac{1}{x}$, so $f'(x) = 1 - \dfrac{1}{x^2} = \dfrac{x^2 - 1}{x^2}$. Thus $f'(x) > 0$ for $x > 1 \quad \Rightarrow \quad f$ is increasing on $[1, \infty)$. Hence for $1 < a < b$, $a + \dfrac{1}{a} = f(a) < f(b) = b + \dfrac{1}{b}$.

32. Let $f(x) = \dfrac{\tan x}{x}$. The $f'(x) = \dfrac{x\sec^2 x - \tan x}{x^2}$. Now by Equation 2.4.2, $\sin x < x$ for $0 < x < \frac{\pi}{2}$. Hence $\tan x < \dfrac{x}{\cos x}$, so $x\sec^2 x - \tan x > x\sec^2 x - x\sec x = x\sec x(\sec x - 1) > 0$ since $\sec x > 1$ for $0 < x < \frac{\pi}{2}$. Hence $f'(x) > 0$ for $0 < x < \frac{\pi}{2}$ so f is increasing on $\left(0, \frac{\pi}{2}\right)$. Thus for $0 < a < b < \frac{\pi}{2}$, we have $\dfrac{\tan a}{a} < \dfrac{\tan b}{b}$ or $\dfrac{b}{a} < \dfrac{\tan b}{\tan a}$.

33. Let $f(x) = 2\sqrt{x} - 3 + \dfrac{1}{x}$. Then $f'(x) = \dfrac{1}{\sqrt{x}} - \dfrac{1}{x^2} > 0$ for $x > 1$ since for $x > 1$, $x^2 > x > \sqrt{x}$. Hence f is increasing, so for $x > 1$, $f(x) > f(1) = 0$ or $2\sqrt{x} - 3 + \dfrac{1}{x} > 0$ for $x > 1$. Hence $2\sqrt{x} > 3 - \dfrac{1}{x}$ for $x > 1$.

34. Let $f(x) = \cos x - 1 + \frac{1}{2}x^2$. Then $f'(x) = -\sin x + x$. Now by Equation 2.4.2, for $0 < x \le \frac{\pi}{2}$, $\sin x < x$, so $x - \sin x > 0$. For $x > \frac{\pi}{2}$, $x - \sin x \ge x - 1 \ge \frac{\pi}{2} - 1 > 0$. Thus f is increasing. So for $x > 0$,
$f(x) > f(0) = 0 \quad \Rightarrow \quad \cos x - 1 + \frac{1}{2}x^2 > 0$ or $\cos x > 1 - \frac{1}{2}x^2$.

35. Let $f(x) = \sin x - x + \frac{1}{6}x^3$. Then $f'(x) = \cos x - 1 + \frac{1}{2}x^2$. By Exercise 34, $f'(x) > 0$ for $x > 0$, so f is increasing for $x > 0$. Thus $f(x) > f(0) = 0$ for $x > 0$.
Therefore $\sin x - x + \frac{1}{6}x^3 > 0$ or $\sin x > x - \frac{1}{6}x^3$ for $x > 0$.

36. $f(x) = \tan x - x$. $f'(x) = \sec^2 x - 1 > 0$ for $0 < x < \frac{\pi}{2}$ since $\sec^2 x > 1$ for $0 < x < \frac{\pi}{2}$. So f is increasing on $\left[0, \frac{\pi}{2}\right)$. Thus $f(x) > f(0) = 0$ for $0 < x < \frac{\pi}{2} \quad \Rightarrow \quad \tan x - x > 0 \quad \Rightarrow \quad \tan x > x$ for $0 < x < \frac{\pi}{2}$.

37. (a) Let $f(x) = x + \dfrac{1}{x}$, so $f'(x) = 1 - \dfrac{1}{x^2} > 0 \quad \Leftrightarrow \quad x^2 < 1 \quad \Leftrightarrow \quad 0 < x < 1$ (since $x > 0$), and $f'(x) > 0$ for $x > 1$. By the First Derivative Test, there is an absolute minimum for $f(x)$ on $(0, \infty)$ where $x = 1$. Thus $f(x) = x + 1/x \ge f(1) = 2$ for $x > 0$.

(b) Let $y = \dfrac{1}{x}$. Then $\left(\sqrt{x} - \sqrt{y}\right)^2 \ge 0 \;\Rightarrow\; \left(\sqrt{x} - \dfrac{1}{\sqrt{x}}\right)^2 \ge 0 \;\Rightarrow\; x - 2 + \dfrac{1}{x} \ge 0 \;\Rightarrow\; x + \dfrac{1}{x} \ge 2$.

38. $f(x) = x^3 + ax^2 + bx + 2 \quad \Rightarrow \quad f'(x) = 3x^2 + 2ax + b$. If $x = -3$ is an extremum, then $f'(-3) = 27 - 6a + b = 0 \quad \Leftrightarrow \quad b = 6a - 27$. If $x = -1$ is an extremum, then $f'(-1) = 3 - 2a + b = 0$ $\Leftrightarrow \quad b = 2a - 3$. So $b = 2a - 3$ and $b = 6a - 27 \quad \Rightarrow \quad b = 9$, $a = 6$. Then
$f'(x) = 3x^2 + 12x + 9 = 3(x+1)(x+3)$ and the First Derivative Test shows that f has a local maximum when $x = -3$ and a local minimum when $x = -1$.

39. $f(x) = ax^3 + bx^2 + cx + d \quad \Rightarrow \quad f(1) = a + b + c + d = 0$ and
$f(-2) = -8a + 4b - 2c + d = 3$.
Also $f'(1) = 3a + 2b + c = 0$ and $f'(-2) = 12a - 4b + c = 0$
by Fermat's Theorem. Solving these four equations, we get
$a = \frac{2}{9}$, $b = \frac{1}{3}$, $c = -\frac{4}{3}$, $d = \frac{7}{9}$, so the function is
$f(x) = \frac{1}{9}(2x^3 + 3x^2 - 12x + 7)$.

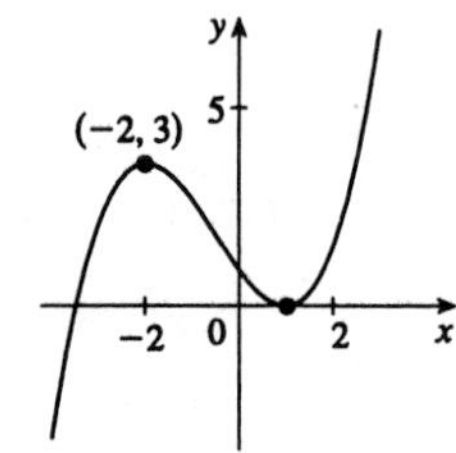

40. $f'(x) = \begin{cases} 1 & \text{if } x < -4 \\ 3x^2 + 6x - 9 & \text{if } -4 < x < 3 \\ -1 & \text{if } x > 3 \end{cases} \quad \Rightarrow \quad f'(x) = 0 \quad \Leftrightarrow \quad 3x^2 + 6x - 9 = 0 \quad \Leftrightarrow$

$(3x - 3)(x + 3) = 0 \quad \Leftrightarrow \quad x = 1$ or -3. Notice that $f'(x) > 0$ for $x < -3$, $f'(x) < 0$ for $-3 < x < 1$, and $f'(x) > 0$ for $1 < x < 3$. Therefore by the First Derivative Test, $f(1) = (1)^3 + 3(1)^2 - 9(1) = -5$ is a local minimum and $f(-3) = (-3)^3 + 3(-3)^2 - 9(-3) = 27$ is a local maximum. Notice also that there are critical points at $x = -4$, and $x = 3$ since at each of these points the left- and right-hand derivatives are not equal, and so $f'(-4)$ and $f'(3)$ do not exist. But $f'(x) > 0$ for $x < -4$ and $f'(x) > 0$ for $-4 < x < -3$, so f' does not change sign at $x = -4$, and there is no maximum or minimum value at $x = -4$. For $x = 3$, $f'(x) > 0$ for $1 < x < 3$ and $f'(x) < 0$ for $x > 3$ so, by the First Derivative Test, there is a local maximum at $x = 3$ where $f(3) = 30 - 3 = 27$.

41. There are many possible functions which satisfy all of the conditions.

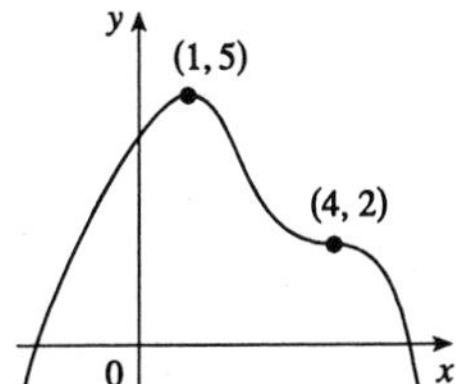

42. There are many possible functions which satisfy all of the conditions.

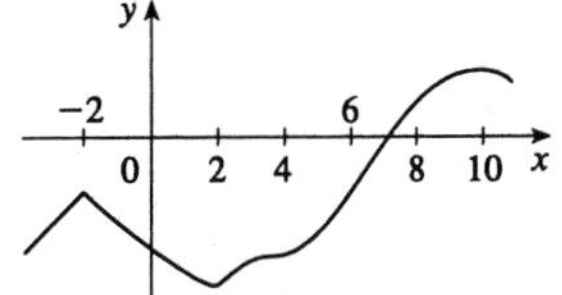

43. There are many possible functions which satisfy all of the conditions.

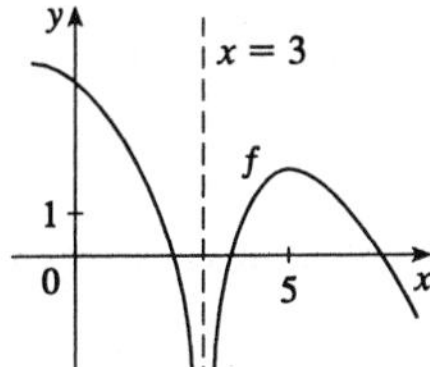

44. There are many possible functions which satisfy all of the conditions.

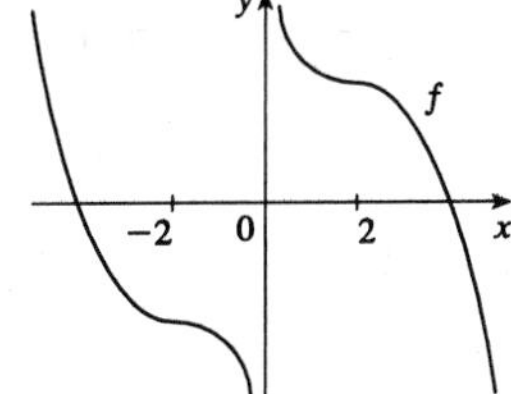

45. Let $x_1, x_2 \in I$ with $x_1 < x_2$. Then $f(x_1) < f(x_2)$ and $g(x_1) < g(x_2)$ (since f and g are increasing on I), so $(f+g)(x_1) = f(x_1) + g(x_1) < f(x_2) + g(x_2) = (f+g)(x_2)$. Therefore $f + g$ is increasing on I.

46. Let $x_1, x_2 \in I$ with $x_1 < x_2$. Then $0 < f(x_1) < f(x_2)$ and $0 < g(x_1) < g(x_2)$ (since f and g are positive and increasing on I). So $(fg)(x_1) = f(x_1)g(x_1) < f(x_2)g(x_1) < f(x_2)g(x_2) = (fg)(x_2)$, which shows that fg is increasing on I.

47. **(a)** Let x_1 and $x_2 \in \mathbb{R}$ with $x_1 < x_2$. Then $g(x_1) < g(x_2)$ since g is increasing on I. So $(f \circ g)(x_1) = f(g(x_1)) < f(g(x_2)) = f \circ g(x_2)$ since f is increasing on $\mathbb{R}$. So for any $x_1 < x_2$ we have $(f \circ g)(x_1) < (f \circ g)(x_2)$, which shows that h is increasing on $\mathbb{R}$.

(b) Let x_1 and $x_2 \in \mathbb{R}$ with $x_1 < x_2$. Then $g(x_1) > g(x_2)$ since g is decreasing on I. So $(f \circ g)(x_1) = f(g(x_1)) < f(g(x_2)) = (f \circ g)(x_2)$ since f is decreasing on $\mathbb{R}$. So for any $x_1 < x_2$ we have $h(x_1) = (f \circ g)(x_1) < (f \circ g)(x_2) = h(x_2)$, which shows that h is increasing on $\mathbb{R}$.

(c) Let x_1 and $x_2 \in \mathbb{R}$ with $x_1 < x_2$. Then $g(x_1) > g(x_2)$ since g is decreasing on $\mathbb{R}$, which implies that $f(g(x_1)) = (f \circ g)(x_1) > (f \circ g)(x_2) = f(g(x_2))$, since f is increasing. This shows that h is decreasing on $\mathbb{R}$.

48. **(a)** By the definition of the derivative, for x between a and b, $f'(x) = \lim_{h\to 0} \dfrac{f(x+h)-f(x)}{h}$. Now for $h > 0$, $f(x+h) > f(x)$ [since f is increasing on (a,b)] and for $h < 0$, $f(x+h) < f(x)$ (for the same reason). In either case $\dfrac{f(x+h)-f(x)}{h} > 0$, and by Theorem 1.3.2, the limit of a positive quantity is either positive or 0. So $f'(x) \geq 0$ for $x \in (a,b)$.

(b) Let $f(x) = x^3$. Then f is increasing and differentiable on $(-1,1)$, but $f'(0) = 3 \cdot 0^2 = 0$.

49. Let x_1 and x_2 be any two numbers in $[a,b]$ with $x_1 < x_2$. Then f is continuous on $[x_1,x_2]$ and differentiable on (x_1,x_2), so by the Mean Value Theorem there is a number c between x_1 and x_2 such that $f(x_2) - f(x_1) = f'(c)(x_2 - x_1)$. Now $f'(c) < 0$ by assumption and $x_2 - x_1 > 0$ because $x_1 < x_2$. Thus $f(x_2) - f(x_1) = f'(c)(x_2 - x_1)$ is negative, so $f(x_2) - f(x_1) < 0$ or $f(x_2) < f(x_1)$. This shows that f is decreasing on $[a,b]$.

50. **(a)** Let $x \in (a,b)$. If $a < x < c$, then $f(x) > f(c)$ since $f' < 0$ implies f is decreasing on $[a,c]$. If $c < x < b$, then $f(x) > f(c)$ since $f' > 0$ implies that f is increasing on $[c,b]$. Therefore $f(c) \leq f(x)$ for all $x \in (a,b)$. Thus, by Definition 3.1.2, f has a local minimum at c.

(b) If f' does not change sign at c, then either $f' > 0$ on some open interval containing c (except at c) or $f' < 0$ on some open interval containing c (except at c). Thus either f is increasing on some interval containing c or f is decreasing on some interval containing c. In either case, f does not have a local extremum at c.

EXERCISES 3.4

1. The derivative f' is increasing when the slopes of the tangent lines are becoming larger as x increases. This seems to be the case on the interval $[2,5]$. The derivative is decreasing when the slopes of the tangent lines are becoming smaller as x increases, and this seems to be the case on $(-\infty,2]$ and $[5,\infty)$. So f' is increasing on $[2,5]$ and decreasing on $(-\infty,2]$ and $[5,\infty)$.

2. **(a)** g is concave upward on $(-\infty,2)$ and $(7,\infty)$.

(b) g is concave downward on $(2,4)$ and $(4,7)$.

(c) The only point of inflection is $(2,2)$. Although the curve is concave down for $4 < x < 7$ and concave up for $x > 7$, the function is not defined at $x = 7$.

3. **(a)** $f(x) = x^3 - x \;\Rightarrow\; f'(x) = 3x^2 - 1 = 0 \;\Leftrightarrow\; x^2 = \frac{1}{3} \;\Leftrightarrow\; x = \pm\frac{1}{\sqrt{3}}$. $f'(x) > 0 \;\Leftrightarrow\; x^2 > \frac{1}{3}$ $\Leftrightarrow\; |x| > \frac{1}{\sqrt{3}} \;\Leftrightarrow\; x > \frac{1}{\sqrt{3}}$ or $x < -\frac{1}{\sqrt{3}}$. $f'(x) < 0 \;\Leftrightarrow\; |x| < \frac{1}{\sqrt{3}} \;\Leftrightarrow\; -\frac{1}{\sqrt{3}} < x < \frac{1}{\sqrt{3}}$. So f is increasing on $\left(-\infty, -\frac{1}{\sqrt{3}}\right]$ and $\left[\frac{1}{\sqrt{3}}, \infty\right)$, and decreasing on $\left[-\frac{1}{\sqrt{3}}, \frac{1}{\sqrt{3}}\right]$.

(b) Local maximum $f\left(-\frac{1}{\sqrt{3}}\right) = \frac{2}{3\sqrt{3}} \approx 0.38$, local minimum $f\left(\frac{1}{\sqrt{3}}\right) = -\frac{2}{3\sqrt{3}} \approx -0.38$.

(c) $f''(x) = 6x \;\Rightarrow\; f''(x) > 0 \;\Leftrightarrow\; x > 0$, so f is CU on $(0, \infty)$ and CD on $(-\infty, 0)$.

(d) Point of inflection at $x = 0$

4. **(a)** $f(x) = 2x^3 + 5x^2 - 4x \;\Rightarrow\; f'(x) = 6x^2 + 10x - 4 = 2(3x-1)(x+2) = 0 \;\Leftrightarrow\; x = \frac{1}{3}$ or -2. $f'(x) > 0 \;\Leftrightarrow\; x < -2$ or $x > \frac{1}{3}$; $f'(x) < 0 \;\Leftrightarrow\; -2 < x < \frac{1}{3}$. So f is increasing on $(-\infty, -2]$, $\left[\frac{1}{3}, \infty\right)$ and decreasing on $\left[-2, \frac{1}{3}\right]$.

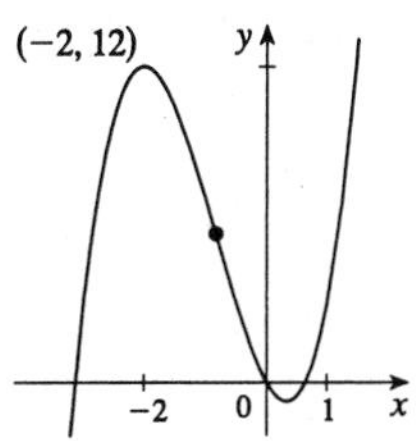

(b) Local maximum $f(-2) = 12$, local minimum $f\left(\frac{1}{3}\right) = -\frac{19}{27}$

(c) $f''(x) = 12x + 10 > 0 \;\Leftrightarrow\; x > -\frac{5}{6}$, so f is CU on $\left(-\frac{5}{6}, \infty\right)$ and CD on $\left(-\infty, -\frac{5}{6}\right)$.

(d) Inflection point at $x = -\frac{5}{6}$

5. **(a)** $f(x) = x^4 - 6x^2 \;\Rightarrow\; f'(x) = 4x^3 - 12x = 4x(x^2 - 3) = 0$ when $x = 0, \pm\sqrt{3}$.

Interval	$4x$	$x^2 - 3$	$f'(x)$	f
$x < -\sqrt{3}$	$-$	$+$	$-$	decreasing on $\left(-\infty, -\sqrt{3}\right]$
$-\sqrt{3} < x < 0$	$-$	$-$	$+$	increasing on $\left[-\sqrt{3}, 0\right]$
$0 < x < \sqrt{3}$	$+$	$-$	$-$	decreasing on $\left[0, \sqrt{3}\right]$
$x > \sqrt{3}$	$+$	$+$	$+$	increasing on $\left[\sqrt{3}, \infty\right)$

(b) Local minima $f\left(\pm\sqrt{3}\right) = -9$, local maximum $f(0) = 0$

(c) $f''(x) = 12x^2 - 12 = 12(x^2 - 1) > 0 \;\Leftrightarrow\; x^2 > 1$ $\Leftrightarrow\; |x| > 1 \;\Leftrightarrow\; x > 1$ or $x < -1$, so f is CU on $(-\infty, -1)$, $(1, \infty)$ and CD on $(-1, 1)$.

(d) Inflection points when $x = \pm 1$

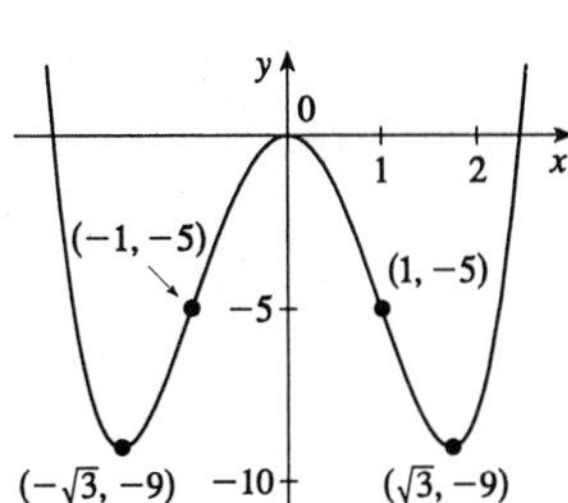

6. **(a)** $g(x) = x^4 - 3x^3 + 3x^2 - x \quad \Rightarrow$
$g'(x) = 4x^3 - 9x^2 + 6x - 1 = (x-1)^2(4x-1) = 0$ when $x = 1$ or $\frac{1}{4}$.
$g'(x) \geq 0 \quad \Leftrightarrow \quad 4x - 1 \geq 0 \quad \Leftrightarrow \quad x \geq \frac{1}{4}$ and $g'(x) \leq 0 \quad \Leftrightarrow$
$x \leq \frac{1}{4}$, so g is increasing on $\left[\frac{1}{4}, \infty\right)$ and decreasing on $\left(-\infty, \frac{1}{4}\right]$.

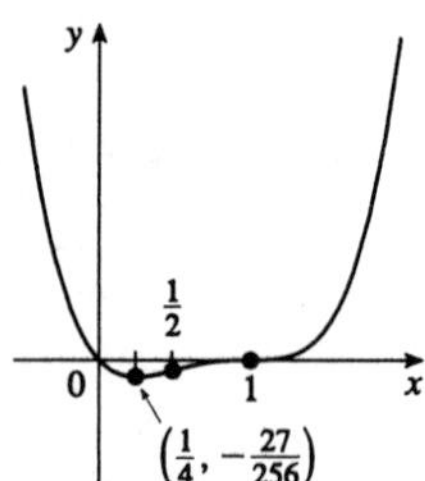

(b) Local minimum $g\left(\frac{1}{4}\right) = -\frac{27}{256}$

(c) $g''(x) = 12x^2 - 18x + 6 = 6(2x-1)(x-1) > 0 \quad \Leftrightarrow$
$x < \frac{1}{2}$ or $x > 1$, $g'(x) < 0 \quad \Leftrightarrow \quad \frac{1}{2} < x < 1$, so g is CU on $\left(-\infty, \frac{1}{2}\right)$ and $(1, \infty)$ and CD on $\left(\frac{1}{2}, 1\right)$.

(d) Inflection points at $x = \frac{1}{2}$ and 1

7. **(a)** $h(x) = 3x^5 - 5x^3 + 3 \quad \Rightarrow \quad h'(x) = 15x^4 - 15x^2 = 15x^2(x^2 - 1) = 0$ when $x = 0, \pm 1$. $h'(x) > 0$
$\Leftrightarrow \quad x^2 > 1 \quad \Leftrightarrow \quad x > 1$ or $x < -1$, so h is increasing on $(-\infty, -1]$ and $[1, \infty)$ and decreasing on $[-1, 1]$.

(b) Local maximum $h(-1) = 5$, local minimum $h(1) = 1$

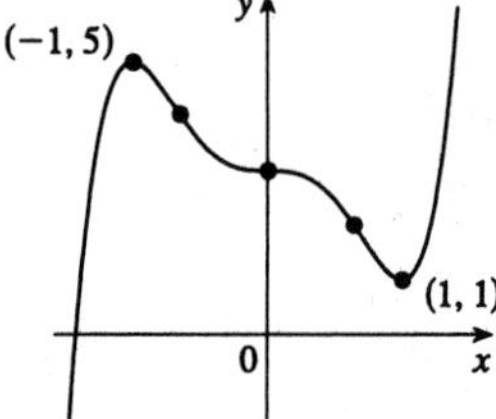

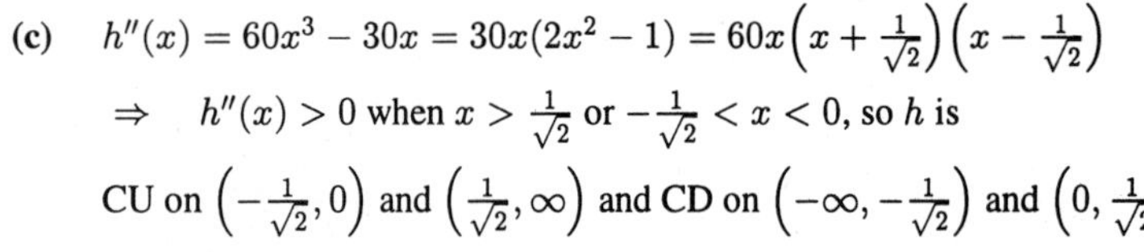

(c) $h''(x) = 60x^3 - 30x = 30x(2x^2 - 1) = 60x\left(x + \frac{1}{\sqrt{2}}\right)\left(x - \frac{1}{\sqrt{2}}\right)$
$\Rightarrow \quad h''(x) > 0$ when $x > \frac{1}{\sqrt{2}}$ or $-\frac{1}{\sqrt{2}} < x < 0$, so h is
CU on $\left(-\frac{1}{\sqrt{2}}, 0\right)$ and $\left(\frac{1}{\sqrt{2}}, \infty\right)$ and CD on $\left(-\infty, -\frac{1}{\sqrt{2}}\right)$ and $\left(0, \frac{1}{\sqrt{2}}\right)$.

(d) Inflection points at $x = \pm\frac{1}{\sqrt{2}}$ and 0

8. **(a)** $h(x) = (x^2 - 1)^3 \quad \Rightarrow \quad h'(x) = 6x(x^2 - 1)^2 \geq 0 \quad \Leftrightarrow \quad x > 0 \ (x \neq 1)$, so h is increasing on $[0, \infty)$ and decreasing on $(-\infty, 0]$.

(b) $h(0) = -1$ is a local minimum.

(c) $h''(x) = 6(x^2 - 1)^2 + 24x^2(x^2 - 1)$
$= 6(x^2 - 1)(5x^2 - 1)$.
The roots ± 1 and $\pm\frac{1}{\sqrt{5}}$ divide $\mathbb{R}$ into five intervals. From the table, we see that h is CU on $(-\infty, -1)$, $\left(-\frac{1}{\sqrt{5}}, \frac{1}{\sqrt{5}}\right)$ and $(1, \infty)$, and CD on $\left(-1, -\frac{1}{\sqrt{5}}\right)$ and $\left(\frac{1}{\sqrt{5}}, 1\right)$.

Interval	$x^2 - 1$	$5x^2 - 1$	$h''(x)$	Concavity
$x < -1$	+	+	+	upward
$-1 < x < -\frac{1}{\sqrt{5}}$	−	+	−	downward
$-\frac{1}{\sqrt{5}} < x < \frac{1}{\sqrt{5}}$	−	−	+	upward
$\frac{1}{\sqrt{5}} < x < 1$	−	+	−	downward
$x > 1$	+	+	+	upward

(d) Inflection points at $x = \pm 1, \pm\frac{1}{\sqrt{5}}$

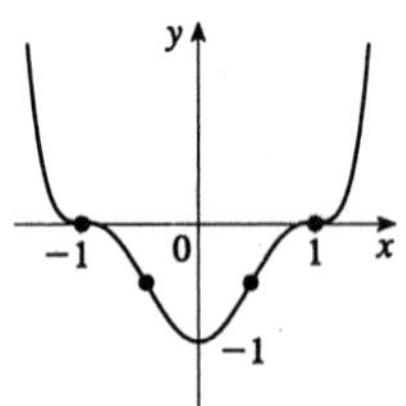

9. **(a)** $P(x) = x\sqrt{x^2+1} \quad\Rightarrow\quad P'(x) = \sqrt{x^2+1} + \dfrac{x^2}{\sqrt{x^2+1}} = \dfrac{2x^2+1}{\sqrt{x^2+1}} > 0$, so P is increasing on $\mathbb{R}$.

(b) No extremum

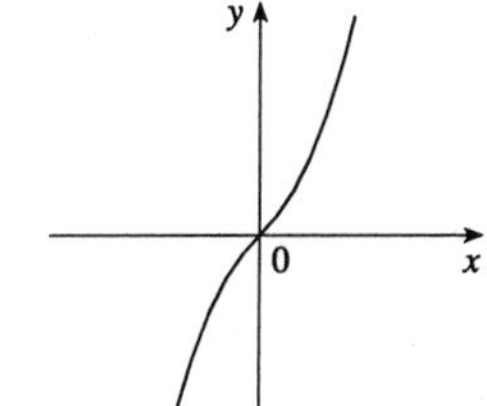

(c) $P''(x) = \dfrac{4x\sqrt{x^2+1} - (2x^2+1)\dfrac{x}{\sqrt{x^2+1}}}{x^2+1} = \dfrac{x(2x^2+3)}{(x^2+1)^{3/2}} > 0$

$\Leftrightarrow \quad x > 0$ so P is CU on $(0,\infty)$ and CD on $(-\infty,0)$.

(d) IP at $x = 0$

10. **(a)** $P(x) = x\sqrt{x+1}$, domain $= [-1,\infty)$, $P'(x) = \sqrt{x+1} + x\dfrac{1}{2\sqrt{x+1}} = \dfrac{3x+2}{2\sqrt{x+1}} > 0$ when $x > -2/3$; $P'(x) < 0$ when $-1 < x < -\frac{2}{3}$, so P is increasing on $\left[-\frac{2}{3},\infty\right)$ and decreasing on $\left[-1,-\frac{2}{3}\right]$.

(b) Local minimum $P\left(-\frac{2}{3}\right) = -\frac{2}{3\sqrt{3}}$

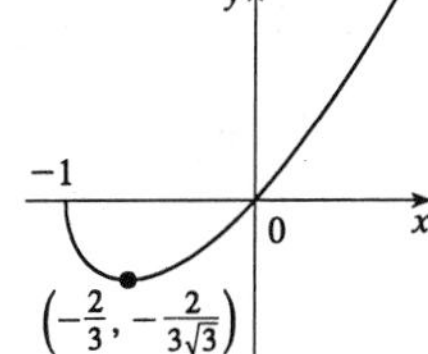

(c) $P''(x) = \dfrac{3\left(2\sqrt{x+1}\right) - (3x+2)\left(1/\sqrt{x+1}\right)}{4(x+1)} = \dfrac{3x+4}{4(x+1)^{3/2}} > 0$

when $x > -1$, so P is CU on $(-1,\infty)$.

(d) No inflection point

11. **(a)** $Q(x) = x^{1/3}(x+3)^{2/3} \quad\Rightarrow\quad Q'(x) = \frac{1}{3}x^{-2/3}(x+3)^{2/3} + x^{1/3}\left(\frac{2}{3}\right)(x+3)^{-1/3} = \dfrac{x+1}{x^{2/3}(x+3)^{1/3}}$. The critical numbers are -3, -1, and 0. Note that $x^{2/3} \geq 0$ for all x. So $Q'(x) > 0$ when $x < -3$ or $x > -1$ and $Q'(x) < 0$ when $-3 < x < -1 \quad\Rightarrow\quad Q$ is increasing on $(-\infty,-3]$ and $[-1,\infty)$ and decreasing on $[-3,-1]$.

(b) $Q(-3) = 0$ is a local maximum and $Q(-1) = -4^{1/3} \approx -1.6$ is a local minimum.

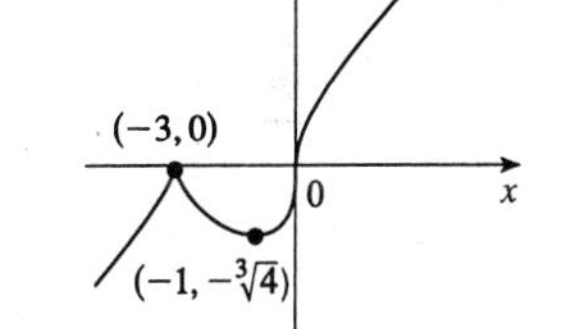

(c) $Q''(x) = -\dfrac{2}{x^{5/3}(x+3)^{4/3}} \quad\Rightarrow\quad Q''(x) > 0$ when $x < 0$,

so Q is CU on $(-\infty,-3)$ and $(-3,0)$ and CD on $(0,\infty)$.

(d) IP at $x = 0$

12. **(a)** $Q(x) = x - 3x^{1/3} \quad\Rightarrow\quad Q'(x) = 1 - \dfrac{1}{x^{2/3}} > 0 \quad\Leftrightarrow\quad x^{2/3} > 1 \quad\Leftrightarrow\quad x^2 > 1 \quad\Leftrightarrow\quad x < -1$ or $x > 1$, so Q is increasing on $(-\infty,-1]$, and $[1,\infty)$, and decreasing on $[-1,1]$.

(b) $Q'(x) = 0 \quad\Leftrightarrow\quad x = \pm 1$; $Q(1) = -2$ is a local minimum, and $Q(-1) = 2$ is a local maximum.

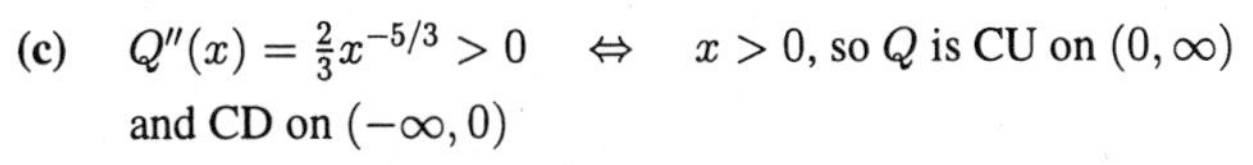

(c) $Q''(x) = \frac{2}{3}x^{-5/3} > 0 \quad\Leftrightarrow\quad x > 0$, so Q is CU on $(0,\infty)$ and CD on $(-\infty,0)$

(d) Inflection point at $(0,0)$

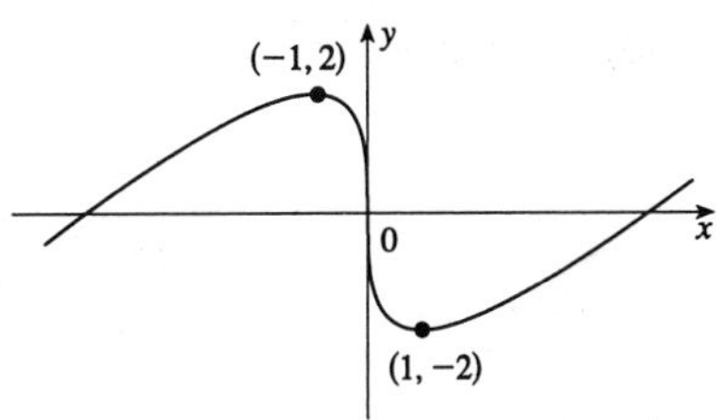

13. **(a)** $f(\theta) = \sin^2\theta \quad \Rightarrow \quad f'(\theta) = 2\sin\theta\cos\theta = \sin 2\theta > 0 \quad \Leftrightarrow \quad 2\theta \in (2n\pi, (2n+1)\pi) \quad \Leftrightarrow$ $\theta \in \left(n\pi, n\pi + \frac{\pi}{2}\right)$, n an integer. So f is increasing on $\left[n\pi, n\pi + \frac{\pi}{2}\right]$ and decreasing on $\left[n\pi + \frac{\pi}{2}, (n+1)\pi\right]$.

(b) Local minima $f(n\pi) = 0$, local maxima $f\left(n\pi + \frac{\pi}{2}\right) = 1$

(c) $f''(\theta) = 2\cos 2\theta > 0 \quad \Leftrightarrow \quad 2\theta \in \left(2n\pi - \frac{\pi}{2}, 2n\pi + \frac{\pi}{2}\right)$ $\Leftrightarrow \quad \theta \in \left(n\pi - \frac{\pi}{4}, n\pi + \frac{\pi}{4}\right)$, so f is CU on these intervals and CD on $\left(n\pi + \frac{\pi}{4}, n\pi + \frac{3\pi}{4}\right)$.

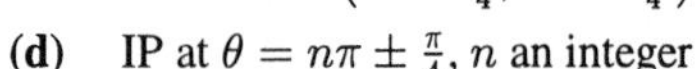

(d) IP at $\theta = n\pi \pm \frac{\pi}{4}$, n an integer

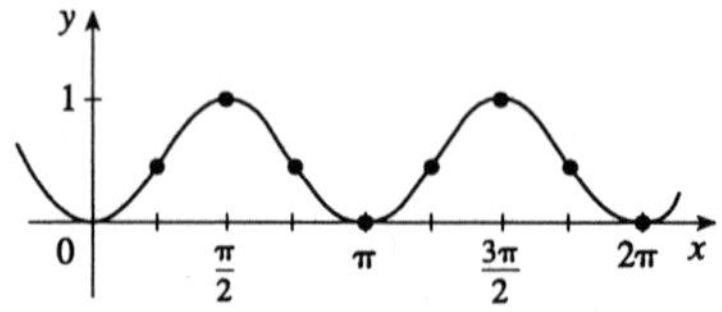

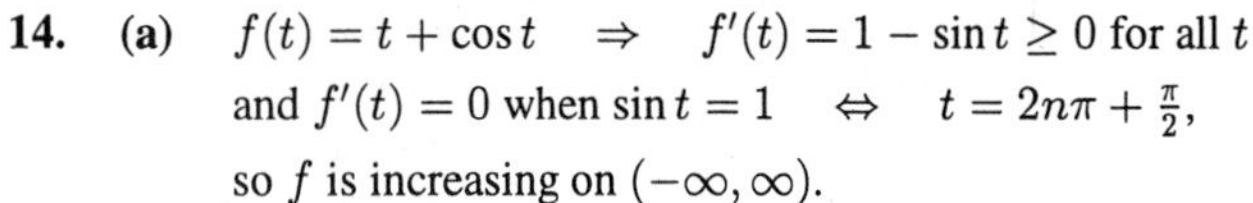

14. **(a)** $f(t) = t + \cos t \quad \Rightarrow \quad f'(t) = 1 - \sin t \geq 0$ for all t and $f'(t) = 0$ when $\sin t = 1 \quad \Leftrightarrow \quad t = 2n\pi + \frac{\pi}{2}$, so f is increasing on $(-\infty, \infty)$.

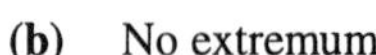

(b) No extremum

(c) $f''(t) = -\cos t > 0 \;\Leftrightarrow\; 2n\pi + \frac{\pi}{2} < t < 2n\pi + \frac{3\pi}{2}$, so f is CU on $\left(2n\pi + \frac{\pi}{2}, 2n\pi + \frac{3\pi}{2}\right)$ and CD on $\left(2n\pi - \frac{\pi}{2}, 2n\pi + \frac{\pi}{2}\right)$.

(d) Points of inflection at $t = 2n\pi \pm \frac{\pi}{2}$, that is, $t = n\pi + \frac{\pi}{2}$

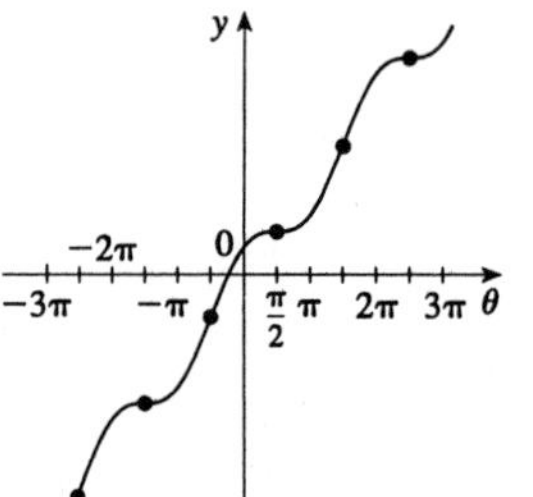

15. $f(x) = 6x^2 - 2x^3 - x^4 \quad \Rightarrow \quad f'(x) = 12x - 6x^2 - 4x^3 \quad \Rightarrow \quad f''(x) = 12 - 12x - 12x^2 = 0 \quad \Leftrightarrow$ $x^2 + x - 1 = 0 \quad \Rightarrow \quad x = \frac{-1 \pm \sqrt{5}}{2}$. For $x < \frac{-1-\sqrt{5}}{2}$, $f''(x) < 0$. For $\frac{-1-\sqrt{5}}{2} < x < \frac{-1+\sqrt{5}}{2}$, $f''(x) > 0$, and if $x > \frac{-1+\sqrt{5}}{2}$ then $f''(x) < 0$. Therefore f is CU on $\left(\frac{-1-\sqrt{5}}{2}, \frac{-1+\sqrt{5}}{2}\right)$.

16. $y = \dfrac{x^2}{\sqrt{1+x}}$, $D = \{x \mid x > -1\} \quad \Rightarrow \quad y' = \dfrac{2x\sqrt{1+x} - \frac{1}{2}(1+x)^{-1/2} \cdot x^2}{1+x} = \dfrac{4x + 3x^2}{2\,(1+x)^{3/2}} \quad \Rightarrow$

$y'' = \dfrac{(4+6x)2(1+x)^{3/2} - 3(1+x)^{1/2}(4x+3x^2)}{4(1+x)^3} = \dfrac{3x^2 + 8x + 8}{4(1+x)^{5/2}} > 0 \quad \Leftrightarrow \quad 3x^2 + 8x + 8 > 0$, which is true for all x since the discriminant is negative, so the function is CU on its domain, which is $(-1, \infty)$.

17. $f(x) = x(1+x)^{-2} \quad \Rightarrow \quad f'(x) = (1+x)^{-2} - 2x(1+x)^{-3} = (1+x)^{-3}(1-x) \quad \Rightarrow$ $f''(x) = -3(1+x)^{-4}(1-x) - (1+x)^{-3} = (1+x)^{-4}(2x-4) > 0 \quad \Leftrightarrow \quad (2x-4) > 0 \quad \Leftrightarrow \quad x > 2$. Therefore f is CU on $(2, \infty)$.

18. $y = \dfrac{x^3}{x^2-3} \quad \Rightarrow \quad y' = \dfrac{x^4 - 9x^2}{(x^2-3)^2} \quad \Rightarrow \quad y'' = \dfrac{(4x^3 - 18x)(x^2-3)^2 - 4x(x^2-3)(x^4 - 9x^2)}{(x^2-3)^4} = \dfrac{6x(x^2+9)}{(x^2-3)^3}$.

Now since $x^2 + 9 > 0$, the quotient is positive $\Leftrightarrow$

$$\frac{x}{x^2-3} = \frac{x}{\left(x-\sqrt{3}\right)\left(x+\sqrt{3}\right)} > 0.$$

So y is concave upward on $\left(-\sqrt{3}, 0\right)$ and $\left(\sqrt{3}, \infty\right)$.

Interval	x	$x+\sqrt{3}$	$x-\sqrt{3}$	$\dfrac{x}{x^2-3}$	Concavity
$x < -\sqrt{3}$	$-$	$-$	$-$	$-$	downward
$-\sqrt{3} < x < 0$	$-$	$+$	$-$	$+$	upward
$0 < x < \sqrt{3}$	$+$	$+$	$-$	$-$	downward
$x > \sqrt{3}$	$+$	$+$	$+$	$+$	upward

19. **(a)**

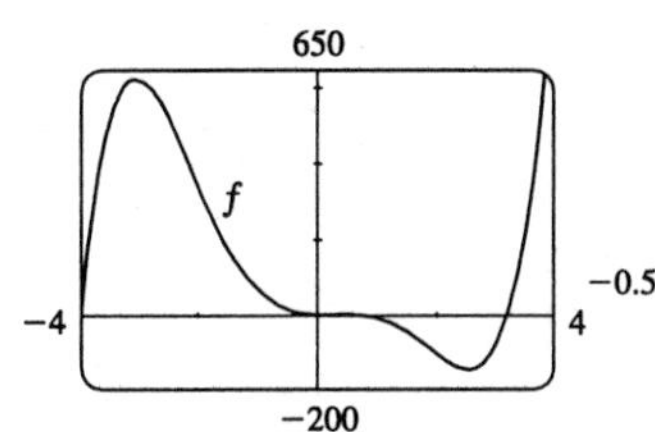

From the graphs of f it seems that f is concave upward on $(-2, 0.25)$ and $(2, \infty)$, and concave downward on $(-\infty, -2)$ and $(0.25, 2)$, with inflection points at about $(-2, 350)$, $(0.25, 1)$, and $(2, -100)$.

(b)

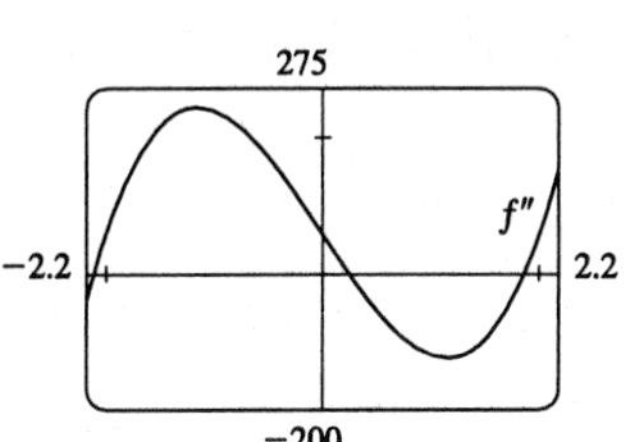

From the graph of f'' it seems that f is CU on $(-2.1, 0.25)$ and $(1.9, \infty)$, and CD on $(-\infty, -2.1)$ and $(0.25, 2)$, with inflection points at about $(-2.1, 386)$, $(0.25, 1.3)$ and $(1.9, -87)$. (We have to check back on the graph of f to find the y-coordinates of the inflection points.)

20. **(a)**

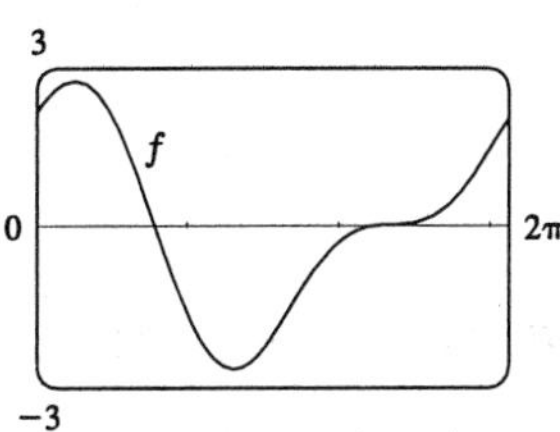

Since the function $f(x) = 2\cos x + \sin 2x$ is periodic with period 2π, we consider it only on the interval $[0, 2\pi]$. From the graphs of f, it seems that f is CU on $(1.5, 3.5)$ and $(4.5, 6.0)$, and CD on $(0, 1.5)$, $(3.5, 4.5)$ and $(6.0, 2\pi)$, with inflection points at about $(1.5, -0.3)$, $(3.5, -1.3)$, $(4.5, 0.0)$ and $(6.0, 1.5)$.

(b)

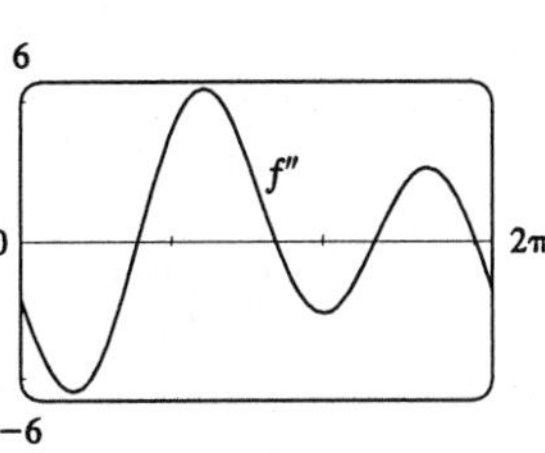

From the graph of f'', it seems that f is CU on $(1.57, 3.39)$ and $(4.71, 6.03)$ and CD on $(0, 1.57)$, $(3.39, 4.71)$ and $(6.03, 2\pi)$, with inflection points at about $(1.57, 0.00)$, $(3.39, -1.45)$, $(4.71, 0.00)$ and $(6.03, 1.45)$.

21. There are many functions which satisfy the given conditions.

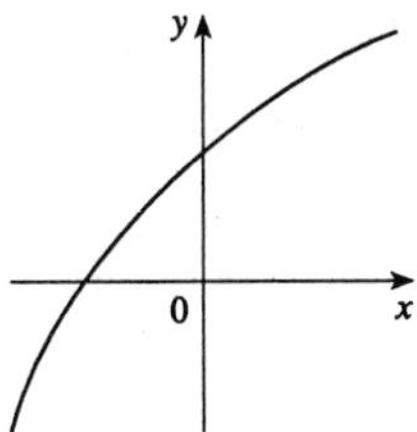

22. There are many functions which satisfy the given conditions.

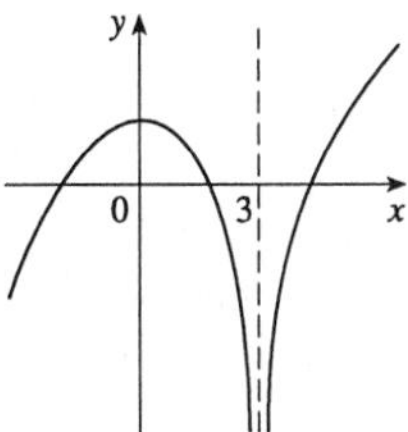

23.

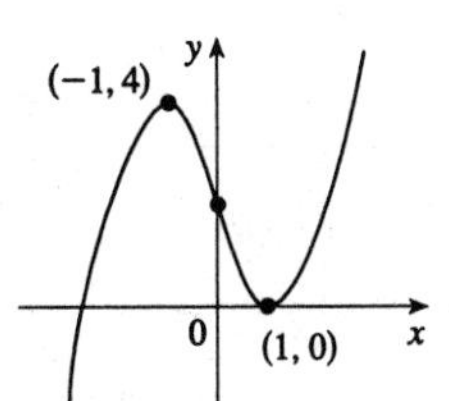

24.

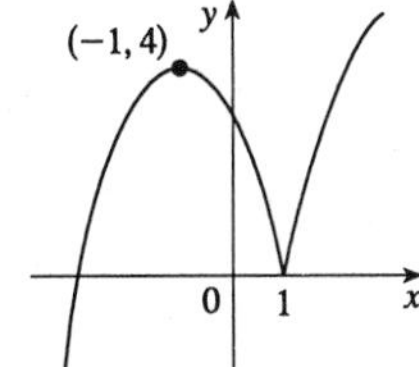

25.

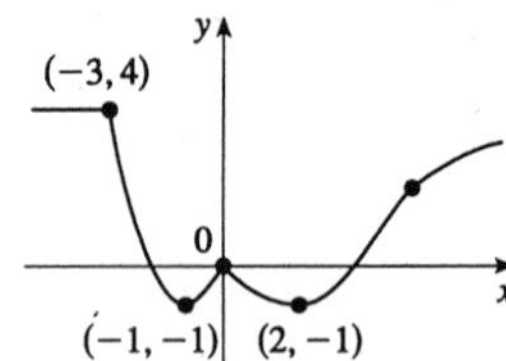

26.

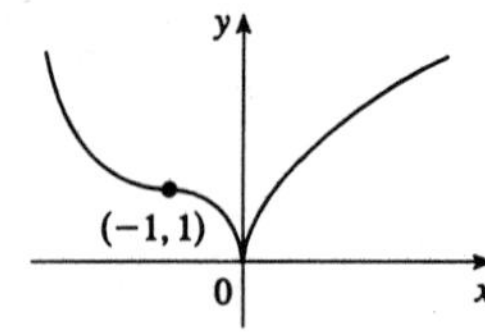

27. (a) f is increasing where f' is positive, that is, on $[0, 2]$, $[4, 6]$, and $[8, \infty)$; and decreasing where f' is negative, that is, on $[2, 4]$ and $[6, 8]$.

(b) f has local maxima where f' changes from positive to negative, at $x = 2$ and at $x = 6$, and a local minimum where f' changes from negative to positive, at $x = 4$.

(c) f is concave upward where f' is increasing, that is, on $(3, 6)$ and $(6, \infty)$, and concave downward where f' is decreasing, that is, on $(0, 3)$.

(d) There is a point of inflection where f changes from being CD to being CU, that is, at $x = 3$.

(e)

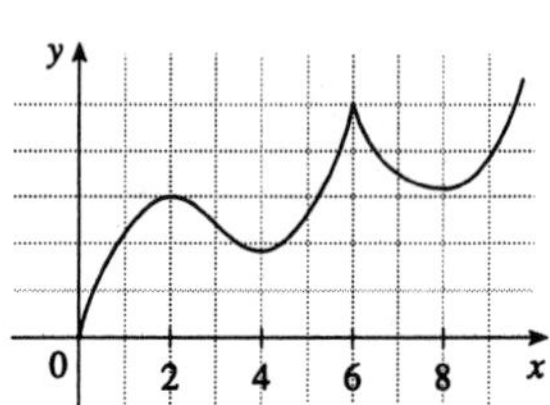

28. (a) f is increasing where f' is positive, on $[1, 6]$ and $[8, \infty)$, and CD where f' is negative, on $(0, 1)$ and $(6, 8)$.

(b) f has a local maximum where f' changes from positive to negative, at $x = 6$, and local minima where f' changes from negative to positive, at $x = 1$ and at $x = 8$.

(c) f is concave upward where f' is increasing, that is, on $(0, 2)$, $(3, 5)$, and $(7, \infty)$, and concave downward where f' is decreasing, that is, on $(2, 3)$ and $(5, 7)$.

(d) There are points of inflection where f changes its direction of concavity, at $x = 2$, $x = 3$, $x = 5$ and $x = 7$.

(e)

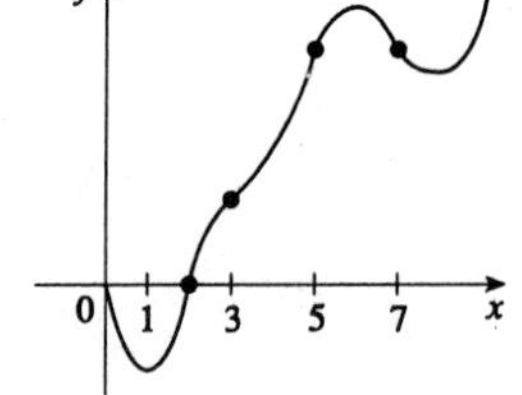

29. In Maple, we define f and then use the command

```
plot(diff(diff(f,x),x),x=-3..3);.
```

In Mathematica, we define f and then use

```
Plot[Dt[Dt[f,x],x],{x,-3,3}].
```

We see that $f'' > 0$ for $x > 0.1$ and $f'' < 0$ for $x < 0.1$. So f is concave up on $(0.1, \infty)$ and concave down on $(-\infty, 0.1)$.

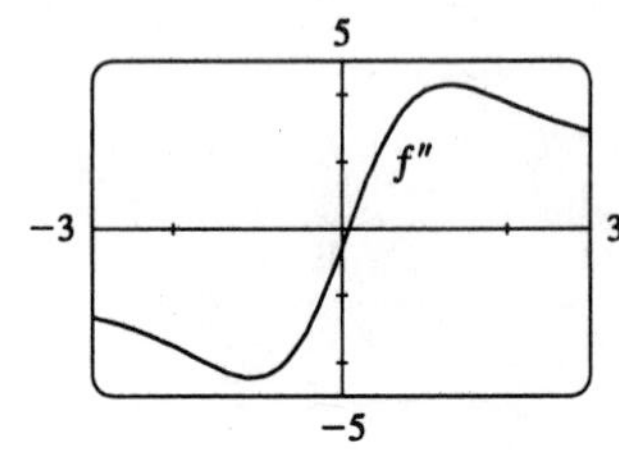

30. It appears that f'' is positive (and thus f is concave up) on $(-1.8, 0.3)$ and $(1.5, \infty)$ and negative (so f is concave down) on $(-\infty, -1.8)$ and $(0.3, 1.5)$.

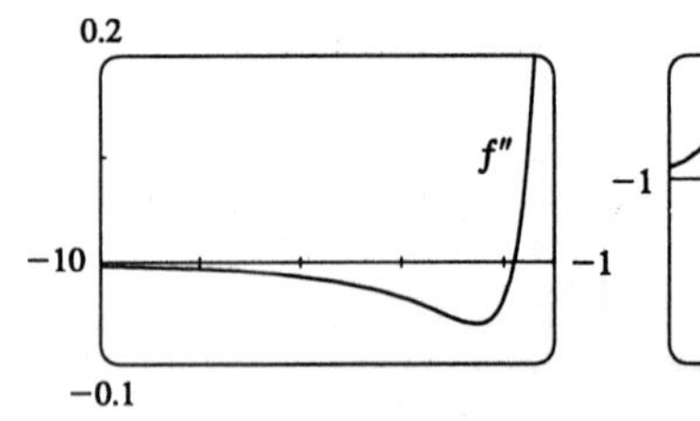

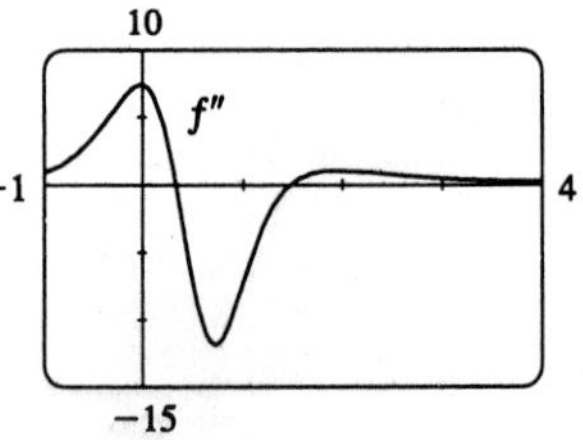

31. By hypothesis $g = f'$ is differentiable on an open interval containing c. Since $(c, f(c))$ is a point of inflection, the concavity changes at $x = c$, so $f'(x)$ changes signs at $x = c$. Hence, by the First Derivative Test, f' has a local extremum at $x = c$. Thus by Fermat's Theorem $f''(c) = 0$.

32. $f(x) = x^4 \quad\Rightarrow\quad f'(x) = 4x^3 \quad\Rightarrow\quad f''(x) = 12x^2 \quad\Rightarrow\quad f''(0) = 0$. For $x < 0$, $f''(x) > 0$, so f is CU on $(-\infty, 0)$; for $x > 0$, $f''(x) > 0$, so f is also CU on $(0, \infty)$. Since f does not change concavity at 0, $(0, 0)$ is not an inflection point.

33. Using the fact that $|x| = \sqrt{x^2}$ (see Exercises 2.5.79-81), we have that $g(x) = x\sqrt{x^2} \quad\Rightarrow$ $g'(x) = \sqrt{x^2} + \sqrt{x^2} = 2\sqrt{x^2} = 2|x| \quad\Rightarrow\quad g''(x) = 2x(x^2)^{-1/2} = \dfrac{2x}{|x|} < 0$ for $x < 0$ and $g''(x) > 0$ for $x > 0$, so $(0, 0)$ is an inflection point. But $g''(0)$ does not exist.

34. $y = x^3 + \cos x \quad\Rightarrow\quad y' = 3x^2 - \sin x \quad\Rightarrow\quad y'' = 6x - \cos x \quad\Rightarrow\quad y''' = 6 + \sin x$. Now to solve $y'' = 0$, try $x_1 = 0$, and then $x_2 = x_1 - \dfrac{y''(x_1)}{y'''(x_1)} \approx 0.1677 \quad\Rightarrow\quad x_3 \approx 0.1643 \quad\Rightarrow\quad x_4 \approx 0.1644$. For $x < 0.1644$, $y'' < 0$, and for $x > 0.1644$, $y'' > 0$. Therefore the point of inflection, correct to three decimal places is $(0.164, f(0.164)) = (0.164, 0.991)$.

35. If f and g are CU on I, then $f'' > 0$ and $g'' > 0$ on I, so $(f + g)'' = f'' + g'' > 0$ on $I \quad\Rightarrow\quad f + g$ is CU on I.

36. Since f is positive and CU on I, $f > 0$ and $f'' > 0$ on I. So $g(x) = [f(x)]^2 \quad\Rightarrow\quad g' = 2ff' \quad\Rightarrow$ $g'' = 2f'f' + 2ff'' = 2(f')^2 + 2ff'' > 0 \quad\Rightarrow\quad g$ is CU on I.

37. Since f and g are positive, increasing, and CU on I, we have $f > 0$, $f' > 0$, $f'' > 0$, $g > 0$, $g' > 0$, $g'' > 0$ on I. Then $(fg)' = f'g + fg' \quad\Rightarrow\quad (fg)'' = f''g + 2f'g' + fg'' > 0 \quad\Rightarrow\quad fg$ is CU on I.

38. Since f and g are CU on $(-\infty, \infty)$, $f'' > 0$ and $g'' > 0$ on $(-\infty, \infty)$.
$h(x) = f(g(x)) \quad\Rightarrow\quad h'(x) = f'(g(x))g'(x) \quad\Rightarrow$
$h''(x) = f''(g(x))g'(x)g'(x) + f'(g(x))g''(x) = f''(g(x))[g'(x)]^2 + f'(g(x))g''(x) > 0$ if $f' > 0$.
So h is CU if f is increasing.

39. Let the cubic function be $f(x) = ax^3 + bx^2 + cx + d \quad\Rightarrow\quad f'(x) = 3ax^2 + 2bx + c \quad\Rightarrow$ $f''(x) = 6ax + 2b$. So f is CU when $6ax + 2b > 0 \quad\Leftrightarrow\quad x > -\dfrac{b}{3a}$, and CD when $x < -\dfrac{b}{3a}$, and so the only point of inflection occurs when $x = -\dfrac{b}{3a}$. If the graph has three x-intercepts x_1, x_2 and x_3, then the equation of $f(x)$ must factor as
$f(x) = a(x - x_1)(x - x_2)(x - x_3) = a[x^3 - (x_1 + x_2 + x_3)x^2 + (x_1x_2 + x_1x_3 + x_2x_3)x - x_1x_2x_3]$.
So $b = -a(x_1 + x_2 + x_3)$. Hence the x-coordinate of the point of inflection is
$-\dfrac{b}{3a} = -\dfrac{-a(x_1 + x_2 + x_3)}{3a} = \dfrac{x_1 + x_2 + x_3}{3}$.

40. $P(x) = x^4 + cx^3 + x^2 \quad \Rightarrow \quad P'(x) = 4x^3 + 3cx^2 + 2x \quad \Rightarrow \quad P''(x) = 12x^2 + 6cx + 2$. The graph of $P''(x)$ is a parabola. If $P''(x)$ has two roots, then it changes sign twice and so has two inflection points. This happens when the discriminant of $P''(x)$ is positive, that is, $(6c)^2 - 4 \cdot 12 \cdot 2 > 0 \quad \Leftrightarrow \quad 36c^2 - 96 > 0 \quad \Leftrightarrow$ $|c| > \frac{2\sqrt{6}}{3}$. If $36c^2 - 96 = 0 \quad \Leftrightarrow \quad c = \pm\frac{2\sqrt{6}}{3}$, $P''(x)$ is 0 at one point, but there are still no inflection points since $P''(x)$ never changes sign, and if $36c^2 - 96 < 0 \quad \Leftrightarrow \quad |c| < \frac{2\sqrt{6}}{3}$, then $P''(x)$ never changes sign, and so there are no inflection points.

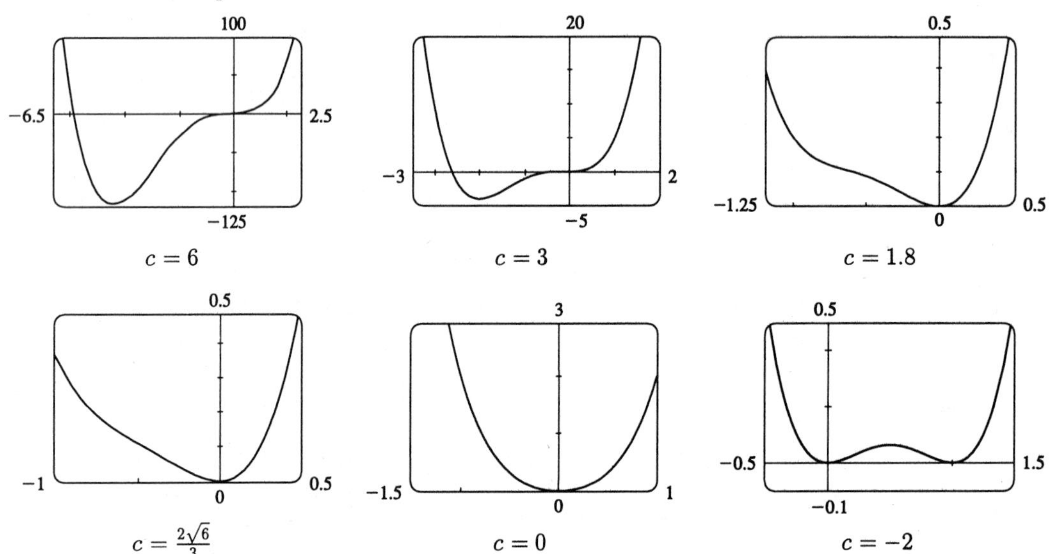

For large positive c, the graph of f has two inflection points and a large dip to the left of the y-axis. As c decreases, the graph of f becomes flatter for $x < 0$, and eventually the dip rises above the x-axis, and then disappears entirely, along with the inflection points. As c continues to decrease, the dip and the inflection points reappear, to the right of the origin.

41. There must exist some interval containing c on which f''' is positive, since $f'''(c)$ is positive and f''' is continuous. On this interval, f'' is increasing (since f''' is positive), so $f'' = (f')'$ changes from negative to positive at c. So by the First Derivative Test, f' has a local minimum at $x = c$ and thus cannot change sign there, so f has no extremum at c. But since f'' changes from negative to positive at c, f has a point of inflection at c (it changes from concave down to concave up).

EXERCISES 3.5

1. $\lim\limits_{x\to\infty} \dfrac{1}{x\sqrt{x}} = \lim\limits_{x\to\infty} \dfrac{1}{x^{3/2}} = 0$ by Theorem 4.

2. $\lim\limits_{x\to\infty} \dfrac{5+2x}{3-x} = \lim\limits_{x\to\infty} \dfrac{\dfrac{5}{x}+2}{\dfrac{3}{x}-1} \overset{(5)}{=} \dfrac{\lim\limits_{x\to\infty}\left[\dfrac{5}{x}+2\right]}{\lim\limits_{x\to\infty}\left[\dfrac{3}{x}-1\right]} \overset{(1,2,3)}{=} \dfrac{5\lim\limits_{x\to\infty}\dfrac{1}{x}+\lim\limits_{x\to\infty}2}{3\lim\limits_{x\to\infty}\dfrac{1}{x}-\lim\limits_{x\to\infty}1} = \dfrac{5(0)+2}{3(0)-1} = -2$ by (7) and Theorem 4.

3. $\lim\limits_{x\to\infty} \dfrac{x+4}{x^2-2x+5} = \lim\limits_{x\to\infty} \dfrac{\dfrac{1}{x}+\dfrac{4}{x^2}}{1-\dfrac{2}{x}+\dfrac{5}{x^2}} \overset{(5)}{=} \dfrac{\lim\limits_{x\to\infty}\left(\dfrac{1}{x}+\dfrac{4}{x^2}\right)}{\lim\limits_{x\to\infty}\left(1-\dfrac{2}{x}+\dfrac{5}{x^2}\right)} \overset{(1,2,3)}{=} \dfrac{\lim\limits_{x\to\infty}\dfrac{1}{x}+4\lim\limits_{x\to\infty}\dfrac{1}{x^2}}{\lim\limits_{x\to\infty}1-2\lim\limits_{x\to\infty}\dfrac{1}{x}+5\lim\limits_{x\to\infty}\dfrac{1}{x^2}}$

$= \dfrac{0+4(0)}{1-2(0)+5(0)} = 0$ by (7) and Theorem 4.

4. $\lim\limits_{t\to\infty} \dfrac{7t^3+4t}{2t^3-t^2+3} = \lim\limits_{t\to\infty} \dfrac{7+\dfrac{4}{t^2}}{2-\dfrac{1}{t}+\dfrac{3}{t^3}} \overset{(5,1,2,3)}{=} \dfrac{\lim\limits_{t\to\infty}7+4\lim\limits_{t\to\infty}\dfrac{1}{t^2}}{\lim\limits_{t\to\infty}2-\lim\limits_{t\to\infty}\dfrac{1}{t}+3\lim\limits_{t\to\infty}\dfrac{1}{t^3}} = \dfrac{7+4(0)}{2-0+3(0)}$

$= \frac{7}{2}$ by (7) and Theorem 4.

5. $\lim\limits_{x\to-\infty} \dfrac{(1-x)(2+x)}{(1+2x)(2-3x)} = \lim\limits_{x\to-\infty} \dfrac{\left[\dfrac{1}{x}-1\right]\left[\dfrac{2}{x}+1\right]}{\left[\dfrac{1}{x}+2\right]\left[\dfrac{2}{x}-3\right]} = \dfrac{\left[\lim\limits_{x\to-\infty}\dfrac{1}{x}-1\right]\left[\lim\limits_{x\to-\infty}\dfrac{2}{x}+1\right]}{\left[\lim\limits_{x\to-\infty}\dfrac{1}{x}+2\right]\left[\lim\limits_{x\to-\infty}\dfrac{2}{x}-3\right]}$

$\overset{(5,4,1,2,7)}{=} \dfrac{(0-1)(0+1)}{(0+2)(0-3)} = \dfrac{1}{6}$

6. $\lim\limits_{x\to\infty} \left[\dfrac{2x^2-1}{x+8x^2}\right]^{1/2} \overset{(11)}{=} \left[\lim\limits_{x\to\infty}\dfrac{2-\dfrac{1}{x^2}}{\dfrac{1}{x}+8}\right]^{1/2} \overset{(5,1,2)}{=} \left[\dfrac{\lim\limits_{x\to\infty}2-\lim\limits_{x\to\infty}\dfrac{1}{x^2}}{\lim\limits_{x\to\infty}\dfrac{1}{x}+\lim\limits_{x\to\infty}8}\right]^{1/2} = \left[\dfrac{2-0}{0+8}\right]^{1/2} = \dfrac{1}{2}$ by (7) and Theorem 4.

7. $\lim\limits_{x\to\infty} \dfrac{1}{3+\sqrt{x}} = \lim\limits_{x\to\infty} \dfrac{1/\sqrt{x}}{(3/\sqrt{x})+1} \overset{(5,1,3)}{=} \dfrac{\lim\limits_{x\to\infty}(1/\sqrt{x})}{3\lim\limits_{x\to\infty}(1/\sqrt{x})+\lim\limits_{x\to\infty}1} = \dfrac{0}{3(0)+1} = 0$ (by Theorem 4 with $r=\frac{1}{2}$.)

Or: Note that $0 < \dfrac{1}{3+\sqrt{x}} < \dfrac{1}{\sqrt{x}}$ and use the Squeeze Theorem.

8. Since $0 \le \sin^2 x \le 1$, we have $0 \le \dfrac{\sin^2 x}{x^2} \le \dfrac{1}{x^2}$ for all $x \ne 0$. By (7) and Theorem 4 we have $\lim\limits_{x\to\infty} 0 = 0$ and $\lim\limits_{x\to\infty} \dfrac{1}{x^2} = 0$. So, by the Squeeze Theorem, $\lim\limits_{x\to\infty} \dfrac{\sin^2 x}{x^2} = 0$.

9. $\lim\limits_{r\to\infty} \dfrac{r^4-r^2+1}{r^5+r^3-r} = \lim\limits_{r\to\infty} \dfrac{\dfrac{1}{r}-\dfrac{1}{r^3}+\dfrac{1}{r^5}}{1+\dfrac{1}{r^2}-\dfrac{1}{r^4}} = \dfrac{\lim\limits_{r\to\infty}\dfrac{1}{r}-\lim\limits_{r\to\infty}\dfrac{1}{r^3}+\lim\limits_{r\to\infty}\dfrac{1}{r^5}}{\lim\limits_{r\to\infty}1+\lim\limits_{r\to\infty}\dfrac{1}{r^2}-\lim\limits_{r\to\infty}\dfrac{1}{r^4}} = \dfrac{0-0+0}{1+0-0} = 0$

10. $\lim_{t\to-\infty}\dfrac{6t^2+5t}{(1-t)(2t-3)}=\lim_{t\to-\infty}\dfrac{6t^2+5t}{-2t^2+5t-3}=\lim_{t\to-\infty}\dfrac{6+5/t}{-2+5/t-3/t^2}$

$$=\frac{\lim_{t\to-\infty}6+5\lim_{t\to-\infty}(1/t)}{\lim_{t\to-\infty}(-2)+5\lim_{t\to-\infty}(1/t)-3\lim_{t\to-\infty}(1/t^2)}=\frac{6+5(0)}{-2+5(0)-3(0)}=-3$$

11. $\lim_{x\to\infty}\dfrac{\sqrt{1+4x^2}}{4+x}=\lim_{x\to\infty}\dfrac{\sqrt{(1/x^2)+4}}{(4/x)+1}=\dfrac{\sqrt{0+4}}{0+1}=2$

12. $\lim_{x\to-\infty}\dfrac{\sqrt{x^2+4x}}{4x+1}=\lim_{x\to-\infty}\dfrac{-\sqrt{1+4/x}}{4+1/x}=\dfrac{-\sqrt{1+0}}{4+0}=-\dfrac{1}{4}.$

Note: In dividing numerator and denominator by x, we used the fact that, for $x<0$, $x=-\sqrt{x^2}$.

13. $\lim_{x\to\infty}\dfrac{1-\sqrt{x}}{1+\sqrt{x}}=\lim_{x\to\infty}\dfrac{(1/\sqrt{x})-1}{(1/\sqrt{x})+1}=\dfrac{0-1}{0+1}=-1$

14. $\lim_{x\to\infty}\left(\sqrt{x^2+3x+1}-x\right)=\lim_{x\to\infty}\left(\sqrt{x^2+3x+1}-x\right)\dfrac{\sqrt{x^2+3x+1}+x}{\sqrt{x^2+3x+1}+x}=\lim_{x\to\infty}\dfrac{x^2+3x+1-x^2}{\sqrt{x^2+3x+1}+x}$

$$=\lim_{x\to\infty}\frac{3x+1}{\sqrt{x^2+3x+1}+x}=\lim_{x\to\infty}\frac{3+1/x}{\sqrt{1+(3/x)+(1/x^2)}+1}=\frac{3+0}{\sqrt{1+3\cdot 0+0}+1}=\frac{3}{2}$$

15. $\lim_{x\to\infty}\left(\sqrt{x^2+1}-\sqrt{x^2-1}\right)=\lim_{x\to\infty}\left(\sqrt{x^2+1}-\sqrt{x^2-1}\right)\dfrac{\sqrt{x^2+1}+\sqrt{x^2-1}}{\sqrt{x^2+1}+\sqrt{x^2-1}}$

$$=\lim_{x\to\infty}\frac{(x^2+1)-(x^2-1)}{\sqrt{x^2+1}+\sqrt{x^2-1}}=\lim_{x\to\infty}\frac{2}{\sqrt{x^2+1}+\sqrt{x^2-1}}$$

$$=\lim_{x\to\infty}\frac{2/x}{\sqrt{1+(1/x^2)}+\sqrt{1-(1/x^2)}}=\frac{0}{\sqrt{1+0}+\sqrt{1-0}}=0$$

16. $\lim_{x\to-\infty}\left(x+\sqrt{x^2+2x}\right)=\lim_{x\to-\infty}\left(x+\sqrt{x^2+2x}\right)\left[\dfrac{x-\sqrt{x^2+2x}}{x-\sqrt{x^2+2x}}\right]=\lim_{x\to-\infty}\dfrac{x^2-(x^2+2x)}{x-\sqrt{x^2+2x}}$

$$=\lim_{x\to-\infty}\frac{-2x}{x-\sqrt{x^2+2x}}=\lim_{x\to-\infty}\frac{-2}{1+\sqrt{1+2/x}}=\frac{-2}{1+\sqrt{1+2\cdot 0}}=-1$$

Note: In dividing numerator and denominator by x, we used the fact that, for $x<0$, $x=-\sqrt{x^2}$.

17. $\lim_{x\to\infty}\left(\sqrt{1+x}-\sqrt{x}\right)=\lim_{x\to\infty}\left(\sqrt{1+x}-\sqrt{x}\right)\left(\dfrac{\sqrt{1+x}+\sqrt{x}}{\sqrt{1+x}+\sqrt{x}}\right)=\lim_{x\to\infty}\dfrac{(1+x)-x}{\sqrt{1+x}+\sqrt{x}}$

$$=\lim_{x\to\infty}\frac{1}{\sqrt{1+x}+\sqrt{x}}=\lim_{x\to\infty}\frac{1/\sqrt{x}}{\sqrt{(1/x)+1}+1}=\frac{0}{\sqrt{0+1}+1}=0$$

18. Using $a^3-b^3=(a-b)(a^2+ab+b^2)$ with $a=\sqrt[3]{1+x}$ and $b=\sqrt[3]{x}$, we have

$$\lim_{x\to\infty}\left(\sqrt[3]{1+x}-\sqrt[3]{x}\right)=\lim_{x\to\infty}\frac{(1+x)-x}{(1+x)^{2/3}+(1+x)^{1/3}x^{1/3}+x^{2/3}}$$

$$=\lim_{x\to\infty}\frac{1}{(1+x)^{2/3}+(1+x)^{1/3}x^{1/3}+x^{2/3}}=0$$

19. $\lim_{x\to-\infty}\left(\sqrt{x^2+x+1}+x\right) = \lim_{x\to-\infty}\left(\sqrt{x^2+x+1}+x\right)\left[\frac{\sqrt{x^2+x+1}-x}{\sqrt{x^2+x+1}-x}\right]$

$= \lim_{x\to-\infty}\frac{x+1}{\left(\sqrt{x^2+x+1}-x\right)} = \lim_{x\to-\infty}\frac{1+(1/x)}{-\sqrt{1+(1/x)+(1/x^2)}-1} = \frac{1+0}{-\sqrt{1+0+0}-1} = -\frac{1}{2}$

20. $\lim_{x\to\infty}\cos x$ does not exist because, as x increases, $\cos x$ does not approach any one value, but oscillates between 1 and -1 forever.

21. $\sqrt{x}$ is large when x is large, so $\lim_{x\to\infty}\sqrt{x}=\infty$.

22. $\sqrt[3]{x}$ is large negative when x is large negative, so $\lim_{x\to-\infty}\sqrt[3]{x}=-\infty$.

23. $\lim_{x\to\infty}(x-\sqrt{x}) = \lim_{x\to\infty}\sqrt{x}(\sqrt{x}-1)=\infty$ since $\sqrt{x}\to\infty$ and $\sqrt{x}-1\to\infty$ as $x\to\infty$.

24. $\lim_{x\to\infty}(x+\sqrt{x})=\infty$ since $x\to\infty$ and $\sqrt{x}\to\infty$.

25. $\lim_{x\to-\infty}(x^3-5x^2)=-\infty$ since $x^3\to-\infty$ and $-5x^2\to-\infty$ as $x\to-\infty$.

Or: $\lim_{x\to-\infty}(x^3-5x^2) = \lim_{x\to-\infty}x^2(x-5)=-\infty$ since $x^2\to\infty$ and $x-5\to-\infty$.

26. $\lim_{x\to\infty}(x^2-x^4)=\lim_{x\to\infty}x^2(1-x^2)=-\infty$ since $x^2\to\infty$ and $1-x^2\to-\infty$.

27. $\lim_{x\to\infty}\frac{x^7-1}{x^6-1}=\lim_{x\to\infty}\frac{1-1/x^7}{(1/x)-(1/x^7)}=\infty$ since $1-\frac{1}{x^7}\to 1$ while $\frac{1}{x}-\frac{1}{x^7}\to 0^+$ as $x\to\infty$.

Or: Divide numerator and denominator by x^6 instead of x^7.

28. $\lim_{x\to\infty}\frac{x^3-1}{x^4+1}=\lim_{x\to\infty}\frac{(1/x)-(1/x^4)}{1+(1/x^4)}=\frac{0-0}{1+0}=0$

29. $\lim_{x\to\infty}\frac{\sqrt{x}+3}{x+3}=\lim_{x\to\infty}\frac{(1/\sqrt{x})+(3/x)}{1+3/x}=\frac{0+0}{1+0}=0$

30. $\lim_{x\to\infty}\frac{x}{\sqrt{x}-1}=\lim_{x\to\infty}\frac{x/\sqrt{x}}{\sqrt{x}-1/\sqrt{x}}=\lim_{x\to\infty}\frac{\sqrt{x}}{\sqrt{1-1/x}}=\infty$ since $\sqrt{x}\to\infty$ and $\sqrt{1-1/x}\to 1$.

Or: Divide numerator and denominator by x instead of $\sqrt{x}$.

31. If $t=1/x$ then $\lim_{x\to\infty}\cos(1/x)=\lim_{t\to0^+}\cos t=\cos 0=1$.

32. If $t=\frac{1}{x}$ then $\lim_{x\to\infty}\left(x-x\cos\frac{1}{x}\right)=\lim_{t\to0^+}\left(\frac{1}{t}-\frac{1}{t}\cos t\right)=\lim_{t\to0^+}\frac{1-\cos t}{t}=0$ by Corollary 2.4.7.

33. If $f(x)=x^2/2^x$, then a calculator gives $f(0)=0$, $f(1)=0.5$, $f(2)=1$, $f(3)=1.125$, $f(4)=1$, $f(5)=0.78125$, $f(6)=0.5625$, $f(7)=0.3828125$, $f(8)=0.25$, $f(9)=0.158203125$, $f(10)=0.09765625$, $f(20)\approx 0.00038147$, $f(50)\approx 2.2204\times10^{-12}$, $f(100)\approx 7.8886\times10^{-27}$. It appears that $\lim_{x\to\infty}(x^2/2^x)=0$.

34. If $f(x)=x^2\sin(5/x^2)$, then a calculator gives the following approximate values: $f(1)=-0.95892$, $f(2)=3.79594$, $f(3)=4.74674$, $f(4)=4.91902$, $f(5)=4.96673$, $f(6)=4.98394$, $f(7)=4.99133$, $f(8)=4.99492$, $f(9)=4.99683$, $f(10)=4.99792$, $f(20)=4.99987$, $f(50)=4.999997$, $f(100)=4.9999998$. It appears that $\lim_{x\to\infty}x^2\sin(5/x^2)=5$.

Let $t=\frac{1}{x^2}$. Then as $x\to\infty$, $t\to0$ and $\lim_{x\to\infty}x^2\sin\left(\frac{5}{x^2}\right)=\lim_{t\to0}\left(\frac{1}{t}\sin 5t\right)=5\lim_{t\to0}\frac{\sin 5t}{5t}=5$.

35. $\lim\limits_{x\to\pm\infty} \dfrac{x}{x+4} = \lim\limits_{x\to\pm\infty} \dfrac{1}{1+4/x} = \dfrac{1}{1+0} = 1$, so $y = 1$ is a horizontal asymptote.

$\lim\limits_{x\to-4^-} \dfrac{x}{x+4} = \infty$ and $\lim\limits_{x\to-4^+} \dfrac{x}{x+4} = -\infty$, so $x = -4$ is a vertical asymptote. The graph confirms these calculations.

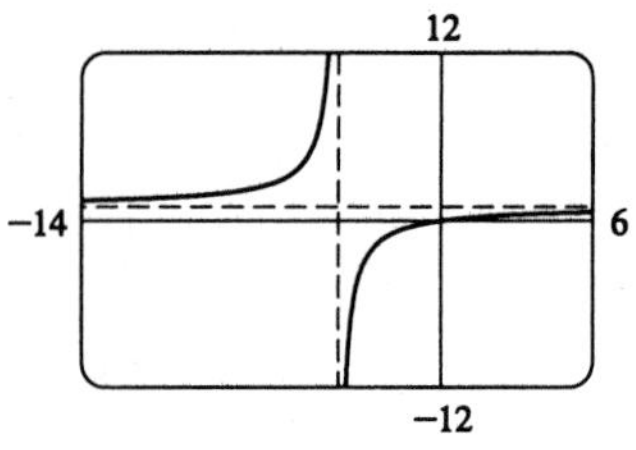

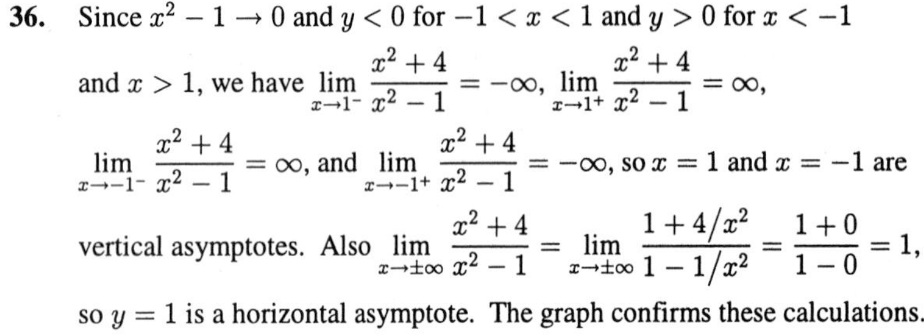

36. Since $x^2 - 1 \to 0$ and $y < 0$ for $-1 < x < 1$ and $y > 0$ for $x < -1$ and $x > 1$, we have $\lim\limits_{x\to1^-} \dfrac{x^2+4}{x^2-1} = -\infty$, $\lim\limits_{x\to1^+} \dfrac{x^2+4}{x^2-1} = \infty$, $\lim\limits_{x\to-1^-} \dfrac{x^2+4}{x^2-1} = \infty$, and $\lim\limits_{x\to-1^+} \dfrac{x^2+4}{x^2-1} = -\infty$, so $x = 1$ and $x = -1$ are vertical asymptotes. Also $\lim\limits_{x\to\pm\infty} \dfrac{x^2+4}{x^2-1} = \lim\limits_{x\to\pm\infty} \dfrac{1+4/x^2}{1-1/x^2} = \dfrac{1+0}{1-0} = 1$, so $y = 1$ is a horizontal asymptote. The graph confirms these calculations.

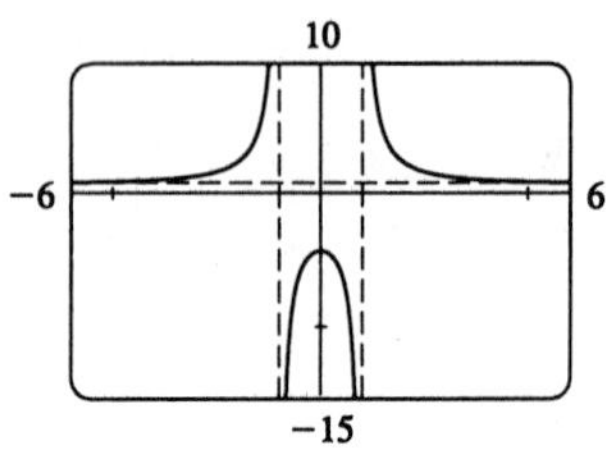

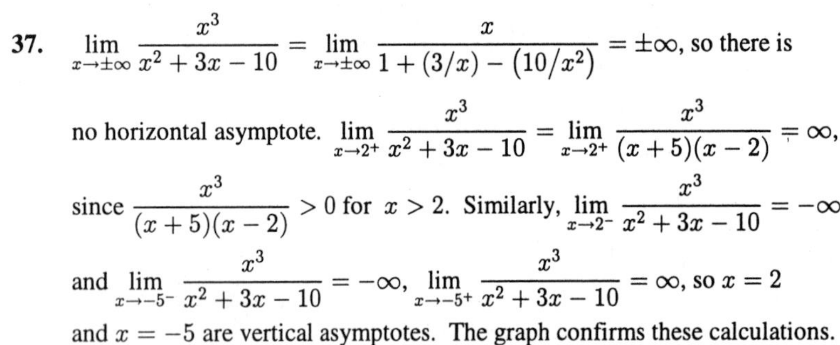

37. $\lim\limits_{x\to\pm\infty} \dfrac{x^3}{x^2+3x-10} = \lim\limits_{x\to\pm\infty} \dfrac{x}{1+(3/x)-(10/x^2)} = \pm\infty$, so there is no horizontal asymptote. $\lim\limits_{x\to2^+} \dfrac{x^3}{x^2+3x-10} = \lim\limits_{x\to2^+} \dfrac{x^3}{(x+5)(x-2)} = \infty$, since $\dfrac{x^3}{(x+5)(x-2)} > 0$ for $x > 2$. Similarly, $\lim\limits_{x\to2^-} \dfrac{x^3}{x^2+3x-10} = -\infty$ and $\lim\limits_{x\to-5^-} \dfrac{x^3}{x^2+3x-10} = -\infty$, $\lim\limits_{x\to-5^+} \dfrac{x^3}{x^2+3x-10} = \infty$, so $x = 2$ and $x = -5$ are vertical asymptotes. The graph confirms these calculations.

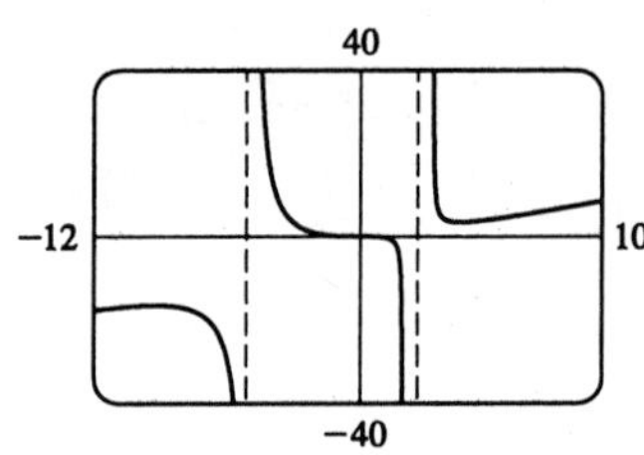

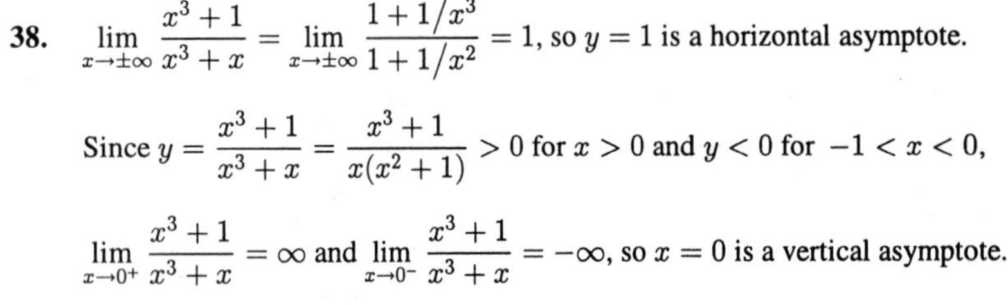

38. $\lim\limits_{x\to\pm\infty} \dfrac{x^3+1}{x^3+x} = \lim\limits_{x\to\pm\infty} \dfrac{1+1/x^3}{1+1/x^2} = 1$, so $y = 1$ is a horizontal asymptote.

Since $y = \dfrac{x^3+1}{x^3+x} = \dfrac{x^3+1}{x(x^2+1)} > 0$ for $x > 0$ and $y < 0$ for $-1 < x < 0$,

$\lim\limits_{x\to0^+} \dfrac{x^3+1}{x^3+x} = \infty$ and $\lim\limits_{x\to0^-} \dfrac{x^3+1}{x^3+x} = -\infty$, so $x = 0$ is a vertical asymptote.

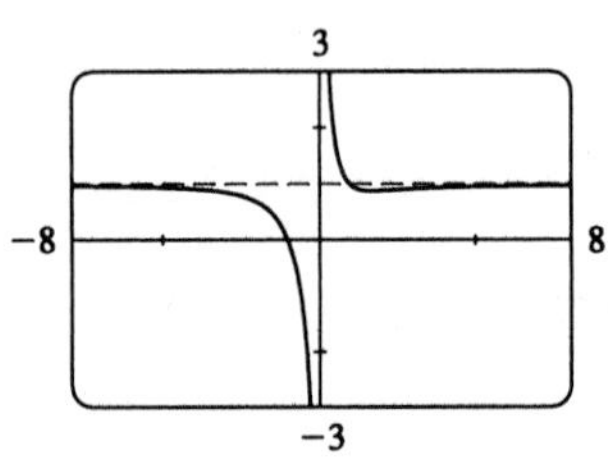

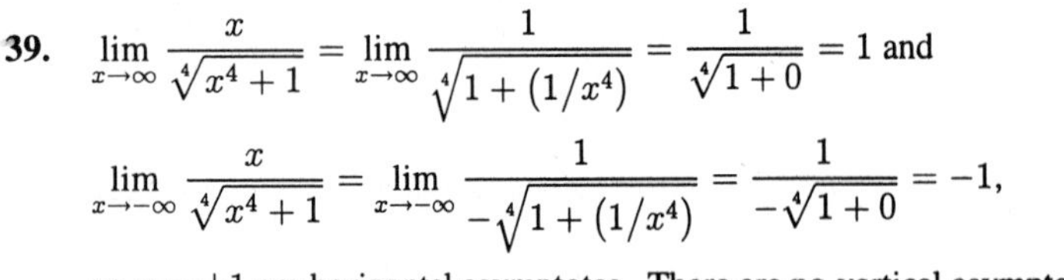

39. $\lim\limits_{x\to\infty} \dfrac{x}{\sqrt[4]{x^4+1}} = \lim\limits_{x\to\infty} \dfrac{1}{\sqrt[4]{1+(1/x^4)}} = \dfrac{1}{\sqrt[4]{1+0}} = 1$ and

$\lim\limits_{x\to-\infty} \dfrac{x}{\sqrt[4]{x^4+1}} = \lim\limits_{x\to-\infty} \dfrac{1}{-\sqrt[4]{1+(1/x^4)}} = \dfrac{1}{-\sqrt[4]{1+0}} = -1$,

so $y = \pm1$ are horizontal asymptotes. There are no vertical asymptotes.

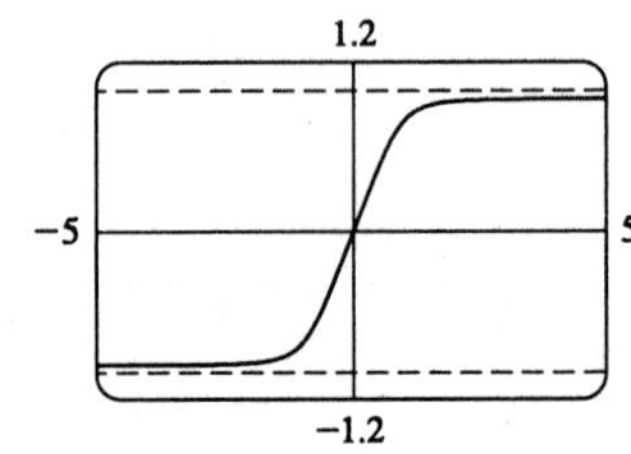

40. $\displaystyle\lim_{x\to\infty}\frac{x-9}{\sqrt{4x^2+3x+2}}=\lim_{x\to\infty}\frac{1-9/x}{\sqrt{4+(3/x)+(2/x^2)}}=\frac{1-0}{\sqrt{4+0+0}}=\frac{1}{2}.$

$\displaystyle\lim_{x\to-\infty}\frac{x-9}{\sqrt{4x^2+3x+2}}=\lim_{x\to-\infty}\frac{1-9/x}{-\sqrt{4+(3/x)+(2/x^2)}}=\frac{1-0}{-\sqrt{4+0+0}}=-\frac{1}{2}.$

The horizontal asymptotes are $y=\pm\frac{1}{2}$. The polynomial $4x^2+3x+1$ is positive for all x, so the denominator never approaches zero, and thus there are no vertical asymptotes.

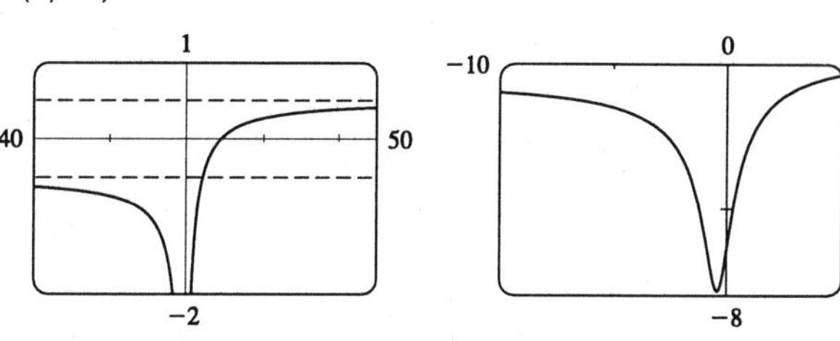

41. $\displaystyle\lim_{x\to 4^+}\frac{4}{x-4}=\infty$ and $\displaystyle\lim_{x\to 4^-}\frac{4}{x-4}=-\infty$, so $x=4$ is a vertical asymptote. $\displaystyle\lim_{x\to\pm\infty}\frac{4}{x-4}=\lim_{x\to\pm\infty}\frac{4/x}{1-4/x}=0$, so $y=0$ is a horizontal asymptote. $y'=-\dfrac{4}{(x-4)^2}<0\ \ (x\neq 4)$ so y is decreasing on $(-\infty,4)$ and $(4,\infty)$. $y''=\dfrac{8}{(x-4)^3}>0$ for $x>4$, so y is CU on $(4,\infty)$ and CD on $(-\infty,4)$.

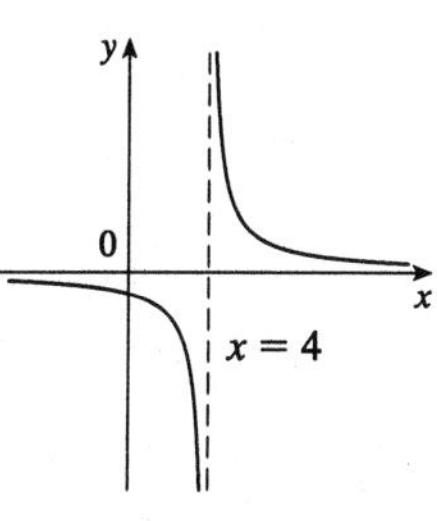

42. $\displaystyle\lim_{x\to\pm\infty}\frac{x-2}{x+2}=\lim_{x\to\pm\infty}\frac{1-2/x}{1+2/x}=\frac{1-0}{1+0}=1$, so $y=1$ is a horizontal asymptote.

$\displaystyle\lim_{x\to-2^+}\frac{x-2}{x+2}=-\infty$ and $\displaystyle\lim_{x\to-2^-}\frac{x-2}{x+2}=\infty$, so $y=-2$ is a vertical asymptote.

$y'=\dfrac{(x+2)\cdot 1-(x-2)\cdot 1}{(x+2)^2}=\dfrac{4}{(x+2)^2}>0$ for $x\neq-2$, so y is increasing on $(-\infty,-2)$ and $(-2,\infty)$. $y''=-\dfrac{8}{(x+2)^3}>0$ for $x<-2$, so y is CU on $(-\infty,-2)$ and CD on $(-2,\infty)$.

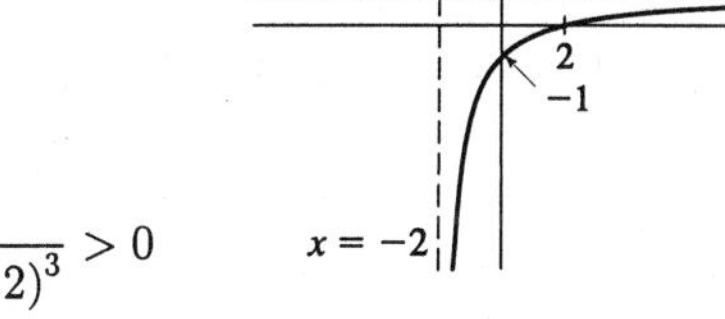

43. $\displaystyle\lim_{x\to\pm\infty}\frac{x}{x^2+1}=\lim_{x\to\pm\infty}\frac{1/x}{1+1/x^2}=\frac{0}{1+0}=0$, so $y=0$ is a horizontal asymptote.

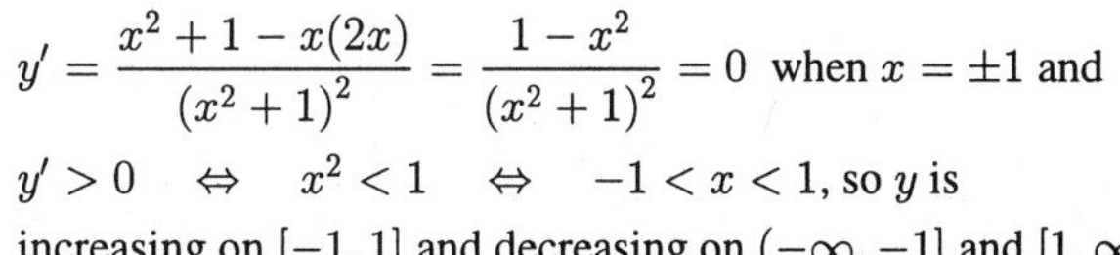

$y'=\dfrac{x^2+1-x(2x)}{(x^2+1)^2}=\dfrac{1-x^2}{(x^2+1)^2}=0$ when $x=\pm1$ and $y'>0 \quad\Leftrightarrow\quad x^2<1 \quad\Leftrightarrow\quad -1<x<1$, so y is increasing on $[-1,1]$ and decreasing on $(-\infty,-1]$ and $[1,\infty)$.

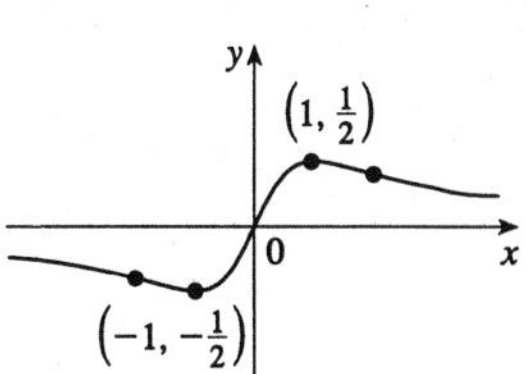

$y''=\dfrac{(1+x^2)^2(-2x)-(1-x^2)\,2\,(x^2+1)\,2x}{(1+x^2)^4}=\dfrac{2x\,(x^2-3)}{(1+x^2)^3}>0 \quad\Leftrightarrow\quad x>\sqrt{3}$ or $-\sqrt{3}<x<0$, so y is CU on $\left(\sqrt{3},\infty\right)$ and $\left(-\sqrt{3},0\right)$ and CD on $\left(-\infty,-\sqrt{3}\right)$ and $\left(0,\sqrt{3}\right)$.

44. $\lim_{x\to\pm\infty} \dfrac{2x^2 - x + 2}{x^2 + 1} = \lim_{x\to\pm\infty} \dfrac{2 - (1/x) + (2/x^2)}{1 + 1/x^2} = \dfrac{2 - 0 + 0}{1 + 0}$

$= 2$, so $y = 2$ is a horizontal asymptote.

$y' = \dfrac{(x^2+1)(4x-1) - (2x^2 - x + 2)(2x)}{(x^2+1)^2} = \dfrac{x^2 - 1}{(x^2+1)^2}$.

$y' > 0 \quad \Leftrightarrow \quad x^2 > 1 \quad \Leftrightarrow \quad x > 1$ or $x < -1$ and

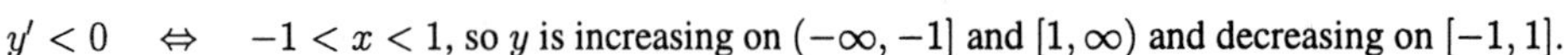

$y' < 0 \quad \Leftrightarrow \quad -1 < x < 1$, so y is increasing on $(-\infty, -1]$ and $[1, \infty)$ and decreasing on $[-1, 1]$.

$y'' = \dfrac{(x^2+1)^2(2x) - (x^2-1)2(x^2+1)(2x)}{(x^2+1)^4} = \dfrac{2x(3 - x^2)}{(x^2+1)^3} > 0 \quad \Leftrightarrow \quad x < -\sqrt{3}$ or $0 < x < \sqrt{3}$, so y is CU on $\left(-\infty, -\sqrt{3}\right)$ and $\left(0, \sqrt{3}\right)$, and CD on $\left(-\sqrt{3}, 0\right)$ and $\left(\sqrt{3}, \infty\right)$.

45. 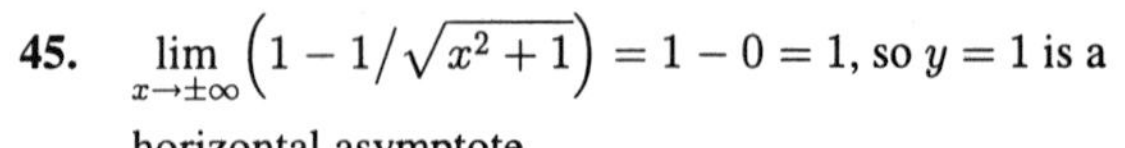

$\lim_{x\to\pm\infty} \left(1 - 1/\sqrt{x^2+1}\right) = 1 - 0 = 1$, so $y = 1$ is a horizontal asymptote.

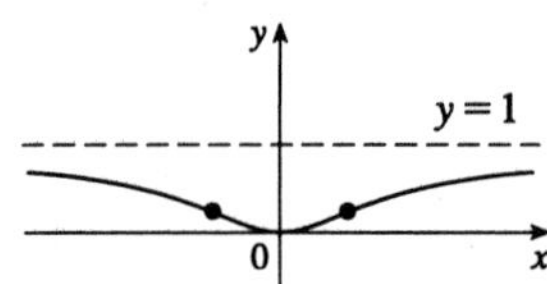

$y' = -\left(-\frac{1}{2}\right)(x^2+1)^{-3/2}(2x) = \dfrac{x}{(x^2+1)^{3/2}} > 0 \quad \Leftrightarrow$

$x > 0$, so y is increasing on $[0, \infty)$ and decreasing on $(-\infty, 0]$.

$y'' = \dfrac{(x^2+1)^{3/2} - x\left(\frac{3}{2}\right)(x^2+1)^{1/2}(2x)}{(x^2+1)^3} = \dfrac{1 - 2x^2}{(x^2+1)^{5/2}} > 0 \quad \Leftrightarrow \quad x^2 < \frac{1}{2} \quad \Leftrightarrow \quad |x| < \frac{1}{\sqrt{2}}$, so y is CU on $\left(-\frac{1}{\sqrt{2}}, \frac{1}{\sqrt{2}}\right)$, and CD on $\left(-\infty, -\frac{1}{\sqrt{2}}\right)$ and $\left(\frac{1}{\sqrt{2}}, \infty\right)$.

46. 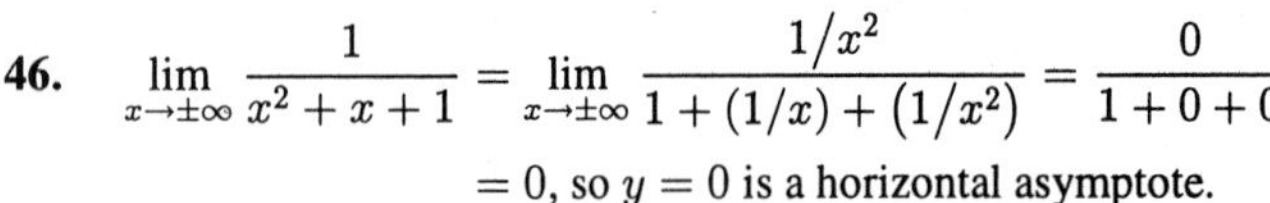

$\lim_{x\to\pm\infty} \dfrac{1}{x^2 + x + 1} = \lim_{x\to\pm\infty} \dfrac{1/x^2}{1 + (1/x) + (1/x^2)} = \dfrac{0}{1 + 0 + 0}$

$= 0$, so $y = 0$ is a horizontal asymptote.

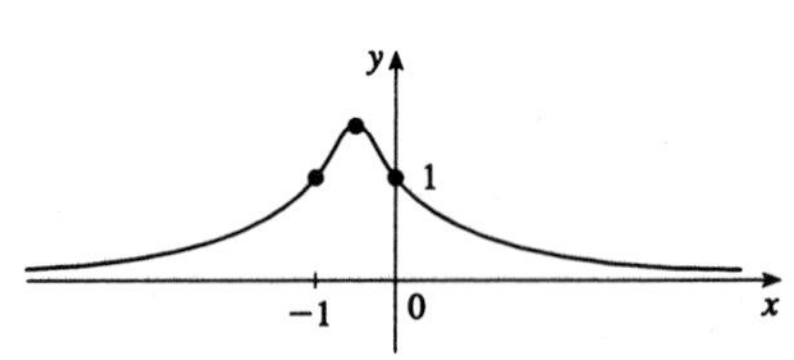

$y' = -\dfrac{2x+1}{(x^2+x+1)^2} > 0 \quad \Leftrightarrow \quad 2x + 1 < 0 \quad \Leftrightarrow \quad x < -\frac{1}{2}$,

and $y' < 0 \quad \Leftrightarrow \quad x > -\frac{1}{2}$, so y is increasing on $\left(-\infty, -\frac{1}{2}\right]$ and decreasing on $\left[-\frac{1}{2}, \infty\right)$.

$y'' = -\dfrac{2(x^2+x+1)^2 - (2x+1)\,2\,(x^2+x+1)(2x+1)}{(x^2+x+1)^4} = \dfrac{(6x)(x+1)}{(x^2+x+1)^3} > 0 \quad \Leftrightarrow \quad x < -1$ or $x > 0$, so y is CU on $(-\infty, -1)$ and $(0, \infty)$ and CD on $(-1, 0)$.

47. **(a)** If $t = \dfrac{1}{x}$ then

$$\lim_{x\to\infty} x \sin\frac{1}{x} = \lim_{t\to 0^+} \frac{1}{t}\sin t = \lim_{t\to 0^+} \frac{\sin t}{t} = 1$$

(b)

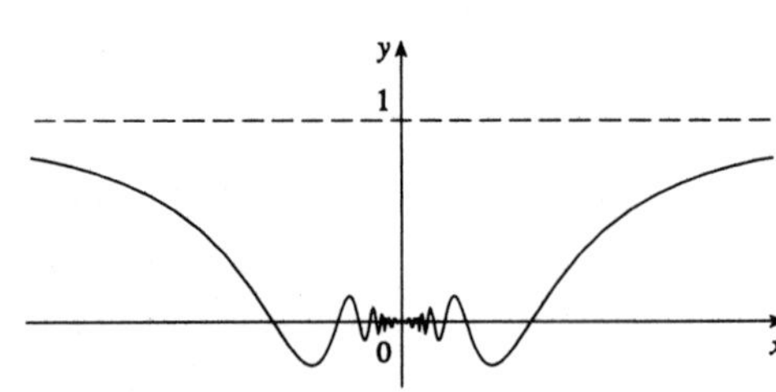

48. **(a)** Since $-1 \le \sin x \le 1$, we have $-\frac{1}{x} \le \frac{\sin x}{x} \le \frac{1}{x}$ for $x > 0$. We know that $\lim_{x\to\infty} \frac{1}{x} = 0 = \lim_{x\to\infty} \frac{1}{x}$, so $\lim_{x\to\infty} \frac{\sin x}{x} = 0$ by the Squeeze Theorem.

(b)

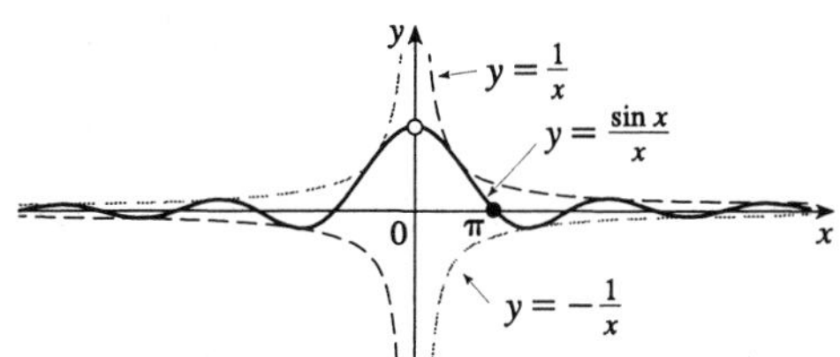

49. $y = f(x) = x^2(x-2)(1-x)$. The y-intercept is $f(0) = 0$, and the x-intercepts occur when $y = 0 \Rightarrow x = 0, 1, 2$. Notice (as in Example 9) that, since x^2 is always positive, the graph does not cross the x-axis at 0, but does cross the x-axis at 1 and 2. $\lim_{x\to\infty} x^2(x-2)(1-x) = -\infty$, since the first two factors are large positive and the third large negative when x is large positive. $\lim_{x\to-\infty} x^2(x-2)(1-x) = -\infty$ because the first and third factors are large positive and the second large negative as $x \to -\infty$.

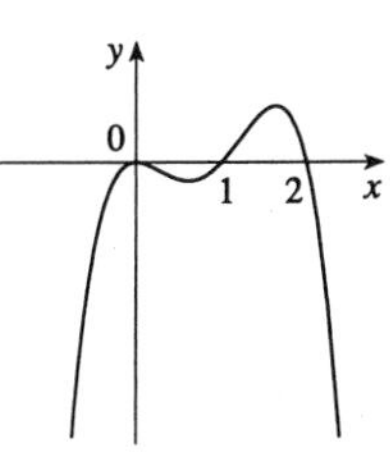

50. $y = (2+x)^3(1-x)(3-x)$. As $x \to \infty$, the first factor is large positive, and the second and third factors are large negative. Therefore $\lim_{x\to\infty} f(x) = \infty$. As $x \to -\infty$, the first factor is large negative, and the second and third factors are large positive. Therefore $\lim_{x\to-\infty} f(x) = -\infty$. Now the y-intercept is $f(0) = (2)^3(1)(3) = 24$ and the x-intercepts are the solutions to $f(x) = 0 \Rightarrow x = -2, 1$ and 3, and the graph crosses the x-axis at all of these points.

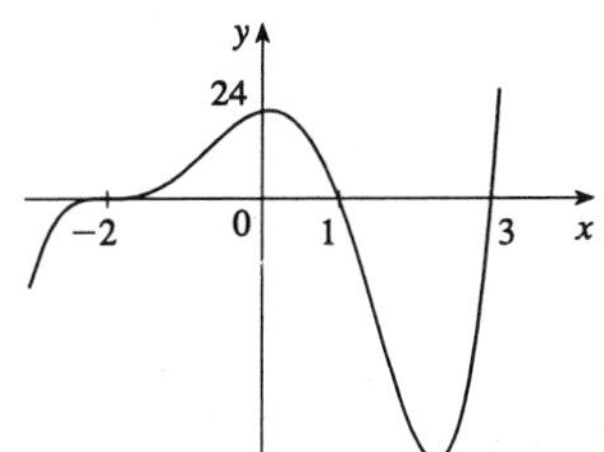

51. $y = f(x) = (x+4)^5(x-3)^4$. The y-intercept is $f(0) = 4^5(-3)^4 = 82{,}944$. The x-intercepts occur when $y = 0 \Rightarrow x = -4, 3$. Notice (as in Example 9) that the graph does not cross the x-axis at 3 because $(x-3)^4$ is always positive, but does cross the x-axis at -4. $\lim_{x\to\infty} (x+4)^5(x-3)^4 = \infty$ since both factors are large positive when x is large positive. $\lim_{x\to-\infty} (x+4)^5(x-3)^4 = -\infty$ since the first factor is large negative and the second factor is large positive when x is large negative.

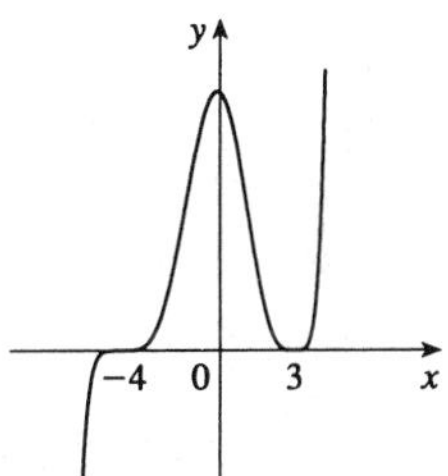

52. $y = (1-x)(x-3)^2(x-5)^2$. As $x \to \infty$, the first factor approaches $-\infty$ while the second and third factors approach ∞. Therefore $\lim_{x\to\infty} (x) = -\infty$. As $x \to -\infty$, the factors all approach ∞. Therefore $\lim_{x\to-\infty} (x) = \infty$. Now the y-intercept is $f(0) = (1)(-3)^2(-5)^2 = 225$ and the x-intercepts are the solutions to $f(x) = 0 \Rightarrow x = 1, 3$, and 5. Notice (as in Example 9) that $f(x)$ does not change sign at $x = 3$ or $x = 5$ because the factors $(x-3)^2$ and $(x-5)^2$ are always positive, so the graph does not cross the x-axis at $x = 3$ or $x = 5$, but does cross the x-axis at $x = 1$.

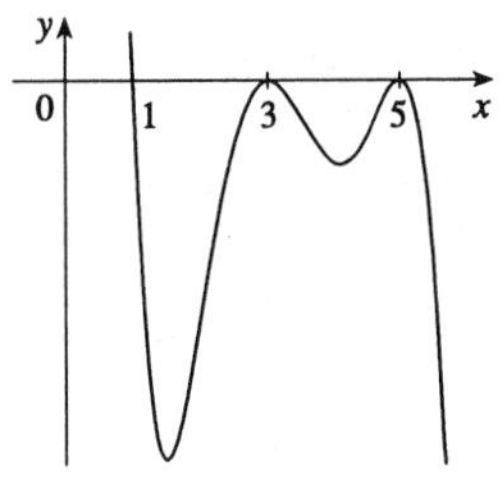

53. First we plot the points which are known to be on the graph: $(2, -1)$ and $(0, 0)$. We can also draw a short line segment of slope 0 at $x = 2$, since we are given that $f'(2) = 0$. Now we know that $f'(x) < 0$ (that is, the function is decreasing) on $(0, 2)$, and that $f''(x) < 0$ on $(0, 1)$ and $f''(x) > 0$ on $(1, 2)$. So we must join the points $(0, 0)$ and $(2, -1)$ in such a way that the curve is concave down on $(0, 1)$ and concave up on $(1, 2)$. The curve must be concave up and increasing on $(2, 4)$, and concave down and increasing on $(4, \infty)$. Now we just need to reflect the curve in the y-axis, since we are given that f is an even function. The diagram shows one possible function satisfying all of the given conditions.

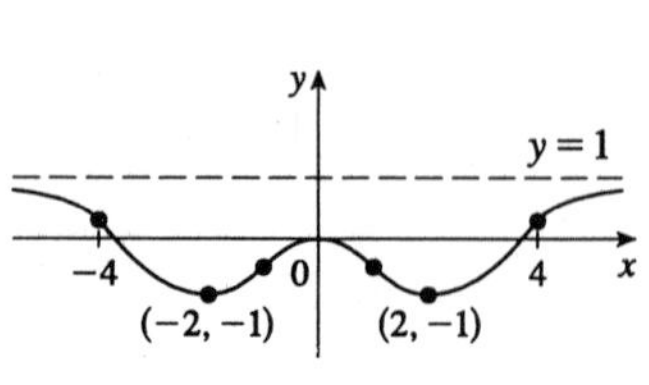

54. The diagram shows one possible function satisfying all of the given conditions.

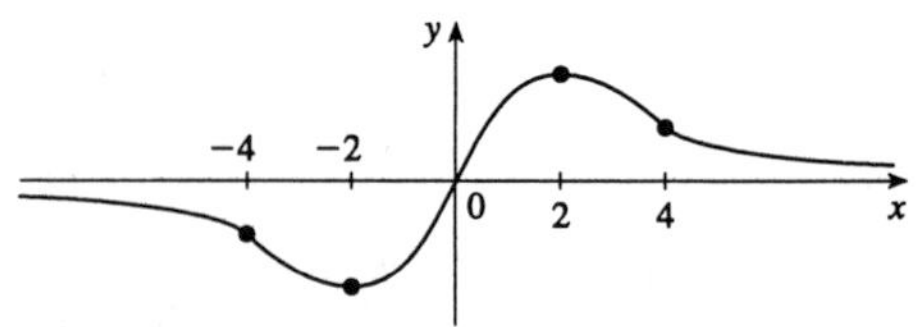

55. We are given that $f(1) = f'(1) = 0$. So we can draw a short horizontal line at the point $(1, 0)$ to represent this situation. We are given that $x = 0$ and $x = 2$ are vertical asymptotes, with $\lim_{x\to 0} f(x) = -\infty$, $\lim_{x\to 2^+} f(x) = \infty$ and $\lim_{x\to 2^-} f(x) = -\infty$, so we can draw the parts of the curve which approach these asymptotes.

On the interval $(-\infty, 0)$, the graph is concave down, and $f(x) \to \infty$ as $x \to -\infty$. Between the asymptotes the graph is concave down. On the interval $(2, \infty)$ the graph is concave up, and $f(x) \to 0$ as $x \to \infty$, so $y = 0$ is a horizontal asymptote. The diagram shows one possible function satisfying all of the given conditions.

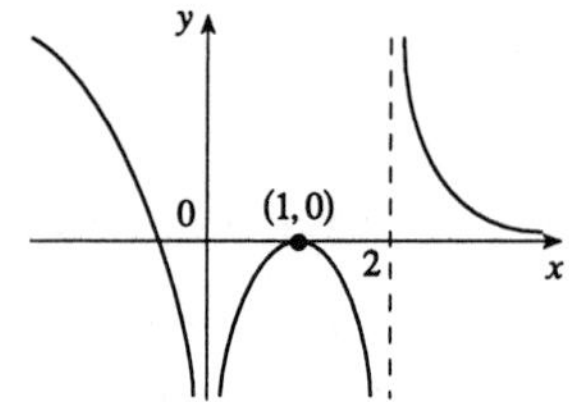

56. The diagram shows one possible function satisfying all of the given conditions.

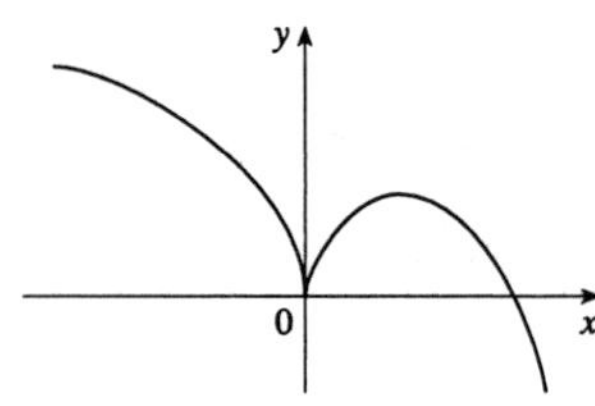

57. Divide numerator and denominator by the highest power of x in $Q(x)$.

(a) If $\deg(P) < \deg(Q)$, then numerator $\to 0$ but denominator doesn't. So $\lim_{x\to\infty} \dfrac{P(x)}{Q(x)} = 0$.

(b) If $\deg(P) > \deg(Q)$, then numerator $\to \pm\infty$ but denominator doesn't, so $\lim_{x\to\infty} \dfrac{P(x)}{Q(x)} = \pm\infty$ (depending on the ratio of the leading coefficients of P and Q.)

58. **(i)** $n = 0$ **(ii)** $n > 0$ (n odd) **(iii)** $n > 0$ (n even) **(iv)** $n < 0$ (n odd) **(v)** $n < 0$ (n even)

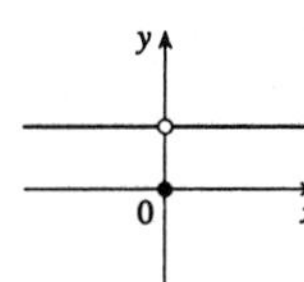

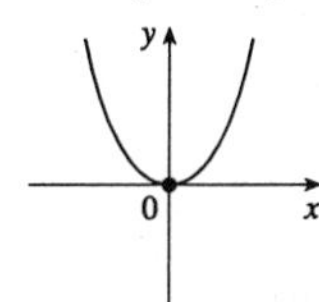

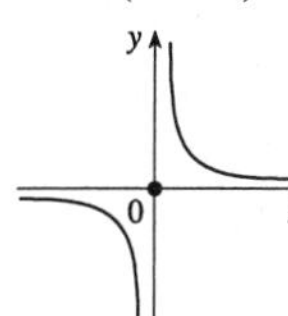

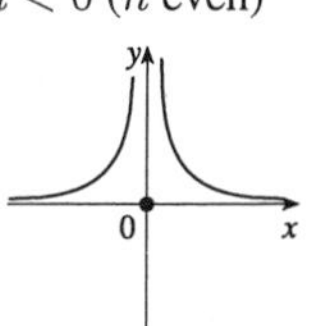

From these sketches we see that

(a) $\displaystyle\lim_{x\to0^+} x^n = \begin{cases} 1 & \text{if } n = 0 \\ 0 & \text{if } n > 0 \\ \infty & \text{if } n < 0 \end{cases}$ **(b)** $\displaystyle\lim_{x\to0^-} x^n = \begin{cases} 1 & \text{if } n = 0 \\ 0 & \text{if } n > 0 \\ -\infty & \text{if } n < 0,\ n \text{ odd} \\ \infty & \text{if } n < 0,\ n \text{ even} \end{cases}$

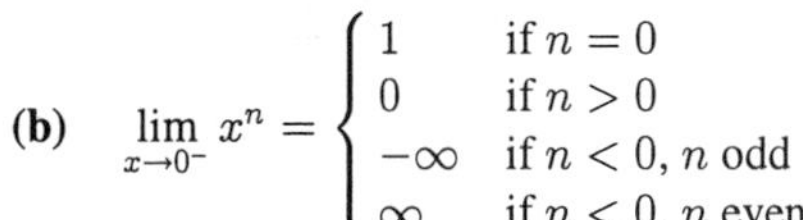

(c) $\displaystyle\lim_{x\to\infty} x^n = \begin{cases} 1 & \text{if } n = 0 \\ \infty & \text{if } n > 0 \\ 0 & \text{if } n < 0 \end{cases}$ **(d)** $\displaystyle\lim_{x\to-\infty} x^n = \begin{cases} 1 & \text{if } n = 0 \\ -\infty & \text{if } n > 0,\ n \text{ odd} \\ \infty & \text{if } n > 0,\ n \text{ even} \\ 0 & \text{if } n < 0 \end{cases}$

59. $\displaystyle\lim_{x\to\infty} \frac{4x-1}{x} = \lim_{x\to\infty}\left(4 - \frac{1}{x}\right) = 4$, and $\displaystyle\lim_{x\to\infty} \frac{4x^2+3x}{x^2} = \lim_{x\to\infty}\left(4 + \frac{3}{x}\right) = 4$. Therefore by the Squeeze Theorem, $\displaystyle\lim_{x\to\infty} f(x) = 4$.

60. **(a)** After t minutes, $25t$ liters of brine with 30 g of salt per liter has been pumped into the tank, so it contains $(5000 + 25t)$ liters of water and $25t \cdot 30 = 750t$ grams of salt. Therefore the salt concentration at time t will be $C(t) = \dfrac{750t}{5000 + 25t} = \dfrac{30t}{200 + t}\,\dfrac{\text{g}}{\text{L}}$.

(b) $\displaystyle\lim_{t\to\infty}(t) = \lim_{t\to\infty}\frac{30t}{200+t} = \lim_{t\to\infty}\frac{30t/t}{200/t + t/t} = \frac{30}{0+1} = 30$. So the salt concentration approaches that of the brine being pumped into the tank.

61. 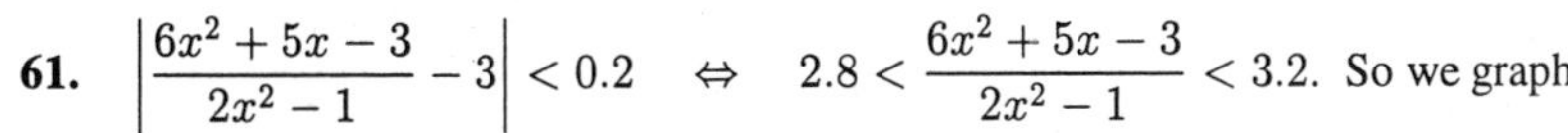
$\left|\dfrac{6x^2+5x-3}{2x^2-1} - 3\right| < 0.2 \quad\Leftrightarrow\quad 2.8 < \dfrac{6x^2+5x-3}{2x^2-1} < 3.2$. So we graph the three parts of this inequality on the same screen, and find that the curve

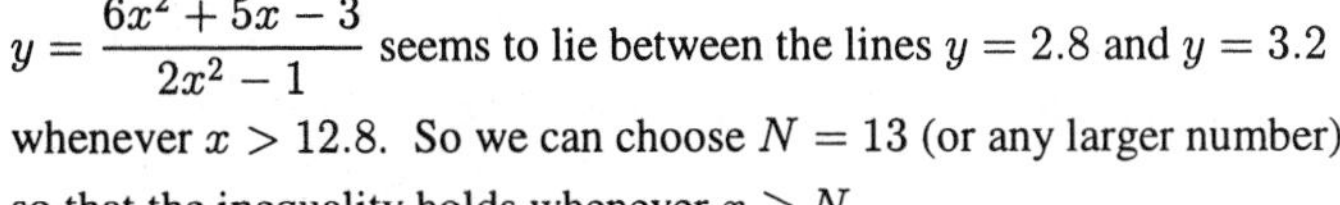
$y = \dfrac{6x^2+5x-3}{2x^2-1}$ seems to lie between the lines $y = 2.8$ and $y = 3.2$ whenever $x > 12.8$. So we can choose $N = 13$ (or any larger number), so that the inequality holds whenever $x \geq N$.

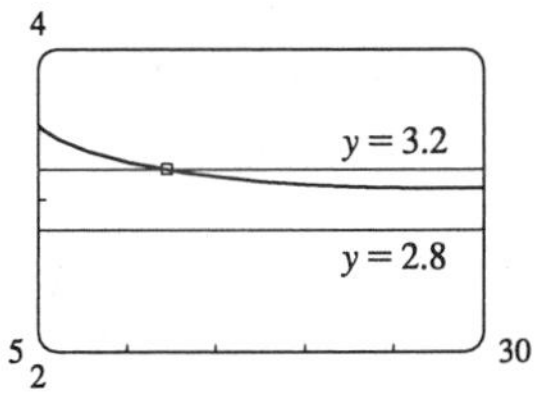

62. For $\epsilon = 0.5$, we must find N such that whenever $x \geq N$, we have $\left|\dfrac{\sqrt{4x^2+1}}{x+1} - 2\right| < 0.5 \quad\Leftrightarrow$

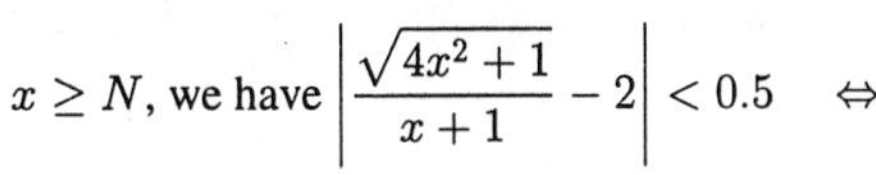
$1.5 < \dfrac{\sqrt{4x^2+1}}{x+1} < 2.5$. We graph the three parts of this inequality on the same screen, and find that it holds whenever $x \geq 3$. So we choose $N = 3$ (or any larger number).

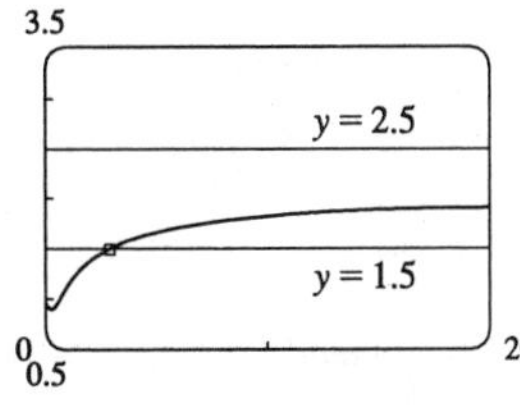

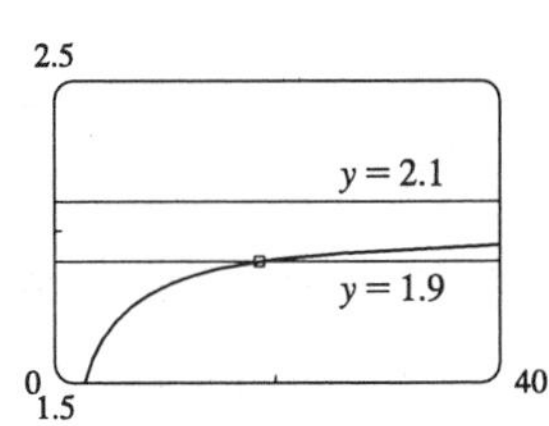

For $\epsilon = 0.1$, we must have $1.9 < \dfrac{\sqrt{4x^2+1}}{x+1} < 2.1$, and the graphs show that this holds whenever $x \geq 19$. So we choose $N = 19$ (or any larger number).

63. For $\epsilon = 0.5$, we need to find N such that

$$\left|\frac{\sqrt{4x^2+1}}{x+1} - (-2)\right| < 0.5 \quad \Leftrightarrow \quad -2.5 < \frac{\sqrt{4x^2+1}}{x+1} < -1.5$$

whenever $x \le N$. We graph the three parts of this inequality on the same screen, and see that the inequality holds for $x \le -6$. So we choose $N = -6$ (or any smaller number).

For $\epsilon = 0.1$, we need $-2.1 < \dfrac{\sqrt{4x^2+1}}{x+1} < -1.9$ whenever $x \le N$. From the graph, it seems that this inequality holds for $x \le -22$. So we choose any $N = -22$ (or any smaller number).

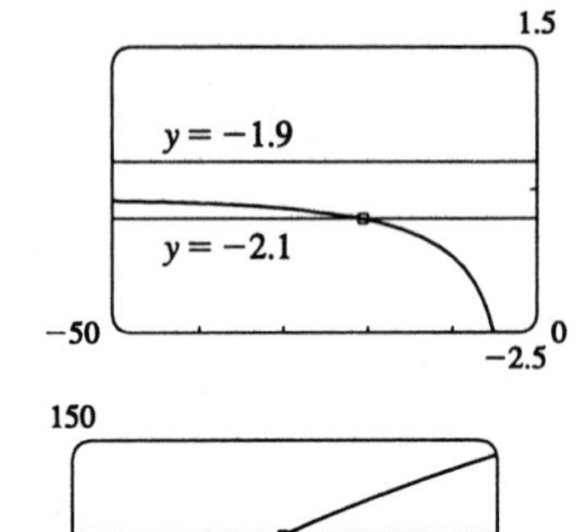

64. We need N such that $\dfrac{2x+1}{\sqrt{x+1}} > 100$ whenever $x \ge N$.

From the graph, we see that this inequality holds for $x \ge 2500$. So we choose $N = 2500$ (or any larger number).

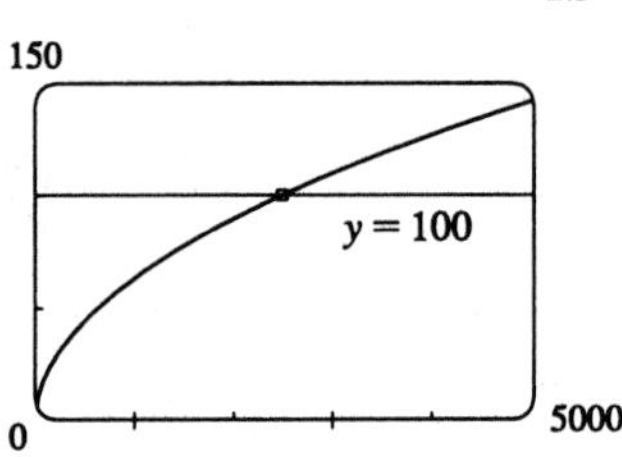

65. **(a)** $1/x^2 < 0.0001 \quad \Leftrightarrow \quad x^2 > 1/0.0001 = 10{,}000 \quad \Leftrightarrow \quad x > 100 \quad (x > 0)$

(b) If $\epsilon > 0$ is given, then $1/x^2 < \epsilon \quad \Leftrightarrow \quad x^2 > 1/\epsilon \quad \Leftrightarrow \quad x > 1/\sqrt{\epsilon}$. Let $N = 1/\sqrt{\epsilon}$.

Then $x > N \quad \Rightarrow \quad x > \dfrac{1}{\sqrt{\epsilon}} \quad \Rightarrow \quad \left|\dfrac{1}{x^2} - 0\right| = \dfrac{1}{x^2} < \epsilon$ so $\displaystyle\lim_{x\to\infty} \frac{1}{x^2} = 0$.

66. **(a)** $1/\sqrt{x} < 0.0001 \quad \Leftrightarrow \quad \sqrt{x} > 1/0.0001 = 10^4 \quad \Leftrightarrow \quad x > 10^8$.

(b) If $\epsilon > 0$ is given, then $1/\sqrt{x} < \epsilon \quad \Leftrightarrow \quad \sqrt{x} > 1/\epsilon \quad \Leftrightarrow \quad x > 1/\epsilon^2$. Let $N = 1/\epsilon^2$.

Then $x > N \quad \Rightarrow \quad x > \dfrac{1}{\epsilon^2} \quad \Rightarrow \quad \left|\dfrac{1}{\sqrt{x}} - 0\right| = \dfrac{1}{\sqrt{x}} < \epsilon$, so $\displaystyle\lim_{x\to\infty} \frac{1}{\sqrt{x}} = 0$.

67. For $x < 0$, $|1/x - 0| = -1/x$. If $\epsilon > 0$ is given, then $-1/x < \epsilon \quad \Leftrightarrow \quad x < -1/\epsilon$.

Take $N = -\dfrac{1}{\epsilon}$. Then $x < N \quad \Rightarrow \quad x < -\dfrac{1}{\epsilon} \quad \Rightarrow \quad \left|\dfrac{1}{x} - 0\right| = -\dfrac{1}{x} < \epsilon$, so $\displaystyle\lim_{x\to-\infty} \frac{1}{x} = 0$.

68. Given $M > 0$, we need $N > 0$ such that $x > N \quad \Rightarrow \quad x^3 > M$. Now $x^3 > M \quad \Leftrightarrow \quad x > \sqrt[3]{M}$, so take $N = \sqrt[3]{M}$. Then $x > N = \sqrt[3]{M} \quad \Rightarrow \quad x^3 > M$, so $\displaystyle\lim_{x\to\infty} x^3 = \infty$.

69. Suppose that $\displaystyle\lim_{x\to\infty} f(x) = L$ and let $\epsilon > 0$ be given. Then there exists $N > 0$ such that $x > N \quad \Rightarrow$ $|f(x) - L| < \epsilon$. Let $\delta = 1/N$. Then $0 < t < \delta \quad \Rightarrow \quad t < 1/N \quad \Rightarrow \quad 1/t > N \quad \Rightarrow \quad |f(1/t) - L| < \epsilon$. So $\displaystyle\lim_{t\to0^+} f(1/t) = L = \lim_{x\to\infty} f(x)$. Now suppose that $\displaystyle\lim_{x\to-\infty} f(x) = L$ and let $\epsilon > 0$ be given. Then there exists $N < 0$ such that $x < N \quad \Rightarrow \quad |f(x) - L| < \epsilon$. Let $\delta = -1/N$. Then

$$-\delta < t < 0 \quad \Rightarrow \quad t > \frac{1}{N} \quad \Rightarrow \quad \frac{1}{t} < N \quad \Rightarrow \quad \left|f\left(\frac{1}{t}\right) - L\right| < \epsilon. \text{ So } \lim_{x\to0^-} f\left(\frac{1}{t}\right) = L = \lim_{x\to-\infty} f(x).$$

EXERCISES 3.6

Abbreviations:	D	the domain of f	VA	vertical asymptote(s)
	HA	horizontal asymptote	IP	inflection point(s)
	CU	concave up	CD	concave down

1. $y = f(x) = 1 - 3x + 5x^2 - x^3$ **A.** $D = \mathbb{R}$ **B.** y-intercept $= f(0) = 1$ **C.** No symmetry **D.** No asymptotes **E.** $f'(x) = -3 + 10x - 3x^2 = -(3x-1)(x-3) > 0 \Leftrightarrow (3x-1)(x-3) < 0 \Leftrightarrow \frac{1}{3} < x < 3$. $f'(x) < 0 \Leftrightarrow x < \frac{1}{3}$ or $x > 3$. So f is increasing on $\left[\frac{1}{3}, 3\right]$ and decreasing on $\left(-\infty, \frac{1}{3}\right]$ and $[3, \infty)$. **F.** The critical numbers occur when $f'(x) = -(3x-1)(x-3) = 0 \Leftrightarrow x = \frac{1}{3}, 3$. The local minimum is $f\left(\frac{1}{3}\right) = \frac{14}{27}$ and the local maximum is $f(3) = 10$. **G.** $f''(x) = 10 - 6x > 0 \Leftrightarrow x < \frac{5}{3}$, so f is CU on $\left(-\infty, \frac{5}{3}\right)$ and CD on $\left(\frac{5}{3}, \infty\right)$. IP $\left(\frac{5}{3}, \frac{142}{27}\right)$

H.

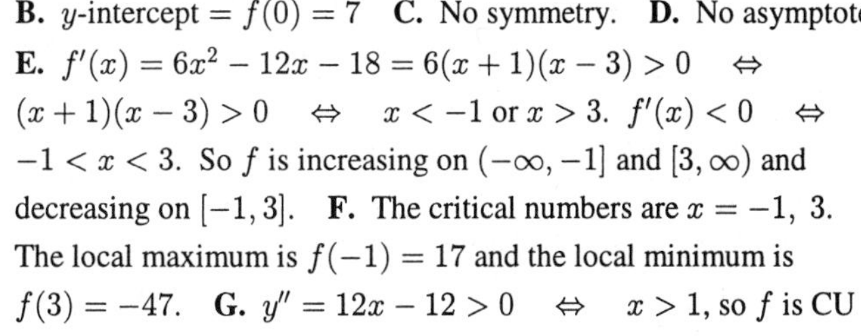

2. $y = f(x) = 2x^3 - 6x^2 - 18x + 7$ **A.** $D = \mathbb{R}$ **B.** y-intercept $= f(0) = 7$ **C.** No symmetry. **D.** No asymptotes. **E.** $f'(x) = 6x^2 - 12x - 18 = 6(x+1)(x-3) > 0 \Leftrightarrow (x+1)(x-3) > 0 \Leftrightarrow x < -1$ or $x > 3$. $f'(x) < 0 \Leftrightarrow -1 < x < 3$. So f is increasing on $(-\infty, -1]$ and $[3, \infty)$ and decreasing on $[-1, 3]$. **F.** The critical numbers are $x = -1,\ 3$. The local maximum is $f(-1) = 17$ and the local minimum is $f(3) = -47$. **G.** $y'' = 12x - 12 > 0 \Leftrightarrow x > 1$, so f is CU on $(1, \infty)$ and CD on $(-\infty, 1)$. IP $(1, -15)$

H.

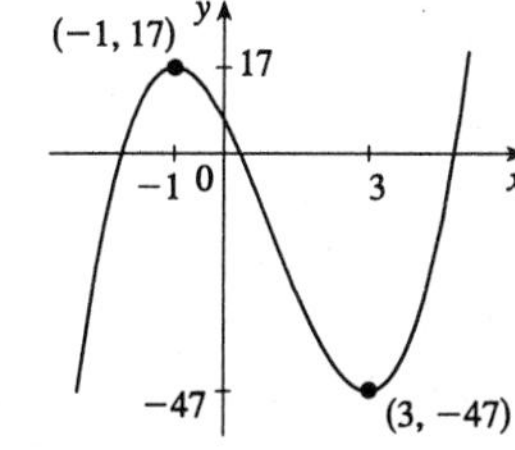

3. $y = f(x) = x^4 - 6x^2$ **A.** $D = \mathbb{R}$ **B.** y-intercept $= f(0) = 0$, x-intercepts occur when $f(x) = 0 \Rightarrow x^4 - 6x^2 = 0 \Leftrightarrow x^2(x^2 - 6) = 0 \Leftrightarrow x = 0, \pm\sqrt{6}$. **C.** Since $f(-x) = (-x)^4 - 6(-x^2) = x^4 - 6x^2 = f(x)$, f is an even function and its graph is symmetric about the y-axis. **D.** No asymptotes. **E.** $f'(x) = 4x^3 - 12x = 4x(x^2 - 3) = 0$ when $x = 0, \pm\sqrt{3}$. $f'(x) < 0$ for $x < -\sqrt{3}$ and $0 < x < \sqrt{3}$. $f'(x) > 0$ for $-\sqrt{3} < x < 0$ and $x > \sqrt{3}$, so that f is increasing on $\left[-\sqrt{3}, 0\right]$ and $\left[\sqrt{3}, \infty\right)$ and decreasing on $\left(-\infty, -\sqrt{3}\right]$ and $\left[0, \sqrt{3}\right]$. **F.** Local minima $f\left(\pm\sqrt{3}\right) = -9$, local maximum $f(0) = 0$. **G.** $f''(x) = 12x^2 - 12 = 12(x^2 - 1) > 0 \Leftrightarrow x^2 > 1 \Leftrightarrow |x| > 1 \Leftrightarrow x > 1$ or $x < -1$, so f is CU on $(-\infty, -1)$, $(1, \infty)$ and CD on $(-1, 1)$. IP $(1, -5)$ and $(-1, -5)$.

H.

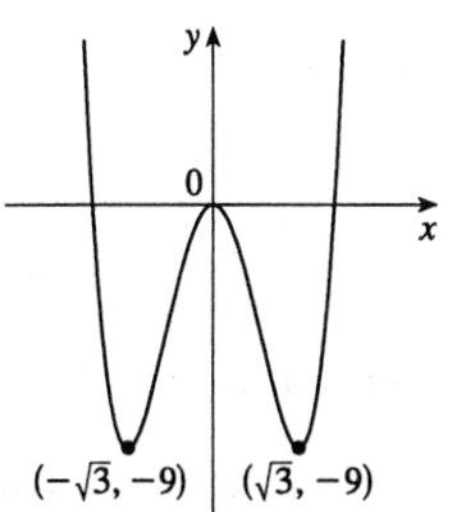

4. $y = f(x) = 4x^3 - x^4$ **A.** $D = \mathbb{R}$ **B.** y-intercept $= f(0) = 0$, x-intercept $\Rightarrow$ $y = 0$ $\Leftrightarrow$ $x^3(4 - x) = 0$ $\Leftrightarrow$ $x = 0, 4$
C. No symmetry **D.** No asymptotes
E. $y' = 12x^2 - 4x^3 = 4x^2(3 - x) > 0$ $\Leftrightarrow$ $x < 3$, so f is increasing on $(-\infty, 3]$ and decreasing on $[3, \infty)$.
F. Local maximum is $f(3) = 27$, no local minimum.
G. $y'' = 12x(2 - x) > 0$ $\Leftrightarrow$ $0 < x < 2$, so f is CU on $(0, 2)$ and CD on $(-\infty, 0)$ and $(2, \infty)$. IP $(0, 0)$ and $(2, 16)$

H.

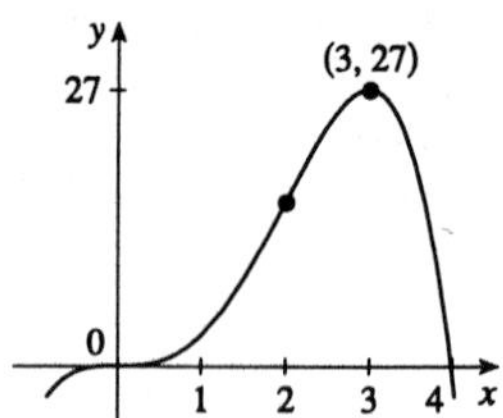

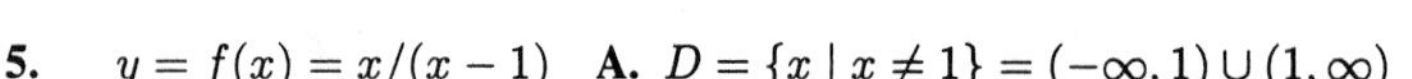

5. $y = f(x) = x/(x - 1)$ **A.** $D = \{x \mid x \neq 1\} = (-\infty, 1) \cup (1, \infty)$
B. x-intercept $= 0$, y-intercept $= f(0) = 0$ **C.** No symmetry
D. $\displaystyle\lim_{x \to \pm\infty} \frac{x}{x - 1} = 1$, so $y = 1$ is a HA. $\displaystyle\lim_{x \to 1^-} \frac{x}{x - 1} = -\infty$, $\displaystyle\lim_{x \to 1^+} \frac{x}{x - 1} = \infty$, so $x = 1$ is a VA.
E. $f'(x) = \dfrac{(x - 1) - x}{(x - 1)^2} = \dfrac{-1}{(x - 1)^2} < 0$ for $x \neq 1$, so f is decreasing on $(-\infty, 1)$ and $(1, \infty)$. **F.** No extremum **G.** $f''(x) = \dfrac{2}{(x - 1)^3} > 0$ $\Leftrightarrow$ $x > 1$, so f is CU on $(1, \infty)$ and CD on $(-\infty, 1)$. No IP

H.

6. $y = x/(x - 1)^2$ **A.** $D = \{x \mid x \neq 1\} = (-\infty, 1) \cup (1, \infty)$ **B.** x-intercept $= 0$, y-intercept $= f(0) = 0$
C. No symmetry **D.** $\displaystyle\lim_{x \to \pm\infty} \frac{x}{(x - 1)^2} = 0$, so $y = 0$ is a HA. $\displaystyle\lim_{x \to 1} \frac{x}{(x - 1)^2} = \infty$, so $x = 1$ is a VA.
E. $f'(x) = \dfrac{(x - 1)^2(1) - x(2)(x - 1)}{(x - 1)^4} = \dfrac{-x - 1}{(x - 1)^3}$. This is negative on $(-\infty, -1)$ and $(1, \infty)$ and positive on $(-1, 1)$, so $f(x)$ is decreasing on $(-\infty, -1]$ and $(1, \infty)$ and increasing on $[-1, 1)$. **F.** Local minimum $f(-1) = -\frac{1}{4}$, no local maximum.
G. $f''(x) = \dfrac{(x - 1)^3(-1) + (x + 1)(3)(x - 1)^2}{(x - 1)^6} = \dfrac{2(x + 2)}{(x - 1)^4}$.
This is negative on $(-\infty, -2)$, and positive on $(-2, 1)$ and $(1, \infty)$. So f is CD on $(-\infty, -2)$ and CU on $(-2, 1)$ and $(1, \infty)$. f has an inflection point at $\left(-2, -\frac{2}{9}\right)$

H.

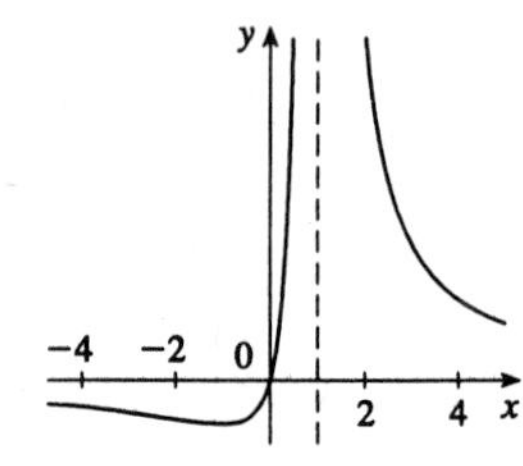

7. $y = f(x) = 1/(x^2 - 9)$ **A.** $D = \{x \mid x \neq \pm 3\} = (-\infty, -3) \cup (-3, 3) \cup (3, \infty)$
B. y-intercept $= f(0) = -\frac{1}{9}$, no x-intercept **C.** $f(-x) = f(x)$ $\Rightarrow$ f is even; the curve is symmetric about the y-axis. **D.** $\displaystyle\lim_{x \to \pm\infty} \frac{1}{x^2 - 9} = 0$, so $y = 0$ is a HA. $\displaystyle\lim_{x \to 3^-} \frac{1}{x^2 - 9} = -\infty$, $\displaystyle\lim_{x \to 3^+} \frac{1}{x^2 - 9} = \infty$, $\displaystyle\lim_{x \to -3^-} \frac{1}{x^2 - 9} = \infty$, $\displaystyle\lim_{x \to -3^+} \frac{1}{x^2 - 9} = -\infty$, so $x = 3$ and $x = -3$ are VA. **E.** $f'(x) = -\dfrac{2x}{(x^2 - 9)^2} > 0$ $\Leftrightarrow$ $x < 0$ $(x \neq -3)$ so f is increasing on $(-\infty, -3)$ and $(-3, 0]$ and decreasing on $[0, 3)$ and $(3, \infty)$.

F. Local maximum $f(0) = -\frac{1}{9}$.

G. $y'' = \dfrac{-2(x^2-9)^2 + (2x)\,2\,(x^2-9)\,(2x)}{(x^2-9)^4}$

$= \dfrac{6(x^2+3)}{(x^2-9)^3} > 0 \quad \Leftrightarrow \quad x^2 > 9 \quad \Leftrightarrow$

$x > 3$ or $x < -3$, so f is CU on $(-\infty, -3)$ and $(3, \infty)$ and CD on $(-3, 3)$. No IP

H.

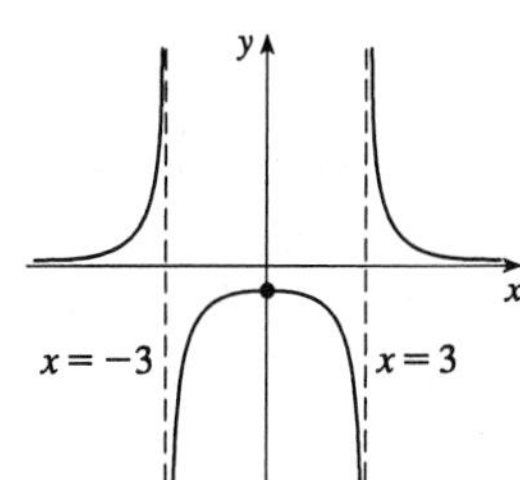

8. $y = f(x) = x/(x^2 - 9)$ **A.** $D = \{x \mid x \neq \pm 3\} = (-\infty, -3) \cup (-3, 3) \cup (3, \infty)$ **B.** x-intercept $= 0$, y-intercept $= f(0) = 0$. **C.** $f(-x) = -f(x)$, so f is odd; the curve is symmetric about the origin.
D. $\lim\limits_{x\to\pm\infty} \dfrac{x}{x^2-9} = 0$, so $y = 0$ is a HA. $\lim\limits_{x\to 3^+} \dfrac{x}{x^2-9} = \infty$, $\lim\limits_{x\to 3^-} \dfrac{x}{x^2-9} = -\infty$, $\lim\limits_{x\to -3^+} \dfrac{x}{x^2-9} = \infty$,

$\lim\limits_{x\to -3^-} \dfrac{x}{x^2-9} = -\infty$, so $x = 3$ and $x = -3$ are VA.

E. $f'(x) = \dfrac{(x^2-9) - x\,(2x)}{(x^2-9)^2} = -\dfrac{x^2+9}{(x^2-9)^2} < 0 \ (x \neq \pm 3)$

so f is decreasing on $(-\infty, -3)$, $(-3, 3)$, and $(3, \infty)$.

F. No extremum

G. $f''(x) = -\dfrac{2x(x^2-9)^2 - (x^2+9)\cdot 2(x^2-9)(2x)}{(x^2-9)^4}$

$= \dfrac{2x\,(x^2+27)}{(x^2-9)^3} > 0$ when $-3 < x < 0$ or $x > 3$,

so f is CU on $(-3, 0)$ and $(3, \infty)$; CD on $(-\infty, -3)$ and $(0, 3)$. IP is $(0, 0)$.

H.

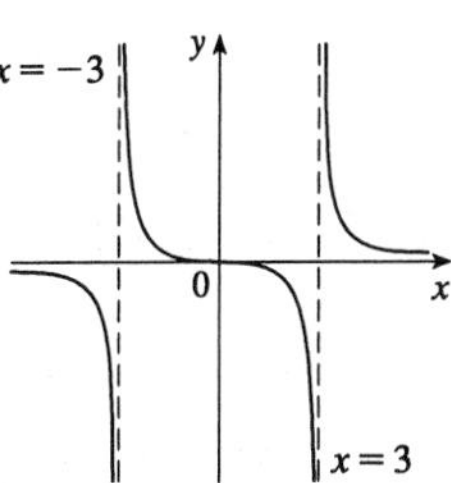

9. $y = f(x) = \dfrac{1}{(x-1)(x+2)}$ **A.** $D = \{x \mid x \neq 1, -2\} = (-\infty, -2) \cup (-2, 1) \cup (1, \infty)$ **B.** No x-intercept, y-intercept $= f(0) = -\frac{1}{2}$ **C.** No symmetry **D.** $\lim\limits_{x\to\pm\infty} \dfrac{1}{(x-1)(x+2)} = 0$, so $y = 0$ is a HA.

$\lim\limits_{x\to 1^-} \dfrac{1}{(x-1)(x+2)} = -\infty$, $\lim\limits_{x\to 1^+} \dfrac{1}{(x-1)(x+2)} = \infty$, $\lim\limits_{x\to -2^-} \dfrac{1}{(x-1)(x+2)} = \infty$,

$\lim\limits_{x\to -2^+} \dfrac{1}{(x-1)(x+2)} = -\infty$. So $x = 1$ and $x = -2$ are VA.

E. $f'(x) = -\dfrac{2x+1}{[(x-1)(x+2)]^2} \quad \Rightarrow \quad f'(x) > 0 \quad \Leftrightarrow$

$x < -\frac{1}{2}$ $(x \neq -2)$, so f is increasing on $(-\infty, -2)$ and $\left(-2, -\frac{1}{2}\right]$ and decreasing on $\left[-\frac{1}{2}, 1\right)$ and $(1, \infty)$. **F.** $f\left(-\frac{1}{2}\right) = -\frac{4}{9}$ is a local maximum. **G.** $f''(x) = \dfrac{6(x^2+x+1)}{[(x-1)(x+2)]^3}$. Now $x^2 + x + 1 > 0$ for all x, so $f''(x) > 0 \ \Leftrightarrow \ (x-1)(x+2) > 0 \ \Leftrightarrow \ x < -2$ or $x > 1$. Thus f is CU on $(-\infty, -2)$ and $(1, \infty)$ and CD on $(-2, 1)$. No IP

H.

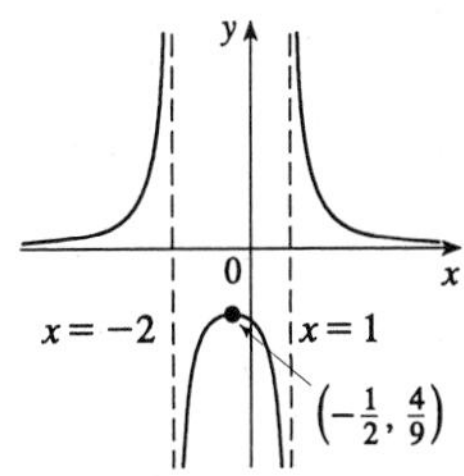

10. $y = f(x) = \dfrac{1}{x^2(x+3)}$ **A.** $D = \{x \mid x \neq 0, -3\} = (-\infty, -3) \cup (-3, 0) \cup (0, \infty)$ **B.** No intercepts

C. No symmetry **D.** $\lim\limits_{x \to \pm\infty} \dfrac{1}{x^2(x+3)} = 0$, so $y = 0$ is a HA. $\lim\limits_{x \to 0} \dfrac{1}{x^2(x+3)} = \infty$ and

$\lim\limits_{x \to -3^+} \dfrac{1}{x^2(x+3)} = \infty$, $\lim\limits_{x \to -3^-} \dfrac{1}{x^2(x+3)} = -\infty$, so $x = 0$ and $x = -3$ are VA. **E.** $f'(x) = -\dfrac{3(x+2)}{x^3(x+3)^2} > 0$

$\Leftrightarrow \quad -2 < x < 0$; $f'(x) < 0 \quad \Leftrightarrow \quad x < -2$ or $x > 0$. So f is increasing on $[-2, 0)$ and decreasing on $(-\infty, -3)$, $(-3, -2]$, and $(0, \infty)$.

F. $f(-2) = \frac{1}{4}$ is a local minimum.

H.

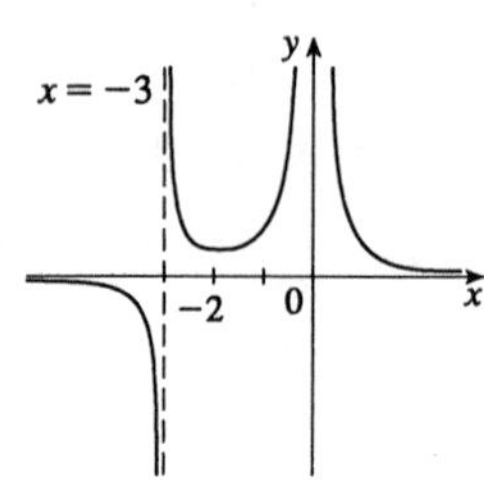

G. $f''(x) = -3\dfrac{x^3(x+3)^2 - (x+2)[3x^2(x+3)^2 + x^3 2(x+3)]}{x^6(x+3)^4}$

$= \dfrac{6(2x^2 + 8x + 9)}{x^4(x+3)^3}$. Since $2x^2 + 8x + 9 > 0$ for all x,

$f''(x) > 0 \quad \Leftrightarrow \quad x > -3 \ (x \neq 0)$, so f is CU on $(-3, 0)$ and $(0, \infty)$, and CD on $(-\infty, -3)$.

11. $y = f(x) = \dfrac{1 + x^2}{1 - x^2} = -1 + \dfrac{2}{1 - x^2}$ **A.** $D = \{x \mid x \neq \pm 1\}$ **B.** No x-intercept, y-intercept $= f(0) = 1$

C. $f(-x) = f(x)$, so f is even and the curve is symmetric about the y-axis.

D. $\lim\limits_{x \to \pm\infty} \dfrac{1 + x^2}{1 - x^2} = \lim\limits_{x \to \pm\infty} \dfrac{(1/x^2) + 1}{(1/x^2) - 1} = -1$, so $y = -1$ is a HA. $\lim\limits_{x \to 1^-} \dfrac{1 + x^2}{1 - x^2} = \infty$, $\lim\limits_{x \to 1^+} \dfrac{1 + x^2}{1 - x^2} = -\infty$,

$\lim\limits_{x \to -1^-} \dfrac{1 + x^2}{1 - x^2} = -\infty$, $\lim\limits_{x \to -1^+} \dfrac{1 + x^2}{1 - x^2} = \infty$. So $x = 1$ and $x = -1$ are VA.

E. $f'(x) = \dfrac{4x}{(1 - x^2)^2} > 0 \quad \Leftrightarrow \quad x > 0 \ (x \neq 1)$, so f increases on $[0, 1)$, $(1, \infty)$ and decreases on $(-\infty, -1)$, $(-1, 0]$.

F. $f(0) = 1$ is a local minimum.

H.

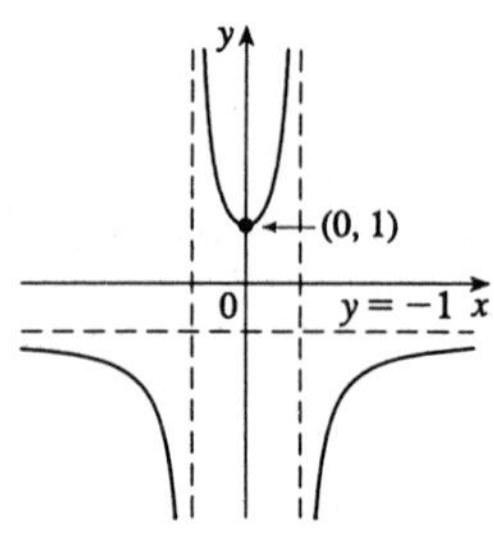

G. $y'' = \dfrac{4(1 - x^2)^2 - 4x \cdot 2(1 - x^2)(-2x)}{(1 - x^2)^4} = \dfrac{4(1 + 3x^2)}{(1 - x^2)^3} > 0 \quad \Leftrightarrow$

$x^2 < 1 \quad \Leftrightarrow \quad -1 < x < 1$, so f is CU on $(-1, 1)$ and CD on $(-\infty, -1)$ and $(1, \infty)$. No IP

12. $y = f(x) = \dfrac{x^3 - 1}{x^3 + 1}$ **A.** $D = \{x \mid x \neq -1\} = (-\infty, -1) \cup (-1, \infty)$ **B.** x-intercept $= 1$,

y-intercept $= f(0) = -1$ **C.** No symmetry **D.** $\lim\limits_{x \to \pm\infty} \dfrac{x^3 - 1}{x^3 + 1} = \lim\limits_{x \to \pm\infty} \dfrac{1 - 1/x^3}{1 + 1/x^3} = 1$, so $y = 1$ is a HA.

$\lim\limits_{x \to -1^-} \dfrac{x^3 - 1}{x^3 + 1} = \infty$ and $\lim\limits_{x \to -1^+} \dfrac{x^3 - 1}{x^3 + 1} = -\infty$, so $x = -1$ is a VA.

E. $f'(x) = \dfrac{(x^3 + 1)(3x^2) - (x^3 - 1)(3x^2)}{(x^3 + 1)^2} = \dfrac{6x^2}{(x^3 + 1)^2} > 0 \ (x \neq -1)$ so f is increasing on $(-\infty, -1)$ and $(-1, \infty)$. **F.** No extremum

G. $y'' = \dfrac{12x(x^3+1)^2 - 6x^2 \cdot 2\,(x^3+1) \cdot 3x^2}{(x^3+1)^4}$

$= \dfrac{12x\,(1-2x^3)}{(x^3+1)^3} > 0 \quad \Leftrightarrow \quad x < -1 \text{ or } 0 < x < \frac{1}{\sqrt[3]{2}},$

so f is CU on $(-\infty, -1)$ and $\left(0, \frac{1}{\sqrt[3]{2}}\right)$

and CD on $(-1, 0)$ and $\left(\frac{1}{\sqrt[3]{2}}, \infty\right)$. IP $(0, -1)$, $\left(\frac{1}{\sqrt[3]{2}}, -\frac{1}{3}\right)$

H.

y

$y=1$

0

$(1,0)$

x

$(0,-1)$

$x=-1$

13. $y = f(x) = \dfrac{1}{x^3 - x} = \dfrac{1}{x(x-1)(x+1)}$ **A.** $D = \{x \mid x \neq 0, \pm 1\}$ **B.** No intercepts **C.** $f(-x) = -f(x)$, symmetric about $(0,0)$ **D.** $\lim\limits_{x\to\pm\infty} \dfrac{1}{x^3-x} = 0$, so $y = 0$ is a HA. $\lim\limits_{x\to 0^-} \dfrac{1}{x^3-x} = \infty$, $\lim\limits_{x\to 0^+} \dfrac{1}{x^3-x} = -\infty$, $\lim\limits_{x\to 1^-} \dfrac{1}{x^3-x} = -\infty$, $\lim\limits_{x\to 1^+} \dfrac{1}{x^3-x} = \infty$, $\lim\limits_{x\to -1^-} \dfrac{1}{x^3-x} = -\infty$, $\lim\limits_{x\to -1^+} \dfrac{1}{x^3-x} = \infty$. So $x = 0$, $x = 1$, and $x = -1$ are VA. **E.** $f'(x) = \dfrac{1-3x^2}{(x^3-x)^2} \quad \Rightarrow \quad f'(x) > 0 \quad \Leftrightarrow \quad x^2 < \frac{1}{3} \quad \Leftrightarrow \quad -\frac{1}{\sqrt{3}} < x < \frac{1}{\sqrt{3}}$ $(x \neq 0)$, so f is increasing on $\left[-\frac{1}{\sqrt{3}}, 0\right)$, $\left(0, \frac{1}{\sqrt{3}}\right]$ and decreasing on $(-\infty, -1)$, $\left(-1, -\frac{1}{\sqrt{3}}\right]$, $\left[\frac{1}{\sqrt{3}}, 1\right)$, and $(1, \infty)$.

F. Local minimum $f\left(-\frac{1}{\sqrt{3}}\right) = \frac{3\sqrt{3}}{2}$, local maximum $f\left(\frac{1}{\sqrt{3}}\right) = -\frac{3\sqrt{3}}{2}$ **G.** $f''(x) = \dfrac{2(6x^4 - 3x^2 + 1)}{(x^3-x)^3}$.

Since $6x^4 - 3x^2 + 1$ has negative discriminant as a quadratic in x^2, it is positive, so $f''(x) > 0 \quad \Leftrightarrow \quad x^3 - x > 0 \quad \Leftrightarrow \quad x > 1$ or $-1 < x < 0$. f is CU on $(-1, 0)$ and $(1, \infty)$, and CD on $(-\infty, -1)$ and $(0, 1)$. No IP

H.

y

$x=-1$

0

$x=1$

x

14. $y = f(x) = \dfrac{1-x^2}{x^3} = \dfrac{1}{x^3} - \dfrac{1}{x}$ **A.** $D = \{x \mid x \neq 0\}$ **B.** x-intercepts ± 1, no y-intercepts **C.** $f(-x) = -f(x)$, so the curve is symmetric about $(0,0)$. **D.** $\lim\limits_{x\to\pm\infty} \dfrac{1-x^2}{x^3} = 0$, so $y = 0$ is a HA. $\lim\limits_{x\to 0^+} \dfrac{1-x^2}{x^3} = \infty$, $\lim\limits_{x\to 0^-} \dfrac{1-x^2}{x^3} = -\infty$, so $x = 0$ is a VA. **E.** $f'(x) = -\dfrac{3}{x^4} + \dfrac{1}{x^2} = \dfrac{x^2-3}{x^4} > 0$ $\Leftrightarrow \quad |x| > \sqrt{3}$, so f is increasing on $\left(-\infty, -\sqrt{3}\right]$, $\left[\sqrt{3}, \infty\right)$ and decreasing on $\left[-\sqrt{3}, 0\right)$ and $\left(0, \sqrt{3}\right]$. **F.** $f\left(\sqrt{3}\right) = -\frac{2}{3\sqrt{3}}$ is a local minimum, $f\left(-\sqrt{3}\right) = \frac{2}{3\sqrt{3}}$ is a local maximum.

G. $f''(x) = \dfrac{12}{x^5} - \dfrac{2}{x^3} = \dfrac{2(6-x^2)}{x^5} > 0 \quad \Leftrightarrow \quad x < -\sqrt{6}$ or $0 < x < \sqrt{6}$, so f is CU on $\left(-\infty, -\sqrt{6}\right)$, $\left(0, \sqrt{6}\right)$ and CD on $\left(-\sqrt{6}, 0\right)$ and $\left(\sqrt{6}, \infty\right)$. IP $\left(\sqrt{6}, -\frac{5}{6\sqrt{6}}\right)$ and $\left(-\sqrt{6}, \frac{5}{6\sqrt{6}}\right)$.

H.

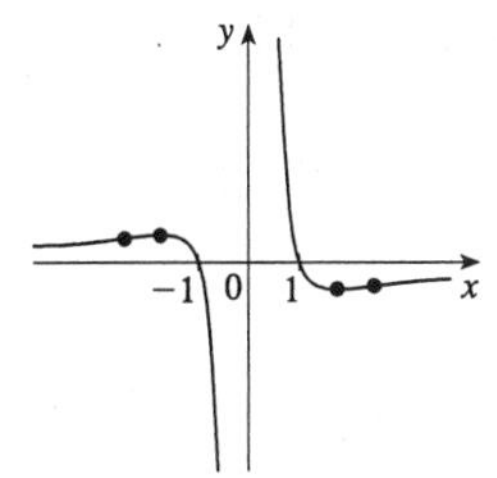

15. $y = f(x) = x\sqrt{x+3}$ **A.** $D = \{x \mid x \geq -3\} = [-3, \infty)$ **B.** x-intercepts $0, -3$, y-intercept $= f(0) = 0$ **C.** No symmetry **D.** $\lim_{x\to\infty} \sqrt{x+3} = \infty$, no asymptotes **E.** $f'(x) = \sqrt{x+3} + \dfrac{x}{2\sqrt{x+3}} = \dfrac{3(x+2)}{2\sqrt{x+3}} > 0$ $\Leftrightarrow$ $x > -2$ and $f'(x) < 0$ $\Leftrightarrow$ $-3 < x < -2$.
So f is increasing on $[-2, \infty)$, decreasing on $[-3, -2]$.
F. $f(-2) = -2$ is a local minimum.
G. $f''(x) = \dfrac{6\sqrt{x+3} - 3(x+2)(1/\sqrt{x+3})}{4(x+3)} = \dfrac{3(x+4)}{4(x+3)^{3/2}} > 0$
for all $x > -3$, so f is CU on $(-3, \infty)$.

H.

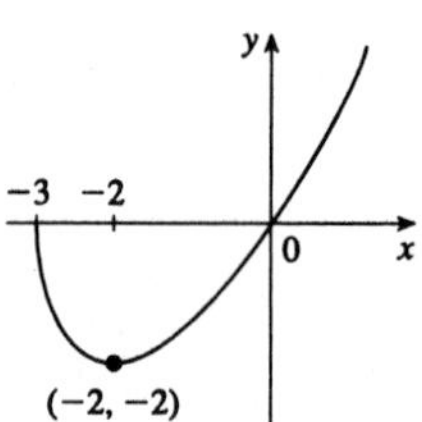

16. $y = f(x) = \sqrt{x} - \sqrt{x-1}$ **A.** $D = \{x \mid x \geq 0 \text{ and } x \geq 1\} = \{x \mid x \geq 1\} = [1, \infty)$ **B.** No intercepts **C.** No symmetry **D.** $\lim_{x\to\infty}\left(\sqrt{x} - \sqrt{x-1}\right) = \lim_{x\to\infty}\left(\sqrt{x} - \sqrt{x-1}\right)\dfrac{\sqrt{x}+\sqrt{x-1}}{\sqrt{x}+\sqrt{x-1}} = \lim_{x\to\infty}\dfrac{1}{\sqrt{x}+\sqrt{x-1}}$ $= 0$, so $y = 0$ is a HA. **E.** $f'(x) = \dfrac{1}{2\sqrt{x}} - \dfrac{1}{2\sqrt{x-1}} < 0$ for all $x > 1$, since $x - 1 < x$ $\Rightarrow$ $\sqrt{x-1} < \sqrt{x}$, so f is decreasing on $[1, \infty)$. **F.** No local extremum.
G. $f''(x) = -\dfrac{1}{4}\left[\dfrac{1}{x^{3/2}} - \dfrac{1}{(x-1)^{3/2}}\right]$ $\Rightarrow$ $f''(x) > 0$ for $x > 1$,
so f is CU on $(1, \infty)$.

H.

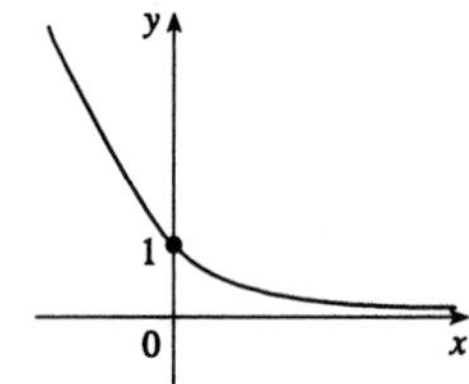

17. $y = f(x) = \sqrt{x^2+1} - x$ **A.** $D = \mathbb{R}$ **B.** No x-intercept, y-intercept $= 1$ **C.** No symmetry
D. $\lim_{x\to-\infty}\left(\sqrt{x^2+1} - x\right) = \infty$ and $\lim_{x\to\infty}\left(\sqrt{x^2+1} - x\right) = \lim_{x\to\infty}\left(\sqrt{x^2+1} - x\right)\dfrac{\sqrt{x^2+1}+x}{\sqrt{x^2+1}+x}$
$= \lim_{x\to\infty}\dfrac{1}{\sqrt{x^2+1}+x} = 0$, so $y = 0$ is a HA.
E. $f'(x) = \dfrac{x}{\sqrt{x^2+1}} - 1 = \dfrac{x - \sqrt{x^2+1}}{\sqrt{x^2+1}}$ $\Rightarrow$ $f'(x) < 0$, so
f is decreasing on $\mathbb{R}$. **F.** No extremum
G. $f''(x) = \dfrac{1}{(x^2+1)^{3/2}} > 0$, so f is CU on $\mathbb{R}$. No IP

H.

18. $y = f(x) = \sqrt{x/(x-5)}$ **A.** $D = \{x \mid x/(x-5) \geq 0\} = (-\infty, 0] \cup (5, \infty)$. **B.** Intercepts are 0
C. No symmetry **D.** $\lim_{x\to\pm\infty}\sqrt{\dfrac{x}{x-5}} = \lim_{x\to\pm\infty}\sqrt{\dfrac{1}{1-5/x}} = 1$, so $y = 1$ is a HA. $\lim_{x\to5^+}\sqrt{\dfrac{x}{x-5}} = \infty$, so $x = 5$ is a VA. **E.** $f'(x) = \dfrac{1}{2}\left(\dfrac{x}{x-5}\right)^{-1/2}\dfrac{(-5)}{(x-5)^2} = -\frac{5}{2}\left[x(x-5)^3\right]^{-1/2} < 0$, so f is decreasing on $(-\infty, 0]$ and $(5, \infty)$. **F.** No local extremum
G. $f''(x) = \frac{5}{4}\left[x(x-5)^3\right]^{-3/2}(x-5)^2(4x-5) > 0$ for $x > 5$,
and $f''(x) < 0$ for $x < 0$, so f is CU on $(5, \infty)$ and
CD on $(-\infty, 0)$. No IP

H.

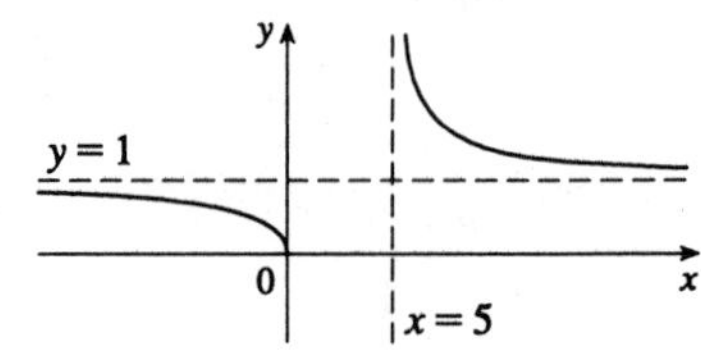

19. $y = f(x) = \sqrt[4]{x^2 - 25}$ **A.** $D = \{x \mid x^2 \geq 25\} = (-\infty, -5] \cup [5, \infty)$ **B.** x-intercepts are ± 5, no y-intercept **C.** $f(-x) = f(x)$, so the curve is symmetric about the y-axis.

D. $\lim\limits_{x \to \pm\infty} \sqrt[4]{x^2 - 25} = \infty$, no asymptotes

E. $f'(x) = \frac{1}{4}(x^2 - 25)^{-3/4}(2x) = \dfrac{x}{2(x^2 - 25)^{3/4}} > 0$ if $x > 5$, so f is increasing on $[5, \infty)$ and decreasing on $(-\infty, -5]$.

F. No local extremum

G. $y'' = \dfrac{2(x^2 - 25)^{3/4} - 3x^2(x^2 - 25)^{-1/4}}{4\,(x^2 - 25)^{3/2}} = -\dfrac{x^2 + 50}{4(x^2 - 25)^{7/4}} < 0$, so f is CD on $(-\infty, -5)$ and $(5, \infty)$. No IP

H.

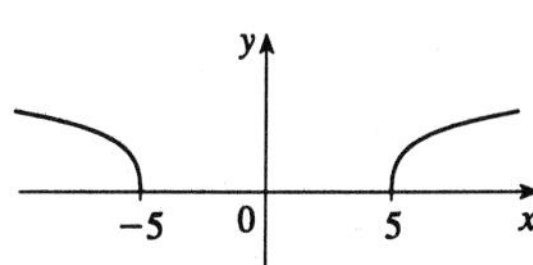

20. $y = f(x) = x\sqrt{x^2 - 9}$ **A.** $D = \{x \mid x^2 \geq 9\} = (-\infty, -3] \cup [3, \infty)$ **B.** x-intercepts are ± 3, no y-intercept. **C.** $f(-x) = -f(x)$, so the curve is symmetric about the origin. **D.** $\lim\limits_{x \to \infty} \sqrt{x^2 - 9} = \infty$, $\lim\limits_{x \to -\infty} \sqrt{x^2 - 9} = -\infty$, no asymptotes

E. $f'(x) = \sqrt{x^2 - 9} + \dfrac{x^2}{\sqrt{x^2 - 9}} > 0$ for $x \in D$, so f is increasing on $(-\infty, -3]$ and $[3, \infty)$. **F.** No extremum

G. $f''(x) = \dfrac{x}{\sqrt{x^2 - 9}} + \dfrac{2x\sqrt{x^2 - 9} - x^2\left(x/\sqrt{x^2 - 9}\right)}{x^2 - 9}$
$= \dfrac{x(2x^2 - 27)}{(x^2 - 9)^{3/2}} > 0 \quad \Leftrightarrow \quad x > 3\sqrt{\frac{3}{2}}$ or $-3\sqrt{\frac{3}{2}} < x < 0$,

so f is CU on $\left(3\sqrt{\frac{3}{2}}, \infty\right)$ and $\left(-3\sqrt{\frac{3}{2}}, -3\right)$ and CD on $\left(-\infty, -3\sqrt{\frac{3}{2}}\right)$ and $\left(3, 3\sqrt{\frac{3}{2}}\right)$. IP $\left(\pm 3\sqrt{\frac{3}{2}}, \pm\frac{9\sqrt{3}}{2}\right)$

H.

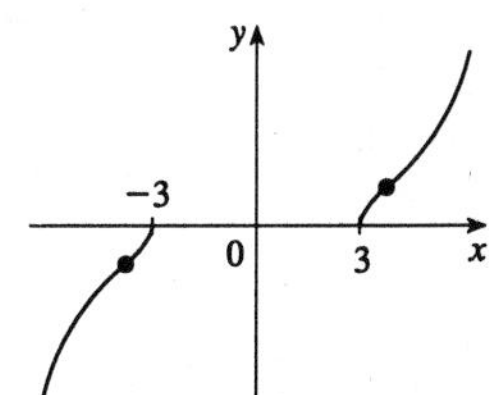

21. $y = f(x) = \dfrac{\sqrt{1 - x^2}}{x}$ **A.** $D = \{x \mid |x| \leq 1, x \neq 0\} = [-1, 0) \cup (0, 1]$ **B.** x-intercepts ± 1, no y-intercept

C. $f(-x) = -f(x)$, so the curve is symmetric about $(0, 0)$. **D.** $\lim\limits_{x \to 0^+} \dfrac{\sqrt{1 - x^2}}{x} = \infty$, $\lim\limits_{x \to 0^-} \dfrac{\sqrt{1 - x^2}}{x} = -\infty$, so $x = 0$ is a VA.

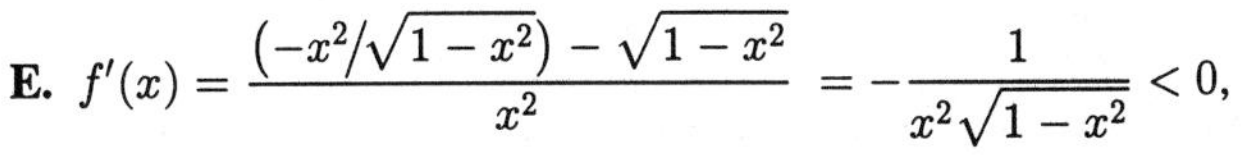

E. $f'(x) = \dfrac{\left(-x^2/\sqrt{1 - x^2}\right) - \sqrt{1 - x^2}}{x^2} = -\dfrac{1}{x^2\sqrt{1 - x^2}} < 0$, so f is decreasing on $[-1, 0)$ and $(0, 1]$. **F.** No extremum

G. $f''(x) = \dfrac{2 - 3x^2}{x^3(1 - x^2)^{3/2}} > 0 \quad \Leftrightarrow \quad -1 < x < -\sqrt{\frac{2}{3}}$ or $0 < x < \sqrt{\frac{2}{3}}$, so f is CU on $\left(-1, -\sqrt{\frac{2}{3}}\right)$ and $\left(0, \sqrt{\frac{2}{3}}\right)$ and CD on $\left(-\sqrt{\frac{2}{3}}, 0\right)$ and $\left(\sqrt{\frac{2}{3}}, 1\right)$. IP $\left(\pm\sqrt{\frac{2}{3}}, \pm\frac{1}{\sqrt{2}}\right)$.

H.

y
−1
0
1
x

22. $y = f(x) = \dfrac{x+1}{\sqrt{x^2+1}}$ **A.** $D = \mathbb{R}$ **B.** x-intercept -1, y-intercept 1 **C.** No symmetry

D. $\lim\limits_{x\to\infty} \dfrac{x+1}{\sqrt{x^2+1}} = 1$, and $\lim\limits_{x\to-\infty} \dfrac{x+1}{\sqrt{x^2+1}} = -1$, so horizontal asymptotes are $y = \pm 1$.

E. $f'(x) = \dfrac{\sqrt{x^2+1} - \dfrac{1}{2\sqrt{x^2+1}}(2x)(x+1)}{(x^2+1)} = \dfrac{1-x}{(x^2+1)^{3/2}} > 0 \quad \Leftrightarrow \quad x < 1$, so f is increasing on $(-\infty, 1]$, and decreasing on $[1, \infty)$. **F.** $f(1) = \sqrt{2}$ is a local maximum.

G. $f''(x) = \dfrac{-1(x^2+1)^{3/2} - \frac{3}{2}(x^2+1)^{1/2}(2x)(1-x)}{(x^2+1)^3}$

$= \dfrac{2x^2 - 3x - 1}{(x^2+1)^{5/2}}.$

H.

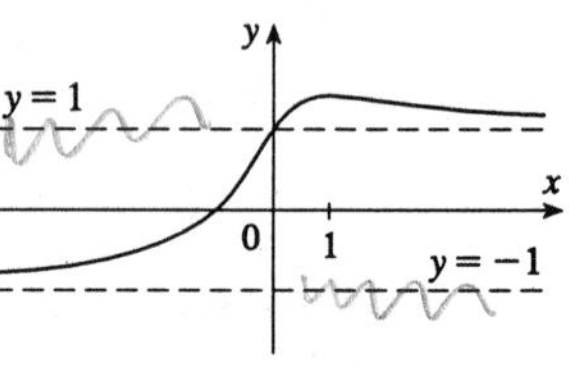

$f''(x) = 0 \quad \Leftrightarrow \quad 2x^2 - 3x - 1 = 0$

$\Leftrightarrow \quad x = \dfrac{3 \pm \sqrt{9 - 4(2)(-1)}}{2(2)} = \dfrac{3 \pm \sqrt{17}}{4}$. $f(x)$ is CU on $\left(-\infty, \frac{3-\sqrt{17}}{4}\right)$ and $\left(\frac{3+\sqrt{17}}{4}, \infty\right)$ and CD on $\left(\frac{3-\sqrt{17}}{4}, \frac{3+\sqrt{17}}{4}\right)$. IP at $x = \frac{3\pm\sqrt{17}}{4}$

23. $y = f(x) = x + 3x^{2/3}$ **A.** $D = \mathbb{R}$ **B.** $y = x + 3x^{2/3} = x^{2/3}(x^{1/3} + 3) = 0$ if $x = 0$ or -27 (x-intercepts), y-intercept $= f(0) = 0$ **C.** No symmetry **D.** $\lim\limits_{x\to\infty} (x + 3x^{2/3}) = \infty$, $\lim\limits_{x\to-\infty} (x + 3x^{2/3}) = \lim\limits_{x\to-\infty} x^{2/3}(x^{1/3} + 3) = -\infty$, no asymptotes

H.

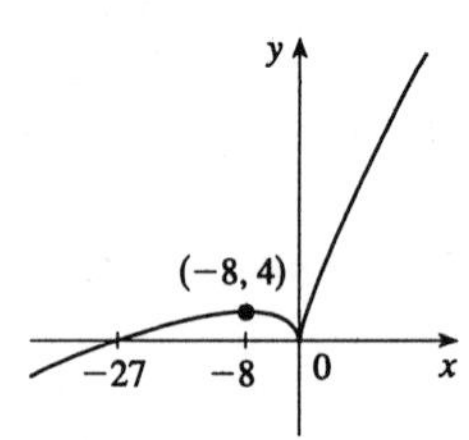

E. $f'(x) = 1 + 2x^{-1/3} = (x^{1/3} + 2)/x^{1/3} > 0$ $\Leftrightarrow \quad x > 0$ or $x < -8$, so f increases on $(-\infty, -8]$, $[0, \infty)$ and decreases on $[-8, 0]$. **F.** Local maximum $f(-8) = 4$, local minimum $f(0) = 0$ **G.** $f''(x) = -\frac{2}{3}x^{-4/3} < 0$ $(x \neq 0)$ so f is CD on $(-\infty, 0)$ and $(0, \infty)$. No IP

24. $y = f(x) = x^{5/3} - 5x^{2/3} = x^{2/3}(x - 5)$ **A.** $D = \mathbb{R}$ **B.** x-intercepts 0, 5, y-intercept 0 **C.** No symmetry **D.** $\lim\limits_{x\to\pm\infty} x^{2/3}(x-5) = \pm\infty$, so there is no asymptote

H.

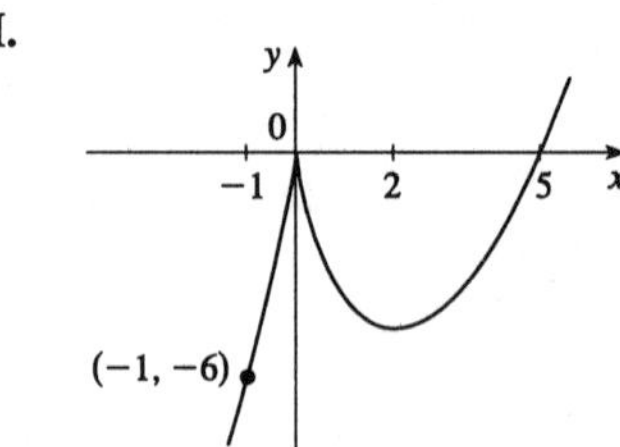

E. $f'(x) = \frac{5}{3}x^{2/3} - \frac{10}{3}x^{-1/3} = \frac{5}{3}x^{-1/3}(x - 2) > 0 \quad \Leftrightarrow$ $x < 0$ or $x > 2$, so f is increasing on $(-\infty, 0]$, $[2, \infty)$ and decreasing on $[0, 2]$. **F.** $f(0) = 0$ is a local maximum. $f(2) = -3\sqrt[3]{4}$ is a local minimum

G. $f''(x) = \frac{10}{9}x^{-1/3} + \frac{10}{9}x^{-4/3} = \frac{10}{9}x^{-4/3}(x + 1) > 0 \quad \Leftrightarrow$ $x > -1$, so f is CU on $(-1, 0)$ and $(0, \infty)$, CD on $(-\infty, -1)$. IP $(-1, -6)$

25. $y = f(x) = x + \sqrt{|x|}$ **A.** $D = \mathbb{R}$ **B.** x-intercepts $= 0, -1$, y-intercept 0 **C.** No symmetry **D.** $\lim\limits_{x\to\infty}\left(x + \sqrt{|x|}\right) = \infty$, $\lim\limits_{x\to-\infty}\left(x + \sqrt{|x|}\right) = -\infty$. No asymptotes **E.** For $x > 0$, $f(x) = x + \sqrt{x}$ $\Rightarrow$ $f'(x) = 1 + \dfrac{1}{2\sqrt{x}} > 0$, so f increases on $[0, \infty)$.

For $x < 0$, $f(x) = x + \sqrt{-x}$ $\Rightarrow$ $f'(x) = 1 - \dfrac{1}{2\sqrt{-x}} > 0$ $\Leftrightarrow$ $2\sqrt{-x} > 1$ $\Leftrightarrow$ $-x > \frac{1}{4}$ $\Leftrightarrow$ $x < -\frac{1}{4}$, so f increases on $\left(-\infty, -\frac{1}{4}\right]$ and decreases on $\left[-\frac{1}{4}, 0\right]$.

F. $f\left(-\frac{1}{4}\right) = \frac{1}{4}$ is a local maximum, $f(0) = 0$ is a local minimum. **G.** For $x > 0$, $f''(x) = -\frac{1}{4}x^{-3/2}$ $\Rightarrow$ $f''(x) < 0$, so f is CD on $(0, \infty)$. For $x < 0$, $f''(x) = -\frac{1}{4}(-x)^{-3/2}$ $\Rightarrow$ $f''(x) < 0$, so f is CD on $(-\infty, 0)$. No IP

H.

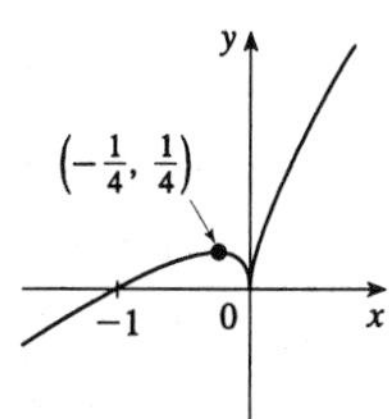

26. $y = f(x) = (x^2 - 1)^{2/3}$ **A.** $D = \mathbb{R}$ **B.** x-intercepts ± 1, y-intercept 1 **C.** $f(-x) = f(x)$, so the curve is symmetric about the y-axis. **D.** $\lim\limits_{x\to\pm\infty}(x^2 - 1)^{2/3} = \infty$, no asymptotes **E.** $f'(x) = \frac{4}{3}x(x^2 - 1)^{-1/3}$ $\Rightarrow$ $f'(x) > 0$ $\Leftrightarrow$ $x > 1$ or $-1 < x < 0$, $f'(x) < 0$ $\Leftrightarrow$ $x < -1$ or $0 < x < 1$. So f is increasing on $[-1, 0]$, $[1, \infty)$ and decreasing on $(-\infty, -1]$, $[0, 1]$.

F. $f(-1) = f(1) = 0$ are local minima, $f(0) = 1$ is a local maximum.

G. $f''(x) = \frac{4}{3}(x^2 - 1)^{-1/3} + \frac{4}{3}x\left(-\frac{1}{3}\right)(x^2 - 1)^{-4/3}(2x)$ $= \frac{4}{9}(x^2 - 3)(x^2 - 1)^{-4/3} > 0$ $\Leftrightarrow$ $|x| > \sqrt{3}$, so f is CU on $\left(-\infty, -\sqrt{3}\right)$, $\left(\sqrt{3}, \infty\right)$ and CD on $\left(-\sqrt{3}, -1\right)$, $(-1, 1)$, $\left(1, \sqrt{3}\right)$. IP $\left(\pm\sqrt{3}, \sqrt[3]{4}\right)$

H.

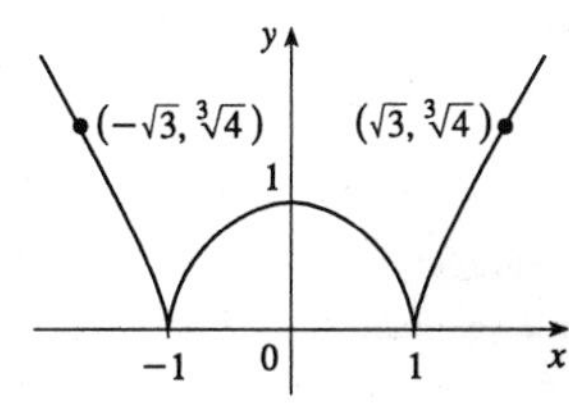

27. $y = f(x) = \cos x - \sin x$ **A.** $D = \mathbb{R}$ **B.** $y = 0$ $\Leftrightarrow$ $\cos x = \sin x$ $\Leftrightarrow$ $x = n\pi + \frac{\pi}{4}$, n an integer (x-intercepts), y-intercept $= f(0) = 1$. **C.** Periodic with period 2π **D.** No asymptotes

E. $f'(x) = -\sin x - \cos x = 0$ $\Leftrightarrow$ $\cos x = -\sin x$ $\Leftrightarrow$ $x = 2n\pi + \frac{3\pi}{4}$ or $2n\pi + \frac{7\pi}{4}$. $f'(x) > 0$ $\Leftrightarrow$ $\cos x < -\sin x$ $\Leftrightarrow$ $2n\pi + \frac{3\pi}{4} < x < 2n\pi + \frac{7\pi}{4}$, so f is increasing on $\left[2n\pi + \frac{3\pi}{4}, 2n\pi + \frac{7\pi}{4}\right]$ and decreasing on $\left[2n\pi - \frac{\pi}{4}, 2n\pi + \frac{3\pi}{4}\right]$.

F. Local maximum $f\left(2n\pi - \frac{\pi}{4}\right) = \sqrt{2}$, local minimum $f\left(2n\pi + \frac{3\pi}{4}\right) = -\sqrt{2}$.

G. $f''(x) = -\cos x + \sin x > 0$ $\Leftrightarrow$ $\sin x > \cos x$ $\Leftrightarrow$ $x \in \left(2n\pi + \frac{\pi}{4}, 2n\pi + \frac{5\pi}{4}\right)$, so f is CU on these intervals and CD on $\left(2n\pi - \frac{3\pi}{4}, 2n\pi + \frac{\pi}{4}\right)$. IP $\left(n\pi + \frac{\pi}{4}, 0\right)$

H.

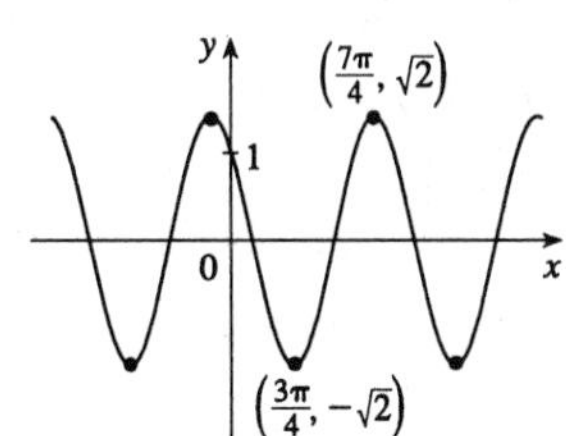

28. $y = f(x) = \sin x - \tan x$ **A.** $D = \{x \mid x \neq (2n+1)\frac{\pi}{2}\}$ **B.** $y = 0 \Leftrightarrow \sin x = \tan x = \dfrac{\sin x}{\cos x} \Leftrightarrow$ $\sin x = 0$ or $\cos x = 1 \Leftrightarrow x = n\pi$ (x-intercepts), y-intercept $= f(0) = 0$ **C.** $f(-x) = -f(x)$, so the curve is symmetric about $(0,0)$. Also periodic with period 2π. **D.** $\lim\limits_{x \to \pi/2^-} (\sin x - \tan x) = -\infty$ and $\lim\limits_{x \to \pi/2^+} (\sin x - \tan x) = \infty$, so $x = n\pi + \frac{\pi}{2}$ are VA.

E. $f'(x) = \cos x - \sec^2 x \leq 0$, so f decreases on each interval in its domain, that is, on $\left((2n-1)\frac{\pi}{2}, (2n+1)\frac{\pi}{2}\right)$.

F. No extremum

G. $f''(x) = -\sin x - 2\sec^2 x \tan x = -\sin x\,(1 + 2\sec^3 x)$. Note that $1 + 2\sec^3 x \neq 0$ since $\sec^3 x \neq -\frac{1}{2}$. $f''(x) > 0$ for $-\frac{\pi}{2} < x < 0$ and $\frac{3\pi}{2} < x < 2\pi$, so f is CU on $\left((n - \frac{1}{2})\pi, n\pi\right)$ and CD on $\left(n\pi, (n + \frac{1}{2})\pi\right)$. IP $(n\pi, 0)$. Note also that $f'(0) = 0$ but $f'(\pi) = -2$.

H.

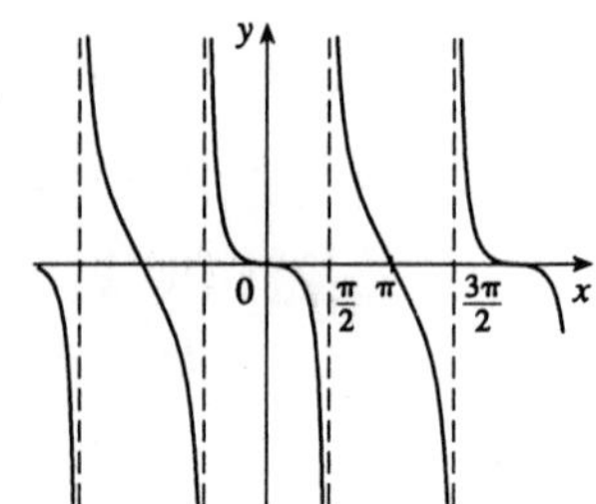

29. $y = f(x) = x \tan x$, $-\frac{\pi}{2} < x < \frac{\pi}{2}$ **A.** $D = \left(-\frac{\pi}{2}, \frac{\pi}{2}\right)$ **B.** Intercepts are 0 **C.** $f(-x) = f(x)$, so the curve is symmetric about the y-axis. **D.** $\lim\limits_{x \to \pi/2^-} x \tan x = \infty$ and $\lim\limits_{x \to -\pi/2^+} x \tan x = \infty$, so $x = \frac{\pi}{2}$ and $x = -\frac{\pi}{2}$ are VA.

E. $f'(x) = \tan x + x \sec^2 x > 0 \Leftrightarrow 0 < x < \frac{\pi}{2}$, so f increases on $\left[0, \frac{\pi}{2}\right)$ and decreases on $\left(-\frac{\pi}{2}, 0\right]$.

F. Absolute minimum $f(0) = 0$.

G. $y'' = 2\sec^2 x + 2x \tan x \sec^2 x > 0$ for $-\frac{\pi}{2} < x < \frac{\pi}{2}$, so f is CU on $\left(-\frac{\pi}{2}, \frac{\pi}{2}\right)$. No IP

H. $x = -\frac{\pi}{2}$, $x = \frac{\pi}{2}$

30. $y = f(x) = 2x + \cot x$, $0 < x < \pi$ **A.** $D = (0, \pi)$. **B.** No y-intercept **C.** No symmetry

D. $\lim\limits_{x \to 0^+} (2x + \cot x) = \infty$, $\lim\limits_{x \to \pi^-} (2x + \cot x) = -\infty$, so $x = 0$ and $x = \pi$ are VA. **E.** $f'(x) = 2 - \csc^2 x > 0$ when $\csc^2 x < 2 \Leftrightarrow \sin x > \frac{1}{\sqrt{2}} \Leftrightarrow \frac{\pi}{4} < x < \frac{3\pi}{4}$, so f is increasing on $\left[\frac{\pi}{4}, \frac{3\pi}{4}\right]$ and decreasing on $\left(0, \frac{\pi}{4}\right]$ and $\left[\frac{3\pi}{4}, \pi\right)$.

F. $f\left(\frac{\pi}{4}\right) = 1 + \frac{\pi}{2}$ is a local minimum, $f\left(\frac{3\pi}{4}\right) = \frac{3\pi}{2} - 1$ is a local maximum. **G.** $f''(x) = -2\csc x\,(-\csc x \cot x) = 2\csc^2 x \cot x > 0$ $\Leftrightarrow \cot x > 0 \Leftrightarrow 0 < x < \frac{\pi}{2}$, so f is CU on $\left(0, \frac{\pi}{2}\right)$, CD on $\left(\frac{\pi}{2}, \pi\right)$. IP $\left(\frac{\pi}{2}, \pi\right)$

H.

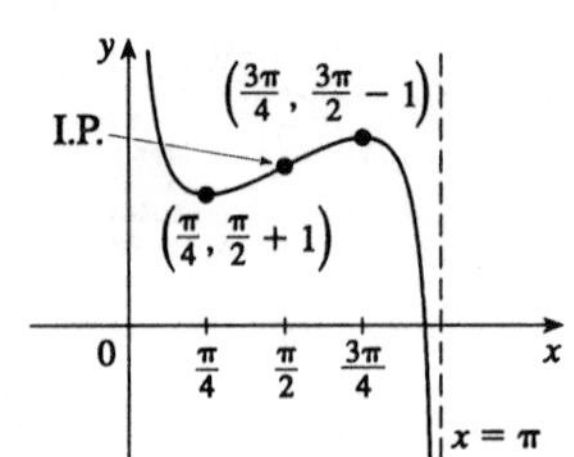

31. $y = f(x) = x/2 - \sin x,\ 0 < x < 3\pi$ **A.** $D = (0, 3\pi)$ **B.** No y-intercept. The x-intercept can be found approximately by Newton's Method (see Exercise 2.10.11). **C.** No symmetry **D.** No asymptotes.
E. $f'(x) = \frac{1}{2} - \cos x > 0 \quad \Leftrightarrow \quad \cos x < \frac{1}{2} \quad \Leftrightarrow \quad \frac{\pi}{3} < x < \frac{5\pi}{3}$ or $\frac{7\pi}{3} < x < 3\pi$, so f is increasing on $\left[\frac{\pi}{3}, \frac{5\pi}{3}\right]$ and $\left[\frac{7\pi}{3}, 3\pi\right)$ and decreasing on $\left(0, \frac{\pi}{3}\right]$ and $\left[\frac{5\pi}{3}, \frac{7\pi}{3}\right]$.
F. $f\left(\frac{\pi}{3}\right) = \frac{\pi}{6} - \frac{\sqrt{3}}{2}$ is a local minimum,
$f\left(\frac{5\pi}{3}\right) = \frac{5\pi}{6} + \frac{\sqrt{3}}{2}$ is a local maximum,
$f\left(\frac{7\pi}{3}\right) = \frac{7\pi}{6} - \frac{\sqrt{3}}{2}$ is a local minimum.
G. $f''(x) = \sin x > 0 \quad \Leftrightarrow \quad 0 < x < \pi$ or $2\pi < x < 3\pi$, so f is CU on $(0, \pi)$ and $(2\pi, 3\pi)$ and CD on $(\pi, 2\pi)$. IP $\left(\pi, \frac{\pi}{2}\right)$ and $(2\pi, \pi)$.

H.

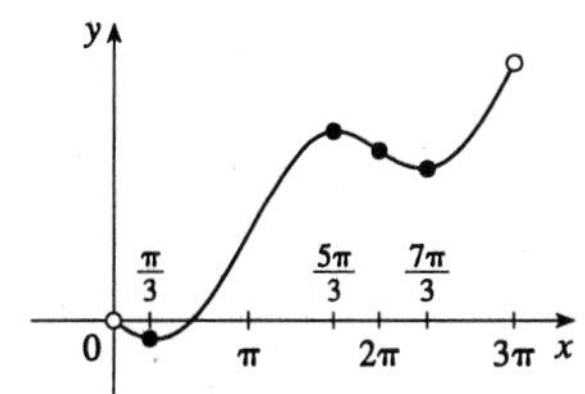

32. $y = f(x) = 2\sin x + \sin^2 x$ **A.** $D = \mathbb{R}$ *Note:* f is periodic with period 2π, so we determine B-G on $[0, 2\pi]$.
B. y-intercept $= 0$, x-intercepts occur when $2\sin x(2 + \sin x) = 0 \quad \Leftrightarrow \quad \sin x = 0 \quad \Leftrightarrow \quad x = 0, \pi, 2\pi$.
C. No symmetry other than periodicity **D.** No asymptotes
E. $f'(x) = 2\cos x + 2\sin x \cos x = 2\cos x(1 + \sin x) > 0 \quad \Leftrightarrow \quad \cos x > 0 \quad \Leftrightarrow \quad 0 < x < \frac{\pi}{2}$ or $\frac{3\pi}{2} < x < 2\pi$, so f is increasing on $\left[0, \frac{\pi}{2}\right]$, $\left[\frac{3\pi}{2}, 2\pi\right]$ and decreasing on $\left[\frac{\pi}{2}, \frac{3\pi}{2}\right]$. **F.** $f\left(\frac{\pi}{2}\right) = 3$ is a local maximum, $f\left(\frac{3\pi}{2}\right) = -1$ is a local minimum.

G. $f''(x) = -2\sin x + 2\cos^2 x - 2\sin^2 x$
$= 2\left(-\sin x + 1 - 2\sin^2 x\right)$
$= 2(1 + \sin x)(1 - 2\sin x) > 0 \quad \Leftrightarrow$
$1 - 2\sin x > 0 \quad \Leftrightarrow \quad \sin x < \frac{1}{2} \quad \Leftrightarrow \quad 0 \le x < \frac{\pi}{6}$ or $\frac{5\pi}{6} < x \le 2\pi$. So f is CU on $\left(0, \frac{\pi}{6}\right)$, $\left(\frac{5\pi}{6}, 2\pi\right)$, and CD on $\left(\frac{\pi}{6}, \frac{5\pi}{6}\right)$. IP $\left(\frac{\pi}{6}, \frac{5}{4}\right)$ and $\left(\frac{5\pi}{6}, \frac{5}{4}\right)$

H.

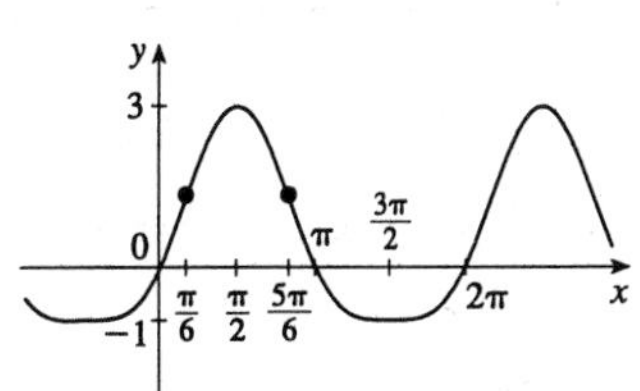

33. $y = f(x) = 2\cos x + \sin^2 x$ **A.** $D = \mathbb{R}$ **B.** y-intercept $= f(0) = 2$ **C.** $f(-x) = f(x)$, so the curve is symmetric about the y-axis. Periodic with period 2π **D.** No asymptotes
E. $f'(x) = -2\sin x + 2\sin x \cos x = 2\sin x\,(\cos x - 1) > 0 \quad \Leftrightarrow \quad \sin x < 0 \quad \Leftrightarrow$ $(2n-1)\pi < x < 2n\pi$, so f is increasing on $[(2n-1)\pi, 2n\pi]$ and decreasing on $[2n\pi, (2n+1)\pi]$.
F. $f(2n\pi) = 2$ is a local maximum.
$f((2n+1)\pi) = -2$ is a local minimum.
G. $f''(x) = -2\cos x + 2\cos 2x = 2\left(2\cos^2 x - \cos x - 1\right)$
$= 2(2\cos x + 1)(\cos x - 1) > 0$
$\Leftrightarrow \quad \cos x < -\frac{1}{2} \quad \Leftrightarrow \quad x \in \left(2n\pi + \frac{2\pi}{3}, 2n\pi + \frac{4\pi}{3}\right)$, so f is CU on these intervals and CD on $\left(2n\pi - \frac{2\pi}{3}, 2n\pi + \frac{2\pi}{3}\right)$.
IP when $x = 2n\pi \pm \frac{2\pi}{3}$

H.

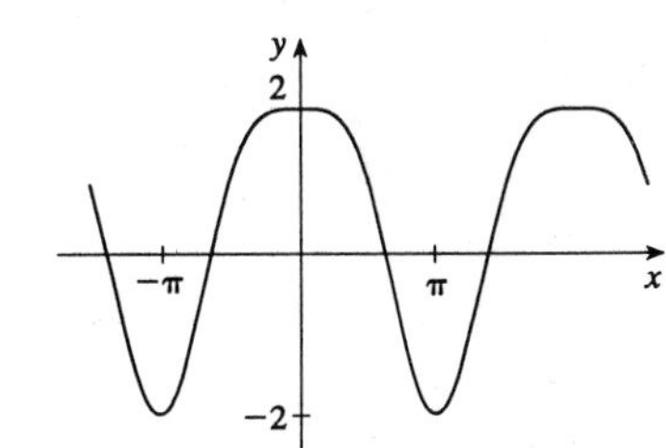

34. $f(x) = \sin x - x$ **A.** $D = \mathbb{R}$ **B.** x-intercept $= 0 =$ y-intercept **C.** $f(-x) = \sin(-x) - (-x) = -(\sin x - x)$ $= -f(-x)$, so f is odd. **D.** No asymptotes **E.** $f'(x) = \cos x - 1 \leq 0$ for all x, so f is decreasing on $(-\infty, \infty)$. **F.** No local extremum **G.** $f''(x) = -\sin x \quad \Rightarrow \quad f''(x) > 0 \quad \Leftrightarrow$ $\sin x < 0 \quad \Leftrightarrow \quad (2n-1)\pi < x < 2n\pi$, so f is CU on $((2n-1)\pi, 2n\pi)$ and CD on $(2n\pi, (2n+1)\pi)$, n an integer. Points of inflection occur when $x = n\pi$.

H.

35. $y = f(x) = \sin 2x - 2\sin x$ **A.** $D = \mathbb{R}$ **B.** y-intercept $= f(0) = 0$. $y = 0 \quad \Leftrightarrow$ $2\sin x = \sin 2x = 2\sin x \cos x \quad \Leftrightarrow \quad \sin x = 0$ or $\cos x = 1 \quad \Leftrightarrow \quad x = n\pi$ (x-intercepts) **C.** $f(-x) = -f(x)$, so the curve is symmetric about $(0, 0)$. *Note:* f is periodic with period 2π, so we determine D-G for $-\pi \leq x \leq \pi$. **D.** No asymptotes **E.** $f'(x) = 2\cos 2x - 2\cos x$. As in Exercise 33G, we see that $f'(x) > 0 \quad \Leftrightarrow \quad -\pi < x < -\frac{2\pi}{3}$ or $\frac{2\pi}{3} < x < \pi$, so f is increasing on $\left[-\pi, -\frac{2\pi}{3}\right]$ and $\left[\frac{2\pi}{3}, \pi\right]$ and decreasing on $\left[-\frac{2\pi}{3}, \frac{2\pi}{3}\right]$. **F.** $f\left(-\frac{2\pi}{3}\right) = \frac{3\sqrt{3}}{2}$ is a local maximum, $f\left(\frac{2\pi}{3}\right) = -\frac{3\sqrt{3}}{2}$ is a local minimum. **G.** $f''(x) = -4\sin 2x + 2\sin x$ $= 2\sin x\,(1 - 4\cos x) = 0$ when $x = 0, \pm\pi$ or $\cos x = \frac{1}{4}$. If $\alpha = \cos^{-1}\frac{1}{4}$, then f is CU on $(-\alpha, 0)$ and (α, π) and CD on $(-\pi, -\alpha)$ and $(0, \alpha)$. IP $(0,0)$,$(\pi, 0)$, $\left(\alpha, -\frac{3\sqrt{15}}{8}\right)$, $\left(-\alpha, \frac{3\sqrt{15}}{8}\right)$.

H.

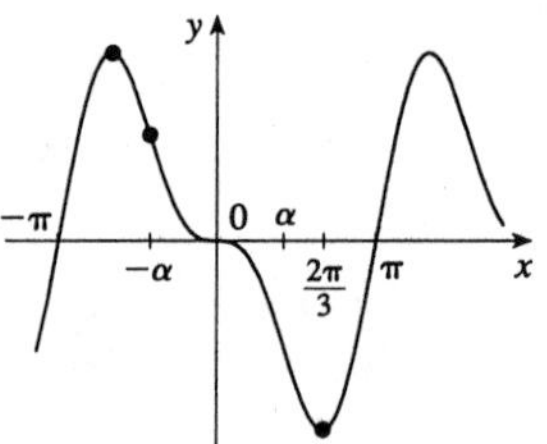

36. $y = f(x) = \cos x/(2 + \sin x)$ **A.** $D = \mathbb{R}$ *Note:* f is periodic with period 2π, so we determine B-G on $[0, 2\pi]$. **B.** x-intercepts $\frac{\pi}{2}, \frac{3\pi}{2}$, y-intercept $= f(0) = \frac{1}{2}$ **C.** No symmetry other than periodicity **D.** No asymptotes **E.** $f'(x) = \dfrac{(2 + \sin x)(-\sin x) - \cos x(\cos x)}{(2 + \sin x)^2} = -\dfrac{2\sin x + 1}{(2 + \sin x)^2}$. $f'(x) > 0 \quad \Leftrightarrow$ $2\sin x + 1 < 0 \quad \Leftrightarrow \quad \sin x < -\frac{1}{2} \quad \Leftrightarrow \quad \frac{7\pi}{6} < x < \frac{11\pi}{6}$, so f is increasing on $\left[\frac{7\pi}{6}, \frac{11\pi}{6}\right]$ and decreasing on $\left[0, \frac{7\pi}{6}\right]$, $\left[\frac{11\pi}{6}, 2\pi\right]$. **F.** $f\left(\frac{7\pi}{6}\right) = -\frac{1}{\sqrt{3}}$ is a local minimum, $f\left(\frac{11\pi}{6}\right) = \frac{1}{\sqrt{3}}$ is a local maximum.

G. $f''(x) = -\dfrac{(2 + \sin x)^2(2\cos x) - (2\sin x + 1)\,2\,(2 + \sin x)\cos x}{(2 + \sin x)^4}$

$= -\dfrac{2\cos x(1 - \sin x)}{(2 + \sin x)^3} > 0 \quad \Leftrightarrow$

$\cos x < 0 \quad \Leftrightarrow \quad \frac{\pi}{2} < x < \frac{3\pi}{2}$,

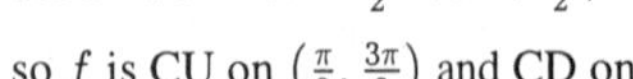

so f is CU on $\left(\frac{\pi}{2}, \frac{3\pi}{2}\right)$ and CD on $\left(0, \frac{\pi}{2}\right)$ and $\left(\frac{3\pi}{2}, 2\pi\right)$. IP $\left(\frac{\pi}{2}, 0\right)$, $\left(\frac{3\pi}{2}, 0\right)$.

H.

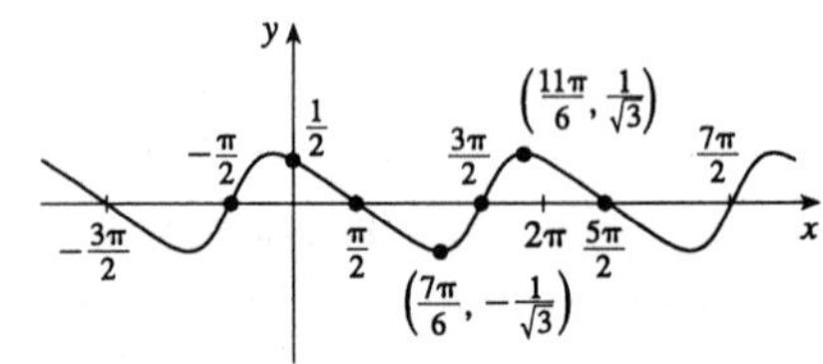

37. $y = f(x) = x^3/(x^2-1)$ **A.** $D = \{x \mid x \neq \pm 1\} = (-\infty,-1) \cup (-1,1) \cup (1,\infty)$ **B.** x-intercept $= 0$, y-intercept $= 0$ **C.** $f(-x) = -f(x)$ $\Rightarrow$ f is odd, so the curve is symmetric about the origin.
D. $\lim\limits_{x\to\infty} \dfrac{x^3}{x^2-1} = \infty$ but long division gives $\dfrac{x^3}{x^2-1} = x + \dfrac{x}{x^2-1}$ so $f(x) - x = \dfrac{x}{x^2-1} \to 0$ as $x \to \pm\infty$ $\Rightarrow$ $y = x$ is a slant asymptote. $\lim\limits_{x\to 1^-} \dfrac{x^3}{x^2-1} = -\infty$, $\lim\limits_{x\to 1^+} \dfrac{x^3}{x^2-1} = \infty$, $\lim\limits_{x\to -1^-} \dfrac{x^3}{x^2-1} = -\infty$, $\lim\limits_{x\to -1^+} \dfrac{x^3}{x^2-1} = \infty$, so $x = 1$ and $x = -1$ are VA. **E.** $f'(x) = \dfrac{3x^2(x^2-1) - x^3(2x)}{(x^2-1)^2} = \dfrac{x^2(x^2-3)}{(x^2-1)^2}$ $\Rightarrow$ $f'(x) > 0$ $\Leftrightarrow$ $x^2 > 3$ $\Leftrightarrow$ $x > \sqrt{3}$ or $x < -\sqrt{3}$, so f is increasing on $\left(-\infty,-\sqrt{3}\right]$ and $\left[\sqrt{3},\infty\right)$ and decreasing on $\left[-\sqrt{3},-1\right)$, $(-1,1)$, and $\left(1,\sqrt{3}\right]$.
F. $f\left(-\sqrt{3}\right) = -\frac{3\sqrt{3}}{2}$ is a local maximum and $f\left(\sqrt{3}\right) = \frac{3\sqrt{3}}{2}$ is a local minimum.
G. $y'' = \dfrac{2x\,(x^2+3)}{(x^2-1)^3} > 0$ $\Leftrightarrow$ $x > 1$ or $-1 < x < 0$, so f is CU on $(-1,0)$ and $(1,\infty)$ and CD on $(-\infty,-1)$ and $(0,1)$. IP $(0,0)$

H.

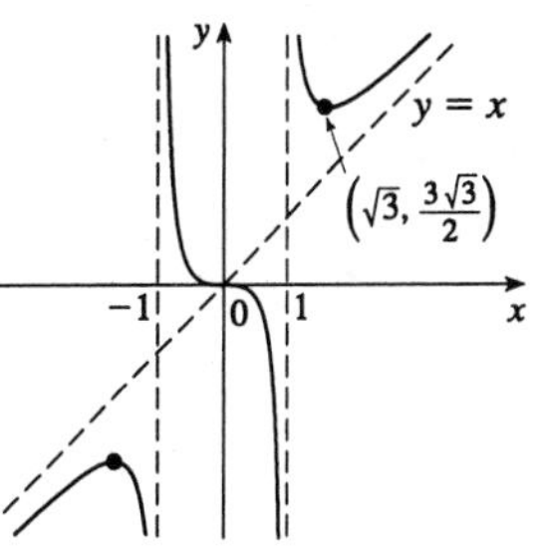

38. $y = f(x) = x - 1/x$ **A.** $D = \{x \mid x \neq 0\} = (-\infty,0) \cup (0,\infty)$ **B.** x-intercepts ± 1, no y-intercept
C. $f(-x) = -f(x)$, so the curve is symmetric about the origin. **D.** $\lim\limits_{x\to\pm\infty} (x - 1/x) = \pm\infty$, so no HA. But $(x - 1/x) - x = -1/x \to 0$ as $x \to \pm\infty$, so $y = x$ is a slant asymptote. Also $\lim\limits_{x\to 0^+} (x - 1/x) = -\infty$ and $\lim\limits_{x\to 0^-} (x - 1/x) = \infty$, so $x = 0$ is a VA.
E. $f'(x) = 1 + 1/x^2 > 0$, so f is increasing on $(-\infty,0)$ and $(0,\infty)$. **F.** No extremum **G.** $f''(x) = -2/x^3$ $\Rightarrow$ $f''(x) > 0$ $\Leftrightarrow$ $x < 0$, so f is CU on $(-\infty,0)$ and CD on$(0,\infty)$. No IP

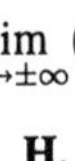

H.

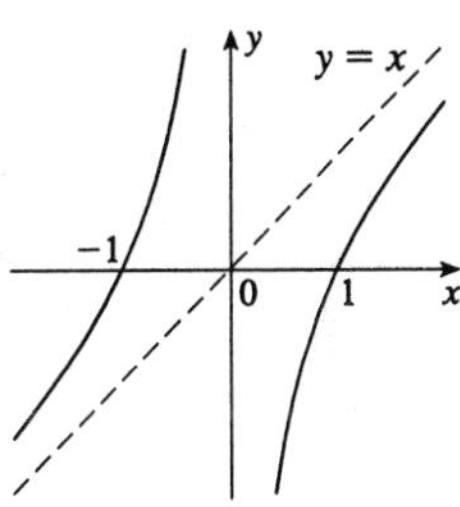

39. $y = f(x) = (x^2+4)/x = x + 4/x$ **A.** $D = \{x \mid x \neq 0\} = (-\infty,0) \cup (0,\infty)$ **B.** No intercept
C. $f(-x) = -f(x)$ $\Rightarrow$ symmetry about the origin **D.** $\lim\limits_{x\to\infty} (x + 4/x) = \infty$ but $f(x) - x = 4/x \to 0$ as $x \to \pm\infty$, so $y = x$ is a slant asymptote. $\lim\limits_{x\to 0^+} (x + 4/x) = \infty$ and $\lim\limits_{x\to 0^-} (x + 4/x) = -\infty$, so $x = 0$ is a VA.
E. $f'(x) = 1 - 4/x^2 > 0$ $\Leftrightarrow$ $x^2 > 4$ $\Leftrightarrow$ $x > 2$ or $x < -2$, so f is increasing on $(-\infty,-2]$ and $[2,\infty)$ and decreasing on $[-2,0)$ and $(0,2]$.
F. $f(-2) = -4$ is a local maximum and $f(2) = 4$ is a local minimum. **G.** $f''(x) = 8/x^3 > 0$ $\Leftrightarrow$ $x > 0$ so f is CU on $(0,\infty)$ and CD on $(-\infty,0)$. No IP

H.

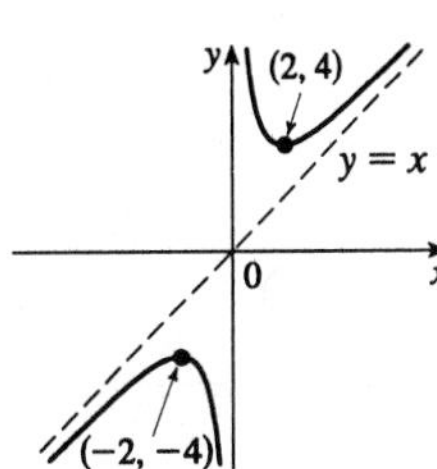

40. $y = f(x) = \dfrac{x^2+x+1}{x} = x + 1 + \dfrac{1}{x}$ **A.** $D = \{x \mid x \neq 0\} = (-\infty, 0) \cup (0, \infty)$ **B.** No intercepts (x-intercepts occur when $x^2 + x + 1 = 0$ but this equation has no real roots since $b^2 - 4ac = -3 < 0$.)
C. No symmetry **D.** $\lim\limits_{x \to \pm\infty} (x + 1 + 1/x) = \pm\infty$, so no HA. But $(x + 1 + 1/x) - (x + 1) = 1/x \to 0$ as $x \to \pm\infty$, so $y = x + 1$ is a slant asymptote. Also $\lim\limits_{x \to 0^+} (x + 1 + 1/x) = \infty$, $\lim\limits_{x \to 0^-} (x + 1 + 1/x) = -\infty$, so $x = 0$ is a VA. **E.** $f'(x) = 1 - 1/x^2 > 0$ when $x^2 > 1$ $\Leftrightarrow$ $x > 1$ or $x < -1$; $f'(x) < 0$ $\Leftrightarrow$ $-1 < x < 1$. So f is increasing on $(-\infty, -1]$, $[1, \infty)$ and decreasing on $[-1, 0)$, $(0, 1]$. **F.** $f(1) = 3$ is a local minimum, $f(-1) = -1$ is a local maximum.
G. $f''(x) = 2/x^3 > 0$ $\Leftrightarrow$ $x > 0$, so f is CU on $(0, \infty)$ and CD on $(-\infty, 0)$. No IP

H.

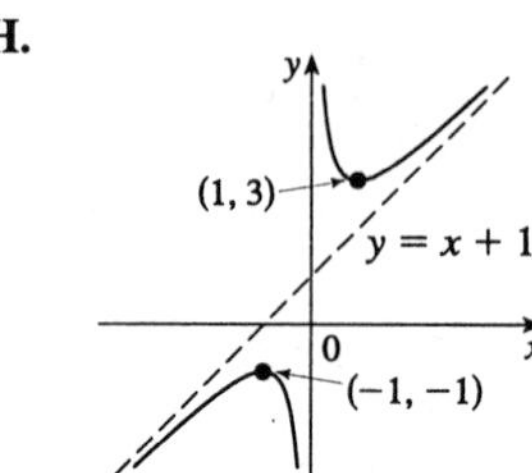

41. $y = \dfrac{1}{x-1} - x$ **A.** $D = \{x \mid x \neq 1\}$ **B.** $y = 0$ $\Leftrightarrow$ $x = \dfrac{1}{x-1}$ $\Leftrightarrow$ $x^2 - x - 1 = 0$ $\Rightarrow$ $x = \frac{1 \pm \sqrt{5}}{2}$ (x-intercepts), y-intercept $= f(0) = -1$ **C.** No symmetry **D.** $y - (-x) = \dfrac{1}{x-1} \to 0$ as $x \to \pm\infty$, so $y = -x$ is a slant asymptote.
$\lim\limits_{x \to 1^+} \left(\dfrac{1}{x-1} - x\right) = \infty$ and $\lim\limits_{x \to 1^-} \left(\dfrac{1}{x-1} - x\right) = -\infty$, so $x = 1$ is a VA. **E.** $f'(x) = -1 - 1/(x-1)^2 < 0$ for all $x \neq 1$, so f is decreasing on $(-\infty, 1)$ and $(1, \infty)$.
F. No local extremum **G.** $f''(x) = \dfrac{2}{(x-1)^3} > 0$ $\Leftrightarrow$ $x > 1$, so f is CU on $(1, \infty)$ and CD on $(-\infty, 1)$. No IP

H.

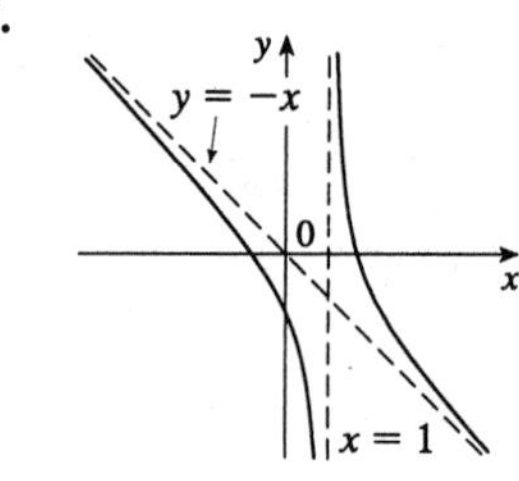

42. $y = f(x) = x^2/(2x + 5)$ **A.** $D = \left\{x \mid x \neq -\frac{5}{2}\right\} = \left(-\infty, -\frac{5}{2}\right) \cup \left(-\frac{5}{2}, \infty\right)$ **B.** Intercepts are 0.
C. No symmetry **D.** $\lim\limits_{x \to \pm\infty} \dfrac{x^2}{2x+5} = \pm\infty$, so no HA. $\lim\limits_{x \to -5/2^+} \dfrac{x^2}{2x+5} = \infty$, $\lim\limits_{x \to -5/2^-} \dfrac{x^2}{2x+5} = -\infty$, so $x = -\frac{5}{2}$ is a VA. By long division, $\dfrac{x^2}{2x+5} = \dfrac{x}{2} - \dfrac{5}{4} + \dfrac{25/4}{2x+5}$, so $\dfrac{x^2}{2x+5} - \left(\dfrac{x}{2} - \dfrac{5}{4}\right) = \dfrac{25/4}{2x+5} \to 0$ as $x \to \pm\infty$, so $y = \frac{1}{2}x - \frac{5}{4}$ is a slant asymptote. **E.** $f'(x) = \dfrac{2x(x+5) - 2x^2}{(2x+5)^2} = \dfrac{2x(x+5)}{(2x+5)^2}$ $\Rightarrow$ $f'(x) > 0$ $\Leftrightarrow$ $x < -5$ or $x > 0$; $f'(x) < 0$ $\Leftrightarrow$ $-5 < x < 0$. So f is increasing on $(-\infty, -5]$ and $[0, \infty)$, decreasing on $\left[-5, -\frac{5}{2}\right)$, $\left(-\frac{5}{2}, 0\right]$. **F.** $f(0) = 0$ is a local minimum, $f(-5) = -5$ is a local maximum.

G. $f''(x) = \dfrac{(4x+10)(2x+5)^2 - (2x^2+10x) \cdot 2(2x+5)(2)}{(2x+5)^4}$

$= \dfrac{50}{(2x+5)^3} > 0$ $\Leftrightarrow$ $x > -\frac{5}{2}$, so f is CU on $\left(-\frac{5}{2}, \infty\right)$ and CD on $\left(-\infty, -\frac{5}{2}\right)$. No IP.

H.

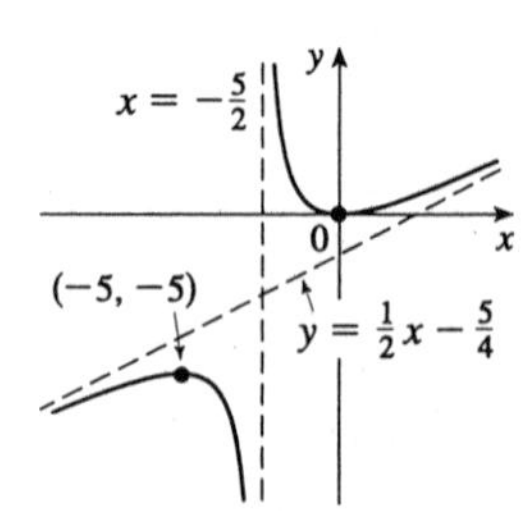

43. $\dfrac{x^2}{a^2} - \dfrac{y^2}{b^2} = 1 \;\Rightarrow\; y = \pm\dfrac{b}{a}\sqrt{x^2 - a^2}$. Now

$$\lim_{x\to\infty}\left[\frac{b}{a}\sqrt{x^2-a^2} - \frac{b}{a}x\right] = \frac{b}{a}\cdot\lim_{x\to\infty}\left(\sqrt{x^2-a^2} - x\right)\frac{\sqrt{x^2-a^2}+x}{\sqrt{x^2-a^2}+x} = \frac{b}{a}\cdot\lim_{x\to\infty}\frac{-a^2}{\sqrt{x^2-a^2}+x} = 0, \text{ which}$$

shows that $y = \dfrac{b}{a}x$ is a slant asymptote.

Similarly $\displaystyle\lim_{x\to\infty}\left[-\frac{b}{a}\sqrt{x^2-a^2} - \left(-\frac{b}{a}x\right)\right] = -\frac{b}{a}\cdot\lim_{x\to\infty}\frac{-a^2}{\sqrt{x^2-a^2}+x} = 0$, which shows that $y = -\dfrac{b}{a}x$ is a slant asymptote.

44. $f(x) - x^2 = \dfrac{x^3+1}{x} - x^2 = \dfrac{x^3+1-x^3}{x} = \dfrac{1}{x}$, and $\displaystyle\lim_{x\to\pm\infty}\frac{1}{x} = 0$. Therefore $\displaystyle\lim_{x\to\pm\infty}\left[f(x) - x^2\right] = 0$, and so the graph of f is asymptotic to that of $y = x^2$. For purposes of differentiation, we will use $f(x) = x^2 + \dfrac{1}{x}$.

A. $D = \{x \mid x \neq 0\}$ **B.** No y-intercept; to find the x-intercept, we set $y = 0 \;\Leftrightarrow\; x = -1$.

C. No symmetry **D.** $\displaystyle\lim_{x\to 0^+}\frac{x^3+1}{x} = \infty$ and $\displaystyle\lim_{x\to 0^-}\frac{x^3+1}{x} = -\infty$, so $x = 0$ is a vertical asymptote.

Also, the graph is asymptotic to the parabola $y = x^2$, as shown above. **E.** $f'(x) = 2x - 1/x^2 > 0 \;\Leftrightarrow\; x > \frac{1}{\sqrt[3]{2}}$, so f is increasing on $\left[\frac{1}{\sqrt[3]{2}}, \infty\right)$ and decreasing on $(-\infty, 0)$ and $\left(0, \frac{1}{\sqrt[3]{2}}\right]$.

F. Local minimum $f\left(\frac{1}{\sqrt[3]{2}}\right) = \frac{3\sqrt[3]{3}}{2}$, no local maximum

G. $f''(x) = 2 + 2/x^3 > 0 \;\Leftrightarrow\; x < -1$ or $x > 0$, so f is CU on $(-\infty, -1)$ and $(0, \infty)$, and CD on $(-1, 0)$. IP $(-1, 0)$.

H.

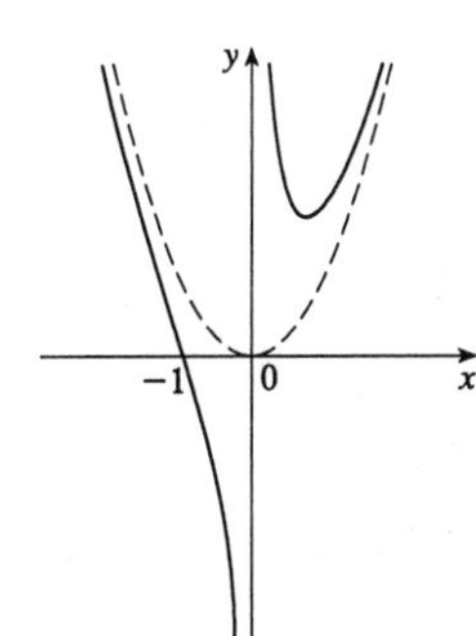

45. $\displaystyle\lim_{x\to\pm\infty}\left[f(x) - x^3\right] = \lim_{x\to\pm\infty}\frac{x^4+1}{x} - \frac{x^4}{x} = \lim_{x\to\pm\infty}\frac{1}{x} = 0$, so the graph of f is asymptotic to that of $y = x^3$.

A. $D = \{x \mid x \neq 0\}$ **B.** No intercepts **C.** f is symmetric about the origin. **D.** $\displaystyle\lim_{x\to 0^-}\left(x^3 + \frac{1}{x}\right) = -\infty$ and $\displaystyle\lim_{x\to 0^+}\left(x^3 + \frac{1}{x}\right) = \infty$, so $x = 0$ is a vertical asymptote, and as shown above, the graph of f is asymptotic to that of $y = x^3$. **E.** $f'(x) = 3x^2 - 1/x^2 > 0$ $\Leftrightarrow\; x^4 > \frac{1}{3} \;\Leftrightarrow |x| > \frac{1}{\sqrt[4]{3}}$, so f is increasing on $\left(-\infty, -\frac{1}{\sqrt[4]{3}}\right)$ and $\left(\frac{1}{\sqrt[4]{3}}, \infty\right)$ and decreasing on $\left(-\frac{1}{\sqrt[4]{3}}, 0\right)$ and $\left(0, \frac{1}{\sqrt[4]{3}}\right)$.

F. Local maximum $f\left(-\frac{1}{\sqrt[4]{3}}\right) = -4\cdot 3^{-5/4}$, local minimum $f\left(\frac{1}{\sqrt[4]{3}}\right) = 4\cdot 3^{-5/4}$ **G.** $f''(x) = 6x + 2/x^3 > 0 \;\Leftrightarrow\; x > 0$, so f is CU on $(0, \infty)$ and CD on $(-\infty, 0)$.

H.

46. $\lim_{x\to\pm\infty}[f(x) - \cos x] = \lim_{x\to\pm\infty} 1/x^2 = 0$, so the graph of f is asymptotic to that of $\cos x$. The intercepts can only be found approximately. $f(x) = f(-x)$, so f is even.

$\lim_{x\to 0}\left(\cos x + \frac{1}{x^2}\right) = \infty$, so $x = 0$ is a vertical asymptote.

We don't need to calculate the derivatives, since we know the asymptotic behavior of the curve.

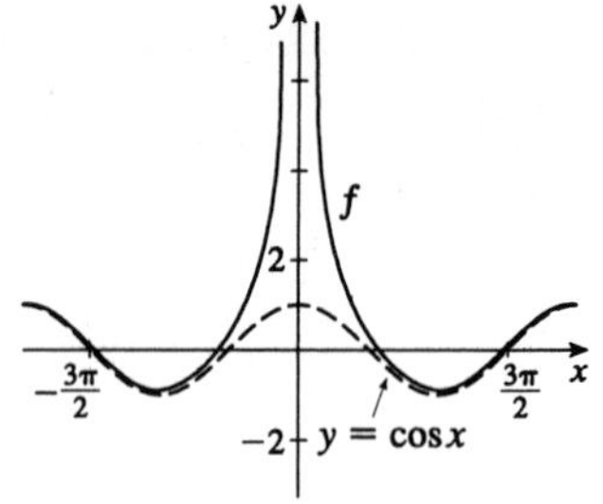

EXERCISES 3.7

Abbreviations:	**HA**	**horizontal asymptote(s)**	**VA**	**vertical asymptote(s)**
	CU	**concave up**	**CD**	**concave down**
	IP	**inflection point(s)**	**FDT**	**First Derivative Test**

1. $f(x) = 4x^4 - 7x^2 + 4x + 6 \quad\Rightarrow\quad f'(x) = 16x^3 - 14x + 4 \quad\Rightarrow\quad f''(x) = 48x^2 - 14$

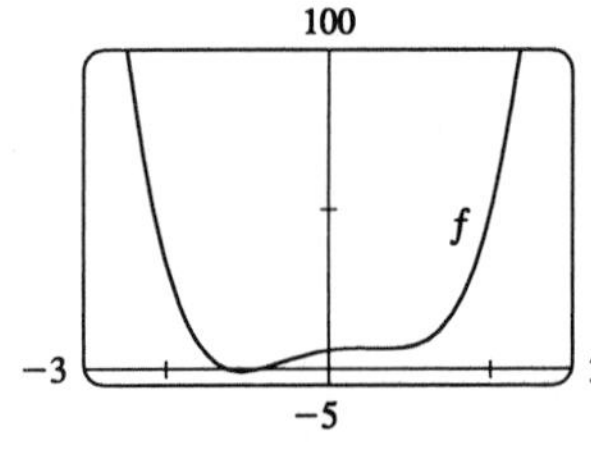

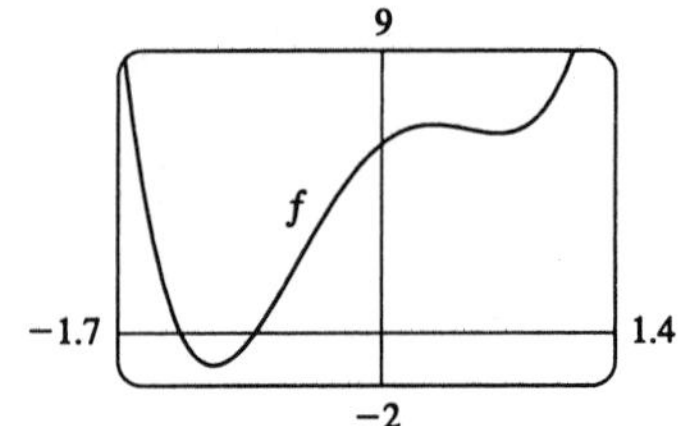

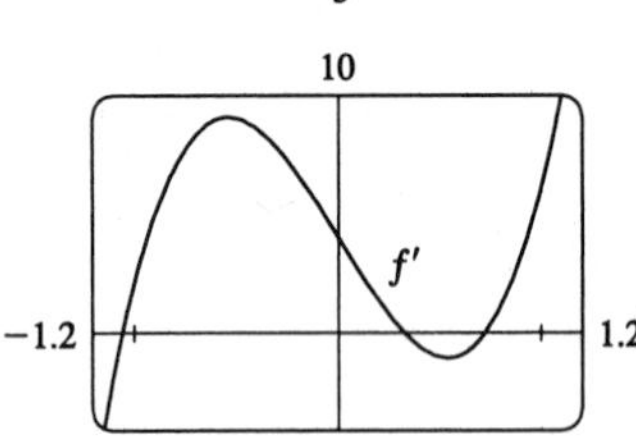

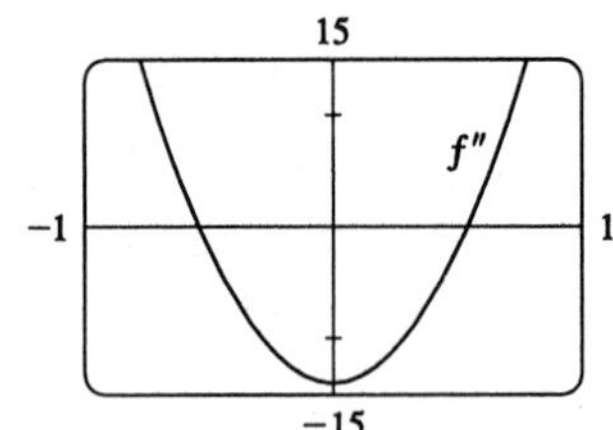

After finding suitable viewing rectangles (by ensuring that we have located all of the x-values where either $f' = 0$ or $f'' = 0$) we estimate from the graph of f' that f is increasing on $[-1.1, 0.3]$ and $[0.7, \infty)$ and decreasing on $(-\infty, -1.1]$ and $[0.3, 0.7]$, with a local maximum of $f(0.3) \approx 6.6$ and minima of $f(-1.1) \approx -1.0$ and $f(0.7) \approx 6.3$. We estimate from the graph of f'' that f is CU on $(-\infty, -0.5)$ and $(0.5, \infty)$ and CD on $(-0.5, 0.5)$, and that f has inflection points at about $(-0.5, 2.0)$ and $(0.5, 6.5)$.

2. $f(x) = 8x^5 + 45x^4 + 80x^3 + 90x^2 + 200x \quad \Rightarrow$

$f'(x) = 40x^4 + 180x^3 + 240x^2 + 180x + 200 \quad \Rightarrow$

$f''(x) = 160x^3 + 540x^2 + 480x + 180$

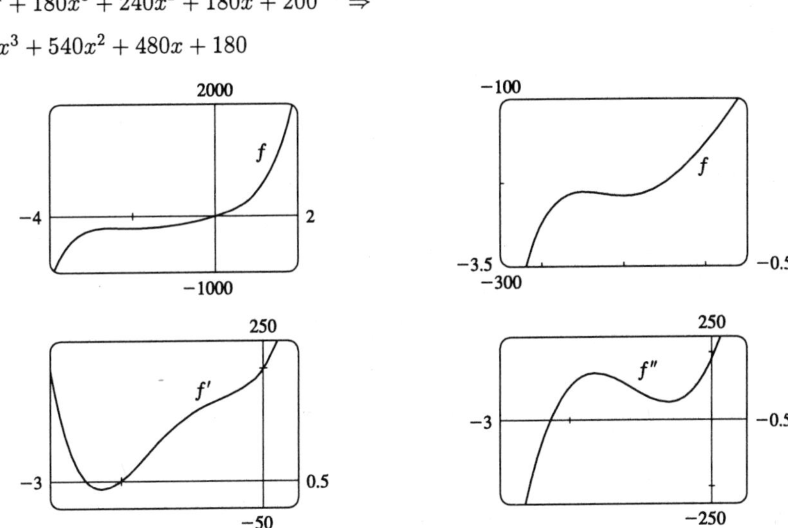

After finding suitable viewing rectangles, we estimate from the graph of f' that f is increasing on $(-\infty, -2.5]$ and $[-2.0, \infty)$ and decreasing on $[-2.5, -2.0]$. Maximum: $f(-2.5) \approx -211$. Minimum: $f(-2) \approx -216$. We estimate from the graph of f'' that f is CU on $(-2.3, \infty)$ and CD on $(-\infty, -2.3)$, and has an IP at $(-2.3, -213)$.

3. $f(x) = \sqrt[3]{x^2 - 3x - 5} \;\Rightarrow\; f'(x) = \dfrac{1}{3}\dfrac{2x-3}{(x^2-3x-5)^{2/3}} \;\Rightarrow\; f''(x) = -\dfrac{2}{9}\dfrac{x^2-3x+24}{(x^2-3x-5)^{5/3}}$

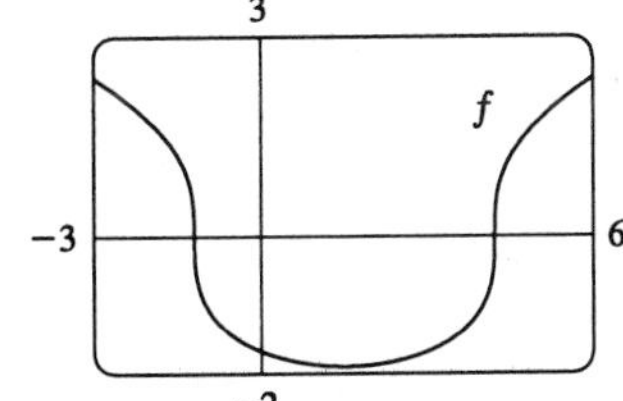

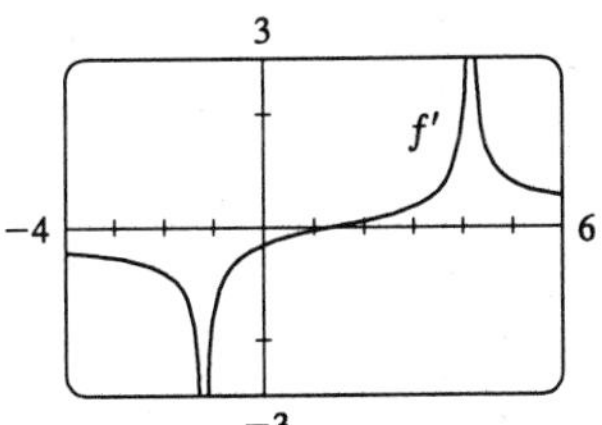

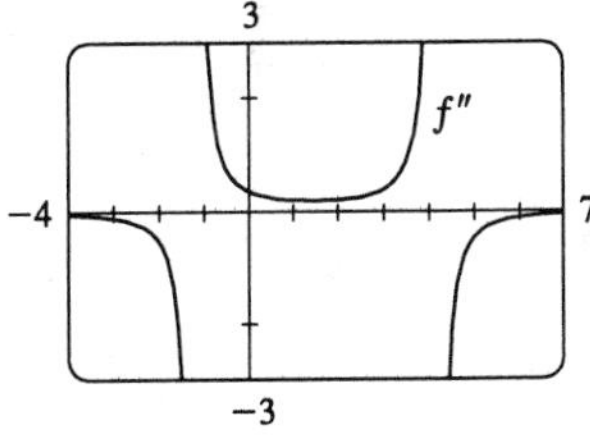

Note: With some CAS's, including Maple, it is necessary to define $f(x) = \dfrac{x^2 - 3x - 5}{|x^2 - 3x - 5|}\left|x^2 - 3x - 5\right|^{1/3}$, since the CAS does not compute real cube roots of negative numbers. (See Example 7 in Section 3 of Review and Preview.) We estimate from the graph of f' that f is increasing on $[1.5, 4.2)$ and $(4.2, \infty)$, and decreasing on $(-\infty, -1.2)$ and $(-1.2, 1.5]$. f has no maximum. Minimum: $f(1.5) \approx -1.9$. From the graph of f'', we estimate that f is CU on $(-1.2, 4.2)$ and CD on $(-\infty, -1.2)$ and $(4.2, \infty)$. IP $(-1.2, 0)$ and $(4.2, 0)$.

4. $f(x) = \dfrac{x^4 + x^3 - 2x^2 + 2}{x^2 + x - 2}$, so $f'(x) = 2\dfrac{x^5 + 2x^4 - 3x^3 - 4x^2 + 2x - 1}{(x^2 + x - 2)^2}$ and $f''(x) = 2\dfrac{x^6 + 3x^5 - 3x^4 - 11x^3 + 12x^2 + 18x - 2}{(x^2 + x - 2)^3}$.

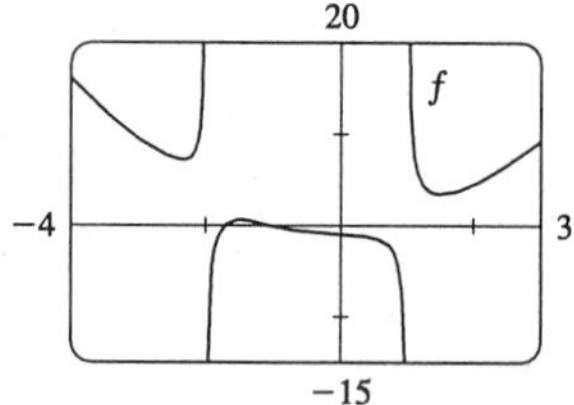

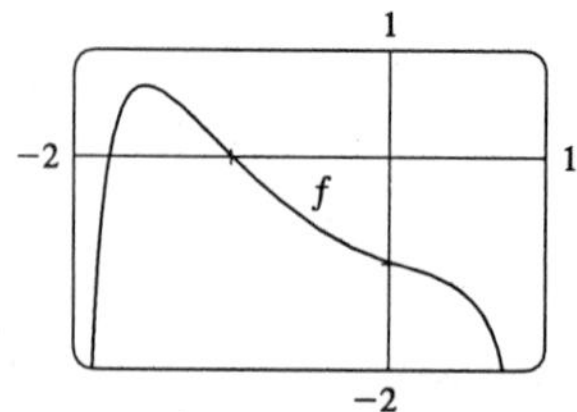

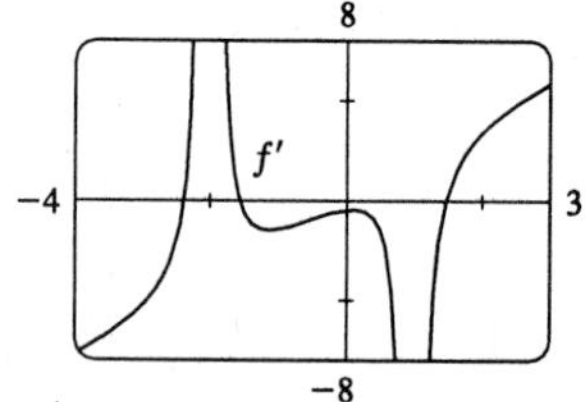

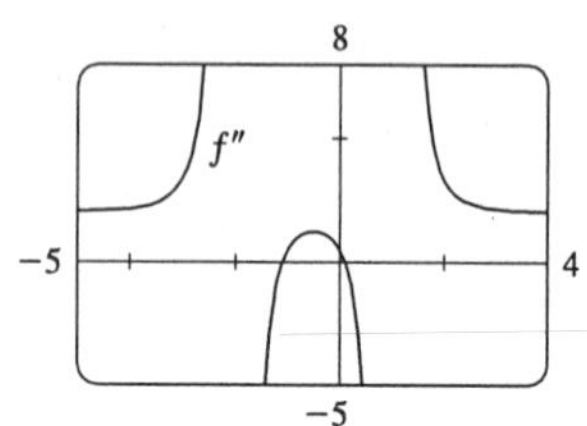

We estimate from the graph of f' that f is increasing on $[-2.4, -2)$, $(-2, -1.5]$ and $[1.5, \infty)$ and decreasing on $(-\infty, -2.4]$, $[-1.5, 1)$ and $(1, 1.5]$. Local maximum: $f(-1.5) \approx 0.7$. Local minima: $f(-2.4) \approx 7.2$, $f(1.5) \approx 3.4$. From the graph of f'', we estimate that f is CU on $(-\infty, -2)$, $(-1.1, 0.1)$ and $(1, \infty)$ and CD on $(-2, -1.1)$ and $(0.1, 1)$. f has IP at $(-1.1, 0.2)$ and $(0.1, -1.1)$.

5. $f(x) = x^2 \sin x \;\Rightarrow\; f'(x) = 2x\sin x + x^2 \cos x \;\Rightarrow\; f''(x) = 2\sin x + 4x\cos x - x^2 \sin x$

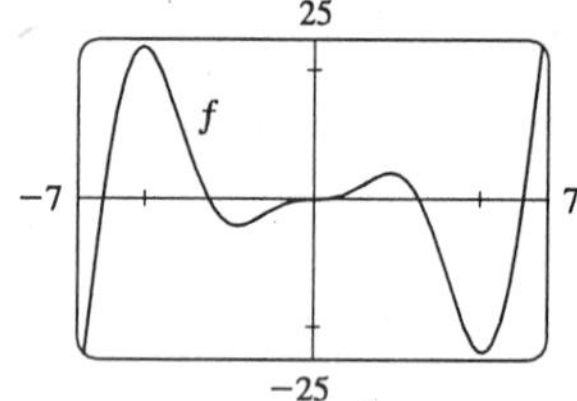

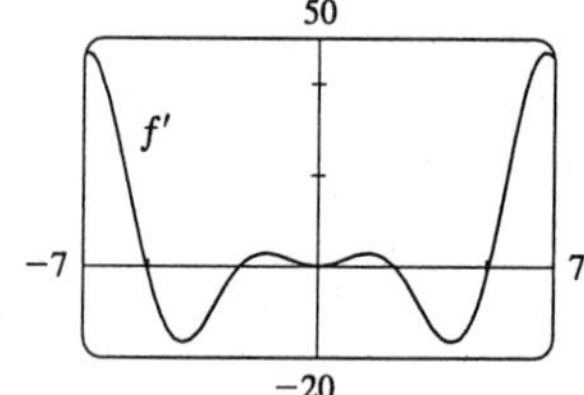

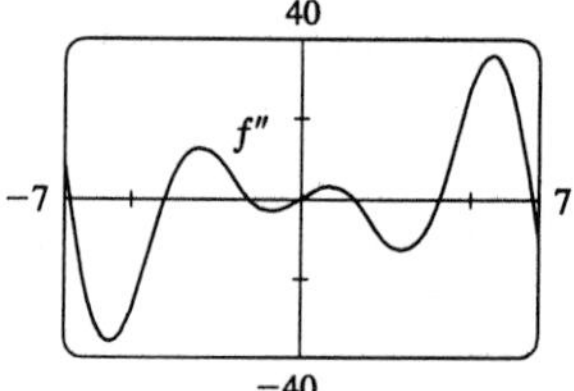

We estimate from the graph of f' that f is increasing on $[-7, -5.1]$, $[-2.3, 2.3]$, and $[5.1, 7]$ and decreasing on $[-5.1, -2.3]$, and $[2.3, 5.1]$. Local maxima: $f(-5.1) \approx 24.1$, $f(2.3) \approx 3.9$. Local minima: $f(-2.3) \approx -3.9$, $f(5.1) \approx -24.1$. From the graph of f'', we estimate that f is CU on $(-7, -6.8)$, $(-4.0, -1.5)$, $(0, 1.5)$, and $(4.0, 6.8)$, and CD on $(-6.8, -4.0)$, $(-1.5, 0)$, $(1.5, 4.0)$, and $(6.8, 7)$. f has IP at $(-6.8, -24.4)$, $(-4, 12.0)$, $(-1.5, -2.3)$, $(0, 0)$, $(1.5, 2.3)$, $(4.0, -12.0)$ and $(6.8, 24.4)$.

6. $f(x) = \sin x + \frac{1}{3}\sin 3x \quad\Rightarrow\quad f'(x) = \cos x + \cos 3x \quad\Rightarrow\quad f''(x) = -\sin x - 3\sin 3x$

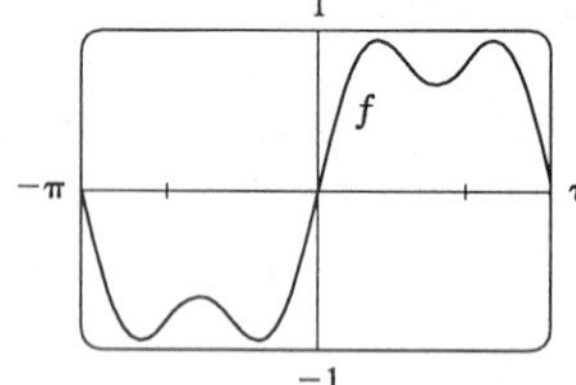

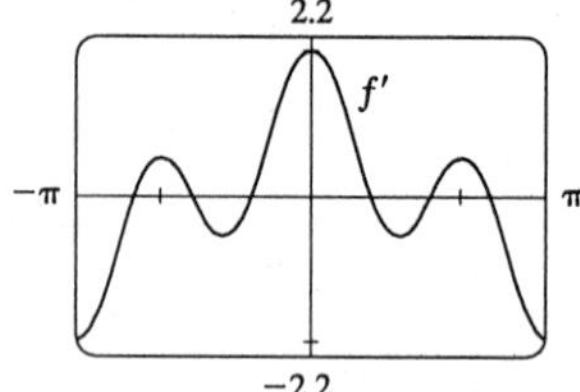

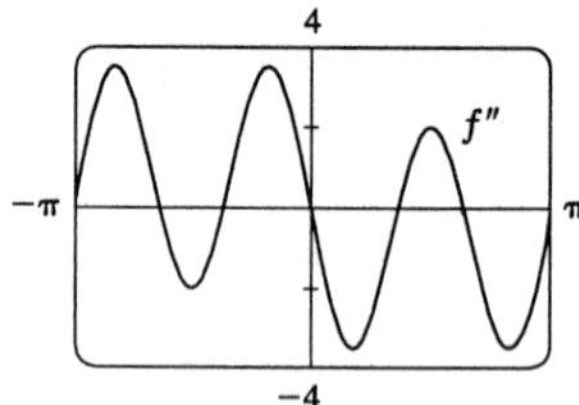

Note that f is periodic with period 2π, so we consider it on the interval $[-\pi, \pi]$. From the graph of f', we estimate that f is increasing on $[-2.4, -1.6]$, $[-0.8, 0.8]$, and $[1.6, 2.4]$ and decreasing on $[-\pi, -2.4]$, $[-1.6, -0.8]$, $[0.8, 1.6]$ and $[2.4, \pi]$. Maxima: $f(-1.6) \approx -0.7$, $f(0.8) \approx 0.9$, $f(2.4) \approx 0.9$. Minima: $f(-2.4) \approx -0.9$, $f(-0.8) \approx -0.9$, $f(1.6) \approx 0.7$. We estimate from the graph of f'' that f is CD on $(-2.0, -1.2)$, $(0, 1.2)$ and $(2.0, \pi)$ and CU on $(-\pi, -2.0)$, $(-1.2, 0)$ and $(1.2, 2)$. f has IP at $(-\pi, 0)$, $(-2.0, -0.8)$, $(-1.2, -0.8)$, $(0, 0)$, $(1.2, 0.8)$, $(2, 0.8)$, and $(\pi, 0)$.

7.

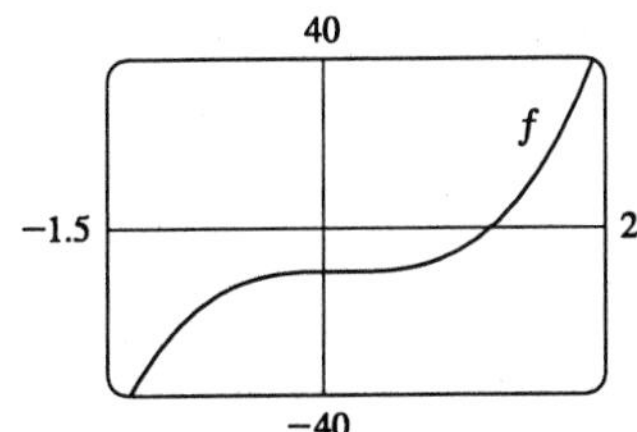

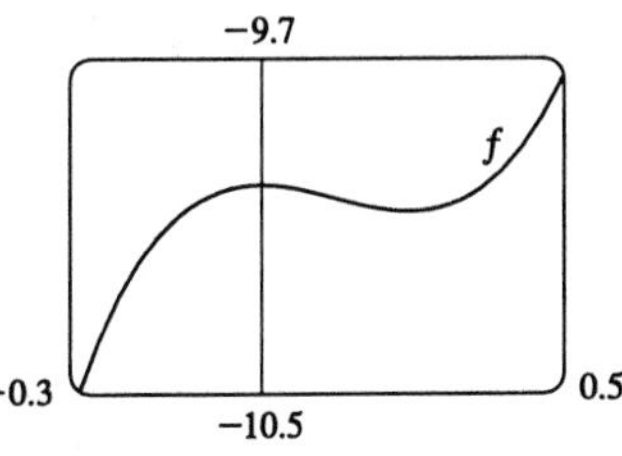

From the graphs, it appears that $f(x) = 8x^3 - 3x^2 - 10$ increases on $(-\infty, 0]$ and $[0.25, \infty)$ and decreases on $[0, 0.25]$; that f has a local maximum of $f(0) = -10.0$ and a local minimum of $f(0.25) \approx -10.1$; that f is CU on $(0.1, \infty)$ and CD on $(-\infty, 0.1)$; and that f has an IP at $(0.1, -10)$. $f(x) = 8x^3 - 3x^2 - 10 \Rightarrow$ $f'(x) = 24x^2 - 6x = 6x(4x - 1)$, which is positive ($f$ is increasing) for $(-\infty, 0]$ and $\left[\frac{1}{4}, \infty\right)$, and negative ($f$ is decreasing) on $\left[0, \frac{1}{4}\right]$. By the FDT, f has a local maximum at $x = 0$: $f(0) = 8(0)^3 - 3(0)^2 - 10 = -10$; and f has a local minimum at $\frac{1}{4}$: $f\left(\frac{1}{4}\right) = \frac{1}{8} - \frac{3}{16} - 10 = -\frac{161}{16}$. $f'(x) = 24x^2 - 6x \Rightarrow$ $f''(x) = 48x - 6 = 6(8x - 1)$, which is positive ($f$ is CU) on $\left(\frac{1}{8}, \infty\right)$, and negative ($f$ is CD) on $\left(-\infty, \frac{1}{8}\right)$. f has an IP at $\left(\frac{1}{8}, f\left(\frac{1}{8}\right)\right) = \left(\frac{1}{8}, -\frac{321}{32}\right)$.

8.

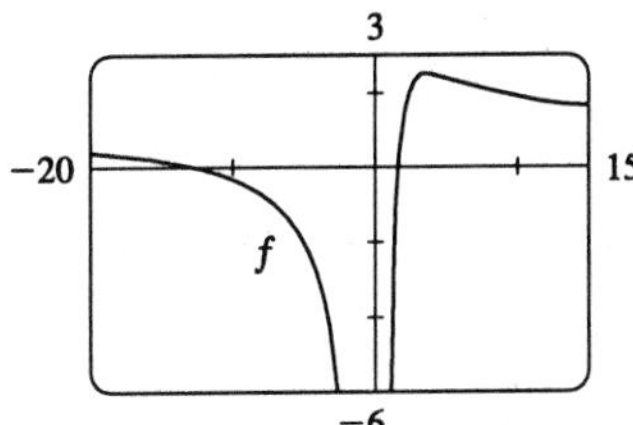

From the graphs, it appears that f increases on $(0, 3.6]$ and decreases on $(-\infty, 0)$ and $[3.6, \infty)$; that f has a local maximum of $f(3.6) \approx 2.5$ and no local minima; that f is CU on $(5.5, \infty)$ and CD on $(-\infty, 0)$ and $(0, 5.5)$; and that f has an IP at $(5.5, 2.3)$. $f(x) = \dfrac{x^2 + 11x - 20}{x^2} = 1 + \dfrac{11}{x} - \dfrac{20}{x^2} \Rightarrow$ $f'(x) = -11x^{-2} + 40x^{-3} = -x^{-3}(11x - 40)$, which is positive ($f$ is increasing) on $\left(0, \frac{40}{11}\right]$, and negative ($f$ is decreasing) on $(-\infty, 0)$ and on $\left[\frac{40}{11}, \infty\right)$. By the FDT, f has a local maximum at $x = \frac{40}{11}$:

$$f\left(\tfrac{40}{11}\right) = \frac{\left(\frac{40}{11}\right)^2 + 11\left(\frac{40}{11}\right) - 20}{\left(\frac{40}{11}\right)^2} = \frac{1600 + 11 \cdot 11 \cdot 40 - 20 \cdot 121}{1600} = \frac{201}{80};$$ and f has no local minima.

$f'(x) = -11x^{-2} + 40x^{-3} \Rightarrow f''(x) = 22x^{-3} - 120x^{-4} = 2x^{-4}(11x - 60)$, which is positive ($f$ is CU) on $\left(\frac{60}{11}, \infty\right)$, and negative ($f$ is CD) on $(-\infty, 0)$ and $\left(0, \frac{60}{11}\right)$. f has an IP at $\left(\frac{60}{11}, f\left(\frac{60}{11}\right)\right) = \left(\frac{60}{11}, \frac{211}{90}\right)$.

9.

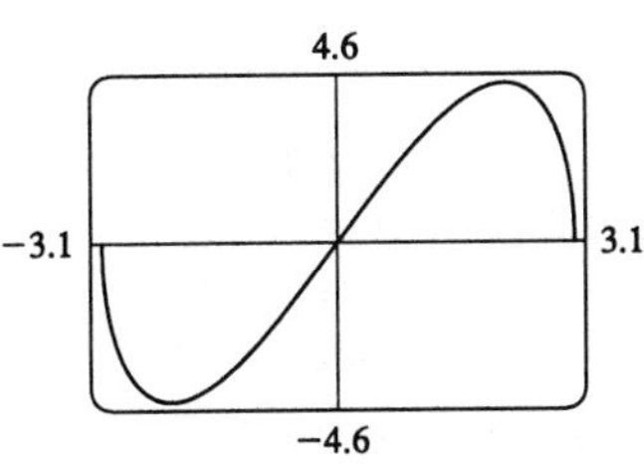

From the graph, it appears that f increases on $[-2.1, 2.1]$ and decreases on $[-3, -2.1]$ and $[2.1, 3]$; that f has a local maximum of $f(2.1) \approx 4.5$ and a local minimum of $f(-2.1) \approx -4.5$; that f is CU on $(-3.0, 0)$ and CD on $(0, 3.0)$, and that f has an IP at $(0, 0)$.

$f(x) = x\sqrt{9-x^2} \quad \Rightarrow \quad f'(x) = \dfrac{-x^2}{\sqrt{9-x^2}} + \sqrt{9-x^2} = \dfrac{9-2x^2}{\sqrt{9-x^2}}$, which is positive ($f$ is increasing) on $\left[\frac{-3\sqrt{2}}{2}, \frac{3\sqrt{2}}{2}\right]$ and negative (f is decreasing) on $\left[-3, \frac{-3\sqrt{2}}{2}\right]$ and $\left[\frac{3\sqrt{2}}{2}, 3\right]$. By the FDT, f has a local maximum of $f\left(\frac{3\sqrt{2}}{2}\right) = \frac{3\sqrt{2}}{2}\sqrt{9-\left(\frac{3\sqrt{2}}{2}\right)^2} = \frac{9}{2}$; and f has a local minimum of $f\left(\frac{-3\sqrt{2}}{2}\right) = -\frac{9}{2}$ (since f is an odd function.) $f'(x) = \dfrac{-x^2}{\sqrt{9-x^2}} + \sqrt{9-x^2} \quad \Rightarrow$

$$f''(x) = \frac{\sqrt{9-x^2}(-2x) + x^2\left(\frac{1}{2}\right)(9-x^2)^{-1/2}(-2x)}{9-x^2} - x(9-x^2)^{-1/2} = \frac{-2x - x^3(9-x^2)^{-1} - x}{\sqrt{9-x^2}}$$

$$= \frac{-3x}{\sqrt{9-x^2}} - \frac{x^3}{(9-x^2)^{3/2}} = \frac{x(2x^2-27)}{(9-x^2)^{3/2}},$$

which is positive (f is CU) on $(-3, 0)$, and negative (f is CD) on $(0, 3)$. f has an IP at $(0, 0)$.

10.

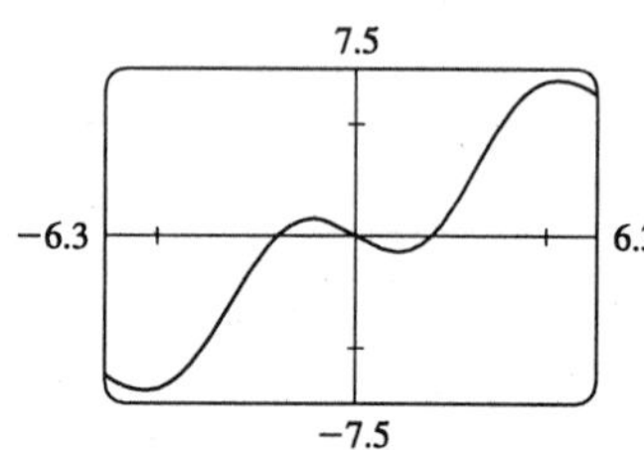

From the graph, it appears that f increases on $[-5.2, -1.0]$ and $[1.0, 5.2]$ and decreases on $[-2\pi, -5.2]$, $[-1.0, 1.0]$, and $[5.2, 2\pi]$; that f has local maxima of $f(-1.0) \approx 0.7$ and $f(5.2) \approx 7.0$ and minima of $f(-5.2) \approx -7.0$ and $f(1.0) \approx -0.7$; that f is CU on $(-2\pi, -3.1)$ and $(0, 3.1)$ and CD on $(-3.1, 0)$ and $(3.1, 2\pi)$, and that f has IP at $(0, 0)$, $(-3.1, -3.1)$ and $(3.1, 3.1)$.

$f(x) = x - 2\sin x \quad \Rightarrow \quad f'(x) = 1 - 2\cos x$, which is positive ($f$ is increasing) when $\cos x < \frac{1}{2}$, that is, on $\left[-\frac{5\pi}{3}, -\frac{\pi}{3}\right]$ and $\left[\frac{\pi}{3}, \frac{5\pi}{3}\right]$, and negative ($f$ is decreasing) on $\left[-2\pi, -\frac{5\pi}{3}\right]$, $\left[-\frac{\pi}{3}, \frac{\pi}{3}\right]$, and $\left[\frac{5\pi}{3}, 2\pi\right]$. By the FDT, f has local maxima of $f\left(-\frac{\pi}{3}\right) = \frac{\pi}{3} + \sqrt{3}$ and $f\left(\frac{5\pi}{3}\right) = \frac{5\pi}{3} + \sqrt{3}$, and local minima of $f\left(-\frac{5\pi}{3}\right) = -\frac{5\pi}{3} - \sqrt{3}$ and $f\left(\frac{\pi}{3}\right) = -\frac{\pi}{3} - \sqrt{3}$.

$f'(x) = 1 - 2\cos x \quad \Rightarrow \quad f''(x) = 2\sin x$, which is positive ($f$ is CU) on $(-2\pi, -\pi)$ and $(0, \pi)$ and negative (f is CD) on $(-\pi, 0)$ and $(\pi, 2\pi)$. f has IP at $(0, 0)$, $(-\pi, -\pi)$ and (π, π).

11.

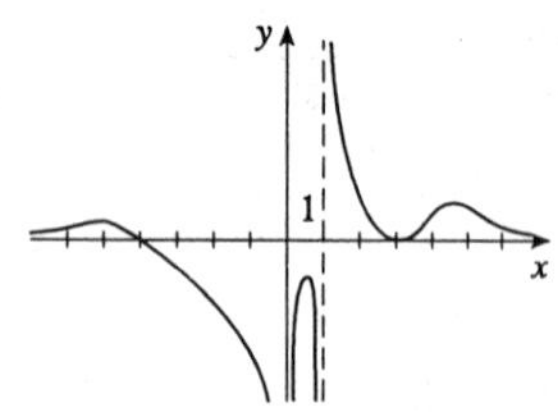

$f(x) = \dfrac{(x+4)(x-3)^2}{x^4(x-1)}$ has VA at $x = 0$ and at $x = 1$ since $\lim\limits_{x\to 0} f(x) = -\infty$, $\lim\limits_{x\to 1^-} f(x) = -\infty$ and $\lim\limits_{x\to 1^+} f(x) = \infty$.

$f(x) = \dfrac{(1 + 4/x)(1 - 3/x)^2}{x(x-1)} \to 0^+$ as $x \to \pm\infty$, so f is asymptotic to the x-axis. Since f is undefined at $x = 0$, it has no y-intercept. $f(x) = 0 \quad \Rightarrow \quad (x+4)(x-3)^2 = 0$ $\Rightarrow \quad x = -4$ or $x = 3$, so f has x-intercepts -4 and 3. Note, however, that the graph of f is only tangent to the x-axis and does not cross it at $x = 3$, since f is positive as $x \to 3^-$ and as $x \to 3^+$.

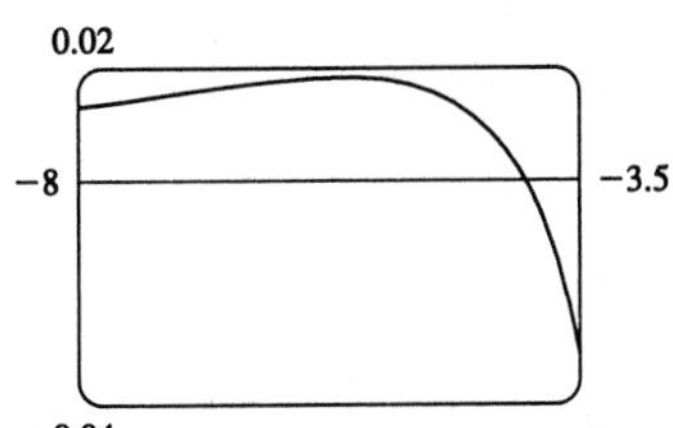

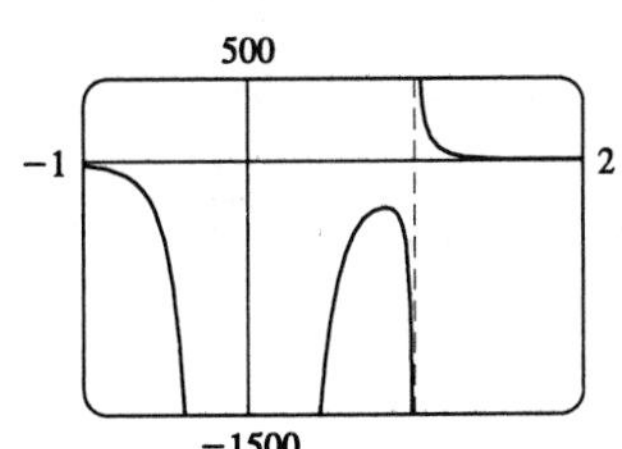

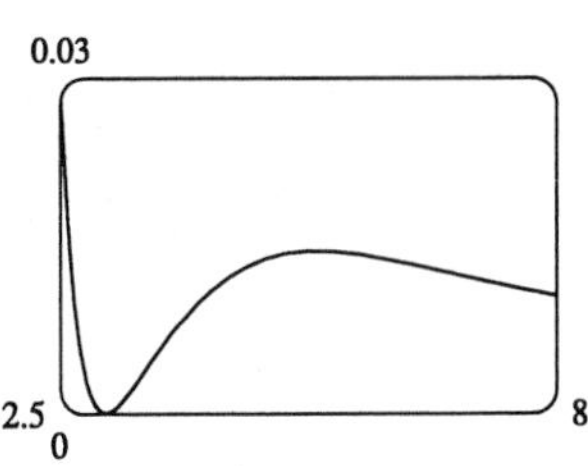

From these graphs, it appears that f has three maxima and one minimum. The maxima are approximately $f(-5.6) = 0.0182$, $f(0.82) = -281.5$ and $f(5.2) = 0.0145$ and we know (since the graph is tangent to the x-axis at $x = 3$) that the minimum is $f(3) = 0$.

12.

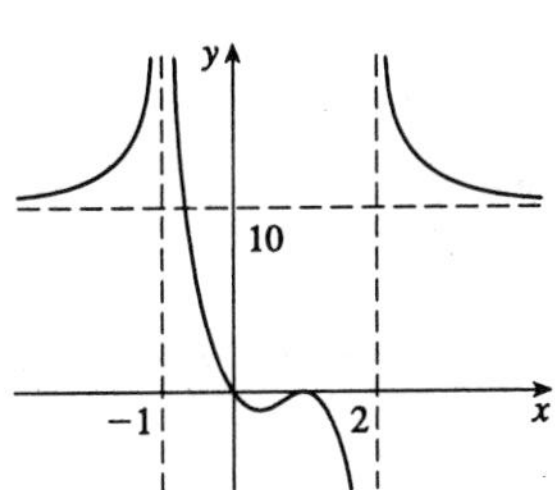

$f(x) = \dfrac{10x(x-1)^4}{(x-2)^3(x+1)^2}$ has VA at $x = -1$ and at $x = 2$ since $\lim\limits_{x\to-1} f(x) = \infty$, $\lim\limits_{x\to2^-} f(x) = -\infty$ and $\lim\limits_{x\to2^+} f(x) = \infty$.

$f(x) = \dfrac{10(1-1/x)^4}{(1-2/x)^3(1+1/x)^2} \to 10$ as $x \to \pm\infty$, so f is asymptotic to the line $y = 10$. $f(0) = 0$, so f has a y-intercept at 0. $f(x) = 0 \;\Rightarrow\; 10x(x-1)^4 = 0 \;\Rightarrow\; x = 0$ or $x = 1$.

So f has x-intercepts 0 and 1. Note, however, that f does not change sign at $x = 1$, so the graph is tangent to the x-axis and does not cross it.

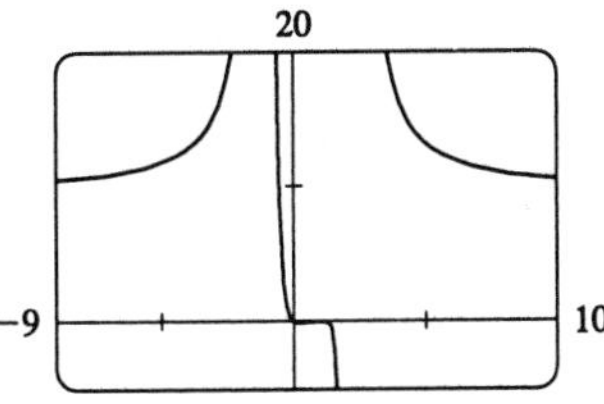

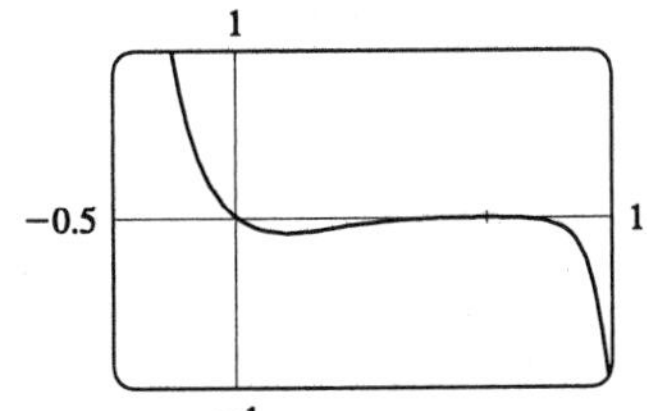

We know (since the graph is tangent to the x-axis at $x = 1$) that the maximum is $f(1) = 0$. From the graphs it appears that the minimum is about $f(0.2) = -0.1$.

13. We use `diff(f,x);` (in Maple) or `Dt[f,x]` (in Mathematica) on the function $f(x) = \dfrac{x^2(x+1)^3}{(x-2)^2(x-4)^4}$, and get $f'(x) = 2\dfrac{x(x+1)^3}{(x-2)^2(x-4)^4} + 3\dfrac{x^2(x+1)^2}{(x-2)^2(x-4)^4} - 2\dfrac{x^2(x+1)^3}{(x-2)^3(x-4)^4} - 4\dfrac{x^2(x+1)^3}{(x-2)^2(x-4)^5}$.

If we then use a CAS to simplify this expression, we get $f'(x) = -\dfrac{x(x+1)^2(x^3+18x^2-44x-16)}{(x-2)^3(x-4)^5}$.

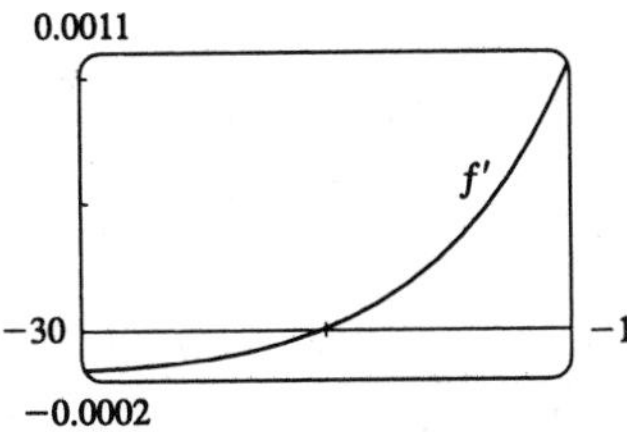

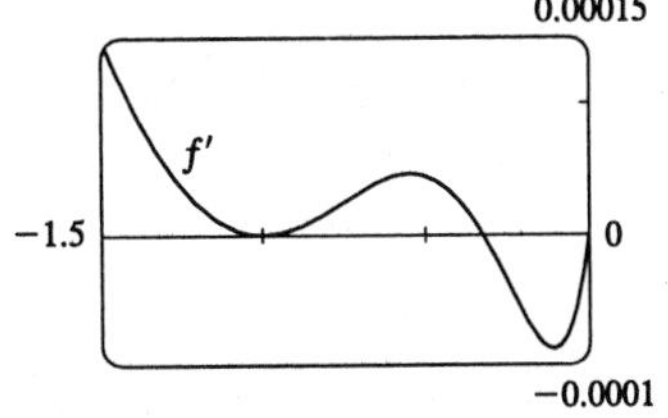

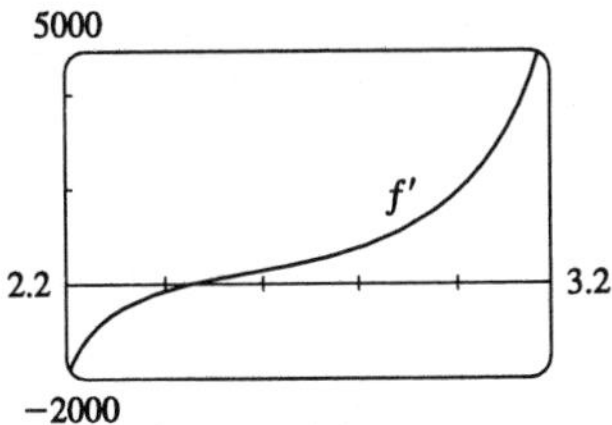

From the graphs of f', it seems that the critical points which indicate extrema occur at $x \approx -20$, -0.3, and 2.5, as estimated in Example 3. (There is another critical point at $x = -1$, but the sign of f' does not change there.)

We differentiate again, and after simplifying, we find that

$$f''(x) = 2\frac{(x+1)(x^6 + 36x^5 + 6x^4 - 628x^3 + 684x^2 + 672x + 64)}{(x-2)^4(x-4)^6}.$$

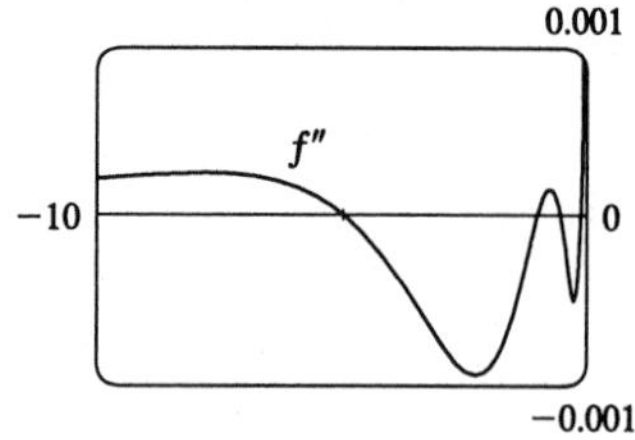

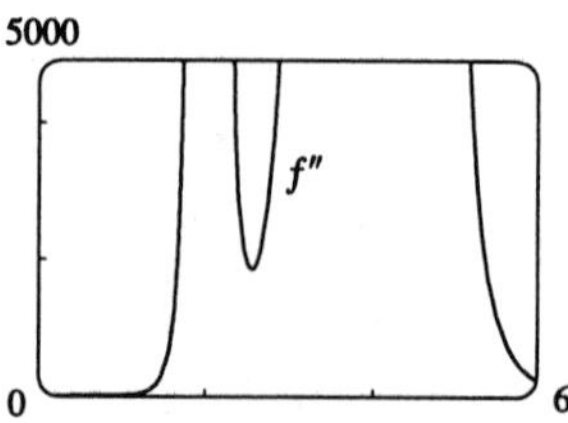

From the graphs of f'', it appears that f is CU on $(-\infty, -5.0)$, $(-1.0, -0.5)$, $(-0.1, 2.0)$, $(2.0, 4.0)$ and $(4.0, \infty)$ and CD on $(-5.0, -1.0)$ and $(-0.5, -0.1)$. We check back on the graphs of f to find the y-coordinates of the inflection points, and find that these points are approximately $(-5, -0.005)$, $(-1, 0)$, $(-0.5, 0.00001)$, and $(-0.1, 0.0000066)$.

14. We use `diff(f,x);` (in Maple) or `Dt[f,x]` (in Mathematica) on the function $f(x) = \dfrac{10x(x-1)^4}{(x-2)^3(x+1)^2}$, and get $f'(x) = 10\dfrac{(x-1)^4}{(x-2)^3(x+1)^2} + 40\dfrac{x(x-1)^3}{(x-2)^3(x+1)^2} - 30\dfrac{x(x-1)^4}{(x-2)^4(x+1)^2} - 20\dfrac{x(x-1)^4}{(x-2)^3(x+1)^3}$. We simplify this expression to get $f'(x) = -20\dfrac{(x-1)^3(5x-1)}{(x-2)^4(x+1)^3}$.

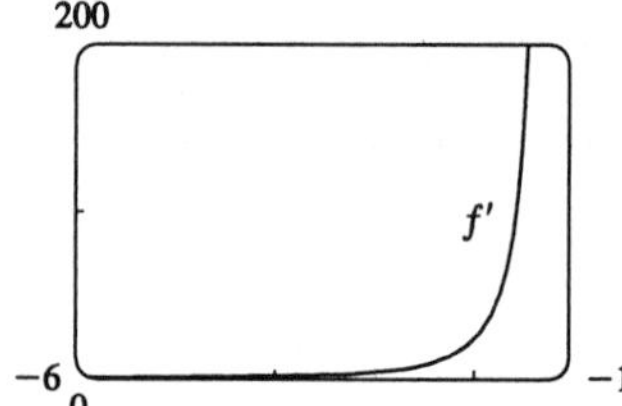

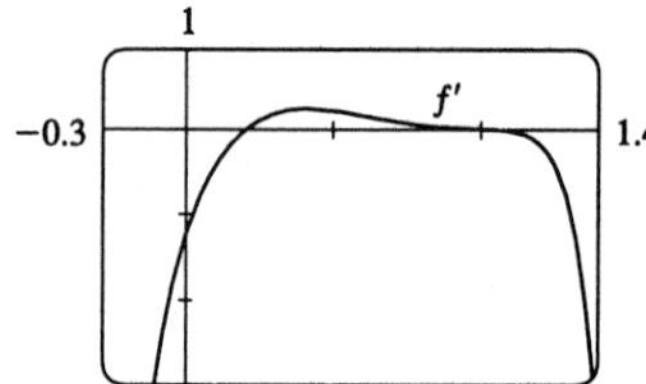

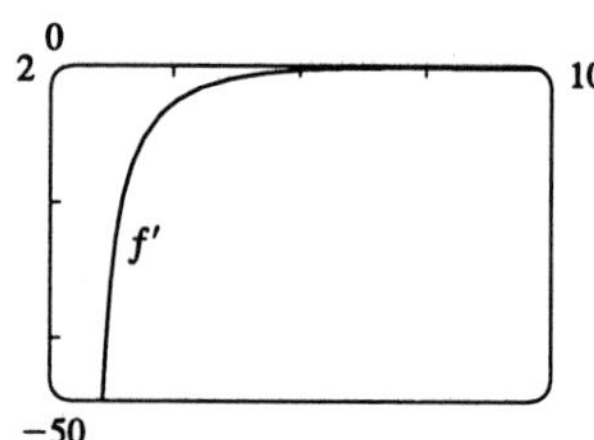

From the graphs of f', we estimate that f is increasing on $(-\infty, -1)$ and $[0.2, 1]$ and decreasing on $(-1, 0.2]$, $(1, 2)$ and $(2, \infty)$. We use the `diff` command on $f'(x)$, and get an unwieldy expression for $f''(x)$, which we simplify to get $f''(x) = 60\dfrac{(x-1)^2(5x^3 - 8x^2 + 17x - 6)}{(x-2)^5(x+1)^4}$.

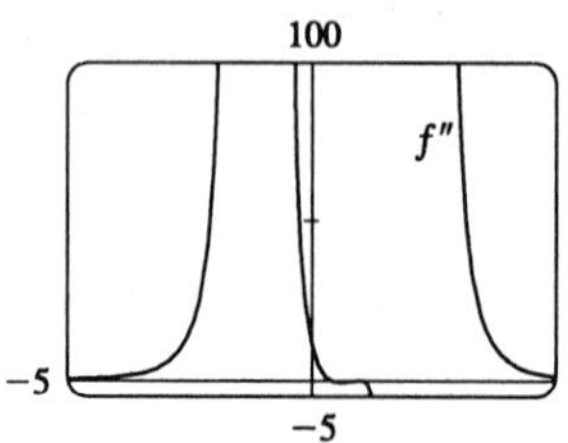

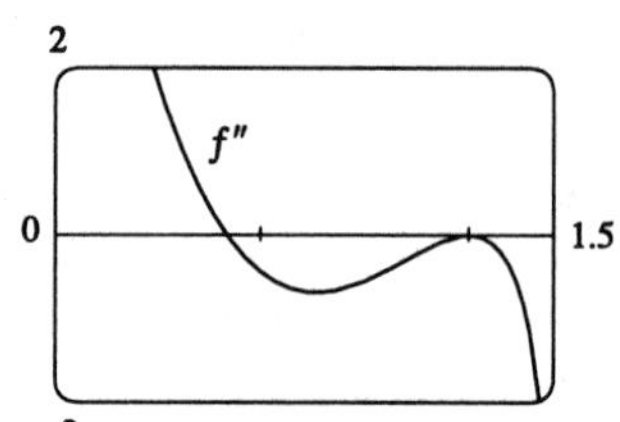

From the graphs of f'', it seems that f is CU on $(-\infty, -1.0)$, $(-1.0, 0.4)$ and $(2.0, \infty)$, and CD on $(0.4, 2)$.

15.

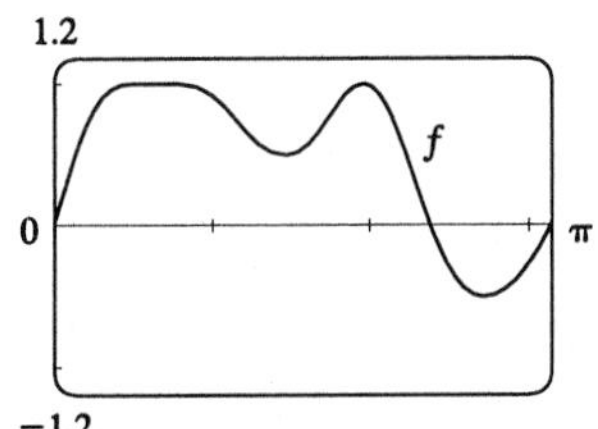

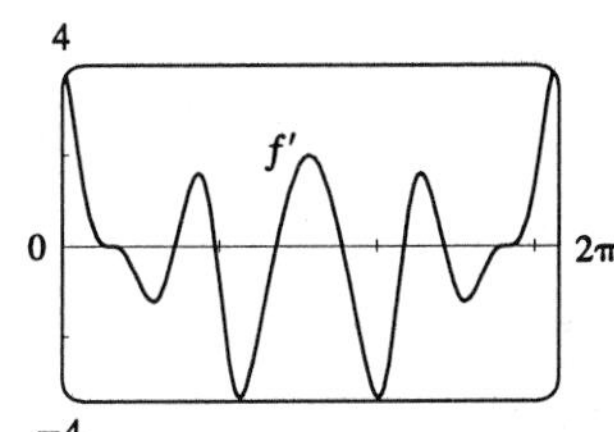

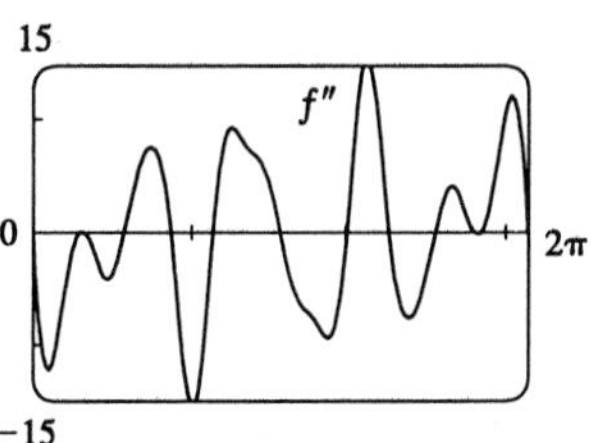

We consider the function only on the interval $[0, \pi]$ and use symmetry to extend. From the graph of f in the viewing rectangle $[0, \pi]$ by $[-1.2, 1.2]$, it looks like f has two maxima and two minima. If we calculate and graph $f'(x) = [\cos(x + \sin 3x)](1 + 3\cos 3x)$ on the same x-interval, we see that the graph of f' appears to be almost tangent to the x-axis at about $x = 0.7$. The graph of $f'' = -[\sin(x + \sin 3x)](1 + 3\cos 3x)^2 + \cos(x + \sin 3x)(-9\sin 3x)$ is even more interesting near this x-value: it seems to just touch the x-axis.

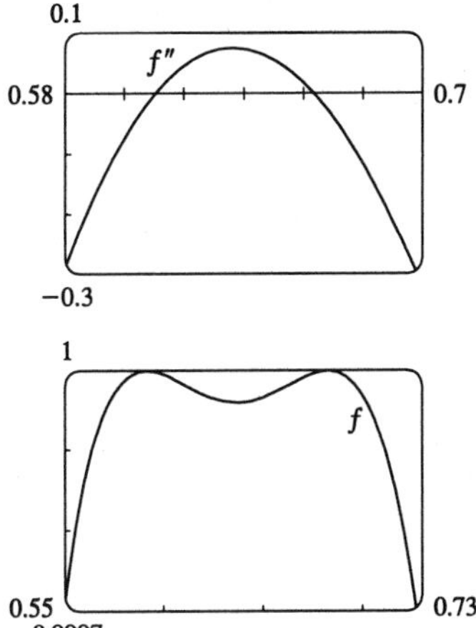

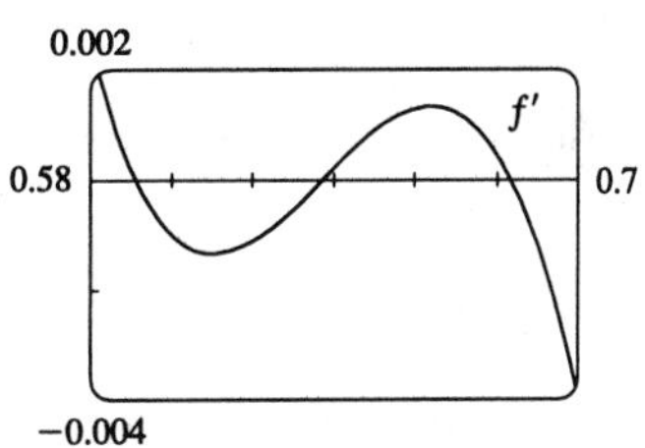

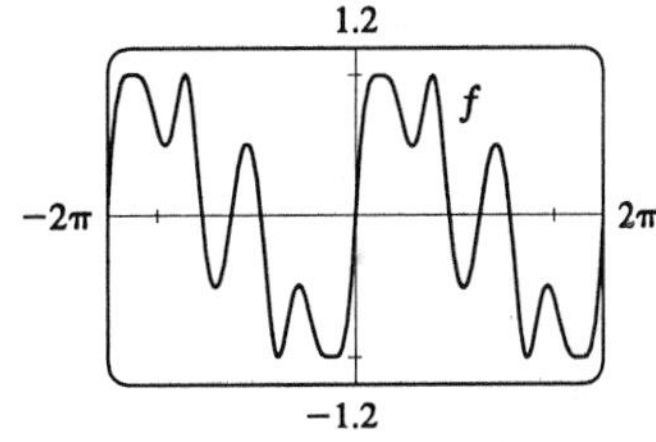

If we zoom in on this place on the graph of f'', we see that f'' actually does cross the axis twice near $x = 0.65$, indicating a change in concavity for a very short interval. If we look at the graph of f' on the same interval, we see that it changes sign three times near $x = 0.65$, indicating that what we had thought was a broad extremum at about $x = 0.7$ actually consists of three extrema (two maxima and a minimum). These maxima are roughly $f(0.59) = 1$ and $f(0.68) = 1$, and the minimum is roughly $f(0.64) = 0.99996$. There are also a maximum of about $f(1.96) = 1$ and minima of about $f(1.46) = 0.49$ and $f(2.73) = -0.51$. The points of inflection are roughly $(0.61, 0.99998)$, $(0.66, 0.99998)$, $(1.17, 0.72)$, $(1.75, 0.77)$, and $(2.28, 0.34)$.

Note that the function is odd and periodic with period 2π, and it is also rotationally symmetric about all points of the form $((2n + 1)\pi, 0)$, n an integer.

16. We need only consider the function $f(x) = x^2\sqrt{c^2 - x^2}$ for $c \geq 0$, because if c is replaced by $-c$, the function is unchanged. For $c = 0$, the graph consists of the single point $(0, 0)$. The domain of f is $[-c, c]$, and the graph of f is symmetric about the y-axis.

$$f'(x) = 2x\sqrt{c^2 - x^2} + x^2 \frac{-2x}{2\sqrt{c^2 - x^2}} = 2x\sqrt{c^2 - x^2} - \frac{x^3}{\sqrt{c^2 - x^2}} = \frac{2x(c^2 - x^2) - x^3}{\sqrt{c^2 - x^2}} = -\frac{3x\left(x^2 - \frac{2}{3}c^2\right)}{\sqrt{c^2 - x^2}}.$$

So we see that all members of the family of curves have horizontal tangents at $x = 0$, since $f'(0) = 0$ for all $c > 0$. Also, the tangents to all the curves become very steep as $x \to \pm c$, since $\lim_{x \to -c^+} f'(x) = \infty$ and $\lim_{x \to c^-} f'(x) = -\infty$. We set $f'(x) = 0 \quad \Leftrightarrow \quad x = 0$ or $x^2 - \frac{2}{3}c^2 = 0$, so the absolute maxima are $f\left(\pm\sqrt{\frac{2}{3}}c\right) = \frac{2}{3\sqrt{3}}c^3$.

$$f''(x) = \frac{(-9x^2 + 2c^2)\sqrt{c^2 - x^2} - (-3x^3 + 2c^2x)\left(-x/\sqrt{c^2 - x^2}\right)}{c^2 - x^2} = \frac{6x^4 - 9c^2x^2 + 2c^4}{(c^2 - x^2)^{3/2}}.$$ Using the quadratic formula, we find that $f''(x) = 0 \quad \Leftrightarrow \quad x^2 = \dfrac{9c^2 \pm c^2\sqrt{33}}{12}$. Since $-c < x < c$, we take $x^2 = \frac{9 - \sqrt{33}}{12}c^2$, so the inflection points are $\left(\pm\sqrt{\frac{9 - \sqrt{33}}{12}}c, \frac{\left(9 - \sqrt{33}\right)\left(\sqrt{33} - 3\right)}{144}c^3\right)$.

From these calculations we can see that the maxima and the points of inflection get both horizontally and vertically further from the origin as c increases. Since all of the functions have two maxima and two inflection points, we see that the basic shape of the curve does not change as c changes.

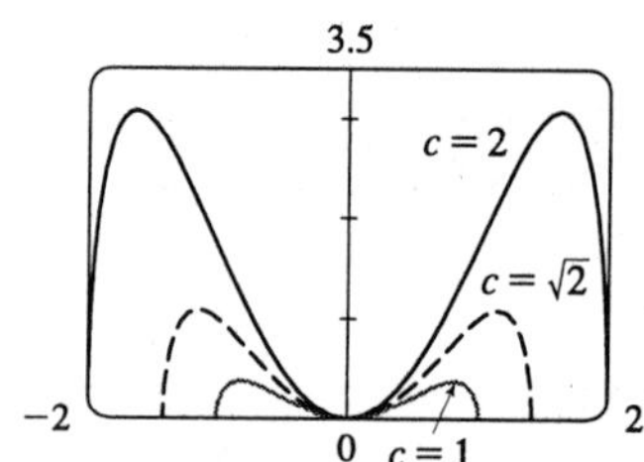

17. Note that $c = 0$ is a transitional value at which the graph consists of the x-axis. Also, we can see that if we substitute $-c$ for c, the function $f(x) = \dfrac{cx}{1 + c^2x^2}$ will be reflected in the x-axis, so we investigate only positive values of c (except $c = -1$, as a demonstration of this reflective property). Also, f is an odd function. $\lim_{x \to \pm\infty} f(x) = 0$, so $y = 0$ is a horizontal asymptote for all c. We calculate

$$f'(x) = \frac{c(1 + c^2x^2) - cx(2c^2x)}{(1 + c^2x^2)^2} = -\frac{c(c^2x^2 - 1)}{(1 + c^2x^2)^2}.$$ So there is an absolute maximum of $f(1/c) = \frac{1}{2}$ and an absolute minimum of $f(-1/c) = -\frac{1}{2}$. These extrema have the same value regardless of c, but the maximum points move closer to the y-axis as c increases.

$$f''(x) = \frac{(-2c^3x)(1 + c^2x^2)^2 - (-c^3x^2 + c)[2(1 + c^2x^2)(2c^2x)]}{(1 + c^2x^2)^4}$$
$$= \frac{(-2cx)(1 + c^2x^2) + (c^3x^2 - c)(4c^2x)}{(1 + c^2x^2)^3} = \frac{2c^3x(c^2x^2 - 3)}{(1 + c^2x^2)^3},$$

so there are inflection points at $(0, 0)$ and at $\left(\pm\sqrt{3}/c, \pm\sqrt{3}/4\right)$. Again, the y-coordinate of the inflection points does not depend on c, but as c increases, both inflection points approach the y-axis.

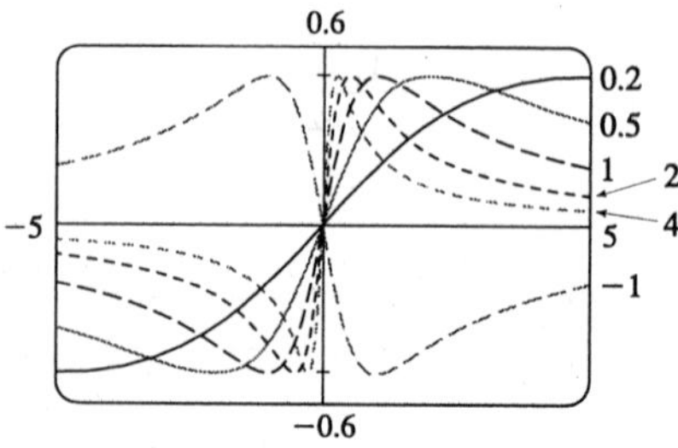

18. Note that f is an even function, and also that $\lim_{x\to\pm\infty} \dfrac{1}{(1-x^2)^2+cx^2} = 0$ for any value of c, so $y=0$ is a horizontal asymptote. We calculate the derivatives: $f'(x) = \dfrac{-4(1-x^2)x+2cx}{\left[(1-x^2)^2+cx^2\right]^2} = \dfrac{4x\left[x^2+\left(\frac{1}{2}c-1\right)\right]}{\left[(1-x^2)^2+cx^2\right]^2}$, and $f''(x) = 2\dfrac{10x^6+(9c-18)x^4+(3c^2-12c+6)x^2+2-c}{\left[x^4+(c-2)x^2+1\right]^2}$. We first consider the case $c>0$. Then the denominator of f' is positive, that is, $(1-x^2)^2+cx^2>0$ for all x, so f has domain $\mathbb{R}$ and furthermore $f>0$. If $\frac{1}{2}c-1\ge 0$, that is, $c\ge 2$, then the only critical point is $f(0)=1$, a maximum. Graphing a few examples for $c\ge 2$ shows that there are two inflection points which approach the y-axis as $c\to\infty$.

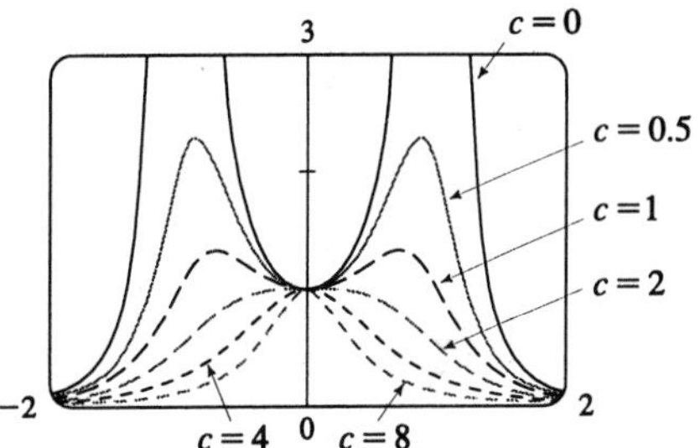

$c=2$ and $c=0$ are transitional values of c at which the shape of the curve changes. For $0<c<2$, there are three critical points: $f(0)=1$, a minimum, and $f\left(\pm\sqrt{1-\frac{1}{2}c}\right) = \dfrac{1}{c(1-c/4)}$, both maxima. As c decreases from 2 to 0, the maximum values get larger and larger, and the x-values at which they occur go from 0 to ± 1. Graphs show that there are four inflection points for $0<c<2$, and that they get farther away from the origin, both vertically and horizontally, as $c\to 0^+$. For $c=0$, the function is simply asymptotic to the x-axis and to the lines $x=\pm 1$, approaching $+\infty$ from both sides of each. The y-intercept is 1, and $(0,1)$ is a local minimum. There are no inflection points. Now if $c<0$, we can write $f(x) = \dfrac{1}{(1-x^2)^2+cx^2} = \dfrac{1}{(1-x^2)^2-\left(\sqrt{-c}x\right)^2} = \dfrac{1}{\left(x^2-\sqrt{-c}x-1\right)\left(x^2+\sqrt{-c}x-1\right)}$. So f has vertical asymptotes where $x^2\pm\sqrt{-c}x-1=0 \quad\Leftrightarrow\quad x = \dfrac{-\sqrt{-c}\pm\sqrt{4-c}}{2}$ or $x = \dfrac{\sqrt{-c}\pm\sqrt{4-c}}{2}$.

As c decreases, the two exterior asymptotes move away from the origin, while the two interior ones move toward it. We graph a few examples to see the behavior of the graph near the asymptotes, and the nature of the critical points $x=0$ and $x=\pm\sqrt{1-\frac{1}{2}c}$:

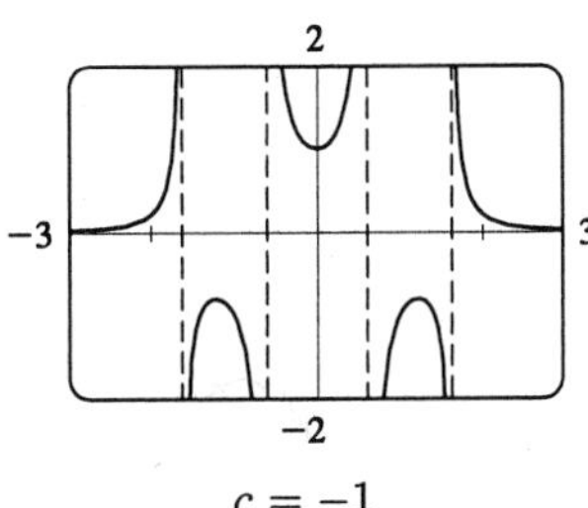

$c=-1$

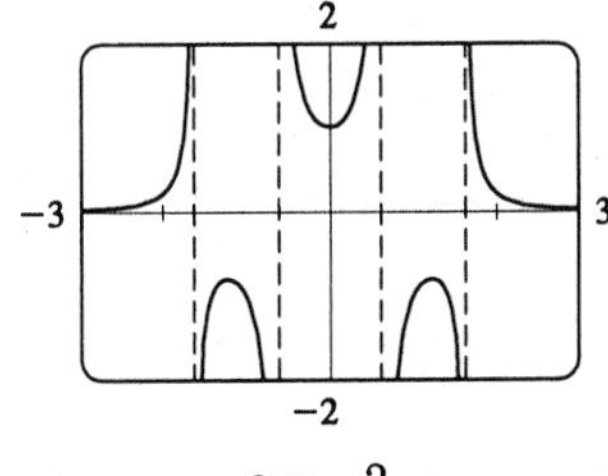

$c=-2$

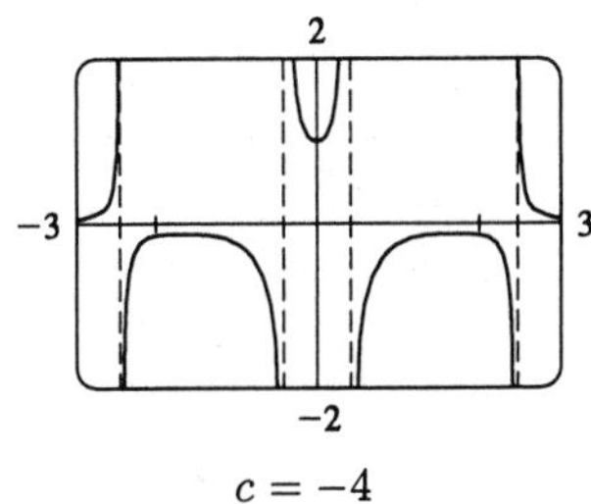

$c=-4$

We see that there is one local minimum, $f(0)=1$, and there are two local maxima, $f\left(\pm\sqrt{1-\frac{1}{2}c}\right) = \dfrac{1}{c(1-c/4)}$ as before. As c decreases, the x-values at which these maxima occur get larger, and the maximum values themselves approach 0, though they are always negative.

19. $f(x) = x^4 + cx^2 = x^2(x^2 + c)$. Note that f is an even function.

For $c \geq 0$, the only x-intercept is the point $(0, 0)$. We calculate $f'(x) = 4x^3 + 2cx = 4x\left(x^2 + \frac{1}{2}c\right) \Rightarrow$ $f''(x) = 12x^2 + 2c$. If $c \geq 0$, $x = 0$ is the only critical point and there are no inflection points. As we can see from the examples, there is no change in the basic shape of the graph for $c \geq 0$; it merely becomes steeper as c increases. For $c = 0$, the graph is the simple curve $y = x^4$.

For $c < 0$, there are x-intercepts at 0 and at $\pm\sqrt{-c}$. Also, there is a maximum at $(0, 0)$, and there are minima at $\left(\pm\sqrt{-\frac{1}{2}c}, -\frac{1}{4}c^2\right)$. As $c \to -\infty$, the x-coordinates of these minima get larger in absolute value, and the minimum points move downward. There are inflection points at $\left(\pm\sqrt{-\frac{1}{6}c}, -\frac{5}{36}c^2\right)$, which also move away from the origin as $c \to -\infty$.

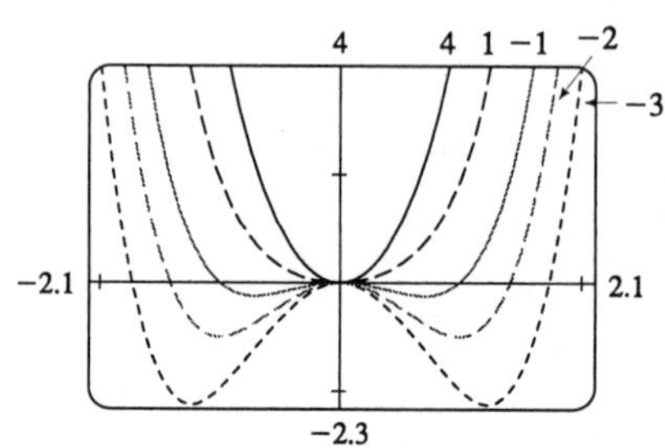

20. For $c = 0$, there are no inflection points; the curve is CU everywhere. If c increases, the curve simply becomes steeper, and there are still no inflection points. If c starts at 0 and decreases, a slight upward bulge appears near $x = 0$, so that there are two inflection points for any $c < 0$. This can be seen algebraically by calculating the second derivative: $f(x) = x^4 + cx^2 + x \Rightarrow f'(x) = 4x^3 + 2cx + 1 \Rightarrow f''(x) = 12x^2 + 2c$. Thus $f''(x) \geq 0$ when $c > 0$. For $c < 0$, there are inflection points when $x = \pm\sqrt{-\frac{1}{6}c}$.

For $c = 0$, the graph has one critical number, at the absolute minimum somewhere around $x = -0.6$. As c increases, the number of critical points does not change. If c instead decreases from 0, we see that the graph eventually sprouts another local minimum, to the right of the origin, somewhere between $x = 1$ and $x = 2$. Consequently, there is also a maximum near $x = 0$.

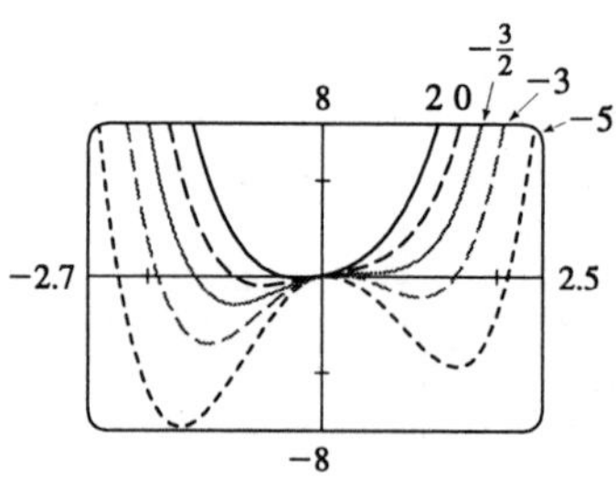

After a bit of experimentation, we find that at $c = -1.5$, there appear to be two critical numbers: the absolute minimum at about $x = -1$, and a horizontal tangent with no extremum at about $x = 0.5$. For any c smaller than this there will be 3 critical points, as shown in the graphs with $c = -3$ and with $c = -5$. To prove this algebraically, we calculate $f'(x) = 4x^3 + 2cx + 1$. Now if we substitute our value of $c = -1.5$, the formula for $f'(x)$ becomes $4x^3 - 3x + 1 = (x + 1)(2x - 1)^2$. This has a double root at $x = \frac{1}{2}$, indicating that the function has two critical points: $x = -1$ and $x = \frac{1}{2}$, just as we had guessed from the graph.

EXERCISES 3.8

1. If x is one number, the other is $100 - x$. Maximize $f(x) = x(100 - x) = 100x - x^2$. $f'(x) = 100 - 2x = 0$ $\Rightarrow$ $x = 50$. Now $f''(x) = -2 < 0$, so there is an absolute maximum at $x = 50$. The numbers are 50 and 50.

2. The two numbers are $x + 100$ and x. Minimize $f(x) = (x + 100)\,x = x^2 + 100x$. $f'(x) = 2x + 100 = 0$ $\Rightarrow$ $x = -50$. Since $f''(x) = 2 > 0$, there is an absolute minimum at $x = -50$. The two numbers are 50 and -50.

3. The two numbers are x and $\dfrac{100}{x}$ where $x > 0$. Minimize $f(x) = x + \dfrac{100}{x}$. $f'(x) = 1 - \dfrac{100}{x^2} = \dfrac{x^2 - 100}{x^2}$. The critical number is $x = 10$. Since $f'(x) < 0$ for $0 < x < 10$ and $f'(x) > 0$ for $x > 10$, there is an absolute minimum at $x = 10$. The numbers are 10 and 10.

4. Let the rectangle have sides x and y and area A, so $A = xy$ or $y = A/x$. The problem is to minimize the perimeter $= 2x + 2y = 2x + 2A/x = P(x)$. Now $P'(x) = 2 - 2A/x^2 = 2(x^2 - A)/x^2$. So the critical number is $x = \sqrt{A}$. Since $P'(x) < 0$ for $0 < x < \sqrt{A}$ and $P'(x) > 0$ for $x > \sqrt{A}$, there is an absolute minimum at $x = \sqrt{A}$. The sides of the rectangle are $\sqrt{A}$ and $A/\sqrt{A} = \sqrt{A}$, so the rectangle is a square.

5. Let p be the perimeter and x and y the lengths of the sides, so $p = 2x + 2y$ $\Rightarrow$ $y = \frac{1}{2}p - x$. The area is $A(x) = x\left(\frac{1}{2}p - x\right) = \frac{1}{2}px - x^2$. Now $0 = A'(x) = \frac{1}{2}p - 2x$ $\Rightarrow$ $x = \frac{1}{4}p$. Since $A''(x) = -2 < 0$, there is an absolute maximum where $x = \frac{1}{4}p$. The sides of the rectangle are $\frac{1}{4}p$ and $\frac{1}{2}p - \frac{1}{4}p = \frac{1}{4}p$, so the rectangle is a square.

6. 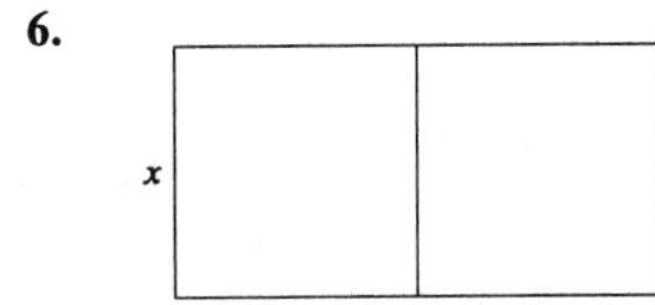

$xy = 1.5 \times 10^6$, so $y = 1.5 \times 10^6/x$. Minimize the amount of fencing, which is $3x + 2y = 3x + 2\left(1.5 \times 10^6/x\right) = 3x + 3 \times 10^6/x = F(x)$. $F'(x) = 3 - 3 \times 10^6/x^2 = 3(x^2 - 10^6)/x^2$. The critical number is $x = 10^3$ and $F'(x) < 0$ for $0 < x < 10^3$ and $F'(x) > 0$ if $x > 10^3$, so the absolute minimum occurs when $x = 10^3$ and $y = 1.5 \times 10^3$.

The field should be 1000 feet by 1500 feet with the middle fence parallel to the short side of the field.

7. 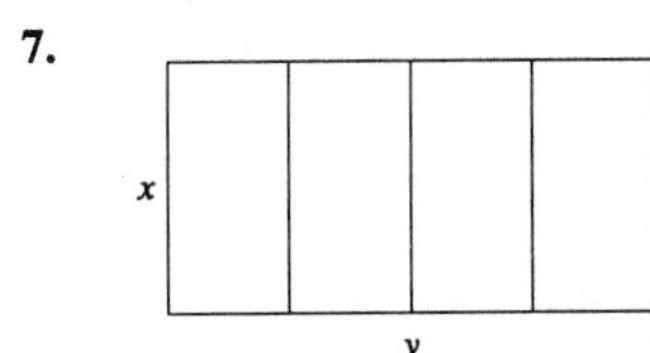

Here $5x + 2y = 750$ so $y = (750 - 5x)/2$. Maximize $A = xy = x(750 - 5x)/2 = 375x - \frac{5}{2}x^2$. Now $A'(x) = 375 - 5x = 0$ $\Rightarrow$ $x = 75$. Since $A''(x) = -5 < 0$ there is an absolute maximum when $x = 75$. Then $y = \frac{375}{2}$. The largest area is $75\left(\frac{375}{2}\right) = 14{,}062.5\text{ ft}^2$.

8. Let b be the area of the base of the box and h be its height, so $32{,}000 = hb^2$ or $h = 32{,}000/b^2$. The surface area of the open box is $b^2 + 4hb = b^2 + 4\left(32{,}000/b^2\right)b = b^2 + 4(32{,}000)/b$. So $V'(b) = 2b - 4(32{,}000)/b^2 = 2(b^3 - 64{,}000)/b^2 = 0$ $\Leftrightarrow$ $b = \sqrt[3]{64{,}000} = 40$. This gives an absolute minimum since $V'(b) < 0$ if $b < 40$ and $V'(b) > 0$ if $b > 40$. The box should be $40 \times 40 \times 20$.

9. Let b be the base of the box and h the height. The surface area is $1200 = b^2 + 4hb \Rightarrow h = (1200 - b^2)/(4b)$. The volume is $V = b^2h = b^2(1200 - b^2)/4b = 300b - b^3/4 \Rightarrow V'(b) = 300 - \frac{3}{4}b^2$. $V'(b) = 0 \Rightarrow b = \sqrt{400} = 20$. Since $V'(b) > 0$ for $0 < b < 20$ and $V'(b) < 0$ for $b > 20$, there is an absolute maximum when $b = 20$. Then $h = 10$, so the largest possible volume is $(20)^2(10) = 4000\text{ cm}^3$.

10. $10 = (2w)(w)\,h = 2w^2h$ so $h = 5/w^2$. The cost is $10(2w^2) + 6[2(2wh) + 2hw] = 20w^2 + 36wh$, so $C(w) = 20w^2 + 36w(5/w^2) = 20w^2 + 180/w$. $C'(w) = 40w - 180/w^2 = 40\left(w^3 - \frac{9}{2}\right)/w^2 \Rightarrow w = \sqrt[3]{\frac{9}{2}}$ is the critical number. There is an absolute minimum for $w = \sqrt[3]{\frac{9}{2}}$ since $C'(w) < 0$ for $0 < w < \sqrt[3]{\frac{9}{2}}$ and $C'(w) > 0$ for $w > \sqrt[3]{\frac{9}{2}}$. $C\left(\sqrt[3]{\frac{9}{2}}\right) = 20\left(\sqrt[3]{\frac{9}{2}}\right)^2 + \frac{180}{\sqrt[3]{9/2}} \approx \$163.54$.

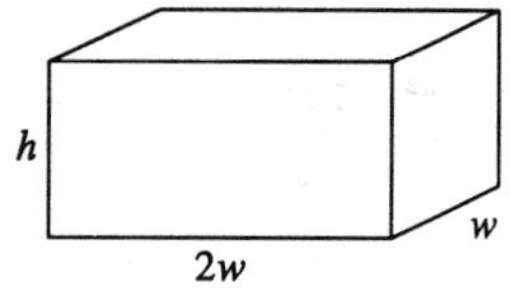

11. $10 = (2w)(w)\,h = 2w^2h$, so $h = 5/w^2$. The cost is $C(w) = 10(2w^2) + 6[2(2wh) + 2hw] + 6(2w^2) = 32w^2 + 36wh = 32w^2 + 180/w$.
$C'(w) = 64w - 180/w^2 = 4(16w^3 - 45)/w^2 \Rightarrow w = \sqrt[3]{\frac{45}{16}}$ is the critical number. $C'(w) < 0$ for $0 < w < \sqrt[3]{\frac{45}{16}}$ and $C'(w) > 0$ for $w > \sqrt[3]{\frac{45}{16}}$. The minimum cost is $C\left(\sqrt[3]{\frac{45}{16}}\right) = 32(2.8125)^{2/3} + 180/\sqrt[3]{2.8125} \approx \$191.28$.

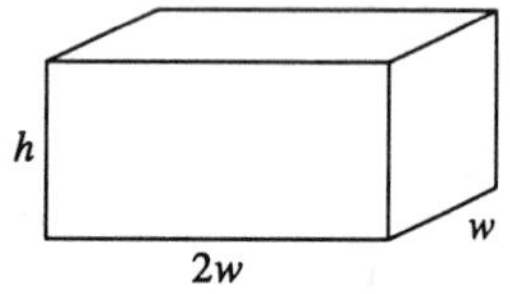

12. $V(x) = x(3 - 2x)^2 = x(4x^2 - 12x + 9) = 4x^3 - 12x^2 + 9x \Rightarrow$
$V'(x) = 12x^2 - 24x + 9 = 3(4x^2 - 8x + 3) = 3(2x - 1)(2x - 3)$, so the critical numbers are $x = \frac{1}{2}$, $x = \frac{3}{2}$. Now $0 \le x \le \frac{3}{2}$ and $V(0) = V\left(\frac{3}{2}\right) = 0$, so the maximum is $V\left(\frac{1}{2}\right) = \left(\frac{1}{2}\right)(2)^2 = 2\text{ ft}^3$.

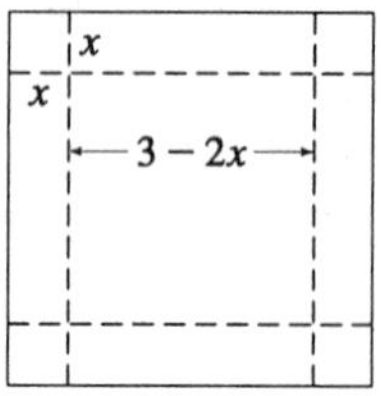

13. For (x, y) on the line $y = 2x - 3$, the distance to the origin is $\sqrt{(x-0)^2 + (2x-3)^2}$. We minimize the square of the distance, that is, $x^2 + (2x - 3)^2 = 5x^2 - 12x + 9 = D(x)$. $D'(x) = 10x - 12 = 0 \Rightarrow x = \frac{6}{5}$. Since there is a point closest to the origin, $x = \frac{6}{5}$ and hence $y = -\frac{3}{5}$. So the point is $\left(\frac{6}{5}, -\frac{3}{5}\right)$.

14. Here $y = -\frac{2}{3}x - \frac{5}{3}$. Let (x, y) be on the line, so the square of its distance from $(-1, -2)$ is $D(x) = (x+1)^2 + \left(-\frac{2}{3}x - \frac{5}{3} + 2\right)^2 = (13x^2 + 14x + 10)/9$. $D'(x) = (26x + 14)/9 = 0 \Rightarrow x = -\frac{7}{13}$. Since there is a point closest to $(-1, -2)$, we must have $x = -\frac{7}{13} \Rightarrow y = -\frac{17}{13}$, so the point is $\left(-\frac{7}{13}, -\frac{17}{13}\right)$.

15. By symmetry, the points are (x, y) and $(x, -y)$, where $y > 0$. The square of the distance is $D(x) = (x-2)^2 + y^2 = (x-2)^2 + 4 + x^2 = 2x^2 - 4x + 8$. So $D'(x) = 4x - 4 = 0 \Rightarrow x = 1$ and $y = \pm\sqrt{4+1} = \pm\sqrt{5}$. The points are $\left(1, \pm\sqrt{5}\right)$.

16. The square of the distance from a point (x, y) on the parabola $x = -y^2$ is $x^2 + (y+3)^2 = y^4 + y^2 + 6y + 9 = D(y)$. Now $D'(y) = 4y^3 + 2y + 6 = 2(y+1)(2y^2 - 2y + 3)$. Since $2y^2 - 2y + 3 = 0$ has no real roots, $y = -1$ is the only critical number. Then $x = -(-1)^2 = -1$, so the point is $(-1, -1)$.

17.

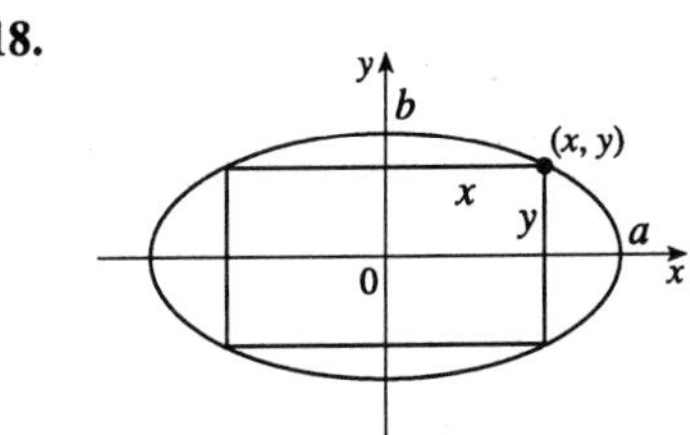

Area of rectangle is $4xy$. Also $r^2 = x^2 + y^2$ so $y = \sqrt{r^2 - x^2}$, so the area is $A(x) = 4x\sqrt{r^2 - x^2}$. Now

$$A'(x) = 4\left(\sqrt{r^2 - x^2} - \frac{x^2}{\sqrt{r^2 - x^2}}\right) = 4\frac{r^2 - 2x^2}{\sqrt{r^2 - x^2}}.$$

The critical number is $x = \frac{1}{\sqrt{2}}r$. Clearly this gives a maximum. The dimensions are $2x = \sqrt{2}r$ and $2y = \sqrt{2}r$.

18.

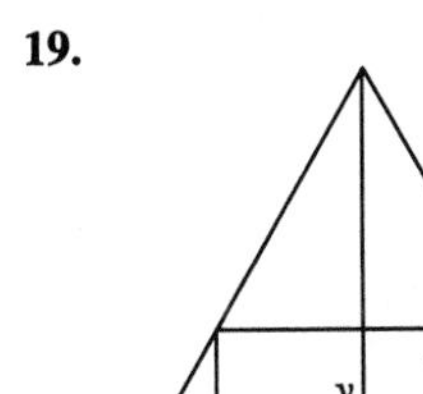

Area is $4xy$. Now the equation of the ellipse gives $y = \frac{b}{a}\sqrt{a^2 - x^2}$, so we maximize $A(x) = 4\frac{b}{a}x\sqrt{a^2 - x^2}$.

$$A'(x) = \frac{4b}{a}\sqrt{a^2 - x^2} + \frac{4bx}{a}\left[-\frac{2x}{2\sqrt{a^2 - x^2}}\right] = \frac{4b}{a\sqrt{a^2 - x^2}}\left[a^2 - 2x^2\right].$$

So the critical number is $x = \frac{1}{\sqrt{2}}a$, and this clearly gives a maximum. Then $y = \frac{1}{\sqrt{2}}b$, so the maximum area is $4\left(\frac{1}{\sqrt{2}}a\right)\left(\frac{1}{\sqrt{2}}b\right) = 2ab$.

19.

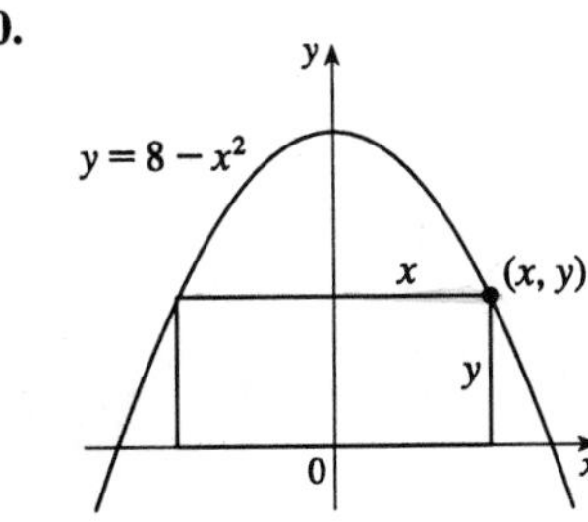

$$\frac{\frac{\sqrt{3}}{2}L - y}{x} = \frac{\frac{\sqrt{3}}{2}L}{L/2} = \sqrt{3} \text{ (similar triangles)} \Rightarrow$$

$\sqrt{3}x = \frac{\sqrt{3}}{2}L - y \quad \Rightarrow \quad y = \frac{\sqrt{3}}{2}(L - 2x)$. The area of the inscribed rectangle is $A(x) = (2x)y = \sqrt{3}x(L - 2x)$ where $0 \le x \le L/2$. Now $0 = A'(x) = \sqrt{3}L - 4\sqrt{3}x \quad \Rightarrow$ $x = \sqrt{3}L/(4\sqrt{3}) = L/4$. Since $A(0) = A(L/2) = 0$, the maximum occurs when $x = L/4$, and $y = \frac{\sqrt{3}}{2}L - \frac{\sqrt{3}}{4}L = \frac{\sqrt{3}}{4}L$, so the dimensions are $L/2$ and $\frac{\sqrt{3}}{4}L$.

20.

The rectangle has area $A(x) = 2xy = 2x(8 - x^2) = 16x - 2x^3$, where $0 \le x \le 2\sqrt{2}$. Now $A'(x) = 16 - 6x^2 = 0 \quad \Rightarrow$ $x = 2\sqrt{\frac{2}{3}}$. Since $A(0) = A\left(2\sqrt{2}\right) = 0$, there is a maximum when $x = 2\sqrt{\frac{2}{3}}$. Then $y = \frac{16}{3}$, so the rectangle has dimensions $4\sqrt{\frac{2}{3}}$ and $\frac{16}{3}$.

21. The area of the triangle is

$A(x) = \frac{1}{2}(2t)(r+x) = t(r+x) = \sqrt{r^2-x^2}(r+x)$. Then

$$0 = A'(x) = r\frac{-2x}{2\sqrt{r^2-x^2}} + \sqrt{r^2-x^2} + x\frac{-2x}{2\sqrt{r^2-x^2}}$$

$$= -\frac{x^2+rx}{\sqrt{r^2-x^2}} + \sqrt{r^2-x^2} \quad \Rightarrow \quad \frac{x^2+rx}{\sqrt{r^2-x^2}} = \sqrt{r^2-x^2}$$

$$\Rightarrow \quad x^2+rx = r^2-x^2 \quad \Rightarrow \quad 0 = 2x^2+rx-r^2 = (2x-r)(x+r)$$

$\Rightarrow \quad x = \frac{1}{2}r$ or $x = -r$. Now $A(r) = 0 = A(-r)$ $\Rightarrow$ the maximum occurs where $x = \frac{1}{2}r$, so the triangle has height $r + \frac{1}{2}r = \frac{3}{2}r$ and base $2\sqrt{r^2 - \left(\frac{1}{2}r\right)^2} = 2\sqrt{\frac{3}{4}r^2} = \sqrt{3}r$.

22. 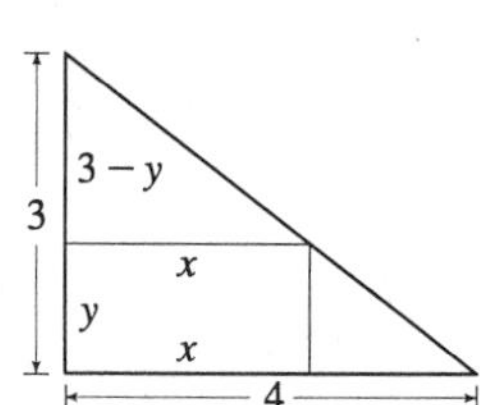

The rectangle has area xy. By similar triangles $\dfrac{3-y}{x} = \dfrac{3}{4}$ $\Rightarrow$ $-4y + 12 = 3x$ or $y = -\frac{3}{4}x + 3$. So the area is $A(x) = x\left[-\frac{3}{4}x + 3\right] = -\frac{3}{4}x^2 + 3x$ where $0 \le x \le 4$. Now $0 = A'(x) = -\frac{3}{2}x + 3 \quad \Rightarrow \quad x = 2$. Since $A(0) = A(4) = 0$, the maximum area is $A(2) = 2\left(\frac{3}{2}\right) = 3\text{ cm}^2$.

23. The cylinder has volume $V = \pi y^2(2x)$. Also $x^2 + y^2 = r^2 \quad \Rightarrow$ $y^2 = r^2 - x^2$, so $V(x) = \pi(r^2-x^2)(2x) = 2\pi\left(r^2x - x^3\right)$, where $0 \le x \le r$. $V'(x) = 2\pi(r^2 - 3x^2) = 0 \quad \Rightarrow \quad x = r/\sqrt{3}$. Now $V(0) = V(r) = 0$, so there is a maximum when $x = r/\sqrt{3}$ and $V\left(r/\sqrt{3}\right) = 4\pi r^3/\left(3\sqrt{3}\right)$.

24. 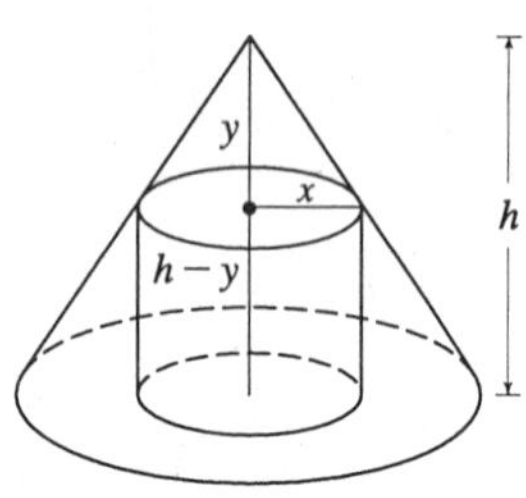

By similar triangles $y/x = h/r$, so $y = hx/r$. The volume of the cylinder is $\pi x^2(h-y) = \pi hx^2 - (\pi h/r)\,x^3 = V(x)$. Now $V'(x) = 2\pi hx - (3\pi h/r)\,x^2 = \pi hx(2 - 3x/r)$. So $0 = V'(x) \quad \Rightarrow \quad x = 0$ or $x = \frac{2}{3}r$. The maximum clearly occurs when $x = \frac{2}{3}r$ and then the volume is $\pi\left(\frac{2}{3}r\right)^2 h\left(1 - \frac{2}{3}\right) = \frac{4}{27}\pi r^2 h$.

25. 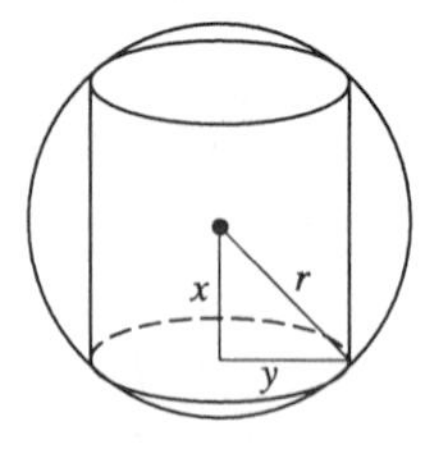

The cylinder has surface area $2\pi y^2 + 2\pi y(2x)$. Now $x^2 + y^2 = r^2 \quad \Rightarrow$ $y = \sqrt{r^2-x^2}$, so the surface area is $S(x) = 2\pi(r^2-x^2) + 4\pi x\sqrt{r^2-x^2}$, $0 \le x \le r$. $S'(x) = -4\pi x + 4\pi\sqrt{r^2-x^2} - 4\pi x^2/\sqrt{r^2-x^2}$

$$= \frac{4\pi\left(r^2 - 2x^2 - x\sqrt{r^2-x^2}\right)}{\sqrt{r^2-x^2}} = 0 \quad \Rightarrow \quad x\sqrt{r^2-x^2} = r^2 - 2x^2 \quad (\bigstar)$$

$$\Rightarrow \quad x^2(r^2-x^2) = r^4 - 4r^2x^2 + 4x^4 \quad \Rightarrow \quad 5x^4 - 5r^2x^2 + r^4 = 0.$$

By the quadratic formula, $x^2 = \frac{5\pm\sqrt{5}}{10}r^2$, but we reject the root with the + sign since it doesn't satisfy ($\bigstar$). So $x = \sqrt{\frac{5-\sqrt{5}}{10}}r$. Since $S(0) = S(r) = 0$, the maximum occurs at the critical number and $x^2 = \frac{5-\sqrt{5}}{10}r^2 \quad \Rightarrow$ $y^2 = \frac{5+\sqrt{5}}{10}r^2 \quad \Rightarrow \quad$ the surface area is $2\pi\left(\frac{5+\sqrt{5}}{10}\right)r^2 + 4\pi\sqrt{\frac{5-\sqrt{5}}{10}}\sqrt{\frac{5+\sqrt{5}}{10}}r^2 = \pi r^2\left(1+\sqrt{5}\right)$.

26.

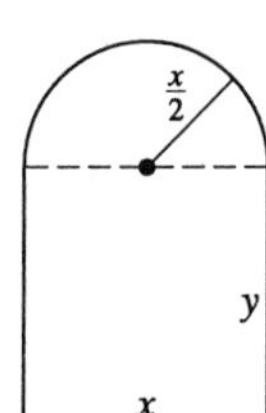

We are given $2y + x + \pi\left(\frac{x}{2}\right) = 30$, so $y = \frac{1}{2}\left[30 - x - \frac{\pi x}{2}\right]$.

The area is $xy + \frac{1}{2}\pi\left(\frac{x}{2}\right)^2$, so

$A(x) = x\left[15 - \frac{x}{2} - \frac{\pi x}{4}\right] + \frac{1}{8}\pi x^2 = 15x - \frac{1}{2}x^2 - \frac{\pi}{8}x^2.$

$A'(x) = 15 - \left(1 + \frac{\pi}{4}\right)x = 0 \quad\Rightarrow\quad x = \dfrac{15}{1 + \pi/4} = \dfrac{60}{4 + \pi}.$

Clearly this gives a maximum, so the dimensions are

$x = \dfrac{60}{4+\pi}$ ft and $y = 15 - \dfrac{30}{4+\pi} - \dfrac{15\pi}{4+\pi} = \dfrac{30}{4+\pi}$ ft.

27.

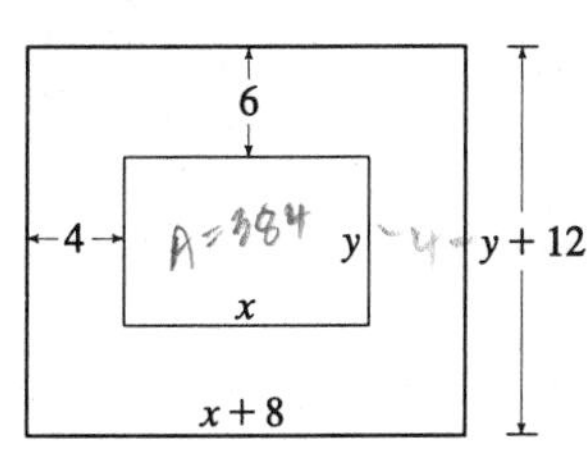

$xy = 384 \quad\Rightarrow\quad y = 384/x$. Total area is

$A(x) = (8 + x)(12 + 384/x) = 12(40 + x + 256/x)$, so

$A'(x) = 12\left(1 - 256/x^2\right) = 0 \quad\Rightarrow\quad x = 16$. There is an absolute minimum when $x = 16$ since $A'(x) < 0$ for $0 < x < 16$ and $A'(x) > 0$ for $x > 16$. When $x = 16$, $y = 384/16 = 24$, so the dimensions are 24 cm and 36 cm.

28.

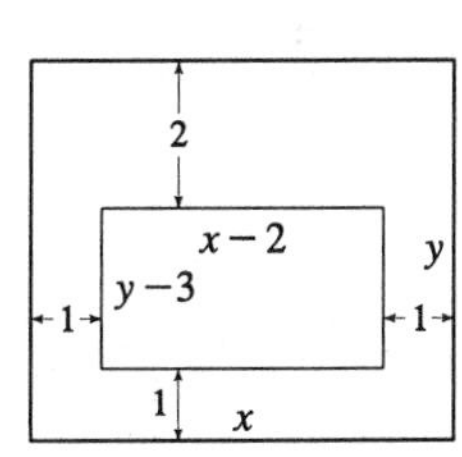

$xy = 180$, so $y = 180/x$. The printed area is

$(x - 2)(y - 3) = (x - 2)(180/x - 3) = 186 - 3x - 360/x = A(x).$

$A'(x) = -3 + 360/x^2 = 0$ when $x^2 = 120 \quad\Rightarrow\quad x = 2\sqrt{30}.$

This gives an absolute maximum since $A'(x) > 0$ for $0 < x < 2\sqrt{30}$ and $A'(x) < 0$ for $x > 2\sqrt{30}$. When $x = 2\sqrt{30}$, $y = 180/\left(2\sqrt{30}\right)$, so the dimensions are $2\sqrt{30}$ in. and $90/\sqrt{30}$ in.

29.

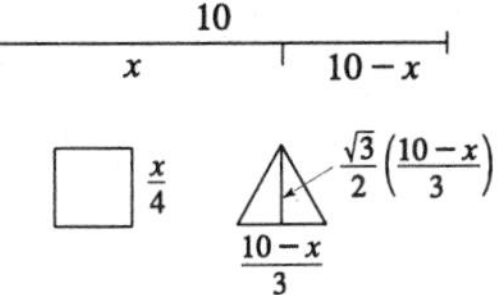

Let x be the length of the wire used for the square. The total area is

$A(x) = \frac{1}{16}x^2 + \frac{\sqrt{3}}{36}(10 - x)^2,\ 0 \le x \le 10.$

$A'(x) = \frac{1}{8}x - \frac{\sqrt{3}}{18}(10 - x) = 0 \quad\Leftrightarrow\quad x = \dfrac{40\sqrt{3}}{9 + 4\sqrt{3}}$. Now

$A(0) = \left(\frac{\sqrt{3}}{36}\right)100 \approx 4.81$, $A(10) = \frac{100}{16} = 6.25$ and $A\left(\frac{40\sqrt{3}}{9+4\sqrt{3}}\right) \approx 2.72$, so

(a) The maximum occurs when $x = 10$ m, and all the wire is used for the square.

(b) The minimum occurs when $x = \frac{40\sqrt{3}}{9+4\sqrt{3}} \approx 4.35$ m.

30.

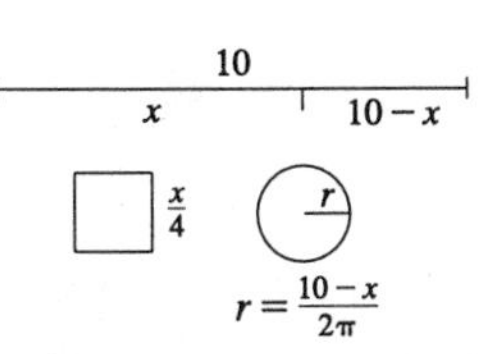

Total area is $A(x) = \left(\dfrac{x}{4}\right)^2 + \pi\left(\dfrac{10 - x}{2\pi}\right)^2 = \dfrac{x^2}{16} + \dfrac{(10 - x)^2}{4\pi}$,

$0 \le x \le 10$. $A'(x) = \dfrac{x}{8} - \dfrac{10 - x}{2\pi} = \left(\dfrac{1}{2\pi} + \dfrac{1}{8}\right)x - \dfrac{5}{\pi} = 0 \quad\Rightarrow$

$x = 40/(4 + \pi)$. $A(0) = 25/\pi \approx 7.96$, $A(10) = 6.25$, and $A(40/(4 + \pi)) \approx 3.5$, so the maximum occurs when $x = 0$ m and the minimum occurs when $x = 40/(4 + \pi)$ m.

31. 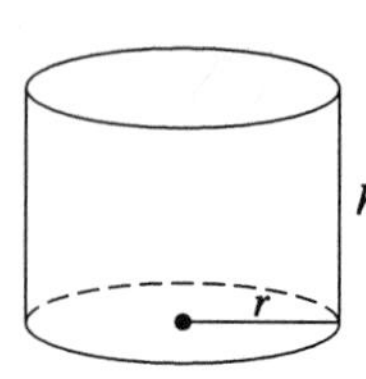

The volume is $V = \pi r^2 h$ and the surface area is

$$S(r) = \pi r^2 + 2\pi rh = \pi r^2 + 2\pi r\left(\frac{V}{\pi r^2}\right) = \pi r^2 + \frac{2V}{r}.$$

$S'(r) = 2\pi r - 2V/r^2 = 0 \quad \Rightarrow \quad 2\pi r^3 = 2V \quad \Rightarrow \quad r = \sqrt[3]{V/\pi}$ cm.

This gives an absolute minimum since $S'(r) < 0$ for $0 < r < \sqrt[3]{V/\pi}$ and $S'(r) > 0$ for $r > \sqrt[3]{\frac{V}{\pi}}$. When $r = \sqrt[3]{\frac{V}{\pi}}$, $h = \frac{V}{\pi r^2} = \frac{V}{\pi (V/\pi)^{2/3}} = \sqrt[3]{\frac{V}{\pi}}$ cm.

32. 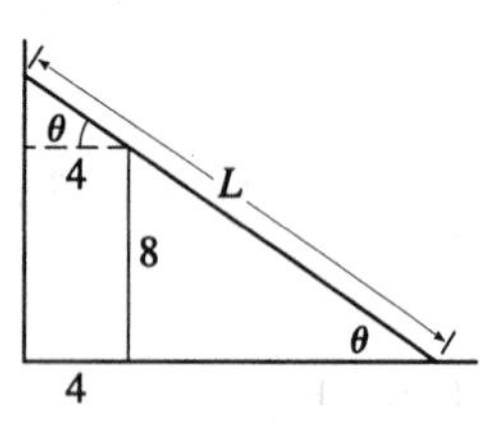

$L = 8\csc\theta + 4\sec\theta$, $0 < \theta < \frac{\pi}{2}$, $\frac{dL}{d\theta} = -8\csc\theta\cot\theta + 4\sec\theta\tan\theta = 0$ when $\sec\theta\tan\theta = 2\csc\theta\cot\theta \quad \Leftrightarrow \quad \tan^3\theta = 2 \quad \Leftrightarrow \quad \tan\theta = \sqrt[3]{2}$ $\Leftrightarrow \quad \theta = \tan^{-1}\sqrt[3]{2}$. $dL/d\theta < 0$ when $0 < \theta < \tan^{-1}\sqrt[3]{2}$, $dL/d\theta > 0$ when $\tan^{-1}\sqrt[3]{2} < \theta < \frac{\pi}{2}$, so L has an absolute minimum when $\theta = \tan^{-1}\sqrt[3]{2}$, so the shortest ladder has length

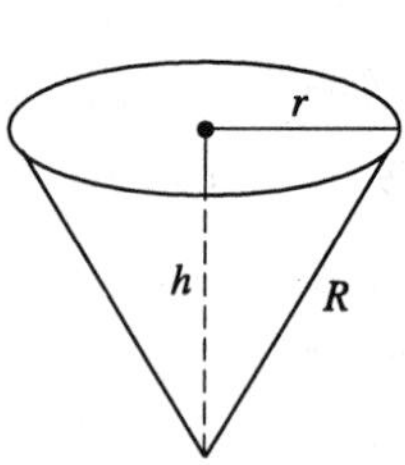

$$L = 8\frac{\sqrt{1 + 2^{2/3}}}{2^{1/3}} + 4\sqrt{1 + 2^{2/3}} \approx 16.65 \text{ ft.}$$

Another Method: Minimize $L^2 = x^2 + (4 + y)^2$, where $\frac{x}{4 + y} = \frac{8}{y}$.

33. 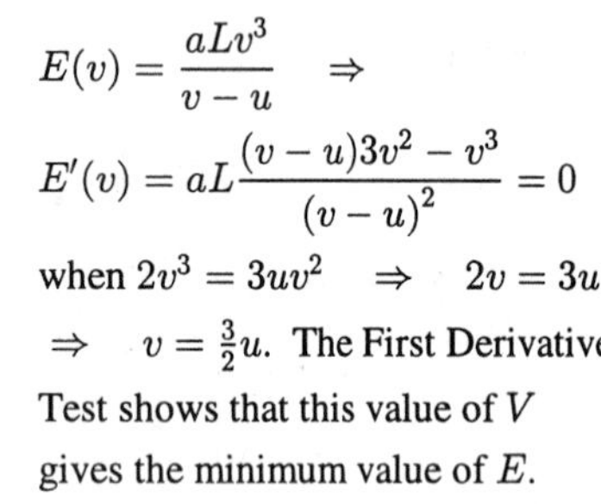

$h^2 + r^2 = R^2 \quad \Rightarrow \quad V = \frac{\pi}{3}r^2 h = \frac{\pi}{3}(R^2 - h^2)h = \frac{\pi}{3}(R^2 h - h^3)$.

$V'(h) = \frac{\pi}{3}(R^2 - 3h^2) = 0$ when $h = \frac{1}{\sqrt{3}}R$. This gives an absolute maximum since $V'(h) > 0$ for $0 < h < \frac{1}{\sqrt{3}}R$ and $V'(h) < 0$ for $h > \frac{1}{\sqrt{3}}R$. Maximum volume is $V\left(\frac{1}{\sqrt{3}}R\right) = \frac{2}{9\sqrt{3}}\pi R^3$.

34. **(a)** $E(v) = \dfrac{aLv^3}{v - u} \quad \Rightarrow$

$$E'(v) = aL\frac{(v - u)3v^2 - v^3}{(v - u)^2} = 0$$

when $2v^3 = 3uv^2 \quad \Rightarrow \quad 2v = 3u$ $\Rightarrow \quad v = \frac{3}{2}u$. The First Derivative Test shows that this value of V gives the minimum value of E.

(b) 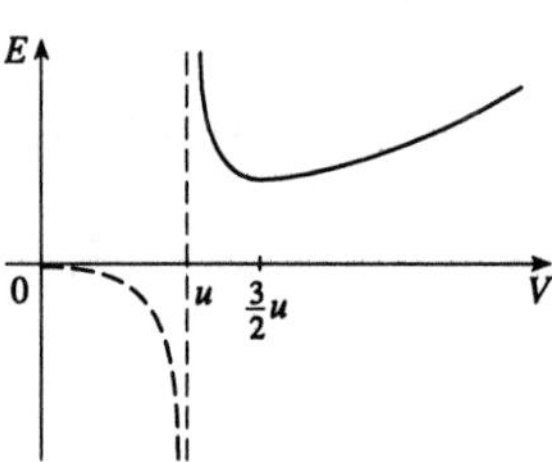

35. $S = 6sh - \frac{3}{2}s^2\cot\theta + 3s^2\frac{\sqrt{3}}{2}\csc\theta$

(a) $\dfrac{dS}{d\theta} = \frac{3}{2}s^2\csc^2\theta - 3s^2\frac{\sqrt{3}}{2}\csc\theta\cot\theta$ or $\frac{3}{2}s^2\csc\theta\left(\csc\theta - \sqrt{3}\cot\theta\right)$.

(b) $\dfrac{dS}{d\theta} = 0$ when $\csc\theta - \sqrt{3}\cot\theta = 0 \quad \Rightarrow \quad \dfrac{1}{\sin\theta} - \sqrt{3}\dfrac{\cos\theta}{\sin\theta} = 0 \quad \Rightarrow \quad \cos\theta = \frac{1}{\sqrt{3}}$. The First Derivative Test shows that the minimum surface area occurs when $\theta = \cos^{-1}\frac{1}{\sqrt{3}} \approx 55°$.

(c)

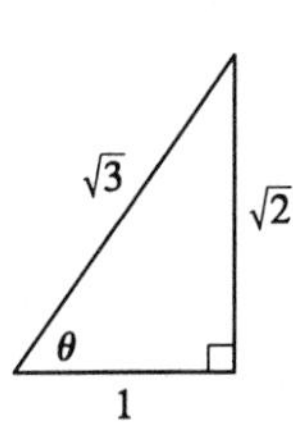

If $\cos\theta = \frac{1}{\sqrt{3}}$, then $\cot\theta = \frac{1}{\sqrt{2}}$ and $\csc\theta = \frac{\sqrt{3}}{\sqrt{2}}$, so the surface area is

$S = 6sh - \frac{3}{2}s^2\frac{1}{\sqrt{2}} + 3s^2\frac{\sqrt{3}}{2}\frac{\sqrt{3}}{\sqrt{2}} = 6sh - \frac{3}{2\sqrt{2}}s^2 + \frac{9}{2\sqrt{2}}s^2 = 6s\left(h + \frac{1}{2\sqrt{2}}s\right)$

36.

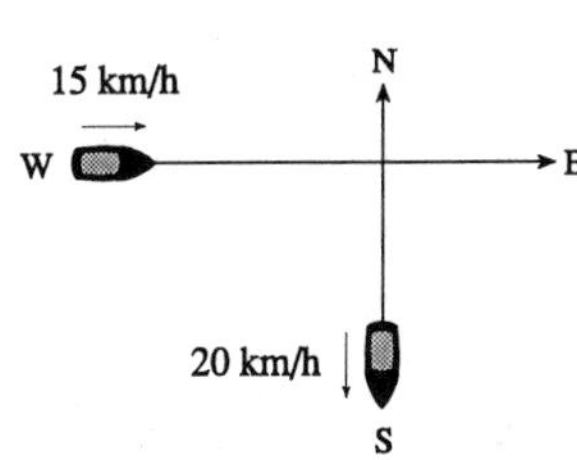

Let t be the time, in hours, after 2:00 P.M.. The position of the boat heading south at time t is $(0, -20t)$. The position of the boat heading east at time t is $(-15 + 15t, 0)$. If $D(t)$ is the distance between the boats at time t, we minimize $f(t) = [D(t)]^2 = 20^2t^2 + 15^2(t-1)^2$.

$f'(t) = 800t + 450(t-1) = 1250t - 450 = 0$ when $t = \frac{450}{1250} = 0.36$ h.

$0.36\text{ h} \times \dfrac{60\text{ min}}{\text{h}} = 21.6\text{ min} = 21\text{ min } 36\text{ s}$. Since $f''(t) > 0$, this gives a minimum, so the boats are closest together at 2:21:36 P.M..

37. Here $T(x) = \dfrac{\sqrt{x^2+25}}{6} + \dfrac{5-x}{8}$, $0 \le x \le 5$, $\Rightarrow$ $T'(x) = \dfrac{x}{6\sqrt{x^2+25}} - \dfrac{1}{8} = 0$ $\Leftrightarrow$ $8x = 6\sqrt{x^2+25}$ $\Leftrightarrow$ $16x^2 = 9(x^2+25)$ $\Leftrightarrow$ $x = \frac{15}{\sqrt{7}}$. But $\frac{15}{\sqrt{7}} > 5$, so T has no critical number. Since $T(0) \approx 1.46$ and $T(5) \approx 1.18$, he should row directly to B.

38.

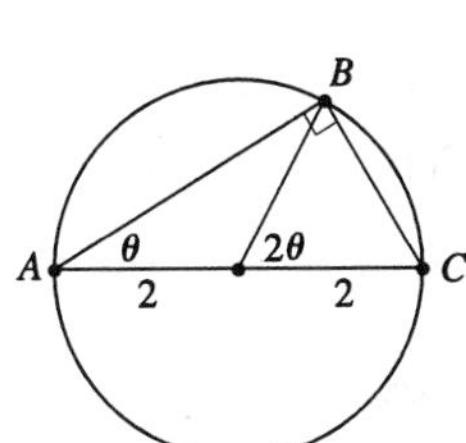

The distance rowed is $4\cos\theta$ while the distance walked is the length of arc $BC = 2(2\theta) = 4\theta$. The time taken is given by

$T(\theta) = \dfrac{4\cos\theta}{2} + \dfrac{4\theta}{4} = 2\cos\theta + \theta$, $0 \le \theta \le \frac{\pi}{2}$. $T'(\theta) = -2\sin\theta + 1 = 0$ $\Leftrightarrow$ $\sin\theta = \frac{1}{2}$ $\Rightarrow$ $\theta = \frac{\pi}{6}$. Check the value of T at $\theta = \frac{\pi}{6}$ and at the endpoints of the domain of T, that is, $\theta = 0$ and $\theta = \frac{\pi}{2}$.

$T(0) = 2\cos 0 + 0 = 2$, $T\left(\frac{\pi}{6}\right) = 2\cos\left(\frac{\pi}{6}\right) + \frac{\pi}{6} = \sqrt{3} + \frac{\pi}{6}$, and $T\left(\frac{\pi}{2}\right) = 2\cos\frac{\pi}{2} + \frac{\pi}{2} = \frac{\pi}{2}$. Therefore the minimum value of T is $\frac{\pi}{2}$ when $\theta = \frac{\pi}{2}$, that is, the woman should walk all the way.

39.

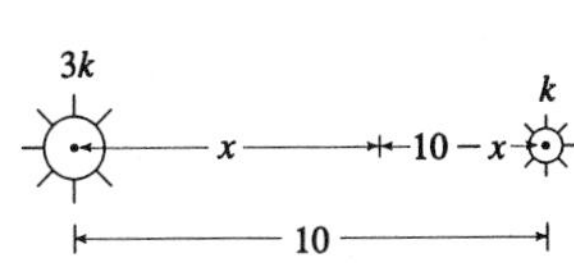

The total illumination is $I(x) = \dfrac{3k}{x^2} + \dfrac{k}{(10-x)^2}$, $0 < x < 10$.

Then $I'(x) = \dfrac{-6k}{x^3} + \dfrac{2k}{(10-x)^3} = 0$ $\Rightarrow$ $6k(10-x)^3 = 2kx^3$

$\Rightarrow$ $\sqrt[3]{3}(10-x) = x$ $\Rightarrow$ $x = 10\sqrt[3]{3}/\left(1+\sqrt[3]{3}\right) \approx 5.9$ ft.

This gives a minimum since there is clearly no maximum.

40.

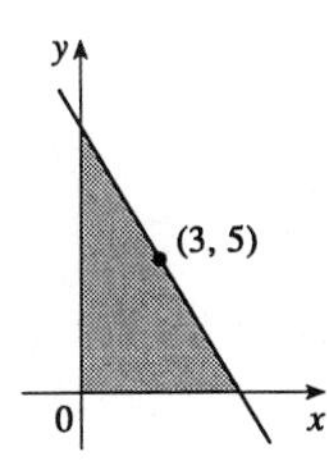

The line with slope m (where $m < 0$) through $(3,5)$ has equation $y - 5 = m(x-3)$ or $y = mx + (5 - 3m)$. The y-intercept is $5 - 3m$ and the x-intercept is $-5/m + 3$. So the triangle has area $A(m) = \frac{1}{2}(5-3m)(-5/m+3) = 15 - 25/(2m) - \frac{9}{2}m$.

Now $A'(m) = \dfrac{25}{2m^2} - \dfrac{9}{2} = 0$ $\Leftrightarrow$ $m = -\frac{5}{3}$ (since $m < 0$). Now $A''(m) = -25/m^3 > 0$, so there is an absolute minimum when $m = -\frac{5}{3}$. Therefore the equation of the line is $y - 5 = -\frac{5}{3}(x-3)$ or $y = -\frac{5}{3}x + 10$.

41. 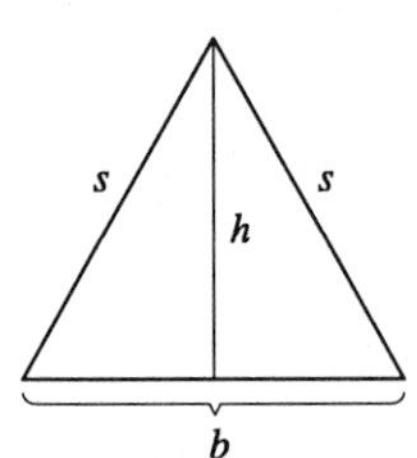

Here $s^2 = h^2 + b^2/4$ so $h^2 = s^2 - b^2/4$. The area is $A = \frac{1}{2}b\sqrt{s^2 - b^2/4}$. Let the perimeter be p, so $2s + b = p$ or $s = (p - b)/2 \Rightarrow$ $A(b) = \frac{1}{2}b\sqrt{(p-b)^2/4 - b^2/4} = b\sqrt{p^2 - 2pb}/4$. Now $A'(b) = \dfrac{\sqrt{p^2 - 2pb}}{4} - \dfrac{bp/4}{\sqrt{p^2 - 2pb}} = \dfrac{-3pb + p^2}{4\sqrt{p^2 - 2pb}}$. Therefore $A'(b) = 0$ $\Rightarrow \quad -3pb + p^2 = 0 \quad \Rightarrow \quad b = p/3$. Since $A'(b) > 0$ for $b < p/3$ and $A'(b) < 0$ for $b > p/3$, there is an absolute maximum when $b = p/3$. But then $2s + p/3 = p$ so $s = p/3 \Rightarrow$ $s = b \quad \Rightarrow \quad$ the triangle is equilateral.

42. The area is given by

$$A(x) = \tfrac{1}{2}\left(2\sqrt{a^2 - x^2}\right)x + \tfrac{1}{2}\left(2\sqrt{a^2 - x^2}\right)\left(\sqrt{x^2 + b^2 - a^2}\right) = \sqrt{a^2 - x^2}\left(x + \sqrt{x^2 + b^2 - a^2}\right) \text{ for}$$

$0 \le x \le a$. Now $A'(x) = \dfrac{-x}{\sqrt{a^2 - x^2}}\left(x + \sqrt{x^2 + b^2 - a^2}\right) + \sqrt{a^2 - x^2}\left(1 + \dfrac{x}{\sqrt{x^2 + b^2 - a^2}}\right) = 0 \quad \Leftrightarrow$

$$\frac{x}{\sqrt{a^2 - x^2}}\left(x + \sqrt{x^2 + b^2 - a^2}\right) = \sqrt{a^2 - x^2}\left(\frac{x + \sqrt{x^2 + b^2 - a^2}}{\sqrt{x^2 + b^2 - a^2}}\right).$$

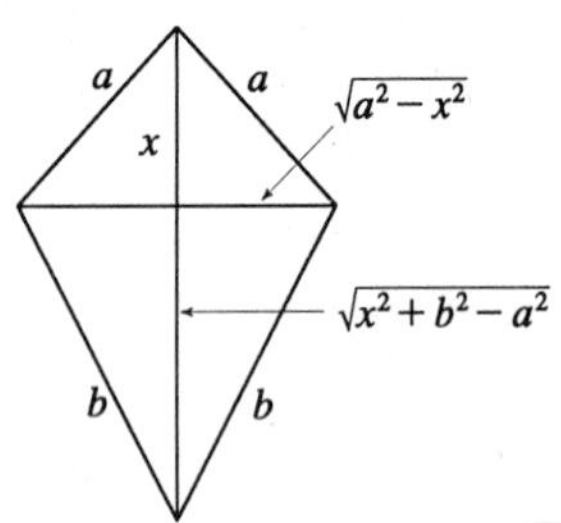

Except for the trivial case where $x = 0$, $a = b$ and $A(x) = 0$, we have that $x + \sqrt{x^2 + b^2 - a^2} > 0$. Hence cancelling this factor gives $\dfrac{x}{\sqrt{a^2 - x^2}} = \dfrac{\sqrt{a^2 - x^2}}{\sqrt{x^2 + b^2 - a^2}}$, so $x\sqrt{x^2 + b^2 - a^2} = a^2 - x^2 \;\Rightarrow\; x^2(x^2 + b^2 - a^2) = a^4 - 2a^2x^2 + x^4$, so $x^4 + x^2(b^2 - a^2) = a^4 - 2a^2x^2 + x^4$ and hence

$$x^2(b^2 + a^2) = a^4, \text{ so } x = \frac{a^2}{\sqrt{a^2 + b^2}}.$$

Now we must check the value of A at this point as well as at the endpoints of the domain to see which gives the maximum value. $A(0) = a\sqrt{b^2 - a^2}$, $A(a) = 0$ and

$$A\left(\frac{a^2}{\sqrt{a^2 + b^2}}\right) = \sqrt{a^2 - \left(\frac{a^2}{\sqrt{a^2 + b^2}}\right)^2}\left[\frac{a^2}{\sqrt{a^2 + b^2}} + \sqrt{\left(\frac{a^2}{\sqrt{a^2 + b^2}}\right)^2 + b^2 - a^2}\right]$$

$$= \frac{ab}{\sqrt{a^2 + b^2}}\left[\frac{a^2}{\sqrt{a^2 + b^2}} + \frac{b^2}{\sqrt{a^2 + b^2}}\right] = \frac{ab(a^2 + b^2)}{a^2 + b^2} = ab.$$

Since $b \ge \sqrt{b^2 - a^2}$, $A\left(a^2/\sqrt{a^2 + b^2}\right) \ge A(0)$. So there is an absolute maximum when $x = \dfrac{a^2}{\sqrt{a^2 + b^2}}$. In this case the horizontal piece should be $\dfrac{2ab}{\sqrt{a^2 + b^2}}$ and the vertical piece should be $\dfrac{a^2 + b^2}{\sqrt{a^2 + b^2}} = \sqrt{a^2 + b^2}$.

43.

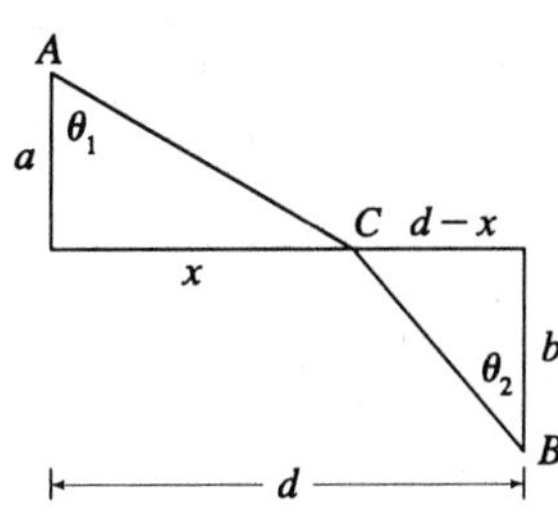

The total time is $T(x) = (\text{time from } A \text{ to } C) + (\text{time from } C \text{ to } B)$

$= \sqrt{a^2 + x^2}/v_1 + \sqrt{b^2 + (d - x)^2}/v_2,\ 0 < x < d.$

$$T'(x) = \frac{x}{v_1\sqrt{a^2 + x^2}} - \frac{d - x}{v_2\sqrt{b^2 + (d - x)^2}} = \frac{\sin\theta_1}{v_1} - \frac{\sin\theta_2}{v_2}.$$

The minimum occurs when $T'(x) = 0 \quad \Rightarrow \quad \dfrac{\sin\theta_1}{v_1} = \dfrac{\sin\theta_2}{v_2}$.

44.

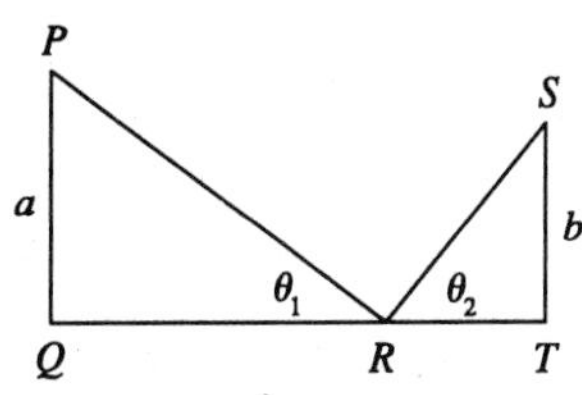

If $d = |QT|$, we minimize $f(\theta_1) = |PR| + |RS| = a\csc\theta_1 + b\csc\theta_2$.

Differentiating with respect to θ_1, and setting $\dfrac{df}{d\theta_1}$ equal to 0, we get

$\dfrac{df}{d\theta_1} = 0 = -a\csc\theta_1\cot\theta_1 - b\csc\theta_2\cot\theta_2\dfrac{d\theta_2}{d\theta_1}$. So we need to find

an expression for $\dfrac{d\theta_2}{d\theta_1}$. We can do this by observing that

$|QT| = \text{constant} = a\cot\theta_1 + b\cot\theta_2$. Differentiating this equation implicitly with respect to θ_1, we get

$-a\csc^2\theta_1 - b\csc^2\theta_2\dfrac{d\theta_2}{d\theta_1} = 0 \quad \Rightarrow \quad \dfrac{d\theta_2}{d\theta_1} = -\dfrac{a\csc^2\theta_1}{b\csc^2\theta_2}$. We substitute this into the expression for $\dfrac{df}{d\theta_1}$ to get

$-a\csc\theta_1\cot\theta_1 - b\csc\theta_2\cot\theta_2\left(-\dfrac{a\csc^2\theta_1}{b\csc^2\theta_2}\right) = 0 \quad \Leftrightarrow \quad -a\csc\theta_1\cot\theta_1 + a\dfrac{\csc^2\theta_1\cot\theta_2}{\csc\theta_2} = 0 \quad \Leftrightarrow$

$\cot\theta_1\csc\theta_2 = \csc\theta_1\cot\theta_2 \quad \Leftrightarrow \quad \dfrac{\cot\theta_1}{\csc\theta_1} = \dfrac{\cot\theta_2}{\csc\theta_2} \quad \Leftrightarrow \quad \cos\theta_1 = \cos\theta_2$. So, since $0 < \theta < \frac{\pi}{2}$ for both angles,

we have $\theta_1 = \theta_2$.

45.

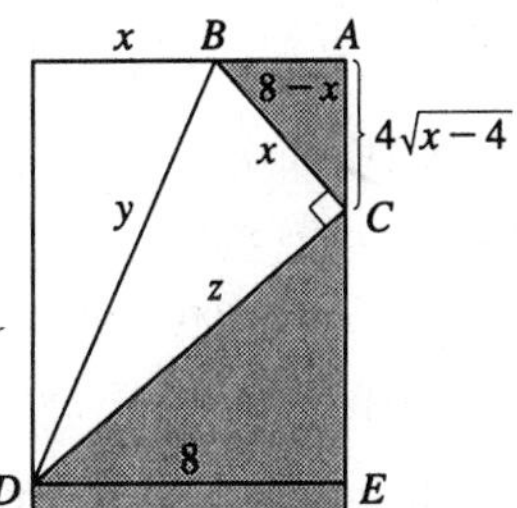

$y^2 = x^2 + z^2$, but triangles CDE and BCA are similar, so

$z/8 = x\big/\left(4\sqrt{x - 4}\right)$. Thus we minimize

$f(x) = y^2 = x^2 + 4x^2/(x - 4) = x^3/(x - 4),\ 4 < x \le 8.$

$$f'(x) = \frac{3x^2(x - 4) - x^3}{(x - 4)^2} = \frac{2x^2(x - 6)}{(x - 4)^2} = 0 \text{ when } x = 6.$$

$f'(x) < 0$ when $x < 6$, $f'(x) > 0$ when $x > 6$, so the minimum occurs when $x = 6$ in.

46.

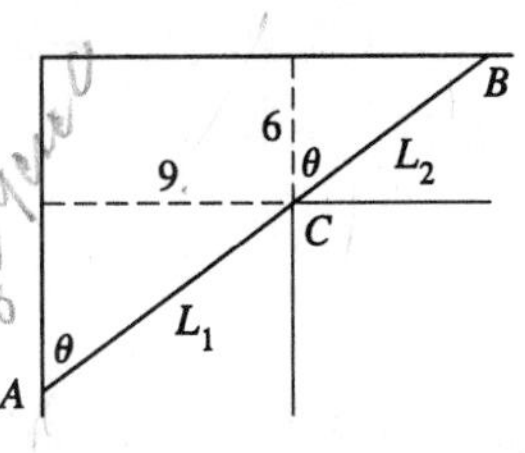

Paradoxically, we solve this maximum problem by solving a minimum problem. Let L be the length of the line ACB going from wall to wall touching the inner corner C. As $\theta \to 0$ or $\theta \to \frac{\pi}{2}$, we have $L \to \infty$ and there will be an angle that makes L a minimum. A pipe of this length will just fit around the corner. From the diagram, $L = L_1 + L_2 = 9\csc\theta + 6\sec\theta \quad \Rightarrow$

$dL/d\theta = -9\csc\theta\cot\theta + 6\sec\theta\tan\theta = 0$ when $6\sec\theta\tan\theta = 9\csc\theta\cot\theta \quad \Leftrightarrow \quad \tan^3\theta = \frac{9}{6} = 1.5 \quad \Leftrightarrow$

$\tan\theta = \sqrt[3]{1.5}$. Then $\csc^2\theta = 1 + \left(\frac{3}{2}\right)^{-2/3}$ and $\sec^2\theta = 1 + \left(\frac{3}{2}\right)^{2/3}$, so the longest pipe has length

$$L = 9\left[1 + \left(\tfrac{3}{2}\right)^{-2/3}\right]^{1/2} + 6\left[1 + \left(\tfrac{3}{2}\right)^{2/3}\right]^{1/2} \approx 21.07 \text{ ft.}$$

Or, use $\theta = \tan^{-1}\left(\sqrt[3]{1.5}\right) \approx 0.852 \quad \Rightarrow \quad L = 9\csc\theta + 6\sec\theta \approx 21.07$ ft.

47. 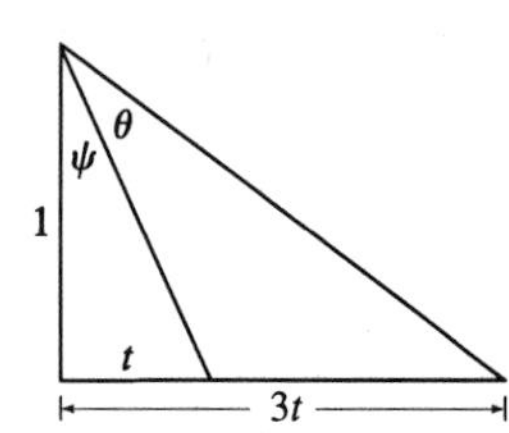

It suffices to maximize $\tan\theta$. Now $\frac{3t}{1} = \tan(\psi+\theta) = \frac{\tan\psi + \tan\theta}{1 - \tan\psi\tan\theta}$ (see endpapers) $= \frac{t+\tan\theta}{1 - t\tan\theta}$. So $3t(1 - t\tan\theta) = t + \tan\theta$ $\Rightarrow$ $2t = (1+3t^2)\tan\theta$ $\Rightarrow$ $\tan\theta = \frac{2t}{1+3t^2}$.

Let $f(t) = \tan\theta = \frac{2t}{1+3t^2}$ $\Rightarrow$
$f'(t) = \frac{2(1+3t^2) - 2t(6t)}{(1+3t^2)^2} = \frac{2(1-3t^2)}{(1+3t^2)^2} = 0 \Leftrightarrow 1 - 3t^2 = 0 \Leftrightarrow t = \frac{1}{\sqrt{3}}$ since $t \geq 0$. Now $f'(t) > 0$ for $0 \leq t < \frac{1}{\sqrt{3}}$ and $f'(t) < 0$ for $t > \frac{1}{\sqrt{3}}$, so f has an absolute maximum when $t = \frac{1}{\sqrt{3}}$ and

$$\tan\theta = \frac{2\left(1/\sqrt{3}\right)}{1+3\left(1/\sqrt{3}\right)^2} = \frac{1}{\sqrt{3}} \quad\Rightarrow\quad \theta = \frac{\pi}{6}.$$

48. 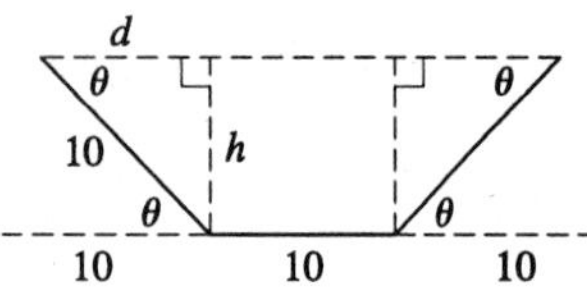

We maximize the cross-sectional area

$$A(\theta) = 10h + 2\left(\tfrac{1}{2}dh\right) = 10h + dh = 10(10\sin\theta) + (10\cos\theta)(10\sin\theta)$$
$$= 100(\sin\theta + \sin\theta\cos\theta),\ 0 \leq \theta \leq \frac{\pi}{2}.$$
$$A'(\theta) = 100\left(\cos\theta + \cos^2\theta - \sin^2\theta\right) = 100\left(\cos\theta + 2\cos^2\theta - 1\right)$$
$$= 100(2\cos\theta - 1)(\cos\theta + 1) = 0 \text{ when } \cos\theta = \tfrac{1}{2} \Leftrightarrow \theta = \tfrac{\pi}{3}.$$

($\cos\theta \neq -1$ since $0 \leq \theta \leq \frac{\pi}{2}$.) Now $A(0) = 0$, $A\left(\frac{\pi}{2}\right) = 100$ and $A\left(\frac{\pi}{3}\right) = 100 \cdot \frac{3\sqrt{3}}{4} > 100$, so the maximum occurs when $\theta = \frac{\pi}{3}$.

49. 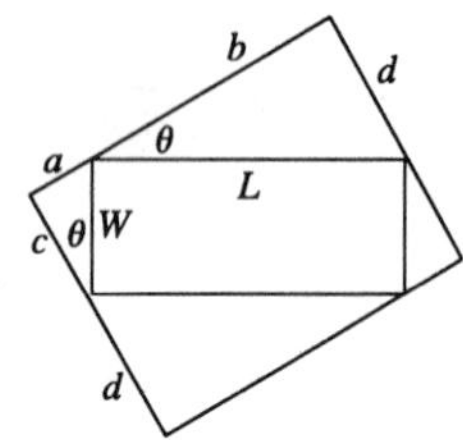

$a = W\sin\theta$, $c = W\cos\theta$, $b = L\cos\theta$, $d = L\sin\theta$, so the area of the circumscribed rectangle is

$$A(\theta) = (a+b)(c+d) = (W\sin\theta + L\cos\theta)(W\cos\theta + L\sin\theta)$$
$$= LW\sin^2\theta + LW\cos^2\theta + \left(L^2 + W^2\right)\sin\theta\cos\theta$$
$$= LW + \tfrac{1}{2}(L^2 + W^2)\sin 2\theta,\ 0 \leq \theta \leq \tfrac{\pi}{2}.$$

This expression shows, without calculus, that the maximum value of $A(\theta)$ occurs when $\sin 2\theta = 1 \Leftrightarrow 2\theta = \frac{\pi}{2} \Leftrightarrow x = \frac{\pi}{4}$. So the maximum area is $A\left(\frac{\pi}{4}\right) = LW + \frac{1}{2}(L^2 + W^2) = \frac{1}{2}(L+W)^2$.

50. (a) $|BC|/b = \cot\theta$, so $|BC| = b\cot\theta \Rightarrow |AB| = a - b\cot\theta$. $|BD|/b = \csc\theta \Rightarrow |BD| = b\csc\theta$.
The total resistance is $R(\theta) = C\frac{|AB|}{r_1^4} + C\frac{|BD|}{r_2^4} = C\left(\frac{a - b\cot\theta}{r_1^4} + \frac{b\csc\theta}{r_2^4}\right)$

(b) $R'(\theta) = C\left(\frac{b\csc^2\theta}{r_1^4} - \frac{b\csc\theta\cot\theta}{r_2^4}\right) = bC\csc\theta\left(\frac{\csc\theta}{r_1^4} - \frac{\cot\theta}{r_2^4}\right)$.

$R'(\theta) = 0 \Leftrightarrow \frac{\csc\theta}{r_1^4} = \frac{\cot\theta}{r_2^4} \Leftrightarrow \frac{r_2^4}{r_1^4} = \frac{\cot\theta}{\csc\theta} = \cos\theta$.

$R'(\theta) > 0 \Leftrightarrow \frac{\csc\theta}{r_1^4} > \frac{\cot\theta}{r_2^4} \Rightarrow \cos\theta < \frac{r_2^4}{r_1^4}$ and $R'(\theta) < 0$ when $\cos\theta < \frac{r_2^4}{r_1^4}$, so there is an absolute minimum when $\cos\theta = r_2^4/r_1^4$.

(c) When $r_2 = \frac{2}{3}r_1$, we have $\cos\theta = \left(\frac{2}{3}\right)^4$, so $\theta = \cos^{-1}\left(\frac{2}{3}\right)^4 \approx 79°$.

51. $L(x) = |AP| + |BP| + |CP| = x + \sqrt{(5-x)^2 + 2^2} + \sqrt{(5-x)^2 + 3^2}$

$= x + \sqrt{x^2 - 10x + 29} + \sqrt{x^2 - 10x + 34} \quad \Rightarrow$

$L'(x) = 1 + \dfrac{x-5}{\sqrt{x^2 - 10x + 29}} + \dfrac{x-5}{\sqrt{x^2 - 10x + 34}}$

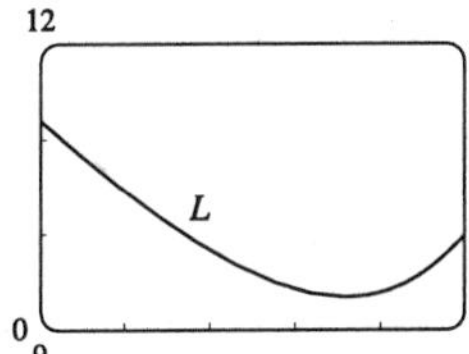

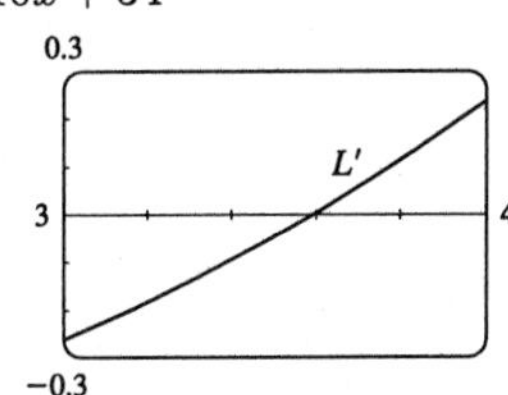

From the graphs of L and L', it seems that the minimum value of L is about $L(3.59) = 9.35$ m.

52. **(a)** $I(x) \propto \dfrac{\text{strength of source}}{\text{distance from source}}$. Adding the intensities from the left and right lightbulbs,

$$I(x) = \frac{k}{\sqrt{x^2 + d^2}} + \frac{k}{\sqrt{(10-x)^2 + d^2}} = \frac{k}{\sqrt{x^2 + d^2}} + \frac{k}{\sqrt{x^2 - 20x + 100 + d^2}}.$$

(b) The magnitude of the constant k won't affect the location of the point of maximum intensity, so for convenience we take $k = 1$. $I'(x) = -\dfrac{x}{(x^2 + d^2)^{3/2}} - \dfrac{x - 10}{(x^2 - 20x + 100 + d^2)^{3/2}}$.

Substituting $d = 5$ into the equations for $I(x)$ and $I'(x)$, we get

$$I_5(x) = \frac{1}{\sqrt{x^2 + 25}} + \frac{1}{\sqrt{x^2 - 20x + 125}} \text{ and } I_5'(x) = -\frac{x}{(x^2 + 25)^{3/2}} - \frac{x - 10}{(x^2 - 20x + 125)^{3/2}}.$$

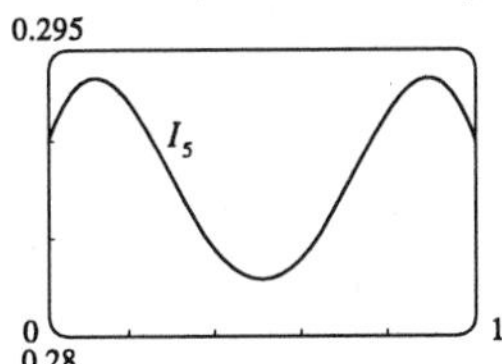

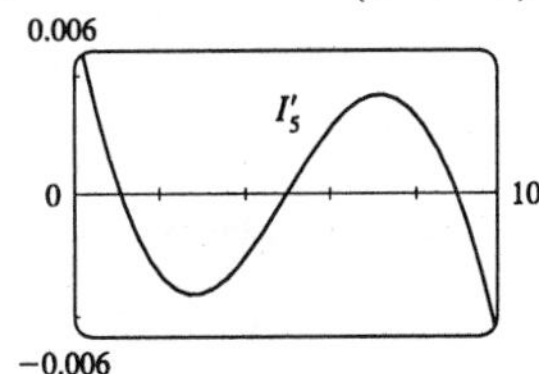

From the graphs, it appears that $I_5(x)$ has a minimum at $x = 5$ m.

(c) Substituting $d = 10$ into $I(x)$ gives $I_{10}(x) = \dfrac{1}{\sqrt{x^2 + 100}} + \dfrac{1}{\sqrt{x^2 - 20x + 100}}$.

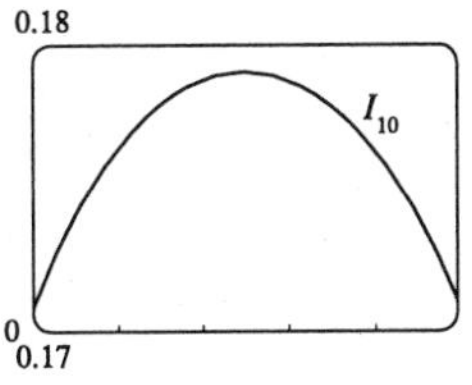

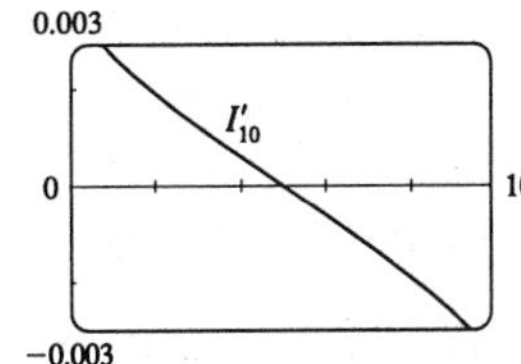

From the graphs, it seems that for $d = 10$, the intensity is minimized at the endpoints, that is, $x = 0$ and $x = 10$. The midpoint is now the most brightly lit point!

(d)

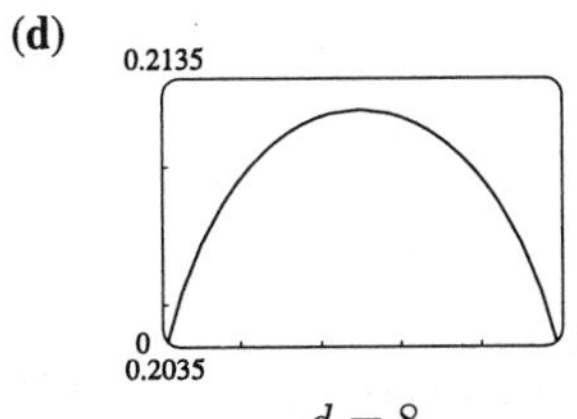

$d = 8$

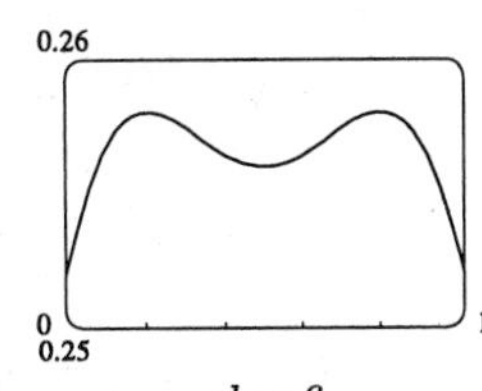

$d = 6$

$d = 5.54$

We plot the graph of $I(x)$ for various values of d. It seems that at $d \approx 5.54$, the point of minimum illumination changes from the midpoint ($x = 5$) to the endpoints ($x = 0$ and $x = 10$).

EXERCISES 3.9

1. **(a)** $C(0)$ represents the fixed costs of production, such as rent, utilities, machinery etc., which are incurred even when nothing is produced.

(b) The inflection point is the point at which $C''(x)$ changes from negative to positive, that is, the marginal cost $C'(x)$ changes from decreasing to increasing. So the marginal cost is minimized.

(c) The marginal cost function is $C'(x)$.

2. **(a)** Profit is maximized when the marginal revenue is equal to the marginal cost, that is, R and C have equal slopes.

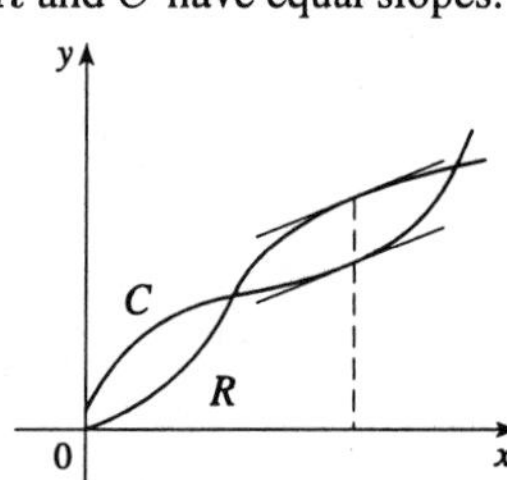

(b) $P(x) = R(x) - C(x)$ is sketched.

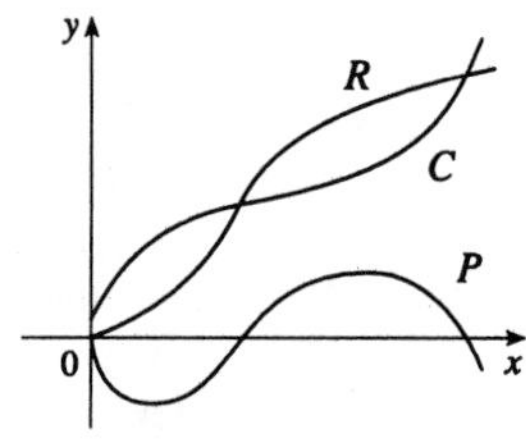

(c) The marginal profit function is defined as $P'(x)$.

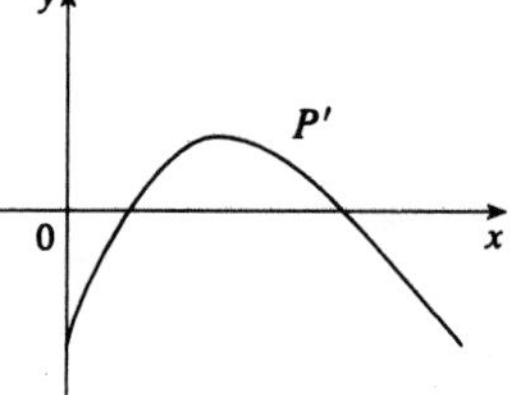

3. **(a)** $C(x) = 10{,}000 + 25x + x^2$, $C(1000) = \$1{,}035{,}000$, $c(x) = \dfrac{C(x)}{x} = \dfrac{10{,}000}{x} + 25 + x$,

$c(1000) = \$1035$. $C'(x) = 25 + 2x$, $C'(1000) = \$2025/\text{unit}$.

(b) We must have $c(x) = C'(x) \quad\Rightarrow\quad 10{,}000/x + 25 + x = 25 + 2x \quad\Rightarrow\quad 10{,}000/x = x \quad\Rightarrow$

$x^2 = 10{,}000 \quad\Rightarrow\quad x = 100$. This is a minimum since $c''(x) = 20{,}000/x^3 > 0$.

(c) The minimum average cost is $c(100) = \$225$.

4. **(a)** $C(x) = 1600 + 8x + 0.01x^2$, $C(1000) = \$19{,}600$. $c(x) = \dfrac{1600}{x} + 8 + 0.01x$, $c(1000) = \$19.60$.

$C'(x) = 8 + 0.02x$, $C'(1000) = \$28$.

(b) We must have $C'(x) = c(x) \quad\Leftrightarrow\quad 8 + 0.02x = \dfrac{1600}{x} + 8 + 0.01x \quad\Leftrightarrow\quad 0.01x = \dfrac{1600}{x} \quad\Leftrightarrow$

$x^2 = \frac{1600}{0.01} = 160{,}000 \quad\Leftrightarrow\quad x = 400$. This is a minimum since $c''(x) = 3200/x^3 > 0$ for $x > 0$.

(c) The minimum average cost is $c(400) = \$16$.

5. **(a)** $C(x) = 45 + \dfrac{x}{2} + \dfrac{x^2}{560}$, $C(1000) = \$2330.71$. $c(x) = \dfrac{45}{x} + \dfrac{1}{2} + \dfrac{x}{560}$, $c(1000) = \$2.33$.

$C'(x) = \dfrac{1}{2} + \dfrac{x}{280}$, $C'(1000) = \$4.07/\text{unit}$

(b) We must have $C'(x) = c(x) \quad \Rightarrow \quad \frac{1}{2} + \frac{x}{280} = \frac{45}{x} + \frac{1}{2} + \frac{x}{560} \quad \Rightarrow \quad \frac{45}{x} = \frac{x}{560} \quad \Rightarrow$ $x^2 = (45)(560) \quad \Rightarrow \quad x = \sqrt{25{,}200} \approx 159$. This is a minimum since $c''(x) = 90/x^2 > 0$.

(c) The minimum average cost is $c(159) = \$1.07$.

6. **(a)** $C(x) = 2000 + 10x + 0.001x^3$. $C(1000) = \$1{,}012{,}000$. $c(x) = \frac{2000}{x} + 10 + 0.001x^2$, $c(1000) = \$1012$. $C'(x) = 10 + 0.003x^2$, $C'(1000) = \$3010$.

(b) We must have $C'(x) = c(x) \quad \Leftrightarrow \quad 10 + 0.003x^2 = \frac{2000}{x} + 10 + 0.001x^2 \quad \Leftrightarrow \quad \frac{2000}{x} = 0.002x^2$ $\Leftrightarrow \quad x^3 = 2000/0.002 = 1{,}000{,}000 \quad \Leftrightarrow \quad x = 100$. This is a minimum since $c''(x) = \frac{4000}{x^3} + 0.002 > 0$ for $x > 0$.

(c) The minimum average cost is $c(100) = \$40$.

7. **(a)** $C(x) = 2\sqrt{x} + \frac{x^2}{8000}$, $C(1000) = \$188.25$. $c(x) = \frac{2}{\sqrt{x}} + \frac{x}{8000}$, $c(1000) = \$0.19$.

$C'(x) = \frac{1}{\sqrt{x}} + \frac{x}{4000}$, $C'(1000) = \$0.28/\text{unit}$.

(b) We must have $C'(x) = c(x) \quad \Rightarrow \quad \frac{1}{\sqrt{x}} + \frac{x}{4000} = \frac{2}{\sqrt{x}} + \frac{x}{8000} \quad \Rightarrow \quad \frac{x}{8000} = \frac{1}{\sqrt{x}} \quad \Rightarrow$ $x^{3/2} = 8000 \quad \Rightarrow \quad x = (8000)^{2/3} = 400$. This is a minimum since $c''(x) = \frac{3}{2}x^{-5/2} > 0$.

(c) The minimum average cost is $c(400) = \$0.15$.

8. **(a)** $C(x) = 1000 + 96x + 2x^{3/2}$, $C(1000) = \$160{,}245.55$. $c(x) = \frac{1000}{x} + 96 + 2\sqrt{x}$, $c(1000) = \$160.25$. $C'(x) = 96 + 3\sqrt{x}$, $C'(1000) = \$190.87$.

(b) We must have $C'(x) = c(x) \quad \Leftrightarrow \quad 96 + 3\sqrt{x} = 1000/x + 96 + 2\sqrt{x} \quad \Leftrightarrow \quad \sqrt{x} = 1000/x \quad \Leftrightarrow$ $x^{3/2} = 1000 \quad \Leftrightarrow \quad x = (1000)^{2/3} = 100$. Since $c'(x) = \left(x^{3/2} - 1000\right)/x^2 < 0$ for $0 < x < 100$ and $c'(x) > 0$ for $x > 100$, there is an absolute minimum at $x = 100$.

(c) The minimum average cost is $c(100) = \$126$.

9. $C(x) = 680 + 4x + 0.01x^2$, $p(x) = 12 \quad \Rightarrow \quad R(x) = xp(x) = 12x$. If the profit is maximum, then $R'(x) = C'(x) \quad \Rightarrow \quad 12 = 4 + 0.02x \quad \Rightarrow \quad 0.02x = 8 \quad \Rightarrow \quad x = 400$. Now $R''(x) = 0 < 0.02 = C''(x)$, so $x = 400$ gives a maximum.

10. $C(x) = 680 + 4x + 0.01x^2$, $p(x) = 12 - x/500$. Then $R(x) = xp(x) = 12x - x^2/500$. If the profit is maximum, then $R'(x) = C'(x) \quad \Leftrightarrow \quad 12 - x/250 = 4 + 0.02x \quad \Leftrightarrow \quad 8 = 0.024x \quad \Leftrightarrow$ $x = 8/0.024 = \frac{1000}{3}$. Now $R''(x) = -\frac{1}{250} < 0.02 = C''(x)$, so $x = \frac{1000}{3}$ gives a maximum.

11. $C(x) = 1200 + 25x - 0.0001x^2$, $p(x) = 55 - x/1000$. Then $R(x) = xp(x) = 55x - x^2/1000$. If the profit is maximum, then $R'(x) = C'(x) \quad \Leftrightarrow \quad 55 - x/500 = 25 - 0.0002x \quad \Rightarrow \quad 30 = 0.0018x \quad \Rightarrow$ $x = 30/0.0018 \approx 16{,}667$. Now $R''(x) = -\frac{1}{500} < -0.0002 = C''(x)$, so $x = 16{,}667$ gives a maximum.

12. $C(x) = 900 + 110x - 0.1x^2 + 0.02x^3$, $p(x) = 260 - 0.1x$. Then $R(x) = xp(x) = 260x - 0.1x^2$. If the profit is maximum, then $C'(x) = R'(x) \quad \Leftrightarrow \quad 110 - 0.2x + 0.06x^2 = 260 - 0.2x \quad \Leftrightarrow \quad 0.06x^2 = 150 \quad \Leftrightarrow$ $x^2 = 150/0.06 = 2500$ so $x = 50$. Now $R''(x) = -0.2 < -0.2 + 0.12x = C''(x)$ for all $x > 0$, so $x = 50$ gives a maximum.

13. $C(x) = 1450 + 36x - x^2 + 0.001x^3$, $p(x) = 60 - 0.01x$. Then $R(x) = xp(x) = 60x - 0.01x^2$. If the profit is maximum, then $R'(x) = C'(x) \quad \Leftrightarrow \quad 60 - 0.02x = 36 - 2x + 0.003x^2 \quad \Rightarrow \quad 0.003x^2 - 1.98x - 24 = 0$.

By the quadratic formula, $x = \dfrac{1.98 \pm \sqrt{(-1.98)^2 + 4(0.003)(24)}}{2(0.003)} = \dfrac{1.98 \pm \sqrt{4.2084}}{0.006}$. Since $x > 0$, $x \approx (1.98 + 2.05)/0.006 \approx 672$. Now $R''(x) = -0.02$ and $C''(x) = -2 + 0.006x \quad \Rightarrow \quad C''(672) = 2.032$ $\Rightarrow \quad R''(672) < C''(672) \quad \Rightarrow \quad$ there is a maximum at $x = 672$.

14. $C(x) = 10{,}000 + 28x - 0.01x^2 + 0.002x^3$, $p(x) = 90 - 0.02x$. Then $R(x) = xp(x) = 90x - 0.02x^2$. If the profit is maximum, then $R'(x) = C'(x) \quad \Leftrightarrow \quad 90 - 0.04x = 28 - 0.02x + 0.006x^2 \quad \Leftrightarrow$ $0.006x^2 + 0.02x - 62 = 0 \quad \Leftrightarrow \quad 3x^2 + 10x - 31{,}000 = 0 \quad \Leftrightarrow \quad (x - 100)(3x + 310) = 0 \quad \Leftrightarrow$ $x = 100$ (since $x > 0$). Now $R''(x) = -0.04 < -0.02 + 0.012x = C''(x)$ for $x > 0$, so there is a maximum at $x = 100$.

15. $C(x) = 0.001x^3 - 0.3x^2 + 6x + 900$. The marginal cost is $C'(x) = 0.003x^2 - 0.6x + 6$. $C'(x)$ is increasing when $C''(x) > 0 \quad \Leftrightarrow \quad 0.006x - 0.6 > 0 \quad \Leftrightarrow \quad x > 0.6/0.006 = 100$. So $C'(x)$ starts to increase when $x = 100$.

16. $C(x) = 0.0002x^3 - 0.25x^2 + 4x + 1500$. The marginal cost is $C'(x) = 0.0006x^2 - 0.50x + 4$. $C'(x)$ is increasing when $C''(x) > 0 \quad \Leftrightarrow \quad 0.0012x - 0.5 > 0 \quad \Leftrightarrow \quad x > 0.5/0.0012 \approx 417$. So $C'(x)$ starts to increase when $x = 417$.

17. (a) We are given that the demand function p is linear and $p(27{,}000) = 10$, $p(33{,}000) = 8$, so the slope is $\dfrac{10 - 8}{27{,}000 - 33{,}000} = -\dfrac{1}{3000}$ and the equation of the graph is $y - 10 = \left(-\frac{1}{3000}\right)(x - 27{,}000) \quad \Rightarrow$ $p(x) = 19 - x/3000$.

(b) The revenue is $R(x) = xp(x) = 19x - x^2/3000 \quad \Rightarrow \quad R'(x) = 19 - x/1500 = 0$ when $x = 28{,}500$. Since $R''(x) = -1/1500 < 0$, the maximum revenue occurs when $x = 28{,}500 \quad \Rightarrow \quad$ the price is $p(28{,}500) = \$9.50$.

18. (a) Let $p(x)$ be the demand function. Then $p(x)$ is linear and $y = p(x)$ passes through $(20, 10)$ and $(18, 11)$, so the slope is $-\frac{1}{2}$ and the equation of the line is y $- 10 = -\frac{1}{2}(x - 10) \quad \Leftrightarrow \quad y = -\frac{1}{2}x + 20$. Thus the demand is $p(x) = -\frac{1}{2}x + 20$ and the revenue is $R(x) = xp(x) = -\frac{1}{2}x^2 + 20x$.

(b) The cost is $C(x) = 6x$, so the profit is $P(x) = R(x) - C(x) = -\frac{1}{2}x^2 + 14x$. Then $0 = P'(x) = -x + 14 \quad \Rightarrow \quad x = 14$. Since $P''(x) = -1 < 0$, the selling price for maximum profit is $p(14) = 20 - \left(\frac{1}{2}\right)14 = \$13$.

19. **(a)** $p(x) = 450 - \frac{1}{10}(x - 1000) = 550 - x/10$.

(b) $R(x) = xp(x) = 500x - x^2/10$. $R'(x) = 550 - x/5 = 0$ when $x = 5(550) = 2750$. $p(2750) = 275$, so the rebate should be $450 - 275 = \$175$.

(c) $P(x) = R(x) - C(x) = 550x - x^2/10 - 6800 - 150x = 400x - x^2/10 - 6800$, $P'(x) = 400 - x/5 = 0$ when $x = 2000$. $p(2000) = 550 - 200 = 350$. Therefore the rebate to maximize profits should be $450 - 350 = \$100$.

20. Let x be the number of \$5 increases in rent. So the price, $p(x) = 400 + 5x$, and the number of units occupied is $100 - x$. Now revenue $= R(x) =$ (rental price per unit) $\times$ (number of units rented) $= (400 + 5x)(100 - x) = -5x^2 + 100x + 40{,}000$, $0 \le x \le 100 \quad \Rightarrow \quad R'(x) = -10x + 100 = 0 \quad \Leftrightarrow$ $x = 10$. This is a maximum since $R''(x) = -10 < 0$ for all x. Now we must check the value of R at $x = 10$ and at the endpoints of the domain to see which gives the maximum value. $R(0) = 40{,}000$, $R(10) = -5(10)^2 + 100(10) + 40{,}000 = 40{,}500$ and $R(100) = -5(100)^2 + 100(100) + 40{,}000 = 0$. Therefore $x = 10$ gives the maximum revenue and the rent should be $400 + 5(10) = \$450$.

EXERCISES 3.10

1. $f(x) = 12x^2 + 6x - 5 \quad \Rightarrow \quad F(x) = 12\left(\frac{1}{3}x^3\right) + 6\left(\frac{1}{2}x^2\right) - 5x + C = 4x^3 + 3x^2 - 5x + C$

2. $f(x) = x^3 - 4x^2 + 17 \quad \Rightarrow \quad F(x) = \frac{1}{4}x^4 - \frac{4}{3}x^3 + 17x + C$

3. $f(x) = 6x^9 - 4x^7 + 3x^2 + 1 \quad \Rightarrow$
$F(x) = 6\left(\frac{1}{10}x^{10}\right) - 4\left(\frac{1}{8}x^8\right) + 3\left(\frac{1}{3}x^3\right) + x + C = \frac{3}{5}x^{10} - \frac{1}{2}x^8 + x^3 + x + C$

4. $f(x) = x^{99} - 2x^{49} - 1 \quad \Rightarrow \quad F(x) = \left(\frac{1}{100}x^{100}\right) - 2\left(\frac{1}{50}x^{50}\right) - x + C = \frac{1}{100}x^{100} - \frac{1}{25}x^{50} - x + C$

5. $f(x) = \sqrt{x} + \sqrt[3]{x} = x^{1/2} + x^{1/3} \quad \Rightarrow \quad F(x) = \frac{1}{3/2}x^{3/2} + \frac{1}{4/3}x^{4/3} + C = \frac{2}{3}x^{3/2} + \frac{3}{4}x^{4/3} + C$

6. $f(x) = \sqrt[3]{x^2} - \sqrt{x^3} = x^{2/3} - x^{3/2} \quad \Rightarrow \quad F(x) = \frac{1}{5/3}x^{5/3} - \frac{1}{5/2}x^{5/2} + C = \frac{3}{5}x^{5/3} - \frac{2}{5}x^{5/2} + C$

7. $f(x) = 6/x^5 = 6x^{-5} \quad \Rightarrow \quad F(x) = \begin{cases} 6x^{-4}/(-4) + C_1 = -3/(2x^4) + C_1 & \text{if } x < 0 \\ -3/(2x^4) + C_2 & \text{if } x > 0 \end{cases}$

8. $f(x) = 3x^{-2} - 5x^{-4}$ has domain $(-\infty, 0) \cup (0, \infty)$, so

$$F(x) = \begin{cases} \dfrac{3x^{-1}}{-1} - \dfrac{5x^{-3}}{-3} + C_1 = -\dfrac{3}{x} + \dfrac{5}{3x^3} + C_1 & \text{if } x < 0 \\ -\dfrac{3}{x} + \dfrac{5}{3x^3} + C_2 & \text{if } x > 0. \end{cases}$$

9. $g(t) = \dfrac{t^3 + 2t^2}{\sqrt{t}} = t^{5/2} + 2t^{3/2} \quad \Rightarrow \quad G(t) = \dfrac{t^{7/2}}{7/2} + \dfrac{2t^{5/2}}{5/2} + C = \frac{2}{7}t^{7/2} + \frac{4}{5}t^{5/2} + C$

10. $f(x) = x^{2/3} + 2x^{-1/3}$ has domain $(-\infty, 0) \cup (0, \infty)$, so

$$F(x) = \begin{cases} \dfrac{x^{5/3}}{5/3} + \dfrac{2x^{2/3}}{2/3} + C_1 = \frac{3}{5}x^{5/3} + 3x^{2/3} + C_1 & \text{if } x > 0 \\ \frac{3}{5}x^{5/3} + 3x^{2/3} + C_2 & \text{if } x < 0. \end{cases}$$

11. $h(x) = \sin x - 2\cos x \quad \Rightarrow \quad H(x) = -\cos x - 2\sin x + C$

12. $f(t) = \sin t - 2\sqrt{t} \quad \Rightarrow \quad F(t) = -\cos t - 2\left(\frac{1}{3/2}\right)t^{3/2} + C = -\cos t - \frac{4}{3}t^{3/2} + C$

13. $f(t) = \sec^2 t + t^2 \quad \Rightarrow \quad F(t) = \tan t + \frac{1}{3}t^3 + C_n$ on the interval $\left(n\pi - \frac{\pi}{2}, n\pi + \frac{\pi}{2}\right)$.

14. $f(\theta) = \theta + \sec\theta\tan\theta \quad \Rightarrow \quad F(\theta) = \frac{1}{2}\theta^2 + \sec\theta + C_n$ on the interval $\left(n\pi - \frac{\pi}{2}, n\pi + \frac{\pi}{2}\right)$.

15. $f''(x) = x^2 + x^3 \quad \Rightarrow \quad f'(x) = \frac{1}{3}x^3 + \frac{1}{4}x^4 + C \quad \Rightarrow \quad f(x) = \frac{1}{12}x^4 + \frac{1}{20}x^5 + Cx + D$

16. $f''(x) = 60x^4 - 45x^2 \quad \Rightarrow \quad f'(x) = 60\left(\frac{1}{5}x^5\right) - 45\left(\frac{1}{3}x^3\right) + C = 12x^5 - 15x^3 + C \quad \Rightarrow$
$f(x) = 12\left(\frac{1}{6}x^6\right) - 15\left(\frac{1}{4}x^4\right) + Cx + D = 2x^6 - \frac{15}{4}x^4 + Cx + D$

17. $f''(x) = 1 \quad \Rightarrow \quad f'(x) = x + C \quad \Rightarrow \quad f(x) = \frac{1}{2}x^2 + Cx + D$

18. $f''(x) = \sin x \quad \Rightarrow \quad f'(x) = -\cos x + C \quad \Rightarrow \quad f(x) = -\sin x + Cx + D$

19. $f'''(x) = 24x \quad \Rightarrow \quad f''(x) = 12x^2 + C \quad \Rightarrow \quad f'(x) = 4x^3 + Cx + D \quad \Rightarrow$
$f(x) = x^4 + \frac{1}{2}Cx^2 + Dx + E$

20. $f'''(x) = x^{1/2} \quad \Rightarrow \quad f''(x) = \frac{2}{3}x^{3/2} + C \quad \Rightarrow \quad f'(x) = \frac{2}{3} \cdot \frac{2}{5}x^{5/2} + Cx + D = \frac{4}{15}x^{5/2} + Cx + D \quad \Rightarrow$
$f(x) = \frac{4}{15} \cdot \frac{2}{7}x^{7/2} + C\left(\frac{1}{2}x^2\right) + Dx + E = \frac{8}{105}x^{7/2} + \frac{1}{2}Cx^2 + Dx + E$

21. $f'(x) = 4x + 3 \quad \Rightarrow \quad f(x) = 2x^2 + 3x + C \quad \Rightarrow \quad -9 = f(0) = C \quad \Rightarrow \quad f(x) = 2x^2 + 3x - 9$

22. $f'(x) = 12x^2 - 24x + 1 \quad \Rightarrow \quad f(x) = 4x^3 - 12x^2 + x + C \quad \Rightarrow \quad f(1) = 4 - 12 + 1 + C = -2 \quad \Rightarrow$
$C = 5$, so $f(x) = 4x^3 - 12x^2 + x + 5$

23. $f'(x) = 3\sqrt{x} - 1/\sqrt{x} = 3x^{1/2} - x^{-1/2} \quad \Rightarrow \quad f(x) = 3\left(\frac{1}{3/2}\right)x^{3/2} - \frac{1}{1/2}x^{1/2} + C \quad \Rightarrow$
$2 = f(1) = 2 - 2 + C = C \quad \Rightarrow \quad f(x) = 2x^{3/2} - 2x^{1/2} + 2$

24. $f'(x) = 1 + x^{-2}, x > 0 \quad \Rightarrow \quad f(x) = x - 1/x + C \quad \Rightarrow \quad f(1) = 1 - 1 + C = 1 \quad \Rightarrow \quad C = 1$, so
$f(x) = 1 + x - 1/x$

25. $f'(x) = 3\cos x + 5\sin x \quad \Rightarrow \quad f(x) = 3\sin x - 5\cos x + C \quad \Rightarrow \quad 4 = f(0) = -5 + C \quad \Rightarrow \quad C = 9$
$\Rightarrow \quad f(x) = 3\sin x - 5\cos x + 9.$

26. $f'(x) = 3x^{-2} \quad \Rightarrow \quad f(x) = \begin{cases} -3/x + C_1 & \text{if } x > 0 \\ -3/x + C_2 & \text{if } x < 0 \end{cases} \quad f(1) = -3 + C_1 = 0 \quad \Rightarrow \quad C_1 = 3,$

$f(-1) = 3 + C_2 = 0 \quad \Rightarrow \quad C_2 = -3.$ So $f(x) = \begin{cases} -3/x + 3 & \text{if } x > 0 \\ -3/x - 3 & \text{if } x < 0 \end{cases}$

27. $f''(x) = x \Rightarrow f'(x) = \frac{1}{2}x^2 + C \Rightarrow 2 = f'(0) = C \Rightarrow f'(x) = \frac{1}{2}x^2 + 2 \Rightarrow$
$f(x) = \frac{1}{6}x^3 + 2x + D \Rightarrow -3 = f(0) = D \Rightarrow f(x) = \frac{1}{6}x^3 + 2x - 3$

28. $f''(x) = 20x^3 - 10 \Rightarrow f'(x) = 5x^4 - 10x + C \Rightarrow -5 = f'(1) = 5 - 10 + C \Rightarrow C = 0 \Rightarrow$
$f'(x) = 5x^4 - 10x \Rightarrow f(x) = x^5 - 5x^2 + D \Rightarrow 1 = f(1) = 1 - 5 + D \Rightarrow D = 5 \Rightarrow$
$f(x) = x^5 - 5x^2 + 5$

29. $f''(x) = x^2 + 3\cos x \Rightarrow f'(x) = \frac{1}{3}x^3 + 3\sin x + C \Rightarrow 3 = f'(0) = C \Rightarrow$
$f'(x) = \frac{1}{3}x^3 + 3\sin x + 3 \Rightarrow f(x) = \frac{1}{12}x^4 - 3\cos x + 3x + D \Rightarrow 2 = f(0) = -3 + D \Rightarrow$
$D = 5 \Rightarrow f(x) = \frac{1}{12}x^4 - 3\cos x + 3x + 5$

30. $f''(x) = x + x^{1/2} \Rightarrow f'(x) = \frac{1}{2}x^2 + \frac{2}{3}x^{3/2} + C \Rightarrow 2 = f'(1) = \frac{1}{2} + \frac{2}{3} + C \Rightarrow C = \frac{5}{6} \Rightarrow$
$f'(x) = \frac{1}{2}x^2 + \frac{2}{3}x^{3/2} + \frac{5}{6} \Rightarrow f(x) = \frac{1}{6}x^3 + \frac{4}{15}x^{5/2} + \frac{5}{6}x + D \Rightarrow 1 = f(1) = \frac{1}{6} + \frac{4}{15} + \frac{5}{6} + D \Rightarrow$
$D = -\frac{4}{15} \Rightarrow f(x) = \frac{1}{6}x^3 + \frac{4}{15}x^{5/2} + \frac{5}{6}x - \frac{4}{15}$

31. $f''(x) = 6x + 6 \Rightarrow f'(x) = 3x^2 + 6x + C \Rightarrow f(x) = x^3 + 3x^2 + Cx + D \Rightarrow 4 = f(0) = D$
and $3 = f(1) = 1 + 3 + C + D = 4 + C + 4 \Rightarrow C = -5 \Rightarrow f(x) = x^3 + 3x^2 - 5x + 4$

32. $f''(x) = 12x^2 - 6x + 2 \Rightarrow f'(x) = 4x^3 - 3x^2 + 2x + C \Rightarrow f(x) = x^4 - x^3 + x^2 + Cx + D$, so
$1 = f(0) = D$ and $11 = f(2) = 16 - 8 + 4 + 2C + D = 13 + 2C \Rightarrow C = -1$, so
$f(x) = x^4 - x^3 + x^2 - x + 1$

33. $f''(x) = x^{-3} \Rightarrow f'(x) = -\frac{1}{2}x^{-2} + C \Rightarrow f(x) = \frac{1}{2}x^{-1} + Cx + D \Rightarrow 0 = f(1) = \frac{1}{2} + C + D$
and $0 = f(2) = \frac{1}{4} + 2C + D$. Solving these equations, we get $C = \frac{1}{4}$, $D = -\frac{3}{4}$, so $f(x) = 1/(2x) + \frac{1}{4}x - \frac{3}{4}$.

34. $f'''(x) = \sin x \Rightarrow f''(x) = -\cos x + C \Rightarrow 1 = f''(0) = -1 + C = 1 \Rightarrow C = 2$, so
$f''(x) = -\cos x + 2 \Rightarrow f'(x) = -\sin x + 2x + D \Rightarrow 1 = f'(0) = D \Rightarrow$
$f'(x) = -\sin x + 2x + 1 \Rightarrow f(x) = \cos x + x^2 + x + E \Rightarrow 1 = f(0) = 1 + E \Rightarrow E = 0$, so
$f(x) = \cos x + x^2 + x$

35. We have that $f'(x) = 2x + 1 \Rightarrow f(x) = x^2 + x + C$. But f passes through $(1, 6)$ so that
$6 = f(1) = 1^2 + 1 + C \Rightarrow C = 4$. Therefore $f(x) = x^2 + x + 4 \Rightarrow f(2) = 2^2 + 2 + 4 = 10$.

36. $f'(x) = x^3 \Rightarrow f(x) = \frac{1}{4}x^4 + C$. $x + y = 0 \Rightarrow y = -x \Rightarrow m = -1$. Now
$m = f'(x) = x^3 = -1 \Leftrightarrow x = -1 \Rightarrow y = 1$ (from the equation of the tangent line), so $(-1, 1)$ is a point on the graph of f, so $1 = (-1)^4/4 + C \Rightarrow C = \frac{3}{4}$. Therefore the function is $f(x) = \frac{1}{4}x^4 + \frac{3}{4}$.

37. b is the antiderivative of f. For small x, f is negative, so the graph of its antiderivative must be decreasing. But both a and c are increasing for small x, so only b can be f's antiderivative. Also, f is positive where b is increasing, which supports our conclusion.

38. We know right away that c cannot be f's antiderivative, since the slope of c is not zero at the x-value where $f = 0$. Now f is positive when a is increasing and negative when a is decreasing, so a is the antiderivative of f.

39. The graph of F will have a minimum at 0 and a maximum at 2, since $f = F'$ goes from negative to positive at $x = 0$, and from positive to negative at $x = 2$.

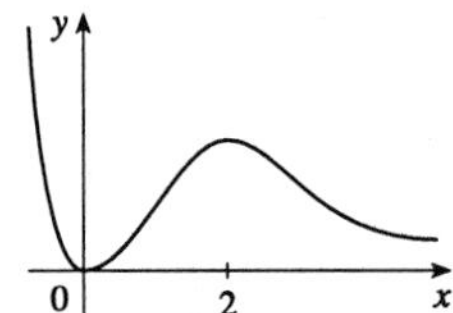

40. The position function is the antiderivative of the velocity function, so its graph will have be horizontal where the velocity function is equal to 0.

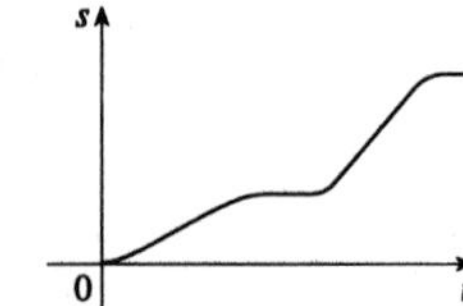

41.

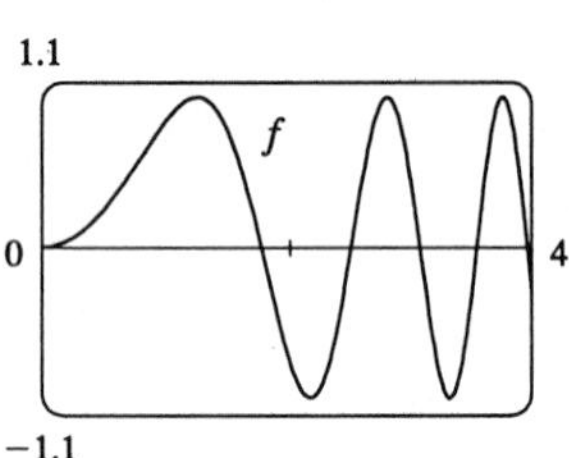

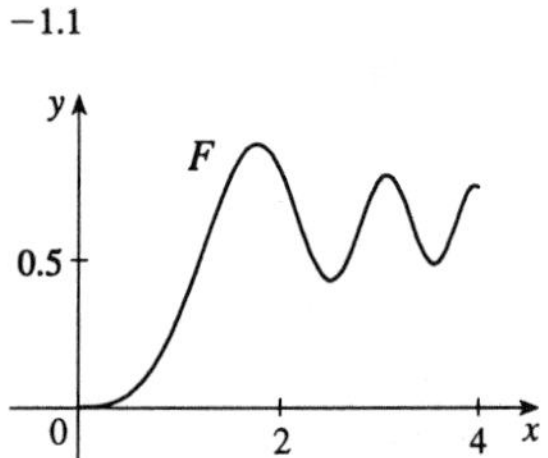

42.

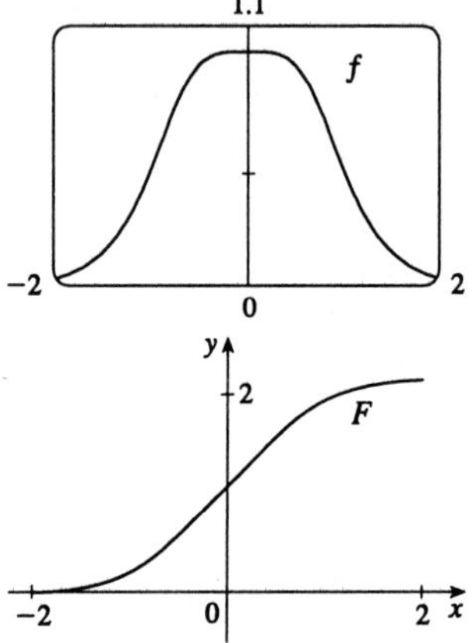

43.

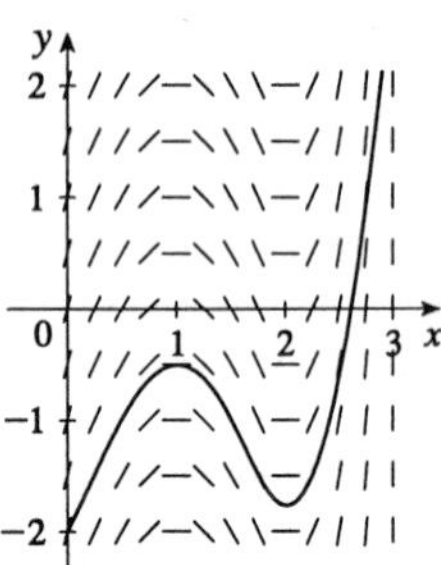

44.

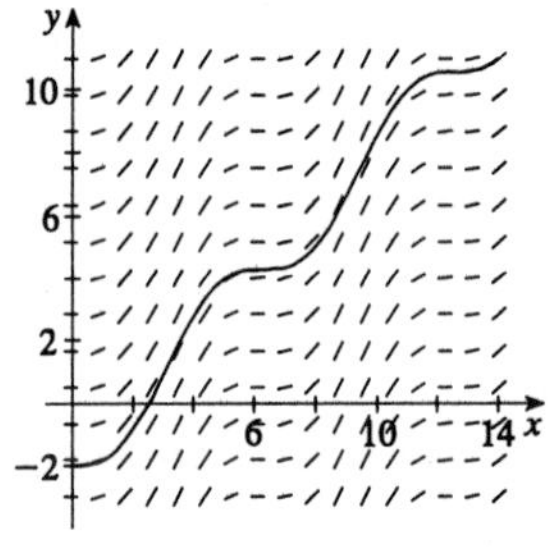

45.

x	$f(x)$
0	1
0.5	0.959
1.0	0.841
1.5	0.665
2.0	0.455
2.5	0.239
3.0	0.470

x	$f(x)$
3.5	−0.100
4.0	−0.189
4.5	−0.217
5.0	−0.192
5.5	−0.128
6.0	−0.047

We compute slopes as in the table and draw a direction field as in Example 6. Then we use the direction field to graph F starting at $(0, 0)$.

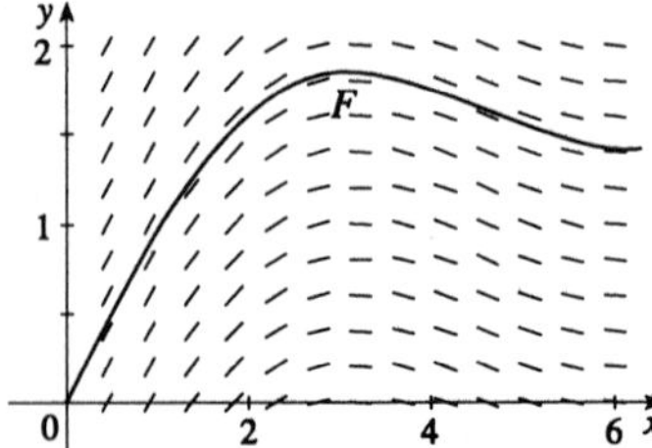

46.

x	$f(x)$
0	0
± 0.2	0.041
± 0.4	0.169
± 0.6	0.410
± 0.8	0.824
± 1.0	1.557
± 1.2	3.087
± 1.4	8.117
± 1.5	21.152

We compute slopes as in the table and draw a direction field as in Example 6. Then we use the direction field to graph F starting at $(0, 0)$ and extending in both directions.

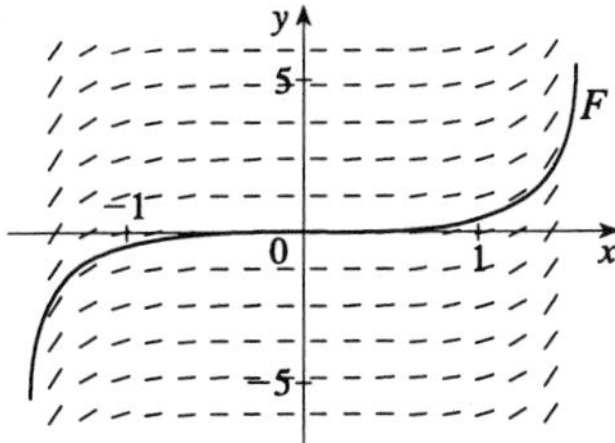

47.

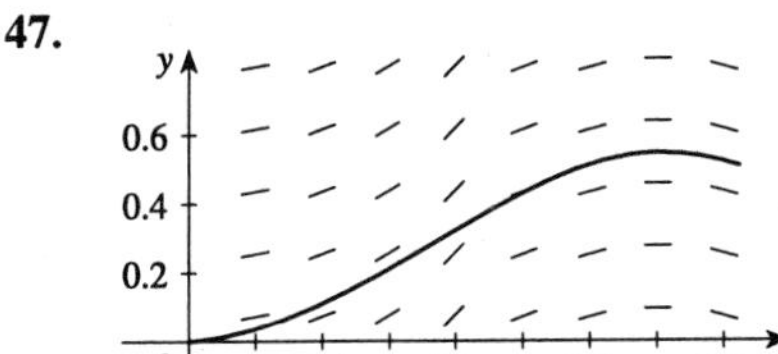

48. **(a)**

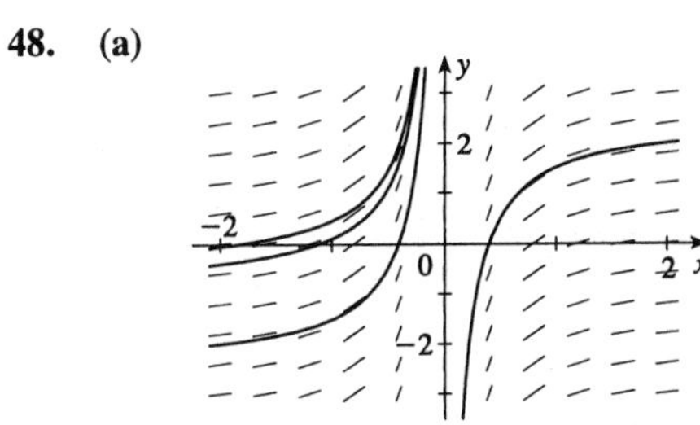

(b) The general antiderivative of $f(x) = x^{-2}$ is

$$F(x) = \begin{cases} -1/x + C_1 & \text{if } x < 0 \\ -1/x + C_2 & \text{if } x > 0 \end{cases}$$

since $f(x)$ is not defined at $x = 0$. The graph of the general antiderivatives of $f(x)$ looks like the graph in part (a), as expected.

49. $v(t) = s'(t) = 3 - 2t \quad \Rightarrow \quad s(t) = 3t - t^2 + C \quad \Rightarrow \quad 4 = s(0) = C \quad \Rightarrow \quad s(t) = 3t - t^2 + 4$

50. $v(t) = s'(t) = 3\sqrt{t} \quad \Rightarrow \quad s(t) = 2t^{3/2} + C \quad \Rightarrow \quad 5 = s(1) = 2 + C \quad \Rightarrow \quad C = 3$, so $s(t) = 2t^{3/2} + 3$

51. $a(t) = v'(t) = 3t + 8 \quad \Rightarrow \quad v(t) = \frac{3}{2}t^2 + 8t + C \quad \Rightarrow \quad -2 = v(0) = C \quad \Rightarrow \quad v(t) = \frac{3}{2}t^2 + 8t - 2 \quad \Rightarrow$
$s(t) = \frac{1}{2}t^3 + 4t^2 - 2t + D \quad \Rightarrow \quad 1 = s(0) = D \quad \Rightarrow \quad s(t) = \frac{1}{2}t^3 + 4t^2 - 2t + 1$

52. $a(t) = v'(t) = \cos t + \sin t \quad \Rightarrow \quad v(t) = \sin t - \cos t + C \quad \Rightarrow \quad 5 = v(0) = -1 + C \quad \Rightarrow \quad C = 6$, so
$v(t) = \sin t - \cos t + 6 \quad \Rightarrow \quad s(t) = -\cos t - \sin t + 6t + D \quad \Rightarrow \quad 0 = s(0) = -1 + D \quad \Rightarrow \quad D = 1$, so
$s(t) = -\cos t - \sin t + 6t + 1$

53. $a(t) = v'(t) = t^2 - t \quad \Rightarrow \quad v(t) = \frac{1}{3}t^3 - \frac{1}{2}t^2 + C \quad \Rightarrow \quad s(t) = \frac{1}{12}t^4 - \frac{1}{6}t^3 + Ct + D \quad \Rightarrow \quad 0 = s(0) = D$
and $12 = s(6) = 108 - 36 + 6C + 0 \quad \Rightarrow \quad C = -10 \quad \Rightarrow \quad s(t) = \frac{1}{12}t^4 - \frac{1}{6}t^3 - 10t$

54. $a(t) = v'(t) = 10 + 3t - 3t^2 \quad \Rightarrow \quad v(t) = 10t + \frac{3}{2}t^2 - t^3 + C \quad \Rightarrow \quad s(t) = 5t^2 + \frac{1}{2}t^3 - \frac{1}{4}t^4 + Ct + D$
$\Rightarrow \quad 0 = s(0) = D$ and $10 = s(2) = 20 + 4 - 4 + 2C \quad \Rightarrow \quad C = -5$, so $s(t) = -5t + 5t^2 + \frac{1}{2}t^3 - \frac{1}{4}t^4$

55. **(a)** $v'(t) = a(t) = -9.8 \Rightarrow v(t) = -9.8t + C$, but $C = v(0) = 0$, so $v(t) = -9.8t \Rightarrow$
$s(t) = -4.9t^2 + D \Rightarrow D = s(0) = 450 \Rightarrow s(t) = 450 - 4.9t^2$

(b) It reaches the ground when $0 = s(t) = 450 - 4.9t^2 \Rightarrow t^2 = 450/4.9 \Rightarrow t = \sqrt{450/4.9} \approx 9.58$ s.

(c) $v = -9.8\sqrt{450/4.9} \approx -93.9$ m/s

56. **(a)** $v'(t) = a(t) = -9.8 \Rightarrow v(t) = -9.8t + C$ and $-5 = v(0) = C \Rightarrow v(t) = -9.8t - 5 \Rightarrow$
$s(t) = -4.9t^2 - 5t + D \Rightarrow 450 = s(0) = D$, so $s(t) = -4.9t^2 - 5t + 450$

(b) $s(t) = 0 \Rightarrow 4.9t^2 + 5t - 450 = 0 \Rightarrow t = \dfrac{-5 + \sqrt{25 + 4(4.9)(450)}}{9.8} \approx 9.1$ s

(c) $v(9.1) \approx -94.05$ m/s

57. **(a)** $v'(t) = -9.8 \Rightarrow v(t) = -9.8t + C \Rightarrow 5 = v(0) = C$, so $v(t) = 5 - 9.8t \Rightarrow$
$s(t) = 5t - 4.9t^2 + D \Rightarrow D = s(0) = 450 \Rightarrow s(t) = 450 + 5t - 4.9t^2$

(b) It reaches the ground when $450 + 5t - 4.9t^2 = 0$. By the quadratic formula, the positive root of this equation is $t = \dfrac{5 + \sqrt{8845}}{9.8} \approx 10.1$ s

(c) $v = 5 - 9.8 \cdot \dfrac{5 + \sqrt{8845}}{9.8} \approx -94.0$ m/s

58. $v'(t) = a(t) = a \Rightarrow v(t) = at + C$ and $v_0 = v(0) = C \Rightarrow v(t) = at + v_0 \Rightarrow$
$s(t) = \frac{1}{2}at^2 + v_0 t + D \Rightarrow s_0 = s(0) = D \Rightarrow s(t) = \frac{1}{2}at^2 + v_0 t + s_0$

59. By Exercise 58, $s(t) = -4.9t^2 + v_0 t + s_0$ and $v(t) = s'(t) = -9.8t + v_0$. So $[v(t)]^2 = (9.8)^2 t^2 - 19.6 v_0 t + v_0^2$ and $v_0^2 - 19.6[s(t) - s_0] = v_0^2 - 19.6[-4.9t^2 + v_0 t] = v_0^2 + (9.8)^2 t^2 - 19.6 v_0 t = [v(t)]^2$

60. For the first ball, $s_1(t) = -16t^2 + 48t + 432$ from Example 8. For the second ball, $a(t) = -32 \Rightarrow$
$v(t) = -32t + C$, but $v(1) = -32(1) + C = 24 \Rightarrow C = 56$, so $v(t) = -32t + 56 \Rightarrow$
$s(t) = -16t^2 + 56t + D$, but $s(1) = -16(1)^2 + 56(1) + D = 432 \Rightarrow D = 392$, and
$s_2(t) = -16t^2 + 56t + 392$. The balls pass each other when $s_1(t) = s_2(t) \Rightarrow$
$-16t^2 + 48t + 432 = -16t^2 + 56t + 392 \Leftrightarrow 8t = 40 \Leftrightarrow t = 5$ s.

61. Marginal cost $= 1.92 - 0.002x = C'(x) \Rightarrow C(x) = 1.92x - 0.001x^2 + K$. But
$C(1) = 1.92 - 0.001 + K = 562 \Rightarrow K = 560.081$. Therefore $C(x) = 1.92x - 0.001x^2 + 560.081 \Rightarrow$
$C(100) = 1.92(100) - 0.001(100)^2 + 560.081 = 742.081$, so the cost of producing 100 items is \$742.08.

62. Let the mass, measured from one end, be $f(x)$. Then $f(0) = 0$ and from Section 2.3, $f'(x) = \rho = \dfrac{dm}{dx} = x^{-1/2}$
$\Rightarrow f(x) = m = 2x^{1/2} + C$ and $f(0) = C = 0$, so $f(x) = 2\sqrt{x}$. Thus the mass of the rod is
$f(100) = 2\sqrt{100} = 20$ g.

63. Taking the upward direction to be positive we have that for $0 \le t \le 10$ (using the subscript 1 to refer to $0 \le t \le 10$), $a_1(t) = -9 + 0.9t = v_1'(t) \quad \Rightarrow \quad v_1(t) = -9t + 0.45t^2 + v_0$, but $v_1(0) = v_0 = -10 \quad \Rightarrow$ $v_1(t) = -9t + 0.45t^2 - 10 = s_1'(t) \quad \Rightarrow \quad s_1(t) = -\frac{9}{2}t^2 + 0.15t^3 - 10t + s_0$. But $s_1(0) = 500 = s_0 \quad \Rightarrow$ $s_1(t) = -\frac{9}{2}t^2 + 0.15t^3 - 10t + 500$. Now for $t > 10$, $a(t) = 0 = v'(t) \quad \Rightarrow$ $v(t) = \text{constant} = v_1(10) = -9(10) + 0.45(10)^2 - 10 = -55 \quad \Rightarrow \quad v(t) = -55 = s'(t) \quad \Rightarrow$ $s(t) = -55t + s_{10}$. But $s(10) = s_1(10) \quad \Rightarrow \quad -55(10) + s_{10} = 100 \quad \Rightarrow \quad s_{10} = 650 \quad \Rightarrow$ $s(t) = -55t + 650$. When the raindrop hits the ground we have that $s(t) = 0 \quad \Rightarrow \quad -55t + 650 = 0 \quad \Rightarrow$ $t = \frac{650}{55} = \frac{130}{11} \approx 11.8$ s.

64. $v'(t) = a(t) = -40$. The initial velocity is 50 mi/h $= \dfrac{50 \cdot 5280}{3600} = \dfrac{220}{3}$ ft/s, so $v(t) = -40t + \dfrac{220}{3}$. The car stops when $v(t) = 0 \quad \Leftrightarrow \quad t = \frac{220}{3 \cdot 40} = \frac{11}{6}$. Since $s(t) = -20t^2 + \frac{220}{3}t$, the distance covered is $s\left(\frac{11}{6}\right) = -20\left(\frac{11}{6}\right)^2 + \frac{220}{3} \cdot \frac{11}{6} \approx 67.2$ ft.

65. $a(t) = a$ and the initial velocity is 30 mi/h $= 30 \cdot \frac{5280}{3600} = 44$ ft/s and final velocity 50 mi/h $= 50 \cdot \frac{5280}{3600} = \frac{220}{3}$ ft/s. So $v(t) = at + 44 \quad \Rightarrow \quad \frac{220}{3} = v(5) = 5a + 44 \quad \Rightarrow \quad a = \frac{88}{15} \approx 5.87\ \text{ft/s}^2$.

66. $a(t) = -40 \quad \Rightarrow \quad v(t) = -40t + v_0$ where v_0 is the car's speed (in ft/s) when the brakes were applied. The car stops when $-40t + v_0 = 0 \quad \Leftrightarrow \quad t = \frac{1}{40}v_0$. Now $s(t) = -20t^2 + v_0 t \quad \Rightarrow$ $160 = s\left(\frac{1}{40}v_0\right) = -20\left(\frac{1}{40}v_0\right)^2 + v_0\left(\frac{1}{40}v_0\right) = \frac{1}{80}v_0^2 \quad \Rightarrow \quad v_0^2 = 12{,}800 \quad \Rightarrow \quad v_0 = 80\sqrt{2} \approx 113$ ft/s.

67. The height at time t is $s(t) = -16t^2 + h$, where $h = s(0)$ is the height of the cliff. $v(t) = -32t = -120$ when $t = 3.75$, so $0 = s(3.75) = -16(3.75)^2 + h \quad \Rightarrow \quad h = 16(3.75)^2 = 225$ ft.

REVIEW EXERCISES FOR CHAPTER 3

1. False. For example, take $f(x) = x^3$, then $f'(x) = 3x^2$ and $f'(0) = 3(0)^2 = 0$, but $f(0) = 0$ is not a maximum or minimum; $(0, 0)$ is an inflection point.

2. False. For example, $f(x) = |x|$ has an absolute minimum at 0, but $f'(0)$ does not exist.

3. False. For example, $f(x) = x$ is continuous on $(0, 1)$ but attains neither a maximum nor a minimum value on $(0, 1)$.

4. True, by Rolle's Theorem, since $|c| < 1 \quad \Leftrightarrow \quad c \in (-1, 1)$.

5. True, by the Test for Monotonic Functions.

6. False. For example, the curve $y = f(x) = 1$ has no inflection points but $f''(c) = 0$ for all c. (See also Exercise 3.4.32.)

7. False. $f(x) = g(x) + C$ by Corollary 3.2.7. For example, $f(x) = x + 2$, $g(x) = x + 1 \quad \Rightarrow$ $f'(x) = g'(x) = 1$, but $f(x) \neq g(x)$.

8. False. Assume there is a function f such that $f(1) = -2$ and $f(3) = 0$. Then by the Mean Value Theorem there exists a number $c \in (1, 3)$ such that $f'(c) = \dfrac{f(3) - f(1)}{3 - 1} = \dfrac{0 - (-2)}{3 - 1} = 1$. But $f'(x) > 1$ for all x, a contradiction.

9. True. The graph of one such function is sketched. [An example is $f(x) = e^{-x}$.]

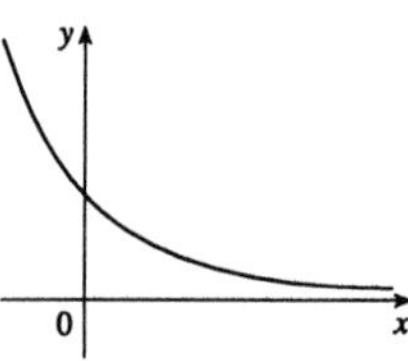

10. False. At any point $(a, f(a))$, we know that $f'(a) < 0$. So since the tangent line at $(a, f(a))$ is not horizontal, it must cross the x-axis — at $x = b$, say. But since $f''(x) > 0$ for all x, the graph of f must lie above all of its tangents; in particular, $f(b) > 0$. But this is a contradiction, since we are given that $f(x) < 0$ for all x.

11. True. Let $x_1 < x_2$ where $x_1, x_2 \in I$. Then $f(x_1) < f(x_2)$ and $g(x_1) < g(x_2)$ (since f and g are increasing on I), so $(f + g)(x_1) = f(x_1) + g(x_1) < f(x_2) + g(x_2) = (f + g)(x_2)$.

12. False. $f(x) = x$ and $g(x) = 2x$ are both increasing on $[0, 1]$, but $f(x) - g(x) = -x$ is not increasing on $[0, 1]$.

13. False. Take $f(x) = x$ and $g(x) = x - 1$. Then both f and g are increasing on $[0, 1]$. But $f(x)g(x) = x(x - 1)$ is not increasing on $[0, 1]$.

14. True. Let $x_1 < x_2$ where $x_1, x_2 \in I$. Then $0 < f(x_1) < f(x_2)$ and $0 < g(x_1) < g(x_2)$ (since f and g are both positive and increasing). Hence $f(x_1)g(x_1) < f(x_2)g(x_1) < f(x_2)g(x_2)$. So fg is increasing on I.

15. True. Let $x_1, x_2 \in I$ and $x_1 < x_2$, then $f(x_1) < f(x_2)$ (f is increasing) $\Rightarrow$ $\dfrac{1}{f(x_1)} > \dfrac{1}{f(x_2)}$ (f is positive) $\Rightarrow \quad g(x_1) > g(x_2) \quad \Rightarrow \quad g(x) = \dfrac{1}{f(x)}$ is decreasing on I.

16. False. The most general antiderivative is $F(x) = -1/x + C_1$ for $x < 0$ and $F(x) = -1/x + C_2$ for $x > 0$ (see Example 1 in Section 3.10.)

17. $f(x) = x^3 - 12x + 5$, $-5 \leq x \leq 3$. $f'(x) = 3x^2 - 12 = 0 \quad \Rightarrow \quad x^2 = 4 \quad \Rightarrow \quad x = \pm 2$. $f''(x) = 6x \quad \Rightarrow$ $f''(-2) = -12 < 0$, so $f(-2) = 21$ is a local maximum, and $f''(2) = 12 > 0$, so $f(2) = -11$ is a local minimum. Also $f(-5) = -60$ and $f(3) = -4$, so $f(-2) = 21$ is the absolute maximum and $f(-5) = -60$ is the absolute minimum.

18. $f(x) = 3x^5 - 25x^3 + 60x$, $-1 \leq x \leq 3$. $f'(x) = 15x^4 - 75x^2 + 60 = 15(x^2 - 4)(x^2 - 1) = 0$ when $x = \pm 1$, ± 2 (but -2 is not in the interval). $f''(x) = 60x^3 - 150x \quad \Rightarrow \quad f''(1) = -90 < 0$, $f''(2) = 180 > 0$, so $f(1) = 38$ is a local maximum, $f(2) = 16$ is a local minimum. Also $f(-1) = -38$ and $f(3) = 234$, so $f(-1) = -38$ is an absolute minimum and $f(3) = 234$ is an absolute maximum.

19. $f(x) = \dfrac{x-2}{x+2}$, $0 \le x \le 4$. $f'(x) = \dfrac{(x+2)-(x-2)}{(x+2)^2} = \dfrac{4}{(x+2)^2} > 0 \quad \Rightarrow \quad f$ is increasing on $[0, 4]$, so f has no local extremum and $f(0) = -1$ is the absolute minimum and $f(4) = \frac{1}{3}$ is the absolute maximum.

20. $f(x) = \sqrt{x^2+4x+8}$, $-3 \le x \le 0$. $f'(x) = (x+2)/\sqrt{x^2+4x+8} = 0$ when $x = -2$, and $f'(x) < 0$ for $x < -2$, $f'(x) > 0$ for $x > -2$. So $f(-2) = 2$ is a local and absolute minimum. Also $f(-3) = \sqrt{5}$, $f(0) = 2\sqrt{2}$, so $f(0) = 2\sqrt{2}$ is an absolute maximum.

21. $f(x) = x - \sqrt{2}\sin x$, $0 \le x \le \pi$. $f'(x) = 1 - \sqrt{2}\cos x = 0 \quad \Rightarrow \quad \cos x = \frac{1}{\sqrt{2}} \quad \Rightarrow \quad x = \frac{\pi}{4}$.
$f''\left(\frac{\pi}{4}\right) = \sqrt{2}\sin\frac{\pi}{4} = 1 > 0$, so $f\left(\frac{\pi}{4}\right) = \frac{\pi}{4} - 1$ is a local minimum. Also $f(0) = 0$ and $f(\pi) = \pi$, so the absolute minimum is $f\left(\frac{\pi}{4}\right) = \frac{\pi}{4} - 1$, the absolute maximum is $f(\pi) = \pi$.

22. $f(x) = 2x + 2\cos x - 4\sin x - \cos 2x$, $0 \le x \le \pi$. $f'(x) = 2 - 2\sin x - 4\cos x + 2\sin 2x$
$= 2(1 - \sin x - 2\cos x + 2\sin x\cos x) = 2(\sin x - 1)(2\cos x - 1) = 0$ when $\sin x = 1$ or $\cos x = \frac{1}{2}$, so $x = \frac{\pi}{2}$ or $\frac{\pi}{3}$. $f'(x) < 0$ for $0 < x < \frac{\pi}{3}$, $f'(x) \ge 0$ for $\frac{\pi}{3} < x < \pi$, so $f\left(\frac{\pi}{3}\right) = \frac{2\pi}{3} + \frac{3}{2} - 2\sqrt{3}$ is a local minimum. Also $f\left(\frac{\pi}{3}\right) \approx 0.13$, $f(0) = 1$, $f(\pi) = 2\pi - 3 \approx 3.28$, so $f\left(\frac{\pi}{3}\right) = \frac{2\pi}{3} + \frac{3}{2} - 2\sqrt{3}$ is an absolute minimum and $f(\pi) = 2\pi - 3$ is an absolute maximum.

23. $\displaystyle\lim_{x\to\infty}\frac{1+2x-x^2}{1-x+2x^2} = \lim_{x\to\infty}\frac{(1/x^2)+(2/x)-1}{(1/x^2)-(1/x)+2} = \frac{0+0-1}{0-0+2} = -\frac{1}{2}$

24. If $t = \dfrac{1}{x}$ then $\displaystyle\lim_{x\to\infty} x\tan\frac{1}{x} = \lim_{t\to 0^+}\frac{\tan t}{t} = \lim_{t\to 0^+}\frac{\sin t}{t}\frac{1}{\cos t} = 1\cdot 1 = 1.$

25. $\displaystyle\lim_{x\to\infty}\frac{\sqrt{x^2-9}}{2x-6} = \lim_{x\to\infty}\frac{\sqrt{1-9/x^2}}{2-6/x} = \frac{\sqrt{1-0}}{2-0} = \frac{1}{2}$

26. $0 \le \cos^2 x \le 1 \quad \Rightarrow \quad 0 \le \dfrac{\cos^2 x}{x^2} \le \dfrac{1}{x^2}$ and $\displaystyle\lim_{x\to-\infty} 0 = 0$, $\displaystyle\lim_{x\to-\infty}\frac{1}{x^2} = 0$, so by the Squeeze Theorem, $\displaystyle\lim_{x\to-\infty}\frac{\cos^2 x}{x^2} = 0.$

27. $\displaystyle\lim_{x\to\infty}\left(\sqrt[3]{x} - \tfrac{1}{3}x\right) = \lim_{x\to\infty}\sqrt[3]{x}\left(1 - \tfrac{1}{3}x^{2/3}\right) = -\infty$, since $\sqrt[3]{x} \to \infty$ and $1 - \frac{1}{3}x^{2/3} \to -\infty$.

28. $\displaystyle\lim_{x\to\infty}\left(\sqrt{x^2+x+1} - \sqrt{x^2-x}\right) = \lim_{x\to\infty}\left(\sqrt{x^2+x+1} - \sqrt{x^2-x}\right)\frac{\sqrt{x^2+x+1}+\sqrt{x^2-x}}{\sqrt{x^2+x+1}+\sqrt{x^2-x}}$

$$= \lim_{x\to\infty}\frac{(x^2+x+1)-(x^2-x)}{\sqrt{x^2+x+1}+\sqrt{x^2-x}} = \lim_{x\to\infty}\frac{2x+1}{\sqrt{x^2+x+1}+\sqrt{x^2-x}}$$

$$= \lim_{x\to\infty}\frac{2+1/x}{\sqrt{1+(1/x)+(1/x^2)}+\sqrt{1-1/x}} = \frac{2+0}{\sqrt{1+0+0}+\sqrt{1-0}} = 1$$

29. $y = f(x) = 1 + x + x^3$ **A.** $D = \mathbb{R}$ **B.** y-intercept $= 1$ **C.** No symmetry **D.** $\lim_{x\to\infty}(1 + x + x^3) = \infty$, $\lim_{x\to-\infty}(1 + x + x^3) = -\infty$, no asymptotes

E. $f'(x) = 1 + 3x^2 \quad\Rightarrow\quad f'(x) > 0$, so f is increasing on $\mathbb{R}$ **F.** No local extremum

G. $f''(x) = 6x \quad\Rightarrow\quad f''(x) > 0$ if $x > 0$ and $f''(x) < 0$ if $x < 0$, so f is CU on $(0, \infty)$ and CD on $(-\infty, 0)$. IP $(0, 1)$

H.

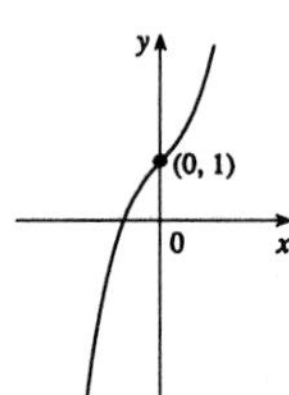

30. $y = f(x) = 3x^4 - 4x^3 - 12x^2 + 2$ **A.** $D = \mathbb{R}$ **B.** y-intercept $= f(0) = 2$ **C.** No symmetry

D. $\lim_{x\to\pm\infty}\left(3x^4 - 4x^3 - 12x^2 + 2\right) = \infty$, no asymptote

E. $f'(x) = 12x^3 - 12x^2 - 24x = 12x(x - 2)(x + 1) = 0$ when $x = -1, 0, 2$

Interval	$12x$	$x - 2$	$x + 1$	$f'(x)$	f
$x < -1$	$-$	$-$	$-$	$-$	decreasing on $(-\infty, -1]$
$-1 < x < 0$	$-$	$-$	$+$	$+$	increasing on $[-1, 0]$
$0 < x < 2$	$+$	$-$	$+$	$-$	decreasing on $[0, 2]$
$x > 2$	$+$	$+$	$+$	$+$	increasing on $[2, \infty)$

F. $f(-1) = -3$ is a local minimum, $f(0) = 2$ is a local maximum, $f(2) = -30$ is a local minimum. **G.** $f''(x) = 12(3x^2 - 2x - 2) = 0$ $\Rightarrow\quad x = \frac{1\pm\sqrt{7}}{3}$. $f''(x) > 0 \quad\Leftrightarrow$ $x > \frac{1+\sqrt{7}}{3}$ or $x < \frac{1-\sqrt{7}}{3}$, so f is CU on $\left(-\infty, \frac{1-\sqrt{7}}{3}\right)$ and $\left(\frac{1+\sqrt{7}}{3}, \infty\right)$ and CD on $\left(\frac{1-\sqrt{7}}{3}, \frac{1+\sqrt{7}}{3}\right)$. IP at $x = \frac{1\pm\sqrt{7}}{3}$

H.

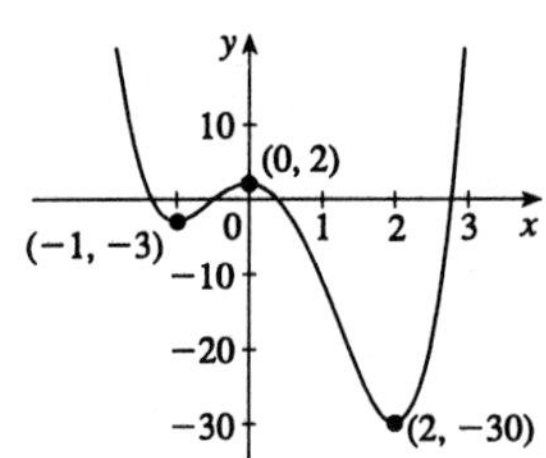

31. $y = f(x) = \dfrac{1}{[x(x-3)]^2}$ **A.** $D = \{x \mid x \neq 0, 3\} = (-\infty, 0) \cup (0, 3) \cup (3, \infty)$

B. No intercepts. **C.** No symmetry. **D.** $\lim_{x\to\pm\infty}\dfrac{1}{x(x-3)^2} = 0$, so $y = 0$ is a HA. $\lim_{x\to0^+}\dfrac{1}{x(x-3)^2} = \infty$, $\lim_{x\to0^-}\dfrac{1}{x(x-3)^2} = -\infty$, $\lim_{x\to3}\dfrac{1}{x(x-3)^2} = \infty$, so $x = 0$ and $x = 3$ are VA.

E. $f'(x) = -\dfrac{(x-3)^2 + 2x(x-3)}{x^2(x-3)^4} = \dfrac{3(1-x)}{x^2(x-3)^3} \quad\Rightarrow\quad f'(x) > 0 \quad\Leftrightarrow\quad 1 < x < 3$, so f is increasing on $[1, 3)$ and decreasing on $(-\infty, 0)$, $(0, 1]$, and $(3, \infty)$. **F.** $f(1) = \frac{1}{4}$ is a local minimum.

G. $f''(x) = \dfrac{6(2x^2 - 4x + 3)}{x^3(x-3)^4}$. Note that $2x^2 - 4x + 3 > 0$ for all x since it has negative discriminant. So $f''(x) > 0$ $\Leftrightarrow\quad x > 0 \quad\Rightarrow\quad f$ is CU on $(0, 3)$ and $(3, \infty)$ and CD on $(-\infty, 0)$. No IP

H.

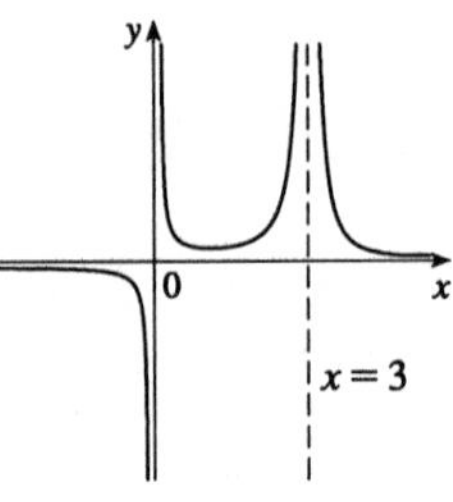

32. $y = f(x) = \dfrac{1}{x^2 - x - 6} = \dfrac{1}{(x-3)(x+2)}$ **A.** $D = \{x \mid x \neq -2, 3\}$.

B. No x-intercept; y-intercept $= f(0) = -\frac{1}{6}$ **C.** No symmetry **D.** $\lim\limits_{x\to\pm\infty} \dfrac{1}{x^2 - x - 6} = 0$, so $y = 0$ is a HA.

$\lim\limits_{x\to 3^+} \dfrac{1}{(x-3)(x+2)} = \infty$, $\lim\limits_{x\to 3^-} \dfrac{1}{(x-3)(x+2)} = -\infty$, $\lim\limits_{x\to -2^+} \dfrac{1}{(x-3)(x+2)} = -\infty$,

$\lim\limits_{x\to -2^-} \dfrac{1}{(x-3)(x+2)} = \infty$, so $x = 3$, $x = -2$ are VA. **E.** $f'(x) = \dfrac{1-2x}{(x^2-x-6)^2} > 0 \quad \Leftrightarrow \quad x < \dfrac{1}{2}$, so f is increasing on $(-\infty, -2)$ and $\left(-2, \frac{1}{2}\right]$, and decreasing on $\left[\frac{1}{2}, 3\right)$ and $(3, \infty)$. **F.** $f\left(\frac{1}{2}\right) = -\frac{4}{25}$ is a local maximum. **G.** $f''(x) = \dfrac{(x^2-x-6)^2(-2) - (1-2x)\,2\,(x^2-x-6)(2x-1)}{(x^2-x-6)^4} = \dfrac{2(3x^2-3x+7)}{(x^2-x-6)^3}$.

Since $3x^2 - 3x + 7 > 0$ for all x, $f''(x) > 0$ $\Leftrightarrow \quad (x-3)(x+2) > 0 \quad \Leftrightarrow$ $x < -2$ or $x > 3$. So f is CU on $(-\infty, -2)$ and $(3, \infty)$, and CD on $(-2, 3)$. No IP

H.

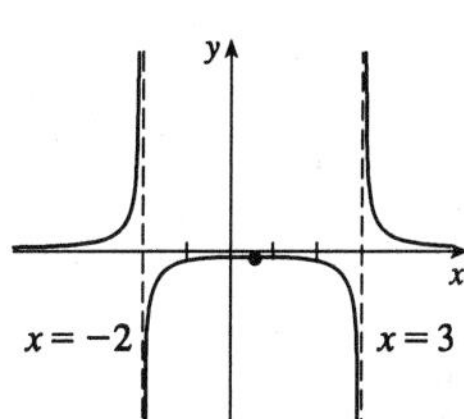

33. $y = f(x) = x\sqrt{5-x}$ **A.** $D = \{x \mid x \leq 5\} = (-\infty, 5]$ **B.** x-intercepts $0, 5$; y-intercept $= f(0) = 0$

C. No symmetry **D.** $\lim\limits_{x\to -\infty} x\sqrt{5-x} = -\infty$, no asymptotes

E. $f'(x) = \sqrt{5-x} - \dfrac{x}{2\sqrt{5-x}} = \dfrac{10-3x}{2\sqrt{5-x}} > 0 \quad \Leftrightarrow \quad x < \frac{10}{3}$. So f is increasing on $\left(-\infty, \frac{10}{3}\right]$ and decreasing on $\left[\frac{10}{3}, 5\right]$. **F.** $f\left(\frac{10}{3}\right) = \frac{10\sqrt{5}}{3\sqrt{3}}$ is a local and absolute maximum.

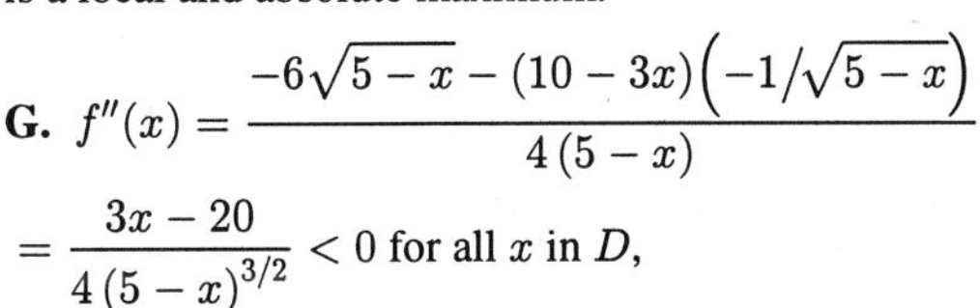

G. $f''(x) = \dfrac{-6\sqrt{5-x} - (10-3x)\left(-1/\sqrt{5-x}\right)}{4\,(5-x)}$

$= \dfrac{3x-20}{4\,(5-x)^{3/2}} < 0$ for all x in D, so f is CD on $(-\infty, 5)$.

H. (graph: y, 0, $\frac{10}{3}$, 5, x)

34. $y = f(x) = \dfrac{1}{x} + \dfrac{1}{x+1} = \dfrac{2x+1}{x(x+1)}$ **A.** $D = \{x \mid x \neq 0, -1\}$ **B.** No y-intercept, x-intercept $= -\frac{1}{2}$

C. No symmetry **D.** $\lim\limits_{x\to\pm\infty} f(x) = 0$, so $y = 0$ is a HA. $\lim\limits_{x\to 0^+} \dfrac{2x+1}{x(x+1)} = \infty$, $\lim\limits_{x\to 0^-} \dfrac{2x+1}{x(x+1)} = -\infty$,

$\lim\limits_{x\to -1^+} \dfrac{2x+1}{x(x+1)} = \infty$, $\lim\limits_{x\to -1^-} \dfrac{2x+1}{x(x+1)} = -\infty$, so $x = 0$, $x = -1$ are VA. **E.** $f'(x) = -\dfrac{1}{x^2} - \dfrac{1}{(x+1)^2} < 0$, so f is decreasing on $(-\infty, -1)$, $(-1, 0)$ and $(0, \infty)$.

F. No extremum **G.** $f''(x) = \dfrac{2}{x^3} + \dfrac{2}{(x+1)^3}$

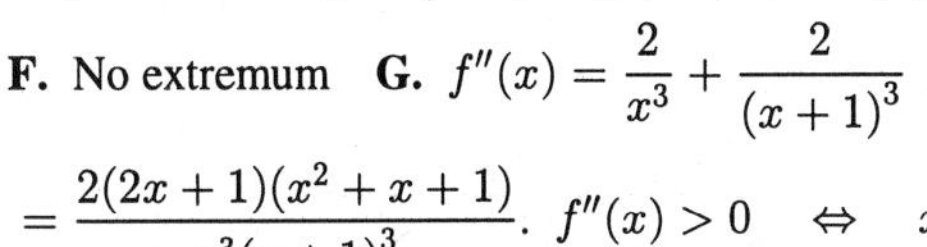

$= \dfrac{2(2x+1)(x^2+x+1)}{x^3(x+1)^3}$. $f''(x) > 0 \quad \Leftrightarrow \quad x > 0$ or $-1 < x < -\frac{1}{2}$, so f is CU on $(0, \infty)$ and $\left(-1, -\frac{1}{2}\right)$ and CD on $(-\infty, -1)$ and $\left(-\frac{1}{2}, 0\right)$. IP $\left(-\frac{1}{2}, 0\right)$

H.

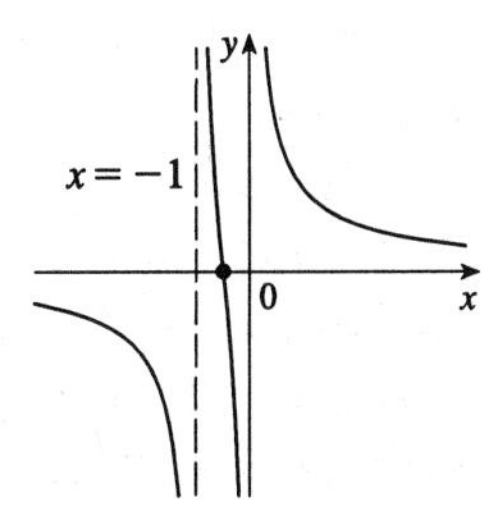

35. $y = f(x) = \dfrac{x^2}{x+8} = x - 8 + \dfrac{64}{x+8}$ **A.** $D = \{x \mid x \neq -8\}$ **B.** Intercepts are 0 **C.** No symmetry

D. $\lim\limits_{x\to\infty} \dfrac{x^2}{x+8} = \infty$, but $f(x) - (x-8) = \dfrac{64}{x+8} \to 0$ as $x \to \infty$, so $y = x - 8$ is a slant asymptote.

$\lim\limits_{x\to-8^+} \dfrac{x^2}{x+8} = \infty$ and $\lim\limits_{x\to-8^-} \dfrac{x^2}{x+8} = -\infty$, so $x = -8$ is a VA. **E.** $f'(x) - 1 - \dfrac{64}{(x+8)^2} = \dfrac{x(x+16)}{(x+8)^2} > 0$

$\Leftrightarrow$ $x > 0$ or $x < -16$, so f is increasing on $(-\infty, -16]$ and $[0, \infty)$ and decreasing on $[-16, -8)$ and $(-8, 0]$. **F.** $f(-16) = -32$ is a local maximum, $f(0) = 0$ is a local minimum.

G. $f''(x) = 128/(x+8)^3 > 0 \quad \Leftrightarrow \quad x > -8$, so f is CU on $(-8, \infty)$ and CD on $(-\infty, -8)$. No IP

H.

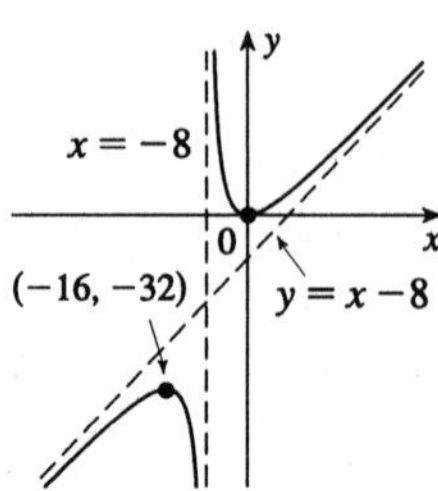

36. $y = f(x) = x + \sqrt{1-x}$ **A.** $D = \{x \mid x \leq 1\} = (-\infty, 1]$ **B.** y-intercept $= 1$; x-intercepts occur when $x + \sqrt{1-x} = 0 \quad \Rightarrow \quad \sqrt{1-x} = -x \quad \Rightarrow \quad 1 - x = x^2 \quad \Rightarrow \quad x^2 + x - 1 = 0 \quad \Rightarrow \quad x = \frac{-1 \pm \sqrt{5}}{2}$, but the larger root is extraneous, so the only x-intercept is $\frac{-1-\sqrt{5}}{2}$. **C.** No symmetry **D.** No asymptotes

E. $f'(x) = 1 - 1/\left(2\sqrt{1-x}\right) = 0 \quad \Leftrightarrow \quad 2\sqrt{1-x} = 1 \quad \Leftrightarrow \quad 1 - x = \frac{1}{4} \quad \Leftrightarrow \quad x = \frac{3}{4}$ and

$f'(x) > 0 \quad \Leftrightarrow \quad x < \frac{3}{4}$, so f is increasing on $\left(-\infty, \frac{3}{4}\right]$, decreasing on $\left[\frac{3}{4}, 1\right]$.

F. $f\left(\frac{3}{4}\right) = \frac{5}{4}$ is a local maximum.

G. $f''(x) = -\dfrac{1}{4(1-x)^{3/2}} < 0 \quad \Leftrightarrow$ $x < 1$, so f is CD on $(-\infty, 1)$. No IP

H.

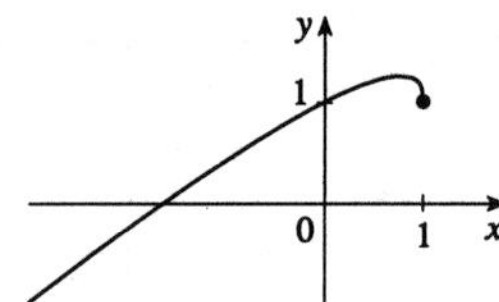

37. $y = f(x) = \sqrt{x} - \sqrt[3]{x}$ **A.** $D = [0, \infty)$ **B.** y-intercept 0, x-intercepts 0, 1 **C.** No symmetry

D. $\lim\limits_{x\to\infty} \left(x^{1/2} - x^{1/3}\right) = \lim\limits_{x\to\infty} \left[x^{1/3}\left(x^{1/6} - 1\right)\right] = \infty$, no asymptotes

E. $f'(x) = \frac{1}{2}x^{-1/2} - \frac{1}{3}x^{-2/3} = \dfrac{3x^{1/6} - 2}{6x^{2/3}} > 0 \quad \Leftrightarrow \quad 3x^{1/6} > 2 \quad \Leftrightarrow \quad x > \left(\frac{2}{3}\right)^6$, so f is increasing on $\left[\left(\frac{2}{3}\right)^6, \infty\right)$ and decreasing on $\left[0, \left(\frac{2}{3}\right)^6\right]$. **F.** $f\left(\left(\frac{2}{3}\right)^6\right) = -\frac{4}{27}$ is a local minimum.

G. $f''(x) = -\frac{1}{4}x^{-3/2} + \frac{2}{9}x^{-5/3} = \dfrac{8 - 9x^{1/6}}{36x^{5/3}} > 0 \quad \Leftrightarrow \quad x^{1/6} < \frac{8}{9} \quad \Leftrightarrow$ $x < \left(\frac{8}{9}\right)^6$, so f is CU on $\left(0, \left(\frac{8}{9}\right)^6\right)$ and CD on $\left(\left(\frac{8}{9}\right)^6, \infty\right)$. IP $\left(\frac{8}{9}, -\frac{64}{729}\right)$

H.

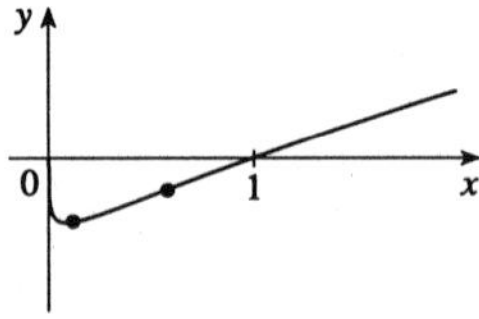

38. $y = f(x) = 4x - \tan x$, $-\frac{\pi}{2} < x < \frac{\pi}{2}$ **A.** $D = \left(-\frac{\pi}{2}, \frac{\pi}{2}\right)$. **B.** y-intercept $= f(0) = 0$ **C.** $f(-x) = -f(x)$, so the curve is symmetric about $(0, 0)$. **D.** $\lim\limits_{x\to\pi/2^-} (4x - \tan x) = -\infty$, $\lim\limits_{x\to-\pi/2^+} (4x - \tan x) = \infty$, so $x = \frac{\pi}{2}$

and $x = -\frac{\pi}{2}$ are VA. **E.** $f'(x) = 4 - \sec^2 x > 0 \quad \Leftrightarrow \quad \sec x < 2 \quad \Leftrightarrow \quad \cos x > \frac{1}{2} \quad \Leftrightarrow \quad -\frac{\pi}{3} < x < \frac{\pi}{3}$, so f is increasing on $\left[-\frac{\pi}{3}, \frac{\pi}{3}\right]$ and decreasing on $\left(-\frac{\pi}{2}, -\frac{\pi}{3}\right]$ and $\left[\frac{\pi}{3}, \frac{\pi}{2}\right)$.
F. $f\left(\frac{\pi}{3}\right) = \frac{4\pi}{3} - \sqrt{3}$ is a local maximum, $f\left(-\frac{\pi}{3}\right) = \sqrt{3} - \frac{4\pi}{3}$ is a local minimum.
G. $f''(x) = -2\sec^2 x \tan x > 0 \quad \Leftrightarrow$ $\tan x < 0 \Leftrightarrow \quad -\frac{\pi}{2} < x < 0$, so f is CU on $\left(-\frac{\pi}{2}, 0\right)$ and CD on $\left(0, \frac{\pi}{2}\right)$. IP $(0, 0)$

H.

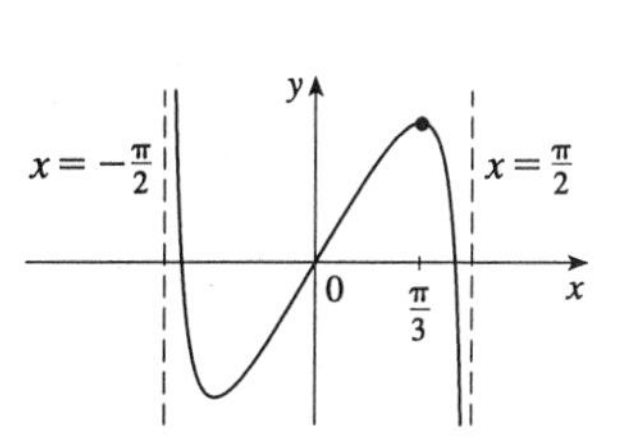

39. $f(x) = \dfrac{x^2 - 1}{x^3} \quad \Rightarrow \quad f'(x) = \dfrac{x^3(2x) - (x^2 - 1)3x^2}{x^6} = \dfrac{3 - x^2}{x^4} \quad \Rightarrow$

$$f''(x) = \frac{x^4(-2x) - (3 - x^2)4x^3}{x^8} = \frac{2x^2 - 12}{x^5}$$

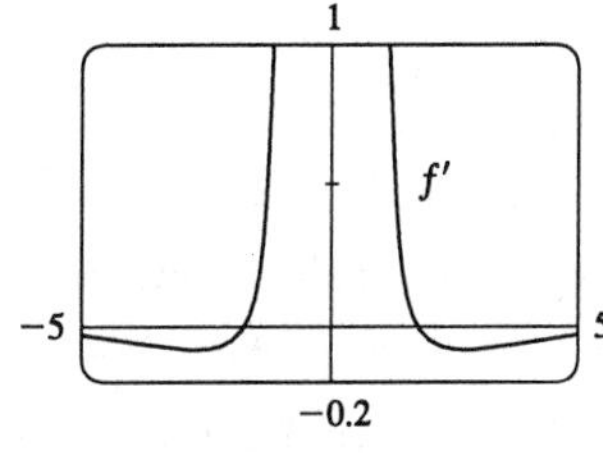

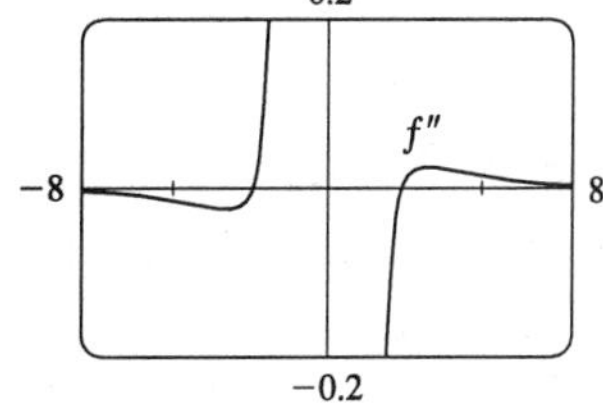

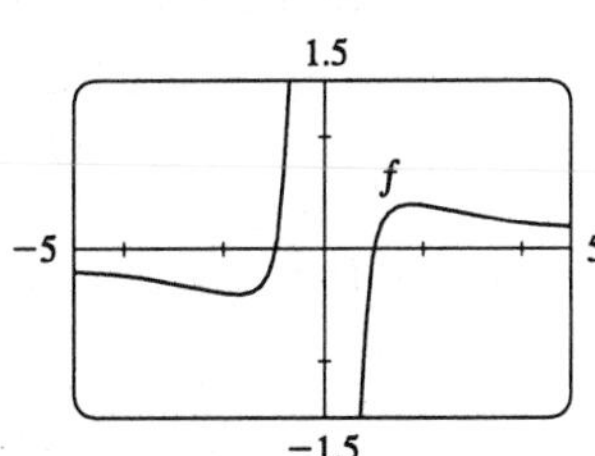

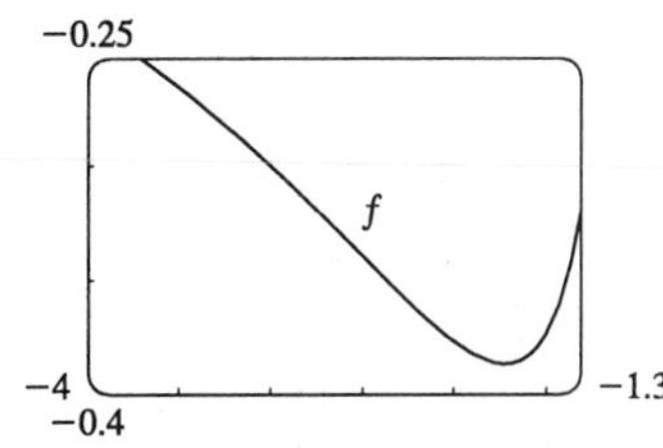

From the graphs of f' and f'', it appears that f is increasing on $[-1.73, 0)$ and $(0, 1.73]$ and decreasing on $(-\infty, -1.73]$ and $[1.73, \infty)$; f has a local maximum of about $f(1.73) = 0.38$ and a local minimum of about $f(-1.7) = -0.38$; f is CU on $(-2.45, 0)$ and $(2.45, \infty)$, and CD on $(-\infty, -2.45)$ and $(0, 2.45)$; and f has inflection points at about $(-2.45, -0.34)$ and $(2.45, 0.34)$. Now $f'(x) = \dfrac{3 - x^2}{x^4}$ is positive for $0 < x^2 < 3$, that is, f is increasing on $\left[-\sqrt{3}, 0\right)$ and $\left(0, \sqrt{3}\right]$; and $f'(x)$ is negative (and so f is decreasing) on $\left(-\infty, -\sqrt{3}\right]$ and $\left[\sqrt{3}, \infty\right)$. $f'(x) = 0$ when $x = \pm\sqrt{3}$. f' goes from positive to negative at $x = \sqrt{3}$, so f has a local maximum of $f\left(\sqrt{3}\right) = \dfrac{\left(\sqrt{3}\right)^2 - 1}{\left(\sqrt{3}\right)^3} = \dfrac{2\sqrt{3}}{9}$; and since f is odd, we know that maxima on the interval $[0, \infty)$ correspond to minima on $(-\infty, 0]$, so f has a local minimum of $f\left(-\sqrt{3}\right) = -\dfrac{2\sqrt{3}}{9}$. Also, $f''(x) = \dfrac{2x^2 - 12}{x^5}$ is positive (so f is CU) on $\left(-\sqrt{6}, 0\right)$ and $\left(\sqrt{6}, \infty\right)$, and negative (so f is CD) on $\left(-\infty, -\sqrt{6}\right)$ and $\left(0, \sqrt{6}\right)$. There are IP at $\left(\sqrt{6}, \frac{5\sqrt{6}}{36}\right)$ and $\left(-\sqrt{6}, -\frac{5\sqrt{6}}{36}\right)$.

40. $f(x) = \dfrac{\sqrt[3]{x}}{1-x} = x^{1/3}(1-x)^{-1} \quad \Rightarrow$

$f'(x) = x^{1/3}(-1)(1-x)^{-2}(-1) + (1-x)^{-1}\left(\frac{1}{3}\right)x^{-2/3} = \frac{1}{3}x^{-2/3}\dfrac{1+2x}{(x-1)^2} \quad \Rightarrow$

$f''(x) = \frac{1}{3}x^{-2/3}\dfrac{(x-1)^2(2)-(1+2x)(2)(x-1)}{(x-1)^4} + \dfrac{1+2x}{(x-1)^2}\left(-\frac{4}{9}x^{-5/3}\right) = -\frac{2}{9}x^{-5/3}\dfrac{5x^2+5x-1}{(x-1)^3}$

From the graphs, it appears that f is increasing on $[-0.50, 1)$ and $(1, \infty)$, with a vertical asymptote at $x = 1$, and decreasing on $(-\infty, -0.50]$; f has no local maximum, but a local minimum of about $f(-0.50) = -0.53$; f is CU on $(-1.17, 0)$ and $(0.17, 1)$ and CD on $(-\infty, -1.17)$, $(0, 0.17)$ and $(1, \infty)$; and f has inflection points at about $(-1.17, -0.49)$, $(0, 0)$ and $(0.17, 0.67)$. Note also that $\lim\limits_{x \to \pm\infty} f(x) = 0$, so $y = 0$ is a horizontal asymptote.

41. $f(x) = 3x^6 - 5x^5 + x^4 - 5x^3 - 2x^2 + 3$, $f'(x) = 18x^5 - 25x^4 + 4x^3 - 15x^2 - 4x$, $f''(x) = 90x^4 - 100x^3 + 12x^2 - 30x - 4$

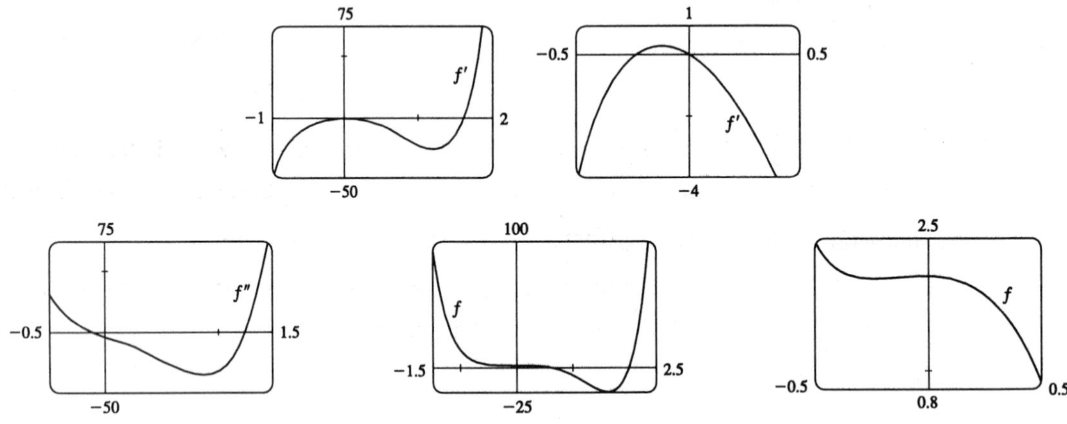

From the graphs of f' and f'', it appears that f is increasing on $[-0.23, 0]$ and $[1.62, \infty)$ and decreasing on $(-\infty, -0.23]$ and $[0, 1.62]$; f has a local maximum of about $f(0) = 2$ and local minima of about $f(-0.23) = 1.96$ and $f(1.62) = -19.2$; f is CU on $(-\infty, -0.12)$ and $(1.24, \infty)$ and CD on $(-0.12, 1.24)$; and f has inflection points at about $(-0.12, 1.98)$ and $(1.2, -12.1)$.

42. $f(x) = \sin x \cos^2 x \quad \Rightarrow \quad f'(x) = \cos^3 x - 2\sin^2 x \cos x \quad \Rightarrow \quad f''(x) = -7\sin x \cos^2 x + 2\sin^3 x$

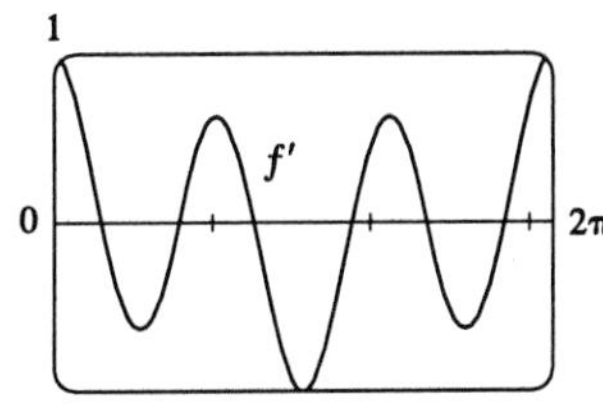

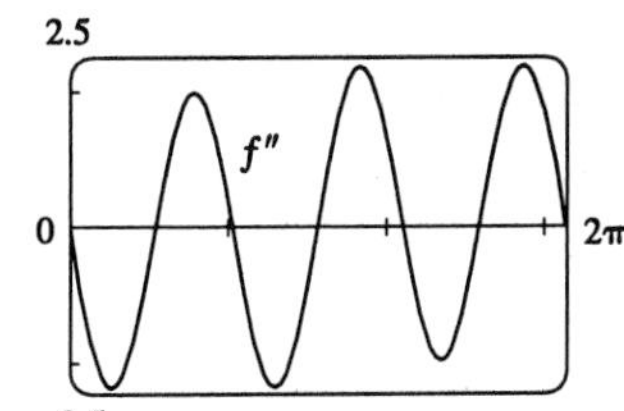

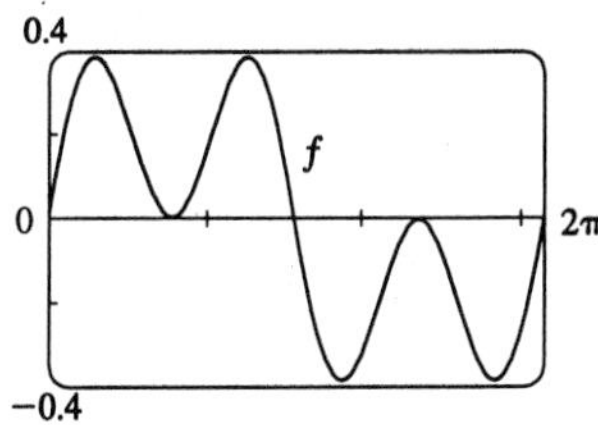

From the graphs of f' and f'', it appears that f is increasing on $(0, 0.62)$, $(1.57, 2.53)$, $(3.76, 4.71)$ and $(5.67, 2\pi)$ and decreasing on $(0.62, 1.57)$, $(2.53, 3.76)$ and $(4.71, 5.67)$; f has local maxima of about $f(0.62) = 0.38$, $f(2.53) = 0.38$ and $f(4.71) = 0$ and local minima of about $f(1.57) = 0$, $f(3.76) = -0.38$ and $f(5.67) = -0.38$; f is CU on $(1.08, 2.06)$, $(3.14, 4.22)$ and $(5.20, 2\pi)$ and CD on $(0, 1.1)$, $(2.06, 3.14)$ and $(4.22, 5.20)$; and f has inflection points at about $(0, 0)$, $(1.08, 0.20)$, $(2.06, 0.20)$, $(3.14, 0)$, $(4.22, -0.20)$, $(5.20, -0.20)$ and $(2\pi, 0)$.

43. $f(x) = x^{101} + x^{51} + x - 1 = 0$. Since f is continuous and $f(0) = -1$ and $f(1) = 2$, the equation has at least one root in $(0, 1)$, by the Intermediate Value Theorem. Suppose the equation has two roots, a and b, with $a < b$. Then $f(a) = 0 = f(b)$, so by Rolle's Theorem, $f'(x) = 0$ has a root in (a, b). But this is impossible since $f'(x) = 101x^{100} + 51x^{50} + 1 \geq 1$ for all x.

44. By the Mean Value Theorem, $f(4) - 1 = f(4) - f(0) = f'(c)(4 - 0) = 4f'(c)$ for some c with $0 < c < 4$. Since $2 \leq f'(c) \leq 5$, we have $4(2) \leq f(4) - 1 \leq 4(5)$ or $8 \leq f(4) - 1 \leq 20$ or $9 \leq f(4) \leq 21$.

45. Since f is continuous on $[32, 33]$ and differentiable on $(32, 33)$, then by the Mean Value Theorem there exists a number c in $(32, 33)$ such that $f'(c) = \frac{1}{5}c^{-4/5} = \dfrac{\sqrt[5]{33} - \sqrt[5]{32}}{33 - 32} = \sqrt[5]{33} - 2$, but $\frac{1}{5}c^{-4/5} > 0 \quad \Rightarrow$ $\sqrt[5]{33} - 2 > 0 \quad \Rightarrow \quad \sqrt[5]{33} > 2$. Also f' is decreasing, so that $f'(c) < f'(32) = \frac{1}{5}(32)^{-4/5} = 0.0125 \quad \Rightarrow$ $0.0125 > f'(c) = \sqrt[5]{33} - 2 \quad \Rightarrow \quad \sqrt[5]{33} < 2.0125$. Therefore $2 < \sqrt[5]{33} < 2.0125$.

46. Let $g(x) = f(x) - x$. So $g'(x) = f'(x) - 1 < 0$ (since $f'(x) < 1$) which means that $g(x)$ is decreasing on $(-\infty, \infty)$. Now if $f(x_0) = x_0$, then $g(x_0) = f(x_0) - x_0 = 0$. But since $g(x)$ is decreasing, there is at most one point x_0 with $g(x_0) = 0$ which means that there is at most one point x_0 with $f(x_0) = x_0$.

Or: Use Rolle's Theorem. If f has two fixed points, say $f(a) = a$ and $f(b) = b$, then $g(a) = 0 = g(b)$, so there exists c with $g'(c) = 0 \quad \Rightarrow \quad f'(c) = 1$, which contradicts $f'(x) < 1$.

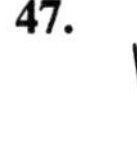

47.

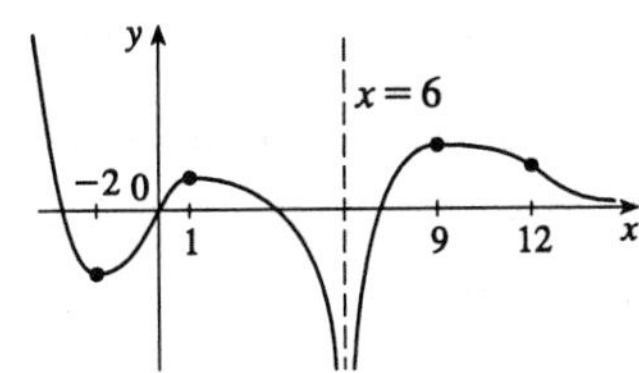

48. For $0 < x < 1$, $f'(x) = 2x$, so $f(x) = x^2 + C$. Since $f(0) = 0$, $f(x) = x^2$ on $[0, 1]$. For $1 < x < 3$, $f'(x) = -1$, so $f(x) = -x + D$. $1 = f(1) = -1 + D \Rightarrow D = 2$, so $f(x) = 2 - x$. For $x > 3$, $f'(x) = 1$, so $f(x) = x + E$. $-1 = f(3) = 3 + E \Rightarrow E = -4$, so $f(x) = x - 4$. Since f is even, its graph is symmetric about the y-axis.

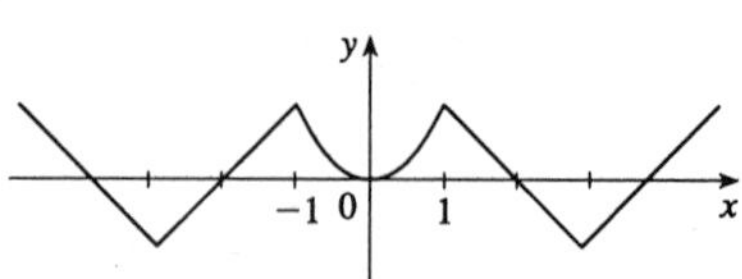

49. For $(1, 6)$ to be on the curve, we have that $6 = y(1) = 1^3 + a(1)^2 + b(1) + 1 = a + b + 2 \Rightarrow b = 4 - a$. Now $y' = 3x^2 + 2ax + b$ and $y'' = 6x + 2a$. Also, for $(1, 6)$ to be an inflection point it must be true that $y''(1) = 6(1) + 2a = 0 \Rightarrow a = -3 \Rightarrow b = 4 - (-3) = 7$.

50. $y = f(x) = x^2 + \sin x$ **A.** $D = \mathbb{R}$ **B.** y-intercept $= f(0) = 0$, x-intercepts occur when $x^2 + \sin x = 0$. $x = 0$ is a solution, and using Newton's Method try $x_1 = -1$ to get $x_2 \approx -0.8914$, $x_3 \approx -0.8770$, $x_4 \approx -0.8767$, $x_5 \approx -0.8767$. So the x-intercepts are $x \approx -0.88$, and $x = 0$. **C.** No symmetry **D.** No asymptotes **E.** $f'(x) = 2x + \cos x$ and to find intervals of increase or decrease solve $2x + \cos x = 0$. Use Newton's Method with $x_1 = -0.5$ to get $x_2 \approx -0.4506$, $x_3 \approx -0.4502$, $x_4 \approx 0.4502$. So the root is $\alpha \approx -0.45$. For $x > \alpha$, $f'(x) > 0$, and for $x < \alpha$, $f'(x) < 0$. So f is increasing on $[\alpha, \infty)$, and decreasing on $(-\infty, \alpha]$. **F.** $f(\alpha) \approx -0.23$ is a local minimum. **G.** $f''(x) = 2 - \sin x > 0$ for all x, so f is CU on $(-\infty, \infty)$. No IP **H.**

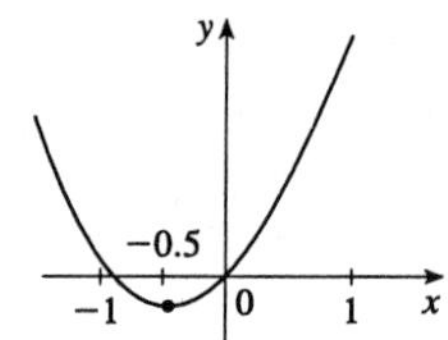

51. **(a)** $g(x) = f(x^2) \Rightarrow g'(x) = 2xf'(x^2)$ by the Chain Rule. Since $f'(x) > 0$ for all $x \neq 0$, we must have $f'(x^2) > 0$ for $x \neq 0$, so $g'(x) = 0 \Leftrightarrow x = 0$. Now $g'(x)$ changes sign (from negative to positive) at $x = 0$, since one of its factors, namely $f'(x^2)$, is positive for all x, and its other factor, namely $2x$, changes from negative to positive at this point, so by the First Derivative Test, f has a local and absolute minimum at $x = 0$.

(b) $g'(x) = 2xf'(x^2) \Rightarrow g''(x) = 2[xf''(x^2)(2x) + f'(x^2)] = 4x^2f''(x^2) + 2f'(x^2)$ by the Product Rule and the Chain Rule. But $x^2 > 0$ for all $x \neq 0$ and thus $f''(x^2) > 0$ (since f is CU for $x > 0$) and $f'(x^2) > 0$ for all $x \neq 0$, so since all of its factors are positive, $g''(x) > 0$ for $x \neq 0$. Whether $g''(0)$ is positive or 0 doesn't matter (since the sign of g'' does not change there); g is concave up on $\mathbb{R}$.

52. **(a)** Using the Test for Monotonic Functions we know that f is increasing on $[2, 0]$ and $[4, \infty)$ because $f' > 0$ on $(2, 0)$ and $(4, \infty)$, and that f is decreasing on $(-\infty, 2]$ and $[0, 4]$ because $f' < 0$ on $(-\infty, 2)$ and $(0, 4)$.

(b) Using the First Derivative Test, we know that f has a local maximum at $x = 0$ because f' changes from positive to negative at $x = 0$, and that f has local minima at $x = 2$ and $x = 4$ because f' changes from negative to positive at these points.

(c)

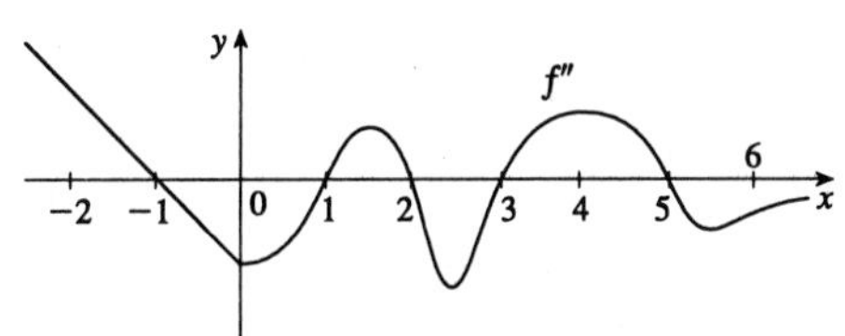

(d)

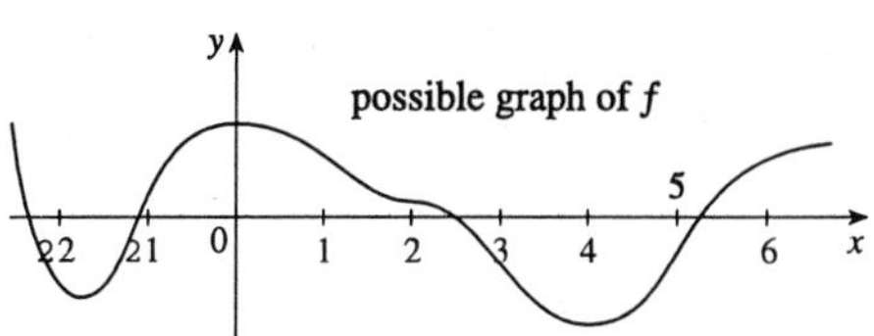

53. If $B = 0$, the line is vertical and the distance from $x = -\frac{C}{A}$ to (x_1, y_1) is $\left|x_1 + \frac{C}{A}\right| = \frac{|Ax_1 + By_1 + C|}{\sqrt{A^2 + B^2}}$, so assume $B \neq 0$. The square of the distance from (x_1, y_1) to the line is $f(x) = (x - x_1)^2 + (y - y_1)^2$ where $Ax + By + C = 0$, so we minimize $f(x) = (x - x_1)^2 + \left(-\frac{A}{B}x - \frac{C}{B} - y_1\right)^2 \Rightarrow$ $f'(x) = 2(x - x_1) + 2\left(-\frac{A}{B}x - \frac{C}{B} - y_1\right)\left(-\frac{A}{B}\right)$. $f'(x) = 0 \Rightarrow x = \frac{B^2x_1 - ABy_1 - AC}{A^2 + B^2}$ and this gives a minimum since $f''(x) = 2\left(1 + \frac{A^2}{B^2}\right) > 0$. Substituting this value of x and simplifying gives $f(x) = \frac{(Ax_1 + By_1 + C)^2}{A^2 + B^2}$, so the minimum distance is $\frac{|Ax_1 + By_1 + C|}{\sqrt{A^2 + B^2}}$.

54. If $d(x)$ is the distance from the point $(x, 8/x)$ on the hyperbola to $(3, 0)$, then $[d(x)]^2 = (x - 3)^2 + 64/x^2 = f(x)$. $f'(x) = 2(x - 3) - 128/x^3 = 0 \Rightarrow x^4 - 3x^3 - 64 = 0 \Rightarrow$ $(x - 4)(x^3 + x^2 + 4x + 16) = 0 \Rightarrow x = 4$ since the solution must have $x > 0$. Then $y = \frac{8}{4} = 2$, so the point is $(4, 2)$.

55.

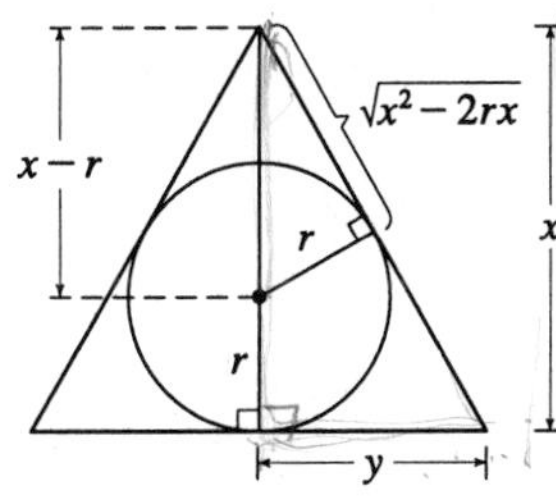

By similar triangles, $\frac{y}{x} = \frac{r}{\sqrt{x^2 - 2rx}}$, so the area of the triangle is

$$A(x) = \tfrac{1}{2}(2y)x = xy = \frac{rx^2}{\sqrt{x^2 - 2rx}} \Rightarrow$$

$$A'(x) = \frac{2rx\sqrt{x^2 - 2rx} - rx^2(x - r)/\sqrt{x^2 - 2rx}}{x^2 - 2rx} = \frac{rx^2(x - 3r)}{(x^2 - 2rx)^{3/2}} = 0$$

when $x = 3r$. $A'(x) < 0$ when $2r < x < 3r$, $A'(x) > 0$ when $x > 3r$. So $x = 3r$ gives a minimum and and $A(3r) = r(9r^2)/\left(\sqrt{3}r\right) = 3\sqrt{3}r^2$.

56.

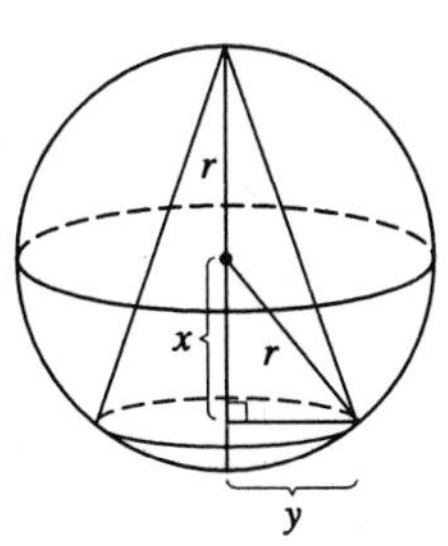

The volume is

$$V = \tfrac{1}{3}\pi y^2(r + x) = \tfrac{1}{3}\pi(r^2 - x^2)(r + x)$$
$$= \tfrac{1}{3}\pi(r^3 + r^2x - rx^2 - x^3),\ -r \le x \le r.$$

$V'(x) = \frac{\pi}{3}(r^2 - 2rx - 3x^2) = -\frac{\pi}{3}(3x - r)(x + r) = 0$ when $x = -r$ or $x = r/3$. Now $V(r) = 0 = V(-r)$, so the maximum occurs at $x = r/3$ and the volume is

$$V\left(\frac{r}{3}\right) = \frac{\pi}{3}\left(r^2 - \frac{r^2}{9}\right)\left(\frac{4r}{3}\right) = \frac{32\pi r^3}{81}.$$

57. 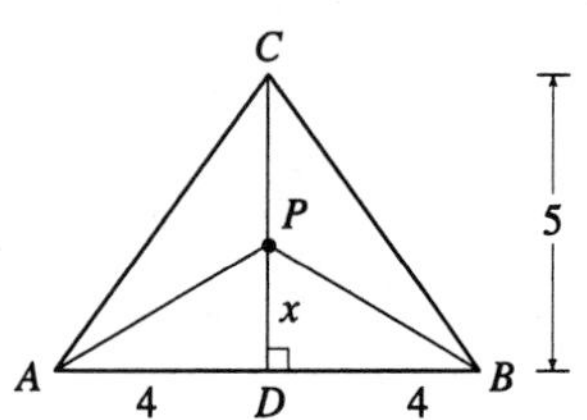

We minimize
$L(x) = |PA| + |PB| + |PC| = 2\sqrt{x^2+16} + (5-x), 0 \le x \le 5.$
$L'(x) = 2x/\sqrt{x^2+16} - 1 = 0 \quad \Leftrightarrow \quad 2x = \sqrt{x^2+16} \quad \Leftrightarrow$
$4x^2 = x^2 + 16 \quad \Leftrightarrow \quad x = \frac{4}{\sqrt{3}}$. $L(0) = 13$, $L\left(\frac{4}{\sqrt{3}}\right) \approx 11.9$,
$L(5) \approx 12.8$, so the minimum occurs when $x = \frac{4}{\sqrt{3}}$.

58. 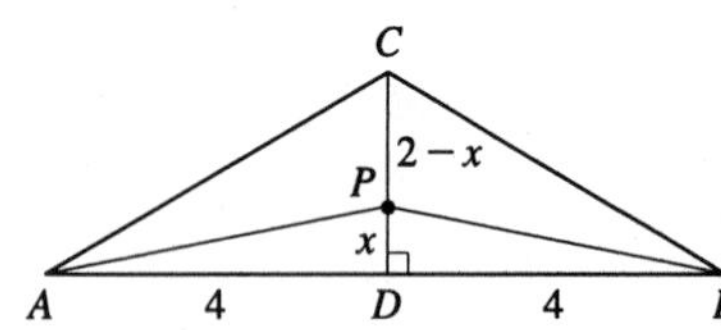

We minimize $L(x) = |PA| + |PB| + |PC| = 2\sqrt{x^2+16} + (2-x)$, $0 \le x \le 2$. $L'(x) = 2x/\sqrt{x^2+16} - 1 = 0 \quad \Leftrightarrow$ $2x = \sqrt{x^2+16} \quad \Leftrightarrow \quad 4x^2 = x^2+16 \quad \Leftrightarrow \quad x = \frac{4}{\sqrt{3}} \approx 2.3$ which isn't in the interval $[0, 2]$. Now $L(0) = 10$ and $L(2) = 2\sqrt{20} = 4\sqrt{5} \approx 8.9$. The minimum occurs when $P = C$.

59. $v = K\sqrt{\dfrac{L}{C} + \dfrac{C}{L}} \quad \Rightarrow \quad \dfrac{dv}{dL} = \dfrac{K}{2\sqrt{(L/C)+(C/L)}}\left(\dfrac{1}{C} - \dfrac{C}{L^2}\right) = 0 \quad \Leftrightarrow \quad \dfrac{1}{C} = \dfrac{C}{L^2} \quad \Leftrightarrow \quad L^2 = C^2$

$\Leftrightarrow \quad L = C$. This gives the minimum velocity since $v' < 0$ for $0 < L < C$ and $v' > 0$ for $L > C$.

60. 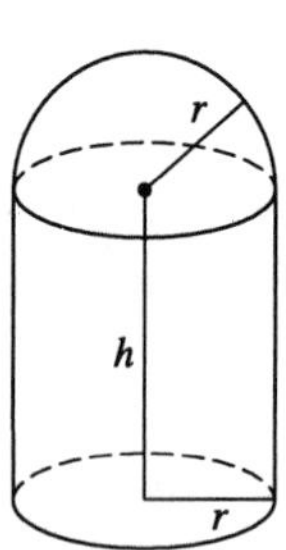

We minimize the surface area
$S = \pi r^2 + 2\pi rh + 2\pi r^2 = 3\pi r^2 + 2\pi rh$. Solving $V = \pi r^2 h + \frac{2}{3}\pi r^3$ for h, we get $h = V/(\pi r^2) - \frac{2}{3}r$, so

$$S(r) = 3\pi r^2 + 2\pi r\left[\frac{V}{\pi r^2} - \frac{2}{3}r\right] = \frac{5}{3}\pi r^2 + \frac{2V}{r}.$$

$$S'(r) = -\frac{2V}{r^2} + \frac{10}{3}\pi r = \frac{\frac{10}{3}\pi r^3 - 2V}{r^2} = 0 \quad \Leftrightarrow$$

$\frac{10}{3}\pi r^3 = 2V \quad \Leftrightarrow \quad r^3 = 3V/(5\pi) \quad \Leftrightarrow \quad r = \sqrt[3]{3V/(5\pi)}$.
This gives an absolute minimum since $S'(r) < 0$ for $0 < r < \sqrt[3]{3V/(5\pi)}$ and $S'(r) > 0$ for $r > \sqrt[3]{3V/(5\pi)}$.
Thus $r = \sqrt[3]{3V/(5\pi)} = h$.

61. Let x = selling price of ticket. Then $12 - x$ is the amount the ticket price has been lowered, so the number of tickets sold is $11{,}000 + 1000(12 - x) = 23{,}000 - 1000x$. The revenue is $R(x) = x(23{,}000 - 1000x) = 23{,}000x - 1000x^2$, so $R'(x) = 23{,}000 - 2000x = 0$ when $x = 11.5$. Since $R''(x) = -2000 < 0$, the maximum revenue occurs when the ticket prices are \$11.50.

62. **(a)** Average cost $= c(x) = \dfrac{C(x)}{x} = \dfrac{250{,}000}{x} + 0.84 + 0.0002x$, for $x > 0$.

$c'(x) = -\dfrac{250{,}000}{x^2} + 0.0002 = 0 \quad \Leftrightarrow \quad x^2 = \dfrac{250{,}000}{0.0002} \quad \Rightarrow \quad x = \sqrt{1.25 \times 10^9}$. For $x < \sqrt{1.25 \times 10^9}$, $c'(x) < 0$, and for $x > \sqrt{1.25 \times 10^9}$, $c'(x) > 0$. Therefore $x = \sqrt{1.25 \times 10^9} \approx 35{,}355.34$ will minimize the average cost.

(b) Profit $= P(x) = R(x) - C(x) = xp(x) - C(x) = 10x - 0.05x^2 - 250{,}000 - 0.84x - 0.0002x^2$
$= -0.0502x^2 + 9.16x - 250{,}000$

$P'(x) = -0.1004x + 9.16 = 0 \quad \Leftrightarrow \quad x \approx 91.235$. This gives a maximum since $P''(x) = -0.1004 < 0$ for all x. Therefore selling 91 units will maximize profits.

63. $f'(x) = x - \sqrt[4]{x} = x - x^{1/4} \quad \Rightarrow \quad f(x) = \frac{1}{2}x^2 - \frac{4}{5}x^{5/4} + C$

64. $f'(x) = 2x^{-5/2} \quad \Rightarrow \quad f(x) = 2\left(\frac{1}{-3/2}\right)x^{-3/2} + C = -\frac{4}{3}x^{-3/2} + C$

65. $f'(x) = (1+x)/\sqrt{x} = x^{-1/2} + x^{1/2} \quad \Rightarrow \quad f(x) = 2x^{1/2} + \frac{2}{3}x^{3/2} + C \quad \Rightarrow \quad 0 = f(1) = 2 + \frac{2}{3} + C \quad \Rightarrow$
$C = -\frac{8}{3} \quad \Rightarrow \quad f(x) = 2x^{1/2} + \frac{2}{3}x^{3/2} - \frac{8}{3}$

66. $f'(x) = 1 + 2\sin x - \cos x \quad \Rightarrow \quad f(x) = x - 2\cos x - \sin x + C \quad \Rightarrow \quad 3 = f(0) = -2 + C \quad \Rightarrow$
$C = 5$, so $f(x) = x - 2\cos x - \sin x + 5$

67. $f''(x) = x^3 + x \quad \Rightarrow \quad f'(x) = \frac{1}{4}x^4 + \frac{1}{2}x^2 + C \quad \Rightarrow \quad 1 = f'(0) = C \quad \Rightarrow \quad f'(x) = \frac{1}{4}x^4 + \frac{1}{2}x^2 + 1 \quad \Rightarrow$
$f(x) = \frac{1}{20}x^5 + \frac{1}{6}x^3 + x + D \quad \Rightarrow \quad -1 = f(0) = D \quad \Rightarrow \quad f(x) = \frac{1}{20}x^5 + \frac{1}{6}x^3 + x - 1$

68. $f''(x) = x^4 - 4x^2 + 3x - 2 \quad \Rightarrow \quad f'(x) = \frac{1}{5}x^5 - \frac{4}{3}x^3 + \frac{3}{2}x^2 - 2x + C \quad \Rightarrow$
$f(x) = \frac{1}{30}x^6 - \frac{1}{3}x^4 + \frac{1}{2}x^3 - x^2 + Cx + D$. $0 = f(0) = D \quad \Rightarrow \quad f(x) = \frac{1}{30}x^6 - \frac{1}{3}x^4 + \frac{1}{2}x^3 - x^2 + Cx$.
$1 = f(1) = \frac{1}{30} - \frac{1}{3} + \frac{1}{2} - 1 + C \quad \Rightarrow \quad C = \frac{9}{5}$, so $f(x) = \frac{1}{30}x^6 - \frac{1}{3}x^4 + \frac{1}{2}x^3 - x^2 + \frac{9}{5}x$

69.

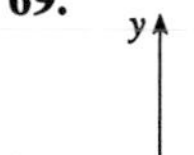
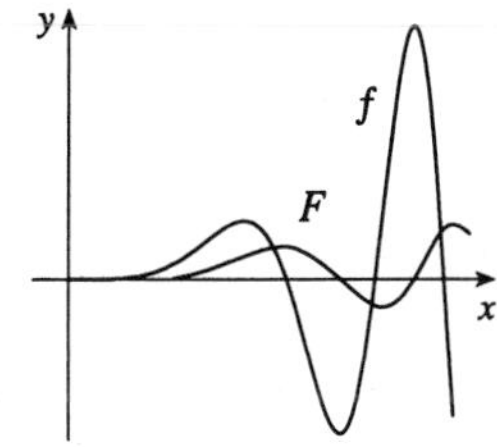

70. We calculate $f(x) = x^4 + x^3 + cx^2 \quad \Rightarrow \quad f'(x) = 4x^3 + 3x^2 + 2cx$. This is 0 when $x(4x^2 + 3x + 2c) = 0$ $\Leftrightarrow \quad x = 0$ or $4x^2 + 3x + 2c = 0$. Using the quadratic formula, we find that the roots of this last equation are $x = \dfrac{-3 \pm \sqrt{9 - 32c}}{8}$. Now if $\sqrt{9 - 32c} < 0 \quad \Leftrightarrow \quad c > \frac{9}{32}$, then $x = 0$ is the only critical point, a minimum. If $c = \frac{9}{32}$, then there are two critical points (a minimum at $x = 0$, and a horizontal tangent with no extremum at $x = -\frac{3}{8}$) and if $c < \frac{9}{32}$, then there are three critical points except when $c = 0$, in which case the root with the $+$ sign coincides with the critical point at $x = 0$. For $0 < c < \frac{9}{32}$, there is a minimum at $x = -\dfrac{3}{8} - \dfrac{\sqrt{9 - 32c}}{8}$, a maximum at $x = -\dfrac{3}{8} - \dfrac{\sqrt{9 - 32c}}{8}$, and a minimum at $x = 0$. For $c = 0$, there is a minimum at $x = -\frac{3}{4}$ and a horizontal tangent with no extremum at $x = 0$, and for $c < 0$, there is a maximum at $x = 0$, and there are minima at $x = -\frac{3}{8} \pm \frac{\sqrt{9-32c}}{8}$. Now we calculate $f''(x) = 12x^2 + 6x + 2c$. The roots of this equation are $x = \frac{-6 \pm \sqrt{36 - 4 \cdot 12 \cdot 2c}}{24}$. So if $36 - 96c \leq 0 \quad \Leftrightarrow \quad c \geq \frac{3}{8}$, then there are no inflection points. If $c < \frac{3}{8}$, then there are two inflection points at $x = -\frac{1}{4} \pm \frac{\sqrt{9-24c}}{12}$.

Value of c	No. of CP	No. of IP
$c < 0$	3	2
$c = 0$	2	2
$0 < c < \frac{9}{32}$	3	2
$c = \frac{9}{32}$	2	2
$\frac{9}{32} < c < \frac{3}{8}$	1	2
$c \geq \frac{3}{8}$	1	0

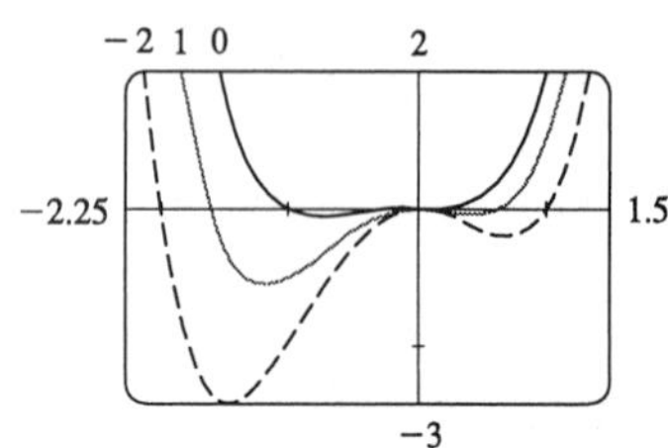

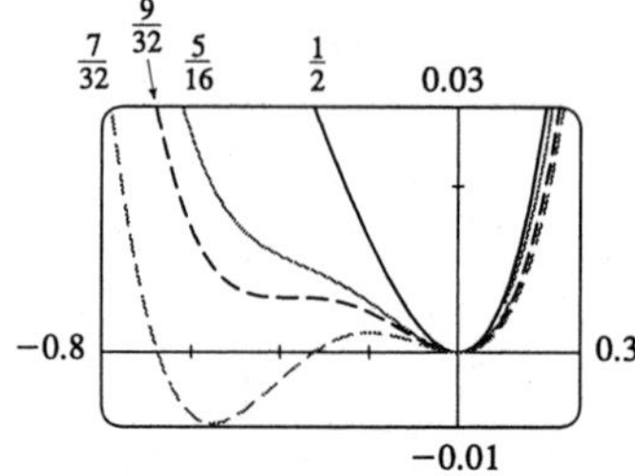

71. Choosing the positive direction to be upward, we have $a(t) = -9.8 \quad \Rightarrow \quad v(t) = -9.8t + v_0$, but $v(0) = 0 = v_0 \quad \Rightarrow \quad v(t) = -9.8t = s'(t) \quad \Rightarrow \quad s(t) = -4.9t^2 + s_0$, but $s(0) = s_0 = 500 \quad \Rightarrow$ $s(t) = -4.9t^2 + 500$. When $s = 0$, $-4.9t^2 + 500 = 0 \quad \Rightarrow \quad t = \sqrt{\frac{500}{4.9}} \quad \Rightarrow$ $v = -9.8\sqrt{\frac{500}{4.9}} \approx -98.995$ m/s. Therefore the canister will not burst.

72. Let $s_A(t)$ and $s_B(t)$ be the position functions for cars A and B and let $f(t) = s_A(t) - s_B(t)$. Since A passed B twice, there must be 3 values of t such that $f(t) = 0$. Then by three applications of Rolle's Theorem [see Exercise 3.2.24(b)] there is a number c such that $f''(c) = 0$. So $s''_A(c) = s''_B(c)$, that is, A and B had equal accelerations at $t = c$.

73. **(a)** The cross-sectional area is

$$A = 2x \cdot 2y = 4xy = 4x\sqrt{100 - x^2}, \; 0 \leq x \leq 10, \text{ so}$$

$$\frac{dA}{dx} = 4x\left(\tfrac{1}{2}\right)\left(100 - x^2\right)^{-1/2}(-2x) + \left(100 - x^2\right)^{1/2} \cdot 4$$

$$= \frac{-4x^2}{(100 - x^2)^{1/2}} + 4\left(100 - x^2\right)^{1/2} = 0 \text{ when}$$

$$-4x^2 + 4(100 - x^2) = 0 \quad \Rightarrow \quad -8x^2 + 400 = 0$$

$\Rightarrow \quad x^2 = 50 \quad \Rightarrow \quad x = \sqrt{50} \quad \Rightarrow \quad y = \sqrt{50}$. And $A(0) = A(10) = 0$. Therefore, the rectangle of maximum area is a square.

(b) $y = \sqrt{100 - x^2}$. The cross-sectional area of each rectangular plank is

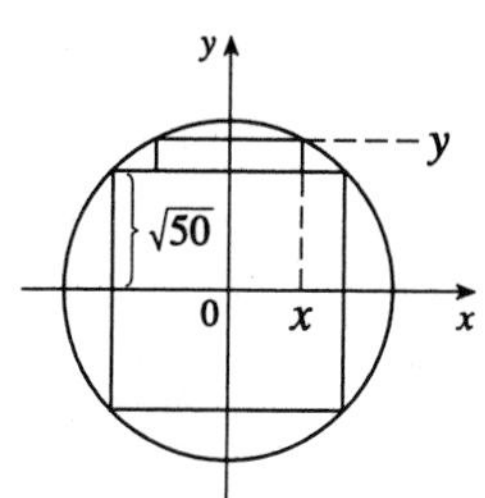

$$A = 2x\left(y - \sqrt{50}\right) = 2x\left[\sqrt{100 - x^2} - \sqrt{50}\right], 0 \le x \le \sqrt{50}, \text{ so}$$

$$\frac{dA}{dx} = 2\left[\sqrt{100 - x^2} - \sqrt{50}\right] + 2x\left(\tfrac{1}{2}\right)\left(100 - x^2\right)^{-1/2}(-2x)$$

$$= 2\left(100 - x^2\right)^{1/2} - 2\sqrt{50} - \frac{2x^2}{\left(100 - x^2\right)^{1/2}}.$$

Set $\dfrac{dA}{dx} = 0$: $\left(100 - x^2\right) - \sqrt{50}\left(100 - x^2\right)^{1/2} - x^2 = 0 \quad\Rightarrow\quad 100 - 2x^2 = \sqrt{50}\left(100 - x^2\right)^{1/2}$

$\Rightarrow \quad 10{,}000 - 400x^2 + 4x^4 = 50(100 - x^2) \quad\Rightarrow\quad 2500 - 175x^2 + 2x^4 = 0 \quad\Rightarrow$

$x^2 = \dfrac{175 \pm \sqrt{10{,}625}}{4} \approx 69.5$ or $17.98 \quad\Rightarrow\quad x \approx 8.3$ or 4.2. But $8.3 > \sqrt{50}$, so $x \approx 4.2 \quad\Rightarrow$

$y \approx \sqrt{100 - (4.2)^2} \approx 2.0$. Each plank should have dimensions about 8.4 inches by 2 inches.

(c) The strength is $S = k\,(2x)(2y)^2 = 8kxy^2 = 8kx(100 - x^2)$, $0 \le x \le 10$.

$\dfrac{dS}{dx} = 800k - 24kx^2 = 0$ when $24kx^2 = 800k \quad\Rightarrow\quad x^2 = \frac{100}{3} \quad\Rightarrow\quad x = \frac{10}{\sqrt{3}} \quad\Rightarrow$

$y = \sqrt{\frac{200}{3}} = \frac{10\sqrt{2}}{\sqrt{3}}$ and $S(0) = S(10) = 0$, so the maximum occurs when $x = \frac{10}{\sqrt{3}}$. The dimensions should be $\frac{20}{\sqrt{3}}$ inches by $\frac{20\sqrt{2}}{\sqrt{3}}$ inches.

74. **(a)**

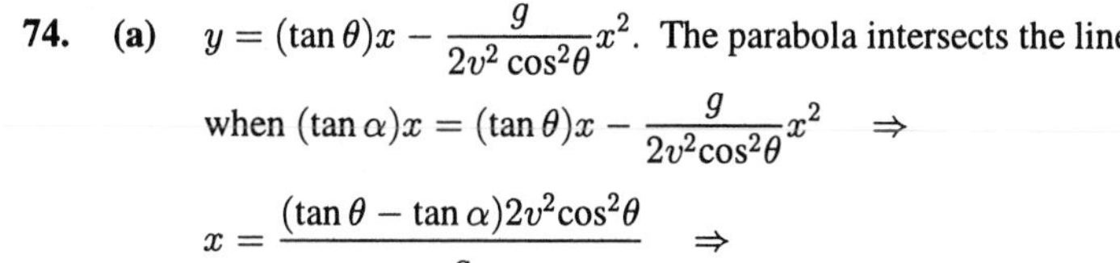

$y = (\tan\theta)x - \dfrac{g}{2v^2\cos^2\theta}x^2$. The parabola intersects the line when $(\tan\alpha)x = (\tan\theta)x - \dfrac{g}{2v^2\cos^2\theta}x^2 \quad\Rightarrow$

$$x = \frac{(\tan\theta - \tan\alpha)2v^2\cos^2\theta}{g} \quad\Rightarrow$$

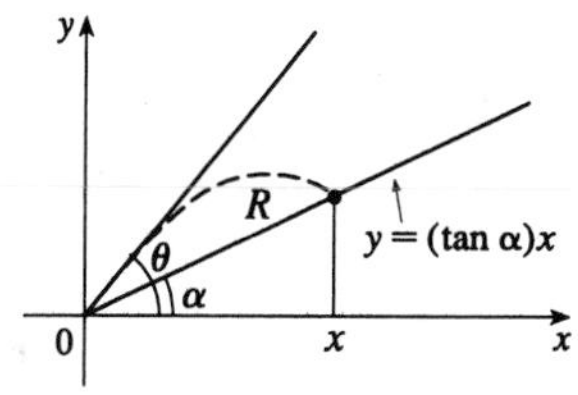

$$R(\theta) = \frac{x}{\cos\alpha} = \left(\frac{\sin\theta}{\cos\theta} - \frac{\sin\alpha}{\cos\alpha}\right)\frac{2v^2\cos\theta}{g\cos\alpha} = \left(\frac{\sin\theta}{\cos\theta} - \frac{\sin\alpha}{\cos\alpha}\right)(\cos\theta\cos\alpha)\frac{2v^2\cos\theta}{g\cos^2\alpha}$$

$$= (\sin\theta\cos\alpha - \sin\alpha\cos\theta)\frac{2v^2\cos\theta}{g\cos^2\alpha} = \sin(\theta - \alpha)\frac{2v^2\cos\theta}{g\cos^2\alpha}.$$

(b) $$R'(\theta) = \frac{2v^2}{g\cos^2\alpha}[\cos\theta\cdot\cos(\theta - \alpha) + \sin(\theta - \alpha)(-\sin\theta)] = \frac{2v^2}{g\cos^2\alpha}\cos[\theta + (\theta - \alpha)]$$

$$= \frac{2v^2}{g\cos^2\alpha}\cos(2\theta - \alpha) = 0 \text{ when } \cos(2\theta - \alpha) = 0 \Rightarrow 2\theta - \alpha = \tfrac{\pi}{2} \Rightarrow \theta = \frac{\pi/2 + \alpha}{2} = \frac{\pi}{4} + \frac{\alpha}{2}.$$

The First Derivative Test shows that this gives a maximum value for $R(\theta)$. [This could be done without calculus by applying the formula for $\sin x\cos y$ (Formula 18a in Appendix D) to $R(\theta)$.]

(c) Replacing α by $-\alpha$ in part (a), we get $R(\theta) = \dfrac{2v^2\cos\theta\sin(\theta + \alpha)}{g\cos^2\alpha}$. Proceeding as in part (b), or simply by replacing α by $-\alpha$ in the result of part (b), we see that $R(\theta)$ is maximized when $\theta = \dfrac{\pi}{4} - \dfrac{\alpha}{2}$ (see the quotation of René Descartes at the beginning of Chapter 6.)

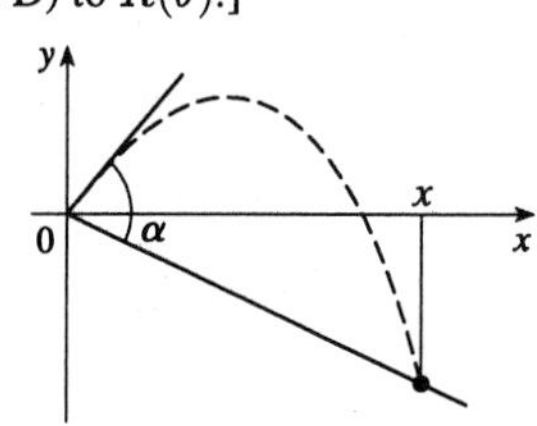

APPLICATIONS PLUS (page 255)

1. **(a)** $I = \dfrac{k\cos\theta}{d^2} = \dfrac{k(h/d)}{d^2} = k\dfrac{h}{d^2} = k\dfrac{h}{\left(\sqrt{1600+h^2}\right)^3} = k\dfrac{h}{(1600+h^2)^{3/2}} \quad\Rightarrow$

$$\frac{dI}{dh} = k\frac{(1600+h^2)^{3/2} - kh\frac{3}{2}(1600+h^2)^{1/2}\cdot 2h}{(1600+h^2)^3} = \frac{k(1600+h^2)^{1/2}(1600+h^2-3h^2)}{(1600+h^2)^3}$$

$$= \frac{k(1600-2h^2)}{(1600+h^2)^{5/2}}. \text{ Set } \frac{dI}{dh} = 0\text{: } 1600 - 2h^2 = 0 \quad\Rightarrow\quad h^2 = 800 \quad\Rightarrow\quad h = \sqrt{800} = 20\sqrt{2}.$$

By the First Derivative Test, I has a relative maximum at $h = 20\sqrt{2} \approx 28$ ft.

(b)

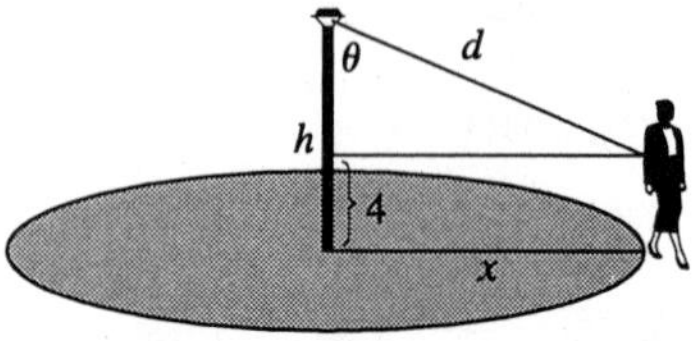

$$\frac{dx}{dt} = 4 \text{ ft/s}$$

$$I = \frac{k\cos\theta}{d^2} = \frac{k[(h-4)/d]}{d^2} = \frac{k\,(h-4)}{d^3} = \frac{k(h-4)}{\left[(h-4)^2+x^2\right]^{3/2}} = k(h-4)\left([h-4]^2+x^2\right)^{-3/2}$$

$$\frac{dI}{dt} = k(h-4)\left(-\tfrac{3}{2}\right)\left[(h-4)^2+x^2\right]^{-5/2}\cdot 2x\cdot\frac{dx}{dt} = k(h-4)(-3x)\left[(h-4)^2+x^2\right]^{-5/2}\cdot 4$$

$$= \frac{-12xk(h-4)}{\left[(h-4)^2+x^2\right]^{5/2}}$$

$$\left.\frac{dI}{dt}\right|_{x=40} = -\frac{480k(h-4)}{\left[(h-4)^2+1600\right]^{5/2}}$$

2. **(a)** Consider $G(x) = T(x+180°) - T(x)$. Fix any number a. If $G(a) = 0$, we are done: Temperature at a = Temperature at $a + 180°$. If $G(a) > 0$, then $G(a+180°) = T(a+360°) - T(a+180°)$ $= T(a) - T(a+180°) = -G(a) < 0$. Also, G is continuous since temperature varies continuously. So, by the Intermediate Value Theorem, G has a zero on the interval $[a, a+180°]$. If $G(a) < 0$ then a similar argument applies.

(b) Yes. The same argument applies.

(c) The same argument applies for quantities that vary continuously, such as barometric pressure. But one could argue that altitude above sea level is sometimes discontinuous, so the result might not always hold for that quantity.

3. We can assume without loss of generality that $\theta = 0$ at time $t = 0$, so that $\theta = 12\pi t$ rad. (The angular velocity of the wheel is 360 rpm $= 360\cdot 2\pi$ rad/60 s $= 12\pi$ rad/s.) Then the position of A as a function of time is $A = (40\cos\theta, 40\sin\theta) = (40\cos 12\pi t, 40\sin 12\pi t)$, so $\sin\alpha = \dfrac{40\sin\theta}{120} = \dfrac{\sin\theta}{3} = \frac{1}{3}\sin 12\pi t$.

(a) Differentiating the expression for $\sin\alpha$, we get $\cos\alpha\cdot\dfrac{d\alpha}{dt}=\frac{1}{3}\cdot 12\pi\cdot\cos 12\pi t=4\pi\cos\theta$. When $\theta=\frac{\pi}{3}$, we have $\sin\alpha=\frac{1}{3}\sin\theta=\frac{\sqrt{3}}{6}$, so $\cos\alpha=\sqrt{1-\left(\frac{\sqrt{3}}{6}\right)^2}=\sqrt{\frac{11}{12}}$ and

$$\frac{d\alpha}{dt}=\frac{4\pi\cos\frac{\pi}{3}}{\cos\alpha}=\frac{2\pi}{\sqrt{11/12}}=\frac{4\pi\sqrt{3}}{\sqrt{11}}\text{ rad/s.}$$

(b) By the Law of Cosines, $|AP|^2=|OA|^2+|OP|^2-2|OA||OP|\cos\theta \Rightarrow$
$120^2=40^2+|OP|^2-2\cdot 40|OP|\cos\theta \Rightarrow |OP|^2-(80\cos\theta)|OP|-12{,}800=0 \Rightarrow$
$|OP|=\frac{1}{2}\left(80\cos\theta\pm\sqrt{6400\cos^2\theta+51{,}200}\right)=40\cos\theta\pm 40\sqrt{\cos^2\theta+8}$
$=40\left(\cos\theta+\sqrt{8+\cos^2\theta}\right)$ cm (since $|OP|>0$).

As a check, note that $|OP|=160$ cm when $\theta=0$ and $|OP|=80\sqrt{2}$ cm when $\theta=\frac{\pi}{2}$.

(c) By part (b), the x-coordinate of P is given by $x=40\left(\cos\theta+\sqrt{8+\cos^2\theta}\right)$, so

$$\frac{dx}{dt}=\frac{dx}{d\theta}\frac{d\theta}{dt}=40\left(-\sin\theta-\frac{2\cos\theta\sin\theta}{2\sqrt{8+\cos^2\theta}}\right)\cdot 12\pi=-480\pi\sin\theta\left(1+\frac{\sin\theta\cos\theta}{\sqrt{8+\cos^2\theta}}\right)\text{cm/s. In}$$

particular, $dx/dt=0$ cm/s when $\theta=0$ and $dx/dt=-480\pi$ cm/s when $\theta=\frac{\pi}{2}$.

4. (a) $v(r)=k(r_0-r)r^2=k\left(r_0r^2-r^3\right) \Rightarrow dv/dr=k\left(2r_0r-3r^2\right)=kr\left(2r_0-3r\right)$.
Set $dv/dr=0$: $kr\left(2r_0-3r\right)=0 \Rightarrow r=\frac{2}{3}r_0$ (since $r>0$). $d^2v/dr^2=k\left(2r_0-6r\right) \Rightarrow$
$d^2v/dr^2\big|_{r=2r_0/3}=k\left(2r_0-6\cdot\frac{2}{3}r_0\right)=-2kr_0<0$

Thus v has a maximum at $r=\frac{2}{3}r_0$ by the Second Derivative Test, and experimental evidence is confirmed.

Alternately, we could use the First Derivative Test or we could argue as follows:
$v\left(\frac{r_0}{2}\right)=k\left(r_0-\frac{r_0}{2}\right)\left(\frac{r_0}{2}\right)^2=k\left(\frac{r_0}{2}\right)^3=\frac{kr_0^3}{8}$, $v\left(\frac{2r_0}{3}\right)=k\left(r_0-\frac{2r_0}{3}\right)\left(\frac{2r_0}{3}\right)^2=\frac{4kr_0^3}{27}$ (absolute maximum), $v(r_0)=k(r_0-r_0)\,r_0=0$

(b) From part (a), the absolute maximum value of f is $\dfrac{4kr_0^3}{27}$.

(c)

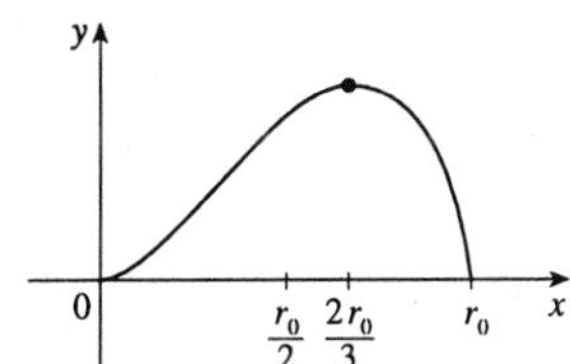

5. (a) $T=2\pi\sqrt{\dfrac{L}{g}}=\dfrac{2\pi}{\sqrt{g}}L^{1/2} \Rightarrow dT=\dfrac{2\pi}{\sqrt{g}}\dfrac{1}{2}L^{-1/2}dL \Rightarrow \dfrac{dT}{2\pi/\sqrt{g}}=\dfrac{dL}{2\sqrt{L}} \Rightarrow$
$\dfrac{dT}{(2\pi/\sqrt{g})\sqrt{L}}=\dfrac{dL}{2L} \Rightarrow \dfrac{dT}{T}=\dfrac{dL}{2L}$

(b) $dL=\dfrac{2L}{T}dT$. Set $dT=-15$ s, $T=3600$ s. Then $dL=\dfrac{2L}{3600}\cdot(-15)=-\frac{30}{3600}L=-\frac{1}{120}L$. Thus, shorten the pendulum by $\frac{1}{120}L$.

(c) $T=\dfrac{2\pi\sqrt{L}}{\sqrt{g}}=2\pi\sqrt{L}g^{-1/2} \Rightarrow dT=2\pi\sqrt{L}\left(-\frac{1}{2}\right)t^{-3/2}dg$. Therefore, $dg=-\dfrac{g\sqrt{g}\,dT}{\pi\sqrt{L}}$.

6. **(a)** If $k =$ energy/km over land, then energy/km over water $= 1.4k$. So the total energy is $E = 1.4k\sqrt{25 + x^2} + k(13 - x)$, $0 \le x \le 13$, and so $\dfrac{dE}{dx} = \dfrac{1.4kx}{(25 + x^2)^{1/2}} - k$.

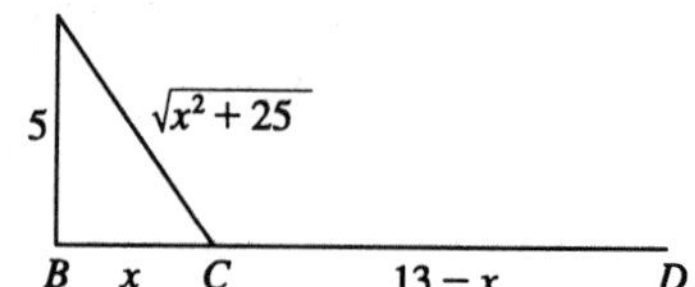

Set $\dfrac{dE}{dx} = 0$: $1.4kx = k(25 + x^2)^{1/2} \Rightarrow 1.96x^2 = x^2 + 25$

$\Rightarrow 0.96x^2 = 25 \Rightarrow x = \dfrac{5}{\sqrt{0.96}} \approx 5.1$. Testing against the value of E at the endpoints: $E(0) = 1.4k(5) + 13k = 20k$, $E(5.1) \approx 17.9k$, $E(13) \approx 19.5k$. Thus, to minimize energy, the bird should fly to a point about 5.1 km from B.

(b) $E = W\sqrt{25 + x^2} + L(13 - x) \Rightarrow \dfrac{dE}{dx} = \dfrac{Wx}{\sqrt{25 + x^2}} - L = 0$ when $\dfrac{W}{L} = \dfrac{\sqrt{25 + x^2}}{x}$. By the same sort of argument as in part (a), this ratio will give the minimal expenditure of energy if the bird heads for the point x km from B.

(c) For flight direct to D, $x = 13$, so from part (b), $\dfrac{W}{L} = \dfrac{\sqrt{25 + 13^2}}{13} \approx 1.07$. There is no value of $\dfrac{W}{L}$ for which the bird should fly directly to B. But note that $\lim\limits_{x \to 0^+} \dfrac{W}{L} = \infty$, so if the point at which E is a minimum is close to B, then $\dfrac{W}{L}$ is large.

7. **(a)** $T_1 = \dfrac{D}{c_1}$, $T_2 = \dfrac{2|PR|}{c_1} + \dfrac{|RS|}{c_2} = \dfrac{2h\sec\theta}{c_1} + \dfrac{D - 2h\tan\theta}{c_2}$, $T_3 = \dfrac{2\sqrt{h^2 + D^2/4}}{c_1} = \dfrac{\sqrt{4h^2 + D^2}}{c_1}$.

(b) $\dfrac{dT_2}{d\theta} = \dfrac{2h}{c_1} \cdot \sec\theta\tan\theta - \dfrac{2h}{c_2}\sec^2\theta = 0$ when $2h\sec\theta\left(\dfrac{1}{c_1}\tan\theta - \dfrac{1}{c_2}\sec\theta\right) = 0 \Rightarrow$

$\dfrac{1}{c_1}\dfrac{\sin\theta}{\cos\theta} - \dfrac{1}{c_2}\dfrac{1}{\cos\theta} = 0 \Rightarrow \sin\theta = \dfrac{c_1}{c_2}$. The First Derivative Test shows that this gives a minimum.

(c) Using part (a), we have $\dfrac{1}{4} = \dfrac{1}{c_1}$. Therefore, $c_1 = 4$. Also $\dfrac{3}{4\sqrt{5}} = \dfrac{\sqrt{4h^2 + 1}}{4} \Rightarrow \sqrt{4h^2 + 1} = \dfrac{3}{\sqrt{5}}$

$\Rightarrow 4h^2 + 1 = \frac{9}{5} \Rightarrow h^2 = \frac{1}{5} \Rightarrow h = \frac{1}{\sqrt{5}}$. From (b), $\sin\theta = \dfrac{c_1}{c_2} = \dfrac{4}{c_2} \Rightarrow$

$\sec\theta = \dfrac{c_2}{\sqrt{c_2^2 - 16}}$ and $\tan\theta = \dfrac{4}{\sqrt{c_2^2 - 16}}$. Thus $\dfrac{1}{3} = \dfrac{\dfrac{2}{\sqrt{5}} \cdot \dfrac{c_2}{\sqrt{c_2^2 - 16}}}{4} + \dfrac{1 - \dfrac{2}{\sqrt{5}} \cdot \dfrac{4}{\sqrt{c_2^2 - 16}}}{c_2} \Rightarrow$

$4c_2 = \dfrac{6c_2^2}{\sqrt{5}\sqrt{c_2^2 - 16}} + 12\left(1 - \dfrac{8}{\sqrt{5}\sqrt{c_2^2 - 16}}\right) \Rightarrow 4c_2 - 12 = \dfrac{6(c_2^2 - 16)}{\sqrt{5}\sqrt{c_2^2 - 16}} = \frac{6}{\sqrt{5}}\sqrt{c_2^2 - 16}$

$\Rightarrow 2c_2 - 6 = \frac{3}{\sqrt{5}}\sqrt{c_2^2 - 16} \Rightarrow 20c_2^2 - 120c_2 + 180 = 9c_2^2 - 144 \Rightarrow$

$11c_2^2 - 120c_2 + 324 = (c_2 - 6)(11c_2 - 54) = 0 \Rightarrow c_2 = 6$ or $\frac{54}{11}$. But the root $\frac{54}{11}$ is inadmissible because if $\tan\theta = \dfrac{4}{\sqrt{\left(\frac{54}{11}\right)^2 - 16}} \approx 1.4$, then $\theta > 45°$, which is impossible (from the diagram). So $c_2 = 6$.

8. **(a)** For $0 \le t \le 3$ we have $a(t) = 60t \quad \Rightarrow \quad v(t) = 30t^2 + C \quad \Rightarrow \quad v(0) = 0 = C \quad \Rightarrow \quad v(t) = 30t^2$, so $s(t) = 10t^3 + C \quad \Rightarrow \quad s(0) = 0 = C \quad \Rightarrow \quad s(t) = 10t^3$.

For $3 \le t \le 17$: $a(t) = -g = -32\text{ ft/s} \quad \Rightarrow \quad v(t) = -32(t-3) + C \quad \Rightarrow \quad v(3) = 270 = C \quad \Rightarrow$ $v(t) = -32(t-3) + 270 \quad \Rightarrow \quad s(t) = -16(t-3)^2 + 270(t-3) + C \quad \Rightarrow \quad s(3) = 270 = C \quad \Rightarrow$ $s(t) = -16(t-3)^2 + 270(t-3) + 270$.

For $17 \le t \le 22$: $v(17) = -32(14) + 270 = -178\text{ ft/s} \quad \Rightarrow$
$s(17) = -16(14)^2 + 270(14) + 270 - 3136 + 3780 + 270 = 914\text{ ft}$.
Also, $v(22) = -18\text{ ft/s}$ and "slope" $= \dfrac{-18-(-178)}{22-17} = \dfrac{160}{5} = 32$, so $v(t) = 32(t-17) - 178 \quad \Rightarrow$
$s(t) = 16(t-17)^2 - 178(t-17) + 914$ and $s(22) = 400 - 890 + 914 = 424\text{ ft}$.

For $t \ge 22$: $v(t) = -18 \quad \Rightarrow \quad s(t) = -18(t-22) + C$. But $s(22) = 424 = C \quad \Rightarrow$
$s(t) = -18(t-22) + 424$.

Therefore, until the rocket lands, we have

$$v(t) = \begin{cases} 30t^2 & \text{if } 0 \le t \le 3 \\ -32(t-3) + 270 & \text{if } 3 \le t \le 17 \\ 32(t-17) - 178 & \text{if } 17 \le t \le 22 \\ -18 & \text{if } t \ge 22. \end{cases} \quad \text{and}$$

$$s(t) = \begin{cases} 10t^3 & \text{if } 0 \le t \le 3 \\ -16(t-3)^2 + 270(t-3) + 270 & 3 \le t \le 17 \\ 16(t-17)^2 - 178(t-17) + 914 & 17 \le t \le 22 \\ -18(t-22) + 424 & t \ge 22. \end{cases}$$

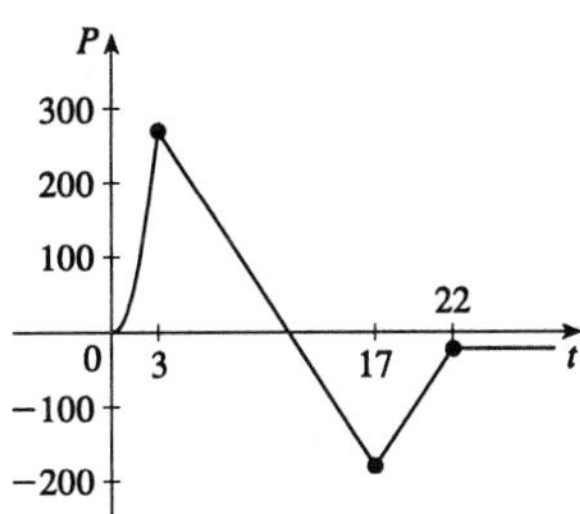

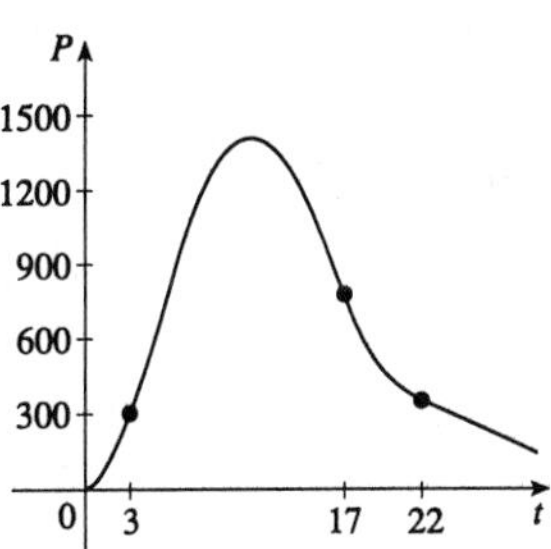

(b) To find the maximum height set $v(t) = -32(t-3) + 270 = 0$. Then $t - 3 = \frac{270}{32} \approx 8.4$, so the time to maximum height is $t \approx 11.4$ s and the maximum height is about
$s(11.4) = -16(8.4)^2 + 270(8.4) + 270 \approx 1409\text{ ft}$.

(c) To find the time to land set $s(t) = -18(t-22) + 424 = 0$. Then $t - 22 = \dfrac{424}{18} = 23.\overline{5}$, so $t \approx 45.6$ s.

9. **(a)** Condition (i) will hold if and only if all of the following four conditions hold: (α) $P(0) = 0$; (β) $P'(0) = 0$ (for a smooth landing); (γ) $P'(\ell) = 0$ (since the plane is cruising horizontally when it begins its descent); and (δ) $P(\ell) = h$.

First of all, condition α implies that $P(0) = d = 0$, so $P(x) = ax^3 + bx^2 + cx \Rightarrow P'(x) = 3ax^2 + 2bx + c$. But $P'(0) = c = 0$ by condition β. So $P'(\ell) = 3a\ell^2 + 2b\ell = \ell(3a\ell + 2b)$. Now by condition γ, $3a\ell + 2b = 0 \Rightarrow a = -\dfrac{2b}{3\ell}$. Therefore $P(x) = -\dfrac{2b}{3\ell}x^3 + bx^2$. Setting $P(\ell) = h$ for condition δ, we get $P(\ell) = -\dfrac{2b}{3\ell}\ell^3 + b\ell^2 = h \Rightarrow b = \dfrac{3h}{\ell^2} \Rightarrow a = -\dfrac{2h}{\ell^3}$. So $P(x) = -\dfrac{2h}{\ell^3}x^3 + \dfrac{3h}{\ell^2}x^2$.

(b) By condition (ii), $\dfrac{dx}{dt} = -v$ for all t, so $x(t) = \ell - vt$. Condition (iii) states that $\left|\dfrac{d^2y}{dt^2}\right| \le k$. By the Chain Rule, we have $\dfrac{dy}{dt} = -\dfrac{2h}{\ell^3}\left(3x^2\right)\dfrac{dx}{dt} + \dfrac{3h}{\ell^2}(2x)\dfrac{dx}{dt} = -\dfrac{6hx^2v}{\ell^3} + \dfrac{6hxv}{\ell^2}$ (for $x \le \ell$) $\Rightarrow$ $\dfrac{d^2y}{dt^2} = -\dfrac{6hv}{\ell^3}(2x)\dfrac{dx}{dt} + \dfrac{6hv}{\ell^2}\dfrac{dx}{dt} = -\dfrac{12hv^2}{\ell^3}x + \dfrac{6hv^2}{\ell^2}$. In particular, when $t = 0$, $x = \ell$ and so $\left.\dfrac{d^2y}{dt^2}\right|_{t=0} = -\dfrac{12hv^2}{\ell^3}\ell + \dfrac{6hv^2}{\ell^2} = -\dfrac{6hv^2}{\ell^2}$. Thus $\left|\dfrac{d^2y}{dt^2}\right|_{t=0} = \dfrac{6hv^2}{\ell^2} \le k$. (This condition also follows from taking $x = 0$.)

(c) We substitute $k = 860\text{ mi}/\text{h}^2$, $h = 35{,}000\text{ ft} \times \dfrac{1\text{ mi}}{5280\text{ ft}}$, and $v = 300\text{ mi/h}$ into the result of part (b): $\dfrac{6\left(35{,}000 \cdot \frac{1}{5280}\right)(300)^2}{\ell^2} \le 860 \Leftrightarrow \ell \ge 300\sqrt{6 \cdot \dfrac{35{,}000}{5280 \cdot 860}} \approx 64.5$ miles.

10. **(a)** In this case, the amount of metal used in the making of each top or bottom is $(2r)^2 = 4r^2$. So the quantity we want to minimize is $A = 2\pi rh + 2(4r^2)$. But $V = \pi r^2 h \Leftrightarrow h = \dfrac{V}{\pi r^2}$. Substituting this expression for h and differentiating A with respect to r, we get $\dfrac{dA}{dr} = -\dfrac{2V}{r^2} + 16r = 0 \Rightarrow 16r^3 = 2V = 2\pi r^2 h \Leftrightarrow \dfrac{h}{r} = \dfrac{8}{\pi}$. This gives a minimum because $\dfrac{d^2A}{dr^2} = 16 + \dfrac{4V}{r^3} > 0$.

(b) We need to find the area of metal used up by each end, that is, the area of each hexagon.

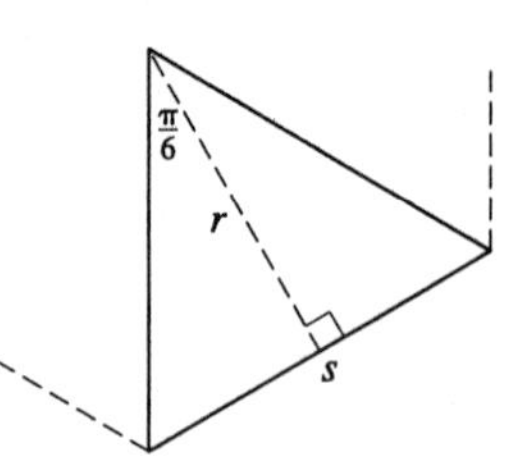

We subdivide the hexagon into six congruent triangles, each sharing one side (s in the diagram) with the hexagon. We calculate the length of $s = 2r\tan\frac{\pi}{6} = \frac{2}{\sqrt{3}}r$, so the area of each triangle is $\frac{1}{2}sr = \frac{1}{\sqrt{3}}r^2$, and the total area of the hexagon is $6 \cdot \frac{1}{\sqrt{3}}r^2 = 2\sqrt{3}r^2$. So the quantity we want to minimize is $A = 2\pi rh + 2 \cdot 2\sqrt{3}r^2$.

Substituting for h as in part (a) and differentiating, we get $\dfrac{dA}{dr} = -\dfrac{2V}{r^2} + 8\sqrt{3}r$. Setting this to 0, we get $8\sqrt{3}r^3 = 2V = 2\pi r^2 h \quad \Rightarrow \quad \dfrac{h}{r} = \dfrac{4\sqrt{3}}{\pi}$. Again this minimizes A because $\dfrac{d^2A}{dr^2} = 8\sqrt{3} + \dfrac{4V}{r^3} > 0$.

(c) Let $C = 4\sqrt{3}r^2 + 2\pi rh + k(4\pi r + h) = 4\sqrt{3}r^2 + 2\pi r\left(\dfrac{V}{\pi r^2}\right) + k\left(4\pi r + \dfrac{V}{\pi r^2}\right)$. Then $\dfrac{dC}{dr} = 8\sqrt{3}r - \dfrac{2V}{r^2} + 4k\pi - \dfrac{2kV}{\pi r^3}$. Setting this equal to 0, dividing by 2 and substituting $\dfrac{V}{r^2} = \pi h$ and $\dfrac{V}{\pi r^3} = \dfrac{h}{r}$ in the second and fourth terms respectively, we get $0 = 4\sqrt{3}r - \pi h + 2k\pi - \dfrac{kh}{r} \quad \Leftrightarrow$ $k\left(2\pi - \dfrac{h}{r}\right) = \pi h - 4\sqrt{3}r \quad \Rightarrow \quad \dfrac{k}{r}\dfrac{2\pi - h/r}{\pi h/r - 4\sqrt{3}} = 1$. We now multiply by $\dfrac{\sqrt[3]{V}}{k}$, noting that $\dfrac{\sqrt[3]{V}}{k}\dfrac{k}{r} = \sqrt[3]{\dfrac{V}{r^3}} = \sqrt[3]{\dfrac{\pi h}{r}}$, and get $\dfrac{\sqrt[3]{V}}{k} = \sqrt[3]{\dfrac{\pi h}{r}} \cdot \dfrac{2\pi - h/r}{\pi h/r - 4\sqrt{3}}$.

(d)

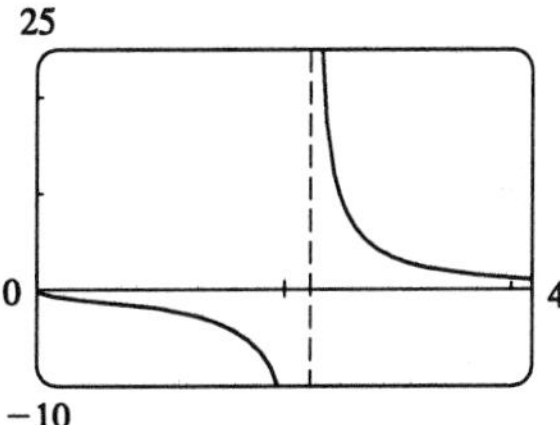

We see from the graph that when the ratio $\dfrac{\sqrt[3]{V}}{k}$ is large, that is, either the volume of the can is large or the cost of joining (proportional to k) is small, the optimum value of $\dfrac{h}{r}$ is about 2.21, but when $\dfrac{\sqrt[3]{V}}{k}$ is small, indicating small volume or expensive joining, the optimum value of $\dfrac{h}{r}$ is larger. (The part of the graph for $\dfrac{\sqrt[3]{V}}{k} < 0$ has no physical meaning, but confirms the location of the asymptote.)

CHAPTER FOUR

EXERCISES 4.1

1. $\sum_{i=1}^{5}\sqrt{i}=\sqrt{1}+\sqrt{2}+\sqrt{3}+\sqrt{4}+\sqrt{5}$

2. $\sum_{i=1}^{6}\frac{1}{i+1}=\frac{1}{2}+\frac{1}{3}+\frac{1}{4}+\frac{1}{5}+\frac{1}{6}+\frac{1}{7}$

3. $\sum_{i=4}^{6}3^i=3^4+3^5+3^6$

4. $\sum_{i=4}^{6}i^3=4^3+5^3+6^3$

5. $\sum_{k=0}^{4}\frac{2k-1}{2k+1}=-1+\frac{1}{3}+\frac{3}{5}+\frac{5}{7}+\frac{7}{9}$

6. $\sum_{k=5}^{8}x^k=x^5+x^6+x^7+x^8$

7. $\sum_{i=1}^{n}i^{10}=1^{10}+2^{10}+3^{10}+\cdots+n^{10}$

8. $\sum_{j=n}^{n+3}j^2=n^2+(n+1)^2+(n+2)^2+(n+3)^2$

9. $\sum_{j=0}^{n-1}(-1)^j=1-1+1-1+\cdots+(-1)^{n-1}$

10. $\sum_{i=1}^{n}f(x_i)\Delta x_i=f(x_1)\Delta x_1+f(x_2)\Delta x_2+f(x_3)\Delta x_3+\cdots+f(x_n)\Delta x_n$

11. $1+2+3+4+\cdots+10=\sum_{i=1}^{10}i$

12. $\sqrt{3}+\sqrt{4}+\sqrt{5}+\sqrt{6}+\sqrt{7}=\sum_{i=3}^{7}\sqrt{i}$

13. $\frac{1}{2}+\frac{2}{3}+\frac{3}{4}+\frac{4}{5}+\cdots+\frac{19}{20}=\sum_{i=1}^{19}\frac{i}{i+1}$

14. $\frac{3}{7}+\frac{4}{8}+\frac{5}{9}+\frac{6}{10}+\cdots+\frac{23}{27}=\sum_{i=3}^{23}\frac{i}{i+4}$

15. $2+4+6+8+\cdots+2n=\sum_{i=1}^{n}2i$

16. $1+3+5+7+\cdots+(2n-1)=\sum_{i=1}^{n}(2i-1)$

17. $1+2+4+8+16+32=\sum_{i=0}^{5}2^i$

18. $\frac{1}{1}+\frac{1}{4}+\frac{1}{9}+\frac{1}{16}+\frac{1}{25}+\frac{1}{36}=\sum_{i=1}^{6}\frac{1}{i^2}$

19. $x+x^2+x^3+\cdots+x^n=\sum_{i=1}^{n}x^i$

20. $1-x+x^2-x^3+\cdots+(-1)^nx^n=\sum_{i=0}^{n}(-1)^ix^i$

21. $\sum_{i=4}^{8}(3i-2)=10+13+16+19+22=80$

22. $\sum_{i=3}^{6}i(i+2)=3\cdot5+4\cdot6+5\cdot7+6\cdot8=15+24+35+48=122$

23. $\sum_{j=1}^{6}3^{j+1}=3^2+3^3+3^4+3^5+3^6+3^7=9+27+81+243+729+2187=3276$

(For a more general method, see Exercise 47.)

24. $\sum_{k=0}^{8}\cos k\pi=\cos 0+\cos\pi+\cos 2\pi+\cos 3\pi+\cos 4\pi+\cos 5\pi+\cos 6\pi+\cos 7\pi+\cos 8\pi$

$=1-1+1-1+1-1+1-1+1=1$

25. $\displaystyle\sum_{n=1}^{20}(-1)^n = -1+1-1+1-1+1-1+1-1+1-1+1-1+1-1+1-1+1-1+1 = 0$

26. $\displaystyle\sum_{i=1}^{100} 4 = \underbrace{4+4+4+\cdots+4}_{\text{100 summands}} = 100 \cdot 4 = 400$

27. $\displaystyle\sum_{i=0}^{4}(2^i + i^2) = (1+0)+(2+1)+(4+4)+(8+9)+(16+16) = 61$

28. $\displaystyle\sum_{i=-2}^{4} 2^{3-i} = 2^5 + 2^4 + 2^3 + 2^2 + 2^1 + 2^0 + 2^{-1} = 63.5$

29. $\displaystyle\sum_{i=1}^{n} 2i = 2\sum_{i=1}^{n} i = n(n+1)$

30. $\displaystyle\sum_{i=1}^{n}(2-5i) = \sum_{i=1}^{n} 2 - \sum_{i=1}^{n} 5i = 2n - 5\sum_{i=1}^{n} i = 2n - \frac{5n(n+1)}{2} = \frac{4n}{2} - \frac{5n^2+5n}{2} = -\frac{n(5n+1)}{2}$

31. $\displaystyle\sum_{i=1}^{n}(i^2+3i+4) = \sum_{i=1}^{n} i^2 + 3\sum_{i=1}^{n} i + \sum_{i=1}^{n} 4 = \frac{n(n+1)(2n+1)}{6} + \frac{3n(n+1)}{2} + 4n$

$\displaystyle = \tfrac{1}{6}[(2n^3+3n^2+n)+(9n^2+9n)+24n] = \tfrac{1}{6}(2n^3+12n^2+34n) = \tfrac{1}{3}n(n^2+6n+17)$

32. $\displaystyle\sum_{i=1}^{n}(3+2i)^2 = \sum_{i=1}^{n}(9+12i+4i^2) = \sum_{i=1}^{n} 9 + 12\sum_{i=1}^{n} i + 4\sum_{i=1}^{n} i^2$

$\displaystyle = 9n + 6n(n+1) + \frac{2n(n+1)(2n+1)}{3} = \frac{27n+18n^2+18n+4n^3+6n^2+2n}{3}$

$\displaystyle = \tfrac{1}{3}(4n^3+24n^2+47n) = \tfrac{1}{3}n(4n^2+24n+47)$

33. $\displaystyle\sum_{i=1}^{n}(i+1)(i+2) = \sum_{i=1}^{n}(i^2+3i+2) = \sum_{i=1}^{n} i^2 + 3\sum_{i=1}^{n} i + \sum_{i=1}^{n} 2$

$\displaystyle = \frac{n(n+1)(2n+1)}{6} + \frac{3n(n+1)}{2} + 2n = \frac{n(n+1)}{6}[(2n+1)+9] + 2n$

$\displaystyle = \frac{n(n+1)}{3}(n+5) + 2n = \frac{n}{3}[(n+1)(n+5)+6] = \frac{n}{3}(n^2+6n+11)$

34. $\displaystyle\sum_{i=1}^{n} i(i+1)(i+2) = \sum_{i=1}^{n}(i^3+3i^2+2i) = \sum_{i=1}^{n} i^3 + 3\sum_{i=1}^{n} i^2 + 2\sum_{i=1}^{n} i$

$\displaystyle = \left[\frac{n(n+1)}{2}\right]^2 + \frac{3n(n+1)(2n+1)}{6} + \frac{2n(n+1)}{2}$

$\displaystyle = n(n+1)\left[\frac{n(n+1)}{4} + \frac{2n+1}{2} + 1\right] = \frac{n(n+1)}{4}(n^2+n+4n+2+4)$

$\displaystyle = \frac{n(n+1)}{4}(n^2+5n+6) = \frac{n(n+1)(n+2)(n+3)}{4}$

35. $\displaystyle\sum_{i=1}^{n}(i^3-i-2) = \sum_{i=1}^{n} i^3 - \sum_{i=1}^{n} i - \sum_{i=1}^{n} 2 = \left[\frac{n(n+1)}{2}\right]^2 - \frac{n(n+1)}{2} - 2n$

$\displaystyle = \tfrac{1}{4}n(n+1)[n(n+1)-2] - 2n = \tfrac{1}{4}n(n+1)(n+2)(n-1) - 2n$

$\displaystyle = \tfrac{1}{4}n[(n+1)(n-1)(n+2)-8] = \tfrac{1}{4}n[(n^2-1)(n+2)-8] = \tfrac{1}{4}n(n^3+2n^2-n-10)$

36. $\sum_{k=1}^{n} k^2(k^2-k+1) = \sum_{k=1}^{n} k^4 - \sum_{k=1}^{n} k^3 + \sum_{k=1}^{n} k^2$

$$= \frac{n(n+1)(2n+1)(3n^2+3n-1)}{30} - \frac{n^2(n+1)^2}{4} + \frac{n(n+1)(2n+1)}{6}$$
$$= \tfrac{1}{60}n(n+1)[2(2n+1)(3n^2+3n-1) - 15n(n+1) + 10(2n+1)]$$
$$= \tfrac{1}{60}n(n+1)(12n^3+3n^2+7n+8)$$

37. By Theorem 2(a) and Example 3, $\sum_{i=1}^{n} c = c\sum_{i=1}^{n} 1 = cn$.

38. Let S_n be the statement that $\sum_{i=1}^{n} i^3 = \left[\frac{n(n+1)}{2}\right]^2$.

1. S_1 is true because $1^3 = \left(\frac{1\cdot 2}{2}\right)^2$.

2. Assume S_k is true. Then $\sum_{i=1}^{k} i^3 = \left[\frac{k(k+1)}{2}\right]^2$, so

$$\sum_{i=1}^{k+1} i^3 = \left[\frac{k(k+1)}{2}\right]^2 + (k+1)^3 = \frac{(k+1)^2}{4}\left[k^2+4(k+1)\right]$$
$$= \frac{(k+1)^2}{4}(k+2)^2 = \left(\frac{(k+1)[(k+1)+1]}{2}\right)^2,$$ showing that S_{k+1} is true.

Therefore S_n is true for all n by mathematical induction.

39. $\sum_{i=1}^{n}\left[(i+1)^4 - i^4\right] = (2^4-1^4) + (3^4-2^4) + (4^4-3^4) + \cdots + \left[(n+1)^4 - n^4\right]$

$$= (n+1)^4 - 1^4 = n^4 + 4n^3 + 6n^2 + 4n$$

On the other hand, $\sum_{i=1}^{n}\left[(i+1)^4 - i^4\right] = \sum_{i=1}^{n}(4i^3+6i^2+4i+1) = 4\sum_{i=1}^{n} i^3 + 6\sum_{i=1}^{n} i^2 + 4\sum_{i=1}^{n} i + \sum_{i=1}^{n} 1$

$$= 4S + n(n+1)(2n+1) + 2n(n+1) + n \quad (\text{where } S = \textstyle\sum_{i=1}^{n} i^3)$$
$$= 4S + 2n^3 + 3n^2 + n + 2n^2 + 2n + n = 4S + 2n^3 + 5n^2 + 4n$$

Thus $n^4 + 4n^3 + 6n^2 + 4n = 4S + 2n^3 + 5n^2 + 4n$, from which it follows that

$4S = n^4 + 2n^3 + n^2 = n^2(n^2+2n+1) = n^2(n+1)^2$ and $S = \left[\frac{n(n+1)}{2}\right]^2$.

40. The area of G_i is

$$\left(\sum_{k=1}^{i} k\right)^2 - \left(\sum_{k=1}^{i-1} k\right)^2 = \left[\frac{i(i+1)}{2}\right]^2 - \left[\frac{(i-1)i}{2}\right]^2$$
$$= \frac{i^2}{4}\left[(i+1)^2 - (i-1)^2\right] = \frac{i^2}{4}\left[(i^2+2i+1) - (i^2-2i+1)\right] = \frac{i^2}{4}(4i) = i^3.$$

Thus the area of $ABCD$ is $\sum_{i=1}^{n} i^3 = \left[\frac{n(n+1)}{2}\right]^2$.

41. **(a)** $\sum_{i=1}^{n}\left(i^4-(i-1)^4\right)=(1^4-0^4)+(2^4-1^4)+(3^4-2^4)+\cdots+\left[n^4-(n-1)^4\right]=n^4-0=n^4$

(b) $\sum_{i=1}^{100}(5^i-5^{i-1})=(5^1-5^0)+(5^2-5^1)+(5^3-5^2)+\cdots+(5^{100}-5^{99})=5^{100}-5^0=5^{100}-1$

(c) $\sum_{i=3}^{99}\left(\frac{1}{i}-\frac{1}{i+1}\right)=\left(\frac{1}{3}-\frac{1}{4}\right)+\left(\frac{1}{4}-\frac{1}{5}\right)+\left(\frac{1}{5}-\frac{1}{6}\right)+\cdots+\left(\frac{1}{99}-\frac{1}{100}\right)=\frac{1}{3}-\frac{1}{100}=\frac{97}{300}$

(d) $\sum_{i=1}^{n}(a_i-a_{i-1})=(a_1-a_0)+(a_2-a_1)+(a_3-a_2)+\cdots+(a_n-a_{n-1})=a_n-a_0$

42. Summing the inequalities $-|a_i|\le a_i\le |a_i|$ for $i=1,2,\ldots,n$, we get $-\sum_{i=1}^{n}|a_i|\le\sum_{i=1}^{n}a_i\le\sum_{i=1}^{n}|a_i|$. Since $|x|\le c \quad\Leftrightarrow\quad -c\le x\le c$, we have $\left|\sum_{i=1}^{n}a_i\right|\le\sum_{i=1}^{n}|a_i|$. *Another method:* Use mathematical induction.

43. $\lim_{n\to\infty}\sum_{i=1}^{n}\frac{1}{n}\left(\frac{i}{n}\right)^2=\lim_{n\to\infty}\frac{1}{n^3}\sum_{i=1}^{n}i^2=\lim_{n\to\infty}\frac{1}{n^3}\frac{n(n+1)(2n+1)}{6}=\lim_{n\to\infty}\frac{1}{6}\left(1+\frac{1}{n}\right)\left(2+\frac{1}{n}\right)=\frac{1}{6}(1)(2)=\frac{1}{3}$

44. $\lim_{n\to\infty}\sum_{i=1}^{n}\frac{1}{n}\left[\left(\frac{i}{n}\right)^3+1\right]=\lim_{n\to\infty}\sum_{i=1}^{n}\left[\frac{i^3}{n^4}+\frac{1}{n}\right]=\lim_{n\to\infty}\left[\frac{1}{n^4}\sum_{i=1}^{n}i^3+\frac{1}{n}\sum_{i=1}^{n}1\right]$

$=\lim_{n\to\infty}\left[\frac{1}{n^4}\left(\frac{n(n+1)}{2}\right)^2+\frac{1}{n}(n)\right]=\lim_{n\to\infty}\frac{1}{4}\left(1+\frac{1}{n}\right)^2+1=\frac{1}{4}+1=\frac{5}{4}$

45. $\lim_{n\to\infty}\sum_{i=1}^{n}\frac{2}{n}\left[\left(\frac{2i}{n}\right)^3+5\left(\frac{2i}{n}\right)\right]=\lim_{n\to\infty}\sum_{i=1}^{n}\left[\frac{16}{n^4}i^3+\frac{20}{n^2}i\right]=\lim_{n\to\infty}\left[\frac{16}{n^4}\sum_{i=1}^{n}i^3+\frac{20}{n^2}\sum_{i=1}^{n}i\right]$

$=\lim_{n\to\infty}\left[\frac{16}{n^4}\frac{n^2(n+1)^2}{4}+\frac{20}{n^2}\frac{n(n+1)}{2}\right]=\lim_{n\to\infty}\left[\frac{4(n+1)^2}{n^2}+\frac{10n(n+1)}{n^2}\right]$

$=\lim_{n\to\infty}\left[4\left(1+\frac{1}{n}\right)^2+10\left(1+\frac{1}{n}\right)\right]=4\cdot 1+10\cdot 1=14$

46. $\lim_{n\to\infty}\sum_{i=1}^{n}\frac{3}{n}\left[\left(1+\frac{3i}{n}\right)^3-2\left(1+\frac{3i}{n}\right)\right]=\lim_{n\to\infty}\sum_{i=1}^{n}\frac{3}{n}\left[1+\frac{9i}{n}+\frac{27i^2}{n^2}+\frac{27i^3}{n^3}-2-\frac{6i}{n}\right]$

$=\lim_{n\to\infty}\sum_{i=1}^{n}\left[\frac{81}{n^4}i^3+\frac{81}{n^3}i^2+\frac{9}{n^2}i-\frac{3}{n}\right]$

$=\lim_{n\to\infty}\left[\frac{81}{n^4}\frac{n^2(n+1)^2}{4}+\frac{81}{n^3}\frac{n(n+1)(2n+1)}{6}+\frac{9}{n^2}\frac{n(n+1)}{2}-\frac{3}{n}n\right]$

$=\lim_{n\to\infty}\left[\frac{81}{4}\left(1+\frac{1}{n}\right)^2+\frac{27}{2}\left(1+\frac{1}{n}\right)\left(2+\frac{1}{n}\right)+\frac{9}{2}\left(1+\frac{1}{n}\right)-3\right]=\frac{81}{4}+\frac{54}{2}+\frac{9}{2}-3=\frac{195}{4}$

47. Let $S=\sum_{i=1}^{n}ar^{i-1}=a+ar+ar^2+\cdots+ar^{n-1}$. Then $rS=ar+ar^2+\cdots+ar^{n-1}+ar^n$. Subtracting the first equation from the second, we find $(r-1)S=ar^n-a=a(r^n-1)$, so $S=\dfrac{a(r^n-1)}{r-1}$.

48. $\sum_{i=1}^{n}\frac{3}{2^{i-1}}=3\sum_{i=1}^{n}\left(\frac{1}{2}\right)^{i-1}=\frac{3\left[\left(\frac{1}{2}\right)^n-1\right]}{\frac{1}{2}-1}$ $\quad$ [using Exercise 47 with $a=3$ and $r=\frac{1}{2}$] $\quad =6\left[1-\left(\frac{1}{2}\right)^n\right]$

49. $\sum_{i=1}^{n}(2i+2^i) = 2\sum_{i=1}^{n} i + \sum_{i=1}^{n} 2\cdot 2^{i-1} = 2\,\frac{n(n+1)}{2} + \frac{2(2^n-1)}{2-1} = 2^{n+1} + n^2 + n - 2.$

For the first sum we have used Theorem 3(c), and for the second, Exercise 47 with $a = r = 2$.

50. $\sum_{i=1}^{m}\left[\sum_{j=1}^{n}(i+j)\right] = \sum_{i=1}^{m}\left(\sum_{j=1}^{n} i + \sum_{j=1}^{n} j\right)$ [Theorem 2(b)] $= \sum_{i=1}^{m}\left(ni + \frac{n(n+1)}{2}\right)$ [Theorem 3(b)]

$$= \sum_{i=1}^{m} ni + \sum_{i=1}^{m}\frac{n(n+1)}{2} = \frac{nm(m+1)}{2} + \frac{nm(n+1)}{2} = \frac{nm}{2}(m+n+2)$$

51. By Theorem 3(c) we have that $\sum_{i=1}^{n} i = \frac{n(n+1)}{2} = 78 \Leftrightarrow n(n+1) = 156 \Leftrightarrow n^2 + n - 156 = 0 \Leftrightarrow$

$(n+13)(n-12) = 0 \Leftrightarrow n = 12$ or -13. But $n = -13$ produces a negative answer for the sum, so $n = 12$.

52. **(a)** From Formula 18a in Appendix D, $2\sin u\cos v = \sin(u+v) + \sin(u-v)$. Taking $u = \frac{1}{2}x$ and $v = ix$, we get $2\sin\left(\frac{1}{2}x\right)\cos ix = \sin\left(\frac{1}{2}x + ix\right) + \sin\left(\frac{1}{2}x - ix\right)$

$$= \sin\left(\left(i+\tfrac{1}{2}\right)x\right) + \sin\left(\left(\tfrac{1}{2}-i\right)x\right) = \sin\left(\left(i+\tfrac{1}{2}\right)x\right) - \sin\left(\left(i-\tfrac{1}{2}\right)x\right)$$

(b) $2\sin\left(\frac{1}{2}x\right)\sum_{i=1}^{n}\cos ix = \sum_{i=1}^{n} 2\sin\left(\frac{1}{2}x\right)\cos ix = \sum_{i=1}^{n}\left[\sin\left(\left(i+\frac{1}{2}\right)x\right) - \sin\left(\left(i-\frac{1}{2}\right)x\right)\right]$

$$= \sin\left(\left(n+\tfrac{1}{2}\right)x\right) - \sin\left(\tfrac{1}{2}x\right) \quad \text{(telescoping sum)}$$

Hence $\sum_{i=1}^{n}\cos ix = \frac{\sin\left(\left(n+\frac{1}{2}\right)x\right) - \sin\left(\frac{1}{2}x\right)}{2\sin\left(\frac{1}{2}x\right)}$. [Note that $\sin\left(\frac{1}{2}x\right) \neq 0$ since x is not an integer multiple of 2π.] Now $\sin\left(\left(n+\frac{1}{2}\right)x\right) = \sin\left(\frac{1}{2}(n+1)x + \frac{1}{2}nx\right)$

$$= \sin\left(\tfrac{1}{2}(n+1)x\right)\cos\left(\tfrac{1}{2}nx\right) + \cos\left(\tfrac{1}{2}(n+1)x\right)\sin\left(\tfrac{1}{2}x\right)$$

and $\sin\left(\frac{1}{2}x\right) = \sin\left(\frac{1}{2}(n+1)x - \frac{1}{2}nx\right)$

$$= \sin\left(\tfrac{1}{2}(n+1)x\right)\cos\left(\tfrac{1}{2}nx\right) - \cos\left(\tfrac{1}{2}(n+1)x\right)\sin\left(\tfrac{1}{2}nx\right)$$

Subtracting, we get $\sin\left(\left(n+\frac{1}{2}\right)x\right) - \sin\left(\frac{1}{2}x\right) = 2\cos\left(\frac{1}{2}(n+1)x\right)\sin\left(\frac{1}{2}nx\right)$.

Thus $\sum_{i=1}^{n}\cos ix = \frac{\cos\left(\frac{1}{2}(n+1)x\right)\sin\left(\frac{1}{2}nx\right)}{\sin\left(\frac{1}{2}x\right)}$.

53. From Formula 18c in Appendix D, $\sin x\sin y = \frac{1}{2}[\cos(x-y) - \cos(x+y)]$, so

$2\sin u\sin v = \cos(u-v) - \cos(u+v)$ (★). Taking $u = \frac{1}{2}x$ and $v = ix$, we get

$2\sin\left(\frac{1}{2}x\right)\sin ix = \cos\left(\left(\frac{1}{2}-i\right)x\right) - \cos\left(\left(\frac{1}{2}+i\right)x\right) = \cos\left(\left(i-\frac{1}{2}\right)x\right) - \cos\left(\left(i+\frac{1}{2}\right)x\right)$. Thus

$2\sin\left(\frac{1}{2}x\right)\sum_{i=1}^{n}\sin ix = \sum_{i=1}^{n} 2\sin\left(\frac{1}{2}x\right)\sin ix = \sum_{i=1}^{n}\left[\cos\left(\left(i-\frac{1}{2}\right)x\right) - \cos\left(\left(i+\frac{1}{2}\right)x\right)\right]$

$$= -\sum_{i=1}^{n}\left[\cos\left(\left(i+\tfrac{1}{2}\right)x\right) - \cos\left(\left(i-\tfrac{1}{2}\right)x\right)\right] = -\left[\cos\left(\left(n+\tfrac{1}{2}\right)x\right) - \cos\left(\tfrac{1}{2}x\right)\right] \quad \text{(telescoping sum)}$$

$$= \cos\left(\tfrac{1}{2}(n+1)x - \tfrac{1}{2}nx\right) - \cos\left(\tfrac{1}{2}(n+1)x + \tfrac{1}{2}nx\right)$$

$$= 2\sin\left(\tfrac{1}{2}(n+1)x\right)\sin\left(\tfrac{1}{2}nx\right) \quad \text{[by (★) with } u = \tfrac{1}{2}(n+1)x \text{ and } v = \tfrac{1}{2}nx]$$

If x is not an integer multiple of 2π, then $\sin\left(\frac{1}{2}x\right) \neq 0$, so we can divide by $2\sin\left(\frac{1}{2}x\right)$ and get

$\sum_{i=1}^{n}\sin ix = \frac{\sin\left(\frac{1}{2}nx\right)\sin\left(\frac{1}{2}(n+1)x\right)}{\sin\left(\frac{1}{2}x\right)}$.

EXERCISES 4.2

1. **(a)** $\|P\| = \max\{1, 1, 1, 1\} = 1$

(b) $$\sum_{i=1}^{n} f(x_i^*)\Delta x_i = \sum_{i=1}^{4} f(i-1)\cdot 1 = 16 + 15 + 12 + 7 = 50$$

(c)

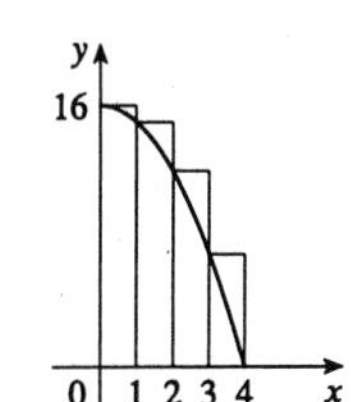

2. **(a)** $\|P\| = \max\{1, 1, 1, 1\} = 1$

(b) $$\sum_{i=1}^{n} f(x_i^*)\Delta x_i = \sum_{i=1}^{4} f(i)\cdot 1 = 15 + 12 + 7 + 0 = 34$$

(c)

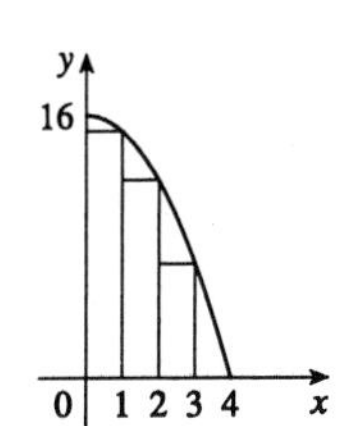

3. **(a)** $\|P\| = \max\{1, 1, 1, 1\} = 1$

(b) $$\sum_{i=1}^{n} f(x_i^*)\Delta x_i = \sum_{i=1}^{4} f\left(i-\tfrac{1}{2}\right)\cdot 1 = 15.75 + 13.75 + 9.75 + 3.75 = 43$$

(c)

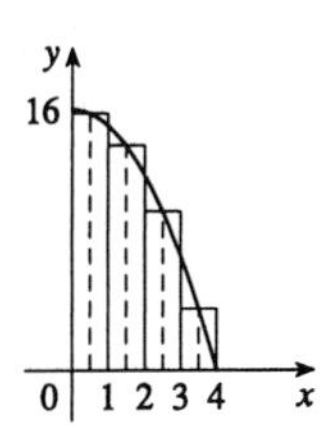

4. **(a)** $\|P\| = \max\{0.5, 0.5, 1, 2\} = 2$

(b)

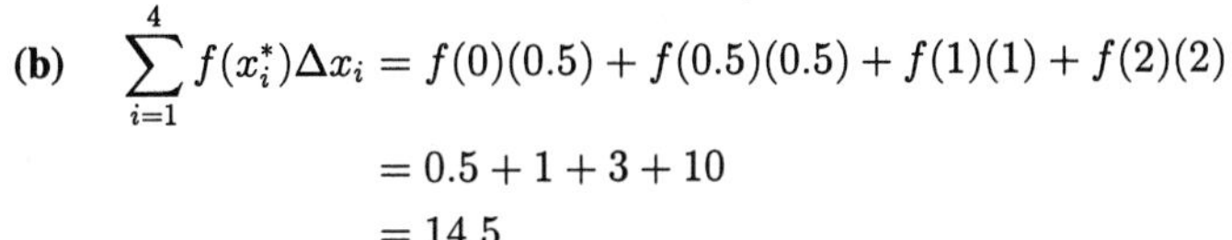

$$\sum_{i=1}^{4} f(x_i^*)\Delta x_i = f(0)(0.5) + f(0.5)(0.5) + f(1)(1) + f(2)(2)$$

$$= 0.5 + 1 + 3 + 10 = 14.5$$

(c)

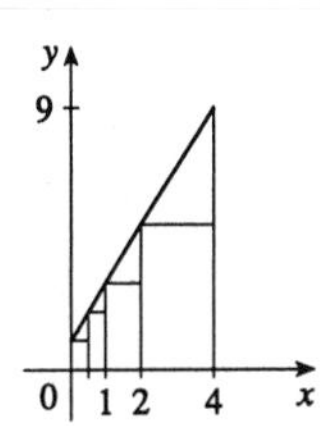

5. **(a)** $\|P\| = \max\{0.5, 0.5, 0.5, 0.5, 0.5, 0.5\} = 0.5$

(b)

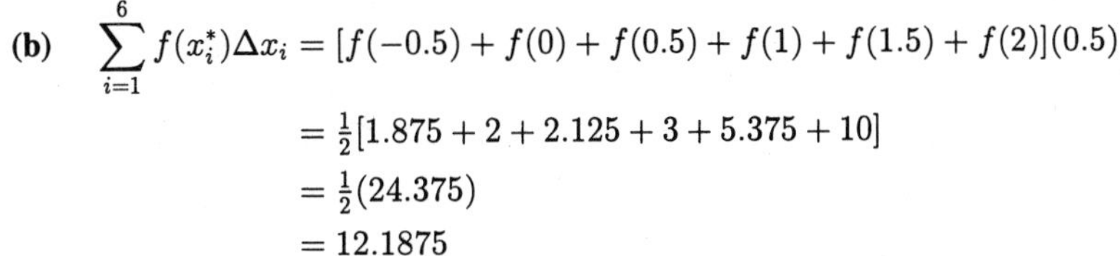

$$\sum_{i=1}^{6} f(x_i^*)\Delta x_i = [f(-0.5) + f(0) + f(0.5) + f(1) + f(1.5) + f(2)](0.5)$$

$$= \tfrac{1}{2}[1.875 + 2 + 2.125 + 3 + 5.375 + 10] = \tfrac{1}{2}(24.375) = 12.1875$$

(c)

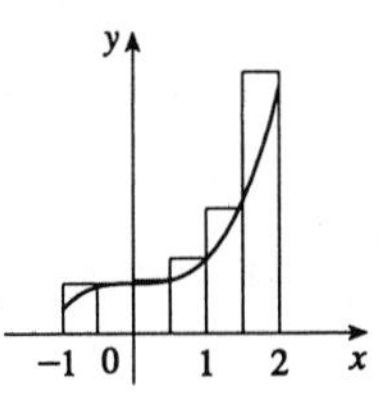

6. **(a)** $\|P\| = \max\{0.5, 0.5, 0.5, 0.5\} = 0.5$

(b)

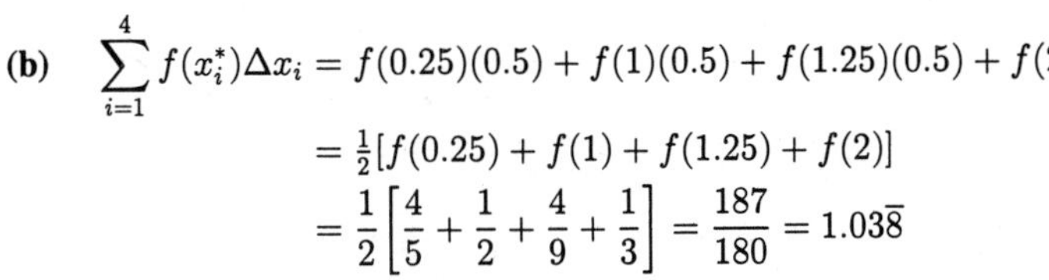

$$\sum_{i=1}^{4} f(x_i^*)\Delta x_i = f(0.25)(0.5) + f(1)(0.5) + f(1.25)(0.5) + f(2)(0.5)$$

$$= \tfrac{1}{2}[f(0.25) + f(1) + f(1.25) + f(2)] = \frac{1}{2}\left[\frac{4}{5} + \frac{1}{2} + \frac{4}{9} + \frac{1}{3}\right] = \frac{187}{180} = 1.03\overline{8}$$

(c)

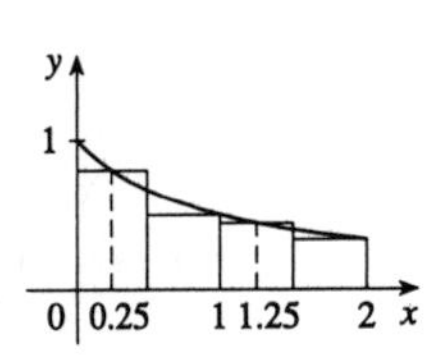

7. **(a)** $\|P\| = \max\{\frac{\pi}{4}, \frac{\pi}{4}, \frac{\pi}{4}, \frac{\pi}{4}\} = \frac{\pi}{4}$

(b) $$\sum_{i=1}^{n} f(x_i^*)\Delta x_i = \sum_{i=1}^{4} f(x_i^*)\tfrac{\pi}{4} = \tfrac{\pi}{4}\left[f\left(\tfrac{\pi}{6}\right) + f\left(\tfrac{\pi}{3}\right) + f\left(\tfrac{2\pi}{3}\right) + f\left(\tfrac{5\pi}{6}\right)\right]$$
$$= \tfrac{\pi}{4}\left[1 + \sqrt{3} + \sqrt{3} + 1\right]$$
$$= \tfrac{\pi}{2}\left(1 + \sqrt{3}\right)$$

(c)

8. **(a)** $\|P\| = \max\{\frac{\pi}{6}, \frac{\pi}{12}, \frac{\pi}{12}, \frac{\pi}{6}\} = \frac{\pi}{6}$

(b) $$\sum_{i=1}^{4} f(x_i^*)\Delta x_i = f(0)\tfrac{\pi}{6} + f\left(\tfrac{\pi}{6}\right)\tfrac{\pi}{12} + f\left(\tfrac{\pi}{4}\right)\tfrac{\pi}{12} + f\left(\tfrac{\pi}{3}\right)\tfrac{\pi}{6}$$
$$= 4 \cdot \tfrac{\pi}{6} + 2\sqrt{3} \cdot \tfrac{\pi}{12} + 2\sqrt{2} \cdot \tfrac{\pi}{12} + 2 \cdot \tfrac{\pi}{6}$$
$$= \tfrac{\pi}{6}\left(6 + \sqrt{3} + \sqrt{2}\right)$$

(c)

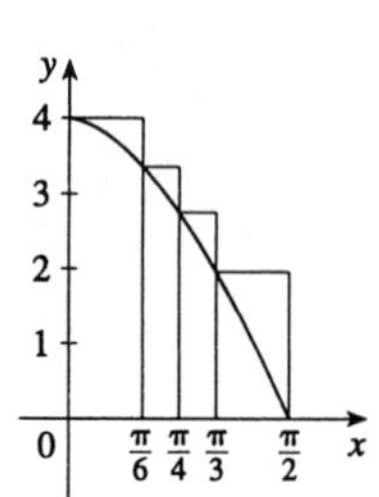

9. **(a)**

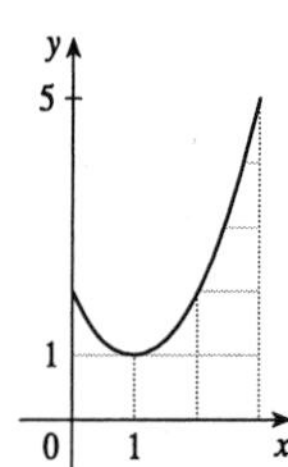

$y = x^2 - 2x + 2 = (x-1)^2 + 1$. By counting squares, we estimate that the area under the curve is between 5 and 7 — perhaps near 6.

(b) $f(x) = y = x^2 - 2x + 2$ on $[0, 3]$ with partition points $x_i = 0 + \dfrac{3i}{n} = \dfrac{3i}{n}$, $\Delta x_i = \dfrac{3}{n}$ and $x_i^* = x_i$, so

$$R_n = \sum_{i=1}^{n} f(x_i^*)\Delta x_i = \frac{3}{n}\sum_{i=1}^{n} f\left(\frac{3i}{n}\right) = \frac{3}{n}\sum_{i=1}^{n}\left[\left(\frac{3i}{n}\right)^2 - 2\left(\frac{3i}{n}\right) + 2\right] = \frac{27}{n^3}\sum_{i=1}^{n} i^2 - \frac{18}{n^2}\sum_{i=1}^{n} i + 6$$
$$= \frac{27}{6}\left[\frac{n(n+1)(2n+1)}{n^3}\right] - \frac{18}{2}\left[\frac{n(n+1)}{n^2}\right] + 6 = \frac{9}{2}\left(2 + \frac{3}{n} + \frac{1}{n^2}\right) - 9\left(1 + \frac{1}{n}\right) + 6$$
$$= 6 + \frac{9}{2n} + \frac{9}{2n^2}.$$

(c) $R_6 = 6 + \dfrac{9}{2 \cdot 6} + \dfrac{9}{2 \cdot 36} = \dfrac{55}{8} = 6.875$, $R_{12} = 6 + \dfrac{9}{2 \cdot 12} + \dfrac{9}{2 \cdot 144} = \dfrac{205}{32} = 6.40625$,

$R_{24} = 6 + \dfrac{9}{2 \cdot 24} + \dfrac{9}{2 \cdot 576} = \dfrac{793}{128} = 6.1953125$

(d) Since $\|P\| \to 0$ as $n \to \infty$, the area is $A = \lim\limits_{n\to\infty} R_n = \lim\limits_{n\to\infty}\left(6 + \dfrac{9}{2n} + \dfrac{9}{2n^2}\right) = 6$.

10. **(a)**

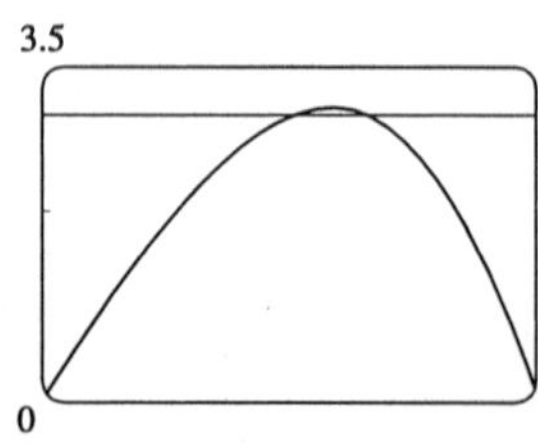

From the graph, it appears that the area under the curve is about two-thirds of the area of a rectangle with height 3 and width 2. So we estimate the area under the curve to be 4.

(b) $f(x) = y = 4x - x^3$ on $[0, 2]$ with partition points $x_i = 0 + \dfrac{2i}{n} = \dfrac{2i}{n}$, $\Delta i = \dfrac{2}{n}$ and $x_i^* = x_i$, so

$$R_n = \sum_{i=1}^{n} f(x_i^*)\Delta x_i = \frac{2}{n}\sum_{i=1}^{n} f\left(\frac{2i}{n}\right) = \frac{2}{n}\sum_{i=1}^{n} 4\frac{2i}{n} - \left(\frac{2i}{n}\right)^3 = \frac{16}{n^2}\sum_{i=1}^{n} i - \frac{16}{n^4}\sum_{i=1}^{n} i^3$$

$$= \frac{16}{n^2}\left[\frac{n(n+1)}{2}\right] - \frac{16}{n^4}\left[\frac{n(n+1)}{2}\right]^2 = 8\left(1+\frac{1}{n}\right) - 4\left(1+\frac{1}{n}\right)^2 = 4 - \frac{4}{n^2}.$$

(c) $R_{10} = 4 - \frac{4}{100} = 3.96$, $R_{20} = 4 - \frac{4}{400} = 3.99$, $R_{30} = 4 - \frac{4}{900} = 3.99\overline{5}$

(d) Since $\|P\| \to 0$ as $n \to \infty$, the area is $A = \lim\limits_{n\to\infty} R_n = \lim\limits_{n\to\infty}\left(4 - \dfrac{4}{n^2}\right) = 4$.

11. $f(x) = x^2 + 1$ on $[0, 2]$ with partition points $x_i = 2i/n$ $(i = 0, 1, 2, \ldots, n)$, so $\Delta x_1 = \Delta x_2 = \cdots = \Delta x_n = \dfrac{2}{n}$.

$\|P\| = \max\{\Delta x_i\} = \dfrac{2}{n}$, so $\|P\| \to 0$ is equivalent to $n \to \infty$. Taking x_i^* to be the midpoint of $[x_{i-1}, x_i] = \left[2\dfrac{i-1}{n}, 2\dfrac{i}{n}\right]$, we get $x_i^* = \dfrac{2i-1}{n}$. Thus

$$A = \lim_{\|P\|\to 0}\sum_{i=1}^{n} f(x_i^*)\Delta x_i = \lim_{n\to\infty}\sum_{i=1}^{n}\left[\left(\frac{2i-1}{n}\right)^2 + 1\right]\frac{2}{n}$$

$$= \lim_{n\to\infty}\sum_{i=1}^{n}\left[\frac{8i^2}{n^3} - \frac{8i}{n^3} + \frac{2}{n^3} + \frac{2}{n}\right]$$

$$= \lim_{n\to\infty}\left[\frac{8}{n^3}\sum_{i=1}^{n} i^2 - \frac{8}{n^3}\sum_{i=1}^{n} i + \left(\frac{2}{n^3} + \frac{2}{n}\right)\sum_{i=1}^{n} 1\right]$$

$$= \lim_{n\to\infty}\left[\frac{8}{n^3}\frac{n(n+1)(2n+1)}{6} - \frac{8}{n^3}\frac{n(n+1)}{2} + \left(\frac{2}{n^3} + \frac{2}{n}\right)n\right]$$

$$= \lim_{n\to\infty}\left[\frac{4}{3}\cdot 1\left(1+\frac{1}{n}\right)\left(2+\frac{1}{n}\right) - \frac{4}{n}\cdot 1\left(1+\frac{1}{n}\right) + \frac{2}{n^2} + 2\right]$$

$$= \left(\tfrac{4}{3}\cdot 1\cdot 1\cdot 2\right) - (0\cdot 1\cdot 1) + 0 + 2 = \tfrac{8}{3} + 2 = \tfrac{14}{3}.$$

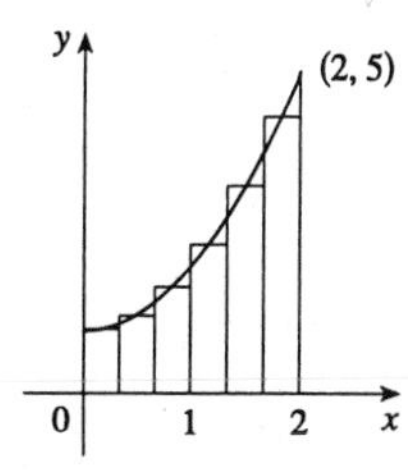

12. $f(x) = y = x^3$ on $[0, 1]$ with partition points $x_i = 0 + \dfrac{1i}{n} = \dfrac{i}{n}$ and $\Delta x_i = \dfrac{1}{n}$.

(a) $x_i^* = x_{i-1}$, so

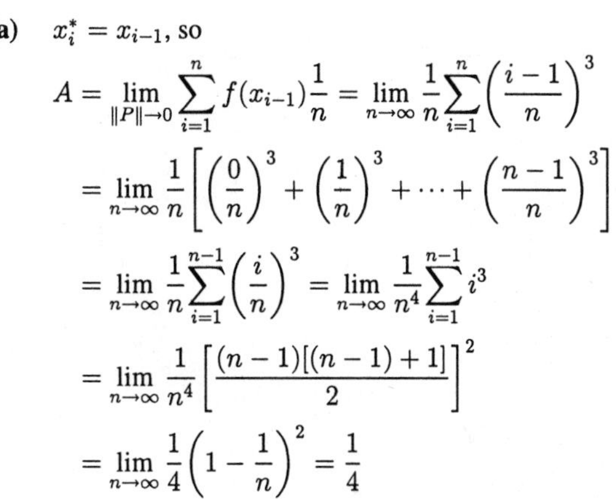

$$A = \lim_{\|P\|\to 0}\sum_{i=1}^{n} f(x_{i-1})\frac{1}{n} = \lim_{n\to\infty}\frac{1}{n}\sum_{i=1}^{n}\left(\frac{i-1}{n}\right)^3$$

$$= \lim_{n\to\infty}\frac{1}{n}\left[\left(\frac{0}{n}\right)^3 + \left(\frac{1}{n}\right)^3 + \cdots + \left(\frac{n-1}{n}\right)^3\right]$$

$$= \lim_{n\to\infty}\frac{1}{n}\sum_{i=1}^{n-1}\left(\frac{i}{n}\right)^3 = \lim_{n\to\infty}\frac{1}{n^4}\sum_{i=1}^{n-1} i^3$$

$$= \lim_{n\to\infty}\frac{1}{n^4}\left[\frac{(n-1)[(n-1)+1]}{2}\right]^2$$

$$= \lim_{n\to\infty}\frac{1}{4}\left(1-\frac{1}{n}\right)^2 = \frac{1}{4}$$

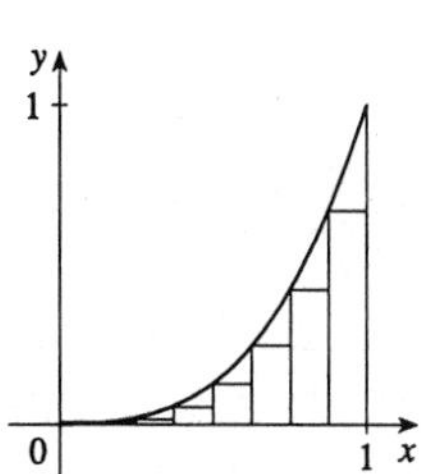

(b) $A = \lim_{\|P\|\to 0} \sum_{i=1}^{n} f(x_i)\frac{1}{n} = \lim_{n\to\infty} \frac{1}{n}\sum_{i=1}^{n}\left(\frac{i}{n}\right)^3$

$= \lim_{n\to\infty} \frac{1}{n^4}\sum_{i=1}^{n} i^3 = \lim_{n\to\infty} \frac{1}{n^4}\left[\frac{n(n+1)}{2}\right]^2$

$= \lim_{n\to\infty} \frac{1}{4}\left(1+\frac{1}{n}\right)^2 = \frac{1}{4}$

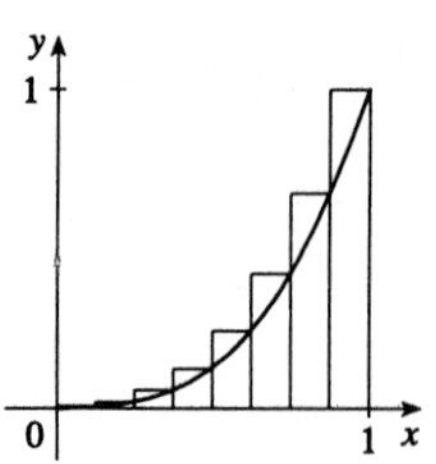

(c) $x_i^* = \dfrac{x_{i-1}+x_i}{2}$, so

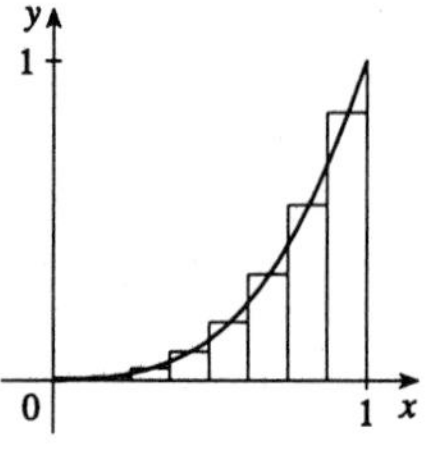

$$A = \lim_{\|P\|\to 0} \sum_{i=1}^{n} f\left(\frac{x_{i-1}+x_i}{2}\right)\frac{1}{n} = \lim_{n\to\infty} \frac{1}{n}\sum_{i=1}^{n}\left[\frac{\frac{i-1}{n}+\frac{i}{n}}{2}\right]^3$$

$$= \lim_{n\to\infty} \frac{1}{n}\sum_{i=1}^{n}\left(\frac{2i-1}{2n}\right)^3 = \lim_{n\to\infty} \frac{1}{8n^4}\sum_{i=1}^{n}(2i-1)^3$$

$$= \lim_{n\to\infty} \frac{1}{8n^4}\sum_{i=1}^{n}(8i^3-12i^2+6i-1)$$

$$= \lim_{n\to\infty} \frac{1}{8n^4}\left[8\left(\frac{n(n+1)}{2}\right)^2 - 12\frac{n(n+1)(2n+1)}{6} + 6\frac{n(n+1)}{2} - 1\right]$$

$$= \lim_{n\to\infty} \left[\frac{1}{4}\left(1+\frac{1}{n}\right)^2 - \frac{1}{4n}\left(1+\frac{1}{n}\right)\left(2+\frac{1}{n}\right) + \frac{3}{8n^2}\left(1+\frac{1}{n}\right) - \frac{1}{8n^4}\right] = \frac{1}{4}$$

13. $f(x) = 2x+1$ on $[0,5]$. $x_i^* = x_i = \dfrac{5i}{n}$ for $i = 1,\ldots,n$ and $\Delta x_i = \dfrac{5}{n}$.

$$A = \lim_{n\to\infty} \sum_{i=1}^{n}\left[2\left(\frac{5i}{n}\right)+1\right]\frac{5}{n} = \lim_{n\to\infty}\left[\frac{50}{n^2}\sum_{i=1}^{n} i + \frac{5}{n}\sum_{i=1}^{n} 1\right]$$

$$= \lim_{n\to\infty}\left[\frac{50}{n^2}\frac{n(n+1)}{2} + \frac{5}{n}n\right] = \lim_{n\to\infty}\left[25\cdot 1\left(1+\frac{1}{n}\right)+5\right]$$

$$= 25+5 = 30$$

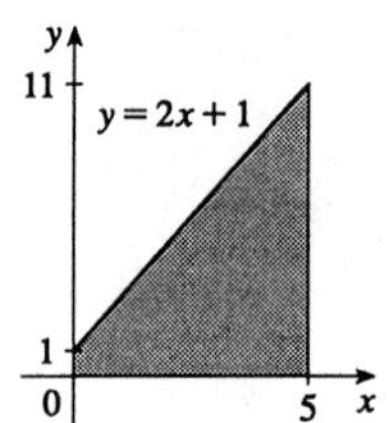

14. $f(x) = x^2+3x-2$ on $[1,4]$. $x_i^* = x_i = 1+\dfrac{3i}{n}$ for $i = 1,2,\ldots,n$. $\Delta x_i = \dfrac{3}{n}$ for all i.

$$A = \lim_{n\to\infty} \sum_{i=1}^{n}\left[\left(1+\frac{3i}{n}\right)^2 + 3\left(1+\frac{3i}{n}\right) - 2\right]\frac{3}{n}$$

$$= \lim_{n\to\infty} \sum_{i=1}^{n}\left[\frac{9i^2}{n^2}+\frac{15i}{n}+2\right]\frac{3}{n} = \lim_{n\to\infty}\left[\frac{27}{n^3}\sum_{i=1}^{n} i^2 + \frac{45}{n^2}\sum_{i=1}^{n} i + \frac{6}{n}\sum_{i=1}^{n} 1\right]$$

$$= \lim_{n\to\infty}\left[\frac{27}{n^3}\frac{n(n+1)(2n+1)}{6} + \frac{45}{n^2}\frac{n(n+1)}{2} + \frac{6}{n}n\right]$$

$$= \lim_{n\to\infty}\left[\frac{9}{2}\cdot 1\left(1+\frac{1}{n}\right)\left(2+\frac{1}{n}\right) + \frac{45}{2}\cdot 1\left(1+\frac{1}{n}\right)+6\right]$$

$$= \left(\tfrac{9}{2}\cdot 1\cdot 1\cdot 2\right) + \left(\tfrac{45}{2}\cdot 1\cdot 1\right) + 6 = 9 + \tfrac{45}{2} + 6 = \tfrac{75}{2} = 37.5$$

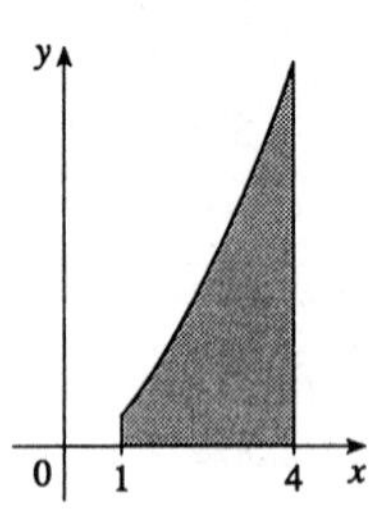

15. $f(x) = 2x^2 - 4x + 5$ on $[-3, 2]$. $x_i^* = x_i = -3 + \frac{5i}{n} \quad (i = 1, 2, \ldots, n)$.

$$A = \lim_{n\to\infty} \sum_{i=1}^{n} \left[2\left(-3 + \frac{5i}{n}\right)^2 - 4\left(-3 + \frac{5i}{n}\right) + 5\right]\frac{5}{n}$$

$$= \lim_{n\to\infty} \sum_{i=1}^{n} \left[\frac{50i^2}{n^2} - \frac{80i}{n} + 35\right]\frac{5}{n}$$

$$= \lim_{n\to\infty} \left[\frac{250}{n^3}\sum_{i=1}^{n} i^2 - \frac{400}{n^2}\sum_{i=1}^{n} i + \frac{175}{n}\sum_{i=1}^{n} 1\right]$$

$$= \lim_{n\to\infty} \left[\frac{250}{n^3}\frac{n(n+1)(2n+1)}{6} - \frac{400}{n^2}\frac{n(n+1)}{2} + \frac{175}{n}n\right]$$

$$= \lim_{n\to\infty} \left[\frac{125}{3} \cdot 1\left(1 + \frac{1}{n}\right)\left(2 + \frac{1}{n}\right) - 200 \cdot 1\left(1 + \frac{1}{n}\right) + 175\right]$$

$$= \left(\tfrac{125}{3} \cdot 1 \cdot 1 \cdot 2\right) - (200 \cdot 1 \cdot 1) + 175 = \tfrac{175}{3}$$

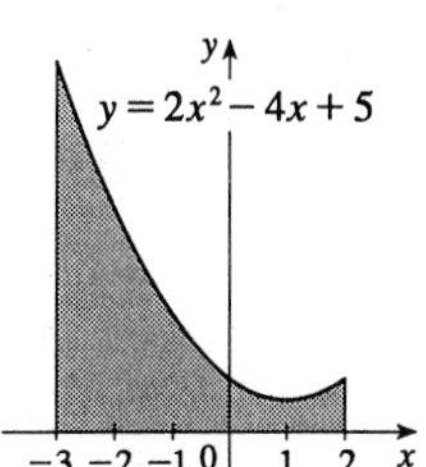

16. $f(x) = x^3 + 2x^2 + x$ on $[0, 1]$. $x_i^* = x_i = \frac{i}{n}$, $\Delta x_i = \frac{1}{n}$.

$$A = \lim_{n\to\infty} \sum_{i=1}^{n} \left[\frac{i^3}{n^3} + 2\frac{i^2}{n^2} + \frac{i}{n}\right]\frac{1}{n} = \lim_{n\to\infty} \left[\frac{1}{n^4}\sum_{i=1}^{n} i^3 + \frac{2}{n^3}\sum_{i=1}^{n} i^2 + \frac{1}{n^2}\sum_{i=1}^{n} i\right]$$

$$= \lim_{n\to\infty} \left[\frac{1}{n^4}\frac{n^2(n+1)^2}{4} + \frac{2}{n^3}\frac{n(n+1)(2n+1)}{6} + \frac{1}{n^2}\frac{n(n+1)}{2}\right]$$

$$= \lim_{n\to\infty} \left[\frac{1}{4}\left(1 + \frac{1}{n}\right)^2 + \frac{1}{3}\left(1 + \frac{1}{n}\right)\left(2 + \frac{1}{n}\right) + \frac{1}{2}\left(1 + \frac{1}{n}\right)\right]$$

$$= \left(\tfrac{1}{4} \cdot 1\right) + \left(\tfrac{1}{3} \cdot 1 \cdot 2\right) + \left(\tfrac{1}{2} \cdot 1\right) = \tfrac{17}{12}$$

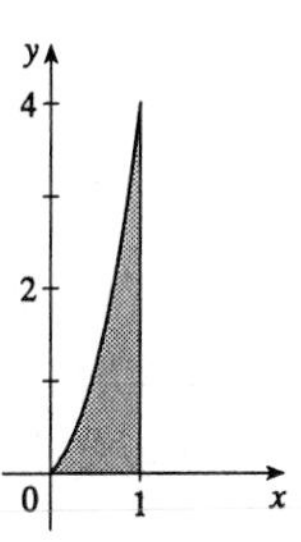

17. $f(x) = x^3 + 2x$ on $[0, 2]$. $x_i^* = x_i = \frac{2i}{n}$ for $i = 1, 2, \ldots, n$.

$$A = \lim_{n\to\infty} \sum_{i=1}^{n} \left[\left(\frac{2i}{n}\right)^3 + \frac{4i}{n}\right]\frac{2}{n} = \lim_{n\to\infty} \left[\frac{16}{n^4}\sum_{i=1}^{n} i^3 + \frac{8}{n^2}\sum_{i=1}^{n} i\right]$$

$$= \lim_{n\to\infty} \left[\frac{16}{n^4}\frac{n^2(n+1)^2}{4} + \frac{8}{n^2}\frac{n(n+1)}{2}\right]$$

$$= \lim_{n\to\infty} \left[4\left(1 + \frac{1}{n}\right)^2 + 4\left(1 + \frac{1}{n}\right)\right] = 4 \cdot 1 + 4 \cdot 1 = 8$$

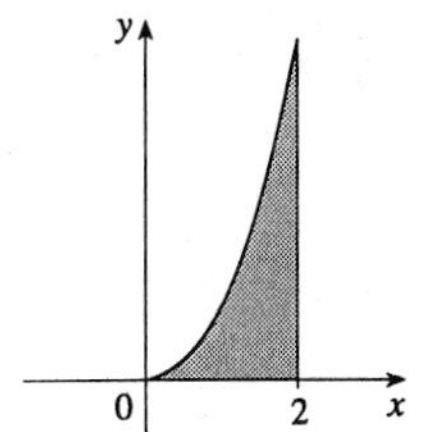

18. $f(x) = x^4 + 3x + 2$ on $[0, 3]$, $x_i^* = x_i = \frac{3i}{n}$, $\Delta x_i = \frac{3}{n}$.

$$A = \lim_{n\to\infty} \sum_{i=1}^{n} \left[\left(\frac{3i}{n}\right)^4 + 3\left(\frac{3i}{n}\right) + 2\right]\frac{3}{n} = \lim_{n\to\infty} \left[\frac{243}{n^5}\sum_{i=1}^{n} i^4 + \frac{27}{n^2}\sum_{i=1}^{n} i + \frac{6}{n}\sum_{i=1}^{n} 1\right]$$

$$= \lim_{n\to\infty} \left[\frac{243}{n^5}\frac{n(n+1)(2n+1)(3n^2+3n-1)}{30} + \frac{27}{n^2}\frac{n(n+1)}{2} + \frac{6}{n}n\right]$$

$$= \lim_{n\to\infty} \left[\frac{81}{10} \cdot 1\left(1 + \frac{1}{n}\right)\left(2 + \frac{1}{n}\right)\left(3 + \frac{1}{n} - \frac{1}{n^2}\right) + \frac{27}{2} \cdot 1\left(1 + \frac{1}{n}\right) + 6\right]$$

$$= \left(\tfrac{81}{10} \cdot 1^2 \cdot 2 \cdot 3\right) + \left(\tfrac{27}{2} \cdot 1^2\right) + 6 = \tfrac{243}{5} + \tfrac{27}{2} + 6 = 68.1.$$

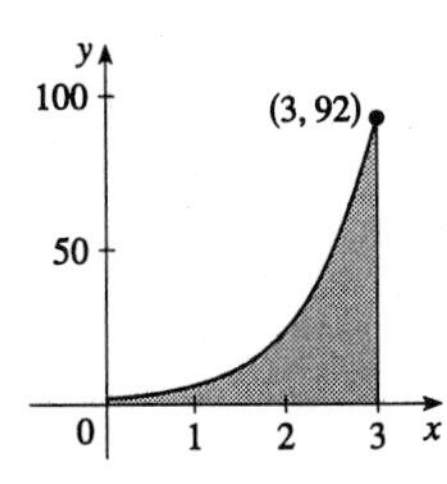

19. Here is one possible algorithm (ordered sequence of operations) for calculating the sums:

1 Let SUM = 0, let X-MIN = 0, let X-MAX = π, let STEP-SIZE = $\frac{\pi}{10}$ (or $\frac{\pi}{30}$ or $\frac{\pi}{50}$, depending on which sum we are calculating), and let RIGHT-ENDPOINT = X-MIN + STEP-SIZE.

2 Repeat steps 2a, 2b in sequence until RIGHT-ENDPOINT > X-MAX.

2a Add sin(RIGHT-ENDPOINT) to SUM.

2b Add STEP-SIZE to RIGHT-ENDPOINT.

At the end of this procedure, the variable SUM is equal to the answer we are looking for. We find that $R_{10} = \frac{\pi}{10}\sum_{i=1}^{10}\sin\left(\frac{i\pi}{10}\right) \approx 1.9835$, $R_{30} = \frac{\pi}{30}\sum_{i=1}^{30}\sin\left(\frac{i\pi}{30}\right) \approx 1.9982$, and $R_{50} = \frac{\pi}{50}\sum_{i=1}^{50}\sin\left(\frac{i\pi}{50}\right) \approx 1.9993$.

It appears that the exact area is 2.

20. See Exercise 19 for a possible algorithm (in which the endpoints and the function must be changed) for calculating the sums. We find that $R_{10} = \frac{1}{10}\sum_{i=1}^{10}\frac{1}{(1+i/10)^2} \approx 0.4640$, $R_{30} = \frac{1}{30}\sum_{i=1}^{30}\frac{1}{(1+i/30)^2} \approx 0.4877$, and $R_{50} = \frac{1}{50}\sum_{i=1}^{50}\frac{1}{(1+i/50)^2} \approx 0.4926$. It appears that the exact area is $\frac{1}{2}$.

21. In Maple, we have to perform a number of steps before getting a numerical answer. After loading the `student` package [command: `with(student);`] we use the command `leftsum(x^(1/2),x=1..4,10` [*or* `30`, *or* `50`]`);` which gives us the expression in summation notation. To get a numerical approximation to the sum, we use `evalf(");`.

Mathematica does not have a special command for these sums, so we must type them in manually. For example, the first left sum is given by `(3/10)*Sum[Sqrt[1+3(i-1)/10],{i,1,10}]`, and we use the `N` command on the resulting output to get a numerical approximation.

In Derive, we use the `LEFT_RIEMANN` command to get the left sums, but must define the right sums ourselves.

(a) The left sums are $L_{10} = \frac{3}{10}\sum_{i=1}^{10}\sqrt{1+\frac{3(i-1)}{10}} \approx 4.5148$, $L_{30} = \frac{3}{30}\sum_{i=1}^{30}\sqrt{1+\frac{3(i-1)}{30}} \approx 4.6165$, $L_{50} = \frac{3}{50}\sum_{i=1}^{50}\sqrt{1+\frac{3(i-1)}{50}} \approx 4.6366$. The right sums are $R_{10} = \frac{3}{10}\sum_{i=1}^{10}\sqrt{1+\frac{3i}{10}} \approx 4.8148$, $R_{30} = \frac{3}{30}\sum_{i=1}^{30}\sqrt{1+\frac{3i}{30}} \approx 4.7165$, $R_{50} = \frac{3}{50}\sum_{i=1}^{50}\sqrt{1+\frac{3i}{50}} \approx 4.6966$.

(b) In Maple, we use the `leftbox` and `rightbox` commands (with the same arguments as `leftsum` and `rightsum` above) to generate the graphs.

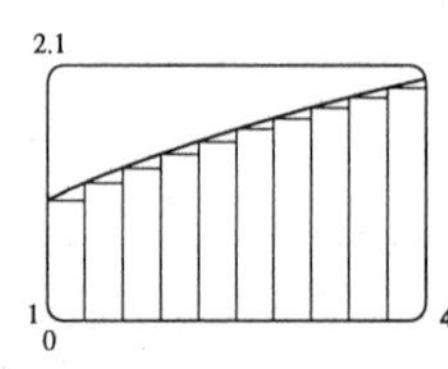

left endpoints, $n = 10$

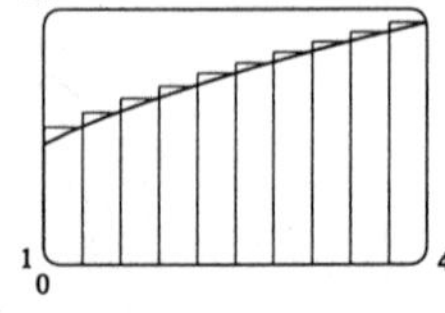

right endpoints, $n = 10$

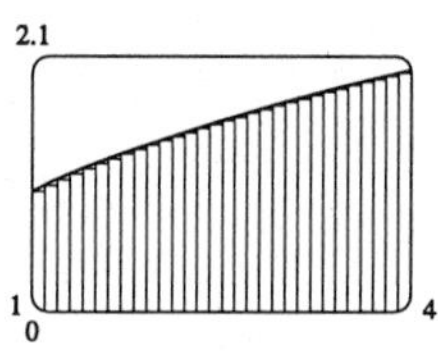

left endpoints, $n = 30$

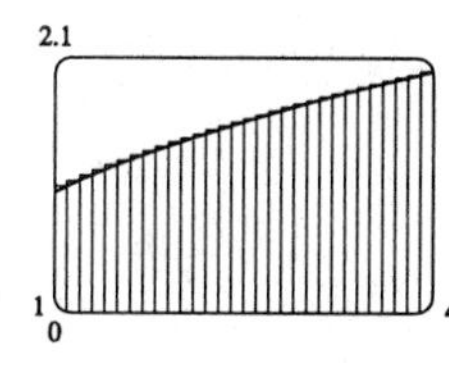

right endpoints, $n = 30$

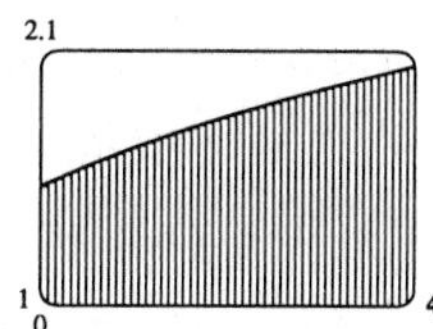

left endpoints, $n = 50$

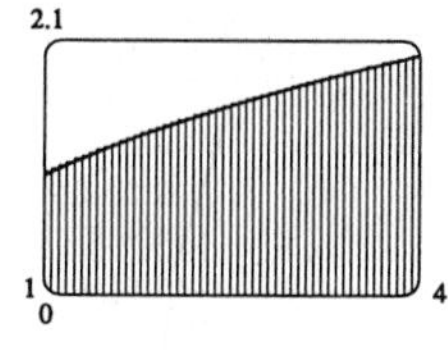

right endpoints, $n = 50$

(c) We know that since $\sqrt{x}$ is an increasing function on $[1, 4]$, all of the left sums are smaller than the actual area, and all of the right sums are larger than the actual area (see Example 3). Since the left sum with $n = 50$ is about $4.637 > 4.6$ and the right sum with $n = 50$ is about $4.697 < 4.7$, we conclude that $4.6 < L_{50} <$ actual area $< R_{50} < 4.7$, so the actual area is between 4.6 and 4.7.

22. See the solution to Exercise 21 for the CAS commands for evaluating the sums.

(a) The left sums are $L_{10} = \frac{1}{10}\left(\frac{\pi}{2}\right)\sum_{i=1}^{10} \sin\left[\sin\left(\frac{i-1}{10} \cdot \frac{\pi}{2}\right)\right] \approx 0.8251$,

$L_{30} = \frac{1}{30}\left(\frac{\pi}{2}\right)\sum_{i=1}^{30} \sin\left[\sin\left(\frac{i-1}{30} \cdot \frac{\pi}{2}\right)\right] \approx 0.8710$, $L_{50} = \frac{1}{50}\left(\frac{\pi}{2}\right)\sum_{i=1}^{50} \sin\left[\sin\left(\frac{i-1}{50} \cdot \frac{\pi}{2}\right)\right] \approx 0.8799$.

The right sums are $R_{10} = \frac{1}{10}\left(\frac{\pi}{2}\right)\sum_{i=1}^{10} \sin\left[\sin\left(\frac{i}{10} \cdot \frac{\pi}{2}\right)\right] \approx 0.9573$,

$R_{30} = \frac{1}{30}\left(\frac{\pi}{2}\right)\sum_{i=1}^{30} \sin\left[\sin\left(\frac{i}{30} \cdot \frac{\pi}{2}\right)\right] \approx 0.9150$, $R_{50} = \frac{1}{50}\left(\frac{\pi}{2}\right)\sum_{i=1}^{50} \sin\left[\sin\left(\frac{i}{50} \cdot \frac{\pi}{2}\right)\right] \approx 0.9064$.

(b) In Maple, we use the `leftbox` and `rightbox` commands (with the same arguments as `leftsum` and `rightsum` above) to generate the graphs.

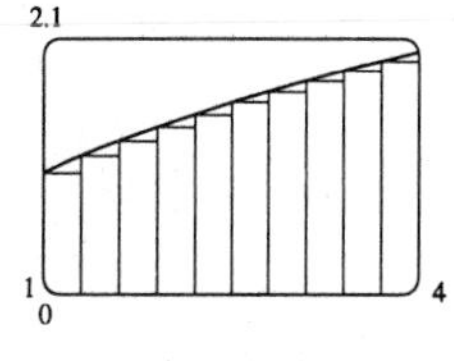

left endpoints, $n = 10$

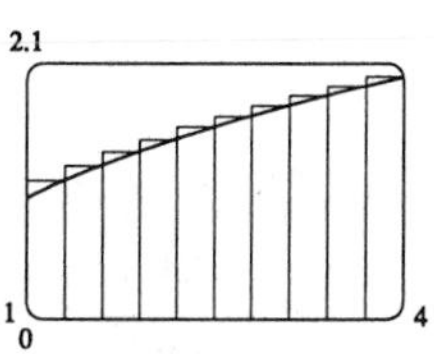

right endpoints, $n = 10$

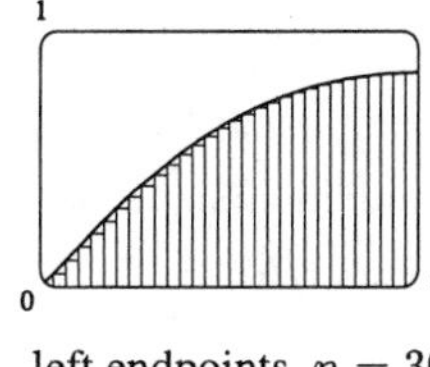

left endpoints, $n = 30$

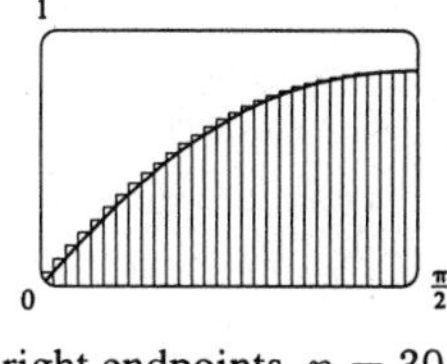

right endpoints, $n = 30$

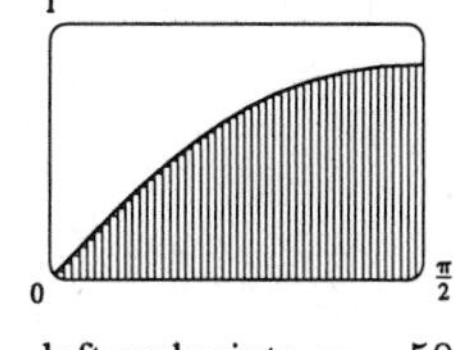

left endpoints, $n = 50$

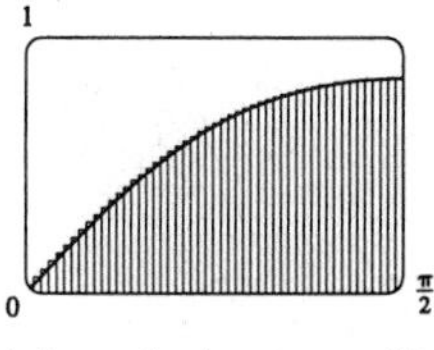

right endpoints, $n = 50$

(c) We know (see Example 3) that since $\sin(\sin x)$ is an increasing function on $\left[0, \frac{\pi}{2}\right]$ [and this is true because its derivative, $-\cos(\sin x)(-\cos x)$, is positive on that interval], all of the left sums are smaller than the actual area, and all of the right sums are larger than the actual area. Since the left sum with $n = 50$ is about $0.8799 > 0.87$ and the right sum with $n = 50$ is about $0.9064 < 0.91$, we conclude that $0.87 < L_{50} <$ actual area $< R_{50} < 0.91$, so the actual area is between 0.87 and 0.91.

23. $\lim_{n\to\infty}\sum_{i=1}^{n}\frac{\pi}{4n}\tan\frac{\pi i}{4n}$ can be interpreted as the area of the region lying under the graph of $y=\tan x$ on the interval $\left[0,\frac{\pi}{4}\right]$, since for $y=\tan x$ on $\left[0,\frac{\pi}{4}\right]$ with partition points $x_i=\left(\frac{\pi}{4}\right)\frac{i}{n}$, $\Delta x=\frac{\pi}{4n}$ and $x_i^*=x_i$, the expression for the area is $A=\lim_{n\to\infty}\sum_{i=1}^{n}f(x_i^*)\Delta x=\lim_{n\to\infty}\sum_{i=1}^{n}\tan\left(\frac{\pi i}{4n}\right)\frac{\pi}{4n}$. Note that this answer is not unique, since the expression for the area is the same for the function $y=\tan(k\pi+x)$ on the interval $\left[k\pi,k\pi+\frac{\pi}{4}\right]$ where k is any integer.

24. The two most obvious ways to interpret $\lim_{n\to\infty}\sum_{i=1}^{n}\frac{3}{n}\sqrt{1+\frac{3i}{n}}$ are: as the area of the region under the graph of $\sqrt{x}$ on the interval $[1,4]$ [see Exercise 21 (a)], or as the area of the region lying under the graph of $\sqrt{x+1}$ on the interval $[0,3]$, since for $y=\sqrt{x+1}$ on $[0,3]$ with partition points $x_i=\frac{i}{3n}$, $\Delta x=\frac{1}{3n}$ and $x_i^*=x_i$, the expression for the area is $A=\lim_{n\to\infty}\sum_{i=1}^{n}f(x_i^*)\Delta x=\lim_{n\to\infty}\sum_{i=1}^{n}\sqrt{1+\frac{3i}{n}}\left(\frac{3}{n}\right)$.

25. $f(x)=\sin x$. $\Delta x_i=\frac{\pi}{n}$ and $x_i^*=x_i=\frac{i\pi}{n}$. So

$$A=\lim_{n\to\infty}\sum_{i=1}^{n}\left[\sin\frac{i\pi}{n}\left(\frac{\pi}{n}\right)\right]=\lim_{n\to\infty}\frac{\pi}{n}\sum_{i=1}^{n}\sin\frac{i\pi}{n}=\lim_{n\to\infty}\frac{\pi}{n}\frac{\sin\left(\frac{n}{2}\cdot\frac{\pi}{n}\right)\sin\left(\frac{n+1}{2}\cdot\frac{\pi}{n}\right)}{\sin\left(\frac{1}{2}\frac{\pi}{n}\right)}\quad\text{(see Exercise 4.1.53)}$$

$$=\lim_{n\to\infty}\frac{\pi}{n}\frac{\sin\left(\frac{\pi}{2}+\frac{\pi}{2n}\right)}{\sin\left(\frac{\pi}{2n}\right)}=\lim_{n\to\infty}2\frac{\frac{\pi}{2n}}{\sin\left(\frac{\pi}{2n}\right)}\cdot\lim_{n\to\infty}\cos\left(\frac{\pi}{2n}\right)=2\cdot1\cdot1=2.$$

Here we have used the identity $\sin\left(\frac{\pi}{2}+x\right)=\cos x$.

26. **(a)**

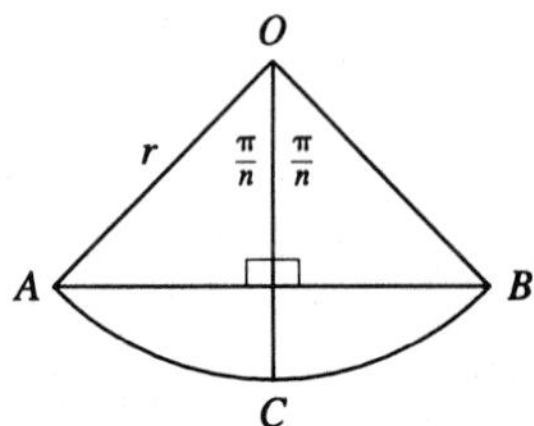

The diagram shows one of the n congruent triangles. O is the center of the circle and AB is one of the sides of the polygon. $\angle AOB=2\pi/n$. Radius OC is drawn so as to bisect the central angle, $\angle AOB$. It follows that OC intersects AB at right angles and bisects AB. Thus $\triangle OAB$ is divided into two right triangles with legs of length $r\sin(\pi/n)$ and $r\cos(\pi/n)$.

$\triangle OAB$ has area $2\cdot\frac{1}{2}[r\sin(\pi/n)][r\cos(\pi/n)]=r^2\sin(\pi/n)\cos(\pi/n)=\frac{1}{2}r^2\sin(2\pi/n)$, so $A_n=n\cdot\text{area}(\triangle OAB)=\frac{1}{2}nr^2\sin(2\pi/n)$.

(b) Using Theorem 2.4.4, $\lim_{n\to\infty}A_n=\lim_{n\to\infty}\frac{1}{2}nr^2\sin(2\pi/n)=\lim_{n\to\infty}\frac{\sin(2\pi/n)}{2\pi/n}\pi r^2=\pi r^2$.

EXERCISES 4.3

1. $f(x)=7-2x$ **(a)** $\|P\|=\max\{0.6,0.6,0.8,1.2,0.8\}=1.2$

(b) $$\sum_{i=1}^{5} f(x_i^*)\Delta x_i = f(1.3)(0.6)+f(1.9)(0.6)+f(2.6)(0.8)+f(3.6)(1.2)+f(4.6)(0.8)$$
$$=(4.4)(0.6)+(3.2)(0.6)+(1.8)(0.8)+(-0.2)(1.2)+(-2.2)(0.8)=4.$$

2. $f(x)=3x-1$ **(a)** $\|P\|=\max\{0.8,0.6,0.6,0.8,0.8,0.4\}=0.8$

(b) $$\sum_{i=1}^{6} f(x_i^*)\Delta x_i = f(-1.6)(0.8)+f(-0.9)(0.6)+f(-0.3)(0.6)+f(0.4)(0.8)$$
$$+f(1.2)(0.8)+f(1.8)(0.4)$$
$$=(-5.8)(0.8)+(-3.7)(0.6)+(-1.9)(0.6)+(0.2)(0.8)+(2.6)(0.8)+(4.4)(0.4)$$
$$=-4$$

3. $f(x)=2-x^2$ **(a)** $\|P\|=\max\{0.6,0.4,1,0.8,0.6,0.6\}=1$

(b) $$\sum_{i=1}^{6} f(x_i^*)\Delta x_i = f(-1.4)(0.6)+f(-1)(0.4)+f(0)(1)+f(0.8)(0.8)+f(1.4)(0.6)+f(2)(0.6)$$
$$=(0.04)(0.6)+(1)(0.4)+(2)(1)+(1.36)(0.8)+(0.04)(0.6)+(-2)(0.6)=2.336$$

4. $f(x)=x+x^2$ **(a)** $\|P\|=\max\{0.5,0.5,0.3,0.3,0.4\}=0.5$

(b) $$\sum_{i=1}^{5} f(x_i^*)\Delta x_i = f(-2)(0.5)+f(-1.5)(0.5)+f(-1)(0.3)+f(-0.7)(0.3)+f(-0.4)(0.4)$$
$$=2(0.5)+(0.75)(0.5)+(0)(0.3)+(-0.21)(0.3)+(-0.24)(0.4)=1.216$$

5. $f(x)=x^3$ **(a)** $\|P\|=\max\{0.5,0.5,0.5,0.5\}=0.5$

(b) $$\sum_{i=1}^{n} f(x_i^*)\Delta x_i = \frac{1}{2}\sum_{i=1}^{4} f(x_i^*) = \tfrac{1}{2}\left[(-1)^3+(-0.4)^3+(0.2)^3+1^3\right]=-0.028$$

6. $f(x)=\sin x$ **(a)** $\|P\|=\max\left\{\frac{\pi}{2}-1,1,1,1,\pi-2\right\}=\pi-2\approx 1.14$

(b) $$\sum_{i=1}^{5} f(x_i^*)\Delta x_i = \sin(-1.5)\left(\tfrac{\pi}{2}-1\right)+\sin(-0.5)1+\sin(0.5)1+\sin(1.5)1+(\sin 3)(\pi-2)\approx 0.589$$

7. **(a)** Using the right endpoints, we calculate
$$\int_0^8 f(x)\,dx \approx \sum_{i=1}^{4} f(x_i)\Delta x_i = 2[f(2)+f(4)+f(6)+f(8)]=2(1+2-2+1)=4$$

(b) Using the left endpoints, we calculate
$$\int_0^8 f(x)\,dx \approx \sum_{i=1}^{4} f(x_{i-1})\Delta x_i = 2[f(0)+f(2)+f(4)+f(6)]=2(2+1+2-2)=6$$

(c) Using the midpoint of each interval, we calculate
$$\int_0^8 f(x)\,dx \approx \sum_{i=1}^{4} f\left(\frac{x_i+x_{i-1}}{2}\right)\Delta x_i = 2[f(1)+f(3)+f(5)+f(7)]=2(3+2+1-1)=10$$

8. **(a)** Using the right endpoints,

$\int_0^6 f(x)\,dx \approx \sum_{i=1}^{3} f(x_i)\Delta x_i = 2[f(2)+f(4)+f(6)] = 2(8.3+2.3-10.5) = 0.2$

(b) Using the left endpoints,

$\int_0^6 f(x)\,dx \approx \sum_{i=1}^{3} f(x_{i-1})\Delta x_i = 2[f(0)+f(2)+f(4)] = 2(9.3+8.3+2.3) = 39.8$

(c) Using the midpoint of each interval, we calculate

$\int_0^6 f(x)\,dx \approx \sum_{i=1}^{3} f\left(\frac{1}{2}(x_i+x_{i-1})\right)\Delta x_i = 2[f(1)+f(3)+f(5)] = 2(9+6.5-7.6) = 15.8.$

The estimate using the right endpoints must be less than $\int_0^6 f(x)dx$, since if we take x_i^* to be the right endpoint x_i of each interval, then $f(x_i) \le f(x)$ for all x on $[x_{i-1}, x_i]$, which implies that $f(x_i)\Delta x_i \le \int_{x_{i-1}}^{x_i} f(x)dx$, and so the sum $\sum_{i=1}^{3}[f(x_i)\Delta x_i] \le \sum_{i=1}^{3}\left[\int_{x_{i-1}}^{x_i} f(x)dx\right] = \int_0^6 f(x)\,dx$. Similarly, if we take x_i^* to be the left endpoint x_{i-1} of each interval then $f(x_{i-1}) \ge f(x)$ for all x on $[x_{i-1}, x_i]$, and so $\sum_{i=1}^{3} f(x_{i-1})\Delta x_i \ge \int_0^6 f(x)dx$. We cannot say anything about the midpoint estimate.

9. The width of the intervals is $\Delta x = (5-0)/5 = 1$ so the partition points are $0, 1, 2, 3, 4, 5$ and the midpoints are $0.5, 1.5, 2.5, 3.5, 4.5$. The Midpoint Rule gives

$\int_0^5 x^3\,dx \approx \sum_{i=1}^{5} f(\overline{x}_i)\Delta x = (0.5)^3 + (1.5)^3 + (2.5)^3 + (3.5)^3 + (4.5)^3 = 153.125.$

10. The width of the interval $\Delta x = (3-1)/4 = 0.5$ so the partition points are $1.0, 1.5, 2.0, 2.5, 3.0$ and the midpoints are $1.25, 1.75, 2.25, 2.75$.

$$\int_1^3 \frac{1}{2x-7}\,dx \approx \sum_{i=1}^{4} f(\overline{x}_i)\Delta x = 0.5\left[\frac{1}{2(1.25)-7} + \frac{1}{2(1.75)-7} + \frac{1}{2(2.25)-7} + \frac{1}{2(2.75)-7}\right] \approx -0.7873$$

11. $\Delta x = (2-1)/10 = 0.1$ so the partition points are $1.0, 1.1, \ldots, 2.0$ and the midpoints are $1.05, 1.15, \ldots, 1.95$.

$$\int_1^2 \sqrt{1+x^2}\,dx \approx \sum_{i=1}^{10} f(\overline{x}_i)\Delta x = 0.1\left[\sqrt{1+(1.05)^2} + \sqrt{1+(1.15)^2} + \cdots + \sqrt{1+(1.95)^2}\right] \approx 1.8100$$

12. $\Delta x = \frac{1}{4}\left(\frac{\pi}{4} - 0\right) = \frac{\pi}{16}$ so the partition points are $0, \frac{\pi}{16}, \frac{2\pi}{16}, \frac{3\pi}{16}, \frac{4\pi}{16}$ and the midpoints are $\frac{\pi}{32}, \frac{3\pi}{32}, \frac{5\pi}{32}, \frac{7\pi}{32}$. The Midpoint Rule gives $\int_0^{\pi/4} \tan x\,dx \approx \left(\frac{\pi}{16}\right)\left(\tan\frac{\pi}{32} + \tan\frac{3\pi}{32} + \tan\frac{5\pi}{32} + \tan\frac{7\pi}{32}\right) \approx 0.3450$

13. In Maple, we use the command `with(student);` to load the `sum` and `box` commands, then `m:=middlesum(sqrt(1+x^2),x=1..2,10);` which gives us the sum in summation notation, then `M:=evalf(m);` which gives $M_{10} \approx 1.81001414$, confirming the result of Exercise 11. The command `middlebox(sqrt(1+x^2),x=1..2,10);` generates the graph. Repeating for $n = 20$ and $n = 30$ gives $M_{20} \approx 1.81007263$ and $M_{30} \approx 1.81008347$.

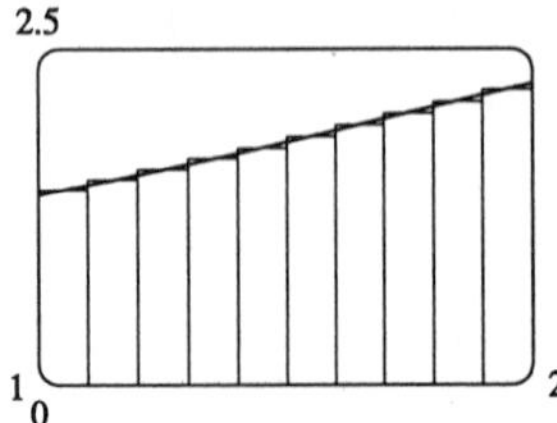

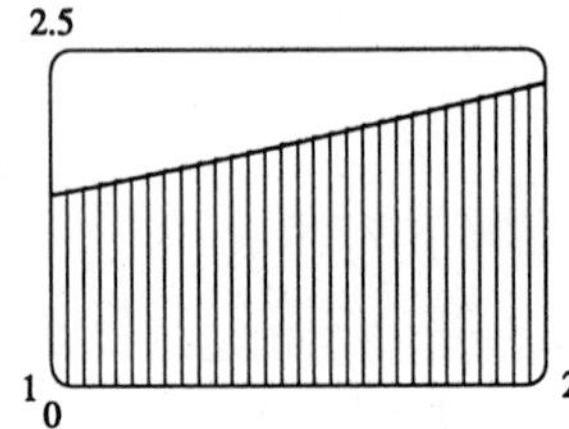

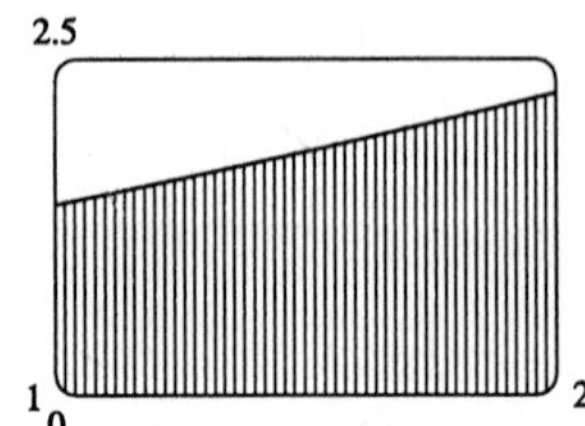

14. See the solution to Exercise 4.2.19 for a possible algorithm to calculate the sums. We calculate that the left Riemann sum is $L_{100} = \sum_{i=1}^{100} \sqrt{1 + (x_{i-1})^2}\,\frac{1}{100} \approx 1.80598$, and the right Riemann sum is $R_{100} = \sum_{i=1}^{100} \sqrt{1 + (x_{i-1})^2}\,\Delta x_i \approx 1.81420$, so since $\sqrt{1+x^2}$ is an increasing function, we must have $L_{100} \le \int_1^2 \sqrt{1+x^2}\,dx \le R_{100}$, so $1.805 < L_{100} \le \int_1^2 \sqrt{1+x^2}\,dx \le R_{100} < 1.815$. Therefore the approximate value 1.8100 in Exercise 11 must be accurate to two decimal places.

15. $\int_a^b c\,dx = \lim_{n\to\infty} \frac{b-a}{n} \sum_{i=1}^{n} c = \lim_{n\to\infty} \frac{b-a}{n} nc$ [By Theorem 4.1.3] $= \lim_{n\to\infty} (b-a)c = (b-a)c$

16. $$\int_{-2}^{7} (6-2x)\,dx = \lim_{n\to\infty} \frac{9}{n} \sum_{i=1}^{n} \left[6 - 2\left(-2 + \frac{9i}{n}\right)\right] = \lim_{n\to\infty} \frac{9}{n} \sum_{i=1}^{n} \left[10 - \frac{18i}{n}\right] = \lim_{n\to\infty} \left[\frac{90}{n} \sum_{i=1}^{n} 1 - \frac{162}{n^2} \sum_{i=1}^{n} i\right]$$
$$= \lim_{n\to\infty} \left[\frac{90}{n} n - \frac{162}{n^2} \frac{n(n+1)}{2}\right] = \lim_{n\to\infty} \left[90 - 81 \cdot 1\left(1 + \frac{1}{n}\right)\right] = 90 - 81 = 9$$

17. $$\int_1^4 (x^2 - 2)\,dx = \lim_{n\to\infty} \frac{3}{n} \sum_{i=1}^{n} \left[\left(1 + \frac{3i}{n}\right)^2 - 2\right] = \lim_{n\to\infty} \frac{3}{n} \sum_{i=1}^{n} \left[\frac{9i^2}{n^2} + \frac{6i}{n} - 1\right]$$
$$= \lim_{n\to\infty} \left[\frac{27}{n^3} \sum_{i=1}^{n} i^2 + \frac{18}{n^2} \sum_{i=1}^{n} i - \frac{3}{n} \sum_{i=1}^{n} 1\right] = \lim_{n\to\infty} \left[\frac{27}{n^3} \frac{n(n+1)(2n+1)}{6} + \frac{18}{n^2} \frac{n(n+1)}{2} - \frac{3}{n} n\right]$$
$$= \lim_{n\to\infty} \left[\frac{9}{2} \cdot 1\left(1 + \frac{1}{n}\right)\left(2 + \frac{1}{n}\right) + 9 \cdot 1\left(1 + \frac{1}{n}\right) - 3\right] = \left(\tfrac{9}{2} \cdot 2\right) + 9 - 3 = 15$$

18. $$\int_1^5 (2 + 3x - x^2)\,dx = \lim_{n\to\infty} \frac{4}{n} \sum_{i=1}^{n} \left[2 + 3\left(1 + \frac{4i}{n}\right) - \left(1 + \frac{4i}{n}\right)^2\right] = \lim_{n\to\infty} \frac{4}{n} \sum_{i=1}^{n} \left[-\frac{16i^2}{n^2} + \frac{4i}{n} + 4\right]$$
$$= \lim_{n\to\infty} \left[-\frac{64}{n^3} \sum_{i=1}^{n} i^2 + \frac{16}{n^2} \sum_{i=1}^{n} i + \frac{16}{n} \sum_{i=1}^{n} 1\right]$$
$$= \lim_{n\to\infty} \left[-\frac{64}{n^3} \frac{n(n+1)(2n+1)}{6} + \frac{16}{n^2} \frac{n(n+1)}{2} + \frac{16}{n} n\right]$$
$$= \lim_{n\to\infty} \left[-\frac{32}{3} \cdot 1\left(1 + \frac{1}{n}\right)\left(2 + \frac{1}{n}\right) + 8 \cdot 1\left(1 + \frac{1}{n}\right) + 16\right] = -\frac{64}{3} + 8 + 16 = \frac{8}{3}$$

19. $$\int_0^b (x^3 + 4x)\,dx = \lim_{n\to\infty} \frac{b}{n} \sum_{i=1}^{n} \left[\left(\frac{bi}{n}\right)^3 + 4\left(\frac{bi}{n}\right)\right] = \lim_{n\to\infty} \left[\frac{b^4}{n^4} \sum_{i=1}^{n} i^3 + 4\frac{b^2}{n^2} \sum_{i=1}^{n} i\right]$$
$$= \lim_{n\to\infty} \left[\frac{b^4}{n^4} \frac{n^2(n+1)^2}{4} + \frac{4b^2}{n^2} \frac{n(n+1)}{2}\right] = \lim_{n\to\infty} \left[\frac{b^4}{4} \cdot 1^2\left(1 + \frac{1}{n}\right)^2 + 2b^2 \cdot 1\left(1 + \frac{1}{n}\right)\right]$$
$$= \frac{b^4}{4} + 2b^2$$

20. $$\int_0^1 (x^3 - 5x^4)\,dx = \lim_{n\to\infty} \frac{1}{n} \sum_{i=1}^{n} \left[\left(\frac{i}{n}\right)^3 - 5\left(\frac{i}{n}\right)^4\right] = \lim_{n\to\infty} \left[\frac{1}{n^4} \sum_{i=1}^{n} i^3 - \frac{5}{n^5} \sum_{i=1}^{n} i^4\right]$$
$$= \lim_{n\to\infty} \left[\frac{n^2(n+1)^2}{4n^4} - \frac{5n(n+1)(2n+1)(3n^2+3n-1)}{30n^5}\right]$$
$$= \lim_{n\to\infty} \left[\frac{1}{4} \cdot 1^2\left(1 + \frac{1}{n}\right)^2 - \frac{1}{6} \cdot 1\left(1 + \frac{1}{n}\right)\left(2 + \frac{1}{n}\right)\left(3 + \frac{3}{n} - \frac{1}{n^2}\right)\right] = \frac{1}{4} - 1 = -\frac{3}{4}$$

21. $\int_a^b x\,dx = \lim_{n\to\infty} \frac{b-a}{n}\sum_{i=1}^{n}\left[a + \frac{b-a}{n}i\right] = \lim_{n\to\infty}\left[\frac{a(b-a)}{n}\sum_{i=1}^{n}1 + \frac{(b-a)^2}{n^2}\sum_{i=1}^{n}i\right]$

$= \lim_{n\to\infty}\left[\frac{a(b-a)}{n}n + \frac{(b-a)^2}{n^2}\cdot\frac{n(n+1)}{2}\right] = a(b-a) + \lim_{n\to\infty}\frac{(b-a)^2}{2}\left(1+\frac{1}{n}\right)$

$= a(b-a) + \frac{1}{2}(b-a)^2 = (b-a)\left(a + \frac{1}{2}b - \frac{1}{2}a\right) = (b-a)\frac{1}{2}(b+a) = \frac{1}{2}(b^2-a^2)$

22. $\int_a^b x^2\,dx = \lim_{n\to\infty}\frac{b-a}{n}\sum_{i=1}^{n}\left[a+\frac{b-a}{n}i\right]^2 = \lim_{n\to\infty}\frac{b-a}{n}\sum_{i=1}^{n}\left[a^2 + 2a\frac{(b-a)}{n}i + \frac{(b-a)^2}{n^2}i^2\right]$

$= \lim_{n\to\infty}\left[\frac{(b-a)^3}{n^3}\sum_{i=1}^{n}i^2 + \frac{2a(b-a)^2}{n^2}\sum_{i=1}^{n}i + \frac{a^2(b-a)}{n}\sum_{i=1}^{n}1\right]$

$= \lim_{n\to\infty}\left[\frac{(b-a)^3}{n^3}\frac{n(n+1)(2n+1)}{6} + \frac{2a(b-a)^2}{n^2}\frac{n(n+1)}{2} + \frac{a^2(b-a)}{n}n\right]$

$= \lim_{n\to\infty}\left[\frac{(b-a)^3}{6}\cdot 1\cdot\left(1+\frac{1}{n}\right)\left(2+\frac{1}{n}\right) + a(b-a)^2\cdot 1\cdot\left(1+\frac{1}{n}\right) + a^2(b-a)\right]$

$= \frac{(b-a)^3}{3} + a(b-a)^2 + a^2(b-a) = \frac{b^3 - 3ab^2 + 3a^2b - a^3}{3} + ab^2 - 2a^2b + a^3 + a^2b - a^3$

$= \frac{b^3}{3} - \frac{a^3}{3} - ab + a^2b + ab^2 - a^2b = \frac{b^3-a^3}{3}$

23. $\int_1^3(1+2x)dx$ can be interpreted as the area under the graph of $f(x) = 1+2x$ between $x=1$ and $x=3$. This is equal to the area of the rectangle plus the area of the triangle (see diagram) so $\int_1^3(1+2x)dx = A = 2\cdot 3 + \frac{1}{2}\cdot 2\cdot 4 = 10.$ *Or:* Use the formula for the area of a trapezoid: $a = \frac{1}{2}(2)(3+7) = 10.$

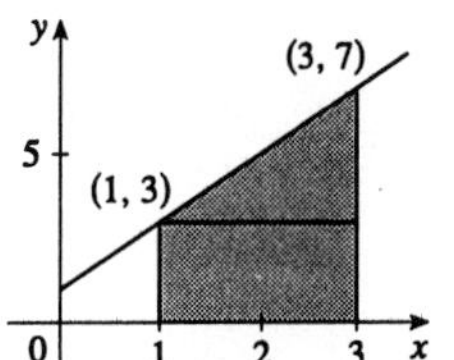

24. $\int_{-2}^{2}\sqrt{4-x^2}\,dx$ can be interpreted as the area under the graph of $f(x) = \sqrt{4-x^2}$ between $x=-2$ and $x=2$. This is equal to half the area of the circle with radius 2, so $\int_1^3\sqrt{4-x^2}\,dx = \frac{\pi\cdot 2^2}{2} = 2\pi.$

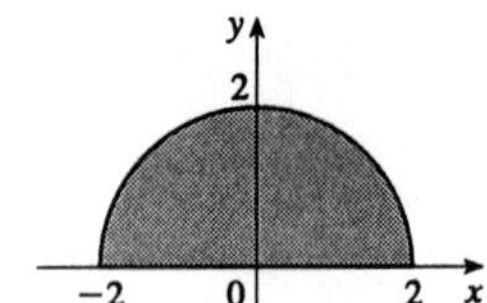

25. $\int_{-3}^{0}\left(1+\sqrt{9-x^2}\right)dx$ can be interpreted as the area under the graph of $f(x) = 1+\sqrt{9-x^2}$ between $x=-3$ and $x=0$. This is equal to one-quarter the area of the circle with radius 3, plus the area of the rectangle (see diagram), so $\int_{-3}^{0}\left(1+\sqrt{9-x^2}\right)dx = \frac{1}{4}\pi 3^2 + 1\cdot 3 = 3 + \frac{9}{4}\pi.$

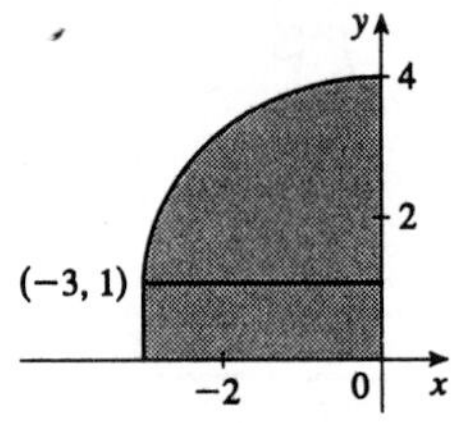

26. $\int_{-1}^{3}(2-x)dx$ can be interpreted as $A_1 - A_2$, where A_1 and A_2 are the areas of the triangles shown. Thus $\int_{-1}^{3}(2-x)dx = \frac{1}{2}\cdot 3\cdot 3 - \frac{1}{2}\cdot 1\cdot 1 = 4.$

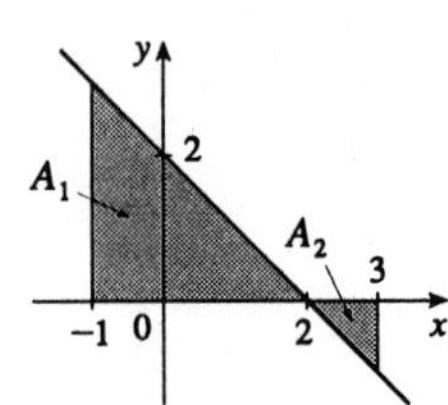

27. $\int_{-2}^{2}(1-|x|)dx$ can be interpreted as the area of the central triangle minus the areas of the outside ones (see diagram), so

$\int_{-2}^{2}(1-|x|)dx = \frac{1}{2}\cdot 2\cdot 1 - 2\cdot\frac{1}{2}\cdot 1\cdot 1 = 0.$

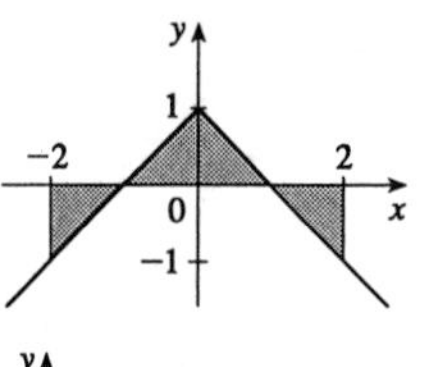

28. $\int_0^3|3x-5|dx$ can be interpreted as the area under the graph of the function $f(x)=|3x-5|$ between $x=0$ and $x=3$. This is equal to the sum of the areas of the two triangles (see diagram), so

$\int_0^3|3x-5|dx = \frac{1}{2}\cdot\frac{5}{3}\cdot 5 + \frac{1}{2}\left(3-\frac{5}{3}\right)4 = \frac{41}{6}.$

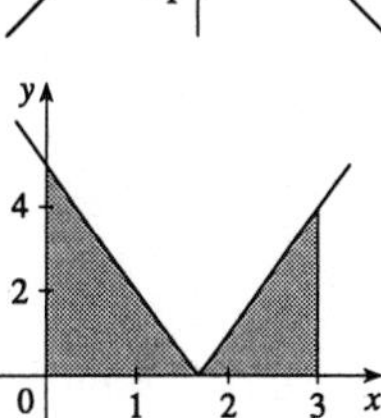

29. $\lim\limits_{\|P\|\to 0}\sum\limits_{i=1}^{n}\left[2(x_i^*)^2-5x_i^*\right]\Delta x_i = \int_0^1(2x^2-5x)\,dx$ by Definition 2.

30. $\lim\limits_{\|P\|\to 0}\sum\limits_{i=1}^{n}\sqrt{x_i^*}\,\Delta x_i = \int_1^4\sqrt{x}\,dx$

31. $\lim\limits_{\|P\|\to 0}\sum\limits_{i=1}^{n}\cos x_i\,\Delta x_i = \int_0^{\pi}\cos x\,dx$

32. $\lim\limits_{\|P\|\to 0}\sum\limits_{i=1}^{n}\frac{\tan x_i}{x_i}\,\Delta x_i = \int_2^4\frac{\tan x}{x}\,dx$

33. $\lim\limits_{n\to\infty}\sum\limits_{i=1}^{n}\frac{i^4}{n^5} = \lim\limits_{n\to\infty}\frac{1}{n}\sum\limits_{i=1}^{n}\left(\frac{i}{n}\right)^4 = \int_0^1 x^4\,dx$

34. $\lim\limits_{n\to\infty}\frac{1}{n}\sum\limits_{i=1}^{n}\frac{1}{1+(i/n)^2} = \int_0^1\frac{dx}{1+x^2}$

35. $\lim\limits_{n\to\infty}\sum\limits_{i=1}^{n}\left[3\left(1+\frac{2i}{n}\right)^5-6\right]\frac{2}{n} = \int_1^3(3x^5-6)\,dx$

Note: To get started, notice that $\frac{2}{n}=\Delta x=\frac{b-a}{n}$ and $1+\frac{2i}{n}=a+\frac{b-a}{n}i$.

36. $\int_1^1 x^2\cos x\,dx = 0$ by the definition in Note 6.

37. By the definition in Note 6, $\int_9^4\sqrt{t}\,dt = -\int_4^9\sqrt{t}\,dt = -\frac{38}{3}$.

38. **(a)** Choose equal partitions, so $\Delta x_i=(4-0)/8=0.5$ and $x_i^*=x_i=0.5i$.

$$\int_0^4(x^2-3x)\,dx \approx \sum_{i=1}^{8} f(x_i^*)\Delta x_i$$
$$= 0.5\{[0.5^2-3(0.5)]+[1.0^2-3(1.0)]+\cdots+[3.5^2-3(3.5)]+[4.0^2-3(4.0)]\}$$
$$= \tfrac{1}{2}\left[-\tfrac{5}{4}-2-\tfrac{9}{4}-2-\tfrac{5}{4}+0+\tfrac{7}{4}+4\right] = -1.5$$

(b)

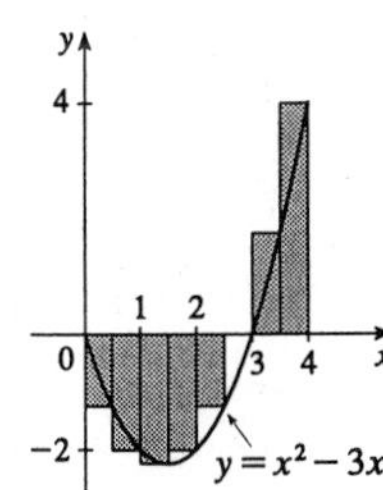

(c)

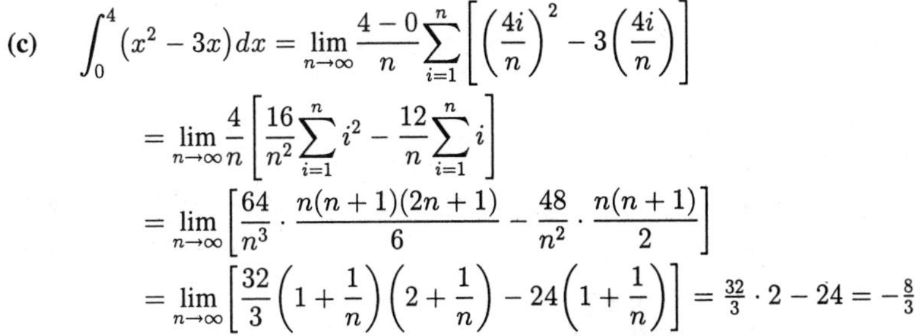

$$\int_0^4(x^2-3x)\,dx = \lim_{n\to\infty}\frac{4-0}{n}\sum_{i=1}^{n}\left[\left(\frac{4i}{n}\right)^2-3\left(\frac{4i}{n}\right)\right]$$
$$= \lim_{n\to\infty}\frac{4}{n}\left[\frac{16}{n^2}\sum_{i=1}^{n}i^2-\frac{12}{n}\sum_{i=1}^{n}i\right]$$
$$= \lim_{n\to\infty}\left[\frac{64}{n^3}\cdot\frac{n(n+1)(2n+1)}{6}-\frac{48}{n^2}\cdot\frac{n(n+1)}{2}\right]$$
$$= \lim_{n\to\infty}\left[\frac{32}{3}\left(1+\frac{1}{n}\right)\left(2+\frac{1}{n}\right)-24\left(1+\frac{1}{n}\right)\right] = \frac{32}{3}\cdot 2-24 = -\frac{8}{3}$$

(d)

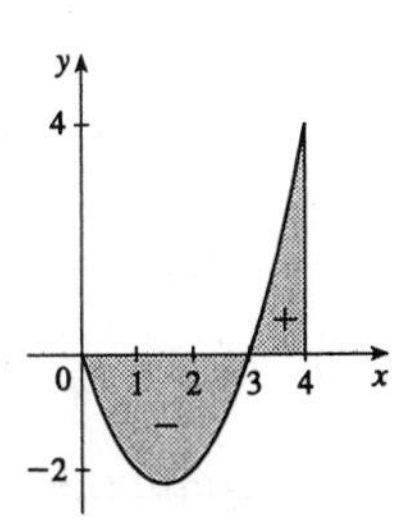

39. $\int_{-4}^{-1} \sqrt{3}\,dx = \sqrt{3}(-1+4) = 3\sqrt{3}$

40. $\int_3^6 (4-7x)dx = \int_3^6 4\,dx - \int_3^6 7x\,dx$ [Property 4] $= 4(6-3) - 7\int_3^6 x\,dx$ [Properties 1 and 3] $= 12 - 7\cdot\frac{1}{2}(6^2-3^2) = 12 - \frac{7}{2}(27) = -\frac{165}{2}$

41. $\int_1^4 (2x^2-3x+1)dx = 2\int_1^4 x^2\,dx - 3\int_1^4 x\,dx + \int_1^4 1\,dx$
$= 2\cdot\frac{1}{3}(4^3-1^3) - 3\cdot\frac{1}{2}(4^2-1^2) + 1(4-1) = \frac{45}{2} = 22.5$

42. $\int_0^1 (5\cos x + 4x)dx = 5\int_0^1 \cos x\,dx + 4\int_0^1 x\,dx = 5\sin 1 + 4\cdot\frac{1}{2}(1^2-0^2) = 5\sin 1 + 2$

43. $\int_{-1}^1 f(x)dx = \int_{-1}^0 f(x)dx + \int_0^1 f(x)dx = \int_{-1}^0(-2x)dx + \int_0^1 3x^2\,dx$
$= -2\int_{-1}^0 x\,dx + 3\int_0^1 x^2\,dx = -2\cdot\frac{1}{2}\left[0^2-(-1)^2\right] + 3\cdot\frac{1}{3}[1^3-0^3] = 2$

44. $\int_3^4 f(x)dx + \int_1^3 f(x)dx + \int_4^1 f(x)dx = \int_1^3 f(x)dx + \int_3^4 f(x)dx - \int_1^4 f(x)dx = \int_1^4 f(x)dx - \int_1^4 f(x)dx = 0$

45. $\int_1^3 f(x)dx + \int_3^6 f(x)dx + \int_6^{12} f(x)dx = \int_1^6 f(x)dx + \int_6^{12} f(x)dx = \int_1^{12} f(x)dx$

46. $\int_5^8 f(x)dx + \int_0^5 f(x)dx = \int_0^5 f(x)dx + \int_5^8 f(x)dx = \int_0^8 f(x)dx$

47. $\int_2^{10} f(x)dx - \int_2^7 f(x)dx = \int_2^7 f(x)dx + \int_7^{10} f(x)dx - \int_2^7 f(x)dx = \int_7^{10} f(x)dx$

48. $\int_{-3}^5 f(x)dx - \int_{-3}^0 f(x)dx + \int_5^6 f(x)dx = \int_{-3}^0 f(x)dx + \int_0^5 f(x)dx - \int_{-3}^0 f(x)dx + \int_5^6 f(x)dx = \int_0^6 f(x)dx$

49. $0 \le \sin x < 1$ on $\left[0,\frac{\pi}{4}\right]$, so $\sin^3 x \le \sin^2 x$ on $\left[0,\frac{\pi}{4}\right]$. Hence $\int_0^{\pi/4}\sin^3 x\,dx \le \int_0^{\pi/4}\sin^2 x\,dx$ (Property 7).

50. $5 - x \ge 3 \ge x+1$ on $[1,2]$, so $\sqrt{5-x} \ge \sqrt{x+1}$ and $\int_1^2 \sqrt{5-x}\,dx \ge \int_1^2 \sqrt{x+1}\,dx$.

51. $x \ge 4 \ge 8-x$ on $[4,6]$, so $\dfrac{1}{x} \le \dfrac{1}{8-x}$ on $[4,6]$, and $\displaystyle\int_4^6 \frac{1}{x}\,dx \le \int_4^6 \frac{1}{8-x}\,dx$.

52. $\frac{1}{2} \le \sin x \le 1$ for $\frac{\pi}{6} \le x \le \frac{\pi}{2}$, so $\frac{1}{2}\left(\frac{\pi}{2}-\frac{\pi}{6}\right) \le \int_{\pi/6}^{\pi/2}\sin x\,dx \le 1\left(\frac{\pi}{2}-\frac{\pi}{6}\right)$ (Property 8); that is,
$\frac{\pi}{6} \le \int_{\pi/6}^{\pi/2}\sin x\,dx \le \frac{\pi}{3}$.

53. If $-1 \le x \le 1$, then $0 \le x^2 \le 1$ and $1 \le 1+x^2 \le 2$, so $1 \le \sqrt{1+x^2} \le \sqrt{2}$ and
$1[1-(-1)] \le \int_{-1}^1 \sqrt{1+x^2}\,dx \le \sqrt{2}[1-(-1)]$ [Property 8]; that is, $2 \le \int_{-1}^1 \sqrt{1+x^2}\,dx \le 2\sqrt{2}$.

54. From Property 9 we know that $\left|\int_0^{2\pi} f(x)\sin 2x\,dx\right| \le \int_0^{2\pi}|f(x)\sin 2x|dx$. But $|\sin 2x| \le 1$ for all x, so $|f(x)\sin 2x| \le |f(x)|$ for all x, so by Property 7 we have $\int_0^{2\pi}|f(x)\sin 2x|dx \le \int_0^{2\pi}|f(x)|dx$. The first and last inequalities together give the result.

55. If $1 \le x \le 2$, then $\dfrac{1}{2} \le \dfrac{1}{x} \le 1$, so $\frac{1}{2}(2-1) \le \displaystyle\int_1^2 \frac{1}{x}\,dx \le 1(2-1)$ or $\dfrac{1}{2} \le \displaystyle\int_1^2 \frac{1}{x}\,dx \le 1$.

56. If $0 \le x \le 2$, then $0 \le x^3 \le 8$, so $1 \le x^3+1 \le 9$ and $1 \le \sqrt{x^3+1} \le 3$. Thus
$1(2-0) \le \int_0^2 \sqrt{x^3+1}\,dx \le 3(2-0)$; that is, $2 \le \int_0^2 \sqrt{x^3+1}\,dx \le 6$.

57. If $f(x) = x^2+2x$, $-3 \le x \le 0$, then $f'(x) = 2x+2 = 0$ when $x = -1$, and $f(-1) = -1$. At the endpoints, $f(-3) = 3$, $f(0) = 0$. Thus the absolute minimum is $m = -1$ and the absolute maximum is $M = 3$. Thus $-1[0-(-3)] \le \int_{-3}^0 (x^2+2x)dx \le 3[0-(-3)]$ or $-3 \le \int_{-3}^0 (x^2+2x)dx \le 9$.

58. If $\frac{\pi}{4} \le x \le \frac{\pi}{3}$, then $\frac{1}{2} \le \cos x \le \frac{\sqrt{2}}{2}$, so $\frac{1}{2}\left(\frac{\pi}{3} - \frac{\pi}{4}\right) \le \int_{\pi/4}^{\pi/3} \cos x\,dx \le \frac{\sqrt{2}}{2}\left(\frac{\pi}{3} - \frac{\pi}{4}\right)$ or $\frac{\pi}{24} \le \int_{\pi/4}^{\pi/3} \cos x\,dx \le \frac{\sqrt{2}\pi}{24}$.

59. For $-1 \le x \le 1$, $0 \le x^4 \le 1$ and $1 \le \sqrt{1+x^4} \le \sqrt{2}$, so $1[1-(-1)] \le \int_{-1}^{1} \sqrt{1+x^4}\,dx \le \sqrt{2}[1-(-1)]$ or $2 \le \int_{-1}^{1} \sqrt{1+x^4}\,dx \le 2\sqrt{2}$.

60. If $\frac{1}{4}\pi \le x \le \frac{3}{4}\pi$, then $\frac{\sqrt{2}}{2} \le \sin x \le 1$ and $\frac{1}{2} \le \sin^2 x \le 1$, so $\frac{1}{2}\left(\frac{3}{4}\pi - \frac{1}{4}\pi\right) \le \int_{\pi/4}^{3\pi/4} \sin^2 x\,dx \le 1\left(\frac{3}{4}\pi - \frac{1}{4}\pi\right)$; that is, $\frac{1}{4}\pi \le \int_{\pi/4}^{3\pi/4} \sin^2 x\,dx \le \frac{1}{2}\pi$.

61. $\sqrt{x^4+1} \ge \sqrt{x^4} = x^2$, so $\int_1^3 \sqrt{x^4+1}\,dx \ge \int_1^3 x^2\,dx = \frac{1}{3}(3^3 - 1^3) = \frac{26}{3}$.

62. $x^2 - 1 < x^2 \quad \Rightarrow \quad \sqrt{x^2-1} < \sqrt{x^2} = |x| = x$ for $x \ge 0$, so $\int_2^5 \sqrt{x^2-1}\,dx \le \int_2^5 x\,dx = \frac{1}{2}(5^2 - 2^2) = 10.5$.

63. $0 \le \sin x \le 1$ for $0 \le x \le \frac{\pi}{2}$, so $x \sin x \le x \quad \Rightarrow \quad \int_0^{\pi/2} x \sin x\,dx \le \int_0^{\pi/2} x\,dx = \frac{1}{2}\left[\left(\frac{\pi}{2}\right)^2 - 0^2\right] = \frac{\pi^2}{8}$.

64. By Properties 9 and 7 we have $\left|\int_0^{\pi} x^2 \cos x\,dx\right| \le \int_0^{\pi} |x^2 \cos x|\,dx \le \int_0^{\pi} x^2\,dx = \frac{1}{3}(\pi^3 - 0^3) = \frac{1}{3}\pi^3$.

65. Using a regular partition and right endpoints as in the proof of Property 2, we calculate

$$\int_a^b cf(x)dx = \lim_{n\to\infty} \sum_{i=1}^{n} cf(x_i)\Delta x_i = \lim_{n\to\infty} c\sum_{i=1}^{n} f(x_i)\Delta x_i = c\lim_{n\to\infty} \sum_{i=1}^{n} f(x_i)\Delta x_i = c\int_a^b f(x)dx.$$

66. As in the proof of Property 2, we write $\displaystyle\int_a^b f(x)dx = \lim_{n\to\infty} \sum_{i=1}^{n} f(x_i)\Delta x$. Now $f(x_i) \ge 0$ and $\Delta x \ge 0$, so $f(x_i)\Delta x \ge 0$ and therefore $\displaystyle\sum_{i=1}^{n} f(x_i)\Delta x \ge 0$. But the limit of nonnegative quantities is nonnegative by Theorem 1.3.2, so $\int_a^b f(x)dx \ge 0$.

67. By Property 7, the inequalities $-|f(x)| \le f(x) \le |f(x)|$ imply that $\int_a^b (-|f(x)|)dx \le \int_a^b f(x)dx \le \int_a^b |f(x)|dx$. By Property 3, the left-hand integral equals $-\int_a^b |f(x)|dx$. Thus $-M \le \int_a^b f(x)dx \le M$, where $M = \int_a^b |f(x)|dx$. [Notice that $M \ge 0$ by Property 6.] It follows that $\left|\int_a^b f(x)dx\right| \le M = \int_a^b |f(x)|dx$.

68. Since f is continuous on $[a,b]$, f attains an absolute minimum value m at some point of $[a,b]$ by the Extreme Value Theorem. Since $f(x) > 0$ for all x in $[a,b]$, it follows that $m > 0$. Now $f(x) \ge m$ for $a \le x \le b$, so $\int_a^b f(x)dx \ge m(b-a)$ by Property 8. Assuming that $a < b$, we see that $m > 0$ and $b - a > 0$, so $m(b-a) > 0$. This proves that $\int_a^b f(x)dx > 0$.

69. **(a)** $f(x) = x^2 \sin x$ is continuous on $[0,2]$ and hence integrable by Theorem 4.

(b) $f(x) = \sec x$ is unbounded on $[0,2]$, so f is not integrable (see the remarks following Theorem 4.)

(c) $f(x)$ is piecewise continuous on $[0,2]$ with a single jump discontinuity at $x = 1$, so f is integrable.

(d) $f(x)$ has an infinite discontinuity at $x = 1$, so f is not integrable on $[0,2]$.

70. **(a)** f is not continuous on $[0,1]$ because $\lim_{x \to 0^+} f(x) = \lim_{x \to 0^+} 1/x$ does not exist.

(b) f is unbounded on $[0,1]$ since if we pick any proposed positive number M to serve as a bound, then for $0 < x < 1/M$, we see that $f(x) = 1/x > M$.

(c) $\int_0^1 f(x)dx$ does not exist since an approximating sum $\sum_{i=1}^n f(x_i^*)\Delta x_i$ can be made arbitrarily large, no matter how small the norm $\|P\|$ is. The first subinterval of the partition P of $[0,1]$ will be of the form $[0, x_1]$. By choosing x_1^* to be positive but sufficiently small, we can make $f(x_1^*)$ as large as we like. This means that we can make $\sum_{i=1}^n f(x_i^*)\Delta x_i$ as large as we like by an appropriate choice of x_1^*. Thus $\sum_{i=1}^n f(x_i^*)\Delta x_i$ cannot approach a limiting value as $\|P\| \to 0$.

71. f is bounded since $|f(x)| \le 1$ for all x in $[a,b]$. To see that f is not integrable on $[a,b]$, notice that $\sum_{i=1}^n f(x_i^*)\Delta x_i = 0$ if $x_1^*, \ldots, x_n^*$ are all chosen to be rational numbers, but $\sum_{i=1}^n f(x_i^*)\Delta x_i = \sum_{i=1}^n \Delta x_i = b - a$ if $x_1^*, \ldots, x_n^*$ are all chosen to be irrational numbers. This is true no matter how small $\|P\|$ is, since every interval $[x_{i-1}, x_i]$ with $x_{i-1} < x_i$ contains both rational and irrational numbers. $\sum_{i=1}^n f(x_i^*)\Delta x_i$ cannot approach both 0 and $b-a$ as $\|P\| \to 0$, so it has no limit as $\|P\| \to 0$.

72. First notice that if $x_i^* = x_i = 2^{i/n}$ for $i = 1, 2, \ldots, n$, then $\Delta x_i = x_i - x_{i-1} = 2^{i/n} - 2^{(i-1)/n}$ and $\|P\| = 2 - 2^{(n-1)/n}$, so $\|P\| \to 2 - 2 = 0$ as $n \to \infty$. Thus

$$\int_1^2 x^3\,dx = \lim_{n\to\infty} \sum_{i=1}^n x_i^3 \Delta x_i = \lim_{n\to\infty} \sum_{i=1}^n 2^{3i/n}\left(2^{i/n} - 2^{(i-1)/n}\right)$$

$$= \lim_{n\to\infty} \sum_{i=1}^n 2^{4i/n}\left(1 - 2^{-1/n}\right) = \lim_{n\to\infty} \left(1 - 2^{-1/n}\right) \sum_{i=1}^n \left(2^{4/n}\right)^i$$

$$= \lim_{n\to\infty} \left(1 - 2^{-1/n}\right) \cdot 2^{4/n} \frac{\left(2^{4/n}\right)^n - 1}{2^{4/n} - 1}$$

(by Exercise 4.1.49, which says in particular that

$$\sum_{i=1}^n r^i = \sum_{i=1}^n r \cdot r^{i-1} = \frac{r(r^n - 1)}{r - 1}.$$ Here $r = 2^{4/n}$.)

$$= \lim_{n\to\infty} \left(2^4 - 1\right) \cdot \frac{2^{4/n} - 2^{3/n}}{2^{4/n} - 1}$$

$$= 15 \cdot \lim_{n\to\infty} \frac{\left(2^{1/n}\right)^4 - \left(2^{1/n}\right)^3}{\left(2^{1/n}\right)^4 - 1}$$

$$= 15 \cdot \lim_{n\to\infty} \frac{\left(2^{1/n}\right)^3 \left(2^{1/n} - 1\right)}{\left(2^{1/n} - 1\right)\left(2^{1/n} + 1\right)\left[\left(2^{1/n}\right)^2 + 1\right]}$$

$$= 15 \cdot \lim_{n\to\infty} \frac{\left(2^{1/n}\right)^3}{\left(2^{1/n} + 1\right)\left(2^{2/n} + 1\right)}$$

$$= 15 \cdot \frac{1}{2 \cdot 2} = \frac{15}{4}.$$

73. Choose $x_i = 1 + \dfrac{i}{n}$ and $x_i^* = \sqrt{x_{i-1}\, x_i} = \sqrt{\left(1 + \dfrac{i-1}{n}\right)\left(1 + \dfrac{i}{n}\right)}$. Then

$$\int_1^2 x^{-2}\,dx = \lim_{n\to\infty} \frac{1}{n} \sum_{i=1}^{n} \frac{1}{\left(1 + \dfrac{i-1}{n}\right)\left(1 + \dfrac{i}{n}\right)} = \lim_{n\to\infty} n \sum_{i=1}^{n} \frac{1}{(n+i-1)(n+i)}$$

$$= \lim_{n\to\infty} n \sum_{i=1}^{n} \left[\frac{1}{n+i-1} - \frac{1}{n+1}\right] \quad \text{(by the hint)}$$

$$= \lim_{n\to\infty} n \left[\sum_{i=0}^{n-1} \frac{1}{n+i} - \sum_{i=1}^{n} \frac{1}{n+i}\right] = \lim_{n\to\infty} n \left[\frac{1}{n} - \frac{1}{2n}\right]$$

$$= \lim_{n\to\infty} \left[1 - \frac{1}{2}\right] = \frac{1}{2}.$$

74. **(a)** $f(x) = \cos(x^2)$

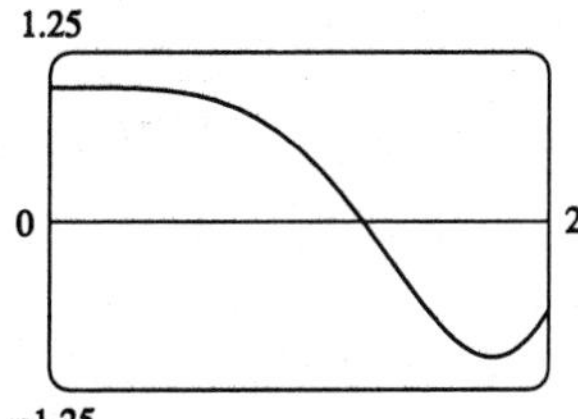

(b) $g(x) = \int_0^x \cos(t^2)dt$. We estimate: $g(0) = 0$, $g(0.2) \approx 0.2$, $g(0.4) \approx 0.4$, $g(0.6) \approx 0.6$, $g(0.8) \approx 0.8$, $g(1.0) \approx 0.9$, $g(1.2) \approx 1.0$, $g(1.4) \approx 0.9$, $g(1.6) \approx 0.8$, $g(1.8) \approx 0.6$, $g(2.0) \approx 0.5$. $g(x)$ starts to decrease at that value of x where $\cos(t^2)$ changes from positive to negative, that is, at about $x = 1.25$.

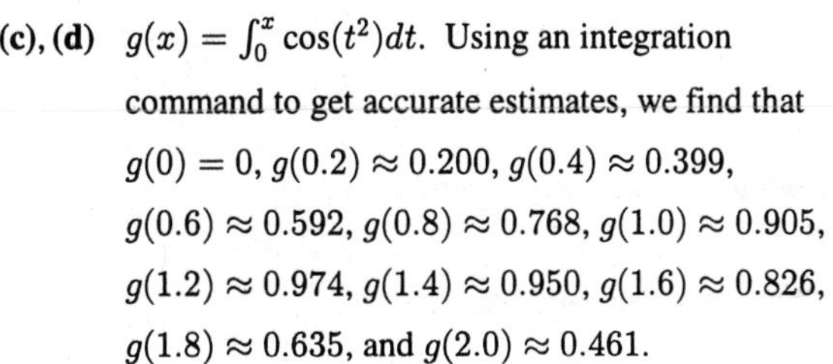

(c), (d) $g(x) = \int_0^x \cos(t^2)dt$. Using an integration command to get accurate estimates, we find that $g(0) = 0$, $g(0.2) \approx 0.200$, $g(0.4) \approx 0.399$, $g(0.6) \approx 0.592$, $g(0.8) \approx 0.768$, $g(1.0) \approx 0.905$, $g(1.2) \approx 0.974$, $g(1.4) \approx 0.950$, $g(1.6) \approx 0.826$, $g(1.8) \approx 0.635$, and $g(2.0) \approx 0.461$.

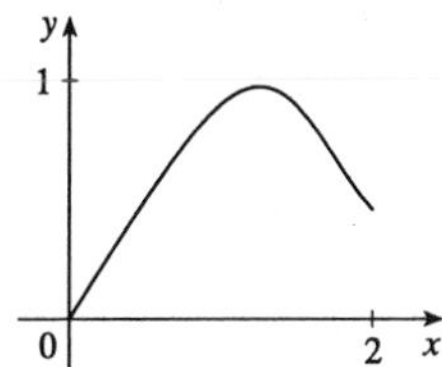

(e) The graphs of $g'(x)$ and $f(x)$ look alike. This makes sense, since as x increases, the "area function" $g(x) = \int_0^x f(t)dt$ increases at a rate proportional (in fact, equal) to the value of $f(x)$: if $f(x)$ is large, the area under its graph increases quickly as x increases; if $f(x)$ is small, the area increases slowly, and so on. This is the basic idea behind the Fundamental Theorem of Calculus, which is discussed in the next section.

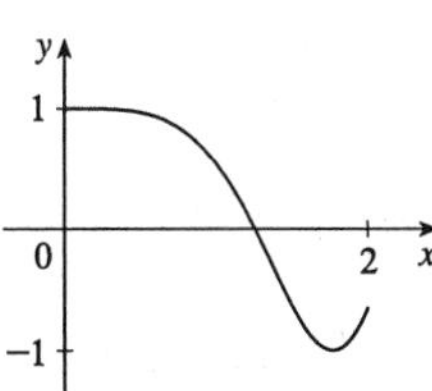

EXERCISES 4.4

1. **(a)** $g(0) = \int_0^0 f(t)dt = 0$, $g(1) = \int_0^1 f(t)dt = 1 \cdot 2 = 2$,

$g(2) = \int_0^2 f(t)dt = \int_0^1 f(t)dt + \int_1^2 f(t)dt = g(1) + \int_1^2 f(t)dt = 2 + 1 \cdot 2 + \frac{1}{2} \cdot 1 \cdot 2 = 5$,

$g(3) = \int_0^3 f(t)dt = g(2) + \int_2^3 f(t)dt = 5 + \frac{1}{2} \cdot 1 \cdot 4 = 7$,

$g(6) = g(3) + \int_3^6 f(t)dt = 7 + \left[-\left(\frac{1}{2} \cdot 2 \cdot 2 + 1 \cdot 2\right)\right] = 7 - 4 = 3$

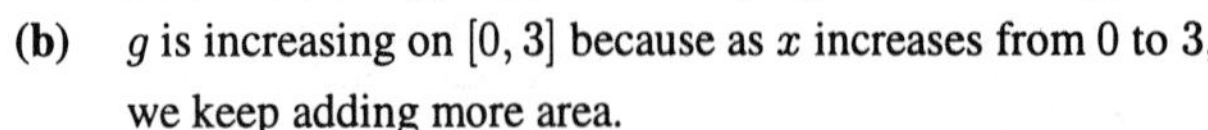

(b) g is increasing on $[0, 3]$ because as x increases from 0 to 3, we keep adding more area.

(c) g has a maximum value when we start subtracting area, that is, at $x = 3$.

(d)

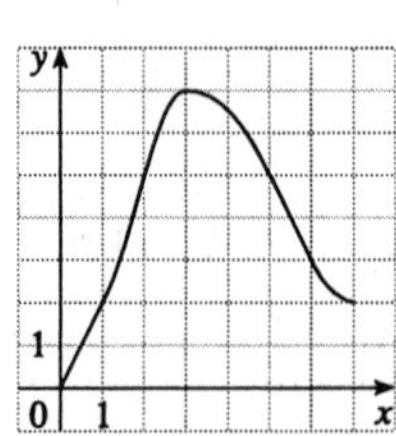

2. **(a)** $g(-3) = \int_{-3}^{-3} f(t)dt = 0$, $g(3) = \int_{-3}^{3} f(t)dt = \int_{-3}^{0} f(t)dt + \int_0^3 f(t)dt = 0$ by symmetry, since the area above the x-axis is the same as the area below the axis.

(b) From the graph, it appears that to the nearest $\frac{1}{2}$, $g(-2) = \int_{-3}^{-2} f(t)dt \approx 1$, $g(-1) = \int_{-3}^{-1} f(t)dt \approx 3\frac{1}{2}$, and $g(0) = \int_{-3}^{0} f(t)dt \approx 5\frac{1}{2}$.

(c) g is increasing on $[-3, 0]$ because as x increases from -3 to 0, we keep adding more area.

(d) g has a maximum value when we start subtracting area, that is, at $x = 0$.

(e) [graph: y, 1, 0, 1, x]

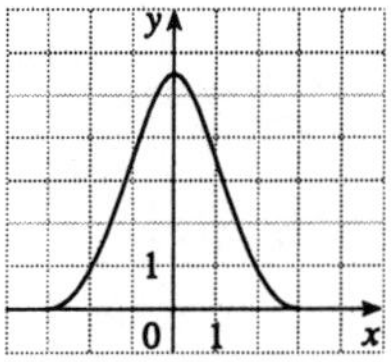

(f) The graph of $g'(x)$ is the same as that of $f(x)$, as indicated by the Fundamental Theorem of Calculus.

3.

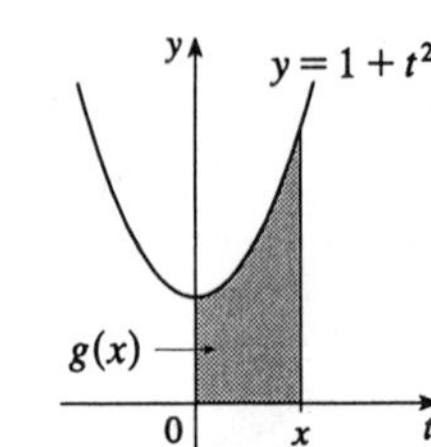

(a) By Part 1 of the Fundamental Theorem, $g(x) = \int_0^x (1 + t^2)dt \;\Rightarrow\; g'(x) = f(x) = 1 + x^2$.

(b) By Part 2 of the Fundamental Theorem,

$$g(x) = \int_0^x (1 + t^2)dt = \left[t + \tfrac{1}{3}t^3\right]_0^x = \left(x + \tfrac{1}{3}x^3\right) - \left(0 + \tfrac{1}{3}0^3\right) = x + \tfrac{1}{3}x^3$$

$\Rightarrow\; g'(x) = 1 + x^2$.

4.

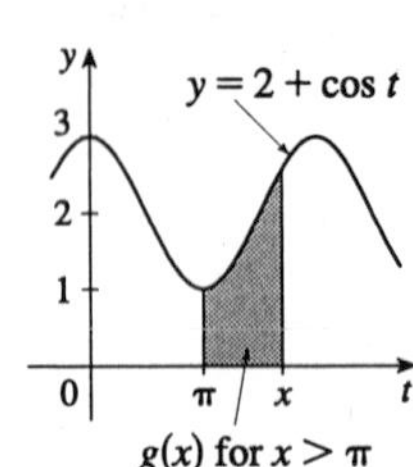

(a) $g(x) = \int_\pi^x (2 + \cos t)dt \;\Rightarrow\; g'(x) = 2 + \cos x$

(b) $g(x) = \int_\pi^x (2 + \cos t)dt = [2t + \sin t]_\pi^x = (2x + \sin x) - (2\pi + 0) = 2x + \sin x - 2\pi$,

so $g'(x) = 2 + \cos x$.

5. $g(x) = \int_1^x (t^2 - 1)^{20} dt \;\Rightarrow\; g'(x) = (x^2 - 1)^{20}$

6. $g(x) = \int_{-1}^x \sqrt{t^3 + 1}\, dt \;\Rightarrow\; g'(x) = \sqrt{x^3 + 1}$

7. $g(u) = \displaystyle\int_\pi^u \frac{1}{1 + t^4}\, dt \;\Rightarrow\; g'(u) = \frac{1}{1 + u^4}$

8. $g(t) = \int_0^t \sin(x^2)dx \;\Rightarrow\; g'(t) = \sin(t^2)$

9. $F(x) = \int_x^2 \cos(t^2)dt = -\int_2^x \cos(t^2)dt \quad \Rightarrow \quad F'(x) = -\cos(x^2)$

10. $F(x) = \int_x^4 (2+\sqrt{u})^8 du = -\int_4^x (2+\sqrt{u})^8 du \quad \Rightarrow \quad F'(x) = -(2+\sqrt{x})^8$

11. Let $u = \dfrac{1}{x}$. Then $\dfrac{du}{dx} = -\dfrac{1}{x^2}$, so $\dfrac{d}{dx}\displaystyle\int_2^{1/x} \sin^4 t\, dt = \dfrac{d}{du}\int_2^u \sin^4 t\, dt \cdot \dfrac{du}{dx} = \sin^4 u \dfrac{du}{dx} = \dfrac{-\sin^4(1/x)}{x^2}$.

12. Let $u = \sqrt{x}$. Then $\dfrac{du}{dx} = \dfrac{1}{2\sqrt{x}}$, so

$$h'(x) = \frac{d}{dx}\int_1^{\sqrt{x}} \frac{s^2}{s^2+1} ds = \frac{d}{du}\int_1^u \frac{s^2}{s^2+1} ds \cdot \frac{du}{dx} = \frac{u^2}{u^2+1}\frac{du}{dx} = \frac{x}{x+1}\frac{1}{2\sqrt{x}} = \frac{\sqrt{x}}{2(x+1)}.$$

13. Let $u = \tan x$. Then $\dfrac{du}{dx} = \sec^2 x$, so

$$\frac{d}{dx}\int_{\tan x}^{17} \sin(t^4)dt = -\frac{d}{dx}\int_{17}^{\tan x} \sin(t^4)dt = -\frac{d}{du}\int_{17}^{u} \sin(t^4)dt \cdot \frac{du}{dx} = -\sin(u^4)\frac{du}{dx} = -\sin(\tan^4 x)\sec^2 x.$$

14. Let $u = x^2$. Then $\dfrac{du}{dx} = 2x$, so $\dfrac{dy}{dx} = \dfrac{d}{dx}\displaystyle\int_{x^2}^{\pi} \frac{\sin t}{t} dt = -\frac{d}{dx}\int_{\pi}^{x^2} \frac{\sin t}{t} dt$

$$= -\frac{d}{du}\int_{\pi}^{u} \frac{\sin t}{t} dt \cdot \frac{du}{dx} = -\frac{\sin u}{u} \cdot \frac{du}{dx} = -\frac{\sin(x^2)}{x^2} \cdot 2x = -\frac{2\sin(x^2)}{x}.$$

15. Let $t = 5x+1$. Then $\dfrac{dt}{dx} = 5$, so

$$\frac{d}{dx}\int_0^{5x+1} \frac{1}{u^2-5} du = \frac{d}{dt}\int_0^t \frac{1}{u^2-5} du \cdot \frac{dt}{dx} = \frac{1}{t^2-5}\frac{dt}{dx} = \frac{5}{25x^2+10x-4}.$$

16. Let $u = \sin x$. Then $\dfrac{du}{dx} = \cos x$, so

$$\frac{dy}{dx} = \frac{dy}{du}\frac{du}{dx} = \frac{d}{du}\int_{-5}^{u} t\cos(t^3)dt \cdot \frac{du}{dx} = u\cos(u^3)\frac{du}{dx} = \sin x\cos(\sin^3 x)\cos x = \sin x \cos x \cos(\sin^3 x).$$

17. $\int_{-2}^4 (3x-5)dx = \left(3 \cdot \tfrac{1}{2}x^2 - 5x\right)\Big|_{-2}^4 = (3\cdot 8 - 5\cdot 4) - [3\cdot 2 - (-10)] = -12$

18. $\int_1^2 x^{-2}\, dx = [-x^{-1}]_1^2 = [-1/x]_1^2 = -\tfrac{1}{2} + 1 = \tfrac{1}{2}$

19. $\int_0^1 (1-2x-3x^2)dx = \left[x - 2\cdot\tfrac{1}{2}x^2 - 3\cdot\tfrac{1}{3}x^3\right]_0^1 = [x - x^2 - x^3]_0^1 = (1-1-1) - 0 = -1$

20. $\int_1^2 (5x^2-4x+3)dx = \left[5\cdot\tfrac{1}{3}x^3 - 4\cdot\tfrac{1}{2}x^2 + 3x\right]_1^2 = 5\cdot\tfrac{8}{3} - 4\cdot 2 + 6 - \left(\tfrac{5}{3} - 2 + 3\right) = \tfrac{26}{3}$

21. $\int_{-3}^0 (5y^4 - 6y^2 + 14)dy = \left[5\left(\tfrac{1}{5}y^5\right) - 6\left(\tfrac{1}{3}y^3\right) + 14y\right]_{-3}^0 = [y^5 - 2y^3 + 14y]_{-3}^0 = 0 - (-243 + 54 - 42) = 231$

22. $\int_0^1 (y^9 - 2y^5 + 3y)dy = \left[\tfrac{1}{10}y^{10} - 2\left(\tfrac{1}{6}y^6\right) + 3\left(\tfrac{1}{2}y^2\right)\right]_0^1 = \left(\tfrac{1}{10} - \tfrac{1}{3} + \tfrac{3}{2}\right) - 0 = \tfrac{19}{15}$

23. $$\int_0^4 \sqrt{x}\, dx = \int_0^4 x^{1/2}\, dx = \left[\frac{x^{3/2}}{3/2}\right]_0^4 = \left[\frac{2x^{3/2}}{3}\right]_0^4 = \frac{2(4)^{3/2}}{3} - 0 = \frac{16}{3}$$

24. $$\int_0^1 x^{3/7}\, dx = \left[\frac{x^{10/7}}{10/7}\right]_0^1 = \left[\tfrac{7}{10}x^{10/7}\right]_0^1 = \tfrac{7}{10} - 0 = \tfrac{7}{10}$$

25. $$\int_1^3 \left[\frac{1}{t^2} - \frac{1}{t^4}\right] dt = \int_1^3 (t^{-2} - t^{-4})dt = \left[\frac{t^{-1}}{-1} - \frac{t^{-3}}{-3}\right]_1^3 = \left[\frac{1}{3t^3} - \frac{1}{t}\right]_1^3 = \left(\frac{1}{81} - \frac{1}{3}\right) - \left(\frac{1}{3} - 1\right) = \frac{28}{81}$$

26. $\int_1^2 \frac{t^6 - t^2}{t^4}\,dt = \int_1^2 (t^2 - t^{-2})\,dt = \left[\frac{t^3}{3} - \frac{t^{-1}}{-1}\right]_1^2 = \left[\frac{t^3}{3} + \frac{1}{t}\right]_1^2 = \left(\frac{8}{3} + \frac{1}{2}\right) - \left(\frac{1}{3} + 1\right) = \frac{11}{6}$

27. $\int_1^2 \frac{x^2+1}{\sqrt{x}}\,dx = \int_1^2 (x^{3/2} + x^{-1/2})\,dx = \left[\frac{x^{5/2}}{5/2} + \frac{x^{1/2}}{1/2}\right]_1^2 = \left[\frac{2}{5}x^{5/2} + 2x^{1/2}\right]_1^2$

$= \left(\frac{2}{5}4\sqrt{2} + 2\sqrt{2}\right) - \left(\frac{2}{5} + 2\right) = \frac{18\sqrt{2} - 12}{5} = \frac{6}{5}\left(3\sqrt{2} - 2\right)$

28. $\int_0^2 (x^3 - 1)^2\,dx = \int_0^2 (x^6 - 2x^3 + 1)\,dx = \left[\frac{1}{7}x^7 - 2\left(\frac{1}{4}x^4\right) + x\right]_0^2 = \left(\frac{128}{7} - 2\cdot 4 + 2\right) - 0 = \frac{86}{7}$

29. $\int_0^1 u\left(\sqrt{u} + \sqrt[3]{u}\right)du = \int_0^1 (u^{3/2} + u^{4/3})\,du = \left[\frac{u^{5/2}}{5/2} + \frac{u^{7/3}}{7/3}\right]_0^1 = \left[\frac{2}{5}u^{5/2} + \frac{3}{7}u^{7/3}\right]_0^1 = \frac{2}{5} + \frac{3}{7} = \frac{29}{35}$

30. $\int_{-1}^1 \frac{3}{t^4}\,dt$ does not exist since $f(t) = \frac{3}{t^4}$ has an infinite discontinuity at 0.

31. $\int_{-2}^3 |x^2 - 1|\,dx = \int_{-2}^{-1} (x^2 - 1)\,dx + \int_{-1}^1 (1 - x^2)\,dx + \int_1^3 (x^2 - 1)\,dx$

$= \left[\frac{x^3}{3} - x\right]_{-2}^{-1} + \left[x - \frac{x^3}{3}\right]_{-1}^1 + \left[\frac{x^3}{3} - x\right]_1^3$

$= \left(-\frac{1}{3} + 1\right) - \left(-\frac{8}{3} + 2\right) + \left(1 - \frac{1}{3}\right) - \left(-1 + \frac{1}{3}\right) + (9 - 3) - \left(\frac{1}{3} - 1\right) = \frac{28}{3}$

32. $\int_1^2 \left(x + \frac{1}{x}\right)^2 dx = \int_1^2 (x^2 + 2 + x^{-2})\,dx = \left[\frac{x^3}{3} + 2x + \frac{x^{-1}}{-1}\right]_1^2 = \left[\frac{x^3}{3} + 2x - \frac{1}{x}\right]_1^2$

$= \left(\frac{8}{3} + 4 - \frac{1}{2}\right) - \left(\frac{1}{3} + 2 - 1\right) = \frac{29}{6}$

33. $\int_3^3 \sqrt{x^5 + 2}\,dx = 0$ by the definition in Note 6 in Section 4.3.

34. $\int_{-1}^2 |x - x^2|\,dx = \int_{-1}^0 (x^2 - x)\,dx + \int_0^1 (x - x^2)\,dx + \int_1^2 (x^2 - x)\,dx$

$= \left[\frac{x^3}{3} - \frac{x^2}{2}\right]_{-1}^0 + \left[\frac{x^2}{2} - \frac{x^3}{3}\right]_0^1 + \left[\frac{x^3}{3} - \frac{x^2}{2}\right]_1^2$

$= 0 - \left(-\frac{1}{3} - \frac{1}{2}\right) + \left(\frac{1}{2} - \frac{1}{3}\right) - 0 + \left(\frac{8}{3} - 2\right) - \left(\frac{1}{3} - \frac{1}{2}\right) = \frac{11}{6}$

35. $\int_{-4}^2 \frac{2}{x^6}\,dx$ does not exist since $f(x) = \frac{2}{x^6}$ has an infinite discontinuity at 0.

36. $\int_1^{-1} (x - 1)(3x + 2)\,dx = -\int_{-1}^1 (3x^2 - x - 2)\,dx = -\left[3\frac{x^3}{3} - \frac{x^2}{2} - 2x\right]_{-1}^1 = \left[-x^3 + \frac{x^2}{2} + 2x\right]_{-1}^1$

$= \left(-1 + \frac{1}{2} + 2\right) - \left(1 + \frac{1}{2} - 2\right) = 2$

37. $\int_1^4 \left(\sqrt{t} - \frac{2}{\sqrt{t}}\right)dt = \int_1^4 (t^{1/2} - 2t^{-1/2})\,dt = \left[\frac{t^{3/2}}{3/2} - 2\frac{t^{1/2}}{1/2}\right]_1^4 = \left[\frac{2}{3}t^{3/2} - 4t^{1/2}\right]_1^4$

$= \left[\frac{2}{3}\cdot 8 - 4\cdot 2\right] - \left[\frac{2}{3} - 4\right] = \frac{2}{3}$

38. $\int_1^8 \left[\sqrt[3]{r} + \frac{1}{\sqrt[3]{r}}\right] dr = \int_1^8 (r^{1/3} + r^{-1/3})\,dr = \left[\frac{r^{4/3}}{4/3} + \frac{r^{2/3}}{2/3}\right]_1^8 = \left[\frac{3}{4}r^{4/3} + \frac{3}{2}r^{2/3}\right]_1^8$

$= \left(\frac{3}{4}\cdot 16 + \frac{3}{2}\cdot 4\right) - \left(\frac{3}{4} + \frac{3}{2}\right) = \frac{63}{4}$

39. $\int_{-1}^{0}(x+1)^3\,dx = \int_{-1}^{0}(x^3+3x^2+3x+1)\,dx = \left[\frac{x^4}{4}+3\frac{x^3}{3}+3\frac{x^2}{2}+x\right]_{-1}^{0} = 0-\left[\frac{1}{4}-1+\frac{3}{2}-1\right]$
$= 2-\frac{7}{4} = \frac{1}{4}$

40. $\int_{-5}^{-2}\frac{x^4-1}{x^2+1}\,dx = \int_{-5}^{-2}(x^2-1)\,dx = \left[\frac{x^3}{3}-x\right]_{-5}^{-2} = \left(-\frac{8}{3}+2\right)-\left(\frac{-125}{3}+5\right) = 36$

41. $\int_{\pi/4}^{\pi/3}\sin t\,dt = [-\cos t]_{\pi/4}^{\pi/3} = -\cos\frac{\pi}{3}+\cos\frac{\pi}{4} = -\frac{1}{2}+\frac{1}{\sqrt{2}} = \frac{\sqrt{2}-1}{2}$

42. $\int_0^{\pi/2}(\cos\theta+2\sin\theta)d\theta = [\sin\theta-2\cos\theta]_0^{\pi/2} = (1-2\cdot 0)-(0-2\cdot 1) = 3$

43. $\int_{\pi/2}^{\pi}\sec x\tan x\,dx$ does not exist since $\sec x\tan x$ has an infinite discontinuity at $\frac{\pi}{2}$.

44. $\int_{\pi/3}^{\pi/2}\csc x\cot x\,dx = [-\csc x]_{\pi/3}^{\pi/2} = -\csc\frac{\pi}{2}+\csc\frac{\pi}{3} = -1+\frac{2}{3}\sqrt{3}$

45. $\int_{\pi/6}^{\pi/3}\csc^2\theta\,d\theta = [-\cot\theta]_{\pi/6}^{\pi/3} = -\cot\frac{\pi}{3}+\cot\frac{\pi}{6} = -\frac{1}{3}\sqrt{3}+\sqrt{3} = \frac{2}{3}\sqrt{3}$

46. $\int_{\pi/4}^{\pi}\sec^2\theta\,d\theta$ does not exist since $\sec^2\theta$ has an infinite discontinuity at $\frac{\pi}{2}$.

47. $\int_0^1\left[\sqrt[4]{x^5}+\sqrt[5]{x^4}\right]dx = \int_0^1(x^{5/4}+x^{4/5})\,dx = \left[\frac{x^{9/4}}{9/4}+\frac{x^{9/5}}{9/5}\right]_0^1 = \left[\frac{4}{9}x^{9/4}+\frac{5}{9}x^{9/5}\right]_0^1 = \frac{4}{9}+\frac{5}{9}-0 = 1$

48. $\int_1^8\frac{x-1}{\sqrt[3]{x^2}}\,dx = \int_1^8(x^{1/3}-x^{-2/3})\,dx = \left[\frac{x^{4/3}}{4/3}-\frac{x^{1/3}}{1/3}\right]_1^8 = \left[\frac{3}{4}x^{4/3}-3x^{1/3}\right]_1^8$
$= \left(\frac{3}{4}\cdot 16-3\cdot 2\right)-\left(\frac{3}{4}-3\right) = \frac{33}{4}$

49. $\int_{-1}^{2}(x-2|x|)dx = \int_{-1}^{0}3x\,dx+\int_0^2(-x)dx = 3\left[\frac{1}{2}x^2\right]_{-1}^{0}-\left[\frac{1}{2}x^2\right]_0^2 = \left(3\cdot 0-3\cdot\frac{1}{2}\right)-(2-0) = -\frac{7}{2} = -3.5$

50. $\int_0^2(x^2-|x-1|)\,dx = \int_0^1(x^2+x-1)\,dx+\int_1^2(x^2-x+1)\,dx = \left[\frac{x^3}{3}+\frac{x^2}{2}-x\right]_0^1+\left[\frac{x^3}{3}-\frac{x^2}{2}+x\right]_1^2$
$= \left(\frac{1}{3}+\frac{1}{2}-1\right)-0+\left(\frac{8}{3}-2+2\right)-\left(\frac{1}{3}-\frac{1}{2}+1\right) = \frac{5}{3}$

51. $\int_0^2 f(x)dx = \int_0^1 x^4dx+\int_1^2 x^5dx = \frac{1}{5}x^5\big|_0^1+\frac{1}{6}x^6\big|_1^2 = \left(\frac{1}{5}-0\right)+\left(\frac{64}{6}-\frac{1}{6}\right) = 10.7$

52. $\int_{-\pi}^{\pi}f(x)dx = \int_{-\pi}^{0}x\,dx+\int_0^{\pi}\sin x\,dx = \frac{1}{2}x^2\big|_{-\pi}^{0}-\cos x\big|_0^{\pi} = \left(0-\frac{\pi^2}{2}\right)-(\cos\pi-\cos 0)$
$= -\frac{\pi^2}{2}-(-1-1) = 2-\frac{\pi^2}{2}$

53. From the graph, it appears that the area is about 60. The actual area is $\int_0^{27}x^{1/3}dx = \left[\frac{3}{4}x^{4/3}\right]_0^{27} = \frac{3}{4}\cdot 81-0$ $= \frac{243}{4} = 60.75$. This is $\frac{3}{4}$ of the area of the viewing rectangle.

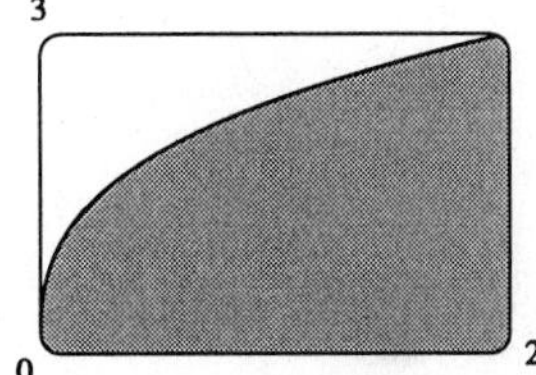

54. From the graph, it appears that the area is about $\frac{1}{3}$. The actual area is

$$\int_1^6 x^{-4}dx = \left[\frac{x^{-3}}{-3}\right]_1^6 = \left[\frac{-1}{3x^3}\right]_1^6$$

$$= -\frac{1}{3\cdot 216} + \frac{1}{3} = \frac{215}{648} \approx 0.3318.$$

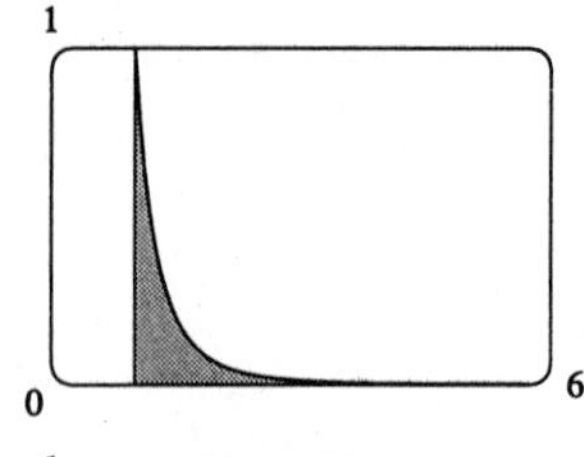

55. It appears that the area under the graph is about $\frac{2}{3}$ of the area of the viewing rectangle, or about $\frac{2}{3}\pi \approx 2.1$. The actual area is

$$\int_0^\pi \sin x\,dx = [-\cos x]_0^\pi = -\cos\pi + \cos 0$$
$$= -(-1) + 1 = 2.$$

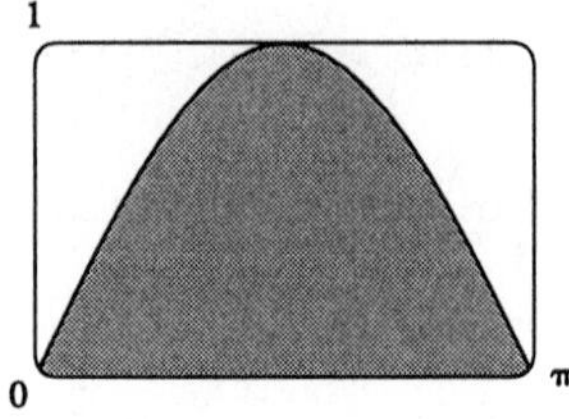

56. Splitting up the region as shown, we estimate that the area under the graph is $\frac{\pi}{3} + \frac{1}{4}\left(3\cdot\frac{\pi}{3}\right) \approx 1.8$. The actual area is

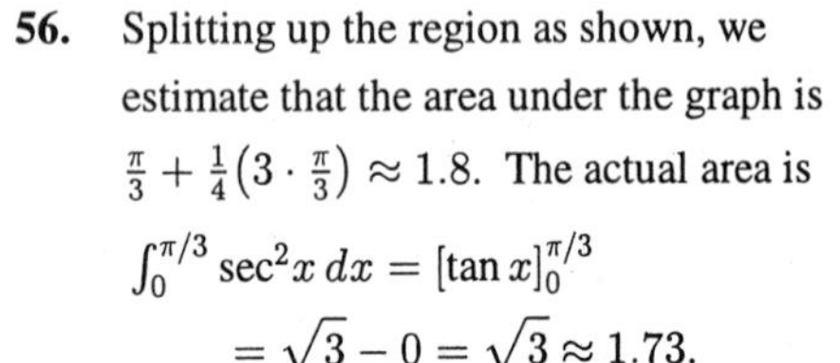

$$\int_0^{\pi/3} \sec^2 x\,dx = [\tan x]_0^{\pi/3}$$
$$= \sqrt{3} - 0 = \sqrt{3} \approx 1.73.$$

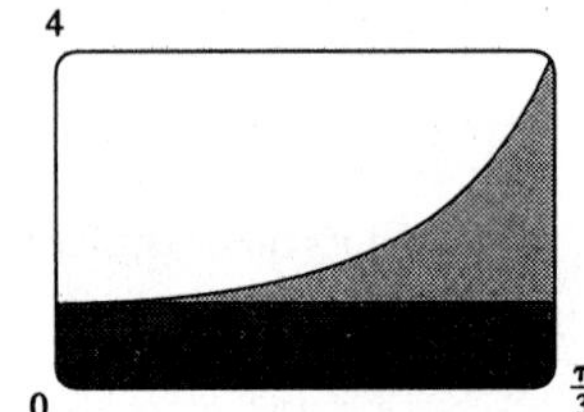

57. By zooming in on the graph of $y = x + x^2 - x^4$, we see that the graph has x-intercepts at $x = 0$ and at $x \approx 1.32$. So the area of the region below the curve and above the y-axis is about

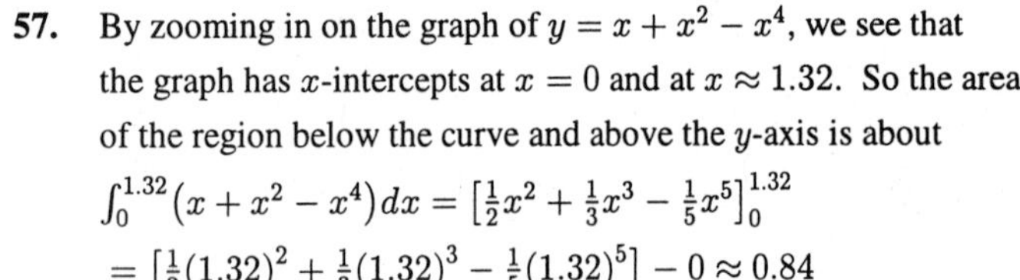

$$\int_0^{1.32}(x + x^2 - x^4)\,dx = \left[\tfrac{1}{2}x^2 + \tfrac{1}{3}x^3 - \tfrac{1}{5}x^5\right]_0^{1.32}$$
$$= \left[\tfrac{1}{2}(1.32)^2 + \tfrac{1}{3}(1.32)^3 - \tfrac{1}{5}(1.32)^5\right] - 0 \approx 0.84$$

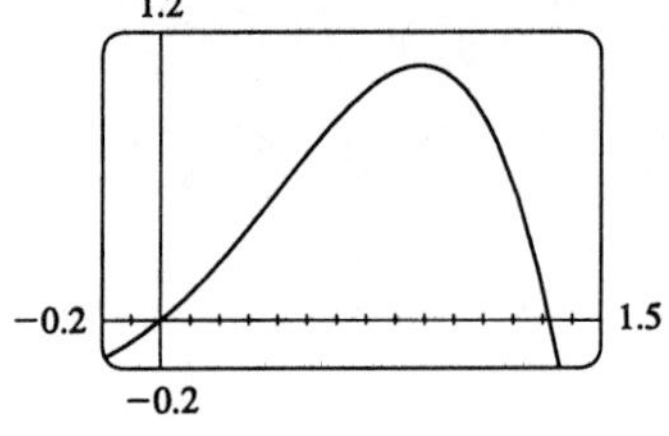

58. By zooming in on the graph of $y = 2x + 3x^4 - 2x^6$, we see that its x-intercepts are $x = 0$ and $x \approx 1.37$. So the area of the region below the curve and above the y-axis is about

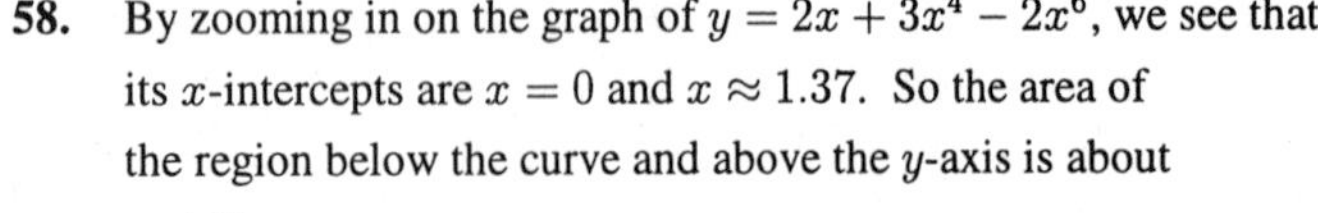

$$\int_0^{1.37}(2x + 3x^4 - 2x^6)\,dx = \left[x^2 + \tfrac{3}{5}x^5 - \tfrac{2}{7}x^7\right]_0^{1.37}$$
$$= \left[(1.37)^2 + \tfrac{3}{5}(1.37)^5 - \tfrac{2}{7}(1.37)^7\right] - 0 \approx 2.18$$

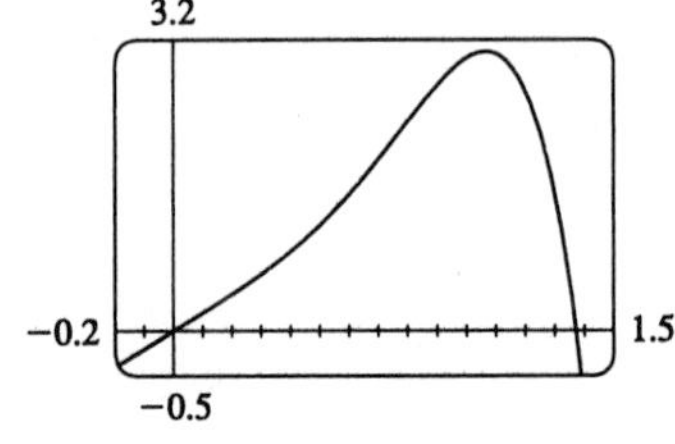

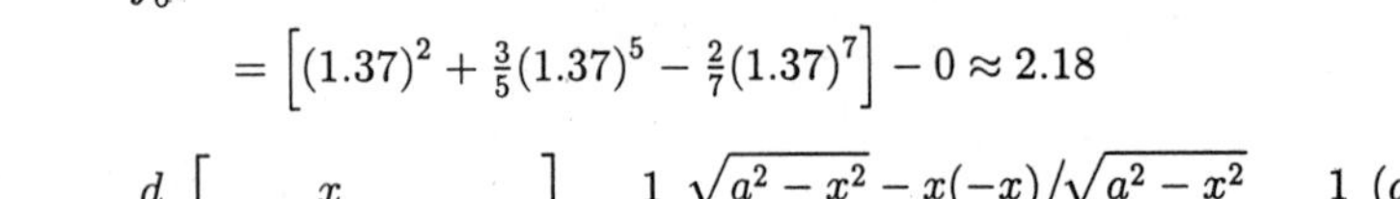

59. $$\frac{d}{dx}\left[\frac{x}{a^2\sqrt{a^2-x^2}} + C\right] = \frac{1}{a^2}\frac{\sqrt{a^2-x^2} - x(-x)/\sqrt{a^2-x^2}}{a^2-x^2} = \frac{1}{a^2}\frac{(a^2-x^2)+x^2}{(a^2-x^2)^{3/2}} = \frac{1}{\sqrt{(a^2-x^2)^3}}$$

60. $$\frac{d}{dx}\left[-\frac{\sqrt{x^2+a^2}}{a^2x} + C\right] = -\frac{1}{a^2}\frac{d}{dx}\left[\frac{\sqrt{x^2+a^2}}{x}\right] = -\frac{x\left(x/\sqrt{x^2+a^2}\right) - \sqrt{x^2+a^2}\cdot 1}{a^2x^2}$$
$$= -\frac{x^2 - (x^2+a^2)}{a^2x^2\sqrt{x^2+a^2}} = \frac{1}{x^2\sqrt{x^2+a^2}}$$

61. $$\frac{d}{dx}\left(\frac{x}{2} - \frac{\sin 2x}{4} + C\right) = \tfrac{1}{2} - \tfrac{1}{4}(\cos 2x)(2) + 0 = \tfrac{1}{2} - \tfrac{1}{2}\cos 2x = \tfrac{1}{2} - \tfrac{1}{2}(1 - 2\sin^2 x) = \sin^2 x$$

62. $\frac{d}{dx}(-x^2\cos x + 2\int x\cos x\,dx) = -x^2(-\sin x) - 2x\cos x + 2x\cos x = x^2\sin x$

63. $\int x\sqrt{x}\,dx = \int x^{3/2}\,dx = \frac{2}{5}x^{5/2} + C$

64. $\int \sqrt{x}(x^2 - 1/x)dx = \int(x^{5/2} - x^{-1/2})dx = \frac{2}{7}x^{7/2} - 2x^{1/2} + C$

65. $\displaystyle\int(2-\sqrt{x})^2dx = \int(4 - 4\sqrt{x} + x)dx = 4x - 4\frac{x^{3/2}}{3/2} + \frac{x^2}{2} + C = 4x - \frac{8}{3}x^{3/2} + \frac{1}{2}x^2 + C$

66. $\int(\cos x - 2\sin x)dx = \sin x + 2\cos x + C$

67. $\int(2x + \sec x\tan x)dx = x^2 + \sec x + C$

68. $\int(x^2 + 1 + 1/x^2)dx = \frac{1}{3}x^3 + x - 1/x + C$

69. **(a)** displacement $= \int_0^3(3t-5)dt = \left[\frac{3}{2}t^2 - 5t\right]_0^3 = \frac{27}{2} - 15 = -\frac{3}{2}$ m

(b) distance traveled $= \int_0^3|3t-5|dt = \int_0^{5/3}(5-3t)dt + \int_{5/3}^3(3t-5)dt = \left[5t - \frac{3}{2}t^2\right]_0^{5/3} + \left[\frac{3}{2}t^2 - 5t\right]_{5/3}^3$
$= \frac{25}{3} - \frac{3}{2}\cdot\frac{25}{9} + \frac{27}{2} - 15 - \left(\frac{3}{2}\cdot\frac{25}{9} - \frac{25}{3}\right) = \frac{41}{6}$ m

70. **(a)** displacement $= \int_1^6(t^2 - 2t - 8)dt = \left[\frac{1}{3}t^3 - t^2 - 8t\right]_1^6 = (72 - 36 - 48) - \left(\frac{1}{3} - 1 - 8\right) = -\frac{10}{3}$ m

(b) distance traveled $= \int_1^6|t^2 - 2t - 8|dt = \int_1^6|(t-4)(t+2)|dt = \int_1^4(-t^2 + 2t + 8)dt + \int_4^6(t^2 - 2t - 8)dt$
$= \left[-\frac{1}{3}t^3 + t^2 + 8t\right]_1^4 + \left[\frac{1}{3}t^3 - t^2 - 8t\right]_4^6$
$= \left(-\frac{64}{3} + 16 + 32\right) - \left(-\frac{1}{3} + 1 + 8\right) + (72 - 36 - 48) - \left(\frac{64}{3} - 16 - 32\right) = \frac{98}{3}$ m

71. **(a)** $v'(t) = a(t) = t + 4 \quad\Rightarrow\quad v(t) = \frac{1}{2}t^2 + 4t + C \quad\Rightarrow\quad 5 = v(0) = C \quad\Rightarrow\quad v(t) = \frac{1}{2}t^2 + 4t + 5$ m/s

Or: $v(t) - v(0) = \int_0^t a(u)du = \int_0^t(u+4)du = \frac{1}{2}u^2 + 4u\Big|_0^t = \frac{1}{2}t^2 + 4t \;\Rightarrow\; v(t) = \frac{1}{2}t^2 + 4t + 5$ m/s.

(b) distance traveled $= \int_0^{10}|v(t)|\,dt = \int_0^{10}\left|\frac{1}{2}t^2 + 4t + 5\right|dt = \int_0^{10}\left(\frac{1}{2}t^2 + 4t + 5\right)dt$
$= \left[\frac{1}{6}t^3 + 2t^2 + 5t\right]_0^{10} = \frac{500}{3} + 200 + 50 = 416\frac{2}{3}$ m

72. **(a)** $v'(t) = 2t + 3 \quad\Rightarrow\quad v(t) = t^2 + 3t + C \quad\Rightarrow\quad C = v(0) = -4 \quad\Rightarrow\quad v(t) = t^2 + 3t - 4$

(b) distance $= \int_0^3|t^2 + 3t - 4|dt = \int_0^3|(t+4)(t-1)|dt = \int_0^1(-t^2 - 3t + 4)dt + \int_1^3(t^2 + 3t - 4)dt$
$= \left[-\frac{1}{3}t^3 - \frac{3}{2}t^2 + 4t\right]_0^1 + \left[\frac{1}{3}t^3 + \frac{3}{2}\cdot 3t^2 - 4t\right]_1^3 = \left(-\frac{1}{3} - \frac{3}{2} + 4\right) + \left(9 + \frac{27}{2} - 12\right) - \left(\frac{1}{3} + \frac{3}{2} - 4\right) = \frac{89}{6}$ m

73. Since $m'(x) = \rho(x)$, $m = \int_0^4\rho(x)dx = \int_0^4(9 + 2\sqrt{x})dx = \left[9x + \frac{4}{3}x^{3/2}\right]_0^4 = 36 + \frac{32}{3} - 0 = \frac{140}{3} = 46\frac{2}{3}$ kg.

74. $n(10) - n(4) = \int_4^{10}(200 + 50t)dt = [200t + 25t^2]_4^{10} = 2000 + 2500 - (800 + 400) = 3300$

75. Let s be the position of the car. We know from Equation 12 that $s(100) - s(0) = \int_0^{100}v(t)dt$. We use the Midpoint Rule for $0 \le t \le 100$ with $n = 5$. Note that the length of each of the five time intervals is 20 seconds $= \frac{1}{180}$ hour. So the distance traveled is

$\int_0^{100}v(t)dt \approx \frac{1}{180}[v(10) + v(30) + v(50) + v(70) + v(90)] = \frac{1}{180}(38 + 58 + 51 + 53 + 47) = \frac{247}{180} \approx 1.4$ miles

76. Let w be the amount of water in the tank. We are given that the rate of water leaving the tank is $r(t) = -dw/dt$. So by the Fundamental Theorem of Calculus, the total loss of water from the tank after four hours is $w(0) - w(4) = \int_0^4 r(t)dt$. We use the Midpoint Rule with $n = 4$ and $\Delta t_i = 1$:

$\int_0^4 r(t)dt \approx \sum_{i=1}^4 r(\bar{t}_i)(1) = r(0.5) + r(1.5) + r(2.5) + r(3.5) \approx 5.9 + 5.4 + 4.7 + 3.6 = 19.6$ L.

77. $g(x) = \int_{2x}^{3x} \frac{u-1}{u+1}\,du = \int_{2x}^{0} \frac{u-1}{u+1}\,du + \int_{0}^{3x} \frac{u-1}{u+1}\,du = -\int_{0}^{2x} \frac{u-1}{u+1}\,du + \int_{0}^{3x} \frac{u-1}{u+1}\,du \Rightarrow$

$g'(x) = -\frac{2x-1}{2x+1} \cdot \frac{d}{dx}(2x) + \frac{3x-1}{3x+1} \cdot \frac{d}{dx}(3x) = -2 \cdot \frac{2x-1}{2x+1} + 3 \cdot \frac{3x-1}{3x+1}$

78. $g(x) = \int_{\tan x}^{x^2} \frac{1}{\sqrt{2+t^4}}\,dt = \int_{\tan x}^{1} \frac{dt}{\sqrt{2+t^4}} + \int_{1}^{x^2} \frac{dt}{\sqrt{2+t^4}} = -\int_{1}^{\tan x} \frac{dt}{\sqrt{2+t^4}} + \int_{1}^{x^2} \frac{dt}{\sqrt{2+t^4}} \Rightarrow$

$g'(x) = \frac{-1}{\sqrt{2+\tan^4 x}} \frac{d}{dx}(\tan x) + \frac{1}{\sqrt{2+x^8}} \frac{d}{dx}(x^2) = -\frac{\sec^2 x}{\sqrt{2+\tan^4 x}} + \frac{2x}{\sqrt{2+x^8}}$

79. $y = \int_{\sqrt{x}}^{x^3} \sqrt{t}\sin t\,dt = \int_{\sqrt{x}}^{1} \sqrt{t}\sin t\,dt + \int_{1}^{x^3} \sqrt{t}\sin t\,dt = -\int_{1}^{\sqrt{x}} \sqrt{t}\sin t\,dt + \int_{1}^{x^3} \sqrt{t}\sin t\,dt \Rightarrow$

$y' = -\sqrt[4]{x}(\sin\sqrt{x}) \cdot \frac{d}{dx}(\sqrt{x}) + x^{3/2}\sin(x^3) \cdot \frac{d}{dx}(x^3) = -\frac{\sqrt[4]{x}\sin\sqrt{x}}{2\sqrt{x}} + x^{3/2}\sin(x^3)(3x^2)$

$= 3x^{7/2}\sin(x^3) - (\sin\sqrt{x})/(2\sqrt[4]{x})$

80. $y = \int_{\cos x}^{5x} \cos(u^2)du = \int_0^{5x} \cos(u^2)du - \int_0^{\cos x} \cos(u^2)du \Rightarrow$

$y' = \cos(25x^2) \cdot \frac{d}{dx}(5x) - \cos(\cos^2 x) \cdot \frac{d}{dx}(\cos x)$

$= \cos(25x^2) \cdot 5 - \cos(\cos^2 x) \cdot (-\sin x) = 5\cos(25x^2) + \sin x \cos(\cos^2 x)$

81. $F(x) = \int_1^x f(t)dt \Rightarrow F'(x) = f(x) = \int_1^{x^2} \frac{\sqrt{1+u^4}}{u}\,du \Rightarrow$

$F''(x) = f'(x) = \frac{\sqrt{1+(x^2)^4}}{x^2} \cdot \frac{d}{dx}(x^2) = \frac{2\sqrt{1+x^8}}{x}$. So $F''(2) = \sqrt{1+2^8} = \sqrt{257}$.

82. For the curve to be concave up, we must have $y'' > 0$. $y = \int_0^x \frac{1}{1+t+t^2}dt \Leftrightarrow y' = \frac{1}{1+x+x^2} \Leftrightarrow$

$y'' = \frac{-(1+2x)}{(1+x+x^2)^2}$. For this expression to be positive we must have $(1+2x) < 0$, since $(1+x+x^2)^2 > 0$ for all x. $(1+2x) < 0 \Leftrightarrow x < -\frac{1}{2}$. Thus the curve is concave up on $(-\infty, -\frac{1}{2})$.

83. **(a)** The Fresnel Function $S(x) = \int_0^x \sin(\frac{\pi}{2}t^2)dt$ has local maximum values where $0 = S'(x) = \sin(\frac{\pi}{2}x^2)$ and S' changes from positive to negative. For $x > 0$, this happens when $\frac{\pi}{2}x^2 = (2n-1)\pi \Leftrightarrow$ $x = \sqrt{2(2n-1)}$, n any positive integer. For $x < 0$, S' changes from positive to negative where $x = -2\sqrt{n}$, since if $x < 0$, then as x increases, x^2 decreases. S' does not change sign at $x = 0$.

(b) S is concave upward on those intervals where $S''(x) \geq 0$. Differentiating our expression for $S'(x)$, we get $S''(x) = \cos(\frac{\pi}{2}x^2)(2\frac{\pi}{2}x) = \pi x\cos(\frac{\pi}{2}x^2)$. For $x > 0$, $S''(x) > 0$ where $\cos(\frac{\pi}{2}x^2) > 0 \Leftrightarrow 0 < \frac{\pi}{2}x^2 < \frac{\pi}{2}$ or $(2n - \frac{1}{2})\pi < \frac{\pi}{2}x^2 < (2n + \frac{1}{2})\pi$, n any integer $\Leftrightarrow$ $0 < x < 1$ or $\sqrt{4n-1} < x < \sqrt{4n+1}$, n any positive integer. For $x < 0$, as x increases, x^2 decreases, so the intervals of upward concavity for $x < 0$ are $\left(-\sqrt{4n-1}, -\sqrt{4n-3}\right)$, n any positive integer. To summarize: S is concave upward on the intervals $(0,1)$, $\left(-\sqrt{3}, -1\right)$, $\left(\sqrt{3}, \sqrt{5}\right)$, $\left(-\sqrt{7}, -\sqrt{5}\right)$, $\left(\sqrt{7}, 3\right)$, ...

(c)

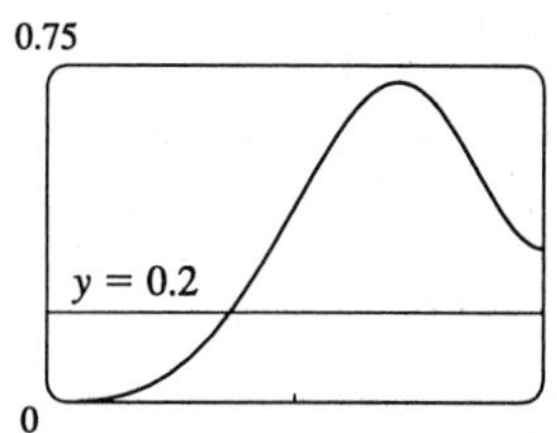

In Maple, we use `plot({int(sin(Pi*t^2/2),t=0..x),0.2},x=0..2);`. Note that Maple recognizes the Fresnel function, calling it `FresnelS(x)`. In Mathematica, we use `Plot[{Integrate[Sin[Pi*t^2/2],{t,0,x}],0.2},{x,0,2}]`. From the graphs, we see that $\int_0^x \sin\left(\frac{\pi}{2}t^2\right)dt = 0.2$ at $x \approx 0.74$.

84. In Maple, we should start by setting `si:=int(sin(t)/t,t=0..x);`. In Mathematica, the command is `si=Integrate[Sin[t]/t,{t,0,x}]`. Note that both systems recognize this function; Maple calls it `Si(x)` and Mathematica calls it `SinIntegral[x]`.

(a) In Maple, the command to generate the graph is `plot(si,x=-4*Pi..4*Pi);`. In Mathematica, it is `Plot[si,{x,-4*Pi,4*Pi}]`.

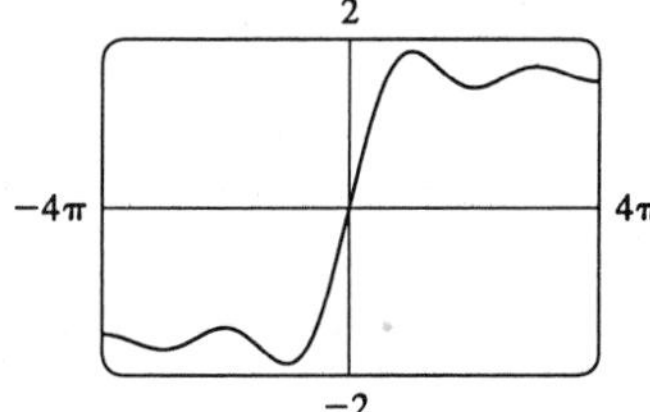

(b) $\text{Si}(x)$ has local maximum values where $\text{Si}'(x)$ changes from positive to negative, passing through 0. From the Fundamental Theorem we know that $\text{Si}'(x) = \dfrac{d}{dx}\displaystyle\int_0^x \frac{\sin t}{t}\,dt = \frac{\sin x}{x}$, so we must have $\sin x = 0$ for an extremum, and for $x > 0$ we must have $x = (2n-1)\pi$, n any positive integer, for Si' to be changing from positive to negative at x. For $x < 0$, we must have $x = 2n\pi$, n any positive integer, for a maximum, since the denominator of $\text{Si}'(x)$ is negative for $x < 0$. Thus the local maxima occur at $x = \pi, -2\pi, 3\pi, -4\pi, 5\pi, -6\pi, \ldots$.

(c) To find the first inflection point, we first find $\text{Si}''(x) = \dfrac{\cos x}{x} - \dfrac{\sin x}{x^2}$. In Maple this is done with the command `DDsi:=diff((diff(si),x),x);` (or use some other name); in Mathematica we use `DDsi=Dt[Dt[si,x],x]`. Then we try to solve the equation $\text{Si}''(x) = 0$. Neither system is able to solve this equation exactly — if we try the `solve` command (in Maple) or `Solve` (in Mathematica), they return nothing — so we use the approximation commands `fsolve(DDsi=0,x,3..5);` (in Maple) or `FindRoot[DDsi==0,{x,4}]` (in Mathematica). The `3..5` in `fsolve` indicates the interval $(3,5)$, since we can see from the graph that the first inflection point lies somewhere between $x = 3$ and $x = 5$. The 4 in `FindRoot` gives Mathematica a clue as to where to find the root. Both systems return the value $x \approx 4.4934$. To find the y-coordinate of the inflection point we evaluate $\text{Si}(4.4934) \approx 1.6556$. So the coordinates of the first inflection point to the right of the origin are about $(4.4934, 1.6556)$. Alternatively, we could graph $S''(x)$ and estimate the first positive x-value at which it changes sign.

(d) It seems from the graph that the function has horizontal asymptotes at $x \approx 1.5$, with $\lim\limits_{x\to\pm\infty} \text{Si}(x) \approx \pm 1.5$ respectively. We can use Maple's `limit` command to see what limit, if any, is approached by $\text{Si}(x)$ as $x \to \pm\infty$: use `limit(si,x=infinity);`. This command returns $\frac{\pi}{2}$, so we have $\lim\limits_{x\to\infty} \text{Si}(x) = \frac{\pi}{2}$. Also `limit(si,x=-infinity);` returns $-\frac{\pi}{2}$ [as we could have predicted, since $\text{Si}(x)$ is an odd function], so $\lim\limits_{x\to-\infty} \text{Si}(x) = -\frac{\pi}{2}$. So $\text{Si}(x)$ has the horizontal asymptotes $y = \pm\frac{\pi}{2}$.

(e) We use the `fsolve` command in Maple (or `FindRoot` in Mathematica) to find that the solution is $x \approx 1.1$. Or, as in Exercise 83(c), we graph $y = \text{Si}(x)$ and $y = 1$ on the same screen to see where they intersect.

85. **(a)** By the Fundamental Theorem of Calculus, $g'(x) = f(x)$. So $g'(x) = 0$ at $x = 1, 3, 5, 7$, and 9. g has local maxima at $x = 1$ and at $x = 5$ (since $f = g'$ changes from positive to negative there) and local minima at $x = 3$ and at $x = 7$. There is no local extremum at $x = 9$, since f is not defined for $x > 9$.

(b) We can see from the graph that $\left|\int_0^1 f\,dt\right| < \left|\int_1^3 f\,dt\right| < \cdots < \left|\int_7^9 f\,dt\right|$. So

$$g(9) = \int_0^9 f\,dt = \left|\int_0^1 f\,dt\right| - \left|\int_1^3 f\,dt\right| + \cdots - \left|\int_5^7 f\,dt\right| + \left|\int_7^9 f\,dt\right|$$

$$> g(5) = \int_0^5 f\,dt = \left|\int_0^1 f\,dt\right| - \cdots + \left|\int_3^5 f\,dt\right|,$$

which in turn is larger than $g(1)$.
So the absolute maximum of $g(x)$ occurs at $x = 9$.

(c) g is concave downward on those intervals where $g'' < 0$.
But $g'(x) = f(x)$, so $g''(x) = f'(x)$, which is negative on $\left(\frac{1}{2}, 2\right)$, $(4, 6)$ and $(8, 9)$. So g is concave down on these intervals.

(d)

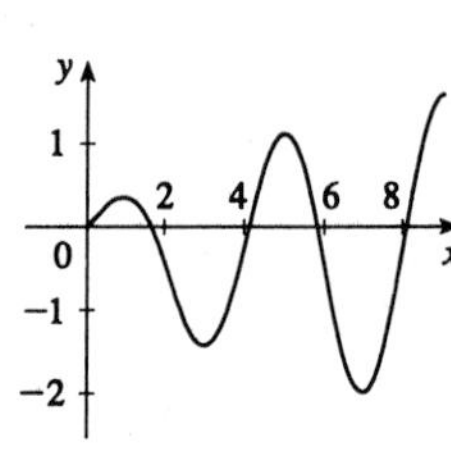

86. **(a)** By the Fundamental Theorem of Calculus, $g'(x) = f(x)$. So $g'(x) = f(x) = 0$ at $x = 2, 4, 6, 8$, and 10. g has maxima at $x = 2, 6$, and 10, since $f = g'$ changes from positive to negative there, and g has minima at $x = 4$ and 8.

(b) We can see from the graph that $\left|\int_0^2 f\,dt\right| > \left|\int_2^4 f\,dt\right| > \cdots > \left|\int_8^{10} f\,dt\right|$. So

$$g(2) = \int_0^2 f\,dt > g(6) = \left|\int_0^2 f\,dt\right| - \left|\int_2^4 f\,dt\right| + \left|\int_4^6 f\,dt\right|,$$

which in turn is larger than $g(10)$. So the absolute maximum of $g(x)$ occurs at $x = 2$.

(c) g is concave downward on those intervals where $g'' < 0$.
But $g'(x) = f(x)$, so $g''(x) = f'(x)$, which is negative on $(1, 3)$, $(5, 7)$ and $(9, 10)$. So g is concave down on these intervals.

(d)

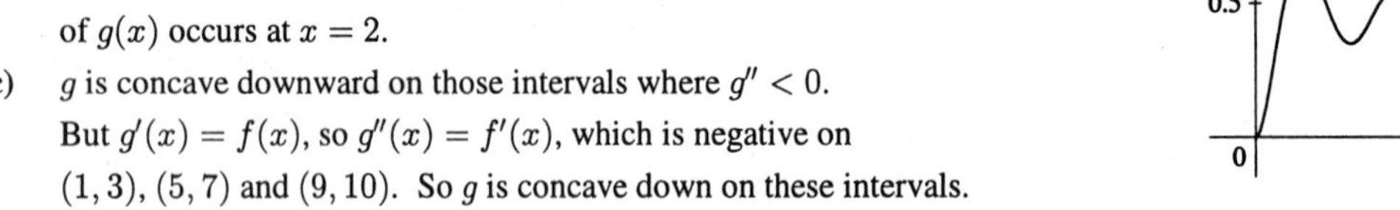

y
g
0.5
0
10 x

87. $$\lim_{n\to\infty} \sum_{i=1}^{n} \frac{i^3}{n^4} = \lim_{n\to\infty} \frac{1-0}{n} \sum_{i=1}^{n} \left(\frac{i}{n}\right)^3 = \int_0^1 x^3\,dx = \left.\frac{x^4}{4}\right|_0^1 = \frac{1}{4}$$

88. $$\lim_{n\to\infty} \frac{1}{n}\left(\sqrt{\frac{1}{n}} + \sqrt{\frac{2}{n}} + \cdots + \sqrt{\frac{n}{n}}\right) = \lim_{n\to\infty} \frac{1-0}{n} \sum_{i=1}^{n} \sqrt{\frac{i}{n}} = \int_0^1 \sqrt{x}\,dx = \left.\frac{2x^{3/2}}{3}\right|_0^1 = \frac{2}{3} - 0 = \frac{2}{3}$$

89. Suppose $h < 0$. Since f is continuous on $[x+h, x]$, the Extreme Value Theorem says that there are numbers u and v in $[x+h, x]$ such that $f(u) = m$ and $f(v) = M$, where m and M are the absolute minimum and maximum values of f on $[x+h, x]$. By Property 8 of integrals, $m(-h) \le \int_{x+h}^{x} f(t)dt \le M(-h)$; that is, $f(u)(-h) \le -\int_{x}^{x+h} f(t)dt \le f(v)(-h)$. Since $-h > 0$, we can divide this inequality by $-h$: $f(u) \le \dfrac{1}{h}\displaystyle\int_{x}^{x+h} f(t)dt \le f(v)$. By Equation 3, $\dfrac{g(x+h)-g(x)}{h} = \dfrac{1}{h}\displaystyle\int_{x}^{x+h} f(t)dt$ for $h \ne 0$, and hence $f(u) \le \dfrac{g(x+h)-g(x)}{h} \le f(v)$ which is Equation 4 in the case where $h < 0$.

90. $\dfrac{d}{dx}\int_{g(x)}^{h(x)} f(t)dt = \dfrac{d}{dx}\left[\int_{g(x)}^{a} f(t)dt + \int_{a}^{h(x)} f(t)dt\right]$ (where a is in the domain of f)

$= \dfrac{d}{dx}\left[-\int_{a}^{g(x)} f(t)dt\right] + \dfrac{d}{dx}\left[\int_{a}^{h(x)} f(t)dt\right] = -f(g(x))g'(x) + f(h(x))h'(x)$

$= f(h(x))h'(x) - f(g(x))g'(x)$

91. **(a)** Let $f(x) = \sqrt{x} \Rightarrow f'(x) = 1/(2\sqrt{x}) > 0$ for $x > 0 \Rightarrow f$ is increasing on $[0, \infty)$. If $x \ge 0$, then $x^3 \ge 0$, so $1 + x^3 \ge 1$ and since f is increasing, this means that $f(1+x^3) \ge f(1) \Rightarrow \sqrt{1+x^3} \ge 1$ for $x \ge 0$. Next let $g(t) = t^2 - t \Rightarrow g'(t) = 2t - 1 \Rightarrow g'(t) > 0$ when $t \ge 1$. Thus g is increasing on $[1, \infty)$. And since $g(1) = 0$, $g(t) \ge 0$ when $t \ge 1$. Now let $t = \sqrt{1+x^3}$, where $x \ge 0$. $\sqrt{1+x^3} \ge 1$ (from above) $\Rightarrow t \ge 1 \Rightarrow g(t) \ge 0 \Rightarrow (1+x^3) - \sqrt{1+x^3} \ge 0$ for $x \ge 0$. Therefore $1 \le \sqrt{1+x^3} \le 1 + x^3$ for $x \ge 0$.

(b) From part (a) and Property 7: $\int_0^1 1\,dx \le \int_0^1 \sqrt{1+x^3}\,dx \le \int_0^1 (1+x^3)dx \Leftrightarrow$ $x\big|_0^1 \le \int_0^1 \sqrt{1+x^3}\,dx \le \left[x + \frac{1}{4}x^4\right]_0^1 \Leftrightarrow 1 \le \int_0^1 \sqrt{1+x^3}\,dx \le 1 + \frac{1}{4} = 1.25$.

92. **(a)** If $x < 0$ then $g(x) = \int_0^x f(t)dt = \int_0^x 0\,dt = 0$. If $0 \le x \le 1$ then $g(x) = \int_0^x t\,dt = \frac{1}{2}t^2\big|_0^x = \frac{1}{2}x^2$.

If $1 < x \le 2$ then $g(x) = \int_0^x f(t)dt = \int_0^1 t\,dt + \int_1^x (2-t)dt = \frac{1}{2}t^2\big|_0^1 + \left[2t - \frac{1}{2}t^2\right]_1^x$

$= \frac{1}{2} + \left(2x - \frac{1}{2}x^2\right) - \left(2 - \frac{1}{2}\right) = 2x - \frac{1}{2}x^2 - 1.$

If $x > 2$, $g(x) = \int_0^x f(t)dt = \int_0^1 t\,dt + \int_1^2 (2-t)dt + \int_2^x 0\,dt = g(2) + 0 = 1$. So

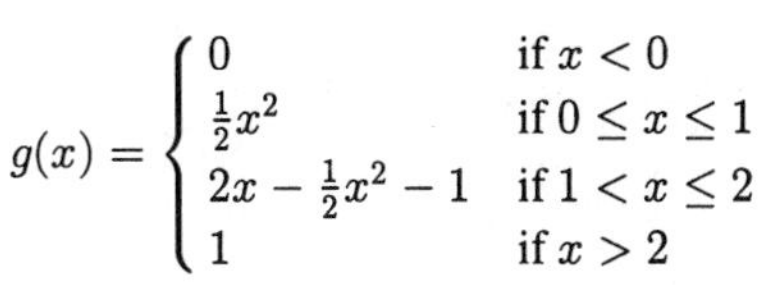

$$g(x) = \begin{cases} 0 & \text{if } x < 0 \\ \frac{1}{2}x^2 & \text{if } 0 \le x \le 1 \\ 2x - \frac{1}{2}x^2 - 1 & \text{if } 1 < x \le 2 \\ 1 & \text{if } x > 2 \end{cases}$$

(b)

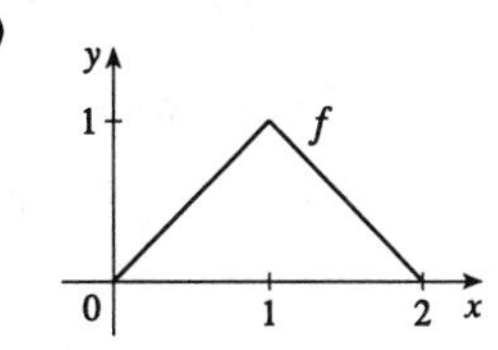

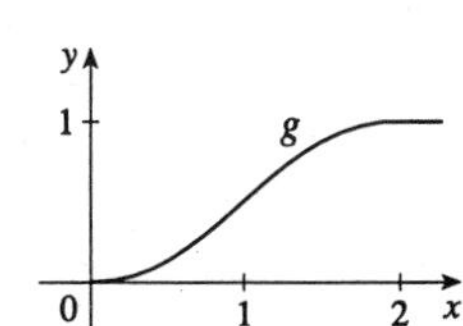

(c) f is differentiable on $(-\infty, 0)$, $(0, 1)$, $(1, 2)$ and $(2, \infty)$. g is differentiable on $(-\infty, \infty)$.

93. If $w'(t)$ is the rate of change of weight in pounds per year, then $w(t)$ represents the weight in pounds of the child at age t. We know from the Fundamental Theorem of Calculus that $\int_5^{10} w'(t)dt = w(10) - w(5)$, so the integral represents the weight gained by the child between the ages of 5 and 10.

94. By Part 2 of the Fundamental Theorem of Calculus, $f(x) = f(1) + \int_1^x f'(t)dt = f(1) + \int_1^x (2^t/t)\,dt$. This integral cannot be expressed in any simpler form. Since we want $f(1) = 0$, we have $f(x) = \int_1^x (2^t/t)\,dt$.

95. We differentiate both sides, using the Fundamental Theorem of Calculus, to get $\dfrac{f(x)}{x^2} = 2\dfrac{1}{2\sqrt{x}}$ $\Leftrightarrow$

$f(x) = x^{3/2}$. To find a, we substitute $x = a$ in the original equation to obtain $6 + \displaystyle\int_a^a \frac{f(t)}{t^2}\,dt = 2\sqrt{a}$ $\Rightarrow$

$6 + 0 = 2\sqrt{a} \quad \Rightarrow \quad a = 9.$

96. By the Fundamental Theorem of Calculus, $\int_1^2 (h')'(u)du = h'(2) - h'(1) = 5 - 2 = 3$. The other information is unnecessary.

97. $\int_4^8 (1/x)dx = [\ln x]_4^8 = \ln 8 - \ln 4 = \ln \frac{8}{4} = \ln 2$

98. $\int_{\ln 3}^{\ln 6} 8e^x\,dx = [8e^x]_{\ln 3}^{\ln 6} = 8(e^{\ln 6} - e^{\ln 3}) = 8(6 - 3) = 24$

99. $\displaystyle\int_8^9 2^t\,dt = \left[\frac{1}{\ln 2}2^t\right]_8^9 = \frac{1}{\ln 2}(2^9 - 2^8) = \frac{2^8}{\ln 2}$

100. $\int_{-e^2}^{-e} (3/x)dx = [3\ln|x|]_{-e^2}^{-e} = 3\ln e - 3\ln(e^2) = 3\cdot 1 - 3\cdot 2 = -3$

101. $\displaystyle\int_1^{\sqrt{3}} \frac{6}{1+x^2}\,dx = 6\left[\tan^{-1} x\right]_1^{\sqrt{3}} = 6\tan^{-1}\sqrt{3} - 6\tan^{-1} 1 = 6\cdot\frac{\pi}{3} - 6\cdot\frac{\pi}{4} = \frac{\pi}{2}$

102. $\displaystyle\int_0^{0.5} \frac{dx}{\sqrt{1-x^2}} = [\sin^{-1} x]_0^{0.5} = \sin^{-1}\frac{1}{2} - \sin^{-1} 0 = \frac{\pi}{6}$

103.
$$\int_1^e \frac{x^2+x+1}{x}\,dx = \int_1^e \left[x + 1 + \frac{1}{x}\right]dx = \left[\tfrac{1}{2}x^2 + x + \ln x\right]_1^e$$
$$= \left[\tfrac{1}{2}e^2 + e + \ln e\right] - \left[\tfrac{1}{2} + 1 + \ln 1\right] = \tfrac{1}{2}e^2 + e - \tfrac{1}{2}$$

104.
$$\int_4^9 \left(\sqrt{x} + \frac{1}{\sqrt{x}}\right)^2 dx = \int_4^9 \left(x + 2 + \frac{1}{x}\right)dx = \left[\tfrac{1}{2}x^2 + 2x + \ln x\right]_4^9$$
$$= \tfrac{81}{2} + 18 + \ln 9 - (8 + 8 + \ln 4) = \tfrac{85}{2} + \ln\tfrac{9}{4}$$

105. $\displaystyle\int \left[x^2 + 1 + \frac{1}{x^2+1}\right]dx = \tfrac{1}{3}x^3 + x + \tan^{-1} x + C$

106. area $= \int_1^2 (1/x)dx = [\ln x]_1^2 = \ln 2 - \ln 1 = \ln 2$

EXERCISES 4.5

1. Let $u = x^2 - 1$. Then $du = 2x\,dx$, so $\int x(x^2-1)^{99}\,dx = \int u^{99}\left(\frac{1}{2}\,du\right) = \frac{1}{2}\frac{u^{100}}{100} + C = \frac{1}{200}(x^2-1)^{100} + C$.

2. $u = 2 + x^3$. Then $du = 3x^2 dx$, so

$$\int \frac{x^2\,dx}{\sqrt{2+x^3}} = \int \frac{(1/3)\,du}{\sqrt{u}} = \frac{1}{3}\int u^{-1/2}\,du = \frac{1}{3}\frac{u^{1/2}}{1/2} + C = \tfrac{2}{3}\sqrt{2+x^3} + C.$$

3. Let $u = 4x$. Then $du = 4\,dx$, so $\int \sin 4x\,dx = \int \sin u\left(\frac{1}{4}\,du\right) = \frac{1}{4}(-\cos u) + C = -\frac{1}{4}\cos 4x + C$

4. Let $u = 2x + 1$. Then $du = 2\,dx$, so

$$\int \frac{dx}{(2x+1)^2} = \int \frac{(1/2)\,du}{u^2} = \frac{1}{2}\int u^{-2}\,du = -\tfrac{1}{2}u^{-1} + C = -\frac{1}{2(2x+1)} + C.$$

5. Let $u = x^2 + 6x$. Then $du = 2(x+3)dx$, so

$$\int \frac{x+3}{(x^2+6x)^2}\,dx = \frac{1}{2}\int \frac{du}{u^2} = \frac{1}{2}\int u^{-2}\,du = -\tfrac{1}{2}u^{-1} + C = -\frac{1}{2(x^2+6x)} + C.$$

6. Let $u = a\theta$. Then $du = a\,d\theta$, so $\int \sec a\theta \tan a\theta\,d\theta = \int \sec u \tan u\left(\frac{1}{a}\right)du = \frac{1}{a}\sec u + C = \frac{\sec a\theta}{a} + C$.

7. Let $u = x^2 + x + 1$. Then $du = (2x+1)dx$, so

$\int (2x+1)(x^2+x+1)^3\,dx = \int u^3\,du = \frac{1}{4}u^4 + C = \frac{1}{4}(x^2+x+1)^4 + C$.

8. Let $u = 1 - x^4$. Then $du = -4x^3\,dx$, so

$\int x^3(1-x^4)^5\,dx = \int u^5\left(-\frac{1}{4}\,du\right) = -\frac{1}{4}\left(\frac{1}{6}u^6\right) + C = -\frac{1}{24}(1-x^4)^6 + C$.

9. Let $u = x - 1$. Then $du = dx$, so $\int \sqrt{x-1}\,dx = \int u^{1/2}\,du = \frac{2}{3}u^{3/2} + C = \frac{2}{3}(x-1)^{3/2} + C$.

10. Let $u = 1 - x$. Then $du = -dx$, so $\int \sqrt[3]{x-1}\,dx = -\int u^{1/3}\,du = -\frac{3}{4}u^{4/3} + C = -\frac{3}{4}(1-x)^{4/3} + C$.

11. Let $u = 2 + x^4$. Then $du = 4x^3\,dx$, so $\int x^3\sqrt{2+x^4}\,dx = \int u^{1/2}\left(\frac{1}{4}\,du\right) = \frac{1}{4}\frac{u^{3/2}}{3/2} + C = \frac{1}{6}(2+x^4)^{3/2} + C$.

12. Let $u = x^2 + 1$. Then $du = 2x\,dx$, so

$$\int x(x^2+1)^{3/2}\,dx = \int u^{3/2}\left(\tfrac{1}{2}\,du\right) = \frac{1}{2}\frac{u^{5/2}}{5/2} + C = \tfrac{1}{5}u^{5/2} + C = \tfrac{1}{5}(x^2+1)^{5/2} + C.$$

13. Let $u = t + 1$. Then $du = dt$, so $\int \frac{2}{(t+1)^6}\,dt = 2\int u^{-6}\,du = -\frac{2}{5}u^{-5} + C = -\frac{2}{5(t+1)^5} + C$.

14. Let $u = 1 - 3t$. Then $du = -3\,dt$, so

$$\int \frac{1}{(1-3t)^4}\,dt = \int u^{-4}\left(-\tfrac{1}{3}\,du\right) = -\frac{1}{3}\left(\frac{u^{-3}}{-3}\right) + C = \frac{1}{9u^3} + C = \frac{1}{9(1-3t)^3} + C.$$

15. Let $u = 1 - 2y$. Then $du = -2\,dy$, so

$$\int (1-2y)^{1.3} dy = \int u^{1.3}\left(-\tfrac{1}{2}\,du\right) = -\frac{1}{2}\left(\frac{u^{2.3}}{2.3}\right) + C = -\frac{(1-2y)^{2.3}}{4.6} + C.$$

16. Let $u = 3 - 5y$. Then $du = -5\,dy$, so

$\int \sqrt[5]{3-5y}\,dy = \int u^{1/5}\left(-\frac{1}{5}\,du\right) = -\frac{1}{5}\cdot\frac{5}{6}\,u^{6/5} + C = -\frac{1}{6}(3-5y)^{6/5} + C.$

17. Let $u = 2\theta$. Then $du = 2\,d\theta$, so $\int \cos 2\theta\,d\theta = \int \cos u\left(\frac{1}{2}\,du\right) = \frac{1}{2}\sin u + C = \frac{1}{2}\sin 2\theta + C.$

18. Let $u = 3\theta$. Then $du = 3\,d\theta$, so $\int \sec^2 3\theta\,d\theta = \int \sec^2 u\left(\frac{1}{3}\,du\right) = \frac{1}{3}\tan u + C = \frac{1}{3}\tan 3\theta + C.$

19. Let $u = x + 2$. Then $du = dx$, so $\displaystyle\int \frac{x}{\sqrt[4]{x+2}}\,dx = \int \frac{u-2}{\sqrt[4]{u}}\,du = \int \left(u^{3/4} - 2u^{-1/4}\right)du$

$= \frac{4}{7}u^{7/4} - 2\cdot\frac{4}{3}u^{3/4} + C = \frac{4}{7}(x+2)^{7/4} - \frac{8}{3}(x+2)^{3/4} + C.$

20. Let $u = 1 - x$. Then $x = 1 - u$ and $dx = -du$, so

$$\int \frac{x^2}{\sqrt{1-x}}\,dx = \int \frac{(1-u)^2}{\sqrt{u}}(-du) = -\int \frac{1-2u+u^2}{\sqrt{u}}\,du = -\int \left(u^{-1/2} - 2u^{1/2} + u^{3/2}\right)du$$

$$= -\left[2u^{1/2} - 2\cdot\tfrac{2}{3}u^{3/2} + \tfrac{2}{5}u^{5/2}\right] + C = -2\sqrt{1-x} + \tfrac{4}{3}(1-x)^{3/2} - \tfrac{2}{5}(1-x)^{5/2} + C$$

21. Let $u = t^2$. Then $du = 2t\,dt$, so $\int t\sin(t^2)dt = \int \sin u\left(\frac{1}{2}\,du\right) = -\frac{1}{2}\cos u + C = -\frac{1}{2}\cos(t^2) + C.$

22. Let $u = 1 + \sqrt{x}$. Then $du = \dfrac{dx}{2\sqrt{x}}$, so $\displaystyle\int \frac{\left(1+\sqrt{x}\right)^9}{\sqrt{x}}\,dx = \int u^9\,2\,du = 2\frac{u^{10}}{10} + C = \frac{\left(1+\sqrt{x}\right)^{10}}{5} + C.$

23. Let $u = 1 - x^2$. Then $x^2 = 1 - u$ and $2x\,dx = -du$, so

$$\int x^3\left(1-x^2\right)^{3/2}dx = \int \left(1-x^2\right)^{3/2}x^2\cdot x\,dx = \int u^{3/2}(1-u)\left(-\tfrac{1}{2}\right)du = \tfrac{1}{2}\int\left(u^{5/2} - u^{3/2}\right)du$$

$$= \tfrac{1}{2}\left[\tfrac{2}{7}u^{7/2} - \tfrac{2}{5}u^{5/2}\right] + C = \tfrac{1}{7}\left(1-x^2\right)^{7/2} - \tfrac{1}{5}\left(1-x^2\right)^{5/2} + C.$$

24. Let $u = 1 - t^3$. Then $du = -3t^2\,dt$, so

$\int t^2\cos(1-t^3)dt = \int \cos u\left(-\frac{1}{3}\,du\right) = -\frac{1}{3}\sin u + C = -\frac{1}{3}\sin(1-t^3) + C.$

25. Let $u = 1 + \sec x$. Then $du = \sec x\tan x\,dx$, so

$\int \sec x\tan x\sqrt{1+\sec x}\,dx = \int u^{1/2}\,du = \frac{2}{3}u^{3/2} + C = \frac{2}{3}(1+\sec x)^{3/2} + C.$

26. Let $u = \sqrt{x}$. Then $du = \dfrac{dx}{2\sqrt{x}}$, so $\displaystyle\int \frac{\cos\sqrt{x}}{\sqrt{x}}\,dx = \int \cos u\cdot 2\,du = 2\sin u + C = 2\sin\sqrt{x} + C.$

27. Let $u = \cos x$. Then $du = -\sin x\,dx$, so $\int \cos^4 x\sin x\,dx = \int u^4(-du) = -\frac{1}{5}u^5 + C = -\frac{1}{5}\cos^5 x + C.$

28. Let $u = ax^2 + 2bx + c$. Then $du = 2(ax+b)dx$, so

$$\int \frac{(ax+b)dx}{\sqrt{ax^2+2bx+c}} = \int \frac{(1/2)\,du}{\sqrt{u}} = \frac{1}{2}\int u^{-1/2}du = u^{1/2} + C = \sqrt{ax^2+2bx+c} + C.$$

29. Let $u = 2x + 3$. Then $du = 2\,dx$, so

$\int \sin(2x+3)dx = \int \sin u\left(\frac{1}{2}\,du\right) = -\frac{1}{2}\cos u + C = -\frac{1}{2}\cos(2x+3) + C.$

30. Let $u = 7 - 3x$. Then $du = -3\,dx$, so
$\int \cos(7-3x)dx = \int \cos u \left(-\frac{1}{3}\,du\right) = -\frac{1}{3}\sin u + C = -\frac{1}{3}\sin(7-3x) + C.$

31. Let $u = 3x$. Then $du = 3\,dx$, so
$\int(\sin 3\alpha - \sin 3x)dx = \int(\sin 3\alpha - \sin u)\frac{1}{3}\,du = \frac{1}{3}[(\sin 3\alpha)\,u + \cos u] + C = (\sin 3\alpha)\,x + \frac{1}{3}\cos 3x + C.$

32. Let $u = x^3 + 1$. Then $x^3 = u - 1$ and $du = 3x^2\,dx$, so

$$\int \sqrt[3]{x^3+1}\,x^5\,dx = \int u^{1/3}(u-1)\tfrac{1}{3}\,du = \tfrac{1}{3}\int\left(u^{4/3} - u^{1/3}\right)du = \tfrac{1}{3}\left[\tfrac{3}{7}u^{7/3} - \tfrac{3}{4}u^{4/3}\right] + C$$
$$= \tfrac{1}{7}(x^3+1)^{7/3} - \tfrac{1}{4}(x^3+1)^{4/3} + C.$$

33. Let $u = b + cx^{a+1}$. Then $du = (a+1)cx^a\,dx$, so

$$\int x^a\sqrt{b + cx^{a+1}}\,dx = \int u^{1/2}\frac{1}{(a+1)c}\,du = \frac{1}{(a+1)c}\left(\tfrac{2}{3}u^{3/2}\right) + C = \frac{2}{3c(a+1)}\left(b + cx^{a+1}\right)^{3/2} + C.$$

34. Let $u = \sin x$. Then $du = \cos x\,dx$, so $\int \cos x\cos(\sin x)dx = \int \cos u\,du = \sin u + C = \sin(\sin x) + C.$

35. $f(x) = \dfrac{3x-1}{(3x^2-2x+1)^4}$. Let $u = 3x^2 - 2x + 1$.

Then $du = (6x-2)dx = 2(3x-1)dx$, so

$$\int \frac{3x-1}{(3x^2-2x+1)^4}\,dx = \int \frac{1}{u^4}\left(\tfrac{1}{2}\,du\right) = \tfrac{1}{2}\int u^{-4}du$$
$$= -\tfrac{1}{6}u^{-3} + C = -\frac{1}{6(3x^2-2x+1)^3} + C.$$

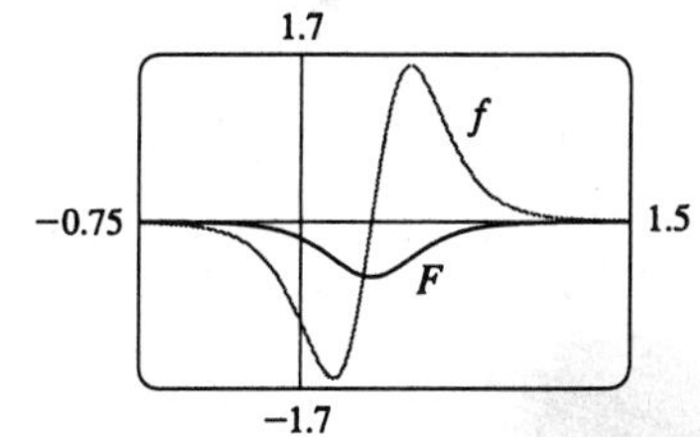

Notice that at $x = \frac{1}{3}$, the integrand goes from negative to positive, and the graph of the integral has a horizontal tangent (a local minimum).

36. $f(x) = \dfrac{x}{\sqrt{x^2+1}}$. Let $u = x^2 + 1$. Then $du = 2x\,dx$, so

$$\int \frac{x}{\sqrt{x^2+1}}\,dx = \int \frac{1}{\sqrt{u}}\left(\tfrac{1}{2}\,du\right) = \tfrac{1}{2}\int u^{-1/2}du$$
$$= u^{1/2} + C = \sqrt{x^2+1} + C.$$

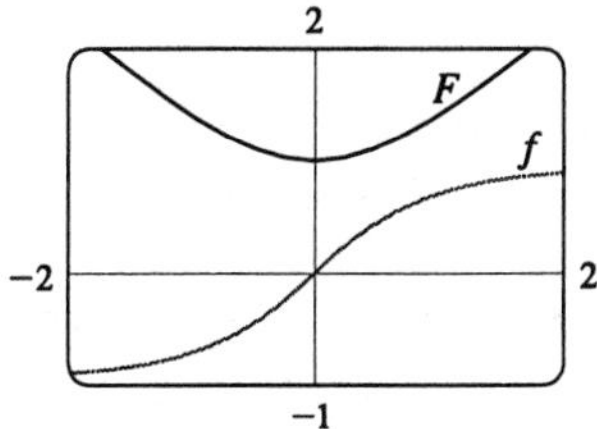

Note that at $x = 0$, the graph of the integrand crosses the x-axis from below, and the graph of the integral has a horizontal tangent (minimum).

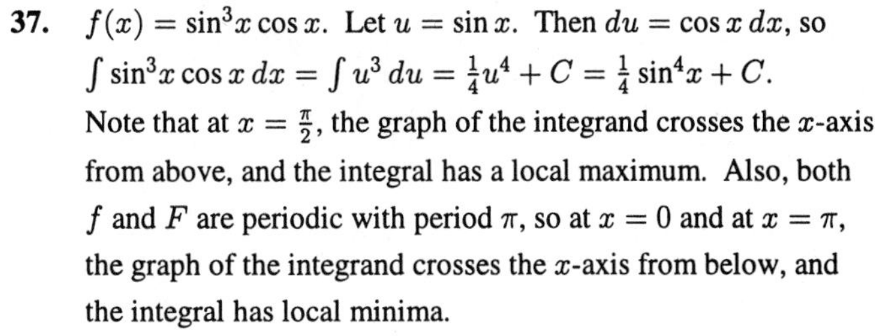

37. $f(x) = \sin^3 x\cos x$. Let $u = \sin x$. Then $du = \cos x\,dx$, so
$\int \sin^3 x\cos x\,dx = \int u^3\,du = \frac{1}{4}u^4 + C = \frac{1}{4}\sin^4 x + C.$
Note that at $x = \frac{\pi}{2}$, the graph of the integrand crosses the x-axis from above, and the integral has a local maximum. Also, both f and F are periodic with period π, so at $x = 0$ and at $x = \pi$, the graph of the integrand crosses the x-axis from below, and the integral has local minima.

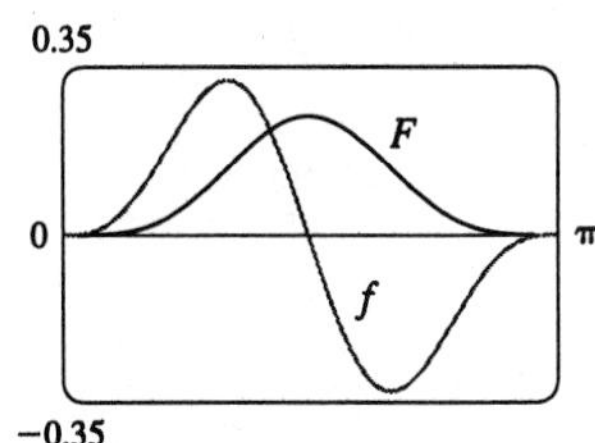

38. $f(x) = \tan^2\theta \sec^2\theta$

Let $u = \tan\theta$. Then $du = \sec^2\theta\, d\theta$, so

$$\int \tan^2\theta \sec^2\theta\, d\theta = \int u^2 du = \tfrac{1}{3}u^3 + C = \tfrac{1}{3}\tan^3\theta + C.$$

Note that f is positive, and F is increasing.

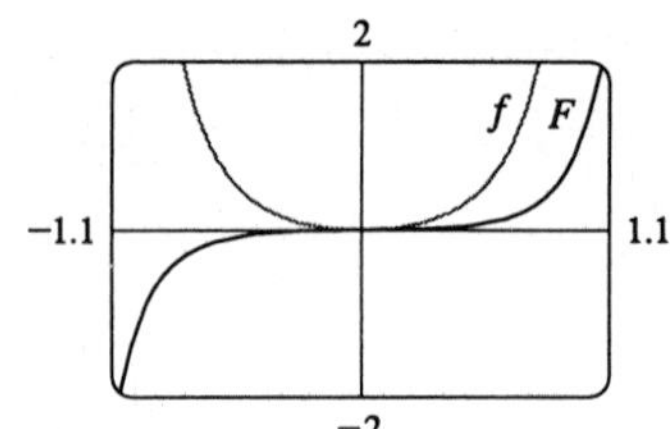

39. Let $u = 2x - 1$. Then $du = 2\,dx$, so $\int_0^1 (2x-1)^{100} dx = \int_{-1}^1 u^{100}\left(\frac{1}{2}\,du\right) = \int_0^1 u^{100}\,du$ [since the integrand is an even function] $= \left[\frac{1}{101}u^{101}\right]_0^1 = \frac{1}{101}$.

40. Let $u = 1 - 2x$. Then $du = -2\,dx$, so

$\int_0^{-4} \sqrt{1-2x}\,dx = \int_1^9 u^{1/2}\left(-\frac{1}{2}\,du\right) = -\frac{1}{2}\cdot\frac{2}{3}u^{3/2}\Big|_1^9 = -\frac{1}{3}(27-1) = -\frac{26}{3}$.

41. Let $u = x^4 + x$. Then $du = (4x^3+1)dx$, so $\displaystyle\int_0^1 (x^4+x)^5(4x^3+1)\,dx = \int_0^2 u^5 du = \left[\frac{u^6}{6}\right]_0^2 = \frac{2^6}{6} = \frac{32}{3}$.

42. Let $u = x^3 - x$. Then $du = (3x^2-1)dx$, so $\displaystyle\int_2^3 \frac{3x^2-1}{(x^3-x)^2}\,dx = \int_6^{24}\frac{du}{u^2} = \left[-\frac{1}{u}\right]_6^{24} = -\frac{1}{24}+\frac{1}{6} = \frac{1}{8}$.

43. Let $u = x - 1$. Then $du = dx$, so

$\int_1^2 x\sqrt{x-1}\,dx = \int_0^1 (u+1)\sqrt{u}\,du = \int_0^1 \left(u^{3/2}+u^{1/2}\right)du = \left[\frac{2}{5}u^{5/2}+\frac{2}{3}u^{3/2}\right]_0^1 = \dfrac{2}{5}+\dfrac{2}{3} = \dfrac{16}{15}$.

44. Let $u = 1 + 2x$. Then $x = \frac{1}{2}(u-1)$ and $du = 2\,dx$, so

$$\int_0^4 \frac{x\,dx}{\sqrt{1+2x}} = \int_1^9 \frac{\frac{1}{2}(u-1)}{\sqrt{u}}\frac{du}{2} = \tfrac{1}{4}\int_1^9\left(u^{1/2}-u^{-1/2}\right)du = \tfrac{1}{4}\left[\tfrac{2}{3}u^{3/2}-2u^{1/2}\right]_1^9$$
$$= \tfrac{1}{2}\left[\tfrac{1}{3}u^{3/2}-u^{1/2}\right]_1^9 = \tfrac{1}{2}\left[(9-3)-\left(\tfrac{1}{3}-1\right)\right] = \tfrac{10}{3}.$$

45. Let $u = \pi t$. Then $du = \pi\,dt$, so $\int_0^1 \cos\pi t\,dt = \int_0^\pi \cos u\left(\frac{1}{\pi}\,du\right) = \frac{1}{\pi}\sin u\Big|_0^\pi = \frac{1}{\pi}(0-0) = 0$.

46. Let $u = 4t$. Then $du = 4\,dt$, so $\int_0^{\pi/4}\sin 4t\,dt = \int_0^\pi \sin u\left(\frac{1}{4}\,du\right) = -\frac{1}{4}\cos u\Big|_0^\pi = \frac{1}{4}-\left(-\frac{1}{4}\right) = \frac{1}{2}$.

47. Let $u = 1 + \dfrac{1}{x}$. Then $du = -\dfrac{dx}{x^2}$, so

$\displaystyle\int_1^4 \frac{1}{x^2}\sqrt{1+\frac{1}{x}}\,dx = \int_2^{5/4} u^{1/2}(-du) = \int_{5/4}^2 u^{1/2}\,du = \left[\tfrac{2}{3}u^{3/2}\right]_{5/4}^2 = \tfrac{2}{3}\left[2\sqrt{2}-\tfrac{5\sqrt{5}}{8}\right] = \tfrac{4\sqrt{2}}{3}-\tfrac{5\sqrt{5}}{12}$.

48. $\displaystyle\int_0^2 \frac{dx}{(2x-3)^2}$ does not exist since $\dfrac{1}{(2x-3)^2}$ has an infinite discontinuity at $x = \frac{3}{2}$.

49. Let $u = \cos\theta$. Then $du = -\sin\theta\,d\theta$, so

$$\int_0^{\pi/3}\frac{\sin\theta}{\cos^2\theta}\,d\theta = \int_1^{1/2}\frac{-du}{u^2} = \int_{1/2}^1 u^{-2}\,du = \left[-\frac{1}{u}\right]_{1/2}^1 = -1+2 = 1.$$

50. $\displaystyle\int_{-\pi/2}^{\pi/2}\frac{x^2\sin x}{1+x^6}\,dx = 0$ by Equation 6, since $f(x) = \dfrac{x^2\sin x}{1+x^6}$ is an odd function.

51. Let $u = 1 + 2x$. Then $du = 2\,dx$, so $\displaystyle\int_0^{13} \frac{dx}{\sqrt[3]{(1+2x)^2}} = \int_1^{27} u^{-2/3}\left(\tfrac{1}{2}\,du\right) = \tfrac{1}{2} \cdot 3u^{1/3}\Big|_1^{27} = \tfrac{3}{2}(3-1) = 3.$

52. $\int_{-\pi/3}^{\pi/3} \sin^5\theta\,d\theta = 0$ since $f(\theta) = \sin^5\theta$ is an odd function.

53. $\displaystyle\int_0^4 \frac{dx}{(x-2)^3}$ does not exist since $\dfrac{1}{(x-2)^3}$ has an infinite discontinuity at $x = 2$.

54. Let $u = a^2 - x^2$. Then $du = -2x\,dx$, so
$\int_0^a x\sqrt{a^2 - x^2}\,dx = \int_{a^2}^0 u^{1/2}\left(-\tfrac{1}{2}\,du\right) = \tfrac{1}{2}\int_0^{a^2} u^{1/2}\,du = \tfrac{1}{2} \cdot \tfrac{2}{3}u^{3/2}\Big|_0^{a^2} = \tfrac{1}{3}a^3.$

55. Let $u = x^2 + a^2$. Then $du = 2x\,dx$, so
$\int_0^a x\sqrt{x^2 + a^2}\,dx = \int_{a^2}^{2a^2} u^{1/2}\left(\tfrac{1}{2}\,du\right) = \tfrac{1}{2}\left[\tfrac{2}{3}\,u^{3/2}\right]_{a^2}^{2a^2} = \left[\tfrac{1}{3}u^{3/2}\right]_{a^2}^{2a^2} = \tfrac{1}{3}\left(2\sqrt{2} - 1\right)a^3.$

56. $\int_{-a}^a x\sqrt{x^2 + a^2}\,dx = 0$ since $f(x) = x\sqrt{x^2 + a^2}$ is an odd function.

57. From the graph, it appears that the area under the curve is about $1 +$ a little more than $\tfrac{1}{2} \cdot 1 \cdot 0.7$, or about 1.4. The exact area is given by $A = \int_0^1 \sqrt{2x+1}\,dx$. Let $u = 2x + 1$, so $du = 2\,dx$, the limits change to $2 \cdot 0 + 1 = 1$ and $2 \cdot 1 + 1 = 3$, and
$A = \int_1^3 \sqrt{u}\left(\tfrac{1}{2}\,du\right) = \tfrac{1}{3}u^{3/2}\Big|_1^3 = \tfrac{1}{3}\left(3\sqrt{3} - 1\right)$
$= \sqrt{3} - \tfrac{1}{3} \approx 1.399.$

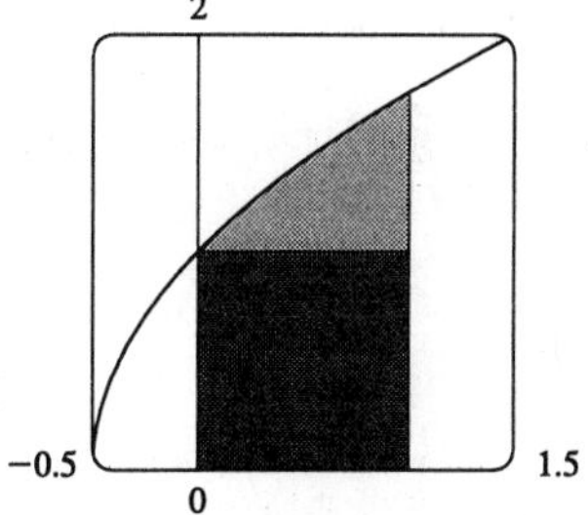

58. From the graph, it appears that the area under the curve is almost $\tfrac{1}{2} \cdot \pi \cdot 2.6$, or about 4. The exact area is given by
$A = \int_0^\pi (2\sin x - \sin 2x)dx = -2\cos x\Big|_0^\pi - \int_0^\pi \sin 2x\,dx.$
But the second integral is 0 by symmetry. So the area is
$A = -2(\cos\pi - \cos 0) = -2(-1-1) = 4.$

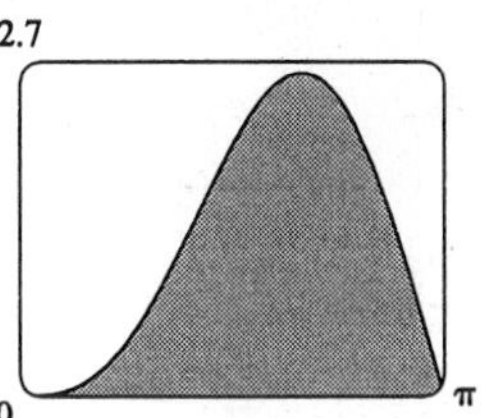

59. We split the integral: $\int_{-2}^2 (x+3)\sqrt{4-x^2}\,dx = \int_{-2}^2 x\sqrt{4-x^2}\,dx + \int_{-2}^2 3\sqrt{4-x^2}\,dx$. The first integral is 0 by Theorem 6, since $f(x) = x\sqrt{4-x^2}$ is an odd function and we are integrating from $x = -2$ to $x = 2$. The second integral we interpret as three times the area of the half-circle with radius 2, so the original integral is equal to $0 + 3 \cdot \tfrac{1}{2}(\pi \cdot 2^2) = 6\pi$.

60. We make the substitution $u = x^2$, so $du = 2x\,dx$. The limits are $0^2 = 0$ and $1^2 = 1$, and the integral becomes $\tfrac{1}{2}\int_0^1 \sqrt{1-u^2}\,du$. But this can be interpreted as one-half the area of the quarter-circle with radius 1. So the integral is $\tfrac{1}{2} \cdot \tfrac{1}{4}(\pi \cdot 1^2) = \tfrac{1}{8}\pi$.

61. The volume of inhaled air in the lungs at time t is
$V(t) = \int_0^t f(u)du = \int_0^t \tfrac{1}{2}\sin\left(\tfrac{2}{5}\pi u\right)du = \int_0^{2\pi t/5} \tfrac{1}{2}\sin v\left(\tfrac{5}{2\pi}\,dv\right)$ [We substitute $v = \tfrac{2\pi}{5}u \quad\Rightarrow\quad dv = \tfrac{2\pi}{5}\,du$]
$= \tfrac{5}{4\pi}(-\cos v)\Big|_0^{2\pi t/5} = \tfrac{5}{4\pi}\left[-\cos\left(\tfrac{2}{5}\pi t\right) + 1\right] = \tfrac{5}{4\pi}\left[1 - \cos\left(\tfrac{2}{5}\pi t\right)\right]$ liters.

62. Number of calculators $= x(4) - x(2) = \int_2^4 5000\left[1 - 100(t+10)^{-2}\right]dt$

$= 5000\left[t + 100(t+10)^{-1}\right]_2^4 = 5000\left[\left(4 + \frac{100}{14}\right) - \left(2 + \frac{100}{12}\right)\right] \approx 4048$

63. We make the substitution $u = 2x$ in $\int_0^2 f(2x)dx$. So $du = 2dx$ and the limits become $u = 2 \cdot 0 = 2$ and $u = 2 \cdot 2 = 4$. Hence $\int_0^4 f(u)\left(\frac{1}{2}\,du\right) = \frac{1}{2}\int_0^4 f(u)\,du = \frac{1}{2}(10) = 5$.

64. We make the substitution $u = x^2$ in $\int_0^3 xf(x^2)dx$. So $du = 2x\,dx$ and the limits become $u = 0^2 = 0$ and $u = 3^2 = 9$. Hence $\int_0^9 f(u)\left(\frac{1}{2}\,du\right) = \frac{1}{2}\int_0^9 f(u)du = \frac{1}{2}(4) = 2$.

65. Let $u = -x$. Then $du = -dx$. When $x = a$, $u = -a$; when $x = b$, $u = -b$. So

$\int_a^b f(-x)dx = \int_{-a}^{-b} f(u)(-du) = \int_{-b}^{-a} f(u)du = \int_{-b}^{-a} f(x)dx$.

From the diagram, we see that the equality follows from the fact that we are reflecting the graph of f, and the limits of integration, about the y-axis.

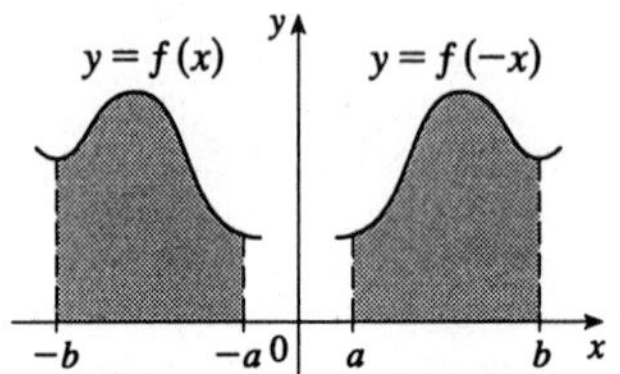

66. Let $u = x + c$. Then $du = dx$. When $x = a$, $u = a + c$; when $x = b$, $u = b + c$. So

$\int_a^b f(x+c)dx = \int_{a+c}^{b+c} f(u)du = \int_{a+c}^{b+c} f(x)dx$.

From the diagram, we see that the equality follows from the fact that we are translating the graph of f, and the limits of integration, by a distance c.

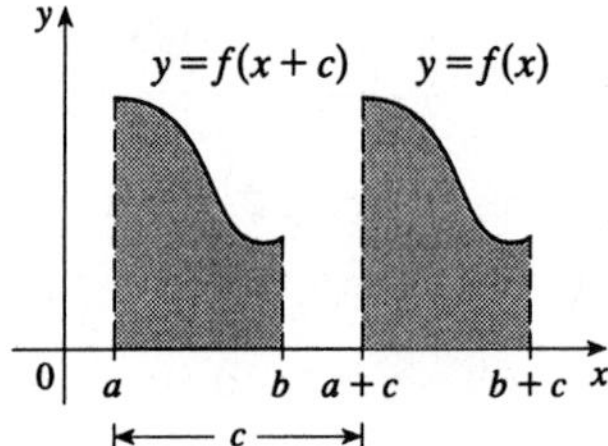

67. Let $u = 1 - x$. Then $du = -dx$. When $x = 1$, $u = 0$ and when $x = 0$, $u = 1$. So

$\int_0^1 x^a(1-x)^b\,dx = -\int_1^0 (1-u)^a u^b\,du = \int_0^1 u^b(1-u)^a\,du = \int_0^1 x^b(1-x)^a\,dx$.

68. Let $u = \pi - x$. Then $du = -dx$. When $x = \pi$, $u = 0$ and when $x = 0$, $u = \pi$. So

$\int_0^\pi xf(\sin x)dx = -\int_\pi^0 (\pi - u)f(\sin(\pi - u))du = \int_0^\pi (\pi - u)f(\sin u)du$

$= \pi\int_0^\pi f(\sin u)du - \int_0^\pi uf(\sin u)du = \pi\int_0^\pi f(\sin x)dx - \int_0^\pi xf(\sin x)dx$

$\Rightarrow \quad 2\int_0^\pi xf(\sin x)dx = \pi\int_0^\pi f(\sin x)dx \quad \Rightarrow \quad \int_0^\pi xf(\sin x)dx = \frac{\pi}{2}\int_0^\pi f(\sin x)dx$.

69. Let $u = 2x - 1$. Then $du = 2\,dx$, so $\displaystyle\int \frac{dx}{2x-1} = \int \frac{\frac{1}{2}du}{u} = \frac{1}{2}\ln|u| + C = \frac{1}{2}\ln|2x-1| + C$.

70. Let $u = x^2 + 1$. Then $du = 2x\,dx$, so $\displaystyle\int \frac{x\,dx}{x^2+1} = \int \frac{\frac{1}{2}\,du}{u} = \frac{1}{2}\ln|u| + C = \frac{1}{2}\ln(x^2+1) + C$ or $\ln\sqrt{x^2+1} + C$.

71. Let $u = \ln x$. Then $du = \dfrac{dx}{x}$, so $\displaystyle\int \frac{(\ln x)^2}{x}dx = \int u^2 du = \frac{1}{3}u^3 + C = \frac{1}{3}(\ln x)^3 + C$.

72. Let $u = x^2$. Then $du = 2x\,dx$, so $\displaystyle\int xe^{x^2}dx = \int e^u\left(\frac{1}{2}\,du\right) = \frac{1}{2}e^u + C = \frac{1}{2}e^{x^2} + C$.

73. Let $u = 1 + e^x$. Then $du = e^x\,dx$, so $\int e^x(1+e^x)^{10}\,dx = \int u^{10}\,du = \frac{1}{11}u^{11} + C = \frac{1}{11}(1+e^x)^{11} + C$.

74. Let $u = \tan^{-1} x$. Then $du = \dfrac{dx}{1+x^2}$, so $\displaystyle\int \frac{\tan^{-1} x}{1+x^2}\,dx = \int u\,du = \frac{u^2}{2} + C = \frac{(\tan^{-1}x)^2}{2} + C$.

75. Let $u = \ln x$. Then $du = \dfrac{dx}{x}$, so $\displaystyle\int \frac{dx}{x\ln x} = \int \frac{du}{u} = \ln|u| + C = \ln|\ln x| + C$.

76. Let $u = e^x$. Then $du = e^x\,dx$, so $\int e^x \sin(e^x)dx = \int \sin u\,du = -\cos u + C = -\cos(e^x) + C$.

77. $\displaystyle\int \frac{e^x+1}{e^x}\,dx = \int (1+e^{-x})dx = x - e^{-x} + C$ [Substitute $u = -x$.]

78. Let $u = e^x + 1$. Then $du = e^x\,dx$, so $\displaystyle\int \frac{e^x}{e^x+1}\,dx = \int \frac{du}{u} = \ln|u| + C = \ln(e^x+1) + C$.

79. Let $u = x^2 + 2x$. Then $du = 2(x+1)dx$, so $\displaystyle\int \frac{x+1}{x^2+2x}\,dx = \int \frac{\frac{1}{2}\,du}{u} = \tfrac{1}{2}\ln|u| + C = \tfrac{1}{2}\ln|x^2+2x| + C$.

80. Let $u = \cos x$. Then $du = -\sin x\,dx$, so $\displaystyle\int \frac{\sin x}{1+\cos^2 x}\,dx = \int \frac{-du}{1+u^2} = -\tan^{-1} u + C = -\tan^{-1}(\cos x) + C$.

81. $\displaystyle\int \frac{1+x}{1+x^2}\,dx = \int \frac{1}{1+x^2}\,dx + \int \frac{x}{1+x^2}\,dx = \tan^{-1} x + \frac{1}{2}\int \frac{2x\,dx}{1+x^2} = \tan^{-1} x + \tfrac{1}{2}\ln(1+x^2) + C$

(At the last step, we evaluate $\int du/u$ where $u = 1 + x^2$.)

82. Let $u = x^2$. Then $du = 2x\,dx$, so $\displaystyle\int \frac{x}{1+x^4}\,dx = \int \frac{\frac{1}{2}\,du}{1+u^2} = \tfrac{1}{2}\tan^{-1} u + C = \tfrac{1}{2}\tan^{-1}(x^2) + C$.

83. Let $u = 2x + 3$. Then $du = 2\,dx$, so

$$\int_0^3 \frac{dx}{2x+3} = \int_3^9 \frac{\frac{1}{2}\,du}{u} = \tfrac{1}{2}\ln u\Big|_3^9 = \tfrac{1}{2}(\ln 9 - \ln 3) = \tfrac{1}{2}(\ln 3^2 - \ln 3) = \tfrac{1}{2}(2\ln 3 - \ln 3) = \tfrac{1}{2}\ln 3 \quad \left(\text{or } \ln\sqrt{3}\right).$$

84. Let $u = -t^3$. Then $du = -3t^2\,dt$, so

$$\int_0^1 t^2 2^{-t^3} dt = \int_0^{-1} 2^u\left(-\tfrac{1}{3}\,du\right) = \frac{1}{3}\int_{-1}^0 2^u\,du = \frac{1}{3}\left[\frac{2^u}{\ln 2}\right]_{-1}^0 = \frac{1}{3\ln 2}\left(1 - \frac{1}{2}\right) = \frac{1}{6\ln 2}.$$

85. Let $u = \ln x$. Then $du = \dfrac{dx}{x}$, so $\displaystyle\int_e^{e^4} \frac{dx}{x\sqrt{\ln x}} = \int_1^4 u^{-1/2}\,du = 2u^{1/2}\Big|_1^4 = 2\cdot 2 - 2\cdot 1 = 2$.

86. Let $u = \sin^{-1}x$. Then $du = \dfrac{dx}{\sqrt{1-x^2}}$ so $\displaystyle\int_0^{1/2} \frac{\sin^{-1}x}{\sqrt{1-x^2}}\,dx = \int_0^{\pi/6} u\,du = \left[\frac{u^2}{2}\right]_0^{\pi/6} = \frac{\pi^2}{72}$.

87. $\dfrac{x\sin x}{1+\cos^2 x} = x \cdot \dfrac{\sin x}{2-\sin^2 x} = xf(\sin x)$, where $f(t) = \dfrac{t}{2-t^2}$. By Exercise 68,

$$\int_0^\pi \frac{x\sin x}{1+\cos^2 x}\,dx = \int_0^\pi xf(\sin x)dx = \frac{\pi}{2}\int_0^\pi f(\sin x)dx = \frac{\pi}{2}\int_0^\pi \frac{\sin x}{1+\cos^2 x}\,dx.$$

Let $u = \cos x$. Then $du = -\sin x\,dx$. When $x = \pi$, $u = -1$ and when $x = 0$, $u = 1$. So

$$\begin{aligned}\frac{\pi}{2}\int_0^\pi \frac{\sin x}{1+\cos^2 x}\,dx &= -\frac{\pi}{2}\int_1^{-1} \frac{du}{1+u^2} = \frac{\pi}{2}\int_{-1}^1 \frac{du}{1+u^2} = \frac{\pi}{2}\left[\tan^{-1} u\right]_{-1}^1 \\ &= \tfrac{\pi}{2}\left[\tan^{-1}1 - \tan^{-1}(-1)\right] = \tfrac{\pi}{2}\left[\tfrac{\pi}{4} - \left(-\tfrac{\pi}{4}\right)\right] = \tfrac{\pi^2}{4}.\end{aligned}$$

REVIEW EXERCISES FOR CHAPTER 4

1. True by Theorem 4.1.2 (b).

2. False. Try $n = 2$, $a_1 = a_2 = b_1 = b_2 = 1$ as a counterexample.

3. True by repeated application of Theorem 2.2.3 (b).

4. True by Theorem 4.1.3 (b).

5. True by Property 2 of Integrals.

6. False. Try $a = 0$, $b = 2$, $f(x) = g(x) = 1$ as a counterexample.

7. False. For example, let $f(x) = x^2$. Then $\int_0^1 \sqrt{x^2}\, dx = \int_0^1 x\, dx = \frac{1}{2}$, but $\sqrt{\int_0^1 x^2\, dx} = \sqrt{\frac{1}{3}} = \frac{1}{\sqrt{3}}$.

8. True by Part 2 of the Fundamental Theorem of Calculus.

9. True by Property 7 of Integrals.

10. False. For example, let $a = 0$, $b = 1$, $f(x) = 3$, $g(x) = x$. $f(x) > g(x)$ for each x in $[0, 1]$, but $f'(x) = 0 < 1 = g'(x)$ for $x \in [0, 1]$.

11. True. The integrand is an odd function that is continuous on $[-1, 1]$, so the result follows from Equation 4.5.6 (b).

12. True by the remarks following Theorem 4.3.4.

13. False. The function $f(x) = 1/x^4$ is not bounded on the interval $[-2, 1]$. It has an infinite discontinuity at $x = 0$, so it is not integrable on the interval. (If the integral were to exist, a positive value would be expected by Property 6 of Integrals.)

14. False. See the remarks following Note 4 in Section 3, and notice that $x - x^3 < 0$ for $x > 1$.

15. False. For example, the function $y = |x|$ is continuous on $\mathbb{R}$, but has no derivative at $x = 0$.

16. True by Part 1 of the Fundamental Theorem of Calculus.

17. First note that either a or b must be the graph of $\int_0^x f(t)dt$, since $\int_0^0 f(t)dt = 0$, and $c(0) \neq 0$. Now notice that $b > 0$ when c is increasing, and that $c > 0$ when a is increasing. It follows that c is the graph of $f(x)$, b is the graph of $f'(x)$, and a is the graph of $\int_0^x f(t)dt$.

18. (a) By Part 2 of the Fundamental Theorem of Calculus, we have

$$\int_0^{\pi/2} \frac{d}{dx}\left(\sin\frac{x}{2}\cos\frac{x}{3}\right)dx = \sin\frac{x}{2}\cos\frac{x}{3}\Big|_0^{\pi/2} = \frac{1}{\sqrt{2}}\cdot\frac{\sqrt{3}}{2} - 0\cdot 1 = \frac{\sqrt{6}}{4}.$$

(b) $\dfrac{d}{dx}\displaystyle\int_0^{\pi/2} \sin\frac{x}{2}\cos\frac{x}{3}\,dx = 0$, since the definite integral is a constant.

(c) $\dfrac{d}{dx}\displaystyle\int_x^{\pi/2} \sin\frac{t}{2}\cos\frac{t}{3}\,dt = \frac{d}{dx}\left(-\int_{\pi/2}^{x} \sin\frac{t}{2}\cos\frac{t}{3}\,dt\right) = -\frac{d}{dx}\int_{\pi/2}^{x} \sin\frac{t}{2}\cos\frac{t}{2}\,dt = -\sin\frac{x}{2}\cos\frac{x}{3}$, by Part 1 of the Fundamental Theorem of Calculus.

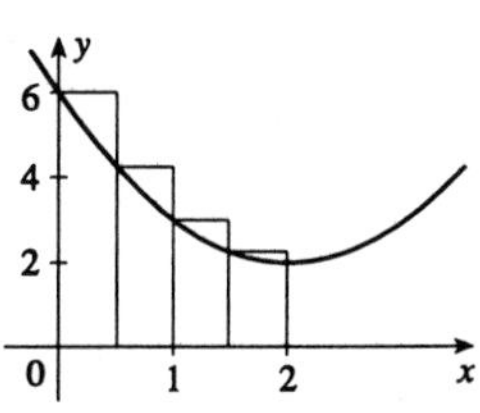

19. $\sum_{i=1}^{n} f(x_i^*)\,\Delta x_i = \sum_{i=1}^{4} f\left(\frac{i-1}{2}\right) \cdot \frac{1}{2} = \frac{1}{2}\left[f(0) + f\left(\frac{1}{2}\right) + f(1) + f\left(\frac{3}{2}\right)\right].$

$f(x) = 2 + (x-2)^2$, so $f(0) = 6$, $f\left(\frac{1}{2}\right) = 4.25$, $f(1) = 3$, and

$f\left(\frac{3}{2}\right) = 2.25$. Thus $\sum_{i=1}^{n} f(x_i^*)\,\Delta x_i = \frac{1}{2}(15.5) = 7.75$.

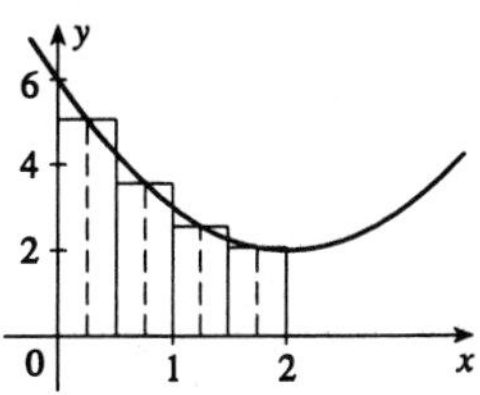

20. $$\sum_{i=1}^{n} f(x_i^*)\,\Delta x_i = \sum_{i=1}^{4} f\left(\frac{2i-1}{4}\right) \cdot \frac{1}{2}$$
$$= \frac{1}{2}\left[f\left(\frac{1}{4}\right) + f\left(\frac{3}{4}\right) + f\left(\frac{5}{4}\right) + f\left(\frac{7}{4}\right)\right]$$
$$= \tfrac{1}{2}[5.0625 + 3.5625 + 2.5625 + 2.0625] = \tfrac{1}{2}(13.25) = 6.625$$

21. By Theorem 4.3.5, $\int_2^4 (3-4x)dx = \lim_{n\to\infty} \frac{2}{n}\sum_{i=1}^{n}\left[3 - 4\left(2 + \frac{2i}{n}\right)\right] = \lim_{n\to\infty} \frac{2}{n}\sum_{i=1}^{n}\left[-5 - \frac{8}{n}i\right]$

$$= \lim_{n\to\infty} \frac{2}{n}\left[-5n - \frac{8}{n}\frac{n(n+1)}{2}\right] = \lim_{n\to\infty}\left[-10 - 8\cdot 1\left(1 + \frac{1}{n}\right)\right] = -10 - 8 = -18.$$

22. By Theorem 4.3.5, $\int_0^5 (x^3 - 2x^2)dx = \lim_{n\to\infty} \frac{5}{n}\sum_{i=1}^{n}\left[\left(\frac{5i}{n}\right)^3 - 2\left(\frac{5i}{n}\right)^2\right]$

$$= \lim_{n\to\infty}\left[\frac{625}{n^4}\sum_{i=1}^{n} i^3 - \frac{250}{n^3}\sum_{i=1}^{n} i^2\right] = \lim_{n\to\infty}\left[\frac{625}{n^4}\frac{n^2(n+1)^2}{4} - \frac{250}{n^3}\frac{n(n+1)(2n+1)}{6}\right]$$

$$= \lim_{n\to\infty}\left[\frac{625}{4}\left(1+\frac{1}{n}\right)^2 - \frac{125}{3}\left(1+\frac{1}{n}\right)\left(2+\frac{1}{n}\right)\right] = \frac{625}{4} - \frac{250}{3} = \frac{875}{12}.$$

23. $\int_0^5 (x^3 - 2x^2)dx = \frac{1}{4}x^4 - \frac{2}{3}x^3\Big|_0^5 = \frac{625}{4} - \frac{250}{3} = \frac{875}{12}$ (We did this integral in Exercise 22.)

24. $\int_0^b (x^3 + 4x - 1)dx = \frac{1}{4}x^4 + 2x^2 - x\Big|_0^b = \frac{1}{4}b^4 + 2b^2 - b$

25. $\int_0^1 (1 - x^9)dx = \left[x - \frac{1}{10}x^{10}\right]_0^1 = 1 - \frac{1}{10} = \frac{9}{10}$

26. Let $u = 1 - x$. Then $du = -dx$, so $\int_0^1 (1-x)^9 dx = \int_1^0 u^9(-du) = \int_0^1 u^9 du = \frac{1}{10}u^{10}\Big|_0^1 = \frac{1}{10}$

27. $\int_1^8 \sqrt[3]{x}(x-1)dx = \int_1^8 (x^{4/3} - x^{1/3})dx = \frac{3}{7}x^{7/3} - \frac{3}{4}x^{4/3}\Big|_1^8 = \left(\frac{3}{7}\cdot 128 - \frac{3}{4}\cdot 16\right) - \left(\frac{3}{7} - \frac{3}{4}\right) = \frac{1209}{28}$

28. $$\int_1^4 \frac{x^2 - x + 1}{\sqrt{x}}\,dx = \int_1^4 (x^{3/2} - x^{1/2} + x^{-1/2})dx = \tfrac{2}{5}x^{5/2} - \tfrac{2}{3}x^{3/2} + 2x^{1/2}\Big|_1^4$$
$$= \left(\tfrac{2}{5}\cdot 32 - \tfrac{2}{3}\cdot 8 + 4\right) - \left(\tfrac{2}{5} - \tfrac{2}{3} + 2\right) = \tfrac{146}{15}$$

29. Let $u = 1 + 2x^3$. Then $du = 6x^2\,dx$, so $\int_0^2 x^2(1+2x^3)^3\,dx = \int_1^{17} u^3\left(\frac{1}{6}\,du\right) = \frac{1}{24}u^4\Big|_1^{17} = \frac{1}{24}(17^4 - 1) = 3480.$

30. Let $u = 16 - 3x$. Then $x = \frac{1}{3}(16-u)$, $dx = -\frac{1}{3}du$, so

$$\int_0^4 x\sqrt{16-3x}\,dx = \int_{16}^4 u^{1/2}\left(\frac{16-u}{3}\right)\left(-\tfrac{1}{3}\,du\right) = \frac{1}{9}\int_4^{16}(16\,u^{1/2} - u^{3/2})du$$
$$= \tfrac{1}{9}\left[16\cdot\tfrac{2}{3}u^{3/2} - \tfrac{2}{5}u^{5/2}\right]_4^{16} = \tfrac{1}{9}\left[\tfrac{32}{3}\cdot 64 - \tfrac{2}{5}\cdot 1024 - \tfrac{32}{3}\cdot 8 + \tfrac{2}{5}\cdot 32\right] = \tfrac{3008}{135}.$$

31. Let $u = 2x + 3$. Then $du = 2\,dx$, so $\displaystyle\int_3^{11} \frac{dx}{\sqrt{2x+3}} = \int_9^{25} u^{-1/2}\left(\tfrac{1}{2}\,du\right) = u^{1/2}\Big|_9^{25} = 5 - 3 = 2.$

32. $\displaystyle\int_0^2 \frac{x\,dx}{(x^2-1)^2}$ does not exist since the integrand has an infinite discontinuity at $x = 1$.

33. $\displaystyle\int_{-2}^{-1} \frac{dx}{(2x+3)^4}$ does not exist since the integrand has an infinite discontinuity at $x = -\frac{3}{2}$.

34. $\displaystyle\int_{-1}^{1} \frac{x + x^3 + x^5}{1 + x^2 + x^4}\,dx = 0$ by Equation 4.5.6, since the integrand is odd.

35. Let $u = 2 + x^5$. Then $du = 5x^4\,dx$, so

$$\int \frac{x^4\,dx}{(2+x^5)^6} = \int u^{-6}\left(\tfrac{1}{5}\,du\right) = \frac{1}{5}\left(\frac{u^{-5}}{-5}\right) + C = -\frac{1}{25u^5} + C = -\frac{1}{25(2+x^5)^5} + C.$$

36. Let $u = 2x - x^2$. Then $du = 2(1-x)dx$, so

$\int (1-x)\sqrt{2x - x^2}\,dx = \int u^{1/2}\left(\tfrac{1}{2}\,du\right) = \tfrac{1}{2} \cdot \tfrac{2}{3}u^{3/2} + C = \tfrac{1}{3}(2x - x^2)^{3/2} + C.$

37. Let $u = \pi x$. Then $du = \pi\,dx$, so $\displaystyle\int \sin \pi x\,dx = \int \frac{\sin u\,du}{\pi} = \frac{-\cos u}{\pi} + C = -\frac{\cos \pi x}{\pi} + C.$

38. Let $u = 3t$. Then $du = 3\,dt$, so $\int \csc^2 3t\,dt = \int \csc^2 u\left(\tfrac{1}{3}\,du\right) = -\tfrac{1}{3}\cot u + C = -\tfrac{1}{3}\cot 3t + C.$

39. Let $u = \dfrac{1}{t}$. Then $du = -\dfrac{1}{t^2}\,dt$, so $\displaystyle\int \frac{\cos(1/t)}{t^2}\,dt = \int \cos u\,(-du) = -\sin u + C = -\sin\left(\frac{1}{t}\right) + C.$

40. Let $u = \cos x$. Then $du = -\sin x\,dx$, so

$\int \sin x \sec^2(\cos x)dx = \int \sec^2 u(-du) = -\tan u + C = -\tan(\cos x) + C.$

41. $\int_0^{2\pi} |\sin x|dx = \int_0^{\pi} \sin x\,dx - \int_{\pi}^{2\pi} \sin x\,dx = 2\int_0^{\pi} \sin x\,dx = -2\cos x\big|_0^{\pi} = -2[(-1) - 1] = 4$

42. $\int_0^8 |x^2 - 6x + 8|dx = \int_0^8 |(x-2)(x-4)|dx$

$= \int_0^2 (x^2 - 6x + 8)dx - \int_2^4 (x^2 - 6x + 8)dx + \int_4^8 (x^2 - 6x + 8)dx$

$= \left[\tfrac{1}{3}x^3 - 3x^2 + 8x\right]_0^2 - \left[\tfrac{1}{3}x^3 - 3x^2 + 8x\right]_2^4 + \left[\tfrac{1}{3}x^3 - 3x^2 + 8x\right]_4^8$

$= \left(\tfrac{8}{3} - 12 + 16\right) - 0 - \left(\tfrac{64}{3} - 48 + 32\right) + \left(\tfrac{8}{3} - 12 + 16\right) + \left(\tfrac{512}{3} - 192 + 64\right) - \left(\tfrac{64}{3} - 48 + 32\right) = \tfrac{136}{3}$

43.

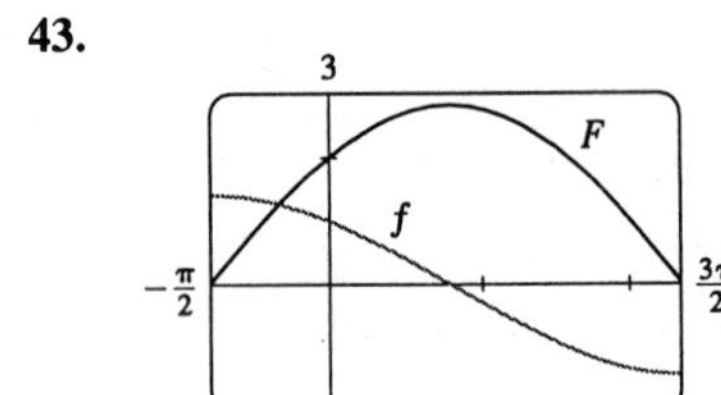

$f(x) = \dfrac{\cos x}{\sqrt{1 + \sin x}}$. Let $u = 1 + \sin x$. Then $du = \cos x\,dx$, so

$$\int \frac{\cos x\,dx}{\sqrt{1+\sin x}} = \int u^{-1/2}\,du = 2u^{1/2} + C = 2\sqrt{1 + \sin x} + C.$$

44.

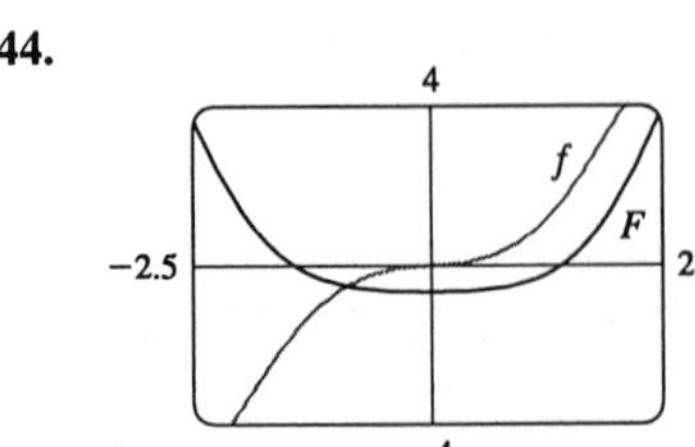

$f(x) = \dfrac{x^3}{\sqrt{x^2+1}}$. Let $u = x^2 + 1$. Then $x^2 = u - 1$ and $x\,dx = \frac{1}{2}\,du$,

so $\displaystyle\int \frac{x^3}{\sqrt{x^2+1}}\,dx = \int \frac{(u-1)}{\sqrt{u}}\left(\tfrac{1}{2}\,du\right) = \frac{1}{2}\int \left(u^{1/2} - u^{-1/2}\right)du$

$= \tfrac{1}{2}\left(\tfrac{2}{3}u^{3/2} - 2u^{1/2}\right) + C = \tfrac{1}{3}(x^2+1)^{3/2} - (x^2+1)^{1/2} + C$

$= \tfrac{1}{3}\sqrt{x^2+1}\,(x^2 - 2) + C.$

45.

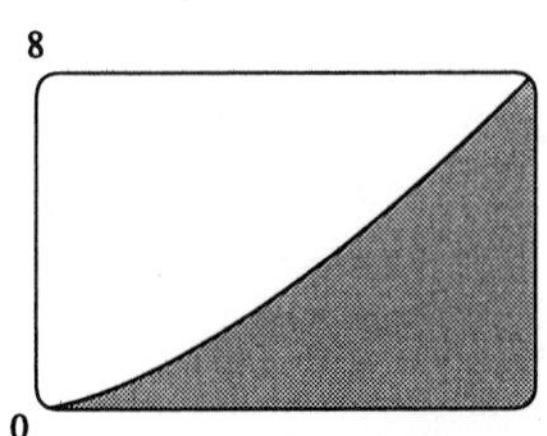

From the graph, it appears that the area under the curve $y = x\sqrt{x}$ between $x = 0$ and $x = 4$ is somewhat less than half the area of an 8×4 rectangle, so perhaps about 13 or 14. To find the exact value, we evaluate

$\int_0^4 x\sqrt{x}\,dx = \int_0^4 x^{3/2}\,dx = \frac{2}{5}x^{5/2}\Big|_0^4 = \frac{2}{5}(4)^{5/2} = \frac{64}{5} = 12.8.$

46.

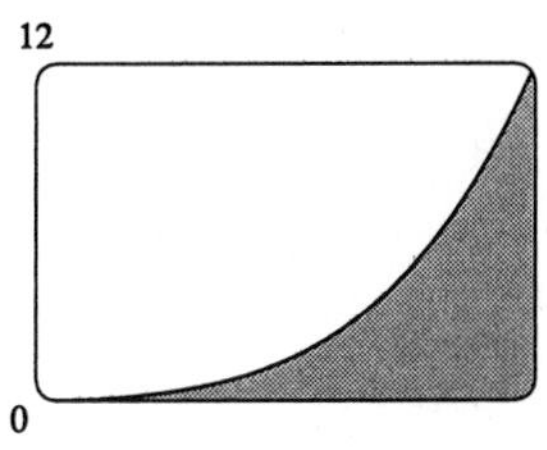

From the graph, it appears that the area under the curve $y = x^2\sqrt{1+x^3}$ between $x = 0$ and $x = 2$ is about one-quarter the area of a 2×12 rectangle, so about 6. To find the exact value, we evaluate $\int_0^2 x^2\sqrt{1+x^3}\,dx$. Let $u = 1 + x^3$, so $du = 3x^2\,dx$, the limits change to $u = 1 + 0^3 = 1$ and $u = 1 + 2^3 = 9$, and the integral becomes

$\int_1^9 \sqrt{u}\left(\frac{1}{3}\,du\right) = \frac{1}{3}\left(\frac{2}{3}u^{3/2}\right)\Big|_1^9 = \frac{2}{9}(27-1) = \frac{52}{9} = 5.\overline{7}.$

47. $F(x) = \int_1^x \sqrt{1+t^4}\,dt \quad\Rightarrow\quad F'(x) = \sqrt{1+x^4}$

48. $F(x) = \int_\pi^x \tan(s^2)ds \quad\Rightarrow\quad F'(x) = \tan(x^2)$

49. $g(x) = \displaystyle\int_0^{x^3} \frac{t\,dt}{\sqrt{1+t^3}}$. Let $y = g(x)$ and $u = x^3$. Then

$$g'(x) = \frac{dy}{dx} = \frac{dy}{du}\frac{du}{dx} = \frac{u}{\sqrt{1+u^3}}3x^2 = \frac{x^3}{\sqrt{1+x^9}}3x^2 = \frac{3x^5}{\sqrt{1+x^9}}.$$

50. $g(x) = \displaystyle\int_1^{\cos x} \sqrt[3]{1-t^2}\,dt$. Let $y = g(x)$ and $u = \cos x$. Then $g'(x) = \dfrac{dy}{dx} = \dfrac{dy}{du}\dfrac{du}{dx}$

$$= \sqrt[3]{1-u^2}(-\sin x) = \sqrt[3]{1-\cos^2 x}(-\sin x) = -\sin x\sqrt[3]{\sin^2 x} = -(\sin x)^{5/3}.$$

51. $y = \displaystyle\int_{\sqrt{x}}^{x} \frac{\cos\theta}{\theta}\,d\theta = \int_1^x \frac{\cos\theta}{\theta}\,d\theta + \int_{\sqrt{x}}^1 \frac{\cos\theta}{\theta}\,d\theta = \int_1^x \frac{\cos\theta}{\theta}\,d\theta - \int_1^{\sqrt{x}} \frac{\cos\theta}{\theta}\,d\theta \quad\Rightarrow$

$$y' = \frac{\cos x}{x} - \frac{\cos\sqrt{x}}{\sqrt{x}}\frac{1}{2\sqrt{x}} = \frac{2\cos x - \cos\sqrt{x}}{2x}$$

52. $y = \displaystyle\int_{2x}^{3x+1} \sin(t^4)\,dt = \int_0^{3x+1} \sin(t^4)\,dt - \int_0^{2x} \sin(t^4)\,dt \quad\Rightarrow\quad y' = 3\sin\left[(3x+1)^4\right] - 2\sin\left[(2x)^4\right].$

53. If $1 \le x \le 3$, then $2 \le \sqrt{x^2+3} \le 2\sqrt{3}$, so $2(3-1) \le \int_1^3 \sqrt{x^2+3}\,dx \le 2\sqrt{3}(3-1)$; that is, $4 \le \int_1^3 \sqrt{x^2+3}\,dx \le 4\sqrt{3}$.

54. If $3 \le x \le 5$, then $4 \le x+1 \le 6$ and $\dfrac{1}{6} \le \dfrac{1}{x+1} \le \dfrac{1}{4}$, so $\frac{1}{6}(5-3) \le \displaystyle\int_3^5 \frac{1}{x+1}\,dx \le \frac{1}{4}(5-3)$; that is,

$$\frac{1}{3} \le \int_3^5 \frac{1}{x+1}dx \le \frac{1}{2}.$$

55. $0 \le x \le 1 \;\Rightarrow\; 0 \le \cos x \le 1 \;\Rightarrow\; x^2\cos x \le x^2 \;\Rightarrow\; \int_0^1 x^2\cos x\,dx \le \int_0^1 x^2\,dx = \frac{1}{3}x^3\Big|_0^1 = \frac{1}{3}$ [Property 7].

56. On the interval $\left[\frac{\pi}{4}, \frac{\pi}{2}\right]$, x is increasing and $\sin x$ is decreasing, so $(\sin x)/x$ is decreasing. Therefore the largest value of $\dfrac{\sin x}{x}$ on $\left[\dfrac{\pi}{4}, \dfrac{\pi}{2}\right]$ is $\dfrac{\sin(\pi/4)}{\pi/4} = \dfrac{\sqrt{2}/2}{\pi/4} = \dfrac{2\sqrt{2}}{\pi}$. By Property 8 with $M = \dfrac{2\sqrt{2}}{\pi}$ we get

$$\int_{\pi/4}^{\pi/2} \frac{\sin x}{x}\,dx \le \frac{2\sqrt{2}}{\pi}\left(\frac{\pi}{2} - \frac{\pi}{4}\right) = \frac{\sqrt{2}}{2}.$$

57. Let $f(x) = \sqrt{1 + x^3}$ on $[0, 1]$. The Midpoint Rule with $n = 5$ gives

$$\int_0^1 \sqrt{1 + x^3}\,dx \approx \tfrac{1}{5}[f(0.1) + f(0.3) + f(0.5) + f(0.7) + f(0.9)]$$
$$= \tfrac{1}{5}\left[\sqrt{1 + (0.1)^3} + \sqrt{1 + (0.3)^3} + \cdots + \sqrt{1 + (0.9)^3}\right] \approx 1.110.$$

58. (a) displacement $= \int_0^5 (t^2 - t)dt = \left[\frac{1}{3}t^3 - \frac{1}{2}t^2\right]_0^5 = \frac{125}{3} - \frac{25}{2} = \frac{175}{6}$.

(b) distance traveled $= \int_0^5 |t^2 - t|dt = \int_0^5 |t(t-1)|dt = \int_0^1 (t - t^2)dt + \int_1^5 (t^2 - t)dt$

$$= \left[\tfrac{1}{2}t^2 - \tfrac{1}{3}t^3\right]_0^1 + \left[\tfrac{1}{3}t^3 - \tfrac{1}{2}t^2\right]_1^5 = \tfrac{1}{2} - \tfrac{1}{3} - 0 + \left(\tfrac{125}{3} - \tfrac{25}{2}\right) - \left(\tfrac{1}{3} - \tfrac{1}{2}\right) = 29.5$$

59. $y = \sqrt{16 - x^2}$ is a semicircle with a radius of 4. So $\int_{-4}^4 \sqrt{16 - x^2}\,dx$ represents the area between the semicircle $y = \sqrt{16 - x^2}$ and the x-axis and is equal to $\frac{1}{2}\pi(4)^2 = 8\pi$.

60. $A_1 = \frac{1}{2}bh = \frac{1}{2}(2)(2) = 2$, $A_2 = \frac{1}{2}bh = \frac{1}{2}(1)(1) = \frac{1}{2}$, and since $y = -\sqrt{1 - x^2}$ for $0 \le x \le 1$ represents a quarter-circle with radius 1, $A_3 = \frac{1}{4}\pi r^2 = \frac{1}{4}\pi(1)^2 = \frac{\pi}{4}$.
So $\int_{-3}^1 f(x)dx = A_1 - A_2 - A_3 = 2 - \frac{1}{2} - \frac{\pi}{4} = \frac{1}{4}(6 - \pi)$.

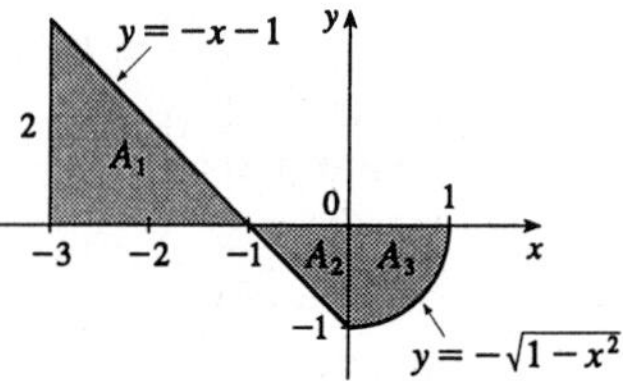

61. By the Fundamental Theorem of Calculus, we know that $F(x) = \int_a^x t^2 \sin(t^2)dt$ is an antiderivative of $f(x) = x^2 \sin(x^2)$. This integral cannot be expressed in any simpler form. Since $\int_a^a f\,dt = 0$ for any a, we can take $a = 1$, and then $F(1) = 0$, as required. So $F(x) = \int_1^x t^2 \sin(t^2)dt$ is the desired function.

62. (a) C is increasing on those intervals where C' is positive. But by the Fundamental Theorem of Calculus, $C'(x) = \dfrac{d}{dx}\left[\int_0^x \cos\left(\frac{\pi}{2}t^2\right)dt\right] = \cos\left(\frac{\pi}{2}x^2\right)$. This is positive when $\frac{\pi}{2}x^2$ is in the interval $\left(\left(2n - \frac{1}{2}\right)\pi, \left(2n + \frac{1}{2}\right)\pi\right)$, n any integer. This implies that $\left(2n - \frac{1}{2}\right)\pi < \frac{\pi}{2}x^2 < \left(2n + \frac{1}{2}\right)\pi \quad \Leftrightarrow$ $0 \le |x| \le 1$ or $\sqrt{4n - 1} < |x| < \sqrt{4n + 1}$, n any positive integer. So C is increasing on the intervals $[-1, 1]$, $\left[\sqrt{3}, \sqrt{5}\right]$, $\left[-\sqrt{5}, -\sqrt{3}\right]$, $\left[\sqrt{7}, 3\right]$, $\left[-3, -\sqrt{7}\right]$, ...

(b) C is concave up on those intervals where $C'' > 0$. We differentiate C' to find C'': $C'(x) = \cos\left(\frac{\pi}{2}x^2\right)$ $\Rightarrow \quad C''(x) = -\sin\left(\frac{\pi}{2}x^2\right)\left(\frac{\pi}{2} \cdot 2x\right) = -\pi x \sin\left(\frac{\pi}{2}x^2\right)$. For $x > 0$, this is positive where $(2n - 1)\pi < \frac{\pi}{2}x^2 < 2n\pi$, n any positive integer $\quad \Leftrightarrow \quad \sqrt{2(2n - 1)} < x < 2\sqrt{n}$, n any positive integer. Since there is a factor of $-x$ in C'', the intervals of upward concavity for $x < 0$ are $\left(-\sqrt{2(2n + 1)}, -2\sqrt{n}\right)$, n any nonnegative integer. That is, C is concave up on $\left(-\sqrt{2}, 0\right)$, $\left(\sqrt{2}, 2\right)$, $\left(-\sqrt{6}, -2\right)$, $\left(\sqrt{6}, 2\sqrt{2}\right)$, ...

(c)

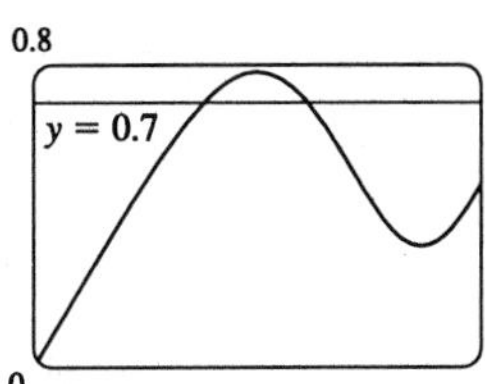

In Maple, we use `plot({Int(cos(Pi*t^2/2),t=0..x),0.7},x=0..2);` and zoom in, with successive `plot` commands, on the intersections. From the graphs, we see that $\int_0^x \cos\left(\frac{\pi}{2}t^2\right)dt = 0.7$ at $x \approx 0.76$ and $x \approx 1.22$.

(d) The graphs of $S(x)$ and $C(x)$ have similar shapes, except that S's flattens out near the origin, while C's does not. Note that for $x > 0$, C is increasing where S is concave up, and C is decreasing where S is concave down. Similarly, S is increasing where C is concave down, and S is decreasing where C is concave up. For $x < 0$, these relationships are reversed; that is, C is increasing where S is concave down, and S is increasing where C is concave up.

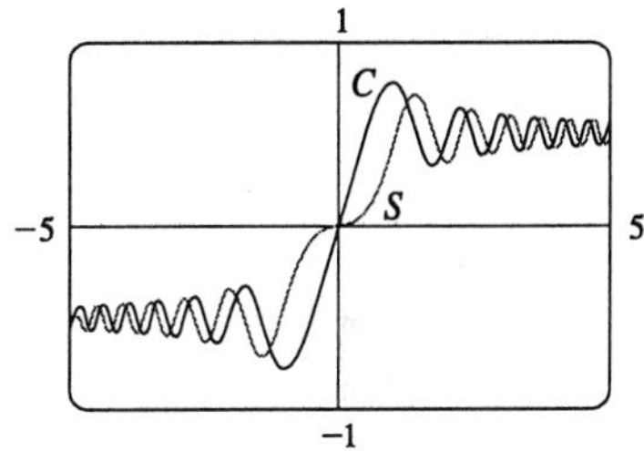

See Example 4.4.2 and Exercise 4.4.83 for a discussion of $S(x)$.

63. Following the hint, we have $\Delta x_i = x_i - x_{i-1} = \dfrac{i^2}{n^2} - \dfrac{(i-1)^2}{n^2} = \dfrac{2i-1}{n^2}$. So

$$\int_0^1 \sqrt{x}\,dx = \lim_{n\to\infty}\sum_{i=1}^n \Delta x_i\, f(x_i^*) = \lim_{n\to\infty}\sum_{i=1}^n\left(\frac{2i-1}{n^2}\right)\sqrt{\frac{i^2}{n^2}} = \lim_{n\to\infty}\sum_{i=1}^n \frac{2i^2-i}{n^3}$$

$$= \lim_{n\to\infty}\frac{1}{n^3}\left[2\sum_{i=1}^n i^2 - \sum_{i=1}^n i\right] = \lim_{n\to\infty}\frac{1}{n^3}\left[\frac{n(n+1)(2n+1)}{3} - \frac{n(n+1)}{2}\right] \text{ [by Theorem 4.1.3(c), (d)]}$$

$$= \lim_{n\to\infty}\left[\frac{(1+1/n)(2+1/n)}{3} - \frac{1+1/n}{2n}\right] = \frac{2}{3} - 0 = \frac{2}{3}.$$

64. From the given equation, $\int_a^x f(t)dt = \sin x - \frac{1}{2}$. Differentiating both sides gives $f(x) = \cos x$, by Part 1 of the Fundamental Theorem of Calculus. We put $x = a$ into the first equation to get $0 = \sin a - \frac{1}{2}$, so $a = \frac{\pi}{6}$, $f(x) = \cos x$ satisfy the given equation.

65. Let $u = f(x)$ so $du = f'(x)dx$. So $2\int_a^b f(x)f'(x)dx = 2\int_{f(a)}^{f(b)} u\,du = u^2\Big|_{f(a)}^{f(b)} = [f(b)]^2 - [f(a)]^2$.

66. $\displaystyle\lim_{h\to 0}\frac{1}{h}\int_2^{2+h}\sqrt{1+t^3}\,dt = \lim_{h\to 0}\frac{F(2+h)-F(2)}{h} = F'(2)$, where $F(x) = \displaystyle\int_2^x \sqrt{1+t^3}dt$. Now $F'(x) = \sqrt{1+x^3}$ by the Fundamental Theorem of Calculus, Part 1, so the required limit is $F'(2) = \sqrt{1+2^3} = 3$.

67. Let $u = 1 - x$, then $du = -dx$, so $\int_0^1 f(1-x)dx = \int_1^0 f(u)(-du) = \int_0^1 f(u)du = \int_0^1 f(x)dx$.

68. $\displaystyle\lim_{n\to\infty}\frac{1}{n}\left[\left(\frac{1}{n}\right)^9 + \left(\frac{2}{n}\right)^9 + \left(\frac{3}{n}\right)^9 + \cdots + \left(\frac{n}{n}\right)^9\right] = \lim_{n\to\infty}\frac{1-0}{n}\sum_{i=1}^n\left(\frac{i}{n}\right)^9 = \int_0^1 x^9\,dx = \frac{x^{10}}{10}\Big|_0^1 = \frac{1}{10}$. The limit is based on Riemann sums using right endpoints and subintervals of equal length. See Theorem 4.3.5.

PROBLEMS PLUS (page 304)

1. Differentiating both sides of the equation $x \sin \pi x = \int_0^{x^2} f(t)dt$ (using Part 1 of the Fundamental Theorem of Calculus for the right side) gives $\sin \pi x + \pi x \cos \pi x = 2x f(x^2)$. Putting $x = 2$, we obtain $\sin 2\pi + 2\pi \cos 2\pi = 4f(4)$, so $f(4) = \frac{1}{4}(0 + 2\pi \cdot 1) = \frac{\pi}{2}$.

2. **(a)** Let $f(x) = ax^2 + bx + c$. $f(0) = f(\pi) = 0$, so we know that $f(0) = 0 \quad \Leftrightarrow \quad c = 0$, and $f(\pi) = 0 \quad \Leftrightarrow \quad a\pi^2 + b\pi = 0 \quad \Leftrightarrow \quad b = -a\pi$. So $f(x) = ax^2 - a\pi x$. Now we want the maximum value of $f(x)$ on $[0, \pi]$ to be the same as that of $\sin x$, that is, 1. So we find the value of x at which f has a maximum by differentiating and setting $f'(x) = 0 \quad \Leftrightarrow \quad 2ax - a\pi = 0 \quad \Leftrightarrow \quad x = \frac{\pi}{2}$. Now $f\left(\frac{\pi}{2}\right) = a\left(\frac{\pi}{2}\right)^2 - a\pi\left(\frac{\pi}{2}\right) = -\frac{1}{4}\pi^2 a$. We set this equal to 1, in order to find a: $-\frac{1}{4}\pi^2 a = 1 \quad \Leftrightarrow \quad a = -\frac{4}{\pi^2} \quad \Rightarrow \quad b = -a\pi = \frac{4}{\pi}$. So the desired function is $f(x) = -\frac{4}{\pi^2}x^2 + \frac{4}{\pi}x$.

(b) Once again, $g(0) = g(\pi) = 0 \quad \Rightarrow \quad g(x) = ax^2 - a\pi x$. We want $g'(0) = \left.\frac{d}{dx}(\sin x)\right|_0 = \cos 0 = 1$. We calculated $g'(x)$ in part (a), so we set $g'(0) = 1 \quad \Leftrightarrow \quad 2a(0) - a\pi = 1 \quad \Leftrightarrow \quad a = -\frac{1}{\pi}$. We also want $g'(\pi) = \left[\frac{d}{dx}(\sin x)\right]_\pi = \cos \pi = -1$, so we make sure that $g'(\pi) = -1$ with $a = -\frac{1}{\pi}$: $g'(\pi) = 2\left(-\frac{1}{\pi}\right)\pi - \left(-\frac{1}{\pi}\right)\pi = -1$. So the desired function is $g(x) = -\frac{1}{\pi}x^2 + x$.

(c) Again, $h(x) = ax^2 - a\pi x$. Now we want the area under the curves of $h(x)$ and $\sin x$ to be the same; that is, $\int_0^\pi h(x)dx = \int_0^\pi \sin x\, dx = -\cos x\big|_0^\pi = (1) - (-1) = 2$. We integrate h between 0 and π and set the result equal to 2: $\int_0^\pi (ax^2 - a\pi x)dx = \left[\frac{1}{3}ax^3 - \frac{1}{2}a\pi x^2\right]_0^\pi = \frac{1}{3}a\pi^3 - \frac{1}{2}a\pi^3 = -\frac{1}{6}a\pi^3 = 2 \Leftrightarrow \quad a = -\frac{12}{\pi^3}$. So the desired function is $h(x) = -\frac{12}{\pi^3}x^2 + \frac{12}{\pi^2}x$.

(d)

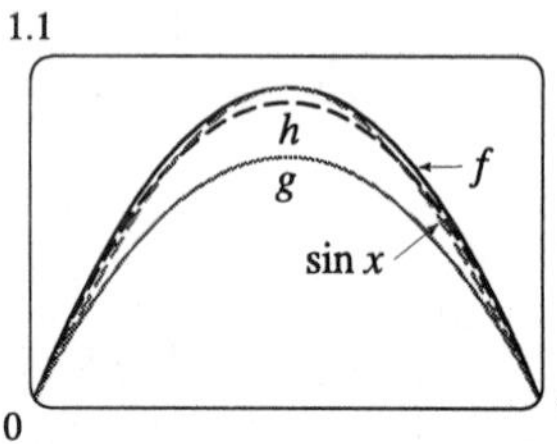

1.05
h
sin x
f
g
0.6
0.55
2.5

3. For $1 \le x \le 2$, we have $x^4 \le 2^4 = 16$, so $1 + x^4 \le 17$ and $\frac{1}{1+x^4} \ge \frac{1}{17}$. Thus $\int_1^2 \frac{1}{1+x^4}dx \ge \int_1^2 \frac{1}{17}dx = \frac{1}{17}$. Also $1 + x^4 > x^4$ for $1 \le x \le 2$, so $\frac{1}{1+x^4} < \frac{1}{x^4}$ and $\int_1^2 \frac{1}{1+x^4}dx < \int_1^2 x^{-4}dx = \left[\frac{x^{-3}}{-3}\right]_1^2 = -\frac{1}{24} + \frac{1}{3} = \frac{7}{24}$. Thus we have the estimate $\frac{1}{17} \le \int_1^2 \frac{1}{1+x^4}dx \le \frac{7}{24}$.

4. Let $P(a, 1-a^2)$ be the point of contact. The equation of the tangent line at P is $y-(1-a^2)=(-2a)(x-a)$ $\Rightarrow$ $y-1+a^2=-2ax+2a^2$ $\Rightarrow$ $y=-2ax+a^2+1$. To find the x-intercept, put $y=0$: $2ax=a^2+1$ $\Rightarrow$ $x=\dfrac{a^2+1}{2a}$. To find the y-intercept, put $x=0$: $y=a^2+1$. Therefore the area of the triangle is $\dfrac{1}{2}\left(\dfrac{a^2+1}{2a}\right)(a^2+1)=\dfrac{(a^2+1)^2}{4a}$. Therefore we minimize the function $A(a)=\dfrac{(a^2+1)^2}{4a}$, $0<a\le 1$. $A'(a)=\dfrac{(4a)2(a^2+1)(2a)-(a^2+1)^2(4)}{16a^2}=\dfrac{(a^2+1)[4a^2-(a^2+1)]}{4a^2}=\dfrac{(a^2+1)(3a^2-1)}{4a^2}$.
$A'(a)=0$ when $3a^2-1=0$ $\Rightarrow$ $a=\frac{1}{\sqrt{3}}$. $A'(a)<0$ for $a<\frac{1}{\sqrt{3}}$, $A'(a)>0$ for $a>\frac{1}{\sqrt{3}}$. So by the First Derivative Test, there is an absolute minimum when $a=\frac{1}{\sqrt{3}}$. The required point is $\left(\frac{1}{\sqrt{3}},\frac{2}{3}\right)$.

5. Differentiating $x^2+xy+y^2=12$ implicitly with respect to x gives $2x+y+x\dfrac{dy}{dx}+2y\dfrac{dy}{dx}=0$, so $\dfrac{dy}{dx}=-\dfrac{2x+y}{x+2y}$. At a highest or lowest point, $\dfrac{dy}{dx}=0$ $\Leftrightarrow$ $y=-2x$. Substituting this into the original equation gives $x^2+x(-2x)+(-2x)^2=12$, so $3x^2=12$ and $x=\pm 2$. If $x=2$, then $y=-2x=-4$, and if $x=-2$ then $y=4$. Thus the highest and lowest points are $(-2,4)$ and $(2,-4)$.

6. By Part 2 of the Fundamental Theorem of Calculus, $\int_0^1 f'(x)dx=f(1)-f(0)=1-0=1$.

7. Such a function cannot exist. $f'(x)>3$ for all x means that f is differentiable (and hence continuous) for all x. So by Part 2 of the Fundamental Theorem, $\int_1^4 f'(x)dx=f(4)-f(1)=7-(-1)=8$. However, if $f'(x)>3$ for all x, then $\int_1^4 f'(x)dx\ge 3\cdot(4-1)=9$ by Property 8 of integrals.

Alternate Solution: By the Mean Value Theorem there exists a number $c\in(1,4)$ such that $f'(c)=\dfrac{f(4)-f(1)}{4-1}=\dfrac{7-(-1)}{3}=\dfrac{8}{3}$ $\Rightarrow$ $8=3f'(c)$. But $f'(x)>3$ $\Rightarrow$ $3f'(c)>9$, so such a function cannot exist.

8. Such a function cannot exist. If $f''(x)>0$ for all x, then, by the Test for Monotonic Functions, f' is increasing on $(-\infty,\infty)$, so $f'(0)>f'(-1)$.

9. $f(x)=\dfrac{1}{1+|x|}+\dfrac{1}{1+|x-2|}$

$$=\begin{cases}\dfrac{1}{1-x}+\dfrac{1}{1-(x-2)} & \text{if } x<0\\ \dfrac{1}{1+x}+\dfrac{1}{1-(x-2)} & \text{if } 0\le x<2\\ \dfrac{1}{1+x}+\dfrac{1}{1+(x-2)} & \text{if } x\ge 2\end{cases}\quad\Rightarrow\quad f'(x)=\begin{cases}\dfrac{1}{(1-x)^2}+\dfrac{1}{(3-x)^2} & \text{if } x<0\\ \dfrac{-1}{(1+x)^2}+\dfrac{1}{(3-x)^2} & \text{if } 0<x<2\\ \dfrac{-1}{(1+x)^2}-\dfrac{1}{(x-1)^2} & \text{if } x>2\end{cases}$$

Clearly $f'(x)>0$ for $x<0$ and $f'(x)<0$ for $x>2$. For $0<x<2$, we have $f'(x)=\dfrac{1}{(3-x)^2}-\dfrac{1}{(x+1)^2}=\dfrac{(x^2+2x+1)-(x^2-6x+9)}{(3-x)^2(x+1)^2}=\dfrac{8(x-1)}{(3-x)^2(x+1)^2}$, so $f'(x)<0$ for $x<1$, $f'(1)=0$ and $f'(x)>0$ for $x>1$. We have shown that $f'(x)>0$ for $x<0$; $f'(x)<0$ for $0<x<1$; $f'(x)>0$ for $1<x<2$; and $f'(x)<0$ for $x>2$. Therefore by the First Derivative Test, the local maxima of f are at $x=0$ and $x=2$, where f takes the value $\frac{4}{3}$. Therefore $\frac{4}{3}$ is the absolute maximum value of f.

10. $y = x^3 - 3x + 4 \;\Rightarrow\; y' = 3x^2 - 3$, and $y = 3(x^2 - x) \;\Rightarrow\; y' = 6x - 3$. The slopes of the tangents of the two curves are equal when $3x^2 - 3 = 6x - 3$, that is, when $x = 0$ or 2. At $x = 0$, both tangents have slope -3, but the curves do not intersect. At $x = 2$, both tangents have slope 9 and the curves intersect at $(2, 6)$. So there is a common tangent line at $(2, 6)$.

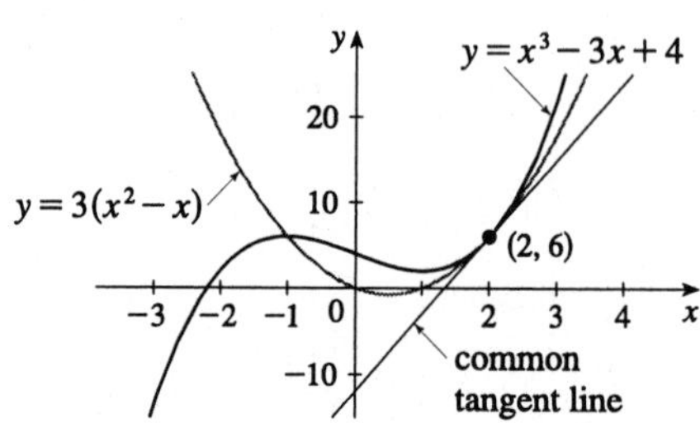

11. We must find a value x_0 such that the normal lines to the parabola $y = x^2$ at $x = \pm x_0$ intersect at a point one unit from the points $(\pm x_0, x_0^2)$. The normals to $y = x^2$ at $x = \pm x_0$ have slopes $-\dfrac{1}{\pm 2x_0}$ and pass through $(\pm x_0, x_0^2)$ respectively, so the normals have the equations $y - x_0^2 = -\dfrac{1}{2x_0}(x - x_0)$ and $y - x_0^2 = \dfrac{1}{2x_0}(x + x_0)$. The common y-intercept is $x_0^2 + \frac{1}{2}$. We want to find the value of x_0 for which the distance from $\left(0, x_0^2 + \frac{1}{2}\right)$ to (x_0, x_0^2) equals 1. The square of the distance is $(x_0 - 0)^2 + \left[x_0^2 - \left(x_0^2 + \frac{1}{2}\right)\right]^2 = x_0^2 + \frac{1}{4} = 1 \;\Leftrightarrow\; x_0 = \pm\frac{\sqrt{3}}{2}$. For these values of x_0, the y-intercept is $x_0^2 + \frac{1}{2} = \frac{5}{4}$, so the center of the circle is at $\left(0, \frac{5}{4}\right)$.

Alternate Solution: Let the center of the circle be $(0, a)$. Then the equation of the circle is $x^2 + (y - a)^2 = 1$. Solving with the equation of the parabola, $y = x^2$, we get $x^2 + \left(x^2 - a\right)^2 = 1 \;\Leftrightarrow\; x^2 + x^4 - 2ax^2 + a^2 = 1 \;\Leftrightarrow\; x^4 + (1 - 2a)x^2 + a^2 - 1 = 0$. The parabola and the circle will be tangent to each other when this quadratic equation in x^2 has equal roots, that is, when the discriminant is 0. Thus $(1 - 2a)^2 - 4(a^2 - a) = 0 \;\Leftrightarrow\; 1 - 4a + 4a^2 - 4a^2 + 4 = 0 \;\Leftrightarrow\; 4a = 5$, so $a = \frac{5}{4}$. The center of the circle is $\left(0, \frac{5}{4}\right)$.

12. To sketch the region $\{(x, y) \mid 2xy \le |x - y| \le x^2 + y^2\}$, we consider two cases.

Case 1: $x \ge y$ This is the case in which (x, y) lies on or below the line $y = x$. The double inequality becomes $2xy \le x - y \le x^2 + y^2$. The right-hand inequality holds if and only if $x^2 - x + y^2 + y \ge 0 \;\Leftrightarrow\; \left(x - \frac{1}{2}\right)^2 + \left(y + \frac{1}{2}\right)^2 \ge \frac{1}{2} \;\Leftrightarrow\; (x, y)$ lies on or outside the circle with radius $\frac{1}{\sqrt{2}}$ centered at $\left(\frac{1}{2}, -\frac{1}{2}\right)$. The left-hand inequality holds if and only if $2xy - x + y \le 0 \;\Leftrightarrow\; \left(x + \frac{1}{2}\right)\left(y - \frac{1}{2}\right) \le -\frac{1}{4} \;\Leftrightarrow\; (x, y)$ lies on or below the hyperbola $\left(x + \frac{1}{2}\right)\left(y - \frac{1}{2}\right) = -\frac{1}{4}$, which passes through the origin and approaches the lines $y = \frac{1}{2}$ and $x = -\frac{1}{2}$ asymptotically.

Case 2: $x \le y$ This is the case in which (x, y) lies on or above the line $y = x$. The double inequality becomes $2xy \le y - x \le x^2 + y^2$. The right-hand inequality holds if and only if $x^2 + x + y^2 - y \ge 0 \;\Leftrightarrow\; \left(x + \frac{1}{2}\right)^2 + \left(y - \frac{1}{2}\right)^2 \ge \frac{1}{2} \;\Leftrightarrow\; (x, y)$ lies on or outside the circle of radius $\frac{1}{\sqrt{2}}$ centered at $\left(\frac{1}{2}, -\frac{1}{2}\right)$. The left-hand inequality holds if and only if $2xy + x - y \le 0 \;\Leftrightarrow\; xy + \frac{1}{2}x - \frac{1}{2}y \le 0 \;\Leftrightarrow\; \left(x - \frac{1}{2}\right)\left(y + \frac{1}{2}\right) \le -\frac{1}{4}$

$\Leftrightarrow$ (x, y) lies on or above the left-hand branch of the hyperbola $\left(x - \frac{1}{2}\right)\left(y + \frac{1}{2}\right) = -\frac{1}{4}$, which passes through the origin and approaches the lines $y = -\frac{1}{2}$ and $x = \frac{1}{2}$ asymptotically.

Therefore the region of interest consists of the points on or above the left branch of the hyperbola $\left(x - \frac{1}{2}\right)\left(y + \frac{1}{2}\right) = -\frac{1}{4}$ that are on or outside the circle $\left(x + \frac{1}{2}\right)^2 + \left(y - \frac{1}{2}\right)^2 = \frac{1}{2}$, together with the points on or below the right branch of the hyperbola $\left(x + \frac{1}{2}\right)\left(y - \frac{1}{2}\right) = -\frac{1}{4}$ that are on or outside the circle $\left(x - \frac{1}{2}\right)^2 + \left(y + \frac{1}{2}\right)^2 = \frac{1}{2}$. Note that the inequalities are unchanged when x and y are interchanged, so the region is symmetric about the line $y = x$. So we need only have analyzed case 1 and then reflected that region about the line $y = x$, instead of considering case 2.

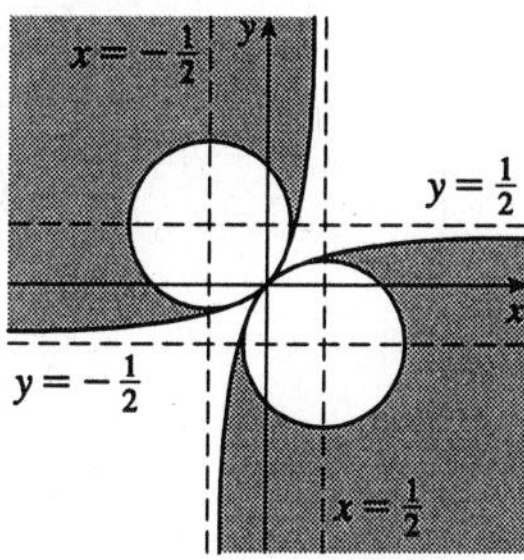

13. $f(x) = \int_0^{g(x)} \frac{1}{\sqrt{1+t^3}}\,dt$, where $g(x) = \int_0^{\cos x} \left[1 + \sin(t^2)\right] dt$. Using Part 1 of the Fundamental Theorem of Calculus and the Chain Rule (twice) we have that

$f'(x) = \dfrac{1}{\sqrt{1 + [g(x)]^3}}\, g'(x) = \dfrac{1}{\sqrt{1 + [g(x)]^3}}\left[1 + \sin(\cos^2 x)\right](-\sin x)$. Now $g\left(\frac{\pi}{2}\right) = \int_0^0 [1 + \sin(t^2)]dt = 0$, so $f'\left(\frac{\pi}{2}\right) = \frac{1}{\sqrt{1+0}}(1 + \sin 0)(-1) = 1 \cdot 1 \cdot (-1) = -1$.

14. If $f'(x) < 0$ for all x, $f''(x) > 0$ for $|x| > 1$, $f''(x) < 0$ for $|x| < 1$, and $\lim\limits_{x \to \pm\infty} [f(x) + x] = 0$, then f is decreasing everywhere, concave up on $(-\infty, -1)$ and $(1, \infty)$, concave down on $(-1, 1)$, and approaches the line $y = -x$ as $x \to \pm\infty$. An example of such a graph is sketched.

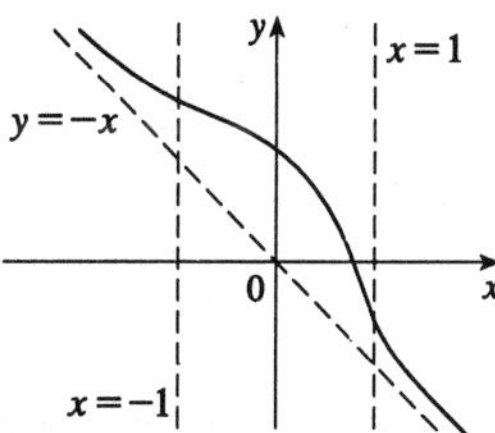

15. $f(x) = [\![x]\!] + \sqrt{x - [\![x]\!]}$. On each interval of the form $[n, n+1)$, where n is an integer, we have $f(x) = n + \sqrt{x - n}$. It is easy to see that this function is continuous and increasing on $[n, n+1)$. Also, the left-hand limit

$$\begin{aligned}\lim_{x \to (n+1)^-} f(x) &= \lim_{x \to (n+1)^-} \left[[\![x]\!] + \sqrt{x - [\![x]\!]}\right] = \lim_{x \to (n+1)^-} [\![x]\!] + \sqrt{\lim_{x \to (n+1)^-} x - \lim_{x \to (n+1)^-} [\![x]\!]} \\ &= n + \sqrt{n + 1 - n} = n + 1 = f(n+1) = \lim_{x \to (n+1)^+} f(x),\end{aligned}$$

so f is continuous and increasing everywhere.

16. Let $h(x) = [\![1/x]\!]$ for $\frac{1}{6} \le |x| \le 2$. Then $f(x) = (-1)^{h(x)}$ and $g(x) = xf(x)$ for $\frac{1}{6} \le |x| \le 2$. Notice that $\frac{1}{6} \le |x| \le 2 \;\Leftrightarrow\; x \in \left[-2, -\frac{1}{6}\right] \cup \left[\frac{1}{6}, 2\right] \;\Leftrightarrow\; 1/x \in \left[-6, -\frac{1}{2}\right] \cup \left[\frac{1}{2}, 6\right] \;\Leftrightarrow\; [\![1/x]\!] \in \{0, \pm 1, \pm 2, \ldots, \pm 6\}$.

More specifically,

$$h(x) = \begin{cases} -1 & \text{if } x \in [-2,-1] \\ -2 & x \in \left(-1, -\frac{1}{2}\right] \\ -3 & x \in \left(-\frac{1}{2}, -\frac{1}{3}\right] \\ -4 & x \in \left(-\frac{1}{3}, -\frac{1}{4}\right] \\ -5 & x \in \left(-\frac{1}{4}, -\frac{1}{5}\right] \\ -6 & x \in \left(-\frac{1}{5}, -\frac{1}{6}\right] \\ 6 & x = \frac{1}{6} \\ 5 & x \in \left(\frac{1}{6}, \frac{1}{5}\right] \\ 4 & x \in \left(\frac{1}{5}, \frac{1}{4}\right] \\ 3 & x \in \left(\frac{1}{4}, \frac{1}{3}\right] \\ 2 & x \in \left(\frac{1}{3}, \frac{1}{2}\right] \\ 1 & x \in \left(\frac{1}{2}, 1\right] \\ 0 & x \in (1, 2] \end{cases} \quad \text{so} \quad \begin{array}{l} \left.\begin{array}{l} f(x) = -1 \\ g(x) = -x \end{array}\right\} \text{ if } \begin{cases} x \in [-2,-1] \\ x \in \left(-\frac{1}{2}, -\frac{1}{3}\right] \\ x \in \left(-\frac{1}{4}, -\frac{1}{5}\right] \\ x \in \left(\frac{1}{6}, \frac{1}{5}\right] \\ x \in \left(\frac{1}{4}, \frac{1}{3}\right] \\ x \in \left(\frac{1}{2}, 1\right] \end{cases} \\ \left.\begin{array}{l} f(x) = 1 \\ g(x) = x \end{array}\right\} \text{ if } \begin{cases} x \in \left(-1, -\frac{1}{2}\right] \\ x \in \left(-\frac{1}{3}, -\frac{1}{4}\right] \\ x \in \left(-\frac{1}{5}, -\frac{1}{6}\right] \\ x = \frac{1}{6} \\ x \in \left(\frac{1}{5}, \frac{1}{4}\right] \\ x \in \left(\frac{1}{3}, \frac{1}{2}\right] \\ x \in (1, 2] \end{cases} \end{array}$$

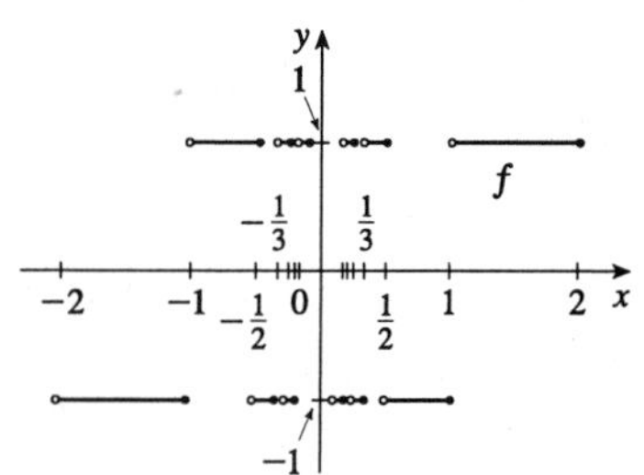

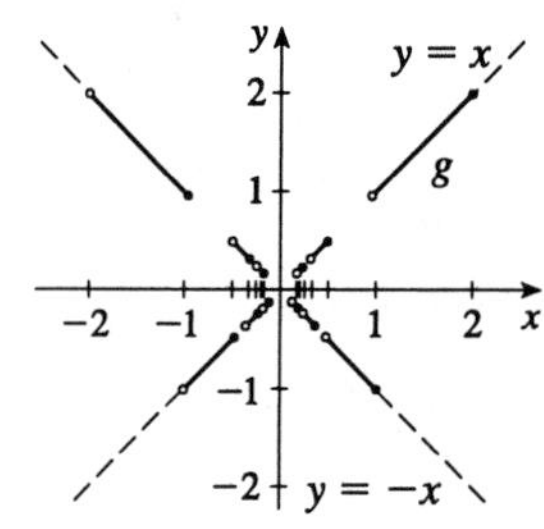

If we were to define $g(x) = x(-1)^{[\![1/x]\!]}$ for $x \ne 0$ and $g(0) = 0$, then g would be continuous at 0. Since $|g(x)| = |x|$ for $x \ne 0$, we would have $|g(x) - g(0)| = |x - 0|$. This equality implies that $g(x) \to g(0)$ as $x \to 0$.

17. $f(x) = 2 + x - x^2 = (-x+2)(x+1)$. So $f(x) = 0 \;\Leftrightarrow\; x = 2$ or $x = -1$, and $f(x) \ge 0$ for $x \in [-1, 2]$ and $f(x) < 0$ everywhere else. The integral $\int_a^b (2 + x - x^2)dx$ has a maximum on the interval where the integrand is positive, which is $[-1, 2]$. So $a = -1$, $b = 2$. (Any larger interval gives a smaller integral since $f(x) < 0$ outside $[-1, 2]$. Any smaller interval also gives a smaller integral since $f(x) \ge 0$ in $[-1, 2]$.)

18. If $f(x) = \int_0^x x^2 \sin(t^2)dt = x^2 \int_0^x \sin(t^2)dt$, then $f'(x) = 2x \int_0^x \sin(t^2)dt + x^2 \sin(x^2)$ by the Product Rule and Part 1 of the Fundamental Theorem.

19. $A = (x_1, x_1^2)$ and $B = (x_2, x_2^2)$, where x_1 and x_2 are the solutions of the quadratic equation $x^2 = mx + b$. Let $P = (x, x^2)$ and set $A_1 = (x_1, 0)$, $B_1 = (x_2, 0)$, and $P_1 = (x, 0)$. Let $f(x)$ denote the area of triangle PAB. Then $f(x)$ can be expressed in terms of the areas of three trapezoids as follows:

$$f(x) = \text{area}(A_1ABB_1) - \text{area}(A_1APP_1) - \text{area}(B_1BPP_1)$$
$$= \tfrac{1}{2}(x_2 - x_1)(x_1^2 + x_2^2) - \tfrac{1}{2}(x - x_1)(x_1^2 + x^2) - \tfrac{1}{2}(x_2 - x)(x^2 + x_2^2).$$

After expansion, canceling of terms, and factoring, we find that $f(x) = \frac{1}{2}(x_2 - x_1)(x - x_1)(x_2 - x)$.

Note: Another way to get an expression for $f(x)$ is to use the formula for an area of a triangle in terms of the coordinates of the vertices: $f(x) = \frac{1}{2}[(x_2x_1^2 - x_1x_2^2) + (x_1x^2 - xx_1^2) + (xx_2^2 - x_2x^2)]$. From our expression for $f(x)$, it follows that $f'(x) = \frac{1}{2}(x_2 - x_1)(x_1 + x_2 - 2x)$ and $f''(x) = -(x_2 - x_1) < 0$. Thus the area $f(x)$ is maximized when $x = \frac{1}{2}(x_1 + x_2)$, and $f\left(\frac{1}{2}(x_1 + x_2)\right) = \frac{1}{2}(x_2 - x_1)\frac{1}{2}(x_2 - x_1)\frac{1}{2}(x_2 - x_1) = \frac{1}{8}(x_2 - x_1)^3$. In terms of m and b, $x_1 = \frac{1}{2}\left(m - \sqrt{m^2 + 4b}\right)$ and $x_2 = \frac{1}{2}\left(m + \sqrt{m^2 + 4b}\right)$, so the maximal area is $\frac{1}{8}(m^2 + 4b)^{3/2}$ and it is attained at the point $P\left(\frac{1}{2}m, \frac{1}{4}m^2\right)$.

20.

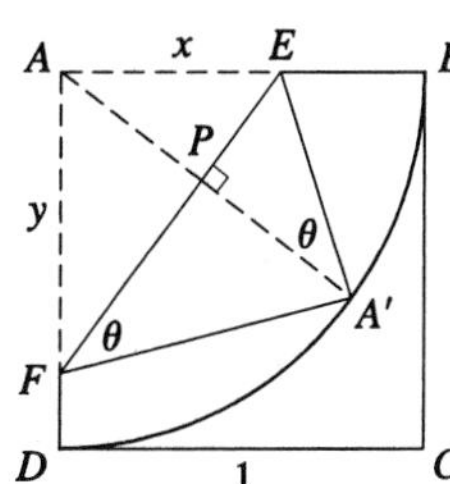

Let $x = |AE|$, $y = |AF|$ as shown. The area of the triangle AEF is therefore $\mathcal{A} = \frac{1}{2}xy$. We need a relationship between x and y, so we can take the derivative $\dfrac{d\mathcal{A}}{dx}$ and find the maxima and minima of $\mathcal{A}$. Now let A' be the point on which A ends up after the fold has been performed, and let P be the intersection of AA' and EF. Note that AA' is perpendicular to EF since we are reflecting A through the line EF to get to A', and that $|AP| = |AP'|$ for the same reason. But $|AA'| = 1$, since AA' is a radius of the circle. So since $|AP| + |AP'| = |AA'|$, we have $AP = \frac{1}{2}$. So another way to express the area of the triangle is $\mathcal{A} = \frac{1}{2}|EF||AP| = \frac{1}{2}\sqrt{x^2 + y^2}\left(\frac{1}{2}\right) = \frac{1}{4}\sqrt{x^2 + y^2}$. Equating the two expressions for $\mathcal{A}$, we get $\frac{1}{2}xy = \frac{1}{4}\sqrt{x^2 + y^2} \quad \Rightarrow \quad x^2y^2 = \frac{1}{4}(x^2 + y^2) \quad \Rightarrow \quad y^2(4x^2 - 1) = x^2 \quad \Rightarrow \quad y = x/\sqrt{4x^2 - 1}$. (Note that we could also have derived this result from the similarity of $\triangle A'PE$ and $\triangle A'FE$.) Now we can substitute and calculate $\dfrac{d\mathcal{A}}{dx}$: $\mathcal{A} = \dfrac{1}{2}\dfrac{x^2}{\sqrt{4x^2 - 1}} \quad \Rightarrow \quad \dfrac{d\mathcal{A}}{dx} = \dfrac{1}{2}\left[\dfrac{\sqrt{4x^2 - 1}(2x) - x^2\left(\frac{1}{2}\right)(4x^2 - 1)^{-1/2}(8x)}{4x^2 - 1}\right]$. This is 0 when $2x\sqrt{4x^2 - 1} - 4x^3(4x^2 - 1)^{-1/2} = 0 \quad \Leftrightarrow \quad 2(4x^2 - 1) - 4x^2 = 0 \quad \Leftrightarrow \quad 4x^2 = 2 \quad \Leftrightarrow \quad x = \frac{1}{\sqrt{2}}$.

So this is one possible value for an extremum. We must also test the endpoints of the interval over which x ranges. The largest value that x can attain is 1, and the smallest value of x occurs when $y = 1 \quad \Leftrightarrow \quad 1 = x/\sqrt{4x^2 - 1} \quad \Leftrightarrow \quad x^2 = 4x^2 - 1 \quad \Leftrightarrow \quad 3x^2 = 1 \quad \Leftrightarrow \quad x = \frac{1}{\sqrt{3}}$. This will give the same value of $\mathcal{A}$ as will $x = 1$, since the geometric situation is the same (reflected through the line $y = x$). We calculate

$$\mathcal{A}\left(\frac{1}{\sqrt{2}}\right) = \frac{1}{2}\frac{(1/\sqrt{2})^2}{\sqrt{4(1/\sqrt{2})^2 - 1}} = \frac{1}{4}, \text{ and } \mathcal{A}(1) = \frac{1}{2}\left[\frac{1^2}{\sqrt{4(1)^2 - 1}}\right] = \frac{1}{2\sqrt{3}}.$$

So the maximum area is $\mathcal{A}(1) = \mathcal{A}\left(\frac{1}{\sqrt{3}}\right) = \frac{1}{2\sqrt{3}}$ and the minimum area is $\mathcal{A}\left(\frac{1}{\sqrt{2}}\right) = \frac{1}{4}$.

Another Method: Use the angle θ (see diagram above) as a variable:

$\mathcal{A} = \frac{1}{2}\left(\frac{1}{2}\sec\theta\right)\left(\frac{1}{2}\csc\theta\right) = \dfrac{1}{8\sin\theta\cos\theta} = \dfrac{1}{4\sin 2\theta}$. $\mathcal{A}$ is minimized when $\sin 2\theta$ is maximal, that is, when $\sin 2\theta = 1 \quad \Rightarrow \quad 2\theta = \frac{\pi}{2} \quad \Rightarrow \quad \theta = \frac{\pi}{4}$. Also note that $A'E = \frac{1}{2}\sec\theta \le 1 \quad \Rightarrow \quad \sec\theta \le 2 \quad \Rightarrow \quad \cos\theta \ge \frac{1}{2} \quad \Rightarrow \quad \theta \le \frac{\pi}{3}$, and similarly $\theta \ge \frac{\pi}{6}$. As above, we find that $\mathcal{A}$ is maximized at these endpoints:

$$\mathcal{A}\left(\tfrac{\pi}{6}\right) = \frac{1}{4\sin\left(\frac{\pi}{3}\right)} = \frac{1}{2\sqrt{3}} = \frac{1}{4\sin\left(\frac{2\pi}{3}\right)} = \mathcal{A}\left(\tfrac{\pi}{3}\right).$$

21. Since $[\![x]\!] \le x < [\![x]\!] + 1$, we have $1 \le \dfrac{x}{[\![x]\!]} \le 1 + \dfrac{1}{[\![x]\!]}$ for $x > 0$. As $x \to \infty$, $[\![x]\!] \to \infty$, so $\dfrac{1}{[\![x]\!]} \to 0$ and $1 + \dfrac{1}{[\![x]\!]} \to 1$. Thus $\lim\limits_{x\to\infty} \dfrac{x}{[\![x]\!]} = 1$ by the Squeeze Theorem.

22. $\dfrac{d}{dx}\displaystyle\int_0^x \left(\int_1^{\sin t} \sqrt{1+u^4}\,du\right) dt = \int_1^{\sin x} \sqrt{1+u^4}\,du$, by Part 1 of the Fundamental Theorem of Calculus.

So $\dfrac{d^2}{dx^2}\displaystyle\int_0^x \left(\int_1^{\sin t} \sqrt{1+u^4}\,du\right) dt = \dfrac{d}{dx}\int_1^{\sin x} \sqrt{1+u^4}\,du = \sqrt{1+\sin^4 x}\cos x$, again by Part 1 of the Fundamental Theorem of Calculus.

23. Differentiating the equation $\int_0^x f(t)dt = [f(x)]^2$ using Part 1 of the Fundamental Theorem gives $f(x) = 2f(x)f'(x) \;\Rightarrow\; f(x)[2f'(x) - 1] = 0$, so $f(x) = 0$ or $f'(x) = \frac{1}{2}$. $f'(x) = \frac{1}{2} \;\Rightarrow$ $f(x) = \frac{1}{2}x + C$. To find C we substitute into the original equation to get $\int_0^x \left(\frac{1}{2}t + C\right)dt = \left(\frac{1}{2}x + C\right)^2 \;\Leftrightarrow$ $\frac{1}{4}x^2 + Cx = \frac{1}{4}x^2 + Cx + C^2$. It follows that $C = 0$ so $f(x) = \frac{1}{2}x$. Therefore $f(x) = 0$ or $f(x) = \frac{1}{2}x$.

24. **(a)**

θ
c
b
a

By Exercise 88 in Appendix D, the area of the triangle is $A = \frac{1}{2}bc\sin\theta$. (This formula can also be found in the endpapers.) But A is a constant, so differentiating this equation with respect to t, we get

$$\frac{dA}{dt} = 0 = \frac{1}{2}\left[bc\cos\theta\,\frac{d\theta}{dt} + b\frac{dc}{dt}\sin\theta + \frac{db}{dt}c\sin\theta\right]$$

$$\Rightarrow \quad bc\cos\theta\,\frac{d\theta}{dt} = -\sin\theta\left[b\frac{dc}{ct} + c\frac{db}{dt}\right] \quad \Rightarrow \quad \frac{d\theta}{dt} = -\tan\theta\left[\frac{1}{c}\frac{dc}{dt} + \frac{1}{b}\frac{db}{dt}\right].$$

(b) We use the Law of Cosines to get the length of side a in terms of those of b and c, and then we differentiate implicitly with respect to t: $a^2 = b^2 + c^2 - 2bc\cos\theta \;\Rightarrow$

$$2a\frac{da}{dt} = 2b\frac{db}{dt} + 2c\frac{dc}{dt} - 2\left[bc(-\sin\theta)\frac{d\theta}{dt} + b\frac{dc}{dt}\cos\theta + \frac{db}{dt}c\cos\theta\right] \quad \Rightarrow$$

$\dfrac{da}{dt} = \dfrac{1}{a}\left[b\dfrac{db}{dt} + c\dfrac{dc}{dt} + bc\sin\theta\,\dfrac{d\theta}{dt} - b\dfrac{dc}{dt}\cos\theta - c\dfrac{db}{dt}\cos\theta\right]$. Now we substitute our value of a from the Law of Cosines and the value of $\dfrac{d\theta}{dt}$ from part (a), and simplify (primes signify differentiation by t):

$$\begin{aligned}\frac{da}{dt} &= \frac{bb' + cc' + bc\sin\theta[-\tan\theta(c'/c + b'/b)] - (bc' + cb')(\cos\theta)}{\sqrt{b^2 + c^2 - 2bc\cos\theta}}\\ &= \frac{bb' + cc' - [\sin^2\theta(bc' + cb') + \cos^2\theta(bc' + cb')]/\cos\theta}{\sqrt{b^2 + c^2 - 2bc\cos\theta}}\\ &= \frac{bb' + cc' - (bc' + cb')\sec\theta}{\sqrt{b^2 + c^2 - 2bc\cos\theta}}.\end{aligned}$$

25. We find the tangent line to $f_n(x) = n\cos nx$ at the point $\left(\frac{\pi}{2n}, 0\right)$:

$\frac{df_n}{dx} = n(-\sin nx)n = -n^2 \sin nx \quad \Rightarrow$

$f_n'\left(\frac{\pi}{2n}\right) = -n^2 \sin\left(n\frac{\pi}{2n}\right) = -n^2(1) = -n^2$. So the equation of the tangent line to f_n is $y = t_n(x) = -n^2\left(x - \frac{\pi}{2n}\right) = \frac{\pi}{2}n - n^2x$.

Note that the tangent line lies above the curve on $\left(0, \frac{\pi}{2n}\right)$.

The area bounded by the graphs of $t_n(x)$ and $f_n(x)$ is

$A_n = \int_0^{\pi/(2n)} [t_n(x) - f_n(x)]dx = \int_0^{\pi/(2n)} \left(\frac{\pi}{2}n - n^2x - n\cos nx\right)dx$

$= \left[\frac{\pi}{2}nx - \frac{1}{2}n^2x^2 - \sin nx\right]_0^{\pi/(2n)} = \frac{\pi}{2}n\left(\frac{\pi}{2n}\right) - \frac{1}{2}n^2\left(\frac{\pi}{2n}\right)^2 - \sin\left[n\left(\frac{\pi}{2n}\right)\right]$

$= \frac{1}{8}\pi^2 - 1$, a constant.

26. **(a)** We can split the integral $\int_0^n [\![x]\!]\,dx$ into the sum $\sum_{i=1}^{n}\left[\int_{i-1}^{i} [\![x]\!]\,dx\right]$. But on each of the intervals $[i-1, i)$ of integration, $[\![x]\!]$ is a constant function, namely $i - 1$. So the ith integral in the sum is equal to $(i-1)[i-(i-1)] = (i-1)$. So the original integral is equal to $\sum_{i=1}^{n}(i-1) = \sum_{i=1}^{n-1} i = \frac{(n-1)n}{2}$.

(b) We can write $\int_a^b [\![x]\!]\,dx = \int_0^b [\![x]\!]\,dx - \int_0^a [\![x]\!]\,dx$.

Now $\int_0^b [\![x]\!]\,dx = \int_0^{[\![b]\!]} [\![x]\!]\,dx + \int_{[\![b]\!]}^b [\![x]\!]\,dx$. The first of these integrals is equal to $\frac{1}{2}([\![b]\!] - 1)[\![b]\!]$, by part (a), and since $[\![x]\!] = [\![b]\!]$ on $[[\![b]\!], b]$, the second integral is just $[\![b]\!](b - [\![b]\!])$. So

$\int_0^b [\![x]\!]\,dx = \frac{1}{2}([\![b]\!] - 1)[\![b]\!] + [\![b]\!](b - [\![b]\!]) = \frac{1}{2}[\![b]\!](2b - [\![b]\!] - 1)$ and similarly

$\int_0^a [\![x]\!]\,dx = \frac{1}{2}[\![a]\!](2a - [\![a]\!] - 1)$. Therefore $\int_a^b [\![x]\!]\,dx = \frac{1}{2}[\![b]\!](2b - [\![b]\!] - 1) - \frac{1}{2}[\![a]\!](2a - [\![a]\!] - 1)$.

27. $f(x) = (a^2 + a - 6)\cos 2x + (a-2)x + \cos 1 \quad \Rightarrow \quad f'(x) = -(a^2 + a - 6)\sin 2x(2) + (a - 2)$.

The derivative exists for all x, so the only possible critical points will occur where $f'(x) = 0 \quad \Leftrightarrow$

$2(a-2)(a+3)\sin 2x = a - 2 \quad \Leftrightarrow \quad$ either $a = 2$ or $2(a+3)\sin 2x = 1$, with the latter implying that

$\sin 2x = \frac{1}{2(a+3)}$. Since the range of $\sin 2x$ is $[-1, 1]$, this equation has no solution whenever either

$\frac{1}{2(a+3)} > 1$ or $\frac{1}{2(a+3)} > 1$. Solving these inequalities, we get $-\frac{7}{2} < a < -\frac{5}{2}$.

28. $\int_0^x f(t)dt = \int_x^1 t^2 f(t)dt + 8x^6 + 6x^8 + C = -\int_1^x t^2 f(t)dt + 8x^6 + 6x^8 + C$. We differentiate the equation implicitly with respect to x, using the Fundamental Theorem of Calculus, to get $f(x) = -x^2 f(x) + 48x^5 + 48x^7$

$\Rightarrow \quad f(x) = \frac{48x^5 + 48x^7}{1 + x^2} = 48x^5$. Now we put $x = 0$ in our first equation to find C:

$\int_0^0 48t^5\,dt = 0 = \int_0^1 t^2(48t^5)dt + C \quad \Rightarrow \quad C = -[6t^8]_0^1 = -6$.

29. **(a)**

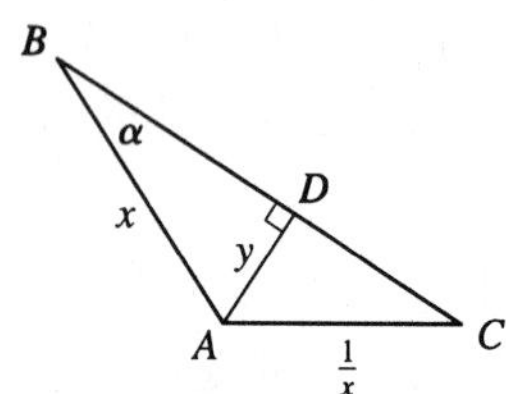

Let $y = AD$. We compute the area of $\triangle ABC$ in two ways.

First, $|AB||AC| = 1$, so $\mathcal{A} = \frac{1}{2}|AB||AC|\sin\frac{2\pi}{3} = \frac{1}{2}\cdot 1\cdot\frac{\sqrt{3}}{2} = \frac{\sqrt{3}}{4}$.

Second, $\mathcal{A} = (\text{area of } \triangle ABD) + (\text{area of } \triangle ACD)$

$$= \tfrac{1}{2}|AB||AD|\sin\tfrac{\pi}{3} + \tfrac{1}{2}|AD||AC|\sin\tfrac{\pi}{3}$$

$$= \tfrac{1}{2}xy\tfrac{\sqrt{3}}{2} + \tfrac{1}{2}y(1/x)\tfrac{\sqrt{3}}{2} = \tfrac{\sqrt{3}}{4}y(x + 1/x).$$

Equating the two expressions for the area, we get $y\left(x + \dfrac{1}{x}\right) = 1$, or $y = \dfrac{1}{x + 1/x} = \dfrac{x}{x^2+1}$, $x > 0$.

Another Method: By using the Law of Sines on the triangles ABD and ABC, we can get $\dfrac{x}{y} = \frac{\sqrt{3}}{2}\cot\alpha + \frac{1}{2}$ and $\frac{\sqrt{3}}{2}\cot\alpha = x^2 + \frac{1}{2}$. Eliminating $\cot\alpha$ gives $\dfrac{x}{y} = \left(x^2 + \frac{1}{2}\right) + \frac{1}{2} \Rightarrow$ $y = \dfrac{x}{x^2+1}$, $x > 0$.

(b) We differentiate our expression for y with respect to x to find the maximum:

$\dfrac{dy}{dx} = \dfrac{(x^2+1) - x(2x)}{(x^2+1)^2} = \dfrac{1-x^2}{(x^2+1)^2} = 0$ when $x = 1$. This indicates a maximum by the First Derivative Test, since $y'(x) > 0$ for $0 < x < 1$ and $y'(x) < 0$ for $x > 1$, so the maximum value of y is $y(1) = \frac{1}{2}$.

30. **(a)**

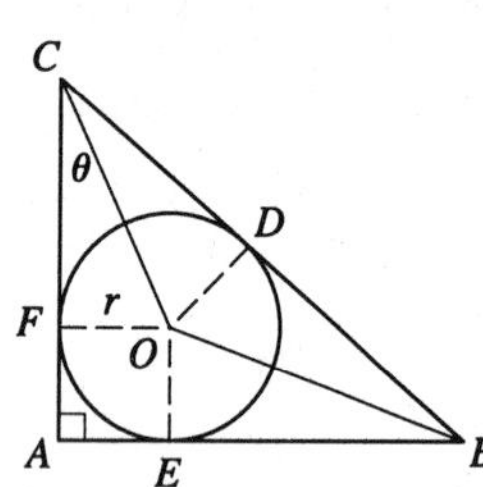

By elementary geometry, two tangents to a circle from a given point have the same length, so $|CF| = |CD|$, $|AE| = |AF|$, and $|BD| = |BE|$. Thus

$$|CD| = \tfrac{1}{2}(|CD| + |CF|) = \tfrac{1}{2}[(|BC| - |BD|) + (|AC| - |AF|)]$$

$$= \tfrac{1}{2}[|AC| + |BC| - (|AF| + |BD|)]$$

$$= \tfrac{1}{2}[|AC| + |BC| - (|AE| + |BE|)] = \tfrac{1}{2}(|AC| + |BC| - |AB|).$$

(b) By part (a), $\dfrac{r}{\tan\theta} = |CD| = \frac{1}{2}(|AC| + |BC| - |AB|) = \frac{1}{2}(a + a\cos 2\theta - a\sin 2\theta) \Leftrightarrow$

$r = \frac{1}{2}a\tan\theta(2\cos^2\theta - 2\sin\theta\cos\theta) = \frac{1}{2}a(2\sin\theta\cos\theta - 2\sin^2\theta) = \frac{1}{2}a(\sin 2\theta + \cos 2\theta - 1)$.

(c) We differentiate r with respect to θ and set $dr/d\theta = 0$ to find the maximum values:

$dr/d\theta = \frac{1}{2}a(2\cos 2\theta - 2\sin 2\theta) = a\cos 2\theta(1 - \tan 2\theta)$. Since $0 < \theta < \frac{\pi}{4}$, $dr/d\theta = 0 \Leftrightarrow \cos 2\theta = 0$ $\Leftrightarrow \theta = \frac{\pi}{8}$. This gives a maximum by the First Derivative Test, since $dr/d\theta > 0$ for $0 < \theta < \frac{\pi}{8}$, and $dr/d\theta > 0$ for $\frac{\pi}{8} < \theta < \frac{\pi}{4}$. The maximum value is

$$r\left(\tfrac{\pi}{8}\right) = \tfrac{1}{2}a\left(\sin\tfrac{\pi}{4} + \cos\tfrac{\pi}{4} - 1\right) = \tfrac{1}{2}\left(\sqrt{2} - 1\right)a \approx 0.207a.$$

31. $$\lim_{n\to\infty}\left(\frac{1}{\sqrt{n}\sqrt{n+1}} + \frac{1}{\sqrt{n}\sqrt{n+2}} + \cdots + \frac{1}{\sqrt{n}\sqrt{n+n}}\right) = \lim_{n\to\infty}\frac{1}{n}\left(\sqrt{\frac{n}{n+1}} + \sqrt{\frac{n}{n+2}} + \cdots + \sqrt{\frac{n}{n+n}}\right)$$

$$= \lim_{n\to\infty}\frac{1}{n}\left(\frac{1}{\sqrt{1+1/n}} + \frac{1}{\sqrt{1+2/n}} + \cdots + \frac{1}{\sqrt{1+1}}\right) = \lim_{n\to\infty}\frac{1}{n}\sum_{i=1}^{n} f\left(\frac{i}{n}\right) \quad \left(\text{where } f(x) = \frac{1}{\sqrt{1+x}}\right)$$

$$= \int_0^1 \frac{1}{\sqrt{1+x}}\,dx = \left[2\sqrt{1+x}\right]_0^1 = 2\left(\sqrt{2} - 1\right)$$

32. Note that the graphs of $(x-c)^2$ and $[(x-c)-2]^2$ intersect when $|x-c| = |x-c-2|$ $\Leftrightarrow$ $c - x = x - c - 2$ $\Leftrightarrow$ $x = c+1$. The integration will proceed differently depending on the value of c.

Case 1: $-2 \le c < -1$ In this case, $f(x) = (x-c-2)^2$ for $x \in [0,1]$, so

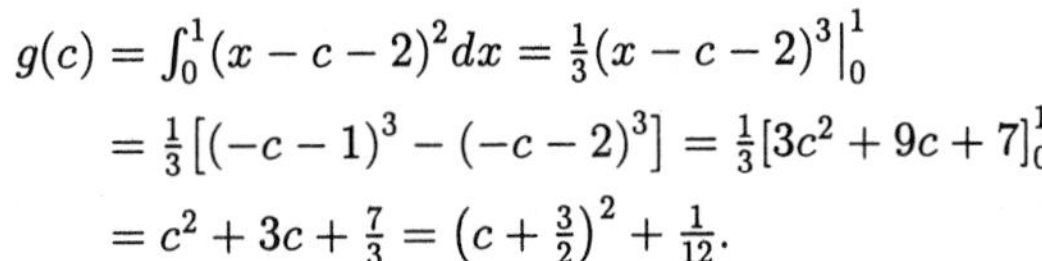

$$\begin{aligned} g(c) &= \int_0^1 (x-c-2)^2 dx = \tfrac{1}{3}(x-c-2)^3\Big|_0^1 \\ &= \tfrac{1}{3}\left[(-c-1)^3 - (-c-2)^3\right] = \tfrac{1}{3}[3c^2+9c+7]_0^1 \\ &= c^2 + 3c + \tfrac{7}{3} = \left(c+\tfrac{3}{2}\right)^2 + \tfrac{1}{12}. \end{aligned}$$

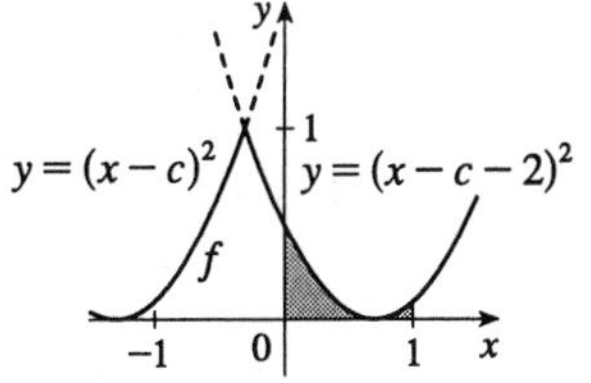

This is a parabola; its maximum for $-2 \le c < -1$ is $g(-2) = g(-1) = \frac{1}{3}$, and its minimum is $g\left(-\frac{3}{2}\right) = \frac{1}{12}$.

Case 2: $-1 \le c < 0$ In this case,

$$f(x) = \begin{cases} (x-c)^2 & \text{if } 0 \le x \le c+1 \\ (x-c-2)^2 & \text{if } c+1 < x \le 1 \end{cases}$$ Therefore

$$\begin{aligned} g(c) &= \int_0^1 f(x)dx = \int_0^{c+1} (x-c)^2 dx + \int_{c+1}^1 (x-c-2)^2 dx \\ &= \tfrac{1}{3}(x-c)^3\Big|_0^{c+1} + \tfrac{1}{3}(x-c-2)^3\Big|_{c+1}^1 \\ &= \tfrac{1}{3}\left[1 + c^3 + (-c-1)^3 - (-1)\right] \\ &= -c^2 - c + \tfrac{1}{3} = -\left(c+\tfrac{1}{2}\right)^2 + \tfrac{7}{12}. \end{aligned}$$

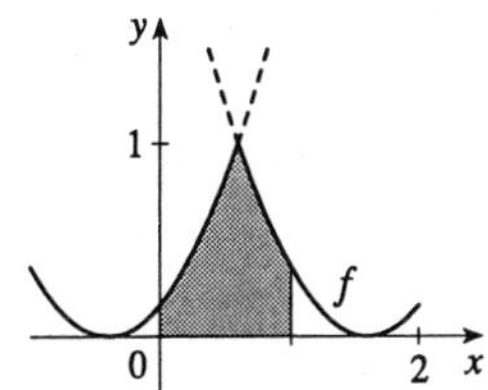

Again, this is a parabola, whose maximum for $-1 \le c < 0$ is $g\left(-\frac{1}{2}\right) = \frac{7}{12}$, and whose minimum on this c-interval is $g(-1) = g(0) = \frac{1}{3}$.

Case 3: $0 \le c \le 2$ In this case, $f(x) = (x-c)^2$ for $x \in [0,1]$, so

$$\begin{aligned} g(c) &= \int_0^1 (x-c)^2 dx = \tfrac{1}{3}(x-c)^3\Big|_0^1 \\ &= \tfrac{1}{3}\left[(1-c)^3 - (-c)^3\right] \\ &= c^2 - c + \tfrac{1}{3} = \left(c-\tfrac{1}{2}\right)^2 + \tfrac{1}{12}. \end{aligned}$$

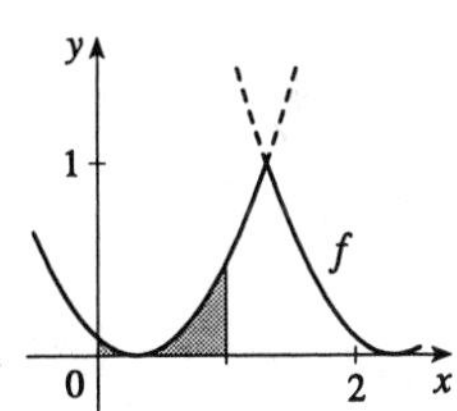

This parabola has a maximum of $g(2) = \frac{7}{3}$ and a minimum of $g\left(\frac{1}{2}\right) = \frac{1}{12}$.

We conclude that $g(c)$ has an absolute maximum of $g(2) = \frac{7}{3}$; and absolute minima of $g\left(-\frac{3}{2}\right) = g\left(\frac{1}{2}\right) = \frac{1}{12}$.

33. Let the roots have common difference d. If we let m be the average of the roots, then we can simplify the calculations by writing the roots as $m-3a$, $m-a$, $m+a$, $m+3a$, where $d = 2a$. The polynomial is then of the form $P(x) = c(x-m+3a)(x-m+a)(x-m-a)(x-m-3a)$. To simplify further, we make the change of variable $u = x - m$. Then $P(x) = c(u+3a)(u+a)(u-a)(u-3a) = c(u^2-a^2)(u^2-9a^2)$.

So $\dfrac{dP}{dx} = \dfrac{dP}{du}\dfrac{du}{dx} = \dfrac{dP}{du} = c\left[2u(u^2-9a^2) + (u^2-a^2)(2u)\right] = 2cu(2u^2-10a^2) = 4cu(u^2-5a^2) = 0$ when $u = 0, \pm\sqrt{5}a$. So the roots of P' are $m - \sqrt{5}a$, m, and $m + \sqrt{5}a$, which form an arithmetic sequence.

34. **(a)** To find $B_1(x)$, we use the fact that $B_1'(x) = B_0(x) \Rightarrow B_1(x) = \int B_0(x)dx = \int 1\,dx = x + C$. Now we impose the condition that $\int_0^1 B_1(x)dx = 0 \Rightarrow 0 = \int_0^1 (x + C)dx = \frac{1}{2}x^2\big|_0^1 + Cx\big|_0^1 = \frac{1}{2} + C \Rightarrow$ $C = -\frac{1}{2}$. So $B_1(x) = x - \frac{1}{2}$. Similarly $B_2(x) = \int B_1(x)dx = \int \left(x - \frac{1}{2}\right)dx = \frac{1}{2}x^2 - \frac{1}{2}x + D$. But $\int_0^1 B_2(x)dx = 0 \Rightarrow 0 = \int_0^1 \left(\frac{1}{2}x^2 - \frac{1}{2}x + D\right)dx = \frac{1}{6} - \frac{1}{4} + D \Rightarrow D = \frac{1}{12}$, so $B_2(x) = \frac{1}{2}x^2 - \frac{1}{2}x + \frac{1}{12}$. $B_3(x) = \int B_2(x)dx = \int \left(\frac{1}{2}x^2 - \frac{1}{2}x + \frac{1}{12}\right)dx = \frac{1}{6}x^3 - \frac{1}{4}x^2 + \frac{1}{12}x + E$. But $\int_0^1 B_3(x)dx = 0 \Rightarrow$ $0 = \int_0^1 \left(\frac{1}{6}x^3 - \frac{1}{4}x^2 + \frac{1}{12}x + E\right)dx = \frac{1}{24} - \frac{1}{12} + \frac{1}{24} + E \Rightarrow E = 0$. So $B_3(x) = \frac{1}{6}x^3 - \frac{1}{4}x^2 + \frac{1}{12}x$. $B_4(x) = \int B_3(x)dx = \int \left(\frac{1}{6}x^3 - \frac{1}{4}x^2 + \frac{1}{12}x\right)dx = \frac{1}{24}x^4 - \frac{1}{12}x^3 + \frac{1}{24}x^2 + F$. But $\int_0^1 B_4(x)dx = 0 \Rightarrow$ $0 = \int_0^1 \left(\frac{1}{24}x^4 - \frac{1}{12}x^3 + \frac{1}{24}x^2 + F\right)dx = \frac{1}{120} - \frac{1}{48} + \frac{1}{72} + F \Rightarrow F = -\frac{1}{720}$. So $B_4(x) = \frac{1}{24}x^4 - \frac{1}{12}x^3 + \frac{1}{24}x^2 - \frac{1}{720}$.

(b) By Part 2 of the Fundamental Theorem of Calculus, $B_n(1) - B_n(0) = \int_0^1 B_n'(x)dx = \int_0^1 B_{n-1}(x)dx = 0$ for $n - 1 \geq 1$, by definition. Thus $B_n(0) = B_n(1)$ for $n \geq 2$.

(c) We know that $B_n(x) = \dfrac{1}{n!}\displaystyle\sum_{k=0}^{n}\binom{n}{k}b_k x^{n-k}$. If we set $x = 1$ in this expression, and use the fact that $B_n(1) = B_n(0) = \dfrac{b_n}{n!}$ for $n \geq 2$, we get $b_n = \displaystyle\sum_{k=0}^{n}\binom{n}{k}b_k$. Now if we expand the right-hand side, we get $b_n = \binom{n}{0}b_0 + \binom{n}{1}b_1 + \cdots + \binom{n}{n-2}b_{n-2} + \binom{n}{n-1}b_{n-1} + \binom{n}{n}b_n$. We cancel the b_n terms, move the b_{n-1} term to the LHS and divide by $-\binom{n}{n-1} = -n$:

$b_{n-1} = -\dfrac{1}{n}\left[\binom{n}{0}b_0 + \binom{n}{1}b_1 + \cdots + \binom{n}{n-2}b_{n-2}\right]$ for $n \geq 2$, as required.

(d) We use mathematical induction. For $n = 0$: $B_0(1 - x) = 1$ and $(-1)^0 B_0(x) = 1$, so the equation holds for $n = 0$ since $b_0 = 1$. Now if $B_k(1 - x) = (-1)^k B_k(x)$, then since $\dfrac{d}{dx}B_{k+1}(1 - x) = B_{k+1}'(1 - x)\dfrac{d}{dx}(1 - x) = -B_k(1 - x)$, we have $\dfrac{d}{dx}B_{k+1}(1 - x) = (-1)(-1)^k B_k(x) = (-1)^{k+1}B_k(x)$. Integrating, we get $B_{k+1}(1 - x) = (-1)^{k+1}B_{k+1}(x) + C$. But the constant of integration must be 0, since if we substitute $x = 0$ in the equation, we get $B_{k+1}(1) = (-1)^{k+1}B_{k+1}(0) + C$, and if we substitute $x = 1$ we get $B_{k+1}(0) = (-1)^{k+1}B_{k+1}(1) + C$, and these two equations together imply that $B_{k+1}(0) = (-1)^{k+1}\left[(-1)^{k+1}B_{k+1}(0) + C\right] + C = B_{k+1}(0) + 2C \Leftrightarrow C = 0$.

So the equation holds for all n, by induction.

Now if the power of -1 is odd, then we have $B_{2n+1}(1 - x) = -B_{2n+1}(x)$. In particular, $B_{2n+1}(1) = -B_{2n+1}(0)$. But from part (b), we know that $B_k(1) = B_k(0)$ for $k > 1$. The only possibility is that $B_{2n+1}(0) = B_{2n+1}(1) = 0$ for all $n > 0$, and this implies that $b_{2n+1} = (2n + 1)!\, B_{2n+1}(0) = 0$ for $n > 0$.

(e) From part (a), we know that $b_0 = 0!\,B_0(0) = 1$, and similarly $b_1 = -\frac{1}{2}$, $b_2 = \frac{1}{6}$, $b_3 = 0$ and $b_4 = -\frac{1}{30}$.

We use the formula to find

$$b_6 = b_{7-1} = -\frac{1}{7}\left[\binom{7}{0}b_0 + \binom{7}{1}b_1 + \binom{7}{2}b_2 + \binom{7}{3}b_3 + \binom{7}{4}b_4 + \binom{7}{5}b_5\right].$$

The b_3 and b_5 terms are 0, so this is equal to

$-\frac{1}{7}\left[1 + 7\left(-\frac{1}{2}\right) + \frac{7\cdot 6}{2\cdot 1}\left(\frac{1}{6}\right) + \frac{7\cdot 6\cdot 5}{3\cdot 2\cdot 1}\left(-\frac{1}{30}\right)\right] = -\frac{1}{7}\left(1 - \frac{7}{2} + \frac{7}{2} - \frac{7}{6}\right) = \frac{1}{42}$. Similarly,

$$b_8 = -\frac{1}{9}\left[\binom{9}{0}b_0 + \binom{9}{1}b_1 + \binom{9}{2}b_2 + \binom{9}{4}b_4 + \binom{9}{6}b_6\right]$$

$$= -\frac{1}{9}\left[1 + 9\left(-\frac{1}{2}\right) + \frac{9\cdot 8}{2\cdot 1}\left(\frac{1}{6}\right) + \frac{9\cdot 8\cdot 7\cdot 6}{4\cdot 3\cdot 2\cdot 1}\left(-\frac{1}{30}\right) + \frac{9\cdot 8\cdot 7}{3\cdot 2\cdot 1}\left(\frac{1}{42}\right)\right] = -\frac{1}{9}\left(1 - \frac{9}{2} + 6 - \frac{21}{5} + 2\right) = -\frac{1}{30}.$$

Now we can calculate

$$B_5(x) = \frac{1}{5!}\sum_{k=0}^{5}\binom{5}{k}b_k x^{5-k} = \frac{1}{120}\left[x^5 + 5\left(-\frac{1}{2}\right)x^4 + \frac{5\cdot 4}{2\cdot 1}\left(\frac{1}{6}\right)x^3 + 5\left(-\frac{1}{30}\right)x\right]$$

$$= \frac{1}{120}\left(x^5 - \frac{5}{2}x^4 + \frac{5}{3}x^3 - \frac{1}{6}x\right)$$

$$B_6(x) = \frac{1}{720}\left[x^6 + 6\left(-\frac{1}{2}\right)x^5 + \frac{6\cdot 5}{2\cdot 1}\left(\frac{1}{6}\right)x^4 + \frac{6\cdot 5}{2\cdot 1}\left(-\frac{1}{30}\right)x^2 + \frac{1}{42}\right] = \frac{1}{720}\left(x^6 - 3x^5 + \frac{5}{2}x^4 - \frac{1}{2}x^2 + \frac{1}{42}\right)$$

$$B_7(x) = \frac{1}{5040}\left[x^7 + 7\left(-\frac{1}{2}\right)x^6 + \frac{7\cdot 6}{2\cdot 1}\left(\frac{1}{6}\right)x^5 + \frac{7\cdot 6\cdot 5}{3\cdot 2\cdot 1}\left(-\frac{1}{30}\right)x^3 + 7\left(\frac{1}{42}\right)x\right]$$

$$= \frac{1}{5040}\left(x^7 - \frac{7}{2}x^6 + \frac{7}{2}x^5 - \frac{7}{6}x^3 + \frac{1}{6}x\right)$$

$$B_8(x) = \frac{1}{40{,}320}\left[x^8 + 8\left(-\frac{1}{2}\right)x^7 + \frac{8\cdot 7}{2\cdot 1}\left(\frac{1}{6}\right)x^6 + \frac{8\cdot 7\cdot 6\cdot 5}{4\cdot 3\cdot 2\cdot 1}\left(-\frac{1}{30}\right)x^4 + \frac{8\cdot 7}{2\cdot 1}\left(\frac{1}{42}\right)x^2 + \left(-\frac{1}{30}\right)\right]$$

$$= \frac{1}{40{,}320}\left(x^8 - 4x^7 + \frac{14}{3}x^6 - \frac{7}{3}x^4 + \frac{2}{3}x^2 - \frac{1}{30}\right)$$

$$B_9(x) = \frac{1}{362{,}880}\left[x^9 + 9\left(-\frac{1}{2}\right)x^8 + \frac{9\cdot 8}{2\cdot 1}\left(\frac{1}{6}\right)x^7 + \frac{9\cdot 8\cdot 7\cdot 6}{4\cdot 3\cdot 2\cdot 1}\left(-\frac{1}{30}\right)x^5 + \frac{9\cdot 8\cdot 7}{3\cdot 2\cdot 1}\left(\frac{1}{42}\right)x^3 + 9\left(-\frac{1}{30}\right)x\right]$$

$$= \frac{1}{362{,}880}\left(x^9 - \frac{9}{2}x^8 + 6x^7 - \frac{21}{5}x^5 + 2x^3 - \frac{3}{10}x\right)$$

(f)

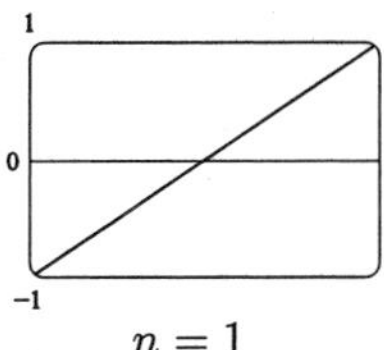

$n = 1$

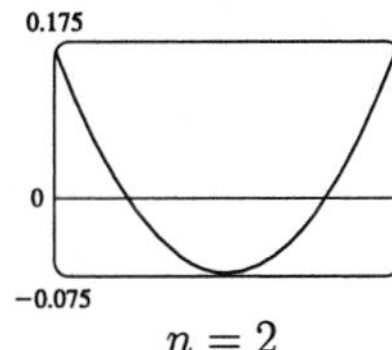

$n = 2$

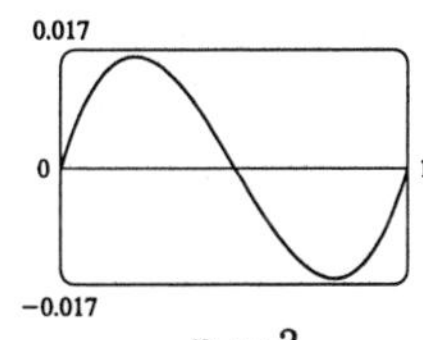

$n = 3$

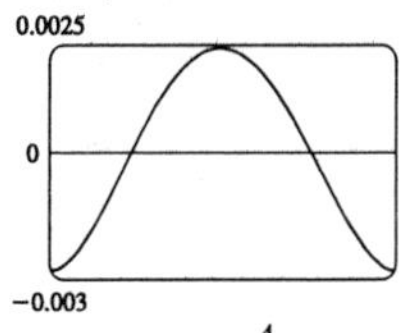

$n = 4$

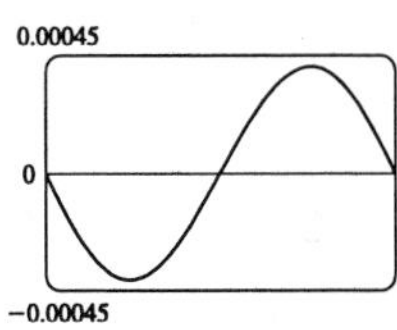

$n = 5$

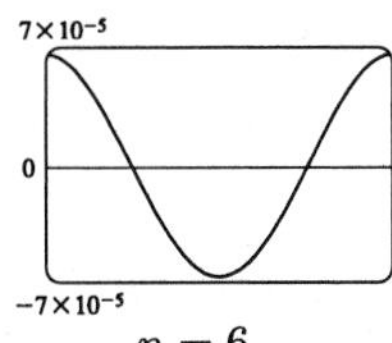

$n = 6$

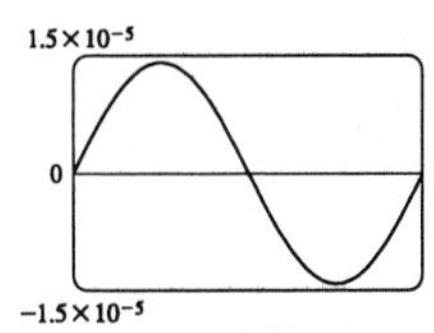

$n = 7$

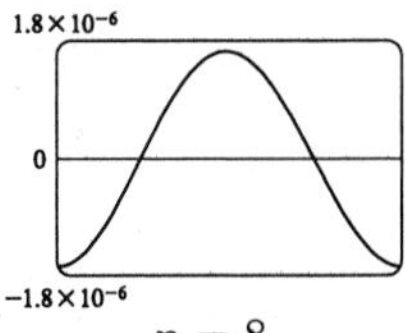

$n = 8$

2.75×10⁻⁷
0
1
-2.75×10⁻⁷
$n = 9$

There are four basic shapes for the graphs of B_n (excluding B_1), and as n increases they repeat in a cycle of four. For $n = 4m$, the shape resembles that of the graph of $-\cos 2\pi x$; For $n = 4m + 1$, that of $-\sin 2\pi x$; for $n = 4m + 2$, that of $\cos 2\pi x$; and for $n = 4m + 3$, that of $\sin 2\pi x$.

(g) For $k=0$: $B_1(x+1)-B_1(x)=x+1-\frac{1}{2}-\left(x-\frac{1}{2}\right)=1$, and $\dfrac{x^0}{0!}=1$, so the equation holds for $k=0$.

We now assume that $B_n(x+1)-B_n(x)=\dfrac{x^{n-1}}{(n-1)!}$. We integrate this equation with respect to x:

$\displaystyle\int[B_n(x+1)-B_n(x)]dx=\int\frac{x^{n-1}}{(n-1)!}\,dx$. But we can evaluate the LHS using the definition

$B_{n+1}(x)=\int B_n(x)dx$, and the RHS is a simple integral. The equation becomes

$B_{n+1}(x+1)-B_{n+1}(x)=\dfrac{1}{(n-1)!}\left(\dfrac{1}{n}x^n\right)=\dfrac{1}{n!}x^n$, since by part (b) $B_{n+1}(1)-B_{n+1}(0)=0$, and so

the constant of integration must vanish. So the equation holds for all k, by induction.

(h) The result from part (g) implies that $p^k=k![B_{k+1}(p+1)-B_{k+1}(p)]$. If we sum both sides of this equation from $p=0$ to $p=n$ (note that k is fixed in this process), we get

$\displaystyle\sum_{p=0}^{n}p^k=k!\sum_{p=0}^{n}[B_{k+1}(p+1)-B_{k+1}(p)]$. But the RHS is just a telescoping sum, so the equation becomes

$1^k+2^k+3^k+\cdots+n^k=k![B_{k+1}(n+1)-B_{k+1}(0)]$. But from the definition of Bernoulli polynomials (and using the Fundamental Theorem of Calculus), the RHS is equal to $k!\int_0^{n+1}B_k(x)dx$.

(i) If we let $k=3$ and then substitute from part (a), the formula in part (h) becomes

$$\begin{aligned}1^3+2^3+\cdots+n^3&=3![B_4(n+1)-B_4(0)]\\&=6\left[\tfrac{1}{24}(n+1)^4-\tfrac{1}{12}(n+1)^3+\tfrac{1}{24}(n+1)^2-\tfrac{1}{720}-\left(\tfrac{1}{24}-\tfrac{1}{12}+\tfrac{1}{24}-\tfrac{1}{720}\right)\right]\\&=\frac{(n+1)^2[1+(n+1)^2-2(n+1)]}{4}=\frac{(n+1)^2[1-(n+1)]^2}{4}=\left[\frac{n(n+1)}{2}\right]^2.\end{aligned}$$

(j) $1^k+2^k+3^k+\cdots+n^k=k!\int_0^{n+1}B_k(x)dx$ [by part (h)]

$$=k!\int_0^{n+1}\frac{1}{k!}\sum_{j=0}^{k}\binom{k}{j}b_jx^{k-j}\,dx=\int_0^{n+1}\sum_{j=0}^{k}\binom{k}{j}b_jx^{k-j}\,dx.$$

Now view $\displaystyle\sum_{j=0}^{k}\binom{k}{j}b_jx^{k-j}$ as $(x+b)^k$, as explained in the problem. Then

$$1^k+2^k+3^k+\cdots+n^k \text{ “=” } \int_0^{n+1}(x+b)^k\,dx=\left[\frac{(x+b)^{k+1}}{k+1}\right]_0^{n+1}=\frac{(n+1+b)^{k+1}-b^{k+1}}{k+1}$$

(k) We expand the RHS of the formula in (j), turning the b^i into b_i, and remembering that $b_{2i+1}=0$ for $i>0$:

$$\begin{aligned}1^5+2^5+\cdots+n^5&=\tfrac{1}{6}\left[(n+1+b)^6-b^6\right]\\&=\tfrac{1}{6}\left[(n+1)^6+6(n+1)^5b_1+\tfrac{6\cdot5}{2\cdot1}(n+1)^4b_2+\tfrac{6\cdot5}{2\cdot1}(n+1)^2b_4\right]\\&=\tfrac{1}{6}\left[(n+1)^6-3(n+1)^5+\tfrac{5}{2}(n+1)^4-\tfrac{1}{2}(n+1)^2\right]\\&=\tfrac{1}{12}(n+1)^2\left[2(n+1)^4-6(n+1)^3+5(n+1)^2-1\right]\\&=\tfrac{1}{12}(n+1)^2[(n+1)-1]^2\left[2(n+1)^2-2(n+1)-1\right]\\&=\tfrac{1}{12}n^2(n+1)^2(2n^2+2n-1)\end{aligned}$$

CHAPTER FIVE

EXERCISES 5.1

1. $A = \int_{-1}^{1}[(x^2+3)-x]dx = \int_{-1}^{1}(x^2-x+3)dx$
$= \left[\frac{1}{3}x^3 - \frac{1}{2}x^2 + 3x\right]_{-1}^{1} = \left(\frac{1}{3}-\frac{1}{2}+3\right) - \left(-\frac{1}{3}-\frac{1}{2}-3\right)$
$= \frac{20}{3}$

2. $A = \int_0^6[2x-(x^2-4x)]dx = \int_0^6(6x-x^2)dx$
$= \left[3x^2 - \frac{1}{3}x^3\right]_0^6 = 108 - 72$
$= 36$

3. $A = \int_{-1}^{1}\left[(1-y^4)-(y^3-y)\right]dy = \int_{-1}^{1}\left(-y^4-y^3+y+1\right)dy$
$= \left[-\frac{1}{5}y^5 - \frac{1}{4}y^4 + \frac{1}{2}y^2 + y\right]_{-1}^{1} = \left(-\frac{1}{5}-\frac{1}{4}+\frac{1}{2}+1\right) - \left(\frac{1}{5}-\frac{1}{4}+\frac{1}{2}-1\right)$
$= \frac{8}{5}$

4. $A = \int_{-1}^{2}[y^2-(y-5)]dy = \left[\frac{1}{3}y^3 - \frac{1}{2}y^2 + 5y\right]_{-1}^{2}$
$= \left(\frac{8}{3}-2+10\right) - \left(-\frac{1}{3}-\frac{1}{2}-5\right)$
$= 16.5$

5. $A = \int_0^1(x-x^2)dx = \left[\frac{1}{2}x^2 - \frac{1}{3}x^3\right]_0^1$
$= \frac{1}{2} - \frac{1}{3} = \frac{1}{6}$

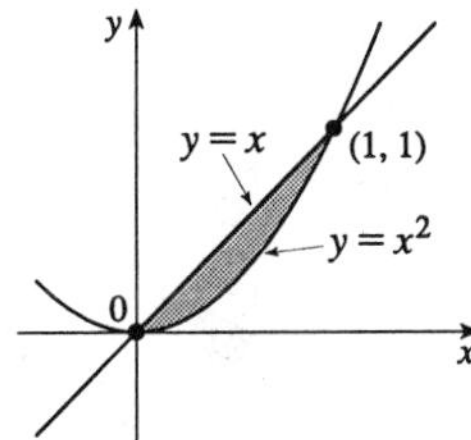

6. $A = \int_{-1}^{0}(x^3-x)dx + \int_0^1(x-x^3)dx$
$= 2\int_0^1(x-x^3)dx = 2\left[\frac{1}{2}x^2 - \frac{1}{4}x^4\right]_0^1$
$= 2\left(\frac{1}{2}-\frac{1}{4}\right) = \frac{1}{2}$

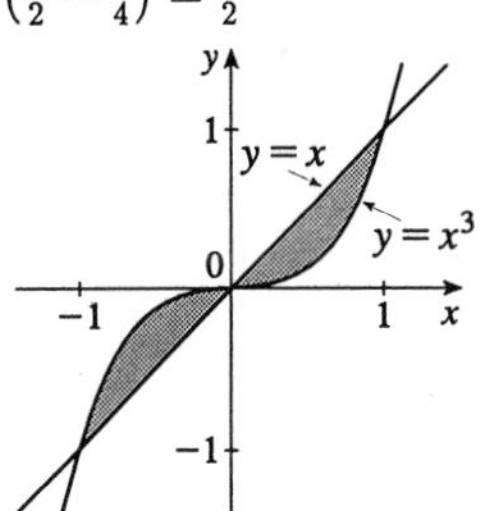

7. $A = \int_0^1\left(\sqrt{x}-x^2\right)dx = \left[\frac{2}{3}x^{3/2} - \frac{1}{3}x^3\right]_0^1$
$= \frac{2}{3} - \frac{1}{3} = \frac{1}{3}$

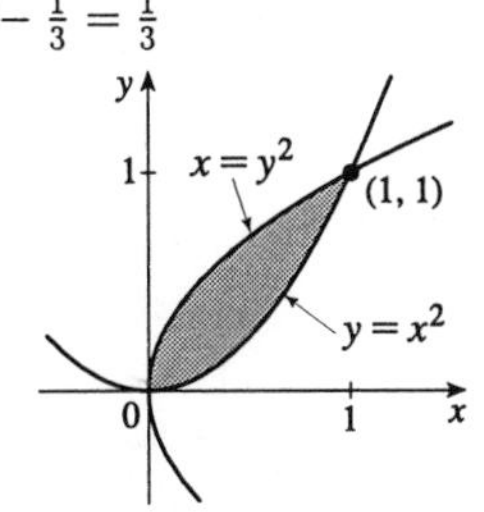

8. $A = \int_{-1}^{1}\left(x^2-x^4\right)dx = 2\int_0^1\left(x^2-x^4\right)dx$
$= 2\left[\frac{1}{3}x^3 - \frac{1}{5}x^5\right]_0^1 = 2\left(\frac{1}{3}-\frac{1}{5}\right) = \frac{4}{15}$

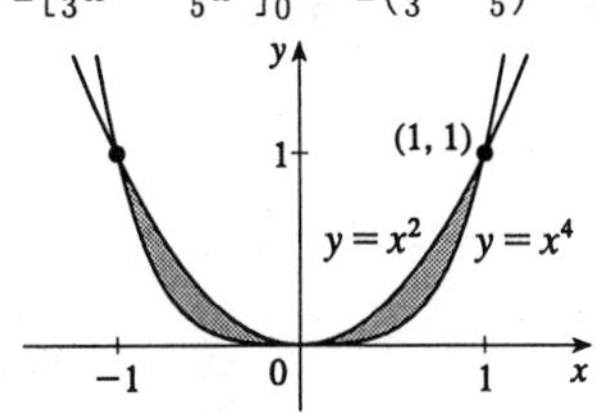

9. $A = \int_0^4 \left(\sqrt{x} - \frac{1}{2}x\right) dx = \left[\frac{2}{3}x^{3/2} - \frac{1}{4}x^2\right]_0^4$
$= \left(\frac{16}{3} - 4\right) - 0 = \frac{4}{3}$

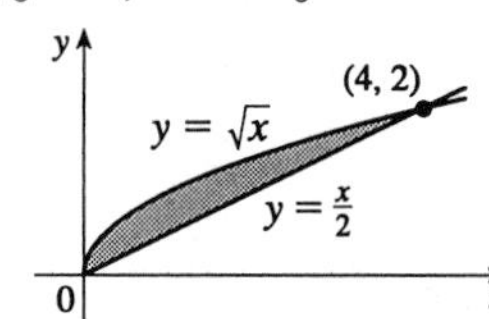

10. $A = \int_2^5 \left(\sqrt{x-1} - \frac{1}{3}x - \frac{1}{3}\right) dx$
$= \left[\frac{2}{3}(x-1)^{3/2} - \frac{1}{6}x^2 - \frac{1}{3}x\right]_2^5$
$= \left(\frac{16}{3} - \frac{25}{6} - \frac{5}{3}\right) - \left(\frac{2}{3} - \frac{4}{6} - \frac{2}{3}\right) = \frac{1}{6}$

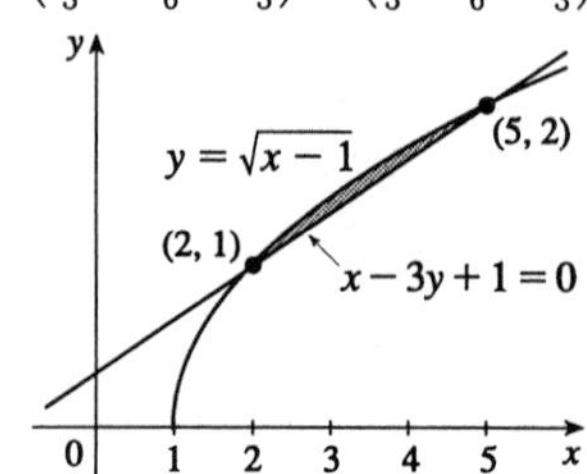

11. $A = \int_{-1}^1 [(x^2 + 3) - 4x^2] dx = 2\int_0^1 (3 - 3x^2) dx$
$= [2(3x - x^3)]_0^1 = 2(3 - 1) - 0 = 4$

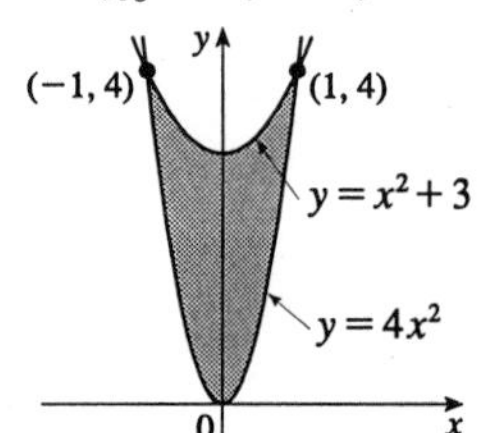

12. $A = \int_{-1}^1 \left[(1 - x^2) - (x^4 - x^2)\right] dx$
$= 2\int_0^1 \left(1 - x^4\right) dx = 2\left[x - \frac{1}{5}x^5\right]_0^1$
$= 2\left(1 - \frac{1}{5}\right) = \frac{8}{5}$

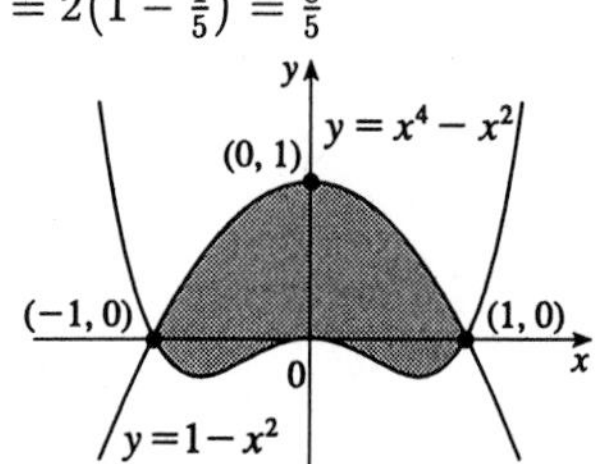

13. $A = \int_0^3 [(2x + 5) - (x^2 + 2)] dx + \int_3^6 [(x^2 + 2) - (2x + 5)] dx$
$= \int_0^3 (-x^2 + 2x + 3) dx + \int_3^6 (x^2 - 2x - 3) dx$
$= \left[-\frac{1}{3}x^3 + x^2 + 3x\right]_0^3 + \left[\frac{1}{3}x^3 - x^2 - 3x\right]_3^6$
$= (-9 + 9 + 9) - 0 + (72 - 36 - 18) - (9 - 9 - 9) = 36$

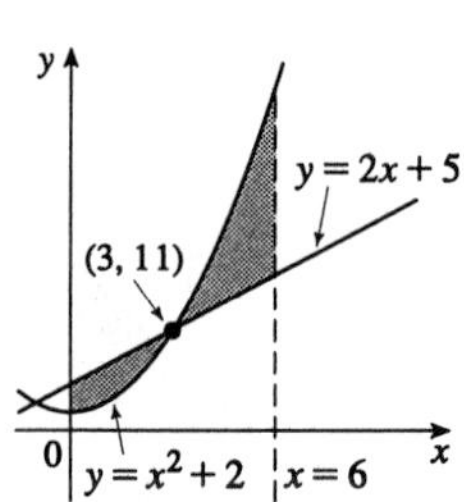

14. $A = \int_{-2}^{-1} [(x^2 + 1) - (3 - x^2)] dx + \int_{-1}^1 [(3 - x^2) - (x^2 + 1)] dx$
$+ \int_1^2 [(x^2 + 1) - (3 - x^2)] dx$
$= \int_{-2}^{-1} (2x^2 - 2) dx + \int_{-1}^1 (2 - 2x^2) dx + \int_1^2 (2x^2 - 2) dx$
$= 2\int_0^1 (2 - 2x^2) dx + 2\int_1^2 (2x^2 - 2) dx$ [by symmetry]
$= 2\left[2x - \frac{2}{3}x^3\right]_0^1 + 2\left[\frac{2}{3}x^3 - 2x\right]_1^2 = 2\left(2 - \frac{2}{3}\right) + 2\left(\frac{16}{3} - 4\right) - 2\left(\frac{2}{3} - 2\right) = 8$

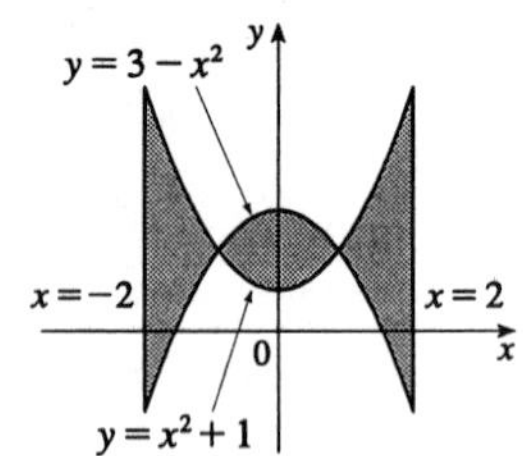

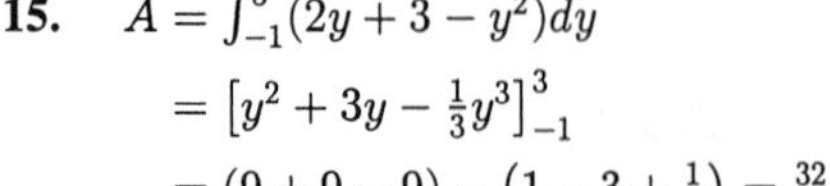

15. $A = \int_{-1}^3 (2y + 3 - y^2) dy$
$= \left[y^2 + 3y - \frac{1}{3}y^3\right]_{-1}^3$
$= (9 + 9 - 9) - \left(1 - 3 + \frac{1}{3}\right) = \frac{32}{3}$

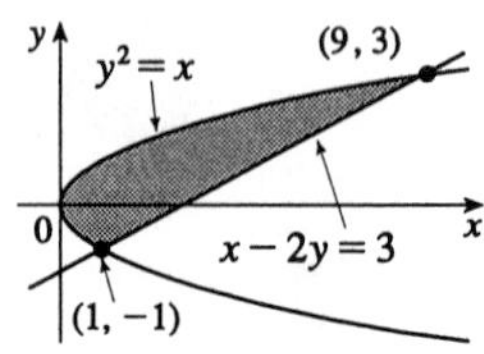

16. $A = \int_{-1}^{2}[2 - y^2 - (-y)]dy$
$= \int_{-1}^{2}(-y^2 + y + 2)dy$
$= \left[-\frac{1}{3}y^3 + \frac{1}{2}y^2 + 2y\right]_{-1}^{2}$
$= \left(-\frac{8}{3} + 2 + 4\right) - \left(\frac{1}{3} + \frac{1}{2} - 2\right) = \frac{9}{2}$

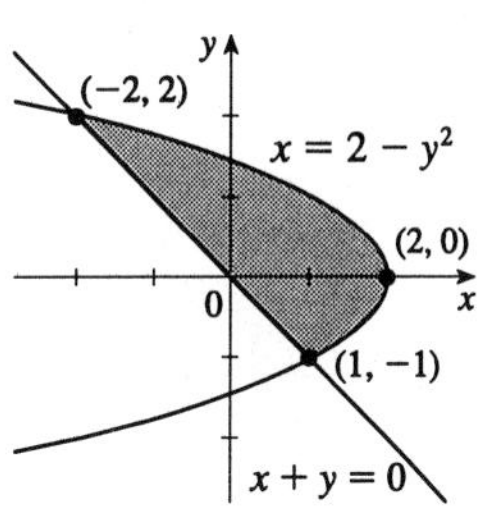

17. $A = \int_{-1}^{1}[(1 - y^2) - (y^2 - 1)]dy$
$= \int_{-1}^{1} 2(1 - y^2)dy = 4\int_{0}^{1}(1 - y^2)dy$
$= 4\left[y - \frac{1}{3}y^3\right]_0^1$
$= 4\left(1 - \frac{1}{3}\right)$
$= \frac{8}{3}$

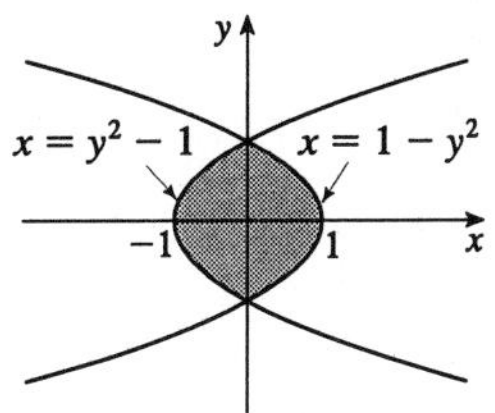

18. $A = \int_0^1[(x^3 - 4x^2 + 3x) - (x^2 - x)]dx$
$+ \int_1^4[x^2 - x - (x^3 - 4x^2 + 3x)]dx$
$= \int_0^1(x^3 - 5x^2 + 4x)dx + \int_1^4(-x^3 + 5x^2 - 4x)dx$
$= \left[\frac{1}{4}x^4 - \frac{5}{3}x^3 + 2x^2\right]_0^1 + \left[-\frac{1}{4}x^4 + \frac{5}{3}x^3 - 2x^2\right]_1^4$
$= \left(\frac{1}{4} - \frac{5}{3} + 2\right) - 0 + \left(-64 + \frac{320}{3} - 32\right) - \left(-\frac{1}{4} + \frac{5}{3} - 2\right)$
$= \frac{71}{6}$

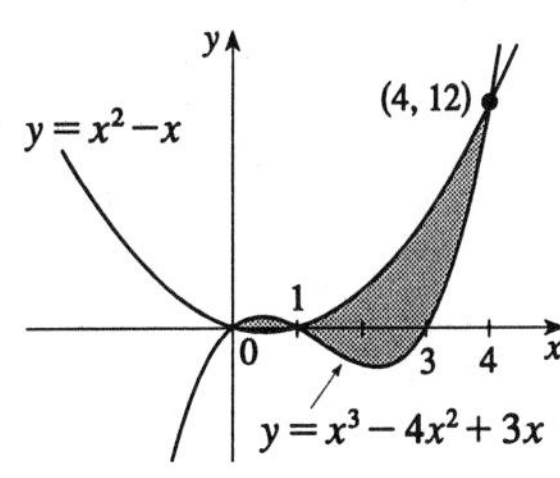

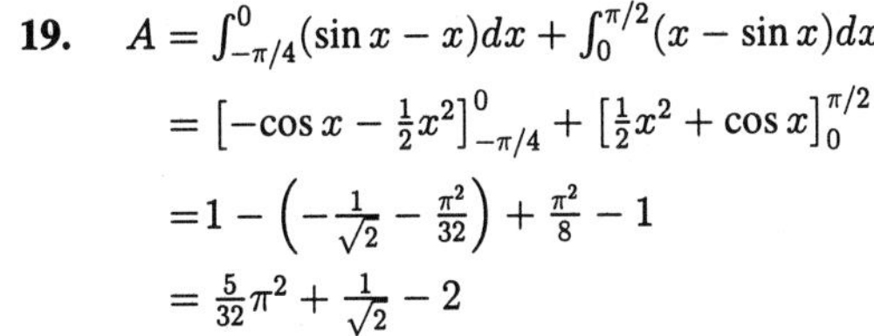

19. $A = \int_{-\pi/4}^{0}(\sin x - x)dx + \int_0^{\pi/2}(x - \sin x)dx$
$= \left[-\cos x - \frac{1}{2}x^2\right]_{-\pi/4}^{0} + \left[\frac{1}{2}x^2 + \cos x\right]_0^{\pi/2}$
$= 1 - \left(-\frac{1}{\sqrt{2}} - \frac{\pi^2}{32}\right) + \frac{\pi^2}{8} - 1$
$= \frac{5}{32}\pi^2 + \frac{1}{\sqrt{2}} - 2$

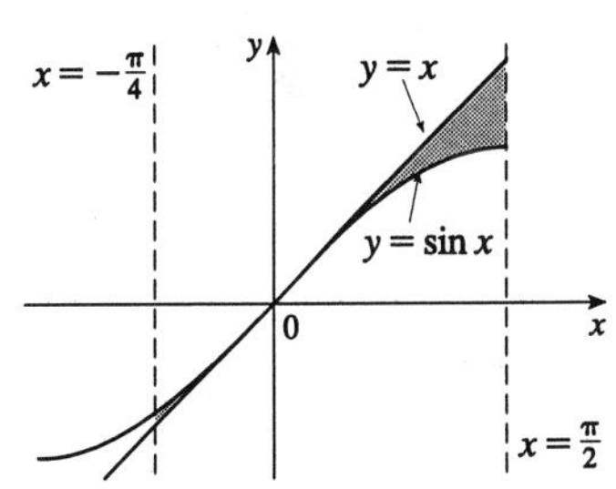

20. $A = 2\int_0^{\pi/4}(\sec^2 x - \cos x)dx$
$= 2[\tan x - \sin x]_0^{\pi/4}$
$= 2\left(1 - \frac{1}{\sqrt{2}}\right) - 0$
$= 2 - \sqrt{2}$

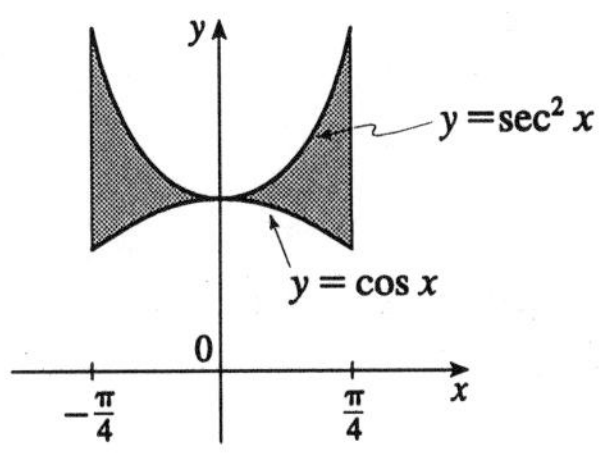

21. Notice that $\cos x = \sin 2x = 2\sin x\cos x \Leftrightarrow$ $2\sin x = 1$ or $\cos x = 0 \Leftrightarrow x = \frac{\pi}{6}$ or $\frac{\pi}{2}$.

$$A = \int_0^{\pi/6}(\cos x - \sin 2x)dx + \int_{\pi/6}^{\pi/2}(\sin 2x - \cos x)dx$$
$$= \left[\sin x + \tfrac{1}{2}\cos 2x\right]_0^{\pi/6} + \left[-\tfrac{1}{2}\cos 2x - \sin x\right]_{\pi/6}^{\pi/2}$$
$$= \tfrac{1}{2} + \tfrac{1}{2}\cdot\tfrac{1}{2} - \left(0 + \tfrac{1}{2}\cdot 1\right) + \left(\tfrac{1}{2} - 1\right) - \left(-\tfrac{1}{2}\cdot\tfrac{1}{2} - \tfrac{1}{2}\right) = \tfrac{1}{2}$$

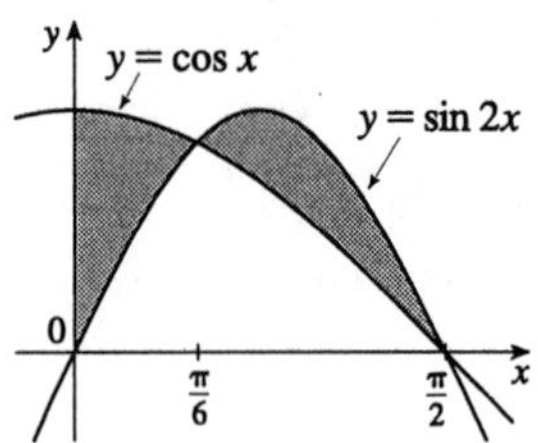

22. $\sin x = \sin 2x = 2\sin x\cos x$ when $\sin x = 0$ and when $\cos x = \frac{1}{2}$; that is, when $x = 0$ or $\frac{\pi}{3}$.

$$A = \int_0^{\pi/3}(\sin 2x - \sin x)dx + \int_{\pi/3}^{\pi/2}(\sin x - \sin 2x)dx$$
$$= \left[-\tfrac{1}{2}\cos 2x + \cos x\right]_0^{\pi/3} + \left[\tfrac{1}{2}\cos 2x - \cos x\right]_{\pi/3}^{\pi/2}$$
$$= \left[-\tfrac{1}{2}\left(-\tfrac{1}{2}\right) + \tfrac{1}{2}\right] - \left(-\tfrac{1}{2} + 1\right) + \left(-\tfrac{1}{2} - 0\right) - \left[\tfrac{1}{2}\left(-\tfrac{1}{2}\right) - \tfrac{1}{2}\right] = \tfrac{1}{2}$$

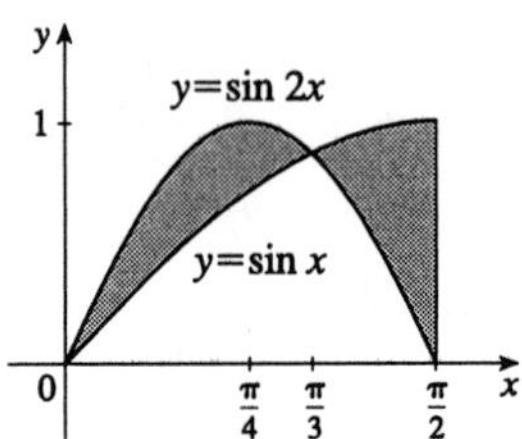

23. $\cos x = \sin 2x = 2\sin x\cos x \Leftrightarrow \cos x = 0$ or $\sin x = \frac{1}{2} \Leftrightarrow x = \frac{\pi}{2}$ or $\frac{5\pi}{6}$.

$$A = \int_{\pi/2}^{5\pi/6}(\cos x - \sin 2x)dx + \int_{5\pi/6}^{\pi}(\sin 2x - \cos x)dx$$
$$= \left[\sin x + \tfrac{1}{2}\cos 2x\right]_{\pi/2}^{5\pi/6} - \left[\sin x + \tfrac{1}{2}\cos 2x\right]_{5\pi/6}^{\pi}$$
$$= \left(\tfrac{1}{2} + \tfrac{1}{2}\cdot\tfrac{1}{2}\right) - \left(1 - \tfrac{1}{2}\right) - \left(0 + \tfrac{1}{2}\right) + \left(\tfrac{1}{2} + \tfrac{1}{2}\cdot\tfrac{1}{2}\right) = \tfrac{1}{2}$$

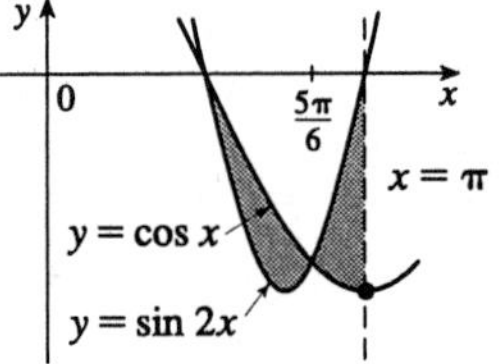

24. $\sin x = \cos 2x = 1 - 2\sin^2 x \Leftrightarrow 2\sin^2 x + \sin x - 1 = 0$ $\Leftrightarrow (2\sin x - 1)(\sin x + 1) = 0 \Leftrightarrow \sin x = \frac{1}{2}$ or -1 $\Leftrightarrow x = \frac{\pi}{6}$.

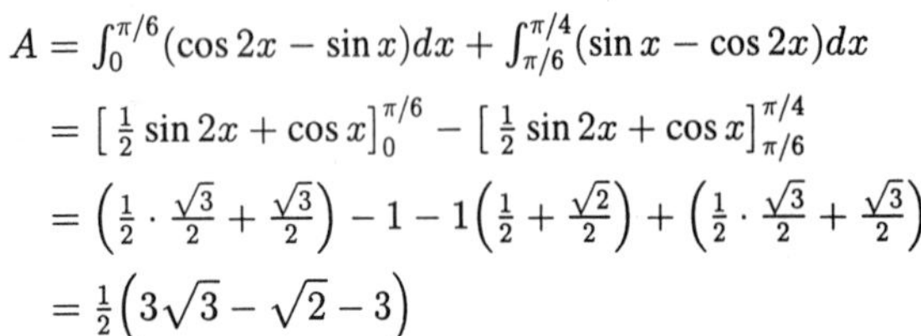

$$A = \int_0^{\pi/6}(\cos 2x - \sin x)dx + \int_{\pi/6}^{\pi/4}(\sin x - \cos 2x)dx$$
$$= \left[\tfrac{1}{2}\sin 2x + \cos x\right]_0^{\pi/6} - \left[\tfrac{1}{2}\sin 2x + \cos x\right]_{\pi/6}^{\pi/4}$$
$$= \left(\tfrac{1}{2}\cdot\tfrac{\sqrt{3}}{2} + \tfrac{\sqrt{3}}{2}\right) - 1 - 1\left(\tfrac{1}{2} + \tfrac{\sqrt{2}}{2}\right) + \left(\tfrac{1}{2}\cdot\tfrac{\sqrt{3}}{2} + \tfrac{\sqrt{3}}{2}\right)$$
$$= \tfrac{1}{2}\left(3\sqrt{3} - \sqrt{2} - 3\right)$$

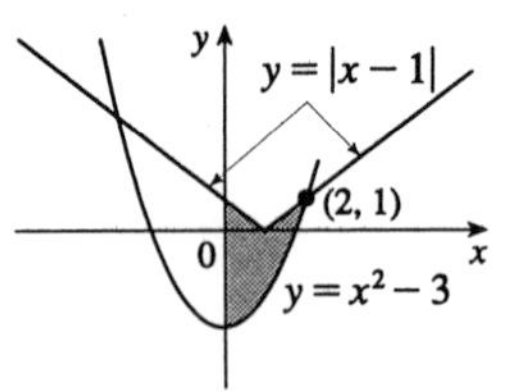

25. $A = \int_{-4}^{0}\left(-x - \left[(x+1)^2 - 7\right]\right)dx + \int_0^2\left(x - \left[(x+1)^2 - 7\right]\right)dx$

$$= \int_{-4}^{0}(-x^2 - 3x + 6)dx + \int_0^2(-x^2 - x + 6)dx$$
$$= \left[-\tfrac{1}{3}x^3 - \tfrac{3}{2}x^2 + 6x\right]_{-4}^{0} + \left[-\tfrac{1}{3}x^3 - \tfrac{1}{2}x^2 + 6x\right]_0^2$$
$$= 0 - \left(\tfrac{64}{3} - 24 - 24\right) + \left(-\tfrac{8}{3} - 2 + 12\right) - 0 = 34$$

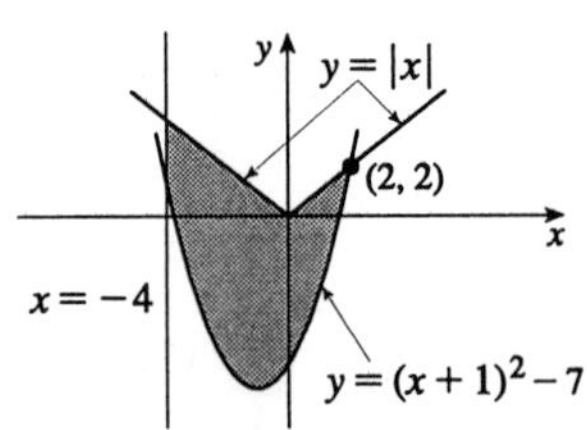

26. $A = \int_0^1 [(1-x)-(x^2-3)]dx + \int_1^2 [(x-1)-(x^2-3)]dx$
$= \left[-\frac{1}{3}x^3 - \frac{1}{2}x^2 + 4x\right]_0^1 + \left[-\frac{1}{3}x^3 + \frac{1}{2}x^2 + 2x\right]_1^2$
$= \left(-\frac{1}{3} - \frac{1}{2} + 4\right) - 0 + \left(-\frac{8}{3} + 2 + 4\right) - \left(-\frac{1}{3} + \frac{1}{2} + 2\right) = \frac{13}{3}$

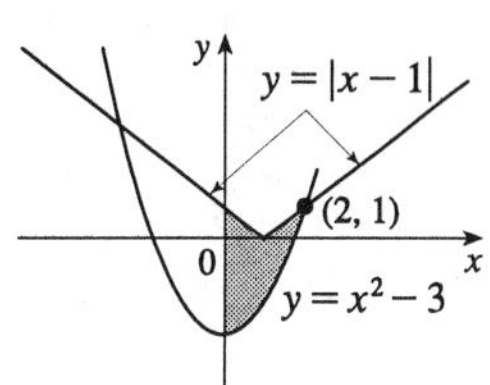

27. $A = \int_0^3 \left[\frac{1}{3}x - (-x)\right]dx + \int_3^6 \left[\left(8 - \frac{7}{3}x\right) - (-x)\right]dx$
$= \int_0^3 \frac{4}{3}x\,dx + \int_3^6 \left(-\frac{4}{3}x + 8\right)dx$
$= \left[\frac{2}{3}x^2\right]_0^3 + \left[-\frac{2}{3}x^2 + 8x\right]_3^6$
$= (6-0) + (24-18) = 12$

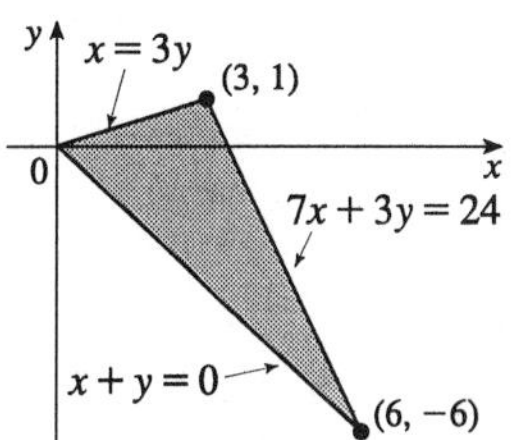

28. $A = \displaystyle\int_{-1}^{1} \left| x\sqrt{1-x^2} - (x - x^3) \right| dx$
$= \displaystyle\int_{-1}^{0} \left(x - x^3 - x\sqrt{1-x^2}\right)dx + \int_0^1 \left(x\sqrt{1-x^2} - x + x^3\right)dx$
$= 2\displaystyle\int_0^1 \left(x\sqrt{1-x^2} - x + x^3\right)dx$ (by symmetry)
$= 2\left[-\frac{1}{3}(1-x^2)^{3/2} - \frac{1}{2}x^2 + \frac{1}{4}x^4\right]_0^1 = 2\left[\left(-\frac{1}{2} + \frac{1}{4}\right) + \frac{1}{3}\right] = \frac{1}{6}$

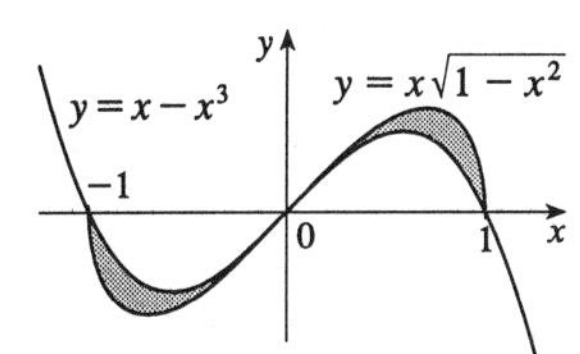

29. **(a)** 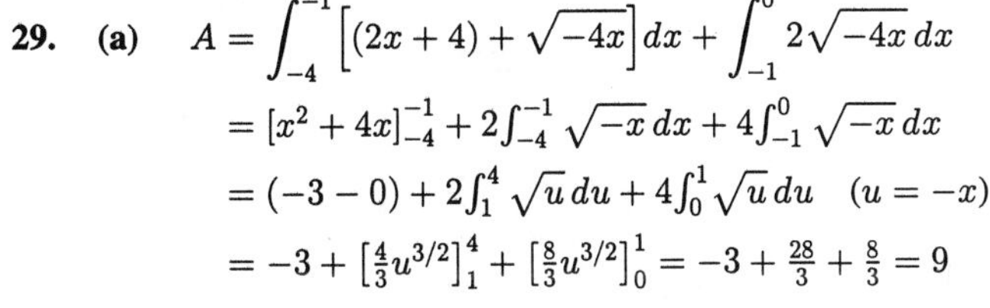
$A = \displaystyle\int_{-4}^{-1} \left[(2x+4) + \sqrt{-4x}\right]dx + \int_{-1}^{0} 2\sqrt{-4x}\,dx$
$= \left[x^2 + 4x\right]_{-4}^{-1} + 2\int_{-4}^{-1} \sqrt{-x}\,dx + 4\int_{-1}^{0} \sqrt{-x}\,dx$
$= (-3-0) + 2\int_1^4 \sqrt{u}\,du + 4\int_0^1 \sqrt{u}\,du \quad (u = -x)$
$= -3 + \left[\frac{4}{3}u^{3/2}\right]_1^4 + \left[\frac{8}{3}u^{3/2}\right]_0^1 = -3 + \frac{28}{3} + \frac{8}{3} = 9$

(b) $A = \int_{-4}^{2} \left[-\frac{1}{4}y^2 - \left(\frac{1}{2}y - 2\right)\right]dy = \left[-\frac{1}{12}y^3 - \frac{1}{4}y^2 + 2y\right]_{-4}^{2}$
$= \left(-\frac{2}{3} - 1 + 4\right) - \left(\frac{16}{3} - 4 - 8\right) = 9$

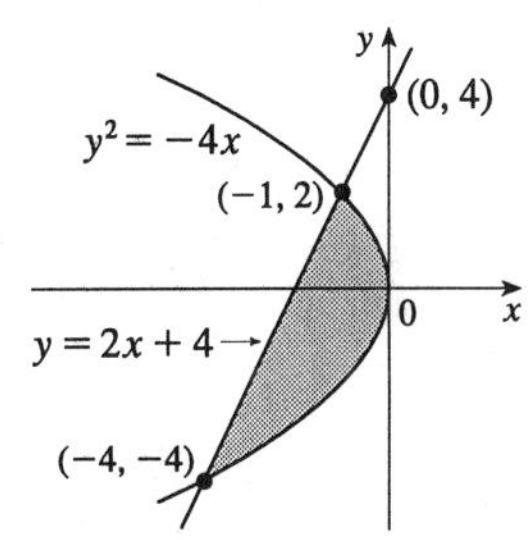

30. **(a)** 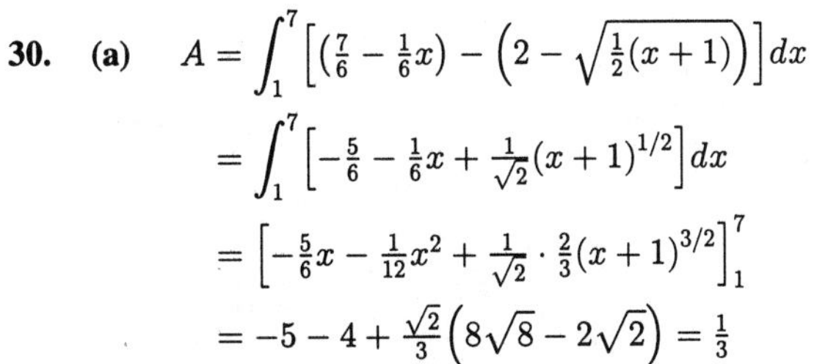
$A = \displaystyle\int_1^7 \left[\left(\frac{7}{6} - \frac{1}{6}x\right) - \left(2 - \sqrt{\frac{1}{2}(x+1)}\right)\right]dx$
$= \displaystyle\int_1^7 \left[-\frac{5}{6} - \frac{1}{6}x + \frac{1}{\sqrt{2}}(x+1)^{1/2}\right]dx$
$= \left[-\frac{5}{6}x - \frac{1}{12}x^2 + \frac{1}{\sqrt{2}} \cdot \frac{2}{3}(x+1)^{3/2}\right]_1^7$
$= -5 - 4 + \frac{\sqrt{2}}{3}\left(8\sqrt{8} - 2\sqrt{2}\right) = \frac{1}{3}$

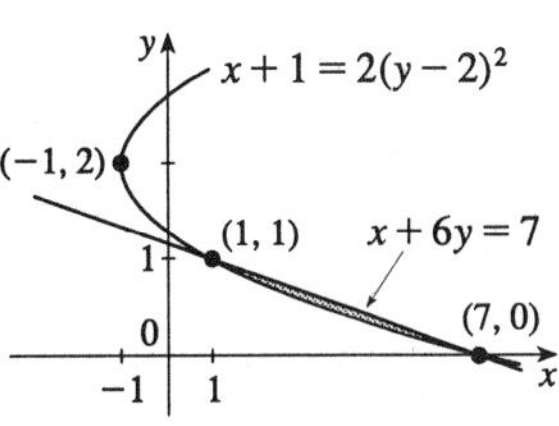

(b) $A = \int_0^1 \left((7-6y) - \left[2(y-2)^2 - 1\right]\right)dy = \int_0^1 (-2y^2 + 2y)dy = \left[-\frac{2}{3}y^3 + y^2\right]_0^1 = -\frac{2}{3} + 1 = \frac{1}{3}$

31. $A = \int_0^1 \left(8x - \frac{3}{4}x\right) dx + \int_1^4 \left[\left(-\frac{5}{3}x + \frac{29}{3}\right) - \frac{3}{4}x\right] dx$
$= \frac{29}{4}\int_0^1 x\,dx + \int_1^4 \left(-\frac{29}{12}x + \frac{29}{3}\right) dx$
$= \frac{29}{4}\left[\frac{1}{2}x^2\right]_0^1 - \frac{29}{12}\left[\frac{1}{2}x^2 - 4x\right]_1^4$
$= \frac{29}{8} - \frac{29}{12}\left(-8 - \frac{1}{2} + 4\right) = 14.5$

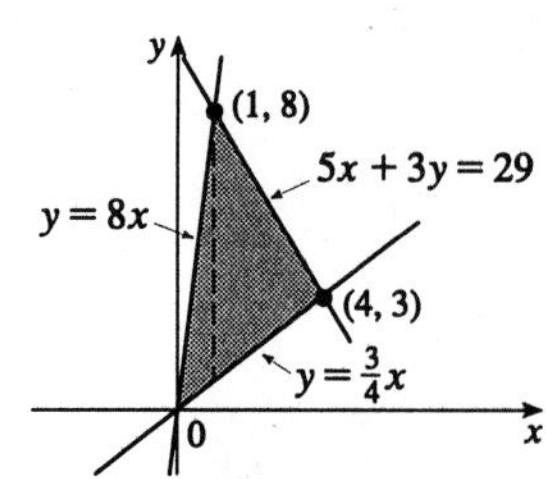

32. $A = \int_{-2}^0 \left[\left(-\frac{3}{7}x + \frac{29}{7}\right) - (-4x - 3)\right] dx$
$+ \int_0^5 \left[\left(-\frac{3}{7}x + \frac{29}{7}\right) - (x - 3)\right] dx$
$= \int_{-2}^0 \left[\frac{25}{7}x + \frac{50}{7}\right] dx + \int_0^5 \left[-\frac{10}{7}x + \frac{50}{7}\right] dx$
$= \left[\frac{25}{7}\left(\frac{1}{2}x^2\right) + \frac{50}{7}x\right]_{-2}^0 + \left[-\frac{5}{7}x^2 + \frac{50}{7}x\right]_0^5$
$= \frac{25}{7}(0 - 2) + \frac{50}{7}(0 + 2) - \frac{5}{7}(25 - 0) + \frac{50}{7}(5 - 0)$
$= 25$

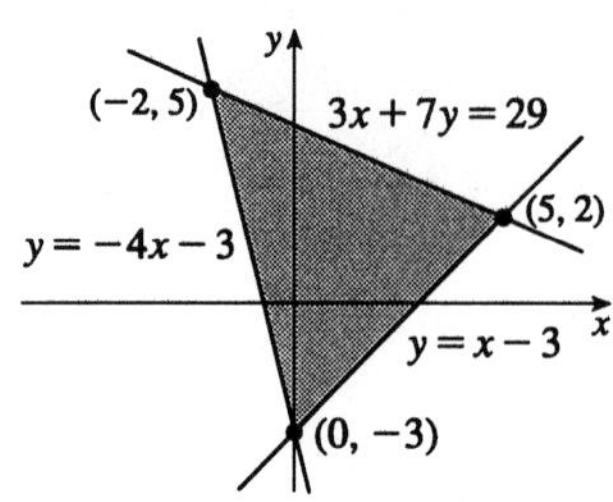

33. $\int_0^2 |x^2 - x^3| dx = \int_0^1 (x^2 - x^3) dx + \int_1^2 (x^3 - x^2) dx$
$= \left[\frac{1}{3}x^3 - \frac{1}{4}x^4\right]_0^1 + \left[\frac{1}{4}x^4 - \frac{1}{3}x^3\right]_1^2$
$= \frac{1}{3} - \frac{1}{4} + \left(4 - \frac{8}{3}\right) - \left(\frac{1}{4} - \frac{1}{3}\right) = 1.5$

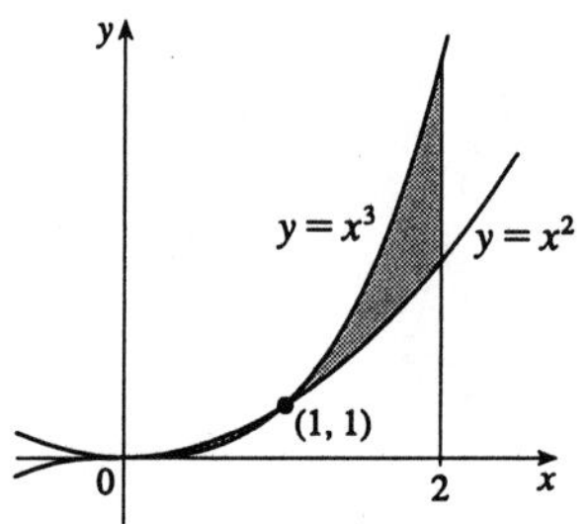

34. $$\int_0^\pi \left|\sin x - \frac{2}{\pi}x\right| dx = \int_0^{\pi/2} \left(\sin x - \frac{2}{\pi}x\right) dx + \int_{\pi/2}^\pi \left(\frac{2}{\pi}x - \sin x\right) dx$$
$$= \left[-\cos x - \frac{x^2}{\pi}\right]_0^{\pi/2} + \left[\frac{x^2}{\pi} + \cos x\right]_{\pi/2}^\pi$$
$$= -\frac{\pi}{4} + 1 + (\pi - 1) - \frac{\pi}{4} = \frac{\pi}{2}$$

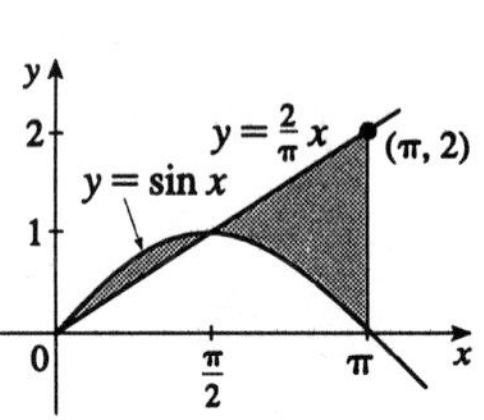

35. $\int_{-1}^2 x^3\,dx = \left[\frac{1}{4}x^4\right]_{-1}^2 = 4 - \frac{1}{4} = \frac{15}{4} = 3.75$

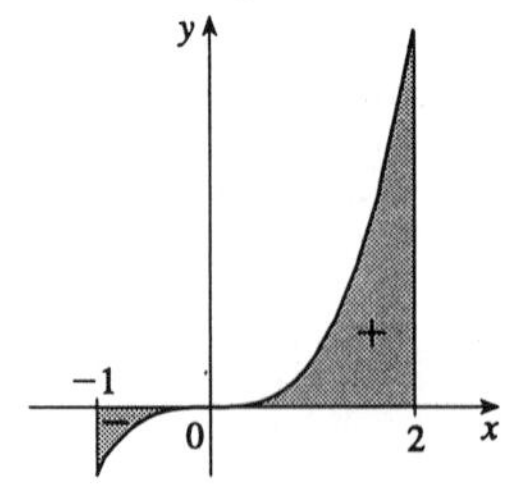

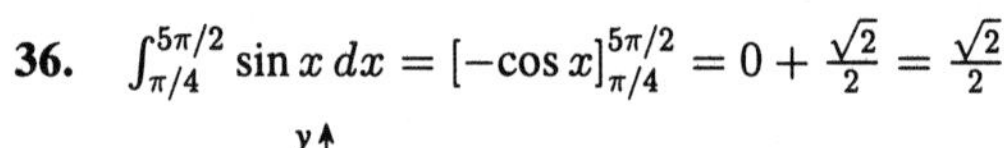

36. $\int_{\pi/4}^{5\pi/2} \sin x\,dx = [-\cos x]_{\pi/4}^{5\pi/2} = 0 + \frac{\sqrt{2}}{2} = \frac{\sqrt{2}}{2}$

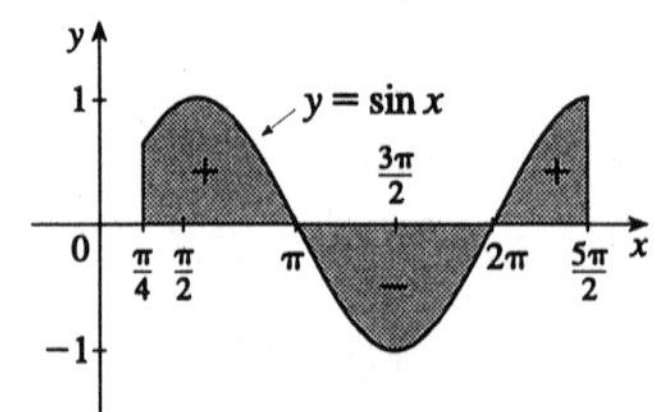

37. Let $f(x) = \sqrt{1+x^3} - (1-x)$, $\Delta x = \dfrac{2-0}{4} = \dfrac{1}{2}$.

$$A = \int_0^2 \left[\sqrt{1+x^3} - (1-x)\right] dx \approx \tfrac{1}{2}\left[f\left(\tfrac{1}{4}\right) + f\left(\tfrac{3}{4}\right) + f\left(\tfrac{5}{4}\right) + f\left(\tfrac{7}{4}\right)\right]$$
$$= \tfrac{1}{2}\left[\left(\tfrac{\sqrt{65}}{8} - \tfrac{3}{4}\right) + \left(\tfrac{\sqrt{91}}{8} - \tfrac{1}{4}\right) + \left(\tfrac{3\sqrt{21}}{8} + \tfrac{1}{4}\right) + \left(\tfrac{\sqrt{407}}{8} + \tfrac{3}{4}\right)\right]$$
$$= \tfrac{1}{16}\left(\sqrt{65} + \sqrt{91} + 3\sqrt{21} + \sqrt{407}\right) \approx 3.22$$

38. Let $f(x) = x - x\tan x$, and $\Delta x = \dfrac{\pi/4 - 0}{4} = \dfrac{\pi}{16}$. Then

$$A = \textstyle\int_0^{\pi/4} (x - x\tan x)dx \approx \tfrac{\pi}{16}\left[f\left(\tfrac{\pi}{32}\right) + f\left(\tfrac{3\pi}{32}\right) + f\left(\tfrac{5\pi}{32}\right) + f\left(\tfrac{7\pi}{32}\right)\right]$$
$$\approx \tfrac{\pi}{16}\left[\tfrac{\pi}{32}\left(1 - \tan\tfrac{\pi}{32}\right) + \tfrac{3\pi}{32}\left(1 - \tan\tfrac{3\pi}{32}\right) + \tfrac{5\pi}{32}\left(1 - \tan\tfrac{5\pi}{32}\right) + \tfrac{7\pi}{32}\left(1 - \tan\tfrac{7\pi}{32}\right)\right]$$
$$= \tfrac{\pi^2}{512}\left[16 - \tan\tfrac{\pi}{32} - 3\tan\tfrac{3\pi}{32} - 5\tan\tfrac{5\pi}{32} - 7\tan\tfrac{7\pi}{32}\right] \approx 0.1267$$

39.

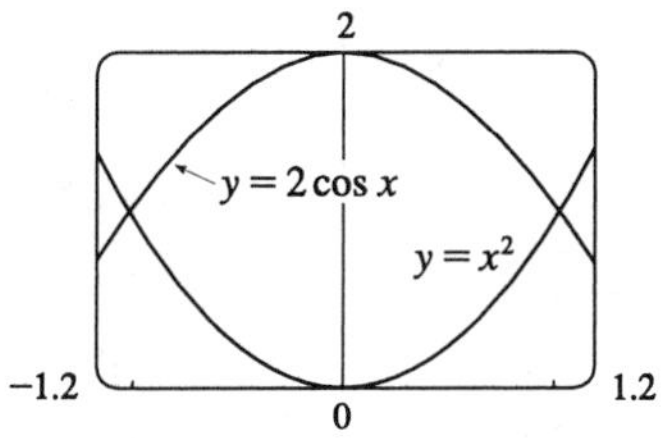

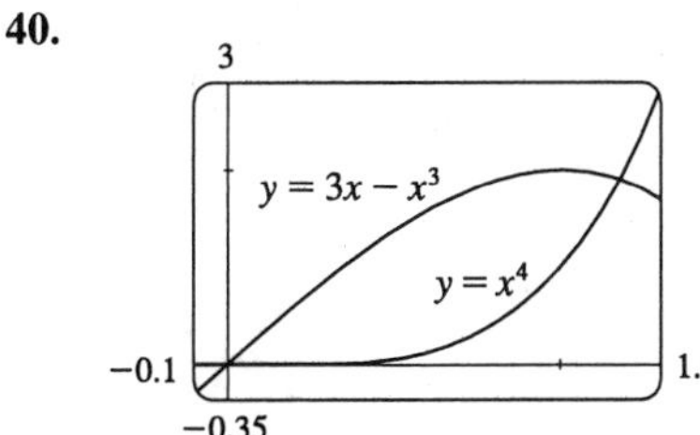

From zooming in on the graph or using a cursor, we see that the curves intersect at $x \approx \pm 1.02$, with $2\cos x > x^2$ on $[-1.02, 1.02]$. So the area between them is

$$A \approx \textstyle\int_{-1.02}^{1.02} (2\cos x - x^2)dx = 2\int_0^{1.02} (2\cos x - x^2)dx$$
$$= 2\left[2\sin x - \tfrac{1}{3}x^3\right]_0^{1.02} \approx 2.70.$$

40.

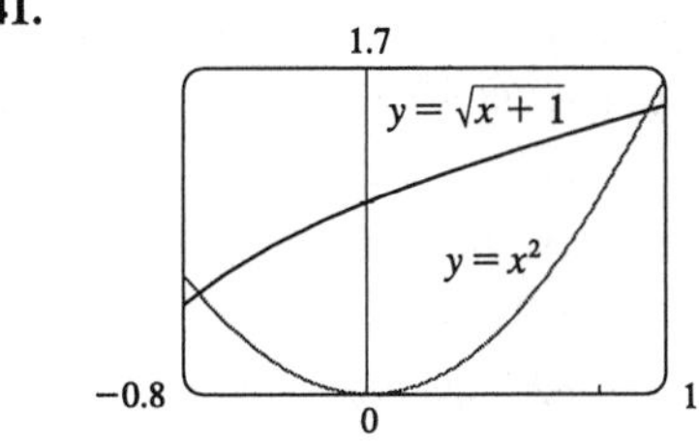

From the graph, we see that the curves intersect at $x = 0$ and at $x \approx 1.17$, with $3x - x^3 > x^4$ on $[0, 1.17]$. So the area between the curves is

$$A \approx \textstyle\int_0^{1.17} \left[(3x - x^3) - x^4\right]dx = \left[\tfrac{3}{2}x^2 - \tfrac{1}{4}x^4 - \tfrac{1}{5}x^5\right]_0^{1.17}$$
$$\approx 1.15.$$

41.

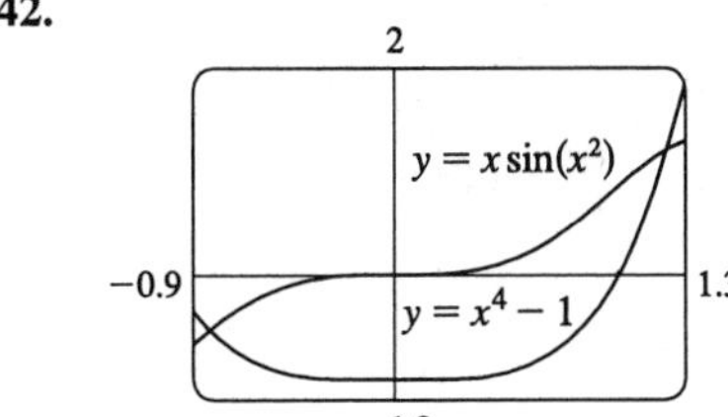

From the graph, we see that the curves intersect at $x \approx -0.72$ and at $x \approx 1.22$, with $\sqrt{x+1} > x^2$ on $[-0.72, 1.22]$. So the area between the curves is

$$A \approx \textstyle\int_{-0.72}^{1.22} \left(\sqrt{x+1} - x^2\right)dx = \left[\tfrac{2}{3}(x+1)^{3/2} - \tfrac{1}{3}x^3\right]_{-0.72}^{1.22}$$
$$\approx 1.38.$$

42.

From the graph, we see that the curves intersect at $x \approx -0.83$ and $x \approx 1.22$, with $x\sin(x^2) > x^4 - 1$ on $[-0.83, 1.22]$. So the area between the curves is

$$A \approx \textstyle\int_{-0.83}^{1.22} \left[x\sin(x^2) - (x^4 - 1)\right]dx$$
$$= \left[-\tfrac{1}{2}\cos\left(x^2\right) - \tfrac{1}{5}x^5 + x\right]_{-0.83}^{1.22} \approx 1.78.$$

43.

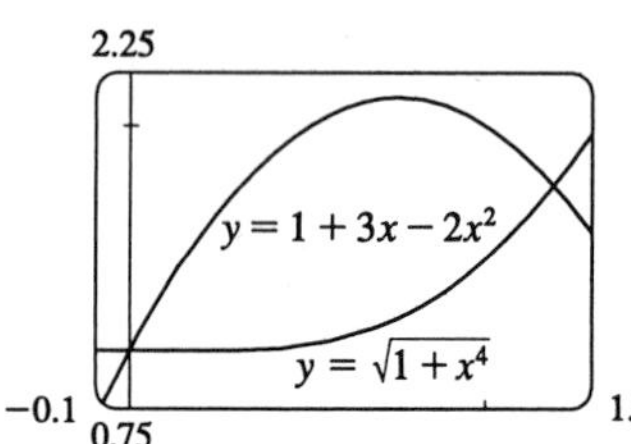

From the graph, we see that the curves intersect at $x = 0$ and at $x \approx 1.19$, with $1 + 3x - 2x^2 > \sqrt{1 + x^4}$ on $[0, 1.19]$. So, using the Midpoint Rule with $f(x) = 1 + 3x - 2x^2 - \sqrt{1 + x^4}$ on $[0, 1.19]$ with $n = 4$, we calculate the approximate area between the curves:

$$A \approx \int_0^{1.19} \left(1 + 3x - 2x^2 - \sqrt{1 + x^4}\right) dx$$
$$\approx \tfrac{1.19}{4}\left[f\left(\tfrac{1.19}{8}\right) + f\left(\tfrac{3 \cdot 1.19}{8}\right) + f\left(\tfrac{5 \cdot 1.19}{8}\right) + f\left(\tfrac{7 \cdot 1.19}{8}\right)\right] \approx 0.83.$$

44.

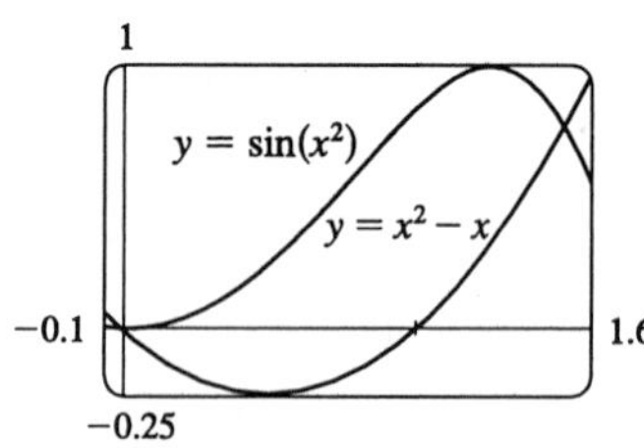

From the graph, we see that the curves intersect at $x = 0$ and at $x \approx 1.5$, with $\sin(x^2) > x^2 - x$ on $[0, 1.51]$. So, using the Midpoint Rule with $f(x) = \sin(x^2) - x^2 + x$ on $[0, 1.51]$ with $n = 4$, we calculate that the area between the curves is

$$A \approx \int_0^{1.51} [\sin(x^2) - (x^2 - x)]dx$$
$$\approx \tfrac{1.51}{4}\left[f\left(\tfrac{1.51}{8}\right) + f\left(\tfrac{3 \cdot 1.51}{8}\right) + f\left(\tfrac{5 \cdot 1.51}{8}\right) + f\left(\tfrac{7 \cdot 1.51}{8}\right)\right] \approx 0.81.$$

45.

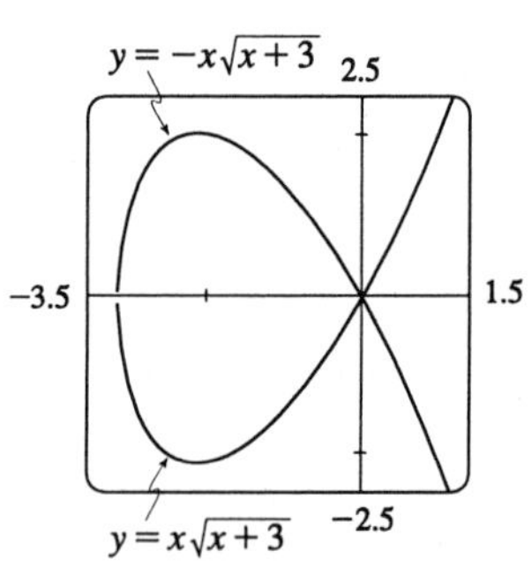

To graph this function, we must first express it as a combination of explicit functions of y; namely, $y = \pm x\sqrt{x + 3}$. We can see from the graph that the loop extends from $x = -3$ to $x = 0$, and that by symmetry, the area we seek is just twice the area under the top half of the curve on this interval, the equation of the top half being $y = -x\sqrt{x + 3}$. So the area is $A = 2\int_{-3}^{0}\left(-x\sqrt{x + 3}\right)dx$. We substitute $u = x + 3$, so $du = dx$ and the limits change to 0 and 3, and we get

$$A = -2\int_0^3 \left[(u - 3)\sqrt{u}\right] du = -2\int_0^3 \left(u^{3/2} - 3u^{1/2}\right) du$$
$$= -2\left[\tfrac{2}{5}u^{5/2} - 2u^{3/2}\right]_0^3 = -2\left[\tfrac{2}{5}\left(3^2\sqrt{3}\right) - 2\left(3\sqrt{3}\right)\right] = \tfrac{24}{5}\sqrt{3}.$$

46.

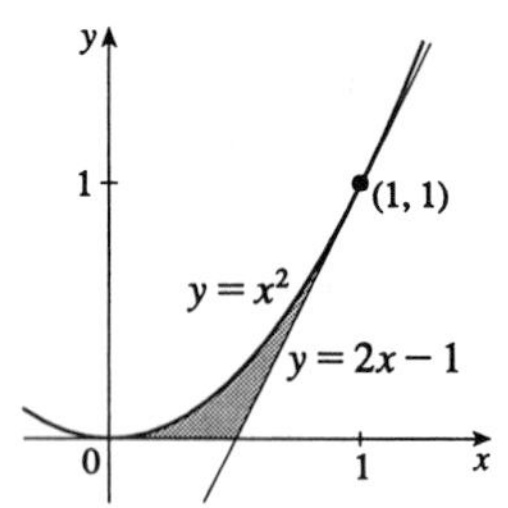

We start by finding the equation of the tangent line to $y = x^2$ at the point $(1, 1)$: $y' = 2x$, so the slope of the tangent is $2(1) = 2$, and its equation is therefore $y - 1 = 2(x - 1) \quad \Leftrightarrow \quad y = 2x - 1$. Now we must find the x-intercept of the tangent, since we only want the area above the x-axis. The x-intercept is given by $0 = 2x - 1 \quad \Leftrightarrow \quad x = \frac{1}{2}$. So on the interval $\left[0, \frac{1}{2}\right]$, we want the area under the graph of $y = x^2$, and on $\left[\frac{1}{2}, 1\right]$, we want the area between the functions $y = x^2$ and $y = 2x - 1$. So we integrate to find the area:

$$A = \int_0^{1/2} x^2\, dx + \int_{1/2}^{1} [x^2 - (2x - 1)]dx = \int_0^{1/2} x^2\, dx + \int_{1/2}^{1} (x - 1)^2\, dx = \left[\tfrac{1}{3}x^3\right]_0^{1/2} + \left[\tfrac{1}{3}(x - 1)^3\right]_{1/2}^{1}$$
$$= \tfrac{1}{3}\left(\tfrac{1}{2}\right)^3 + \left[-\tfrac{1}{3}\left(-\tfrac{1}{2}\right)^3\right] = \tfrac{1}{24} + \tfrac{1}{24} = \tfrac{1}{12}.$$

Alternate Method: Integrating with respect to y instead of x, we have

$$A = \int_0^1 \left[\tfrac{1}{2}(y + 1) - \sqrt{y}\right] dy = \left[\tfrac{1}{4}y^2 + \tfrac{1}{2}y - \tfrac{2}{3}y^{3/2}\right]_0^1 = \tfrac{1}{4} + \tfrac{1}{2} - \tfrac{2}{3} = \tfrac{1}{12}.$$

47. We first assume that $c > 0$, since c can be replaced by $-c$ in both equations without changing the graphs, and if $c = 0$ the curves do not enclose a region. We see from the graph that the enclosed area lies between $x = -c$ and $x = c$, and by symmetry, it is equal to twice the area under the top half of the graph (whose equation is $y = c^2 - x^2$). The enclosed area is

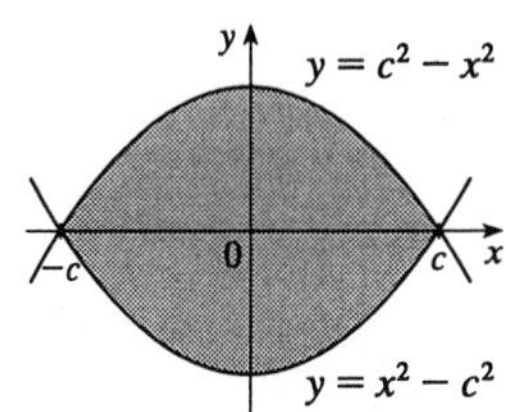

$$2\int_{-c}^{c}(c^2 - x^2)dx = 2\left[c^2x + \tfrac{1}{3}x^3\right]_{-c}^{c} = 2\left(\left[c^3 + \tfrac{1}{3}c^3\right] - \left[-c^3 + \tfrac{1}{3}(-c)^3\right]\right)$$
$$= \tfrac{8}{3}c^3, \text{ which is equal to 576 when } c = \sqrt[3]{216} = 6.$$

Note that $c = -6$ is another solution, since the graphs are the same.

48. The area under the graph of f from 0 to t is equal to $\int_0^t f(x)\,dx$, so the requirement is that $\int_0^t f(x)\,dx = t^3$ for all t. We differentiate this equation with respect to t (with the help of Part 1 of the Fundamental Theorem of Calculus) to get $f(t) = 3t^2$. This function is positive and continuous, as required.

49. By the symmetry of the problem, we consider only the first quadrant where $y = x^2 \Rightarrow x = \sqrt{y}$. We are looking for a number b such that $\int_0^4 x\,dy = 2\int_0^b x\,dy \Rightarrow \int_0^4 \sqrt{y}\,dy = 2\int_0^b \sqrt{y}\,dy \Rightarrow$ $\tfrac{2}{3}\left[y^{3/2}\right]_0^4 = \tfrac{4}{3}\left[y^{3/2}\right]_0^b \Rightarrow \tfrac{2}{3}(8 - 0) = \tfrac{4}{3}(b^{3/2} - 0) \Rightarrow b^{3/2} = 4 \Rightarrow b = 4^{2/3}$.

50. **(a)** We want to choose a so that $\displaystyle\int_1^a \frac{1}{x^2}\,dx = \int_a^4 \frac{1}{x^2}\,dx \Rightarrow \left[\frac{-1}{x}\right]_1^a = \left[\frac{-1}{x}\right]_a^4 \Rightarrow 1 - \frac{1}{a} = \frac{1}{a} - \frac{1}{4}$ $\displaystyle\Rightarrow \frac{2}{a} = \frac{5}{4} \Rightarrow a = \frac{8}{5}$.

(b) The area under the curve $y = \dfrac{1}{x^2}$ from $x = 1$ to $x = 4$ is $\frac{3}{4}$ [take $a = 4$ in the first integral in part (a)], so that b must be greater than $\frac{1}{16}$, since the area under the line $y = \frac{1}{16}$ from $x = 1$ to $x = 4$ is only $\frac{3}{16}$, which is less than half of $\frac{3}{4}$. We want to choose b so that the upper area in the diagram is half of the total area under the curve $y = \dfrac{1}{x^2}$ from $x = 1$ to $x = 4$. This implies that $\displaystyle\int_b^1 \left(\frac{1}{\sqrt{y}} - 1\right)dy = \frac{1}{2}\int_1^4 \frac{1}{x^2}\,dx \Rightarrow$

$\displaystyle\left[2\sqrt{y} - y\right]_b^1 = \frac{1}{2}\left[-\frac{1}{x}\right]_1^4 \Rightarrow 1 - 2\sqrt{b} + b = \tfrac{3}{8} \Rightarrow b - 2\sqrt{b} + \tfrac{5}{8} = 0$. Letting $c = \sqrt{b}$, we get $c^2 - 2c + \frac{5}{8} = 0 \Rightarrow 8c^2 - 16c + 5 = 0$. Thus

$$c = \frac{16 \pm \sqrt{256 - 160}}{16} = 1 \pm \tfrac{\sqrt{6}}{4}. \text{ But}$$

$c = \sqrt{b} < 1 \Rightarrow c = 1 - \frac{\sqrt{6}}{4} \Rightarrow$

$b = c^2 = 1 + \frac{3}{8} - \frac{\sqrt{6}}{2} = \frac{1}{8}\left(11 - 4\sqrt{6}\right) \approx 0.1503$.

51. We know that the area under curve A between $t = 0$ and $t = x$ is $\int_0^x v_A(t)\,dt = s_A(x)$, where $v_A(t)$ is the velocity of car A and s_A is its displacement. Similarly, the area under curve B between $t = 0$ and $t = x$ is $\int_0^x v_B(t)\,dt = s_B(x)$.

(a) After one minute, the area under curve A is greater than the area under curve B. So A is ahead after one minute.

(b) After two minutes, car B is traveling faster than car A and has gained some ground, but the area under curve A from $t = 0$ to $t = 2$ is still greater than the corresponding area for curve B, so car A is still ahead.

(c) From the graph, it appears that the area between curves A and B for $0 \le t \le 1$ (when car A is going faster), which corresponds to the distance by which car A is ahead, seems to be about 3 squares. Therefore the cars will be side by side at the time x where the area between the curves for $1 \le t \le x$ (when car B is going faster) is the same as the area for $0 \le t \le 1$. From the graph, it appears that this time is $x \approx 2.2$. So the cars are side by side when $t \approx 2.2$ minutes.

52.

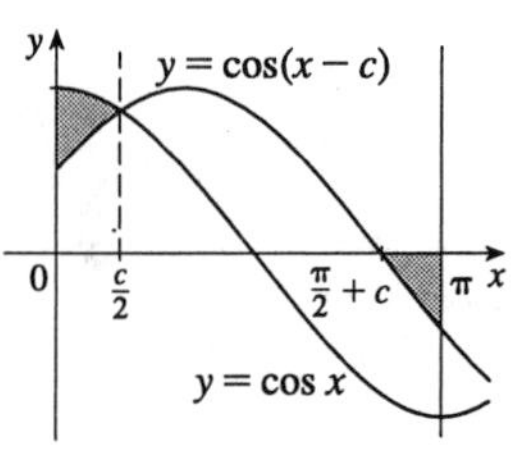

We see something close to the desired situation in the diagram. The point of intersection of $y = \cos x$ and $y = \cos(x - c)$ occurs where $x = c/2$ [since $\cos(c/2) = \cos(c/2 - c)$ by the evenness of the cosine function] and the point where $\cos(x - c)$ crosses the x-axis is $x = \frac{\pi}{2} + c$, since $\cos\left[\left(\frac{\pi}{2} + c\right) - c\right] = 0$. So we require that $\int_0^{c/2}[\cos x - \cos(x - c)]dx = -\int_{\pi/2+c}^{\pi}\cos(x - c)dx$ (the negative sign on the RHS is needed since the second area is beneath the x-axis) $\Leftrightarrow$ $[\sin x - \sin(x - c)]_0^{c/2} = -[\sin(x - c)]_{\pi/2+c}^{\pi}$ $\Rightarrow$ $[\sin(c/2) - \sin(-c/2)] - [-\sin(-c)] = \sin\left[\left(\frac{\pi}{2} + c\right) - c\right] - \sin(\pi - c)$ $\Leftrightarrow$ $2\sin(c/2) - \sin c = 1 - \sin c$. [Here we have used the oddness of the sine function, and the fact that $\sin(\pi - c) = \sin c$]. So $2\sin(c/2) = 1$ $\Leftrightarrow$ $c/2 = \frac{\pi}{6}$ $\Leftrightarrow$ $c = \frac{\pi}{3}$.

53. $A = \displaystyle\int_1^2 \left(\frac{1}{x} - \frac{1}{x^2}\right)dx = \left[\ln x + \frac{1}{x}\right]_1^2$
$= \left(\ln 2 + \frac{1}{2}\right) - (\ln 1 + 1) = \ln 2 - \frac{1}{2}$

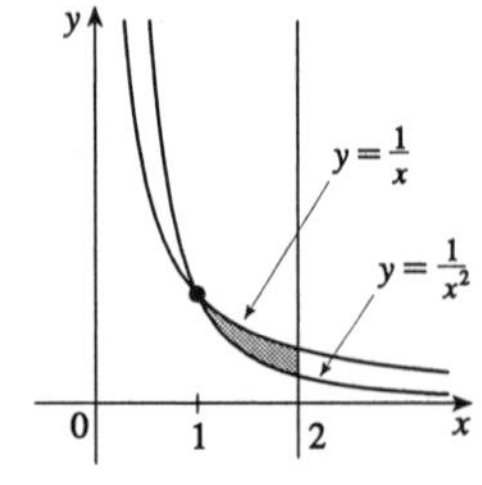

54. $A = \int_1^2 (1/y)dy = [\ln y]_1^2$
$= \ln 2 - \ln 1 = \ln 2$

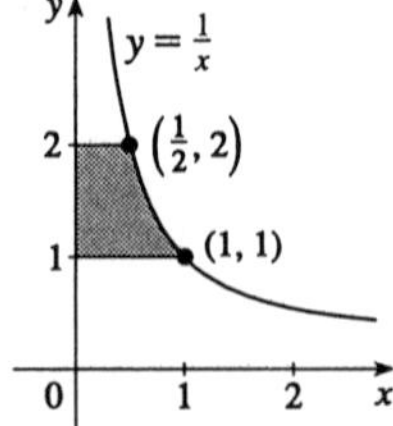

55. $A = 2\int_0^1 \left(\dfrac{2}{x^2+1} - x^2\right)dx = \left[4\tan^{-1}x - \frac{2}{3}x^3\right]_0^1 = 4 \cdot \frac{\pi}{4} - \frac{2}{3}$
$= \pi - \frac{2}{3}$

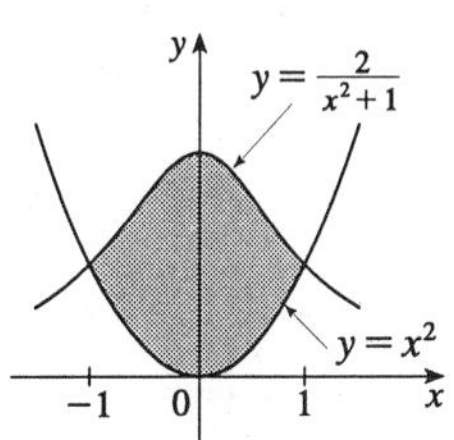

56. $A = \int_{-1}^0 (2^x - 5^x)dx + \int_0^1 (5^x - 2^x)dx$

$$= \left[\frac{2^x}{\ln 2} - \frac{5^x}{\ln 5}\right]_{-1}^0 + \left[\frac{5^x}{\ln 5} - \frac{2^x}{\ln 2}\right]_0^1$$

$$= \left(\frac{1}{\ln 2} - \frac{1}{\ln 5}\right) - \left(\frac{1/2}{\ln 2} - \frac{1/5}{\ln 5}\right) + \left(\frac{5}{\ln 5} - \frac{2}{\ln 2}\right) - \left(\frac{1}{\ln 5} - \frac{1}{\ln 2}\right)$$

$$= \frac{16}{5\ln 5} - \frac{1}{2\ln 2}$$

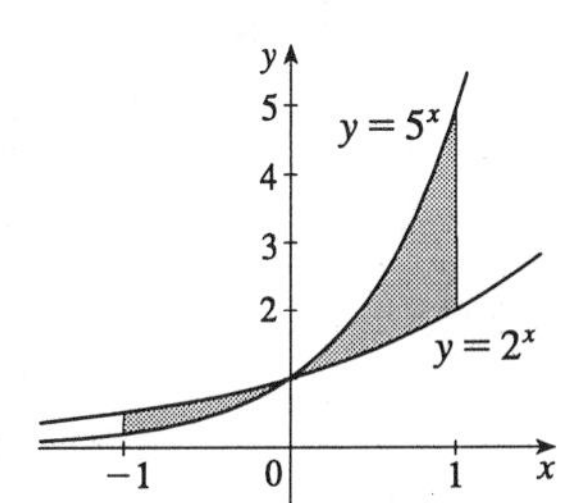

57. $A = \int_0^1 (e^{3x} - e^x)dx = \left[\frac{1}{3}e^{3x} - e^x\right]_0^1 = \left(\frac{1}{3}e^3 - e\right) - \left(\frac{1}{3} - 1\right)$
$= \frac{1}{3}e^3 - e + \frac{2}{3}$

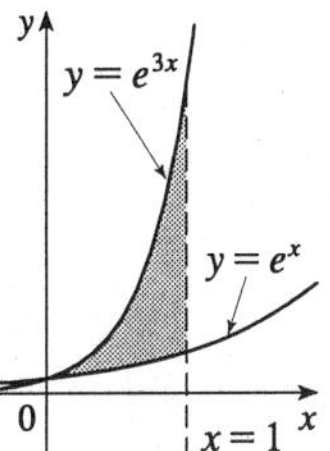

58. The curve and the line will determine a region when they intersect at two or more points. So we solve the equation $x/(x^2+1) = mx$ $\Rightarrow$ $x = 0$ or $mx^2 + m - 1 = 0$ $\Rightarrow$

$$x = 0 \text{ or } x = \frac{\pm\sqrt{-4(m)(m-1)}}{2m} = \pm\sqrt{\frac{1}{m} - 1}.$$

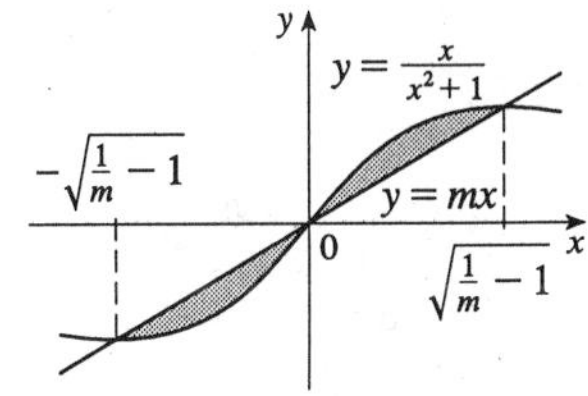

Note that if $m = 1$, this has only the solution $x = 0$, and no region is determined. But if $1/m - 1 > 0$ $\Leftrightarrow$ $1/m > 1$ $\Leftrightarrow$ $0 < m < 1$, then there are two solutions. [Another way of seeing this is to observe that the slope of the tangent to $y = x/(x^2+1)$ at the origin is $y' = 1$ and therefore we must have $0 < m < 1$.] Note that we cannot just integrate between the positive and negative roots, since the curve and the line cross at the origin. Since mx and $x/(x^2+1)$ are both odd functions, the total area is twice the area between the curves on the interval $\left[0, \sqrt{1/m - 1}\right]$. So the total area enclosed is

$$2\int_0^{\sqrt{1/m-1}} \left[\frac{x}{x^2+1} - mx\right]dx = 2\left[\tfrac{1}{2}\ln(x^2+1) - \tfrac{1}{2}mx^2\right]_0^{\sqrt{1/m-1}}$$

$$= \left[\ln\left(\frac{1}{m} - 1 + 1\right) - m\left(\frac{1}{m} - 1\right)\right] - (\ln 1 - 0) = \ln\left(\frac{1}{m}\right) + m - 1 = m - \ln m - 1.$$

EXERCISES 5.2

1. $V = \int_0^1 \pi(x^2)^2\,dx = \pi\int_0^1 x^4\,dx = \pi\left[\frac{1}{5}x^5\right]_0^1 = \frac{\pi}{5}$

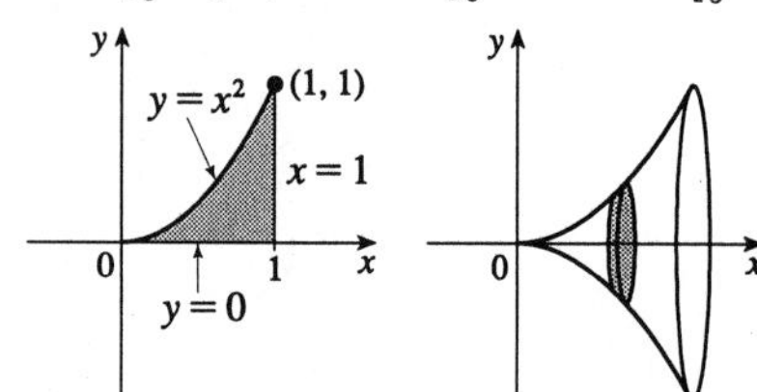

2. $V = \int_0^4 \pi\left(x^{3/2}\right)^2 dx = \pi\left[\frac{1}{4}x^4\right]_0^4 = 64\pi$

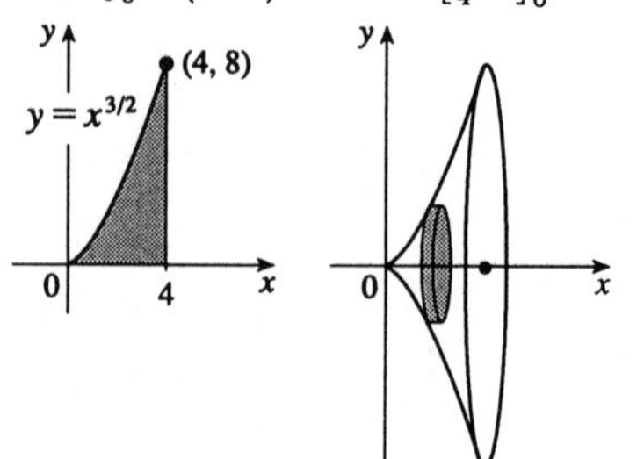

3. $V = \int_0^1 \pi(-x+1)^2\,dx = \pi\int_0^1 (x^2 - 2x + 1)dx$
$= \pi\left[\frac{1}{3}x^3 - x^2 + x\right]_0^1 = \pi\left(\frac{1}{3} - 1 + 1\right)$
$= \frac{\pi}{3}$

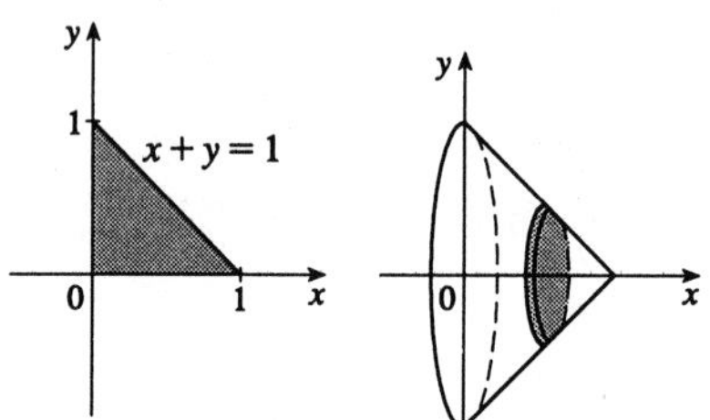

4. $V = \int_2^5 \pi(x-1)dx = \pi\left[\frac{1}{2}x^2 - x\right]_2^5$
$= \pi\left(\frac{25}{2} - 5 - \frac{4}{2} + 2\right) = \frac{15}{2}\pi$

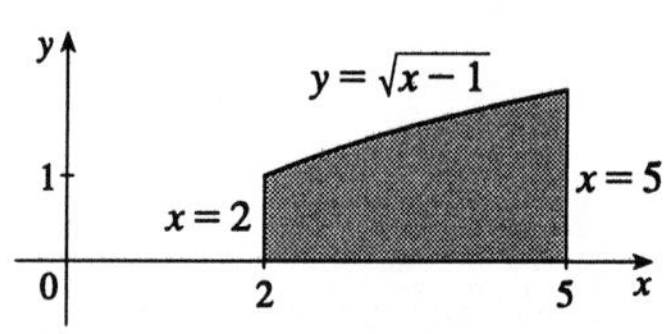

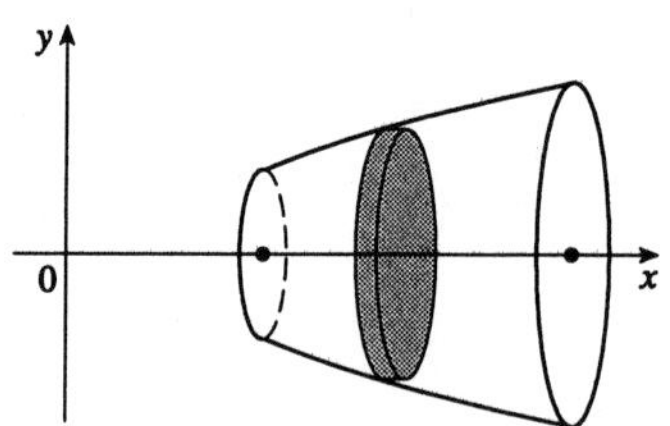

5. $V = \int_0^4 \pi\left(\sqrt{y}\right)^2 dy = \pi\int_0^4 y\,dy = \pi\left[\frac{1}{2}y^2\right]_0^4$
$= 8\pi$

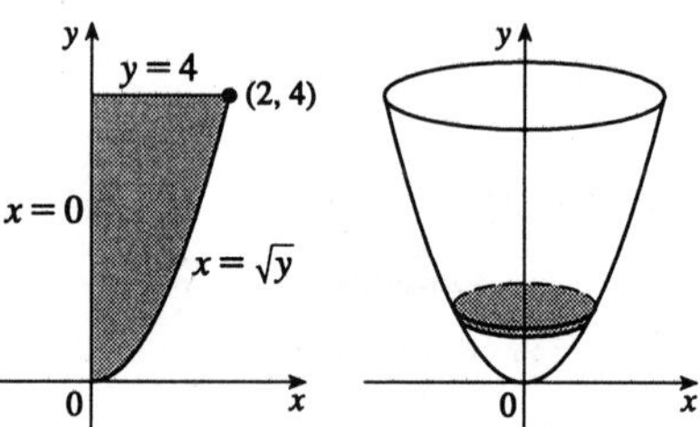

6. $V = \int_0^1 \pi(y - y^2)^2 dy = \pi\int_0^1 (y^4 - 2y^3 + y^2)\,dy$
$= \pi\left[\frac{1}{5}y^5 - \frac{1}{2}y^4 + \frac{1}{3}y^3\right]_0^1 = \pi\left(\frac{1}{5} - \frac{1}{2} + \frac{1}{3}\right) = \frac{\pi}{30}$

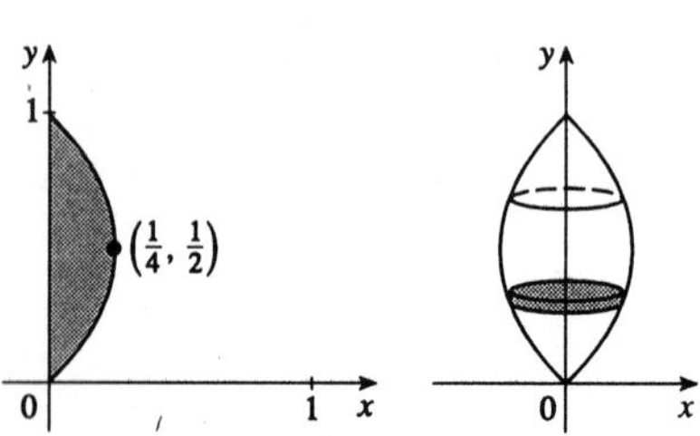

7. $V = \pi\int_0^1 \left[\left(\sqrt{x}\right)^2 - \left(x^2\right)^2\right]dx = \pi\int_0^1 \left(x - x^4\right)dx$
$= \pi\left[\frac{1}{2}x^2 - \frac{1}{5}x^5\right]_0^1 = \pi\left(\frac{1}{2} - \frac{1}{5}\right) = \frac{3\pi}{10}$

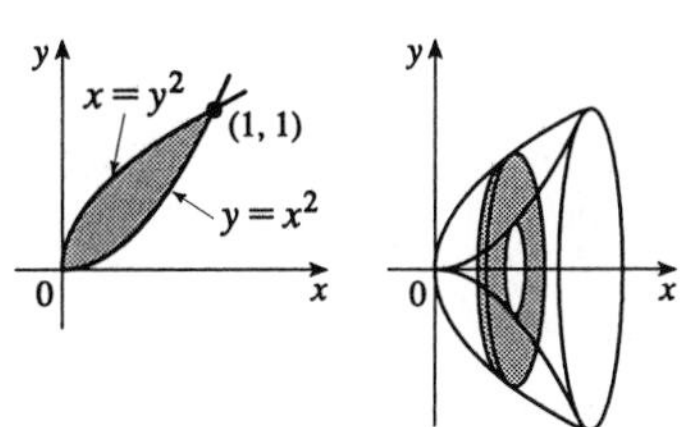

8. $V = \pi\int_{-1}^1 \left[\left(3 - x^2\right)^2 - \left(x^2 + 1\right)^2\right]dx$
$= \pi\int_{-1}^1 (8 - 8x^2)dx = 2\pi\int_0^1 (8 - 8x^2)dx$
$= 2\pi\left[8x - \frac{8}{3}x^3\right]_0^1 = 2\pi\left(8 - \frac{8}{3}\right) = \frac{32}{3}\pi$

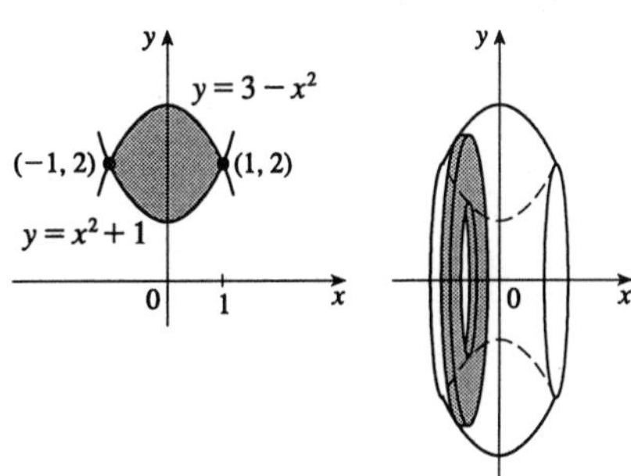

9. $V = \pi\int_0^2 \left[(2y)^2 - \left(y^2\right)^2\right]dy = \pi\int_0^2 \left(4y^2 - y^4\right)dy$
$= \pi\left[\frac{4}{3}y^3 - \frac{1}{5}y^5\right]_0^2 = \pi\left(\frac{32}{3} - \frac{32}{5}\right) = \frac{64\pi}{15}$

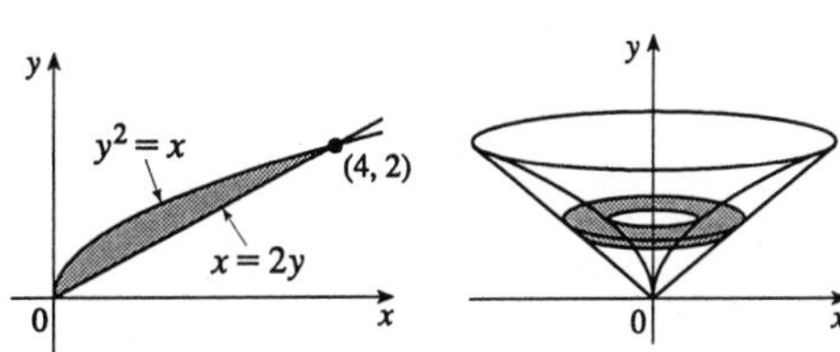

10. $V = \pi\int_0^1 \left[1^2 - \left(1 - \sqrt{1-y}\right)^2\right]dy$
$= \pi\int_0^1 \left(2\sqrt{1-y} - 1 + y\right)dy$
$= \pi\left[-\frac{4}{3}(1-y)^{3/2} - y + \frac{1}{2}y^2\right]_0^1$
$= \pi\left[\left(0 - 1 + \frac{1}{2}\right) - \left(-\frac{4}{3} - 0 + 0\right)\right] = \frac{5}{6}\pi$

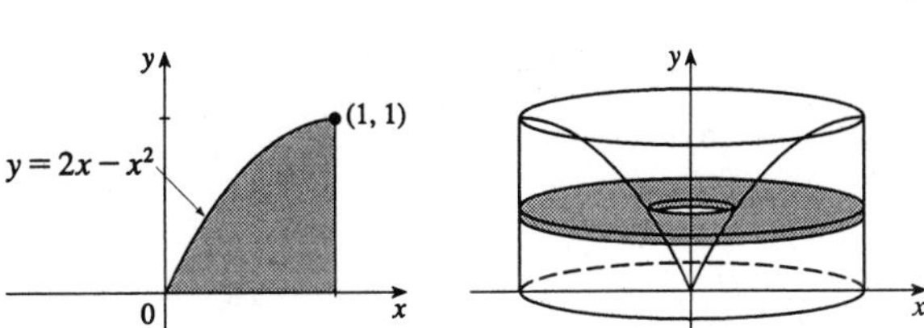

11. $V = \pi\int_{-1}^1 \left[\left(2 - x^4\right)^2 - 1^2\right]dx$
$= 2\pi\int_0^1 \left(3 - 4x^4 + x^8\right)dx$
$= 2\pi\left[3x - \frac{4}{5}x^5 + \frac{1}{9}x^9\right]_0^1$
$= 2\pi\left(3 - \frac{4}{5} + \frac{1}{9}\right) = \frac{208}{45}\pi$

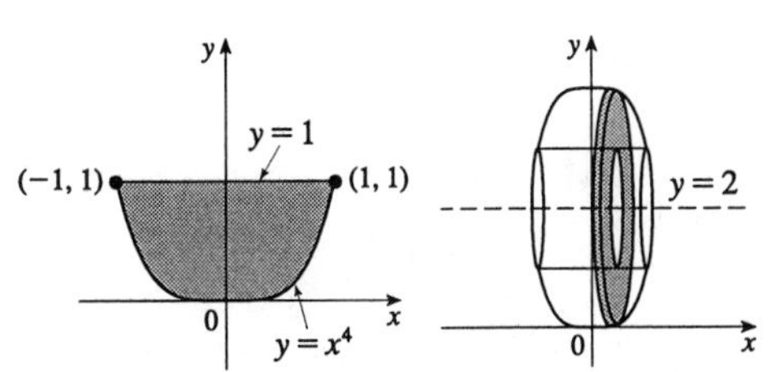

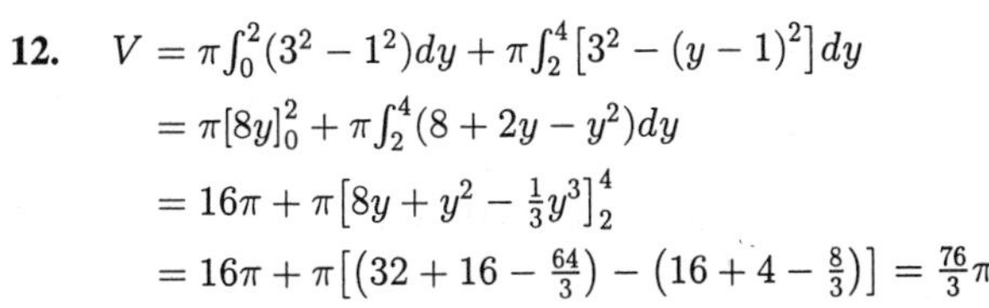

12. $V = \pi\int_0^2 (3^2 - 1^2)dy + \pi\int_2^4 \left[3^2 - (y-1)^2\right]dy$
$= \pi[8y]_0^2 + \pi\int_2^4 (8 + 2y - y^2)dy$
$= 16\pi + \pi\left[8y + y^2 - \frac{1}{3}y^3\right]_2^4$
$= 16\pi + \pi\left[\left(32 + 16 - \frac{64}{3}\right) - \left(16 + 4 - \frac{8}{3}\right)\right] = \frac{76}{3}\pi$

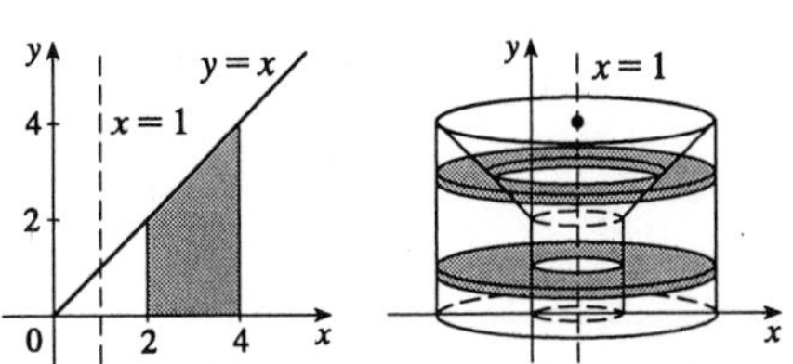

13. $V = \pi\int_0^8 \left(\frac{1}{4}x\right)^2 dx = \frac{\pi}{16}\left[\frac{1}{3}x^3\right]_0^8 = \frac{32}{3}\pi$

14. $V = \pi\int_0^2 \left[8^2 - (4y)^2\right] dy = \pi\left[64y - \frac{16}{3}y^3\right]_0^2 = \pi\left(128 - \frac{128}{3}\right) = \frac{256}{3}\pi$

15. $V = \pi\int_0^2 (8-4y)^2 dy = \pi\left[64y - 32y^2 + \frac{16}{3}y^3\right]_0^2 = \pi\left(128 - 128 + \frac{128}{3}\right) = \frac{128}{3}\pi$

16. $V = \pi\int_0^8 \left[2^2 - \left(2 - \frac{1}{4}x\right)^2\right] dx = \pi\int_0^8 \left(x - \frac{1}{16}x^2\right) dx = \pi\left[\frac{1}{2}x^2 - \frac{1}{48}x^3\right]_0^8 = \pi\left(32 - \frac{32}{3}\right) = \frac{64}{3}\pi$

17. $V = \pi\int_0^8 \left[\left(\sqrt[3]{x}\right)^2 - \left(\frac{1}{4}x\right)^2\right] = \pi\int_0^8 \left(x^{2/3} - \frac{1}{16}x^2\right) dx = \pi\left[\frac{3}{5}x^{5/3} - \frac{1}{48}x^3\right]_0^8 = \pi\left(\frac{96}{5} - \frac{32}{3}\right) = \frac{128}{15}\pi$

18. $V = \pi\int_0^2 \left[(4y)^2 - \left(y^3\right)^2\right] dy = \pi\int_0^2 (16y^2 - y^6) dy = \pi\left[\frac{16}{3}y^3 - \frac{1}{7}y^7\right]_0^2 = \pi\left(\frac{128}{3} - \frac{128}{7}\right) = \frac{512}{21}\pi$

19. $V = \pi\int_0^8 \left[\left(2 - \frac{1}{4}x\right)^2 - \left(2 - \sqrt[3]{x}\right)^2\right] dx = \pi\int_0^8 \left(-x + \frac{1}{16}x^2 + 4x^{1/3} - x^{2/3}\right) dx$

$= \pi\left[-\frac{1}{2}x^2 + \frac{1}{48}x^3 + 3x^{4/3} - \frac{3}{5}x^{5/3}\right]_0^8 = \pi\left(-32 + \frac{32}{3} + 48 - \frac{96}{5}\right) = \frac{112}{15}\pi$

20. $V = \pi\int_0^2 \left[\left(8 - y^3\right)^2 - (8 - 4y)^2\right] dy = \pi\int_0^2 (-16y^3 + y^6 + 64y - 16y^2) dy$

$= \pi\left[-4y^4 + \frac{1}{7}y^7 + 32y^2 - \frac{16}{3}y^3\right]_0^2 = \pi\left(-64 + \frac{128}{7} + 128 - \frac{128}{3}\right) = \frac{832}{21}\pi$

21. $V = \pi\int_0^8 \left(2^2 - x^{2/3}\right) dx = \pi\left[4x - \frac{3}{5}x^{5/3}\right]_0^8 = \pi\left(32 - \frac{96}{5}\right) = \frac{64}{5}\pi$

22. $V = \pi\int_0^2 \left(y^3\right)^2 dy = \pi\left[\frac{1}{7}y^7\right]_0^2 = \frac{128}{7}\pi$

23. $V = \pi\int_0^8 \left(2 - \sqrt[3]{x}\right)^2 dx = \pi\int_0^8 \left(4 - 4x^{1/3} + x^{2/3}\right) dx$

$= \pi\left[4x - 3x^{4/3} + \frac{3}{5}x^{5/3}\right]_0^8 = \pi\left(32 - 48 + \frac{96}{5}\right) = \frac{16}{5}\pi$

24. $V = \pi\int_0^2 \left[8^2 - \left(8 - y^3\right)^2\right] dy = \pi\int_0^2 (16y^3 - y^6) dy = \pi\left[4y^4 - \frac{1}{7}y^7\right]_0^2 = \pi\left(64 - \frac{128}{7}\right) = \frac{320}{7}\pi$

25. $V = \pi\int_0^2 \left(x^2 - 1\right)^2 dx = \pi\int_0^2 \left(x^4 - 2x^2 + 1\right) dx = \pi\left[\frac{1}{5}x^5 - \frac{2}{3}x^3 + x\right]_0^2 = \pi\left(\frac{32}{5} - \frac{16}{3} + 2\right) = \frac{46}{15}\pi$

26. $V = \pi\displaystyle\int_1^3 \left[\frac{1}{x}\right]^2 dx = \pi\left[\frac{-1}{x}\right]_1^3 = \pi\left(-\frac{1}{3} + 1\right) = \frac{2}{3}\pi$

27. $V = \pi\int_{-1}^1 (\sec^2 x - 1^2) dx = \pi[\tan x - x]_{-1}^1 = \pi[(\tan 1 - 1) - (-\tan 1 + 1)] = 2\pi(\tan 1 - 1)$

28. $V = \pi\int_0^{\pi/4} (\cos^2 x - \sin^2 x) dx = \frac{\pi}{2}\int_0^{\pi/4} \cos 2x\,(2\,dx) = \frac{\pi}{2}[\sin 2x]_0^{\pi/4} = \frac{\pi}{2}(1 - 0) = \frac{\pi}{2}$

29. $V = \pi\int_{-3}^{-2} (-x-2)^2 dx + \pi\int_{-2}^0 (x+2)^2 dx = \pi\int_{-3}^0 (x+2)^2 dx = \left[\frac{\pi}{3}(x+2)^3\right]_{-3}^0 = \frac{\pi}{3}[8 - (-1)] = 3\pi$

30. $V = \pi\int_1^2 1^2\,dx + \pi\int_2^3 2^2\,dx + \pi\int_3^4 3^2\,dx + \pi\int_4^5 4^2\,dx + \pi\int_5^6 5^2\,dx$

$= \pi\cdot 1 + \pi\cdot 4 + \pi\cdot 9 + \pi\cdot 16 + \pi\cdot 25 = 55\pi$

31. $V = \pi\int_0^{\pi/4} [1^2 - \tan^2 x] dx$

32. $V = \pi\int_0^2 \left[5^2 - \left(y^2 + 1\right)^2\right] dy = \pi\int_0^2 \left(24 - y^4 - 2y^2\right) dy$

33. $x - 1 = (x-4)^2 + 1 \quad\Leftrightarrow\quad x^2 - 9x + 18 = 0 \quad\Leftrightarrow\quad x = 3$ or 6, so

$$V = \pi\int_3^6 \left[\left[6 - (x-4)^2\right]^2 - (8-x)^2\right] dx = \pi\int_3^6 \left(x^4 - 16x^3 + 83x^2 - 144x + 36\right) dx.$$

34. $V = \pi \int_0^{\pi/2} \left[1^2 - (1 - \cos x)^2\right] dx = \pi \int_0^{\pi/2} (2\cos x - \cos^2 x)\,dx$

35. $V = \pi \int_0^{\pi/2} \left[(1 + \cos x)^2 - 1^2\right] dx = \pi \int_0^{\pi/2} (2\cos x + \cos^2 x)\,dx$

36. The points of intersection of the two curves are $(3, 0)$ and $\left(-\frac{9}{4}, \frac{7}{2}\right)$. Therefore
$V = \pi \int_0^{7/2} \left[\left[4 - (y-1)^2 + 5\right]^2 - \left(3 - \frac{3}{2}y + 5\right)^2\right] dy = \pi \int_0^{7/2} \left[\left[9 - (y-1)^2\right]^2 - \left(8 - \frac{3}{2}y\right)^2\right] dy.$

37. We see from the graph in Exercise 5.1.41 that the x-coordinates of the points of intersection are $x \approx -0.72$ and $x \approx 1.22$, with $\sqrt{x+1} > x^2$ on $[-0.72, 1.22]$, so the volume of revolution is about

$$\pi \int_{-0.72}^{1.22} \left[\left(\sqrt{x+1}\right)^2 - \left(x^2\right)^2\right] dx = \pi \int_{-0.72}^{1.22} (x + 1 - x^4)\,dx = \pi \left[\tfrac{1}{2}x^2 + x - \tfrac{1}{5}x^5\right]_{-0.72}^{1.22} \approx 5.80.$$

38. We see from the graph in Exercise 5.1.40 that the x-coordinates of the points of intersection are $x = 0$ and $x \approx 1.17$, with $3x - x^3 > x^4$ on $[0, 1.17]$, so the volume of revolution is about

$$\pi \int_0^{1.17} \left[\left(3x - x^3\right)^2 - \left(x^4\right)^2\right] dx = \pi \int_0^{1.17} \left[9x^2 - 6x^4 + x^6 - x^8\right] dx = \pi \left[3x^3 - \tfrac{6}{5}x^5 + \tfrac{1}{7}x^7 - \tfrac{1}{9}x^9\right]_0^{1.17} \approx 6.74.$$

39. $V = \pi \int_0^1 3^2\,dx + \pi \int_1^4 1^2\,dx + \pi \int_4^5 3^2\,dx$
$= 9\pi + 3\pi + 9\pi = 21\pi$

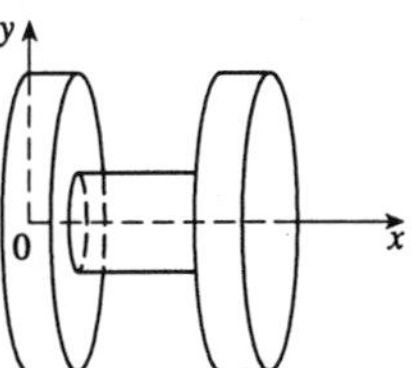

40. $V = \pi \int_0^1 \left(\frac{1}{2}\right)^2 dx + \pi \int_1^2 \left(x^2 - 2x + 2\right)^2 dx$
$= \frac{\pi}{4} + \pi \int_1^2 \left(x^4 - 4x^3 + 8x^2 - 8x + 4\right) dx$
$= \frac{\pi}{4} + \pi \left[\frac{1}{5}x^5 - x^4 + \frac{8}{3}x^3 - 4x^2 + 4x\right]_1^2$
$= \frac{\pi}{4} + \pi \left[\left(\frac{32}{5} - 16 + \frac{64}{3} - 16 + 8\right) - \left(\frac{1}{5} - 1 + \frac{8}{3} - 4 + 4\right)\right]$
$= \frac{127\pi}{60}$

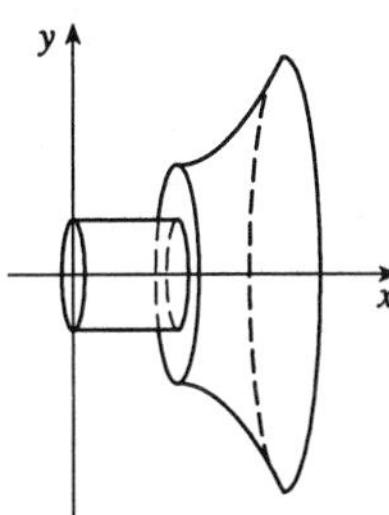

41. The solid is obtained by rotating the region under the curve $y = \tan x$, from $x = 0$ to $x = \frac{\pi}{4}$, about the x-axis.

42. The solid is obtained by rotating the region bounded by the curve $x = y^3$ and the lines $y = 1$, $y = 2$, and $x = 0$ about the y-axis.

43. The solid is obtained by rotating the region between the curves $x = y$ and $x = \sqrt{y}$ about the y-axis.

44. The solid is obtained by rotating the region bounded by the curve $y = (x - 2)^2$ and the line $y = 4$ about the x-axis.

45. The solid is obtained by rotating the region between the curves $y = 5 - 2x^2$ and $y = 5 - 2x$ about the x-axis.
Or: The solid is obtained by rotating the region bounded by the curves $y = 2x$ and $y = 2x^2$ about the line $y = 5$.

46. The solid is obtained by rotating the region bounded by the curves $y = 2 + \cos x$ and $y = 2 + \sin x$ and the line $x = \frac{\pi}{2}$ about the x-axis.

47. $V = \pi \int_0^h \left(-\frac{r}{h}y + r\right)^2 dy$

$= \pi \int_0^h \left[\frac{r^2}{h^2}y^2 - \frac{2r^2}{h}y + r^2\right] dy$

$= \pi \left[\frac{r^2}{3h^2}y^3 - \frac{r^2}{h}y^2 + r^2 y\right]_0^h = \frac{1}{3}\pi r^2 h$

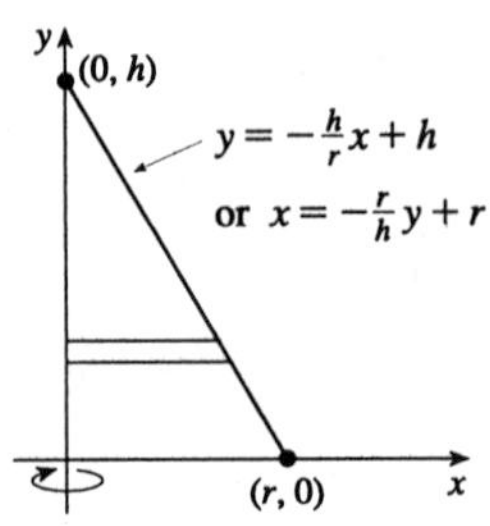

48. $V = \pi \int_0^h \left(R - \frac{R-r}{h}y\right)^2 dy$

$= \pi \int_0^h \left[R^2 - \frac{2R(R-r)}{h}y + \left(\frac{R-r}{h}\right)^2 y^2\right] dy$

$= \pi \left[R^2 y - \frac{R(R-r)}{h}y^2 + \frac{1}{3}\left(\frac{R-r}{h}\right)^2 y^3\right]_0^h$

$= \pi\left[R^2 h - R(R-r)h + \frac{1}{3}(R-r)^2 h\right]$

$= \pi\left[Rrh + \frac{1}{3}(R^2 - 2Rr + r^2)h\right] = \frac{1}{3}\pi h(R^2 + Rr + r^2)$

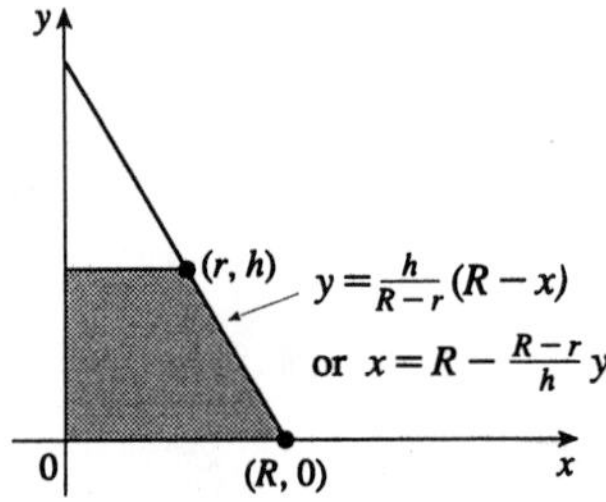

Alternate Solution: $\frac{H}{R} = \frac{H-h}{r}$ by similar triangles. Therefore $Hr = HR - hR$ and $hR = H(R-r)$, so

$H = \frac{hR}{R-r}$. Now

$V = \frac{1}{3}\pi R^2 H - \frac{1}{3}\pi r^2 (H - h)$ (by Exercise 47)

$= \frac{1}{3}\pi R^2 \frac{hR}{R-r} - \frac{1}{3}\pi r^2 \frac{rh}{R-r} = \frac{\pi h}{3}\frac{R^3 - r^3}{R-r}$

$= \frac{1}{3}\pi h(R^2 + Rr + r^2) = \frac{1}{3}\left[\pi R^2 + \pi r^2 + \sqrt{(\pi R^2)(\pi r^2)}\right]h$

$= \frac{1}{3}\left(A_1 + A_2 + \sqrt{A_1 A_2}\right)h,$

where A_1 and A_2 are the areas of the bases of the frustum.

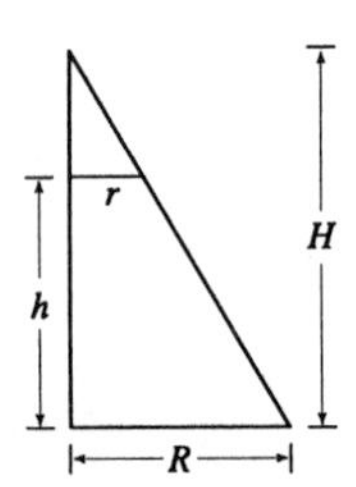

49. $V = \pi \int_{r-h}^{r} (r^2 - y^2)\,dy = \pi\left[r^2 y - \frac{y^3}{3}\right]_{r-h}^{r}$

$= \pi\left[\left(r^3 - \frac{r^3}{3}\right) - \left(r^2(r-h) - \frac{(r-h)^3}{3}\right)\right]$

$= \pi h^2\left(r - \frac{h}{3}\right)$

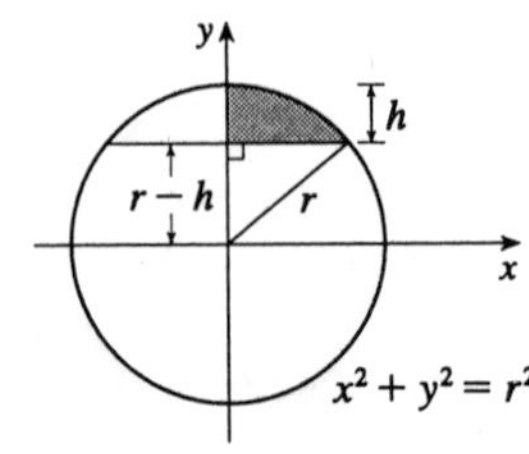

50. $V = \int_0^h A(y)\,dy = \int_0^h \left[\frac{a-b}{h}y + b\right]^2 dy$

$= \int_0^h \left[\frac{(a-b)^2}{h^2}y^2 + \frac{2b(a-b)}{h}y + b^2\right] dy$

$= \left[\frac{(a-b)^2}{3h^2}y^3 + \frac{b(a-b)}{h}y^2 + b^2y\right]_0^h = \frac{1}{3}(a-b)^2h + b(a-b)\,h + b^2h$

$= \frac{1}{3}(a^2 - 2ab + b^2 + 3ab)h = \frac{1}{3}(a^2 + ab + b^2)h$

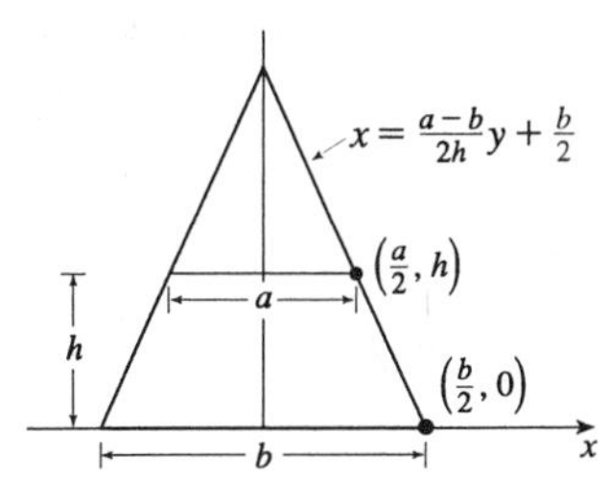

[Note that this can be written as $\frac{1}{3}\left(A_1 + A_2 + \sqrt{A_1A_2}\right)h$ as in Exercise 48.]

51. For a cross-section at height y, we see from similar triangles that $\frac{\alpha/2}{b/2} = \frac{h-y}{h}$, so $\alpha = b\left(1 - \frac{y}{h}\right)$. Similarly, $\beta = 2b\left(1 - \frac{y}{h}\right)$. So

$V = \int_0^h A(y)\,dy = \int_0^h 2b^2\left(1 - \frac{y}{h}\right)^2 dy = 2b^2\int_0^h \left(1 - \frac{2y}{h} + \frac{y^2}{h^2}\right) dy$

$= 2b^2\left[y - \frac{y^2}{h} + \frac{y^3}{3h^2}\right]_0^h = 2b^2\left[h - h + \frac{1}{3}h\right] = \frac{2}{3}b^2h$

$\left(= \frac{1}{3}Bh\right.$, where B is the area of the base, as with any pyramid.$\left.\right)$

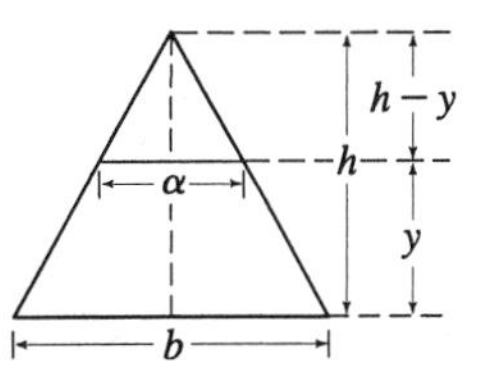

52. Consider the triangle consisting of two vertices of the base and the center of the base. This triangle is similar to the corresponding triangle at a height y, so $a/b = \alpha/\beta \quad\Rightarrow\quad \alpha = a\beta/b$. Also by similar triangles, $b/h = \beta/(h-y) \quad\Rightarrow\quad \beta = b(h-y)/h$. These two equations imply that $\alpha = a(1 - y/h)$, and since the cross-section is an equilateral triangle, it has area

$A(y) = \frac{\alpha}{2} \cdot \frac{\sqrt{3}\alpha}{2} = \frac{a^2(1-y/h)^2}{4}\sqrt{3}$, so

$V = \int_0^h A(y)\,dy = \frac{a^2\sqrt{3}}{4}\int_0^h \left(1 - \frac{y}{h}\right)^2 dy$

$= \frac{a^2\sqrt{3}}{4}\left[-\frac{h}{3}\left(1 - \frac{y}{h}\right)^3\right]_0^h$

$= -\frac{1}{12}\sqrt{3}a^2h(-1) = \frac{\sqrt{3}}{12}a^2h.$

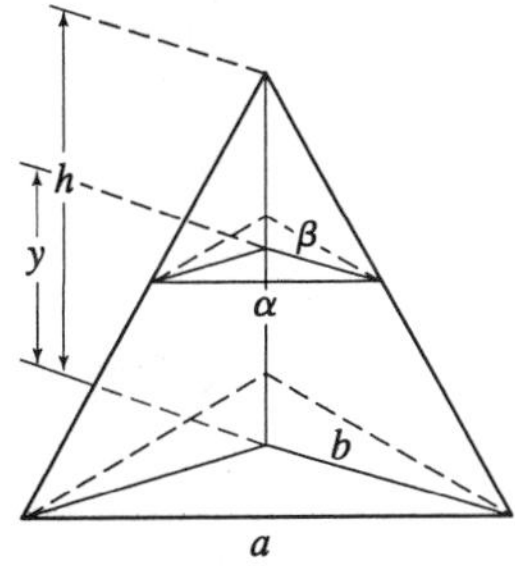

53. A cross-section at height z is a triangle similar to the base, so its area is

$A(z) = \frac{1}{2} \cdot 3\left(\frac{5-z}{5}\right) \cdot 4\left(\frac{5-z}{5}\right) = 6\left(1 - \frac{z}{5}\right)^2$, so

$V = \int_0^5 A(z)\,dz = 6\int_0^5 \left(1 - \frac{z}{5}\right)^2 dz$

$= 6\left[(-5)\frac{1}{3}\left(1 - \frac{1}{5}z\right)^3\right]_0^5 = -10(-1) = 10\text{ cm}^3.$

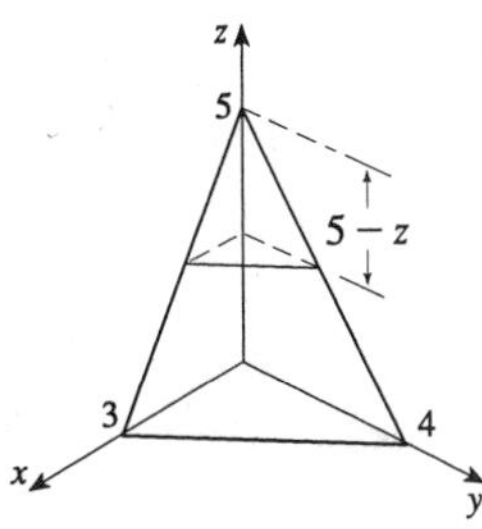

54. A cross-section is shaded in the diagram.

$A(x) = y^2 = \left(2\sqrt{r^2 - x^2}\right)^2$, so

$V = \int_{-r}^{r} A(x)dx = 2\int_0^r 4(r^2 - x^2)dx$

$= 8\left[r^2 x - \frac{1}{3}x^3\right]_0^r = \frac{16}{3}r^3$.

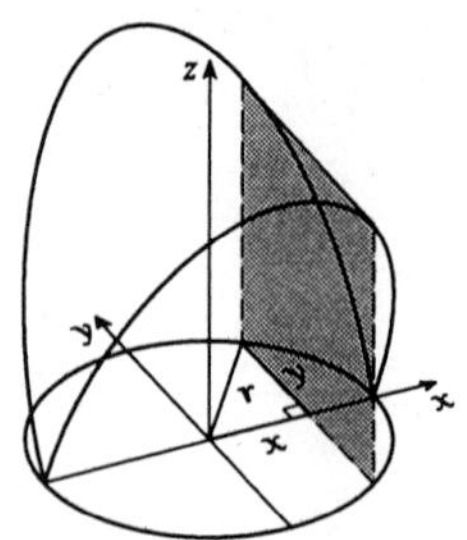

55. $V = \int_{-2}^{2} A(x)\,dx = 2\int_0^2 A(x)\,dx$

$= 2\int_0^2 \frac{1}{2}\left(\sqrt{2}y\right)^2 dx = 2\int_0^2 y^2\,dx$

$= \frac{1}{2}\int_0^2 (36 - 9x^2)\,dx = \frac{9}{2}\int_0^2 (4 - x^2)\,dx$

$= \frac{9}{2}\left[4x - \frac{x^3}{3}\right]_0^2 = \frac{9}{2}\left(8 - \frac{8}{3}\right) = 24$

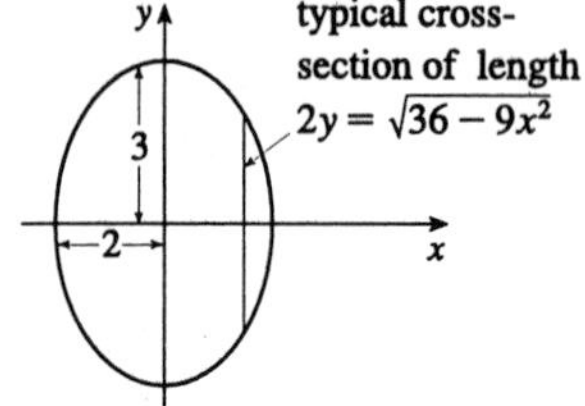

56. The cross-section of the base corresponding to the coordinate y has length $2x = 2\sqrt{y}$. The corresponding equilateral triangle has area $A(y) = \left(2\sqrt{y}\right)^2 \sqrt{3}/4 = y\sqrt{3}$. Therefore, $V = \int_0^1 A(y)\,dy = \int_0^1 y\sqrt{3}\,dy = \sqrt{3}\left[\frac{1}{2}y^2\right]_0^1 = \frac{\sqrt{3}}{2}$.

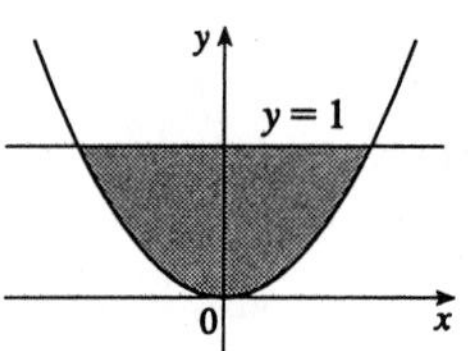

57. The cross section of the base corresponding to the coordinate y has length $2x = 2\sqrt{y}$, so

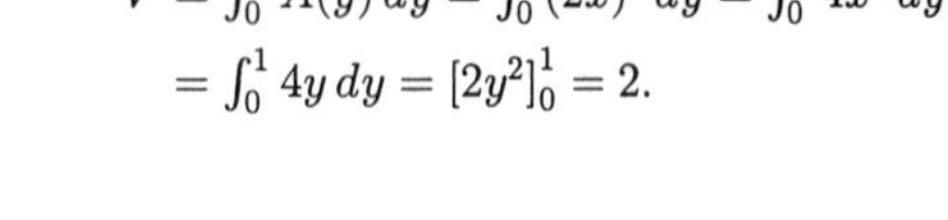

$V = \int_0^1 A(y)\,dy = \int_0^1 (2x)^2\,dy = \int_0^1 4x^2\,dy$

$= \int_0^1 4y\,dy = \left[2y^2\right]_0^1 = 2$.

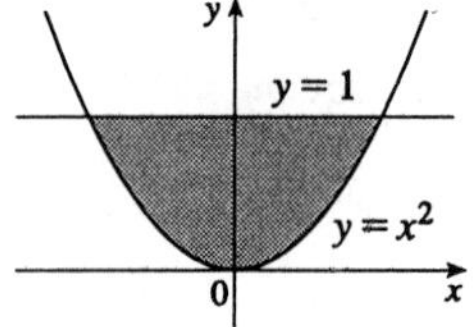

58. Since the area of a semicircle of diameter y is $\frac{\pi y^2}{8}$, we have

$V = \int_0^2 A(x)\,dx = \int_0^2 \frac{\pi}{8}y^2\,dx = \frac{\pi}{8}\int_0^2 \left(1 - \frac{1}{2}x\right)^2 dx$

$= \frac{\pi}{4}\int_0^2 \left(\frac{1}{2}x - 1\right)^2 \frac{1}{2}\,dx = \frac{\pi}{4}\left[\frac{1}{3}\left(\frac{1}{2}x - 1\right)^3\right]_0^2$

$= \frac{\pi}{12}[0 - (-1)] = \frac{\pi}{12}$.

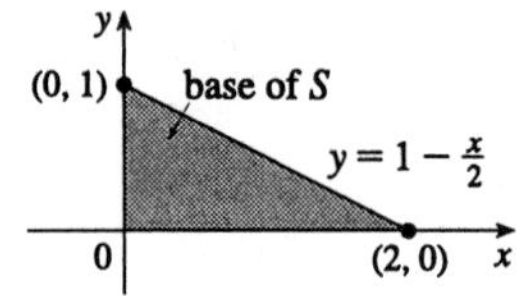

59. Assume that the base of each isosceles triangle lies in the base of S. Then its area is $A(x) = \frac{1}{2}bh = \frac{1}{2}\left(1 - \frac{1}{2}x\right)\left(1 - \frac{1}{2}x\right) = \frac{1}{2}\left(1 - \frac{1}{2}x\right)^2$, and the volume is $V = \int_0^2 A(x)\,dx = \int_0^2 \frac{1}{2}y^2\,dx = \frac{1}{2}\int_0^2 \left(1 - \frac{1}{2}x\right)^2 dx = \frac{1}{2}\left[\frac{2}{3}\left(\frac{1}{2}x - 1\right)^3\right]_0^2 = \frac{1}{3}$.

60. **(a)** $V = \int_{-r}^{r} A(x)\,dx = 2\int_0^r A(x)\,dx = 2\int_0^r \frac{1}{2}h\left(2\sqrt{r^2 - x^2}\right)dx = 2h\int_0^r \sqrt{r^2 - x^2}\,dx$

(b) Observe that the integral represents one quarter of the area of a circle of radius r, so $V = 2h\frac{1}{4}\pi r^2 = \frac{1}{2}\pi h r^2$.

61. **(a)** The torus is obtained by rotating the circle $(x - R)^2 + y^2 = r^2$ about the y-axis. Solving for y, we see that the right half of the circle is given by $x = R + \sqrt{r^2 - y^2} = f(y)$ and the left half by $x = R - \sqrt{r^2 - y^2} = g(y)$.

So $V = \pi \int_{-r}^{r} \left([f(y)]^2 - [g(y)]^2\right) dy$
$= 2\pi \int_0^r 4R\sqrt{r^2 - y^2}\, dy$
$= 8\pi R \int_0^r \sqrt{r^2 - y^2}\, dy.$

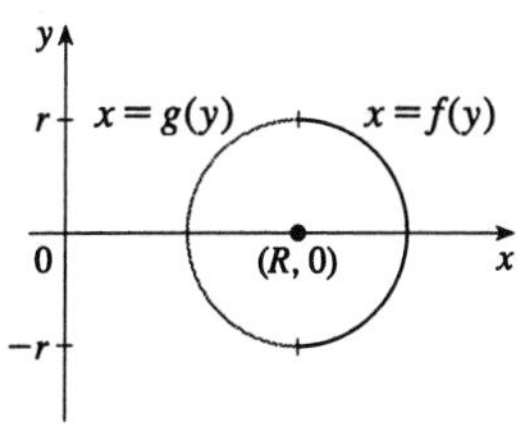

(b) Observe that the integral represents a quarter of the area of a circle with radius r, so $8\pi R \int_0^r \sqrt{r^2 - y^2}\, dy = 8\pi R \frac{1}{4}(\pi r^2) = 2\pi^2 r^2 R.$

62. Each cross-section of the solid S in a plane perpendicular to the x-axis is a square (since the edges of the cut lie on the cylinders, which are perpendicular). One-quarter of this square and one-eighth of S are shown. The area of this quarter-square is $|PQ|^2 = r^2 - x^2$. Therefore $A(x) = 4(r^2 - x^2)$ and the volume of S is

$V = \int_{-r}^{r} A(x)\, dx = 4\int_{-r}^{r} (r^2 - x^2) dx$
$= 8\int_0^r (r^2 - x^2) dx = 8\left[r^2 x - \frac{1}{3}x^3\right]_0^r = \frac{16}{3} r^3.$

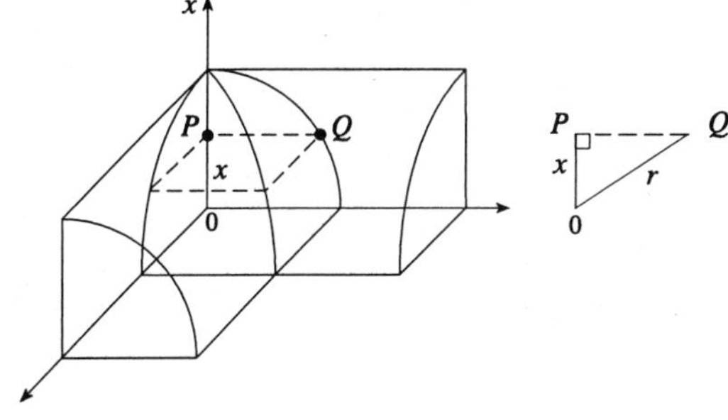

63. The volume is obtained by rotating the area common to two circles of radius r, as shown. The volume of the right half is

$$V_{\text{right}} = \pi \int_0^{r/2} \left[r^2 - \left(\tfrac{1}{2}r + x\right)^2\right] dx$$
$$= \pi\left[r^2 x - \tfrac{1}{3}\left(\tfrac{1}{2}r + x\right)^3\right]_0^{r/2}$$
$$= \pi\left[\left(\tfrac{1}{2}r^3 - \tfrac{1}{3}r^3\right) - \left(0 - \tfrac{1}{24}r^3\right)\right]$$
$$= \tfrac{5}{24}\pi r^3.$$

So by symmetry, the total volume is twice this, or $\frac{5}{12}\pi r^3$.

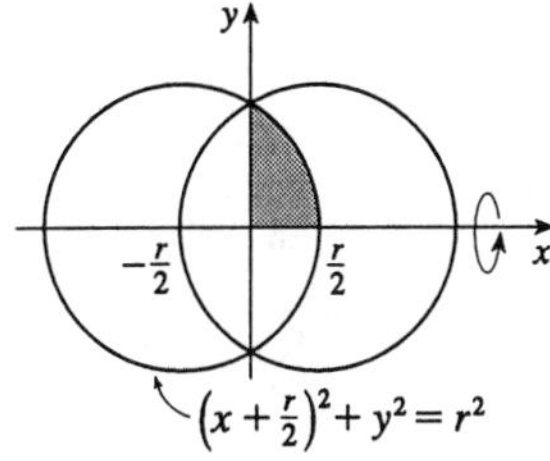

Alternate Solution: We observe that the volume is the twice the volume of a cap of a sphere, so we can use the formula from Exercise 49 with $h = \frac{1}{2}r$: $V = 2\pi\left(\frac{1}{2}r\right)^2\left(r - \dfrac{r/2}{3}\right) = \frac{5}{12}\pi r^3$.

64. We consider two cases: one in which the ball is not completely submerged and the other in which it is.

Case 1: $0 \le h \le 10$ The ball will not be completely submerged, and so a cross-section of the water parallel to the surface will be the shaded area shown in the first diagram. We can find the area of the cross-section at height x above the bottom of the bowl by

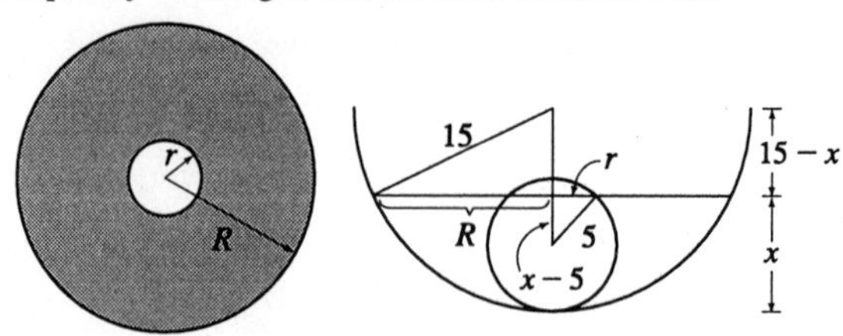

using the Pythagorean Theorem: $R^2 = 15^2 - (15-x)^2$ and $r^2 = 5^2 - (5-x)^2$, so $A(x) = \pi(R^2 - r^2) = 20\pi x$. The volume of water when it has depth h is then $V(h) = \int_0^h A(x)\,dx = \int_0^h 20\pi x\,dx = \left[10\pi x^2\right]_0^h = 10\pi h^2$, $0 \le h \le 10$.

Case 2: $10 < h \le 15$ In this case we can find the volume by simply subtracting the volume displaced by the ball from the total volume inside the bowl underneath the surface of the water. The total volume underneath the surface is just the volume of a cap of the bowl, so we use the formula from Exercise 49: $V_{\text{cap}}(h) = \pi h^2\left(15 - \frac{1}{3}h\right)$. The volume of the small sphere is $V_{\text{ball}} = \frac{4}{3}\pi(5)^3 = \frac{500}{3}\pi$, so the total volume is $V_{\text{cap}} - V_{\text{ball}} = \pi\left(15h^2 - \frac{1}{3}h^3 - \frac{500}{3}\right)\text{cm}^3$.

65. The cross-sections perpendicular to the y-axis in Figure 17 are rectangles. The rectangle corresponding to the coordinate y has a base of length $2\sqrt{16-y^2}$ in the xy-plane and a height of $\frac{1}{\sqrt{3}}y$, since $\angle BAC = 30°$ and $|BC| = \frac{1}{\sqrt{3}}|AB|$. Thus $A(y) = \frac{2}{\sqrt{3}}y\sqrt{16-y^2}$ and

$$V = \int_0^4 A(y)\,dy = \int_0^4 A(y)\,dy = \frac{2}{\sqrt{3}}\int_0^4 \sqrt{16-y^2}\,y\,dy$$
$$= \frac{2}{\sqrt{3}}\int_{16}^0 u^{1/2}\left(-\tfrac{1}{2}\,du\right) \quad [\text{Put } u = 16 - y^2, \text{ so } du = -2y\,dy]$$
$$= \frac{1}{\sqrt{3}}\int_0^{16} u^{1/2}\,du = \frac{1}{\sqrt{3}}\frac{2}{3}\left[u^{3/2}\right]_0^{16} = \frac{2}{3\sqrt{3}}(64) = \frac{128}{3\sqrt{3}}.$$

66. If the angle is 45°, then the rectangular cross-section perpendicular to the y-axis at coordinate y has height y (since $|BC| = |AB| = y$ in Figure 17), so $A(y) = 2y\sqrt{16-y^2}$ and $V = \int_0^4 A(y)dy = \int_0^4 2y\sqrt{16-y^2}\,dy = -\frac{2}{3}\left[\left(16-y^2\right)^{3/2}\right]_0^4 = \frac{128}{3}$. More generally, if the angle is θ, where $0 \le \theta < 90°$, then $|BC| = |AB|\tan\theta$, so $A(y) = (y\tan\theta)\sqrt{16-y^2}$ and $V = \int_0^4 A(y)\,dy = \frac{128}{3}\tan\theta$.

67. Take the x-axis to be the axis of the cylindrical hole of radius r. A quarter of the cross-section through y, perpendicular to the y-axis, is the rectangle shown. Using Pythagoras twice, we see that the dimensions of this rectangle are

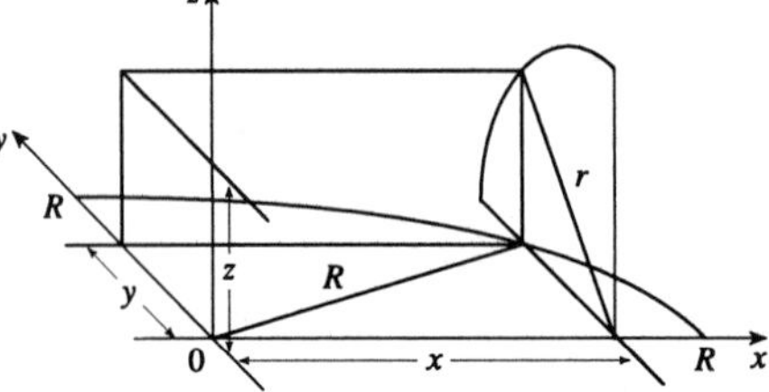

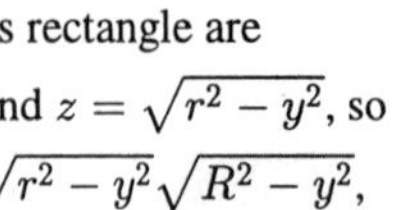

$x = \sqrt{R^2 - y^2}$ and $z = \sqrt{r^2 - y^2}$, so $\frac{1}{4}A(y) = xz = \sqrt{r^2-y^2}\sqrt{R^2-y^2}$,

and $V = \int_{-r}^{r} A(y)\,dy = \int_{-r}^{r} 4\sqrt{r^2-y^2}\sqrt{R^2-y^2}\,dy = 8\int_0^r \sqrt{r^2-y^2}\sqrt{R^2-y^2}\,dy$.

68. $V = 2\int_0^{\sqrt{R^2-r^2}} A(z)dz$

$= 2\int_0^{\sqrt{R^2-r^2}} [\pi(R^2 - z^2) - \pi r^2]dz$

$= 2\left[\pi R^2 z - \frac{1}{3}\pi z^3 - \pi r^2 z\right]_0^{\sqrt{R^2-r^2}}$

$= 2\pi(R^2 - r^2)\sqrt{R^2 - r^2} - \frac{2}{3}\pi(R^2 - r^2)^{3/2}$

$= \frac{4}{3}\pi(R^2 - r^2)^{3/2}$

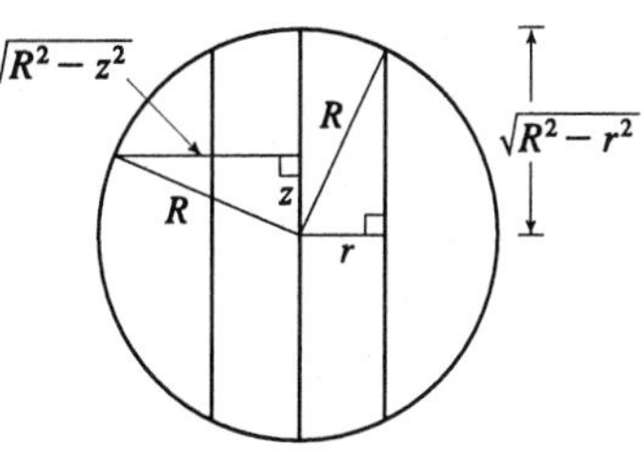

[Note that this is the same as the volume of a sphere whose radius is half the height of the solid obtained by drilling a hole of radius r vertically through the center of a sphere with radius R.]

69. **(a)** Volume$(S_1) = \int_0^h A(z)dz =$ Volume(S_2) since the cross-sectional area $A(z)$ at height z is the same for both solids.

(b) By Cavalieri's Principle, the volume of the cylinder in the figure is the same as that of a right circular cylinder with radius r and height h, that is, $\pi r^2 h$.

70. Use the Midpoint Rule with $\Delta x = 2$, $n = 5$ and $x_i^* = (2i + 1)$, where $i = 0, 1, 2, 3, 4$:
$V \approx 2(0.65 + 0.61 + 0.59 + 0.55 + 0.50) = 5.80\,\text{m}^3$.

71. **(a)** The radius of the barrel is the same at each end by symmetry, since the function $y = R - cx^2$ is even. Since the barrel is obtained by rotating the function y about the x-axis, this radius is equal to the value of y at $x = \frac{1}{2}h$, which is $R - c\left(\frac{1}{2}h\right)^2 = R - d = r$.

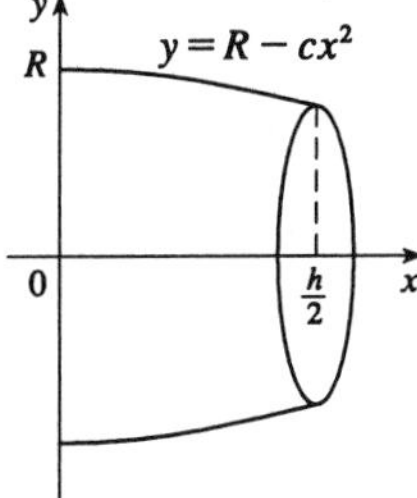

(b) The barrel is symmetric about the y-axis, so its volume is twice the volume of that part of the barrel for $x > 0$. Also, the barrel is a volume of rotation, so we use Formula 2:

$V = 2\int_0^{h/2} \pi(y^2)dx = 2\pi\int_0^{h/2}\left(R - cx^2\right)^2 dx = 2\pi\left[R^2x - \frac{2}{3}Rcx^3 + \frac{1}{5}c^2x^5\right]_0^{h/2}$

$= 2\pi\left(\frac{1}{2}R^2h - \frac{1}{12}Rch^3 + \frac{1}{160}c^2h^5\right)$.

Trying to make this look more like the expression we want, we rewrite it as
$V = \frac{1}{3}\pi h\left[2R^2 + \left(R^2 - \frac{1}{2}Rch^2 + \frac{3}{80}c^2h^4\right)\right]$. But
$R^2 - \frac{1}{2}Rch^2 + \frac{3}{80}c^2h^4 = \left(R - \frac{1}{4}ch^2\right)^2 - \frac{1}{40}c^2h^4 = (R - d)^2 - \frac{2}{5}\left(\frac{1}{4}ch^2\right)^2 = r^2 - \frac{2}{5}d^2$. Substituting this back into V, we see that $V = \frac{1}{3}\pi h\left(2R^2 + r^2 - \frac{2}{5}d^2\right)$, as required.

72. It suffices to consider the case where $\mathcal{R}$ is bounded by the curves $y = f(x)$ and $y = g(x)$ for $a \le x \le b$, where $g(x) \le f(x)$ for all x in $[a, b]$, since other regions can be decomposed into subregions of this type. We are concerned with the volume obtained when $\mathcal{R}$ is rotated about the line $y = -k$. By Formula 4, this is equal to
$V_2 = \pi\int_a^b\left([f(x) + k]^2 - [g(x) + k]^2\right)dx = \pi\int_a^b\left([f(x)]^2 - [g(x)]^2\right)dx + 2\pi k\int_a^b[f(x) - g(x)]dx = V_1 + 2\pi kA$.

73. We are given that the rate of change of the volume of water is $dV/dt = -kA(x)$, where k is some positive constant and $A(x)$ is the area of the surface when the water has depth x. Now we are concerned with the rate of change of the depth of the water with respect to time, that is, $\frac{dx}{dt}$. But by the Chain Rule, $\frac{dV}{dt} = \frac{dV}{dx}\frac{dx}{dt}$, so the first equation can be written $\frac{dV}{dx}\frac{dx}{dt} = -kA(x)$ (★). Also, we know that the total volume of water up to a depth x is $V(x) = \int_0^x A(s)\,ds$, where $A(s)$ is the area of a cross-section of the water at a depth s. Differentiating this equation with respect to x, we get $dV/dx = A(x)$. Substituting this into equation ★, we get $A(x)(dx/dt) = -kA(x) \quad \Rightarrow \quad dx/dt = -k$, a constant.

EXERCISES 5.3

1. $V = \int_1^2 2\pi x \cdot x^2\,dx = 2\pi \int_1^2 x^3\,dx = 2\pi\left[\frac{1}{4}x^4\right]_1^2 = 2\pi\left(\frac{15}{4}\right) = \frac{15}{2}\pi$

2. $V = \int_1^{10} 2\pi x(1/x)dx = 2\pi[x]_1^{10} = 20\pi - 2\pi = 18\pi$

3. $V = \int_0^4 2\pi x\sqrt{4+x^2}\,dx = \pi\int_0^4 \sqrt{x^2+4}\,2x\,dx = \left[\pi\,\frac{2}{3}(x^2+4)^{3/2}\right]_0^4$
$= \left(\frac{2}{3}\pi\right)\left(20\sqrt{20} - 8\right) = \frac{16}{3}\pi\left(5\sqrt{5} - 1\right)$

4. $V = \int_0^{\sqrt{\pi}} 2\pi x\sin(x^2)dx = \pi[-\cos(x^2)]_0^{\sqrt{\pi}} = \pi[1 - (-1)] = 2\pi$

5. $V = \int_0^2 2\pi x(4 - x^2)dx = 2\pi\int_0^2(4x - x^3)dx = 2\pi\left[2x^2 - \frac{1}{4}x^4\right]_0^2 = 2\pi(8 - 4) = 8\pi$

Note: If we integrated from -2 to 2, we would be generating the volume twice.

6. The curves $y^2 = x$ and $x = 2y$ intersect at $(0,0)$ and $(4,2)$. $V = \int_0^4 2\pi x\left(\sqrt{x} - \frac{1}{2}x\right)dx$
$= 2\pi\int_0^4 x^{3/2}\,dx - \pi\int_0^4 x^2\,dx = 2\pi\left[\frac{2}{5}x^{5/2}\right]_0^4 - \pi\left[\frac{1}{3}x^3\right]_0^4 = \frac{4}{5}\pi(32) - \frac{64}{3}\pi = \frac{64}{15}\pi$

7. $V = \int_0^1 2\pi x(x^2 - x^3)dx = 2\pi\int_0^1(x^3 - x^4)dx = 2\pi\left[\frac{1}{4}x^4 - \frac{1}{5}x^5\right]_0^1 = 2\pi\left(\frac{1}{4} - \frac{1}{5}\right) = \frac{1}{10}\pi$

8. The two curves intersect at $(2,2)$ and $(4,2)$. For $2 < x < 4$, $-x^2 + 6x - 6 > x^2 - 6x + 10$. [To see this, just notice that $(x^2 - 6x + 10) - (-x^2 + 6x - 6) = 2x^2 - 12x + 16 = 2(x-2)(x-4) < 0$ for $2 < x < 4$.] Thus
$V = \int_2^4 2\pi x[(-x^2 + 6x - 6) - (x^2 - 6x + 10)]dx = 2\pi\int_2^4 x(-2x^2 + 12x - 16)dx$
$= 4\pi\int_2^4(-x^3 + 6x^2 - 8x)dx = 4\pi\left[-\frac{1}{4}x^4 + 2x^3 - 4x^2\right]_2^4$
$= 4\pi[(-64 + 128 - 64) - (-4 + 16 - 16)] = 16\pi.$

9. $V = \int_0^{16} 2\pi y\sqrt[4]{y}\,dy = 2\pi\int_0^{16} y^{5/4}\,dy = 2\pi\left[\frac{4}{9}y^{9/4}\right]_0^{16} = \frac{8}{9}\pi(512 - 0) = \frac{4096}{9}\pi$

10. $V = \int_2^5 2\pi y \cdot y^2\,dy = 2\pi\left[\frac{1}{4}y^4\right]_2^5 = \frac{\pi}{2}(625 - 16) = \frac{609}{2}\pi$

11. $V = \int_0^9 2\pi y \cdot 2\sqrt{y}\,dy = 4\pi\int_0^9 y^{3/2}\,dy = 4\pi\left[\frac{2}{5}y^{5/2}\right]_0^9 = \frac{8}{5}\pi(243 - 0) = \frac{1944}{5}\pi$

12. The two curves intersect at $(0,0)$ and $(0,6)$, so
$V=\int_0^6 2\pi y(-y^2+6y)dy=2\pi\left[-\frac{1}{4}y^4+2y^3\right]_0^6=2\pi(-324+432)=216\pi.$

13. $V=\int_0^1 2\pi y[(2-y)-y^2]dy=2\pi\left[y^2-\frac{1}{3}y^3-\frac{1}{4}y^4\right]_0^1=2\pi\left(1-\frac{1}{3}-\frac{1}{4}\right)=\frac{5}{6}\pi$

14. $V=\int_0^1 2\pi y[(2-y)-y]dy=4\pi\int_0^1 y(1-y)dy=4\pi\left[\frac{1}{2}y^2-\frac{1}{3}y^3\right]_0^1=4\pi\left(\frac{1}{6}\right)=\frac{2}{3}\pi$

15. $V=\int_1^4 2\pi x\sqrt{x}\,dx=2\pi\int_1^4 x^{3/2}\,dx$
$=2\pi\left[\frac{2}{5}x^{5/2}\right]_1^4=\frac{4}{5}\pi(32-1)=\frac{124}{5}\pi$

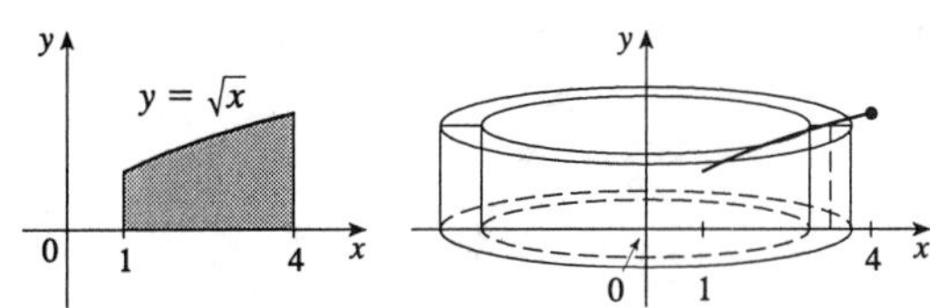

16. $V=\int_{-2}^{-1} 2\pi(-x)\cdot x^2\,dx=2\pi\left[-\frac{1}{4}x^4\right]_{-2}^{-1}$
$=2\pi\left[\left(-\frac{1}{4}\right)-(-4)\right]=\frac{15}{2}\pi$

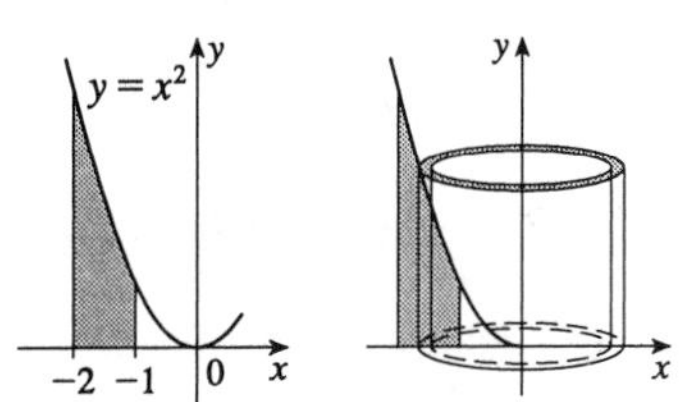

17. $V=\int_1^2 2\pi(x-1)\,x^2\,dx=2\pi\left[\frac{1}{4}x^4-\frac{1}{3}x^3\right]_1^2$
$=2\pi\left[\left(4-\frac{8}{3}\right)-\left(\frac{1}{4}-\frac{1}{3}\right)\right]=\frac{17}{6}\pi$

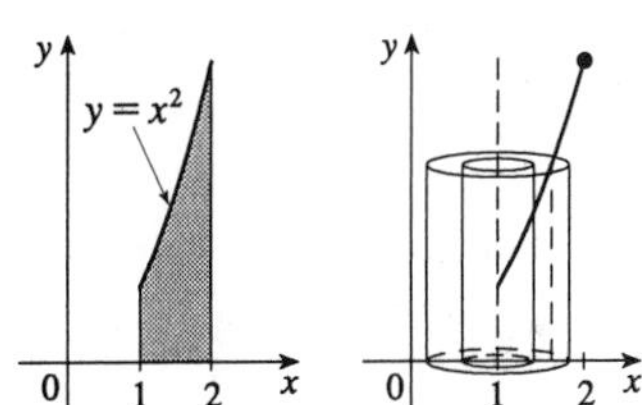

18. $V=\int_1^2 2\pi(4-x)x^2\,dx=2\pi\left[\frac{4}{3}x^3-\frac{1}{4}x^4\right]_1^2$

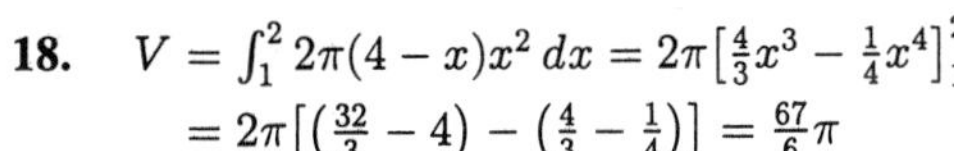
$=2\pi\left[\left(\frac{32}{3}-4\right)-\left(\frac{4}{3}-\frac{1}{4}\right)\right]=\frac{67}{6}\pi$

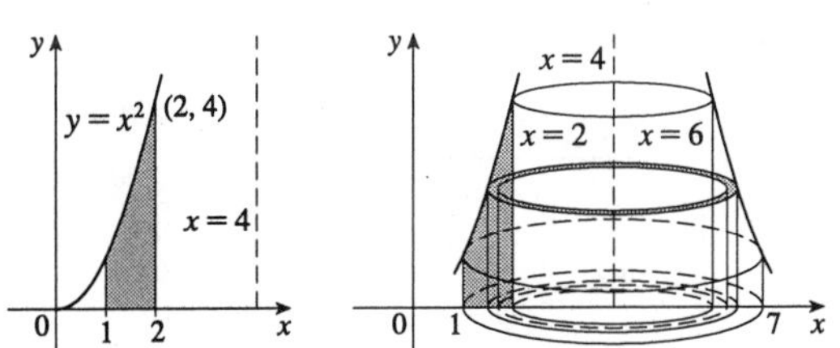

19. $V=\int_0^2 2\pi(3-y)(5-x)dy$
$=\int_0^2 2\pi(3-y)(5-y^2-1)dy$
$=\int_0^2 2\pi(12-4y-3y^2+y^3)\,dy$
$=2\pi\left[12y-2y^2-y^3+\frac{1}{4}y^4\right]_0^2$
$=2\pi(24-8-8+4)=24\pi$

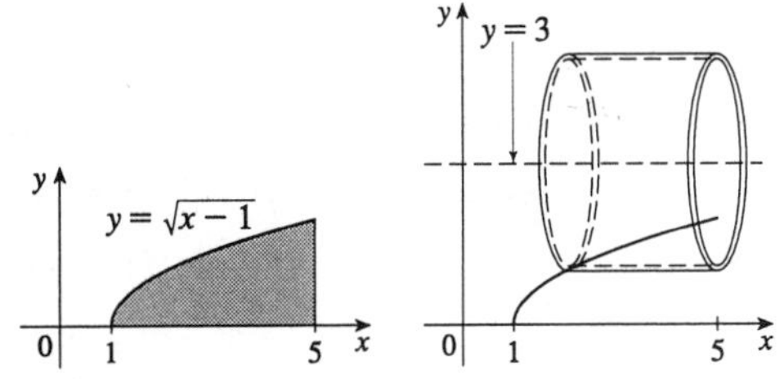

20. $V = \int_0^4 2\pi(2+x)[(8x-2x^2)-(4x-x^2)]dx$
$= \int_0^4 2\pi(2+x)(4x-x^2)dx$
$= 2\pi\int_0^4 (8x+2x^2-x^3)dx$
$= 2\pi\left[4x^2+\frac{2}{3}x^3-\frac{1}{4}x^4\right]_0^4$
$= 2\pi\left(64+\frac{128}{3}-64\right) = \frac{256}{3}\pi$

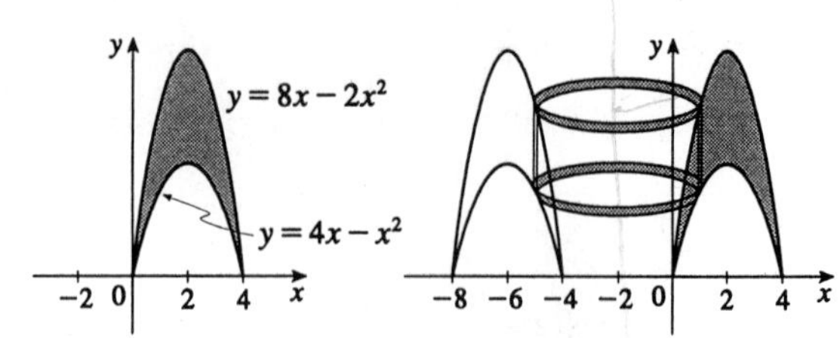

21. $V = \int_{2\pi}^{3\pi} 2\pi x \sin x\, dx$

22. $V = \int_0^3 2\pi x \dfrac{1}{1+x^2}\, dx = 2\pi \int_0^3 \dfrac{x}{1+x^2}\, dx$

23. $V = \int_0^{\pi/4} 2\pi y \cos y\; dy$

24. $-x^2+7x-10 = x-2 \;\Leftrightarrow\; x^2-6x+8=0 \;\Leftrightarrow\; x=2$ or $4 \;\Leftrightarrow\; (x,y)=(2,0)$ or $(4,2)$. Use washers:
$$V = \pi\int_2^4 \left[\left(-x^2+7x-10\right)^2-(x-2)^2\right]dx = \pi\int_2^4 \left(x^4-14x^3+68x^2-136x+96\right)dx$$

25. $V = \int_0^1 2\pi(x+1)\left(\sin\frac{\pi}{2}x - x^4\right)dx$

26. $V = \int_{-2}^2 2\pi(5-y)[(8-2y^2)-(4-y^2)]dy = \int_{-2}^2 2\pi(5-y)(4-y^2)dy$

27. The solid is obtained by rotating the region bounded by the curve $y=\cos x$ and the line $y=0$, from $x=0$ to $x=\frac{\pi}{2}$, about the y-axis.

28. The solid is obtained by rotating the region bounded by the curve $x=\sqrt{y}$ and the lines $y=9$ and $x=0$ about the x-axis.

29. The solid is obtained by rotating the region in the first quadrant bounded by the curves $y=x^2$ and $y=x^6$ about the y-axis.

30. The solid is obtained by rotating the region under the curve $y=\sin^4 x$, above $y=0$, from $x=0$ to $x=\pi$, about the line $x=4$.

31.

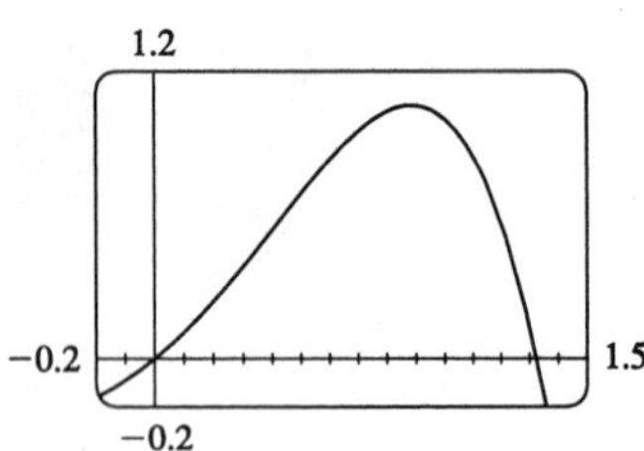

From the graph, it appears that the curves intersect at $x=0$ and at $x\approx 1.32$, with $x+x^2-x^4>0$ on $(0,1.32)$. So the volume of the solid obtained by rotating the region about the y-axis is

$V \approx 2\pi\int_0^{1.32} x\left(x+x^2-x^4\right)dx$
$= 2\pi\left[\frac{1}{3}x^3+\frac{1}{4}x^4-\frac{1}{6}x^6\right]_0^{1.32} \approx 4.05.$

32.

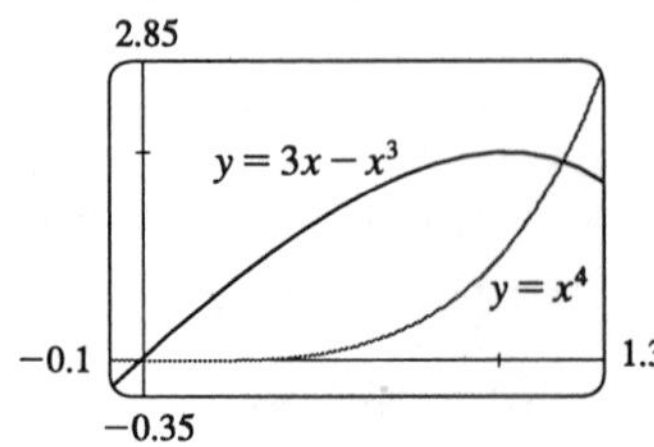

From the graph, it appears that the curves intersect at $x=0$ and at $x\approx 1.17$, with $3x-x^3>x^4$ on $(0,1.17)$. So the volume of the solid obtained by rotation about the y-axis is

$V \approx 2\pi\int_0^{1.17} x\left[(3x-x^3)-x^4\right]dx$
$= 2\pi\left[x^3-\frac{1}{5}x^5-\frac{1}{6}x^6\right]_0^{1.17} \approx 4.62.$

33. Use disks: $V = \int_{-2}^{1} \pi(x^2+x-2)^2 dx = \pi\int_{-2}^{1}(x^4+2x^3-3x^2-4x+4)\,dx$
$= \pi\left[\frac{1}{5}x^5+\frac{1}{2}x^4-x^3-2x^2+4x\right]_{-2}^{1} = \pi\left[\left(\frac{1}{5}+\frac{1}{2}-1-2+4\right)-\left(-\frac{32}{5}+8+8-8-8\right)\right]$
$= \pi\left(\frac{33}{5}+\frac{3}{2}\right) = \frac{81}{10}\pi$

34. Use shells: $V = \int_1^2 2\pi x(-x^2+3x-2)dx = 2\pi\int_1^2(-x^3+3x^2-2x)dx$
$= 2\pi\left[-\frac{1}{4}x^4+x^3-x^2\right]_1^2 = 2\pi\left[(-4+8-4)-\left(-\frac{1}{4}+1-1\right)\right] = \frac{1}{2}\pi$

35. Use disks: $V = \pi\int_{-1}^{1}(1-y^2)^2dy = 2\pi\int_0^1(y^4-2y^2+1)\,dy = 2\pi\left[\frac{1}{5}y^5-\frac{2}{3}y^3+y\right]_0^1 = 2\pi\left(\frac{1}{5}-\frac{2}{3}+1\right) = \frac{16}{15}\pi$

36. Use shells: $V = \displaystyle\int_0^2 2\pi x^2\sqrt{1+x^3}\,dx = 2\pi\left[\frac{1}{3}\cdot\frac{2}{3}(1+x^3)^{3/2}\right]_0^2 = \left(\frac{4}{9}\pi\right)(27-1) = \frac{104}{9}\pi$

37. Use disks: $V = \pi\displaystyle\int_0^2\left[\sqrt{1-(y-1)^2}\right]^2dy = \pi\int_0^2(2y-y^2)dy = \pi\left[y^2-\frac{1}{3}y^3\right]_0^2 = \pi\left(4-\frac{8}{3}\right) = \frac{4}{3}\pi$

38. Using shells, we have $V = 2\pi\displaystyle\int_0^2 y\cdot 2\sqrt{1-(y-1)^2}\,dy = 4\pi\int_{-1}^{1}(u+1)\sqrt{1-u^2}\,du$ [Put $u = y-1$]
$= 4\pi\int_{-1}^{1}\sqrt{1-u^2}\,du - 4\pi\int_{-1}^{1}u\sqrt{1-u^2}\,du$. The first definite integral is the area of a semicircle of radius 1, that is, $\frac{\pi}{2}$. The second equals zero because its integrand is an odd function. Thus $V = 4\pi\frac{\pi}{2}-4\pi\cdot 0 = 2\pi^2$.

39. $V = 2\int_0^r 2\pi x\sqrt{r^2-x^2}\,dx = -2\pi\int_0^r(r^2-x^2)^{1/2}(-2x)dx = \left[-2\pi\cdot\frac{2}{3}(r^2-x^2)^{3/2}\right]_0^r = -\frac{4}{3}\pi(0-r^3) = \frac{4}{3}\pi r^3$

40. $V = \displaystyle\int_{R-r}^{R+r} 2\pi x\cdot 2\sqrt{r^2-(x-R)^2}\,dx$
$= \displaystyle\int_{-r}^{r} 4\pi(u+R)\sqrt{r^2-u^2}\,du$ [Put $u = x-R$]
$= 4\pi R\displaystyle\int_{-r}^{r}\sqrt{r^2-u^2}\,du + 4\pi\int_{-r}^{r}u\sqrt{r^2-u^2}\,du$

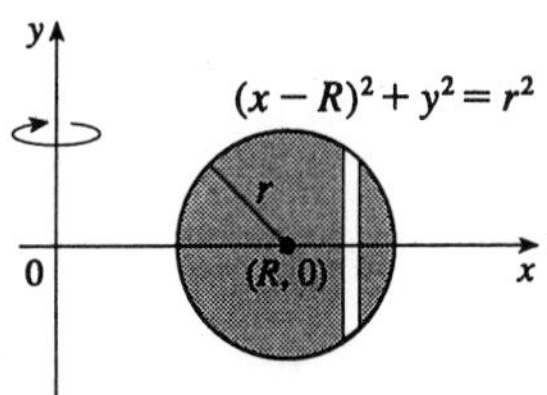

The first integral is the area of a semicircle of radius r, that is, $\frac{1}{2}\pi r^2$, and the second is zero since the integrand is an odd function. Thus $V = 4\pi R\left(\frac{1}{2}\pi r^2\right)+4\pi\cdot 0 = 2\pi Rr^2$.

41. $V = 2\pi\displaystyle\int_0^r x\left(-\frac{h}{r}x+h\right)dx = 2\pi h\int_0^r\left(-\frac{x^2}{r}+x\right)dx = 2\pi h\left[-\frac{x^3}{3r}+\frac{x^2}{2}\right]_0^r = 2\pi h\frac{r^2}{6} = \frac{\pi r^2 h}{3}$

42. **(b)** By symmetry, the volume of a napkin ring obtained by drilling a hole of radius r through a sphere with radius R is twice the volume obtained by rotating the area above the x-axis and below the curve $y = \sqrt{R^2-x^2}$ (the equation of the top half of the cross-section of the sphere), between $x = r$ and $x = R$, about the y-axis. By Formula 2, this is equal to $2\cdot 2\pi\displaystyle\int_r^R x\sqrt{R^2-x^2}\,dx = 4\pi\left[-\frac{1}{3}(R^2-x^2)^{3/2}\right]_r^R$ (Put $u = R^2-x^2$) $= \frac{4}{3}\pi(R^2-r^2)^{3/2}$. But by the Pythagorean Theorem, $R^2-r^2 = \left(\frac{1}{2}h\right)^2$, so the volume of the napkin ring is $\frac{4}{3}\pi\left(\frac{1}{2}h\right)^3 = \frac{1}{6}\pi h^3$, which is independent of both R and r; that is, the amount of wood in a napkin ring of height h is the same regardless of the size of the sphere used. Note that most of this calculation has been done already, but with more difficulty, in Exercise 5.2.68.

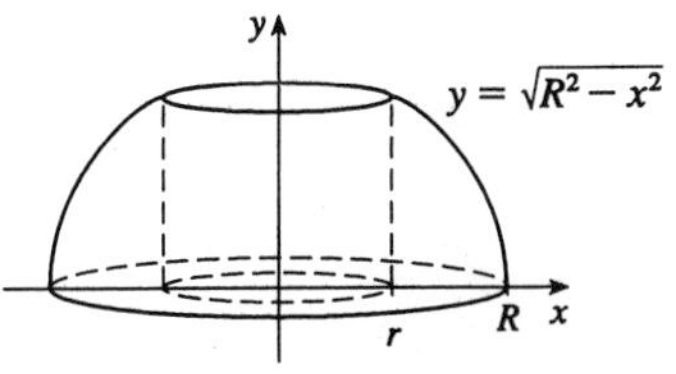

43. If $a < b \le 0$, then a typical cylindrical shell has radius $-x$ and height $f(x)$, so
$V = \int_a^b 2\pi(-x)f(x)dx = -\int_a^b 2\pi x f(x)dx$.

44. 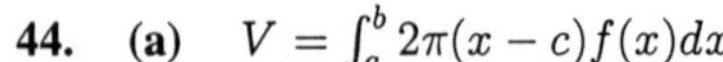
(a) $V = \int_a^b 2\pi(x-c)f(x)dx$ **(b)** $V = \int_a^b 2\pi(c-x)f(x)dx$

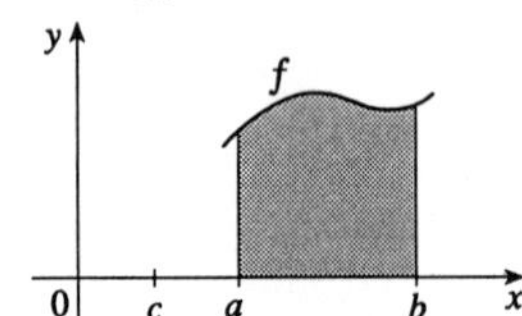

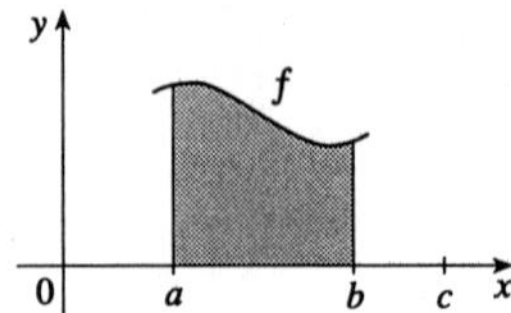

45. $\Delta x = \dfrac{\pi/4 - 0}{4} = \dfrac{\pi}{16}$.
$V = \int_0^{\pi/4} 2\pi x \tan x\, dx \approx 2\pi \cdot \frac{\pi}{16}\left(\frac{\pi}{32}\tan\frac{\pi}{32} + \frac{3\pi}{32}\tan\frac{3\pi}{32} + \frac{5\pi}{32}\tan\frac{5\pi}{32} + \frac{7\pi}{32}\tan\frac{7\pi}{32}\right) \approx 1.142$

46. $\Delta x = \dfrac{12-2}{5} = 2$, $n = 5$ and $x_i^* = 2 + (2i+1)$, where $i = 0, 1, 2, 3, 4$. The values of $f(x)$ are taken directly from the diagram.
$V = \int_2^{12} 2\pi x\, f(x)\, dx \approx 2\pi[3f(3) + 5f(5) + 7f(7) + 9f(9) + 11f(11)] \cdot 2$
$\approx 2\pi[3(2) + 5(4) + 7(4) + 9(2) + 11(1)]2 = 332\pi$

EXERCISES 5.4

1. By Equation 2, $W = Fd = (900)(8) = 7200\,\text{J}$.

2. $F = mg = (60)(9.8) = 588\,\text{N}$; $W = Fd = 588 \cdot 2 = 1176\,\text{J}$

3. By Equation 4, $W = \int_a^b f(x)\,dx = \int_0^{10}(5x^2+1)dx = \left[\frac{5}{3}x^3 + x\right]_0^{10} = \frac{5000}{3} + 10 = \frac{5030}{3}$ ft-lb.

4. $W = \int_1^2 \cos\left(\frac{1}{3}\pi x\right)dx = \frac{3}{\pi}\int_{\pi/3}^{2\pi/3} \cos u\,du$ $\left[\text{Put } u = \frac{1}{3}\pi x,\ du = \frac{1}{3}\pi\,dx\right] = \frac{3}{\pi}[\sin u]_{\pi/3}^{2\pi/3} = \frac{3}{\pi}\left(\frac{\sqrt{3}}{2} - \frac{\sqrt{3}}{2}\right) = 0\,\text{J}$.

Interpretation: From $x = 1$ to $x = \frac{3}{2}$, the force does work equal to $\int_1^{3/2}\cos\left(\frac{1}{3}\pi x\right)dx = \frac{3}{\pi}\left(1 - \frac{\sqrt{3}}{2}\right)$ J in accelerating the particle and increasing its kinetic energy. From $x = \frac{3}{2}$ to $x = 2$, the force opposes the motion of the particle, decreasing its kinetic energy. This is negative work, equal in magnitude but opposite in sign to the work done from $x = 1$ to $x = \frac{3}{2}$.

5. $10 = f(x) = kx = \frac{1}{3}k$ (4 inches $= \frac{1}{3}$ foot), so $k = 30$ (The units for k are pounds per foot.)
Now 6 inches $= \frac{1}{2}$ foot, so $W = \int_0^{1/2} 30x\,dx = [15x^2]_0^{1/2} = \frac{15}{4}$ ft-lb.

6. $25 = f(x) = kx = k(0.1)$ (10 cm $= 0.1$ m), so $k = 250\,\text{N/m}$. Thus $f(x) = 250x$ and the work required is
$W = \int_0^{0.05} 250x\,dx = [125x^2]_0^{0.05} = 125(0.0025) = 0.3125 \approx 0.31\,\text{J}$.

7. If $\int_0^{0.12} kx\,dx = 2\,\text{J}$, then $2 = \left[\frac{1}{2}kx^2\right]_0^{0.12} = \frac{1}{2}k(0.0144) = 0.0072k$ and $k = \frac{2}{0.0072} = \frac{2500}{9} \approx 277.78$. Thus the work needed to stretch the spring from 35 cm to 40 cm is

$\int_{0.05}^{0.10} \frac{2500}{9}x\,dx = \left[\frac{1250}{9}x^2\right]_{1/20}^{1/10} = \frac{1250}{9}\left(\frac{1}{100} - \frac{1}{400}\right) = \frac{25}{24} \approx 1.04\,\text{J}$.

8. If $12 = \int_0^1 kx\,dx = \left[\frac{1}{2}kx^2\right]_0^1 = \frac{1}{2}k$, then $k = 24$ and the work required is

$\int_0^{3/4} 24x\,dx = \left[12x^2\right]_0^{3/4} = 12 \cdot \frac{9}{16} = \frac{27}{4} = 6.75$ ft-lb.

9. $f(x) = kx$, so $30 = \frac{2500}{9}x$ and $x = \frac{270}{2500}\,\text{m} = 10.8\,\text{cm}$.

10. Let L be the natural length of the spring in meters. Then

$6 = \int_{0.10-L}^{0.12-L} kx\,dx = \left[\frac{1}{2}kx^2\right]_{0.10-L}^{0.12-L} = \frac{1}{2}k\left[(0.12-L)^2 - (0.10-L)^2\right]$ and

$10 = \int_{0.12-L}^{0.14-L} kx\,dx = \left[\frac{1}{2}kx^2\right]_{0.12-L}^{0.14-L} = \frac{1}{2}k\left[(0.14-L)^2 - (0.12-L)^2\right]$. In other words,

$12 = k(0.0044 - 0.04L)$ and $20 = k(0.0052 - 0.04L)$. Subtracting the first equation from the second gives $8 = 0.0008k$, so $k = 10{,}000$. Now the second equation becomes $20 = 52 - 400L$, so $L = \frac{32}{400}\,\text{m} = 8\,\text{cm}$.

11. First notice that the exact height of the building does not matter. The portion of the rope from x ft to $(x + \Delta x)$ft below the top of the building weighs $\frac{1}{2}\Delta x$ lb and must be lifted x ft (approximately), so its contribution to the total work is $\frac{1}{2}x\Delta x$ ft-lb. The total work is $W = \int_0^{50} \frac{1}{2}x\,dx = \left[\frac{1}{4}x^2\right]_0^{50} = \frac{2500}{4} = 625$ ft-lb.

12. The cable weighs 1.5 lb/ft. Each part of the top 10 ft of cable is lifted a distance equal to its distance from the top. The remaining 30 ft of cable is lifted 10 ft. Thus

$W = \int_0^{10} \frac{3}{2}x\,dx + \int_{10}^{40} \frac{3}{2} \cdot 10\,dx = \left[\frac{3}{4}x^2\right]_0^{10} + \left[15x\right]_{10}^{40} = \frac{3}{4}(100) + 15(30) = 75 + 450 = 525$ ft-lb.

13. The work needed to lift the cable is $\int_0^{500} 2x\,dx = x^2\big|_0^{500} = 250{,}000$ ft-lb. The work needed to lift the coal is $800\,\text{lb} \cdot 500\,\text{ft} = 400{,}000$ ft-lb. Thus the total work required is $250{,}000 + 400{,}000 = 650{,}000$ ft-lb.

14. The work needed to lift the bucket itself is $4\,\text{lb} \cdot 80\,\text{ft} = 320$ ft-lb. At time t (in seconds) the bucket is $2t$ ft above its original 80 ft depth, but it now holds only $(40 - 0.2t)$ lb of water. In terms of distance, the bucket holds $\left(40 - \frac{1}{10}x\right)$ lb of water when it is x ft above its original 80 ft depth. Moving this amount of water a distance Δx requires $\left(40 - \frac{1}{10}x\right)\Delta x$ ft-lb of work. Thus the work needed to lift the water is

$\int_0^{80}\left(40 - \frac{1}{10}x\right)dx = \left[40x - \frac{1}{20}x^2\right]_0^{80} = (3200 - 320)$ ft-lb. Adding in the work of lifting the bucket gives a total of 3200 ft-lb of work.

15. A "slice" of water Δx m thick and lying at a depth of x m (where $0 \le x \le \frac{1}{2}$) has volume $2\Delta x\,\text{m}^3$, a mass of $2000\Delta x$ kg, weighs about $(9.8)(2000\Delta x) = 19{,}600\,\Delta x$ N, and thus requires about $19{,}600x\Delta x$ J of work for its removal. So $W \approx \int_0^{1/2} 19{,}600x\,dx = 9800x^2\big|_0^{1/2} = 2450\,\text{J}$.

16. A horizontal cylindrical slice of water Δx ft thick has a volume of $\pi r^2 h = \pi \cdot 12^2 \cdot \Delta x\,\text{ft}^3$ and weighs about $\left(62.5\,\text{lb}/\text{ft}^3\right)\left(144\pi\Delta x\,\text{ft}^3\right) = 9000\pi\Delta x$ lb. If the slice lies x ft below the edge of the pool (where $1 \le x \le 5$), then the work needed to pump it out is about $9000\pi x\Delta x$. Thus

$W \approx \int_1^5 9000\pi x\,dx = \left[4500\pi x^2\right]_1^5 = 4500\pi(25 - 1) = 108{,}000\pi$ ft-lb.

17. A "slice" of water Δx m thick and lying x ft above the bottom has volume $8x\Delta x$ m^3 and weighs about $(9.8\times 10^3)(8x\Delta x)$ N. It must be lifted $(5-x)$ m by the pump, so the work needed is about $(9.8\times 10^3)(5-x)(8x\Delta x)$ J. The total work required is

$$\begin{aligned} W &\approx \textstyle\int_0^3 (9.8\times 10^3)(5-x)8x\,dx = (9.8\times 10^3)\int_0^3 (40x-8x^2)dx \\ &= (9.8\times 10^3)\left[20x^2-\tfrac{8}{3}x^3\right]_0^3 = (9.8\times 10^3)(180-72) = (9.8\times 10^3)(108) = 1058.4\times 10^3 \\ &\approx 1.06\times 10^6 \text{ J.} \end{aligned}$$

18. For convenience, measure depth x from the middle of the tank, so that $-1.5\le x\le 1.5$ m. Lifting a slice of water of thickness Δx at depth x requires a work contribution of

$$\Delta W \approx (9.8\times 10^3)\left(2\sqrt{(1.5)^2-x^2}\right)(6\Delta x)(2.5+x), \text{ so}$$

$$\begin{aligned} W &\approx \textstyle\int_{-1.5}^{1.5}(9.8\times 10^3)12\sqrt{2.25-x^2}(2.5+x)dx \\ &= (9.8\times 10^3)\left[60\int_0^{3/2}\sqrt{\tfrac{9}{4}-x^2}\,dx + 12\int_{-3/2}^{3/2} x\sqrt{\tfrac{9}{4}-x^2}\,dx\right] \end{aligned}$$

The second integral vanishes because its integrand is an odd function, and the first integral represents the area of a quarter-circle of radius $\frac{3}{2}$. Therefore

$$W \approx (9.8\times 10^3)60\textstyle\int_0^{3/2}\sqrt{\tfrac{9}{4}-x^2}\,dx = (9.8\times 10^3)(60)\left(\tfrac{9}{8}\right)\left(\tfrac{\pi}{2}\right) = 330{,}750\pi \approx 1.04\times 10^6 \text{ J.}$$

19. Measure depth x downward from the flat top of the tank, so that $0\le x\le 2$ ft. Then

$$\Delta W = (62.5)\left(2\sqrt{4-x^2}\right)(8\Delta x)(x+1) \text{ ft-lb, so}$$

$$\begin{aligned} W &\approx (62.5)(16)\textstyle\int_0^2 (x+1)\sqrt{4-x^2}\,dx = 1000\left(\int_0^2 x\sqrt{4-x^2}\,dx + \int_0^2\sqrt{4-x^2}\,dx\right) \\ &= 1000\left[\textstyle\int_0^4 u^{1/2}\left(\tfrac{1}{2}\,du\right) + \tfrac{1}{4}\pi(2^2)\right] \qquad (\text{Put } u = 4-x^2, \text{ so } du = -2x\,dx) \\ &= 1000\left(\tfrac{1}{2}\cdot\tfrac{2}{3}u^{3/2}\Big|_0^4 + \pi\right) = 1000\left(\tfrac{8}{3}+\pi\right) \approx 5.8\times 10^3 \text{ ft-lb.} \end{aligned}$$

Note: The second integral represents the area of a quarter-circle of radius 2.

20. Let x be depth in feet, so that $0\le x\le 5$. Then $\Delta W = (62.5)\pi\left(\sqrt{5^2-x^2}\right)^2\Delta x\cdot x$ ft-lb and

$$W \approx 62.5\pi\textstyle\int_0^5 x(25-x^2)dx = 62.5\pi\left[\tfrac{25}{2}x^2-\tfrac{1}{4}x^4\right]_0^5 = 62.5\pi\left(\tfrac{625}{2}-\tfrac{625}{4}\right) = 62.5\pi\left(\tfrac{625}{4}\right) \approx 3.07\times 10^4 \text{ ft-lb.}$$

21. If only 4.7×10^5 J of work is done, then only the water above a certain level (call it h) will be pumped out. So we use the same formula as in Exercise 17, except that the work is fixed, and we are trying to find the lower limit of integration: $4.7\times 10^5 \approx \int_h^3 (9.8\times 10^3)(5-x)8x\,dx = (9.8\times 10^3)\left[20x^2-\frac{8}{3}x^3\right]_h^3 \Leftrightarrow$

$\frac{4.7}{9.8}\times 10^2 \approx 48 = \left(20\cdot 3^2 - \frac{8}{3}\cdot 3^3\right) - \left(20h^2 - \frac{8}{3}h^3\right) \quad\Leftrightarrow\quad 2h^3-15h^2+45=0.$

To find the solution of this equation, we plot $2h^3-15h^2+45$ between $h=0$ and $h=3$. We see that the equation is satisfied for $h\approx 2.0$. So the depth of water remaining in the tank is about 2.0 m.

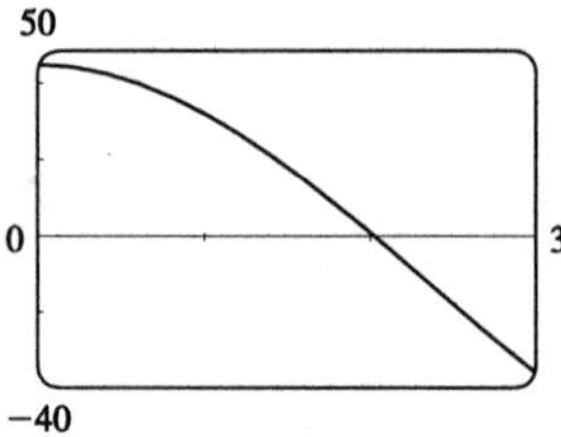

22. $W \approx (9.8 \times 920)\int_0^{3/2} 12\sqrt{\frac{9}{4} - x^2}\left(\frac{5}{2} + x\right)dx = 9016\left[30\int_0^{3/2}\sqrt{\frac{9}{4} - x^2}\,dx + 12\int_0^{3/2} x\sqrt{\frac{9}{4} - x^2}\,dx\right]$

Here $\displaystyle\int_0^{3/2}\sqrt{\tfrac{9}{4} - x^2}\,dx = \tfrac{1}{4}\pi\left(\tfrac{3}{2}\right)^2 = \tfrac{9\pi}{16}$ and $\displaystyle\int_0^{3/2} x\sqrt{\tfrac{9}{4} - x^2}\,dx = \int_0^{9/4} \tfrac{1}{2}u^{1/2}\,du$ (where $u = \frac{9}{4} - x^2$, so $du = -2x\,dx$) $= \left[\frac{1}{3}u^{3/2}\right]_0^{9/4} = \frac{1}{3}\left(\frac{27}{8}\right) = \frac{9}{8}$, so

$W \approx 9016\left[30 \cdot \frac{9}{16}\pi + 12 \cdot \frac{9}{8}\right] = 9016\left(\frac{135}{8}\pi + \frac{27}{2}\right) \approx 6.00 \times 10^5$ J.

23. $V = \pi r^2 x$, so V is a function of x and P can also be regarded as a function of x. If $V_1 = \pi r^2 x_1$ and $V_2 = \pi r^2 x_2$, then $W = \int_{x_1}^{x_2} F(x)\,dx = \int_{x_1}^{x_2} \pi r^2 P(V(x))dx = \int_{x_1}^{x_2} P(V(x))dV(x)$ [Put $V(x) = \pi r^2 x$, so $dV(x) = \pi r^2\,dx$] $= \int_{V_1}^{V_2} P(V)dV$ by the Substitution Rule.

24. $160\,\text{lb/in}^2 = 160 \cdot 144\,\text{lb/ft}^2$, $100\,\text{in}^3 = \frac{100}{1728}\,\text{ft}^3$, and $800\,\text{in}^3 = \frac{800}{1728}\,\text{ft}^3$.

$k = PV^{1.4} = (160 \cdot 144)\left(\frac{100}{1728}\right)^{1.4} = 23{,}040\left(\frac{25}{432}\right)^{1.4} \approx 426.5$. Therefore $P \approx 426.5V^{-1.4}$ and

$W = \int_{100/1728}^{800/1728} 426.5V^{-1.4}\,dV = 426.5\left[\frac{1}{-0.4}V^{-0.4}\right]_{25/432}^{25/54} = (426.5)(2.5)\left[\left(\frac{432}{25}\right)^{0.4} - \left(\frac{54}{25}\right)^{0.4}\right]$

$\approx 1.88 \times 10^3$ ft-lb.

25. $$W = \int_a^b F(r)dr = \int_a^b G\frac{m_1 m_2}{r^2}\,dr = Gm_1m_2\left[\frac{-1}{r}\right]_a^b = Gm_1m_2\left(\frac{1}{a} - \frac{1}{b}\right)$$

26. By Exercise 25, $W = GMm\left(\dfrac{1}{R} - \dfrac{1}{R + 1{,}000{,}000}\right)$ where M = mass of earth in kg, R = radius of earth in m, m = mass of satellite in kg. (Note 1000 km = 1,000,000 m.) Thus

$$W = (6.67 \times 10^{-11})(5.98 \times 10^{24})(1000) \times \left(\frac{1}{6.37 \times 10^6} - \frac{1}{7.37 \times 10^6}\right) \approx 8.50 \times 10^9 \text{ J}.$$

EXERCISES 5.5

1. $f_{\text{ave}} = \frac{1}{3-0}\int_0^3 (x^2 - 2x)dx = \frac{1}{3}\left[\frac{1}{3}x^3 - x^2\right]_0^3 = \frac{1}{3}(9 - 9) = 0$

2. $f_{\text{ave}} = \frac{1}{\pi - 0}\int_0^\pi \sin x\,dx = \frac{1}{\pi}(-\cos x)\Big|_0^\pi = \frac{1}{\pi}(1 + 1) = \frac{2}{\pi}$

3. $f_{\text{ave}} = \frac{1}{1-(-1)}\int_{-1}^1 x^4\,dx = \frac{1}{2} \cdot 2\int_0^1 x^4\,dx = \left[\frac{1}{5}x^5\right]_0^1 = \frac{1}{5}$

4. $f_{\text{ave}} = \frac{1}{3-1}\int_1^3 (x^3 - x)dx = \frac{1}{2}\left[\frac{1}{4}x^4 - \frac{1}{2}x^2\right]_1^3 = \frac{1}{2}\left[\left(\frac{81}{4} - \frac{9}{2}\right) - \left(\frac{1}{4} - \frac{1}{2}\right)\right] = \frac{1}{2} \cdot 16 = 8$

5. $$f_{\text{ave}} = \frac{1}{\frac{\pi}{4} - \left(-\frac{\pi}{2}\right)}\int_{-\pi/2}^{\pi/4} \sin^2 x \cos x\,dx = \frac{4}{3\pi}\int_{-\pi/2}^{\pi/4} \sin^2 x \cos x\,dx$$

$= \frac{4}{3\pi}\int_{-1}^{1/\sqrt{2}} u^2\,du$ [Put $u = \sin x$, so $du = \cos x\,dx$] $= \frac{4}{3\pi}\left[\frac{1}{3}u^3\right]_{-1}^{1/\sqrt{2}}$

$= \frac{4}{9\pi}\left(\frac{1}{2\sqrt{2}} + 1\right) = \frac{4}{9\pi}\left(\frac{\sqrt{2}}{4} + 1\right) = \frac{\sqrt{2}+4}{9\pi}$

6. $f_{\text{ave}} = \frac{1}{9-4}\int_4^9 \sqrt{x}\,dx = \frac{1}{5} \cdot \frac{2}{3}x^{3/2}\Big|_4^9 = \frac{2}{15}(27 - 8) = \frac{38}{15}$

7. **(a)** $f_{\text{ave}} = \dfrac{1}{2-0}\displaystyle\int_0^2 (4 - x^2)\,dx$

$= \frac{1}{2}\left[4x - \frac{1}{3}x^3\right]_0^2 = \frac{1}{2}\left[\left(8 - \frac{8}{3}\right) - 0\right]$

$= \frac{8}{3}$

(b) $f_{\text{ave}} = f(c) \quad \Leftrightarrow \quad \frac{8}{3} = 4 - c^2$

$\Leftrightarrow \quad c^2 = \frac{4}{3} \quad \Leftrightarrow \quad c = \frac{2}{\sqrt{3}}$

(c)

8. **(a)** $f_{\text{ave}} = \frac{1}{3-0}\int_0^3 (4x - x^2)dx$

$= \frac{1}{3}\left[2x^2 - \frac{1}{3}x^3\right]_0^3 = \frac{1}{3}(18 - 9)$

$= 3$

(b) $f_{\text{ave}} = f(c) \quad \Leftrightarrow \quad 3 = 4c - c^2$

$\Leftrightarrow \quad c^2 - 4c + 3 = 0 \quad \Leftrightarrow$

$c = 1$ or 3

(c)

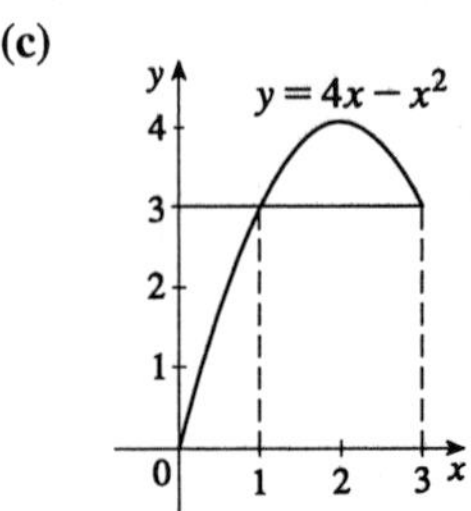

9. **(a)** $f_{\text{ave}} = \frac{1}{2-0}\int_0^2 (x^3 - x + 1)dx$

$= \frac{1}{2}\left[\frac{1}{4}x^4 - \frac{1}{2}x^2 + x\right]_0^2$

$= \frac{1}{2}(4 - 2 + 2) = 2$

(b) From the graph, it appears that $f(x) = 2$ at $x \approx 1.32$.

(c)

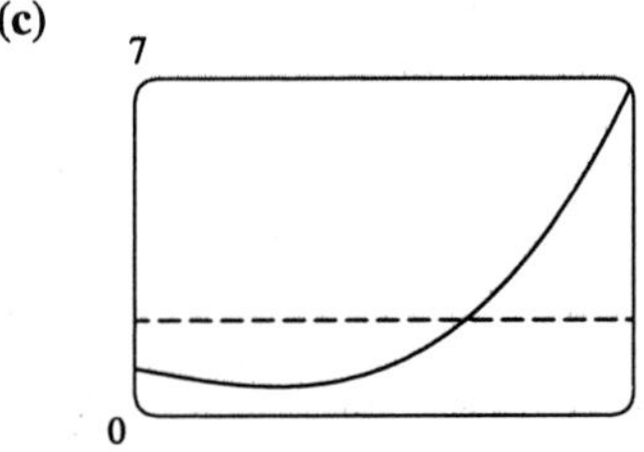

10. **(a)** $f_{\text{ave}} = \dfrac{1}{\sqrt{\pi} - 0}\displaystyle\int_0^{\sqrt{\pi}} \left[x \sin(x^2)\right]dx$

$= \frac{1}{\sqrt{\pi}}\left[-\frac{1}{2}\cos(x^2)\right]_0^{\sqrt{\pi}}$

$= -\frac{1}{2\sqrt{\pi}}(\cos \pi - \cos 0) = 1/\sqrt{\pi}$

(b) From the graph, it appears that $f(x) = 1/\sqrt{\pi}$ at $x \approx 0.85$ and at $x \approx 1.67$.

(c)

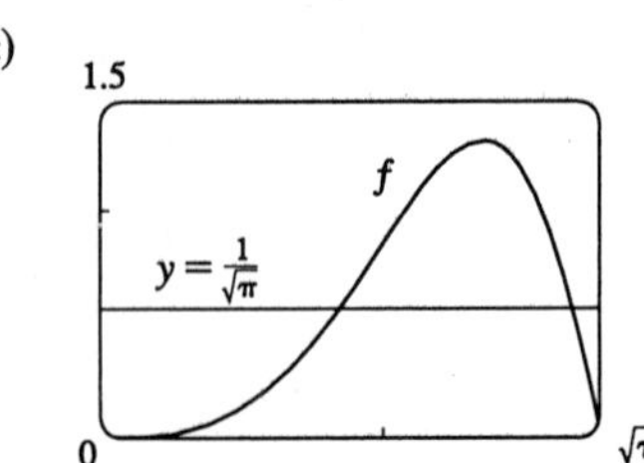

11. Since f is continuous on $[1, 3]$, by the Mean Value Theorem for Integrals there exists a number c in $[1, 3]$ such that $\int_1^3 f(x)\,dx = 8 = f(c)(3 - 1) = 2f(c)$; that is, there is a number c such that $f(c) = \frac{8}{2} = 4$.

12. The requirement is that $\dfrac{1}{b-0}\displaystyle\int_0^b f(x)\,dx = 3$. The LHS of this equation is equal to

$\dfrac{1}{b}\displaystyle\int_0^b (2 + 6x - 3x^2)\,dx = \dfrac{1}{b}\left[2x + 3x^2 - x^3\right]_0^b = 2 + 3b - b^2$, so we solve the equation $2 + 3b - b^2 = 3 \quad \Leftrightarrow$

$b^2 - 3b + 1 = 0 \quad \Leftrightarrow \quad b = \dfrac{3 \pm \sqrt{(-3)^2 - 4 \cdot 1 \cdot 1}}{2 \cdot 1} = \dfrac{3 \pm \sqrt{5}}{2}$. Both roots are valid since they are both positive.

13. $T_{\text{ave}} = \frac{1}{12}\int_0^{12}\left[50 + 14\sin\frac{1}{12}\pi t\right]dt = \frac{1}{12}\left[50t - 14\cdot\frac{12}{\pi}\cos\frac{1}{12}\pi t\right]_0^{12}$
$= \frac{1}{12}\left[50\cdot 12 + 14\cdot\frac{12}{\pi} + 14\cdot\frac{12}{\pi}\right] = \left(50 + \frac{28}{\pi}\right)^\circ\text{F} \approx 59^\circ\text{F}$

14. $T_{\text{ave}} = \frac{1}{5}\int_0^5 4x\,dx = \frac{2}{5}x^2\Big|_0^5 = 10^\circ\text{C}$

15. $\rho_{\text{ave}} = \dfrac{1}{8}\displaystyle\int_0^8 \frac{12}{\sqrt{x+1}}\,dx = \frac{3}{2}\int_0^8 (x+1)^{-1/2}\,dx = \left[3\sqrt{x+1}\right]_0^8 = 9 - 3 = 6\,\text{kg/m}$

16. $s(t) = \frac{1}{2}gt^2 \quad\Rightarrow\quad v(t) = s'(t) = gt \quad\Rightarrow\quad v_T = v(T) = gT$, and $s(T) = \frac{1}{2}gT^2$. Also, $s'(t) = gt = \sqrt{2gs} = v(s)$.

The average of the velocities with respect to time during the interval $[0, T]$ is

$$v_{\text{ave}} = \frac{1}{T}\int_0^T v(t)dt = \frac{1}{T}[s(T) - s(0)] \quad \text{(by the Fundamental Theorem of Calculus)}$$
$$= \frac{1}{2}\cdot\frac{1}{T}gT^2 = \tfrac{1}{2}gT = \tfrac{1}{2}v_T.$$

But with respect to s,

$$v_{\text{ave}} = \frac{1}{s(T)}\int_0^{s(T)} v(s)\,ds = \frac{1}{s(T)}\int_0^{s(T)}\sqrt{2gs}\,ds = \frac{2}{gT^2}\sqrt{2g}\int_0^{s(T)} s^{1/2}\,ds$$
$$= \frac{2\sqrt{2}}{\sqrt{gT^2}}\left(\tfrac{2}{3}\right)\left[s^{3/2}\right]_0^{gT^2/2} = \frac{4\sqrt{2}}{3\sqrt{gT^2}}\left(\tfrac{1}{2}gT^2\right)^{3/2} = \tfrac{2}{3}gT = \tfrac{2}{3}v_T.$$

17. $V_{\text{ave}} = \frac{1}{5}\int_0^5 V(t)dt = \frac{1}{5}\int_0^5 \frac{5}{4\pi}\left[1 - \cos\left(\frac{2}{5}\pi t\right)\right]dt = \frac{1}{4\pi}\int_0^5\left[1 - \cos\left(\frac{2}{5}\pi t\right)\right]dt$
$= \frac{1}{4\pi}\left[t - \frac{5}{2\pi}\sin\left(\frac{2}{5}\pi t\right)\right]_0^5 = \frac{1}{4\pi}[(5-0) - 0] = \frac{5}{4\pi} \approx 0.4\,\text{L}$

18. $v_{\text{ave}} = \dfrac{1}{R-0}\displaystyle\int_0^R v(r)\,dr = \frac{1}{R}\int_0^R \frac{P}{4\eta l}\left(R^2 - r^2\right)dr = \frac{P}{4\eta lR}\left[R^2 r - \tfrac{1}{3}r^3\right]_0^R = \frac{P}{4\eta lR}\left(\tfrac{2}{3}\right)R^3 = \frac{PR^2}{6\eta l}$. Since $v(r)$ is decreasing on $[0, R]$, $v_{\max} = v(0) = \dfrac{PR^2}{4\eta l}$. Thus $v_{\text{ave}} = \frac{2}{3}v_{\max}$.

19. Let $F(x) = \int_a^x f(t)\,dt$ for x in $[a, b]$. Then F is continuous on $[a, b]$ and differentiable on (a, b), so by the Mean Value Theorem there is a number c in (a, b) such that $F(b) - F(a) = F'(c)(b - a)$. But $F'(x) = f(x)$ by the Fundamental Theorem of Calculus. Therefore $\int_a^b f(t)dt - 0 = f(c)(b - a)$.

20. $$f_{\text{ave}}[a, b] = \frac{1}{b-a}\int_a^b f(x)\,dx$$
$$= \frac{1}{b-a}\int_a^c f(x)\,dx + \frac{1}{b-a}\int_c^b f(x)\,dx$$
$$= \frac{c-a}{b-a}\left[\frac{1}{c-a}\int_a^c f(x)\,dx\right] + \frac{b-c}{b-a}\left[\frac{1}{b-c}\int_c^b f(x)\,dx\right]$$
$$= \frac{c-a}{b-a}f_{\text{ave}}[a, c] + \frac{b-c}{b-a}f_{\text{ave}}[c, b]$$

REVIEW EXERCISES FOR CHAPTER 5

1. (a) $A = \int_a^b [f(x) - g(x)]dx$ (b) $A = \int_c^d [u(y) - v(y)]dy$

   (c) Here we use disks: $V = \pi \int_a^b \left([f(x)]^2 - [g(x)]^2\right)dx$

   (d) Here we use cylindrical shells: $V = 2\pi \int_a^b x[f(x) - g(x)]dx$

   (e) Use shells: $V = 2\pi \int_c^d y[u(y) - v(y)]dy$

   (f) Use disks: $V = \pi \int_c^d \left([u(y)]^2 - [v(y)]^2\right)dy$

2. $A = \int_{-1}^4 (4 + 3x - x^2)dx = \left[4x + \frac{3}{2}x^2 - \frac{1}{3}x^3\right]_{-1}^4 = \left(16 + 24 - \frac{64}{3}\right) - \left(-4 + \frac{3}{2} + \frac{1}{3}\right) = \frac{125}{6}$

3. $A = \int_0^6 [(12x - 2x^2) - (x^2 - 6x)]dx = \int_0^6 (18x - 3x^2)dx = [9x^2 - x^3]_0^6 = 9 \cdot 36 - 216 = 108$

4. $12 - x^2 = x^2 - 6 \quad \Leftrightarrow \quad x^2 = 9 \quad \Leftrightarrow \quad x = \pm 3$. By symmetry,
   $A = 2\int_0^5 |(12 - x^2) - (x^2 - 6)|dx = 4\int_0^5 |9 - x^2|dx = 4\int_0^3 (9 - x^2)dx + 4\int_3^5 (x^2 - 9)dx$
   $= 4\left(\left[9x - \frac{1}{3}x^3\right]_0^3 + \left[\frac{1}{3}x^3 - 9x\right]_3^5\right) = 4\left[\left(27 - \frac{27}{3}\right) - 0 + \left(\frac{125}{3} - 45\right) - \left(\frac{27}{3} - 27\right)\right] = \frac{392}{3}$

5. By symmetry, $A = 2\int_0^1 \left(x^{1/3} - x^3\right)dx = 2\left[\frac{3}{4}x^{4/3} - \frac{1}{4}x^4\right]_0^1 = 2\left(\frac{3}{4} - \frac{1}{4}\right) = 1$.

6. $A = \int_1^7 [(2y - 7) - (y^2 - 6y)]dy = \int_1^7 (-y^2 + 8y - 7)dy$
   $= \left[-\frac{1}{3}y^3 + 4y^2 - 7y\right]_1^7 = -\frac{343}{3} + 196 - 49 - \left(-\frac{1}{3} + 4 - 7\right) = 36$

7. $A = \int_0^\pi |\sin x - (-\cos x)|dx = \int_0^{3\pi/4} (\sin x + \cos x)dx - \int_{3\pi/4}^\pi (\sin x + \cos x)dx$
   $= [\sin x - \cos x]_0^{3\pi/4} - [-\cos x + \sin x]_{3\pi/4}^\pi$
   $= \left(\frac{1}{\sqrt{2}} + \frac{1}{\sqrt{2}}\right) - (0 - 1) - (1 + 0) + \left(\frac{1}{\sqrt{2}} + \frac{1}{\sqrt{2}}\right) = \sqrt{2} + 1 - 1 + \sqrt{2} = 2\sqrt{2}$

8. The curves intersect at $(1, 1)$, so the area is
   $A = \int_0^2 |x^3 - (x^2 - 4x + 4)|dx = \int_0^1 (-x^3 + x^2 - 4x + 4)dx + \int_1^2 (x^3 - x^2 + 4x - 4)dx$
   $= \left[-\frac{1}{4}x^4 + \frac{1}{3}x^3 - 2x^2 + 4x\right]_0^1 + \left[\frac{1}{4}x^4 - \frac{1}{3}x^3 + 2x^2 - 4x\right]_1^2$
   $= -\frac{1}{4} + \frac{1}{3} - 2 + 4 + 4 - \frac{8}{3} + 8 - 8 - \frac{1}{4} + \frac{1}{3} - 2 + 4 = 5.5.$

9. $V = \displaystyle\int_1^3 \pi\left(\sqrt{x - 1}\right)^2 dx = \pi \int_1^3 (x - 1)dx = \pi\left[\frac{1}{2}x^2 - x\right]_1^3 = \pi\left[\left(\frac{9}{2} - 3\right) - \left(\frac{1}{2} - 1\right)\right] = 2\pi$

10. $V = \displaystyle\int_0^1 \pi\left[\left(x^2\right)^2 - \left(x^3\right)^2\right]dx = \pi \int_0^1 \left(x^4 - x^6\right)dx = \pi\left[\frac{1}{5}x^5 - \frac{1}{7}x^7\right]_0^1 = \pi\left(\frac{1}{5} - \frac{1}{7}\right) = \frac{2\pi}{35}$

11. $V = \int_1^3 2\pi y(-y^2 + 4y - 3)dy = 2\pi\int_1^3 (-y^3 + 4y^2 - 3y)dy = 2\pi\left[-\frac{1}{4}y^4 + \frac{4}{3}y^3 - \frac{3}{2}y^2\right]_1^3$
    $= 2\pi\left[\left(-\frac{81}{4} + 36 - \frac{27}{2}\right) - \left(-\frac{1}{4} + \frac{4}{3} - \frac{3}{2}\right)\right] = \frac{16\pi}{3}$

12. $V = \int_0^8 \pi\left(y^{1/3}\right)^2 dy = \pi\int_0^8 y^{2/3}\,dy = \pi\left[\frac{3}{5}y^{5/3}\right]_0^8 = \frac{96\pi}{5}$

13. $V = \int_a^{a+h} 2\pi x \cdot 2\sqrt{x^2 - a^2}\,dx = 2\pi\int_0^{2ah+h^2} u^{1/2}\,du$ (Put $u = x^2 - a^2$, so $du = 2x\,dx$)
    $= 2\pi\left[\frac{2}{3}u^{3/2}\right]_0^{2ah+h^2} = \frac{4}{3}\pi(2ah + h^2)^{3/2}$

14. $V = \int_{3\pi/2}^{5\pi/2} 2\pi x \cos x \, dx$ (by the method of cylindrical shells)

15. $V = \int_0^1 \pi\left[\left(1 - x^3\right)^2 - \left(1 - x^2\right)^2\right]dx$

16. $V = \int_0^2 2\pi(8 - x^3)(2 - x)dx$

17. **(a)** $V = \int_0^1 \pi(x^2 - x^4)\,dx = \pi\left[\frac{1}{3}x^3 - \frac{1}{5}x^5\right]_0^1 = \pi\left[\frac{1}{3} - \frac{1}{5}\right] = \frac{2\pi}{15}$

Or: $V = \int_0^1 2\pi y(\sqrt{y} - y)\,dy = 2\pi\left[\frac{2}{5}y^{5/2} - \frac{1}{3}y^3\right]_0^1 = \frac{2\pi}{15}$

(b) $V = \int_0^1 \pi\left[\left(\sqrt{y}\right)^2 - y^2\right]dy = \pi\left[\frac{1}{2}y^2 - \frac{1}{3}y^3\right]_0^1 = \pi\left[\frac{1}{2} - \frac{1}{3}\right] = \frac{\pi}{6}$

Or: $V = \int_0^1 2\pi x(x - x^2)dx = 2\pi\left[\frac{1}{3}x^3 - \frac{1}{4}x^4\right]_0^1 = \frac{\pi}{6}$

(c) $V = \int_0^1 \pi\left[\left(2 - x^2\right)^2 - (2 - x)^2\right]dx = \int_0^1 \pi\left(x^4 - 5x^2 + 4x\right)dx$

$= \pi\left[\frac{1}{5}x^5 - \frac{5}{3}x^3 + 2x^2\right]_0^1 = \pi\left[\frac{1}{5} - \frac{5}{3} + 2\right] = \frac{8\pi}{15}$

Or: $V = \int_0^1 2\pi(2 - y)\left(\sqrt{y} - y\right)dy = 2\pi\int_0^1 \left(y^2 - y^{3/2} - 2y + 2y^{1/2}\right)dy$

$= 2\pi\left[\frac{1}{3}y^3 - \frac{2}{5}y^{5/2} - y^2 + \frac{4}{3}y^{3/2}\right]_0^1 = \frac{8\pi}{15}$

18. **(a)** $A = \int_0^1 (2x - x^2 - x^3)dx = \left[x^2 - \frac{1}{3}x^3 - \frac{1}{4}x^4\right]_0^1 = 1 - \frac{1}{3} - \frac{1}{4} = \frac{5}{12}$

(b) $V = \int_0^1 \pi\left[\left(2x - x^2\right)^2 - x^6\right]dx = \int_0^1 \pi\left(4x^2 - 4x^3 + x^4 - x^6\right)dx$

$= \pi\left[\frac{4}{3}x^3 - x^4 + \frac{1}{5}x^5 - \frac{1}{7}x^7\right]_0^1 = \pi\left(\frac{4}{3} - 1 + \frac{1}{5} - \frac{1}{7}\right) = \frac{41\pi}{105}$

(c) $V = \int_0^1 2\pi x(2x - x^2 - x^3)dx = \int_0^1 2\pi\left(2x^2 - x^3 - x^4\right)dx = 2\pi\left[\frac{2}{3}x^3 - \frac{1}{4}x^4 - \frac{1}{5}x^5\right]_0^1$

$= 2\pi\left(\frac{2}{3} - \frac{1}{4} - \frac{1}{5}\right) = \frac{13\pi}{30}$ (by the method of cylindrical shells)

19. **(a)** Using the Midpoint Rule on $[0, 1]$ with $f(x) = \tan(x^2)$ and $n = 4$, we estimate

$A = \int_0^1 \tan(x^2)dx \approx \frac{1}{4}\left[\tan\left(\left(\frac{1}{8}\right)^2\right) + \tan\left(\left(\frac{3}{8}\right)^2\right) + \tan\left(\left(\frac{5}{8}\right)^2\right) + \tan\left(\left(\frac{7}{8}\right)^2\right)\right] \approx 0.38.$

(b) Using the Midpoint Rule on $[0, 1]$ with $f(x) = \pi\tan^2(x^2)$ (for disks) and $n = 4$, we estimate

$V = \int_0^1 f(x)dx \approx \frac{1}{4}\pi\left[\tan^2\left(\left(\frac{1}{8}\right)^2\right) + \tan^2\left(\left(\frac{3}{8}\right)^2\right) + \tan^2\left(\left(\frac{5}{8}\right)^2\right) + \tan^2\left(\left(\frac{7}{8}\right)^2\right)\right] \approx 0.87.$

20. **(a)** From the graph, it appears that the curves intersect at $x = 0$ and at $x \approx 0.75$, with $1 - x^2 > x^6 - x + 1$ on $[0, 0.75]$.

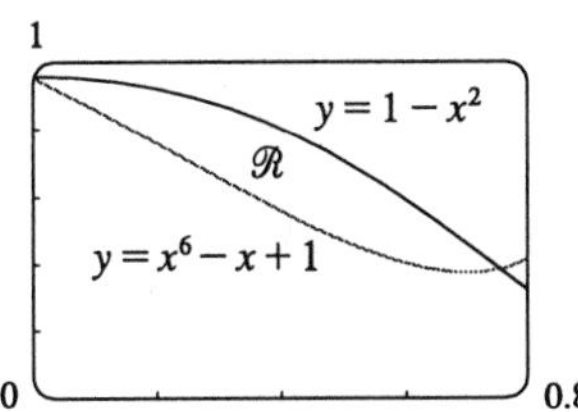

(b) We estimate

$A \approx \int_0^{0.75}[(1 - x^2) - (x^6 - x + 1)]dx$

$= \left[-\frac{1}{3}x^3 - \frac{1}{7}x^7 + \frac{1}{2}x^2\right]_0^{0.75} \approx 0.12.$

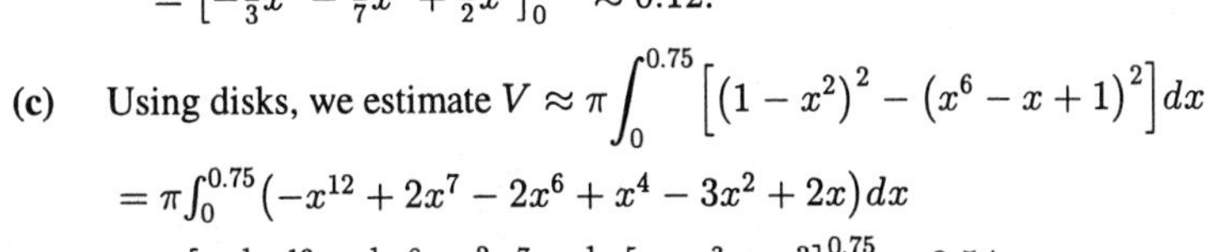

(c) Using disks, we estimate $V \approx \pi\int_0^{0.75}\left[\left(1 - x^2\right)^2 - \left(x^6 - x + 1\right)^2\right]dx$

$= \pi\int_0^{0.75}\left(-x^{12} + 2x^7 - 2x^6 + x^4 - 3x^2 + 2x\right)dx$

$= \pi\left[-\frac{1}{13}x^{13} + \frac{1}{4}x^8 - \frac{2}{7}x^7 + \frac{1}{5}x^5 - x^3 + x^2\right]_0^{0.75} \approx 0.54.$

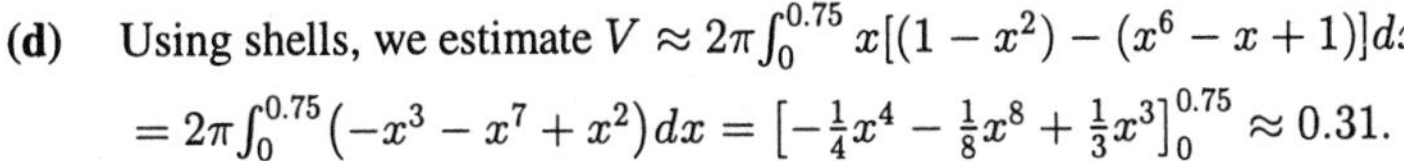

(d) Using shells, we estimate $V \approx 2\pi\int_0^{0.75} x[(1 - x^2) - (x^6 - x + 1)]dx$

$= 2\pi\int_0^{0.75}\left(-x^3 - x^7 + x^2\right)dx = \left[-\frac{1}{4}x^4 - \frac{1}{8}x^8 + \frac{1}{3}x^3\right]_0^{0.75} \approx 0.31.$

21. The solid is obtained by rotating the region under the curve $y = \sin x$, above $y = 0$, from $x = 0$ to $x = \pi$, about the x-axis.

22. The solid is obtained by rotating the region under the curve $y = \sin x$, above $y = 0$, from $x = 0$ to $x = \pi$, about the y-axis.

23. The solid is obtained by rotating the region in the first quadrant bounded by the curve $x = 4 - y^2$ and the coordinate axes about the x-axis.

24. The solid is obtained by rotating the region between the curves $y = 2 - x^2$ and $y = 2 - \sqrt{x}$ about the x-axis. *Or:* It is obtained by rotating the region between $y = x^2$ and $y = \sqrt{x}$ about the line $y = 2$.

25. Take the base to be the disk $x^2 + y^2 \leq 9$. Then $V = \int_{-3}^{3} A(x)dx$, where $A(x_0)$ is the area of the isosceles right triangle whose hypotenuse lies along the line $x = x_0$ in the xy-plane. $A(x) = \frac{1}{4}\left(2\sqrt{9 - x^2}\right)^2 = 9 - x^2$, so $V = 2\int_0^3 A(x)dx = 2\int_0^3 (9 - x^2)dx = 2\left[9x - \frac{1}{3}x^3\right]_0^3 = 2(27 - 9) = 36$.

26. $V = \int_{-1}^{1} A(x)dx = 2\int_0^1 A(x)dx = 2\int_0^1 \left[(2 - x^2) - x^2\right]^2 dx = 2\int_0^1 \left[2(1 - x^2)\right]^2 dx$
$= 8\int_0^1 (1 - 2x^2 + x^4)dx = 8\left[x - \frac{2}{3}x^3 + \frac{1}{5}x^5\right]_0^1 = 8\left(1 - \frac{2}{3} + \frac{1}{5}\right) = \frac{64}{15}$

27. Equilateral triangles with sides measuring $\frac{1}{4}x$ meters have height $\frac{1}{4}x \sin 60° = \frac{\sqrt{3}}{8}x$. Therefore, $A(x) = \frac{1}{2} \cdot \frac{1}{4}x \cdot \frac{\sqrt{3}}{8}x = \frac{\sqrt{3}}{64}x^2$. $V = \int_0^{20} A(x)dx = \frac{\sqrt{3}}{64}\int_0^{20} x^2\,dx = \frac{\sqrt{3}}{64}\left[\frac{1}{3}x^3\right]_0^{20} = \frac{1000\sqrt{3}}{24} = \frac{125\sqrt{3}}{3}$ m^3

28. **(a)** By the symmetry of the problem, we consider only the solid to the right of the origin. The semicircular cross-sections perpendicular to the x-axis have radius $1 - x$, so $A(x) = \frac{1}{2}\pi(1 - x)^2$. Now we can calculate $V = 2\int_0^1 A(x)dx = 2\int_0^1 \frac{1}{2}\pi(1 - x)^2\,dx = \int_0^1 \pi(1 - x)^2\,dx = -\frac{\pi}{3}\left[(1 - x)^3\right]_0^1 = \frac{\pi}{3}$

(b) Cut the solid with a plane perpendicular to the x-axis and passing through the y-axis. Fold the half of the solid in the region $x \leq 0$ under the xy-plane so that the point $(-1, 0)$ comes around and touches the point $(1, 0)$. The resulting solid is a right circular cone of radius 1 with vertex at $(1, 0, 0)$ and with its base in the yz-plane, centered at the origin. The volume of this cone is $\frac{1}{3}r^2h = \frac{1}{3}\pi \cdot 1^2 \cdot 1 = \frac{\pi}{3}$.

29. $30\,\text{N} = f(x) = kx = k(0.03\,\text{m})$, so $k = 30/0.03 = 1000\,\text{N/m}$.
$W = \int_0^{0.08} kx\,dx = 1000\int_0^{0.08} x\,dx = 500\left[x^2\right]_0^{0.08} = 500(0.08)^2 = 3.2\,\text{J}$.

30. The work needed to raise the elevator alone is 1600 lb $\times$ 30 ft $=$ 48,000 ft-lb. The work needed to raise the bottom 170 ft of cable is170 ft $\times$ 10 lb/ft $\times$ 30 ft $=$ 51,000 ft-lb. The work needed to raise the top 30 ft of cable is $\int_0^{30} 10x\,dx = \left[5x^2\right]_0^{30} = 5 \cdot 900 = 4500$ ft-lb. Adding these, we see that the total work needed is $48{,}000 + 51{,}000 + 4{,}500 = 103{,}500$ ft-lb.

31. **(a)** $W = \int_0^4 \pi\left(2\sqrt{y}\right)^2 62.5(4-y)dy = 250\pi \int_h^4 y(4-y)dy = 250\pi\left[2y^2 - \frac{1}{3}y^3\right]_0^4$
$= 250\pi\left(32 - \frac{64}{3}\right) = \frac{8000\pi}{3}$ ft-lb

(b) In part (a) we knew the final water level (0) but not the amount of work done. Here we use the same equation, except with the work fixed, and the lower limit of integration (that is, the final water level — call it h) unknown:

$W = 4000 = \int_h^4 \pi\left(2\sqrt{y}\right)^2 62.5(4-y)dy = 250\pi \int_h^4 y(4-y)dy$
$= 250\pi\left[2y^2 - \frac{1}{3}y^3\right]_h^4 = 250\pi\left[\left(2\cdot 16 - \frac{1}{3}\cdot 64\right) - \left(2h^2 - \frac{1}{3}h^3\right)\right]$
$\Leftrightarrow \quad h^3 - 6h^2 + 32 - \frac{48}{\pi} = 0$. We plot the graph of the function $f(h) = h^3 - 6h^2 + 32 - \frac{48}{\pi}$ on the interval $[0, 4]$ to see where it is 0.

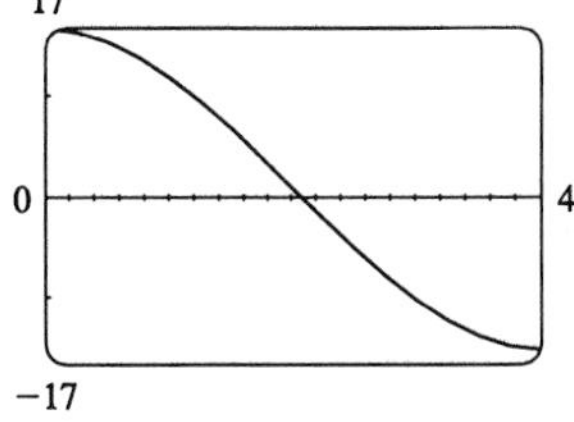

From the graph, it appears that $f(h) = 0$ for $h \approx 2.1$.
So the depth of water remaining is about 2.1 ft.

32. $f_{\text{ave}} = \frac{1}{4-2}\int_2^4 x^3\,dx = \frac{1}{2}\left[\frac{1}{4}x^4\right]_2^4 = \frac{1}{8}\left(4^4 - 2^4\right) = 30$

33. $\lim\limits_{h\to 0} f_{\text{ave}} = \lim\limits_{h\to 0} \dfrac{1}{h}\displaystyle\int_x^{x+h} f(t)dt = \lim\limits_{h\to 0} \dfrac{F(x+h) - F(x)}{h}$, where $F(x) = \int_a^x f(t)dt$.

But we recognize this limit as being $F'(x)$ by the definition of the derivative. Therefore $\lim\limits_{h\to 0} f_{\text{ave}} = F'(x) = f(x)$ by Part 1 of the Fundamental Theorem of Calculus.

34. **(a)** R_1 is the region below the graph of $y = x^2$ and above the x-axis between $x = 0$ and $x = b$, and R_2 is the region to the left of the graph of $x = \sqrt{y}$ and to the right of the y-axis between $y = 0$ and $y = b^2$. So the area of R_1 is $A_1 = \int_0^b x^2\,dx = \left[\frac{1}{3}x^3\right]_0^b = \frac{1}{3}b^3$, and the area of R_2 is $A_2 = \int_0^{b^2} \sqrt{y}\,dy = \left[\frac{2}{3}y^{3/2}\right]_0^{b^2} = \frac{2}{3}b^3$. So there are no solutions to $A_1 = A_2$ for $b \neq 0$.

(b) Using disks, we calculate the volume of rotation of R_1 about the x-axis to be $V_{1x} = \pi\int_0^b \left(x^2\right)^2 dx = \frac{1}{5}\pi b^5$. Using cylindrical shells, we calculate the volume of rotation of R_1 about the y-axis to be $V_{1y} = 2\pi\int_0^b x(x^2)dx = 2\pi\left[\frac{1}{4}x^4\right]_0^b = \frac{1}{2}\pi b^4$. So $V_{1x} = V_{1y} \quad\Leftrightarrow\quad \frac{1}{5}\pi b^5 = \frac{1}{2}\pi b^4 \quad\Leftrightarrow\quad 2b = 5 \quad\Leftrightarrow$ $b = \frac{5}{2}$. So the volumes of rotation about the x- and y-axes are the same for $b = \frac{5}{2}$.

(c) We use cylindrical shells to calculate the volume of rotation of R_2 about the x-axis: $R_{2x} = 2\pi\int_0^{b^2} y\left(\sqrt{y}\right)dy = 2\pi\left[\frac{2}{5}y^{5/2}\right]_0^{b^2} = \frac{4}{5}\pi b^5$. We already know the volume of rotation of R_1 about the x-axis from part (b), and $R_{1x} = R_{2x} \quad\Leftrightarrow\quad \frac{1}{5}\pi b^5 = \frac{4}{5}\pi b^5$, which has no solution for $b \neq 0$.

(d) We use disks to calculate the volume of rotation of R_2 about the y-axis: $R_{2y} = \pi\int_0^{b^2} \left(\sqrt{y}\right)^2 dy = \pi\left[\frac{1}{2}y^2\right]_0^{b^2} = \frac{1}{2}\pi b^4$. We know the volume of rotation of R_1 about the y-axis from part (b), and $R_{1y} = R_{2y} \quad\Leftrightarrow\quad \frac{1}{2}\pi b^4 = \frac{1}{2}\pi b^4$. But this equation is true for all b, so the volumes of rotation about the y-axis are equal for all values of b.

APPLICATIONS PLUS (page 340)

1. (a) By Formula 5.2.3, $V = \int_0^h \pi[f(y)]^2 dy$.

(b) $\dfrac{dV}{dt} = \dfrac{dV}{dh} \cdot \dfrac{dh}{dt} = \pi[f(h)]^2 \dfrac{dh}{dt}$

(c) $kA\sqrt{h} = \pi[f(h)]^2 \dfrac{dh}{dt}$. Set $\dfrac{dh}{dt} = C$: $\pi[f(h)]^2 C = kA\sqrt{h} \quad \Rightarrow \quad [f(h)]^2 = \dfrac{kA}{\pi C}\sqrt{h} \quad \Rightarrow$ $f(h) = \sqrt{\dfrac{kA}{\pi C}} h^{1/4}$, that is, $f(y) = \sqrt{\dfrac{kA}{\pi C}}\, y^{1/4}$. The advantage of having $\dfrac{dh}{dt} = C$ is that the markings on the container will be equally spaced.

2. We note that since c is the consumption in gallons per hour, and v is the velocity in miles per hour, then $\dfrac{c}{v} = \dfrac{\text{gallons/hour}}{\text{hours/mile}} = \dfrac{\text{gallons}}{\text{mile}}$ gives us the consumption in gallons per mile, that is, the quantity G. To find the minimum, we calculate $\dfrac{dG}{dv} = \dfrac{d}{dv}\left(\dfrac{c}{v}\right) = \dfrac{v\dfrac{dc}{dv} - c\dfrac{dv}{dv}}{v^2} = \dfrac{v\dfrac{dc}{dv} - c}{v^2}$. This is 0 when $v\dfrac{dc}{dv} - c = 0 \iff \dfrac{dc}{dv} = \dfrac{c}{v}$. This implies that the tangent line of $c(v)$ passes through the origin, and this seems to be the case when $v \approx 53$ mi/h. Note that the slope of the secant line through the origin and a point $(v, c(v))$ on the graph is equal to $G(v)$, and it is intuitively clear that G is minimized in the case where the secant is in fact a tangent.

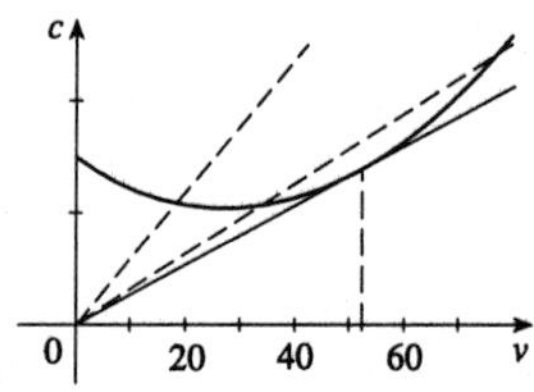

3. (a) First note that $90 \text{ mi/h} = 90 \times \frac{5280}{3600} \text{ ft/s} = 132 \text{ ft/s}$. Then $a(t) = 4 \text{ ft/s}^2 \quad \Rightarrow \quad v(t) = 4t = 132$ when $t = \frac{132}{4} = 33$ s. It takes 33 s to reach 132 ft/s. Therefore, taking $s(0) = 0$, we have $s(t) = 2t^2$, $0 \le t \le 33$. So $s(33) = 2178$ ft.

For $33 \le t \le 933$ we have $v(t) = 132$ ft/s $\quad \Rightarrow \quad s(t) = 132(t - 33) + C$ and $s(33) = 2178 \quad \Rightarrow$ $C = 2178$, so $s(t) = 132(t - 33) + 2178$, $33 \le t \le 933$.

Therefore $s(933) = 132(933) + 2178 = 120{,}978$ ft $= 22.9125$ mi.

(b) As in part (a), the train accelerates for 33 s and travels 2178 ft while doing so. Similarly, it decelerates for 33 s and travels 2178 ft at the end of its trip. During the remaining $900 - 66 = 834$ s it travels at 132 ft/s, so the distance traveled is $132 \cdot 834 = 110{,}088$ ft. Thus the total distance is

$2178 + 110{,}088 + 2178 = 114{,}444 \text{ ft} = 21.675 \text{ mi}$.

4. **(a)** $C(t) = \dfrac{1}{t}\displaystyle\int_0^t [f(s) + g(s)]ds$. Using Part 1 of the Fundamental Theorem of Calculus, we have

$C'(t) = \dfrac{1}{t}[f(t) + g(t)] - \dfrac{1}{t^2}\displaystyle\int_0^t [f(s) + g(s)]ds$. Set $C'(t) = 0$:

$$\frac{1}{t}[f(t) + g(t)] - \frac{1}{t^2}\int_0^t [f(s) + g(s)]ds = 0 \quad \Rightarrow \quad [f(t) + g(t)] - \frac{1}{t}\int_0^t [f(s) + g(s)]ds = 0 \quad \Rightarrow$$

$[f(t) + g(t)] - C(t) = 0$ or $C(t) = f(t) + g(t)$.

(b) For $0 \le t \le 30$, we have $D(t) = \displaystyle\int_0^t \left(\frac{V}{15} - \frac{V}{450}s\right)ds = \left(\frac{V}{15}s - \frac{V}{900}s^2\right)\Big|_0^t = \frac{V}{15}t - \frac{V}{900}t^2$. So

$$D(t) = V \;\Rightarrow\; 60t - t^2 = 900 \;\Rightarrow\; t^2 - 60t + 900 = 0 \;\Rightarrow\; (t - 30)^2 = 0 \;\Rightarrow\; T = 30.$$

(c) $$C(t) = \frac{1}{t}\int_0^t \left(\frac{V}{15} - \frac{V}{450}s + \frac{V}{12{,}900}s^2\right)ds = \frac{1}{t}\left[\frac{V}{15}s - \frac{V}{900}s^2 + \frac{V}{38{,}700}s^3\right]_0^t$$

$$= \frac{1}{t}\left(\frac{V}{15}t - \frac{V}{900}t^2 + \frac{Vt^3}{38{,}700}\right) = \frac{V}{15} - \frac{V}{900}t + \frac{V}{38{,}700}t^2 \quad \Rightarrow$$

$$C'(t) = -\frac{V}{900} + \frac{V}{19{,}350}t = 0 \text{ when } \frac{1}{19{,}350}t = \frac{1}{900} \quad \Rightarrow \quad t = 21.5.$$

$C(21.5) = \dfrac{V}{15} - \dfrac{V}{900}(21.5) + \dfrac{V}{38{,}700}(21.5)^2 \approx 0.05472V$, $C(0) = \dfrac{V}{15}$, and

$C(30) = \dfrac{V}{15} - \dfrac{V}{900}(30) + \dfrac{V}{38{,}700}(30)^2 \approx 0.05659V$, so the absolute minimum is $C(21.5) \approx 0.05472V$.

(d) As in part (c) we have $C(t) = \dfrac{V}{15} - \dfrac{V}{900}t + \dfrac{V}{38{,}700}t^2$, so $C(t) = f(t) + g(t) \quad \Leftrightarrow$

$$\frac{V}{15} - \frac{V}{900}t + \frac{V}{38{,}700}t^2 = \frac{V}{15} - \frac{V}{450}t + \frac{V}{12{,}900}t^2$$

$$\Leftrightarrow \quad t^2\left(\frac{1}{12{,}900} - \frac{1}{38{,}700}\right) = t\left(\frac{1}{450} - \frac{1}{900}\right)$$

$$\Leftrightarrow \quad t = \frac{1/900}{2/38{,}700} = \frac{43}{2} = 21.5.$$ This is the value of t that we obtained as the critical number of C in part (c), so we have verified the result of (a) in this case.

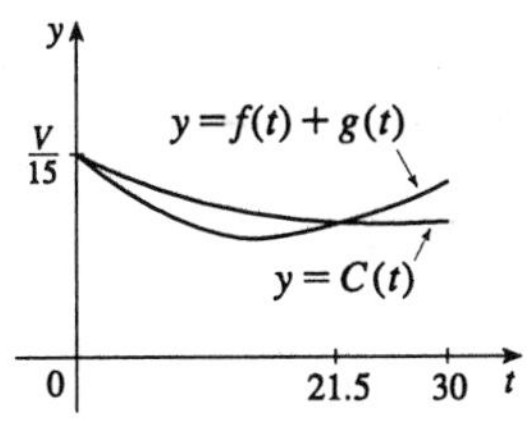

5. **(a)** Let $F(t) = \int_0^t f(s)ds$. Then, by Part 1 of the Fundamental Theorem of Calculus,

$F'(t) = f(t) =$ rate of depreciation, so $F(t)$ represents the loss in value over the interval $[0, t]$.

(b) $C(t) = [A + F(t)]/t$ represents the average expenditure over the interval $[0, t]$. The company wants to minimize average expenditure.

(c) $C(t) = \dfrac{1}{t}\left(A + \displaystyle\int_0^t f(s)ds\right)$. Using Part 1 of the Fundamental Theorem of Calculus, we have

$$C'(t) = -\frac{1}{t^2}\left(A + \int_0^t f(s)ds\right) + \frac{1}{t}f(t) = 0 \text{ when } tf(t) = A + \int_0^t f(s)ds \quad \Rightarrow$$

$$f(t) = \frac{1}{t}\left(A + \int_0^t f(s)ds\right) = C(t).$$

6. **(a)** We first use the cylindrical shell method to express the volume V in terms of h, r, and ω:

$$V = \int_0^r 2\pi xy\,dx = \int_0^r 2\pi x\left[h + \frac{\omega^2 x^2}{2g}\right]dx = 2\pi\int_0^r \left(hx + \frac{\omega^2 x^3}{2g}\right)dx$$

$$= 2\pi\left[\frac{hx^2}{2} + \frac{\omega^2 x^4}{8g}\right]_0^r = 2\pi\left[\frac{hr^2}{2} + \frac{\omega^2 r^4}{8g}\right] = \pi hr^2 + \frac{\omega^2 r^4}{4g} \quad\Rightarrow$$

$$h = \frac{V - (\pi\omega^2 r^4)/(4g)}{\pi r^2} = \frac{4gV - \pi\omega^2 r^4}{4\pi g r^2}$$

(b) The surface touches the bottom when $h = 0 \quad\Rightarrow\quad 4gV - \pi\omega^2 r^4 = 0 \quad\Rightarrow\quad \omega^2 = \dfrac{4gV}{\pi r^4} \quad\Rightarrow$

$\omega = \dfrac{2\sqrt{gV}}{\sqrt{\pi r^2}}$. To spill over the top, $y(r) > L \quad\Leftrightarrow$

$$L > h + \frac{\omega^2 r^2}{2g} = \frac{4gV - \pi\omega^2 r^4}{4\pi g r^2} + \frac{\omega^2 r^2}{2g} = \frac{4gV}{4\pi g r^2} - \frac{\pi\omega^2 r^2}{4\pi g r^2} + \frac{\omega^2 r^2}{2g}$$

$$= \frac{V}{\pi r^2} - \frac{\omega^2 r^2}{4g} + \frac{\omega^2 r^2}{2g} = \frac{V}{\pi r^2} + \frac{\omega^2 r^2}{4g} \quad\Leftrightarrow\quad \frac{\omega^2 r^2}{4g} > L - \frac{V}{\pi r^2} = \frac{\pi r^2 L - V}{\pi r^2} \quad\Leftrightarrow$$

$\omega^2 > \dfrac{4g(\pi r^2 L - V)}{\pi r^4}$. So for spillage, the angular speed should be $\omega > \dfrac{2\sqrt{g(\pi r^2 L - V)}}{r^2\sqrt{\pi}}$.

(c) **(i)** Here we have $r = 2$, $L = 7$, $h = 7 - 5 = 2$. When $x = 1$, $y = 7 - 4 = 3$. Therefore

$$3 = 2 + \frac{\omega^2 \cdot 1^2}{2\cdot 32} \quad\Rightarrow\quad 1 = \frac{\omega^2}{2\cdot 32} \quad\Rightarrow\quad \omega^2 = 64 \quad\Rightarrow\quad \omega = 8\text{ rad/s}.$$

$$V = \pi(2)(2)^2 + \frac{\pi\cdot 8^2\cdot 2^4}{4g} = 8\pi + 8\pi = 16\pi\text{ ft}^2.$$

(ii) At the wall, $x = 2$, so $y = 2 + \dfrac{8^2\cdot 2^2}{2\cdot 32} = 6$ and the surface is $7 - 6 = 1$ ft below the top of the tank.

7. **(a)** $P = \dfrac{\text{area under } y = L\sin\theta}{\text{area of rectangle}} = \dfrac{\int_0^\pi L\sin\theta\,d\theta}{\pi L} = \dfrac{-L\cos\theta|_0^\pi}{\pi L} = \dfrac{-(-1)+1}{\pi} = \dfrac{2}{\pi}$

(b) $P = \dfrac{\text{area under } y = \frac{1}{2}L\sin\theta}{\text{area of rectangle}} = \dfrac{\int_0^\pi \frac{1}{2}L\sin\theta\,d\theta}{\pi L} = \dfrac{\int_0^\pi \sin\theta\,d\theta}{2\pi} = \dfrac{-\cos\theta|_0^\pi}{2\pi} = \dfrac{2}{2\pi} = \dfrac{1}{\pi}$

(c) $P = \dfrac{\text{area under } y = \frac{1}{5}L\sin\theta}{\text{area of rectangle}} = \dfrac{\int_0^\pi \frac{1}{5}L\sin\theta\,d\theta}{\pi L} = \dfrac{\int_0^\pi \sin\theta\,d\theta}{5\pi} = \dfrac{2}{5\pi}$

8. **(a)** $W = \displaystyle\int_{s_0}^{s_1} F(s)ds$, where $F(s) = m\dfrac{dv}{dt} = m\dfrac{dv}{ds}\dfrac{ds}{dt} = mv\dfrac{dv}{ds}$ and so, by the Substitution Rule,

$$W = \int_{s_0}^{s_1} F(s)ds = \int_{s_0}^{s_1} mv\frac{dv}{ds}\,ds = \int_{v_0}^{v_1} mv\,dv = \tfrac{1}{2}mv^2\big|_{v_0}^{v_1} = \tfrac{1}{2}mv_1^2 - \tfrac{1}{2}mv_0^2$$

(b) First we note that 90 mi/h $= \frac{90(5280)}{(60)^2} = 132$ ft/s. Assume $v_0 = v(s_0) = 0$ and $v_1 = v(s_1) = 132$ ft/s. The mass of the baseball is $m = \dfrac{w}{g} = \dfrac{5/16}{32} = \dfrac{5}{512}$, so the work done is

$$W = \tfrac{1}{2}mv_1^2 - \tfrac{1}{2}mv_0^2 = \tfrac{1}{2}\cdot\tfrac{5}{512}\cdot(132)^2 = \tfrac{87{,}120}{1024} \approx 85\text{ ft-lb}.$$

9. **(a)** $F = ma = m\dfrac{dv}{dt}$, so by the Substitution Rule we have

$$\int_{t_0}^{t_1} F(t)dt = \int_{t_0}^{t_1} m\left(\frac{dv}{dt}\right)dt = m\int_{v_0}^{v_1} dv = mv\big|_{v_0}^{v_1} = mv_1 - mv_0 = p(t_1) - p(t_0).$$

(b) **(i)** We have $v_1 = 110\text{ mi/h} = \frac{110(5280)}{3600}\text{ ft/s} = 161.\overline{3}\text{ ft/s}$, $v_0 = -90\text{ mi/h} = -132\text{ ft/s}$, and $m = \frac{5/16}{32} = \frac{5}{512}$. So the change in momentum is

$p(t_1) - p(t_0) = mv_1 - mv_0 = \frac{5}{512}\left[161.\overline{3} - (-132)\right] = \frac{1466.\overline{6}}{512} \approx 2.86$ slug-ft/s.

(ii) From part (a) and part (b)(i) we have $\int_0^{0.01} F(t)dt = p(0.01) - p(0) \approx 2.86$, so the average force over the interval $[0, 0.01] = \frac{1}{0.01}\int_0^{0.01} F(t)dt \approx \frac{1}{0.01}(2.86) = 286$ lb.

10. **(a)** $F = ma = m\dfrac{dv}{dt} \quad\Rightarrow\quad \displaystyle\int_{t_0}^{t} F(u)v(u)du = \int_{t_0}^{t} mv(u)\frac{dv(u)}{du}du = \tfrac{1}{2}mv^2(t) - \tfrac{1}{2}mv^2(t_0) = W(t)$ by Problem 8. Thus $P = \dfrac{dW}{dt} = \dfrac{d}{dt}\displaystyle\int_{t_0}^{t} F(u)v(u)du = F(t)v(t)$.

(b) Note that $60\text{ mi/h} = 60 \cdot \dfrac{5280}{(60)^2}\text{ ft/s} = 88\text{ ft/s}$. Assume constant acceleration: $a = \dfrac{88-0}{10} = 8.8\text{ ft/s}^2$.

Then $s(t) = 8.8\left(\frac{1}{2}t^2\right) \quad\Rightarrow\quad s(10) = 440$, so average velocity $= \dfrac{440-0}{10-0} = 44$ ft/s. Therefore

$P = Fv = mav = \frac{3000}{32}(44)(8.8) = 36{,}300$ ft-lb/s $= \frac{36{,}300}{550}$ horsepower $= 66$ horsepower.

(c) If θ is the angle of inclination, then $\tan\theta = 0.04 \quad\Rightarrow\quad \sin\theta = \dfrac{0.04}{\sqrt{1.0016}} \approx 0.04$. So

$F = mg\sin\theta \approx 3000(0.04) = 120\text{ lb} \quad\Rightarrow\quad P = Fv = 120(88) = 10{,}560\text{ ft-lb/s} = \frac{10{,}560}{550}\text{ hp} = 19.2\text{ hp}$.

11. **(a)** The segment is obtained by rotating the part of the circle $x^2 + y^2 = r^2$ given by $r - h \le y \le r$ about the y-axis. (Take the y-axis pointing downward.) So

$$\begin{aligned} V &= \textstyle\int_{r-h}^{r} \pi x^2\,dy = \int_{r-h}^{r} \pi(r^2 - y^2)dy = \pi\left[r^2y - \frac{1}{3}y^3\right]_{r-h}^{r} = \pi\left[\left(r^3 - \frac{1}{3}r^3\right) - \left(r^2[r-h] - \frac{1}{3}[r-h]^3\right)\right] \\ &= \pi\left[\tfrac{2}{3}r^3 - \left(r^3 - r^2h - \tfrac{1}{3}r^3 + r^2h - rh^2 + \tfrac{1}{3}h^3\right)\right] = \pi\left[\tfrac{2}{3}r^3 - \left(\tfrac{2}{3}r^3 - rh^2 + h^3\right)\right] = \pi\left(rh^2 - \tfrac{1}{3}h^3\right) \\ &= \pi h^2\left(r - \tfrac{h}{3}\right) = \tfrac{1}{3}\pi h^2(3r - h) \end{aligned}$$

(b) The smaller segment has height $h = 1 - x$ and so by part (a) its volume is $V = \frac{1}{3}\pi(1-x)^2[3 - (1-x)] = \frac{1}{3}\pi(x-1)^2(x+2)$. This volume must be $\frac{1}{3}$ of the total volume of the sphere, which is $\frac{4}{3}\pi(1)^3$. So $\frac{1}{3}\pi(x-1)^2(x+2) = \frac{1}{3}\left(\frac{4}{3}\pi\right) \quad\Rightarrow\quad (x^2 - 2x + 1)(x+2) = \frac{4}{3} \quad\Rightarrow$ $x^3 - 3x + 2 = \frac{4}{3} \quad\Rightarrow\quad 3x^3 - 9x + 2 = 0$. Using Newton's method with $f(x) = 3x^3 - 9x + 2$, $f'(x) = 9x^2 - 9$, we get $x_{n+1} = x_n - \dfrac{3x_n^3 - 9x_n + 2}{9x_n^2 - 9}$. Taking $x_1 = 0$, we get $x_2 \approx 0.2222$, $x_3 \approx 0.2261$, and $x_4 \approx 0.2261$, so, correct to four decimal places, $x \approx 0.2261$.

(c) With $r = 0.5$ and $s = 0.75$, the given equation becomes $x^3 - 3(0.5)x^2 + 4(0.5)^3(0.75) = 0 \quad\Rightarrow$ $x^3 - \frac{3}{2}x^2 + 4\left(\frac{1}{8}\right)\frac{3}{4} = 0 \quad\Rightarrow\quad 8x^3 - 12x^2 + 3 = 0$. We use Newton's method with $f(x) = 8x^3 - 12x^2 + 3$, $f'(x) = 24x^2 - 24x$, so $x_{n+1} = x_n - \dfrac{8x_n^3 - 12x_n^2 + 3}{24x_n^2 - 24x_n}$. Take $x_1 = 0.5$. Then $x_2 \approx 0.6667$, $x_3 \approx 0.6736$, and $x_4 \approx 0.6736$. So to four decimals the depth is 0.6736 m.

(d) **(i)** From part (a), the volume of water in the bowl is $V = \frac{1}{3}\pi h^2(3r - h) = \frac{1}{3}\pi h^2(15 - h) = 5\pi h^2 - \frac{1}{3}\pi h^3$. We are given that $\dfrac{dV}{dt} = 0.2\,\text{m}^3/\text{s}$ and we want to find $\dfrac{dh}{dt}$ when $h = 3$. Now $\dfrac{dV}{dt} = 10\pi h\dfrac{dh}{dt} - \pi h^2\dfrac{dh}{dt}$, so $\dfrac{dh}{dt} = \dfrac{0.2}{\pi(10h - h^2)}$. When $h = 3$, we have $\dfrac{dh}{dt} = \dfrac{0.2}{\pi(10 \cdot 3 - 3^2)} = \dfrac{1}{105\pi} \approx 0.003\,\text{in/s}$.

(ii) From part (a), the volume of water required to fill the bowl is $V = \frac{1}{2} \cdot \frac{4}{3}\pi(5)^3 - \frac{1}{3}\pi(4)^2(15 - 4) = \frac{2}{3} \cdot 125\pi - \frac{16}{3} \cdot 11\pi = \frac{74}{3}\pi$. To find the time required to fill the bowl we divide this volume by the rate: Time $= \dfrac{74\pi/3}{0.2} = \dfrac{370\pi}{3} \approx 387\,\text{s} \approx 6.5\,\text{min}$

12. **(a)** The volume above the surface is $\int_0^{L-h} A(y)dy = \int_{-h}^{L-h} A(y)dy - \int_{-h}^{0} A(y)dy$. So the proportion of volume above the surface is $\dfrac{\int_0^{L-h} A(y)dy}{\int_{-h}^{L-h} A(y)dy} = \dfrac{\int_{-h}^{L-h} A(y)dy - \int_{-h}^{0} A(y)dy}{\int_{-h}^{L-h} A(y)dy}$. Now by Archimedes' Principle, we have $\rho_f g \int_{-h}^{0} A(y)dy = \rho_0 g \int_{-h}^{L-h} A(y)dy$, so $\int_{-h}^{0} A(y)dy = \dfrac{\rho_0}{\rho_f}\int_{-h}^{L-h} A(y)dy$. Therefore

$$\frac{\int_0^{L-h} A(y)dy}{\int_{-h}^{L-h} A(y)dy} = \frac{\int_{-h}^{L-h} A(y)dy - \frac{\rho_0}{\rho_f}\int_{-h}^{L-h} A(y)dy}{\int_{-h}^{L-h} A(y)dy} = \frac{\rho_f - \rho_0}{\rho_f},$$ so the proportion of volume above the surface is $100\left(\dfrac{\rho_f - \rho_0}{\rho_f}\right)\%$.

(b) For an iceberg, the percentage of volume above the surface is $100\left(\dfrac{1030 - 917}{1030}\right)\% \approx 11\%$.

(c) No, the water does not overflow. Let V_i be the volume of the ice cube, and let V_w be the volume of the water which results from the melting. Then by the formula derived in part (a), the volume of ice above the surface of the water is $\dfrac{\rho_f - \rho_0}{\rho_f}V_i$, and so the volume below the surface is $V_i - \dfrac{\rho_f - \rho_0}{\rho_f}V_i = \dfrac{\rho_0}{\rho_f}V_i$. Now the mass of the ice cube is the same as the mass of the water which is created when it melts, namely $m = \rho_0 V_i = \rho_f V_w \quad \Rightarrow \quad V_w = \dfrac{\rho_0}{\rho_f}V_i$. So when the ice cube melts, the volume of the resulting water is the same as the underwater volume of the ice cube, and so the water does not overflow.

(d) The figure shows the instant when the height of the exposed part of the ball is y. Using the formula in (a) with $r = 0.4$ and $h = 0.8 - y$, we see that the volume of the submerged part of the sphere is $\frac{1}{3}\pi(0.8 - y)^2[1.2 - (0.8 - y)]$, so its weight is $\frac{1000}{3}\pi g(0.8 - y)^2(0.4 + y)$.

y
0.4
y
x
0.8 − y

Let $s = 0.8 - y$. Then the work done to submerge the sphere is

$$W = \int_0^{0.8} g\tfrac{1000}{3}\pi s^2(1.2 - s)ds = \tfrac{1000}{3}\pi g\int_0^{0.8}(1.2s^2 - s^3)ds = \tfrac{1000}{3}\pi g\left[\tfrac{1.2}{3}s^3 - \tfrac{1}{4}s^4\right]_0^{0.8}$$
$$= g\tfrac{1000}{3}\pi\left[0.4s^3 - \tfrac{1}{4}s^4\right]_0^{0.8} = g\tfrac{1000}{3}\pi(0.2048 - 0.1024) = 9.8\tfrac{1000}{3}\pi(0.1024) \approx 1.05 \times 10^3\,\text{joules}.$$

CHAPTER SIX

EXERCISES 6.1

1. The diagram shows that there is a horizontal line which intersects the graph more than once, so the function is not one-to-one.

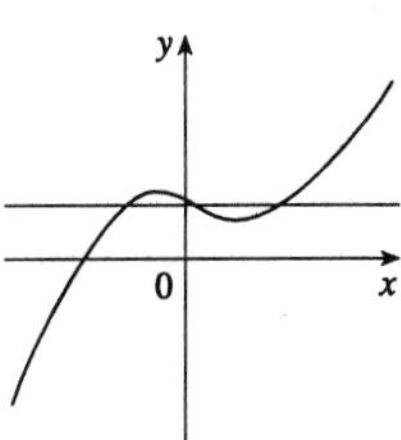

2. The function is one-to-one because no horizontal line intersects the graph more than once.

3. The function is one-to-one because no horizontal line intersects the graph more than once.

4. The diagram shows that there is a horizontal line which intersects the graph more than once, so the function is not one-to-one.

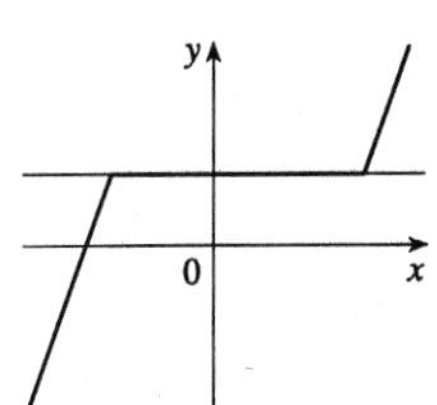

5. The diagram shows that there is a horizontal line which intersects the graph more than once, so the function is not one-to-one.

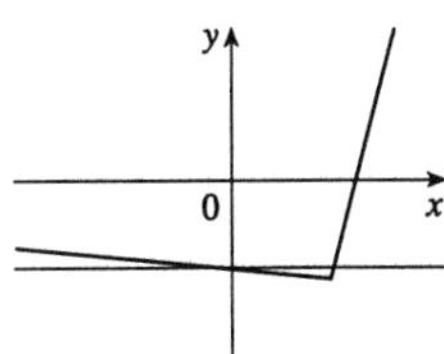

6. The function is one-to-one because no horizontal line intersects the graph more than once.

7. $x_1 \neq x_2 \quad\Rightarrow\quad 7x_1 \neq 7x_2 \quad\Rightarrow\quad 7x_1 - 3 \neq 7x_2 - 3 \quad\Rightarrow\quad f(x_1) \neq f(x_2)$, so f is 1-1.

8. $f(x) = x^2 - 2x + 5 \quad\Rightarrow\quad f(0) = 5 = f(2)$, so f is not one-to-one.

9. $x_1 \neq x_2 \quad\Rightarrow\quad \sqrt{x_1} \neq \sqrt{x_2} \quad\Rightarrow\quad g(x_1) \neq g(x_2)$, so g is 1-1.

10. $g(x) = |x| \quad\Rightarrow\quad g(-1) = 1 = g(1)$, so g is not one-to-one.

11. $h(x) = x^4 + 5 \quad\Rightarrow\quad h(1) = 6 = h(-1)$, so h is not 1-1.

12. $x_1 \neq x_2 \;\Rightarrow\; x_1^4 \neq x_2^4$ (since $x \geq 0$) $\Rightarrow\; x_1^4 + 5 \neq x_2^4 + 5 \;\Rightarrow\; h(x_1) \neq h(x_2)$, so h is 1-1.

13. $x_1 \neq x_2 \quad\Rightarrow\quad 4x_1 \neq 4x_2 \quad\Rightarrow\quad 4x_1 + 7 \neq 4x_2 + 7 \quad\Rightarrow\quad f(x_1) \neq f(x_2)$, so f is 1-1. $y = 4x + 7 \quad\Rightarrow$ $4x = y - 7 \quad\Rightarrow\quad x = (y - 7)/4$. Interchange x and y: $y = (x - 7)/4$. So $f^{-1}(x) = (x - 7)/4$.

14. $f(x) = \dfrac{x-2}{x+2}$. If $f(x_1) = f(x_2)$, then $\dfrac{x_1 - 2}{x_1 + 2} = \dfrac{x_2 - 2}{x_2 + 2} \quad\Rightarrow$

$x_1x_2 + 2x_1 - 2x_2 - 4 = x_1x_2 - 2x_1 + 2x_2 - 4 \quad\Rightarrow\quad 4x_1 = 4x_2 \quad\Rightarrow\quad x_1 = x_2$, so f is 1-1. $y = \dfrac{x-2}{x+2}$

$\Rightarrow \quad xy + 2y = x - 2 \quad\Rightarrow\quad x(1-y) = 2(y+1) \quad\Rightarrow\quad x = \dfrac{2(1+y)}{1-y}$.

Interchange x and y: $y = \dfrac{2(1+x)}{1-x}$. So $f^{-1}(x) = \dfrac{2(1+x)}{1-x}$.

15. $f(x) = \dfrac{1+3x}{5-2x}$. If $f(x_1) = f(x_2)$, then $\dfrac{1+3x_1}{5-2x_1} = \dfrac{1+3x_2}{5-2x_2} \quad\Rightarrow$

$5 + 15x_1 - 2x_2 - 6x_1x_2 = 5 - 2x_1 + 15x_2 - 6x_1x_2 \quad\Rightarrow\quad 17x_1 = 17x_2 \quad\Rightarrow\quad x_1 = x_2$, so f is one-to-one.

$y = \dfrac{1+3x}{5-2x} \quad\Rightarrow\quad 5y - 2xy = 1 + 3x \quad\Rightarrow\quad x(3+2y) = 5y - 1 \quad\Rightarrow\quad x = \dfrac{5y-1}{2y+3}$.

Interchange x and y: $y = \dfrac{5x-1}{2x+3}$. So $f^{-1}(x) = \dfrac{5x-1}{2x+3}$.

16. $x_1 \neq x_2 \quad\Rightarrow\quad x_1^3 \neq x_2^3 \quad\Rightarrow\quad -4x_1^3 \neq -4x_2^3 \quad\Rightarrow\quad 5 - 4x_1^3 \neq 5 - 4x_2^3 \quad\Rightarrow\quad f(x_1) \neq f(x_2)$, so f is one-to-one. $y = 5 - 4x^3 \quad\Rightarrow\quad 4x^3 = 5 - y \quad\Rightarrow\quad x^3 = (5-y)/4 \quad\Rightarrow\quad x = \left(\dfrac{5-y}{4}\right)^{1/3}$.

Interchange x and y: $y = \left(\dfrac{5-x}{4}\right)^{1/3}$. So $f^{-1}(x) = \left(\dfrac{5-x}{4}\right)^{1/3}$.

17. $x_1 \neq x_2 \quad\Rightarrow\quad 5x_1 \neq 5x_2 \quad\Rightarrow\quad 2 + 5x_1 \neq 2 + 5x_2 \quad\Rightarrow\quad \sqrt{2+5x_1} \neq \sqrt{2+5x_2} \quad\Rightarrow$

$f(x_1) \neq f(x_2)$, so f is 1-1. $y = \sqrt{2+5x} \quad\Rightarrow\quad y^2 = 2 + 5x$ and $y \geq 0 \quad\Rightarrow\quad 5x = y^2 - 2 \quad\Rightarrow$

$x = \dfrac{y^2-2}{5}$, $y \geq 0$. Interchange x and y: $y = \dfrac{x^2-2}{5}$, $x \geq 0$. So $f^{-1}(x) = \dfrac{x^2-2}{5}$, $x \geq 0$.

18. $f(x) = x^2 + x$, $x \geq -\frac{1}{2}$. $f'(x) = 2x + 1 > 0$ for $x > -\frac{1}{2}$, so f is increasing on $\left[-\frac{1}{2}, \infty\right)$ and hence one-to-one. (Or use the Horizontal Line Test.) $y = x^2 + x \quad\Rightarrow\quad x^2 + x - y = 0 \quad\Rightarrow\quad x = \frac{1}{2}\left(-1 \pm \sqrt{1+4y}\right)$ by the quadratic formula. But $x \geq -\frac{1}{2} \quad\Rightarrow\quad x = \frac{1}{2}\left(-1 + \sqrt{1+4y}\right)$.

Interchange x and y: $y = \frac{1}{2}\left(-1 + \sqrt{1+4x}\right)$. So $f^{-1}(x) = \frac{1}{2}\left(-1 + \sqrt{1+4x}\right)$.

19. **(a)** $x_1 \neq x_2 \Rightarrow 2x_1 \neq 2x_2 \Rightarrow 2x_1 + 1 \neq 2x_2 + 1 \Rightarrow f(x_1) \neq f(x_2)$, so f is 1-1.

(b) $f(1) = 3 \quad\Rightarrow\quad g(3) = 1$. Also $f'(x) = 2$, so $g'(3) = 1/f'(3) = \frac{1}{2}$.

(c) $y = 2x + 1 \quad\Rightarrow\quad x = \frac{1}{2}(y-1)$.
Interchanging x and y gives
$y = \frac{1}{2}(x-1)$, so $f^{-1}(x) = \frac{1}{2}(x-1)$.
Domain(g) = range$(f) = \mathbb{R}$.
Range(g) = domain$(f) = \mathbb{R}$.

(d) $g(x) = \frac{1}{2}(x-1) \quad\Rightarrow\quad g'(x) = \frac{1}{2}$
$\Rightarrow \quad g'(3) = \frac{1}{2}$ as in (b).

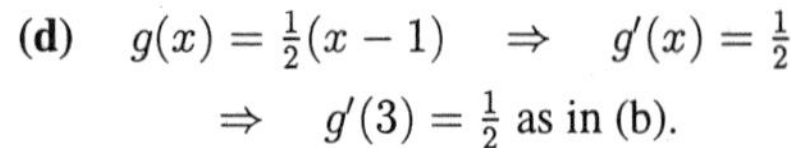

(e)

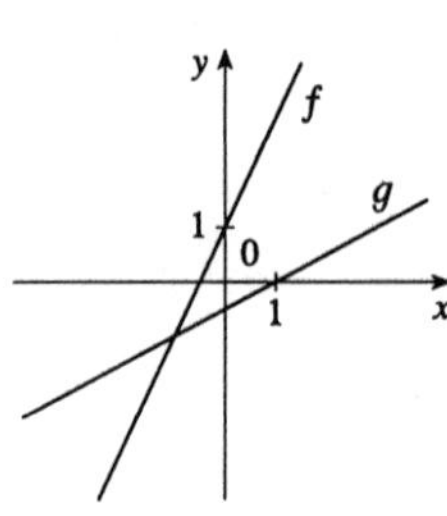

20. **(a)** $x_1 \neq x_2 \Rightarrow -x_1 \neq -x_2 \Rightarrow 6 - x_1 \neq 6 - x_2 \Rightarrow f(x_1) \neq f(x_2)$, so f is 1-1.

(b) $f'(x) = -1$ and $g(2) = 4$ since $f(4) = 2$, so by Theorem 8, $g'(2) = \dfrac{1}{f'(4)} = \dfrac{1}{-1} = -1$.

(c) $y = 6 - x \quad \Rightarrow \quad x = 6 - y$. Interchange x and y: $y = 6 - x$. So $g(x) = 6 - x$. Domain $= \mathbb{R} =$ Range.

(d) $g'(x) = -1$, so $g'(2) = -1$.

(e)

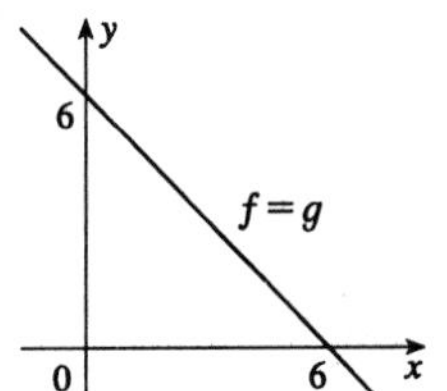

21. **(a)** $x_1 \neq x_2 \quad \Rightarrow \quad x_1^3 \neq x_2^3 \quad \Rightarrow \quad f(x_1) \neq f(x_2)$, so f is one-to-one.

(b) $f'(x) = 3x^2$ and $f(2) = 8 \quad \Rightarrow \quad g(8) = 2$, so $g'(8) = 1/f'(g(8)) = 1/f'(2) = \frac{1}{12}$.

(c) $y = x^3 \quad \Rightarrow \quad x = y^{1/3}$. Interchanging x and y gives $y = x^{1/3}$, so $f^{-1}(x) = x^{1/3}$.
Domain$(g) =$ range$(f) = \mathbb{R}$.
Range$(g) =$ domain$(f) = \mathbb{R}$.

(d) $g(x) = x^{1/3} \quad \Rightarrow$
$g'(x) = \frac{1}{3}x^{-2/3} \quad \Rightarrow$
$g'(8) = \frac{1}{3}\left(\frac{1}{4}\right) = \frac{1}{12}$ as in part (b).

(e)

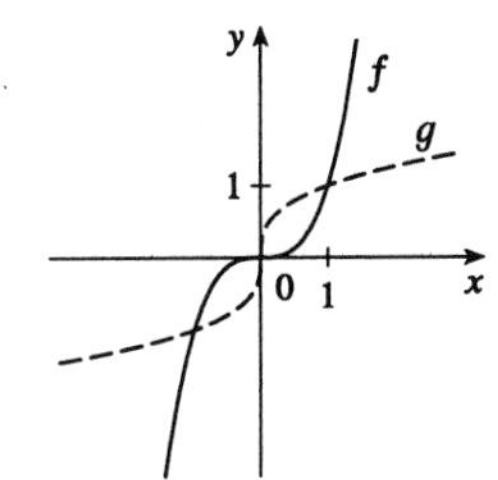

22. **(a)** $x_1 \neq x_2 \quad \Rightarrow \quad x_1 - 2 \neq x_2 - 2 \quad \Rightarrow \quad \sqrt{x_1 - 2} \neq \sqrt{x_2 - 2} \quad \Rightarrow \quad f(x_1) \neq f(x_2)$, so f is 1-1.

(b) $f(6) = 2$, so $g(2) = 6$. Also $f'(x) = \dfrac{1}{2\sqrt{x - 2}}$, so

$$g'(2) = \frac{1}{f'(g(2))} = \frac{1}{f'(6)} = \frac{1}{1/4} = 4.$$

(c) $y = \sqrt{x - 2} \quad \Rightarrow \quad y^2 = x - 2$
$\Rightarrow \quad x = y^2 + 2$. Interchange x and y:
$y = x^2 + 2$. So $g(x) = x^2 + 2$.

(d) Domain $= [0, \infty)$, range $= [2, \infty)$.
$g'(x) = 2x \quad \Rightarrow \quad g'(2) = 4$.

(e)

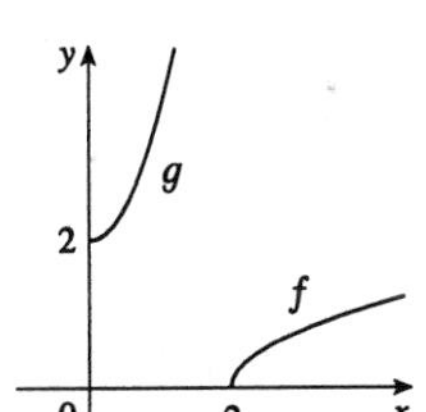

23. **(a)** Since $x \geq 0$, $x_1 \neq x_2 \Rightarrow x_1^2 \neq x_2^2 \Rightarrow 9 - x_1^2 \neq 9 - x_2^2 \Rightarrow f(x_1) \neq f(x_2)$, so f is 1-1.

(b) $f'(x) = -2x$ and $f(1) = 8 \Rightarrow g(8) = 1$, so $g'(8) = \dfrac{1}{f'(g(8))} = \dfrac{1}{f'(1)} = \dfrac{1}{(-2)} = -\dfrac{1}{2}$.

(c) $y = 9 - x^2 \quad \Rightarrow \quad x^2 = 9 - y \quad \Rightarrow$

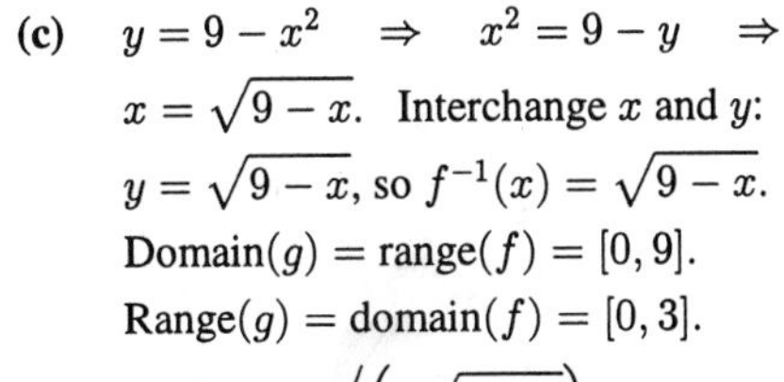

$x = \sqrt{9 - x}$. Interchange x and y:
$y = \sqrt{9 - x}$, so $f^{-1}(x) = \sqrt{9 - x}$.
Domain$(g) =$ range$(f) = [0, 9]$.
Range$(g) =$ domain$(f) = [0, 3]$.

(d) $g'(x) = -1\Big/\left(2\sqrt{9 - x}\right) \quad \Rightarrow$
$g'(8) = -\frac{1}{2}$ as in (b).

(e)

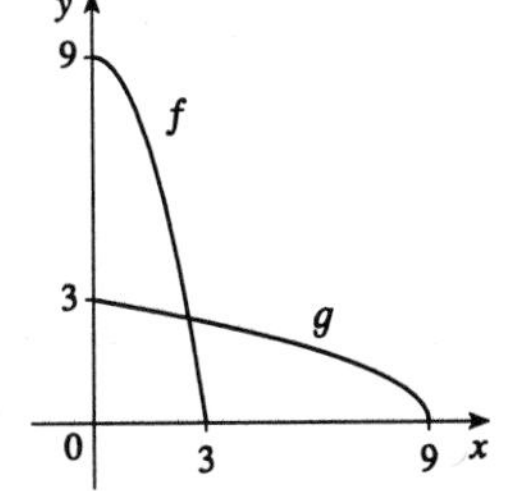

24. **(a)** $x_1 \neq x_2 \Rightarrow x_1 - 1 \neq x_2 - 1 \Rightarrow \dfrac{1}{x_1 - 1} \neq \dfrac{1}{x_2 - 1} \Rightarrow f(x_1) \neq f(x_2)$, so f is 1-1.

(b) $g(2) = \frac{3}{2}$ since $f\left(\frac{3}{2}\right) = 2$. Also $f'(x) = -\dfrac{1}{(x-1)^2}$, so $g'(2) = \dfrac{1}{f'\left(\frac{3}{2}\right)} = \dfrac{1}{-4} = -\dfrac{1}{4}$.

(c) $y = \dfrac{1}{x-1} \quad\Rightarrow\quad x - 1 = \dfrac{1}{y} \quad\Rightarrow$

$x = 1 + \dfrac{1}{y}$. Interchange x and y: $y = 1 + \dfrac{1}{x}$.

So $g(x) = 1 + 1/x$, $x > 0$ (since $y > 1$)

Domain $= (0, \infty)$, range $= (1, \infty)$.

(d) $g'(x) = -1/x^2$, so $g'(2) = -\frac{1}{4}$.

(e)

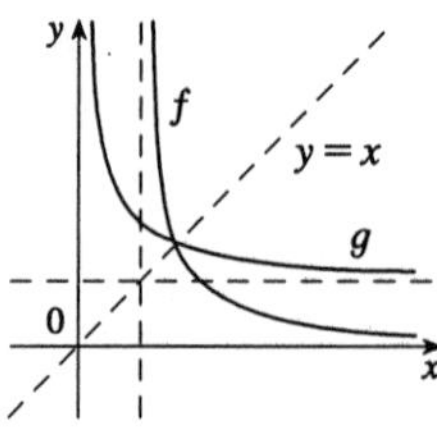

25. $f(0) = 1 \;\Rightarrow\; g(1) = 0$, and $f'(x) = 3x^2 + 1 \;\Rightarrow\; f'(0) = 1$. Therefore $g'(1) = \dfrac{1}{f'(g(1))} = \dfrac{1}{f'(0)} = \dfrac{1}{1} = 1$.

26. $f(1) = 2 \;\Rightarrow\; g(2) = 1$, and $f'(x) = 5x^4 - 3x^2 + 2 \;\Rightarrow\; f'(1) = 4$. Thus $g'(2) = \dfrac{1}{f'(g(2))} = \dfrac{1}{f'(1)} = \dfrac{1}{4}$.

27. $f(0) = 3 \quad\Rightarrow\quad g(3) = 0$, and $f'(x) = 2x + \frac{\pi}{2}\sec^2(\pi x/2) \quad\Rightarrow\quad f'(0) = 1 \cdot \frac{\pi}{2} = \frac{\pi}{2}$. Thus

$g'(3) = \dfrac{1}{f'(g(3))} = \dfrac{1}{f'(0)} = \dfrac{2}{\pi}$.

28. $f(1) = 2 \quad\Rightarrow\quad g(2) = 1$, and $f'(x) = \dfrac{3x^2 + 2x + 1}{2\sqrt{x^3 + x^2 + x + 1}} \quad\Rightarrow\quad f'(1) = \dfrac{3 + 2 + 1}{2\sqrt{1 + 1 + 1 + 1}} = \dfrac{3}{2}$.

Therefore, $g'(2) = \dfrac{1}{f'(g(2))} = \dfrac{2}{3}$.

29. $f(4) = 5 \quad\Rightarrow\quad g(5) = 4$. Therefore, $g'(5) = \dfrac{1}{f'(g(5))} = \dfrac{1}{f'(4)} = \dfrac{1}{2/3} = \dfrac{3}{2}$.

30. $f(3) = 2 \quad\Rightarrow\quad g(2) = 3$. Thus $g'(2) = \dfrac{1}{f'(g(2))} = \dfrac{1}{f'(3)} = 9$. Thus, $G(x) = \dfrac{1}{g(x)} \quad\Rightarrow$

$G'(x) = -\dfrac{g'(x)}{[g(x)]^2} \quad\Rightarrow\quad G'(2) = -\dfrac{g'(2)}{[g(2)]^2} = -\dfrac{9}{(3)^2} = -1$.

31. $y = 1 - 2/x^2 \quad\Rightarrow\quad 1 - y = 2/x^2$

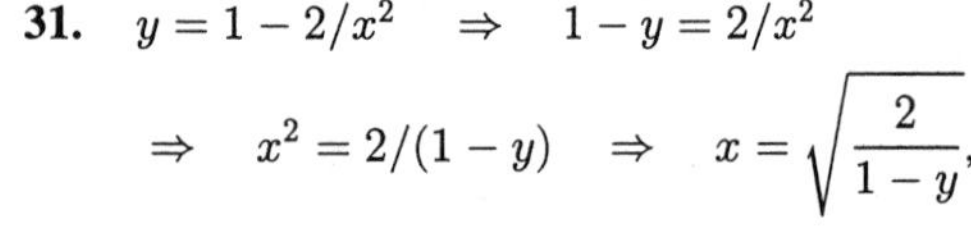

$\Rightarrow\quad x^2 = 2/(1-y) \quad\Rightarrow\quad x = \sqrt{\dfrac{2}{1-y}}$,

since $x > 0$. Interchange x and y:

$y = \sqrt{\dfrac{2}{1-x}}$. So $f^{-1}(x) = \sqrt{\dfrac{2}{1-x}}$.

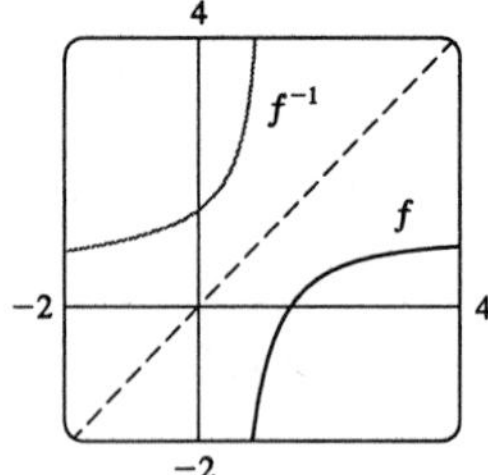

32. $y = \sqrt{x^2 + 2x}$, $x > 0 \quad\Rightarrow\quad y > 0$ and $y^2 = x^2 + 2x$

$\Rightarrow\quad x^2 + 2x - y^2 = 0$. Now we use the quadratic

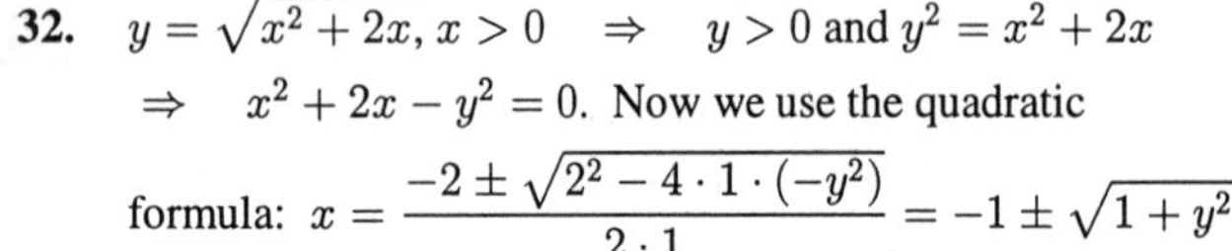

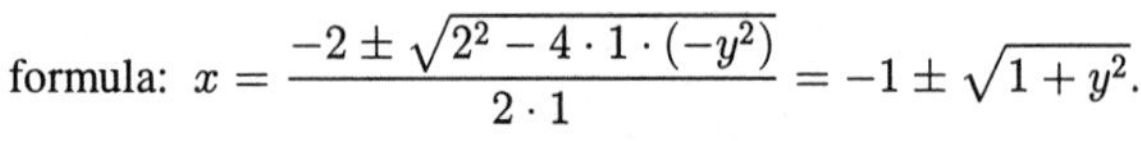

formula: $x = \dfrac{-2 \pm \sqrt{2^2 - 4 \cdot 1 \cdot (-y^2)}}{2 \cdot 1} = -1 \pm \sqrt{1 + y^2}$.

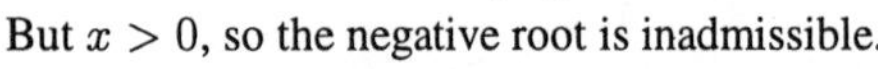

But $x > 0$, so the negative root is inadmissible.

Interchange x and y: $y = -1 + \sqrt{1 + x^2}$.

So $f^{-1}(x) = -1 + \sqrt{1 + x^2}$, $x > 0$.

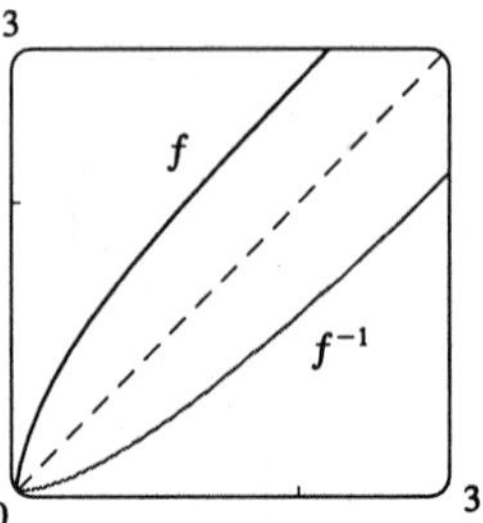

33. Since $f'(x) = \dfrac{2x}{2\sqrt{x^2+1}} - 1 = \dfrac{x - \sqrt{x^2+1}}{\sqrt{x^2+1}}$ is negative for all x, we know that f is a decreasing function on $\mathbb{R}$, and hence is 1-1. We could also use the Horizontal Line Test to show that f is 1-1.
The parametric equations for the graph of f are $x = t$, $y = \sqrt{t^2+1} - t$; for the graph of f^{-1} they are $x = \sqrt{t^2+1} - t$, $y = t$.

34. Since $f'(x) = 1 + \cos x \geq 0$ for all x, f is increasing and is therefore one-to-one.
We can also use the Horizontal Line Test to show that f is 1-1. The parametric equations for the graph of f are $x = t$, $y = t + \sin t$; for the graph of f^{-1} they are $x = t + \sin t$, $y = t$.

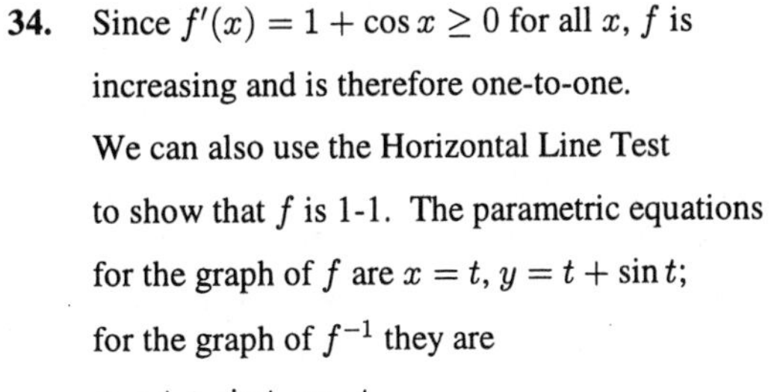

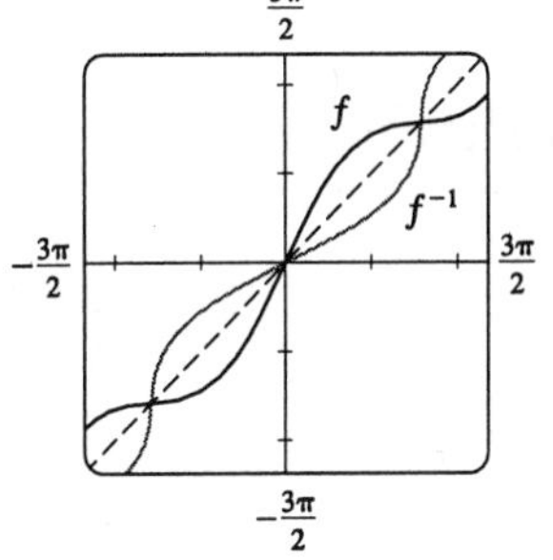

35. **(a)** $\sqrt[5]{x} - \sqrt[5]{y} = y \;\Rightarrow\; \sqrt[5]{x} = y + \sqrt[5]{y} \;\Rightarrow$ $x = \left(y + \sqrt[5]{y}\right)^5$. Interchange x and y: $y = \left(x + \sqrt[5]{x}\right)^5$. So $f^{-1}(x) = \left(x + \sqrt[5]{x}\right)^5$.

(b) The parametric equations for the graph of f^{-1} are $x = t$, $y = \left(t + \sqrt[5]{t}\right)^5$.
So the parametric equations for the graph of $f = \left(f^{-1}\right)^{-1}$ are $x = \left(t + \sqrt[5]{t}\right)^5$, $y = t$.

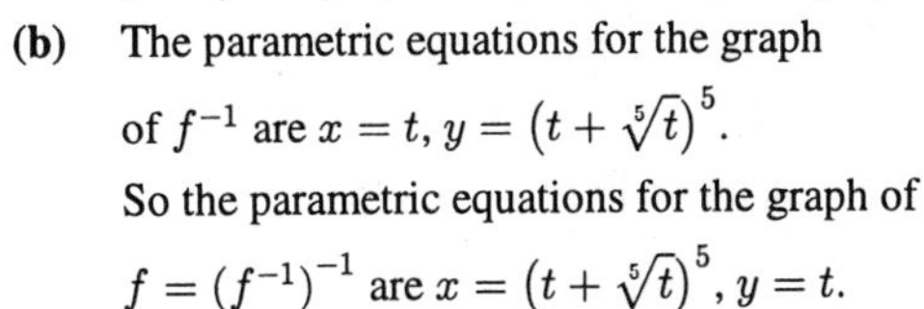

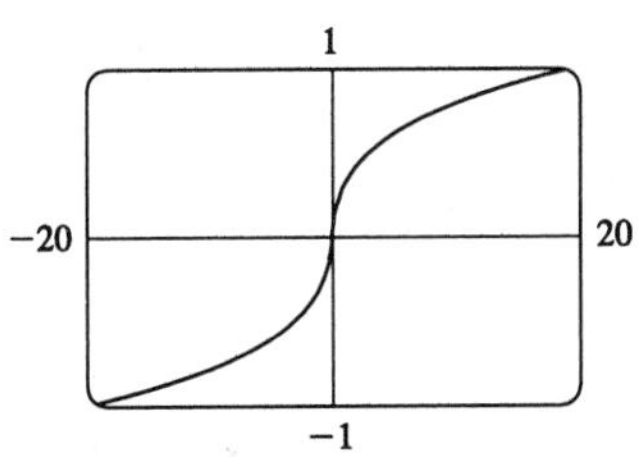

36. In this case $f'(x) = \dfrac{3x^2 + 2x + 1}{2\sqrt{x^3 + x^2 + x + 1}}$, so the parametric equations for the graph of $\left(f^{-1}\right)'$ are

$$x = f(t) = \sqrt{t^3 + t^2 + t + 1},\; y = \frac{1}{f'(t)} = \frac{2\sqrt{t^3 + t^2 + t + 1}}{3t^2 + 2t + 1}.$$

The graph looks like it represents the derivative of the function f^{-1} considered in Example 7, as expected. [Notice that $(f^{-1})'$ changes from increasing to decreasing where f' appears to have an inflection point in Figure 13.]

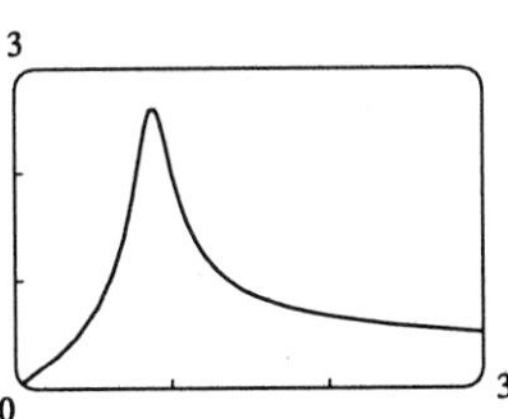

37. $y = \sqrt[n]{x} \;\Rightarrow\; y^n = x \;\Rightarrow\; ny^{n-1}y' = 1 \;\Rightarrow\; y' = \dfrac{1}{ny^{n-1}} = \dfrac{1}{n\left(\sqrt[n]{x}\right)^{n-1}} = \dfrac{1}{n}x^{(1/n)-1}$

38. See Section 6.6.

39. Suppose that f is increasing. If $x_1 \neq x_2$, then either $x_1 < x_2$ or $x_2 < x_1$. If $x_1 < x_2$, then $f(x_1) < f(x_2)$. If $x_2 < x_1$, then $f(x_2) < f(x_1)$. In either case, $x_1 \neq x_2 \Rightarrow f(x_1) \neq f(x_2)$, so f is one-to-one.

40. **(a)** We know that $g'(x) = \dfrac{1}{f'(g(x))}$, so . Thus

$$g''(x) = -\frac{g'(x)f''(g(x))}{[f'(g(x))]^2} = -\frac{f''(g(x))}{f'(g(x))[f'(g(x))]^2} = -\frac{f''(g(x))}{[f'(g(x))]^3}.$$

(b) If f is increasing and concave upward, then $f'(x) > 0$ and $f''(x) > 0$. Thus $g''(x) = -\dfrac{f''(g(x))}{[f'(g(x))]^2}$ is negative, and so g is concave downward.

EXERCISES 6.2

1.

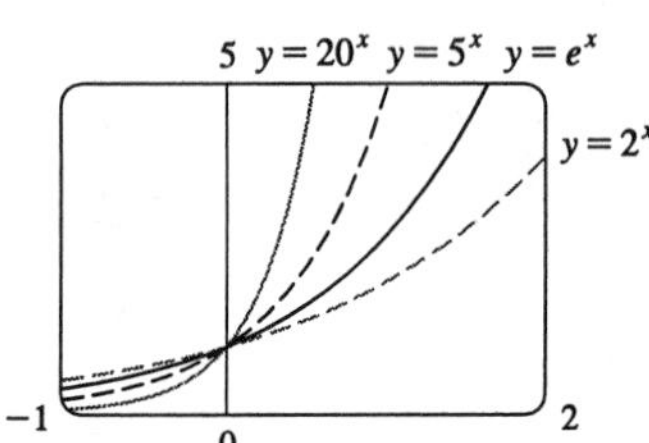

All of these graphs approach 0 as $x \to -\infty$, all of them pass through the point $(0, 1)$, and all of them are increasing and approach ∞ as $x \to \infty$. The larger the base, the faster the function increases for $x > 0$, and the faster it approaches 0 as $x \to -\infty$.

2.

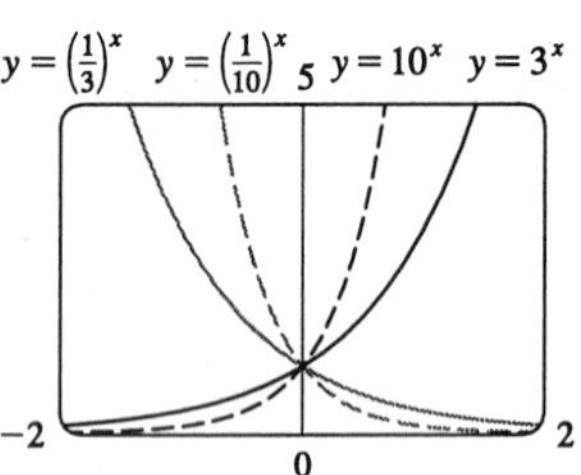

The graph of e^{-x} is the reflection of the graph of e^x in the y-axis, and the graph of 8^{-x} is the reflection of that of 8^x in the y-axis. The graph of 8^x increases more quickly than that of e^x for $x > 0$, and approaches 0 faster as $x \to -\infty$.

3.

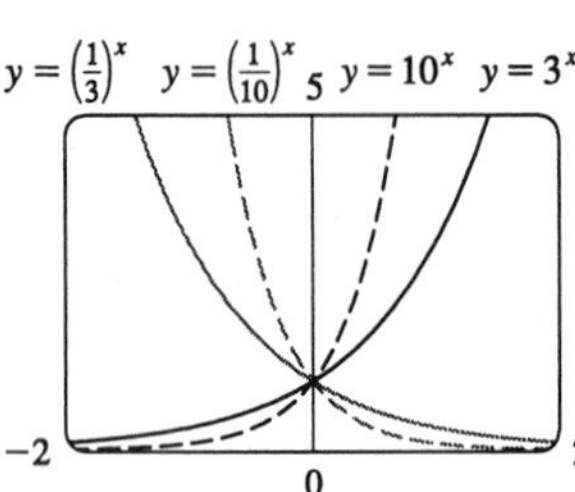

The functions with bases greater than 1 (3^x and 10^x) are increasing, while those with bases less than 1 $\left[\left(\frac{1}{3}\right)^x \text{ and } \left(\frac{1}{10}\right)^x\right]$ are decreasing. The graph of $\left(\frac{1}{3}\right)^x$ is the reflection of that of 3^x about the y-axis, and the graph of $\left(\frac{1}{10}\right)^x$ is the reflection of that of 10^x about the y-axis. The graph of 10^x increases more quickly than that of 3^x for $x > 0$, and approaches 0 faster as $x \to -\infty$.

4.

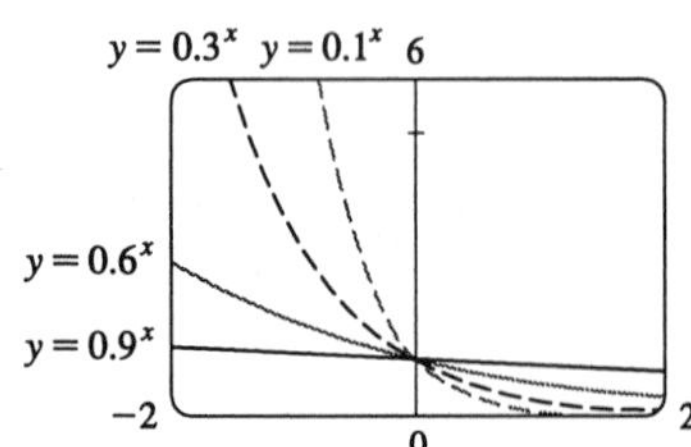

Each of the graphs approaches ∞ as $x \to -\infty$, and each approaches 0 as $x \to \infty$. The smaller the base, the faster the function grows as $x \to -\infty$, and the faster it approaches 0 as $x \to \infty$.

5. $y = 2^x$ $\quad$ $y = 2^x + 1$

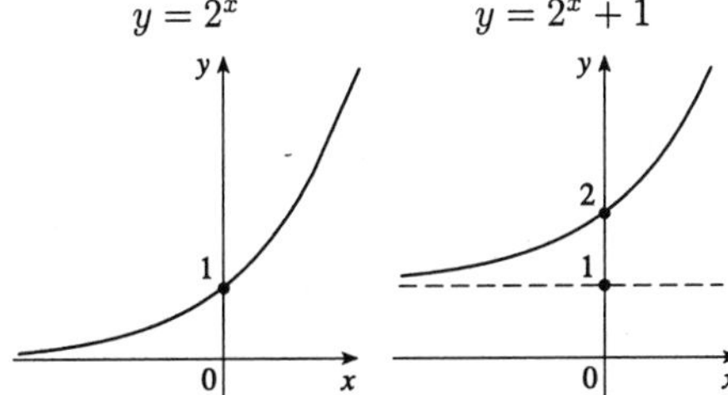

6. $y = 2^x$ $\quad$ $y = 2^{x+1}$

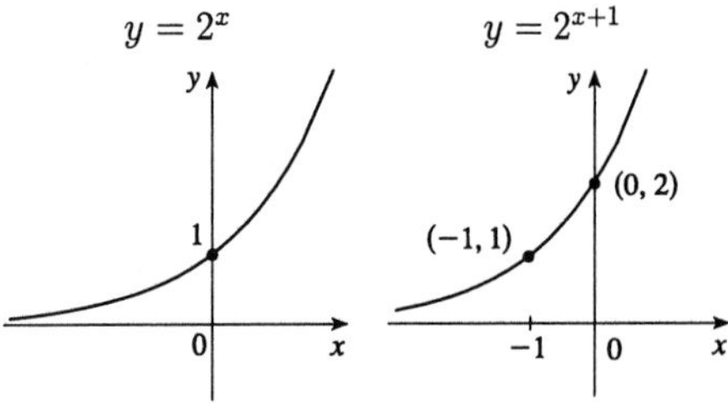

7. $y = 3^x$ $\quad$ $y = 3^{-x}$

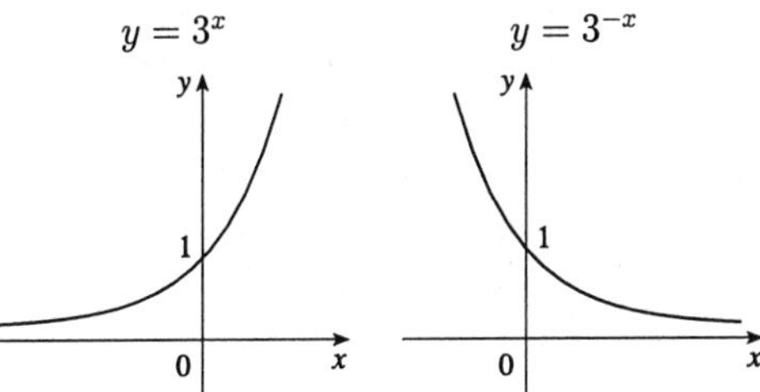

8. $y = 3^x$ $\quad$ $y = -3^x$

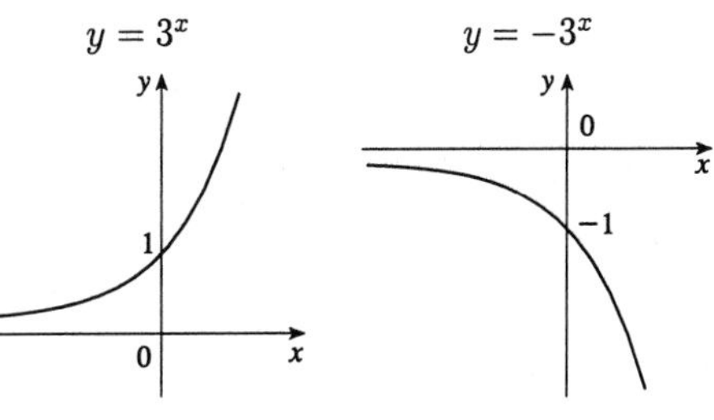

9. $y = -3^{-x}$

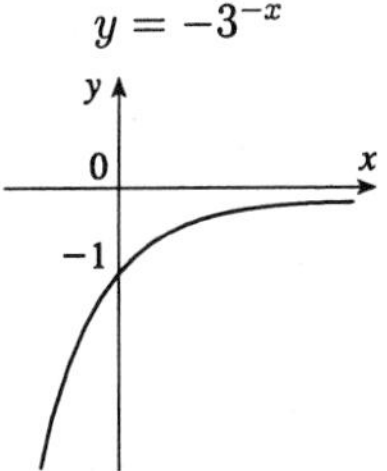

10. $y = 2^x$ $\quad$ $y = 2^{|x|}$

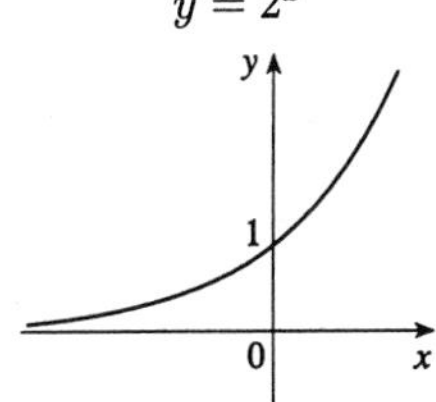

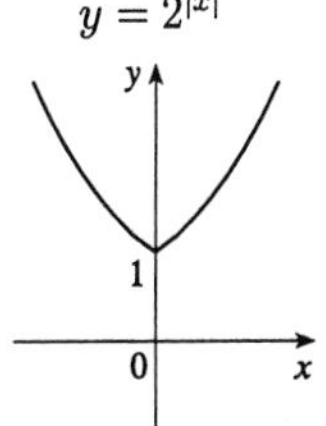

11. $y = -e^x$ $\quad$ $y = 3 - e^x$

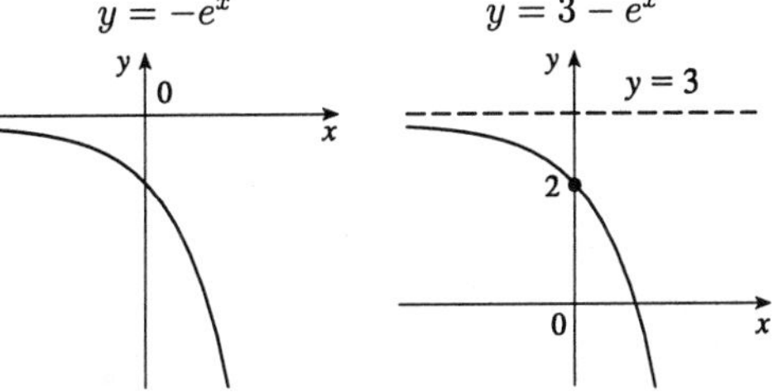

12. $y = -10^{-x}$ $\quad$ $y = 2 + 5(1 - 10^{-x})$

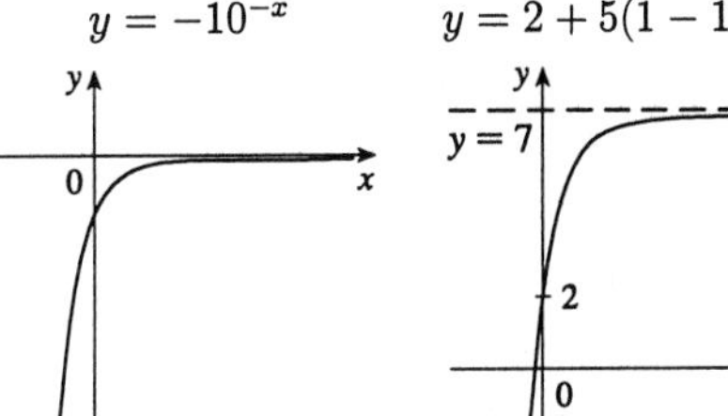

13. $\lim_{x\to\infty} (1.1)^x = \infty$ by Equation 3 since $1.1 > 1$.

14. $\lim_{x\to-\infty} (1.1)^x = 0$ by Equation 3 since $1.1 > 1$.

15. $\lim_{x\to-\infty} \left(\frac{\pi}{4}\right)^x = \infty$ since $0 < \frac{\pi}{4} < 1$.

16. $\lim_{x\to\infty} \left(\frac{2\pi}{7}\right)^x = 0$ since $0 < \frac{2\pi}{7} < 1$.

17. Divide numerator and denominator by e^{3x}: $\displaystyle\lim_{x\to\infty} \frac{e^{3x} - e^{-3x}}{e^{3x} + e^{-3x}} = \lim_{x\to\infty} \frac{1 - e^{-6x}}{1 + e^{-6x}} = \frac{1-0}{1+0} = 1$

18. Divide numerator and denominator by e^{-3x}: $\displaystyle\lim_{x\to-\infty} \frac{e^{3x} - e^{-3x}}{e^{3x} + e^{-3x}} = \lim_{x\to-\infty} \frac{e^{6x} - 1}{e^{6x} + 1} = \frac{0-1}{0+1} = -1$

19. $\displaystyle\lim_{x\to1^-} e^{2/(x-1)} = 0$ since $\displaystyle\frac{2}{x-1} \to -\infty$ as $x \to 1^-$.

20. $\displaystyle\lim_{x\to1^+} e^{2/(x-1)} = \infty$ since $\displaystyle\frac{2}{x-1} \to \infty$ as $x \to 1^+$.

21. $\displaystyle\lim_{x\to\pi/2^-} \frac{2}{1 + e^{\tan x}} = 0$ since $\tan x \to \infty \quad \Rightarrow \quad e^{\tan x} \to \infty$.

22. As $x \to 0^-$, $\cot x = \dfrac{\cos x}{\sin x} \to -\infty$, so $e^{\cot x} \to 0$ and $\lim\limits_{x\to 0} \dfrac{2}{1+e^{\cot x}} = \dfrac{2}{1+0} = 2$.

23. $2\text{ ft} = 24\text{ in}$, $f(24) = 24^2\text{ in} = 576\text{ in} = 48\text{ ft}$. $g(24) = 2^{24}\text{ in} = 2^{24}/(12 \cdot 5280)\text{mi} \approx 265\text{ mi}$

24.

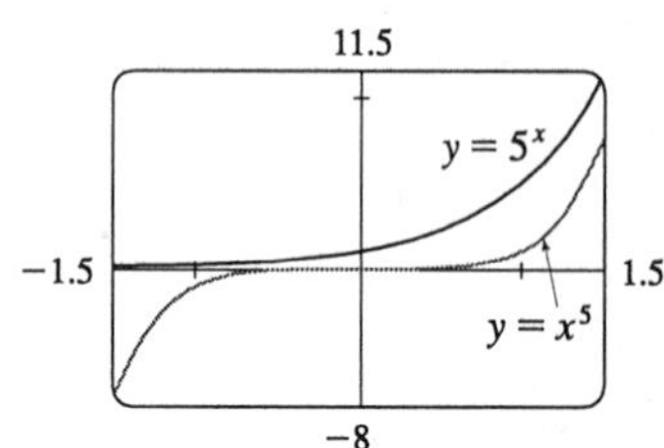

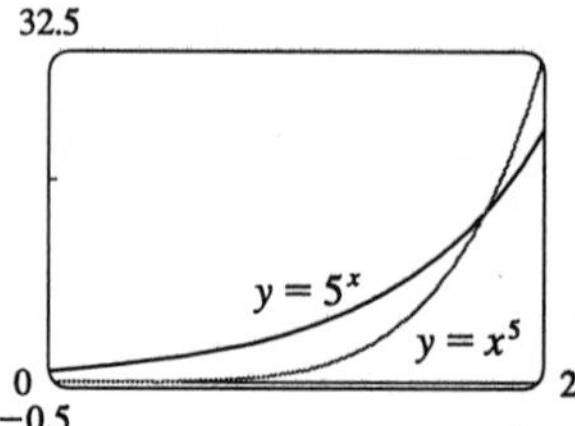

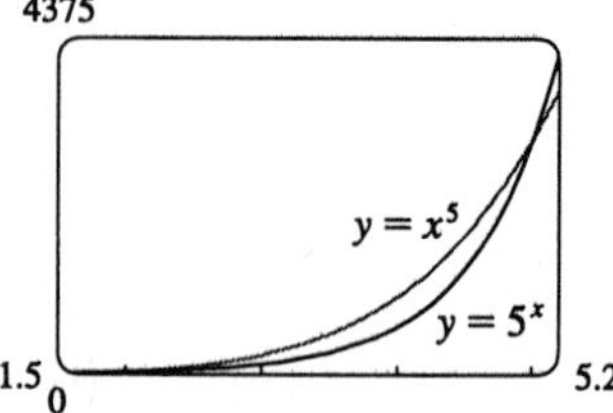

50,000
y = 5^x
y = x^5
0
7

We see from the graphs that for x less than about 1.8, $g(x) > f(x)$, and then near the point $(1.8, 17.1)$ the curves intersect. Then $f(x) > g(x)$ from $x \approx 1.8$ until $x = 5$. At $(5, 3125)$ there is another point of intersection, and for $x > 5$ we see that $g(x) > f(x)$. In fact, g increases much more rapidly than f beyond that point.

25. **(a)** Let $f(h) = \dfrac{4^h - 1}{h}$. Then $f(0.1) \approx 1.487$, $f(0.01) \approx 1.396$, $f(0.001) \approx 1.387$, and $f(0.0001) \approx 1.386$.

These quantities represent the slopes of secant lines to the curve $y = 4^x$, through the points $(0, 1)$ and $(h, 4^h)$ $\left(\text{since they are of the form } \dfrac{4^h - 4^0}{h - 0}\right)$.

(b) The value of the limit $\lim\limits_{h\to 0} \dfrac{4^h - 1}{h}$ is about 1.39, judging from the calculations in part (a).

(c) The limit in part (b) represents the slope of the tangent line to the curve $y = 4^x$ at $(0, 1)$.

26. **(a)**

h	$\dfrac{2.7^h - 1}{h}$
0.1	1.044
0.01	0.998
0.001	0.994
0.0001	0.993
0.00001	0.993

From these calculations, it seems that $\lim\limits_{h\to 0} \dfrac{2.7^h - 1}{h} \approx 0.99$.

(b)

h	$\dfrac{2.8^h - 1}{h}$
0.1	1.084
0.01	1.035
0.001	1.030
0.0001	1.030
0.00001	1.030

From these calculations, it seems that $\lim\limits_{h\to 0} \dfrac{2.8^h - 1}{h} \approx 1.03$.

Since e is defined such that $\lim\limits_{h\to 0} \dfrac{e^h - 1}{h} = 1$, and the limit in part (a) is slightly smaller than 1, and that in part (b) slightly larger, these limits indicate that $2.7 < e < 2.8$. In fact, they suggest that e is closer to 2.7, since the limit in part (a) is closer to 1 than that in part (b).

27. $f(x) = e^{\sqrt{x}} \quad \Rightarrow \quad f'(x) = e^{\sqrt{x}} \big/ (2\sqrt{x})$

28. $f(x) = xe^{-x^2} \quad \Rightarrow \quad f'(x) = e^{-x^2} + xe^{-x^2}(-2x) = e^{-x^2}(1 - 2x^2)$

29. $y = xe^{2x} \quad \Rightarrow \quad y' = e^{2x} + xe^{2x}(2) = e^{2x}(1 + 2x)$

30. $g(x) = e^{-5x} \cos 3x \quad \Rightarrow \quad g'(x) = -5e^{-5x} \cos 3x - 3e^{-5x} \sin 3x$

31. $h(t) = \sqrt{1 - e^t} \quad \Rightarrow \quad h'(t) = -e^t \Big/ \left(2\sqrt{1 - e^t}\right)$

32. $h(\theta) = e^{\sin 5\theta} \quad \Rightarrow \quad h'(\theta) = 5\cos(5\theta)e^{\sin 5\theta}$

33. $y = e^{x \cos x} \quad \Rightarrow \quad y' = e^{x \cos x}(\cos x - x \sin x)$

34. $y = \dfrac{e^{-x^2}}{x} \quad \Rightarrow \quad y' = \dfrac{xe^{-x^2}(-2x) - e^{-x^2}}{x^2} = \dfrac{e^{-x^2}(-2x^2 - 1)}{x^2}$

35. $y = e^{-1/x} \quad \Rightarrow \quad y' = e^{-1/x}/x^2$

36. $y = e^{x + e^x} \quad \Rightarrow \quad y' = e^{x + e^x}(1 + e^x)$

37. $y = \tan(e^{3x-2}) \quad \Rightarrow \quad y' = 3e^{3x-2} \sec^2(e^{3x-2})$

38. $y = (2x + e^{3x})^{1/3} \quad \Rightarrow \quad y' = \frac{1}{3}(2 + 3e^{3x})(2x + e^{3x})^{-2/3}$

39. $y = \dfrac{e^{3x}}{1 + e^x} \quad \Rightarrow \quad y' = \dfrac{3e^{3x}(1 + e^x) - e^{3x}(e^x)}{(1 + e^x)^2} = \dfrac{3e^{3x} + 3e^{4x} - e^{4x}}{(1 + e^x)^2} = \dfrac{3e^{3x} + 2e^{4x}}{(1 + e^x)^2}$

40. $y = \dfrac{e^x + e^{-x}}{e^x - e^{-x}} \quad \Rightarrow$

$$y' = \frac{(e^x - e^{-x})(e^x - e^{-x}) - (e^x + e^{-x})(e^x + e^{-x})}{(e^x - e^{-x})^2} = \frac{(e^{2x} - 2 + e^{-2x}) - (e^{2x} + 2 + e^{-2x})}{(e^x - e^{-x})^2} = -\frac{4}{(e^x - e^{-x})^2}$$

41. $y = x^e \quad \Rightarrow \quad y' = ex^{e-1}$

42. $y = \sec\left(e^{\tan x^2}\right) \quad \Rightarrow \quad y' = \sec\left(e^{\tan x^2}\right)\tan\left(e^{\tan x^2}\right)\left(e^{\tan x^2}\right)[\sec^2(x^2)](2x)$

43. $y = f(x) = e^{-x} \sin x \quad \Rightarrow \quad f'(x) = -e^{-x} \sin x + e^{-x} \cos x \quad \Rightarrow \quad f'(\pi) = e^{-\pi}(\cos \pi - \sin \pi) = -e^{-\pi}$, so the equation of the tangent at $(\pi, 0)$ is $y - 0 = -e^{-\pi}(x - \pi)$ or $x + e^{\pi}y = \pi$.

44. $y' = 2xe^{-x} - x^2e^{-x}$. At $(1, 1/e)$, $y' = 2e^{-1} - e^{-1} = 1/e$. So the equation of the tangent line is $y - \dfrac{1}{e} = \dfrac{1}{e}(x - 1) \quad \Rightarrow \quad y = \dfrac{x}{e}$.

45. $\cos(x - y) = xe^x \quad \Rightarrow \quad -\sin(x - y)(1 - y') = e^x + xe^x \quad \Rightarrow \quad y' = 1 + \dfrac{e^x(1 + x)}{\sin(x - y)}$

46. Using implicit differentiation, $2e^{xy} = x + y \quad \Rightarrow \quad (y + xy')2e^{xy} = 1 + y' \quad \Rightarrow \quad y'(2xe^{xy} - 1) = 1 - 2ye^{xy}$ $\Rightarrow \quad y' = (1 - 2ye^{xy})/(2xe^{xy} - 1)$. So at $(0, 2)$, $m = y' = 3$, and the equation of the tangent line is $y - 2 = 3(x - 0) \quad \Rightarrow \quad y = 3x + 2$.

47. $y = e^{2x} + e^{-3x} \quad \Rightarrow \quad y' = 2e^{2x} - 3e^{-3x} \quad \Rightarrow \quad y'' = 4e^{2x} + 9e^{-3x}$, so
$y'' + y' - 6y = (4e^{2x} + 9e^{-3x}) + (2e^{2x} - 3e^{-3x}) - 6(e^{2x} + e^{-3x}) = 0$.

48. $y = Ae^{-x} + Bxe^{-x} \quad \Rightarrow \quad y' = -Ae^{-x} + Be^{-x} - Bxe^{-x} = (B - A)e^{-x} - Bxe^{-x} \quad \Rightarrow$
$y'' = (A - B)e^{-x} - Be^{-x} + Bxe^{-x} = (A - 2B)e^{-x} + Bxe^{-x}$, so
$y'' + 2y' + y = (A - 2B)e^{-x} + Bxe^{-x} + 2[(B - A)e^{-x} - Bxe^{-x}] + Ae^{-x} + Bxe^{-x} = 0$.

49. $y = e^{rx} \quad \Rightarrow \quad y' = re^{rx} \quad \Rightarrow \quad y'' = r^2e^{rx}$, so $y'' + 5y' - 6y = r^2e^{rx} + 5re^{rx} - 6e^{rx}$
$= e^{rx}(r^2 + 5r - 6) = e^{rx}(r + 6)(r - 1) = 0 \quad \Rightarrow \quad (r + 6)(r - 1) = 0 \quad \Rightarrow \quad r = 1$ or -6.

50. $y = e^{\lambda x} \quad \Rightarrow \quad y' = \lambda e^{\lambda x} \quad \Rightarrow \quad y'' = \lambda^2 e^{\lambda x}$. Thus, $y + y' = y'' \quad \Leftrightarrow \quad e^{\lambda x} + \lambda e^{\lambda x} = \lambda^2 e^{\lambda x} \quad \Leftrightarrow$
$e^{\lambda x}(\lambda^2 - \lambda - 1) = 0 \quad \Leftrightarrow \quad \lambda = \frac{1 \pm \sqrt{5}}{2}$, since $e^{\lambda x} \neq 0$.

51. $f(x) = e^{-2x} \quad \Rightarrow \quad f'(x) = -2e^{-2x} \quad \Rightarrow \quad f''(x) = (-2)^2 e^{-2x} \quad \Rightarrow \quad f'''(x) = (-2)^3 e^{-2x} \quad \Rightarrow \quad \cdots$
$\Rightarrow \quad f^{(8)}(x) = (-2)^8\, e^{-2x} = 256e^{-2x}$

52. $f(x) = xe^{-x}$, $f'(x) = e^{-x} - xe^{-x} = (1 - x)e^{-x}$, $f''(x) = -e^{-x} + (1 - x)(-e^{-x}) = (x - 2)e^{-x}$. Similarly,
$f'''(x) = (3 - x)e^{-x}$, $f^{(4)}(x) = (x - 4)e^{-x}, \ldots, f^{(1000)}(x) = (x - 1000)e^{-x}$.

53. **(a)** $f(x) = e^x + x$ is continuous on $\mathbb{R}$ and $f(-1) = e^{-1} - 1 < 0 < 1 = f(0)$, so by the Intermediate Value Theorem, $e^x + x = 0$ has a root in $(-1, 0)$.

(b) $f(x) = e^x + x \quad \Rightarrow \quad f'(x) = e^x + 1$, so $x_{n+1} = x_n - \dfrac{e^{x_n} + x_n}{e^{x_n} + 1}$. From Exercise 31 we know that there is a root between -1 and 0, so we take $x_1 = -0.5$. Then $x_2 \approx -0.566311$, $x_3 \approx -0.567143$, and $x_4 \approx -0.567143$, so the root is -0.567143 to six decimal places.

54. From the graph, it appears that the curves intersect at about $x \approx 1.2$ or 1.3. We use Newton's Method with $f(x) = x^3 + x - 3 - e^{-x^2}$, so $f'(x) = 3x^2 + 1 - 2xe^{-x^2}$, and the formula is $x_{n+1} = x_n - f(x_n)/f'(x_n)$. We take $x_1 = 1.2$, and the formula gives $x_2 \approx 1.252462$, $x_3 \approx 1.251045$, and $x_4 \approx x_5 \approx 1.251044$. So the root of the equation, correct to six decimal places, is $x = 1.251044$.

55. **(a)** $\displaystyle\lim_{t\to\infty} p(t) = \lim_{t\to\infty} \frac{1}{1 + ae^{-kt}} = \frac{1}{1 + a \cdot 0} = 1$, since $k > 0 \quad \Rightarrow \quad -kt \to -\infty$.

(b) $\dfrac{dp}{dt} = -\left(1 + ae^{-kt}\right)^{-2}\left(-kae^{-kt}\right) = \dfrac{kae^{-kt}}{(1 + ae^{-kt})^2}$

(c) From the graph, it seems that $p(t) = 0.8$ (indicating that 80% of the population has heard the rumor) when $t \approx 7.4$ hours.

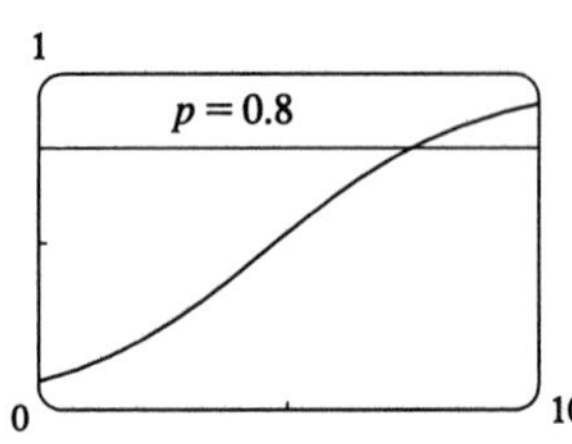

56. **(a)**

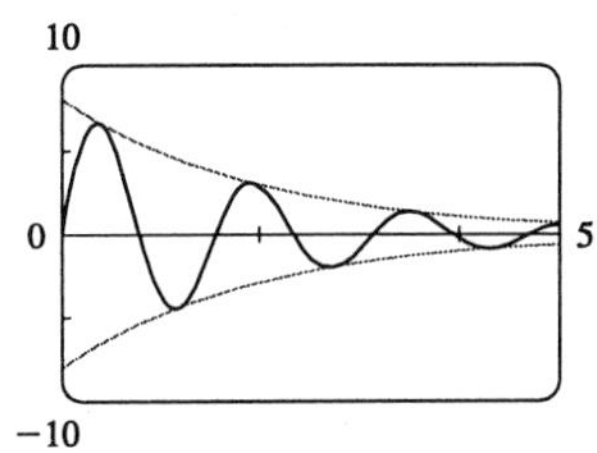

The displacement function is squeezed between the other two functions. This is because $-1 \le \sin 4t \le 1 \quad \Rightarrow$ $-8e^{-t/2} \le 8e^{-t/2} \sin 4t \le 8e^{-t/2}$.

(b) The maximum value of the displacement is about 6.6 cm, occurring at $t \approx 0.36$ s. It occurs just before the graph of the displacement function touches the graph of $8e^{-t/2}$ (when $t = \frac{\pi}{8} \approx 0.39$).

(c) The velocity of the object is the derivative of its displacement function, that is,

$\dfrac{d}{dt}\left(8e^{-t/2}\sin 4t\right) = 8\left(e^{-t/2}\cos 4t(4) + \sin 4t\left(-\frac{1}{2}\right)e^{-t/2}\right).$

If the displacement is zero, then we must have $\sin 4t = 0$ (since the exponential term in the displacement function is always positive). The first time that $\sin 4t = 0$ after $t = 0$ occurs at $t = \frac{\pi}{4}$. Substituting this into our expression for the velocity, and noting that the second term vanishes, we get $v\left(\frac{\pi}{4}\right) = 8e^{-\pi/8}\cos\left(4 \cdot \frac{\pi}{4}\right) \cdot 4 = -32e^{-\pi/8} \approx -21.6$ cm/s.

(d) The graph indicates that the displacement is less than 2 cm from equilibrium whenever t is larger than about 2.8.

57. $x - e^x \quad \Rightarrow \quad f'(x) = 1 - e^x = 0 \quad \Leftrightarrow \quad e^x = 1 \quad \Leftrightarrow \quad x = 0$. Now $f'(x) > 0$ for all $x < 0$ and $f'(x) < 0$ for all $x > 0$, so the absolute maximum value is $f(0) = 0 - 1 = -1$.

58. $g(x) = \dfrac{e^x}{x} \quad \Rightarrow \quad g'(x) = \dfrac{xe^x - e^x}{x^2} = 0 \quad \Leftrightarrow \quad e^x(x-1) = 0 \quad \Rightarrow \quad x = 1$. Now $g'(x) > 0 \quad \Leftrightarrow$ $\dfrac{xe^x - e^x}{x^2} > 0 \quad \Leftrightarrow \quad x - 1 > 0 \quad \Leftrightarrow \quad x > 1$ and $g'(x) < 0 \quad \Leftrightarrow \quad \dfrac{xe^x - e^x}{x^2} < 0 \quad \Leftrightarrow \quad x - 1 < 0 \quad \Leftrightarrow$ $x < 1$. Thus there is an absolute minimum value of $g(1) = e$ at $x = 1$.

59. **(a)** $f(x) = xe^x \quad \Rightarrow \quad f'(x) = e^x + xe^x = e^x(1+x) > 0 \quad \Leftrightarrow \quad 1 + x > 0 \quad \Leftrightarrow \quad x > -1$, so f is increasing on $[-1, \infty)$ and decreasing on $(-\infty, -1]$.

(b) $f''(x) = e^x(1+x) + e^x = e^x(2+x) > 0 \quad \Leftrightarrow \quad 2 + x > 0 \quad \Leftrightarrow \quad x > -2$, so f is CU on $(-2, \infty)$ and CD on $(-\infty, -2)$.

(c) f has an inflection point at $(-2, -2e^{-2})$.

60. **(a)** $f(x) = x^2e^x \quad \Rightarrow \quad f'(x) = 2xe^x + x^2e^x = (x^2 + 2x)e^x$. $f'(x) > 0 \quad \Leftrightarrow \quad x(x+2) > 0 \quad \Leftrightarrow$ $x < -2$ or $x > 0$, $f'(x) < 0 \quad \Leftrightarrow \quad -2 < x < 0$, so f is increasing on $(-\infty, -2]$ and $[0, \infty)$ and decreasing on $[-2, 0]$.

(b) $f''(x) = (2x+2)e^x + (x^2 + 2x)e^x = (x^2 + 4x + 2)e^x = 0 \quad \Leftrightarrow \quad x^2 + 4x + 2 = 0 \quad \Leftrightarrow$ $x = -2 \pm \sqrt{2}$. $f''(x) > 0$ when $x > -2 + \sqrt{2}$ or $x < -2 - \sqrt{2}$, so f is CU on $\left(-\infty, -2 - \sqrt{2}\right)$ and $\left(-2 + \sqrt{2}, \infty\right)$ and CD on $\left(-2 - \sqrt{2}, -2 + \sqrt{2}\right)$.

(c) f has inflection points at $\left(-2 + \sqrt{2}, \left(6 - 4\sqrt{2}\right)e^{-2+\sqrt{2}}\right)$ and $\left(-2 - \sqrt{2}, \left(6 + 4\sqrt{2}\right)e^{-2-\sqrt{2}}\right)$.

61. $y = f(x) = e^{-1/(x+1)}$ **A.** $D = \{x \mid x \neq -1\} = (-\infty, -1) \cup (-1, \infty)$ **B.** No x-intercept; y-intercept $= f(0) = e^{-1}$ **C.** No symmetry **D.** $\lim_{x \to \pm\infty} e^{-1/(x+1)} = 1$ since $-1/(x+1) \to 0$, so $y = 1$ is a HA. $\lim_{x \to -1^+} e^{-1/(x+1)} = 0$ since $-1/(x+1) \to -\infty$, $\lim_{x \to -1^-} e^{-1/(x+1)} = \infty$ since $-1/(x+1) \to \infty$, so $x = -1$ is a VA. **E.** $f'(x) = e^{-1/(x+1)}/(x+1)^2 \Rightarrow f'(x) > 0$ for all x except 1, so f is increasing on $(-\infty, -1)$ and $(-1, \infty)$. **F.** No extrema

G. $f''(x) = \dfrac{e^{-1/(x+1)}}{(x+1)^4} + \dfrac{e^{-1/(x+1)}(-2)}{(x+1)^3} = -\dfrac{e^{-1/(x+1)}(2x+1)}{(x+1)^4}$

$\Rightarrow \quad f''(x) > 0 \quad \Leftrightarrow \quad 2x + 1 < 0 \quad \Leftrightarrow \quad x < -\frac{1}{2}$, so f is CU on $(-\infty, -1)$ and $\left(-1, -\frac{1}{2}\right)$, and CD on $\left(-\frac{1}{2}, \infty\right)$. f has an IP at $\left(-\frac{1}{2}, e^{-2}\right)$.

H.

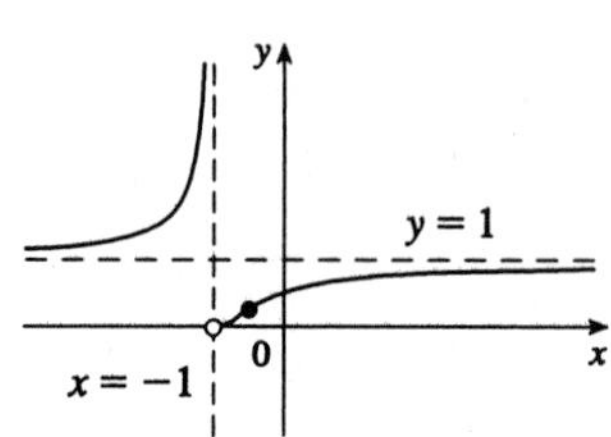

62. $y = f(x) = xe^{x^2}$ **A.** $D = \mathbb{R}$ **B.** Both intercepts are 0. **C.** $f(-x) = -f(x)$, so the curve is symmetric about the origin. **D.** $\lim_{x \to \infty} xe^{x^2} = \infty$, $\lim_{x \to -\infty} xe^{x^2} = -\infty$, no asymptotes **E.** $f'(x) = e^{x^2} + xe^{x^2}(2x) = e^{x^2}(1 + 2x^2) > 0$, so f is increasing on $\mathbb{R}$. **F.** No extrema

G. $f''(x) = e^{x^2}(2x)(1 + 2x^2) + e^{x^2}(4x)$
$= e^{x^2}(2x)(3 + 2x^2) > 0$
$\Leftrightarrow \quad x > 0$, so f is CU on $(0, \infty)$ and CD on $(-\infty, 0)$. f has an inflection point at $(0, 0)$.

H.

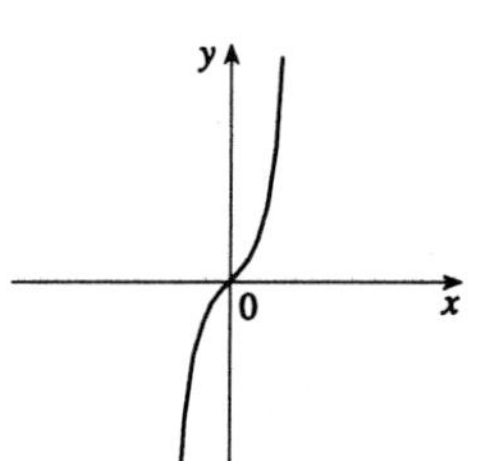

63. $y = 1/(1 + e^{-x})$ **A.** $D = \mathbb{R}$ **B.** No x-intercepts; y-intercept $= f(0) = \frac{1}{2}$. **C.** No symmetry **D.** $\lim_{x \to \infty} 1/(1 + e^{-x}) = \frac{1}{1+0} = 1$ and $\lim_{x \to -\infty} 1/(1 + e^{-x}) = 0$ $\left(\text{since } \lim_{x \to -\infty} e^{-x} = \infty\right)$, so f has horizontal asymptotes $y = 0$ and $y = 1$. **E.** $f'(x) = -(1 + e^{-x})^{-2}(-e^{-x}) = e^{-x}/(1 + e^{-x})^2$. This is positive for all x, so f is increasing on $\mathbb{R}$. **F.** No extrema

G. $f''(x) = \dfrac{(1 + e^{-x})^2(-e^{-x}) - e^{-x}(2)(1 + e^{-x})(-e^{-x})}{(1 + e^{-x})^4}$

$= \dfrac{e^{-x}(e^{-x} - 1)}{(1 + e^{-x})^3}$. The second factor in the numerator is negative for $x > 0$ and positive for $x < 0$, and the other factors are always positive, so f is CU on $(-\infty, 0)$ and CD on $(0, \infty)$. f has an inflection point at $\left(0, \frac{1}{2}\right)$.

H.

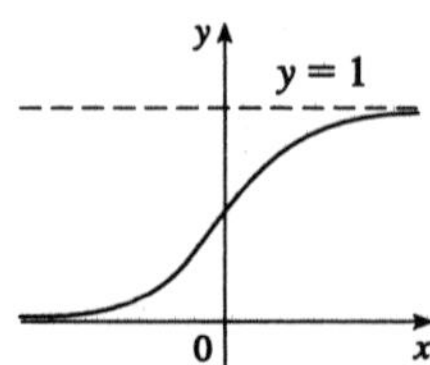

64.

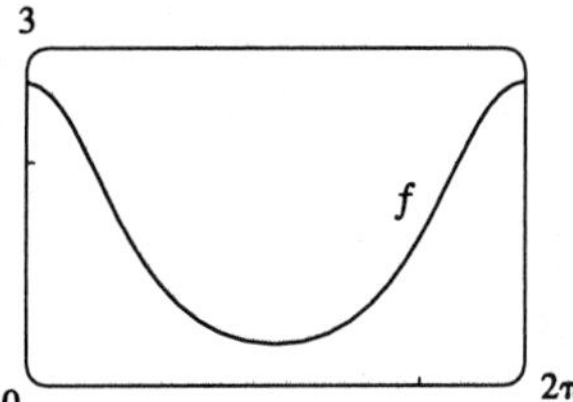

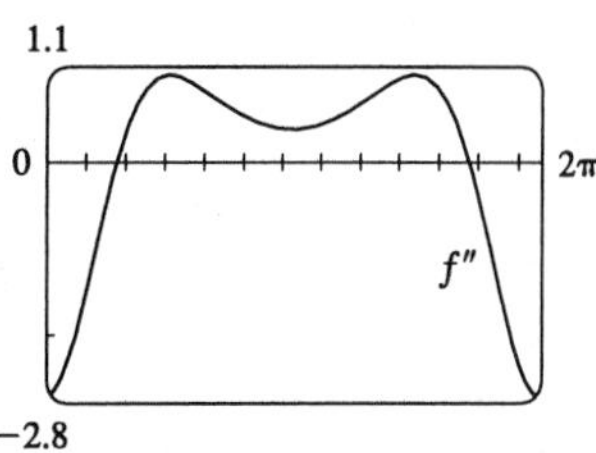

The function $f(x) = e^{\cos x}$ is periodic with period 2π, so we consider it only on the interval $[0, 2\pi]$. We see that it has local maxima of about $f(0) \approx 2.72$ and $f(2\pi) \approx 2.72$, and a local minimum of about $f(3.14) \approx 0.37$. To find the exact values, we calculate $f'(x) = -\sin x\, e^{\cos x}$. This is 0 when $-\sin x = 0 \quad \Leftrightarrow \quad x = 0, \pi$ or 2π (since we are only considering $x \in [0, 2\pi]$). Also $f'(x) > 0 \quad \Leftrightarrow \quad \sin x < 0 \quad \Leftrightarrow \quad 0 < x < \pi$. So $f(0) = f(2\pi) = e$ (both maxima) and $f(\pi) = e^{\cos \pi} = 1/e$ (minimum). To find the inflection points, we calculate and graph $f''(x) = \dfrac{d}{dx}(-\sin x\, e^{\cos x}) = -\cos x\, e^{\cos x} - \sin x(e^{\cos x})(-\sin x) = e^{\cos x}\left(\sin^2 x - \cos x\right)$. From the graph of $f''(x)$, we see that f has inflection points at $x \approx 0.90$ and at $x \approx 5.38$. These x-coordinates correspond to inflection points $(0.90, 1.86)$ and $(5.38, 1.86)$.

65.

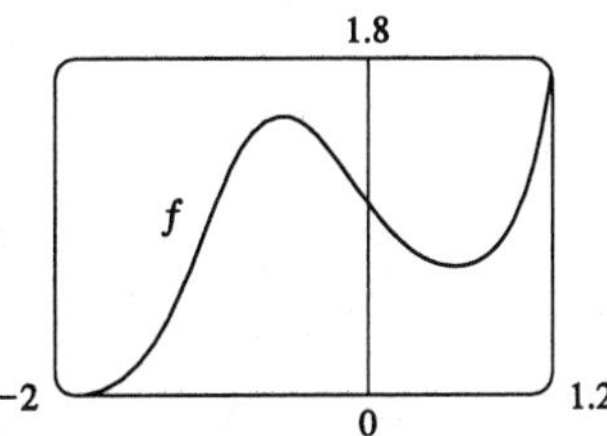

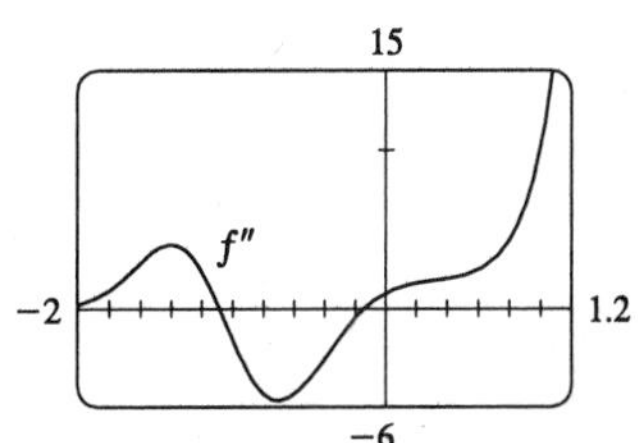

$f(x) = e^{x^3 - x} \to 0$ as $x \to -\infty$, and $f(x) \to \infty$ as $x \to \infty$. From the graph, it appears that f has a local minimum of about $f(0.58) = 0.68$, and a local maximum of about $f(-0.58) = 1.47$. To find the exact values, we calculate $f'(x) = (3x^2 - 1)e^{x^3 - x}$, which is 0 when $3x^2 - 1 = 0 \quad \Leftrightarrow \quad x = \pm\frac{1}{\sqrt{3}}$. The negative root corresponds to the local maximum $f\left(-\frac{1}{\sqrt{3}}\right) = e^{(-1/\sqrt{3})^3 - (-1/\sqrt{3})} = e^{2\sqrt{3}/9}$, and the positive root corresponds to the local minimum $f\left(\frac{1}{\sqrt{3}}\right) = e^{(1/\sqrt{3})^3 - (1/\sqrt{3})} = e^{-2\sqrt{3}/9}$. To estimate the inflection points, we calculate and graph $f''(x) = \dfrac{d}{dx}\left[(3x^2 - 1)e^{x^3 - x}\right] = (3x^2 - 1)e^{x^3 - x}(3x^2 - 1) + e^{x^3 - x}(6x) = e^{x^3 - x}\left(9x^4 - 6x^2 + 6x + 1\right)$. From the graph, it appears that $f''(x)$ changes sign (and thus f has inflection points) at $x \approx -0.15$ and $x \approx -1.09$. From the graph of f, we see that these x-values correspond to inflection points at about $(-0.15, 1.15)$ and $(-1.09, 0.82)$.

66. To find the asymptotes of f, we note that since $b > 0$, $\lim_{x\to\pm\infty} e^{-x^2/b} = 0$, so $y = 0$ is a horizontal asymptote. To find the maxima and minima we differentiate: $f(x) = e^{-x^2/b} \Rightarrow f'(x) = -\dfrac{2x}{b}e^{-x^2/b}$. This is 0 only at $x = 0$, and at that point $f'(x)$ changes from positive to negative, so there is an absolute maximum of $f(0) = e^0 = 1$. To find the inflection points we differentiate again:
$f''(x) = -\dfrac{2}{b}\left[x\left(-\dfrac{2x}{b}\right)e^{-x^2/b} + e^{-x^2/b}(1)\right] = \dfrac{2}{b^2}e^{-x^2/b}\left[2x^2 - b\right]$. This is 0 when $2x^2 - b = 0 \Leftrightarrow$ $x = \pm\sqrt{b/2}$, and $f''(x)$ changes sign at these values.
Also $f\left(\pm\sqrt{b/2}\right) = e^{-\left(\pm\sqrt{b/2}\right)^2/b} = e^{-1/2} = \sqrt{1/e}$, so the inflection points are $\left(\pm\sqrt{b/2}, \sqrt{1/e}\right)$. The value of b affects the graph by stretching or shrinking it in the x-direction: as b increases, the inflection points move away from the y-axis and the graph of f becomes more spread out.

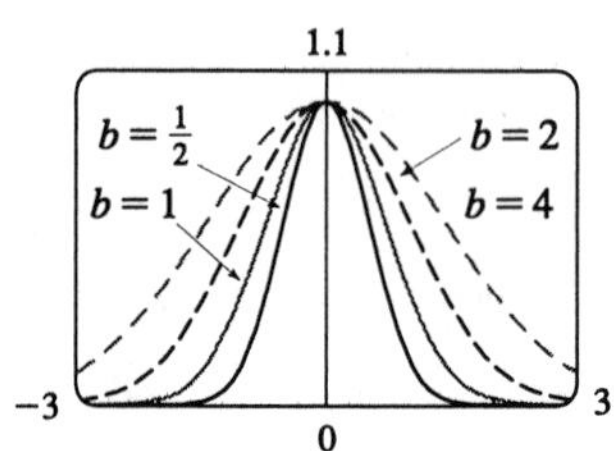

67. $u = -6x \Rightarrow du = -6\,dx$, so $\int e^{-6x}\,dx = -\frac{1}{6}\int e^u\,du = -\frac{1}{6}e^u + C = -\frac{1}{6}e^{-6x} + C$.

68. Let $u = x^2$. Then $du = 2x\,dx \Rightarrow \int xe^{x^2}dx = \frac{1}{2}\int e^u\,du = \frac{1}{2}e^u + C = \frac{1}{2}e^{x^2} + C$.

69. Let $u = 1 + e^x$. Then $du = e^x\,dx$, so $\int e^x(1 + e^x)^{10}\,dx = \int u^{10}\,du = \frac{1}{11}u^{11} + C = \frac{1}{11}(1 + e^x)^{11} + C$.

70. Let $u = \tan x$. Then $du = \sec^2 x\,dx$, so $\int \sec^2 x\, e^{\tan x}\,dx = \int e^u\,du = e^u + C = e^{\tan x} + C$.

71. $\displaystyle\int \frac{e^x + 1}{e^x}\,dx = \int (1 + e^{-x})dx = x - e^{-x} + C$

72. Let $u = 2 - x$. Then $du = -dx$ so $\int_2^3 e^{2-x}\,dx = \int_0^{-1}(-e^u)du = -e^u\big|_0^{-1} = 1 - 1/e$.

73. Let $u = -x^2$. Then $du = -2x\,dx$, so $\int xe^{-x^2}dx = -\frac{1}{2}\int e^u\,du = -\frac{1}{2}e^u + C = -\frac{1}{2}e^{-x^2} + C$.

74. Let $u = \dfrac{1}{x}$. Then $du = -\dfrac{1}{x^2}\,dx$, so $\displaystyle\int \frac{e^{1/x}}{x^2}\,dx = -\int e^u\,du = -e^u + C = -e^{1/x} + C$.

75. Let $u = x^2 - 4x - 3$. Then $du = 2(x - 2)dx$, so
$\int (x - 2)e^{x^2-4x-3}\,dx = \frac{1}{2}\int e^u\,du = \frac{1}{2}e^u + C = \frac{1}{2}e^{x^2-4x-3} + C$.

76. Let $u = e^x$. Then $du = e^x\,dx$, so $\int e^x\sin(e^x)dx = \int \sin u\,du = -\cos u + C = -\cos(e^x) + C$.

77. Area $= \int_0^1 (e^{3x} - e^x)dx = \left[\frac{1}{3}e^{3x} - e^x\right]_0^1 = \left(\frac{1}{3}e^3 - e\right) - \left(\frac{1}{3} - 1\right) = \frac{1}{3}e^3 - e + \frac{2}{3} \approx 4.644$

78. $f''(x) = 3e^x + 5\sin x \Rightarrow f'(x) = 3e^x - 5\cos x + C \Rightarrow 2 = f'(0) = 3 - 5 + C \Rightarrow C = 4$, so $f'(x) = 3e^x - 5\cos x + 4 \Rightarrow f(x) = 3e^x - 5\sin x + 4x + D \Rightarrow 1 = f(0) = 3 + D \Rightarrow$ $D = -2$, so $f(x) = 3e^x - 5\sin x + 4x - 2$.

79. $V = \int_0^1 \pi(e^x)^2\,dx = \int_0^1 \pi e^{2x}\,dx = \frac{1}{2}\left[\pi e^{2x}\right]_0^1 = \frac{\pi}{2}(e^2 - 1)$

80. $V = \int_0^1 2\pi xe^{-x^2}dx$. Let $u = x^2$. Thus $du = 2xdx$, so $V = \pi\int_0^1 e^{-u}\,du = \pi[-e^{-u}]_0^1 = \pi(1 - 1/e)$.

81. We use Theorem 6.1.8. Note that $f(0) = 3 + 0 + e^0 = 4$, so $f^{-1}(4) = 0$. Also $f'(x) = 1 + e^x$. Therefore, $(f^{-1})'(4) = \dfrac{1}{f'(f^{-1}(4))} = \dfrac{1}{f'(0)} = \dfrac{1}{1 + e^0} = \dfrac{1}{2}$.

82. We recognize this limit as the definition of the derivative of the function $f(x) = e^{\sin x}$ at $x = \pi$, since it is of the form $\lim\limits_{x\to\pi} \dfrac{f(x) - f(\pi)}{x - \pi}$. Therefore, the limit is equal to $f'(\pi) = (\cos\pi)e^{\sin\pi} = -1 \cdot e^0 = -1$.

83. **(a)** Let $f(x) = e^x - 1 - x$. Now $f(0) = e^0 - 1 = 0$, and for $x \ge 0$, we have $f'(x) = e^x - 1 \ge 0$. Now, since $f(0) = 0$ and f is increasing on $[0, \infty)$, $f(x) \ge 0$ for $x \ge 0 \;\Rightarrow\; e^x - 1 - x \ge 0 \;\Rightarrow\; e^x \ge 1 + x$.

(b) For $0 \le x \le 1$, $x^2 \le x$, so $e^{x^2} \le e^x$ (since e^x is increasing.) Hence [from (a)] $1 + x^2 \le e^{x^2} \le e^x$. So $\frac{4}{3} = \int_0^1 (1 + x^2)dx \le \int_0^1 e^{x^2}dx \le \int_0^1 e^x\,dx = e - 1 < e \;\Rightarrow\; \frac{4}{3} \le \int_0^1 e^{x^2}dx \le e$.

84. **(a)** Let $f(x) = e^x - 1 - x - \frac{1}{2}x^2$. Thus, $f'(x) = e^x - 1 - x$, which is positive for $x \ge 0$ by Exercise 83(a). Thus $f(x)$ is increasing on $[0, \infty)$, so on that interval, $0 = f(0) \le f(x) = e^x - 1 - x - \frac{1}{2}x^2 \;\Rightarrow\; e^x \ge 1 + x + \frac{1}{2}x^2$.

(b) Using the same argument as in Exercise 83(b), from part (a) we have $1 + x^2 + \frac{1}{2}x^4 \le e^{x^2} \le e^x$ (for $0 \le x \le 1$) $\Rightarrow\; \int_0^1 (1 + x^2 + \frac{1}{2}x^4)dx \le \int_0^1 e^{x^2}dx \le \int_0^1 e^x\,dx \;\Rightarrow\; \frac{43}{30} \le \int_0^1 e^{x^2}dx \le e - 1$.

85. **(a)** By Exercise 83(a), the result holds for $n = 1$. Suppose that $e^x \ge 1 + x + \dfrac{x^2}{2!} + \cdots + \dfrac{x^k}{k!}$ for $x \ge 0$. Let $f(x) = e^x - 1 - x - \dfrac{x^2}{2!} - \cdots - \dfrac{x^k}{k!} - \dfrac{x^{k+1}}{(k+1)!}$. Then $f'(x) = e^x - 1 - x - \cdots - \dfrac{x^k}{k!} \ge 0$ by assumption. Hence $f(x)$ is increasing on $[0, \infty)$. So $0 \le x$ implies that $0 = f(0) \le f(x) = e^x - 1 - x - \cdots - \dfrac{x^k}{k!} - \dfrac{x^{k+1}}{(k+1)!}$, and hence $e^x \ge 1 + x + \cdots + \dfrac{x^k}{k!} + \dfrac{x^{k+1}}{(k+1)!}$ for $x \ge 0$. Therefore, for $x \ge 0$, $e^x \ge 1 + x + \dfrac{x^2}{2!} + \cdots + \dfrac{x^n}{n!}$ for every positive integer n, by mathematical induction.

(b) Taking $n = 4$ and $x = 1$ in (a), we have $e = e^1 \ge 1 + \dfrac{1}{2} + \dfrac{1}{6} + \dfrac{1}{24} = 2.708\overline{3} > 2.7$.

(c) $e^x \ge 1 + x + \cdots + \dfrac{x^k}{k!} + \dfrac{x^{k+1}}{(k+1)!} \;\Rightarrow\; \dfrac{e^x}{x^k} \ge \dfrac{1}{x^k} + \dfrac{1}{x^{k-1}} + \cdots + \dfrac{1}{k!} + \dfrac{x}{(k+1)!} \ge \dfrac{x}{(k+1)!}$. But $\lim\limits_{x\to\infty} \dfrac{x}{(k+1)!} = \infty$, so $\lim\limits_{x\to\infty} \dfrac{e^x}{x^k} = \infty$.

86. **(a)** The graph of g finally surpasses that of f at $x \approx 35.8$.

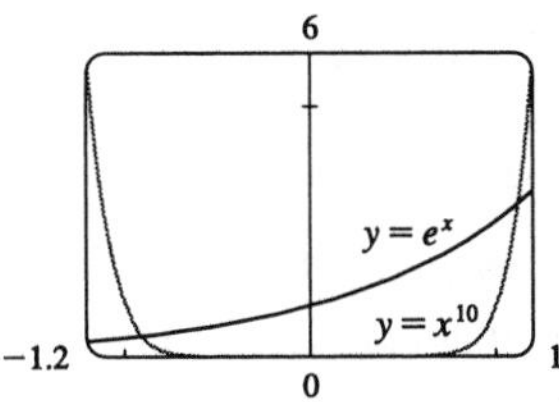

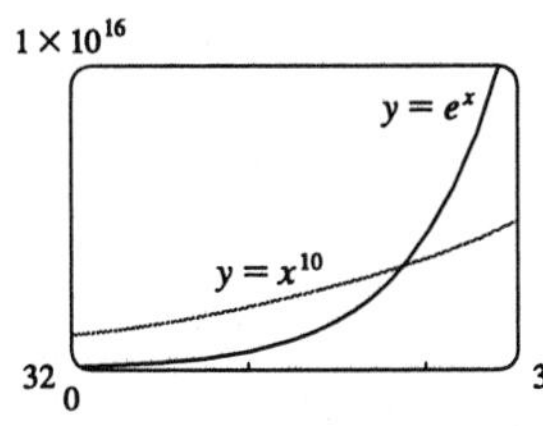

(b)

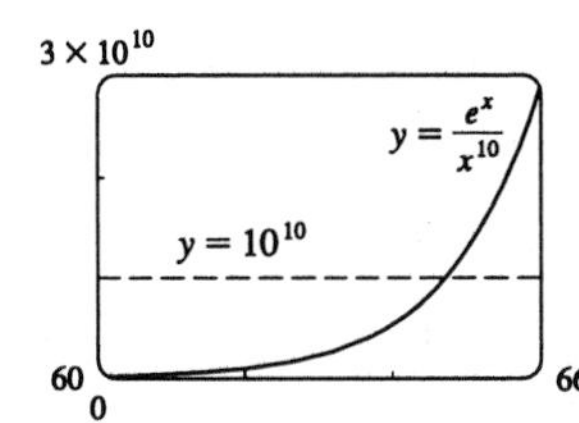

(c) From the graph in (b), it seems that $\dfrac{e^x}{x^{10}} > 10^{10}$ whenever $x > 65$, approximately. So we can take $N \ge 65$.

EXERCISES 6.3

1. $\log_2 64 = 6$ since $2^6 = 64$.

2. $\log_6 \frac{1}{36} = -2$ since $6^{-2} = \frac{1}{36}$.

3. $\log_8 2 = \frac{1}{3}$ since $8^{1/3} = 2$.

4. $\log_8 4 = \frac{2}{3}$ since $8^{2/3} = 4$.

5. $\log_3 \frac{1}{27} = -3$ since $3^{-3} = \frac{1}{27}$.

6. $e^{\ln 6} = 6$

7. $\ln e^{\sqrt{2}} = \sqrt{2}$

8. $\log_3 3^{\sqrt{5}} = \sqrt{5}$

9. $\log_{10} 1.25 + \log_{10} 80 = \log_{10}(1.25 \cdot 80) = \log_{10} 100 = 2$

10. $\log_3 108 - \log_3 4 = \log_3 \frac{108}{4} = \log_3 27 = 3$

11. $\log_8 6 - \log_8 3 + \log_8 4 = \log_8 \frac{6 \cdot 4}{3} = \log_8 8 = 1$

12. $\log_5 10 + \log_5 20 - 3\log_5 2 = \log_5 \dfrac{10 \cdot 20}{2^3} = \log_5 25 = 2$

13. $2^{(\log_2 3 + \log_2 5)} = 2^{\log_2 15} = 15$

14. $e^{3\ln 2} = e^{\ln(2^3)} = e^{\ln 8} = 8 \quad \left[\textit{Or: } e^{3\ln 2} = \left(e^{\ln 2}\right)^3 = 2^3 = 8\right]$

15. $\log_5 a + \log_5 b - \log_5 c = \log_5(ab/c)$

16. $\log_2 x + 5\log_2(x+1) + \frac{1}{2}\log_2(x-1) = \log_2 x + \log_2(x+1)^5 + \log_2 \sqrt{x-1} = \log_2\left(x(x+1)^5\sqrt{x-1}\right)$

17. $2\ln 4 - \ln 2 = \ln 4^2 - \ln 2 = \ln 16 - \ln 2 = \ln \frac{16}{2} = \ln 8$

Or: $2\ln 4 - \ln 2 = 2\ln 2^2 - \ln 2 = 4\ln 2 - \ln 2 = 3\ln 2$

18. $\ln 10 + \frac{1}{2}\ln 9 = \ln 10 + \ln 9^{1/2} = \ln 10 + \ln 3 = \ln 30$

19. $\frac{1}{3}\ln x - 4\ln(2x+3) = \ln(x^{1/3}) - \ln(2x+3)^4 = \ln\left(x^{1/3}/(2x+3)^4\right)$

20. $\ln x + a\ln y - b\ln z = \ln x + \ln y^a - \ln z^b = \ln(xy^a/z^b)$

21. **(a)** $\log_2 5 = \dfrac{\ln 5}{\ln 2} \approx 2.321928$

(b) $\log_5 26.05 = \dfrac{\ln 26.05}{\ln 5} \approx 2.025563$

(c) $\log_3 e = \dfrac{1}{\ln 3} \approx 0.910239$

(d) $\log_{0.7} 14 = \dfrac{\ln 14}{\ln 0.7} \approx -7.399054$

22.

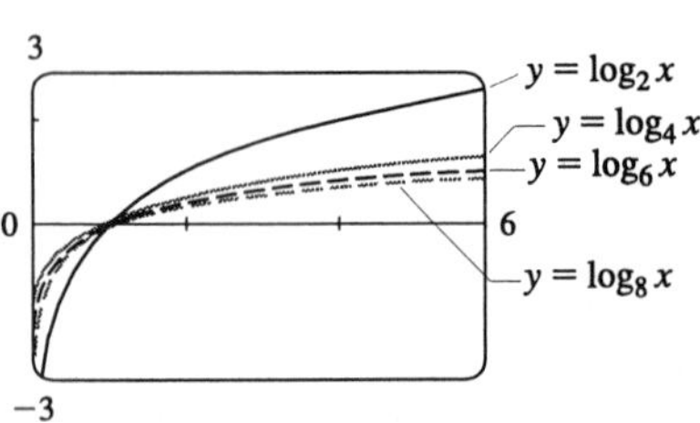

To graph the functions, we use $\log_2 x = \dfrac{\ln x}{\ln 2}$, $\log_4 x = \dfrac{\ln x}{\ln 4}$, etc. These graphs all approach $-\infty$ as $x \to 0^+$, and they all pass through the point $(1, 0)$. Also, they are all increasing, and all approach ∞ as $x \to \infty$. The smaller the base, the larger the rate of increase of the function (for $x > 1$) and the closer the approach to the y-axis $\left(\text{as } x \to 0^+\right)$.

23.

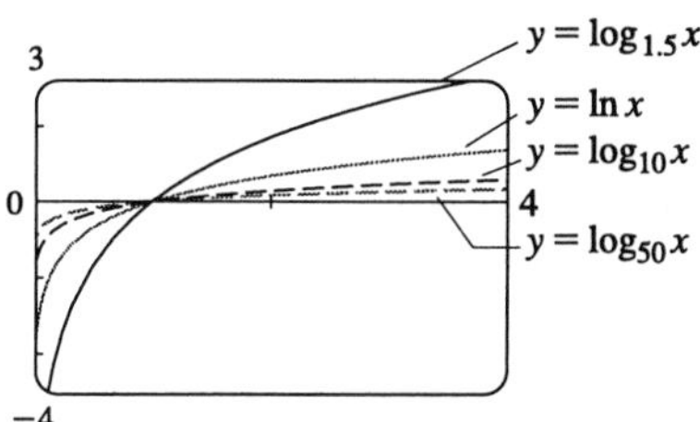

To graph these functions, we use $\log_{1.5} x = \dfrac{\ln x}{\ln 1.5}$ and $\log_{50} x = \dfrac{\ln x}{\ln 50}$. These graphs all approach $-\infty$ as $x \to 0^+$, and they all pass through the point $(1, 0)$. Also, they are all increasing, and all approach ∞ as $x \to \infty$. The functions with larger bases increase extremely slowly, and the ones with smaller bases do so somewhat more quickly. The functions with large bases approach the y-axis more closely as $x \to 0^+$.

24.

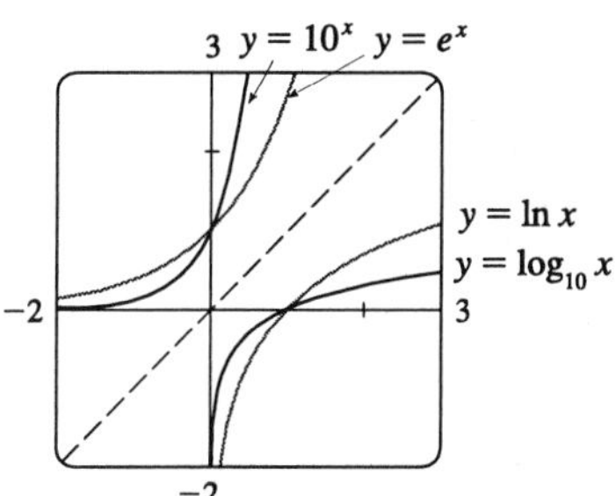

We see that the graph of $\ln x$ is the reflection of the graph of e^x about the line $y = x$, and that the graph of $\log_{10} x$ is the reflection of the graph of 10^x about the same line.
The graph of 10^x increases more quickly than that of e^x. Also note that $\log_{10} x \to \infty$ as $x \to \infty$ more slowly than $\ln x$.

25. $y = \log_{10} x$ $\quad$ $y = \log_{10}(x + 5)$

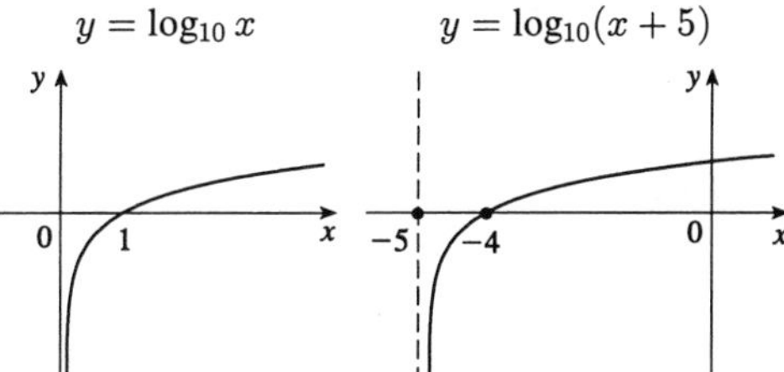

26. $y = \log_5 x$ $\quad$ $y = 1 + \log_5(x - 1)$

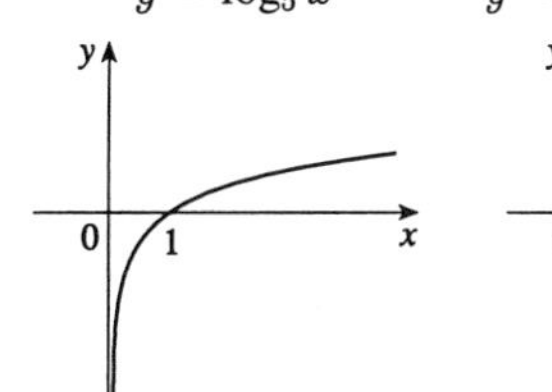

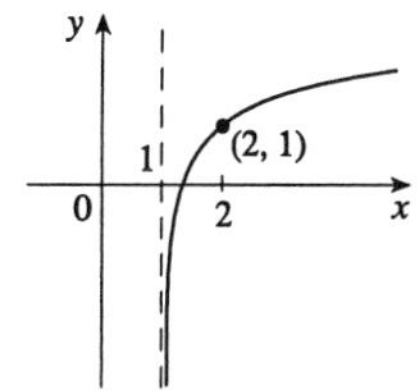

27. $y = \ln x$ $\quad$ $y = -\ln x$

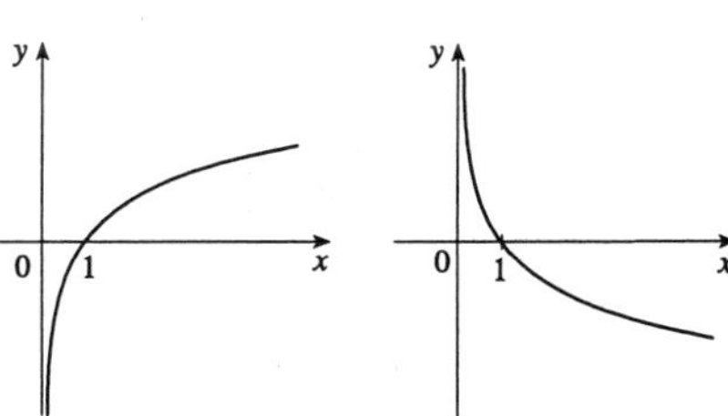

28. $y = \ln x$ $\quad$ $y = \ln(-x)$

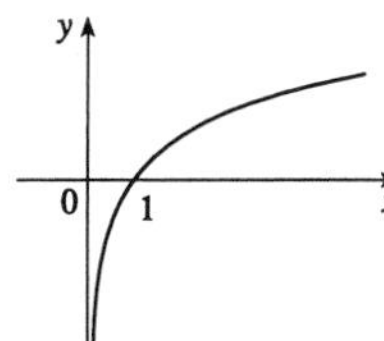

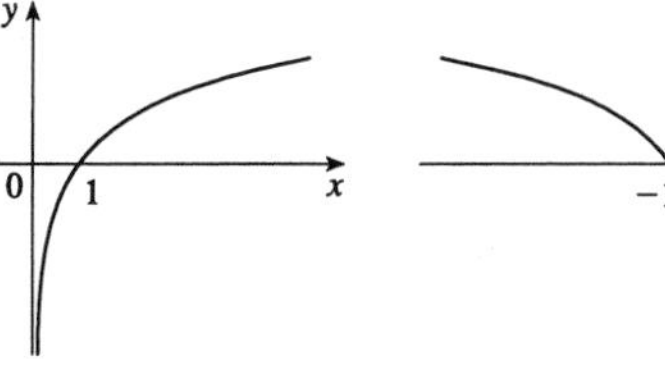

29. $y = \ln(-x)$ $\quad$ $y = -\ln(-x)$

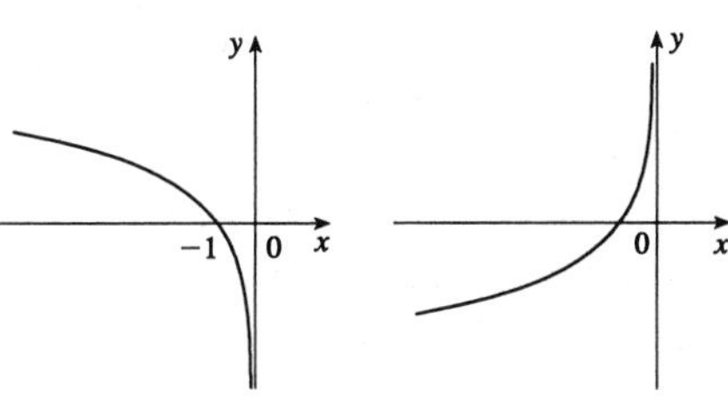

30. $y = \ln|x|$

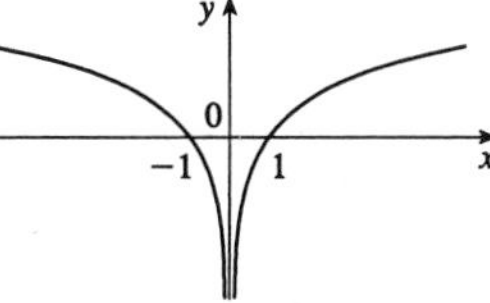

31. $y = \ln(x^2) = 2\ln|x|$

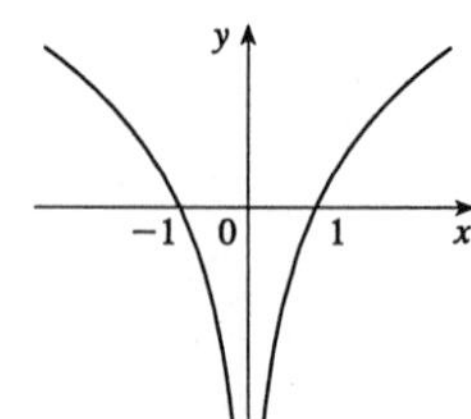

32. $y = \ln x$ $\qquad y = \ln 1/x = -\ln x$

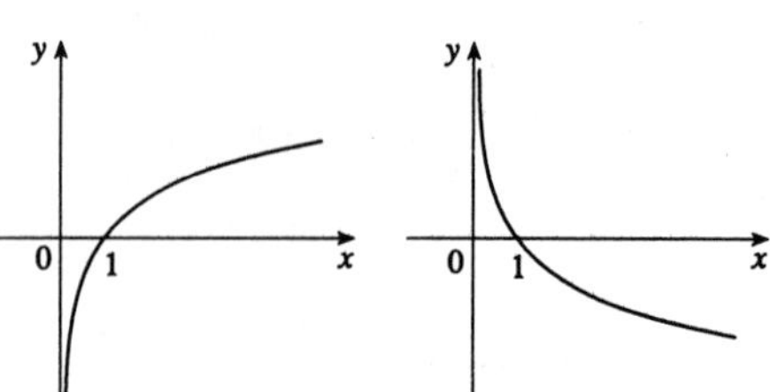

33. $y = \ln x$ $\qquad y = \ln(x+3)$

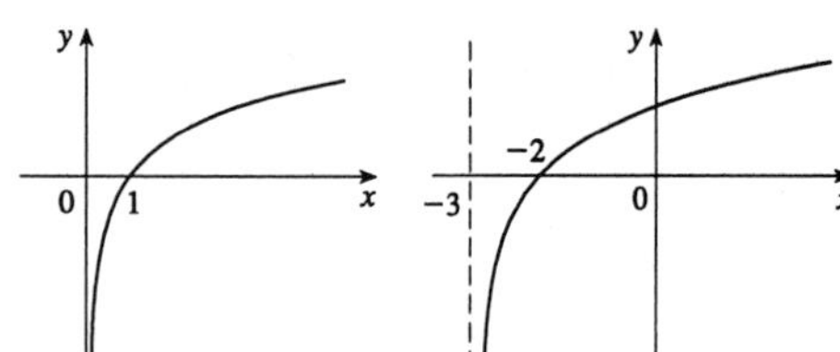

34. $y = \ln|x+3|$

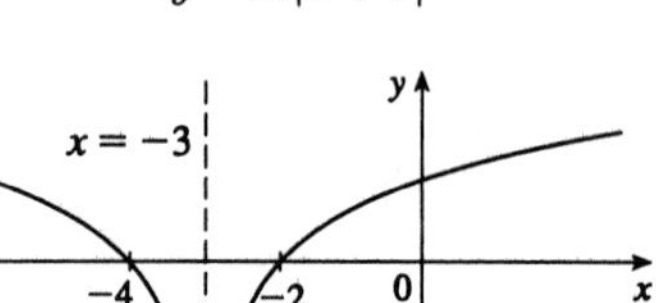

35. $\log_2 x = 3 \quad \Leftrightarrow \quad x = 2^3 = 8$

36. $2^{x-5} = 3 \quad \Leftrightarrow \quad \log_2 3 = x - 5 \quad \Leftrightarrow \quad x = 5 + \log_2 3$. *Or:* $2^{x-5} = 3 \quad \Leftrightarrow \quad \ln(2^{x-5}) = \ln 3 \quad \Leftrightarrow$
$(x-5)\ln 2 = \ln 3 \quad \Leftrightarrow \quad x - 5 = \dfrac{\ln 3}{\ln 2} \quad \Leftrightarrow \quad x = 5 + \dfrac{\ln 3}{\ln 2}$

37. $e^x = 16 \quad \Leftrightarrow \quad \ln e^x = \ln 16 \quad \Leftrightarrow \quad x = \ln 16 \quad \Leftrightarrow \quad x = 4\ln 2$

38. $\ln x = -1 \quad \Leftrightarrow \quad e^{\ln x} = e^{-1} \quad \Leftrightarrow \quad x = 1/e$

39. $\ln(2x-1) = 3 \quad \Leftrightarrow \quad e^{\ln(2x-1)} = e^3 \quad \Leftrightarrow \quad 2x - 1 = e^3 \quad \Leftrightarrow \quad x = \frac{1}{2}(e^3+1)$

40. $e^{3x-4} = 2 \quad \Leftrightarrow \quad \ln\left(e^{3x-4}\right) = \ln 2 \quad \Leftrightarrow \quad 3x - 4 = \ln 2 \quad \Leftrightarrow \quad x = \frac{1}{3}(\ln 2 + 4)$

41. $3^{x+2} = m \quad \Leftrightarrow \quad \log_3 m = x + 2 \quad \Leftrightarrow \quad x = \log_3 m - 2$

42. $6 = 5^{\log_5(2x)} = 2x \quad \Leftrightarrow \quad x = 3$

43. $\ln x = \ln 5 + \ln 8 = \ln 40 \quad \Leftrightarrow \quad x = 40$

44. $\ln x^2 = 2\ln 4 - 4\ln 2 = 4\ln 2 - 4\ln 2 = 0 \quad \Leftrightarrow \quad x^2 = 1 \quad \Leftrightarrow \quad x = \pm 1$

45. $5 = \ln(e^{2x-1}) = 2x - 1 \quad \Leftrightarrow \quad x = 3$

46. $\ln x + \ln(x-1) = \ln[x(x-1)] = 1 \quad \Leftrightarrow \quad x(x-1) = e \quad \Leftrightarrow \quad x^2 - x - e = 0$. The quadratic formula gives $x = \frac{1}{2}\left(1 \pm \sqrt{1+4e}\,\right)$, but we reject the negative root since the natural logarithm is not defined for $x < 0$. So $x = \frac{1}{2}\left(1 + \sqrt{1+4e}\,\right)$.

47. $\ln(\ln x) = 1 \quad \Leftrightarrow \quad e^{\ln(\ln x)} = e^1 \quad \Leftrightarrow \quad \ln x = e^1 = e \quad \Leftrightarrow \quad e^{\ln x} = e^e \quad \Leftrightarrow \quad x = e^e$

48. $e^{e^x} = 10 \quad \Leftrightarrow \quad \ln\left(e^{e^x}\right) = \ln 10 \quad \Leftrightarrow \quad e^x \ln e = e^x = \ln 10 \quad \Leftrightarrow \quad \ln e^x = \ln(\ln 10) \quad \Leftrightarrow \quad x = \ln(\ln 10)$

49. $2^{3^x} = 5 \quad\Leftrightarrow\quad 3^x = \log_2 5 \quad\Leftrightarrow\quad \log_3(\log_2 5) = x$. *Or:* $2^{3^x} = 5 \quad\Leftrightarrow\quad \ln 2^{3^x} = \ln 5 \quad\Leftrightarrow\quad 3^x \ln 2 = \ln 5$ $\quad\Leftrightarrow\quad 3^x = \dfrac{\ln 5}{\ln 2}$. Hence $\ln 3^x = x \ln 3 = \ln\left(\dfrac{\ln 5}{\ln 2}\right) \quad\Leftrightarrow\quad x = \dfrac{\ln(\ln 5/\ln 2)}{\ln 3}$.

50. $\log_2[\log_3(\log_4 x)] = C \quad\Leftrightarrow\quad 2^C = \log_3(\log_4 x) \quad\Leftrightarrow\quad 3^{2^C} = \log_4 x \quad\Leftrightarrow\quad 4^{3^{2^C}} = x$

51. $\ln(x+6) + \ln(x-3) = \ln 5 + \ln 2 \quad\Leftrightarrow\quad \ln[(x+6)(x-3)] = \ln 10 \quad\Leftrightarrow\quad (x+6)(x-3) = 10 \quad\Leftrightarrow$ $x^2 + 3x - 18 = 10 \quad\Leftrightarrow\quad x^2 + 3x - 28 = 0 \quad\Leftrightarrow\quad (x+7)(x-4) = 0 \quad\Leftrightarrow\quad x = -7$ or 4. However, $x = -7$ is not a solution since $\ln(-7+6)$ is not defined. So $x = 4$ is the only solution.

52. $\ln\left(\dfrac{x-2}{x-1}\right) = 1 + \ln\left(\dfrac{x-3}{x-1}\right) \quad\Leftrightarrow\quad \ln\left(\dfrac{x-2}{x-1}\right) - \ln\left(\dfrac{x-3}{x-1}\right) = 1 \quad\Leftrightarrow\quad \ln\left(\dfrac{x-2}{x-3}\right) = 1 \quad\Leftrightarrow$ $\dfrac{x-2}{x-3} = e^1 = e$

53. $e^{ax} = Ce^{bx} \quad\Leftrightarrow\quad \ln e^{ax} = \ln(Ce^{bx}) \quad\Leftrightarrow\quad ax = \ln C + bx \quad\Leftrightarrow\quad (a-b)x = \ln C \quad\Leftrightarrow\quad x = \dfrac{\ln C}{a-b}$

54. $7e^x - e^{2x} = 12 \quad\Leftrightarrow\quad (e^x)^2 - 7e^x + 12 = 0 \quad\Leftrightarrow\quad (e^x - 3)(e^x - 4) = 0$, so we have either $e^x = 3 \quad\Leftrightarrow$ $x = \ln 3$, or $e^x = 4 \quad\Leftrightarrow\quad x = \ln 4$.

55. $\ln(x-5) = 3 \quad\Rightarrow\quad x - 5 = e^3 \quad\Rightarrow\quad x = e^3 + 5 \approx 25.0855$

56. $e^{5x-1} = 12 \quad\Rightarrow\quad 5x - 1 = \ln 12 \quad\Rightarrow\quad x = \frac{1}{5}(\ln 12 + 1) \approx 0.6970$

57. $e^{2-3x} = 20 \quad\Rightarrow\quad 2 - 3x = \ln 20 \quad\Rightarrow\quad x = \frac{1}{3}(2 - \ln 20) \approx -0.3319$

58. $2^{-x} = 5 \quad\Rightarrow\quad -x \ln 2 = \ln 5 \quad\Rightarrow\quad x = -\ln 5/\ln 2 \approx -2.3219$

59. 3 ft = 36 in, so we need x such that $\log_2 x = 36 \quad\Leftrightarrow\quad x = 2^{36} = 68{,}719{,}476{,}736$. In miles, this is $68{,}719{,}476{,}736 \text{ in} \cdot \dfrac{1 \text{ ft}}{12 \text{ in}} \cdot \dfrac{1 \text{ mi}}{5280 \text{ ft}} \approx 1{,}084{,}587.7$ mi.

60. **(a)** $v(t) = ce^{-kt} \quad\Rightarrow\quad a(t) = v'(t) = -kce^{-kt} = -kv(t)$

(b) $v(0) = ce^0 = c$, so c is the initial velocity.

(c) $v(t) = ce^{-kt} = c/2 \quad\Rightarrow\quad e^{-kt} = \frac{1}{2} \quad\Rightarrow\quad -kt = \ln\frac{1}{2} = -\ln 2 \quad\Rightarrow\quad t = (\ln 2)/k$

61. If I is the intensity of the 1989 San Francisco earthquake, then $\log_{10}(I/S) = 7.1 \quad\Rightarrow$ $\log_{10}(16I/S) = \log_{10}16 + \log_{10}(I/S) = \log_{10}16 + 7.1 \approx 8.3$.

62. Let I_1 and I_2 be the intensities of the music and the mower. Then $10\log_{10}\left(\dfrac{I_1}{I_0}\right) = 120$ and $10\log_{10}\left(\dfrac{I_2}{I_0}\right) = 106$, so $\log_{10}\left(\dfrac{I_1}{I_2}\right) = \log_{10}\left(\dfrac{I_1/I_0}{I_2/I_0}\right) = \log_{10}\left(\dfrac{I_1}{I_0}\right) - \log_{10}\left(\dfrac{I_2}{I_0}\right) = 12 - 10.6 = 1.4 \quad\Rightarrow$ $\dfrac{I_1}{I_2} = 10^{1.4} \approx 25$.

63. $\lim\limits_{x\to5^+} \ln(x-5) = -\infty$ since $x - 5 \to 0^+$ as $x \to 5^+$.

64. $\lim\limits_{x\to0^+} \log_{10}(4x) = -\infty$ since $4x \to 0^+$ as $x \to 0^+$.

65. $\lim_{x\to\infty} \log_2(x^2 - x) = \infty$ since $x^2 - x \to \infty$ as $x \to \infty$.

66. $\lim_{x\to 0^+} \ln(\sin x) = -\infty$ since $\sin x \to 0^+$ as $x \to 0^+$.

67. $\lim_{x\to\pi/2^-} \log_{10}(\cos x) = -\infty$ since $\cos x \to 0^+$ as $x \to \frac{\pi}{2}^-$.

68. $\lim_{x\to\infty} \dfrac{\ln x}{1+\ln x} = \lim_{x\to\infty} \dfrac{1}{(1/\ln x)+1} = \dfrac{1}{0+1} = 1$

69. $\lim_{x\to\infty} \ln(1+e^{-x^2}) = \ln\left(1+\lim_{x\to\infty} e^{-x^2}\right) = \ln(1+0) = 0$

70. $\lim_{x\to\infty} [\ln(2+x) - \ln(1+x)] = \lim_{x\to\infty} \ln\left(\dfrac{2+x}{1+x}\right) = \lim_{x\to\infty} \ln\left(\dfrac{2/x+1}{1/x+1}\right) = \ln\dfrac{1}{1} = \ln 1 = 0$

71. $f(x) = \log_{10}(1-x)$ Domain$(f) = \{x \mid 1-x > 0\} = \{x \mid x < 1\} = (-\infty, 1)$. Range$(f) = \mathbb{R}$.

72. $g(x) = \ln(4-x^2)$ Domain$(g) = \{x \mid 4-x^2 > 0\} = \{x \mid |x| < 2\} = (-2, 2)$. Since $4 - x^2 \le 4$, we have $\ln(4-x^2) \le \ln 4$. Also $\lim_{x\to 2^-} g(x) = -\infty$, so range$(g) = (-\infty, \ln 4]$.

73. $F(t) = \sqrt{t}\ln(t^2-1)$ Domain$(F) = \{t \mid t \ge 0 \text{ and } t^2 - 1 > 0\} = \{t \mid t > 1\} = (1, \infty)$. Range$(F) = \mathbb{R}$.

74. $G(t) = \ln(t^3 - t)$ Domain$(G) = \{t \mid t^3 - t > 0\} = \{t \mid t(t^2-1) > 0\}$
$= \{t \mid t > 1 \text{ or } -1 < t < 0\} = (-1, 0) \cup (1, \infty)$. Range$(G) = \mathbb{R}$.

75. $y = \ln(x+3) \Rightarrow e^y = e^{\ln(x+3)} = x + 3 \Rightarrow x = e^y - 3$.
Interchange x and y: the inverse function is $y = e^x - 3$.

76. $y = 2^{10^x} \Rightarrow \log_2 y = 10^x \Rightarrow \log_{10}(\log_2 y) = x$.
Interchange x and y: $y = \log_{10}(\log_2 x)$ is the inverse function.

77. $y = e^{\sqrt{x}} \Rightarrow \ln y = \ln e^{\sqrt{x}} = \sqrt{x} \Rightarrow x = (\ln y)^2$. Also note that $\sqrt{x} \ge 0 \Rightarrow y = e^{\sqrt{x}} \ge 1$.
Interchange x and y: the inverse function is $y = (\ln x)^2$, $x \ge 1$.

78. $y = (\ln x)^2$, $x \ge 1$, $\ln x = \sqrt{y} \Rightarrow x = e^{\sqrt{y}}$. Interchange x and y: $y = e^{\sqrt{x}}$ is the inverse function.

79. $y = \dfrac{10^x}{10^x+1} \Rightarrow 10^x y + y = 10^x \Rightarrow 10^x(1-y) = y \Rightarrow 10^x = \dfrac{y}{1-y} \Rightarrow$
$x = \log_{10}\left(\dfrac{y}{1-y}\right)$. Interchange x and y: $y = \log_{10}\left(\dfrac{x}{1-x}\right)$ is the inverse function.

80. $y = \dfrac{1+e^x}{1-e^x} \Rightarrow y - ye^x = 1 + e^x \Rightarrow e^x(y+1) = y - 1 \Rightarrow e^x = \dfrac{y-1}{y+1} \Rightarrow x = \ln\left(\dfrac{y-1}{y+1}\right)$.
Interchange x and y: $y = \ln\left(\dfrac{x-1}{x+1}\right)$ is the inverse function.

81. $y = e^x - 2e^{-x}$, so $y' = e^x + 2e^{-x}$, $y'' = e^x - 2e^{-x}$. $y'' > 0 \Leftrightarrow e^x - 2e^{-x} > 0 \Leftrightarrow e^x > 2e^{-x} \Leftrightarrow$
$e^{2x} > 2 \Leftrightarrow 2x > \ln 2 \Leftrightarrow x > \frac{1}{2}\ln 2$. Therefore, y is concave upward on $\left(\frac{1}{2}\ln 2, \infty\right)$.

82. $f(x) = e^x + e^{-2x}$, $f'(x) = e^x - 2e^{-2x} > 0 \Leftrightarrow e^x > 2e^{-2x} \Leftrightarrow e^{3x} > 2 \Leftrightarrow 3x > \ln 2 \Leftrightarrow$
$x > \frac{1}{3}\ln 2$. Thus f is increasing on $\left[\frac{1}{3}\ln 2, \infty\right)$.

83. **(a)** We have to show that $-f(x) = f(-x)$.

$$-f(x) = -\ln\left(x+\sqrt{x^2+1}\right) = \ln\left[\left(x+\sqrt{x^2+1}\right)^{-1}\right] = \ln\frac{1}{x+\sqrt{x^2+1}}$$

$$= \ln\left(\frac{1}{x+\sqrt{x^2+1}}\cdot\frac{x-\sqrt{x^2+1}}{x-\sqrt{x^2+1}}\right) = \ln\frac{x-\sqrt{x^2+1}}{x^2-x^2-1}$$

$$= \ln\left(\sqrt{x^2+1}-x\right) = f(-x).$$ Thus, f is an odd function.

(b) Let $y = \ln\left(x+\sqrt{x^2+1}\right)$, then $e^y = x+\sqrt{x^2+1} \quad\Leftrightarrow\quad (e^y - x)^2 = x^2+1 \quad\Leftrightarrow$

$e^{2y} - 2xe^y + x^2 = x^2 + 1 \quad\Leftrightarrow\quad 2xe^y = e^{2y} - 1 \quad\Leftrightarrow\quad x = \dfrac{e^{2y}-1}{2e^y} = \frac{1}{2}(e^y - e^{-y})$. Thus, the inverse function is $f^{-1}(x) = \frac{1}{2}(e^x - e^{-x})$.

84. Let (a, e^{-a}) be the point where the tangent meets the curve. The tangent has slope $-e^{-a}$ and is perpendicular to the line $2x - y = 8$, which has slope 2. So $-e^{-a} = -\frac{1}{2} \quad\Rightarrow\quad e^{-a} = \frac{1}{2} \quad\Rightarrow\quad e^a = 2 \quad\Rightarrow$ $a = \ln(e^a) = \ln 2$. Thus the point on the curve is $\left(\ln 2, \frac{1}{2}\right)$ and the equation of the tangent is $y - \frac{1}{2} = -\frac{1}{2}(x - \ln 2)$ or $x + 2y = 1 + \ln 2$.

85. Let $x = \log_{10} 99$, $y = \log_9 82$. Then $10^x = 99 < 10^2 \quad\Rightarrow\quad x < 2$, and $9^y = 82 > 9^2 \quad\Rightarrow\quad y > 2$. Therefore $y = \log_9 82$ is larger.

86. **(a)** $\lim\limits_{x\to\infty} x^{\ln x} = \lim\limits_{x\to\infty} \left(e^{\ln x}\right)^{\ln x} = \lim\limits_{x\to\infty} e^{(\ln x)^2} = \infty$ since $(\ln x)^2 \to \infty$.

(b) $\lim\limits_{x\to 0^+} x^{-\ln x} = \lim\limits_{x\to 0^+} \left(e^{\ln x}\right)^{-\ln x} = \lim\limits_{x\to 0^+} e^{-(\ln x)^2} = 0$ since $-(\ln x)^2 \to -\infty$ as $x \to 0^+$.

(c) $\lim\limits_{x\to 0^+} x^{1/x} = \lim\limits_{x\to 0^+} \left(e^{\ln x}\right)^{1/x} = \lim\limits_{x\to 0^+} e^{(\ln x/x)} = 0$ since $\dfrac{\ln x}{x} \to -\infty$ as $x \to 0^+$.

(d) $\lim\limits_{x\to\infty} (\ln 2x)^{-\ln x} = \lim\limits_{x\to\infty} \left[e^{\ln(\ln 2x)}\right]^{-\ln x} = \lim\limits_{x\to\infty} e^{-\ln x \ln(\ln 2x)} = 0$ since $-\ln x \ln(\ln 2x) \to -\infty$ as $x \to \infty$.

87. **(a)** Let $\epsilon > 0$ be given. We need N such that $|a^x - 0| < \epsilon$ when $x < N$. But $a^x < \epsilon \quad\Leftrightarrow\quad x < \log_a \epsilon$. Let $N = \log_a \epsilon$. Then $x < N \quad\Rightarrow\quad x < \log_a \epsilon \quad\Rightarrow\quad |a^x - 0| = a^x < \epsilon$, so $\lim\limits_{x\to-\infty} a^x = 0$.

(b) Let $M > 0$ be given. We need N such that $a^x > M$ when $x > N$. But $a^x > M \quad\Leftrightarrow\quad x > \log_a M$. Let $N = \log_a M$. Then $x > N \quad\Rightarrow\quad x > \log_a M \quad\Rightarrow\quad a^x > M$, so $\lim\limits_{x\to\infty} a^x = \infty$.

88. **(a)**

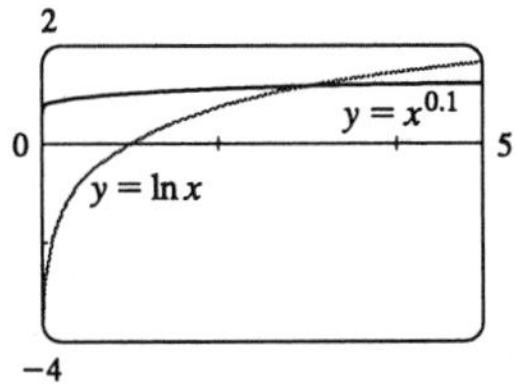

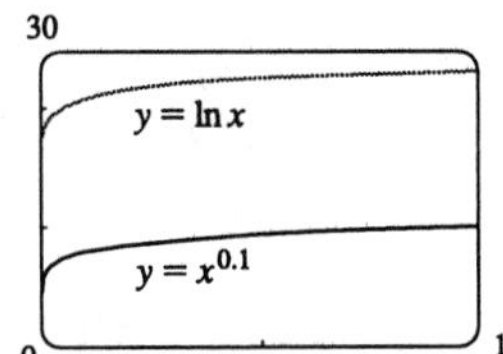

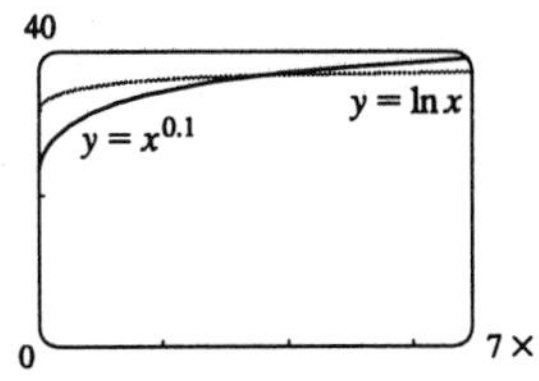

From the graphs, we see that $f(x) > g(x)$ for approximately $0 < x < 3$, and then $g(x) > f(x)$ for $3 < x < 3.5 \times 10^{15}$ (approximately). At that point, the graph of f finally surpasses the graph of g for good.

(b)

0.2

$y = \frac{\ln x}{x^{0.1}}$

$y = 0.1$

0

5×10^{28}

(c) From the graph at left, it seems that $\dfrac{\ln x}{x^{0.1}} < 0.1$ whenever $x > 1.3 \times 10^{28}$ (approximately). So we can take $N = 1.3 \times 10^{28}$, or any larger number.

89. $\ln(x^2 - 2x - 2) \le 0 \quad \Rightarrow \quad 0 < x^2 - 2x - 2 \le 1$. Now $x^2 - 2x - 2 \le 1$ gives $x^2 - 2x - 3 \le 0$ and hence $(x-3)(x+1) \le 0$. So $-1 \le x \le 3$. Now $0 < x^2 - 2x - 2 \quad \Rightarrow \quad x < 1 - \sqrt{3}$ or $x > 1 + \sqrt{3}$. Therefore $\ln(x^2 - 2x - 2) \le 0 \quad \Leftrightarrow \quad -1 \le x < 1 - \sqrt{3}$ or $1 + \sqrt{3} < x \le 3$.

90. (a) The primes less than 25 are: 2, 3, 5, 7, 11, 13, 17, 19, and 23. There are 9 of them, so $\pi(25) = 9$. We use the sieve of Eratosthenes, and arrive at the figure at right. There are 25 numbers left over, so $\pi(100) = 25$.

	2	3	~~4~~	5	~~6~~	7	~~8~~	~~9~~	~~10~~
11	~~12~~	13	~~14~~	~~15~~	~~16~~	17	~~18~~	19	~~20~~
~~21~~	~~22~~	23	~~24~~	~~25~~	~~26~~	~~27~~	~~28~~	29	~~30~~
31	~~32~~	~~33~~	~~34~~	~~35~~	~~36~~	37	~~38~~	~~39~~	~~40~~
41	~~42~~	43	~~44~~	~~45~~	~~46~~	47	~~48~~	~~49~~	~~50~~
~~51~~	~~52~~	53	~~54~~	~~55~~	~~56~~	57	~~58~~	59	~~60~~
61	~~62~~	~~63~~	~~64~~	~~65~~	~~66~~	67	~~68~~	~~69~~	~~70~~
71	~~72~~	73	~~74~~	~~75~~	~~76~~	~~77~~	~~78~~	~~79~~	~~80~~
~~81~~	~~82~~	83	~~84~~	~~35~~	~~86~~	~~87~~	~~88~~	89	~~90~~
91	~~92~~	~~93~~	~~94~~	~~95~~	~~96~~	97	~~98~~	~~99~~	~~100~~

(b) Let $f(n) = \dfrac{\pi(n)}{n/\ln n}$. Then we compute $f(100) = \dfrac{25}{100/\ln 100} \approx 1.15$, $f(1000) \approx 1.16$, $f(10^4) \approx 1.13$, $f(10^5) \approx 1.10$, $f(10^6) \approx 1.08$, and $f(10^7) \approx 1.07$.

(c) By the Prime Number Theorem, the number of primes less than a billion, that is, $\pi(10^9)$, should be close to $10^9/\ln 10^9 \approx 48{,}254{,}942$. In fact, $\pi(10^9) = 50{,}847{,}543$, so our estimate is off by about 5.1%. Do not attempt this calculation at home.

EXERCISES 6.4

1. $f(x) = \ln(x+1) \quad \Rightarrow \quad f'(x) = 1/(x+1)$, $\text{Dom}(f) = \text{Dom}(f') = \{x \mid x + 1 > 0\}$
$= \{x \mid x > -1\} = (-1, \infty)$ [Note that, in general, $\text{Dom}(f') \subset \text{Dom}(f)$.]

2. $f(x) = \cos(\ln x) \quad \Rightarrow \quad f'(x) = -\sin(\ln x)/x$, $\text{Dom}(f) = \text{Dom}(f') = (0, \infty)$

3. $f(x) = x^2 \ln(1 - x^2) \quad \Rightarrow \quad f'(x) = 2x \ln(1 - x^2) + \dfrac{x^2(-2x)}{1 - x^2} = 2x \ln(1 - x^2) - \dfrac{2x^3}{1 - x^2}$,
$\text{Dom}(f) = \text{Dom}(f') = \{x \mid 1 - x^2 > 0\} = \{x \mid |x| < 1\} = (-1, 1)$

4. $f(x) = \ln \ln \ln x \quad \Rightarrow \quad f'(x) = \dfrac{1}{\ln \ln x} \cdot \dfrac{1}{\ln x} \cdot \dfrac{1}{x}$,
$\text{Dom}(f) = \{x \mid \ln \ln x > 0\} = \{x \mid \ln x > 1\} = \{x \mid x > e\} = (e, \infty)$. $\text{Dom}(f') = (e, \infty)$

5. $f(x) = \log_3(x^2 - 4) \quad \Rightarrow \quad f'(x) = \dfrac{2x}{(x^2-4)\ln 3}$,

$\text{Dom}(f) = \text{Dom}(f') = \{x \mid x^2 - 4 > 0\} = \{x \mid |x| > 2\} = (-\infty, -2) \cup (2, \infty)$

6. $f(x) = \sqrt{3 - 2^x} \quad \Rightarrow \quad f'(x) = \dfrac{1}{2\sqrt{3-2^x}}(-2^x \ln 2) = -\ln 2\, \dfrac{2^{x-1}}{\sqrt{3-2^x}}$,

$\text{Dom}(f) = \{x \mid 2^x \le 3\} = \{x \mid x \le \log_2 3\} = (-\infty, \log_2 3]$, $\text{Dom}(f') = \{x \mid 2^x < 3\} = (-\infty, \log_2 3)$

7. $y = x \ln x \quad \Rightarrow \quad y' = \ln x + x(1/x) = \ln x + 1 \quad \Rightarrow \quad y'' = 1/x$

8. $y = \ln(ax) \quad \Rightarrow \quad y' = \dfrac{a}{ax} = \dfrac{1}{x} \quad \Rightarrow \quad y'' = -\dfrac{1}{x^2}$

9. $y = \log_{10} x \quad \Rightarrow \quad y' = \dfrac{1}{x \ln 10} \quad \Rightarrow \quad y'' = -\dfrac{1}{x^2 \ln 10}$

10. $y = \ln(\sec x + \tan x) \quad \Rightarrow \quad y' = \dfrac{\sec x \tan x + \sec^2 x}{\sec x + \tan x} = \sec x \quad \Rightarrow \quad y'' = \sec x \tan x$

11. $f(x) = \sqrt{x} \ln x \quad \Rightarrow \quad f'(x) = \dfrac{1}{2\sqrt{x}} \ln x + \sqrt{x}\left(\dfrac{1}{x}\right) = \dfrac{\ln x + 2}{2\sqrt{x}}$

12. $f(x) = \log_{10}\left(\dfrac{x}{x-1}\right) = \log_{10} x - \log_{10}(x-1) \quad \Rightarrow \quad f'(x) = \dfrac{1}{x \ln 10} - \dfrac{1}{(x-1)\ln 10}$ or $-\dfrac{1}{x(x-1)\ln 10}$

13. $g(x) = \ln \dfrac{a-x}{a+x} = \ln(a-x) - \ln(a+x) \quad \Rightarrow \quad g'(x) = \dfrac{-1}{a-x} - \dfrac{1}{a+x} = \dfrac{-2a}{a^2 - x^2}$

14. $h(x) = \ln\left(x + \sqrt{x^2-1}\right) \quad \Rightarrow \quad h'(x) = \dfrac{1}{x + \sqrt{x^2-1}}\left[1 + \dfrac{x}{\sqrt{x^2-1}}\right] = \dfrac{1}{\sqrt{x^2-1}}$

15. $F(x) = \ln \sqrt{x} = \frac{1}{2} \ln x \quad \Rightarrow \quad F'(x) = \dfrac{1}{2}\left(\dfrac{1}{x}\right) = \dfrac{1}{2x}$

16. $G(x) = \sqrt{\ln x} \quad \Rightarrow \quad G'(x) = \dfrac{1}{2\sqrt{\ln x}}\left(\dfrac{1}{x}\right) = \dfrac{1}{2x\sqrt{\ln x}}$

17. $f(t) = \log_2(t^4 - t^2 + 1) \quad \Rightarrow \quad f'(t) = \dfrac{4t^3 - 2t}{(t^4 - t^2 + 1)\ln 2}$

18. $h(y) = \ln(y^3 \sin y) = 3 \ln y + \ln(\sin y) \quad \Rightarrow \quad h'(y) = \dfrac{3}{y} + \dfrac{1}{\sin y}(\cos y) = \dfrac{3}{y} + \cot y$

19. $g(u) = \dfrac{1 - \ln u}{1 + \ln u} \quad \Rightarrow \quad g'(u) = \dfrac{(1 + \ln u)(-1/u) - (1 - \ln u)(1/u)}{(1 + \ln u)^2} = -\dfrac{2}{u(1 + \ln u)^2}$

20. $G(u) = \ln \sqrt{\dfrac{3u+2}{3u-2}} = \frac{1}{2}[\ln(3u+2) - \ln(3u-2)] \quad \Rightarrow \quad G'(u) = \dfrac{1}{2}\left(\dfrac{3}{3u+2} - \dfrac{3}{3u-2}\right) = \dfrac{-6}{9u^2 - 4}$

21. $y = (\ln \sin x)^3 \quad \Rightarrow \quad y' = 3(\ln \sin x)^2 \dfrac{\cos x}{\sin x} = 3(\ln \sin x)^2 \cot x$

22. $y = \ln(x + \ln x) \quad \Rightarrow \quad y' = \dfrac{1}{x + \ln x}\left(1 + \dfrac{1}{x}\right)$

23. $y = \dfrac{\ln x}{1 + x^2} \quad \Rightarrow \quad y' = \dfrac{(1 + x^2)(1/x) - 2x \ln x}{(1 + x^2)^2} = \dfrac{1 + x^2 - 2x^2 \ln x}{x(1 + x^2)^2}$

24. $y = \ln\left(x\sqrt{1 - x^2} \sin x\right) = \ln x + \frac{1}{2} \ln(1 - x^2) + \ln \sin x \quad \Rightarrow$

$y' = \dfrac{1}{x} + \dfrac{1}{2}\left(\dfrac{-2x}{1 - x^2}\right) + \dfrac{\cos x}{\sin x} = \dfrac{1}{x} - \dfrac{x}{1 - x^2} + \cot x$

25. $y = \ln|x^3 - x^2| \quad\Rightarrow\quad y' = \dfrac{1}{x^3 - x^2}(3x^2 - 2x) = \dfrac{x(3x-2)}{x^2(x-1)} = \dfrac{3x-2}{x(x-1)}$

26. $y = \ln|\tan 2x| \quad\Rightarrow\quad y' = \dfrac{2\sec^2 2x}{\tan 2x}$

27. $F(x) = e^x \ln x \quad\Rightarrow\quad F'(x) = e^x \ln x + e^x\left(\dfrac{1}{x}\right) = e^x\left(\ln x + \dfrac{1}{x}\right)$

28. $G(x) = 5^{\tan x} \quad\Rightarrow\quad G'(x) = 5^{\tan x}(\ln 5)\sec^2 x$

29. $f(t) = \pi^{-t} \quad\Rightarrow\quad f'(t) = \pi^{-t}(\ln\pi)(-1) = -\pi^{-t}\ln\pi$

30. $g(x) = 1.6^x + x^{1.6} \quad\Rightarrow\quad g'(x) = 1.6^x \ln(1.6) + 1.6\,x^{0.6}$

31. $h(t) = t^3 - 3^t \quad\Rightarrow\quad h'(t) = 3t^2 - 3^t \ln 3$

32. $y = 2^{3^x} \quad\Rightarrow\quad y' = 2^{3^x}(\ln 2)3^x \ln 3 = (\ln 2)(\ln 3)3^x 2^{3^x}$

33. $y = \ln[e^{-x}(1+x)] = \ln(e^{-x}) + \ln(1+x) = -x + \ln(1+x) \quad\Rightarrow\quad y' = -1 + \dfrac{1}{1+x} = -\dfrac{x}{1+x}$

34. $y = x^x \quad\Rightarrow\quad \ln y = x\ln x \quad\Rightarrow\quad \dfrac{y'}{y} = \ln x + x\left(\dfrac{1}{x}\right) \quad\Rightarrow\quad y' = x^x(\ln x + 1)$

35. $y = x^{\sin x} \quad\Rightarrow\quad \ln y = \sin x \ln x \quad\Rightarrow\quad \dfrac{y'}{y} = \cos x \ln x + \dfrac{\sin x}{x} \quad\Rightarrow\quad y' = x^{\sin x}\left(\cos x \ln x + \dfrac{\sin x}{x}\right)$

36. $y = (\sin x)^x \quad\Rightarrow\quad \ln y = x\ln(\sin x) \quad\Rightarrow\quad \dfrac{y'}{y} = \ln(\sin x) + x\dfrac{\cos x}{\sin x} \quad\Rightarrow$

$y' = (\sin x)^x[\ln(\sin x) + x\cot x]$

37. $y = x^{e^x} \quad\Rightarrow\quad \ln y = e^x \ln x \quad\Rightarrow\quad \dfrac{y'}{y} = e^x \ln x + \dfrac{e^x}{x} \quad\Rightarrow\quad y' = x^{e^x} e^x\left(\ln x + \dfrac{1}{x}\right)$

38. $y = x^{1/x} \quad\Rightarrow\quad \ln y = \dfrac{1}{x}\ln x \quad\Rightarrow\quad \dfrac{y'}{y} = -\dfrac{1}{x^2}\ln x + \dfrac{1}{x}\left(\dfrac{1}{x}\right) \quad\Rightarrow\quad y' = x^{1/x}\dfrac{(1-\ln x)}{x^2}$

39. $y = (\ln x)^x \quad\Rightarrow\quad \ln y = x\ln\ln x \quad\Rightarrow\quad \dfrac{y'}{y} = \ln\ln x + x\cdot\dfrac{1}{\ln x}\cdot\dfrac{1}{x} \quad\Rightarrow\quad y' = (\ln x)^x\left(\ln\ln x + \dfrac{1}{\ln x}\right)$

40. $y = x^{\ln x} \quad\Rightarrow\quad \ln y = \ln x \ln x = (\ln x)^2 \quad\Rightarrow\quad \dfrac{y'}{y} = 2\ln x\left(\dfrac{1}{x}\right) \quad\Rightarrow\quad y' = x^{\ln x}\left(\dfrac{2\ln x}{x}\right)$

41. $y = x^{1/\ln x} \quad\Rightarrow\quad \ln y = \left(\dfrac{1}{\ln x}\right)\ln x = 1 \quad\Rightarrow\quad y = e \quad\Rightarrow\quad y' = 0$

42. $y = (\sin x)^{\cos x} \quad\Rightarrow\quad \ln y = \cos x \ln(\sin x) \quad\Rightarrow\quad \dfrac{y'}{y} = -\sin x \ln\sin x + \cos x\left(\dfrac{\cos x}{\sin x}\right) \quad\Rightarrow$

$y' = (\sin x)^{\cos x}(-\sin x \ln\sin x + \cos x \cot x)$

43. $y = \cos\left(x^{\sqrt{x}}\right) \quad\Rightarrow\quad y' = -\sin\left(x^{\sqrt{x}}\right)x^{\sqrt{x}}\left(\dfrac{\ln x + 2}{2\sqrt{x}}\right)$ by Example 16

44. $y = x^{x^x} \quad\Rightarrow\quad \ln y = x^x \ln x \quad\Rightarrow\quad \dfrac{y'}{y} = x^x(\ln x + 1)\ln x + x^x\left(\dfrac{1}{x}\right)$ $\Big[$since $z = x^x \quad\Rightarrow\quad \ln z = x\ln x$

$\Rightarrow\quad \dfrac{z'}{z} = \ln x + x\left(\dfrac{1}{x}\right) \quad\Rightarrow\quad z' = x^x(\ln x + 1)\Big] \quad\Rightarrow\quad y' = x^{x^x}\left[x^x(\ln x + 1)\ln x + x^{x-1}\right]$

45. $f(x) = \dfrac{x}{\ln x} \quad\Rightarrow\quad f'(x) = \dfrac{\ln x - x(1/x)}{(\ln x)^2} = \dfrac{\ln x - 1}{(\ln x)^2} \quad\Rightarrow\quad f'(e) = \dfrac{1-1}{1^2} = 0$

46. $f(x) = x^2 \ln x \;\Rightarrow\; f'(x) = 2x\ln x + x^2\left(\dfrac{1}{x}\right) = 2x\ln x + x \;\Rightarrow\; f'(1) = 2\ln 1 + 1 = 1$

47. $f(x) = \sin x + \ln x \quad\Rightarrow$

$f'(x) = \cos x + \dfrac{1}{x}$

This is reasonable, because the graph shows that f increases when $f'(x)$ is positive.

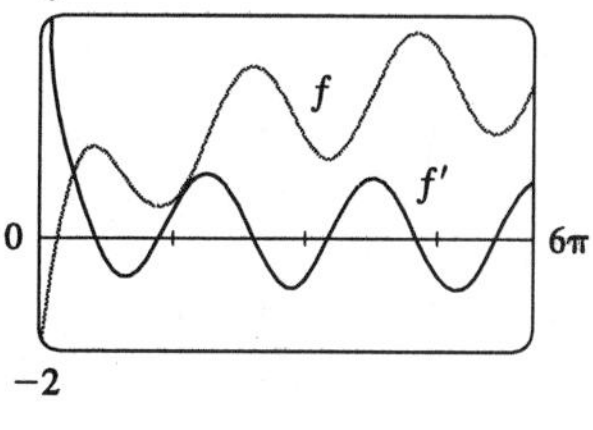

48. $f(x) = x^{\cos x} = e^{\ln x \cos x} \quad\Rightarrow$

$$f'(x) = e^{\ln x \cos x}\left[\ln x(-\sin x) + \cos x\left(\frac{1}{x}\right)\right] = x^{\cos x}\left[\frac{\cos x}{x} - \sin x \ln x\right]$$

This is reasonable, because the graph shows that f increases when $f'(x)$ is positive.

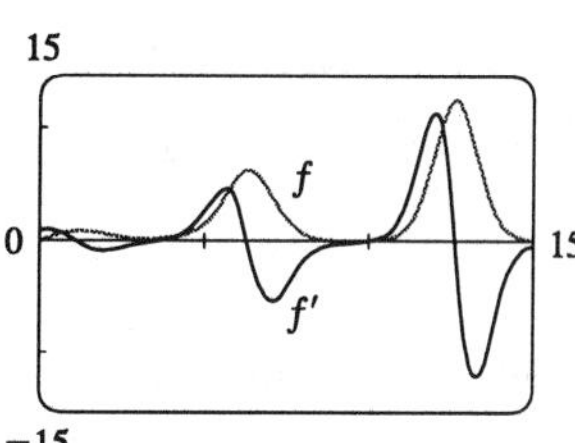

49. $y = f(x) = \ln\ln x \quad\Rightarrow\quad f'(x) = \dfrac{1}{\ln x}\left(\dfrac{1}{x}\right) \quad\Rightarrow\quad f'(e) = \dfrac{1}{e}$, so the equation of the tangent at $(e, 0)$ is $y - 0 = \dfrac{1}{e}(x - e)$ or $x - ey = e$.

50. $f(x) = 10^x \quad\Rightarrow\quad f'(x) = 10^x \ln 10$, so the slope of the tangent at $(1, 10)$ is $f'(1) = 10\ln 10$ and the equation is $y - 10 = 10\ln 10(x-1)$ or $y = 10[(x-1)\ln 10 + 1]$.

51. $y = \ln(x^2 + y^2) \;\Rightarrow\; y' = \dfrac{2x + 2yy'}{x^2 + y^2} \;\Rightarrow\; x^2y' + y^2y' = 2x + 2yy' \;\Rightarrow\; y' = \dfrac{2x}{x^2 + y^2 - 2y}$

52. $x^y = y^x \quad\Rightarrow\quad y\ln x = x\ln y \quad\Rightarrow\quad y'\ln x + \dfrac{y}{x} = \ln y + x\dfrac{y'}{y} \quad\Rightarrow\quad y' = \dfrac{\ln y - y/x}{\ln x - x/y}$

53. $f(x) = \ln(x-1) \quad\Rightarrow\quad f'(x) = 1/(x-1) = (x-1)^{-1} \quad\Rightarrow\quad f''(x) = -(x-1)^{-2} \quad\Rightarrow$

$f'''(x) = 2(x-1)^{-3} \quad\Rightarrow\quad f^{(4)}(x) = -2\cdot 3(x-1)^{-4} \quad\Rightarrow\quad \cdots \quad\Rightarrow$

$f^{(n)}(x) = (-1)^{n-1}\cdot 2\cdot 3\cdot 4\cdot\cdots\cdot(n-1)(x-1)^{-n} = (-1)^{n-1}\dfrac{(n-1)!}{(x-1)^n}$

54. $y = x^8 \ln x$, so $D^9 y = D^8\left(8x^7\ln x + x^7\right) = D^8\left(8x^7\ln x\right) = D^7\left(8\cdot 7x^6\ln x + 8x^6\right)$

$= D^7\left(8\cdot 7x^6 \ln x\right) = D^6\left(8\cdot 7\cdot 6x^5\ln x\right) = \cdots = D\left(8!\,x^0\ln x\right) = 8!/x$

55.

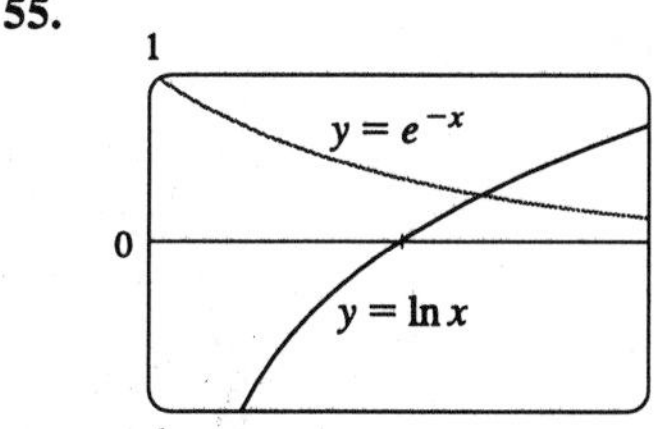

From the graph, it appears that the only root of the equation occurs at about $x = 1.3$. So we use Newton's Method with this as our initial approximation, and with $f(x) = \ln x - e^{-x} \quad\Rightarrow\quad f'(x) = 1/x - e^{-x}$. The formula is $x_{n+1} = x_n - f(x_n)/f'(x_n)$, and we calculate $x_1 = 1.3$, $x_2 \approx 1.309760$, $x_3 \approx x_4 \approx 1.309800$. So, correct to six decimal places, the root of the equation $\ln x = e^{-x}$ is $x = 1.309800$.

56. 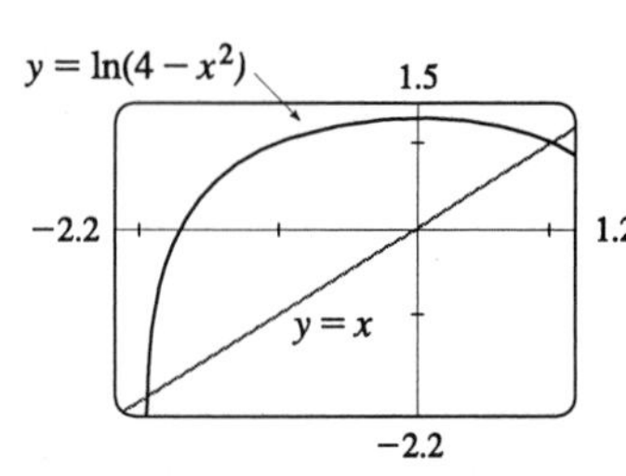

We use Newton's Method with $f(x) = \ln(4 - x^2) - x$ and $f'(x) = \dfrac{1}{4 - x^2}(-2x) - 1 = 1 - \dfrac{2x}{4 - x^2}$. The formula is $x_{n+1} = x_n - f(x_n)/f'(x_n)$. From the graphs it seems that the roots occur at approximately $x = -1.9$ and $x = 1.1$. However, if we use $x_1 = -1.9$ as an initial approximation to the first root, we get $x_2 \approx -2.009611$, and $f(x) = \ln(x-2)^2 - x$ is undefined at this point, making it impossible to calculate x_3. We must use a more accurate first estimate for this root, such as $x_1 = -1.95$. With this approximation, we get $x_1 = -1.95$, $x_2 \approx -1.1967495$, $x_3 \approx -1.964760$, $x_4 \approx x_5 \approx -1.964636$. Calculating the second root gives $x_1 = 1.1$, $x_2 \approx 1.058649$, $x_3 \approx 1.058007$, $x_4 \approx x_5 \approx 1.058006$. So, correct to six decimal places, the two roots of the equation $\ln(4 - x^2) = x$ are $x = -1.964636$ and $x = 1.058006$.

57. $f(x) = \dfrac{\ln x}{\sqrt{x}} \quad \Rightarrow \quad f'(x) = \dfrac{\sqrt{x}(1/x) - (\ln x)\left[1/(2\sqrt{x})\right]}{x} = \dfrac{2 - \ln x}{2x^{3/2}} \quad \Rightarrow$

$f''(x) = \dfrac{2x^{3/2}(-1/x) - (2 - \ln x)\left(3x^{1/2}\right)}{4x^3} = \dfrac{3\ln x - 8}{4x^{5/2}} > 0 \quad \Leftrightarrow \quad \ln x > \frac{8}{3} \quad \Leftrightarrow \quad x > e^{8/3}$, so f is CU on $\left(e^{8/3}, \infty\right)$ and CD on $\left(0, e^{8/3}\right)$. The inflection point is $\left(e^{8/3}, \frac{8}{3}e^{-4/3}\right)$.

58. $f(x) = x\ln x$, $f'(x) = \ln x + 1 = 0$ when $\ln x = -1 \quad \Leftrightarrow \quad x = e^{-1}$. $f'(x) > 0 \quad \Leftrightarrow \quad \ln x + 1 > 0 \quad \Leftrightarrow \quad \ln x > -1 \quad \Leftrightarrow \quad x > 1/e$. $f'(x) < 0 \quad \Leftrightarrow \quad \ln x + 1 < 0 \quad \Leftrightarrow \quad x < 1/e$. Therefore, there is an absolute minimum value of $f(1/e) = (1/e)\ln(1/e) = -1/e$.

59. $y = f(x) = \ln(\cos x)$ **A.** $D = \{x \mid \cos x > 0\} = \left(-\frac{\pi}{2}, \frac{\pi}{2}\right) \cup \left(\frac{3\pi}{2}, \frac{5\pi}{2}\right) \cup \cdots$ $= \left\{x \mid 2n\pi - \frac{\pi}{2} < x < 2n\pi + \frac{\pi}{2}, n = 0, \pm 1, \pm 2, \ldots\right\}$ **B.** x-intercepts occur when $\ln(\cos x) = 0 \quad \Leftrightarrow \quad \cos x = 1 \quad \Leftrightarrow \quad x = 2n\pi$, y-intercept $= f(0) = 0$. **C.** $f(-x) = f(x)$, so the curve is symmetric about the y-axis. $f(x + 2\pi) = f(x)$, f has period 2π, so in parts D-G we consider only $-\frac{\pi}{2} < x < \frac{\pi}{2}$.

D. $\lim\limits_{x \to \pi/2^-} \ln(\cos x) = -\infty$ and $\lim\limits_{x \to -\pi/2^+} \ln(\cos x) = -\infty$, so $x = \frac{\pi}{2}$ and $x = -\frac{\pi}{2}$ are VA. No HA.
E. $f'(x) = (1/\cos x)(-\sin x) = -\tan x > 0 \quad \Leftrightarrow \quad -\frac{\pi}{2} < x < 0$, so f is increasing on $\left(-\frac{\pi}{2}, 0\right]$ and decreasing on $\left[0, \frac{\pi}{2}\right)$. **F.** $f(0) = 0$ is a local maximum.
G. $f''(x) = -\sec^2 x < 0 \quad \Rightarrow \quad f$ is CD on $\left(-\frac{\pi}{2}, \frac{\pi}{2}\right)$. No IP.

H.

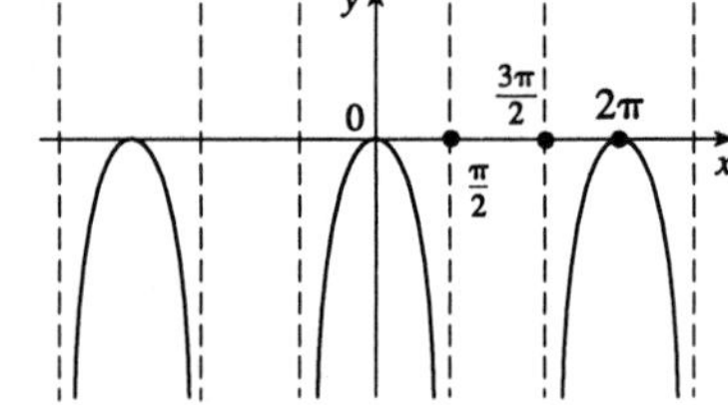

60. $y = f(x) = x^2 + \ln x$ **A.** $D = (0, \infty)$ **B.** No y-intercept **C.** No symmetry

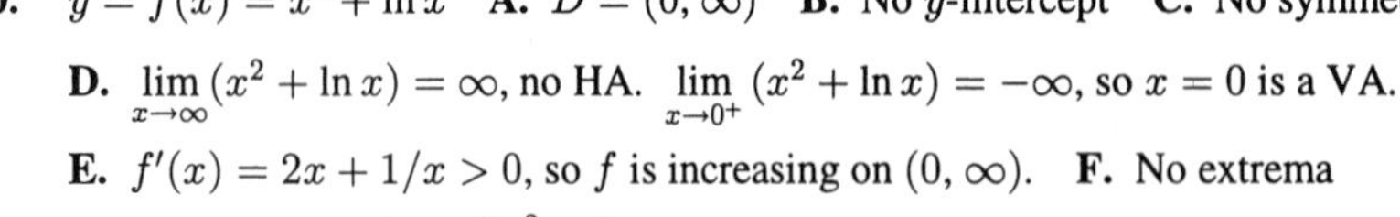

D. $\lim\limits_{x \to \infty}(x^2 + \ln x) = \infty$, no HA. $\lim\limits_{x \to 0^+}(x^2 + \ln x) = -\infty$, so $x = 0$ is a VA.
E. $f'(x) = 2x + 1/x > 0$, so f is increasing on $(0, \infty)$. **F.** No extrema
G. $f''(x) = 2 - \dfrac{1}{x^2} = \dfrac{2x^2 - 1}{x^2} > 0 \quad \Leftrightarrow \quad 2x^2 > 1 \quad \Leftrightarrow \quad x > \frac{1}{\sqrt{2}}$, so f is CU on $\left(\frac{1}{\sqrt{2}}, \infty\right)$ and CD on $\left(0, \frac{1}{\sqrt{2}}\right)$. IP $\left(\frac{1}{\sqrt{2}}, \frac{1}{2}(1 - \ln 2)\right)$.

H.

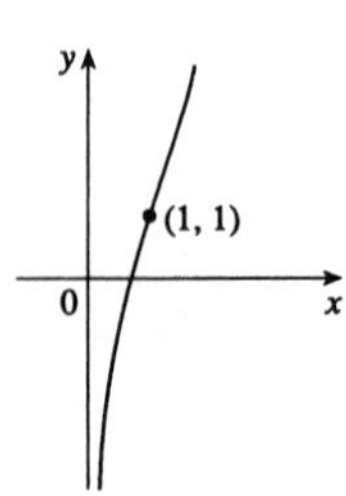

61. $y = f(x) = \ln(1 + x^2)$ **A.** $D = \mathbb{R}$ **B.** Both intercepts are 0. **C.** $f(-x) = f(x)$, so the curve is symmetric about the y-axis. **D.** $\lim\limits_{x\to\pm\infty} \ln(1 + x^2) = \infty$, no asymptotes. **E.** $f'(x) = \dfrac{2x}{1 + x^2} > 0 \quad \Leftrightarrow \quad x > 0$, so f is increasing on $[0, \infty)$ and decreasing on $(-\infty, 0]$.

F. $f(0) = 0$ is a local and absolute minimum.

G. $f''(x) = \dfrac{2(1 + x^2) - 2x(2x)}{(1 + x^2)^2} = \dfrac{2(1 - x^2)}{(1 + x^2)^2} > 0$

$\Leftrightarrow \quad |x| < 1$, so f is CU on $(-1, 1)$, CD on $(-\infty, -1)$ and $(1, \infty)$. IP $(1, \ln 2)$ and $(-1, \ln 2)$.

H.

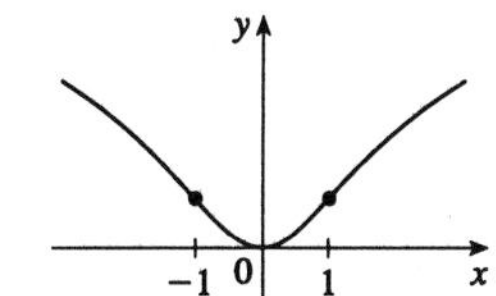

62. $y = \ln(\tan^2 x)$ **A.** $D = \{x \mid x \neq n\pi/2\}$ **B.** x-intercepts $n\pi + \frac{\pi}{4}$, no y-intercept. **C.** $f(-x) = f(x)$, so the curve is symmetric about the y-axis. Also $f(x + \pi) = f(x)$, so f is periodic with period π, and we consider parts D-G only for $-\frac{\pi}{2} < x < \frac{\pi}{2}$. **D.** $\lim\limits_{x\to 0} \ln(\tan^2 x) = -\infty$ and $\lim\limits_{x\to\pi/2^-} \ln(\tan^2 x) = \infty$, $\lim\limits_{x\to-\pi/2^+} \ln(\tan^2 x) = \infty$, so $x = 0$, $x = \pm\frac{\pi}{2}$ are VA. **E.** $f'(x) = \dfrac{2\tan x \sec^2 x}{\tan^2 x} = 2\dfrac{\sec^2 x}{\tan x} > 0 \quad \Leftrightarrow \quad \tan x > 0 \quad \Leftrightarrow \quad 0 < x < \frac{\pi}{2}$, so f is increasing on $\left(0, \frac{\pi}{2}\right)$ and decreasing on $\left(-\frac{\pi}{2}, 0\right)$.

F. No extrema **G.** $f'(x) = \dfrac{2}{\sin x \cos x} = \dfrac{4}{\sin 2x}$

$\Rightarrow \quad f''(x) = \dfrac{-8\cos 2x}{\sin^2 2x} < 0 \quad \Leftrightarrow \quad \cos 2x > 0$

$\Leftrightarrow \quad -\frac{\pi}{4} < x < \frac{\pi}{4}$, so f is CD on $\left(-\frac{\pi}{4}, 0\right)$ and $\left(0, \frac{\pi}{4}\right)$ and CU on $\left(-\frac{\pi}{2}, -\frac{\pi}{4}\right)$ and $\left(\frac{\pi}{4}, \frac{\pi}{2}\right)$. IP are $\left(\pm\frac{\pi}{4}, 0\right)$.

H.

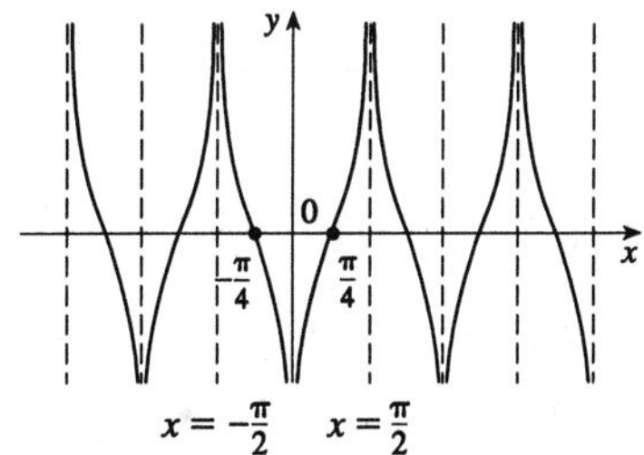

63. $y = f(x) = \ln(x^2 - x)$ **A.** $\{x \mid x^2 - x > 0\} = \{x \mid x < 0 \text{ or } x > 1\} = (-\infty, 0) \cup (1, \infty)$. **B.** x-intercepts occur when $x^2 - x = 1 \quad \Leftrightarrow \quad x^2 - x - 1 = 0 \quad \Leftrightarrow \quad x = \frac{1}{2}\left(1 \pm \sqrt{5}\right)$. No y-intercept **C.** No symmetry **D.** $\lim\limits_{x\to\infty} \ln(x^2 - x) = \infty$, no HA. $\lim\limits_{x\to 0^-} \ln(x^2 - x) = -\infty$, $\lim\limits_{x\to 1^+} \ln(x^2 - x) = -\infty$, so $x = 0$ and $x = 1$ are VA. **E.** $f'(x) = \dfrac{2x - 1}{x^2 - x} > 0$ when $x > 1$ and $f'(x) < 0$ when $x < 0$, so f is increasing on $(1, \infty)$ and decreasing on $(-\infty, 0)$. **F.** No extrema

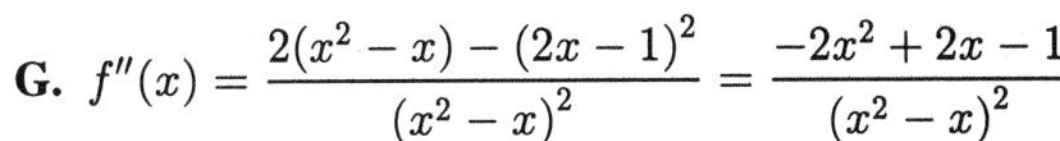

G. $f''(x) = \dfrac{2(x^2 - x) - (2x - 1)^2}{(x^2 - x)^2} = \dfrac{-2x^2 + 2x - 1}{(x^2 - x)^2}$

$\Rightarrow \quad f''(x) < 0$ for all x since $-2x^2 + 2x - 1$ has a negative discriminant. So f is CD on $(-\infty, 0)$ and $(1, \infty)$. No IP.

H.

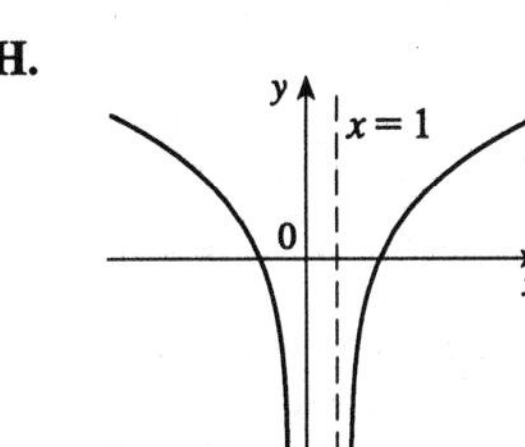

64. $y = x^{-\ln x}$ **A.** The domain is $(0, \infty)$ **B.** No intercepts **C.** No symmetry

D. $\lim\limits_{x \to \infty} x^{-\ln x} = \lim\limits_{x \to \infty} \left(e^{\ln x}\right)^{-\ln x} = \lim\limits_{x \to \infty} e^{-(\ln x)^2} = 0$ since $-(\ln x)^2 \to 0$ as $x \to \infty$. Thus $y = 0$ is a HA.

$\lim\limits_{x \to 0^+} x^{-\ln x} = \lim\limits_{x \to 0^+} e^{-(\ln x)^2} = 0$ since $-(\ln x)^2 \to -\infty$ as $x \to 0^+$. **E.** Using logarithmic differentiation, we obtain $f'(x) = \dfrac{x^{-\ln x}(-2\ln x)}{x}$. $f'(x) > 0 \Leftrightarrow \ln x < 0 \Leftrightarrow 0 < x < 1$ and $f'(x) < 0 \Leftrightarrow \ln x > 0$ $\Leftrightarrow x > 1$. Thus f is increasing on $(0, 1]$ and decreasing on $[1, \infty)$. **F.** Since $f'(1) = 0$ and f' changes from positive to negative at 1, $f(1) = 1$ is a local maximum by the First Derivative Test.

G. $f''(x) = \dfrac{2x^{-\ln x}}{x^2}(2\ln x - 1)(\ln x + 1)$.

$2\ln x - 1 = 0 \Leftrightarrow \ln x = \frac{1}{2} \Leftrightarrow$ $x = e^{1/2} = \sqrt{e}$ and $\ln x + 1 = 0 \Leftrightarrow \ln x = -x$ $\Leftrightarrow x = e^{-1} = 1/e$. Therefore f is CU on $(0, 1/e)$ and $(\sqrt{e}, \infty)$ and CD on $(1/e, \sqrt{e})$. The IP are $(1/e, 1/e)$ and $(\sqrt{e}, e^{-1/4})$.

H.

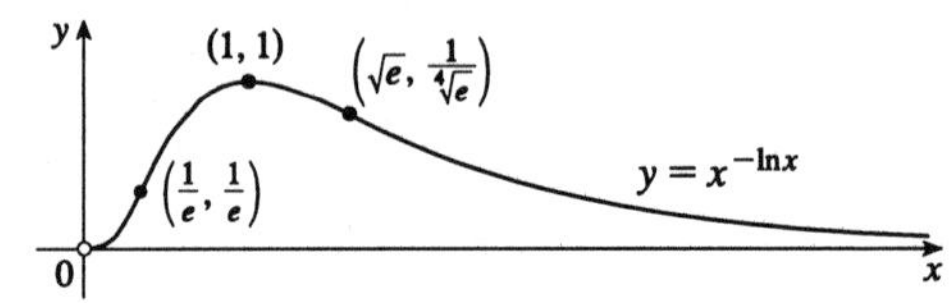

65.

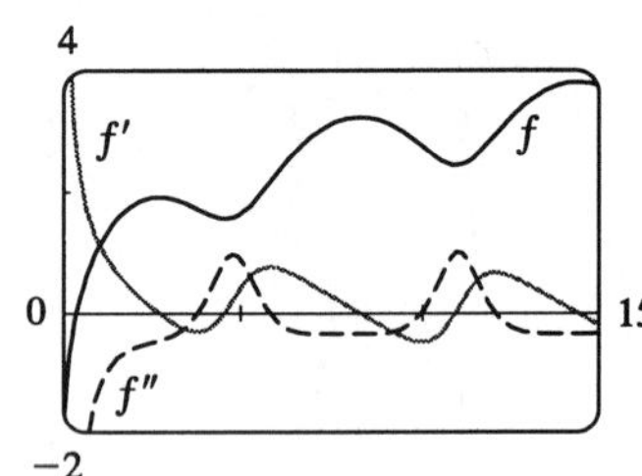

We use the CAS to calculate $f'(x) = \dfrac{2 + \sin x + x\cos x}{2x + x\sin x}$ and $f''(x) = \dfrac{2x^2\sin x + 4\sin x - \cos^2 x + x^2 + 5}{x^2(\cos^2 x - 4\sin x - 5)}$. From the graphs, it seems that $f' > 0$ (and so f is increasing) on approximately the intervals $[0, 2.7]$, $[4.5, 8.2]$ and $[10.9, 14.3]$. It seems that f'' changes sign (indicating inflection points) at $x \approx 3.8, 5.7$, 10.0 and 12.0. Looking back at the graph of f, this implies that the inflection points have approximate coordinates $(3.8, 1.7)$, $(5.7, 2.1)$, $(10.0, 2.7)$, and $(12.0, 2.9)$.

66. We see that if $c \le 0$, $f(x) = \ln(x^2 + c)$ is only defined for $x^2 > -c \Rightarrow |x| > \sqrt{-c}$, and $\lim\limits_{x \to \sqrt{-c}^+} f(x) = \lim\limits_{x \to -\sqrt{-c}^-} f(x) = -\infty$, since $\ln y \to -\infty$ as $y \to 0$. Thus, for $c < 0$, there are vertical asymptotes at $x = \pm\sqrt{c}$, and as c decreases (that is, $|c|$ increases), the asymptotes get further apart. For $c = 0$, $\lim\limits_{x \to 0} f(x) = -\infty$, so there is a vertical asymptote at $x = 0$. If $c > 0$, there are no asymptotes. To find the extrema and inflection points, we differentiate: $f(x) = \ln(x^2 + c) \Rightarrow f'(x) = \dfrac{1}{x^2 + c}(2x)$, so by the First Derivative Test there is a local and absolute minimum at $x = 0$. Differentiating again, we get

$$f''(x) = \frac{1}{x^2 + c}(2) + 2x\left[-(x^2 + c)^{-2}(2x)\right] = \frac{2(c - x^2)}{(x^2 + c)^2}.$$

Now if $c \le 0$, this is always negative, so f is concave down on both of the intervals on which it is defined. If $c > 0$, then f'' changes sign when $c = x^2 \Leftrightarrow x = \pm\sqrt{c}$. So for $x > 0$ there are inflection points at $\pm\sqrt{c}$, and as c increases, the inflection points get further apart.

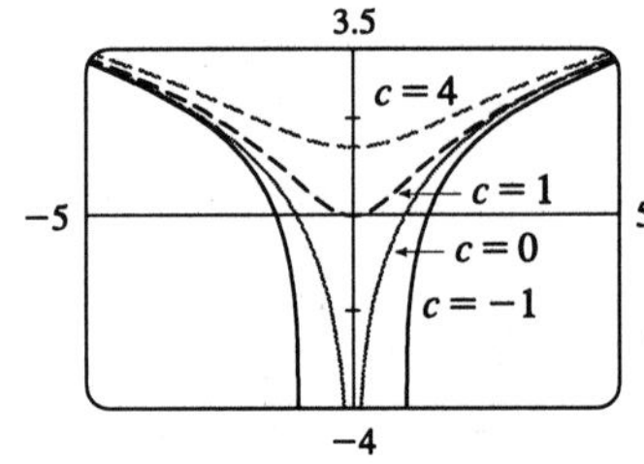

67. $\displaystyle\int_4^8 \frac{1}{x}\,dx = \ln x\big|_4^8 = \ln 8 - \ln 4 = \ln\tfrac{8}{4} = \ln 2$

68. $\displaystyle\int_{-e^2}^{-e} \frac{3}{x}\,dx = 3\ln|x|\big|_{-e^2}^{-e} = 3\ln e - 3\ln(e^2) = 3 - 6 = -3$

69. $\displaystyle\int_1^e \frac{x^2+x+1}{x}dx = \int_1^e \left(x+1+\frac{1}{x}\right)dx = \left[\tfrac{1}{2}x^2 + x + \ln x\right]_1^e = \left(\tfrac{1}{2}e^2 + e + 1\right) - \left(\tfrac{1}{2} + 1 + 0\right) = \tfrac{1}{2}e^2 + e - \tfrac{1}{2}$

70. $\displaystyle\int_4^9 \left[\sqrt{x} + \frac{1}{\sqrt{x}}\right]^2 dx = \int_4^9 \left(x + 2 + \frac{1}{x}\right)dx = \left[\tfrac{1}{2}x^2 + 2x + \ln x\right]_4^9 = \tfrac{81}{2} + 18 + \ln 9 - (8 + 8 + \ln 4)$

$\displaystyle = \tfrac{85}{2} + \ln\tfrac{9}{4}$

71. Let $u = 2x - 1$. Then $du = 2\,dx$, so $\displaystyle\int \frac{dx}{2x-1} = \frac{1}{2}\int\frac{du}{u} = \tfrac{1}{2}\ln|u| + C = \tfrac{1}{2}\ln|2x-1| + C.$

72. Let $u = x^3 + 3x + 1$. Then $du = 3(x^2+1)dx$, so

$\displaystyle\int \frac{x^2+1}{x^3+3x+1}\,dx = \frac{1}{3}\int\frac{du}{u} = \tfrac{1}{3}\ln|u| + C = \tfrac{1}{3}\ln|x^3+3x+1| + C.$

73. Let $u = x^2 + 2x$. Then $du = 2(x+1)dx$, so

$\displaystyle\int \frac{x+1}{x^2+2x}\,dx = \frac{1}{2}\int\frac{du}{u} = \tfrac{1}{2}\ln|u| + C = \tfrac{1}{2}\ln|x^2+2x| + C.$

74. Let $u = \ln x$. Then $du = \dfrac{1}{x}\,dx$, so $\displaystyle\int\frac{dx}{x\ln x} = \int\frac{du}{u} = \ln|u| + C = \ln|\ln x| + C.$

75. Let $u = \ln x$. Then $du = \dfrac{dx}{x}$ $\Rightarrow$ $\displaystyle\int\frac{(\ln x)^2}{x}\,dx = \int u^2\,du = \tfrac{1}{3}u^3 + C = \tfrac{1}{3}(\ln x)^3 + C.$

76. Let $u = 2 - \tan x$. Then $du = -\sec^2 x\,dx$, so $\displaystyle\int\frac{\sec^2 x}{2-\tan x}\,dx = -\int\frac{du}{u} = -\ln|u| + C = -\ln|2-\tan x| + C.$

77. Let $u = 1 + \cos x$. Then $du = -\sin x\,dx$, so

$\displaystyle\int\frac{\sin x}{1+\cos x}\,dx = -\int\frac{du}{u} = -\ln|u| + C = -\ln(1+\cos x) + C.$

78. Let $u = 1 + \ln x$. Then $du = \dfrac{dx}{x}$, so $\displaystyle\int\frac{(1+\ln x)^4}{x}\,dx = \int u^4\,du = \frac{u^5}{5} + C = \frac{(1+\ln x)^5}{5} + C.$

79. $\displaystyle\int_3^4 5^t\,dt = \left[\frac{5^t}{\ln 5}\right]_3^4 = \frac{5^4 - 5^3}{\ln 5} = \frac{500}{\ln 5}$

80. Let $u = \sqrt{x}$. Then $du = \dfrac{dx}{2\sqrt{x}}$ so $\displaystyle\int\frac{10^{\sqrt{x}}}{\sqrt{x}}\,dx = 2\int 10^u\,du = 2\frac{10^u}{\ln 10} + C = \frac{2}{\ln 10}10^{\sqrt{x}} + C.$

81. **(a)** $\displaystyle\frac{d}{dx}(\ln|\sin x| + C) = \frac{1}{\sin x}\cos x = \cot x$

(b) Let $u = \sin x$. Then $du = \cos x\,dx$, so $\displaystyle\int\cot x\,dx = \int\frac{\cos x}{\sin x}\,dx = \int\frac{du}{u} = \ln|u| + C = \ln|\sin x| + C.$

82. Let $u = x - 2$. Then the area is

$\displaystyle A = -\int_{-4}^{-1}\frac{2}{x-2}\,dx = -2\int_{-6}^{-3}\frac{du}{u} = \left[-2\ln|u|\right]_{-6}^{-3} = -2\ln 3 + 2\ln 6 = 2\ln 2 \approx 1.386.$

83. The cross-sectional area is $\pi\left(1/\sqrt{x+1}\right)^2 = \pi/(x+1)$. Therefore, the volume is
$\displaystyle\int_0^1 \frac{\pi}{x+1}\,dx = \pi[\ln(x+1)]_0^1 = \pi\ln 2 - \ln 1 = \pi\ln 2$

84. Using cylindrical shells, we get $V = \displaystyle\int_0^3 \frac{2\pi x}{x^2+1}\,dx = \pi\left[\ln\left(1+x^2\right)\right]_0^3 = \pi\ln 10$.

85. $y = (3x-7)^4\left(8x^2-1\right)^3 \;\Rightarrow\; \ln|y| = 4\ln|3x-7| + 3\ln|8x^2-1| \;\Rightarrow\; \dfrac{y'}{y} = \dfrac{12}{3x-7} + \dfrac{48x}{8x^2-1} \;\Rightarrow$
$y' = (3x-7)^4\left(8x^2-1\right)^3\left(\dfrac{12}{3x-7} + \dfrac{48x}{8x^2-1}\right)$

86. $y = x^{2/5}\left(x^2+8\right)^4 e^{x^2+x} \;\Rightarrow\; \ln|y| = \frac{2}{5}\ln|x| + 4\ln(x^2+8) + x^2 + x \;\Rightarrow$
$\dfrac{y'}{y} = \dfrac{2}{5}\cdot\dfrac{1}{x} + 4\dfrac{2x}{x^2+8} + 2x + 1 \;\Rightarrow\; y' = x^{2/5}\left(x^2+8\right)^4 e^{x^2+x}\left[\dfrac{2}{5x} + \dfrac{8x}{x^2+8} + 2x + 1\right]$

87. $y = \dfrac{(x+1)^4(x-5)^3}{(x-3)^8} \;\Rightarrow\; \ln|y| = 4\ln|x+1| + 3\ln|x-5| - 8\ln|x-3| \;\Rightarrow$
$\dfrac{y'}{y} = \dfrac{4}{x+1} + \dfrac{3}{x-5} - \dfrac{8}{x-3} \;\Rightarrow\; y' = \dfrac{(x+1)^4(x-5)^3}{(x-3)^8}\left(\dfrac{4}{x+1} + \dfrac{3}{x-5} - \dfrac{8}{x-3}\right)$

88. $y = \sqrt{\dfrac{x^2+1}{x+1}} \;\Rightarrow\; \ln y = \frac{1}{2}[\ln(x^2+1) - \ln(x+1)] \;\Rightarrow\; \dfrac{y'}{y} = \dfrac{1}{2}\left(\dfrac{2x}{x^2+1} - \dfrac{1}{x+1}\right) \;\Rightarrow$
$y' = \sqrt{\dfrac{x^2+1}{x+1}}\left[\dfrac{x}{x^2+1} - \dfrac{1}{2(x+1)}\right]$

89. $y = \dfrac{e^x\sqrt{x^5+2}}{(x+1)^4(x^2+3)^2} \;\Rightarrow\; \ln y = x + \frac{1}{2}\ln(x^5+2) - 4\ln|x+1| - 2\ln(x^2+3) \;\Rightarrow$
$\dfrac{y'}{y} = 1 + \dfrac{5x^4}{2(x^5+2)} - \dfrac{4}{x+1} - \dfrac{4x}{x^2+3}$. So $y' = \dfrac{e^x\sqrt{x^5+2}}{(x+1)^4(x^2+3)^2}\left[1 + \dfrac{5x^4}{2(x^5+2)} - \dfrac{4}{x+1} - \dfrac{4x}{x^2+3}\right]$.

90. $y = \dfrac{\left(x^3+1\right)^4\sin^2 x}{x^{1/3}} \;\Rightarrow\; \ln|y| = 4\ln\left|x^3+1\right| + 2\ln|\sin x| - \frac{1}{3}\ln|x|$. So $\dfrac{y'}{y} = 4\dfrac{3x^2}{x^3+1} + 2\dfrac{\cos x}{\sin x} - \dfrac{1}{3x}$
$\Rightarrow\; y' = \dfrac{\left(x^3+1\right)^4\sin^2 x}{x^{1/3}}\left(\dfrac{12x^2}{x^3+1} + 2\cot x - \dfrac{1}{3x}\right)$.

91. The domain of $f(x) = 1/x$ is $(-\infty,0)\cup(0,\infty)$, so its general antiderivative is $F(x) = \begin{cases}\ln x + C_1 & \text{if } x > 0\\ \ln|x| + C_2 & \text{if } x < 0\end{cases}$

92. $f''(x) = x^{-2}, x > 0 \;\Rightarrow\; f'(x) = -1/x + C \;\Rightarrow\; f(x) = -\ln x + Cx + D$. $0 = f(1) = C + D$ and
$0 = f(2) = -\ln 2 + 2C + D = -\ln 2 + 2C - C = -\ln 2 + C \;\Rightarrow\; C = \ln 2$ and $D = -\ln 2$. So
$f(x) = -\ln x + (\ln 2)x - \ln 2$.

93. $f(x) = 2x + \ln x \;\Rightarrow\; f'(x) = 2 + 1/x$. If $g = f^{-1}$, then $f(1) = 2 \;\Rightarrow\; g(2) = 1$, so
$g'(2) = 1/f'(g(2)) = 1/f'(1) = \frac{1}{3}$.

94. $f(x) = e^x + \ln x \;\Rightarrow\; f'(x) = e^x + 1/x$. $h = f^{-1}$ and $f(1) = e \;\Rightarrow\; h(e) = 1$, so
$h'(e) = 1/f'(1) = 1/(e+1)$.

95. The curve and the line will determine a region when they intersect at two or more points. So we solve the equation $x/(x^2+1) = mx$ $\Rightarrow$ $x = 0$ or $mx^2 + m - 1 = 0$ $\Rightarrow$ $x = 0$ or $x = \dfrac{\pm\sqrt{-4(m)(m-1)}}{2m} = \pm\sqrt{\dfrac{1}{m} - 1}$. Note that if $m = 1$, this has only the solution $x = 0$, and no region is determined. But if $1/m - 1 > 0$ $\Leftrightarrow$ $1/m > 1$ $\Leftrightarrow$ $0 < m < 1$, then there are two solutions.

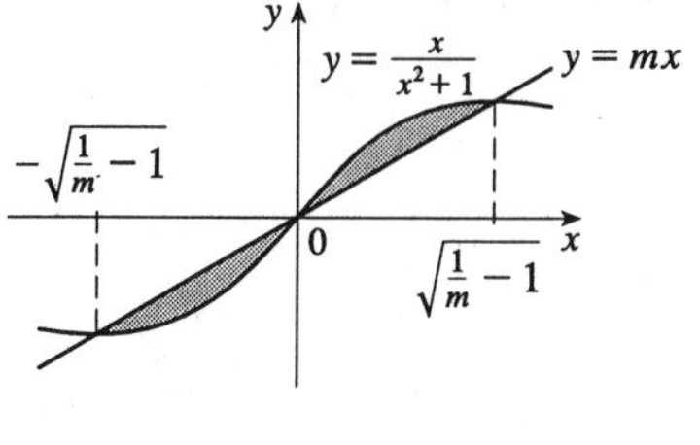

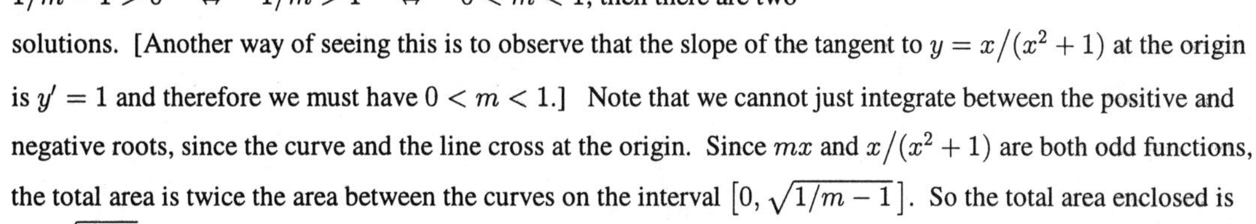

[Another way of seeing this is to observe that the slope of the tangent to $y = x/(x^2+1)$ at the origin is $y' = 1$ and therefore we must have $0 < m < 1$.] Note that we cannot just integrate between the positive and negative roots, since the curve and the line cross at the origin. Since mx and $x/(x^2+1)$ are both odd functions, the total area is twice the area between the curves on the interval $\left[0, \sqrt{1/m - 1}\right]$. So the total area enclosed is

$$2\int_0^{\sqrt{1/m-1}} \left[\frac{x}{x^2+1} - mx\right] dx = 2\left[\tfrac{1}{2}\ln(x^2+1) - \tfrac{1}{2}mx^2\right]_0^{\sqrt{1/m-1}}$$

$$= \left[\ln\left(\frac{1}{m} - 1 + 1\right) - m\left(\frac{1}{m} - 1\right)\right] - (\ln 1 - 0) = \ln\left(\frac{1}{m}\right) + m - 1 = m - \ln m - 1.$$

96. **(a)** Let $f(x) = \ln x$ $\Rightarrow$ $f'(x) = 1/x$ $\Rightarrow$ $f''(x) = -1/x^2$. Then from Equation 2.9.5, the linear approximation to $\ln x$ near 1 is

$\ln x \approx f(1) + f'(1)(x-1) = \ln 1 + \frac{1}{1}(x-1)$

$= x - 1$, and from Equation 2.9.10, the quadratic approximation is

$\ln x \approx f(1) + f'(1)(x-1) + \frac{1}{2}f''(1)(x-1)^2$

$= \ln 1 + \frac{1}{1}(x-1) + \frac{1}{2}\left(\frac{-1}{1^2}\right)(x-1)^2$

$= x - 1 - \frac{1}{2}(x-1)^2$.

(b)

1, L, f, P, 0, 2, −1.25

(c)

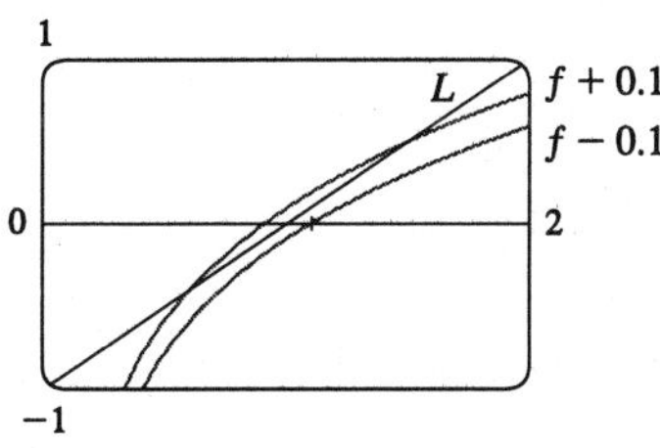

From the graph, it appears that the linear approximation is accurate to within 0.1 for x between about 0.62 and 1.51.

(d)

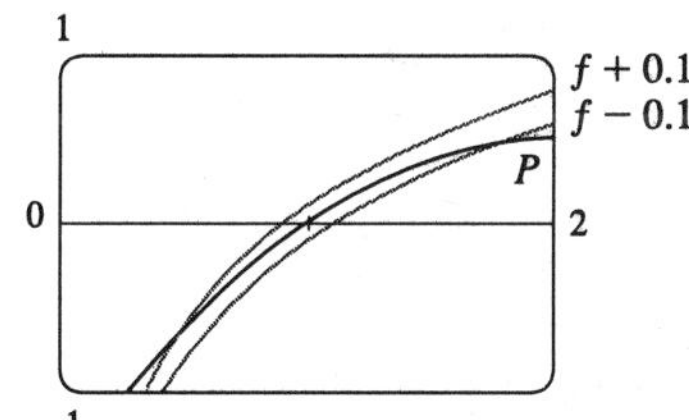

It appears that the quadratic approximation is accurate to within 0.1 for x between about 0.45 and 1.77.

97. If $f(x) = \ln(1+x)$, then $f'(x) = 1/(1+x)$, so $f'(0) = 1$. Thus

$$\lim_{x\to 0}\frac{\ln(1+x)}{x} = \lim_{x\to 0}\frac{f(x)}{x} = \lim_{x\to 0}\frac{f(x) - f(0)}{x - 0} = f'(0) = 1.$$

98. Let $m = n/x$. Then $n = xm$, and as $n \to \infty$, $m \to \infty$. Therefore,

$$\lim_{n\to\infty}\left(1 + \frac{x}{n}\right)^n = \lim_{m\to\infty}\left(1 + \frac{1}{m}\right)^{mx} = \left[\lim_{m\to\infty}\left(1 + \frac{1}{m}\right)^m\right]^x = e^x \text{ by Equation 9.}$$

EXERCISES 6.2*

1. $\ln \dfrac{ab^2}{c} = \ln ab^2 - \ln c = \ln a + \ln b^2 - \ln c = \ln a + 2\ln b - \ln c$

2. $\ln x(x^2+1)^3 = \ln x + \ln(x^2+1)^3 = \ln x + 3\ln(x^2+1)$

3. $\ln \sqrt[3]{2xy} = \ln(2xy)^{1/3} = \frac{1}{3}\ln(2xy) = \frac{1}{3}(\ln 2 + \ln x + \ln y)$

4. $\ln x\sqrt{\dfrac{y}{z}} = \ln x + \ln\sqrt{\dfrac{y}{z}} = \ln x + \ln\left(\dfrac{y}{z}\right)^{1/2} = \ln x + \frac{1}{2}\ln\left(\dfrac{y}{z}\right) = \ln x + \frac{1}{2}(\ln y - \ln z)$

5. $2\ln 4 - \ln 2 = \ln 4^2 - \ln 2 = \ln 16 - \ln 2 = \ln\frac{16}{2} = \ln 8$

Or: $2\ln 4 - \ln 2 = 2\ln 2^2 - \ln 2 = 4\ln 2 - \ln 2 = 3\ln 2$

6. $\ln 10 + \frac{1}{2}\ln 9 = \ln 10 + \ln 9^{1/2} = \ln 10 + \ln 3 = \ln 30$

7. $\frac{1}{3}\ln x - 4\ln(2x+3) = \ln\left(x^{1/3}\right) - \ln(2x+3)^4 = \ln\left[\sqrt[3]{x}/(2x+3)^4\right]$

8. $\ln x + a\ln y - b\ln z = \ln x + \ln y^a - \ln z^b = \ln\left(xy^a/z^b\right)$

9. $y = \ln x$

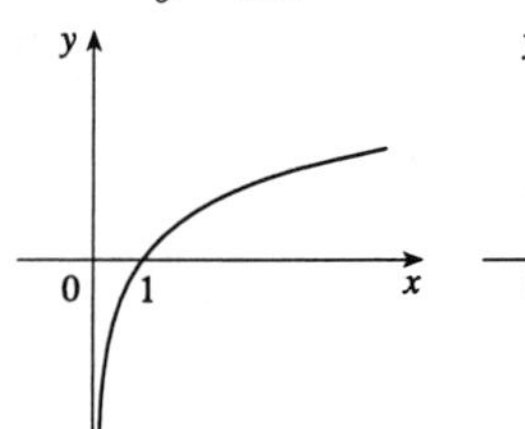

$y = -\ln x$

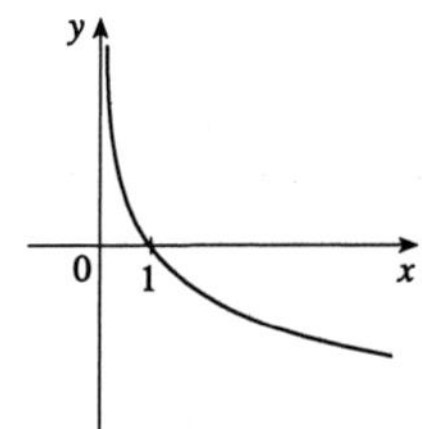

10. $y = \ln|x|$

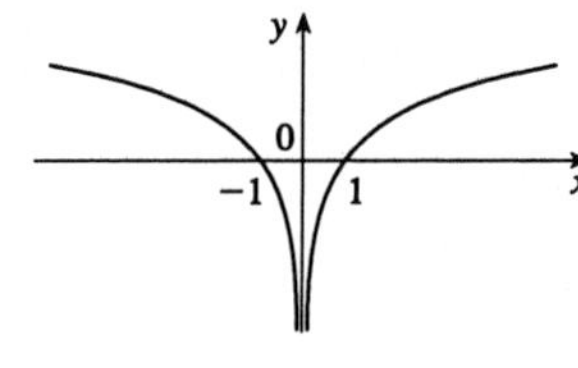

11. $y = \ln x$

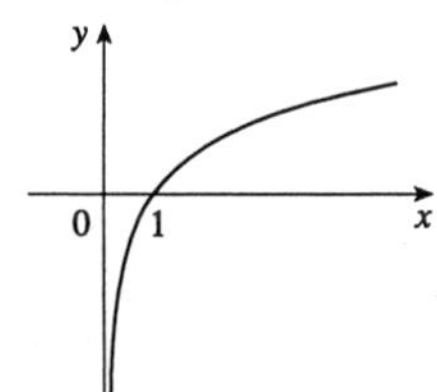

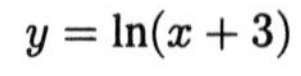
$y = \ln(x+3)$

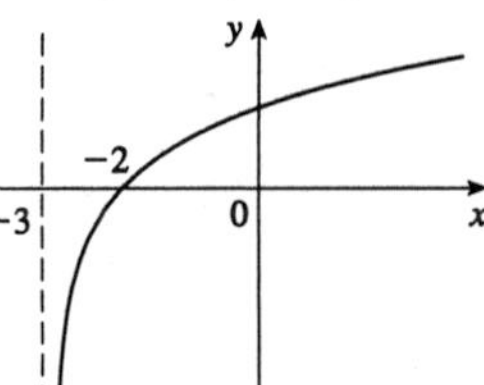

12. $y = \ln(x-2)$

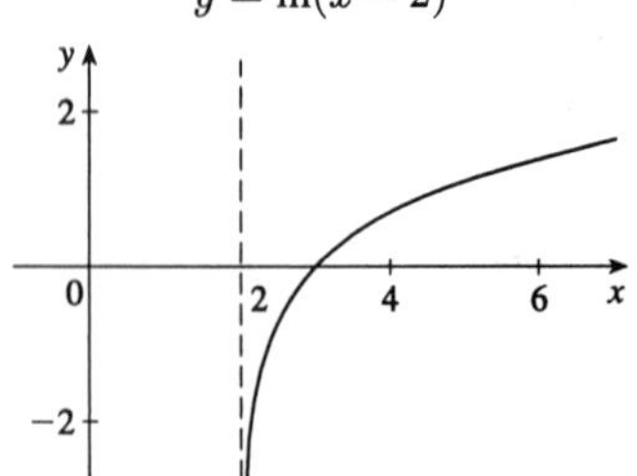

$y = 1 + \ln(x-2)$

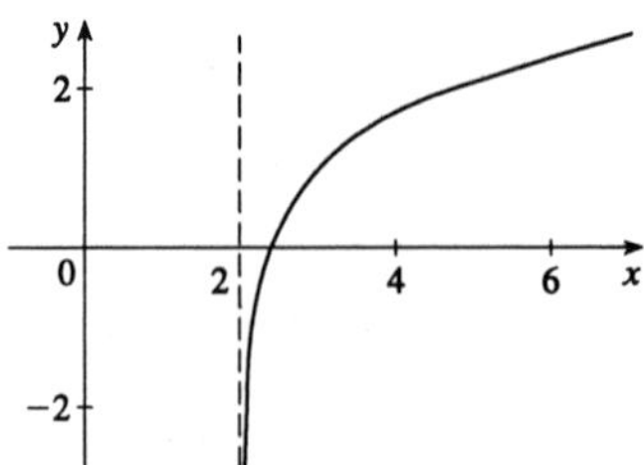

13. $f(x) = \ln(x+1) \quad \Rightarrow \quad f'(x) = 1/(x+1)$, $\text{Dom}(f) = \text{Dom}(f') = \{x \mid x+1 > 0\}$
$= \{x \mid x > -1\} = (-1, \infty)$. [Note that, in general, $\text{Dom}(f') \subset \text{Dom}(f)$.]

14. $f(x) = \cos(\ln x) \quad \Rightarrow \quad f'(x) = -\sin(\ln x)/x$, $\text{Dom}(f) = \text{Dom}(f') = (0, \infty)$

15. $f(x) = x^2 \ln(1-x^2) \quad \Rightarrow \quad f'(x) = 2x\ln(1-x^2) + \dfrac{x^2(-2x)}{1-x^2} = 2x\ln(1-x^2) - \dfrac{2x^3}{1-x^2}$,
$\text{Dom}(f) = \text{Dom}(f') = \{x \mid 1 - x^2 > 0\} = \{x \mid |x| < 1\} = (-1, 1)$

16. $f(x) = \ln\ln\ln x \quad \Rightarrow \quad f'(x) = \dfrac{1}{\ln\ln x} \cdot \dfrac{1}{\ln x} \cdot \dfrac{1}{x}$,
$\text{Dom}(f) = \{x \mid \ln\ln x > 0\} = \{x \mid \ln x > 1\} = \{x \mid x > e\} = (e, \infty)$, $\text{Dom}(f') = (e, \infty)$

17. $y = x\ln x \quad \Rightarrow \quad y' = \ln x + x(1/x) = \ln x + 1 \quad \Rightarrow \quad y'' = 1/x$

18. $y = \ln(\sec x + \tan x) \quad \Rightarrow \quad y' = \dfrac{1}{\sec x + \tan x}(\sec x \tan x + \sec^2 x) = \sec x \quad \Rightarrow \quad y'' = \sec x \tan x$

19. $f(x) = \sqrt{x}\ln x \quad \Rightarrow \quad f'(x) = \dfrac{1}{2\sqrt{x}}\ln x + \sqrt{x}\left(\dfrac{1}{x}\right) = \dfrac{\ln x + 2}{2\sqrt{x}}$

20. $g(t) = \sin(\ln t) \quad \Rightarrow \quad g'(t) = \cos(\ln t)/t$

21. $g(x) = \ln\dfrac{a-x}{a+x} = \ln(a-x) - \ln(a+x) \quad \Rightarrow \quad g'(x) = \dfrac{-1}{a-x} - \dfrac{1}{a+x} = \dfrac{-2a}{a^2 - x^2}$

22. $h(x) = \ln\left(x + \sqrt{x^2-1}\right) \quad \Rightarrow \quad h'(x) = \dfrac{1}{x + \sqrt{x^2-1}}\left(1 + \dfrac{x}{\sqrt{x^2-1}}\right) = \dfrac{1}{\sqrt{x^2-1}}$

23. $F(x) = \ln\sqrt{x} = \frac{1}{2}\ln x \quad \Rightarrow \quad F'(x) = \dfrac{1}{2}\left(\dfrac{1}{x}\right) = \dfrac{1}{2x}$

24. $G(x) = \sqrt{\ln x} \quad \Rightarrow \quad G'(x) = \dfrac{1}{2\sqrt{\ln x}}\left(\dfrac{1}{x}\right) = \dfrac{1}{2x\sqrt{\ln x}}$

25. $h(y) = \ln(y^3 \sin y) = 3\ln y + \ln(\sin y) \quad \Rightarrow \quad h'(y) = \dfrac{3}{y} + \dfrac{1}{\sin y}(\cos y) = \dfrac{3}{y} + \cot y$

26. $k(r) = r\sin r \ln r \quad \Rightarrow \quad k'(r) = \sin r \ln r + r\cos r \ln r + \sin r$

27. $g(u) = \dfrac{1 - \ln u}{1 + \ln u} \quad \Rightarrow \quad g'(u) = \dfrac{(1+\ln u)(-1/u) - (1 - \ln u)(1/u)}{(1+\ln u)^2} = -\dfrac{2}{u(1+\ln u)^2}$

28. $G(u) = \ln\sqrt{\dfrac{3u+2}{3u-2}} = \frac{1}{2}[\ln(3u+2) - \ln(3u-2)] \quad \Rightarrow \quad G'(u) = \dfrac{1}{2}\left[\dfrac{3}{3u+2} - \dfrac{3}{3u-2}\right] = \dfrac{-6}{9u^2-4}$

29. $y = (\ln\sin x)^3 \quad \Rightarrow \quad y' = 3(\ln\sin x)^2 \dfrac{\cos x}{\sin x} = 3(\ln\sin x)^2 \cot x$

30. $y = \ln(x + \ln x) \quad \Rightarrow \quad y' = \dfrac{1}{x + \ln x}\left(1 + \dfrac{1}{x}\right)$

31. $y = \dfrac{\ln x}{1+x^2} \quad \Rightarrow \quad y' = \dfrac{(1+x^2)(1/x) - 2x\ln x}{(1+x^2)^2} = \dfrac{1 + x^2 - 2x^2\ln x}{x(1+x^2)^2}$

32. $y = \ln\left(x\sqrt{1-x^2}\sin x\right) = \ln x + \frac{1}{2}\ln(1-x^2) + \ln\sin x \quad \Rightarrow$
$y' = \dfrac{1}{x} + \dfrac{1}{2}\left(\dfrac{-2x}{1-x^2}\right) + \dfrac{\cos x}{\sin x} = \dfrac{1}{x} - \dfrac{x}{1-x^2} + \cot x$

33. $y = \ln\left(\dfrac{x+1}{x-1}\right)^{3/5} = \frac{3}{5}[\ln(x+1) - \ln(x-1)] \quad \Rightarrow \quad y' = \dfrac{3}{5}\left(\dfrac{1}{x+1} - \dfrac{1}{x-1}\right) = \dfrac{-6}{5(x^2-1)}$

34. $y = \ln|\tan 2x| \quad \Rightarrow \quad y' = \dfrac{2\sec^2 2x}{\tan 2x}$

35. $y = \ln|x^3 - x^2| \quad \Rightarrow \quad y' = \dfrac{1}{x^3 - x^2}(3x^2 - 2x) = \dfrac{x(3x-2)}{x^2(x-1)} = \dfrac{3x-2}{x(x-1)}$

36. $y = \tan[\ln(ax+b)] \quad \Rightarrow \quad y' = \sec^2(\ln(ax+b))\dfrac{a}{ax+b}$

37. $f(x) = \dfrac{x}{\ln x} \quad \Rightarrow \quad f'(x) = \dfrac{\ln x - x(1/x)}{(\ln x)^2} = \dfrac{\ln x - 1}{(\ln x)^2} \quad \Rightarrow \quad f'(e) = \dfrac{1-1}{1^2} = 0$

38. $f(x) = x^2 \ln x \;\Rightarrow\; f'(x) = 2x\ln x + x^2\left(\dfrac{1}{x}\right) = 2x\ln x + x \;\Rightarrow\; f'(1) = 2\ln 1 + 1 = 1$

39. $f(x) = \sin x + \ln x \quad \Rightarrow$

$f'(x) = \cos x + \dfrac{1}{x}$

This is reasonable, because the graph shows that f increases when $f'(x)$ is positive.

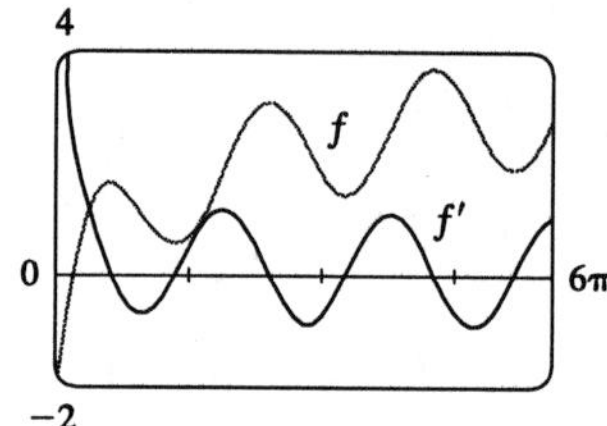

40. $f(x) = \ln(x^2 + x + 1) \quad \Rightarrow$

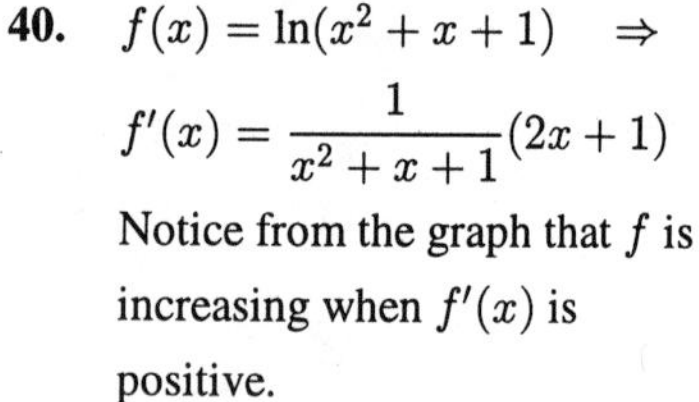

$f'(x) = \dfrac{1}{x^2+x+1}(2x+1)$

Notice from the graph that f is increasing when $f'(x)$ is positive.

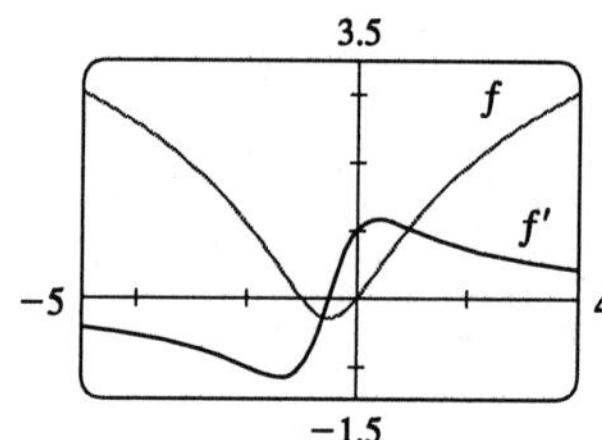

41. $y = f(x) = \ln\ln x \quad \Rightarrow \quad f'(x) = \dfrac{1}{\ln x}\left(\dfrac{1}{x}\right) \quad \Rightarrow \quad f'(e) = \dfrac{1}{e}$, so the equation of the tangent at $(e, 0)$ is $y - 0 = \dfrac{1}{e}(x - e)$ or $x - ey = e$.

42. $y = f(x) = \sin(\ln x) \quad \Rightarrow \quad f'(x) = \cos(\ln x)(1/x) \quad \Rightarrow \quad f'(1) = (\cos 0)\left(\frac{1}{1}\right) = 1$, so the equation of the tangent at $(1, 0)$ is $y - 0 = 1(x-1) \quad \Leftrightarrow \quad y = x - 1$.

43. $y = \ln(x^2 + y^2) \;\Rightarrow\; y' = \dfrac{2x + 2yy'}{x^2 + y^2} \;\Rightarrow\; x^2y' + y^2y' = 2x + 2yy' \;\Rightarrow\; y' = \dfrac{2x}{x^2 + y^2 - 2y}$

44. $\ln xy = \ln x + \ln y = y\sin x \quad \Rightarrow \quad 1/x + y'/y = y\cos x + y'\sin x \quad \Rightarrow \quad y'(1/y - \sin x) = y\cos x - 1/x$

$\Rightarrow \quad y' = \dfrac{y\cos x - 1/x}{1/y - \sin x} = \left(\dfrac{y}{x}\right)\dfrac{xy\cos x - 1}{1 - y\sin x}$

45. $f(x) = \ln(x-1) \quad \Rightarrow \quad f'(x) = 1/(x-1) = (x-1)^{-1} \quad \Rightarrow \quad f''(x) = -(x-1)^{-2} \quad \Rightarrow$

$f'''(x) = 2(x-1)^{-3} \quad \Rightarrow \quad f^{(4)}(x) = -2\cdot 3(x-1)^{-4} \quad \Rightarrow \quad \cdots \quad \Rightarrow$

$f^{(n)}(x) = (-1)^{n-1}\cdot 2\cdot 3\cdot 4\cdot\cdots\cdot(n-1)(x-1)^{-n} = (-1)^{n-1}\dfrac{(n-1)!}{(x-1)^n}$

46. $y = x^8 \ln x$, so $D^9 y = D^8\left(8x^7\ln x + x^7\right) = D^8\left(8x^7\ln x\right) = D^7\left(8\cdot 7x^6\ln x + 8x^6\right)$

$= D^7\left(8\cdot 7x^6\ln x\right) = D^6\left(8\cdot 7\cdot 6x^5\ln x\right) = \cdots = D\left(8!\,x^0\ln x\right) = 8!/x$

47.

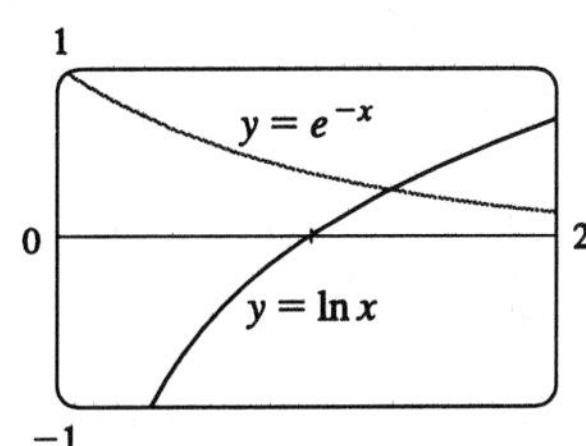

From the graph, it appears that the only root of the equation occurs at about $x = 1.3$. So we use Newton's Method with this as our initial approximation, and with $f(x) = \ln x - e^{-x} \Rightarrow f'(x) = 1/x - e^{-x}$. The formula is $x_{n+1} = x_n - f(x_n)/f'(x_n)$, and we calculate $x_1 = 1.3$, $x_2 \approx 1.309760$, $x_3 \approx x_4 \approx 1.309800$. So, correct to six decimal places, the root of the equation $\ln x = e^{-x}$ is $x = 1.309800$.

48.

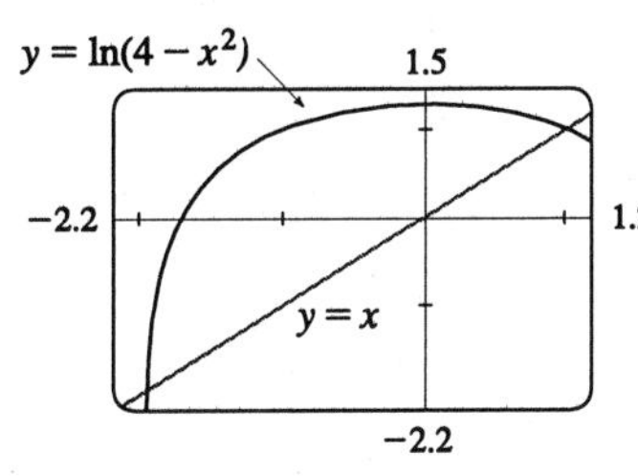

We use Newton's Method with $f(x) = \ln(4 - x^2) - x$ and $f'(x) = \dfrac{1}{4 - x^2}(-2x) - 1 = 1 - \dfrac{2x}{4 - x^2}$. The formula is $x_{n+1} = x_n - f(x_n)/f'(x_n)$. From the graphs it seems that the roots occur at approximately $x = -1.9$ and $x = 1.1$. However, if we use $x_1 = -1.9$ as an initial approximation to the first root, we get $x_2 \approx -2.009611$, and $f(x) = \ln(x-2)^2 - x$ is undefined at this point, making it impossible to calculate x_3. We must use a more accurate first estimate for this root, such as $x_1 = -1.95$. With this approximation, we get $x_1 = -1.95$, $x_2 \approx -1.1967495$, $x_3 \approx -1.964760$, $x_4 \approx x_5 \approx -1.964636$. Calculating the second root gives $x_1 = 1.1$, $x_2 \approx 1.058649$, $x_3 \approx 1.058007$, $x_4 \approx x_5 \approx 1.058006$. So, correct to six decimal places, the two roots of the equation $\ln(4 - x^2) = x$ are $x = -1.964636$ and $x = 1.058006$.

49. $y = f(x) = \ln(\cos x)$ **A.** $D = \{x \mid \cos x > 0\} = \left(-\frac{\pi}{2}, \frac{\pi}{2}\right) \cup \left(\frac{3\pi}{2}, \frac{5\pi}{2}\right) \cup \cdots$ $= \left\{x \mid 2n\pi - \frac{\pi}{2} < x < 2n\pi + \frac{\pi}{2}, n = 0, \pm 1, \pm 2, \dots\right\}$ **B.** x-intercepts occur when $\ln(\cos x) = 0 \Leftrightarrow \cos x = 1 \Leftrightarrow x = 2n\pi$, y-intercept $= f(0) = 0$. **C.** $f(-x) = f(x)$, so the curve is symmetric about the y-axis. $f(x + 2\pi) = f(x)$, f has period 2π, so in parts D-G we consider only $-\frac{\pi}{2} < x < \frac{\pi}{2}$.

D. $\lim\limits_{x \to \pi/2^-} \ln(\cos x) = -\infty$ and $\lim\limits_{x \to -\pi/2^+} \ln(\cos x) = -\infty$, so $x = \frac{\pi}{2}$ and $x = -\frac{\pi}{2}$ are VA. No HA.

E. $f'(x) = (1/\cos x)(-\sin x) = -\tan x > 0 \Leftrightarrow -\frac{\pi}{2} < x < 0$, so f is increasing on $\left(-\frac{\pi}{2}, 0\right]$ and decreasing on $\left[0, \frac{\pi}{2}\right)$. **F.** $f(0) = 0$ is a local maximum.

G. $f''(x) = -\sec^2 x < 0 \Rightarrow f$ is CD on $\left(-\frac{\pi}{2}, \frac{\pi}{2}\right)$. No IP.

H.

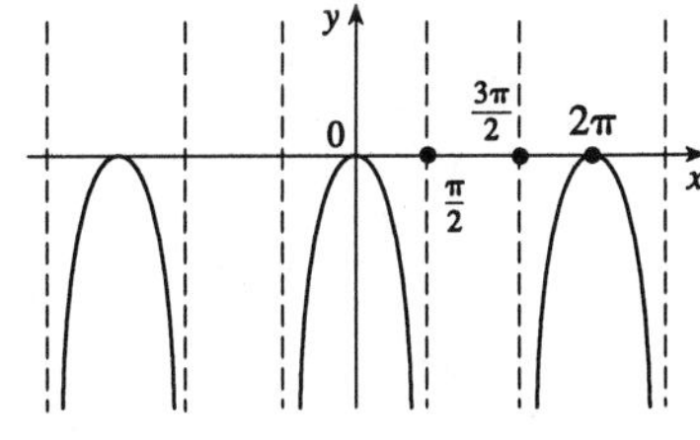

50. $y = f(x) = x^2 + \ln x$ **A.** $D = (0, \infty)$ **B.** No y-intercept **C.** No symmetry

D. $\lim\limits_{x \to \infty} (x^2 + \ln x) = \infty$, no HA. $\lim\limits_{x \to 0^+} (x^2 + \ln x) = -\infty$, so $x = 0$ is a VA.

E. $f'(x) = 2x + 1/x > 0$, so f is increasing on $(0, \infty)$. **F.** No extrema

G. $f''(x) = 2 - \dfrac{1}{x^2} = \dfrac{2x^2 - 1}{x^2} > 0 \Leftrightarrow 2x^2 > 1 \Leftrightarrow x > \frac{1}{\sqrt{2}}$, so f is CU on $\left(\frac{1}{\sqrt{2}}, \infty\right)$ and CD on $\left(0, \frac{1}{\sqrt{2}}\right)$. IP $\left(\frac{1}{\sqrt{2}}, \frac{1}{2}(1 - \ln 2)\right)$.

H.

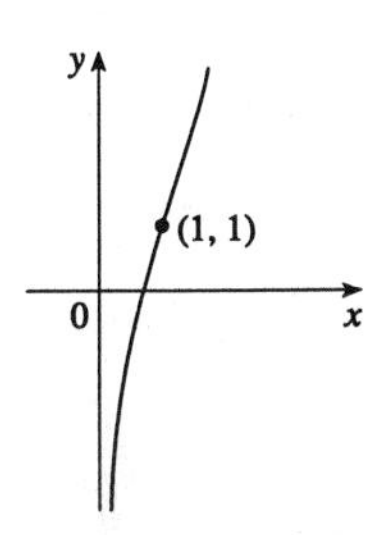

51. $y = f(x) = \ln\left(x + \sqrt{1+x^2}\right)$ **A.** $x + \sqrt{1+x^2} > 0$ for all x since $1 + x^2 > x^2 \Rightarrow \sqrt{1+x^2} > |x|$, so $D = \mathbb{R}$. **B.** y-intercept $= f(0) = 0$, x–intercept occurs when $x + \sqrt{1+x^2} = 1 \Rightarrow \sqrt{1+x^2} = 1 - x \Rightarrow 1 + x^2 = 1 - 2x + x^2 \Rightarrow x = 0$. **C.** $\ln\left(-x + \sqrt{1+x^2}\right) = -\ln\left(x + \sqrt{1+x^2}\right)$ since $\left(\sqrt{1+x^2} - x\right)\left(\sqrt{1+x^2} + x\right) = 1$, so the curve is symmetric about the origin.

D. $\lim\limits_{x\to\infty} \ln\left(x + \sqrt{1+x^2}\right) = \infty$, $\lim\limits_{x\to-\infty} \ln\left(x + \sqrt{1+x^2}\right) = \lim\limits_{x\to-\infty} \ln \dfrac{1}{\sqrt{1+x^2} - x} = -\infty$, no HA.

E. $f'(x) = \dfrac{1}{x + \sqrt{1+x^2}}\left(1 + \dfrac{x}{\sqrt{1+x^2}}\right) = \dfrac{1}{\sqrt{1+x^2}} > 0$, so f is increasing on $(-\infty, \infty)$.

F. No extrema.

G. $f''(x) = -\dfrac{x}{(1+x^2)^{3/2}}$ $\Rightarrow$ $f''(x) > 0 \Leftrightarrow x < 0$, so f is CU on $(-\infty, 0)$ and CD on $(0, \infty)$, and there is an IP at $(0, 0)$.

H.

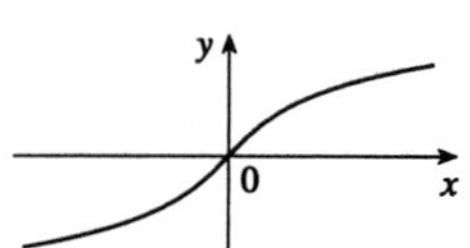

52. $y = \ln(\tan^2 x)$ **A.** $D = \{x \mid x \neq n\pi/2\}$ **B.** x-intercepts $n\pi + \frac{\pi}{4}$, no y-intercept. **C.** $f(-x) = f(x)$, so the curve is symmetric about the y-axis. Also $f(x + \pi) = f(x)$, so f is periodic with period π, and we consider parts D-G only for $-\frac{\pi}{2} < x < \frac{\pi}{2}$. **D.** $\lim\limits_{x\to0} \ln(\tan^2 x) = -\infty$ and $\lim\limits_{x\to\pi/2^-} \ln(\tan^2 x) = \infty$, $\lim\limits_{x\to-\pi/2^+} \ln(\tan^2 x) = \infty$, so $x = 0$, $x = \pm\frac{\pi}{2}$ are VA. **E.** $f'(x) = \dfrac{2\tan x \sec^2 x}{\tan^2 x} = 2\dfrac{\sec^2 x}{\tan x} > 0 \Leftrightarrow \tan x > 0 \Leftrightarrow 0 < x < \frac{\pi}{2}$, so f is increasing on $\left(0, \frac{\pi}{2}\right)$ and decreasing on $\left(-\frac{\pi}{2}, 0\right)$.

F. No extrema **G.** $f'(x) = \dfrac{2}{\sin x \cos x} = \dfrac{4}{\sin 2x}$

$\Rightarrow f''(x) = \dfrac{-8\cos 2x}{\sin^2 2x} < 0 \Leftrightarrow \cos 2x > 0$

$\Leftrightarrow -\frac{\pi}{4} < x < \frac{\pi}{4}$, so f is CD on $\left(-\frac{\pi}{4}, 0\right)$ and $\left(0, \frac{\pi}{4}\right)$ and CU on $\left(-\frac{\pi}{2}, -\frac{\pi}{4}\right)$ and $\left(\frac{\pi}{4}, \frac{\pi}{2}\right)$. IP are $\left(\pm\frac{\pi}{4}, 0\right)$.

H.

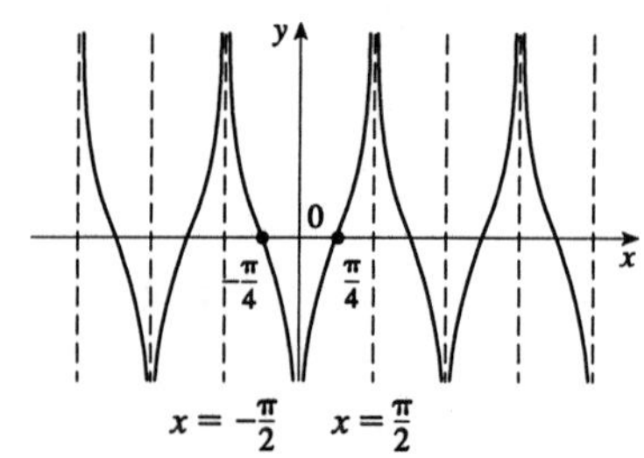

53. $y = f(x) = \ln(1 + x^2)$ **A.** $D = \mathbb{R}$ **B.** Both intercepts are 0. **C.** $f(-x) = f(x)$, so the curve is symmetric about the y-axis. **D.** $\lim\limits_{x\to\pm\infty} \ln(1 + x^2) = \infty$, no asymptotes. **E.** $f'(x) = \dfrac{2x}{1+x^2} > 0 \Leftrightarrow x > 0$, so f is increasing on $[0, \infty)$ and decreasing on $(-\infty, 0]$.

F. $f(0) = 0$ is a local and absolute minimum.

G. $f''(x) = \dfrac{2(1+x^2) - 2x(2x)}{(1+x^2)^2} = \dfrac{2(1-x^2)}{(1+x^2)^2} > 0$

$\Leftrightarrow |x| < 1$, so f is CU on $(-1, 1)$, CD on $(-\infty, -1)$ and $(1, \infty)$. IP $(1, \ln 2)$ and $(-1, \ln 2)$.

H.

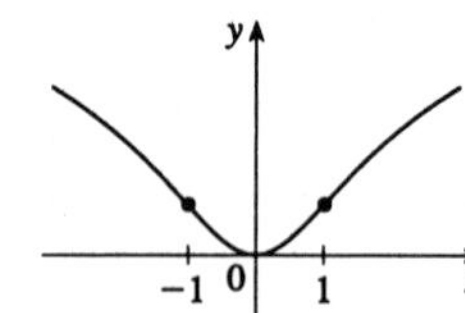

54. $y = f(x) = \ln(x^2 - x)$ **A.** $\{x \mid x^2 - x > 0\} = \{x \mid x < 0 \text{ or } x > 1\} = (-\infty, 0) \cup (1, \infty)$. **B.** x-intercepts occur when $x^2 - x = 1 \quad \Leftrightarrow \quad x^2 - x - 1 = 0 \quad \Leftrightarrow \quad x = \frac{1}{2}\left(1 \pm \sqrt{5}\right)$. No y-intercept **C.** No symmetry **D.** $\lim\limits_{x\to\infty} \ln(x^2 - x) = \infty$, no HA. $\lim\limits_{x\to 0^-} \ln(x^2 - x) = -\infty$, $\lim\limits_{x\to 1^+} \ln(x^2 - x) = -\infty$, so $x = 0$ and $x = 1$ are VA. **E.** $f'(x) = \dfrac{2x - 1}{x^2 - x} > 0$ when $x > 1$ and $f'(x) < 0$ when $x < 0$, so f is increasing on $(1, \infty)$ and decreasing on $(-\infty, 0)$. **F.** No extrema

G. $f''(x) = \dfrac{2(x^2 - x) - (2x - 1)^2}{(x^2 - x)^2} = \dfrac{-2x^2 + 2x - 1}{(x^2 - x)^2}$

$\Rightarrow \quad f''(x) < 0$ for all x since $-2x^2 + 2x - 1$ has a negative discriminant. So f is CD on $(-\infty, 0)$ and $(1, \infty)$. No IP.

H.

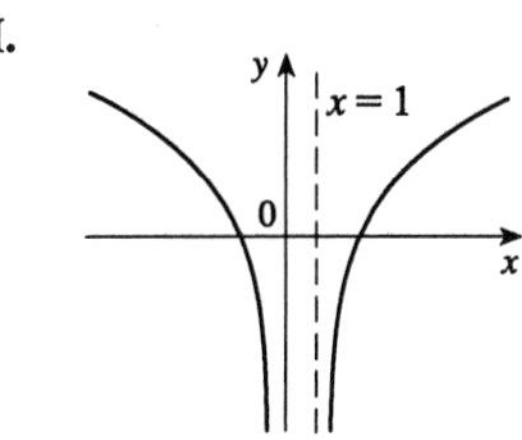

55.

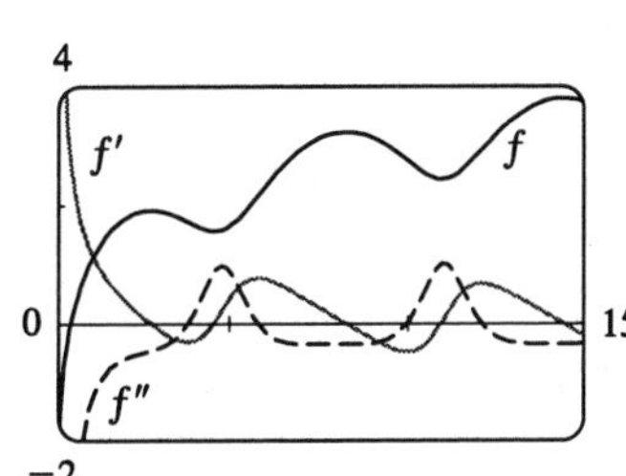

We use the CAS to calculate $f'(x) = \dfrac{2 + \sin x + x\cos x}{2x + x\sin x}$ and $f''(x) = \dfrac{2x^2\sin x + 4\sin x - \cos^2 x + x^2 + 5}{x^2(\cos^2 x - 4\sin x - 5)}$. From the graphs, it seems that $f' > 0$ (and so f is increasing) on approximately the intervals $[0, 2.7]$, $[4.5, 8.2]$ and $[10.9, 14.3]$. It seems that f'' changes sign (indicating inflection points) at $x \approx 3.8, 5.7$, 10.0 and 12.0. Looking back at the graph of f, this implies that the inflection points have approximate coordinates $(3.8, 1.7)$, $(5.7, 2.1)$, $(10.0, 2.7)$, and $(12.0, 2.9)$.

56. We see that if $c \le 0$, $f(x) = \ln(x^2 + c)$ is only defined for $x^2 > -c \quad \Rightarrow \quad |x| > \sqrt{-c}$, and $\lim\limits_{x\to\sqrt{-c}^+} f(x) = \lim\limits_{x\to-\sqrt{-c}^-} f(x) = -\infty$, since $\ln y \to -\infty$ as $y \to 0$. Thus, for $c < 0$, there are vertical asymptotes at $x = \pm\sqrt{c}$, and as c decreases (that is, $|c|$ increases), the asymptotes get further apart. For $c = 0$, $\lim\limits_{x\to 0} f(x) = -\infty$, so there is a vertical asymptote at $x = 0$. If $c > 0$, there are no asymptotes. To find the extrema and inflection points, we differentiate: $f(x) = \ln\left(x^2 + c\right) \quad \Rightarrow \quad f'(x) = \dfrac{1}{x^2 + c}(2x)$, so by the First Derivative Test there is a local and absolute minimum at $x = 0$. Differentiating again, we get

$$f''(x) = \frac{1}{x^2 + c}(2) + 2x\left[-\left(x^2 + c\right)^{-2}(2x)\right] = \frac{2(c - x^2)}{(x^2 + c)^2}.$$

Now if $c \le 0$, this is always negative, so f is concave down on both of the intervals on which it is defined. If $c > 0$, then f'' changes sign when $c = x^2 \quad \Leftrightarrow \quad x = \pm\sqrt{c}$. So for $x > 0$ there are inflection points at $\pm\sqrt{c}$, and as c increases, the inflection points get further apart.

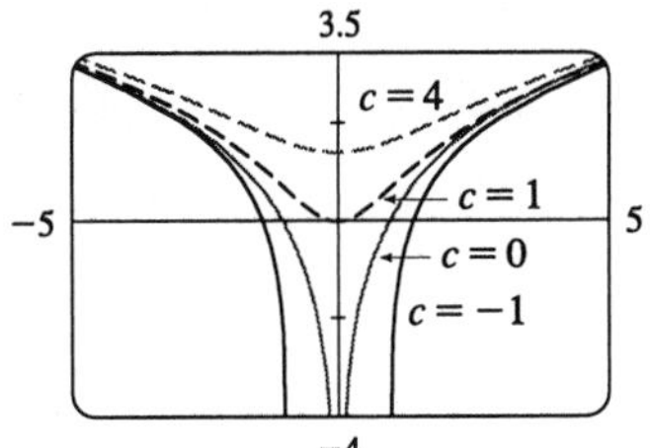

57. $\displaystyle\int_4^8 \frac{1}{x}\,dx = \ln x\big|_4^8 = \ln 8 - \ln 4 = \ln\tfrac{8}{4} = \ln 2$

58. $\displaystyle\int_{-e^2}^{-e} \frac{3}{x}\,dx = 3\ln|x|\big|_{-e^2}^{-e} = 3\ln e - 3\ln\left(e^2\right) = 3 - 6 = -3$

59. $\displaystyle\int_1^e \frac{x^2+x+1}{x}dx = \int_1^e \left(x+1+\frac{1}{x}\right)dx = \left[\tfrac{1}{2}x^2+x+\ln x\right]_1^e = \left(\tfrac{1}{2}e^2+e+1\right)-\left(\tfrac{1}{2}+1+0\right) = \tfrac{1}{2}e^2+e-\tfrac{1}{2}$

60. $\displaystyle\int_4^9 \left[\sqrt{x}+\frac{1}{\sqrt{x}}\right]^2 dx = \int_4^9 \left(x+2+\frac{1}{x}\right)dx = \left[\tfrac{1}{2}x^2+2x+\ln x\right]_4^9$

$= \frac{81}{2}+18+\ln 9-(8+8+\ln 4) = \frac{85}{2}+\ln\frac{9}{4}$

61. Let $u=2x-1$. Then $du = 2\,dx$, so $\displaystyle\int \frac{dx}{2x-1} = \frac{1}{2}\int \frac{du}{u} = \tfrac{1}{2}\ln|u|+C = \tfrac{1}{2}\ln|2x-1|+C.$

62. Let $u=x^3+3x+1$. Then $du = 3(x^2+1)dx$, so

$\displaystyle\int \frac{x^2+1}{x^3+3x+1}dx = \tfrac{1}{3}\int\frac{du}{u} = \tfrac{1}{3}\ln|u|+C = \tfrac{1}{3}\ln|x^3+3x+1|+C.$

63. Let $u=x^2+2x$. Then $du = 2(x+1)dx$, so

$\displaystyle\int \frac{x+1}{x^2+2x}dx = \tfrac{1}{2}\int\frac{du}{u} = \tfrac{1}{2}\ln|u|+C = \tfrac{1}{2}\ln|x^2+2x|+C.$

64. Let $u=\ln x$. Then $du = \dfrac{1}{x}\,dx$, so $\displaystyle\int\frac{dx}{x\ln x} = \int\frac{du}{u} = \ln|u|+C = \ln|\ln x|+C.$

65. Let $u=\ln x$. Then $du = \dfrac{dx}{x}$ $\Rightarrow$ $\displaystyle\int\frac{(\ln x)^2}{x}dx = \int u^2\,du = \tfrac{1}{3}u^3+C = \tfrac{1}{3}(\ln x)^3+C.$

66. Let $u = 2-\tan x$. Then $du = -\sec^2 x\,dx$, so $\displaystyle\int\frac{\sec^2 x}{2-\tan x}dx = -\int\frac{du}{u} = -\ln|u|+C = -\ln|2-\tan x|+C.$

67. Let $u=1+\cos x$. Then $du = -\sin x\,dx$, so

$\displaystyle\int\frac{\sin x}{1+\cos x}dx = -\int\frac{du}{u} = -\ln|u|+C = -\ln(1+\cos x)+C.$

68. Let $u = 1+\ln x$. Then $du = \dfrac{dx}{x}$, so $\displaystyle\int\frac{(1+\ln x)^4}{x}dx = \int u^4\,du = \frac{u^5}{5}+C = \frac{(1+\ln x)^5}{5}+C.$

69. (a) $\dfrac{d}{dx}(\ln|\sin x|+C) = \dfrac{1}{\sin x}\cos x = \cot x$

(b) Let $u=\sin x$. Then $du = \cos x\,dx$, so $\displaystyle\int\cot x\,dx = \int\frac{\cos x}{\sin x}dx = \int\frac{du}{u} = \ln|u|+C = \ln|\sin x|+C.$

70. Let $u = x-2$. Then the area is

$\displaystyle A = -\int_{-4}^{-1}\frac{2}{x-2}dx = -2\int_{-6}^{-3}\frac{du}{u} = \left[-2\ln|u|\right]_{-6}^{-3} = -2\ln 3+2\ln 6 = 2\ln 2\approx 1.386.$

71. The cross-sectional area is $\pi\left(1/\sqrt{x+1}\right)^2 = \pi/(x+1)$. Therefore, the volume is

$\displaystyle\int_0^1\frac{\pi}{x+1}dx = \pi[\ln(x+1)]_0^1 = \pi\ln 2-\ln 1 = \pi\ln 2$

72. Using cylindrical shells, we get $\displaystyle V = \int_0^3\frac{2\pi x}{x^2+1}dx = \pi\left[\ln(1+x^2)\right]_0^3 = \pi\ln 10.$

73. $y = (3x-7)^4(8x^2-1)^3$ $\Rightarrow$ $\ln|y| = 4\ln|3x-7|+3\ln|8x^2-1|$ $\Rightarrow$ $\dfrac{y'}{y} = \dfrac{12}{3x-7}+\dfrac{48x}{8x^2-1}$ $\Rightarrow$

$y' = (3x-7)^4(8x^2-1)^3\left(\dfrac{12}{3x-7}+\dfrac{48x}{8x^2-1}\right)$

74. $y = \dfrac{(x+1)^4(x-5)^3}{(x-3)^8} \quad \Rightarrow \quad \ln|y| = 4\ln|x+1| + 3\ln|x-5| - 8\ln|x-3| \quad \Rightarrow$

$\dfrac{y'}{y} = \dfrac{4}{x+1} + \dfrac{3}{x-5} - \dfrac{8}{x-3} \quad \Rightarrow \quad y' = \dfrac{(x+1)^4(x-5)^3}{(x-3)^8}\left(\dfrac{4}{x+1} + \dfrac{3}{x-5} - \dfrac{8}{x-3}\right)$

75. $y = \sqrt{\dfrac{x^2+1}{x+1}} \quad \Rightarrow \quad \ln y = \frac{1}{2}[\ln(x^2+1) - \ln(x+1)] \quad \Rightarrow \quad \dfrac{y'}{y} = \dfrac{1}{2}\left(\dfrac{2x}{x^2+1} - \dfrac{1}{x+1}\right) \quad \Rightarrow$

$y' = \sqrt{\dfrac{x^2+1}{x+1}}\left[\dfrac{x}{x^2+1} - \dfrac{1}{2(x+1)}\right]$

76. $y = \dfrac{(x^3+1)^4 \sin^2 x}{x^{1/3}} \quad \Rightarrow \quad \ln|y| = 4\ln|x^3+1| + 2\ln|\sin x| - \frac{1}{3}\ln|x|$. So $\dfrac{y'}{y} = 4\dfrac{3x^2}{x^3+1} + 2\dfrac{\cos x}{\sin x} - \dfrac{1}{3x}$

$\Rightarrow \quad y' = \dfrac{(x^3+1)^4 \sin^2 x}{x^{1/3}}\left(\dfrac{12x^2}{x^3+1} + 2\cot x - \dfrac{1}{3x}\right).$

77. The domain of $f(x) = \dfrac{1}{x}$ is $(-\infty, 0) \cup (0, \infty)$, so its general antiderivative is $F(x) = \begin{cases} \ln x + C_1 & \text{if } x > 0 \\ \ln|x| + C_2 & \text{if } x < 0. \end{cases}$

78. $f''(x) = x^{-2}, x > 0 \quad \Rightarrow \quad f'(x) = -1/x + C \quad \Rightarrow \quad f(x) = -\ln x + Cx + D$. $0 = f(1) = C + D$ and $0 = f(2) = -\ln 2 + 2C + D = -\ln 2 + 2C - C = -\ln 2 + C \quad \Rightarrow \quad C = \ln 2$ and $D = -\ln 2$. So $f(x) = -\ln x + (\ln 2)x - \ln 2$.

79. $f(x) = 2x + \ln x \quad \Rightarrow \quad f'(x) = 2 + 1/x$. If $g = f^{-1}$, then $f(1) = 2 \quad \Rightarrow \quad g(2) = 1$, so $g'(2) = 1/f'(g(2)) = 1/f'(1) = \frac{1}{3}$.

80. **(a)** Let $f(x) = \ln x \quad \Rightarrow \quad f'(x) = 1/x \quad \Rightarrow \quad f''(x) = -1/x^2$. Then from Equation 2.9.5, the linear approximation to $\ln x$ near 1 is

$\ln x \approx f(1) + f'(x)(x-1) = \ln 1 + \frac{1}{1}(x-1)$
$= x - 1$, and from Equation 2.9.10,

the quadratic approximation is

$\ln x \approx f(1) + f'(1)(x-1) + \frac{1}{2}f''(1)(x-1)^2$
$= \ln 1 + \frac{1}{1}(x-1) + \frac{1}{2}\left(\frac{-1}{1^2}\right)(x-1)^2$
$= x - 1 - \frac{1}{2}(x-1)^2.$

(b)

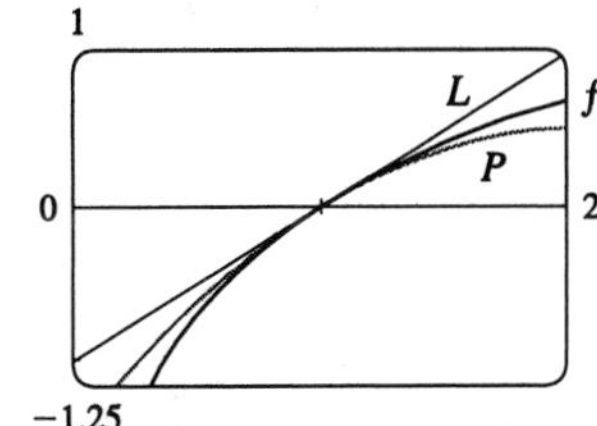

(c)

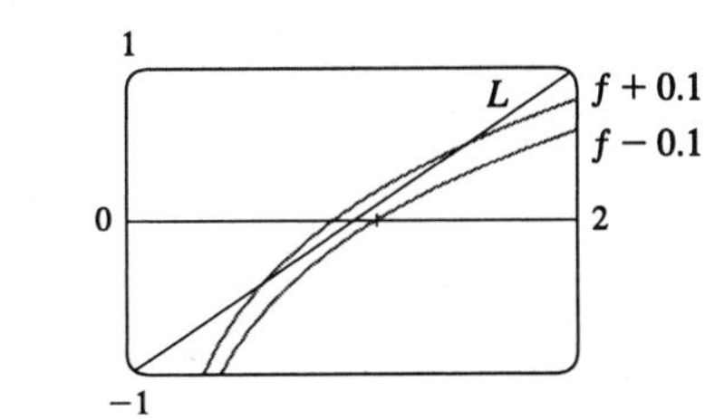

From the graph, it appears that the linear approximation is accurate to within 0.1 for x between about 0.62 and 1.51.

(d)

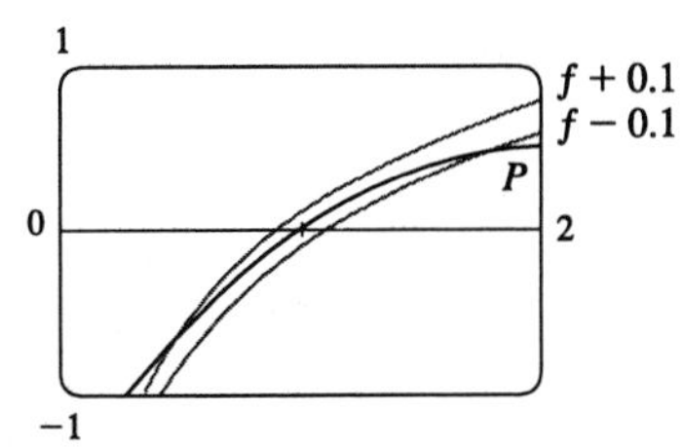

It appears that the quadratic approximation is accurate to within 0.1 for x between about 0.45 and 1.77.

81. **(a)**

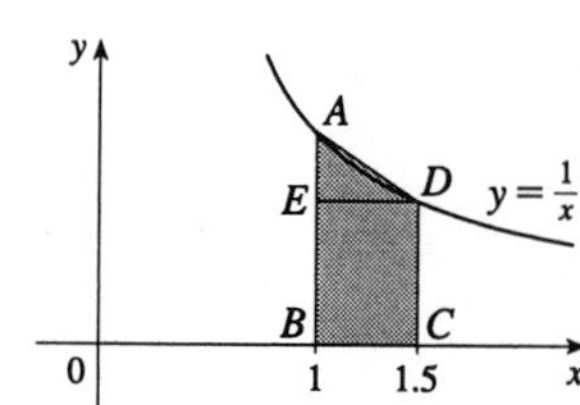

We interpret $\ln 1.5$ as the area under the curve $y = 1/x$ from $x = 1$ to $x = 1.5$. The area of the square $BCDE$ is $\frac{1}{2} \cdot \frac{2}{3} = \frac{1}{3}$. The area of the trapezoid $ABCD$ is $\frac{1}{2} \cdot \frac{1}{2}\left(1 + \frac{2}{3}\right) = \frac{5}{12}$. Thus, by comparing areas, we observe that $\frac{1}{3} < \ln 1.5 < \frac{5}{12}$.

(b) With $f(t) = 1/t$, $n = 10$, and $\Delta x = 0.05$, we have

$$\ln 1.5 = \int_1^{1.5} (1/t)dt \approx (0.05)[f(1.025) + f(1.075) + \cdots + f(1.475)]$$

$$= (0.05)\left[\frac{1}{1.025} + \frac{1}{1.075} + \cdots + \frac{1}{1.475}\right] \approx 0.4054.$$

82. **(a)** $y = \dfrac{1}{t}$, $y' = -\dfrac{1}{t^2}$. The slope of AD is $\dfrac{1/2 - 1}{2 - 1} = -\dfrac{1}{2}$. Let c be the t-coordinate of the point on $y = \dfrac{1}{t}$ with slope $-\frac{1}{2}$. Then $-\frac{1}{c^2} = -\frac{1}{2} \quad \Rightarrow \quad c^2 = 2 \quad \Rightarrow \quad c = \sqrt{2}$ since $c > 0$. Therefore the tangent line is given by $y - \frac{1}{\sqrt{2}} = -\frac{1}{2}\left(t - \sqrt{2}\right) \quad \Rightarrow \quad y = -\frac{1}{2}t + \sqrt{2}$.

(b)

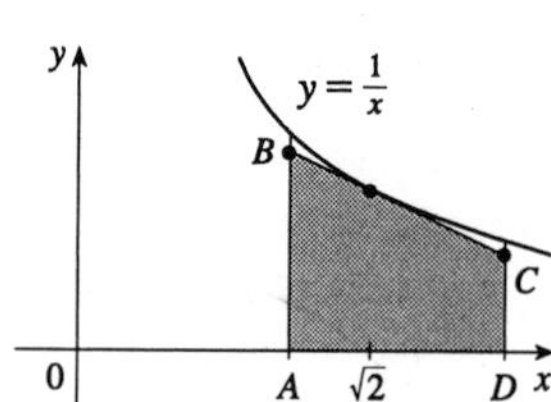

Since the graph of $y = 1/t$ is concave upward, the graph lies above the tangent line, that is, above the line segment BC. Now $|AB| = -\frac{1}{2} + \sqrt{2}$ and $|CD| = -1 + \sqrt{2}$. So the area of the trapezoid $ABCD$ is

$$\tfrac{1}{2}\left[\left(-\tfrac{1}{2} + \sqrt{2}\right) + \left(-1 + \sqrt{2}\right)1\right] = -\tfrac{3}{4} + \sqrt{2} \approx 0.6642.$$

So $\ln 2 >$ area of trapezoid $ABCD > 0.66$.

83.

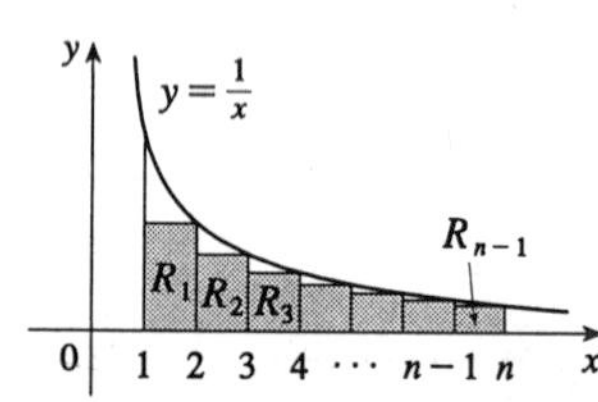

The area of R_i is $\dfrac{1}{i+1}$ and so

$$\frac{1}{2} + \frac{1}{3} + \cdots + \frac{1}{n} < \int_1^n \frac{1}{t}\,dt = \ln n.$$

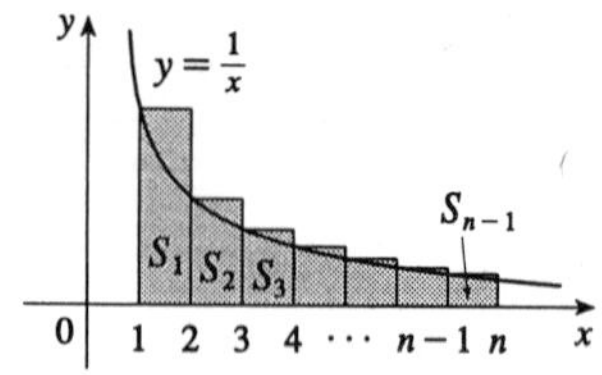

The area of S_i is $\dfrac{1}{i}$ and so

$$1 + \frac{1}{2} + \cdots + \frac{1}{n-1} > \int_1^n \frac{1}{t}\,dt = \ln n.$$

84. If $f(x) = \ln(x^r)$, then $f'(x) = (1/x^r)(rx^{r-1}) = r/x$. But if $g(x) = r\ln x$, then $g'(x) = r/x$. So f and g must differ by a constant: $\ln(x^r) = r\ln x + C$. Put $x = 1$: $\ln(1^r) = r\ln 1 + C \quad \Rightarrow \quad C = 0$, so $\ln(x^r) = r\ln x$.

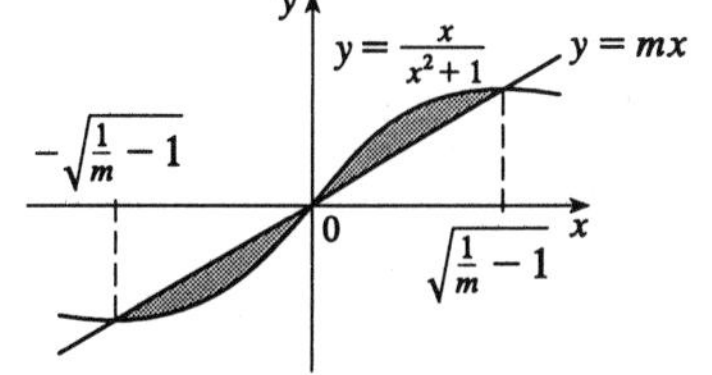

85. The curve and the line will determine a region when they intersect at two or more points. So we solve the equation $x/(x^2+1) = mx$ $\Rightarrow$ $x = 0$ or $mx^2 + m - 1 = 0$ $\Rightarrow$ $x = 0$ or $x = \dfrac{\pm\sqrt{-4(m)(m-1)}}{2m} = \pm\sqrt{\dfrac{1}{m} - 1}$. Note that if $m = 1$, this has only the solution $x = 0$, and no region is determined. But if $1/m - 1 > 0$ $\Leftrightarrow$ $1/m > 1$ $\Leftrightarrow$ $0 < m < 1$, then there are two solutions. [Another way of seeing this is to observe that the slope of the tangent to $y = x/(x^2+1)$ at the origin is $y' = 1$ and therefore we must have $0 < m < 1$.] Note that we cannot just integrate between the positive and negative roots, since the curve and the line cross at the origin. Since mx and $x/(x^2+1)$ are both odd functions, the total area is twice the area between the curves on the interval $\left[0, \sqrt{1/m - 1}\right]$. So the total area enclosed is

$$2\int_0^{\sqrt{1/m-1}} \left[\frac{x}{x^2+1} - mx\right] dx = 2\left[\tfrac{1}{2}\ln(x^2+1) - \tfrac{1}{2}mx^2\right]_0^{\sqrt{1/m-1}}$$
$$= \left[\ln\left(\frac{1}{m} - 1 + 1\right) - m\left(\frac{1}{m} - 1\right)\right] - (\ln 1 - 0) = \ln\left(\frac{1}{m}\right) + m - 1 = m - \ln m - 1.$$

86. $\displaystyle\lim_{x\to\infty} [\ln(2+x) - \ln(1+x)] = \lim_{x\to\infty} \ln\left(\frac{2+x}{1+x}\right) = \lim_{x\to\infty} \ln\left(\frac{2/x+1}{1/x+1}\right) = \ln\frac{1}{1} = \ln 1 = 0$

87. If $f(x) = \ln(1+x)$, then $f'(x) = 1/(1+x)$, so $f'(0) = 1$. Thus
$\displaystyle\lim_{x\to 0}\frac{\ln(1+x)}{x} = \lim_{x\to 0}\frac{f(x)}{x} = \lim_{x\to 0}\frac{f(x)-f(0)}{x-0} = f'(0) = 1.$

88. **(a)**

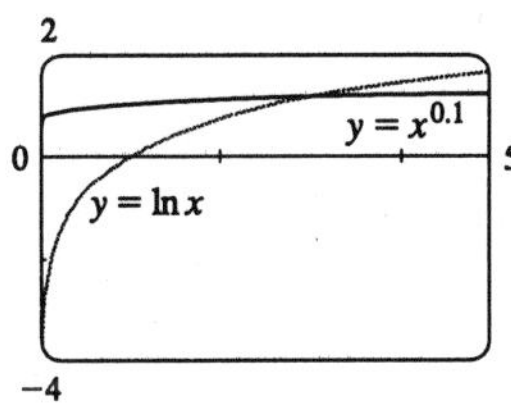

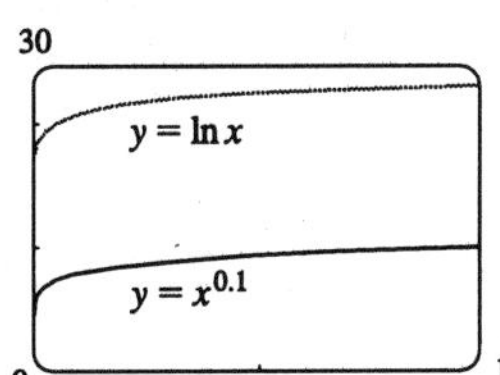

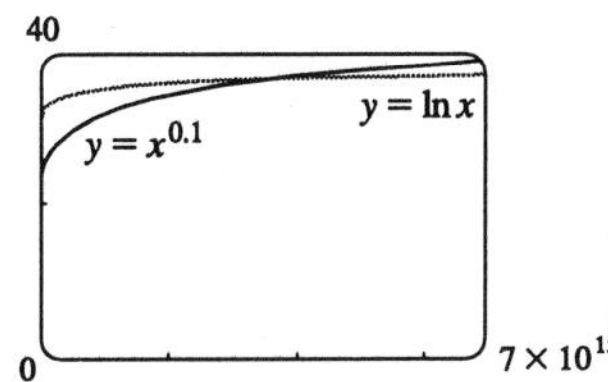

From the graphs, we see that $f(x) > g(x)$ for approximately $0 < x < 3$, and then $g(x) > f(x)$ for $3 < x < 3.5 \times 10^{15}$ (approximately). At that point, the graph of f finally surpasses the graph of g for good.

(b)

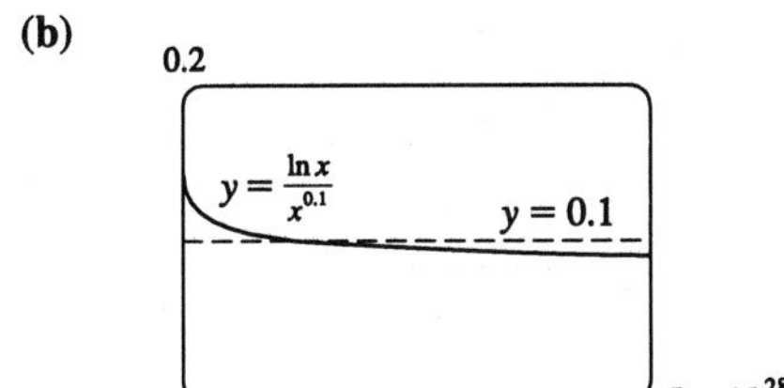

(c) From the graph at left, it seems that $\dfrac{\ln x}{x^{0.1}} < 0.1$ whenever $x > 1.3 \times 10^{28}$ (approximately). So we can take $N = 1.3 \times 10^{28}$, or any larger number.

EXERCISES 6.3*

1. $\ln e^{\sqrt{2}} = \sqrt{2}$ **2.** $e^{\ln 6} = 6$ **3.** $e^{3\ln 2} = \left(e^{\ln 2}\right)^3 = 2^3 = 8$

4. $\ln \sqrt{e} = \ln\left(e^{1/2}\right) = \frac{1}{2}$ **5.** $\ln e^{\sin x} = \sin x$ **6.** $e^{x+\ln x} = e^x e^{\ln x} = xe^x$

7. $e^x = 16 \quad\Leftrightarrow\quad \ln e^x = \ln 16 \quad\Leftrightarrow\quad x = \ln 16 \quad\Leftrightarrow\quad x = 4\ln 2$

8. $\ln x = -1 \quad\Leftrightarrow\quad e^{\ln x} = e^{-1} \quad\Leftrightarrow\quad x = 1/e$

9. $\ln(2x-1) = 3 \quad\Leftrightarrow\quad e^{\ln(2x-1)} = e^3 \quad\Leftrightarrow\quad 2x - 1 = e^3 \quad\Leftrightarrow\quad x = \frac{1}{2}(e^3+1)$

10. $e^{3x-4} = 2 \quad\Leftrightarrow\quad \ln\left(e^{3x-4}\right) = \ln 2 \quad\Leftrightarrow\quad 3x - 4 = \ln 2 \quad\Leftrightarrow\quad x = \frac{1}{3}(\ln 2 + 4)$

11. $\ln(\ln x) = 1 \quad\Leftrightarrow\quad e^{\ln(\ln x)} = e^1 \quad\Leftrightarrow\quad \ln x = e^1 = e \quad\Leftrightarrow\quad e^{\ln x} = e^e \quad\Leftrightarrow\quad x = e^e$

12. $e^{e^x} = 10 \Leftrightarrow \ln\left(e^{e^x}\right) = \ln 10 \Leftrightarrow e^x \ln e = e^x = \ln 10 \Leftrightarrow \ln e^x = \ln(\ln 10) \Leftrightarrow x = \ln(\ln 10)$

13. $e^{ax} = Ce^{bx} \Leftrightarrow \ln e^{ax} = \ln\left(Ce^{bx}\right) \Leftrightarrow ax = \ln C + bx \Leftrightarrow (a-b)x = \ln C \Leftrightarrow x = \dfrac{\ln C}{a-b}$

14. $7e^x - e^{2x} = 12 \quad\Leftrightarrow\quad (e^x)^2 - 7e^x + 12 = 0 \quad\Leftrightarrow\quad (e^x - 3)(e^x - 4) = 0$, so we have either $e^x = 3 \quad\Leftrightarrow\quad x = \ln 3$, or $e^x = 4 \quad\Leftrightarrow\quad x = \ln 4$.

15. $\ln(x-5) = 3 \quad\Rightarrow\quad x - 5 = e^3 \quad\Rightarrow\quad x = e^3 + 5 \approx 25.0855$

16. $e^{5x-1} = 12 \quad\Rightarrow\quad 5x - 1 = \ln 12 \quad\Rightarrow\quad x = \frac{1}{5}(\ln 12 + 1) \approx 0.6970$

17. $y = e^x$

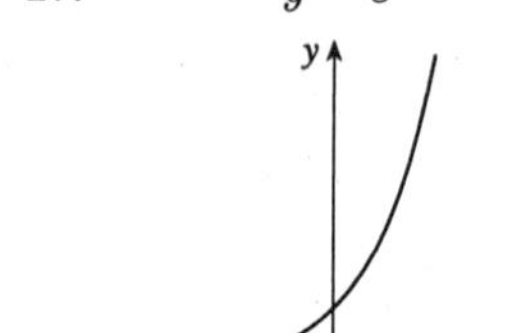

$y = e^{-x}$

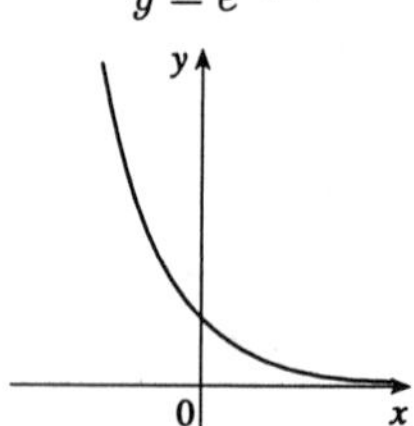

18. $y = -e^x$

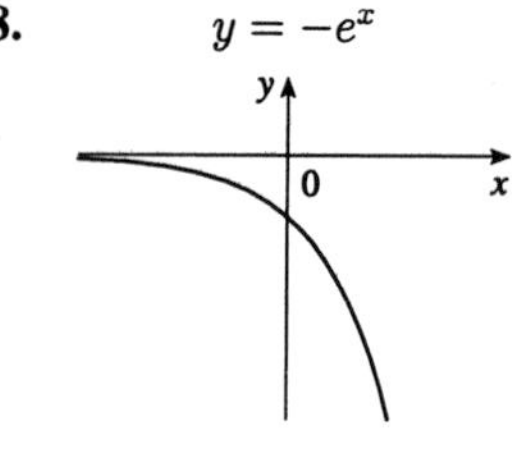

19. $y = -e^x$ $\qquad$ $y = 3 - e^x$

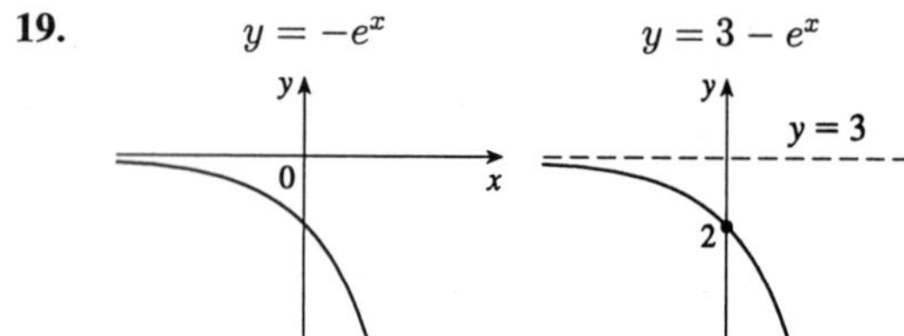

20. $y = e^x$ $\qquad$ $y = e^{x-3}$

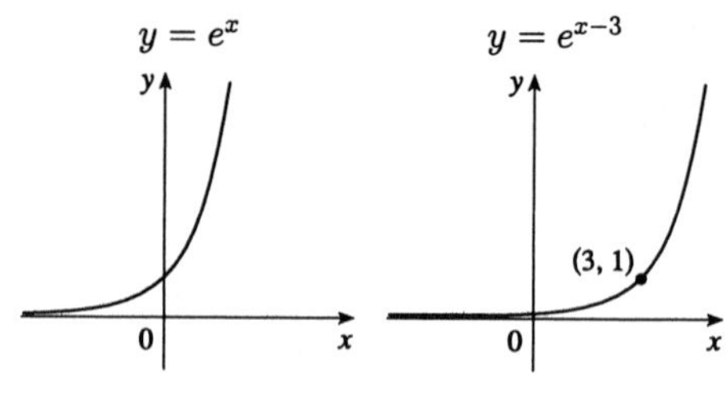

21. Divide numerator and denominator by e^{3x}: $\displaystyle\lim_{x\to\infty} \frac{e^{3x} - e^{-3x}}{e^{3x} + e^{-3x}} = \lim_{x\to\infty} \frac{1 - e^{-6x}}{1 + e^{-6x}} = \frac{1-0}{1+0} = 1.$

22. Divide numerator and denominator by e^{-3x}: $\displaystyle\lim_{x\to-\infty} \frac{e^{3x} - e^{-3x}}{e^{3x} + e^{-3x}} = \lim_{x\to-\infty} \frac{e^{6x} - 1}{e^{6x} + 1} = \frac{0-1}{0+1} = -1.$

23. $\lim\limits_{x\to 1^-} e^{2/(x-1)} = 0$ since $\dfrac{2}{x-1} \to -\infty$ as $x \to 1^-$.

24. $\lim\limits_{x\to 1^+} e^{2/(x-1)} = \infty$ since $\dfrac{2}{x-1} \to \infty$ as $x \to 1^+$.

25. $f(x) = e^{\sqrt{x}} \quad\Rightarrow\quad f'(x) = \dfrac{e^{\sqrt{x}}}{2\sqrt{x}}$

26. $f(x) = xe^{-x^2} \quad\Rightarrow\quad f'(x) = e^{-x^2} + xe^{-x^2}(-2x) = e^{-x^2}(1-2x^2)$

27. $y = xe^{2x} \quad\Rightarrow\quad y' = e^{2x} + xe^{2x}(2) = e^{2x}(1+2x)$

28. $g(x) = e^{-5x}\cos 3x \quad\Rightarrow\quad g'(x) = -5e^{-5x}\cos 3x - 3e^{-5x}\sin 3x$

29. $h(t) = \sqrt{1-e^t} \quad\Rightarrow\quad h'(t) = -\dfrac{e^t}{2\sqrt{1-e^t}}$

30. $h(\theta) = e^{\sin 5\theta} \quad\Rightarrow\quad h'(\theta) = 5\cos(5\theta)e^{\sin 5\theta}$

31. $y = e^{x\cos x} \quad\Rightarrow\quad y' = e^{x\cos x}(\cos x - x\sin x)$

32. $y = \dfrac{e^{-x^2}}{x} \quad\Rightarrow\quad y' = \dfrac{xe^{-x^2}(-2x) - e^{-x^2}}{x^2} = \dfrac{e^{-x^2}(-2x^2-1)}{x^2}$

33. $y = e^{-1/x} \quad\Rightarrow\quad y' = \dfrac{e^{-1/x}}{x^2}$

34. $y = e^{x+e^x} \quad\Rightarrow\quad y' = e^{x+e^x}(1+e^x)$

35. $y = \tan(e^{3x-2}) \quad\Rightarrow\quad y' = 3e^{3x-2}\sec^2(e^{3x-2})$

36. $y = (2x+e^{3x})^{1/3} \quad\Rightarrow\quad y' = \frac{1}{3}(2+3e^{3x})(2x+e^{3x})^{-2/3}$

37. $y = \dfrac{e^{3x}}{1+e^x} \quad\Rightarrow\quad y' = \dfrac{3e^{3x}(1+e^x) - e^{3x}(e^x)}{(1+e^x)^2} = \dfrac{3e^{3x}+3e^{4x}-e^{4x}}{(1+e^x)^2} = \dfrac{3e^{3x}+2e^{4x}}{(1+e^x)^2}$

38. $y = \dfrac{e^x+e^{-x}}{e^x-e^{-x}} \quad\Rightarrow$

$$y' = \frac{(e^x-e^{-x})(e^x-e^{-x}) - (e^x+e^{-x})(e^x+e^{-x})}{(e^x-e^{-x})^2} = \frac{(e^{2x}-2+e^{-2x}) - (e^{2x}+2+e^{-2x})}{(e^x-e^{-x})^2} = -\frac{4}{(e^x-e^{-x})^2}.$$

39. $y = e^x\ln x \quad\Rightarrow\quad y' = e^x\left(\dfrac{1}{x}\right) + (\ln x)(e^x) = e^x\left(\ln x + \dfrac{1}{x}\right)$

40. $y = \sec\left(e^{\tan x^2}\right) \quad\Rightarrow\quad y' = \sec\left(e^{\tan x^2}\right)\tan\left(e^{\tan x^2}\right)\left(e^{\tan x^2}\right)[\sec^2(x^2)](2x)$

41. $y = f(x) = e^{-x}\sin x \quad\Rightarrow\quad f'(x) = -e^{-x}\sin x + e^{-x}\cos x \quad\Rightarrow\quad f'(\pi) = e^{-\pi}(\cos\pi - \sin\pi) = -e^{-\pi}$, so the equation of the tangent at $(\pi, 0)$ is $y - 0 = -e^{-\pi}(x-\pi)$ or $x + e^{\pi}y = \pi$.

42. $y' = 2xe^{-x} - x^2e^{-x}$. At $(1, 1/e)$, $y' = 2e^{-1} - e^{-1} = 1/e$. So the equation of the tangent line is $y - \dfrac{1}{e} = \dfrac{1}{e}(x-1) \quad\Rightarrow\quad y = \dfrac{x}{e}$.

43. $\cos(x-y) = xe^x \quad\Rightarrow\quad -\sin(x-y)(1-y') = e^x + xe^x \quad\Rightarrow\quad y' = 1 + \dfrac{e^x(1+x)}{\sin(x-y)}$

44. $y = Ae^{-x} + Bxe^{-x} \quad \Rightarrow \quad y' = -Ae^{-x} + Be^{-x} - Bxe^{-x} = (B - A)e^{-x} - Bxe^{-x} \quad \Rightarrow$
$y'' = (A - B)e^{-x} - Be^{-x} + Bxe^{-x} = (A - 2B)e^{-x} + Bxe^{-x}$, so
$y'' + 2y' + y = (A - 2B)e^{-x} + Bxe^{-x} + 2[(B - A)e^{-x} - Bxe^{-x}] + Ae^{-x} + Bxe^{-x} = 0.$

45. $y = e^{rx} \quad \Rightarrow \quad y' = re^{rx} \quad \Rightarrow \quad y'' = r^2e^{rx}$, so $y'' + 5y' - 6y = r^2e^{rx} + 5re^{rx} - 6e^{rx}$
$= e^{rx}(r^2 + 5r - 6) = e^{rx}(r + 6)(r - 1) = 0 \quad \Rightarrow \quad (r + 6)(r - 1) = 0 \quad \Rightarrow \quad r = 1$ or -6.

46. $y = e^{\lambda x} \quad \Rightarrow \quad y' = \lambda e^{\lambda x} \quad \Rightarrow \quad y'' = \lambda^2 e^{\lambda x}$. Thus, $y + y' = y'' \quad \Leftrightarrow \quad e^{\lambda x} + \lambda e^{\lambda x} = \lambda^2 e^{\lambda x} \quad \Leftrightarrow$
$e^{\lambda x}(\lambda^2 - \lambda - 1) = 0 \quad \Leftrightarrow \quad \lambda = \frac{1 \pm \sqrt{5}}{2}$, since $e^{\lambda x} \neq 0$.

47. $f(x) = e^{-2x} \quad \Rightarrow \quad f'(x) = -2e^{-2x} \quad \Rightarrow \quad f''(x) = (-2)^2 e^{-2x} \quad \Rightarrow \quad f'''(x) = (-2)^3 e^{-2x} \quad \Rightarrow \quad \cdots$
$\Rightarrow \quad f^{(8)}(x) = (-2)^8 e^{-2x} = 256e^{-2x}$

48. $f(x) = xe^{-x}$, $f'(x) = e^{-x} - xe^{-x} = (1 - x)e^{-x}$, $f''(x) = -e^{-x} + (1 - x)(-e^{-x}) = (x - 2)e^{-x}$. Similarly,
$f'''(x) = (3 - x)e^{-x}$, $f^{(4)}(x) = (x - 4)e^{-x}, \ldots, f^{(1000)}(x) = (x - 1000)e^{-x}$.

49. (a) $f(x) = e^x + x$ is continuous on $\mathbb{R}$ and $f(-1) = e^{-1} - 1 < 0 < 1 = f(0)$, so by the Intermediate Value Theorem, $e^x + x = 0$ has a root in $(-1, 0)$.

(b) $f(x) = e^x + x \quad \Rightarrow \quad f'(x) = e^x + 1$, so $x_{n+1} = x_n - \dfrac{e^{x_n} + x_n}{e^{x_n} + 1}$. From Exercise 31 we know that there is a root between -1 and 0, so we take $x_1 = -0.5$. Then $x_2 \approx -0.566311$, $x_3 \approx -0.567143$, and $x_4 \approx -0.567143$, so the root is -0.567143 to six decimal places.

50.

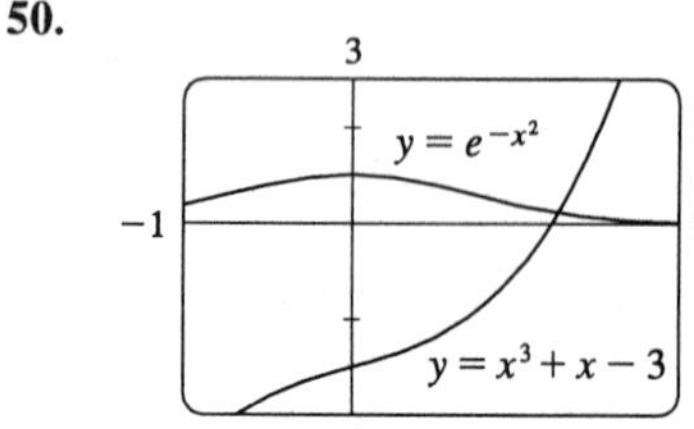

From the graph, it appears that the curves intersect at about $x \approx 1.2$ or 1.3. We use Newton's Method with $f(x) = x^3 + x - 3 - e^{-x^2}$, so $f'(x) = 3x^2 + 1 - 2xe^{-x^2}$, and the formula is $x_{n+1} = x_n - f(x_n)/f'(x_n)$.
We take $x_1 = 1.2$, and the formula gives $x_2 \approx 1.252462$, $x_3 \approx 1.251045$, and $x_4 \approx x_5 \approx 1.251044$. So the root of the equation, correct to six decimal places, is $x = 1.251044$.

51. (a) 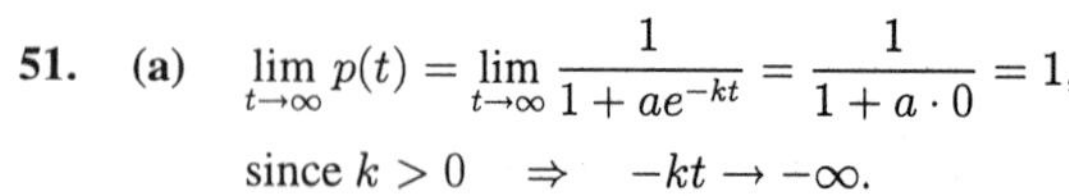
$\displaystyle\lim_{t\to\infty} p(t) = \lim_{t\to\infty} \frac{1}{1 + ae^{-kt}} = \frac{1}{1 + a \cdot 0} = 1,$
since $k > 0 \quad \Rightarrow \quad -kt \to -\infty$.

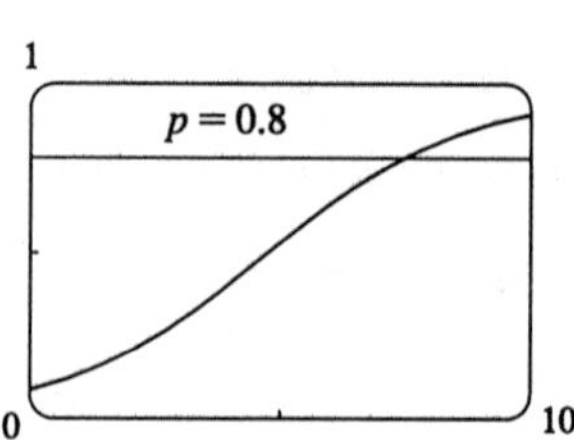

(b) 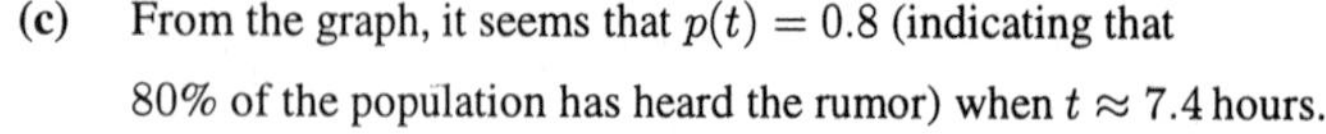
$\dfrac{dp}{dt} = -(1 + ae^{-kt})^{-2}(-kae^{-kt}) = \dfrac{kae^{-kt}}{(1 + ae^{-kt})^2}$

(c) From the graph, it seems that $p(t) = 0.8$ (indicating that 80% of the population has heard the rumor) when $t \approx 7.4$ hours.

52. **(a)**

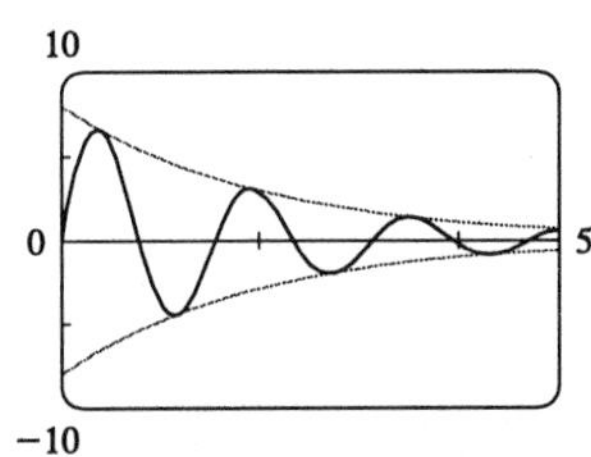

The displacement function is squeezed between the other two functions. This is because $-1 \le \sin 4t \le 1 \quad \Rightarrow$ $-8e^{-t/2} \le 8e^{-t/2}\sin 4t \le 8e^{-t/2}$.

(b) The maximum value of the displacement is about 6.6 cm, occurring at $t \approx 0.36$ s. It occurs just before the graph of the displacement function touches the graph of $8e^{-t/2}$ (when $t = \frac{\pi}{8} \approx 0.39$).

(c) The velocity of the object is the derivative of its displacement function, that is,

$$\frac{d}{dt}\left(8e^{-t/2}\sin 4t\right) = 8\left(e^{-t/2}\cos 4t(4) + \sin 4t\left(-\tfrac{1}{2}\right)e^{-t/2}\right).$$

If the displacement is zero, then we must have $\sin 4t = 0$ (since the exponential term in the displacement function is always positive). The first time that $\sin 4t = 0$ after $t = 0$ occurs at $t = \frac{\pi}{4}$. Substituting this into our expression for the velocity, and noting that the second term vanishes, we get $v\left(\frac{\pi}{4}\right) = 8e^{-\pi/8}\cos\left(4 \cdot \frac{\pi}{4}\right) \cdot 4 = -32e^{-\pi/8} \approx -21.6$ cm/s.

(d)

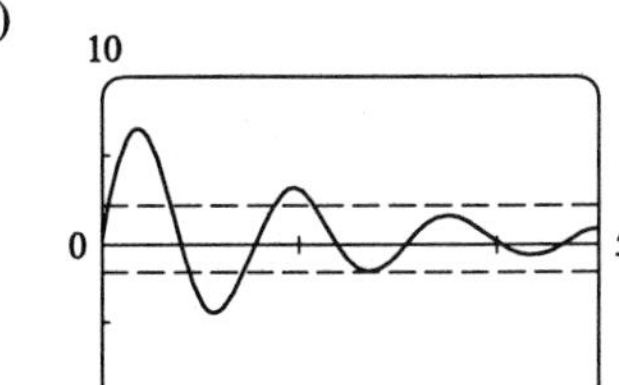

The graph indicates that the displacement is less than 2 cm from equilibrium whenever t is larger than about 2.8.

53. $x - e^x \quad \Rightarrow \quad f'(x) = 1 - e^x = 0 \quad \Leftrightarrow \quad e^x = 1 \quad \Leftrightarrow \quad x = 0$. Now $f'(x) > 0$ for all $x < 0$ and $f'(x) < 0$ for all $x > 0$, so the absolute maximum value is $f(0) = 0 - 1 = -1$.

54. $g(x) = \dfrac{e^x}{x} \quad \Rightarrow \quad g'(x) = \dfrac{xe^x - e^x}{x^2} = 0 \quad \Leftrightarrow \quad e^x(x-1) = 0 \quad \Rightarrow \quad x = 1$. Now $g'(x) > 0 \quad \Leftrightarrow$ $\dfrac{xe^x - e^x}{x^2} > 0 \quad \Leftrightarrow \quad x - 1 > 0 \quad \Leftrightarrow \quad x > 1$ and $g'(x) < 0 \quad \Leftrightarrow \quad \dfrac{xe^x - e^x}{x^2} < 0 \quad \Leftrightarrow \quad x - 1 < 0 \quad \Leftrightarrow$ $x < 1$. Thus there is an absolute minimum value of $g(1) = e$ at $x = 1$.

55. $y = e^x - 2e^{-x}$, so $y' = e^x + 2e^{-x}$, $y'' = e^x - 2e^{-x}$. $y'' > 0 \quad \Leftrightarrow \quad e^x - 2e^{-x} > 0 \quad \Leftrightarrow \quad e^x > 2e^{-x} \quad \Leftrightarrow$ $e^{2x} > 2 \quad \Leftrightarrow \quad 2x > \ln 2 \quad \Leftrightarrow \quad x > \frac{1}{2}\ln 2$. Therefore, y is concave upward on $\left(\frac{1}{2}\ln 2, \infty\right)$.

56. $f(x) = e^x + e^{-2x}$, $f'(x) = e^x - 2e^{-2x} > 0 \quad \Leftrightarrow \quad e^x > 2e^{-2x} \quad \Leftrightarrow \quad e^{3x} > 2 \quad \Leftrightarrow \quad 3x > \ln 2 \quad \Leftrightarrow$ $x > \frac{1}{3}\ln 2$. Thus f is increasing on $\left[\frac{1}{3}\ln 2, \infty\right)$.

57. $y = f(x) = e^{-1/(x+1)}$ **A.** $D = \{x \mid x \ne -1\} = (-\infty, -1) \cup (-1, \infty)$ **B.** No x-intercept; y-intercept $= f(0) = e^{-1}$ **C.** No symmetry **D.** $\lim\limits_{x\to\pm\infty} e^{-1/(x+1)} = 1$ since $-1/(x+1) \to 0$, so $y = 1$ is a HA. $\lim\limits_{x\to -1^+} e^{-1/(x+1)} = 0$ since $-1/(x+1) \to -\infty$, $\lim\limits_{x\to -1^-} e^{-1/(x+1)} = \infty$ since $-1/(x+1) \to \infty$, so $x = -1$ is a VA. **E.** $f'(x) = e^{-1/(x+1)}/(x+1)^2 \quad \Rightarrow \quad f'(x) > 0$ for all x except 1, so f is increasing on $(-\infty, -1)$ and $(-1, \infty)$. **F.** No extrema

G. $f''(x) = \dfrac{e^{-1/(x+1)}}{(x+1)^4} + \dfrac{e^{-1/(x+1)}(-2)}{(x+1)^3} = -\dfrac{e^{-1/(x+1)}(2x+1)}{(x+1)^4} \quad \Rightarrow$

$f''(x) > 0 \quad \Leftrightarrow \quad 2x + 1 < 0 \quad \Leftrightarrow \quad x < -\frac{1}{2}$, so f is CU on $(-\infty, -1)$ and $\left(-1, -\frac{1}{2}\right)$, and CD on $\left(-\frac{1}{2}, \infty\right)$. IP at $\left(-\frac{1}{2}, e^{-2}\right)$.

H.

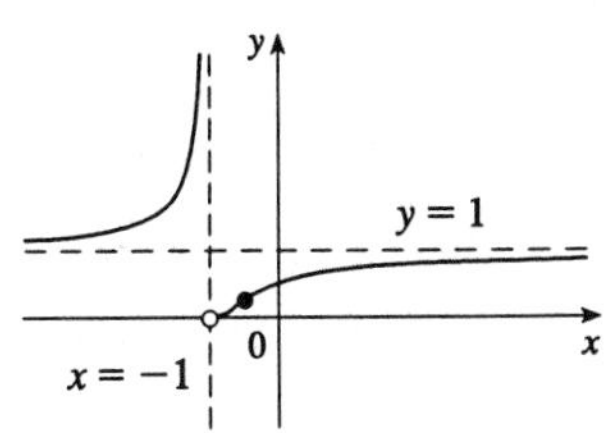

58. $y = f(x) = xe^{x^2}$ **A.** $D = \mathbb{R}$ **B.** Both intercepts are 0. **C.** $f(-x) = -f(x)$, so the curve is symmetric about the origin. **D.** $\lim\limits_{x\to\infty} xe^{x^2} = \infty$, $\lim\limits_{x\to-\infty} xe^{x^2} = -\infty$, no asymptotes **E.** $f'(x) = e^{x^2} + xe^{x^2}(2x) = e^{x^2}(1+2x^2) > 0$, so f is increasing on $\mathbb{R}$. **F.** No extrema **G.** $f''(x) = e^{x^2}(2x)(1+2x^2) + e^{x^2}(4x) = e^{x^2}(2x)(3+2x^2) > 0$ $\Leftrightarrow$ $x > 0$, so f is CU on $(0,\infty)$ and CD on $(-\infty, 0)$. f has an inflection point at $(0,0)$.

H.

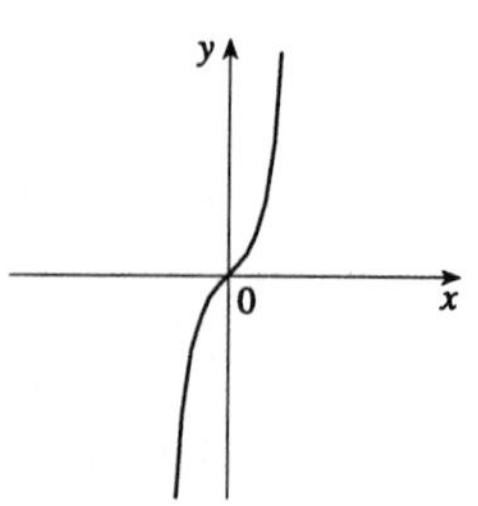

59. $y = 1/(1+e^{-x})$ **A.** $D = \mathbb{R}$ **B.** No x-intercepts; y-intercept $= f(0) = \frac{1}{2}$. **C.** No symmetry **D.** $\lim\limits_{x\to\infty} 1/(1+e^{-x}) = \frac{1}{1+0} = 1$ and $\lim\limits_{x\to-\infty} 1/(1+e^{-x}) = 0$ $\left(\text{since } \lim\limits_{x\to-\infty} e^{-x} = \infty\right)$, so f has horizontal asymptotes $y = 0$ and $y = 1$. **E.** $f'(x) = -(1+e^{-x})^{-2}(-e^{-x}) = e^{-x}/(1+e^{-x})^2$. This is positive for all x, so f is increasing on $\mathbb{R}$. **F.** No extrema

G. $f''(x) = \dfrac{(1+e^{-x})^2(-e^{-x}) - e^{-x}(2)(1+e^{-x})(-e^{-x})}{(1+e^{-x})^4}$

$= \dfrac{e^{-x}(e^{-x}-1)}{(1+e^{-x})^3}$. The second factor in the numerator is negative for $x > 0$ and positive for $x < 0$, and the other factors are always positive, so f is CU on $(-\infty, 0)$ and CD on $(0,\infty)$. f has an inflection point at $\left(0, \frac{1}{2}\right)$.

H.

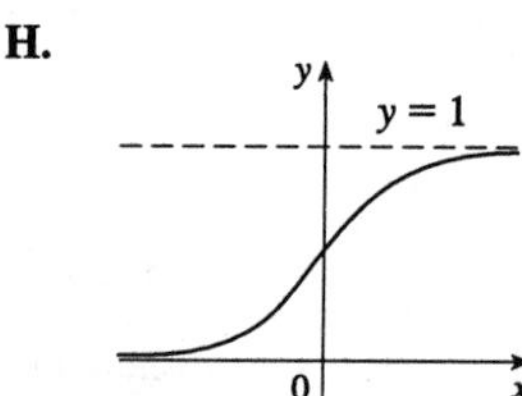

60.

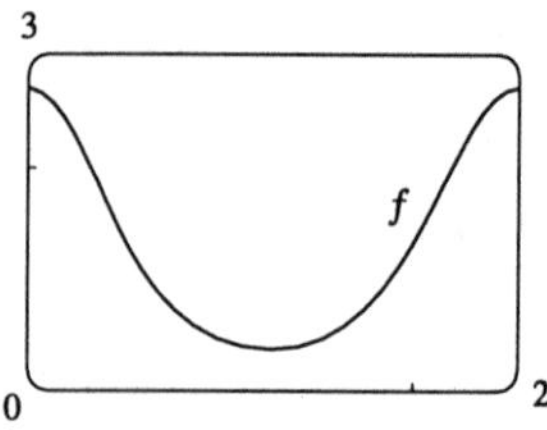

1.1
0
2π
f''
−2.8

The function $f(x) = e^{\cos x}$ is periodic with period 2π, so we consider it only on the interval $[0, 2\pi]$. We see that it has local maxima of about $f(0) \approx 2.72$ and $f(2\pi) \approx 2.72$, and a local minimum of about $f(3.14) \approx 0.37$. To find the exact values, we calculate $f'(x) = -\sin x\, e^{\cos x}$. This is 0 when $-\sin x = 0$ $\Leftrightarrow$ $x = 0, \pi$ or 2π $\left(\text{since we are only considering } x \in [0, 2\pi]\right)$. Also $f'(x) > 0$ $\Leftrightarrow$ $\sin x < 0$ $\Leftrightarrow$ $0 < x < \pi$. So $f(0) = f(2\pi) = e$ (both maxima) and $f(\pi) = e^{\cos\pi} = 1/e$ (minimum). To find the inflection points, we calculate and graph $f''(x) = \dfrac{d}{dx}(-\sin x\, e^{\cos x}) = -\cos x\, e^{\cos x} - \sin x(e^{\cos x})(-\sin x) = e^{\cos x}\left(\sin^2 x - \cos x\right)$. From the graph of $f''(x)$, we see that f has inflection points at $x \approx 0.90$ and at $x \approx 5.38$. These x-coordinates correspond to inflection points $(0.90, 1.86)$ and $(5.38, 1.86)$.

61.

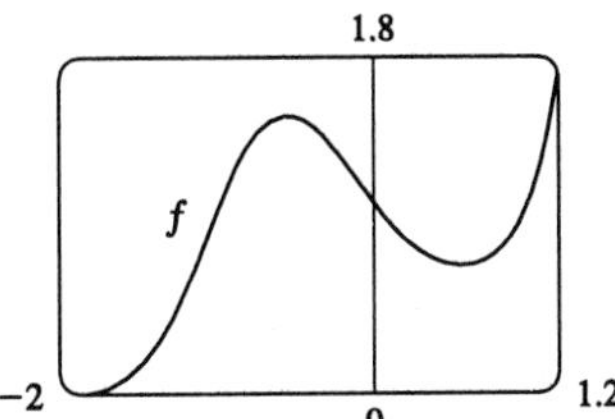

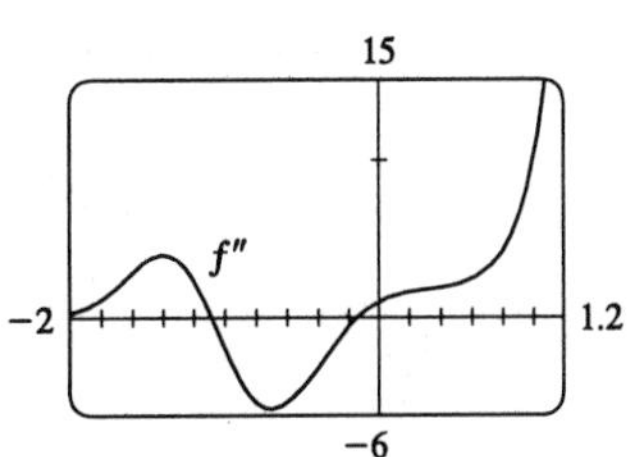

$f(x)=e^{x^3-x}\to 0$ as $x\to-\infty$, and $f(x)\to\infty$ as $x\to\infty$. From the graph, it appears that f has a local minimum of about $f(0.58)=0.68$, and a local maximum of about $f(-0.58)=1.47$. To find the exact values, we calculate $f'(x)=(3x^2-1)e^{x^3-x}$, which is 0 when $3x^2-1=0 \quad\Leftrightarrow\quad x=\pm\frac{1}{\sqrt{3}}$. The negative root corresponds to the local maximum $f\left(-\frac{1}{\sqrt{3}}\right)=e^{(-1/\sqrt{3})^3-(-1/\sqrt{3})}=e^{2\sqrt{3}/9}$, and the positive root corresponds to the local minimum $f\left(\frac{1}{\sqrt{3}}\right)=e^{(1/\sqrt{3})^3-(1/\sqrt{3})}=e^{-2\sqrt{3}/9}$. To estimate the inflection points, we calculate and graph $f''(x)=\dfrac{d}{dx}\left[(3x^2-1)e^{x^3-x}\right]=(3x^2-1)e^{x^3-x}(3x^2-1)+e^{x^3-x}(6x)=e^{x^3-x}(9x^4-6x^2+6x+1)$. From the graph, it appears that $f''(x)$ changes sign (and thus f has inflection points) at $x\approx-0.15$ and $x\approx-1.09$. From the graph of f, we see that these x-values correspond to inflection points at about $(-0.15,1.15)$ and $(-1.09,0.82)$.

62. To find the asymptotes of f, we note that since $b>0$, $\lim\limits_{x\to\pm\infty}e^{-x^2/b}=0$, so $y=0$ is a horizontal asymptote. To find the maxima and minima we differentiate: $f(x)=e^{-x^2/b} \quad\Rightarrow\quad f'(x)=-\dfrac{2x}{b}e^{-x^2/b}$. This is 0 only at $x=0$, and at that point $f'(x)$ changes from positive to negative, so there is an absolute maximum of $f(0)=e^0=1$. To find the inflection points we differentiate again:
$f''(x)=-\dfrac{2}{b}\left[x\left(-\dfrac{2x}{b}\right)e^{-x^2/b}+e^{-x^2/b}(1)\right]=\dfrac{2}{b^2}e^{-x^2/b}\left[2x^2-b\right]$. This is 0 when $2x^2-b=0 \quad\Leftrightarrow$
$x=\pm\sqrt{b/2}$, and $f''(x)$ changes sign at these values.
Also $f\left(\pm\sqrt{b/2}\right)=e^{-(\pm\sqrt{b/2})^2/b}=e^{-1/2}=\sqrt{1/e}$, so the inflection points are $\left(\pm\sqrt{b/2},\sqrt{1/e}\right)$. The value of b affects the graph by stretching or shrinking it in the x-direction: as b increases, the inflection points move away from the y-axis and the graph of f becomes more spread out.

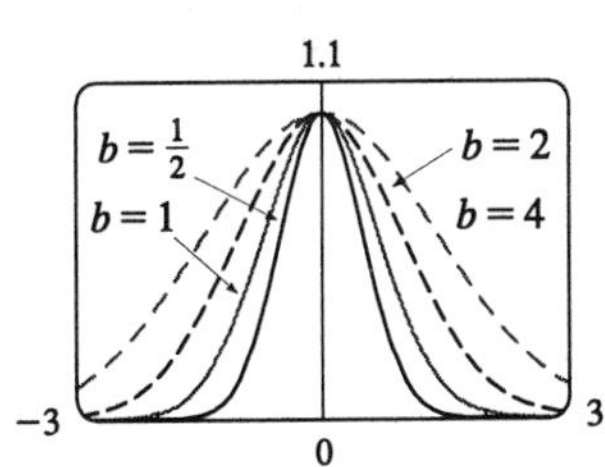

63. $u=-6x \quad\Rightarrow\quad du=-6\,dx$, so $\int e^{-6x}\,dx=-\frac{1}{6}\int e^u\,du=-\frac{1}{6}e^u+C=-\frac{1}{6}e^{-6x}+C$.

64. Let $u=x^2$. Then $du=2x\,dx \quad\Rightarrow\quad \int xe^{x^2}dx=\frac{1}{2}\int e^u\,du=\frac{1}{2}e^u+C=\frac{1}{2}e^{x^2}+C$.

65. Let $u=1+e^x$. Then $du=e^x\,dx$, so $\int e^x(1+e^x)^{10}\,dx=\int u^{10}\,du=\frac{1}{11}u^{11}+C=\frac{1}{11}(1+e^x)^{11}+C$.

66. Let $u=\tan x$. Then $du=\sec^2x\,dx$, so $\int\sec^2x\,e^{\tan x}\,dx=\int e^u\,du=e^u+C=e^{\tan x}+C$.

67. $\displaystyle\int\frac{e^x+1}{e^x}\,dx=\int(1+e^{-x})dx=x-e^{-x}+C$

68. Let $u=2-x$. Then $du=-dx$ so $\int_2^3 e^{2-x}\,dx=\int_0^{-1}(-e^u)du=-e^u\big|_0^{-1}=1-1/e$.

69. Let $u = -x^2$. Then $du = -2x\,dx$, so $\int xe^{-x^2}dx = -\frac{1}{2}\int e^u\,du = -\frac{1}{2}e^u + C = -\frac{1}{2}e^{-x^2} + C$.

70. Let $u = \dfrac{1}{x}$. Then $du = -\dfrac{1}{x^2}\,dx$, so $\displaystyle\int \frac{e^{1/x}}{x^2}\,dx = -\int e^u\,du = -e^u + C = -e^{1/x} + C$.

71. Let $u = x^2 - 4x - 3$. Then $du = 2(x-2)dx$, so
$\int (x-2)e^{x^2-4x-3}\,dx = \frac{1}{2}\int e^u\,du = \frac{1}{2}e^u + C = \frac{1}{2}e^{x^2-4x-3} + C$.

72. Let $u = e^x$. Then $du = e^x\,dx$, so $\int e^x \sin(e^x)dx = \int \sin u\,du = -\cos u + C = -\cos(e^x) + C$.

73. Area $= \int_0^1 (e^{3x} - e^x)dx = \left[\frac{1}{3}e^{3x} - e^x\right]_0^1 = \left(\frac{1}{3}e^3 - e\right) - \left(\frac{1}{3} - 1\right) = \frac{1}{3}e^3 - e + \frac{2}{3} \approx 4.644$

74. $f''(x) = 3e^x + 5\sin x \quad\Rightarrow\quad f'(x) = 3e^x - 5\cos x + C \quad\Rightarrow\quad 2 = f'(0) = 3 - 5 + C \quad\Rightarrow\quad C = 4$, so
$f'(x) = 3e^x - 5\cos x + 4 \quad\Rightarrow\quad f(x) = 3e^x - 5\sin x + 4x + D \quad\Rightarrow\quad 1 = f(0) = 3 + D \quad\Rightarrow$
$D = -2$, so $f(x) = 3e^x - 5\sin x + 4x - 2$.

75. $V = \int_0^1 \pi(e^x)^2\,dx = \int_0^1 \pi e^{2x}\,dx = \frac{1}{2}[\pi e^{2x}]_0^1 = \frac{\pi}{2}(e^2 - 1)$

76. $V = \int_0^1 2\pi x e^{-x^2}dx$. Let $u = x^2$. Thus $du = 2x dx$, so $V = \pi\int_0^1 e^{-u}\,du = \pi[-e^{-u}]_0^1 = \pi(1 - 1/e)$.

77. $y = e^{\sqrt{x}} \quad\Rightarrow\quad \ln y = \ln e^{\sqrt{x}} = \sqrt{x}$
$\Rightarrow\quad x = (\ln y)^2$. Interchange x and y:
the inverse function is $y = (\ln x)^2$. The domain of the inverse function is the range of the original function $y = e^{\sqrt{x}}$, that is, $[1, \infty)$.

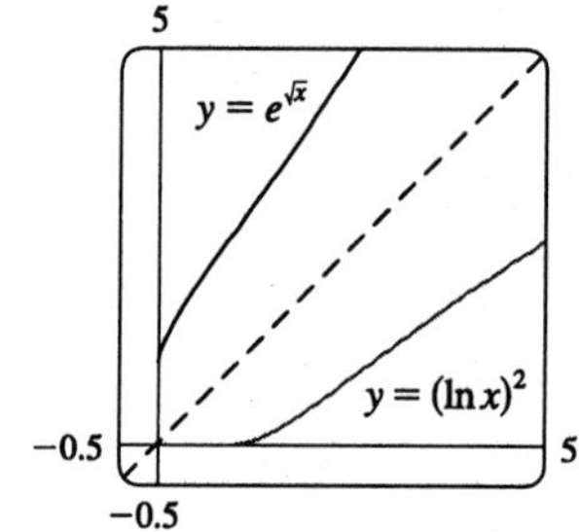

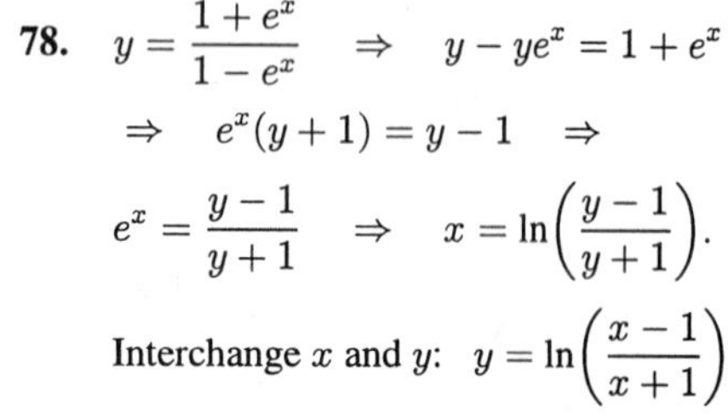

78. $y = \dfrac{1+e^x}{1-e^x} \quad\Rightarrow\quad y - ye^x = 1 + e^x$
$\Rightarrow\quad e^x(y+1) = y - 1 \quad\Rightarrow$
$e^x = \dfrac{y-1}{y+1} \quad\Rightarrow\quad x = \ln\left(\dfrac{y-1}{y+1}\right)$.

Interchange x and y: $y = \ln\left(\dfrac{x-1}{x+1}\right)$ is the inverse function.

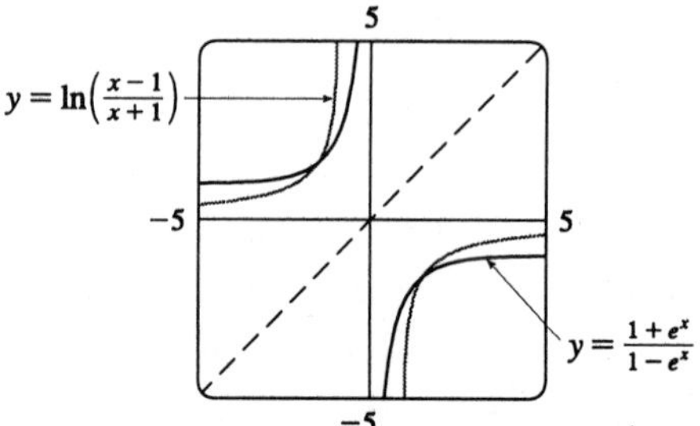

79. We use Theorem 6.1.8. Note that $f(0) = 3 + 0 + e^0 = 4$, so $f^{-1}(4) = 0$. Also $f'(x) = 1 + e^x$. Therefore,
$(f^{-1})'(4) = \dfrac{1}{f'(f^{-1}(4))} = \dfrac{1}{f'(0)} = \dfrac{1}{1+e^0} = \dfrac{1}{2}$.

80. We recognize this limit as the definition of the derivative of the function $f(x) = e^{\sin x}$ at $x = \pi$, since it is of the form $\displaystyle\lim_{x\to\pi} \frac{f(x) - f(\pi)}{x - \pi}$. Therefore, the limit is equal to $f'(\pi) = (\cos\pi)e^{\sin\pi} = -1\cdot e^0 = -1$.

81. Using the second law of logarithms and Equation 5, we have $\ln(e^x/e^y) = \ln e^x - \ln e^y = x - y = \ln(e^{x-y})$. Since ln is a one-to-one function, it follows that $e^x/e^y = e^{x-y}$.

82. Using the third law of logarithms and Equation 5, we have $\ln e^{rx} = rx = r \ln e^x = \ln(e^x)^r$. Since ln is a one-to-one function, it follows that $e^{rx} = (e^x)^r$.

83. **(a)** Let $f(x) = e^x - 1 - x$. Now $f(0) = e^0 - 1 = 0$, and for $x \geq 0$, we have $f'(x) = e^x - 1 \geq 0$. Now, since $f(0) = 0$ and f is increasing on $[0, \infty)$, $f(x) \geq 0$ for $x \geq 0 \Rightarrow e^x - 1 - x \geq 0 \Rightarrow e^x \geq 1 + x$.

(b) For $0 \leq x \leq 1$, $x^2 \leq x$, so $e^{x^2} \leq e^x$ (since e^x is increasing). Hence [from (a)] $1 + x^2 \leq e^{x^2} \leq e^x$. So $\frac{4}{3} = \int_0^1 (1 + x^2)dx \leq \int_0^1 e^{x^2} dx \leq \int_0^1 e^x\,dx = e - 1 < e \Rightarrow \frac{4}{3} \leq \int_0^1 e^{x^2} dx \leq e$.

84. **(a)** Let $f(x) = e^x - 1 - x - \frac{1}{2}x^2$. Thus, $f'(x) = e^x - 1 - x$, which is positive for $x \geq 0$ by Exercise 83(a). Thus $f(x)$ is increasing on $[0, \infty)$, so on that interval, $0 = f(0) \leq f(x) = e^x - 1 - x - \frac{1}{2}x^2 \Rightarrow e^x \geq 1 + x + \frac{1}{2}x^2$.

(b) Using the same argument as in Exercise 83(b), from part (a) we have $1 + x^2 + \frac{1}{2}x^4 \leq e^{x^2} \leq e^x$ (for $0 \leq x \leq 1$) $\Rightarrow \int_0^1 (1 + x^2 + \frac{1}{2}x^4)dx \leq \int_0^1 e^{x^2} dx \leq \int_0^1 e^x\,dx \Rightarrow \frac{43}{30} \leq \int_0^1 e^{x^2} dx \leq e - 1$.

85. **(a)** By Exercise 83(a), the result holds for $n = 1$. Suppose that $e^x \geq 1 + x + \dfrac{x^2}{2!} + \cdots + \dfrac{x^k}{k!}$ for $x \geq 0$. Let $f(x) = e^x - 1 - x - \dfrac{x^2}{2!} - \cdots - \dfrac{x^k}{k!} - \dfrac{x^{k+1}}{(k+1)!}$. Then $f'(x) = e^x - 1 - x - \cdots - \dfrac{x^k}{k!} \geq 0$ by assumption. Hence $f(x)$ is increasing on $[0, \infty)$. So $0 \leq x$ implies that $0 = f(0) \leq f(x) = e^x - 1 - x - \cdots - \dfrac{x^k}{k!} - \dfrac{x^{k+1}}{(k+1)!}$, and hence $e^x \geq 1 + x + \cdots + \dfrac{x^k}{k!} + \dfrac{x^{k+1}}{(k+1)!}$ for $x \geq 0$. Therefore, for $x \geq 0$, $e^x \geq 1 + x + \dfrac{x^2}{2!} + \cdots + \dfrac{x^n}{n!}$ for every positive integer n, by mathematical induction.

(b) Taking $n = 4$ and $x = 1$ in (a), we have $e = e^1 \geq 1 + \dfrac{1}{2} + \dfrac{1}{6} + \dfrac{1}{24} = 2.708\overline{3} > 2.7$.

(c) $e^x \geq 1 + x + \cdots + \dfrac{x^k}{k!} + \dfrac{x^{k+1}}{(k+1)!} \Rightarrow \dfrac{e^x}{x^k} \geq \dfrac{1}{x^k} + \dfrac{1}{x^{k-1}} + \cdots + \dfrac{1}{k!} + \dfrac{x}{(k+1)!} \geq \dfrac{x}{(k+1)!}$. But $\lim\limits_{x\to\infty} \dfrac{x}{(k+1)!} = \infty$, so $\lim\limits_{x\to\infty} \dfrac{e^x}{x^k} = \infty$.

86. **(a)** The graph of g finally surpasses that of f at $x \approx 35.8$.

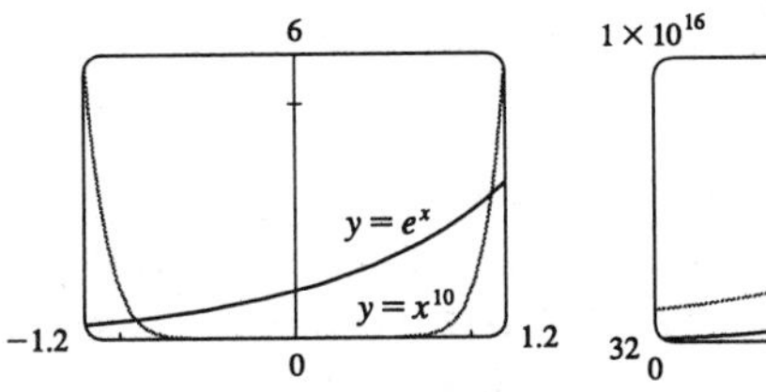

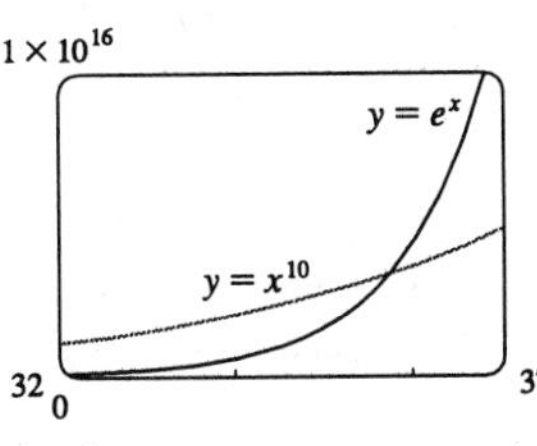

(b)

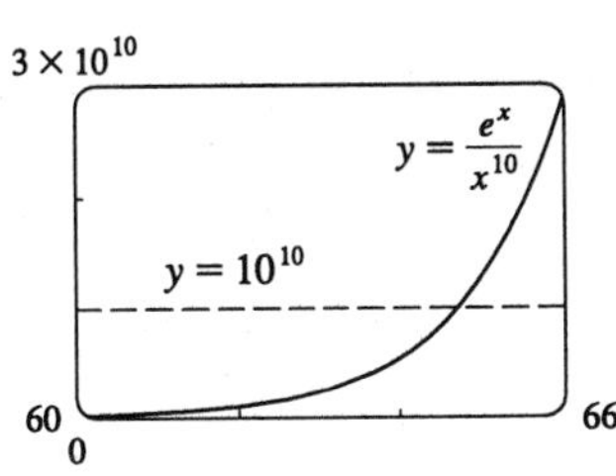

(c) From the graph in (b), it seems that $\dfrac{e^x}{x^{10}} > 10^{10}$ whenever $x > 65$, approximately. So we can take $N \geq 65$.

EXERCISES 6.4*

1. $10^{\pi} = e^{\pi \ln 10}$

2. $x^{\sqrt{2}} = e^{\sqrt{2}\ln x}$

3. $2^{\cos x} = e^{(\cos x)\ln 2}$

4. $(\sin x)^{\ln x} = e^{(\ln x)\ln(\sin x)}$

5. $\log_2 64 = 6$ since $2^6 = 64$.

6. $\log_6 \frac{1}{36} = -2$ since $6^{-2} = \frac{1}{36}$.

7. $2^{(\log_2 3 + \log_2 5)} = 2^{\log_2 15} = 15$

8. $\log_3 3^{\sqrt{5}} = \sqrt{5}$

9.

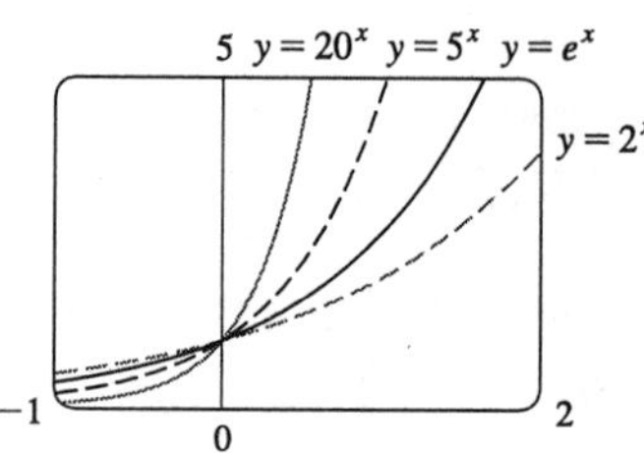

All of these graphs approach 0 as $x \to -\infty$, all of them pass through the point $(0, 1)$, and all of them are increasing and approach ∞ as $x \to \infty$. The larger the base, the faster the function increases for $x > 0$, and the faster it approaches 0 as $x \to -\infty$.

10.

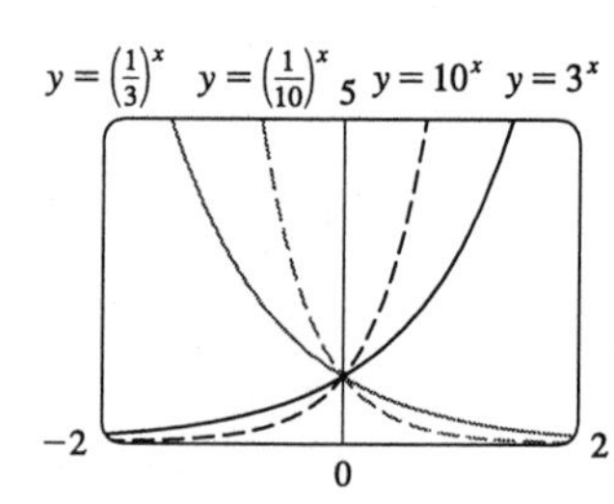

The functions with bases greater than 1 (3^x and 10^x) are increasing, while those with bases less than 1 $\left[\left(\frac{1}{3}\right)^x \text{ and } \left(\frac{1}{10}\right)^x\right]$ are decreasing. The graph of $\left(\frac{1}{3}\right)^x$ is the reflection of that of 3^x about the y-axis, and the graph of $\left(\frac{1}{10}\right)^x$ is the reflection of that of 10^x about the y-axis. The graph of 10^x increases more quickly than that of 3^x for $x > 0$, and approaches 0 faster as $x \to -\infty$.

11. **(a)** $\log_2 5 = \dfrac{\ln 5}{\ln 2} \approx 2.321928$ **(b)** $\log_5 26.05 = \dfrac{\ln 26.05}{\ln 5} \approx 2.025563$

(c) $\log_3 e = \dfrac{1}{\ln 3} \approx 0.910239$ **(d)** $\log_{0.7} 14 = \dfrac{\ln 14}{\ln 0.7} \approx -7.399054$

12.

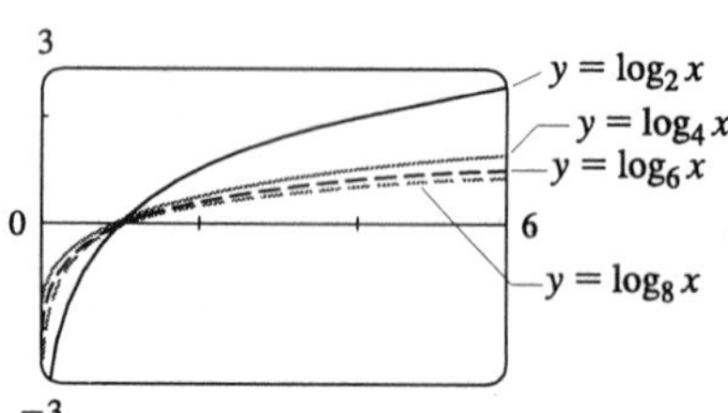

To graph the functions, we use $\log_2 x = \dfrac{\ln x}{\ln 2}$, $\log_4 x = \dfrac{\ln x}{\ln 4}$, etc. These graphs all approach $-\infty$ as $x \to 0^+$, and they all pass through the point $(1, 0)$. Also, they are all increasing, and all approach ∞ as $x \to \infty$. The smaller the base, the larger the rate of increase of the function (for $x > 1$) and the closer the approach to the y-axis $\left(\text{as } x \to 0^+\right)$.

13.

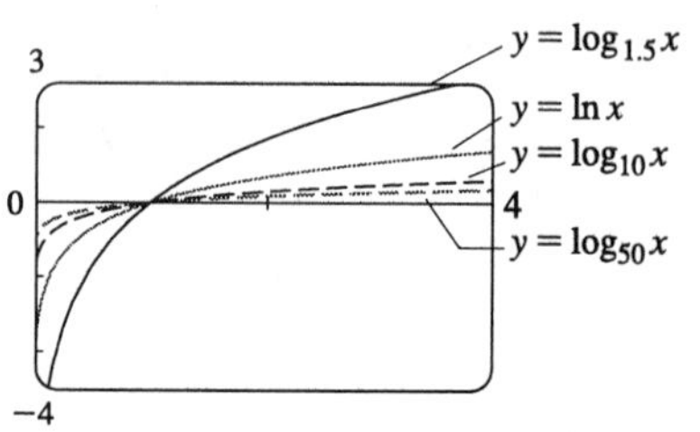

To graph these functions, we use $\log_{1.5} x = \dfrac{\ln x}{\ln 1.5}$ and $\log_{50} x = \dfrac{\ln x}{\ln 50}$. These graphs all approach $-\infty$ as $x \to 0^+$, and they all pass through the point $(1, 0)$. Also, they are all increasing, and all approach ∞ as $x \to \infty$. The functions with larger bases increase extremely slowly, and the ones with smaller bases do so somewhat more quickly. The functions with large bases approach the y-axis more closely as $x \to 0^+$.

14. 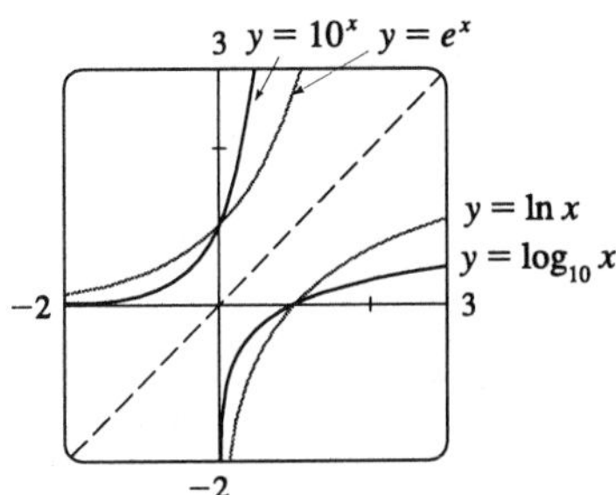

We see that the graph of $\ln x$ is the reflection of the graph of e^x about the line $y = x$, and that the graph of $\log_{10} x$ is the reflection of the graph of 10^x about the same line.
The graph of 10^x increases more quickly than that of e^x. Also note that as $x \to \infty$, $\log_{10} x \to \infty$ more slowly than $\ln x$.

15. **(a)** $2\text{ ft} = 24\text{ in}$, so $f(24) = 24^2\text{ in} = 576\text{ in} = 48\text{ ft}$, and $g(24) = 2^{24}\text{ in} = 2^{24}/(12 \cdot 5280)\text{mi} \approx 265\text{ mi}$.

(b) $3\text{ ft} = 36\text{ in}$, so we need x such that $\log_2 x = 36 \quad \Leftrightarrow \quad x = 2^{36} = 68{,}719{,}476{,}736$. In miles, this is
$68{,}719{,}476{,}736\text{ in} \cdot \dfrac{1\text{ mi}}{5280\text{ ft}} \cdot \dfrac{1\text{ ft}}{12\text{ in}} \approx 1{,}084{,}587.7\text{ mi}$.

16. 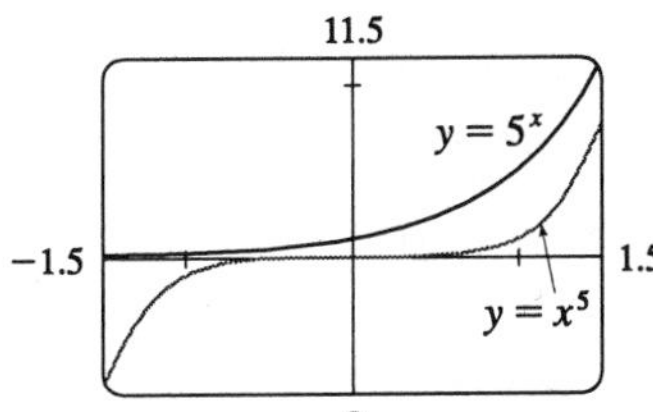

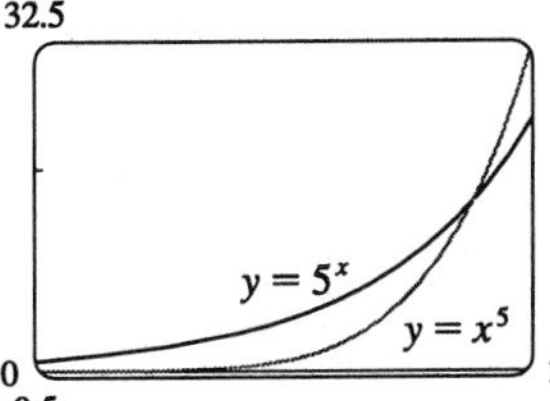

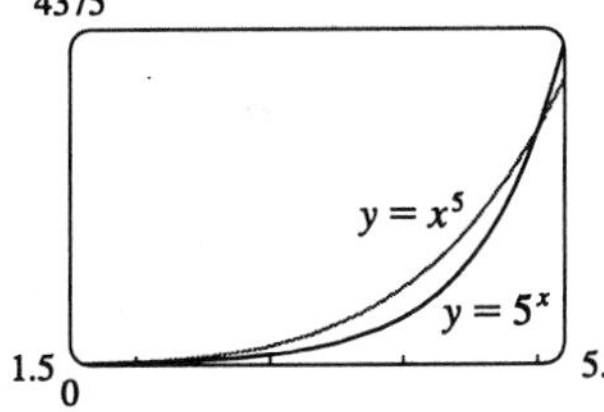

50,000
y = 5^x
y = x^5
0
7

We see from the graphs that for x less than about 1.8, $g(x) > f(x)$, and then near the point $(1.8, 17.1)$ the curves intersect. Then $f(x) > g(x)$ from $x \approx 1.8$ until $x = 5$. At $(5, 3125)$ there is another point of intersection, and for $x > 5$ we see that $g(x) > f(x)$. In fact, g increases much more rapidly than f beyond that point.

17. $\lim\limits_{x \to \pi/2^-} \log_{10}(\cos x) = -\infty$ since $\cos x \to 0^+$ as $x \to \frac{\pi}{2}^-$.

18. Since $(1.1)^x$ is continuous, we know that $\ln\left(\lim\limits_{x \to -\infty} (1.1)^x\right) = \lim\limits_{x \to -\infty} \ln((1.1)^x) = \lim\limits_{x \to -\infty} x \ln 1.1 = -\infty$.
Therefore $\lim\limits_{x \to -\infty} (1.1)^x = \lim\limits_{y \to -\infty} e^{\ln y} = 0$.

19. $h(t) = t^3 - 3^t \quad \Rightarrow \quad h'(t) = 3t^2 - 3^t \ln 3$

20. $g(x) = 1.6^x + x^{1.6} \quad \Rightarrow \quad g'(x) = 1.6^x \ln(1.6) + 1.6\, x^{0.6}$

21. $f(t) = \pi^{-t} \quad \Rightarrow \quad f'(t) = \pi^{-t}(\ln \pi)(-1) = -\pi^{-t} \ln \pi$

22. $G(x) = 5^{\tan x} \quad \Rightarrow \quad G'(x) = 5^{\tan x}(\ln 5)\sec^2 x$

23. $f(x) = \log_3(x^2 - 4) \quad \Rightarrow \quad f'(x) = \dfrac{2x}{(\ln 3)(x^2 - 4)}$

24. $f(x) = \sqrt{3 - 2^x} \quad \Rightarrow \quad f'(x) = \dfrac{1}{2\sqrt{3 - 2^x}}(-2^x \ln 2) = -\ln 2\, \dfrac{2^{x-1}}{\sqrt{3 - 2^x}}$

25. $f(t) = \log_2(t^4 - t^2 + 1) \quad \Rightarrow \quad f'(t) = \dfrac{4t^3 - 2t}{(\ln 2)(t^4 - t^2 + 1)}$

26. $f(x) = \log_{10}\left(\dfrac{x}{x-1}\right) = \log_{10} x - \log_{10}(x-1) \quad \Rightarrow \quad f'(x) = \dfrac{1}{x \ln 10} - \dfrac{1}{(x-1)\ln 10}$ or $-\dfrac{1}{x(x-1)\ln 10}$

27. $y = 2^{3^x} \quad \Rightarrow \quad y' = 2^{3^x}(\ln 2)3^x \ln 3 = (\ln 2)(\ln 3)3^x 2^{3^x}$

28. $y = x^x \quad \Rightarrow \quad \ln y = x \ln x \quad \Rightarrow \quad \dfrac{y'}{y} = \ln x + x\left(\dfrac{1}{x}\right) \quad \Rightarrow \quad y' = x^x(\ln x + 1)$

29. $y = x^{\sin x} \Rightarrow \ln y = \sin x \ln x \Rightarrow \dfrac{y'}{y} = \cos x \ln x + \dfrac{\sin x}{x} \Rightarrow y' = x^{\sin x}\left[\cos x \ln x + \dfrac{\sin x}{x}\right]$

30. $y = (\sin x)^x \Rightarrow \ln y = x \ln(\sin x) \Rightarrow \dfrac{y'}{y} = \ln(\sin x) + x\dfrac{\cos x}{\sin x} \Rightarrow y' = (\sin x)^x[\ln(\sin x) + x \cot x]$

31. $y = x^{e^x} \quad \Rightarrow \quad \ln y = e^x \ln x \quad \Rightarrow \quad \dfrac{y'}{y} = e^x \ln x + \dfrac{e^x}{x} \quad \Rightarrow \quad y' = x^{e^x} e^x \left(\ln x + \dfrac{1}{x}\right)$

32. $y = x^{1/x} \quad \Rightarrow \quad \ln y = \dfrac{1}{x} \ln x \quad \Rightarrow \quad \dfrac{y'}{y} = -\dfrac{1}{x^2} \ln x + \dfrac{1}{x}\left(\dfrac{1}{x}\right) \quad \Rightarrow \quad y' = x^{1/x}\dfrac{(1 - \ln x)}{x^2}$

33. $y = (\ln x)^x \quad \Rightarrow \quad \ln y = x \ln \ln x \quad \Rightarrow \quad \dfrac{y'}{y} = \ln \ln x + x \cdot \dfrac{1}{\ln x} \cdot \dfrac{1}{x} \quad \Rightarrow \quad y' = (\ln x)^x \left(\ln \ln x + \dfrac{1}{\ln x}\right)$

34. $y = x^{\ln x} \quad \Rightarrow \quad \ln y = \ln x \ln x = (\ln x)^2 \quad \Rightarrow \quad \dfrac{y'}{y} = 2 \ln x \left(\dfrac{1}{x}\right) \quad \Rightarrow \quad y' = x^{\ln x}\left(\dfrac{2 \ln x}{x}\right)$

35. $y = x^{1/\ln x} \quad \Rightarrow \quad \ln y = \left(\dfrac{1}{\ln x}\right)\ln x = 1 \quad \Rightarrow \quad y = e \quad \Rightarrow \quad y' = 0$

36. $y = (\sin x)^{\cos x} \quad \Rightarrow \quad \ln y = \cos x \ln(\sin x) \quad \Rightarrow \quad \dfrac{y'}{y} = -\sin x \ln \sin x + \cos x\left(\dfrac{\cos x}{\sin x}\right) \quad \Rightarrow$

$y' = (\sin x)^{\cos x}(-\sin x \ln \sin x + \cos x \cot x)$

37. $y = \cos\left(x^{\sqrt{x}}\right) \quad \Rightarrow \quad y' = -\sin\left(x^{\sqrt{x}}\right)x^{\sqrt{x}}\left(\dfrac{\ln x + 2}{2\sqrt{x}}\right)$ by Example ~~16~~ 4

38. $y = x^{x^x} \quad \Rightarrow \quad \ln y = x^x \ln x \quad \Rightarrow \quad \dfrac{y'}{y} = x^x(\ln x + 1)\ln x + x^x\left(\dfrac{1}{x}\right) \quad \left[\text{since } z = x^x \quad \Rightarrow \quad \ln z = x \ln x\right.$

$\left.\Rightarrow \quad \dfrac{z'}{z} = \ln x + x\left(\dfrac{1}{x}\right) \quad \Rightarrow \quad z' = x^x(\ln x + 1)\right] \quad \Rightarrow \quad y' = x^{x^x}\left[x^x(\ln x + 1)\ln x + x^{x-1}\right]$

39. $y = 10^x \quad \Rightarrow \quad y' = 10^x \ln 10$, so at $(1, 10)$, the slope of the tangent line is $10^1 \ln 10 = 10 \ln 10$, and its equation is $y - 10 = 10 \ln 10(x - 1)$, or $y = (10 \ln 10)x + 10(1 - \ln 10)$.

40. $f(x) = x^{\cos x} = e^{\ln x \cos x} \quad \Rightarrow$

$$f'(x) = e^{\ln x \cos x}\left[\ln x(-\sin x) + \cos x\left(\frac{1}{x}\right)\right] = x^{\cos x}\left[\frac{\cos x}{x} - \sin x \ln x\right]$$

This is reasonable, because the graph shows that f increases when $f'(x)$ is positive.

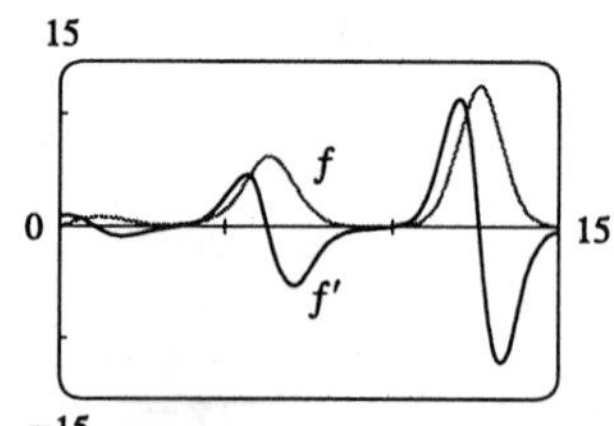

41. $\displaystyle\int_3^4 5^t\,dt = \left[\frac{5^t}{\ln 5}\right]_3^4 = \frac{5^4 - 5^3}{\ln 5} = \frac{500}{\ln 5}$

42. Let $u = \sqrt{x}$. Then $du = \dfrac{dx}{2\sqrt{x}}$ so $\displaystyle\int \frac{10^{\sqrt{x}}}{\sqrt{x}}\,dx = 2\int 10^u\,du = 2\frac{10^u}{\ln 10} + C = \frac{2}{\ln 10}10^{\sqrt{x}} + C$.

43. $\displaystyle\int \frac{\log_{10} x}{x}\,dx = \int \frac{(\ln x)/(\ln 10)}{x}\,dx = \frac{1}{\ln 10}\int \frac{\ln x}{x}\,dx$. Now put $u = \ln x$, so $du = \dfrac{1}{x}\,dx$, and the expression becomes $\displaystyle\frac{1}{\ln 10}\int u\,du = \frac{1}{\ln 10}\left(\tfrac{1}{2}u^2 + C_1\right) = \frac{1}{2\ln 10}(\ln x)^2 + C$.

Or: The substitution $u = \log_{10} x$ gives $du = \dfrac{dx}{x\ln 10}$ and we get $\displaystyle\int \frac{\log_{10} x}{x}\,dx = \tfrac{1}{2}\ln 10(\log_{10} x)^2 + C$.

44. $\displaystyle\int (x^\pi - \pi^x)dx = \int x^\pi\,dx - \int \pi^x\,dx = \frac{x^{\pi+1}}{\pi+1} - \frac{\pi^x}{\ln \pi} + C$

45.
$$\begin{aligned}
A &= \int_{-1}^{0}(2^x - 5^x)dx + \int_0^1 (5^x - 2^x)dx \\
&= \left[\frac{2^x}{\ln 2} - \frac{5^x}{\ln 5}\right]_{-1}^{0} + \left[\frac{5^x}{\ln 5} - \frac{2^x}{\ln 2}\right]_0^1 \\
&= \left(\frac{1}{\ln 2} - \frac{1}{\ln 5}\right) - \left(\frac{1/2}{\ln 2} - \frac{1/5}{\ln 5}\right) + \left(\frac{5}{\ln 5} - \frac{2}{\ln 2}\right) - \left(\frac{1}{\ln 5} - \frac{1}{\ln 2}\right) \\
&= \frac{16}{5\ln 5} - \frac{1}{2\ln 2}
\end{aligned}$$

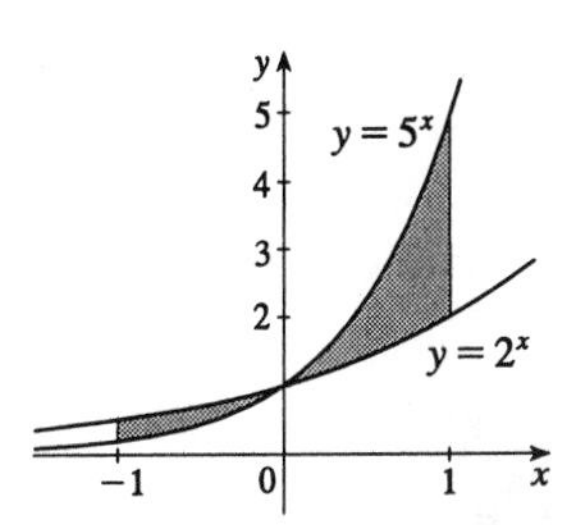

46. We use the formula for disks (Equation 5.2.2). The volume is $V = \int_0^1 \pi[10^{-x}]^2\,dx = \pi\int_0^1 10^{-2x}\,dx$. To evaluate the integral, we let $u = -2x \Rightarrow du = -2\,dx$, $x = 0 \Rightarrow u = 0$, and $x = 1 \Rightarrow u = -2$, so we have $\displaystyle V = -\frac{\pi}{2}\int_0^{-2} 10^u\,du = -\frac{\pi}{2}\left[\frac{1}{\ln 10}10^u\right]_0^{-2} = -\frac{\pi}{2\ln 10}\left(10^{-2} - 1\right) = \frac{99\pi}{200\ln 10}$.

47. We see that the graphs of $y = 2^x$ and $y = 1 + 3^{-x}$ intersect at $x \approx 0.6$. We let $f(x) = 2^x - 1 - 3^{-x}$ and calculate $f'(x) = 2^x \ln 2 + 3^{-x}\ln 3$, and using the formula $x_{n+1} = x_n - f(x_n)/f'(x_n)$ (Newton's Method), we get $x_1 = 0.6$, $x_2 \approx x_3 \approx 0.600967$. So, correct to six decimal places, the root occurs at $x = 0.600967$.

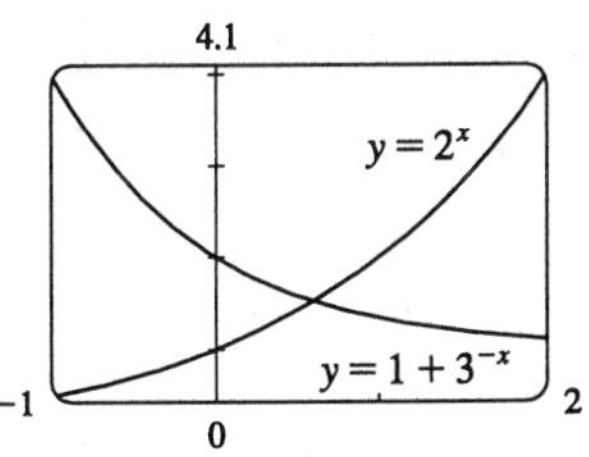

48. $x^y = y^x \Rightarrow y\ln x = x\ln y \Rightarrow y'\ln x + y/x = \ln y + xy'/y \Rightarrow y' = \dfrac{\ln y - y/x}{\ln x - x/y}$

49. $y = \dfrac{10^x}{10^x + 1} \Leftrightarrow (10^x + 1)y = 10^x \Leftrightarrow y = 10^x(1 - y) \Leftrightarrow 10^x = \dfrac{y}{1 - y} \Leftrightarrow$
$\log_{10} 10^x = \log_{10}\left(\dfrac{y}{1-y}\right) \Leftrightarrow x = \log_{10} y - \log_{10}(1 - y)$. Interchange x and y:
$y = \log_{10} x - \log_{10}(1 - x)$ is the inverse function.

50. $\displaystyle\lim_{x\to 0^+} x^{-\ln x} = \lim_{x\to 0^+}\left(e^{\ln x}\right)^{-\ln x} = \lim_{x\to 0^+} e^{-(\ln x)^2} = 0$ since $-(\ln x)^2 \to -\infty$ as $x \to 0^+$.

51. If I is the intensity of the 1989 San Francisco earthquake, then $\log_{10}\dfrac{I}{S} = 7.1 \quad \Rightarrow$

$\log_{10}\dfrac{16I}{S} = \log_{10}16 + \log_{10}\dfrac{I}{S} = \log_{10}16 + 7.1 \approx 8.3.$

52. Let I_1 and I_2 be the intensities of the music and the mower. Then $10\log_{10}\left(\dfrac{I_1}{I_0}\right) = 120$ and

$10\log_{10}\left(\dfrac{I_2}{I_0}\right) = 106$, so $\log_{10}\left(\dfrac{I_1}{I_2}\right) = \log_{10}\left(\dfrac{I_1/I_0}{I_2/I_0}\right) = \log_{10}\left(\dfrac{I_1}{I_0}\right) - \log_{10}\left(\dfrac{I_2}{I_0}\right) = 12 - 10.6 = 1.4 \quad \Rightarrow$

$\dfrac{I_1}{I_2} = 10^{1.4} \approx 25.$

53. We find I with the loudness formula from Exercise 52, substituting $I_0 = 10^{-12}$ and $L = 50$: $50 = 10\log_{10}\dfrac{I}{10^{-12}}$

$\Leftrightarrow \quad 5 = \log_{10}\dfrac{I}{10^{-12}} \quad \Leftrightarrow \quad 10^5 = \dfrac{I}{10^{-12}} \quad \Leftrightarrow \quad I = 10^{-7}\text{ watt}/\text{m}^2$. Now we differentiate L with respect

to I: $L = 10\log_{10}\dfrac{I}{I_0} \quad \Rightarrow \quad \dfrac{dL}{dI} = 10\dfrac{1}{(I/I_0)\ln 10}\left(\dfrac{1}{I_0}\right) = \dfrac{10}{\ln 10}\left(\dfrac{1}{I}\right)$. Substituting $I = 10^{-7}$, we get

$L'(50) = \dfrac{10}{\ln 10}\left(\dfrac{1}{10^{-7}}\right) = \dfrac{10^8}{\ln 10} \approx 4.34 \times 10^7 \dfrac{\text{dB}}{\text{watt}/\text{m}^2}.$

54. **(a)** $I(x) = I_0 a^x \quad \Rightarrow \quad I'(x) = I_0(\ln a)a^x = (I_0 a^x)\ln a = I(x)\ln a$

(b) We substitute $I_0 = 8$, $a = 0.38$ and $x = 20$ into the first expression for $I'(x)$ above:

$I'(20) = 8(\ln 0.38)(0.38)^{20} \approx -3.05 \times 10^{-8}$

(c) The average value of the function $I(x)$ between $x = 0$ and $x = 20$ is

$$\frac{\int_0^{20} I(x)dx}{20 - 0} = \frac{1}{20}\int_0^{20} 8(0.38)^x\,dx = \frac{2}{5}\left[\frac{(0.38)^x}{\ln 0.38}\right]_0^{20} = \frac{2(0.38^{20} - 1)}{5\ln 0.38} \approx 0.41.$$

55. Using Definition 1 and the second law of exponents for e^x, we have

$a^{x-y} = e^{(x-y)\ln a} = e^{x\ln a - y\ln a} = \dfrac{e^{x\ln a}}{e^{y\ln a}} = \dfrac{a^x}{a^y}.$

56. Using Definition 1, the first law of logarithms, and the first law of exponents for e^x, we have

$(ab)^x = e^{x\ln(ab)} = e^{x(\ln a + \ln b)} = e^{x\ln a + x\ln b} = e^{x\ln a}e^{x\ln b} = a^x b^x.$

57. Let $\log_a x = r$ and $\log_a y = s$. Then $a^r = x$ and $a^s = y$.

(a) $xy = a^r a^s = a^{r+s} \quad \Rightarrow \quad \log_a(xy) = r + s = \log_a x + \log_a y$

(b) $\dfrac{x}{y} = \dfrac{a^r}{a^s} = a^{r-s} \quad \Rightarrow \quad \log_a\dfrac{x}{y} = r - s = \log_a x - \log_a y$

(c) $x^y = (a^r)^y = a^{ry} \quad \Rightarrow \quad \log_a(x^y) = ry = y\log_a x$

58. Let $m = \dfrac{n}{x}$. Then $n = xm$, and as $n \to \infty$, $m \to \infty$. Therefore,

$\lim\limits_{n\to\infty}\left(1 + \dfrac{x}{n}\right)^n = \lim\limits_{m\to\infty}\left(1 + \dfrac{1}{m}\right)^{mx} = \left[\lim\limits_{m\to\infty}\left(1 + \dfrac{1}{m}\right)^m\right]^x = e^x$ by Equation 8.

EXERCISES 6.5

1. **(a)** By Theorem 2, $y(t) = y(0)e^{kt} = 100e^{kt} \Rightarrow y(\frac{1}{3}) = 100e^{k/3} = 200 \Rightarrow$ $k/3 = \ln(200/100) = \ln 2 \Rightarrow k = 3\ln 2$. So $y(t) = 100e^{(3\ln 2)t} = 100 \cdot 2^{3t}$.

(b) $y(10) = 100 \cdot 2^{30} \approx 1.07 \times 10^{11}$ cells

(c) $y(t) = 100 \cdot 2^{3t} = 10{,}000 \Rightarrow 2^{3t} = 100 \Rightarrow 3t\ln 2 = \ln 100 \Rightarrow t = (\ln 100)/(3\ln 2) \approx 2.2\text{ h}$

2. **(a)** By Theorem 2, $y(t) = y(0)e^{kt} = 4000\,e^{kt} \Rightarrow y(\frac{1}{2}) = 4000e^{k/2} = 12{,}000 \Rightarrow e^{k/2} = 3 \Rightarrow$ $k/2 = \ln 3 \Rightarrow k = 2\ln 3$, so $y(t) = 4000e^{(2\ln 3)t} = 4000 \cdot 9^t$.

(b) $y(\frac{1}{3}) = 4000 \cdot 9^{1/3} \approx 8320$

(c) $4000 \cdot 9^t = 20{,}000 \Rightarrow 9^t = 5 \Rightarrow t\ln 9 = \ln 5 \Rightarrow t = (\ln 5)/(\ln 9) \approx 0.73\text{ h} \approx 44\text{ min}$

3. **(a)** $y(t) = y(0)e^{kt} = 500\,e^{kt} \Rightarrow y(3) = 500e^{3k} = 8000 \Rightarrow e^{3k} = 16 \Rightarrow 3k = \ln 16 \Rightarrow$ $y(t) = 500e^{(\ln 16)t/3} = 500 \cdot 16^{t/3}$

(b) $y(4) = 500 \cdot 16^{4/3} \approx 20{,}159$

(c) $y(t) = 500 \cdot 16^{t/3} = 30{,}000 \Rightarrow 16^{t/3} = 60 \Rightarrow \frac{1}{3}t\ln 16 = \ln 60 \Rightarrow$ $t = 3(\ln 60)/(\ln 16) \approx 4.4\text{ h}$

4. **(a)** $y(t) = y(0)e^{kt} \Rightarrow y(2) = y(0)e^{2k} = 400$, $y(6) = y(0)e^{6k} = 25{,}600$. Dividing these equations, we get $e^{6k}/e^{2k} = 25{,}600/400 \Rightarrow e^{4k} = 64 \Rightarrow 4k = \ln 64 = 6\ln 2 \Rightarrow k = \frac{3}{2}\ln 2 = \frac{1}{2}\ln 8$. Thus $y(0) = 400/e^{2k} = 400/e^{\ln 8} = \frac{400}{8} = 50$.

(b) $y(t) = y(0)\,e^{kt} = 50\,e^{(\ln 8)t/2}$

(c) $y(t) = 50\,e^{(3\ln 2)t/2} = 100 \Leftrightarrow e^{(3\ln 2)t/2} = 2 \Leftrightarrow (3\ln 2)t/2 = \ln 2 \Leftrightarrow t = 2/3\text{ h} = 40\text{ min}$

(d) $50\,e^{(\ln 8)t/2} = 100{,}000 \Leftrightarrow e^{(\ln 8)t/2} = 2000 \Leftrightarrow (\ln 8)t/2 = \ln 2000 \Leftrightarrow$ $t = (2\ln 2000)/\ln 8 \approx 7.3\text{ h}$.

5. **(a)** Let the population (in millions) in the year t be $P(t)$. Since the initial time is the year 1750, we substitute $t - 1750$ for t in Theorem 2, so the exponential model gives $P(t) = P(1750)e^{k(t-1750)}$. Then $P(1800) = 906 = 728e^{k(1800-1750)} \Rightarrow \ln\frac{906}{728} = k(50) \Rightarrow k = \frac{1}{50}\ln\frac{906}{728} \approx 0.0043748$. So with this model, we estimate $P(1900) \approx P(1750)e^{k(1900-1750)} \approx 728e^{150(0.0043748)} \approx 1403$ million, and $P(1950) \approx 728e^{200(0.0043748)} \approx 1746$ million. Both of these estimates are much too low.

(b) In this case, the exponential model gives $P(t) = P(1850)e^{k(t-1850)} \Rightarrow$ $P(1900) = 1608 = 1171e^{k(1900-1850)} \Rightarrow \ln\frac{1608}{1171} = k(50) \Rightarrow k = \frac{1}{50}\ln\frac{1608}{1171} \approx 0.006343$. So with this model, we estimate $P(1950) \approx 1171e^{100(0.006343)} \approx 2208$ million. This is still too low, but closer than the estimate of $P(1950)$ in part (a).

(c) The exponential model gives $P(t) = P(1900)e^{k(t-1900)} \quad \Rightarrow \quad P(1950) = 2517 = 1608e^{k(1950-1900)}$ $\Rightarrow \quad \ln \frac{2517}{1608} = k(50) \quad \Rightarrow \quad k = \frac{1}{50} \ln \frac{2517}{1608} \approx 0.008962$. With this model, we estimate $P(1992) \approx 1608e^{0.008962(1992-1900)} \approx 3667$ million. This is much too low.

The discrepancy is explained by the fact that the world birth rate (average yearly number of births per person) is about the same as always, whereas the mortality rate (especially the infant mortality rate) is much lower, owing mostly to advances in medical science and to the wars in the first part of the twentieth century. The exponential model assumes, among other things, that the birth and mortality rates will remain constant.

6. **(a)** Let $P(t)$ be the population (in millions) in the year t. Since the initial time is the year 1990, we substitute $t - 1900$ for t in Theorem 2, and find that the exponential model gives $P(t) = P(1900)e^{k(t-1900)} \quad \Rightarrow$ $P(1910) = 92 = 76e^{k(1910-1900)} \quad \Rightarrow \quad k = \frac{1}{10} \ln \frac{92}{76} \approx 0.0191$. With this model, we estimate $P(1990) \approx 76e^{0.0191(1990-1900)} \approx 424$ million. This estimate is much too high.

The discrepancy is explained by the fact that, between the years 1900 and 1910, an enormous number of immigrants (compared to the total population) came to the United States. Since that time, immigration (as a proportion of total population) has been much lower. Also, the birth rate in the United States has declined since the turn of the century. So our calculation of the constant k was based partly on factors which no longer exist.

(b) Substituting $t - 1970$ for t in Theorem 2, we find that the exponential model gives $P(t) = P(1970)e^{k(t-1970)} \quad \Rightarrow \quad P(1980) = 227 = 203e^{k(1990-1980)} \quad \Rightarrow \quad k = \frac{1}{10} \ln \frac{227}{203} \approx 0.01117$. With this model, we estimate $P(1990) \approx 203e^{0.01117(1990-1970)} \approx 254$ million. This is quite accurate. The further estimates are $P(2000) \approx 203e^{0.01117(30)} \approx 284$ million and $P(2010) \approx 203e^{0.01117(40)} \approx 317$ million.

(c)

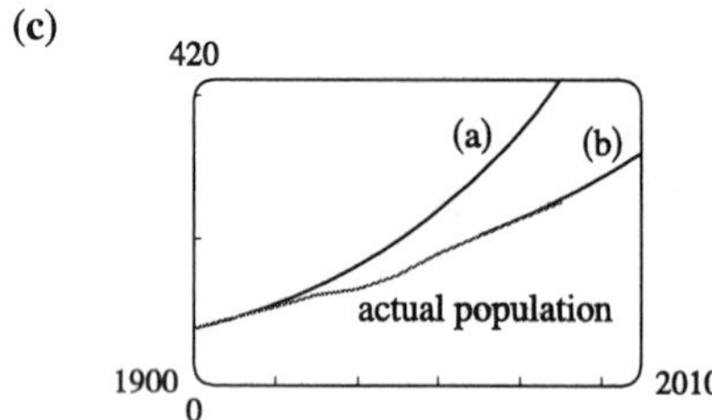

The model in part (a) is quite inaccurate after 1910 (off by 5 million in 1920 and 28 million in 1930). The model in part (b) is more accurate (which is unsurprising, since it is based on more recent information).

7. **(a)** If $y = [N_2O_5]$ then $\dfrac{dy}{dt} = -0.0005y \quad \Rightarrow \quad y(t) = y(0)e^{-0.0005t} = Ce^{-0.0005t}$.

(b) $y(t) = Ce^{-0.0005t} = 0.9C \quad \Rightarrow \quad e^{-0.0005t} = 0.9 \quad \Rightarrow \quad -0.0005t = \ln 0.9 \quad \Rightarrow$ $t = -2000 \ln 0.9 \approx 211$ s

8. **(a)** The mass remaining after t days is $y(t) = y(0)e^{kt} = 200e^{kt}$. Since the half-life is 140 days,
$y(140) = 200\,e^{140k} = 100 \quad \Rightarrow \quad e^{140k} = \frac{1}{2} \quad \Rightarrow \quad 140k = \ln\frac{1}{2} \quad \Rightarrow \quad k = -(\ln 2)/140$, so
$y(t) = 200e^{-(\ln 2)t/140} = 200 \cdot 2^{-t/140}$.

(b) $y(100) = 200 \cdot 2^{-100/140} \approx 121.9$ mg

(c) $200e^{-(\ln 2)t/140} = 10 \quad \Leftrightarrow$

$$-\ln 2\frac{t}{140} = \ln\frac{1}{20} = -\ln 20 \quad \Leftrightarrow$$

$$t = 140\frac{\ln 20}{\ln 2} \approx 605 \text{ days}$$

(d)

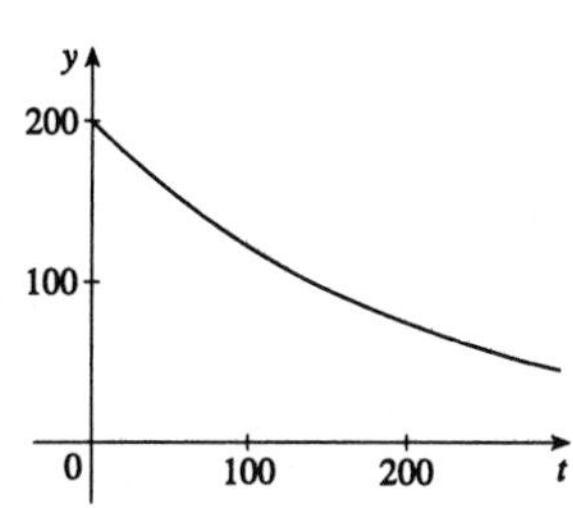

9. **(a)** If $y(t)$ is the mass remaining after t days, then $y(t) = y(0)e^{kt} = 50e^{kt} \quad \Rightarrow$
$y(0.00014) = 50e^{0.00014k} = 25 \quad \Rightarrow \quad e^{0.00014k} = \frac{1}{2} \quad \Rightarrow \quad k = -(\ln 2)/0.00014 \quad \Rightarrow \quad y(t) = 50$
$e^{-(\ln 2)t/0.00014} = 50 \cdot 2^{-t/0.00014}$

(b) $y(0.01) = 50 \cdot 2^{-0.01/0.00014} \approx 1.57 \times 10^{-20}$ mg

(c) $50e^{-(\ln 2)t/0.00014} = 40 \quad \Rightarrow \quad -(\ln 2)t/0.00014 = \ln 0.8 \quad \Rightarrow \quad t = -0.00014\dfrac{\ln 0.8}{\ln 2} \approx 4.5 \times 10^{-5}$ s

10. **(a)** If $y(t)$ is the mass after t days and $y(0) = A$, then $y(t) = Ae^{kt} \quad \Rightarrow \quad y(3) = Ae^{3k} = 0.58A \quad \Rightarrow$
$e^{3k} = 0.58 \quad \Rightarrow \quad k = \frac{1}{3}\ln 0.58$. Then
$Ae^{\ln(0.58)t/3} = \dfrac{A}{2} \quad \Leftrightarrow \quad \dfrac{\ln(0.58)t}{3} = \ln\frac{1}{2}$, so the half-life is $t = -\dfrac{3\ln 2}{\ln 0.58} \approx 3.82$ days.

(b) $Ae^{\ln(0.58)t/3} = \dfrac{A}{10} \quad \Rightarrow \quad \dfrac{\ln(0.58)t}{3} = \ln\dfrac{1}{10} \quad \Leftrightarrow \quad t = -3\dfrac{\ln 10}{\ln 0.58} \approx 12.68$ days

11. Let $y(t)$ be the level of radioactivity. Thus, $y(t) = y(0)e^{-kt}$ and k is determined by using the half-life:
$\frac{1}{2} = e^{-5730k} \quad \Rightarrow \quad k = -\dfrac{\ln\frac{1}{2}}{5730} = \dfrac{\ln 2}{5730}$. If 0.74 of the ^{14}C remains, then we know that $0.74 = e^{-t(\ln 2)/5730} \quad \Rightarrow$
$\ln 0.74 = -\dfrac{t\ln 2}{5730} \quad \Rightarrow \quad t = -\dfrac{5730(\ln 0.74)}{\ln 2} \approx 2489 \approx 2500$ years.

12. From the information given, we know that $\dfrac{dy}{dx} = 2y \quad \Rightarrow \quad y = Ce^{2x}$ by Theorem 2. To calculate C we use the point $(0, 5)$: $5 = Ce^{2(0)} \quad \Rightarrow \quad C = 5$. Thus, the equation of the curve is $y = 5e^{2x}$.

13. Let $y(t)$ = temperature after t minutes. Then $\dfrac{dy}{dt} = -\frac{1}{10}[y(t) - 21]$. If $u(t) = y(t) - 21$, then $\dfrac{du}{dt} = -\dfrac{u}{10}$
$\Rightarrow \quad u(t) = u(0)\,e^{-t/10} = 12\,e^{-t/10} \quad \Rightarrow \quad y(t) = 21 + u(t) = 21 + 12\,e^{-t/10}$.

14. **(a)** Let $y(t)$ = temperature after t minutes. Newton's Law of Cooling implies that $\dfrac{dy}{dt} = k(y - 5)$. Let
$u(t) = y(t) - 5$. Then $\dfrac{du}{dt} = ku$, so $u(t) = u(0)e^{kt} = 15e^{kt} \quad \Rightarrow \quad y(t) = 5 + 15e^{kt} \quad \Rightarrow$
$y(1) = 5 + 15e^{k} = 12 \quad \Rightarrow \quad e^{k} = \frac{7}{15} \quad \Rightarrow \quad k = \ln\frac{7}{15}$, so $y(t) = 5 + 15e^{\ln(7/15)t}$ and
$y(2) = 5 + 15e^{2\ln(7/15)} \approx 8.3°$.

(b) $5 + 15e^{\ln(7/15)t} = 6$ when $e^{\ln(7/15)t} = \frac{1}{15} \quad \Rightarrow \quad \ln\left(\frac{7}{15}\right)t = \ln\frac{1}{15} = -\ln 15 \quad \Rightarrow \quad t = \dfrac{-\ln 15}{\ln(7/15)} \approx 3.6$ min.

15. **(a)** Let $y(t) =$ temperature after t minutes. Newton's Law of Cooling implies that $\frac{dy}{dt} = k(y - 75)$. Let $u(t) = y(t) - 75$. Then $\frac{du}{dt} = ku$, so $u(t) = u(0)e^{kt} = 110e^{kt} \Rightarrow y(t) = 75 + 110e^{kt} \Rightarrow$ $y(30) = 75 + 110e^{30k} = 150 \Rightarrow e^{30k} = \frac{75}{110} = \frac{15}{22} \Rightarrow k = \frac{1}{30}\ln\frac{15}{22}$, so $y(t) = 75 + 110e^{\frac{1}{30}t\ln\left(\frac{15}{22}\right)}$ and $y(45) = 75 + 110e^{\frac{45}{30}\ln\left(\frac{15}{22}\right)} \approx 137\,°\text{F}$.

(b) $y(t) = 75 + 110e^{\frac{1}{30}t\ln\left(\frac{15}{22}\right)} = 100 \Rightarrow e^{\frac{1}{30}t\ln\left(\frac{15}{22}\right)} = \frac{25}{110} \Rightarrow \frac{1}{30}t\ln\frac{15}{22} = \ln\frac{25}{110} \Rightarrow t = \dfrac{30\ln\frac{25}{110}}{\ln\frac{15}{22}} \approx 116\text{ min}$

16. **(a)** Let $y(t) =$ temperature after t minutes. Newton's Law of Cooling implies that $\frac{dy}{dt} = k(y - a)$ where a is the surrounding temperature. Let $u(t) = y(t) - a$. Then $\frac{du}{dt} = ku$, so $u(t) = u(0)e^{kt} = (21 - a)e^{kt}$ and so $y(t) = a + (21 - a)e^{kt}$. Using the knowledge of $y(t)$ at $t = 1$ and $t = 2$ we have:
$27 = y(1) = a + (21 - a)e^{k}$ and $30 = y(2) = a + (21 - a)e^{2k}$. Rearranged, these become
$27 - a = (21 - a)e^{k}$ and $30 - a = (21 - a)e^{2k}$.
To determine a we must eliminate k. To do so, we divide the square of the first equation by the second and get:

$$\frac{(27 - a)^2}{30 - a} = \frac{(21 - a)^2 e^{2k}}{(21 - a)e^{2k}} \quad \Rightarrow \quad \frac{(27 - a)^2}{30 - a} = 21 - a \quad \Rightarrow$$

$(27 - a)^2 = (30 - a)(21 - a)$
$\Leftrightarrow \quad 729 - 54a + a^2 = 630 - 51a + a^2 \quad \Rightarrow \quad -3a = -99$
$\Rightarrow \quad a = 33$. Hence, the outdoor temperature is $33°$ C.

(b)

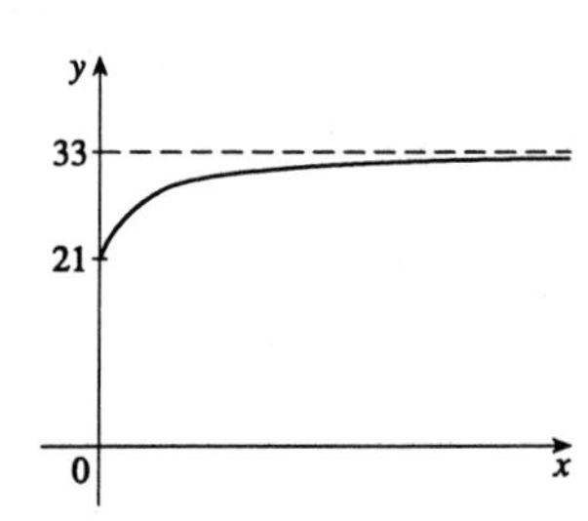

17. **(a)** Let $P(h)$ be the pressure at altitude h. Then $dP/dh = kP \Rightarrow P(h) = P(0)e^{kh} = 101.3e^{kh} \Rightarrow$ $P(1000) = 101.3e^{1000k} = 87.14 \Rightarrow 1000k = \ln\left(\frac{87.14}{101.3}\right) \Rightarrow P(h) = 101.3\,e^{\frac{1}{1000}h\ln\left(\frac{87.14}{101.3}\right)}$, so $P(3000) = 101.3e^{3\ln\left(\frac{87.14}{101.3}\right)} \approx 64.5\text{ kPa}$.

(b) $P(6187) = 101.3\,e^{\frac{6187}{1000}\ln\left(\frac{87.14}{101.3}\right)} \approx 39.9\text{ kPa}$

18. With the notation of Example 4, $A_0 = 500$, $i = 0.14$, and $t = 2$.

(a) $n = 1$: $A = 500(1.14)^2 = \$649.80$ **(b)** $n = 4$: $A = 500\left(1 + \frac{0.14}{4}\right)^8 = \$658.40$

(c) $n = 12$: $A = 500\left(1 + \frac{0.14}{12}\right)^{24} = \$660.49$ **(d)** $n = 365$: $A = 500\left(1 + \frac{0.14}{365}\right)^{2\cdot365} = \$661.53$

(e) $n = 365 \cdot 24$: $A = 500\left(1 + \frac{0.14}{365\cdot24}\right)^{2\cdot365\cdot24} = \$661.56$

(f) continuously: $A = 500e^{(0.14)2} = \$661.56$

19. With the notation of Example 4, $A_0 = 3000$, $i = 0.05$, and $t = 5$.

(a) $n = 1$: $A = 3000(1.05)^5 = \$3828.84$ **(b)** $n = 2$: $A = 3000\left(1 + \frac{0.05}{2}\right)^{10} = \$3840.25$

(c) $n = 12$: $A = 3000\left(1 + \frac{0.05}{12}\right)^{60} = \$3850.08$ **(d)** $n = 52$: $A = 3000\left(1 + \frac{0.05}{52}\right)^{5\cdot52} = \$3851.61$

(e) $n = 365$: $A = 3000\left(1 + \frac{0.05}{365}\right)^{5\cdot365} = \$3852.01$ **(f)** continuously: $A = 3000e^{(0.05)5} = \$3852.08$

20. $A_0 e^{0.06t} = 2A_0 \quad \Leftrightarrow \quad e^{0.06t} = 2 \quad \Leftrightarrow \quad 0.06t = \ln 2 \quad \Leftrightarrow \quad t = \frac{50}{3}\ln 2 \approx 11.55$, so the investment will double in about 11.55 years.

21. **(a)** If $y(t)$ is the amount of salt at time t, then $y(0) = 1500(0.3) = 450\,\text{kg}$. The rate of change of y is $\frac{dy}{dt} = -\left(\frac{y(t)}{1500}\frac{\text{kg}}{\text{L}}\right)\left(20\,\frac{\text{L}}{\text{min}}\right) = -\frac{1}{75}y(t)\,\frac{\text{kg}}{\text{min}}$, so $y(t) = y(0)e^{-t/75} = 450e^{-t/75} \quad \Rightarrow$ $y(30) = 450e^{-0.4} \approx 301.6\,\text{kg}$.

(b) When the concentration is 0.2 kg/L, the amount of salt is $1500(0.2) = 300\,\text{kg}$. So $y(t) = 450e^{-t/75} = 300$ $\Rightarrow \quad e^{-t/75} = \frac{2}{3} \quad \Rightarrow \quad -t/75 = \ln\frac{2}{3} \quad \Rightarrow \quad t = -75\ln\frac{2}{3} \approx 30.41$ min.

22. **(a)** If $y(t)$ is the amount of salt at time t, then $y(0) = 1500(0.3) = 450$ kg. The rate of change of y is $\frac{dy}{dt} = \left(0.1\,\frac{\text{kg}}{\text{L}}\right)\left(20\,\frac{\text{L}}{\text{min}}\right) - \left(\frac{y(t)}{1500}\frac{\text{kg}}{\text{L}}\right)\left(20\,\frac{\text{L}}{\text{min}}\right) = 2 - \frac{1}{75}\,y(t) = -\frac{1}{75}[y(t) - 150]$. Let $u(t) = y(t) - 150$. Then $du/dt = -\frac{1}{75}u(t) \quad \Rightarrow \quad u(t) = u(0)\,e^{-t/75} = 300e^{-t/75} \quad \Rightarrow \quad 150 + 300$ $e^{-t/75}$, so $y(30) = 150 + 300e^{-0.4} \approx 351$ kg.

(b) When the concentration is 0.2 kg/L, the amount of salt is $1500(0.2) = 300$ kg. So $y(t) = 150 + 300$ $e^{-t/75} = 300 \quad \Rightarrow \quad 300\,e^{-t/75} = 150 \quad \Rightarrow \quad e^{-t/75} = \frac{1}{2} \quad \Rightarrow \quad -t/75 = -\ln 2 \quad \Rightarrow$ $t = 75\ln 2 \approx 52$ min.

EXERCISES 6.6

1. $\cos^{-1}(-1) = \pi$ since $\cos\pi = -1$.

2. $\sin^{-1}(0.5) = \frac{\pi}{6}$ since $\sin\frac{\pi}{6} = 0.5$.

3. $\tan^{-1}\sqrt{3} = \frac{\pi}{3}$ since $\tan\frac{\pi}{3} = \sqrt{3}$.

4. $\arctan(-1) = -\frac{\pi}{4}$ since $\tan\left(-\frac{\pi}{4}\right) = -1$.

5. $\csc^{-1}\sqrt{2} = \frac{\pi}{4}$ since $\csc\frac{\pi}{4} = \sqrt{2}$.

6. $\arcsin 1 = \frac{\pi}{2}$ since $\sin\frac{\pi}{2} = 1$.

7. $\cot^{-1}\left(-\sqrt{3}\right) = \frac{5\pi}{6}$ since $\cot\frac{5\pi}{6} = -\sqrt{3}$.

8. $\sec^{-1}2 = \frac{\pi}{3}$ since $\sec\frac{\pi}{3} = 2$.

9. $\sin(\sin^{-1}0.7) = 0.7$

10. $\sin^{-1}(\sin 1) = 1$ since $-\frac{\pi}{2} \le 1 \le \frac{\pi}{2}$.

11. $\tan^{-1}\left(\tan\frac{4\pi}{3}\right) = \tan^{-1}\sqrt{3} = \frac{\pi}{3}$

12. $\tan(\cos^{-1}0.5) = \tan\frac{\pi}{3} = \sqrt{3}$

13. Let $\theta = \cos^{-1}\frac{4}{5}$, so $\cos\theta = \frac{4}{5}$. Then $\sin\left(\cos^{-1}\frac{4}{5}\right) = \sin\theta = \sqrt{1 - \left(\frac{4}{5}\right)^2} = \sqrt{\frac{9}{25}} = \frac{3}{5}$.

14. Let $\theta = \arctan 2$, so $\tan\theta = 2 \quad \Rightarrow \quad \sec^2\theta = 1 + \tan^2\theta = 1 + 4 = 5 \quad \Rightarrow \quad \sec\theta = \sqrt{5} \quad \Rightarrow$ $\sec(\arctan 2) = \sec\theta = \sqrt{5}$.

15. $\arcsin\left(\sin\frac{5\pi}{4}\right) = \arcsin\left(-\frac{1}{\sqrt{2}}\right) = -\frac{\pi}{4}$

16. Let $\theta = \sin^{-1}\frac{3}{5}$. Then $\sin\theta = \frac{3}{5} \quad \Rightarrow \quad \cos\theta = \sqrt{1 - \left(\frac{3}{5}\right)^2} = \frac{4}{5}$, so $\sin\left(2\sin^{-1}\frac{3}{5}\right) = \sin 2\theta = 2\sin\theta\cos\theta = 2\cdot\frac{3}{5}\cdot\frac{4}{5} = \frac{24}{25}$.

17. Let $\theta = \sin^{-1}\frac{5}{13}$. Then $\sin\theta = \frac{5}{13}$, so $\cos\left(2\sin^{-1}\frac{5}{13}\right) = \cos 2\theta = 1 - 2\sin^2\theta = 1 - 2\left(\frac{5}{13}\right)^2 = \frac{119}{169}$.

18. Let $x = \sin^{-1}\frac{1}{3}$ and $y = \sin^{-1}\frac{2}{3}$. Then $\sin x = \frac{1}{3}$, $\cos x = \sqrt{1 - \left(\frac{1}{3}\right)^2} = \frac{2\sqrt{2}}{3}$, $\sin y = \frac{2}{3}$,
$\cos y = \sqrt{1 - \left(\frac{2}{3}\right)^2} = \frac{\sqrt{5}}{3}$, so
$\sin\left(\sin^{-1}\frac{1}{3} + \sin^{-1}\frac{2}{3}\right) = \sin(x+y) = \sin x\cos y + \cos x\sin y = \frac{1}{3}\left(\frac{\sqrt{5}}{3}\right) + \frac{2\sqrt{2}}{3}\left(\frac{2}{3}\right) = \frac{1}{9}\left(\sqrt{5} + 4\sqrt{2}\right)$

19. Let $y = \sin^{-1}x$. Then $-\frac{\pi}{2} \le y \le \frac{\pi}{2} \Rightarrow \cos y \ge 0$, so $\cos(\sin^{-1}x) = \cos y = \sqrt{1 - \sin^2 y} = \sqrt{1 - x^2}$

20. Let $y = \sin^{-1}x$. Then $\sin y = x$, so from the triangle we see that $\tan\left(\sin^{-1}x\right) = \tan y = \dfrac{x}{\sqrt{1-x^2}}$.

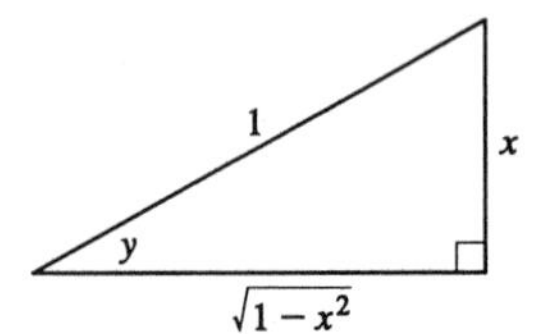

21. Let $y = \tan^{-1}x$. Then $\tan y = x$, so from the triangle we see that $\sin\left(\tan^{-1}x\right) = \sin y = \dfrac{x}{\sqrt{1+x^2}}$.

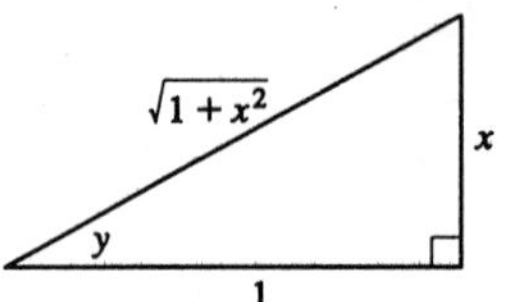

22. Let $y = \cos^{-1}x$. Then $\cos y = x \Rightarrow \sin y = \sqrt{1-x^2}$ since $0 \le y \le \pi$. So
$\sin(2\cos^{-1}x) = \sin 2y = 2\sin y\cos y = 2x\sqrt{1-x^2}$.

23.

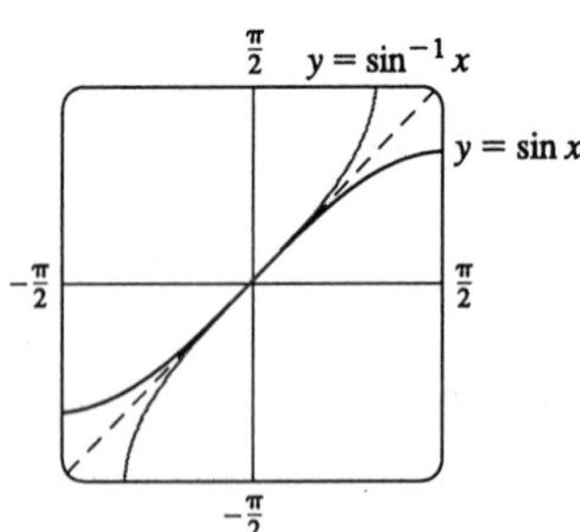

The graph of $\sin^{-1}x$ is the reflection of the graph of $\sin x$ about the line $y = x$.

24.

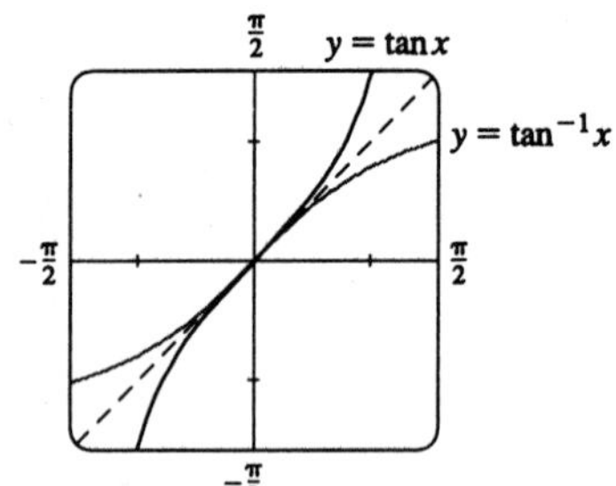

The graph of $\tan^{-1}x$ is the reflection of the graph of $\tan x$ about the line $y = x$.

25. Let $y = \cos^{-1}x$. Then $\cos y = x$ and $0 \le y \le \pi \Rightarrow -\sin y\dfrac{dy}{dx} = 1 \Rightarrow$
$\dfrac{dy}{dx} = -\dfrac{1}{\sin y} = -\dfrac{1}{\sqrt{1-\cos^2 y}} = -\dfrac{1}{\sqrt{1-x^2}}$ (Note that $\sin y \ge 0$ for $0 \le y \le \pi$.)

26. **(a)** Let $a = \sin^{-1}x$ and $b = \cos^{-1}x$. Then $\cos a = \sqrt{1-\sin^2 a} = \sqrt{1-x^2}$ since $\cos a \ge 0$ for $-\frac{\pi}{2} \le a \le \frac{\pi}{2}$.
Similarly, $\sin b = \sqrt{1-x^2}$. So $\sin(\sin^{-1}x + \cos^{-1}x) = \sin(a+b)$
$= \sin a\cos b + \cos a\sin b = x \cdot x + \sqrt{1-x^2}\sqrt{1-x^2} = x^2 + (1-x^2) = 1$. But
$\frac{\pi}{2} \le \sin^{-1}x + \cos^{-1}x \le \frac{\pi}{2}$, and so $\sin^{-1}x + \cos^{-1}x = \frac{\pi}{2}$.

(b) We differentiate the result of part (a) with respect to x, and get $\dfrac{1}{\sqrt{1-x^2}} + \dfrac{d}{dx}\cos^{-1}x = 0 \Rightarrow$
$\dfrac{d}{dx}\cos^{-1}x = -\dfrac{1}{\sqrt{1-x^2}}$.

27. Let $y = \cot^{-1}x$. Then $\cot y = x \quad\Rightarrow\quad -\csc^2 y\,\dfrac{dy}{dx} = 1 \quad\Rightarrow\quad \dfrac{dy}{dx} = -\dfrac{1}{\csc^2 y} = -\dfrac{1}{1+\cot^2 y} = -\dfrac{1}{1+x^2}$.

28. Let $y = \sec^{-1}x$. Then $\sec y = x$ and $y \in \left(0, \frac{\pi}{2}\right] \cup \left(\pi, \frac{3\pi}{2}\right]$. Differentiate with respect to x:
$\sec y \tan y\left(\dfrac{dy}{dx}\right) = 1 \quad\Rightarrow\quad \dfrac{dy}{dx} = \dfrac{1}{\sec y \tan y} = \dfrac{1}{\sec y\sqrt{\sec^2 y - 1}} = \dfrac{1}{x\sqrt{x^2-1}}$. Note that
$\tan^2 y = \sec^2 y - 1 \;\Rightarrow\; \tan y = \sqrt{\sec^2 y - 1}$ since $\tan y > 0$ when $0 < y < \frac{\pi}{2}$ or $\pi < y < \frac{3\pi}{2}$.

29. Let $y = \csc^{-1}x$. Then $\csc y = x \quad\Rightarrow\quad -\csc y \cot y\,\dfrac{dy}{dx} = 1 \quad\Rightarrow$
$\dfrac{dy}{dx} = -\dfrac{1}{\csc y \cot y} = -\dfrac{1}{\csc y\sqrt{\csc^2 y - 1}} = -\dfrac{1}{x\sqrt{x^2-1}}$. Note that $\cot y \geq 0$ on the domain of $\csc^{-1}x$.

30. $f(x) = \sin^{-1}(2x-1) \quad\Rightarrow\quad f'(x) = \dfrac{1}{\sqrt{1-(2x-1)^2}}(2) = \dfrac{1}{\sqrt{x-x^2}}$

31. $g(x) = \tan^{-1}(x^3) \quad\Rightarrow\quad g'(x) = \dfrac{1}{1+(x^3)^2}(3x^2) = \dfrac{3x^2}{1+x^6}$

32. $y = \left(\sin^{-1}x\right)^2 \quad\Rightarrow\quad y' = \dfrac{2\sin^{-1}x}{\sqrt{1-x^2}}$

33. $y = \sin^{-1}(x^2) \quad\Rightarrow\quad y' = \dfrac{1}{\sqrt{1-(x^2)^2}}(2x) = \dfrac{2x}{\sqrt{1-x^4}}$

34. $h(x) = \arcsin x \ln x \quad\Rightarrow\quad h'(x) = \dfrac{\ln x}{\sqrt{1-x^2}} + \dfrac{\arcsin x}{x}$

35. $H(x) = (1+x^2)\arctan x \quad\Rightarrow\quad H'(x) = (2x)\arctan x + (1+x^2)\dfrac{1}{1+x^2} = 1 + 2x\arctan x$

36. $f(t) = \dfrac{\cos^{-1}t}{t} \quad\Rightarrow\quad f'(t) = \dfrac{t\left(-1/\sqrt{1-t^2}\right) - \cos^{-1}t}{t^2} = -\dfrac{\cos^{-1}t}{t^2} - \dfrac{1}{t\sqrt{1-t^2}}$

37. $g(t) = \sin^{-1}\left(\dfrac{4}{t}\right) \quad\Rightarrow\quad g'(t) = \dfrac{1}{\sqrt{1-(4/t)^2}}\left(-\dfrac{4}{t^2}\right) = -\dfrac{4}{\sqrt{t^4-16t^2}}$

38. $F(t) = \sqrt{1-t^2} + \sin^{-1}t \quad\Rightarrow\quad F'(t) = \dfrac{-2t}{2\sqrt{1-t^2}} + \dfrac{1}{\sqrt{1-t^2}} = \dfrac{1-t}{\sqrt{1-t^2}}$

39. $G(t) = \cos^{-1}\sqrt{2t-1} \quad\Rightarrow\quad G'(t) = -\dfrac{1}{\sqrt{1-(2t-1)}}\dfrac{2}{2\sqrt{2t-1}} = -\dfrac{1}{\sqrt{2(-2t^2+3t-1)}}$

40. $y = \tan^{-1}\dfrac{x}{a} + \frac{1}{2}\ln(x-a) - \frac{1}{2}\ln(x+a) \quad\Rightarrow$
$y' = \dfrac{a}{x^2+a^2} + \dfrac{1/2}{x-a} - \dfrac{1/2}{x+a} = \dfrac{a}{x^2+a^2} + \dfrac{a}{x^2-a^2} = \dfrac{2ax^2}{x^4-a^4}$

41. $y = \sec^{-1}\sqrt{1+x^2} \quad\Rightarrow\quad y' = \left[\dfrac{1}{\sqrt{1+x^2}\sqrt{(1+x^2)-1}}\right]\left[\dfrac{2x}{2\sqrt{1+x^2}}\right] = \dfrac{x}{(1+x^2)\sqrt{x^2}} = \dfrac{x}{(1+x^2)|x|}$

42. $y = x\cos^{-1}x - \sqrt{1-x^2} \quad\Rightarrow\quad y' = \cos^{-1}x - \dfrac{x}{\sqrt{1-x^2}} + \dfrac{x}{\sqrt{1-x^2}} = \cos^{-1}x$

43. $y = \tan^{-1}(\sin x) \quad\Rightarrow\quad y' = \dfrac{\cos x}{1+\sin^2 x}$

44. $y = \sin^{-1}\left(\dfrac{\cos x}{1+\sin x}\right) \quad \Rightarrow$

$$y' = \frac{1}{\sqrt{1-[\cos x/(1+\sin x)]^2}} \cdot \frac{-\sin x(1+\sin x) - \cos^2 x}{(1+\sin x)^2}$$

$$= \frac{1}{\sqrt{(1+2\sin x+\sin^2 x - \cos^2 x)/(1+\sin x)^2}} \cdot \frac{-(1+\sin x)}{(1+\sin x)^2} = \frac{-1}{\sqrt{2\sin x + 2\sin^2 x}}$$

45. $y = (\tan^{-1}x)^{-1} \quad \Rightarrow \quad y' = -(\tan^{-1}x)^{-2}\left(\dfrac{1}{1+x^2}\right) = -\dfrac{1}{(1+x^2)(\tan^{-1}x)^2}$

46. $y = \tan^{-1}\left(x - \sqrt{x^2+1}\right) \quad \Rightarrow \quad y' = \dfrac{1}{1+\left(x-\sqrt{x^2+1}\right)^2}\left(1 - \dfrac{x}{\sqrt{x^2+1}}\right)$

$$= \frac{\sqrt{x^2+1}-x}{2\left(1+x^2-x\sqrt{x^2+1}\right)\sqrt{x^2+1}} = \frac{\sqrt{x^2+1}-x}{2(1+x^2)\left(\sqrt{x^2+1}-x\right)} = \frac{1}{2(1+x^2)}$$

47. $y = x^2\cot^{-1}(3x) \quad \Rightarrow \quad y' = 2x\cot^{-1}(3x) + x^2\left[-\dfrac{1}{1+(3x)^2}\right](3) = 2x\cot^{-1}(3x) - \dfrac{3x^2}{1+9x^2}$

48. $y = x\sin x\csc^{-1}x \quad \Rightarrow$

$$y' = \sin x\csc^{-1}x + x\cos x\csc^{-1}x - \frac{x\sin x}{x\sqrt{x^2-1}} = \sin x\csc^{-1}x + x\cos x\csc^{-1}x - \frac{\sin x}{\sqrt{x^2-1}}$$

49. $y = \arccos\left(\dfrac{b+a\cos x}{a+b\cos x}\right) \quad \Rightarrow$

$$y' = -\frac{1}{\sqrt{1-\left(\dfrac{b+a\cos x}{a+b\cos x}\right)^2}}\,\frac{(a+b\cos x)(-a\sin x)-(b+a\cos x)(-b\sin x)}{(a+b\cos x)^2}$$

$$= \frac{1}{\sqrt{a^2+b^2\cos^2 x-b^2-a^2\cos^2 x}}\,\frac{(a^2-b^2)\sin x}{|a+b\cos x|}$$

$$= \frac{1}{\sqrt{a^2-b^2}\sqrt{1-\cos^2 x}}\,\frac{(a^2-b^2)\sin x}{|a+b\cos x|} = \frac{\sqrt{a^2-b^2}}{|a+b\cos x|}\,\frac{\sin x}{|\sin x|}$$

But $0 \le x \le \pi$, so $|\sin x| = \sin x$. Also $a > b > 0 \quad \Rightarrow \quad b\cos x \ge -b > -a$, so $a + b\cos x > 0$.

Thus $y' = \dfrac{\sqrt{a^2-b^2}}{a+b\cos x}$.

50. $f(x) = \cos^{-1}(\sin^{-1}x) \quad \Rightarrow \quad f'(x) = -\dfrac{1}{\sqrt{1-(\sin^{-1}x)^2}} \cdot \dfrac{1}{\sqrt{1-x^2}}$,

$\text{Dom}(f) = \{x \mid -1 \le \sin^{-1}x \le 1\} = \{x \mid \sin(-1) \le x \le \sin 1\} = [-\sin 1, \sin 1]$,

$\text{Dom}(f') = \{x \mid -1 < \sin^{-1}x < 1\} = (-\sin 1, \sin 1)$

51. $g(x) = \sin^{-1}(3x+1) \quad \Rightarrow \quad g'(x) = \dfrac{3}{\sqrt{1-(3x+1)^2}} = \dfrac{3}{\sqrt{-9x^2-6x}}$,

$\text{Dom}(g) = \{x \mid -1 \le 3x+1 \le 1\} = \left\{x \mid -\frac{2}{3} \le x \le 0\right\} = \left[-\frac{2}{3}, 0\right]$,

$\text{Dom}(g') = \{x \mid -1 < 3x+1 < 1\} = \left(-\frac{2}{3}, 0\right)$

52. $F(x) = \sqrt{\sin^{-1}(2/x)} \quad \Rightarrow \quad F'(x) = \dfrac{1}{2\sqrt{\sin^{-1}(2/x)}} \cdot \dfrac{1}{\sqrt{1-(2/x)^2}}\left[-\dfrac{2}{x^2}\right] = -\dfrac{1}{x^2\sqrt{\sin^{-1}(2/x)}\sqrt{1-4/x^2}}$

$\text{Dom}(F) = \{x \mid -1 \le 2/x \le 1 \text{ and } \sin^{-1}(2/x) \ge 0\} = \{x \mid 0 < 2/x \le 1\} = \{x \mid x \ge 2\} = [2, \infty)$

$\text{Dom}(F') = \{x \mid x > 2\} = (2, \infty)$

53. $S(x) = \sin^{-1}(\tan^{-1}x) \quad \Rightarrow \quad S'(x) = \left[\sqrt{1-(\tan^{-1}x)^2}(1+x^2)\right]^{-1}$,

$\text{Dom}(S) = \{x \mid -1 \le \tan^{-1}x \le 1\} = \{x \mid \tan(-1) \le x \le \tan 1\} = [-\tan 1, \tan 1]$,

$\text{Dom}(S') = \{x \mid -1 < \tan^{-1}x < 1\} = (-\tan 1, \tan 1)$

54. $R(t) = \arcsin 2^t \quad \Rightarrow \quad R'(t) = \dfrac{1}{\sqrt{1-(2^t)^2}}(2^t \ln 2) = \dfrac{2^t \ln 2}{\sqrt{1-4^t}}$,

$\text{Dom}(R) = \{t \mid -1 \le 2^t \le 1\} = \{t \mid t \le 0\} = (-\infty, 0]$, $\text{Dom}(R') = (-\infty, 0)$

55. $U(t) = 2^{\arctan t} \quad \Rightarrow \quad U'(t) = 2^{\arctan t}(\ln 2)/(1+t^2)$, $\text{Dom}(U) = \text{Dom}(U') = \mathbb{R}$

56. $f(x) = x\tan^{-1}x \quad \Rightarrow \quad f'(x) = \tan^{-1}x + \dfrac{x}{1+x^2} \quad \Rightarrow \quad f'(1) = \frac{\pi}{4} + \frac{1}{2}$

57. $g(x) = x\sin^{-1}\left(\dfrac{x}{4}\right) + \sqrt{16-x^2} \quad \Rightarrow \quad g'(x) = \sin^{-1}\left(\dfrac{x}{4}\right) + \dfrac{x}{4\sqrt{1-(x/4)^2}} - \dfrac{x}{\sqrt{16-x^2}} = \sin^{-1}\left(\dfrac{x}{4}\right)$

$\Rightarrow \quad g'(2) = \sin^{-1}\frac{1}{2} = \frac{\pi}{6}$

58. $h(x) = (3\tan^{-1}x)^4 \quad \Rightarrow \quad h'(x) = 4(3\tan^{-1}x)^3\left(\dfrac{3}{1+x^2}\right) \quad \Rightarrow \quad h'(3) = 4(3\tan^{-1}3)^3\left(\frac{3}{10}\right) = \frac{162}{5}(\tan^{-1}3)^3$

59. $f(x) = e^x - x^2\arctan x \quad \Rightarrow$

$$f'(x) = e^x - \left[x^2\left(\frac{1}{1+x^2}\right) + 2x\arctan x\right]$$

$$= e^x - \frac{x^2}{1+x^2} - 2x\arctan x$$

This is reasonable because the graphs show that f is increasing when $f'(x)$ is positive.

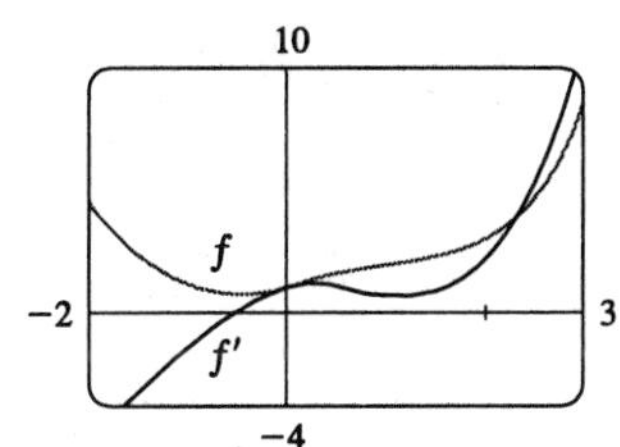

60. $f(x) = x\arcsin(1-x^2) \quad \Rightarrow$

$$f'(x) = \arcsin(1-x^2) + x\left[\frac{-2x}{\sqrt{1-(1-x^2)^2}}\right]$$

$$= \arcsin(1-x^2) - \frac{2x^2}{\sqrt{2x^2-x^4}}$$

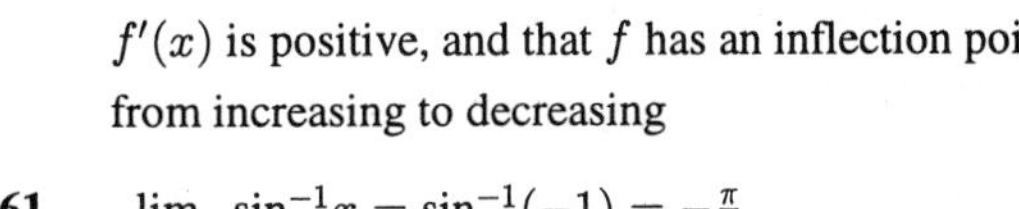
This is reasonable because the graphs show that f is increasing when $f'(x)$ is positive, and that f has an inflection point when f' changes from increasing to decreasing

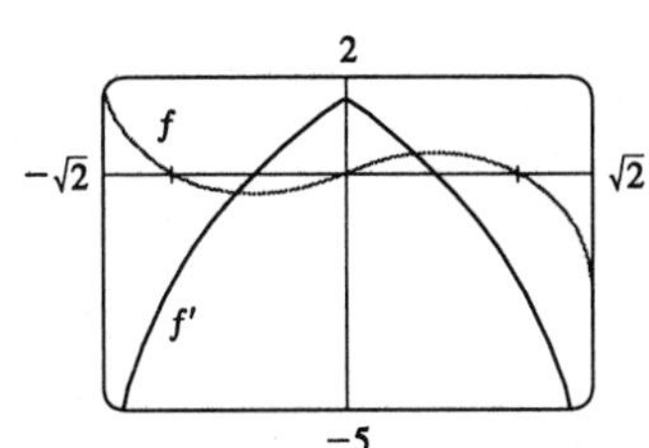

61. $\lim\limits_{x\to -1^+} \sin^{-1}x = \sin^{-1}(-1) = -\frac{\pi}{2}$

62. $\lim\limits_{x\to\infty} \sin^{-1}\left(\dfrac{x+1}{2x+1}\right) = \sin^{-1}\left(\lim\limits_{x\to\infty} \dfrac{x+1}{2x+1}\right) = \sin^{-1}\frac{1}{2} = \frac{\pi}{6}$

63. $\lim\limits_{x\to\infty} \tan^{-1}(x^2) = \frac{\pi}{2}$ since $x^2 \to \infty$ as $x \to \infty$.

64. $\lim\limits_{x\to\infty} \tan^{-1}(x - x^2) = -\frac{\pi}{2}$ since $x - x^2 = x(1-x) \to -\infty$ as $x \to \infty$.

65. $x = 5\cot\alpha,\ 3 - x = 2\cot\beta \quad \Rightarrow$

$$\theta = \pi - \cot^{-1}\left(\frac{x}{5}\right) - \cot^{-1}\left(\frac{3-x}{2}\right) \quad \Rightarrow$$

$$\frac{d\theta}{dx} = \frac{1}{1+\left(\frac{x}{5}\right)^2}\left(\frac{1}{5}\right) + \frac{1}{1+\left(\frac{3-x}{2}\right)^2}\left(-\frac{1}{2}\right) = 0 \quad \Rightarrow$$

$$5\left(1+\frac{x^2}{25}\right) = 2\left(1+\frac{9-6x+x^2}{4}\right)$$

$\Rightarrow \quad 50 + 2x^2 = 65 - 30x + 5x^2 \quad \Rightarrow \quad x^2 - 10x + 5 = 0 \quad \Rightarrow \quad x = 5 \pm 2\sqrt{5}$. We reject the root with the $+$ sign, since it is larger than 3. $d\theta/dx > 0$ for $x < 5 - 2\sqrt{5}$ and $d\theta/dx < 0$ for $x > 5 - 2\sqrt{5}$, so θ is maximized when $|AP| = x = 5 - 2\sqrt{5}$.

66. Let x be the distance from the observer to the wall. Then, from the given figure,

$$\theta = \tan^{-1}\left(\frac{h+d}{x}\right) - \tan^{-1}\left(\frac{d}{x}\right),\ x > 0 \quad \Rightarrow$$

$$\frac{d\theta}{dx} = \frac{1}{1+[(h+d)/x]^2}\left[-\frac{h+d}{x^2}\right] - \frac{1}{1+(d/x)^2}\left[-\frac{d}{x^2}\right] = -\frac{h+d}{x^2+(h+d)^2} + \frac{d}{x^2+d^2}$$

$$= \frac{d[x^2+(h+d)^2] - (h+d)(x^2+d^2)}{[x^2+(h+d)^2](x^2+d^2)} = \frac{h^2d + hd^2 - hx^2}{[x^2+(h+d)^2](x^2+d^2)} = 0 \quad \Leftrightarrow$$

$hx^2 = h^2d + hd^2 \quad \Leftrightarrow \quad x^2 = hd + d^2 \quad \Leftrightarrow \quad x = \sqrt{d(h+d)}$. Since $d\theta/dx > 0$ for all $x < \sqrt{d(h+d)}$ and $d\theta/dx < 0$ for all $x > \sqrt{d(h+d)}$, the absolute maximum occurs when $x = \sqrt{d(h+d)}$.

67.

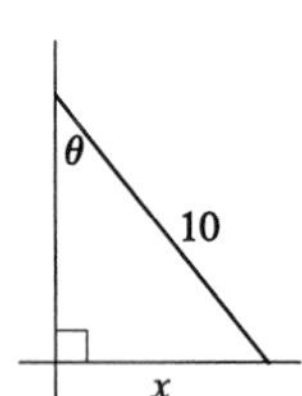

$\dfrac{dx}{dt} = 2\text{ ft/s},\ \sin\theta = \dfrac{x}{10} \quad \Rightarrow \quad \theta = \sin^{-1}\left(\dfrac{x}{10}\right)$,

$$\frac{d\theta}{dx} = \frac{1/10}{\sqrt{1-(x/10)^2}},\ \frac{d\theta}{dt} = \frac{d\theta}{dx}\frac{dx}{dt} = \frac{1/10}{\sqrt{1-(x/10)^2}}(2)\text{ rad/s},$$

$$\left.\frac{d\theta}{dt}\right|_{x=6} = \frac{2/10}{\sqrt{1-(6/10)^2}}\text{ rad/s} = \frac{1}{4}\text{ rad/s}$$

68.

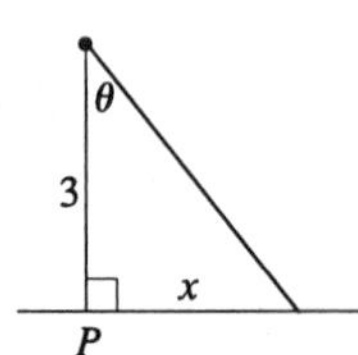

$\dfrac{d\theta}{dt} = 4\text{ rev/min} = 8\pi\cdot 60\text{ rad/h}$. From the diagram, we see that $\tan\theta = \dfrac{x}{3}$

$\Rightarrow \quad \theta = \tan^{-1}\left(\dfrac{x}{3}\right)$. Thus, $8\pi\cdot 60 = \dfrac{d\theta}{dt} = \dfrac{d\theta}{dx}\dfrac{dx}{dt} = \dfrac{1/3}{1+(x/3)^2}\dfrac{dx}{dt}$.

So $\dfrac{dx}{dt} = 8\pi\cdot 60\cdot 3\left[1+\left(\dfrac{x}{3}\right)^2\right]$ km/h, and at $x = 1$,

$$\frac{dx}{dt} = 8\pi\cdot 60\cdot 3\left[1+\tfrac{1}{9}\right]\text{ km/h} = 1600\pi\text{ km/h}.$$

69. By reflecting the graph of $y = \sec x$ (see Figure 11) about the line $y = x$, we get the graph of $y = \sec^{-1} x$.

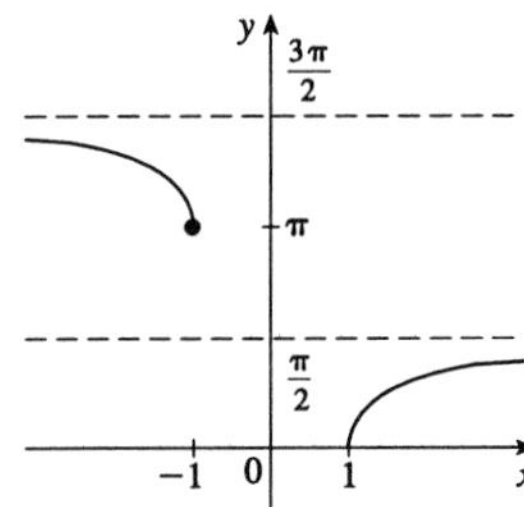

70. $y = f(x) = \arctan(\tan x)$. Since tan has period π, so does f. For $-\frac{\pi}{2} < x < \frac{\pi}{2}$, $f(x) = x$. These facts enable us to draw the graph.

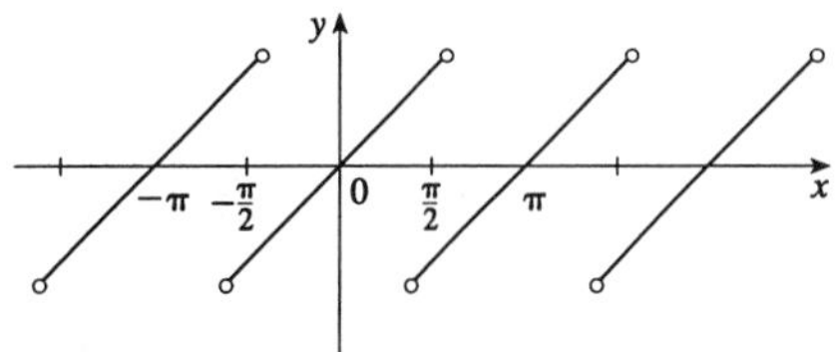

71. $y = f(x) = \sin^{-1}[x/(x+1)]$ **A.** $D = \{x \mid -1 \le x/(x+1) \le 1\}$. For $x > -1$ we have $-x - 1 \le x \le x + 1 \quad \Leftrightarrow \quad 2x \ge -1 \quad \Leftrightarrow \quad x \ge -\frac{1}{2}$, so $D = \left[-\frac{1}{2}, \infty\right)$. **B.** Intercepts are 0 **C.** No symmetry **D.** $\lim\limits_{x\to\infty} \sin^{-1}\left(\dfrac{x}{x+1}\right) = \lim\limits_{x\to\infty} \sin^{-1}\left(\dfrac{1}{1+1/x}\right) = \sin^{-1} 1 = \frac{\pi}{2}$, so $y = \frac{\pi}{2}$ is a HA.

E. $f'(x) = \dfrac{1}{\sqrt{1 - [x/(x+1)]^2}} \dfrac{(x+1) - x}{(x+1)^2} = \dfrac{1}{(x+1)\sqrt{2x+1}} > 0$, so f is increasing on $\left[-\frac{1}{2}, \infty\right)$. **F.** No local maximum or minimum, $f\left(-\frac{1}{2}\right) = \sin^{-1}(-1) = -\frac{\pi}{2}$ is an absolute minimum **G.** $f''(x) = \dfrac{\sqrt{2x+1} + (x+1)/\sqrt{2x+1}}{(x+1)^2(2x+1)}$

$= -\dfrac{3x+2}{(x+1)^2(2x+1)^{3/2}} < 0$ on D, so f is CD on $\left(-\frac{1}{2}, \infty\right)$.

H.

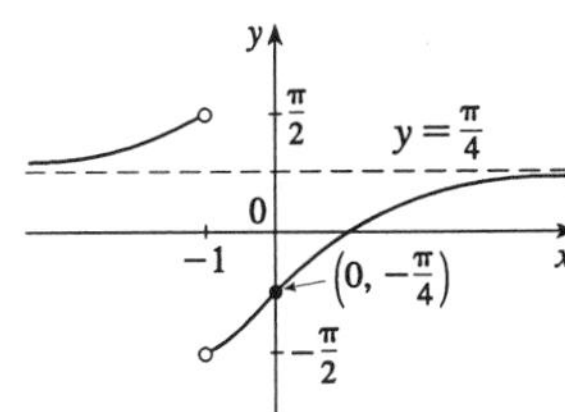

72. $y = f(x) = \tan^{-1}\left(\dfrac{x-1}{x+1}\right)$ **A.** $D = \{x \mid x \ne -1\}$ **B.** x-intercept $= 1$, y-intercept $= f(0) = \tan^{-1}(-1) = -\frac{\pi}{4}$ **C.** No symmetry

D. $\lim\limits_{x\to\pm\infty} \tan^{-1}\left(\dfrac{x-1}{x+1}\right) = \lim\limits_{x\to\pm\infty} \tan^{-1}\left(\dfrac{1-1/x}{1+1/x}\right) = \tan^{-1} 1 = \frac{\pi}{4}$, so $y = \frac{\pi}{4}$ is a HA. Also $\lim\limits_{x\to -1^+} \tan^{-1}\left(\dfrac{x-1}{x+1}\right) = -\dfrac{\pi}{2}$ and $\lim\limits_{x\to -1^-} \tan^{-1}\left(\dfrac{x-1}{x+1}\right) = \dfrac{\pi}{2}$.

E. $f'(x) = \dfrac{1}{1 + \left[(x-1)/(x+1)\right]^2} \dfrac{(x+1) - (x-1)}{(x+1)^2}$

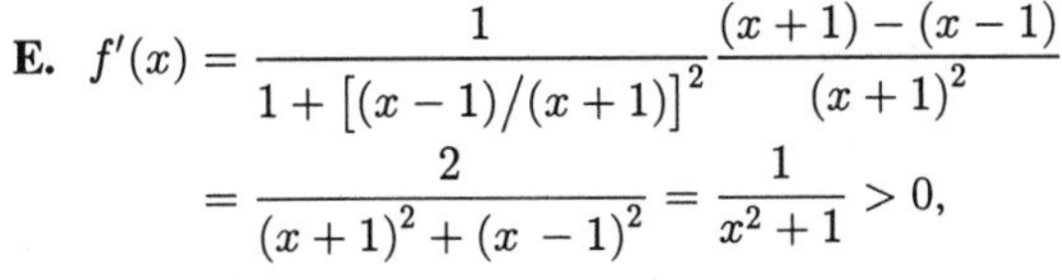

$= \dfrac{2}{(x+1)^2 + (x-1)^2} = \dfrac{1}{x^2+1} > 0$,

so f is increasing on $(-\infty, -1)$ and $(-1, \infty)$.

F. No extrema **G.** $f''(x) = -2x/(x^2+1)^2 > 0 \quad \Leftrightarrow \quad x < 0$, so f is CU on $(-\infty, -1)$ and $(-1, 0)$, and CD on $(0, \infty)$. IP is $\left(0, -\frac{\pi}{4}\right)$.

H.

73. $y = f(x) = x - \tan^{-1}x$ **A.** $D = \mathbb{R}$ **B.** Intercepts are 0 **C.** $f(-x) = -f(x)$, so the curve is symmetric about the origin. **D.** $\lim\limits_{x\to\infty}(x - \tan^{-1}x) = \infty$ and $\lim\limits_{x\to-\infty}(x - \tan^{-1}x) = -\infty$, no HA.
But $f(x) - \left(x - \frac{\pi}{2}\right) = -\tan^{-1}x + \frac{\pi}{2} \to 0$ as $x \to \infty$, and

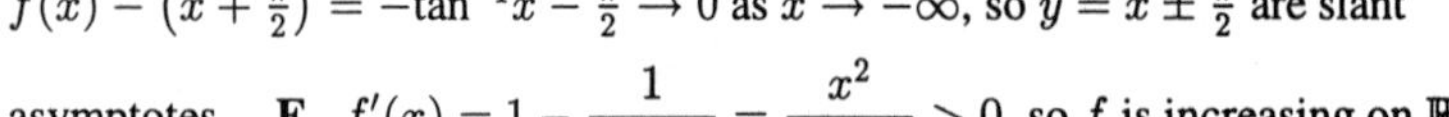

$f(x) - \left(x + \frac{\pi}{2}\right) = -\tan^{-1}x - \frac{\pi}{2} \to 0$ as $x \to -\infty$, so $y = x \pm \frac{\pi}{2}$ are slant asymptotes. **E.** $f'(x) = 1 - \dfrac{1}{x^2+1} = \dfrac{x^2}{x^2+1} > 0$, so f is increasing on $\mathbb{R}$.

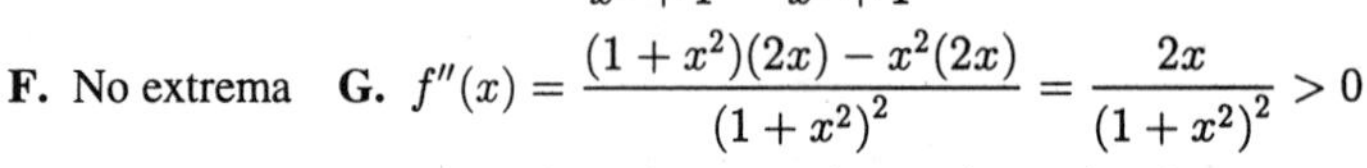

F. No extrema **G.** $f''(x) = \dfrac{(1+x^2)(2x) - x^2(2x)}{(1+x^2)^2} = \dfrac{2x}{(1+x^2)^2} > 0$ $\Leftrightarrow$ $x > 0$, so f is CU on $(0, \infty)$, CD on $(-\infty, 0)$. IP $(0, 0)$

H.

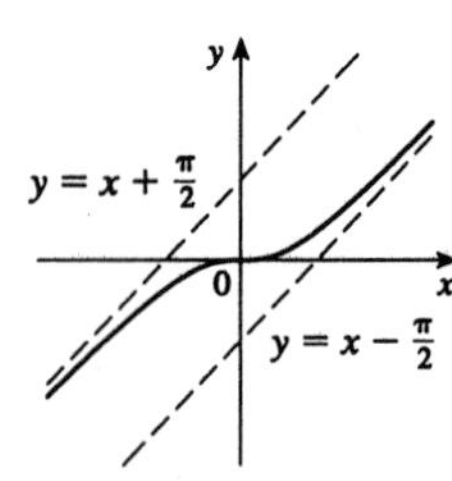

74. $y = \tan^{-1}(\ln x)$ **A.** $D = (0, \infty)$ **B.** No y-intercept, x-intercept when $\tan^{-1}(\ln x) = 0$ $\Leftrightarrow$ $\ln x = 0$ $\Leftrightarrow$ $x = 1$. **C.** No symmetry **D.** $\lim\limits_{x\to\infty}\tan^{-1}(\ln x) = \frac{\pi}{2}$, so $y = \frac{\pi}{2}$ is a HA. Also $\lim\limits_{x\to0^+}\tan^{-1}(\ln x) = -\frac{\pi}{2}$. **E.** $f'(x) = \dfrac{1}{x\left[1 + (\ln x)^2\right]} > 0$, so f is increasing on $(0, \infty)$. **F.** No extrema

G. $f''(x) = \dfrac{-\left[1 + (\ln x)^2 + x(2\ln x/x)\right]}{x^2\left[1 + (\ln x)^2\right]^2} = -\dfrac{(1 + \ln x)^2}{x^2[1 + (\ln x)^2]^2} < 0$, so f is CD on $(0, \infty)$.

H.

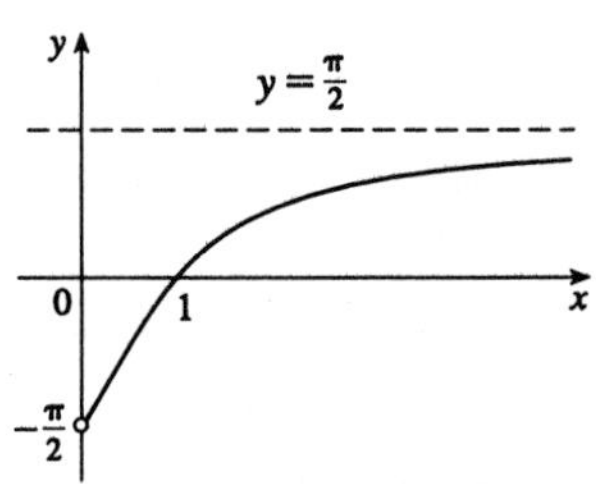

75. $f(x) = \arctan[\cos(3\arcsin x)]$. We use a CAS to compute f' and f'', and to graph f, f', and f'':

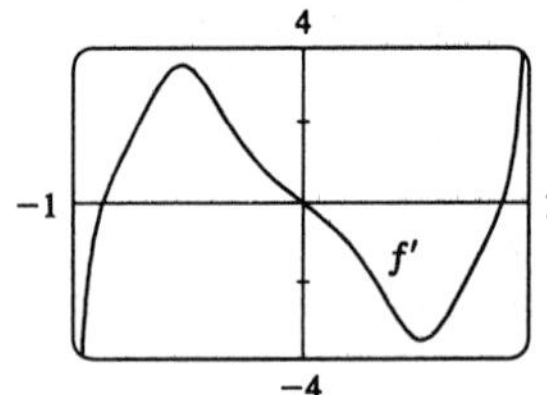

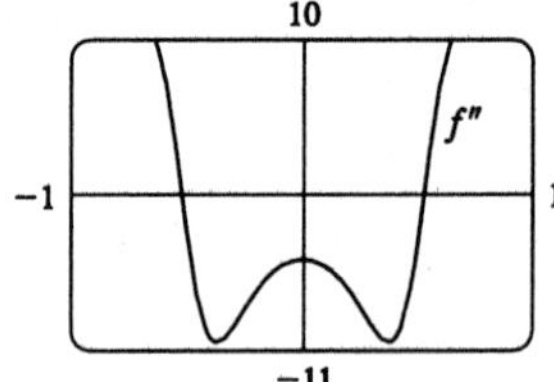

From the graph of f', it appears that the only maximum occurs at $x = 0$ and there are minima at $x = \pm0.87$. From the graph of f'', it appears that there are inflection points at $x = \pm0.52$.

76. First note that the function $f(x) = x - c\sin^{-1}x$ is only defined on the interval $[-1, 1]$, since $\sin^{-1}$ is only defined on that interval. We differentiate to get $f'(x) = 1 - c/\sqrt{1-x^2}$. Now if $c \le 0$, then $f'(x) \ge 1$, so there are no extrema and f is increasing on its domain. If $c > 1$, then $f'(x) < 0$, so there are no local extrema and f is decreasing on its domain, and if $c = 1$, then there are still no extrema, since $f'(x)$ does not change sign at $x = 0$. So we can only have local extrema if $0 < c < 1$. In this case, f is increasing where $f'(x) > 0$ $\Leftrightarrow$ $\sqrt{1-x^2} > c$ $\Leftrightarrow$ $|x| < \sqrt{1-c^2}$, and decreasing where $\sqrt{1-c^2} < |x| \le 1$. f has a maximum at $x = \sqrt{1-c^2}$ and a minimum at $x = -\sqrt{1-c^2}$.

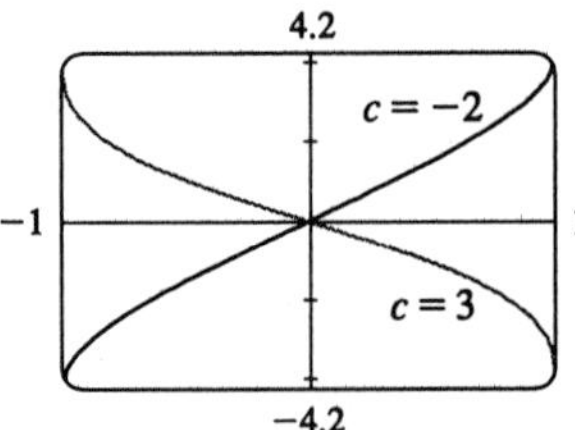

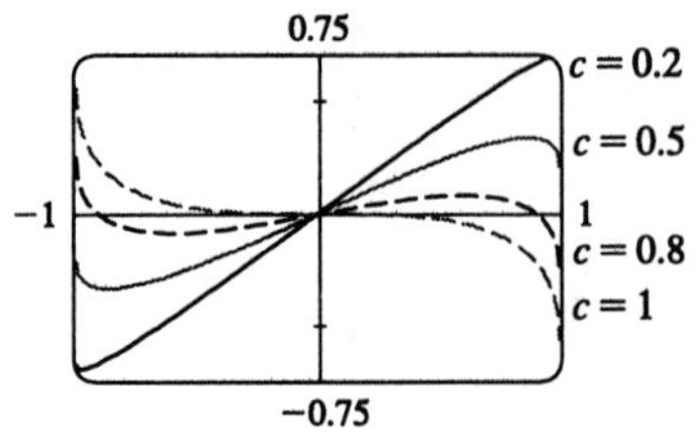

77. $f(x) = 2x + 5/\sqrt{1-x^2} \quad \Rightarrow \quad F(x) = x^2 + 5\sin^{-1}x + C$

78. $f'(x) = 4 - 3(1+x^2)^{-1} \quad \Rightarrow \quad f(x) = 4x - 3\tan^{-1}x + C \quad \Rightarrow \quad f(\frac{\pi}{4}) = \pi - 3 + C = 0 \quad \Rightarrow$

$C = 3 - \pi$, so $f(x) = 4x - 3\tan^{-1}x + 3 - \pi$.

79. $\displaystyle\int_1^{\sqrt{3}} \frac{6}{1+x^2}\,dx = \left[6\tan^{-1}x\right]_1^{\sqrt{3}} = 6\left(\tan^{-1}\sqrt{3} - \tan^{-1}1\right) = 6\left(\tfrac{\pi}{3} - \tfrac{\pi}{4}\right) = \tfrac{\pi}{2}$

80. $\displaystyle\int_0^{0.5} \frac{dx}{\sqrt{1-x^2}} = \left[\sin^{-1}x\right]_0^{0.5} = \sin^{-1}\tfrac{1}{2} - \sin^{-1}0 = \tfrac{\pi}{6}$

81. Let $u = x^3$. Then $du = 3x^2\,dx$, so $\displaystyle\int \frac{x^2}{\sqrt{1-x^6}}\,dx = \frac{1}{3}\int \frac{1}{\sqrt{1-u^2}}\,du = \tfrac{1}{3}\sin^{-1}u + C = \tfrac{1}{3}\sin^{-1}(x^3) + C.$

82. Let $u = \tan^{-1}x$. Then $du = dx/(1+x^2)$, so $\displaystyle\int \frac{\tan^{-1}x}{1+x^2}\,dx = \int u\,du = \tfrac{1}{2}u^2 + C = \tfrac{1}{2}(\tan^{-1}x)^2 + C.$

83. $\displaystyle\int \frac{x+9}{x^2+9}\,dx = \int \frac{x}{x^2+9}\,dx + 9\int \frac{1}{x^2+9}\,dx = \tfrac{1}{2}\ln(x^2+9) + 3\tan^{-1}\frac{x}{3} + C$

(Let $u = x^2 + 9$ in the first integral; use Equation 14 in the second.)

84. Let $u = \cos x$. Then $du = -\sin x\,dx$, so

$\displaystyle\int \frac{\sin x}{1+\cos^2 x}\,dx = -\int \frac{1}{1+u^2}\,du = -\tan^{-1}u + C = -\tan^{-1}(\cos x) + C.$

85. Let $u = 3x$. Then $du = 3\,dx$, so $\displaystyle\int \frac{dx}{1+9x^2} = \frac{1}{3}\int \frac{du}{1+u^2} = \tfrac{1}{3}\tan^{-1}u + C = \tfrac{1}{3}\tan^{-1}(3x) + C.$

86. Let $u = \frac{1}{2}x$. Then $du = \frac{1}{2}\,dx \Rightarrow \displaystyle\int \frac{dx}{x\sqrt{x^2-4}} = \int \frac{dx}{2x\sqrt{(x/2)^2-1}} = \int \frac{2\,du}{4u\sqrt{u^2-1}} = \frac{1}{2}\int \frac{du}{u\sqrt{u^2-1}}$

$= \tfrac{1}{2}\sec^{-1}u + C = \tfrac{1}{2}\sec^{-1}\left(\tfrac{1}{2}x\right) + C.$

87. Let $u = e^x$. Then $du = e^x\,dx$, so $\displaystyle\int \frac{e^x\,dx}{e^{2x}+1} = \int \frac{du}{u^2+1} = \tan^{-1}u + C = \tan^{-1}(e^x) + C.$

88. Let $u = e^{2x}$. Then $du = 2e^{2x}\,dx \quad \Rightarrow \quad \displaystyle\int \frac{e^{2x}\,dx}{\sqrt{1-e^{4x}}} = \frac{1}{2}\int \frac{du}{\sqrt{1-u^2}} = \tfrac{1}{2}\sin^{-1}u + C = \tfrac{1}{2}\sin^{-1}(e^{2x}) + C$

89. Let $u = \sin^{-1}x$. Then $du = \dfrac{1}{\sqrt{1-x^2}}\,dx$, so $\displaystyle\int_0^{1/2} \frac{\sin^{-1}x}{\sqrt{1-x^2}}\,dx = \int_0^{\pi/6} u\,du = \frac{u^2}{2}\bigg|_0^{\pi/6} = \frac{1}{2}\left(\frac{\pi}{6}\right)^2 = \frac{\pi^2}{72}.$

90. Let $u = \ln x$. Then $du = (1/x)dx \quad \Rightarrow$

$\displaystyle\int \frac{dx}{x\left[4 + (\ln x)^2\right]} = \int \frac{du}{4+u^2} = \tfrac{1}{2}\tan^{-1}\left(\frac{u}{2}\right) + C = \tfrac{1}{2}\tan^{-1}\left(\tfrac{1}{2}\ln x\right) + C.$

91. Let $u = x/a$. Then $du = dx/a$, so

$\displaystyle\int \frac{dx}{\sqrt{a^2-x^2}} = \int \frac{dx}{a\sqrt{1-(x/a)^2}} = \int \frac{du}{\sqrt{1-u^2}} = \sin^{-1}u + C = \sin^{-1}(x/a) + C.$

92. We use the disk method: $A = \displaystyle\int_0^2 \pi\left[\frac{1}{\sqrt{x^2+4}}\right]^2 dx = \pi\int_0^2 \frac{1}{x^2+4}\,dx$. By Formula 14, this is equal to

$\pi\left[\tfrac{1}{2}\tan^{-1}(x/2)\right]_0^2 = \tfrac{\pi}{2}\left(\tfrac{\pi}{4} - 0\right) = \tfrac{\pi^2}{8}.$

93. 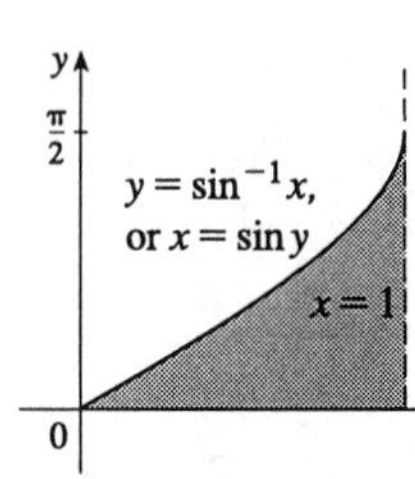

The integral represents the area below the curve $y = \sin^{-1}x$ on the interval $x \in [0, 1]$. The bounding curves are $y = \sin^{-1}x \quad \Leftrightarrow \quad x = \sin y$, $y = 0$ and $x = 1$. We see that y ranges between $\sin^{-1}0 = 0$ and $\sin^{-1}1 = \frac{\pi}{2}$. So we have to integrate the function $x = 1 - \sin y$ between $y = 0$ and $y = \frac{\pi}{2}$:

$\int_0^1 \sin^{-1}x\,dx = \int_0^{\pi/2}(1 - \sin y)dy = \left(\frac{\pi}{2} + \cos\frac{\pi}{2}\right) - (0 + \cos 0) = \frac{\pi}{2} - 1.$

94. Let $a = \arctan x$ and $b = \arctan y$. Then by the addition formula for the tangent (see endpapers),

$$\tan(a + b) = \frac{\tan a + \tan b}{1 - (\tan a)(\tan b)} = \frac{\tan(\arctan x) + \tan(\arctan y)}{1 - \tan(\arctan x)\tan(\arctan y)} \quad \Rightarrow \quad \tan(a + b) = \frac{x + y}{1 - xy} \quad \Rightarrow$$

$\arctan x + \arctan y = a + b = \arctan\left(\dfrac{x + y}{1 - xy}\right)$, since $-\frac{\pi}{2} < \arctan x + \arctan y < \frac{\pi}{2}$.

95. **(a)** $\arctan\frac{1}{2} + \arctan\frac{1}{3} = \arctan\left(\dfrac{\frac{1}{2} + \frac{1}{3}}{1 - \frac{1}{2}\cdot\frac{1}{3}}\right) = \arctan 1 = \dfrac{\pi}{4}$

(b) $2\arctan\frac{1}{3} + \arctan\frac{1}{7} = \left(\arctan\frac{1}{3} + \arctan\frac{1}{3}\right) + \arctan\frac{1}{7} = \arctan\left(\dfrac{\frac{1}{3} + \frac{1}{3}}{1 - \frac{1}{3}\cdot\frac{1}{3}}\right) + \arctan\frac{1}{7}$

$$= \arctan\tfrac{3}{4} + \arctan\tfrac{1}{7} = \arctan\left(\frac{\frac{3}{4} + \frac{1}{7}}{1 - \frac{3}{4}\cdot\frac{1}{7}}\right) = \arctan 1 = \frac{\pi}{4}$$

96. **(a)** $f(x) = \sin(\sin^{-1}x)$

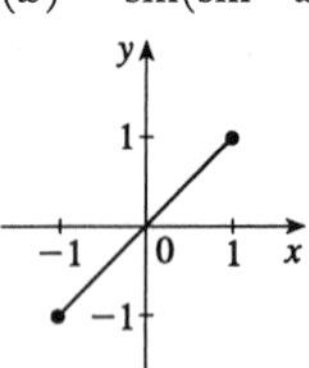

(b) $g(x) = \sin^{-1}(\sin x)$

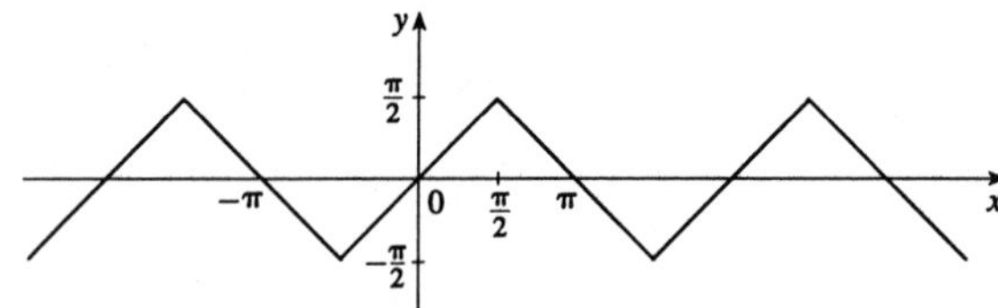

(c) $g'(x) = \dfrac{d}{dx}\sin^{-1}(\sin x) = \dfrac{1}{\sqrt{1 - \sin^2 x}}\cos x = \dfrac{\cos x}{\sqrt{\cos^2 x}} = \dfrac{\cos x}{|\cos x|}$

(d) $h(x) = \cos^{-1}(\sin x)$, so

$h'(x) = -\dfrac{\cos x}{\sqrt{1 - \sin^2 x}} = \dfrac{\cos x}{|\cos x|}$.

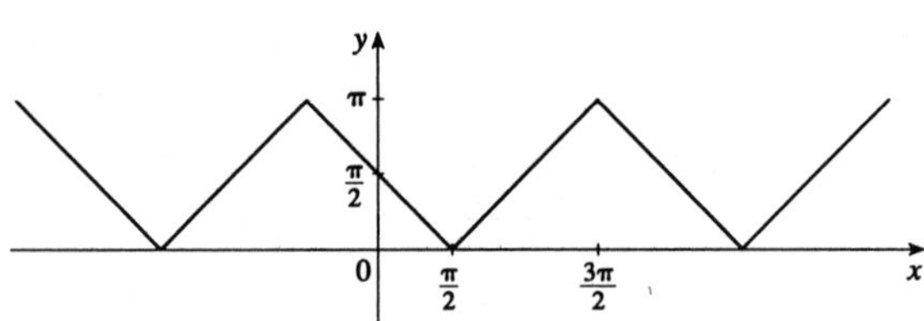

97. Let $f(x) = 2\sin^{-1}x - \cos^{-1}(1 - 2x^2)$. Then

$f'(x) = \dfrac{2}{\sqrt{1 - x^2}} - \dfrac{4x}{\sqrt{1 - (1 - 2x^2)^2}} = \dfrac{2}{\sqrt{1 - x^2}} - \dfrac{4x}{2x\sqrt{1 - x^2}} = 0$ (since $x \geq 0$). Thus $f'(x) = 0$ for all $x \in [0, 1)$. Thus $f(x) = C$. To find C let $x = 0$. Thus $2\sin^{-1}(0) - \cos^{-1}(1) = 0 = C$. Therefore we see that $f(x) = 2\sin^{-1}x - \cos^{-1}(1 - 2x^2) = 0 \quad \Rightarrow \quad 2\sin^{-1}x = \cos^{-1}(1 - 2x^2)$.

98. Let $f(x) = \sin^{-1}\left(\dfrac{x-1}{x+1}\right) - 2\tan^{-1}\sqrt{x} + \frac{\pi}{2}$. Note that the domain of f is $[0, \infty)$. Thus

$$f'(x) = \frac{1}{\sqrt{1-\left(\dfrac{x-1}{x+1}\right)^2}}\,\frac{(x+1)-(x-1)}{(x+1)^2} - \frac{2}{1+x}\cdot\frac{1}{2\sqrt{x}} = \frac{1}{\sqrt{x}(x+1)} - \frac{1}{\sqrt{x}(x+1)} = 0.$$ Then

$f(x) = C$. To find C, we let $x = 0 \Rightarrow \sin^{-1}(-1) - 2\tan^{-1}(0) + \frac{\pi}{2} = C \Rightarrow -\frac{\pi}{2} - 0 + \frac{\pi}{2} = 0 = C$.

Thus, $f(x) = 0 \Rightarrow \sin^{-1}\left(\dfrac{x-1}{x+1}\right) = 2\tan^{-1}\sqrt{x} - \frac{\pi}{2}$.

99. $y = \sec^{-1}x \Rightarrow \sec y = x \Rightarrow \sec y \tan y\,\dfrac{dy}{dx} = 1 \Rightarrow \dfrac{dy}{dx} = \dfrac{1}{\sec y \tan y}$. Now

$\tan^2 y = \sec^2 y - 1 = x^2 - 1$, so $\tan y = \pm\sqrt{x^2-1}$. For $y \in \left[0, \frac{\pi}{2}\right)$, $x \geq 1$, so $\sec y = x = |x|$ and $\tan y \geq 0$

$\Rightarrow \dfrac{dy}{dx} = \dfrac{1}{x\sqrt{x^2-1}} = \dfrac{1}{|x|\sqrt{x^2-1}}$. For $y \in \left(\frac{\pi}{2}, \pi\right]$, $x \leq -1$, so $|x| = -x$ and $\tan y = -\sqrt{x^2-1} \Rightarrow$

$$\frac{dy}{dx} = \frac{1}{\sec y \tan y} = \frac{1}{x\left(-\sqrt{x^2-1}\right)} = \frac{1}{(-x)\sqrt{x^2-1}} = \frac{1}{|x|\sqrt{x^2-1}}.$$

100. (a) Since $|\arctan(1/x)| < \frac{\pi}{2}$, we have $0 \leq |x \arctan(1/x)| \leq \frac{\pi}{2}|x| \to 0$ as $x \to 0$. So, by the Squeeze Theorem, $\lim\limits_{x\to 0} f(x) = 0 = f(0)$, so f is continuous at 0.

(b) Here $\dfrac{f(x)-f(0)}{x-0} = \dfrac{x\arctan(1/x) - 0}{x} = \arctan\left(\dfrac{1}{x}\right)$. So (see Exercise 57 in Section 2.1 for a discussion of left- and right-hand derivatives)

$f'_-(0) = \lim\limits_{x\to 0^-}\dfrac{f(x)-f(0)}{x-0} = \lim\limits_{x\to 0^-}\arctan\left(\dfrac{1}{x}\right) = \lim\limits_{y\to -\infty}\arctan y = -\dfrac{\pi}{2}$, while

$f'_+(0) = \lim\limits_{x\to 0^+}\dfrac{f(x)-f(0)}{x-0} = \lim\limits_{x\to 0^+}\arctan\left(\dfrac{1}{x}\right) = \lim\limits_{y\to \infty}\arctan y = \dfrac{\pi}{2}$. So $f'(0)$ does not exist.

EXERCISES 6.7

1. (a) $\sinh 0 = \frac{1}{2}(e^0 - e^0) = 0$ **(b)** $\cosh 0 = \frac{1}{2}(e^0 + e^0) = \frac{1}{2}(1+1) = 1$

2. (a) $\tanh 0 = \dfrac{(e^0 - e^{-0})/2}{(e^0 + e^{-0})/2} = 0$ **(b)** $\tanh 1 = \dfrac{e^1 - e^{-1}}{e^1 + e^{-1}} = \dfrac{e^2-1}{e^2+1} \approx 0.76159$.

3. (a) $\sinh(\ln 2) = \dfrac{e^{\ln 2} - e^{-\ln 2}}{2} = \dfrac{2 - \frac{1}{2}}{2} = \dfrac{3}{4}$ **(b)** $\sinh 2 = \frac{1}{2}(e^2 - e^{-2}) \approx 3.62686$

4. (a) $\cosh 3 = \frac{1}{2}(e^3 + e^{-3}) \approx 10.06766$ **(b)** $\cosh(\ln 3) = \dfrac{e^{\ln 3} + e^{-\ln 3}}{2} = \dfrac{3 + \frac{1}{3}}{2} = \dfrac{5}{3}$

5. (a) $\operatorname{sech} 0 = \dfrac{1}{\cosh 0} = \dfrac{1}{1} = 1$ **(b)** $\cosh^{-1}1 = 0$ because $\cosh 0 = 1$.

6. (a) $\sinh 1 = \frac{1}{2}(e^1 - e^{-1}) \approx 1.17520$

(b) Using Equation 3, we have $\sinh^{-1}1 = \ln\left(1 + \sqrt{1^2+1}\right) = \ln\left(1+\sqrt{2}\right) \approx 0.88137$.

7. $\sinh(-x) = \frac{1}{2}\left[e^{-x} - e^{-(-x)}\right] = \frac{1}{2}(e^{-x} - e^{x}) = -\frac{1}{2}(e^{x} - e^{-x}) = -\sinh x$

8. $\cosh(-x) = \frac{1}{2}\left[e^{-x} + e^{-(-x)}\right] = \frac{1}{2}(e^{-x} + e^{x}) = \frac{1}{2}(e^{x} + e^{-x}) = \cosh x$

9. $\cosh x + \sinh x = \frac{1}{2}(e^{x} + e^{-x}) + \frac{1}{2}(e^{x} - e^{-x}) = \frac{1}{2}(2e^{x}) = e^{x}$

10. $\cosh x - \sinh x = \frac{1}{2}(e^{x} + e^{-x}) - \frac{1}{2}(e^{x} - e^{-x}) = \frac{1}{2}(2e^{-x}) = e^{-x}$

11. $\sinh x \cosh y + \cosh x \sinh y = \left[\frac{1}{2}(e^{x} - e^{-x})\right]\left[\frac{1}{2}(e^{y} + e^{-y})\right] + \left[\frac{1}{2}(e^{x} + e^{-x})\right]\left[\frac{1}{2}(e^{y} - e^{-y})\right]$

$$= \tfrac{1}{4}[(e^{x+y} + e^{x-y} - e^{-x+y} - e^{-x-y}) + (e^{x+y} - e^{x-y} + e^{-x+y} - e^{-x-y})]$$
$$= \tfrac{1}{4}(2e^{x+y} - 2e^{-x-y}) = \tfrac{1}{2}\left[e^{x+y} - e^{-(x+y)}\right] = \sinh(x+y)$$

12. $\cosh x \cosh y + \sinh x \sinh y = \left[\frac{1}{2}(e^{x} + e^{-x})\right]\left[\frac{1}{2}(e^{y} + e^{-y})\right] + \left[\frac{1}{2}(e^{x} - e^{-x})\right]\left[\frac{1}{2}(e^{y} - e^{-y})\right]$

$$= \tfrac{1}{4}[(e^{x+y} + e^{x-y} + e^{-x+y} + e^{-x-y}) + (e^{x+y} - e^{x-y} - e^{-x+y} + e^{-x-y})]$$
$$= \tfrac{1}{4}(2e^{x+y} + 2e^{-x-y}) = \tfrac{1}{2}\left[e^{x+y} + e^{-(x+y)}\right] = \cosh(x+y)$$

13. Divide both sides of the identity $\cosh^2 x - \sinh^2 x = 1$ by $\sinh^2 x$:

$$\frac{\cosh^2 x}{\sinh^2 x} - 1 = \frac{1}{\sinh^2 x} \quad \Leftrightarrow \quad \coth^2 x - 1 = \operatorname{csch}^2 x.$$

14. $$\tanh(x+y) = \frac{\sinh(x+y)}{\cosh(x+y)} = \frac{\sinh x \cosh y + \cosh x \sinh y}{\cosh x \cosh y + \sinh x \sinh y} = \frac{\dfrac{\sinh x \cosh y}{\cosh x \cosh y} + \dfrac{\cosh x \sinh y}{\cosh x \cosh y}}{\dfrac{\cosh x \cosh y}{\cosh x \cosh y} + \dfrac{\sinh x \sinh y}{\cosh x \cosh y}}$$
$$= \frac{\tanh x + \tanh y}{1 + \tanh x \tanh y}$$

15. By Exercise 11, $\sinh 2x = \sinh(x+x) = \sinh x \cosh x + \cosh x \sinh x = 2 \sinh x \cosh x$.

16. Putting $y = x$ in the result from Exercise 12, we have

$\cosh 2x = \cosh(x+x) = \cosh x \cosh x + \sinh x \sinh x = \cosh^2 x + \sinh^2 x$.

17. $$\tanh(\ln x) = \frac{\sinh(\ln x)}{\cosh(\ln x)} = \frac{(e^{\ln x} - e^{-\ln x})/2}{(e^{\ln x} + e^{-\ln x})/2} = \frac{x - 1/x}{x + 1/x} = \frac{x^2 - 1}{x^2 + 1}$$

18. $$\frac{1 + (\sinh x)/\cosh x}{1 - (\sinh x)/\cosh x} = \frac{\cosh x + \sinh x}{\cosh x - \sinh x} = \frac{\frac{1}{2}(e^{x} + e^{-x}) + \frac{1}{2}(e^{x} - e^{-x})}{\frac{1}{2}(e^{x} + e^{-x}) - \frac{1}{2}(e^{x} - e^{-x})}$$
$$= \frac{e^{x} + e^{-x} + e^{x} - e^{-x}}{e^{x} + e^{-x} - e^{x} + e^{-x}} = \frac{2e^{x}}{2e^{-x}} = e^{2x}$$

19. By Exercise 9, $(\cosh x + \sinh x)^n = (e^{x})^n = e^{nx} = \cosh nx + \sinh nx$.

20. $\sinh x = \frac{3}{4} \;\Rightarrow\; \operatorname{csch} x = 1/\sinh x = \frac{4}{3}$. $\cosh^2 x = \sinh^2 x + 1 = \frac{9}{16} + 1 = \frac{25}{16} \;\Rightarrow\; \cosh x = \frac{5}{4}$ (since $\cosh x > 0$). $\operatorname{sech} x = 1/\cosh x = \frac{4}{5}$, $\tanh x = \sinh x/\cosh x = \frac{3/4}{5/4} = \frac{3}{5}$, and $\coth x = 1/\tanh x = \frac{5}{3}$.

21. $\tanh x = \frac{4}{5} > 0$, so $x > 0$. $\coth x = 1/\tanh x = \frac{5}{4}$, $\operatorname{sech}^2 x = 1 - \tanh^2 x = 1 - \left(\frac{4}{5}\right)^2 = \frac{9}{25} \;\Rightarrow\; \operatorname{sech} x = \frac{3}{5}$ (since $\operatorname{sech} x > 0$), $\cosh x = 1/\operatorname{sech} x = \frac{5}{3}$, $\sinh x = \tanh x \cosh x = \frac{4}{5} \cdot \frac{5}{3} = \frac{4}{3}$, and $\operatorname{csch} x = 1/\sinh x = \frac{3}{4}$.

22.

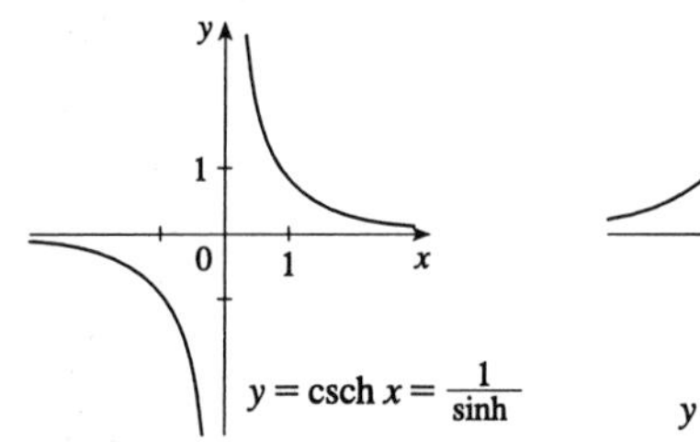

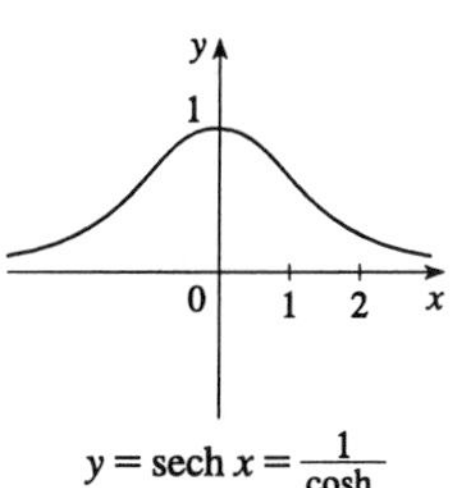

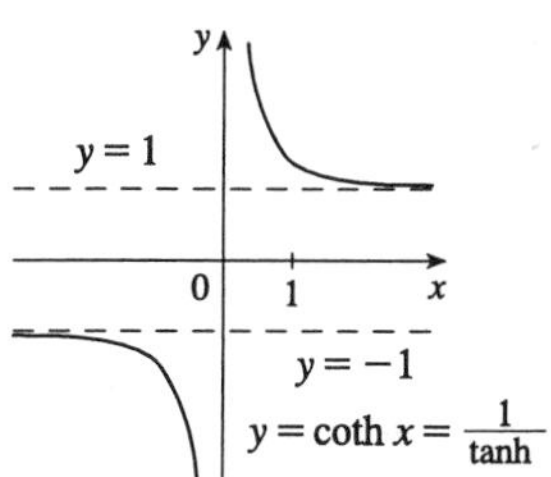

23. **(a)** $\displaystyle\lim_{x\to\infty} \tanh x = \lim_{x\to\infty} \frac{e^x - e^{-x}}{e^x + e^{-x}} = \lim_{x\to\infty} \frac{1 - e^{-2x}}{1 + e^{-2x}} = \frac{1-0}{1+0} = 1$

(b) $\displaystyle\lim_{x\to-\infty} \tanh x = \lim_{x\to-\infty} \frac{e^x - e^{-x}}{e^x + e^{-x}} = \lim_{x\to-\infty} \frac{e^{2x} - 1}{e^{2x} + 1} = \frac{0-1}{0+1} = -1$

(c) $\displaystyle\lim_{x\to\infty} \sinh x = \lim_{x\to\infty} \frac{e^x - e^{-x}}{2} = \infty$

(d) $\displaystyle\lim_{x\to-\infty} \sinh x = \lim_{x\to-\infty} \frac{e^x - e^{-x}}{2} = -\infty$

(e) $\displaystyle\lim_{x\to\infty} \operatorname{sech} x = \lim_{x\to\infty} \frac{2}{e^x + e^{-x}} = 0$

(f) $\displaystyle\lim_{x\to\infty} \coth x = \lim_{x\to\infty} \frac{e^x + e^{-x}}{e^x - e^{-x}} = \lim_{x\to\infty} \frac{1 + e^{-2x}}{1 - e^{-2x}} = \frac{1+0}{1-0} = 1$ [*Or:* Use part (a)]

(g) $\displaystyle\lim_{x\to0^+} \coth x = \lim_{x\to0^+} \frac{\cosh x}{\sinh x} = \infty$, since $\sinh x \to 0$ and $\coth x > 0$.

(h) $\displaystyle\lim_{x\to0^-} \coth x = \lim_{x\to0^-} \frac{\cosh x}{\sinh x} = -\infty$, since $\sinh x \to 0$ and $\coth x < 0$.

(i) $\displaystyle\lim_{x\to-\infty} \operatorname{csch} x = \lim_{x\to-\infty} \frac{2}{e^x - e^{-x}} = 0$

24. **(a)** $\displaystyle\frac{d}{dx}\cosh x = \frac{d}{dx}\left[\tfrac{1}{2}(e^x + e^{-x})\right] = \tfrac{1}{2}(e^x - e^{-x}) = \sinh x$

(b) $\displaystyle\frac{d}{dx}\tanh x = \frac{d}{dx}\left[\frac{\sinh x}{\cosh x}\right] = \frac{\cosh x \cosh x - \sinh x \sinh x}{\cosh^2 x} = \frac{\cosh^2 x - \sinh^2 x}{\cosh^2 x} = \frac{1}{\cosh^2 x} = \operatorname{sech}^2 x$

(c) $\displaystyle\frac{d}{dx}\operatorname{csch} x = \frac{d}{dx}\left[\frac{1}{\sinh x}\right] = -\frac{\cosh x}{\sinh^2 x} = -\frac{1}{\sinh x}\cdot\frac{\cosh x}{\sinh x} = -\operatorname{csch} x \coth x$

(d) $\displaystyle\frac{d}{dx}\operatorname{sech} x = \frac{d}{dx}\left[\frac{1}{\cosh x}\right] = -\frac{\sinh x}{\cosh^2 x} = -\frac{1}{\cosh x}\cdot\frac{\sinh x}{\cosh x} = -\operatorname{sech} x \tanh x$

(e) $\displaystyle\frac{d}{dx}\coth x = \frac{d}{dx}\left[\frac{\cosh x}{\sinh x}\right] = \frac{\sinh x \sinh x - \cosh x \cosh x}{\sinh^2 x} = \frac{\sinh^2 x - \cosh^2 x}{\sinh^2 x} = -\frac{1}{\sinh^2 x} = -\operatorname{csch}^2 x$

25. Let $y = \sinh^{-1} x$. Then $\sinh y = x$ and, by Example 1(a), $\cosh y = \sqrt{1 + \sinh^2 y} = \sqrt{1 + x^2}$. So by Exercise 9, $e^y = \sinh y + \cosh y = x + \sqrt{1+x^2} \quad\Rightarrow\quad y = \ln\!\left(x + \sqrt{1+x^2}\,\right)$.

26. Let $y = \cosh^{-1} x$. Then $\cosh y = x$ and $y \geq 0$, so $\sinh y = \sqrt{\cosh^2 y - 1} = \sqrt{x^2 - 1}$. So, by Exercise 9, $e^y = \cosh y + \sinh y = x + \sqrt{x^2 - 1} \quad\Rightarrow\quad y = \ln\!\left(x + \sqrt{x^2-1}\,\right)$.

Another method: Write $x = \cosh y = \frac{1}{2}(e^y + e^{-y})$ and solve a quadratic, as in Example 3.

27. **(a)** Let $y = \tanh^{-1} x$. Then $x = \tanh y = \dfrac{e^y - e^{-y}}{e^y + e^{-y}} = \dfrac{e^{2y} - 1}{e^{2y} + 1} \quad \Rightarrow \quad xe^{2y} + x = e^{2y} - 1 \quad \Rightarrow$

$$e^{2y} = \frac{1+x}{1-x} \quad \Rightarrow \quad 2y = \ln\left(\frac{1+x}{1-x}\right) \quad \Rightarrow \quad y = \frac{1}{2}\ln\left(\frac{1+x}{1-x}\right).$$

(b) Let $y = \tanh^{-1} x$. Then $x = \tanh y$, so from Exercise 18 we have $e^{2y} = \dfrac{1 + \tanh y}{1 - \tanh y} = \dfrac{1+x}{1-x} \quad \Rightarrow$

$$2y = \ln\left(\frac{1+x}{1-x}\right) \quad \Rightarrow \quad y = \frac{1}{2}\ln\left(\frac{1+x}{1-x}\right).$$

28. **(a)** **(i)** $y = \operatorname{csch}^{-1} x \quad \Leftrightarrow \quad \operatorname{csch} y = x \quad (x \neq 0)$

(ii) We sketch the graph of csch^{-1} by reflecting the graph of csch (see Exercise 22) about the line $y = x$.

(iii) Let $y = \operatorname{csch}^{-1} x$. Then $x = \operatorname{csch} y = \dfrac{2}{e^y - e^{-y}} \quad \Rightarrow \quad xe^y - xe^{-y} = 2$

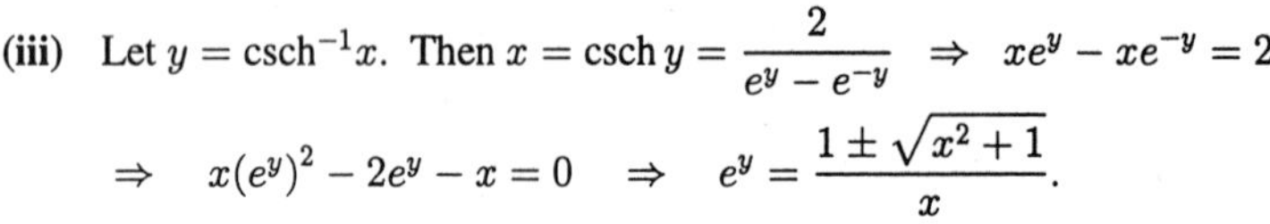

$$\Rightarrow \quad x(e^y)^2 - 2e^y - x = 0 \quad \Rightarrow \quad e^y = \frac{1 \pm \sqrt{x^2+1}}{x}.$$

But $e^y > 0$, so for $x > 0$, $e^y = \dfrac{1 + \sqrt{x^2+1}}{x}$ and for $x < 0$, $e^y = \dfrac{1 - \sqrt{x^2+1}}{x}$.

Thus $\operatorname{csch}^{-1} x = \ln\left(\dfrac{1}{x} + \dfrac{\sqrt{x^2+1}}{|x|}\right)$.

(b) **(i)** $y = \operatorname{sech}^{-1} x \quad \Leftrightarrow \quad \operatorname{sech} y = x$ and $y > 0$.

(ii) We sketch the graph of sech^{-1} by reflecting the graph of sech (see Exercise 22) about the line $y = x$.

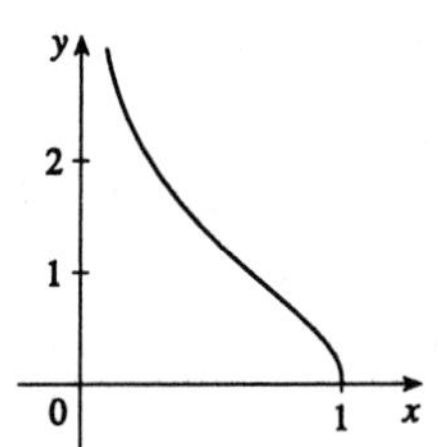

(iii) Let $y = \operatorname{sech}^{-1} x$, so $x = \operatorname{sech} y = \dfrac{2}{e^y + e^{-y}} \quad \Rightarrow \quad xe^y + xe^{-y} = 2$

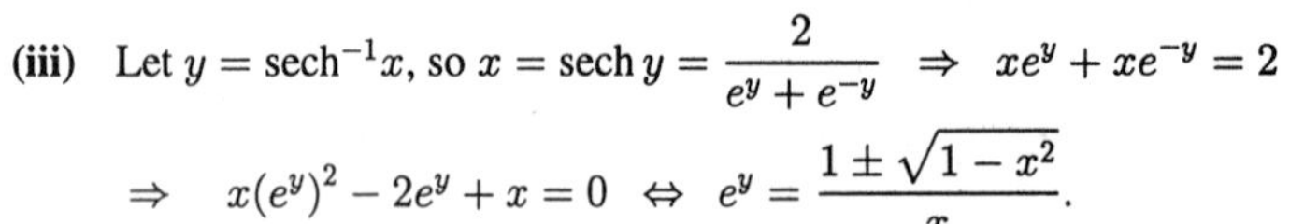

$$\Rightarrow \quad x(e^y)^2 - 2e^y + x = 0 \quad \Leftrightarrow \quad e^y = \frac{1 \pm \sqrt{1-x^2}}{x}.$$

But $y > 0 \;\Rightarrow\; e^y > 1$. This rules out the minus sign because $\dfrac{1 - \sqrt{1-x^2}}{x} > 1 \quad \Leftrightarrow$

$1 - \sqrt{1-x^2} > x \quad \Leftrightarrow \quad 1 - x > \sqrt{1-x^2} \quad \Leftrightarrow \quad 1 - 2x + x^2 > 1 - x^2 \quad \Leftrightarrow \quad x^2 > x \quad \Leftrightarrow$

$x > 1$, but $x = \sec y \leq 1$. Thus $e^y = \dfrac{1 + \sqrt{1-x^2}}{x} \quad \Rightarrow \quad \operatorname{sech}^{-1} x = \ln\left(\dfrac{1 + \sqrt{1-x^2}}{x}\right)$.

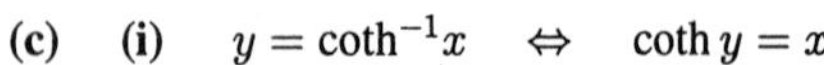

(c) **(i)** $y = \coth^{-1} x \quad \Leftrightarrow \quad \coth y = x$

(ii) We sketch the graph of $\coth^{-1}$ by reflecting the graph of coth (see Exercise 22) about the line $y = x$.

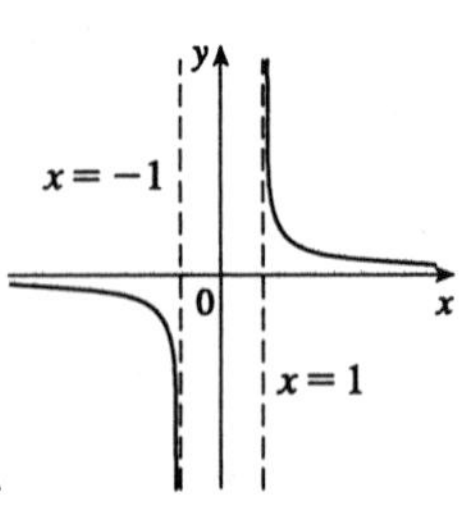

(iii) Let $y = \coth^{-1} x$. Then $x = \coth y = \dfrac{e^y + e^{-y}}{e^y - e^{-y}} \quad \Rightarrow$

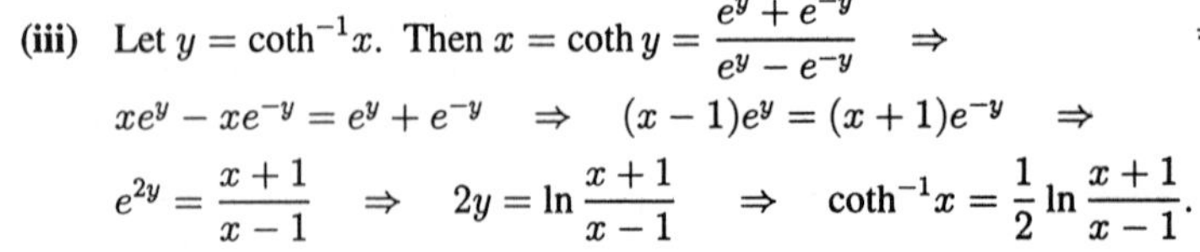

$xe^y - xe^{-y} = e^y + e^{-y} \quad \Rightarrow \quad (x-1)e^y = (x+1)e^{-y} \quad \Rightarrow$

$$e^{2y} = \frac{x+1}{x-1} \quad \Rightarrow \quad 2y = \ln\frac{x+1}{x-1} \quad \Rightarrow \quad \coth^{-1} x = \frac{1}{2}\ln\frac{x+1}{x-1}.$$

29. **(a)** Let $y = \cosh^{-1}x$. Then $\cosh y = x$ and $y \geq 0 \quad \Rightarrow \quad \sinh y\,\dfrac{dy}{dx} = 1 \quad \Rightarrow$

$\dfrac{dy}{dx} = \dfrac{1}{\sinh y} = \dfrac{1}{\sqrt{\cosh^2 y - 1}} = \dfrac{1}{\sqrt{x^2 - 1}}$ (since $\sinh y \geq 0$ for $y \geq 0$). *Or:* Use Formula 4.

(b) Let $y = \tanh^{-1}x$. Then $\tanh y = x \Rightarrow \operatorname{sech}^2 y\,\dfrac{dy}{dx} = 1 \Rightarrow \dfrac{dy}{dx} = \dfrac{1}{\operatorname{sech}^2 y} = \dfrac{1}{1 - \tanh^2 y} = \dfrac{1}{1 - x^2}$.

Or: Use Formula 5.

(c) Let $y = \operatorname{csch}^{-1}x$. Then $\operatorname{csch} y = x \quad \Rightarrow \quad -\operatorname{csch} y \coth y\,\dfrac{dy}{dx} = 1 \quad \Rightarrow \quad \dfrac{dy}{dx} = -\dfrac{1}{\operatorname{csch} y \coth y}$.

By Exercise 13, $\coth y = \pm\sqrt{\operatorname{csch}^2 y + 1} = \pm\sqrt{x^2 + 1}$. If $x > 0$, then $\coth y > 0$, so $\coth y = \sqrt{x^2 + 1}$. If $x < 0$, then $\coth y < 0$, so $\coth y = -\sqrt{x^2 + 1}$. In either case we have

$\dfrac{dy}{dx} = -\dfrac{1}{\operatorname{csch} y \coth y} = -\dfrac{1}{|x|\sqrt{x^2 + 1}}$.

(d) Let $y = \operatorname{sech}^{-1}x$. Then $\operatorname{sech} y = x \quad \Rightarrow \quad -\operatorname{sech} y \tanh y\,\dfrac{dy}{dx} = 1 \quad \Rightarrow$

$\dfrac{dy}{dx} = -\dfrac{1}{\operatorname{sech} y \tanh y} = -\dfrac{1}{\operatorname{sech} y\,\sqrt{1 - \operatorname{sech}^2 y}} = -\dfrac{1}{x\sqrt{1 - x^2}}$. (Note that $y > 0$ and so $\tanh y > 0$.)

(e) Let $y = \coth^{-1}x$. Then $\coth y = x \Rightarrow -\operatorname{csch}^2 y\,\dfrac{dy}{dx} = 1 \Rightarrow \dfrac{dy}{dx} = -\dfrac{1}{\operatorname{csch}^2 y} = \dfrac{1}{1 - \coth^2 y} = \dfrac{1}{1 - x^2}$

by Exercise 13.

30. $f(x) = e^x \sinh x \quad \Rightarrow \quad f'(x) = e^x \sinh x + e^x \cosh x$

31. $f(x) = \tanh 3x \quad \Rightarrow \quad f'(x) = 3\operatorname{sech}^2 3x$

32. $g(x) = \cosh^4 x \quad \Rightarrow \quad g'(x) = 4\cosh^3 x \sinh x$

33. $h(x) = \cosh(x^4) \quad \Rightarrow \quad h'(x) = \sinh(x^4)4x^3 = 4x^3 \sinh(x^4)$

34. $F(x) = e^{\coth 2x} \quad \Rightarrow \quad F'(x) = e^{\coth 2x}(-\operatorname{csch}^2 2x)(2) = -2e^{\coth 2x}\operatorname{csch}^2 2x$

35. $G(x) = x^2 \operatorname{sech} x \quad \Rightarrow \quad G'(x) = 2x \operatorname{sech} x - x^2 \operatorname{sech} x \tanh x$

36. $f(t) = \ln(\sinh t) \quad \Rightarrow \quad f'(t) = \dfrac{1}{\sinh t}\cosh t = \coth t$

37. $H(t) = \tanh(e^t) \quad \Rightarrow \quad H'(t) = \operatorname{sech}^2(e^t)[e^t] = e^t \operatorname{sech}^2(e^t)$

38. $y = \cos(\sinh x) \quad \Rightarrow \quad y' = -\sin(\sinh x)\cosh x$

39. $y = x^{\cosh x} \quad \Rightarrow \ln y = \cosh x \ln x \quad \Rightarrow \quad \dfrac{y'}{y} = \sinh x \ln x + \dfrac{\cosh x}{x} \quad \Rightarrow \quad y' = x^{\cosh x}\left(\sinh x \ln x + \dfrac{\cosh x}{x}\right)$

40. $y = e^{\tanh x}\cosh(\cosh x) \quad \Rightarrow \quad y' = e^{\tanh x}\operatorname{sech}^2 x \cosh(\cosh x) + e^{\tanh x}\sinh(\cosh x)\sinh x$

41. $y = \cosh^{-1}(x^2) \quad \Rightarrow \quad y' = \left(1\Big/\sqrt{(x^2)^2 - 1}\right)(2x) = 2x/\sqrt{x^4 - 1}$

42. $y = \sqrt{x}\sinh^{-1}\sqrt{x} \quad \Rightarrow \quad y' = \dfrac{1}{2\sqrt{x}}\sinh^{-1}\sqrt{x} + \sqrt{x}\dfrac{1}{\sqrt{1 + (\sqrt{x})^2}}\dfrac{1}{2\sqrt{x}} = \dfrac{1}{2\sqrt{x}}\sinh^{-1}\sqrt{x} + \dfrac{1}{2\sqrt{1 + x}}$

43. $y = x\ln(\operatorname{sech} 4x) \quad\Rightarrow\quad y' = \ln(\operatorname{sech} 4x) + x\,\dfrac{-\operatorname{sech} 4x\tanh 4x}{\operatorname{sech} 4x}(4) = \ln(\operatorname{sech} 4x) - 4x\tanh 4x$

44. $y = x\tanh^{-1}x + \ln\sqrt{1-x^2} = x\tanh^{-1}x + \frac{1}{2}\ln(1-x^2) \quad\Rightarrow$

$$y' = \tanh^{-1}x + \frac{x}{1-x^2} + \frac{1}{2}\left(\frac{1}{1-x^2}\right)(-2x) = \tanh^{-1}x$$

45. $y = x\sinh^{-1}(x/3) - \sqrt{9+x^2} \quad\Rightarrow$

$$y' = \sinh^{-1}\left(\frac{x}{3}\right) + x\frac{1/3}{\sqrt{1+(x/3)^2}} - \frac{2x}{2\sqrt{9+x^2}} = \sinh^{-1}\left(\frac{x}{3}\right) + \frac{x}{\sqrt{9+x^2}} - \frac{x}{\sqrt{9+x^2}} = \sinh^{-1}\left(\frac{x}{3}\right)$$

46. $y = \operatorname{sech}^{-1}\sqrt{1-x^2} \quad\Rightarrow\quad y' = -\dfrac{1}{\sqrt{1-x^2}\sqrt{1-(1-x^2)}}\dfrac{-2x}{2\sqrt{1-x^2}} = \dfrac{x}{(1-x^2)|x|}$

47. $y = \coth^{-1}\sqrt{x^2+1} \quad\Rightarrow\quad y' = \dfrac{1}{1-(x^2+1)}\dfrac{2x}{2\sqrt{x^2+1}} = -\dfrac{1}{x\sqrt{x^2+1}}$

48. **(a)** $f(x) = \sinh x - (x-1)\cosh x \quad\Rightarrow\quad f'(x) = \cosh x - [(x-1)\sinh x + \cosh x(1)] = (1-x)\sinh x.$ This is 0 when $\sinh x = 0$ or $1 - x = 0$, that is, $x = 0$ or 1. We see that f' changes from negative to positive at $x = 0$, then back to negative at $x = 1$, so there is a local minimum of $f(0) = \sinh 0 - (0-1)\cosh 0 = 1$, and a local maximum of $f(1) = \sinh 1 - (1-1)\cosh 1 = \sinh 1 = \frac{1}{2}(e - 1/e)$.

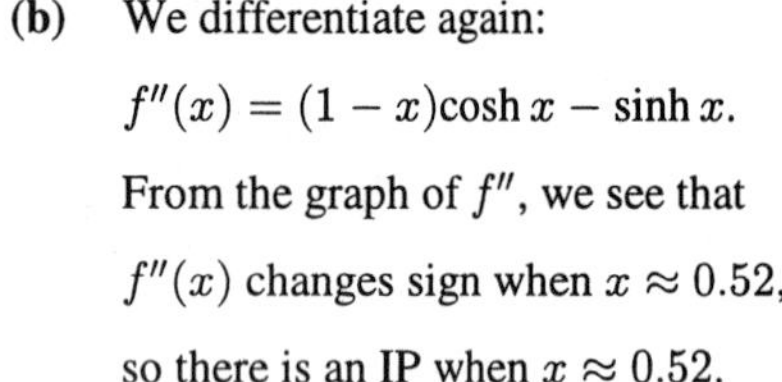

(b) We differentiate again: $f''(x) = (1-x)\cosh x - \sinh x.$ From the graph of f'', we see that $f''(x)$ changes sign when $x \approx 0.52$, so there is an IP when $x \approx 0.52$.

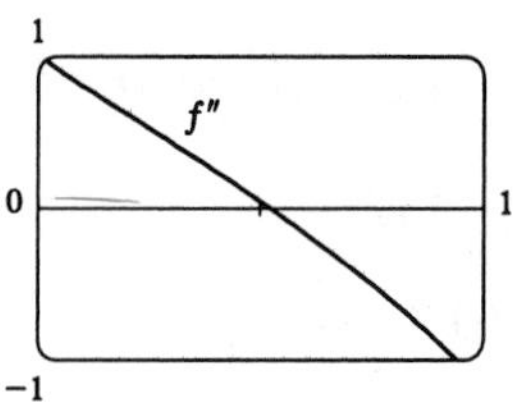

(c)

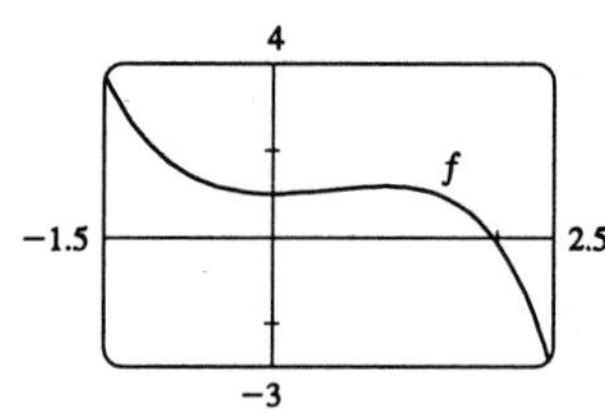

49. The tangent to $y = \cosh x$ has slope 1 when $y' = \sinh x = 1 \quad\Rightarrow\quad x = \sinh^{-1}1 = \ln\left(1+\sqrt{2}\right)$, by Equation 3. Since $\sinh x = 1$ and $y = \cosh x = \sqrt{1+\sinh^2 x}$, we have $\cosh x = \sqrt{2}$. The point is $\left(\ln\left(1+\sqrt{2}\right), \sqrt{2}\right)$.

50. $\int \operatorname{sech}^2 x\,dx = \tanh x + C$

51. $u = 2x \quad\Rightarrow\quad du = 2\,dx$, so $\int \sinh 2x\,dx = \frac{1}{2}\int \sinh u\,du = \frac{1}{2}\cosh u + C = \frac{1}{2}\cosh 2x + C$.

52. Let $u = \cosh x$. Then $du = \sinh x\,dx$, and

$$\int \tanh x\,dx = \int \frac{\sinh x}{\cosh x}\,dx = \int \frac{du}{u} = \ln|u| + C = \ln(\cosh x) + C.$$

53. Let $u = \sinh x$, so $du = \cosh x\,dx$, and $\displaystyle\int \coth x\,dx = \int \frac{\cosh x}{\sinh x}\,dx = \int \frac{du}{u} = \ln|u| + C = \ln|\sinh x| + C$.

54. Let $u = 1 + \cosh x$. Then $du = \sinh x\,dx \quad\Rightarrow\quad \displaystyle\int \frac{\sinh x}{1+\cosh x}\,dx = \int \frac{du}{u} = \ln|u| + C = \ln(1+\cosh x) + C$.

55. Let $u = x/2$. Then $du = \frac{1}{2}\,dx \quad\Rightarrow$

$$\int \frac{1}{\sqrt{4+x^2}}\,dx = \frac{1}{2}\int \frac{1}{\sqrt{1+(x/2)^2}}\,dx = \int \frac{1}{\sqrt{1+u^2}}\,du = \sinh^{-1}u + C = \sinh^{-1}\left(\frac{x}{2}\right) + C.$$

56. $\displaystyle\int_2^3 \frac{1}{\sqrt{x^2-1}}\,dx = \left[\cosh^{-1}x\right]_2^3 = \cosh^{-1}3 - \cosh^{-1}2$. Using Equation 4, we could write this as $\ln\!\left(3+2\sqrt{2}\right) - \ln\!\left(2+\sqrt{3}\right) = \ln\!\left[\left(3+2\sqrt{2}\right)/\left(2+\sqrt{3}\right)\right]$.

57. $\displaystyle\int_0^{1/2} \frac{1}{1-x^2}\,dx = \left[\tanh^{-1}x\right]_0^{1/2} = \tanh^{-1}\tfrac{1}{2} = \tfrac{1}{2}\ln\!\left(\frac{1+1/2}{1-1/2}\right)$ (from Equation 5) $= \tfrac{1}{2}\ln 3$.

58. We want $\int_0^1 \sinh cx\,dx = 1$. To calculate the integral, we put $u = cx$, so $du = c\,dx$, the upper limit becomes c, and the equation becomes $\displaystyle\frac{1}{c}\int_0^c \sinh u\,du = 1 \quad\Leftrightarrow\quad \frac{1}{c}[\cosh c - 1] = 1$ $\Leftrightarrow\quad \cosh c - 1 = c$. We plot the function $f(c) = \cosh c - c - 1$, and see that its positive root lies at approximately $c = 1.62$. So the equation $\int_0^1 \sinh cx\,dx = 1$ holds for $c \approx 1.62$.

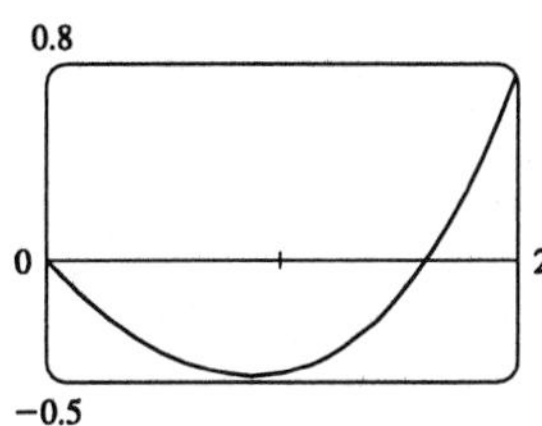

59. **(a)** From the graphs, we estimate that the two curves $y = \cosh 2x$ and $y = 1 + \sinh x$ intersect at $x = 0$ and at $x \approx 0.481$.

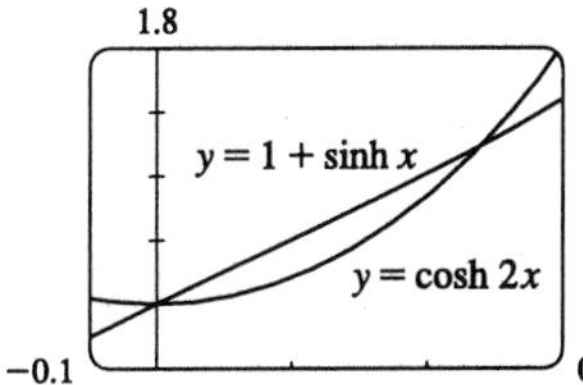

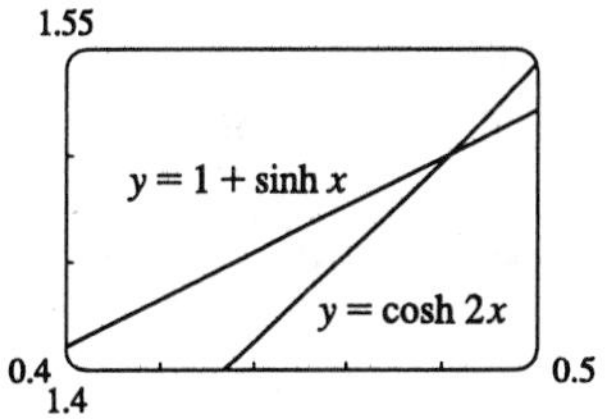

(b) We have found the two roots of the equation $\cosh 2x = 1 + \sinh x$ to be $x = 0$ and $x \approx 0.481$. Note from the first graph that $1 + \sinh x > \cosh 2x$ on the interval $(0, 0.481)$, so the area between the two curves is

$$A \approx \int_0^{0.481}(1 + \sinh x - \cosh 2x)dx = \left[x + \cosh x - \tfrac{1}{2}\sinh 2x\right]_0^{0.481}$$
$$= \left[0.481 + \cosh 0.481 - \tfrac{1}{2}\sinh(2\cdot 0.481)\right] - \left[0 + \cosh 0 - \tfrac{1}{2}\sinh(2\cdot 0)\right] \approx 0.0402.$$

60. $\displaystyle\lim_{x\to\infty}\frac{\sinh x}{e^x} = \lim_{x\to\infty}\frac{e^x - e^{-x}}{2e^x} = \lim_{x\to\infty}\frac{1 - e^{-2x}}{2} = \frac{1-0}{2} = \frac{1}{2}$

61. **(a)** $y = A\sinh mx + B\cosh mx \quad\Rightarrow\quad y' = mA\cosh mx + mB\sinh mx \quad\Rightarrow$
$y'' = m^2A\sinh mx + m^2B\cosh mx = m^2y$

(b) From part (a), a solution of $y'' = 9y$ is $y(x) = A\sinh 3x + B\cosh 3x$. So $-4 = y(0) = A\sinh 0 + B\cosh 0 = B$, so $B = -4$. Now $y'(x) = 3A\cosh 3x - 12\sinh 3x \quad\Rightarrow$ $6 = y'(0) = 3A \quad\Rightarrow\quad A = 2$, so $y = 2\sinh 3x - 4\cosh 3x$.

62. We differentiate the function twice, then substitute into the differential equation: $\displaystyle y = \frac{T}{\rho g}\cosh\frac{\rho g x}{T} \quad\Rightarrow$

$$\frac{dy}{dx} = \frac{T}{\rho g}\sinh\!\left(\frac{\rho g x}{T}\right)\frac{\rho g}{T} = \sinh\frac{\rho g x}{T} \quad\Rightarrow\quad \frac{d^2y}{dx^2} = \cosh\!\left(\frac{\rho g x}{T}\right)\frac{\rho g}{T} = \frac{\rho g}{T}\cosh\frac{\rho g x}{T}.$$

We evaluate the two sides separately: $\displaystyle \text{LHS} = \frac{d^2y}{dx^2} = \frac{\rho g}{T}\cosh\frac{\rho g x}{T}$,

$\displaystyle \text{RHS} = \frac{\rho g}{T}\sqrt{1 + \left(\frac{dy}{dx}\right)^2} = \frac{\rho g}{T}\sqrt{1 + \sinh^2\frac{\rho g x}{T}} = \frac{\rho g}{T}\cosh\frac{\rho g x}{T}$, by the identity proved in Example 1(a).

63. $\cosh x = \cosh[\ln(\sec\theta + \tan\theta)] = \frac{1}{2}\left[e^{\ln(\sec\theta+\tan\theta)} + e^{-\ln(\sec\theta+\tan\theta)}\right]$

$$= \frac{1}{2}\left[\sec\theta + \tan\theta + \frac{1}{\sec\theta + \tan\theta}\right] = \frac{1}{2}\left[\sec\theta + \tan\theta + \frac{\sec\theta - \tan\theta}{(\sec\theta + \tan\theta)(\sec\theta - \tan\theta)}\right]$$

$$= \frac{1}{2}\left[\sec\theta + \tan\theta + \frac{\sec\theta - \tan\theta}{\sec^2\theta - \tan^2\theta}\right] = \tfrac{1}{2}(\sec\theta + \tan\theta + \sec\theta - \tan\theta) = \sec\theta.$$

64. The area of the triangle with vertices O, P, and $(\cosh t, 0)$ is $\frac{1}{2}\sinh t\cosh t$, and the area under the curve $x^2 - y^2 = 1$, from $x = 1$ to $x = \cosh t$, is $\int_1^{\cosh t}\sqrt{x^2-1}\,dx$. Therefore, the area of the shaded region is $A(t) = \frac{1}{2}\sinh t\cosh t - \int_1^{\cosh t}\sqrt{x^2-1}\,dx$. So, by Part 1 of the Fundamental Theorem of Calculus,

$$A'(t) = \tfrac{1}{2}(\cosh^2 t + \sinh^2 t) - \sqrt{\cosh^2 t - 1}\,\sinh t = \tfrac{1}{2}(\cosh^2 t + \sinh^2 t) - \sqrt{\sinh^2 t}\,\sinh t$$

$$= \tfrac{1}{2}(\cosh^2 t + \sinh^2 t) - \sinh^2 t = \tfrac{1}{2}(\cosh^2 t - \sinh^2 t) = \tfrac{1}{2}(1) = \tfrac{1}{2}.$$

Thus $A(t) = \frac{1}{2}t + C$ since $A'(t) = \frac{1}{2}$. To calculate C, we let $t = 0$. Thus,

$A(0) = \frac{1}{2}\sinh 0\cosh 0 - \int_1^{\cosh 0}\sqrt{x^2-1}\,dx = \frac{1}{2}(0) + C \quad\Rightarrow\quad C = 0$. Thus $A(t) = \frac{1}{2}t$.

EXERCISES 6.8

NOTE: The use of l'Hospital's Rule is indicated by an H above the equal sign: $\overset{\text{H}}{=}$

1. $\displaystyle\lim_{x\to 2}\frac{x-2}{x^2-4} = \lim_{x\to 2}\frac{x-2}{(x-2)(x+2)} = \lim_{x\to 2}\frac{1}{x+2} = \frac{1}{4}$

2. $\displaystyle\lim_{x\to 1}\frac{x^2+3x-4}{x-1} = \lim_{x\to 1}\frac{(x-1)(x+4)}{x-1} = \lim_{x\to 1}(x+4) = 5$

3. $\displaystyle\lim_{x\to -1}\frac{x^6-1}{x^4-1} \overset{\text{H}}{=} \lim_{x\to -1}\frac{6x^5}{4x^3} = \frac{-6}{-4} = \frac{3}{2}$

4. $\displaystyle\lim_{x\to 1}\frac{x^a-1}{x^b-1} \overset{\text{H}}{=} \lim_{x\to 1}\frac{ax^{a-1}}{bx^{b-1}} = \frac{a}{b}$

5. $\displaystyle\lim_{x\to 0}\frac{e^x-1}{\sin x} \overset{\text{H}}{=} \lim_{x\to 0}\frac{e^x}{\cos x} = \frac{1}{1} = 1$

6. $\displaystyle\lim_{x\to 1}\frac{\ln x}{x-1} \overset{\text{H}}{=} \lim_{x\to 1}\frac{1/x}{1} = 1$

7. $\displaystyle\lim_{x\to 0}\frac{\sin x}{x^3} \overset{\text{H}}{=} \lim_{x\to 0}\frac{\cos x}{3x^2} = \infty$

8. $\displaystyle\lim_{x\to \pi}\frac{\tan x}{x} = \frac{\tan\pi}{\pi} = \frac{0}{\pi} = 0$

9. $\displaystyle\lim_{x\to 0}\frac{\tan x}{x+\sin x} \overset{\text{H}}{=} \lim_{x\to 0}\frac{\sec^2 x}{1+\cos x} = \frac{1}{1+1} = \frac{1}{2}$

10. $\displaystyle\lim_{x\to 3\pi/2}\frac{\cos x}{x-3\pi/2} \overset{\text{H}}{=} \lim_{x\to 3\pi/2}\frac{-\sin x}{1} = -\sin\tfrac{3\pi}{2} = 1$

11. $\displaystyle\lim_{x\to \infty}\frac{\ln x}{x} \overset{\text{H}}{=} \lim_{x\to \infty}\frac{1/x}{1} = 0$

12. $\displaystyle\lim_{x\to 0^+}\frac{\ln x}{\sqrt{x}} = -\infty$ since $\ln x \to -\infty$ and $\sqrt{x} \to 0^+$.

13. $\displaystyle\lim_{x\to \infty}\frac{e^x}{x^3} \overset{\text{H}}{=} \lim_{x\to \infty}\frac{e^x}{3x^2} \overset{\text{H}}{=} \lim_{x\to \infty}\frac{e^x}{6x} \overset{\text{H}}{=} \lim_{x\to \infty}\frac{e^x}{6} = \infty$

14. $\displaystyle\lim_{x\to \infty}\frac{(\ln x)^3}{x^2} \overset{\text{H}}{=} \lim_{x\to \infty}\frac{3(\ln x)^2(1/x)}{2x} = \lim_{x\to \infty}\frac{3(\ln x)^2}{2x^2} \overset{\text{H}}{=} \lim_{x\to \infty}\frac{6(\ln x)(1/x)}{4x}$

$$= \lim_{x\to \infty}\frac{3\ln x}{2x^2} \overset{\text{H}}{=} \lim_{x\to \infty}\frac{3/x}{4x} = \lim_{x\to \infty}\frac{3}{4x^2} = 0$$

15. $\displaystyle\lim_{x\to a}\frac{x^{1/3}-a^{1/3}}{x-a} \overset{\text{H}}{=} \lim_{x\to a}\frac{(1/3)x^{-2/3}}{1} = \frac{1}{3a^{2/3}}$

16. $\lim_{x\to 0}\frac{6^x-2^x}{x}\overset{\text{H}}{=}\lim_{x\to 0}\frac{6^x(\ln 6)-2^x(\ln 2)}{1}=\ln 6-\ln 2=\ln\frac{6}{2}=\ln 3$

17. $\lim_{x\to 0}\frac{e^x-1-x}{x^2}\overset{\text{H}}{=}\lim_{x\to 0}\frac{e^x-1}{2x}\overset{\text{H}}{=}\lim_{x\to 0}\frac{e^x}{2}=\frac{1}{2}$

18. $\lim_{x\to 0}\frac{e^x-1-x-x^2/2}{x^3}\overset{\text{H}}{=}\lim_{x\to 0}\frac{e^x-1-x}{3x^2}\overset{\text{H}}{=}\lim_{x\to 0}\frac{e^x-1}{6x}\overset{\text{H}}{=}\lim_{x\to 0}\frac{e^x}{6}=\frac{1}{6}$

19. $\lim_{x\to 0}\frac{\sin x}{e^x}=\frac{0}{1}=0$

20. $\lim_{x\to 0}\frac{\sin^2 x}{\tan(x^2)}\overset{\text{H}}{=}\lim_{x\to 0}\frac{2\sin x\cos x}{2x\sec^2(x^2)}=\lim_{x\to 0}\frac{\sin x}{x}\lim_{x\to 0}\frac{\cos x}{\sec^2(x^2)}=1\cdot 1=1$

21. $\lim_{x\to 0}\frac{1-\cos x}{x^2}\overset{\text{H}}{=}\lim_{x\to 0}\frac{\sin x}{2x}\overset{\text{H}}{=}\lim_{x\to 0}\frac{\cos x}{2}=\frac{1}{2}$

22. $\lim_{x\to 0}\frac{\sin x-x}{x^3}\overset{\text{H}}{=}\lim_{x\to 0}\frac{\cos x-1}{3x^2}\overset{\text{H}}{=}\lim_{x\to 0}\frac{-\sin x}{6x}\overset{\text{H}}{=}\lim_{x\to 0}\frac{-\cos x}{6}=-\frac{1}{6}$

23. $\lim_{x\to 2^-}\frac{\ln x}{\sqrt{2-x}}=\infty$ since $\sqrt{2-x}\to 0$ but $\ln x\to\ln 2$

24. $\lim_{x\to 0}\frac{\sin x}{\sinh x}\overset{\text{H}}{=}\lim_{x\to 0}\frac{\cos x}{\cosh x}=\frac{1}{1}=1$

25. $\lim_{x\to\infty}\frac{\ln\ln x}{\sqrt{x}}\overset{\text{H}}{=}\lim_{x\to\infty}\frac{1/(x\ln x)}{1/(2\sqrt{x})}=\lim_{x\to\infty}\frac{2}{\sqrt{x}\ln x}=0$

26. $\lim_{x\to\infty}\frac{\ln(1+e^x)}{5x}\overset{\text{H}}{=}\lim_{x\to\infty}\frac{e^x/(1+e^x)}{5}=\lim_{x\to\infty}\frac{e^x}{5(1+e^x)}\overset{\text{H}}{=}\lim_{x\to\infty}\frac{e^x}{5e^x}=\frac{1}{5}$

27. $\lim_{x\to 0}\frac{\tan^{-1}(2x)}{3x}\overset{\text{H}}{=}\lim_{x\to 0}\frac{2/(1+4x^2)}{3}=\frac{2}{3}$

28. $\lim_{x\to 0}\frac{x}{\sin^{-1}(3x)}\overset{\text{H}}{=}\lim_{x\to 0}\frac{1}{3\big/\sqrt{1-(3x)^2}}=\lim_{x\to 0}\frac{1}{3}\sqrt{1-9x^2}=\frac{1}{3}$

29. $\lim_{x\to 0}\frac{\tan\alpha x}{x}\overset{\text{H}}{=}\lim_{x\to 0}\frac{\alpha\sec^2\alpha x}{1}=\alpha$

30. $\lim_{x\to 0}\frac{\sin mx}{\sin nx}\overset{\text{H}}{=}\lim_{x\to 0}\frac{m\cos mx}{n\cos nx}=\frac{m}{n}$

31. $\lim_{x\to 0}\frac{\tan 2x}{\tanh 3x}\overset{\text{H}}{=}\lim_{x\to 0}\frac{2\sec^2 2x}{3\,\text{sech}^2 3x}=\frac{2}{3}$

32. $\lim_{x\to 0}\frac{\sin^{10}x}{\sin(x^{10})}\overset{\text{H}}{=}\lim_{x\to 0}\frac{10\sin^9 x\cos x}{10x^9\cos(x^{10})}=\left[\lim_{x\to 0}\frac{\sin x}{x}\right]^9\lim_{x\to 0}\frac{\cos x}{\cos(x^{10})}=1^9\cdot 1=1$

33. $\lim_{x\to 0}\frac{x+\sin 3x}{x-\sin 3x}\overset{\text{H}}{=}\lim_{x\to 0}\frac{1+3\cos 3x}{1-3\cos 3x}=\frac{1+3}{1-3}=-2$

34. $\lim_{x\to 0}\frac{2x-\sin^{-1}x}{2x+\cos^{-1}x}=\frac{2(0)-0}{2(0)+\pi/2}=0$

35. $\lim_{x\to 0}\frac{e^{4x}-1}{\cos x}=\frac{0}{1}=0$

36. $\lim_{x\to 0}\frac{2x-\sin^{-1}x}{2x+\tan^{-1}x}\overset{\text{H}}{=}\lim_{x\to 0}\frac{2-1/\sqrt{1-x^2}}{2+1/(1+x^2)}=\frac{2-1}{2+1}=\frac{1}{3}$

37. $\lim_{x\to 0}\frac{\tan x-\sin x}{x^3}\overset{\text{H}}{=}\lim_{x\to 0}\frac{\sec^2 x-\cos x}{3x^2}\overset{\text{H}}{=}\lim_{x\to 0}\frac{2\sec^2 x\tan x+\sin x}{6x}$

$\overset{\text{H}}{=}\lim_{x\to 0}\frac{4\sec^2 x\tan^2 x+2\sec^4 x+\cos x}{6}=\frac{0+2+1}{6}=\frac{1}{2}$

38. $\lim_{x\to 0}\frac{\cos mx-\cos nx}{x^2}\overset{\text{H}}{=}\lim_{x\to 0}\frac{-m\sin mx+n\sin nx}{2x}=\lim_{x\to 0}\frac{-m^2\cos mx+n^2\cos nx}{2}=\frac{1}{2}(n^2-m^2)$

39. $\lim_{x\to 0^+}\sqrt{x}\ln x=\lim_{x\to 0^+}\frac{\ln x}{x^{-1/2}}\overset{\text{H}}{=}\lim_{x\to 0^+}\frac{1/x}{-\frac{1}{2}x^{-3/2}}=\lim_{x\to 0^+}\left(-2\sqrt{x}\right)=0$

40. $\lim_{x\to -\infty}xe^x=\lim_{x\to -\infty}\frac{x}{e^{-x}}\overset{\text{H}}{=}\lim_{x\to -\infty}\frac{1}{-e^{-x}}=\lim_{x\to -\infty}-e^x=0$

41. $\lim_{x\to\infty}e^{-x}\ln x=\lim_{x\to\infty}\frac{\ln x}{e^x}\overset{\text{H}}{=}\lim_{x\to\infty}\frac{1/x}{e^x}=\lim_{x\to\infty}\frac{1}{xe^x}=0$

42. $\lim_{x\to\pi/2^-}\sec 7x\cos 3x=\lim_{x\to\pi/2^-}\frac{\cos 3x}{\cos 7x}=\lim_{x\to\pi/2^-}\frac{-3\sin 3x}{-7\sin 7x}=\frac{3(-1)}{7(-1)}=\frac{3}{7}$

43. $\lim_{x\to\infty}x^3e^{-x^2}=\lim_{x\to\infty}\frac{x^3}{e^{x^2}}\overset{\text{H}}{=}\lim_{x\to\infty}\frac{3x^2}{2xe^{x^2}}=\lim_{x\to\infty}\frac{3x}{2e^{x^2}}\overset{\text{H}}{=}\lim_{x\to\infty}\frac{3}{4xe^{x^2}}=0$

44. $\lim_{x\to 0^+}\sqrt{x}\sec x=0\cdot 1=0$

45. $\lim_{x\to\pi}(x-\pi)\cot x=\lim_{x\to\pi}\frac{x-\pi}{\tan x}\overset{\text{H}}{=}\lim_{x\to\pi}\frac{1}{\sec^2 x}=\frac{1}{(-1)^2}=1$

46. $\lim_{x\to 1^+}(x-1)\tan(\pi x/2)=\lim_{x\to 1^+}\frac{x-1}{\cot(\pi x/2)}\overset{\text{H}}{=}\lim_{x\to 1^+}\frac{1}{-\csc^2(\pi x/2)\frac{\pi}{2}}=-\frac{2}{\pi}$

47. $\lim_{x\to 0}\left(\frac{1}{x^4}-\frac{1}{x^2}\right)=\lim_{x\to 0}\frac{1-x^2}{x^4}=\infty$

48. $\lim_{x\to 0}(\csc x-\cot x)=\lim_{x\to 0}\left(\frac{1}{\sin x}-\frac{\cos x}{\sin x}\right)=\lim_{x\to 0}\frac{1-\cos x}{\sin x}\overset{\text{H}}{=}\lim_{x\to 0}\frac{\sin x}{\cos x}=0$

49. $\lim_{x\to 0}\left(\frac{1}{x}-\csc x\right)=\lim_{x\to 0}\left(\frac{1}{x}-\frac{1}{\sin x}\right)=\lim_{x\to 0}\frac{\sin x-x}{x\sin x}$

$$\overset{\text{H}}{=}\lim_{x\to 0}\frac{\cos x-1}{\sin x+x\cos x}\overset{\text{H}}{=}\lim_{x\to 0}\frac{-\sin x}{2\cos x-x\sin x}=\frac{0}{2}=0$$

50. $\lim_{x\to 1}\left(\frac{1}{\ln x}-\frac{1}{x-1}\right)=\lim_{x\to 1}\frac{x-1-\ln x}{(x-1)\ln x}\overset{\text{H}}{=}\lim_{x\to 1}\frac{1-1/x}{\ln x+(x-1)(1/x)}$

$$=\lim_{x\to 1}\frac{x-1}{x\ln x+x-1}\overset{\text{H}}{=}\lim_{x\to 1}\frac{1}{\ln x+1+1}=\frac{1}{0+2}=\frac{1}{2}$$

51. $\lim_{x\to\infty}\left(x-\sqrt{x^2-1}\right)=\lim_{x\to\infty}\left(x-\sqrt{x^2-1}\right)\frac{x+\sqrt{x^2-1}}{x+\sqrt{x^2-1}}=\lim_{x\to\infty}\frac{x^2-(x^2-1)}{x+\sqrt{x^2-1}}=\lim_{x\to\infty}\frac{1}{x+\sqrt{x^2-1}}=0$

52. $\lim_{x\to\infty}\left(\sqrt{x^2+x+1}-\sqrt{x^2-x}\right)=\lim_{x\to\infty}\left(\sqrt{x^2+x+1}-\sqrt{x^2-x}\right)\frac{\sqrt{x^2+x+1}+\sqrt{x^2-x}}{\sqrt{x^2+x+1}+\sqrt{x^2-x}}$

$$=\lim_{x\to\infty}\frac{(x^2+x+1)-(x^2-x)}{\sqrt{x^2+x+1}+\sqrt{x^2-x}}=\lim_{x\to\infty}\frac{2x+1}{\sqrt{x^2+x+1}+\sqrt{x^2-x}}$$

$$=\lim_{x\to\infty}\frac{2+1/x}{\sqrt{1+1/x+1/x^2}+\sqrt{1-1/x}}=\frac{2}{1+1}=1$$

53. $\lim_{x\to\infty}\left(\frac{x^3}{x^2-1}-\frac{x^3}{x^2+1}\right)=\lim_{x\to\infty}\frac{x^3(x^2+1)-x^3(x^2-1)}{(x^2-1)(x^2+1)}=\lim_{x\to\infty}\frac{2x^3}{x^4-1}=\lim_{x\to\infty}\frac{2/x}{1-1/x^4}=0$

54. $\lim_{x\to\infty}\left(xe^{1/x}-x\right)=\lim_{x\to\infty}x\left(e^{1/x}-1\right)=\lim_{x\to\infty}\frac{e^{1/x}-1}{1/x}\overset{\text{H}}{=}\lim_{x\to\infty}\frac{e^{1/x}(-1/x^2)}{-1/x^2}=\lim_{x\to\infty}e^{1/x}=e^0=1$

55. $y = x^{\sin x} \quad \Rightarrow \quad \ln y = \sin x \ln x$, so $\lim\limits_{x\to 0^+} \ln y = \lim\limits_{x\to 0^+} \sin x \ln x = \lim\limits_{x\to 0^+} \dfrac{\ln x}{\csc x} \overset{\text{H}}{=} \lim\limits_{x\to 0^+} \dfrac{1/x}{-\csc x \cot x}$

$= -\left(\lim\limits_{x\to 0^+} \dfrac{\sin x}{x}\right)\left(\lim\limits_{x\to 0^+} \tan x\right) = -1 \cdot 0 = 0 \quad \Rightarrow \quad \lim\limits_{x\to 0^+} x^{\sin x} = \lim\limits_{x\to 0^+} e^{\ln y} = e^0 = 1.$

56. Let $y = (\sin x)^{\tan x}$. Then $\ln y = \tan x \ln(\sin x) \quad \Rightarrow \quad \lim\limits_{x\to 0^+} \ln y = \lim\limits_{x\to 0^+} \tan x \ln(\sin x)$

$= \lim\limits_{x\to 0^+} \dfrac{\ln(\sin x)}{\cot x} \overset{\text{H}}{=} \lim\limits_{x\to 0^+} \dfrac{(\cos x)/\sin x}{-\csc^2 x} = \lim\limits_{x\to 0^+} (-\sin x \cos x) = 0$, so $\lim\limits_{x\to 0^+} (\sin x)^{\tan x} = \lim\limits_{x\to 0^+} e^{\ln y} = e^0 = 1.$

57. $y = (1-2x)^{1/x} \quad \Rightarrow \quad \ln y = \dfrac{1}{x}\ln(1-2x) \quad \Rightarrow \quad \lim\limits_{x\to 0} \ln y = \lim\limits_{x\to 0} \dfrac{\ln(1-2x)}{x} \overset{\text{H}}{=} \lim\limits_{x\to 0} \dfrac{-2/(1-2x)}{1} = -2$

$\Rightarrow \quad \lim\limits_{x\to 0} (1-2x)^{1/x} = \lim\limits_{x\to 0} e^{\ln y} = e^{-2}$

58. Let $y = (1 + a/x)^{bx}$. Then $\ln y = bx \ln\left(1 + \dfrac{a}{x}\right) \quad \Rightarrow \quad \lim\limits_{x\to\infty} \ln y = \lim\limits_{x\to\infty} \dfrac{b\ln(1+a/x)}{1/x}$

$\overset{\text{H}}{=} \lim\limits_{x\to\infty} \dfrac{b\left(\dfrac{1}{1+a/x}\right)\left(-\dfrac{a}{x^2}\right)}{-1/x^2} = \lim\limits_{x\to\infty} \dfrac{ab}{1+a/x} = ab$, so $\lim\limits_{x\to\infty} \left[1 + \dfrac{a}{x}\right]^{bx} = \lim\limits_{x\to\infty} e^{\ln y} = e^{ab}$

59. $y = \left(1 + \dfrac{3}{x} + \dfrac{5}{x^2}\right)^x \quad \Rightarrow \quad \ln y = x \ln\left(1 + \dfrac{3}{x} + \dfrac{5}{x^2}\right) \quad \Rightarrow$

$\lim\limits_{x\to\infty} \ln y = \lim\limits_{x\to\infty} \dfrac{\ln\left(1 + \dfrac{3}{x} + \dfrac{5}{x^2}\right)}{1/x} \overset{\text{H}}{=} \lim\limits_{x\to\infty} \dfrac{\left(-\dfrac{3}{x^2} - \dfrac{10}{x^3}\right)\Big/\left(1 + \dfrac{3}{x} + \dfrac{5}{x^2}\right)}{-1/x^2} = \lim\limits_{x\to\infty} \dfrac{3 + 10/x}{1 + 3/x + 5/x^2} = 3$, so

$\lim\limits_{x\to\infty} \left(1 + \dfrac{3}{x} + \dfrac{5}{x^2}\right)^x = \lim\limits_{x\to\infty} e^{\ln y} = e^3.$

60. Let $y = \left(1 + \dfrac{1}{x^2}\right)^x$. Then $\ln y = x \ln\left(1 + \dfrac{1}{x^2}\right) \quad \Rightarrow \quad \lim\limits_{x\to\infty} \ln y = \lim\limits_{x\to\infty} x \ln\left(1 + \dfrac{1}{x^2}\right)$

$= \lim\limits_{x\to\infty} \dfrac{\ln\left(1 + \dfrac{1}{x^2}\right)}{1/x} \overset{\text{H}}{=} \lim\limits_{x\to\infty} \dfrac{\left(-\dfrac{2}{x^3}\right)\Big/\left(1 + \dfrac{1}{x^2}\right)}{-1/x^2} = \lim\limits_{x\to\infty} \dfrac{2/x}{1 + 1/x^2} = 0$, so

$\lim\limits_{x\to\infty} (1 + 1/x^2)^x = \lim\limits_{x\to\infty} e^{\ln y} = e^0 = 1.$

61. $y = x^{1/x} \quad \Rightarrow \quad \ln y = (1/x)\ln x \quad \Rightarrow \quad \lim\limits_{x\to\infty} \ln y = \lim\limits_{x\to\infty} \dfrac{\ln x}{x} \overset{\text{H}}{=} \lim\limits_{x\to\infty} \dfrac{1/x}{1} = 0 \quad \Rightarrow$

$\lim\limits_{x\to\infty} x^{1/x} = \lim\limits_{x\to\infty} e^{\ln y} = e^0 = 1$

62. Let $y = (e^x + x)^{1/x}$. Then $\ln y = \dfrac{1}{x}\ln(e^x + x) \quad \Rightarrow$

$\lim\limits_{x\to\infty} \ln y = \lim\limits_{x\to\infty} \dfrac{\ln(e^x + x)}{x} \overset{\text{H}}{=} \lim\limits_{x\to\infty} \dfrac{e^x + 1}{e^x + x} \overset{\text{H}}{=} \lim\limits_{x\to\infty} \dfrac{e^x}{e^x + 1} \overset{\text{H}}{=} \lim\limits_{x\to\infty} \dfrac{e^x}{e^x} = 1 \quad \Rightarrow$

$\lim\limits_{x\to\infty} (e^x + x)^{1/x} = \lim\limits_{x\to\infty} e^{\ln y} = e^1 = e.$

63. $y = (\cot x)^{\sin x} \quad \Rightarrow \quad \ln y = \sin x \ln(\cot x) \quad \Rightarrow$

$\lim\limits_{x\to 0^+} \ln y = \lim\limits_{x\to 0^+} \dfrac{\ln(\cot x)}{\csc x} \overset{\text{H}}{=} \lim\limits_{x\to 0^+} \dfrac{(-\csc^2 x)/\cot x}{-\csc x \cot x} = \lim\limits_{x\to 0^+} \dfrac{\csc x}{\cot^2 x}$

$= \lim\limits_{x\to 0^+} \dfrac{\sin x}{\cos^2 x} = 0$, so $\lim\limits_{x\to 0^+} (\cot x)^{\sin x} = \lim\limits_{x\to 0^+} e^{\ln y} = e^0 = 1.$

64. Let $y = (1 + 1/x)^{x^2}$. Then $\ln y = x^2 \ln(1 + 1/x)$ $\Rightarrow$

$$\lim_{x\to\infty} \ln y = \lim_{x\to\infty} x^2 \ln(1 + 1/x) = \lim_{x\to\infty} \frac{\ln(1 + 1/x)}{1/x^2} \overset{\text{H}}{=} \lim_{x\to\infty} \frac{(-1/x^2)/(1 + 1/x)}{-2/x^3}$$

$$= \lim_{x\to\infty} \frac{x}{2(1 + 1/x)} = \infty \quad \Rightarrow \quad \lim_{x\to\infty} (1 + 1/x)^{x^2} = \lim_{x\to\infty} e^{\ln y} = \infty.$$

65. $y = \left(\dfrac{x}{x+1}\right)^x \quad \Rightarrow \quad \ln y = x \ln\left(\dfrac{x}{x+1}\right) \quad \Rightarrow$

$$\lim_{x\to\infty} \ln y = \lim_{x\to\infty} x \ln\left(\frac{x}{x+1}\right) = \lim_{x\to\infty} \frac{\ln x - \ln(x+1)}{1/x} \overset{\text{H}}{=} \lim_{x\to\infty} \frac{1/x - 1/(x+1)}{-1/x^2}$$

$$= \lim_{x\to\infty} \left(-x + \frac{x^2}{x+1}\right) = \lim_{x\to\infty} \frac{-x}{x+1} = -1, \text{ so } \lim_{x\to\infty} \left(\frac{x}{x+1}\right)^x = \lim_{x\to\infty} e^{\ln y} = e^{-1}$$

Or: $\displaystyle \lim_{x\to\infty} \left(\frac{x}{x+1}\right)^x = \lim_{x\to\infty} \left[\left(\frac{x+1}{x}\right)^{-1}\right]^x = \left[\lim_{x\to\infty} \left(1 + \frac{1}{x}\right)^x\right]^{-1} = e^{-1}$

66. Let $y = (\cos 3x)^{5/x}$. Then $\ln y = \dfrac{5}{x} \ln(\cos 3x) \quad \Rightarrow \quad \displaystyle \lim_{x\to 0} \ln y = 5 \lim_{x\to 0} \frac{\ln(\cos 3x)}{x} \overset{\text{H}}{=} 5 \lim_{x\to 0} \frac{-3\tan 3x}{1} = 0$, so $\displaystyle \lim_{x\to 0} (\cos 3x)^{5/x} = e^0 = 1$.

67. Let $y = (-\ln x)^x$. Then $\ln y = x \ln(-\ln x) \quad \Rightarrow \quad \displaystyle \lim_{x\to 0^+} \ln y = \lim_{x\to 0^+} x \ln(-\ln x) = \lim_{x\to 0^+} \frac{\ln(-\ln x)}{1/x}$

$$\overset{\text{H}}{=} \lim_{x\to 0^+} \frac{(1/-\ln x)(-1/x)}{-1/x^2} = \lim_{x\to 0^+} \frac{-x}{\ln x} = 0 \quad \Rightarrow \quad \lim_{x\to 0^+} (-\ln x)^x = e^0 = 1.$$

68. Let $y = \left(\dfrac{2x-3}{2x+5}\right)^{2x+1}$. Then $\ln y = (2x+1)\ln\left(\dfrac{2x-3}{2x+5}\right) \quad \Rightarrow$

$$\lim_{x\to\infty} \ln y = \lim_{x\to\infty} \frac{\ln(2x-3) - \ln(2x+5)}{1/(2x+1)} \overset{\text{H}}{=} \lim_{x\to\infty} \frac{2/(2x-3) - 2/(2x+5)}{-2/(2x+1)^2} = \lim_{x\to\infty} \frac{-8(2x+1)^2}{(2x-3)(2x+5)}$$

$$= \lim_{x\to\infty} \frac{-8(2 + 1/x)^2}{(2 - 3/x)(2 + 5/x)} = -8 \quad \Rightarrow \quad \lim_{x\to\infty} \left(\frac{2x-3}{2x+5}\right)^{2x+1} = e^{-8}.$$

69.

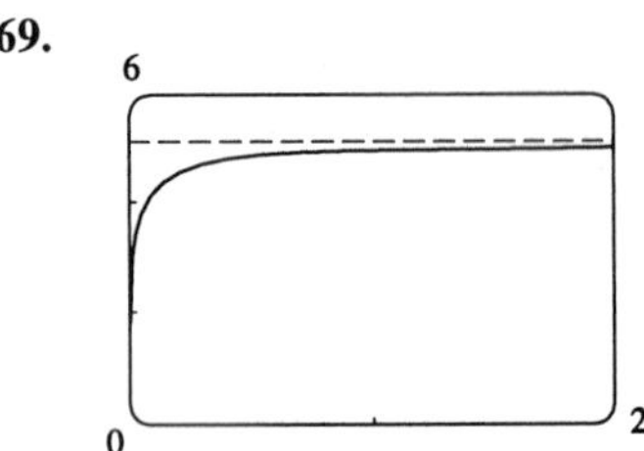

From the graph, it appears that $\displaystyle \lim_{x\to\infty} x[\ln(x+5) - \ln x] = 5$.

Now $\displaystyle \lim_{x\to\infty} x[\ln(x+5) - \ln x] = \lim_{x\to\infty} \frac{\ln(x+5) - \ln x}{1/x}$

$$\overset{\text{H}}{=} \lim_{x\to\infty} \frac{1/(x+5) - 1/x}{-1/x^2} = \lim_{x\to\infty} \frac{5x^2}{x(x+5)} = 5.$$

70.

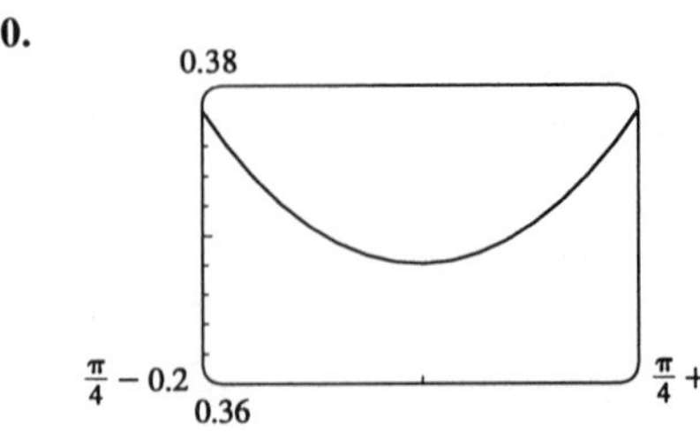

From the graph, it appears that $\displaystyle \lim_{x\to\pi/4} (\tan x)^{\tan 2x} \approx 0.368$.

(Note that $\frac{\pi}{4} \approx 0.785$.) Let $y = (\tan x)^{\tan 2x}$, so $\ln y = \tan 2x \ln(\tan x)$. Then by l'Hospital's Rule,

$$\lim_{x\to\pi/4} \ln y = \lim_{x\to\pi/4} \frac{\ln(\tan x)}{\cot 2x} = \lim_{x\to\pi/4} \frac{\sec^2 x/\tan x}{-2\csc^2 2x} = \frac{2/1}{-2(1)}$$

$$= -1, \text{ so } \lim_{x\to\pi/4} (\tan x)^{\tan 2x} = \lim_{x\to\pi/4} e^{\ln y} = e^{-1} = 1/e \approx 0.3679.$$

71.

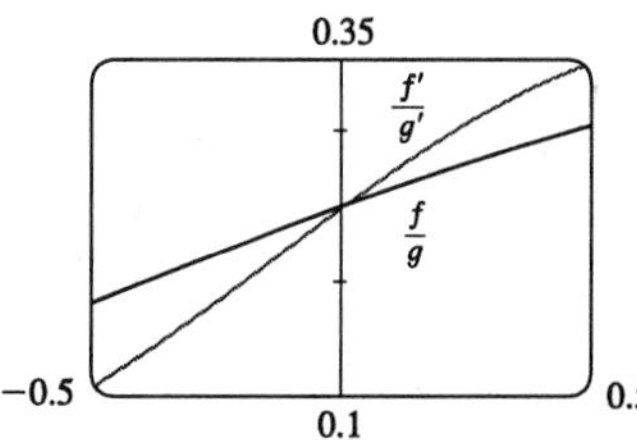

From the graph, it appears that $\lim\limits_{x\to 0}\dfrac{f(x)}{g(x)}=\lim\limits_{x\to 0}\dfrac{f'(x)}{g'(x)}\approx 0.25$.

We calculate $\lim\limits_{x\to 0}\dfrac{f(x)}{g(x)}=\lim\limits_{x\to 0}\dfrac{e^x-1}{x^3+4x}$

$\overset{\text{H}}{=}\lim\limits_{x\to 0}\dfrac{e^x}{3x^2+4}=\dfrac{1}{4}$.

72.

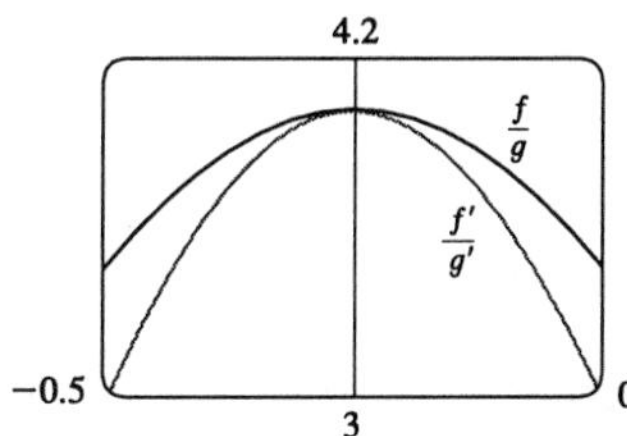

From the graph, it appears that $\lim\limits_{x\to 0}\dfrac{f(x)}{g(x)}=\lim\limits_{x\to 0}\dfrac{f'(x)}{g'(x)}=4$. We calculate

$\lim\limits_{x\to 0}\dfrac{f(x)}{g(x)}=\lim\limits_{x\to 0}\dfrac{2\,x\sin x}{\sec x-1}\overset{\text{H}}{=}\lim\limits_{x\to 0}\dfrac{2(x\cos x+\sin x)}{\sec x\tan x}$

$\overset{\text{H}}{=}\lim\limits_{x\to 0}\dfrac{2(-x\sin x+\cos x+\cos x)}{\sec x(\sec^2 x)+\tan x(\sec x\tan x)}=\dfrac{4}{1}=4.$

73. $y=f(x)=xe^{-x}$ **A.** $D=\mathbb{R}$ **B.** Intercepts are 0 **C.** No symmetry

D. $\lim\limits_{x\to\infty}xe^{-x}=\lim\limits_{x\to\infty}\dfrac{x}{e^x}\overset{\text{H}}{=}\lim\limits_{x\to\infty}\dfrac{1}{e^x}=0$, so $y=0$ is a HA. $\lim\limits_{x\to-\infty}xe^{-x}=-\infty$

E. $f'(x)=e^{-x}-xe^{-x}=e^{-x}(1-x)>0$ $\Leftrightarrow$ $x<1$, so f is increasing on $(-\infty,1]$ and decreasing on $[1,\infty)$.

F. Absolute maximum $f(1)=1/e$.

G. $f''(x)=e^{-x}(x-2)>0$ $\Leftrightarrow$ $x>2$, so f is CU on $(2,\infty)$ and CD on $(-\infty,2)$. IP is $(2,2/e^2)$.

H.

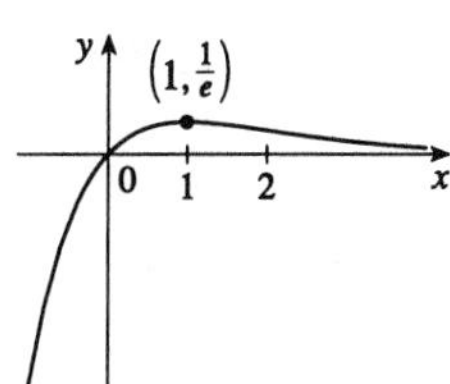

74. $y=f(x)=x^2e^{-x}$ **A.** $D=\mathbb{R}$ **B.** Intercepts are 0 **C.** No symmetry

D. $\lim\limits_{x\to\infty}x^2e^{-x}=\lim\limits_{x\to\infty}\dfrac{x^2}{e^x}\overset{\text{H}}{=}\lim\limits_{x\to\infty}\dfrac{2x}{e^x}\overset{\text{H}}{=}\lim\limits_{x\to\infty}\dfrac{2}{e^x}=0$, so $y=0$ is a HA. Also $\lim\limits_{x\to-\infty}x^2e^{-x}=\infty$.

E. $f'(x)=2xe^{-x}-x^2e^{-x}=x(2-x)e^{-x}>0$ when $0<x<2$, so f is increasing on $[0,2]$ and decreasing on $(-\infty,0]$ and $[2,\infty)$. **F.** $f(0)=0$ is a local minimum, $f(2)=4e^{-2}$ is a local maximum.

G. $f''(x)=(2-2x)e^{-x}-(2x-x^2)e^{-x}$ $=(x^2-4x+2)e^{-x}=0$ when $x^2-4x+2=0$ $\Leftrightarrow$ $x=2\pm\sqrt{2}$. $f''(x)>0$ $\Leftrightarrow$ $x<2-\sqrt{2}$ or $x>2+\sqrt{2}$, so f is CU on $\left(-\infty,2-\sqrt{2}\right)$ and $\left(2+\sqrt{2},\infty\right)$ and CD on $\left(2-\sqrt{2},2+\sqrt{2}\right)$. IP $\left(2\pm\sqrt{2},\left(6\pm4\sqrt{2}\right)e^{\sqrt{2}\pm2}\right)$

H.

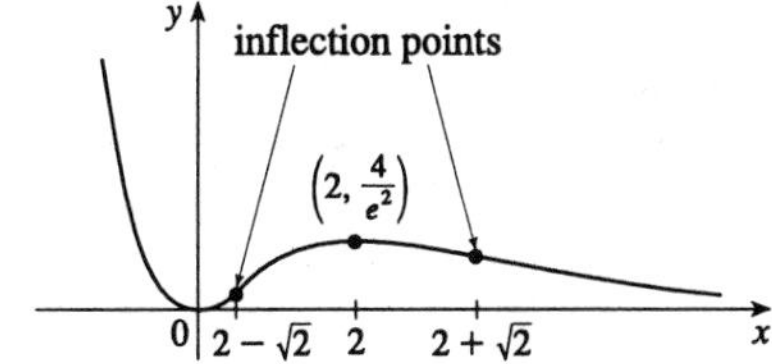

75. $y = f(x) = x \ln x$ **A.** $D = (0, \infty)$ **B.** x-intercept when $\ln x = 0 \Leftrightarrow x = 1$, no y-intercept **C.** No symmetry **D.** $\lim_{x\to\infty} x \ln x = \infty$, $\lim_{x\to 0^+} x \ln x = \lim_{x\to 0^+} \dfrac{\ln x}{1/x} \overset{\text{H}}{=} \lim_{x\to 0^+} \dfrac{1/x}{-1/x^2} = \lim_{x\to 0^+} (-x) = 0$, no asymptotes. **E.** $f'(x) = \ln x + 1 = 0$ when $\ln x = -1$ $\Leftrightarrow$ $x = e^{-1}$. $f'(x) > 0$ $\Leftrightarrow$ $\ln x > -1$ $\Leftrightarrow$ $x > e^{-1}$, so f is increasing on $[1/e, \infty)$ and decreasing on $(0, 1/e]$. **F.** $f(1/e) = -1/e$ is an absolute and local minimum. **G.** $f''(x) = 1/x > 0$, so f is CU on $(0, \infty)$. No IP

H.

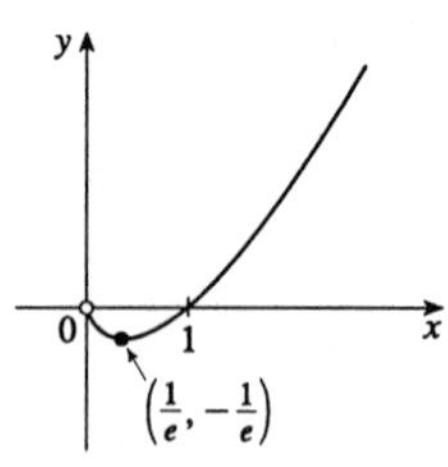

76. $f(x) = (\ln x)/x$ **A.** $D = (0, \infty)$ **B.** x-intercept $= 1$ **C.** No symmetry **D.** $\lim_{x\to\infty} \dfrac{\ln x}{x} \overset{\text{H}}{=} \lim_{x\to\infty} \dfrac{1/x}{1} = 0$, so $y = 0$ is a horizontal asymptote. Also $\lim_{x\to 0^+} \dfrac{\ln x}{x} = -\infty$ since $\ln x \to -\infty$ and $x \to 0^+$, so $x = 0$ is a vertical asymptote. **E.** $f'(x) = \dfrac{1 - \ln x}{x^2} = 0$ when $\ln x = 1$ $\Leftrightarrow$ $x = e$. $f'(x) > 0$ $\Leftrightarrow$ $1 - \ln x > 0$ $\Leftrightarrow$ $\ln x < 1$ $\Leftrightarrow$ $0 < x < e$. $f'(x) < 0$ $\Leftrightarrow$ $x > e$. So f is increasing on $(0, e]$ and decreasing on $[e, \infty)$.
F. Thus $f(e) = 1/e$ is a local (and absolute) maximum.

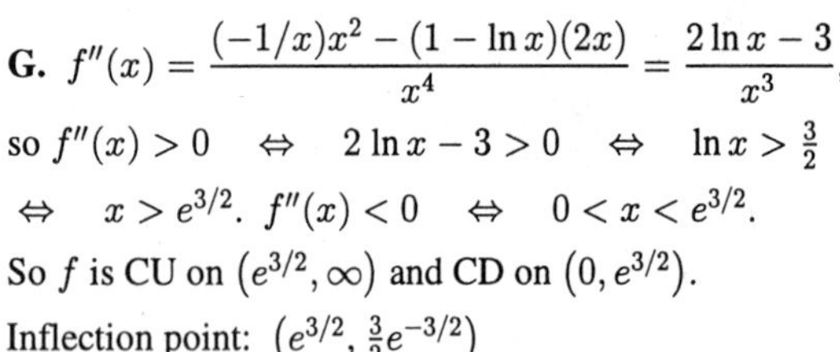

G. $f''(x) = \dfrac{(-1/x)x^2 - (1 - \ln x)(2x)}{x^4} = \dfrac{2\ln x - 3}{x^3}$,
so $f''(x) > 0$ $\Leftrightarrow$ $2\ln x - 3 > 0$ $\Leftrightarrow$ $\ln x > \frac{3}{2}$ $\Leftrightarrow$ $x > e^{3/2}$. $f''(x) < 0$ $\Leftrightarrow$ $0 < x < e^{3/2}$.
So f is CU on $\left(e^{3/2}, \infty\right)$ and CD on $\left(0, e^{3/2}\right)$.
Inflection point: $\left(e^{3/2}, \frac{3}{2}e^{-3/2}\right)$

H.

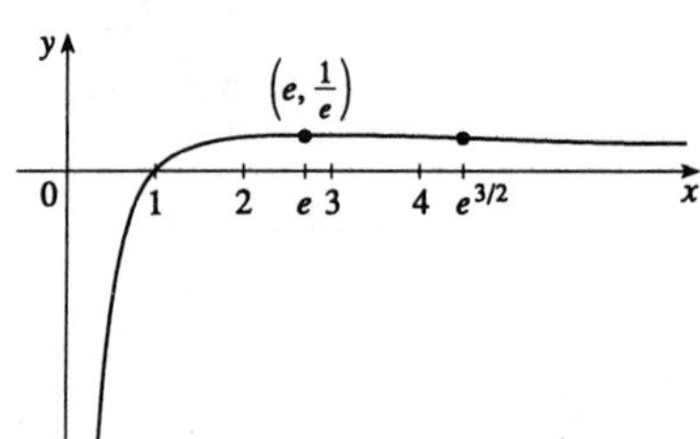

77. $y = f(x) = x^2 \ln x$ **A.** $D = (0, \infty)$ **B.** x-intercept when $\ln x = 0 \Leftrightarrow x = 1$, no y-intercept **C.** No symmetry **D.** $\lim_{x\to\infty} x^2 \ln x = \infty$, $\lim_{x\to 0^+} x^2 \ln x = \lim_{x\to 0^+} \dfrac{\ln x}{1/x^2} \overset{\text{H}}{=} \lim_{x\to 0^+} \dfrac{1/x}{-2/x^3} = \lim_{x\to 0^+} \left(-\dfrac{x^2}{2}\right) = 0$, no asymptote
E. $f'(x) = 2x \ln x + x = x(2\ln x + 1) > 0$ $\Leftrightarrow$ $\ln x > -\frac{1}{2}$ $\Leftrightarrow$ $x > e^{-1/2}$, so f is increasing on $\left[1/\sqrt{e}, \infty\right)$, decreasing on $\left(0, 1/\sqrt{e}\,\right]$.
F. $f\left(1/\sqrt{e}\right) = -1/(2e)$ is an absolute minimum.
G. $f''(x) = 2\ln x + 3 > 0$ $\Leftrightarrow$ $\ln x > -\frac{3}{2}$ $\Leftrightarrow$ $x > e^{-3/2}$, so f is CU on $\left(e^{-3/2}, \infty\right)$ and CD on $\left(0, e^{-3/2}\right)$. IP is $\left(e^{-3/2}, -3/(2e^3)\right)$

H.

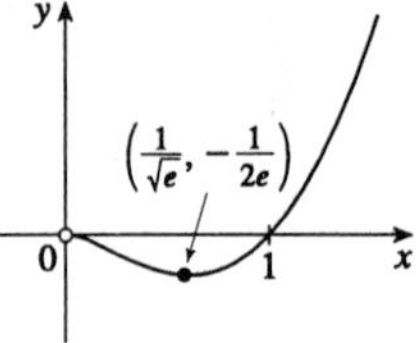

78. $y = f(x) = x(\ln x)^2$ **A.** $D = (0, \infty)$ **B.** x-intercept $= 1$, no y-intercept **C.** No symmetry
D. $\lim_{x\to\infty} x(\ln x)^2 = \infty$, $\lim_{x\to 0^+} x(\ln x)^2 = \lim_{x\to 0^+} \dfrac{(\ln x)^2}{1/x} \overset{\text{H}}{=} \lim_{x\to 0^+} \dfrac{2(\ln x)(1/x)}{-1/x^2} = \lim_{x\to 0^+} \dfrac{2\ln x}{-1/x} \overset{\text{H}}{=} \lim_{x\to 0^+} \dfrac{2/x}{1/x^2}$ $= \lim_{x\to 0^+} 2x = 0$, no asymptote **E.** $f'(x) = (\ln x)^2 + 2\ln x = (\ln x)(\ln x + 2) = 0$ when $\ln x = 0$ $\Leftrightarrow$ $x = 1$ and when $\ln x = -2$ $\Leftrightarrow$ $x = e^{-2}$. $f'(x) > 0$ when $0 < x < e^{-2}$ and when $x > 1$, so f is increasing on $\left(0, e^{-2}\right]$ and $[1, \infty)$ and decreasing on $\left[e^{-2}, 1\right]$. **F.** $f(e^{-2}) = 4e^{-2}$ is a local maximum,

$f(1) = 0$ is a local minimum.

G. $f''(x) = 2\,(\ln x)(1/x) + 2/x$
$= (2/x)(\ln x + 1) = 0$ when
$\ln x = -1 \quad \Leftrightarrow \quad x = e^{-1}$. $f''(x) > 0$
$\Leftrightarrow \quad x > 1/e$, so f is CU on $(1/e, \infty)$,
CD on $(0, 1/e)$. IP $(1/e, 1/e)$

H.

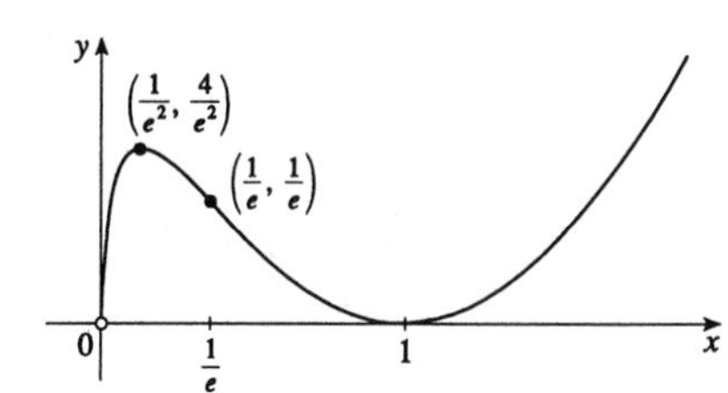

79. $y = f(x) = xe^{-x^2}$ **A.** $D = \mathbb{R}$ **B.** Intercepts are 0 **C.** $f(-x) = -f(x)$, so the curve is symmetric about the origin. **D.** $\lim\limits_{x\to\pm\infty} xe^{-x^2} = \lim\limits_{x\to\pm\infty} \dfrac{x}{e^{x^2}} \overset{\text{H}}{=} \lim\limits_{x\to\pm\infty} \dfrac{1}{2xe^{x^2}} = 0$, so $y = 0$ is a HA.

E. $f'(x) = e^{-x^2} - 2x^2\,e^{-x^2} = e^{-x^2}(1 - 2x^2) > 0 \quad \Leftrightarrow \quad x^2 < \frac{1}{2} \quad \Leftrightarrow \quad |x| < \frac{1}{\sqrt{2}}$, so f is increasing on $\left[-\frac{1}{\sqrt{2}}, \frac{1}{\sqrt{2}}\right]$ and decreasing on $\left(-\infty, -\frac{1}{\sqrt{2}}\right]$ and $\left[\frac{1}{\sqrt{2}}, \infty\right)$. **F.** $f\left(\frac{1}{\sqrt{2}}\right) = 1/\sqrt{2e}$ is a local maximum, $f\left(-\frac{1}{\sqrt{2}}\right) = -1/\sqrt{2e}$ is a local minimum.

G. $f''(x) = -2xe^{-x^2}(1 - 2x^2) - 4xe^{-x^2}$
$= 2xe^{-x^2}(2x^2 - 3) > 0 \quad \Leftrightarrow$
$x > \sqrt{\frac{3}{2}}$ or $-\sqrt{\frac{3}{2}} < x < 0$, so f is CU on $\left(\sqrt{\frac{3}{2}}, \infty\right)$ and $\left(-\sqrt{\frac{3}{2}}, 0\right)$ and CD on $\left(-\infty, -\sqrt{\frac{3}{2}}\right)$ and $\left(0, \sqrt{\frac{3}{2}}\right)$.
IP are $(0, 0)$ and $\left(\pm\sqrt{\frac{3}{2}}, \pm\sqrt{\frac{3}{2}}e^{-3/2}\right)$.

H.

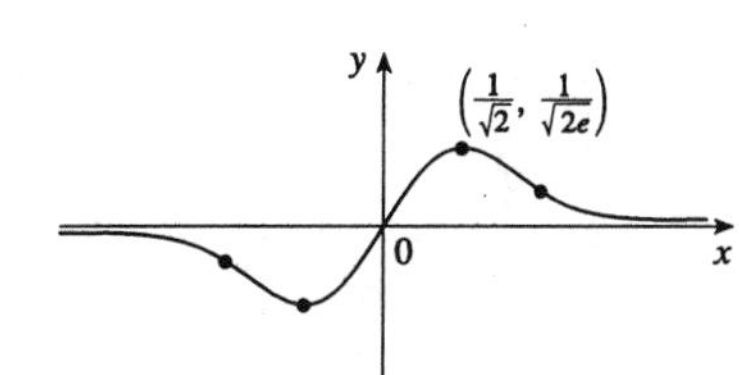

80. $y = f(x) = e^x/x$ **A.** $D = \{x \mid x \neq 0\}$ **B.** No intercepts **C.** No symmetry **D.** $\lim\limits_{x\to\infty} \dfrac{e^x}{x} \overset{\text{H}}{=} \lim\limits_{x\to\infty} \dfrac{e^x}{1} = \infty$, $\lim\limits_{x\to-\infty} \dfrac{e^x}{x} = 0$, so $y = 0$ is a HA. $\lim\limits_{x\to0^+} \dfrac{e^x}{x} = \infty$, $\lim\limits_{x\to0^-} \dfrac{e^x}{x} = -\infty$, so $x = 0$ is a VA. **E.** $f'(x) = \dfrac{xe^x - e^x}{x^2} > 0$ $\Leftrightarrow \quad (x - 1)e^x > 0 \quad \Leftrightarrow \quad x > 1$, so f is increasing on $[1, \infty)$, and decreasing on $(-\infty, 0)$ and $(0, 1]$. **F.** $f(1) = e$ is a local minimum.

G. $f''(x) = \dfrac{x^2(xe^x) - 2x(xe^x - e^x)}{x^4} = \dfrac{e^x(x^2 - 2x + 2)}{x^3} > 0$
$\Leftrightarrow \quad x > 0$ since $x^2 - 2x + 2 > 0$ for all x. So f is CU on $(0, \infty)$ and CD on $(-\infty, 0)$. No IP.

H.

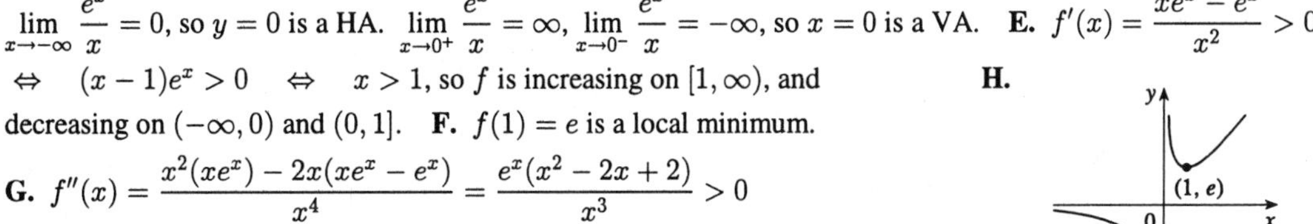

81. $y = f(x) = xe^{1/x}$ **A.** $D = \{x \mid x \neq 0\}$ **B.** No intercepts **C.** No symmetry **D.** $\lim\limits_{x\to\infty} xe^{1/x} = \infty$,

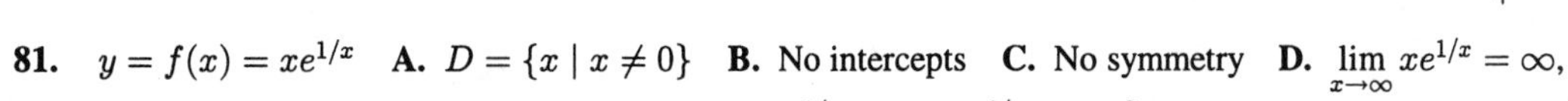

$\lim\limits_{x\to-\infty} xe^{1/x} = -\infty$, no HA. $\lim\limits_{x\to0^+} xe^{1/x} = \lim\limits_{x\to0^+} \dfrac{e^{1/x}}{1/x} \overset{\text{H}}{=} \lim\limits_{x\to0^+} \dfrac{e^{1/x}(-1/x^2)}{-1/x^2} = \lim\limits_{x\to0^+} e^{1/x} = \infty$, so $x = 0$ is a VA.

Also $\lim\limits_{x\to0^-} xe^{1/x} = 0$ since $\dfrac{1}{x} \to -\infty \quad \Rightarrow \quad e^{1/x} \to 0$.

E. $f'(x) = e^{1/x} + xe^{1/x}\left(-\dfrac{1}{x^2}\right) = e^{1/x}\left(1 - \dfrac{1}{x}\right) > 0 \quad \Leftrightarrow \quad \dfrac{1}{x} < 1$
$\Leftrightarrow \quad x < 0$ or $x > 1$, so f is increasing on $(-\infty, 0)$ and $[1, \infty)$, decreasing on $(0, 1]$. **F.** $f(1) = e$ is a local minimum.

G. $f''(x) = e^{1/x}(-1/x^2)(1 - 1/x) + e^{1/x}(1/x^2) = e^{1/x}/x^3 > 0 \quad \Leftrightarrow$
$x > 0$, so f is CU on $(0, \infty)$ and CD on $(-\infty, 0)$. No IP

H.

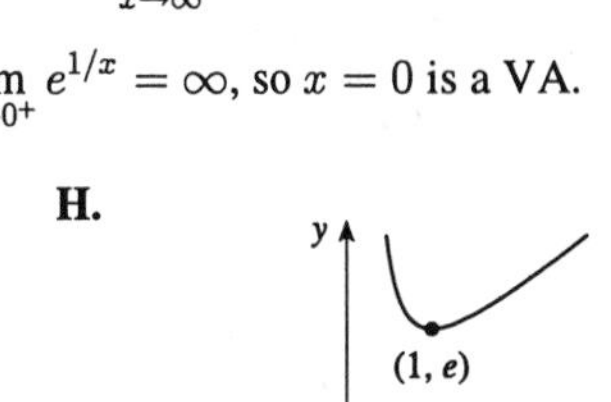
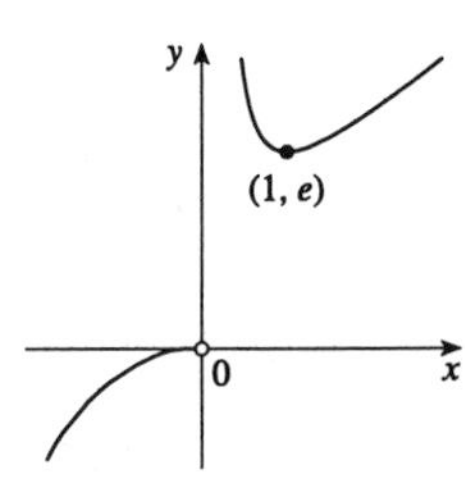

82. $y = f(x) = e^x - x$ **A.** $D = \mathbb{R}$ **B.** No x-intercepts; y-intercept $= 1$ **C.** No symmetry
D. $\lim\limits_{x\to-\infty}(e^x - x) = \infty$, $\lim\limits_{x\to\infty}(e^x - x) = \lim\limits_{x\to\infty} x\left(\dfrac{e^x}{x} - 1\right) = \infty$ since $\lim\limits_{x\to\infty}\dfrac{e^x}{x} \overset{\text{H}}{=} \lim\limits_{x\to\infty}\dfrac{e^x}{1} = \infty$. $y = -x$ is a slant asymptote since $(e^x - x) - (-x) = e^x \to 0$ as $x \to -\infty$. **E.** $f'(x) = e^x - 1 > 0 \Leftrightarrow e^x > 1 \Leftrightarrow x > 0$, so f is increasing on $[0, \infty)$ and decreasing on $(-\infty, 0]$. **F.** $f(0) = 1$ is a local and absolute minimum. **G.** $f''(x) = e^x > 0$ for all x, so f is CU on $\mathbb{R}$.

H.

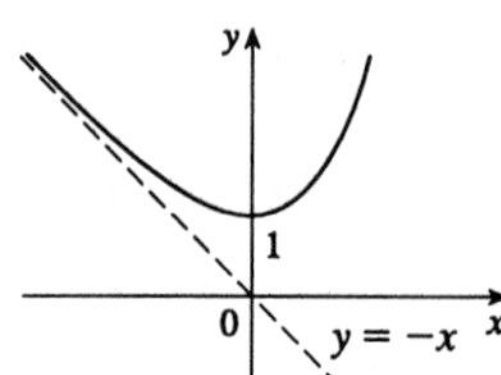

83. $y = f(x) = x - \ln(1 + x)$ **A.** $D = \{x \mid x > -1\} = (-1, \infty)$ **B.** Intercepts are 0 **C.** No symmetry
D. $\lim\limits_{x\to-1^+}[x - \ln(1 + x)] = \infty$, so $x = -1$ is a VA. $\lim\limits_{x\to\infty}[x - \ln(1 + x)] = \lim\limits_{x\to\infty} x\left(1 - \dfrac{\ln(1 + x)}{x}\right) = \infty$, since $\lim\limits_{x\to\infty}\dfrac{\ln(1 + x)}{x} \overset{\text{H}}{=} \lim\limits_{x\to\infty}\dfrac{1/(1 + x)}{1} = 0$.
E. $f'(x) = 1 - \dfrac{1}{1 + x} = \dfrac{x}{1 + x} > 0 \Leftrightarrow x > 0$ since $x + 1 > 0$. So f is increasing on $[0, \infty)$ and decreasing on $(-1, 0]$. **F.** $f(0) = 0$ is an absolute minimum. **G.** $f''(x) = 1/(1 + x)^2 > 0$, so f is CU on $(-1, \infty)$.

H.

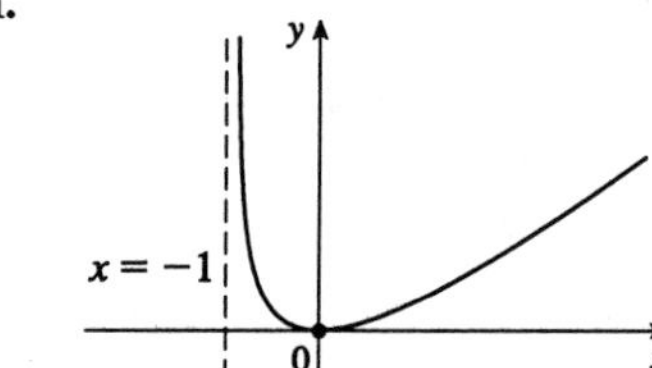

84. $y = f(x) = e^x - 3e^{-x} - 4x$ **A.** $D = \mathbb{R}$ **B.** y-intercept $= -2$ **C.** No symmetry
D. $\lim\limits_{x\to\infty}(e^x - 3e^{-x} - 4x) = \lim\limits_{x\to\infty} x\left(\dfrac{e^x}{x} - 3\dfrac{e^{-x}}{x} - 4\right) = \infty$, since $\lim\limits_{x\to\infty}\dfrac{e^x}{x} \overset{\text{H}}{=} \lim\limits_{x\to\infty}\dfrac{e^x}{1} = \infty$. Similarly, $\lim\limits_{x\to-\infty}(e^x - 3e^{-x} - 4x) = -\infty$. **E.** $f'(x) = e^x + 3e^{-x} - 4 = e^{-x}(e^{2x} - 4e^x + 3) = e^{-x}(e^x - 3)(e^x - 1) > 0$ $\Leftrightarrow e^x > 3$ or $e^x < 1 \Leftrightarrow x > \ln 3$ or $x < 0$. So f is increasing on $(-\infty, 0]$ and $[\ln 3, \infty)$ and decreasing on $[0, \ln 3]$. **F.** $f(0) = -2$ is a local maximum and $f(\ln 3) = 2 - 4\ln 3$ is a local minimum. **G.** $f''(x) = e^x - 3e^{-x} = e^{-x}(e^{2x} - 3) > 0 \Leftrightarrow e^{2x} > 3 \Leftrightarrow x > \frac{1}{2}\ln 3$, so f is CU on $\left(\frac{1}{2}\ln 3, \infty\right)$ and CD on $\left(-\infty, \frac{1}{2}\ln 3\right)$. IP at $x = \frac{1}{2}\ln 3$.

H.

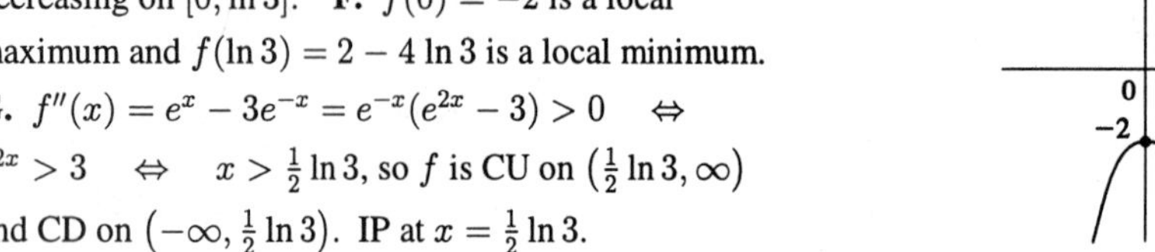

85. **(a)**

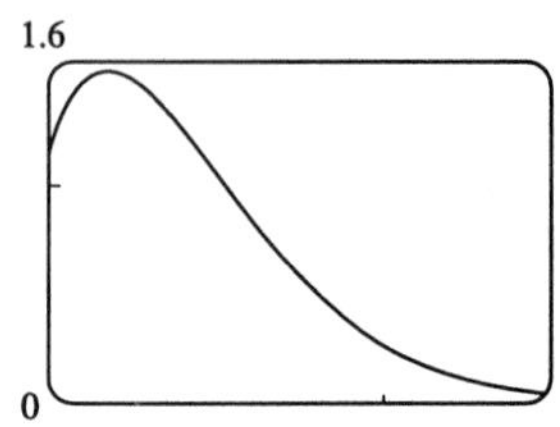

(b) We note that

$\ln f(x) = \ln x^{-x} = -x \ln x = -\dfrac{\ln x}{1/x}$, so

$\lim\limits_{x\to0^+}\ln f(x) \overset{\text{H}}{=} \lim\limits_{x\to0^+} -\dfrac{1/x}{-x^{-2}} = \lim\limits_{x\to0^+} x = 0$.

Thus $\lim\limits_{x\to0^+} f(x) = \lim\limits_{x\to0^+} e^{\ln f(x)} = e^0 = 1$.

(c) From the graph, it appears that there is a local and absolute maximum of about $f(0.37) \approx 1.44$. To find the exact value, we differentiate: $f(x) = x^{-x} = e^{-x\ln x} \quad \Rightarrow$

$f'(x) = e^{-x\ln x}\left[-x\left(\dfrac{1}{x}\right) + \ln x(-1)\right] = -x^{-x}(1 + \ln x)$. This is 0 only when $1 + \ln x = 0 \quad \Leftrightarrow$

$x = e^{-1}$. Also $f'(x)$ changes from positive to negative at e^{-1}.

So the maximum value is $f(1/e) = (1/e)^{-1/e} = e^{1/e}$.

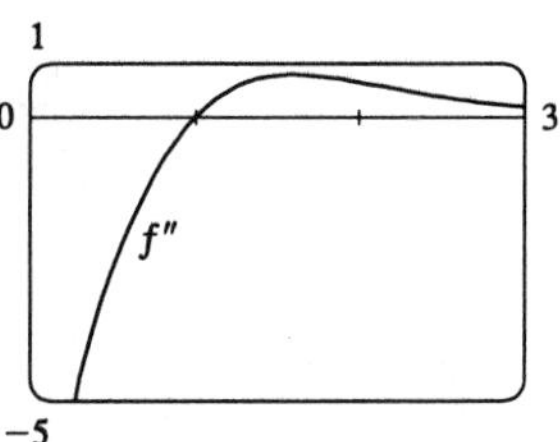

(d) We differentiate again to get

$f''(x) = -x^{-x}(1/x) + (1 + \ln x)^2(x^{-x}) = x^{-x}\left[(1 + \ln x)^2 - 1/x\right]$.

From the graph of $f''(x)$, it seems that $f''(x)$ changes from negative to positive at $x = 1$, so f has an

86. **(a)**

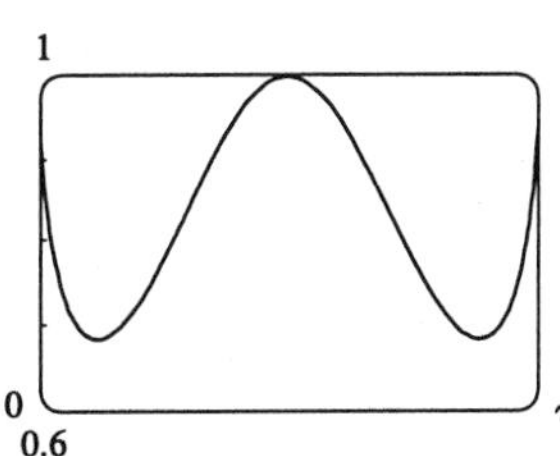

Note that the function is only defined for $\sin x > 0$, and is periodic with period 2π, so we consider it only on $[0, \pi]$.

(b) $\ln f(x) = \ln(\sin x)^{\sin x} = \sin x \ln(\sin x) = \dfrac{\ln(\sin x)}{\csc x}$, so

$\lim\limits_{x\to 0^+} \ln f(x) \overset{\text{H}}{=} \lim\limits_{x\to 0^+} \dfrac{\cot x}{-\csc x \cot x} = \lim\limits_{x\to 0^+} -\sin x = 0$.

Thus $\lim\limits_{x\to 0^+} f(x) = \lim\limits_{x\to 0^+} e^{\ln f(x)} = e^0 = 1$.

(c) From the graph, it seems that there are local minima at about $f(0.38) = f(2.76) \approx 0.69$, and a local maximum of about $f(1.57) = 1$. To find the exact values, we differentiate:

$f(x) = (\sin x)^{\sin x} = e^{\sin x \ln(\sin x)} \quad \Rightarrow$

$f'(x) = e^{\sin x\ln(\sin x)}\left[\sin x\left(\dfrac{1}{\sin x}\right)\cos x + \ln(\sin x)\cos x\right] = \cos x[1 + \ln(\sin x)](\sin x)^{\sin x}$. This is 0 when

$\cos x = 0 \quad \Leftrightarrow \quad x = \frac{\pi}{2}$ and when $1 + \ln(\sin x) = 0 \quad \Leftrightarrow \quad \ln(\sin x) = -1 \quad \Leftrightarrow \quad \sin x = 1/e$. This occurs at $x = \sin^{-1} 1/e$ and at $x = \pi - \sin^{-1} 1/e$ [since $\sin x = \sin(\pi - x)$]. So the local maximum is $f\left(\frac{\pi}{2}\right) = \left(\sin\frac{\pi}{2}\right)^{\sin(\pi/2)} = 1^1 = 1$, and the local minima are

$f(\sin^{-1}(1/e)) = f(\pi - \sin^{-1}(1/e)) = (1/e)^{1/e} = e^{-1/e}$.

(d) We differentiate again to get

$$\begin{aligned} f''(x) = {} & [\cos x(1 + \ln \sin x)]^2(\sin x)^{\sin x} \\ & + \cos x(\csc x \cos x)(\sin x)^{\sin x} \\ & - \sin x\,[1 + \ln(\sin x)](\sin x)^{\sin x}. \end{aligned}$$

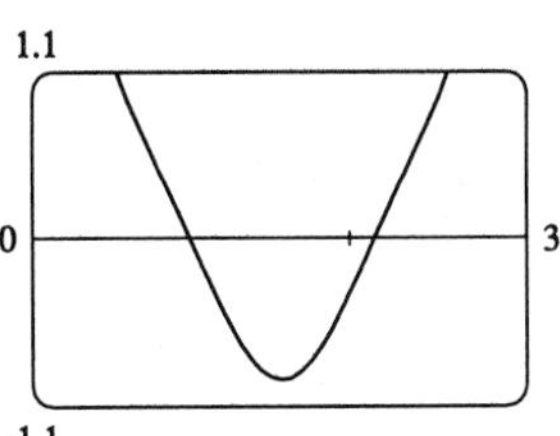

From the graph of $f''(x)$, it seems that $f''(x)$ changes sign at $x \approx 0.94$ and at $x \approx 2.20$, and so f has inflection points at approximately those x-values.

87. (a)

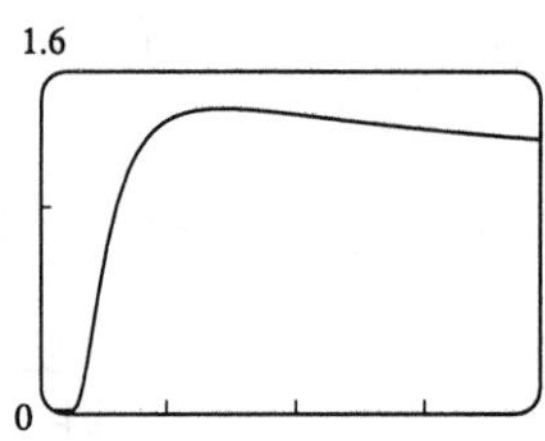

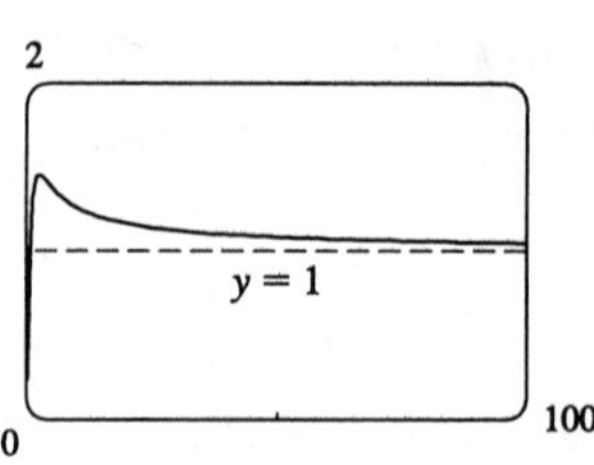

(b) $\ln f(x) = \ln x^{1/x} = \dfrac{1}{x}\ln x$, so $\lim\limits_{x\to\infty} \ln f(x) \overset{\text{H}}{=} \lim\limits_{x\to\infty} \dfrac{1/x}{1} = 0$. Therefore $\lim\limits_{x\to\infty} f(x) = \lim\limits_{x\to\infty} e^{\ln f(x)} = 1$.

Also $\lim\limits_{x\to 0^+} \ln f(x) = \lim\limits_{x\to 0^+} \left(\dfrac{1}{x}\ln x\right) = -\infty$. So $\lim\limits_{x\to 0^+} f(x) = \lim\limits_{x\to 0^+} e^{\ln f(x)} = 0$.

(c) From the graph, it appears that f has a local maximum at about $f(2.7) \approx 1.44$. To find the exact value, we differentiate: $f(x) = x^{1/x} = e^{(1/x)\ln x}$ $\Rightarrow$

$f'(x) = e^{(1/x)\ln x}\left[\dfrac{1}{x}\left(\dfrac{1}{x}\right) + \ln x\left(-x^{-2}\right)\right] = (1 - \ln x)x^{1/x}x^{-2}$. This is 0 only when $1 - \ln x = 0$ $\Rightarrow$

$x = e$, and $f'(x)$ changes from positive to negative there. So the local maximum value is $f(e) = e^{1/e}$.

(d) We differentiate again to get $f''(x) = (1 - \ln x)x^{1/x}(-2x^{-3}) + (1 - \ln x)^2 x^{1/x}x^{-4} + (-1/x)x^{1/x}x^{-2}$.

From the graphs it appears that $f''(x)$ changes sign at $x \approx 0.58$ and at $x \approx 4.4$, so f has inflection points there.

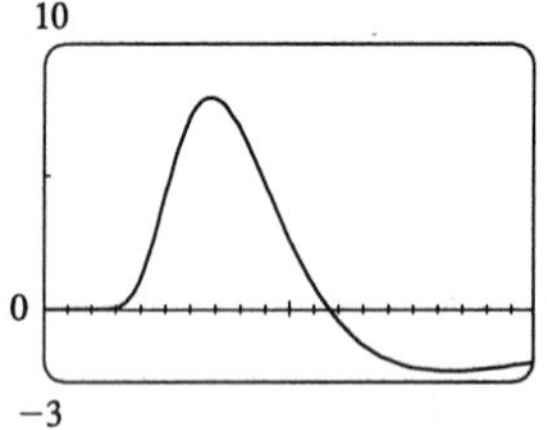

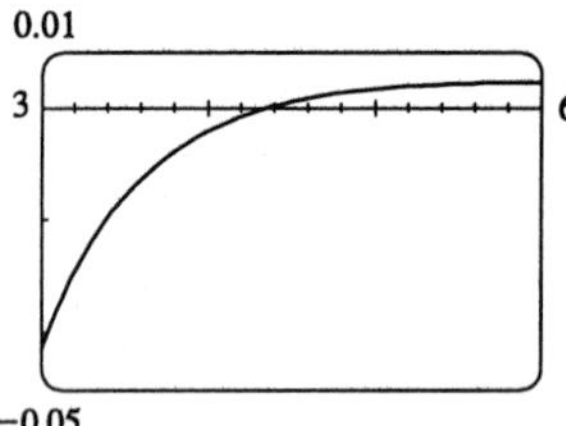

88. Note that for all n, we have $f(0) = 0$ and $\lim\limits_{x\to\infty} f(x) \overset{\text{H}}{=} \lim\limits_{x\to\infty} \dfrac{nx^{n-1}}{e^x} \overset{\text{H}}{=} \cdots \overset{\text{H}}{=} \lim\limits_{x\to\infty} \dfrac{n!}{e^x} = 0$.

To find the maxima and minima, we differentiate: $f(x) = x^n e^{-x}$ $\Rightarrow$

$f'(x) = x^n(-e^{-x}) + e^{-x}(nx^{n-1}) = e^{-x}x^{n-1}(n - x)$. This is 0 at $x = 0$ and at $x = n$. Now if n is even, then $f(0)$ will be a minimum, since $f(0) = 0$, and $f(x) > 0$ for $x \neq 0$. But if n is odd, then there is no extremum at $x = 0$ since $f(x) < 0$ for $x < 0$ and $f(x) > 0$ for $x > 0$. In either case, there is a maximum at $x = n$ by the First Derivative Test. So as n gets larger, so does the x-coordinate of the maximum. The maximum value itself also gets larger as n gets larger, since $f(n) = n^n e^{-n} = (n/e)^n$, which increases as n increases.

To find the points of inflection, we differentiate again: $f'(x) = e^{-x}x^{n-1}(n - x)$ $\Rightarrow$

$$\begin{aligned} f''(x) &= e^{-x}x^{n-1}(-1) + e^{-x}(n-1)x^{n-2}(n - x) + (-e^{-x})x^{n-1}(n - x) \\ &= e^{-x}x^{n-2}[-x + (n-1)(n-x) - x(n-x)] = e^{-x}x^{n-2}\left(x^2 - 2nx + n^2 - n\right). \end{aligned}$$

Now for $n = 1$, $f''(x) = \dfrac{e^{-x}}{x}(x^2 - 2x) = e^{-x}(x - 2)$, which changes sign only at $x = 2$. If $n = 2$, then

$f''(x) = e^{-x}(x^2 - 4x + 2)$, which changes sign only at $x = 2 \pm \sqrt{2}$. If $n \geq 3$, then $f''(x) = 0$ at $x = 0$ and when $x^2 - 2nx + n^2 - n = 0 \quad \Leftrightarrow$

$$x = \frac{2n \pm \sqrt{4n^2 - 4(n^2 - n)}}{2} = n \pm \sqrt{n}.$$

If n is even, there is no IP at $x = 0$, since $f''(x) > 0$ on both sides of the origin. If n is odd (except for $n = 1$), there is an IP at $x = 0$. In either case, there are IP at $x = n \pm \sqrt{n}$, so as n increases, these two inflection points get further from the origin, and further from each other.

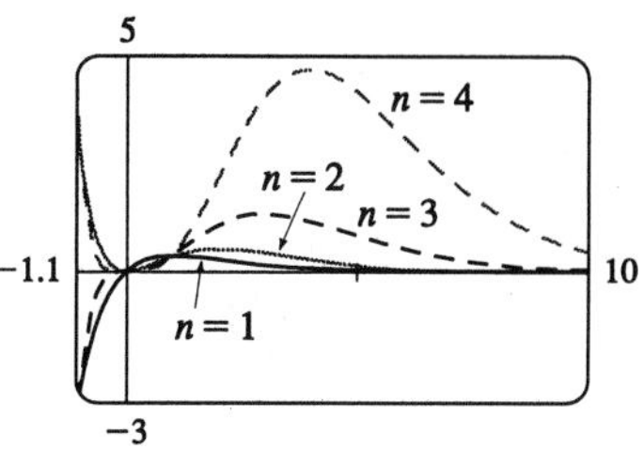

89. If $c < 0$, then $\lim_{x \to -\infty} f(x) = \lim_{x \to -\infty} \frac{x}{e^{cx}} \overset{\text{H}}{=} \lim_{x \to -\infty} \frac{1}{ce^{cx}} = 0$, and $\lim_{x \to \infty} f(x) = \infty$.

If $c > 0$, then $\lim_{x \to -\infty} f(x) = -\infty$, and $\lim_{x \to \infty} f(x) \overset{\text{H}}{=} \lim_{x \to \infty} \frac{1}{ce^{cx}} = 0$.

If $c = 0$, then $f(x) = x$, so $\lim_{x \to \pm\infty} f(x) = \pm\infty$ respectively.

So we see that $c = 0$ is a transitional value. We now exclude the case $c = 0$, since we know how the function behaves in that case.

To find the maxima and minima of f, we differentiate: $f(x) = xe^{-cx} \quad \Rightarrow$

$f'(x) = x(-ce^{-cx}) + e^{-cx} = (1 - cx)e^{-cx}$. This is 0 when $1 - cx = 0 \quad \Leftrightarrow \quad x = 1/c$. If $c < 0$ then this represents a minimum of $f(1/c) = 1/(ce)$, since $f'(x)$ changes from negative to positive at $x = 1/c$; and if $c > 0$, it represents a maximum. As $|c|$ increases, the extremum gets closer to the origin.

To find the inflection points, we differentiate again:

$f'(x) = e^{-cx}(1 - cx) \quad \Rightarrow$

$f''(x) = e^{-cx}(-c) + (1 - cx)(-ce^{-cx}) = (cx - 2)ce^{-cx}$.

This changes sign when $cx - 2 = 0 \quad \Leftrightarrow \quad x = 2/c$.

So as $|c|$ increases, the points of inflection get closer to the origin.

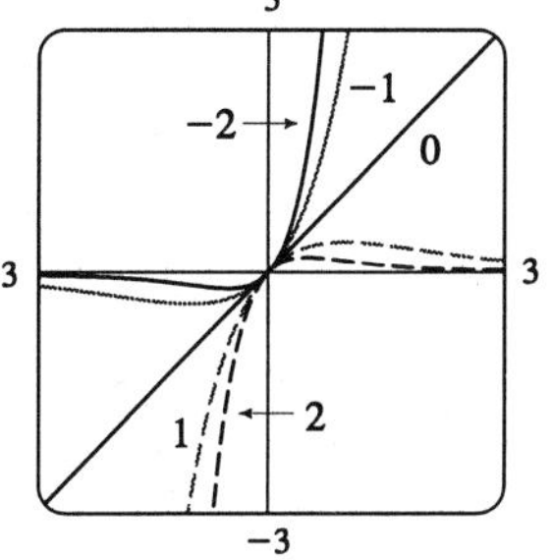

90. We see that both numerator and denominator approach 0, so we can use l'Hospital's Rule:

$$\lim_{x \to a} \frac{\sqrt{2a^3x - x^4} - a\sqrt[3]{aax}}{a - \sqrt[4]{ax^3}} \overset{\text{H}}{=} \lim_{x \to a} \frac{\frac{1}{2}\left(2a^3x - x^4\right)^{-1/2}\left(2a^3 - 4x^3\right) - a\left(\frac{1}{3}\right)(aax)^{-2/3}a^2}{-\frac{1}{4}\left(ax^3\right)^{-3/4}\left(3ax^2\right)}$$

$$= \frac{\frac{1}{2}\left(2a^3a - a^4\right)^{-1/2}\left(2a^3 - 4a^3\right) - \frac{1}{3}a^3\left(a^2a\right)^{-2/3}}{-\frac{1}{4}\left(aa^3\right)^{-3/4}\left(3aa^2\right)}$$

$$= \frac{\left(a^4\right)^{-1/2}\left(-a^3\right) - \frac{1}{3}a^3\left(a^3\right)^{-2/3}}{-\frac{3}{4}a^3\left(a^4\right)^{-3/4}} = \frac{-a - \frac{1}{3}a}{-3/4} = \frac{4}{3}\left(\frac{4}{3}a\right) = \frac{16}{9}a$$

91. Both numerator and denominator approach 0 as $x \to 0$, so we use l'Hospital's Rule (and the Fundamental Theorem of Calculus, Part 1):

$$\lim_{x \to 0} \frac{S(x)}{x^3} = \lim_{x \to 0} \frac{\int_0^x \sin\left(\pi t^2/2\right) dt}{x^3} \overset{\text{H}}{=} \lim_{x \to 0} \frac{\sin\left(\pi x^2/2\right)}{3x^2} \overset{\text{H}}{=} \lim_{x \to 0} \frac{\pi x \cos\left(\pi x^2/2\right)}{6x} = \frac{\pi}{6} \cdot \cos 0 = \frac{\pi}{6}$$

92. Both numerator and denominator approach 0 as $a \to 0$, so we use l'Hospital's Rule. (Note that we are differentiating *with respect to a,* since that is the quantity which is changing.) We also use the Fundamental Theorem of Calculus, Part 1: $\lim_{a\to 0} T(x,t) = \lim_{a\to 0} \dfrac{C\int_0^a e^{-(x-u)^2/(4kt)}\,du}{a\sqrt{4\pi kt}} \overset{\text{H}}{=} \lim_{a\to 0} \dfrac{Ce^{-(x-a)^2/(4kt)}}{\sqrt{4\pi kt}} = \dfrac{Ce^{-x^2/(4kt)}}{\sqrt{4\pi kt}}$.

93. Since $\lim_{h\to 0} [f(x+h) - f(x-h)] = f(x) - f(x) = 0$ (f is differentiable and hence continuous) and $\lim_{h\to 0} 2h = 0$, we use l'Hospital's Rule:

$$\lim_{h\to 0} \frac{f(x+h)-f(x-h)}{2h} \overset{\text{H}}{=} \lim_{h\to 0} \frac{f'(x+h)-f'(x-h)(-1)}{2} = \frac{f'(x)+f'(x)}{2} = \frac{2f'(x)}{2} = f'(x)$$

94. Since $\lim_{h\to 0}[f(x+h) - 2f(x) + f(x-h)] = f(x) - 2f(x) + f(x) = 0$ (f is differentiable and hence continuous) and $\lim_{h\to 0} h^2 = 0$, we can apply l'Hospital's Rule:

$$\lim_{h\to 0} \frac{f(x+h)-2f(x)+f(x-h)}{h^2} \overset{\text{H}}{=} \lim_{h\to 0} \frac{f'(x+h)-f'(x-h)}{2h} = f''(x)$$

At the last step, we have applied the result of Exercise 93 to $f'(x)$.

95. $\lim_{x\to\infty} \dfrac{e^x}{x^n} \overset{\text{H}}{=} \lim_{x\to\infty} \dfrac{e^x}{n\,x^{n-1}} \overset{\text{H}}{=} \lim_{x\to\infty} \dfrac{e^x}{n(n-1)x^{n-2}} \overset{\text{H}}{=} \cdots \overset{\text{H}}{=} \lim_{x\to\infty} \dfrac{e^x}{n!} = \infty$

96. $\lim_{x\to\infty} \dfrac{\ln x}{x^p} \overset{\text{H}}{=} \lim_{x\to\infty} \dfrac{1/x}{px^{p-1}} = \lim_{x\to\infty} \dfrac{1}{px^p} = 0$ since $p > 0$.

97. $\lim_{x\to 0^+} x^\alpha \ln x = \lim_{x\to 0^+} \dfrac{\ln x}{x^{-\alpha}} \overset{\text{H}}{=} \lim_{x\to 0^+} \dfrac{1/x}{-\alpha x^{-\alpha-1}} = \lim_{x\to 0^+} \dfrac{x^\alpha}{-\alpha} = 0$ since $\alpha > 0$.

98. Using l'Hospital's Rule and the Fundamental Theorem of Calculus, Part 1, we have

$$\lim_{x\to 0} \frac{\int_0^x \sin(t^2)dt}{x^3} = \lim_{x\to 0} \frac{\frac{d}{dx}\int_0^x \sin(t^2)dt}{3x^2} = \lim_{x\to 0} \frac{\sin(x^2)}{3x^2} = \frac{1}{3}\lim_{x\to 0} \frac{\sin(x^2)}{x^2} = \frac{1}{3}.$$

99. Let the radius of the circle be r. We see that $A(\theta)$ is just the area of the whole figure (a sector of the circle with radius 1), minus the area of $\triangle OPR$. But the area of the sector of the circle is $\frac{1}{2}r^2\theta$ (see endpapers), and the area of the triangle is $\frac{1}{2}r|PQ| = \frac{1}{2}r(r\sin\theta) = \frac{1}{2}r^2\sin\theta$. So we have $A(\theta) = \frac{1}{2}r^2\theta - \frac{1}{2}r^2\sin\theta = \frac{1}{2}r^2(\theta - \sin\theta)$. Now by elementary trigonometry, $B(\theta) = \frac{1}{2}|QR||PQ| = \frac{1}{2}r(1-\cos\theta)(r\sin\theta)$. So the limit we want is

$$\lim_{\theta\to 0^+} \frac{A(\theta)}{B(\theta)} = \lim_{\theta\to 0^+} \frac{\frac{1}{2}r^2(\theta-\sin\theta)}{\frac{1}{2}r^2(1-\cos\theta)\sin\theta} \overset{\text{H}}{=} \lim_{\theta\to 0^+} \frac{1-\cos\theta}{(1-\cos\theta)\cos\theta + \sin\theta(\sin\theta)}$$

$$= \lim_{\theta\to 0^+} \frac{1-\cos\theta}{\cos\theta - \cos^2\theta + \sin^2\theta} \overset{\text{H}}{=} \lim_{\theta\to 0^+} \frac{\sin\theta}{-\sin\theta + 4\sin\theta(\cos\theta)} = \frac{1}{-1+4\cos 0} = \frac{1}{3}.$$

100. The area $A(t) = \int_0^t \sin(x^2)dx$, and the area $B(t) = \frac{1}{2}t\sin(t^2)$. Since $\lim_{t\to 0^+} A(t) = 0 = \lim_{t\to 0^+} B(t)$, we can use l'Hospital's Rule:

$$\lim_{t\to 0^+} \frac{A(t)}{B(t)} \overset{\text{H}}{=} \lim_{t\to 0^+} \frac{\sin(t^2)}{\frac{1}{2}\sin(t^2) + \frac{1}{2}t[2t\cos(t^2)]} \quad \text{(by the Fundamental Theorem of Calculus, Part 1)}$$

$$\overset{\text{H}}{=} \lim_{t\to 0^+} \frac{2t\cos(t^2)}{t\cos(t^2) - 2t^3\sin(t^2) + 2t\cos(t^2)} = \lim_{t\to 0^+} \frac{2\cos(t^2)}{3\cos(t^2) - 2t^2\sin(t^2)} = \frac{2}{3-0} = \frac{2}{3}.$$

101. **(a)** We show that $\lim\limits_{x\to 0} \dfrac{f(x)}{x^n} = 0$ for every integer $n \ge 0$. Let $y = \dfrac{1}{x^2}$. Then

$$\lim_{x\to 0} \frac{f(x)}{x^{2n}} = \lim_{x\to 0} \frac{e^{-1/x^2}}{(x^2)^n} = \lim_{y\to\infty} \frac{y^n}{e^y} \overset{\text{H}}{=} \lim_{y\to\infty} \frac{ny^{n-1}}{e^y} \overset{\text{H}}{=} \cdots \overset{\text{H}}{=} \lim_{y\to\infty} \frac{n!}{e^y} = 0 \quad \Rightarrow$$

$$\lim_{x\to 0} \frac{f(x)}{x^n} = \lim_{x\to 0} x^n \frac{f(x)}{x^{2n}} = \lim_{x\to 0} x^n \lim_{x\to 0} \frac{f(x)}{x^{2n}} = 0. \text{ Thus } f'(0) = \lim_{x\to 0} \frac{f(x) - f(0)}{x - 0} = \lim_{x\to 0} \frac{f(x)}{x} = 0.$$

(b) Using the Chain Rule and the Quotient Rule we see that $f^{(n)}(x)$ exists for $x \ne 0$. In fact, we prove by induction that for each $n \ge 0$, there is a polynomial p_n and a non-negative integer k_n with $f^{(n)}(x) = p_n(x) f(x)/x^{k_n}$ for $x \ne 0$. This is true for $n = 0$; suppose it is true for the nth derivative. Then

$$\begin{aligned} f^{(n+1)}(x) &= \left[x^{k_n} [p_n'(x) f(x) + p_n(x) f'(x)] - k_n x^{k_n - 1} p_n(x) f(x)\right] x^{-2k_n} \\ &= \left[x^{k_n} p_n'(x) + p_n(x)(2/x^3) - k_n x^{k_n-1} p_n(x)\right] f(x) x^{-2k_n} \\ &= \left[x^{k_n+3} p_n'(x) + 2\, p_n(x) - k_n x^{k_n+2} p_n(x)\right] f(x) x^{-(2k_n+3)}, \end{aligned}$$

which has the desired form.

Now we show by induction that $f^{(n)}(0) = 0$ for all n. By (a), $f'(0) = 0$. Suppose that $f^{(n)}(0) = 0$. Then

$$\begin{aligned} f^{(n+1)}(0) &= \lim_{x\to 0} \frac{f^{(n)}(x) - f^{(n)}(0)}{x - 0} = \lim_{x\to 0} \frac{f^{(n)}(x)}{x} \\ &= \lim_{x\to 0} \frac{p_n(x) f(x)/x^{k_n}}{x} = \lim_{x\to 0} \frac{p_n(x) f(x)}{x^{k_n+1}} \\ &= \lim_{x\to 0} p_n(x) \lim_{x\to 0} \frac{f(x)}{x^{k_n+1}} = p_n(0) \cdot 0 = 0. \end{aligned}$$

102. **(a)** For f to be continuous, we need $\lim\limits_{x\to 0} f(x) = f(0) = 1$. We note that for $x \ne 0$, $\ln f(x) = \ln|x|^x = x \ln|x|$. So $\lim\limits_{x\to 0} \ln f(x) = \lim\limits_{x\to 0} \dfrac{\ln|x|}{1/x} \overset{\text{H}}{=} \lim\limits_{x\to 0} \dfrac{1/x}{-x^{-2}} = 0$. Therefore, $\lim\limits_{x\to 0} f(x) = \lim\limits_{x\to 0} e^{\ln f(x)} = e^0 = 1$. So f is continuous at 0.

(b) From the graphs, it seems that $f(x)$ is differentiable at 0.

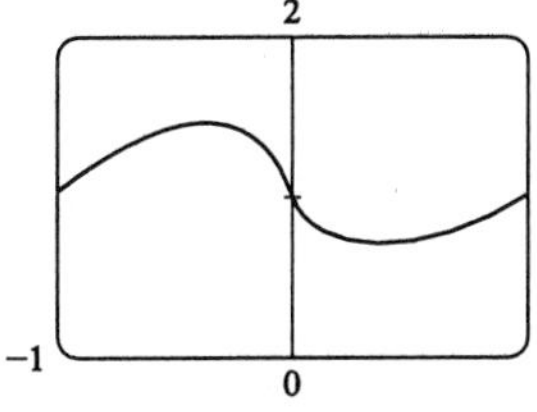

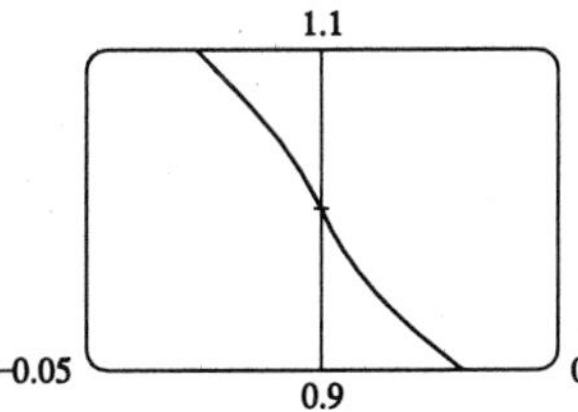

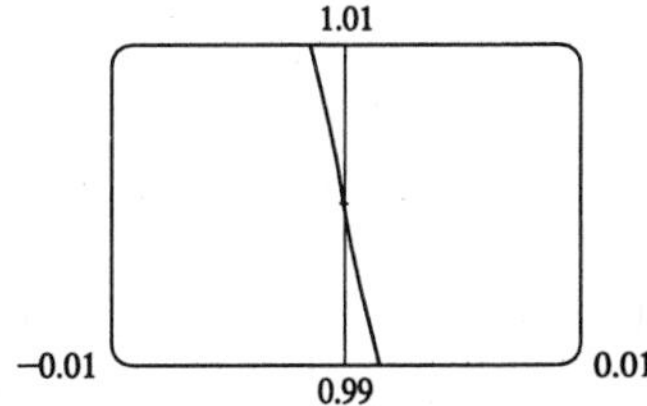

(c) To find f', we use logarithmic differentiation: $\ln f(x) = x \ln|x| \quad \Rightarrow \quad \dfrac{f'(x)}{f(x)} = x\left(\dfrac{1}{x}\right) + \ln|x| \quad \Rightarrow$ $f'(x) = f(x)(1 + \ln|x|) = |x|^x (1 + \ln|x|)$, $x \ne 0$. Now $f'(x) \to -\infty$ as $x \to 0$ [since $|x|^x \to 1$ and $(1 + \ln|x|) \to -\infty$], so the curve has a vertical tangent at $(0, 1)$ and is therefore not differentiable there. The fact cannot be seen in the graphs in part (b) because $\ln|x| \to -\infty$ very slowly as $x \to 0$.

REVIEW EXERCISES FOR CHAPTER 6

1. False. For example, $\cos\frac{\pi}{2} = \cos\left(-\frac{\pi}{2}\right) = 0$, so $\cos x$ is not 1-1.

2. False, since the range of $\tan^{-1}$ is $\left(-\frac{\pi}{2}, \frac{\pi}{2}\right)$, so $\tan^{-1}(-1) = -\frac{\pi}{4}$.

3. True, since $\ln x$ is an increasing function on $(0, \infty)$.

4. True, by Equation 6.3.6 or by Definition 6.4*.1.

5. True, since $e^x \neq 0$ for all x.

6. False. For example, $\ln(1+1) = \ln 2$, but $\ln 1 + \ln 1 = 0$. In fact $\ln a + \ln b = \ln(ab)$.

7. False. For example, $(\ln e)^6 = 1^6 = 1$, but $6\ln e = 6$. In fact $\ln(x^6) = 6\ln x$.

8. False. $(d/dx)10^x = 10^x \ln 10$ by Equation 6.4.7 or by Equation 6.4*.3.

9. False. $\ln 10$ is a constant, so its derivative is 0.

10. True. $y = e^{3x} \quad\Rightarrow\quad \ln y = 3x \quad\Rightarrow\quad x = \frac{1}{3}\ln y \quad\Rightarrow\quad$ the inverse function is $y = \frac{1}{3}\ln x$.

11. False. The "-1" is not an exponent; it is an indication of an inverse function. See Equation 6.6.4.

12. False. For example, $\tan^{-1}20$ is defined; $\sin^{-1}20$ and $\cos^{-1}20$ are not.

13. True. See Figure 2 in Section 6.7.

14. True. $\ln\frac{1}{10} = -\ln 10 = -\int_1^{10}(1/x)dx$ by Equation 6.4.4 or by Definition 6.2*.1.

15. True. $\int_2^{16}(1/x)dx = \ln x\big|_2^{16} = \ln 16 - \ln 2 = \ln\frac{16}{2} = \ln 8 = \ln 2^3 = 3\ln 2$.

16. False. We cannot use l'Hospital's Rule because the denominator does not approach 0 as $x \to \pi^-$. In fact

$$\lim_{x\to\pi^-}\frac{\tan x}{1-\cos x} = \frac{\tan\pi}{1-\cos\pi} = \frac{0}{1-(-1)} = 0.$$

17. $y = e^x$

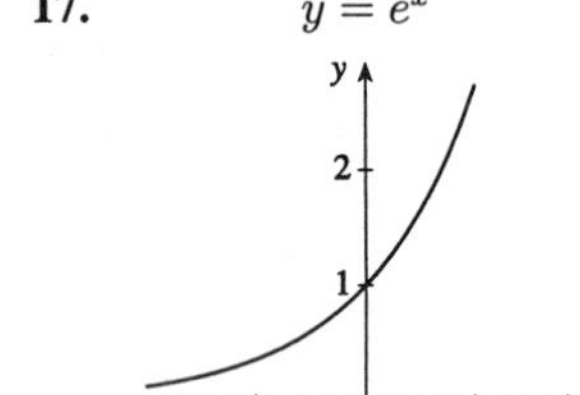

18. $y = e^{-x}$

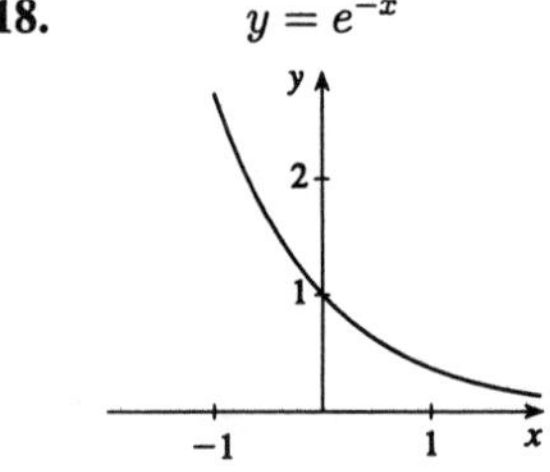

19. $y = e^{-x}$

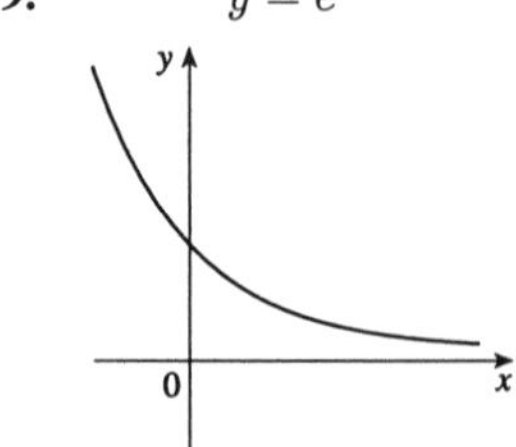

$y = -e^{-x}$

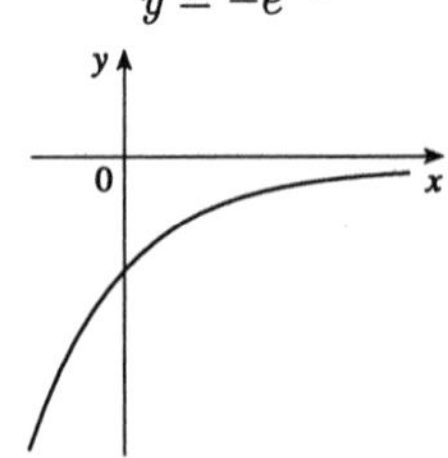

20. $y = 7^x$ $y = 1 + 7^x$

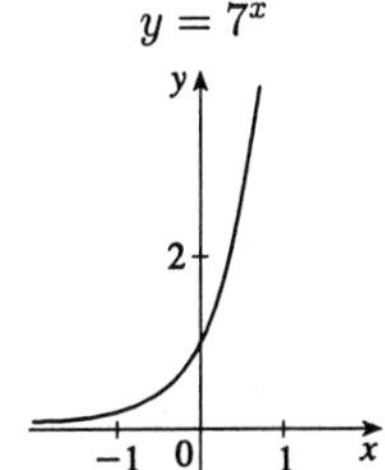

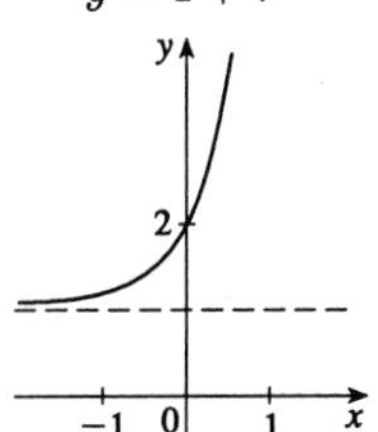

21. $y = \ln x$

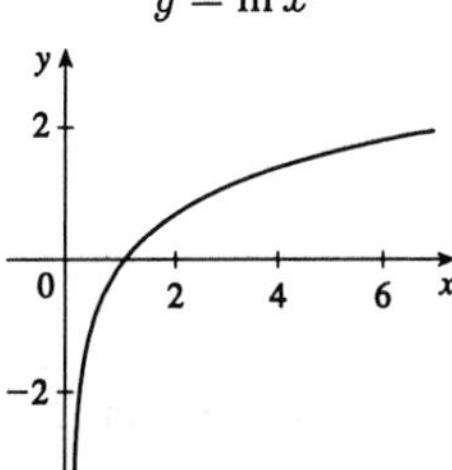

22. $y = \ln(x - 1)$

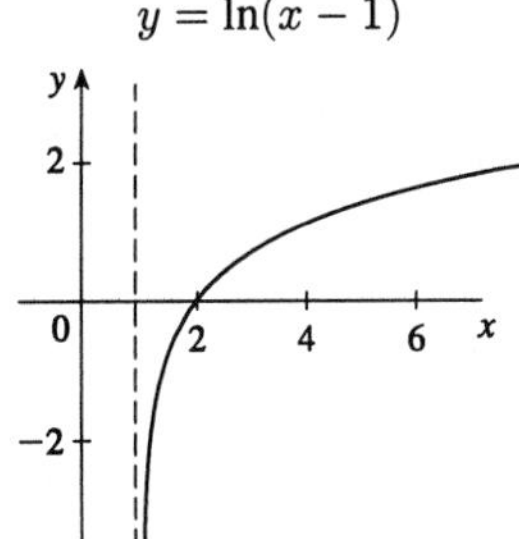

23. $y = 2 - \ln x$

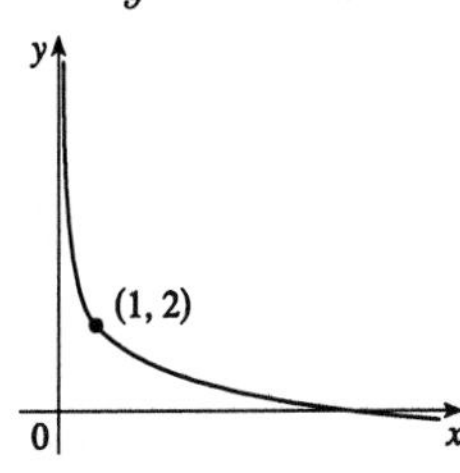

24. $y = e^x \cos x$

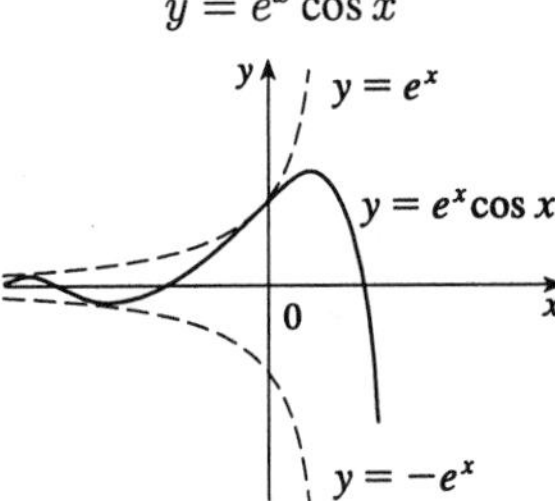

25. $y = \tan^{-1} x$

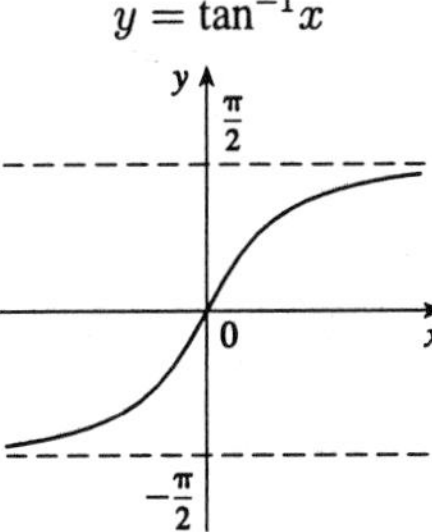

26. We have seen that if $a > 1$, then $a^x > x^a$ for sufficiently large x. [For $a = 2$, see Figures 4 and 5 in Section 6.2. In general, we could show that $\lim_{x\to\infty} (a^x/x^a) = \infty$ by using l'Hospital's Rule repeatedly.]

Also, $\log_a x$ increases much more slowly than either x^a or a^x. [Compare the graph of $\log_a x$ with those of x^a and a^x, or use l'Hospital's Rule to show that $\lim_{x\to\infty} [(\log_a x)/x^a] = 0$.] So for large x, $\log_a x < x^a < a^x$.

27. $e^x = 5 \quad \Rightarrow \quad x = \ln(e^x) = \ln 5$

28. $\ln x = 2 \quad \Rightarrow \quad x = e^{\ln x} = e^2$

29. $\log_{10}(e^x) = 1 \quad \Rightarrow \quad e^x = 10 \quad \Rightarrow \quad x = \ln(e^x) = \ln 10$

Or: $1 = \log_{10}(e^x) = x \log_{10} e \quad \Rightarrow \quad x = 1/\log_{10} e = \ln 10$

30. $e^{e^x} = 2 \quad \Rightarrow \quad \ln 2 = \ln e^{e^x} = e^x \quad \Rightarrow \quad x = \ln e^x = \ln \ln 2$

31. $2 = \ln(x^\pi) = \pi \ln x \quad \Rightarrow \quad \ln x = 2/\pi \quad \Rightarrow \quad x = e^{2/\pi}$

32. $1 = \ln(x + 1) - \ln(x) = \ln\left(\dfrac{x+1}{x}\right) \quad \Rightarrow \quad \dfrac{x+1}{x} = e \;\Rightarrow\; ex = x + 1 \;\Rightarrow\; x = \dfrac{1}{e-1}$

33. $\tan x = 4 \quad \Rightarrow \quad x = \tan^{-1} 4 + n\pi = \arctan 4 + n\pi$, n an integer

34. $\sin^{-1}x = 1 \quad \Rightarrow \quad x = \sin 1$

35. $y = \log_{10}(x^2 - x) \quad \Rightarrow \quad y' = \dfrac{1}{x^2 - x}(\log_{10} e)(2x - 1) = \dfrac{2x - 1}{(\ln 10)(x^2 - x)}$

36. $y = \sqrt{2}^{\,x} \quad \Rightarrow \quad y' = \sqrt{2}^{\,x} \ln \sqrt{2} = \sqrt{2}^{\,x}(\ln 2)/2$

37. $y = \dfrac{\sqrt{x+1}(2-x)^5}{(x+3)^7} \quad \Rightarrow \quad \ln|y| = \frac{1}{2}\ln(x+1) + 5\ln|2-x| - 7\ln(x+3) \quad \Rightarrow$

$\dfrac{y'}{y} = \dfrac{1}{2(x+1)} + \dfrac{-5}{2-x} - \dfrac{7}{x+3} \quad \Rightarrow y' = \dfrac{\sqrt{x+1}(2-x)^5}{(x+3)^7}\left[\dfrac{1}{2(x+1)} - \dfrac{5}{2-x} - \dfrac{7}{x+3}\right]$

38. $y = \ln(\csc 5x) \quad \Rightarrow \quad y' = \dfrac{-5 \csc 5x \cot 5x}{\csc 5x} = -5 \cot 5x$

39. $y = e^{cx}(c \sin x - \cos x) \quad \Rightarrow \quad y' = ce^{cx}(c \sin x - \cos x) + e^{cx}(c \cos x + \sin x) = (c^2 + 1)e^{cx} \sin x$

40. $y = \sin^{-1}(e^x) \quad \Rightarrow \quad y' = e^x/\sqrt{1 - e^{2x}}$

41. $y = \ln(\sec^2 x) = 2\ln|\sec x| \quad \Rightarrow \quad y' = (2/\sec x)(\sec x \tan x) = 2 \tan x$

42. $y = \ln(x^2 e^x) = 2\ln|x| + x \quad \Rightarrow \quad y' = 2/x + 1$

43. $y = xe^{-1/x} \quad \Rightarrow \quad y' = e^{-1/x} + xe^{-1/x}(1/x^2) = e^{-1/x}(1 + 1/x)$

44. $y = \ln|\csc 3x + \cot 3x| \quad \Rightarrow \quad y' = \dfrac{-3 \csc 3x \cot 3x - 3 \csc^2 3x}{\csc 3x + \cot 3x} = -3 \csc 3x$

45. $y = (\cos^{-1}x)^{\sin^{-1}x} \quad \Rightarrow \quad \ln y = \sin^{-1}x \ln(\cos^{-1}x) \quad \Rightarrow$

$\dfrac{y'}{y} = \dfrac{1}{\sqrt{1-x^2}} \ln(\cos^{-1}x) + (\sin^{-1}x)\left(\dfrac{1}{\cos^{-1}x}\right)\left(-\dfrac{1}{\sqrt{1-x^2}}\right) \quad \Rightarrow$

$y' = (\cos^{-1}x)^{\sin^{-1}x - 1}\left[\dfrac{\cos^{-1}x \ln(\cos^{-1}x) - \sin^{-1}x}{\sqrt{1-x^2}}\right]$

46. $y = x^r e^{sx} \quad \Rightarrow \quad y' = rx^{r-1}e^{sx} + sx^r e^{sx}$

47. $y = e^{e^x} \quad \Rightarrow \quad y' = e^{e^x} e^x = e^{x + e^x}$

48. $y = 5^{x \tan x} \quad \Rightarrow \quad y' = 5^{x \tan x}(\ln 5)(\tan x + x \sec^2 x)$

49. $y = \ln \dfrac{1}{x} + \dfrac{1}{\ln x} = -\ln x + (\ln x)^{-1} \quad \Rightarrow \quad y' = -\dfrac{1}{x} - \dfrac{1}{x(\ln x)^2}$

50. $xe^y = y - 1 \quad \Rightarrow \quad e^y + xe^y y' = y' \quad \Rightarrow \quad y' = e^y/(1 - xe^y)$

51. $y = 7^{\sqrt{2x}} \quad \Rightarrow \quad y' = 7^{\sqrt{2x}}(\ln 7)\left[1/(2\sqrt{2x})\right](2) = 7^{\sqrt{2x}}(\ln 7)/\sqrt{2x}$

52. $y = e^{\cos x} + \cos(e^x) \quad \Rightarrow \quad y' = -\sin x\, e^{\cos x} - e^x \sin(e^x)$

53. $y = \ln(\cosh 3x) \quad \Rightarrow \quad y' = (1/\cosh 3x)(\sinh 3x)(3) = 3 \tanh 3x$

54. $y = \ln|x^2 - 4| - \ln|2x + 5| \quad \Rightarrow \quad y' = \dfrac{2x}{x^2 - 4} - \dfrac{2}{2x + 5}$

55. $y = \cosh^{-1}(\sinh x) \quad \Rightarrow \quad y' = (\cosh x)/\sqrt{\sinh^2 x - 1}$

56. $y = x \tanh^{-1}\sqrt{x} \quad \Rightarrow \quad y' = \tanh^{-1}\sqrt{x} + x\dfrac{1}{1-(\sqrt{x})^2}\dfrac{1}{2\sqrt{x}} = \tanh^{-1}\sqrt{x} + \dfrac{\sqrt{x}}{2(1-x)}$

57. $y = \ln\sin x - \frac{1}{2}\sin^2 x \quad \Rightarrow \quad y' = \dfrac{\cos x}{\sin x} - \sin x\cos x = \cot x - \sin x \cos x$

58. $y = \left(\dfrac{c}{x}\right)^x \quad \Rightarrow \quad \ln y = x\ln\left(\dfrac{c}{x}\right) = x\ln c - x\ln x \quad \Rightarrow \quad \dfrac{y'}{y} = \ln c - \ln x - 1 \quad \Rightarrow \quad y' = \left(\dfrac{c}{x}\right)^x\left[\ln\left(\dfrac{c}{x}\right) - 1\right]$

59. $y = \sin^{-1}\left(\dfrac{x-1}{x+1}\right) \quad \Rightarrow$

$$y' = \frac{1}{\sqrt{1-[(x-1)/(x+1)]^2}}\frac{(x+1)-(x-1)}{(x+1)^2} = \frac{1}{\sqrt{(x+1)^2-(x-1)^2}}\left(\frac{2}{x+1}\right)$$

$$= \frac{2}{\sqrt{4x}(x+1)} = \frac{1}{\sqrt{x}(x+1)}.$$ [Note that the domain of y is $x \geq 0$.]

60. $y = \arctan(\arcsin\sqrt{x}) \quad \Rightarrow \quad y' = \dfrac{1}{1+(\arcsin\sqrt{x})^2}\cdot\dfrac{1}{\sqrt{1-x}}\cdot\dfrac{1}{2\sqrt{x}}$

61. $y = \frac{1}{4}\left[\ln(x^2+x+1) - \ln(x^2-x+1)\right] + \dfrac{1}{2\sqrt{3}}\left[\tan^{-1}\left(\dfrac{2x+1}{\sqrt{3}}\right) + \tan^{-1}\left(\dfrac{2x-1}{\sqrt{3}}\right)\right] \quad \Rightarrow$

$$y' = \frac{1}{4}\left[\frac{2x+1}{x^2+x+1} - \frac{2x-1}{x^2-x+1}\right] + \frac{1}{2\sqrt{3}}\left[\frac{2/\sqrt{3}}{1+\left[(2x+1)/\sqrt{3}\right]^2} + \frac{2/\sqrt{3}}{1+\left[(2x-1)/\sqrt{3}\right]^2}\right]$$

$$= \frac{1}{4}\left[\frac{2x+1}{x^2+x+1} - \frac{2x-1}{x^2-x+1}\right] + \frac{1}{4(x^2+x+1)} + \frac{1}{4(x^2-x+1)}$$

$$= \frac{1}{2}\left[\frac{x+1}{x^2+x+1} - \frac{x-1}{x^2-x+1}\right] = \frac{1}{x^4+x^2+1}$$

62. $y = \dfrac{x}{\sqrt{a^2-1}} - \dfrac{2}{\sqrt{a^2-1}}\arctan\dfrac{\sin x}{a+\sqrt{a^2-1}+\cos x}$. Let $k = a + \sqrt{a^2-1}$. Then

$$y' = \frac{1}{\sqrt{a^2-1}} - \frac{2}{\sqrt{a^2-1}}\cdot\frac{1}{1+\sin^2 x/(k+\cos x)^2}\cdot\frac{\cos x(k+\cos x)+\sin^2 x}{(k+\cos x)^2}$$

$$= \frac{1}{\sqrt{a^2-1}} - \frac{2}{\sqrt{a^2-1}}\cdot\frac{k\cos x+\cos^2 x+\sin^2 x}{(k+\cos x)^2+\sin^2 x} = \frac{1}{\sqrt{a^2-1}} - \frac{2}{\sqrt{a^2-1}}\cdot\frac{k\cos x+1}{k^2+2k\cos x+1}$$

$$= \frac{k^2+2k\cos x+1-2k\cos x-2}{\sqrt{a^2-1}(k^2+2k\cos x+1)} = \frac{k^2-1}{\sqrt{a^2-1}(k^2+2k\cos x+1)}.$$

But $k^2 = 2a^2 + 2a\sqrt{a^2-1} - 1 = 2a\left(a+\sqrt{a^2-1}\right) - 1 = 2ak-1$, so $k^2+1 = 2ak$, and

$k^2 - 1 = 2(ak-1)$. So $y' = \dfrac{2(ak-1)}{\sqrt{a^2-1}\,(2ak+2k\cos x)} = \dfrac{ak-1}{\sqrt{a^2-1}\,k(a+\cos x)}$. But

$ak - 1 = a^2 + a\sqrt{a^2-1} - 1 = k\sqrt{a^2-1}$, so $y' = 1/(a+\cos x)$.

63. $f(x) = 2^x \quad \Rightarrow \quad f'(x) = 2^x\ln 2 \quad \Rightarrow \quad f''(x) = 2^x(\ln 2)^2 \quad \Rightarrow \quad \cdots \quad \Rightarrow \quad f^{(n)}(x) = 2^x(\ln 2)^n$

64. $f(x) = \ln(2x) = \ln 2 + \ln x \quad \Rightarrow \quad f'(x) = x^{-1}$, $f''(x) = -x^{-2}$, $f'''(x) = 2x^{-3}$, $f^{(4)}(x) = -2\cdot 3x^{-4}, \ldots,$
$f^{(n)}(x) = (-1)^{n-1}(n-1)!\,x^{-n}$.

65. We first show it is true for $n = 1$: $f'(x) = e^x + xe^x = (x+1)e^x$. We now assume it is true for $n = k$: $f^{(k)}(x) = (x+k)e^x$. With this assumption, we must show it is true for $n = k+1$:
$f^{(k+1)}(x) = \dfrac{d}{dx}\left[f^{(k)}(x)\right] = \dfrac{d}{dx}[(x+k)e^x] = e^x + (x+k)e^x = [x + (k+1)]e^x$.
Therefore $f^{(n)}(x) = (x+n)e^x$ by mathematical induction.

66. Using implicit differentiation, $y = x + \arctan y \quad\Rightarrow\quad y' = 1 + \dfrac{1}{1+y^2}y' \quad\Rightarrow\quad y'\left(1 - \dfrac{1}{1+y^2}\right) = 1 \quad\Rightarrow$
$y'\left(\dfrac{y^2}{1+y^2}\right) = 1 \quad\Rightarrow\quad y' = \dfrac{1+y^2}{y^2} = \dfrac{1}{y^2} + 1.$

67. $y = f(x) = \ln(e^x + e^{2x}) \quad\Rightarrow\quad f'(x) = \dfrac{e^x + 2e^{2x}}{e^x + e^{2x}} \quad\Rightarrow\quad f'(0) = \frac{3}{2}$, so the tangent line at $(0, \ln 2)$ is
$y - \ln 2 = \frac{3}{2}x$ or $3x - 2y + \ln 4 = 0$.

68. $y = f(x) = x \ln x \quad\Rightarrow\quad f'(x) = \ln x + 1$, so the slope of the tangent at (e, e) is $f'(e) = 2$ and the equation is $y - e = 2(x - e)$ or $y = 2x - e$.

69. $f(x) = \sqrt{x} - \sqrt{x-1}$ has domain $[1, \infty)$. To see that f is 1-1, we can either graph the function and use the Horizontal Line Test, or we can calculate $f'(x) = \dfrac{1}{2\sqrt{x}} - \dfrac{1}{2\sqrt{x-1}} = \dfrac{\sqrt{x-1} - \sqrt{x}}{2\sqrt{x(x-1)}} < 0$, so f is decreasing and hence 1-1. The parametric equations of the graph of f are $x = t$, $y = \sqrt{t} - \sqrt{t-1}$, and so the parametric equations of the graph of f^{-1} are $x = \sqrt{t} - \sqrt{t-1}$, $y = t$.

70. $f(x) = xe^{\sin x} \quad\Rightarrow\quad f'(x) = x\left[e^{\sin x}(\cos x)\right] + e^{\sin x}(1) = e^{\sin x}(x\cos x + 1)$. As a check on our work, we notice from the graphs that $f'(x) > 0$ when f is increasing. Also, we see in the larger viewing rectangle a certain similarity in the graphs of f and f': the sizes of the oscillations of f and f' are linked.

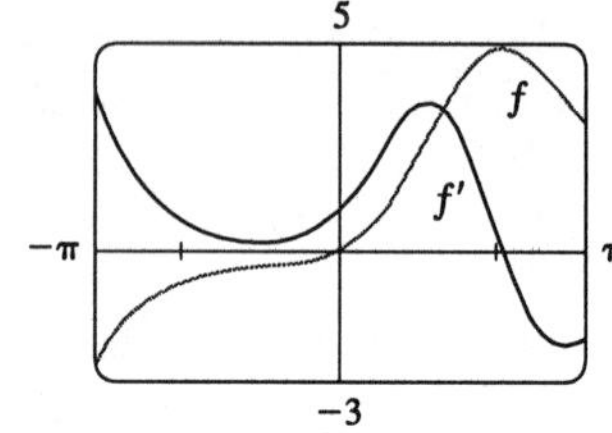

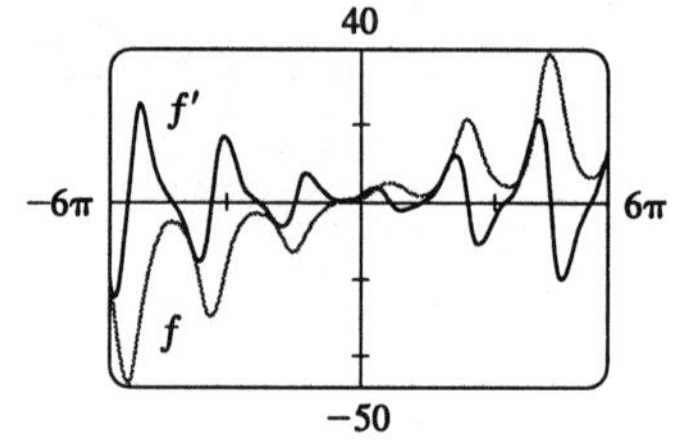

71. $y = [\ln(x+4)]^2 \quad\Rightarrow\quad y' = 2\dfrac{\ln(x+4)}{x+4} = 0 \quad\Leftrightarrow\quad \ln(x+4) = 0 \quad\Leftrightarrow\quad x + 4 = 1 \quad\Leftrightarrow\quad x = -3$, so the tangent is horizontal at $(-3, 0)$.

72. The line $x - 4y = 1$ has slope $\frac{1}{4}$. A tangent to $y = e^x$ has slope $\frac{1}{4}$ when $y' = e^x = \frac{1}{4} \quad\Rightarrow\quad x = \ln\frac{1}{4} = -\ln 4$, so the equation is $y - \frac{1}{4} = \frac{1}{4}(x + \ln 4)$ or $x - 4y + 1 + \ln 4 = 0$.

73. The slope of the tangent at the point (a, e^a) is $\left[\frac{d}{dx}e^x\right]_{x=a} = e^a$. The equation of the tangent line is thus $y - e^a = e^a(x - a)$. We substitute $x = 0$, $y = 0$ into this equation, since we want the line to pass through the origin: $0 - e^a = e^a(0 - a) \quad \Leftrightarrow \quad -e^a = e^a(-a) \quad \Leftrightarrow \quad a = 1$. So the equation of the tangent is $y - e = e(x - 1)$, or $y = ex$.

74. **(a)** $\lim_{t\to\infty} C(t) = \lim_{t\to\infty} K\left(e^{-at} - e^{-bt}\right) = K\lim_{t\to\infty} e^{-at} - K\lim_{t\to\infty} e^{-bt} = 0$ since $a > 0$ and $b > 0$.

(b) $C'(t) = K\left(-ae^{-at} + be^{-bt}\right)$

(c) $C'(t) = 0 \quad\Rightarrow\quad ae^{-at} = be^{-bt} \quad\Rightarrow\quad \frac{b}{a} = \frac{e^{-at}}{e^{-bt}} = e^{(b-a)t} \quad\Rightarrow\quad \ln\left(\frac{b}{a}\right) = \ln\left(e^{(b-a)t}\right) = (b-a)t$

$\Rightarrow \quad t = \frac{\ln(b/a)}{b-a}$

75. $\lim_{x\to-\infty} 10^{-x} = \infty$ since $-x \to \infty$ as $x \to -\infty$.

76. $\lim_{x\to\infty} \frac{4^x}{4^x - 1} = \lim_{x\to\infty} \frac{1}{1 - 4^{-x}} = \frac{1}{1-0} = 1$

77. $\lim_{x\to0^+} \ln(\tan x) = -\infty$ since $\tan x \to 0^+$ as $x \to 0^+$.

78. $-e^{x/2} \le e^{x/2}\cos x \le e^{x/2}$ and $\lim_{x\to-\infty}\left(-e^{x/2}\right) = 0 = \lim_{x\to-\infty} e^{x/2}$, so $\lim_{x\to-\infty} e^{x/2}\cos x = 0$ by the Squeeze Theorem.

79. $\lim_{x\to-4^+} e^{1/(x+4)} = \infty$ since $\frac{1}{x+4} \to \infty$ as $x \to -4^+$.

80. $\lim_{x\to-1^+} e^{\tanh^{-1}x} = 0$ since $\tanh^{-1}x \to -\infty$ as $x \to -1^+$.

81. $\lim_{x\to\infty} \frac{e^x}{e^{2x} + e^{-x}} = \lim_{x\to\infty} \frac{e^{-x}}{1 + e^{-3x}} = \frac{0}{1+0} = 0$

82. $\lim_{x\to0}(1+x)^{2/x} = \lim_{x\to0}\left[(1+x)^{1/x}\right]^2 = \left[\lim_{x\to0}(1+x)^{1/x}\right]^2 = e^2$

83. $\lim_{x\to1} \cos^{-1}\left(\frac{x}{x+1}\right) = \cos^{-1}\frac{1}{2} = \frac{\pi}{3}$

84. $\lim_{x\to-\infty} \tan^{-1}\left(x^4\right) = \frac{\pi}{2}$ since $x^4 \to \infty$ as $x \to -\infty$.

85. $\lim_{x\to\pi} \frac{\sin x}{x^2 - \pi^2} \overset{\text{H}}{=} \lim_{x\to\pi} \frac{\cos x}{2x} = -\frac{1}{2\pi}$

86. $\lim_{x\to0} \frac{e^{ax} - e^{bx}}{x} \overset{\text{H}}{=} \lim_{x\to0} \frac{ae^{ax} - be^{bx}}{1} = a - b$

87. $\lim_{x\to\infty} \frac{\ln(\ln x)}{\ln x} \overset{\text{H}}{=} \lim_{x\to\infty} \frac{1/(x\ln x)}{1/x} = \lim_{x\to\infty} \frac{1}{\ln x} = 0$

88. $\lim_{x\to0} \frac{1 + \sin x - \cos x}{1 - \sin x - \cos x} \overset{\text{H}}{=} \lim_{x\to0} \frac{\cos x + \sin x}{-\cos x + \sin x} = \frac{1+0}{-1+0} = -1$

89. $\lim_{x\to0} \frac{\ln(1-x) + x + \frac{1}{2}x^2}{x^3} \overset{\text{H}}{=} \lim_{x\to0} \frac{-\frac{1}{1-x} + 1 + x}{3x^2} \overset{\text{H}}{=} \lim_{x\to0} \frac{-\frac{1}{(1-x)^2} + 1}{6x} \overset{\text{H}}{=} \lim_{x\to0} \frac{-\frac{2}{(1-x)^3}}{6} = -\frac{2}{6} = -\frac{1}{3}$

90. $\lim_{x\to\pi/2} \left(\frac{\pi}{2} - x\right)\tan x = \lim_{x\to\pi/2} \frac{\pi/2 - x}{\cot x} \overset{\text{H}}{=} \lim_{x\to\pi/2} \frac{-1}{-\csc^2 x} = \lim_{x\to\pi/2} \sin^2 x = 1$

91. $\lim_{x\to0^+} \sin x(\ln x)^2 = \lim_{x\to0^+} \frac{(\ln x)^2}{\csc x} \overset{\text{H}}{=} \lim_{x\to0^+} \frac{2\ln x/x}{-\csc x\cot x} = -2\lim_{x\to0} \frac{\sin x}{x} \lim_{x\to0} \frac{\ln x}{\cot x} = -2\lim_{x\to0} \frac{\ln x}{\cot x}$

$\overset{\text{H}}{=} -2\lim_{x\to0} \frac{1/x}{-\csc^2 x} = 2\lim_{x\to0} \frac{\sin^2 x}{x} = 2\lim_{x\to0} \frac{\sin x}{x} \lim_{x\to0} \sin x = 2 \cdot 1 \cdot 0 = 0$

92. $\lim\limits_{x\to 0}\left(\csc^2 x - x^{-2}\right) = \lim\limits_{x\to 0}\left[\dfrac{1}{\sin^2 x} - \dfrac{1}{x^2}\right] = \lim\limits_{x\to 0}\dfrac{x^2 - \sin^2 x}{x^2\sin^2 x} \overset{\text{H}}{=} \lim\limits_{x\to 0}\dfrac{2x - \sin 2x}{2x\sin^2 x + x^2\sin 2x}$

$\overset{\text{H}}{=} \lim\limits_{x\to 0}\dfrac{2 - 2\cos 2x}{2\sin^2 x + 4x\sin 2x + 2x^2\cos 2x} \overset{\text{H}}{=} \lim\limits_{x\to 0}\dfrac{4\sin 2x}{6\sin 2x + 12x\cos 2x - 4x^2\sin 2x}$

$\overset{\text{H}}{=} \lim\limits_{x\to 0}\dfrac{8\cos 2x}{24\cos 2x - 32x\sin 2x - 8x^2\cos 2x} = \dfrac{8}{24} = \dfrac{1}{3}$

93. $\lim\limits_{x\to 1}(\ln x)^{\sin x} = (\ln 1)^{\sin 1} = 0^{\sin 1} = 0$

94. Let $y = x^{1/(1-x)}$. Then $\ln y = \dfrac{\ln x}{1-x}$, so $\lim\limits_{x\to 1}\ln y = \lim\limits_{x\to 1}\dfrac{\ln x}{1-x} \overset{\text{H}}{=} \lim\limits_{x\to 1}\dfrac{1/x}{-1} = -1 \quad\Rightarrow\quad \lim\limits_{x\to 1}x^{1/(1-x)} = e^{-1}$.

95. $\lim\limits_{x\to 0^+}\dfrac{x^{1/3}-1}{x^{1/4}-1} = \dfrac{0-1}{0-1} = 1$

96. $\lim\limits_{x\to\infty}\dfrac{\sqrt{x}}{\ln x} \overset{\text{H}}{=} \lim\limits_{x\to\infty}\dfrac{1/(2\sqrt{x})}{1/x} = \lim\limits_{x\to\infty}\dfrac{\sqrt{x}}{2} = \infty$, so $\lim\limits_{x\to\infty}\tan^{-1}\left(\dfrac{\sqrt{x}}{\ln x}\right) = \dfrac{\pi}{2}$

97. $y = f(x) = \tan^{-1}(1/x)$ **A.** $D = \{x \mid x \neq 0\}$ **B.** No intercepts **C.** $f(-x) = -f(x)$, so the curve is symmetric about the origin. **D.** $\lim\limits_{x\to\pm\infty}\tan^{-1}(1/x) = \tan^{-1}0 = 0$, so $y = 0$ is a HA. $\lim\limits_{x\to 0^+}\tan^{-1}(1/x) = \frac{\pi}{2}$ and $\lim\limits_{x\to 0^-}\tan^{-1}(1/x) = -\frac{\pi}{2}$ since $\dfrac{1}{x} \to \pm\infty$ as $x \to 0^\pm$.

E. $f'(x) = \dfrac{1}{1+(1/x)^2}\left(-1/x^2\right) = \dfrac{-1}{x^2+1} \quad\Rightarrow\quad f'(x) < 0$, so f is decreasing on $(-\infty, 0)$ and $(0,\infty)$. **F.** No extrema

G. $f''(x) = \dfrac{2x}{(x^2+1)^2} > 0 \quad\Leftrightarrow\quad x > 0$, so f is CU on $(0,\infty)$ and CD on $(-\infty, 0)$.

H.

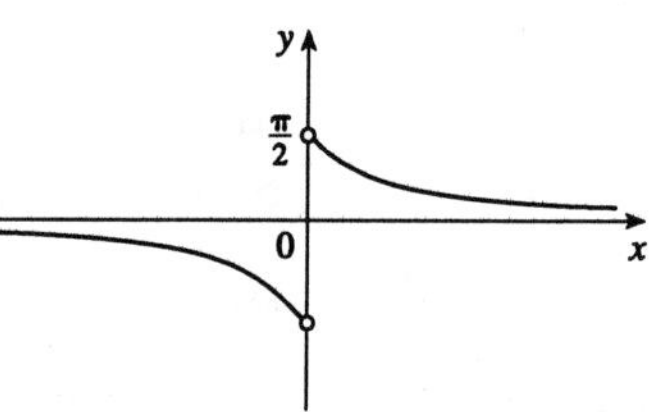

98. $y = f(x) = \sin^{-1}(1/x)$ **A.** $D = \{x \mid -1 \le 1/x \le 1\} = (-\infty,-1] \cup [1,\infty)$. **B.** No intercepts

C. $f(-x) = -f(x)$, symmetric about the origin **D.** $\lim\limits_{x\to\pm\infty}\sin^{-1}(1/x) = \sin^{-1}(0) = 0$, so $y = 0$ is a HA.

E. $f'(x) = \dfrac{1}{\sqrt{1-(1/x)^2}}\left(-\dfrac{1}{x^2}\right) = \dfrac{-1}{\sqrt{x^4-x^2}} < 0$, so f is decreasing on $(-\infty,-1]$ and $[1,\infty)$.

F. No local extrema, but $f(1) = \frac{\pi}{2}$ is the absolute maximum and $f(-1) = -\frac{\pi}{2}$ is the absolute minimum.

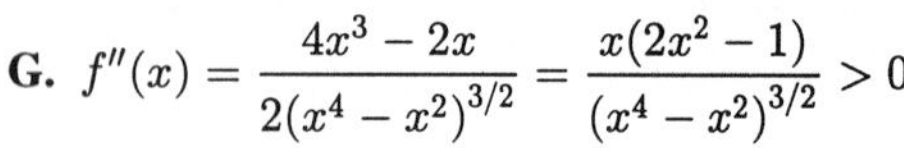

G. $f''(x) = \dfrac{4x^3 - 2x}{2(x^4-x^2)^{3/2}} = \dfrac{x(2x^2-1)}{(x^4-x^2)^{3/2}} > 0$ for $x > 1$ and $f''(x) < 0$ for $x < -1$, so f is CU on $(1,\infty)$ and CD on $(-\infty,-1)$.

H.

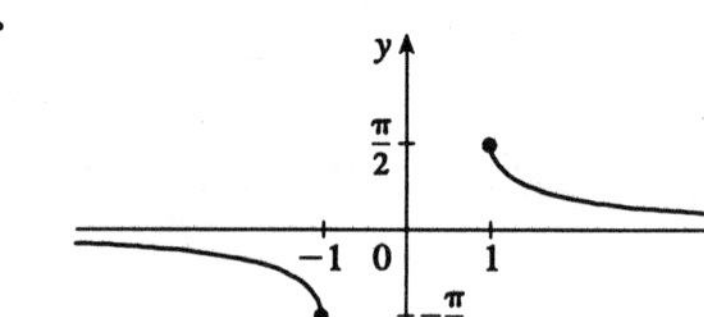

99. $y = f(x) = 2^{1/(x-1)}$ **A.** $D = \{x \mid x \neq 1\}$ **B.** No x-intercepts; y-intercept $= f(0) = 1/2$. **C.** No symmetry **D.** $\lim\limits_{x\to\pm\infty}2^{1/(x-1)} = 2^0 = 1$, so $y = 1$ is a HA. $\lim\limits_{x\to 1^+}2^{1/(x-1)} = \infty$, so $x = 1$ is a VA. Also $\lim\limits_{x\to 1^-}2^{1/(x-1)} = 0$.

E. $f'(x) = 2^{1/(x-1)}(-\ln 2)/(x-1)^2 < 0$, so f is decreasing on $(-\infty, 1)$ and $(1,\infty)$. **F.** No extrema

G. $y'' = \dfrac{2^{1/(x-1)}(\ln 2)(2x - 2 + \ln 2)}{(x-1)^4} > 0$

$\Leftrightarrow \quad 2x - 2 + \ln 2 > 0 \quad \Leftrightarrow \quad x > 1 - \frac{1}{2}\ln 2,$

so f is CU on $\left(1 - \ln\sqrt{2}, 1\right)$ and $(1, \infty)$ and

CD on $\left(-\infty, 1 - \ln\sqrt{2}\right)$. IP at $x = 1 - \ln\sqrt{2}$.

H.

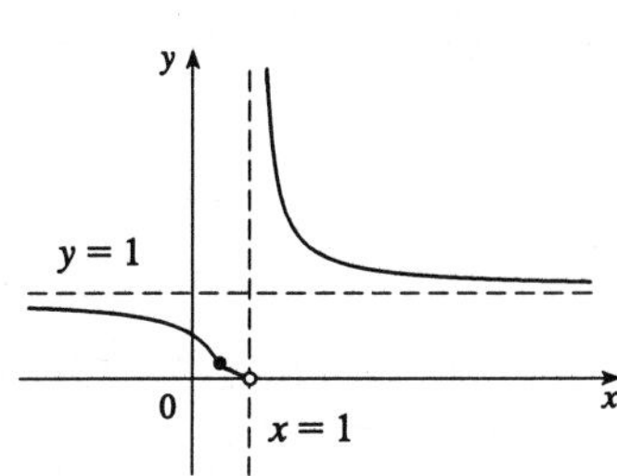

100. $y = f(x) = e^{2x-x^2}$ **A.** $D = \mathbb{R}$ **B.** No x-intercepts; y-intercept $= 1$ **C.** No symmetry

D. $\lim\limits_{x\to\pm\infty} e^{2x-x^2} = 0$, so $y = 0$ is a HA. **E.** $f'(x) = 2(1-x)e^{2x-x^2} > 0 \quad \Leftrightarrow \quad x < 1$, so f is increasing on $(-\infty, 1]$ and decreasing on $[1, \infty)$.

F. $f(1) = e$ is a local and absolute maximum.

G. $f''(x) = 2(2x^2 - 4x + 1)e^{2x-x^2} = 0 \quad \Leftrightarrow$

$x = 1 \pm \frac{\sqrt{2}}{2}$. $f''(x) > 0 \quad \Leftrightarrow \quad x < 1 - \frac{\sqrt{2}}{2}$ or $x > 1 + \frac{\sqrt{2}}{2}$,

so f is CU on $\left(-\infty, 1 - \frac{\sqrt{2}}{2}\right)$ and $\left(1 + \frac{\sqrt{2}}{2}, \infty\right)$, and

CD on $\left(1 - \frac{\sqrt{2}}{2}, 1 + \frac{\sqrt{2}}{2}\right)$. IP at $x = 1 \pm \frac{\sqrt{2}}{2}$

H.

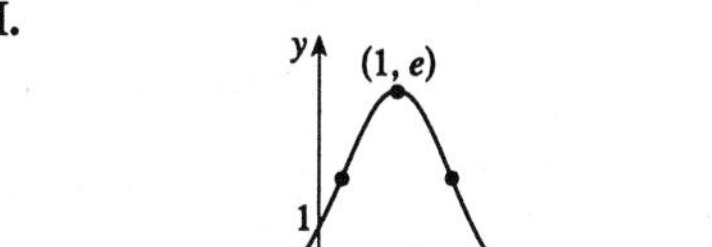

101. $y = f(x) = e^x + e^{-3x}$ **A.** $D = \mathbb{R}$ **B.** No x-intercepts; y-intercept $= f(0) = 2$ **C.** No symmetry

D. $\lim\limits_{x\to\pm\infty} (e^x + e^{-3x}) = \infty$, no asymptote

E. $f'(x) = e^x - 3e^{-3x} = e^{-3x}(e^{4x} - 3) > 0 \quad \Leftrightarrow$

$e^{4x} > 3 \quad \Leftrightarrow \quad 4x > \ln 3 \quad \Leftrightarrow \quad x > \frac{1}{4}\ln 3$, so f is

increasing on $\left[\frac{1}{4}\ln 3, \infty\right)$ and decreasing on $\left(-\infty, \frac{1}{4}\ln 3\right]$.

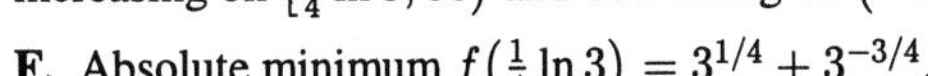

F. Absolute minimum $f\left(\frac{1}{4}\ln 3\right) = 3^{1/4} + 3^{-3/4}$.

G. $f''(x) = e^x + 9e^{-3x} > 0$, so f is CU on $(-\infty, \infty)$.

H.

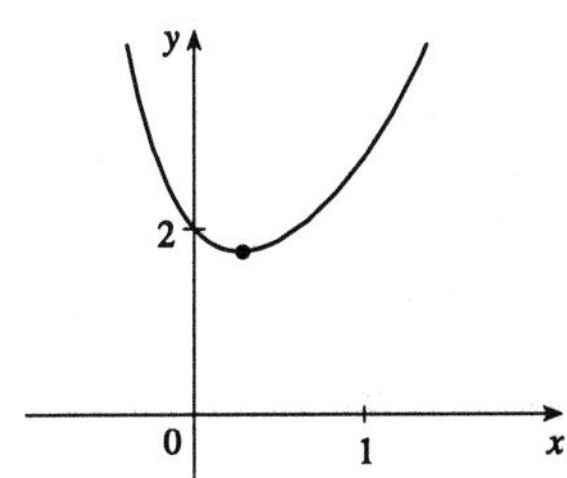

102. $y = f(x) = \ln(x^2 - 1)$ **A.** $D = \{x \mid x^2 > 1\} = (-\infty, -1) \cup (1, \infty)$ **B.** No y-intercept, x-intercepts occur when $x^2 - 1 = 1 \quad \Leftrightarrow \quad x = \pm\sqrt{2}$.

C. $f(-x) = f(x)$, so the graph is symmetric about the y-axis.

D. $\lim\limits_{x\to\pm\infty} \ln(x^2 - 1) = \infty$, $\lim\limits_{x\to 1^+} \ln(x^2 - 1) = -\infty$,

$\lim\limits_{x\to -1^-} \ln(x^2 - 1) = -\infty$, so $x = 1$ and $x = -1$ are VA.

E. $f'(x) = \dfrac{2x}{x^2 - 1} > 0$ for $x > 1$ and $f'(x) < 0$ for $x < -1$, so f is increasing on $(1, \infty)$ and decreasing on $(-\infty, -1)$. **F.** No extrema

G. $f''(x) = -2\dfrac{x^2 + 1}{(x^2 - 1)^2} < 0$, so f is CD on $(-\infty, -1)$ and $(1, \infty)$.

H.

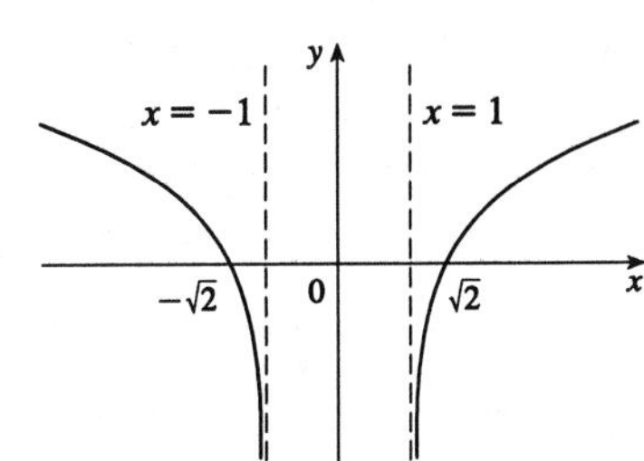

103. 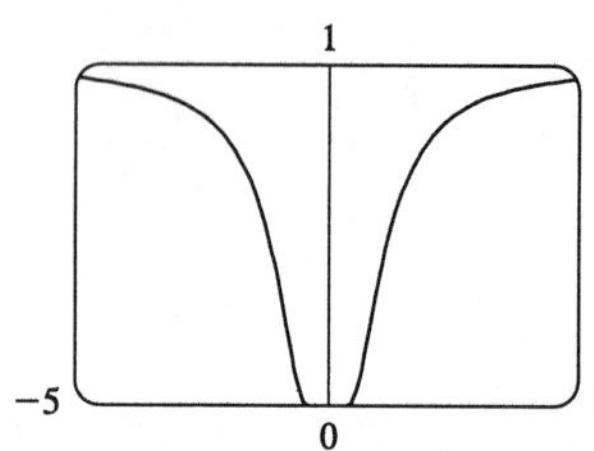

From the graph, we estimate the points of inflection to be about $(\pm 0.8, 0.2)$.

$f(x) = e^{-1/x^2} \quad\Rightarrow\quad f'(x) = 2x^{-3}e^{-1/x^2} \quad\Rightarrow$

$f''(x) = 2\left[x^{-3}(2x^{-3})e^{-1/x^2} + e^{-1/x^2}(-3x^{-4})\right]$

$= 2x^{-6}e^{-1/x^2}(2 - 3x^2)$. This is 0 when $2 - 3x^2 = 0$

$\Leftrightarrow \quad x = \pm\sqrt{\frac{2}{3}}$, so the inflection points are $\left(\pm\sqrt{\frac{2}{3}}, e^{-3/2}\right)$.

104. We exclude the case $c = 0$, since in that case $f(x) = 0$ for all x.

To find the maxima and minima, we differentiate: $f(x) = cxe^{-cx^2} \quad\Rightarrow$

$f'(x) = c\left[xe^{-cx^2}(-2cx) + e^{-cx^2}(1)\right] = ce^{-cx^2}(-2cx^2 + 1)$. This is 0 where $-2cx^2 + 1 = 0 \quad\Leftrightarrow$

$x = \pm 1/\sqrt{2c}$. So if $c > 0$, there are two extrema, whose x-coordinates approach 0 as c increases. The negative root gives a minimum and the positive root gives a maximum, by the First Derivative Test. By substituting back into the equation, we see that $f\left(\pm 1/\sqrt{2c}\right) = c\left(\pm 1/\sqrt{2c}\right)e^{-c(\pm 1/\sqrt{2c})^2} = \pm\sqrt{c/2e}$ respectively. So as c increases, the extreme points become more pronounced. Note that if $c > 0$, then $\lim\limits_{x\to\pm\infty} f(x) = 0$. If $c < 0$, then there are no extrema, and $\lim\limits_{x\to\pm\infty} f(x) = \mp\infty$ respectively.

To find the points of inflection, we differentiate again: $f'(x) = ce^{-cx^2}(-2cx^2 + 1) \quad\Rightarrow$

$f''(x) = c\left[e^{-cx^2}(-4cx) + (-2cx^2 + 1)(-2cxe^{-cx^2})\right] = -2c^2xe^{-cx^2}(3 - 2cx^2)$. This is 0 at $x = 0$ and where $3 - 2cx^2 = 0 \quad\Leftrightarrow\quad x = \pm\sqrt{3/(2c)}$. If $c > 0$ there are three inflection points, and as c increases, the x-coordinates of the nonzero inflection points approach 0. If $c < 0$, there is only one inflection point, namely the origin.

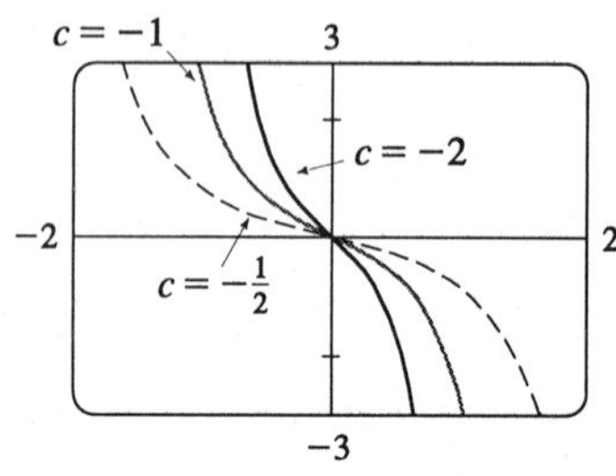

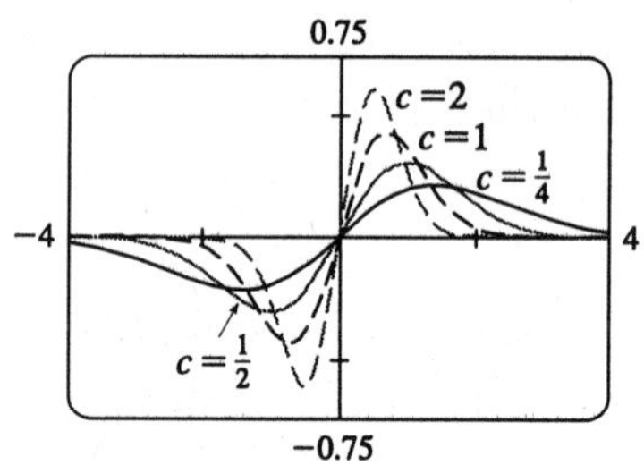

105. **(a)** $y(t) = y(0)e^{kt} = 1000e^{kt} \quad\Rightarrow\quad y(2) = 1000e^{2k} = 9000 \quad\Rightarrow\quad e^{2k} = 9 \quad\Rightarrow\quad 2k = \ln 9 \quad\Rightarrow$

$k = \frac{1}{2}\ln 9 = \ln 3 \quad\Rightarrow\quad y(t) = 1000e^{(\ln 3)t} = 1000 \cdot 3^t$

(b) $y(3) = 1000 \cdot 3^3 = 27{,}000$

(c) $1000 \cdot 3^t = 2000 \quad\Rightarrow\quad 3^t = 2 \quad\Rightarrow\quad t\ln 3 = \ln 2 \quad\Rightarrow\quad t = (\ln 2)/(\ln 3) \approx 0.63\text{ h}$

106. **(a)** If $y(t)$ is the mass remaining after t years, then $y(t) = y(0)e^{kt} = 18e^{kt} \quad\Rightarrow\quad y(25) = 18e^{25k} = 9 \quad\Rightarrow$

$e^{25k} = \frac{1}{2} \quad\Rightarrow\quad 25k = -\ln 2 \quad\Rightarrow\quad k = -\frac{1}{25}\ln 2 \quad\Rightarrow\quad y(t) = 18\,e^{-(\ln 2)t/25} = 18 \cdot 2^{-t/25}$.

(b) $18 \cdot 2^{-t/25} = 2 \;\Rightarrow\; 2^{-t/25} = \frac{1}{9} \;\Rightarrow\; -\frac{1}{25}t\ln 2 = -\ln 9 \;\Rightarrow\; t = 25\dfrac{\ln 9}{\ln 2} \approx 79$ years

107. Using the formula in Example 6.5.4, $A(t) = A_0\left(1+\dfrac{i}{n}\right)^{nt}$, where $A_0 = 10{,}000.00$ and $i = 0.06$ then for:

(a) $n = 1$: $A(4) = 10{,}000(1+0.06)^{1\cdot 4} = \$12{,}624.77$

(b) $n = 2$: $A(4) = 10{,}000\left(1+\frac{0.06}{2}\right)^{2\cdot 4} = \$12{,}667.70$

(c) $n = 4$: $A(4) = 10{,}000\left(1+\frac{0.06}{4}\right)^{4\cdot 4} = \$12{,}689.86$

(d) $n = 12$: $A(4) = 10{,}000\left(1+\frac{0.06}{12}\right)^{12\cdot 4} = \$12{,}704.89$

(e) $n = 365$: $A(4) = 10{,}000\left(1+\frac{0.06}{365}\right)^{365\cdot 4} = \$12{,}712.24$

(f) Using the formula for continuous interest, $A(t) = A_0e^{it}$, we have $A(4) = 10{,}000 \cdot e^{0.06\cdot 4} = \$12{,}712.49$.

108. (a) Let $y(t)$ = temperature after t minutes. Newton's Law of Cooling implies that $\dfrac{dy}{dt} = k(y-70)$. Let $u(t) = y(t) - 70$. Then $\dfrac{du}{dt} = ku$, so $u(t) = u(0)e^{kt} = 130e^{kt} \;\Rightarrow\; y(t) = 70 + 130e^{kt} \;\Rightarrow$ $y(10) = 70 + 130e^{10k} = 150 \;\Rightarrow\; e^{10k} = \frac{8}{13} \;\Rightarrow\; k = \frac{1}{10}\ln\left(\frac{8}{13}\right)$, so $y(t) = 70 + 130e^{\frac{1}{10}t\ln\left(\frac{8}{13}\right)}$ and $y(15) = 70 + 130e^{\frac{1}{10}\cdot 15\ln\left(\frac{8}{13}\right)} \approx 133°\,\text{F}$.

(b) $70 + 130e^{\frac{1}{10}t\ln\left(\frac{8}{13}\right)} = 100$ when $e^{\frac{1}{10}t\ln\left(\frac{8}{13}\right)} = \frac{30}{130} \;\Rightarrow\; \ln\left(\frac{8}{13}\right)t = 10\ln\frac{3}{13} \;\Rightarrow\; t = \dfrac{10\ln(3/13)}{\ln(8/13)} \approx 30$ min.

109. (a) $C'(t) = -kC(t) \;\Rightarrow\; C(t) = C(0)e^{-kt}$ by Theorem 9.5.4. But $C(0) = C_0$. Thus $C(t) = C_0e^{-kt}$.

(b) $C(30) = \frac{1}{2}C_0$ since the concentration is reduced by half. Thus, $\frac{1}{2}C_0 = C_0e^{-30k} \;\Rightarrow\; \ln\frac{1}{2} = -30k \;\Rightarrow$ $k = -\frac{1}{30}\ln\frac{1}{2} = \frac{1}{30}\ln 2$. Since 10% of the original concentration remains if 90% is eliminated, we want the value of t such that $C(t) = \frac{1}{10}C_0$. Therefore, $\frac{1}{10}C_0 = C_0e^{-t(\ln 2)/30} \;\Rightarrow\; t = -\dfrac{30}{\ln 2}\ln 0.1 \approx 100$ h.

110. (a) Let $f(x) = \ln x + x - 3$. Then $f'(x) = 1/x + 1 > 0$ (for $x > 0$) and $f(2) \approx -0.307$ and $f(e) \approx 0.718$. f is differentiable on $(2, e)$, continuous on $[2, e]$ and $f(2) < 0$, $f(e) > 0$. Therefore by the Intermediate Value Theorem there exists a number c in $(2, e)$ such that $f(c) = 0$. Thus, there is one root. But $f'(x) > 0$ for $x \in [2, e]$ so f is increasing on $[2, e]$, which means that there is exactly one root.

(b) We use Newton's Method with $f(x) = \ln x + x - 3$, $f'(x) = 1/x + 1$, and $x_1 = 2$.

$$x_2 = x_1 - \frac{\ln x_1 + x_1 - 3}{1/x_1 + 1} = 2 - \frac{\ln 2 + 2 - 3}{1/2 + 1} \approx 2.20457.$$ Similarly, $x_3 \approx 2.20794$, $x_4 = 2.20794$.

Thus the root of the equation, correct to four decimal places, is 2.2079.

111. $s(t) = Ae^{-ct}\cos(\omega t + \delta) \;\Rightarrow$

$$\begin{aligned}
v(t) &= s'(t) = -cAe^{-ct}\cos(\omega t+\delta) + Ae^{-ct}[-\omega\sin(\omega t+\delta)] = -Ae^{-ct}[c\cos(\omega t+\delta) + \omega\sin(\omega t+\delta)] \;\Rightarrow \\
a(t) &= v'(t) = cAe^{-ct}[c\cos(\omega t+\delta) + \omega\sin(\omega t+\delta)] = -Ae^{-ct}\left[-\omega c\sin(\omega t+\delta) + \omega^2\cos(\omega t+\delta)\right] \\
&= Ae^{-ct}\left[(c^2-\omega^2)\cos(\omega t+\delta) + 2c\omega\sin(\omega t+\delta)\right]
\end{aligned}$$

112. $f(x) = g(e^x) \Rightarrow f'(x) = g'(e^x)e^x$

113. $f(x) = e^{g(x)} \Rightarrow f'(x) = e^{g(x)}g'(x)$

114. $f(x) = g(\ln x) \Rightarrow f'(x) = g'(\ln x)/x$

115. $f(x) = \ln|g(x)| \Rightarrow f'(x) = g'(x)/g(x)$

116. $f(x) = xe^{g(\sqrt{x})} \Rightarrow f'(x) = e^{g(\sqrt{x})} + xe^{g(\sqrt{x})}g'(x)\dfrac{1}{2\sqrt{x}} = e^{g(\sqrt{x})}\left[1 + \frac{1}{2}\sqrt{x}\,g'(x)\right]$

117. $f(x) = \ln g(e^x) \Rightarrow f'(x) = \dfrac{1}{g(e^x)}\,g'(e^x)e^x$

118. Let $u = 2 - 3x$. Then $du = -3\,dx \Rightarrow \displaystyle\int_1^2 \frac{1}{2-3x}\,dx = -\frac{1}{3}\int_{-1}^{-4}\frac{du}{u} = \left[-\frac{\ln|u|}{3}\right]_{-1}^{-4} = -\frac{\ln 4}{3}$

119. $\displaystyle\int_0^{2\sqrt{3}} \frac{1}{x^2+4}\,dx = \left[\tfrac{1}{2}\tan^{-1}(x/2)\right]_0^{2\sqrt{3}} = \tfrac{1}{2}\left(\tan^{-1}\sqrt{3} - \tan^{-1}0\right) = \tfrac{1}{2}\cdot\tfrac{\pi}{3} = \tfrac{\pi}{6}$

120. $\int_0^1 e^{\pi t}\,dt = \left[\frac{1}{\pi}e^{\pi t}\right]_0^1 = \frac{1}{\pi}(e^\pi - 1)$

121. $\displaystyle\int_2^4 \frac{1+x-x^2}{x^2}\,dx = \int_2^4\left(x^{-2} + \frac{1}{x} - 1\right)dx = \left[-\frac{1}{x} + \ln x - x\right]_2^4$

$= \left(-\frac{1}{4} + \ln 4 - 4\right) - \left(-\frac{1}{2} + \ln 2 - 2\right) = \ln 2 - \frac{7}{4}$

122. $\int_{\ln 3}^{\ln 6} 8e^x\,dx = [8e^x]_{\ln 3}^{\ln 6} = 8(e^{\ln 6} - e^{\ln 3}) = 8(6-3) = 24$

123. Let $u = e^x + 1$. Then $du = e^x\,dx$, so $\displaystyle\int\frac{e^x}{e^x+1}\,dx = \int\frac{du}{u} = \ln|u| + C = \ln(e^x+1) + C.$

124. Let $u = \ln x$. Then $du = \dfrac{dx}{x} \Rightarrow \displaystyle\int\frac{\cos(\ln x)}{x}\,dx = \int\cos u\,du = \sin u + C = \sin(\ln x) + C.$

125. Let $u = \sqrt{x}$. Then $du = \dfrac{dx}{2\sqrt{x}} \Rightarrow \displaystyle\int\frac{e^{\sqrt{x}}}{\sqrt{x}}\,dx = 2\int e^u\,du = 2e^u + C = 2e^{\sqrt{x}} + C.$

126. Let $u = x^2$. Then $du = 2x\,dx \Rightarrow \displaystyle\int\frac{x}{\sqrt{1-x^4}}\,dx = \frac{1}{2}\int\frac{du}{\sqrt{1-u^2}} = \tfrac{1}{2}\sin^{-1}u + C = \tfrac{1}{2}\sin^{-1}(x^2) + C.$

127. Let $u = \ln(\cos x)$. Then $du = \dfrac{-\sin x}{\cos x}\,dx = -\tan x\,dx \Rightarrow$

$\int\tan x\ln(\cos x)dx = -\int u\,du = -\frac{1}{2}u^2 + C = -\frac{1}{2}[\ln(\cos x)]^2 + C.$

128. Let $u = \ln(e^x + 1)$. Then $du = \left[e^x/(e^x+1)\right]dx \Rightarrow$

$\displaystyle\int\frac{e^x}{(e^x+1)\ln(e^x+1)}\,dx = \int\frac{du}{u} = \ln|u| + C = \ln\ln(e^x+1) + C.$

129. Let $u = 1 + x^4$. Then $du = 4x^3\,dx \Rightarrow \displaystyle\int\frac{x^3}{1+x^4}\,dx = \frac{1}{4}\int\frac{1}{u}\,du = \tfrac{1}{4}\ln|u| + C = \tfrac{1}{4}\ln(1+x^4) + C.$

130. Let $u = \sqrt{x}$, so $du = dx/(2\sqrt{x}) \Rightarrow \displaystyle\int\frac{1}{\sqrt{x}(1+x)}\,dx = 2\int\frac{du}{1+u^2} = 2\tan^{-1}u + C = 2\tan^{-1}\sqrt{x} + C.$

131. Let $u = 1 + \sec\theta$, so $du = \sec\theta\tan\theta\,d\theta \Rightarrow \displaystyle\int\frac{\sec\theta\tan\theta}{1+\sec\theta}\,d\theta = \int\frac{1}{u}\,du = \ln|u| + C = \ln|1+\sec\theta| + C.$

132. Let $u = -x^3$. Then $du = -3x^2\,dx \quad \Rightarrow \quad \int x^2 2^{-x^3}\,dx = -\frac{1}{3}\int 2^u\,du = -\frac{1}{3}\frac{2^u}{\ln 2} + C = -\frac{1}{3\ln 2}2^{-x^3} + C.$

133. $u = 3t \quad \Rightarrow \quad \int \cosh 3t\,dt = \frac{1}{3}\int \cosh u\,du = \frac{1}{3}\sinh u + C = \frac{1}{3}\sinh 3t + C$

134. $1 + e^{2x} > e^{2x} \quad \Rightarrow \quad \sqrt{1+e^{2x}} > \sqrt{e^{2x}} = e^x \quad \Rightarrow \quad \int_0^1 \sqrt{1+e^{2x}}\,dx \geq \int_0^1 e^x\,dx = [e^x]_0^1 = e - 1$

135. $\cos x \leq 1 \quad \Rightarrow \quad e^x \cos x \leq e^x \quad \Rightarrow \quad \int_0^1 e^x \cos x\,dx \leq \int_0^1 e^x\,dx = [e^x]_0^1 = e - 1$

136. For $0 \leq x \leq 1$, $0 \leq \sin^{-1}x \leq \frac{\pi}{2}$, so $\int_0^1 x\sin^{-1}x\,dx \leq \int_0^1 x\left(\frac{\pi}{2}\right)dx = \left[\frac{\pi}{4}x^2\right]_0^1 = \frac{\pi}{4}$.

137. $f'(x) = \dfrac{d}{dx}\displaystyle\int_1^{\sqrt{x}} \frac{e^s}{s}\,ds = \frac{e^{\sqrt{x}}}{\sqrt{x}}\frac{d}{dx}\sqrt{x} = \frac{e^{\sqrt{x}}}{\sqrt{x}}\frac{1}{2\sqrt{x}} = \frac{e^{\sqrt{x}}}{2x}$

138. $f'(x) = \dfrac{d}{dx}\displaystyle\int_{\ln x}^{2x} e^{-t^2}\,dt = -\frac{d}{dx}\int_0^{\ln x} e^{-t^2}\,dt + \frac{d}{dx}\int_0^{2x} e^{-t^2}\,dt = -e^{-(\ln x)^2}\left(\frac{1}{x}\right) + e^{-(2x)^2}(2) = -\frac{e^{-(\ln x)^2}}{x} + 2e^{-4x^2}$

139. $f_{\text{ave}} = \dfrac{1}{4-1}\displaystyle\int_1^4 \frac{1}{x}\,dx = \left[\tfrac{1}{3}\ln x\right]_1^4 = \tfrac{1}{3}\ln 4$

140. $A = \int_{-2}^{0}(e^{-x} - e^x)dx + \int_0^1 (e^x + e^{-x})dx = [-e^{-x} - e^x]_{-2}^{0} + [e^x + e^{-x}]_0^1$
$= (-1-1) - (-e^2 - e^{-2}) + (e + e^{-1}) - (1+1) = e^2 + e + e^{-1} + e^{-2} - 4$

141. $V = \displaystyle\int_0^1 \frac{2\pi x}{1+x^4}\,dx$ by cylindrical shells. Let $u = x^2 \quad \Rightarrow \quad du = 2x\,dx$. Then
$V = \displaystyle\int_0^1 \frac{\pi}{1+u^2}\,du = \pi\left[\tan^{-1}u\right]_0^1 = \pi\left(\tan^{-1}1 - \tan^{-1}0\right) = \pi\left(\frac{\pi}{4}\right) = \frac{\pi^2}{4}.$

142. $f(x) = x + x^2 + e^x \quad \Rightarrow \quad f'(x) = 1 + 2x + e^x$ and $f(0) = 1 \quad \Rightarrow \quad g(1) = 0$, so
$g'(1) = \dfrac{1}{f'(g(1))} = \dfrac{1}{f'(0)} = \dfrac{1}{2}.$

143. $f(x) = \ln x + \tan^{-1}x \quad \Rightarrow \quad f(1) = \ln 1 + \tan^{-1}1 = \frac{\pi}{4} \quad \Rightarrow \quad g\left(\frac{\pi}{4}\right) = 1.$
$f'(x) = \dfrac{1}{x} + \dfrac{1}{1+x^2}$, so $g'\left(\frac{\pi}{4}\right) = \dfrac{1}{f'(1)} = \dfrac{1}{3/2} = \dfrac{2}{3}.$

144.

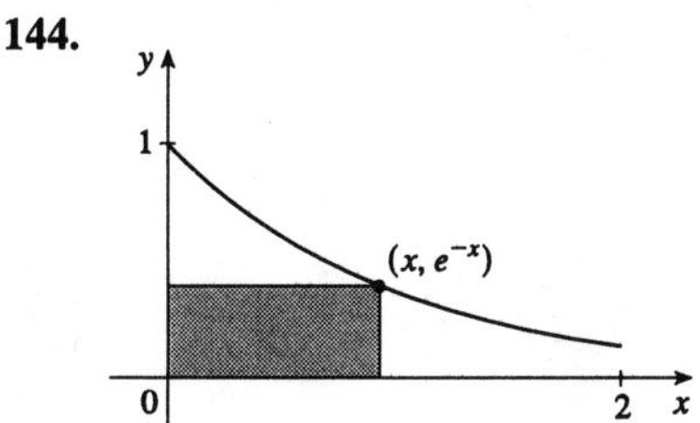

The area of such a rectangle is just the product of its sides, that is, $A(x) = x \cdot e^{-x}$. We want to find the maximum of this function, so we differentiate: $A'(x) = x(-e^{-x}) + e^{-x}(1) = e^{-x}(1-x)$. This is 0 only at $x = 1$, and changes from positive to negative there, so by the First Derivative Test this gives a local maximum. So the largest area is $A(1) = 1/e$.

145. 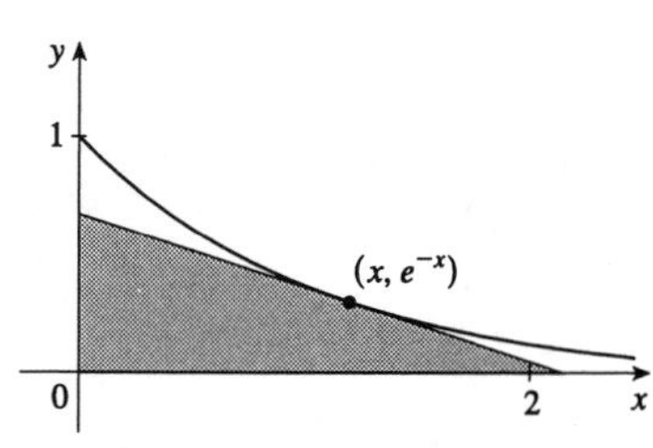

We find the equation of a tangent to the curve $y = e^{-x}$, so that we can find the x- and y-intercepts of this tangent, and then we can find the area of the triangle. The slope of the tangent at the point (a, e^{-a}) is given by $\left.\frac{d}{dx}e^{-x}\right|_{x=a} = -e^{-a}$, and so the equation of the tangent is $y - e^{-a} = -e^{-a}(x - a) \quad \Leftrightarrow$ $y = e^{-a}(a - x + 1)$. The y-intercept of this line is $y = e^{-a}(a - 0 + 1) = e^{-a}(a + 1)$. To find the x-intercept we set $y = 0 \quad \Rightarrow \quad e^{-a}(a - x + 1) = 0 \quad \Rightarrow$ $x = a + 1$. So the area of the triangle is $A(a) = \frac{1}{2}[e^{-a}(a + 1)](a + 1) = \frac{1}{2}e^{-a}(a + 1)^2$. We differentiate this with respect to a: $A'(a) = \frac{1}{2}\left[e^{-a}(2)(a + 1) + (a + 1)^2 e^{-a}(-1)\right] = \frac{1}{2}e^{-a}(1 - a^2)$. This is 0 at $a = \pm 1$, and the root $a = 1$ gives a maximum, by the First Derivative Test. So the maximum area of the triangle is $A(1) = \frac{1}{2}e^{-1}(1 + 1)^2 = 2e^{-1} = 2/e$.

146. Using Theorem 4.3.5 with $a = 0$ and $b = 1$, we have $\int_0^1 e^x\,dx = \lim_{n\to\infty}\frac{1}{n}\sum_{i=1}^{n} e^{i/n}$. This series is a geometric series with $a = r = e^{1/n}$, so $\sum_{i=1}^{n} e^{i/n} = e^{1/n}\frac{e^{n/n} - 1}{e^{1/n} - 1} = e^{1/n}\frac{e - 1}{e^{1/n} - 1} \quad \Rightarrow$

$\int_0^1 e^x\,dx = \lim_{n\to\infty}\frac{1}{n}\sum_{i=1}^{n} e^{i/n} = \lim_{n\to\infty}(e - 1)e^{1/n}\frac{1/n}{e^{1/n} - 1}$. As $n \to \infty$, $1/n \to 0^+$, so $e^{1/n} \to e^0 = 1$. Let $t = 1/n$. Then $e^{1/n} - 1 = e^t - 1 \to 0^+$, so l'Hospital's Rule gives $\lim_{t\to 0}\frac{t}{e^t - 1} = \lim_{t\to 0}\frac{1}{e^t} = 1$ and we have $\int_0^1 e^x\,dx = \left[\lim_{t\to 0^+}(e - 1)e^t\right]\left[\lim_{t\to 0^+}\frac{t}{e^t - 1}\right] = e - 1$.

147. $\lim_{x\to -1} F(x) = \lim_{x\to -1}\frac{b^{x+1} - a^{x+1}}{x + 1} \overset{H}{=} \lim_{x\to -1}\frac{b^{x+1}\ln b - a^{x+1}\ln a}{1} = \ln b - \ln a = F(-1)$, so F is continuous at -1.

148. Let $\theta_1 = \operatorname{arccot} x$, so $\cot\theta_1 = x = x/1$. So $\sin(\operatorname{arccot} x) = \sin\theta_1 = \dfrac{1}{\sqrt{x^2 + 1}}$.

Let $\theta_2 = \arctan\left[\frac{1}{\sqrt{x^2+1}}\right]$, so $\tan\theta_2 = \dfrac{1}{\sqrt{x^2 + 1}}$.

Hence $\cos(\arctan[\sin(\operatorname{arccot} x)]) = \cos\theta_2 = \dfrac{\sqrt{x^2 + 1}}{\sqrt{x^2 + 2}} = \sqrt{\dfrac{x^2 + 1}{x^2 + 2}}$.

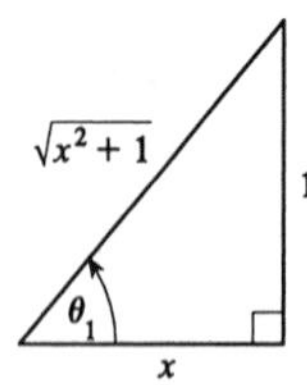

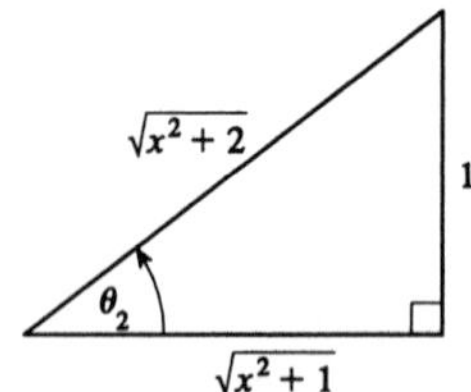

149. Differentiating both sides of the given equation, using the Fundamental Theorem for each side, gives $f(x) = e^{2x} + 2xe^{2x} + e^{-x}f(x)$. So $f(x)(1 - e^{-x}) = e^{2x} + 2xe^{2x}$. Hence $f(x) = \dfrac{e^{2x}(1 + 2x)}{1 - e^{-x}}$.

150. **(a)** Let $f(x) = x - \ln x - 1$, so $f'(x) = 1 - \frac{1}{x} = \frac{x-1}{x}$. Since $x > 0$, $f'(x) < 0$ for $0 < x < 1$ and $f'(x) > 0$ for $x > 1$. So there is an absolute minimum at $x = 1$ with $f(1) = 0$.

So for $x > 0$, $x \neq 1$, $x - \ln x - 1 = f(x) > f(1) = 0$, and hence $\ln x < x - 1$.

(b) Here let $f(x) = \ln x - \frac{x-1}{x} = \ln x - 1 + \frac{1}{x}$. So $f'(x) = \frac{1}{x} - \frac{1}{x^2} = \frac{x-1}{x^2}$. As in (a), we see that there is an absolute minimum value at $x = 1$ and that $f(1) = 0$. So for $x > 0$, $x \neq 1$,

$\ln x - \frac{x-1}{x} = f(x) > f(1) = 0$ and hence $\frac{x-1}{x} < \ln x$.

(c) Let $b > a > 0$, so $b/a > 1$. Letting $x = b/a$ in the inequalities in (a) and (b) gives

$\frac{b-a}{b} = \frac{b/a - 1}{b/a} < \ln \frac{b}{a} < \frac{b}{a} - 1 = \frac{b-a}{a}$. Noting that $\ln \frac{b}{a} = \ln b - \ln a$, the result follows after dividing through by $b - a$.

(d) Let $f(x) = \ln x$. From the given diagram, we see that

(slope of tangent at $x = b$) $<$ (slope of secant line) $<$ (slope of tangent at $x = a$). Since $f'(x) = \frac{1}{x}$, we therefore have $\frac{1}{b} < \frac{\ln b - \ln a}{b - a} < \frac{1}{a}$.

To make this geometric argument more rigorous, we could use the Mean Value Theorem: For any a and b with $0 < a < b$, there exists some $c \in (a, b)$ for which $f'(c) = \frac{1}{c} = \frac{\ln b - \ln a}{b - a}$. But $\frac{1}{x}$ is a decreasing function on $(0, \infty)$, so $\frac{1}{b} < \frac{1}{c} = \frac{\ln b - \ln a}{b - a} < \ln \frac{1}{a}$.

(e) Since $\frac{1}{b} < \frac{1}{x} < \frac{1}{a}$ for $a < x < b$, Property 8 says that $\frac{1}{b}(b - a) < \int_a^b \frac{1}{x}\,dx < \frac{1}{a}(b - a) \quad \Rightarrow$

$\frac{1}{b}(b - a) < \ln b - \ln a < \frac{1}{a}(b - a) \quad \Rightarrow \quad \frac{1}{b} < \frac{\ln b - \ln a}{b - a} < \frac{1}{a}$. (Note from the proof of Property 8 that we are justified in making all of the inequalities strict.)

151. Let $y = \tan^{-1} x$. Then $\tan y = x$, so from the triangle we see that $\sin(\tan^{-1} x) = \sin y = \frac{x}{\sqrt{1 + x^2}}$.

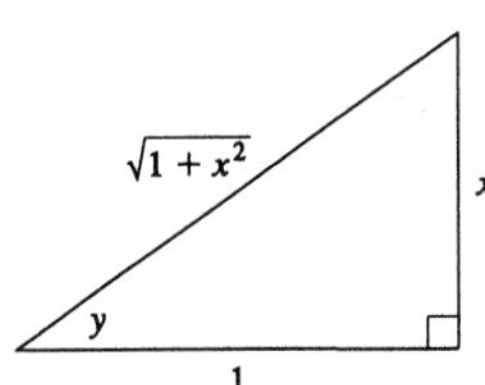

Using this fact we have that

$\sin(\tan^{-1}(\sinh x)) = \frac{\sinh x}{\sqrt{1 + \sinh^2 x}} = \frac{\sinh x}{\cosh x} = \tanh x$.

Hence $\sin^{-1}(\tanh x) = \sin^{-1}(\sin(\tan^{-1}(\sinh x))) = \tan^{-1}(\sinh x)$.

PROBLEMS PLUS (page 433)

1. Let $y = f(x) = e^{-x^2}$. The area of the rectangle under the curve from $-x$ to x is $A(x) = 2xe^{-x^2}$ where $x \geq 0$. We maximize $A(x)$: $A'(x) = 2e^{-x^2} - 4x^2e^{-x^2} = 2e^{-x^2}(1 - 2x^2) = 0 \quad \Rightarrow \quad x = \frac{1}{\sqrt{2}}$. This gives a maximum since $A'(x) > 0$ for $0 \leq x < \frac{1}{\sqrt{2}}$ and $A'(x) < 0$ for $x > \frac{1}{\sqrt{2}}$. We next determine the points of inflection of $f(x)$. Now $f'(x) = -2xe^{-x^2} = -A(x)$. So $f''(x) = -A'(x)$. So $f''(x) < 0$ for $-\frac{1}{\sqrt{2}} < x < \frac{1}{\sqrt{2}}$ and $f''(x) > 0$ for $x < -\frac{1}{\sqrt{2}}$ and $x > \frac{1}{\sqrt{2}}$. So $f(x)$ changes concavity at $x = \pm\frac{1}{\sqrt{2}}$. So the two vertices of the rectangle of largest area are at the inflection points.

2. The total region bounded by the parabola and the x-axis is $\int_0^1 (x - x^2)dx = \left[\frac{1}{2}x^2 - \frac{1}{3}x^3\right]_0^1 = \frac{1}{6}$.

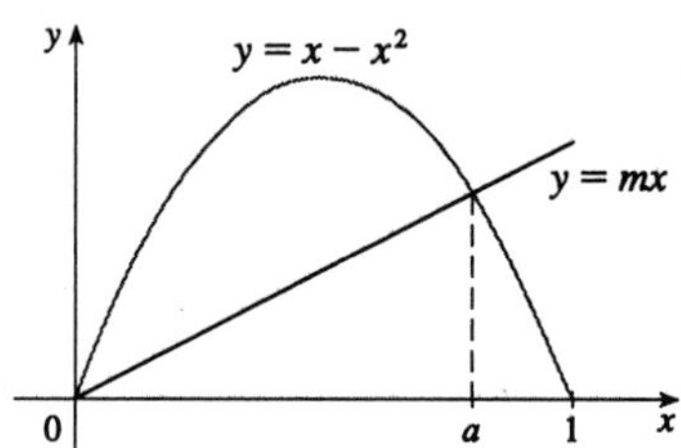

Let the slope of the line we are looking for be m. Then the area above this line but below the parabola is $\int_0^a [(x - x^2) - mx]dx$, where a is the x-coordinate of the point of intersection of the line and the parabola. We find the point of intersection by solving the equation $x - x^2 = mx \quad \Leftrightarrow \quad 1 - x = m \quad \Leftrightarrow$ $x = 1 - m$. So the value of a is $1 - m$, and

$\int_0^{1-m} [(x - x^2) - mx]dx = \int_0^{1-m} [(1 - m)x - x^2]dx = \left[\frac{1}{2}(1 - m)x^2 - \frac{1}{3}x^3\right]_0^{1-m}$
$= \frac{1}{2}(1 - m)(1 - m)^2 - \frac{1}{3}(1 - m)^3 = \frac{1}{6}(1 - m)^3$. We want this to be half of $\frac{1}{6}$, so
$\frac{1}{6}(1 - m)^3 = \frac{1}{12} \quad \Rightarrow \quad m = 1 - \frac{1}{\sqrt[3]{2}}$ So the slope of the required line is $1 - \frac{1}{\sqrt[3]{2}} \approx 0.206$.

3. We use proof by contradiction. Suppose that $\log_2 5$ is a rational number. Then $\log_2 5 = m/n$ where m and n are positive integers $\quad \Rightarrow \quad 2^{m/n} = 5 \quad \Rightarrow \quad 2^m = 5^n$. But this is impossible since 2^m is even and 5^n is odd. So $\log_2 5$ is irrational.

4. **(a)**

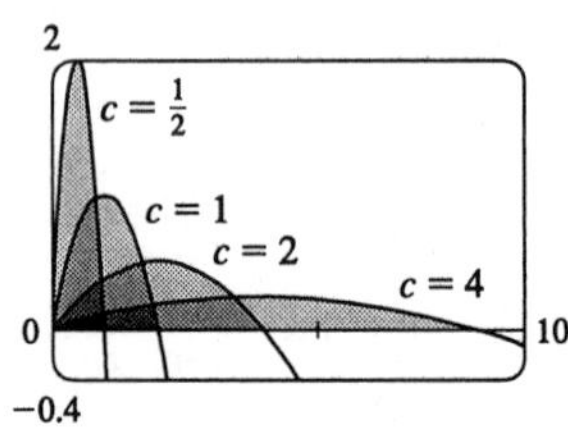

From the graph, it appears that the areas are equal; that is, the area enclosed is independent of c.

(b) We first find the x-intercepts of the curve, to determine the limits of integration: $y = 0 \quad \Leftrightarrow$ $2cx - x^2 = 0 \quad \Leftrightarrow \quad x = 0$ or $x = 2c$. Now we integrate the function between these limits to find the enclosed area:

$$A = \int_0^{2c} \frac{2cx - x^2}{c^3}\,dx = \frac{1}{c^3}\left[cx^2 - \tfrac{1}{3}x^3\right]_0^{2c} = \frac{1}{c^3}\left[c(2c)^2 - \tfrac{1}{3}(2c)^3\right] = \frac{1}{c^3}\left[4c^3 - \tfrac{8}{3}c^3\right] = \frac{4}{3}, \text{ a constant.}$$

(c)

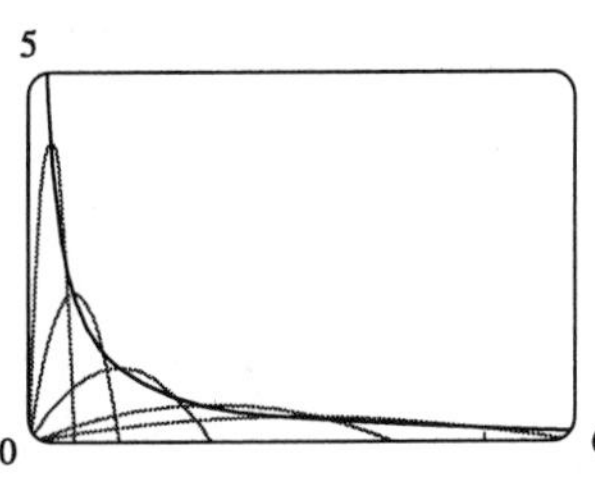

The vertices of the family of parabolas seem to determine a branch of a hyperbola.

(d) For a particular c, the vertex is the point where the maximum occurs. We have seen that the x-intercepts are 0 and $2c$, so by symmetry, the maximum occurs at $x = c$, and its value is $\dfrac{2c(c) - c^2}{c^3} = \dfrac{1}{c}$. So we are interested in the curve consisting of all points of the form $\left(c, \dfrac{1}{c}\right)$, $c > 0$. This is the part of the hyperbola $y = 1/x$ lying in the first quadrant.

5. First notice that if we can prove the simpler inequality $\dfrac{x^2+1}{x} \geq 2$ for $x > 0$, then the desired inequality follows because $\dfrac{(x^2+1)(y^2+1)(z^2+1)}{xyz} = \left(\dfrac{x^2+1}{x}\right)\left(\dfrac{y^2+1}{y}\right)\left(\dfrac{z^2+1}{z}\right) \geq 2 \cdot 2 \cdot 2 = 8$. So we let $f(x) = \dfrac{x^2+1}{x} = x + \dfrac{1}{x}$, $x > 0$. Then $f'(x) = 1 - \dfrac{1}{x^2} = 0$ if $x = 1$, and $f'(x) < 0$ for $0 < x < 1$, $f'(x) > 0$ for $x > 1$. Thus the absolute minimum value of $f(x)$ for $x > 0$ is $f(1) = 2$. Therefore $\dfrac{x^2+1}{x} \geq 2$ for all positive x. $\Big($Or, without calculus, $\dfrac{x^2+1}{x} \geq 2 \quad \Leftrightarrow \quad x^2 + 1 \geq 2x \quad \Leftrightarrow \quad x^2 - 2x + 1 \geq 0 \quad \Leftrightarrow$ $(x-1)^2 \geq 0$, which is true.$\Big)$

6. Let the circle have radius r, so $|OP| = |OQ| = r$, where O is the center of the circle. Now $\angle POR$ has measure $\frac{1}{2}\theta$, and $\angle OPR$ is a right angle, so $\dfrac{|PR|}{r} = \tan\frac{1}{2}\theta$. So the area of $\triangle OPR$ is $\frac{1}{2}r^2\tan\frac{1}{2}\theta$. The area of the sector cut by OP and OR is $\frac{1}{2}\left(\frac{1}{2}\theta\right)r^2 = \frac{1}{4}\theta r^2$. Let S be the intersection of PQ and OR. Then $\dfrac{|PS|}{r} = \sin\frac{1}{2}\theta$ and $\dfrac{|OS|}{r} = \cos\frac{1}{2}\theta$. So the area of $\triangle OSP$ is $\frac{1}{2}|OS||PS| = \frac{1}{2}r\cos\frac{1}{2}\theta\, r\sin\frac{1}{2}\theta = \frac{1}{2}r^2\sin\frac{1}{2}\theta\cos\frac{1}{2}\theta = \frac{1}{4}r^2\sin\theta$.
So $B(\theta) = 2\left(\frac{1}{2}r^2\tan\frac{1}{2}\theta - \frac{1}{4}\theta r^2\right) = r^2\left(\tan\frac{1}{2}\theta - \frac{1}{2}\theta\right)$ and $A(\theta) = 2\left(\frac{1}{4}\theta r^2 - \frac{1}{4}r^2\sin\theta\right) = \frac{1}{2}r^2(\theta - \sin\theta) \quad \Rightarrow$
$\dfrac{A(\theta)}{B(\theta)} = \dfrac{\frac{1}{2}r^2(\theta - \sin\theta)}{r^2\left(\tan\frac{1}{2}\theta - \frac{1}{2}\theta\right)} = \dfrac{\theta - \sin\theta}{2\left(\tan\frac{1}{2}\theta - \frac{1}{2}\theta\right)}$. Now $\lim\limits_{\theta\to 0^+} A(\theta) = 0 = \lim\limits_{\theta\to 0^+} B(\theta)$, so by l'Hospital's Rule,

$$\lim_{\theta\to 0^+}\frac{A(\theta)}{B(\theta)} = \lim_{\theta\to 0^+}\frac{1-\cos\theta}{2\left(\frac{1}{2}\sec^2\frac{1}{2}\theta - \frac{1}{2}\right)} = \lim_{\theta\to 0^+}\frac{1-\cos\theta}{\sec^2\frac{1}{2}\theta - 1} = \lim_{\theta\to 0^+}\frac{1-\cos\theta}{\tan^2\frac{1}{2}\theta} \overset{\text{H}}{=} \lim_{\theta\to 0^+}\frac{\sin\theta}{2\left(\tan\frac{1}{2}\theta\right)\left(\sec^2\frac{1}{2}\theta\right)\frac{1}{2}}$$

$$= \lim_{\theta\to 0^+}\frac{\sin\theta\cos^3\frac{1}{2}\theta}{\sin\frac{1}{2}\theta} = \lim_{\theta\to 0^+}\frac{2\sin\frac{1}{2}\theta\cos\frac{1}{2}\theta\cos^3\frac{1}{2}\theta}{\sin\frac{1}{2}\theta} = 2\lim_{\theta\to 0^+}\cos^4\tfrac{1}{2}\theta = 2.$$

7. Consider the statement that $\dfrac{d^n}{dx^n}(e^{ax}\sin bx) = r^n e^{ax}\sin(bx + n\theta)$. For $n = 1$,

$$\frac{d}{dx}(e^{ax}\sin bx) = ae^{ax}\sin bx + be^{ax}\cos bx, \text{ and}$$

$$re^{ax}\sin(bx+\theta) = re^{ax}[\sin bx\cos\theta + \cos bx\sin\theta] = re^{ax}\left(\frac{a}{r}\sin bx + \frac{b}{r}\cos bx\right)$$
$$= ae^{ax}\sin bx + be^{ax}\cos bx, \text{ since } \tan\theta = b/a \quad\Rightarrow\quad \sin\theta = b/r \text{ and } \cos\theta = a/r.$$

So the statement is true for $n = 1$. Assume it is true for $n = k$. Then

$$\frac{d^{k+1}}{dx^{k+1}}(e^{ax}\sin bx) = \frac{d}{dx}\left[r^k e^{ax}\sin(bx+k\theta)\right] = r^k ae^{ax}\sin(bx+k\theta) + r^k e^{ax}b\cos(bx+k\theta)$$
$$= r^k e^{ax}[a\sin(bx+k\theta) + b\cos(bx+k\theta)]. \text{ But}$$

$$\sin[bx+(k+1)\theta] = \sin[(bx+k\theta)+\theta] = \sin(bx+k\theta)\cos\theta + \sin\theta\cos(bx+k\theta)$$
$$= \frac{a}{r}\sin(bx+k\theta) + \frac{b}{r}\cos(bx+k\theta).$$

Hence $a\sin(bx+k\theta) + b\cos(bx+k\theta) = r\sin[bx+(k+1)\theta]$. So

$$\frac{d^{k+1}}{dx^{k+1}}(e^{ax}\sin bx) = r^k e^{ax}[a\sin(bx+k\theta) + b\sin(bx+k\theta)] = r^k e^{ax}[r\sin(bx+(k+1)\theta)]$$
$$= r^{k+1}e^{ax}[\sin(bx+(k+1)\theta)].$$

Therefore the statement is true for all n by mathematical induction.

8. Differentiating both sides of the equation $\int_0^x f(t)dt = 3f(x) - 2$ using the Fundamental Theorem of Calculus gives $f(x) = 3f'(x) \quad\Rightarrow\quad f'(x) = \frac{1}{3}f(x)$. So $f(x) = f(0)e^{x/3}$ by Theorem 6.5.2. Put $x = 0$ in the given equation: $0 = 3f(0) - 2 \quad\Rightarrow\quad f(0) = \frac{2}{3}$, so $f(x) = \frac{2}{3}e^{x/3}$.

9. The volume generated from $x = 0$ to $x = b$ is $\int_0^b \pi[f(x)]^2\,dx$. Hence we are given that $b^2 = \int_0^b \pi[f(x)]^2\,dx$ for all $b > 0$. Differentiating both sides of this equation using the Fundamental Theorem of Calculus gives $2b = \pi[f(b)]^2 \quad\Rightarrow\quad f(b) = \sqrt{2b/\pi}$, since f is positive. Therefore $f(x) = \sqrt{2x/\pi}$.

10. Using Theorem 4.3.5 with $a = 0$, $b = 10{,}000$, and $f(x) = \sqrt{x}$, we have

$$\int_0^{10{,}000}\sqrt{x}\,dx = \lim_{n\to\infty}\frac{10{,}000}{n}\sum_{i=1}^{n}\sqrt{\frac{10{,}000i}{n}}.$$ We use the approximation with $n = 10{,}000$:

$$\int_0^{10{,}000}\sqrt{x}\,dx \approx \sum_{i=1}^{10{,}000}\sqrt{i}. \text{ So } \sum_{i=1}^{10{,}000}\sqrt{i} \approx \left[\tfrac{2}{3}x^{3/2}\right]_0^{10{,}000} = \tfrac{2}{3}(1{,}000{,}000) \approx 666{,}667.$$

Or: We can use graphical methods as follows: from the figure we see that $\int_{i-1}^{i}\sqrt{x}\,dx < \sqrt{i} < \int_i^{i+1}\sqrt{x}\,dx$, so $\int_0^{10{,}000}\sqrt{x}\,dx < \sum_{i=1}^{10{,}000}\sqrt{i} < \int_1^{10{,}001}\sqrt{x}\,dx$. Since $\int\sqrt{x}\,dx = \frac{2}{3}x^{3/2} + C$, we get $\int_0^{10{,}000}\sqrt{x}\,dx = 666{,}666.\overline{6}$ and $\int_1^{10{,}001}\sqrt{x}\,dx = \frac{2}{3}\left[(10{,}001)^{3/2} - 1\right] \approx 666{,}766$.

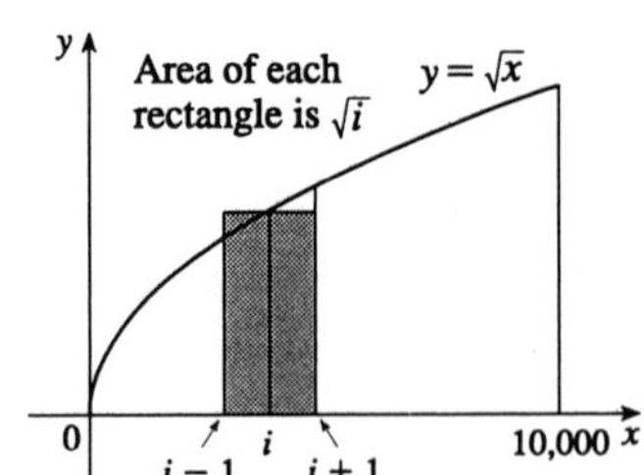

Hence $666{,}666.\overline{6} < \sum_{i=1}^{10{,}000}\sqrt{i} < 666{,}766$.

11. Let the line through A and B have equation $y = mx + b$. Now $x^2 = mx + b$ gives $x^2 - mx - b = 0$ and hence $x = \dfrac{m \pm \sqrt{m^2 + 4b}}{2}$. So A has x-coordinate $x_1 = \dfrac{1}{2}\left(m - \sqrt{m^2+4b}\right)$ and B has x-coordinate $x_2 = \frac{1}{2}\left(m + \sqrt{m^2+4b}\right)$. So the parabolic segment has area

$$\begin{aligned}\int_{x_1}^{x_2}[(mx+b) - x^2]dx &= \left[\tfrac{1}{2}mx^2 + bx - \tfrac{1}{3}x^3\right]_{x_1}^{x_2} = \tfrac{1}{2}m(x_2^2 - x_1^2) + b(x_2 - x_1) - \tfrac{1}{3}(x_2^3 - x_1^3)\\ &= (x_2 - x_1)\left[\tfrac{1}{2}m(x_2 + x_1) + b - \tfrac{1}{3}(x_2^2 + x_1x_2 + x_1^2)\right]\\ &= \sqrt{m^2+4b}\left(\tfrac{1}{2}m^2 + b - \tfrac{1}{3}(mx_2 + b - b + mx_1 + b)\right)\\ &= \sqrt{m^2+4b}\left[\tfrac{1}{2}m^2 + b - \tfrac{1}{3}(m^2+b)\right]\\ &= \tfrac{1}{6}(m^2+4b)^{3/2}.\end{aligned}$$

Now since the line through C has slope m, we see that if C has x-coordinate c, then $2c = m$ $\left[\text{since } (x^2)' = 2x\right]$ and hence $c = m/2$. So C has coordinates $\left(\frac{1}{2}m, \frac{1}{4}m^2\right)$. The line through AC has slope $\dfrac{\frac{1}{4}m^2 - x_1^2}{\frac{1}{2}m - x_1} = \dfrac{m}{2} + x_1$ and equation $y - x_1^2 = \left(\frac{1}{2}m + x_1\right)(x - x_1)$ or $y = \left(\frac{1}{2}m + x_1\right)x - \frac{1}{2}mx_1$. Similarly, the equation of the line through BC is $y = \left(\frac{1}{2}m + x_2\right)x - \frac{1}{2}mx_2$. So the area of the triangular region is

$$\begin{aligned}&\int_{x_1}^{m/2}\left[(mx+b) - \left[\left(\tfrac{1}{2}m + x_1\right)x - \tfrac{1}{2}mx_1\right]\right]dx + \int_{m/2}^{x_2}\left[(mx+b) - \left[\left(\tfrac{1}{2}m + x_2\right)x - \tfrac{1}{2}mx_2\right]\right]dx\\ &\quad= \int_{x_1}^{m/2}\left[\left(\tfrac{1}{2}m - x_1\right)x + \left(b + \tfrac{1}{2}mx_1\right)\right]dx + \int_{m/2}^{x_2}\left[\left(\tfrac{1}{2}m - x_2\right)x + \left(b + \tfrac{1}{2}mx_2\right)\right]dx\\ &\quad= \left[\tfrac{1}{2}\left(\tfrac{1}{2}m - x_1\right)x^2 + \left(b + \tfrac{1}{2}mx_1\right)x\right]_{x_1}^{m/2} + \left[\tfrac{1}{2}\left(\tfrac{1}{2}m - x_2\right)x^2 + \left(b + \tfrac{1}{2}mx_2\right)x\right]_{m/2}^{x_2}\\ &\quad= \left[\tfrac{1}{16}m^3 - \tfrac{1}{8}m^2x_1 + \tfrac{1}{2}bm + \tfrac{1}{4}m^2x_1\right] - \left[\tfrac{1}{4}mx_1^2 - \tfrac{1}{2}x_1^3 + bx_1 + \tfrac{1}{2}mx_1^2\right]\\ &\qquad\quad + \left[\tfrac{1}{4}mx_2^2 - \tfrac{1}{2}x_2^3 + bx_2 + \tfrac{1}{2}mx_2^2\right] - \left[\tfrac{1}{16}m^3 - \tfrac{1}{8}m^2x_2 + \tfrac{1}{2}bm + \tfrac{1}{4}m^2x_2\right]\\ &\quad= \left(b - \tfrac{1}{8}m^2\right)(x_2 - x_1) + \tfrac{3}{4}m(x_2^2 - x_1^2) - \tfrac{1}{2}(x_2^3 - x_1^3)\\ &\quad= (x_2 - x_1)\left[\left(b - \tfrac{1}{8}m^2\right) + \tfrac{3}{4}m(x_2 + x_1) - \tfrac{1}{2}(x_2^2 + x_2x_1 + x_1^2)\right]\\ &\quad= \sqrt{m^2+4b}\left[\left(b - \tfrac{1}{8}m^2\right) + \tfrac{3}{4}m^2 - \tfrac{1}{2}(m^2+b)\right] = \tfrac{1}{8}\sqrt{m^2+4b}(m^2+4b) = \tfrac{1}{8}(m^2+4b)^{3/2}.\end{aligned}$$

The result follows since $\frac{4}{3}\left[\frac{1}{8}(m^2+4b)^{3/2}\right] = \frac{1}{6}(m^2+4b)^{3/2}$, the area of the parabolic segment.

Alternate Solution: Let $A = (a, a^2)$, $B = (b, b^2)$. Then $m_{AB} = (b^2 - a^2)/(b - a) = a + b$, so the equation of AB is $y - a^2 = (a+b)(x - a)$, or $y = (a+b)x - ab$, and the area of the parabolic segment is $\int_a^b[(a+b)x - ab - x^2]dx = \left[(a+b)\frac{1}{2}x^2 - abx - \frac{1}{3}x^3\right]_a^b = \frac{1}{2}(a+b)(b^2 - a^2) - ab(b-a) - \frac{1}{3}(b^3 - a^3)$ $= \frac{1}{6}(b-a)^3$. At C, $y' = 2x = b + a$, so the x-coordinate of C is $\frac{1}{2}(a+b)$. If we calculate the area of triangle ABC as in Problems Plus #19 after Chapter 4 (by subtracting areas of trapezoids) we find that the area is $\frac{1}{2}(b-a)\left[\frac{1}{2}(a+b) - a\right]\left[b - \frac{1}{2}(a+b)\right] = \frac{1}{2}(b-a)\frac{1}{2}(b-a)\frac{1}{2}(b-a) = \frac{1}{8}(b-a)^3$. This is $\frac{3}{4}$ of the area of the parabolic segment calculated above.

12. Let $g(x) = f''(x)$. Then $g'(x) = g(x)$, so $g(x) = Ae^x$ by Theorem 6.5.2. Now $f''(x) = Ae^x \quad\Rightarrow$ $f'(x) = \int Ae^x\,dx = Ae^x + C \quad\Rightarrow\quad f(x) = \int(Ae^x + C)dx = Ae^x + Cx + D$.

13. By l'Hospital's Rule and the Fundamental Theorem, using the notation $\exp(y) = e^y$,

$$\lim_{x\to 0} \frac{\int_0^x (1-\tan 2t)^{1/t}\,dt}{x} \overset{\text{H}}{=} \lim_{x\to 0} \frac{(1-\tan 2x)^{1/x}}{1} = \exp\left[\lim_{x\to 0} \frac{\ln(1-\tan 2x)}{x}\right]$$

$$\overset{\text{H}}{=} \exp\left(\lim_{x\to 0} \frac{-2\sec^2 2x}{1-\tan 2x}\right) = \exp\left(\frac{-2\cdot 1^2}{1-0}\right) = e^{-2}.$$

14. The cone has volume $\frac{1}{3}\pi r^2$. Suppose the cuts to be made are distances a and b respectively from the top. By similar triangles, the respective radii are ar and br. The top cone has volume $\frac{1}{3}\pi a(ar)^2 = \frac{1}{3}\pi a^3 r^2$. Since $\frac{1}{3}\pi a^3 r^2 = \frac{1}{3}\left(\frac{1}{3}\pi r^2\right)$, we get $a^3 = \frac{1}{3}$, so $a = \frac{1}{\sqrt[3]{3}}$. The middle cone has volume $\frac{1}{3}\pi b^3 r^2 = \frac{2}{3}\left(\frac{1}{3}\pi r^2\right)$. So $b^3 = \frac{2}{3}$ and hence $b = \sqrt[3]{\frac{2}{3}}$.

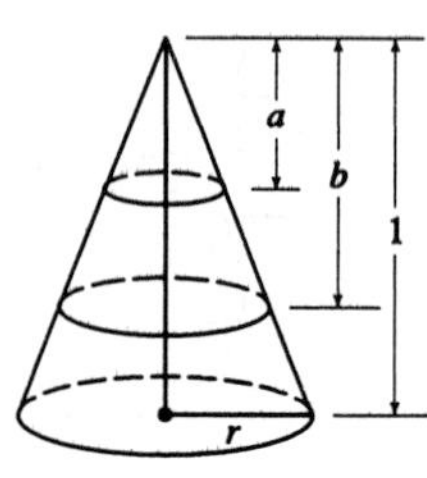

15. We first show that $\frac{x}{1+x^2} < \tan^{-1}x$ for $x > 0$. Let $f(x) = \tan^{-1}x - \frac{x}{1+x^2}$. Then

$f'(x) = \frac{1}{1+x^2} - \frac{1(1+x^2) - x(2x)}{(1+x^2)^2} = \frac{(1+x^2)-(1-x^2)}{(1+x^2)^2} = \frac{2x^2}{(1+x^2)^2} > 0$ for $x > 0$. So $f(x)$ is increasing on $[0,\infty)$. Hence $0 < x \quad\Rightarrow\quad 0 = f(0) < f(x) = \tan^{-1}x - \frac{x}{1+x^2}$. So $\frac{x}{1+x^2} < \tan^{-1}x$ for $0 < x$. We next show that $\tan^{-1}x < x$ for $x > 0$. Let $h(x) = x - \tan^{-1}x$. Then

$h'(x) = 1 - \frac{1}{1+x^2} = \frac{x^2}{1+x^2} > 0$. Hence $h(x)$ is increasing on $[0,\infty)$. So for $0 < x$,

$0 = h(0) < h(x) = x - \tan^{-1}x$. Hence $\tan^{-1}x < x$ for $x > 0$.

16. The shaded region has area $\int_0^1 f(x)dx = \frac{1}{3}$. The integral $\int_0^1 f^{-1}(y)dy$ gives the area of the unshaded region, which we know to be $1 - \frac{1}{3} = \frac{2}{3}$. So $\int_0^1 f^{-1}(y)dy = \frac{2}{3}$.

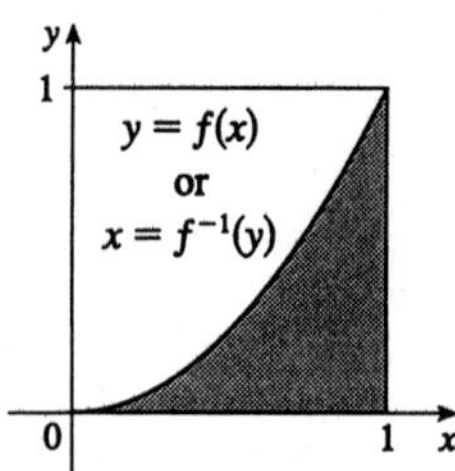

17. By the Fundamental Theorem of Calculus, $f'(x) = \sqrt{1+x^3} > 0$ for $x > -1$. So f is increasing on $[-1,\infty)$ and hence is one-to-one. Note that $f(1) = 0$, so $f^{-1}(1) = 0 \quad\Rightarrow\quad (f^{-1})'(0) = 1/f'(1) = \frac{1}{\sqrt{2}}$.

18.

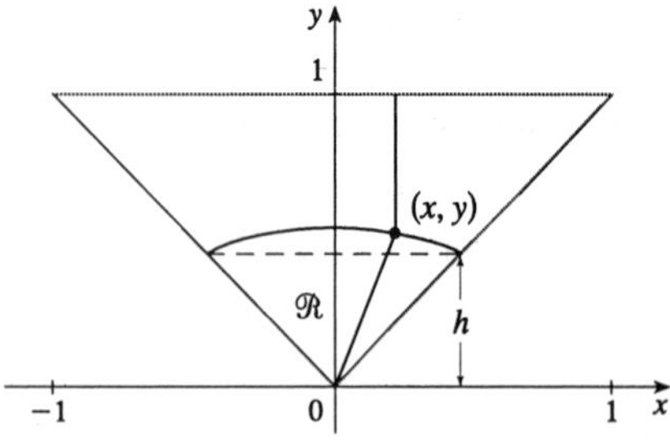

We first assume that $L = 2$ for simplicity in calculation. Afterwards, we will multiply by $\left(\frac{1}{2}L\right)^2$ to find the actual area. Also, we restrict our attention to the triangle shown. A point in this triangle is closer to the side shown than to any other side, so if we find the area of the region $\mathfrak{R}$ consisting of all points in the triangle that are closer to the center than to that side, we can multiply this area by 4 to find the total area.

We find the equation of the set of points which are equidistant from the center and the side: the distance of the point (x, y) from the side is $1 - y$, and its distance from the centre is $\sqrt{x^2 + y^2}$. So the distances are equal if $\sqrt{x^2 + y^2} = 1 - y \quad \Leftrightarrow \quad x^2 + y^2 = 1 + y^2 - 2y \quad \Leftrightarrow \quad y = \frac{1}{2}(1 - x^2)$. Note that the area we are interested in is equal to the area of a triangle plus a crescent-shaped area. To find these areas, we have to find the y-coordinate of the horizontal line separating them. Let this be called h. From the diagram, $1 - h = \sqrt{2}h \quad \Leftrightarrow$ $h = \frac{1}{1+\sqrt{2}} = \sqrt{2} - 1$. We calculate the areas in terms of h, and substitute afterward.

The area of the triangle is h^2, and the area of the crescent-shaped section is $\int_{-h}^{h} \left[\frac{1}{2}(1 - x^2) - h\right] dx = 2\left[\left(\frac{1}{2} - h\right)x - \frac{1}{6}x^3\right]_0^h = h - 2h^2 - \frac{1}{3}h^3$. So the area of the whole region is $4\left[\left(h - 2h^2 - \frac{1}{3}h^3\right) + h^2\right] = 4\left(\sqrt{2} - 1\right)\left[1 - \left(\sqrt{2} - 1\right) - \frac{1}{3}\left(\sqrt{2} - 1\right)^2\right] = 4\left(\sqrt{2} - 1\right)\left(1 - \frac{1}{3}\sqrt{2}\right)$ $= \frac{4}{3}\left(4\sqrt{2} - 5\right)$. Now we must multiply by $\left(\frac{1}{2}L\right)^2$ to account for the size of the square, so the total area of the region, in a square with side L, is $\frac{1}{3}\left(4\sqrt{2} - 5\right)L^2$.

19. Let $L = \lim_{x\to\infty}\left(\frac{x+a}{x-a}\right)^x$, so $\ln L = \lim_{x\to\infty} \ln\left(\frac{x+a}{x-a}\right)^x = \lim_{x\to\infty} x \ln\left(\frac{x+a}{x-a}\right) = \lim_{x\to\infty} \frac{\ln(x+a) - \ln(x-a)}{1/x}$

$$\overset{\text{H}}{=} \lim_{x\to\infty} \frac{\frac{1}{x+a} - \frac{1}{x-a}}{-1/x^2} = -\lim_{x\to\infty} \frac{(x-a)x^2 - (x+a)x^2}{(x+a)(x-a)} = -\lim_{x\to\infty} \frac{-2ax^2}{x^2 - a^2} = \lim_{x\to\infty} \frac{2a}{1 - a^2/x^2} = 2a.$$

Hence $\ln L = 2a$, so $L = e^{2a}$. Hence $L = e^1 \quad \Rightarrow \quad 2a = 1 \quad \Rightarrow \quad a = \frac{1}{2}$.

20. *Case (i):* For $x + y \geq 0$, that is, $y \geq -x$, $x + y = |x + y| \leq e^x \quad \Rightarrow \quad y \leq e^x - x$. Note that $y = e^x - x$ is always above the line $y = -x$ and that $y = -x$ is a slant asymptote. (See Exercise 6.8.62.)

Case (ii): For $x + y < 0$, $-x - y = |x + y| \leq e^x \quad \Rightarrow \quad -x - e^x \leq y$. Note that $-x - e^x$ is always below the line $y = -x$ and $y = -x$ is a slant asymptote. Putting the two pieces together gives the graph:

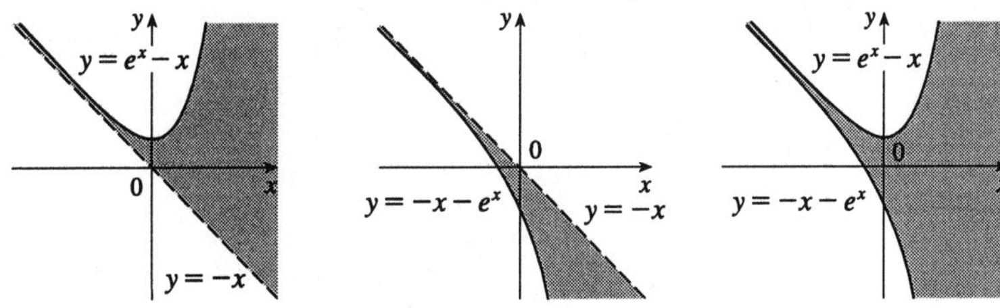

21. Note that $\frac{d}{dx}\left(\int_0^x \left[\int_0^u f(t)dt\right] du\right) = \int_0^x f(t)dt$ by Part 1 of the Fundamental Theorem of Calculus, while

$$\frac{d}{dx}\left[\int_0^x f(u)(x-u)du\right] = \frac{d}{dx}\left[x\int_0^x f(u)du\right] - \frac{d}{dx}\left[\int_0^x f(u)u\,du\right] = \int_0^x f(u)du + xf(x) - f(x)x$$
$$= \int_0^x f(u)du.$$

Hence $\int_0^x f(u)(x-u)du = \int_0^x \left[\int_0^u f(t)dt\right] du + C$. Setting $x = 0$ gives $C = 0$.

22. Let x be the distance between the center of the disk and the surface of the liquid. The wetted circular region has area $\pi r^2 - \pi x^2$ while the unexposed wetted region has area $2\int_x^r \sqrt{r^2 - t^2}\,dt$, so the exposed wetted region has area $A(x) = \pi r^2 - \pi x^2 - 2\int_x^r \sqrt{r^2 - t^2}\,dt$, $0 \le x \le r$.

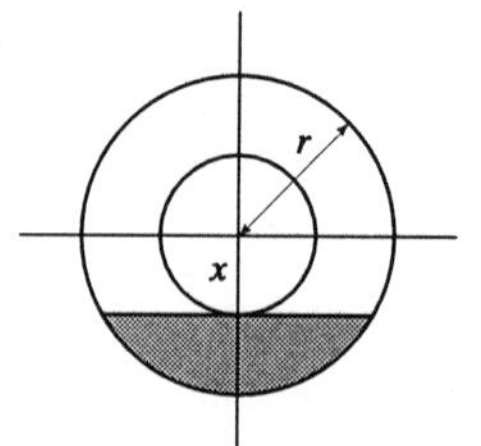

By Part 1 of the Fundamental Theorem, we have

$A'(x) = -2\pi x + 2\sqrt{r^2 - x^2}$, so $A'(x) = 0$ when $\pi x = \sqrt{r^2 - x^2} \quad \Rightarrow \quad \pi^2 x^2 = r^2 - x^2$ and hence $(1 + \pi^2)x^2 = r^2$, so $x = \dfrac{r}{\sqrt{1 + \pi^2}}$. Now since $A(0) = A(r) = 0$, while $A\left(r/\sqrt{1 + \pi^2}\right) > 0$, there is an absolute maximum when $x = \dfrac{r}{\sqrt{1 + \pi^2}}$.

23. $\displaystyle\lim_{x\to\infty} x^c e^{-2x} \int_0^x e^{2t}\sqrt{t^2 + 1}\,dt = \lim_{x\to\infty} \frac{\int_0^x e^{2t}\sqrt{t^2 + 1}\,dt}{x^{-c}e^{2x}}$. By l'Hospital's Rule, this is equal to

$\displaystyle\lim_{x\to\infty} \frac{e^{2x}\sqrt{x^2 + 1}}{-cx^{-c-1}e^{2x} + 2x^{-c}e^{2x}} = \lim_{x\to\infty} \frac{\sqrt{x^2 + 1}}{2x^{-c} - cx^{-c-1}} = \lim_{x\to\infty} \frac{x^{c+1}\sqrt{x^2 + 1}}{2x - c}$. This limit is finite only for $c \le -1$, and is zero for $c < -1$. If $c = -1$, the limit is $\displaystyle\lim_{x\to\infty} \frac{\sqrt{x^2 + 1}}{2x + 1} = \frac{1}{2}$.

24.

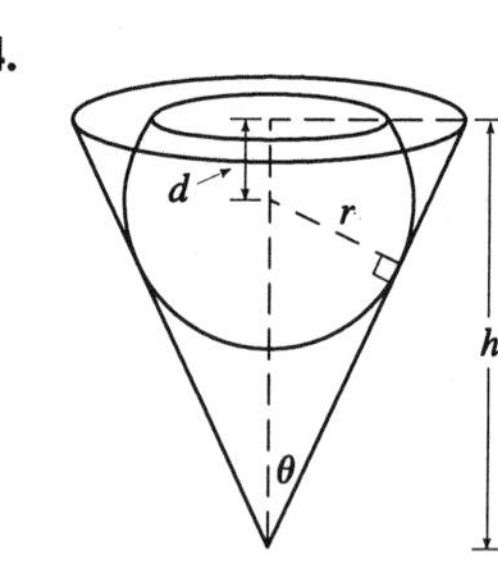

We want to find the volume of that part of the sphere which is below the surface of the water. As we can see from the diagram, this region is a cap of a sphere with radius r and height $r + d$. If we can find an expression for d in terms of h, r and θ, then we can determine the volume of the region (see Exercise 49 in Section 5.2), and then differentiate with respect to r to find the maximum.

We see that $\sin\theta = \dfrac{r}{h - d} \quad \Leftrightarrow \quad h - d = \dfrac{r}{\sin\theta} \quad \Leftrightarrow \quad d = h - r\csc\theta$.

Now we can use the formula from Exercise 5.2.49 to find the volume of water displaced:

$$V = \pi(r + d)^2\left(r - \frac{r + d}{3}\right) = \pi[r + (h - r\csc\theta)]^2\left[\frac{2r}{3} - \left(\frac{h}{3} - \frac{r}{3}\csc\theta\right)\right]$$

$= \frac{\pi}{3}[r(1 - \csc\theta) + h]^2[r(2 + \csc\theta) - h]$. Now we differentiate with respect to r:

$$\frac{dV}{dr} = \tfrac{\pi}{3}\left([r(1 - \csc\theta) + h]^2(2 + \csc\theta) + 2[r(1 - \csc\theta) + h](1 - \csc\theta)[r(2 + \csc\theta) + h]\right)$$

$$= \tfrac{\pi}{3}[r(1 - \csc\theta) + h]\left([r(1 - \csc\theta) + h](2 + \csc\theta) + 2(1 - \csc\theta)[r(2 + \csc\theta) - h]\right)$$

$$= \tfrac{\pi}{3}[r(1 - \csc\theta) + h](3(2 + \csc\theta)(1 - \csc\theta)r + [(2 + \csc\theta) - 2(1 - \csc\theta)]h)$$

$= \frac{\pi}{3}[r(1 - \csc\theta) + h][3(2 + \csc\theta)(1 - \csc\theta)r + 3h\csc\theta]$. This is 0 when

$r = \dfrac{h}{\csc\theta - 1}$ and when $r = \dfrac{h\csc\theta}{(\csc\theta + 2)(\csc\theta - 1)}$. Now since $V\left(\dfrac{h}{\csc\theta - 1}\right) = 0$ (the first factor vanishes; this corresponds to $d = -r$), the maximum volume of water is displaced when $r = \dfrac{h\csc\theta}{(\csc\theta - 1)(\csc\theta + 2)}$.

(Our intuition tells that a maximum value does exist, and it must occur at a critical number.) Multiplying numerator and denominator by $\sin^2\theta$, we get an alternative form of the answer: $r = \dfrac{h\sin\theta}{\sin\theta + \cos 2\theta}$.

25. Both sides of the inequality are positive, so $\cosh(\sinh x) < \sinh(\cosh x) \iff \cosh^2(\sinh x) < \sinh^2(\cosh x)$

$\iff \sinh^2(\sinh x) + 1 < \sinh^2(\cosh x) \iff 1 < [\sinh(\cosh x) - \sinh(\sinh x)][\sinh(\cosh x) + \sinh(\sinh x)]$

$\iff 1 < \left[\sinh\left(\frac{e^x + e^{-x}}{2}\right) - \sinh\left(\frac{e^x - e^{-x}}{2}\right)\right]\left[\sinh\left(\frac{e^x + e^{-x}}{2}\right) + \sinh\left(\frac{e^x - e^{-x}}{2}\right)\right]$

$\iff 1 < \left[2\cosh(e^x/2)\sinh(e^{-x}/2)\right]\left[2\sinh(e^x/2)\cosh(e^{-x}/2)\right]$ (use the addition formulas and cancel)

$\iff 1 < \left[2\sinh(e^x/2)\cosh(e^x/2)\right]\left[2\sinh(e^{-x}/2)\cosh(e^{-x}/2)\right] \iff 1 < \sinh e^x \sinh e^{-x}$, by the half-angle formula. Now both e^x and e^{-x} are positive, and $\sinh y > y$ for $y > 0$, since $\sinh 0 = 0$ and $(\sinh y - y)' = \cosh y - 1 > 0$ for $x > 0$, so $1 = e^x e^{-x} < \sinh e^x \sinh e^{-x}$. So, following this chain of reasoning backward, we arrive at the desired result.

Another Method: Using Formula 6.7.3, we have

$\sinh^{-1}(\cosh(\sinh x)) = \ln\left(\cosh(\sinh x) + \sqrt{1 + \cosh^2(\sinh x)}\right) = \ln(\cosh(\sinh x) + \sinh(\cosh x))$

$= \ln\left(e^{\sinh x}\right) = \sinh x$. But $\sinh x < \cosh x$, so $\sinh^{-1}(\cosh(\sinh x)) < \cosh x$. Since sinh is an increasing function, we can apply it to both sides of the inequality and get $\cosh(\sinh x) < \sinh(\cosh x)$.

26. We assume that P lies in the region of positive x. Since $y = x^3$ is an odd function, this assumption will not affect the result of the calculation.

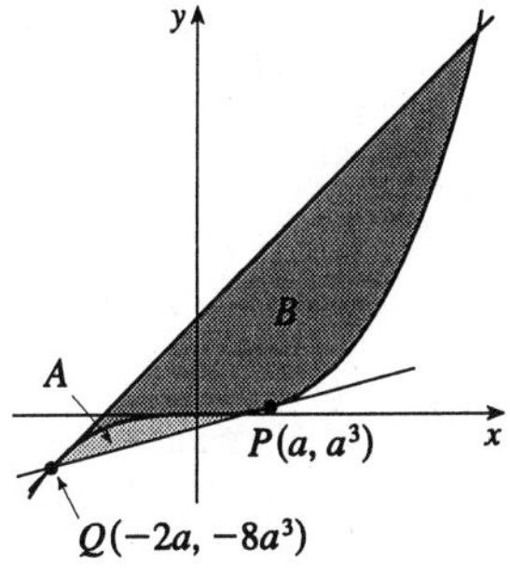

Let $P = (a, a^3)$. The slope of the tangent to the curve $y = x^3$ at P is $3a^2$, and so the equation of the tangent is $y - a^3 = 3a^2(x - a) \iff y = 3a^2 x - 2a^3$. We solve this simultaneously with $y = x^3$ to find the other point of intersection: $x^3 = 3a^2 x - 2a^3 \iff (x - a)^2(x + 2a) = 0$. So $Q = (-2a, -8a^3)$ is the other point of intersection. (See Example 1 in Problems Plus after Chapter 2 for a more arduous method.) The equation of the tangent at Q is

$y - (-8a^3) = 12a^2[x - (-2a)] \iff y = 12a^2 x + 16a^3$. By symmetry, this tangent will intersect the curve again at $x = -2(-2a) = 4a$. The curve lies above the first tangent, and below the second, so we are looking for a relationship between $A = \int_{-2a}^{a}[x^3 - (3a^2 x - 2a^3)]dx$ and $B = \int_{-2a}^{4a}[(12a^2 x + 16a^3) - x^3]dx$. We calculate $A = \left[\frac{1}{4}x^4 - \frac{3}{2}a^2 x^2 + 2a^3 x\right]_{-2a}^{a} = \frac{3}{4}a^4 - (-6a^4) = \frac{27}{4}a^4$, and

$B = \left[6a^2 x^2 + 16a^3 x - \frac{1}{4}x^4\right]_{-2a}^{4a} = 96a^4 - (-8a^4) = 108a^4$. We see that $B = 16A = 2^4 A$. This is because our calculation of area B was essentially the same as that of area A, with a replaced by $-2a$, so if we replace a with $-2a$ in our expression for A, we get $\frac{27}{4}(-2a)^4 = 108a^4 = B$.

27. We must find expressions for the areas A and B, and then set them equal and see what this says about the curve C. If $P = (a, 2a^2)$, then area A is just $\int_0^a (2x^2 - x^2)dx = \int_0^a x^2\,dx = \frac{1}{3}a^3$. To find area B, we use y as the variable of integration. So we find the equation of the middle curve as a function of y: $y = 2x^2 \iff x = \sqrt{y/2}$, since we are concerned with the first quadrant only. We can express area B as

$\int_0^{2a^2}\left[\sqrt{y/2} - C(y)\right]dy = \left[\frac{4}{3}(y/2)^{3/2}\right]_0^{2a^2} - \int_0^{2a^2} C(y)dy = \frac{4}{3}a^3 - \int_0^{2a^2} C(y)dy$, where $C(y)$ is the function with graph C. Setting $A = B$, we get $\frac{1}{3}a^3 = \frac{4}{3}a^3 - \int_0^{2a^2} C(y)dy \iff \int_0^{2a^2} C(y)dy = a^3$. Now we differentiate this equation with respect to a using the Chain Rule and the Fundamental Theorem: $C(2a^2)(4a) = 3a^2 \Rightarrow C(y) = \frac{3}{4}\sqrt{y/2}$, where $y = 2a^2$. Now we can solve for y: $x = \frac{3}{4}\sqrt{y/2} \Rightarrow x^2 = \frac{9}{16}(y/2) \Rightarrow y = \frac{32}{9}x^2$.

28. 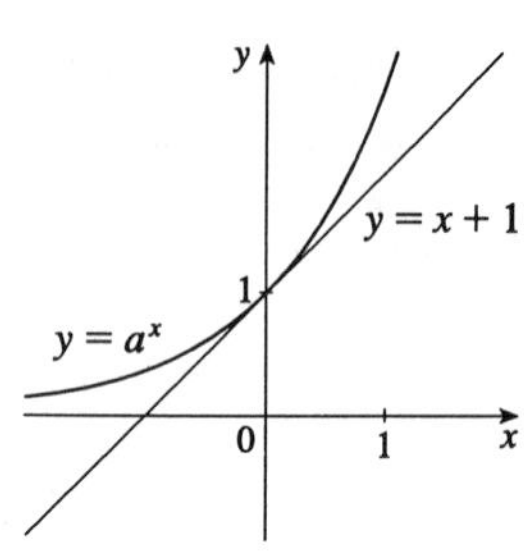

We see that at $x = 0$, $f(x) = a^x = 1 + x = 1$, so if $y = a^x$ is to lie above $y = 1 + x$, the two curves must just touch at $(0, 1)$, that is, we must have $f'(0) = 1$. [To see this analytically, note that $a^x \geq 1 + x \quad \Rightarrow$
$a^x - 1 \geq x \quad \Rightarrow \quad \dfrac{a^x - 1}{x} \geq 1$ for $x > 0$, so $f'(0) = \displaystyle\lim_{x \to 0^+} \frac{a^x - 1}{x} \geq 1$.
Similarly, for $x < 0$, $\dfrac{a^x - 1}{x} \leq 1$, so $f'(0) = \displaystyle\lim_{x \to 0^-} \frac{a^x - 1}{x} \leq 1$.
Since $1 \leq f'(0) \leq 1$, we must have $f'(0) = 1$.]
But $f'(x) = a^x \ln a \;\Rightarrow\; f'(0) = \ln a$, so we have $\ln a = 1 \;\Leftrightarrow\; a = e$.

Another Method: The inequality certainly holds for $x \leq -1$, so consider $x > -1$, $x \neq 0$. Then $a^x \geq 1 + x$ $\Rightarrow \quad a \geq (1 + x)^{1/x}$ for $x > 0 \quad \Rightarrow \quad a \geq \displaystyle\lim_{x \to 0^+} (1 + x)^{1/x} = e$, by Equation 6.4.8 or Equation 6.4*.7. Also, $a^x \geq 1 + x \quad \Rightarrow \quad a \leq (1 + x)^{1/x}$ for $x < 0 \quad \Rightarrow \quad a \leq \displaystyle\lim_{x \to 0^-} (1 + x)^{1/x} = e$. So since $e \leq a \leq e$, we must have $a = e$.

29. Suppose that the curve $y = a^x$ intersects the line $y = x$. Then $a^{x_0} = x_0$ for some $x_0 > 0$, and hence $a = x_0^{1/x_0}$. We find the maximum value of $g(x) = x^{1/x}$, > 0, because if a is larger than the maximum value of this function, then the curve $y = a^x$ does not intersect the line $y = x$.
$g'(x) = e^{(1/x)\ln x}\left(-\dfrac{1}{x^2}\ln x + \dfrac{1}{x}\cdot\dfrac{1}{x}\right) = x^{1/x}\left(\dfrac{1}{x^2}\right)(1 - \ln x)$. This is 0 only where $x = e$, and for $0 < x < e$, $f'(x) > 0$, while for $x > e$, $f'(x) < 0$, so g has an absolute maximum of $g(e) = e^{1/e}$. So if $y = a^x$ intersects $y = x$, we must have $0 < a \leq e^{1/e}$. Conversely, suppose that $0 < a \leq e^{1/e}$. Then $a^e \leq e$, so the graph of $y = a^x$ lies below or touches the graph of $y = x$ at $x = e$. Also $a^0 = 1 > 0$, so the graph of $y = a^x$ lies above that of $y = x$ at $x = 0$. Therefore, by the Intermediate Value Theorem, the graphs of $y = a^x$ and $y = x$ must intersect somewhere between $x = 0$ and $x = e$.

30.

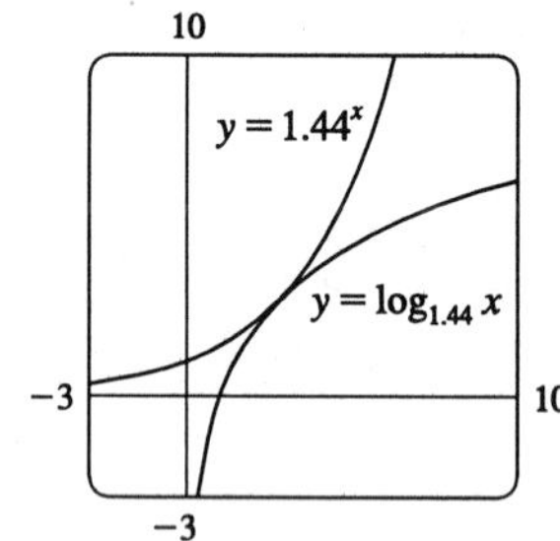

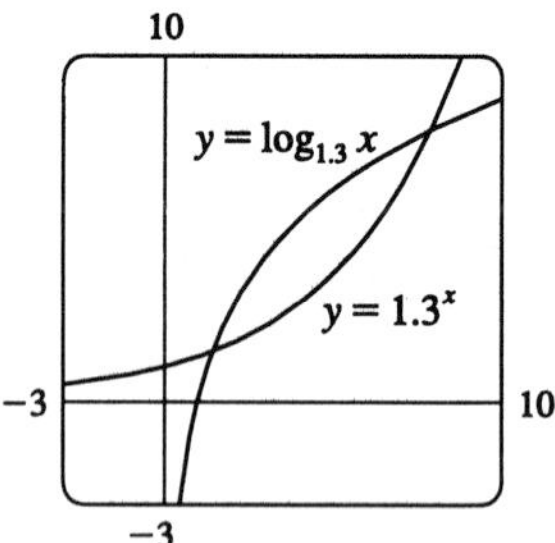

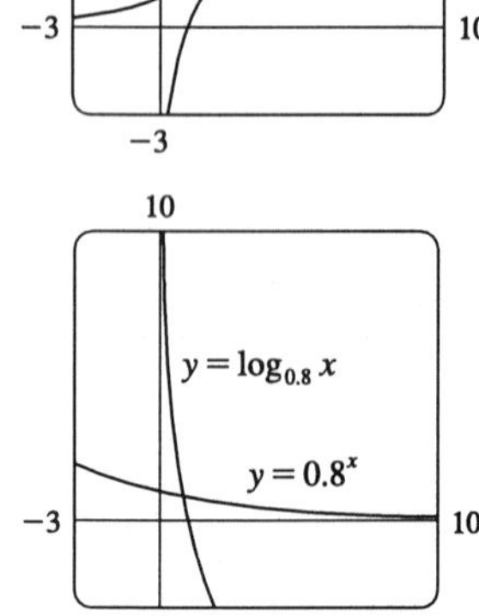

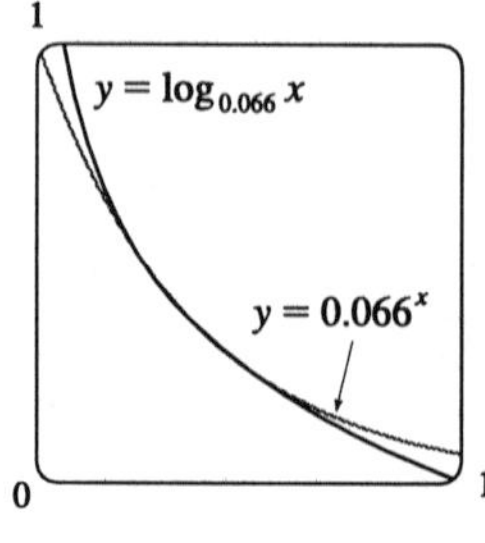

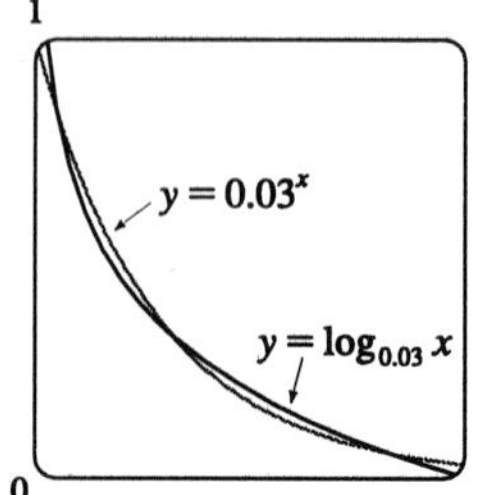

From the graphs, it appears that the number of points of intersection is: none when $a > 1.44$, one when $a \approx 1.44$, two when $1 < a < 1.44$, one when $0.066 < a < 1.44$, and three when $0 < a < 0.066$. For the proof, note first that $y = \log_a x \Leftrightarrow x = a^y$, so we look for points (x, y) which satisfy both $x = a^y$ and $y = a^x$. If there are any solutions to this system of two equations, then there is one on the line $y = x$: $x = a^x$.

Case 1: $a > e^{1/e}$ There are no solutions; see Exercise 29.

Case 2: $a = e^{1/e}$ In this case, $x = e$ is the only solution, since $a^x = x \Leftrightarrow a = x^{1/x}$, and the function $g(x) = x^{1/x}$ discussed in Exercise 29 takes its maximum value of $e^{1/e}$ only once.

Case 3: $1 < a < e^{1/e}$ There is at least one point for which $x > a^x$, namely $x = e$, since $e = \left(e^{1/e}\right)^e > a^e$. So using the Intermediate Value Theorem twice, we see that $f(x) = a^x - x$ must have two roots, since $f(0) = a > 0$, $f(e) < 0$, and $\lim\limits_{x\to\infty} f(x) = \infty$. Furthermore, there can be only two roots, since $f''(x) = a^x(\ln a)^2 > 0$ for all x, and so the graph of f is everywhere concave up, and thus cannot intersect the x-axis more than twice.

Case 4: $0 < a < 1$ Let $f(x) = \log_a x - a^x$. Then $f'(x) = \dfrac{1}{x\ln a} - a^x \ln a = \dfrac{1 - xa^x(\ln a)^2}{x \ln a}$ (★). $f'(x) \le 0 \Leftrightarrow xa^x \le \dfrac{1}{(\ln a)^2}$. We can verify that the function $y = xa^x$ has a maximum value of $-\dfrac{1}{e\ln a}$ for $x = -\dfrac{1}{\ln a}$, so $f'(x) \le 0$ if $-\dfrac{1}{e\ln a} \le \dfrac{1}{(\ln a)^2} \Leftrightarrow (\ln a)^2 \le -e\ln a$ (note that $\ln a < 0$ because $0 < a < 1$) $\Leftrightarrow \ln a \ge -e \Leftrightarrow a \ge e^{-e}$. This means that f is decreasing when $e^{-e} \le a < 1$. We also know that $f(x) > 0$ when x is close to 0 (because $\lim\limits_{x\to 0^+} f(x) = \infty$ for $a < 1$) and $f(1) = -1$. Thus f has exactly one zero and so the curves $y = a^x$ and $y = \log_a x$ intersect exactly once when $e^{-e} \le a < 1$.

For $0 < a < e^{-e}$ we have $-\dfrac{1}{e\ln a} > \dfrac{1}{(\ln a)^2}$ and so (★) shows that $f'(x) < 0 \Leftrightarrow x < \alpha$ or $x > \beta$, where α and β are the roots of the equation $xa^x = 1/(\ln a)^2$. (Notice that $\alpha < -1/\ln a < \beta$.) By considering the function $g(x) = a^x - x$, we see that the equation $a^x = x$ has exactly one root; call it x_2. It is also a solution of $a^x = \log_a x$, that is, a zero of f. To show that $\alpha < x_2 < \beta$, note that $0 < \alpha < -\dfrac{1}{\ln a} < \beta \Rightarrow \alpha^2 < \dfrac{1}{(\ln a)^2} < \beta^2$. But, by definition of α and β, $\alpha a^\alpha = \dfrac{1}{(\ln a)^2} = \beta a^\beta$, so $\alpha^2 < \alpha a^\alpha \Rightarrow \alpha < a^\alpha$ and $\beta a^\beta < \beta^2 \Rightarrow a^\beta < \beta$. Thus $g(\alpha) > 0$ and $g(\beta) < 0 \Rightarrow \alpha < x_2 < \beta$. Since f decreases on $(0, \alpha)$ and changes from positive to negative on this interval, we conclude that it has a unique zero in $(0, \alpha)$; call it x_1. Similarly, f has a unique zero x_3 in (β, ∞). So the curves $y = a^x$ and $\log_a x$ intersect three times for $0 < a < e^{-e}$, namely, at $x = x_1$, x_2, and x_3, where $x_1 < \alpha < x_2 < \beta < x_3$. (You can verify that $x_3 = a^{x_1}$.)

31. Note that $f(0) = 0$, so for $x \neq 0$, $\left|\dfrac{f(x) - f(0)}{x - 0}\right| = \left|\dfrac{f(x)}{x}\right| = \dfrac{|f(x)|}{|x|} \leq \dfrac{|\sin x|}{|x|} = \dfrac{\sin x}{x}$. Therefore $|f'(0)| = \left|\lim\limits_{x\to 0} \dfrac{f(x) - f(0)}{x - 0}\right| = \lim\limits_{x\to 0}\left|\dfrac{f(x) - f(0)}{x - 0}\right| \leq \lim\limits_{x\to 0} \dfrac{\sin x}{x} = 1$. But $f'(x) = a_1 \cos x + 2a_2 \cos 2x + \cdots + na_n \cos nx$, so $|f'(0)| = |a_1 + 2a_2 + \cdots + na_n| \leq 1$.

Another Solution: We are given that $\left|\sum\limits_{k=1}^{n} a_k \sin kx\right| \leq |\sin x|$. So for x close to 0, and $x \neq 0$, we have $\left|\sum\limits_{k=1}^{n} a_k \dfrac{\sin kx}{\sin x}\right| \leq 1 \;\Rightarrow\; \lim\limits_{x\to 0}\left|\sum\limits_{k=1}^{n} a_k \dfrac{\sin kx}{\sin x}\right| \leq 1 \;\Rightarrow\; \left|\sum\limits_{k=1}^{n} a_k \lim\limits_{x\to 0} \dfrac{\sin kx}{\sin x}\right| \leq 1$. But by l'Hospital's Rule, $\lim\limits_{x\to 0} \dfrac{\sin kx}{\sin x} = \lim\limits_{x\to 0} \dfrac{k \cos kx}{\cos x} = k$, so $\left|\sum\limits_{k=1}^{n} ka_k\right| \leq 1$.

32. We split up the integral, and use (twice) the fact that $\int_p^q |f - g|\,dx \geq \int_p^q (f - g)\,dx$ for $q \geq p$:

$$\begin{aligned}
\int_0^a |e^x - c|\,dx &= \int_0^{a/2} |c - e^x|\,dx + \int_{a/2}^a |e^x - c|\,dx \\
&\geq \int_0^{a/2} (c - e^x)\,dx + \int_{a/2}^a (e^x - c)\,dx = \left[cx - e^x\right]_0^{a/2} + \left[e^x - cx\right]_{a/2}^a \\
&= \left[\left(\tfrac{1}{2}ca - e^{a/2}\right) - (0 - 1)\right] + \left[(e^a - ca) - \left(e^{a/2} - \tfrac{1}{2}ca\right)\right] = e^a + 1 - 2e^{a/2} \\
&= \left(e^{a/2} - 1\right)^2.
\end{aligned}$$

Another Method: Since $e^x - c \geq 0 \;\Leftrightarrow\; x \geq \ln c$, we can write $I(c) = \int_0^a |e^x - c|\,dx = \int_0^{\ln c} (c - e^x)\,dx + \int_{\ln c}^a (e^x - c)\,dx$. Evaluation gives $I(c) = 2c \ln c - 2c - ac + e^a + 1$. Then we can show that the minimum value of the function $I(c)$ is $\left(e^{a/2} - 1\right)^2$.

33. The volume is $\int_0^{\sqrt{2}} \pi r^2\,ds$, where s is measured along the line $y = x$ from the origin to P. From the figure we have $r^2 + s^2 = d^2 = x^2 + x^4$, and from the distance formula we have

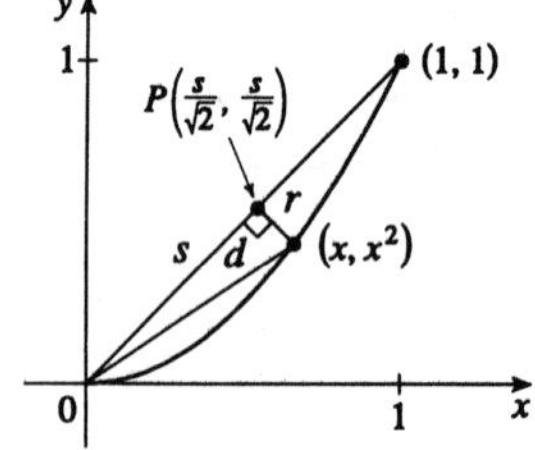

$$\begin{aligned}
r^2 &= \left(x - \tfrac{1}{\sqrt{2}}s\right)^2 + \left(x^2 - \tfrac{1}{\sqrt{2}}s\right)^2 \\
&= x^2 + x^4 + s^2 - \sqrt{2}s\left(x + x^2\right)
\end{aligned}$$

$\Rightarrow\quad x^2 + x^4 - s^2 = x^2 + x^4 + s^2 - \sqrt{2}s(x + x^2) \quad\Rightarrow\quad 2s^2 = \sqrt{2}s(x + x^2) \quad\Rightarrow\quad s = (x + x^2)/\sqrt{2}$

$\Rightarrow\quad r^2 = x^2 + x^4 - s^2 = x^2 + x^4 - \frac{1}{2}\left(x^2 + 2x^3 + x^4\right) = \frac{1}{2}x^4 - x^3 + \frac{1}{2}x^2$. Also, $ds = \frac{1}{\sqrt{2}}(1 + 2x)\,dx$, so

$$\begin{aligned}
V &= \int_0^{\sqrt{2}} \pi r^2\,ds = \pi \int_0^1 \left(\tfrac{1}{2}x^4 - x^3 + \tfrac{1}{2}x^2\right) \tfrac{1}{\sqrt{2}}(1 + 2x)\,dx \\
&= \tfrac{\pi}{2\sqrt{2}} \int_0^1 \left(2x^5 - 3x^4 + x^2\right) dx = \tfrac{\pi}{2\sqrt{2}}\left(\tfrac{1}{3} - \tfrac{3}{5} + \tfrac{1}{3}\right) \\
&= \tfrac{\pi}{30\sqrt{2}}.
\end{aligned}$$

For a more general method, see Problems Plus 22 after Chapter 8.

CHAPTER SEVEN

EXERCISES 7.1

1. Let $u = x$, $dv = e^{2x}\,dx \quad \Rightarrow \quad du = dx$, $v = \frac{1}{2}e^{2x}$. Then by Equation 2,
$\int xe^{2x}\,dx = \frac{1}{2}xe^{2x} - \int \frac{1}{2}e^{2x}\,dx = \frac{1}{2}xe^{2x} - \frac{1}{4}e^{2x} + C$.

2. Let $u = x$, $dv = \cos x\,dx \quad \Rightarrow \quad du = dx$, $v = \sin x$. Then by Equation 2,
$\int x\cos x\,dx = x\sin x - \int \sin x\,dx = x\sin x + \cos x + C$.

3. Let $u = x$, $dv = \sin 4x\,dx \quad \Rightarrow \quad du = dx$, $v = -\frac{1}{4}\cos 4x$. Then
$\int x\sin 4x\,dx = -\frac{1}{4}x\cos 4x - \int\left(-\frac{1}{4}\cos 4x\right)dx = -\frac{1}{4}x\cos 4x + \frac{1}{16}\sin 4x + C$.

4. Let $u = \ln x$, $dv = x\,dx \quad \Rightarrow \quad du = dx/x$, $v = \frac{1}{2}x^2$. Then
$\int x\ln x\,dx = \frac{1}{2}x^2\ln x - \int \frac{1}{2}x^2(dx/x) = \frac{1}{2}x^2\ln x - \frac{1}{4}x^2 + C$.

5. Let $u = x^2$, $dv = \cos 3x\,dx \quad \Rightarrow \quad du = 2x\,dx$, $v = \frac{1}{3}\sin 3x$. Then
$I = \int x^2\cos 3x\,dx = \frac{1}{3}x^2\sin 3x - \frac{2}{3}\int x\sin 3x\,dx$ by Equation 2. Next let $U = x$, $dV = \sin 3x\,dx \quad \Rightarrow$
$dU = dx$, $V = -\frac{1}{3}\cos 3x$ to get $\int x\sin 3x\,dx = -\frac{1}{3}x\cos 3x + \frac{1}{3}\int \cos 3x\,dx = -\frac{1}{3}x\cos 3x + \frac{1}{9}\sin 3x + C_1$.
Substituting for $\int x\sin 3x\,dx$, we get $I = \frac{1}{3}x^2\sin 3x - \frac{2}{3}\left(-\frac{1}{3}x\cos 3x + \frac{1}{9}\sin 3x + C_1\right)$
$= \frac{1}{3}x^2\sin 3x + \frac{2}{9}x\cos 3x - \frac{2}{27}\sin 3x + C$, where $C = -\frac{2}{3}C_1$.

6. Let $u = x^2$, $dv = \sin 2x\,dx \quad \Rightarrow \quad du = 2x\,dx$, $v = -\frac{1}{2}\cos 2x$. Then
$I = \int x^2\sin 2x\,dx = -\frac{1}{2}x^2\cos 2x - \int\left(-\frac{1}{2}\cos 2x \cdot 2x\right)dx = -\frac{1}{2}x^2\cos 2x + \int x\cos 2x\,dx$. Next let $U = x$,
$dV = \cos 2x\,dx \quad \Rightarrow \quad dU = dx$, $V = \frac{1}{2}\sin 2x$ to get
$\int x\cos 2x\,dx = \frac{1}{2}x\sin 2x - \frac{1}{2}\int \sin 2x\,dx = \frac{1}{2}x\sin 2x + \frac{1}{4}\cos 2x + C$. Substituting into the previous formula,
we obtain $I = -\frac{1}{2}x^2\cos 2x + \frac{1}{2}x\sin 2x + \frac{1}{4}\cos 2x + C$.

7. Let $u = (\ln x)^2$, $dv = dx \quad \Rightarrow \quad du = 2\ln x \cdot \frac{1}{x}\,dx$, $v = x$. Then $I = \int (\ln x)^2 dx = x(\ln x)^2 - 2\int \ln x\,dx$.
Taking $U = \ln x$, $dV = dx \quad \Rightarrow \quad dU = 1/x\,dx$, $V = x$, we find that
$\int \ln x\,dx = x\ln x - \displaystyle\int x \cdot \frac{1}{x}\,dx = x\ln x - x + C_1$. Thus $I = x(\ln x)^2 - 2x\ln x + 2x + C$, where $C = -2C_1$.

8. Let $u = \sin^{-1}x$, $dv = dx \quad \Rightarrow \quad du = dx/\sqrt{1-x^2}$, $v = x$. Then $\int \sin^{-1}x\,dx = x\sin^{-1}x - \displaystyle\int \frac{x}{\sqrt{1-x^2}}\,dx$.
Setting $t = 1 - x^2$, we get $dt = -2x\,dx$, so $-\displaystyle\int \frac{x\,dx}{\sqrt{1-x^2}} = \int t^{-1/2} \cdot \frac{1}{2}\,dt = t^{1/2} + C = \sqrt{1-x^2} + C$.
Hence $\int \sin^{-1}x\,dx = x\sin^{-1}x + \sqrt{1-x^2} + C$.

9. $I = \int \theta\sin\theta\cos\theta\,d\theta = \frac{1}{4}\int 2\theta\sin 2\theta\,d\theta = \frac{1}{8}\int t\sin t\,dt$ (Put $t = 2\theta \quad \Rightarrow \quad dt = d\theta/2$.)
Let $u = t$, $dv = \sin t\,dt \quad \Rightarrow \quad du = dt$, $v = -\cos t$. Then
$I = \frac{1}{8}\left(-t\cos t + \int \cos t\,dt\right) = \frac{1}{8}(-t\cos t + \sin t) + C = \frac{1}{8}(\sin 2\theta - 2\theta\cos 2\theta) + C$.

10. Let $u = \theta$, $dv = \sec^2\theta\, d\theta \Rightarrow du = d\theta$, $v = \tan\theta$. Then
$\int \theta \sec^2\theta\, d\theta = \theta \tan\theta - \int \tan\theta\, d\theta = \theta\tan\theta - \ln|\sec\theta| + C$.

11. Let $u = \ln t$, $dv = t^2\, dt \Rightarrow du = dt/t$, $v = \frac{1}{3}t^3$. Then
$\int t^2 \ln t\, dt = \frac{1}{3}t^3 \ln t - \int \frac{1}{3}t^3(1/t)\, dt = \frac{1}{3}t^3\ln t - \frac{1}{9}t^3 + C = \frac{1}{9}t^3(3\ln t - 1) + C$.

12. Let $u = t^3$, $dv = e^t\, dt \Rightarrow du = 3t^2$, $v = e^t$. Then $I = \int t^3 e^t\, dt = t^3e^t - \int 3t^2 e^t\, dt$. Integrate by parts twice more with $dv = e^t\, dt$. $I = t^3e^t - (3t^2e^t - \int 6te^t\, dt) = t^3e^t - 3t^2e^t + 6te^t - \int 6e^t\, dt$
$= t^3e^t - 3t^2e^t + 6te^t - 6e^t + C = (t^3 - 3t^2 + 6t - 6)e^t + C$. More generally, if $p(t)$ is a polynomial of degree n in t, then repeated integration by parts shows that
$\int p(t)e^t\, dt = \left[p(t) - p'(t) + p''(t) - p'''(t) + \cdots + (-1)^n p^{(n)}(t)\right]e^t + C$.

13. First let $u = \sin 3\theta$, $dv = e^{2\theta}d\theta \Rightarrow du = 3\cos 3\theta\, d\theta$, $v = \frac{1}{2}e^{2\theta}$. Then
$I = \int e^{2\theta}\sin 3\theta\, d\theta = \frac{1}{2}e^{2\theta}\sin 3\theta - \frac{3}{2}\int e^{2\theta}\cos 3\theta\, d\theta$. Next let $U = \cos 3\theta$, $dU = -3\sin 3\theta\, d\theta$, $dV = e^{2\theta}d\theta$, $v = \frac{1}{2}e^{2\theta}$ to get $\int e^{2\theta}\cos 3\theta\, d\theta = \frac{1}{2}e^{2\theta}\cos 3\theta + \frac{3}{2}\int e^{2\theta}\sin 3\theta\, d\theta$. Substituting in the previous formula gives $I = \frac{1}{2}e^{2\theta}\sin 3\theta - \frac{3}{4}e^{2\theta}\cos 3\theta - \frac{9}{4}\int e^{2\theta}\sin 3\theta\, d\theta$ or $\frac{13}{4}\int e^{2\theta}\sin 3\theta\, d\theta = \frac{1}{2}e^{2\theta}\sin 3\theta - \frac{3}{4}e^{2\theta}\cos 3\theta + C_1$. Hence $\int e^{2\theta}\sin 3\theta\, d\theta = \frac{1}{13}e^{2\theta}(2\sin 3\theta - 3\cos 3\theta) + C$, where $C = \frac{4}{13}C_1$.

14. Let $u = \cos 3\theta$, $dv = e^{-\theta}\, d\theta \Rightarrow du = -3\sin 3\theta\, d\theta$, $v = -e^{-\theta}$. Then
$I = \int e^{-\theta}\cos 3\theta\, d\theta = -e^{-\theta}\cos 3\theta - 3\int e^{-\theta}\sin 3\theta\, d\theta$. Integrate by parts again:
$I = -e^{-\theta}\cos 3\theta + 3e^{-\theta}\sin 3\theta - \int e^{-\theta}9\cos 3\theta\, d\theta$, so $10\int e^{-\theta}\cos 3\theta\, d\theta = e^{-\theta}(3\sin 3\theta - \cos 3\theta) + C_1$ and
$I = \frac{1}{10}e^{-\theta}(3\sin 3\theta - \cos 3\theta) + C$, where $C = C_1/10$.

15. Let $u = y$, $dv = \sinh y\, dy \Rightarrow du = dy$, $v = \cosh y$. Then
$\int y\sinh y\, dy = y\cosh y - \int \cosh y\, dy = y\cosh y - \sinh y + C$.

16. Let $u = y$, $dv = \cosh ay\, dy \Rightarrow du = dy$, $v = \dfrac{\sinh ay}{a}$. Then

$$\int y\cosh ay\, dy = \frac{y\sinh ay}{a} - \frac{1}{a}\int \sinh ay\, dy = \frac{y\sinh ay}{a} - \frac{\cosh ay}{a^2} + C.$$

17. Let $u = t$, $dv = e^{-t}\, dt \Rightarrow du = dt$, $v = -e^{-t}$. Then Formula 6 says $\int_0^1 te^{-t}\, dt = [-te^{-t}]_0^1 + \int_0^1$ $e^{-t}\, dt = -1/e + [-e^{-t}]_0^1 = -1/e - 1/e + 1 = 1 - 2/e$.

18. Let $u = \ln t$, $dv = \sqrt{t}\, dt \Rightarrow du = dt/t$, $v = \frac{2}{3}t^{3/2}$. By Formula 6,
$\int_1^4 \sqrt{t}\ln t\, dt = \left[\frac{2}{3}t^{3/2}\ln t\right]_1^4 - \frac{2}{3}\int_1^4 \sqrt{t}\, dt = \frac{2}{3}\cdot 8\cdot \ln 4 - 0 - \left[\frac{2}{3}\cdot\frac{2}{3}t^{3/2}\right]_1^4 = \frac{16}{3}\ln 4 - \frac{4}{9}(8-1) = \frac{32}{3}\ln 2 - \frac{28}{9}$.

19. Let $u = x$, $dv = \cos 2x\, dx \Rightarrow du = dx$, $v = \frac{1}{2}\sin 2x\, dx$. Then
$\int_0^{\pi/2} x\cos 2x\, dx = \left[\frac{1}{2}x\sin 2x\right]_0^{\pi/2} - \frac{1}{2}\int_0^{\pi/2}\sin 2x\, dx = 0 + \left[\frac{1}{4}\cos 2x\right]_0^{\pi/2} = \frac{1}{4}(-1-1) = -\frac{1}{2}$.

20. Let $u = x^2$, $dv = e^{-x}\, dx \Rightarrow du = 2x\, dx$, $v = -e^{-x}$. Then
$I = \int_0^1 x^2e^{-x}\, dx = [-x^2e^{-x}]_0^1 + \int_0^1 2xe^{-x}\, dx = -1/e + \int_0^1 2xe^{-x}\, dx$.

Now use parts again with $u = 2x$, $dv = e^{-x}$. Then
$I = -1/e - [2xe^{-x}]_0^1 + \int_0^1 2e^{-x}\, dx = -1/e - 2/e - [2e^{-x}]_0^1 = -3/e - 2/e + 2 = 2 - 5/e$.

21. Let $u = \cos^{-1}x$, $dv = dx \quad\Rightarrow\quad du = -\dfrac{dx}{\sqrt{1-x^2}}$, $v = x$. Then

$I = \displaystyle\int_0^{1/2} \cos^{-1}x\,dx = \left[x\cos^{-1}x\right]_0^{1/2} + \int_0^{1/2} \frac{x\,dx}{\sqrt{1-x^2}} = \frac{1}{2}\cdot\frac{\pi}{3} + \int_1^{3/4} t^{-1/2}\left[-\frac{1}{2}\,dt\right]$, where $t = 1 - x^2 \quad\Rightarrow$

$dt = -2x\,dx$. Thus $I = \frac{\pi}{6} + \frac{1}{2}\int_{3/4}^{1} t^{-1/2}\,dt = \left[\sqrt{t}\right]_{3/4}^{1} = \frac{\pi}{6} + 1 - \frac{\sqrt{3}}{2} = \frac{1}{6}\left(\pi + 6 - 3\sqrt{3}\right)$.

22. Let $u = x$, $dv = \csc^2 x\,dx \quad\Rightarrow\quad du = dx$, $v = -\cot x$. Then

$\int_{\pi/4}^{\pi/2} x\csc^2 x\,dx = \left[-x\cot x\right]_{\pi/4}^{\pi/2} + \int_{\pi/4}^{\pi/2} \cot x\,dx = -\frac{\pi}{2}\cdot 0 + \frac{\pi}{4}\cdot 1 + \left[\ln|\sin x|\right]_{\pi/4}^{\pi/2}$

$= \frac{\pi}{4} + \ln 1 - \ln\frac{1}{\sqrt{2}} = \frac{\pi}{4} + \frac{1}{2}\ln 2$.

23. Let $u = \ln(\sin x)$, $dv = \cos x\,dx \quad\Rightarrow\quad du = \dfrac{\cos x}{\sin x}\,dx$, $v = \sin x$. Then

$I = \int \cos x \ln(\sin x)dx = \sin x\ln(\sin x) - \int \cos x\,dx = \sin x\ln(\sin x) - \sin x + C$.

Another Method: Substitute $t = \sin x$, so $dt = \cos x\,dx$. Then $I = \int \ln t\,dt = t\ln t - t + C$ (see Example 2) and so $I = \sin x(\ln\sin x - 1) + C$.

24. Substitute $t = x^2 \quad\Rightarrow\quad dt = 2x\,dx$. Then use parts with $u = t$, $dv = e^t\,dt \quad\Rightarrow\quad du = dt$, $v = e^t$. Thus

$\int x^3 e^{x^2}\,dx = \frac{1}{2}\int te^t\,dt = \frac{1}{2}te^t - \frac{1}{2}\int e^t\,dt = \frac{1}{2}te^t - \frac{1}{2}e^t + C = \frac{1}{2}e^{x^2}(x^2-1) + C$.

25. Let $u = 2x + 3$, $dv = e^x\,dx \quad\Rightarrow\quad du = 2\,dx$, $v = e^x$. Then

$\int(2x+3)e^x\,dx = (2x+3)e^x - \int e^x\cdot 2\,dx = (2x+3)e^x - 2e^x + C = (2x+1)e^x + C$.

26. Let $u = x$, $dv = 5^x\,dx \quad\Rightarrow\quad du = dx$, $v = \left(5^x/\ln 5\right)dx$. Then

$\displaystyle\int x5^x\,dx = \frac{x5^x - \int 5^x\,dx}{\ln 5} = \frac{1}{\ln 5}\left(x5^x - \frac{5^x}{\ln 5}\right) + C = \frac{5^x}{\ln 5}\left(x - \frac{1}{\ln 5}\right) + C$.

27. Let $w = \ln x \quad\Rightarrow\quad dw = dx/x$. Then $x = e^w$ and $dx = e^w\,dw$, so

$\int \cos(\ln x)dx = \int e^w \cos w\,dw = \frac{1}{2}e^w(\sin w + \cos w) + C$ (by the method of Example 4)

$= \frac{1}{2}x[\sin(\ln x) + \cos(\ln x)] + C$.

28. Let $u = \tan^{-1}x$, $dv = x\,dx \quad\Rightarrow\quad du = dx/(1+x^2)$, $v = \frac{1}{2}x^2$. Then

$\int x\tan^{-1}x\,dx = \frac{1}{2}x^2\tan^{-1}x - \dfrac{1}{2}\displaystyle\int \frac{x^2\,dx}{1+x^2}$. But

$\displaystyle\int \frac{x^2\,dx}{1+x^2} = \int \frac{(1+x^2)-1}{1+x^2}\,dx = \int 1\,dx - \int \frac{dx}{1+x^2} = x - \tan^{-1}x + C_1 \quad\Rightarrow$

$\int x\tan^{-1}x\,dx = \frac{1}{2}x^2\tan^{-1}x + \frac{1}{2}\tan^{-1}x - \frac{1}{2}x + C$.

29. $I = \int_1^4 \ln\sqrt{x}\,dx = \frac{1}{2}\int_1^4 \ln x\,dx = \frac{1}{2}[x\ln x - x]_1^4$ as in Example 2. So

$I = \frac{1}{2}[(4\ln 4 - 4) - (0 - 1)] = 4\ln 2 - \frac{3}{2}$.

30. Let $w = \ln x$, so that $x = e^w$ and $dx = e^w\,dw$. Then

$\int \sin(\ln x)dx = \int e^w \sin w\,dw = \frac{1}{2}e^w(\sin w - \cos w) + C$ (by Example 4) $= \frac{1}{2}x[\sin(\ln x) - \cos(\ln x)] + C$.

31. Let $w = \sqrt{x}$, so that $x = w^2$ and $dx = 2w\,dw$. Then use $u = 2w$, $dv = \sin w\,dw$. Thus
$\int \sin\sqrt{x}\,dx = \int 2w \sin w\,dw = -2w\cos w + \int 2\cos w\,dw = -2w\cos w + 2\sin w + C$
$= -2\sqrt{x}\cos\sqrt{x} + 2\sin\sqrt{x} + C.$

32. Substitute $t = x^3 \Rightarrow dt = 3x^2\,dx$. Then use parts with $u = t$, $dv = \cos t\,dt$. Thus
$\int x^5\cos(x^3)dx = \frac{1}{3}\int x^3\cos(x^3)\cdot 3x^2\,dx = \frac{1}{3}\int t\cos t\,dt = \frac{1}{3}t\sin t - \frac{1}{3}\int \sin t\,dt$
$= \frac{1}{3}t\sin t + \frac{1}{3}\cos t + C = \frac{1}{3}x^3\sin(x^3) + \frac{1}{3}\cos(x^3) + C.$

33. $\int x^5 e^{x^2}\,dx = \int (x^2)^2 e^{x^2} x\,dx = \int t^2 e^t \frac{1}{2}\,dt$ (where $t = x^2 \Rightarrow \frac{1}{2}\,dt = x\,dx$)
$= \frac{1}{2}(t^2 - 2t + 2)e^t + C$ (by Example 3) $= \frac{1}{2}(x^4 - 2x^2 + 2)e^{x^2} + C.$

34. Let $w = \sqrt{x}$, so that $x = w^2$ and $dx = 2w\,dw$. Then use $u = 2w$, $dv = e^w\,dw$. So
$\int_1^4 e^{\sqrt{x}}\,dx = \int_1^2 e^w\,2w\,dw = [2we^w]_1^2 - 2\int_1^2 e^w\,dw = 4e^2 - 2e - 2(e^2 - e) = 2e^2.$

35. Let $u = x$, $dv = \cos\pi x\,dx \Rightarrow du = dx$,

$$v = \int \cos\pi x\,dx = \frac{\sin\pi x}{\pi}.$$ Thus

$$\int x\cos\pi x\,dx = x\cdot\frac{\sin\pi x}{\pi} - \int\frac{\sin\pi x}{\pi}\,dx = \frac{x\sin\pi x}{\pi} + \frac{\cos\pi x}{\pi^2} + C.$$

We see from the graph that this is reasonable, since the antiderivative has extrema where the original function is 0.

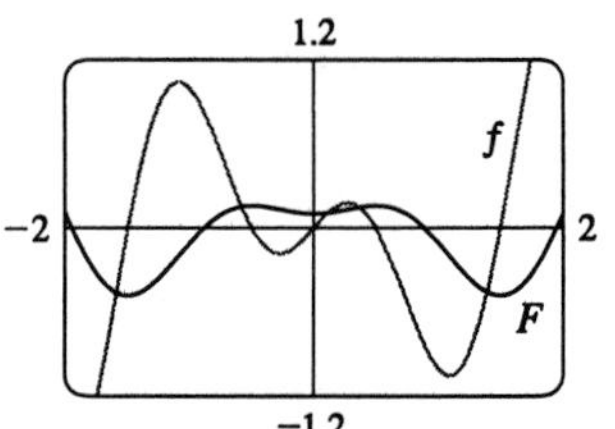

36. Let $u = \ln x$, $dv = \sqrt{x}\,dx \Rightarrow du = dx/x$,

$v = \int\sqrt{x}\,dx = \frac{2}{3}x^{3/2}$. Thus

$\int\sqrt{x}\ln x\,dx = \frac{2}{3}x^{3/2}\ln x - \int\frac{2}{3}x^{3/2}(1/x)dx = \frac{2}{3}x^{3/2}\ln x - \frac{4}{9}x^{3/2} + C.$

We see from the graph that this is reasonable, since the antiderivative is increasing where the original function is positive.

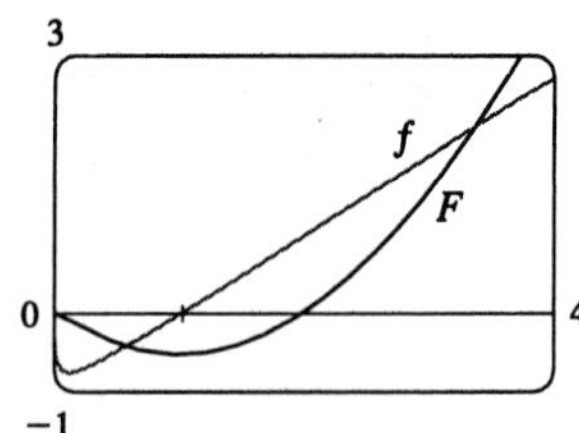

37. **(a)** Take $n = 2$ in Example 6 to get $\int\sin^2 x\,dx = -\frac{1}{2}\cos x\sin x + \frac{1}{2}\int 1\,dx = \dfrac{x}{2} - \dfrac{\sin 2x}{4} + C.$

(b) $\int\sin^4 x\,dx = -\frac{1}{4}\cos x\sin^3 x + \frac{3}{4}\int\sin^2 x\,dx = -\frac{1}{4}\cos x\sin^3 x + \frac{3}{8}x - \frac{3}{16}\sin 2x + C.$

38. **(a)** Let $u = \cos^{n-1}x$, $dv = \cos x\,dx \Rightarrow du = -(n-1)\cos^{n-2}x\sin x\,dx$, $v = \sin x$ in (2):
$\int\cos^n x\,dx = \cos^{n-1}x\sin x + (n-1)\int\cos^{n-2}x\sin^2 x\,dx$
$= \cos^{n-1}x\sin x + (n-1)\int\cos^{n-2}x\,(1 - \cos^2 x)dx$
$= \cos^{n-1}x\sin x + (n-1)\int\cos^{n-2}x\,dx - (n-1)\int\cos^n x\,dx.$

Rearranging terms gives $n\int\cos^n x\,dx = \cos^{n-1}x\sin x + (n-1)\int\cos^{n-2}x\,dx$ or

$$\int\cos^n x\,dx = \frac{1}{n}\cos^{n-1}x\sin x + \frac{n-1}{n}\int\cos^{n-2}x\,dx.$$

(b) Take $n = 2$ in (a) to get $\int\cos^2 x\,dx = \frac{1}{2}\cos x\sin x + \frac{1}{2}\int 1\,dx = \dfrac{x}{2} + \dfrac{\sin 2x}{4} + C.$

(c) $\int\cos^4 x\,dx = \frac{1}{4}\cos^3 x\sin x + \frac{3}{4}\int\cos^2 x\,dx = \frac{1}{4}\cos^3 x\sin x + \frac{3}{8}x + \frac{3}{16}\sin 2x + C$

39. **(a)** $\int_0^{\pi/2} \sin^n x\,dx = \left[-\dfrac{\cos x \sin^{n-1} x}{n}\right]_0^{\pi/2} + \dfrac{n-1}{n}\displaystyle\int_0^{\pi/2} \sin^{n-2} x\,dx = \dfrac{n-1}{n}\displaystyle\int_0^{\pi/2} \sin^{n-2} x\,dx$

(b) $\int_0^{\pi/2} \sin^3 x\,dx = \frac{2}{3}\int_0^{\pi/2} \sin x\,dx = \left[-\frac{2}{3}\cos x\right]_0^{\pi/2} = \frac{2}{3}$; $\int_0^{\pi/2} \sin^5 x\,dx = \frac{4}{5}\int_0^{\pi/2} \sin^3 x\,dx = \frac{4}{5}\cdot\frac{2}{3} = \frac{8}{15}$

(c) The formula holds for $n = 1$ (that is, $2n+1 = 3$) by (b). Assume it holds for some $k \geq 1$. Then $\displaystyle\int_0^{\pi/2} \sin^{2k+1} x\,dx = \frac{2\cdot 4\cdot 6\cdot\cdots\cdot(2k)}{3\cdot 5\cdot 7\cdot\cdots\cdot(2k+1)}$.

By Example 6, $\displaystyle\int_0^{\pi/2} \sin^{2k+3} x\,dx = \frac{2k+2}{2k+3}\int_0^{\pi/2} \sin^{2k+1} x\,dx = \frac{2\cdot 4\cdot 6\cdot\cdots\cdot[2(k+1)]}{2\cdot 4\cdot 6\cdot\cdots\cdot[2(k+1)+1]}$ as desired.

By induction, the formula holds for all $n \geq 1$.

40. The formula holds for $n = 1$ because $\int_0^{\pi/2} \sin^2 x\,dx = \frac{1}{2}\left[x - \frac{1}{2}\sin 2x\right]_0^{\pi/2} = \dfrac{\pi}{4} = \dfrac{1}{2}\cdot\dfrac{\pi}{2}$.

Now assume it holds for some $k \geq 1$. Then by the reduction formula in Example 6, $\displaystyle\int_0^{\pi/2} \sin^{2(k+1)} x\,dx = \frac{2k+1}{2k+2}\int_0^{\pi/2} \sin^{2k} x\,dx = \frac{1\cdot 3\cdot 5\cdot\cdots\cdot(2k+1)}{2\cdot 4\cdot 6\cdot\cdots\cdot(2k+2)}\cdot\frac{\pi}{2}$, so the formula holds for $k+1$. By induction, the formula holds for all $n \geq 1$.

41. Let $u = (\ln x)^n$, $dv = dx \Rightarrow du = n(\ln x)^{n-1}(dx/x)$, $v = x$. Then $\int (\ln x)^n\,dx = x(\ln x)^n - n\int(\ln x)^{n-1}\,dx$, by Equation 2.

42. Let $u = x^n$, $dv = e^x\,dx \Rightarrow du = nx^{n-1}\,dx$, $v = e^x$. Then $\int x^n e^x\,dx = x^n e^x - n\int x^{n-1}e^x\,dx$, by Equation 2.

43. Let $u = (x^2+a^2)^n$, $dv = dx \Rightarrow du = n(x^2+a^2)^{n-1}2x\,dx$, $v = x$. Then

$$\begin{aligned}\int (x^2+a^2)^n\,dx &= x(x^2+a^2)^n - 2n\int x^2(x^2+a^2)^{n-1}\,dx\\ &= x(x^2+a^2)^n - 2n\left[\int (x^2+a^2)^n\,dx - a^2\int (x^2+a^2)^{n-1}\,dx\right] \quad [\text{since } x^2 = (x^2+a^2) - a^2]\end{aligned}$$

$\Rightarrow (2n+1)\int(x^2+a^2)^n\,dx = x(x^2+a^2)^n + 2na^2\int(x^2+a^2)^{n-1}\,dx$, and

$$\int (x^2+a^2)^n\,dx = \frac{x(x^2+a^2)^n}{2n+1} + \frac{2na^2}{2n+1}\int (x^2+a^2)^{n-1}\,dx \text{ (provided } 2n+1 \neq 0).$$

44. Let $u = \sec^{n-2}x$, $dv = \sec^2 x\,dx \Rightarrow du = (n-2)\sec^{n-3}x\sec x\tan x\,dx$, $v = \tan x$. Then by Equation 2,

$$\begin{aligned}\int \sec^n x\,dx &= \tan x\sec^{n-2}x - (n-2)\int \sec^{n-2}x\tan^2 x\,dx\\ &= \tan x\sec^{n-2}x - (n-2)\int \sec^{n-2}x(\sec^2 x - 1)dx\\ &= \tan x\sec^{n-2}x - (n-2)\int \sec^n x\,dx + (n-2)\int \sec^{n-2}x\,dx,\end{aligned}$$

so $(n-1)\int \sec^n x\,dx = \tan x\sec^{n-2}x + (n-2)\int\sec^{n-2}x\,dx$. If $n - 1 \neq 0$, then

$$\int \sec^n x\,dx = \frac{\tan x\sec^{n-2}x}{n-1} + \frac{n-2}{n-1}\int \sec^{n-2}x\,dx.$$

45. Take $n = 3$ in Exercise 41 to get

$\int(\ln x)^3\,dx = x(\ln x)^3 - 3\int(\ln x)^2\,dx = x(\ln x)^3 - 3x(\ln x)^2 + 6x\ln x - 6x + C$ (by Exercise 7).

46. Take $n = 4$ in Exercise 42 to get

$\int x^4 e^x\,dx = x^4 e^x - 4\int x^3 e^x\,dx = x^4 e^x - 4(x^3 - 3x^2 + 6x - 6)e^x + C$ (by Exercise 12)

$= e^x(x^4 - 4x^3 + 12x^2 - 24x + 24) + C.$

Or: Instead of using Exercise 12, apply Exercise 42 with $n = 3$, then $n = 2$, etc.

47. Let $u = \sin^{-1}x$, $dv = dx$ $\Rightarrow$ $du = \dfrac{dx}{\sqrt{1-x^2}}$, $v = x$. Then

$$\text{area} = \int_0^{1/2} \sin^{-1}x\,dx = \left[x\sin^{-1}x\right]_0^{1/2} - \int_0^{1/2} \frac{x}{\sqrt{1-x^2}}\,dx = \tfrac{1}{2}\left(\tfrac{\pi}{6}\right) + \left[\sqrt{1-x^2}\right]_0^{1/2}$$

$$= \tfrac{\pi}{12} + \tfrac{\sqrt{3}}{2} - 1 = \tfrac{1}{12}\left(\pi + 6\sqrt{3} - 12\right).$$

48. The curves intersect when $(x-5)\ln x = 0$; that is, when $x = 1$ or $x = 5$. For $1 < x < 5$, we have $5\ln x > x\ln x$ since $\ln x > 0$. Thus area $= \int_1^5 (5\ln x - x\ln x)dx$. Let $u = \ln x$, $dv = (5-x)dx$ $\Rightarrow$ $du = dx/x$, $v = 5x - \frac{1}{2}x^2$. Then

$$\text{area} = (\ln x)\left[5x - \tfrac{1}{2}x^2\right]_1^5 - \int_1^5 \left(5x - \tfrac{1}{2}x^2\right)\frac{1}{x}\,dx = (\ln 5)\left(\tfrac{25}{2}\right) - 0 - \int_1^5 \left[5 - \tfrac{1}{2}x\right]dx$$

$$= \tfrac{25}{2}\ln 5 - \left[5x - \tfrac{1}{4}x^2\right]_1^5 = \tfrac{25}{2}\ln 5 - \left[\left(25 - \tfrac{25}{4}\right) - \left(5 - \tfrac{1}{4}\right)\right] = \tfrac{25}{2}\ln 5 - 14.$$

49.

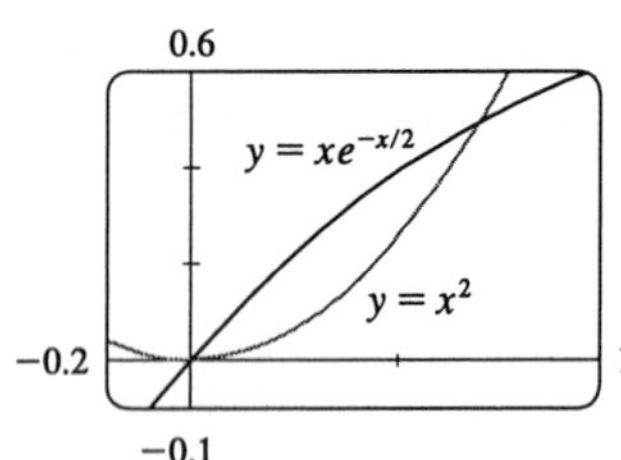

From the graph, we see that the curves intersect at approximately $x = 0$ and $x = 0.70$, with $xe^{-x/2} > x^2$ on $(0, 0.70)$. So the area bounded by the curves is approximately $A = \int_0^{0.70} (xe^{-x/2} - x^2)dx$. We separate this into two integrals, and evaluate the first one by parts with $u = x$, $dv = e^{-x/2}\,dx$ $\Rightarrow$ $du = dx$, $v = -2e^{-x/2}$:

$A = \left[-2xe^{-x/2}\right]_0^{0.70} - \int_0^{0.70}(-2e^{-x/2})dx - \left[\frac{1}{3}x^3\right]_0^{0.70} = [-2(0.70)e^{-0.35} - 0] - \left[4e^{-x/2}\right]_0^{0.70} - \frac{1}{3}[0.70^3 - 0]$
$\approx 0.080.$

50.

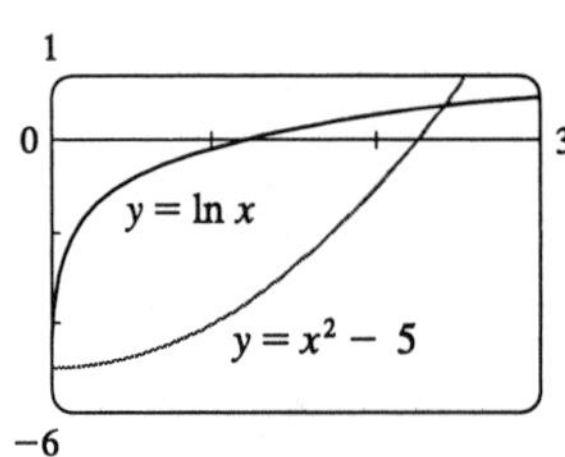

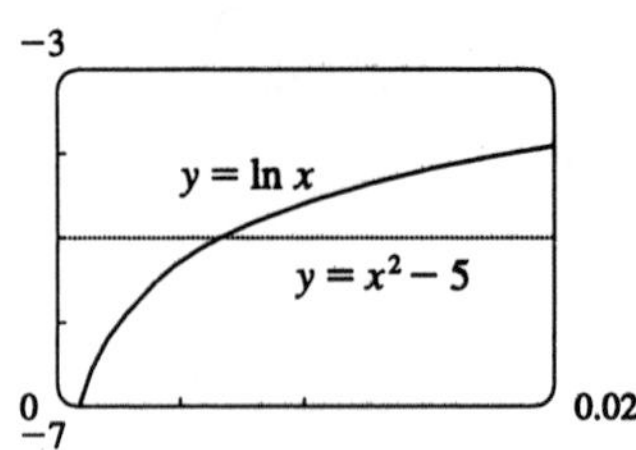

From the graphs, we see that the curves intersect at approximately $x = 0.0067$ and $x = 2.43$, with $\ln x > x^2 - 5$ on $(0.0067, 2.43)$. So the area bounded by the curves is about

$A = \int_{0.0067}^{2.43} [\ln x - (x^2 - 5)]dx = \int_{0.0067}^{2.43} (\ln x - x^2 + 5)dx$

$= \left[(x\ln x - x) - \frac{1}{3}x^3 + 5x\right]_{0.0067}^{2.43}$ (see Example 2) $\approx 7.10.$

51. Volume $= \int_{2\pi}^{3\pi} 2\pi x \sin x\, dx$. Let $u = x$, $dv = \sin x\, dx \Rightarrow du = dx$, $v = -\cos x \Rightarrow$
$V = 2\pi[-x\cos x + \sin x]_{2\pi}^{3\pi} = 2\pi[(3\pi + 0) - (-2\pi + 0)] = 2\pi(5\pi) = 10\pi^2$.

52. Volume $= \int_0^1 2\pi x(e^x - e^{-x})dx = 2\pi\int_0^1 (xe^x - xe^{-x})dx = 2\pi\left[\int_0^1 xe^x\, dx - \int_0^1 xe^{-x}\, dx\right]$
(both integrals by parts) $= 2\pi[(xe^x - e^x) - (-xe^{-x} - e^{-x})]_0^1 = 2\pi[2/e - 0] = 4\pi/e$

53. Volume $= \int_{-1}^{0} 2\pi(1-x)e^{-x}\, dx$. Let $u = 1 - x$, $dv = e^{-x}\, dx \Rightarrow du = -\, dx$, $v = -e^{-x} \Rightarrow$
$V = 2\pi[xe^{-x}]_{-1}^{0} = 2\pi(0 + e) = 2\pi e$

54. Volume $= \int_1^{\pi} 2\pi y \cdot \ln y\, dy = 2\pi\left[\frac{1}{2}y^2\ln y - \frac{1}{4}y^2\right]_1^{\pi} = 2\pi\left[\frac{1}{4}y^2(2\ln y - 1)\right]_1^{\pi}$ (by parts)

$$= 2\pi\left[\frac{\pi^2(2\ln\pi - 1)}{4} - \frac{(0-1)}{4}\right] = \pi^3\ln\pi - \frac{\pi^3}{2} + \frac{\pi}{2}$$

55. Since $v(t) > 0$ for all t, the desired distance $s(t) = \int_0^t v(w)dw = \int_0^t w^2e^{-w}\, dw$. Let $u = w^2$, $dv = e^{-w}\, dw$ $\Rightarrow du = 2w\, dw$, $v = -e^{-w}$. Then $s(t) = [-w^2e^{-w}]_0^t + 2\int_0^t we^{-w}\, dw$.

Now let $U = w$, $dV = e^{-w}\, dw \Rightarrow dU = dw$, $V = -e^{-w}$. Then
$s(t) = -t^2e^{-t} + 2\left([-we^{-w}]_0^t + \int_0^t e^{-w}\, dw\right) = -t^2e^{-t} - 2te^{-t} - 2e^{-t} + 2 = 2 - e^{-t}(t^2 + 2t + 2)$ meters.

56. Suppose $f(0) = g(0) = 0$ and put $u = f(x)$, $dv = g''(x)dx \Rightarrow du = f'(x)dx$, $v = g'(x)$. Then
$\int_0^a f(x)g''(x)dx = [f(x)g'(x)]_0^a - \int_0^a f'(x)g'(x)dx = f(a)g'(a) - \int_0^a f'(x)g'(x)dx$.

Now put $U = f'(x)$, $dV = g'(x)dx \Rightarrow dU = f''(x)dx$ and $V = g(x)$, so
$\int_0^a f'(x)g'(x)dx = [f'(x)g(x)]_0^a - \int_0^a f''(x)g(x)dx = f'(a)g(a) - \int_0^a f''(x)g(x)dx$.

Combining the two results, we get $\int_0^a f(x)g''(x)dx = f(a)g'(a) - f'(a)g(a) + \int_0^a f''(x)g(x)dx$.

57. Take $g(x) = x$ in Equation 1.

58. By Exercise 57, $\int_a^b f(x)dx = bf(b) - a\,f(a) - \int_a^b x\,f'(x)dx$. Now let $y = f(x)$, so that $x = g(y)$ and $dy = f'(x)dx$. Then $\int_a^b x\,f'(x)dx = \int_{f(a)}^{f(b)} g(y)dy$. The result follows.

59. By Exercise 58, $\int_1^e \ln x\, dx = e\ln e - 1\ln 1 - \int_{\ln 1}^{\ln e} e^y\, dy = e - \int_0^1 e^y\, dy = e - [e^y]_0^1 = e - (e-1) = 1$.

60. Exercise 58 says that the area of region $ABFC$ is
(area of rectangle $OBFE$) − (area of rectangle $OACD$) − (area of region $DCFE$).

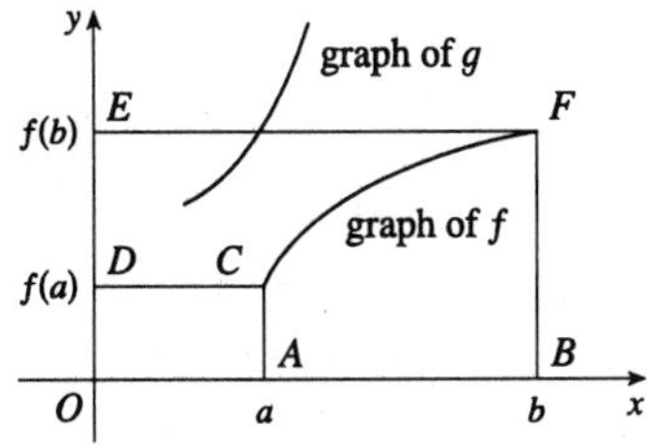

51. Volume $= \int_{2\pi}^{3\pi} 2\pi x \sin x\,dx$. Let $u = x$, $dv = \sin x\,dx \quad \Rightarrow \quad du = dx$, $v = -\cos x \quad \Rightarrow$
$V = 2\pi[-x\cos x + \sin x]_{2\pi}^{3\pi} = 2\pi[(3\pi + 0) - (-2\pi + 0)] = 2\pi(5\pi) = 10\pi^2$.

52. Volume $= \int_0^1 2\pi x(e^x - e^{-x})dx = 2\pi \int_0^1 (xe^x - xe^{-x})dx = 2\pi\left[\int_0^1 xe^x\,dx - \int_0^1 xe^{-x}\,dx\right]$
(both integrals by parts) $= 2\pi[(xe^x - e^x) - (-xe^{-x} - e^{-x})]_0^1 = 2\pi[2/e - 0] = 4\pi/e$

53. Volume $= \int_{-1}^0 2\pi(1-x)e^{-x}\,dx$. Let $u = 1 - x$, $dv = e^{-x}\,dx \quad \Rightarrow \quad du = -\,dx$, $v = -e^{-x} \quad \Rightarrow$
$V = 2\pi[xe^{-x}]_{-1}^0 = 2\pi(0 + e) = 2\pi e$

54. Volume $= \int_1^\pi 2\pi y \cdot \ln y\,dy = 2\pi\left[\frac{1}{2}y^2 \ln y - \frac{1}{4}y^2\right]_1^\pi = 2\pi\left[\frac{1}{4}y^2(2\ln y - 1)\right]_1^\pi$ (by parts)
$= 2\pi\left[\frac{\pi^2(2\ln\pi - 1)}{4} - \frac{(0-1)}{4}\right] = \pi^3 \ln \pi - \frac{\pi^3}{2} + \frac{\pi}{2}$

55. Since $v(t) > 0$ for all t, the desired distance $s(t) = \int_0^t v(w)dw = \int_0^t w^2 e^{-w}\,dw$. Let $u = w^2$, $dv = e^{-w}\,dw$
$\Rightarrow \quad du = 2w\,dw$, $v = -e^{-w}$. Then $s(t) = [-w^2 e^{-w}]_0^t + 2\int_0^t we^{-w}\,dw$.

Now let $U = w$, $dV = e^{-w}\,dw \quad \Rightarrow \quad dU = dw$, $V = -e^{-w}$. Then
$s(t) = -t^2 e^{-t} + 2\left([-we^{-w}]_0^t + \int_0^t e^{-w}\,dw\right) = -t^2 e^{-t} - 2te^{-t} - 2e^{-t} + 2 = 2 - e^{-t}(t^2 + 2t + 2)$ meters.

56. Suppose $f(0) = g(0) = 0$ and put $u = f(x)$, $dv = g''(x)dx \quad \Rightarrow \quad du = f'(x)dx$, $v = g'(x)$. Then
$\int_0^a f(x)g''(x)dx = [f(x)g'(x)]_0^a - \int_0^a f'(x)g'(x)dx = f(a)g'(a) - \int_0^a f'(x)g'(x)dx$.

Now put $U = f'(x)$, $dV = g'(x)dx \quad \Rightarrow \quad dU = f''(x)dx$ and $V = g(x)$, so
$\int_0^a f'(x)g'(x)dx = [f'(x)g(x)]_0^a - \int_0^a f''(x)g(x)dx = f'(a)g(a) - \int_0^a f''(x)g(x)dx$.

Combining the two results, we get $\int_0^a f(x)g''(x)dx = f(a)g'(a) - f'(a)g(a) + \int_0^a f''(x)g(x)dx$.

57. Take $g(x) = x$ in Equation 1.

58. By Exercise 57, $\int_a^b f(x)dx = bf(b) - a\,f(a) - \int_a^b x\,f'(x)dx$. Now let $y = f(x)$, so that $x = g(y)$ and $dy = f'(x)dx$. Then $\int_a^b x\,f'(x)dx = \int_{f(a)}^{f(b)} g(y)dy$. The result follows.

59. By Exercise 58, $\int_1^e \ln x\,dx = e\ln e - 1\ln 1 - \int_{\ln 1}^{\ln e} e^y\,dy = e - \int_0^1 e^y\,dy = e - [e^y]_0^1 = e - (e-1) = 1$.

60. Exercise 58 says that the area of region $ABFC$ is
(area of rectangle $OBFE$) – (area of rectangle $OACD$) – (area of region $DCFE$).

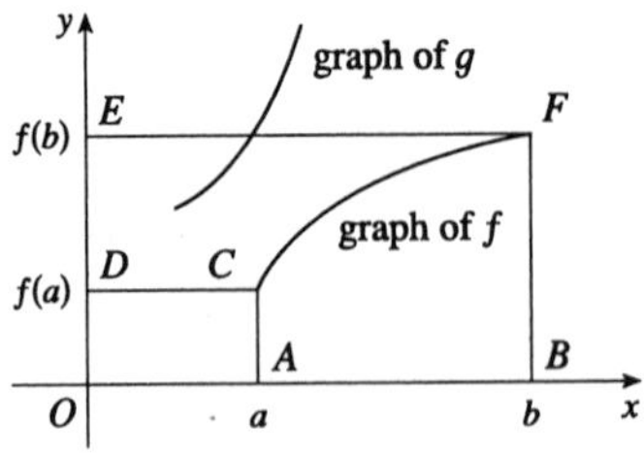

61. Using the formula for volumes of rotation (Equation 5.2.3) and the figure, we see that
Volume $= \int_0^d \pi b^2\,dy - \int_0^c \pi a^2\,dy - \int_c^d \pi[g(y)]^2\,dy = \pi b^2 d - \pi a^2 c - \int_c^d \pi[g(y)]^2\,dy$.
Let $y = f(x)$, which gives $dy = f'(x)dx$ and $g(y) = x$, so that $V = \pi b^2 d - \pi a^2 c - \pi\int_a^b x^2 f'(x)dx$. Now integrate by parts with $u = x^2$, and $dv = f'(x)dx \quad\Rightarrow\quad du = 2x\,dx$, $v = f(x)$, and
$\int_a^b x^2 f'(x)dx = \left[x^2 f(x)\right]_a^b - \int_a^b 2x\,f(x)dx = b^2 f(b) - a^2 f(a) - \int_a^b 2x\,f(x)dx$, but $f(a) = c$ and $f(b) = d$
$\Rightarrow \quad V = \pi b^2 d - \pi a^2 c - \pi\left[b^2 d - a^2 c - \int_a^b 2x f(x)dx\right] = \int_a^b 2\pi x f(x)dx$.

62. **(a)** We note that for $0 \le x \le \frac{\pi}{2}$, $0 \le \sin x \le 1$, so $\sin^{2n+2}x \le \sin^{2n+1}x \le \sin^{2n}x$. So by Property 7 of integrals, $I_{2n+2} \le I_{2n+1} \le I_{2n}$.

(b) Substituting directly into the result from Exercise 40, we get

$$\frac{I_{2n+2}}{I_{2n}} = \frac{\dfrac{1\cdot 3\cdot 5\cdot\cdots\cdot[2(n+1)-1]}{2\cdot 4\cdot 6\cdot\cdots\cdot[2(n+1)]}\dfrac{\pi}{2}}{\dfrac{1\cdot 3\cdot 5\cdot\cdots\cdot(2n-1)}{2\cdot 4\cdot 6\cdot\cdots\cdot(2n)}\dfrac{\pi}{2}} = \frac{2(n+1)-1}{2(n+1)} = \frac{2n+1}{2n+2}.$$

(c) We divide the result from part (a) by I_{2n}. The inequalities are preserved since I_{2n} is positive: $\dfrac{I_{2n+2}}{I_{2n}} \le \dfrac{I_{2n+1}}{I_{2n}} \le \dfrac{I_{2n}}{I_{2n}}$. Now from part (b), the left term is equal to $\dfrac{2n+1}{2n+2}$, so the expression becomes $\dfrac{2n+1}{2n+2} \le \dfrac{I_{2n+1}}{I_{2n}} \le 1$. Now $\lim\limits_{n\to\infty}\dfrac{2n+1}{2n+2} = \lim\limits_{n\to\infty} 1 = 1$, so by the Squeeze Theorem, $\lim\limits_{n\to\infty}\dfrac{I_{2n+1}}{I_{2n}} = 1$.

(d) We substitute the results from Exercises 39 and 40 into the result from part (c):

$$1 = \lim_{n\to\infty}\frac{I_{2n+1}}{I_{2n}} = \lim_{n\to\infty}\frac{\dfrac{2\cdot 4\cdot 6\cdot\cdots\cdot(2n)}{3\cdot 5\cdot 7\cdot\cdots\cdot(2n+1)}}{\dfrac{1\cdot 3\cdot 5\cdot\cdots\cdot(2n-1)}{2\cdot 4\cdot 6\cdot\cdots\cdot(2n)}\dfrac{\pi}{2}}$$

$$= \lim_{n\to\infty}\left[\frac{2\cdot 4\cdot 6\cdot\cdots\cdot(2n)}{3\cdot 5\cdot 7\cdot\cdots\cdot(2n+1)}\right]\left[\frac{2\cdot 4\cdot 6\cdot\cdots\cdot(2n)}{1\cdot 3\cdot 5\cdot\cdots\cdot(2n-1)}\left(\frac{2}{\pi}\right)\right].$$

Rearranging the terms and multiplying by $\frac{\pi}{2}$, we get $\lim\limits_{n\to\infty}\dfrac{2}{1}\cdot\dfrac{2}{3}\cdot\dfrac{4}{3}\cdot\dfrac{4}{5}\cdot\dfrac{6}{5}\cdot\dfrac{6}{7}\cdot\cdots\cdot\dfrac{2n}{2n-1}\cdot\dfrac{2n}{2n+1} = \dfrac{\pi}{2}$, as required.

(e) The area of the kth rectangle is k. At the $2n$th step, the area is increased from $2n-1$ to $2n$ by multiplying the length by $\dfrac{2n}{2n-1}$, and at the $(2n+1)$th step, the area is increased from $2n$ to $2n+1$ by multiplying the height by $\dfrac{2n+1}{2n}$. These two steps multiply the ratio of length to height by $\dfrac{2n}{2n-1}$ and $\dfrac{1}{(2n+1)/(2n)} = \dfrac{2n}{2n+1}$ respectively. So, by part (d), the limiting ratio is $\dfrac{2}{1}\cdot\dfrac{2}{3}\cdot\dfrac{4}{3}\cdot\dfrac{4}{5}\cdot\dfrac{6}{5}\cdot\dfrac{6}{7}\cdot\cdots = \dfrac{\pi}{2}$.

13. Let $u = \cos x$, $du = -\sin x\,dx$. Then $\int \sin^3 x\,\sqrt{\cos x}\,dx = \int (1-\cos^2 x)\sqrt{\cos x}\,\sin x\,dx$
$= \int (1-u^2)u^{1/2}(-du) = \int \left(u^{5/2} - u^{1/2}\right)du = \frac{2}{7}u^{7/2} - \frac{2}{3}u^{3/2} + C$
$= \frac{2}{7}(\cos x)^{7/2} - \frac{2}{3}(\cos x)^{3/2} + C = \left[\frac{2}{7}\cos^3 x - \frac{2}{3}\cos x\right]\sqrt{\cos x} + C.$

14. Let $u = x^2 \Rightarrow du = 2x\,dx$. Then $\int x\sin^3(x^2)dx = \int \sin^3 u \cdot \frac{1}{2}\,du = \frac{1}{2}\left(-\cos u + \frac{1}{3}\cos^3 u\right) + C$ (by Exercise 4) $= -\frac{1}{2}\cos(x^2) + \frac{1}{6}\cos^3(x^2) + C.$

15. Let $u = \cos x \Rightarrow du = -\sin x\,dx$. Then $\displaystyle\int \cos^2 x\tan^3 x\,dx = \int \frac{\sin^3 x}{\cos x}\,dx$

$$= \int \frac{(1-u^2)(-du)}{u} = \int \left[\frac{-1}{u} + u\right]du = -\ln|u| + \tfrac{1}{2}u^2 + C = \tfrac{1}{2}\cos^2 x - \ln|\cos x| + C.$$

16. Let $u = \sin x \Rightarrow du = \cos x\,dx$. Then

$$\int \cot^5 x\sin^2 x\,dx = \int \frac{\cos^5 x}{\sin^3 x}\,dx = \int \frac{\left(1-u^2\right)^2}{u^3}\,du = \int \left(u^{-3} - 2u^{-1} + u\right)du$$
$$= -\tfrac{1}{2}u^{-2} - 2\ln|u| + \tfrac{1}{2}u^2 + C = \tfrac{1}{2}\sin^2 x - \tfrac{1}{2}\csc^2 x - 2\ln|\sin x| + C.$$

17. $\displaystyle\int \frac{1-\sin x}{\cos x}\,dx = \int(\sec x - \tan x)dx = \ln|\sec x + \tan x| - \ln|\sec x| + C$ (by Example 8)

$$= \ln|(\sec x + \tan x)\cos x| + C = \ln|1 + \sin x| + C = \ln(1 + \sin x) + C,$$

since $1 + \sin x \geq 0$.

Or: $\displaystyle\int \frac{1-\sin x}{\cos x}\,dx = \int \frac{1-\sin x}{\cos x}\cdot\frac{1+\sin x}{1+\sin x}\,dx = \int \frac{(1-\sin^2 x)dx}{\cos x(1+\sin x)} = \int \frac{\cos x\,dx}{1+\sin x} = \int \frac{dw}{w}$

(where $w = 1 + \sin x$, $dw = \cos x\,dx$) $= \ln|w| + C = \ln|1 + \sin x| + C = \ln(1 + \sin x) + C.$

18. $\displaystyle\int \frac{dx}{1-\sin x} = \int \frac{1+\sin x}{1-\sin^2 x}\,dx = \int \frac{1+\sin x}{\cos^2 x}\,dx = \int \left(\sec^2 x + \sec x\tan x\right)dx = \tan x + \sec x + C$

19. $\int \tan^2 x\,dx = \int(\sec^2 x - 1)dx = \tan x - x + C.$

20. $\int \tan^4 x\,dx = \int \tan^2 x(\sec^2 x - 1)dx = \int \tan^2 x\sec^2 x\,dx - \int \tan^2 x\,dx = \frac{1}{3}\tan^3 x - \tan x + x + C$

(Set $u = \tan x$ in the first integral and use Exercise 25 for the second.)

21. $\int \sec^4 x\,dx = \int(\tan^2 x + 1)\sec^2 x\,dx = \int \tan^2 x\sec^2 x\,dx + \int \sec^2 x\,dx = \frac{1}{3}\tan^3 x + \tan x + C$

22. $\int \sec^6 x\,dx = \int\left(\tan^2 x + 1\right)^2\sec^2 x\,dx = \int \tan^4 x\sec^2 x\,dx + 2\int \tan^2 x\sec^2 x\,dx + \int \sec^2 x\,dx$
$= \frac{1}{5}\tan^5 x + \frac{2}{3}\tan^3 x + \tan x + C$ (Set $u = \tan x$ in the first two integrals.)

23. Let $u = \tan x \Rightarrow du = \sec^2 x\,dx$. Then $\int_0^{\pi/4}\tan^4 x\sec^2 x\,dx = \int_0^1 u^4\,du = \left[\frac{1}{5}u^5\right]_0^1 = \frac{1}{5}.$

24. Let $u = \tan x \Rightarrow du = \sec^2 x\,dx$. Then
$\int_0^{\pi/4}\tan^2 x\sec^4 x\,dx = \int_0^1 u^2(u^2+1)du = \int_0^1\left(u^4 + u^2\right)du = \left[\frac{1}{5}u^5 + \frac{1}{3}u^3\right]_0^1 = \frac{1}{5} + \frac{1}{3} = \frac{8}{15}.$

25. Let $u = \sec x \Rightarrow du = \sec x\tan x\,dx$. Then
$\int \tan x\sec^3 x\,dx = \int \sec^2 x\sec x\tan x\,dx = \int u^2\,du = \frac{1}{3}u^3 + C = \frac{1}{3}\sec^3 x + C.$

26. Let $u = \sec x \quad \Rightarrow \quad du = \sec x \tan x\,dx$. Then

$$\int \tan^3 x \sec^3 x\,dx = \int \sec^2 x \tan^2 x \sec x \tan x\,dx = \int u^2(u^2 - 1)du$$
$$= \int (u^4 - u^2)du = \tfrac{1}{5}u^5 - \tfrac{1}{3}u^3 + C = \tfrac{1}{5}\sec^5 x - \tfrac{1}{3}\sec^3 x + C.$$

27. $\int \tan^5 x\,dx = \int (\sec^2 x - 1)^2 \tan x\,dx = \int \sec^4 x \tan x\,dx - 2\int \sec^2 x \tan x\,dx + \int \tan x\,dx$

$$= \int \sec^3 x \sec x \tan x\,dx - 2\int \tan x \sec^2 x\,dx + \int \tan x\,dx$$
$$= \tfrac{1}{4}\sec^4 x - \tan^2 x + \ln|\sec x| + C. \text{ } \textit{Or: } \tfrac{1}{4}\sec^4 x - \sec^2 x + \ln|\sec x| + C.$$

28. $\int \tan^6 x\,dx = \int \tan^4 x(\sec^2 x - 1)dx = \int \tan^4 x \sec^2 x\,dx - \int \tan^4 x\,dx$

$$= \tfrac{1}{5}\tan^5 x - \int \tan^2 x(\sec^2 x - 1)dx = \tfrac{1}{5}\tan^5 x - \int \tan^2 x \sec^2 x\,dx + \int (\sec^2 x - 1)dx$$
$$= \tfrac{1}{5}\tan^5 x - \tfrac{1}{3}\tan^3 x + \tan x - x + C.$$ (Set $u = \tan x$ in the first two integrals.)

29. Let $u = \sec x \quad \Rightarrow \quad du = \sec x \tan x\,dx$. Then

$$\int_0^{\pi/3} \tan^5 x \sec x\,dx = \int_0^{\pi/3} (\sec^2 x - 1)^2 \sec x \tan x\,dx = \int_1^2 (u^2 - 1)^2\,du$$
$$= \int_1^2 (u^4 - 2u^2 + 1)du = \left[\tfrac{1}{5}u^5 - \tfrac{2}{3}u^3 + u\right]_1^2 = \left[\tfrac{32}{5} - \tfrac{16}{3} + 2\right] - \left[\tfrac{1}{5} - \tfrac{2}{3} + 1\right] = \tfrac{38}{15}.$$

30. Let $u = \sec x \quad \Rightarrow \quad du = \sec x \tan x\,dx$. Then

$$\int_0^{\pi/3} \tan^5 x \sec^3 x\,dx = \int_0^{\pi/3} (\sec^2 x - 1)^2 \sec^2 x \sec x \tan x\,dx = \int_1^2 (u^2 - 1)^2 u^2\,du$$
$$= \int_1^2 (u^6 - 2u^4 + u^2)du = \left[\tfrac{1}{7}u^7 - \tfrac{2}{5}u^5 + \tfrac{1}{3}u^3\right]_1^2 = \left[\tfrac{128}{7} - \tfrac{64}{5} + \tfrac{8}{3}\right] - \left[\tfrac{1}{7} - \tfrac{2}{5} + \tfrac{1}{3}\right] = \tfrac{848}{105}.$$

31. Let $u = \tan x \quad \Rightarrow \quad du = \sec^2 x\,dx$. Then

$$\int \frac{\sec^2 x}{\cot x}\,dx = \int \tan x \sec^2 x\,dx = \int u\,du = \tfrac{1}{2}u^2 + C = \tfrac{1}{2}\tan^2 x + C.$$

32. $\int \tan^2 x \sec x\,dx = \int (\sec^2 x - 1)\sec x\,dx = \int \sec^3 x\,dx - \int \sec x\,dx$

$$= \tfrac{1}{2}(\sec x \tan x + \ln|\sec x + \tan x|) - \ln|\sec x + \tan x| + C$$ (by Examples 9 and 8)
$$= \tfrac{1}{2}(\sec x \tan x - \ln|\sec x + \tan x|) + C$$

33. $\int_{\pi/6}^{\pi/2} \cot^2 x\,dx = \int_{\pi/6}^{\pi/2} (\csc^2 x - 1)dx = [-\cot x - x]_{\pi/6}^{\pi/2} = \left(0 - \tfrac{\pi}{2}\right) - \left(-\sqrt{3} - \tfrac{\pi}{6}\right) = \sqrt{3} - \tfrac{\pi}{3}$

34. $\displaystyle\int_{\pi/4}^{\pi/2} \cot^3 x\,dx = \int_{\pi/4}^{\pi/2} \cot x(\csc^2 x - 1)dx = \int_{\pi/4}^{\pi/2} \cot x \csc^2 x\,dx - \int_{\pi/4}^{\pi/2} \frac{\cos x}{\sin x}\,dx$

$$= \left[-\tfrac{1}{2}\cot^2 x - \ln|\sin x|\right]_{\pi/4}^{\pi/2} = (0 - \ln 1) - \left[-\tfrac{1}{2} - \ln \tfrac{1}{\sqrt{2}}\right] = \tfrac{1}{2} + \ln \tfrac{1}{\sqrt{2}} = \tfrac{1}{2}(1 - \ln 2)$$

35. Let $u = \cot x \quad \Rightarrow \quad du = -\csc^2 x\,dx$. Then

$$\int \cot^4 x \csc^4 x\,dx = \int u^4(u^2 + 1)(-du) = -\int (u^6 + u^4)du = -\tfrac{1}{7}u^7 - \tfrac{1}{5}u^5 + C = -\tfrac{1}{7}\cot^7 x - \tfrac{1}{5}\cot^5 x + C.$$

36. Let $u = \cot x \quad \Rightarrow \quad du = -\csc^2 x\,dx$. Then $\int \cot^3 x \csc^4 x\,dx = \int \cot^3 x(\cot^2 x + 1)\csc^2 x\,dx$

$$= \int u^3(u^2 + 1)(-du) = -\tfrac{1}{6}u^6 - \tfrac{1}{4}u^4 + C = -\tfrac{1}{6}\cot^6 x - \tfrac{1}{4}\cot^4 x + C.$$

37. $\displaystyle I = \int \csc x\,dx = \int \frac{\csc x(\csc x - \cot x)}{\csc x - \cot x}\,dx = \int \frac{-\csc x \cot x + \csc^2 x}{\csc x - \cot x}\,dx$. Let $u = \csc x - \cot x \quad \Rightarrow$

$du = (-\csc x \cot x + \csc^2 x)dx$. Then $I = \int du/u = \ln|u| = \ln|\csc x - \cot x| + C$.

38. Let $u = \csc x$, $dv = \csc^2 x\, dx$. Then $du = -\csc x \cot x\, dx$, $v = -\cot x \quad \Rightarrow$

$$\int \csc^3 x\, dx = -\csc x \cot x - \int \csc x \cot^2 x\, dx = -\csc x \cot x - \int \csc x(\csc^2 x - 1)dx$$
$$= -\csc x \cot x + \int \csc x\, dx - \int \csc^3 x\, dx.$$

Solving for $\int \csc^3 x\, dx$ and using Exercise 45, we get

$$\int \csc^3 x\, dx = -\tfrac{1}{2}\csc x \cot x + \tfrac{1}{2}\int \csc x\, dx = -\tfrac{1}{2}\csc x \cot x + \tfrac{1}{2}\ln|\csc x - \cot x| + C.$$

39. $\int \sin 5x \sin 2x\, dx = \int \frac{1}{2}[\cos(5x - 2x) - \cos(5x + 2x)]dx = \frac{1}{2}\int(\cos 3x - \cos 7x)dx$
$= \frac{1}{6}\sin 3x - \frac{1}{14}\sin 7x + C$

40. $\int \sin 3x \cos x\, dx = \int \frac{1}{2}[\sin(3x + x) + \sin(3x - x)]dx = \frac{1}{2}\int(\sin 4x + \sin 2x)dx = -\frac{1}{8}\cos 4x - \frac{1}{4}\cos 2x + C$

41. $\int \cos 3x \cos 4x\, dx = \int \frac{1}{2}[\cos(3x - 4x) + \cos(3x + 4x)]dx = \frac{1}{2}\int(\cos x + \cos 7x)dx = \frac{1}{2}\sin x + \frac{1}{14}\sin 7x + C$

42. $\int \sin 3x \sin 6x\, dx = \int \frac{1}{2}[\cos(3x - 6x) - \cos(3x + 6x)]dx = \frac{1}{2}\int(\cos 3x - \cos 9x)dx$
$= \frac{1}{6}\sin 3x - \frac{1}{18}\sin 9x + C$

43. $\displaystyle\int \frac{1 - \tan^2 x}{\sec^2 x}dx = \int(\cos^2 x - \sin^2 x)dx = \int \cos 2x\, dx = \tfrac{1}{2}\sin 2x + C$

44. $\displaystyle\int \frac{\cos x + \sin x}{\sin 2x}dx = \frac{1}{2}\int \frac{\cos x + \sin x}{\sin x \cos x}dx = \frac{1}{2}\int(\csc x + \sec x)dx$
$= \frac{1}{2}(\ln|\csc x - \cot x| + \ln|\sec x + \tan x|) + C$

45. Let $u = \cos x \quad \Rightarrow \quad du = -\sin x\, dx$. Then

$$\int \sin^5 x\, dx = \int(1 - \cos^2 x)^2 \sin x\, dx = \int(1 - u^2)^2(-du)$$
$$= \int(-1 + 2u^2 - u^4)du = -\tfrac{1}{5}u^5 + \tfrac{2}{3}u^3 - u + C$$
$$= -\tfrac{1}{5}\cos^5 x + \tfrac{2}{3}\cos^3 x - \cos x + C.$$

Notice that F is increasing when $f(x) > 0$, so the graphs serve as a check on our work.

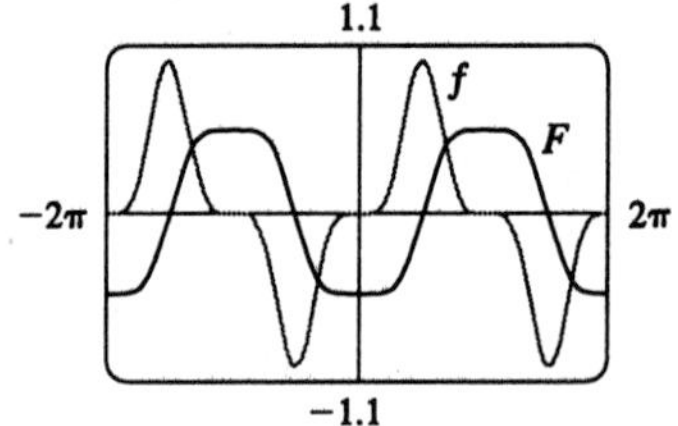

46.

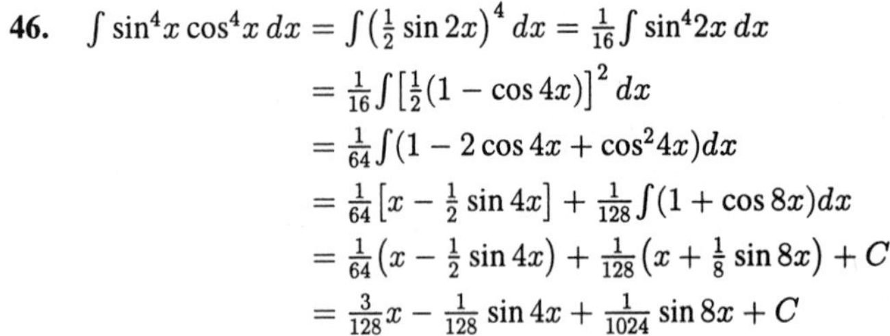

$$\int \sin^4 x \cos^4 x\, dx = \int\left(\tfrac{1}{2}\sin 2x\right)^4 dx = \tfrac{1}{16}\int \sin^4 2x\, dx$$
$$= \tfrac{1}{16}\int\left[\tfrac{1}{2}(1 - \cos 4x)\right]^2 dx$$
$$= \tfrac{1}{64}\int(1 - 2\cos 4x + \cos^2 4x)dx$$
$$= \tfrac{1}{64}\left[x - \tfrac{1}{2}\sin 4x\right] + \tfrac{1}{128}\int(1 + \cos 8x)dx$$
$$= \tfrac{1}{64}\left(x - \tfrac{1}{2}\sin 4x\right) + \tfrac{1}{128}\left(x + \tfrac{1}{8}\sin 8x\right) + C$$
$$= \tfrac{3}{128}x - \tfrac{1}{128}\sin 4x + \tfrac{1}{1024}\sin 8x + C$$

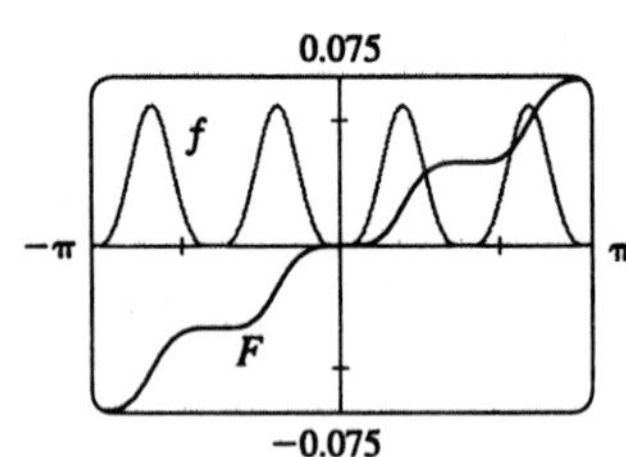

Notice that $f(x) = 0$ whenever F has a horizontal tangent.

47. $f_{\text{ave}} = \frac{1}{2\pi}\int_{-\pi}^{\pi}\sin^2 x \cos^3 x\, dx = \frac{1}{2\pi}\int_{-\pi}^{\pi}\sin^2 x(1 - \sin^2 x)\cos x\, dx = \frac{1}{2\pi}\int_0^0 u^2(1 - u^2)du$ (where $u = \sin x$) $= 0$

48. (a) Let $u = \cos x$. Then $du = -\sin x\, dx \Rightarrow \int \sin x \cos x\, dx = \int u(-du) = -\frac{1}{2}u^2 + C = -\frac{1}{2}\cos^2 x + C_1$.

(b) Let $u = \sin x$. Then $du = \cos x\, dx \Rightarrow \int \sin x \cos x\, dx = \int u\, du = \frac{1}{2}u^2 + C = \frac{1}{2}\sin^2 x + C_2$.

(c) $\int \sin x \cos x\, dx = \int \frac{1}{2}\sin 2x\, dx = -\frac{1}{4}\cos 2x + C_3$

(d) Let $u = \sin x$, $dv = \cos x\,dx$. Then $du = \cos x\,dx$, $v = \sin x$, so
$\int \sin x \cos x\,dx = \sin^2 x - \int \sin x \cos x\,dx$, by Equation 2, so $\int \sin x \cos x\,dx = \frac{1}{2}\sin^2 x + C_4$.

The answers differ from one another by constants.

Since $\cos 2x = 1 - 2\sin^2 x = 2\cos^2 x - 1$, we find that $-\frac{1}{4}\cos 2x = \frac{1}{2}\sin^2 x - \frac{1}{4} = -\frac{1}{2}\cos^2 x + \frac{1}{4}$.

49. For $0 < x < \frac{\pi}{2}$, we have $0 < \sin x < 1$, so $\sin^3 x < \sin x$. Hence the area is
$\int_0^{\pi/2}(\sin x - \sin^3 x)dx = \int_0^{\pi/2} \sin x(1 - \sin^2 x)dx = \int_0^{\pi/2} \cos^2 x \sin x\,dx$. Now let $u = \cos x \Rightarrow$
$du = -\sin x\,dx$. Then area $= \int_1^0 u^2(-du) = \int_0^1 u^2\,du = \left[\frac{1}{3}u^3\right]_0^1 = \frac{1}{3}$.

50. $\sin x > 0$ for $0 < x < \frac{\pi}{2}$, so the sign of $2\sin^2 x - \sin x$ $\left[\text{which equals } 2\sin x\left(\sin x - \frac{1}{2}\right)\right]$ is the same as that of $\sin x - \frac{1}{2}$. Thus $2\sin^2 x - \sin x$ is positive on $\left(\frac{\pi}{6}, \frac{\pi}{2}\right)$ and negative on $\left(0, \frac{\pi}{6}\right)$. The desired area is

$$\int_0^{\pi/6}(\sin x - 2\sin^2 x)dx + \int_{\pi/6}^{\pi/2}(2\sin^2 x - \sin x)dx = \int_0^{\pi/6}(\sin x - 1 + \cos 2x)dx + \int_{\pi/6}^{\pi/2}(1 - \cos 2x - \sin x)dx$$

$$= \left[-\cos x - x + \tfrac{1}{2}\sin 2x\right]_0^{\pi/6} + \left[x - \tfrac{1}{2}\sin 2x + \cos x\right]_{\pi/6}^{\pi/2}$$

$$= -\tfrac{\sqrt{3}}{2} - \tfrac{\pi}{6} + \tfrac{\sqrt{3}}{4} - (-1) + \tfrac{\pi}{2} - \left(\tfrac{\pi}{6} - \tfrac{\sqrt{3}}{4} + \tfrac{\sqrt{3}}{2}\right) = 1 + \tfrac{\pi}{6} - \tfrac{\sqrt{3}}{2}.$$

51.

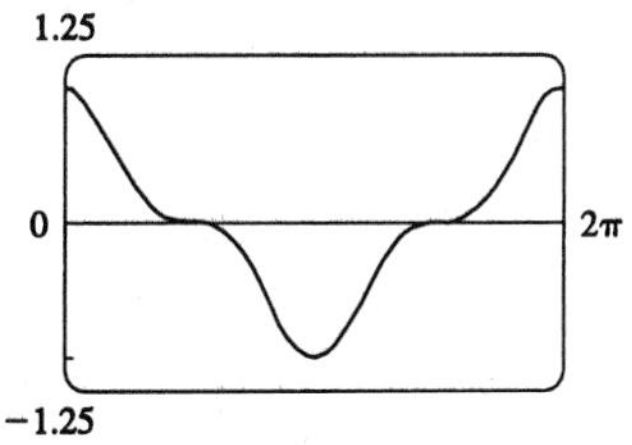

It seems from the graph that $\int_0^{2\pi} \cos^3 x\,dx = 0$, since the area below the x-axis and above the graph looks about equal to the area above the axis and below the graph. By Example 1, the integral is $\left[\sin x - \frac{1}{3}\sin^3 x\right]_0^{2\pi} = 0$. Note that due to symmetry, the integral of any odd power of $\sin x$ or $\cos x$ between limits which differ by $2n\pi$ (n any integer) is 0.

52.

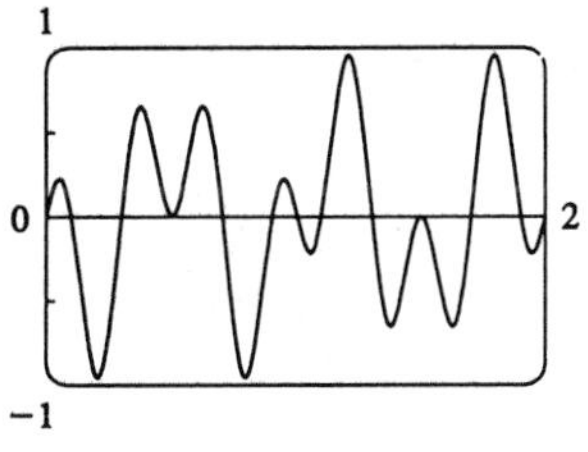

It seems from the graph that $\int_0^2 \sin 2\pi x \cos 5\pi x\,dx = 0$, since each bulge above the x-axis seems to have a corresponding depression below the x-axis. To evaluate the integral, we use a trigonometric identity:

$$\int_0^1 \sin 2\pi x \cos 5\pi x\,dx = \tfrac{1}{2}\int_0^2[\sin(2\pi x - 5\pi x) + \sin(2\pi x + 5\pi x)]dx$$

$$= \tfrac{1}{2}\int_0^2[\sin(-3\pi x) + \sin 7\pi x]dx = \tfrac{1}{2}\left[\tfrac{1}{3\pi}\cos(-3\pi x) - \tfrac{1}{7\pi}\cos 7\pi x\right]_0^2$$

$$= \tfrac{1}{2}\left[\tfrac{1}{3\pi}(1 - 1) - \tfrac{1}{8\pi}(1 - 1)\right] = 0.$$

53. $V = \int_{\pi/2}^{\pi} \pi \sin^2 x\,dx = \pi\int_{\pi/2}^{\pi} \frac{1}{2}(1 - \cos 2x)dx = \pi\left[\frac{1}{2}x - \frac{1}{4}\sin 2x\right]_{\pi/2}^{\pi} = \pi\left(\frac{\pi}{2} - 0 - \frac{\pi}{4} + 0\right) = \frac{\pi^2}{4}$

54. Volume $= \int_0^{\pi/4} \pi(\tan^2 x)^2\,dx = \pi\int_0^{\pi/4}\tan^2 x(\sec^2 x - 1)dx = \pi\int_0^{\pi/4}\tan^2 x \sec^2 x\,dx - \pi\int_0^{\pi/4}\tan^2 x\,dx$

$$= \pi\int_0^{\pi/4} u^2\,du - \pi\int_0^{\pi/4}(\sec^2 x - 1)dx \quad [\text{where } u = \tan x, \text{ and } du = \sec^2 x\,dx]$$

$$= \pi\left[\tfrac{1}{3}u^3\right]_0^{\pi/4} - \pi[\tan x - x]_0^{\pi/4} = \pi\left[\tfrac{1}{3}\tan^3 x - \tan x + x\right]_0^{\pi/4} = \pi\left[\tfrac{1}{3} - 1 + \tfrac{\pi}{4}\right] = \pi\left(\tfrac{\pi}{4} - \tfrac{2}{3}\right)$$

55. Volume $= \pi\int_0^{\pi/2}\left[(1 + \cos x)^2 - 1^2\right]dx = \pi\int_0^{\pi/2}(2\cos x + \cos^2 x)dx$

$$= \pi\left[2\sin x + \tfrac{1}{2}x + \tfrac{1}{4}\sin 2x\right]_0^{\pi/2} = \pi\left(2 + \tfrac{\pi}{4}\right) = 2\pi + \tfrac{\pi^2}{4}$$

(d) Let $u = \sin x$, $dv = \cos x\,dx$. Then $du = \cos x\,dx$, $v = \sin x$, so
$\int \sin x \cos x\,dx = \sin^2 x - \int \sin x \cos x\,dx$, by Equation 2, so $\int \sin x \cos x\,dx = \frac{1}{2}\sin^2 x + C_4$.

The answers differ from one another by constants.

Since $\cos 2x = 1 - 2\sin^2 x = 2\cos^2 x - 1$, we find that $-\frac{1}{4}\cos 2x = \frac{1}{2}\sin^2 x - \frac{1}{4} = -\frac{1}{2}\cos^2 x + \frac{1}{4}$.

49. For $0 < x < \frac{\pi}{2}$, we have $0 < \sin x < 1$, so $\sin^3 x < \sin x$. Hence the area is
$\int_0^{\pi/2}(\sin x - \sin^3 x)dx = \int_0^{\pi/2}\sin x(1 - \sin^2 x)dx = \int_0^{\pi/2}\cos^2 x \sin x\,dx$. Now let $u = \cos x \Rightarrow$
$du = -\sin x\,dx$. Then area $= \int_1^0 u^2(-du) = \int_0^1 u^2\,du = \left[\frac{1}{3}u^3\right]_0^1 = \frac{1}{3}$.

50. $\sin x > 0$ for $0 < x < \frac{\pi}{2}$, so the sign of $2\sin^2 x - \sin x$ $\left[\text{which equals } 2\sin x\left(\sin x - \frac{1}{2}\right)\right]$ is the same as that of $\sin x - \frac{1}{2}$. Thus $2\sin^2 x - \sin x$ is positive on $\left(\frac{\pi}{6}, \frac{\pi}{2}\right)$ and negative on $\left(0, \frac{\pi}{6}\right)$. The desired area is

$$\int_0^{\pi/6}(\sin x - 2\sin^2 x)dx + \int_{\pi/6}^{\pi/2}(2\sin^2 x - \sin x)dx = \int_0^{\pi/6}(\sin x - 1 + \cos 2x)dx + \int_{\pi/6}^{\pi/2}(1 - \cos 2x - \sin x)dx$$
$$= \left[-\cos x - x + \tfrac{1}{2}\sin 2x\right]_0^{\pi/6} + \left[x - \tfrac{1}{2}\sin 2x + \cos x\right]_{\pi/6}^{\pi/2}$$
$$= -\tfrac{\sqrt{3}}{2} - \tfrac{\pi}{6} + \tfrac{\sqrt{3}}{4} - (-1) + \tfrac{\pi}{2} - \left(\tfrac{\pi}{6} - \tfrac{\sqrt{3}}{4} + \tfrac{\sqrt{3}}{2}\right) = 1 + \tfrac{\pi}{6} - \tfrac{\sqrt{3}}{2}.$$

51.

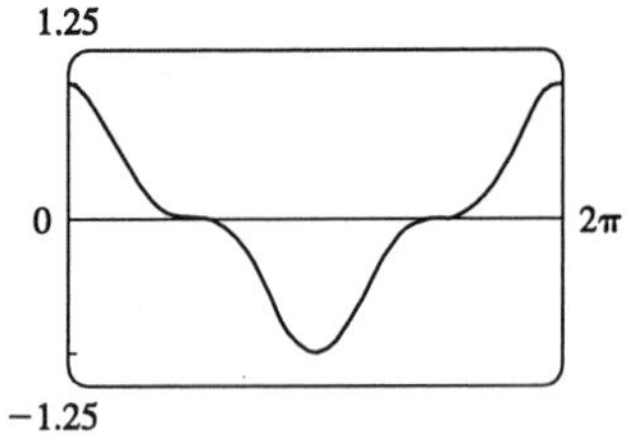

1.25
0
2π
−1.25

It seems from the graph that $\int_0^{2\pi}\cos^3 x\,dx = 0$, since the area below the x-axis and above the graph looks about equal to the area above the axis and below the graph. By Example 1, the integral is $\left[\sin x - \frac{1}{3}\sin^3 x\right]_0^{2\pi} = 0$. Note that due to symmetry, the integral of any odd power of $\sin x$ or $\cos x$ between limits which differ by $2n\pi$ (n any integer) is 0.

52.

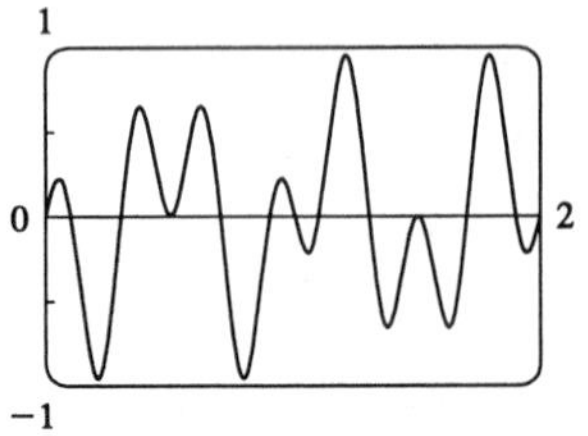

1
0
2
−1

It seems from the graph that $\int_0^2 \sin 2\pi x \cos 5\pi x\,dx = 0$, since each bulge above the x-axis seems to have a corresponding depression below the x-axis. To evaluate the integral, we use a trigonometric identity:

$$\int_0^1 \sin 2\pi x \cos 5\pi x\,dx = \tfrac{1}{2}\int_0^2[\sin(2\pi x - 5\pi x) + \sin(2\pi x + 5\pi x)]dx$$
$$= \tfrac{1}{2}\int_0^2[\sin(-3\pi x) + \sin 7\pi x]dx = \tfrac{1}{2}\left[\tfrac{1}{3\pi}\cos(-3\pi x) - \tfrac{1}{7\pi}\cos 7\pi x\right]_0^2$$
$$= \tfrac{1}{2}\left[\tfrac{1}{3\pi}(1 - 1) - \tfrac{1}{8\pi}(1 - 1)\right] = 0.$$

53. $V = \int_{\pi/2}^{\pi}\pi\sin^2 x\,dx = \pi\int_{\pi/2}^{\pi}\frac{1}{2}(1 - \cos 2x)dx = \pi\left[\frac{1}{2}x - \frac{1}{4}\sin 2x\right]_{\pi/2}^{\pi} = \pi\left(\frac{\pi}{2} - 0 - \frac{\pi}{4} + 0\right) = \frac{\pi^2}{4}$

54. Volume $= \int_0^{\pi/4}\pi(\tan^2 x)^2\,dx = \pi\int_0^{\pi/4}\tan^2 x(\sec^2 x - 1)dx = \pi\int_0^{\pi/4}\tan^2 x\sec^2 x\,dx - \pi\int_0^{\pi/4}\tan^2 x\,dx$
$= \pi\int_0^{\pi/4}u^2\,du - \pi\int_0^{\pi/4}(\sec^2 x - 1)dx$ [where $u = \tan x$, and $du = \sec^2 x\,dx$]
$= \pi\left[\frac{1}{3}u^3\right]_0^{\pi/4} - \pi[\tan x - x]_0^{\pi/4} = \pi\left[\frac{1}{3}\tan^3 x - \tan x + x\right]_0^{\pi/4} = \pi\left[\frac{1}{3} - 1 + \frac{\pi}{4}\right] = \pi\left(\frac{\pi}{4} - \frac{2}{3}\right)$

55. Volume $= \pi\int_0^{\pi/2}\left[(1 + \cos x)^2 - 1^2\right]dx = \pi\int_0^{\pi/2}(2\cos x + \cos^2 x)dx$
$= \pi\left[2\sin x + \frac{1}{2}x + \frac{1}{4}\sin 2x\right]_0^{\pi/2} = \pi\left(2 + \frac{\pi}{4}\right) = 2\pi + \frac{\pi^2}{4}$

EXERCISES 7.3

1. Let $x = \sin\theta$, where $-\frac{\pi}{2} \le \theta \le \frac{\pi}{2}$. Then $dx = \cos\theta\, d\theta$ and $\sqrt{1-x^2} = |\cos\theta| = \cos\theta$ (since $\cos\theta > 0$ for θ in $\left[-\frac{\pi}{2}, \frac{\pi}{2}\right]$). Thus

$$\int_{1/2}^{\sqrt{3}/2} \frac{dx}{x^2\sqrt{1-x^2}} = \int_{\pi/6}^{\pi/3} \frac{\cos\theta\, d\theta}{\sin^2\theta\cos\theta} = \int_{\pi/6}^{\pi/3} \csc^2\theta\, d\theta = [-\cot\theta]_{\pi/6}^{\pi/3}$$
$$= -\frac{1}{\sqrt{3}} - \left(-\sqrt{3}\right) = \frac{3}{\sqrt{3}} - \frac{1}{\sqrt{3}} = \frac{2}{\sqrt{3}}.$$

2. Let $x = 2\sin\theta$, $-\frac{\pi}{2} \le \theta \le \frac{\pi}{2}$. Then $dx = 2\cos\theta\, d\theta$ and $\sqrt{4-x^2} = |2\cos\theta| = 2\cos\theta$, so

$$\int_0^2 x^3\sqrt{4-x^2}\, dx = \int_0^{\pi/2} 8\sin^3\theta(2\cos\theta)(2\cos\theta)d\theta$$
$$= 32\int_0^{\pi/2} \cos^2\theta(1-\cos^2\theta)\sin\theta\, d\theta = 32\int_1^0 u^2(1-u^2)(-du) \text{ (where } u = \cos\theta)$$
$$= 32\int_0^1 (u^2-u^4)du = 32\left[\tfrac{1}{3}u^3 - \tfrac{1}{5}u^5\right]_0^1 = 32\left(\tfrac{1}{3}-\tfrac{1}{5}\right) = \tfrac{64}{15}.$$

3. Let $u = 1-x^2$. Then $du = -2x\, dx$, so $\displaystyle\int \frac{x}{\sqrt{1-x^2}}dx = -\frac{1}{2}\int\frac{du}{\sqrt{u}} = -\sqrt{u} + C = -\sqrt{1-x^2} + C.$

4. Let $u = 4-x^2$. Then $du = -2x\, dx \quad\Rightarrow$

$$\int x\sqrt{4-x^2}\, dx = -\tfrac{1}{2}\int\sqrt{u}\, du = -\tfrac{1}{2}\cdot\tfrac{2}{3}u^{3/2} + C = -\tfrac{1}{3}(4-x^2)^{3/2} + C.$$

5. Let $2x = \sin\theta$, where $-\frac{\pi}{2} \le \theta \le \frac{\pi}{2}$. Then $x = \frac{1}{2}\sin\theta$, $dx = \frac{1}{2}\cos\theta\, d\theta$, and $\sqrt{1-4x^2} = \sqrt{1-(2x)^2} = \cos\theta$.

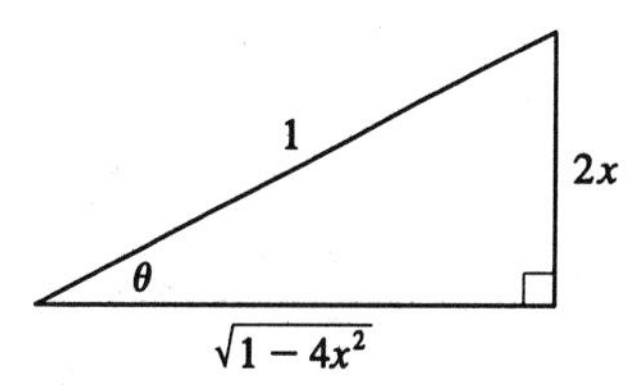

$$\int\sqrt{1-4x^2}\, dx = \int\cos\theta\left(\tfrac{1}{2}\cos\theta\right)d\theta = \tfrac{1}{4}\int(1+\cos 2\theta)d\theta$$
$$= \tfrac{1}{4}\left(\theta + \tfrac{1}{2}\sin 2\theta\right) + C = \tfrac{1}{4}(\theta + \sin\theta\cos\theta) + C$$
$$= \tfrac{1}{4}\left[\sin^{-1}(2x) + 2x\sqrt{1-4x^2}\right] + C$$

6. Let $x = 2\tan\theta$, where $-\frac{\pi}{2} < \theta < \frac{\pi}{2}$. Then $dx = 2\sec^2\theta\, d\theta$ and $\sqrt{x^2+4} = 2\sec\theta$, so

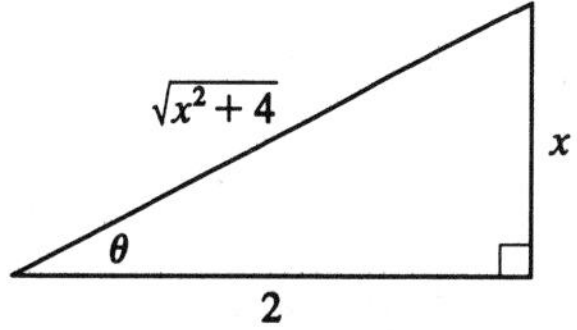

$$\int_0^2 \frac{x^3}{\sqrt{x^2+4}}dx = \int_0^{\pi/4}\frac{8\tan^3\theta}{2\sec\theta}2\sec^2\theta\, d\theta$$
$$= 8\int_0^{\pi/4}(\sec^2\theta - 1)\sec\theta\tan\theta\, d\theta = 8\left[\tfrac{1}{3}\sec^3\theta - \sec\theta\right]_0^{\pi/4}$$
$$= 8\left[\tfrac{1}{3}\cdot 2\sqrt{2} - \sqrt{2}\right] - 8\left[\tfrac{1}{3} - 1\right] = \tfrac{8}{3}\left(2-\sqrt{2}\right).$$

7. Let $x = 3\tan\theta$, where $-\frac{\pi}{2} < \theta < \frac{\pi}{2}$. Then $dx = 3\sec^2\theta\, d\theta$ and $\sqrt{9+x^2} = 3\sec\theta$.

$$\int_0^3 \frac{dx}{\sqrt{9+x^2}} = \int_0^{\pi/4}\frac{3\sec^2\theta\, d\theta}{3\sec\theta} = \int_0^{\pi/4}\sec\theta\, d\theta = [\ln|\sec\theta + \tan\theta|]_0^{\pi/4} = \ln\left(\sqrt{2}+1\right) - \ln 1 = \ln\left(\sqrt{2}+1\right)$$

8. Let $x = \tan\theta$, where $-\frac{\pi}{2} < \theta < \frac{\pi}{2}$. Then $dx = \sec^2\theta\,d\theta$, $\sqrt{x^2+1} = \sec\theta$ and $x = 0 \Rightarrow \theta = 0$, $x = 1 \Rightarrow \theta = \frac{\pi}{4}$, so

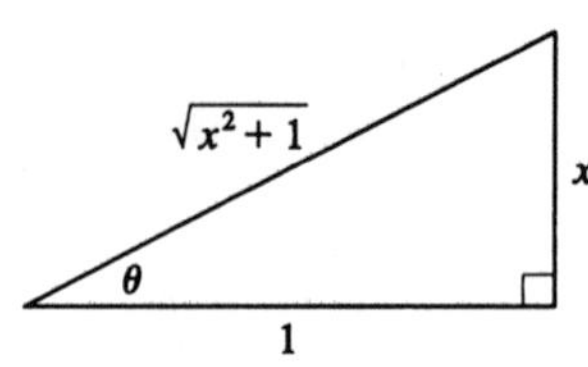

$$\int_0^1 \sqrt{x^2+1}\,dx = \int_0^{\pi/4} \sec\theta\sec^2\theta\,d\theta = \int_0^{\pi/4}\sec^3\theta\,d\theta$$
$$= \tfrac{1}{2}\left[\sec\theta\tan\theta + \ln|\sec\theta+\tan\theta|\right]_0^{\pi/4} \quad \text{(by Example 7.2.9)}$$
$$= \tfrac{1}{2}\left[\sqrt{2}\cdot 1 + \ln\left(1+\sqrt{2}\right) - 0 - \ln(1+0)\right]$$
$$= \tfrac{1}{2}\left[\sqrt{2} + \ln\left(1+\sqrt{2}\right)\right].$$

9. Let $x = 4\sec\theta$, where $0 \le \theta < \frac{\pi}{2}$ or $\pi \le \theta < \frac{3\pi}{2}$. Then $dx = 4\sec\theta\tan\theta\,d\theta$ and $\sqrt{x^2-16} = 4|\tan\theta| = 4\tan\theta$. Thus

$$\int \frac{dx}{x^3\sqrt{x^2-16}} = \int \frac{4\sec\theta\tan\theta\,d\theta}{64\sec^3\theta\cdot 4\tan\theta} = \tfrac{1}{64}\int\cos^2\theta\,d\theta = \tfrac{1}{128}\int(1+\cos 2\theta)d\theta$$
$$= \tfrac{1}{128}\left(\theta + \tfrac{1}{2}\sin 2\theta\right) + C = \tfrac{1}{128}(\theta + \sin\theta\cos\theta) + C = \tfrac{1}{128}\left(\sec^{-1}\frac{x}{4} + \frac{4\sqrt{x^2-16}}{x^2}\right) + C$$

by the diagrams for $0 \le \theta < \frac{\pi}{2}$ and $\pi \le \theta < \frac{3\pi}{2}$, where the labels of the legs in the second diagram indicate the x-and y-coordinates of P rather than the lengths of those sides. Henceforth we omit the second diagram from our solutions.

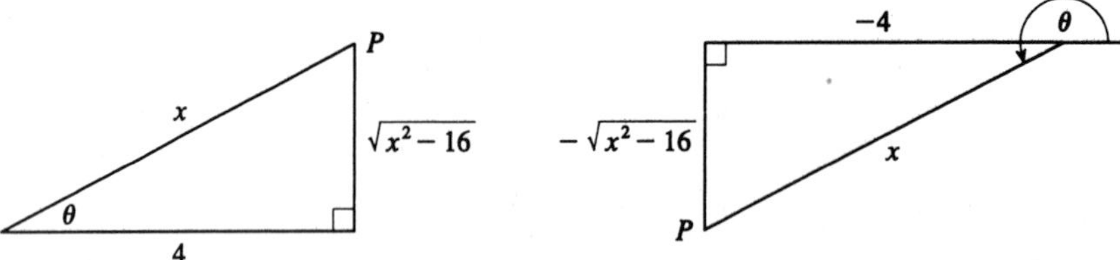

10. Let $x = a\sec\theta$, where $0 \le \theta < \frac{\pi}{2}$ or $\pi \le \theta < \frac{3\pi}{2}$. Then $dx = a\sec\theta\tan\theta\,d\theta$ and $\sqrt{x^2-a^2} = a\tan\theta$, so

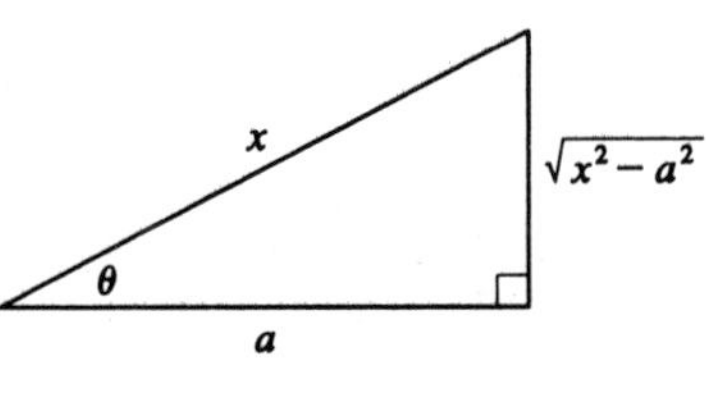

$$\int \frac{\sqrt{x^2-a^2}}{x^4}\,dx = \int \frac{a\tan\theta}{a^4\sec^4\theta}\,a\sec\theta\tan\theta\,d\theta$$
$$= \frac{1}{a^2}\int \sin^2\theta\cos\theta\,d\theta = \frac{1}{3a^2}\sin^3\theta + C = \frac{(x^2-a^2)^{3/2}}{3a^2x^3} + C$$

11. $9x^2 - 4 = (3x)^2 - 4$, so let $3x = 2\sec\theta$, where $0 \le \theta < \frac{\pi}{2}$ or $\pi \le \theta < \frac{3\pi}{2}$. Then $dx = \frac{2}{3}\sec\theta\tan\theta\,d\theta$ and $\sqrt{9x^2-4} = 2\tan\theta$.

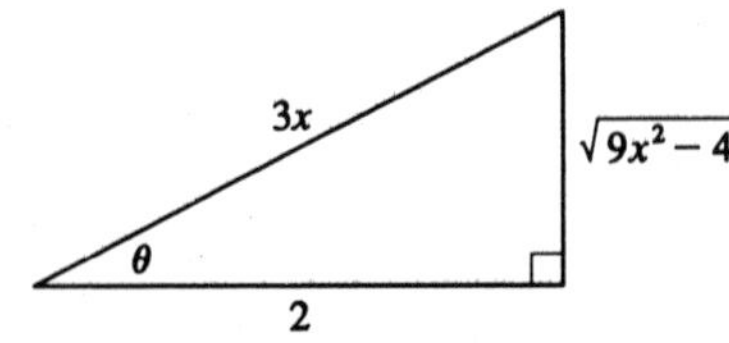

$$\int \frac{\sqrt{9x^2-4}}{x}\,dx = \int \frac{2\tan\theta}{\frac{2}{3}\sec\theta}\cdot\tfrac{2}{3}\sec\theta\tan\theta\,d\theta$$
$$= 2\int\tan^2\theta\,d\theta = 2\int(\sec^2\theta - 1)d\theta = 2(\tan\theta - \theta) + C$$
$$= \sqrt{9x^2-4} - 2\sec^{-1}\left(\frac{3x}{2}\right) + C$$

8. Let $x = \tan\theta$, where $-\frac{\pi}{2} < \theta < \frac{\pi}{2}$. Then $dx = \sec^2\theta\, d\theta$, $\sqrt{x^2+1} = \sec\theta$ and $x = 0 \;\Rightarrow\; \theta = 0$, $x = 1 \;\Rightarrow\; \theta = \frac{\pi}{4}$, so

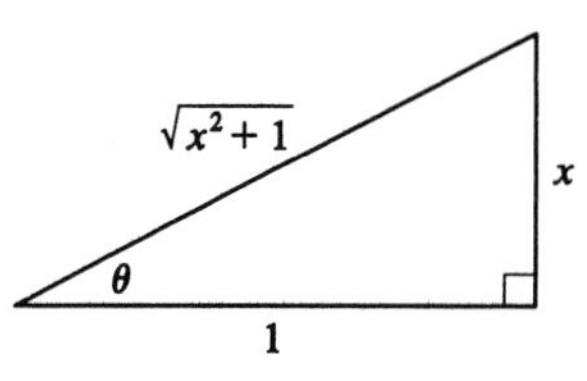

$$\begin{aligned}\int_0^1 \sqrt{x^2+1}\,dx &= \int_0^{\pi/4} \sec\theta\sec^2\theta\,d\theta = \int_0^{\pi/4}\sec^3\theta\,d\theta \\ &= \tfrac{1}{2}\left[\sec\theta\tan\theta + \ln|\sec\theta+\tan\theta|\right]_0^{\pi/4} \quad \text{(by Example 7.2.9)} \\ &= \tfrac{1}{2}\left[\sqrt{2}\cdot 1 + \ln\left(1+\sqrt{2}\right) - 0 - \ln(1+0)\right] \\ &= \tfrac{1}{2}\left[\sqrt{2} + \ln\left(1+\sqrt{2}\right)\right].\end{aligned}$$

9. Let $x = 4\sec\theta$, where $0 \le \theta < \frac{\pi}{2}$ or $\pi \le \theta < \frac{3\pi}{2}$. Then $dx = 4\sec\theta\tan\theta\,d\theta$ and $\sqrt{x^2-16} = 4|\tan\theta| = 4\tan\theta$. Thus

$$\begin{aligned}\int \frac{dx}{x^3\sqrt{x^2-16}} &= \int \frac{4\sec\theta\tan\theta\,d\theta}{64\sec^3\theta\cdot 4\tan\theta} = \tfrac{1}{64}\int\cos^2\theta\,d\theta = \tfrac{1}{128}\int(1+\cos 2\theta)d\theta \\ &= \tfrac{1}{128}\left(\theta + \tfrac{1}{2}\sin 2\theta\right) + C = \tfrac{1}{128}(\theta + \sin\theta\cos\theta) + C = \tfrac{1}{128}\left(\sec^{-1}\frac{x}{4} + \frac{4\sqrt{x^2-16}}{x^2}\right) + C\end{aligned}$$

by the diagrams for $0 \le \theta < \frac{\pi}{2}$ and $\pi \le \theta < \frac{3\pi}{2}$, where the labels of the legs in the second diagram indicate the x-and y-coordinates of P rather than the lengths of those sides. Henceforth we omit the second diagram from our solutions.

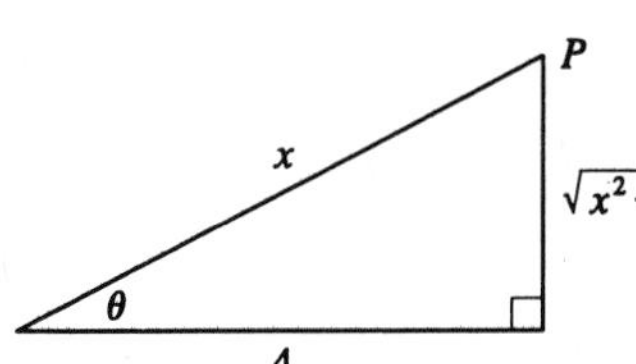

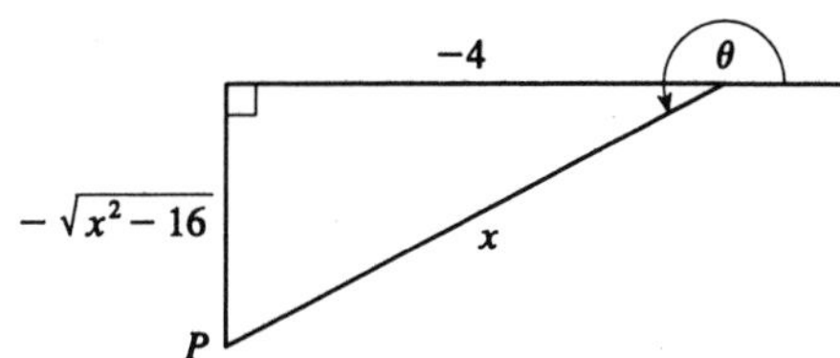

10. Let $x = a\sec\theta$, where $0 \le \theta < \frac{\pi}{2}$ or $\pi \le \theta < \frac{3\pi}{2}$. Then $dx = a\sec\theta\tan\theta\,d\theta$ and $\sqrt{x^2-a^2} = a\tan\theta$, so

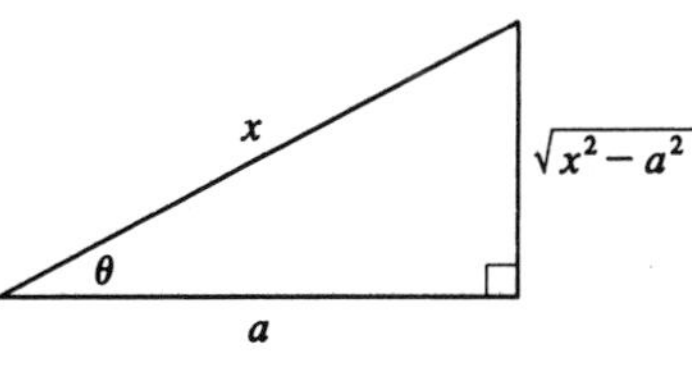

$$\begin{aligned}\int \frac{\sqrt{x^2-a^2}}{x^4}\,dx &= \int \frac{a\tan\theta}{a^4\sec^4\theta}\,a\sec\theta\tan\theta\,d\theta \\ &= \frac{1}{a^2}\int \sin^2\theta\cos\theta\,d\theta = \frac{1}{3a^2}\sin^3\theta + C = \frac{(x^2-a^2)^{3/2}}{3a^2x^3} + C\end{aligned}$$

11. $9x^2 - 4 = (3x)^2 - 4$, so let $3x = 2\sec\theta$, where $0 \le \theta < \frac{\pi}{2}$ or $\pi \le \theta < \frac{3\pi}{2}$. Then $dx = \frac{2}{3}\sec\theta\tan\theta\,d\theta$ and $\sqrt{9x^2-4} = 2\tan\theta$.

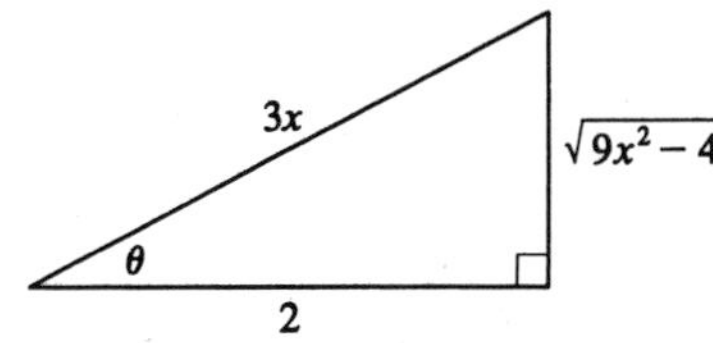

$$\begin{aligned}\int \frac{\sqrt{9x^2-4}}{x}\,dx &= \int \frac{2\tan\theta}{\frac{2}{3}\sec\theta}\cdot\tfrac{2}{3}\sec\theta\tan\theta\,d\theta \\ &= 2\int\tan^2\theta\,d\theta = 2\int(\sec^2\theta - 1)d\theta = 2(\tan\theta - \theta) + C \\ &= \sqrt{9x^2-4} - 2\sec^{-1}\left(\frac{3x}{2}\right) + C\end{aligned}$$

18. Let $x = 3\sin\theta$, where $-\frac{\pi}{2} \le \theta \le \frac{\pi}{2}$. Then

$\int_0^3 x^2\sqrt{9-x^2}\,dx = \int_0^{\pi/2} 9\sin^2\theta(3\cos\theta)(3\cos\theta)d\theta = \frac{81}{4}\int_0^{\pi/2}\sin^2 2\theta\,d\theta$

$= \frac{81}{8}\int_0^{\pi/2}(1-\cos 4\theta)d\theta = \frac{81}{8}\left[\theta - \frac{1}{4}\sin 4\theta\right]_0^{\pi/2} = \frac{81\pi}{16}$.

19. Let $u = 1 + x^2$, $du = 2x\,dx$. Then $\int 5x\sqrt{1+x^2}\,dx = \frac{5}{2}\int u^{1/2}\,du = \frac{5}{3}u^{3/2} + C = \frac{5}{3}(1+x^2)^{3/2} + C$.

20. Let $2x = 5\sec\theta$, where $0 \le \theta < \frac{\pi}{2}$ or $\pi \le \theta < \frac{3\pi}{2}$. Then

$$\int \frac{dx}{(4x^2-25)^{3/2}} = \int \frac{\frac{5}{2}\sec\theta\tan\theta\,d\theta}{125\tan^3\theta} = \frac{1}{50}\int \frac{\cos\theta}{\sin^2\theta}\,d\theta$$

$$= -\frac{1}{50\sin\theta} + C \quad (\text{Set } u = \sin\theta)$$

$$= -\frac{1}{50}\cdot\frac{2x}{\sqrt{4x^2-25}} + C = -\frac{x}{25\sqrt{4x^2-25}} + C.$$

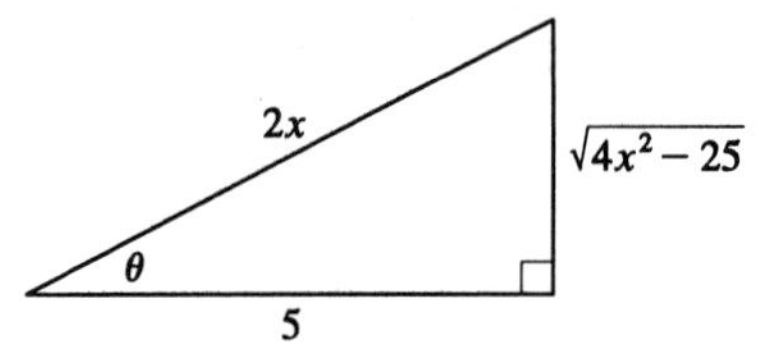

21. $2x - x^2 = -(x^2 - 2x + 1) + 1 = 1 - (x-1)^2$. Let $u = x - 1$. Then $du = dx$ and

$\int\sqrt{2x-x^2}\,dx = \int\sqrt{1-u^2}\,du = \int\cos^2\theta\,d\theta$ (where $u = \sin\theta$, $-\frac{\pi}{2} \le \theta \le \frac{\pi}{2}$)

$$= \tfrac{1}{2}\int(1+\cos 2\theta)d\theta = \tfrac{1}{2}\left(\theta + \tfrac{1}{2}\sin 2\theta\right) + C = \tfrac{1}{2}\left(\sin^{-1}u + u\sqrt{1-u^2}\right) + C$$

$$= \tfrac{1}{2}\left[\sin^{-1}(x-1) + (x-1)\sqrt{2x-x^2}\right] + C.$$

22. $x^2 + 4x + 8 = (x+2)^2 + 4$. Let $u = x + 2 \quad\Rightarrow\quad du = dx$. Then let $u = 2\tan\theta$.

$$\int \frac{dx}{\sqrt{x^2+4x+8}} = \int \frac{du}{\sqrt{u^2+4}} = \int \frac{2\sec^2\theta\,d\theta}{2\sec\theta}$$

$$= \int \sec\theta\,d\theta = \ln|\sec\theta + \tan\theta| + C_1 = \ln\left(\frac{\sqrt{u^2+4}+u}{2}\right) + C_1$$

$$= \ln\left(\sqrt{u^2+4}+u\right) + C = \ln\left(\sqrt{x^2+4x+8}+x+2\right) + C$$

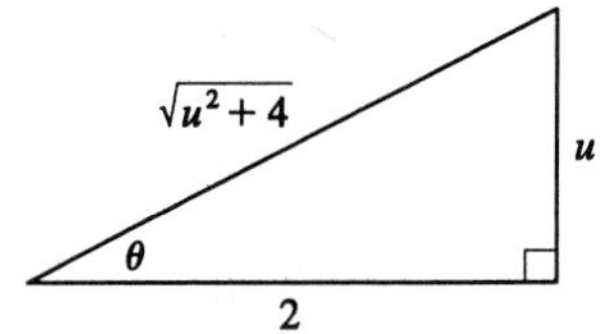

23. $9x^2 + 6x - 8 = (3x+1)^2 - 9$, so let $u = 3x + 1$, $du = 3\,dx$. Then $\displaystyle\int \frac{dx}{\sqrt{9x^2+6x-8}} = \int \frac{\frac{1}{3}\,du}{\sqrt{u^2-9}}$. Now let $u = 3\sec\theta$, where $0 \le \theta < \frac{\pi}{2}$ or $\pi \le \theta < \frac{3\pi}{2}$. Then $du = 3\sec\theta\tan\theta\,d\theta$ and $\sqrt{u^2-9} = 3\tan\theta$, so

$$\int \frac{\frac{1}{3}\,du}{\sqrt{u^2-9}} = \int \frac{\sec\theta\tan\theta\,d\theta}{3\tan\theta} = \tfrac{1}{3}\int\sec\theta\,d\theta = \tfrac{1}{3}\ln|\sec\theta + \tan\theta| + C_1$$

$$= \frac{1}{3}\ln\left|\frac{u+\sqrt{u^2-9}}{3}\right| + C_1 = \tfrac{1}{3}\ln\left|u+\sqrt{u^2-9}\right| + C = \tfrac{1}{3}\ln\left|3x+1+\sqrt{9x^2+6x-8}\right| + C.$$

24. $4x - x^2 = -(x^2 - 4x + 4) + 4 = 4 - (x-2)^2$, so let $u = x - 2$. Then $x = u + 2$ and $dx = du$, so

$$\int \frac{x^2\,dx}{\sqrt{4x-x^2}} = \int \frac{(u+2)^2\,du}{\sqrt{4-u^2}} = \int \frac{(2\sin\theta+2)^2}{2\cos\theta}2\cos\theta\,d\theta \ (\text{Put } u = 2\sin\theta)$$

$$= 4\int(\sin^2\theta + 2\sin\theta + 1)d\theta = 2\int(1-\cos 2\theta)d\theta + 8\int\sin\theta\,d\theta + 4\int d\theta$$

$$= 2\theta - \sin 2\theta - 8\cos\theta + 4\theta + C = 6\theta - 8\cos\theta - 2\sin\theta\cos\theta + C$$

$$= 6\sin^{-1}\left(\tfrac{1}{2}u\right) - 4\sqrt{4-u^2} - \tfrac{1}{2}u\sqrt{4-u^2} + C$$

$$= 6\sin^{-1}\left(\frac{x-2}{2}\right) - 4\sqrt{4x-x^2} - \left(\frac{x-2}{2}\right)\sqrt{4x-x^2} + C.$$

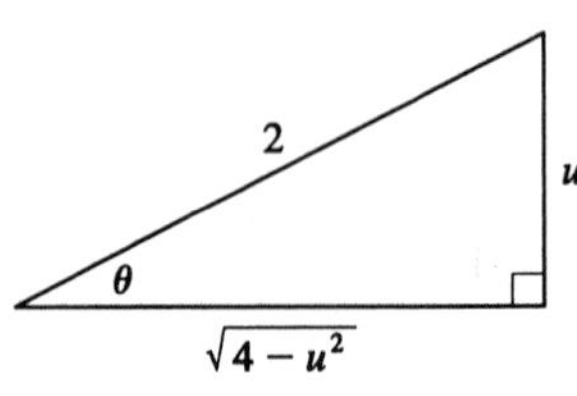

25. $x^2+2x+2=(x+1)^2+1$. Let $u=x+1$, $du=dx$. Then

$$\int\frac{dx}{(x^2+2x+2)^2}=\int\frac{du}{(u^2+1)^2}=\int\frac{\sec^2\theta\,d\theta}{\sec^4\theta}\quad\left(\begin{array}{r}\text{where } u=\tan\theta,\ du=\sec^2\theta\,d\theta,\\ \text{and } u^2+1=\sec^2\theta\end{array}\right)$$

$$=\int\cos^2\theta\,d\theta=\tfrac{1}{2}(\theta+\sin\theta\cos\theta)+C\quad\text{(as in Exercise 21)}$$

$$=\frac{1}{2}\left[\tan^{-1}u+\frac{u}{1+u^2}\right]+C=\frac{1}{2}\left[\tan^{-1}(x+1)+\frac{x+1}{x^2+2x+2}\right]+C.$$

26. $5-4x-x^2=-(x^2+4x+4)+9=9-(x+2)^2$. Let $u=x+2\quad\Rightarrow\quad du=dx$. Then

$$\int\frac{dx}{(5-4x-x^2)^{5/2}}=\int\frac{du}{(9-u^2)^{5/2}}=\int\frac{3\cos\theta\,d\theta}{(3\cos\theta)^5}\quad\left(\begin{array}{r}\text{where } u=3\sin\theta,\ du=3\cos\theta\,d\theta,\\ \text{and } \sqrt{9-u^2}=3\cos\theta\end{array}\right)$$

$$=\tfrac{1}{81}\int\sec^4\theta\,d\theta=\tfrac{1}{81}\int(\tan^2\theta+1)\sec^2\theta\,d\theta=\tfrac{1}{81}\left[\tfrac{1}{3}\tan^3\theta+\tan\theta\right]+C$$

$$=\frac{1}{243}\left[\frac{u^3}{(9-u^2)^{3/2}}+\frac{3u}{\sqrt{9-u^2}}\right]+C=\frac{1}{243}\left[\frac{(x+2)^3}{(5-4x-x^2)^{3/2}}+\frac{3(x+2)}{\sqrt{5-4x-x^2}}\right]+C.$$

27. Let $u=e^t\quad\Rightarrow\quad du=e^t\,dt$. Then $\int e^t\sqrt{9-e^{2t}}\,dt=\int\sqrt{9-u^2}\,du=\int(3\cos\theta)3\cos\theta\,d\theta$ (where $u=3\sin\theta$, $-\frac{\pi}{2}\le\theta\le\frac{\pi}{2}$) $=9\int\cos^2\theta\,d\theta=\frac{9}{2}(\theta+\sin\theta\cos\theta)+C$ (as in Exercise 21)

$$=\frac{9}{2}\left[\sin^{-1}\left(\frac{u}{3}\right)+\frac{u}{3}\cdot\frac{\sqrt{9-u^2}}{3}\right]+C=\tfrac{9}{2}\sin^{-1}\left(\tfrac{1}{3}e^t\right)+\tfrac{1}{2}e^t\sqrt{9-e^{2t}}+C.$$

28. Let $u=e^t$. Then $t=\ln u$ and $dt=du/u$. Hence $I=\int\sqrt{e^{2t}-9}\,dt=\int\left(\sqrt{u^2-9}/u\right)du$. Now let $u=3\sec\theta$, where $0\le\theta<\frac{\pi}{2}$ or $\pi\le\theta<\frac{3\pi}{2}$. Then $\sqrt{u^2-9}=3\tan\theta$ and $du=3\sec\theta\tan\theta\,d\theta$, so

$$I=\int\frac{3\tan\theta}{3\sec\theta}3\sec\theta\tan\theta\,d\theta=3\int\tan^2\theta\,d\theta=3\int(\sec^2\theta-1)d\theta=3(\tan\theta-\theta)+C$$

$$=3\left[\tfrac{1}{3}\sqrt{u^2-9}-\sec^{-1}\left(\tfrac{1}{3}u\right)\right]+C=\sqrt{e^{2t}-9}-3\sec^{-1}\left(\tfrac{1}{3}e^t\right)+C.$$

29. **(a)** Let $x=a\tan\theta$, where $-\frac{\pi}{2}<\theta<\frac{\pi}{2}$. Then $\sqrt{x^2+a^2}=a\sec\theta$ and

$$\int\frac{dx}{\sqrt{x^2+a^2}}=\int\frac{a\sec^2\theta\,d\theta}{a\sec\theta}=\int\sec\theta\,d\theta=\ln|\sec\theta+\tan\theta|+C_1=\ln\left|\frac{\sqrt{x^2+a^2}}{a}+\frac{x}{a}\right|+C_1$$

$$=\ln\left(x+\sqrt{x^2+a^2}\right)+C,\ \text{where } C=C_1-\ln|a|$$

(b) Let $x=a\sinh t$, so that $dx=a\cosh t\,dt$ and $\sqrt{x^2+a^2}=a\cosh t$. Then

$$\int\frac{dx}{\sqrt{x^2+a^2}}=\int\frac{a\cosh t\,dt}{a\cosh t}=t+C=\sinh^{-1}(x/a)+C.$$

30. **(a)** Let $x=a\tan\theta$, $-\frac{\pi}{2}<\theta<\frac{\pi}{2}$. Then $I=\displaystyle\int\frac{x^2}{(x^2+a^2)^{3/2}}\,dx=\int\frac{a^2\tan^2\theta}{a^3\sec^3\theta}a\sec^2\theta\,d\theta$

$$=\int\frac{\tan^2\theta}{\sec\theta}\,d\theta=\int\frac{\sec^2\theta-1}{\sec\theta}\,d\theta=\int(\sec\theta-\cos\theta)d\theta=\ln|\sec\theta+\tan\theta|-\sin\theta+C$$

$$=\ln\left|\frac{\sqrt{x^2+a^2}}{a}+\frac{x}{a}\right|-\frac{x}{\sqrt{x^2+a^2}}+C=\ln\left(x+\sqrt{x^2+a^2}\right)-\frac{x}{\sqrt{x^2+a^2}}+C_1.$$

(b) Let $x=a\sinh t$. Then $I=\displaystyle\int\frac{a^2\sinh^2t}{a^3\cosh^3t}a\cosh t\,dt=\int\tanh^2t\,dt=\int(1-\operatorname{sech}^2t)dt$

$$=t-\tanh t+C=\sinh^{-1}(x/a)-x/\sqrt{a^2+x^2}+C.$$

31. Area of $\triangle POQ = \frac{1}{2}(r\cos\theta)(r\sin\theta) = \frac{1}{2}r^2\sin\theta\cos\theta$. Area of region $PQR = \int_{r\cos\theta}^{r}\sqrt{r^2-x^2}\,dx$.

Let $x = r\cos u \quad\Rightarrow\quad dx = -r\sin u\,du$ for $\theta \le u \le \frac{\pi}{2}$. Then we obtain

$\int\sqrt{r^2-x^2}\,dx = \int r\sin u(-r\sin u)du = -r^2\int \sin^2 u\,du$

$\qquad = -\frac{1}{2}r^2(u - \sin u\cos u) + C = -\frac{1}{2}r^2\cos^{-1}(x/r) + \frac{1}{2}x\sqrt{r^2-x^2} + C$. So

area of region $PQR = \frac{1}{2}\left[-r^2\cos^{-1}(x/r) + x\sqrt{r^2-x^2}\right]_{r\cos\theta}^{r} = \frac{1}{2}[0 - (-r^2\theta + r\cos\theta\, r\sin\theta)]$

$\qquad = \frac{1}{2}r^2\theta - \frac{1}{2}r^2\sin\theta\cos\theta$, so (area of sector POR) = (area of $\triangle POQ$) + (area of region PQR) $= \frac{1}{2}r^2\theta$.

32. $9x^2 - 4y^2 = 36 \quad\Rightarrow\quad y = \pm\frac{3}{2}\sqrt{x^2-4} \quad\Rightarrow$

area $= 2\int_2^3 \frac{3}{2}\sqrt{x^2-4}\,dx = 3\int_2^3\sqrt{x^2-4}\,dx$

$$= 3\int_0^\alpha 2\tan\theta\, 2\sec\theta\tan\theta\,d\theta \quad \begin{bmatrix} \text{where } x = 2\sec\theta, \\ dx = 2\sec\theta\tan\theta\,d\theta, \\ \alpha = \sec^{-1}\frac{3}{2} \end{bmatrix}$$

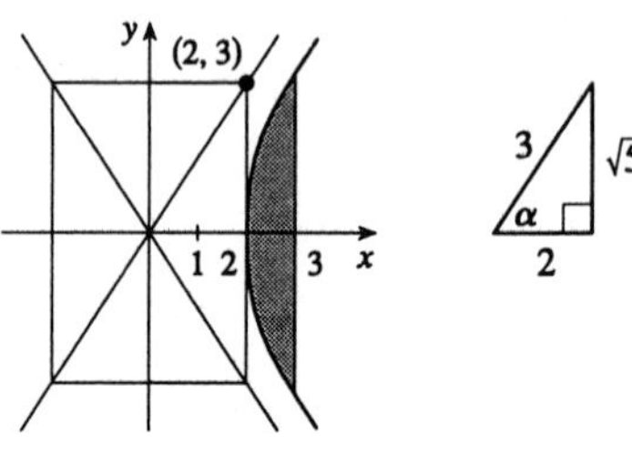

$= 12\int_0^\alpha(\sec^2\theta - 1)\sec\theta\,d\theta = 12\int_0^\alpha(\sec^3\theta - \sec\theta)d\theta$

$= 12\left[\frac{1}{2}(\sec\theta\tan\theta + \ln|\sec\theta+\tan\theta|) - \ln|\sec\theta+\tan\theta|\right]_0^\alpha$

$= 6[\sec\theta\tan\theta - \ln|\sec\theta+\tan\theta|]_0^\alpha$

$= 6\left[\frac{3\sqrt{5}}{4} - \ln\left(\frac{3}{2}+\frac{\sqrt{5}}{2}\right)\right] = \frac{9\sqrt{5}}{2} - 6\ln\left(\frac{3+\sqrt{5}}{2}\right)$

33. From the graph, it appears that the curve $y = x^2\sqrt{4-x^2}$ and the line $y = 2 - x$ intersect at about $x = 0.81$ and $x = 2$, with $x^2\sqrt{4-x^2} > 2 - x$ on $(0.81, 2)$. So the area bounded by the curve and the line is

$$A \approx \int_{0.81}^{2}\left[x^2\sqrt{4-x^2} - (2-x)\right]dx = \int_{0.81}^2 x^2\sqrt{4-x^2}\,dx - \left[2x - \tfrac{1}{2}x^2\right]_{0.81}^2.$$

To evaluate the integral, we put $x = 2\sin\theta$, where $-\frac{\pi}{2} \le \theta \le \frac{\pi}{2}$. Then $dx = 2\cos\theta\,d\theta$, $x = 2 \quad\Rightarrow$ $\theta = \sin^{-1}1 = \frac{\pi}{2}$, and $x = 0.81 \quad\Rightarrow\quad \theta = \sin^{-1}0.405 \approx 0.417$. So

$\int_{0.81}^2 x^2\sqrt{4-x^2}\,dx \approx \int_{0.417}^{\pi/2} 4\sin^2\theta(2\cos\theta)(2\cos\theta\,d\theta) = 4\int_{0.417}^{\pi/2}\sin^2 2\theta\,d\theta$

$= 4\int_{0.417}^{\pi/2}\frac{1}{2}(1-\cos 4\theta)d\theta = 2\left[\theta - \frac{1}{4}\sin 4\theta\right]_{0.417}^{\pi/2} = 2\left(\left[\frac{\pi}{2} - 0\right] - \left[0.417 - \frac{1}{4}(0.995)\right]\right) \approx 2.81$. So

$A \approx 2.81 - \left[\left(2\cdot 2 - \frac{1}{2}\cdot 2^2\right) - \left(2\cdot 0.81 - \frac{1}{2}\cdot 0.81^2\right)\right] \approx 2.10.$

34. Let $x = \sqrt{2}\sec\theta$, where $0 \le \theta < \frac{\pi}{2}$ or $\pi \le \theta < \frac{3\pi}{2}$, so $dx = \sqrt{2}\sec\theta\tan\theta\,d\theta$. Then

$$\int\frac{dx}{x^4\sqrt{x^2-2}} = \int\frac{\sqrt{2}\sec\theta\tan\theta\,d\theta}{4\sec^4\theta\sqrt{2}\tan\theta}$$

$= \frac{1}{4}\int\cos^3\theta\,d\theta = \frac{1}{4}\int(1-\sin^2\theta)\cos\theta\,d\theta$

$= \frac{1}{4}\left[\sin\theta - \frac{1}{3}\sin^3\theta\right] + C$ (substitute $u = \sin\theta$)

$$= \frac{1}{4}\left[\frac{\sqrt{x^2-2}}{x} - \frac{(x^2-2)^{3/2}}{3x^3}\right] + C.$$

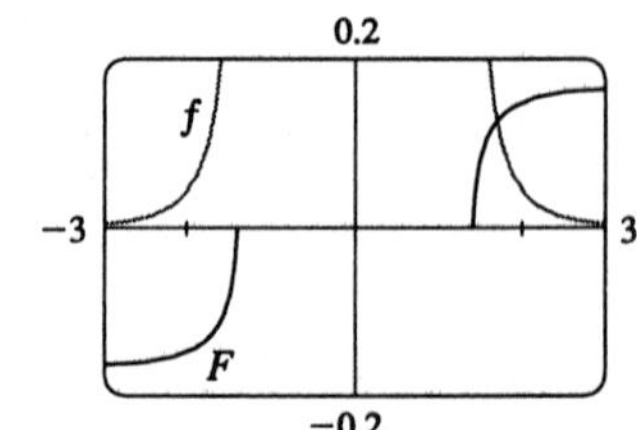

From the graph, it appears that our answer is reasonable. [Notice that $f(x)$ is large when F increases rapidly and small when F levels out.]

35. Let the equation of the large circle be $x^2 + y^2 = R^2$. Then the equation of the small circle is $x^2 + (y-b)^2 = r^2$, where $b = \sqrt{R^2 - r^2}$ is the distance between the centers of the circles. The desired area is

$$A = \int_{-r}^{r}\left[\left(b + \sqrt{r^2 - x^2}\right) - \sqrt{R^2 - x^2}\right]dx = 2\int_0^r\left(b + \sqrt{r^2 - x^2} - \sqrt{R^2 - x^2}\right)dx$$
$$= 2\int_0^r b\,dx + 2\int_0^r \sqrt{r^2 - x^2}\,dx - 2\int_0^r \sqrt{R^2 - x^2}\,dx.$$

The first integral is just $2br = 2r\sqrt{R^2 - r^2}$. To evaluate the other two integrals, note that

$$\int \sqrt{a^2 - x^2}\,dx = \int a^2\cos^2\theta\,d\theta \quad (x = a\sin\theta,\ dx = a\cos\theta\,d\theta)$$
$$= \tfrac{1}{2}\left(\tfrac{1}{2}a^2\right)\int(1 + \cos 2\theta) = \tfrac{1}{2}a^2\left(\theta + \tfrac{1}{2}\sin 2\theta\right) + C = \tfrac{1}{2}a^2(\theta + \sin\theta\cos\theta) + C$$
$$= \frac{a^2}{2}\arcsin\left(\frac{x}{a}\right) + \frac{a^2}{2}\left(\frac{x}{a}\right)\frac{\sqrt{a^2 - x^2}}{a} + C = \frac{a^2}{2}\arcsin\left(\frac{x}{a}\right) + \frac{x}{2}\sqrt{a^2 - x^2} + C,$$ so the desired area is

$$A = 2r\sqrt{R^2 - r^2} + \left[r^2\arcsin(x/r) + x\sqrt{r^2 - x^2}\right]_0^r - \left[R^2\arcsin(x/R) + x\sqrt{R^2 - x^2}\right]_0^r$$
$$= 2r\sqrt{R^2 - r^2} + r^2\left(\tfrac{\pi}{2}\right) - \left[R^2\arcsin(r/R) + r\sqrt{R^2 - r^2}\right] = r\sqrt{R^2 - r^2} + \frac{\pi}{2}r^2 - R^2\arcsin(r/R).$$

36. Note that the circular cross-sections of the tank are the same everywhere, so the percentage of the total capacity that is being used is equal to the percentage of any cross-section that is under water. The underwater area is

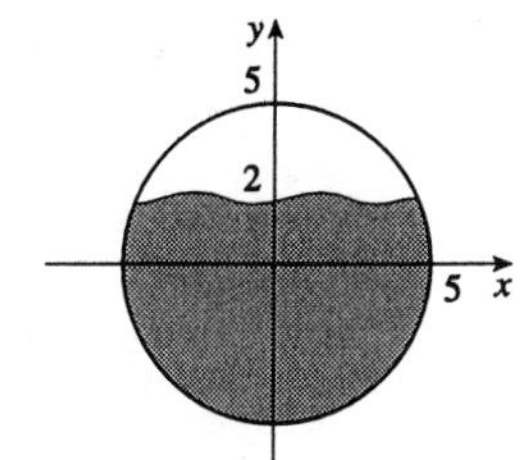

$$A = 2\int_{-5}^{2}\sqrt{25 - y^2}\,dy = \left[25\arcsin(y/5) + y\sqrt{25 - y^2}\right]_{-5}^{2}$$
(substitute $y = 5\sin\theta$) $\quad = 25\arcsin\frac{2}{5} + 2\sqrt{21} + \frac{25}{2}\pi \approx 58.72\text{ ft}^2$,

so the fraction of the total capacity in use is $\dfrac{A}{\pi(5)^2} \approx \dfrac{58.72}{25\pi} \approx 0.748$ or 74.8%.

37. We use cylindrical shells and assume that $R > r$. $x^2 = r^2 - (y-R)^2 \Rightarrow x = \pm\sqrt{r^2 - (y-R)^2}$, so $g(y) = 2\sqrt{r^2 - (y-R)^2}$ in Formula 5.3.3, and

$$V = \int_{R-r}^{R+r} 2\pi y \cdot 2\sqrt{r^2 - (y-R)^2}\,dy = \int_{-r}^{r} 4\pi(u + R)\sqrt{r^2 - u^2}\,du \quad (\text{where } u = y - R)$$
$$= 4\pi\int_{-r}^{r} u\sqrt{r^2 - u^2}\,du + 4\pi R\int_{-r}^{r}\sqrt{r^2 - u^2}\,du \quad \left(\begin{array}{c}\text{where } u = r\sin\theta,\ du = r\cos\theta\,d\theta \\ \text{in the second integral}\end{array}\right)$$
$$= 4\pi\left[-\tfrac{1}{3}(r^2 - u^2)^{3/2}\right]_{-r}^{r} + 4\pi R\int_{-\pi/2}^{\pi/2} r^2\cos^2\theta\,d\theta = -\tfrac{4\pi}{3}(0 - 0) + 4\pi R r^2\int_{-\pi/2}^{\pi/2}\cos^2 d\theta$$
$$= 2\pi R r^2\int_{-\pi/2}^{\pi/2}(1 + \cos 2\theta)d\theta = 2\pi R r^2\left[\theta + \tfrac{1}{2}\sin 2\theta\right]_{-\pi/2}^{\pi/2} = 2\pi^2 R r^2.$$

Another Method: Use washers instead of shells, so $V = 8\pi R\int_0^r\sqrt{r^2 - y^2}\,dy$ as in Exercise 5.2.61(a), but evaluate the integral using $y = r\sin\theta$.

38. For each k satisfying $-c < k < c$, the cross-section of the ellipsoid in the plane $z = k$ is the ellipse $\dfrac{x^2}{a^2} + \dfrac{y^2}{b^2} = 1 - \dfrac{k^2}{c^2}$, whose interior has area $\pi ab\left(1 - \dfrac{k^2}{c^2}\right)$ (see Example 2). Thus the volume enclosed by the ellipsoid is $V = \displaystyle\int_{-c}^{c} A(z)dz = 2\int_0^c A(z)dz = 2\pi ab\int_0^c\left(1 - \frac{z^2}{c^2}\right)dz = 2\pi ab\left[z - \frac{z^3}{3c^2}\right]_0^c = \tfrac{4}{3}\pi abc.$

EXERCISES 7.4

1. $\dfrac{1}{(x-1)(x+2)} = \dfrac{A}{x-1} + \dfrac{B}{x+2}$

2. $\dfrac{7}{(2x-3)(x+4)} = \dfrac{A}{2x-3} + \dfrac{B}{x+4}$

3. $\dfrac{x^2+3x-4}{(2x-1)^2(2x+3)} = \dfrac{A}{2x-1} + \dfrac{B}{(2x-1)^2} + \dfrac{C}{2x+3}$

4. $\dfrac{x^3-x^2}{(x-6)(5x+3)^3} = \dfrac{A}{x-6} + \dfrac{B}{5x+3} + \dfrac{C}{(5x+3)^2} + \dfrac{D}{(5x+3)^3}$

5. $\dfrac{1}{x^4-x^3} = \dfrac{1}{x^3(x-1)} = \dfrac{A}{x} + \dfrac{B}{x^2} + \dfrac{C}{x^3} + \dfrac{D}{x-1}$

6. $\dfrac{1+x+x^2}{(x+1)(x+2)^2(x+3)^3} = \dfrac{A}{x+1} + \dfrac{B}{x+2} + \dfrac{C}{(x+2)^2} + \dfrac{D}{x+3} + \dfrac{E}{(x+3)^2} + \dfrac{F}{(x+3)^3}$

7. $\dfrac{x^2+1}{x^2-1} = 1 + \dfrac{2}{(x-1)(x+1)} = 1 + \dfrac{A}{x-1} + \dfrac{B}{x+1}$

8. $\dfrac{x^4+x^3-x^2-x+1}{x^3-x} = x+1+\dfrac{1}{x(x+1)(x-1)} = x+1+\dfrac{A}{x} + \dfrac{B}{x+1} + \dfrac{C}{x-1}$

9. $\dfrac{x^2-2}{x(x^2+2)} = \dfrac{A}{x} + \dfrac{Bx+C}{x^2+2}$

10. $\dfrac{x^3-4x^2+2}{(x^2+1)(x^2+2)} = \dfrac{Ax+B}{x^2+1} + \dfrac{Cx+D}{x^2+2}$

11. $\dfrac{x^4+x^2+1}{(x^2+1)(x^2+4)^2} = \dfrac{Ax+B}{x^2+1} + \dfrac{Cx+D}{x^2+4} + \dfrac{Ex+F}{(x^2+4)^2}$

12. $\dfrac{1+16x}{(2x-3)(x+5)^2(x^2+x+1)} = \dfrac{A}{2x-3} + \dfrac{B}{x+5} + \dfrac{C}{(x+5)^2} + \dfrac{Dx+E}{x^2+x+1}$

13. $\dfrac{x^4}{(x^2+9)^3} = \dfrac{Ax+B}{x^2+9} + \dfrac{Cx+D}{(x^2+9)^2} + \dfrac{Ex+F}{(x^2+9)^3}$

14. $\dfrac{19x}{(x-1)^3(4x^2+5x+3)^2} = \dfrac{A}{x-1} + \dfrac{B}{(x-1)^2} + \dfrac{C}{(x-1)^3} + \dfrac{Dx+E}{4x^2+5x+3} + \dfrac{Fx+G}{(4x^2+5x+3)^2}$

15. $\dfrac{x^3+x^2+1}{x^4+x^3+2x^2} = \dfrac{x^3+x^2+1}{x^2(x^2+x+2)} = \dfrac{A}{x} + \dfrac{B}{x^2} + \dfrac{Cx+D}{x^2+x+2}$

16. $\dfrac{1}{x^6-x^3} = \dfrac{1}{x^3(x^3-1)} = \dfrac{1}{x^3(x-1)(x^2+x+1)} = \dfrac{A}{x} + \dfrac{B}{x^2} + \dfrac{C}{x^3} + \dfrac{D}{x-1} + \dfrac{Ex+F}{x^2+x+1}$

17. $\displaystyle\int \frac{x^2}{x+1}\,dx = \int\left(x-1+\frac{1}{x+1}\right)dx = \tfrac{1}{2}x^2 - x + \ln|x+1| + C$

18. $\displaystyle\int \frac{x}{x-5}\,dx = \int \frac{(x-5)+5}{x-5}\,dx = \int\left(1+\frac{5}{x-5}\right)dx = x + 5\ln|x-5| + C$

19. $\dfrac{4x-1}{(x-1)(x+2)} = \dfrac{A}{x-1} + \dfrac{B}{x+2} \quad\Rightarrow\quad 4x-1 = A(x+2)+B(x-1)$ Take $x=1$ to get $3=3A$, then $x=-2$ to get $-9=-3B \quad\Rightarrow\quad A=1,\ B=3$. Now

$$\int_2^4 \frac{4x-1}{(x-1)(x+2)}\,dx = \int_2^4\left[\frac{1}{x-1}+\frac{3}{x+2}\right]dx = [\ln(x-1)+3\ln(x+2)]_2^4$$
$$= \ln 3 + 3\ln 6 - \ln 1 - 3\ln 4 = 4\ln 3 - 3\ln 2 = \ln\tfrac{81}{8}.$$

20. $\dfrac{1}{(x+1)(x-2)} = \dfrac{A}{x+1} + \dfrac{B}{x-2} \quad\Rightarrow\quad 1 = A(x-2) + B(x+1)$. Taking $x = -1$, then $x = 2$, gives $A = -\frac{1}{3}$, $B = \frac{1}{3}$. Hence

$$\int_3^7 \frac{dx}{(x+1)(x-2)} = \frac{1}{3}\int_3^7 \left[\frac{1}{x-2} - \frac{1}{x+1}\right] dx = \tfrac{1}{3}[\ln|x-2| - \ln|x+1|]_3^7$$
$$= \tfrac{1}{3}(\ln 5 - \ln 8 - \ln 1 + \ln 4) = \tfrac{1}{3}\ln\tfrac{5}{2}.$$

21. $\displaystyle\int \frac{6x-5}{2x+3}\,dx = \int \left[3 - \frac{14}{2x+3}\right] dx = 3x - 7\ln|2x+3| + C$

22. If $a \neq b$, $\dfrac{1}{(x+a)(x+b)} = \dfrac{1}{b-a}\left(\dfrac{1}{x+a} - \dfrac{1}{x+b}\right)$, so if $a \neq b$, then

$$\int \frac{dx}{(x+a)(x+b)} = \frac{1}{b-a}(\ln|x+a| - \ln|x+b|) + C = \frac{1}{b-a}\ln\left|\frac{x+a}{x+b}\right| + C.$$

If $a = b$, then $\displaystyle\int \frac{dx}{(x+a)^2} = -\frac{1}{x+a} + C.$

23. $\dfrac{x^2+1}{x^2-x} = 1 + \dfrac{x+1}{x(x-1)} = 1 - \dfrac{1}{x} + \dfrac{2}{x-1}$, so

$$\int \frac{x^2+1}{x^2-x}\,dx = x - \ln|x| + 2\ln|x-1| + C = x + \ln\frac{(x-1)^2}{|x|} + C.$$

24. $\dfrac{x^3+x^2-12x+1}{x^2+x-12} = x + \dfrac{1}{x^2+x-12} = x + \dfrac{1}{(x-3)(x+4)} = x + \dfrac{1}{7}\left(\dfrac{1}{x-3} - \dfrac{1}{x+4}\right)$. So

$$\int_0^2 \frac{x^3+x^2-12x+1}{x^2+x-12}\,dx = \left[\tfrac{1}{2}x^2 + \tfrac{1}{7}(\ln|x-3| - \ln|x+4|)\right]_0^2 = 2 + \tfrac{1}{7}\ln\tfrac{2}{9}.$$

25. $\dfrac{2x+3}{(x+1)^2} = \dfrac{A}{x+1} + \dfrac{B}{(x+1)^2} \quad\Rightarrow\quad 2x+3 = A(x+1) + B$. Take $x = -1$ to get $B = 1$, and equate coefficients of x to get $A = 2$. Now

$$\int_0^1 \frac{2x+3}{(x+1)^2}\,dx = \int_0^1 \left[\frac{2}{x+1} + \frac{1}{(x+1)^2}\right] dx = \left[2\ln(x+1) - \frac{1}{x+1}\right]_0^1$$
$$= 2\ln 2 - \tfrac{1}{2} - (2\ln 1 - 1) = 2\ln 2 + \tfrac{1}{2}.$$

26. $\dfrac{1}{x(x+1)(2x+3)} = \dfrac{A}{x} + \dfrac{B}{x+1} + \dfrac{C}{2x+3} \quad\Rightarrow\quad 1 = A(x+1)(2x+3) + B(x)(2x+3) + C(x)(x+1)$.

Set $x = 0$ to get $A = \frac{1}{3}$, take $x = -1$ to get $B = -1$, and finally set $x = -\frac{3}{2}$, giving $C = \frac{4}{3}$. Now

$$\int \frac{dx}{x(x+1)(2x+3)} = \int \left[\frac{1/3}{x} - \frac{1}{x+1} + \frac{4/3}{2x+3}\right] dx = \tfrac{1}{3}\ln|x| - \ln|x+1| + \tfrac{2}{3}\ln|2x+3| + C.$$

27. $\dfrac{6x^2+5x-3}{x^3+2x^2-3x} = \dfrac{A}{x} + \dfrac{B}{x+3} + \dfrac{C}{x-1} \quad\Rightarrow$

$6x^2 + 5x - 3 = A(x+3)(x-1) + B(x)(x-1) + C(x)(x+3)$.

Set $x = 0$ to get $A = 1$, then take $x = -3$ to get $B = 3$, then set $x = 1$ to get $C = 2$:

$$\int_2^3 \frac{6x^2+5x-3}{x^3+2x^2-3x}\,dx = \int_2^3 \left[\frac{1}{x} + \frac{3}{x+3} + \frac{2}{x-1}\right] dx = [\ln x + 3\ln(x+3) + 2\ln(x-1)]_2^3$$
$$= (\ln 3 + 3\ln 6 + 2\ln 2) - (\ln 2 + 3\ln 5) = 4\ln 6 - 3\ln 5.$$

28. $\dfrac{x}{x^2+4x+4} = \dfrac{A}{x+2} + \dfrac{B}{(x+2)^2} \quad\Rightarrow\quad x = A(x+2)+B$. Set $x=-2$ to get $B=-2$ and equate coefficients of x to get $A=1$. Then

$$\int_0^1 \frac{x\,dx}{x^2+4x+4} = \int_0^1 \left[\frac{1}{x+2} - \frac{2}{(x+2)^2}\right]dx = \left[\ln(x+2) + \frac{2}{x+2}\right]_0^1 = \ln 3 + \frac{2}{3} - (\ln 2 + 1) = \ln\frac{3}{2} - \frac{1}{3}.$$

29. $\dfrac{1}{(x-1)^2(x+4)} = \dfrac{A}{x-1} + \dfrac{B}{(x-1)^2} + \dfrac{C}{x+4} \quad\Rightarrow\quad 1 = A(x-1)(x+4) + B(x+4) + C(x-1)^2$. Set $x=1$ to get $B=\frac{1}{5}$ and take $x=-4$ to get $C=\frac{1}{25}$. Now equating the coefficients of x^2, we get $0 = Ax^2 + Cx^2$ or $A = -C = -\frac{1}{25} \quad\Rightarrow$

$$\int \frac{dx}{(x-1)^2(x+4)} = \int\left[\frac{-1/25}{x-1} + \frac{1/5}{(x-1)^2} + \frac{1/25}{x+4}\right]dx = -\frac{1}{25}\ln|x-1| - \frac{1}{5}\cdot\frac{1}{x-1} + \frac{1}{25}\ln|x+4| + C$$

$$= \frac{1}{25}\left[\ln\left|\frac{x+4}{x-1}\right| - \frac{5}{x-1}\right] + C.$$

30. $\dfrac{x^2}{(x-3)(x+2)^2} = \dfrac{A}{x-3} + \dfrac{B}{x+2} + \dfrac{C}{(x+2)^2} \quad\Rightarrow\quad x^2 = A(x+2)^2 + B(x-3)(x+2) + C(x-3)$. Setting $x=3$ gives $A=\frac{9}{25}$. Take $x=-2$ to get $C=-\frac{4}{5}$, and equate the coefficients of x^2 to get $1=A+B$ $\Rightarrow\quad B=\frac{16}{25}$. Then

$$\int \frac{x^2}{(x-3)(x+2)^2}\,dx = \int\left[\frac{9/25}{x-3} + \frac{16/25}{x+2} - \frac{4/5}{(x+2)^2}\right]dx = \tfrac{9}{25}\ln|x-3| + \tfrac{16}{25}\ln|x+2| + \frac{4}{5(x+2)} + C.$$

31. $\dfrac{5x^2+3x-2}{x^3+2x^2} = \dfrac{5x^2+3x-2}{x^2(x+2)} = \dfrac{A}{x} + \dfrac{B}{x^2} + \dfrac{C}{x+2}$. Multiply by $x^2(x+2)$ to get $5x^2+3x-2 = Ax(x+2) + B(x+2) + Cx^2$. Set $x=-2$ to get $C=3$, and take $x=0$ to get $B=-1$. Equating the coefficients of x^2 gives $5x^2 = Ax^2 + Cx^2$ or $A=2$. So

$$\int \frac{5x^2+3x-2}{x^3+2x^2}\,dx = \int\left[\frac{2}{x} - \frac{1}{x^2} + \frac{3}{x+2}\right]dx = 2\ln|x| + \frac{1}{x} + 3\ln|x+2| + C.$$

32. $\dfrac{18-2x-4x^2}{x^3+4x^2+x-6} = \dfrac{18-2x-4x^2}{(x-1)(x+2)(x+3)} = \dfrac{A}{x-1} + \dfrac{B}{x+2} + \dfrac{C}{x+3} \quad\Rightarrow$ $18-2x-4x^2 = A(x+2)(x+3) + B(x-1)(x+3) + C(x-1)(x+2)$. Set $x=1$ to get $A=1$. Now setting $x=-2$ gives $B=-2$, and setting $x=-3$ gives $C=-3$. Then

$$\int \frac{18-2x-4x^2}{x^3+4x^2+x-6}\,dx = \int\left(\frac{1}{x-1} - \frac{2}{x+2} - \frac{3}{x+3}\right)dx = \ln|x-1| - 2\ln|x+2| - 3\ln|x+3| + C.$$

33. Let $u = x^3+3x^2+4$. Then $du = 3(x^2+2x)dx \Rightarrow \displaystyle\int \frac{x^2+2x}{x^3+3x^2+4}\,dx = \frac{1}{3}\int\frac{du}{u} = \tfrac{1}{3}\ln|x^3+3x^2+4| + C.$

34. $\dfrac{1}{x^2(x-1)^2} = \dfrac{A}{x} + \dfrac{B}{x^2} + \dfrac{C}{x-1} + \dfrac{D}{(x-1)^2} \quad\Rightarrow\quad 1 = Ax(x-1)^2 + B(x-1)^2 + Cx^2(x-1) + Dx^2$. Set $x=0$, giving $B=1$. Then set $x=1$ to get $D=1$. Equate the coefficients of x^3 to get $0=A+C$ or $A=-C$, and finally set $x=2$ to get $1 = 2A+1-4A+4$ or $A=2$. Now

$$\int \frac{dx}{x^2(x-1)^2} = \int\left[\frac{2}{x} + \frac{1}{x^2} - \frac{2}{x-1} + \frac{1}{(x-1)^2}\right]dx = 2\ln|x| - \frac{1}{x} - 2\ln|x-1| - \frac{1}{x-1} + C.$$

35. $\dfrac{x^2}{(x+1)^3} = \dfrac{A}{x+1} + \dfrac{B}{(x+1)^2} + \dfrac{C}{(x+1)^3}$. Multiply by $(x+1)^3$ to get $x^2 = A(x+1)^2 + B(x+1) + C$.

Setting $x = -1$ gives $C = 1$. Equating the coefficients of x^2 gives $A = 1$, and setting $x = 0$ gives $B = -2$.

Now $\displaystyle\int \frac{x^2\,dx}{(x+1)^3} = \int \left[\frac{1}{x+1} - \frac{2}{(x+1)^2} + \frac{1}{(x+1)^3}\right] dx = \ln|x+1| + \frac{2}{x+1} - \frac{1}{2(x+1)^2} + C.$

36. $\dfrac{x}{x+1} = \dfrac{(x+1)-1}{x+1} = 1 - \dfrac{1}{x+1}$, so $\dfrac{x^3}{(x+1)^3} = \left[1 - \dfrac{1}{x+1}\right]^3 = 1 - \dfrac{3}{x+1} + \dfrac{3}{(x+1)^2} - \dfrac{1}{(x+1)^3}$. Thus

$$\int \frac{x^3}{(x+1)^3}\,dx = \int \left[1 - \frac{3}{x+1} + \frac{3}{(x+1)^2} - \frac{1}{(x+1)^3}\right] dx = x - 3\ln|x+1| - \frac{3}{x+1} + \frac{1}{2(x+1)^2} + C.$$

37. $\dfrac{1}{x^4 - x^2} = \dfrac{1}{x^2(x-1)(x+1)} = \dfrac{A}{x} + \dfrac{B}{x^2} + \dfrac{C}{x-1} + \dfrac{D}{x+1}$. Multiply by $x^2(x-1)(x+1)$ to get

$1 = Ax(x-1)(x+1) + B(x-1)(x+1) + Cx^2(x+1) + Dx^2(x-1)$. Setting $x = 1$ gives $C = \frac{1}{2}$, taking $x = -1$ gives $D = -\frac{1}{2}$. Equating the coefficients of x^3 gives $0 = A + C + D = A$. Finally, setting $x = 0$ yields $B = -1$. Now $\displaystyle\int \frac{dx}{x^4 - x^2} = \int \left[\frac{-1}{x^2} + \frac{1/2}{x-1} - \frac{1/2}{x+1}\right] dx = \frac{1}{x} + \frac{1}{2}\ln\left|\frac{x-1}{x+1}\right| + C.$

38. Let $u = x^4 - x^2 + 1$. Then $du = (4x^3 - 2x)dx \quad \Rightarrow$

$$\int \frac{2x^3 - x}{x^4 - x^2 + 1}\,dx = \int \frac{\frac{1}{2}\,du}{u} = \tfrac{1}{2}\ln|x^4 - x^2 + 1| + C = \tfrac{1}{2}\ln(x^4 - x^2 + 1) + C.$$

39. $\dfrac{x^3}{x^2+1} = \dfrac{(x^3+x) - x}{x^2+1} = x - \dfrac{x}{x^2+1}$, so $\displaystyle\int_0^1 \frac{x^3}{x^2+1}\,dx = \int_0^1 x\,dx - \int_0^1 \frac{x\,dx}{x^2+1}$

$= \left[\frac{1}{2}x^2\right]_0^1 - \dfrac{1}{2}\displaystyle\int_1^2 \frac{1}{u}\,du$ (where $u = x^2+1$, $du = 2x\,dx$) $= \frac{1}{2} - \left[\frac{1}{2}\ln u\right]_1^2 = \frac{1}{2} - \frac{1}{2}\ln 2 = \frac{1}{2}(1 - \ln 2)$

40. $\displaystyle\int_0^1 \frac{x-1}{x^2+2x+2}\,dx = \int_0^1 \frac{x+1}{x^2+2x+2}\,dx - \int_0^1 \frac{2}{x^2+2x+2}\,dx$

$= \left[\frac{1}{2}\ln(x^2+2x+2)\right]_0^1 - 2\displaystyle\int_0 \frac{dx}{(x+1)^2+1}$ $\quad\left[\begin{array}{c}\text{set } u = x^2+2x+2,\ du = 2(x+1)dx \\ \text{in the first integral}\end{array}\right]$

$= \frac{1}{2}(\ln 5 - \ln 2) - 2\left[\tan^{-1}(x+1)\right]_0^1 = \frac{1}{2}\ln\frac{5}{2} - 2\tan^{-1}2 + \frac{\pi}{2}.$

Or: Complete the square and let $u = x + 1$.

41. Complete the square: $x^2 + x + 1 = \left(x + \frac{1}{2}\right)^2 + \frac{3}{4}$ and let $u = x + \frac{1}{2}$. Then

$$\int_0^1 \frac{x}{x^2+x+1}\,dx = \int_{1/2}^{3/2} \frac{u - 1/2}{u^2 + 3/4}\,du = \int_{1/2}^{3/2} \frac{u}{u^2+3/4}\,du - \frac{1}{2}\int_{1/2}^{3/2} \frac{1}{u^2+3/4}\,du$$

$$= \tfrac{1}{2}\ln\left(u^2 + \tfrac{3}{4}\right) - \tfrac{1}{2}\tfrac{1}{\sqrt{3}/2}\left[\tan^{-1}\left(\tfrac{2}{\sqrt{3}}u\right)\right]_{1/2}^{3/2} = \tfrac{1}{2}\ln 3 - \tfrac{1}{\sqrt{3}}\left(\tfrac{\pi}{3} - \tfrac{\pi}{6}\right) = \ln\sqrt{3} - \tfrac{\pi}{6\sqrt{3}}.$$

42. $\displaystyle\int_{-1/2}^{1/2} \frac{4x^2+5x+7}{4x^2+4x+5}\,dx = \int_{-1/2}^{1/2}\left[1+\frac{x+2}{4x^2+4x+5}\right]dx$

$$= [x]_{-1/2}^{1/2} + \int_{-1/2}^{1/2}\frac{x+1/2}{4x^2+4x+5}\,dx + \int_{-1/2}^{1/2}\frac{3/2}{4x^2+4x+5}\,dx$$

$$= 1 + \left[\tfrac{1}{8}\ln(4x^2+4x+5)\right]_{-1/2}^{1/2} + \frac{3}{2}\int_{-1/2}^{1/2}\frac{dx}{(2x+1)^2+4} \quad \left[\begin{array}{c}\text{set } u = 4x^2+4x+5,\ du = 8\left(x+\tfrac{1}{2}\right)dx\\ \text{in the first integral}\end{array}\right]$$

$$= 1 + \tfrac{1}{8}(\ln 8 - \ln 4) + \frac{3}{2}\int_0^2\frac{(1/2)du}{u^2+4} \quad (\text{set } u = 2x+1, \text{ so } du = 2\,dx)$$

$$= 1 + \tfrac{1}{8}\ln 2 + \tfrac{3}{4}\left[\tfrac{1}{2}\tan^{-1}\left(\tfrac{1}{2}u\right)\right]_0^2 = 1 + \tfrac{1}{8}\ln 2 + \tfrac{3\pi}{32}$$

43. $\displaystyle\frac{3x^2-4x+5}{(x-1)(x^2+1)} = \frac{A}{x-1}+\frac{Bx+C}{x^2+1} \quad\Rightarrow\quad 3x^2-4x+5 = A(x^2+1)+(Bx+C)(x-1)$. Take $x=1$ to get $4=2A$ or $A=2$. Now $(Bx+C)(x-1) = 3x^2-4x+5-2(x^2+1) = x^2-4x+3$. Equating coefficients of x^2 and then comparing the constant terms, we get $B=1$ and $C=-3$. Hence

$$\int\frac{3x^2-4x+5}{(x-1)(x^2+1)}\,dx = \int\left[\frac{2}{x-1}+\frac{x-3}{x^2+1}\right]dx = 2\ln|x-1| + \int\frac{x\,dx}{x^2+1} - 3\int\frac{dx}{x^2+1}$$

$$= 2\ln|x-1| + \tfrac{1}{2}\ln(x^2+1) - 3\tan^{-1}x + C = \ln(x-1)^2 + \ln\sqrt{x^2+1} - 3\tan^{-1}x + C.$$

44. $\displaystyle\frac{2x+3}{x(x^2+3)} = \frac{A}{x}+\frac{Bx+C}{x^2+3} \quad\Rightarrow\quad 2x+3 = A(x^2+3)+(Bx+C)x = (A+B)x^2+Cx+3A \quad\Rightarrow$

$A=1, C=2, B=-A=-1$. Hence $\displaystyle\int\frac{2x+3}{x^3+3x}\,dx = \int\left[\frac{1}{x}+\frac{-x+2}{x^2+3}\right]dx$

$$= \ln|x| - \frac{1}{2}\int\frac{2x\,dx}{x^2+3} + 2\int\frac{dx}{x^2+3} = \ln|x| - \tfrac{1}{2}\ln(x^2+3) + \tfrac{2}{\sqrt{3}}\tan^{-1}\left(\tfrac{1}{\sqrt{3}}x\right) + C.$$

45. $\displaystyle\frac{1}{x^3-1} = \frac{1}{(x-1)(x^2+x+1)} = \frac{A}{x-1}+\frac{Bx+C}{x^2+x+1} \quad\Rightarrow\quad 1 = A(x^2+x+1)+(Bx+C)(x-1)$. Take $x=1$ to get $A=\frac{1}{3}$. Equate coefficients of x^2 and 1 to get $0=\frac{1}{3}+B$, $1=\frac{1}{3}-C$, so $B=-\frac{1}{3}$, $C=-\frac{2}{3} \quad\Rightarrow$

$$\int\frac{dx}{x^3-1} = \int\frac{1/3}{x-1}\,dx + \int\frac{(-1/3)x-2/3}{x^2+x+1}\,dx = \tfrac{1}{3}\ln|x-1| - \frac{1}{3}\int\frac{x+2}{x^2+x+1}\,dx$$

$$= \tfrac{1}{3}\ln|x-1| - \frac{1}{3}\int\frac{x+1/2}{x^2+x+1}\,dx - \frac{1}{3}\int\frac{(3/2)dx}{(x+1/2)^2+3/4}$$

$$= \tfrac{1}{3}\ln|x-1| - \tfrac{1}{6}\ln(x^2+x+1) - \tfrac{1}{2}\left(\tfrac{2}{\sqrt{3}}\right)\tan^{-1}\left[\left(x+\tfrac{1}{2}\right)\Big/\left(\tfrac{\sqrt{3}}{2}\right)\right] + K$$

$$= \tfrac{1}{3}\ln|x-1| - \tfrac{1}{6}\ln(x^2+x+1) - \tfrac{1}{\sqrt{3}}\tan^{-1}\left[\tfrac{1}{\sqrt{3}}(2x+1)\right] + K.$$

46. $\displaystyle\frac{x^3}{x^3+1} = \frac{(x^3+1)-1}{x^3+1} = 1-\frac{1}{x^3+1} = 1-\left(\frac{A}{x+1}+\frac{Bx+C}{x^2-x+1}\right) \quad\Rightarrow$

$1 = A(x^2-x+1)+(Bx+C)(x+1)$. Equate the terms of degree 2, 1 and 0 to get $0=A+B$, $0=-A+B+C$, $1=A+C$. Solve the three equations to get $A=\frac{1}{3}$, $B=-\frac{1}{3}$, and $C=\frac{2}{3}$. So

$$\int\frac{x^3}{x^3+1}\,dx = \int\left[1-\frac{1/3}{x+1}+\frac{\frac{1}{3}x-\frac{2}{3}}{x^2-x+1}\right]dx$$

$$= x - \tfrac{1}{3}\ln|x+1| + \frac{1}{6}\int\frac{2x-1}{x^2-x+1}\,dx - \frac{1}{2}\int\frac{dx}{\left(x-\frac{1}{2}\right)^2+\frac{3}{4}}$$

$$= x - \tfrac{1}{3}\ln|x+1| + \tfrac{1}{6}\ln(x^2-x+1) - \tfrac{1}{\sqrt{3}}\tan^{-1}\left[\tfrac{1}{\sqrt{3}}(2x-1)\right] + K.$$

47. $\dfrac{x^2-2x-1}{(x-1)^2(x^2+1)} = \dfrac{A}{x-1} + \dfrac{B}{(x-1)^2} + \dfrac{Cx+D}{x^2+1} \quad \Rightarrow$

$x^2-2x-1 = A(x-1)(x^2+1) + B(x^2+1) + (Cx+D)(x-1)^2$. Setting $x=1$ gives $B=-1$. Equating the coefficients of x^3 gives $A=-C$. Equating the constant terms gives $-1=-A-1+D$, so $D=A$, and setting $x=2$ gives $-1=5A-5-2A+A$ or $A=1$. We have

$$\int \frac{x^2-2x-1}{(x-1)^2(x^2+1)}\,dx = \int \left[\frac{1}{x-1} - \frac{1}{(x-1)^2} - \frac{x-1}{x^2+1}\right] dx$$
$$= \ln|x-1| + \frac{1}{x-1} - \tfrac{1}{2}\ln(x^2+1) + \tan^{-1}x + C.$$

48. $\dfrac{x^4}{x^4-1} = 1 + \dfrac{1}{x^4-1}$ and $\dfrac{1}{x^4-1} = \dfrac{1}{(x-1)(x+1)(x^2+1)} = \dfrac{A}{x-1} + \dfrac{B}{x+1} + \dfrac{Cx+D}{x^2+1} \quad \Rightarrow$

$1 = A(x+1)(x^2+1) + B(x-1)(x^2+1) + (Cx+D)(x-1)(x+1)$. Set $x=1$ to get $A=\frac{1}{4}$, and set $x=-1$ to get $B=-\frac{1}{4}$. Now take $x=0$ to get $1=A-B-D=-D+\frac{1}{2}$, so that $D=-\frac{1}{2}$. Finally equate the coefficients of x^4 to get $C=0$. Now

$$\int \frac{x^4\,dx}{x^4-1} = \int \left[1 + \frac{1/4}{x-1} - \frac{1/4}{x+1} - \frac{1/2}{x^2+1}\right] dx = x + \frac{1}{4}\ln\left|\frac{x-1}{x+1}\right| - \tfrac{1}{2}\tan^{-1}x + C.$$

49. $\dfrac{3x^3-x^2+6x-4}{(x^2+1)(x^2+2)} = \dfrac{Ax+B}{x^2+1} + \dfrac{Cx+D}{x^2+2} \quad \Rightarrow$

$3x^3-x^2+6x-4 = (Ax+B)(x^2+2) + (Cx+D)(x^2+1)$. Equating the coefficients gives $A+C=3$, $B+D=-1$, $2A+C=6$, and $2B+D=-4 \quad \Rightarrow \quad A=3$, $C=0$, $B=-3$, and $D=2$. Now

$$\int \frac{3x^3-x^2+6x-4}{(x^2+1)(x^2+2)}\,dx = 3\int \frac{x-1}{x^2+1}\,dx + 2\int \frac{dx}{x^2+2} = \tfrac{3}{2}\ln(x^2+1) - 3\tan^{-1}x + \sqrt{2}\tan^{-1}\left(\frac{x}{\sqrt{2}}\right) + C.$$

50. $\dfrac{x^3-2x^2+x+1}{x^4+5x^2+4} = \dfrac{x^3-2x^2+x+1}{(x^2+1)(x^2+4)} = \dfrac{Ax+B}{x^2+1} + \dfrac{Cx+D}{x^2+4} \quad \Rightarrow$

$x^3-2x^2+x+1 = (Ax+B)(x^2+4) + (Cx+D)(x^2+1)$. Equating coefficients gives $A+C=1$, $B+D=-2$, $4A+C=1$, $4B+D=1 \quad \Rightarrow \quad A=0$, $C=1$, $B=1$, $D=-3$. Now

$$\int \frac{x^3-2x^2+x+1}{x^4+5x^2+4}\,dx = \int \frac{dx}{x^2+1} + \int \frac{x-3}{x^2+4}\,dx = \tan^{-1}x + \tfrac{1}{2}\ln(x^2+4) - \tfrac{3}{2}\tan^{-1}(x/2) + C.$$

51. $$\int \frac{x-3}{(x^2+2x+4)^2}\,dx = \int \frac{x-3}{[(x+1)^2+3]^2}\,dx = \int \frac{u-4}{(u^2+3)^2}\,du \quad (\text{with } u = x+1)$$

$$= \int \frac{u\,du}{(u^2+3)^2} - 4\int \frac{du}{(u^2+3)^2} = \frac{1}{2}\int \frac{dv}{v^2} - 4\int \frac{\sqrt{3}\sec^2\theta\,d\theta}{9\sec^4\theta} \quad \left[\begin{array}{c} v = u^2+3 \text{ in the first integral;} \\ u = \sqrt{3}\tan\theta \text{ in the second} \end{array}\right]$$

$$= \frac{-1}{(2v)} - \frac{4\sqrt{3}}{9}\int \cos^2\theta\,d\theta = \frac{-1}{2(u^2+3)} - \tfrac{2\sqrt{3}}{9}(\theta + \sin\theta\cos\theta) + C$$

$$= \frac{-1}{2(x^2+2x+4)} - \frac{2\sqrt{3}}{9}\left[\tan^{-1}\left(\frac{x+1}{\sqrt{3}}\right) + \frac{\sqrt{3}(x+1)}{x^2+2x+4}\right] + C$$

$$= \frac{-1}{2(x^2+2x+4)} - \frac{2\sqrt{3}}{9}\tan^{-1}\left(\frac{x+1}{\sqrt{3}}\right) - \frac{2(x+1)}{3(x^2+2x+4)} + C$$

52. $\dfrac{x^4+1}{x(x^2+1)^2} = \dfrac{A}{x} + \dfrac{Bx+C}{x^2+1} + \dfrac{Dx+E}{(x^2+1)^2} \quad \Rightarrow \quad x^4+1 = A(x^2+1)^2 + (Bx+C)x(x^2+1) + (Dx+E)x.$

Setting $x=0$ gives $A=1$, and equating the coefficients of x^4 gives $1=A+B$, so $B=0$. Now

$$\frac{C}{x^2+1} + \frac{Dx+E}{(x^2+1)^2} = \frac{x^4+1}{x(x^2+1)^2} - \frac{1}{x} = \frac{1}{x}\left[\frac{x^4+1-(x^4+2x^2+1)}{(x^2+1)^2}\right] = \frac{-2x}{(x^2+1)^2},$$ so we can take $C=0$,

$D=-2$, and $E=0$. Hence $\displaystyle\int \frac{x^4+1}{x(x^2+1)^2}\,dx = \int\left[\frac{1}{x} - \frac{2x}{(x^2+1)^2}\right]dx = \ln|x| + \frac{1}{x^2+1} + C.$

53. Let $u = \sin^2 x - 3\sin x + 2$. Then $du = (2\sin x\cos x - 3\cos x)dx$, so

$$\int \frac{(2\sin x - 3)\cos x}{\sin^2 x - 3\sin x + 2}\,dx = \int \frac{du}{u} = \ln|u| + C = \ln\left|\sin^2 x - 3\sin x + 2\right| + C.$$

54. Let $u = \cos x$, then $du = -\sin x\,dx \quad \Rightarrow \quad \displaystyle\int \frac{\sin x\cos^2 x}{5+\cos^2 x}\,dx = \int \frac{-u^2\,du}{5+u^2} = -\int\left[1 - \frac{5}{u^2+5}\right]du$

$$= -u + \tfrac{5}{\sqrt{5}}\tan^{-1}\left(\tfrac{1}{\sqrt{5}}u\right) + C = -\cos x + \sqrt{5}\tan^{-1}\left(\tfrac{1}{\sqrt{5}}\cos x\right) + C.$$

55.

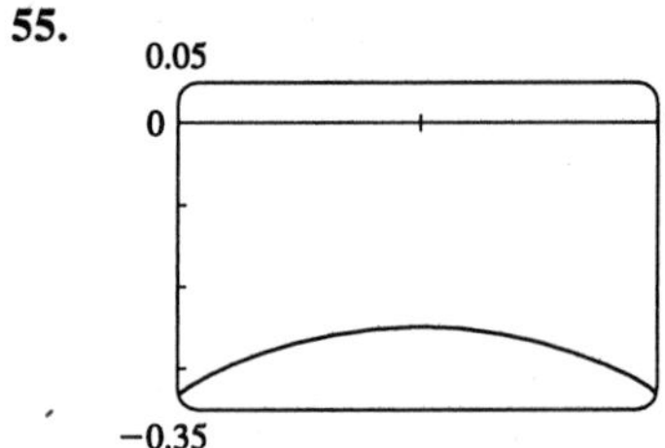

From the graph, we see that the integral will be negative, and we guess that the area is about the same as that of a rectangle with width 2 and height 0.3, so we estimate the integral to be $-(2\cdot 0.3) = -0.6$.

Now $\dfrac{1}{x^2-2x-3} = \dfrac{1}{(x-3)(x+1)} = \dfrac{A}{x-3} + \dfrac{B}{x+1} \quad \Leftrightarrow$

$1 = (A+B)x + A - 3B$, so $A = -B$ and $A - 3B = 1 \quad \Leftrightarrow$

$A = \frac{1}{4}$ and $B = -\frac{1}{4}$, so the integral becomes

$$\int_0^2 \frac{dx}{x^2-2x-3} = \frac{1}{4}\int_0^2 \frac{dx}{x-3} - \frac{1}{4}\int_0^2 \frac{dx}{x+1} = \tfrac{1}{4}\left[\ln|x-3| - \ln|x+1|\right]_0^2 = \frac{1}{4}\left[\ln\left|\frac{x-3}{x+1}\right|\right]_0^2$$

$$= \tfrac{1}{4}\left(\ln\tfrac{1}{3} - \ln 3\right) = -\tfrac{1}{2}\ln 3 \approx -0.55.$$

56. $\dfrac{1}{x^3-2x^2} = \dfrac{1}{x^2(x-2)} = \dfrac{A}{x} + \dfrac{B}{x^2} + \dfrac{C}{x-2} \quad \Rightarrow \quad 1 = (A+C)x^2 + (B-2A)x - 2B$, so $A+C = B-2A$

$= 0$ and $-2B = 1 \quad \Rightarrow \quad B = -\frac{1}{2}$, $A = -\frac{1}{4}$, and $C = \frac{1}{4}$. So the general antiderivative of $\dfrac{1}{x^3-2x^2}$ is

$$\int \frac{dx}{x^3-2x^2} = -\frac{1}{4}\int\frac{dx}{x} - \frac{1}{2}\int\frac{dx}{x^2} + \frac{1}{4}\int\frac{dx}{x-2}$$

$$= -\tfrac{1}{4}\ln|x| - \frac{1}{2}\left(-\frac{1}{x}\right) + \tfrac{1}{4}\ln|x-2| + C$$

$$= \frac{1}{4}\ln\left|\frac{x-2}{x}\right| + \frac{1}{2x} + C.$$

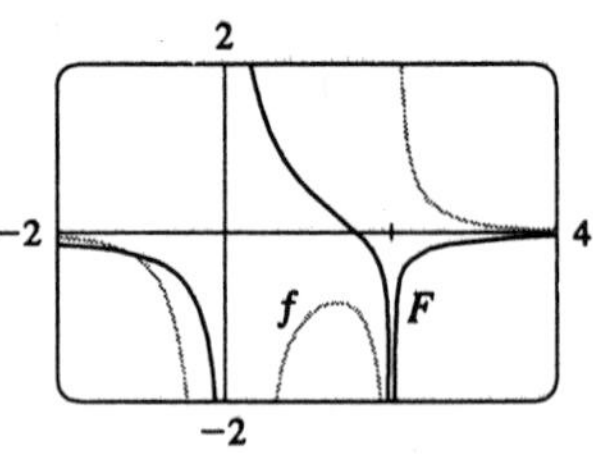

We plot this function with $C=0$ on the same screen as $y = \dfrac{1}{x^3-2x^2}$.

57. If $|x| < a$, then $\displaystyle\int \frac{dx}{a^2-x^2} = \int \frac{a\,\text{sech}^2 u\,du}{a^2\,\text{sech}^2 u}$ (put $x = a\tanh u$) $= \dfrac{u}{a} + C = \dfrac{1}{a}\tanh^{-1}\left(\dfrac{x}{a}\right) + C.$

If $|x| > a$, then $\displaystyle\int \frac{dx}{a^2-x^2} = \int \frac{-a\,\text{csch}^2 u\,du}{-a^2\,\text{csch}^2 u}$ (put $x = a\coth u$) $= \dfrac{u}{a} + C = \dfrac{1}{a}\coth^{-1}\left(\dfrac{x}{a}\right) + C.$

58. $\displaystyle\int \frac{dx}{x^2+2x-3} = \int \frac{dx}{(x+1)^2-4} = \int \frac{du}{u^2-4}$ [substitute $u = x+1$]

$\displaystyle= \frac{1}{4}\ln\left|\frac{u-2}{u+2}\right| + C$ (by Equation 6) $\displaystyle= \frac{1}{4}\ln\left|\frac{x-1}{x+3}\right| + C$

59. $\displaystyle\int \frac{dx}{x^2-2x} = \int \frac{dx}{(x-1)^2-1} = \int \frac{du}{u^2-1}$ (put $u = x-1$)

$\displaystyle= \frac{1}{2}\ln\left|\frac{u-1}{u+1}\right| + C$ (by Equation 6) $\displaystyle= \frac{1}{2}\ln\left|\frac{x-2}{x}\right| + C$

60. $\displaystyle\int \frac{(2x+1)dx}{4x^2+12x-7} = \frac{1}{4}\int \frac{(8x+12)dx}{4x^2+12x-7} - \int \frac{2\,dx}{(2x+3)^2-16}$

$\displaystyle= \tfrac{1}{4}\ln|4x^2+12x-7| - \int \frac{du}{u^2-16}$ [put $u = 2x+3$]

$= \tfrac{1}{4}\ln|4x^2+12x-7| - \tfrac{1}{8}\ln|(u-4)/(u+4)| + C$ (by Equation 6)

$= \tfrac{1}{4}\ln|4x^2+12x-7| - \tfrac{1}{8}\ln|(2x-1)/(2x+7)| + C$

61. $\displaystyle\int \frac{x\,dx}{x^2+x-1} = \frac{1}{2}\int \frac{(2x+1)dx}{x^2+x-1} - \frac{1}{2}\int \frac{dx}{\left(x+\frac{1}{2}\right)^2-\frac{5}{4}} = \tfrac{1}{2}\ln|x^2+x-1| - \frac{1}{2}\int \frac{du}{u^2-\left(\frac{\sqrt{5}}{2}\right)^2}$

(where $u = x+\frac{1}{2}$) $\displaystyle= \tfrac{1}{2}\ln|x^2+x-1| - \frac{1}{2\sqrt{5}}\ln\left|\frac{u-\sqrt{5}/2}{u+\sqrt{5}/2}\right| + C$

$\displaystyle= \tfrac{1}{2}\ln|x^2+x-1| - \frac{1}{2\sqrt{5}}\ln\left|\frac{2x+1-\sqrt{5}}{2x+1+\sqrt{5}}\right| + C$

62. $x^2-6x+8 = (x-3)^2-1$ is positive for $5 \le x \le 10$, so

area $\displaystyle= \int_5^{10} \frac{dx}{(x-3)^2-1} = \int_2^7 \frac{du}{u^2-1}$ (put $u = x-3$) $\displaystyle= \frac{1}{2}\ln\left|\frac{u-1}{u+1}\right|\Bigg]_2^7$

$= \tfrac{1}{2}\ln\tfrac{3}{4} - \tfrac{1}{2}\ln\tfrac{1}{3} = \tfrac{1}{2}(\ln 3 - 2\ln 2 + \ln 3) = \ln 3 - \ln 2 = \ln\tfrac{3}{2}$.

63. $\displaystyle\frac{x+1}{x-1} = 1 + \frac{2}{x-1} > 0$ for $2 \le x \le 3$, so

area $\displaystyle= \int_2^3 \left[1 + \frac{2}{x-1}\right]dx = [x + 2\ln|x-1|]_2^3 = (3+2\ln 2) - (2+2\ln 1) = 1 + 2\ln 2$.

64. We use disks, so the volume is $\displaystyle V = \pi\int_0^1 \left[\frac{1}{x^2+3x+2}\right]^2 dx = \pi\int_0^1 \frac{dx}{(x+1)^2(x+2)^2}$. To evaluate the integral, we use partial fractions: $\displaystyle\frac{1}{(x+1)^2(x+2)^2} = \frac{A}{x+1} + \frac{B}{(x+1)^2} + \frac{C}{x+2} + \frac{D}{(x+2)^2} \Rightarrow$ $1 = A(x+1)(x+2)^2 + B(x+2)^2 + C(x+1)^2(x+2) + D(x+1)^2$. We set $x = -1$, giving $B = 1$, then set $x = -2$, giving $D = 1$. Now equating coefficients of x^3 gives $A = -C$, and then equating constants gives $1 = 4A + 4 + 2(-A) + 1 \Rightarrow A = -2 \Rightarrow C = 2$. So the expression becomes

$\displaystyle V = \pi\int_0^1 \left[\frac{-2}{x+1} + \frac{1}{(x+1)^2} + \frac{2}{(x+2)} + \frac{1}{(x+2)^2}\right]dx = \pi\left[2\ln\left|\frac{x+2}{x+1}\right| - \frac{1}{x+1} - \frac{1}{x+2}\right]_0^1$

$= \pi\left[\left(2\ln\tfrac{3}{2} - \tfrac{1}{2} - \tfrac{1}{3}\right) - \left(2\ln 2 - 1 - \tfrac{1}{2}\right)\right] = \pi\left(2\ln\tfrac{3/2}{2} + \tfrac{2}{3}\right) = \pi\left(\tfrac{2}{3} + \ln\tfrac{9}{16}\right)$.

65. In this case, we use cylindrical shells, so the volume is $V = 2\pi \int_0^1 \frac{x\,dx}{x^2+3x+2} = 2\pi \int_0^1 \frac{x}{(x+1)(x+2)}$. We use partial fractions to simplify the integrand: $\frac{x}{(x+1)(x+2)} = \frac{A}{x+1} + \frac{B}{x+2} \Rightarrow$

$x = (A+B)x + 2A + B$. So $A + B = 1$ and $2A + B = 0 \Rightarrow A = -1$ and $B = 2$. So the volume is

$$2\pi \int_0^1 \left[\frac{-1}{x+1} + \frac{2}{x+2}\right] dx = 2\pi\left[-\ln|x+1| + 2\ln|x+2|\right]_0^1$$
$$= 2\pi(-\ln 2 + 2\ln 3 + \ln 1 - 2\ln 2) = 2\pi(2\ln 3 - 3\ln 2) = 2\pi \ln \tfrac{9}{8}.$$

66. If we subtract and add $2x^2$, we get

$$x^4 + 1 = x^4 + 2x^2 + 1 - 2x^2 = \left(x^2+1\right)^2 - 2x^2 = \left(x^2+1\right)^2 - \left(\sqrt{2}x\right)^2$$
$$= \left[\left(x^2+1\right) - \sqrt{2}x\right]\left[\left(x^2+1\right) + \sqrt{2}x\right] = \left(x^2 - \sqrt{2}x + 1\right)\left(x^2 + \sqrt{2}x + 1\right).$$

So we can decompose $\frac{1}{x^4+1} = \frac{Ax+B}{x^2+\sqrt{2}x+1} + \frac{Cx+D}{x^2-\sqrt{2}x+1} \Rightarrow$

$1 = (Ax+B)\left(x^2 - \sqrt{2}x + 1\right) + (Cx+D)\left(x^2 + \sqrt{2}x + 1\right)$. Setting the constant terms equal gives

$B + D = 1$, then from the coefficients of x^3 we get $A + C = 0$. Now from the coefficients of x we get $A + C + (B - D)\sqrt{2} = 0 \Leftrightarrow [(1-D) - D]\sqrt{2} = 0 \Rightarrow D = \frac{1}{2} \Rightarrow B = \frac{1}{2}$, and finally, from the coefficients of x^2 we get $\sqrt{2}(C - A) + B + D = 0 \Rightarrow C - A = -\frac{1}{\sqrt{2}} \Rightarrow C = -\frac{\sqrt{2}}{4}$ and $A = \frac{\sqrt{2}}{4}$.

So we rewrite the integrand, splitting the terms into forms which we know how to integrate:

$$\frac{1}{x^4+1} = \frac{\frac{\sqrt{2}}{4}x + \frac{1}{2}}{x^2+\sqrt{2}x+1} + \frac{-\frac{\sqrt{2}}{4}x + \frac{1}{2}}{x^2-\sqrt{2}x+1} = \frac{1}{4\sqrt{2}}\left[\frac{2x+2\sqrt{2}}{x^2+\sqrt{2}x+1} - \frac{2x-2\sqrt{2}}{x^2-\sqrt{2}x+1}\right]$$
$$= \frac{\sqrt{2}}{8}\left[\frac{2x+\sqrt{2}}{x^2+\sqrt{2}x+1} - \frac{2x-\sqrt{2}}{x^2-\sqrt{2}x+1}\right] + \frac{1}{4}\left[\frac{1}{\left(x+\frac{1}{\sqrt{2}}\right)^2 + \frac{1}{2}} + \frac{1}{\left(x-\frac{1}{\sqrt{2}}\right)^2 + \frac{1}{2}}\right].$$

Now we integrate: $\int \frac{dx}{x^4+1} = \frac{\sqrt{2}}{8} \ln\left(\frac{x^2+\sqrt{2}x+1}{x^2-\sqrt{2}x+1}\right) + \frac{\sqrt{2}}{4}\left[\tan^{-1}\left(\sqrt{2}x+1\right) + \tan^{-1}\left(\sqrt{2}x-1\right)\right] + C.$

67. **(a)** In Maple, we define $f(x)$, and then use `convert(f,parfrac,x);` to obtain

$f(x) = \frac{24{,}110/4879}{5x+2} - \frac{668/323}{2x+1} - \frac{9438/80{,}155}{3x-7} + \frac{(22{,}098x + 48{,}935)/260{,}015}{x^2+x+5}$. In Mathematica, we use the command `Apart`, and in Derive, we use `Expand`.

(b) $\int f(x)dx = \frac{24{,}110}{4879} \cdot \frac{1}{5}\ln|5x+2| - \frac{668}{323} \cdot 2\ln|2x+1| - \frac{9438}{80{,}155} \cdot \frac{1}{3}\ln|3x-7|$

$$+ \frac{1}{260{,}015}\int \frac{22{,}098\left(x+\frac{1}{2}\right) + 37{,}886}{\left(x+\frac{1}{2}\right)^2 + \frac{19}{4}}\,dx + C$$
$$= \frac{24{,}110}{4879} \cdot \frac{1}{5}\ln|5x+2| - \frac{668}{323} \cdot \frac{1}{2}\ln|2x+1| - \frac{9438}{80{,}155} \cdot \frac{1}{3}\ln|3x-7|$$
$$+ \frac{1}{260{,}015}\left[22{,}098 \cdot \frac{1}{2}\ln\left|x^2+x+5\right| + 37{,}886 \cdot \sqrt{\frac{4}{19}}\tan^{-1}\left(\frac{1}{\sqrt{19/4}}\left(x+\frac{1}{2}\right)\right)\right] + C$$
$$= \frac{4822}{4879}\ln|5x+2| - \frac{334}{323}\ln|2x+1| - \frac{3146}{80{,}155}\ln|3x-7| + \frac{11{,}049}{260{,}015}\ln\left|x^2+x+5\right|$$
$$+ \frac{75{,}772}{260{,}015\sqrt{19}}\tan^{-1}\left[\frac{1}{\sqrt{19}}(2x+1)\right] + C.$$

If we tell Maple to `integrate(f,x);` we get

$$\frac{4822\ln(5x+2)}{4879} - \frac{334\ln(2x+1)}{323} - \frac{3146\ln(3x-7)}{80{,}155} + \frac{11{,}049\ln(x^2+x+5)}{260{,}015} + \frac{3988\sqrt{19}}{260{,}115}\tan^{-1}\left[\tfrac{\sqrt{19}}{19}(2x+1)\right].$$

The main difference in Maple's answer is that the absolute value signs and the constant of integration have been omitted. Also, the fractions have been reduced and the denominators rationalized.

68. **(a)** In Maple, we define $f(x)$, and then use `convert(f,parfrac,x);` to get

$$f(x) = \frac{5828/1815}{(5x-2)^2} - \frac{59{,}096/19{,}965}{5x-2} + \frac{2(2843x+816)/3993}{2x^2+1} + \frac{(313x-251)/363}{(2x^2+1)^2}.$$

In Mathematica, we use the command `Apart`, and in Derive, we use `Expand`.

(b) As we saw in Exercise 67, computer algebra systems omit the absolute value signs in $\int 1/y\,dy = \ln|y|$. So we use the CAS to integrate the expression in part (a) and add the necessary absolute value signs and constant of integration to get

$$\int f(x)dx = -\frac{5828}{9075(5x-2)} - \frac{59{,}096\ln|5x-2|}{99{,}825} + \frac{2843\ln(2x^2+1)}{7986} + \frac{503}{15{,}972}\sqrt{2}\tan^{-1}(\sqrt{2}x) - \frac{1}{2904}\frac{1004x+626}{2x^2+1} + C.$$

(c) From the graph, we see that f goes from negative to positive at $x \approx -0.78$, then back to negative at $x \approx 0.8$, and finally back to positive at $x = 1$. Also $\lim_{x\to 0.4} f(x) = \infty$. So we see (by the First Derivative Test) that $\int f(x)dx$ has minima at $x \approx -0.78$ and $x = 1$, and a maximum at $x \approx 0.80$, and that $\int f(x)dx$ is unbounded as $x \to 0.4$. Note also that just to the right of $x = 0.4$, f has large values, so $\int f(x)dx$ increases rapidly, but slows down as f drops toward 0. $\int f(x)dx$ decreases from about 0.8 to 1, then increases slowly since f stays small and positive.

69. There are only finitely many values of x where $Q(x) = 0$ (assuming that Q is not the zero polynomial). At all other values of x, $F(x)/Q(x) = G(x)/Q(x)$, so $F(x) = G(x)$. In other words, the values of F and G agree at all except perhaps finitely many values of x. By continuity of F and G, the polynomials F and G must agree at those values of x too.

More explicitly: if a is a value of x such that $Q(a) = 0$, then $Q(x) \neq 0$ for all x sufficiently close to a. Thus

$$F(a) = \lim_{x\to a} F(x) \begin{bmatrix}\text{by continuity}\\ \text{of } F\end{bmatrix} = \lim_{x\to a} G(x) \begin{bmatrix}\text{since } F(x) = G(x)\\ \text{whenever } Q(x) \neq 0\end{bmatrix} = G(a) \begin{bmatrix}\text{by continuity}\\ \text{of } G\end{bmatrix}.$$

70. Let $f(x) = ax^2 + bx + c$. We calculate the partial fraction decomposition of $\dfrac{f(x)}{x^2(x+1)^3}$. Since $f(0) = 1$, we must have $c = 1$, so $\dfrac{f(x)}{x^2(x+1)^3} = \dfrac{ax^2+bx+1}{x^2(x+1)^3} = \dfrac{A}{x} + \dfrac{B}{x^2} + \dfrac{C}{x+1} + \dfrac{D}{(x+1)^2} + \dfrac{E}{(x+1)^3}$. Now in order for the integral not to contain any logarithms, we must have $A = C = 0$, so $ax^2 + bx + 1 = B(x+1)^3 + Dx^2(x+1) + Ex^2$. Equating constant terms gives $B = 1$, then equating coefficients of x gives $3B = b \quad\Rightarrow\quad b = 3$. This is the quantity we are looking for, since $f'(0) = b$.

EXERCISES 7.5

1. Let $u = \sqrt{x}$. Then $x = u^2$, $dx = 2u\,du$ $\Rightarrow$

$$\int_0^1 \frac{dx}{1+\sqrt{x}} = \int_0^1 \frac{2u\,du}{1+u} = 2\int_0^1 \left[1 - \frac{1}{1+u}\right] du = 2[u - \ln(1+u)]_0^1 = 2(1 - \ln 2).$$

2. Let $u = \sqrt[3]{x}$. Then $x = u^3$, $dx = 3u^2\,du$ $\Rightarrow$

$$\int_0^1 \frac{x}{1+\sqrt[3]{x}}\,dx = \int_0^1 \frac{3u^2\,du}{1+u} = \int_0^1 \left(3u - 3 + \frac{3}{1+u}\right) du = \left[\tfrac{3}{2}u^2 - 3u + 3\ln(1+u)\right]_0^1 = 3\left(\ln 2 - \tfrac{1}{2}\right).$$

3. Let $u = \sqrt{x}$. Then $x = u^2$, $dx = 2u\,du$ $\Rightarrow$

$$\int \frac{\sqrt{x}\,dx}{x+1} = \int \frac{u \cdot 2u\,du}{u^2+1} = 2\int \left[1 - \frac{1}{u^2+1}\right] du = 2(u - \tan^{-1} u) + C = 2\left(\sqrt{x} - \tan^{-1}\sqrt{x}\right) + C.$$

4. Let $u = \sqrt{x+1}$. Then $x = u^2 - 1$, $dx = 2u\,du$ $\Rightarrow$

$$\int \frac{dx}{x\sqrt{x+1}} = \int \frac{2u\,du}{(u^2-1)u} = 2\int \frac{du}{u^2-1} = \ln\left|\frac{u-1}{u+1}\right| + C = \ln\left|\frac{\sqrt{x+1}-1}{\sqrt{x+1}+1}\right| + C.$$

5. Let $u = \sqrt[3]{x}$. Then $x = u^3$, $dx = 3u^2\,du$ $\Rightarrow$

$$\int \frac{dx}{x - \sqrt[3]{x}} = \int \frac{3u^2\,du}{u^3 - u} = 3\int \frac{u\,du}{u^2-1} = \tfrac{3}{2}\ln\left|u^2 - 1\right| + C = \tfrac{3}{2}\ln\left|x^{2/3} - 1\right| + C.$$

6. Let $u = \sqrt{x+2}$. Then $x = u^2 - 2$, $dx = 2u\,du$ $\Rightarrow$

$$I = \int \frac{dx}{x - \sqrt{x+2}} = \int \frac{2u\,du}{u^2 - 2 - u} = 2\int \frac{u\,du}{u^2 - u - 2} \text{ and } \frac{u}{u^2-u-2} = \frac{A}{u-2} + \frac{B}{u+1} \quad \Rightarrow$$

$u = A(u+1) + B(u-2)$ $\Rightarrow$ $A = \frac{2}{3}$, $B = \frac{1}{3}$, so

$$I = \frac{2}{3}\int \left[\frac{2}{u-2} + \frac{1}{u+1}\right] du = \tfrac{2}{3}(2\ln|u-2| + \ln|u+1|) + C$$

$$= \tfrac{2}{3}\left[2\ln\left|\sqrt{x+2} - 2\right| + \ln\left(\sqrt{x+2} + 1\right)\right] + C.$$

7. Let $u = \sqrt{x-1}$. Then $x = u^2 + 1$, $dx = 2u\,du$ $\Rightarrow$ $\displaystyle\int_5^{10} \frac{x^2\,dx}{\sqrt{x-1}} = \int_2^3 \frac{(u^2+1)^2\,2u\,du}{u}$

$$= 2\textstyle\int_2^3 (u^4 + 2u^2 + 1)\,du = 2\left[\tfrac{1}{5}u^5 + \tfrac{2}{3}u^3 + u\right]_2^3 = 2\left(\tfrac{243}{5} + 18 + 3\right) - 2\left(\tfrac{32}{5} + \tfrac{16}{3} + 2\right) = \tfrac{1676}{15}.$$

8. Let $u = \sqrt{x-1}$. Then $x = u^2 + 1$, $dx = 2u\,du$ $\Rightarrow$

$$\int_1^3 \frac{\sqrt{x-1}}{x+1}\,dx = \int_0^{\sqrt{2}} \frac{u}{u^2+2}\,2u\,du = 2\int_0^{\sqrt{2}} \left[1 - \frac{2}{u^2+2}\right] du = \left[2u - \frac{4}{\sqrt{2}}\tan^{-1}\frac{u}{\sqrt{2}}\right]_0^{\sqrt{2}}$$

$$= 2\sqrt{2} - 2\sqrt{2}\tan^{-1}1 = 2\sqrt{2}\left(1 - \tfrac{\pi}{4}\right).$$

9. Let $u = \sqrt{x}$. Then $x = u^2$, $dx = 2u\,du$ $\Rightarrow$

$$\int \frac{dx}{\sqrt{1+\sqrt{x}}} = \int \frac{2u\,du}{\sqrt{1+u}} = 2\int \frac{(v^2-1)2v\,dv}{v} \quad \left(\text{put } v = \sqrt{1+u},\ u = v^2 - 1,\ du = 2v\,dv\right)$$

$$= 4\textstyle\int (v^2 - 1)\,dv = \tfrac{4}{3}v^3 - 4v + C = \tfrac{4}{3}\left(1 + \sqrt{x}\right)^{3/2} - 4\sqrt{1+\sqrt{x}} + C.$$

10. Let $u = \sqrt{x}$. Then $x = u^2$, $dx = 2u\,du \quad \Rightarrow$

$$\int_{1/3}^{3} \frac{\sqrt{x}}{x^2 + x}\,dx = \int_{1/\sqrt{3}}^{\sqrt{3}} \frac{u \cdot 2u\,du}{u^4 + u^2} = 2\int_{1/\sqrt{3}}^{\sqrt{3}} \frac{du}{u^2+1} = 2\left[\tan^{-1}u\right]_{1/\sqrt{3}}^{\sqrt{3}} = 2\left(\tfrac{\pi}{3} - \tfrac{\pi}{6}\right) = \tfrac{\pi}{3}.$$

11. Let $u = \sqrt{x}$. Then $x = u^2$, $dx = 2u\,du \quad \Rightarrow$

$$\int \frac{\sqrt{x}+1}{\sqrt{x}-1}\,dx = \int \frac{u+1}{u-1}\,2u\,du = 2\int \frac{u^2+u}{u-1}\,du = 2\int \left[u + 2 + \frac{2}{u-1}\right]du$$
$$= u^2 + 4u + 4\ln|u-1| + C = x + 4\sqrt{x} + 4\ln\left|\sqrt{x} - 1\right| + C.$$

12. Let $u = \sqrt[3]{x}$. Then $x = u^3$, $dx = 3u^2\,du \quad \Rightarrow$

$$\int \frac{\sqrt[3]{x}+1}{\sqrt[3]{x}-1}\,dx = \int \frac{u+1}{u-1}\,3u^2\,du = 3\int \left(u^2 + 2u + 2 + \frac{2}{u-1}\right)du$$
$$= u^3 + 3u^2 + 6u + 6\ln|u-1| + C = x + 3x^{2/3} + 6\sqrt[3]{x} + 6\ln\left|\sqrt[3]{x} - 1\right| + C.$$

13. Let $u = \sqrt[3]{x^2+1}$. Then $x^2 = u^3 - 1$, $2x\,dx = 3u^2\,du \quad \Rightarrow$

$$\int \frac{x^3\,dx}{\sqrt[3]{x^2+1}} = \int \frac{(u^3-1)\frac{3}{2}u^2\,du}{u} = \tfrac{3}{2}\int (u^4 - u)\,du$$
$$= \tfrac{3}{10}u^5 - \tfrac{3}{4}u^2 + C = \tfrac{3}{10}(x^2+1)^{5/3} - \tfrac{3}{4}(x^2+1)^{2/3} + C.$$

14. Let $u = \sqrt[6]{x}$. Then $x = u^6$, $dx = 6u^5\,du$, $\sqrt{x} = u^3$, and $\sqrt[3]{x} = u^2$, so

$$\int \frac{\sqrt{x}\,dx}{\sqrt{x} - \sqrt[3]{x}} = \int \frac{u^3 \cdot 6u^5\,du}{u^3 - u^2} = 6\int \frac{u^6\,du}{u-1} = 6\int \left(u^5 + u^4 + u^3 + u^2 + u + 1 + \frac{1}{u-1}\right)du$$
$$= u^6 + \tfrac{6}{5}u^5 + \tfrac{3}{2}u^4 + 2u^3 + 3u^2 + 6u + 6\ln|u-1| + C$$
$$= x + \tfrac{6}{5}x^{5/6} + \tfrac{3}{2}x^{2/3} + 2\sqrt{x} + 3\sqrt[3]{x} + 6\sqrt[6]{x} + 6\ln\left|\sqrt[6]{x} - 1\right| + C.$$

15. Let $u = \sqrt[4]{x}$. Then $x = u^4$, $dx = 4u^3\,du \quad \Rightarrow$

$$\int \frac{dx}{\sqrt{x} + \sqrt[4]{x}} = \int \frac{4u^3\,du}{u^2+u} = 4\int \frac{u^2\,du}{u+1} = 4\int \left[u - 1 + \frac{1}{u+1}\right]du$$
$$= 2u^2 - 4u + 4\ln|u+1| + C = 2\sqrt{x} - 4\sqrt[4]{x} + 4\ln\left(\sqrt[4]{x} + 1\right) + C.$$

16. Let $u = \sqrt[12]{x}$. Then $x = u^{12}$, $dx = 12u^{11}\,du \quad \Rightarrow$

$$\int \frac{dx}{\sqrt[3]{x} + \sqrt[4]{x}} = \int \frac{12u^{11}\,du}{u^4 + u^3} = 12\int \frac{u^8\,du}{u+1} = 12\int \left(u^7 - u^6 + u^5 - u^4 + u^3 - u^2 + u - 1 + \frac{1}{u+1}\right)du$$
$$= \tfrac{3}{2}u^8 - \tfrac{12}{7}u^7 + 2u^6 - \tfrac{12}{5}u^5 + 3u^4 - 4u^3 + 6u^2 - 12u + 12\ln|u+1| + C$$
$$= \tfrac{3}{2}x^{2/3} - \tfrac{12}{7}x^{7/12} + 2\sqrt{x} - \tfrac{12}{5}x^{5/12} + 3\sqrt[3]{x} - 4\sqrt[4]{x} + 6\sqrt[6]{x} - 12\sqrt[12]{x} + 12\ln\left(\sqrt[12]{x} + 1\right) + C.$$

17. Let $u = \sqrt{x}$. Then $x = u^2$, $dx = 2u\,du \quad \Rightarrow$

$$\int \sqrt{\frac{1-x}{x}}\,dx = \int \frac{\sqrt{1-u^2}}{u}\,2u\,du = 2\int \sqrt{1-u^2}\,du = 2\int \cos^2\theta\,d\theta \quad \text{(put } u = \sin\theta\text{)}$$
$$= \theta + \sin\theta\cos\theta + C = \sin^{-1}\sqrt{x} + \sqrt{x(1-x)} + C.$$

Or: Let $u = \sqrt{\dfrac{1-x}{x}}$. This gives $I = \sqrt{x(1-x)} - \tan^{-1}\sqrt{\dfrac{1-x}{x}} + C.$

18. Let $u = \sin x$. Then $du = \cos x\,dx \quad \Rightarrow$

$$\int \frac{\cos x\,dx}{\sin^2 x + \sin x} = \int \frac{du}{u^2+u} = \int \frac{du}{u(u+1)} = \int \left[\frac{1}{u} - \frac{1}{u+1}\right] du = \ln\left|\frac{u}{u+1}\right| + C = \ln\left|\frac{\sin x}{1+\sin x}\right| + C.$$

19. Let $u = e^x$. Then $x = \ln u$, $dx = \dfrac{du}{u} \quad \Rightarrow$

$$\int \frac{e^{2x}\,dx}{e^{2x}+3e^x+2} = \int \frac{u^2(du/u)}{u^2+3u+2} = \int \frac{u\,du}{(u+1)(u+2)} = \int \left[\frac{-1}{u+1} + \frac{2}{u+2}\right] du$$
$$= 2\ln|u+2| - \ln|u+1| + C = \ln\left[(e^x+2)^2/(e^x+1)\right] + C.$$

20. Let $u = e^x$. Then $x = \ln u$, $dx = \dfrac{du}{u} \quad \Rightarrow$

$$\int \frac{dx}{\sqrt{1+e^x}} = \int \frac{du/u}{\sqrt{1+u}} = \int \frac{2v\,dv}{v(v^2-1)} \quad (\text{put } v = \sqrt{1+u},\ u = v^2-1,\ du = 2v\,dv)$$
$$= 2\int \frac{dv}{v^2-1} = \ln\left|\frac{v-1}{v+1}\right| + C = \ln\left|\frac{\sqrt{1+e^x}-1}{\sqrt{1+e^x}+1}\right| + C.$$

21. Let $u = e^x$. Then $x = \ln u$, $dx = \dfrac{du}{u} \quad \Rightarrow$

$$\int \sqrt{1-e^x}\,dx = \int \sqrt{1-u}\left(\frac{du}{u}\right) = \int \frac{\sqrt{1-u}\,du}{u} = \int \frac{v(-2v)dv}{1-v^2} \quad \left[\begin{array}{l}\text{where } v = \sqrt{1-u},\ u = 1-v^2, \\ \quad du = -2v\,dv)\end{array}\right]$$
$$= 2\int \left[1 + \frac{1}{v^2-1}\right] dv = 2\left[v + \frac{1}{2}\ln\left|\frac{v-1}{v+1}\right|\right] + C = 2\sqrt{1-e^x} + \ln\left(\frac{1-\sqrt{1-e^x}}{1+\sqrt{1-e^x}}\right) + C.$$

22. Let $t = \tan\left(\dfrac{x}{2}\right)$. Then $dx = \dfrac{2}{1+t^2}\,dt$, $\sin x = \dfrac{2t}{1+t^2} \quad \Rightarrow$

$$\int \frac{dx}{3-5\sin x} = \int \frac{2\,dt/(1+t^2)}{3-10t/(1+t^2)} = \int \frac{2\,dt}{3(1+t^2)-10t} = 2\int \frac{dt}{3t^2-10t+3}$$
$$= \frac{1}{4}\int \left[\frac{1}{t-3} - \frac{3}{3t-1}\right] dt = \tfrac{1}{4}(\ln|t-3| - \ln|3t-1|) + C = \frac{1}{4}\ln\left|\frac{\tan(x/2)-3}{3\tan(x-2)-1}\right| + C.$$

23. Let $t = \tan\left(\dfrac{x}{2}\right)$. Then, by Equation 1,

$$\int_0^{\pi/2} \frac{dx}{\sin x + \cos x} = \int_0^1 \frac{2\,dt}{2t+1-t^2} = -2\int_0^1 \frac{dt}{t^2-2t-1} = -2\int_0^1 \frac{dt}{(t-1)^2-2}$$
$$= -\frac{1}{\sqrt{2}}\ln\left|\frac{t-1-\sqrt{2}}{t-1+\sqrt{2}}\right|\Bigg|_0^1 = -\frac{1}{\sqrt{2}}\left[\ln 1 - \ln\frac{\sqrt{2}+1}{\sqrt{2}-1}\right] = \frac{1}{\sqrt{2}}\ln\left(\sqrt{2}+1\right)^2$$
$$= \sqrt{2}\ln\left(\sqrt{2}+1\right) \text{ or } -\sqrt{2}\ln\left(\sqrt{2}-1\right) \text{ since } \sqrt{2}+1 = \tfrac{1}{\sqrt{2}-1}$$
$$\text{or } \tfrac{1}{\sqrt{2}}\ln\left(3+2\sqrt{2}\right) \text{ since } \left(\sqrt{2}+1\right)^2 = 3+2\sqrt{2}.$$

24. Let $t = \tan\left(\dfrac{x}{2}\right)$. Then, by Equation 1,

$$\int_{\pi/3}^{\pi/2} \frac{dx}{1+\sin x - \cos x} = \int_{1/\sqrt{3}}^1 \frac{2\,dt/(1+t^2)}{1+2t/(1+t^2)-(1-t^2)/(1+t^2)} = \int_{1/\sqrt{3}}^1 \frac{2\,dt}{1+t^2+2t-1+t^2}$$
$$= \int_{1/\sqrt{3}}^1 \left[\frac{1}{t} - \frac{1}{t+1}\right] dt = \left[\ln t - \ln(t+1)\right]_{1/\sqrt{3}}^1 = \ln\tfrac{1}{2} - \ln\tfrac{1}{\sqrt{3}+1} = \ln\tfrac{\sqrt{3}+1}{2}.$$

25. Let $t = \tan(x/2)$. Then, by Equation 1,

$$\int \frac{dx}{3\sin x + 4\cos x} = \int \frac{2\,dt}{6t + 4(1-t^2)} = \int \frac{-\,dt}{2t^2 - 3t - 2} = -\int \left[\frac{-2/5}{2t+1} + \frac{1/5}{t-2}\right] dt$$

$$= \tfrac{1}{5}\ln\left|\frac{2t+1}{t-2}\right| + C = \tfrac{1}{5}\ln\left|\frac{2\tan(x/2)+1}{\tan(x/2)-2}\right| + C.$$

26. Let $t = \tan(x/2)$. Then

$$\int \frac{dx}{\sin x + \tan x} = \int \frac{2\,dt/(1+t^2)}{2t/(1+t^2) + 2t/(1-t^2)} = \int \frac{2(1-t^2)dt}{2t(1-t^2) + 2t(1+t^2)} = \int \frac{1-t^2}{2t}\,dt$$

$$= \frac{1}{2}\int\left[\frac{1}{t} - t\right]dt = \tfrac{1}{2}\left(\ln|t| - \tfrac{1}{2}t^2\right) + C = \tfrac{1}{2}\ln\left|\tan\frac{x}{2}\right| - \tfrac{1}{4}\tan^2\frac{x}{2} + C.$$

27. Let $t = \tan\left(\frac{x}{2}\right)$. Then $\displaystyle\int \frac{dx}{2\sin x + \sin 2x} = \frac{1}{2}\int \frac{dx}{\sin x + \sin x\cos x}$

$$= \frac{1}{2}\int \frac{2\,dt/(1+t^2)}{2t/(1+t^2) + 2t(1-t^2)/(1+t^2)^2} = \frac{1}{2}\int \frac{(1+t^2)dt}{t(1+t^2) + t(1-t^2)}$$

$$= \frac{1}{4}\int \frac{(1+t^2)dt}{t} = \frac{1}{4}\int\left[\frac{1}{t} + t\right]dt = \tfrac{1}{4}\ln|t| + \tfrac{1}{8}t^2 + C = \frac{1}{4}\ln\left|\tan\frac{x}{2}\right| + \frac{1}{8}\tan^2\frac{x}{2} + C.$$

28. Let $t = \tan\left(\frac{x}{2}\right)$. Then $\displaystyle\int \frac{\sec x\,dx}{1+\sin x} = \int \frac{(1+t^2)/(1-t^2)}{1+2t/(1+t^2)}\frac{2\,dt}{1+t^2} = \int \frac{2/(1-t^2)dt}{1+2t/(1+t^2)}$

$$= \int \frac{2(1+t^2)dt}{(1-t^2)(1+t^2+2t)} = 2\int \frac{(t^2+1)dt}{(1-t)(1+t)^3} = 2\int\left[\frac{1/4}{1-t} + \frac{1/4}{1+t} - \frac{1/2}{(1+t)^2} + \frac{1}{(1+t)^3}\right]dt$$

$$= -\tfrac{1}{2}\ln|1-t| + \tfrac{1}{2}\ln|1+t| + \frac{1}{1+t} - \frac{1}{(1+t)^2} + C$$

$$= \frac{1}{2}\ln\left|\frac{1+\tan(x/2)}{1-\tan(x/2)}\right| + \frac{1}{1+\tan(x/2)} - \frac{1}{[1+\tan(x/2)]^2} + C.$$

29. Let $t = \tan(x/2)$. Then

$$\int \frac{dx}{a\sin x + b\cos x} = \int \frac{2\,dt}{a(2t) + b(1-t^2)} = -\frac{2}{b}\int \frac{dt}{t^2 - 2(a/b)t - 1}$$

$$= -\frac{2}{b}\int \frac{dt}{(t-a/b)^2 - (1+a^2/b^2)} = -\frac{1}{b}\frac{b}{\sqrt{a^2+b^2}}\ln\left|\frac{t - a/b - \sqrt{a^2+b^2}/b}{t - a/b + \sqrt{a^2+b^2}/b}\right| + C$$

$$= \frac{1}{\sqrt{a^2+b^2}}\ln\left|\frac{b\tan(x/2) - a + \sqrt{a^2+b^2}}{b\tan(x/2) - a - \sqrt{a^2+b^2}}\right| + C.$$

30. $\displaystyle\int \frac{dx}{a^2\sin^2 x + b^2\cos^2 x} = \int \frac{dx}{(a^2/2)(1-\cos 2x) + (b^2/2)(1+\cos 2x)} = \int \frac{dx}{\frac{1}{2}(b^2+a^2) + \frac{1}{2}(b^2-a^2)\cos 2x}$

$$= \int \frac{dX}{(b^2+a^2) + (b^2-a^2)\cos X}\ \text{(where } X = 2x\text{)}\ = \int \frac{2\,dt/(1+t^2)}{(b^2+a^2) + (b^2-a^2)(1-t^2)/(1+t^2)}$$

$$[\text{where } t = \tan(X/2)]\ = \int \frac{2\,dt}{(b^2+a^2)(1+t^2) + (b^2-a^2)(1-t^2)} = \int \frac{2\,dt}{2a^2t^2 + 2b^2}$$

$$= \int \frac{dt}{a^2t^2 + b^2} = \frac{1}{a}\int \frac{du}{u^2+b^2}\ [\text{put } u = at,\ dt = du/a]$$

$$= \frac{1}{ab}\tan^{-1}\frac{u}{b} + C = \frac{1}{ab}\tan^{-1}\frac{at}{b} + C = \frac{1}{ab}\tan^{-1}\frac{a\tan x}{b} + C.$$

31. **(a)** Let $t = \tan\left(\frac{x}{2}\right)$. Then $\int \sec x\,dx = \int \frac{dx}{\cos x} = \int \frac{2\,dt}{1-t^2} = \int \left[\frac{1}{1-t} + \frac{1}{1+t}\right] dt$

$$= \ln|1+t| - \ln|1-t| + C = \ln\left|\frac{1+t}{1-t}\right| + C = \ln\left|\frac{1+\tan(x/2)}{1-\tan(x/2)}\right| + C.$$

(b) $\tan\left(\frac{\pi}{4} + \frac{x}{2}\right) = \frac{\tan(\pi/4) + \tan(x/2)}{1 - \tan(\pi/4)\tan(x/2)} = \frac{1+\tan(x/2)}{1-\tan(x/2)}$. Substituting in the formula from part (a), we get $\int \sec x\,dx = \ln\left|\tan\left(\frac{1}{4}\pi + \frac{1}{2}x\right)\right| + C$.

32. Let $t = \tan\left(\frac{x}{2}\right)$. Then $\int \csc x\,dx = \int \frac{dx}{\sin x} = \int \frac{2\,dt}{2t} = \ln|t| + C = \ln\left|\tan\frac{x}{2}\right| + C$.

33. According to Equation 7.2.1, $\int \sec x\,dx = \ln|\sec x + \tan x| + C$. Now

$$\frac{1+\tan(x/2)}{1-\tan(x/2)} = \frac{1+\sin(x/2)/\cos(x/2)}{1-\sin(x/2)/\cos(x/2)} = \frac{\cos(x/2)+\sin(x/2)}{\cos(x/2)-\sin(x/2)}$$

$$= \frac{[\cos(x/2)+\sin(x/2)]^2}{[\cos(x/2)-\sin(x/2)][\cos(x/2)+\sin(x/2)]} = \frac{1+2\cos(x/2)\sin(x/2)}{\cos^2(x/2)-\sin^2(x/2)}$$

$$= \frac{1+\sin x}{\cos x} \quad \text{(using identities from the endpapers)} \quad = \sec x + \tan x,$$

so $\ln\left|\frac{1+\tan(x/2)}{1-\tan(x/2)}\right| = \ln|\sec x + \tan x|$, and the formula in Exercise 31(a) agrees with (7.2.1).

34. Let $u = \tan x$ and $du = \sec^2 x\,dx \quad \Rightarrow$

$$\int \frac{dx}{a\sin^2 x + b\sin x\cos x + c\cos^2 x} = \int \frac{\sec^2 x\,dx}{a\tan^2 x + b\tan x + c} = \int \frac{du}{au^2+bu+c}$$

$$= \int \frac{du}{a[u+b/(2a)]^2 + [c - b^2/(4a)]} \quad \text{(complete the square)} \quad = \frac{1}{a}\int \frac{du}{[u+b/(2a)]^2 + (4ac-b^2)/(4a^2)}$$

$$= \frac{1}{a}\int \frac{dv}{v^2 - (b^2-4ac)/(4a^2)} \quad \left(\text{where } v = u + \frac{b}{2a} \text{ and } dv = du\right).$$

Case 1: $b^2 - 4ac > 0$ Let $k = \sqrt{b^2-4ac}/2a$. The integral becomes

$$\frac{1}{a}\int \frac{dv}{v^2-k^2} = \frac{1}{a}\int \left[\frac{1/(2k)}{v-k} - \frac{1/(2k)}{v+k}\right] dv \quad \text{(by partial fractions)}$$

$$= \frac{1}{2ak}\int \left[\frac{1}{v-k} - \frac{1}{v+k}\right] dv = \frac{1}{2ak}\ln\left|\frac{v-k}{v+k}\right| + C$$

$$= \frac{1}{\sqrt{b^2-4ac}}\ln\left|\frac{2a\tan x + b - \sqrt{b^2-4ac}}{2a\tan x + b + \sqrt{b^2-4ac}}\right| + C.$$

Case 2: $b^2 - 4ac = 0$ The integral becomes $\frac{1}{a}\int \frac{dv}{v^2} = \frac{-1}{av} + C = \frac{-1}{a\tan x + b/2} + C$.

Case 3: $b^2 - 4ac < 0$ Let $k = \sqrt{4ac-b^2}/2a$. The integral becomes

$$\frac{1}{a}\int \frac{dv}{v^2+k^2} = \frac{1}{a}\int \frac{d\theta}{k} \quad (\text{where } v = k\tan\theta,\ dv = k\sec^2\theta\,d\theta) \quad = \frac{\theta}{ak} + C$$

$$= \frac{1}{ak}\tan^{-1}\frac{v}{k} + C = \frac{2}{\sqrt{4ac-b^2}}\tan^{-1}\left(\frac{2a\tan x + b}{\sqrt{4ac-b^2}}\right) + C.$$

31. **(a)** Let $t = \tan\left(\frac{x}{2}\right)$. Then $\int \sec x\,dx = \int \frac{dx}{\cos x} = \int \frac{2\,dt}{1-t^2} = \int \left[\frac{1}{1-t} + \frac{1}{1+t}\right] dt$

$= \ln|1+t| - \ln|1-t| + C = \ln\left|\frac{1+t}{1-t}\right| + C = \ln\left|\frac{1+\tan(x/2)}{1-\tan(x/2)}\right| + C.$

(b) $\tan\left(\frac{\pi}{4} + \frac{x}{2}\right) = \frac{\tan(\pi/4) + \tan(x/2)}{1 - \tan(\pi/4)\tan(x/2)} = \frac{1+\tan(x/2)}{1-\tan(x/2)}$. Substituting in the formula from part (a), we get $\int \sec x\,dx = \ln\left|\tan\left(\frac{1}{4}\pi + \frac{1}{2}x\right)\right| + C.$

32. Let $t = \tan\left(\frac{x}{2}\right)$. Then $\int \csc x\,dx = \int \frac{dx}{\sin x} = \int \frac{2\,dt}{2t} = \ln|t| + C = \ln\left|\tan\frac{x}{2}\right| + C.$

33. According to Equation 7.2.1, $\int \sec x\,dx = \ln|\sec x + \tan x| + C$. Now

$$\frac{1+\tan(x/2)}{1-\tan(x/2)} = \frac{1+\sin(x/2)/\cos(x/2)}{1-\sin(x/2)/\cos(x/2)} = \frac{\cos(x/2)+\sin(x/2)}{\cos(x/2)-\sin(x/2)}$$

$$= \frac{[\cos(x/2)+\sin(x/2)]^2}{[\cos(x/2)-\sin(x/2)][\cos(x/2)+\sin(x/2)]} = \frac{1+2\cos(x/2)\sin(x/2)}{\cos^2(x/2)-\sin^2(x/2)}$$

$$= \frac{1+\sin x}{\cos x} \quad \text{(using identities from the endpapers)} \quad = \sec x + \tan x,$$

so $\ln\left|\frac{1+\tan(x/2)}{1-\tan(x/2)}\right| = \ln|\sec x + \tan x|$, and the formula in Exercise 31(a) agrees with (7.2.1).

34. Let $u = \tan x$ and $du = \sec^2 x\,dx \quad \Rightarrow$

$$\int \frac{dx}{a\sin^2 x + b\sin x\cos x + c\cos^2 x} = \int \frac{\sec^2 x\,dx}{a\tan^2 x + b\tan x + c} = \int \frac{du}{au^2 + bu + c}$$

$$= \int \frac{du}{a[u+b/(2a)]^2 + [c - b^2/(4a)]} \quad \text{(complete the square)} \quad = \frac{1}{a}\int \frac{du}{[u+b/(2a)]^2 + (4ac-b^2)/(4a^2)}$$

$$= \frac{1}{a}\int \frac{dv}{v^2 - (b^2-4ac)/(4a^2)} \quad \left(\text{where } v = u + \frac{b}{2a} \text{ and } dv = du\right).$$

Case 1: $b^2 - 4ac > 0$ Let $k = \sqrt{b^2-4ac}/2a$. The integral becomes

$$\frac{1}{a}\int \frac{dv}{v^2-k^2} = \frac{1}{a}\int \left[\frac{1/(2k)}{v-k} - \frac{1/(2k)}{v+k}\right] dv \quad \text{(by partial fractions)}$$

$$= \frac{1}{2ak}\int \left[\frac{1}{v-k} - \frac{1}{v+k}\right] dv = \frac{1}{2ak}\ln\left|\frac{v-k}{v+k}\right| + C$$

$$= \frac{1}{\sqrt{b^2-4ac}}\ln\left|\frac{2a\tan x + b - \sqrt{b^2-4ac}}{2a\tan x + b + \sqrt{b^2-4ac}}\right| + C.$$

Case 2: $b^2 - 4ac = 0$ The integral becomes $\frac{1}{a}\int \frac{dv}{v^2} = \frac{-1}{av} + C = \frac{-1}{a\tan x + b/2} + C.$

Case 3: $b^2 - 4ac < 0$ Let $k = \sqrt{4ac-b^2}/2a$. The integral becomes

$$\frac{1}{a}\int \frac{dv}{v^2+k^2} = \frac{1}{a}\int \frac{d\theta}{k} \quad (\text{where } v = k\tan\theta,\ dv = k\sec^2\theta\,d\theta) \quad = \frac{\theta}{ak} + C$$

$$= \frac{1}{ak}\tan^{-1}\frac{v}{k} + C = \frac{2}{\sqrt{4ac-b^2}}\tan^{-1}\left(\frac{2a\tan x + b}{\sqrt{4ac-b^2}}\right) + C.$$

14. Integrate by parts: $u = \sin^{-1}x$, $dv = x\,dx \quad \Rightarrow \quad du = \left(1/\sqrt{1-x^2}\right)dx$, $v = \frac{1}{2}x^2$, so

$$\int x\sin^{-1}x\,dx = \tfrac{1}{2}x^2\sin^{-1}x - \frac{1}{2}\int\frac{x^2\,dx}{\sqrt{1-x^2}} = \tfrac{1}{2}x^2\sin^{-1}x - \frac{1}{2}\int\frac{\sin^2\theta\cos\theta\,d\theta}{\cos\theta} \quad \begin{bmatrix}\text{where } x = \sin\theta \\ \text{for } -\frac{\pi}{2} \le \theta \le \frac{\pi}{2}\end{bmatrix}$$

$$= \tfrac{1}{2}x^2\sin^{-1}x - \tfrac{1}{4}\textstyle\int(1-\cos 2\theta)d\theta = \tfrac{1}{2}x^2\sin^{-1}x - \tfrac{1}{4}(\theta - \sin\theta\cos\theta) + C$$

$$= \tfrac{1}{2}x^2\sin^{-1}x - \tfrac{1}{4}\left[\sin^{-1}x - x\sqrt{1-x^2}\right] + C = \tfrac{1}{4}\left[(2x^2-1)\sin^{-1}x + x\sqrt{1-x^2}\right] + C.$$

15. Let $u = \sqrt{9-x^2}$. Then $u^2 = 9 - x^2$, $u\,du = -x\,dx \quad \Rightarrow$

$$\int\frac{\sqrt{9-x^2}}{x}\,dx = \int\frac{\sqrt{9-x^2}}{x^2}\,x\,dx = \int\frac{u}{9-u^2}(-u)du = \int\left[1 - \frac{9}{9-u^2}\right]du$$

$$= u + 9\int\frac{du}{u^2-9} = u + \frac{9}{2\cdot 3}\ln\left|\frac{u-3}{u+3}\right| + C = \sqrt{9-x^2} + \frac{3}{2}\ln\left|\frac{\sqrt{9-x^2}-3}{\sqrt{9-x^2}+3}\right| + C$$

$$= \sqrt{9-x^2} + \frac{3}{2}\ln\frac{\left(\sqrt{9-x^2}-3\right)^2}{x^2} + C = \sqrt{9-x^2} + 3\ln\left|\frac{3-\sqrt{9-x^2}}{x}\right| + C.$$

Or: Put $x = 3\sin\theta$.

16. $$\int\frac{x}{x^2+3x+2}\,dx = \int\left[\frac{-1}{x+1} + \frac{2}{x+2}\right]dx = \ln\left[\frac{(x+2)^2}{|x+1|}\right] + C$$

17. Integrate by parts: $u = x^2$, $dv = \cosh x\,dx \quad \Rightarrow \quad du = 2x\,dx$, $v = \sinh x$, so

$I = \int x^2\cosh x\,dx = x^2\sinh x - \int 2x\sinh x\,dx$. Now let $U = x$, $dV = \sinh x\,dx \Rightarrow dU = dx$, $V = \cosh x$.
So $I = x^2\sinh x - 2(x\cosh x - \int\cosh x\,dx)$

$= x^2\sinh x - 2[x\cosh x - \sinh x] = (x^2+2)\sinh x - 2x\cosh x + C$.

18. $$\int\frac{x^3+x+1}{x^4+2x^2+4x}\,dx = \tfrac{1}{4}\ln\left|x^4+2x^2+4x\right| + C$$

19. Let $u = \sin x$. Then $\displaystyle\int\frac{\cos x\,dx}{1+\sin^2 x} = \int\frac{du}{1+u^2} = \tan^{-1}u + C = \tan^{-1}(\sin x) + C$

20. Let $u = \sqrt{x}$. Then $x = u^2$, $dx = 2u\,du \quad \Rightarrow$

$\int\cos\sqrt{x}\,dx = \int\cos u\cdot 2u\,du = 2u\sin u - \int 2\sin u\,du$ (by parts)

$= 2u\sin u + 2\cos u + C = 2\left(\sqrt{x}\sin\sqrt{x} + \cos\sqrt{x}\right) + C$.

21. $\int_0^1\cos\pi x\tan\pi x\,dx = \int_0^1\sin\pi x\,dx = -\frac{1}{\pi}\int_0^1(-\pi\sin\pi x)dx = -\frac{1}{\pi}[\cos\pi x]_0^1 = -\frac{1}{\pi}(-1-1) = \frac{2}{\pi}$.

22. Let $u = e^x$. Then $x = \ln u$, $dx = du/u \quad \Rightarrow \quad \displaystyle\int\frac{e^{2x}}{1+e^x}\,dx = \int\frac{u^2}{1+u}\frac{du}{u} = \int\frac{u}{1+u}\,du$

$$= \int\left(1 - \frac{1}{1+u}\right)du = u - \ln|1+u| + C = e^x - \ln(1+e^x) + C.$$

23. Integrate by parts twice, first with $u = e^{3x}$, $dv = \cos 5x\,dx$:

$\int e^{3x}\cos 5x\,dx = \frac{1}{5}e^{3x}\sin 5x - \int\frac{3}{5}e^{3x}\sin 5x\,dx = \frac{1}{5}e^{3x}\sin 5x + \frac{3}{25}e^{3x}\cos 5x - \frac{9}{25}\int e^{3x}\cos 5x\,dx$, so
$\frac{34}{25}\int e^{3x}\cos 5x\,dx = \frac{1}{25}e^{3x}(5\sin 5x + 3\cos 5x) + C_1$ and $\int e^{3x}\cos 5x\,dx = \frac{1}{34}e^{3x}(5\sin 5x + 3\cos 5x) + C$.

24. $\int\cos 3x\cos 5x\,dx = \int\frac{1}{2}[\cos 8x + \cos 2x]dx = \frac{1}{16}\sin 8x + \frac{1}{4}\sin 2x + C$. *Or:* Use integration by parts.

14. Integrate by parts: $u = \sin^{-1}x$, $dv = x\,dx \quad \Rightarrow \quad du = \left(1/\sqrt{1-x^2}\right)dx$, $v = \frac{1}{2}x^2$, so

$$\int x\sin^{-1}x\,dx = \tfrac{1}{2}x^2\sin^{-1}x - \frac{1}{2}\int\frac{x^2\,dx}{\sqrt{1-x^2}} = \tfrac{1}{2}x^2\sin^{-1}x - \frac{1}{2}\int\frac{\sin^2\theta\cos\theta\,d\theta}{\cos\theta} \quad \begin{bmatrix}\text{where } x=\sin\theta \\ \text{for } -\frac{\pi}{2}\le\theta\le\frac{\pi}{2}\end{bmatrix}$$

$$= \tfrac{1}{2}x^2\sin^{-1}x - \tfrac{1}{4}\int(1-\cos 2\theta)d\theta = \tfrac{1}{2}x^2\sin^{-1}x - \tfrac{1}{4}(\theta - \sin\theta\cos\theta) + C$$

$$= \tfrac{1}{2}x^2\sin^{-1}x - \tfrac{1}{4}\left[\sin^{-1}x - x\sqrt{1-x^2}\right] + C = \tfrac{1}{4}\left[(2x^2-1)\sin^{-1}x + x\sqrt{1-x^2}\right] + C.$$

15. Let $u = \sqrt{9-x^2}$. Then $u^2 = 9 - x^2$, $u\,du = -x\,dx \quad \Rightarrow$

$$\int\frac{\sqrt{9-x^2}}{x}\,dx = \int\frac{\sqrt{9-x^2}}{x^2}\,x\,dx = \int\frac{u}{9-u^2}(-u)du = \int\left[1 - \frac{9}{9-u^2}\right]du$$

$$= u + 9\int\frac{du}{u^2-9} = u + \frac{9}{2\cdot 3}\ln\left|\frac{u-3}{u+3}\right| + C = \sqrt{9-x^2} + \frac{3}{2}\ln\left|\frac{\sqrt{9-x^2}-3}{\sqrt{9-x^2}+3}\right| + C$$

$$= \sqrt{9-x^2} + \frac{3}{2}\ln\frac{\left(\sqrt{9-x^2}-3\right)^2}{x^2} + C = \sqrt{9-x^2} + 3\ln\left|\frac{3-\sqrt{9-x^2}}{x}\right| + C.$$

Or: Put $x = 3\sin\theta$.

16. $$\int\frac{x}{x^2+3x+2}\,dx = \int\left[\frac{-1}{x+1} + \frac{2}{x+2}\right]dx = \ln\left[\frac{(x+2)^2}{|x+1|}\right] + C$$

17. Integrate by parts: $u = x^2$, $dv = \cosh x\,dx \quad \Rightarrow \quad du = 2x\,dx$, $v = \sinh x$, so

$I = \int x^2\cosh x\,dx = x^2\sinh x - \int 2x\sinh x\,dx$. Now let $U = x$, $dV = \sinh x\,dx \Rightarrow dU = dx$, $V = \cosh x$.
So $I = x^2\sinh x - 2(x\cosh x - \int\cosh x\,dx)$

$= x^2\sinh x - 2[x\cosh x - \sinh x] = (x^2+2)\sinh x - 2x\cosh x + C$.

18. $$\int\frac{x^3+x+1}{x^4+2x^2+4x}\,dx = \tfrac{1}{4}\ln\left|x^4+2x^2+4x\right| + C$$

19. Let $u = \sin x$. Then $\displaystyle\int\frac{\cos x\,dx}{1+\sin^2 x} = \int\frac{du}{1+u^2} = \tan^{-1}u + C = \tan^{-1}(\sin x) + C$

20. Let $u = \sqrt{x}$. Then $x = u^2$, $dx = 2u\,du \quad \Rightarrow$

$\int\cos\sqrt{x}\,dx = \int\cos u\cdot 2u\,du = 2u\sin u - \int 2\sin u\,du$ (by parts)

$= 2u\sin u + 2\cos u + C = 2\left(\sqrt{x}\sin\sqrt{x} + \cos\sqrt{x}\right) + C$.

21. $\int_0^1\cos\pi x\tan\pi x\,dx = \int_0^1\sin\pi x\,dx = -\frac{1}{\pi}\int_0^1(-\pi\sin\pi x)dx = -\frac{1}{\pi}[\cos\pi x]_0^1 = -\frac{1}{\pi}(-1-1) = \frac{2}{\pi}$.

22. Let $u = e^x$. Then $x = \ln u$, $dx = du/u \quad \Rightarrow \quad \displaystyle\int\frac{e^{2x}}{1+e^x}\,dx = \int\frac{u^2}{1+u}\frac{du}{u} = \int\frac{u}{1+u}\,du$

$$= \int\left(1 - \frac{1}{1+u}\right)du = u - \ln|1+u| + C = e^x - \ln(1+e^x) + C.$$

23. Integrate by parts twice, first with $u = e^{3x}$, $dv = \cos 5x\,dx$:

$\int e^{3x}\cos 5x\,dx = \frac{1}{5}e^{3x}\sin 5x - \int\frac{3}{5}e^{3x}\sin 5x\,dx = \frac{1}{5}e^{3x}\sin 5x + \frac{3}{25}e^{3x}\cos 5x - \frac{9}{25}\int e^{3x}\cos 5x\,dx$, so

$\frac{34}{25}\int e^{3x}\cos 5x\,dx = \frac{1}{25}e^{3x}(5\sin 5x + 3\cos 5x) + C_1$ and $\int e^{3x}\cos 5x\,dx = \frac{1}{34}e^{3x}(5\sin 5x + 3\cos 5x) + C$.

24. $\int\cos 3x\cos 5x\,dx = \int\frac{1}{2}[\cos 8x + \cos 2x]dx = \frac{1}{16}\sin 8x + \frac{1}{4}\sin 2x + C$. *Or:* Use integration by parts.

36. Let $u = x + 2$. Then $du = dx \quad \Rightarrow$

$$\int \frac{dx}{\sqrt{5 - 4x - x^2}} = \int \frac{dx}{\sqrt{9 - (x+2)^2}} = \int \frac{du}{\sqrt{9 - u^2}} = \sin^{-1}\left(\frac{u}{3}\right) + C = \sin^{-1}\left(\frac{x+2}{3}\right) + C.$$

37. Let $u = 1 - x^2$. Then $du = -2x\,dx \quad \Rightarrow$

$$\int \frac{x\,dx}{1 - x^2 + \sqrt{1 - x^2}} = -\frac{1}{2}\int \frac{du}{u + \sqrt{u}} = -\int \frac{v\,dv}{v^2 + v} \quad \text{(put } v = \sqrt{u},\, u = v^2,\, du = 2v\,dv\text{)}$$

$$= -\int \frac{dv}{v+1} = -\ln|v + 1| + C = -\ln\left(\sqrt{1 - x^2} + 1\right) + C.$$

38. $\displaystyle\int \frac{1 + \cos x}{\sin x}\,dx = \int (\csc x + \cot x)dx = \ln|\csc x - \cot x| + \ln|\sin x| + C = \ln|1 - \cos x| + C.$

Or: $\displaystyle\int \frac{1 + \cos x}{\sin x}\,dx = \int \frac{1 - \cos^2 x}{\sin x(1 - \cos x)}\,dx = \int \frac{\sin x\,dx}{1 - \cos x} = \ln|1 - \cos x| + C.$

39. Let $u = e^x$. Then $x = \ln u$, $dx = du/u \quad \Rightarrow$

$$\int \frac{e^x\,dx}{e^{2x} - 1} = \int \frac{u(du/u)}{u^2 - 1} = \int \frac{du}{u^2 - 1} = \frac{1}{2}\ln\left|\frac{u-1}{u+1}\right| + C = \tfrac{1}{2}\ln\left|\frac{e^x - 1}{e^x + 1}\right| + C.$$

40. $\displaystyle\int \frac{dx}{x^3 - 8} = \int \left[\frac{1/12}{x-2} - \frac{x/12 + 1/3}{x^2 + 2x + 4}\right]dx = \frac{1}{12}\int \left[\frac{1}{x-2} - \frac{x+4}{x^2 + 2x + 4}\right]dx$

$$= \tfrac{1}{12}\ln|x - 2| - \frac{1}{24}\int \frac{2x+2}{x^2 + 2x + 4}\,dx - \frac{1}{4}\int \frac{dx}{(x+1)^2 + 3}$$

$$= \tfrac{1}{12}\ln|x - 2| - \tfrac{1}{24}\ln(x^2 + 2x + 4) - \tfrac{1}{4\sqrt{3}}\tan^{-1}\left[\tfrac{1}{\sqrt{3}}(x+1)\right] + C$$

41. $\displaystyle\int_{-1}^{1} x^5 \cosh x\,dx = 0$ by Theorem 4.5.6, since $x^5 \cosh x$ is odd.

42. Let $u = \tan x$. Then

$$\int_{\pi/4}^{\pi/3} \frac{\ln(\tan x)dx}{\sin x \cos x} = \int_{\pi/4}^{\pi/3} \frac{\ln(\tan x)}{\tan x}\sec^2 x\,dx = \int_1^{\sqrt{3}} \frac{\ln u}{u}\,du = \left[\tfrac{1}{2}(\ln u)^2\right]_1^{\sqrt{3}} = \tfrac{1}{2}\left(\ln\sqrt{3}\right)^2 = \tfrac{1}{8}(\ln 3)^2.$$

43. $\int_{-3}^{3}|x^3 + x^2 - 2x|dx = \int_{-3}^{3}|(x+2)x(x-1)|dx$

$= -\int_{-3}^{-2}(x^3 + x^2 - 2x)dx + \int_{-2}^{0}(x^3 + x^2 - 2x)dx - \int_0^1(x^3 + x^2 - 2x)dx + \int_1^3(x^3 + x^2 - 2x)dx.$

Let $f(x) = \frac{1}{4}x^4 + \frac{1}{3}x^3 - x^2$. Then $f'(x) = x^3 + x^2 - 2x$, so

$\int_{-3}^{3}|x^3 + x^2 - 2x|dx = -f(-2) + f(-3) + f(0) - f(-2) - f(1) + f(0) + f(3) - f(1)$

$= f(-3) - 2f(-2) + 2f(0) - 2f(1) + f(3) = \frac{9}{4} - 2\left(-\frac{8}{3}\right) + 2 \cdot 0 - 2\left(-\frac{5}{12}\right) + \frac{81}{4} = \frac{86}{3}.$

44. Let $u = \sin\theta$. Then $\int_0^{\pi/4}\cos^5\theta\,d\theta = \int_0^{\pi/4}(1 - \sin^2\theta)^2\cos\theta\,d\theta = \int_0^{1/\sqrt{2}}(1 - u^2)^2\,du$

$= \int_0^{1/\sqrt{2}}(u^4 - 2u^2 + 1)du = \left[\frac{1}{5}u^5 - \frac{2}{3}u^3 + u\right]_0^{1/\sqrt{2}} = \frac{1}{20\sqrt{2}} - \frac{1}{3\sqrt{2}} + \frac{1}{\sqrt{2}} = \frac{43\sqrt{2}}{120}.$

45. Let $u = \ln(\sin x)$. Then $du = \cot x\,dx \quad \Rightarrow \quad \int \cot x \ln(\sin x)dx = \int u\,du = \frac{1}{2}u^2 + C = \frac{1}{2}[\ln(\sin x)]^2 + C.$

46. Let $u = e^x$. Then $x = \ln u$, $dx = du/u \quad \Rightarrow$

$$\int \frac{1 + e^x}{1 - e^x}\,dx = \int \frac{(1+u)du}{(1-u)u} = -\int \frac{(u+1)du}{(u-1)u} = -\int \left[\frac{2}{u-1} - \frac{1}{u}\right]du$$

$$= \ln|u| - 2\ln|u - 1| + C = \ln e^x - 2\ln|e^x - 1| + C = x - 2\ln|e^x - 1| + C.$$

36. Let $u = x + 2$. Then $du = dx \quad \Rightarrow$

$$\int \frac{dx}{\sqrt{5 - 4x - x^2}} = \int \frac{dx}{\sqrt{9 - (x+2)^2}} = \int \frac{du}{\sqrt{9 - u^2}} = \sin^{-1}\left(\frac{u}{3}\right) + C = \sin^{-1}\left(\frac{x+2}{3}\right) + C.$$

37. Let $u = 1 - x^2$. Then $du = -2x\,dx \quad \Rightarrow$

$$\int \frac{x\,dx}{1 - x^2 + \sqrt{1 - x^2}} = -\frac{1}{2}\int \frac{du}{u + \sqrt{u}} = -\int \frac{v\,dv}{v^2 + v} \quad (\text{put } v = \sqrt{u},\, u = v^2,\, du = 2v\,dv)$$

$$= -\int \frac{dv}{v+1} = -\ln|v+1| + C = -\ln\left(\sqrt{1-x^2} + 1\right) + C.$$

38. $\displaystyle\int \frac{1 + \cos x}{\sin x}\,dx = \int (\csc x + \cot x)dx = \ln|\csc x - \cot x| + \ln|\sin x| + C = \ln|1 - \cos x| + C.$

Or: $\displaystyle\int \frac{1 + \cos x}{\sin x}\,dx = \int \frac{1 - \cos^2 x}{\sin x(1 - \cos x)}\,dx = \int \frac{\sin x\,dx}{1 - \cos x} = \ln|1 - \cos x| + C.$

39. Let $u = e^x$. Then $x = \ln u$, $dx = du/u \quad \Rightarrow$

$$\int \frac{e^x\,dx}{e^{2x} - 1} = \int \frac{u(du/u)}{u^2 - 1} = \int \frac{du}{u^2 - 1} = \frac{1}{2}\ln\left|\frac{u-1}{u+1}\right| + C = \tfrac{1}{2}\ln\left|\frac{e^x - 1}{e^x + 1}\right| + C.$$

40. $\displaystyle\int \frac{dx}{x^3 - 8} = \int \left[\frac{1/12}{x-2} - \frac{x/12 + 1/3}{x^2 + 2x + 4}\right]dx = \frac{1}{12}\int \left[\frac{1}{x-2} - \frac{x+4}{x^2+2x+4}\right]dx$

$$= \tfrac{1}{12}\ln|x-2| - \frac{1}{24}\int \frac{2x+2}{x^2+2x+4}\,dx - \frac{1}{4}\int \frac{dx}{(x+1)^2 + 3}$$

$$= \tfrac{1}{12}\ln|x-2| - \tfrac{1}{24}\ln(x^2+2x+4) - \tfrac{1}{4\sqrt{3}}\tan^{-1}\left[\tfrac{1}{\sqrt{3}}(x+1)\right] + C$$

41. $\displaystyle\int_{-1}^{1} x^5\cosh x\,dx = 0$ by Theorem 4.5.6, since $x^5\cosh x$ is odd.

42. Let $u = \tan x$. Then

$$\int_{\pi/4}^{\pi/3} \frac{\ln(\tan x)dx}{\sin x\cos x} = \int_{\pi/4}^{\pi/3} \frac{\ln(\tan x)}{\tan x}\sec^2 x\,dx = \int_1^{\sqrt{3}} \frac{\ln u}{u}\,du = \left[\tfrac{1}{2}(\ln u)^2\right]_1^{\sqrt{3}} = \tfrac{1}{2}\left(\ln\sqrt{3}\right)^2 = \tfrac{1}{8}(\ln 3)^2.$$

43. $\int_{-3}^{3}|x^3 + x^2 - 2x|dx = \int_{-3}^{3}|(x+2)x(x-1)|dx$

$= -\int_{-3}^{-2}(x^3 + x^2 - 2x)dx + \int_{-2}^{0}(x^3 + x^2 - 2x)dx - \int_0^1(x^3 + x^2 - 2x)dx + \int_1^3(x^3 + x^2 - 2x)dx.$

Let $f(x) = \frac{1}{4}x^4 + \frac{1}{3}x^3 - x^2$. Then $f'(x) = x^3 + x^2 - 2x$, so

$\int_{-3}^{3}|x^3 + x^2 - 2x|dx = -f(-2) + f(-3) + f(0) - f(-2) - f(1) + f(0) + f(3) - f(1)$

$= f(-3) - 2f(-2) + 2f(0) - 2f(1) + f(3) = \frac{9}{4} - 2\left(-\frac{8}{3}\right) + 2\cdot 0 - 2\left(-\frac{5}{12}\right) + \frac{81}{4} = \frac{86}{3}.$

44. Let $u = \sin\theta$. Then $\int_0^{\pi/4}\cos^5\theta\,d\theta = \int_0^{\pi/4}(1 - \sin^2\theta)^2\cos\theta\,d\theta = \int_0^{1/\sqrt{2}}(1 - u^2)^2\,du$

$= \int_0^{1/\sqrt{2}}(u^4 - 2u^2 + 1)du = \left[\frac{1}{5}u^5 - \frac{2}{3}u^3 + u\right]_0^{1/\sqrt{2}} = \frac{1}{20\sqrt{2}} - \frac{1}{3\sqrt{2}} + \frac{1}{\sqrt{2}} = \frac{43\sqrt{2}}{120}.$

45. Let $u = \ln(\sin x)$. Then $du = \cot x\,dx \quad \Rightarrow \quad \int \cot x\ln(\sin x)dx = \int u\,du = \frac{1}{2}u^2 + C = \frac{1}{2}[\ln(\sin x)]^2 + C.$

46. Let $u = e^x$. Then $x = \ln u$, $dx = du/u \quad \Rightarrow$

$$\int \frac{1 + e^x}{1 - e^x}\,dx = \int \frac{(1+u)du}{(1-u)u} = -\int \frac{(u+1)du}{(u-1)u} = -\int \left[\frac{2}{u-1} - \frac{1}{u}\right]du$$

$$= \ln|u| - 2\ln|u-1| + C = \ln e^x - 2\ln|e^x - 1| + C = x - 2\ln|e^x - 1| + C.$$

58. $\int (x+\sin x)^2\,dx = \int (x^2 + 2x\sin x + \sin^2 x)dx = \frac{1}{3}x^3 + 2(\sin x - x\cos x) + \frac{1}{2}(x - \sin x\cos x) + C$
$= \frac{1}{3}x^3 + \frac{1}{2}x + 2\sin x - \frac{1}{2}\sin x\cos x - 2x\cos x + C$

59. Let $u = \arctan x$. Then $du = \dfrac{dx}{1+x^2} \Rightarrow \displaystyle\int \frac{e^{\arctan x}}{1+x^2}\,dx = \int e^u\,du = e^u + C = e^{\arctan x} + C.$

60. Let $u = x^2$. Then $du = 2x\,dx \Rightarrow \displaystyle\int \frac{dx}{x(x^4+1)} = \int \frac{x\,dx}{x^2(x^4+1)} = \frac{1}{2}\int \frac{du}{u(u^2+1)}$
$= \dfrac{1}{2}\displaystyle\int \left[\frac{1}{u} - \frac{u}{u^2+1}\right]du = \frac{1}{2}\ln|u| - \frac{1}{4}\ln(u^2+1) + C = \frac{1}{2}\ln(x^2) - \frac{1}{4}\ln(x^4+1) + C$
$= \frac{1}{4}\left[\ln(x^4) - \ln(x^4+1)\right] + C = \frac{1}{4}\ln\left(\dfrac{x^4}{x^4+1}\right) + C.$ *Or:* Write $I = \displaystyle\int \frac{x^3\,dx}{x^4(x^4+1)}$ and let $u = x^4$.

61. Integrate by parts three times, first with $u = t^3$, $dv = e^{-2t}\,dt$:
$\int t^3e^{-2t}\,dt = -\frac{1}{2}t^3e^{-2t} + \frac{1}{2}\int 3t^2e^{-2t}\,dt = -\frac{1}{2}t^3e^{-2t} - \frac{3}{4}t^2e^{-2t} + \frac{1}{2}\int 3te^{-2t}\,dt$
$= -e^{-2t}\left[\frac{1}{2}t^3 + \frac{3}{4}t^2\right] - \frac{3}{4}te^{-2t} + \frac{3}{4}\int e^{-2t}\,dt = -e^{-2t}\left[\frac{1}{2}t^3 + \frac{3}{4}t^2 + \frac{3}{4}t + \frac{3}{8}\right] + C$
$= -\frac{1}{8}e^{-2t}(4t^3 + 6t^2 + 6t + 3) + C.$

62. Let $u = \sqrt[6]{t}$. Then $t = u^6$, $dt = 6u^5\,du \Rightarrow$
$\displaystyle\int \frac{\sqrt{t}\,dt}{1+\sqrt[3]{t}} = \int \frac{u^3\cdot 6u^5\,du}{1+u^2} = 6\int \frac{u^8}{u^2+1}\,du = 6\int\left(u^6 - u^4 + u^2 - 1 + \frac{1}{u^2+1}\right)du$
$= 6\left(\frac{1}{7}u^7 - \frac{1}{5}u^5 + \frac{1}{3}u^3 - u + \tan^{-1}u\right) + C = 6\left(\frac{1}{7}t^{7/6} - \frac{1}{5}t^{5/6} + \frac{1}{3}t^{1/2} - t^{1/6} + \tan^{-1}t^{1/6}\right) + C.$

63. $\int \sin x\sin 2x\sin 3x\,dx = \int \sin x\cdot\frac{1}{2}[\cos(2x-3x) - \cos(2x+3x)]dx$
$= \frac{1}{2}\int(\sin x\cos x - \sin x\cos 5x)dx = \frac{1}{4}\int \sin 2x\,dx - \frac{1}{2}\int \frac{1}{2}[\sin(x+5x) + \sin(x-5x)]dx$
$= -\frac{1}{8}\cos 2x - \frac{1}{4}\int(\sin 6x - \sin 4x)dx = -\frac{1}{8}\cos 2x + \frac{1}{24}\cos 6x - \frac{1}{16}\cos 4x + C$

64. $\int_1^3 |\ln(x/2)|dx = -\int_1^2 \ln(x/2)\,dx + \int_2^3 \ln(x/2)\,dx$. Let $u = x/2$. Then $x = 2u$, $dx = 2\,du \Rightarrow$
$\int \ln(x/2)\,dx = 2\int \ln u\,du = 2(u\ln u - u) + C = x\ln(x/2) - x + C$, so
$\int_1^3 |\ln(x/2)|dx = [x - x\ln(x/2)]_1^2 + [x\ln(x/2) - x]_2^3$
$= 2 - \left[1 - \ln\frac{1}{2}\right] + \left[3\ln\frac{3}{2} - 3\right] - (-2) = 3\ln 3 - 4\ln 2 = \ln\frac{27}{16}.$

65. As in Example 5, $\displaystyle\int\sqrt{\frac{1+x}{1-x}}\,dx = \int \frac{1+x}{\sqrt{1-x^2}}\,dx = \int \frac{dx}{\sqrt{1-x^2}} + \int \frac{x\,dx}{\sqrt{1-x^2}} = \sin^{-1}x - \sqrt{1-x^2} + C.$

Another Method: Substitute $u = \sqrt{(1+x)/(1-x)}$.

66. Let $t = \sqrt{x^2-1}$. Then $dt = \left(x/\sqrt{x^2-1}\right)dx$, $x^2 - 1 = t^2$, $x = \sqrt{t^2+1}$, so
$I = \displaystyle\int \frac{x\ln x}{\sqrt{x^2-1}}\,dx = \int \ln\sqrt{t^2+1}\,dt = \frac{1}{2}\int \ln(t^2+1)\,dt.$ Now use parts with $u = \ln(t^2+1)$, $dv = dt$:
$I = \frac{1}{2}t\ln(t^2+1) - \displaystyle\int \frac{t^2}{t^2+1}\,dt = \frac{1}{2}t\ln(t^2+1) - \int\left[1 - \frac{1}{t^2+1}\right]dt$
$= \frac{1}{2}t\ln(t^2+1) - t + \tan^{-1}t + C = \sqrt{x^2-1}\ln x - \sqrt{x^2-1} + \tan^{-1}\sqrt{x^2-1} + C.$

Another Method: First integrate by parts with $u = \ln x$, $dv = \left(x/\sqrt{x^2-1}\right)dx$ and then use substitution $\left(x = \sec\theta \text{ or } u = \sqrt{x^2-1}\right)$.

67. $\displaystyle\int \frac{x+a}{x^2+a^2}\,dx = \frac{1}{2}\int \frac{2x\,dx}{x^2+a^2} + a\int \frac{dx}{x^2+a^2} = \tfrac{1}{2}\ln(x^2+a^2) + a\cdot\frac{1}{a}\tan^{-1}\!\left(\frac{x}{a}\right) + C$

$\displaystyle = \ln\sqrt{x^2+a^2} + \tan^{-1}(x/a) + C.$

68. Let $x - \frac{1}{2} = \frac{\sqrt{5}}{2}u$. Then $dx = \frac{\sqrt{5}}{2}\,du$, $u = \frac{1}{\sqrt{5}}(2x-1) \quad\Rightarrow$

$$\int \sqrt{1+x-x^2}\,dx = \int \sqrt{\tfrac{5}{4} - \left(x-\tfrac{1}{2}\right)^2}\,dx = \frac{5}{4}\int \sqrt{1-u^2}\,du = \frac{5}{4}\int \cos\theta\cos\theta\,d\theta$$

$$= \tfrac{5}{8}(\theta + \sin\theta\cos\theta) + C = \frac{5}{8}\left[\sin^{-1}\!\left(\frac{2x-1}{\sqrt{5}}\right) + \frac{2x-1}{\sqrt{5}}\frac{2}{\sqrt{5}}\sqrt{1+x-x^2}\right] + C$$

$$= \tfrac{5}{8}\sin^{-1}\!\left[\tfrac{1}{\sqrt{5}}(2x-1)\right] + \tfrac{1}{4}(2x-1)\sqrt{1+x-x^2} + C.$$

69. Let $u = x^5$. Then $du = 5x^4\,dx \;\Rightarrow\; \displaystyle\int \frac{x^4\,dx}{x^{10}+16} = \int \frac{\frac{1}{5}\,du}{u^2+16} = \tfrac{1}{5}\cdot\tfrac{1}{4}\tan^{-1}\!\left(\tfrac{1}{4}u\right) + C = \tfrac{1}{20}\tan^{-1}\!\left(\tfrac{1}{4}x^5\right) + C.$

70. $\displaystyle\int \frac{x+2}{x^2+x+2}\,dx = \frac{1}{2}\int \frac{2x+1}{x^2+x+2}\,dx + \frac{3}{2}\int \frac{dx}{(x+1/2)^2+7/4}$

$$= \tfrac{1}{2}\ln(x^2+x+2) + \tfrac{3}{2}\tfrac{2}{\sqrt{7}}\tan^{-1}\!\left[\tfrac{2}{\sqrt{7}}\left(x+\tfrac{1}{2}\right)\right] + C = \ln\sqrt{x^2+x+2} + \tfrac{3}{\sqrt{7}}\tan^{-1}\!\left[\tfrac{1}{\sqrt{7}}(2x+1)\right] + C$$

71. Integrate by parts with $u = x$, $dv = \sec x\tan x\,dx \quad\Rightarrow\quad du = dx$, $v = \sec x$:

$\int x\sec x\tan x\,dx = x\sec x - \int \sec x\,dx = x\sec x - \ln|\sec x + \tan x| + C.$

72. Let $u = x^2$. Then $du = 2x\,dx \Rightarrow \displaystyle\int \frac{x\,dx}{x^4-a^4} = \int \frac{(1/2)du}{u^2-(a^2)^2} = \frac{1}{4a^2}\ln\left|\frac{u-a^2}{u+a^2}\right| + C = \frac{1}{4a^2}\ln\left|\frac{x^2-a^2}{x^2+a^2}\right| + C.$

73. $\displaystyle\int \frac{dx}{\sqrt{x+1}+\sqrt{x}} = \int \left(\sqrt{x+1}-\sqrt{x}\right)dx = \frac{2}{3}\left[(x+1)^{3/2} - x^{3/2}\right] + C.$

74. Let $u = e^x$. Then $x = \ln u$, $dx = du/u \quad\Rightarrow$

$$\int \frac{dx}{1+2e^x-e^{-x}} = \int \frac{du/u}{1+2u-1/u} = \int \frac{du}{2u^2+u-1} = \int \left[\frac{2/3}{2u-1} - \frac{1/3}{u+1}\right]du$$

$$= \tfrac{1}{3}\ln|2u-1| - \tfrac{1}{3}\ln|u+1| + C = \tfrac{1}{3}\ln|(2e^x-1)/(e^x+1)| + C.$$

75. Let $u = \sqrt{x}$. Then $du = dx/(2\sqrt{x}) \quad\Rightarrow$

$$\int \frac{\arctan\sqrt{x}}{\sqrt{x}}\,dx = \int \tan^{-1}u\;2\,du = 2u\tan^{-1}u - \int \frac{2u\,du}{1+u^2} \quad \text{(by parts)}$$

$$= 2u\tan^{-1}u - \ln(1+u^2) + C = 2\sqrt{x}\tan^{-1}\sqrt{x} - \ln(1+x) + C.$$

76. Use parts with $u = \ln(x+1)$, $dv = dx/x^2$:

$$\int \frac{\ln(x+1)}{x^2}\,dx = -\frac{1}{x}\ln(x+1) + \int \frac{dx}{x(x+1)} = -\frac{1}{x}\ln(x+1) + \int \left[\frac{1}{x} - \frac{1}{x+1}\right]dx$$

$$= -\frac{1}{x}\ln(x+1) + \ln|x| - \ln(x+1) + C = -\left(1+\frac{1}{x}\right)\ln(x+1) + \ln|x| + C.$$

77. Let $u = e^x$. Then $x = \ln u$, $dx = du/u \quad\Rightarrow$

$$\int \frac{dx}{e^{3x}-e^x} = \int \frac{du/u}{u^3-u} = \int \frac{du}{(u-1)u^2(u+1)} = \int \left[\frac{1/2}{u-1} - \frac{1}{u^2} - \frac{1/2}{u+1}\right]du = \frac{1}{u} + \frac{1}{2}\ln\left|\frac{u-1}{u+1}\right| + C$$

$$= e^{-x} + \tfrac{1}{2}\ln|(e^x-1)/(e^x+1)| + C.$$

78. $\int \frac{1+\cos^2 x}{1-\cos^2 x}\,dx = \int \frac{1+\cos^2 x}{\sin^2 x}\,dx = \int (\csc^2 x + \cot^2 x)\,dx = \int (2\csc^2 x - 1)\,dx = -2\cot x - x + C$

79. Let $u = \sqrt{2x-25} \Rightarrow u^2 = 2x - 25 \Rightarrow 2u\,du = 2\,dx \Rightarrow$
$\int \frac{dx}{x\sqrt{2x-25}} = \int \frac{u\,du}{\frac{1}{2}(u^2+25)\cdot u} = 2\int \frac{du}{u^2+25} = \frac{2}{5}\tan^{-1}\left(\frac{1}{5}u\right) + C = \frac{2}{5}\tan^{-1}\left(\frac{1}{5}\sqrt{2x-25}\right) + C.$

80. Let $u = \cos^2 x \Rightarrow du = -2\cos x \sin x\,dx = -\sin 2x\,dx \Rightarrow$
$\int \frac{\sin 2x}{\sqrt{9-\cos^4 x}}\,dx = \int \frac{-du}{\sqrt{9-u^2}} = -\sin^{-1}\left(\frac{1}{3}u\right) + C = -\sin^{-1}\left(\frac{1}{3}\cos^2 x\right) + C.$

EXERCISES 7.7

1. By Formula 99,
$\int e^{-3x}\cos 4x\,dx = \frac{e^{-3x}}{(-3)^2+4^2}(-3\cos 4x + 4\sin 4x) + C = \frac{e^{-3x}}{25}(-3\cos 4x + 4\sin 4x) + C.$

2. Let $u = x/2$ and use Formula 72:
$\int \csc^3(x/2)\,dx = 2\int \csc^3 u\,du = -\csc u \cot u + \ln|\csc u - \cot u| + C$
$= -\csc(x/2)\cot(x/2) + \ln|\csc(x/2) - \cot(x/2)| + C.$

3. Let $u = 3x$. Then $du = 3\,dx$, so $\int \frac{\sqrt{9x^2-1}}{x^2}\,dx = \int \frac{\sqrt{u^2-1}}{u^2/9}\,\frac{du}{3} = 3\int \frac{\sqrt{u^2-1}}{u^2}\,du$
$= -\frac{3\sqrt{u^2-1}}{u} + 3\ln\left|u + \sqrt{u^2-1}\right| + C$ (by Formula 42) $= -\frac{\sqrt{9x^2-1}}{x} + 3\ln\left|3x + \sqrt{9x^2-1}\right| + C.$

4. By Formula 32, $\int \frac{\sqrt{4-3x^2}}{x}\,dx = \int \frac{\sqrt{4-u^2}}{u/\sqrt{3}}\,\frac{du}{\sqrt{3}} \quad \left(u = \sqrt{3}x\right)$
$= \int \frac{\sqrt{4-u^2}}{u}\,du = \sqrt{4-u^2} - 2\ln\left|\frac{2+\sqrt{4-u^2}}{u}\right| + C_1 = \sqrt{4-3x^2} - 2\ln\left|\frac{2+\sqrt{4-3x^2}}{\sqrt{3}x}\right| + C_1$
$= \sqrt{4-3x^2} - 2\ln\left|\frac{2+\sqrt{4-3x^2}}{x}\right| + C.$

5. $\int x^2 e^{3x}\,dx = \frac{1}{3}x^2 e^{3x} - \frac{2}{3}\int x e^{3x}\,dx$ (Formula 97) $= \frac{1}{3}x^2 e^{3x} - \frac{2}{3}\left[\frac{1}{9}(3x-1)e^{3x}\right] + C$ (Formula 96)
$= \frac{1}{27}(9x^2 - 6x + 2)e^{3x} + C$

6. Let $u = \sin x$. Then $du = \cos x\,dx$, so $\int \frac{\sin x \cos x}{\sqrt{1+\sin x}}\,dx = \int \frac{u\,du}{\sqrt{1+u}} = \frac{2}{3}(u-2)\sqrt{1+u} + C$ (Formula 55) $= -\frac{2}{3}(2-\sin x)\sqrt{1+\sin x} + C.$

7. Let $u = x^2$. Then $du = 2x\,dx$, so $\int x\sin^{-1}(x^2)dx = \frac{1}{2}\int \sin^{-1}u\,du = \frac{1}{2}\left(u\sin^{-1}u + \sqrt{1-u^2}\right) + C$
(Formula 87) $= \frac{1}{2}\left(x^2\sin^{-1}(x^2) + \sqrt{1-x^4}\right) + C.$

8. Let $u = x^2$. Then $du = 2x\,dx$, so $\int x^3 \sin^{-1}(x^2)dx = \frac{1}{2}\int u \sin^{-1}u\,du = \dfrac{2u^2-1}{8}\sin^{-1}u + \dfrac{u\sqrt{1-u^2}}{8} + C$ (Formula 90) $= \dfrac{2x^4-1}{8}\sin^{-1}(x^2) + \dfrac{x^2\sqrt{1-x^4}}{8} + C$.

9. Let $u = e^x$. Then $du = e^x\,dx$, so $\int e^x \operatorname{sech}(e^x)dx = \int \operatorname{sech} u\,du = \tan^{-1}|\sinh u| + C$ (Formula 107) $= \tan^{-1}[\sinh(e^x)] + C$.

10. Let $u = 3x$. Then $du = 3\,dx$, so $\int x^2 \cos 3x\,dx = \frac{1}{27}\int u^2 \cos u\,du = \frac{1}{27}(u^2 \sin u - 2\int u \sin u\,du)$ (Formula 85) $= \frac{1}{3}x^2 \sin 3x - \frac{2}{27}(\sin 3x - 3x\cos 3x) + C$ (Formula 82) $= \frac{1}{27}[(9x^2-2)\sin 3x + 6x\cos 3x] + C$.

11. Let $u = x + 2$. Then $\displaystyle\int \sqrt{5-4x-x^2}\,dx = \int \sqrt{9-(x+2)^2}\,dx = \int \sqrt{9-u^2}\,du$ $= \dfrac{u}{2}\sqrt{9-u^2} + \dfrac{9}{2}\sin^{-1}\dfrac{u}{3} + C$ (Formula 30) $= \dfrac{x+2}{2}\sqrt{5-4x-x^2} + \dfrac{9}{2}\sin^{-1}\dfrac{x+2}{3} + C$.

12. Let $u = x^2$. Then $du = 2x\,dx$, so by Formula 48, $\displaystyle\int \frac{x^5\,dx}{x^2+\sqrt{2}} = \frac{1}{2}\int \frac{u^2}{u+\sqrt{2}}\,du$

$= \frac{1}{2}\cdot\frac{1}{2}\left[\left(u+\sqrt{2}\right)^2 - 4\sqrt{2}\left(u+\sqrt{2}\right) + 4\ln\left|u+\sqrt{2}\right|\right] + C$

$= \frac{1}{4}\left[\left(x^2+\sqrt{2}\right)^2 - 4\sqrt{2}\left(x^2+\sqrt{2}\right) + 4\ln\left(x^2+\sqrt{2}\right)\right] + C = \frac{1}{4}x^4 - \frac{1}{\sqrt{2}}x^2 + \ln\left(x^2+\sqrt{2}\right) + K$.

Or: Let $u = x^2 + \sqrt{2}$.

13. $\int \sec^5 x\,dx = \frac{1}{4}\tan x \sec^3 x + \frac{3}{4}\int \sec^3 x\,dx$ (Formula 77)

$= \frac{1}{4}\tan x\sec^3 x + \frac{3}{4}\left(\frac{1}{2}\tan x \sec x + \frac{1}{2}\int \sec x\,dx\right)$ (Formula 77 again)

$= \frac{1}{4}\tan x \sec^3 x + \frac{3}{8}\tan x \sec x + \frac{3}{8}\ln|\sec x + \tan x| + C$ (Formula 14)

14. Let $u = 2x$. Then $du = 2\,dx$, so

$\int \sin^6 2x\,dx = \frac{1}{2}\int \sin^6 u\,du = \frac{1}{2}\left(-\frac{1}{6}\sin^5 u \cos u + \frac{5}{6}\int \sin^4 u\,du\right)$ (Formula 73)

$= -\frac{1}{12}\sin^5 u\cos u + \frac{5}{12}\left(-\frac{1}{4}\sin^3 u \cos u + \frac{3}{4}\int \sin^2 u\,du\right)$ (Formula 73)

$= -\frac{1}{12}\sin^5 u \cos u - \frac{5}{48}\sin^3 u\cos u + \frac{5}{16}\left(\frac{1}{2}u - \frac{1}{4}\sin 2u\right) + C$ (Formula 63)

$= -\frac{1}{12}\sin^5 2x\cos 2x - \frac{5}{48}\sin^3 2x \cos 2x - \frac{5}{64}\sin 4x + \frac{5}{16}x + C$.

15. Let $u = \sin x$. Then $du = \cos x\,dx$, so $\int \sin^2 x \cos x \ln(\sin x)dx = \int u^2 \ln u\,du$ $= \frac{1}{9}u^3(3\ln u - 1) + C$ (Formula 101) $= \frac{1}{9}\sin^3 x[3\ln(\sin x) - 1] + C$.

16. Let $u = e^x$. Then $x = \ln u$, $dx = du/u$, so $\displaystyle\int \frac{dx}{e^x(1+2e^x)} = \int \frac{du/u}{u(1+2u)} = \int \frac{du}{u^2(1+2u)}$ $= -\dfrac{1}{u} + 2\ln\left|\dfrac{1+2u}{u}\right| + C$ (Formula 50) $= -e^{-x} + 2\ln(e^{-x}+2) + C$.

17. $\displaystyle\int \sqrt{2+3\cos x}\tan x\,dx = -\int \frac{\sqrt{2+3\cos x}}{\cos x}(-\sin x\,dx) = -\int \frac{\sqrt{2+3u}}{u}\,du$ (where $u = \cos x$)

$\displaystyle = -2\sqrt{2+3u} - 2\int \frac{du}{u\sqrt{2+3u}}$ (Formula 58) $\displaystyle = -2\sqrt{2+3u} - 2\cdot\frac{1}{\sqrt{2}}\ln\left|\frac{\sqrt{2+3u}-\sqrt{2}}{\sqrt{2+3u}+\sqrt{2}}\right| + C$

(Formula 57) $\displaystyle = -2\sqrt{2+3\cos x} - \sqrt{2}\ln\left|\frac{\sqrt{2+3\cos x}-\sqrt{2}}{\sqrt{2+3\cos x}+\sqrt{2}}\right| + C$

18. Let $u = x - 2$. Then $\displaystyle\int \frac{x\,dx}{\sqrt{x^2-4x}} = \int \frac{(x-2)+2}{\sqrt{(x-2)^2-4}}\,dx$

$$= \int \frac{u\,du}{\sqrt{u^2-4}} + 2\int \frac{du}{\sqrt{u^2-4}} = \frac{1}{2}\int v^{-1/2}\,dv + 2\int \frac{du}{\sqrt{u^2-4}} \quad (\text{put } v = u^2 - 4)$$

$$= v^{1/2} + 2\ln\left|u + \sqrt{u^2-4}\right| + C \quad (\text{Formulas 2 and 43}) = \sqrt{x^2-4x} + 2\ln\left|x - 2 + \sqrt{x^2-4x}\right| + C.$$

19. $\int_0^{\pi/2} \cos^5 x\,dx = \frac{1}{5}\left[\cos^4 x \sin x\right]_0^{\pi/2} + \frac{4}{5}\int_0^{\pi/2}\cos^3 x\,dx$ (Formula 74)

$$= 0 + \tfrac{4}{5}\left[\tfrac{1}{3}(2+\cos^2 x)\sin x\right]_0^{\pi/2} \quad (\text{Formula 68}) \quad = \tfrac{4}{15}(2-0) = \tfrac{8}{15}$$

20. Since $\int x^4 e^{-x}\,dx = -x^4 e^{-x} + 4\int x^3 e^{-x}\,dx$ (Formula 97)

$$= -x^4 e^{-x} + 4\left(-x^3 e^{-x} + 3\int x^2 e^{-x}\,dx\right) \quad (\text{Formula 97})$$
$$= -\left(x^4 + 4x^3\right)e^{-x} + 12\left(-x^2 e^{-x} + 2\int x e^{-x}\,dx\right) \quad (\text{Formula 97})$$
$$= -\left(x^4 + 4x^3 + 12x^2\right)e^{-x} + 24[(-x-1)e^{-x}] + C \quad (\text{Formula 96})$$
$$= -\left(x^4 + 4x^3 + 12x^2 + 24x + 24\right)e^{-x} + C,$$

we find that $\int_0^1 x^4 e^{-x}\,dx = -(1+4+12+24+24)e^{-1} + 24e^0 = 24 - 65e^{-1}$.

21. Let $u = x^5$, $du = 5x^4\,dx$.

$$\int \frac{x^4\,dx}{\sqrt{x^{10}-2}} = \frac{1}{5}\int \frac{du}{\sqrt{u^2-2}} = \tfrac{1}{5}\ln\left|u + \sqrt{u^2-2}\right| + C \quad (\text{Formula 43}) \quad = \tfrac{1}{5}\ln\left|x^5 + \sqrt{x^{10}-2}\right| + C.$$

22. Let $u = 3x + 4$. Then $\int e^x \cos(3x+4)dx = \frac{1}{3}\int e^{(u-4)/3}\cos u\,du = \frac{1}{3}e^{-4/3}\int e^{u/3}\cos u\,du$

$$= \tfrac{1}{3}e^{-4/3}\frac{e^{u/3}}{(1/3)^2 + 1^2}\left(\tfrac{1}{3}\cos u + \sin u\right) + C \quad (\text{Formula 99 with } a = \tfrac{1}{3},\ b = 1)$$

$$= \tfrac{3}{10}e^x\left[\tfrac{1}{3}\cos(3x+4) + \sin(3x+4)\right] + C.$$

23. Let $u = 1 + e^x$, so $du = e^x\,dx$. Then $\int e^x \ln(1+e^x)dx = \int \ln u\,du = u\ln u - u + C$ (Formula 100)

$$= (1+e^x)\ln(1+e^x) - e^x - 1 + C = (1+e^x)\ln(1+e^x) - e^x + C_1.$$

24. Using Formula 95 with $n = 2$,

$$\int x^2 \tan^{-1}x\,dx = \frac{1}{3}\left[x^3\tan^{-1}x - \int \frac{x^3\,dx}{1+x^2}\right] = \frac{x^3}{3}\tan^{-1}x - \frac{1}{3}\int\left(x - \frac{x}{x^2+1}\right)dx$$

$$= \frac{x^3}{3}\tan^{-1}x - \frac{1}{3}\frac{x^2}{2} + \frac{1}{6}\int \frac{du}{u} \quad \left[\text{put } u = x^2+1, \text{ so } du = 2x\,dx\right] = \tfrac{1}{3}x^3\tan^{-1}x - \tfrac{1}{6}x^2 + \tfrac{1}{6}\ln(1+x^2) + C.$$

25. Let $u = e^x \quad\Rightarrow\quad \ln u = x \quad\Rightarrow\quad dx = \dfrac{du}{u}$. Then $\displaystyle\int \sqrt{e^{2x}-1}\,dx = \int \frac{\sqrt{u^2-1}}{u}\,du$

$$= \sqrt{u^2-1} - \cos^{-1}(1/u) + C \quad (\text{Formula 41}) \quad = \sqrt{e^{2x}-1} - \cos^{-1}(e^{-x}) + C.$$

Or: Let $u = \sqrt{e^{2x}-1}$.

26. $\int e^{\sin x}\sin 2x\,dx = 2\int e^{\sin x}\sin x\cos x\,dx = 2\int ue^u\,du$ (put $u = \sin x$, $du = \cos x\,dx$)

$$= 2(u-1)e^u + C \quad (\text{Formula 96}) \quad = 2(\sin x - 1)e^{\sin x} + C$$

27. Volume $\displaystyle = \int_0^1 \frac{2\pi x}{(1+5x)^2}\,dx = 2\pi\left[\frac{1}{25(1+5x)} + \tfrac{1}{25}\ln|1+5x|\right]_0^1$ (Formula 51)

$$= \tfrac{2\pi}{25}\left(\tfrac{1}{6} + \ln 6 - 1 - \ln 1\right) = \tfrac{2\pi}{25}\left(\ln 6 - \tfrac{5}{6}\right)$$

28. Volume $= \int_0^{\pi/4} \pi \tan^4 x\, dx = \pi\left(\left[\frac{1}{3}\tan^3 x\right]_0^{\pi/4} - \int_0^{\pi/4} \tan^2 x\, dx\right)$ (Formula 75)

$= \pi\left[\frac{1}{3}\tan^3 x - \tan x + x\right]_0^{\pi/4}$ (Formula 65) $= \pi\left[\frac{1}{3} - 1 + \frac{\pi}{4}\right] = \pi\left[\frac{\pi}{4} - \frac{2}{3}\right]$

29. **(a)** $\dfrac{d}{du}\left[\dfrac{1}{b^3}\left(a + bu - \dfrac{a^2}{a+bu} - 2a\ln|a+bu|\right) + C\right] = \dfrac{1}{b^3}\left[b + \dfrac{ba^2}{(a+bu)^2} - \dfrac{2ab}{(a+bu)}\right]$

$$= \frac{1}{b^3}\left[\frac{b(a+bu)^2 + ba^2 - (a+bu)2ab}{(a+bu)^2}\right] = \frac{1}{b^3}\left[\frac{b^3u^2}{(a+bu)^2}\right] = \frac{u^2}{(a+bu)^2}$$

(b) Let $t = a + bu \;\Rightarrow\; dt = b\,du$.

$$\int \frac{u^2\,du}{(a+bu)^2} = \frac{1}{b^3}\int \frac{(t-a)^2}{t^2}\,dt = \frac{1}{b^3}\int\left(1 - \frac{2a}{t} + \frac{a^2}{t^2}\right)dt = \frac{1}{b^3}\left(t - 2a\ln|t| - \frac{a^2}{t}\right) + C$$

$$= \frac{1}{b^3}\left(a + bu - \frac{a^2}{a+bu} - 2a\ln|a+bu|\right) + C$$

30. **(a)** $\dfrac{d}{du}\left[\dfrac{u}{8}(2u^2 - a^2)\sqrt{a^2-u^2} + \dfrac{a^4}{8}\sin^{-1}\dfrac{u}{a} + C\right]$

$$= \sqrt{a^2-u^2}\left[\tfrac{1}{8}(2u^2-a^2) + \frac{u}{8}(4u)\right] + \frac{u}{8}(2u^2-a^2)\frac{-u}{\sqrt{a^2-u^2}} + \frac{a^4}{8}\frac{1/a}{\sqrt{1-u^2/a^2}}$$

$$= \sqrt{a^2-u^2}\left[\frac{2u^2-a^2}{8} + \frac{u^2}{2}\right] - \frac{u^2(2u^2-a^2)}{8\sqrt{a^2-u^2}} + \frac{a^4}{8\sqrt{a^2-u^2}}$$

$$= \tfrac{1}{2}(a^2-u^2)^{-1/2}\left[\tfrac{1}{4}(a^2-u^2)(2u^2-a^2) + u^2(a^2-u^2) - \frac{u^2}{4}(2u^2-a^2) + \frac{a^4}{4}\right]$$

$$= \tfrac{1}{2}(a^2-u^2)^{-1/2}\left[2u^2a^2 - 2u^4\right] = \frac{u^2(a^2-u^2)}{\sqrt{a^2-u^2}} = u^2\sqrt{a^2-u^2}$$

(b) Let $u = a\sin\theta \;\Rightarrow\; du = a\cos\theta\,d\theta$. Then

$\int u^2\sqrt{a^2-u^2}\,du = \int a^2\sin^2\theta\, a\sqrt{1-\sin^2\theta}\, a\cos\theta\,d\theta$

$= a^4\int \sin^2\theta\cos^2\theta\,d\theta = a^4\int \frac{1}{2}(1+\cos 2\theta)\frac{1}{2}(1-\cos 2\theta)d\theta = \frac{1}{4}a^4\int(1-\cos^2 2\theta)d\theta$

$= \frac{1}{4}a^4\int\left[1 - \frac{1}{2}(1+\cos 4\theta)\right]d\theta = \frac{1}{4}a^4\left(\frac{1}{2}\theta - \frac{1}{8}\sin 4\theta\right) + C = \frac{1}{4}a^4\left(\frac{1}{2}\theta - \frac{1}{8}2\sin 2\theta\cos 2\theta\right) + C$

$= \frac{1}{4}a^4\left[\frac{1}{2}\theta - \frac{1}{2}\sin\theta\cos\theta(1-2\sin^2\theta)\right] + C = \dfrac{a^4}{8}\left[\sin^{-1}\dfrac{u}{a} - \dfrac{u}{a}\dfrac{\sqrt{a^2-u^2}}{a}\left(1 - \dfrac{2u^2}{a^2}\right)\right] + C$

$= \frac{1}{8}a^4\sin^{-1}(u/a) + \frac{1}{8}u\sqrt{a^2-u^2}(2u^2-a^2) + C$

31. Maple, Mathematica and Derive all give $\int x^2\sqrt{5-x^2}\,dx = -\frac{1}{4}x(5-x^2)^{3/2} + \frac{5}{8}x\sqrt{5-x^2} + \frac{25}{8}\sin^{-1}\left(\frac{1}{\sqrt{5}}x\right)$. Using Formula 31, we get $\int x^2\sqrt{5-x^2}\,dx = \frac{1}{8}x(2x^2-5)\sqrt{5-x^2} + \frac{1}{8}(5^2)\sin^{-1}\left(\frac{1}{\sqrt{5}}x\right) + C$. But $-\frac{1}{4}x(5-x^2)^{3/2} + \frac{5}{8}x\sqrt{5-x^2} = \frac{1}{8}x\sqrt{5-x^2}[5 - 2(5-x^2)] = \frac{1}{8}x(2x^2-5)\sqrt{5-x^2}$, and the $\sin^{-1}$ terms are the same in each expression, so the answers are equivalent.

32. Maple and Mathematica both give $\int x^2(1+x^3)^4\,dx = \frac{1}{15}x^{15} + \frac{1}{3}x^{12} + \frac{2}{3}x^9 + \frac{2}{3}x^6 + \frac{1}{3}x^3$, while Derive gives $\int x^2(1+x^3)^4\,dx = \frac{1}{15}(x^3+1)^5$. Using the substitution $u = 1 + x^3 \;\Rightarrow\; du = 3x^2\,dx$, we get $\int x^2(1+x^3)^4\,dx = \int u^4\left(\frac{1}{3}\,du\right) = \frac{1}{15}u^5 + C = \frac{1}{15}(1+x^3)^5 + C$. We can use the Binomial Theorem or a CAS to expand this expression, and we get $\frac{1}{15}(1+x^3)^5 + C = \frac{1}{15} + \frac{1}{3}x^3 + \frac{2}{3}x^6 + \frac{2}{3}x^9 + \frac{1}{3}x^{12} + \frac{1}{15}x^{15} + C$.

28. Volume $= \int_0^{\pi/4} \pi \tan^4 x\,dx = \pi\left(\left[\frac{1}{3}\tan^3 x\right]_0^{\pi/4} - \int_0^{\pi/4} \tan^2 x\,dx\right)$ (Formula 75)

$= \pi\left[\frac{1}{3}\tan^3 x - \tan x + x\right]_0^{\pi/4}$ (Formula 65) $= \pi\left[\frac{1}{3} - 1 + \frac{\pi}{4}\right] = \pi\left[\frac{\pi}{4} - \frac{2}{3}\right]$

29. **(a)** $$\frac{d}{du}\left[\frac{1}{b^3}\left(a + bu - \frac{a^2}{a+bu} - 2a\ln|a+bu|\right) + C\right] = \frac{1}{b^3}\left[b + \frac{ba^2}{(a+bu)^2} - \frac{2ab}{(a+bu)}\right]$$

$$= \frac{1}{b^3}\left[\frac{b(a+bu)^2 + ba^2 - (a+bu)2ab}{(a+bu)^2}\right] = \frac{1}{b^3}\left[\frac{b^3u^2}{(a+bu)^2}\right] = \frac{u^2}{(a+bu)^2}$$

(b) Let $t = a + bu \Rightarrow dt = b\,du$.

$$\int \frac{u^2\,du}{(a+bu)^2} = \frac{1}{b^3}\int \frac{(t-a)^2}{t^2}\,dt = \frac{1}{b^3}\int\left(1 - \frac{2a}{t} + \frac{a^2}{t^2}\right)dt = \frac{1}{b^3}\left(t - 2a\ln|t| - \frac{a^2}{t}\right) + C$$

$$= \frac{1}{b^3}\left(a + bu - \frac{a^2}{a+bu} - 2a\ln|a+bu|\right) + C$$

30. **(a)** $$\frac{d}{du}\left[\frac{u}{8}(2u^2 - a^2)\sqrt{a^2-u^2} + \frac{a^4}{8}\sin^{-1}\frac{u}{a} + C\right]$$

$$= \sqrt{a^2-u^2}\left[\frac{1}{8}(2u^2-a^2) + \frac{u}{8}(4u)\right] + \frac{u}{8}(2u^2-a^2)\frac{-u}{\sqrt{a^2-u^2}} + \frac{a^4}{8}\frac{1/a}{\sqrt{1-u^2/a^2}}$$

$$= \sqrt{a^2-u^2}\left[\frac{2u^2-a^2}{8} + \frac{u^2}{2}\right] - \frac{u^2(2u^2-a^2)}{8\sqrt{a^2-u^2}} + \frac{a^4}{8\sqrt{a^2-u^2}}$$

$$= \frac{1}{2}(a^2-u^2)^{-1/2}\left[\frac{1}{4}(a^2-u^2)(2u^2-a^2) + u^2(a^2-u^2) - \frac{u^2}{4}(2u^2-a^2) + \frac{a^4}{4}\right]$$

$$= \frac{1}{2}(a^2-u^2)^{-1/2}\left[2u^2a^2 - 2u^4\right] = \frac{u^2(a^2-u^2)}{\sqrt{a^2-u^2}} = u^2\sqrt{a^2-u^2}$$

(b) Let $u = a\sin\theta \Rightarrow du = a\cos\theta\,d\theta$. Then

$\int u^2\sqrt{a^2-u^2}\,du = \int a^2\sin^2\theta\, a\sqrt{1-\sin^2\theta}\, a\cos\theta\,d\theta$

$= a^4\int \sin^2\theta\cos^2\theta\,d\theta = a^4\int \frac{1}{2}(1+\cos 2\theta)\frac{1}{2}(1-\cos 2\theta)d\theta = \frac{1}{4}a^4\int(1-\cos^2 2\theta)d\theta$

$= \frac{1}{4}a^4\int\left[1 - \frac{1}{2}(1+\cos 4\theta)\right]d\theta = \frac{1}{4}a^4\left(\frac{1}{2}\theta - \frac{1}{8}\sin 4\theta\right) + C = \frac{1}{4}a^4\left(\frac{1}{2}\theta - \frac{1}{8}2\sin 2\theta\cos 2\theta\right) + C$

$= \frac{1}{4}a^4\left[\frac{1}{2}\theta - \frac{1}{2}\sin\theta\cos\theta(1 - 2\sin^2\theta)\right] + C = \frac{a^4}{8}\left[\sin^{-1}\frac{u}{a} - \frac{u}{a}\frac{\sqrt{a^2-u^2}}{a}\left(1 - \frac{2u^2}{a^2}\right)\right] + C$

$= \frac{1}{8}a^4\sin^{-1}(u/a) + \frac{1}{8}u\sqrt{a^2-u^2}(2u^2-a^2) + C$

31. Maple, Mathematica and Derive all give $\int x^2\sqrt{5-x^2}\,dx = -\frac{1}{4}x(5-x^2)^{3/2} + \frac{5}{8}x\sqrt{5-x^2} + \frac{25}{8}\sin^{-1}\left(\frac{1}{\sqrt{5}}x\right)$. Using Formula 31, we get $\int x^2\sqrt{5-x^2}\,dx = \frac{1}{8}x(2x^2-5)\sqrt{5-x^2} + \frac{1}{8}(5^2)\sin^{-1}\left(\frac{1}{\sqrt{5}}x\right) + C$. But $-\frac{1}{4}x(5-x^2)^{3/2} + \frac{5}{8}x\sqrt{5-x^2} = \frac{1}{8}x\sqrt{5-x^2}[5 - 2(5-x^2)] = \frac{1}{8}x(2x^2-5)\sqrt{5-x^2}$, and the $\sin^{-1}$ terms are the same in each expression, so the answers are equivalent.

32. Maple and Mathematica both give $\int x^2(1+x^3)^4\,dx = \frac{1}{15}x^{15} + \frac{1}{3}x^{12} + \frac{2}{3}x^9 + \frac{2}{3}x^6 + \frac{1}{3}x^3$, while Derive gives $\int x^2(1+x^3)^4\,dx = \frac{1}{15}(x^3+1)^5$. Using the substitution $u = 1 + x^3 \Rightarrow du = 3x^2\,dx$, we get $\int x^2(1+x^3)^4\,dx = \int u^4\left(\frac{1}{3}\,du\right) = \frac{1}{15}u^5 + C = \frac{1}{15}(1+x^3)^5 + C$. We can use the Binomial Theorem or a CAS to expand this expression, and we get $\frac{1}{15}(1+x^3)^5 + C = \frac{1}{15} + \frac{1}{3}x^3 + \frac{2}{3}x^6 + \frac{2}{3}x^9 + \frac{1}{3}x^{12} + \frac{1}{15}x^{15} + C$.

If we substitute $u = \sqrt{x^2+1} \quad \Rightarrow \quad x^4 = (u^2-1)^2$, $x\,dx = u\,du$, then the integral becomes

$\int (u^2-1)^2 u(u\,du) = \int (u^4 - 2u^2 + 1)u^2\,du = \frac{1}{7}u^7 - \frac{2}{5}u^5 + \frac{1}{3}u^3 + C$

$= (x^2+1)^{3/2}\left[\frac{1}{7}(x^2+1)^2 - \frac{2}{5}(x^2+1) + \frac{1}{3}\right] + C = \frac{1}{105}(x^2+1)^{3/2}\left[15(x^2+1)^2 - 42(x^2+1) + 35\right] + C$

$= \frac{1}{105}(x^2+1)^{3/2}(15x^4 - 12x^2 + 8) + C.$

39. Maple gives the antiderivative

$$F(x) = \int \frac{x^2-1}{x^4+x^2+1}\,dx = -\tfrac{1}{2}\ln(x^2+x+1) + \tfrac{1}{2}\ln(x^2-x+1).$$

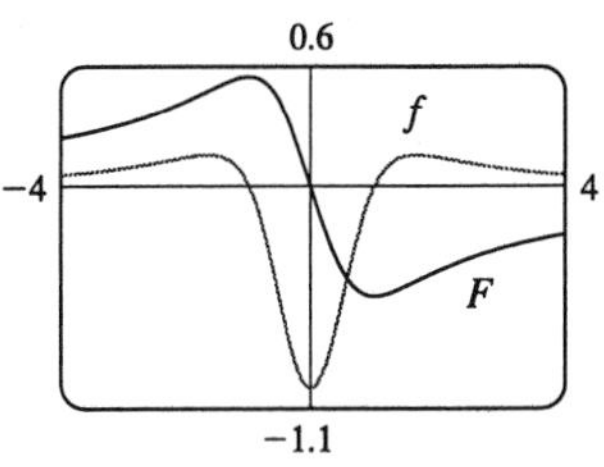

We can see that at 0, this antiderivative is 0. From the graphs, it appears that F has a maximum at $x = -1$ and a minimum at $x = 1$ [since $F'(x) = f(x)$ changes sign at these x-values], and that F has inflection points at $x \approx -1.7$, $x = 0$ and $x \approx 1.7$ [since $f(x)$ has extrema at these x-values].

40. Maple gives the antiderivative which, after we use the `simplify` command, becomes $\int xe^{-x}\sin x\,dx = -\frac{1}{2}e^{-x}(\cos x + x\cos x + x\sin x)$. We calculate that at $x = 0$, this antiderivative takes the value $-\frac{1}{2}$, so we take $F(x) = -\frac{1}{2}e^{-x}(\cos x + x\cos x + x\sin x) + \frac{1}{2}$, so that $F(0) = 0$.

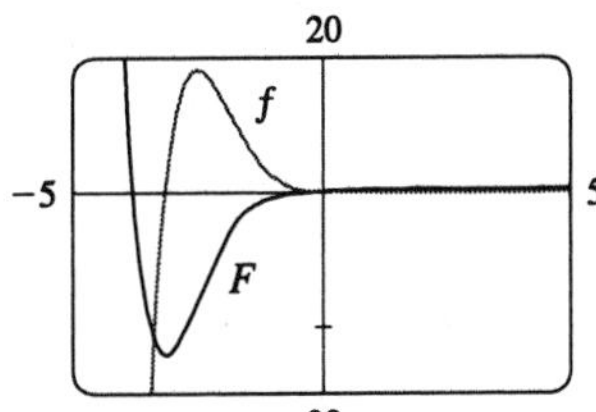

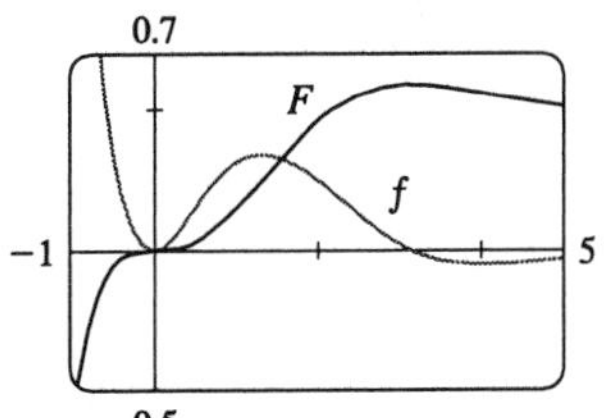

From the graphs, it appears that F has a minimum at $x \approx -3.1$ and a maximum at $x \approx 3.1$ [Note that $f(x) = 0$ at $x = \pm\pi$], and that F has inflection points where f' changes sign, at $x \approx -2.5$, $x = 0$, $x \approx 1.3$ and $x \approx 4.1$.

41. Since f is everywhere positive, we know that its antiderivative F is increasing. The antiderivative given by Maple is

$$\int \sin^4 x\cos^6 x\,dx = -\tfrac{1}{10}\sin^3 x\cos^7 x - \tfrac{3}{80}\sin x\cos^7 x + \tfrac{1}{160}\cos^5 x\sin x + \tfrac{1}{128}\cos^3 x\sin x + \tfrac{3}{256}\cos x\sin x + \tfrac{3}{256}x,$$

and this is 0 at $x = 0$.

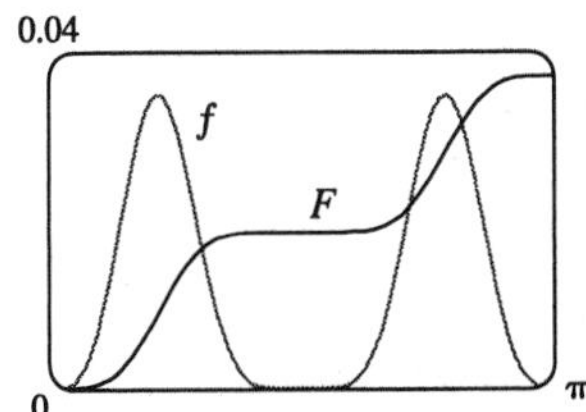

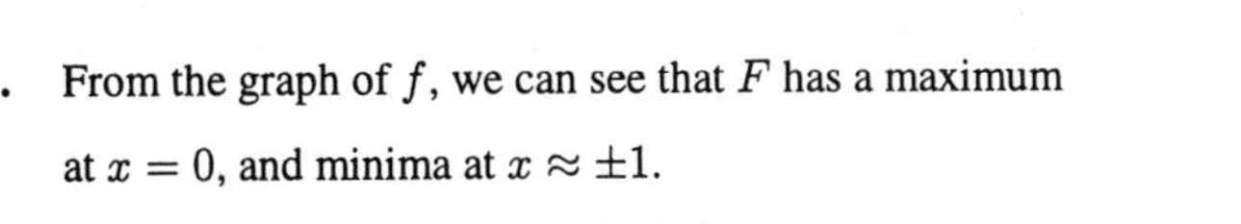

42. From the graph of f, we can see that F has a maximum at $x = 0$, and minima at $x \approx \pm 1$.

The antiderivative given by Maple is

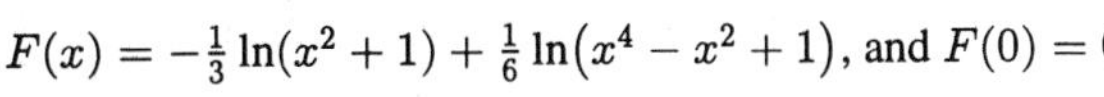

$F(x) = -\frac{1}{3}\ln(x^2+1) + \frac{1}{6}\ln(x^4 - x^2 + 1)$, and $F(0) = 0$.

Note that f is odd, and the integral function F is even.

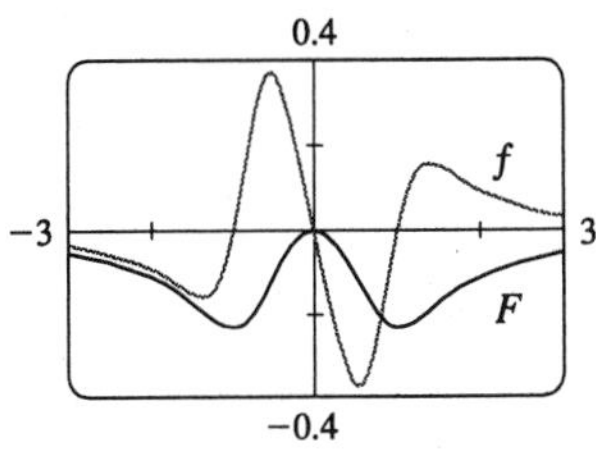

EXERCISES 7.8

1. **(a)** $L_2 = \sum_{i=1}^{2} f(x_{i-1})\Delta x = 2\,f(0) + 2\,f(2) = 2(0.5) + 2(2.5) = 6$

$R_2 = \sum_{i=1}^{2} f(x_i)\Delta x = 2\,f(2) + 2\,f(4) = 2(2.5) + 2(3.5) = 12$

$M_2 = \sum_{i=1}^{2} f(\overline{x}_i)\Delta x = 2\,f(1) + 2\,f(3) \approx 2(1.7) + 2(3.2) = 9.8$

(b) 

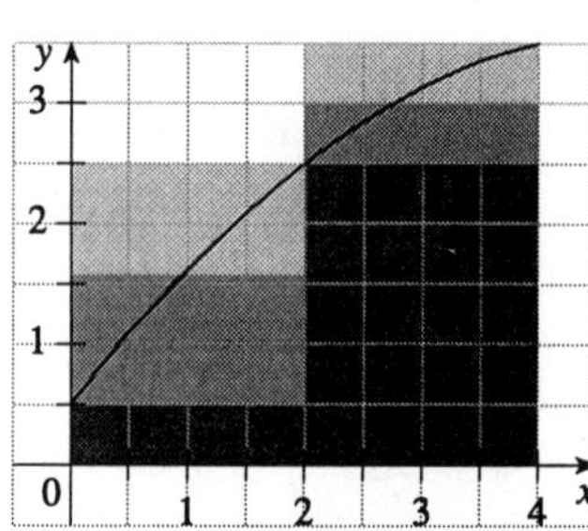

L_2 is an underestimate, since the area under the small rectangles is less than the area under the curve, and R_2 is an overestimate, since the area under the large rectangles is greater than the area under the curve. It appears that M_2 is an overestimate, though it is fairly close to I. See the solution to Exercise 37 for a proof of the fact that if f is concave down on $[a, b]$, then the Midpoint Rule is an overestimate of $\int_a^b f(x)dx$.

(c) $T_2 = \left(\frac{1}{2}\Delta x\right)[f(x_0) + 2f(x_1) + f(x_2)] = \frac{2}{2}[f(0) + 2\,f(2) + f(4)] = 0.5 + 2(2.5) + 3.5 = 9.$

This approximation is an underestimate, since the graph is concave down. See the solution to Exercise 37 for a general proof of this conclusion.

(d) For any n, we will have $L_n < T_n < I < M_n < R_n$.

2.

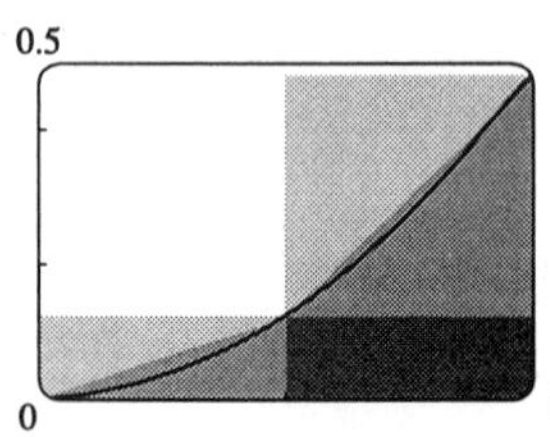

(a) Since f is increasing on $[0, 1]$, L_2 will underestimate I (since the area under the small rectangles is less than the area under the curve), and R_2 will overestimate I. Since f is concave upward on $[0, 1]$, M_2 will underestimate I and T_2 will overestimate I (the area under the straight line segments is greater than the area under the curve).

(b) For any n, we will have $L_n < M_n < I < T_n < R_n$.

(c) $L_5 = \sum_{i=1}^{5} f(x_{i-1})\Delta x = \frac{1}{5}[f(0.0) + f(0.2) + f(0.4) + f(0.6) + f(0.8)] \approx 0.1187$

$R_5 = \sum_{i=1}^{5} f(x_i)\Delta x = \frac{1}{5}[f(0.2) + f(0.4) + f(0.6) + f(0.8) + f(1)] \approx 0.2146$

$M_5 = \sum_{i=1}^{5} f(\overline{x}_i)\Delta x = \frac{1}{5}[f(0.1) + f(0.3) + f(0.5) + f(0.7) + f(0.9)] \approx 0.1622$

$T_5 = \left(\frac{1}{2}\Delta x\right)[f(0) + 2f(0.2) + 2f(0.4) + 2f(0.6) + 2f(0.8) + f(1)] \approx 0.1666$

From the graph, it appears that the Midpoint Rule gives the best approximation. (This is in fact the case, since $I \approx 0.16371405$.)

3. $f(x)=\sqrt{1+x^3}$, $\Delta x=\dfrac{1-(-1)}{8}=\dfrac{1}{4}$

(a) $T_8=\frac{0.25}{2}\left[f(-1)+2f\left(-\frac{3}{4}\right)+2f\left(-\frac{1}{2}\right)+\cdots+2f\left(\frac{1}{2}\right)+2f\left(\frac{3}{4}\right)+f(1)\right]\approx 1.913972$

(b) $S_8=\frac{0.25}{3}\left[f(-1)+4f\left(-\frac{3}{4}\right)+2f\left(-\frac{1}{2}\right)+4f\left(-\frac{1}{4}\right)+2f(0)+4f\left(\frac{1}{4}\right)+2f\left(\frac{1}{2}\right)+4f\left(\frac{3}{4}\right)+f(1)\right]$
≈ 1.934766

4. $f(x)=\cos(x^2)$, $\Delta x=\dfrac{1-0}{4}=\dfrac{1}{4}$

(a) $T_4=\frac{0.25}{2}\left[f(0)+2f\left(\frac{1}{4}\right)+2f\left(\frac{1}{2}\right)+2f\left(\frac{3}{4}\right)+f(1)\right]\approx 0.895759$

(b) $S_4=\frac{0.25}{3}\left[f(0)+4f\left(\frac{1}{4}\right)+2f\left(\frac{1}{2}\right)+4f\left(\frac{3}{4}\right)+f(1)\right]\approx 0.904501$

5. $f(x)=\dfrac{\sin x}{x}$, $\Delta x=\dfrac{\pi-\pi/2}{6}=\dfrac{\pi}{12}$

(a) $T_6=\frac{\pi}{24}\left[f\left(\frac{\pi}{2}\right)+2f\left(\frac{7\pi}{12}\right)+2f\left(\frac{2\pi}{3}\right)+2f\left(\frac{3\pi}{4}\right)+2f\left(\frac{5\pi}{6}\right)+2f\left(\frac{11\pi}{12}\right)+f(\pi)\right]\approx 0.481672$

(b) $S_6=\frac{\pi}{36}\left[f\left(\frac{\pi}{2}\right)+4f\left(\frac{7\pi}{12}\right)+2f\left(\frac{2\pi}{3}\right)+4f\left(\frac{3\pi}{4}\right)+2f\left(\frac{5\pi}{6}\right)+4f\left(\frac{11\pi}{12}\right)+f(\pi)\right]\approx 0.481172$

6. $f(x)=x\tan x$, $\Delta x=\dfrac{\pi/4-0}{6}=\dfrac{\pi}{24}$

(a) $T_6=\frac{\pi}{48}\left[f(0)+2f\left(\frac{\pi}{24}\right)+2f\left(\frac{\pi}{12}\right)+\cdots+2f\left(\frac{5\pi}{24}\right)+f\left(\frac{\pi}{4}\right)\right]\approx 0.189445$

(b) $S_6=\frac{\pi}{72}\left[f(0)+4f\left(\frac{\pi}{24}\right)+2f\left(\frac{\pi}{12}\right)+4f\left(\frac{\pi}{8}\right)+2f\left(\frac{\pi}{6}\right)+4f\left(\frac{5\pi}{24}\right)+f\left(\frac{\pi}{4}\right)\right]\approx 0.185822$

7. $f(x)=e^{-x^2}$, $\Delta x=\dfrac{1-0}{10}=0.1$

(a) $T_{10}=\frac{0.1}{2}[f(0)+2f(0.1)+2f(0.2)+\cdots+2f(0.8)+2f(0.9)+f(1)]\approx 0.746211$

(b) $M_{10}=0.1\left[f(0.05)+f(0.15)+f(0.25)+\cdots+f(0.75)+f(0.85)+f(0.95)\right]\approx 0.747131$

(c) $S_{10}=\frac{0.1}{3}\left[f(0)+4f(0.1)+2f(0.2)+4f(0.3)+2f(0.4)+4f(0.5)\right.$
$\left.+2f(0.6)+4f(0.7)+2f(0.8)+4f(0.9)+f(1)\right]\approx 0.746825$

8. $f(x)=\dfrac{1}{\sqrt{1+x^3}}$, $\Delta x=\dfrac{2-0}{10}=\dfrac{1}{5}$

(a) $T_{10}=\frac{0.2}{2}[f(0)+2f(0.2)+2f(0.4)+\cdots+2f(1.6)+2f(1.8)+f(2)]\approx 1.401435$

(b) $M_{10}=0.2[f(0.1)+f(0.3)+f(0.5)+\cdots+f(1.7)+f(1.9)]\approx 1.402556$

(c) $S_{10}=\frac{0.2}{3}\left[f(0)+4f(0.2)+2f(0.4)+4f(0.6)+2f(0.8)+4f(1)\right.$
$\left.+2f(1.2)+4f(1.4)+2f(1.6)+4f(1.8)+f(2)\right]\approx 1.402206$

9. $f(x)=\cos(e^x)$, $\Delta x=\dfrac{1/2-0}{8}=\dfrac{1}{16}$

(a) $T_8=\frac{1}{32}\left[f(0)+2f\left(\frac{1}{16}\right)+2f\left(\frac{1}{8}\right)+\cdots+2f\left(\frac{7}{16}\right)+f\left(\frac{1}{2}\right)\right]\approx 0.132465$

(b) $M_8=\frac{1}{16}\left[f\left(\frac{1}{32}\right)+f\left(\frac{3}{32}\right)+f\left(\frac{5}{32}\right)+\cdots+f\left(\frac{15}{32}\right)\right]\approx 0.132857$

(c) $S_8=\frac{1}{48}\left[f(0)+4f\left(\frac{1}{16}\right)+2f\left(\frac{1}{8}\right)+4f\left(\frac{3}{16}\right)+2f\left(\frac{1}{4}\right)+4f\left(\frac{5}{16}\right)+2f\left(\frac{3}{8}\right)+4f\left(\frac{7}{16}\right)+f\left(\frac{1}{2}\right)\right]$
≈ 0.132727

10. $f(x) = \dfrac{1}{\ln x}$, $\Delta x = \dfrac{3-2}{10} = \dfrac{1}{10}$

(a) $T_{10} = \frac{0.1}{2}[f(2) + 2f(2.1) + 2f(2.2) + \cdots + 2f(2.9) + f(3)] \approx 1.119061$

(b) $M_{10} = 0.1[f(2.05) + f(2.15) + f(2.25) + \cdots + f(2.95)] = 1.118107$

(c) $S_{10} = \frac{0.1}{3}[f(2) + 4f(2.1) + 2f(2.2) + 4f(2.3) + 2f(2.4) + 4f(2.5)$
$+ 2f(2.6) + 4f(2.7) + 2f(2.8) + 4f(2.9) + f(3)] \approx 1.118428$

11. $f(x) = x^5 e^x$, $\Delta x = \dfrac{1-0}{10} = \dfrac{1}{10}$

(a) $T_{10} = \frac{0.1}{2}[f(0) + 2f(0.1) + 2f(0.2) + \cdots + 2f(0.9) + f(1)] \approx 0.409140$

(b) $M_{10} = 0.1[f(0.05) + f(0.15) + f(0.25) + \cdots + f(0.95)] \approx 0.388849$

(c) $S_{10} = \frac{0.1}{3}[f(0) + 4f(0.1) + 2f(0.2) + 4f(0.3) + 2f(0.4) + 4f(0.5)$
$+ 2f(0.6) + 4f(0.7) + 2f(0.8) + 4f(0.9) + f(1)] \approx 0.395802$

12. $f(x) = \ln(1 + e^x)$, $\Delta x = \dfrac{1-0}{8} = \dfrac{1}{8}$

(a) $T_8 = \frac{1}{8 \cdot 2}[f(0) + 2f(\frac{1}{8}) + 2f(\frac{1}{4}) + 2f(\frac{3}{8}) + 2f(\frac{1}{2}) + 2f(\frac{5}{8}) + 2f(\frac{3}{4}) + 2f(\frac{7}{8}) + f(1)] \approx 0.984120$

(b) $M_8 = \frac{1}{8}[f(\frac{1}{16}) + f(\frac{3}{16}) + f(\frac{5}{16}) + f(\frac{7}{16}) + \cdots + f(\frac{15}{16})] \approx 0.983669$

(c) $S_8 = \frac{1}{8 \cdot 3}[f(0) + 4f(\frac{1}{8}) + 2f(\frac{1}{4}) + 4f(\frac{3}{8}) + 2f(\frac{1}{2}) + 4f(\frac{5}{8}) + 2f(\frac{3}{4}) + 4f(\frac{7}{8}) + f(1)] \approx 0.983819$

13. $f(x) = e^{1/x}$, $\Delta x = \dfrac{2-1}{4} = \dfrac{1}{4}$

(a) $T_4 = \frac{1}{4 \cdot 2}[f(1) + 2f(1.25) + 2f(1.5) + 2f(1.75) + f(2)] \approx 2.031893$

(b) $M_4 = \frac{1}{4}[f(1.125) + f(1.375) + f(1.625) + f(1.875)] \approx 2.014207$

(c) $S_4 = \frac{1}{4 \cdot 3}[f(1) + 4f(1.25) + 2f(1.5) + 4f(1.75) + f(2)] \approx 2.020651$

14. $f(x) = \sqrt{x}\sin x$, $\Delta x = \dfrac{4-0}{8} = \dfrac{1}{2}$

(a) $T_8 = \frac{1}{2 \cdot 2}\{f(0) + 2[f(\frac{1}{2}) + f(1) + f(\frac{3}{2}) + f(2) + f(\frac{5}{2}) + f(3) + f(\frac{7}{2})] + f(4)\} \approx 1.732865$

(b) $M_8 = \frac{1}{2}[f(\frac{1}{4}) + f(\frac{3}{4}) + f(\frac{5}{4}) + f(\frac{7}{4}) + \cdots + f(\frac{15}{4})] \approx 1.787427$

(c) $S_8 = \frac{1}{2 \cdot 3}[f(0) + 4f(\frac{1}{2}) + 2f(1) + 4f(\frac{3}{2}) + 2f(2) + 4f(2.5) + 2f(3) + 4f(3.5) + f(4)] \approx 1.772142$

15. $f(x) = \dfrac{1}{1+x^4}$, $\Delta x = \dfrac{3-0}{6} = \dfrac{1}{2}$

(a) $T_6 = \frac{1}{2 \cdot 2}[f(0) + 2f(0.5) + 2f(1) + 2f(1.5) + 2f(2) + 2f(2.5) + f(3)] \approx 1.098004$

(b) $M_6 = \frac{1}{2}[f(0.25) + f(0.75) + f(1.25) + f(1.75) + f(2.25) + f(2.75)] \approx 1.098709$

(c) $S_6 = \frac{1}{2 \cdot 3}[f(0) + 4f(0.5) + 2f(1) + 4f(1.5) + 2f(2) + 4f(2.5) + f(3)] = 1.109031$

16. $f(x) = \dfrac{e^x}{x}$, $\Delta x = \dfrac{4-2}{10} = \dfrac{1}{5}$

(a) $T_{10} = \frac{1}{5\cdot 2}[f(2) + 2(f(2.2) + f(2.4) + f(2.6) + \cdots + f(3.8)) + f(4)] \approx 14.704592$

(b) $M_{10} = \frac{1}{5}[f(2.1) + f(2.3) + f(2.5) + f(2.7) + \cdots + f(3.7) + f(3.9)] \approx 14.662669$

(c) $S_{10} = \frac{1}{5\cdot 3}[f(2) + 4f(2.2) + 2f(2.4) + 4f(2.6) + \cdots + 2f(3.6) + 4f(3.8) + f(4)] \approx 14.676696$

17. $f(x) = e^{-x^2}$, $\Delta x = \dfrac{2-0}{10} = \dfrac{1}{5}$

(a) $T_{10} = \frac{1}{5\cdot 2}[f(0) + 2(f(0.2) + f(0.4) + \cdots + f(1.8)) + f(2)] \approx 0.881839$

$M_{10} = \frac{1}{5}[f(0.1) + f(0.3) + f(0.5) + \cdots + f(1.7) + f(1.9)] \approx 0.882202$

(b) $f(x) = e^{-x^2}$, $f'(x) = -2xe^{-x^2}$, $f''(x) = (4x^2 - 2)e^{-x^2}$, $f'''(x) = 4x(3 - 2x^2)e^{-x^2}$.
$f'''(x) = 0 \quad \Leftrightarrow \quad x = 0$ or $x = \pm\sqrt{\frac{3}{2}}$. So to find the maximum value of $|f''(x)|$ on $[0, 2]$, we need only consider its values at $x = 0$, $x = 2$, and $x = \sqrt{\frac{3}{2}}$. $|f''(0)| = 2$, $|f''(2)| \approx 0.2564$ and $\left|f''\left(\sqrt{\frac{3}{2}}\right)\right| = 4e^{-3/2} \approx 0.8925$. Thus, taking $K = 2$, $a = 0$, $b = 2$, and $n = 10$ in Theorem 5, we get $|E_T| \le 2 \cdot 2^3/\left[12(10)^2\right] = \frac{1}{75} = 0.01\overline{3}$, and $|E_M| \le 2 \cdot 2^3/\left[24(10)^2\right] = 0.00\overline{6}$.

18. **(a)** $T_4 = \frac{1}{4\cdot 2}\left[f(0) + 2f\left(\frac{1}{4}\right) + 2f\left(\frac{1}{2}\right) + 2f\left(\frac{3}{4}\right) + f(1)\right] \approx 0.895759$

$M_4 = \frac{1}{4}\left[f\left(\frac{1}{8}\right) + f\left(\frac{3}{8}\right) + f\left(\frac{5}{8}\right) + f\left(\frac{7}{8}\right)\right] \approx 0.908907$

$T_8 = \frac{1}{8\cdot 2}\left[f(0) + 2\left(f\left(\frac{1}{8}\right) + f\left(\frac{2}{8}\right) + \cdots + f\left(\frac{7}{8}\right)\right) + f(1)\right] \approx 0.902333$

$M_8 = \frac{1}{8}\left[f\left(\frac{1}{16}\right) + f\left(\frac{3}{16}\right) + f\left(\frac{5}{16}\right) + \cdots + f\left(\frac{15}{16}\right)\right] = 0.905620$

(b) $f(x) = \cos(x^2)$, $f'(x) = -2x\sin(x^2)$, $f''(x) = -2\sin(x^2) - 4x^2\cos(x^2)$. For $0 \le x \le 1$, sin and cos are positive, so $|f''(x)| = 2\sin(x^2) + 4x^2\cos(x^2) \le 2\cdot 1 + 4\cdot 1\cdot 1 = 6$ since $\sin(x^2) \le 1$ and $\cos(x^2) \le 1$ for all x, and $x^2 \le 1$ for $0 \le x \le 1$. So for $n = 4$, we take $K = 6$, $a = 0$, and $b = 1$ in Theorem 5, to get $|E_T| \le 6 \cdot 1^3/\left[12(4)^2\right] = \frac{1}{32} = 0.03125$, and $|E_M| \le 6 \cdot 1^3/\left[24(4)^2\right] = \frac{1}{64} = 0.015625$. And for $n = 8$, we take $K = 6$, $a = 0$, and $b = 1$ in Theorem 5, to get $|E_T| \le 6 \cdot 1^3/\left[12(8)^2\right] = \frac{1}{128} = 0.0078125$ and $|E_M| \le 6 \cdot 1^3/\left[24(8)^2\right] = \frac{1}{256} = 0.00390625$. [A slightly better estimate is obtained by noting that $0 \le x \le 1 \quad \Rightarrow \quad \sin(x^2) \le \sin 1$, so we can take $K = 4 + 2\sin 1$ in Theorem 5.]

19. **(a)** $T_{10} = \frac{1}{10\cdot 2}[f(0) + 2(f(0.1) + f(0.2) + \cdots + f(0.9)) + f(1)] \approx 1.719713$

$S_{10} = \frac{1}{10\cdot 3}[f(0) + 4f(0.1) + 2f(0.2) + 4f(0.3) + \cdots + 4f(0.9) + f(1)] \approx 1.7182828$

Since $\int_0^1 e^x\,dx = \left[e^x\right]_0^1 = e - 1 \approx 1.71828183$, $E_T \approx -0.00143166$ and $E_S \approx -0.00000095$.

(b) $f(x) = e^x \quad \Rightarrow \quad f''(x) = e^x \le e$ for $0 \le x \le 1$. Taking $K = e$, $a = 0$, $b = 1$, and $n = 10$ in Theorem 5, we get $|E_T| \le \dfrac{e(1)^3}{12(10)^2} \approx 0.002265 > 0.00143166$ [actual $|E_T|$ from (a)]. $f^{(4)}(x) = e^x < e$ for $0 \le x \le 1$. Using Theorem 7, we have $|E_S| \le e(1)^5/\left[180(10)^4\right] \approx 0.0000015 > 0.00000095$ [actual $|E_S|$ from (a)]. We see that the actual errors are about two-thirds the size of the error estimates.

20. From Exercise 19, we take $K = e$ to get $|E_T| \le \dfrac{K(b-a)^3}{12n^2} \le 0.00001 \quad \Rightarrow \quad n^2 \ge \dfrac{e(1^3)}{12(0.00001)} \quad \Rightarrow$ $n \ge 150.5$. Take $n = 151$ for T_n. Now $|E_M| \le \dfrac{K(b-a)^3}{24n^2} \le 0.00001 \quad \Rightarrow \quad n \ge 106.4$. Take $n = 107$ for M_n. Finally, $|E_S| \le \dfrac{K(b-a)^5}{180n^4} \le 0.00001 \quad \Rightarrow \quad n^4 \ge \dfrac{e}{180(0.00001)} \quad \Rightarrow \quad n \ge 6.23$. Take $n = 8$ for S_n (since n has to be even for Simpson's Rule).

21. Take $K = 2$ (as in Exercise 17) in Theorem 5. $|E_T| \le \dfrac{K(b-a)^3}{12n^2} \le 10^{-5} \quad \Leftrightarrow \quad \dfrac{1}{6n^2} \le 10^{-5} \quad \Leftrightarrow$ $6n^2 \ge 10^5 \quad \Leftrightarrow \quad n \ge 129.099\ldots \quad \Leftrightarrow \quad n \ge 130$. Take $n = 130$ in the trapezoidal method. For E_M, again take $K = 2$ in Theorem 5 to get $|E_M| \le 2(1)^3/(24n^2) \le 10^{-5} \quad \Leftrightarrow \quad n^2 \ge 2/[24(10^{-5})] \quad \Leftrightarrow \quad n \ge 91.3$ $\Rightarrow \quad n \ge 92$. Take $n = 92$ for M_n.

22. From Example 6(b) we take $K = 76e$ to get $|E_S| \le 76e(1)^5/(180n^4) \le 0.00001 \quad \Rightarrow$ $n^4 \ge 76e/[180(0.00001)] \quad \Rightarrow \quad n \ge 18.4$ Take $n = 20$ (since n must be even.)

23. **(a)** Using the CAS, we differentiate $f(x) = e^{\cos x}$ twice, and find that $f''(x) = e^{\cos x}(\sin^2 x - \cos x)$.

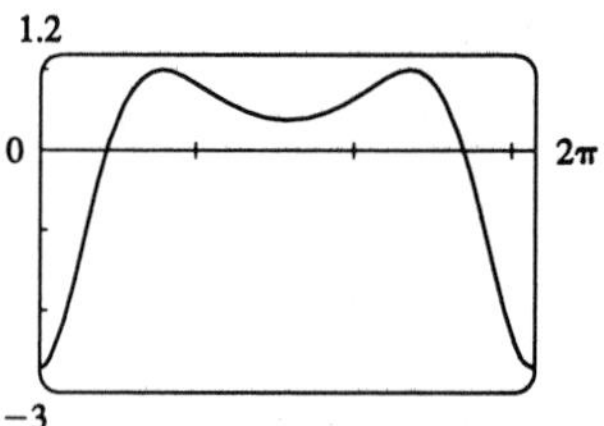

From the graph, we see that $|f''(x)| < 2.8$ on $[0, 2\pi]$. Other possible upper bounds for $|f''(x)|$ are $K = 3$ or $K = e$ (the actual maximum value.)

(b) A CAS gives $M_{10} \approx 7.954926518$. (In Maple, use `student[middlesum]`.)

(c) Using Theorem 5 for the Midpoint Rule, with $K = 2.8$, we get $|E_M| \le \dfrac{2.8(2\pi - 0)^3}{24 \cdot 10^2} \approx 0.287$.

(d) A CAS gives $I \approx 7.954926521$.

(e) The actual error is only about 3×10^{-9}, much less than the estimate in part (c).

(f) We use the CAS to differentiate twice more: $f^{(4)}(x) = e^{\cos x}\left(\sin^4 x - 6\sin^2 x \cos x + 3 - 7\sin^2 x + \cos x\right)$

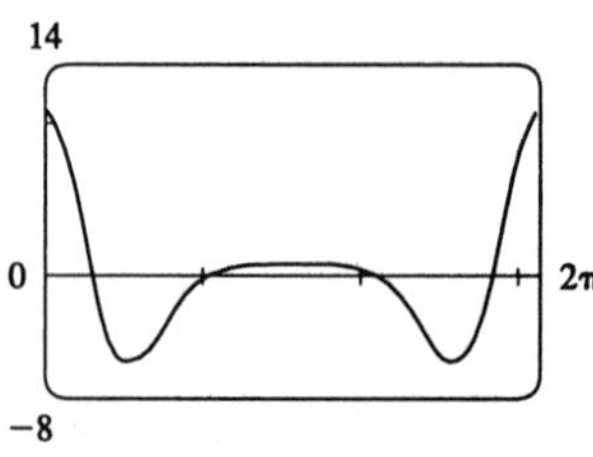

From the graph, it appears that $\left|f^{(4)}(x)\right| < 10.9$ on $[0, 2\pi]$. Another possible upper bound for $\left|f^{(4)}(x)\right|$ is $4e$ (the actual maximum value.)

(g) A CAS gives $S_{10} \approx 7.953789422$. (In Maple, use `student[simpson]`.)

(h) Using Theorem 7 with $K = 10.9$, we get $|E_S| \le \dfrac{10.9(2\pi - 0)^5}{180 \cdot 10^4} \approx 0.0593$.

(i) The actual error is about $7.954926521 - 7.953789427 \approx 0.00114$. This is quite a bit smaller than the estimate in part (h), though the difference is not nearly as great as it was in the case of the Midpoint Rule.

(j) To ensure that $|E_S| \le 0.0001$, we use Theorem 7: $|E_S| \le \dfrac{10.9(2\pi)^5}{180 \cdot n^4} \le 0.0001 \quad \Leftrightarrow \quad \dfrac{10.9(2\pi)^5}{180 \cdot 0.0001} \le n^4$

$\Leftrightarrow \quad n^4 \ge 5{,}929{,}981 \quad \Leftrightarrow \quad n \ge 49.4$. So we must take $n \ge 50$ to ensure that $|I - S_n| \le 0.0001$.

24. (a) Using the CAS, we differentiate $f(x) = \sqrt{4 - x^3}$ twice, and find that

$f''(x) = -\dfrac{9}{4}\dfrac{x^4}{(4 - x^3)^{3/2}} - 3\dfrac{x}{(4 - x^3)^{1/2}}$.

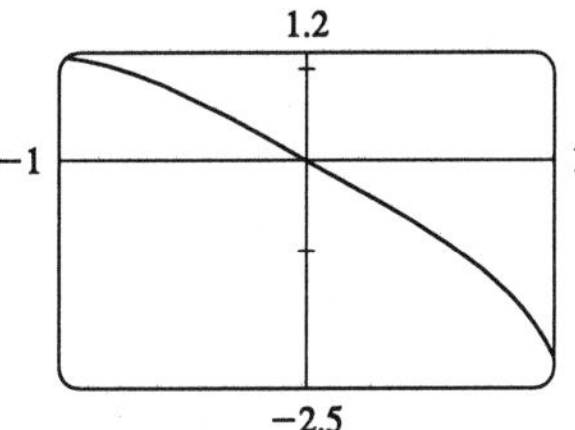

From the graph, it seems that $|f''(x)| < 2.2$ on $[-1, 1]$.

(b) A CAS gives $M_{10} \approx 3.995804152$. (In Maple, use `student[middlesum]`.)

(c) Using Theorem 5 for the Midpoint Rule, with $K = 2.2$, we get $|E_M| \le \dfrac{2.2[1 - (-1)]^3}{24 \cdot 10^2} \approx 0.00733$.

(d) A CAS gives $I \approx 3.995487677$.

(e) The actual error is about -0.0003165, so $|E_M|$ is much smaller than the estimate in (c).

(f) We use the CAS to differentiate twice more, and then graph $f^{(4)}(x) = \dfrac{9}{16}\dfrac{x^2(x^6 - 224x^3 - 1280)}{(4 - x^3)^{7/2}}$.

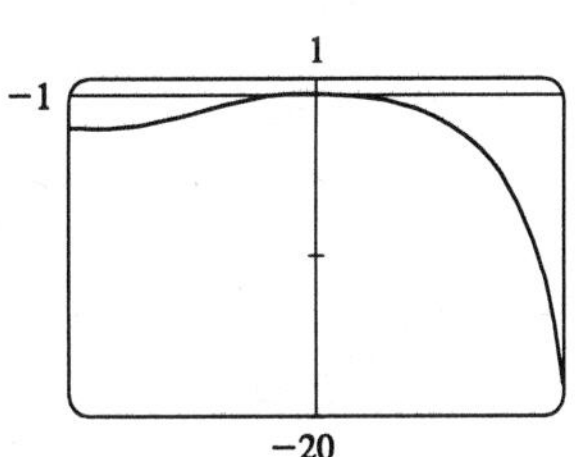

From the graph, we see that $\left|f^{(4)}(x)\right| < 18.1$ on $[-1, 1]$.

(g) A CAS gives $S_{10} \approx 3.995449790$. (In Maple, use `student[simpson]`.)

(h) Using Theorem 7 with $K = 18.1$, we get $|E_S| \le 18.1[1 - (-1)]^5/(180 \cdot 10^4) \approx 0.000322$.

(i) The actual error is about $3.995487677 - 3.995449790 \approx 0.0000379$. This is quite a bit smaller than the estimate in part (h).

(j) To ensure that $|E_S| \le 0.0001$, we use Theorem 7: $|E_S| \le 18.1(2)^5/(180 \cdot n^4) \le 0.0001 \quad \Leftrightarrow$ $18.1(2)^5/(180 \cdot 0.0001) \le n^4 \quad \Leftrightarrow \quad n^4 \ge 32{,}178 \quad \Leftrightarrow \quad n \ge 13.4$. So we must take $n \ge 14$.

25. $\int_0^1 x^3\,dx = \left[\frac{1}{4}x^4\right]_0^1 = 0.25.\ f(x) = x^3.$

$n = 4$: $L_4 = \frac{1}{4}\left[0^3 + \left(\frac{1}{4}\right)^3 + \left(\frac{1}{2}\right)^3 + \left(\frac{3}{4}\right)^3\right] = 0.140625$

$R_4 = \frac{1}{4}\left[\left(\frac{1}{4}\right)^3 + \left(\frac{1}{2}\right)^3 + \left(\frac{3}{4}\right)^3 + 1^3\right] = 0.390625,$

$T_4 = \frac{1}{4\cdot 2}\left[0^3 + 2\left(\frac{1}{4}\right)^3 + 2\left(\frac{1}{2}\right)^3 + 2\left(\frac{3}{4}\right)^3 + 1^3\right] = \frac{17}{64} = 0.265625,$

$M_4 = \frac{1}{4}\left[\left(\frac{1}{8}\right)^3 + \left(\frac{3}{8}\right)^3 + \left(\frac{5}{8}\right)^3 + \left(\frac{7}{8}\right)^3\right] = 0.2421875$

$E_L = \int_0^1 x^3\,dx - L_4 = \frac{1}{4} - 0.140625 = 0.109375,\ E_R = \frac{1}{4} - 0.390625 = -0.140625,$

$E_T = \frac{1}{4} - 0.265625 = -0.015625,\ E_M = \frac{1}{4} - 0.2421875 = 0.0078125$

$n = 8$: $L_8 = \frac{1}{8}\left[f(0) + f\left(\frac{1}{8}\right) + f\left(\frac{2}{8}\right) + \cdots + f\left(\frac{7}{8}\right)\right] \approx 0.191406$

$R_8 = \frac{1}{8}\left[f\left(\frac{1}{8}\right) + f\left(\frac{2}{8}\right) + \cdots + f\left(\frac{7}{8}\right) + f(1)\right] \approx 0.316406$

$T_8 = \frac{1}{8\cdot 2}\left[f(0) + 2\left(f\left(\frac{1}{8}\right) + f\left(\frac{2}{8}\right) + \cdots + f\left(\frac{7}{8}\right)\right) + f(1)\right] \approx 0.253906$

$M_8 = \frac{1}{8}\left[f\left(\frac{1}{16}\right) + f\left(\frac{3}{16}\right) + \cdots + f\left(\frac{13}{16}\right) + f\left(\frac{15}{16}\right)\right] = 0.248047$

$E_L \approx \frac{1}{4} - 0.191406 \approx 0.058594,\ E_R \approx \frac{1}{4} - 0.316406 \approx -0.066406,$

$E_T \approx \frac{1}{4} - 0.253906 \approx -0.003906,\ E_M \approx \frac{1}{4} - 0.248047 \approx 0.001953.$

$n = 16$: $L_{16} = \frac{1}{16}\left[f(0) + f\left(\frac{1}{16}\right) + f\left(\frac{2}{16}\right) + \cdots + f\left(\frac{15}{16}\right)\right] \approx 0.219727$

$R_{16} = \frac{1}{16}\left[f\left(\frac{1}{16}\right) + f\left(\frac{2}{16}\right) + \cdots + f\left(\frac{15}{16}\right) + f(1)\right] \approx 0.282227$

$T_{16} = \frac{1}{16\cdot 2}\left(f(0) + 2\left[f\left(\frac{1}{16}\right) + f\left(\frac{2}{16}\right) + \cdots + f\left(\frac{15}{16}\right)\right] + f(1)\right) \approx 0.250977$

$M_{16} = \frac{1}{16}\left[f\left(\frac{1}{32}\right) + f\left(\frac{3}{32}\right) + \cdots + f\left(\frac{31}{32}\right)\right] \approx 0.249512$

$E_L \approx \frac{1}{4} - 0.219727 \approx 0.030273,\ E_R \approx \frac{1}{4} - 0.282227 \approx -0.032227,$

$E_T \approx \frac{1}{4} - 0.250977 \approx -0.000977,\ E_M \approx \frac{1}{4} - 0.249512 \approx 0.000488.$

n	L_n	R_n	T_n	M_n
4	0.140625	0.390625	0.265625	0.242188
8	0.191406	0.316406	0.253906	0.248047
16	0.219727	0.282227	0.250977	0.249512

n	E_L	E_R	E_T	E_M
4	0.109375	−0.140625	−0.015625	0.007813
8	0.058594	−0.066406	−0.003906	0.001953
16	0.030273	−0.032227	−0.000977	0.000488

Observations:

1. E_L and E_R are always opposite in sign, as are E_T and E_M.
2. As n is doubled, E_L and E_R are decreased by about a factor of 2, and E_T and E_M are decreased by a factor of about 4.
3. The Midpoint approximation is about twice as accurate as the Trapezoidal approximation.
4. All the approximations become more accurate as the value of n increases.
5. The Midpoint and Trapezoidal approximations are much more accurate than the endpoint approximations.

26. $\int_0^2 e^x\,dx = \left[e^x\right]_0^2 = e^2 - 1 \approx 6.389056.\;\; f(x) = e^x$

$n = 4$: $\Delta x = (2-0)/4 = \frac{1}{2}$

$L_4 = \frac{1}{2}\left[e^0 + e^{1/2} + e^1 + e^{3/2}\right] \approx 4.924346$

$R_4 = \frac{1}{2}\left[e^{1/2} + e^1 + e^{3/2} + e^2\right] \approx 8.118874$

$T_4 = \frac{1}{2\cdot 2}\left[e^0 + 2e^{1/2} + 2e^1 + 2e^{3/2} + e^2\right] \approx 6.521610$

$M_4 = \frac{1}{2}\left[e^{1/4} + e^{3/4} + e^{5/4} + e^{7/4}\right] \approx 6.322986.$

$E_L \approx 6.389056 - 4.924346 \approx 1.464710,\; E_R \approx 6.389056 - 8.118874 = -1.729818,$

$E_T \approx 6.389056 - 6.521610 \approx -0.132554,\; E_M \approx 6.389056 - 6.322986 = 0.0660706.$

$n = 8$: $\Delta x = (2-0)/8 = \frac{1}{4}$

$L_8 = \frac{1}{4}\left[e^0 + e^{1/4} + e^{1/2} + e^{3/4} + e^1 + e^{5/4} + e^{3/2} + e^{7/4}\right] \approx 5.623666$

$R_8 = \frac{1}{4}\left[e^{1/4} + e^{1/2} + e^{3/4} + e^1 + e^{5/4} + e^{3/2} + e^{7/4} + e^2\right] \approx 7.220930$

$T_8 = \frac{1}{4\cdot 2}\left[e^0 + 2e^{1/4} + 2e^{1/2} + 2e^{3/4} + 2e^1 + 2e^{5/4} + 2e^{3/2} + 2e^{7/4} + e^2\right] \approx 6.422298$

$M_8 = \frac{1}{4}\left[e^{1/8} + e^{3/8} + e^{5/8} + e^{7/8} + e^{9/8} + e^{11/8} + e^{13/8} + e^{15/8}\right] \approx 6.372448$

$E_L \approx 6.389056 - 5.623666 \approx 0.765390,\; E_R \approx 6.389056 - 7.220930 \approx -0.831874,$

$E_T \approx 6.389056 - 6.422298 \approx -0.033242,\; E_M \approx 6.389056 - 6.372448 \approx 0.016608.$

$n = 16$: $\Delta x = (2-0)/16 = \frac{1}{8}$

$L_{16} = \frac{1}{8}\left[f(0) + f\left(\frac{1}{8}\right) + f\left(\frac{2}{8}\right) + \cdots + f\left(\frac{14}{8}\right) + f\left(\frac{15}{8}\right)\right] \approx 5.998057$

$R_{16} = \frac{1}{8}\left[f\left(\frac{1}{8}\right) + f\left(\frac{2}{8}\right) + f\left(\frac{3}{8}\right) + \cdots + f\left(\frac{15}{8}\right) + f(2)\right] \approx 6.796689$

$T_{16} = \frac{1}{8\cdot 2}\left(f(0) + 2\left[f\left(\frac{1}{8}\right) + f\left(\frac{2}{8}\right) + f\left(\frac{3}{8}\right) + \cdots + f\left(\frac{15}{8}\right)\right] + f(2)\right) \approx 6.397373$

$M_{16} = \frac{1}{8}\left[f\left(\frac{1}{16}\right) + f\left(\frac{3}{16}\right) + f\left(\frac{5}{16}\right) + \cdots + f\left(\frac{29}{16}\right) + f\left(\frac{31}{16}\right)\right] \approx 6.384899$

$E_L \approx 6.389056 - 5.998057 \approx 0.390999,\; E_R \approx 6.389056 - 6.796689 \approx -0.407633,$

$E_T \approx 6.389056 - 6.397373 \approx -0.008317,\; E_M \approx 6.389056 - 6.384899 \approx 0.004158.$

n	E_L	E_R	E_T	E_M
4	1.464710	−1.729818	−0.132554	0.066071
8	0.765390	−0.831874	−0.033242	0.016608
16	0.390999	−0.407633	−0.008317	0.004158

Observations:

1. E_L and E_R are always opposite in sign, as are E_T and E_M.
2. As n is doubled, E_L and E_R are decreased by a factor of about 2, and E_T and E_M are decreased by a factor of about 4.
3. The Midpoint approximation is about twice as accurate as the Trapezoidal approximation.
4. All the approximations become more accurate as the value of n increases.
5. The Midpoint and Trapezoidal approximations are much more accurate than the endpoint approximations.

27. $\int_1^4 \sqrt{x}\,dx = \left[\frac{2}{3}x^{3/2}\right]_1^4 = \frac{2}{3}(8-1) = \frac{14}{3} \approx 4.666667$

$n = 6$: $\Delta x = (4-1)/6 = \frac{1}{2}$

$$T_6 = \frac{1}{2\cdot 2}\left[\sqrt{1} + 2\sqrt{1.5} + 2\sqrt{2} + 2\sqrt{2.5} + 2\sqrt{3} + 2\sqrt{3.5} + \sqrt{4}\right] \approx 4.661488$$

$$M_6 = \frac{1}{2}\left[\sqrt{1.25} + \sqrt{1.75} + \sqrt{2.25} + \sqrt{2.75} + \sqrt{3.25} + \sqrt{3.75}\right] \approx 4.669245$$

$$S_6 = \frac{1}{2\cdot 3}\left[\sqrt{1} + 4\sqrt{1.5} + 2\sqrt{2} + 4\sqrt{2.5} + 2\sqrt{3} + 4\sqrt{3.5} + \sqrt{4}\right] \approx 4.666563$$

$E_T \approx \frac{14}{3} - 4.661488 \approx 0.005178$, $E_M \approx \frac{14}{3} - 4.669245 \approx -0.002578$,

$E_S \approx \frac{14}{3} - 4.666563 \approx 0.000104$.

$n = 12$: $\Delta x = (4-1)/12 = \frac{1}{4}$

$$T_{12} = \frac{1}{4\cdot 2}(f(1) + 2[f(1.25) + f(1.5) + \cdots + f(3.5) + f(3.75)] + f(4)) \approx 4.665367$$

$$M_{12} = \frac{1}{4}[f(1.125) + f(1.375) + f(1.625) + \cdots + f(3.875)] \approx 4.667316$$

$$S_{12} = \frac{1}{4\cdot 3}[f(1) + 4\,f(1.25) + 2\,f(1.5) + 4\,f(1.75) + \cdots + 4\,f(3.75) + f(4)] \approx 4.666659$$

$E_T \approx \frac{14}{3} - 4.665367 \approx 0.001300$, $E_M \approx \frac{14}{3} - 4.667316 \approx -0.000649$,

$E_S \approx \frac{14}{3} - 4.666659 \approx 0.000007$.

Note: These errors were computed more precisely and then rounded to six places. That is, they were not computed by comparing the rounded values of T_n, M_n, and S_n with the rounded value of the actual definite integral.

n	T_n	M_n	S_n
6	4.661488	4.669245	4.666563
12	4.665367	4.667316	4.666659

n	E_T	E_M	E_S
6	0.005178	−0.002578	0.000104
12	0.001300	−0.000649	0.000007

Observations:

1. E_T and E_M are opposite in sign and decrease by a factor of about 4 as n is doubled.
2. The Simpson's approximation is much more accurate than the Midpoint and Trapezoidal approximations, and seems to decrease by a factor of about 16 as n is doubled.

28. $\int_{-1}^2 xe^x\,dx = [xe^x - e^x]_{-1}^2 = [(x-1)e^x]_{-1}^2 = e^2 - (-2e^{-1}) = e^2 + 2/e \approx 8.124815$

$n = 6$: $\Delta x = [2-(-1)]/6 = \frac{1}{2}$

$$T_6 = \frac{1}{2\cdot 2}\left[-1e^{-1} + 2(-0.5e^{-0.5}) + 2\cdot 0 + 2(0.5e^{0.5}) + 2(e) + 2(1.5e^{1.5}) + 2e^2\right] \approx 8.583514$$

$$M_6 = \frac{1}{2}\left[-0.75e^{-0.75} + (-0.25e^{-0.25}) + 0.25e^{0.25} + 0.75e^{0.75} + 1.25e^{1.25} + 1.75e^{1.75}\right] \approx 7.896632$$

$$S_6 = \frac{1}{2\cdot 3}\left[-1e^{-1} + 4(-0.5e^{-0.5}) + 2\cdot 0 + 4(0.5e^{0.5}) + 2e + 4(1.5e^{1.5}) + 2e^2\right] \approx 8.136885$$

$E_T \approx 8.124815 - 8.583514 \approx -0.458699$, $E_M \approx 8.124815 - 7.896632 \approx 0.228183$,

$E_S \approx 8.124815 - 8.136885 \approx -0.012070$.

$n = 12$: $\Delta x = [2 - (-1)]/12 = \frac{1}{4}$

$$T_{12} \approx \frac{1}{4 \cdot 2}\big[-e^{-1} + 2\big(-0.75e^{-0.75}\big) + 2\big(-0.5e^{-0.5}\big) + 2\big(-0.25e^{-0.25}\big) + 2 \cdot 0 + 2\big(0.25e^{0.25}\big) + 2\big(0.5e^{0.5}\big) + 2\big(0.75e^{0.75}\big) + 2e + 2\big(1.25e^{1.25}\big) + 2\big(1.5e^{1.5}\big) + 2\big(1.75e^{1.75}\big) + 2e^2\big] \approx 8.240073$$

$$M_{12} \approx \frac{1}{4}\big[f\big(-\tfrac{7}{8}\big) + f\big(-\tfrac{5}{8}\big) + f\big(-\tfrac{3}{8}\big) + f\big(-\tfrac{1}{8}\big) + f\big(\tfrac{1}{8}\big) + f\big(\tfrac{3}{8}\big) + \cdots + f\big(\tfrac{13}{8}\big) + f\big(\tfrac{15}{8}\big)\big] \approx 8.067259$$

$$S_{12} \approx \frac{1}{4 \cdot 3}\Big[-e^{-1} + 4\big(-0.75e^{-0.75}\big) + 2\big(-0.5e^{-0.5}\big) + 4\big(-0.25e^{-0.25}\big) + 2 \cdot 0 + 4\big(0.25e^{0.25}\big) + 2\big(0.5e^{0.5}\big) + 4\big(0.75e^{0.75}\big) + 2e + 4\big(1.25e^{1.25}\big) + 2\big(1.5e^{1.5}\big) + 4\big(1.75e^{1.75}\big) + 2e^2\Big] \approx 8.125593$$

$E_T \approx 8.124815 - 8.240073 \approx -0.115258$, $E_M \approx 8.124815 - 8.067259 \approx 0.057555$,

$E_S \approx 8.124815 - 8.125593 \approx -0.000778$

n	T_n	M_n	S_n
6	8.583514	7.896632	8.136885
12	8.240073	8.067259	8.125593

n	E_T	E_M	E_S
6	− 0.458699	0.228183	− 0.012070
12	− 0.115258	0.057555	− 0.000778

Observations:

1. E_T and E_M are opposite in sign and decrease by a factor of about 4 as n is doubled.
2. The Simpson's approximation is much more accurate than the Midpoint and Trapezoidal approximations, and seems to decrease by a factor of about 16 as n is doubled.

29. $\int_1^{3.2} y\,dx \approx \frac{0.2}{2}\big[4.9 + 2(5.4) + 2(5.8) + 2(6.2) + 2(6.7) + 2(7.0) + 2(7.3) + 2(7.5) + 2(8.0) + 2(8.2) + 2(8.3) + 8.3\big] = 15.4$

30. $\int_2^6 y\,dx \approx \frac{0.5}{3}\big[9.22 + 4(9.01) + 2(8.76) + 4(8.30) + 2(7.52) + 4(6.83) + 2(7.32) + 4(7.69) + 7.91\big] \approx 31.94$

31. $\Delta t = 1\text{ min} = \frac{1}{60}\text{ h}$, so distance $= \int_0^{1/6} v(t)dt \approx \frac{1/60}{3}\big[40 + 4(42) + 2(45) + 4(49) + 2(52) + 4(54) + 2(56) + 4(57) + 2(57) + 4(55) + 56\big] \approx 8.6\text{ mi}$.

32. If $x =$ distance from left end of pool and $w = w(x) =$ width at x, then Simpson's Rule with $n = 8$ and $\Delta x = 2$ gives Area $= \int_0^{16} w\,dx \approx \frac{2}{3}\big[0 + 4(6.2) + 2(7.2) + 4(6.8) + 2(5.6) + 4(5.0) + 2(4.8) + 4(4.8) + 0\big] \approx 84\text{ m}^2$.

33. $\Delta x = (4 - 0)/4 = 1$

(a) $T_4 = \frac{1}{2}[f(0) + 2f(1) + 2f(2) + 2f(3) + f(4)] \approx \frac{1}{2}[0 + 2(3) + 2(5) + 2(3) + 1] = 11.5$

(b) $M_4 = 1 \cdot [f(.5) + f(1.5) + f(2.5) + f(3.5)] \approx 1 + 4.5 + 4.5 + 2 = 12$

(c) $S_4 = \frac{1}{3}[f(0) + 4f(1) + 2f(2) + 4f(3) + f(4)] \approx \frac{1}{3}[0 + 4(3) + 2(5) + 4(3) + 1] = 11.\overline{6}$

34. By Simpson's Rule,

$$\text{Volume} = \int_0^{10} A(x)dx \approx \frac{1}{3}\big[A(0) + 4A(1) + 2A(2) + 4A(3) + 2A(4) + 4A(5) + 2A(6) + 4A(7) + 2A(8) + 4A(9) + A(10)\big] = 5.7\overline{6} \approx 5.8\text{ m}^3.$$

35. Volume $= \pi \int_0^2 \left(\sqrt[3]{1+x^3}\right)^2 dx = \pi \int_0^2 \left(1+x^3\right)^{2/3} dx$. $V \approx \pi \cdot S_{10}$ where $f(x) = \left(1+x^3\right)^{2/3}$ and $\Delta x = (2-0)/10 = \frac{1}{5}$. Therefore

$$V \approx \pi \cdot S_{10} = \pi \tfrac{1}{5 \cdot 3}[f(0) + 4f(0.2) + 2f(0.4) + 4f(0.6) + 2f(0.8) + 4f(1) + 2f(1.2) + 4f(1.4) + 2f(1.6) + 4f(1.8) + f(2)] \approx 12.325078.$$

36. Using Simpson's Rule with $n = 10$, $\Delta x = \frac{\pi/2}{10}$, $L = 1$, $\theta_0 = 42° \approx 0.733$ radians, $g \approx 9.8\,\text{m/s}^2$, and $k^2 = \sin^2\left(\frac{1}{2}\theta_0\right) \approx \sin^2\left(\frac{0.733}{2}\right) \approx 0.12843$, we get

$$T = 4\sqrt{\frac{L}{g}} \int_0^{\pi/2} \frac{dx}{\sqrt{1-k^2\sin^2 x}}$$

$$\approx 4\sqrt{\frac{1}{9.8}}\left(\frac{\pi/2}{10 \cdot 3}\right)\left(\frac{1}{\sqrt{1-0.1284\sin^2 0}} + \frac{4}{\sqrt{1-0.1284\sin^2\frac{\pi}{20}}} + \frac{2}{\sqrt{1-0.1284\sin^2\frac{\pi}{10}}} + \cdots + \frac{2}{\sqrt{1-0.1284\sin^2\frac{4\pi}{10}}} + \frac{4}{\sqrt{1-0.1284\sin^2\frac{9\pi}{20}}} + \frac{1}{\sqrt{1-0.1284\sin^2\frac{\pi}{2}}}\right) \approx 2.07665.$$

37. Since the Trapezoidal and Midpoint approximations on the interval $[a, b]$ are the sums of the Trapezoidal and Midpoint approximations on the subintervals $[x_{i-1}, x_i]$, $i = 1, 2, \ldots, n$, we can focus our attention on one such interval. The condition $f''(x) < 0$ for $a \le x \le b$ means that the graph of f is concave down as in Figure 5. In that figure, T_n is the area of the trapezoid $AQRD$, $\int_a^b f(x)dx$ is the area of the region $AQPRD$, and M_n is the area of the trapezoid $ABCD$, so $T_n < \int_a^b f(x)dx < M_n$. In general, the condition $f'' < 0$ implies that the graph of f on $[a, b]$ lies above the chord joining the points $(a, f(a))$ and $(b, f(b))$. Thus $\int_a^b f(x)dx > T_n$. Since M_n is the area under a tangent to the graph, and since $f'' < 0$ implies that the tangent lies above the graph, we also have $M_n > \int_a^b f(x)dx$. Thus $T_n < \int_a^b f(x)dx < M_n$.

38. Let f be a polynomial of degree ≤ 3; say $f(x) = Ax^3 + Bx^2 + Cx + D$. It will suffice to show that Simpson's estimate is exact when there are two intervals ($n = 2$), because for a larger even number of intervals the sum of exact estimates is exact. As in the derivation of Simpson's Rule, we can assume that $x_0 = -h$, $x_1 = 0$, and $x_2 = h$. Then Simpson's approximation is

$$\begin{aligned}\int_{-h}^h f(x)dx &\approx \tfrac{1}{3}h[f(-h) + 4f(0) + f(h)] \\ &= \tfrac{1}{3}h[(-Ah^3 + Bh^2 - Ch + D) + 4D + (Ah^3 + Bh^2 + Ch + D)] \\ &= \tfrac{1}{3}h[2Bh^2 + 6D] = \tfrac{2}{3}Bh^3 + 2Dh.\end{aligned}$$

The exact value of the integral is

$$\int_{-h}^h (Ax^3 + Bx^2 + Cx + D)dx = \left[\tfrac{1}{4}Ax^4 + \tfrac{1}{3}Bx^3 + \tfrac{1}{2}Cx^2 + Dx\right]_{-h}^h = \tfrac{2}{3}Bh^3 + 2Dh.$$

Thus Simpson's Rule is exact.

39. $T_n = \frac{1}{2}\Delta x[f(x_0) + 2f(x_1) + \cdots + 2f(x_{n-1}) + f(x_n)]$ and $M_n = \Delta x[f(\overline{x}_1) + f(\overline{x}_2) + \cdots + f(\overline{x}_{n-1}) + f(\overline{x}_n)]$, where $\overline{x}_i = \frac{1}{2}(x_{i-1} + x_i)$. Now

$$T_{2n} = \tfrac{1}{2}\left(\tfrac{1}{2}\Delta x\right)[f(x_0) + 2f(\overline{x}_1) + 2f(x_1) + 2f(\overline{x}_2) + 2f(x_2) + \cdots + 2f(\overline{x}_{n-1}) + 2f(x_{n-1}) + 2f(\overline{x}_n) + f(x_n)], \text{ so}$$

$$\tfrac{1}{2}(T_n + M_n) = \tfrac{1}{4}\Delta x[f(x_0) + 2f(x_1) + \cdots + 2f(x_{n-1}) + f(x_n)] + \tfrac{1}{4}\Delta x[2f(\overline{x}_1) + 2f(\overline{x}_2) + \cdots + 2f(\overline{x}_{n-1}) + 2f(\overline{x}_n)] = T_{2n}.$$

40. $T_n = \dfrac{\Delta x}{2}\left[f(x_0) + 2\displaystyle\sum_{i=1}^{n-1} f(x_i) + f(x_n)\right]$ and $M_n = \Delta x \displaystyle\sum_{i=1}^{n} f\left(x_i - \frac{\Delta x}{2}\right)$, so

$\frac{1}{3}T_n + \frac{2}{3}M_n = \frac{1}{3}(T_n + 2M_n) = \dfrac{\Delta x}{3}\left[\dfrac{f(x_0)}{2} + \displaystyle\sum_{i=1}^{n-1} f(x_i) + \dfrac{f(x_n)}{2} + 2\displaystyle\sum_{i=1}^{n} f\left(x_i - \frac{\Delta x}{2}\right)\right]$, where $\Delta x = \dfrac{b-a}{n}$.

Let $\delta x = \dfrac{b-a}{2n}$. Then $\Delta x = 2\delta x$, so

$$\tfrac{1}{3}T_n + \tfrac{2}{3}M_n = \frac{\delta x}{3}\left[f(x_0) + 2\sum_{i=1}^{n-1} f(x_i) + f(x_n) + 4\sum_{i=1}^{n} f(x_i - \delta x)\right]$$

$$= \tfrac{1}{3}\delta x\left[f(x_0) + 4f(x_1 - \delta x) + 2f(x_1) + 4f(x_2 - \delta x) + 2f(x_2) + \cdots + 2f(x_{n-1}) + 4f(x_n - \delta x) + f(x_n)\right].$$

Since $x_0, x_1 - \delta x, x_1, x_2 - \delta x, x_2, \ldots, x_{n-1}, x_n - \delta x, x_n$ are the partition points for S_{2n}, and since $\delta x = \dfrac{b-a}{2n}$ is the width of the subintervals for S_{2n}, the last expression for $\frac{1}{3}T_n + \frac{2}{3}M_n$ is the usual expression for S_{2n}. Therefore $\frac{1}{3}T_n + \frac{2}{3}M_n = S_{2n}$.

EXERCISES 7.9

1. The area under the graph of $y = 1/x^3 = x^{-3}$ between $x = 1$ and $x = t$ is $A(t) = \int_1^t x^{-3}\,dx = \left[-\frac{1}{2}x^{-2}\right]_1^t = \frac{1}{2} - 1/(2t^2)$. So the area for $0 \le x \le 10$ is $A(10) = 0.5 - 0.005 = 0.495$, the area for $0 \le x \le 100$ is $A(100) = 0.5 - 0.00005 = 0.49995$, and the area for $0 \le x \le 1000$ is $A(1000) = 0.5 - 0.0000005 = 0.4999995$. The total area under the curve for $x \ge 1$ is $\int_1^\infty x^{-3}\,dx = \lim\limits_{t\to\infty}\left[-\frac{1}{2}t^{-2} - \left(-\frac{1}{2}\right)\right] = 0 - \left(-\frac{1}{2}\right) = \frac{1}{2}$.

2. **(a)**

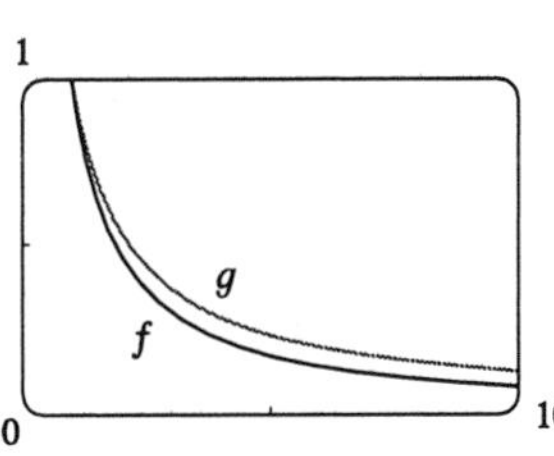

(b) The area under the graph of f from $x = 1$ to $x = t$ is

$F(t) = \int_1^t f(x)\,dx = \int_1^t x^{-1.1}\,dx$
$= \left[-\frac{1}{0.1}x^{-0.1}\right]_1^t = -10(t^{-0.1} - 1)$
$= 10(1 - t^{-0.1})$,

and the area under the graph of g is

$G(t) = \int_1^t g(x)\,dx = \int_1^t x^{-0.9}\,dx = \left[\frac{1}{0.1}x^{0.1}\right]_1^t$
$= 10(t^{0.1} - 1)$.

t	$F(t)$	$G(t)$
10	2.06	2.59
100	3.69	5.85
10^4	6.02	15.12
10^6	7.49	29.81
10^{10}	9	90
10^{20}	9.9	990

(c) The total area under the graph of f is $\lim\limits_{t\to\infty} F(t) = \lim\limits_{t\to\infty} 10(1 - t^{-0.1}) = 10$.

The total area under the graph of g does not exist, since $\lim\limits_{t\to\infty} G(t) = \lim\limits_{t\to\infty} 10(t^{0.1} - 1) = \infty$.

3. $\int_2^\infty \frac{dx}{\sqrt{x+3}} = \lim_{t\to\infty}\int_2^t \frac{dx}{\sqrt{x+3}} = \lim_{t\to\infty}\left[2\sqrt{x+3}\right]_2^t = \lim_{t\to\infty}\left(2\sqrt{t+3}-2\sqrt{5}\right) = \infty$. Divergent

4. $\int_2^\infty \frac{dx}{(x+3)^{3/2}} = \lim_{t\to\infty}\int_2^t \frac{dx}{(x+3)^{3/2}} = \lim_{t\to\infty}(-2)\left[(x+3)^{-1/2}\right]_2^t = \lim_{t\to\infty}\left[\frac{-2}{\sqrt{t+3}}+\frac{2}{\sqrt{5}}\right] = \frac{2}{\sqrt{5}}$

5. $\int_{-\infty}^1 \frac{dx}{(2x-3)^2} = \lim_{t\to-\infty}\frac{1}{2}\int_t^1 \frac{2\,dx}{(2x-3)^2} = \lim_{t\to-\infty}\frac{1}{2}\left[-\frac{1}{2x-3}\right]_t^1 = \lim_{t\to-\infty}\left[\frac{1}{2}+\frac{1}{2(2t-3)}\right] = \frac{1}{2}$

6. $\int_{-\infty}^{-1} \frac{dx}{\sqrt[3]{x-1}} = \lim_{t\to-\infty}\int_t^{-1}(x-1)^{-1/3}\,dx = \lim_{t\to-\infty}\left[\frac{3}{2}(x-1)^{2/3}\right]_t^{-1} = \lim_{t\to-\infty}\left[\frac{3}{2}\sqrt[3]{4}-\frac{3}{2}(t-1)^{2/3}\right] = -\infty$.

Divergent

7. $\int_{-\infty}^\infty x\,dx = \int_{-\infty}^0 x\,dx + \int_0^\infty x\,dx$. $\int_{-\infty}^0 x\,dx = \lim_{t\to-\infty}\left[\frac{1}{2}x^2\right]_t^0 = \lim_{t\to-\infty}\left(-\frac{1}{2}t^2\right) = -\infty$. Divergent

8. $\int_{-\infty}^\infty (2x^2-x+3)dx = \int_{-\infty}^0 (2x^2-x+3)dx + \int_0^\infty (2x^2-x+3)dx$.
$\int_{-\infty}^0 (2x^2-x+3)dx = \lim_{t\to-\infty}\left[\frac{2}{3}x^3-\frac{1}{2}x^2+3x\right]_t^0 = \lim_{t\to-\infty}\left[-\frac{2}{3}t^3+\frac{1}{2}t^2-3t\right] = \infty$. Divergent

9. $\int_0^\infty e^{-x}\,dx = \lim_{t\to\infty}\int_0^t e^{-x}\,dx = \lim_{t\to\infty}[-e^{-x}]_0^t = \lim_{t\to\infty}(-e^{-t}+1) = 1$

10. $\int_{-\infty}^0 e^{3x}\,dx = \lim_{t\to-\infty}\int_t^0 e^{3x}\,dx = \lim_{t\to-\infty}\left[\frac{1}{3}e^{3x}\right]_t^0 = \lim_{t\to-\infty}\left[\frac{1}{3}-\frac{1}{3}e^{3t}\right] = \frac{1}{3}$

11. $\int_{-\infty}^\infty xe^{-x^2}\,dx = \int_{-\infty}^0 xe^{-x^2}\,dx + \int_0^\infty xe^{-x^2}\,dx$, $\int_{-\infty}^0 xe^{-x^2}\,dx = \lim_{t\to-\infty}-\frac{1}{2}\left[e^{-x^2}\right]_0^t = \lim_{t\to-\infty}-\frac{1}{2}\left(1-e^{-t^2}\right) = -\frac{1}{2}$,
and $\int_0^\infty xe^{-x^2}\,dx = \lim_{t\to\infty}-\frac{1}{2}\left[e^{-x^2}\right]_t^0 = \lim_{t\to\infty}-\frac{1}{2}\left(e^{-t^2}-1\right) = \frac{1}{2}$. Therefore $\int_{-\infty}^\infty xe^{-x^2}\,dx = -\frac{1}{2}+\frac{1}{2} = 0$.

12. $\int_{-\infty}^\infty x^2e^{-x^3}\,dx = \int_{-\infty}^0 x^2e^{-x^3}\,dx + \int_0^\infty x^2e^{-x^3}\,dx$, and
$\int_{-\infty}^0 x^2e^{-x^3}\,dx = \lim_{t\to-\infty}\left[-\frac{1}{3}e^{-x^3}\right]_t^0 = -\frac{1}{3}+\frac{1}{3}\left(\lim_{t\to-\infty}e^{-t^3}\right) = \infty$. Divergent

13. $\int_0^\infty \frac{dx}{(x+2)(x+3)} = \lim_{t\to\infty}\int_0^t\left[\frac{1}{x+2}-\frac{1}{x+3}\right]dx = \lim_{t\to\infty}\left[\ln\left(\frac{x+2}{x+3}\right)\right]_0^t = \lim_{t\to\infty}\left[\ln\left(\frac{t+2}{t+3}\right)-\ln\frac{2}{3}\right]$
$= \ln 1 - \ln\frac{2}{3} = -\ln\frac{2}{3}$

14. $\int_0^\infty \frac{x\,dx}{(x+2)(x+3)} = \lim_{t\to\infty}\int_0^t\left[\frac{-2}{x+2}+\frac{3}{x+3}\right]dx = \lim_{t\to\infty}[3\ln(x+3)-2\ln(x+2)]_0^t$
$= \lim_{t\to\infty}\left[\ln\frac{(t+3)^3}{(t+2)^2}-\ln\frac{27}{4}\right] = \infty$. Divergent

15. $\int_0^\infty \cos x\,dx = \lim_{t\to\infty}[\sin x]_0^t = \lim_{t\to\infty}\sin t$, which does not exist. Divergent

16. $\int_1^\infty \sin\pi x\,dx = \lim_{t\to\infty}-\frac{1}{\pi}[\cos\pi x]_1^t = -\frac{1}{\pi}\lim_{t\to\infty}(\cos\pi t+1)$, which does not exist. Divergent

17. $\int_0^\infty \frac{5\,dx}{2x+3} = \frac{5}{2}\lim_{t\to\infty}\int_0^t \frac{2\,dx}{2x+3} = \frac{5}{2}\lim_{t\to\infty}[\ln(2x+3)]_0^t = \frac{5}{2}\lim_{t\to\infty}[\ln(2t+3)-\ln 3] = \infty$. Divergent

18. $\int_{-\infty}^3 \frac{dx}{x^2+9} = \lim_{t\to-\infty}\left[\frac{1}{3}\tan^{-1}\left(\frac{1}{3}x\right)\right]_t^3 = \lim_{t\to-\infty}\frac{1}{3}\left[\frac{\pi}{4}-\tan^{-1}\left(\frac{1}{3}t\right)\right] = \frac{1}{3}\left(\frac{\pi}{4}+\frac{\pi}{2}\right) = \frac{\pi}{4}$

19. $\int_{-\infty}^{1} xe^{2x}\,dx = \lim_{t\to-\infty}\int_t^1 xe^{2x}\,dx = \lim_{t\to-\infty}\left[\frac{1}{2}xe^{2x} - \frac{1}{4}e^{2x}\right]_t^1$ (by parts)

$= \lim_{t\to-\infty}\left[\frac{1}{2}e^2 - \frac{1}{4}e^2 - \frac{1}{2}te^{2t} + \frac{1}{4}e^{2t}\right] = \frac{1}{4}e^2 - 0 + 0 = \frac{1}{4}e^2,$

since $\lim_{t\to-\infty} te^{2t} = \lim_{t\to-\infty}\frac{t}{e^{-2t}} \overset{\text{H}}{=} \lim_{t\to-\infty}\frac{1}{-2e^{-2t}} = \lim_{t\to-\infty} -\frac{1}{2}e^{2t} = 0.$

20. $\displaystyle\int_0^\infty xe^{-x}\,dx = \lim_{t\to\infty}\left[-xe^{-x} - e^{-x}\right]_0^t = \lim_{t\to\infty}\left[1 - (t+1)e^{-t}\right] = 1 - \lim_{t\to\infty}\frac{t+1}{e^t} \overset{\text{H}}{=} 1 - \lim_{t\to\infty}\frac{1}{e^t} = 1 - 0 = 1$

21. $\displaystyle\int_1^\infty \frac{\ln x}{x}\,dx = \lim_{t\to\infty}\left[\frac{(\ln x)^2}{2}\right]_1^t = \lim_{t\to\infty}\frac{(\ln t)^2}{2} = \infty.$ Divergent

22. $\displaystyle\int_e^\infty \frac{dx}{x(\ln x)^2} = \lim_{t\to\infty}\int_e^t \frac{dx}{x(\ln x)^2} = \lim_{t\to\infty}\left[-\frac{1}{\ln x}\right]_e^t = \lim_{t\to\infty}\left[1 - \frac{1}{\ln t}\right] = 1$

23. $\displaystyle\int_{-\infty}^\infty \frac{x\,dx}{1+x^2} = \int_{-\infty}^0 \frac{x\,dx}{1+x^2} + \int_0^\infty \frac{x\,dx}{1+x^2}$ and

$\displaystyle\int_{-\infty}^0 \frac{x\,dx}{1+x^2} = \lim_{t\to-\infty}\left[\frac{1}{2}\ln(1+x^2)\right]_t^0 = \lim_{t\to-\infty}\left[0 - \frac{1}{2}\ln(1+t^2)\right] = -\infty.$ Divergent

24. $\int_{-\infty}^\infty e^{-|x|}\,dx = \int_{-\infty}^0 e^x\,dx + \int_0^\infty e^{-x}\,dx$, $\int_{-\infty}^0 e^x\,dx = \lim_{t\to-\infty}\left[e^x\right]_t^0 = \lim_{t\to-\infty}(1 - e^t) = 1$, and $\int_0^\infty e^{-x}\,dx = \lim_{t\to\infty}\left[-e^{-x}\right]_0^t = \lim_{t\to\infty}(1 - e^{-t}) = 1$. Therefore $\int_{-\infty}^\infty e^{-|x|}\,dx = 1 + 1 = 2$.

25. Integrate by parts with $u = \ln x$, $dv = dx/x^2 \quad\Rightarrow\quad du = dx/x$, $v = -1/x$.

$\displaystyle\int_1^\infty \frac{\ln x}{x^2}\,dx = \lim_{t\to\infty}\int_1^t \frac{\ln x}{x^2}\,dx = \lim_{t\to\infty}\left[-\frac{\ln x}{x} - \frac{1}{x}\right]_1^t = \lim_{t\to\infty}\left[-\frac{\ln t}{t} - \frac{1}{t} + 0 + 1\right]$

$\displaystyle = -0 - 0 + 0 + 1 = 1,$ since $\displaystyle\lim_{t\to\infty}\frac{\ln t}{t} \overset{\text{H}}{=} \lim_{t\to\infty}\frac{1/t}{1} = 0.$

26. Integrate by parts with $u = \ln x$, $dv = dx/x^3$, $du = dx/x$, $v = -1/(2x^2)$:

$\displaystyle\int_1^\infty \frac{\ln x}{x^3}\,dx = \lim_{t\to\infty}\int_1^t \frac{\ln x}{x^3} = \lim_{t\to\infty}\left(\left[-\frac{1}{2x^2}\ln x\right]_1^t + \frac{1}{2}\int_1^t \frac{1}{x^3}\,dx\right)$

$\displaystyle = \lim_{t\to\infty}\left(-\frac{1}{2}\frac{\ln t}{t^2} + 0 - \frac{1}{4t^2} + \frac{1}{4}\right) = \frac{1}{4},$

since, by l'Hospital's Rule, $\displaystyle\lim_{t\to\infty}\frac{\ln t}{t^2} = \lim_{t\to\infty}\frac{1/t}{2t} = \lim_{t\to\infty}\frac{1}{2t^2} = 0.$

27. $\displaystyle\int_0^3 \frac{dx}{\sqrt{x}} = \lim_{t\to0^+}\int_t^3 \frac{dx}{\sqrt{x}} = \lim_{t\to0^+}\left[2\sqrt{x}\right]_t^3 = \lim_{t\to0^+}\left(2\sqrt{3} - 2\sqrt{t}\right) = 2\sqrt{3}$

28. $\displaystyle\int_0^3 \frac{dx}{x\sqrt{x}} = \lim_{t\to0^+}\int_t^3 \frac{dx}{x^{3/2}} = \lim_{t\to0^+}\left[\frac{-2}{\sqrt{x}}\right]_t^3 = \frac{-2}{\sqrt{3}} + \lim_{t\to0^+}\frac{2}{\sqrt{t}} = \infty.$ Divergent

29. $\displaystyle\int_{-1}^0 \frac{dx}{x^2} = \lim_{t\to0^-}\int_{-1}^t \frac{dx}{x^2} = \lim_{t\to0^-}\left[\frac{-1}{x}\right]_{-1}^t = \lim_{t\to0^-}\left[-\frac{1}{t} + \frac{1}{-1}\right] = \infty.$ Divergent

30. $\displaystyle\int_1^9 \frac{dx}{\sqrt[3]{x-9}} = \lim_{t\to9^-}\int_1^t \frac{dx}{\sqrt[3]{x-9}} = \lim_{t\to9^-}\left[\frac{3}{2}(x-9)^{2/3}\right]_1^t = \lim_{t\to9^-}\left[\frac{3}{2}(t-9)^{2/3} - 6\right] = -6$

31. $\displaystyle\int_{-2}^3 \frac{dx}{x^4} = \int_{-2}^0 \frac{dx}{x^4} + \int_0^3 \frac{dx}{x^4}$ and $\displaystyle\int_{-2}^0 \frac{dx}{x^4} = \lim_{t\to0^-}\left[-\frac{1}{3}x^{-3}\right]_{-2}^t = \lim_{t\to0^-}\left[-\frac{1}{3t^3} - \frac{1}{24}\right] = \infty.$ Divergent

32. $\int_0^2 \frac{dx}{4x-5} = \int_0^{5/4} \frac{dx}{4x-5} + \int_{5/4}^2 \frac{dx}{4x-5}$. $\int_0^{5/4} \frac{dx}{4x-5} = \lim_{t\to 5/4^-} \left[\frac{1}{4}\ln|4x-5|\right]_0^t$

$= \lim_{t\to 5/4^-} \frac{1}{4}[\ln|4t-5| - \ln 5] = -\infty$. Divergent

33. $\int_4^5 \frac{dx}{(5-x)^{2/5}} = \lim_{t\to 5^-}\left[-\frac{5}{3}(5-x)^{3/5}\right]_4^t = \lim_{t\to 5^-}\left[-\frac{5}{3}(5-t)^{3/5} + \frac{5}{3}\right] = 0 + \frac{5}{3} = \frac{5}{3}$

34. $\int_{\pi/4}^{\pi/2} \sec^2 x\, dx = \lim_{t\to\pi/2^-} [\tan x]_{\pi/4}^t = \lim_{t\to\pi/2^-} (\tan t - 1) = \infty$. Divergent

35. $\int_{\pi/4}^{\pi/2} \tan^2 x\, dx = \lim_{t\to\pi/2^-} \int_{\pi/4}^t (\sec^2 x - 1)dx = \lim_{t\to\pi/2^-} [\tan x - x]_{\pi/4}^t$

$= \frac{\pi}{4} - 1 + \lim_{t\to\pi/2^-}(\tan t - t) = \infty$. Divergent

36. $\int_0^{\pi/4} \frac{\cos x\, dx}{\sqrt{\sin x}} = \lim_{t\to 0^+} \int_t^{\pi/4} \frac{\cos x\, dx}{\sqrt{\sin x}} = \lim_{t\to 0^+}\left[2\sqrt{\sin x}\right]_t^{\pi/4}$

$= \lim_{t\to 0^+}\left(2\sqrt{\frac{1}{\sqrt{2}}} - 2\sqrt{\sin t}\right) = 2\sqrt{\frac{1}{\sqrt{2}}} = \frac{2}{2^{1/4}} = 2^{3/4}$

37. $\int_0^\pi \sec x\, dx = \int_0^{\pi/2} \sec x\, dx + \int_{\pi/2}^\pi \sec x\, dx$. $\int_0^{\pi/2} \sec x\, dx = \lim_{t\to\pi/2^-} \int_0^t \sec x\, dx$

$= \lim_{t\to\pi/2^-} [\ln|\sec x + \tan x|]_0^t = \lim_{t\to\pi/2^-} \ln|\sec t + \tan t| = \infty$. Divergent

38. $\int_0^4 \frac{dx}{x^2+x-6} = \int_0^4 \frac{dx}{(x+3)(x-2)} = \int_0^2 \frac{dx}{(x-2)(x+3)} + \int_2^4 \frac{dx}{(x-2)(x+3)}$, and

$\int_0^2 \frac{dx}{(x-2)(x+3)} = \lim_{t\to 2^-} \int_0^t \left[\frac{1/5}{x-2} - \frac{1/5}{x+3}\right]dx = \lim_{t\to 2^-}\left[\frac{1}{5}\ln\left|\frac{x-2}{x+3}\right|\right]_0^t = \lim_{t\to 2^-}\frac{1}{5}\left[\ln\left|\frac{t-2}{t+3}\right| - \ln\frac{2}{3}\right]$

$= -\infty$. Divergent

39. $\int_{-2}^2 \frac{dx}{x^2-1} = \int_{-2}^{-1} \frac{dx}{x^2-1} + \int_{-1}^0 \frac{dx}{x^2-1} + \int_0^1 \frac{dx}{x^2-1} + \int_1^2 \frac{dx}{x^2-1}$, and

$\int \frac{dx}{x^2-1} = \int \frac{dx}{(x-1)(x+1)} = \frac{1}{2}\ln\left|\frac{x-1}{x+1}\right| + C$, so

$\int_0^1 \frac{dx}{x^2-1} = \lim_{t\to 1^-}\left[\frac{1}{2}\ln\left|\frac{x-1}{x+1}\right|\right]_0^t = \lim_{t\to 1^-}\frac{1}{2}\ln\left|\frac{t-1}{t+1}\right| = -\infty$. Divergent

40. $\int_0^2 \frac{x-3}{2x-3}dx = \int_0^{3/2} \frac{x-3}{2x-3}dx + \int_{3/2}^2 \frac{x-3}{2x-3}dx$ and

$\int \frac{x-3}{2x-3}dx = \frac{1}{2}\int \frac{2x-6}{2x-3}dx = \frac{1}{2}\int\left[1 - \frac{3}{2x-3}\right]dx = \frac{1}{2}x - \frac{3}{4}\ln|2x-3| + C$, so

$\int_0^{3/2} \frac{x-3}{2x-3}dx = \lim_{t\to 3/2^-} \frac{1}{4}[2x - 3\ln|2x-3|]_0^t = \infty$. Divergent

41. Integrate by parts with $u = \ln x$, $dv = x\, dx$:

$\int_0^1 x\ln x\, dx = \lim_{t\to 0^+}\int_t^1 x\ln x\, dx = \lim_{t\to 0^+}\left[\frac{1}{2}x^2\ln x - \frac{1}{4}x^2\right]_t^1 = -\frac{1}{4} - \lim_{t\to 0^+}\frac{1}{2}t^2\ln t$

$= -\frac{1}{4} - \frac{1}{2}\lim_{t\to 0^+}\frac{\ln t}{1/t^2} \overset{H}{=} -\frac{1}{4} - \frac{1}{2}\lim_{t\to 0^+}\frac{1/t}{-2/t^3} = -\frac{1}{4} + \frac{1}{4}\lim_{t\to 0^+} t^2 = -\frac{1}{4}$

42. Integrate by parts with $u = \ln x$, $dv = dx/\sqrt{x} \quad \Rightarrow \quad du = dx/x$, $v = 2\sqrt{x}$:

$$\int_0^1 \frac{\ln x}{\sqrt{x}}\,dx = \lim_{t\to0^+}\int_t^1 \frac{\ln x}{\sqrt{x}}\,dx = \lim_{t\to0^+}\left(\left[2\sqrt{x}\ln x\right]_t^1 - 2\int_t^1 \frac{dx}{\sqrt{x}}\right)$$
$$= \lim_{t\to0^+}\left(-2\sqrt{t}\ln t - 4\left[\sqrt{x}\right]_t^1\right) = \lim_{t\to0^+}\left(-2\sqrt{t}\ln t - 4 + 4\sqrt{t}\right) = -4,$$

since, by l'Hospital's Rule, $\lim\limits_{t\to0^+}\sqrt{t}\ln t = \lim\limits_{t\to0^+}\dfrac{\ln t}{t^{-1/2}} = \lim\limits_{t\to0^+}\dfrac{1/t}{-t^{-3/2}/2} = \lim\limits_{t\to0^+}\left(-2\sqrt{t}\right) = 0.$

43.

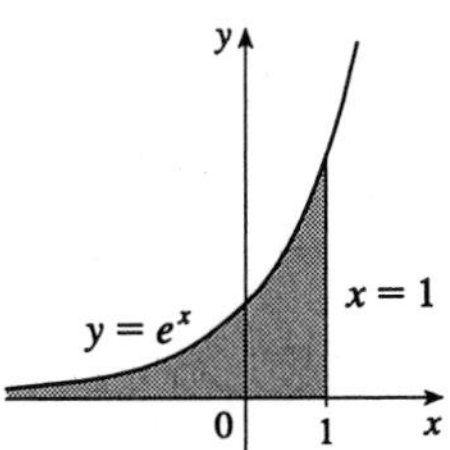

Area $= \int_{-\infty}^{1} e^x\,dx = \lim\limits_{t\to-\infty}\left[e^x\right]_t^1 = e - \lim\limits_{t\to-\infty} e^t = e$

44.

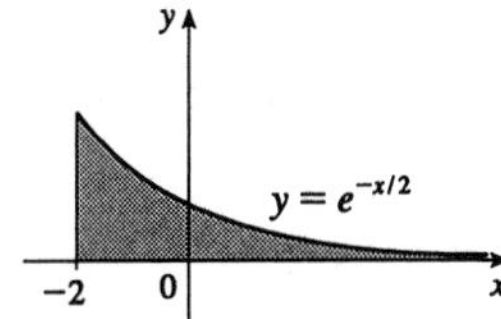

Area $= \int_{-2}^{\infty} e^{-x/2}\,dx = -2\lim\limits_{t\to\infty}\left[e^{-x/2}\right]_{-2}^{t}$
$= -2\lim\limits_{t\to\infty} e^{-t/2} + 2e = 2e$

45.

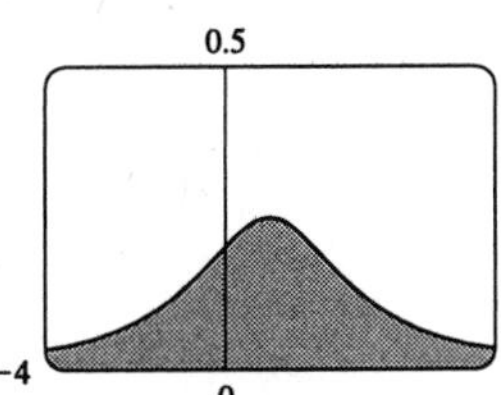

$$\text{Area} = \int_{-\infty}^{\infty}\frac{dx}{x^2-2x+5} = \int_{-\infty}^{0}\frac{dx}{(x-1)^2+4} + \int_0^{\infty}\frac{dx}{(x-1)^2+4}$$
$$= \lim_{t\to-\infty}\left[\frac{1}{2}\tan^{-1}\left(\frac{x-1}{2}\right)\right]_t^0 + \lim_{t\to\infty}\left[\frac{1}{2}\tan^{-1}\left(\frac{x-1}{2}\right)\right]_0^t$$
$$= \tfrac{1}{2}\tan^{-1}\left(-\tfrac{1}{2}\right) - \tfrac{1}{2}\left(-\tfrac{\pi}{2}\right) + \tfrac{1}{2}\left(\tfrac{\pi}{2}\right) - \tfrac{1}{2}\tan^{-1}\left(-\tfrac{1}{2}\right) = \tfrac{\pi}{2}.$$

46.

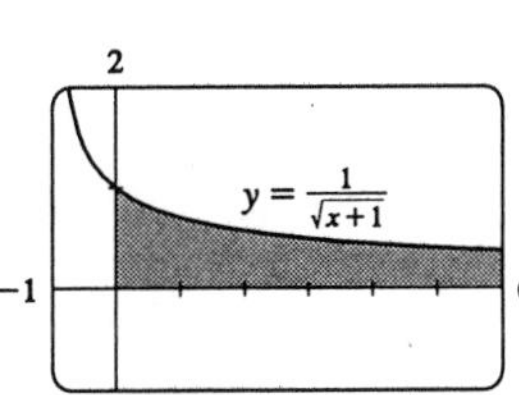

$$\text{Area} = \int_0^{\infty}\frac{dx}{\sqrt{x+1}} = \lim_{t\to\infty}\left[2\sqrt{x+1}\right]_0^t = \infty,$$

so the area is infinite.

47.

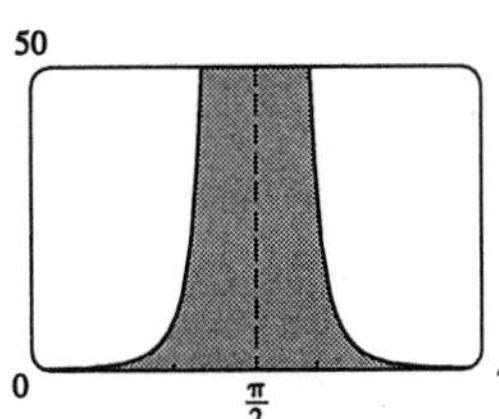

Area $= \int_0^{\pi}\tan^2 x\sec^2 x\,dx$
$= \int_0^{\pi/2}\tan^2 x\sec^2 x\,dx + \int_{\pi/2}^{\pi}\tan^2 x\sec^2 x\,dx.$

But $\int_0^{\pi/2}\tan^2 x\sec^2 x\,dx = \lim\limits_{t\to\pi/2^-}\left[\tfrac{1}{3}\tan^3 x\right]_0^t = \infty,$

so the area is infinite.

48.

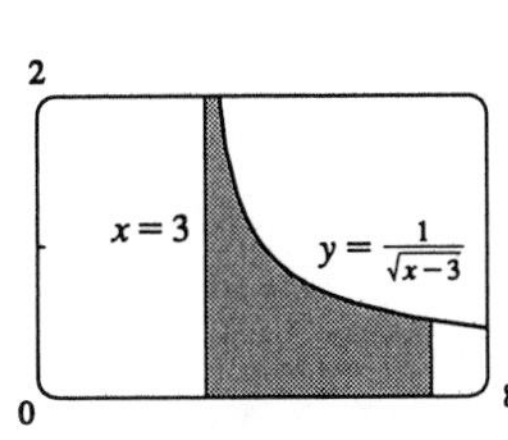

$$\text{Area} = \int_3^7\frac{dx}{\sqrt{x-3}} = \lim_{t\to3^+}\left[2\sqrt{x-3}\right]_t^7$$
$$= 4 - 2\lim_{t\to3^+} 2\sqrt{t-3} = 4 - 0 = 4$$

49. $\dfrac{\sin^2 x}{x^2} \le \dfrac{1}{x^2}$ on $[1, \infty)$. $\displaystyle\int_1^\infty \frac{dx}{x^2}$ is convergent by Example 4, so $\displaystyle\int_1^\infty \frac{\sin^2 x}{x^2}\,dx$ is convergent by the Comparison Theorem.

50. $\dfrac{\sqrt{1+\sqrt{x}}}{\sqrt{x}} > \dfrac{1}{\sqrt{x}}$ on $[1, \infty)$. $\displaystyle\int_1^\infty \frac{dx}{\sqrt{x}}$ is divergent by Example 4, so $\displaystyle\int_1^\infty \frac{\sqrt{1+\sqrt{x}}}{\sqrt{x}}\,dx$ is divergent by the Comparison Theorem.

51. For $x \ge 1$, $x + e^{2x} > e^{2x} > 0 \quad \Rightarrow \quad \dfrac{1}{x+e^{2x}} \le \dfrac{1}{e^{2x}} = e^{-2x}$ on $[1, \infty)$.
$\int_1^\infty e^{-2x}\,dx = \lim\limits_{t\to\infty}\left[-\frac{1}{2}e^{-2x}\right]_1^t = \lim\limits_{t\to\infty}\left[-\frac{1}{2}e^{-2t} + \frac{1}{2}e^{-2}\right] = \frac{1}{2}e^{-2}$. Therefore $\int_1^\infty e^{-2x}\,dx$ is convergent, and by the Comparison Theorem, $\displaystyle\int_1^\infty \frac{dx}{x+e^{2x}}$ is also convergent.

52. $\dfrac{1}{\sqrt{x^3+1}} \le \dfrac{1}{x^{3/2}}$ on $[1, \infty)$. $\displaystyle\int_1^\infty \frac{dx}{x^{3/2}}$ converges by Example 4, so $\displaystyle\int_1^\infty \frac{dx}{\sqrt{x^3+1}}$ converges by the Comparison Theorem.

53. $\dfrac{1}{x\sin x} \ge \dfrac{1}{x}$ on $\left(0, \frac{\pi}{2}\right]$ since $0 \le \sin x \le 1$. $\displaystyle\int_0^{\pi/2} \frac{dx}{x} = \lim_{t\to 0^+}\int_t^{\pi/2}\frac{dx}{x} = \lim_{t\to 0^+}\left[\ln x\right]_t^{\pi/2}$. But $\ln t \to -\infty$ as $t \to 0^+$, so $\displaystyle\int_0^{\pi/2}\frac{dx}{x}$ is divergent, and by the Comparison Theorem, $\displaystyle\int_0^{\pi/2}\frac{dx}{x\sin x}$ is also divergent.

54. $e^{-x} \le 1 \quad \Rightarrow \quad \dfrac{e^{-x}}{\sqrt{x}} \le \dfrac{1}{\sqrt{x}}$ for $0 \le x \le 1$, and $\displaystyle\int_0^1 \frac{1}{\sqrt{x}}\,dx = \lim_{t\to 0^+}\int_t^1 \frac{1}{\sqrt{x}}\,dx = \lim_{t\to 0^+}\left(2 - 2\sqrt{t}\right) = 2$ is convergent. Therefore $\displaystyle\int_0^1 \frac{e^{-x}}{\sqrt{x}}\,dx$ is convergent, by the Comparison Theorem.

55.
$$\int_0^\infty \frac{dx}{\sqrt{x}(1+x)} = \int_0^1 \frac{dx}{\sqrt{x}(1+x)} + \int_1^\infty \frac{dx}{\sqrt{x}(1+x)} = \lim_{t\to 0^+}\int_t^1 \frac{dx}{\sqrt{x}(1+x)} + \lim_{t\to\infty}\int_1^t \frac{dx}{\sqrt{x}(1+x)}$$
$$= \lim_{t\to 0^+}\int_{\sqrt{t}}^1 \frac{2\,du}{1+u^2} + \lim_{t\to\infty}\int_1^{\sqrt{t}} \frac{2\,du}{1+u^2} \quad \left[\text{put } u = \sqrt{x},\ x = u^2\right] = \lim_{t\to 0^+}\left[2\tan^{-1}u\right]_{\sqrt{t}}^1 + \lim_{t\to\infty}\left[2\tan^{-1}u\right]_1^{\sqrt{t}}$$
$$= \lim_{t\to 0^+}\left[2\left(\tfrac{\pi}{4}\right) - 2\tan^{-1}\sqrt{t}\right] + \lim_{t\to\infty}\left[2\tan^{-1}\sqrt{t} - 2\left(\tfrac{\pi}{4}\right)\right] = \tfrac{\pi}{2} - 0 + 2\left(\tfrac{\pi}{2}\right) - \tfrac{\pi}{2} = \pi$$

56. $\displaystyle\int_2^\infty \frac{dx}{x\sqrt{x^2-4}} = \int_2^3 \frac{dx}{x\sqrt{x^2-4}} + \int_3^\infty \frac{dx}{x\sqrt{x^2-4}} = \lim_{t\to 2^+}\int_t^3 \frac{dx}{x\sqrt{x^2-4}} + \lim_{t\to\infty}\int_3^t \frac{dx}{x\sqrt{x^2-4}}$, and
$\displaystyle\int \frac{dx}{x\sqrt{x^2-4}} = \int \frac{2\sec\theta\tan\theta\,d\theta}{2\sec\theta\,2\tan\theta}$ (put $x = 2\sec\theta$) $= \frac{1}{2}\theta + C = \frac{1}{2}\sec^{-1}\left(\frac{1}{2}x\right) + C$, so
$$\int_2^\infty \frac{dx}{x\sqrt{x^2-4}} = \lim_{t\to 2^+}\left[\tfrac{1}{2}\sec^{-1}\left(\tfrac{1}{2}x\right)\right]_t^3 + \lim_{t\to\infty}\left[\tfrac{1}{2}\sec^{-1}\left(\tfrac{1}{2}x\right)\right]_3^t = \tfrac{1}{2}\sec^{-1}\left(\tfrac{3}{2}\right) - 0 + \tfrac{1}{2}\left(\tfrac{\pi}{2}\right) - \tfrac{1}{2}\sec^{-1}\left(\tfrac{3}{2}\right) = \tfrac{\pi}{4}.$$

57. If $p = 1$, then $\displaystyle\int_0^1 \frac{dx}{x^p} = \lim_{t\to 0^+}\left[\ln x\right]_t^1 = \infty$. Divergent. If $p \ne 1$, then $\displaystyle\int_0^1 \frac{dx}{x^p} = \lim_{t\to 0^+}\int_t^1 \frac{dx}{x^p}$ (Note that the integral is not improper if $p < 0$) $\displaystyle= \lim_{t\to 0^+}\left[\frac{x^{-p+1}}{-p+1}\right]_t^1 = \lim_{t\to 0^+}\frac{1}{1-p}\left[1 - \frac{1}{t^{p-1}}\right]$. If $p > 1$, then $p - 1 > 0$, so $\dfrac{1}{t^{p-1}} \to \infty$ as $t \to 0^+$, and the integral diverges.

Finally, if $p < 1$, then $\displaystyle\int_0^1 \frac{dx}{x^p} = \frac{1}{1-p}\left[\lim_{t\to 0^+}\left(1 - t^{1-p}\right)\right] = \frac{1}{1-p}$.

Thus the integral converges if and only if $p < 1$, and in that case its value is $\dfrac{1}{1-p}$.

58. Let $u = \ln x$. Then $du = dx/x \quad\Rightarrow\quad \displaystyle\int_e^\infty \frac{dx}{x(\ln x)^p} = \int_1^\infty \frac{du}{u^p}$. By Example 4, this converges to $\dfrac{1}{p-1}$ if $p > 1$ and diverges otherwise.

59. First suppose $p = -1$. Then

$\displaystyle\int_0^1 x^p \ln x\,dx = \int_0^1 \frac{\ln x}{x}\,dx = \lim_{t\to 0^+}\int_t^1 \frac{\ln x}{x}\,dx = \lim_{t\to 0^+}\left[\tfrac{1}{2}(\ln x)^2\right]_t^1 = -\tfrac{1}{2}\lim_{t\to 0^+}(\ln t)^2 = -\infty$, so the integral diverges. Now suppose $p \neq -1$. Then integration by parts gives

$\displaystyle\int x^p \ln x\,dx = \frac{x^{p+1}}{p+1}\ln x - \int \frac{x^p}{p+1}\,dx = \frac{x^{p+1}}{p+1}\ln x - \frac{x^{p+1}}{(p+1)^2} + C$. If $p < -1$, then $p + 1 < 0$, so

$$\int_0^1 x^p \ln x\,dx = \lim_{t\to 0^+}\left[\frac{x^{p+1}}{p+1}\ln x - \frac{x^{p+1}}{(p+1)^2}\right]_t^1 = \frac{-1}{(p+1)^2} - \left(\frac{1}{p+1}\right)\lim_{t\to 0^+}\left[t^{p+1}\left(\ln t - \frac{1}{p+1}\right)\right] = \infty.$$

If $p > -1$, then $p + 1 > 0$ and

$$\begin{aligned}\int_0^1 x^p \ln x\,dx &= \frac{-1}{(p+1)^2} - \left(\frac{1}{p+1}\right)\lim_{t\to 0^+}\frac{\ln t - 1/(p+1)}{t^{-(p+1)}}\\ &\overset{\text{H}}{=} \frac{-1}{(p+1)^2} - \left(\frac{1}{p+1}\right)\lim_{t\to 0^+}\frac{1/t}{-(p+1)t^{-(p+2)}} = \frac{-1}{(p+1)^2} + \frac{1}{(p+1)^2}\lim_{t\to 0^+} t^{p+1} = \frac{-1}{(p+1)^2}.\end{aligned}$$

Thus the integral converges to $-\dfrac{1}{(p+1)^2}$ if $p > -1$ and diverges otherwise.

60. For n a nonnegative integer, integration by parts with $u = x^{n+1}$, $dv = e^{-x}\,dx$, gives

$\int x^{n+1}e^{-x}\,dx = -x^{n+1}e^{-x} + (n+1)\int x^n e^{-x}\,dx$, so

$$\begin{aligned}\textstyle\int_0^\infty x^{n+1}e^{-x}\,dx &= \lim_{t\to\infty}\textstyle\int_0^t x^{n+1}e^{-x}\,dx = \lim_{t\to\infty}\left[-x^{n+1}e^{-x}\right]_0^t + (n+1)\int_0^\infty x^n e^{-x}\,dx\\ &= \lim_{t\to\infty}\frac{-t^{n+1}}{e^t} + (n+1)\int_0^\infty x^n e^{-x}\,dx = (n+1)\int_0^\infty x^n e^{-x}\,dx.\end{aligned}$$

Now $\int_0^\infty x^0 e^{-x}\,dx = \lim_{t\to\infty}\int_0^t e^{-x}\,dx = \lim_{t\to\infty}\left[-e^{-x}\right]_0^t = 1$, so $\int_0^\infty x^1 e^{-x}\,dx = 1\cdot 1 = 1$, $\int_0^\infty x^2 e^{-x}\,dx = 2\cdot 1 = 2$, and $\int_0^\infty x^3 e^{-x}\,dx = 3\cdot 2 = 6$. In general, we guess that $\int_0^\infty x^n e^{-x}\,dx = n! = 1\cdot 2\cdot 3\cdot\cdots\cdot n$, when n is a positive integer. (Since $0! = 1$, our guess holds for $n = 0$ too.) Our guess works for $n \leq 3$. Suppose that $\int_0^\infty x^k e^{-x}\,dx = k!$ for some positive integer k. Then $\int_0^\infty x^{k+1}e^{-x}\,dx = (k+1)k! = (k+1)!$, so the formula holds for $k+1$. By induction, the formula holds for all integers $n \geq 0$.

61. **(a)** $\int_{-\infty}^\infty x\,dx = \int_{-\infty}^0 x\,dx + \int_0^\infty x\,dx$, and $\int_0^\infty x\,dx = \lim_{t\to\infty}\int_0^t x\,dx = \lim_{t\to\infty}\left[\frac{1}{2}t^2 - \frac{1}{2}(0^2)\right] = \infty$, so the integral is divergent.

(b) $\int_{-t}^t x\,dx = \left[\frac{1}{2}x^2\right]_{-t}^t = \frac{1}{2}t^2 - \frac{1}{2}t^2 = 0$, so $\lim_{t\to\infty}\int_{-t}^t x\,dx = 0$. Therefore $\int_{-\infty}^\infty x\,dx \neq \lim_{t\to\infty}\int_{-t}^t x\,dx$.

62. Assume without loss of generality that $a < b$. Then

$\int_{-\infty}^{a} f(x)dx + \int_{a}^{\infty} f(x)dx = \lim_{t\to-\infty} \int_{t}^{a} f(x)dx + \lim_{u\to\infty} \int_{a}^{u} f(x)dx$

$= \lim_{t\to-\infty} \int_{t}^{a} f(x)dx + \lim_{u\to\infty} \left[\int_{a}^{b} f(x)dx + \int_{b}^{u} f(x)dx\right] = \lim_{t\to-\infty} \int_{t}^{a} f(x)dx + \int_{a}^{b} f(x)dx + \lim_{u\to\infty} \int_{b}^{u} f(x)dx$

$= \lim_{t\to-\infty} \left[\int_{t}^{a} f(x)dx + \int_{a}^{b} f(x)dx\right] + \int_{b}^{\infty} f(x)dx = \lim_{t\to-\infty} \int_{t}^{b} f(x)dx + \int_{b}^{\infty} f(x)dx$

$= \int_{-\infty}^{b} f(x)dx + \int_{b}^{\infty} f(x)dx.$

63. Volume $= \displaystyle\int_1^\infty \pi\left(\frac{1}{x}\right)^2 dx = \pi \lim_{t\to\infty} \int_1^t \frac{dx}{x^2} = \pi \lim_{t\to\infty}\left[-\frac{1}{x}\right]_1^t = \pi \lim_{t\to\infty}\left(1-\frac{1}{t}\right) = \pi < \infty.$

64. Work $= \displaystyle\int_R^\infty \frac{GMm}{r^2}\,dr = \lim_{t\to\infty}\int_R^t \frac{GMm}{r^2}\,dr = \lim_{t\to\infty} GMm\left[\frac{-1}{r}\right]_R^t = GMm \lim_{t\to\infty}\left(\frac{-1}{t}+\frac{1}{R}\right) = \frac{GMm}{R},$

where M = mass of earth $= 5.98\times10^{24}$ kg, m = mass of satellite $= 10^3$ kg,

R = radius of earth $= 6.37\times10^6$ m, and G = gravitational constant $= 6.67\times10^{-11}\ \mathrm{N\cdot m^2/kg}$.

Therefore, Work $= \dfrac{6.67\times10^{-11}\cdot 5.98\times10^{24}\cdot 10^3}{6.37\times10^6} \approx 6.26\times10^{10}$ J.

65. Work $= \displaystyle\int_R^\infty F\,dr = \lim_{t\to\infty}\int_R^t \frac{GmM}{r^2}\,dr = \lim_{t\to\infty} GmM\left(\frac{1}{R}-\frac{1}{t}\right) = \frac{GmM}{R}$. The initial kinetic energy provides the work, so $\frac{1}{2}mv_0^2 = \dfrac{GmM}{R} \quad\Rightarrow\quad v_0 = \sqrt{\dfrac{2GM}{R}}$.

66. **(a)** $\int_{-\infty}^{\infty} f(x)dx = \int_{-\infty}^{0} f(x)dx + \int_{0}^{\infty} f(x)dx = \int_{-\infty}^{0} 0\,dx + \int_{0}^{\infty} ce^{-cx}\,dx$

$= 0 + \lim_{t\to\infty}\left[-e^{-cx}\right]_0^t = \lim_{t\to\infty}(1-e^{-ct}) = 1$

(b) $\mu = \int_{-\infty}^{\infty} xf(x)dx = \int_{-\infty}^{0} xf(x)dx + \int_{0}^{\infty} xf(x)dx = 0 + \lim_{t\to\infty}\int_0^t cxe^{-cx}\,dx.$

Let $u = x,\ dv = e^{-cx}\,dx \quad\Rightarrow\quad du = dx,\ v = -e^{-cx}/c$. Then

$\mu = c\cdot\lim_{t\to\infty}\left[-\dfrac{x}{c}e^{-cx} - \dfrac{1}{c^2}e^{-cx}\right]_0^t = c\left(-\dfrac{t}{c}e^{-ct} - \dfrac{1}{c^2}e^{-ct} + 0 + \dfrac{1}{c^2}\right) = \dfrac{c}{c^2} = \dfrac{1}{c}$, since

$\lim_{t\to\infty}\dfrac{t}{e^{ct}} \overset{H}{=} \lim_{t\to\infty}\dfrac{1}{ce^{ct}} = 0.$

(c) $\sigma^2 = \int_{-\infty}^{\infty}(x-\mu)^2 f(x)dx = \int_{-\infty}^{0}(x-\mu)^2 f(x)dx + \int_{0}^{\infty}(x-\mu)^2 f(x)dx$

$= 0 + \lim_{t\to\infty} c\int_0^t (x-\mu)^2 e^{-cx}\,dx = c\cdot\lim_{t\to\infty}\int_0^t (x^2e^{-cx} - 2x\mu e^{-cx} + \mu^2 e^{-cx})dx$

Now use parts for the first integral, with $u = x^2,\ dv = e^{-cx}\,dx \quad\Rightarrow\quad du = 2x\,dx,\ v = -e^{-cx}/c$; and use part (b) for the second integral to get

$\sigma^2 = c\cdot\lim_{t\to\infty}\left(\left[-\dfrac{x^2}{c}e^{-cx}\right]_0^t + \displaystyle\int_0^t \frac{2x}{c}e^{-cx}\,dx - \frac{2\mu}{c^2} - \left[\frac{\mu^2}{c}e^{-cx}\right]_0^t\right)$

$= c\left[\dfrac{2}{c^3} - \dfrac{2}{c^3} + \dfrac{1}{c^3}\right]$ $\quad$ [using part (b) again, and the fact that $\mu = 1/c$, for the remaining integral] $\quad = \dfrac{1}{c^2} \quad\Rightarrow\quad \sigma = \dfrac{1}{c}$.

67. **(a)** $F(s) = \displaystyle\int_0^\infty f(t)e^{-st}\,dt = \int_0^\infty e^{-st}\,dt = \lim_{n\to\infty}\left[-\frac{e^{-st}}{s}\right]_0^n = \lim_{n\to\infty}\left(\frac{e^{-sn}}{-s}+\frac{1}{s}\right)$. This converges to $\dfrac{1}{s}$ only if $s > 0$. Therefore $F(s) = \dfrac{1}{s}$ with domain $\{s \mid s > 0\}$.

(b) $F(s) = \int_0^\infty f(t)e^{-st}\,dt = \int_0^\infty e^t e^{-st}\,dt = \lim_{n\to\infty}\int_0^n e^{t(1-s)}\,dt$

$= \lim_{n\to\infty}\left[\dfrac{1}{1-s}e^{t(1-s)}\right]_0^n = \lim_{n\to\infty}\left(\dfrac{e^{(1-s)n}}{1-s} - \dfrac{1}{1-s}\right).$

This converges only if $1 - s < 0 \Rightarrow s > 1$, in which case $F(s) = \dfrac{1}{s-1}$ with domain $\{s \mid s > 1\}$.

(c) $F(s) = \int_0^\infty f(t)e^{-st}\,dt = \lim_{n\to\infty}\int_0^n te^{-st}\,dt$. Use integration by parts: let $u = t$, $dv = e^{-st}\,dt \Rightarrow$ $du = dt$, $v = -\dfrac{e^{-st}}{s}$. Then $F(s) = \lim_{n\to\infty}\left[-\dfrac{t}{s}e^{-st} - \dfrac{1}{s^2}e^{-st}\right]_0^n = \lim_{n\to\infty}\left(\dfrac{-n}{se^{sn}} - \dfrac{1}{s^2e^{sn}} + 0 + \dfrac{1}{s^2}\right) = \dfrac{1}{s^2}$ only if $s > 0$. Therefore $F(s) = \dfrac{1}{s^2}$ and the domain of F is $\{s \mid s > 0\}$.

68. $0 \le f(t) \le Me^{at} \Rightarrow 0 \le f(t)e^{-st} \le Me^{at}e^{-st}$ for $t \ge 0$. Now use the Comparison Theorem:

$$\int_0^\infty Me^{at}e^{-st}\,dt = \lim_{n\to\infty} M\int_0^n e^{t(a-s)}\,dt = M \cdot \lim_{n\to\infty}\left[\frac{1}{a-s}e^{t(a-s)}\right]_0^n = M \cdot \lim_{n\to\infty}\frac{1}{a-s}\left[e^{n(a-s)} - 1\right].$$

This is convergent only when $a - s < 0 \Rightarrow s > a$. Therefore, by the Comparison Theorem, $F(s) = \int_0^\infty f(t)e^{-st}\,dt$ is also convergent for $s > a$.

69. $G(s) = \int_0^\infty f'(t)e^{-st}\,dt$. Integrate by parts with $u = e^{-st}$, $dv = f'(t)dt \Rightarrow du = -se^{-st}$, $v = f(t)$:
$G(s) = \lim_{n\to\infty}\left[f(t)e^{-st}\right]_0^n + s\int_0^\infty f(t)e^{-st}\,dt = \lim_{n\to\infty} f(n)e^{-sn} - f(0) + sF(s)$.
But $0 \le f(t) \le Me^{at} \Rightarrow 0 \le f(t)e^{-st} \le Me^{at}e^{-st}$ and $\lim_{t\to\infty} Me^{t(a-s)} = 0$ for $s > a$. So by the Squeeze Theorem, $\lim_{t\to\infty} f(t)e^{-st} = 0$ for $s > a \Rightarrow G(s) = 0 - f(0) + sF(s) = sF(s) - f(0)$ for $s > a$.

70. $\int_0^\infty e^{-x^2}\,dx = \int_0^4 e^{-x^2}\,dx + \int_4^\infty e^{-x^2}\,dx$. Now

$$\int_0^4 e^{-x^2}\,dx \approx S_8 = \tfrac{4}{8\cdot 3}[f(0) + 4f(0.5) + 2f(1) + 4f(1.5) + 2f(2) + 4f(2.5) + 2f(3) + 4f(3.5) + f(4)] \approx 0.886196.$$

Also, for $x \ge 4$, $x^2 \ge 4x \Rightarrow -x^2 \le -4x \Rightarrow e^{-x^2} \le e^{-4x} \Rightarrow$
$\int_4^\infty e^{-x^2}\,dx \le \int_4^\infty e^{-4x}\,dx = \lim_{t\to\infty}\left[-\tfrac{1}{4}e^{-4x}\right]_4^t = \tfrac{1}{4}e^{-16} < 0.0000001 \Rightarrow \int_4^\infty e^{-x^2}\,dx < 0.0000001$.
So $\int_0^\infty e^{-x^2}\,dx \approx 0.89$.

71. Use integration by parts: let $u = x$, $dv = xe^{-x^2}\,dx \Rightarrow du = dx$, $v = -\tfrac{1}{2}e^{-x^2}$. So

$$\int_0^\infty x^2e^{-x^2}\,dx = \lim_{t\to\infty}\left[-\frac{x}{2}e^{-x^2}\right]_0^t + \tfrac{1}{2}\int_0^\infty e^{-x^2}\,dx = \lim_{t\to\infty} -\frac{t}{2e^{t^2}} + \tfrac{1}{2}\int_0^\infty e^{-x^2}\,dx = \tfrac{1}{2}\int_0^\infty e^{-x^2}\,dx.$$

(The limit is 0 by l'Hospital's Rule.)

72. $\int_0^\infty e^{-x^2}\,dx$ is the area under the curve $y = e^{-x^2}$ for $0 \le x < \infty$, and $\int_0^1 \sqrt{-\ln y}\,dy$ is the area to the left of the curve $x = \sqrt{-\ln y}$ for $0 < y \le 1$. But $x = \sqrt{-\ln y} \Rightarrow x^2 = -\ln y \Rightarrow e^{-x^2} = y$, and $0 < y \le 1 \Rightarrow$ $0 \le x < \infty$. Therefore each integral represents the same area, so the integrals are equal.

73. For the first part of the integral, let $x = 2\tan\theta \quad \Rightarrow \quad dx = 2\sec^2\theta\, d\theta$.

$\displaystyle\int \frac{1}{\sqrt{x^2+4}}\,dx = \int \sec\theta = \ln|\sec\theta + \tan\theta|$. But $\tan\theta = \frac{1}{2}x$, and

$\sec\theta = \sqrt{1+\tan^2\theta} = \sqrt{1+\frac{1}{4}x^2} = \frac{1}{2}\sqrt{x^2+4}$. So

$$\int_0^\infty \left(\frac{1}{\sqrt{x^2+4}} - \frac{C}{x+2}\right)dx = \lim_{t\to\infty}\left[\ln\left|\frac{\sqrt{x^2+4}}{2} + \frac{x}{2}\right| - C\ln|x+2|\right]_0^t$$

$$= \lim_{t\to\infty} \ln\left(\frac{\sqrt{t^2+4}+t}{2(t+2)^C}\right) - (\ln 1 - C\ln 2) = \ln\left(\lim_{t\to\infty}\frac{t+\sqrt{t^2+4}}{(t+2)^C}\right) + \ln 2^{C-1}.$$

By l'Hospital's Rule, $\displaystyle\lim_{t\to\infty}\frac{t+\sqrt{t^2+4}}{(t+2)^C} = \lim_{t\to\infty}\frac{1+t/\sqrt{t^2+4}}{C(t+2)^{C-1}} = \frac{2}{C\lim\limits_{t\to\infty}(t+2)^{C-1}}$.

If $C < 1$, we get ∞ and the interval diverges. If $C = 1$, we get 2, so the original integral converges to $\ln 2 + \ln 2^0 = \ln 2$. If $C > 1$, we get 0, so the original integral diverges to $-\infty$.

74. $\displaystyle\int_0^\infty \left(\frac{x}{x^2+1} - \frac{C}{3x+1}\right)dx = \lim_{t\to\infty}\left[\tfrac{1}{2}\ln(x^2+1) - \tfrac{1}{3}C\ln(3x+1)\right]_0^t$

$$= \lim_{t\to\infty}\left[\ln(t^2+1)^{1/2} - \ln(3t+1)^{C/3}\right] = \lim_{t\to\infty}\left(\ln\frac{(t^2+1)^{1/2}}{(3t+1)^{C/3}}\right) = \ln\left(\lim_{t\to\infty}\frac{\sqrt{t^2+1}}{(3t+1)^{C/3}}\right)$$

Clearly the integral diverges for $C \le 0$. For $C > 0$, we use l'Hospital's Rule and get

$\displaystyle\ln\left[\lim_{t\to\infty}\frac{t/\sqrt{t^2+1}}{C(3t+1)^{(C/3)-1}}\right] = \ln\left[\frac{1}{C}\lim_{t\to\infty}\frac{1}{(3t+1)^{(C/3)-1}}\right]$. For $\dfrac{C}{3} < 1$, the integral diverges. For $C = 3$,

$\ln\left(\frac{1}{3}\lim\limits_{t\to\infty}\frac{1}{1}\right) = \ln\frac{1}{3}$. For $C > 3$, the limit is 0 so the integral diverges to $-\infty$.

75. We integrate by parts with $u = \dfrac{1}{\ln(1+x+t)}$, $dv = \sin t\,dt$, so $du = \dfrac{-1}{(1+x+t)[\ln(1+x+t)]^2}$ and

$v = -\cos t$. The integral becomes

$$I = \int_0^\infty \frac{\sin t\,dt}{\ln(1+x+t)} = \lim_{b\to\infty}\left(\frac{-\cos t}{\ln(1+x+t)}\bigg|_0^b - \int_0^b \frac{\cos t\,dt}{(1+x+t)[\ln(1+x+t)]^2}\right)$$

$$= \lim_{b\to\infty}\frac{-\cos b}{\ln(1+x+b)} + \frac{1}{\ln(1+x)} + \int_0^\infty \frac{-\cos t\,dt}{(1+x+t)[\ln(1+x+t)]^2}$$

$$= \frac{1}{\ln(1+x)} + J, \text{ where } J = \int_0^\infty \frac{-\cos t\,dt}{(1+x+t)[\ln(1+x+t)]^2}.$$

Now $-1 \le -\cos t \le 1$ for all t; in fact, the inequality is strict except at isolated points. So

$$-\int_0^\infty \frac{dt}{(1+x+t)[\ln(1+x+t)]^2} < J < \int_0^\infty \frac{dt}{(1+x+t)[\ln(1+x+t)]^2} \quad\Leftrightarrow$$

$$-\frac{1}{\ln(1+x)} < J < \frac{1}{\ln(1+x)} \quad\Leftrightarrow\quad 0 < I < \frac{2}{\ln(1+x)}.$$

REVIEW EXERCISES FOR CHAPTER 7

1. False. Since the numerator has a higher degree than the denominator,

$$\frac{x(x^2+4)}{x^2-4} = x + \frac{8x}{x^2-4} = x + \frac{A}{x+2} + \frac{B}{x-2}.$$

2. True. In fact, $A = -1$, $B = C = 1$.

3. False. It can be put in the form $\frac{A}{x} + \frac{B}{x^2} + \frac{C}{x-4}$.

4. False. The form is $\frac{A}{x} + \frac{Bx+C}{x^2+4}$.

5. False. This is an improper integral, since the denominator vanishes at $x = 1$.

$$\int_0^4 \frac{x}{x^2-1}\,dx = \int_0^1 \frac{x}{x^2-1}\,dx + \int_1^4 \frac{x}{x^2-1}\,dx \text{ and } \int_0^1 \frac{x}{x^2-1} = \lim_{t\to 1^-}\int_0^t \frac{x}{x^2-1}\,dx$$

$$= \lim_{t\to 1^-}\left[\tfrac{1}{2}\ln|x^2-1|\right]_0^t = \lim_{t\to 1^-}\tfrac{1}{2}\ln|t^2-1| = \infty. \text{ So the integral diverges.}$$

6. True by Theorem 7.9.2 with $p = \sqrt{2} > 1$.

7. False. See Exercise 61 in Section 7.9.

8. False. For example, with $n = 1$ the Trapezoidal Rule is much more accurate than the Midpoint Rule for the function in the diagram.

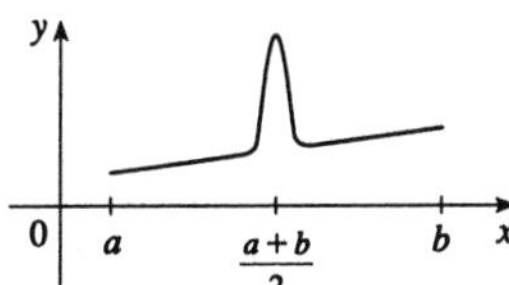

9. $\int \frac{x-1}{x+1}\,dx = \int\left[1 - \frac{2}{x+1}\right]dx = x - 2\ln|x+1| + C$

10. Let $u = \cos x$. Then $\int \frac{\sin^3 x}{\cos x}\,dx = \int \frac{(1-\cos^2 x)\sin x}{\cos x}\,dx = -\int \frac{1-u^2}{u}\,du$

$$= \int\left(u - \frac{1}{u}\right)du = \frac{u^2}{2} - \ln|u| + C = \tfrac{1}{2}\cos^2 x - \ln|\cos x| + C.$$

11. Let $u = \arctan x$. Then $du = dx/(1+x^2)$, so $\int \frac{(\arctan x)^5}{1+x^2}\,dx = \int u^5\,du = \frac{1}{6}u^6 + C = \frac{1}{6}(\arctan x)^6 + C.$

12. Integrate by parts twice, first with $u = x^2$, $dv = e^{-3x}\,dx \quad\Rightarrow\quad du = 2x\,dx$, $v = -\frac{1}{3}e^{-3x}$:

$\int x^2e^{-3x}\,dx = -\frac{1}{3}x^2e^{-3x} + \frac{2}{3}\int xe^{-3x}\,dx = -\frac{1}{3}x^2e^{-3x}\,dx + \frac{2}{3}\left(-\frac{1}{3}xe^{-3x} + \frac{1}{3}\int e^{-3x}\,dx\right)$

$= -\left(\frac{1}{3}x^2 + \frac{2}{9}x + \frac{2}{27}\right)e^{-3x} + C.$

13. Let $u = \sin x$. Then $\int \frac{\cos x\,dx}{e^{\sin x}} = \int e^{-u}\,du = -e^{-u} + C = -\frac{1}{e^{\sin x}} + C.$

14. $\int \frac{x^2+1}{x-1}\,dx = \int\left(x + 1 + \frac{2}{x-1}\right)dx = \frac{1}{2}x^2 + x + 2\ln|x-1| + C$

15. Use integration by parts with $u = \ln x$, $dv = x^4\,dx \quad\Rightarrow\quad du = dx/x$, $v = x^5/5$:

$\int x^4 \ln x\,dx = \frac{1}{5}x^5\ln x - \frac{1}{5}\int x^4\,dx = \frac{1}{5}x^5\ln x - \frac{1}{25}x^5 + C = \frac{1}{25}x^5(5\ln x - 1) + C.$

16. Let $u = \tan\theta$. Then $\displaystyle\int \frac{\sec^2\theta\, d\theta}{1-\tan\theta} = \int \frac{du}{1-u} = -\ln|1-u| + C = -\ln|1-\tan\theta| + C.$

17. Let $u = x^2$. Then $du = 2x\,dx$, so $\int x\sin(x^2)dx = \frac{1}{2}\int \sin u\, du = -\frac{1}{2}\cos u + C = -\frac{1}{2}\cos(x^2) + C.$

18. Let $u = 2x$. Then $\int x\sin^2 x\, dx = \frac{1}{2}\int x(1-\cos 2x)dx = \frac{1}{4}x^2 - \frac{1}{8}\int 2x\cos 2x\, 2\, dx$
$= \frac{1}{4}x^2 - \frac{1}{8}\int u\cos u\, du = \frac{1}{4}x^2 - \frac{1}{8}(u\sin u + \cos u) + C = \frac{1}{4}x^2 - \frac{1}{4}x\sin 2x - \frac{1}{8}\cos 2x + C.$

19. $\displaystyle\int \frac{dx}{2x^2-5x+2} = \int\left[\frac{-2/3}{2x-1} + \frac{1/3}{x-2}\right]dx = -\tfrac{1}{3}\ln|2x-1| + \tfrac{1}{3}\ln|x-2| + C = \tfrac{1}{3}\ln\left|\frac{x-2}{2x-1}\right| + C.$

20. $\displaystyle\int \frac{dt}{\sin^2 t + \cos 2t} = \int \frac{dt}{\sin^2 t + (\cos^2 t - \sin^2 t)} = \int \frac{dt}{\cos^2 t} = \int \sec^2 t\, dt = \tan t + C$

21. Let $u = \sec x$. Then $du = \sec x\tan x\, dx$, so $\int \tan^7 x\sec^3 x\, dx = \int \tan^6 x\sec^2 x\sec x\tan x\, dx$
$= \int (u^2-1)^3 u^2\, du = \int (u^8 - 3u^6 + 3u^4 - u^2)du = \frac{1}{9}u^9 - \frac{3}{7}u^7 + \frac{3}{5}u^5 - \frac{1}{3}u^3 + C$
$= \frac{1}{9}\sec^9 x - \frac{3}{7}\sec^7 x + \frac{3}{5}\sec^5 x - \frac{1}{3}\sec^3 x + C.$

22. Let $u = x - 1$. Then $\displaystyle\int \frac{dx}{\sqrt{8+2x-x^2}} = \int \frac{dx}{\sqrt{9-(x-1)^2}}$
$\displaystyle= \int \frac{du}{\sqrt{9-u^2}} = \sin^{-1}\left(\frac{u}{3}\right) + C = \sin^{-1}\left(\frac{x-1}{3}\right) + C.$

23. Let $u = \sqrt{1+2x}$. Then $x = \frac{1}{2}(u^2-1)$, $dx = u\, du$, so $\displaystyle\int \frac{dx}{\sqrt{1+2x}+3} = \int \frac{u\, du}{u+3}$
$\displaystyle= \int\left[1 - \frac{3}{u+3}\right]du = u - 3\ln|u+3| + C = \sqrt{1+2x} - 3\ln\left(\sqrt{1+2x}+3\right) + C.$

24. Let $u = (\tan^{-1}x)^2$, $dv = x\, dx \;\Rightarrow\; du = 2(\tan^{-1}x)/(1+x^2)dx$, $v = \frac{1}{2}x^2$. Then
$\displaystyle I = \int x(\tan^{-1}x)^2\, dx = \tfrac{1}{2}x^2(\tan^{-1}x)^2 - \int \frac{x^2\tan^{-1}x}{1+x^2}\, dx.$

Now let $w = \tan^{-1}x$, $dw = 1/(1+x^2)dx$, and $x^2 = \tan^2 w$. So
$I = \frac{1}{2}x^2(\tan^{-1}x)^2 - \int w\tan^2 w\, dw = \frac{1}{2}x^2(\tan^{-1}x)^2 - \int w\sec^2 w\, dw + \int w\, dw$
$= \frac{1}{2}x^2(\tan^{-1}x)^2 - \left[x\tan^{-1}x - \ln\sqrt{x^2+1}\right] + \frac{1}{2}(\tan^{-1}x)^2$ [use parts with $u = w$, $dv = \sec^2 w\, dw$]
$= \frac{1}{2}(x^2+1)(\tan^{-1}x)^2 - x\tan^{-1}x + \ln\sqrt{x^2+1} + C$
or $\frac{1}{2}(x^2+1)(\tan^{-1}x)^2 - x\tan^{-1}x + \frac{1}{2}\ln(x^2+1) + C.$

25. $\displaystyle u = \sqrt{x} \;\Rightarrow\; du = \frac{dx}{2\sqrt{x}} \;\Rightarrow\; \int \frac{e^{\sqrt{x}}\, dx}{\sqrt{x}} = 2\int e^u\, du = 2e^u + C = 2e^{\sqrt{x}} + C$

26. $\displaystyle\int \frac{dx}{x^3-2x^2+x} = \int \frac{dx}{(x-1)^2 x} = \int\left[\frac{-1}{x-1} + \frac{1}{(x-1)^2} + \frac{1}{x}\right]dx = -\frac{1}{x-1} + \ln\left|\frac{x}{x-1}\right| + C$

27. Let $x = \sec\theta$. Then
$\displaystyle\int \frac{dx}{(x^2-1)^{3/2}} = \int \frac{\sec\theta\tan\theta}{\tan^3\theta}\, d\theta = \int \frac{\sec\theta}{\tan^2\theta}\, d\theta = \int \frac{\cos\theta\, d\theta}{\sin^2\theta} = -\frac{1}{\sin\theta} + C = -\frac{x}{\sqrt{x^2-1}} + C.$

28. Let $x = \tan\theta$, $-\frac{\pi}{2} < \theta < \frac{\pi}{2}$. Then $\displaystyle\int \frac{dx}{x^2\sqrt{1+x^2}} = \int \frac{\sec^2\theta\, d\theta}{\tan^2\theta \sec\theta} = \int \frac{\sec\theta\, d\theta}{\tan^2\theta} = \int \frac{\cos\theta\, d\theta}{\sin^2\theta} = \int \frac{du}{u^2}$

(put $u = \sin\theta$) $\displaystyle = -\frac{1}{u} + C = -\frac{1}{\sin\theta} + C = -\frac{\sqrt{1+x^2}}{x} + C.$

29. $\displaystyle\int \frac{dx}{x^3+x} = \int \left(\frac{1}{x} - \frac{x}{x^2+1}\right) dx = \ln|x| - \tfrac{1}{2}\ln(x^2+1) + C$

30. Let $u = e^x$. Then $x = \ln u$, $dx = \dfrac{du}{u}$, so $\displaystyle\int \frac{dx}{1+e^x} = \int \frac{du/u}{1+u} = \int \left[\frac{1}{u} - \frac{1}{u+1}\right] du$

$= \ln u - \ln(u+1) + C = \ln e^x - \ln(1+e^x) + C = x - \ln(1+e^x) + C$

31. $\int \cot^2 x\, dx = \int (\csc^2 x - 1) dx = -\cot x - x + C$

32. Let $t = \tan\left(\dfrac{x}{2}\right)$. Then $\displaystyle\int \frac{dx}{5 - 3\cos x} = \int \frac{2\,dt/(1+t^2)}{5 - 3(1-t^2)/(1+t^2)}$

$\displaystyle = \int \frac{2\,dt}{5(1+t^2) - 3(1-t^2)} = \int \frac{2\,dt}{8t^2+2} = \frac{1}{2}\int \frac{2\,dt}{(2t)^2+1} = \frac{1}{2}\int \frac{du}{u^2+1} \quad (u = 2t)$

$= \frac{1}{2}\tan^{-1}u + C = \frac{1}{2}\tan^{-1}(2t) + C = \frac{1}{2}\tan^{-1}[2\tan(x/2)] + C.$

33. $\displaystyle\int \frac{2x^2+3x+11}{x^3+x^2+3x-5}\, dx = \int \left(\frac{2}{x-1} - \frac{1}{x^2+2x+5}\right) dx = 2\ln|x-1| - \int \frac{dx}{(x+1)^2+4}$

$\displaystyle = 2\ln|x-1| - \frac{1}{2}\tan^{-1}\left(\frac{x+1}{2}\right) + C$

34. Let $u = x+1$. Then $\displaystyle\int \frac{x^3}{(x+1)^{10}}\, dx = \int \frac{(u-1)^3}{u^{10}}\, dx = \int \frac{u^3 - 3u^2 + 3u - 1}{u^{10}}\, du$

$= \int (u^{-7} - 3u^{-8} + 3u^{-9} - u^{-10})\, du = -\frac{1}{6}u^{-6} + \frac{3}{7}u^{-7} - \frac{3}{8}u^{-8} + \frac{1}{9}u^{-9} + C$

$\displaystyle = \frac{-1}{6(x+1)^6} + \frac{3}{7(x+1)^7} - \frac{3}{8(x+1)^8} + \frac{1}{9(x+1)^9} + C.$

35. Let $u = \cot 4x$. Then $du = -4\csc^2 4x\, dx \quad\Rightarrow\quad \displaystyle\int \csc^4 4x\, dx = \int (\cot^2 4x + 1)\csc^2 4x\, dx$

$= \int (u^2+1)\left(-\frac{1}{4}\,du\right) = -\frac{1}{4}\left(\frac{1}{3}u^3 + u\right) + C = -\frac{1}{12}(\cot^3 4x + 3\cot 4x) + C.$

36. Integrate by parts twice, first with $u = (\arcsin x)^2$, $dv = dx$:

$\displaystyle I = \int (\arcsin x)^2\, dx = x(\arcsin x)^2 - \int 2x \arcsin x \left(\frac{dx}{\sqrt{1-x^2}}\right).$

Now let $U = \arcsin x$, $dV = \dfrac{x}{\sqrt{1-x^2}}\, dx \quad\Rightarrow\quad dU = \dfrac{1}{\sqrt{1-x^2}}\, dx$, $V = -\sqrt{1-x^2}$. So

$I = x(\arcsin x)^2 - 2\left[\arcsin x\left(-\sqrt{1-x^2}\right) + \int dx\right] = x(\arcsin x)^2 + 2\sqrt{1-x^2}\arcsin x - 2x + C.$

37. Let $u = \ln x$. Then $\displaystyle\int \frac{\ln(\ln x)}{x}\, dx = \int \ln u\, du$. Now use parts with $w = \ln u$, $dv = du \quad\Rightarrow\quad dw = du/u$,

$v = u \quad\Rightarrow\quad \int \ln u\, du = u\ln u - u + C = (\ln x)[\ln(\ln x) - 1] + C.$

38. $\displaystyle\int \frac{\sin x}{1+\sin x}\,dx = \int \frac{\sin x(1-\sin x)}{1-\sin^2 x}\,dx = \int \frac{\sin x - \sin^2 x}{\cos^2 x}\,dx = \int(\sec x \tan x - \tan^2 x)\,dx$

$= \sec x - \int(\sec^2 x - 1)dx = \sec x - \tan x + x + C$

39. Let $u = 2x+1$. Then $du = 2\,dx \quad \Rightarrow$

$$\int \frac{dx}{\sqrt{4x^2+4x+5}} = \int \frac{(1/2)du}{\sqrt{u^2+4}} = \frac{1}{2}\int \frac{2\sec^2\theta\,d\theta}{2\sec\theta} \quad \text{(put } u = 2\tan\theta,\ du = 2\sec^2\theta\,d\theta\text{)}$$

$$= \tfrac{1}{2}\int \sec\theta\,d\theta = \tfrac{1}{2}\ln|\sec\theta + \tan\theta| + C_1 = \frac{1}{2}\ln\left|\frac{\sqrt{u^2+4}}{2} + \frac{u}{2}\right| + C_1$$

$$= \tfrac{1}{2}\ln\left(u + \sqrt{u^2+4}\right) + C = \tfrac{1}{2}\ln\left(2x+1+\sqrt{4x^2+4x+5}\right) + C.$$

40. $\int e^{-x}\sinh x\,dx = \int e^{-x}\cdot\frac{1}{2}(e^x - e^{-x})dx = \int\left[\frac{1}{2} - \frac{1}{2}e^{-2x}\right]dx = \frac{x}{2} + \frac{1}{4}e^{-2x} + C$

41. $\int(\cos x + \sin x)^2\cos 2x\,dx = \int(\cos^2 x + 2\sin x\cos x + \sin^2 x)\cos 2x\,dx$

$= \int(1+\sin 2x)\cos 2x\,dx = \int\cos 2x\,dx + \frac{1}{2}\int\sin 4x\,dx = \frac{1}{2}\sin 2x - \frac{1}{8}\cos 4x + C.$

Or: $\int(\cos x + \sin x)^2\cos 2x\,dx = \int(\cos x + \sin x)^2(\cos^2 x - \sin^2 x)dx$

$= \int(\cos x + \sin x)^3(\cos x - \sin x)dx = \frac{1}{4}(\cos x + \sin x)^4 + C_2.$

42. Let $u = e^{2x} \quad \Rightarrow \quad du = 2e^{2x}\,dx.$

$$\int \frac{e^{2x}}{e^{4x}-7}\,dx = \frac{1}{2}\int\frac{du}{u^2-7} = \frac{1}{2}\int\left[\frac{1}{2\sqrt{7}(u-\sqrt{7})} - \frac{1}{2\sqrt{7}(u+\sqrt{7})}\right]du$$

$$= \frac{1}{4\sqrt{7}}\ln\left|\frac{u-\sqrt{7}}{u+\sqrt{7}}\right| + C = \frac{1}{4\sqrt{7}}\ln\left|\frac{e^{2x}-\sqrt{7}}{e^{2x}+\sqrt{7}}\right| + C.$$

43. $\int_0^{\pi/2}\cos^3 x\sin 2x\,dx = \int_0^{\pi/2} 2\cos^4 x\sin x\,dx = \left[-\frac{2}{5}\cos^5 x\right]_0^{\pi/2} = \frac{2}{5}$

44. $\displaystyle\int_{-1}^{1}\frac{dx}{2x+1} = \int_{-1}^{-1/2}\frac{dx}{2x+1} + \int_{-1/2}^{1}\frac{dx}{2x+1}.$

$$\int_{-1/2}^{1}\frac{dx}{2x+1} = \lim_{t\to -1/2^+}\int_t^1\frac{dx}{2x+1} = \lim_{t\to -1/2^+}\left[\tfrac{1}{2}\ln|2x+10|\right]_t^1 = \infty, \text{ so } \int_{-1}^{1}\frac{dx}{2x+1} \text{ diverges.}$$

45. $\displaystyle\int_0^3\frac{dx}{x^2-x-2} = \int_0^3\frac{dx}{(x+1)(x-2)} = \int_0^2\frac{dx}{(x+1)(x-2)} + \int_2^3\frac{dx}{(x+1)(x-2)}$, and

$$\int_2^3\frac{dx}{x^2-x-2} = \lim_{t\to 2^+}\int_t^3\left[\frac{-1/3}{x+1} + \frac{1/3}{x-2}\right]dx = \lim_{t\to 2^+}\left[\frac{1}{3}\ln\left|\frac{x-2}{x+1}\right|\right]_t^3 = \lim_{t\to 2^+}\left[\tfrac{1}{3}\ln\tfrac{1}{4} - \frac{1}{3}\ln\left|\frac{t-2}{t+1}\right|\right] = \infty.$$

Divergent

46. Let $x = 2\theta$. Then $\int_0^{\pi/4}\cos^5(2\theta)d\theta = \frac{1}{2}\int_0^{\pi/2}\cos^5 x\,dx = \frac{1}{2}\int_0^{\pi/2}(1-\sin^2 x)^2\cos x\,dx = \frac{1}{2}\int_0^1(1-u^2)^2\,du$

$(u = \sin x) \quad = \frac{1}{2}\int_0^1(u^4 - 2u^2 + 1)du = \frac{1}{2}\left[\frac{1}{5}u^5 - \frac{2}{3}u^3 + u\right]_0^1 = \frac{1}{2}\left[\frac{1}{5} - \frac{2}{3} + 1\right] = \frac{4}{15}.$

47. $\displaystyle\int_0^1\frac{t^2-1}{t^2+1}\,dt = \int_0^1\left[1 - \frac{2}{t^2+1}\right]dt = \left[t - 2\tan^{-1}t\right]_0^1 = \left(1 - 2\cdot\tfrac{\pi}{4}\right) - 0 = 1 - \tfrac{\pi}{2}$

48. Let $u = \sqrt{y-2}$. Then $y = u^2 + 2$, so $\displaystyle\int \frac{y\,dy}{\sqrt{y-2}} = \int \frac{(u^2+2)2u\,du}{u} = 2\int (u^2+2)\,du$

$$= 2\left[\tfrac{1}{3}u^3 + 2u\right] + C, \text{ so } \int_2^6 \frac{y\,dy}{\sqrt{y-2}} = \lim_{t\to 2^+}\int_t^6 \frac{y\,dy}{\sqrt{y-2}} = \lim_{t\to 2^+}\left[\tfrac{2}{3}(y-2)^{3/2} + 4\sqrt{y-2}\right]_t^6$$

$$= \lim_{t\to 2^+}\left[\tfrac{16}{3} + 8 - \tfrac{2}{3}(t-2)^{3/2} - 4\sqrt{t-2}\right] = \tfrac{40}{3}.$$

49. $$\int_0^\infty \frac{dx}{(x+2)^4} = \lim_{t\to\infty}\left[\frac{-1}{3(x+2)^3}\right]_0^t = \lim_{t\to\infty}\left[\frac{1}{3\cdot 2^3} - \frac{1}{3(t+2)^3}\right] = \frac{1}{24}$$

50. Let $u = \dfrac{1}{x}$. Then $du = -\dfrac{dx}{x^2}$, so $\displaystyle\int_1^4 \frac{e^{1/x}}{x^2}\,dx = \int_{1/4}^1 e^u\,du = \left[e^u\right]_{1/4}^1 = e - e^{1/4}$.

51. Let $u = \ln x$. Then

$$\int_1^e \frac{dx}{x\sqrt{\ln x}} = \lim_{t\to 1^+}\int_t^e \frac{dx}{x\sqrt{\ln x}} = \lim_{t\to 1^+}\int_{\ln t}^1 \frac{du}{\sqrt{u}} = \lim_{t\to 1^+}\left[2\sqrt{u}\right]_{\ln t}^1 = \lim_{t\to 1^+}\left(2 - 2\sqrt{\ln t}\right) = 2.$$

52. $$\int_0^\infty \frac{dx}{(x+1)^2(x+2)} = \lim_{t\to\infty}\int_0^t\left[\frac{1}{x+2} - \frac{1}{x+1} + \frac{1}{(x+1)^2}\right]dx = \lim_{t\to\infty}\left[\ln\left(\frac{x+2}{x+1}\right) - \frac{1}{x+1}\right]_0^t = 1 - \ln 2$$

53. Let $u = \sqrt{x} + 2$. Then $x = (u-2)^2$, $dx = 2(u-2)du$, so

$$\int_1^4 \frac{\sqrt{x}\,dx}{\sqrt{x}+2} = \int_3^4 \frac{2(u-2)^2\,du}{u} = \int_3^4\left[2u - 8 + \frac{8}{u}\right]du = \left[u^2 - 8u + 8\ln u\right]_3^4$$

$$= (16 - 32 + 8\ln 4) - (9 - 24 + 8\ln 3) = -1 + 8\ln 4 - 8\ln 3 = 8\ln\tfrac{4}{3} - 1.$$

54. $\int_{-3}^3 x\sqrt{1+x^4}\,dx = 0$ since the integrand is an odd function.

55. Let $u = 2x+1$. Then $\displaystyle\int_{-\infty}^\infty \frac{dx}{4x^2+4x+5} = \int_{-\infty}^\infty \frac{\frac{1}{2}\,du}{u^2+4} = \frac{1}{2}\int_{-\infty}^0 \frac{du}{u^2+4} + \frac{1}{2}\int_0^\infty \frac{du}{u^2+4}$

$$= \tfrac{1}{2}\lim_{t\to-\infty}\left[\tfrac{1}{2}\tan^{-1}\left(\tfrac{1}{2}u\right)\right]_t^0 + \tfrac{1}{2}\lim_{t\to\infty}\left[\tfrac{1}{2}\tan^{-1}\left(\tfrac{1}{2}u\right)\right]_0^t = \tfrac{1}{4}\left[0 - \left(-\tfrac{\pi}{2}\right)\right] + \tfrac{1}{4}\left[\tfrac{\pi}{2} - 0\right] = \tfrac{\pi}{4}.$$

56. Let $u = \ln x$. Then $du = \dfrac{dx}{x}$, so $\displaystyle\int_1^e \frac{dx}{x[1+(\ln x)^2]} = \int_0^1 \frac{du}{1+u^2} = \left[\tan^{-1}u\right]_0^1 = \tfrac{\pi}{4} - 0 = \tfrac{\pi}{4}$.

57. Let $x = \sec\theta$. Then $\displaystyle\int_1^2 \frac{\sqrt{x^2-1}}{x}\,dx = \int_0^{\pi/3}\frac{\tan\theta}{\sec\theta}\sec\theta\tan\theta\,d\theta = \int_0^{\pi/3}\tan^2\theta\,d\theta$

$$= \int_0^{\pi/3}(\sec^2\theta - 1)d\theta = \left[\tan\theta - \theta\right]_0^{\pi/3} = \sqrt{3} - \tfrac{\pi}{3}.$$

58. $\int_{-1}^1 \left(x^{-1/3} + x^{-4/3}\right)dx = \int_{-1}^0 \left(x^{-1/3} + x^{-4/3}\right)dx + \int_0^1 \left(x^{-1/3} + x^{-4/3}\right)dx.$

$$\int_0^1 \left(x^{-1/3} + x^{-4/3}\right)dx = \lim_{t\to 0^+}\int_t^1\left(x^{-1/3} + x^{-4/3}\right)dx = \lim_{t\to 0^+}\left[\tfrac{3}{2}x^{2/3} - 3x^{-1/3}\right]_t^1$$

$$= \lim_{t\to 0^+}\left[\tfrac{3}{2} - 3 - \tfrac{3}{2}t^{2/3} + 3t^{-1/3}\right] = \infty. \text{ Divergent}$$

59. $\int_0^\infty e^{ax}\cos bx\,dx = \lim\limits_{t\to\infty}\int_0^t e^{ax}\cos bx\,dx$. Integrate by parts twice:

$$\int e^{ax}\cos bx\,dx = \frac{1}{b}e^{ax}\sin bx - \frac{a}{b}\int e^{ax}\sin bx\,dx = \frac{1}{b}e^{ax}\sin bx + \frac{a}{b^2}e^{ax}\cos bx - \frac{a^2}{b^2}\int e^{ax}\cos bx\,dx, \text{ so}$$

$$\left(1+\frac{a^2}{b^2}\right)\int e^{ax}\cos bx\,dx = \frac{1}{b}e^{ax}\sin bx + \frac{a}{b^2}e^{ax}\cos bx + C_1. \text{ Thus}$$

$$\int e^{ax}\cos bx\,dx = \frac{e^{ax}}{a^2+b^2}(b\sin bx + a\cos bx) + C. \text{ Now}$$

$$\int_0^\infty e^{ax}\cos bx\,dx = \lim_{t\to\infty}\left[\frac{e^{ax}}{a^2+b^2}(a\cos bx + b\sin bx)\right]_0^t = \lim_{t\to\infty}\frac{e^{at}}{a^2+b^2}(a\cos bt + b\sin bt) - \frac{a}{a^2+b^2}.$$

If $a \geq 0$, the limit does not exist and the integral is divergent. If $a < 0$, the limit is 0 (since $|e^{at}\cos bt| \leq e^{at}$ and $|e^{at}\sin bt| \leq e^{at}$), so the integral converges to $-a/(a^2+b^2)$.

60. $\displaystyle\int_1^\infty \frac{\tan^{-1}x}{x^2}\,dx = \lim_{t\to\infty}\int_1^t \frac{\tan^{-1}x}{x^2}\,dx$. Integrate by parts:

$$\int \frac{\tan^{-1}x}{x^2}\,dx = \frac{-\tan^{-1}x}{x} + \int \frac{1}{x}\frac{dx}{1+x^2} = \frac{-\tan^{-1}x}{x} + \int\left[\frac{1}{x} - \frac{x}{x^2+1}\right]dx$$

$$= \frac{-\tan^{-1}x}{x} + \ln|x| - \tfrac{1}{2}\ln(x^2+1) + C = \frac{-\tan^{-1}x}{x} + \frac{1}{2}\ln\frac{x^2}{x^2+1} + C, \text{ so}$$

$$\int_1^\infty \frac{\tan^{-1}x}{x^2}\,dx = \lim_{t\to\infty}\left[-\frac{\tan^{-1}x}{x} + \frac{1}{2}\ln\frac{x^2}{x^2+1}\right]_1^t = \lim_{t\to\infty}\left[-\frac{\tan^{-1}t}{t} + \frac{1}{2}\ln\frac{t^2}{t^2+1} + \tfrac{\pi}{4} - \tfrac{1}{2}\ln\tfrac{1}{2}\right]$$

$$= 0 + \tfrac{1}{2}\ln 1 + \tfrac{\pi}{4} + \tfrac{1}{2}\ln 2 = \tfrac{\pi}{4} + \tfrac{1}{2}\ln 2.$$

61. We first make the substitution $t = x + 1$, so $\ln(x^2+2x+2) = \ln\left[(x+1)^2+1\right] = \ln(t^2+1)$. Then we use parts with $u = \ln(t^2+1)$, $dv = dt$:

$$\int \ln(t^2+1)\,dt = t\ln(t^2+1) - \int\frac{t(2t)dt}{t^2+1} = t\ln(t^2+1) - 2\int\frac{t^2\,dt}{t^2+1}$$

$$= t\ln(t^2+1) - 2\int\left(1 - \frac{1}{t^2+1}\right)dt = t\ln(t^2+1) - 2t + 2\arctan t + C$$

$$= (x+1)\ln(x^2+2x+2) - 2x + 2\arctan(x+1) + K.$$

[Alternately, we could have integrated by parts immediately with $u = \ln(x^2+2x+2)$]
Notice from the graph that $f = 0$ where F has a horizontal tangent. Also, F is always increasing, and $f \geq 0$.

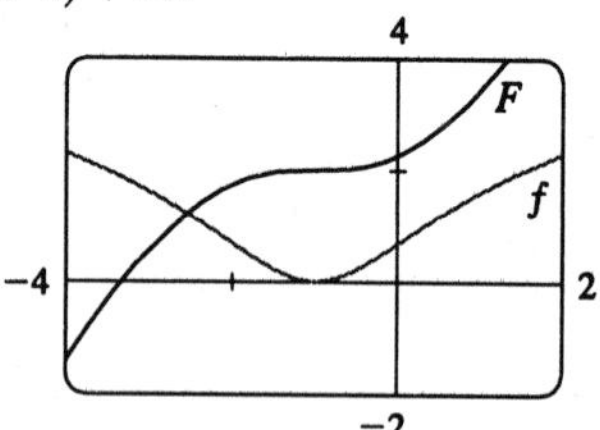

62. From the graph, it seems as though $\int_0^{2\pi}\cos^2x\sin^3x\,dx = 0$. To evaluate the integral, we use the procedure outlined in Section 7.2, and the integral becomes $\int_0^{2\pi}\cos^2x(1-\cos^2x)\sin x\,dx$
$= \left[\tfrac{1}{3}\cos^3x - \tfrac{1}{5}\cos^5x\right]_0^{2\pi} = \left(\tfrac{1}{3}-\tfrac{1}{5}\right) - \left(\tfrac{1}{3}-\tfrac{1}{5}\right) = 0.$

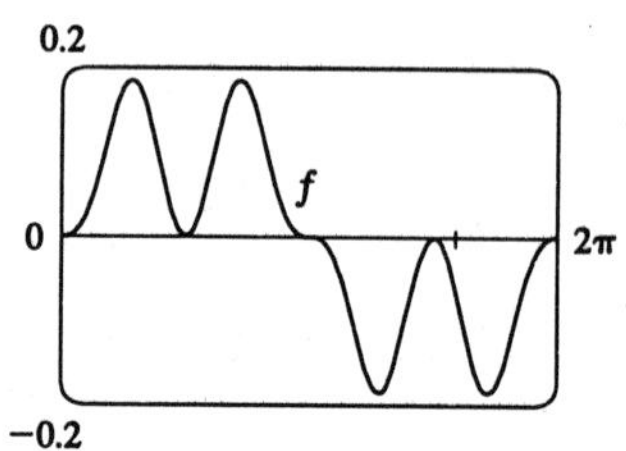

63. Let $u = e^x$. Then $du = e^x\,dx$, so $\int e^x\sqrt{1-e^{2x}}\,dx = \int\sqrt{1-u^2}\,du = \frac{1}{2}u\sqrt{1-u^2} + \frac{1}{2}\sin^{-1}u + C$

(Formula 30) $= \frac{1}{2}\left[e^x\sqrt{1-e^{2x}} + \sin^{-1}(e^x)\right] + C.$

64. $\int \tan^5 x\,dx = \frac{1}{4}\tan^4 x - \int \tan^3 x\,dx$ (Formula 75 with $n = 5$)

$= \frac{1}{4}\tan^4 x - \frac{1}{2}\tan^2 x - \ln|\cos x| + C$ (Formula 69)

65. Let $u = x + \frac{1}{2}$. Then $du = dx$, so $\displaystyle\int\sqrt{x^2+x+1}\,dx = \int\sqrt{\left(x+\tfrac{1}{2}\right)^2 + \tfrac{3}{4}}\,dx$

$$= \int\sqrt{u^2 + \left(\tfrac{\sqrt{3}}{2}\right)^2}\,du = \tfrac{1}{2}u\sqrt{u^2+\tfrac{3}{4}} + \tfrac{3}{8}\ln\left|u + \sqrt{u^2+\tfrac{3}{4}}\right| + C \quad \text{(Formula 21)}$$

$$= \frac{2x+1}{4}\sqrt{x^2+x+1} + \tfrac{3}{8}\ln\left|x + \tfrac{1}{2} + \sqrt{x^2+x+1}\right| + C.$$

66. Let $u = \sin x$. Then $du = \cos x\,dx$, so $\displaystyle\int\frac{\cot x\,dx}{\sqrt{1+2\sin x}} = \int\frac{du}{u\sqrt{1+2u}} = \ln\left|\frac{\sqrt{1+2u}-1}{\sqrt{1+2u}+1}\right| + C$

(Formula 57 with $a = 1$ and $b = 2$) $\displaystyle = \ln\left|\frac{\sqrt{1+2\sin x}-1}{\sqrt{1+2\sin x}+1}\right| + C.$

67. **(a)** $\displaystyle\frac{d}{du}\left[-\frac{1}{u}\sqrt{a^2-u^2} - \sin^{-1}\left(\frac{u}{a}\right) + C\right] = \frac{1}{u^2}\sqrt{a^2-u^2} + \frac{1}{\sqrt{a^2-u^2}} - \frac{1}{\sqrt{1-u^2/a^2}}\cdot\frac{1}{a}$

$$= \left(a^2-u^2\right)^{-1/2}\left[\frac{1}{u^2}\left(a^2-u^2\right) + 1 - 1\right] = \frac{\sqrt{a^2-u^2}}{u^2}.$$

(b) Let $u = a\sin\theta \quad\Rightarrow\quad du = a\cos\theta\,d\theta,\ a^2 - u^2 = a^2(1-\sin^2\theta) = a^2\cos^2\theta.$

$$\int\frac{\sqrt{a^2-u^2}}{u^2}\,du = \int\frac{a^2\cos^2\theta}{a^2\sin^2\theta}\,d\theta = \int\frac{1-\sin^2\theta}{\sin^2\theta}\,d\theta$$

$$= \int\left(\csc^2\theta - 1\right)d\theta = -\cot\theta - \theta + C = -\frac{\sqrt{a^2-u^2}}{u} - \sin^{-1}\left(\frac{u}{a}\right) + C.$$

68. Work backward, and use integration by parts with $U = u^{-(n-1)}$ and $dV = (a+bu)^{-1/2}\,du \quad\Rightarrow$

$\displaystyle dU = \frac{-(n-1)du}{u^n}$ and $V = \frac{2}{b}\sqrt{a+bu}$, to get

$$\int\frac{du}{u^{n-1}\sqrt{a+bu}} = \int U\,dV = UV - \int V\,dU = \frac{2\sqrt{a+bu}}{bu^{n-1}} + \frac{2(n-1)}{b}\int\frac{\sqrt{a+bu}}{u^n}\,du$$

$$= \frac{2\sqrt{a+bu}}{bu^{n-1}} + \frac{2(n-1)}{b}\int\frac{a+bu}{u^n\sqrt{a+bu}}\,du$$

$\displaystyle = \frac{2\sqrt{a+bu}}{bu^{n-1}} + 2(n-1)\int\frac{du}{u^{n-1}\sqrt{a+bu}} + \frac{2a(n-1)}{b}\int\frac{du}{u^n\sqrt{a+bu}}$. Rearranging

the equation gives $\displaystyle\frac{2a(n-1)}{b}\int\frac{du}{u^n\sqrt{a+bu}} = -\frac{2\sqrt{a+bu}}{bu^{n-1}} - (2n-3)\int\frac{du}{u^{n-1}\sqrt{a+bu}} \quad\Rightarrow$

$$\int\frac{du}{u^n\sqrt{a+bu}} = \frac{-\sqrt{a+bu}}{a(n-1)u^{n-1}} - \frac{b(2n-3)}{2a(n-1)}\int\frac{du}{u^{n-1}\sqrt{a+bu}}.$$

69. $f(x) = \sqrt{1+x^4}$, $\Delta x = \dfrac{b-a}{n} = \dfrac{1-0}{10} = \dfrac{1}{10}$

(a) $T_{10} = \frac{0.1}{2}[f(0) + 2\,f(0.1) + 2\,f(0.2) + \cdots + 2\,f(0.8) + 2\,f(0.9) + f(1)] \approx 1.090608$

(b) $M_{10} = 0.1\left[f\left(\frac{1}{20}\right) + f\left(\frac{3}{20}\right) + f\left(\frac{5}{20}\right) + \cdots + f\left(\frac{19}{20}\right)\right] \approx 1.088840$

(c) $S_{10} = \frac{0.1}{3}\left[f(0) + 4\,f(0.1) + 2\,f(0.2) + 4\,f(0.3) + 2\,f(0.4) + 4\,f(0.5)\right.$
$\left.+\,2\,f(0.6) + 4\,f(0.7) + 2\,f(0.8) + 4\,f(0.9) + f(1)\right] \approx 1.089429$

70. $f(x) = \sqrt{\sin x}$, $\Delta x = \dfrac{\frac{\pi}{2} - 0}{10} = \dfrac{\pi}{20}$

(a) $T_{10} = \frac{\pi}{20 \cdot 2}\left[f(0) + 2\,f\left(\frac{\pi}{20}\right) + 2\,f\left(\frac{\pi}{10}\right) + 2\,f\left(\frac{3\pi}{20}\right) + \cdots + 2\,f\left(\frac{2\pi}{5}\right) + 2\,f\left(\frac{9\pi}{20}\right) + f\left(\frac{\pi}{2}\right)\right] \approx 1.185197$

(b) $M_{10} = \frac{\pi}{20}\left[f\left(\frac{\pi}{40}\right) + f\left(\frac{3\pi}{40}\right) + f\left(\frac{5\pi}{40}\right) + \cdots + f\left(\frac{17\pi}{40}\right) + f\left(\frac{19\pi}{40}\right)\right] \approx 1.201932$

(c) $S_{10} = \frac{\pi}{20 \cdot 3}\left[f(0) + 4\,f\left(\frac{\pi}{20}\right) + 2\,f\left(\frac{\pi}{10}\right) + 4\,f\left(\frac{3\pi}{20}\right) + 2\,f\left(\frac{\pi}{5}\right) + 4\,f\left(\frac{\pi}{4}\right)\right.$
$\left.+\,2\,f\left(\frac{3\pi}{10}\right) + 4\,f\left(\frac{7\pi}{20}\right) + 2\,f\left(\frac{2\pi}{5}\right) + 4\,f\left(\frac{9\pi}{20}\right) + f\left(\frac{\pi}{2}\right)\right] \approx 1.193089$

71. $f(x) = \left(1+x^4\right)^{1/2}$, $f'(x) = \frac{1}{2}\left(1+x^4\right)^{-1/2}(4x^3) = 2x^3\left(1+x^4\right)^{-1/2}$, $f''(x) = (2x^6 + 6x^2)\left(1+x^4\right)^{-3/2}$. Thus $|f''(x)| \le 8 \cdot 1^{-3/2} = 8$ on $[0,1]$. By taking $K = 8$, we find that the error in Exercise 69(a) is bounded by $K\dfrac{(b-a)^3}{12n^2} = \dfrac{8}{1200} = \dfrac{1}{150} < 0.0067$, and in (b) by $K\dfrac{(b-a)^3}{24n^2} = \dfrac{1}{300} = 0.00\overline{3}$.

72. $\displaystyle\int_1^4 \frac{e^x}{x}\,dx \approx S_6 = \frac{(4-1)/6}{3}\left[f(1) + 4\,f(1.5) + 2\,f(2) + 4\,f(2.5) + 2\,f(3) + 4\,f(3.5) + f(4)\right] \approx 17.74.$

73. **(a)** $f(x) = \sin(\sin x)$. A CAS gives

$f^{(4)}(x) = \sin(\sin x)\cos^4 x + 6\cos(\sin x)\cos^2 x \sin x$
$+\,3\sin(\sin x) + \sin(\sin x)\cos^2 x + \cos(\sin x)\sin x.$

From the graph, we see that $f^{(4)}(x) < 3.8$ for $x \in [0, \pi]$.

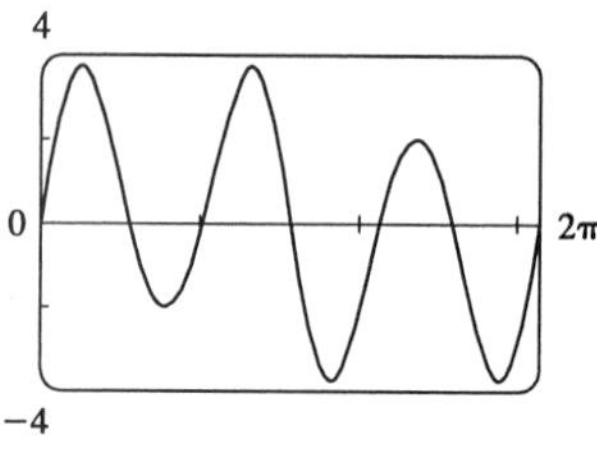

(b) We use Simpson's Rule with $f(x) = \sin(\sin x)$ and $\Delta x = \frac{\pi}{10}$:

$\int_0^\pi f(x)dx \approx \frac{\pi}{10 \cdot 3}\left[f(0) + 4f\left(\frac{\pi}{10}\right) + 2f\left(\frac{2\pi}{10}\right) + \cdots\right.$
$\left.+\,2f\left(\frac{8\pi}{10}\right) + 4f\left(\frac{9\pi}{10}\right) + f(\pi)\right] \approx 1.7867$

From part (a), we know that $f^{(4)}(x) < 3.8$ on $[0, \pi]$, so we use Theorem 7.8.7 with $K = 3.8$, and estimate the error as $|E_S| \le \dfrac{3.8(\pi - 0)^5}{180(10)^4} \approx 0.000646$.

(c) If we want the error to be less than 0.00001, we must have $|E_S| \le \dfrac{3.8\pi^5}{180n^4} \le 0.00001$, so $n^4 \ge \dfrac{3.8\pi^5}{180(0.00001)} \approx 646{,}041.5 \quad \Rightarrow \quad n \ge 28.35$. Since n must be even for Simpson's Rule, we must have $n \ge 30$ to ensure the desired accuracy.

74. **(a)** To evaluate $\int x^5 e^{-2x}\,dx$ by hand, we would integrate by parts repeatedly, always taking $dv = e^{-2x}$ and starting with $u = x^5$. Each time we would reduce by 1 the degree of the x-factor.

(b) To evaluate the integral using tables, we would use Formula 97 (which is proved using integration by parts) until the exponent of x was reduced to 1, and then we would use Formula 96.

(d)

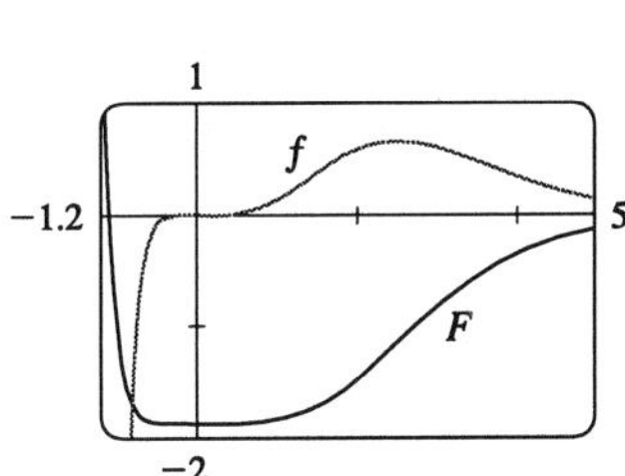

(c) $\int x^5 e^{-2x}\,dx = -\frac{15}{4}xe^{-2x} - \frac{15}{8}e^{-2x} - \frac{5}{4}x^4e^{-2x} - \frac{15}{4}x^2e^{-2x} - \frac{5}{2}x^3e^{-2x} - \frac{1}{2}x^5e^{-2x} + C$

$= -\frac{1}{8}e^{-2x}\left(4x^5 + 10x^4 + 20x^3 + 30x^2 + 30x + 15\right) + C$

75. $\dfrac{x^3}{x^5+2} \le \dfrac{x^3}{x^5} = \dfrac{1}{x^2}$ for x in $[1,\infty)$. $\displaystyle\int_1^\infty \frac{1}{x^2}\,dx$ is convergent by (7.9.2) with $p = 2 > 1$. Therefore $\displaystyle\int_1^\infty \frac{x^3}{x^5+2}\,dx$ is convergent by the Comparison Theorem.

76. The line $y = 3$ intersects the hyperbola $y^2 - x^2 = 1$ at two points on its upper branch, namely $\left(-2\sqrt{2}, 3\right)$ and $\left(2\sqrt{2}, 3\right)$. The desired area is

$A = \int_{-2\sqrt{2}}^{2\sqrt{2}}\left(3 - \sqrt{x^2+1}\right)dx = 2\int_0^{2\sqrt{2}}\left(3 - \sqrt{x^2+1}\right)dx$

$= 2\left[3x - \frac{1}{2}x\sqrt{x^2+1} - \frac{1}{2}\ln\left(x + \sqrt{x^2+1}\right)\right]_0^{2\sqrt{2}}$ (Formula 21)

$= \left[6x - x\sqrt{x^2+1} - \ln\left(x + \sqrt{x^2+1}\right)\right]_0^{2\sqrt{2}} = 12\sqrt{2} - 2\sqrt{2}\cdot 3 - \ln\left(2\sqrt{2}+3\right)$

$= 6\sqrt{2} - \ln\left(3 + 2\sqrt{2}\right)$.

Another Method: $A = 2\int_1^3 \sqrt{y^2-1}\,dy$ and use Formula 39.

77. For x in $\left[0, \frac{\pi}{2}\right]$, $0 \le \cos^2 x \le \cos x$. For x in $\left[\frac{\pi}{2}, \pi\right]$, $\cos x \le 0 \le \cos^2 x$. Thus

area $= \int_0^{\pi/2}(\cos x - \cos^2 x)dx + \int_{\pi/2}^{\pi}(\cos^2 x - \cos x)dx$

$= \left[\sin x - \frac{1}{2}x - \frac{1}{4}\sin 2x\right]_0^{\pi/2} + \left[\frac{1}{2}x + \frac{1}{4}\sin 2x - \sin x\right]_{\pi/2}^{\pi} = \left[\left(1 - \frac{\pi}{4}\right) - 0\right] + \left[\frac{\pi}{2} - \left(\frac{\pi}{4} - 1\right)\right] = 2.$

78. The curves $y = \dfrac{1}{2 \pm \sqrt{x}}$ are defined for $x \ge 0$. For $x > 0$, $\dfrac{1}{2-\sqrt{x}} > \dfrac{1}{2+\sqrt{x}}$. Thus the required area is

$$\int_0^1\left[\frac{1}{2-\sqrt{x}} - \frac{1}{2+\sqrt{x}}\right]dx = \int_0^1\left[\frac{1}{2-u} - \frac{1}{2+u}\right]2u\,du \;\left(\text{put } u = \sqrt{x}\right) \;= 2\int_0^1\left[-\frac{u}{u-2} - \frac{u}{u+2}\right]du$$

$$= 2\int_0^1\left[-1 - \frac{2}{u-2} - 1 + \frac{2}{u+2}\right]du = 2\left[2\ln\left|\frac{u+2}{u-2}\right| - 2u\right]_0^1 = 4\ln 3 - 4.$$

79. Using the formula for disks, the volume is

$V = \int_0^{\pi/2}\pi[f(x)]^2\,dx = \pi\int_0^{\pi/2}\left(\cos^2 x\right)^2 dx = \pi\int_0^{\pi/2}\left[\frac{1}{2}(1+\cos 2x)\right]^2 dx = \frac{\pi}{4}\int_0^{\pi/2}(1 + \cos^2 2x + 2\cos 2x)dx$

$= \frac{\pi}{4}\int_0^{\pi/2}\left[1 + \frac{1}{2}(1+\cos 4x) + 2\cos 2x\right]dx = \frac{\pi}{4}\left[\frac{3}{2}x + \frac{1}{2}\left(\frac{1}{4}\sin 4x\right) + 2\left(\frac{1}{2}\sin 2x\right)\right]_0^{\pi/2}$

$= \frac{\pi}{4}\left[\left(\frac{3\pi}{4} + \frac{1}{8}\cdot 0 + 0\right) - 0\right] = \frac{3\pi^2}{16}.$

80. Using the formula for cylindrical shells, the volume is

$V = \int_0^{\pi/2} 2\pi x\, f(x)dx = 2\pi \int_0^{\pi/2} x\cos^2 x\, dx = 2\pi \int_0^{\pi/2} x\left[\frac{1}{2}(1+\cos 2x)\right]dx$

$= 2\left(\frac{1}{2}\right)\pi \int_0^{\pi/2}(x + x\cos 2x)dx = \pi\left(\left[\frac{1}{2}x^2\right]_0^{\pi/2} + \left[x\left(\frac{1}{2}\sin 2x\right)\right]_0^{\pi/2} - \int_0^{\pi/2}\frac{1}{2}\sin 2x\, dx\right)$

(parts with $u = x$, $dv = \cos 2x$)

$= \pi\left[\frac{1}{2}\left(\frac{\pi}{2}\right)^2 + 0 - \frac{1}{2}\left[-\frac{1}{2}\cos 2x\right]_0^{\pi/2}\right] = \frac{\pi^3}{8} + \frac{\pi}{4}(-1-1) = \frac{1}{8}(\pi^3 - 4\pi).$

81. For $n \geq 0$, $\int_0^\infty x^n\, dx = \lim\limits_{t\to\infty}\left[x^{n+1}/(n+1)\right]_0^t = \infty$. For $n < 0$, $\int_0^\infty x^n\, dx = \int_0^1 x^n\, dx + \int_1^\infty x^n\, dx$. Both integrals are improper. By (7.9.2), the second integral diverges if $-1 \leq n < 0$. By Exercise 7.9.57, the first integral diverges if $n \leq -1$. Thus $\int_0^\infty x^n\, dx$ is divergent for all values of n.

82. n is a positive integer, so

$\int(\ln x)^n\, dx = x(\ln x)^n - \int x \cdot n(\ln x)^{n-1}(dx/x)$ (by parts) $= x(\ln x)^n - n\int(\ln x)^{n-1}\, dx$, so

$\int_0^1(\ln x)^n\, dx = \lim\limits_{t\to 0^+}\int_t^1(\ln x)^n\, dx = \lim\limits_{t\to 0^+}\left[x(\ln x)^n\right]_t^1 - n\lim\limits_{t\to 0^+}\int_t^1(\ln x)^{n-1}\, dx$

$$= -\lim_{t\to 0^+}\frac{(\ln t)^n}{1/t} - n\int_0^1(\ln x)^{n-1}\, dx = -n\int_0^1(\ln x)^{n-1}\, dx,$$

by repeated application of l'Hospital's Rule. We want to prove that $\int_0^1(\ln x)^n\, dx = (-1)^n n!$ for every positive integer n. For $n = 1$, we have $\int_0^1(\ln x)^1\, dx = (-1)\int_0^1(\ln x)^0\, dx = -\int_0^1 dx = -1$ $\left(\text{or } \int_0^1 \ln x\, dx = \lim\limits_{t\to 0^+}[x\ln x - x]_t^1 = -1\right)$. Assuming that the formula holds for n, we find that

$\int_0^1(\ln x)^{n+1}\, dx = -(n+1)\int_0^1(\ln x)^n\, dx = -(n+1)(-1)^n n! = (-1)^{n+1}(n+1)!$.

This is the formula for $n+1$. Thus the formula holds for all positive integers n by induction.

83. By the Fundamental Theorem of Calculus,

$\int_0^\infty f'(x)dx = \lim\limits_{t\to\infty}\int_0^t f'(x)dx = \lim\limits_{t\to\infty}\left[f(t) - f(0)\right] = \lim\limits_{t\to\infty} f(t) - f(0) = 0 - f(0) = -f(0).$

84. If the distance between P and the point charge is d, then the potential V at P is

$$V = W = \int_\infty^d F\, dr = \int_\infty^d \frac{q}{4\pi\epsilon_0 r^2}\, dr = \lim_{t\to\infty}\frac{q}{4\pi\epsilon_0}\left[-\frac{1}{r}\right]_t^d = \frac{q}{4\pi\epsilon_0}\lim_{t\to\infty}\left(-\frac{1}{d} + \frac{1}{t}\right) = -\frac{q}{4\pi\epsilon_0 d}.$$

85. Let $u = 1/x \quad\Rightarrow\quad x = 1/u \quad\Rightarrow\quad dx = -(1/u^2)du.$

$$\int_0^\infty \frac{\ln x}{1+x^2}\, dx = \int_\infty^0 \frac{\ln(1/u)}{1+1/u^2}\left(-\frac{du}{u^2}\right) = \int_\infty^0 \frac{-\ln u}{u^2+1}(-du) = \int_\infty^0 \frac{\ln u}{1+u^2}\, du$$

$$= -\int_0^\infty \frac{\ln u}{1+u^2}\, du. \text{ Therefore } \int_0^\infty \frac{\ln x}{1+x^2}\, dx = -\int_0^\infty \frac{\ln x}{1+x^2}\, dx = 0.$$

APPLICATIONS PLUS (page 499)

1. **(a)** Coefficient of inequality $= \dfrac{\text{area between Lorenz curve and straight line}}{\text{area under straight line}}$

$$= \frac{\int_0^1 [x - L(x)]dx}{\int_0^1 x\,dx} = \frac{\int_0^1 [x - L(x)]dx}{\left[x^2/2\right]_0^1} = \frac{\int_0^1 [x - L(x)]dx}{1/2} = 2\int_0^1 [x - L(x)]dx$$

(b) $L(x) = \frac{5}{12}x^2 + \frac{7}{12}x \quad \Rightarrow \quad L\left(\frac{1}{2}\right) = \frac{5}{48} + \frac{7}{24} = \frac{19}{48} = 0.3958\overline{3}$, so the bottom 50% of the households receive about 40% of the income.

Coefficient of inequality $= 2\int_0^1 \left[x - \frac{5}{12}x^2 - \frac{7}{12}x\right]dx = 2\int_0^1 \frac{5}{12}(x - x^2)dx = \frac{5}{6}\left(\frac{1}{2}x^2 - \frac{1}{3}x^3\right)\Big|_0^1 = \frac{5}{36}$

(c) Coefficient of inequality $= 2\displaystyle\int_0^1 [x - L(x)]dx = 2\int_0^1 \left(x - \frac{5x^3}{4 + x^2}\right)dx$

$$= 2\int_0^1 \left[x - \left(5x - \frac{20x}{x^2 + 4}\right)\right]dx = 2\int_0^1 \left(-4x + \frac{20x}{x^2 + 4}\right)dx$$

$$= 2\left[-2x^2 + 10\ln(x^2 + 4)\right]_0^1 = 2(-2 + 10\ln 5 - 10\ln 4)$$

$$= -4 + 20\ln\tfrac{5}{4} \approx 0.46$$

2. **(a)** Take slices perpendicular to the line through the center C of the bottom of the glass and the point P where the top surface of the water meets the bottom of the glass.

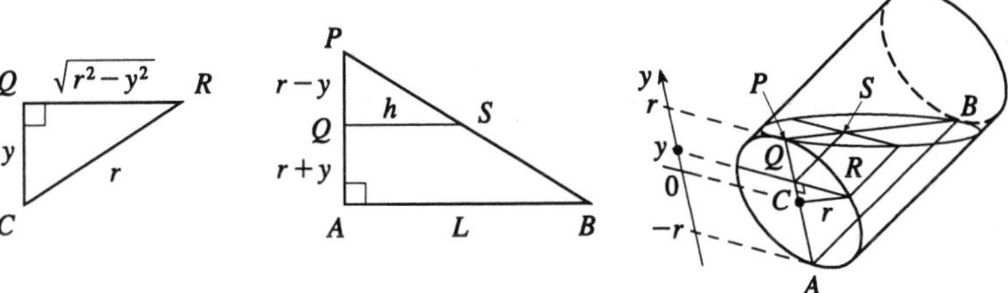

A typical rectangular cross-section y units above the axis of the graph has width $2|QR| = 2\sqrt{r^2 - y^2}$ and length $h = |QS| = \dfrac{L}{2r}(r - y)$. $\left(\text{Triangles } PQS \text{ and } PAB \text{ are similar, so } \dfrac{h}{L} = \dfrac{|PQ|}{|PA|} = \dfrac{r - y}{2r}\right)$. Thus

$$V = \int_{-r}^{r} 2\sqrt{r^2 - y^2}\cdot\frac{L}{2r}(r - y)dy = \int_{-r}^{r} L\left(1 - \frac{y}{r}\right)\sqrt{r^2 - y^2}\,dy$$

$$= L\int_{-r}^{r} \sqrt{r^2 - y^2}\,dy + \frac{L}{2r}\int_{-r}^{r} (-2y)\sqrt{r^2 - y^2}\,dy$$

$$\overset{\bigstar}{=} L\cdot\frac{\pi r^2}{2} + \frac{L}{2r}\cdot\frac{2}{3}\left[\left(r^2 - y^2\right)^{2/3}\right]_{-r}^{r} = \frac{\pi r^2 L}{2} + \frac{L}{3r}(0 - 0) = \frac{\pi r^2 L}{2}$$

★ The first integral is the area of a semicircle of radius r.

(b) Slice parallel to the plane through the axis of the glass and the point of contact P. (This is the plane determined by P, B, and C in the figure.) $STUV$ is a typical trapezoidal slice. With respect to an x-axis with origin at C as shown, if S and V have coordinate x, then $|SV| = 2\sqrt{r^2 - x^2}$. Projecting the trapezoid $STUV$ onto the plane of the triangle PAB, we see that $|AP| = 2r$, $|SV| = 2\sqrt{r^2 - x^2}$, and $|SP| = |VA| = \frac{1}{2}(|AP| - |SV|) = r - \sqrt{r^2 - x^2}$.

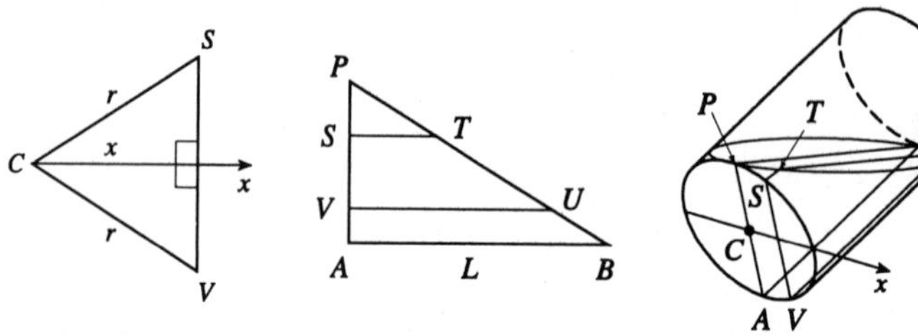

By similar triangles, $\dfrac{|ST|}{|SP|} = \dfrac{|AB|}{|AP|}$, so $|ST| = \left(r - \sqrt{r^2 - x^2}\right) \cdot \dfrac{L}{2r}$. In the same way, we find that $\dfrac{|VU|}{|VP|} = \dfrac{|AB|}{|AP|}$, so $|VU| = |VP| \cdot \dfrac{L}{2r} = (|AP| - |VA|) \cdot \dfrac{L}{2r} = \left(r + \sqrt{r^2 - x^2}\right) \cdot \dfrac{L}{2r}$.

The area $A(x)$ of the trapezoid $STUV$ is $\frac{1}{2}|SV| \cdot (|ST| + |VU|)$; that is,

$$A(x) = \tfrac{1}{2} \cdot 2\sqrt{r^2 - x^2} \cdot \left[\left(r - \sqrt{r^2 - x^2}\right) \cdot \frac{L}{2r} + \left(r + \sqrt{r^2 - x^2}\right) \cdot \frac{L}{2r}\right]$$

$$= L\sqrt{r^2 - x^2}. \text{ Thus } V = \int_{-r}^{r} A(x)dx = L\int_{-r}^{r} \sqrt{r^2 - x^2}\, dx = L \cdot \frac{\pi r^2}{2} = \frac{\pi r^2 L}{2}.$$

(c) See the computation of V in (a) or (b).

(d) The volume of the water is exactly half the volume of the cylindrical glass, so $V = \frac{1}{2}\pi r^2 L$.

(e) Choose x-, y-, and z-axes as shown in the figure. Then slices perpendicular to the x-axis are triangular, slices perpendicular to the y-axis are rectangular, and slices perpendicular to the z-axis are segments of circles. Using triangular slices, we find that the area $A(x)$ of a typical slice DEF, where D has coordinate x, is given by

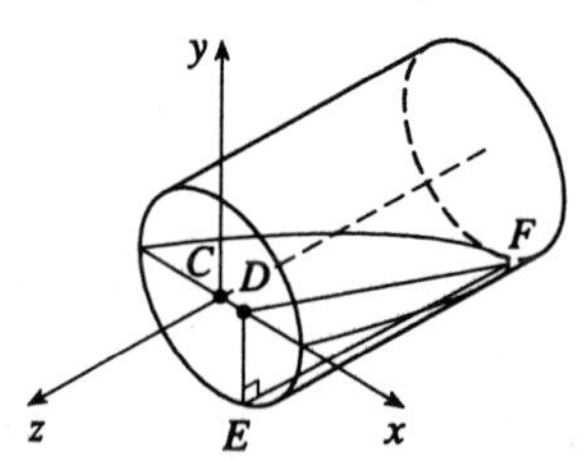

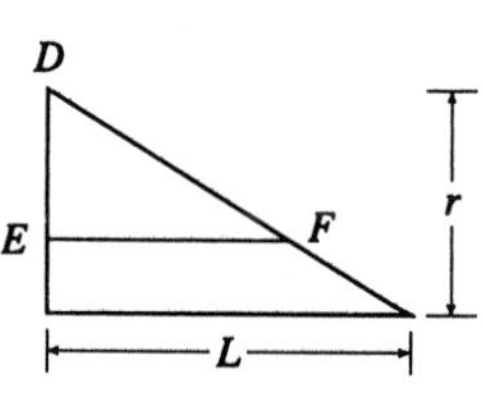

$$A(x) = \tfrac{1}{2}|DE| \cdot |EF| = \tfrac{1}{2}|DE| \cdot \left(\frac{L}{r}|DE|\right) = \frac{L}{2r}|DE|^2 = \frac{L}{2r}(r^2 - x^2). \text{ Thus}$$

$$V = \int_{-r}^{r} A(x)dx = \frac{L}{2r}\int_{-r}^{r} (r^2 - x^2)dx = \frac{L}{2r}\left[r^2 x - \frac{x^3}{3}\right]_{-r}^{r}$$

$$= \frac{L}{2r}\left[\left(r^3 - \frac{r^3}{3}\right) - \left(-r^3 + \frac{r^3}{3}\right)\right] = \frac{L}{2r} \cdot \tfrac{4}{3}r^3 = \tfrac{2}{3}r^2 L.$$

3. **(a)** The tangent to the curve $y = f(x)$ at $x = x_0$ has the equation $y - f(x_0) = f'(x_0)(x - x_0)$.

The y-intercept of this tangent line is $f(x_0) - f'(x_0)x_0$. Thus L is the distance from the point $(0, f(x_0) - f'(x_0)x_0)$ to the point $(x_0, f'(x_0))$. That is, $L^2 = x_0^2 + [f'(x_0)]^2 x_0^2$, so $[f'(x_0)]^2 = \dfrac{L^2 - x_0^2}{x_0^2}$ and $f'(x_0) = -\dfrac{\sqrt{L^2 - x_0^2}}{x_0}$ for each $0 < x_0 < L$.

(b) $\dfrac{dy}{dx} = -\dfrac{\sqrt{L^2 - x^2}}{x} \quad \Rightarrow$

$$y = \int -\frac{\sqrt{L^2 - x^2}}{x}\,dx = \int \frac{-L\cos\theta\, L\cos\theta\, d\theta}{L\sin\theta} \quad (\text{where } x = L\sin\theta)$$

$$= L\int \frac{\sin^2\theta - 1}{\sin\theta}\,d\theta = L\int(\sin\theta - \csc\theta)d\theta = -L\cos\theta + L\ln|\csc\theta + \cot\theta| + C$$

$$= -\sqrt{L^2 - x^2} + L\ln\left(\frac{L}{x} + \frac{\sqrt{L^2 - x^2}}{x}\right) + C$$

When $x = L, 0 = y = -0 + L\ln(1 + 0) + C$, so $C = 0$. Therefore

$$y = -\sqrt{L^2 - x^2} + L\ln\left(\frac{L + \sqrt{L^2 - x^2}}{x}\right).$$

4. **(a)** $p(r) = \dfrac{4}{a_0^3} r^2 e^{-2r/a_0}$ for $r \geq 0$, where $a_0 \approx 5.59 \times 10^{-11}$ m.

$p'(r) = \dfrac{4}{a_0^3}\left[r^2\left(\dfrac{-2}{a_0}\right)e^{-2r/a_0} + 2re^{-2r/a_0}\right] = \dfrac{8r}{a_0^4}e^{-2r/a_0}[a_0 - r]$. Therefore $p'(r) = 0 \quad \Leftrightarrow \quad r = 0$ (the right-hand derivative of p vanishes there) or $r = a_0$. Also $p'(r) > 0$ for $0 < r < a_0$ and $p'(r) < 0$ for $r > a_0$. So p has a local maximum (in fact, an absolute maximum) at $r = a_0$. We find $p(a_0) = \dfrac{4}{a_0}e^{-2} = \dfrac{4}{a_0 e^2}$. Also, $p(r) \geq 0$ for all $r \geq 0$, $\lim\limits_{r\to 0^+} p(r) = p(0) = 0$, and

$$\lim_{r\to\infty} p(r) = \left(\frac{4}{a_0^3}\right)\lim_{r\to\infty}\frac{r^2}{e^{2r/a_0}} \overset{\text{H}}{=} \left(\frac{4}{a_0^3}\right)\lim_{r\to\infty}\frac{2r}{2e^{2r/a_0}/a_0} = \left(\frac{4}{a_0^2}\right)\lim_{r\to\infty}\frac{r}{e^{2r/a_0}} \overset{\text{H}}{=} \left(\frac{4}{a_0^2}\right)\lim_{r\to\infty}\frac{1}{2e^{2r/a_0}/a_0} = 0,$$

where each H marks an application of the ∞/∞ form of l'Hospital's rule. Thus p has an absolute minimum at $r = 0$.

$$p''(r) = \frac{4}{a_0^3}\left[r^2\left(\frac{-2}{a_0}\right)^2 e^{-2r/a_0} + 2r\left(\frac{-2}{a_0}\right)e^{-2r/a_0} + 2re^{-2r/a_0} + 2e^{-2r/a_0}\right]$$

$$= \frac{4}{a_0^3}e^{-2r/a_0}\left[\frac{4}{a_0^2}r^2 - \frac{8}{a_0}r + 2\right] = \frac{8}{a_0^5}e^{-2r/a_0}\left(2r^2 - 4a_0 r + a_0^2\right)$$

Therefore $p''(r) = 0 \quad \Leftrightarrow \quad r = \left(1 \pm \frac{\sqrt{2}}{2}\right)a_0$.

In other words, the graph of p has points of inflection at $r = \left(1 \pm \frac{\sqrt{2}}{2}\right)a_0$.

(b) Probability that electron is within $4a_0$ of the nucleus $= P(4a_0) = \int_0^{4a_0} \frac{4}{a_0^3} s^2 e^{-2s/a_0}\, ds.$

Let $t = -2s/a_0$. Then $s = -a_0 t/2$ and $dt = -\frac{2}{a_0}\, ds$, $ds = -\frac{a_0}{2}\, dt$. So

$$P(4a_0) = \int_0^{-8} \frac{4}{a_0^3}\frac{a_0^2}{4} t^2 e^t \left(\frac{-a_0}{2}\right) dt = \frac{1}{2}\int_{-8}^{0} t^2 e^t\, dt = \tfrac{1}{2}\left[t^2 e^t - 2te^t + 2e^t\right]_{-8}^{0} \quad \left[\begin{array}{l}\text{see Example 3}\\ \text{in Section 7.1}\end{array}\right]$$

$$= \tfrac{1}{2}\left[2 - \left(64e^{-8} + 16e^{-8} + 2e^{-8}\right)\right] = 1 - 41e^{-8} \approx 0.986.$$

(c) $E = \int_0^\infty r\,p(r)dr = \frac{4}{a_0^3}\int_0^\infty r^3 e^{-2r/a_0}\, dr$. Let $t = -2r/a_0$. Then $dt = \frac{-2}{a_0}\, dr$, so

$$E = \frac{4}{a_0^3}\int_0^{-\infty} \frac{a_0^3}{-8} t^3 e^t \frac{a_0}{-2}\, dt = \frac{a_0}{4}\int_0^{-\infty} t^3 e^t\, dt = -\frac{a_0}{4}\int_{-\infty}^{0} t^3 e^t\, dt$$

$$= \left(\frac{-a_0}{4}\right)\lim_{b\to-\infty}\int_b^0 t^3 e^t\, dt \overset{\bigstar}{=} -\left(\frac{a_0}{4}\right)\lim_{b\to-\infty}\left[t^3 e^t - 3t^2 e^t + 6te^t - 6e^t\right]_b^0$$

$$= -\left(\frac{a_0}{4}\right)\lim_{b\to-\infty}\left[-6 - e^b\left(b^3 - 3b^2 + 6b - 6\right)\right] = \frac{-a_0}{4}\cdot(-6) = \tfrac{3}{2}a_0$$

★ Here we use the result of Example 3 in Section 7.1, and integration by parts with $u = t^3$ and $dv = e^t\, dt$.

5. **(a)** $\frac{dC}{dt} = r - kC = k\left(\frac{r}{k} - C\right)$. Let $u(t) = \frac{r}{k} - C(t)$. Then $\frac{du}{dt} = -\frac{dC}{dt}$. Therefore $\frac{du}{dt} = -ku$, and by Theorem 6.5.2, the solution to this equation is $u(t) = u(0)e^{-kt} = \left[\frac{r}{k} - C(0)\right]e^{-kt}$. Therefore

$$\frac{r}{k} - C(t) = \left(\frac{r}{k} - C_0\right)e^{-kt} \quad\Rightarrow\quad C(t) = \frac{r}{k} - \left(\frac{r}{k} - C_0\right)e^{-kt} = C_0 e^{-kt} + \frac{r}{k}\left(1 - e^{-kt}\right).$$

(b) If $C_0 < r/k$, then the first formula for $C(t)$ shows that $C(t)$ increases monotonically and $\lim_{t\to\infty} C(t) = r/k$. The second expression for $C(t)$ shows how the role of C_0 steadily diminishes as that of r/k increases.

6. **(a)** Let $S(t)$ be sales (in \$ millions) at time t. Then $\frac{dS}{dt} = k(25 - S)$ for some constant k.

(b) $\frac{dS}{dt} = k(25 - S)$. Let $\sigma = 25 - S$, so $\frac{d\sigma}{dt} = -\frac{dS}{dt}$ and the differential equation becomes $\frac{d\sigma}{dt} = -k\sigma$.
By Theorem 6.5.2, the solution to this equation is $\sigma(t) = \sigma(0)e^{-kt} \quad\Rightarrow$
$25 - S(t) = [25 - S(0)]e^{-kt} = 25e^{-kt} \quad\Rightarrow\quad S(t) = 25\left(1 - e^{-kt}\right)$. To find k, we substitute
$S(3) = 3.5 = 25\left(1 - e^{-3k}\right) \quad\Rightarrow\quad -3k = \ln 0.86 \quad\Rightarrow\quad k = -\frac{1}{3}\ln 0.86$. So
$S(t) = 25\left[1 - (0.86)^{t/3}\right]$.

(c) $S(8) = 25\left(1 - 0.86^{8/3}\right) \approx \$8.28$ (million)

(d) $12.5 = S = 25\left[1 - (0.86)^{t/3}\right] \quad\Rightarrow\quad \frac{1}{2} = 1 - (0.86)^{t/3} \quad\Rightarrow\quad (0.86)^{8/3} = \frac{1}{2} \quad\Rightarrow\quad (0.86)^t = \frac{1}{8}$

$\Rightarrow\quad t\ln(0.86) = -\ln 8 \quad\Rightarrow\quad t = \frac{\ln 8}{-\ln(0.86)} \approx 13.8$ years

7. **(a)** Here we have a differential equation of the form $\dfrac{dv}{dt} = kv$, so by Theorem 6.5.2, the solution is $v(t) = v(0)e^{kt}$. In this case $k = -\frac{1}{10}$ and $v(0) = 100$ ft/s, so $v(t) = 100e^{-t/10}$. We are interested in the time that the ball takes to travel 280 ft, so we find the distance function

$s(t) = \int_0^t v(x)dx = \int_0^t 100e^{-x/10}\,dx = 100\big[-10e^{-x/10}\big]_0^t = -1000\big(e^{-t/10} - 1\big) = 1000\big(1 - e^{-t/10}\big)$.

Now we set $s(t) = 280$ and solve for t: $280 = 1000\big(1 - e^{-t/10}\big) \quad\Rightarrow\quad 1 - e^{-t/10} = \frac{7}{25} \quad\Rightarrow$ $-\frac{1}{10}t = \ln\big(1 - \frac{7}{25}\big) \approx -0.3285$, so $t \approx 3.285$ seconds.

(b) Let x be the distance of the shortstop from home plate. We calculate the time for the ball to reach home plate as a function of x, then differentiate with respect to x to find the value of x which corresponds to the minimum time.

The total time that it takes the ball to reach home is the sum of the times of the two throws, plus the relay time $\left(\frac{1}{2}\text{ s}\right)$. The distance from the fielder to the shortstop is $280 - x$, so to find the time t_1 taken by the first throw, we solve the equation $s_1(t_1) = 280 - x \;\Leftrightarrow\; 1 - e^{-t_1/10} = \dfrac{280 - x}{1000} \;\Leftrightarrow\; t_1 = -10\ln\dfrac{720 + x}{1000}$.

We find the time t_2 taken by the second throw if the shortstop throws with velocity w, since we see that this velocity varies in the rest of the problem. We use $v = we^{-t/10}$ and isolate t_2 in the equation

$s(t_2) = 10w\big(1 - e^{-t_2/10}\big) = x \quad\Leftrightarrow\quad e^{-t_2/10} = 1 - \dfrac{x}{10w} \quad\Leftrightarrow\quad t_2 = -10\ln\dfrac{10w - x}{10w}$, so the total time is $t_w(x) = \dfrac{1}{2} - 10\left[\ln\dfrac{720 + x}{1000} + \ln\dfrac{10w - x}{10w}\right]$. To find the minimum, we differentiate:

$\dfrac{dt_w}{dx} = -10\left[\dfrac{1}{720 + x} - \dfrac{1}{10w - x}\right]$, which changes from negative to positive when $720 + x = 10w - x$ $\Leftrightarrow \quad x = 5w - 360$. So by the First Derivative Test, t_w has a minimum at this distance from the shortstop to home plate. So if the shortstop throws at $w = 105$ ft/s, the minimum time is

$t_{105}(165) = -10\ln\frac{720 + 165}{1000} + \frac{1}{2} - 10\ln\frac{1050 - 165}{1050} \approx 3.431$ seconds. This is longer than the time taken in part (a), so in this case the manager should encourage a direct throw.

If $w = 115$ ft/s, the minimum time is $t_{115}(215) = -10\ln\frac{720 + 215}{1000} + \frac{1}{2} - 10\ln\frac{1150 - 215}{1150} \approx 3.242$ seconds. This is less than the time taken in part (a), so in this case, the manager should encourage a relayed throw.

(c) In general, the minimum time is

$t_w(5w - 360) = \dfrac{1}{2} - 10\left[\ln\dfrac{360 + 5w}{1000} + \ln\dfrac{360 + 5w}{10w}\right] = \dfrac{1}{2} - 10\ln\dfrac{(w + 72)^2}{400w}$.

We want to find out when this is about 3.285 seconds, the same time as the direct throw. From the graph, we estimate that this is the case for $w \approx 112.8$ ft/s. So if the shortstop can throw the ball with this velocity, then a relayed throw takes the same time as a direct throw.

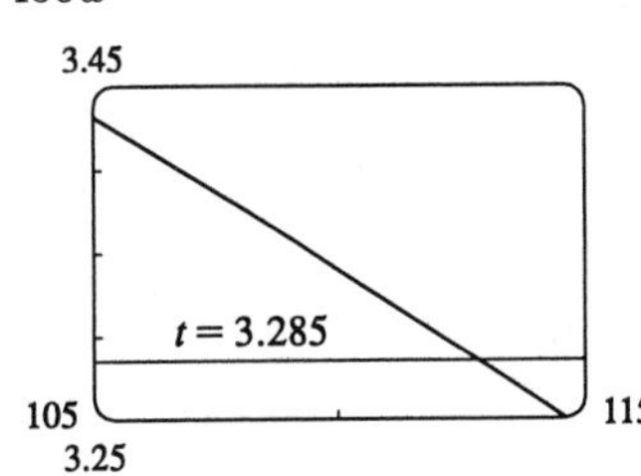

8. **(a)** If we write $f(\lambda) = \dfrac{a\lambda^{-5}}{e^{b/(\lambda T)} - 1}$, then as $\lambda \to 0$ it is of the form $\dfrac{\infty}{\infty}$, and as $\lambda \to \infty$ it is of the form $\dfrac{0}{0}$, so in either case we can use l'Hospital's Rule. First of all,

$$\lim_{\lambda\to\infty} f(\lambda) \stackrel{\text{H}}{=} \lim_{\lambda\to\infty} \frac{a(-5\lambda^{-6})}{-\dfrac{bT}{(\lambda T)^2}e^{b/(\lambda T)}} = 5\frac{aT}{b}\lim_{\lambda\to\infty}\frac{\lambda^2\lambda^{-6}}{e^{b/(\lambda T)}} = 5\frac{aT}{b}\lim_{\lambda\to\infty}\frac{\lambda^{-4}}{e^{b/(\lambda T)}} = 0.$$ Also

$$\lim_{\lambda\to 0} f(\lambda) \stackrel{\text{H}}{=} 5\frac{aT}{b}\lim_{\lambda\to 0}\frac{\lambda^{-4}}{e^{b/(\lambda T)}} \stackrel{\text{H}}{=} 5\frac{aT}{b}\lim_{\lambda\to 0}\frac{-4\lambda^{-5}}{-\dfrac{bT}{(\lambda T)^2}e^{b/(\lambda T)}} = -20\left(\frac{aT}{b}\right)^2\lim_{\lambda\to 0}\frac{\lambda^{-3}}{e^{b/(\lambda T)}}.$$ This is still

indeterminate, but note that each time we use l'Hospital's Rule, we gain a factor of λ in the numerator, as well as a constant factor, and the denominator is unchanged. So if we use l'Hospital's Rule three more times, the exponent of λ in the numerator will become 0. That is, for some $\{k_i\}$, all constant,

$$\lim_{\lambda\to 0} f(\lambda) \stackrel{\text{H}}{=} k_1\lim_{\lambda\to 0}\frac{\lambda^{-3}}{e^{b/(\lambda T)}} \stackrel{\text{H}}{=} k_2\lim_{\lambda\to 0}\frac{\lambda^{-2}}{e^{b/(\lambda T)}} \stackrel{\text{H}}{=} k_3\lim_{\lambda\to 0}\frac{\lambda^{-1}}{e^{b/(\lambda T)}} \stackrel{\text{H}}{=} k_4\lim_{\lambda\to 0}\frac{1}{e^{b/(\lambda T)}} = 0.$$

(b) We have graphed the function $\dfrac{f(\lambda)}{a}$ so as to have a more convenient scale, as suggested in the problem. From the graph, it seems that $f(\lambda)$ has a maximum at $\lambda \approx 0.51\ \mu\text{m}$.

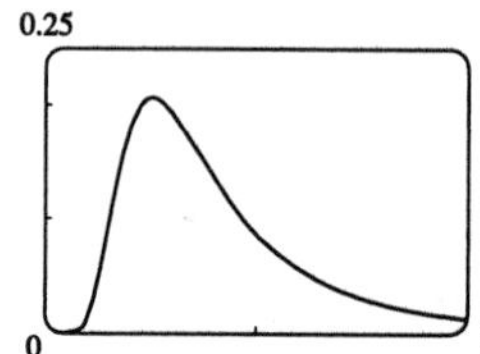

(c) We differentiate $f(\lambda) = a\lambda^{-5}\left(e^{b/(\lambda T)} - 1\right)^{-1}$ to find

$$f'(\lambda) = a\left[-\lambda^{-5}\left(e^{b/(\lambda T)}-1\right)^{-2}\left(\frac{-b}{(\lambda T)^2}Te^{b/(\lambda T)}\right) + \left(e^{b/(\lambda T)}-1\right)^{-1}\left(-5\lambda^{-6}\right)\right]$$

$$= a\left[\frac{be^{b/(\lambda T)}}{T\lambda^7\left[e^{b/(\lambda T)}-1\right]^2} - \frac{5}{\lambda^6\left[e^{b/(\lambda T)}-1\right]}\right] = \frac{a}{T\lambda^7\left[e^{b/(\lambda T)}-1\right]^2}\left[be^{b/(\lambda T)} - 5\lambda T\left(e^{b/(\lambda T)}-1\right)\right]$$

$$= \frac{a}{T\lambda^7\left[e^{b/(\lambda T)}-1\right]^2}\left[5\lambda T + e^{b/(\lambda T)}(b - 5\lambda T)\right].$$

This is 0 when $g(\lambda) = 5\lambda T + e^{b/(\lambda T)}(b - 5\lambda T) = 0$, so we find

$g'(\lambda) = 5T - \dfrac{be^{b/(\lambda T)}}{\lambda^2 T}(b - 5\lambda T) - 5Te^{b/(\lambda T)}$ and use Newton's method: $\lambda_1 = 0.51$, $\lambda_2 \approx 0.50834$, $\lambda_3 \approx 0.50837 \approx \lambda_4$. So, correct to five decimal places, the root of the equation $f'(\lambda) = 0$ is $\lambda = 0.50837$.

(d)

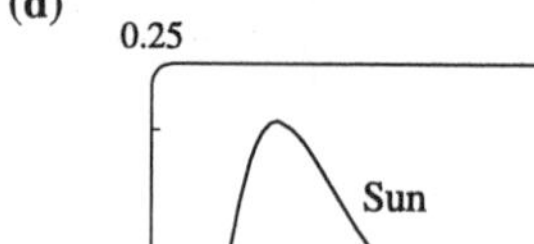

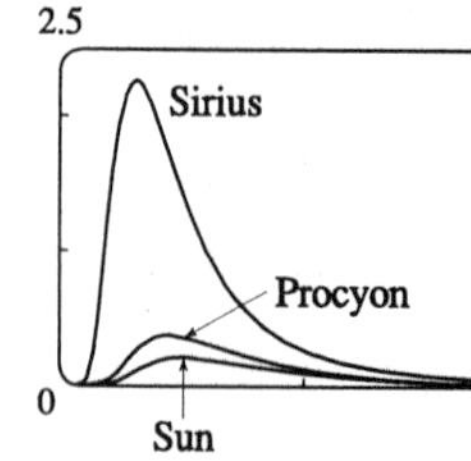

As T gets larger, the total area under the curve increases, as we would expect: the hotter the star, the more energy it emits. Also, as T increases, the λ-value of the maximum decreases, so the higher the temperature, the shorter the peak wavelength (and consequently the average average wavelength) of light emitted. This is why Sirius is a blue star and Betelgeuse is a red star: most of Sirius's light is of a fairly short wavelength, that is, a higher frequency, toward the blue end of the spectrum, whereas most of Betelgeuse's light is of a lower frequency, toward the red end of the spectrum.

9. **(a)** $|VP| = 9 + x\cos\alpha$, $|PT| = 35 - (4 + x\sin\alpha) = 31 - x\sin\alpha$, and $|PB| = (4 + x\sin\alpha) - 10 = x\sin\alpha - 6$.

So using the Pythagorean Theorem, we have

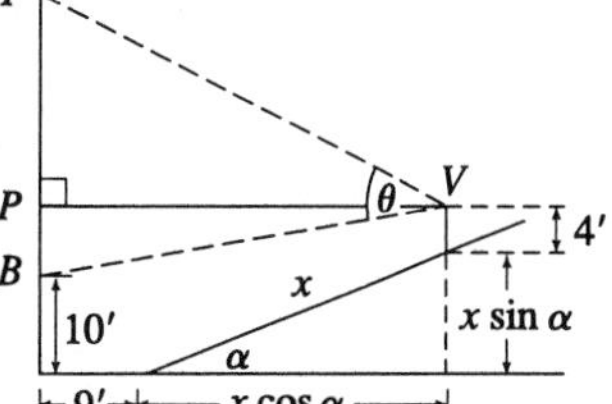

$$|VT| = \sqrt{|VP|^2 + |PT|^2} = \sqrt{(9 + x\cos\alpha)^2 + (31 - x\sin\alpha)^2} = a, \text{ and}$$

$|VB| = \sqrt{|VP|^2 + |PB|^2} = \sqrt{(9 + x\cos\alpha)^2 + (x\sin\alpha - 6)^2} = b$. Using the Law of Cosines on $\triangle VBT$, we get $25^2 = a^2 + b^2 - 2ab\cos\theta \quad \Leftrightarrow$

$$\cos\theta = \frac{a^2 + b^2 - 625}{2ab} \quad \Leftrightarrow \quad \theta = \arccos\left(\frac{a^2 + b^2 - 625}{2ab}\right), \text{ as required.}$$

(b) From the graph, it appears that the value of x which maximizes θ is $x \approx 8.25$ ft. The row closest to this value of x is the fourth row, at $x = 9$ ft, and from the graph, the viewing angle in this row seems to be about 0.85 radians, or about 49°.

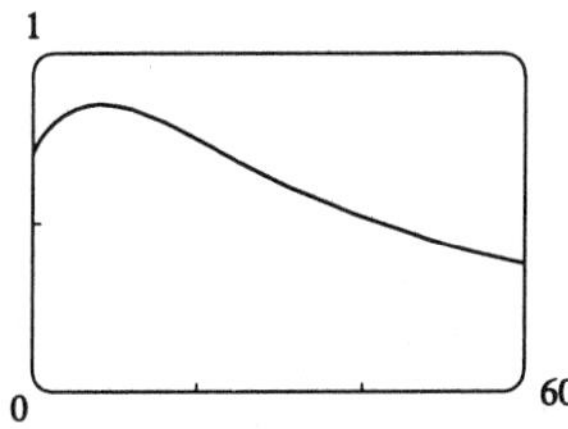

(c) With a CAS, we type in the definition of θ (calling it T), substitute in the proper values of a and b in terms of x and $\alpha = 20° = \frac{\pi}{9}$ radians, and then use the differentiation command (`diff` in Maple) to find the derivative. We use a numerical root finder (`fsolve` in Maple) and find that the root of the equation $d\theta/dx = 0$ is $x \approx 8.25306209$, as approximated above.

(d) From the graph in part (b), it seems that the average value of the function on the interval $[0, 60]$ is about 0.6. We can use a CAS to approximate $\frac{1}{60}\int_0^{60} \theta(x)dx \approx 0.625 \approx 36°$. (The calculation is faster if we reduce the number of digits of accuracy required.) The minimum value is $\theta(60) \approx 0.38$ and, from part (b), the maximum value is about 0.85.

10. **(a)** From Snell's Law, we have $\sin\alpha = k\sin\beta \approx \frac{4}{3}\sin\beta \quad \Leftrightarrow \quad \beta \approx \arcsin\left(\frac{3}{4}\sin\alpha\right)$. We substitute this into $D(\alpha) = \pi + 2\alpha - 4\beta = \pi + 2\alpha - 4\arcsin\left(\frac{3}{4}\sin\alpha\right)$, and then differentiate to find the minimum:

$$D'(\alpha) = 2 - 4\left[1 - \left(\tfrac{3}{4}\sin\alpha\right)^2\right]^{-1/2}\left(\tfrac{3}{4}\cos\alpha\right) = 2 - \frac{3\cos\alpha}{\sqrt{1 - \frac{9}{16}\sin^2\alpha}}. \text{ This is 0 when } \frac{3\cos\alpha}{\sqrt{1 - \frac{9}{16}\sin^2\alpha}} = 2$$

$$\Leftrightarrow \quad \tfrac{9}{4}\cos^2\alpha = 1 - \tfrac{9}{16}\sin^2\alpha \quad \Leftrightarrow \quad \tfrac{27}{16}\cos^2\alpha = \tfrac{7}{16} \quad \Leftrightarrow \quad \cos\alpha = \sqrt{\tfrac{7}{27}} \quad \Leftrightarrow$$

$\alpha = \arccos\sqrt{\frac{7}{27}} \approx 59.4°$, and so the extremum is $D(59.4°) \approx 2.4$ radians $\approx 138°$. To see that this is a minimum, we check the endpoints, which in this case are $\alpha = 0$ and $\alpha = \frac{\pi}{2}$: $D(0) = \pi$ radians $= 180°$, and $D\left(\frac{\pi}{2}\right) \approx 165°$.

Another Method: We first calculate $\sin\alpha = \frac{4}{3}\sin\beta \quad \Leftrightarrow \quad \cos\alpha = \dfrac{4}{3}\cos\beta\,\dfrac{d\beta}{d\alpha} \quad \Leftrightarrow \quad \dfrac{d\beta}{d\alpha} = \dfrac{3\cos\alpha}{4\cos\beta}$,

so since $D'(\alpha) = 2 - 4\dfrac{d\beta}{d\alpha} = 0 \quad \Leftrightarrow \quad \dfrac{d\beta}{d\alpha} = \dfrac{1}{2}$, the minimum occurs when $3\cos\alpha = 2\cos\beta$. Now we square both sides and substitute $\sin\alpha = \frac{4}{3}\sin\beta$, leading to the same result.

(b) If we repeat part (a) with k in place of $\frac{4}{3}$, we get $D(\alpha) = \pi + 2\alpha - 4\arcsin\left(\frac{1}{k}\sin\alpha\right) \Rightarrow$

$D'(\alpha) = 2 - \dfrac{4\cos\alpha}{k\sqrt{1 - [(\sin\alpha)/k]^2}}$, which is 0 when $\dfrac{2\cos\alpha}{k} = \sqrt{1 - \left(\dfrac{\sin\alpha}{k}\right)^2} \Leftrightarrow$

$\left(\dfrac{2\cos\alpha}{k}\right)^2 = 1 - \left(\dfrac{\sin\alpha}{k}\right)^2 \quad \Leftrightarrow \quad \cos^2\alpha + \sin^2\alpha + 3\cos^2\alpha = k^2 \quad \Leftrightarrow \quad 3\cos^2\alpha = k^2 - 1 \quad \Leftrightarrow$

$\theta = \arccos\sqrt{\dfrac{k^2 - 1}{3}}$. So for $k \approx 1.3318$ (red light) the minimum occurs at

$\theta \approx \arccos\sqrt{\dfrac{(1.3318)^2 - 1}{3}} \approx 1.038$ radians, and so the rainbow angle is about

$\pi - D(1.038) \approx \pi - \left[\pi + 2(1.038) - 4\arcsin\left(\frac{1}{1.3318}\sin 1.038\right)\right] \approx 42.3°$.

For $k \approx 1.3435$ (violet light) the minimum occurs at $\theta \approx \arccos\sqrt{\dfrac{(1.3435)^2 - 1}{3}} \approx 1.026$ radians, and so the rainbow angle is about $\pi - D(1.026) \approx \pi - \left[\pi + 2(1.026) - 4\arcsin\left(\frac{1}{1.3435}\sin 1.026\right)\right] \approx 40.6°$.

Another Method: As in part (a), we can instead find $D'(\alpha)$ in terms of $\dfrac{d\beta}{d\alpha}$, and then substitute $\dfrac{d\beta}{d\alpha} = \dfrac{\cos\alpha}{k\cos\beta}$.

(c) At each reflection or refraction, the light is bent in a counterclockwise direction: the bend at A is $\alpha - \beta$, the bend at B is $\pi - 2\beta$, the bend at C is again $\pi - 2\beta$, and the bend at D is $\alpha - \beta$. So the total bend is $2(\alpha - \beta) + 2(\pi - 2\beta) = 2\alpha - 6\beta + 2\pi$, as required. We substitute $\beta = \arcsin\left(\dfrac{\sin\alpha}{k}\right)$ and differentiate,

to get $D'(\alpha) = 2 - \dfrac{6\cos\alpha}{k\sqrt{1 - [(\sin\alpha)/k]^2}}$, which is 0 when $\dfrac{3\cos\alpha}{k} = \sqrt{1 - \left(\dfrac{\sin\alpha}{k}\right)^2} \Leftrightarrow$

$9\cos^2\alpha + \sin^2\alpha = k^2 \quad \Leftrightarrow \quad \cos\alpha = \sqrt{\dfrac{k^2 - 1}{8}}$. If $k = \dfrac{4}{3}$, then the maximum occurs at

$\alpha = \arccos\sqrt{\dfrac{\left(\frac{4}{3}\right)^2 - 1}{8}} \approx 1.254$ radians, and so the rainbow angle is about

$\pi - D(1.254) \approx \pi - \left[6\arcsin\left(\frac{3}{4}\sin 1.254\right) - 2 \cdot 1.254\right] \approx 51°$.

(d) In the primary rainbow, the rainbow angle gets smaller as k gets larger (for α fixed), as we found in part (b), so the colors appear from top to bottom in order of increasing k.

But in the secondary rainbow, the rainbow angle $\pi - D(\alpha)$ gets larger as k gets larger (for α fixed). This is because as k increases, $\dfrac{\sin\alpha}{k}$ decreases, so $\arcsin\left(\dfrac{\sin\alpha}{k}\right)$ decreases, so $D(\alpha)$ decreases. Consequently, the rainbow angle is larger for colors with higher indices of refraction, and the colors appear from bottom to top in order of increasing k, the reverse of their order in the primary rainbow.

CHAPTER EIGHT

EXERCISES 8.1

1. $\dfrac{dy}{dx} = y^2 \;\Rightarrow\; \dfrac{dy}{y^2} = dx\ (y \neq 0) \;\Rightarrow\; \displaystyle\int \frac{dy}{y^2} = \int dx \;\Rightarrow\; -\frac{1}{y} = x + C \;\Rightarrow\; -y = \frac{1}{x+C} \;\Rightarrow\; y = \frac{-1}{x+C}$, and $y = 0$ is also a solution.

2. $\dfrac{dy}{dx} = \dfrac{x + \sin x}{3y^2} \;\Rightarrow\; \displaystyle\int 3y^2\,dy = \int (x + \sin x)dx \;\Rightarrow\; y^3 = \frac{x^2}{2} - \cos x + C \;\Rightarrow\; y = \sqrt[3]{\tfrac{1}{2}x^2 - \cos x + C}$

3. $yy' = x \;\Rightarrow\; \int y\,dy = \int x\,dx \;\Rightarrow\; \dfrac{y^2}{2} = \dfrac{x^2}{2} + C_1 \;\Rightarrow\; y^2 = x^2 + 2C_1 \;\Rightarrow\; x^2 - y^2 = C$ (where $C = -2C_1$). This represents a family of hyperbolas.

4. $y' = xy \;\Rightarrow\; \displaystyle\int \frac{dy}{y} = \int x\,dx\ (y \neq 0) \;\Rightarrow\; \ln|y| = \frac{x^2}{2} + C \;\Rightarrow\; |y| = e^C e^{x^2/2} \;\Rightarrow\; y = Ke^{x^2/2}$, where $K = \pm e^C$ is a constant. (In our derivation, K was nonzero, but we can restore the excluded case $y = 0$ by allowing K to be zero.)

5. $x^2 y' + y = 0 \;\Rightarrow\; \dfrac{dy}{dx} = -\dfrac{y}{x^2} \;\Rightarrow\; \displaystyle\int \frac{dy}{y} = \int \frac{-dx}{x^2}\ (y \neq 0) \;\Rightarrow\; \ln|y| = \frac{1}{x} + K \;\Rightarrow$
$|y| = e^K e^{1/x} \;\Rightarrow\; y = Ce^{1/x}$, where now we allow C to be any constant.

6. $y' = \dfrac{\ln x}{xy + xy^3} = \dfrac{\ln x}{x(y + y^3)} \;\Rightarrow\; \displaystyle\int (y + y^3)dy = \int \frac{\ln x}{x}\,dx \;\Rightarrow\; \frac{y^2}{2} + \frac{y^4}{4} = \tfrac{1}{2}(\ln x)^2 + C_1 \;\Rightarrow$
$y^4 + 2y^2 = 2(\ln x)^2 + 2C_1 \;\Rightarrow\; (y^2 + 1)^2 = 2(\ln x)^2 + K$ (where $K = 2C_1 + 1$) $\Rightarrow$
$y^2 + 1 = \sqrt{2(\ln x)^2 + K}$

7. $\dfrac{du}{dt} = e^{u+2t} = e^u e^{2t} \;\Rightarrow\; \int e^{-u}\,du = \int e^{2t}\,dt \;\Rightarrow\; -e^{-u} = \tfrac{1}{2}e^{2t} + C_1 \;\Rightarrow\; e^{-u} = -\tfrac{1}{2}e^{2t} + C$ (where $C = -C_1$ and the right-hand side is positive, since $e^{-u} > 0$) $\Rightarrow\; -u = \ln\left(C - \tfrac{1}{2}e^{2t}\right) \;\Rightarrow$
$u = -\ln\left(C - \tfrac{1}{2}e^{2t}\right)$

8. $\dfrac{dx}{dt} = 1 + t - x - tx = (1+t)(1-x) \;\Rightarrow\; \displaystyle\int \frac{dx}{1-x} = \int (1+t)dt\ (x \neq 1) \;\Rightarrow$
$-\ln|1 - x| = \tfrac{1}{2}t^2 + t + C \;\Rightarrow\; |1 - x| = e^{-(t^2/2 + t + C)} \;\Rightarrow\; 1 - x = \pm e^{-(t^2/2+t+C)} \;\Rightarrow$
$x = 1 + Ae^{-(t^2/2+t)}$ (where $A = \pm e^C$ or 0)

9. $e^y y' = \dfrac{3x^2}{1+y}$, $y(2) = 0$. $\displaystyle\int e^y(1+y)dy = \int 3x^2\,dx \;\Rightarrow\; ye^y = x^3 + C$. $y(2) = 0$, so $0 = 2^3 + C$ and $C = -8$. Thus $ye^y = x^3 - 8$.

10. $\dfrac{dy}{dx} = \dfrac{1+x}{xy}$, $x > 0$, $y(1) = -4$. $\displaystyle\int y\,dy = \int \frac{1+x}{x}\,dx = \int \left(\frac{1}{x} + 1\right)dx \;\Rightarrow$
$\tfrac{1}{2}y^2 = \ln|x| + x + C = \ln x + x + C$ (since $x > 0$). $y(1) = -4 \;\Rightarrow\; \frac{(-4)^2}{2} = \ln 1 + 1 + C \;\Rightarrow$
$8 = 0 + 1 + C \;\Rightarrow\; C = 7$, so $y^2 = 2\ln x + 2x + 14$.

11. $xe^{-t}\dfrac{dx}{dt}=t,\ x(0)=1.\ \int x\,dx=\int te^t dt \quad\Rightarrow\quad \frac{1}{2}x^2=(t-1)e^t+C.\ x(0)=1$, so $\frac{1}{2}=(0-1)e^0+C$ and $C=\frac{3}{2}$. Thus $x^2=2(t-1)e^t+3 \quad\Rightarrow\quad x=\sqrt{2(t-1)e^t+3}$.

12. $x\,dx+2y\sqrt{x^2+1}\,dy=0,\ y(0)=1.\ \int 2y\,dy=-\displaystyle\int\frac{x\,dx}{\sqrt{x^2+1}} \quad\Rightarrow\quad y^2=-\sqrt{x^2+1}+C.\ y(0)=1 \quad\Rightarrow$ $1=-1+C \quad\Rightarrow\quad C=2$, so $y^2=2-\sqrt{x^2+1}$.

13. $\dfrac{du}{dt}=\dfrac{2t+1}{2(u-1)},\ u(0)=-1.\ \int 2(u-1)du=\int(2t+1)dt \quad\Rightarrow\quad u^2-2u=t^2+t+C.\ u(0)=-1$ so $(-1)^2-2(-1)=0^2+0+C$ and $C=3$. Thus $u^2-2u=t^2+t+3$; the quadratic formula gives $u=1-\sqrt{t^2+t+4}$.

14. $\dfrac{dy}{dt}=\dfrac{ty+3t}{t^2+1}=\dfrac{t(y+3)}{t^2+1},\ y(2)=2.\ \displaystyle\int\frac{dy}{y+3}=\int\frac{t\,dt}{t^2+1} \quad\Rightarrow\quad \ln|y+3|=\frac{1}{2}\ln(t^2+1)+C \quad\Rightarrow$ $y+3=A\sqrt{t^2+1}.\ y(2)=2 \quad\Rightarrow\quad 5=A\sqrt{5} \quad\Rightarrow\quad A=\sqrt{5} \quad\Rightarrow\quad y=-3+\sqrt{5t^2+5}$.

15. Let $y=f(x)$. Then $\dfrac{dy}{dx}=x^3y$ and $y(0)=1$. $\dfrac{dy}{y}=x^3\,dx$ (if $y\neq 0$), so $\displaystyle\int\frac{dy}{y}=\int x^3\,dx$ and $\ln|y|=\frac{1}{4}x^4+C;\ y(0)=1 \quad\Rightarrow\quad C=0$, so $\ln|y|=\frac{1}{4}x^4$, $|y|=e^{x^4/4}$ and $y=f(x)=e^{x^4/4}$ [since $y(0)=1$].

16. Let $y=g(x)$. Then $\dfrac{dy}{dx}=y(1+y)$ and $y(0)=1$. $\displaystyle\int\frac{dy}{y(1+y)}=\int dx \Rightarrow \int\left(\frac{1}{y}-\frac{1}{1+y}\right)dy=\int dx$ $\Rightarrow\quad \ln|y|-\ln|1+y|=x+C \quad\Rightarrow\quad \left|\dfrac{y}{1+y}\right|=e^Ce^x \quad\Rightarrow\quad \dfrac{y}{1+y}=Ae^x.\ y(0)=1 \quad\Rightarrow\quad \dfrac{1}{2}=$ A, so $\dfrac{y}{1+y}=\dfrac{e^x}{2}$. Solve for y: $y=\dfrac{e^x}{2-e^x}$.

17. $\dfrac{dy}{dx}=4x^3y,\ y(0)=7.\ \dfrac{dy}{y}=4x^3\,dx\quad$ (if $y\neq 0$) $\quad\Rightarrow\quad \displaystyle\int\frac{dy}{y}=\int 4x^3\,dx \quad\Rightarrow\quad \ln|y|=x^4+C \quad\Rightarrow$ $y=Ae^{x^4};\ y(0)=7 \quad\Rightarrow\quad A=7 \quad\Rightarrow\quad y=7e^{x^4}$.

18. $\dfrac{dy}{dx}=\dfrac{y^2}{x^3},\ y(1)=1.\ \displaystyle\int\frac{dy}{y^2}=\int\frac{dx}{x^3} \quad\Rightarrow\quad -\frac{1}{y}=-\frac{1}{2x^2}+C.\ y(1)=1 \quad\Rightarrow\quad -1=-\frac{1}{2}+C \quad\Rightarrow$ $C=-\frac{1}{2}$. So $\dfrac{1}{y}=\dfrac{1}{2x^2}+\dfrac{1}{2} \quad\Rightarrow\quad y=\dfrac{2x^2}{x^2+1}$.

19.

3

−3 0 3

$y'=e^{x-y},\ y(0)=1.$

So $\dfrac{dy}{dx}=e^xe^{-y} \quad\Leftrightarrow\quad \displaystyle\int e^y dy=\int e^x dx \quad\Leftrightarrow$

$e^y=e^x+C$. From the initial condition, we must have

$e^1=e^0+C \quad\Rightarrow C=e-1$. So the solution is

$e^y=e^x+e-1\Rightarrow\quad y=\ln(e^x+e-1)$.

20. $e^{-y}y' + \cos x = 0 \quad \Leftrightarrow \quad \int e^{-y}\,dy = -\int \cos x\,dx \quad \Leftrightarrow \quad -e^{-y} = -\sin x + C_1 \quad \Leftrightarrow \quad y = -\ln(\sin x + C)$.
The solution is periodic, with period 2π. Note that for $C > 1$, the domain of the solution is $\mathbb{R}$, but for $-1 < C \leq 1$ it is only defined on the intervals where $\sin x + C > 0$, and it is meaningless for $C \leq -1$, since then $\sin x + C \leq 0$, and the logarithm is undefined.

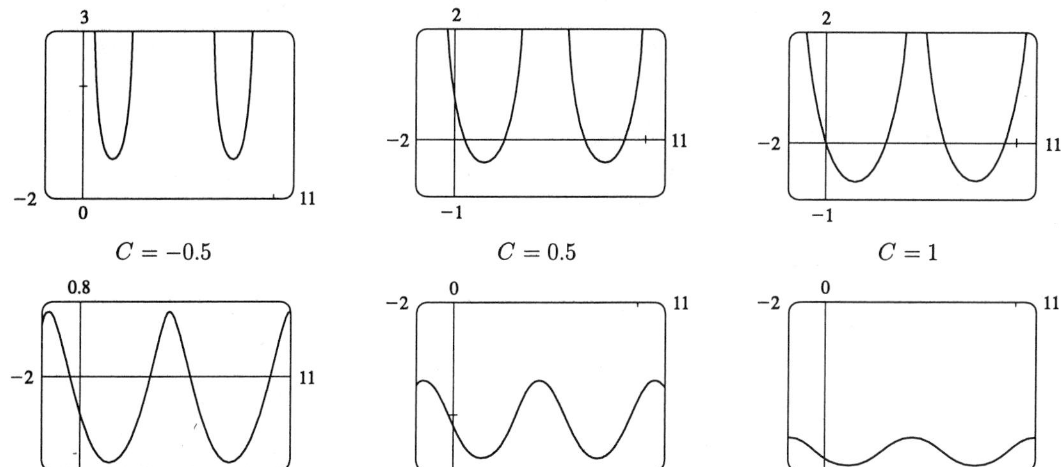

For $-1 < C < 1$, the solution curve consists of concave-up pieces separated by intervals on which the solution is not defined (where $\sin x + C \leq 0$). For $C = 1$, the solution curve consists of concave-up pieces separated by vertical asymptotes at the points where $\sin x + C = 0 \quad \Leftrightarrow \quad \sin x = -1$. For $C > 1$, the curve is continuous, and as C increases, the graph moves downward, and the amplitude of the oscillations decreases.

21. $\dfrac{dy}{dx} = \dfrac{\sin x}{\sin y}$, $y(0) = \dfrac{\pi}{2}$. So $\int \sin y\,dy = \int \sin x\,dx \quad \Leftrightarrow \quad -\cos y = -\cos x + C \quad \Leftrightarrow \quad \cos y = \cos x - C$.
From the initial condition, we need $\cos \frac{\pi}{2} = \cos 0 - C \quad \Rightarrow \quad 0 = 1 - C \quad \Rightarrow \quad C = 1$, so the solution is $\cos y = \cos x - 1$. Note that we cannot take $\cos^{-1}$ of both sides, since that would unnecessarily restrict the solution to the case where $-1 \leq \cos x - 1 \quad \Leftrightarrow \quad 0 \leq \cos x$, since $\cos^{-1}$ is defined only on $[-1, 1]$. Instead we plot the graph using Maple's `plots[implicitplot]` or Mathematica's `Plot[Evaluate[···]]`.

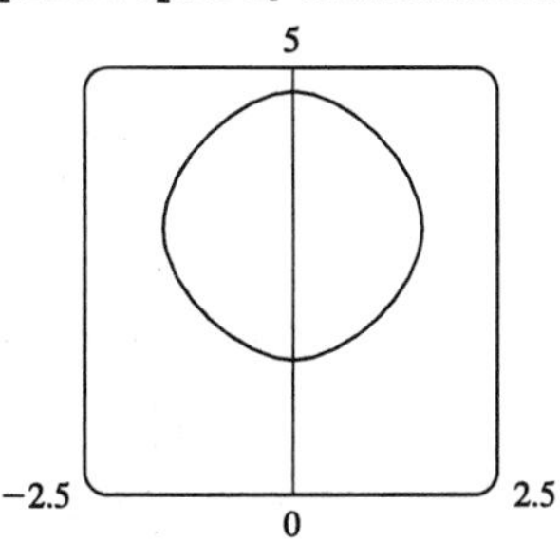

22. $\dfrac{dy}{dx} = \dfrac{x\sqrt{x^2+1}}{ye^y} \quad \Leftrightarrow \quad \int ye^y\,dy = \int x\sqrt{x^2+1}\,dx$. We use parts on the LHS with $u = y$, $dv = e^y\,dy$, and on the RHS we use the substitution $z = x^2+1$, so $dz = \frac{1}{2}x\,dx$. The equation becomes $ye^y - \int e^y\,dy = \frac{1}{2}\int\sqrt{z}\,dz \quad \Leftrightarrow \quad e^y(y-1) = \frac{1}{3}(x^2+1)^{3/2} + C$, so we see that the curves are symmetric about the y-axis. Every point (x, y) in the plane lies on one of the curves, namely the one for which $C = (y-1)e^y - \frac{1}{3}(x^2+1)^{3/2}$. For example, along the y-axis, $C = (y-1)e^y - \frac{1}{3}$, so the origin lies on the curve with $C = -\frac{4}{3}$. We use Maple's `plots[implicitplot]` command or `Plot[Evaluate[···]]` in Mathematica to plot the solution curves for various values of C.

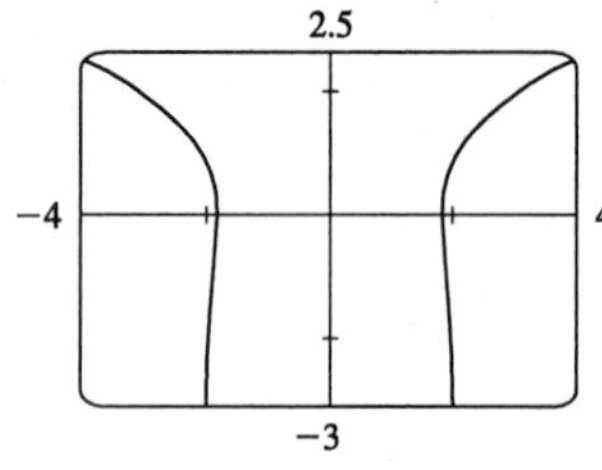

$C = -4$

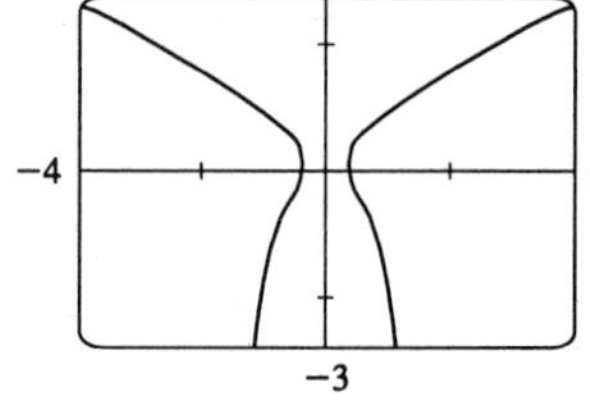

$C = -1.4$

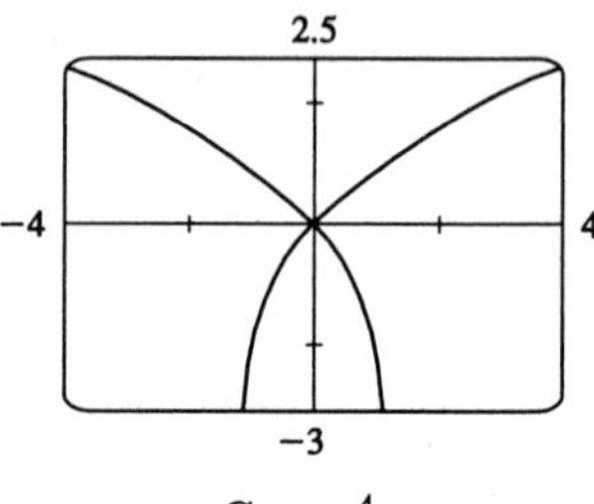

$C = -\frac{4}{3}$

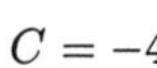

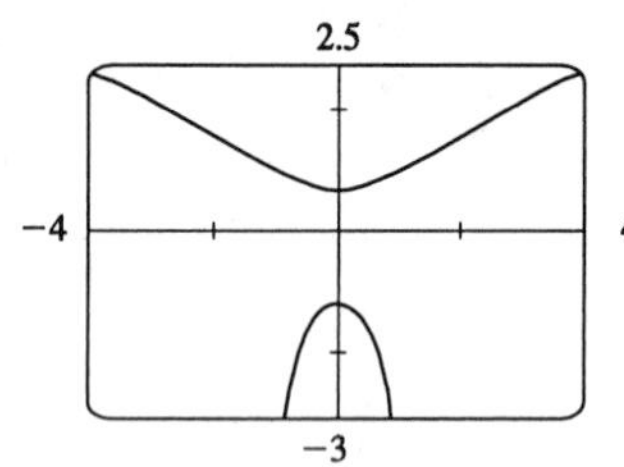

$C = -1$

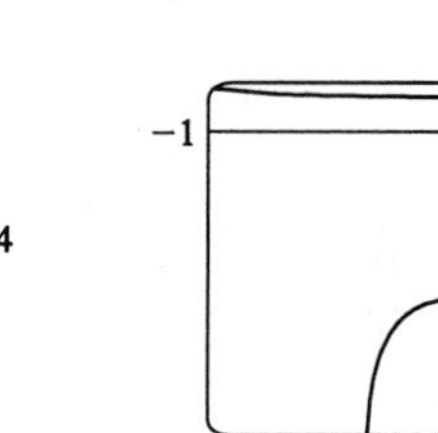

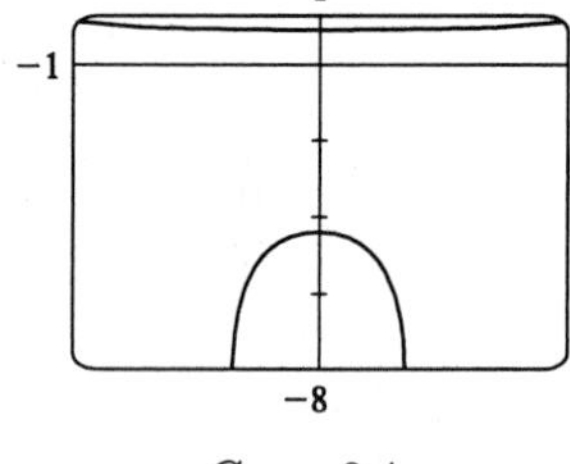

$C = -0.4$

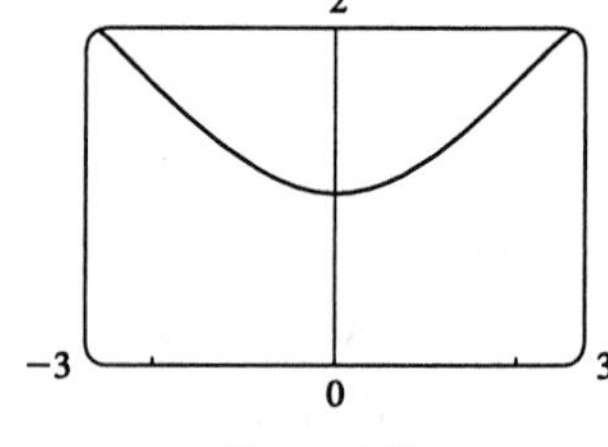

$C = -1/3$

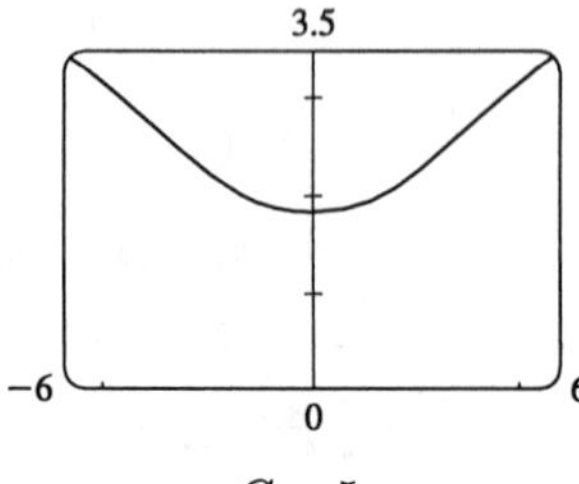

$C = 5$

It seems that the transitional values of C are $-\frac{4}{3}$ and $-\frac{1}{3}$. For $C < -\frac{4}{3}$, the graph consists of left and right branches. At $C = -\frac{4}{3}$, the two branches become connected at the origin, and as C increases, the graph splits into top and bottom branches. At $C = -\frac{1}{3}$, the bottom half disappears. As C increases further, the graph moves upward, but doesn't change shape much.

23. (a)

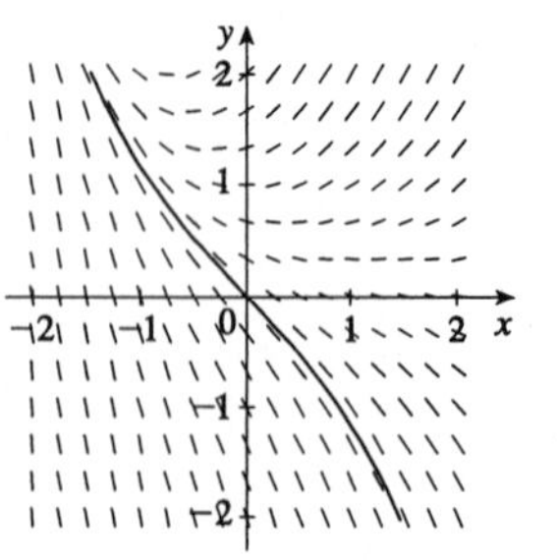

(b)

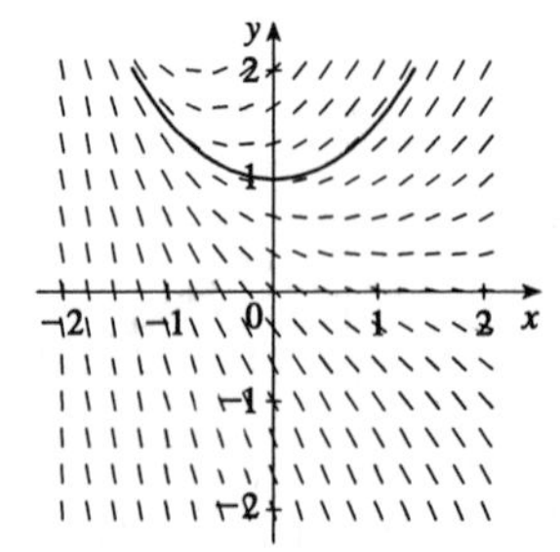

(c)

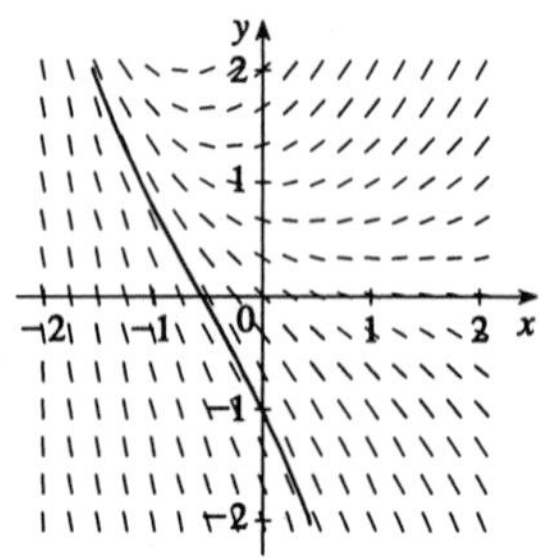

24. **(a)**

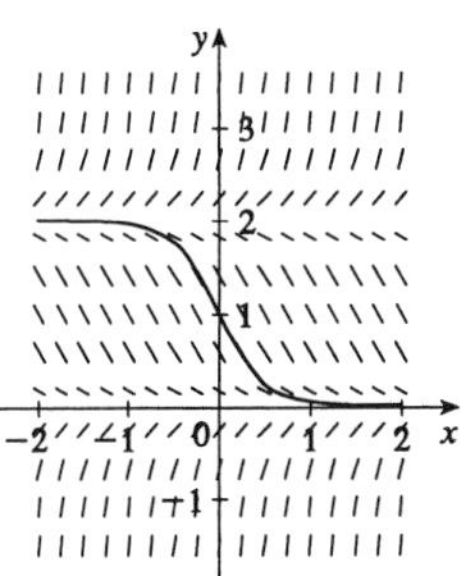

(b)

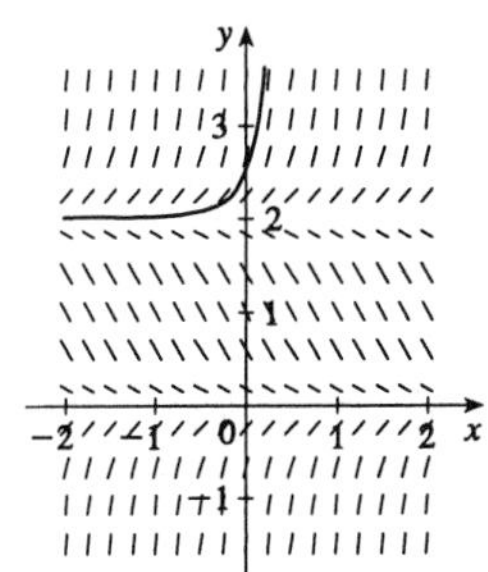

(c)

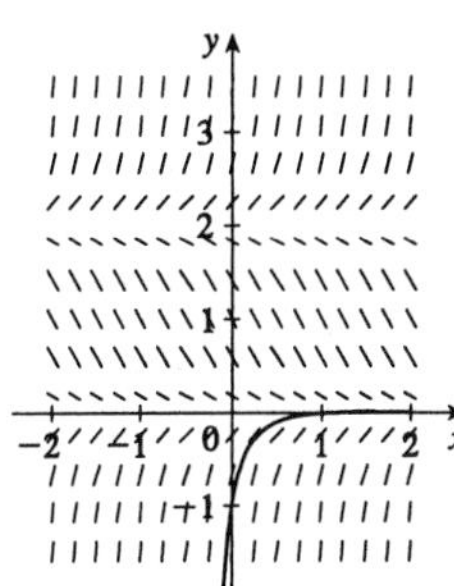

25. $y' = x - y$

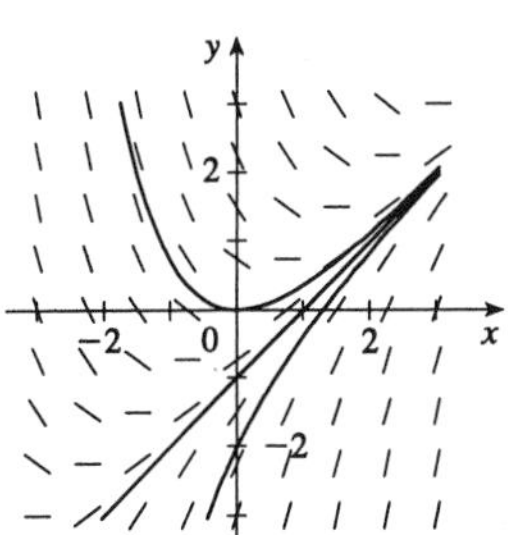

26. $y' = xy + y^2$

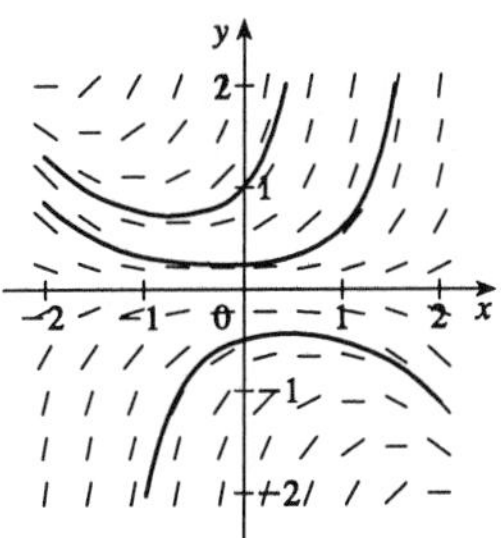

27. In Maple, we can use either `directionfield` (in Maple's share library) or `plots[fieldplot]` to plot the direction field. To plot the solution, we can either use the initial-value option in `directionfield`, or actually solve the equation. In *Mathematica*, we use `PlotVectorField` for the direction field, and the `Plot[Evaluate[···]]` construction to plot the solution, which is $y = e^{(1-\cos 2x)/2}$.

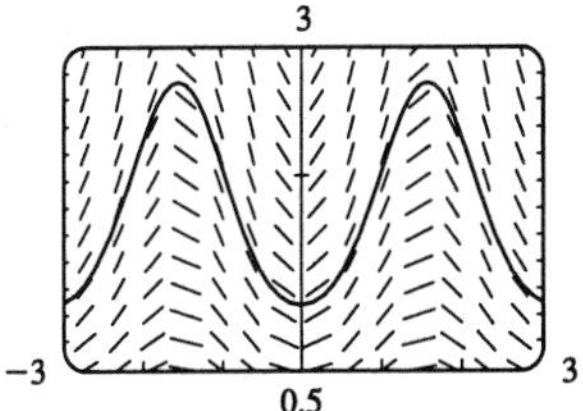

28. In Maple, we can use either `directionfield` (in Maple's share library) or `plots[fieldplot]` to plot the direction field. To plot the solution, we can either use the initial-value option in `directionfield`, or actually solve the equation. In *Mathematica*, we use `PlotVectorField` for the direction field, and the `Plot[Evaluate[...]]` construction to plot the solution, which is $y = -x - 2\arctan\dfrac{2 + x - 2/(1 + \tan\frac{1}{2})}{x - 2/(1 + \tan\frac{1}{2})}$.

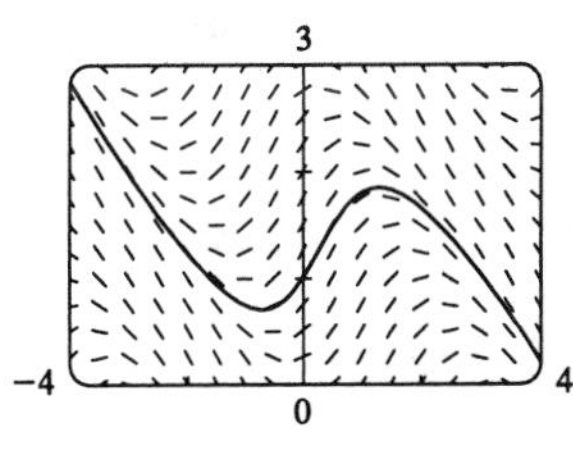

29. **(a)** Let $y(t)$ be the amount of salt (in kg) after t minutes. Then $y(0) = 15$. The amount of liquid in the tank is 1000 L at all times, so the concentration at time t (in minutes) is $y(t)/1000$ kg/L and

$$\frac{dy}{dt} = -\left[\frac{y(t)}{1000}\,\frac{\text{kg}}{\text{L}}\right]\left(10\,\frac{\text{L}}{\text{min}}\right) = -\frac{y(t)}{100}\,\frac{\text{kg}}{\text{min}}.\quad \int\frac{dy}{y} = -\frac{1}{100}\int dt \quad\Rightarrow\quad \ln y = -\frac{t}{100} + C, \text{ and}$$

$$y(0) = 15 \quad\Rightarrow\quad \ln 15 = C, \text{ so } \ln y = \ln 15 - \frac{t}{100}. \text{ It follows that } \ln\left(\frac{y}{15}\right) = -\frac{t}{100} \text{ and } \frac{y}{15} = e^{-t/100},$$

so $y = 15e^{-t/100}$ kg.

(b) After 20 minutes, $y = 15e^{-20/100} = 15e^{-0.2} \approx 12.3$ kg.

30. **(a)** If $y(t)$ is the amount of salt (in kg) after t minutes, then $y(0)=0$ and the total amount of liquid in the tank remains constant at 1000 L.

$$\frac{dy}{dt}=\left(0.05\,\frac{\text{kg}}{\text{L}}\right)\left(5\,\frac{\text{L}}{\text{min}}\right)+\left(0.04\,\frac{\text{kg}}{\text{L}}\right)\left(10\,\frac{\text{L}}{\text{min}}\right)-\left(\frac{y(t)}{1000}\,\frac{\text{kg}}{\text{L}}\right)\left(15\,\frac{\text{L}}{\text{min}}\right)$$

$$=0.25+0.40-0.015y=0.65-0.015y=\frac{130-3y}{200}\,\frac{\text{kg}}{\text{min}}\text{, so}$$

$\displaystyle\int\frac{dy}{130-3y}=\int\frac{dt}{200}$ and $-\frac{1}{3}\ln|130-3y|=\frac{1}{200}t+C$; since $y(0)=0$, we have $-\frac{1}{3}\ln 130=C$, so $-\frac{1}{3}\ln|130-3y|=\frac{1}{200}t-\frac{1}{3}\ln 130$, $\ln|130-3y|=-\frac{3}{200}t+\ln 130=\ln\left(130e^{-3t/200}\right)$, and $|130-3y|=130e^{-3t/200}$.

Since y is continuous, $y(0)=0$, and the right-hand side is never zero, we deduce that $130-3y$ is always positive. Thus $130-3y=130e^{-3t/200}$ and $y=\frac{130}{3}\left(1-e^{-3t/200}\right)$kg.

(b) After an hour, $y=\frac{130}{3}\left(1-e^{-180/200}\right)=\frac{130}{3}\left(1-e^{-0.9}\right)\approx 25.7$ kg.

Note: As $t\to\infty$, $y(t)\to\frac{130}{3}=43\frac{1}{3}$ kg.

31. $\displaystyle\frac{dx}{dt}=k(a-x)(b-x)$, $a\neq b$. $\displaystyle\int\frac{dx}{(a-x)(b-x)}=\int k\,dt \quad\Rightarrow\quad \frac{1}{b-a}\int\left(\frac{1}{a-x}-\frac{1}{b-x}\right)dx=\int k\,dt$

$\displaystyle\Rightarrow\quad \frac{1}{b-a}(-\ln|a-x|+\ln|b-x|)=kt+C \quad\Rightarrow\quad \ln\left|\frac{b-x}{a-x}\right|=(b-a)(kt+C)$. Here the concentrations $[A]=a-x$ and $[B]=b-x$ cannot be negative, so $\displaystyle\frac{b-x}{a-x}\geq 0$ and $\displaystyle\left|\frac{b-x}{a-x}\right|=\frac{b-x}{a-x}$. We now have $\displaystyle\ln\left(\frac{b-x}{a-x}\right)=(b-a)(kt+C)$. Since $x(0)=0$, $\displaystyle\ln\left(\frac{b}{a}\right)=(b-a)C$. Hence

$$\ln\left(\frac{b-x}{a-x}\right)=(b-a)kt+\ln\left(\frac{b}{a}\right),\ \frac{b-x}{a-x}=\frac{b}{a}e^{(b-a)kt},\text{ and }x=\frac{b\left[e^{(b-a)kt}-1\right]}{be^{(b-a)kt}/a-1}=\frac{ab\left[e^{(b-a)kt}-1\right]}{be^{(b-a)kt}-a}\text{ moles/L.}$$

32. **(a)** If $b=a$, then $\displaystyle\frac{dx}{dt}=k(a-x)^2$, so $\displaystyle\int\frac{dx}{(a-x)^2}=\int k\,dt$ and $\displaystyle\frac{1}{a-x}=kt+C$. Taking $x(0)=0$, we get $\displaystyle C=\frac{1}{a}$. Thus $\displaystyle a-x=\frac{1}{kt+1/a}$ and $\displaystyle x=a-\frac{a}{akt+1}=\frac{a^2kt}{akt+1}\,\frac{\text{moles}}{\text{L}}$.

(b) Suppose $[C]=a/2$ after 20 seconds. If t is measured in seconds then $x(20)=a/2$, so $\displaystyle\frac{a}{2}=\frac{20a^2k}{20ak+1}$ and $40a^2k=20a^2k+a$. Thus $20a^2k=a$ and $\displaystyle k=\frac{1}{20a}$, so $\displaystyle x=\frac{a^2t/(20a)}{1+at/(20a)}=\frac{at/20}{1+t/20}=\frac{at}{t+20}\,\frac{\text{moles}}{\text{L}}$.

33. **(a)** Let $P(t)$ be the world population in the year t. Then $dP/dt=0.02P$, so $\int(1/P)dP=\int 0.02\,dt$ and $\ln P=0.02t+C \quad\Rightarrow\quad P(t)=Ae^{0.02t}$. $P(1986)=5\times 10^9 \quad\Rightarrow\quad P(t)=5\times 10^9e^{0.02(t-1986)}$.

(b) **(i)** The predicted population in 2000 is $P(2000)=5e^{0.28}\times 10^9\approx 6.6$ billion.

(ii) The predicted population in 2100 is $P(2100)=5e^{2.28}\times 10^9\approx 49$ billion.

(iii) The predicted population in 2500 is $P(2500)=5e^{10.28}\times 10^9\approx 146$ trillion.

(c) According to this model, in 2000 the area per person will be $\dfrac{1.8\times10^{15}}{6.6\times10^{9}}\approx 270{,}000\text{ ft}^2$. In 2100 it will be $\dfrac{1.8\times10^{15}}{49\times10^{9}}\approx 37{,}000\text{ ft}^2$, and in 2500 it will be $\dfrac{1.8\times10^{15}}{146\times10^{12}}\approx 12\text{ ft}^2$. (!)

34. Let $y(t)$ be the world population (in billions) t years after 1986. Assuming that $\dfrac{dy}{dt}=ky(M-y)$ with $M=100$ and $\left.\dfrac{dy}{dt}\right|_{t=0}=0.02y(0)=0.02(5)=0.10$, we get $0.10=ky(0)[M-y(0)]=5k(100-5)=475k$, so $k=\dfrac{0.10}{475}=\dfrac{1}{4750}$. The solution to Equation 10 is $y(t)=\dfrac{y_0M}{y_0+(M-y_0)e^{-kMt}}=\dfrac{500}{5+95e^{-2t/95}}$, from which we compute $y(14)\approx 6.6$ billion, $y(114)\approx 36.7$ billion, and $y(514)\approx 100$ billion. According to this model, the population initially grows at the same rate as before, but then levels off at $M=100$ billion.

35. **(a)** Our assumption is that $\dfrac{dy}{dt}=ky(1-y)$, where y is the fraction of the population that has heard the rumor.

(b) Take $M=1$ in (11) to get $y=\dfrac{y_0}{y_0+(1-y_0)e^{-kt}}$.

(c) Let t be the number of hours since 8 A.M. Then $y_0=y(0)=\dfrac{80}{1000}=0.08$ and $y(4)=\dfrac{1}{2}$, so $\frac{1}{2}=y(4)=\dfrac{0.08}{0.08+0.92e^{-4k}}$. Thus $0.08+0.92e^{-4k}=0.16$, $e^{-4k}=\dfrac{0.08}{0.92}=\dfrac{2}{23}$, and $e^{-k}=\left(\dfrac{2}{23}\right)^{1/4}$, so $y=\dfrac{0.08}{0.08+0.92(2/23)^{t/4}}=\dfrac{2}{2+23(2/23)^{t/4}}$ and $\left(\dfrac{2}{23}\right)^{t/4}=\dfrac{2}{23}\cdot\dfrac{1-y}{y}$ or $\left(\dfrac{2}{23}\right)^{t/4-1}=\dfrac{1-y}{y}$. It follows that $\dfrac{t}{4}-1=\dfrac{\ln[(1-y)/y]}{\ln(2/23)}$, so $t=4\left[1+\dfrac{\ln[(1-y)/y]}{\ln(2/23)}\right]$. When $y=0.9$, $\dfrac{1-y}{y}=\dfrac{1}{9}$, so $t=4\left(1-\dfrac{\ln 9}{\ln 23}\right)\approx 7.6$ h or 7 h 36 min. Thus 90% of the population will have heard the rumor by 3:36 P.M..

36. **(a)** $y(0)=y_0=400$, $y(1)=1200$ and $M=10{,}000$. From the solution to the logistic differential equation, $y=\dfrac{400(10{,}000)}{400+(9600)e^{-10{,}000kt}}=\dfrac{10{,}000}{1+24e^{-10{,}000kt}}$. $y(1)=1200 \Rightarrow 1+24e^{-10{,}000k}=\dfrac{100}{12} \Rightarrow$ $e^{10{,}000k}=\frac{288}{88} \Rightarrow k=\frac{1}{10{,}000}\ln\frac{36}{11}$. So $y=\dfrac{10{,}000}{1+24e^{-t\ln(36/11)}}=\dfrac{10{,}000}{1+24\cdot(11/36)^t}$.

(b) $5000=\dfrac{10{,}000}{1+24(11/36)^t} \Rightarrow 24\left(\frac{11}{36}\right)^t=1 \Rightarrow t\ln\frac{11}{36}=\ln\frac{1}{24} \Rightarrow t\approx 2.68$ years.

37. y increases most rapidly when y' is maximal, that is, when $y''=0$. But $y'=ky(M-y) \Rightarrow$ $y''=ky'(M-y)+ky(-y')=ky'(M-2y)=k^2y(M-y)(M-2y)$. Since $0<y<M$, we see that $y''=0$ $\Leftrightarrow$ $y=M/2$.

38.

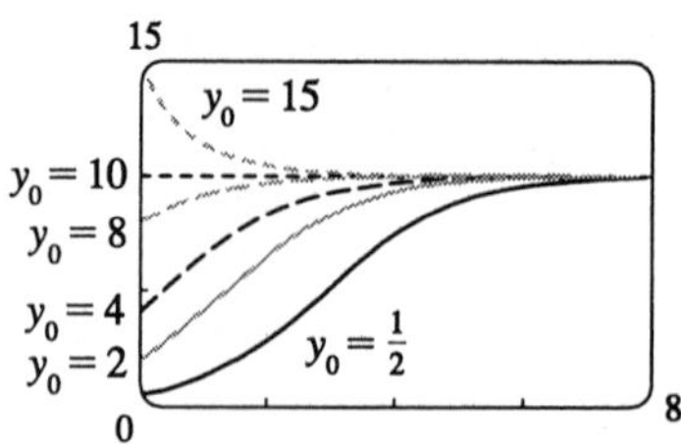

First we keep k constant (at 0.1, say) and change y_0 in the function $y = \dfrac{10y_0}{y_0 + (10 - y_0)e^{-t}}$. (Notice that y_0 is the y-intercept.) If $y_0 = 0$, the function is 0 everywhere. For $0 < y_0 < 5$, the curve has an inflection point, which moves to the right as y_0 decreases. If $5 < y_0 < 10$, the graph is concave down everywhere. (We are considering only $t \geq 0$.) If $y_0 = 10$, the function is the constant function $y = 10$, and if $y_0 > 10$, the function decreases. For all $y_0 \neq 0$, $\lim\limits_{t \to \infty} y = 10$.

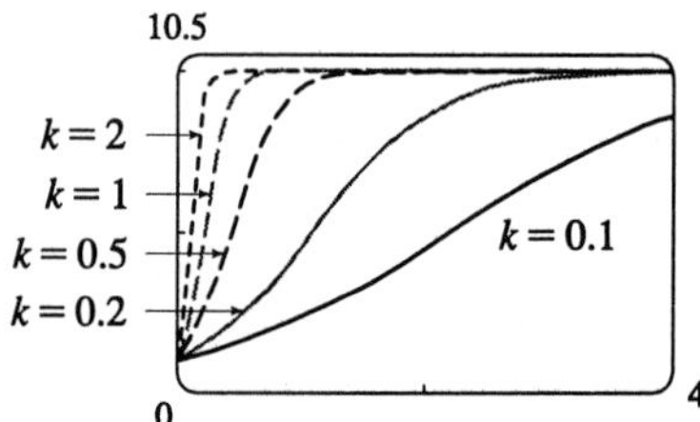

Now we instead keep y_0 constant (at $y_0 = 1$) and change k in the function $y = \dfrac{10}{1 + 9e^{-10kt}}$. It seems that as k increases, the graph approaches the line $y = 10$ more and more quickly. (Note that the only difference in the shape shape of the curves is in the horizontal scaling; if we choose suitable x-scales, the graphs all look the same.)

39. At $t = 0$, the exponential model $y = e^{0.1t}$ has derivative $y' = 0.1e^0 = 0.1$. From the original differential equation, the logistic model has derivative $y' = ky(M - y)$. At $t = 0$, this is equal to $ky_0(M - y_0) = 9k$. So the two derivatives are equal at $t = 0$ if $9k = 0.1 \quad \Leftrightarrow \quad k = 0.1/9 = \frac{1}{90}$.

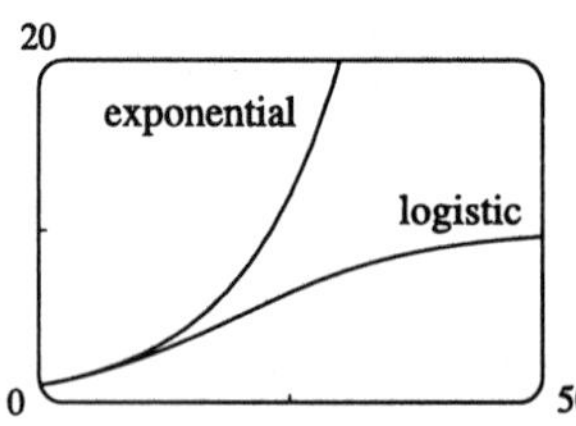

We graph both models, and see that for small values of t they agree closely, but for large values of t, the exponential model increases rapidly, while the logistic model levels off and approaches the line $y = 10$.

40. **(a)** $\dfrac{dy}{dt} = c\ln\left(\dfrac{M}{y}\right)y \quad\Rightarrow\quad \displaystyle\int \frac{dy}{y\ln(M/y)} = \int c\,dt$. Let $u = \ln\dfrac{M}{y} = \ln M - \ln y$. Then $du = -\dfrac{dy}{y}$, so

$-\int du/u = ct + K \quad\Rightarrow\quad -\ln|u| = ct + K \quad\Rightarrow\quad |u| = e^{-(ct+K)} \quad\Rightarrow\quad |\ln(M/y)| = e^{-(ct+K)} \quad\Rightarrow$

$\ln(M/y) = e^{-(ct+K)}$ [We know that $\ln(M/y) \geq 0$ since $M \geq y$] $\quad\Rightarrow\quad M/y = e^{e^{-(ct+K)}} \quad\Rightarrow$

$y(t) = Me^{-e^{-(ct+K)}}$. Taking $t = 0$, we get $y_0 = Me^{-e^{-K}}$, so $e^{-K} = \ln(M/y_0)$ and

$e^{-(ct+K)} = e^{-K}e^{-ct} = \ln(M/y_0)e^{-ct}$. Thus $y(t) = Me^{-\ln(M/y_0)e^{-ct}}$, $c \neq 0$.

(b) $\lim\limits_{t\to\infty} y(t) = M$ since $\lim\limits_{t\to\infty} e^{-ct} = 0$.

(c) Let $b = \ln(M/y_0)$. Then $b > 0$ since $M > y_0$, and $y(t) = Me^{-be^{-ct}}$. We compute

$y'(t) = Me^{-be^{-ct}}\dfrac{d}{dt}\left(-be^{-ct}\right) = Me^{-be^{-ct}}\left(bce^{-ct}\right) = Mbce^{-ct-be^{-ct}}$;

$y''(t) = Mbce^{-ct-be^{-ct}}\dfrac{d}{dt}\left(-ct - be^{-ct}\right) = Mbce^{-ct-be^{-ct}}\left(-c + bce^{-ct}\right)$.

y' is never zero, so there is no local extremum. $y'' = 0 \quad\Leftrightarrow\quad -c + bce^{-ct} = 0 \quad\Leftrightarrow\quad be^{-ct} = 1 \quad\Leftrightarrow$ $e^{ct} = b \quad\Leftrightarrow\quad t = (\ln b)/c$. If $b \leq 1$ (that is, if $M/y_0 \leq e$,) then $-c + bce^{-ct} = c(be^{-ct} - 1) < 0$, so the graph is concave down on $(0, \infty)$ with no inflection points there. If $b > 1$, then the graph is concave up on $(0, (\ln b)/c)$, has an inflection point at $t = (\ln b)/c$, and is concave down on $((\ln b)/c, \infty)$.

Note that $t = (\ln b)/c \quad\Rightarrow\quad e^{-ct} = e^{-\ln b} = 1/b \quad\Rightarrow\quad y = Me^{-b(1/b)} = M/e$.

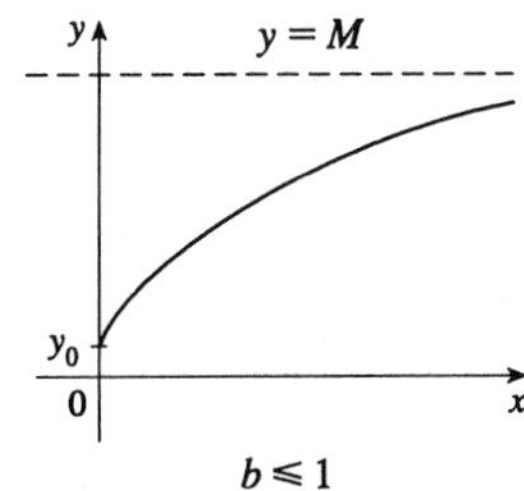

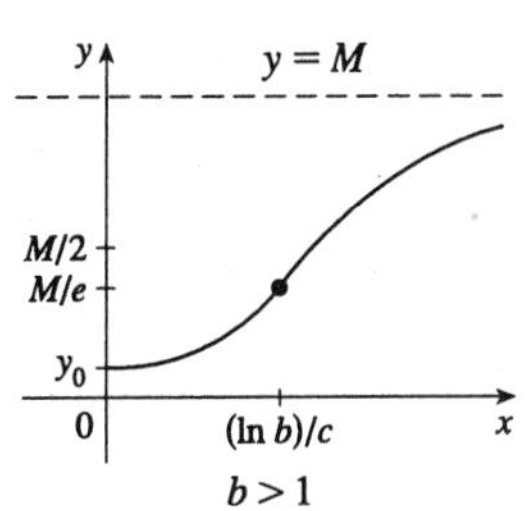

41. **(a)** The rate of growth of the area is jointly proportional to $\sqrt{A(t)}$ and $M - A(t)$; that is, the rate is proportional to the product of those two quantities. So for some constant k, $dA/dt = k\sqrt{A}(M - A)$. We are interested in the maximum of the function dA/dt, so we differentiate, using the Chain Rule and then substituting for dA/dt from the differential equation:

$$\frac{d}{dt}\left(\frac{dA}{dt}\right) = k\left[\tfrac{1}{2}A^{-1/2}(M - A)\frac{dA}{dt} + \sqrt{A}(-1)\frac{dA}{dt}\right] = \tfrac{1}{2}kA^{-1/2}\frac{dA}{dt}[(M - A) - 2A]$$

$= \tfrac{1}{2}k^2(M - A)(M - 3A)$. This is 0 when $M - A = 0$ [this situation never actually occurs, since the graph of $A(t)$ is asymptotic to the line $y = M$, as in the logistic model] and when $M - 3A = 0$ $\Leftrightarrow\quad A(t) = M/3$. This represents a maximum by the First Derivative Test, since $\dfrac{d}{dt}\left(\dfrac{dA}{dt}\right)$ goes from positive to negative when $A(t) = M/3$.

(b) To solve the differential equation, we separate variables in our expression for dA/dt and integrate: $\int \frac{dA}{\sqrt{A}(M-A)} = \int k\,dt = kt + C_1$. To evaluate the LHS, we make the substitution $x = \sqrt{A}$, so $A = x^2$ and $dA = 2x\,dx$. The LHS becomes

$$2\int \frac{dx}{M-x^2} = \int \left(\frac{1/\sqrt{M}}{\sqrt{M}+x} + \frac{1/\sqrt{M}}{\sqrt{M}-x} \right) dx \quad \text{(difference of squares; partial fractions)}$$

$$= \frac{1}{\sqrt{M}}\left(\ln\left|\sqrt{M}+x\right| - \ln\left|\sqrt{M}-x\right|\right) = \frac{1}{\sqrt{M}} \ln \frac{\sqrt{M}+\sqrt{A}}{\sqrt{M}-\sqrt{A}}.$$

So, multiplying by $\sqrt{M}$ and exponentiating both sides of the equation, we get $\frac{\sqrt{M}+\sqrt{A}}{\sqrt{M}-\sqrt{A}} = Ce^{\sqrt{M}kt}$, where $C = e^{\sqrt{M}C_1}$. Solving for A: $\sqrt{M}+\sqrt{A} = Ce^{\sqrt{M}kt}\left(\sqrt{M}-\sqrt{A}\right) \quad \Leftrightarrow$

$$\sqrt{A}\left(Ce^{\sqrt{M}kt}+1\right) = \sqrt{M}\left(Ce^{\sqrt{M}kt}-1\right) \quad \Leftrightarrow \quad A = M\left(\frac{Ce^{\sqrt{M}kt}-1}{Ce^{\sqrt{M}kt}+1}\right)^2.$$

To get C in terms of the initial area A_0 and the maximum area M, we substitute $t=0$ and $A = A_0$:

$A_0 = M\left(\frac{C-1}{C+1}\right)^2 \quad \Leftrightarrow \quad (C+1)\sqrt{A_0} = (C-1)\sqrt{M} \quad \Leftrightarrow \quad C = \frac{\sqrt{M}+\sqrt{A_0}}{\sqrt{M}-\sqrt{A_0}}$. (Notice that if $A_0 = 0$, then $C = 1$.)

42. Assume that the raindrop begins at rest, so that $v(0) = 0$. $dm/dt = km$ and $(mv)' = gm \quad \Rightarrow$ $m'v + mv' = gm \quad \Rightarrow \quad (km)v + mv' = gm \quad \Rightarrow \quad v' = g - kv \quad \Rightarrow \quad \int \frac{dv}{g-kv} = \int dt \quad \Rightarrow$ $-(1/k)\ln|g-kv| = t + C \quad \Rightarrow \quad g - kv = Ae^{-kt}$. $v(0) = 0 \quad \Rightarrow \quad A = g$. So $v = (g/k)\left(1-e^{-kt}\right)$. Since $k > 0$, as $t \to \infty$, $e^{-kt} \to 0$ and therefore $\lim\limits_{t\to\infty} v(t) = g/k$.

43. $RI + LI'(t) = V$, $I(0) = 0$. $LI' = V - RI \;\Rightarrow\; L\frac{dI}{dt} = V - RI \;\Rightarrow\; \int \frac{L\,dI}{V-RI} = \int dt \;\Rightarrow$ $-\frac{L}{R}\ln|V-RI| = t + C \quad \Rightarrow \quad V - RI = Ae^{-Rt/L} \quad \Rightarrow \quad I = \frac{V}{R} - \frac{A}{R}e^{-Rt/L}$.

$I(0) = 0 \quad \Rightarrow \quad 0 = \frac{V}{R} - \frac{A}{R}\cdot e^0 \quad \Rightarrow \quad A = V$. So $I = \frac{V}{R}\left(1-e^{-Rt/L}\right)$.

44. **(a)** According to the hint we use the Chain Rule: $m\frac{dv}{dt} = m\frac{dv}{dx}\cdot\frac{dx}{dt} = mv\frac{dv}{dx} = -\frac{mgR^2}{(x+R)^2} \quad \Rightarrow$

$\int v\,dv = \int \frac{-gR^2\,dx}{(x+R)^2} \quad \Rightarrow \quad \frac{v^2}{2} = \frac{gR^2}{x+R} + C$. When $x = 0$, $v = v_0$, so $\frac{v_0^2}{2} = \frac{gR^2}{0+R} \quad \Rightarrow$ $C = \frac{1}{2}v_0^2 - gR \quad \Rightarrow \quad \frac{1}{2}v^2 - \frac{1}{2}v_0^2 = gR^2/(x+R) - gR$. Now at the top of its flight, the rocket's velocity will be 0, and its height will be $x = h$. Solving for v_0: $-\frac{1}{2}v_0^2 = gR^2/(h+R) - gR \quad \Rightarrow$

$$\frac{v_0^2}{2} = g\left[-\frac{R^2}{R+h} + \frac{R(R+h)}{R+h}\right] = \frac{gRh}{R+h} \quad \Rightarrow \quad v_0 = \sqrt{\frac{2gRh}{R+h}}.$$

(b) $v_e = \lim\limits_{h\to\infty} v_0 = \lim\limits_{h\to\infty}\sqrt{\frac{2gRh}{R+h}} = \lim\limits_{h\to 0}\sqrt{\frac{2gR}{(R/h)+1}} = \sqrt{2gR}$

(c) $v_e = \sqrt{2\cdot 32\text{ ft/s}^2\cdot 3960\text{ mi}\cdot 5280\text{ ft/mi}} \approx 36{,}600\text{ ft/s} \approx 6.93\text{ mi/s}$

EXERCISES 8.2

1. $L = \int_{-1}^{3} \sqrt{1 + \left(\frac{dy}{dx}\right)^2}\, dx = \int_{-1}^{3} \sqrt{1 + 2^2}\, dx = \sqrt{5}[3 - (-1)] = 4\sqrt{5}.$

The arc length can be calculated using the distance formula, since the curve is a straight line, so $L = [\text{distance from } (-1, -1) \text{ to } (3, 7)] = \sqrt{[3 - (-1)]^2 + [7 - (-1)]^2} = \sqrt{80} = 4\sqrt{5}.$

2. Using the arc length formula with $y = \sqrt{4 - x^2} \Rightarrow \frac{dy}{dx} = -\frac{x}{\sqrt{4 - x^2}}$, we get

$L = \int_0^2 \sqrt{1 + \left(\frac{dy}{dx}\right)^2}\, dx = \int_0^2 \sqrt{1 + \frac{x^2}{4 - x^2}}\, dx = \int_0^2 \frac{2\, dx}{\sqrt{4 - x^2}} = \int_0^{\pi/2} \frac{4 \cos\theta\, d\theta}{2 \cos\theta}$ $\begin{bmatrix} x = 2\sin\theta, \\ dx = 2\cos\theta\, d\theta \end{bmatrix}$

$= 2\int_0^{\pi/2} d\theta = \pi.$

The curve is a quarter of a circle with radius 2, so the length of the arc is $\frac{1}{4}(2\pi \cdot 2) = \pi$, as above.

3. $y^2 = (x - 1)^3,\ y = (x - 1)^{3/2} \Rightarrow \frac{dy}{dx} = \frac{3}{2}(x - 1)^{1/2} \Rightarrow 1 + \left(\frac{dy}{dx}\right)^2 = 1 + \frac{9}{4}(x - 1).$

So $L = \int_1^2 \sqrt{1 + \frac{9}{4}(x - 1)}\, dx = \int_1^2 \sqrt{\frac{9}{4}x - \frac{5}{4}}\, dx = \left[\frac{4}{9} \cdot \frac{2}{3}\left(\frac{9}{4}x - \frac{5}{4}\right)^{3/2}\right]_1^2 = \frac{13\sqrt{13} - 8}{27}.$

4. $12xy = 4y^4 + 3,\ x = \frac{y^3}{3} + \frac{y^{-1}}{4} \Rightarrow \frac{dx}{dy} = y^2 - \frac{y^{-2}}{4}$, so $\left(\frac{dx}{dy}\right)^2 = y^4 - \frac{1}{2} + \frac{y^{-4}}{16} \Rightarrow$

$1 + \left(\frac{dx}{dy}\right)^2 = y^4 + \frac{1}{2} + \frac{y^{-4}}{16} \Rightarrow \sqrt{1 + \left(\frac{dx}{dy}\right)^2} = y^2 + \frac{y^{-2}}{4}.$

So $L = \int_1^2 \left(y^2 + \frac{y^{-2}}{4}\right) dy = \left[\frac{y^3}{3} - \frac{1}{4y}\right]_1^2 = \left(\frac{8}{3} - \frac{1}{8}\right) - \left(\frac{1}{3} - \frac{1}{4}\right) = \frac{59}{24}.$

5. $y = \frac{1}{3}(x^2 + 2)^{3/2} \Rightarrow dy/dx = \frac{1}{2}(x^2 + 2)^{1/2}(2x) = x\sqrt{x^2 + 2} \Rightarrow$
$1 + (dy/dx)^2 = 1 + x^2(x^2 + 2) = (x^2 + 1)^2.$ So $L = \int_0^1 (x^2 + 1) dx = \left[\frac{1}{3}x^3 + x\right]_0^1 = \frac{4}{3}.$

6. $y = \frac{x^3}{6} + \frac{1}{2x} \Rightarrow \frac{dy}{dx} = \frac{x^2}{2} - \frac{x^{-2}}{2} \Rightarrow 1 + \left(\frac{dy}{dx}\right)^2 = \frac{x^4}{4} + \frac{1}{2} + \frac{x^{-4}}{4}.$

So $L = \int_1^2 \left(\frac{x^2}{2} + \frac{x^{-2}}{2}\right) dx = \frac{1}{2}\left[\frac{x^3}{3} - \frac{1}{x}\right]_1^2 = \frac{1}{2}\left[\left(\frac{8}{3} - \frac{1}{2}\right) - \left(\frac{1}{3} - 1\right)\right] = \frac{17}{12}.$

7. $y = \frac{x^4}{4} + \frac{1}{8x^2} \Rightarrow \frac{dy}{dx} = x^3 - \frac{1}{4x^3} \Rightarrow 1 + \left(\frac{dy}{dx}\right)^2 = 1 + x^6 - \frac{1}{2} + \frac{1}{16x^6} = x^6 + \frac{1}{2} + \frac{1}{16x^6}.$

So $L = \int_1^3 \left(x^3 + \frac{1}{4}x^{-3}\right) dx = \left[\frac{1}{4}x^4 - \frac{1}{8}x^{-2}\right]_1^3 = \left(\frac{81}{4} - \frac{1}{72}\right) - \left(\frac{1}{4} - \frac{1}{8}\right) = \frac{181}{9}.$

8. $y = \frac{x^2}{2} - \frac{\ln x}{4} \Rightarrow \frac{dy}{dx} = x - \frac{1}{4x} \Rightarrow 1 + \left(\frac{dy}{dx}\right)^2 = x^2 + \frac{1}{2} + \frac{1}{16x^2}.$

So $L = \int_2^4 \left(x + \frac{1}{4x}\right) dx = \left[\frac{x^2}{2} + \frac{\ln x}{4}\right]_2^4 = \left(8 + \frac{2\ln 2}{4}\right) - \left(2 + \frac{\ln 2}{4}\right) = 6 + \frac{\ln 2}{4}.$

9. $y = \ln(\cos x) \quad \Rightarrow \quad y' = \dfrac{1}{\cos x}(-\sin x) = -\tan x \quad \Rightarrow \quad 1 + (y')^2 = 1 + \tan^2 x = \sec^2 x.$

So $L = \int_0^{\pi/4} \sec x\,dx = \ln(\sec x + \tan x)\big|_0^{\pi/4} = \ln\left(\sqrt{2} + 1\right).$

10. $y = \ln(\sin x) \quad \Rightarrow \quad \dfrac{dy}{dx} = \dfrac{\cos x}{\sin x} = \cot x \quad \Rightarrow \quad 1 + \left(\dfrac{dy}{dx}\right)^2 = 1 + \cot^2 x = \csc^2 x.$ So

$L = \int_{\pi/6}^{\pi/3} \csc x\,dx = [\ln(\csc x - \cot x)]_{\pi/6}^{\pi/3} = \ln\left(\frac{2}{\sqrt{3}} - \frac{1}{\sqrt{3}}\right) - \ln\left(2 - \sqrt{3}\right)$

$= \ln \frac{1}{\sqrt{3}\left(2-\sqrt{3}\right)} = \ln \frac{2+\sqrt{3}}{\sqrt{3}} = \ln\left(1 + \frac{2}{\sqrt{3}}\right).$

11. $y = \ln\left(1 - x^2\right) \quad \Rightarrow \quad \dfrac{dy}{dx} = \dfrac{-2x}{1 - x^2} \quad \Rightarrow \quad 1 + \left(\dfrac{dy}{dx}\right)^2 = 1 + \dfrac{4x^2}{\left(1 - x^2\right)^2} = \dfrac{\left(1 + x^2\right)^2}{\left(1 - x^2\right)^2}.$ So

$$L = \int_0^{1/2} \frac{1 + x^2}{1 - x^2}\,dx = \int_0^{1/2} \left[-1 + \frac{2}{(1 - x)(1 + x)}\right] dx = \int_0^{1/2} \left[-1 + \frac{1}{1 + x} + \frac{1}{1 - x}\right] dx$$

$= [-x + \ln(1 + x) - \ln(1 - x)]_0^{1/2} = -\frac{1}{2} + \ln\frac{3}{2} - \ln\frac{1}{2} - 0 = \ln 3 - \frac{1}{2}.$

12. $y = \ln\left(\dfrac{e^x + 1}{e^x - 1}\right) = \ln(e^x + 1) - \ln(e^x - 1) \quad \Rightarrow \quad y' = \dfrac{e^x}{e^x + 1} - \dfrac{e^x}{e^x - 1} = \dfrac{-2e^x}{e^{2x} - 1}$

$\Rightarrow \quad 1 + (y')^2 = 1 + \dfrac{4e^{2x}}{\left(e^{2x} - 1\right)^2} = \dfrac{\left(e^{2x} + 1\right)^2}{\left(e^{2x} - 1\right)^2} \quad \Rightarrow \quad \sqrt{1 + (y')^2} = \dfrac{e^{2x} + 1}{e^{2x} - 1} = \dfrac{e^x + e^{-x}}{e^x - e^{-x}} = \dfrac{\cosh x}{\sinh x}.$ So

$$L = \int_a^b \frac{\cosh x}{\sinh x}\,dx = \ln \sinh x\big|_a^b = \ln\left(\frac{\sinh b}{\sinh a}\right) = \ln\left(\frac{e^b - e^{-b}}{e^a - e^{-a}}\right).$$

13. $y = e^x \quad \Rightarrow \quad y' = e^x \quad \Rightarrow \quad 1 + (y')^2 = 1 + e^{2x}.$ So

$$L = \int_0^1 \sqrt{1 + e^{2x}}\,dx = \int_1^e \sqrt{1 + u^2}\,\frac{du}{u} \quad [\text{where } u = e^x, \text{ so } x = \ln u,\ dx = du/u]$$

$$= \int_1^e \frac{\sqrt{1 + u^2}}{u^2}\,u\,du = \int_{\sqrt{2}}^{\sqrt{1+e^2}} \frac{v}{v^2 - 1}\,v\,dv \quad \left[\text{where } v = \sqrt{1 + u^2}, \text{ so } v^2 = 1 + u^2,\ v\,dv = u\,du\right]$$

$$= \int_{\sqrt{2}}^{\sqrt{1+e^2}} \left(1 + \frac{1/2}{v - 1} - \frac{1/2}{v + 1}\right) dv = \left[v + \tfrac{1}{2}\ln\frac{v - 1}{v + 1}\right]_{\sqrt{2}}^{\sqrt{1+e^2}}$$

$$= \sqrt{1 + e^2} - \sqrt{2} + \tfrac{1}{2}\ln\frac{\sqrt{1 + e^2} - 1}{\sqrt{1 + e^2} + 1} - \tfrac{1}{2}\ln\frac{\sqrt{2} - 1}{\sqrt{2} + 1}$$

$$= \sqrt{1 + e^2} - \sqrt{2} + \ln\left(\sqrt{1 + e^2} - 1\right) - 1 - \ln\left(\sqrt{2} - 1\right)$$

Or: Use Formula 23 for $\int \left(\sqrt{1 + u^2}/u\right) du$, or substitute $u = \tan\theta$.

14. $y = \ln x \quad \Rightarrow \quad \dfrac{dy}{dx} = \dfrac{1}{x} \quad \Rightarrow \quad \sqrt{1 + \left(\dfrac{dy}{dx}\right)^2} = \sqrt{1 + \left(\dfrac{1}{x}\right)^2} = \dfrac{\sqrt{1 + x^2}}{x}.$

So $L = \displaystyle\int_1^{\sqrt{3}} \frac{\sqrt{1 + x^2}}{x}\,dx \quad \left[\text{put } v = \sqrt{1 + x^2},\ v^2 = 1 + x^2,\ v\,dv = x\,dx.\right]$ Thus

$$L = \int_{\sqrt{2}}^2 \frac{v}{v^2 - 1}\,v\,dv = \int_{\sqrt{2}}^2 \left(1 + \frac{1/2}{v - 1} - \frac{1/2}{v + 1}\right) dv = \left[v + \tfrac{1}{2}\ln|v - 1| - \tfrac{1}{2}\ln|v + 1|\right]_{\sqrt{2}}^2$$

$$= \left[v - \tfrac{1}{2}\ln\left|\frac{v + 1}{v - 1}\right|\right]_{\sqrt{2}}^2 = 2 - \sqrt{2} - \tfrac{1}{2}\ln 3 + \tfrac{1}{2}\ln\left(\frac{\sqrt{2} + 1}{\sqrt{2} - 1}\right) = 2 - \sqrt{2} + \ln\left(\sqrt{2} + 1\right) - \tfrac{1}{2}\ln 3.$$

Or: Use Formula 23.

15. $y = \cosh x \quad \Rightarrow \quad y' = \sinh x \quad \Rightarrow \quad 1 + (y')^2 = 1 + \sinh^2 x = \cosh^2 x.$

So $L = \int_0^1 \cosh x \, dx = [\sinh x]_0^1 = \sinh 1 = \frac{1}{2}(e - 1/e).$

16. $y^2 = 4x,\ x = \frac{1}{4}y^2 \quad \Rightarrow \quad \dfrac{dy}{dx} = \frac{1}{2}y \quad \Rightarrow \quad 1 + \left(\dfrac{dy}{dx}\right)^2 = 1 + \dfrac{y^2}{4}.$ So

$L = \int_0^2 \sqrt{1 + \frac{1}{4}y^2}\, dy = \int_0^1 \sqrt{1 + u^2} \cdot 2\, du$ (put $u = \frac{1}{2}y \quad \Rightarrow \quad dy = 2\, du$)

$= \left[u\sqrt{1+u^2} + \ln\left|u + \sqrt{1+u^2}\right|\right]_0^1$ (Formula 21 or let $u = \tan\theta$) $= \sqrt{2} + \ln\left(1 + \sqrt{2}\right).$

17. $y = x^3 \quad \Rightarrow \quad y' = 3x^2 \quad \Rightarrow \quad 1 + (y')^2 = 1 + 9x^4.$ So $L = \int_0^1 \sqrt{1 + 9x^4}\, dx.$

18. $y = \tan x \quad \Rightarrow \quad 1 + (y')^2 = 1 + \sec^4 x.$ So $L = \int_0^{\pi/4} \sqrt{1 + \sec^4 x}\, dx.$

19. $y = e^x \cos x \quad \Rightarrow \quad y' = e^x(\cos x - \sin x) \quad \Rightarrow$

$1 + (y')^2 = 1 + e^{2x}(\cos^2 x - 2\cos x \sin x + \sin^2 x) = 1 + e^{2x}(1 - \sin 2x).$ So

$L = \int_0^{\pi/2} \sqrt{1 + e^{2x}(1 - \sin 2x)}\, dx.$

20. $\dfrac{x^2}{a^2} + \dfrac{y^2}{b^2} = 1,\ y = \pm b\sqrt{1 - x^2/a^2} = \pm\dfrac{b}{a}\sqrt{a^2 - x^2}.\ y = \dfrac{b}{a}\sqrt{a^2 - x^2} \quad \Rightarrow \quad \dfrac{dy}{dx} = \dfrac{-bx}{a\sqrt{a^2 - x^2}} \quad \Rightarrow$

$\left(\dfrac{dy}{dx}\right)^2 = \dfrac{b^2x^2}{a^2(a^2 - x^2)}.$ So $L = 2\displaystyle\int_{-a}^{a}\left[1 + \dfrac{b^2x^2}{a^2(a^2 - x^2)}\right]^{1/2} dx = \dfrac{4}{a}\displaystyle\int_0^a \left[\dfrac{(b^2 - a^2)x^2 + a^4}{a^2 - x^2}\right]^{1/2} dx.$

21. $y = x^3 \quad \Rightarrow \quad 1 + (y')^2 = 1 + (3x^2)^2 = 1 + 9x^4.$ So $L = \displaystyle\int_0^1 \sqrt{1 + 9x^4}\, dx.$ Let $f(x) = \sqrt{1 + 9x^4}.$

Then by Simpson's Rule with $n = 10$,

$L \approx \frac{1/10}{3}[f(0) + 4f(0.1) + 2f(0.2) + 4f(0.3) + 2f(0.4)$

$+ 4f(0.5) + 2f(0.6) + 4f(0.7) + 2f(0.8) + 4f(0.9) + f(1)] \approx 1.548.$

22. $y = x^4 \quad \Rightarrow \quad 1 + (y')^2 = 1 + (4x^3)^2 = 1 + 16x^6.$ So $L = \int_0^2 \sqrt{1 + 16x^6}\, dx.$ Let $f(x) = \sqrt{1 + 16x^6}.$ Then

$L \approx \frac{1/5}{3}[f(0) + 4f(0.2) + 2f(0.4) + 4f(0.6) + 2f(0.8) + 4f(1)$

$+ 2f(1.2) + 4f(1.4) + 2f(1.6) + 4f(1.8) + f(2)] \approx 16.65.$

23. $y = \sin x,\ 1 + (dy/dx)^2 = 1 + \cos^2 x,\ L = \int_0^\pi \sqrt{1 + \cos^2 x}\, dx.$ Let $g(x) = \sqrt{1 + \cos^2 x}.$ Then

$L \approx \frac{\pi/10}{3}\left[g(0) + 4g\left(\frac{\pi}{10}\right) + 2g\left(\frac{\pi}{5}\right) + 4g\left(\frac{3\pi}{10}\right) + 2g\left(\frac{2\pi}{5}\right) + 4g\left(\frac{\pi}{2}\right)\right.$

$\left. + 2g\left(\frac{3\pi}{5}\right) + 4g\left(\frac{7\pi}{10}\right) + 2g\left(\frac{4\pi}{5}\right) + 4g\left(\frac{9\pi}{10}\right) + g(\pi)\right] \approx 3.820.$

24. $y = \tan x \quad \Rightarrow \quad 1 + (y')^2 = 1 + \sec^4 x.$ So $L = \int_0^{\pi/4} \sqrt{1 + \sec^4 x}\, dx.$ Let $g(x) = \sqrt{1 + \sec^4 x}.$ Then

$L \approx \frac{\pi/40}{3}\left[g(0) + 4g\left(\frac{\pi}{40}\right) + 2g\left(\frac{2\pi}{40}\right) + 4g\left(\frac{3\pi}{40}\right) + 2g\left(\frac{4\pi}{40}\right)\right.$

$\left. + 4g\left(\frac{5\pi}{40}\right) + 2g\left(\frac{6\pi}{40}\right) + 4g\left(\frac{7\pi}{40}\right) + 2g\left(\frac{8\pi}{40}\right) + 4g\left(\frac{9\pi}{40}\right) + g\left(\frac{\pi}{4}\right)\right] \approx 1.278.$

25. **(a)**

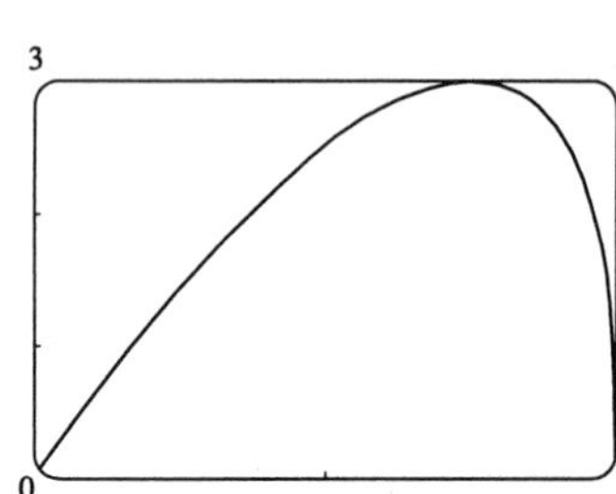

(b)

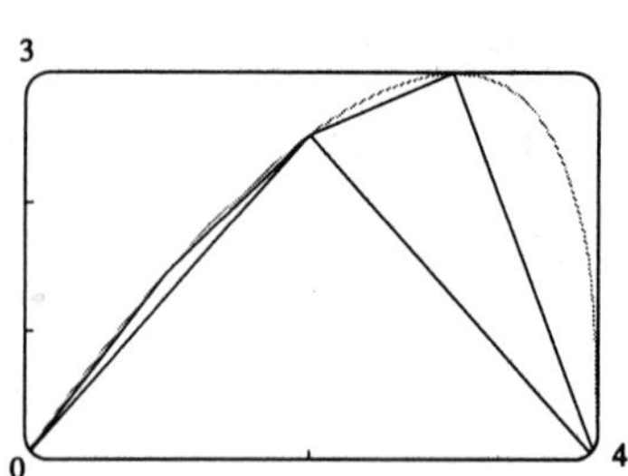

Let $f(x) = y = x\sqrt[3]{4-x}$. The polygon with one side is just the line segment joining the points $(0, f(0)) = (0, 0)$ and $(4, f(4)) = (4, 0)$, and its length is 4. The polygon with two sides joins the points $(0, 0)$, $(2, f(2)) = \left(2, 2\sqrt[3]{2}\right)$ and $(4, 0)$. Its length is

$\sqrt{(2-0)^2 + \left(2\sqrt[3]{2} - 0\right)^2} + \sqrt{(4-2)^2 + \left(0 - 2\sqrt[3]{2}\right)^2} = 2\sqrt{4 + 2^{8/3}} \approx 6.43$. Similarly, the inscribed polygon with four sides joins the points $(0, 0)$, $\left(1, \sqrt[3]{3}\right)$, $\left(2, 2\sqrt[3]{2}\right)$, $(3, 3)$, and $(4, 0)$, so its length is

$\sqrt{1 + \left(\sqrt[3]{3}\right)^2} + \sqrt{1 + \left(2\sqrt[3]{2} - \sqrt[3]{3}\right)^2} + \sqrt{1 + \left(3 - 2\sqrt[3]{2}\right)^2} + \sqrt{1 + 9} \approx 7.50.$

(c) Using the arc length formula with $\dfrac{dy}{dx} = x\left[\frac{1}{3}(4-x)^{-2/3}(-1)\right] + \sqrt[3]{4-x} = \dfrac{12 - 4x}{3(4-x)^{2/3}}$, the length of the curve is $L = \displaystyle\int_0^4 \sqrt{1 + \left(\frac{dy}{dx}\right)^2}\, dx = \int_0^4 \sqrt{1 + \left[\frac{12 - 4x}{3(4-x)^{2/3}}\right]^2}\, dx.$

(d) According to a CAS, the length of the curve is $L \approx 7.7988$. The actual value is larger than any of the approximations in part (b). This is always true, since any approximating straight line between two points on the curve is shorter than the length of the curve between the two points.

26. **(a)**

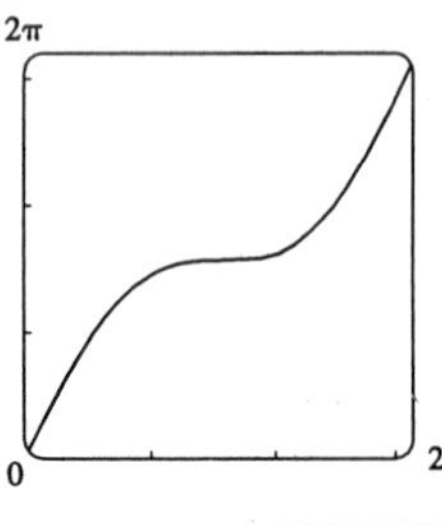

(b)

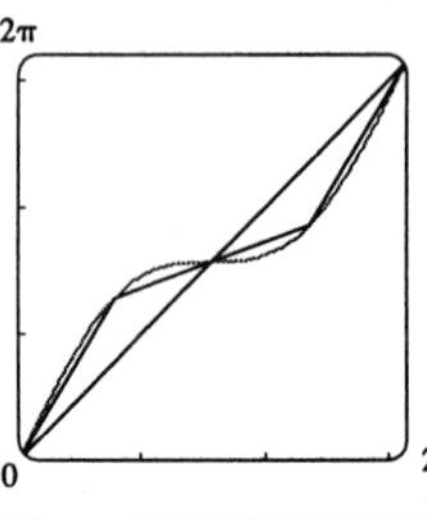

Let $f(x) = y = x + \sin x$. The polygon with one side is just the line segment joining the points $(0, f(0)) = (0, 0)$ and $(2\pi, f(2\pi)) = (2\pi, 2\pi)$, and its length is $\sqrt{(2\pi - 0)^2 + (2\pi - 0)^2} = 2\sqrt{2}\pi \approx 8.9$. The polygon with two sides joins the points $(0, 0)$, $(\pi, f(\pi)) = (\pi, \pi)$, and $(2\pi, 2\pi)$.

Its length is $\sqrt{(\pi - 0)^2 + (\pi - 0)^2} + \sqrt{(2\pi - \pi)^2 + (2\pi - \pi)^2} = \sqrt{2}\pi + \sqrt{2}\pi = 2\sqrt{2}\pi \approx 8.9$. Note from the diagram that the two approximations are the same because the sides of the 2-sided polygon are in fact on the same line, since $f(\pi) = \pi = \frac{1}{2}f(2\pi)$. The four-sided polygon joins the points $(0, 0)$, $\left(\frac{\pi}{2}, \frac{\pi}{2} + 1\right)$, (π, π), $\left(\frac{3\pi}{2}, \frac{3\pi}{2} - 1\right)$, and $(2\pi, 2\pi)$, so its length is

$\sqrt{\left(\frac{\pi}{2}\right)^2 + \left(\frac{\pi}{2} + 1\right)^2} + \sqrt{\left(\frac{\pi}{2}\right)^2 + \left(\frac{\pi}{2} - 1\right)^2} + \sqrt{\left(\frac{\pi}{2}\right)^2 + \left(\frac{\pi}{2} - 1\right)^2} + \sqrt{\left(\frac{\pi}{2}\right)^2 + \left(\frac{\pi}{2} + 1\right)^2} \approx 9.4.$

(c) Using the arc length formula with $dy/dx = 1 + \cos x$, the length of the curve is

$L = \int_0^{2\pi} \sqrt{1 + (1 + \cos x)^2}\, dx = \int_0^{2\pi} \sqrt{2 + 2\cos x + \cos^2 x}\, dx.$

(d) Maple approximates the integral as 9.5076. The actual length is larger than the approximations in part (b).

27. $y = 2x^{3/2} \Rightarrow y' = 3x^{1/2} \Rightarrow 1 + (y')^2 = 1 + 9x$. The arc length function with starting point $P_0(1, 2)$ is $s(x) = \int_1^x \sqrt{1+9t}\,dt = \left[\frac{2}{27}(1+9t)^{3/2}\right]_1^x = \frac{2}{27}\left[(1+9x)^{3/2} - 10\sqrt{10}\right]$.

28. **(a)**

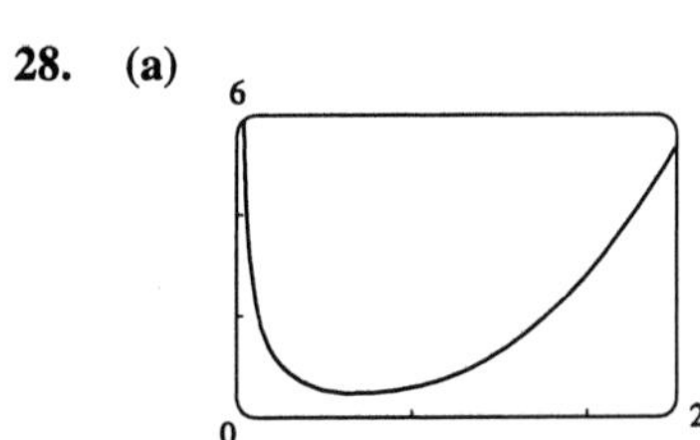

(b) $1 + \left(\frac{dy}{dx}\right)^2 = x^4 + \frac{1}{2} + \frac{1}{16x^4}$,

$$\begin{aligned} s(x) &= \int_1^x [t^2 + 1/(4t^2)]dt \\ &= \left[\tfrac{1}{3}t^3 - 1/(4t)\right]_1^x \\ &= \tfrac{1}{3}x^3 - 1/(4x) - \left(\tfrac{1}{3} - \tfrac{1}{4}\right) \\ &= \tfrac{1}{3}x^3 - 1/(4x) - \tfrac{1}{12} \text{ for } x > 0. \end{aligned}$$

(c)

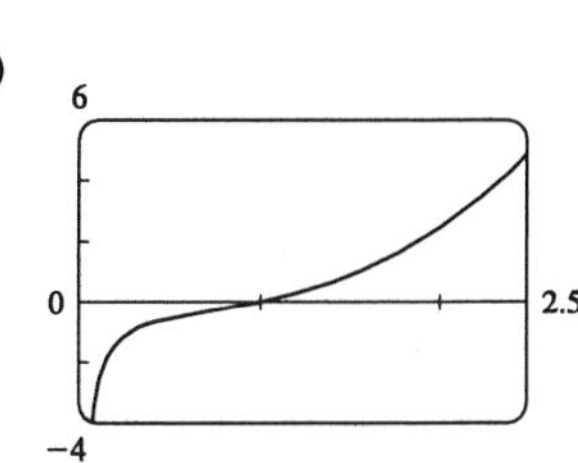

29. $y^{2/3} = 1 - x^{2/3} \Rightarrow y = \left(1 - x^{2/3}\right)^{3/2} \Rightarrow$

$\frac{dy}{dx} = \frac{3}{2}\left(1 - x^{2/3}\right)^{1/2}\left(-\frac{2}{3}x^{-1/3}\right) = -x^{-1/3}\left(1 - x^{2/3}\right)^{1/2} \Rightarrow$

$\left(\frac{dy}{dx}\right)^2 = x^{-2/3}\left(1 - x^{2/3}\right) = x^{-2/3} - 1$. Thus

$L = 4\int_0^1 \sqrt{1 + (x^{-2/3} - 1)}\,dx = 4\int_0^1 x^{-1/3}\,dx = 4\lim_{t\to 0^+}\left[\frac{3}{2}x^{2/3}\right]_t^1 = 6.$

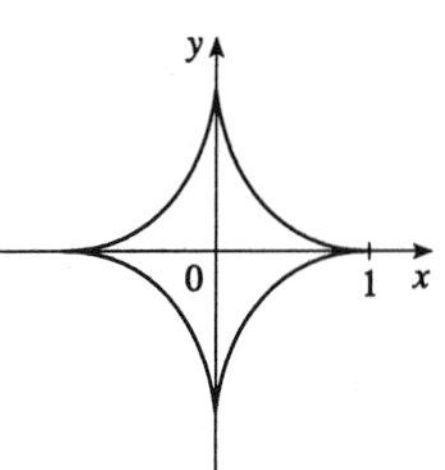

30. **(a)**

(b) $y = x^{2/3} \Rightarrow 1 + \left(\frac{dy}{dx}\right)^2 = 1 + \left(\frac{2}{3}x^{-1/3}\right)^2 = 1 + \frac{4}{9}x^{-2/3}$.

So $L = \int_0^1 \sqrt{1 + \frac{4}{9}x^{-2/3}}\,dx$ (an improper integral).

$x = y^{3/2} \Rightarrow 1 + \left(\frac{dx}{dy}\right)^2 = 1 + \left(\frac{3}{2}y^{1/2}\right)^2 = 1 + \frac{9}{4}y$.

So $L = \int_0^1 \sqrt{1 + \frac{9}{4}y}\,dy$. The second integral equals $\frac{4}{9}\cdot\frac{2}{3}\left[\left(1 + \frac{9}{4}y\right)^{3/2}\right]_0^1 = \frac{8}{27}\left(\frac{13\sqrt{13}}{8} - 1\right) = \frac{13\sqrt{13} - 8}{27}$.

The first integral can be evaluated as follows:

$$\int_0^1 \sqrt{1 + \tfrac{4}{9}x^{-2/3}}\,dx = \int_0^1 \frac{\sqrt{9x^{2/3} + 4}}{3x^{1/3}}\,dx = \int_0^9 \frac{\sqrt{u+4}}{18}\,du \quad \left(\text{put } u = 9x^{2/3} \Rightarrow du = 6x^{-1/3}\,dx\right)$$

$$= \lim_{t\to 0^+} \tfrac{1}{18}\cdot\left[\tfrac{2}{3}(u+4)^{3/2}\right]_t^9 = \tfrac{13\sqrt{13} - 8}{27}.$$

(c) $L = \int_0^1 \sqrt{1 + \frac{9}{4}y}\,dy + \int_0^8 \sqrt{1 + \frac{9}{4}y}\,dy = \frac{13\sqrt{13} - 8}{27} + \frac{8}{27}\left[\left(1 + \frac{9}{4}y\right)^{3/2}\right]_0^8$

$= \frac{13\sqrt{13} - 8}{27} + \frac{8}{27}\left(19\sqrt{19} - 1\right) = \frac{13\sqrt{13} + 152\sqrt{19} - 16}{27}$.

31. $y = 4500 - \frac{x^2}{8000} \Rightarrow \frac{dy}{dx} = -\frac{x}{4000} \Rightarrow \left(\frac{dy}{dx}\right)^2 = \frac{x^2}{16{,}000{,}000}$. When $y = 4500$ m, $x = 0$ m. When $y = 0$ m, $x = 6000$ m. Therefore

$L = \int_0^{6000} \sqrt{1 + (x/4000)^2}\,dx = \int_0^{3/2} \sqrt{1 + u^2}\,4000\,du$ (put $u = \frac{1}{4000}x$)

$= 4000\left[\frac{1}{2}u\sqrt{1+u^2} + \frac{1}{2}\ln\left(u + \sqrt{1+u^2}\right)\right]_0^{3/2}$ (Formula 21 or $u = \tan\theta$)

$= 4000\left[\frac{3}{4}\sqrt{\frac{13}{4}} + \frac{1}{2}\ln\left(\frac{3}{2} + \sqrt{\frac{13}{4}}\right)\right] = 1500\sqrt{13} + 2000\ln\frac{3+\sqrt{13}}{2} \approx 7798$ m.

32. **(a)** $y = a\cosh\left(\dfrac{x}{a}\right) \Rightarrow y' = \sinh\left(\dfrac{x}{a}\right) \Rightarrow 1 + (y')^2 = 1 + \sinh^2\left(\dfrac{x}{a}\right) = \cosh^2\left(\dfrac{x}{a}\right)$.

So $L = \displaystyle\int_{-b}^{b} \cosh\left(\frac{x}{a}\right) dx = \left[a\sinh\left(\frac{x}{a}\right)\right]_{-b}^{b} = 2a\sinh\left(\frac{b}{a}\right)$.

(b) From part (a), we have an expression involving the distance between the poles ($2b$), the height of the lowest point of the wire (a), and the length of the wire (L). For this problem, we substitute $L = 56$ and $b = \frac{1}{2}(50) = 25$ into that expression: $56 = 2a\sinh(25/a)$.

It would be difficult to solve this equation exactly, so we use a machine to graph the line $y = 56$ and the curve $y = 2x\sinh(25/x)$, and estimate the x-coordinate of the point of intersection.

From the graph, it appears that the root of the equation is $x \approx 30$ ft. So the lowest point of the wire is about 30 ft above the ground.

33. The sine wave has amplitude 1 and period 14, since it goes through two periods in a distance of 28 in., so its equation is $y = \sin\left(\frac{2\pi}{14}x\right) = \sin\left(\frac{\pi}{7}x\right)$. The width w of the flat metal sheet needed to make the panel is the arc length of the sine curve from $x = 0$ to $x = 28$. We set up the integral to evaluate w using the arc length formula with $\dfrac{dy}{dx} = \frac{\pi}{7}\cos\left(\frac{\pi}{7}x\right)$: $L = \displaystyle\int_0^{28} \sqrt{1 + \left[\tfrac{\pi}{7}\cos\left(\tfrac{\pi}{7}x\right)\right]^2}\,dx = 2\int_0^{14} \sqrt{1 + \left[\tfrac{\pi}{7}\cos\left(\tfrac{\pi}{7}x\right)\right]^2}\,dx$. This integral would be very difficult to evaluate exactly, so we use a CAS, and find that $L \approx 29.36$ inches.

34.

By symmetry, the length of the curve in each quadrant is the same, so we find the length in the first quadrant and multiply by 4.

$x^{2k} + y^{2k} = 1 \Rightarrow y^{2k} = 1 - x^{2k} \Rightarrow y = \left(1 - x^{2k}\right)^{1/(2k)}$

(in the first quadrant), so we use the arc length formula with

$$\frac{dy}{dx} = \frac{1}{2k}\left(1 - x^{2k}\right)^{1/(2k)-1}\left(-2kx^{2k-1}\right) = -x^{2k-1}\left(1 - x^{2k}\right)^{1/(2k)-1}.$$

The total length is therefore

$$L_{2k} = 4\int_0^1 \sqrt{1 + \left[-x^{2k-1}(1 - x^{2k})^{1/(2k)-1}\right]^2}\,dx$$
$$= 4\int_0^1 \sqrt{1 + x^{2(2k-1)}(1 - x^{2k})^{1/k-2}}\,dx$$

Now from the graph, we see that as k increases, the "corners" of these fat circles get closer to the points $(\pm1, \pm1)$, and the "edges" of the fat circles approach the lines joining these four points. It seems plausible that as $k \to \infty$, the total length of the fat circle with $n = 2k$ will approach the length of the perimeter of the square with sides of length 2. This is supported by taking the limit as $k \to \infty$ of the equation of the fat circle in the first quadrant: $\lim\limits_{k\to\infty}\left(1 - x^{2k}\right)^{1/(2k)} = 1$ for $0 \le x < 1$. So we guess that $\lim\limits_{k\to\infty} L_{2k} = 4 \cdot 2 = 8$.

35. $y = \displaystyle\int_1^x \sqrt{t^3 - 1}\,dt \Rightarrow \frac{dy}{dx} = \sqrt{x^3 - 1}$ (by the Fundamental Theorem, Part 1) $\Rightarrow 1 + \left(\dfrac{dy}{dx}\right)^2 = x^3$

$\Rightarrow L = \int_1^4 x^{3/2}\,dx = \frac{2}{5}\left[x^{5/2}\right]_1^4 = \frac{2}{5}(32 - 1) = \frac{62}{5} = 12.4$.

36. **(a)** From Formula 21, $\int_0^1 \sqrt{1+4x^2}\,dx = 2\int_0^1 \sqrt{\frac{1}{4}+x^2}\,dx = \left[2 \cdot \frac{1}{2}x\sqrt{\frac{1}{4}+x^2} + 2 \cdot \frac{1/4}{2}\ln\left|x+\sqrt{\frac{1}{4}+x^2}\right|\right]_0^1$

$$= \tfrac{\sqrt{5}}{2} + \tfrac{1}{4}\ln\left(1+\tfrac{\sqrt{5}}{2}\right) - \tfrac{1}{4}\ln\left(\tfrac{1}{2}\right) = \tfrac{\sqrt{5}}{2} + \tfrac{1}{4}\ln\left(2+\sqrt{5}\right).$$

Or: Substitute $x = \frac{1}{2}\tan\theta$.

(b) From the arc length formula, one recognizes the above integral to be the length of a curve from $x = 0$ to $x = 1$, with $(dy/dx)^2 = 4x^2 \quad\Rightarrow\quad dy/dx = \pm 2x$. Therefore the equation of the curve is $y = \pm x^2 + C$, and hence the integral gives the length of a segment of a parabola.

(c) For convenience take $y = x^2$ as the parabola being investigated in (b). Then $y(0) = 0$ and $y(1) = 1$. A straight line joining these points has length $\sqrt{1^2+1^2} = \sqrt{2}$, by the distance formula. Since we know that the shortest distance between two points is a straight line, $\sqrt{2} < \frac{\sqrt{5}}{2} + \frac{1}{4}\ln\left(2+\sqrt{5}\right)$ must be true.

EXERCISES 8.3

1. $y = \sqrt{x} \quad\Rightarrow\quad 1 + \left(\dfrac{dy}{dx}\right)^2 = 1 + \left(\dfrac{1}{2\sqrt{x}}\right)^2 = 1 + \dfrac{1}{4x}$. So

$$S = \int_4^9 2\pi y\sqrt{1+\left(\frac{dy}{dx}\right)^2}\,dx = \int_4^9 2\pi\sqrt{x}\sqrt{1+\frac{1}{4x}}\,dx = 2\pi\int_4^9 \sqrt{x+\tfrac{1}{4}}\,dx$$
$$= 2\pi\left[\tfrac{2}{3}\left(x+\tfrac{1}{4}\right)^{3/2}\right]_4^9 = \tfrac{4\pi}{3}\left[\tfrac{1}{8}(4x+1)^{3/2}\right]_4^9 = \tfrac{\pi}{6}\left(37\sqrt{37}-17\sqrt{17}\right).$$

2. The curve $y^2 = 4x+4$ is symmetric about the x-axis, which is the axis of rotation, so we need only consider the upper half of the curve, given by $y = \sqrt{4x+4} = 2\sqrt{x+1}$.

$\dfrac{dy}{dx} = \dfrac{1}{\sqrt{x+1}} \quad\Rightarrow\quad \sqrt{1+\left(\dfrac{dy}{dx}\right)^2} = \sqrt{1+\dfrac{1}{x+1}}$. So

$$S = 2\pi\int_0^8 2\sqrt{x+1}\sqrt{1+\frac{1}{x+1}}\,dx = 4\pi\int_0^8 \sqrt{x+2}\,dx = 4\pi\left[\tfrac{2}{3}(x+2)^{3/2}\right]_0^8 = \tfrac{8\pi}{3}\left(10\sqrt{10}-2\sqrt{2}\right).$$

Another Method: Use $S = \displaystyle\int_2^6 2\pi y\sqrt{1+\left(\frac{dx}{dy}\right)^2}\,dy$, where $x = \dfrac{y^2}{4} - 1$.

3. $y = x^3 \quad\Rightarrow\quad y' = 3x^2$. So $S = \displaystyle\int_0^2 2\pi y\sqrt{1+(y')^2}\,dx = 2\pi\int_0^2 x^3\sqrt{1+9x^4}\,dx$ (Let $u = 1+9x^4$, so $du = 36x^3\,dx$) $= \frac{2\pi}{36}\int_1^{145}\sqrt{u}\,du = \frac{\pi}{18}\left[\frac{2}{3}u^{3/2}\right]_1^{145} = \frac{\pi}{27}\left(145\sqrt{145}-1\right)$.

4. $y = \dfrac{x^2}{4} - \dfrac{\ln x}{2} \quad\Rightarrow\quad \dfrac{dy}{dx} = \dfrac{x}{2} - \dfrac{1}{2x} \quad\Rightarrow\quad 1 + \left(\dfrac{dy}{dx}\right)^2 = \dfrac{x^2}{4} + \dfrac{1}{2} + \dfrac{1}{4x^2}$. So

$$S = 2\pi \int_1^4 \left(\frac{x^2}{4} - \frac{\ln x}{2}\right)\left(\frac{x}{2} + \frac{1}{2x}\right)dx = \frac{\pi}{2}\int_1^4 \left(\frac{x^2}{2} - \ln x\right)\left(x + \frac{1}{x}\right)dx$$
$$= \frac{\pi}{2}\int_1^4 \left(\frac{x^3}{2} + \frac{x}{2} - x\ln x - \frac{\ln x}{x}\right)dx = \frac{\pi}{2}\left[\frac{x^4}{8} + \frac{x^2}{4} - \frac{x^2}{2}\ln x + \frac{x^2}{4} - \tfrac{1}{2}(\ln x)^2\right]_1^4$$
$$= \tfrac{\pi}{2}\left[\left(32 + 4 - 8\ln 4 + 4 - \tfrac{1}{2}(\ln 4)^2\right) - \left(\tfrac{1}{8} + \tfrac{1}{4} - 0 + \tfrac{1}{4} - 0\right)\right] = \pi\left[\tfrac{315}{16} - 8\ln 2 - (\ln 2)^2\right].$$

5. $y = \sin x \quad\Rightarrow\quad 1 + \left(\dfrac{dy}{dx}\right)^2 = 1 + \cos^2 x$. So

$$S = 2\pi\textstyle\int_0^\pi \sin x\sqrt{1+\cos^2 x}\,dx = 2\pi\int_{-1}^1 \sqrt{1+u^2}\,du \quad (\text{put } u = -\cos x \Rightarrow du = \sin x\,dx)$$
$$= 4\pi\textstyle\int_0^1 \sqrt{1+u^2}\,du = 4\pi\int_0^{\pi/4}\sec^3\theta\,d\theta \quad (\text{put } u = \tan\theta \quad\Rightarrow\quad du = \sec^2\theta\,d\theta)$$
$$= 2\pi[\sec\theta\tan\theta + \ln|\sec\theta + \tan\theta|]_0^{\pi/4} = 2\pi\left[\sqrt{2} + \ln\left(\sqrt{2}+1\right)\right].$$

6. $y = \cos x \quad\Rightarrow\quad \sqrt{1 + \left(\dfrac{dy}{dx}\right)^2} = \sqrt{1+\sin^2 x}$. So

$$S = 2\pi\textstyle\int_0^{\pi/3}\cos x\sqrt{1+\sin^2 x}\,dx = 2\pi\int_0^{\sqrt{3}/2}\sqrt{1+u^2}\,du \quad (\text{put } u = \sin x)$$
$$= 2\pi\textstyle\int_0^\alpha \sec^3\theta\,d\theta \quad (\text{put } u = \tan\theta,\ du = \sec^2\theta\,d\theta)$$
$$= \pi[\sec\theta\tan\theta + \ln|\sec\theta + \tan\theta|]_0^\alpha$$
$$= \pi\left[\tfrac{\sqrt{7}}{2}\cdot\tfrac{\sqrt{3}}{2} + \ln\left(\tfrac{\sqrt{7}}{2} + \tfrac{\sqrt{3}}{2}\right)\right] = \pi\left[\tfrac{\sqrt{21}}{4} + \ln\left(\tfrac{\sqrt{7}+\sqrt{3}}{2}\right)\right].$$

$\alpha = \tan^{-1}\left(\frac{\sqrt{3}}{2}\right)$

7. $y = \cosh x \quad\Rightarrow\quad 1 + \left(\dfrac{dy}{dx}\right)^2 = 1 + \sinh^2 x = \cosh^2 x$. So $S = 2\pi\int_0^1 \cosh x\cosh x\,dx$

$$= 2\pi\textstyle\int_0^1 \tfrac{1}{2}(1+\cosh 2x)dx = \pi\left[x + \tfrac{1}{2}\sinh 2x\right]_0^1 = \pi\left(1 + \tfrac{1}{2}\sinh 2\right) \text{ or } \pi\left[1 + \tfrac{1}{4}(e^2 - e^{-2})\right].$$

8. $2y = 3x^{2/3},\ y = \tfrac{3}{2}x^{2/3} \quad\Rightarrow\quad \dfrac{dy}{dx} = x^{-1/3} \quad\Rightarrow\quad 1 + \left(\dfrac{dy}{dx}\right)^2 = 1 + x^{-2/3}$. So

$$S = 2\pi\int_1^8 \tfrac{3}{2}x^{2/3}\sqrt{1+x^{-2/3}}\,dx = 3\pi\int_1^2 u^2\sqrt{1+1/u^2}\,3u^2\,du \quad (\text{put } u = x^{1/3},\ x = u^3,\ dx = 3u^2\,du)$$
$$= 9\pi\textstyle\int_1^2 u^3\sqrt{u^2+1}\,du = 9\pi\int_{\pi/4}^{\tan^{-1}2}\tan^3\theta\sec^3\theta\,d\theta \quad (\text{put } u = \tan\theta,\ du = \sec^2\theta\,d\theta)$$
$$= 9\pi\textstyle\int_{\pi/4}^{\tan^{-1}2}\sec^2\theta(\sec^2\theta - 1)\sec\theta\tan\theta\,d\theta = 9\pi\int_{\sqrt{2}}^{\sqrt{5}} v^2(v^2-1)dv \quad (\text{put } v = \sec\theta)$$
$$= 9\pi\left[\tfrac{1}{5}v^5 - \tfrac{1}{3}v^3\right]_{\sqrt{2}}^{\sqrt{5}} = 9\pi\left[5\sqrt{5} - \tfrac{5\sqrt{5}}{3} - \tfrac{4\sqrt{2}}{5} + \tfrac{2\sqrt{2}}{3}\right] = \tfrac{3\pi}{5}\left(50\sqrt{5} - 2\sqrt{2}\right).$$

9. $x = \tfrac{1}{3}(y^2+2)^{3/2} \quad\Rightarrow\quad dx/dy = \tfrac{1}{2}(y^2+2)^{1/2}(2y) = y\sqrt{y^2+2} \quad\Rightarrow$
$1 + (dx/dy)^2 = 1 + y^2(y^2+2) = (y^2+1)^2$. So
$S = 2\pi\int_1^2 y(y^2+1)dy = 2\pi\left[\tfrac{1}{4}y^4 + \tfrac{1}{2}y^2\right]_1^2 = 2\pi\left(4 + 2 - \tfrac{1}{4} - \tfrac{1}{2}\right) = \tfrac{21\pi}{2}$.

10. $x = 1 + 2y^2 \quad\Rightarrow\quad 1 + (dx/dy)^2 = 1 + (4y)^2 = 1 + 16y^2$. So $S = 2\pi\int_1^2 y\sqrt{1+16y^2}\,dy$

$$= \tfrac{\pi}{16}\textstyle\int_1^2 (16y^2+1)^{1/2}32y\,dy = \tfrac{\pi}{16}\left[\tfrac{2}{3}(16y^2+1)^{3/2}\right]_1^2 = \tfrac{\pi}{24}\left(65\sqrt{65} - 17\sqrt{17}\right).$$

11. $y = \sqrt[3]{x} \quad \Rightarrow \quad x = y^3 \quad \Rightarrow \quad 1 + (dx/dy)^2 = 1 + 9y^4$. So

$$S = 2\pi \int_1^2 x\sqrt{1 + (dx/dy)^2}\, dy = 2\pi \int_1^2 y^3 \sqrt{1 + 9y^4}\, dy = \frac{2\pi}{36} \int_1^2 \sqrt{1 + 9y^4}\, 36y^3\, dy$$

$$= \tfrac{\pi}{18}\left[\tfrac{2}{3}\left(1 + 9y^4\right)^{3/2}\right]_1^2 = \tfrac{\pi}{27}\left(145\sqrt{145} - 10\sqrt{10}\right).$$

12. $x = \sqrt{2y - y^2} \quad \Rightarrow \quad \dfrac{dx}{dy} = \dfrac{1 - y}{\sqrt{2y - y^2}} \quad \Rightarrow \quad 1 + \left(\dfrac{dx}{dy}\right)^2 = 1 + \dfrac{1 - 2y + y^2}{2y - y^2} = \dfrac{1}{2y - y^2}$. So

$$S = 2\pi \int_0^1 \sqrt{2y - y^2}\left(\frac{1}{\sqrt{2y - y^2}}\right) dy = 2\pi \textstyle\int_0^1 dy = 2\pi.$$

13. $x = e^{2y} \quad \Rightarrow \quad 1 + (dx/dy)^2 = 1 + 4e^{4y}$. So

$$S = 2\pi \int_0^{1/2} e^{2y}\sqrt{1 + (2e^{2y})^2}\, dy = 2\pi \textstyle\int_2^{2e} \sqrt{1 + u^2}\, \frac{1}{4}\, du \quad [\text{put } u = 2e^{2y},\ du = 4e^{2y}\, dy]$$

$$= \tfrac{\pi}{2}\textstyle\int_2^{2e} \sqrt{1 + u^2}\, du = \tfrac{\pi}{2}\left[\tfrac{1}{2}u\sqrt{1 + u^2} + \tfrac{1}{2}\ln\left|u + \sqrt{1 + u^2}\right|\right]_2^{2e} \quad [\text{put } u = \tan\theta \text{ or use Formula 21}]$$

$$= \tfrac{\pi}{2}\left[e\sqrt{1 + 4e^2} + \tfrac{1}{2}\ln\left(2e + \sqrt{1 + 4e^2}\right) - \sqrt{5} - \tfrac{1}{2}\ln\left(2 + \sqrt{5}\right)\right]$$

$$= \frac{\pi}{4}\left[2e\sqrt{1 + 4e^2} - 2\sqrt{5} + \ln\left(\frac{2e + \sqrt{1 + 4e^2}}{2 + \sqrt{5}}\right)\right].$$

14. $y = 1 - x^2 \quad \Rightarrow \quad 1 + (dy/dx)^2 = 1 + 4x^2 \quad \Rightarrow$

$$S = 2\pi \textstyle\int_0^1 x\sqrt{1 + 4x^2}\, dx = \frac{\pi}{4}\int_0^1 8x\sqrt{4x^2 + 1}\, dx = \frac{\pi}{4}\left[\frac{2}{3}(4x^2 + 1)^{3/2}\right]_0^1 = \frac{\pi}{6}\left(5\sqrt{5} - 1\right).$$

15. $x = \frac{1}{2\sqrt{2}}(y^2 - \ln y) \quad \Rightarrow \quad \dfrac{dx}{dy} = \dfrac{1}{2\sqrt{2}}\left(2y - \dfrac{1}{y}\right) \quad \Rightarrow \quad 1 + \left(\dfrac{dx}{dy}\right)^2 = 1 + \dfrac{1}{8}\left(2y - \dfrac{1}{y}\right)^2$

$$= 1 + \frac{1}{8}\left(4y^2 - 4 + \frac{1}{y^2}\right) = \frac{1}{8}\left(4y^2 + 4 + \frac{1}{y^2}\right) = \left[\frac{1}{2\sqrt{2}}\left(2y + \frac{1}{y}\right)\right]^2. \text{ So}$$

$$S = 2\pi \int_1^2 \frac{1}{2\sqrt{2}}(y^2 - \ln y)\frac{1}{2\sqrt{2}}\left(2y + \frac{1}{y}\right) dy = \frac{\pi}{4}\int_1^2 \left(2y^3 + y - 2y\ln y - \frac{\ln y}{y}\right) dy$$

$$= \tfrac{\pi}{4}\left[\tfrac{1}{2}y^4 + \tfrac{1}{2}y^2 - y^2\ln y + \tfrac{1}{2}y^2 - \tfrac{1}{2}(\ln y)^2\right]_1^2 = \tfrac{\pi}{8}\left[y^4 + 2y^2 - 2y^2\ln y - (\ln y)^2\right]_1^2$$

$$= \tfrac{\pi}{8}\left[16 + 8 - 8\ln 2 - (\ln 2)^2 - 1 - 2\right] = \tfrac{\pi}{8}\left[21 - 8\ln 2 - (\ln 2)^2\right].$$

16. $x = a\cosh(y/a) \quad \Rightarrow \quad 1 + (dx/dy)^2 = 1 + \sinh^2(y/a) = \cosh^2(y/a)$. So

$$S = 2\pi \int_{-a}^{a} a\cosh\left(\frac{y}{a}\right)\cosh\left(\frac{y}{a}\right) dy = 4\pi a \int_0^a \cosh^2\left(\frac{y}{a}\right) dy = 2\pi a \int_0^a \left[1 + \cosh\left(\frac{2y}{a}\right)\right] dy$$

$$= 2\pi a\left[y + \frac{a}{2}\sinh\left(\frac{2y}{a}\right)\right]_0^a = 2\pi a\left[a + \frac{a}{2}\sinh 2\right] = 2\pi a^2\left[1 + \tfrac{1}{2}\sinh 2\right] \text{ or } \frac{\pi a^2(e^2 + 4 - e^{-2})}{2}.$$

17. $S = 2\pi \int_0^1 x^4\sqrt{1 + (4x^3)^2}\, dx = 2\pi \int_0^1 x^4\sqrt{16x^6 + 1}\, dx$

$$\approx 2\pi\tfrac{1/10}{3}\left[f(0) + 4f(0.1) + 2f(0.2) + 4f(0.3) + 2f(0.4) + 4f(0.5) + 2f(0.6)\right.$$
$$\left.+ 4f(0.7) + 2f(0.8) + 4f(0.9) + f(1)\right] \approx 3.44. \text{ Here } f(x) = x^4\sqrt{16x^6 + 1}.$$

18. $S = 2\pi \int_0^{\pi/4} \tan x \sqrt{1 + (\sec^2 x)^2}\, dx$

$\approx 2\pi \frac{\pi/40}{3} \left[f(0) + 4f\left(\frac{\pi}{40}\right) + 2f\left(\frac{\pi}{20}\right) + 4f\left(\frac{3\pi}{40}\right) + 2f\left(\frac{\pi}{10}\right) \right.$
$\left. + 4f\left(\frac{\pi}{8}\right) + 2f\left(\frac{3\pi}{20}\right) + 4f\left(\frac{7\pi}{40}\right) + 2f\left(\frac{\pi}{5}\right) + 4f\left(\frac{9\pi}{40}\right) + f\left(\frac{\pi}{4}\right) \right] \approx 3.84.$

Here $f(x) = \tan x\, \sqrt{1 + \sec^4 x}$.

19. The curve $8y^2 = x^2(1 - x^2)$ actually consists of two loops in the region described by the inequalities $|x| \le 1$, $|y| \le \frac{\sqrt{2}}{8}$. $\left(\text{The maximum value of } |y| \text{ is attained when } |x| = \frac{1}{\sqrt{2}}.\right)$ If we consider the loop in the region $x \ge 0$, the surface area S it generates when rotated about the x-axis is calculated as follows: $16y\frac{dy}{dx} = 2x - 4x^3$, so $\left(\frac{dy}{dx}\right)^2 = \left(\frac{x - 2x^3}{8y}\right)^2 = \frac{x^2(1 - 2x^2)^2}{64y^2} = \frac{x^2(1 - 2x^2)^2}{8x^2(1 - x^2)} = \frac{(1 - 2x^2)^2}{8(1 - x^2)}$ for $x \ne 0, \pm 1$. The formula also holds for $x = 0$ by continuity. $1 + \left(\frac{dy}{dx}\right)^2 = 1 + \frac{(1 - 2x^2)^2}{8(1 - x^2)} = \frac{9 - 12x^2 + 4x^4}{8(1 - x^2)} = \frac{(3 - 2x^2)^2}{8(1 - x^2)}$. So

$$S = 2\pi \int_0^1 \frac{\sqrt{x^2(1 - x^2)}}{2\sqrt{2}} \cdot \frac{3 - 2x^2}{2\sqrt{2}\sqrt{1 - x^2}}\, dx = \frac{\pi}{4} \int_0^1 x(3 - 2x^2)dx = \frac{\pi}{4}\left[\frac{3}{2}x^2 - \frac{1}{2}x^4\right]_0^1 = \frac{\pi}{4}\left(\frac{3}{2} - \frac{1}{2}\right) = \frac{\pi}{4}.$$

20. $S = 2\pi \int_0^\infty y\sqrt{1 + (dy/dx)^2}\, dx = 2\pi \int_0^\infty e^{-x}\sqrt{1 + e^{-2x}}\, dx$

$= 2\pi \int_0^1 \sqrt{1 + u^2}\, du$ [where $u = e^{-x}$, $du = -e^{-x}\, dx$]

$= 2\pi \left[\frac{1}{2}u\sqrt{1 + u^2} + \frac{1}{2}\ln\left|u + \sqrt{1 + u^2}\right|\right]_0^1$ [Now put $u = \tan\theta$ or $u = \sinh t$, or use Formula 21]

$= 2\pi\left[\frac{1}{2}\sqrt{2} + \frac{1}{2}\ln\left(1 + \sqrt{2}\right)\right] = \pi\left[\sqrt{2} + \ln\left(1 + \sqrt{2}\right)\right].$

21. $S = 2\pi \int_1^\infty y\sqrt{1 + \left(\frac{dy}{dx}\right)^2}\, dx = 2\pi \int_1^\infty \frac{1}{x}\sqrt{1 + \frac{1}{x^4}}\, dx = 2\pi \int_1^\infty \frac{\sqrt{x^4 + 1}}{x^3}\, dx$

$> 2\pi \int_1^\infty \frac{x^2}{x^3}\, dx = 2\pi \int_1^\infty \frac{dx}{x} = 2\pi \lim_{t\to\infty} [\ln x]_1^t = 2\pi \lim_{t\to\infty} \ln t = \infty.$

22. The upper half of the torus is generated by rotating the curve $(x - R)^2 + y^2 = r^2$, $y > 0$, about the y-axis. $y\frac{dy}{dx} = -(x - R) \;\Rightarrow\; 1 + \left(\frac{dy}{dx}\right)^2 = 1 + \frac{(x - R)^2}{y^2} = \frac{y^2 + (x - R)^2}{y^2} = \frac{r^2}{r^2 - (x - R)^2}$. Thus

$S = 2\int_{R-r}^{R+r} 2\pi x\sqrt{1 + \left(\frac{dy}{dx}\right)^2}\, dx = 4\pi \int_{R-r}^{R+r} \frac{rx}{\sqrt{r^2 - (x - R)^2}}\, dx = 4\pi r \int_{-r}^{r} \frac{u + R}{\sqrt{r^2 - u^2}}\, du$ (put $u = x - R$)

$= 4\pi r \int_{-r}^{r} \frac{u\, du}{\sqrt{r^2 - u^2}} + 4\pi Rr \int_{-r}^{r} \frac{du}{\sqrt{r^2 - u^2}}$

$= 4\pi r \cdot 0 + 8\pi Rr \int_0^r \frac{du}{\sqrt{r^2 - u^2}}$ [since the first integrand is odd and the second is even]

$= 8\pi Rr\left[\sin^{-1}(u/r)\right]_0^r = 8\pi Rr\left(\frac{\pi}{2}\right) = 4\pi^2 Rr.$

23. $\frac{x^2}{a^2} + \frac{y^2}{b^2} = 1 \quad \Rightarrow \quad \frac{y(dy/dx)}{b^2} = -\frac{x}{a^2} \quad \Rightarrow \quad \frac{dy}{dx} = -\frac{b^2 x}{a^2 y} \quad \Rightarrow$

$$1 + \left(\frac{dy}{dx}\right)^2 = 1 + \frac{b^4 x^2}{a^4 y^2} = \frac{b^4 x^2 + a^4 y^2}{a^4 y^2} = \frac{b^4 x^2 + a^4 b^2 (1 - x^2/a^2)}{a^4 b^2 (1 - x^2/a^2)}$$

$$= \frac{a^4 b^2 + b^4 x^2 - a^2 b^2 x^2}{a^4 b^2 - a^2 b^2 x^2} = \frac{a^4 + b^2 x^2 - a^2 x^2}{a^4 - a^2 x^2} = \frac{a^4 - (a^2 - b^2)x^2}{a^2(a^2 - x^2)}.$$

The ellipsoid's surface area is twice the area generated by rotating the first quadrant portion of the ellipse about the x-axis. Thus

$$S = 2\int_0^a 2\pi y \sqrt{1 + \left(\frac{dy}{dx}\right)^2}\, dx = 4\pi \int_0^a \frac{b}{a}\sqrt{a^2 - x^2}\,\frac{\sqrt{a^4 - (a^2 - b^2)x^2}}{a\sqrt{a^2 - x^2}}\, dx = \frac{4\pi b}{a^2}\int_0^a \sqrt{a^4 - (a^2 - b^2)x^2}\, dx$$

$$= \frac{4\pi b}{a^2}\int_0^{a\sqrt{a^2-b^2}} \sqrt{a^4 - u^2}\,\frac{du}{\sqrt{a^2 - b^2}} \quad \text{(put } u = \sqrt{a^2 - b^2}\, x\text{)}$$

$$= \frac{4\pi b}{a^2\sqrt{a^2 - b^2}}\left[\frac{u}{2}\sqrt{a^4 - u^2} + \frac{a^4}{2}\sin^{-1}\frac{u}{a^2}\right]_0^{a\sqrt{a^2-b^2}} \quad \text{(Formula 30)}$$

$$= \frac{4\pi b}{a^2\sqrt{a^2 - b^2}}\left[\frac{a\sqrt{a^2 - b^2}}{2}\sqrt{a^4 - a^2(a^2 - b^2)} + \frac{a^4}{2}\sin^{-1}\frac{\sqrt{a^2 - b^2}}{a}\right] = 2\pi\left[b^2 + \frac{a^2 b \sin^{-1}\dfrac{\sqrt{a^2 - b^2}}{a}}{\sqrt{a^2 - b^2}}\right]$$

24. Take the sphere $x^2 + y^2 + z^2 = \frac{1}{4}d^2$ and let the intersecting planes be $y = c$ and $y = c + h$, where $-\frac{1}{2}d \le c \le \frac{1}{2}d - h$. The sphere intersects the xy-plane in the circle $x^2 + y^2 = \frac{1}{4}d^2$. From this equation, we get $x\frac{dx}{dy} + y = 0$, so $\frac{dx}{dy} = -\frac{y}{x}$. The desired surface area is

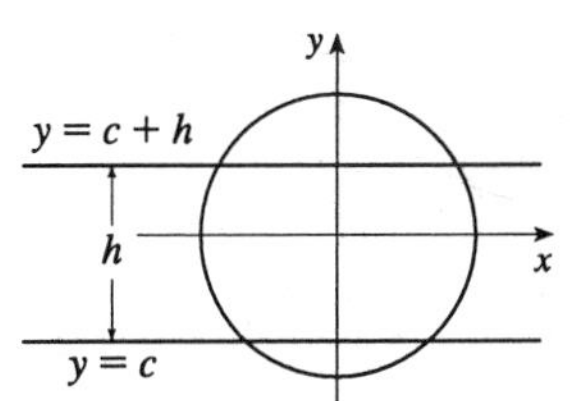

$$S = 2\pi\int x\, ds = 2\pi\int_c^{c+h} x\sqrt{1 + (dx/dy)^2}\, dy = 2\pi\int_c^{c+h} x\sqrt{1 + y^2/x^2}\, dy = 2\pi\int_c^{c+h}\sqrt{x^2 + y^2}\, dy$$

$$= 2\pi\int_c^{c+h}\tfrac{1}{2}d\, dy = \pi d\int_c^{c+h} dy = \pi d h.$$

25. In the derivation of (4), we computed a typical contribution to the surface area to be $2\pi\frac{y_{i-1} + y_i}{2}|P_{i-1}P_i|$, the area of a frustum of a cone. When $f(x)$ is not necessarily positive, the approximations $y_i = f(x_i) \approx f(x_i^*)$ and $y_{i-1} = f(x_{i-1}) \approx f(x_i^*)$ must be replaced by $y_i = |f(x_i)| \approx |f(x_i^*)|$ and $y_{i-1} = |f(x_{i-1})| \approx |f(x_i^*)|$. Thus $2\pi\frac{y_{i-1} + y_i}{2}|P_{i-1}P_i| \approx 2\pi|f(x_i^*)|\sqrt{1 + [f'(x_i^*)]^2}\,\Delta x_i$. Continuing with the rest of the derivation as before, we obtain $S = \int_a^b 2\pi|f(x)|\sqrt{1 + [f'(x)]^2}\, dx$.

26. The analogue of $f(x_i^*)$ in the derivation of (4) is now $c - f(x_i^*)$, so

$$S = \lim_{\|P\|\to 0}\sum_{i=1}^{n} 2\pi[c - f(x_i^*)]\sqrt{1 + [f'(x_i^*)]^2}\,\Delta x_i = \int_a^b 2\pi[c - f(x)]\sqrt{1 + [f'(x)]^2}\, dx.$$

27. For the upper semicircle, $f(x)=\sqrt{r^2-x^2}$, $f'(x)=-x/\sqrt{r^2-x^2}$. The surface area generated is

$$S_1=\int_{-r}^{r}2\pi\left(r-\sqrt{r^2-x^2}\right)\sqrt{1+\frac{x^2}{r^2-x^2}}\,dx=4\pi\int_0^r\left(r-\sqrt{r^2-x^2}\right)\frac{r}{\sqrt{r^2-x^2}}dx$$
$$=4\pi\int_0^r\left(\frac{r^2}{\sqrt{r^2-x^2}}-r\right)dx.$$

For the lower semicircle, $f(x)=-\sqrt{r^2-x^2}$ and $f'(x)=\dfrac{x}{\sqrt{r^2-x^2}}$, so $S_2=4\pi\displaystyle\int_0^r\left(\frac{r^2}{\sqrt{r^2-x^2}}+r\right)dx$.

Thus the total area is $S=S_1+S_2=8\pi\displaystyle\int_0^r\left(\frac{r^2}{\sqrt{r^2-x^2}}\right)dx=8\pi\left[r^2\sin^{-1}\left(\frac{x}{r}\right)\right]_0^r=8\pi r^2\left(\frac{\pi}{2}\right)=4\pi^2r^2$.

28. Since $g(x)=f(x)+c$, we have $g'(x)=f'(x)$. Thus

$$S_g=\int_a^b 2\pi g(x)\sqrt{1+[g'(x)]^2}\,dx=\int_a^b 2\pi[f(x)+c]\sqrt{1+[f'(x)]^2}\,dx$$
$$=\int_a^b 2\pi f(x)\sqrt{1+[f'(x)]^2}\,dx+2\pi c\int_a^b\sqrt{1+[f'(x)]^2}\,dx=S_f+2\pi cL.$$

EXERCISES 8.4

1. $m_1=4$, $m_2=8$; $P_1(-1,2)$, $P_2(2,4)$. $m=m_1+m_2=12$. $M_x=4\cdot 2+8\cdot 4=40$;
$M_y=4\cdot(-1)+8\cdot 2=12$; $\overline{x}=M_y/m=1$ and $\overline{y}=M_x/m=\frac{10}{3}$, so the center of mass is $(\overline{x},\overline{y})=\left(1,\frac{10}{3}\right)$.

2. $M_x=2\cdot 1+3\cdot(-2)+1\cdot 4=0$; $M_y=2\cdot 5+3\cdot 3+1\cdot(-2)=17$, $m=2+3+1=6$, $(\overline{x},\overline{y})=\left(\frac{17}{6},0\right)$.

3. $m=m_1+m_2+m_3=4+2+5=11$. $M_x=4\cdot(-2)+2\cdot 4+5\cdot(-3)=-15$;
$M_y=4\cdot(-1)+2\cdot(-2)+5\cdot 5=17$, $(\overline{x},\overline{y})=\left(\frac{17}{11},-\frac{15}{11}\right)$.

4. $M_x=3\cdot 0+3\cdot 8+8\cdot(-4)+6\cdot(-5)=-38$; $M_y=3\cdot 0+3\cdot 1+8\cdot 3+6\cdot(-6)=-9$;
$m=3+3+8+6=20$, $(\overline{x},\overline{y})=\left(-\frac{9}{20},-\frac{19}{20}\right)$.

5. $A=\int_0^2 x^2\,dx=\left[\frac{1}{3}x^3\right]_0^2=\frac{8}{3}$, $\overline{x}=A^{-1}\int_0^2 x\cdot x^2\,dx=\frac{3}{8}\left[\frac{1}{4}x^4\right]_0^2=\frac{3}{8}\cdot 4=\frac{3}{2}$,
$\overline{y}=A^{-1}\int_0^2\frac{1}{2}(x^2)^2\,dx=\frac{3}{8}\cdot\frac{1}{2}\left[\frac{1}{5}x^5\right]_0^2=\frac{3}{16}\cdot\frac{32}{5}=\frac{6}{5}$. Centroid $(\overline{x},\overline{y})=\left(\frac{3}{2},\frac{6}{5}\right)=(1.5,1.2)$.

6. By symmetry, $\overline{x}=0$ and $A=2\int_0^1(1-x^2)dx=2\left[x-\frac{1}{3}x^3\right]_0^1=\frac{4}{3}$; $\overline{y}=A^{-1}\int_{-1}^1\frac{1}{2}(1-x^2)^2\,dx$
$=A^{-1}\int_0^1(1-x^2)^2\,dx=\frac{3}{4}\left[x-\frac{2}{3}x^3+\frac{1}{5}x^5\right]_0^1=\frac{3}{4}\left(1-\frac{2}{3}+\frac{1}{5}\right)=\frac{2}{5}$. $(\overline{x},\overline{y})=\left(0,\frac{2}{5}\right)$.

7. $A=\int_{-1}^2(3x+5)dx=\left[\frac{3}{2}x^2+5x\right]_{-1}^2=(6+10)-\left(\frac{3}{2}-5\right)=16+\frac{7}{2}=\frac{39}{2}$,
$\overline{x}=A^{-1}\int_{-1}^2 x(3x+5)dx=\frac{2}{39}\int_{-1}^2(3x^2+5x)dx=\frac{2}{39}\left[x^3+\frac{5}{2}x^2\right]_{-1}^2$
$=\frac{2}{39}\left[(8+10)-\left(-1+\frac{5}{2}\right)\right]=\frac{2}{39}\left(\frac{36-3}{2}\right)=\frac{11}{13}$,
$\overline{y}=A^{-1}\int_{-1}^2\frac{1}{2}(3x+5)^2\,dx=\frac{1}{39}\int_{-1}^2(9x^2+30x+25)dx=\frac{1}{39}[3x^3+15x^2+25x]_{-1}^2$
$=\frac{1}{39}[(24+60+50)-(-3+15-25)]=\frac{147}{39}=\frac{49}{13}$. $(\overline{x},\overline{y})=\left(\frac{11}{13},\frac{49}{13}\right)$.

8. $A = \int_0^4 \sqrt{x}\,dx = \left[\frac{2}{3}x^{3/2}\right]_0^4 = \frac{2}{3}\cdot 8 = \frac{16}{3}$,
$\overline{x} = A^{-1}\int_0^4 x\sqrt{x}\,dx = \frac{3}{16}\left[\frac{2}{5}x^{5/2}\right]_0^4 = \frac{3}{16}\cdot\frac{2}{5}\cdot 32 = \frac{12}{5}$,
$\overline{y} = A^{-1}\int_0^4 \frac{1}{2}\left(\sqrt{x}\right)^2 dx = \frac{3}{16}\cdot\frac{1}{2}\int_0^4 x\,dx = \frac{3}{32}\left[\frac{1}{2}x^2\right]_0^4 = \frac{3}{32}\cdot 8 = \frac{3}{4}$. $(\overline{x},\overline{y}) = \left(\frac{12}{5},\frac{3}{4}\right)$.

9. By symmetry, $\overline{x} = 0$ and $A = 2\int_0^{\pi/4}\cos 2x\,dx = \sin 2x\big|_0^{\pi/4} = 1$,
$\overline{y} = A^{-1}\int_{-\pi/4}^{\pi/4}\frac{1}{2}\cos^2 2x\,dx = \int_0^{\pi/4}\cos^2 2x\,dx = \frac{1}{2}\int_0^{\pi/4}(1+\cos 4x)dx = \frac{1}{2}\left[x + \frac{1}{4}\sin 4x\right]_0^{\pi/4}$
$= \frac{1}{2}\left(\frac{\pi}{4} + \frac{1}{4}\cdot 0\right) = \frac{\pi}{8}$. $(\overline{x},\overline{y}) = \left(0,\frac{\pi}{8}\right)$.

10. $A = \int_0^{\pi/2}\sin x\,dx = -\cos x\big|_0^{\pi/2} = 0-(-1) = 1$,
$\overline{x} = A^{-1}\int_0^{\pi/2} x\sin x\,dx = \int_0^{\pi/2} x\sin x\,dx = [-x\cos x + \sin x]_0^{\pi/2} = 1$,
$\overline{y} = A^{-1}\int_0^{\pi/2}\frac{1}{2}\sin^2 x\,dx = \frac{1}{4}\int_0^{\pi/2}(1-\cos 2x)dx = \frac{1}{4}\left[x - \frac{1}{2}\sin 2x\right]_0^{\pi/2} = \frac{1}{4}\cdot\frac{\pi}{2} = \frac{\pi}{8}$. $(\overline{x},\overline{y}) = \left(1,\frac{\pi}{8}\right)$.

11. $A = \int_0^1 e^x\,dx = [e^x]_0^1 = e-1$,

$$\overline{x} = \frac{1}{A}\int_0^1 xe^x\,dx = \frac{1}{e-1}[xe^x - e^x]_0^1 \quad \text{(integration by parts)} \quad = \frac{1}{e-1}[0-(-1)] = \frac{1}{e-1},$$
$$\overline{y} = \frac{1}{A}\int_0^1 \frac{(e^x)^2}{2}\,dx = \frac{1}{e-1}\cdot\tfrac{1}{4}[e^{2x}]_0^1 = \frac{1}{4(e-1)}(e^2-1) = \frac{e+1}{4}. \quad (\overline{x},\overline{y}) = \left(\frac{1}{e-1},\frac{e+1}{4}\right).$$

12. $A = \int_1^e \ln x\,dx = [x\ln x - x]_1^e = 0-(-1) = 1$,
$$\overline{x} = \frac{1}{A}\int_1^e x\ln x\,dx = \left[\tfrac{1}{2}x^2\ln x - \tfrac{1}{4}x^2\right]_1^e = \left(\tfrac{1}{2}e^2 - \tfrac{1}{4}e^2\right) - \left(-\tfrac{1}{4}\right) = \frac{e^2+1}{4},$$
$\overline{y} = \dfrac{1}{A}\displaystyle\int_1^e \frac{(\ln x)^2}{2}\,dx = \frac{1}{2}\int_1^e (\ln x)^2\,dx$. To evaluate $\int (\ln x)^2\,dx$, take $u = \ln x$ and $dv = \ln x\,dx$, so that $du = 1/x\,dx$ and $v = x\ln x - x$. Then
$$\int(\ln x)^2\,dx = x(\ln x)^2 - x(\ln x) - \int(x\ln x - x)\frac{1}{x}\,dx = x(\ln x)^2 - x(\ln x) - \int(\ln x - 1)dx$$
$$= x(\ln x)^2 - x\ln x - x\ln x + x + x + C = x(\ln x)^2 - 2x\ln x + 2x + C.$$ Thus
$$\overline{y} = \tfrac{1}{2}\left[x(\ln x)^2 - 2x\ln x + 2x\right]_1^e = \tfrac{1}{2}[(e-2e+2e)-(0-0+2)] = \frac{e-2}{2}. \quad (\overline{x},\overline{y}) = \left(\frac{e^2+1}{4},\frac{e-2}{2}\right).$$

13. $A = \int_0^1\left(\sqrt{x}-x\right)dx = \left[\frac{2}{3}x^{3/2} - \frac{1}{2}x^2\right]_0^1 = \frac{2}{3} - \frac{1}{2} = \frac{1}{6}$,
$\overline{x} = A^{-1}\int_0^1 x\left(\sqrt{x}-x\right)dx = 6\int_0^1\left(x^{3/2}-x^2\right)dx = 6\left[\frac{2}{5}x^{5/2} - \frac{1}{3}x^3\right]_0^1 = 6\left(\frac{2}{5}-\frac{1}{3}\right) = \frac{2}{5}$,
$\overline{y} = A^{-1}\int_0^1 \frac{1}{2}\left[\left(\sqrt{x}\right)^2 - x^2\right]dx = 3\int_0^1(x-x^2)dx = 3\left[\frac{1}{2}x^2 - \frac{1}{3}x^3\right]_0^1 = 3\left(\frac{1}{2}-\frac{1}{3}\right) = \frac{1}{2}$.
$(\overline{x},\overline{y}) = \left(\frac{2}{5},\frac{1}{2}\right) = (0.4, 0.5)$.

14. By symmetry, $\overline{x} = 0$ and $A = 2\int_0^2[(8-x^2)-x^2]dx = 2\int_0^2(8-2x^2)dx = 2\left[8x - \frac{2}{3}x^3\right]_0^2 = 2\left(16-\frac{16}{3}\right) = \frac{64}{3}$,
$$\overline{y} = \frac{1}{A}\int_{-2}^2 \frac{1}{2}\left[\left(8-x^2\right)^2 - \left(x^2\right)^2\right]dx = \frac{1}{A}\int_0^2(64-16x^2)dx = \tfrac{3}{64}\cdot 16\left[4x - \tfrac{1}{3}x^3\right]_0^2 = \tfrac{3}{4}\left(8-\tfrac{8}{3}\right) = \tfrac{3}{4}\cdot\tfrac{16}{3} = 4.$$
This result could have been predicted by symmetry. $(\overline{x},\overline{y}) = (0,4)$.

15. $A=\int_0^{\pi/4}(\cos x-\sin x)dx=[\sin x+\cos x]_0^{\pi/4}=\sqrt{2}-1$,

$$\overline{x}=A^{-1}\int_0^{\pi/4}x(\cos x-\sin x)dx=A^{-1}[x(\sin x+\cos x)+\cos x-\sin x]_0^{\pi/4}\ \text{[integration by parts]}$$

$$=A^{-1}\left(\tfrac{\pi}{4}\sqrt{2}-1\right)=\frac{\frac{1}{4}\pi\sqrt{2}-1}{\sqrt{2}-1},$$

$$\overline{y}=A^{-1}\int_0^{\pi/4}\frac{1}{2}(\cos^2x-\sin^2x)\,dx=\frac{1}{2A}\int_0^{\pi/4}\cos 2x\,dx=\frac{1}{4A}[\sin 2x]_0^{\pi/4}=\frac{1}{4A}=\frac{1}{4(\sqrt{2}-1)}.$$

$$(\overline{x},\overline{y})=\left(\frac{\pi\sqrt{2}-4}{4(\sqrt{2}-1)},\frac{1}{4(\sqrt{2}-1)}\right).$$

16. $A=\int_0^1 x\,dx+\displaystyle\int_1^2\frac{1}{x}\,dx=\left[\tfrac{1}{2}x^2\right]_0^1+[\ln x]_1^2=\tfrac{1}{2}+\ln 2$,

$$\overline{x}=\frac{1}{A}\left[\int_0^1 x^2\,dx+\int_1^2 1\,dx\right]=\frac{1}{A}\left(\left[\tfrac{1}{3}x^3\right]_0^1+[x]_1^2\right)=\frac{1}{A}\left(\tfrac{1}{3}+1\right)=\frac{2}{1+2\ln 2}\cdot\frac{4}{3}=\frac{8}{3(1+2\ln 2)},$$

$$\overline{y}=\frac{1}{A}\left[\int_0^1\tfrac{1}{2}x^2\,dx+\int_1^2\frac{1}{2x^2}\,dx\right]=\frac{1}{2A}\left(\left[\tfrac{1}{3}x^3\right]_0^1+\left[-\frac{1}{x}\right]_1^2\right)=\frac{1}{2A}\left(\frac{1}{3}+\frac{1}{2}\right)=\frac{5}{12A}=\frac{5}{6+12\ln 2}.$$

$$(\overline{x},\overline{y})=\left(\frac{8}{3(1+2\ln 2)},\frac{5}{6(1+2\ln 2)}\right).$$

Remark: The principle used in this problem is stated after Example 2: the moment of the union of two non-overlapping regions is the sum of the moments of the individual regions.

17. By symmetry, $M_y=0$ and $\overline{x}=0$. $A=\frac{1}{2}bh=\frac{1}{2}\cdot 2\cdot 2=2$.

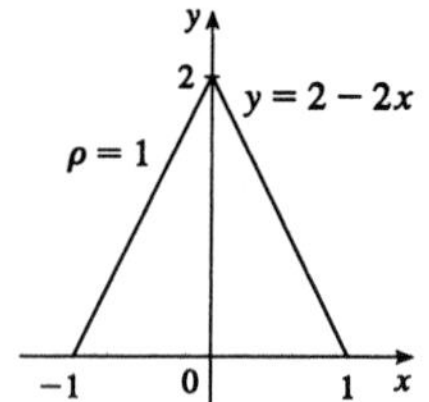

$$M_x=2\rho\int_0^1\tfrac{1}{2}(2-2x)^2\,dx=4\int_0^1(1-x)^2\,dx$$

$$=4\left[-\tfrac{1}{3}(1-x)^3\right]_0^1=4\cdot\tfrac{1}{3}=\tfrac{4}{3}.$$

$$\overline{y}=\frac{1}{\rho A}M_x=\frac{2}{3}.\quad(\overline{x},\overline{y})=\left(0,\tfrac{2}{3}\right).$$

18. By symmetry about the line $y=x$, we expect that $\overline{x}=\overline{y}$, but we will calculate both anyway. $A=\frac{1}{4}\pi r^2$, so $m=\rho A=\frac{1}{2}\pi r^2$. $M_x=\rho\int_0^r\frac{1}{2}\left(\sqrt{r^2-x^2}\right)^2dx=\int_0^r(r^2-x^2)dx=\left[r^2x-\frac{1}{3}x^3\right]_0^r=\frac{2}{3}r^3$;
$M_y=\rho\int_0^r x\sqrt{r^2-x^2}\,dx=\int_0^r(r^2-x^2)^{1/2}2x\,dx=\int_0^{r^2}u^{1/2}\,du$ (put $u=r^2-x^2$) $=\left[\frac{2}{3}u^{3/2}\right]_0^{r^2}=\frac{2}{3}r^3$;
$\overline{x}=\dfrac{1}{m}M_y=\dfrac{2}{\pi r^2}\left(\frac{2}{3}r^3\right)=\frac{4}{3\pi}r$; $\overline{y}=\dfrac{1}{m}M_x=\dfrac{2}{\pi r^2}\left(\frac{2}{3}r^3\right)=\frac{4}{3\pi}r$. $(\overline{x},\overline{y})=\left(\frac{4}{3\pi}r,\frac{4}{3\pi}r\right)$.

19. By symmetry, $M_y=0$ and $\overline{x}=0$. $A=$ area of triangle $+$ area of square $=1+4=5$, so $m=\rho A=4\cdot 5=20$.
$M_x=\rho\cdot 2\int_0^1\frac{1}{2}\left[(1-x)^2-(-2)^2\right]dx=4\int_0^1(x^2-2x-3)dx$
$=4\left[\frac{1}{3}x^3-x^2-3x\right]_0^1=4\left(\frac{1}{3}-1-3\right)=4\left(-\frac{11}{3}\right)=-\frac{44}{3}$. $\overline{y}=M_x/m=\frac{1}{20}\left(-\frac{44}{3}\right)=-\frac{11}{15}$. $(\overline{x},\overline{y})=\left(0,-\frac{11}{15}\right)$.

20. By symmetry, $M_y=0$ and $\overline{x}=0$; $A=\frac{1}{2}\pi\cdot 1^2+4$, so $m=\rho A=5\left(\frac{\pi}{2}+4\right)=\frac{5}{2}(\pi+8)$;

$$M_x=\rho\cdot 2\int_0^1\frac{1}{2}\left[\left(\sqrt{1-x^2}\right)^2-(-2)^2\right]dx=5\int_0^1(-x^2-3)dx=-5\left[\tfrac{1}{3}x^3+3x\right]_0^1=-5\cdot\tfrac{10}{3}=-\tfrac{50}{3};$$

$$\overline{y}=\frac{1}{m}M_x=\frac{2}{5(\pi+8)}\cdot\frac{-50}{3}=\frac{-20}{3(\pi+8)}.\quad(\overline{x},\overline{y})=\left(0,\frac{-20}{3(\pi+8)}\right).$$

21. Choose x- and y-axes so that the base (one side of the triangle) lies along the x-axis with the other vertex along the positive y-axis as shown. From geometry, we know the medians intersect at a point $\frac{2}{3}$ of the way from each vertex (along the median) to the opposite side.

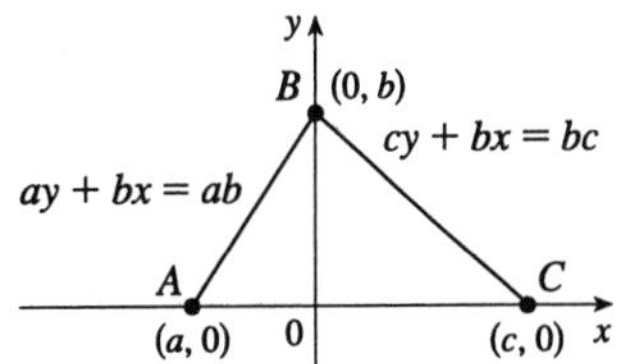

The median from B goes to the midpoint $\left(\frac{1}{2}(a+c), 0\right)$ of side AC, so the point of intersection of the medians is $\left(\frac{2}{3}\cdot\frac{1}{2}(a+c), \frac{1}{3}b\right) = \left(\frac{1}{3}(a+c), \frac{1}{3}b\right)$. This can also be verified by finding the equations of two medians, and solving them simultaneously to find their point of intersection. Now let us compute the location of the centroid of the triangle. The area is $A = \frac{1}{2}(c-a)b$.

$$\overline{x} = \frac{1}{A}\left[\int_a^0 x\cdot\frac{b}{a}(a-x)dx + \int_0^c x\cdot\frac{b}{c}(c-x)dx\right] = \frac{1}{A}\left[\frac{b}{a}\int_a^0 \left(ax-x^2\right)dx + \frac{b}{c}\int_0^c \left(cx-x^2\right)dx\right]$$

$$= \frac{b}{Aa}\left[\tfrac{1}{2}ax^2 - \tfrac{1}{3}x^3\right]_a^0 + \frac{b}{Ac}\left[\tfrac{1}{2}cx^2 - \tfrac{1}{3}x^3\right]_0^c = \frac{b}{Aa}\left[-\tfrac{1}{2}a^3 + \tfrac{1}{3}a^3\right] + \frac{b}{Ac}\left[\tfrac{1}{2}c^3 - \tfrac{1}{3}c^3\right]$$

$$= \frac{2}{a(c-a)}\cdot\frac{-a^3}{6} + \frac{2}{c(c-a)}\cdot\frac{c^3}{6} = \frac{1}{3(c-a)}\left(c^2-a^2\right) = \frac{a+c}{3}, \text{ and}$$

$$\overline{y} = \frac{1}{A}\left[\int_a^0 \tfrac{1}{2}\left(\frac{b}{a}(a-x)\right)^2 dx + \int_0^c \tfrac{1}{2}\left(\frac{b}{c}(c-x)\right)^2 dx\right]$$

$$= \frac{1}{A}\left[\frac{b^2}{2a^2}\int_a^0 \left(a^2-2ax+x^2\right)dx + \frac{b^2}{2c^2}\int_0^c \left(c^2-2cx+x^2\right)dx\right]$$

$$= \frac{1}{A}\left[\frac{b^2}{2a^2}\left[a^2x - ax^2 + \tfrac{1}{3}x^3\right]_a^0 + \frac{b^2}{2c^2}\left[c^2x - cx^2 + \tfrac{1}{3}x^3\right]_0^c\right]$$

$$= \frac{1}{A}\left[\frac{b^2}{2a^2}\left(-a^3+a^3-\tfrac{1}{3}a^3\right) + \frac{b^2}{2c^2}\left(c^3-c^3+\tfrac{1}{3}c^3\right)\right] = \frac{1}{A}\left[\frac{b^2}{6}(-a+c)\right] = \frac{2}{(c-a)b}\cdot\frac{(c-a)b^2}{6} = \frac{b}{3}.$$

Thus $(\overline{x}, \overline{y}) = \left(\dfrac{a+c}{3}, \dfrac{b}{3}\right)$ as claimed.

Remarks: Actually the computation of $\overline{y}$ is all that is needed. By considering each side of the triangle in turn to be the base, we see that the centroid is $\frac{1}{3}$ of the way from each side to the opposite vertex and must therefore be the intersection of the medians.

The computation of $\overline{y}$ in this problem (and many others) can be simplified by using horizontal rather than vertical approximating rectangles. If the length of a thin rectangle at coordinate y is $\ell(y)$, then its area is $\ell(y)\Delta y$, its mass is $\rho\ell(y)\Delta y$, and its moment about the x-axis is $\Delta M_x = \rho y\ell(y)\Delta y$. Thus $M_x = \int \rho y\ell(y)dy$ and

$$\overline{y} = \frac{\int \rho y\ell(y)dy}{\rho A} = \frac{1}{A}\int y\ell(y)dy.$$ In this problem,

$$\ell(y) = \frac{c-a}{b}(b-y)$$ by similar triangles, so

$$\overline{y} = \frac{1}{A}\int_0^b \frac{c-a}{b}y(b-y)dy = \frac{2}{b^2}\int_0^b \left(by-y^2\right)dy = \frac{2}{b^2}\left[\tfrac{1}{2}by^2 - \tfrac{1}{3}y^3\right]_0^b = \frac{2}{b^2}\cdot\frac{b^3}{6} = \frac{b}{3}.$$

Notice that only one integral is needed when this method is used.

Since the position of a centroid is independent of density when the density is constant, we will assume for convenience that $\rho = 1$ in Exercises 22-25.

22. Divide the lamina into three rectangles with masses 2, 2 and 6, with centroids $\left(-\frac{3}{2}, 1\right)$, $\left(0, \frac{1}{2}\right)$ and $\left(2, \frac{3}{2}\right)$ respectively. The total mass of the lamina is 10. So, using Formulas 5, 6, and 7, we have

$\overline{x} = \dfrac{\sum m_i x_i}{m} = \frac{2}{10}\left(-\frac{3}{2}\right) + \frac{2}{10}(0) + \frac{6}{10}(2) = \frac{9}{10}$, and $\overline{y} = \dfrac{\sum m_i y_i}{m} = \frac{2}{10}(1) + \frac{2}{10}\left(\frac{1}{2}\right) + \frac{6}{10}\left(\frac{3}{2}\right) = \frac{6}{5}$.

Therefore $(\overline{x}, \overline{y}) = \left(\frac{9}{10}, \frac{6}{5}\right)$.

23. Divide the lamina into two triangles and one rectangle with respective masses of 2, 2 and 4, so that the total mass is 8. Using the result of Exercise 21, the triangles have centroids $\left(-1, \frac{2}{3}\right)$ and $\left(1, \frac{2}{3}\right)$. The centroid of the rectangle (its center) is $\left(0, -\frac{1}{2}\right)$. So, using Formulas 5 and 7, we have

$\overline{y} = \dfrac{\sum m_i y_i}{m} = \frac{2}{8}\left(\frac{2}{3}\right) + \frac{2}{8}\left(\frac{2}{3}\right) + \frac{4}{8}\left(-\frac{1}{2}\right) = \frac{1}{12}$, and $\overline{x} = 0$, since the lamina is symmetric about the line $x = 0$.

Therefore $(\overline{x}, \overline{y}) = \left(0, \frac{1}{12}\right)$.

24. The two triangles each have mass $\frac{1}{2}$, so that the total mass of the lamina is 1. Using the result of Exercise 21, the centroid of the triangle in the second quadrant is $\overline{x} = -1 + \left(\frac{2}{3}\right)\frac{\sqrt{2}}{2}\cos 45° = -\frac{2}{3}$, and

$\overline{y} = 0 + \left(\frac{2}{3}\right)\frac{\sqrt{2}}{2}\sin 45° = \frac{1}{3}$. Similarly, the centroid of the triangle in the first quadrant is

$\overline{x} = \left(\frac{2}{3}\right)\frac{\sqrt{2}}{2}\cos 45° = \frac{1}{3}$, and $\overline{y} = 0 + \left(\frac{2}{3}\right)\frac{\sqrt{2}}{2}\sin 45° = \frac{1}{3}$. For the entire lamina,

$\overline{x} = \dfrac{\sum m_i x_i}{m} = \frac{1}{2}\left(-\frac{2}{3}\right) + \frac{1}{2}\left(\frac{1}{3}\right) = -\frac{1}{6}$, and $\overline{y} = \dfrac{\sum m_i y_i}{m} = \frac{1}{2}\left(\frac{1}{3}\right) + \frac{1}{2}\left(\frac{1}{3}\right) = \frac{1}{3}$. Therefore, $(\overline{x}, \overline{y}) = \left(-\frac{1}{6}, \frac{1}{3}\right)$.

25. Suppose first that the large rectangle were complete, so that its mass would be $6 \cdot 3 = 18$. Its centroid would be $\left(1, \frac{3}{2}\right)$. The mass removed from this object to create the one being studied is 3. The centroid of the cut-out piece is $\left(\frac{3}{2}, \frac{3}{2}\right)$. Therefore, for the actual lamina, whose mass is 15, $\overline{x} = \frac{18}{15}(1) - \frac{3}{15}\left(\frac{3}{2}\right) = \frac{9}{10}$, and $\overline{y} = \frac{3}{2}$, since the lamina is symmetric about the line $y = \frac{3}{2}$. Therefore $(\overline{x}, \overline{y}) = \left(\frac{9}{10}, \frac{3}{2}\right)$.

26. A sphere can be generated by rotating a semicircle about its diameter. By Example 3, the center of mass travels a distance $2\pi\overline{y} = 2\pi[4r/(3\pi)] = 8r/3$, so by the Theorem of Pappus, the volume of the sphere is

$V = Ad = \dfrac{\pi r^2}{2} \cdot \dfrac{8r}{3} = \frac{4}{3}\pi r^3$.

27. A cone of height h and radius r can be generated by rotating a right triangle about one of its legs as shown. By Exercise 21, $\overline{x} = \frac{1}{3}r$, so by the Theorem of Pappus, the volume of the cone is

$V = Ad = \frac{1}{2}rh \cdot 2\pi\left(\frac{1}{3}r\right) = \frac{1}{3}\pi r^2 h$.

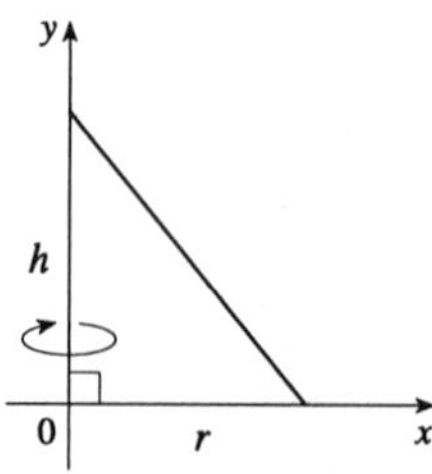

28. The quadrilateral is a parallelogram. By symmetry, the centroid is its center $\left(\frac{7}{2}, 2\right)$, the midpoint of both diagonals. Its area is $A = bh = 6 \cdot 4 = 24$. The Theorem of Pappus implies that the volume of the solid is $V = Ad = A \cdot 2\pi\overline{x} = 24 \cdot 2\pi\left(\frac{7}{2}\right) = 168\pi$.

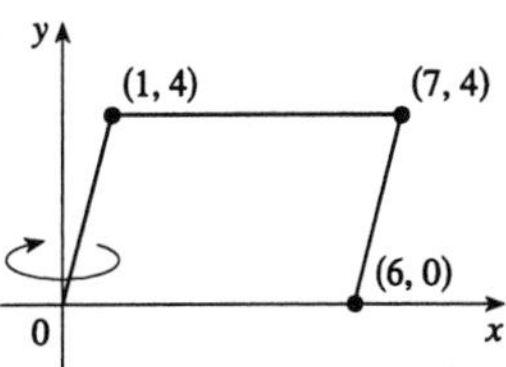

29. Suppose the region lies between two curves $y = f(x)$ and $y = g(x)$ where $f(x) \geq g(x)$, as illustrated in Figure 9. Take a partition P by points x_i with $a = x_0 < x_1 < \cdots < x_n = b$ and choose x_i^* to be the midpoint of the ith subinterval; that is, $x_i^* = \frac{1}{2}(x_{i-1} + x_i)$. Then the centroid of the ith approximating rectangle R_i is its center $C_i = \left(x_i^*, \frac{1}{2}[f(x_i^*) + g(x_i^*)]\right)$. Its area is $[f(x_i^*) - g(x_i^*)]\Delta x_i$, so its mass is $\rho[f(x_i^*) - g(x_i^*)]\Delta x_i$. Thus

$M_y(R_i) = \rho[f(x_i^*) - g(x_i^*)]\Delta x_i \cdot x_i^* = \rho x_i^*[f(x_i^*) - g(x_i^*)]\Delta x_i$ and

$M_x(R_i) = \rho[f(x_i^*) - g(x_i^*)]\Delta x_i \cdot \frac{1}{2}[f(x_i^*) + g(x_i^*)] = \rho \cdot \frac{1}{2}\left[f(x_i^*)^2 - g(x_i^*)^2\right]\Delta x_i$.

Summing over i and taking the limit as $\|P\| \to 0$, we get

$M_y = \lim\limits_{\|P\| \to 0} \sum\limits_i \rho x_i^*[f(x_i^*) - g(x_i^*)]\Delta x_i = \rho\int_a^b x[f(x) - g(x)]dx$ and

$M_x = \lim\limits_{\|P\| \to 0} \sum\limits_i \rho \cdot \frac{1}{2}\left[f(x_i^*)^2 - g(x_i^*)^2\right]\Delta x_i = \rho\int_a^b \frac{1}{2}\left[f(x)^2 - g(x)^2\right]dx$. Thus

$$\overline{x} = \frac{M_y}{m} = \frac{M_y}{\rho A} = \frac{1}{A}\int_a^b x[f(x) - g(x)]dx \text{ and } \overline{y} = \frac{M_x}{m} = \frac{M_x}{\rho A} = \frac{1}{A}\int_a^b \tfrac{1}{2}\left[f(x)^2 - g(x)^2\right]dx.$$

30. **(a)**

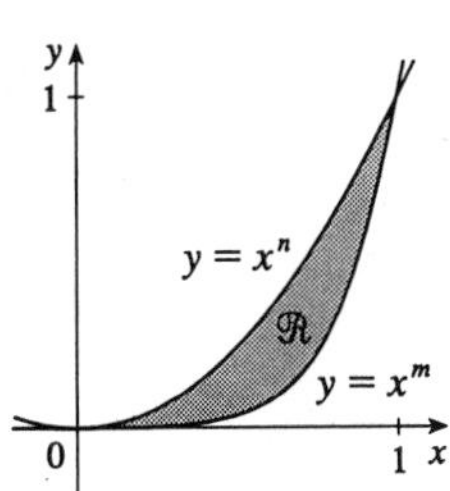

(b) Using Formula 11 and the fact that the area of $\Re$ is

$$A = \int_0^1 (x^n - x^m)dx = \frac{1}{n+1} - \frac{1}{m+1} = \frac{m-n}{(n+1)(m+1)}, \text{ we get}$$

$$\overline{x} = \frac{(n+1)(m+1)}{m-n}\int_0^1 x[x^n - x^m]dx$$

$$= \frac{(n+1)(m+1)}{m-n}\left[\frac{1}{n+2} - \frac{1}{m+2}\right] = \frac{(n+1)(m+1)}{(n+2)(m+2)} \text{ and}$$

$$\overline{y} = \frac{(n+1)(m+1)}{m-n}\int_0^1 \tfrac{1}{2}\left[(x^n)^2 - (x^m)^2\right]dx = \frac{(n+1)(m+1)}{2(m-n)}\left[\frac{1}{2n+1} - \frac{1}{2m+1}\right]$$

$$= \frac{(n+1)(m+1)}{(2n+1)(2m+1)}.$$

(c) If we take $n = 3$ and $m = 4$, then $(\overline{x}, \overline{y}) = \left(\frac{4 \cdot 5}{5 \cdot 6}, \frac{4 \cdot 5}{7 \cdot 9}\right) = \left(\frac{2}{3}, \frac{20}{63}\right)$, which lies outside $\Re$ since $\left(\frac{2}{3}\right)^3 = \frac{8}{27} < \frac{20}{63}$.

(This is the simplest of many possibilities.)

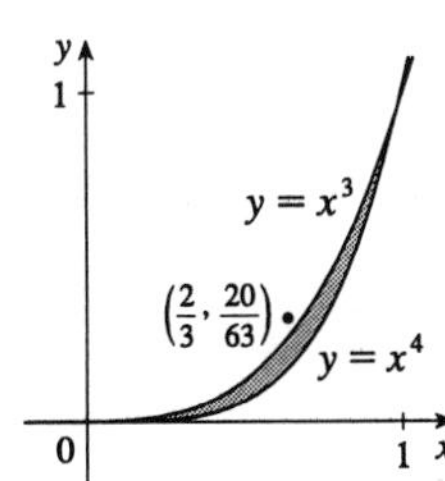

EXERCISES 8.5

1. **(a)** $P = \rho g d = (1000\,\text{kg/m}^3)(9.8\,\text{m/s}^2)(1\,\text{m}) = 9800\,\text{Pa} = 9.8\,\text{kPa}$

(b) $F = \rho g d A = PA = (9800\,\text{N/m}^2)(2\,\text{m}^2) = 1.96 \times 10^4\,\text{N}$

(c) $F = \int_0^1 \rho g x \cdot 1\,dx = 9800 \int_0^1 x\,dx = 4900x^2\big|_0^1 = 4.90 \times 10^3\,\text{N}$

2. **(a)** $P = \rho g d = 1030(9.8)(2.5) = 25{,}235 \approx 2.52 \times 10^4\,\text{Pa} = 25.2\,\text{kPa}$

(b) $F = PA \approx \left(2.52 \times 10^4\,\text{N/m}^2\right)\left(50\,\text{m}^2\right) = 1.26 \times 10^6\,\text{N}$

(c) $F = \int_0^{2.5} \rho g x \cdot 5\,dx = (1030)(9.8)(5)\int_0^{2.5} x\,dx \approx 2.52 \times 10^4 [x^2]_0^{2.5} \approx 1.58 \times 10^5\,\text{N}$

3. $F = \int_0^{10} \rho g x \cdot 2\sqrt{100 - x^2}\,dx = 9.8 \times 10^3 \int_0^{10} \sqrt{100 - x^2}\,2x\,dx$

$= 9.8 \times 10^3 \int_{100}^0 u^{1/2}(-du) \quad (\text{put } u = 100 - x^2)$

$= 9.8 \times 10^3 \int_0^{100} u^{1/2}\,du = 9.8 \times 10^3 \left[\frac{2}{3}u^{3/2}\right]_0^{100}$

$= \frac{2}{3} \cdot 9.8 \times 10^6 \approx 6.5 \times 10^6\,\text{N}$

4. $F = \int_5^{10} \rho g(x - 5) \cdot 2\sqrt{100 - x^2}\,dx = \rho g \int_5^{10} 2x\sqrt{100 - x^2}\,dx - 10\rho g \int_5^{10} \sqrt{100 - x^2}\,dx$

$= -\rho g\left[\frac{2}{3}\left(100 - x^2\right)^{3/2}\right]_5^{10} - 10\rho g\left[\frac{1}{2}x\sqrt{100 - x^2} + 50\sin^{-1}(x/10)\right]_5^{10}$

$= \frac{2}{3}\rho g(75)^{3/2} - 10\rho g\left[50\left(\frac{\pi}{2}\right) - \frac{5}{2}\sqrt{75} - 50\left(\frac{\pi}{6}\right)\right] = 250\rho g\left(\frac{3\sqrt{3}}{2} - \frac{2\pi}{3}\right) \approx 1.23 \times 10^6\,\text{N}$

5. $F = \displaystyle\int_{-r}^{r} \rho g(x + r) \cdot 2\sqrt{r^2 - x^2}\,dx = \rho g \int_{-r}^{r} \sqrt{r^2 - x^2}\,2x\,dx + 2\rho g r \int_{-r}^{r} \sqrt{r^2 - x^2}\,dx$. The first integral is 0 because the integrand is an odd function. The second integral can be interpreted as the area of a semicircular disk with radius r, or we could make the trigonometric substitution $x = r\sin\theta$. Continuing:
$F = \rho g \cdot 0 + 2\rho g r \cdot \frac{1}{2}\pi r^2 = \rho g \pi r^3 = 1000 g\pi r^3$ N (SI units assumed).

6. $F = \displaystyle\int_{-10}^{0} \rho g(x + 10)2\sqrt{100 - x^2}\,dx = \rho g \int_{-10}^{0} \sqrt{100 - x^2}\,2x\,dx + 20\rho g \int_{-10}^{0} \sqrt{100 - x^2}\,dx$

$= \rho g \displaystyle\int_0^{100} u^{1/2}(-du) + 20\rho g \cdot \frac{1}{4}\pi(10)^2 = -\frac{2}{3}\rho g\left[u^{3/2}\right]_0^{100} + 500\rho g\pi = -\frac{2000}{3}\rho g + 500\rho g\pi$

$= 1000\rho g\left(\frac{\pi}{2} - \frac{2}{3}\right) \approx 8.86 \times 10^6\,\text{N}$. *Another Method:* $F = \int_0^{10} \rho g(10 - x)2\sqrt{100 - x^2}\,dx$.

7. $F = \displaystyle\int_0^6 \delta x \cdot \frac{2x}{3}\,dx = \left[\frac{2}{9}\delta x^3\right]_0^6$

$= 48\delta \approx 48 \times 62.5 = 3000\,\text{lb}$

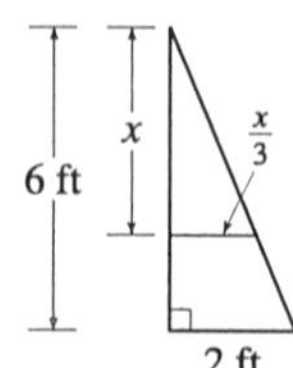

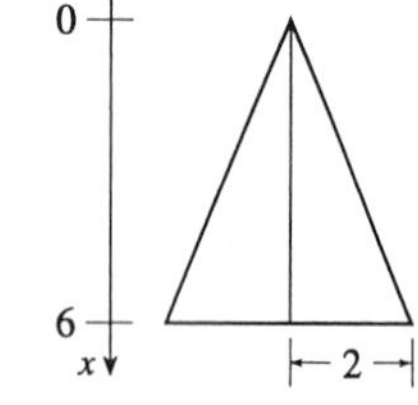

8. $F = \displaystyle\int_0^h \rho g x \cdot \frac{b}{h}x\,dx = \left[\frac{\rho g b}{3h}x^3\right]_0^h = \frac{\rho g b h^2}{3}$

$= \frac{1000}{3} g b h^2$ (SI units assumed)

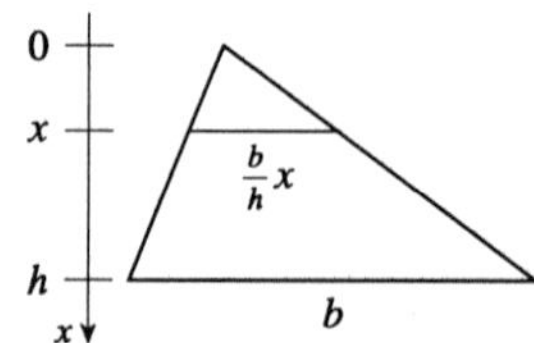

9. $F = \int_2^6 \delta(x-2)\frac{2}{3}x\,dx = \frac{2}{3}\delta\int_2^6 (x^2-2x)dx = \frac{2}{3}\delta\left[\frac{1}{3}x^3 - x^2\right]_2^6 = \frac{2}{3}\delta\left[36 - \left(-\frac{4}{3}\right)\right] = \frac{224}{9}\delta \approx 1.56\times 10^3\text{ lb}$

10. $$F = \int_0^h \rho g(h-x)\frac{b}{h}x\,dx = \frac{\rho g b}{h}\int_0^h (hx - x^2)\,dx = \frac{\rho g b}{h}\left[\frac{hx^2}{2} - \frac{x^3}{3}\right]_0^h = \frac{\rho g b}{h}\frac{h^3}{6} = \frac{\rho g b h^2}{6} = \frac{1000 g b h^2}{6}$$

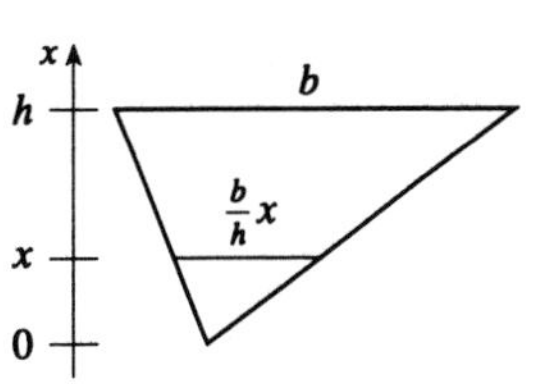

11. $$F = \int_0^8 \delta x\cdot(12+x)dx = \delta\int_0^8 (12x + x^2)dx = \delta\left[6x^2 + \frac{x^3}{3}\right]_0^8 = \delta\left(384 + \tfrac{512}{3}\right) = (62.5)\tfrac{1664}{3} \approx 3.47\times 10^4\text{ lb}$$

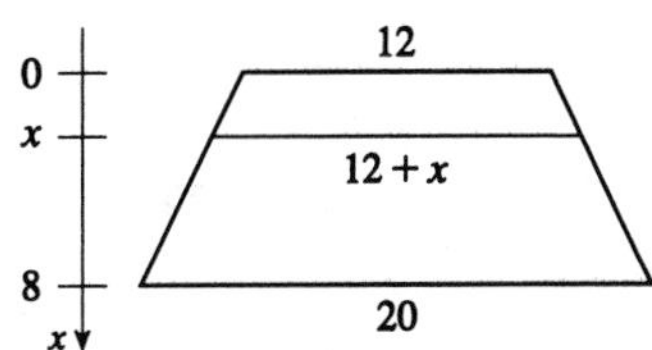

12. $$F = \int_0^h \rho g x\left[a + \tfrac{x}{h}(b-a)\right]dx = \rho g a\int_0^h x\,dx + \frac{\rho g(b-a)}{h}\int_0^h x^2\,dx = \rho g a\frac{h^2}{2} + \rho g\frac{b-a}{h}\frac{h^3}{3} = \rho g h^2\left(\frac{a}{2} + \frac{b-a}{3}\right) = \rho g h^2\frac{a+2b}{6} \approx \tfrac{500}{3}gh^2(a+2b)\text{N}$$

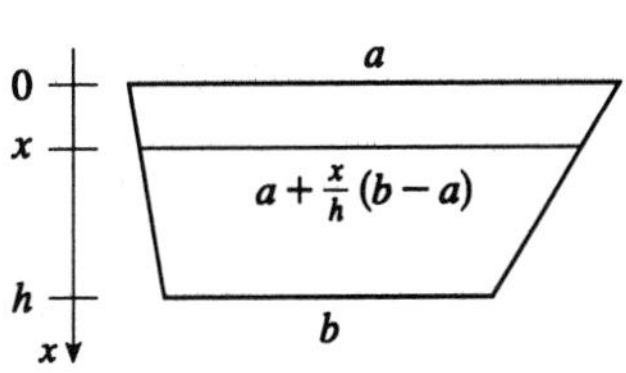

13. $$F = \int_0^{4\sqrt{3}} \rho g\left(4\sqrt{3} - x\right)\frac{2x}{\sqrt{3}}\,dx = 8\rho g\int_0^{4\sqrt{3}} x\,dx - \frac{2\rho g}{\sqrt{3}}\int_0^{4\sqrt{3}} x^2\,dx = 4\rho g\left[x^2\right]_0^{4\sqrt{3}} - \frac{2\rho g}{3\sqrt{3}}\left[x^3\right]_0^{4\sqrt{3}} = 192\rho g - \frac{2\rho g}{3\sqrt{3}}64\cdot 3\sqrt{3} = 192\rho g - 128\rho g = 64\rho g \approx 64(840)(9.8) \approx 5.27\times 10^5\text{ N}$$

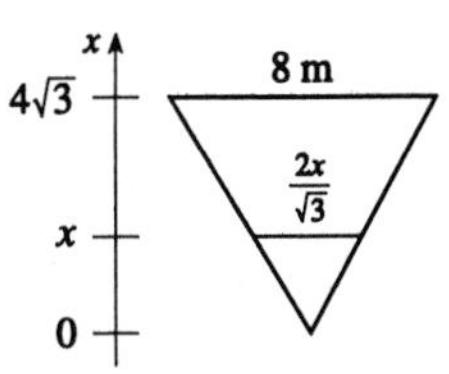

14. $$F = \int_0^4 \rho g(4-x)\cdot\tfrac{2}{\sqrt{3}}x\,dx = \left(\tfrac{8}{\sqrt{3}}\rho g\right)\int_0^4 x\,dx - \left(\tfrac{2}{\sqrt{3}}\rho g\right)\int_0^4 x^2\,dx = \left(\tfrac{8}{\sqrt{3}}\rho g\right)\left[\tfrac{1}{2}x^2\right]_0^4 - \left(\tfrac{2}{\sqrt{3}}\rho g\right)\left[\tfrac{1}{3}x^3\right]_0^4 = \left(\tfrac{2}{\sqrt{3}}\rho g\right)\left(32 - \tfrac{64}{3}\right) = \tfrac{64}{3\sqrt{3}}\rho g \approx 1.01\times 10^5\text{ N}$$

15. **(a)** $F = \rho g d A \approx (1000)(9.8)(0.8)(0.2)^2 \approx 314\text{ N}$

(b) $F = \int_{0.8}^1 \rho g x(0.2)dx = 0.2\rho g\left[\tfrac{1}{2}x^2\right]_{0.8}^1 = (0.2\rho g)(0.18) = 0.036\rho g \approx 353\text{ N}$

16. $$F = \int_0^2 \rho g(10-x)2\sqrt{4-x^2}\,dx = 20\rho g\int_0^2 \sqrt{4-x^2}\,dx - \rho g\int_0^2 \sqrt{4-x^2}\,2x\,dx = 20\rho g\tfrac{1}{4}\pi(2^2) - \rho g\int_0^4 u^{1/2}\,du \quad \begin{bmatrix}\text{put } u = 4-x^2,\\ du = -2x\,dx\end{bmatrix} = 20\pi\rho g - \tfrac{2}{3}\rho g\left[u^{3/2}\right]_0^4 = 20\pi\rho g - \tfrac{16}{3}\rho g = \rho g\left(20\pi - \tfrac{16}{3}\right) \approx 5.63\times 10^5\text{ N}$$

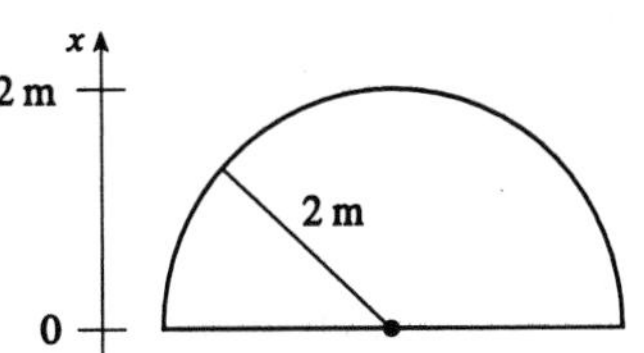

17. $F = \int_0^2 \rho g x \cdot 3 \cdot \sqrt{2}\,dx = 3\sqrt{2}\rho g \int_0^2 x\,dx$
$= 3\sqrt{2}\rho g\left[\frac{1}{2}x^2\right]_0^2 = 6\sqrt{2}\rho g$
$\approx 8.32 \times 10^4$ N

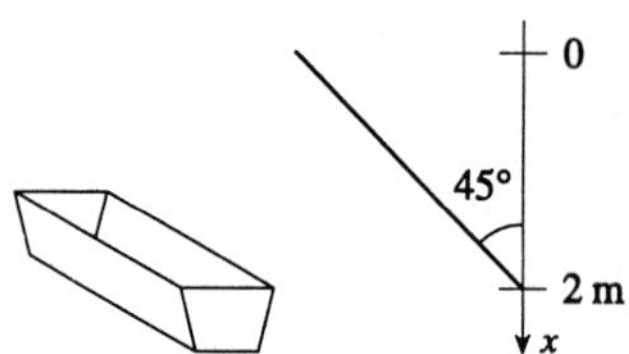

18. The height of the dam is $h = 15\sqrt{19}\left(\frac{\sqrt{3}}{2}\right)$, so

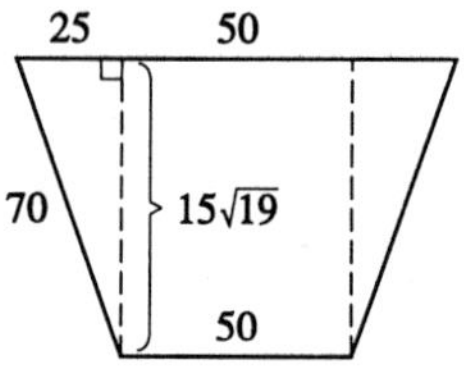

$$F = \int_0^h \delta x\left(100 - \frac{50x}{h}\right)\frac{2}{\sqrt{3}}\,dx = \frac{200\delta}{\sqrt{3}}\int_0^h x\,dx - \frac{100\delta}{h\sqrt{3}}\int_0^h x^2\,dx$$
$$= \frac{200\delta}{\sqrt{3}}\frac{h^2}{2} - \frac{100\delta}{h\sqrt{3}}\frac{h^3}{3} = \frac{200\delta h^2}{3\sqrt{3}} = \frac{200(62.5)}{3\sqrt{3}} \cdot \frac{12{,}825}{4} \approx 7.71 \times 10^6 \text{ lb.}$$

19. Assume that the pool is filled with water.

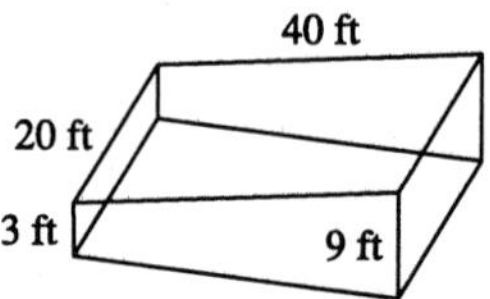

(a) $F = \int_0^3 \delta x\, 20\,dx = 20\delta\left[\frac{1}{2}x^2\right]_0^3 = 20\delta \cdot \frac{9}{2} = 90\delta \approx 5625 \text{ lb} \approx 5.63 \times 10^3 \text{ lb}$

(b) $F = \int_0^9 \delta x 20\,dx = 20\delta\left[\frac{1}{2}x^2\right]_0^9 = 810\delta \approx 50625 \text{ lb} \approx 5.06 \times 10^4 \text{ lb.}$

(c) $$F = \int_0^3 \delta x\, 40\,dx + \int_3^9 \delta x(40)\frac{9-x}{6}\,dx = 40\delta\left[\tfrac{1}{2}x^2\right]_0^3 + \tfrac{20}{3}\delta\int_3^9 (9x - x^2)dx$$
$$= 180\delta + \tfrac{20}{3}\delta\left[\tfrac{9}{2}x^2 - \tfrac{1}{3}x^3\right]_3^9 = 180\delta + \tfrac{20}{3}\delta\left[\left(\tfrac{729}{2} - 243\right) - \left(\tfrac{81}{2} - 9\right)\right]$$
$$= 780\delta \approx 4.88 \times 10^4 \text{ lb}$$

(d) $$F = \int_3^9 \delta x\, 20\frac{\sqrt{409}}{3}\,dx$$
$$= \tfrac{1}{3}\left(20\sqrt{409}\right)\delta\left[\tfrac{1}{2}x^2\right]_3^9$$
$$= \tfrac{1}{3} \cdot 10\sqrt{409}\delta(81 - 9)$$
$$\approx 3.03 \times 10^5 \text{ lb}$$

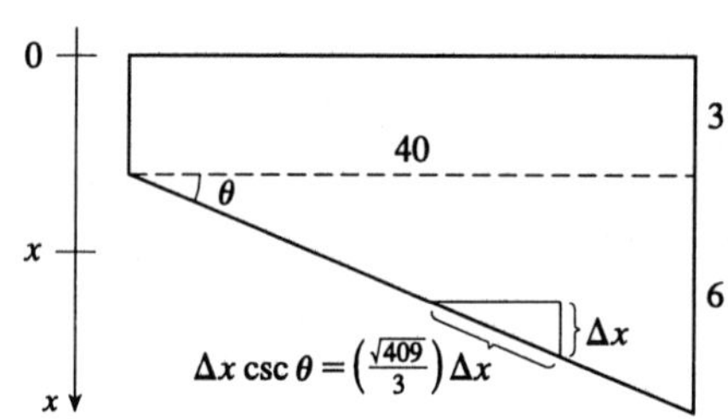

20. Partition the interval $[a, b]$ by points x_i as usual and choose $x_i^* \in [x_{i-1}, x_i]$ for each i. The ith horizontal strip of the immersed plate is approximated by a rectangle of height Δx_i and width $w(x_i^*)$, so its area is $A_i \approx w(x_i^*)\Delta x_i$. For small Δx_i, the pressure P_i on the ith strip is almost constant and $P_i \approx \rho g x_i^*$ by Equation 1. The hydrostatic force F_i acting on the ith strip is $F_i = P_i A_i \approx \rho g x_i^* w(x_i^*)\Delta x_i$. Adding these forces and taking the limit as $\|P\| \to 0$, we obtain the hydrostatic force on the immersed plate:
$F = \lim_{\|P\|\to 0}\sum_{i=1}^n F_i = \lim_{\|P\|\to 0}\sum_{i=1}^n \rho g x_i^* w(x_i^*)\Delta x_i = \int_a^b \rho g x w(x)dx$

21. $\overline{x} = A^{-1}\int_a^b x w(x)dx$ (Equation 1) $\Rightarrow$ $A\overline{x} = \int_a^b x w(x)dx$ $\Rightarrow$ $(\rho g\overline{x})A = \int_a^b \rho g x w(x)dx = F$ by Exercise 20.

22. $F = (\rho g\overline{x})A = (\rho g r)\pi r^2 = \rho g \pi r^3$. Note that $\overline{x} = r$ because the centroid of a circle is its center, which in this case is at a depth of r meters.

EXERCISES 8.6

1. $C(2000) = C(0) + \int_0^{2000} C'(x)\,dx = 1{,}500{,}000 + \int_0^{2000} (0.006x^2 - 1.5x + 8)\,dx$

$= 1{,}500{,}000 + \left[0.002x^3 - 0.75x^2 + 8x\right]_0^{2000} = \$14{,}516{,}000$

2. $R'(x) = 90 - 0.02x$ and $R(100) = \$8800$, so $R(200) = R(100) + \int_{100}^{200} R'(x)\,dx$

$= 8800 + \int_{100}^{200} (90 - 0.02x)\,dx = 8800 + [90x - 0.01x^2]_{100}^{200}$

$= 8800 + (18{,}000 - 400) - (9000 - 100) = \$17{,}500$

3. $C(5000) - C(3000) = \int_{3000}^{5000} (140 - 0.5x + 0.012x^2)\,dx = [140x - 0.25x^2 + 0.004x^3]_{3000}^{5000}$

$= 494{,}450{,}000 - 106{,}170{,}000 = \$388{,}280{,}000$

4. Consumer surplus $= \displaystyle\int_0^{30} [p(x) - p(30)]\,dx$

$= \displaystyle\int_0^{30} \left[5 - \frac{x}{10} - \left(5 - \tfrac{30}{10}\right)\right] dx$

$= \left[3x - \tfrac{1}{20}x^2\right]_0^{30}$

$= 90 - 45 = \$45$

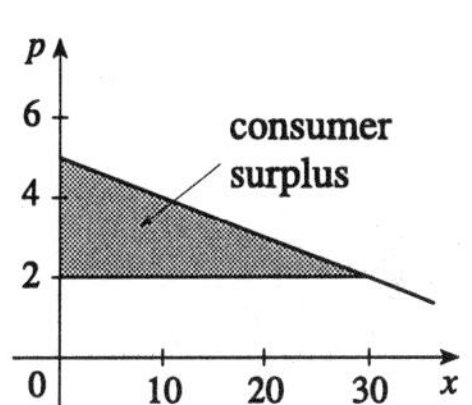

5. $p(x) = 20 = \dfrac{1000}{x + 20} \quad \Rightarrow \quad x + 20 = 50 \quad \Rightarrow \quad x = 30.$

Consumer surplus $= \displaystyle\int_0^{30} [p(x) - 20]\,dx = \int_0^{30} \left(\frac{1000}{x+20} - 20\right) dx = [1000\ln(x+20) - 20x]_0^{30}$

$= 1000\ln\left(\tfrac{50}{20}\right) - 600 = 1000\ln\left(\tfrac{5}{2}\right) - 600 \approx \$316.29.$

6. $P = p_s(10) = 3 + 1 = 4.$

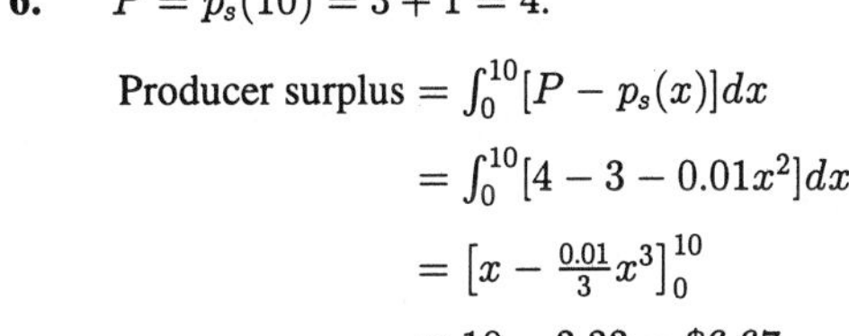

Producer surplus $= \int_0^{10} [P - p_s(x)]\,dx$

$= \int_0^{10} [4 - 3 - 0.01x^2]\,dx$

$= \left[x - \tfrac{0.01}{3}x^3\right]_0^{10}$

$\approx 10 - 3.33 = \$6.67$

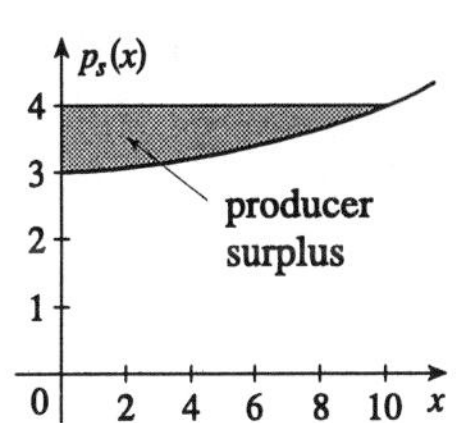

7. $P = p(x) = 10 = 5 + \frac{1}{10}\sqrt{x} \quad \Rightarrow \quad 50 = \sqrt{x} \quad \Rightarrow \quad x = 2500.$

Producer surplus $= \int_0^{2500} [P - p(x)]\,dx = \int_0^{2500} \left(10 - 5 - \frac{1}{10}\sqrt{x}\right) dx = \left[5x - \tfrac{1}{15}x^{3/2}\right]_0^{2500} \approx \$4166.67$

8. $p = 50 - \dfrac{x}{20}$ and $p = 20 + \dfrac{x}{10}$ intersect at $p = 40$ and $x = 200.$

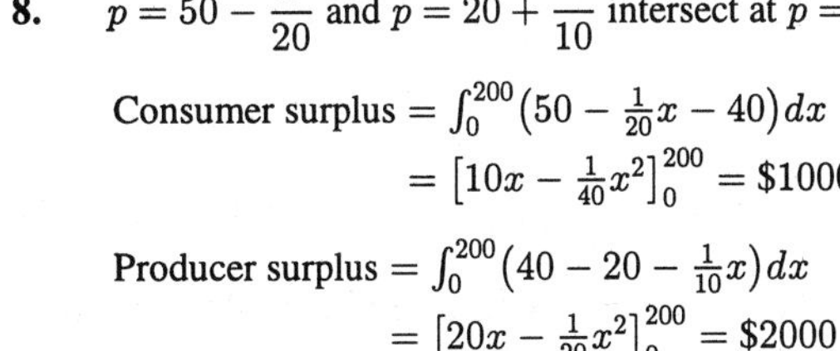

Consumer surplus $= \int_0^{200} \left(50 - \tfrac{1}{20}x - 40\right) dx$

$= \left[10x - \tfrac{1}{40}x^2\right]_0^{200} = \$1000$

Producer surplus $= \int_0^{200} \left(40 - 20 - \tfrac{1}{10}x\right) dx$

$= \left[20x - \tfrac{1}{20}x^2\right]_0^{200} = \$2000$

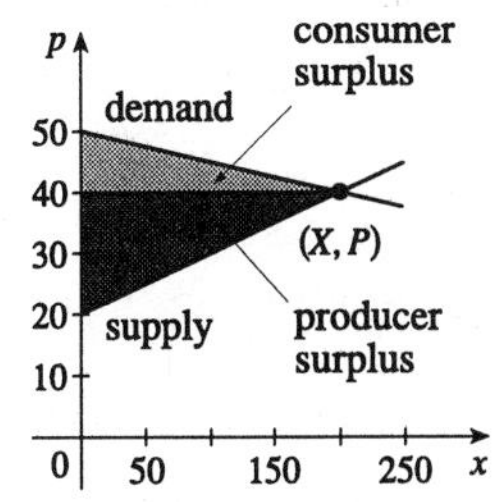

9. The demand function is linear, with slope $\frac{-10}{100}$ and $p(1000) = 450$. So its equation is $p - 450 = -\frac{1}{10}(x - 1000)$ or $p = -\frac{1}{10}x + 550$. A selling price of \$400 $\Rightarrow$ $400 = -\frac{1}{10}x + 550$ $\Rightarrow$ $x = 1500$.

Consumer surplus $= \int_0^{1500} \left(550 - \frac{1}{10}x - 400\right) dx = \left[150x - \frac{1}{20}x^2\right]_0^{1500} = \$112{,}500$

10. Present value $= \int_0^5 2000e^{-0.07t}\,dt = -\frac{2000}{0.07}\left[e^{-0.07t}\right]_0^5 = \frac{2000}{0.07}\left(1 - e^{-0.35}\right) \approx \$8437.48$

11. Pretend that it is five years later. Then the fund will start in five years and continue for 15 years, so the present value (five years from now) is $\int_5^{20} 12{,}000e^{-0.06t}\,dt = -\frac{12{,}000}{0.06}\left[e^{-0.06t}\right]_5^{20} = \frac{12{,}000}{0.06}\left(e^{-0.3} - e^{-1.2}\right) \approx \$87{,}924.80$.

12. Present value $= \int_0^4 f(t)e^{-0.06t}\,dt = \int_0^4 \left(1 + \frac{1}{2}t\right)e^{-0.06t}\,dt = \int_0^4 e^{-0.06t}\,dt + \int_0^4 \frac{1}{2}te^{-0.06t}\,dt$. The second integral can be evaluated by parts using $u = \frac{1}{2}t$, $dv = e^{-0.06t}\,dt$ $\Rightarrow$ $du = \frac{1}{2}\,dt$, $v = -\frac{1}{0.06}e^{-0.06t}$. Continuing:

present value $= -\frac{1}{0.06}\left[e^{-0.06t}\right]_0^4 + \left(-\frac{1}{2 \cdot 0.06}te^{-0.06t}\Big|_0^4 + \frac{1}{2 \cdot 0.06}\int_0^4 e^{-0.06t}\,dt\right)$

$= \frac{1}{0.06}\left(1 - e^{-0.24}\right) - \frac{1}{2 \cdot 0.06}\left(4e^{-0.24}\right) + \frac{1}{2 \cdot (0.06)^2}\left(1 - e^{-0.24}\right) \approx 6.97$ million.

13. **(a)** $f(t) = A$, so present value $= \int_0^\infty Ae^{-rt}\,dt = \lim_{x\to\infty} \int_0^x Ae^{-rt}\,dt = \lim_{x\to\infty} -(A/r)\left[e^{-rt}\right]_0^x$

$= \lim_{x\to\infty} -(A/r)\left[e^{-rx} - 1\right] = A/r$ (since $r > 0$, $e^{-rx} \to 0$ as $x \to \infty$.)

(b) $r = 0.08$, $A = 5000$, so present value $= \frac{5000}{0.08} = \$62{,}500$ [by part (a)].

14. If r is the annual interest rate, then

present value $= \int_0^\infty f(t)e^{-rt}\,dt = \lim_{x\to\infty} \int_0^x (5000 + 1000t)e^{-rt}\,dt = \lim_{x\to\infty} \int_0^x 5000e^{-rt}\,dt + \lim_{x\to\infty} \int_0^x 1000te^{-rt}\,dt$.

We evaluate the second integral by parts with $u = 1000t$, $dv = e^{-rt}\,dt$ $\Rightarrow$ $du = 1000\,dt$, $v = -e^{-rt}/r$:

$$\text{present value} = \lim_{x\to\infty} -\frac{5000}{r}\left[e^{-rt}\right]_0^x + \lim_{x\to\infty}\left[-\frac{1000t}{r}e^{-rt}\Big|_0^x + \int_0^x \frac{1000}{r}e^{-rt}\,dt\right]$$

$$= \lim_{x\to\infty} -\frac{5000}{r}\left[e^{-rx} - 1\right] + \lim_{x\to\infty}\left[-\frac{1000}{r}e^{-rt}\left(t + \frac{1}{r}\right)\right]_0^x = \frac{5000}{r} + \frac{1000}{r^2} = \frac{1000}{r^2}(5r + 1)$$

(since $r > 0$, as $x \to \infty$, $e^{-rx} \to 0$ and $xe^{-rx} \to 0$ by l'Hospital's Rule).

15. $f(8) - f(4) = \int_4^8 f'(t)dt = \int_4^8 \sqrt{t}\,dt = \frac{2}{3}t^{3/2}\Big|_4^8 = \frac{2}{3}\left(16\sqrt{2} - 8\right) = \frac{16(2\sqrt{2}-1)}{3} \approx \$9.75$ million

16. $n(10) - n(4) = \int_4^{10}(200 + 50t)dt = \left[200t + 25t^2\right]_4^{10} = 2000 + 2500 - (800 + 400) = 3300$

17. $F = \dfrac{\pi PR^4}{8\eta\ell} = \dfrac{\pi(4000)(0.008)^4}{8(0.027)(2)} \approx 1.19 \times 10^{-4}\,\text{cm}^3/\text{s}$

18. If the flux remains constant, then $\dfrac{\pi P_0R_0^4}{8\eta\ell} = \dfrac{\pi PR^4}{8\eta\ell}$ $\Rightarrow$ $\dfrac{P}{P_0} = \left(\dfrac{R_0}{R}\right)^4$.

$R = \frac{3}{4}R_0$ $\Rightarrow$ $P = P_0\left(\frac{4}{3}\right)^4 \approx 3.1605 > 3P_0$.

19. $\int_0^{12} c(t)dt = \int_0^{12} \frac{1}{4}t(12 - t)dt = \left[\frac{3}{2}t^2 - \frac{1}{12}t^3\right]_0^{12} = \frac{144}{2} = 72\,\text{mg}\cdot\text{s/L}$. Therefore,
$F = A/72 = \frac{8}{72} = \frac{1}{9}\,\text{L/s} = \frac{60}{9}\,\text{L/min}$.

20. As in Example 3, we will estimate the cardiac output using Simpson's Rule with $\Delta t = 2$, to approximate $\int_0^{20} c(t)dt \approx \frac{2}{3}\left[0 + 4(2.1) + 2(4.5) + 4(7.3) + 2(5.8) + 4(3.6) + 2(2.8) + 4(1.4) + 2(0.6) + 4(0.2) + 0\right]$

$= 57.2\,\text{mg}\cdot\text{s/L}$. Therefore, $F \approx A/57.2 = \dfrac{6}{57.2} \approx 0.1049$ L/s or $6.29\,\text{L/min}$.

REVIEW EXERCISES FOR CHAPTER 8

1. $y^2\dfrac{dy}{dx} = x + \sin x \quad\Rightarrow\quad \displaystyle\int y^2\,dy = \int (x + \sin x)dx \quad\Rightarrow\quad \frac{y^3}{3} = \frac{x^2}{2} - \cos x + C \quad\Rightarrow$
$y^3 = \frac{3}{2}x^2 - 3\cos x + K$ (where $K = 3C$) $\quad\Rightarrow\quad y = \sqrt[3]{\frac{3}{2}x^2 - 3\cos x + K}$

2. $\dfrac{dy}{dx} = \dfrac{y^2+1}{xy},\ x > 0 \quad\Rightarrow\quad \displaystyle\int \frac{y\,dy}{y^2+1} = \int \frac{dx}{x} \quad\Rightarrow\quad \tfrac{1}{2}\ln(y^2+1) = \ln x + C \quad\Rightarrow$
$\ln\sqrt{y^2+1} = \ln(e^C x) \quad\Rightarrow\quad \sqrt{y^2+1} = e^C x \quad\Rightarrow\quad y^2 = Kx^2 - 1$, where $K = e^{2C} > 0$.

3. $y' = \dfrac{1}{x^2y - 2x^2 + y - 2} \quad\Rightarrow\quad \dfrac{dy}{dx} = \dfrac{1}{(x^2+1)(y-2)} \quad\Rightarrow\quad \displaystyle\int (y-2)dy = \int \frac{dx}{x^2+1} \quad\Rightarrow$
$\tfrac{1}{2}y^2 - 2y = \tan^{-1}x + K \quad\Rightarrow\quad y = 2 \pm \sqrt{2\tan^{-1}x + C}$, where $C = 4 + 2K$.

4. $2yy' = xe^x \quad\Rightarrow\quad \int 2y\,dy = \int xe^x\,dx \quad\Rightarrow$
$y^2 = xe^x - \int e^x\,dx$ (by parts) $= (x-1)e^x + C$.
We substitute the initial condition: $1^2 = (0-1)e^0 + C$
$\Rightarrow\quad C = 2$. So the solution is $y = \sqrt{(x-1)e^x + 2}$.
The negative square root is inadmissible due to the initial condition.

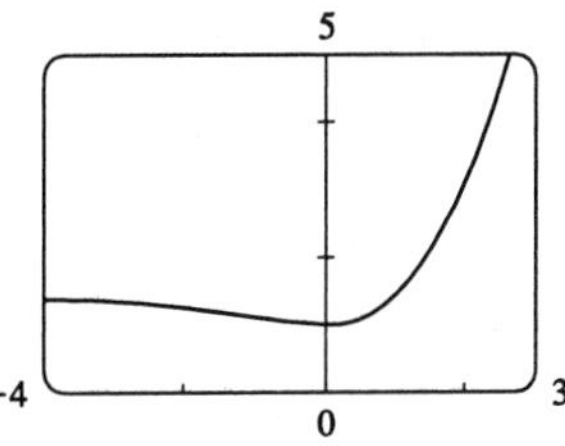

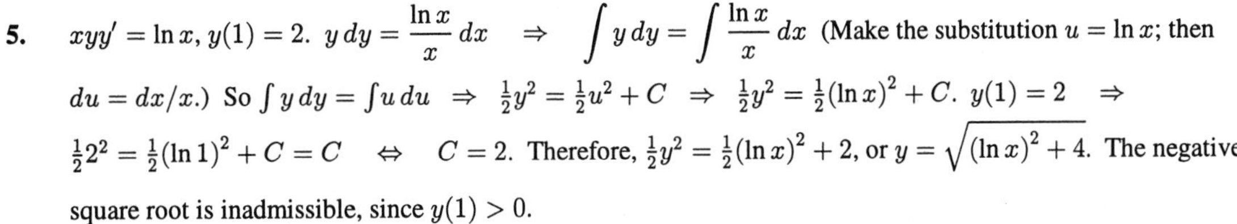

5. $xyy' = \ln x,\ y(1) = 2.\ y\,dy = \dfrac{\ln x}{x}\,dx \quad\Rightarrow\quad \displaystyle\int y\,dy = \int \frac{\ln x}{x}\,dx$ (Make the substitution $u = \ln x$; then $du = dx/x$.) So $\int y\,dy = \int u\,du \quad\Rightarrow\quad \frac{1}{2}y^2 = \frac{1}{2}u^2 + C \quad\Rightarrow\quad \frac{1}{2}y^2 = \frac{1}{2}(\ln x)^2 + C.\ y(1) = 2 \quad\Rightarrow$
$\frac{1}{2}2^2 = \frac{1}{2}(\ln 1)^2 + C = C \quad\Leftrightarrow\quad C = 2$. Therefore, $\frac{1}{2}y^2 = \frac{1}{2}(\ln x)^2 + 2$, or $y = \sqrt{(\ln x)^2 + 4}$. The negative square root is inadmissible, since $y(1) > 0$.

6. We sketch the direction field and four solution curves, as shown. Note that the slope $y' = x/y$ is not defined on the line $y = 0$. (As a check on our work, we solve the equation: $y' = x/y \quad\Leftrightarrow$ $y\,dy = x\,dx \quad\Leftrightarrow\quad y^2 = x^2 + C$. For $C = 0$ this is the pair of lines $y = \pm x$. For $C \neq 0$ it is a hyperbola.)

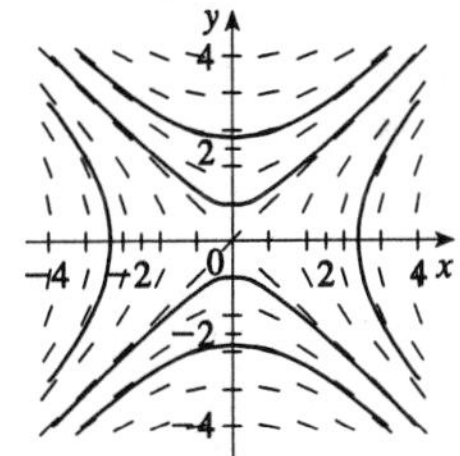

7. $3x = 2(y-1)^{3/2},\ 2 \le y \le 5.\ x = \frac{2}{3}(y-1)^{3/2}$, so $dx/dy = (y-1)^{1/2}$ and the arc length formula gives
$$L = \int_2^5 \sqrt{1 + (dx/dy)^2}\,dy = \int_2^5 \sqrt{1 + (y-1)}\,dy = \int_2^5 \sqrt{y}\,dy = \left[\tfrac{2}{3}y^{3/2}\right]_2^5 = \tfrac{2}{3}\left(5\sqrt{5} - 2\sqrt{2}\right).$$

8. $y = \sqrt{x - x^2} + \sin^{-1}\sqrt{x}$ is defined when $x - x^2 \ge 0$ and $\sqrt{x} \le 1$, that is, for x in $[0, 1]$.
$\dfrac{dy}{dx} = \dfrac{1-2x}{2\sqrt{x-x^2}} + \dfrac{1}{2\sqrt{x}\sqrt{1-x}} = \dfrac{2-2x}{2\sqrt{x-x^2}} = \sqrt{\dfrac{1-x}{x}}$, so $1 + \left(\dfrac{dy}{dx}\right)^2 = 1 + \dfrac{1-x}{x} = \dfrac{1}{x}$. Thus
$$L = \int_0^1 \sqrt{1 + (dy/dx)^2}\,dx = \int_0^1 \sqrt{1/x}\,dx = \int_0^1 x^{-1/2}\,dx = \left[2x^{1/2}\right]_0^1 = 2.$$

9. **(a)** $y = \frac{1}{6}x^3 + \dfrac{1}{2x}, 1 \le x \le 2 \quad\Rightarrow\quad y' = \dfrac{1}{2}\left(x^2 - \dfrac{1}{x^2}\right) \quad\Rightarrow\quad (y')^2 = \dfrac{1}{4}\left(x^4 - 2 + \dfrac{1}{x^4}\right) \quad\Rightarrow$

$1 + (y')^2 = \dfrac{1}{4}\left(x^4 + 2 + \dfrac{1}{x^4}\right) = \dfrac{1}{4}\left(x^2 + \dfrac{1}{x^2}\right)^2 \quad\Rightarrow$

$$L = \int_1^2 \sqrt{1 + (y')^2}\, dy = \frac{1}{2}\int_1^2 \left(x^2 + \frac{1}{x^2}\right) dx = \frac{1}{2}\left[\frac{x^3}{3} - \frac{1}{x}\right]_1^2 = \frac{1}{2}\left(\frac{17}{6}\right) = \frac{17}{12}.$$

(b) $S = \displaystyle\int_1^2 2\pi y \sqrt{1 + \left(\frac{dy}{dx}\right)^2}\, dx = 2\pi \int_1^2 \left(\frac{x^3}{6} + \frac{1}{2x}\right)\frac{1}{2}\left(x^2 + \frac{1}{x^2}\right) dx$

$= \pi \int_1^2 \left(\frac{1}{6}x^5 + \frac{2}{3}x + \frac{1}{2}x^{-3}\right) dx = \pi\left[\frac{1}{36}x^6 + \frac{1}{3}x^2 - \frac{1}{4}x^{-2}\right]_1^2$

$= \pi\left[\left(\frac{64}{36} + \frac{4}{3} - \frac{1}{16}\right) - \left(\frac{1}{36} + \frac{1}{3} - \frac{1}{4}\right)\right] = \frac{47\pi}{16}$

10. **(a)** $y = x^2 \quad\Rightarrow\quad 1 + (y')^2 = 1 + 4x^2 \quad\Rightarrow\quad S = \int_0^1 2\pi x\sqrt{1 + 4x^2}\, dx = \int_1^5 \frac{\pi}{4}\sqrt{u}\, du$ (put $u = 1 + 4x^2$)
$= \frac{\pi}{6}\left[u^{3/2}\right]_1^5 = \frac{\pi}{6}\left(5^{3/2} - 1\right).$

(b) $y = x^2 \quad\Rightarrow\quad 1 + (y')^2 = 1 + 4x^2$. So $S = 2\pi \displaystyle\int_0^1 x^2\sqrt{1 + 4x^2}\, dx = 2\pi \int_0^2 \frac{1}{4}u^2\sqrt{1 + u^2}\,\frac{1}{2}\, du$ (put
$u = 2x$) $= \frac{\pi}{4}\int_0^2 u^2\sqrt{1 + u^2}\, du = \frac{\pi}{4}\left[\frac{1}{8}u(1 + 2u^2)\sqrt{1 + u^2} - \frac{1}{8}\ln\left|u + \sqrt{1 + u^2}\right|\right]_0^2$ (put $u = \tan\theta$ or
use Formula 22) $= \frac{\pi}{4}\left[\frac{1}{4}(9)\sqrt{5} - \frac{1}{8}\ln\left(2 + \sqrt{5}\right) - 0\right] = \frac{\pi}{32}\left[18\sqrt{5} - \ln\left(2 + \sqrt{5}\right)\right].$

11. $y = \dfrac{1}{x^2}, 1 \le x \le 2.$ $\dfrac{dy}{dx} = -\dfrac{2}{x^3}$, so $1 + \left(\dfrac{dy}{dx}\right)^2 = 1 + \dfrac{4}{x^6}$. $L = \displaystyle\int_1^2 \sqrt{1 + \frac{4}{x^6}}\, dx$. By Simpson's Rule with
$n = 10$, $L \approx \frac{1/10}{3}[f(1) + 4f(1.1) + 2f(1.2) + 4f(1.3) + 2f(1.4) + 4f(1.5)$
$+ 2f(1.6) + 4f(1.7) + 2f(1.8) + 4f(1.9) + f(2)] \approx 1.297$. Here $f(x) = \sqrt{1 + 4/x^6}$.

12. $y = \dfrac{1}{x^2}, 1 \le x \le 2 \quad\Rightarrow\quad S = \displaystyle\int_1^2 2\pi y\sqrt{1 + \left(\frac{dy}{dx}\right)^2}\, dx = 2\pi \int_1^2 \frac{1}{x^2}\sqrt{1 + \left(\frac{2}{x^3}\right)^2}\, dx$. By Simpson's Rule
with $n = 10$, $S \approx 2\pi \cdot \frac{1/10}{3}[g(1) + 4g(1.1) + 2g(1.2) + 4g(1.3) + 2g(1.4)$
$+ 4g(1.5) + 2g(1.6) + 4g(1.7) + 2g(1.8) + 4g(1.9) + g(2)]$

where $g(x) = \dfrac{1}{x^2}\sqrt{1 + \left(\dfrac{2}{x^3}\right)^2}$. Thus

$S \approx \frac{2\pi}{30}[2.236 + 4(1.492) + 2(1.062) + 4(0.800) + 2(0.631) + 4(0.517)$
$+ 2(0.435) + 4(0.374) + 2(0.326) + 4(0.289) + 0.258] \approx \frac{2\pi}{30}(21.285) \approx 4.46.$

13. The loop lies between $x = 0$ and $x = 3a$ and is symmetric about the x-axis. We can assume without loss of generality that $a > 0$, since if $a = 0$, the graph is the parallel lines $x = 0$ and $x = 3a$, so there is no loop. The upper half of the loop is given by $y = \dfrac{1}{3\sqrt{a}}\sqrt{x}(3a - x) = \sqrt{a}x^{1/2} - \dfrac{x^{3/2}}{3\sqrt{a}}$, $0 \le x \le 3a$. The desired surface area is twice the area generated by the upper half of the loop, that is, $S = 2(2\pi)\displaystyle\int_0^{3a} x\sqrt{1 + \left(\frac{dy}{dx}\right)^2}\, dx.$

$\frac{dy}{dx} = \frac{\sqrt{a}}{2}x^{-1/2} - \frac{x^{1/2}}{2\sqrt{a}} \quad \Rightarrow \quad 1 + \left(\frac{dy}{dx}\right)^2 = \frac{a}{4x} + \frac{1}{2} + \frac{x}{4a}$. Therefore

$$S = 2(2\pi)\int_0^{3a} x\left(\frac{\sqrt{a}}{2}x^{-1/2} + \frac{x^{1/2}}{2\sqrt{a}}\right)dx = 2\pi\int_0^{3a}\left(\sqrt{a}x^{1/2} + \frac{x^{3/2}}{\sqrt{a}}\right)dx$$

$$= 2\pi\left[\frac{2\sqrt{a}}{3}x^{3/2} + \frac{2}{5\sqrt{a}}x^{5/2}\right]_0^{3a} = 2\pi\left[\frac{2\sqrt{a}}{3}3a\sqrt{3a} + \frac{2}{5\sqrt{a}}9a^2\sqrt{3a}\right] = \frac{56\sqrt{3}\pi a^2}{5}.$$

14. $y^2 = ax - \frac{2}{3}x^2 + \frac{x^3}{9a} = \left(\sqrt{ax} - \frac{x^{3/2}}{3\sqrt{a}}\right)^2, y = \sqrt{ax} - \frac{x^{3/2}}{3\sqrt{a}} \quad \Rightarrow$

$1 + \left(\frac{dy}{dx}\right)^2 = \left(\frac{\sqrt{a}}{2\sqrt{x}} - \frac{\sqrt{x}}{2\sqrt{a}}\right)^2 + 1 = \left(\frac{\sqrt{a}}{2\sqrt{x}} + \frac{\sqrt{x}}{2\sqrt{a}}\right)^2$. So

$$S = 2\pi\int_0^{3a}\left(\sqrt{ax} - \frac{x^{3/2}}{3\sqrt{a}}\right)\left(\frac{\sqrt{a}}{2\sqrt{x}} + \frac{\sqrt{x}}{2\sqrt{a}}\right)dx = 2\pi\int_0^{3a}\left(\frac{a}{2} + \frac{x}{2} - \frac{x}{6} - \frac{x^2}{6a}\right)dx$$

$$= 2\pi\left[\frac{a}{2}x + \frac{x^2}{6} - \frac{x^3}{18a}\right]_0^{3a} = 2\pi\left(\frac{3a^2}{2} + \frac{9a^2}{6} - \frac{27a^3}{18a}\right) = 3\pi a^2.$$

15. $A = \int_{-2}^{1}[(4 - x^2) - (x + 2)]dx = \int_{-2}^{1}(2 - x - x^2)dx = \left[2x - \frac{1}{2}x^2 - \frac{1}{3}x^3\right]_{-2}^{1}$

$= \left(2 - \frac{1}{2} - \frac{1}{3}\right) - \left(-4 - 2 + \frac{8}{3}\right) = \frac{9}{2} \quad \Rightarrow$

$\overline{x} = A^{-1}\int_{-2}^{1} x(2 - x - x^2)dx = \frac{2}{9}\int_{-2}^{1}(2x - x^2 - x^3)dx = \frac{2}{9}\left[x^2 - \frac{1}{3}x^3 - \frac{1}{4}x^4\right]_{-2}^{1}$

$= \frac{2}{9}\left[\left(1 - \frac{1}{3} - \frac{1}{4}\right) - \left(4 + \frac{8}{3} - 4\right)\right] = -\frac{1}{2}$ and

$\overline{y} = A^{-1}\int_{-2}^{1}\frac{1}{2}\left[(4 - x^2)^2 - (x + 2)^2\right]dx = \frac{1}{9}\int_{-2}^{1}(x^4 - 9x^2 - 4x + 12)dx$

$= \frac{1}{9}\left[\frac{1}{5}x^5 - 3x^3 - 2x^2 + 12x\right]_{-2}^{1} = \frac{1}{9}\left[\left(\frac{1}{5} - 3 - 2 + 12\right) - \left(-\frac{32}{5} + 24 - 8 - 24\right)\right] = \frac{12}{5}$.

So $(\overline{x}, \overline{y}) = \left(-\frac{1}{2}, \frac{12}{5}\right)$.

16. The region is symmetric about the line $x = 2$, so $\overline{x} = 2$. $A = \int_0^4(4x - x^2)dx = \left[2x^2 - \frac{1}{3}x^3\right]_0^4 = 32 - \frac{64}{3} = \frac{32}{3}$;

$\overline{y} = A^{-1}\int_0^4 \frac{1}{2}(4x - x^2)^2\,dx = \frac{3}{64}\int_0^4(16x^2 - 8x^3 + x^4)dx$

$= \frac{3}{64}\left[\frac{16}{3}x^3 - 2x^4 + \frac{1}{5}x^5\right]_0^4 = \frac{3}{64}\left[\frac{16}{3}\cdot 64 - 2\cdot 256 + \frac{16\cdot 64}{5}\right] = \frac{8}{5}$. $(\overline{x}, \overline{y}) = \left(2, \frac{8}{5}\right)$.

17. The equation of the line passing through $(0, 0)$ and $(3, 2)$ is $y = \frac{2}{3}x$. $A = \frac{1}{2}\cdot 3\cdot 2 = 3$. Therefore,

$\overline{x} = \frac{1}{3}\int_0^3 x\left(\frac{2}{3}x\right)dx = \frac{2}{27}\left[x^3\right]_0^3 = 2$, and $\overline{y} = \frac{1}{3}\int_0^3\frac{1}{2}\left(\frac{2}{3}x\right)^2dx = \frac{2}{81}\left[x^3\right]_0^3 = \frac{2}{3}$. $(\overline{x}, \overline{y}) = \left(2, \frac{2}{3}\right)$.

Or: Use Exercise 8.4.21.

18. Total area $= 14$. If the lamina were a regular rectangle of area 20, its centroid and center would be $\left(\frac{1}{2}, 0\right)$. Divide the area removed into three squares, one in each of the first, third and fourth quadrants, with areas of 1, 1 and 4, respectively. The centers of these rectangles, in the same order, are $\left(\frac{3}{2}, \frac{1}{2}\right)$, $\left(-\frac{1}{2}, -\frac{3}{2}\right)$ and $(1, -1)$. Therefore $\overline{x} = \frac{1}{14}\left[20\left(\frac{1}{2}\right) - 1\left(\frac{3}{2}\right) - 1\left(-\frac{1}{2}\right) - 4(1)\right] = \frac{5}{14}$ and $\overline{y} = \frac{1}{14}\left[20(0) - 1\left(\frac{1}{2}\right) - 1\left(-\frac{3}{2}\right) - 4(-1)\right] = \frac{5}{14}$.
So $(\overline{x}, \overline{y}) = \left(\frac{5}{14}, \frac{5}{14}\right)$.

19. The centroid of this circle, $(1, 0)$, travels a distance $2\pi(1)$ when the lamina is rotated about the y-axis. The area of the circle is $\pi(1)^2$. So by the Theorem of Pappus, $V = A2\pi\overline{x} = \pi(1)^2 2\pi(1) = 2\pi^2$.

20. The semicircular region has an area of $\frac{1}{2}\pi r^2$, and sweeps out a sphere of radius r when rotated about the x-axis. $\overline{x} = 0$ because of symmetry about the line $x = 0$. And by the Theorem of Pappus, Volume $= \frac{4}{3}\pi r^3 = \frac{1}{2}\pi r^2(2\pi\overline{y})$ $\Rightarrow$ $\overline{y} = \frac{4}{3\pi}r$. Therefore, $(\overline{x}, \overline{y}) = \left(0, \frac{4}{3\pi}r\right)$.

21. As in Example 1 of Section 8.5, $F = \int_0^2 \rho g x(5-x)\,dx = \rho g\left[\frac{5}{2}x^2 - \frac{1}{3}x^3\right]_0^2 = \rho g\frac{22}{3} = \frac{22}{3}\delta \approx \frac{22}{3}\cdot 62.5 \approx 458$ lb.

22. 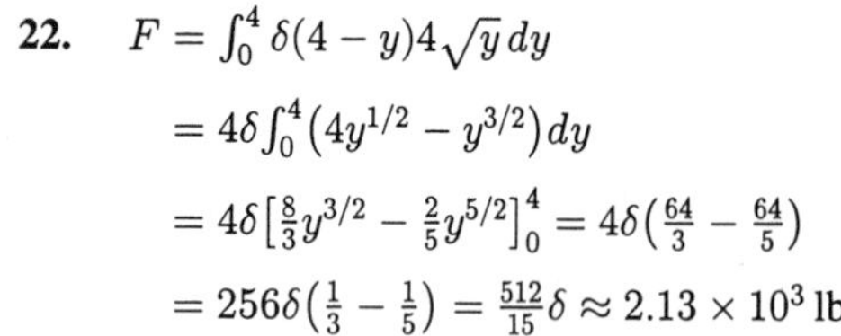

$F = \int_0^4 \delta(4-y)4\sqrt{y}\,dy$

$= 4\delta\int_0^4 \left(4y^{1/2} - y^{3/2}\right)dy$

$= 4\delta\left[\frac{8}{3}y^{3/2} - \frac{2}{5}y^{5/2}\right]_0^4 = 4\delta\left(\frac{64}{3} - \frac{64}{5}\right)$

$= 256\delta\left(\frac{1}{3} - \frac{1}{5}\right) = \frac{512}{15}\delta \approx 2.13 \times 10^3$ lb

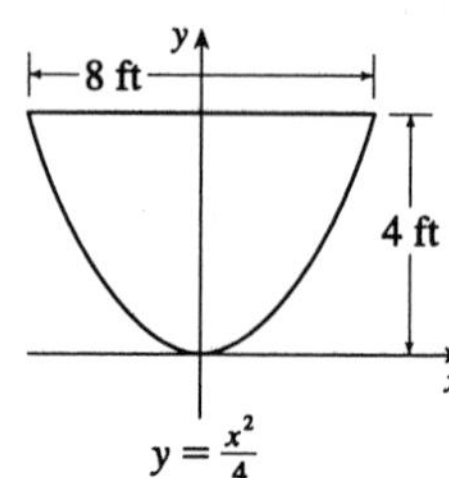

23. $x = 100 \quad \Rightarrow \quad P = 2000 - 0.1(100) - 0.01(100)^2 = 1890$

Consumer surplus $= \int_0^{100}[p(x) - P]\,dx = \int_0^{100}(2000 - 0.1x - 0.01x^2 - 1890)\,dx$

$= \left[110x - 0.05x^2 - \frac{0.01}{3}x^3\right]_0^{100} = 11{,}000 - 500 - \frac{10{,}000}{3} \approx \$7166.67$

24. **(a)** $\int_8^{15} 10{,}000e^{-0.08t}\,dt = -\frac{10{,}000}{0.08}\left[e^{-0.08t}\right]_8^{15} = -\frac{10{,}000}{0.08}\left[e^{-1.2} - e^{-0.64}\right] \approx \$28{,}262.28$

(b) $\int_3^{10} 10{,}000e^{-0.08t}\,dt = -\frac{10{,}000}{0.08}\left[e^{-0.08t}\right]_3^{10} = -\frac{10{,}000}{0.08}\left[e^{-0.8} - e^{-0.24}\right] \approx \$42{,}162.36$

(c) $\int_8^{\infty} 10{,}000e^{-0.08t}\,dt = \lim_{x\to\infty}\int_8^x 10{,}000e^{-0.08t}\,dt = \lim_{x\to\infty}\left(-\frac{10{,}000}{0.08}\left[e^{-0.08t}\right]_{12}^{x}\right)$

$= \lim_{x\to\infty} -\frac{10{,}000}{0.08}\left[e^{-0.08x} - e^{-1.2}\right] = \frac{10{,}000}{0.08}e^{-0.64} \approx \$65{,}911.55$

25. **(a)** $\frac{dL}{dt} \propto L_\infty - L \quad \Rightarrow \quad \frac{dL}{dt} = k(L_\infty - L) \quad \Rightarrow \quad \int \frac{dL}{L_\infty - L} = \int k\,dt \quad \Rightarrow \quad -\ln|L_\infty - L| = kt + C$

$\Rightarrow \quad L_\infty - L = Ae^{-kt} \quad \Rightarrow \quad L = L_\infty - Ae^{-kt}$. At $t = 0$, $L = L(0) = L_\infty - A \quad \Rightarrow$

$A = L_\infty - L(0) \quad \Rightarrow \quad L(t) = L_\infty - [L_\infty - L(0)]e^{-kt}$

(b) $L_\infty = 53$ cm, $L(0) = 10$ cm and $k = 0.2$. So $L(t) = 53 - (53 - 10)e^{-0.2t} = 53 - 43e^{-0.2t}$.

26. Denote the amount of salt in the tank (in kg) by y. $y(0) = 0$ since initially there is only water in the tank. The rate at which y increases is equal to the rate at which salt flows into the tank minus the rate at which it flows out. That rate is $\frac{dy}{dt} = 0.1\,\frac{\text{kg}}{\text{L}} \times 10\,\frac{\text{L}}{\text{min}} - \frac{y}{100}\,\frac{\text{kg}}{\text{L}} \times 10\,\frac{\text{L}}{\text{min}} = 1 - \frac{y}{10}\,\frac{\text{kg}}{\text{min}} \quad \Rightarrow \quad \int\frac{dy}{10 - y} = \int\frac{1}{10}\,dt \quad \Rightarrow$ $-\ln|10 - y| = \frac{1}{10}t + C \quad \Rightarrow \quad 10 - y = Ae^{-t/10}$. $y(0) = 0 \quad \Rightarrow \quad 10 = A \quad \Rightarrow \quad y = 10\left(1 - e^{-t/10}\right)$. At $t = 6$ minutes, $y = 10\left(1 - e^{-6/10}\right) \approx 4.512$ kg.

27. Let P be the population and I be the number of infected people. The rate of spread dI/dt is jointly proportional to I and to $P - I$, so for some constant k, $\frac{dI}{dt} = kI(P - I) \quad \Rightarrow \quad I = \frac{I_0 P}{I_0 + (P - I_0)e^{-kPt}}$ (from the discussion of logistic growth in Section 8.1).

Now, measuring t in days, we substitute $t = 7$, $P = 5000$, $I_0 = 160$ and $I(7) = 1200$ to find k:
$1200 = \dfrac{160 \cdot 5000}{160 + (5000 - 160)e^{-5000 \cdot 7 \cdot k}} \quad \Leftrightarrow \quad k \approx 0.00006448$. So, putting $I = 5000 \times 80\% = 4000$, we solve
for t: $4000 = \dfrac{160 \cdot 5000}{160 + (5000 - 160)e^{-0.00006448 \cdot 5000 \cdot t}} \quad \Leftrightarrow \quad 160 + 4840e^{-0.3224t} = 200 \quad \Leftrightarrow$
$-0.3224t = \ln \dfrac{40}{4840} \quad \Leftrightarrow \quad t \approx 14.9$. So it takes about 15 days for 80% of the population to be infected.

28. First note that, in this question, "weighs" is used in the informal sense, so what we really require is Barbara's mass m in kg as a function of t. Barbara's net intake of calories per day at time t (measured in days) is $c(t) = 1600 - 850 - 15m(t) = 750 - 15m(t)$, where $m(t)$ is her mass at time t. We are given that $m(0) = 60$ kg and $\dfrac{dm}{dt} = \dfrac{c(t)}{10{,}000}$, so $\dfrac{dm}{dt} = \dfrac{750 - 15m}{10{,}000} = \dfrac{150 - 3m}{2000} = \dfrac{-3(m - 50)}{2000}$ with $m(0) = 60$.
From $\displaystyle\int \frac{dm}{m - 50} = \int \frac{-3\,dt}{2000}$, we get $\ln|m - 50| = -\frac{3}{2000}t + C$. Since $m(0) = 60$, $C = \ln 10$.

Now $\ln(|m - 50|/10) = -3t/2000$, so $|m - 50| = 10e^{-3t/2000}$. The quantity $m - 50$ is continuous and initially positive; the right-hand side is never zero. Thus $m - 50$ is positive for all t, and $m(t) = 50 + 10e^{-3t/2000}$ kg. As $t \to \infty$, $m(t) \to 50$ kg. Thus Barbara's mass gradually settles down to 50 kg.

29. **(a)** We are given that $V = \frac{1}{3}\pi r^2 h$, $dV/dt = 60{,}000\pi \text{ ft}^3/\text{h}$, and $r = 1.5h = \frac{3}{2}h$. So $V = \frac{1}{3}\pi\left(\frac{3}{2}h\right)^2 \Rightarrow$
$\dfrac{dV}{dt} = \frac{3}{4}\pi \cdot 3h^2 \dfrac{dh}{dt} = \frac{9}{4}\pi h^2 \dfrac{dh}{dt}$. Therefore, $\dfrac{dh}{dt} = \dfrac{4(dV/dt)}{9\pi h^2} = \dfrac{240{,}000\pi}{9\pi h^2} = \dfrac{80{,}000}{3h^2}$ (★) $\Rightarrow$
$\int 3h^2\,dh = \int 80{,}000\,dt \;\Rightarrow\; h^3 = 80{,}000t + C$. When $t = 0$, $h = 60$. Therefore, $C = 60^3 = 216{,}000$, so $h^3 = 80{,}000t + 216{,}000$. Let $h = 100$. Then $100^3 = 1{,}000{,}000 = 80{,}000t + 216{,}000 \;\Rightarrow$
$80{,}000t = 784{,}000 \;\Rightarrow\; t = 9.8$, so the time required is 9.8 hours.

(b) The floor area of the silo is $F = \pi \cdot 200^2 = 40{,}000\pi \text{ ft}^2$, and the area of the base of the pile is $A = \pi r^2 = \pi\left(\frac{3}{2}h\right)^2 = \frac{9\pi}{4}h^2 = 8100\pi \text{ ft}^2$. So the area of the floor which is not covered is $F - A = 31{,}900\pi \approx 100{,}000 \text{ ft}^2$.

Now $A = \frac{9\pi}{4}h^2 \;\Rightarrow\; \dfrac{dA}{dt} = \dfrac{9\pi}{4} \cdot 2h\dfrac{dh}{dt}$, and from (★) in part (a) we know that when $h = 60$,
$\dfrac{dh}{dt} = \dfrac{80{,}000}{3(60)^2} = \dfrac{200}{27}\dfrac{\text{ft}}{\text{h}}$. Therefore $\dfrac{dA}{dt} = \dfrac{9\pi}{4}(2)(60)\left(\dfrac{200}{27}\right) \approx 6283\,\dfrac{\text{ft}^2}{\text{h}}$.

(c) At $h = 90$ ft, $\dfrac{dV}{dt} = 60{,}000\pi - 20{,}000\pi = 40{,}000\pi \text{ ft}^3/\text{h}$. From (★) in (a),
$\dfrac{dh}{dt} = \dfrac{4(dV/dt)}{9\pi h^2} = \dfrac{4(40{,}000\pi)}{9\pi h^2} = \dfrac{160{,}000}{9h^2} \;\Rightarrow\; \int 9h^2\,dh = \int 160{,}000\,dt \;\Rightarrow$
$3h^3 = 160{,}000t + C$. When $t = 0$, $h = 90$; therefore, $C = 3 \cdot 729{,}000 = 2{,}187{,}000$. So
$3h^3 = 160{,}000t + 2{,}187{,}000$. At the top, $h = 100$, so $3(100)^3 = 160{,}000t + 2{,}187{,}000 \;\Rightarrow$
$t = \dfrac{813{,}000}{160{,}000} \approx 5.1$. The pile reaches the top after about 5.1 h.

PROBLEMS PLUS (page 544)

1. By symmetry, the problem can be reduced to finding the line $x = c$ such that the shaded area is one-third of the area of the quarter-circle. The equation of the circle is $y = \sqrt{49 - x^2}$, so we require that $\int_0^c \sqrt{49 - x^2}\,dx = \frac{1}{3} \cdot \frac{1}{4}\pi(7)^2 \quad \Leftrightarrow$ $\left[\frac{1}{2}x\sqrt{49 - x^2} + \frac{49}{2}\sin^{-1}(x/7)\right]_0^c = \frac{49}{12}\pi$ (Formula 30) $\Leftrightarrow$ $\frac{1}{2}c\sqrt{49 - c^2} + \frac{49}{2}\sin^{-1}(c/7) = \frac{49}{12}\pi$.

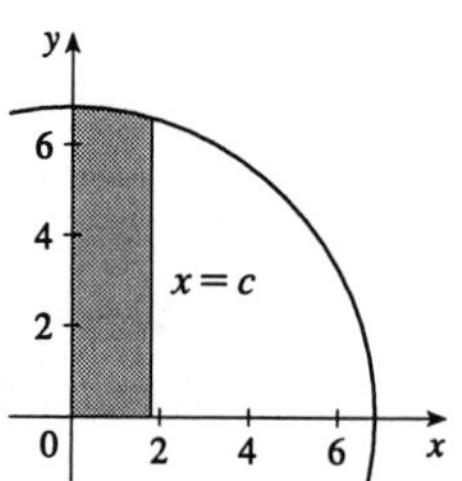

This equation would be difficult to solve exactly, so we plot the left-hand side as a function of c, and find that the equation holds for $c \approx 1.85$. So the cuts should be made at distances of about 1.85 inches from the center of the pizza.

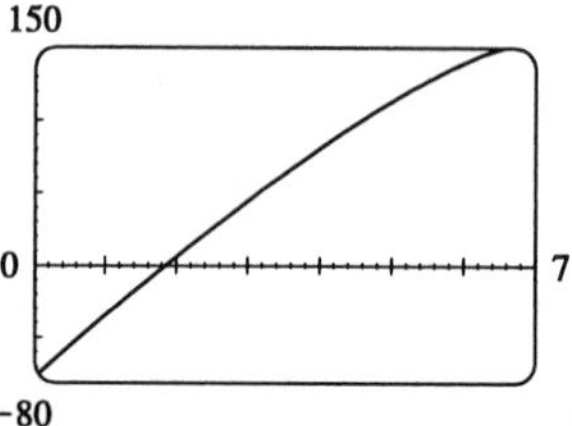

2. $(fg)' = f'g'$, where $f(x) = e^{x^2} \quad \Rightarrow \quad \left(e^{x^2}g\right)' = 2xe^{x^2}g'$. Since the student's mistake did not affect the answer, $\left(e^{x^2}g\right)' = 2xe^{x^2}g + e^{x^2}g' = 2xe^{x^2}g'$. So $(2x - 1)g' = 2xg$, or $\dfrac{g'}{g} = \dfrac{2x}{2x-1} = 1 + \dfrac{1}{2x-1} \quad \Rightarrow$ $\ln|g(x)| = x + \frac{1}{2}\ln(2x - 1) + C \Rightarrow g(x) = Ae^x\sqrt{2x - 1}$.

3.

$$\begin{aligned} f'(x) &= \lim_{h\to 0} \frac{f(x+h) - f(x)}{h} \\ &= \lim_{h\to 0} \frac{f(x)[f(h) - 1]}{h} \quad [\text{since } f(x+h) = f(x)f(h)] \\ &= f(x) \lim_{h\to 0} \frac{f(h) - 1}{h} \\ &= f(x) \lim_{h\to 0} \frac{f(h) - f(0)}{h - 0} \\ &= f(x)f'(0) \\ &= f(x) \end{aligned}$$

Therefore, $f'(x) = f(x)$ for all x. In Leibniz notation, we can either solve the differential equation $\dfrac{dy}{dx} = y$ $\Rightarrow \quad \displaystyle\int \frac{dy}{y} = \int dx \quad \Rightarrow \quad \ln|y| = x + C \quad \Rightarrow \quad y = Ae^x$, or we can use Theorem 6.5.2 to get the same result. Now $f(0) = 1 \quad \Rightarrow \quad A = 1 \quad \Rightarrow \quad f(x) = e^x$.

4. $y = \pm\sqrt{x^3 - x^4} \quad \Rightarrow$ The loop of the curve is symmetric about $y = 0$, and therefore $\overline{y} = 0$. At each point x where $0 \le x \le 1$, the lamina has a vertical length of $\sqrt{x^3 - x^4} - \left(-\sqrt{x^3 - x^4}\right) = 2\sqrt{x^3 - x^4}$. Therefore,

$$\overline{x} = \frac{\int_0^1 x2\sqrt{x^3 - x^4}\,dx}{\int_0^1 2\sqrt{x^3 - x^4}\,dx} = \frac{\int_0^1 x\sqrt{x^3 - x^4}\,dx}{\int_0^1 \sqrt{x^3 - x^4}\,dx}.$$ We evaluate the integrals separately:

$\int_0^1 x\sqrt{x^3 - x^4}\,dx = \int_0^1 x^{5/2}\sqrt{1-x}\,dx$

$= \displaystyle\int_0^{\pi/2} 2\sin^6\theta\cos\theta\sqrt{1-\sin^2\theta}\,d\theta \quad \begin{bmatrix}\text{where } \sin\theta = \sqrt{x};\ \cos\theta\,d\theta = dx/(2\sqrt{x}) \\ \Rightarrow \quad 2\sin\theta\cos\theta\,d\theta = dx\end{bmatrix}$

$= \int_0^{\pi/2} 2\sin^6\theta\cos^2\theta\,d\theta = \int_0^{\pi/2} 2\left[\frac{1}{2}(1-\cos 2\theta)\right]^3 \frac{1}{2}(1+\cos 2\theta)d\theta$

$= \int_0^{\pi/2} \frac{1}{8}\left(1 - 2\cos 2\theta + 2\cos^3 2\theta - \cos^4 2\theta\right)d\theta$

$= \int_0^{\pi/2} \frac{1}{8}\left[1 - 2\cos 2\theta + 2\cos 2\theta(1-\sin^2 2\theta) - \frac{1}{4}(1+\cos 4\theta)^2\right]d\theta$

$= \frac{1}{8}\left[\theta - \frac{1}{3}\sin^3 2\theta\right]_0^{\pi/2} - \frac{1}{32}\int_0^{\pi/2}(1 + 2\cos 4\theta + \cos^2 4\theta)d\theta$

$= \frac{\pi}{16} - \frac{1}{32}\left[\theta + \frac{1}{2}\sin 4\theta\right]_0^{\pi/2} - \frac{1}{64}\int_0^{\pi/2}(1+\cos 8\theta)d\theta$

$= \frac{3\pi}{64} - \frac{1}{64}\left[\theta + \frac{1}{8}\sin 8\theta\right]_0^{\pi/2} = \frac{5\pi}{128}$

$\int_0^1 \sqrt{x^3 - x^4}\,dx = \int_0^1 x^{3/2}\sqrt{1-x}\,dx = \int_0^{\pi/2} 2\sin^4\theta\cos\theta\sqrt{1-\sin^2\theta}\,d\theta \quad (\text{where } \sin\theta = \sqrt{x})$

$= \int_0^{\pi/2} 2\sin^4\theta\cos^2\theta\,d\theta = \int_0^{\pi/2} 2\cdot\frac{1}{4}(1-\cos 2\theta)^2\cdot\frac{1}{2}(1+\cos 2\theta)d\theta$

$= \int_0^{\pi/2} \frac{1}{4}(1 - \cos 2\theta - \cos^2 2\theta + \cos^3 2\theta)d\theta$

$= \int_0^{\pi/2} \frac{1}{4}\left[1 - \cos 2\theta - \frac{1}{2}(1+\cos 4\theta) + \cos 2\theta(1-\sin^2 2\theta)\right]d\theta$

$= \frac{1}{4}\left[\frac{\theta}{2} - \frac{1}{8}\sin 4\theta - \frac{1}{6}\sin^3 2\theta\right]_0^{\pi/2} = \frac{\pi}{16}$

Or: Use Formula 114.

Therefore, $\overline{x} = \dfrac{5\pi/128}{\pi/16} = \dfrac{5}{8}$, and $(\overline{x}, \overline{y}) = \left(\dfrac{5}{8}, 0\right)$.

5. First we show that $x(1-x) \le \frac{1}{4}$ for all x. Let $f(x) = x(1-x) = x - x^2$. Then $f'(x) = 1 - 2x$. This is 0 when $x = \frac{1}{2}$ and $f'(x) > 0$ for $x < \frac{1}{2}$, $f'(x) < 0$ for $x > \frac{1}{2}$, so the absolute maximum of f is $f\left(\frac{1}{2}\right) = \frac{1}{4}$. Thus $x(1-x) \le \frac{1}{4}$ for all x.

Now suppose that the given assertion is false, that is, $a(1-b) > \frac{1}{4}$ and $b(1-a) > \frac{1}{4}$. Multiply these inequalities: $a(1-b)b(1-a) > \frac{1}{16} \quad\Rightarrow\quad [a(1-a)][b(1-b)] > \frac{1}{16}$. But we know that $a(1-a) \le \frac{1}{4}$ and $b(1-b) \le \frac{1}{4} \quad\Rightarrow\quad [a(1-a)][b(1-b)] \le \frac{1}{16}$. Thus we have a contradiction, so the given assertion is proved.

6. $\displaystyle\int \frac{1}{x^7 - x}\,dx = \int \frac{dx}{x(x^6-1)} = \int \frac{x^5}{x^6(x^6-1)}\,dx = \frac{1}{6}\int \frac{du}{u(u-1)}\,du \quad \left[u = x^6\right]$

$\displaystyle = \frac{1}{6}\int\left(\frac{1}{u-1} - \frac{1}{u}\right)du = \tfrac{1}{6}(\ln|u-1| - \ln|u|) + C = \tfrac{1}{6}\ln\left|\frac{u-1}{u}\right| + C = \tfrac{1}{6}\ln\left|\frac{x^6-1}{x^6}\right| + C$

Alternate Method: $\displaystyle\int \frac{1}{x^7 - x}\,dx = \int \frac{x^{-7}}{1 - x^{-6}}\,dx \quad \left[\text{put } u = 1 - x^{-6},\ du = 6x^{-7}\,dx\right]$

$\displaystyle = \frac{1}{6}\int \frac{du}{u} = \tfrac{1}{6}\ln|u| + C = \tfrac{1}{6}\ln|1 - x^{-6}| + C$

Other Methods: Substitute $u = x^3$ or $x^3 = \sec\theta$.

7. Let $F(x) = \int_a^x f(t)dt - \int_x^b f(t)dt$. Then $F(a) = -\int_a^b f(t)dt$ and $F(b) = \int_a^b f(t)dt$. Also F is continuous by the Fundamental Theorem. So by the Intermediate Value Theorem, there is a number c in $[a, b]$ such that $F(c) = 0$. For that number c, we have $\int_a^c f(t)dt = \int_c^b f(t)dt$.

8. **(a)** Since the right triangles OAT and OBT are similar, we have $\dfrac{H+r}{r}=\dfrac{r}{a}$ $\Rightarrow$ $a=\dfrac{r^2}{H+r}$. The surface area visible from B is $S=\int_a^r 2\pi x\sqrt{1+(dx/dy)^2}\,dy$.

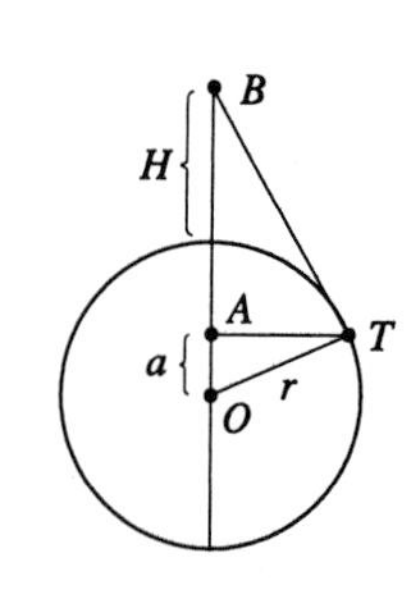

From $x^2+y^2=r^2$, we get $\dfrac{dy}{dx}=-\dfrac{y}{x}$ and $1+\left(\dfrac{dx}{dy}\right)^2=\dfrac{x^2+y^2}{x^2}=\dfrac{r^2}{x^2}$. Thus

$$S=\int_a^r 2\pi x\cdot\frac{r}{x}\,dy=2\pi r(r-a)=2\pi r\left(r-\frac{r^2}{H+r}\right)=2\pi r^2\left(1-\frac{r}{H+r}\right)$$

$$=2\pi r^2\cdot\frac{H}{H+R}=\frac{2\pi r^2H}{r+H}.$$

(b) If a light is placed at point L, at a distance x from the center of the sphere of radius r, then from (a) we find that the total illuminated area on the two spheres is $A=\dfrac{2\pi r^2(x-R)}{x}+\dfrac{2\pi R^2(d-x-R)}{d-x}$.

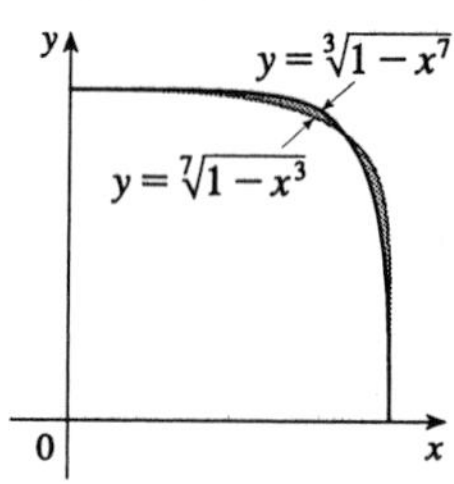

$\dfrac{A}{2\pi}=r^2\left(1-\dfrac{r}{x}\right)+R^2\left(1-\dfrac{R}{d-x}\right)$ $\Rightarrow$ $\dfrac{dA}{dx}=0$ $\Leftrightarrow$

$0=r^2\cdot\dfrac{r}{x^2}+R^2\cdot\dfrac{-R}{(d-x)^2}$ $\Leftrightarrow$ $0=\dfrac{r^3}{x^2}-\dfrac{R^3}{(d-x)^2}$ $\Leftrightarrow$ $R^3x^2=r^3\left[d^2-(2d)x+x^2\right]$ $\Leftrightarrow$

$(R^3-r^3)x^2+(2r^3d)x-r^3d^2=0$ $\Leftrightarrow$ $\left[(R/r)^3-1\right]x^2+(2d)x-d^2=0$. Assume, without loss of generality, that $R=\lambda r$, where $\lambda\ge 1$. Then $dA/dx=0$ $\Leftrightarrow$ $(\lambda^3-1)x^2+(2d)x-d^2=0$ $\Leftrightarrow$

$x=\dfrac{-2d\pm 2d\sqrt{\lambda^3}}{2(\lambda^3-1)}$ $\Leftrightarrow$ $x=\dfrac{\lambda^{3/2}-1}{\lambda^3-1}d$ (since $x>0$) $\Leftrightarrow$ $x=\dfrac{d}{\lambda^{3/2}-1}$.

9. The given integral represents the difference of the shaded areas, which appears to be 0. It can be calculated by integrating with respect to either x or y, so we find x in terms of y for each curve:

$y=\sqrt[3]{1-x^7}$ $\Rightarrow$ $x=\sqrt[7]{1-y^3}$ and $y=\sqrt[7]{1-x^3}$ $\Rightarrow$ $x=\sqrt[3]{1-y^7}$, so

$$\int_0^1\left(\sqrt[3]{1-y^7}-\sqrt[7]{1-y^3}\right)dy=\int_0^1\left(\sqrt[7]{1-x^3}-\sqrt[3]{1-x^7}\right)dx.$$

But this equation is of the form $z=-z$. So $\int_0^1\left(\sqrt[3]{1-x^7}-\sqrt[7]{1-x^3}\right)dx=0$.

10. Let b be the number of hours before noon that it began to snow, t the time measured in hours after noon, and $x=x(t)=$ distance traveled by the plow at time t. Then $dx/dt=$ speed of plow. Since the snow falls steadily, the height at time t is $h(t)=k(t+b)$, where k is a constant. We are given that the rate of removal is constant, say R (in m^3/h). If the width of the path is w, then

$R=\text{height}\times\text{width}\times\text{speed}=h(t)\times w\times\dfrac{dx}{dt}=k(t+b)w\dfrac{dx}{dt}$. Thus $\dfrac{dx}{dt}=\dfrac{C}{t+b}$ where $C=\dfrac{R}{kw}$ is a constant. This is a separable equation. $\int dx=C\int\dfrac{dt}{t+b}$ $\Rightarrow$ $x=C\ln(t+b)+K$.

Put $t=0$: $0=C\ln b+K$ $\Rightarrow$ $K=-C\ln b$, so $x(t)=C\ln(t+b)-C\ln b=C\ln(1+t/b)$.

Put $t=1$: $6000=C\ln(1+1/b)$. Put $t=2$: $9000=C\ln(1+2/b)$.

Solve for b: $\dfrac{\ln(1+1/b)}{6000} = \dfrac{\ln(1+2/b)}{9000} \;\Rightarrow\; 3\ln\left(1+\frac{1}{b}\right) = 2\ln\left(1+\frac{2}{b}\right) \;\Rightarrow$
$\left(1+\frac{1}{b}\right)^3 = \left(1+\frac{2}{b}\right)^2 \;\Rightarrow\; 1+\frac{3}{b}+\frac{3}{b^2}+\frac{1}{b^3} = 1+\frac{4}{b}+\frac{4}{b^2} \;\Rightarrow\; \frac{1}{b}+\frac{1}{b^2}-\frac{1}{b^3}=0 \;\Rightarrow$
$b^2+b-1=0 \;\Rightarrow\; b=\frac{-1\pm\sqrt{5}}{2}$. But $b>0$, so $b=\frac{-1+\sqrt{5}}{2}$. The snow began to fall $\frac{\sqrt{5}-1}{2}$ hours before noon, that is, at about 11:23 A.M..

11. First we find the domain explicitly. $5x^2 \geq x^4+4 \;\Leftrightarrow\; x^4-5x^2+4 \leq 0 \;\Leftrightarrow\; (x^2-4)(x^2-1)\leq 0$ $\Leftrightarrow\; (x-2)(x-1)(x+1)(x+2)\leq 0 \;\Leftrightarrow\; x\in[-2,-1]$ or $[1,2]$. Therefore, the domain is $\{x \mid 5x^2 \geq x^4+4\} = [-2,-1]\cup[1,2]$. Now f is decreasing on its domain, since $f(x)=3x-2x^3 \;\Rightarrow$ $f'(x)=3-6x^2 \leq -3$. So f's maximum value is either $f(-2)$ or $f(1)$. $f(-2)=10$ and $f(1)=1$, so the maximum value of f is 10.

12. $\left[\int f(x)dx\right]\left[\int \frac{dx}{f(x)}\right] = -1 \;\Rightarrow\; \int \frac{dx}{f(x)} = \frac{-1}{\int f(x)dx} \;\Rightarrow\; \frac{1}{f(x)} = \frac{f(x)}{\left[\int f(x)dx\right]^2}$ (after differentiating)
$\Rightarrow\; \int f(x)dx = \pm f(x)$ (after taking square roots) $\Rightarrow\; f(x)=\pm f'(x)$ (after differentiating again)
$\Rightarrow\; \int \frac{dy}{y} = \pm\int dx \;\Rightarrow\; \ln|y| = \pm x + C \;\Rightarrow\; y=Ae^x$ or $y=Ae^{-x}$. Therefore, $f(x)=Ae^x$ or $f(x)=Ae^{-x}$, for all non-zero constants A, are the functions satisfying the original equation.

13. Recall that $\cos A\cos B = \frac{1}{2}[\cos(A+B)+\cos(A-B)]$. So

$$\begin{aligned} f(x) &= \int_0^\pi \cos t\cos(x-t)dt = \tfrac{1}{2}\int_0^\pi[\cos(t+x-t)+\cos(t-x+t)]dt \\ &= \tfrac{1}{2}\int_0^\pi[\cos x+\cos(2t-x)]dt = \tfrac{1}{2}\left[t\cos x + \tfrac{1}{2}\sin(2t-x)\right]_0^\pi \\ &= \tfrac{\pi}{2}\cos x + \tfrac{1}{4}\sin(2\pi - x) - \tfrac{1}{4}\sin(-x) = \tfrac{\pi}{2}\cos x + \tfrac{1}{4}\sin(-x) - \tfrac{1}{4}\sin(-x) \\ &= \tfrac{\pi}{2}\cos x. \end{aligned}$$

The minimum of $\cos x$ on this domain is -1, so the minimum value of $f(x)$ is $f(\pi) = -\frac{\pi}{2}$.

14. The problem can be reduced to finding the line which minimizes the shaded area in the diagram. The equation of the circle in the first quadrant is $y=\sqrt{1-x^2}$, so if the equation of the line is $y=h$, then the circle and the line intersect where $h=\sqrt{1-x^2} \Rightarrow x=\sqrt{1-h^2}$. So the shaded area is

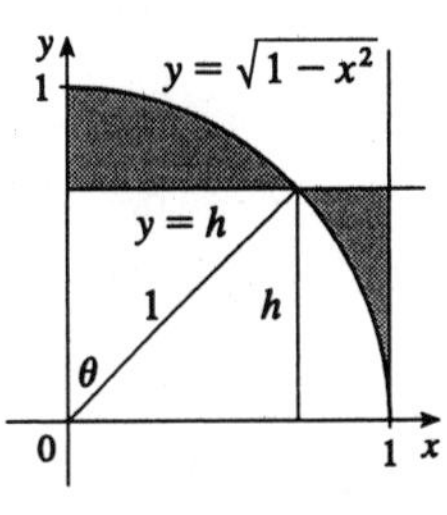

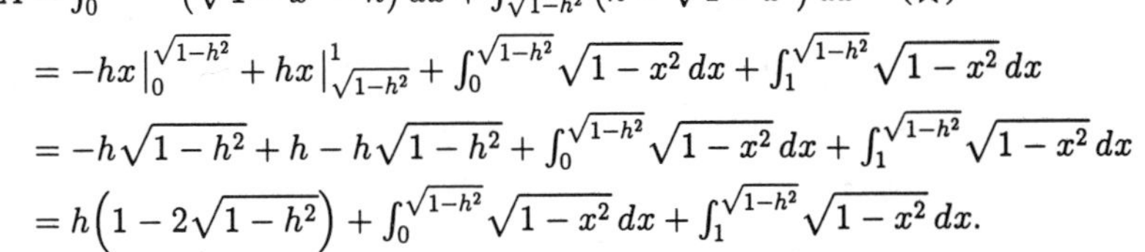

$$\begin{aligned} A &= \int_0^{\sqrt{1-h^2}}\left(\sqrt{1-x^2}-h\right)dx + \int_{\sqrt{1-h^2}}^1\left(h-\sqrt{1-x^2}\right)dx \quad (\bigstar) \\ &= -hx\Big|_0^{\sqrt{1-h^2}} + hx\Big|_{\sqrt{1-h^2}}^1 + \int_0^{\sqrt{1-h^2}}\sqrt{1-x^2}\,dx + \int_1^{\sqrt{1-h^2}}\sqrt{1-x^2}\,dx \\ &= -h\sqrt{1-h^2} + h - h\sqrt{1-h^2} + \int_0^{\sqrt{1-h^2}}\sqrt{1-x^2}\,dx + \int_1^{\sqrt{1-h^2}}\sqrt{1-x^2}\,dx \\ &= h\left(1-2\sqrt{1-h^2}\right) + \int_0^{\sqrt{1-h^2}}\sqrt{1-x^2}\,dx + \int_1^{\sqrt{1-h^2}}\sqrt{1-x^2}\,dx. \end{aligned}$$

Note that at the second line, we reversed the limits of integration and changed the sign in the last integral.

We are interested in the minimum of $A(h) = h\left(1 - 2\sqrt{1-h^2}\right) + \int_0^{\sqrt{1-h^2}} \sqrt{1-x^2}\,dx + \int_1^{\sqrt{1-h^2}} \sqrt{1-x^2}\,dx$, so we find dA/dh using the Fundamental Theorem of Calculus, Part 1, and the Chain Rule:

$$\frac{dA}{dh} = h\left(-2\frac{-h}{\sqrt{1-h^2}}\right) + \left(1 - 2\sqrt{1-h^2}\right) + 2\left[\sqrt{1 - \left(\sqrt{1-h^2}\right)^2}\right]\frac{d}{dh}\left(\sqrt{1-h^2}\right)$$
$$= \frac{1}{\sqrt{1-h^2}}\left[2h^2 + \sqrt{1-h^2} - 2\left(1-h^2\right)\right] + 2h\frac{-h}{\sqrt{1-h^2}}.$$

This is 0 when $\sqrt{1-h^2} - 2(1-h^2) = 0 \quad \Leftrightarrow \quad u - 2u^2 = 0 \ \left(\text{where } u = \sqrt{1-h^2}\right) \quad \Leftrightarrow \quad u = 0 \text{ or } \frac{1}{2}$ $\Leftrightarrow \quad h = 1 \text{ or } \frac{\sqrt{3}}{2}$. By the First Derivative Test, $h = \frac{\sqrt{3}}{2}$ represents a minimum for $A(h)$, since $A'(h) = 1 - \dfrac{2}{\sqrt{1-h^2}}$ goes from negative to positive at $h = \frac{\sqrt{3}}{2}$.

Another Method: Use Part 2 of the Fundamental Theorem to evaluate all of the integrals before differentiating.

Note: Another strategy is to use the angle θ as the variable (see diagram above) and show that $A = \theta + \cos\theta - \frac{\pi}{4} - \frac{1}{2}\sin 2\theta$, which is minimized when $\theta = \frac{\pi}{3}$.

15. $0 < a < b$. $\displaystyle\int_0^1 [bx + a(1-x)]^t\,dx = \int_a^b \frac{u^t}{(b-a)}\,du$ [put $u = bx + a(1-x)$]

$$= \left[\frac{u^{t+1}}{(t+1)(b-a)}\right]_a^b = \frac{b^{t+1} - a^{t+1}}{(t+1)(b-a)}.\ \text{Now let } y = \lim_{t\to 0}\left[\frac{b^{t+1} - a^{t+1}}{(t+1)(b-a)}\right]^{1/t}.$$

Then $\ln y = \displaystyle\lim_{t\to 0}\left[\frac{1}{t}\ln\frac{b^{t+1} - a^{t+1}}{(t+1)(b-a)}\right]$. This limit is of the form $\frac{0}{0}$, so we can apply l'Hospital's Rule to get

$$\ln y = \lim_{t\to 0}\left[\frac{b^{t+1}\ln b - a^{t+1}\ln a}{b^{t+1} - a^{t+1}} - \frac{1}{t+1}\right] = \frac{b\ln b - a\ln a}{b-a} - 1 = \frac{b\ln b}{b-a} - \frac{a\ln a}{b-a} - \ln e = \ln\frac{b^{b/(b-a)}}{ea^{a/(b-a)}}.$$

Therefore, $y = e^{-1}\left(\dfrac{b^b}{a^a}\right)^{1/(b-a)}$.

16.

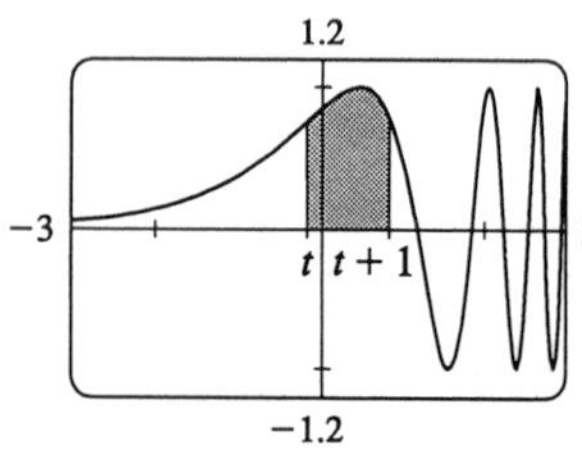

From the graph, it appears that the area under the graph of $f(x) = \sin(e^x)$ on the interval $[t, t+1]$ is greatest when $t \approx -0.2$. To find the exact value, we write the integral as $I = \int_t^{t+1} f(x)dx = \int_0^{t+1} f(x)dx - \int_0^t f(x)dx$, and use the Fundamental Theorem of Calculus to find $\dfrac{dI}{dt} = f(t+1) - f(t) = \sin\left(e^{t+1}\right) - \sin\left(e^t\right) = 0$ when $\sin(e^{t+1}) = \sin(e^t)$. Now we have $\sin x = \sin y$ whenever $x - y = 2k\pi$ and also whenever x and y are the same distance from $\left(k + \frac{1}{2}\right)\pi$, k any integer, since $\sin x$ is symmetric about the line $x = \left(k + \frac{1}{2}\right)\pi$. The first possibility is the more obvious one, but if we calculate $e^{t+1} - e^t = 2k\pi$, we get $t = \ln(2k\pi/(e-1))$, which is about 1.3 for $k = 1$ (the least possible value of k). From the graph, this looks unlikely to give the maximum we are looking for. So instead we set $e^{t+1} - \left(k + \frac{1}{2}\right)\pi = \left(k + \frac{1}{2}\right)\pi - e^t \quad \Leftrightarrow \quad e^{t+1} + e^t = (2k+1)\pi \quad \Leftrightarrow$ $e^t(e+1) = (2k+1)\pi \ \Leftrightarrow \ t = \ln((2k+1)\pi/(e+1))$. Now $k = 0 \quad \Rightarrow \quad t = \ln(\pi/(e+1)) \approx -0.16853$, which does give the maximum value, as we have seen from the graph of f.

17. In accordance with the hint, we let $I_k = \int_0^1 (1-x^2)^k\,dx$, and we find an expression for I_{k+1} in terms of I_k. We integrate I_{k+1} by parts with $u = (1-x^2)^{k+1} \Rightarrow du = (k+1)(1-x^2)^k(-2x)$, $dv = dx \Rightarrow v = x$, and then split the remaining integral into identifiable quantities:

$$I_{k+1} = x(1-x^2)^{k+1}\Big|_0^1 + 2(k+1)\int_0^1 x^2(1-x^2)^k\,dx = (2k+2)\int_0^1 (1-x^2)^k[1-(1-x^2)]dx$$

$= (2k+2)(I_k - I_{k+1})$. So $I_{k+1}[1+(2k+2)] = (2k+2)I_k$

$\Rightarrow \quad I_{k+1} = \dfrac{2k+2}{2k+3}I_k$. Now to complete the proof, we use induction: $I_0 = 1 = \dfrac{2^0(0!)^2}{1!}$, so the formula holds for $n=0$. Now suppose it holds for $n=k$. Then

$$I_{k+1} = \frac{2k+2}{2k+3}I_k = \frac{2k+2}{2k+3}\left[\frac{2^{2k}(k!)^2}{(2k+1)!}\right] = \frac{2(k+1)2^{2k}(k!)^2}{(2k+3)(2k+1)!} = \frac{2(k+1)}{2k+2}\cdot\frac{2(k+1)2^{2k}(k!)^2}{(2k+3)(2k+1)!}$$

$$= \frac{[2(k+1)]^2 2^{2k}(k!)^2}{(2k+3)(2k+2)(2k+1)!} = \frac{2^{2(k+1)}[(k+1)!]^2}{[2(k+1)+1]!}.$$

So by induction, the formula holds for all integers $n \geq 0$.

18. **(a)** Since $-1 \leq \sin \leq 1$, we have $-f(x) \leq f(x)\sin nx \leq f(x)$, and the graph of $y = f(x)\sin nx$ oscillates between $f(x)$ and $-f(x)$. (The diagram shows the case $f(x) = e^x$ and $n = 10$.) As $n \to \infty$, the graph oscillates more and more frequently; see the graphs in part (b).

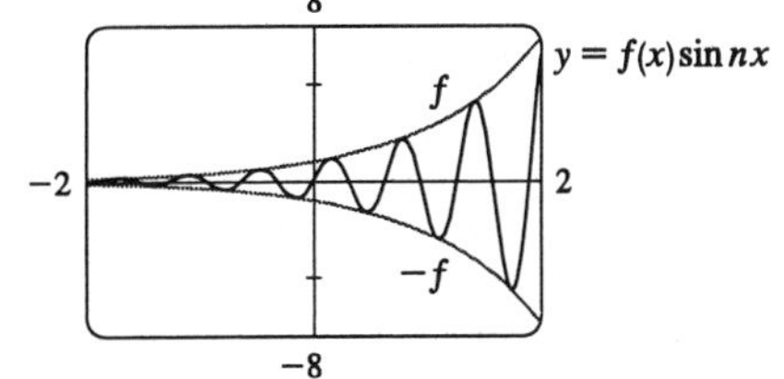

(b) From the graphs of the integrand, it seems that $\lim\limits_{n\to\infty} \int_0^1 f(x)\sin nx\,dx = 0$, since as n increases, the integrand oscillates more and more rapidly, and thus (since f' is continuous) it makes sense that the areas above the x-axis and below it during each oscillation approach equality.

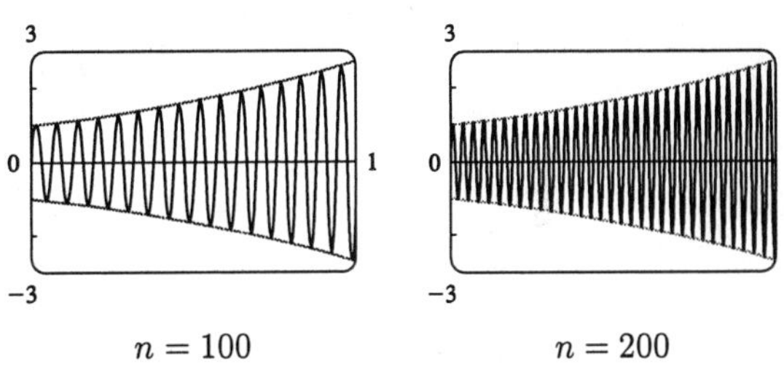

(c) We integrate by parts with $u = f(x) \Rightarrow du = f'(x)dx$, $dv = \sin nx\,dx \Rightarrow v = -\dfrac{\cos nx}{n}$:

$$\int_0^1 f(x)\sin nx\,dx = -\frac{f(x)\cos nx}{n}\Big|_0^1 + \int_0^1 \frac{\cos nx}{n}f'(x)dx = \frac{1}{n}\left[\int_0^1 \cos nx\,f'(x)dx - f(x)\cos nx\Big|_0^1\right]$$

$= \dfrac{1}{n}\left[\int_0^1 \cos nx\,f'(x)dx + f(0) - f(1)\cos n\right]$. Taking absolute values of the first and last terms in this equality, and using the facts that $|\alpha \pm \beta| \leq |\alpha| + |\beta|$, $\int_0^1 f(x)dx \leq \int_0^1 |f(x)|dx$, $|f(0)| = f(0)$ (f is positive), $|f'(x)| \leq M$ for $0 \leq x \leq 1$, and $|\cos nx| \leq 1$, we get

$\left|\int_0^1 f(x)\sin nx\,dx\right| \leq \dfrac{1}{n}\left[\left|\int_0^1 M\,dx\right| + |f(0)| + |f(1)|\right] = \dfrac{1}{n}[M + |f(0)| + |f(1)|]$, which approaches 0 as $n \to \infty$. The result follows by the Squeeze Theorem.

19. We use the Fundamental Theorem of Calculus to differentiate the given equation:

$[f(x)]^2 = 100 + \int_0^x \left([f(t)]^2 + [f'(t)]^2\right) dt \quad \Rightarrow \quad 2f(x)f'(x) = [f(x)]^2 + [f'(x)]^2 \quad \Rightarrow$

$[f(x)]^2 + [f'(x)]^2 - 2f(x)f'(x) = [f(x) - f'(x)]^2 = 0 \quad \Leftrightarrow \quad f(x) = f'(x)$. We can solve this as a separable equation, or else use Theorem 6.5.2, which says that the only solutions are $f(x) = Ce^x$. Now $[f(0)]^2 = 100$, so $f(0) = C = \pm 10$, and hence $f(x) = \pm 10e^x$ are the only functions satisfying the given equation.

20. **(a)** $T_n(x) = \cos(n \arccos x)$. The domain of arccos is $[-1, 1]$, and the domain of cos is $\mathbb{R}$, so the domain of $T_n(x)$ is $[-1, 1]$. As for the range, $T_0(x) = \cos 0 = 1$, so the range of $T_0(x)$ is $\{1\}$. But since the range of $n \arccos x$ is at least $[0, \pi]$ for $n > 0$, and since $\cos y$ takes on all values in $[-1, 1]$ for $y \in [0, \pi]$, the range of $T_n(x)$ is $[-1, 1]$ for $n > 0$.

(b) Using the usual trigonometric identities, $T_2(x) = \cos(2 \arccos x) = 2[\cos(\arccos x)]^2 - 1 = 2x^2 - 1$, and

$$\begin{aligned} T_3(x) &= \cos(3 \arccos x) = \cos(\arccos x + 2 \arccos x) \\ &= \cos(\arccos x)\cos(2 \arccos x) - \sin(\arccos x)\sin(2 \arccos x) \\ &= x(2x^2 - 1) - \sin(\arccos x)[2 \sin(\arccos x)\cos(\arccos x)] \\ &= 2x^3 - x - 2\left[\sin^2(\arccos x)\right]x = 2x^3 - x - 2x\left[1 - \cos^2(\arccos x)\right] \\ &= 2x^3 - x - 2x\left(1 - x^2\right) = 4x^3 - 3x. \end{aligned}$$

(c) Let $y = \arccos x$. Then

$$\begin{aligned} T_{n+1}(x) &= \cos[(n+1)y] = \cos(y + ny) = \cos y \cos ny - \sin y \sin ny \\ &= 2 \cos y \cos ny - (\cos y \cos ny + \sin y \sin ny) = 2xT_n(x) - \cos(ny - y) = 2xT_n(x) - T_{n-1}(x) \end{aligned}$$

(d) Here we use induction. $T_0(x) = 1$, a polynomial of degree 0. Now assume that $T_k(x)$ is a polynomial of degree k. Then $T_{k+1}(x) = 2xT_k(x) - T_{k-1}(x)$. By assumption, the leading term of T_k is $a_k x^k$, say, so the leading term of T_{k+1} is $2xa_k x^k = 2a_k x^{k+1}$, and so T_{k+1} has degree $k+1$.

(e) $T_4(x) = 2xT_3(x) - T_2(x) = 2x(4x^3 - 3x) - (2x^2 - 1) = 8x^4 - 8x^2 + 1$,

$T_5(x) = 2xT_4(x) - T_3(x) = 2x(8x^4 - 8x^2 + 1) - (4x^3 - 3x) = 16x^5 - 20x^3 + 5x$,

$T_6(x) = 2xT_5(x) - T_4(x) = 2x(16x^5 - 20x^3 + 5x) - (8x^4 - 8x^2 + 1) = 32x^6 - 48x^4 + 18x^2 - 1$,

$$\begin{aligned} T_7(x) &= 2xT_6(x) - T_5(x) = 2x(32x^6 - 48x^4 + 18x^2 - 1) - (16x^5 - 20x^3 + 5x) \\ &= 64x^7 - 112x^5 + 56x^3 - 7x \end{aligned}$$

(f) The zeros of $T_n(x) = \cos(n \arccos x)$ occur where $n \arccos x = k\pi + \frac{\pi}{2}$ for some integer k, since then $\cos(n \arccos x) = \cos\left(k\pi + \frac{\pi}{2}\right) = 0$. Note that there will be restrictions on k, since $0 \le \arccos x \le \pi$. We continue: $n \arccos x = k\pi + \frac{\pi}{2} \quad \Leftrightarrow \quad \arccos x = \dfrac{k\pi + \frac{\pi}{2}}{n}$. This only has solutions for $0 \le \dfrac{k\pi + \frac{\pi}{2}}{n} \le \pi$ $\quad \Leftrightarrow \quad 0 < k\pi + \frac{\pi}{2} < n\pi \quad \Leftrightarrow \quad 0 \le k < n$. [This makes sense, because then $T_n(x)$ has n zeros, and it is a polynomial of degree n.] So, taking cosines of both sides of the last equation, we find that the zeros of $T_n(x)$ occur at $x = \cos \dfrac{k\pi + \pi/2}{n}$, k an integer with $0 \le k < n$. To find the values of x at which $T_n(x)$ has local extrema, we set $0 = T_n'(x) = -\sin(n \arccos x)\dfrac{-n}{\sqrt{1 - x^2}} = \dfrac{n \sin(n \arccos x)}{\sqrt{1 - x^2}} \quad \Leftrightarrow$

$\sin(n \arccos x) = 0 \quad \Leftrightarrow \quad n \arccos x = k\pi$, k some integer $\quad \Leftrightarrow \quad \arccos x = k\pi/n$.

This has solutions for $0 \le k \le n$, but we disallow the cases $k = 0$ and $k = n$, since these give $x = 1$ and $x = -1$ respectively. So the local extrema of $T_n(x)$ occur at $x = \cos(k\pi/n)$, k an integer with $0 < k < n$. [Again, this seems reasonable, since a polynomial of degree n has at most $(n-1)$ extrema.] By the First Derivative Test, the cases where k is even give maxima of $T_n(x)$, since then $n \arccos[\cos(k\pi/n)] = k\pi$ is an even multiple of π, so $\sin(n \arccos x)$ goes from negative to positive at $x = \cos(k\pi/n)$. Similarly, the cases where k is odd represent minima of $T_n(x)$.

(g)

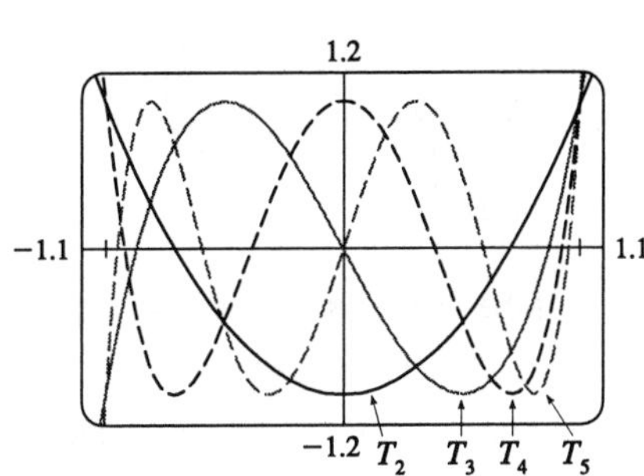

(h)

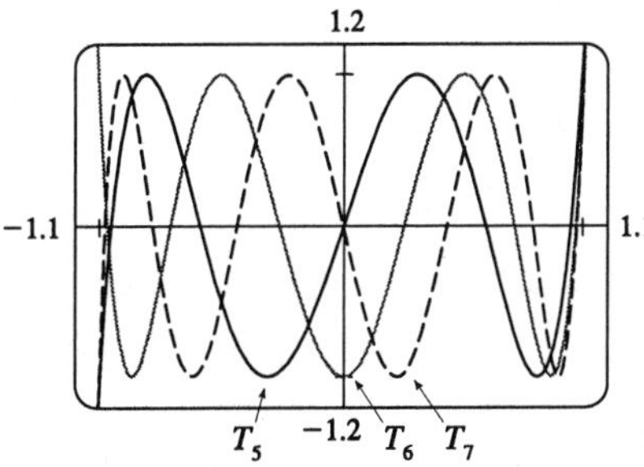

(i) From the graphs, it seems that the zeros of T_n and T_{n+1} alternate; that is, between two adjacent zeros of T_n, there is a zero of T_{n+1}, and vice versa. The same is true of the x-coordinates of the extrema of T_n and T_{n+1}: between the x-coordinates of any two adjacent extrema of one, there is the x-coordinate of an extremum of the other.

(j) When n is odd, the function $T_n(x)$ is odd, since all of its terms have odd degree, and so $\int_{-1}^{1} T_n(x)dx = 0$. When n is even, $T_n(x)$ is even, and it appears that the integral is negative, but decreases in absolute value as n gets larger.

(k) $\int_{-1}^{1} T_n(x)dx = \int_{-1}^{1} \cos(n \arccos x)dx$. We substitute $u = \arccos x \Rightarrow x = \cos u \Rightarrow dx = -\sin u\,du$, $x = -1 \Rightarrow u = \pi$, and $x = 1 \Rightarrow u = 0$. So the integral becomes

$$\int_0^\pi \cos(nu)\sin u\,du = \int_0^\pi \tfrac{1}{2}[\sin(u - nu) + \sin(u + nu)]du = \frac{1}{2}\left[\frac{\cos[(1-n)u]}{n-1} - \frac{\cos[(1+n)u]}{n+1}\right]_0^\pi$$

$$= \begin{cases} \dfrac{1}{2}\left[\left(\dfrac{-1}{n-1} - \dfrac{-1}{n+1}\right) - \left(\dfrac{1}{n-1} - \dfrac{1}{n+1}\right)\right] & n \text{ even} \\ \dfrac{1}{2}\left[\left(\dfrac{1}{n-1} - \dfrac{1}{n+1}\right) - \left(\dfrac{1}{n-1} - \dfrac{1}{n+1}\right)\right] & n \text{ odd} \end{cases} = \begin{cases} -2/(n^2-1) & n \text{ even} \\ 0 & n \text{ odd} \end{cases}$$

(l) From the graph, we see that as c increases through an integer, the graph of f gains a local extremum, which starts at $x = -1$ and moves rightward, compressing the graph of f as c continues to increase.

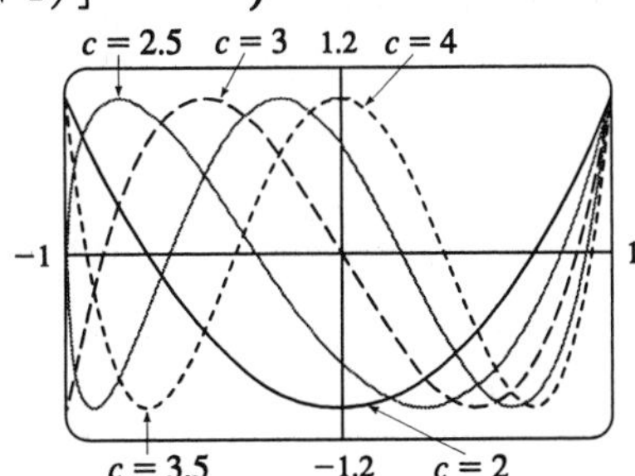

21. To find the height of the pyramid, we use similar triangles. The first figure shows a cross-section of the pyramid passing through the top and through two opposite corners of the square base. Now $|BD| = b$, since it is a radius of the sphere, which has diameter $2b$ since it is tangent to the opposite sides of the square base. Also, $|AD| = b$ since $\triangle ADB$ is isosceles. So the height is $|AB| = \sqrt{b^2 + b^2} = \sqrt{2}b$.

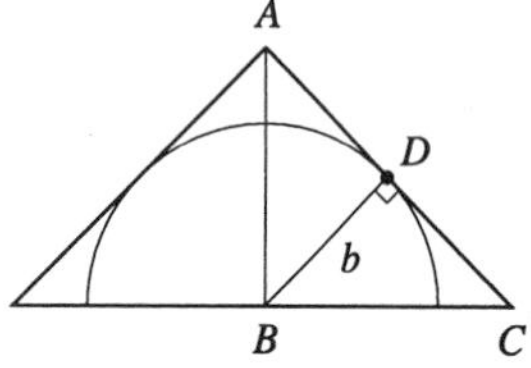

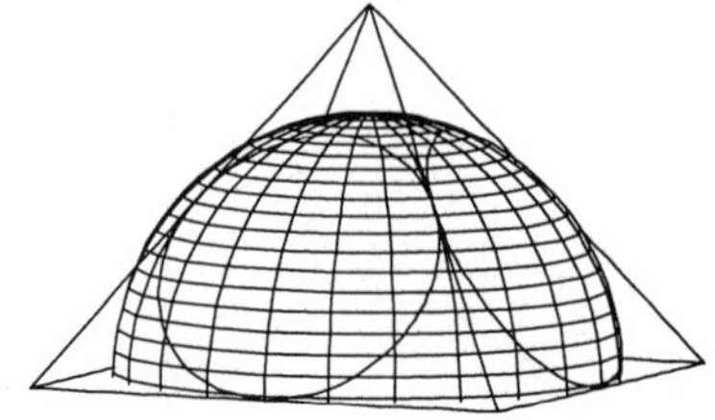
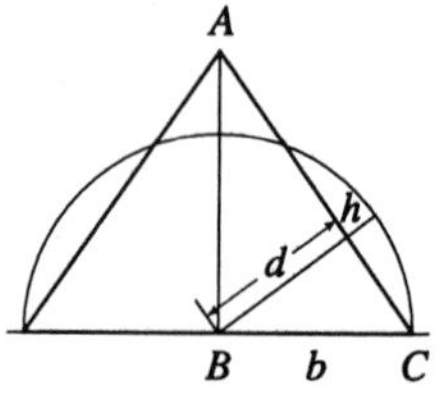

We observe that the shared volume is equal to half the volume of the sphere, minus the sum of the four equal volumes (caps of the sphere) cut off by the triangular faces of the pyramid. See Exercise 5.2.49 for a derivation of the formula for the volume of a cap of a sphere. To use the formula, we need to find the perpendicular distance h of each triangular face from the surface of the sphere. We first find the distance d from the center of the sphere to one of the triangular faces. The third figure shows a cross-section of the pyramid through the top and through the midpoints of opposite sides of the square base. From similar triangles we find that

$\dfrac{d}{b} = \dfrac{|AB|}{|AC|} = \dfrac{\sqrt{2}b}{\sqrt{b^2 + \left(\sqrt{2}b\right)^2}} \quad \Rightarrow \quad d = \dfrac{\sqrt{2}b^2}{\sqrt{3b^2}} = \frac{\sqrt{6}}{3}b$. So $h = b - \frac{\sqrt{6}}{3}b = \frac{3-\sqrt{6}}{3}b$. So, using the formula from Exercise 5.2.49 with $r = b$, we find that the volume of each of the caps is

$\pi\left(\frac{3-\sqrt{6}}{3}b\right)^2\left(b - \frac{3-\sqrt{6}}{3\cdot 3}b\right) = \frac{15-6\sqrt{6}}{9}\cdot\frac{6+\sqrt{6}}{9}\pi b^3 = \left(\frac{2}{3} - \frac{7}{27}\sqrt{6}\right)\pi b^3$. So, using our first observation, the shared volume is $V = \frac{1}{2}\left(\frac{4}{3}\pi b^3\right) - 4\left(\frac{2}{3} - \frac{7}{27}\sqrt{6}\right)\pi b^3 = \left(\frac{28}{27}\sqrt{6} - 2\right)\pi b^3$.

22. In the second figure, the segment lying above the interval $[x_i - \Delta x, x_i]$ along the tangent to C has length $\Delta x \sec\alpha = \Delta x\sqrt{1 + \tan^2\alpha} = \sqrt{1 + [f'(x_i)]^2}\,\Delta x$. The segment from $(x_i, f(x_i))$ drawn perpendicular to the line $y = mx + b$ has length $g(x_i) = [f(x_i) - mx_i - b]\cos\beta = \dfrac{f(x_i) - mx_i - b}{\sqrt{1+m^2}}$. Also

$$\Delta u = \Delta x \sec\alpha\cos(\beta - \alpha) = \Delta x\,\frac{\cos\beta\cos\alpha + \sin\beta\sin\alpha}{\cos\alpha} = \Delta x(\cos\beta + \sin\beta\tan\alpha)$$

$$= \Delta x\left[\frac{1}{\sqrt{1+m^2}} + \frac{m}{\sqrt{1+m^2}}f'(x_i)\right] = \frac{1 + mf'(x_i)}{\sqrt{1+m^2}}\,\Delta x.$$

(a) $\displaystyle \text{Area}(\Re) = \lim_{\|P\|\to 0}\sum_{i=1}^{n} g(x_i)\Delta u = \lim_{\|P\|\to 0}\sum_{i=1}^{n}\frac{f(x_i) - mx_i - b}{\sqrt{1+m^2}}\cdot\frac{1 + mf'(x_i)}{\sqrt{1+m^2}}\,\Delta x$

$$= \frac{1}{1+m^2}\int_p^q [f(x) - mx - b][1 + mf'(x)]\,dx$$

(b) $\displaystyle V = \lim_{\|P\|\to 0}\sum_{i=1}^{n}\pi[g(x_i)]^2\,\Delta u = \frac{\pi}{(1+m^2)^{3/2}}\int_p^q [f(x) - mx - b]^2[1 + mf'(x)]\,dx$

(c) $\displaystyle S = \int_p^q 2\pi g(x)\sqrt{1 + [f'(x)]^2}\,dx = \frac{2\pi}{\sqrt{1+m^2}}\int_p^q [f(x) - mx - b]\sqrt{1 + [f'(x)]^2}\,dx$

CHAPTER NINE

EXERCISES 9.1

1. **(a)**

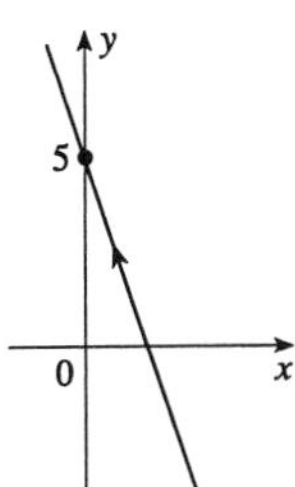

(b) $x = 1 - t, y = 2 + 3t$

$y = 2 + 3(1 - x) = 5 - 3x$, so

$3x + y = 5$

2. **(a)**

t	x	y
-3	-7	5
-2	-5	4
-1	-3	3
0	-1	2
1	1	1
2	3	0

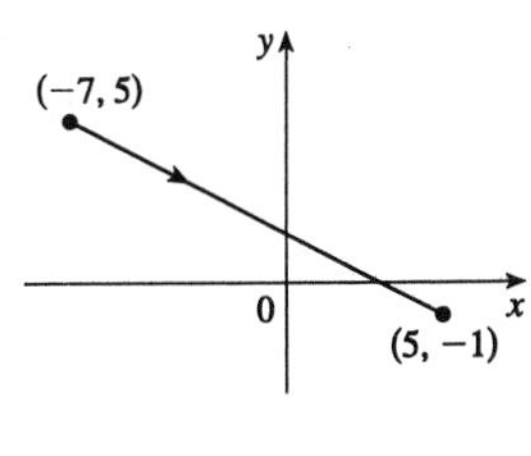

(b) $x = 2t - 1, y = 2 - t, -3 \le t \le 3$

$x = 2(2 - y) - 1 = 3 - 2y$, so

$x + 2y = 3$, with $-7 \le x \le 5$

3. **(a)**

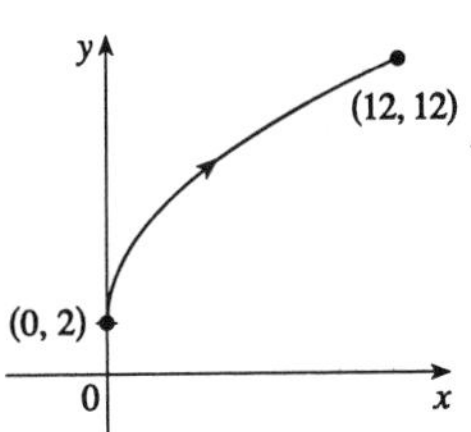

(b) $x = 3t^2, y = 2 + 5t, 0 \le t \le 2$

$$x = 3\left(\frac{y-2}{5}\right)^2 = \tfrac{3}{25}(y-2)^2,$$

$2 \le y \le 12$

4. **(a)**

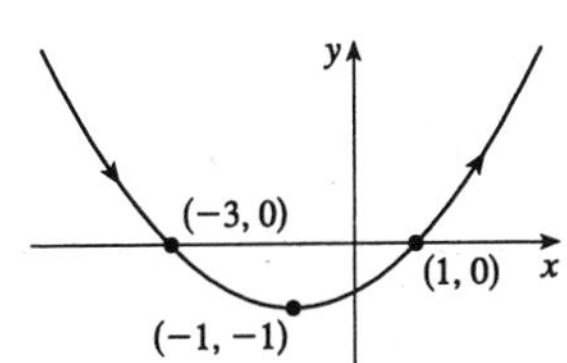

(b) $x = 2t - 1, y = t^2 - 1$

$$y = \left(\frac{x+1}{2}\right)^2 - 1, \text{ so}$$

$y + 1 = \frac{1}{4}(x + 1)^2$

5. **(a)**

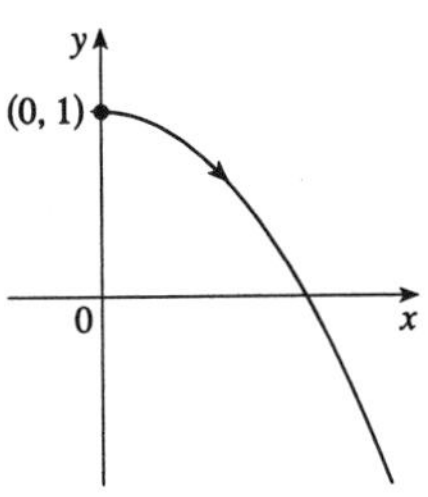

(b) $x = \sqrt{t}, y = 1 - t$

$y = 1 - t = 1 - x^2, x \ge 0$

6. **(a)**

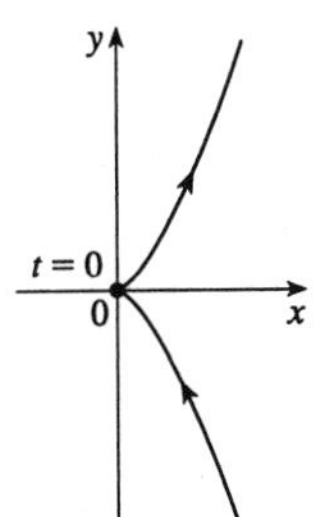

(b) $x = t^2, y = t^3$

$x = t^2 = \sqrt[3]{y^2}$, or $y = \pm\sqrt{x^3}$

7. **(a)**

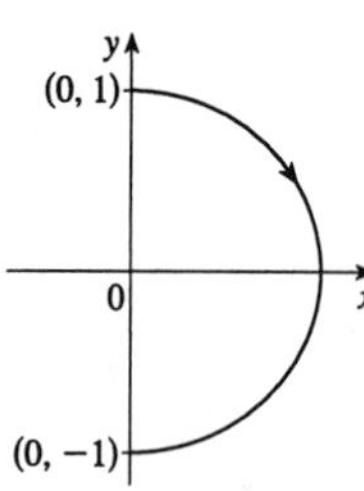

(b) $x = \sin\theta,\ y = \cos\theta,\ 0 \le \theta \le \pi$

$x^2 + y^2 = \sin^2\theta + \cos^2\theta = 1,\ 0 \le x \le 1$

8. **(a)**

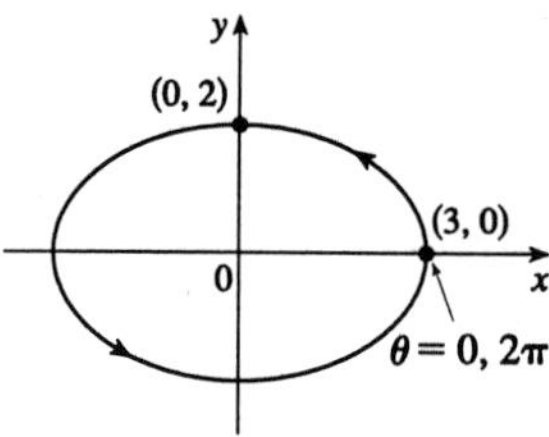

(b) $x = 3\cos\theta,\ y = 2\sin\theta,\ 0 \le \theta \le 2\pi$

$\left(\frac{x}{3}\right)^2 + \left(\frac{y}{2}\right)^2 = \cos^2\theta + \sin^2\theta = 1$, or

$\frac{1}{9}x^2 + \frac{1}{4}y^2 = 1$

9. **(a)**

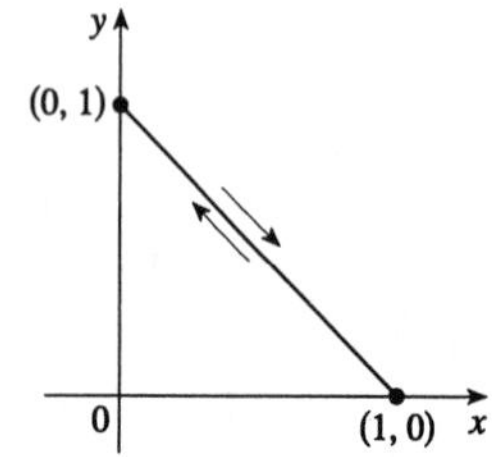

(b) $x = \sin^2\theta,\ y = \cos^2\theta$

$x + y = \sin^2\theta + \cos^2\theta = 1,\ 0 \le x \le 1$

10. **(a)**

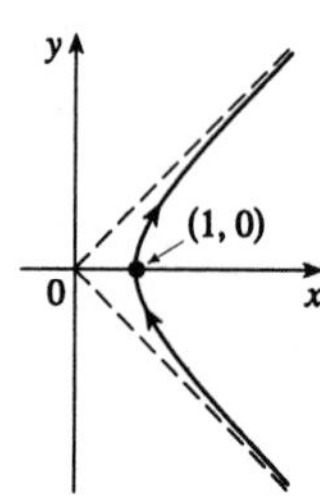

(b) $x = \sec\theta,\ y = \tan\theta,\ -\frac{\pi}{2} < \theta < \frac{\pi}{2}$

$x^2 - y^2 = \sec^2\theta - \tan^2\theta = 1,\ x \ge 1$,

or $x = \sqrt{y^2 + 1}$

11. **(a)**

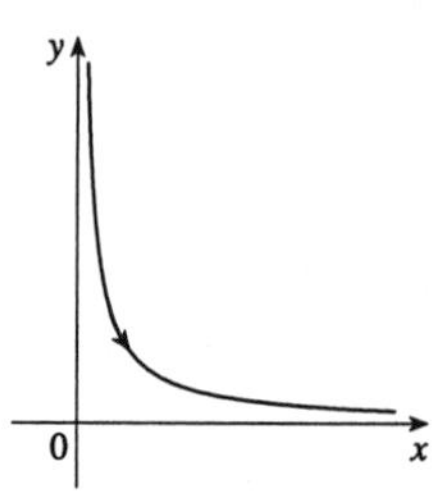

(b) $x = e^t,\ y = e^{-t}$

$y = 1/x,\ x > 0$

12. **(a)**

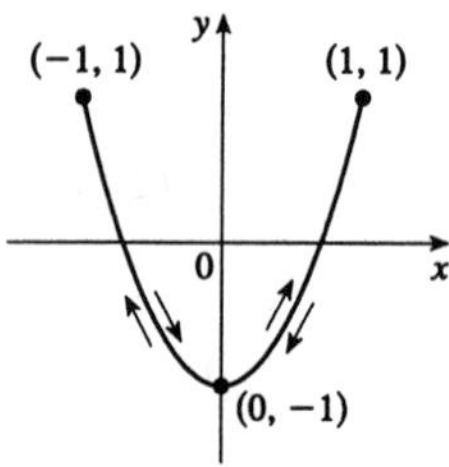

(b) $x = \cos t,\ y = \cos 2t$

$y = \cos 2t = 2\cos^2 t - 1 = 2x^2 - 1$,

so $y + 1 = 2x^2,\ -1 \le x \le 1$

13. **(a)**

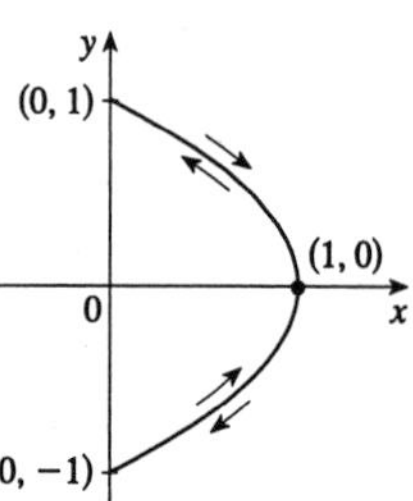

(b) $x = \cos^2\theta,\ y = \sin\theta$

$x + y^2 = \cos^2\theta + \sin^2\theta = 1,\ -1 \le y \le 1$

14. **(a)**

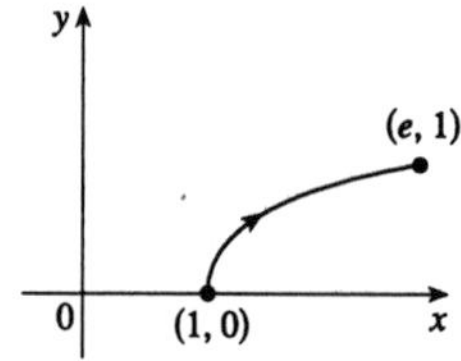

(b) $x = e^t,\ y = \sqrt{t}$

$x = e^{y^2},\ 0 \le y \le 1$

Or: $y = \sqrt{\ln x},\ 1 \le x \le e$

15. **(a)**

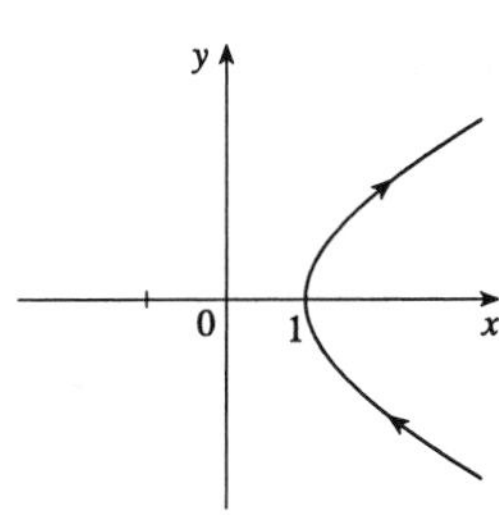

(b) $x = \cosh t$, $y = \sinh t$

$x^2 - y^2 = \cosh^2 t - \sinh^2 t = 1$, $x \geq 1$

16. **(a)**

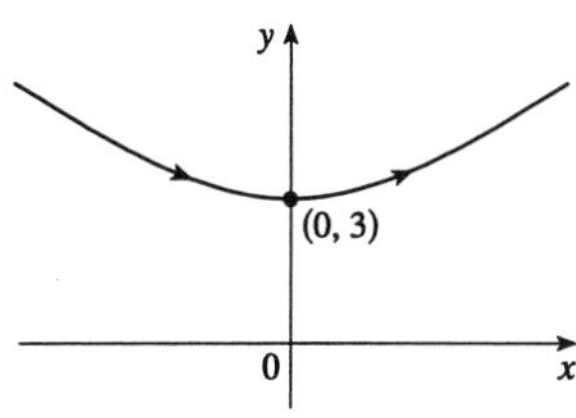

(b) $x = 4\sinh t$, $y = 3\cosh t$

$(y/3)^2 - (x/4)^2 = \cosh^2 t - \sinh^2 t = 1$

with $y \geq 3$, so $\dfrac{y^2}{9} - \dfrac{x^2}{16} = 1$, $y \geq 3$

17. $x^2 + y^2 = \cos^2 \pi t + \sin^2 \pi t = 1$, $1 \leq t \leq 2$, so the particle moves counterclockwise along the circle $x^2 + y^2 = 1$ from $(-1, 0)$ to $(1, 0)$, along the lower half of the circle.

18. $(x - 2)^2 + (y - 3)^2 = \cos^2 t + \sin^2 t = 1$, so the motion takes place on a unit circle centered at $(2, 3)$. As t goes from 0 to 2π, the particle makes one complete counterclockwise rotation around the circle, starting and ending at $(3, 3)$.

19. $x = 8t - 3$, $y = 2 - t$, $0 \leq t \leq 1 \quad \Rightarrow \quad x = 8(2 - y) - 3 = 13 - 8y$, so the particle moves along the line $x + 8y = 13$ from $(-3, 2)$ to $(5, 1)$.

20. $x = \cos^2 t = y^2$, so the particle moves along the parabola $x = y^2$. As t goes from 0 to 4π, the particle moves from $(1, 1)$ down to $(1, -1)$ (at $t = \pi$), back up to $(1, 1)$ again (at $t = 2\pi$), and then repeats this entire cycle between $t = 2\pi$ and $t = 4\pi$.

21. $\left(\frac{1}{2}x\right)^2 + \left(\frac{1}{3}y\right)^2 = \sin^2 t + \cos^2 t = 1$, so the particle moves once clockwise along the ellipse $\frac{1}{4}x^2 + \frac{1}{9}y^2 = 1$, starting and ending at $(0, 3)$.

22. $y = \csc t = 1/\sin t = 1/x$. The particle slides down the first quadrant branch of the hyperbola $xy = 1$ from $\left(\frac{1}{2}, 2\right)$ to $(\sin 1, \csc 1) \approx (0.84147, 1.1884)$ as t goes from $\frac{\pi}{6}$ to 1.

23. From the graphs, it seems that as $t \to -\infty$, $x \to \infty$ and $y \to -\infty$. So the point $(x(t), y(t))$ will move from far out in the fourth quadrant as t increases. At $t \approx -1.7$, both x and y are 0, so the graph passes through the origin. After that the graph passes through the second quadrant (x is negative, y is positive), then intersects the x-axis at $x \approx -9$ when $t = 0$. After this, the graph passes through the third quadrant, going through the origin again at $t \approx 1.7$, and then as $t \to \infty$, $x \to \infty$ and $y \to \infty$. Note that for every point $(x(t), y(t)) = (3(t^2 - 3), t^3 - 3t)$, we can substitute $-t$ to get the corresponding point $(x(-t), y(-t))$ $= \left(3\left[(-t)^2 - 3\right], (-t)^3 - 3(-t)\right)$ $= (x(t), -y(t))$, and so the graph is symmetric about the x-axis.

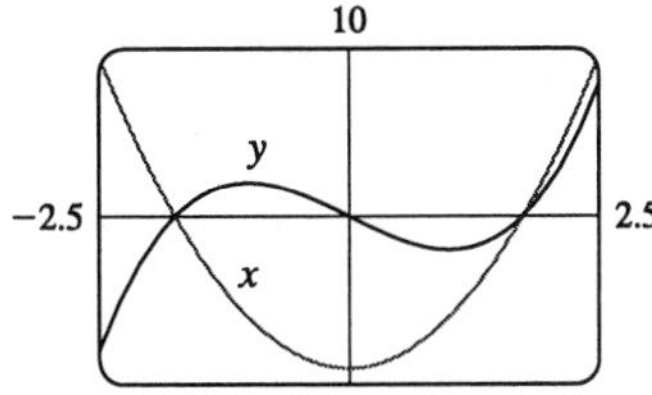

24. As $t \to -\infty$, $y \to -\frac{\pi}{2}$ and x oscillates between 1 and -1. Then, as t increases through 0, y increases while x continues to oscillate, and the graph passes through the origin. Then, as $t \to \infty$, $y \to \frac{\pi}{2}$ as x oscillates.

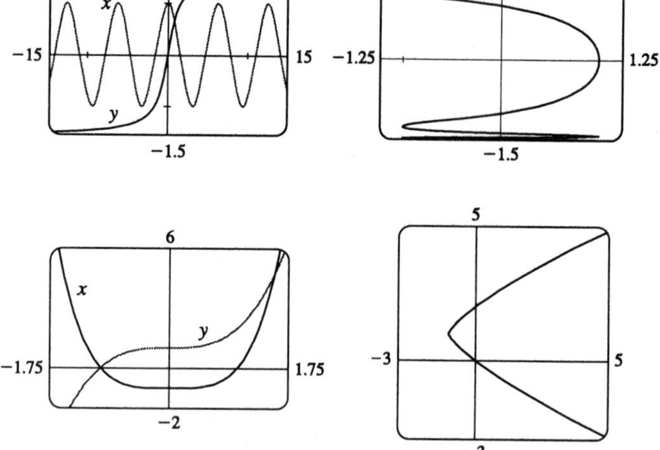

25. As $t \to -\infty$, $x \to \infty$ and $y \to -\infty$. The graph passes through the origin at $t = -1$, and then goes through the second quadrant (x negative, y positive), passing through the point $(-1, 1)$ at $t = 0$. As t increases, the graph passes through the point $(0, 2)$ at $t = 1$, and then as $t \to \infty$, both x and y approach ∞.

26. **(a)** Note that as $t \to -\infty$, we have $x \to -\infty$ and $y \to \infty$, whereas when $t \to \infty$, both x and $y \to \infty$. This description fits only IV. [But also note that $x(t)$ increases, then decreases, then increases again.]

(b) This curve is only defined for $t > 0$. As $t \to 0^+$, $x \to 0$ and $y \to -\infty$. This is only the case with VI.

(c) If $t = 0$, then $(x, y) = (\sin 0, \sin 0) = (0, 0)$. Also, $|x| = |\sin 3t| \le 1$ for all t, and $|y| = |\sin 4t| \le 1$ for all t. The only graph which includes the point $(0, 0)$ and which has $|x| \le 1$ and $|y| \le 1$, is V.

(d) Note that as $t \to -\infty$, both x and $y \to -\infty$, and as $t \to \infty$, both x and $y \to \infty$. This description fits only III. (Also note that, since $\sin 2t$ and $\sin 3t$ lie between -1 and 1, the curve never strays very far from the line $y = x$.)

(e) Note that both $x(t)$ and $y(t)$ are periodic with period 2π and satisfy $|x| \le 1$ and $|y| \le 1$. Now the only y-intercepts occur when $x = \sin(t + \sin t) = 0 \quad \Leftrightarrow \quad t = 0$ or π. So there should be two y-intercepts: $y(0) = \cos 1 \approx 0.54$ and $y(\pi) = \cos(\pi - 1) \approx -0.54$. Similarly, there should be two x-intercepts: $x\left(\frac{\pi}{2}\right) = \sin\left(\frac{\pi}{2} + 1\right) \approx 0.54$ and $x\left(\frac{3\pi}{2}\right) = \sin\left(\frac{3\pi}{2} - 1\right) \approx -0.54$. The only curve with exactly these x- and y-intercepts is I.

(f) Note that $x(t)$ is periodic with period 2π, so the only y-intercepts occur when $x = \cos t = 0 \quad \Leftrightarrow \quad t = \frac{\pi}{2}$ or $\frac{3\pi}{2}$. Also, the graph is symmetric about the x-axis, since

$y(-t) = \sin(-t + \sin 5(-t)) = \sin(-t - \sin 5t) = -\sin(t + \sin 5t) = -y(t)$, and

$x(-t) = \cos(-t) = \cos t = x(t)$. The only graph which has only two y-intercepts, and is symmetric about the x-axis, is II.

27. Clearly the curve passes through (x_1, y_1) when $t = 0$ and through (x_2, y_2) when $t = 1$. For $0 < t < 1$, x is strictly between x_1 and x_2 and y is strictly between y_1 and y_2. For every value of t, x and y satisfy the relation $y - y_1 = \dfrac{y_2 - y_1}{x_2 - x_1}(x - x_1)$, which is the equation of the straight line through (x_1, y_1) and (x_2, y_2).

Finally, any point (x, y) on that line satisfies $\dfrac{y-y_1}{y_2-y_1}=\dfrac{x-x_1}{x_2-x_1}$; if we call that common value t, then the given parametric equations yield the point (x, y); and any (x, y) on the line between (x_1, y_1) and (x_2, y_2) yields a value of t in $[0, 1]$. So the given parametric equations exactly specify the line segment from (x_1, y_1) to (x_2, y_2).

28. **(a)** If $\alpha = 30°$ and $v_0 = 500$ m/s, then the equations become $x = 250\sqrt{3}\,t$ and $y = 250t - 4.9t^2$. $y = 0$ when $t = 0$ (when the gun is fired) and again when $t = \frac{250}{4.9} \approx 51$ s. Then $x = \left(250\sqrt{3}\right)\left(\frac{250}{4.9}\right) \approx 22{,}092$ m.

(b) $y = -4.9\left(t^2 - \frac{250}{4.9}t\right) = -4.9\left(t - \frac{125}{4.9}\right)^2 + \frac{125^2}{4.9} \le \frac{125^2}{4.9}$ with equality when $t = \frac{125}{4.9}$ s, so the maximum height attained is $\frac{125^2}{4.9} \approx 3189$ m.

(c) $t = \dfrac{x}{v_0 \cos\alpha}$, so $y = (v_0 \sin\alpha)\dfrac{x}{v_0\cos\alpha} - \dfrac{g}{2}\left(\dfrac{x}{v_0\cos\alpha}\right)^2 = (\tan\alpha)x - \left(\dfrac{g}{2v_0^2\cos^2\!\alpha}\right)^2 x^2$, which is the equation of a parabola.

29. The case $\frac{\pi}{2} < \theta < \pi$ is illustrated. C has coordinates $(r\theta, r)$ as before, and Q has coordinates $(r\theta, r + r\cos(\pi - \theta)) = (r\theta, r(1 - \cos\theta))$, so P has coordinates $(r\theta - r\sin(\pi - \theta), r(1 - \cos\theta)) = (r(\theta - \sin\theta), r(1 - \cos\theta))$. Again we have the parametric equations $x = r(\theta - \sin\theta)$, $y = r(1 - \cos\theta)$.

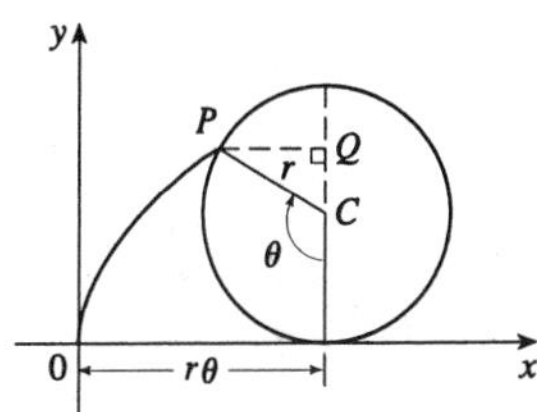

30. The first two diagrams depict the case $\pi < \theta < \frac{3\pi}{2}$, $d < r$. As in Exercise 29, C has coordinates $(r\theta, r)$. Now Q (in the second diagram) has coordinates $(r\theta, r + d\cos(\theta - \pi)) = (r\theta, r - d\cos\theta)$, so a typical point P of the trochoid has coordinates $(r\theta + d\sin(\theta - \pi), r - d\cos\theta)$. That is, P has coordinates (x, y), where $x = r\theta - d\sin\theta$ and $y = r - d\cos\theta$. When $d = r$, these equations agree with those of the cycloid.

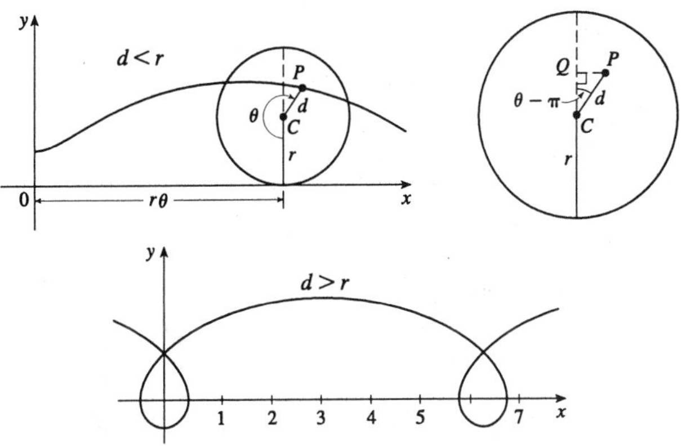

31. It is apparent that $x = |OQ|$ and $y = |QP| = |ST|$. From the diagram, $x = |OQ| = a\cos\theta$ and $y = |ST| = b\sin\theta$. Thus the parametric equations are $x = a\cos\theta$ and $y = b\sin\theta$. To eliminate θ we rearrange: $\sin\theta = y/b \;\Rightarrow\; \sin^2\!\theta = (y/b)^2$ and $\cos\theta = x/a \;\Rightarrow\; \cos^2\!\theta = (x/a)^2$. Adding the two equations: $\sin^2\!\theta + \cos^2\!\theta = 1 = x^2/a^2 + y^2/b^2$. Thus we have an ellipse.

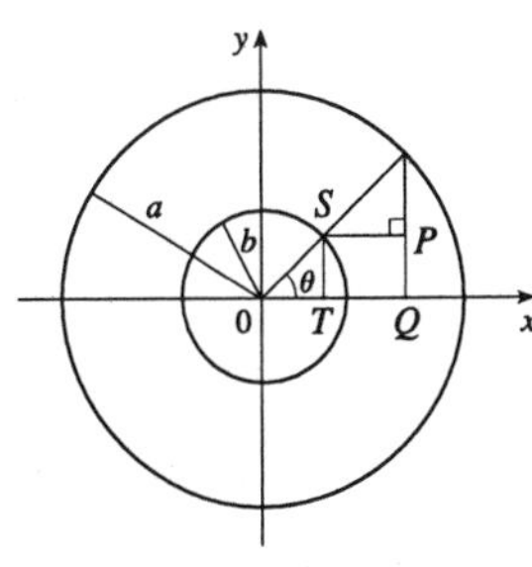

32. A has coordinates $(a\cos\theta, a\sin\theta)$. Since OA is perpendicular to AB, $\triangle OAB$ is a right triangle and B has coordinates $(a\sec\theta, 0)$. It follows that P has coordinates $(a\sec\theta, b\sin\theta)$. Thus the parametric equations are $x = a\sec\theta$ and $y = b\sin\theta$.

33. **(a)** The center Q of the smaller circle has coordinates $((a-b)\cos\theta, (a-b)\sin\theta)$. Arc PS on circle C has length $a\theta$ since it is equal in length to arc AS (the smaller circle rolls without slipping against the larger). Thus $\angle PQS = \frac{a}{b}\theta$ and $\angle PQT = \frac{a}{b}\theta - \theta$, so P has coordinates

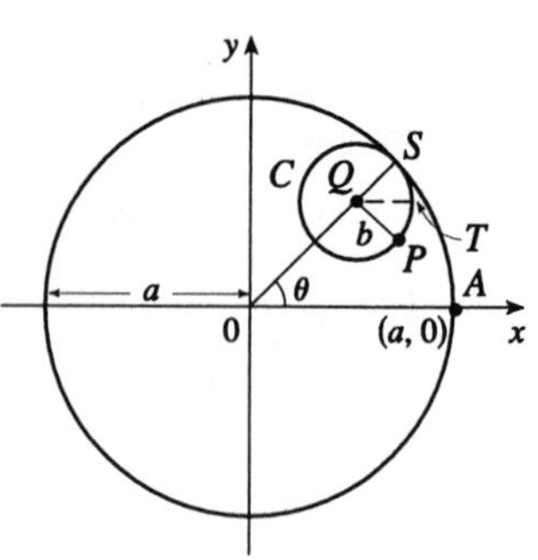

$x = (a-b)\cos\theta + b\cos(\angle PQT) = (a-b)\cos\theta + b\cos\left(\frac{a-b}{b}\theta\right)$,

and $y = (a-b)\sin\theta - b\sin(\angle PQT) = (a-b)\sin\theta - b\sin\left(\frac{a-b}{b}\theta\right)$.

(b) If $b = \frac{a}{4}$, then $a - b = \frac{3a}{4}$ and $\frac{a-b}{b} = 3$, so

$x = \frac{3a}{4}\cos\theta + \frac{a}{4}\cos 3\theta = \frac{3a}{4}\cos\theta + \frac{a}{4}\left(4\cos^3\theta - 3\cos\theta\right) = a\cos^3\theta$ and

$y = \frac{3a}{4}\sin\theta - \frac{a}{4}\sin 3\theta = \frac{3a}{4}\sin\theta - \frac{a}{4}\left(3\sin\theta - 4\sin^3\theta\right) = a\sin^3\theta$.

The curve is symmetric about the origin.

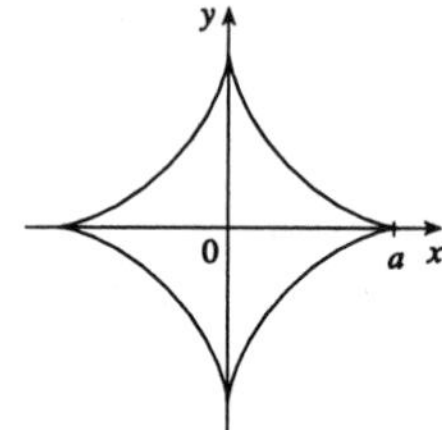

(c)

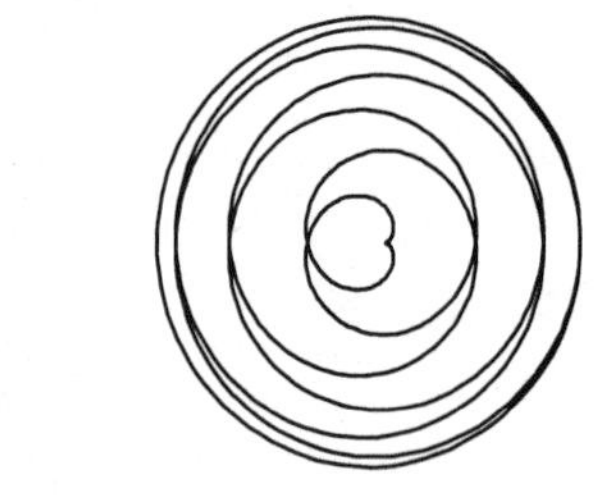

$a/b = \frac{1}{8}$

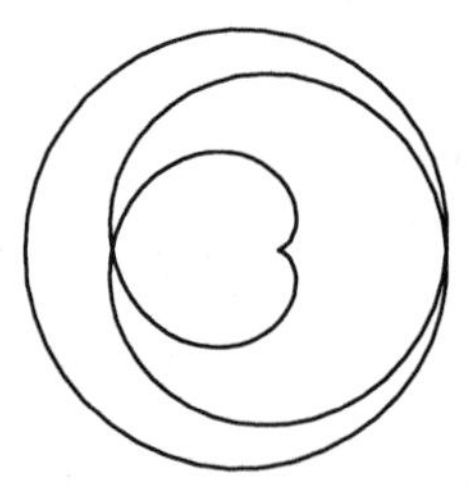

$a/b = \frac{1}{4}$

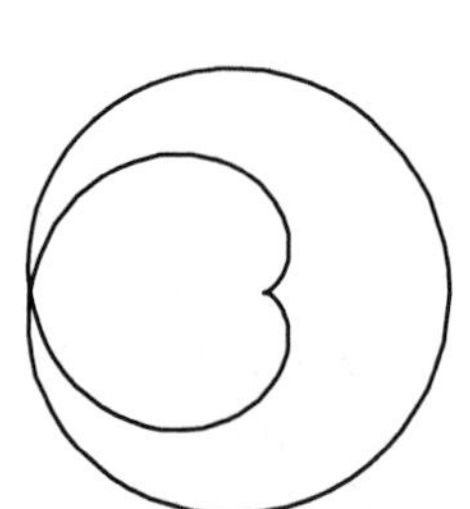

$a/b = \frac{1}{3}$

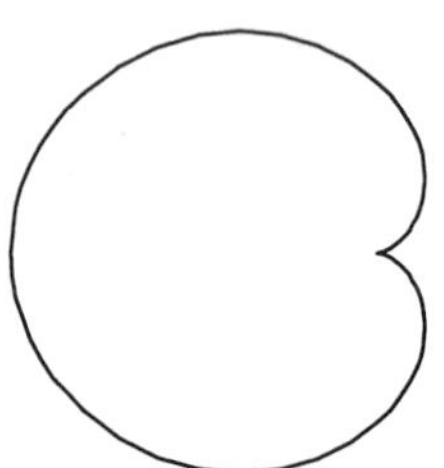

$a/b = \frac{1}{2}$

$a/b = e - 2, 0 \le t \le 446$

$a/b = \frac{7}{5}$

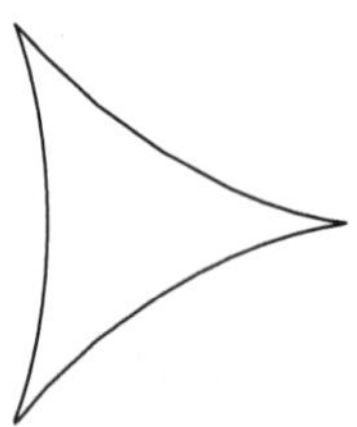

$a/b = 3$

$a/b = \frac{11}{3}$

$a/b = 23$

34. **(a)** The center Q of the smaller circle has coordinates $((a+b)\cos\theta, (a+b)\sin\theta)$. Arc PS has length $a\theta$ (as in Exercise 33), so that $\angle PQS = a\theta/b$, $\angle PQR = \pi - a\theta/b$, and $\angle PQT = \pi - \dfrac{a\theta}{b} - \theta = \pi - \left(\dfrac{a+b}{b}\right)\theta$ since $\angle RQT = \theta$. Thus the coordinates of P are

$$x = (a+b)\cos\theta + b\cos\left(\pi - \frac{a+b}{b}\theta\right) = (a+b)\cos\theta - b\cos\left(\frac{a+b}{b}\theta\right)$$

and $y = (a+b)\sin\theta - b\sin\left(\pi - \dfrac{a+b}{b}\theta\right) = (a+b)\sin\theta - b\sin\left(\dfrac{a+b}{b}\theta\right)$.

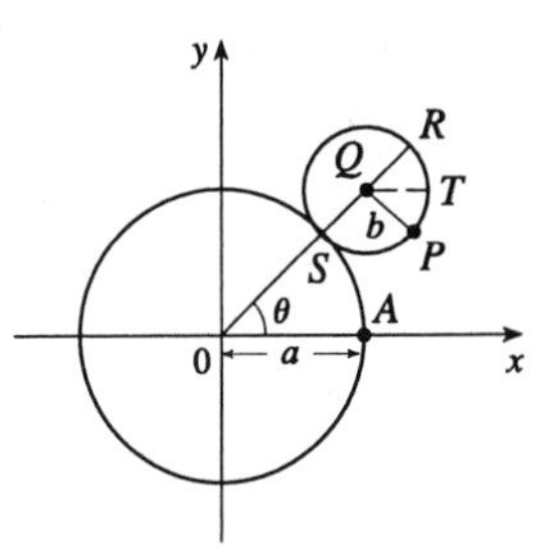

(b)

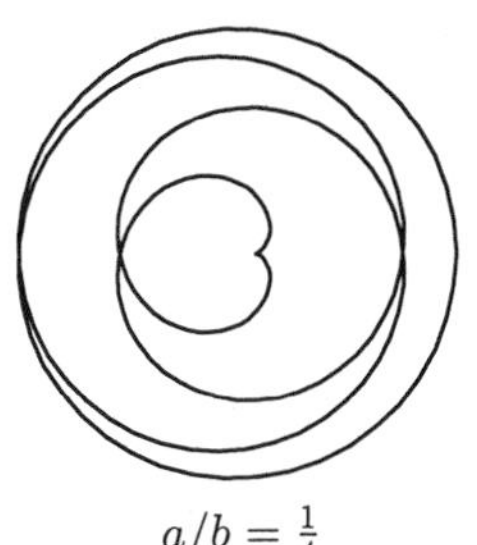

$a/b = \frac{1}{4}$

$a/b = \frac{2}{5}$

$a/b = \frac{3}{7}$

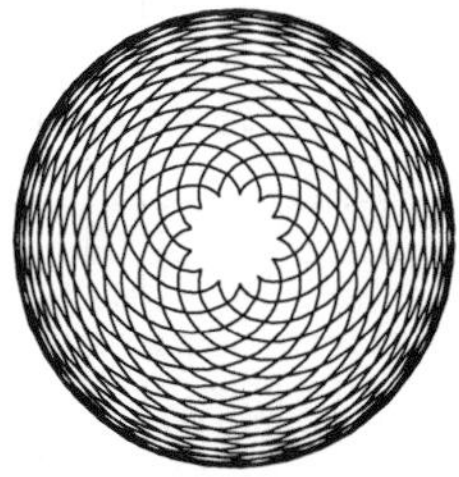

$a/b = \frac{13}{27}$

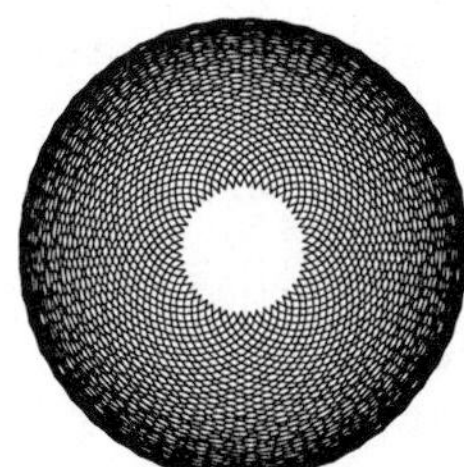

$a/b = e - 2, 0 \le t \le 446$

$a/b = \frac{7}{5}$

$a/b = 3$

$a/b = \frac{11}{3}$

$a/b = 37$

35. $C = (2a\cot\theta, 2a)$, so the x-coordinate of P is $x = 2a\cot\theta$. Let $B = (0, 2a)$. Then $\angle OAB$ is a right angle and $\angle OBA = \theta$, so $|OA| = 2a\sin\theta$ and $A = (2a\sin\theta\cos\theta, 2a\sin^2\theta)$. Thus the y-coordinate of P is $y = 2a\sin^2\theta$.

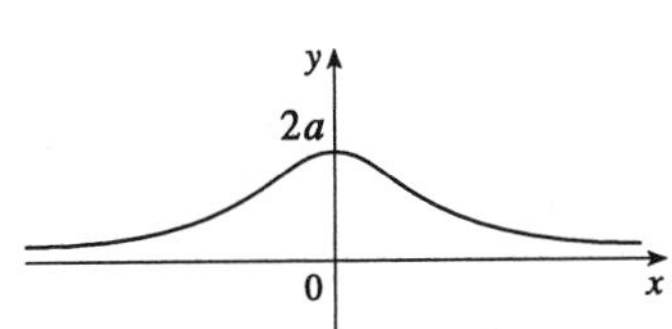

36. Let θ be the angle of inclination of segment OP. Then $|OB| = \dfrac{2a}{\cos\theta}$. Let $C = (2a, 0)$. Then by use of right triangle OAC we see that $|OA| = 2a\cos\theta$. Now

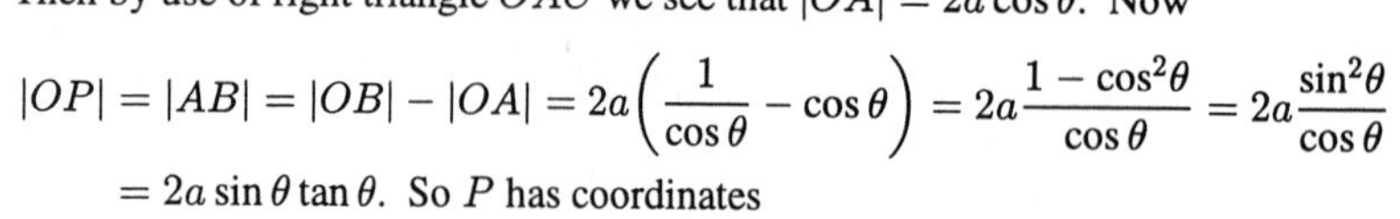

$$|OP| = |AB| = |OB| - |OA| = 2a\left(\frac{1}{\cos\theta} - \cos\theta\right) = 2a\frac{1-\cos^2\theta}{\cos\theta} = 2a\frac{\sin^2\theta}{\cos\theta}$$

$= 2a\sin\theta\tan\theta$. So P has coordinates

$x = 2a\sin\theta\tan\theta\cdot\cos\theta = 2a\sin^2\theta$ and $y = 2a\sin\theta\tan\theta\cdot\sin\theta = 2a\sin^2\theta\tan\theta$

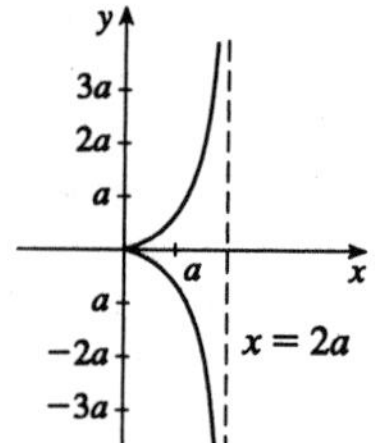

37. $x = t^2, y = t^3 - ct$. We use a graphing device to produce the graphs for various values of c. Note that all the members of the family are symmetric about the x-axis. For $c < 0$, the graph does not cross itself, but for $c = 0$ it has a cusp at $(0,0)$ and for $c > 0$ the graph crosses itself at $x = c$, so the loop grows larger as c increases.

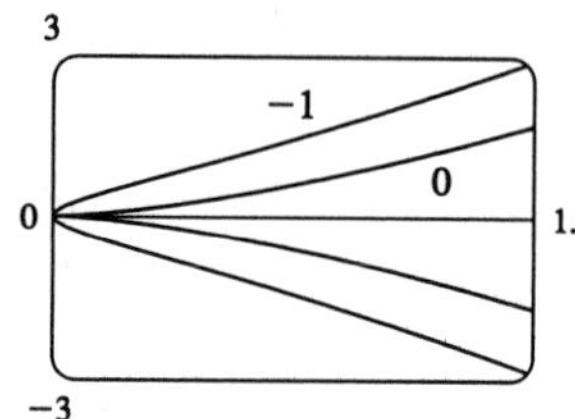

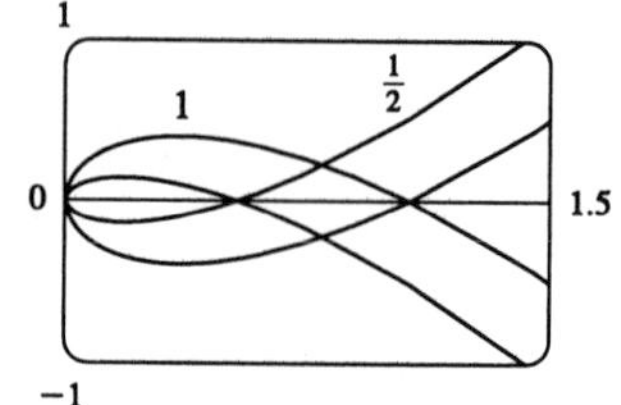

38. $x = 2ct - 4t^3, y = -ct^2 + 3t^4$. We use a graphing device to produce the graphs for various values of c. Note that all the members of the family are symmetric about the y-axis. When $c < 0$, the graph resembles that of a polynomial of even degree, but when $c = 0$ there is a cusp at the origin, and when $c > 0$, the graph crosses itself at the origin, and has two cusps below the x-axis. The size of the "swallowtail" increases as c increases.

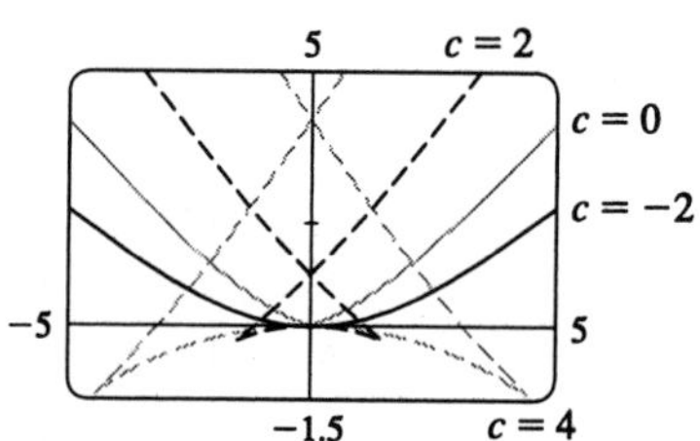

39. Note that all the Lissajous figures are symmetric about the x-axis. The parameters a and b simply stretch the graph in the x- and y-directions respectively. For $a = b = n = 1$ the graph is simply a circle with radius 1. For $n = 2$ the graph crosses itself at the origin and there are loops above and below the x-axis. In general, the figures have $n - 1$ points of intersection, all of which are on the y-axis, and a total of n closed loops.

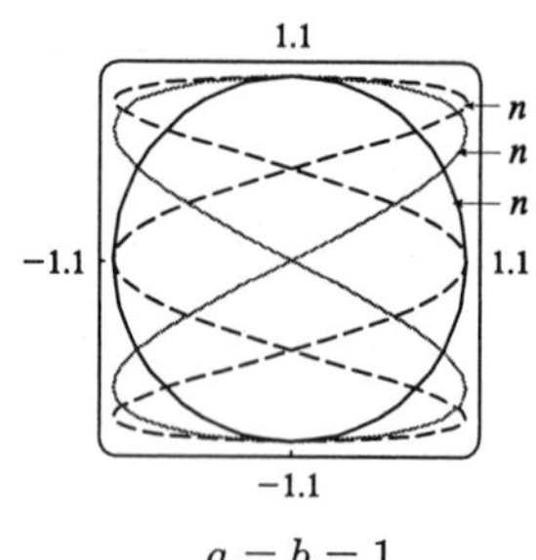

$a = b = 1$

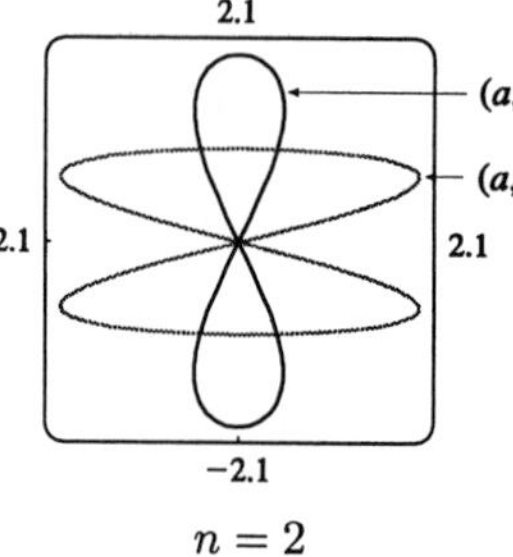

$n = 2$

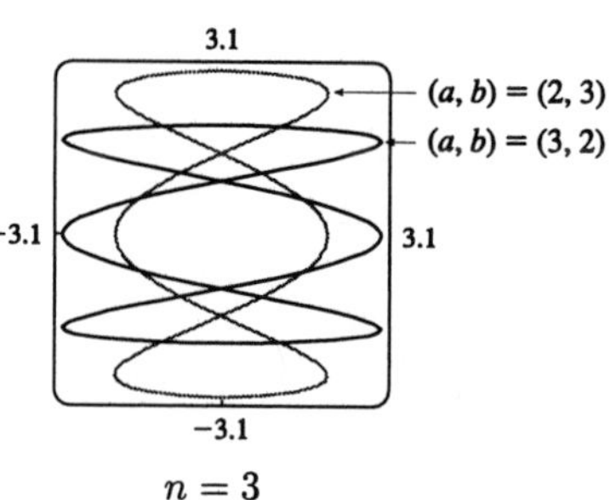

$n = 3$

40. Note that for all values of a, b, c, and d, the graph is symmetric about the x-axis. Also, in all cases we consider only $t \in (-\pi, \pi)$, since for larger t-intervals the graphs get too cluttered.

If you have a CAS, you can use its animation command (`animate` in Maple, `Animate` in Mathematica) to observe the effects of changing one of the parameters.

Note that if either a or b varies, only the x-value is altered for any particular t-value, so the graph is either stretched or shrunk in the x-direction only. Similarly, if c or d varies, the graph is altered only in the y-direction.

First we hold b, c, and d constant and allow a to vary. As a changes, the graph shrinks or expands in the x-direction. The points nearest the x-axis seem to move the fastest as a changes.

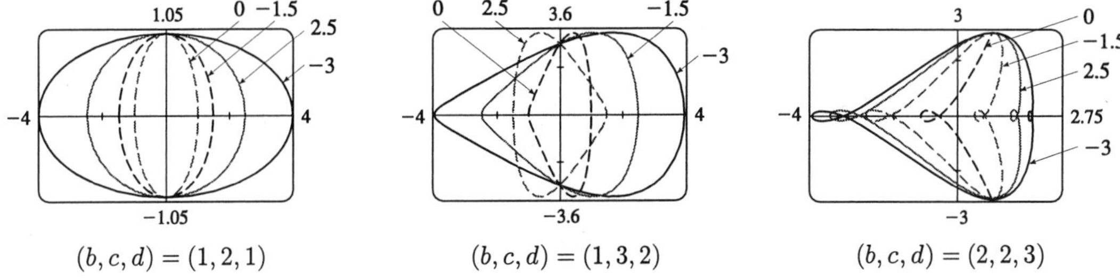

$(b, c, d) = (1, 2, 1)$ $(b, c, d) = (1, 3, 2)$ $(b, c, d) = (2, 2, 3)$

Next, we hold a, c, and d constant, and vary b. It seems that b affects the number of horizontal oscillations.

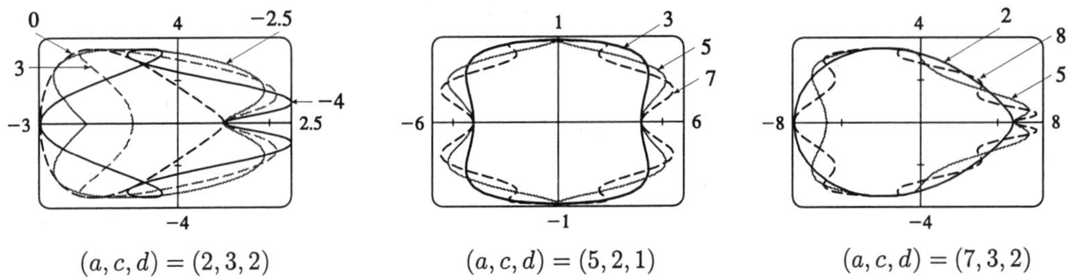

$(a, c, d) = (2, 3, 2)$ $(a, c, d) = (5, 2, 1)$ $(a, c, d) = (7, 3, 2)$

Next, we hold a, b, and d constant, and allow c to vary. This is like the case where a varies, except that the graph changes in the y-direction rather than the x-direction. Two of the x-intercepts remain constant as c changes.

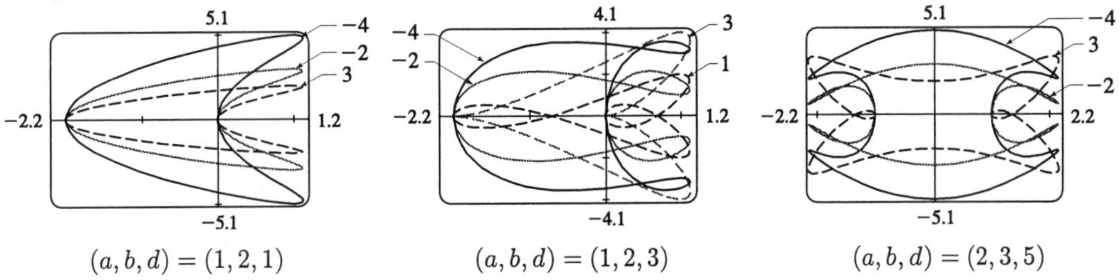

$(a, b, d) = (1, 2, 1)$ $(a, b, d) = (1, 2, 3)$ $(a, b, d) = (2, 3, 5)$

Finally, we fix a, b, and c, and allow d to vary. d seems to affect the number of vertical oscillations in the graph.

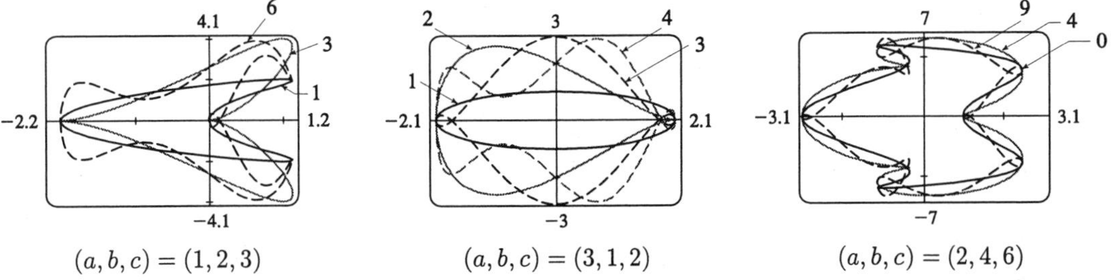

$(a, b, c) = (1, 2, 3)$ $(a, b, c) = (3, 1, 2)$ $(a, b, c) = (2, 4, 6)$

Note to Instructor: This exercise is extremely complex, and there are an enormous number of observations which can be made about the family of curves. The main point of the exercise is to encourage speculation and investigation rather than exhaustive analysis, and this should be considered when evaluating students' work on this exercise.

EXERCISES 9.2

1. $x = t^2 + t$, $y = t^2 - t$; $t = 0$. $\dfrac{dy}{dt} = 2t - 1$, $\dfrac{dx}{dt} = 2t + 1$, so $\dfrac{dy}{dx} = \dfrac{dy/dt}{dx/dt} = \dfrac{2t-1}{2t+1}$. When $t = 0$, $x = y = 0$ and $\dfrac{dy}{dx} = -1$. The tangent is $y - 0 = (-1)(x - 0)$ or $y = -x$.

2. $x = 1 - t^3$, $y = t^2 - 3t + 1$; $t = 1$. $\dfrac{dy}{dt} = 2t - 3$, $\dfrac{dx}{dt} = -3t^2$; $\dfrac{dy}{dx} = \dfrac{dy/dt}{dx/dt} = \dfrac{2t-3}{-3t^2}$. When $t = 1$, $\dfrac{dy}{dx} = \dfrac{-1}{-3} = \dfrac{1}{3}$ and $(x, y) = (0, -1)$, so the equation of the tangent is $y + 1 = \frac{1}{3}(x - 0)$, or $x - 3y = 3$.

3. $x = \ln t$, $y = te^t$; $t = 1$. $\dfrac{dy}{dt} = (t+1)e^t$, $\dfrac{dx}{dt} = \dfrac{1}{t}$, $\dfrac{dy}{dx} = \dfrac{dy/dt}{dx/dt} = t(t+1)e^t$. When $t = 1$, $(x, y) = (0, e)$ and $\dfrac{dy}{dx} = 2e$, so the equation of the tangent is $y - e = 2e(x - 0)$ or $2ex - y + e = 0$.

4. $x = t\sin t$, $y = t\cos t$; $t = \pi$. $\dfrac{dy}{dt} = \cos t - t\sin t$, $\dfrac{dx}{dt} = \sin t + t\cos t$, and $\dfrac{dy}{dx} = \dfrac{dy/dt}{dx/dt} = \dfrac{\cos t - t\sin t}{\sin t + t\cos t}$. When $t = \pi$, $(x, y) = (0, -\pi)$ and $\dfrac{dy}{dx} = \dfrac{-1}{-\pi} = \dfrac{1}{\pi}$, so the equation of the tangent is $y + \pi = \dfrac{1}{\pi}(x - 0)$ or $x - \pi y - \pi^2 = 0$.

5. **(a)** $x = 2t + 3$, $y = t^2 + 2t$; $(5, 3)$. $\dfrac{dy}{dt} = 2t + 2$, $\dfrac{dx}{dt} = 2$, and $\dfrac{dy}{dx} = \dfrac{dy/dt}{dx/dt} = t + 1$.

At $(5, 3)$, $t = 1$ and $\dfrac{dy}{dx} = 2$, so the tangent is $y - 3 = 2(x - 5)$ or $2x - y - 7 = 0$.

(b) $y = t^2 + 2t = \left(\dfrac{x-3}{2}\right)^2 + 2\left(\dfrac{x-3}{2}\right) = \dfrac{(x-3)^2}{4} + x - 3$, so $\dfrac{dy}{dx} = \dfrac{x-3}{2} + 1$.

When $x = 5$, $\dfrac{dy}{dx} = 2$, so the equation of the tangent is $2x - y - 7 = 0$, as before.

6. **(a)** $x = 5\cos t$, $y = 5\sin t$; $(3, 4)$. $\dfrac{dy}{dt} = 5\cos t$, $\dfrac{dx}{dt} = -5\sin t$, $\dfrac{dy}{dx} = \dfrac{dy/dt}{dx/dt} = -\cot t$.

At $(3, 4)$, $t = \tan^{-1}\dfrac{y}{x} = \tan^{-1}\dfrac{4}{3}$, so $\dfrac{dy}{dx} = -\dfrac{3}{4}$, and the tangent is $y - 4 = -\frac{3}{4}(x - 3)$, or $3x + 4y = 25$.

(b) $x^2 + y^2 = 25$, so $2x + 2y\dfrac{dy}{dx} = 0$, or $\dfrac{dy}{dx} = -\dfrac{x}{y}$. At $(3, 4)$, $\dfrac{dy}{dx} = -\dfrac{3}{4}$, and as in (a) the tangent is $3x + 4y = 25$.

7. $x = 2\sin 2t$, $y = 2\sin t$; $\left(\sqrt{3}, 1\right)$.

$\dfrac{dy}{dx} = \dfrac{dy/dt}{dx/dt} = \dfrac{2\cos t}{2 \cdot 2\cos 2t} = \dfrac{\cos t}{2\cos 2t}$. The point $\left(\sqrt{3}, 1\right)$ corresponds only to $t = \frac{\pi}{6}$, so the slope of the tangent at that point is $\dfrac{\cos\frac{\pi}{6}}{2\cos\frac{\pi}{3}} = \dfrac{\sqrt{3}/2}{2 \cdot \frac{1}{2}} = \dfrac{\sqrt{3}}{2}$. The equation of the tangent is therefore $(y - 1) = \frac{\sqrt{3}}{2}\left(x - \sqrt{3}\right)$ or $\sqrt{3}x - 2y - 1 = 0$.

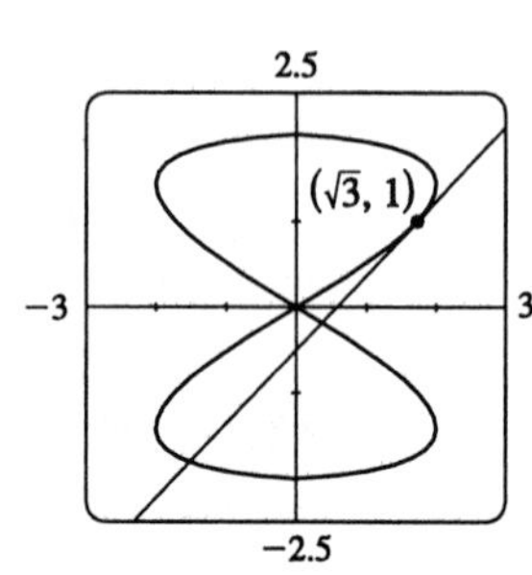

8. $x = \sin t$, $y = \sin(t + \sin t)$; $(0,0)$. $\dfrac{dy}{dx} = \dfrac{dy/dt}{dx/dt} = \dfrac{\cos(t + \sin t)(1 + \cos t)}{\cos t} = (\sec t + 1)\cos(t + \sin t)$.

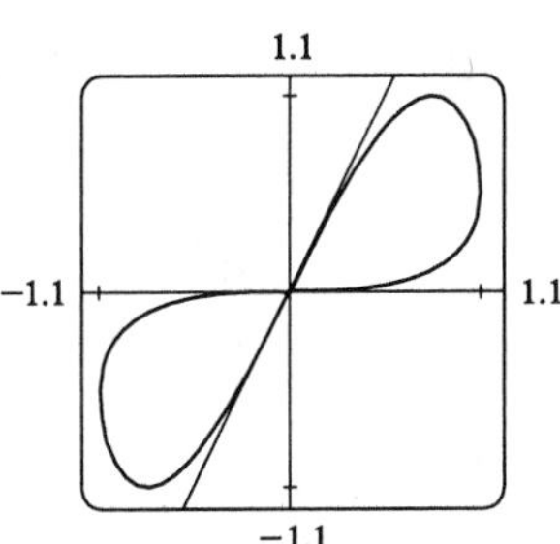

Note that there are two tangents at the point $(0,0)$, since both $t = 0$ and $t = \pi$ correspond to the origin. The tangent corresponding to $t = 0$ has slope $(\sec 0 + 1)\cos(0 + \sin 0) = 2\cos 0 = 2$, and its equation is $y = 2x$. The tangent corresponding to $t = \pi$ has slope $(\sec \pi + 1)\cos(\pi + \sin \pi) = 0$, so it is the x-axis.

9. $x = t^2 + t$, $y = t^2 + 1$. $\dfrac{dy}{dx} = \dfrac{dy/dt}{dx/dt} = \dfrac{2t}{2t+1} = 1 - \dfrac{1}{2t+1}$; $\dfrac{d}{dt}\left(\dfrac{dy}{dx}\right) = \dfrac{2}{(2t+1)^2}$;

$$\frac{d^2y}{dx^2} = \frac{d}{dx}\left(\frac{dy}{dx}\right) = \frac{d(dy/dx)/dt}{dx/dt} = \frac{2}{(2t+1)^3}$$

10. $x = t^3 + t^2 + 1$, $y = 1 - t^2$. $\dfrac{dy}{dt} = -2t$, $\dfrac{dx}{dt} = 3t^2 + 2t$; $\dfrac{dy}{dx} = \dfrac{dy/dt}{dx/dt} = \dfrac{-2t}{3t^2 + 2t} = -\dfrac{2}{3t+2}$;

$$\frac{d}{dt}\left(\frac{dy}{dx}\right) = \frac{6}{(3t+2)^2};\ \frac{d^2y}{dx^2} = \frac{d(dy/dx)/dt}{dx/dt} = \frac{6}{t(3t+2)^3}$$

11. $x = \sin \pi t$, $y = \cos \pi t$. $\dfrac{dy}{dx} = \dfrac{dy/dt}{dx/dt} = \dfrac{-\pi \sin \pi t}{\pi \cos \pi t} = -\tan \pi t$;

$$\frac{d^2y}{dx^2} = \frac{d}{dx}\left(\frac{dy}{dx}\right) = \frac{d(dy/dx)/dt}{dx/dt} = \frac{-\pi \sec^2 \pi t}{\pi \cos \pi t} = -\sec^3 \pi t$$

12. $x = t + 2\cos t$, $y = \sin 2t$. $\dfrac{dy}{dx} = \dfrac{dy/dt}{dx/dt} = \dfrac{2\cos 2t}{1 - 2\sin t}$;

$$\frac{d}{dt}\left(\frac{dy}{dx}\right) = \frac{(1 - 2\sin t)(-4\sin 2t) - 2\cos 2t(-2\cos t)}{(1 - 2\sin t)^2} = \frac{4(\cos t - \sin 2t + \sin t \sin 2t)}{(1 - 2\sin t)^2};$$

$$\frac{d^2y}{dx^2} = \frac{d(dy/dx)/dt}{dx/dt} = \frac{4(\cos t - \sin 2t + \sin t \sin 2t)}{(1 - 2\sin t)^3}$$

13. $x = e^{-t}$, $y = te^{2t}$. $\dfrac{dy}{dx} = \dfrac{dy/dt}{dx/dt} = \dfrac{(2t+1)e^{2t}}{-e^{-t}} = -(2t+1)e^{3t}$;

$$\frac{d}{dt}\left(\frac{dy}{dx}\right) = -3(2t+1)e^{3t} - 2e^{3t} = -(6t+5)e^{3t};$$

$$\frac{d^2y}{dx^2} = \frac{d}{dx}\left(\frac{dy}{dx}\right) = \frac{d(dy/dx)/dt}{dx/dt} = \frac{-(6t+5)e^{3t}}{-e^{-t}} = (6t+5)e^{4t}$$

14. $x = 1 + t^2$, $y = t \ln t$. $\dfrac{dy}{dx} = \dfrac{dy/dt}{dx/dt} = \dfrac{1 + \ln t}{2t}$; $\dfrac{d}{dt}\left(\dfrac{dy}{dx}\right) = \dfrac{2t(1/t) - (1 + \ln t)2}{(2t)^2} = -\dfrac{\ln t}{2t^2}$;

$$\frac{d^2y}{dx^2} = \frac{d(dy/dx)/dt}{dx/dt} = -\frac{\ln t}{4t^3}$$

15. $x = t(t^2 - 3) = t^3 - 3t$, $y = 3(t^2 - 3)$. $\frac{dx}{dt} = 3t^2 - 3 = 3(t-1)(t+1)$; $\frac{dy}{dt} = 6t$. $\frac{dy}{dt} = 0 \quad \Leftrightarrow \quad t = 0$ $\Leftrightarrow \quad (x, y) = (0, -9)$. $\frac{dx}{dt} = 0 \quad \Leftrightarrow \quad t = \pm 1 \quad \Leftrightarrow \quad (x, y) = (-2, -6)$ or $(2, -6)$. So there is a horizontal tangent at $(0, -9)$ and there are vertical tangents at $(-2, -6)$ and $(2, -6)$.

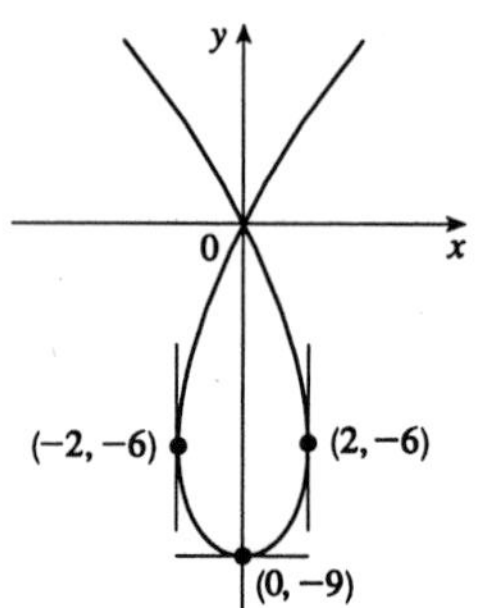

	$t < -1$	$-1 < t < 0$	$0 < t < 1$	$t > 1$
dx/dt	+	−	−	+
dy/dt	−	−	+	+
x	→	←	←	→
y	↓	↓	↑	↑
curve	↘	↙	↖	↗

16. $x = t^3 - 3t^2$, $y = t^3 - 3t$. $\frac{dx}{dt} = 3t^2 - 6t = 3t(t-2)$,
$\frac{dy}{dt} = 3t^2 - 3 = 3(t-1)(t+1)$. $\frac{dy}{dt} = 0 \quad \Leftrightarrow \quad t = +1$ or -1
$\Leftrightarrow \quad (x, y) = (-2, -2)$ or $(-4, 2)$. $\frac{dx}{dt} = 0 \quad \Leftrightarrow \quad t = 0$ or 2
$\Leftrightarrow \quad (x, y) = (0, 0)$ or $(-4, 2)$. So the tangent is horizontal at $(-2, -2)$ and vertical at $(0, 0)$. At $(-4, 2)$ the curve crosses itself and there are two tangents, one horizontal and one vertical.

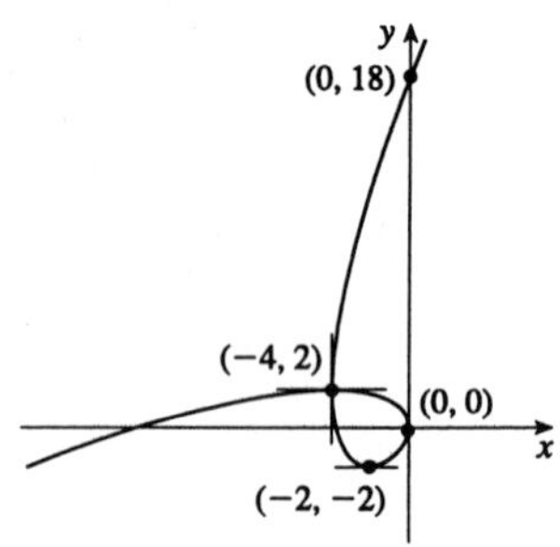

	$t < -1$	$-1 < t < 0$	$0 < t < 1$	$1 < t < 2$	$t > 2$
dx/dt	+	+	−	−	+
dy/dt	+	−	−	+	+
x	→	→	←	←	→
y	↑	↓	↓	↑	↑
curve	↗	↘	↙	↖	↗

17. $x = \frac{3t}{1+t^3}$, $y = \frac{3t^2}{1+t^3}$. $\frac{dx}{dt} = \frac{(1+t^3)3 - 3t(3t^2)}{(1+t^3)^2} = \frac{3 - 6t^3}{(1+t^3)^2}$,
$\frac{dy}{dt} = \frac{(1+t^3)(6t) - 3t^2(3t^2)}{(1+t^3)^2} = \frac{6t - 3t^4}{(1+t^3)^2} = \frac{3t(2-t^3)}{(1+t^3)^2}$. $\frac{dy}{dt} = 0 \quad \Leftrightarrow \quad t = 0$ or $\sqrt[3]{2} \quad \Leftrightarrow$
$(x, y) = (0, 0)$ or $\left(\sqrt[3]{2}, \sqrt[3]{4}\right)$. $\frac{dx}{dt} = 0 \quad \Leftrightarrow \quad t^3 = \frac{1}{2} \quad \Leftrightarrow \quad t = 2^{-1/3} \quad \Leftrightarrow$
$(x, y) = \left(\sqrt[3]{4}, \sqrt[3]{2}\right)$. There are horizontal tangents at $(0, 0)$ and $\left(\sqrt[3]{2}, \sqrt[3]{4}\right)$, and there are vertical tangents at $\left(\sqrt[3]{4}, \sqrt[3]{2}\right)$ and $(0, 0)$. [The vertical tangent at $(0, 0)$ is undetectable by the methods of this section because that tangent corresponds to the limiting position of the point (x, y) as $t \to \pm\infty$.]
In the following table, $\alpha = \sqrt[3]{2}$.

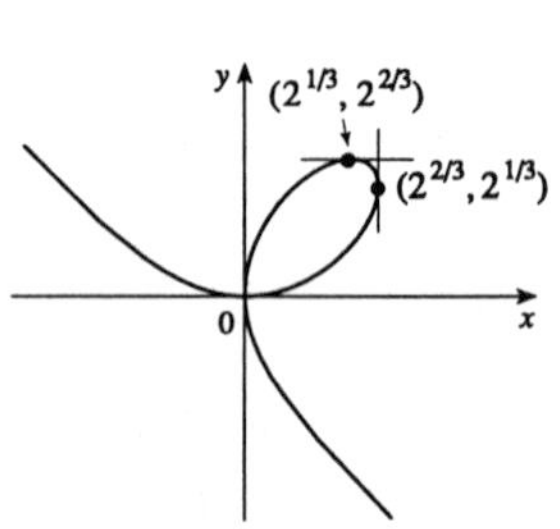

	$t<-1$	$-1<t<0$	$0<t<1/\alpha$	$1/\alpha<t<\alpha$	$t>\alpha$
dx/dt	+	+	+	−	−
dy/dt	−	−	+	+	−
x	→	→	→	←	←
y	↓	↓	↑	↑	↓
curve	↘	↘	↗	↖	↙

18. $x = a(\cos\theta - \cos^2\theta)$, $y = a(\sin\theta - \sin\theta\cos\theta)$. $\dfrac{dx}{d\theta} = a(-\sin\theta + 2\cos\theta\sin\theta)$,

$\dfrac{dy}{d\theta} = a(\cos\theta + \sin^2\theta - \cos^2\theta) = a(\cos\theta + 1 - 2\cos^2\theta)$, $\dfrac{dy}{d\theta} = 0 \quad\Leftrightarrow$
$0 = 2\cos^2\theta - \cos\theta - 1 = (2\cos\theta + 1)(\cos\theta - 1) \quad\Leftrightarrow\quad \cos\theta = -\frac{1}{2}$ or $1 \quad\Leftrightarrow\quad (x,y) = \left(-\frac{3}{4}a, \pm\frac{3\sqrt{3}}{4}a\right)$

or $(0,0)$. $\dfrac{dx}{d\theta} = 0 \quad\Leftrightarrow\quad (2\cos\theta - 1)\sin\theta = 0 \quad\Leftrightarrow\quad \cos\theta = \frac{1}{2}$ or $\sin\theta = 0 \quad\Leftrightarrow\quad (x,y) = (0,0)$ or $\left(\frac{1}{4}a, \pm\frac{\sqrt{3}}{4}a\right)$ or $(-2a, 0)$. The curve has horizontal tangents at $\left(-\frac{3}{4}a, \pm\frac{3\sqrt{3}}{4}a\right)$ and vertical tangents at $(-2a,0)$ and $\left(\frac{1}{4}a, \pm\frac{\sqrt{3}}{4}a\right)$. Since $\dfrac{dy}{dx} = \dfrac{dy/d\theta}{dx/d\theta} = \dfrac{(2\cos\theta + 1)(1 - \cos\theta)}{(2\cos\theta - 1)\sin\theta}$,

we see that $\displaystyle\lim_{\theta\to 0}\frac{dy}{dx} = \lim_{\theta\to 0}\frac{2\cos\theta + 1}{2\cos\theta - 1}\cdot\lim_{\theta\to 0}\frac{1 - \cos\theta}{\sin\theta} = 3\cdot 0 = 0$

(using l'Hospital's Rule). Thus the curve has a horizontal tangent at $(0,0)$, where both $\dfrac{dx}{d\theta}$ and $\dfrac{dy}{d\theta}$ are 0. [The curve is the cardioid $r = a(1 - \cos\theta)$; see Section 9.4.]

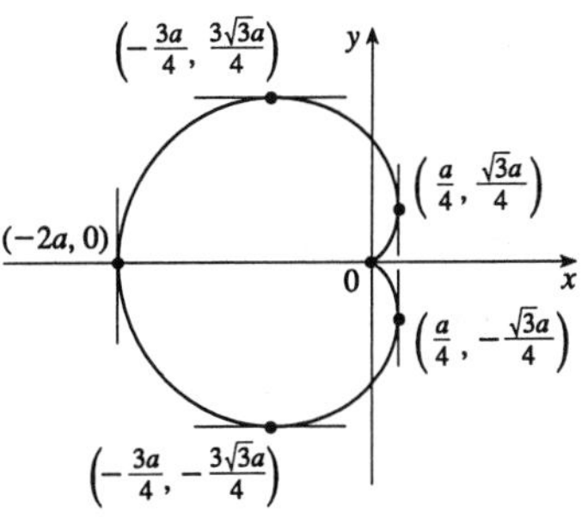

	$0<t<\frac{\pi}{3}$	$\frac{\pi}{3}<t<\frac{2\pi}{3}$	$\frac{2\pi}{3}<t<\pi$	$\pi<t<\frac{4\pi}{3}$	$\frac{4\pi}{3}<t<\frac{5\pi}{3}$	$\frac{5\pi}{3}<t<2\pi$
dx/dt	+	−	−	+	+	−
dy/dt	+	+	−	−	+	+
x	→	←	←	→	→	←
y	↑	↑	↓	↓	↑	↑
curve	↗	↖	↙	↘	↗	↖

19. From the graph, it appears that the leftmost point on the curve $x = t^4 - t^2$, $y = t + \ln t$ is about $(-0.25, 0.36)$. To find the exact coordinates, we find the value of t for which the graph has a vertical tangent, that is, $0 = dx/dt = 4t^3 - 2t$
$\Leftrightarrow\quad 2t(2t^2 - 1) = 0 \quad\Leftrightarrow\quad 2t\left(\sqrt{2}t + 1\right)\left(\sqrt{2}t - 1\right) = 0$
$\Leftrightarrow\quad t = 0$ or $\pm\frac{1}{\sqrt{2}}$. The negative and 0 roots are inadmissible since $y(t)$ is only defined for $t > 0$, so the leftmost point must be

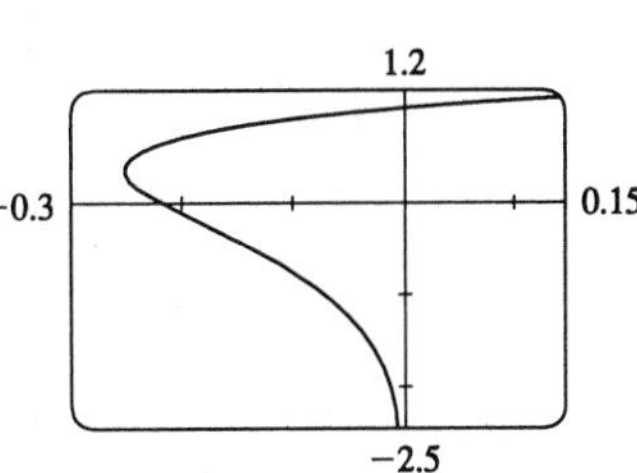

$$\left(x\left(\tfrac{1}{\sqrt{2}}\right), y\left(\tfrac{1}{\sqrt{2}}\right)\right) = \left(\left(\tfrac{1}{\sqrt{2}}\right)^4 - \left(\tfrac{1}{\sqrt{2}}\right)^2, \tfrac{1}{\sqrt{2}} + \ln\tfrac{1}{\sqrt{2}}\right) = \left(-\tfrac{1}{4}, \tfrac{1}{\sqrt{2}} - \tfrac{1}{2}\ln 2\right).$$

20. The curve is symmetric about the line $y = -x$, so if we can find the highest point (x_h, y_h), then the leftmost point is $(x_l, y_l) = (-y_h, -x_h)$. After carefully zooming in, we estimate that the highest point on the curve $x = te^t$, $y = te^{-t}$ is about $(2.7, 0.37)$.

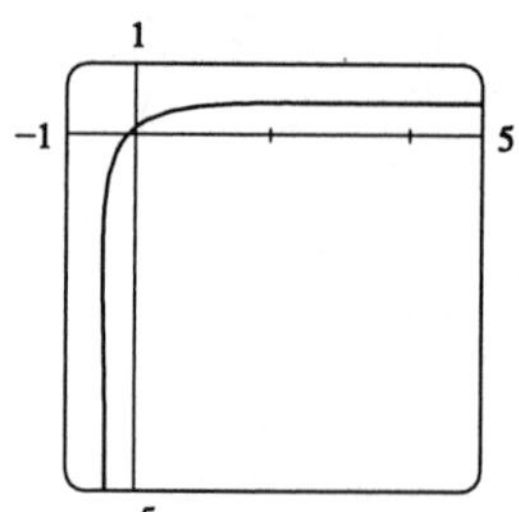

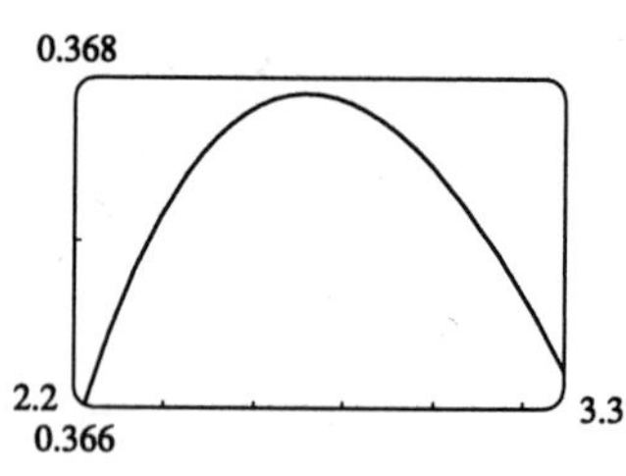

To find the exact coordinates of the highest point, we find the value of t for which the curve has a horizontal tangent, that is, $dy/dt = 0 \quad \Leftrightarrow \quad t(-e^{-t}) + e^{-t} = 0 \quad \Leftrightarrow \quad (1-t)e^{-t} = 0 \quad \Leftrightarrow \quad t = 1$. This corresponds to the point $(x(1), y(1)) = (e, 1/e)$. To find the leftmost point, we find the value of t for which $0 = dx/dt = te^t + e^t \quad \Leftrightarrow \quad (1+t)e^t = 0 \quad \Leftrightarrow \quad t = -1$. This corresponds to the point $(x(-1), y(-1)) = (-1/e, -e)$. As $t \to -\infty$, $x(t) = te^t \to 0^-$ by l'Hospital's Rule and $y(t) = te^{-t} \to -\infty$, so the y-axis is an asymptote. As $t \to \infty$, $x(t) \to \infty$ and $y(t) \to 0^+$, so the x-axis is the other asymptote. The asymptotes can also be determined from the graph, if we use a larger t-interval.

21. We graph the curve $x = t^4 - 2t^3 - 2t^2$, $y = t^3 - t$ in the viewing rectangle $[-2, 1]$ by $[-0.4, 0.4]$. This rectangle corresponds approximately to $t \in [-1, 0.8]$. We estimate that the curve has horizontal tangents at about $(-1, -0.4)$ and $(-0.17, 0.39)$ and vertical tangents at about $(0, 0)$ and $(-0.19, 0.37)$.We calculate $\dfrac{dy}{dx} = \dfrac{dy/dt}{dx/dt} = \dfrac{3t^2 - 1}{4t^3 - 6t^2 - 4t}$. The horizontal tangents occur when $dy/dt = 3t^2 - 1 = 0 \quad \Leftrightarrow \quad t = \pm\frac{1}{\sqrt{3}}$, so both horizontal tangents are shown in our graph. The vertical tangents occur when $dx/dt = 2t(2t^2 - 3t - 2) = 0 \quad \Leftrightarrow \quad 2t(2t+1)(t-2) = 0 \quad \Leftrightarrow \quad t = 0, -\frac{1}{2}$ or 2. It seems that we have missed one vertical tangent, and indeed if we plot the curve on the t-interval $[-1.2, 2.2]$ we see that there is another vertical tangent at $(-8, 6)$.

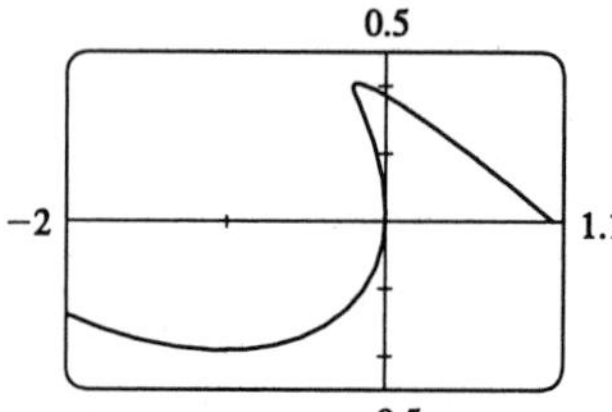

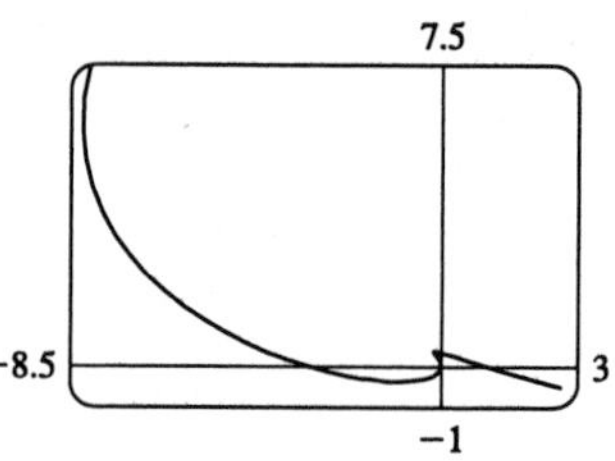

22. We graph the curve $x = t^4 + 4t^3 - 8t^2$, $y = 2t^2 - t$ in the viewing rectangle $[-5, 0]$ by $[-0.2, 2]$. It appears that there is a horizontal tangent at about $(-0.4, -0.1)$, and vertical tangents at about $(-3, 1)$ and $(0, 0)$. We calculate $\dfrac{dy}{dx} = \dfrac{dy/dt}{dx/dt} = \dfrac{4t - 1}{4t^3 + 12t^2 - 16t}$, so there is a horizontal tangent where $dy/dt = 4t - 1 = 0 \quad \Leftrightarrow \quad t = \frac{1}{4}$. This point (the lowest point) is shown in the first graph. There are vertical tangents where $dx/dt = 4t^3 + 12t^2 - 16t = 0 \quad \Leftrightarrow \quad 4t(t^2 + 3t - 4) = 0 \quad \Leftrightarrow \quad 4t(t+4)(t-1) = 0$. We have missed one vertical tangent corresponding to $t = -4$, and if we plot the graph for $t \in [-5, 3]$, we see that the curve has another vertical tangent line at approximately $(-128, 36)$.

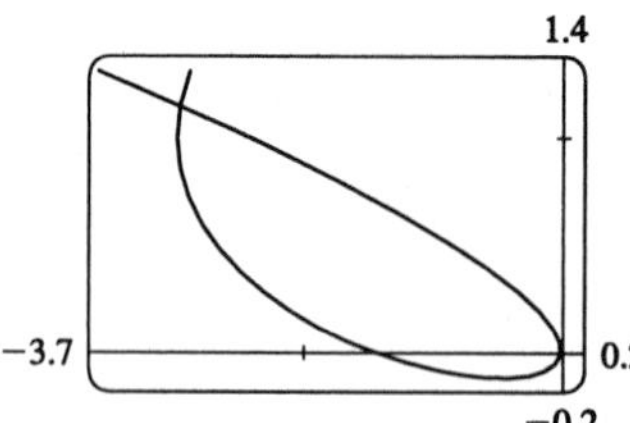

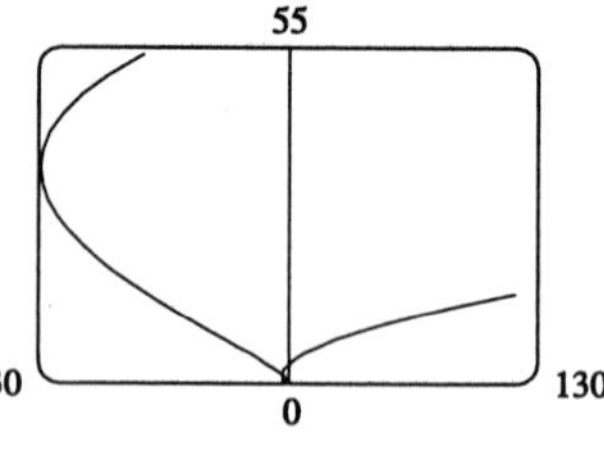

23. $x = \cos t$, $y = \sin t \cos t$. $\dfrac{dx}{dt} = -\sin t$,

$\dfrac{dy}{dt} = -\sin^2 t + \cos^2 t = \cos 2t$. $(x, y) = (0, 0) \Leftrightarrow \quad \cos t = 0$

$\Leftrightarrow \quad t$ is an odd multiple of $\frac{\pi}{2}$.

When $t = \dfrac{\pi}{2}$, $\dfrac{dx}{dt} = -1$ and $\dfrac{dy}{dt} = -1$, so $\dfrac{dy}{dx} = 1$.

When $t = \dfrac{3\pi}{2}$, $\dfrac{dx}{dt} = 1$ and $\dfrac{dy}{dt} = -1$. So $\dfrac{dy}{dx} = -1$.

Thus $y = x$ and $y = -x$ are both tangent to the curve at $(0, 0)$.

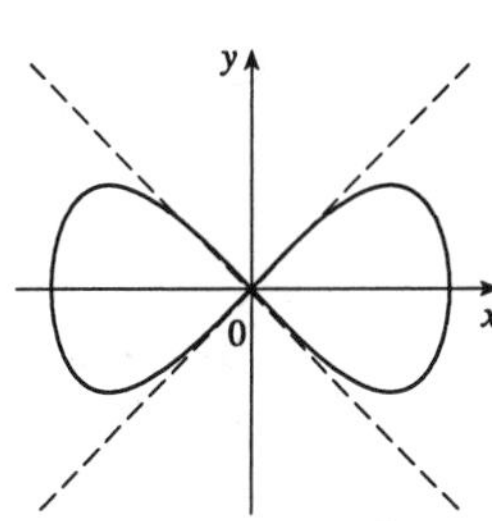

24. $x = 1 - 2\cos^2 t = -\cos 2t$, $y = (\tan t)(1 - 2\cos^2 t) = -(\tan t)\cos 2t$. To find a point where the curve crosses itself, we look for two values of t that give the same point (x, y). Call these values t_1 and t_2. Then $\cos^2 t_1 = \cos^2 t_2$ (from the equation for x) and either $\tan t_1 = \tan t_2$ or $\cos^2 t_1 = \cos^2 t_2 = \frac{1}{2}$ (from the equation for y). We can satisfy $\cos^2 t_1 = \cos^2 t_2$ and $\tan t_1 = \tan t_2$ by choosing t_1 arbitrarily and taking $t_2 = t_1 + \pi$, so evidently the whole curve is retraced every time t traverses an interval of length π. Thus we can restrict our attention to the interval $\left(-\frac{\pi}{2}, \frac{\pi}{2}\right)$. If $t_2 = -t_1$, then $\cos^2 t_2 = \cos^2 t_1$, but $\tan t_2 = -\tan t_1$. This suggests that we try to satisfy the condition $\cos^2 t_1 = \cos^2 t_2 = \frac{1}{2}$. Taking $t_1 = \frac{\pi}{4}$ and $t_2 = -\frac{\pi}{4}$gives $(x, y) = (0, 0)$ for both values of t.

$\dfrac{dx}{dt} = 2\sin 2t$,and $\dfrac{dy}{dt} = 2\sin 2t \tan t - \cos 2t \sec^2 t$.

When $t = \dfrac{\pi}{4}$, $\dfrac{dx}{dt} = 2$ and $\dfrac{dy}{dt} = 2$, so $\dfrac{dy}{dx} = 1$.

When $t = -\dfrac{\pi}{4}$, $\dfrac{dx}{dt} = -2$ and $\dfrac{dy}{dt} = 2$, so $\dfrac{dy}{dx} = -1$.

Thus the equations of the two tangents at $(0, 0)$ are $y = x$ and $y = -x$.

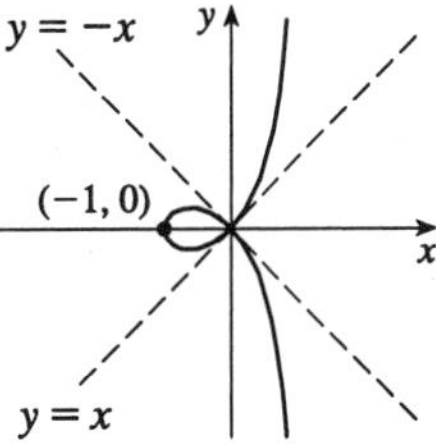

25. (a) $x = r\theta - d\sin\theta$, $y = r - d\cos\theta$; $\dfrac{dx}{d\theta} = r - d\cos\theta$, $\dfrac{dy}{d\theta} = d\sin\theta$. So $\dfrac{dy}{dx} = \dfrac{d\sin\theta}{r - d\cos\theta}$.

(b) If $0 < d < r$, then $|d\cos\theta| \le d < r$, so $r - d\cos\theta \ge r - d > 0$. This shows that $dx/d\theta$ never vanishes, so the trochoid can have no vertical tangents if $d < r$.

26. $x = a\cos^3\theta$, $y = a\sin^3\theta$.

(a) $\dfrac{dx}{d\theta} = -3a\cos^2\theta\sin\theta$, $\dfrac{dy}{d\theta} = 3a\sin^2\theta\cos\theta$, so $\dfrac{dy}{dx} = -\dfrac{\sin\theta}{\cos\theta} = -\tan\theta$.

(b) The tangent is horizontal $\Leftrightarrow$ $dy/dx = 0$ $\Leftrightarrow$ $\tan\theta = 0$ $\Leftrightarrow$ $\theta = n\pi$ $\Leftrightarrow$ $(x, y) = (\pm a, 0)$.
The tangent is vertical $\Leftrightarrow$ $\cos\theta = 0$ $\Leftrightarrow$ θ is an odd multiple of $\frac{\pi}{2}$ $\Leftrightarrow$ $(x, y) = (0, \pm a)$.

(c) $dy/dx = \pm 1$ $\Leftrightarrow$ $\tan\theta = \pm 1$ $\Leftrightarrow$ θ is an odd multiple of $\frac{\pi}{4}$ $\Leftrightarrow$ $(x, y) = \left(\pm\frac{\sqrt{2}}{4}a, \pm\frac{\sqrt{2}}{4}a\right)$

(All sign choices are valid.)

27. The line with parametric equations $x = -7t$, $y = 12t - 5$ is $y = 12\left(-\frac{1}{7}x\right) - 5$, which has slope $-\frac{12}{7}$. The curve $x = t^3 + 4t$, $y = 6t^2$ has slope $\dfrac{dy}{dx} = \dfrac{dy/dt}{dx/dt} = \dfrac{12t}{3t^2 + 4}$. This equals $-\dfrac{12}{7}$ $\Leftrightarrow$ $3t^2 + 4 = -7t$ $\Leftrightarrow$ $(3t + 4)(t + 1) = 0$ $\Leftrightarrow$ $t = -1$ or $t = -\frac{4}{3}$ $\Leftrightarrow$ $(x, y) = (-5, 6)$ or $\left(-\frac{208}{27}, \frac{32}{3}\right)$.

28. $x = 3t^2 + 1$, $y = 2t^3 + 1$, $\dfrac{dx}{dt} = 6t$, $\dfrac{dy}{dt} = 6t^2$, so $\dfrac{dy}{dx} = \dfrac{6t^2}{6t} = t$ (even where $t = 0$).

So at the point corresponding to parameter value t, the equation of the tangent line is $y - (2t^3 + 1) = t[x - (3t^2 + 1)]$. If this line is to pass through $(4, 3)$, we must have $3 - (2t^3 + 1) = t[4 - (3t^2 + 1)] \Leftrightarrow 2t^3 - 2 = 3t^3 - 3t \Leftrightarrow t^3 - 3t + 2 = 0 \Leftrightarrow (t-1)^2(t+2) = 0 \Leftrightarrow$ $t = 1$ or -2. Hence the desired equations are $y - 3 = x - 4$, or $x - y - 1 = 0$, tangent to the curve at $(4, 3)$, and $y - (-15) = -2(x - 13)$, or $2x + y - 11 = 0$, tangent to the curve at $(13, -15)$.

29. By symmetry of the ellipse about the x- and y-axes,

$A = 4\int_0^a y\,dx = 4\int_{\pi/2}^0 b\sin\theta(-a\sin\theta)d\theta = 4ab\int_0^{\pi/2}\sin^2\theta\,d\theta = 4ab\int_0^{\pi/2}\frac{1}{2}(1 - \cos 2\theta)d\theta$

$= 2ab\left[\theta - \frac{1}{2}\sin 2\theta\right]_0^{\pi/2} = 2ab\left(\frac{\pi}{2}\right) = \pi ab.$

30. $t + 1/t = 2.5 \quad\Leftrightarrow\quad t = \frac{1}{2}$ or 2, and for $\frac{1}{2} < t < 2$, we have $t + 1/t < 2.5$.

$x = -\frac{3}{2}$ when $t = \frac{1}{2}$ and $x = \frac{3}{2}$ when $t = 2$.

$A = \int_{-3/2}^{3/2}(2.5 - y)dx = \int_{1/2}^2\left(\frac{5}{2} - t - 1/t\right)\left(1 + 1/t^2\right)dt \quad [x = t - 1/t \quad \Rightarrow dx = (1 + 1/t^2)dt]$

$$= \int_{1/2}^2\left(-t + \tfrac{5}{2} - 2t^{-1} + \tfrac{5}{2}t^{-2} - t^{-3}\right)dt = \left[\frac{-t^2}{2} + \frac{5t}{2} - 2\ln|t| - \frac{5}{2t} + \frac{1}{2t^2}\right]_{1/2}^2$$

$= \left(-2 + 5 - 2\ln 2 - \frac{5}{4} + \frac{1}{8}\right) - \left(-\frac{1}{8} + \frac{5}{4} + 2\ln 2 - 5 + 2\right) = \frac{15}{4} - 4\ln 2$

31. $A = \int_0^1(y - 1)dx = \int_{\pi/2}^0(e^t - 1)(-\sin t)dt = \int_0^{\pi/2}(e^t\sin t - \sin t)dt = \left[\frac{1}{2}e^t(\sin t - \cos t) + \cos t\right]_0^{\pi/2}$

$= \frac{1}{2}\left(e^{\pi/2} - 1\right)$ (Formula 98)

32. By symmetry, $A = 4\int_0^a y\,dx = 4\int_{\pi/2}^0 a\sin^3\theta\,(-3a\cos^2\theta\sin\theta)d\theta = 12a^2\int_0^{\pi/2}\sin^4\theta\cos^2\theta\,d\theta.$

Now $\int\sin^4\theta\cos^2\theta\,d\theta = \int\sin^2\theta\left(\frac{1}{4}\sin^2 2\theta\right)d\theta = \frac{1}{8}\int(1 - \cos 2\theta)\sin^2 2\theta\,d\theta$

$= \frac{1}{8}\int\left(\frac{1}{2}(1 - \cos 4\theta) - \sin^2 2\theta\cos 2\theta\right)d\theta = \frac{1}{16}\theta - \frac{1}{64}\sin 4\theta - \frac{1}{48}\sin^3 2\theta + C$, so

$\int_0^{\pi/2}\sin^4\theta\cos^2\theta\,d\theta = \left[\frac{1}{16}\theta - \frac{1}{64}\sin 4\theta - \frac{1}{48}\sin^3 2\theta\right]_0^{\pi/2} = \frac{\pi}{32}$. Thus $A = 12a^2\left(\frac{\pi}{32}\right) = \frac{3}{8}\pi a^2$.

33. $A = \int_0^{2\pi r} y\,dx = \int_0^{2\pi}(r - d\cos\theta)(r - d\cos\theta)d\theta = \int_0^{2\pi}(r^2 - 2dr\cos\theta + d^2\cos^2\theta)d\theta$

$= \left[r^2\theta - 2dr\sin\theta + \frac{1}{2}d^2\left(\theta + \frac{1}{2}\sin 2\theta\right)\right]_0^{2\pi} = 2\pi r^2 + \pi d^2$

34. **(a)** By symmetry, the area of $\Re$ is twice the area inside $\Re$ above the x-axis. The top half of the loop is described by $x = t^2$, $y = t^3 - 3t$, $-\sqrt{3} \le t \le 0$, so, using the Substitution Rule with $y = t^3 - 3t$ and $dx = 2t\,dt$, we find that

area $= 2\int_0^3 y\,dx = 2\int_0^{-\sqrt{3}}(t^3 - 3t)2t\,dt = 2\int_0^{-\sqrt{3}}(2t^4 - 6t^2)dt = 2\left[\frac{2}{5}t^5 - 2t^3\right]_0^{-\sqrt{3}}$

$= 2\left[\frac{2}{5}\left(-3^{1/2}\right)^5 - 2\left(-3^{1/2}\right)^3\right] = 2\left[\frac{2}{5}\left(-9\sqrt{3}\right) - 2\left(-3\sqrt{3}\right)\right] = \frac{24}{5}\sqrt{3}.$

(b) Here we use the formula for disks and use the Substitution Rule as in part (a):

volume $= \pi\int_0^3 y^2\,dx = \pi\int_0^{-\sqrt{3}}(t^3 - 3t)^2 2t\,dt = 2\pi\int_0^{-\sqrt{3}}(t^6 - 6t^4 + 9t^2)t\,dt = 2\pi\left[\frac{1}{8}t^8 - t^6 + \frac{9}{4}t^4\right]_0^{-\sqrt{3}}$

$= 2\pi\left[\frac{1}{8}\left(-3^{1/2}\right)^8 - \left(-3^{1/2}\right)^6 + \frac{9}{4}\left(-3^{1/2}\right)^4\right] = 2\pi\left[\frac{81}{8} - 27 + \frac{81}{4}\right] = \frac{27}{4}\pi$

(c) By symmetry, the y-coordinate of the centroid is 0. To find the x-coordinate, we note that it is the same as the x-coordinate of the centroid of the top half of $\Re$, the area of which is $\frac{1}{2}\cdot\frac{24}{5}\sqrt{3} = \frac{12}{5}\sqrt{3}$.

So, using Formula 8.4.10 with $A = \frac{12}{5}\sqrt{3}$, we get

$$\overline{x} = \frac{5}{12\sqrt{3}}\int_0^3 xy\,dx = \frac{5}{12\sqrt{3}}\int_0^{-\sqrt{3}} t^2(t^3-3t)2t\,dt = \frac{5}{6\sqrt{3}}\left[\frac{1}{7}t^7 - \frac{3}{5}t^5\right]_0^{-\sqrt{3}} = \frac{5}{6\sqrt{3}}\left[\frac{1}{7}\left(-3^{1/2}\right)^7 - \frac{3}{5}\left(-3^{1/2}\right)^5\right]$$

$= \frac{5}{6\sqrt{3}}\left[-\frac{27}{7}\sqrt{3} + \frac{27}{5}\sqrt{3}\right] = \frac{9}{7}$. So the coordinates of the centroid of $\Re$ are $(x, y) = \left(\frac{9}{7}, 0\right)$.

35. We plot the curve $x = t^3 - 12t$, $y = 3t^2 + 2t + 5$ in the parameter interval $t \in [-4, 3.5]$. In order to find the area of the loop, we need to estimate the two t-values corresponding to the point at which the curve crosses itself. By zooming in, we estimate the y-coordinate of the point of intersection to be 39.633, and so the two t-values at the point of intersection are approximately the two solutions of the equation $y = 3t^2 + 2t + 5 = 39.633$, which are $t = -\frac{1}{3} \pm \frac{\sqrt{419.596}}{6} \approx -3.7473$ or 3.0807. Now as in Example 4, we can evaluate the area of the loop simply by integrating $y\,dx$ between these two t-values, since this integral represents the area under the upper part of the loop for t between the first t-value and t_A, minus the area under the bottom part between t_A and t_B, plus the area under the top part between t_B and the final t-value. So since $dx = (3t^2 - 12)dt$, the area of the loop is $A \approx \int_{-3.7473}^{3.0807} (3t^2 + 2t + 5)(3t^2 - 12)dt \approx 741$.

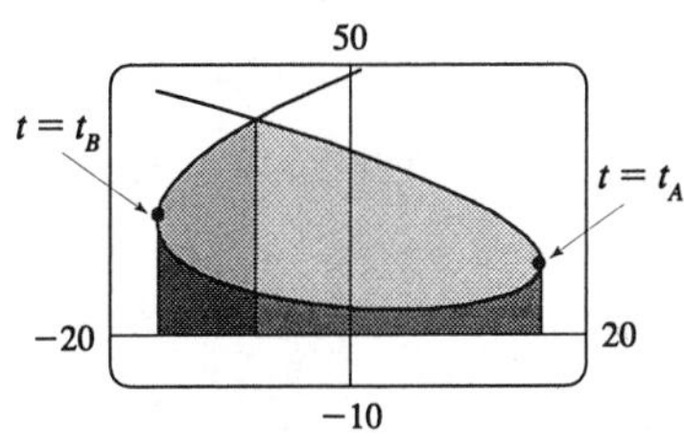

36. If f' is continuous and $f'(t) \neq 0$ for $a \le t \le b$, then either $f'(t) > 0$ for all t in $[a, b]$ or $f'(t) < 0$ for all t in $[a, b]$. Thus f is monotonic (in fact, strictly increasing or strictly decreasing) on $[a, b]$. It follows from Theorem 6.1.2 that f is invertible. Set $F = g \circ f^{-1}$, that is, define F by $F(x) = g(f^{-1}(x))$. Then $x = f(t) \Rightarrow f^{-1}(x) = t$, so $y = g(t) = g(f^{-1}(x)) = F(x)$.

37. **(a)** The equations for line segment P_0P_1 are $x = x_0 + (x_1 - x_0)t$, $y = y_0 + (y_1 - y_0)t$, and those of the other line segments are similar, as shown in Exercise 9.1.27.

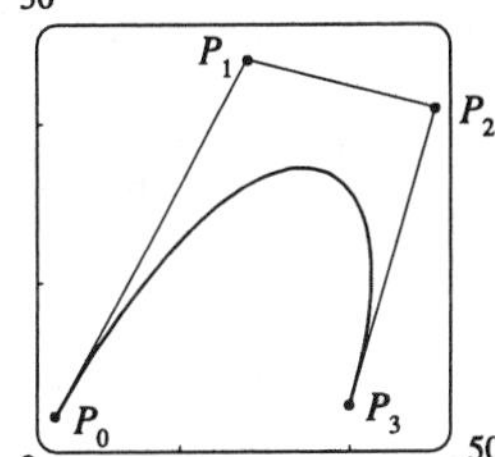

(b) It suffices to show that the slope of the tangent at P_0 is the same as that of line segment P_0P_1, namely $\dfrac{y_1 - y_0}{x_1 - x_0}$. We calculate the slope of the tangent to the Bézier curve:

$$\frac{dy/dt}{dx/dt} = \frac{-3y_0(1-t)^2 + 3y_1\left[-2t(1-t) + (1-t)^2\right] + 3y_2[-t^2 + (2t)(1-t)] + 3y_3t^2}{-3x_0(1-t)^2 + 3x_1\left[-2t(1-t) + (1-t)^2\right] + 3x_2[-t^2 + (2t)(1-t)] + 3x_3t^2}.$$

At point P_0, $t = 0$, so the slope of the tangent is $\dfrac{-3y_0 + 3y_1}{-3x_0 + 3x_1} = \dfrac{y_1 - y_0}{x_1 - x_0}$. So the tangent to the curve at P_0 passes through P_1. Similarly, the slope of the tangent at point P_3 (where $t = 1$) is $\dfrac{-3y_2 + 3y_3}{-3x_2 + 3x_3} = \dfrac{y_3 - y_2}{x_3 - x_2}$, which is also the slope of line P_2P_3.

(c) It appears that if P_1 were to the right of P_2, a loop would appear. We try setting $P_1 = (110, 30)$, and the resulting curve does indeed have a loop.

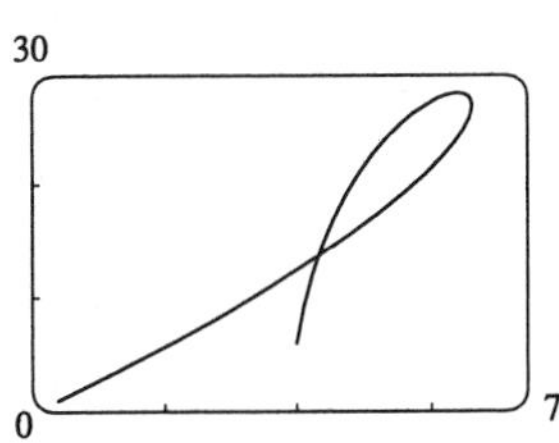

38. **(a)** Based on the behavior of the Bézier curve in Exercise 37, we suspect that the four control points should be in an exaggerated c shape.
We try $P_0(10,12)$, $P_1(4,15)$, $P_2(4,5)$, and $P_3(10,8)$, and these produce a decent c.
If you are using a CAS, it may be necessary to instruct it to make the x- and y-scales the same so as not to distort the figure (this is called a "constrained projection" in Maple.)

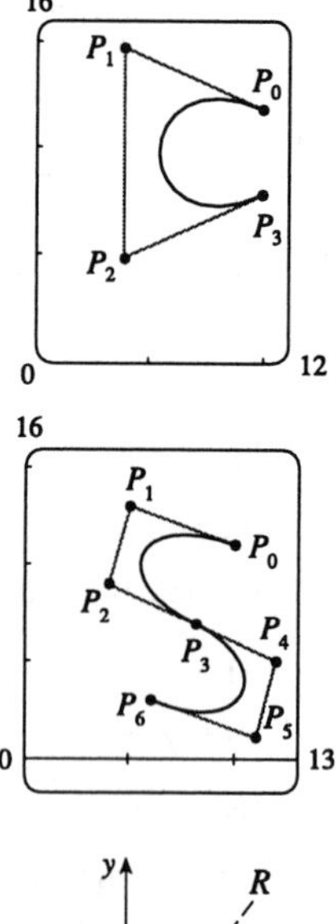

(b) We use the same P_0 and P_1 as in part (a), and use part of our c as the top of an S. To prevent the center line from slanting up too much, we move P_2 up to $(4,6)$ and P_3 down and to the left, to $(8,7)$. In order to have a smooth joint between the top and bottom halves of the S (and a symmetric S), we determine points P_4, P_5, and P_6 by rotating points P_2, P_1, and P_0 about the center of the letter (point P_3). The points are therefore $P_4(12,8)$, $P_5(12,-1)$, and $P_6(6,2)$.

39. The coordinates of T are $(r\cos\theta, r\sin\theta)$. Since TP was unwound from arc TA, TP has length $r\theta$. Also $\angle PTQ = \angle PTR - \angle QTR = \frac{1}{2}\pi - \theta$, so P has coordinates $x = r\cos\theta + r\theta\cos\left(\frac{1}{2}\pi - \theta\right) = r(\cos\theta + \theta\sin\theta)$, $y = r\sin\theta - r\theta\sin\left(\frac{1}{2}\pi - \theta\right) = r(\sin\theta - \theta\cos\theta)$.

40. If the cow walks with the rope taut, it traces out the portion of the involute in Exercise 39 corresponding to the range $0 \le \theta \le \pi$, arriving at the point $(-r, \pi r)$ when $\theta = \pi$. With the rope now fully extended, the cow walks in a semicircle of radius πr, arriving at $(-r, -\pi r)$. Finally, the cow traces out another portion of the involute, namely the reflection about the x-axis of the initial involute path. (This corresponds to the range $-\pi \le \theta \le 0$.)

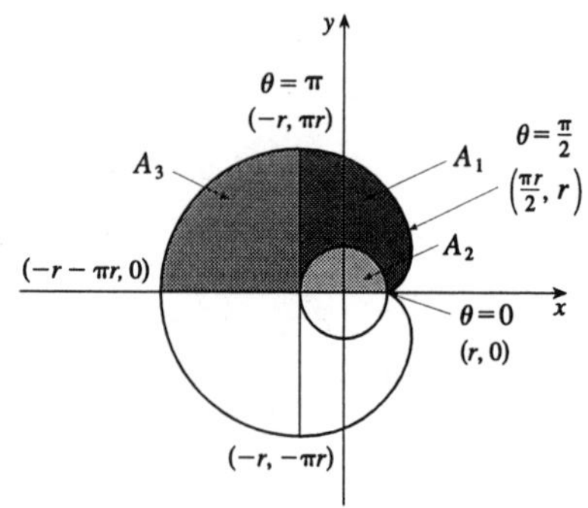

Referring to the figure, we see that the total grazing area is $2(A_1 + A_3)$. A_3 is $\frac{1}{4}$ of the area of a circle of radius πr, so $A_3 = \frac{1}{4}\pi(\pi r)^2 = \frac{1}{4}\pi^3 r^2$. A_1 can be computed directly by using the method of Section 9.5. Instead we will compute $A_1 + A_2$ and then subtract A_2 $\left(= \frac{1}{2}\pi r^2\right)$ to obtain A_1. To find $A_1 + A_2$, first note that the rightmost point of the involute is $\left(\frac{\pi}{2}r, r\right)$. $\left[\text{To see this, note that}\right.$ $dx/d\theta = r\theta\cos\theta$, which vanishes when $\theta = 0$ or $\frac{\pi}{2}$. $\theta = 0$ corresponds to the cusp at $(r, 0)$ and $\theta = \frac{\pi}{2}$ corresponds to $\left(\frac{\pi}{2}r, r\right)$.$\left.\right]$ Thus $A_1 + A_2 = \int_{\theta=\pi}^{\pi/2} y\,dx - \int_{\theta=0}^{\pi/2} y\,dx = \int_{\theta=\pi}^{0} y\,dx$. Now $y\,dx = r(\sin\theta - \theta\cos\theta)r\theta\cos\theta\,d\theta = r^2(\theta\sin\theta\cos\theta - \theta^2\cos^2\theta)d\theta$. Integrate: $(1/r^2)\int y\,dx = -\theta\cos^2\theta - \frac{1}{2}(\theta^2 - 1)\sin\theta\cos\theta - \frac{1}{6}\theta^3 + \frac{1}{2}\theta + C$. This enables us to compute $A_1 + A_2 = r^2\left[-\theta\cos^2\theta - \frac{1}{2}(\theta^2 - 1)\sin\theta\cos\theta - \frac{1}{6}\theta^3 + \frac{1}{2}\theta\right]_{\pi}^{0} = r^2\left[0 - \left(-\pi - \frac{\pi^3}{6} + \frac{\pi}{2}\right)\right] = r^2\left(\frac{\pi}{2} + \frac{\pi^3}{6}\right)$.
Therefore $A_1 = (A_1 + A_2) - A_2 = \frac{1}{6}\pi^3 r^2$, so the grazing area is $2(A_1 + A_3) = 2\left(\frac{1}{6}\pi^3 r^2 + \frac{1}{4}\pi^3 r^2\right) = \frac{5}{6}\pi^3 r^2$.

EXERCISES 9.3

1. $L = \int_0^1 \sqrt{(dx/dt)^2 + (dy/dt)^2}\,dt$ and $dx/dt = 3t^2,\ dy/dt = 4t^3 \quad \Rightarrow$
$L = \int_0^1 \sqrt{9t^4 + 16t^6}\,dt = \int_0^1 t^2\sqrt{9 + 16t^2}\,dt$

2. $L = \int_0^2 \sqrt{(dx/dt)^2 + (dy/dt)^2}\,dt$ and $dx/dt = 2t,\ dy/dt = 4$, so $L = \int_0^2 \sqrt{4t^2 + 16}\,dt = 2\int_0^2 \sqrt{t^2 + 4}\,dt$.

3. $\dfrac{dx}{dt} = \sin t + t\cos t$ and $\dfrac{dy}{dt} = \cos t - t\sin t \quad \Rightarrow$
$L = \int_0^{\pi/2} \sqrt{(\sin t + t\cos t)^2 + (\cos t - t\sin t)^2}\,dt = \int_0^{\pi/2} \sqrt{1 + t^2}\,dt$

4. $dx/dt = -e^{-t}$ and $dy/dt = e^{2t} + 2te^{2t} = e^{2t}(1 + 2t)$ so $L = \int_{-1}^{1} \sqrt{e^{-2t} + e^{4t}(1 + 2t)^2}\,dt$

5. $x = t^3,\ y = t^2,\ 0 \le t \le 4.\ (dx/dt)^2 + (dy/dt)^2 = (3t^2)^2 + (2t)^2 = 9t^4 + 4t^2$
$L = \int_0^4 \sqrt{(dx/dt)^2 + (dy/dt)^2}\,dt = \int_0^4 \sqrt{9t^4 + 4t^2}\,dt = \int_0^4 t\sqrt{9t^2 + 4}\,dt = \frac{1}{18}\int_4^{148} \sqrt{u}\,du$
(where $u = 9t^2 + 4$). So $L = \frac{1}{18}\left(\frac{2}{3}\right)\left[u^{3/2}\right]_4^{148} = \frac{1}{27}\left(148^{3/2} - 4^{3/2}\right) = \frac{8}{27}\left(37^{3/2} - 1\right)$.

6. $x = a(\cos\theta + \theta\sin\theta),\ y = a(\sin\theta - \theta\cos\theta),\ 0 \le \theta \le \pi.$
$(dx/d\theta)^2 + (dy/d\theta)^2 = a^2\left[(-\sin\theta + \sin\theta + \theta\cos\theta)^2 + (\cos\theta - \cos\theta + \theta\sin\theta)^2\right]$
$= a^2\theta^2\left(\cos^2\theta + \sin^2\theta\right) = (a\theta)^2.\ L = \int_0^{\pi} a\theta\,d\theta = \frac{1}{2}\pi^2 a.$

7. $x = 2 - 3\sin^2\theta,\ y = \cos 2\theta,\ 0 \le \theta \le \frac{\pi}{2}.$
$(dx/d\theta)^2 + (dy/d\theta)^2 = (-6\sin\theta\cos\theta)^2 + (-2\sin 2\theta)^2 = (-3\sin 2\theta)^2 + (-2\sin 2\theta)^2 = 13\sin^2 2\theta \quad \Rightarrow$
$L = \int_0^{\pi/2} \sqrt{13}\sin 2\theta\,d\theta = \left[-\frac{\sqrt{13}}{2}\cos 2\theta\right]_0^{\pi/2} = -\frac{\sqrt{13}}{2}(-1 - 1) = \sqrt{13}$

8. $x = e^t - t,\ y = 4e^{t/2},\ 0 \le t \le 1.\ (dx/dt)^2 + (dy/dt)^2 = \left(e^t - 1\right)^2 + \left(2e^{t/2}\right)^2 = e^{2t} + 2e^t + 1 = \left(e^t + 1\right)^2.$
$L = \int_0^1 (e^t + 1)dt = \left[e^t + t\right]_0^1 = (e + 1) - 1 = e$

9. $x = e^t\cos t,\ y = e^t\sin t,\ 0 \le t \le \pi.$

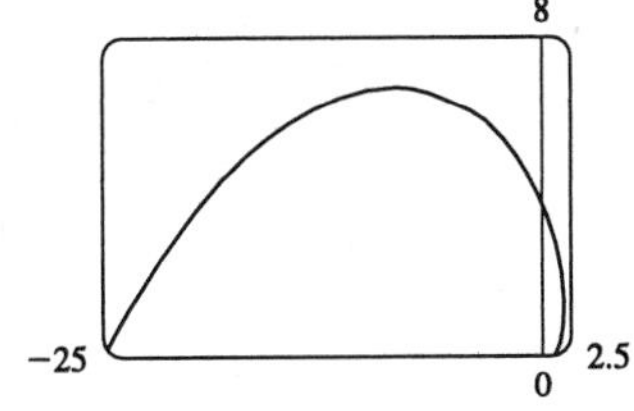

$$\left(\frac{dx}{dt}\right)^2 + \left(\frac{dy}{dt}\right)^2 = \left[e^t(\cos t - \sin t)\right]^2 + \left[e^t(\sin t + \cos t)\right]^2$$
$$= e^{2t}\left(2\cos^2 t + 2\sin^2 t\right) = 2e^{2t} \quad \Rightarrow$$
$L = \int_0^{\pi} \sqrt{2}\,e^t\,dt = \sqrt{2}(e^{\pi} - 1)$

10. $x = 3t - t^3,\ y = 3t^2,\ 0 \le t \le 2.$

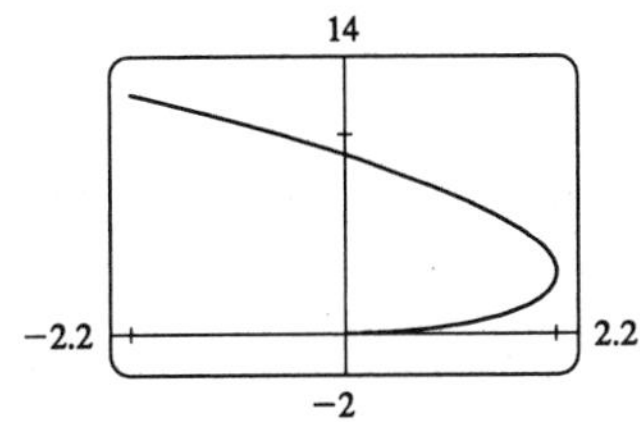

$$\left(\frac{dx}{dt}\right)^2 + \left(\frac{dy}{dt}\right)^2 = \left(3 - 3t^2\right)^2 + (6t)^2$$
$$= 9\left(1 + 2t^2 + t^4\right) = \left[3\left(1 + t^2\right)\right]^2.$$
$L = \int_0^2 3(1 + t^2)dt = \left[3t + t^3\right]_0^2 = 14$

11. $x = \ln t$ and $y = e^{-t} \Rightarrow \dfrac{dx}{dt} = \dfrac{1}{t}$ and $\dfrac{dy}{dt} = -e^{-t} \Rightarrow L = \displaystyle\int_1^2 \sqrt{t^{-2} + e^{-2t}}\,dt$. Using Simpson's Rule with $n = 10$, $\Delta x = (2-1)/10 = 0.1$ and $f(t) = \sqrt{t^{-2} + e^{-2t}}$ we get
$L \approx \frac{0.1}{3}[f(1.0) + 4f(1.1) + 2f(1.2) + \cdots + 2f(1.8) + 4f(1.9) + f(2.0)] \approx 0.7314$.

12. $x = 2a\cot\theta \Rightarrow dx/dt = -2a\csc^2\theta$ and $y = 2a\sin^2\theta \Rightarrow dy/dt = 4a\sin\theta\cos\theta = 2a\sin 2\theta$. So
$L = \int_{\pi/4}^{\pi/2} \sqrt{4a^2\csc^4\theta + 4a^2\sin^2 2\theta}\,d\theta = 2a\int_{\pi/4}^{\pi/2} \sqrt{\csc^4\theta + \sin^2 2\theta}\,d\theta$.

Using Simpson's Rule with $n = 4$, $\Delta x = \frac{\pi/4}{4} = \frac{\pi}{16}$ and $f(\theta) = \sqrt{\csc^4\theta + \sin^2 2\theta}$, we get
$L \approx 2a \cdot S_4 = (2a)\frac{\pi}{16\cdot 3}\left[f\left(\frac{\pi}{4}\right) + 4f\left(\frac{5\pi}{16}\right) + 2f\left(\frac{3\pi}{8}\right) + 4f\left(\frac{7\pi}{16}\right) + f\left(\frac{\pi}{2}\right)\right] \approx 2.2605a$.

13. $x = \sin^2\theta$, $y = \cos^2\theta$, $0 \le \theta \le 3\pi$.

$$\left(\frac{dx}{d\theta}\right)^2 + \left(\frac{dy}{d\theta}\right)^2 = (2\sin\theta\cos\theta)^2 + (-2\cos\theta\sin\theta)^2 = 8\sin^2\theta\cos^2\theta = 2\sin^2 2\theta \quad\Rightarrow$$

$$\text{Distance} = \int_0^{3\pi} \sqrt{2}\,|\sin 2\theta|\,d\theta = 6\sqrt{2}\int_0^{\pi/2} \sin 2\theta\,d\theta \quad \text{(by symmetry)}$$
$$= \left[-3\sqrt{2}\cos 2\theta\right]_0^{\pi/2} = -3\sqrt{2}(-1-1) = 6\sqrt{2}$$

The full curve is traversed as θ goes from 0 to $\frac{\pi}{2}$, because the curve is the segment of $x + y = 1$ that lies in the first quadrant (since $x, y \ge 0$), and this segment is completely traversed as θ goes from 0 to $\frac{\pi}{2}$.
Thus $L = \int_0^{\pi/2} \sin 2\theta\,d\theta = \sqrt{2}$, as above.

14. $x = \cos^2 t$, $y = \cos t$, $0 \le t \le 4\pi$. $\left(\dfrac{dx}{dt}\right)^2 + \left(\dfrac{dy}{dt}\right)^2 = (-2\cos t\sin t)^2 + (-\sin t)^2 = \sin^2 t\left(4\cos^2 t + 1\right)$

$$\text{Distance} = \int_0^{4\pi} |\sin t|\,\sqrt{4\cos^2 t + 1}\,dt = 4\int_0^{\pi} \sin t\,\sqrt{4\cos^2 t + 1}\,dt$$
$$= -4\int_1^{-1} \sqrt{4u^2 + 1}\,du \quad [u = \cos t,\ du = -\sin t\,dt]$$
$$= 4\int_{-1}^{1} \sqrt{4u^2 + 1}du = 8\int_0^1 \sqrt{4u^2 + 1}du = 8\int_0^{\tan^{-1}2} \sec\theta\,\tfrac{1}{2}\sec^2\theta\,d\theta$$
$$= 4\int_0^{\tan^{-1}2} \sec^3\theta\,d\theta = [2\sec\theta\tan\theta + 2\ln|\sec\theta + \tan\theta|]_0^{\tan^{-1}2} \quad \text{(Formula 71)}$$
$$= 4\sqrt{5} + 2\ln\left(\sqrt{5} + 2\right).$$

$$L = \int_0^{\pi} |\sin t|\,\sqrt{4\cos^2 t + 1}\,dt = \sqrt{5} + \tfrac{1}{2}\ln\left(\sqrt{5} + 2\right)$$

15. $x = a\sin\theta$, $y = b\cos\theta$, $0 \le \theta \le 2\pi$.

$$\left(\frac{dx}{d\theta}\right)^2 + \left(\frac{dy}{d\theta}\right)^2 = (a\cos\theta)^2 + (-b\sin\theta)^2 = a^2\cos^2\theta + b^2\sin^2\theta$$
$$= a^2\left(1 - \sin^2\theta\right) + b^2\sin^2\theta = a^2 - \left(a^2 - b^2\right)\sin^2\theta$$
$$= a^2 - c^2\sin^2\theta = a^2\left(1 - \frac{c^2}{a^2}\sin^2\theta\right) = a^2\left(1 - e^2\sin^2\theta\right)$$

So $L = 4\int_0^{\pi/2} \sqrt{a^2(1 - e^2\sin^2\theta)}\,d\theta$ (by symmetry) $= 4a\int_0^{\pi/2} \sqrt{1 - e^2\sin^2\theta}\,d\theta$

16. $x = a\cos^3\theta$, $y = a\sin^3\theta$. $\left(\dfrac{dx}{d\theta}\right)^2 + \left(\dfrac{dy}{d\theta}\right)^2 = \left(-3a\cos^2\theta\sin\theta\right)^2 + \left(3a\sin^2\theta\cos\theta\right)^2 = 9a^2\sin^2\theta\cos^2\theta$.
$L = 4\int_0^{\pi/2} 3a\sin\theta\cos\theta\,d\theta = \left[12a\tfrac{1}{2}\sin^2\theta\right]_0^{\pi/2} = 6a$

17. **(a)** Notice that $0 \le t \le 2\pi$ does not give the complete curve because $x(0) \neq x(2\pi)$. In fact, we must take $t \in [0, 4\pi]$ in order to obtain the complete curve, since the first term in each of the parametric equations has period 2π and the second has period $\frac{4\pi}{11}$, and the least common integer multiple of these two numbers is 4π.

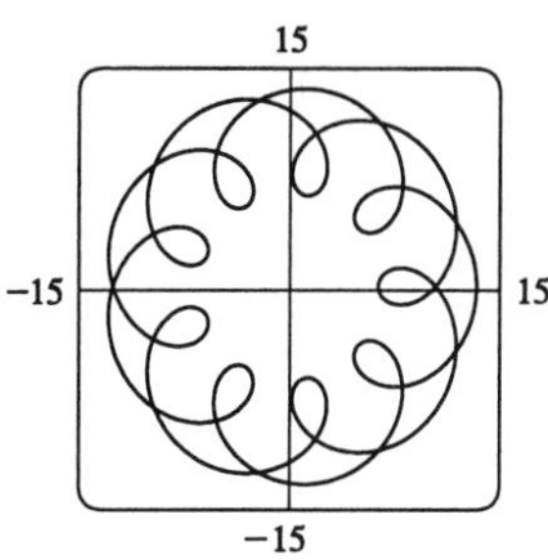

(b) We use the CAS to find the derivatives dx/dt and dy/dt, and then use Theorem 4 to find the arc length. Maple cannot do the integral exactly, so we use the command `evalf(Int(sqrt(diff(x,t)^2+diff(y,t)^2),t=0..4*Pi),4);` to estimate the length to be about 294. The 4 in the Maple command indicates that we want only four digits of accuracy; this speeds up the otherwise glacial calculation.

18. **(a)** It appears that as $t \to \infty$, $(x, y) \to \left(\frac{1}{2}, \frac{1}{2}\right)$, and as $t \to -\infty$, $(x, y) \to \left(-\frac{1}{2}, -\frac{1}{2}\right)$.

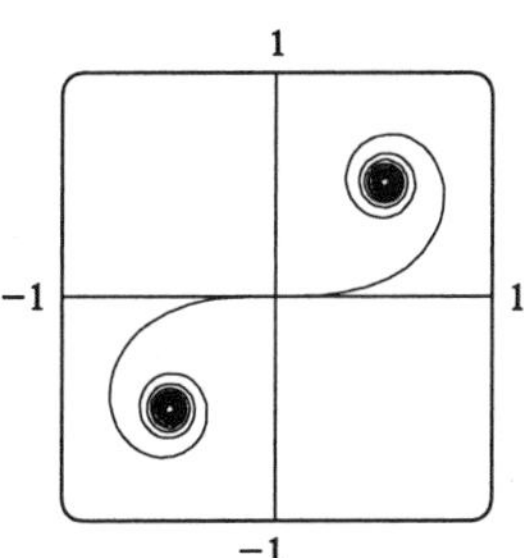

(b) By the Fundamental Theorem of Calculus, $dx/dt = \cos\left(\frac{\pi}{2}t^2\right)$ and $dy/dt = \sin\left(\frac{\pi}{2}t^2\right)$, so by Theorem 4, the length of the curve from the origin to the point with parameter value t is

$$L = \int_0^t \sqrt{(dx/du)^2 + (dy/du)^2}\, du = \int_0^t \sqrt{\cos^2\left(\tfrac{\pi}{2}u^2\right) + \sin^2\left(\tfrac{\pi}{2}u^2\right)}\, du = \textstyle\int_0^t 1\, du = t \quad (\text{or } -t \text{ if } t < 0).$$

We have used u as the dummy variable so as not to confuse it with the upper limit of integration.

19. $x = t^3$ and $y = t^4 \quad \Rightarrow \quad dx/dt = 3t^2$ and $dy/dt = 4t^3$. So $S = \int_0^1 2\pi t^4 \sqrt{9t^4 + 16t^6}\, dt = \int_0^1 2\pi t^6 \sqrt{9 + 16t^2}\, dt$.

20. $S = \int_{\pi/4}^{\pi/2} 2\pi \cdot 2a \sin^2\theta \sqrt{\csc^4\theta + \sin^2 2\theta}\, dt = 4\pi a \int_{\pi/4}^{\pi/2} \sin^2\theta \sqrt{\csc^4\theta + \sin^2 2\theta}\, d\theta$. Using Simpson's Rule with $n = 4$, $\Delta x = \frac{\pi}{16}$ and $f(\theta) = \sin^2\theta \sqrt{\csc^4\theta + \sin^2 2\theta}$, we get $S \approx (4\pi a)\frac{\pi}{16 \cdot 3}\left[f\left(\frac{\pi}{4}\right) + 4f\left(\frac{5\pi}{16}\right) + 2f\left(\frac{3\pi}{8}\right) + 4f\left(\frac{7\pi}{16}\right) + f\left(\frac{\pi}{2}\right)\right] \approx 11.0893a$.

21. $x = t^3, y = t^2, 0 \le t \le 1$. $\left(\dfrac{dx}{dt}\right)^2 + \left(\dfrac{dy}{dt}\right)^2 = (3t^2)^2 + (2t)^2 = 9t^4 + 4t^2$.

$$S = \int_0^1 2\pi y \sqrt{\left(\frac{dx}{dt}\right)^2 + \left(\frac{dy}{dt}\right)^2}\, dt = \int_0^1 2\pi t^2 \sqrt{9t^4 + 4t^2}\, dt$$

$$= 2\pi \int_4^{13} \frac{u - 4}{9} \sqrt{u}\left(\tfrac{1}{18} du\right) \ (\text{where } u = 9t^2 + 4) = \tfrac{\pi}{81}\left[\tfrac{2}{5}u^{5/2} - \tfrac{8}{3}u^{3/2}\right]_4^{13} = \tfrac{2\pi}{1215}\left(247\sqrt{13} + 64\right)$$

22. $x = 3t - t^3, y = 3t^2, 0 \le t \le 1$. $\left(\dfrac{dx}{dt}\right)^2 + \left(\dfrac{dy}{dt}\right)^2 = (3 - 3t^2)^2 + (6t)^2 = 9(1 + 2t^2 + t^4) = \left[3(1 + t^2)\right]^2$.

$S = \int_0^1 2\pi 3t^2 3(1 + t^2) dt = 18\pi \int_0^1 (t^2 + t^4)\, dt = 18\pi\left[\frac{1}{3}t^3 + \frac{1}{5}t^5\right]_0^1 = \frac{48}{5}\pi$

23. $x = a\cos^3\theta,\ y = a\sin^3\theta,\ 0 \le \theta \le \frac{\pi}{2}$

$$\left(\frac{dx}{d\theta}\right)^2 + \left(\frac{dy}{d\theta}\right)^2 = \left(-3a\cos^2\theta\sin\theta\right)^2 + \left(3a\sin^2\theta\cos\theta\right)^2 = 9a^2\sin^2\theta\cos^2\theta$$

$$S = \int_0^{\pi/2} 2\pi a\sin^3\theta\, 3a\sin\theta\cos\theta\, d\theta = 6\pi a^2\int_0^{\pi/2}\sin^4\theta\cos\theta\, d\theta = \tfrac{6}{5}\pi a^2[\sin^5\theta]_0^{\pi/2} = \tfrac{6}{5}\pi a^2$$

24. $$\left(\frac{dx}{d\theta}\right)^2 + \left(\frac{dy}{d\theta}\right)^2 = (-2\sin\theta + 2\sin 2\theta)^2 + (2\cos\theta - 2\cos 2\theta)^2$$

$$= 4\left[\left(\sin^2\theta - 2\sin\theta\sin 2\theta + \sin^2 2\theta\right) + \left(\cos^2\theta - 2\cos\theta\cos 2\theta + \cos^2 2\theta\right)\right]$$

$$= 4[1 + 1 - 2(\cos 2\theta\cos\theta + \sin 2\theta\sin\theta)] = 8[1 - \cos(2\theta - \theta)] = 8(1 - \cos\theta)$$

We plot the graph with parameter interval $[0, 2\pi]$, and see that we should only integrate between 0 and π. (If the interval $[0, 2\pi]$ were taken, the surface of revolution would be generated twice.) Also note that $y = 2\sin\theta - \sin 2\theta = 2\sin\theta(1 - \cos\theta)$. So

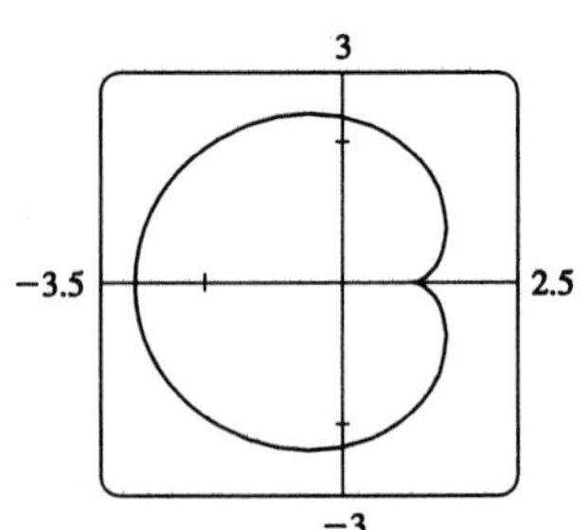

$$S = \int_0^{\pi} 2\pi 2\sin\theta(1 - \cos\theta)2\sqrt{2}\sqrt{1 - \cos\theta}\, d\theta$$

$$= 8\sqrt{2}\pi\int_0^{\pi}(1 - \cos\theta)^{3/2}\sin\theta\, d\theta = 8\sqrt{2}\pi\int_0^2\sqrt{u^3}\, du$$

$$[\text{where } u = 1 - \cos\theta,\ du = \sin\theta\, d\theta] = \left[8\sqrt{2}\pi\left(\tfrac{2}{5}\right)u^{5/2}\right]_0^2 = \tfrac{128}{5}\pi.$$

25. $$\left(\frac{dx}{dt}\right)^2 + \left(\frac{dy}{dt}\right)^2 = (6t)^2 + \left(6t^2\right)^2 = 36t^2\left(1 + t^2\right) \quad\Rightarrow$$

$$S = \int_0^5 2\pi x\sqrt{(dx/dt)^2 + (dy/dt)^2}\, dt = \int_0^5 2\pi\left(3t^2\right)6t\sqrt{1 + t^2}\, dt = 18\pi\int_0^5 t^2\sqrt{1 + t^2}\, 2t\, dt$$

$$= 18\pi\int_1^{26}(u - 1)\sqrt{u}\, du \quad (\text{where } u = 1 + t^2) \quad = 18\pi\int_1^{26}\left(u^{3/2} - u^{1/2}\right)du = 18\pi\left[\tfrac{2}{5}u^{5/2} - \tfrac{2}{3}u^{3/2}\right]_1^{26}$$

$$= 18\pi\left[\left(\tfrac{2}{5}\cdot 676\sqrt{26} - \tfrac{2}{3}\cdot 26\sqrt{26}\right) - \left(\tfrac{2}{5} - \tfrac{2}{3}\right)\right] = \tfrac{24}{5}\pi\left(949\sqrt{26} + 1\right)$$

26. $x = e^t - t,\ y = 4e^{t/2},\ 0 \le t \le 1.$ $\left(\frac{dx}{dt}\right)^2 + \left(\frac{dy}{dt}\right)^2 = \left(e^t - 1\right)^2 + \left(2e^{t/2}\right)^2 = e^{2t} + 2e^t + 1 = \left(e^t + 1\right)^2.$

$$S = \int_0^1 2\pi\left(e^t - t\right)\sqrt{\left(e^t - 1\right)^2 + \left(2e^{t/2}\right)^2}\, dt = \int_0^1 2\pi\left(e^t - t\right)\left(e^t + 1\right)dt$$

$$= 2\pi\left[\tfrac{1}{2}e^{2t} + e^t - (t - 1)e^t - \tfrac{1}{2}t^2\right]_0^1 = \pi\left(e^2 + 2e - 6\right)$$

27. $x = a\cos\theta,\ y = b\sin\theta,\ 0 \le \theta \le 2\pi.$

$$(dx/d\theta)^2 + (dy/d\theta)^2 = (-a\sin\theta)^2 + (b\cos\theta)^2 = a^2\sin^2\theta + b^2\cos^2\theta = a^2\left(1 - \cos^2\theta\right) + b^2\cos^2\theta$$

$$= a^2 - \left(a^2 - b^2\right)\cos^2\theta = a^2 - c^2\cos^2\theta = a^2\left(1 - \frac{c^2}{a^2}\cos^2\theta\right) = a^2\left(1 - e^2\cos^2\theta\right)$$

(a) $$S = \int_0^{\pi} 2\pi b\sin\theta\, a\sqrt{1 - e^2\cos^2\theta}\, d\theta = 2\pi ab\int_{-e}^{e}\sqrt{1 - u^2}\left(\frac{1}{e}\right)du \ [\text{where } u = -e\cos\theta,\ du = e\sin\theta\, d\theta]$$

$$= \frac{4\pi ab}{e}\int_0^e\left(1 - u^2\right)^{1/2}du = \frac{4\pi ab}{e}\int_0^{\sin^{-1}e}\cos^2 v\, dv \ (\text{where } u = \sin v) \ = \frac{2\pi ab}{e}\int_0^{\sin^{-1}e}(1 + \cos 2v)dv$$

$$= \frac{2\pi ab}{e}\left[v + \tfrac{1}{2}\sin 2v\right]_0^{\sin^{-1}e} = \frac{2\pi ab}{e}[v + \sin v\cos v]_0^{\sin^{-1}e} = \frac{2\pi ab}{e}\left(\sin^{-1}e + e\sqrt{1 - e^2}\right).$$

But $\sqrt{1 - e^2} = \sqrt{1 - \dfrac{c^2}{a^2}} = \sqrt{\dfrac{a^2 - c^2}{a^2}} = \sqrt{\dfrac{b^2}{a^2}} = \dfrac{b}{a}$, so $S = \dfrac{2\pi ab}{e}\sin^{-1}e + 2\pi b^2.$

(b) $S = \int_{-\pi/2}^{\pi/2} 2\pi a\cos\theta\, a\sqrt{1-e^2\cos^2\theta}\,d\theta$

$$= 4\pi a^2 \int_0^{\pi/2} \cos\theta\,\sqrt{(1-e^2)+e^2\sin^2\theta}\,d\theta$$

$$= \frac{4\pi a^2(1-e^2)}{e}\int_0^{\pi/2} \frac{e}{\sqrt{1-e^2}}\cos\theta\sqrt{1+\left(\frac{e\sin\theta}{\sqrt{1-e^2}}\right)^2}\,d\theta$$

$$= \frac{4\pi a^2(1-e^2)}{e}\int_0^{e/\sqrt{1-e^2}} \sqrt{1+u^2}\,du \quad \left(\text{where } u = \frac{e\sin\theta}{\sqrt{1-e^2}}\right)$$

$$= \frac{4\pi a^2(1-e^2)}{e}\int_0^{\sin^{-1}e} \sec^3 v\,dv \quad (\text{where } u = \tan v,\ du = \sec^2 v\,dv)$$

$$= \frac{2\pi a^2(1-e^2)}{e}\left[\sec v\tan v + \ln|\sec v+\tan v|\right]_0^{\sin^{-1}e}$$

$$= \frac{2\pi a^2(1-e^2)}{e}\left[\frac{1}{\sqrt{1-e^2}}\frac{e}{\sqrt{1-e^2}} + \ln\left|\frac{1}{\sqrt{1-e^2}}+\frac{e}{\sqrt{1-e^2}}\right|\right]$$

$$= 2\pi a^2 + \frac{2\pi a^2(1-e^2)}{e}\ln\sqrt{\frac{1+e}{1-e}} = 2\pi a^2 + \frac{2\pi b^2}{e}\frac{1}{2}\ln\left(\frac{1+e}{1-e}\right) \quad \left(\text{since } 1-e^2 = \frac{b^2}{a^2}\right)$$

$$= 2\pi\left[a^2 + \frac{b^2}{2e}\ln\frac{1+e}{1-e}\right]$$

28. By Formula 8.3.4, $S = \displaystyle\int_a^b 2\pi F(x)\sqrt{1+F'(x)^2}\,dx$. Now

$1+F'(x)^2 = 1+\left(\dfrac{dy/dt}{dx/dt}\right)^2 = \dfrac{(dx/dt)^2+(dy/dt)^2}{(dx/dt)^2}$. Using the Substitution Rule with $x = x(t)$ $\Rightarrow$

$dx = \dfrac{dx}{dt}\,dt$, we have $S = \displaystyle\int_\alpha^\beta 2\pi y\sqrt{\frac{(dx/dt)^2+(dy/dt)^2}{(dx/dt)^2}}\,\frac{dx}{dt}\,dt = \int_\alpha^\beta 2\pi y\sqrt{(dx/dt)^2+(dy/dt)^2}\,dt.$

29. **(a)** $\phi = \tan^{-1}\left(\dfrac{dy}{dx}\right) \Rightarrow \dfrac{d\phi}{dt} = \dfrac{d}{dt}\tan^{-1}\left(\dfrac{dy}{dx}\right) = \dfrac{1}{1+(dy/dx)^2}\left[\dfrac{d}{dt}\left(\dfrac{dy}{dx}\right)\right]$. But $\dfrac{dy}{dx} = \dfrac{dy/dt}{dx/dt} = \dfrac{\dot y}{\dot x}$

$\Rightarrow \dfrac{d}{dt}\left(\dfrac{dy}{dx}\right) = \dfrac{d}{dt}\left(\dfrac{\dot y}{\dot x}\right) = \dfrac{\ddot y\dot x - \ddot x\dot y}{\dot x^2} \Rightarrow \dfrac{d\phi}{dt} = \dfrac{1}{1+(\dot y/\dot x)^2}\left[\dfrac{\ddot y\dot x - \ddot x\dot y}{\dot x^2}\right] = \dfrac{\dot x\ddot y - \ddot x\dot y}{\dot x^2+\dot y^2}.$

Using the Chain Rule, and the fact that $s = \displaystyle\int_0^t \sqrt{\left(\frac{dx}{dt}\right)^2+\left(\frac{dy}{dt}\right)^2}\,dt \Rightarrow$

$\dfrac{ds}{dt} = \sqrt{\left(\dfrac{dx}{dt}\right)^2+\left(\dfrac{dy}{dt}\right)^2} = \left(\dot x^2+\dot y^2\right)^{1/2}$, we have that

$\dfrac{d\phi}{ds} = \dfrac{d\phi/dt}{ds/dt} = \left(\dfrac{\dot x\ddot y - \ddot x\dot y}{\dot x^2+\dot y^2}\right)\dfrac{1}{\left(\dot x^2+\dot y^2\right)^{1/2}} = \dfrac{\dot x\ddot y - \ddot x\dot y}{\left(\dot x^2+\dot y^2\right)^{3/2}}$. So

$\kappa = \left|\dfrac{d\phi}{ds}\right| = \left|\dfrac{\dot x\ddot y - \ddot x\dot y}{\left(\dot x^2+\dot y^2\right)^{3/2}}\right| = \dfrac{|\dot x\ddot y - \ddot x\dot y|}{\left(\dot x^2+\dot y^2\right)^{3/2}}.$

(b) $x = x$ and $y = f(x) \Rightarrow \dot x = 1$, $\ddot x = 0$ and $\dot y = \dfrac{dy}{dx}$, $\ddot y = \dfrac{d^2y}{dx^2}$.

So $\kappa = \dfrac{|1\cdot(d^2y/dx^2) - 0\cdot(dy/dx)|}{\left[1+(dy/dx)^2\right]^{3/2}} = \dfrac{|d^2y/dx^2|}{\left[1+(dy/dx)^2\right]^{3/2}}.$

30. **(a)** $y = x^2 \Rightarrow \dfrac{dy}{dx} = 2x \Rightarrow \dfrac{d^2y}{dx^2} = 2$. So $\kappa = \dfrac{|d^2y/dx^2|}{\left[1 + (dy/dx)^2\right]^{3/2}} = \dfrac{2}{(1+4x^2)^{3/2}}$, and at $(1,1)$,

$$\kappa = \frac{2}{5^{3/2}} = \frac{2}{5\sqrt{5}}.$$

(b) $\kappa' = \dfrac{d\kappa}{dx} = -3\left(1+4x^2\right)^{-5/2}(8x) = 0 \Leftrightarrow x = 0 \Rightarrow y = 0$. This is a maximum since $\kappa' > 0$ for $x < 0$ and $\kappa' < 0$ for $x > 0$. So the parabola $y = x^2$ has maximum curvature at the origin.

31. $x = \theta - \sin\theta \Rightarrow \dot{x} = 1 - \cos\theta \Rightarrow \ddot{x} = \sin\theta$, and $y = 1 - \cos\theta \Rightarrow \dot{y} = \sin\theta \Rightarrow \ddot{y} = \cos\theta$. Therefore $\kappa = \dfrac{|\cos\theta - \cos^2\theta - \sin^2\theta|}{\left[(1-\cos\theta)^2 + \sin^2\theta\right]^{3/2}} = \dfrac{|\cos\theta - (\cos^2\theta + \sin^2\theta)|}{(1 - 2\cos\theta + \cos^2\theta + \sin^2\theta)^{3/2}} = \dfrac{|\cos\theta - 1|}{(2 - 2\cos\theta)^{3/2}}$. The top of the arch is characterized by a horizontal tangent, and from Example 1 of Section 9.2 the tangent is horizontal when $\theta = (2n-1)\pi$, so take $n = 1$ and substitute $\theta = \pi$ into the expression for κ:

$$\kappa = \frac{|\cos\pi - 1|}{(2 - 2\cos\pi)^{3/2}} = \frac{|-1-1|}{[2 - 2(-1)]^{3/2}} = \frac{1}{4}.$$

32. **(a)** Every straight line has parametrizations of the form $x = a + vt$, $y = b + wt$, where a, b are arbitrary and $v, w \neq 0$. For example, a straight line passing through distinct points (a,b) and (c,d) can be described as the parametrized curve $x = a + (c-a)t$, $y = b + (d-b)t$. Starting with $x = a + vt$, $y = b + wt$, we compute $\dot{x} = v$, $\dot{y} = w$, $\ddot{x} = \ddot{y} = 0$, and $\kappa = \dfrac{|v \cdot 0 - w \cdot 0|}{(v^2 + w^2)^{3/2}} = 0$.

(b) Parametric equations for a circle of radius r are $x = r\cos\theta$ and $y = r\sin\theta$. We can take the center to be the origin. So $\dot{x} = -r\sin\theta \Rightarrow \ddot{x} = -r\cos\theta$ and $\dot{y} = r\cos\theta \Rightarrow \ddot{y} = -r\sin\theta$. Therefore $\kappa = \dfrac{|r^2\sin^2\theta + r^2\cos^2\theta|}{(r^2\sin^2\theta + r^2\cos^2\theta)^{3/2}} = \dfrac{r^2}{r^3} = \dfrac{1}{r}$. And so for any θ (and thus any point), $\kappa = \dfrac{1}{r}$.

EXERCISES 9.4

1. $\left(1, \frac{\pi}{2}\right)$

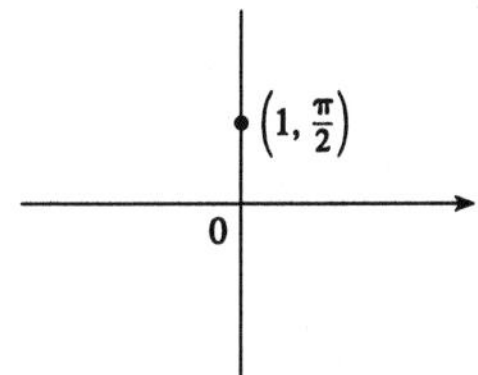

$\left(1, \frac{5\pi}{2}\right), \left(-1, \frac{3\pi}{2}\right)$

2. $(3, 0)$

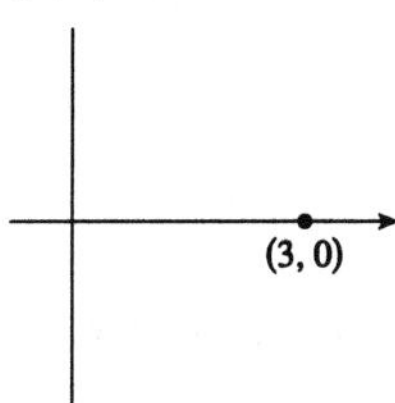

$(3, 2\pi), (-3, \pi)$

3. $\left(-1, \frac{\pi}{5}\right)$

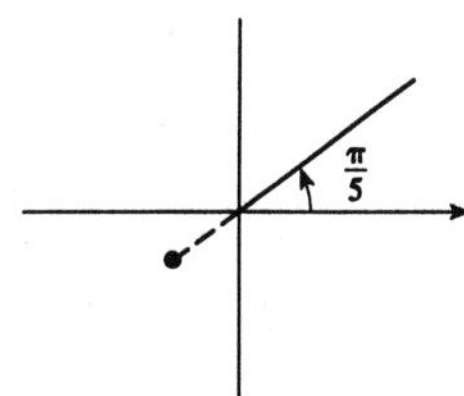

$\left(1, \frac{6\pi}{5}\right), \left(-1, \frac{11\pi}{5}\right)$

4. $\left(2, -\frac{\pi}{7}\right)$

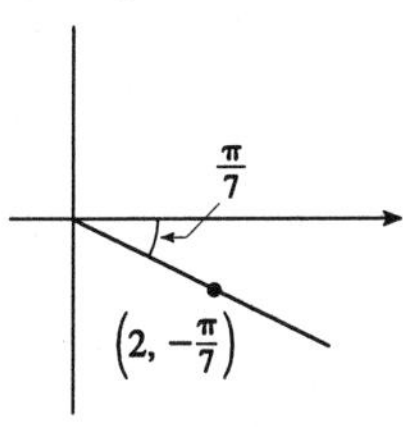

$\left(2, \frac{13\pi}{7}\right), \left(-2, \frac{6\pi}{7}\right)$

5. $(3, 2)$

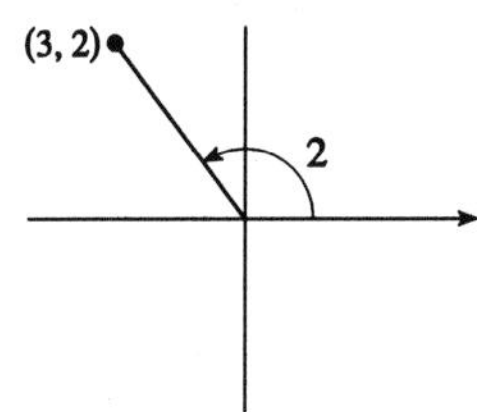

$(3, 2 + 2\pi), (-3, 2 + \pi)$

6. $(-1, \pi)$

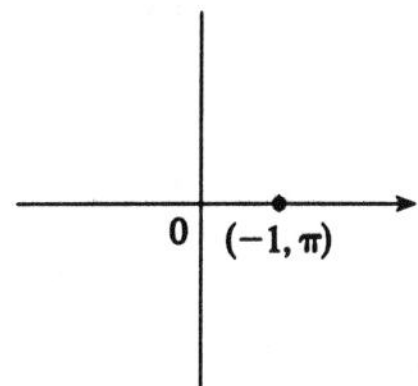

$(-1, 3\pi), (1, 0)$

7.

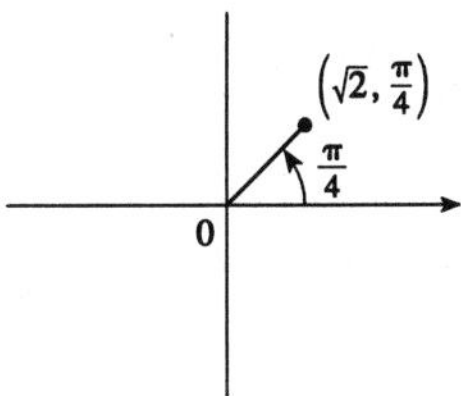

$x = \sqrt{2}\cos\frac{\pi}{4} = 1,$

$y = \sqrt{2}\sin\frac{\pi}{4} = 1$

8.

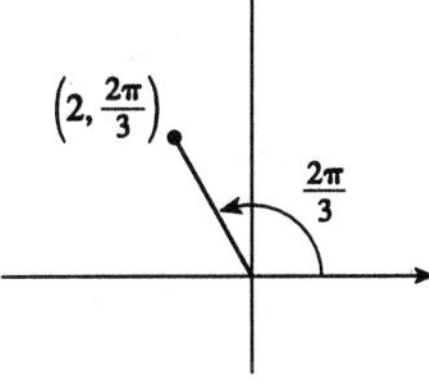

$x = 2\cos\frac{2\pi}{3} = -1$

$y = 2\sin\frac{2\pi}{3} = \sqrt{3}$

9.

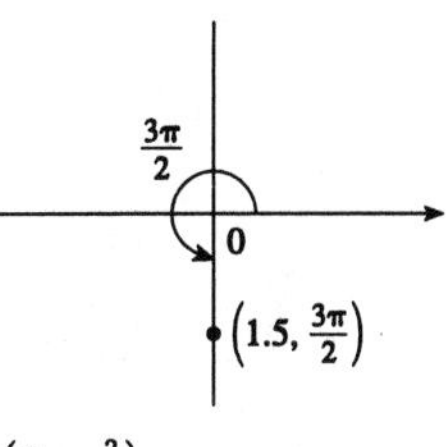

$\left(0, -\frac{3}{2}\right)$

10.

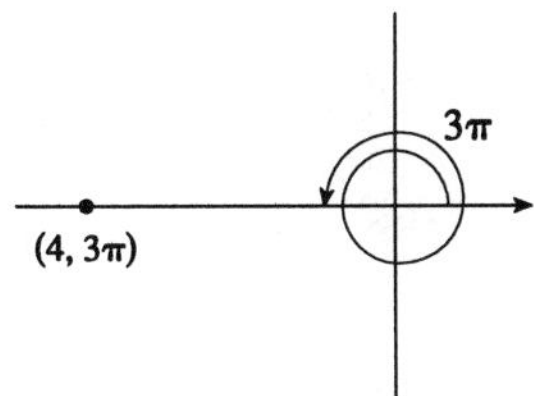

$(-4, 0)$

11.

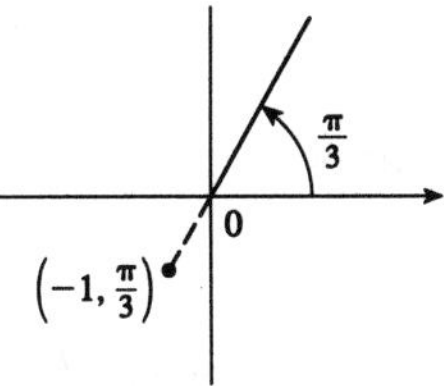

$x = -\cos\frac{\pi}{3} = -\frac{1}{2}$

$y = -\sin\frac{\pi}{3} = -\frac{\sqrt{3}}{2}$

12.

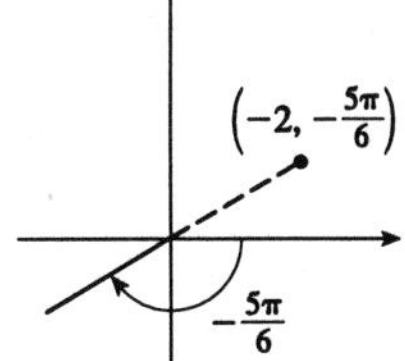

$x = -2\cos\left(-\frac{5\pi}{6}\right) = \sqrt{3}$

$y = -2\sin\left(-\frac{5\pi}{6}\right) = 1$

13. $(x,y)=(-1,1)$, $r=\sqrt{(-1)^2+1^2}=\sqrt{2}$, $\tan\theta=y/x=-1$ and (x,y) is in quadrant II, so $\theta=\frac{3\pi}{4}$. Coordinates $\left(\sqrt{2},\frac{3\pi}{4}\right)$.

14. $(x,y)=\left(-1,-\sqrt{3}\right)$. $r=\sqrt{1+3}=2$, $\tan\theta=y/x=\sqrt{3}\ \Rightarrow\ \theta=\frac{4\pi}{3}$. Coordinates $\left(2,\frac{4\pi}{3}\right)$.

15. $(x,y)=\left(2\sqrt{3},-2\right)$. $r=\sqrt{12+4}=4$, $\tan\theta=y/x=-\frac{1}{\sqrt{3}}\ \Rightarrow\ (x,y)$ is in quadrant IV, so $\theta=\frac{11\pi}{6}$. The polar coordinates are $\left(4,\frac{11\pi}{6}\right)$.

16. $(x,y)=(3,4)$, $r=\sqrt{9+16}=5$, $\tan\theta=y/x=\frac{4}{3}$, so $\theta=\tan^{-1}\frac{4}{3}$. $\left(5,\tan^{-1}\frac{4}{3}\right)$.

17. $r>1$

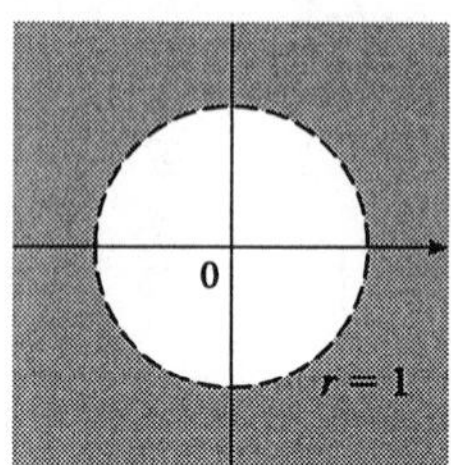

18. $0\le\theta\le\frac{\pi}{3}$

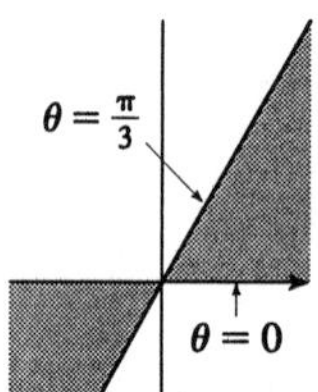

19. $0\le r\le 2,\ \frac{\pi}{2}\le\theta\le\pi$

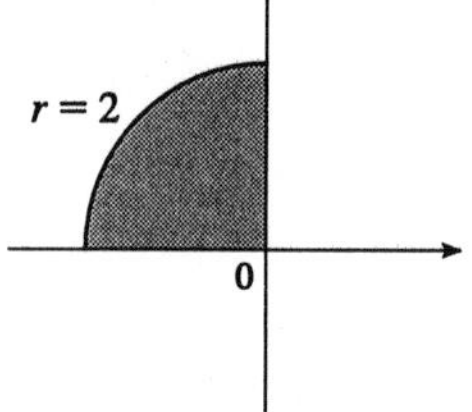

20. $1\le r<3,\ -\frac{\pi}{4}\le\theta\le\frac{\pi}{4}$

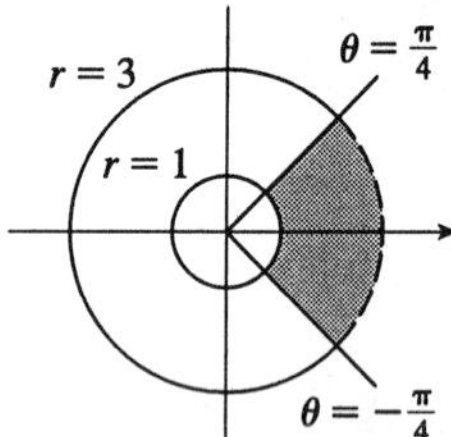

21. $3<r<4,\ -\frac{\pi}{2}\le\theta\le\pi$

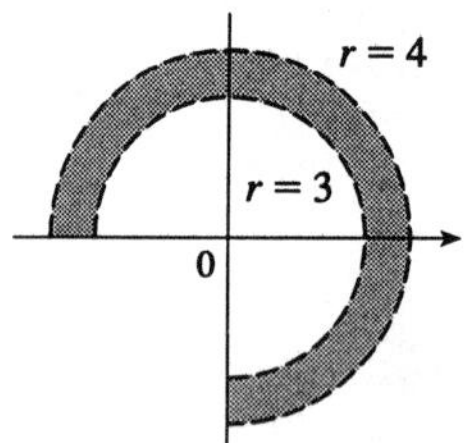

22. $-1\le r\le 1,\ \frac{\pi}{4}\le\theta\le\frac{3\pi}{4}$

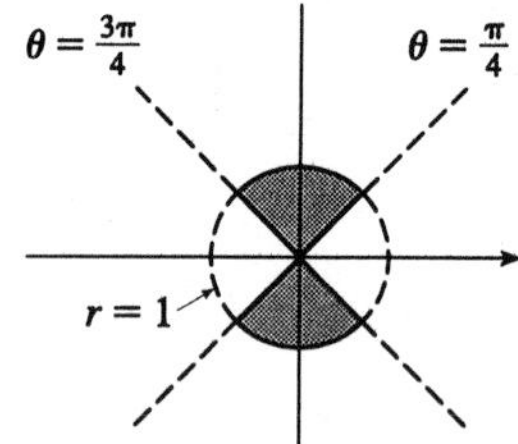

23. $\left(1,\frac{\pi}{6}\right)$ is $\left(\frac{\sqrt{3}}{2},\frac{1}{2}\right)$ Cartesian and $\left(3,\frac{3\pi}{4}\right)$ is $\left(-\frac{3}{\sqrt{2}},\frac{3}{\sqrt{2}}\right)$ Cartesian. The square of the distance between them is $\left(\frac{\sqrt{3}}{2}+\frac{\sqrt{3}}{2}\right)^2+\left(\frac{1}{2}-\frac{3}{\sqrt{2}}\right)^2=\frac{1}{4}\left(40+6\sqrt{6}-6\sqrt{2}\right)$, so the distance is $\frac{1}{2}\sqrt{40+6\sqrt{6}-6\sqrt{2}}$.

24. The points in Cartesian coordinates are $(r_1\cos\theta_1, r_1\sin\theta_1)$ and $(r_2\cos\theta_2, r_2\sin\theta_2)$ respectively. So the square of the distance between them is $(r_2\cos\theta_2-r_1\cos\theta_1)^2+(r_2\sin\theta_2-r_1\sin\theta_1)^2=r_1^2-2r_1r_2\cos(\theta_1-\theta_2)+r_2^2$, and the distance is $\sqrt{r_1^2-2r_1r_2\cos(\theta_1-\theta_2)+r_2^2}$.

25. Since $y=r\sin\theta$, the equation $r\sin\theta=2$ becomes $y=2$.

26. $r=2\sin\theta\ \Rightarrow\ r^2=2r\sin\theta\ \Rightarrow\ x^2+y^2=2y$

27. $r=\dfrac{1}{1-\cos\theta}\ \Leftrightarrow\ r-r\cos\theta=1\ \Leftrightarrow\ r=1+r\cos\theta\ \Leftrightarrow\ r^2=(1+r\cos\theta)^2\ \Leftrightarrow$
$x^2+y^2=(1+x)^2=1+2x+x^2\ \Leftrightarrow\ y^2=1+2x$

28. $r=\dfrac{5}{3-4\sin\theta}\ \Rightarrow\ 3r-4r\sin\theta=5\ \Rightarrow\ 3r=5+4r\sin\theta\ \Rightarrow\ 9r^2=(5+4r\sin\theta)^2\ \Rightarrow$
$9(x^2+y^2)=(5+4y)^2\ \Rightarrow\ 9x^2=7y^2+40y+25$

29. $r^2 = \sin 2\theta = 2\sin\theta\cos\theta \quad \Leftrightarrow \quad r^4 = 2r\sin\theta\, r\cos\theta \quad \Leftrightarrow \quad (x^2+y^2)^2 = 2yx$

30. $r^2 = \theta \quad \Rightarrow \quad \tan(r^2) = \tan\theta \quad \Rightarrow \quad \tan(x^2+y^2) = \dfrac{y}{x}$

31. $y = 5 \quad \Leftrightarrow \quad r\sin\theta = 5$

32. $y = x+1 \quad \Leftrightarrow \quad r\sin\theta = r\cos\theta + 1 \quad \Leftrightarrow \quad r(\sin\theta - \cos\theta) = 1$

33. $x^2+y^2 = 25 \quad \Leftrightarrow \quad r^2 = 25 \quad \Leftrightarrow \quad r = 5$

34. $x^2 = 4y \quad \Leftrightarrow \quad r^2\cos^2\theta = 4r\sin\theta \quad \Leftrightarrow \quad r\cos^2\theta = 4\sin\theta \quad \Leftrightarrow \quad r = 4\tan\theta\sec\theta$

35. $2xy = 1 \quad \Leftrightarrow \quad 2r\cos\theta\, r\sin\theta = 1 \quad \Leftrightarrow \quad r^2\sin 2\theta = 1 \quad \Leftrightarrow \quad r^2 = \csc 2\theta$

36. $x^2 - y^2 = 1 \quad \Leftrightarrow \quad r^2(\cos^2\theta - \sin^2\theta) = 1 \quad \Leftrightarrow \quad r^2\cos 2\theta = 1 \quad \Leftrightarrow \quad r^2 = \sec 2\theta$

37. $r = 5$

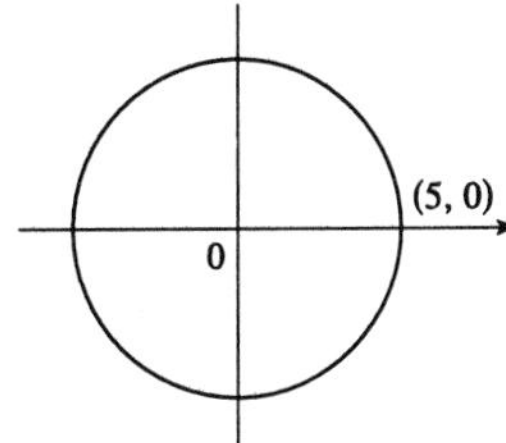

38. $\theta = \frac{3\pi}{4}$

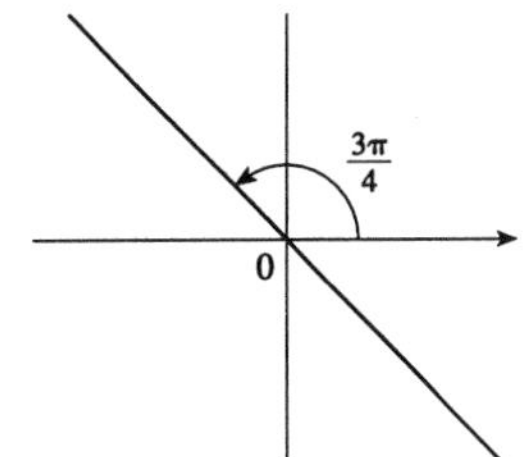

39. $r = 2\sin\theta \quad \Leftrightarrow \quad r^2 = 2r\sin\theta \quad \Leftrightarrow$
$x^2+y^2 = 2y \quad \Leftrightarrow \quad x^2 + (y-1)^2 = 1$

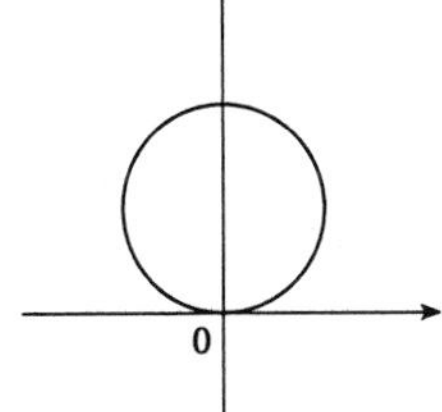

40. $r = -4\sin\theta \quad \Leftrightarrow \quad r^2 = -4r\sin\theta \quad \Leftrightarrow$
$x^2+y^2 = -4y \quad \Leftrightarrow \quad x^2 + (y+2)^2 = 4$

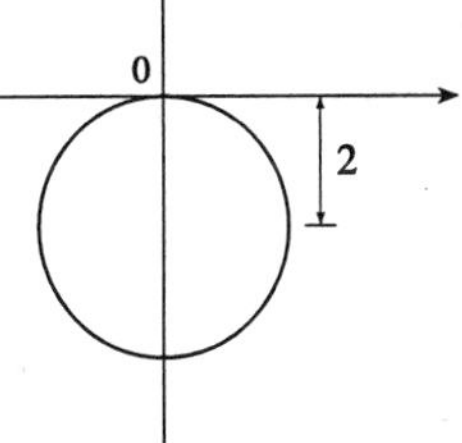

41. $r = -\cos\theta \quad \Leftrightarrow \quad r^2 = -r\cos\theta$
$\Leftrightarrow \; x^2+y^2 = -x \; \Leftrightarrow \; \left(x+\frac{1}{2}\right)^2 + y^2 = \frac{1}{4}$

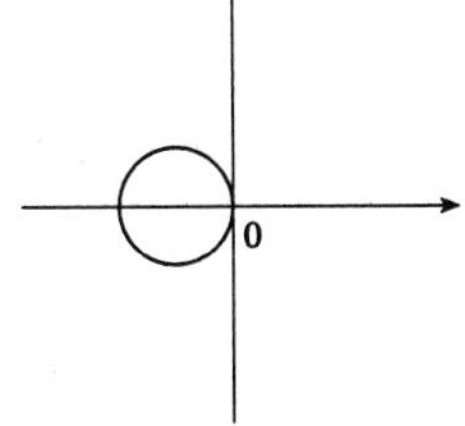

42. $r = 2\sin\theta + 2\cos\theta \; \Leftrightarrow \; r^2 = 2r\sin\theta + 2r\cos\theta$
$x^2+y^2 = 2y + 2x \; \Leftrightarrow \; (x-1)^2 + (y-1)^2 = 2$

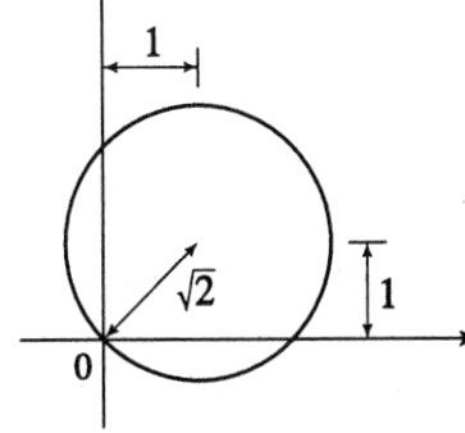

43. $r = 3(1 - \cos\theta)$

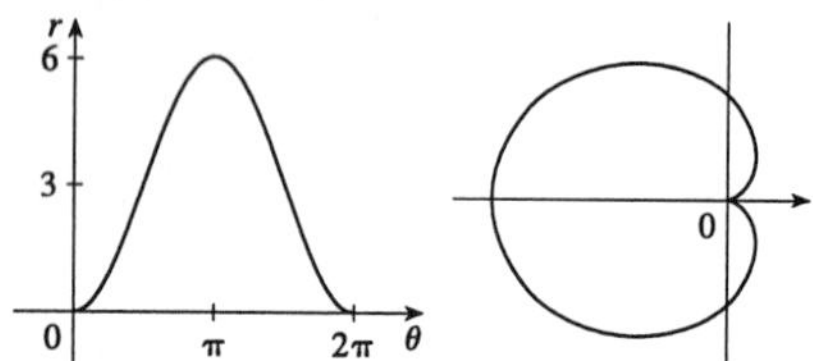

44. $r = 1 + \cos\theta$

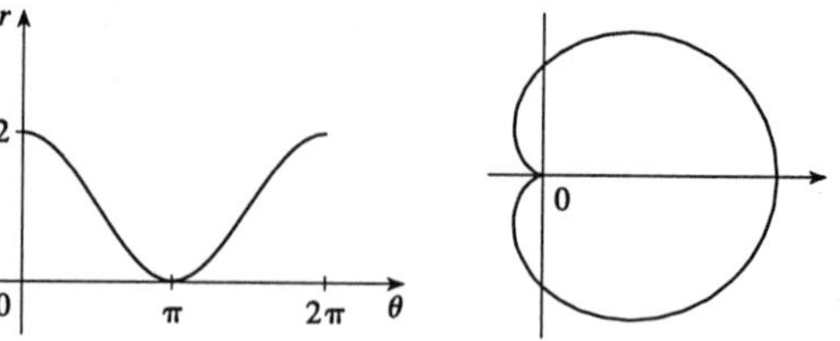

45. $r = \theta, \theta \geq 0$

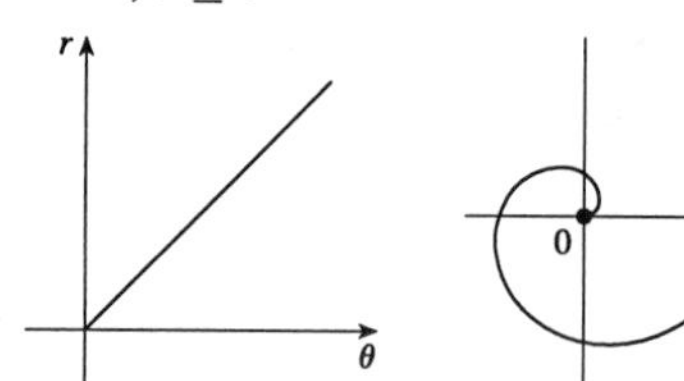

46. $r = \theta/2, -4\pi \leq \theta \leq 4\pi$

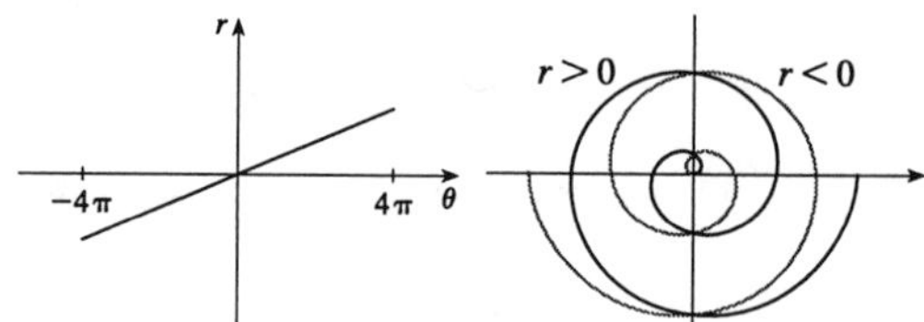

47. $r = 1/\theta$

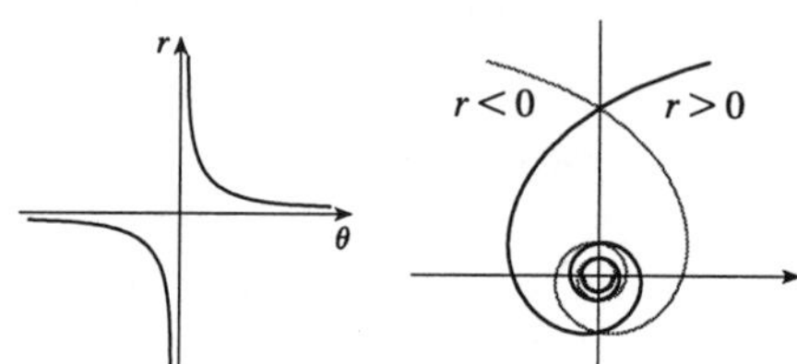

48. $r = e^{\theta}$

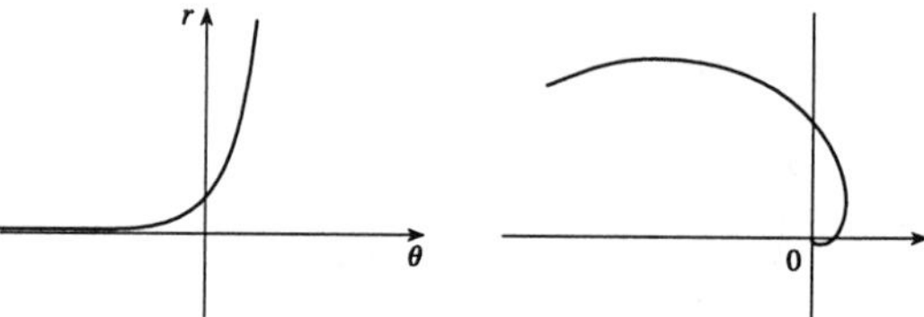

49. $r = 1 - 2\cos\theta$

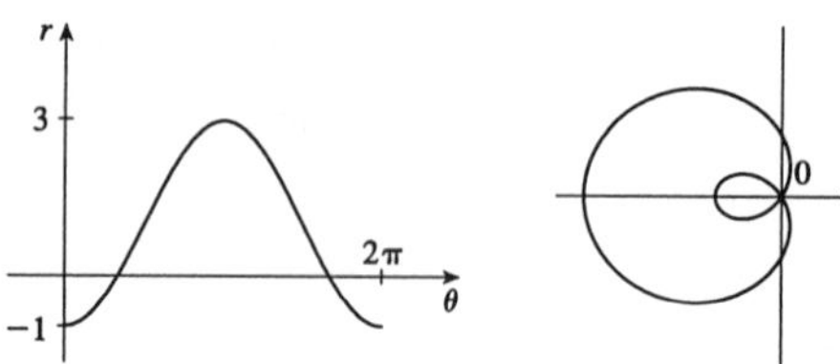

50. $r = 2 + \cos\theta$

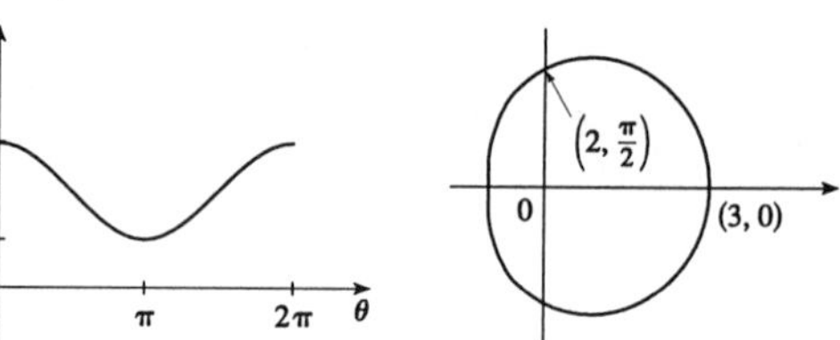

51. $r = \sin 2\theta$

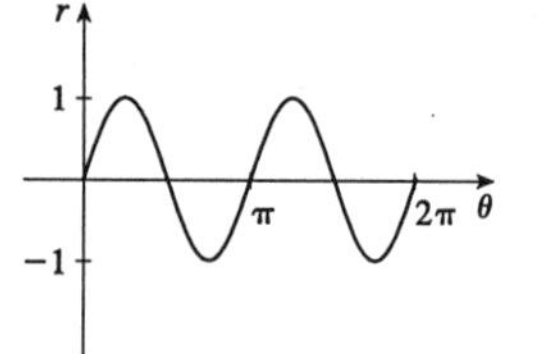

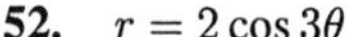
52. $r = 2\cos 3\theta$

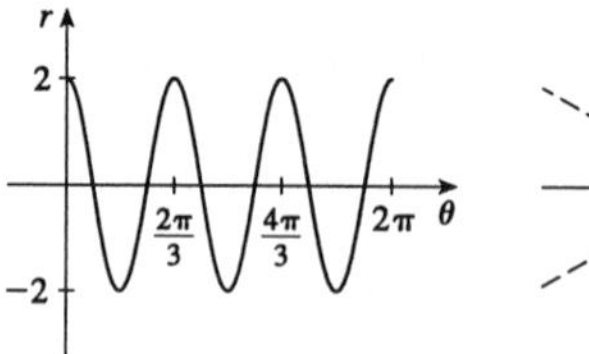

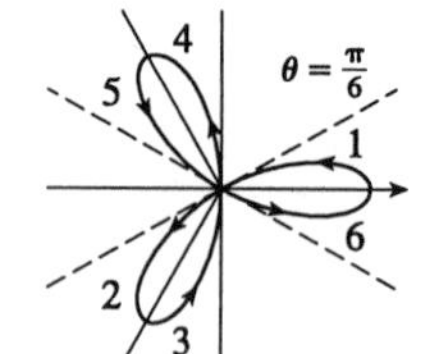

53. $r = 2\cos 4\theta$

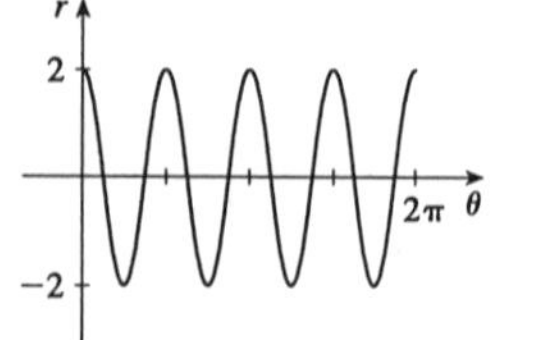

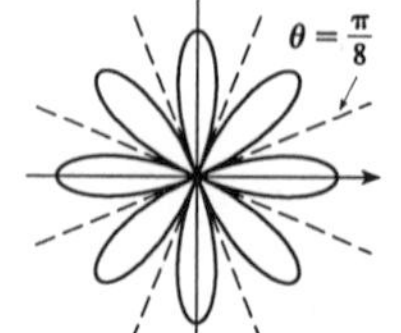

54. $r = \sin 5\theta$

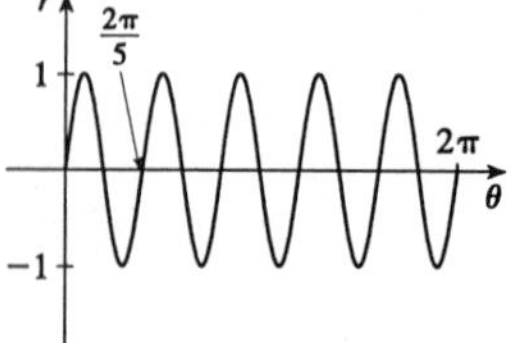

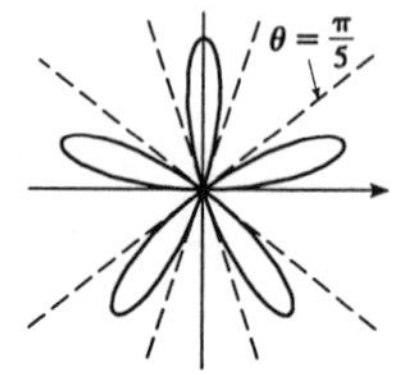

55. $r^2 = 4\cos 2\theta$

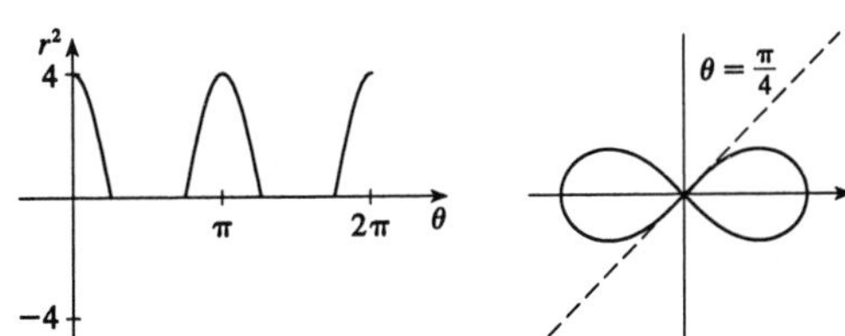

56. $r^2 = \sin 2\theta$

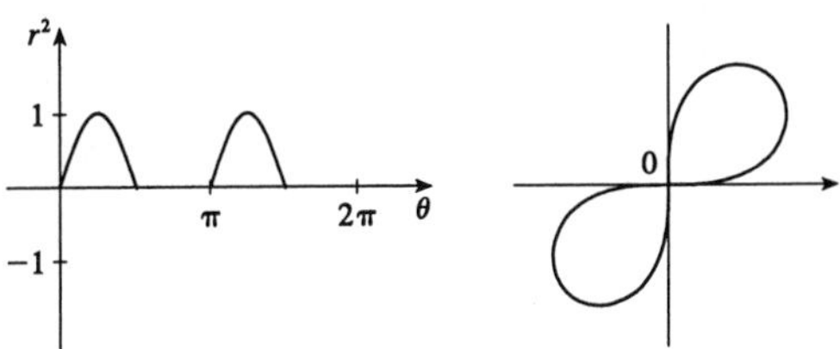

57. $r = 2\cos\left(\frac{3}{2}\theta\right)$

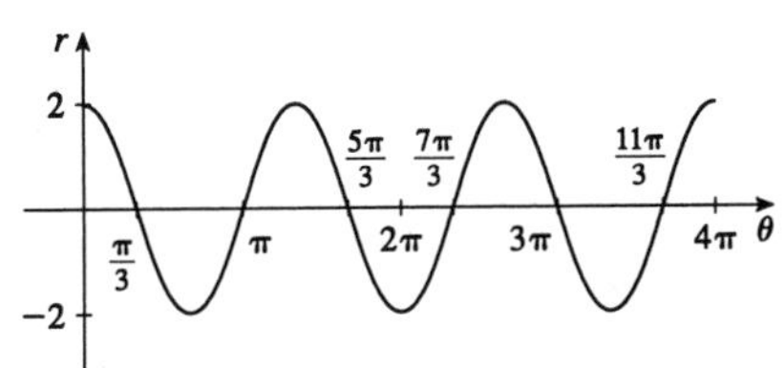

58. $r^2\theta = 1 \quad \Leftrightarrow \quad r = \pm 1/\sqrt{\theta}$ for $\theta > 0$

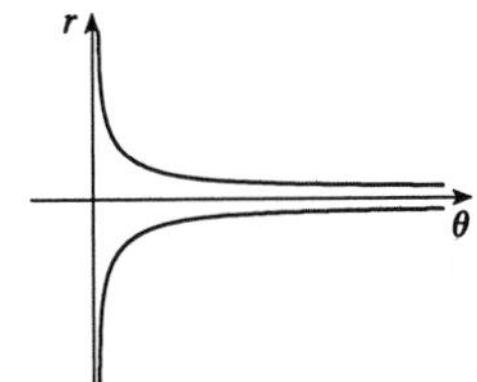

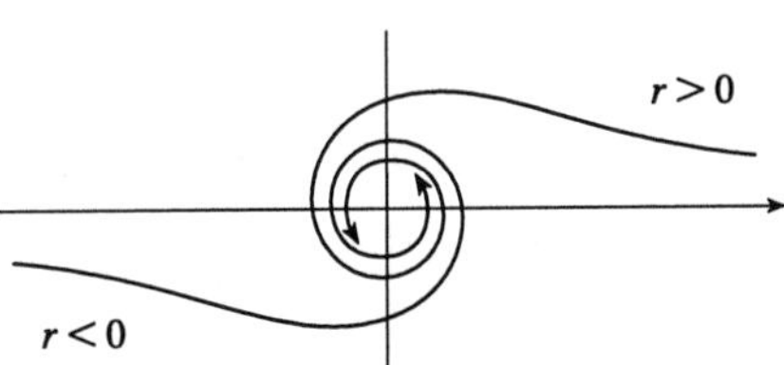

59. $x = r\cos\theta = 4\cos\theta + 2\sec\theta\cos\theta = 4\cos\theta + 2$.
Now, $r \to \infty \;\Rightarrow\; (4 + 2\sec\theta) \to \infty \;\Rightarrow\; \theta \to \frac{\pi}{2}^-$ (since we need only consider $0 \le \theta < 2\pi$), so $\lim\limits_{r\to\infty} x = \lim\limits_{\theta\to\pi/2^-} (4\cos\theta + 2) = 2$. Also, $r \to -\infty \;\Rightarrow\; (4 + 2\sec\theta) \to -\infty \;\Rightarrow\; \theta \to \frac{\pi}{2}^+$, so $\lim\limits_{r\to-\infty} x = \lim\limits_{\theta\to\pi/2^+} (4\cos\theta + 2) = 2$. Therefore $\lim\limits_{r\to\pm\infty} x = 2 \;\Rightarrow\; x = 2$ is a vertical asymptote.

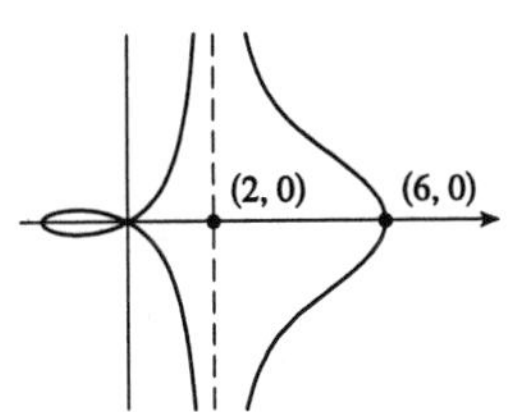

60. $y = r\sin\theta = 2\sin\theta - \csc\theta\sin\theta = 2\sin\theta - 1$.
$r \to \infty \;\Rightarrow\; (2 - \csc\theta) \to \infty \;\Rightarrow\; \csc\theta \to -\infty$
$\Rightarrow\; \theta \to \pi^+$ (since we need only consider $0 \le \theta < 2\pi$)and so $\lim\limits_{r\to\infty} y = \lim\limits_{\theta\to\pi^+} 2\sin\theta - 1 = -1$.
Also $r \to -\infty \;\Rightarrow\; (2 - \csc\theta) \to -\infty \;\Rightarrow$ $\csc\theta \to \infty \;\Rightarrow\; \theta \to \pi^-$ and so $\lim\limits_{r\to-\infty} x = \lim\limits_{\theta\to\pi^-} 2\sin\theta - 1 = -1$. Therefore $\lim\limits_{r\to\pm\infty} y = -1 \;\Rightarrow\; y = -1$ is a horizontal asymptote.

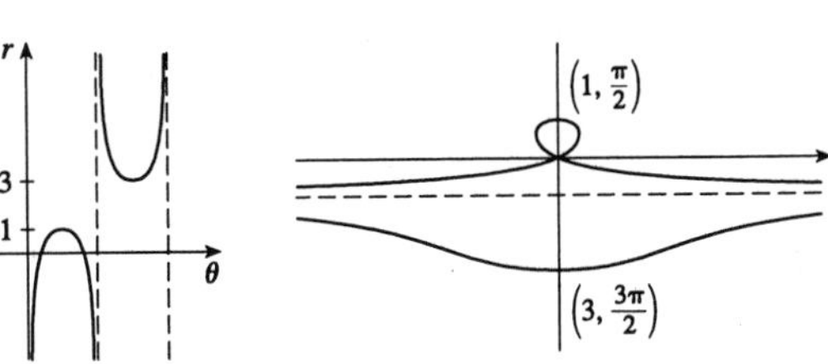

61. To show that $x = 1$ is an asymptote we must prove $\lim\limits_{r\to\pm\infty} x = 1$. $x = r\cos\theta = \sin\theta\tan\theta\cos\theta = \sin^2\theta$. Now, $r \to \infty \;\Rightarrow\; \sin\theta\tan\theta \to \infty \;\Rightarrow\; \theta \to \frac{\pi}{2}^-$, so $\lim\limits_{r\to\infty} x = \lim\limits_{\theta\to\pi/2^-} \sin^2\theta = 1$. Also, $r \to -\infty \;\Rightarrow$ $\sin\theta\tan\theta \to -\infty \Rightarrow\; \theta \to \pi/2^+$, so $\lim\limits_{r\to-\infty} x = \lim\limits_{\theta\to\pi/2^+} \sin^2\theta = 1$.

Therefore $\lim\limits_{r\to\pm\infty} x = 1 \;\Rightarrow\; x = 1$ is a vertical asymptote.

Also notice that $x = \sin^2\theta \geq 0$ for all θ, and $x = \sin^2\theta \leq 1$ for all θ.
And $x \neq 1$, since the curve is not defined at odd multiples of $\frac{\pi}{2}$.
Therefore the curve lies entirely within the vertical strip $0 \leq x < 1$.

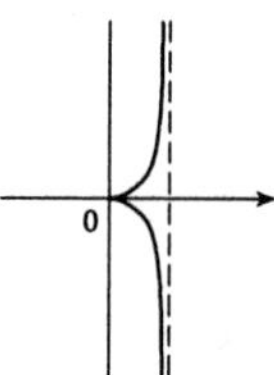

62. The equation is $(x^2+y^2)^3 = 4x^2y^2$, but using polar coordinates we know that $x^2+y^2 = r^2$ and $x = r\cos\theta$ and $y = r\sin\theta$.
Substituting into the given equation:
$r^6 = 4\,r^2\cos^2\theta\, r^2\sin^2\theta \;\Rightarrow\; r^2 = 4\cos^2\theta\sin^2\theta \;\Rightarrow$
$r = \pm 2\cos\theta\sin\theta = \pm\sin 2\theta$.
$r = \pm\sin 2\theta$ is sketched at right.

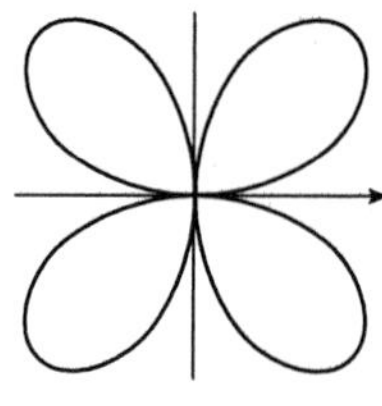

63. $$\frac{dy}{dx} = \frac{dy/d\theta}{dx/d\theta} = \frac{(dr/d\theta)(\sin\theta) + r\cos\theta}{(dr/d\theta)(\cos\theta) - r\sin\theta} = \frac{-3\sin\theta\sin\theta + 3\cos\theta\cos\theta}{-3\sin\theta\cos\theta - 3\cos\theta\sin\theta} = \frac{3\cos 2\theta}{-3\sin 2\theta} = -\cot 2\theta$$
$$= \frac{1}{\sqrt{3}} \text{ when } \theta = \frac{\pi}{3}$$

Alternate Solution: $r = 3\cos\theta \;\Rightarrow\; x = r\cos\theta = 3\cos^2\theta,\ y = r\sin\theta = 3\sin\theta\cos\theta \;\Rightarrow$
$$\frac{dy}{dx} = \frac{dy/d\theta}{dx/d\theta} = \frac{-3\sin^2\theta + 3\cos^2\theta}{-6\cos\theta\sin\theta} = \frac{\cos 2\theta}{-\sin 2\theta} = -\cot 2\theta = \frac{1}{\sqrt{3}} \text{ when } \theta = \frac{\pi}{3}$$

64. $$\frac{dy}{dx} = \frac{(dr/d\theta)\sin\theta + r\cos\theta}{(dr/d\theta)\cos\theta - r\sin\theta} = \frac{(-\sin\theta+\cos\theta)\sin\theta + (\cos\theta+\sin\theta)\cos\theta}{(-\sin\theta+\cos\theta)\cos\theta - (\cos\theta+\sin\theta)\sin\theta} = -1 \text{ when } \theta = \frac{\pi}{4}$$

Alternate Solution: $r = \cos\theta + \sin\theta \;\Rightarrow\; x = r\cos\theta = (\cos\theta+\sin\theta)\cos\theta,\ y = r\sin\theta = (\cos\theta+\sin\theta)\sin\theta$
$$\Rightarrow\; \frac{dy}{dx} = \frac{dy/d\theta}{dx/d\theta} = \frac{\sin\theta(-\sin\theta+\cos\theta) + (\cos\theta+\sin\theta)\cos\theta}{\cos\theta(-\sin\theta+\cos\theta) - (\cos\theta+\sin\theta)\sin\theta} = -1 \text{ when } \theta = \frac{\pi}{4}$$

65. $r = \theta \;\Rightarrow\; x = r\cos\theta = \theta\cos\theta,\ y = r\sin\theta = \theta\sin\theta \;\Rightarrow\; \dfrac{dy}{dx} = \dfrac{dy/d\theta}{dx/d\theta} = \dfrac{\sin\theta + \theta\cos\theta}{\cos\theta - \theta\sin\theta} = -\dfrac{2}{\pi}$ when $\theta = \dfrac{\pi}{2}$

66. $r = \ln\theta \;\Rightarrow\; x = r\cos\theta = \ln\theta\cos\theta,\ y = r\sin\theta = \ln\theta\sin\theta \;\Rightarrow$
$$\frac{dy}{dx} = \frac{dy/d\theta}{dx/d\theta} = \frac{\sin\theta(1/\theta) + \ln\theta\cos\theta}{\cos\theta(1/\theta) - \ln\theta\sin\theta} = \frac{\sin e + e\cos e}{\cos e - e\sin e} \text{ when } \theta = e$$

67. $r = 1 + \cos\theta \;\Rightarrow\; x = r\cos\theta = \cos\theta + \cos^2\theta,\ y = r\sin\theta = \sin\theta + \sin\theta\cos\theta \;\Rightarrow$
$$\frac{dy}{dx} = \frac{dy/d\theta}{dx/d\theta} = \frac{\cos\theta + \cos^2\theta - \sin^2\theta}{-\sin\theta - 2\cos\theta\sin\theta} = \frac{\cos\theta + \cos 2\theta}{-\sin\theta - \sin 2\theta} = -1 \text{ when } \theta = \frac{\pi}{6}$$

68. $r = \sin 3\theta \;\Rightarrow\; x = r\cos\theta = \sin 3\theta\cos\theta,\ y = r\sin\theta = \sin 3\theta\sin\theta \;\Rightarrow$
$$\frac{dy}{dx} = \frac{dy/d\theta}{dx/d\theta} = \frac{3\cos 3\theta\sin\theta + \sin 3\theta\cos\theta}{3\cos 3\theta\cos\theta - \sin 3\theta\sin\theta} = -\sqrt{3} \text{ when } \theta = \frac{\pi}{6}$$

69. $r = 3\cos\theta \;\Rightarrow\; x = r\cos\theta = 3\cos\theta\cos\theta,\ y = r\sin\theta = 3\cos\theta\sin\theta \;\Rightarrow$
$dy/d\theta = -3\sin^2\theta + 3\cos^2\theta = 3\cos 2\theta = 0 \;\Rightarrow\; 2\theta = \frac{\pi}{2}$ or $\frac{3\pi}{2} \;\Leftrightarrow\; \theta = \frac{\pi}{4}$ or $\frac{3\pi}{4}$. So the tangent is horizontal at $\left(\frac{3}{\sqrt{2}}, \frac{\pi}{4}\right)$ and $\left(-\frac{3}{\sqrt{2}}, \frac{3\pi}{4}\right)$ $\left[\text{same as } \left(\frac{3}{\sqrt{2}}, -\frac{\pi}{4}\right)\right]$. $dx/d\theta = -6\sin\theta\cos\theta = -3\sin 2\theta = 0 \;\Rightarrow$ $2\theta = 0$ or $\pi \;\Leftrightarrow\; \theta = 0$ or $\frac{\pi}{2}$. So the tangent is vertical at $(3, 0)$ and $\left(0, \frac{\pi}{2}\right)$.

70. $y = r\sin\theta = \cos\theta\sin\theta + \sin^2\theta = \frac{1}{2}\sin 2\theta + \sin^2\theta \quad\Rightarrow\quad dy/d\theta = \cos 2\theta + \sin 2\theta = 0 \quad\Rightarrow\quad \tan 2\theta = -1$ $\Rightarrow\quad 2\theta = \frac{3\pi}{4}$ or $\frac{7\pi}{4} \quad\Leftrightarrow\quad \theta = \frac{3\pi}{8}$ or $\frac{7\pi}{8} \quad\Rightarrow\quad$ horizontal tangents at $\left(\cos\frac{3\pi}{8} + \sin\frac{3\pi}{8}, \frac{3\pi}{8}\right)$ and $\left(\cos\frac{7\pi}{8} + \sin\frac{7\pi}{8}, \frac{7\pi}{8}\right)$. $x = r\cos\theta = \cos^2\theta + \cos\theta\sin\theta \;\Rightarrow\; dx/d\theta = -\sin 2\theta + \cos 2\theta = 0 \;\Rightarrow\; \tan 2\theta = 1$ $\Rightarrow\; 2\theta = \frac{\pi}{4}$ or $\frac{5\pi}{4} \;\Leftrightarrow\; \theta = \frac{\pi}{8}$ or $\frac{5\pi}{8} \;\Rightarrow\;$ vertical tangents at $\left(\cos\frac{\pi}{8} + \sin\frac{\pi}{8}, \frac{\pi}{8}\right)$ and $\left(\cos\frac{5\pi}{8} + \sin\frac{5\pi}{8}, \frac{5\pi}{8}\right)$.

Note: These expressions can be simplified using trigonometric identities. For example, $\cos\frac{\pi}{8} + \sin\frac{\pi}{8} = \frac{1}{2}\sqrt{4 + 2\sqrt{2}}$.

71. $r = \cos 2\theta \quad\Rightarrow\quad x = r\cos\theta = \cos 2\theta\cos\theta,\; y = r\sin\theta = \cos 2\theta\sin\theta \quad\Rightarrow$

$$dy/d\theta = -2\sin 2\theta\sin\theta + \cos 2\theta\cos\theta = -4\sin^2\theta\cos\theta + \left(\cos^3\theta - \sin^2\theta\cos\theta\right)$$
$$= \cos\theta(\cos^2\theta - 5\sin^2\theta) = \cos\theta(1 - 6\sin^2\theta) = 0 \quad\Rightarrow$$

$\cos\theta = 0$ or $\sin\theta = \pm\frac{1}{\sqrt{6}} \quad\Rightarrow\quad \theta = \frac{\pi}{2}, \frac{3\pi}{2}, \alpha, \pi - \alpha, \pi + \alpha$, or $2\pi - \alpha \quad \left[\text{where } \alpha = \sin^{-1}\frac{1}{\sqrt{6}}\right]$.

So the tangent is horizontal at $\left(1, \frac{3\pi}{2}\right)$, $\left(1, \frac{\pi}{2}\right)$, $\left(\frac{2}{3}, \alpha\right)$, $\left(\frac{2}{3}, \pi - \alpha\right)$, $\left(\frac{2}{3}, \pi + \alpha\right)$, and $\left(\frac{2}{3}, 2\pi - \alpha\right)$.

$dx/d\theta = -2\sin 2\theta\cos\theta - \cos 2\theta\sin\theta = -4\sin\theta\cos^2\theta - (2\cos^2\theta - 1)\sin\theta = \sin\theta(1 - 6\cos^2\theta) = 0 \quad\Rightarrow$ $\sin\theta = 0$ or $\cos\theta = \pm\frac{1}{\sqrt{6}} \quad\Rightarrow\quad \theta = 0, \pi, \frac{\pi}{2} - \alpha, \frac{\pi}{2} + \alpha, \frac{3\pi}{2} - \alpha$, or $\frac{3\pi}{2} + \alpha$. So the tangent is vertical at $(1, 0)$, $(1, \pi)$, $\left(\frac{2}{3}, \frac{3\pi}{2} - \alpha\right)$, $\left(\frac{2}{3}, \frac{3\pi}{2} + \alpha\right)$, $\left(\frac{2}{3}, \frac{\pi}{2} - \alpha\right)$, and $\left(\frac{2}{3}, \frac{\pi}{2} + \alpha\right)$.

72. $dr/d\theta = (1/r)\cos 2\theta$ (by differentiating implicitly), so

$$\frac{dy}{d\theta} = \frac{1}{r}\cos 2\theta\sin\theta + r\cos\theta = \frac{1}{r}\left(\cos 2\theta\sin\theta + r^2\cos\theta\right) = \frac{1}{r}(\cos 2\theta\sin\theta + \sin 2\theta\cos\theta) = \frac{1}{r}\sin 3\theta.$$

This is 0 when $\sin 3\theta = 0 \quad\Rightarrow\quad \theta = 0, \frac{\pi}{3}$ or $\frac{4\pi}{3}$ (restricting θ to the domain of the lemniscate), so there are horizontal tangents at $\left(\sqrt[4]{\frac{3}{4}}, \frac{\pi}{3}\right)$, $\left(\sqrt[4]{\frac{3}{4}}, \frac{4\pi}{3}\right)$ and $(0, 0)$. Similarly, $dx/d\theta = (1/r)\cos 3\theta = 0$ when $\theta = \frac{\pi}{6}$ or $\frac{7\pi}{6}$, so there are vertical tangents at $\left(\sqrt[4]{\frac{3}{4}}, \frac{\pi}{6}\right)$ and $\left(\sqrt[4]{\frac{3}{4}}, \frac{7\pi}{6}\right)$ [and $(0, 0)$]. See the sketch in Exercise 56.

73. $r = 1 + \cos\theta \quad\Rightarrow\quad x = r\cos\theta = \cos\theta(1 + \cos\theta),\; y = r\sin\theta = \sin\theta(1 + \cos\theta) \quad\Rightarrow$ $dy/d\theta = (1 + \cos\theta)\cos\theta - \sin^2\theta = 2\cos^2\theta + \cos\theta - 1 = (2\cos\theta - 1)(\cos\theta + 1) = 0 \quad\Rightarrow\quad \cos\theta = \frac{1}{2}$ or $-1 \quad\Rightarrow\quad \theta = \frac{\pi}{3}, \pi$, or $\frac{5\pi}{3} \quad\Rightarrow\quad$ horizontal tangent at $\left(\frac{3}{2}, \frac{\pi}{3}\right)$, $(0, \pi)$, and $\left(\frac{3}{2}, \frac{5\pi}{3}\right)$.

$dx/d\theta = -(1 + \cos\theta)\sin\theta - \cos\theta\sin\theta = -\sin\theta(1 + 2\cos\theta) = 0 \quad\Rightarrow\quad \sin\theta = 0$ or $\cos\theta = -\frac{1}{2} \quad\Rightarrow$ $\theta = 0, \pi, \frac{2\pi}{3}$, or $\frac{4\pi}{3} \quad\Rightarrow\quad$ vertical tangent at $(2, 0)$, $\left(\frac{1}{2}, \frac{2\pi}{3}\right)$, and $\left(\frac{1}{2}, \frac{4\pi}{3}\right)$. Note that the tangent is horizontal, not vertical when $\theta = \pi$, since $\displaystyle\lim_{\theta\to\pi}\frac{dy/d\theta}{dx/d\theta} = 0$.

74. $\dfrac{dy}{d\theta} = e^\theta\sin\theta + e^\theta\cos\theta = e^\theta(\sin\theta + \cos\theta) = 0 \;\Rightarrow\; \sin\theta = -\cos\theta \;\Rightarrow\; \tan\theta = -1 \;\Rightarrow\; \theta = -\dfrac{1}{4}\pi + n\pi$ (n any integer) $\Rightarrow$ horizontal tangents at $\left(e^{\pi(n-1/4)}, \pi\left(n - \frac{1}{4}\right)\right)$.

$\dfrac{dx}{d\theta} = e^\theta\cos\theta - e^\theta\sin\theta = e^\theta(\cos\theta - \sin\theta) = 0 \;\Rightarrow\; \sin\theta = \cos\theta \;\Rightarrow\; \tan\theta = 1 \;\Rightarrow\; \theta = \dfrac{1}{4}\pi + n\pi$ (n any integer) $\Rightarrow$ vertical tangents at $\left(e^{\pi(n+1/4)}, \pi\left(n + \frac{1}{4}\right)\right)$.

75. $r = a\sin\theta + b\cos\theta \quad\Rightarrow\quad r^2 = ar\sin\theta + br\cos\theta \quad\Rightarrow\quad x^2 + y^2 = ay + bx \quad\Rightarrow$ $\left(x - \frac{1}{2}b\right)^2 + \left(y - \frac{1}{2}a\right)^2 = \frac{1}{4}(a^2 + b^2)$, and this is a circle with center $\left(\frac{1}{2}b, \frac{1}{2}a\right)$ and radius $\frac{1}{2}\sqrt{a^2 + b^2}$.

76. These curves are circles which intersect at the origin and at $\left(\frac{1}{\sqrt{2}}a, \frac{\pi}{4}\right)$. At the origin, the first circle has a horizontal tangent and the second a vertical one, so the tangents are perpendicular here. For the first circle, $dy/d\theta = a\cos\theta\sin\theta + a\sin\theta\cos\theta = a\sin 2\theta = a$ at $\theta = \frac{\pi}{4}$ and $dx/d\theta = a\cos^2\theta - a\sin^2\theta = a\cos 2\theta = 0$ at $\theta = \frac{\pi}{4}$, so the tangent here is vertical. Similarly, for the second circle, $dy/d\theta = a\cos 2\theta = 0$ and $dx/d\theta = -a\sin 2\theta = -a$ at $\theta = \frac{\pi}{4}$, so the tangent is horizontal, and again the tangents are perpendicular.

77. **(a)** We see that the curve crosses itself at the origin, where $r = 0$ (in fact the inner loop corresponds to negative r-values,) so we solve the equation of the limaçon for $r = 0 \quad\Leftrightarrow\quad c\sin\theta = -1 \quad\Leftrightarrow$ $\sin\theta = -1/c$. Now if $|c| < 1$, then this equation has no solution and hence there is no inner loop. But if $c < -1$, then on the interval $(0, 2\pi)$ the equation has the two solutions $\theta = \sin^{-1}(-1/c)$ and $\theta = \pi - \sin^{-1}(-1/c)$, and if $c > 1$, the solutions are $\theta = \pi + \sin^{-1}(1/c)$ and $\theta = 2\pi - \sin^{-1}(1/c)$. In each case, $r < 0$ for θ between the two solutions, indicating a loop.

(b) For $0 < c < 1$, the dimple (if it exists) is characterized by the fact that y has a local maximum at $\theta = \frac{3\pi}{2}$. So we determine for what c-values $\dfrac{d^2y}{d\theta^2}$ is negative at $\theta = \frac{3\pi}{2}$, since by the Second Derivative Test this indicates a maximum: $y = r\sin\theta = \sin\theta + c\sin^2\theta \;\Rightarrow\; \dfrac{dy}{d\theta} = \cos\theta + 2c\sin\theta\cos\theta = \cos\theta + c\sin 2\theta$ $\Rightarrow\; \dfrac{d^2y}{d\theta^2} = -\sin\theta + 2c\cos 2\theta$. At $\theta = \frac{3\pi}{2}$, this is equal to $-(-1) + 2c(-1) = 1 - 2c$, which is negative only for $c > \frac{1}{2}$. A similar argument shows that for $-1 < c < 0$, y only has a local *minimum* at $\theta = \frac{\pi}{2}$ (indicating a dimple) for $c < -\frac{1}{2}$.

78. **(a)** $r = \sin(\theta/2)$. This equation must correspond to one of II, III or VI, since these are the only graphs which are bounded. In fact it must be VI, since this is the only graph which is completed after a rotation of exactly 4π.

(b) $r = \sin(\theta/4)$. This equation must correspond to III, since this is the only graph which is completed after a rotation of exactly 8π.

(c) $r = \sec(3\theta)$. This must correspond to IV, since the graph is unbounded at $\theta = \frac{\pi}{6}, \frac{\pi}{2}, \frac{2\pi}{3}$, and so on.

(d) $r = \theta\sin\theta$. This must correspond to V. Note that $r = 0$ whenever θ is a multiple of π. This graph is unbounded, and each time θ moves through an interval of 2π, the same basic shape is repeated (because of the periodic $\sin\theta$ factor) but it gets larger each time (since θ increases each time we go around.)

(e) $r = 1 + 4\cos 5\theta$. This corresponds to II, since it is bounded, has fivefold rotational symmetry, and takes only one takes only one rotation through 2π to be complete.

(f) $r = 1/\sqrt{\theta}$. This corresponds to I, since it is unbounded at $\theta = 0$, and r decreases as θ increases; in fact $r \to 0$ as $\theta \to \infty$.

Note for Exercises 79-82: Maple is able to plot polar curves using the `polarplot` command, or using the `coords=polar` option in a regular `plot` command. In Mathematica, use `PolarPlot`. If your graphing device cannot plot polar equations, you must convert to parametric equations. For example, in Exercise 79, $x = r\cos\theta = [1 + 2\sin(\theta/2)]\cos\theta$, $y = r\sin\theta = [1 + 2\sin(\theta/2)]\sin\theta$.

79. $r = 1 + 2\sin(\theta/2)$

The correct parameter interval is $[0, 4\pi]$.

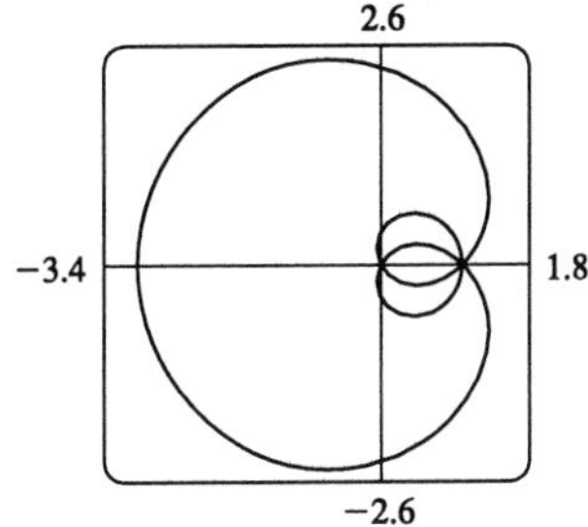

80. $r = \sqrt{1 - 0.8\sin^2\theta}$

The correct parameter interval is $[0, 2\pi]$.

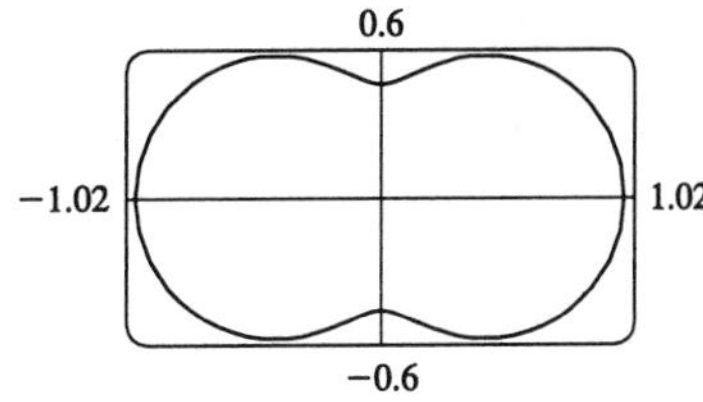

81. $r = \sin(9\theta/4)$

The correct parameter interval is $[0, 8\pi]$.

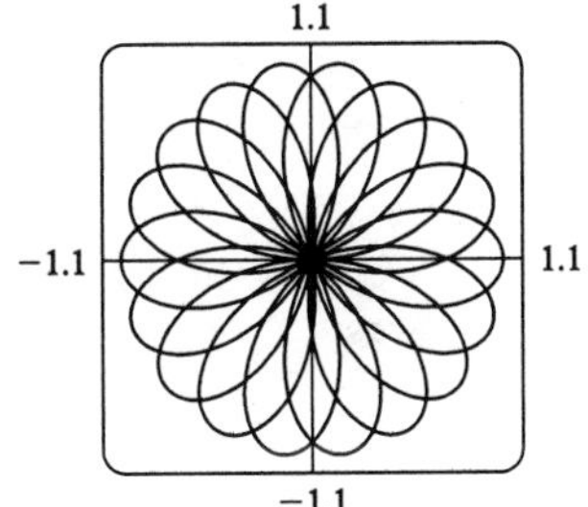

82. $r = 1 + 4\cos(\theta/3)$

The correct parameter interval is $[0, 6\pi]$.

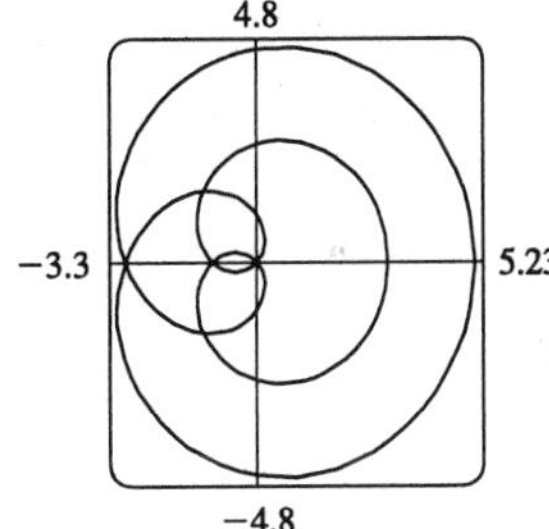

83.

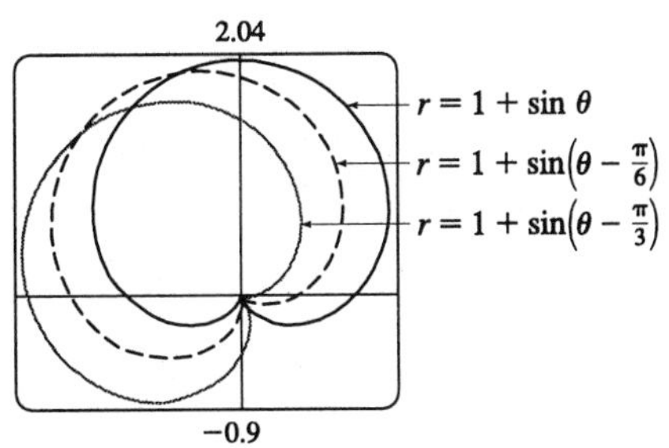

It appears that the graph of $r = 1 + \sin\left(\theta - \frac{\pi}{6}\right)$ is the same shape as the graph of $r = 1 + \sin\theta$, but rotated counterclockwise about the origin by $\frac{\pi}{6}$. Similarly, the graph of $r = 1 + \sin\left(\theta - \frac{\pi}{3}\right)$ is rotated by $\frac{\pi}{3}$. In general, the graph of $r = f(\theta - \alpha)$ is the same shape as that of $r = f(\theta)$, but rotated counterclockwise through α about the origin.

That is, for any point (r_0, θ_0) on the curve $r = f(\theta)$, the point $(r_0, \theta_0 + \alpha)$ is on the curve the curve $r = f(\theta - \alpha)$, since $r_0 = f(\theta_0) = f((\theta_0 + \alpha) - \alpha)$.

84.

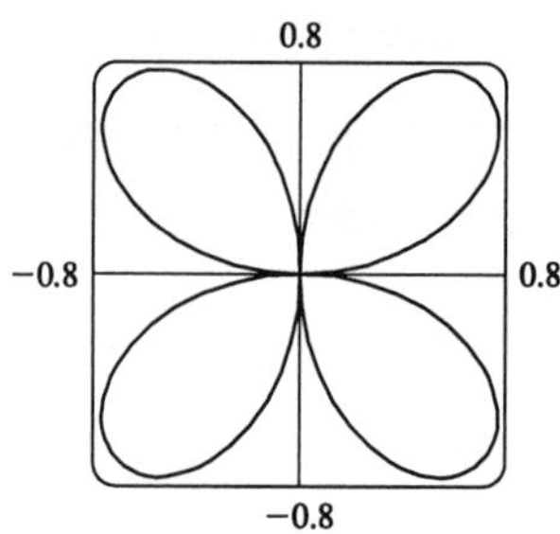

From the graph, the highest points seem to have $y \approx \frac{3}{4}$.

To find the exact value, we set $\dfrac{dy}{d\theta} = 0$. $y = r\sin\theta = \sin\theta\sin 2\theta \quad\Rightarrow$

$\dfrac{dy}{d\theta} = 2\sin\theta\cos 2\theta + \cos\theta\sin 2\theta = 2\sin\theta\left(2\cos^2\theta - 1\right) + \cos\theta(2\sin\theta\cos\theta) = 2\sin\theta\left(3\cos^2\theta - 1\right)$. In the first quadrant, this is 0 when $\cos\theta = \frac{1}{\sqrt{3}} \quad\Leftrightarrow\quad \sin\theta = \sqrt{\frac{2}{3}} \quad\Leftrightarrow$

$y = 2\sin^2\theta\cos\theta = 2 \cdot \frac{2}{3} \cdot \frac{1}{\sqrt{3}} = \frac{4\sqrt{3}}{9} \approx 0.77$.

85. **(a)** $r = \sin n\theta$. From the graphs, it seems that when n is even, the number of loops in the curve (called a rose) is $2n$, and when n is odd, the number of loops is simply n. This is because in the case of n odd, every point on the graph is traversed twice, due to the fact that

$$r(\theta + \pi) = \sin[n(\theta + \pi)] = \sin n\theta\cos n\pi + \cos n\theta\sin n\pi = \begin{cases} \sin n\theta & n \text{ even} \\ -\sin n\theta & n \text{ odd.} \end{cases}$$

$n = 2$

$n = 3$

$n = 4$

$n = 5$

(b) The graph of $r = |\sin n\theta|$ has $2n$ loops whether n is odd or even, since $r(\theta + \pi) = r(\theta)$.

$n = 2$

$n = 3$

$n = 4$

$n = 5$

86. $r = 1 + c\sin n\theta$. We vary n while keeping c constant at 2. As n changes, the curves change in the same way as those in Exercise 85: the number of loops increases. Note that if n is even, the smaller loops are outside the larger ones; if n is odd, they are inside.

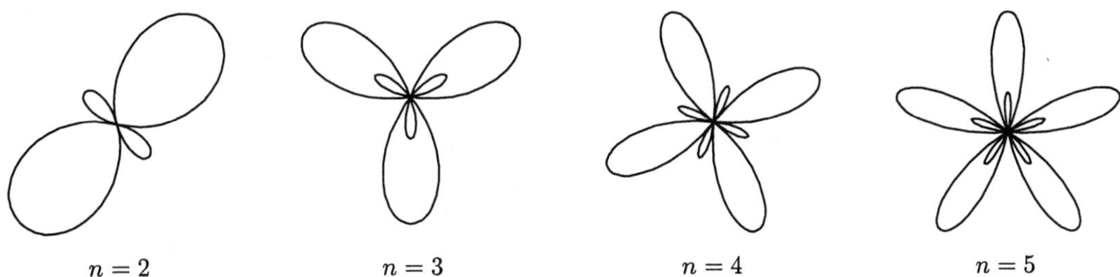

Now we vary c while keeping $n = 3$. As c increases toward 0, the entire graph gets smaller (the graphs below are not to scale) and the smaller loops shrink in relation to the large ones. At $n = -1$, the small loops disappear entirely, and for $-1 < c < 1$, the graph is a simple, closed curve (at $c = 0$ it is a circle.) As n continues to increase, the same changes are seen, but in reverse order, since $1 + (-c)\sin n\theta = 1 + c\sin n(\theta + \pi)$, so the graph for $c = c_0$ is the same as that for $c = -c_0$, with a rotation through π. As $c \to \infty$, the smaller loops get relatively closer in size to the large ones. Note that the distance between the end of an inner loop and the corresponding outer loop is always 1. Maple's `animate` command (or Mathematica's `Animate`) is very useful for seeing the changes that occur as c varies.

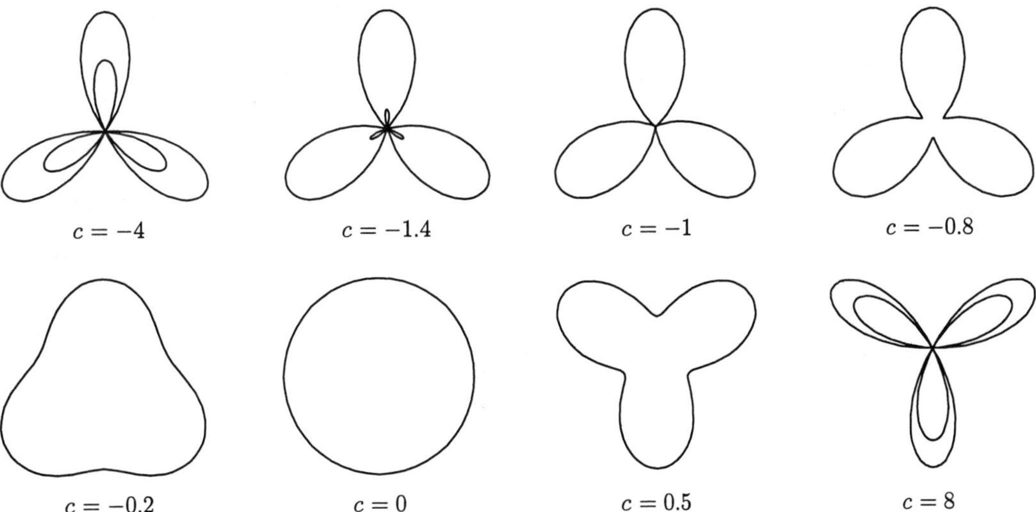

87. $r = \dfrac{1 - a\cos\theta}{1 + a\cos\theta}$. We start with $a = 0$, since in this case the curve is simply the circle $r = 1$. As a increases, the graph moves to the left, and its right side becomes flattened. As a increases through about 0.4, the right side seems to grow a dimple, which upon closer investigation (with narrower θ-ranges) seems to appear at $a \approx 0.42$ (the actual value is $\sqrt{2} - 1$.) As $a \to 1$, this dimple becomes more pronounced, and the curve begins to stretch out horizontally, until at $a = 1$ the denominator vanishes at $\theta = \pi$, and the dimple becomes an actual cusp. For $a > 1$ we must choose our parameter interval carefully, since $r \to \infty$ as $1 + a\cos\theta \to 0 \quad \Leftrightarrow$ $\theta \to \pm\cos^{-1}(-1/a)$. As a increases from 1, the curve splits into two parts. The left part has a loop, which grows larger as a increases, and the right part grows broader vertically, and its left tip develops a dimple when $a \approx 2.42$ (actually, $\sqrt{2} + 1$). As a increases, the dimple grows more and more pronounced.

If $a < 0$, we get the same graph as we do for the corresponding positive a-value, but with a rotation through π about the pole, as happened when c was replaced with $-c$ in Exercise 86.

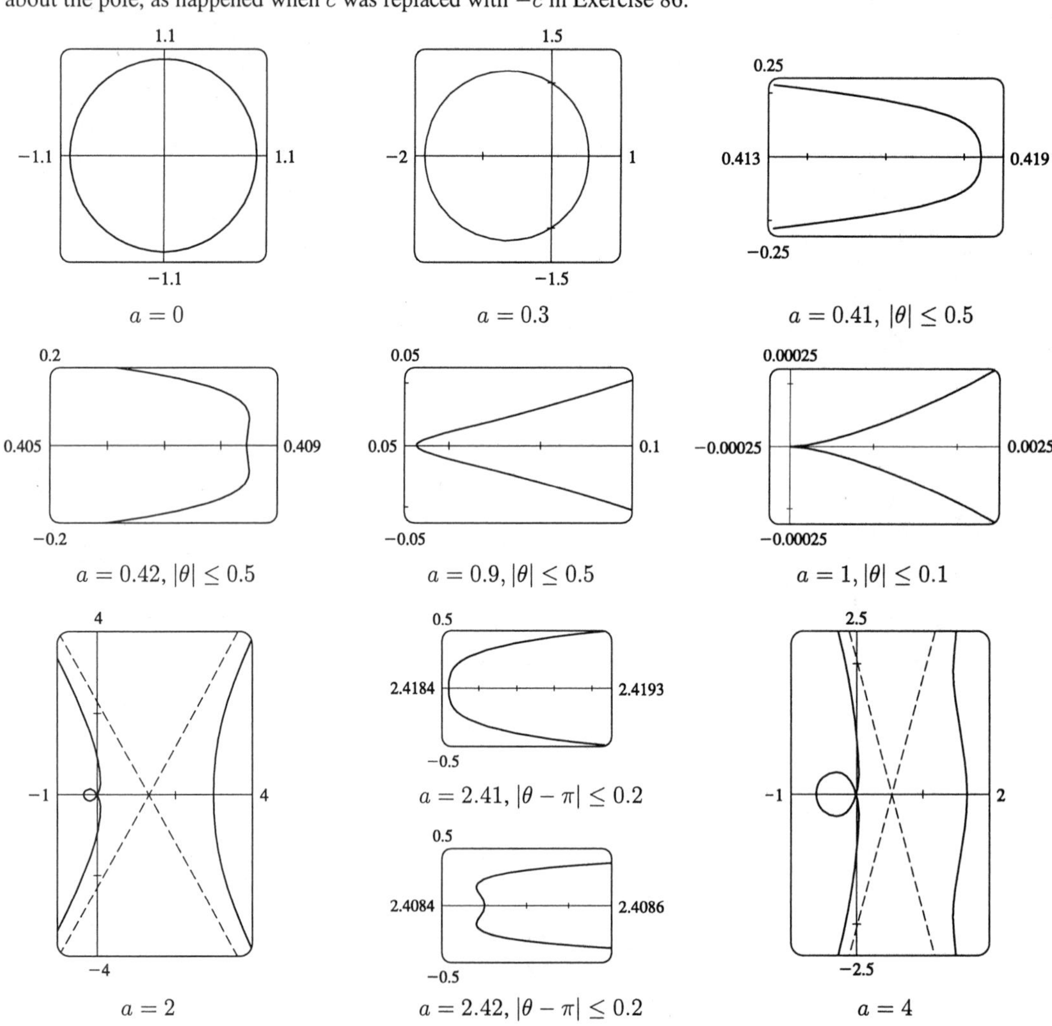

$a = 0$

$a = 0.3$

$a = 0.41,\ |\theta| \le 0.5$

$a = 0.42,\ |\theta| \le 0.5$

$a = 0.9,\ |\theta| \le 0.5$

$a = 1,\ |\theta| \le 0.1$

$a = 2$

$a = 2.41,\ |\theta - \pi| \le 0.2$

$a = 2.42,\ |\theta - \pi| \le 0.2$

$a = 4$

88. Most graphing devices cannot plot implicit polar equations, so we must first find an explicit expression (or expressions) for r in terms of θ, a, and c. We note that the given equation is a quadratic in r^2, so we use the quadratic formula and find that $r^2 = c^2 \cos 2\theta \pm \sqrt{4c^4 \cos^2 2\theta - (4c^4 - a^4)} = c^2 \cos 2\theta \pm 2\sqrt{a^4 - c^4 \sin^2 2\theta}$, so $r = \pm\sqrt{c^2 \cos 2\theta \pm 2\sqrt{a^4 - c^4 \sin^2 2\theta}}$. So for each graph, we must plot four curves to be sure of plotting all the points which satisfy the given equation. Note that all four functions have period π.

We start with the case $a = c = 1$, and the resulting curve resembles the symbol for infinity. If we let a decrease, the curve splits into two symmetric parts, and as a decreases further, the parts become smaller, further apart, and rounder. If instead we let a increase from 1, the two lobes of the curve join together, and as a increases further they continue to merge, until at $a \approx 1.4$, the graph no longer has dimples, and has an oval shape. As $a \to \infty$, the oval becomes larger and rounder, since the c^2 and c^4 terms lose their significance.

Note that the shape of the graph seems to depend only on the *ratio* $\dfrac{c}{a}$, while the *size* of the graph varies as c and a jointly increase.

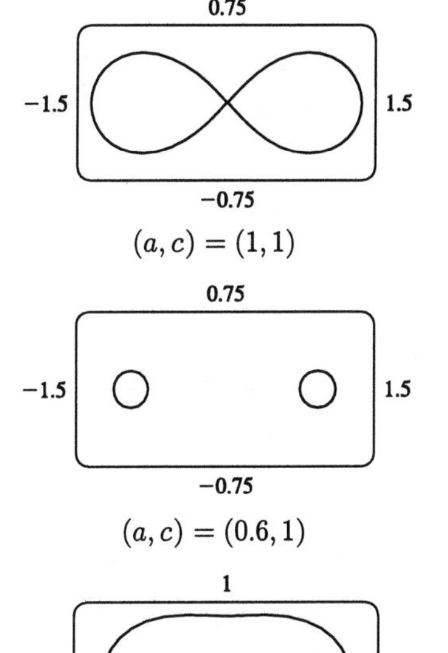

$(a, c) = (1, 1)$

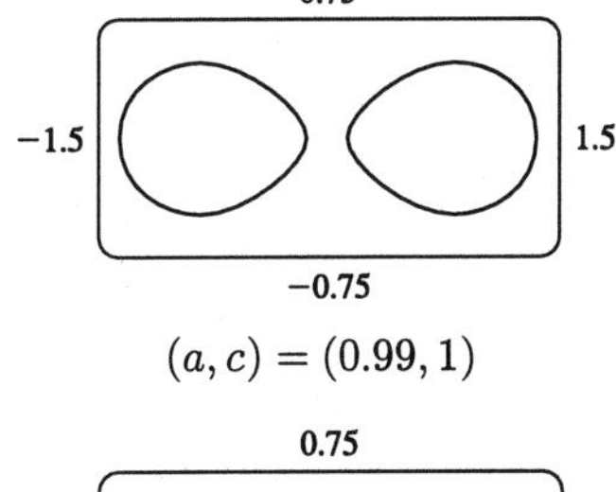

$(a, c) = (0.99, 1)$

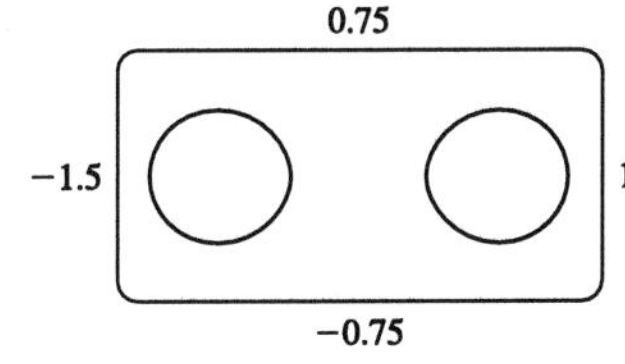

$(a, c) = (0.9, 1)$

$(a, c) = (0.6, 1)$

$(a, c) = (1.01, 1)$

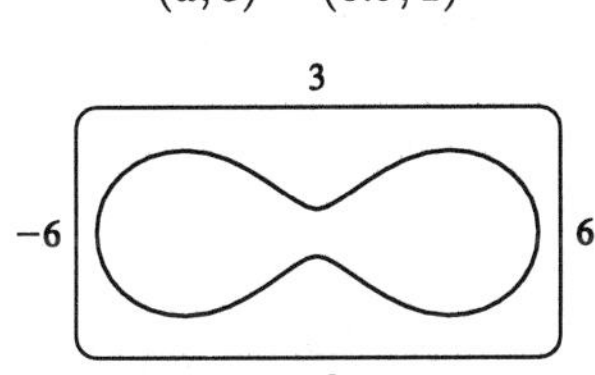

$(a, c) = (4.04, 4)$

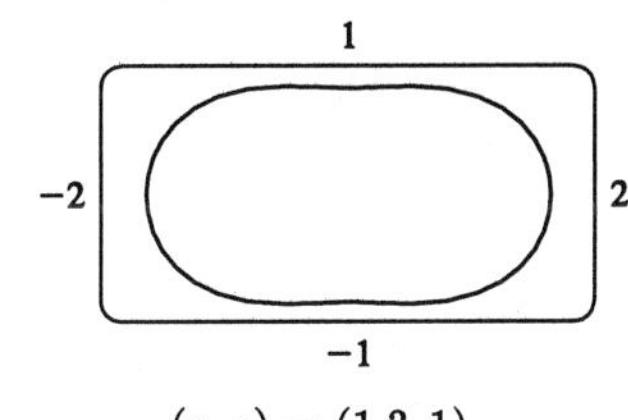

$(a, c) = (1.3, 1)$

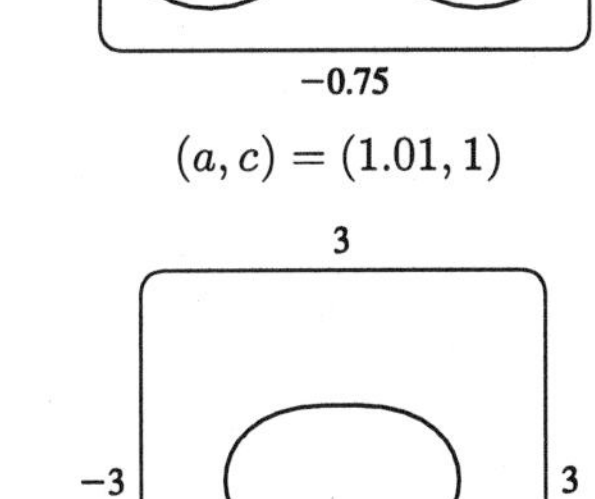

$(a, c) = (1.5, 1)$

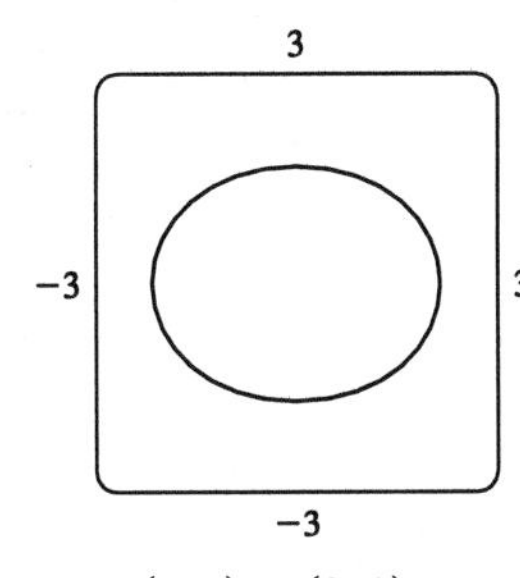

$(a, c) = (2, 1)$

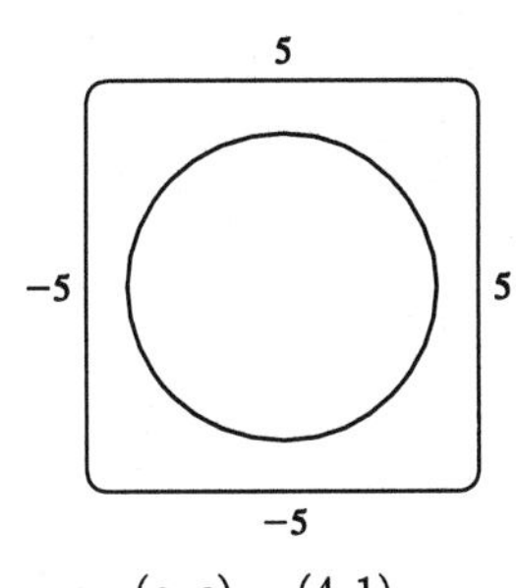

$(a, c) = (4, 1)$

89. $\tan\psi = \tan(\phi - \theta) = \dfrac{\tan\phi - \tan\theta}{1 + \tan\phi\tan\theta} = \dfrac{\dfrac{dy}{dx} - \tan\theta}{1 + \dfrac{dy}{dx}\tan\theta} = \dfrac{\dfrac{dy/d\theta}{dx/d\theta} - \tan\theta}{1 + \dfrac{dy/d\theta}{dx/d\theta}\tan\theta} = \dfrac{\dfrac{dy}{d\theta} - \dfrac{dx}{d\theta}\tan\theta}{\dfrac{dx}{d\theta} + \dfrac{dy}{d\theta}\tan\theta}$

$$= \frac{\left(\dfrac{dr}{d\theta}\sin\theta + r\cos\theta\right) - \tan\theta\left(\dfrac{dr}{d\theta}\cos\theta - r\sin\theta\right)}{\left(\dfrac{dr}{d\theta}\cos\theta - r\sin\theta\right) + \tan\theta\left(\dfrac{dr}{d\theta}\sin\theta + r\cos\theta\right)} = \frac{r\cos^2\theta + r\sin^2\theta}{\dfrac{dr}{d\theta}\cos^2\theta + \dfrac{dr}{d\theta}\sin^2\theta} = \frac{r}{dr/d\theta}$$

90. **(a)** $r = e^\theta \Rightarrow dr/d\theta = e^\theta$, so by Exercise 89, $\tan\psi = r/e^\theta = 1 \Rightarrow \psi = \arctan 1 = \frac{\pi}{4}$.

(b) The Cartesian equation of the tangent line at $(1, 0)$ is $y = x - 1$, and that of the tangent line at $(0, e^{\pi/2})$ is $y = e^{\pi/2} - x$.

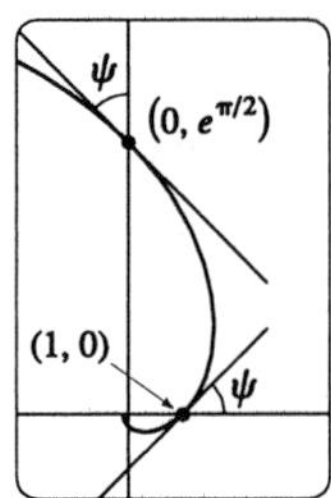

(c) Let a be the tangent of the angle between the tangent and radial lines. Then by Exercise 89, $a = \dfrac{r}{dr/d\theta}$. Now we can either solve this as a separable equation, or we can simply use Theorem 6.5.2, which states that the only solutions are $r = Ce^{k\theta}$, where $k = 1/a$.

EXERCISES 9.5

1. $A = \int_0^\pi \frac{1}{2}r^2\,d\theta = \int_0^\pi \frac{1}{2}\theta^2\,d\theta = \left[\frac{1}{6}\theta^3\right]_0^\pi = \frac{1}{6}\pi^3$

2. $A = \int_{-\pi/2}^{\pi/2} \frac{1}{2}e^{2\theta}\,d\theta = \left[\frac{1}{4}e^{2\theta}\right]_{-\pi/2}^{\pi/2} = \frac{1}{4}(e^\pi - e^{-\pi})$

3. $A = \int_0^{\pi/6} \frac{1}{2}(2\cos\theta)^2\,d\theta = \int_0^{\pi/6}(1 + \cos 2\theta)d\theta = \left[\theta + \frac{1}{2}\sin 2\theta\right]_0^{\pi/6} = \frac{\pi}{6} + \frac{\sqrt{3}}{4}$

4. $A = \int_{\pi/6}^{5\pi/6} \frac{1}{2}(1/\theta)^2\,d\theta = \left[-1/(2\theta)\right]_{\pi/6}^{5\pi/6} = \frac{12}{5\pi}$

5. $A = \int_0^{\pi/6} \frac{1}{2}\sin^2 2\theta\,d\theta = \frac{1}{4}\int_0^{\pi/6}(1 - \cos 4\theta)d\theta = \left[\frac{1}{4}\theta - \frac{1}{16}\sin 4\theta\right]_0^{\pi/6} = \frac{4\pi - 3\sqrt{3}}{96}$

6. $A = 2\int_0^{\pi/12} \frac{1}{2}\cos^2 3\theta\,d\theta = \frac{1}{2}\int_0^{\pi/12}(1 + \cos 6\theta)d\theta = \frac{1}{2}\left[\theta + \frac{1}{6}\sin 6\theta\right]_0^{\pi/12} = \frac{1}{24}(\pi + 2)$

7. $A = \int_0^\pi \frac{1}{2}(5\sin\theta)^2\,d\theta$

$= \frac{25}{4}\int_0^\pi (1 - \cos 2\theta)d\theta$

$= \frac{25}{4}\left[\theta - \frac{1}{2}\sin 2\theta\right]_0^\pi$

$= \frac{25}{4}\pi$

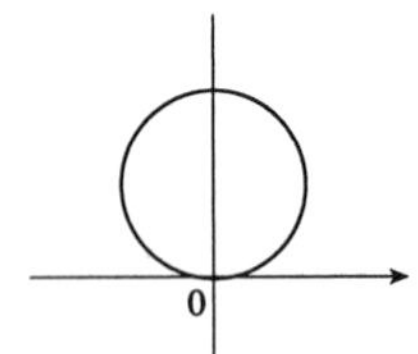

8. $A = 2\int_0^\pi \frac{1}{2}[4(1 - \cos\theta)]^2\,d\theta$

$= 16\int_0^\pi (1 - 2\cos\theta + \cos^2\theta)d\theta$

$= 8\int_0^\pi (3 - 4\cos\theta + \cos 2\theta)d\theta$

$= 4[6\theta - 8\sin\theta + \sin 2\theta]_0^\pi = 24\pi$

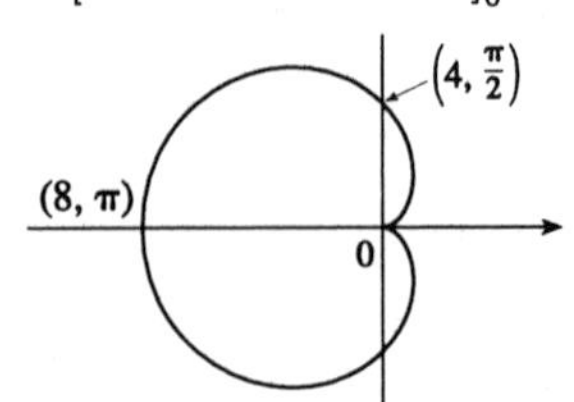

9. $A = 4\int_0^{\pi/4} \frac{1}{2}r^2\,d\theta$
$= 8\int_0^{\pi/4} \cos 2\theta\,d\theta$
$= [4\sin 2\theta]_0^{\pi/4} = 4$

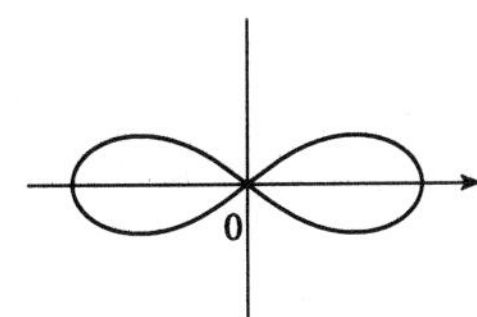

10. 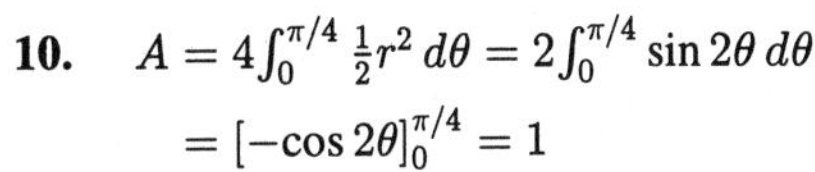
$A = 4\int_0^{\pi/4} \frac{1}{2}r^2\,d\theta = 2\int_0^{\pi/4} \sin 2\theta\,d\theta$
$= [-\cos 2\theta]_0^{\pi/4} = 1$

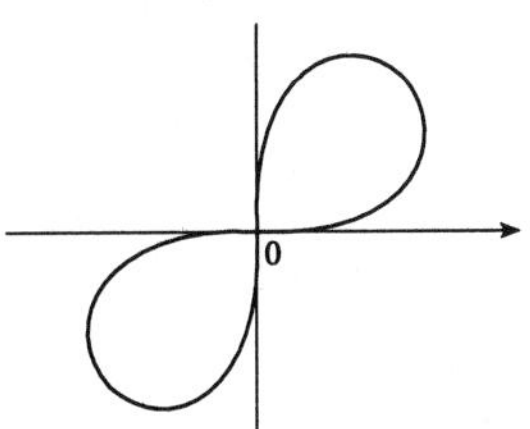

11. $A = 2\int_{-\pi/2}^{\pi/2} \frac{1}{2}(4 - \sin\theta)^2\,d\theta$
$= \int_{-\pi/2}^{\pi/2}(16 - 8\sin\theta + \sin^2\theta)d\theta$
$= 16\pi + 0 + \int_{-\pi/2}^{\pi/2} \sin^2\theta\,d\theta = \frac{33\pi}{2}$

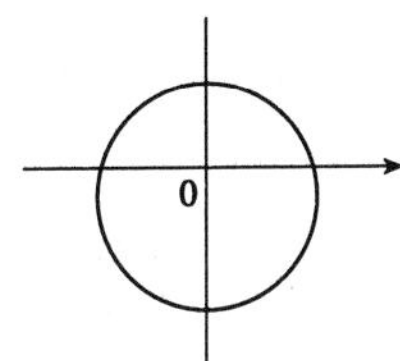

12. $A = 6\int_0^{\pi/6} \frac{1}{2}\sin^2 3\theta\,d\theta = \frac{3}{2}\left[\theta - \frac{1}{6}\sin 6\theta\right]_0^{\pi/6} = \frac{\pi}{4}$

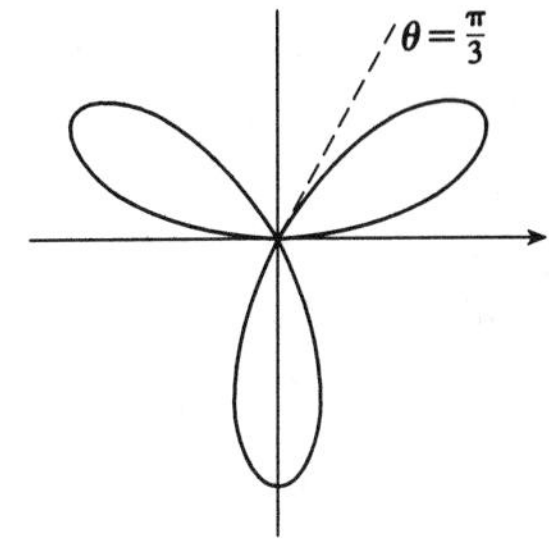

13. By symmetry, the total area is twice the area enclosed above the polar axis, so

$A = 2\int_0^{\pi} \frac{1}{2}r^2\,d\theta = \int_0^{\pi} [2 + \cos 6\theta]^2\,d\theta$
$= \left[4\theta + 4\left(\frac{1}{6}\sin 6\theta\right) + \left(\frac{1}{24}\sin 12\theta + \frac{1}{2}\theta\right)\right]_0^{\pi}$
$= 4\pi + \frac{\pi}{2} = \frac{9\pi}{2}.$

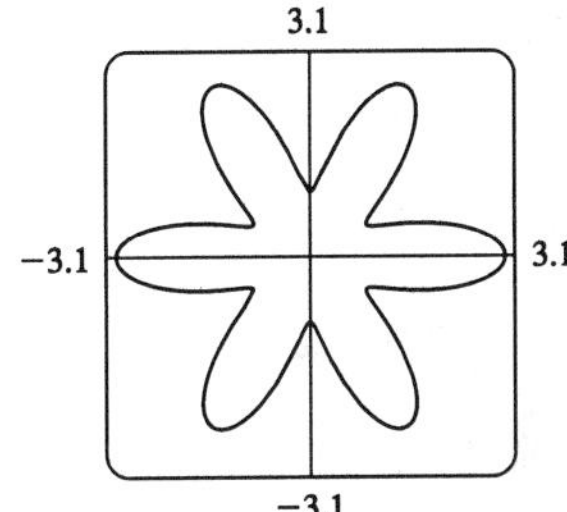

14. Note that the entire curve is generated by $\theta \in [0, \pi]$. The radius is positive on this interval, so the area enclosed is

$A = \int_0^{\pi} \frac{1}{2}r^2\,d\theta = \int_0^{\pi} \frac{1}{2}(2\sin\theta\cos^2\theta)^2 d\theta$
$= 2\int_0^{\pi} \sin^2\theta\cos^4\theta\,d\theta = 2\int_0^{\pi} \left(\frac{1}{2}\sin 2\theta\right)^2 \cos^2\theta\,d\theta$
$= \frac{1}{2}\int_0^{\pi} \sin^2 2\theta\,(\cos 2\theta + 1)d\theta = \frac{1}{2}\left[\int_0^{\pi} \sin^2 2\theta\cos 2\theta\,d\theta + \int_0^{\pi} \sin^2 2\theta\,d\theta\right]$
$= \frac{1}{4}\left[\frac{1}{2}\theta - \frac{1}{4}\sin 4\theta\right]_0^{\pi}$ (the first integral vanishes) $= \frac{\pi}{8}.$

15. $A = 2\int_0^{\pi/6} \frac{1}{2}\cos^2 3\theta\,d\theta = \frac{1}{2}\int_0^{\pi/6}(1 + \cos 6\theta)d\theta = \frac{1}{2}\left[\theta + \frac{1}{6}\sin 6\theta\right]_0^{\pi/6} = \frac{\pi}{12}$

16. $A = 2\int_0^{\pi/4} \frac{1}{2}(3\sin 2\theta)^2\,d\theta = \frac{9}{2}\int_0^{\pi/4}(1 - \cos 4\theta)d\theta = \frac{9}{2}\left[\theta - \frac{1}{4}\sin 4\theta\right]_0^{\pi/4} = \frac{9\pi}{8}$

17. $A = \int_0^{\pi/5} \frac{1}{2}\sin^2 5\theta\,d\theta = \frac{1}{4}\int_0^{\pi/5}(1 - \cos 10\theta)d\theta = \frac{1}{4}\left[\theta - \frac{1}{10}\sin 10\theta\right]_0^{\pi/5} = \frac{\pi}{20}$

18. $A = 2\int_0^{\pi/8} \frac{1}{2}(2\cos 4\theta)^2\,d\theta = 2\int_0^{\pi/8}(1 + \cos 8\theta)d\theta = 2\left[\theta + \frac{1}{8}\sin 8\theta\right]_0^{\pi/8} = \frac{\pi}{4}$

19. This is a limaçon, with inner loop traced out between $\theta = \frac{7\pi}{6}$ and $\frac{11\pi}{6}$.

$$A = 2\int_{7\pi/6}^{3\pi/2} \tfrac{1}{2}(1 + 2\sin\theta)^2\, d\theta = \int_{7\pi/6}^{3\pi/2}(1 + 4\sin\theta + 4\sin^2\theta)d\theta$$
$$= \left[\theta - 4\cos\theta + 2\theta - \sin 2\theta\right]_{7\pi/6}^{3\pi/2} = \pi - \tfrac{3\sqrt{3}}{2}$$

20. $2 + 3\cos\theta = 0 \quad\Rightarrow\quad \cos\theta = -\frac{2}{3} \quad\Rightarrow$

$\theta = \cos^{-1}\left(-\frac{2}{3}\right)\ (=\alpha)$ or $2\pi - \cos^{-1}\left(-\frac{2}{3}\right) \quad\Rightarrow$

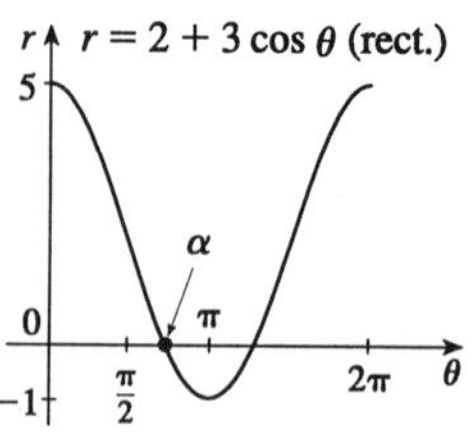

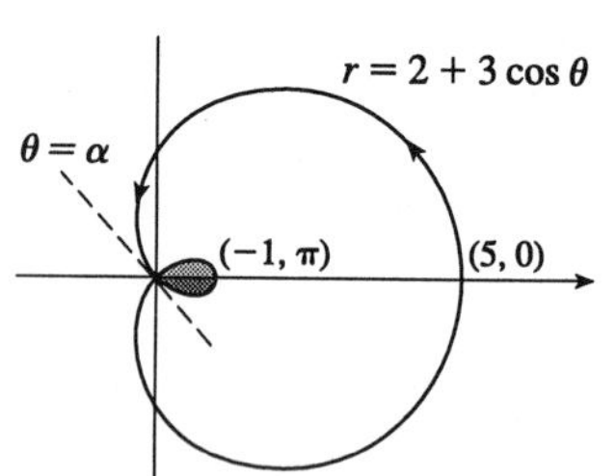

$$A = 2\int_{\alpha}^{\pi} \tfrac{1}{2}(2 + 3\cos\theta)^2\, d\theta$$
$$= \int_{\alpha}^{\pi}(4 + 12\cos\theta + 9\cos^2\theta)d\theta$$
$$= \int_{\alpha}^{\pi}\left(\tfrac{17}{2} + 12\cos\theta + \tfrac{9}{2}\cos 2\theta\right)d\theta$$
$$= \left[\tfrac{17}{2}\theta + 12\sin\theta + \tfrac{9}{4}\sin 2\theta\right]_{\alpha}^{\pi}$$
$$= \tfrac{17}{2}(\pi - \alpha) - 12\sin\alpha - \tfrac{9}{2}\sin\alpha\cos\alpha$$
$$= \tfrac{17}{2}\left[\pi - \cos^{-1}\left(-\tfrac{2}{3}\right)\right] - 12\left(\tfrac{\sqrt{5}}{3}\right) - \tfrac{9}{2}\left(\tfrac{\sqrt{5}}{3}\right)\left(-\tfrac{2}{3}\right) = \tfrac{17}{2}\cos^{-1}\tfrac{2}{3} - 3\sqrt{5}$$

21. $1 - \cos\theta = \frac{3}{2} \quad\Leftrightarrow\quad \cos\theta = -\frac{1}{2} \quad\Rightarrow\quad \theta = \frac{2\pi}{3}$ or $\frac{4\pi}{3} \quad\Rightarrow$

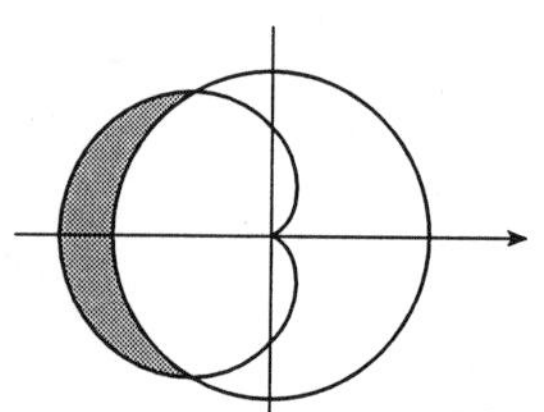

$$A = \int_{2\pi/3}^{4\pi/3} \tfrac{1}{2}\left[(1 - \cos\theta)^2 - \left(\tfrac{3}{2}\right)^2\right]d\theta$$
$$= \tfrac{1}{2}\int_{2\pi/3}^{4\pi/3}\left(-\tfrac{5}{4} - 2\cos\theta + \cos^2\theta\right)d\theta$$
$$= \tfrac{1}{2}\left[-\tfrac{5}{4}\theta - 2\sin\theta\right]_{2\pi/3}^{4\pi/3} + \tfrac{1}{2}\int_{2\pi/3}^{4\pi/3}\frac{1 + \cos 2\theta}{2}\,d\theta$$
$$= -\tfrac{5}{12}\pi + \sqrt{3} + \tfrac{1}{4}\left[\theta + \tfrac{1}{2}\sin 2\theta\right]_{2\pi/3}^{4\pi/3} = \tfrac{9\sqrt{3}}{8} - \tfrac{1}{4}\pi$$

22. 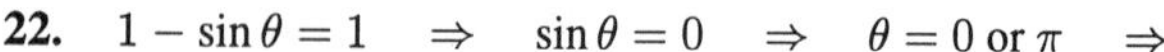
$1 - \sin\theta = 1 \quad\Rightarrow\quad \sin\theta = 0 \quad\Rightarrow\quad \theta = 0$ or $\pi \quad\Rightarrow$

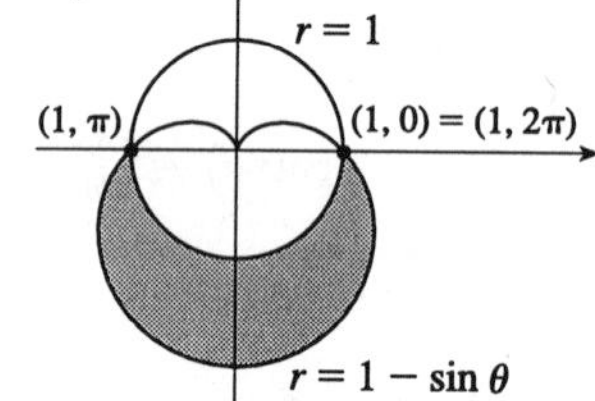

$$A = \int_{\pi}^{2\pi} \tfrac{1}{2}\left[(1 - \sin\theta)^2 - 1\right]d\theta = \tfrac{1}{2}\int_{\pi}^{2\pi}(\sin^2\theta - 2\sin\theta)d\theta$$
$$= \tfrac{1}{4}\int_{\pi}^{2\pi}(1 - \cos 2\theta - 4\sin\theta)d\theta = \tfrac{1}{4}\left[\theta - \tfrac{1}{2}\sin 2\theta + 4\cos\theta\right]_{\pi}^{2\pi}$$
$$= \tfrac{1}{4}\pi + 2$$

23. 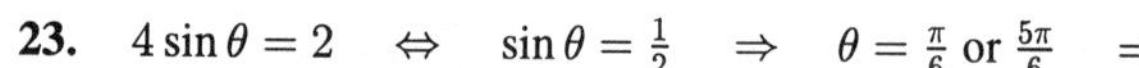
$4\sin\theta = 2 \quad\Leftrightarrow\quad \sin\theta = \frac{1}{2} \quad\Rightarrow\quad \theta = \frac{\pi}{6}$ or $\frac{5\pi}{6} \quad\Rightarrow$

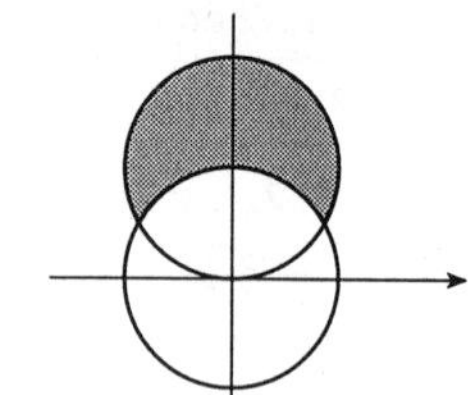

$$A = 2\int_{\pi/6}^{\pi/2} \tfrac{1}{2}\left[(4\sin\theta)^2 - 2^2\right]d\theta$$
$$= \int_{\pi/6}^{\pi/2}(16\sin^2\theta - 4)d\theta = \int_{\pi/6}^{\pi/2}[8(1 - \cos 2\theta) - 4]d\theta$$
$$= \left[4\theta - 4\sin 2\theta\right]_{\pi/6}^{\pi/2} = \tfrac{4}{3}\pi + 2\sqrt{3}$$

24. 
$3\cos\theta = 2 - \cos\theta \quad\Rightarrow\quad \cos\theta = \frac{1}{2} \quad\Rightarrow\quad \theta = \pm\frac{\pi}{3} \quad\Rightarrow$

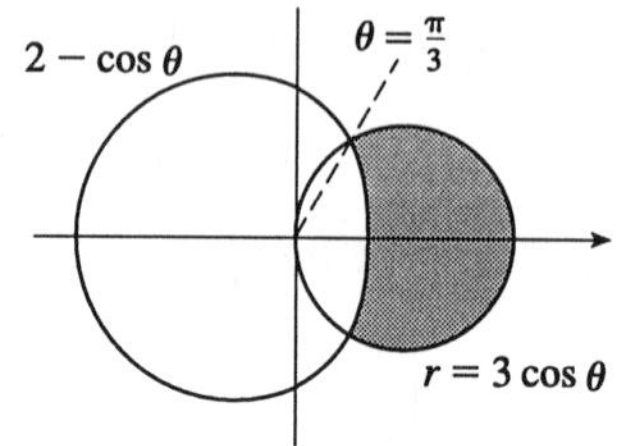

$$A = 2\int_{0}^{\pi/3} \tfrac{1}{2}\left[(3\cos\theta)^2 - (2 - \cos\theta)^2\right]d\theta$$
$$= \int_{0}^{\pi/3}(8\cos^2\theta + 4\cos\theta - 4)d\theta = \int_{0}^{\pi/3}(4\cos 2\theta + 4\cos\theta)d\theta$$
$$= \left[2\sin 2\theta + 4\sin\theta\right]_{0}^{\pi/3} = 3\sqrt{3}$$

25. $3\cos\theta = 1+\cos\theta \quad\Leftrightarrow\quad \cos\theta = \frac{1}{2} \quad\Rightarrow\quad \theta = \frac{\pi}{3} \text{ or } -\frac{\pi}{3}.$

$$\begin{aligned} A &= 2\int_0^{\pi/3} \tfrac{1}{2}\left[(3\cos\theta)^2 - (1+\cos\theta)^2\right]d\theta \\ &= \int_0^{\pi/3}(8\cos^2\theta - 2\cos\theta - 1)d\theta \\ &= \int_0^{\pi/3}[4(1+\cos 2\theta) - 2\cos\theta - 1]d\theta \\ &= [3\theta + 2\sin 2\theta - 2\sin\theta]_0^{\pi/3} = \pi + \sqrt{3} - \sqrt{3} = \pi \end{aligned}$$

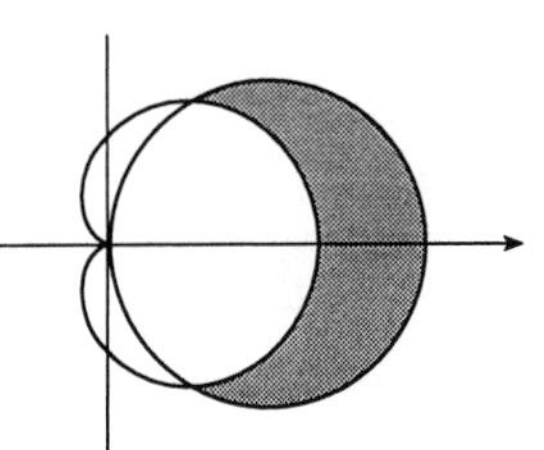

26. $$\begin{aligned} A &= 2\int_{\pi/3}^{\pi/2} \tfrac{1}{2}\left[(1+\cos\theta)^2 - (3\cos\theta)^2\right]d\theta + 2\int_{\pi/2}^{\pi} \tfrac{1}{2}(1+\cos\theta)^2\,d\theta \\ &= \int_{\pi/3}^{\pi}(1+\cos\theta)^2\,d\theta - \int_{\pi/3}^{\pi/2} 9\cos^2\theta\,d\theta \\ &= \left[\theta + 2\sin\theta + \tfrac{1}{2}\left(\theta + \tfrac{1}{2}\sin 2\theta\right)\right]_{\pi/3}^{\pi} - \tfrac{9}{2}\left[\theta + \tfrac{1}{2}\sin 2\theta\right]_{\pi/3}^{\pi/2} = \tfrac{\pi}{4} \end{aligned}$$

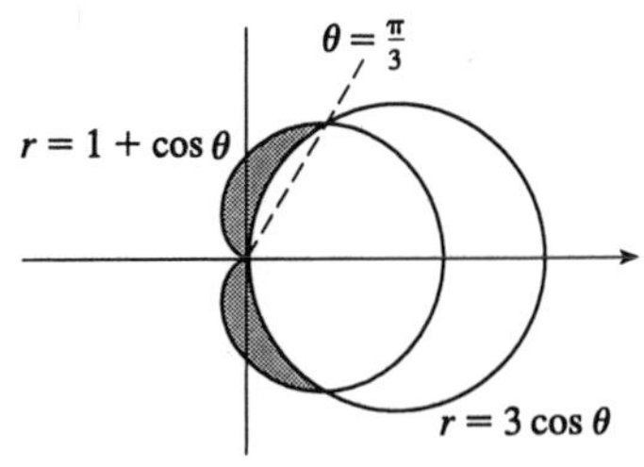

27. $$\begin{aligned} A &= 2\int_0^{\pi/4} \tfrac{1}{2}\sin^2\theta\,d\theta = \int_0^{\pi/4} \frac{1-\cos 2\theta}{2}\,d\theta \\ &= \left[\tfrac{1}{2}\theta - \tfrac{1}{4}\sin 2\theta\right]_0^{\pi/4} = \tfrac{1}{8}\pi - \tfrac{1}{4} \end{aligned}$$

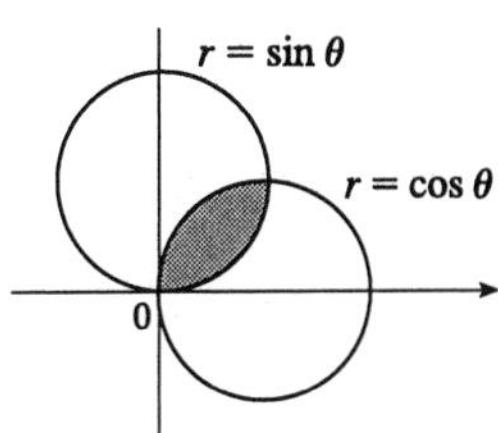

28. $\sin\theta = \pm\sin 2\theta = \pm 2\sin\theta\cos\theta \quad\Rightarrow\quad \sin\theta(1 \pm 2\cos\theta) = 0$

From the figure we can see that the intersections occur where $\cos\theta = \pm\frac{1}{2}$, or $\theta = \frac{\pi}{3}$ and $\frac{2\pi}{3}$.

$$\begin{aligned} A &= 2\left[\int_0^{\pi/3} \tfrac{1}{2}\sin^2\theta\,d\theta + \int_{\pi/3}^{\pi/2} \tfrac{1}{2}\sin^2 2\theta\,d\theta\right] \\ &= \tfrac{1}{2}\left[\theta - \tfrac{1}{2}\sin 2\theta\right]_0^{\pi/3} + \tfrac{1}{2}\left[\theta - \tfrac{1}{4}\sin 4\theta\right]_{\pi/3}^{\pi/2} = \frac{4\pi - 3\sqrt{3}}{16} \end{aligned}$$

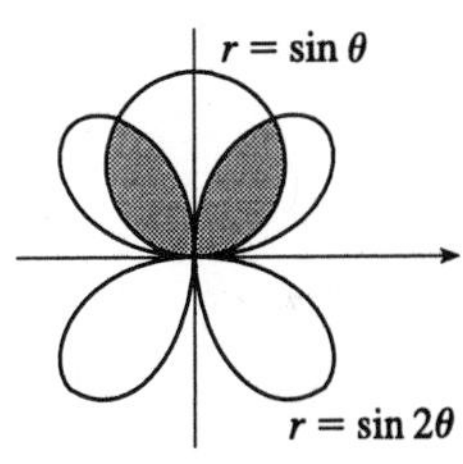

29. $\sin 2\theta = \cos 2\theta \quad\Rightarrow\quad \tan 2\theta = 1 \quad\Rightarrow\quad 2\theta = \frac{\pi}{4} \quad\Rightarrow\quad \theta = \frac{\pi}{8}$

$$\begin{aligned} \Rightarrow\quad A &= 16\int_0^{\pi/8} \tfrac{1}{2}\sin^2 2\theta\,d\theta \\ &= 4\int_0^{\pi/8}(1-\cos 4\theta)d\theta = 4\left[\theta - \tfrac{1}{4}\sin 4\theta\right]_0^{\pi/8} \\ &= \tfrac{1}{2}\pi - 1 \end{aligned}$$

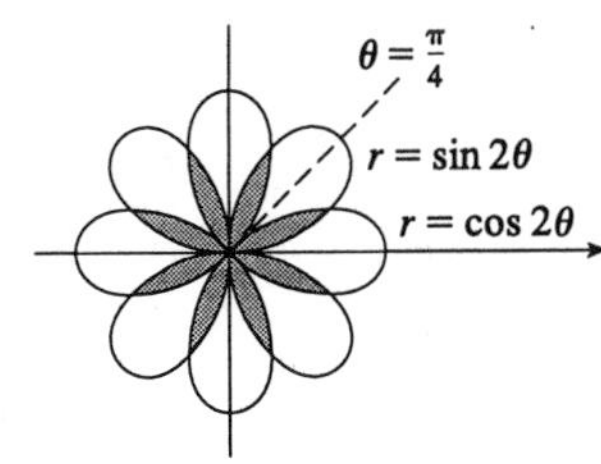

30. $2\sin 2\theta = 1^2 \quad\Rightarrow\quad \sin 2\theta = \frac{1}{2} \quad\Rightarrow$

$2\theta = \frac{\pi}{6} \text{ or } \frac{5\pi}{6} \quad\Rightarrow\quad \theta = \frac{\pi}{12} \text{ or } \frac{5\pi}{12}$

$$\begin{aligned} A &= 4\left[\int_0^{\pi/12} \sin 2\theta\,d\theta + \int_{\pi/12}^{\pi/4} \tfrac{1}{2}\,d\theta\right] \\ &= [-2\cos 2\theta]_0^{\pi/12} + 2\left(\tfrac{1}{4}\pi - \tfrac{1}{12}\pi\right) = 2 - \sqrt{3} + \tfrac{\pi}{3} \end{aligned}$$

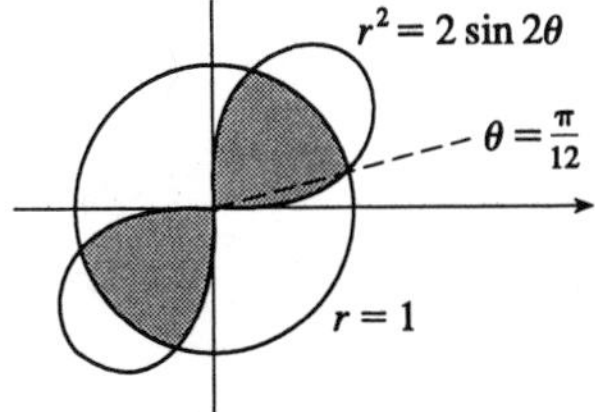

31. $A = 2\left[\int_{-\pi/2}^{-\pi/6} \frac{1}{2}(3+2\sin\theta)^2\,d\theta + \int_{-\pi/6}^{\pi/2} \frac{1}{2}2^2\,d\theta\right]$

$= \int_{-\pi/2}^{-\pi/6}(9+12\sin\theta+4\sin^2\theta)d\theta + [4\theta]_{-\pi/6}^{\pi/2}$

$= [9\theta - 12\cos\theta + 2\theta - \sin 2\theta]_{-\pi/2}^{-\pi/6} + \frac{8\pi}{3}$

$= \frac{19\pi}{3} - \frac{11\sqrt{3}}{2}$

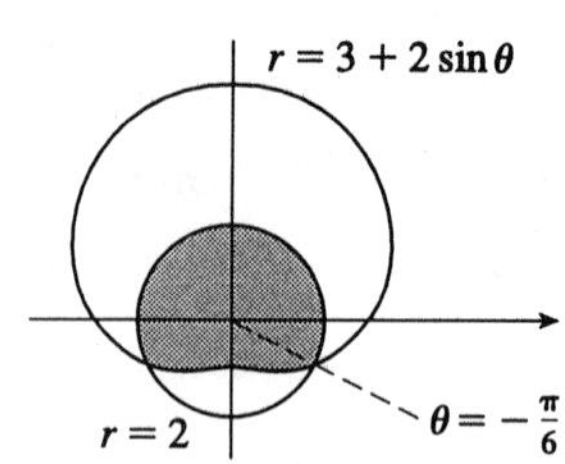

32. Let $\alpha = \tan^{-1}(b/a)$. Then

$A = \int_0^\alpha \frac{1}{2}(a\sin\theta)^2\,d\theta + \int_\alpha^{\pi/2}\frac{1}{2}(b\cos\theta)^2\,d\theta$

$= \frac{1}{4}a^2\left[\theta - \frac{1}{2}\sin 2\theta\right]_0^\alpha + \frac{1}{4}b^2\left[\theta + \frac{1}{2}\sin 2\theta\right]_\alpha^{\pi/2}$

$= \frac{1}{4}\alpha(a^2-b^2) + \frac{1}{8}\pi b^2 - \frac{1}{4}(a^2+b^2)(\sin\alpha\cos\alpha)$

$= \frac{1}{4}(a^2-b^2)\tan^{-1}(b/a) + \frac{1}{8}\pi b^2 - \frac{1}{4}ab$

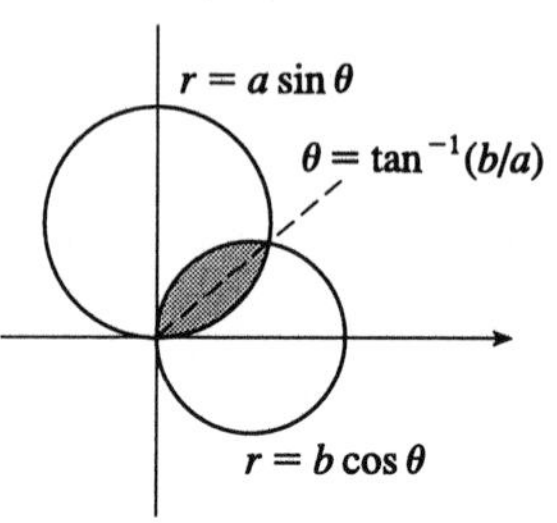

33. $A = 2\left[\int_0^{2\pi/3}\frac{1}{2}\left(\frac{1}{2}+\cos\theta\right)^2 d\theta - \int_{2\pi/3}^{\pi}\frac{1}{2}\left(\frac{1}{2}+\cos\theta\right)^2 d\theta\right]$

$= \left[\frac{\theta}{4} + \sin\theta + \frac{\theta}{2} + \frac{\sin 2\theta}{4}\right]_0^{2\pi/3} - \left[\frac{\theta}{4} + \sin\theta + \frac{\theta}{2} + \frac{\sin 2\theta}{4}\right]_{2\pi/3}^{\pi}$

$= \frac{1}{4}\left(\pi + 3\sqrt{3}\right)$

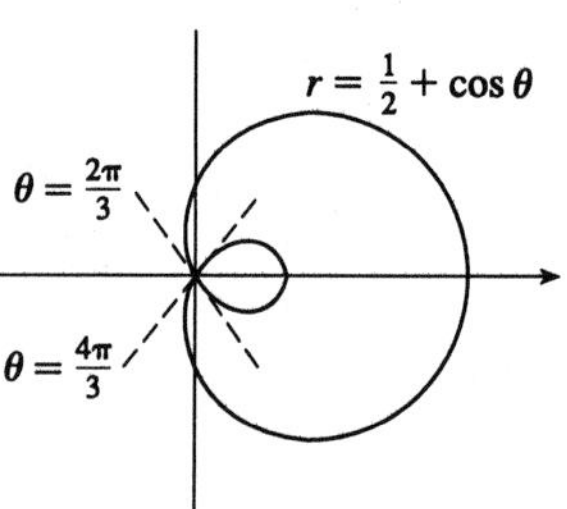

34. The points of intersection occur where $\sqrt{1-0.8\sin^2\theta} = \sin\theta \quad\Leftrightarrow$ $1.8\sin^2\theta = 1 \quad\Leftrightarrow\quad \theta = \arcsin\sqrt{\frac{5}{6}}\ (=\alpha)$. So the area is

$A = 2\int_0^\alpha \frac{1}{2}\sin^2\theta\,d\theta + 2\int_\alpha^{\pi/2}\frac{1}{2}\left(\sqrt{1-0.8\sin^2\theta}\right)^2 d\theta$

$= \left[\frac{1}{2}\theta - \frac{1}{4}\sin 2\theta\right]_0^\alpha + \left[\theta - 0.8\left(\frac{1}{2}\theta - \frac{1}{4}\sin 2\theta\right)\right]_\alpha^{\pi/2}$

$= \frac{1}{2}\alpha - \frac{1}{4}(2\sin\alpha\cos\alpha) + 0.6\cdot\frac{\pi}{2} - [0.6\alpha + 0.2(2\sin\alpha\cos\alpha)]$

$= \frac{1}{2}\arcsin\frac{\sqrt{5}}{3} - \frac{1}{2}\frac{\sqrt{5}}{3}\sqrt{1-\frac{5}{9}} + 0.3\pi - 0.6\arcsin\frac{\sqrt{5}}{3} - 0.4\cdot\frac{\sqrt{5}}{3}\sqrt{1-\frac{5}{9}} \approx 0.411.$

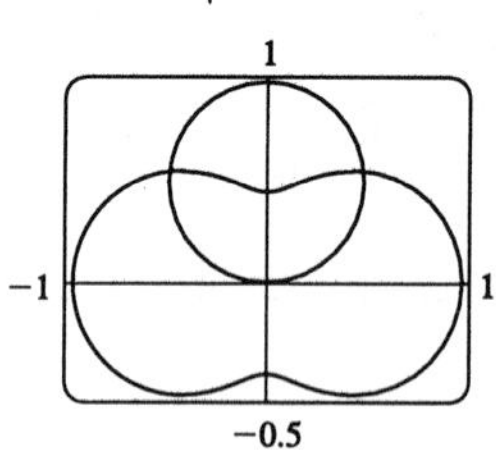

35. The two circles intersect at the pole since $(0,0)$ satisfies the first equation and $\left(0,\frac{\pi}{2}\right)$ the second. The other intersection point $\left(\frac{1}{\sqrt{2}},\frac{\pi}{4}\right)$ occurs where $\sin\theta = \cos\theta$.

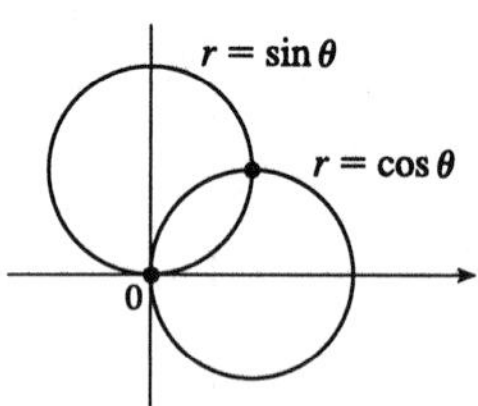

36. $2\cos 2\theta = \pm 2 \quad\Rightarrow\quad \cos 2\theta = \pm 1 \quad\Rightarrow$ $\theta = 0, \frac{\pi}{2}, \pi$, or $\frac{3\pi}{2}$, so the points are $(2,0)$, $\left(2,\frac{\pi}{2}\right)$, $(2,\pi)$, $\left(2,\frac{3\pi}{2}\right)$.

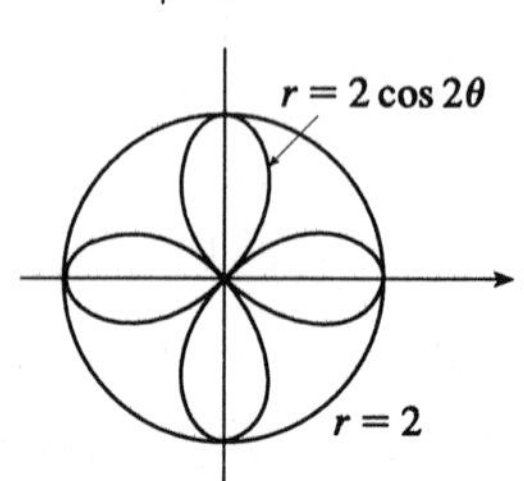

37. The curves intersect at the pole since $\left(0, \frac{\pi}{2}\right)$ satisfies the first equation and $(0, 0)$ the second.
$\cos\theta = 1 - \cos\theta \quad \Rightarrow \quad \cos\theta = \frac{1}{2} \quad \Rightarrow$
$\theta = \frac{\pi}{3}$ or $\frac{5\pi}{3} \quad \Rightarrow$ the other intersection points are $\left(\frac{1}{2}, \frac{\pi}{3}\right)$ and $\left(\frac{1}{2}, \frac{5\pi}{3}\right)$.

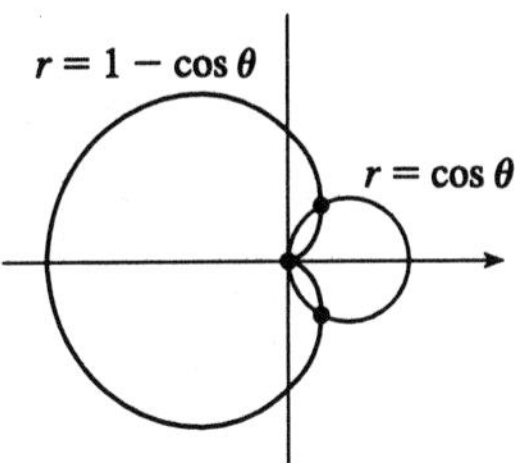

38. Clearly the pole lies on both curves. $\sin 3\theta = \cos 3\theta$
$\Rightarrow \quad \tan 3\theta = 1 \quad \Rightarrow \quad 3\theta = \frac{1}{4}\pi + n\pi \quad$ (n any integer)
$\Rightarrow \quad \theta = \frac{\pi}{12}, \frac{5\pi}{12}, \frac{3\pi}{4}$, so the three remaining intersection points are $\left(\frac{1}{\sqrt{2}}, \frac{\pi}{12}\right), \left(-\frac{1}{\sqrt{2}}, \frac{5\pi}{12}\right), \left(\frac{1}{\sqrt{2}}, \frac{3\pi}{4}\right)$
[The second of these is the same as $\left(\frac{1}{\sqrt{2}}, \frac{17\pi}{12}\right)$].

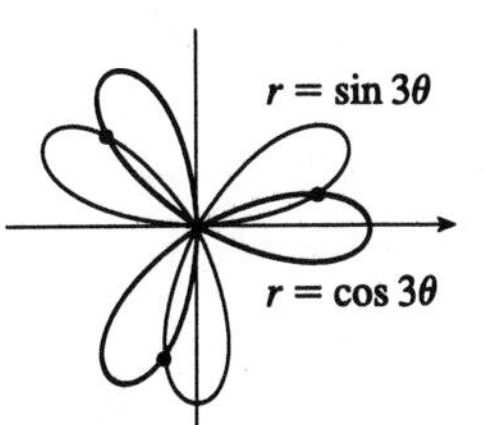

39. The pole is a point of intersection.
$\sin\theta = \sin 2\theta = 2\sin\theta\cos\theta \quad \Leftrightarrow$
$\sin\theta(1 - 2\cos\theta) = 0 \quad \Leftrightarrow \quad \sin\theta = 0$ or $\cos\theta = \frac{1}{2}$
$\Rightarrow \quad \theta = 0, \pi, \frac{\pi}{3}, -\frac{\pi}{3} \quad \Rightarrow \quad \left(\frac{\sqrt{3}}{2}, \frac{\pi}{3}\right)$ and $\left(\frac{\sqrt{3}}{2}, \frac{2\pi}{3}\right)$ are the other intersection points.

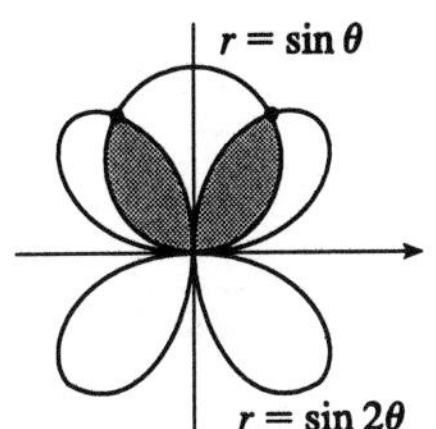

40. Clearly the pole is a point of intersection. $\sin 2\theta = \cos 2\theta$
$\Rightarrow \quad \tan 2\theta = 1 \quad \Rightarrow \quad 2\theta = \frac{\pi}{4} + n\pi$
$\Rightarrow \quad \theta = \frac{\pi}{8}$ or $\frac{9\pi}{8}$ (since $\sin 2\theta$ and $\cos 2\theta$ must be positive in the equations.) So the curves also intersect at $\left(\frac{1}{\sqrt[4]{2}}, \frac{\pi}{8}\right)$ and $\left(\frac{1}{\sqrt[4]{2}}, \frac{9\pi}{8}\right)$.

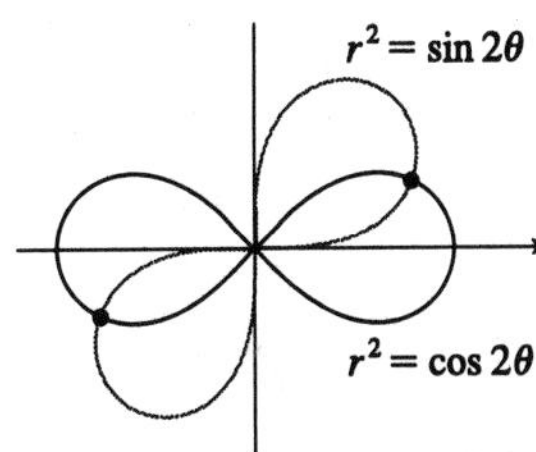

41.

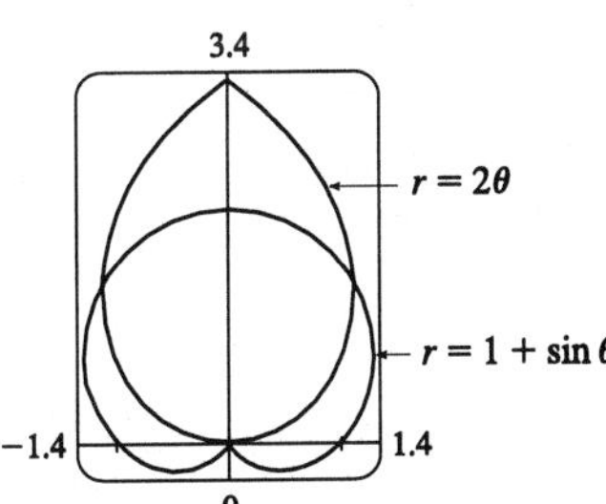

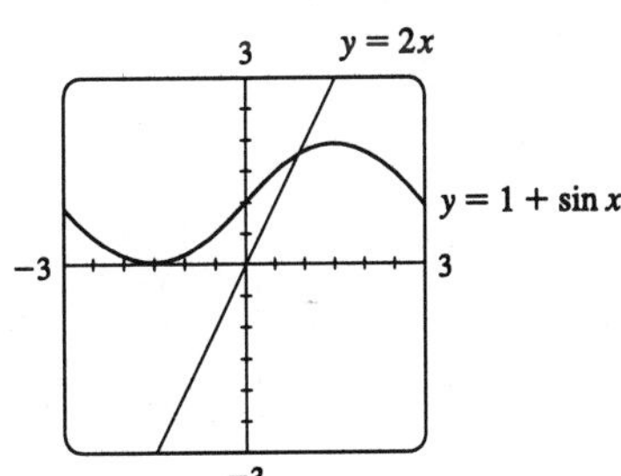

From the first graph, we see that the pole is one point of intersection. By zooming in or using the cursor, we estimate the θ-values of the intersection points to be about 0.89 and 2.25. (The first of these values may be more easily estimated by plotting $y = 1 + \sin x$ and $y = 2x$ in rectangular coordinates; see the second graph.)
Note that the other point of intersection happens when $-2\theta = 1 + \sin(\theta + \pi)$ for $\theta \approx -0.89$. By symmetry, the total area contained is twice the area contained in the first quadrant, that is,
$A \approx \int_0^{0.89} (2\theta)^2 d\theta + \int_{0.89}^{\pi/2} (1 + \sin\theta)^2 d\theta = \left[\frac{4}{3}\theta^3\right]_0^{0.89} + \left[\theta - 2\cos\theta + \left(\frac{1}{2}\theta - \frac{1}{4}\sin 2\theta\right)\right]_{0.89}^{\pi/2} \approx 3.46.$

42.

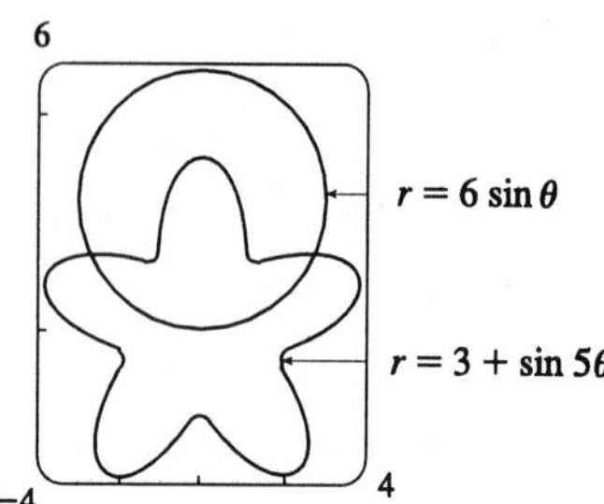

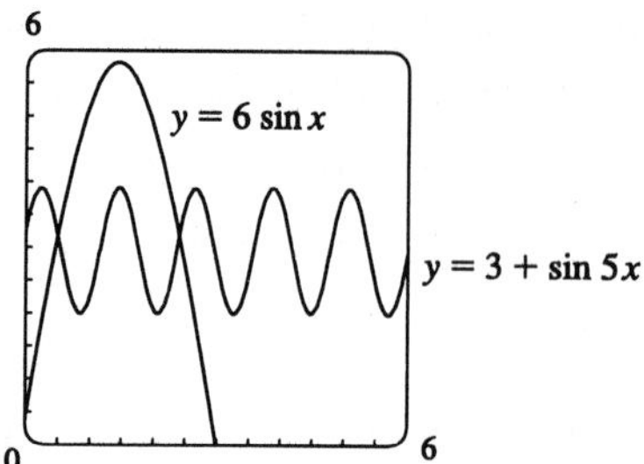

From the first graph, it appears that the θ-values of the points of intersection are about 0.58 and 2.57. (These values may be more easily estimated by plotting $y = 3 + \sin 5x$ and $y = 6\sin x$ in rectangular coordinates; see the second graph.) By symmetry, the total area enclosed in both curves is

$$A \approx \int_0^{0.58} (6\sin\theta)^2 d\theta + \int_{0.58}^{\pi/2} (3 + \sin 5\theta)^2 d\theta$$

$$= \left[36\left(\tfrac{1}{2}\theta - \tfrac{1}{4}\sin 2\theta\right)\right]_0^{0.58} + \left[9\theta - \tfrac{6}{5}\cos 5\theta + \tfrac{1}{5}\left(\tfrac{5}{2}\theta - \tfrac{1}{4}\sin 10\theta\right)\right]_{0.58}^{\pi/2} \approx 10.41$$

43. $L = \displaystyle\int_0^{3\pi/4} \sqrt{r^2 + (dr/d\theta)^2}\, d\theta = \int_0^{3\pi/4} \sqrt{(5\cos\theta)^2 + (-5\sin\theta)^2}\, d\theta$

$= 5\displaystyle\int_0^{3\pi/4} \sqrt{\cos^2\theta + \sin^2\theta}\, d\theta = 5\int_0^{3\pi/4} d\theta = \tfrac{15}{4}\pi$

44. $L = \displaystyle\int_0^{3\pi} \sqrt{(e^{-\theta})^2 + (-e^{-\theta})^2}\, d\theta = \sqrt{2}\int_0^{3\pi} e^{-\theta}\, d\theta = \sqrt{2}(1 - e^{-3\pi})$

45. $L = \displaystyle\int_0^{2\pi} \sqrt{(2^\theta)^2 + [(\ln 2)2^\theta]^2}\, d\theta = \int_0^{2\pi} 2^\theta\sqrt{1 + \ln^2 2}\, d\theta = \left[\sqrt{1 + \ln^2 2}\left(\frac{2^\theta}{\ln 2}\right)\right]_0^{2\pi} = \frac{\sqrt{1 + \ln^2 2}\,(2^{2\pi} - 1)}{\ln 2}$

46. $L = \displaystyle\int_0^{2\pi} \sqrt{\theta^2 + 1}\, d\theta = \left[\frac{\theta}{2}\sqrt{\theta^2 + 1} + \tfrac{1}{2}\ln\left(\theta + \sqrt{\theta^2 + 1}\right)\right]_0^{2\pi}$ (Formula 21)

$= \pi\sqrt{4\pi^2 + 1} + \tfrac{1}{2}\ln\left(2\pi + \sqrt{4\pi^2 + 1}\right)$

47. $L = \displaystyle\int_0^{2\pi} \sqrt{(\theta^2)^2 + (2\theta)^2}\, d\theta = \int_0^{2\pi} \theta\sqrt{\theta^2 + 4}\, d\theta = \tfrac{1}{2}\cdot\tfrac{2}{3}\left[(\theta^2 + 4)^{3/2}\right]_0^{2\pi} = \tfrac{8}{3}\left[(\pi^2 + 1)^{3/2} - 1\right]$

48. $L = 2\displaystyle\int_0^{\pi} \sqrt{(1 + \cos\theta)^2 + (-\sin\theta)^2}\, d\theta = 2\sqrt{2}\int_0^{\pi} \sqrt{1 + \cos\theta}\, d\theta$

$= 2\sqrt{2}\int_0^{\pi} \sqrt{2\cos^2(\theta/2)}\, d\theta = [8\sin(\theta/2)]_0^{\pi} = 8$

49. $L = 2\int_0^{2\pi} \sqrt{\cos^8(\theta/4) + \cos^6(\theta/4)\sin^2(\theta/4)}\, d\theta$

$= 2\int_0^{2\pi} |\cos^3(\theta/4)|\sqrt{\cos^2(\theta/4) + \sin^2(\theta/4)}\, d\theta$

$= 2\int_0^{2\pi} |\cos^3(\theta/4)| d\theta = 8\int_0^{\pi/2} \cos^3 u\, du$ (where $u = \tfrac{1}{4}\theta$)

$= 8\left[\sin u - \tfrac{1}{3}\sin^3 u\right]_0^{\pi/2} = \tfrac{16}{3}$

Note that the curve is retraced after every interval of length 4π.

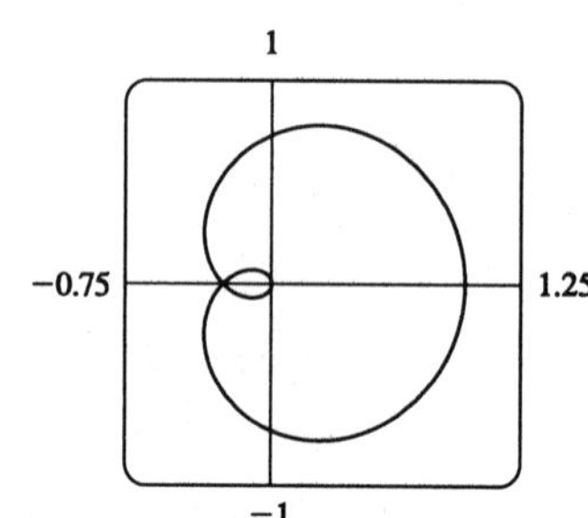

50. $L = 2\int_0^{\pi} \sqrt{\left[\cos^2\left(\frac{1}{2}\theta\right)\right]^2 + \left[-\cos\left(\frac{1}{2}\theta\right)\sin\left(\frac{1}{2}\theta\right)\right]^2}\, d\theta$

$= 2\int_0^{\pi} \cos\left(\frac{1}{2}\theta\right) d\theta = 4\left[\sin\left(\frac{1}{2}\theta\right)\right]_0^{\pi} = 4$

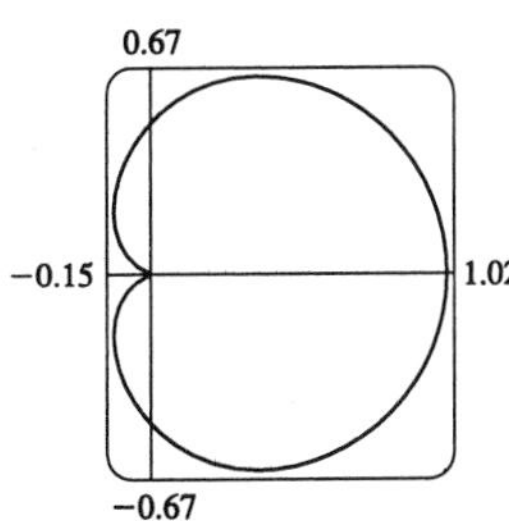

51. From Figure 4 it is apparent that one loop lies between $\theta = -\frac{\pi}{4}$ and $\theta = \frac{\pi}{4}$. Therefore $L = \int_{-\pi/4}^{\pi/4} \sqrt{\cos^2 2\theta + (-2\sin 2\theta)^2}\, d\theta = 2\int_0^{\pi/4} \sqrt{\cos^2 2\theta + 4\sin^2 2\theta}\, d\theta = 2\int_0^{\pi/4} \sqrt{1 + 3\sin^2 2\theta}\, d\theta$. Using Simpson's Rule with $n = 4$, $\Delta x = \frac{\pi/4}{4} = \frac{\pi}{16}$ and $f(\theta) = \sqrt{1 + 3\sin^2 2\theta}$, we get $L \approx 2 \cdot S_4 = 2\frac{\pi}{16 \cdot 3}\left[f(0) + 4f\left(\frac{\pi}{16}\right) + 2f\left(\frac{\pi}{8}\right) + 4f\left(\frac{3\pi}{16}\right) + f\left(\frac{\pi}{4}\right)\right] \approx 2.4228$. Therefore the length of one loop of the four-leaved rose is approximately 2.42.

52. From Exercise 9.4.59 it is apparent that the loop occurs when $4 + 2\sec\theta \geq 0$ and $\frac{\pi}{2} \leq \theta \leq \frac{3\pi}{2}$ so solve $4 + 2\sec\theta = 0 \Rightarrow \sec\theta = -2 \Rightarrow \theta = \frac{2\pi}{3}$ or $\frac{4\pi}{3}$. So the length of the loop is

$L = \int_{2\pi/3}^{4\pi/3} \sqrt{(4 + 2\sec\theta)^2 + (2\sec\theta\tan\theta)^2}\, d\theta = \int_{2\pi/3}^{4\pi/3} \sqrt{16 + 16\sec\theta + 4\sec^2\theta + 4\sec^2\theta\tan^2\theta}\, d\theta$

$= 2\int_{2\pi/3}^{4\pi/3} \sqrt{4 + 4\sec\theta + \sec^2\theta\,(1 + \tan^2\theta)}\, d\theta = 2\int_{2\pi/3}^{4\pi/3} \sqrt{4 + 4\sec\theta + \sec^4\theta}\, d\theta$.

Using Simpson's Rule with $n = 4$, $\Delta x = \frac{1}{4}\left(\frac{4\pi}{3} - \frac{2\pi}{3}\right) = \frac{\pi}{6}$ and $f(\theta) = \sqrt{4 + 4\sec\theta + \sec^4\theta}$, we get $L \approx 2 \cdot S_4 = 2\frac{\pi}{6 \cdot 3}\left[f\left(\frac{2\pi}{3}\right) + 4f\left(\frac{5\pi}{6}\right) + 2f(\pi) + 4f\left(\frac{7\pi}{6}\right) + f\left(\frac{4\pi}{3}\right)\right] \approx 6.12$, so the loop has a length of about 6.1.

53. **(a)** From (9.3.5), $S = \int_a^b 2\pi y\sqrt{(dx/d\theta)^2 + (dy/d\theta)^2}\, d\theta = \int_a^b 2\pi y\sqrt{r^2 + (dr/d\theta)^2}\, d\theta$ (see the derivation of Equation 5) $= \int_a^b 2\pi r\sin\theta\sqrt{r^2 + (dr/d\theta)^2}\, d\theta$.

(b) $r^2 = \cos 2\theta \Rightarrow 2r\dfrac{dr}{d\theta} = -2\sin 2\theta \Rightarrow \left(\dfrac{dr}{d\theta}\right)^2 = \dfrac{\sin^2 2\theta}{r^2} = \dfrac{\sin^2 2\theta}{\cos 2\theta}$

$S = 2\int_0^{\pi/4} 2\pi\sqrt{\cos 2\theta}\sin\theta\sqrt{\cos 2\theta + (\sin^2 2\theta)/\cos 2\theta}\, d\theta = 4\pi\int_0^{\pi/4} \sin\theta\, d\theta$

$= \left[-4\pi\cos\theta\right]_0^{\pi/4} = -4\pi\left(\frac{1}{\sqrt{2}} - 1\right) = 2\pi\left(2 - \sqrt{2}\right)$.

54. **(a)** Rotation around $\theta = \frac{\pi}{2}$ is the same as rotation around the y-axis, that is, $S = \int_a^b 2\pi x\, ds$ where $ds = \sqrt{(dx/dt)^2 + (dy/dt)^2}\, dt$ for a parametric equation, and for the special case of a polar equation, $x = r\cos\theta$ and $ds = \sqrt{(dx/d\theta)^2 + (dy/d\theta)^2}\, d\theta = \sqrt{r^2 + (dr/d\theta)^2}\, d\theta$ (see the derivation of Equation 5.) Therefore, for a polar equation, rotated around $\theta = \frac{\pi}{2}$, $S = \int_a^b 2\pi r\cos\theta\sqrt{r^2 + (dr/d\theta)^2}\, d\theta$.

(b) In the case of the lemniscate we are concerned with $-\frac{\pi}{4} \leq \theta \leq \frac{\pi}{4}$ and $r^2 = \cos 2\theta \Rightarrow 2r\, dr/d\theta = -2\sin 2\theta \Rightarrow (dr/d\theta)^2 = (\sin^2 2\theta)/r^2 = (\sin^2 2\theta)/\cos 2\theta$. Therefore

$S = \int_{-\pi/4}^{\pi/4} 2\pi\sqrt{\cos 2\theta}\cos\theta\sqrt{\cos 2\theta + (\sin^2 2\theta)/\cos 2\theta}\, d\theta$

$= 4\pi\int_0^{\pi/4} \cos\theta\sqrt{\cos 2\theta}\sqrt{1/\cos 2\theta}\, d\theta = 4\pi\int_0^{\pi/4} \cos\theta\, d\theta = 2\sqrt{2}\pi$.

EXERCISES 9.6

1. $x^2 = -8y$. $4p = -8$, so $p = -2$. The vertex is $(0,0)$, the focus is $(0,-2)$, and the directrix is $y = 2$.

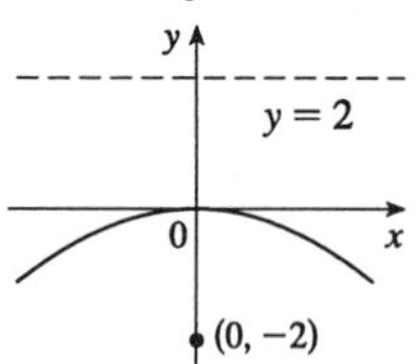

2. $x = -5y^2 \;\Rightarrow\; y^2 = -\frac{1}{5}x \;\Rightarrow\; 4p = -\frac{1}{5}$ $\Rightarrow\; p = -\frac{1}{20} \;\Rightarrow$ vertex $(0,0)$, focus $\left(-\frac{1}{20}, 0\right)$, directrix $x = \frac{1}{20}$

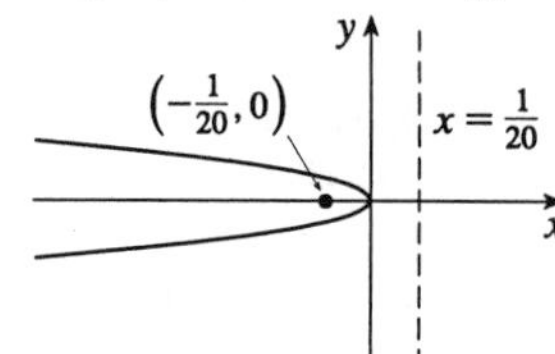

3. $y^2 = x$. $p = \frac{1}{4}$ and the vertex is $(0,0)$, so the focus is $\left(\frac{1}{4}, 0\right)$, and the directrix is $x = -\frac{1}{4}$.

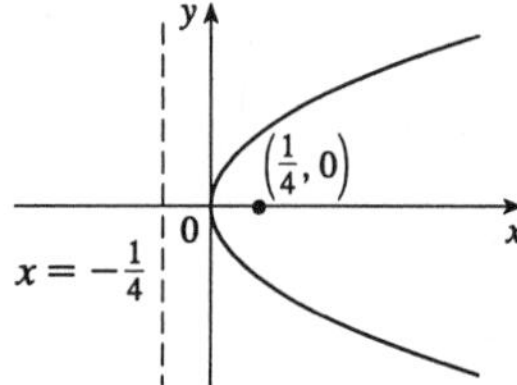

4. $x^2 = \frac{1}{2}y \;\Rightarrow\; p = \frac{1}{8} \;\Rightarrow$ vertex $(0,0)$, focus $\left(0, \frac{1}{8}\right)$, directrix $y = -\frac{1}{8}$

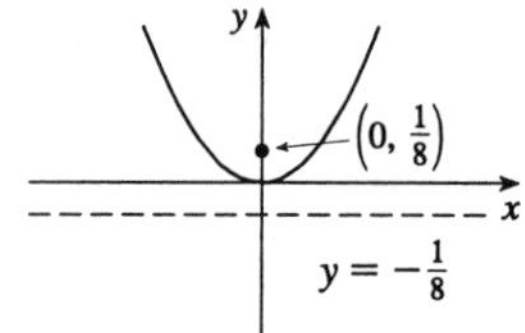

5. $x + 1 = 2(y-3)^2 \;\Rightarrow$ $(y-3)^2 = \frac{1}{2}(x+1) \;\Rightarrow\; p = \frac{1}{8} \;\Rightarrow$ vertex $(-1,3)$, focus $\left(-\frac{7}{8}, 3\right)$, directrix $x = -\frac{9}{8}$

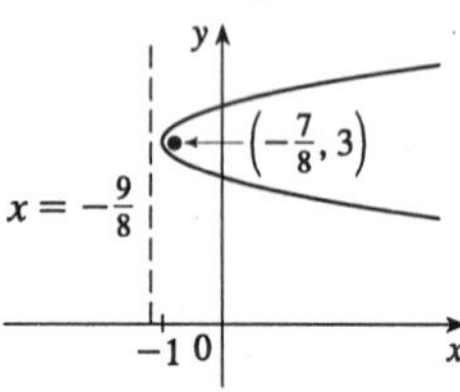

6. $x^2 - 6x + 8y = 7 \;\Leftrightarrow$ $(x-3)^2 = -8y + 16 = -8(y-2) \;\Rightarrow\; p = -2$ $\Rightarrow$ vertex $(3,2)$, focus $(3,0)$, directrix $y = 4$

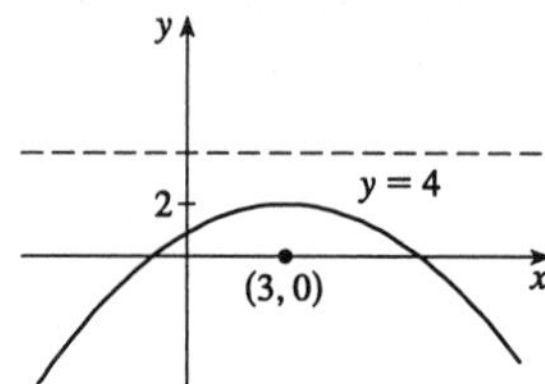

7. $2x + y^2 - 8y + 12 = 0 \;\Rightarrow$ $(y-4)^2 = -2(x-2) \;\Rightarrow$ $p = -\frac{1}{2} \;\Rightarrow$ vertex $(2,4)$, focus $\left(\frac{3}{2}, 4\right)$, directrix $x = \frac{5}{2}$

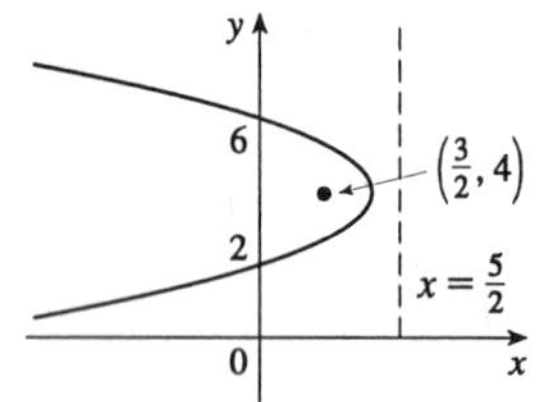

8. $x^2 + 12x - y + 39 = 0 \;\Leftrightarrow$ $(x+6)^2 = y - 3 \;\Rightarrow\; p = \frac{1}{4}$ $\Rightarrow$ vertex $(-6,3)$, focus $\left(-6, \frac{13}{4}\right)$, directrix $y = \frac{11}{4}$

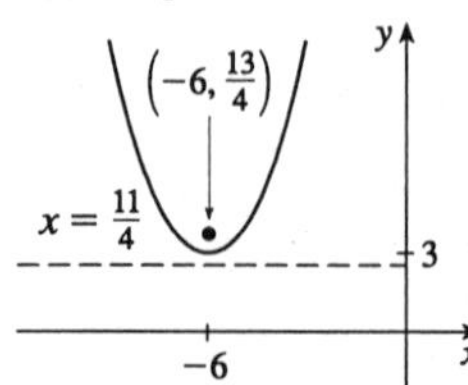

9. $x^2/16 + y^2/4 = 1 \quad \Rightarrow \quad a = 4, b = 2,$
$c = \sqrt{16-4} = 2\sqrt{3} \;\Rightarrow\;$ center $(0, 0)$,
vertices $(\pm 4, 0)$, foci $\left(\pm 2\sqrt{3}, 0\right)$

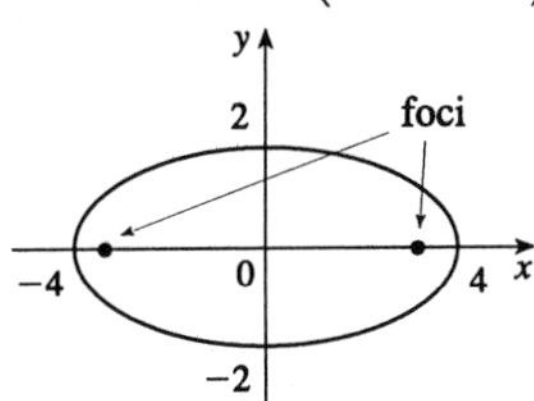

10. $x^2/4 + y^2/25 = 1 \quad \Rightarrow \quad a = 5, b = 2,$
$c = \sqrt{25-4} = \sqrt{21} \;\Rightarrow\;$ center $(0, 0)$,
vertices $(0, \pm 5)$, foci $\left(0, \pm \sqrt{21}\right)$

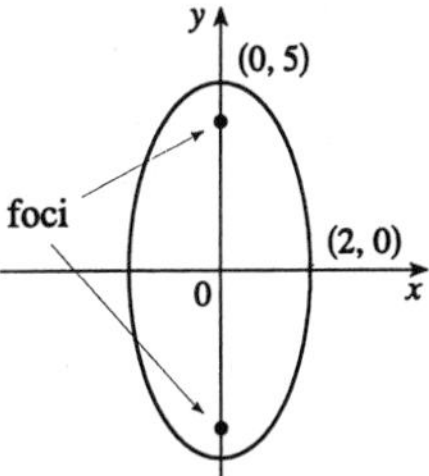

11. $25x^2 + 9y^2 = 225 \quad \Leftrightarrow \quad \frac{1}{9}x^2 + \frac{1}{25}y^2 = 1$
$\Rightarrow \; a = 5, b = 3, c = 4 \quad \Rightarrow$
center $(0, 0)$, vertices $(0, \pm 5)$, foci $(0, \pm 4)$

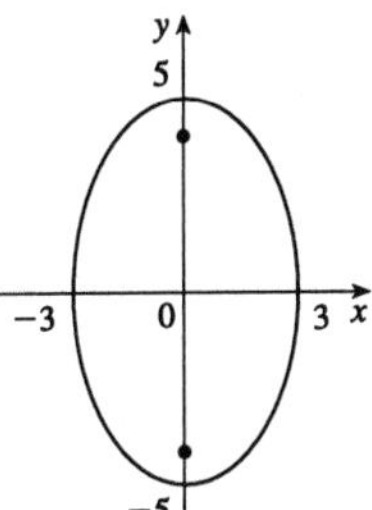

12. $x^2 + 4y^2 = 4 \quad \Leftrightarrow \quad \frac{1}{4}x^2 + y^2 = 1 \quad \Rightarrow$
$a = 2, b = 1, c = \sqrt{3} \quad \Rightarrow \quad$ center $(0, 0)$,
vertices $(\pm 2, 0)$, foci $\left(\pm \sqrt{3}, 0\right)$

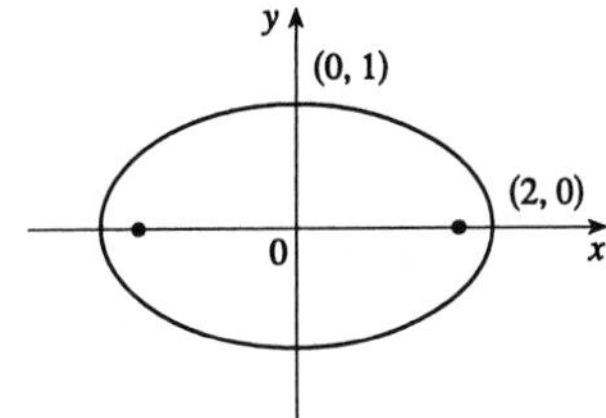

13. $\dfrac{x^2}{144} - \dfrac{y^2}{25} = 1 \quad \Rightarrow \quad a = 12, b = 5,$
$c = \sqrt{144 + 25} = 13 \;\Rightarrow\;$ center $(0, 0)$, vertices
$(\pm 12, 0)$, foci $(\pm 13, 0)$, asymptotes $y = \pm \frac{5}{12}x$

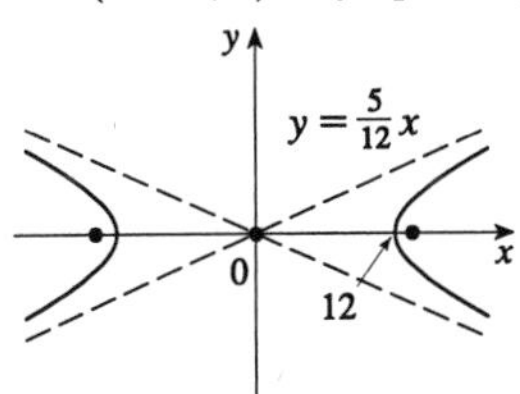

14. $\dfrac{y^2}{25} - \dfrac{x^2}{144} = 1 \;\Rightarrow\; a = 5, b = 12, c = 13 \;\Rightarrow$
center $(0, 0)$, vertices $(0, \pm 5)$, foci $(0, \pm 13)$,
asymptotes $y = \pm \frac{5}{12}x$

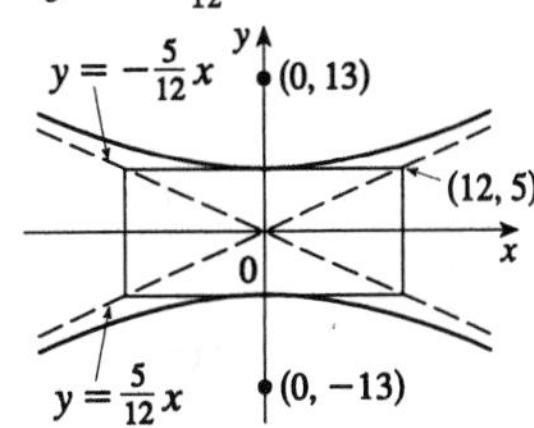

15. $9y^2 - x^2 = 9 \;\Rightarrow\; y^2 - \frac{1}{9}x^2 = 1 \;\Rightarrow\; a = 1,$
$b = 3, c = \sqrt{10} \;\Rightarrow\;$ center $(0, 0)$, vertices
$(0, \pm 1)$, foci $\left(0, \pm \sqrt{10}\right)$, asymptotes $y = \pm \frac{1}{3}x$

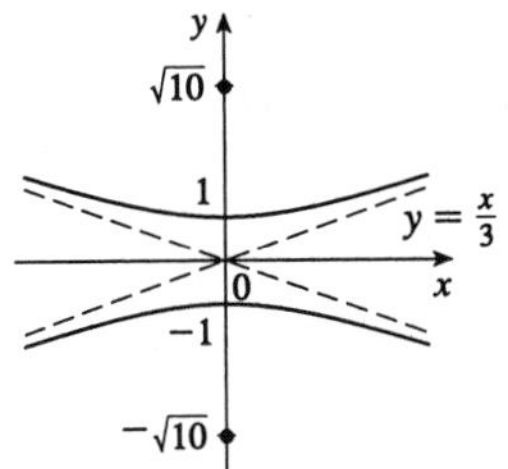

16. $x^2 - y^2 = 1 \quad \Rightarrow \quad a = b = 1, c = \sqrt{2}$
$\Rightarrow \;$ center $(0, 0)$, vertices $(\pm 1, 0)$,
foci $\left(\pm \sqrt{2}, 0\right)$, asymptotes $y = \pm x$

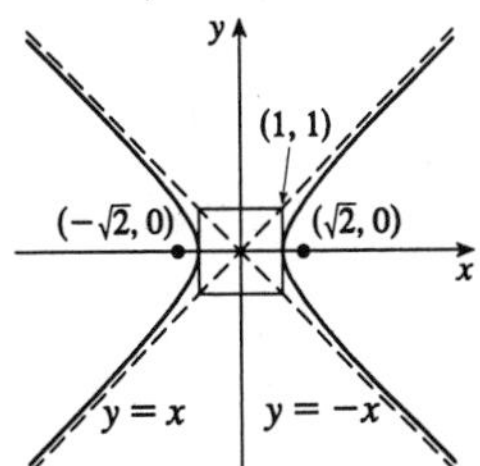

17. $9x^2 - 18x + 4y^2 = 27 \iff \dfrac{(x-1)^2}{4} + \dfrac{y^2}{9} = 1 \Rightarrow a = 3, b = 2, c = \sqrt{5} \Rightarrow$ center $(1, 0)$, vertices $(1, \pm 3)$, foci $\left(1, \pm\sqrt{5}\right)$

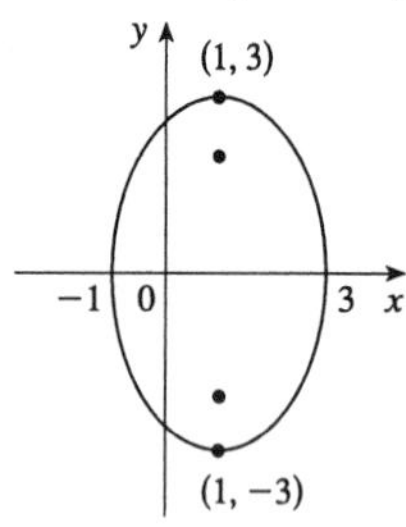

18. $16x^2 + 64x - 9y^2 - 90y = 305 \iff \dfrac{(x+2)^2}{9} - \dfrac{(y+5)^2}{16} = 1 \Rightarrow a = 3, b = 4, c = 5 \Rightarrow$ center $(-2, -5)$, vertices $(-5, -5)$ and $(1, -5)$, foci $(-7, -5)$ and $(3, -5)$, asymptotes $y + 5 = \pm\frac{4}{3}(x + 2)$

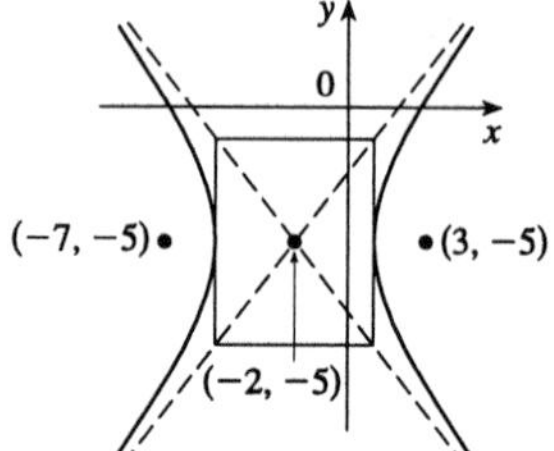

19. $2y^2 - 4y - 3x^2 + 12x = -8 \iff \dfrac{(x-2)^2}{6} - \dfrac{(y-1)^2}{9} = 1 \Rightarrow a = \sqrt{6}, b = 3, c = \sqrt{15} \Rightarrow$ center $(2, 1)$, vertices $\left(2 \pm \sqrt{6}, 1\right)$, foci $\left(2 \pm \sqrt{15}, 1\right)$, asymptotes $y - 1 = \pm\frac{3}{\sqrt{6}}(x - 2)$

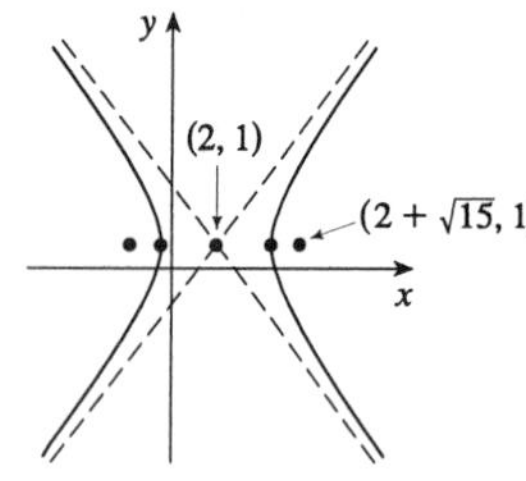

20. $x^2 - 6x + 2y^2 + 4y = -7 \iff \dfrac{(x-3)^2}{4} + \dfrac{(y+1)^2}{2} = 1 \Rightarrow a = 2, b = \sqrt{2} = c \Rightarrow$ center $(3, -1)$, vertices $(1, -1)$ and $(5, -1)$, foci $\left(3 \pm \sqrt{2}, -1\right)$

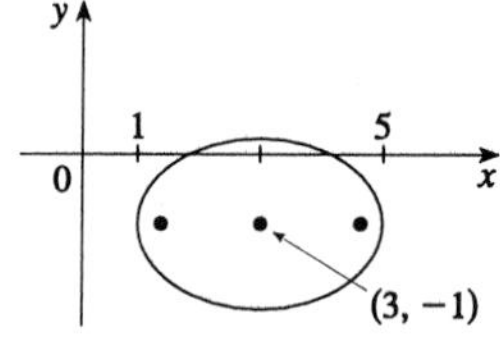

21. Vertex at $(0, 0)$, $p = 3$, opens upward $\Rightarrow x^2 = 4py = 12y$

22. Vertex $(0, 0)$, parabola opens to the left $\Rightarrow p = -2 \Rightarrow y^2 = 4px = -8x$

23. Vertex at $(2, 0)$, $p = 1$, opens to right $\Rightarrow y^2 = 4p(x - 2) = 4(x - 2)$

24. Vertex $(1, 2)$, parabola opens down $\Rightarrow p = -3 \Rightarrow (x-1)^2 = 4p(y-2) = -12(y-2) \iff x^2 - 2x + 12y - 23 = 0$

25. The parabola must have equation $y^2 = 4px$, so $(-4)^2 = 4p(1) \Rightarrow p = 4 \Rightarrow y^2 = 16x$.

26. Vertical axis $\Rightarrow (x-h)^2 = 4p(y-k)$. Substituting $(-2, 3)$ and $(0, 3)$ gives $(-2-h)^2 = 4p(3-k)$ and $(-h)^2 = 4p(3-k) \Rightarrow (-2-h)^2 = (-h)^2 \Rightarrow 4 + 4h + h^2 = h^2 \Rightarrow h = -1 \Rightarrow 1 = 4p(3-k)$. Substituting $(1, 9)$ gives $[1 - (-1)]^2 = 4p(9-k) \Rightarrow 4 = 4p(9-k)$. Solving for p from these equations gives $p = \dfrac{1}{4(3-k)} = \dfrac{1}{9-k} \Rightarrow 4(3-k) = 9-k \Rightarrow k = 1 \Rightarrow p = \frac{1}{8} \Rightarrow (x+1)^2 = \frac{1}{2}(y-1) \Rightarrow 2x^2 + 4x - y + 3 = 0$.

27. Center $(0,0)$, $c=1$, $a=2 \quad\Rightarrow\quad b=\sqrt{2^2-1^2}=\sqrt{3} \quad\Rightarrow\quad \frac{1}{4}x^2+\frac{1}{3}y^2=1$

28. Center $(0,0)$, $c=4$, $a=5$, major axis vertical $\quad\Rightarrow\quad b=3$ and $\frac{1}{9}x^2+\frac{1}{25}y^2=1$

29. Center $(3,0)$, $c=1$, $a=3 \quad\Rightarrow\quad b=\sqrt{8}=2\sqrt{2} \quad\Rightarrow\quad \frac{1}{8}(x-3)^2+\frac{1}{9}y^2=1$

30. Center $(0,2)$, $c=1$, $a=3$, major axis horizontal $\quad\Rightarrow\quad b=2\sqrt{2}$ and $\frac{1}{9}x^2+\frac{1}{8}(y-2)^2=1$

31. Center $(2,2)$, $c=2$, $a=3 \quad\Rightarrow\quad b=\sqrt{5} \quad\Rightarrow\quad \frac{1}{9}(x-2)^2+\frac{1}{5}(y-2)^2=1$

32. Center $(0,0)$, $c=2$, major axis horizontal $\quad\Rightarrow\quad \dfrac{x^2}{a^2}+\dfrac{y^2}{b^2}=1$ and $b^2=a^2-c^2=a^2-4$. Since the ellipse passes through $(2,1)$, we have $2a=|PF_1|+|PF_2|=\sqrt{17}+1 \quad\Rightarrow\quad a^2=\frac{9+\sqrt{17}}{2}$ and $b^2=\frac{1+\sqrt{17}}{2}$, so the ellipse has equation $\dfrac{2x^2}{9+\sqrt{17}}+\dfrac{2y^2}{1+\sqrt{17}}=1.$

33. Center $(0,0)$, vertical axis, $c=3$, $a=1 \quad\Rightarrow\quad b=\sqrt{8}=2\sqrt{2} \quad\Rightarrow\quad y^2-\frac{1}{8}x^2=1$

34. Center $(0,0)$, horizontal axis, $c=6$, $a=4 \quad\Rightarrow\quad b=2\sqrt{5} \quad\Rightarrow\quad \frac{1}{16}x^2-\frac{1}{20}y^2=1$

35. Center $(4,3)$, horizontal axis, $c=3$, $a=2 \quad\Rightarrow\quad b=\sqrt{5} \quad\Rightarrow\quad \frac{1}{4}(x-4)^2-\frac{1}{5}(y-3)^2=1$

36. Center $(2,3)$, vertical axis, $c=5$, $a=3 \quad\Rightarrow\quad b=4 \quad\Rightarrow\quad \frac{1}{9}(y-3)^2-\frac{1}{16}(x-2)^2=1$

37. Center $(0,0)$, horizontal axis, $a=3$, $\frac{b}{a}=2 \quad\Rightarrow\quad b=6 \quad\Rightarrow\quad \frac{1}{9}x^2-\frac{1}{36}y^2=1$

38. Center $(4,2)$, horizontal axis, asymptotes $y-2=\pm(x-4) \quad\Rightarrow\quad c=2,\ b/a=1 \quad\Rightarrow\quad a=b \quad\Rightarrow$
$c^2=4=a^2+b^2=2a^2 \quad\Rightarrow\quad a^2=2 \quad\Rightarrow\quad \frac{1}{2}(x-4)^2-\frac{1}{2}(y-2)^2=1$

39. In Figure 8, we see that the point on the ellipse closest to a focus is the closer vertex (which is a distance $a-c$ from it) while the farthest point is the other vertex (at a distance of $a+c$). So for this lunar orbit, $(a-c)+(a+c)=2a=(1728+110)+(1728+314)$, or $a=1940$; and $(a+c)-(a-c)=2c=314-110$, or $c=102$. Thus $b^2=a^2-c^2=3{,}753{,}196$, and the equation is
$$\frac{x^2}{3{,}763{,}600}+\frac{y^2}{3{,}753{,}196}=1.$$

40. **(a)** Choose V to be the origin, with x-axis through V and F. Then F is $(p,0)$, A is $(p,5)$, so substituting A into the equation $y^2=4px$ gives $25=4p^2$ so $p=\frac{5}{2}$ and $y^2=10x$.

(b) $x=11 \quad\Rightarrow\quad y=\sqrt{110} \quad\Rightarrow\quad |CD|=2\sqrt{110}$

41. **(a)** Set up the coordinate system so that A is $(-200,0)$ and B is $(200,0)$.
$|PA|-|PB|=(1200)(980)=1{,}176{,}000\text{ ft}=\frac{2450}{11}\text{ mi}=2a \quad\Rightarrow\quad a=\frac{1225}{11}$, and $c=200$ so
$$b^2=c^2-a^2=\frac{3{,}339{,}375}{121} \quad\Rightarrow\quad \frac{121x^2}{1{,}500{,}625}-\frac{121y^2}{3{,}339{,}375}=1.$$

(b) Due north of $B \quad\Rightarrow\quad x=200 \quad\Rightarrow\quad \dfrac{(121)(200)^2}{1{,}500{,}625}-\dfrac{121y^2}{3{,}339{,}375}=1 \quad\Rightarrow\quad y=\dfrac{133{,}575}{539}\approx 248$ mi

42. $|PF_1| - |PF_2| = \pm 2a \quad \Leftrightarrow \quad \sqrt{(x+c)^2 + y^2} - \sqrt{(x-c)^2 + y^2} = \pm 2a$

$\Leftrightarrow \quad \sqrt{(x+c)^2 + y^2} = \sqrt{(x-c)^2 + y^2} \pm 2a$

$\Leftrightarrow \quad (x+c)^2 + y^2 = (x-c)^2 + y^2 + 4a^2 \pm 4a\sqrt{(x-c)^2 + y^2}$

$\Leftrightarrow \quad 4cx - 4a^2 = \pm 4a\sqrt{(x-c)^2 + y^2} \quad \Leftrightarrow \quad c^2x^2 - 2a^2cx + a^4 = a^2(x^2 - 2cx + c^2 + y^2)$

$\Leftrightarrow \quad (c^2 - a^2)x^2 - a^2y^2 = a^2(c^2 - a^2) \;\Leftrightarrow\; b^2x^2 - a^2y^2 = a^2b^2$ (where $b^2 = c^2 - a^2$) $\Leftrightarrow \; \dfrac{x^2}{a^2} - \dfrac{y^2}{b^2} = 1$

43. The function whose graph is the upper branch of this hyperbola is concave upward. The function is

$y = f(x) = a\sqrt{1 + \dfrac{x^2}{b^2}} = \dfrac{a}{b}\sqrt{b^2 + x^2}$, so $y' = \dfrac{a}{b}x\left(b^2 + x^2\right)^{-1/2}$ and

$y'' = \dfrac{a}{b}\left[\left(b^2 + x^2\right)^{-1/2} - x^2\left(b^2 + x^2\right)^{-3/2}\right] = ab\left(b^2 + x^2\right)^{-3/2} > 0$ for all x, and so f is concave upward.

44. We can follow exactly the same sequence of steps as in the derivation of Formula 4, except we use the points $(1, 1)$ and $(-1, -1)$ in the distance formula (first equation of that derivation) so $\sqrt{(x-1)^2 + (y-1)^2} + \sqrt{(x+1)^2 + (y+1)^2} = 4$ will lead to $3x^2 - 2xy + 3y^2 = 8$.

45. **(a)** ellipse

(b) hyperbola

(c) empty graph (no curve)

(d) In case (a), $a^2 = k$, $b^2 = k - 16$, and $c^2 = a^2 - b^2 = 16$, so the foci are at $(\pm 4, 0)$. In case (b), $k - 16 < 0$, so $a^2 = k$, $b^2 = 16 - k$, and $c^2 = a^2 + b^2 = 16$, and so again the foci are at $(\pm 4, 0)$.

46. **(a)** $y^2 = 4px \quad \Rightarrow \quad 2yy' = 4p \quad \Rightarrow \quad y' = \dfrac{2p}{y}$, so the tangent line is $y - y_0 = \dfrac{2p}{y_0}(x - x_0) \quad \Rightarrow$

$yy_0 - y_0^2 = 2p(x - x_0) \quad \Leftrightarrow \quad yy_0 - 4px_0 = 2px - 2px_0 \;\Rightarrow\; yy_0 = 2p(x + x_0)$.

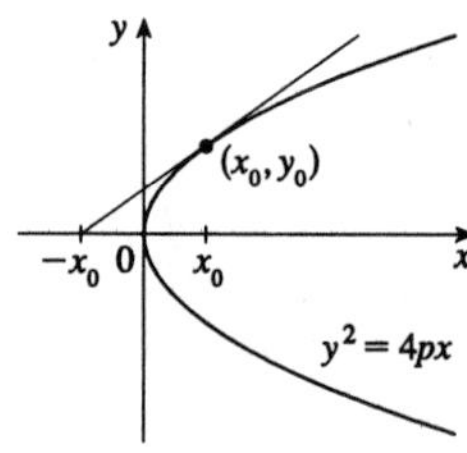

(b) The x-intercept is $-x_0$.

47. Use the parametrization $x = 2\cos t$, $y = \sin t$, $0 \le t \le 2\pi$ to get

$L = 4\displaystyle\int_0^{\pi/2} \sqrt{(dx/dt)^2 + (dy/dt)^2}\,dt = 4\int_0^{\pi/2} \sqrt{4\sin^2 t + \cos^2 t}\,dt = 4\int_0^{\pi/2} \sqrt{3\sin^2 t + 1}\,dt$. Using

Simpson's Rule with $n = 10$, $L \approx \frac{4}{3}\left(\frac{\pi}{20}\right)\left[f(0) + 4f\left(\frac{\pi}{20}\right) + 2f\left(\frac{\pi}{10}\right) + \cdots + 2f\left(\frac{2\pi}{5}\right) + 4f\left(\frac{9\pi}{20}\right) + f\left(\frac{\pi}{2}\right)\right]$, with

$f(t) = \sqrt{3\sin^2 t + 1}$, so $L \approx 9.69$.

48. The length of the major axis is $2a$, so $a = \frac{1}{2}(1.18 \times 10^{10}) = 5.9 \times 10^9$. The length of the minor axis is $2b$, so $b = \frac{1}{2}(1.14 \times 10^{10}) = 5.7 \times 10^9$. Therefore the equation of the ellipse is $\dfrac{x^2}{(5.9 \times 10^9)^2} + \dfrac{y^2}{(5.7 \times 10^9)^2} = 1$.

Converting into parametric equations, $x = 5.9 \times 10^9 \cos\theta$ and $y = 5.7 \times 10^9 \sin\theta$. So

$$L = 4\int_0^{\pi/2} \sqrt{(dx/d\theta)^2 + (dy/d\theta)^2}\,d\theta = 4\int_0^{\pi/2} \sqrt{3.48 \times 10^{19}\sin^2\theta + 3.249 \times 10^{19}\cos^2\theta}\,d\theta$$

$= 4\sqrt{3.249 \times 10^{19}}\int_0^{\pi/2} \sqrt{1.0714\sin^2\theta + \cos^2\theta\,d\theta}$. Using Simpson's Rule with $n = 10$, $\Delta x = \frac{\pi/2}{10} = \frac{\pi}{20}$ and $f(\theta) = \sqrt{1.0714\sin^2\theta + \cos^2\theta}$ we get

$$L \approx 4(5.7 \times 10^9) \cdot S_{10} = 4(5.7 \times 10^9)\tfrac{\pi}{20\cdot 3}\left[f(0) + 4f\left(\tfrac{\pi}{20}\right) + 2f\left(\tfrac{\pi}{10}\right) + \cdots + 2f\left(\tfrac{2\pi}{5}\right) + 4f\left(\tfrac{9\pi}{20}\right) + f\left(\tfrac{\pi}{2}\right)\right]$$

$$\approx \tfrac{\pi}{15}(5.7 \times 10^9)\,(30.529) \approx 3.64 \times 10^{10}\text{ km}.$$

49. $\dfrac{x^2}{a^2} + \dfrac{y^2}{b^2} = 1 \;\Rightarrow\; \dfrac{2x}{a^2} + \dfrac{2yy'}{b^2} = 0 \;\Rightarrow\; y' = -\dfrac{b^2x}{a^2y}$ $(y \neq 0)$. Thus the slope of the tangent line at P is $-\dfrac{b^2x_1}{a^2y_1}$. The slope of F_1P is $\dfrac{y_1}{x_1+c}$ and of F_2P is $\dfrac{y_1}{x_1-c}$. By the formula from Problem 23 of Problems Plus after Chapter 2, we have

$$\tan\alpha = \frac{\dfrac{y_1}{x_1+c} + \dfrac{b^2x_1}{a^2y_1}}{1 - \dfrac{b^2x_1y_1}{a^2y_1(x_1+c)}} = \frac{a^2y_1^2 + b^2x_1(x_1+c)}{a^2y_1(x_1+c) - b^2x_1y_1} = \frac{a^2b^2 + b^2cx_1}{c^2x_1y_1 + a^2cy_1} \quad \begin{bmatrix}\text{using } b^2x_1^2 + a^2y_1^2 = a^2b^2\\ \text{and } a^2 - b^2 = c^2\end{bmatrix}$$

$$= \frac{b^2(cx_1 + a^2)}{cy_1(cx_1 + a^2)} = \frac{b^2}{cy_1}, \text{ and}$$

$$\tan\beta = \frac{-\dfrac{y_1}{x_1-c} - \dfrac{b^2x_1}{a^2y_1}}{1 - \dfrac{b^2x_1y_1}{a^2y_1(x_1-c)}} = \frac{-a^2y_1^2 - b^2x_1(x_1-c)}{a^2y_1(x_1-c) - b^2x_1y_1} = \frac{-a^2b^2 + b^2cx_1}{c^2x_1y_1 - a^2cy_1} = \frac{b^2(cx_1 - a^2)}{cy_1(cx_1 - a^2)} = \frac{b^2}{cy_1}.$$

So $\alpha = \beta$.

50. The slopes of the line segments F_1P and F_2P are $\dfrac{y_1}{x_1+c}$ and $\dfrac{y_1}{x_1-c}$, where P is (x_1, y_1). Differentiating implicitly, $\dfrac{2x}{a^2} - \dfrac{2yy'}{b^2} = 0 \;\Rightarrow\; y' = \dfrac{b^2x}{a^2y} \;\Rightarrow\;$ the slope of the tangent at P is $\dfrac{b^2x_1}{a^2y_1}$, so by the formula from Problem 23 of Problems Plus after Chapter 2,

$$\tan\alpha = \frac{\dfrac{b^2x_1}{a^2y_1} - \dfrac{y_1}{x_1+c}}{1 + \dfrac{b^2x_1y_1}{a^2y_1(x_1+c)}} = \frac{b^2x_1(x_1+c) - a^2y_1^2}{a^2y_1(x_1+c) + b^2x_1y_1} = \frac{b^2(cx_1 + a^2)}{cy_1(cx_1 + a^2)} \quad \begin{bmatrix}\text{using } x_1^2/a^2 - y_1^2/b^2 = 1\\ \text{and } a^2 + b^2 = c^2\end{bmatrix} = \frac{b^2}{cy_1},$$

$$\text{and } \tan\beta = \frac{-\dfrac{b^2x_1}{a^2y_1} + \dfrac{y_1}{x_1-c}}{1 + \dfrac{b^2x_1y_1}{a^2y_1(x_1-c)}} = \frac{-b^2x_1(x_1-c) + a^2y_1^2}{a^2y_1(x_1-c) + b^2x_1y_1} = \frac{b^2(cx_1 - a^2)}{cy_1(cx_1 - a^2)} = \frac{b^2}{cy_1}.$$

So $\alpha = \beta$.

EXERCISES 9.7

1. $r = \dfrac{ed}{1 + e\cos\theta} = \dfrac{\frac{2}{3}\cdot 3}{1 + \frac{2}{3}\cos\theta} = \dfrac{6}{3 + 2\cos\theta}$

2. $r = \dfrac{ed}{1 - e\cos\theta} = \dfrac{\frac{4}{3}\cdot 3}{1 - \frac{4}{3}\cos\theta} = \dfrac{12}{3 - 4\cos\theta}$

3. $r = \dfrac{ed}{1 + e\sin\theta} = \dfrac{1\cdot 2}{1 + \sin\theta} = \dfrac{2}{1 + \sin\theta}$

4. $r = \dfrac{ed}{1 - e\sin\theta} = \dfrac{\frac{1}{2}\cdot 4}{1 - \frac{1}{2}\sin\theta} = \dfrac{4}{2 - \sin\theta}$

5. $r = 5\sec\theta \quad \Leftrightarrow \quad x = r\cos\theta = 5$, so $r = \dfrac{ed}{1 + e\cos\theta} = \dfrac{4\cdot 5}{1 + 4\cos\theta} = \dfrac{20}{1 + 4\cos\theta}$

6. $r = 2\csc\theta \quad \Leftrightarrow \quad y = r\sin\theta = 2$, so $r = \dfrac{ed}{1 + e\sin\theta} = \dfrac{\frac{3}{5}\cdot 2}{1 + \frac{3}{5}\sin\theta} = \dfrac{6}{5 + 3\sin\theta}$

7. Focus $(0, 0)$, vertex $\left(5, \frac{\pi}{2}\right) \quad \Rightarrow \quad$ directrix $y = 10 \quad \Rightarrow \quad r = \dfrac{ed}{1 + e\sin\theta} = \dfrac{10}{1 + \sin\theta}$

8. Directrix $x = 4$ so $r = \dfrac{ed}{1 + e\cos\theta} = \dfrac{\frac{2}{5}\cdot 4}{1 + \frac{2}{5}\cos\theta} = \dfrac{8}{5 + 2\cos\theta}$

9. $e = 3 \quad \Rightarrow \quad$ hyperbola; $ed = 4 \quad \Rightarrow \quad d = \frac{4}{3} \quad \Rightarrow$ directrix $x = \frac{4}{3}$; vertices $(1, 0)$ and $(-2, \pi) = (2, 0)$; center $\left(\frac{3}{2}, 0\right)$; asymptotes parallel to $\theta = \pm\cos^{-1}\left(-\frac{1}{3}\right)$

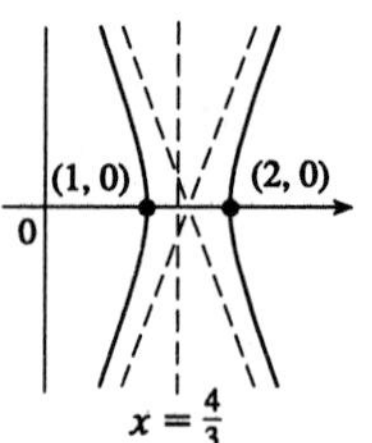

10. $r = \dfrac{8/3}{1 + \cos\theta} \quad \Rightarrow \quad e = 1 \quad \Rightarrow$ parabola; $ed = \frac{8}{3} \quad \Rightarrow \quad d = \frac{8}{3}$ $\Rightarrow \quad$ directrix $x = \frac{8}{3}$; vertex $\left(\frac{4}{3}, 0\right)$

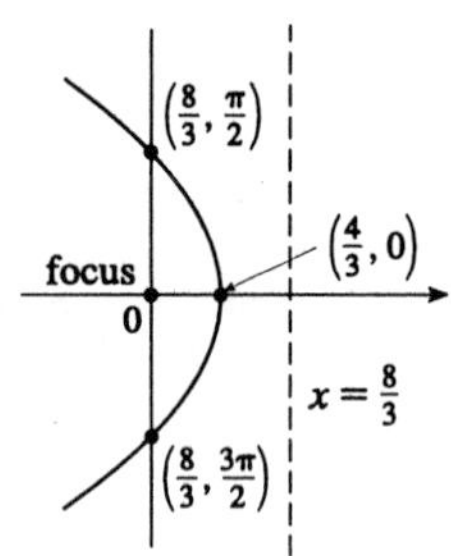

11. $e = 1 \quad \Rightarrow \quad$ parabola; $ed = 2 \quad \Rightarrow \quad d = 2$ $\Rightarrow \quad$ directrix $x = -2$; vertex $(-1, 0) = (1, \pi)$

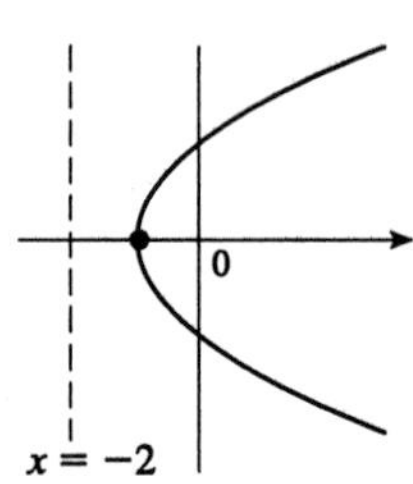

12. $r = \dfrac{10/3}{1 - \frac{2}{3}\sin\theta} \quad \Rightarrow \quad e = \frac{2}{3} \quad \Rightarrow \quad$ ellipse;

$ed = \frac{10}{3} \quad \Rightarrow \quad d = 5 \quad \Rightarrow \quad$ directrix $y = -5$;

vertices $\left(10, \frac{\pi}{2}\right)$ and $\left(2, \frac{3\pi}{2}\right)$; center $\left(4, \frac{\pi}{2}\right)$

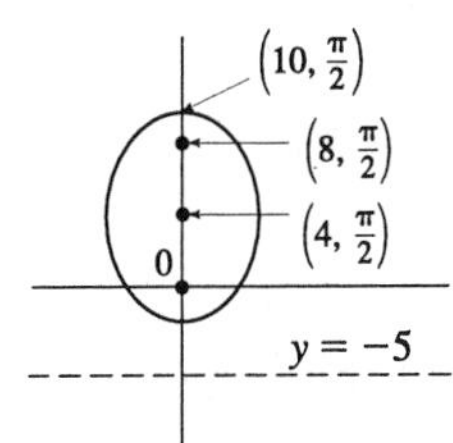

13. $r = \dfrac{3}{1 + \frac{1}{2}\sin\theta} \quad \Rightarrow \quad e = \frac{1}{2} \quad \Rightarrow \quad$ ellipse;

$ed = 3 \quad \Rightarrow \quad d = 6 \quad \Rightarrow \quad$ directrix $y = 6$;

vertices $\left(2, \frac{\pi}{2}\right)$ and $\left(6, \frac{3\pi}{2}\right)$; center $\left(2, \frac{3\pi}{2}\right)$

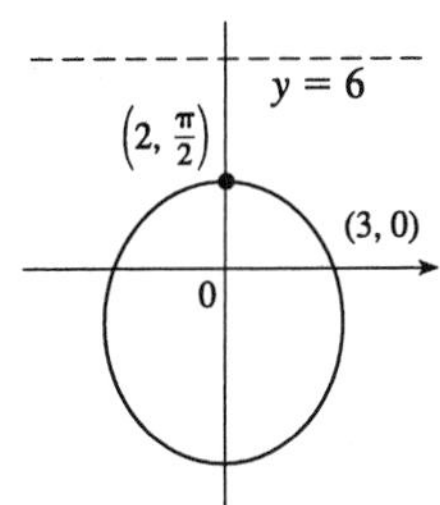

14. $r = \dfrac{5/2}{1 - \frac{3}{2}\sin\theta} \quad \Rightarrow \quad e = \frac{3}{2} \quad \Rightarrow \quad$ hyperbola;

$ed = \frac{5}{2} \quad \Rightarrow \quad d = \frac{5}{3} \quad \Rightarrow \quad$ directrix $y = -\frac{5}{3}$;

vertices $\left(-5, \frac{\pi}{2}\right) = \left(5, \frac{3\pi}{2}\right)$ and $\left(1, \frac{3\pi}{2}\right)$;

center $\left(3, \frac{3\pi}{2}\right)$; foci $(0, 0)$ and $\left(6, \frac{3\pi}{2}\right)$.

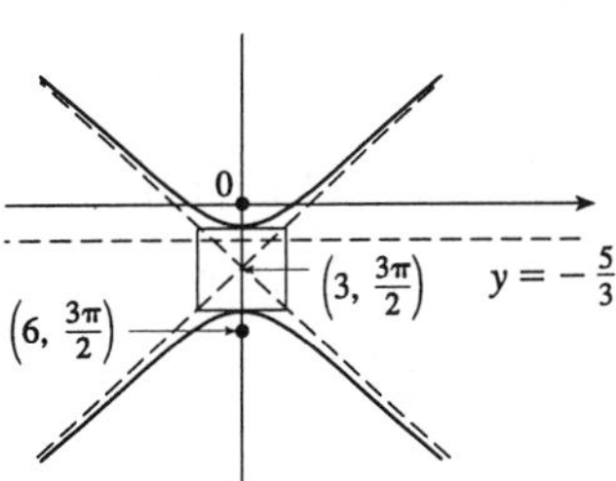

15. $r = \dfrac{7/2}{1 - \frac{5}{2}\sin\theta} \quad \Rightarrow \quad e = \frac{5}{2} \quad \Rightarrow \quad$ hyperbola;

$ed = \frac{7}{2} \quad \Rightarrow \quad d = \frac{7}{5} \quad \Rightarrow \quad$ directrix $y = -\frac{7}{5}$;

center $\left(\frac{5}{3}, \frac{3\pi}{2}\right)$; vertices $\left(-\frac{7}{3}, \frac{\pi}{2}\right) = \left(\frac{7}{3}, \frac{3\pi}{2}\right)$

and $\left(1, \frac{3\pi}{2}\right)$.

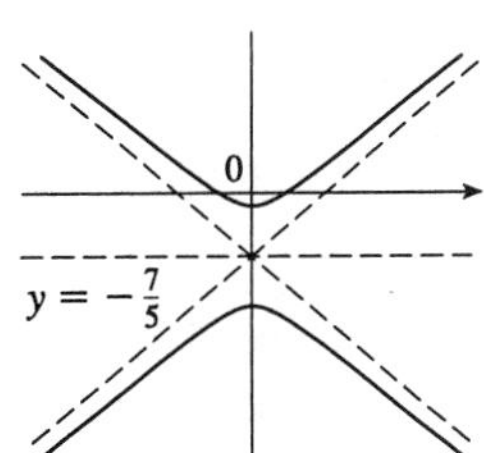

16. $r = \dfrac{8/3}{1 + \frac{1}{3}\cos\theta} \quad \Rightarrow \quad e = \frac{1}{3} \quad \Rightarrow \quad$ ellipse;

$ed = \frac{8}{3} \quad \Rightarrow \quad d = 8 \quad \Rightarrow \quad$ directrix $x = 8$;

vertices $(2, 0)$ and $(4, \pi)$; center $(-1, 0)$.

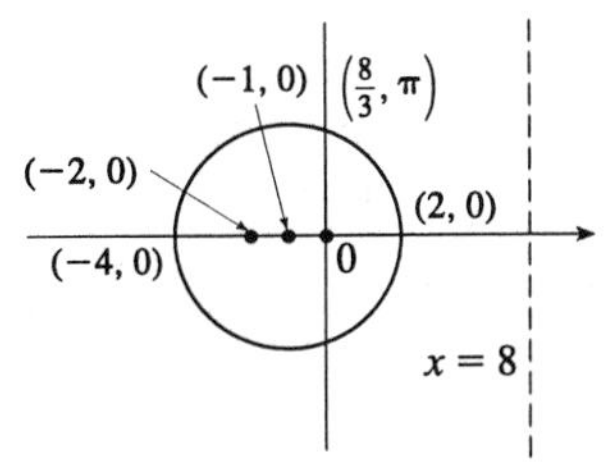

17. **(a)** The equation is $r = \dfrac{1}{4 - 3\cos\theta} = \dfrac{1/4}{1 - \frac{3}{4}\cos\theta}$, so $e = \frac{3}{4}$ and $ed = \frac{1}{4} \Rightarrow d = \frac{1}{3}$. The conic is an ellipse, and the equation of its directrix is $x = r\cos\theta = -\frac{1}{3} \Rightarrow r = -\dfrac{1}{3\cos\theta}$. We must be careful in our choice of parameter values in this equation ($-1 \le \theta \le 1$ works well.)

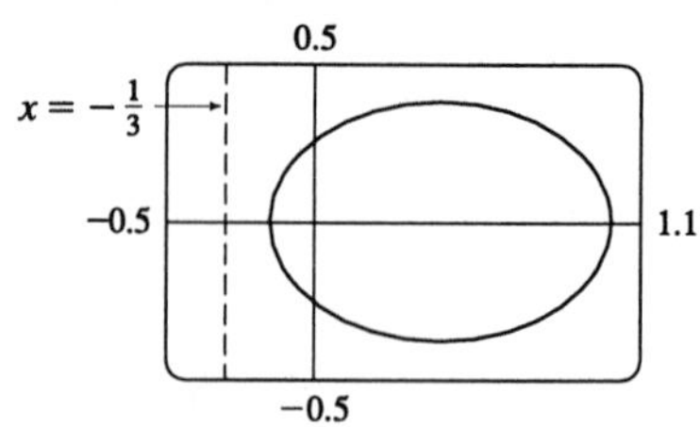

(b) The equation is obtained by replacing θ with $\theta - \frac{\pi}{3}$ in the equation of the original conic (see Example 4), so $r = \dfrac{1}{4 - 3\cos\left(\theta - \frac{\pi}{3}\right)}$.

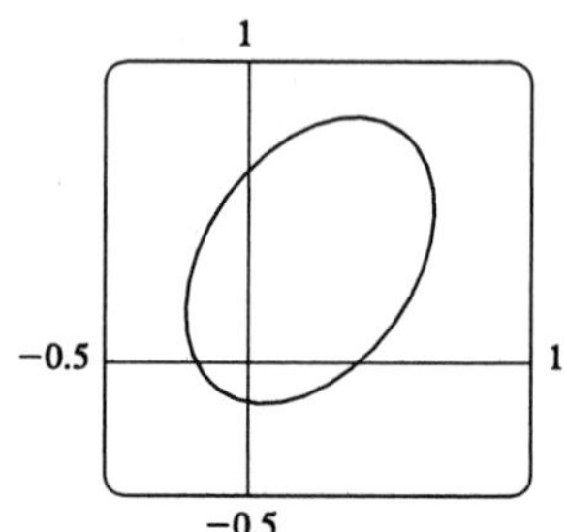

18. $r = \dfrac{5}{2 + 2\sin\theta} = \dfrac{5/2}{1 + \sin\theta}$, so $e = 1$ and $d = \frac{5}{2}$. The equation of the directrix is $y = r\sin\theta = \frac{5}{2} \Rightarrow r = \dfrac{5}{2\sin\theta}$. If the parabola is rotated about its focus (the origin) through $\frac{\pi}{6}$, its equation is the same as that of the original, with θ replaced by $\theta - \frac{\pi}{6}$ (see Example 4), so $r = \dfrac{5}{2 + 2\sin\left(\theta - \frac{\pi}{6}\right)}$.

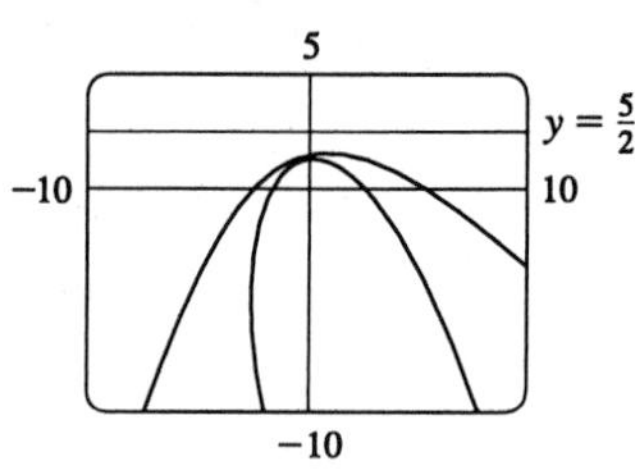

In graphing each of these curves, we must be careful to select parameter ranges which prevent the denominator from vanishing while still showing enough of the curve.

19. For $e < 1$ the curve is an ellipse. It is nearly circular when e is close to 0. As e increases, the graph is stretched out to the right, and grows larger (that is, its right-hand focus moves to the right while its left-hand focus remains at the origin.) At $e = 1$, the curve becomes a parabola with focus at the origin.

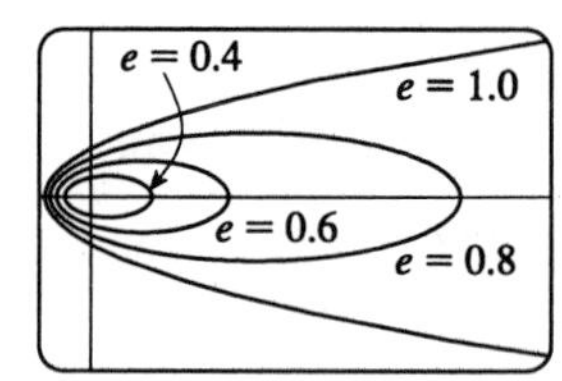

20. **(a)** The value of d does not seem to affect the shape of the conic (a parabola) at all, just its size, position, and orientation (for $d < 0$ it opens upward, for $d > 0$ it opens downward.)

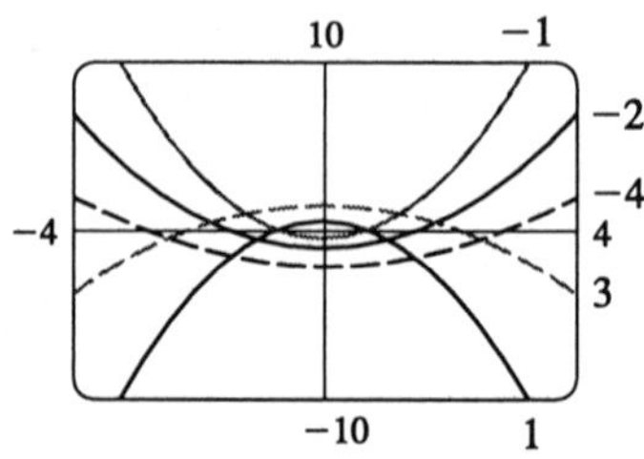

(b) We consider only positive values of e.
When $0 < e < 1$, the conic is an ellipse. As $e \to 0^+$, the graph approaches perfect roundness and zero size. As e increases, the ellipse becomes more elongated, until at $e = 1$ it turns into a parabola. For $e > 1$, the conic is a hyperbola, which moves downward and gets broader as e continues to increase.

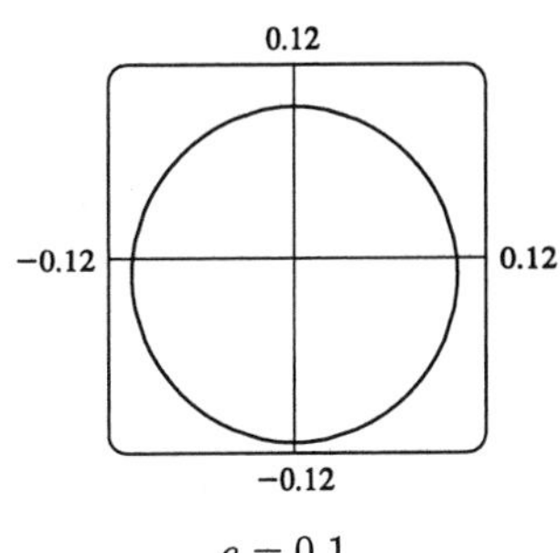

$e = 0.1$

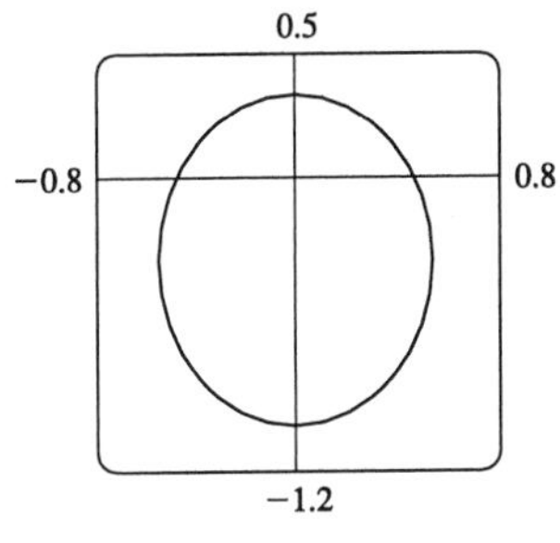

$e = 0.5$

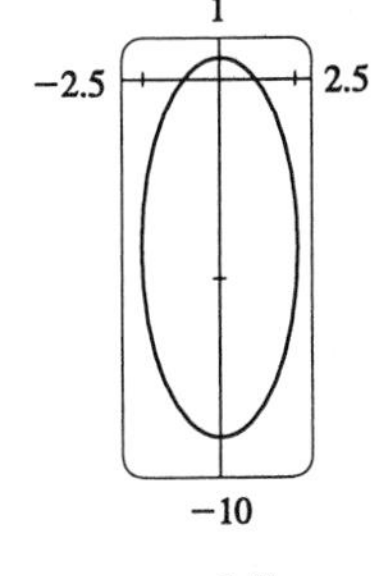

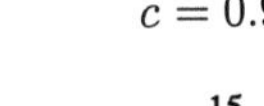

$c = 0.9$

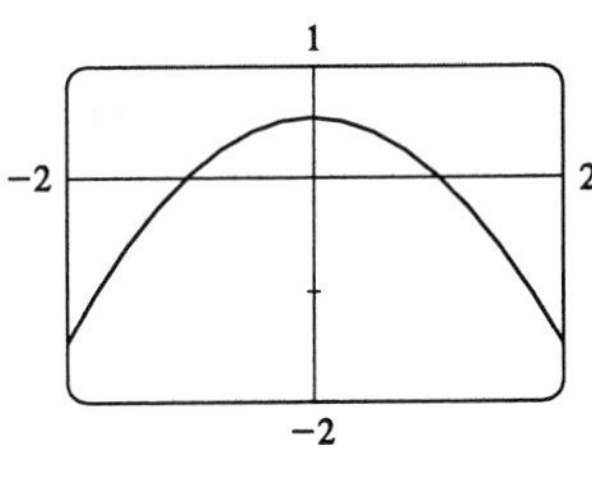

$e = 1$

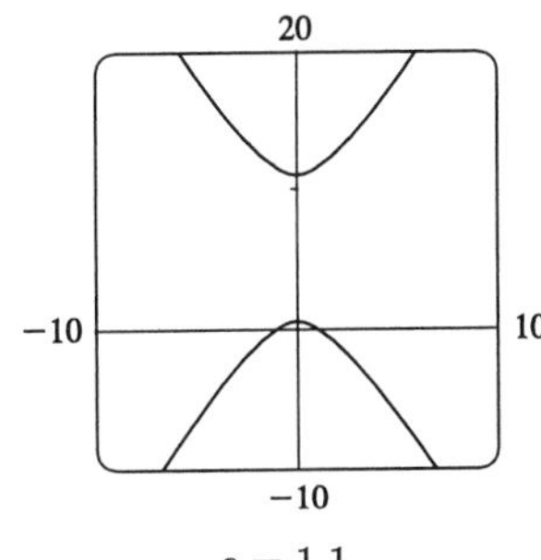

$e = 1.1$

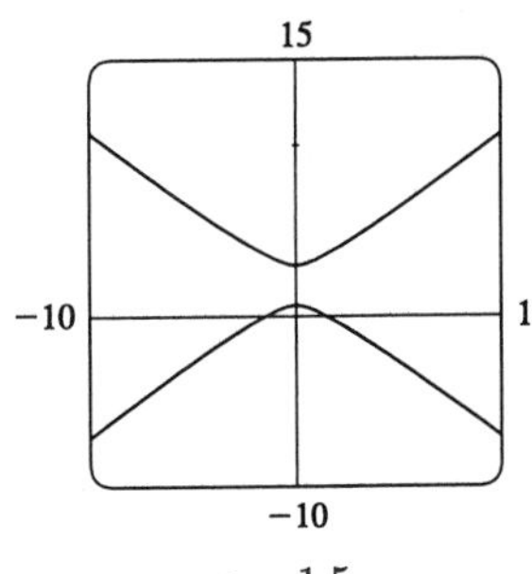

$e = 1.5$

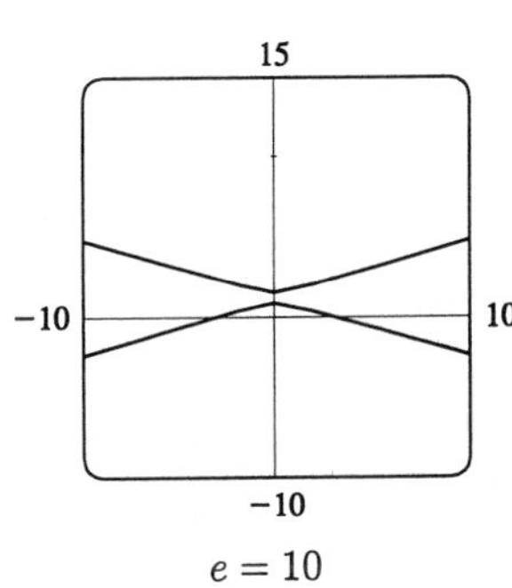

$e = 10$

21. $|PF| = e|Pl| \quad \Rightarrow$

$r = e[d - r\cos(\pi - \theta)] = e(d + r\cos\theta) \quad \Rightarrow$

$r(1 - e\cos\theta) = ed \quad \Rightarrow \quad r = \dfrac{ed}{1 - e\cos\theta}$

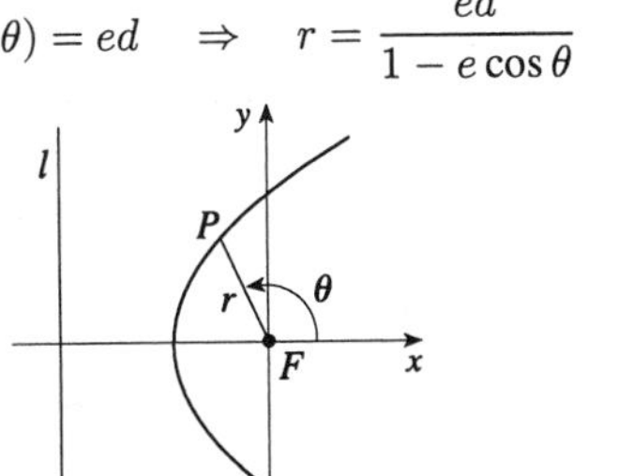

22. $|PF| = e|Pl| \quad \Rightarrow \quad r = e[d - r\sin\theta] \quad \Rightarrow$

$r(1 + e\sin\theta) = ed \quad \Rightarrow \quad r = \dfrac{ed}{1 + e\sin\theta}$

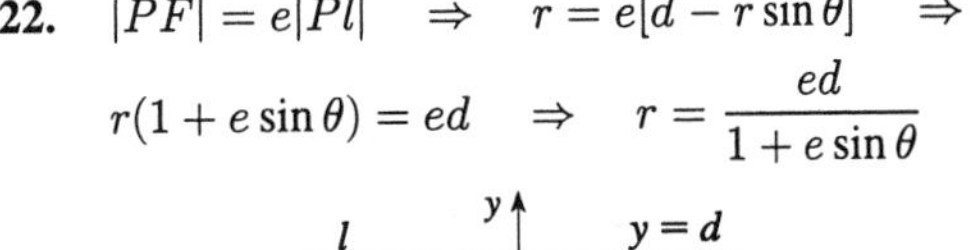

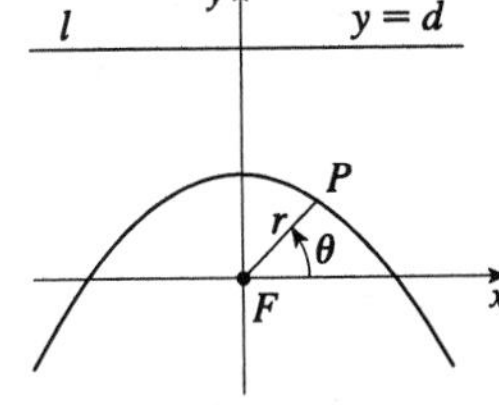

23. $|PF| = e|Pl| \quad \Rightarrow$

$r = e[d - r\sin(\theta - \pi)] = e(d + r\sin\theta)$

$\Rightarrow \quad r(1 - e\sin\theta) = ed \quad \Rightarrow \quad r = \dfrac{ed}{1 - e\sin\theta}$

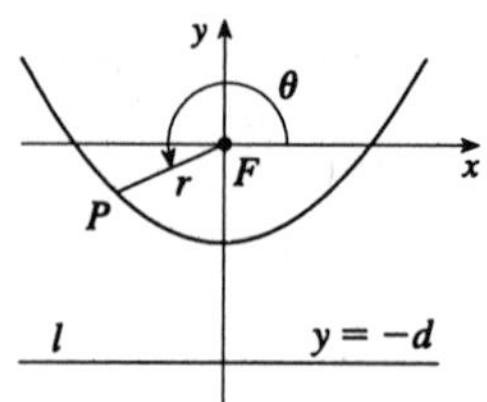

24. The parabolas intersect at the two points where $\dfrac{c}{1+\cos\theta} = \dfrac{d}{1-\cos\theta} \quad\Rightarrow\quad \cos\theta = \dfrac{c-d}{c+d} \quad\Rightarrow\quad r = \dfrac{c+d}{2}$.

For the first parabola, $\dfrac{dr}{d\theta} = \dfrac{c\sin\theta}{(1+\cos\theta)^2}$, so $\dfrac{dy}{dx} = \dfrac{(dr/d\theta)\sin\theta + r\cos\theta}{(dr/d\theta)\cos\theta - r\sin\theta} = \dfrac{c\sin^2\theta + c\cos\theta(1+\cos\theta)}{c\sin\theta\cos\theta - c\sin\theta(1+\cos\theta)}$

$= \dfrac{1+\cos\theta}{-\sin\theta}$, and similarly for the second, $\dfrac{dy}{dx} = \dfrac{1-\cos\theta}{\sin\theta} = \dfrac{\sin\theta}{1+\cos\theta}$. Since the product of these slopes is -1, the parabolas intersect at right angles.

25. **(a)** If the directrix is $x = -d$, then $r = \dfrac{ed}{1 - e\cos\theta}$ [see Figure 8(b)], and, from (4), $a^2 = \dfrac{e^2d^2}{(1-e^2)^2} \quad\Rightarrow$

$ed = a(1-e^2)$. Therefore, $r = \dfrac{a(1-e^2)}{1-e\cos\theta}$.

(b) $e = 0.017$ and the major axis $= 2a = 2.99\times10^8 \quad\Rightarrow\quad a = 1.495\times10^8$.

Therefore $r = \dfrac{1.495\times10^8\left[1-(0.017)^2\right]}{1-0.017\cos\theta} \approx \dfrac{1.49\times10^8}{1-0.017\cos\theta}$.

26. **(a)** At perihelion, $\theta = \pi$, so $r = \dfrac{a(1-e^2)}{1-e\cos\pi} = \dfrac{a(1-e^2)}{1-e(-1)} = \dfrac{a(1-e)(1+e)}{1+e} = a(1-e)$.

At aphelion, $\theta = 0$, so $r = \dfrac{a(1-e^2)}{1-e\cos 0} = \dfrac{a(1-e)(1+e)}{1-e} = a(1+e)$.

(b) At perihelion, $r = a(1-e) \approx (1.495\times10^8)(1-0.017) \approx 1.47\times10^8$ km.

At aphelion, $r = a(1+e) \approx (1.495\times10^8)(1+0.017) \approx 1.52\times10^8$ km.

27. The minimum distance is at perihelion where $4.6\times10^7 = r = a(1-e) = a(1-0.206) = a(0.794) \quad\Rightarrow$ $a = 4.6\times10^7/0.794$. So the maximum distance, which is at aphelion, is $r = a(1+e) = \left(4.6\times10^7/0.794\right)\times10^7(1.206) \approx 7.0\times10^7$ km.

28. At perihelion, $r = a(1-e) = 4.43\times10^9$, and at aphelion, $r = a(1+e) = 7.37\times10^9$. Adding, we get $2a = 11.80\times10^9$, so $a = 5.90\times10^9$ km. Therefore $1+e = a(1+e)/e = \frac{7.37}{5.90} \approx 1.249$ and $e \approx 0.249$.

29. From Exercise 27, we have $e = 0.206$ and $a(1-e) = 4.6\times10^7$ km. Thus $a = 4.6\times10^7/0.794$. From Exercise 25, we can write the equation of Mercury's orbit as $r = a\dfrac{1-e^2}{1-e\cos\theta}$. So since $\dfrac{dr}{d\theta} = \dfrac{-a(1-e^2)e\sin\theta}{(1-e\cos\theta)^2} \quad\Rightarrow$

$r^2 + \left(\dfrac{dr}{d\theta}\right)^2 = \dfrac{a^2(1-e^2)^2}{(1-e\cos\theta)^2} + \dfrac{a^2(1-e^2)^2e^2\sin^2\theta}{(1-e\cos\theta)^4} = \dfrac{a^2(1-e^2)^2}{(1-e\cos\theta)^4}\left(1 - 2e\cos\theta + e^2\right)$, the length of the orbit is $L = \displaystyle\int_0^{2\pi}\sqrt{r^2+(dr/d\theta)^2}\,d\theta = a\left(1-e^2\right)\int_0^{2\pi}\frac{\sqrt{1+e^2-2e\cos\theta}}{(1-e\cos\theta)^2}\,d\theta \approx 3.6\times10^8$ km.

This seems reasonable, since Mercury's orbit is nearly circular, and the circumference of a circle of radius a is $2\pi a \approx 3.6\times10^8$ km.

REVIEW EXERCISES FOR CHAPTER 9

1. $x = 1 - t^2, y = 1 - t \quad (-1 \le t \le 1)$
$x = 1 - (1 - y)^2 = 2y - y^2 \ (0 \le y \le 2)$

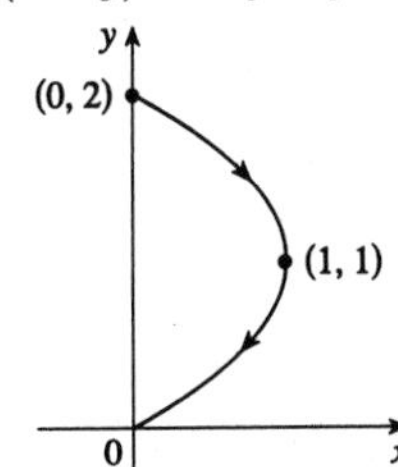

2. $x = t^2 + 1, y = t^2 - 1$
$\Rightarrow \quad y = x - 2, x \ge 1$

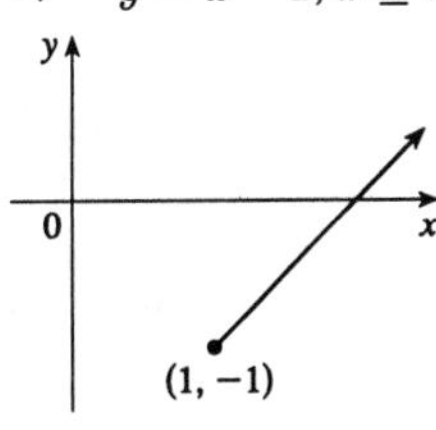

3. $x = 1 + \sin t, y = 2 + \cos t \quad \Rightarrow$
$(x - 1)^2 + (y - 2)^2 = \sin^2 t + \cos^2 t = 1$

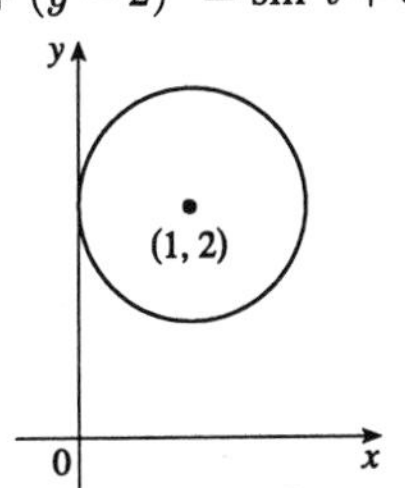

4. $x = 1 + \cos t, y = 1 + \sin^2 t \quad \Rightarrow$
$(x - 1)^2 + (y - 1) = 1 \quad \Leftrightarrow$
$(x - 1)^2 = -(y - 2), 0 \le x \le 2$

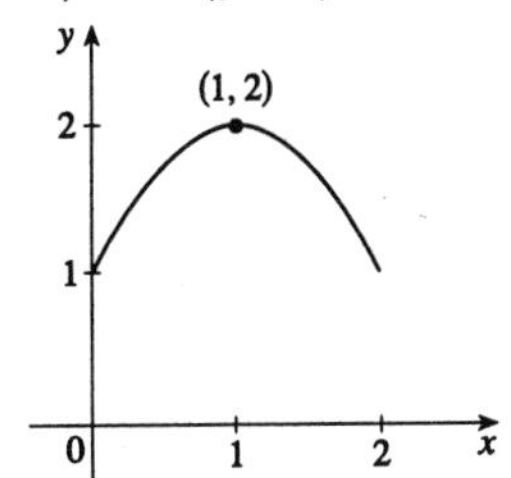

5. $r = 1 + 3\cos\theta$

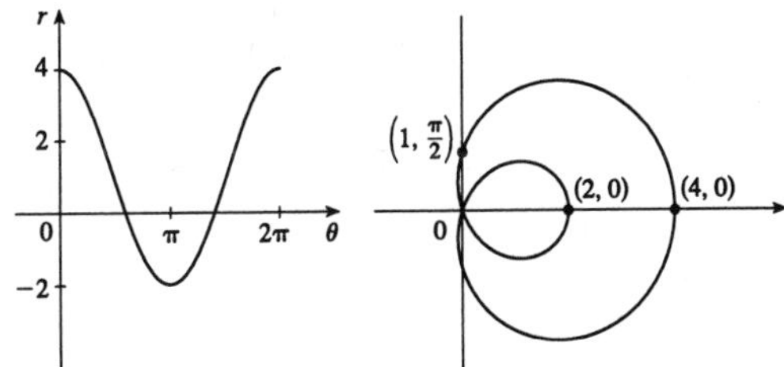

6. $r = 3 - \sin\theta$

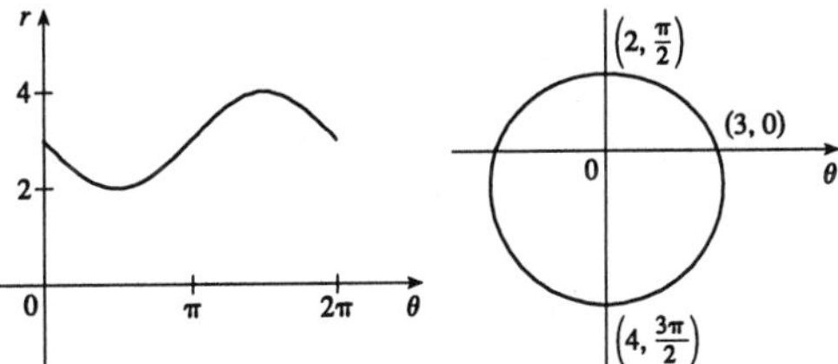

7. $r^2 = \sec 2\theta \quad \Rightarrow \quad r^2 \cos 2\theta = 1 \quad \Rightarrow$
$r^2(\cos^2\theta - \sin^2\theta) = 1 \quad \Rightarrow \quad x^2 - y^2 = 1,$
a hyperbola

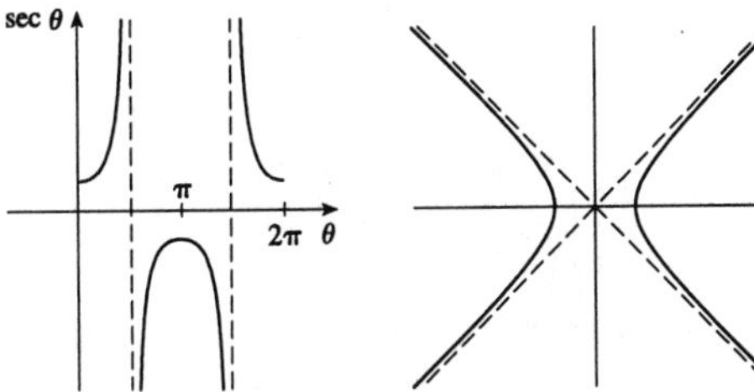

8. $r = \tan\theta \quad \Rightarrow \quad r\cos\theta = \sin\theta \quad \Rightarrow$
$x = \sin\theta$, so $|x| \le 1$.
The curve is also symmetric about the axes.

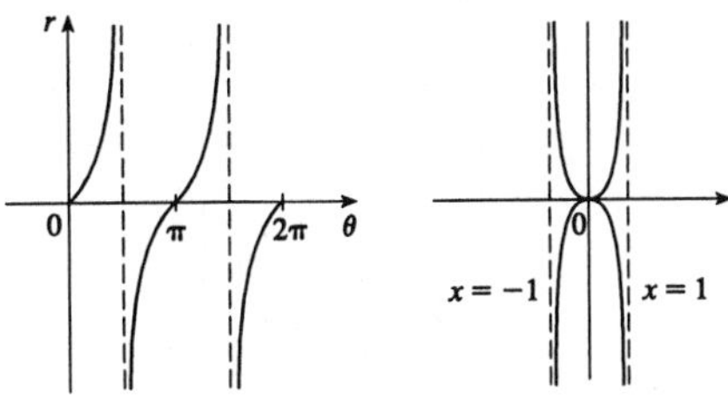

9. $r = 2\cos^2(\theta/2) = 1 + \cos\theta$

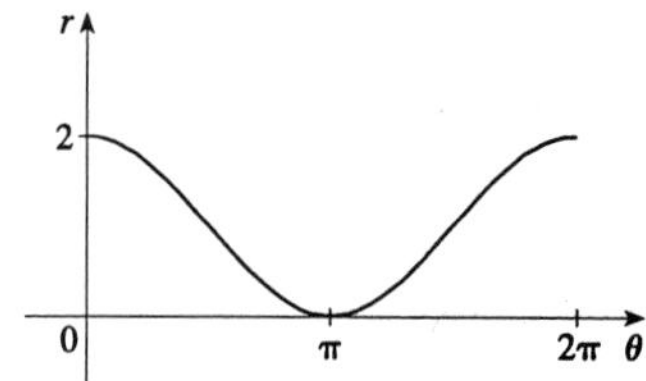

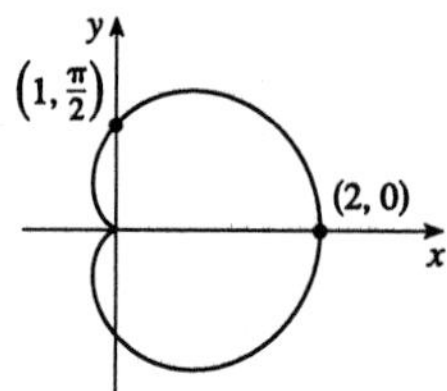

10. $r = 2\cos(\theta/2)$ The curve is symmetric about the pole and both the horizontal and vertical axes.

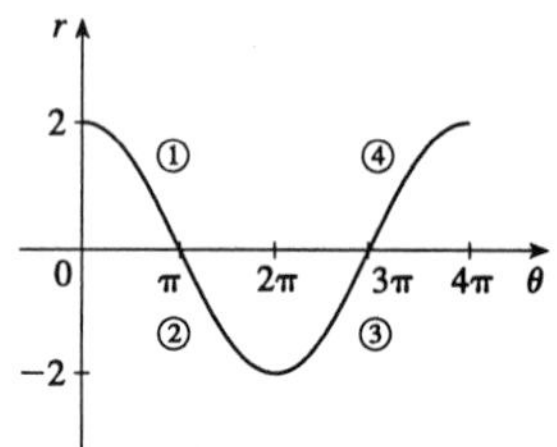

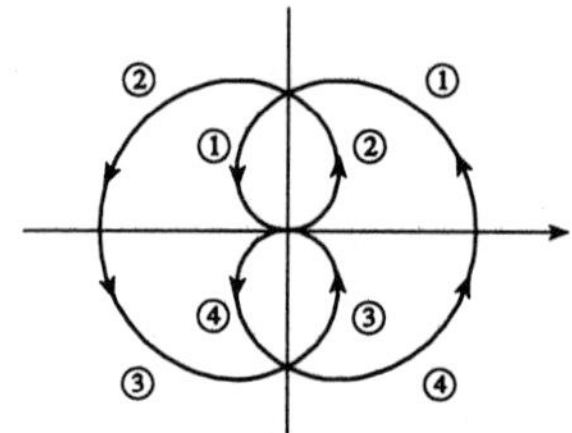

11. $r = \dfrac{1}{1+\cos\theta} \;\Rightarrow\; e = 1 \;\Rightarrow\;$ parabola; $d = 1 \;\Rightarrow\;$ directrix $x = 1$ and vertex $\left(\frac{1}{2}, 0\right)$; y-intercepts are $\left(1, \frac{\pi}{2}\right)$ and $\left(1, \frac{3\pi}{2}\right)$.

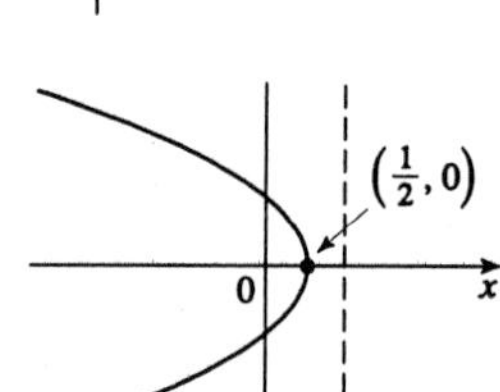

12. $r = \dfrac{5}{1 - 3\sin\theta} \;\Rightarrow\; e = 3 \;\Rightarrow\;$ hyperbola; $p = \frac{5}{3}$ $\;\Rightarrow\;$ directrix $y = -\frac{5}{3}$; vertices $\left(-\frac{5}{2}, \frac{1}{2}\pi\right) = \left(\frac{5}{2}, \frac{3}{2}\pi\right)$ and $\left(\frac{5}{4}, \frac{3}{2}\pi\right) \;\Rightarrow\;$ center $\left(\frac{15}{8}, \frac{3\pi}{2}\right)$, $a = \frac{5}{8}$, $c = \frac{15}{8}$ $\;\Rightarrow\; b = \frac{5}{2\sqrt{2}}$ and foci $(0, 0)$ and $\left(\frac{15}{4}, \frac{3}{2}\pi\right)$.

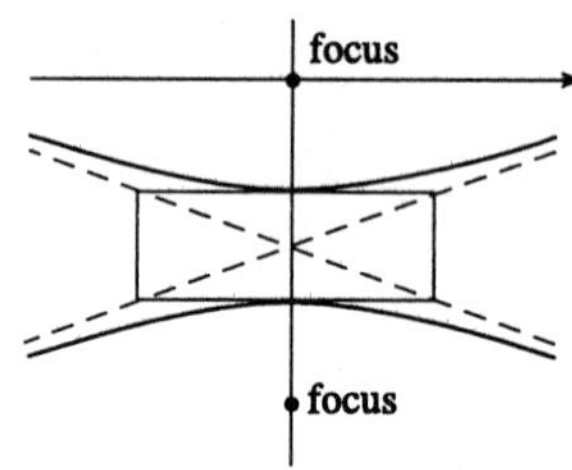

13. $x^2 + y^2 = 4x \;\Leftrightarrow\; r^2 = 4r\cos\theta \;\Leftrightarrow\; r = 4\cos\theta$

14. $x + y^2 = 0 \;\Leftrightarrow\; r\cos\theta + r^2\sin^2\theta = 0 \;\Leftrightarrow\; r = -\dfrac{\cos\theta}{\sin^2\theta}$ or $r = 0 \;\Leftrightarrow\; r = -\cot\theta\csc\theta$ (since $r = 0$ when $\theta = \frac{\pi}{2}$.)

15. $r = (\sin\theta)/\theta$. As $\theta \to \pm\infty$, $r \to 0$.

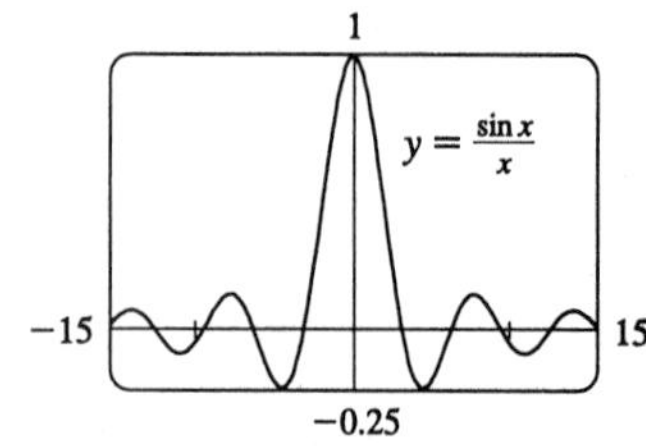

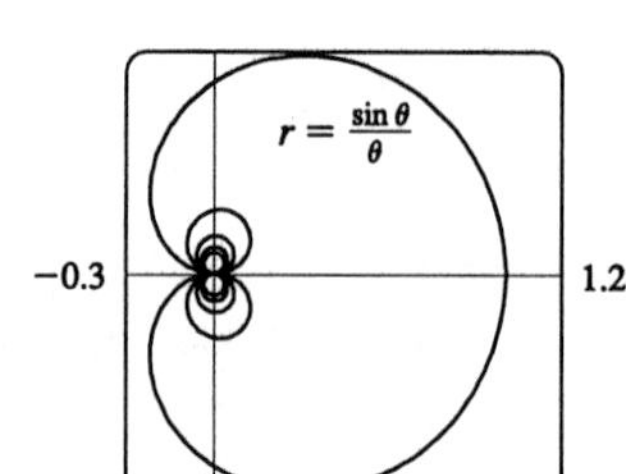

16. $r = \dfrac{2}{4 - 3\cos\theta} = \dfrac{1/2}{1 - \frac{3}{2}\cos\theta} \quad \Rightarrow \quad e = \frac{3}{2}$ and $d = \frac{2}{3}$.

The equation of the directrix is $x = r\cos\theta = -\frac{2}{3} \quad \Rightarrow$ $r = -2/(3\cos\theta)$. To obtain the equation of the rotated ellipse, we replace θ in the original equation with $\theta - \frac{2\pi}{3}$, and get $r = \dfrac{2}{4 - 3\cos\left(\theta - \frac{2\pi}{3}\right)}$.

17. $x = t^2 + 2t,\ y = t^3 - t.\ \dfrac{dy}{dx} = \dfrac{dy/dt}{dx/dt} = \dfrac{3t^2 - 1}{2t + 2} = \dfrac{1}{2}$ when $t = 1$

18. $x = te^t,\ y = 1 + \sqrt{1+t}.\ \dfrac{dy}{dx} = \dfrac{dy/dt}{dx/dt} = \dfrac{1/\left(2\sqrt{1+t}\right)}{(1+t)e^t} = \dfrac{1}{2}$ when $t = 0$

19. $\dfrac{dy}{dx} = \dfrac{(dr/d\theta)\sin\theta + r\cos\theta}{(dr/d\theta)\cos\theta - r\sin\theta} = \dfrac{\sin\theta + \theta\cos\theta}{\cos\theta - \theta\sin\theta} = \dfrac{\frac{1}{\sqrt{2}} + \frac{\pi}{4}\cdot\frac{1}{\sqrt{2}}}{\frac{1}{\sqrt{2}} - \frac{\pi}{4}\cdot\frac{1}{\sqrt{2}}} = \dfrac{4+\pi}{4-\pi}$ when $\theta = \dfrac{\pi}{4}$

20. $\dfrac{dy}{dx} = \dfrac{(dr/d\theta)\sin\theta + r\cos\theta}{(dr/d\theta)\cos\theta - r\sin\theta} = \dfrac{-2\cos\theta\sin\theta + (3 - 2\sin\theta)\cos\theta}{-2\cos^2\theta - (3 - 2\sin\theta)\sin\theta} = \dfrac{3\cos\theta - 2\sin 2\theta}{-3\sin\theta - 2\cos 2\theta} = 0$ when $\theta = \dfrac{\pi}{2}$

21. $\dfrac{dy}{dx} = \dfrac{dy/dt}{dx/dt} = \dfrac{t\cos t + \sin t}{-t\sin t + \cos t}.\ \dfrac{d^2y}{dx^2} = \dfrac{\frac{d}{dt}\left(\frac{dy}{dx}\right)}{dx/dt}$, where

$$\frac{d}{dt}\left(\frac{dy}{dx}\right) = \frac{(-t\sin t + \cos t)(-t\sin t + 2\cos t) - (t\cos t + \sin t)(-t\cos t - 2\sin t)}{(-t\sin t + \cos t)^2} = \frac{t^2+2}{(-t\sin t + \cos t)^2}$$

$$\Rightarrow \quad \frac{d^2y}{dx^2} = \frac{t^2+2}{(-t\sin t + \cos t)^3}$$

22. $x = t^6 + t^3,\ y = t^4 + t^2.\ \dfrac{dy}{dx} = \dfrac{dy/dt}{dx/dt} = \dfrac{4t^3 + 2t}{6t^5 + 3t^2} = \dfrac{4t^2 + 2}{6t^4 + 3t}$,

$$\frac{d}{dt}\left(\frac{dy}{dx}\right) = \frac{(6t^4 + 3t)8t - (4t^2 + 2)(24t^3 + 3)}{(6t^4 + 3t)^2} = \frac{-48t^5 - 48t^3 + 12t^2 - 6}{(6t^4 + 3t)^2},$$

$$\frac{d^2y}{dx^2} = \frac{\frac{d}{dt}\left(\frac{dy}{dx}\right)}{dx/dt} = \frac{-48t^5 - 48t^3 + 12t^2 - 6}{(6t^4 + 3t)^2(6t^5 + 3t^2)} = \frac{-16t^5 - 16t^3 + 4t^2 - 2}{9t^4(2t^3 + 1)^3}$$

23. We graph the curve for $-2.2 \le t \le 1.2$. By zooming in or using a cursor, we find that the lowest point is about $(1.4, 0.75)$. To find the exact values, we find the t-value at which $dy/dt = 2t + 1 = 0 \quad \Leftrightarrow$ $t = -\frac{1}{2} \quad \Leftrightarrow \quad (x, y) = \left(\frac{11}{8}, \frac{3}{4}\right)$.

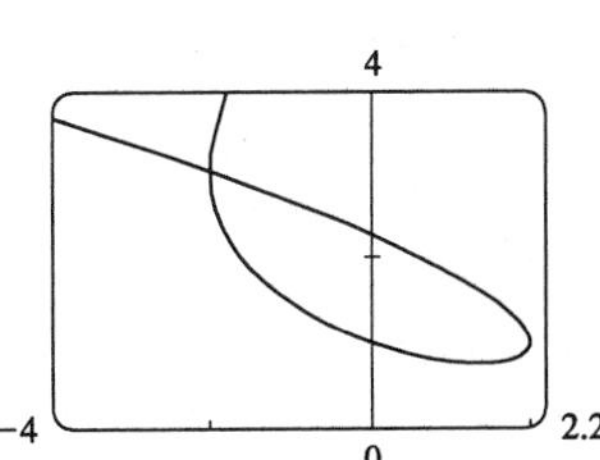

24. We estimate the coordinates of the point of intersection to be $(-2, 3)$. In fact this is exact, since both $t = -2$ and $t = 1$ give the point $(-2, 3)$. So the area enclosed by the loop is

$$\int_{t=-2}^{1} y\,dx = \int_{-2}^{1}(t^2 + t + 1)(3t^2 - 3)dt = \int_{-2}^{1}\left(3t^4 + 3t^3 - 3t - 3\right)dt$$

$$= \left[\tfrac{3}{5}t^5 + \tfrac{3}{4}t^4 - \tfrac{3}{2}t^2 - 3t\right]_{-2}^{1} = \left(\tfrac{3}{5} + \tfrac{3}{4} - \tfrac{3}{2} - 3\right) - \left[-\tfrac{96}{5} + 12 - 6 - (-6)\right] = \tfrac{81}{20}.$$

25. $dx/dt = -2a\sin t + 2a\sin 2t = 2a\sin t(2\cos t - 1) = 0 \iff \sin t = 0$ or $\cos t = \frac{1}{2} \Rightarrow t = 0, \frac{\pi}{3}, \pi$, or $\frac{5\pi}{3}$.

$dy/dt = 2a\cos t - 2a\cos 2t = 2a(1 + \cos t - 2\cos^2 t) = 2a(1 - \cos t)(1 + 2\cos t) = 0 \Rightarrow t = 0, \frac{2\pi}{3}$, or $\frac{4\pi}{3}$.

Thus the graph has vertical tangents where $t = \frac{\pi}{3}$, π and $\frac{5\pi}{3}$, and horizontal tangents where $t = \frac{2\pi}{3}$ and $\frac{4\pi}{3}$. To determine what the slope is where $t = 0$, we use l'Hospital's Rule to evaluate $\lim_{t\to 0} \dfrac{dy/dt}{dx/dt} = 0$, so there is a horizontal tangent there.

t	x	y
0	a	0
$\frac{\pi}{3}$	$\frac{3}{2}a$	$\frac{\sqrt{3}}{2}a$
$\frac{2\pi}{3}$	$-\frac{1}{2}a$	$\frac{3\sqrt{3}}{2}a$
π	$-3a$	0
$\frac{4\pi}{3}$	$-\frac{1}{2}a$	$-\frac{3\sqrt{3}}{2}a$
$\frac{5\pi}{3}$	$\frac{3}{2}a$	$-\frac{\sqrt{3}}{2}a$

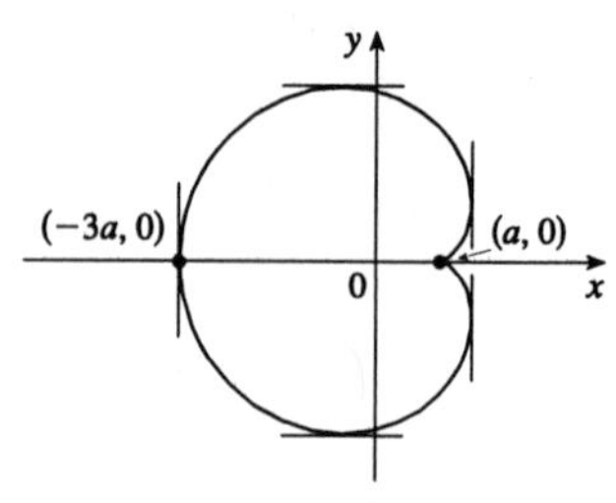

26. From Exercise 25, $x = 2a\cos t - a\cos 2t$, $y = 2a\sin t - a\sin 2t \Rightarrow$

$$A = 2\int_\pi^0 (2a\sin t - a\sin 2t)(-2a\sin t + 2a\sin 2t)\,dt = 4a^2\int_0^\pi (2\sin^2 t + \sin^2 2t - 3\sin t\sin 2t)\,dt$$
$$= 4a^2\int_0^\pi \left[(1 - \cos 2t) + \tfrac{1}{2}(1 - \cos 4t) - 6\sin^2 t\cos t\right]dt = 4a^2\left[t - \tfrac{1}{2}\sin 2t + \tfrac{1}{2}t - \tfrac{1}{8}\sin 4t - 2\sin^3 t\right]_0^\pi$$
$$= 4a^2\left(\tfrac{3}{2}\right)\pi = 6\pi a^2$$

27. This curve has 10 "petals." For instance, for $-\frac{\pi}{10} \le \theta \le \frac{\pi}{10}$, there are two petals, one with $r > 0$ and one with $r < 0$. $A = 10\int_{-\pi/10}^{\pi/10} \frac{1}{2}r^2\,d\theta = 5\int_{-\pi/10}^{\pi/10} 9\cos 5\theta\,d\theta = 90\int_0^{\pi/10}\cos 5\theta\,d\theta = [18\sin 5\theta]_0^{\pi/10} = 18$

28. $r = 1 - 3\sin\theta$. The inner loop is traced out as θ goes from $\alpha = \sin^{-1}\frac{1}{3}$ to $\pi - \alpha$, so

$$A = \int_\alpha^{\pi-\alpha} \tfrac{1}{2}r^2\,d\theta = \int_\alpha^{\pi/2}(1 - 3\sin\theta)^2\,d\theta = \int_\alpha^{\pi/2}\left[1 - 6\sin\theta + \tfrac{9}{2}(1 - \cos 2\theta)\right]d\theta$$
$$= \left[\tfrac{11}{2}\theta + 6\cos\theta - \tfrac{9}{4}\sin 2\theta\right]_\alpha^{\pi/2} = \tfrac{11}{4}\pi - \tfrac{11}{2}\sin^{-1}\tfrac{1}{3} - 3\sqrt{2}.$$

29. The curves intersect where $4\cos\theta = 2$; that is, at $\left(2, \frac{\pi}{3}\right)$ and $\left(2, -\frac{\pi}{3}\right)$.

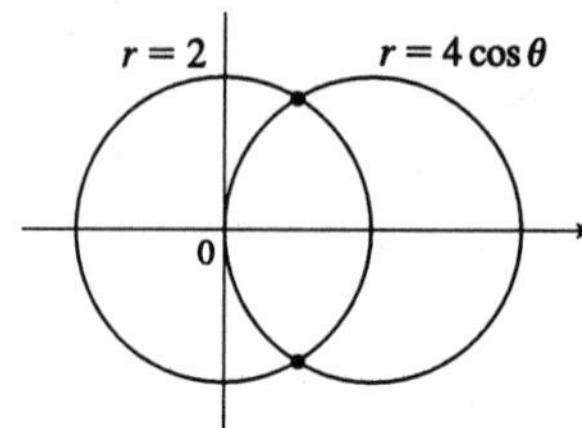

30. The two curves clearly both contain the pole. For other points of intersection, $\cot\theta = 2\cos(\theta + 2n\pi)$ or $-2\cos(\theta + \pi + 2n\pi)$, both of which reduce to $\cot\theta = 2\cos\theta \iff \cos\theta = 2\sin\theta\cos\theta \iff \cos\theta(1 - 2\sin\theta) = 0 \Rightarrow \cos\theta = 0$ or $\sin\theta = \frac{1}{2} \Rightarrow \theta = \frac{\pi}{6}, \frac{\pi}{2}, \frac{5\pi}{6}$ or $\frac{3\pi}{2} \Rightarrow$ intersection points are $\left(0, \frac{\pi}{2}\right)$, $\left(\sqrt{3}, \frac{\pi}{6}\right)$ and $\left(\sqrt{3}, \frac{11\pi}{6}\right)$.

31. The curves intersect where $2\sin\theta = \sin\theta + \cos\theta \Rightarrow \sin\theta = \cos\theta \Rightarrow \theta = \frac{\pi}{4}$, and also at the origin (at which $\theta = \frac{3\pi}{4}$ on the second curve.)

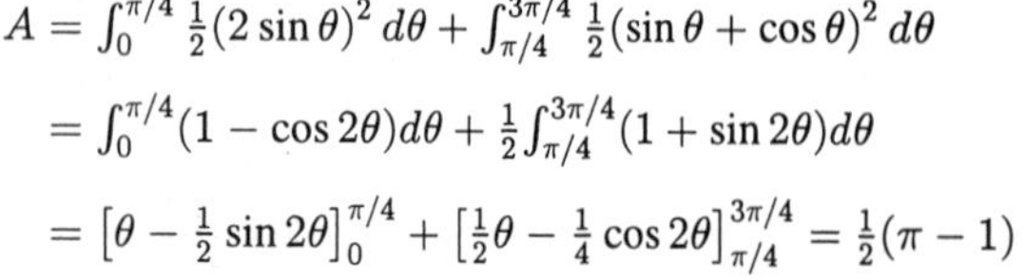

$$A = \int_0^{\pi/4} \tfrac{1}{2}(2\sin\theta)^2\,d\theta + \int_{\pi/4}^{3\pi/4} \tfrac{1}{2}(\sin\theta + \cos\theta)^2\,d\theta$$
$$= \int_0^{\pi/4}(1 - \cos 2\theta)\,d\theta + \tfrac{1}{2}\int_{\pi/4}^{3\pi/4}(1 + \sin 2\theta)\,d\theta$$
$$= \left[\theta - \tfrac{1}{2}\sin 2\theta\right]_0^{\pi/4} + \left[\tfrac{1}{2}\theta - \tfrac{1}{4}\cos 2\theta\right]_{\pi/4}^{3\pi/4} = \tfrac{1}{2}(\pi - 1)$$

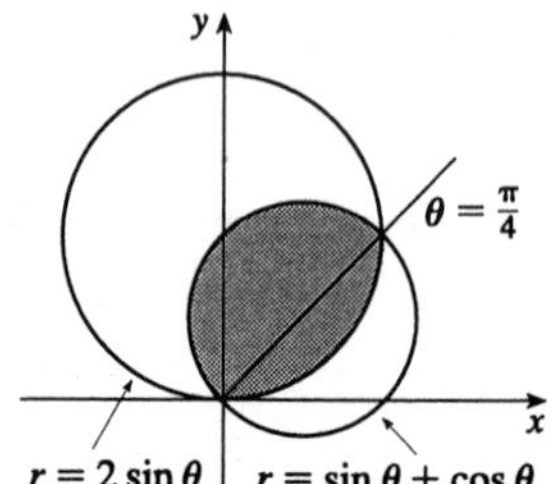

32. $A = 2\int_{-\pi/2}^{\pi/6} \frac{1}{2}\left[(2+\cos 2\theta)^2 - (2+\sin\theta)^2\right]d\theta$

$= \int_{-\pi/2}^{\pi/6}[4\cos 2\theta + \cos^2 2\theta - 4\sin\theta - \sin^2\theta]\,d\theta$

$= \left[2\sin 2\theta + \frac{1}{2}\theta + \frac{1}{8}\sin 4\theta + 4\cos\theta - \frac{1}{2}\theta + \frac{1}{4}\sin 2\theta\right]_{-\pi/2}^{\pi/6}$

$= \frac{51}{16}\sqrt{3}$

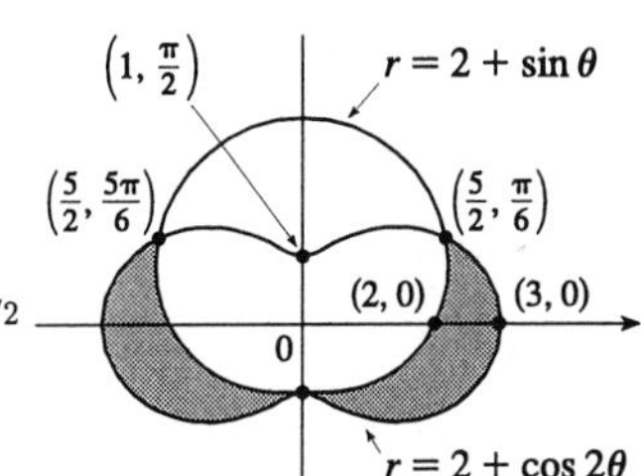

33. $x = 3t^2,\ y = 2t^3.\ L = \int_0^2 \sqrt{(dx/dt)^2 + (dy/dt)^2}\,dt = \int_0^2 \sqrt{(6t)^2 + (6t^2)^2}\,dt$

$= 6\int_0^2 t\sqrt{1+t^2}\,dt = \left[2(1+t^2)^{3/2}\right]_0^2 = 2\left(5\sqrt{5} - 1\right)$

34. $\left(\frac{dx}{dt}\right)^2 + \left(\frac{dy}{dt}\right)^2 = \left[-\sin t + \frac{\frac{1}{2}\sec^2(t/2)}{\tan(t/2)}\right]^2 + \cos^2 t = \left[-\sin t + \frac{1}{2\sin(t/2)\cos(t/2)}\right]^2 + \cos^2 t$

$= \left(-\sin t + \frac{1}{\sin t}\right)^2 + \cos^2 t = \csc^2 t - 1 = \cot^2 t$

$\Rightarrow \quad L = \int_{\pi/2}^{3\pi/4} |\cot t|\,dt = -\int_{\pi/2}^{3\pi/4} \cot t\,dt = \left[-\ln|\sin t|\right]_{\pi/2}^{3\pi/4} = \ln\sqrt{2}$

35. $L = \int_\pi^{2\pi} \sqrt{r^2 + (dr/d\theta)^2}\,d\theta = \int_\pi^{2\pi} \sqrt{(1/\theta)^2 + (-1/\theta^2)^2}\,d\theta = \int_\pi^{2\pi} \frac{\sqrt{\theta^2+1}}{\theta^2}\,d\theta$

$= \left[-\frac{\sqrt{\theta^2+1}}{\theta} + \ln\left|\theta + \sqrt{\theta^2+1}\right|\right]_\pi^{2\pi}$ (Formula 24) $= \frac{\sqrt{\pi^2+1}}{\pi} - \frac{\sqrt{4\pi^2+1}}{2\pi} + \ln\left|\frac{2\pi + \sqrt{4\pi^2+1}}{\pi + \sqrt{\pi^2+1}}\right|$

36. $L = \int_0^\pi \sqrt{r^2 + (dr/d\theta)^2}\,d\theta = \int_0^\pi \sqrt{\sin^6\left(\frac{1}{3}\theta\right) + \sin^4\left(\frac{1}{3}\theta\right)\cos^2\left(\frac{1}{3}\theta\right)}\,d\theta$

$= \int_0^\pi \sin^2\left(\frac{1}{3}\theta\right)d\theta = \left[\frac{1}{2}\left(\theta - \frac{3}{2}\sin\left(\frac{2}{3}\theta\right)\right)\right]_0^\pi = \frac{1}{2}\pi - \frac{3}{8}\sqrt{3}$

37. $S = \int_1^4 2\pi y\sqrt{(dx/dt)^2 + (dy/dt)^2}\,dt = \int_1^4 2\pi\left(\frac{1}{3}t^3 + \frac{1}{2}t^{-2}\right)\sqrt{\left(2/\sqrt{t}\right)^2 + (t^2 - t^{-3})^2}\,dt$

$= 2\pi\int_1^4 \left(\frac{1}{3}t^3 + \frac{1}{2}t^{-2}\right)\sqrt{(t^2 + t^{-3})^2}\,dt = 2\pi\int_1^4 \left(\frac{1}{3}t^5 + \frac{5}{6} + \frac{1}{2}t^{-5}\right)dt = 2\pi\left[\frac{1}{18}t^6 + \frac{5}{6}t - \frac{1}{8}t^{-4}\right]_1^4 = \frac{471{,}295}{1024}\pi$

38. From Exercise 34, we find that $S = \int_{\pi/2}^{3\pi/4} 2\pi\sin t|\cot t|\,dt = -2\pi\int_{\pi/2}^{3\pi/4}\cos t\,dt = \pi\left(2 - \sqrt{2}\right)$.

39. For all c except -1, the curve is asymptotic to the line $x = 1$. For $c < -1$, the curve bulges to the right near $y = 0$. As c increases, the bulge becomes smaller, until at $c = -1$ the curve is the straight line $x = 1$. As c continues to increase, the curve bulges to the left, until at $c = 0$ there is a cusp at the origin. For $c > 0$, there is a loop to the left of the origin, whose size and roundness increase as c increases. Note that the x-intercept of the curve is always $-c$.

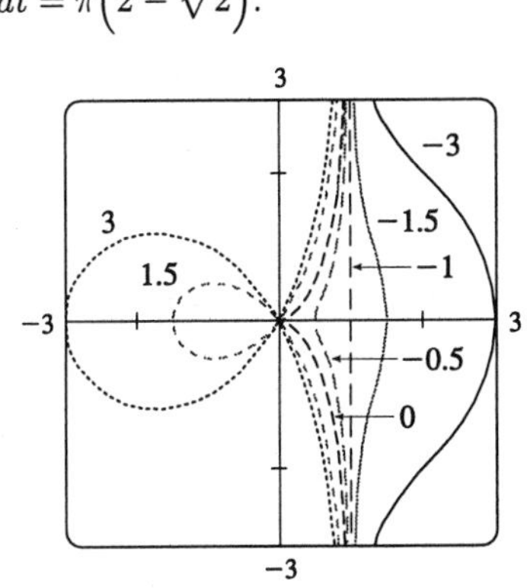

40. For a close to 0, the graph consists of four thin petals. As a increases, the petals get fatter, until as $a \to \infty$, each petal occupies almost its entire quarter-circle.

$a = 0.01$

$a = 0.1$

$a = 1$

$a = 5$

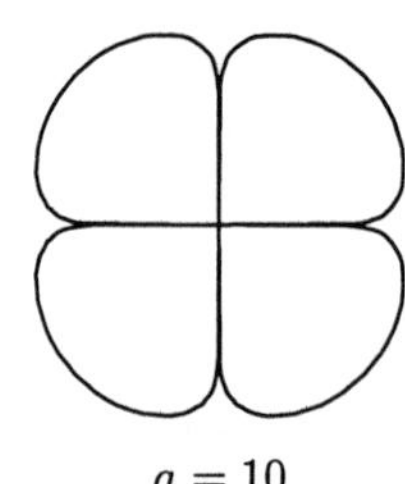

$a = 10$

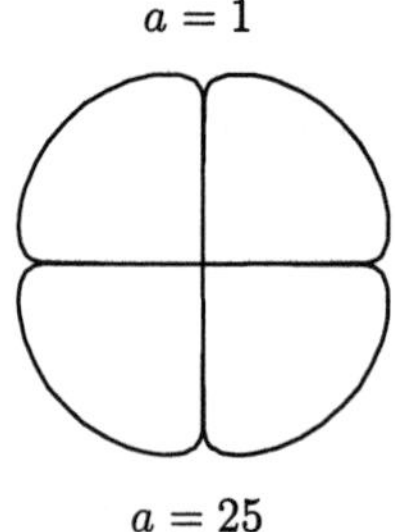

$a = 25$

41. Ellipse, center $(0, 0)$, $a = 3$, $b = 2\sqrt{2}$, $c = 1 \quad \Rightarrow \quad$ foci $(\pm 1, 0)$, vertices $(\pm 3, 0)$

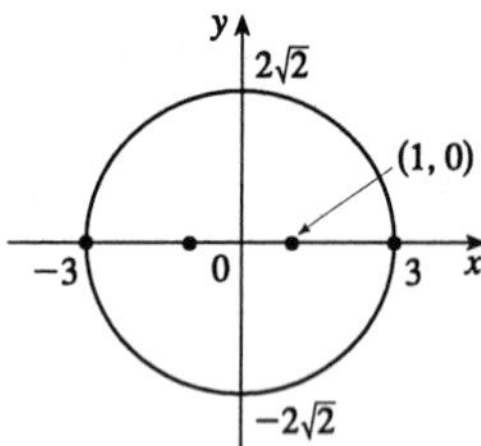

42. $\dfrac{x^2}{4} - \dfrac{y^2}{16} = 1$ is a hyperbola with center $(0, 0)$, vertices $(\pm 2, 0)$, $a = 2$, $b = 4$, $c = \sqrt{16 + 4} = 2\sqrt{5}$, foci $\left(\pm 2\sqrt{5}, 0\right)$ and asymptotes $y = \pm 2x$.

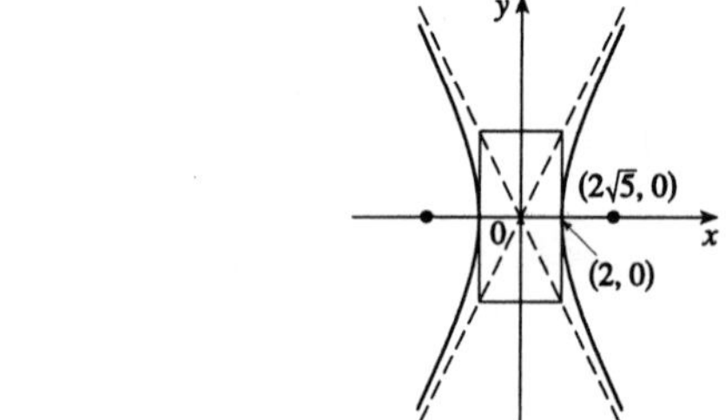

43. $6(y^2 - 6y + 9) = -(x + 1) \quad \Leftrightarrow$

$(y - 3)^2 = -\frac{1}{6}(x + 1)$, a parabola with vertex $(-1, 3)$, opening to the left, $p = -\frac{1}{24} \quad \Rightarrow \quad$ focus $\left(-\frac{25}{24}, 3\right)$ and directrix $x = -\frac{23}{24}$.

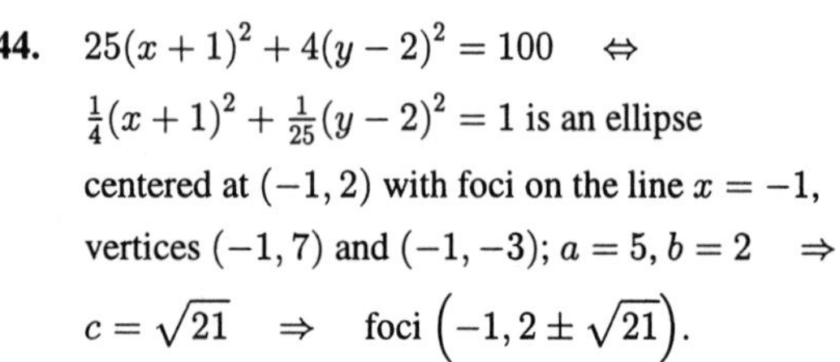

44. $25(x + 1)^2 + 4(y - 2)^2 = 100 \quad \Leftrightarrow$

$\frac{1}{4}(x + 1)^2 + \frac{1}{25}(y - 2)^2 = 1$ is an ellipse centered at $(-1, 2)$ with foci on the line $x = -1$, vertices $(-1, 7)$ and $(-1, -3)$; $a = 5$, $b = 2 \quad \Rightarrow$

$c = \sqrt{21} \quad \Rightarrow \quad$ foci $\left(-1, 2 \pm \sqrt{21}\right)$.

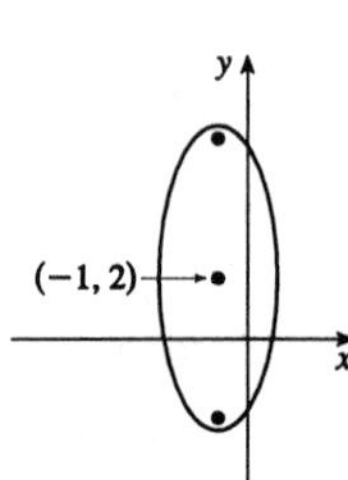

45. The parabola opens upward with vertex $(0, 4)$ and $p = 2$, so its equation is $(x - 0)^2 = 4 \cdot 2(y - 4) \quad \Leftrightarrow$
$x^2 = 8(y - 4)$.

46. Center is $(0, 0)$, and $c = 5$, $a = 2 \Rightarrow b = \sqrt{21}$; foci on y-axis $\Rightarrow$ equation of the hyperbola is $\dfrac{y^2}{4} - \dfrac{x^2}{21} = 1$.

47. The hyperbola has center $(0, 0)$ and foci on the x-axis. $c = 3$ and $b/a = \frac{1}{2}$ (from the asymptotes) $\Rightarrow$
$9 = c^2 = a^2 + b^2 = (2b)^2 + b^2 = 5b^2 \quad \Rightarrow \quad b = \frac{3}{\sqrt{5}} \quad \Rightarrow \quad a = \frac{6}{\sqrt{5}} \quad \Rightarrow$ the equation is $\dfrac{x^2}{36/5} - \dfrac{y^2}{9/5} = 1$
$\Leftrightarrow \quad 5x^2 - 20y^2 = 36$.

48. Center is $(3, 0)$, and $a = \frac{8}{2} = 4$, $c = 2 \quad \Leftrightarrow \quad b = \sqrt{4^2 - 2^2} = 2\sqrt{3} \quad \Rightarrow$ the equation of the ellipse is
$\dfrac{(x-3)^2}{12} + \dfrac{y^2}{16} = 1$.

49. $x^2 = -y + 100$ has its vertex at $(0, 100)$, so one of the vertices of the ellipse is $(0, 100)$. Another form of the equation of a parabola is $x^2 = 4p(y - 100)$ so $4p(y - 100) = -y + 100 \Rightarrow 4py - 4p(100) = 100 - y \Rightarrow$
$4p = \dfrac{100 - y}{y - 100} \Rightarrow p = -\frac{1}{4}$. Therefore the shared focus is found at $\left(0, \frac{399}{4}\right)$ so $2c = \frac{399}{4} - 0 \Rightarrow c = \frac{399}{8}$
and the center of the ellipse is $\left(0, \frac{399}{8}\right)$. So $a = 100 - \frac{399}{8} = \frac{401}{8}$ and $b^2 = a^2 - c^2 = \frac{401^2 - 399^2}{8^2} = 25$. So the equation of the ellipse is $\dfrac{x^2}{b^2} + \dfrac{\left(y - \frac{399}{8}\right)^2}{a^2} = 1 \quad \Rightarrow \quad \dfrac{x^2}{25} + \dfrac{\left(y - \frac{399}{8}\right)^2}{\left(\frac{401}{8}\right)^2} = 1$ or $\dfrac{x^2}{25} + \dfrac{(8y - 399)^2}{160{,}801} = 1$.

50. $\dfrac{x^2}{a^2} + \dfrac{y^2}{b^2} = 1 \quad \Rightarrow \quad \dfrac{2x}{a^2} + \dfrac{2y}{b^2}\dfrac{dy}{dx} = 0 \quad \Rightarrow \quad \dfrac{dy}{dx} = -\dfrac{b^2}{a^2}\dfrac{x}{y}$. Therefore $\dfrac{dy}{dx} = m \quad \Leftrightarrow \quad y = -\dfrac{b^2}{a^2}\dfrac{x}{m}$.
Combining this condition with $\dfrac{x^2}{a^2} + \dfrac{y^2}{b^2} = 1$, we find that $x = \pm\dfrac{a^2 m}{\sqrt{a^2m^2 + b^2}}$. In other words, the two points on the ellipse where the tangent has slope m are $\left(\pm\dfrac{a^2 m}{\sqrt{a^2m^2 + b^2}}, \mp\dfrac{b^2}{\sqrt{a^2m^2 + b^2}}\right)$. The tangent lines at these points have the equations $y \pm \dfrac{b^2}{\sqrt{a^2m^2 + b^2}} = m\left(x \mp \dfrac{a^2 m}{\sqrt{a^2m^2 + b^2}}\right)$ or
$y = mx \mp \dfrac{a^2m^2}{\sqrt{a^2m^2 + b^2}} \mp \dfrac{b^2}{\sqrt{a^2m^2 + b^2}} = mx \pm \sqrt{a^2m^2 + b^2}$.

51. Directrix $x = 4 \quad \Rightarrow \quad d = 4$, so $e = \frac{1}{3} \quad \Rightarrow \quad r = \dfrac{ed}{1 + e\cos\theta} = \dfrac{4}{3 + \cos\theta}$.

52. The asymptotes have slopes $\pm b/a = \pm\sqrt{e^2 - 1}$ (Equations 9.7.4), so the angles they make with the polar axis are $\pm\tan^{-1}\left[\sqrt{e^2 - 1}\right] = \cos^{-1}(\pm 1/e)$.

53. In polar coordinates an equation for the circle is $r = 2a\sin\theta$. Thus the coordinates of Q are
$x = r\cos\theta = 2a\sin\theta\cos\theta$ and $y = r\sin\theta = 2a\sin^2\theta$. The coordinates of R are $x = 2a\cot\theta$ and $y = 2a$.
Since P is the midpoint of QR, we use the midpoint formula to get $x = a(\sin\theta\cos\theta + \cot\theta)$ and
$y = a(1 + \sin^2\theta)$.

APPLICATIONS PLUS (page 593)

1. **(a)** While running from $(L,0)$ to (x,y), the dog travels a distance $s=\int_x^L\sqrt{1+(dy/dx)^2}\,dx$. The dog and rabbit run at the same speed, so the rabbit's position when the dog has traveled a distance s is $(0,s)$. Since the dog runs straight for the rabbit, $\dfrac{dy}{dx}=\dfrac{s-y}{0-x}$ (see the figure). Thus $s=y-x\dfrac{dy}{dx}$ and

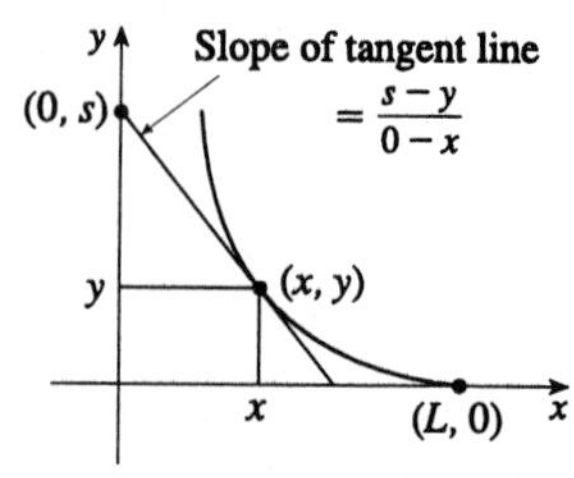

$$-\sqrt{1+\left(\frac{dy}{dx}\right)^2}=\frac{ds}{dx}=\frac{dy}{dx}-x\frac{d^2y}{dx^2}-1\frac{dy}{dx}=-x\frac{d^2y}{dx^2}.\text{ Hence }x\frac{d^2y}{dx^2}=\sqrt{1+\left(\frac{dy}{dx}\right)^2},\text{ as claimed.}$$

(b) Letting $z=\dfrac{dy}{dx}$, we obtain the differential equation $x\dfrac{dz}{dx}=\sqrt{1+z^2}$, or $\dfrac{dz}{\sqrt{1+z^2}}=\dfrac{dx}{x}$. Integrating:

$$\ln x=\int\frac{dz}{\sqrt{1+z^2}}\quad\begin{bmatrix}z=\tan\theta\\ dz=\sec^2\theta\,d\theta\end{bmatrix}\quad=\int(\sec\theta)d\theta=\ln|\sec\theta+\tan\theta|+C=\ln\left|\sqrt{1+z^2}+z\right|+C.$$

When $x=L$, $z=dy/dx=0$, so $\ln L=\ln 1+C$. Therefore $C=\ln L$, so

$\ln x=\ln\left(\sqrt{1+z^2}+z\right)+\ln L=\ln\left[L\left(\sqrt{1+z^2}+z\right)\right]$ and $x=L\left(\sqrt{1+z^2}+z\right)$. $\sqrt{1+z^2}=\dfrac{x}{L}-z$

$$\Rightarrow\quad 1+z^2=\left(\frac{x}{L}\right)^2-\frac{2xz}{L}+z^2\quad\Rightarrow\quad\left(\frac{x}{L}\right)^2-2z\left(\frac{x}{L}\right)-1=0\quad\Rightarrow$$

$z=\dfrac{(x/L)^2-1}{2(x/L)}=\dfrac{x^2-L^2}{2Lx}=\dfrac{x}{2L}-\dfrac{L}{2}\dfrac{1}{x}$ (for $x>0$). Since $z=\dfrac{dy}{dx}$, $y=\dfrac{x^2}{4L}-\dfrac{L}{2}\ln x+C'$. But $y=0$ when $x=L$. Therefore $0=\dfrac{L}{4}-\dfrac{L}{2}\ln L+C'$ and $C'=\dfrac{L}{2}\ln L-\dfrac{L}{4}$. Therefore

$$y=\frac{x^2}{4L}-\frac{L}{2}\ln x+\frac{L}{2}\ln L-\frac{L}{4}=\frac{x^2-L^2}{4L}-\frac{L}{2}\ln\left(\frac{x}{L}\right).$$

(c) As $x\to 0$, $y\to\infty$, so the dog never catches the rabbit.

2. **(a)** If the dog runs twice as fast as the rabbit, then the rabbit's position when the dog has traveled a distance s is $(0,s/2)$. Since the dog runs straight toward the rabbit, the tangent line to the dog's path has slope $\dfrac{dy}{dx}=\dfrac{s/2-y}{0-x}$. Thus $s=2y-2x\dfrac{dy}{dx}$ and $-\sqrt{1+\left(\dfrac{dy}{dx}\right)^2}=\dfrac{ds}{dx}=2\dfrac{dy}{dx}-2x\dfrac{d^2y}{dx^2}-2\dfrac{dy}{dx}=-2x\dfrac{d^2y}{dx^2}$.

Hence $2x\dfrac{d^2y}{dx^2}=\sqrt{1+\left(\dfrac{dy}{dx}\right)^2}$.

(b) Letting $z=\dfrac{dy}{dx}$, we obtain the differential equation $2x\dfrac{dz}{dx}=\sqrt{1+z^2}$, or $\dfrac{2\,dz}{\sqrt{1+z^2}}=\dfrac{dx}{x}$. Integrating, we get $\ln x=\int\dfrac{2\,dz}{\sqrt{1+z^2}}=2\ln\left|\sqrt{1+z^2}+z\right|+C$. [See Exercise 1(b).] When $x=L$, $z=\dfrac{dy}{dx}=0$, so $\ln L=2\ln 1+C=C$. Thus $\ln x=2\ln\left(\sqrt{1+z^2}+z\right)+\ln L=\ln\left[L\left(\sqrt{1+z^2}+z\right)^2\right]$ and $x=L\left(\sqrt{1+z^2}+z\right)^2$.

Therefore $\sqrt{1+z^2} = \sqrt{\frac{x}{L}} - z \quad\Rightarrow\quad 1+z^2 = \frac{x}{L} - 2\sqrt{\frac{x}{L}}z + z^2 \quad\Rightarrow\quad 2\sqrt{\frac{x}{L}}z = \frac{x}{L} - 1 \quad\Rightarrow$

$\frac{dy}{dx} = z = \frac{1}{2}\sqrt{\frac{x}{L}} - \frac{1}{2\sqrt{x/L}} = \frac{1}{2\sqrt{L}}x^{1/2} - \frac{\sqrt{L}}{2}x^{-1/2} \quad\Rightarrow\quad y = \frac{1}{3\sqrt{L}}x^{3/2} - \sqrt{L}x^{1/2} + C'$.

When $x = L$, $y = 0$, so $0 = \frac{1}{3\sqrt{L}}L^{3/2} - \sqrt{L}L^{1/2} + C' = \frac{L}{3} - L + C' = C' - \frac{2}{3}L$. Therefore $C' = \frac{2}{3}L$

and $y = \frac{x^{3/2}}{3\sqrt{L}} - \sqrt{L}x^{1/2} + \frac{2}{3}L$. As $x \to 0$, $y \to \frac{2}{3}L$, so the dog catches the rabbit when the rabbit is at

$\left(0, \frac{2}{3}L\right)$. (At that point, the dog has traveled a distance $\frac{4}{3}L$, twice as far as the rabbit has run.)

3. **(a)**

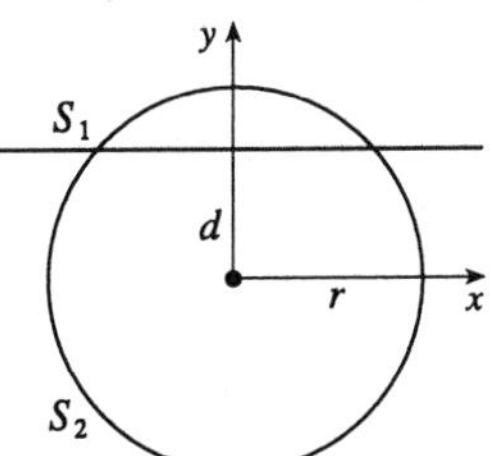

The two spherical zones, whose surface areas we will call S_1 and S_2, are generated by rotation about the y-axis of circular arcs, as indicated in the figure. The arcs are the upper and lower portions of the circle $x^2 + y^2 = r^2$ that are obtained when the circle is cut with the line $y = d$. The portion of the upper arc in the first quadrant is sufficient to generate the upper spherical zone. That portion of the arc can be described by the relation $x = \sqrt{r^2 - y^2}$ for $d \le y \le r$. Thus $\frac{dy}{dx} = \frac{-y}{\sqrt{r^2-y^2}}$ and

$ds = \sqrt{1 + \left(\frac{dx}{dy}\right)^2}\,dy = \sqrt{1 + \frac{y^2}{r^2-y^2}}\,dy = \sqrt{\frac{r^2}{r^2-y^2}}\,dy = \frac{r\,dy}{\sqrt{r^2-y^2}}$. From Formula 8.3.8 we have

$S_1 = \int_d^r 2\pi x\sqrt{1+\left(\frac{dx}{dy}\right)^2}\,dy = \int_d^r 2\pi\sqrt{r^2-y^2}\frac{r\,dy}{\sqrt{r^2-y^2}} = \int_d^r 2\pi r\,dy = 2\pi r(r-d)$. Similarly, we can compute $S_2 = \int_{-r}^d 2\pi x\sqrt{1+\left(\frac{dx}{dy}\right)^2}\,dy = \int_{-r}^d 2\pi r\,dy = 2\pi r(r+d)$. Note that $S_1 + S_2 = 4\pi r^2$, the surface area of the entire sphere.

(b)

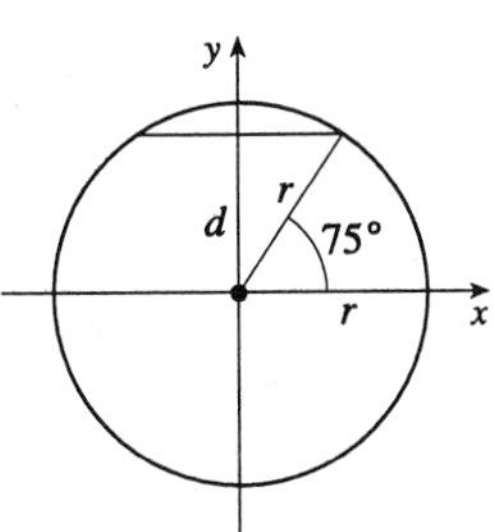

$r = 3960$ mi and $d = r(\sin 75°) \approx 3825$ mi, so the surface area of the Arctic Ocean is about $2\pi r(r-d) \approx 2\pi(3960)(135) \approx 3.36 \times 10^6\ \text{mi}^2$.

(c) The area on the sphere lies between planes $y = y_1$ and $y = y_2$, where $y_2 - y_1 = h$. Thus we compute the surface area on the sphere to be $S = \int_{y_1}^{y_2} 2\pi x \sqrt{1 + \left(\frac{dx}{dy}\right)^2}\, dy = \int_{y_1}^{y_2} 2\pi r\, dy = 2\pi r(y_2 - y_1) = 2\pi rh$.

This equals the lateral area of a cylinder of radius r and height h, since such a cylinder is obtained by rotating the line $x = r$ about the y-axis, so the surface area of the cylinder between the planes $y = y_1$ and $y = y_2$ is

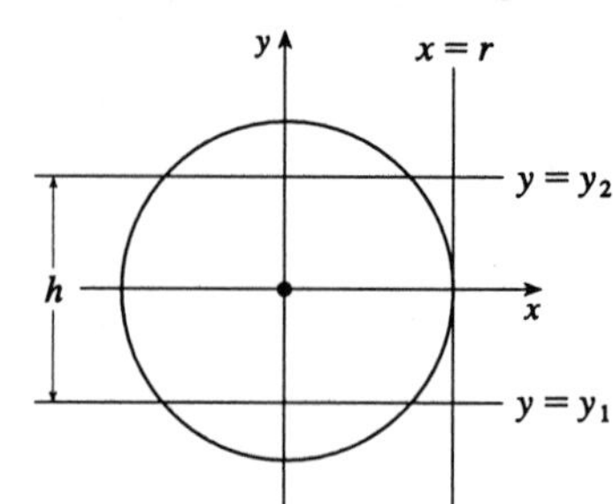

$$A = \int_{y_1}^{y_2} 2\pi x \sqrt{1 + \left(\frac{dx}{dy}\right)^2}\, dy = \int_{y_1}^{y_2} 2\pi r \sqrt{1 + 0^2}\, dy$$
$$= 2\pi r y\Big|_{y=y_1}^{y_2} = 2\pi r(y_2 - y_1) = 2\pi rh.$$

(d) $h = 2r \sin 23.45° \approx 3152$ mi, so the surface area of the Torrid Zone is $2\pi rh \approx 2\pi(3960)(3152) \approx 7.84 \times 10^7\ \text{mi}^2$.

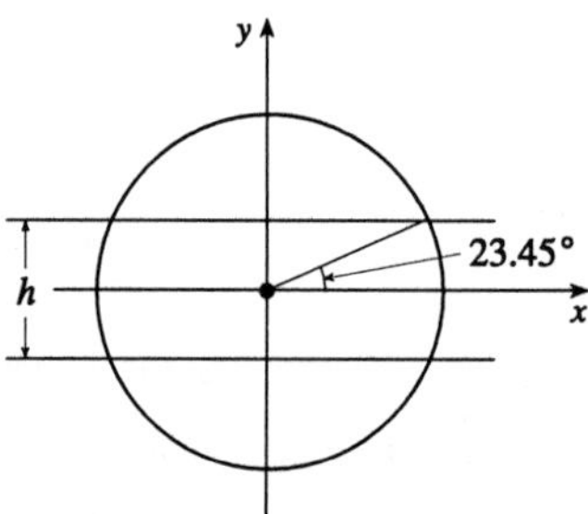

4. (a) $\frac{dx}{dt} = k(a - x)(b - x) - rx^2$, where $k > r$. Thus $\frac{dx}{dt} = (k - r)x^2 - k(a + b)x + kab$. Call the right-hand quadratic polynomial $q(x)$. The roots of $q(x)$ are the distinct positive numbers $\frac{k(a + b) \pm \sqrt{k^2(a + b)^2 - 4(k - r)kab}}{2(k - r)}$. Notice that

$$k^2(a + b)^2 - 4(k - r)kab = k^2(a + b)^2 - 4k^2ab + 4rkab = k^2\left[(a + b)^2 - 4ab\right] + 4rkab$$
$$= k^2\left(a^2 + 2ab + b^2 - 4ab\right) + 4rkab = k^2\left(a^2 - 2ab + b^2\right) + 4rkab = k^2(a - b)^2 + 4rkab > 0,$$

and that $k(a + b) = \sqrt{k^2(a + b)^2} > \sqrt{k^2(a + b)^2 - 4(k - r)kab}$, so the roots are real, distinct, and positive. Call the roots c_1 and c_2, where $0 < c_1 < c_2$. Then $q(x) = (k - r)(x - c_1)(x - c_2)$. Now

$$\frac{dx}{dt} = (k - r)(x - c_1)(x - c_2) \quad\Rightarrow\quad \frac{dx}{(x - c_1)(x - c_2)} = (k - r)dt \quad\Rightarrow$$

$$\int \frac{dx}{(x - c_1)(x - c_2)} = (k - r)t + C \quad\Rightarrow\quad \int \left[\frac{1/(c_1 - c_2)}{x - c_1} + \frac{1/(c_2 - c_1)}{x - c_2}\right] dx = (k - r)t + C \quad\Rightarrow$$

$$\frac{1}{c_2 - c_1} \int \left(\frac{1}{x - c_2} - \frac{1}{x - c_1}\right) dx = (k - r)t + C \quad\Rightarrow$$

$$\frac{1}{c_2 - c_1}\left[\ln|x - c_2| - \ln|x - c_1|\right] = (k - r)t + C \quad\Rightarrow\quad \frac{1}{c_2 - c_1} \ln\left|\frac{x - c_2}{x - c_1}\right| = (k - r)t + C \quad\Rightarrow$$

$$\ln\left|\frac{x - c_2}{x - c_1}\right| = (k - r)(c_2 - c_1)t + C' \quad [\text{where } C' = (c_2 - c_1)C] \quad\Rightarrow\quad \left|\frac{x - c_2}{x - c_1}\right| = M' e^{(k-r)(c_2-c_1)t}$$

$$[\text{where } M' = e^{C'}] \quad\Rightarrow\quad \frac{x - c_2}{x - c_1} = Me^{(k-r)(c_2-c_1)t} \quad [\text{where } M = \pm M'].$$

Now $x(0)=0 \quad\Rightarrow\quad \dfrac{c_2}{c_1}=Me^0=M \quad\Rightarrow\quad x-c_2=(x-c_1)\dfrac{c_2}{c_1}e^{(k-r)(c_2-c_1)t}$. Let's temporarily write $E=e^{(k-r)(c_2-c_1)t}$ for simplicity. Then $x-c_2=(x-c_1)\dfrac{c_2}{c_1}E=\dfrac{c_2}{c_1}xE-c_2E \quad\Rightarrow$

$$x\left(1-\frac{c_2}{c_1}E\right)=c_2(1-E) \quad\Rightarrow$$

$$x=\frac{c_2(1-E)}{1-(c_2/c_1)E}=\frac{c_1c_2(1-E)}{c_1-c_2E}=c_1\frac{c_2-c_2E}{c_1-c_2E}$$

$$=c_1\frac{(c_2-c_1)+(c_1-c_2E)}{c_1-c_2E}=c_1\left(\frac{c_2-c_1}{c_1-c_2E}+1\right)=c_1+\frac{c_1(c_2-c_1)}{c_1-c_2E}$$

$$=c_1-\frac{c_1(c_2-c_1)}{c_2E-c_1}.$$

We have shown that $x(t)=c_1-\dfrac{c_1(c_2-c_1)}{c_2e^{(k-r)(c_2-c_1)t}-c_1}$. [Check that $x(0)=0$.]

(b) As $t\to\infty$, $c_2e^{(k-r)(c_2-c_1)t}\to\infty$, so $\lim\limits_{t\to\infty}x(t)=c_1$. Reminder:

$$c_1=\frac{1}{2(k-r)}\left[k(a+b)-\sqrt{k^2(a+b)^2-4(k-r)kab}\right] \text{ and}$$

$$c_2=\frac{1}{2(k-r)}\left[k(a+b)+\sqrt{k^2(a+b)^2-4(k-r)kab}\right], \text{ so}$$

$c_2-c_1=\dfrac{1}{k-r}\sqrt{k^2(a+b)^2-4(k-r)kab}$. We have avoided use of these expressions in order to make the formula for $x(t)$ as readable as possible.

5. **(a)** $\dfrac{dy}{dt}=ky^{1+\varepsilon} \quad\Rightarrow\quad y^{-1-\varepsilon}\,dy=k\,dt \quad\Rightarrow\quad \dfrac{y^{-\varepsilon}}{-\varepsilon}=kt+C$. Since $y(0)=y_0$, we have $C=\dfrac{y_0^{-\varepsilon}}{-\varepsilon}$. Thus $\dfrac{y^{-\varepsilon}}{-\varepsilon}=kt+\dfrac{y_0^{-\varepsilon}}{-\varepsilon}$, or $y^{-\varepsilon}=y_0^{-\varepsilon}-\varepsilon kt$. So $y^\varepsilon=\dfrac{1}{y_0^{-\varepsilon}-\varepsilon kt}=\dfrac{y_0^\varepsilon}{1-\varepsilon y_0^\varepsilon kt}$ and $y(t)=\dfrac{y_0}{(1-\varepsilon y_0^\varepsilon kt)^{1/\varepsilon}}$.

(b) $y(t)\to\infty$ as $1-\varepsilon y_0^\varepsilon kt\to 0$, that is, as $t\to\dfrac{1}{\varepsilon y_0^\varepsilon k}$. Define $T=\dfrac{1}{\varepsilon y_0^\varepsilon k}$. Then $\lim\limits_{t\to T^-}y(t)=\infty$.

(c) According to the data given, we have $\varepsilon=0.01$, $y(0)=2$, and $y(3)=16$, where the time t is given in months. Thus $y_0=2$ and $16=y(3)=\dfrac{y_0}{(1-\varepsilon y_0^\varepsilon k\cdot 3)^{1/\varepsilon}}$. We could solve for k, but it is easier and more helpful to solve for $\varepsilon y_0^\varepsilon k$. $\left(k \text{ turns out to be } \dfrac{1-8^{-0.01}}{(0.03)(2^{0.01})}\approx 0.68125.\right)$

$16=\dfrac{2}{(1-3\varepsilon y_0^\varepsilon k)^{100}}$, so $1-3\varepsilon y_0^\varepsilon k=\left(\frac{1}{8}\right)^{0.01}=8^{-0.01}$ and $\varepsilon y_0^\varepsilon k=\frac{1}{3}(1-8^{-0.01})$. Thus doomsday occurs when $t=T=\dfrac{1}{\varepsilon y_0^\varepsilon k}=\dfrac{3}{1-8^{-0.01}}\approx 145.77$ months or 12.15 years.

6. **(a)** $\dfrac{dy}{dt} = k\cos(rt-\phi)y \quad\Rightarrow\quad \dfrac{dy}{y} = k\cos(rt-\phi)dt \quad\Rightarrow$

$\ln y = k\int\cos(rt-\phi)dt = \dfrac{k}{r}\sin(rt-\phi)+C$. (Since this is a growth model, $y>0$ and we can write $\ln y$ instead of $\ln|y|$.) Since $y(0)=y_0$, we obtain $\ln y_0 = \dfrac{k}{r}\sin(-\phi)+C = -\dfrac{k}{r}\sin\phi + C$ or $C = \ln y_0 + \dfrac{k}{r}\sin\phi$. Thus $\ln y = \dfrac{k}{r}\sin(rt-\phi)+\ln y_0 + \dfrac{k}{r}\sin\phi$, which we can rewrite as $\ln\dfrac{y}{y_0} = \dfrac{k}{r}[\sin(rt-\phi)+\sin\phi]$ or, after exponentiation, $y(t) = y_0e^{(k/r)[\sin(rt-\phi)+\sin\phi]}$. Thus $y(t)$ oscillates between $y_0e^{(k/r)(1+\sin\phi)}$ and $y_0e^{(k/r)(-1+\sin\phi)}$ (the extrema are attained when $rt-\phi$ is an odd multiple of $\frac{\pi}{2}$], so $\lim_{t\to\infty}y(t)$ does not exist.

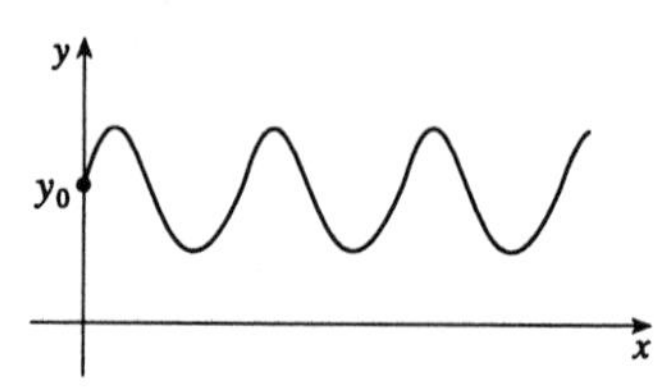

(b) $dy/dt = k\cos^2(rt-\phi)y \quad\Rightarrow\quad dy/y = k\cos^2(rt-\phi)dt \quad\Rightarrow$

$$\ln y = k\int\cos^2(rt-\phi)dt = k\int\frac{1+\cos[2(rt-\phi)]}{2}\,dt = \frac{k}{2}t + \frac{k}{4r}\sin[2(rt-\phi)]+C.$$

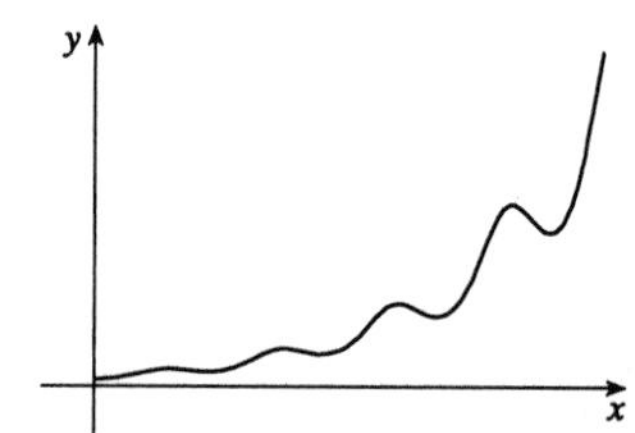

From $y(0)=y_0$, we get $\ln y_0 = \dfrac{k}{4r}\sin(-2\phi)+C = C - \dfrac{k}{4r}\sin 2\phi$, so $C = \ln y_0 + \dfrac{k}{4r}\sin 2\phi$ and $\ln y = \dfrac{k}{2}t + \dfrac{k}{4r}\sin[2(rt-\phi)] + \ln y_0 + \dfrac{k}{4r}\sin 2\phi$. Simplifying, we get $\ln\dfrac{y}{y_0} = \dfrac{k}{2}t + \dfrac{k}{4r}[\sin 2(rt-\phi)+\sin 2\phi]$, or $y(t) = y_0e^{kt/2+(k/4r)[\sin 2(rt-\phi)+\sin 2\phi]}$. Let $f(t) = \dfrac{k}{2}t + \dfrac{k}{4r}[\sin 2(rt-\phi)+\sin 2\phi]$. Then $f'(t) = \dfrac{k}{2}t + \dfrac{k}{4r}[2r\cos 2(rt-\phi)]$ $= \dfrac{k}{2}[1+\cos 2(rt-\phi)] \ge 0$. Since $y(t) = y_0e^{f(t)}$, we have $y'(t) = y_0f'(t)e^{f(t)} \ge 0$, with equality only when $\cos 2(rt-\phi) = -1$, that is, when $rt-\phi$ is an odd multiple of $\frac{\pi}{2}$. Therefore $y(t)$ is an increasing function on $[0,\infty)$. $y(t) = y_0e^{kt/2}e^{(k/4r)[\sin 2(rt-\phi)+\sin 2\phi]}$. The second exponential oscillates between $e^{(k/4r)(1+\sin 2\phi)}$ and $e^{(k/4r)(-1+\sin 2\phi)}$, while the first one, $e^{kt/2}$, grows without bound. So $\lim_{t\to\infty}y(t) = \infty$.

7. **(a)** If $\tan\theta = \sqrt{\dfrac{y}{C-y}}$, then $\tan^2\theta = \dfrac{y}{C-y}$, so $C\tan^2\theta - y\tan^2\theta = y$ and

$$y = \frac{C\tan^2\theta}{1+\tan^2\theta} = \frac{C\tan^2\theta}{\sec^2\theta} = C\tan^2\theta\cos^2\theta = C\sin^2\theta = \frac{C}{2}(1-\cos 2\theta).\ \text{Now}$$

$$dx = \sqrt{\frac{y}{C-y}}\,dy = \tan\theta\cdot\frac{C}{2}\cdot 2\sin 2\theta\,d\theta = C\tan\theta\cdot 2\sin\theta\cos\theta\,d\theta = 2C\sin^2\theta\,d\theta = C(1-\cos 2\theta)d\theta.$$

Thus $x = C\left(\theta - \frac{1}{2}\sin 2\theta\right) + K$ for some constant K. When $\theta = 0$, we have $y = 0$. We require that $x = 0$ when $\theta = 0$ so that the curve passes through the origin when $\theta = 0$. This yields $K = 0$. We now have $x = \frac{1}{2}C(2\theta - \sin 2\theta)$, $y = \frac{1}{2}C(1-\cos 2\theta)$.

(b) Setting $\phi = 2\theta$ and $r = \frac{1}{2}C$, we get $x = r(\phi - \sin\phi)$, $y = r(1-\cos\phi)$. Comparison with Equations 9.1.1 shows that the curve is a cycloid.

8. **(a)** Since the smaller circle rolls without slipping around C, the amount of arc traversed on C ($2r\theta$ in the figure) must equal the amount of arc of the smaller circle that has been in contact with C. Since the smaller circle has radius r, it must have turned through an angle of $2r\theta/r = 2\theta$. In addition to turning through an angle 2θ, the little circle has rolled through an angle θ against C. Thus P has turned through an angle of 3θ as shown in the figure.

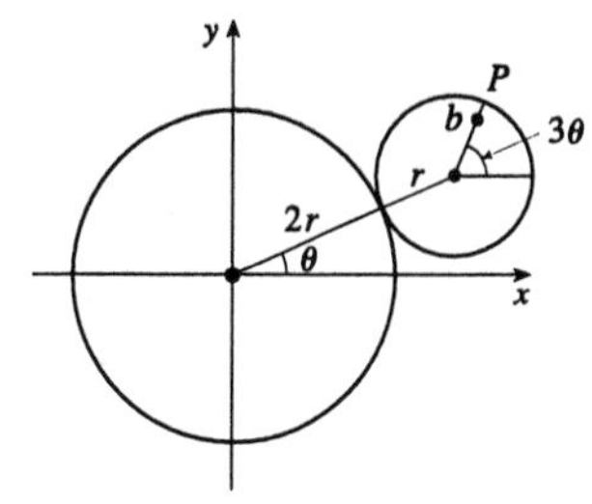

(If the little circle had turned through an angle of 2θ with its center pinned to the x-axis, then P would have turned only 2θ instead of 3θ. The movement of the little circle around C adds θ to the angle.) From the figure, we see that the center of the small circle has coordinates $(3r\cos\theta, 3r\sin\theta)$. Thus P has coordinates (x, y), where $x = 3r\cos\theta + b\cos 3\theta$ and $y = 3r\sin\theta + b\sin 3\theta$.

(b)

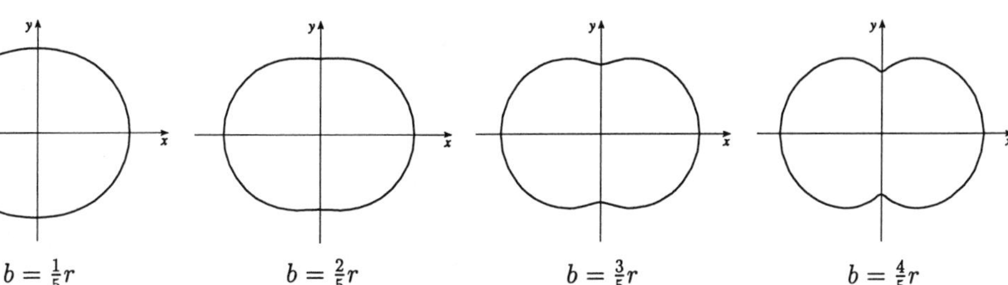

(c) The diagram gives an alternate description of point P on the epitrochoid. Q moves around a circle of radius b, and P rotates one-third as fast with respect to Q at a distance of $3r$. Place an equilateral triangle with sides of length $3\sqrt{3}r$ so that its centroid is at Q and one vertex is at P. (The distance from the centroid to a vertex is $\frac{1}{\sqrt{3}}$ times the length of a side of the equilateral triangle.)

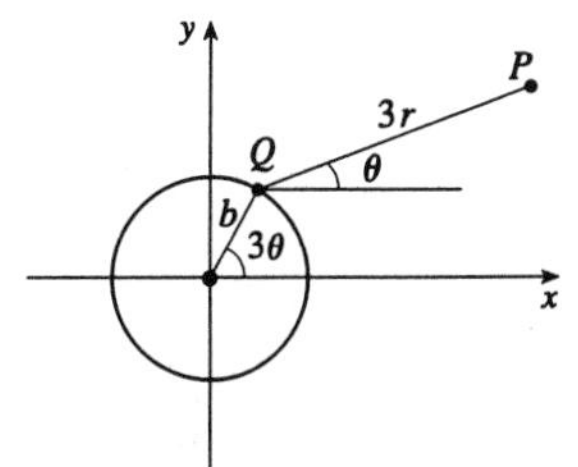

As θ increases by $\frac{2\pi}{3}$, the point Q travels once around the circle of radius b, returning to its original position. At the same time, P (and the rest of the triangle) rotate through an angle of $\frac{2\pi}{3}$ about Q, so P's position is occupied by another vertex. In this way, we see that the epitrochoid traced out by P is simultaneously traced out by the other two vertices as well. The whole equilateral triangle sits inside the epitrochoid (touching it only with its vertices) and each vertex traces out the curve once while the centroid moves around the circle three times.

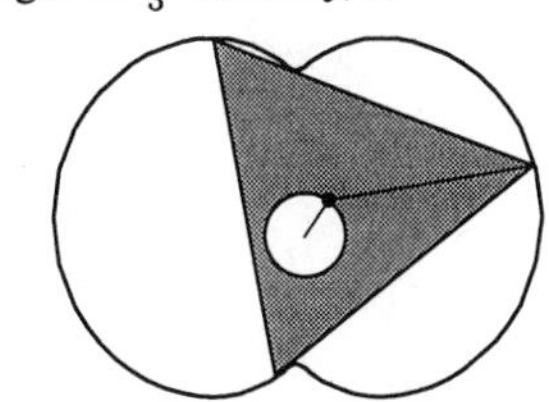

(d) We view the epitrochoid as being traced out in the same way as in part (c), by a rotor for which the distance from its center to each vertex is $3r$, so it has radius $6r$. To show that the rotor fits inside the epitrochoid, it suffices to show that for any position of the tracing point P, there are no points on the opposite side of the rotor which are outside the epitrochoid. But the most likely case of intersection is when P is on the y-axis, so as long as the diameter of the rotor (which is $3\sqrt{3}r$) is less than the distance between the y-intercepts, the rotor will fit. The y-intercepts occur when $\theta = \frac{\pi}{2}$ or $\theta = \frac{3\pi}{2} \Rightarrow y = \pm(3r - b)$, so the distance between the intercepts is $6r - 2b$, and the rotor will fit if $3\sqrt{3}r \le 6r - 2b \Leftrightarrow b \le \frac{3(2-\sqrt{3})}{2}r$.

9. **(a)** Choose a vertical x-axis pointing downward with its origin at the surface. In order to calculate the pressure at depth z, consider a partition P of the interval $[0, z]$ by points x_i and choose a point $x_i^* \in [x_{i-1}, x_i]$ for each i. The thin layer of water lying between depth x_{i-1} and depth x_i has a density of approximately $\rho(x_i^*)$, so the weight of a piece of that layer with unit cross-sectional area would be $\rho(x_i^*)g\,\Delta x_i$, where $\Delta x_i = x_i - x_{i-1}$. The total weight of a column of water extending from the surface to depth z (with unit cross-sectional area) would be approximately $\sum_i \rho(x_i^*)g\,\Delta x_i$. The estimate becomes exact if we take the limit as $\|P\| \to 0$: weight (or force) per unit area at depth z is $W = \lim\limits_{\|P\|\to 0} \sum_i \rho(x_i^*)g\,\Delta x_i$. In other words, $P(z) = \int_0^z \rho(x)g\,dx$. More generally, if we make no assumptions about the location of the origin, then $P(z) = P_0 + \int_0^z \rho(x)g\,dx$, where P_0 is the pressure at $x = 0$. Differentiating, we get $dP/dz = \rho(z)g$.

(b)
$$\begin{aligned} F &= \int_{-r}^{r} P(L+x)\cdot 2\sqrt{r^2-x^2}\,dx \\ &= \int_{-r}^{r}\left(P_0 + \int_0^{L+x} \rho_0 e^{z/H} g\,dz\right)\cdot 2\sqrt{r^2-x^2}\,dx \\ &= P_0\int_{-r}^{r} 2\sqrt{r^2-x^2}\,dx + \rho_0 gH\int_{-r}^{r}\left(e^{(L+x)/H}-1\right)\cdot 2\sqrt{r^2-x^2}\,dx \\ &= (P_0-\rho_0 gH)\int_{-r}^{r} 2\sqrt{r^2-x^2}\,dx + \rho_0 gH\int_{-r}^{r} e^{(L+x)/H}\cdot 2\sqrt{r^2-x^2}\,dx \\ &= (P_0-\rho_0 gH)(\pi r^2) + \rho_0 gHe^{L/H}\int_{-r}^{r} e^{x/H}\cdot 2\sqrt{r^2-x^2}\,dx \end{aligned}$$

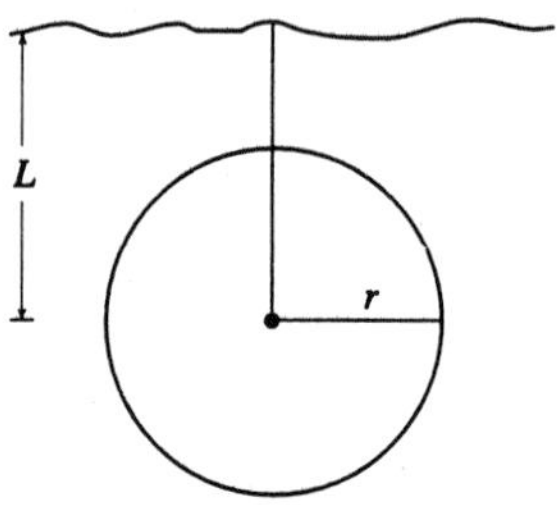

Notice that the result of Exercise 8.5.21 does not apply here since pressure does not increase linearly with depth the way it does for a fluid of constant density.

10. **(a)** $m\dfrac{dv}{dt} = -kv \;\Rightarrow\; \dfrac{dv}{v} = -\dfrac{k}{m}\,dt \;\Rightarrow\; \ln|v| = -\dfrac{k}{m}t + C$. Since $v(0) = v_0$, $\ln|v_0| = C$. Therefore $\ln\left|\dfrac{v}{v_0}\right| = -\dfrac{k}{m}t$ and $\left|\dfrac{v}{v_0}\right| = e^{-kt/m}$. Therefore $v(t) = \pm\, v_0e^{-kt/m}$. The sign is $+$ when $t = 0$, and we assume v is continuous, so that the sign is $+$ for all t. Thus $v(t) = v_0e^{-kt/m}$. $ds/dt = v_0e^{-kt/m} \;\Rightarrow$ $s(t) = -\dfrac{mv_0}{k}e^{-kt/m} + C'$. From $s(0) = s_0$, we get $s_0 = -\dfrac{mv_0}{k} + C'$, so $C' = s_0 + \dfrac{mv_0}{k}$ and $s(t) = s_0 + \dfrac{mv_0}{k}\left(1 - e^{-kt/m}\right)$. The distance traveled from time 0 to time t is $s(t) - s_0$, so the total distance traveled is $\lim\limits_{t\to\infty}[s(t) - s_0] = \dfrac{mv_0}{k}$.

(*Note:* In finding the limit, we use the fact that $k > 0$ to conclude that $\lim\limits_{t\to\infty} e^{-kt/m} = 0$.)

(b) $m\dfrac{dv}{dt} = -kv^2 \;\Rightarrow\; \dfrac{dv}{v^2} = -\dfrac{k}{m}\,dt \;\Rightarrow\; \dfrac{-1}{v} = -\dfrac{kt}{m} + C \;\Rightarrow\; \dfrac{1}{v} = \dfrac{kt}{m} - C$ since $v(0) = v_0$, $C = -\dfrac{1}{v_0}$ and $\dfrac{1}{v} = \dfrac{kt}{m} + \dfrac{1}{v_0}$. Therefore $v(t) = \dfrac{1}{kt/m + 1/v_0} = \dfrac{mv_0}{kv_0t + m}$. $\dfrac{ds}{dt} = \dfrac{mv_0}{kv_0t + m} \;\Rightarrow$ $s(t) = \dfrac{m}{k}\displaystyle\int \dfrac{kv_0\,dt}{kv_0t + m} = \dfrac{m}{k}\ln|kv_0t + m| + C'$. Since $s(0) = s_0$, we get $s_0 = \dfrac{m}{k}\ln m + C'$. Thus $C' = s_0 - \dfrac{m}{k}\ln m$ and $s(t) = s_0 + \dfrac{m}{k}(\ln|kv_0t + m| - \ln m) = s_0 + \dfrac{m}{k}\ln\left|\dfrac{kv_0t + m}{m}\right|$. We can rewrite the formulas for $v(t)$ and $s(t)$ as $v(t) = \dfrac{v_0}{1 + (kv_0/m)t}$ and $s(t) = s_0 + \dfrac{m}{k}\ln\left|1 + \dfrac{kv_0}{m}t\right|$.

Remarks: This model of horizontal motion through a resistive medium was designed to handle the case in which $v_0 > 0$. Then the term $-kv^2$ representing the resisting force causes the object to decelerate. The absolute value in the expression for $s(t)$ is unnecessary (since k, v_0, and m are all positive), and $\lim_{t\to\infty} s(t) = \infty$. In other words, the object travels infinitely far. However, $\lim_{t\to\infty} v(t) = 0$.

When $v_0 < 0$, the term $-kv^2$ increases the size of the object's negative velocity. According to the formula for $s(t)$, the position of the object approaches $-\infty$ as t approaches $m/k(-v_0)$: $\lim_{t\to -m/(kv_0)} s(t) = -\infty$.

Again the object travels infinitely far, but this time the feat is accomplished in a finite amount of time. Notice also that $\lim_{t\to -m/(kv_0)} v(t) = -\infty$ when $v_0 < 0$, showing that the speed of the object increases without limit.

11. **(a)** $\dfrac{d^2y}{dx^2} = k\sqrt{1 + \left(\dfrac{dy}{dx}\right)^2}$. Setting $z = \dfrac{dy}{dx}$, we get $\dfrac{dz}{dx} = k\sqrt{1+z^2} \quad\Rightarrow\quad \dfrac{dz}{\sqrt{1+z^2}} = k\,dx$. Using Formula 25 gives $\ln\left(z + \sqrt{1+z^2}\right) = kx + c \quad\Rightarrow\quad z + \sqrt{1+z^2} = Ce^{kx}$ (where $C = e^c$) $\Rightarrow$ $\sqrt{1+z^2} = Ce^{kx} - z \quad\Rightarrow\quad 1 + z^2 = C^2e^{2kx} - 2Ce^{kx}z + z^2 \quad\Rightarrow\quad 2Ce^{kx}z = C^2e^{2kx} - 1 \quad\Rightarrow$ $z = \dfrac{C}{2}e^{kx} - \dfrac{1}{2C}e^{-kx}$. Now $\dfrac{dy}{dx} = \dfrac{C}{2}e^{kx} - \dfrac{1}{2C}e^{-kx} \quad\Rightarrow\quad y = \dfrac{C}{2k}e^{kx} + \dfrac{1}{2Ck}e^{-kx} + C'$. From the diagram in the text, we see that $y(0) = a$ and $y(\pm b) = h$. $a = y(0) = \dfrac{C}{2k} + \dfrac{1}{2Ck} + C' \quad\Rightarrow$ $C' = a - \dfrac{C}{2k} - \dfrac{1}{2Ck} \quad\Rightarrow\quad y = \dfrac{C}{2k}\left(e^{kx} - 1\right) + \dfrac{1}{2Ck}\left(e^{-kx} - 1\right) + a$. From $h = y(\pm b)$, we find $h = \dfrac{C}{2k}\left(e^{kb} - 1\right) + \dfrac{1}{2Ck}\left(e^{-kb} - 1\right) + a$ and $h = \dfrac{C}{2k}\left(e^{-kb} - 1\right) + \dfrac{1}{2Ck}\left(e^{kb} - 1\right) + a$. Subtracting the second equation from the first, we get $0 = \dfrac{C}{k}\dfrac{e^{kb} - e^{-kb}}{2} - \dfrac{1}{Ck}\dfrac{e^{kb} - e^{-kb}}{2} = \dfrac{1}{k}\left(C - \dfrac{1}{C}\right)\sinh(kb)$. Now $k > 0$ and $b > 0$, so $\sinh(kb) > 0$ and $C = \pm 1$. If $C = 1$, then $y = \dfrac{1}{2k}\left(e^{kx} - 1\right) + \dfrac{1}{2k}\left(e^{-kx} - 1\right) + a = \dfrac{1}{k}\dfrac{e^{kx} + e^{-kx}}{2} - \dfrac{1}{k} + a = a + \dfrac{1}{k}(\cosh kx - 1)$. If $C = -1$, then $y = \dfrac{-1}{2k}\left(e^{kx} - 1\right) - \dfrac{1}{2k}\left(e^{-kx} - 1\right) + a = \dfrac{-1}{k}\dfrac{e^{kx} + e^{-kx}}{2} + \dfrac{1}{k} + a = a - \dfrac{1}{k}(\cosh kx - 1)$.

Since $k > 0$, $\cosh kx \geq 1$, and $y \geq a$, we conclude that $C = 1$ and $y = a + \dfrac{1}{k}(\cosh kx - 1)$, where $h = y(b) = a + \dfrac{1}{k}(\cosh kb - 1)$. Since $\cosh(kb) = \cosh(-kb)$, there is no further information to extract from the condition that $y(b) = y(-b)$. However, we could replace a with the expression $h - \dfrac{1}{k}(\cosh kb - 1)$, obtaining $y = h + \dfrac{1}{k}(\cosh kx - \cosh kb)$. It would be better still to keep a in the expression for y, and use the expression for h to solve for k in terms of a, b, and h. That would enable us to express y in terms of x and the given parameters a, b, and h. Sadly, it is not possible to solve for k in closed form. That would have to be done by numerical methods when specific parameter values are given.

(b) The length of the cable is $L = \displaystyle\int_{-b}^{b} \sqrt{1 + (dy/dx)^2}\,dx = \int_{-b}^{b} \sqrt{1 + \sinh^2 kx}\,dx$

$$= \int_{-b}^{b} \cosh kx\,dx = \frac{1}{k}\sinh kx\Big|_{-b}^{b} = \frac{1}{k}[\sinh(kb) - \sinh(-kb)] = \frac{2}{k}\sinh(kb).$$

12. **(a)** The predator-prey equations are $\frac{dx}{dt} = kx - axy$ and $\frac{dy}{dt} = -ry + bxy$. If $x(t) = \frac{r}{b}$ and $y(t) = \frac{k}{a}$, then $\frac{dx}{dt} = \frac{dy}{dt} = 0$, $kx - axy - \frac{kr}{b} - a\frac{r}{b}\frac{k}{a} = 0$, and $-ry + bxy = -r\frac{k}{a} - b\frac{r}{b}\frac{k}{a} = 0$, so the equations are satisfied.

(b) $\frac{dy}{dx} = \frac{y(bx - r)}{x(k - ay)} \quad \Rightarrow \quad \frac{k - ay}{y}\,dy = \frac{bx - r}{x}\,dx$ (We can assume $x, y > 0$) $\Rightarrow$

$k \ln y - ay = bx - r \ln x + C \quad \Rightarrow \quad r \ln x + k \ln y = bx + ay + C \quad \Rightarrow \quad \ln(x^r y^k) = bx + ay + C$

$\Rightarrow \; x^r y^k = M e^{bx} e^{ay}$ (where $M = e^C > 0$) or $\frac{x^r}{e^{bx}} \frac{y^k}{e^{ay}} = M$.

(c) We are told that a, b, r, and k are positive constants. Since $M > 0$, neither x nor y can be 0. Thus for every time t, $x(t) > 0$ and $y(t) > 0$. This means that the curve traced out by the point $(x(t), y(t))$ as t varies must stay strictly within the first quadrant.

Now consider the behavior of the functions $f(x) = x^r e^{-bx}$ $(x > 0)$ and $g(y) = y^k e^{-ay}$ $(y > 0)$. $\Big($The solution to the differential equation $\frac{dy}{dx} = \frac{y(bx - r)}{x(k - ay)}$ is a curve whose points (x, y) satisfy the condition $f(x)g(y) = M > 0.\Big)$ It is easy to check that $\lim_{x \to 0} f(x) = \lim_{x \to \infty} f(x) = 0$ by repeatedly using l'Hospital's Rule.

Also $f(x) > 0$ for $x > 0$ and $f'(x) = \left(\frac{r}{x} - b\right) f(x)$. From this it follows that f has an absolute maximum at $x = \frac{r}{b}$. Similarly, we find that $g(y) > 0$ for $y > 0$, $\lim_{y \to 0} g(y) = \lim_{y \to \infty} g(y) = 0$, and g has an absolute maximum at $y = \frac{k}{a}$.

If $x \to \infty$ along the solution curve, then $f(x) \to 0$, so $g(y) \to \infty$. [If $g(y)$ were to remain bounded, then $f(x)g(y)$ would approach 0 instead of maintaining the constant value M.] Thus x remains bounded, Similarly, y is bounded, since if $y \to \infty$, then $g(y) \to 0$, forcing $f(x) \to \infty$, which is impossible.

CHAPTER TEN

EXERCISES 10.1

1. $a_n = \dfrac{n}{2n+1}$, so the sequence is $\left\{\dfrac{1}{3}, \dfrac{2}{5}, \dfrac{3}{7}, \dfrac{4}{9}, \dfrac{5}{11}, \ldots\right\}$.

2. $a_n = \left(-\dfrac{2}{3}\right)^n$, so the sequence is $\left\{-\dfrac{2}{3}, \dfrac{4}{9}, -\dfrac{8}{27}, \dfrac{16}{81}, -\dfrac{32}{243}, \ldots\right\}$.

3. $a_n = \dfrac{1 \cdot 3 \cdot 5 \cdot \cdots \cdot (2n-1)}{n!}$, so the sequence is $\left\{1, \dfrac{3}{2}, \dfrac{5}{2}, \dfrac{35}{8}, \dfrac{63}{8}, \ldots\right\}$.

4. $a_n = \dfrac{(-7)^{n+1}}{n!}$, so the sequence is $\left\{49, -\dfrac{343}{2}, \dfrac{2401}{6}, -\dfrac{16{,}807}{24}, \dfrac{117{,}649}{120}, \ldots\right\}$.

5. $a_n = \sin\dfrac{n\pi}{2}$, so the sequence is $\{1, 0, -1, 0, 1, \ldots\}$.

6. $a_1 = 1$, $a_{n+1} = \dfrac{1}{1+a_n}$, so the sequence is $\left\{1, \dfrac{1}{2}, \dfrac{2}{3}, \dfrac{3}{5}, \dfrac{5}{8}, \ldots\right\}$.

7. $a_n = \dfrac{1}{2^n}$ **8.** $a_n = \dfrac{1}{2n}$ **9.** $a_n = 3n - 2$

10. $a_n = \dfrac{n+2}{(n+3)^2}$ **11.** $a_n = (-1)^{n+1}\left(\dfrac{3}{2}\right)^n$ **12.** $a_n = 1 - (-1)^n$

13. $\lim\limits_{n\to\infty} \dfrac{1}{4n^2} = \dfrac{1}{4} \lim\limits_{n\to\infty} \dfrac{1}{n^2} = \dfrac{1}{4} \cdot 0 = 0$. Convergent

14. $\{4\sqrt{n}\}$ clearly diverges since $\sqrt{n} \to \infty$ as $n \to \infty$.

15. $\lim\limits_{n\to\infty} \dfrac{n^2-1}{n^2+1} = \lim\limits_{n\to\infty} \dfrac{1-1/n^2}{1+1/n^2} = 1$. Convergent

16. $\lim\limits_{n\to\infty} \dfrac{4n-3}{3n+4} = \lim\limits_{n\to\infty} \dfrac{4-3/n}{3+4/n} = \dfrac{4}{3}$. Convergent

17. $\{a_n\}$ diverges since $\dfrac{n^2}{n+1} = \dfrac{n}{1+1/n} \to \infty$ as $n \to \infty$.

18. $\lim\limits_{n\to\infty} \dfrac{n^{1/3}+n^{1/4}}{n^{1/2}+n^{1/5}} = \lim\limits_{n\to\infty} \dfrac{1/n^{1/6}+1/n^{1/4}}{1+1/n^{3/10}} = \dfrac{0}{1} = 0$ so the sequence converges.

19. $\lim\limits_{n\to\infty} |a_n| = \lim\limits_{n\to\infty} \dfrac{n^2}{1+n^3} = \lim\limits_{n\to\infty} \dfrac{1/n}{(1/n^3)+1} = 0$, so by Theorem 5, $\lim\limits_{n\to\infty} (-1)^n\left(\dfrac{n^2}{1+n^3}\right) = 0$.

20. $\lim\limits_{n\to\infty} \dfrac{1}{5^n} = \lim\limits_{n\to\infty} \left(\dfrac{1}{5}\right)^n = 0$, by Equation 7 with $r = \dfrac{1}{5}$.

21. $\{a_n\} = \{0, -1, 0, 1, 0, -1, 0, 1, \ldots\}$. This sequence oscillates among 0, -1, and 1 and so diverges.

22. $\{a_n\} = \{1, 0, -1, 0, 1, 0, -1, \ldots\}$. This sequence oscillates among 1, 0, and -1 and so the sequence diverges.

23. $a_n = \left(\frac{\pi}{3}\right)^n$ so $\{a_n\}$ diverges by Equation 7 with $r = \frac{\pi}{3} > 1$.

24. $\lim\limits_{n\to\infty} \arctan 2n = \frac{\pi}{2}$ by (6.6.8) since $2n \to \infty$ as $n \to \infty$. Convergent

25. $0 < \dfrac{3+(-1)^n}{n^2} \le \dfrac{4}{n^2}$ and $\lim\limits_{n\to\infty} \dfrac{4}{n^2} = 0$, so $\left\{\dfrac{3+(-1)^n}{n^2}\right\}$ converges to 0 by the Squeeze Theorem.

26. $\lim\limits_{n\to\infty} \dfrac{n!}{(n+2)!} = \lim\limits_{n\to\infty} \dfrac{1\cdot 2\cdot 3\cdot\cdots\cdot n}{1\cdot 2\cdot 3\cdot\cdots\cdot n(n+1)(n+2)} = \lim\limits_{n\to\infty} \dfrac{1}{(n+2)(n+1)} = 0$. Convergent.

27. $\lim\limits_{x\to\infty} \dfrac{\ln(x^2)}{x} = \lim\limits_{x\to\infty} \dfrac{2\ln x}{x} \overset{\text{H}}{=} \lim\limits_{x\to\infty} \dfrac{2/x}{1} = 0$, so by Theorem 2, $\left\{\dfrac{\ln(n^2)}{n}\right\}$ converges to 0.

28. $\lim\limits_{n\to\infty} \sin(1^4/n) = \sin 0 = 0$ since $1/n \to 0$ as $n \to \infty$, so by Theorem 5, $\{(-1)^n \sin(1/n)\}$ converges to 0.

29. $\sqrt{n+2} - \sqrt{n} = \left(\sqrt{n+2} - \sqrt{n}\right)\dfrac{\sqrt{n+2}+\sqrt{n}}{\sqrt{n+2}+\sqrt{n}} = \dfrac{2}{\sqrt{n+2}+\sqrt{n}} < \dfrac{2}{2\sqrt{n}} = \dfrac{1}{\sqrt{n}} \to 0$ as $n \to \infty$. So by the Squeeze Theorem $\left\{\sqrt{n+2} - \sqrt{n}\right\}$ converges to 0.

30. $\lim\limits_{x\to\infty} \dfrac{\ln(2+e^x)}{3x} \overset{\text{H}}{=} \lim\limits_{x\to\infty} \dfrac{e^x/(2+e^x)}{3} = \lim\limits_{x\to\infty} \dfrac{1}{6e^{-x}+3} = \dfrac{1}{3}$, so by (2), $\lim\limits_{n\to\infty} \dfrac{\ln(2+e^n)}{3n} = \dfrac{1}{3}$. Convergent.

31. $\lim\limits_{x\to\infty} \dfrac{x}{2^x} \overset{\text{H}}{=} \lim\limits_{x\to\infty} \dfrac{1}{(\ln 2)2^x} = 0$, so by Theorem 2 $\{a_n\}$ converges to 0.

32. $a_n = \ln(n+1) - \ln(n) = \ln\left(\dfrac{n+1}{n}\right) = \ln\left(1+\dfrac{1}{n}\right) \to \ln(1) = 0$ as $n \to \infty$. Convergent.

33. Let $y = x^{-1/x}$. Then $\ln y = -(\ln x)/x$ and $\lim\limits_{x\to\infty} (\ln y) \overset{\text{H}}{=} \lim\limits_{x\to\infty} -(1/x)/1 = 0$, so $\lim\limits_{x\to\infty} y = e^0 = 1$, and so $\{a_n\}$ converges to 1.

34. $y = (1+3x)^{1/x} \quad\Rightarrow\quad \ln(y) = (1/x)\ln(1+3x) \quad\Rightarrow\quad \lim\limits_{x\to\infty} \ln y = \lim\limits_{x\to\infty} \dfrac{\ln(1+3x)}{x} \overset{\text{H}}{=} \lim\limits_{x\to\infty} \dfrac{3/(1+3x)}{1} = 0$

$\Rightarrow\quad \lim\limits_{x\to\infty} y = e^0 = 1$, so by Theorem 2, $\left\{(1+3n)^{1/n}\right\}$ converges to 1.

35. $0 \le \dfrac{\cos^2 n}{2^n} \le \dfrac{1}{2^n}$ $\left[\text{since } 0 \le \cos^2 n \le 1\right]$, so since $\lim\limits_{n\to\infty} \dfrac{1}{2^n} = 0$, $\{a_n\}$ converges to 0 by the Squeeze Theorem.

36. $0 \le |a_n| = \dfrac{n|\cos n|}{n^2+1} \le \dfrac{n}{n^2+1} = \dfrac{1}{n+1/n} \to 0$ as $n \to \infty$, so by the Squeeze Theorem and Theorem 5, $\{a_n\}$ converges to 0.

37. The series converges, since $a_n = \dfrac{1+2+3+\cdots+n}{n^2} = \dfrac{n(n+1)/2}{n^2}$ [Theorem 3]

$= \dfrac{n+1}{2n} = \dfrac{1+1/n}{2} \to \dfrac{1}{2}$ as $n \to \infty$.

38. $a_n = \left(\sqrt{n+1} - \sqrt{n}\right)\sqrt{n+\frac{1}{2}} = \left(\sqrt{n+1} - \sqrt{n}\right)\left(\dfrac{\sqrt{n+1}+\sqrt{n}}{\sqrt{n+1}+\sqrt{n}}\right)\sqrt{n+\frac{1}{2}}$

$= \dfrac{\sqrt{n+1/2}}{\sqrt{n+1}+\sqrt{n}} = \dfrac{\sqrt{1+1/(2n)}}{\sqrt{1+1/n}+1} \to \dfrac{1}{2}$ as $n \to \infty$. Convergent.

39. $a_n = \dfrac{1}{2}\cdot\dfrac{2}{2}\cdot\dfrac{3}{2}\cdot\cdots\cdot\dfrac{(n-1)}{2}\cdot\dfrac{n}{2} \ge \dfrac{1}{2}\cdot\dfrac{n}{2} = \dfrac{n}{4} \to \infty$ as $n \to \infty$, so $\{a_n\}$ diverges.

40. $0 < |a_n| = \dfrac{3^n}{n!} = \dfrac{3}{1}\cdot\dfrac{3}{2}\cdot\dfrac{3}{3}\cdot\cdots\cdot\dfrac{3}{(n-1)}\cdot\dfrac{3}{n} \le 3\cdot\dfrac{3}{2}\cdot\dfrac{3}{n} = \dfrac{27}{2n} \to 0$ as $n \to \infty$, so by the Squeeze Theorem and Theorem 5, $\{a_n\}$ converges to 0.

41. From the graph, we see that the sequence is divergent, since it oscillates between 1 and -1 (approximately).

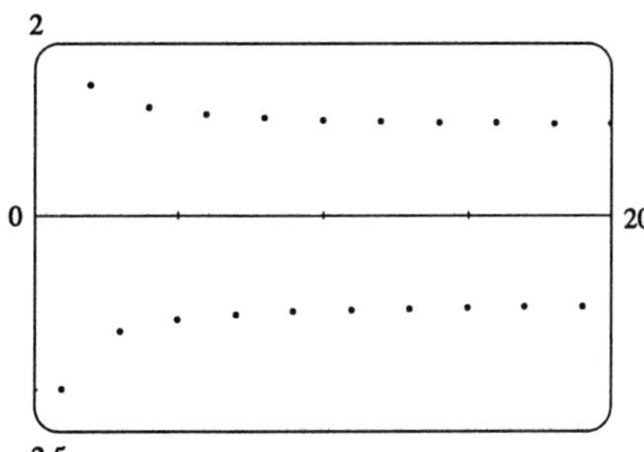

42. From the graph, it appears that the sequence converges to 2. $\left\{\left(-\frac{2}{\pi}\right)^n\right\}$ converges to 0 by (7), and hence $\left\{2+\left(-\frac{2}{\pi}\right)^n\right\}$ converges to $2+0=2$.

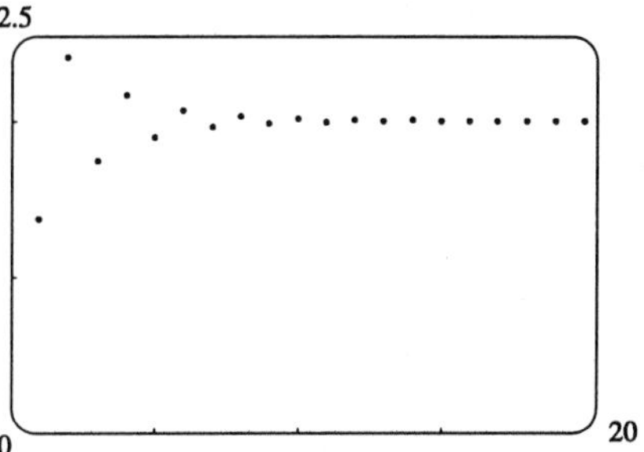

43.

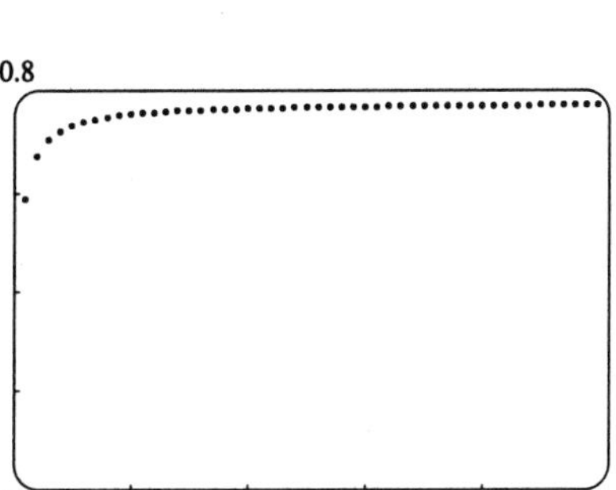

From the graph, it appears that the sequence converges to about 0.78.

$\lim\limits_{n\to\infty}\dfrac{2n}{2n+1}=\lim\limits_{n\to\infty}\dfrac{2}{2+1/n}=1$, so

$\lim\limits_{n\to\infty}\arctan\left(\dfrac{2n}{2n+1}\right)=\arctan 1=\dfrac{\pi}{4}$.

44.

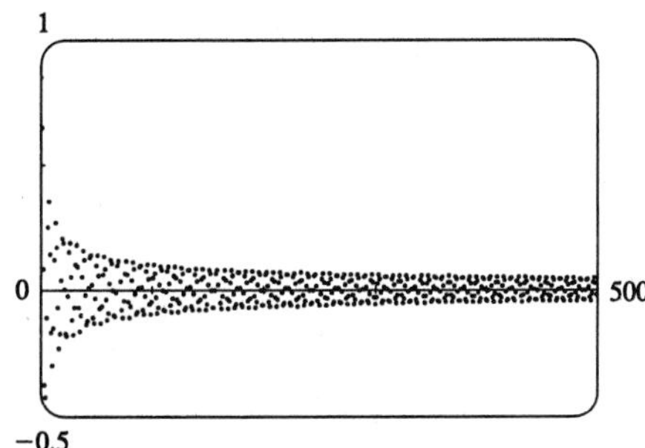

From the graph, it appears that the sequence converges (slowly) to 0.

$0\le\dfrac{|\sin n|}{\sqrt{n}}\le\dfrac{1}{\sqrt{n}}\to 0$ as $n\to\infty$, so by the Squeeze Theorem and Theorem 5,

$\left\{\dfrac{\sin n}{\sqrt{n}}\right\}$ converges to 0.

45.

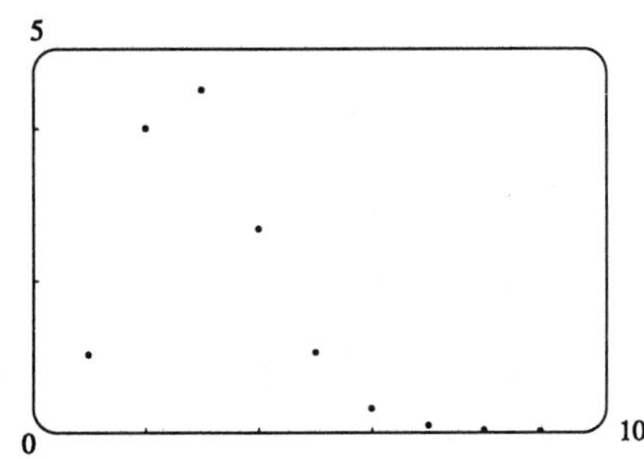

From the graph, it appears that the sequence converges to 0. $0<a_n=\dfrac{n^3}{n!}$

$=\dfrac{n}{n}\cdot\dfrac{n}{(n-1)}\cdot\dfrac{n}{(n-2)}\cdot\dfrac{1}{(n-3)}\cdot\cdots\cdot\dfrac{1}{3}\cdot\dfrac{1}{2}\cdot\dfrac{1}{1}$

$\le\dfrac{n^2}{(n-1)(n-2)(n-3)}$ (for $n\ge 4$)

$=\dfrac{1/n}{(1-1/n)(1-2/n)(1-3/n)}\to 0$ as $n\to\infty$,

so by the Squeeze Theorem, $\{a_n\}$ converges to 0.

46. 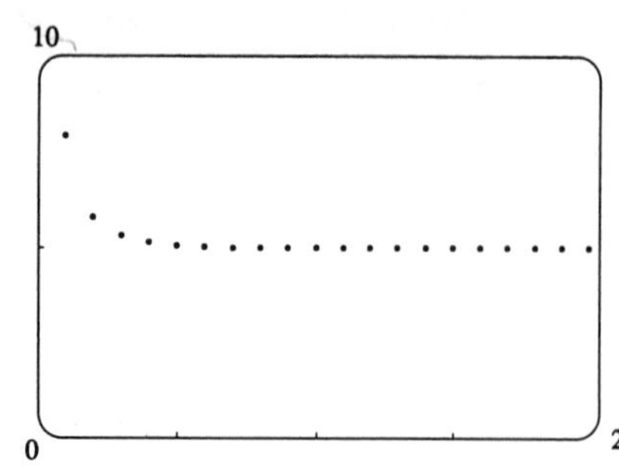

From the graph, it appears that the sequence converges to 5. $5 = \sqrt[n]{5^n} \leq \sqrt[n]{3^n + 5^n} \leq \sqrt[n]{5^n + 5^n} = \sqrt[n]{2}\sqrt[n]{5^n} = \sqrt[n]{2} \cdot 5 \to 5$ as $n \to \infty$. Hence $a_n \to 5$ by the Squeeze Theorem.

Alternate Solution: Let $y = (3^x + 5^x)^{1/x}$. Then

$$\lim_{x\to\infty} \ln y = \lim_{x\to\infty} \frac{\ln(3^x + 5^x)}{x} \overset{\text{H}}{=} \lim_{x\to\infty} \frac{3^x \ln 3 + 5^x \ln 5}{3^x + 5^x}$$

$$= \lim_{x\to\infty} \frac{\left(\frac{3}{5}\right)^x \ln 3 + \ln 5}{\left(\frac{3}{5}\right)^x + 1} = \ln 5, \text{ so } \lim_{x\to\infty} y = e^{\ln 5} = 5,$$

and so $\left\{\sqrt[n]{3^n + 5^n}\right\}$ converges to 5.

47. 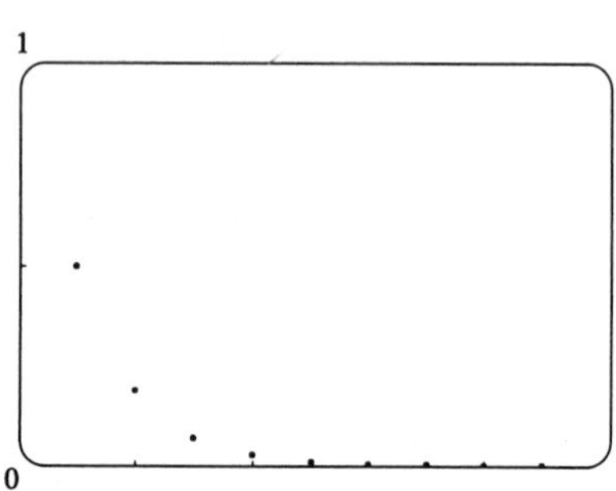

From the graph, it appears that the sequence approaches 0.

$$0 < a_n = \frac{1 \cdot 3 \cdot 5 \cdot \cdots \cdot (2n-1)}{(2n)^n}$$
$$= \frac{1}{2n} \cdot \frac{3}{2n} \cdot \frac{5}{2n} \cdot \cdots \cdot \frac{2n-1}{2n}$$
$$\leq \frac{1}{2n} \cdot (1) \cdot (1) \cdot \cdots \cdot (1) = \frac{1}{2n} \to 0 \text{ as}$$

$n \to \infty$, so by the Squeeze Theorem $\{a_n\}$ converges to 0.

48.

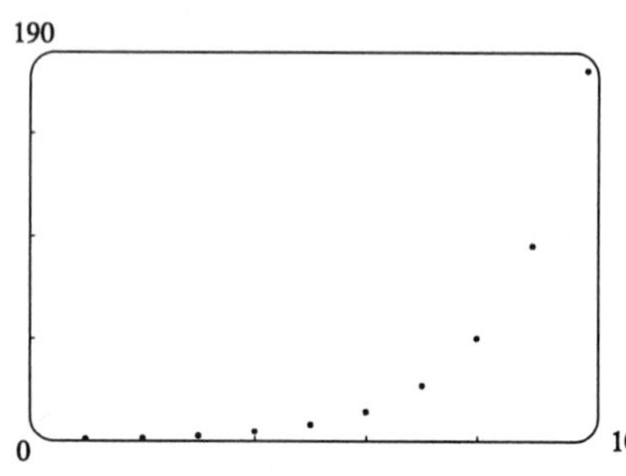

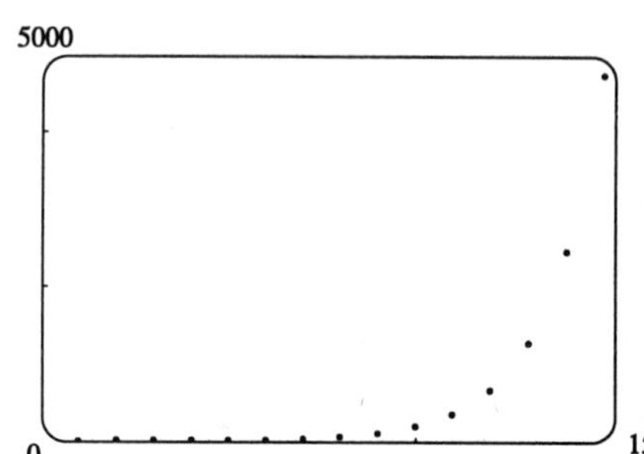

From the graphs, it seems that the sequence diverges.

$a_n = \dfrac{1 \cdot 3 \cdot 5 \cdot \cdots \cdot (2n-1)}{n!}$. We first prove by induction that $a_n \geq \left(\dfrac{3}{2}\right)^{n-1}$ for all n. This is clearly true for $n = 1$, so let $P(n)$ be the statement that the above is true for n. We must show it is then true for $n+1$.

$a_{n+1} = a_n \cdot \dfrac{2n+1}{n+1} \geq \left(\dfrac{3}{2}\right)^{n-1} \cdot \dfrac{2n+1}{n+1}$ (induction hypothesis). But $\dfrac{2n+1}{n+1} \geq \dfrac{3}{2}$ [since $2(2n+1) \geq 3(n+1) \quad \Leftrightarrow \quad 4n+2 \geq 3n+3 \quad \Leftrightarrow \quad n \geq 1$], and so we get that $a_{n+1} \geq \left(\frac{3}{2}\right)^{n-1} \cdot \frac{3}{2} = \left(\frac{3}{2}\right)^n$ which is $P(n+1)$. Thus we have proved our first assertion, so since $\left\{\left(\frac{3}{2}\right)^{n-1}\right\}$ diverges (by Equation 7), so does the given sequence $\{a_n\}$.

49. If $|r| \geq 1$, then $\{r^n\}$ diverges by (7), so $\{nr^n\}$ diverges also since $|nr^n| = n|r^n| \geq |r^n|$. If $|r| < 1$ then $\lim\limits_{x\to\infty} xr^x = \lim\limits_{x\to\infty} \dfrac{x}{r^{-x}} \overset{\text{H}}{=} \lim\limits_{x\to\infty} \dfrac{1}{(-\ln r)r^{-x}} = \lim\limits_{x\to\infty} \dfrac{r^x}{-\ln r} = 0$, so $\lim\limits_{n\to\infty} nr^n = 0$, and hence $\{nr^n\}$ converges whenever $|r| < 1$.

50. **(a)** Let $\lim\limits_{n\to\infty} a_n = L$. By Definition 1 this means that for every $\epsilon > 0$ there is an integer N such that $|a_n - L| < \epsilon$ whenever $n > N$. Thus $|a_{n+1} - L| < \epsilon$ whenever $n + 1 > N \quad \Leftrightarrow \quad n > N - 1$. It follows that $\lim\limits_{n\to\infty} a_{n+1} = L$ and so $\lim\limits_{n\to\infty} a_n = \lim\limits_{n\to\infty} a_{n+1}$.

(b) If $L = \lim\limits_{n\to\infty} a_n$ then $\lim\limits_{n\to\infty} a_{n+1} = L$ also, so L must satisfy $L = 1/(1+L) \quad \Rightarrow \quad L^2 + L - 1 = 0 \quad \Rightarrow$ $L = \frac{-1+\sqrt{5}}{2}$ (since L has to be non-negative if it exists).

51. $3(n+1) + 5 > 3n + 5$ so $\dfrac{1}{3(n+1)+5} < \dfrac{1}{3n+5} \quad \Leftrightarrow \quad a_{n+1} < a_n$ so $\{a_n\}$ is decreasing.

52. $\left\{\dfrac{1}{5^n}\right\}$ is decreasing, since $a_{n+1} \leq a_n \;\Leftrightarrow\; \dfrac{1}{5^{n+1}} \leq \dfrac{1}{5^n} \;\Leftrightarrow\; 5^{n+1} \geq 5^n \;\Leftrightarrow\; 5 \geq 1$, which is obviously true.

53. $\left\{\dfrac{n-2}{n+2}\right\}$ is increasing since $a_n < a_{n+1} \quad \Leftrightarrow \quad \dfrac{n-2}{n+2} < \dfrac{(n+1)-2}{(n+1)+2} \;\Leftrightarrow\; (n-2)(n+3) < (n+2)(n-1)$ $\Leftrightarrow \quad n^2 + n - 6 < n^2 + n - 2 \quad \Leftrightarrow \quad -6 < -2$, which is of course true.

54. $\left\{\dfrac{3n+4}{2n+5}\right\}$ is increasing since $a_{n+1} \geq a_n \quad \Leftrightarrow \quad \dfrac{3(n+1)+4}{2(n+1)+5} \geq \dfrac{3n+4}{2n+5} \quad \Leftrightarrow$ $(3n+7)(2n+5) \geq (3n+4)(2n+7) \;\Leftrightarrow\; 6n^2 + 29n + 35 \geq 6n^2 + 29n + 28 \;\Leftrightarrow\; 35 \geq 28$.

55. $a_1 = 0 > a_2 = -1 < a_3 = 0$, so the sequence is not monotonic.

56. $\left\{3 + \dfrac{(-1)^n}{n}\right\} = \left\{2, \dfrac{7}{2}, \dfrac{8}{3}, \ldots\right\}$ is not monotonic since $2 < \dfrac{7}{2} > \dfrac{8}{3}$.

57. $\left\{\dfrac{n}{n^2+n-1}\right\}$ is decreasing since $a_{n+1} < a_n \quad \Leftrightarrow \quad \dfrac{n+1}{(n+1)^2 + (n+1) - 1} < \dfrac{n}{n^2+n-1} \quad \Leftrightarrow$
$(n+1)(n^2+n-1) < n(n^2+3n+1) \quad \Leftrightarrow \quad n^3 + 2n^2 - 1 < n^3 + 3n^2 + n \quad \Leftrightarrow$
$0 < n^2 + n + 1 = \left(n + \frac{1}{2}\right)^2 + \frac{3}{4}$, which is obviously true.

58. Let $f(x) = \dfrac{\sqrt{x+1}}{5x+3}$. Then $f'(x) = \dfrac{(5x+3)/\left(2\sqrt{x+1}\right) - 5\sqrt{x+1}}{(5x+3)^2} = \dfrac{-(5x+7)}{2(5x+3)^2\sqrt{x+1}}$, which is clearly negative for all $x \geq 1$. So f is decreasing for $x \geq 1$ and $\left\{\dfrac{\sqrt{n+1}}{5n+3}\right\}$ is a decreasing sequence.

59. $a_1 = 2^{1/2}, a_2 = 2^{3/4}, a_3 = 2^{7/8}, \ldots, a_n = 2^{(2^n-1)/2^n} = 2^{(1-1/2^n)}$. $\lim\limits_{n\to\infty} a_n = \lim\limits_{n\to\infty} 2^{(1-1/2^n)} = 2^1 = 2$.

Alternate Solution: Let $L = \lim\limits_{n\to\infty} a_n$ (We could show the limit exists by showing that $\{a_n\}$ is bounded and increasing.) So L must satisfy $L = \sqrt{2 \cdot L} \;\Rightarrow\; L^2 = 2L \;\Rightarrow\; L(L-2) = 0 \quad (L \neq 0$ since the sequence increases) so $L = 2$.

60. **(a)** Let $P(n)$ be the statement that $a_{n+1} \geq a_n$ and $a_n \leq 3$. $P(1)$ is obviously true. We will assume $P(n)$ is true and then show that as a consequence $P(n+1)$ must also be true. $a_{n+2} \geq a_{n+1} \quad \Leftrightarrow$ $\sqrt{2 + a_{n+1}} \geq \sqrt{2 + a_n} \quad \Leftrightarrow \quad 2 + a_{n+1} \geq 2 + a_n \quad \Leftrightarrow \quad a_{n+1} \geq a_n$ which is the induction hypothesis. $a_{n+1} \leq 3 \quad \Leftrightarrow \quad \sqrt{2 + a_n} \leq 3 \quad \Leftrightarrow \quad 2 + a_n \leq 9 \quad \Leftrightarrow \quad a_n \leq 7$, which is certainly true because we are assuming $a_n \leq 3$. So $P(n)$ is true for all n, and so $a_1 \leq a_n \leq 3$ (the sequence is bounded), and hence by Theorem 10, $\lim_{n \to \infty} a_n$ exists.

(b) If $L = \lim_{n \to \infty} a_n$ then $\lim_{n \to \infty} a_{n+1} = L$ also, so $L = \sqrt{2 + L} \quad \Rightarrow \quad L^2 - L - 2 = (L+1)(L-2) = 0$, so $L = 2$ (since L can't be negative).

61. We show by induction that $\{a_n\}$ is increasing and bounded above by 3.
Let $P(n)$ be the proposition that $a_{n+1} > a_n$ and $0 < a_n < 3$. Clearly $P(1)$ is true. Assume $P(n)$ is true. Then

$$a_{n+1} > a_n \quad \Rightarrow \quad \frac{1}{a_{n+1}} < \frac{1}{a_n} \quad \Rightarrow \quad -\frac{1}{a_{n+1}} > -\frac{1}{a_n} \quad \Rightarrow \quad a_{n+2} = 3 - \frac{1}{a_{n+1}} > 3 - \frac{1}{a_n} = a_{n+1} \quad \Leftrightarrow$$

$P(n+1)$. This proves that $\{a_n\}$ is increasing and bounded above by 3, so $1 = a_1 < a_n < 3$, that is, $\{a_n\}$ is bounded, and hence convergent by Theorem 10.
If $L = \lim_{n \to \infty} a_n$, then $\lim_{n \to \infty} a_{n+1} = L$ also, so L must satisfy $L = 3 - 1/L$, so $L^2 - 3L + 1 = 0$ and the quadratic formula gives $L = \frac{3 \pm \sqrt{5}}{2}$. But $L > 1$, so $L = \frac{3 + \sqrt{5}}{2}$.

62. We use induction. Let $P(n)$ be the statement that $0 < a_{n+1} \leq a_n \leq 2$. Clearly $P(1)$ is true, since $a_2 = 1/(3-1) = \frac{1}{2}$. Now assume $P(n)$ is true. Then $a_{n+1} \leq a_n \quad \Rightarrow \quad -a_{n+1} \geq -a_n \quad \Rightarrow$

$$3 - a_{n+1} \geq 3 - a_n \quad \Rightarrow \quad a_{n+2} = \frac{1}{3 - a_{n+1}} \leq \frac{1}{3 - a_n} = a_{n+1}.$$

Also $a_{n+2} > 0$ (since $3 - a_{n+1}$ is positive) and $a_{n+1} \leq 2$ by the induction hypothesis, so $P(n+1)$ is true.
To find the limit, we use the fact that $\lim_{n \to \infty} a_n = \lim_{n \to \infty} a_{n+1} \quad \Rightarrow \quad L = \dfrac{1}{3 - L} \quad \Rightarrow \quad L^2 - 3L + 1 = 0 \quad \Rightarrow$ $L = \frac{3 \pm \sqrt{5}}{2}$. But $L \leq 2$, so we must have $L = \frac{3 - \sqrt{5}}{2}$.

63. **(a)** Let a_n be the number of rabbit pairs in the nth month. Clearly $a_1 = 1 = a_2$. In the nth month, each pair that is 2 or more months old (that is, a_{n-2} pairs) will have a pair of children to add to the a_{n-1} pairs already present. Thus $a_n = a_{n-1} + a_{n-2}$, so that $\{a_n\} = \{f_n\}$, the Fibonacci sequence.

(b) $a_{n-1} = \dfrac{f_n}{f_{n-1}} = \dfrac{f_{n-1} + f_{n-2}}{f_{n-1}} = 1 + \dfrac{f_{n-2}}{f_{n-1}} = 1 + \dfrac{1}{a_{n-2}}$. If $L = \lim_{n \to \infty} a_n$, then L must satisfy $L = 1 + \dfrac{1}{L}$ or $L^2 - L - 1 = 0$, so $L = \frac{1 + \sqrt{5}}{2}$ (since L must be positive.)

64. **(a)** If f is continuous, then $f(L) = f\left(\lim_{n \to \infty} a_n\right) = \lim_{n \to \infty} f(a_n) = \lim_{n \to \infty} a_{n+1} = L$ by Exercise 50(a).

(b) By repeatedly pressing the cosine key on the calculator until the displayed value stabilizes, we see that $L \approx 0.73909$.

65. **(a)**

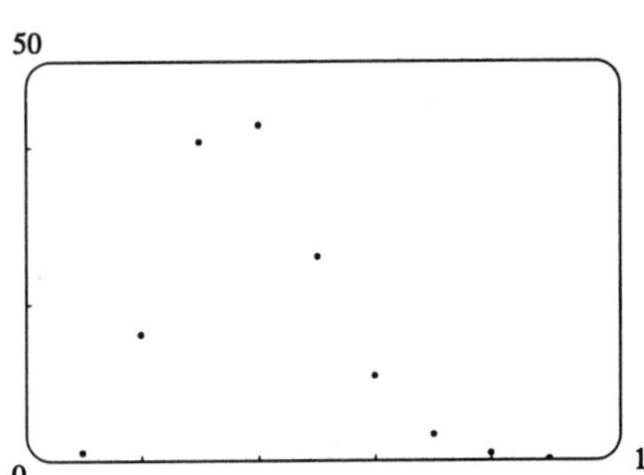

From the graph, it appears that the sequence $\left\{\dfrac{n^5}{n!}\right\}$ converges to 0, that is, $\lim\limits_{n\to\infty} \dfrac{n^5}{n!} = 0$.

(b)

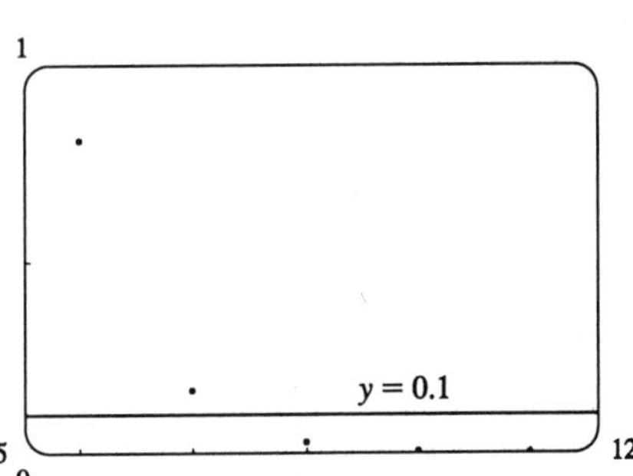

0.03

$y = 0.001$

9.5

0

15.5

From the first graph, it seems that the smallest possible value of N corresponding to $\epsilon = 0.1$ is 9, since $n^5/n! < 0.1$ whenever $n \geq 10$, but $9^5/9! > 0.1$. From the second graph, it seems that for $\epsilon = 0.001$, the smallest possible value for N is 11.

66. Let $\epsilon > 0$ and let N be any positive integer larger than $\ln(\epsilon)/\ln|r|$. If $n > N$ then $n > \ln(\epsilon)/\ln|r| \Rightarrow n\ln|r| < \ln\epsilon$ [since $|r| < 1 \Rightarrow \ln|r| < 0$] $\Rightarrow \ln(|r|^n) < \ln\epsilon \Rightarrow |r|^n < \epsilon \Rightarrow |r^n - 0| < \epsilon$, and so by Definition 1, $\lim\limits_{n\to\infty} r^n = 0$.

67. If $\lim\limits_{n\to\infty} |a_n| = 0$ then $\lim\limits_{n\to\infty} -|a_n| = 0$, and since $-|a_n| \leq a_n \leq |a_n|$, we have that $\lim\limits_{n\to\infty} a_n = 0$ by the Squeeze Theorem.

68. **(a)** $$\frac{b^{n+1} - a^{n+1}}{b-a} = b^n + b^{n-1}a + b^{n-2}a^2 + b^{n-3}a^3 + \cdots + ba^{n-1} + a^n$$
$$< b^n + b^{n-1}b + b^{n-2}b^2 + b^{n-3}b^3 + \cdots + bb^{n-1} + b^n = (n+1)b^n$$

(b) Since $b - a > 0$, we have $b^{n+1} - a^{n+1} < (n+1)b^n(b-a) \Rightarrow b^{n+1} - (n+1)b^n(b-a) < a^{n+1} \Rightarrow b^n[(n+1)a - nb] < a^{n+1}$.

(c) With this substitution, $(n+1)a - nb = 1$, and so $b^n = \left(1 + \dfrac{1}{n}\right)^n < a^{n+1} = \left(1 + \dfrac{1}{n+1}\right)^{n+1}$.

(d) With this substitution, we get $\left(1 + \dfrac{1}{2n}\right)^n \left(\dfrac{1}{2}\right) < 1 \Rightarrow \left(1 + \dfrac{1}{2n}\right)^n < 2 \Rightarrow \left(1 + \dfrac{1}{2n}\right)^{2n} < 4$.

(e) $a_n < a_{2n}$ since $\{a_n\}$ is increasing, so $a_n < a_{2n} < 4$.

(f) Since $\{a_n\}$ is increasing and bounded above by 4, $a_1 \leq a_n \leq 4$, and so $\{a_n\}$ is bounded and monotonic, and hence has a limit by Theorem 10.

69. **(a)** First we show that $a > a_1 > b_1 > b$.

$a_1 - b_1 = \dfrac{a+b}{2} - \sqrt{ab} = \frac{1}{2}\left(a - 2\sqrt{ab} + b\right) = \frac{1}{2}\left(\sqrt{a} - \sqrt{b}\right)^2 > 0 \quad (\text{since } a > b) \quad \Rightarrow \quad a_1 > b_1.$

Also $a - a_1 = a - \frac{1}{2}(a+b) = \frac{1}{2}(a-b) > 0$ and $b - b_1 = b - \sqrt{ab} = \sqrt{b}\left(\sqrt{b} - \sqrt{a}\right) < 0$, so $a > a_1 > b_1 > b$. In the same way we can show that $a_1 > a_2 > b_2 > b_1$ and so the given assertion is true for $n = 1$. Suppose it is true for $n = k$, that is, $a_k > a_{k+1} > b_{k+1} > b_k$. Then

$a_{k+2} - b_{k+2} = \frac{1}{2}(a_{k+1} + b_{k+1}) - \sqrt{a_{k+1}b_{k+1}} = \frac{1}{2}\left(a_{k+1} - 2\sqrt{a_{k+1}b_{k+1}} + b_{k+1}\right)$

$= \frac{1}{2}\left(\sqrt{a_{k+1}} - \sqrt{b_{k+1}}\right)^2 > 0$ and $a_{k+1} - a_{k+2} = a_{k+1} - \frac{1}{2}(a_{k+1} + b_{k+1}) = \frac{1}{2}(a_{k+1} - b_{k+1}) > 0,$

$b_{k+1} - b_{k+2} = b_{k+1} - \sqrt{a_{k+1}b_{k+1}} = \sqrt{b_{k+1}}\left(\sqrt{b_{k+1}} - \sqrt{a_{k+1}}\right) < 0 \quad \Rightarrow \quad a_{k+1} > a_{k+2} > b_{k+2} > b_{k+1},$

so the assertion is true for $n = k + 1$. Thus it is true for all n by mathematical induction.

(b) From part (a) we have $a > a_n > a_{n+1} > b_{n+1} > b_n > b$, which shows that both sequences are monotonic and bounded. So they are both convergent by Theorem 10.

(c) Let $\lim\limits_{n\to\infty} a_n = \alpha$ and $\lim\limits_{n\to\infty} b_n = \beta$. Then $\lim\limits_{n\to\infty} a_{n+1} = \lim\limits_{n\to\infty} \dfrac{a_n + b_n}{2} \quad \Rightarrow \quad \alpha = \dfrac{\alpha + \beta}{2} \quad \Rightarrow$ $2\alpha = \alpha + \beta \quad \Rightarrow \quad \alpha = \beta$.

70. **(a)** Let $\epsilon > 0$. Since $\lim\limits_{n\to\infty} a_{2n} = L$, there exists N_1 such that $|a_{2n} - L| < \epsilon$ for $n > N_1$. Since $\lim\limits_{n\to\infty} a_{2n+1} = L$, there exists N_2 such that $|a_{2n+1} - L| < \epsilon$ for $n > N_2$. Let $N = \max\{2N_1, 2N_2 + 1\}$ and let $n > N$. If n is even, then $n = 2m$ where $m > N_1$, so $|a_n - L| = |a_{2m} - L| < \epsilon$. If n is odd, then $n = 2m + 1$, where $m > N_2$, so $|a_n - L| = |a_{2m+1} - L| < \epsilon$. Therefore $\lim\limits_{n\to\infty} a_n = L$.

(b) $a_1 = 1$, $a_2 = 1 + \frac{1}{1+1} = \frac{3}{2} = 1.5$, $a_3 = 1 + \frac{1}{5/2} = \frac{7}{5} = 1.4$, $a_4 = 1 + \frac{1}{12/5} = \frac{17}{12} = 1.41\overline{6}$,

$a_5 = 1 + \frac{1}{29/12} = \frac{41}{29} \approx 1.413793$, $a_6 = 1 + \frac{1}{70/29} = \frac{99}{70} \approx 1.414286$, $a_7 = 1 + \frac{1}{169/70} = \frac{239}{169} \approx 1.414201$,

$a_8 = 1 + \frac{1}{408/169} = \frac{577}{408} \approx 1.414216$. Notice that $a_1 < a_3 < a_5 < a_7$ and $a_2 > a_4 > a_6 > a_8$. It appears that the odd terms are increasing and the even terms are decreasing. Let's prove that $a_{2n-2} > a_{2n}$ and $a_{2n-1} < a_{2n+1}$ by mathematical induction. Suppose that $a_{2k-2} > a_{2k}$. Then $1 + a_{2k-2} > 1 + a_{2k} \quad \Rightarrow$

$\dfrac{1}{1 + a_{2k-2}} < \dfrac{1}{1 + a_{2k}} \quad \Rightarrow \quad 1 + \dfrac{1}{1 + a_{2k-2}} < 1 + \dfrac{1}{1 + a_{2k}} \quad \Rightarrow \quad a_{2k-1} < a_{2k+1} \quad \Rightarrow$

$1 + a_{2k-1} < 1 + a_{2k+1} \quad \Rightarrow \quad \dfrac{1}{1 + a_{2k-1}} > \dfrac{1}{1 + a_{2k+1}} \quad \Rightarrow \quad 1 + \dfrac{1}{1 + a_{2k-1}} > 1 + \dfrac{1}{1 + a_{2k+1}} \quad \Rightarrow$

$a_{2k} > a_{2k+2}$. We have thus shown, by induction, that the odd terms are increasing and the even terms are decreasing. Also all terms lie between 1 and 2, so both $\{a_n\}$ and $\{b_n\}$ are bounded monotonic sequences and therefore convergent by Theorem 10. Let $\lim\limits_{n\to\infty} a_{2n} = L$. Then $\lim\limits_{n\to\infty} a_{2n+2} = L$ also. We have

$a_{n+2} = 1 + \dfrac{1}{1 + 1 + 1/(1 + a_n)} = 1 + \dfrac{1}{(3 + 2a_n)/(1 + a_n)} = \dfrac{4 + 3a_n}{3 + 2a_n}$, so $a_{2n+2} = \dfrac{4 + 3a_{2n}}{3 + 2a_{2n}}$. Taking limits of both sides, we get $L = \dfrac{4 + 3L}{3 + 2L} \quad \Rightarrow \quad 3L + 2L^2 = 4 + 3L \quad \Rightarrow \quad L^2 = 2 \quad \Rightarrow \quad L = \sqrt{2}$ (since $L > 0$). Thus $\lim\limits_{n\to\infty} a_{2n} = \sqrt{2}$.

Similarly we find that $\lim\limits_{n\to\infty} a_{2n+1} = \sqrt{2}$. So, by part (a), $\lim\limits_{n\to\infty} a_n = \sqrt{2}$.

71. **(a)** $2\cos\theta - 1 = \dfrac{1+2\cos 2\theta}{1+2\cos\theta} \quad\Leftrightarrow\quad (2\cos\theta+1)(2\cos\theta-1) = 1+2\cos 2\theta$ (provided $\cos\theta \neq -1$), and this is certainly true since the LHS $= 4\cos^2\theta - 1$ and the RHS $= 1 + 2(2\cos^2\theta - 1) = 4\cos^2\theta - 1$.

(b) By part (a), we can write each a_k as $2\cos(\theta/2^k) - 1 = \dfrac{1+2\cos(\theta/2^{k-1})}{1+2\cos(\theta/2^k)}$, so we get

$$b_n = \frac{1+2\cos\theta}{1+2\cos(\theta/2)} \cdot \frac{1+2\cos(\theta/2)}{1+2\cos(\theta/4)} \cdot \cdots \cdot \frac{1+2\cos(\theta/2^{n-2})}{1+2\cos(\theta/2^{n-1})} \cdot \frac{1+2\cos(\theta/2^{n-1})}{1+2\cos(\theta/2^n)} = \frac{1+2\cos\theta}{1+2\cos(\theta/2^n)}$$

(telescoping product). So $\lim\limits_{n\to\infty} b_n = \lim\limits_{n\to\infty} \dfrac{1+2\cos\theta}{1+2\cos(\theta/2^n)} = \dfrac{1+2\cos\theta}{1+2\cos 0} = \frac{1}{3}(1+2\cos\theta)$.

72. To write such a program in Maple it is best to calculate all the points first and then graph them. One possible sequence of commands [taking $p_0 = \frac{1}{2}$ and $k = 1.5$ for part (a)] is

```
p(0):=1/2;k:=1.5;
for j from 1 to 20 do p(j):=k*p(j-1)*(1-p(j-1)) od;
plot({[t,p(t)] $t=0..20},style=point);
```

In Mathematica, we can use the program

```
p[0]=1/2
k=1.5
p[j_]:=k*p[j-1]*(1-p[j-1])
P=Table[p[t],{t,20}]
ListPlot[P]
```

(a) With $p_0 = \frac{1}{2}$ and $k = 1.5$:

n	p_n	n	p_n	n	p_n
0	0.5	7	0.3338465076	14	0.3333373304
1	0.375	8	0.3335895255	15	0.3333353319
2	0.3515625	9	0.3334613310	16	0.3333343326
3	0.3419494629	10	0.3333973076	17	0.3333338330
4	0.3375300416	11	0.3333653144	18	0.3333335832
5	0.3354052689	12	0.3333493224	19	0.3333334583
6	0.3343628618	13	0.3333413276	20	0.3333333959

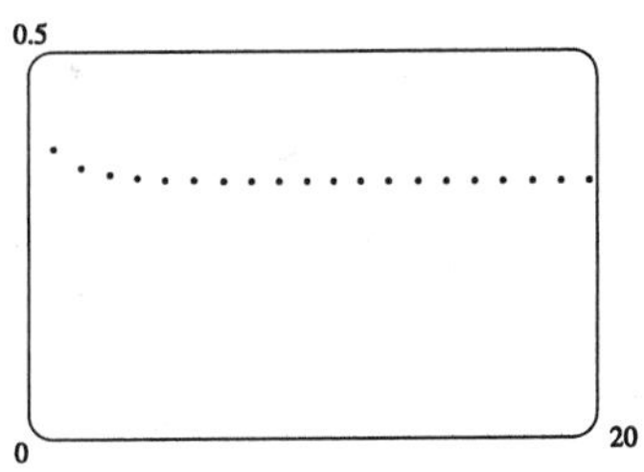

With $p_0 = \frac{1}{2}$ and $k = 2.5$:

n	p_n	n	p_n	n	p_n
0	0.5	7	0.6004164790	14	0.5999967418
1	0.625	8	0.5997913268	15	0.6000016290
2	0.5859375	9	0.6001042278	16	0.5999991855
3	0.6065368653	10	0.5999478590	17	0.6000004073
4	0.5966247408	11	0.6000260638	18	0.5999997963
5	0.6016591488	12	0.5999869665	19	0.6000001018
6	0.5991635438	13	0.6000065163	20	0.5999999490

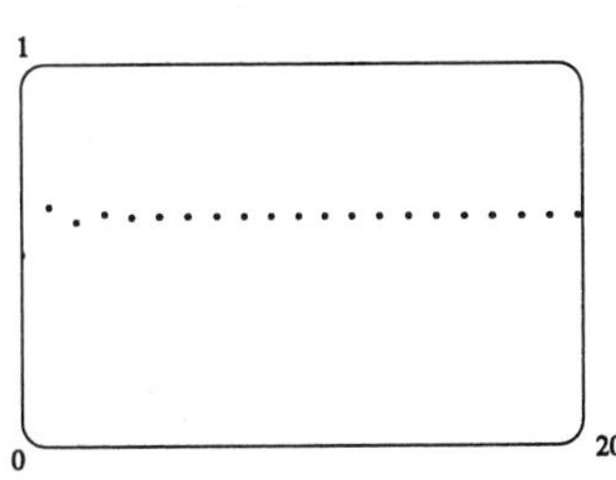

Both of these sequences seem to converge (the first to about $\frac{1}{3}$, the second to about 0.60).

With $p_0 = \frac{7}{8}$ and $k = 1.5$:

n	p_n	n	p_n	n	p_n
0	0.875	7	0.3239166555	14	0.3332554830
1	0.1640625000	8	0.3284919837	15	0.3332943992
2	0.2057189942	9	0.3308775005	16	0.3333138641
3	0.2450980344	10	0.3320963703	17	0.3333235982
4	0.2775374819	11	0.3327125567	18	0.3333284657
5	0.3007656420	12	0.3330223670	19	0.3333308996
6	0.3154585059	13	0.3331777052	20	0.3333321165

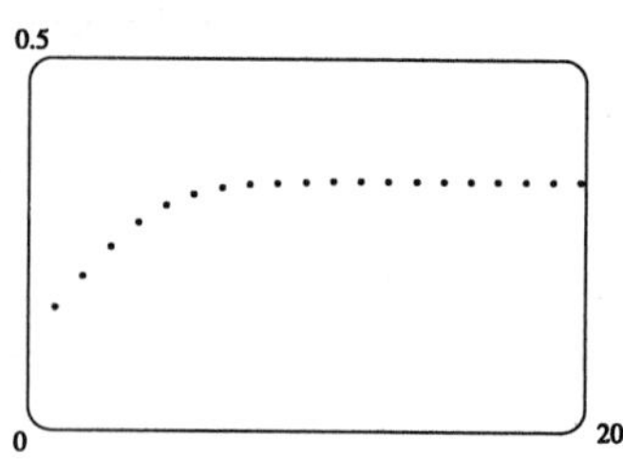

With $p_0 = \frac{7}{8}$ and $k = 2.5$:

n	p_n	n	p_n	n	p_n
0	0.875	7	0.6016572368	14	0.5999869815
1	0.2734375000	8	0.5991645155	15	0.6000065088
2	0.4966735840	9	0.6004159973	16	0.5999967455
3	0.6249723375	10	0.5997915688	17	0.6000016273
4	0.5859547873	11	0.6001041070	18	0.5999991863
5	0.6065294363	12	0.5999479195	19	0.6000004068
6	0.5966286980	13	0.6000260335	20	0.5999997965

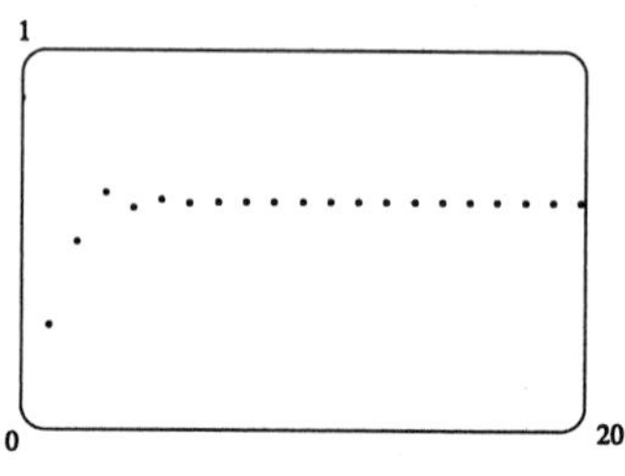

The limit of the sequence seems to depend on k, but not on p_0.

(b) With $p_0 = \frac{7}{8}$ and $k = 3.2$:

n	p_n	n	p_n	n	p_n
0	0.875	7	0.5830728499	14	0.7990633827
1	0.35	8	0.7779164851	15	0.5137954979
2	0.728	9	0.5528397674	16	0.7993909894
3	0.6336512	10	0.7910654688	17	0.5131681136
4	0.7428395414	11	0.5288988573	18	0.7994451226
5	0.6112926627	12	0.7973275392	19	0.5130643795
6	0.7603646182	13	0.5171082701	20	0.7994538304

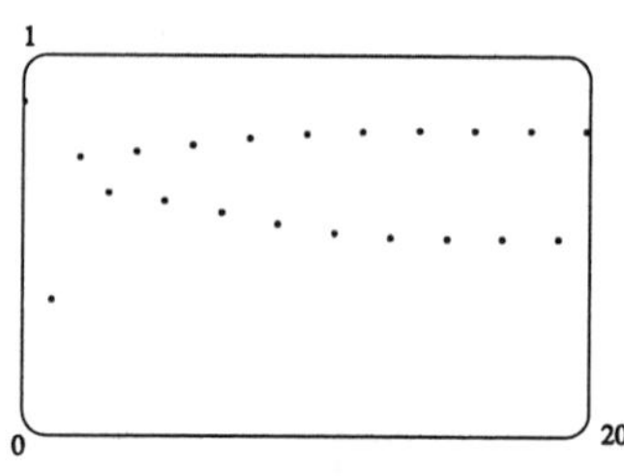

It seems that eventually the terms fluctuate between two values (about 0.5 and 0.8 in this case).

(c) With $p_0 = \frac{7}{8}$ and $k = 3.42$:

n	p_n	n	p_n	n	p_n
0	0.875	7	0.4523028595	14	0.8442074954
1	0.3740625	8	0.8472194412	15	0.4498025043
2	0.8007579317	9	0.4426802162	16	0.8463823230
3	0.5456427594	10	0.8437633930	17	0.4446659591
4	0.8478752457	11	0.4508474152	18	0.8445284521
5	0.4411212218	12	0.8467373600	19	0.4490464983
6	0.8431438501	13	0.4438243549	20	0.8461207932

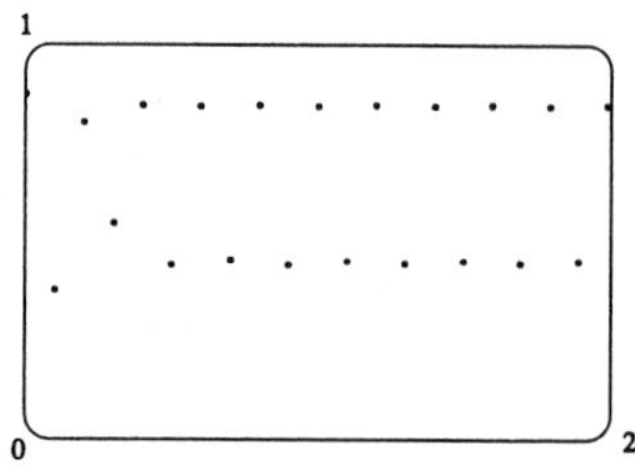

With $p_0 = \frac{7}{8}$ and $k = 3.45$:

n	p_n	n	p_n	n	p_n
0	0.875	7	0.4670259173	14	0.8403376119
1	0.37734375	8	0.8587488492	15	0.4628875692
2	0.8105962828	9	0.4184824580	16	0.8577482029
3	0.5296783244	10	0.8395743715	17	0.4209559704
4	0.8594612300	11	0.4646778994	18	0.8409445428
5	0.4167173031	12	0.8581956047	19	0.4614610245
6	0.8385707738	13	0.4198508854	20	0.8573758785

From the graphs, it seems that for k between 3.4 and 3.5, the terms eventually fluctuate between four values. In the graph at right, the pattern followed by the terms is $0.395, 0.832, 0.487, 0.869, 0.395, \ldots$. Note that even for $k = 3.42$ (as in the first graph), there are four distinct "branches;" even after 1000 terms, the first and third terms in the pattern differ by about 2×10^{-9}, while the first and fifth terms differ by only 2×10^{-10}.

With $p_0 = \frac{7}{8}$ and $k = 3.48$:

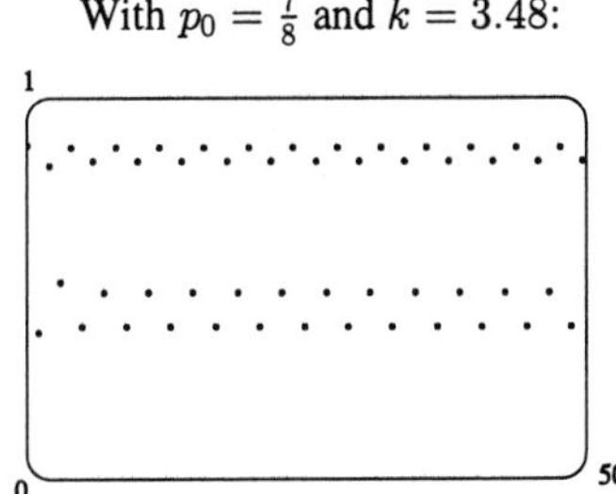

(d)

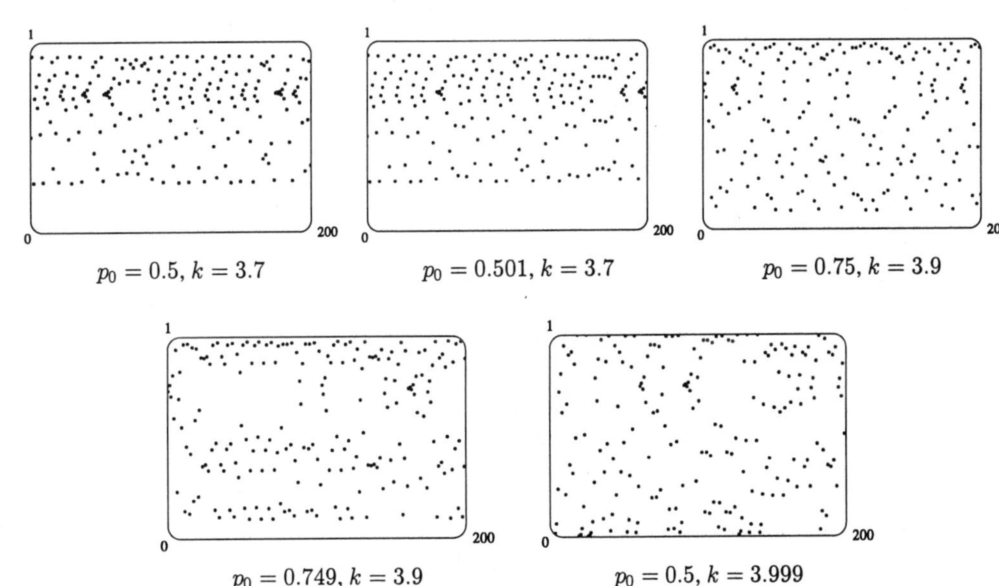

$p_0 = 0.5$, $k = 3.7$ $p_0 = 0.501$, $k = 3.7$ $p_0 = 0.75$, $k = 3.9$

$p_0 = 0.749$, $k = 3.9$ $p_0 = 0.5$, $k = 3.999$

From the graphs, it seems that if p_0 is changed by 0.01, the whole graph changes completely. (Note, however, that this might be partially due to accumulated round-off error in the CAS. These graphs were generated by Maple with 100-digit accuracy, and different degrees of accuracy give different graphs.) There seem to be some some fleeting patterns in these graphs, but on the whole they are certainly very chaotic. As k increases, the graph spreads out vertically, with more extreme values close to 0 or 1.

EXERCISES 10.2

1.

n	s_n
1	3.33333
2	4.44444
3	4.81481
4	4.93827
5	4.97942
6	4.99314
7	4.99771
8	4.99924
9	4.99975
10	4.99992
11	4.99997
12	4.99999

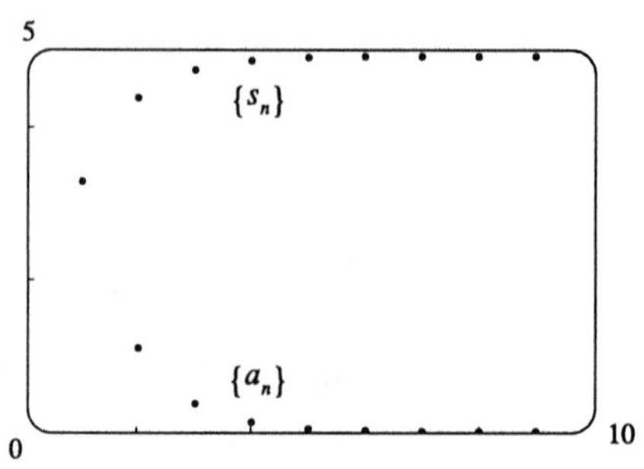

From the graph, it seems that the series converges. In fact, it is a geometric series with $a = \frac{10}{3}$ and $r = \frac{1}{3}$, so its sum is

$$\sum_{n=1}^{\infty} \frac{10}{3^n} = \frac{10/3}{1 - 1/3} = 5.$$

2.

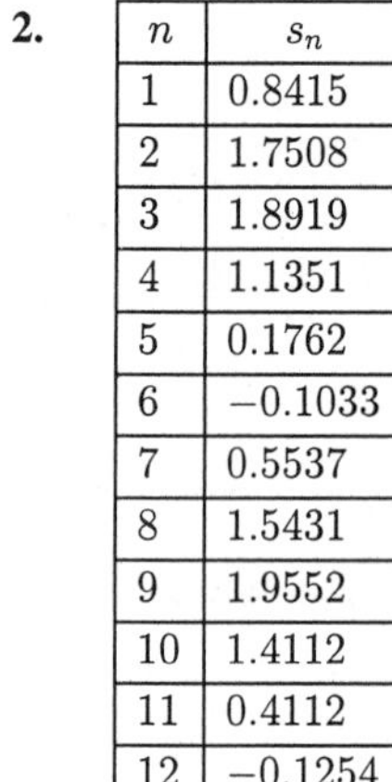

n	s_n
1	0.8415
2	1.7508
3	1.8919
4	1.1351
5	0.1762
6	−0.1033
7	0.5537
8	1.5431
9	1.9552
10	1.4112
11	0.4112
12	−0.1254

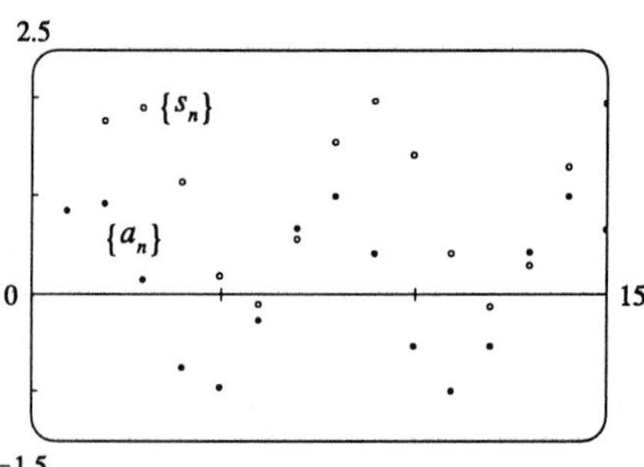

The series diverges, since its terms do not approach 0.

3.

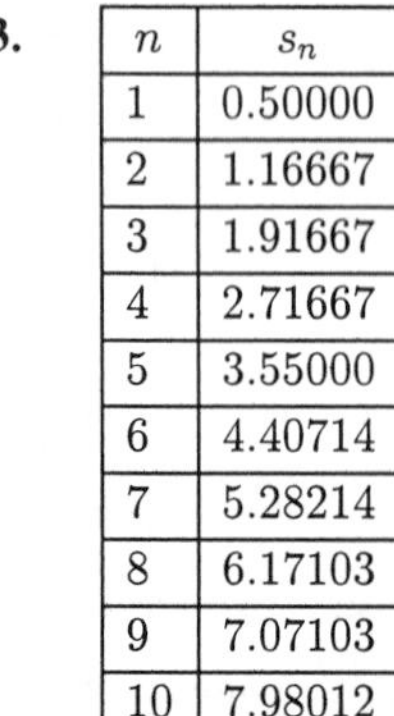

n	s_n
1	0.50000
2	1.16667
3	1.91667
4	2.71667
5	3.55000
6	4.40714
7	5.28214
8	6.17103
9	7.07103
10	7.98012

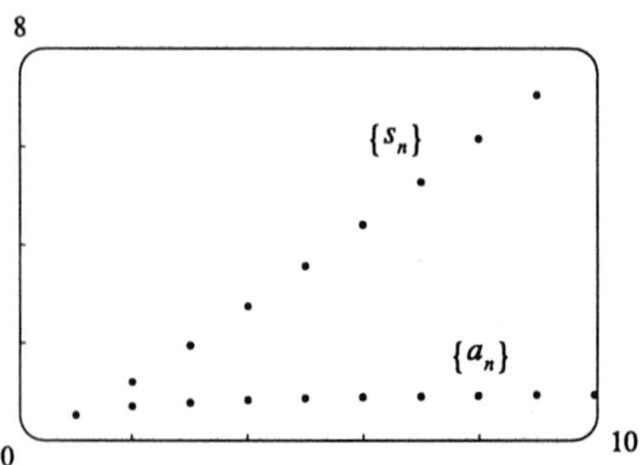

The series diverges, since its terms do not approach 0.

4.

n	s_n
4	0.25000
5	0.40000
6	0.50000
7	0.57143
8	0.62500
9	0.66667
10	0.70000
11	0.72727
12	0.75000
13	0.76923
...	...
99	0.96970
100	0.97000

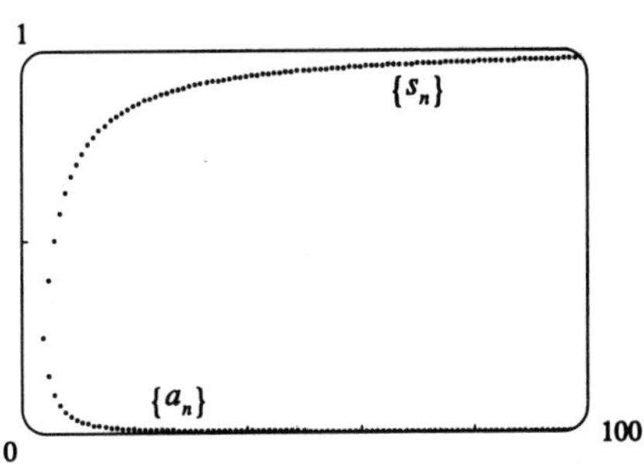

From the graph, it seems that the series converges to about 1. To find the sum, we proceed as in Example 6: since $\dfrac{3}{i(i-1)} = \dfrac{3}{i-1} - \dfrac{3}{i}$, the partial sums are

$$s_n = \sum_{i=4}^{n} \left(\frac{3}{i-1} - \frac{3}{i} \right) = \left(\frac{3}{3} - \frac{3}{4} \right) + \left(\frac{3}{4} - \frac{3}{5} \right) + \cdots + \left(\frac{3}{n-2} - \frac{3}{n-1} \right) + \left(\frac{3}{n-1} - \frac{3}{n} \right) = 1 - \frac{3}{n},$$

and so the sum is $\lim_{n\to\infty} s_n = 1$.

5.

n	s_n
1	0.6464
2	0.8075
3	0.8750
4	0.9106
5	0.9320
6	0.9460
7	0.9558
8	0.9630
9	0.9684
10	0.9726

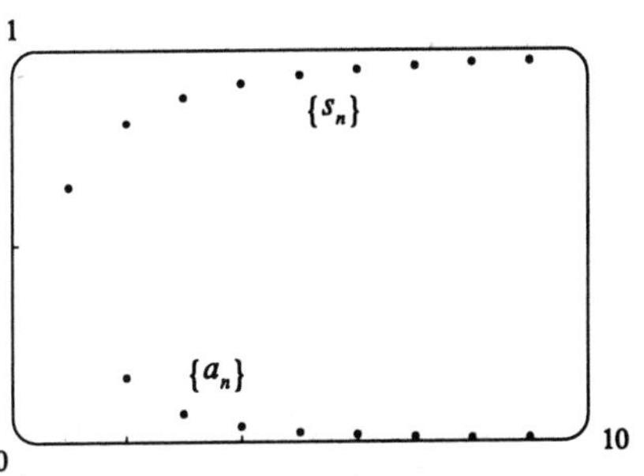

From the graph, it seems that the series converges to 1. To find the sum, we write

$$s_n = \sum_{i=1}^{n} \left(\frac{1}{i^{1.5}} - \frac{1}{(i+1)^{1.5}} \right) = \left(1 - \frac{1}{2^{1.5}} \right) + \left(\frac{1}{2^{1.5}} - \frac{1}{3^{1.5}} \right) + \left(\frac{1}{3^{1.5}} - \frac{1}{4^{1.5}} \right) + \cdots + \left(\frac{1}{n^{1.5}} - \frac{1}{(n+1)^{1.5}} \right)$$

$= 1 - 1/(n+1)^{1.5}$. So the sum is $\lim_{n\to\infty} s_n = 1$.

6.

n	s_n
1	1.000000
2	0.714286
3	0.795918
4	0.772595
5	0.779259
6	0.777355
7	0.777899
8	0.777743
9	0.777788
10	0.777775
11	0.777779
12	0.777778

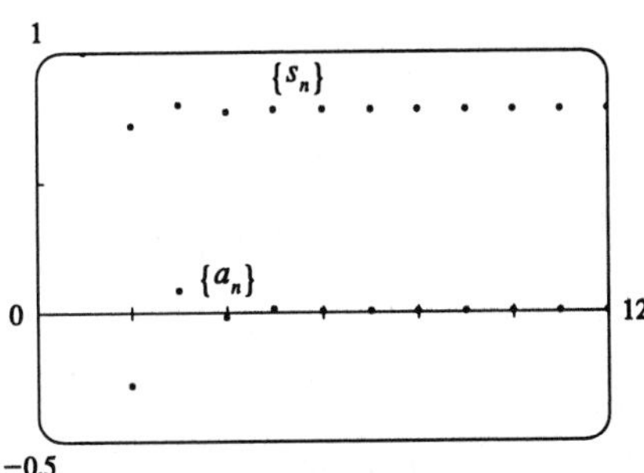

From the graph, it seems that the series converges to about 0.8. In fact, it is a geometric series with $a = 1$ and $r = -2/7$, so its sum is $\displaystyle\sum_{n=1}^{\infty} \left(-\frac{2}{7} \right)^{n-1} = \frac{1}{1-(-2/7)} = \frac{7}{9}$.

7. $\sum_{n=1}^{\infty} 4\left(\frac{2}{5}\right)^{n-1}$ is a geometric series with $a = 4$, $r = \frac{2}{5}$, so it converges to $\frac{4}{1-2/5} = \frac{20}{3}$.

8. $a = 1$, $|r| = \left|-\frac{1}{2}\right| < 1$ so the series converges with sum $\frac{1}{1-(-1/2)} = \frac{2}{3}$.

9. $\sum_{n=1}^{\infty} \frac{2}{3}\left(-\frac{1}{3}\right)^{n-1}$ is geometric with $a = \frac{2}{3}$, $r = -\frac{1}{3}$, so it converges to $\frac{2/3}{1-(-1/3)} = \frac{1}{2}$.

10. $a = -\frac{81}{100}$, $|r| = \left|-\frac{10}{9}\right| > 1$, so the series diverges.

11. $a = 2$, $r = \frac{3}{4} < 1$, so the series converges to $\frac{2}{1-3/4} = 8$.

12. $a = 1$, $|r| = \left|-\frac{3}{\pi}\right| < 1$, so the series converges to $\frac{1}{1-(-3/\pi)} = \frac{\pi}{\pi+3}$.

13. $a = \frac{5e}{3}$, $r = \frac{e}{3} < 1$, so the series converges to $\frac{5e/3}{1-e/3} = \frac{5e}{3-e}$.

14. $\sum_{n=1}^{\infty} \left(\frac{1}{e^2}\right)^n \Rightarrow a = \frac{1}{e^2} = |r| < 1$ so the series converges to $\frac{1/e^2}{1-1/e^2} = \frac{1}{e^2-1}$.

15. $a = 1$, $r = \frac{5}{8} < 1$, so the series converges to $\frac{1}{1-5/8} = \frac{8}{3}$.

16. $\sum_{n=0}^{\infty} 4\left(\frac{4}{5}\right)^n \Rightarrow a = 4$, $|r| = \frac{4}{5} < 1$ so the series converges to $\frac{4}{1-4/5} = 20$.

17. $a = \frac{64}{3}$, $r = \frac{8}{3} > 1$, so the series diverges.

18. $\sum_{n=1}^{\infty} (-1)^{n-1} \frac{3^{2n}}{2^{3n+1}} = \sum_{n=1}^{\infty} -\frac{1}{2}\left(-\frac{9}{8}\right)^n$, $|r| = \frac{9}{8} > 1$ so the series diverges.

19. This series diverges, since if it converged, so would $2 \cdot \sum_{n=1}^{\infty} \frac{1}{2n} = \sum_{n=1}^{\infty} \frac{1}{n}$ [by Theorem 8(a)], which we know diverges (Example 7).

20. $\lim_{n\to\infty} \frac{n^2}{3(n+1)(n+2)} = \lim_{n\to\infty} \frac{1}{3(1+1/n)(1+2/n)} = \frac{1}{3} \neq 0$ so the series diverges by the Test for Divergence.

21. Converges. $s_n = \sum_{i=1}^{n} \frac{1}{(3i-2)(3i+1)} = \sum_{i=1}^{n} \left[\frac{1/3}{3i-2} - \frac{1/3}{3i+1}\right]$ (partial fractions)

$$= \left[\frac{1}{3} \cdot 1 - \frac{1}{3} \cdot \frac{1}{4}\right] + \left[\frac{1}{3} \cdot \frac{1}{4} - \frac{1}{3} \cdot \frac{1}{7}\right] + \left[\frac{1}{3} \cdot \frac{1}{7} - \frac{1}{3} \cdot \frac{1}{10}\right] + \cdots$$

$$+ \left[\frac{1}{3} \cdot \frac{1}{3n-2} - \frac{1}{3} \cdot \frac{1}{3n+1}\right] = \frac{1}{3} - \frac{1}{3(3n+1)} \quad \text{(telescoping series)}$$

$$\Rightarrow \lim_{n\to\infty} s_n = \frac{1}{3} \Rightarrow \sum_{n=1}^{\infty} \frac{1}{(3n-2)(3n+1)} = \frac{1}{3}$$

22. $\sum_{n=1}^{\infty} \left(\frac{1}{2^{n-1}} + \frac{2}{3^{n-1}}\right) = \sum_{n=1}^{\infty} \frac{1}{2^{n-1}} + 2\sum_{n=1}^{\infty} \frac{1}{3^{n-1}} = \frac{1}{1-1/2} + 2\left(\frac{1}{1-1/3}\right) = 5$

23. Converges by Theorem 8.

$$\sum_{n=1}^{\infty}[2(0.1)^n + (0.2)^n] = 2\sum_{n=1}^{\infty}(0.1)^n + \sum_{n=1}^{\infty}(0.2)^n = 2\left(\frac{0.1}{1-0.1}\right) + \frac{0.2}{1-0.2} = \frac{2}{9} + \frac{1}{4} = \frac{17}{36}$$

24. $\lim_{n\to\infty} a_n = \lim_{n\to\infty}\left(\frac{1}{n} + 2^n\right)$ does not exist, so the series diverges by the Test for Divergence.

25. Diverges by the Test for Divergence. $\lim_{n\to\infty}\frac{n}{\sqrt{1+n^2}} = \lim_{n\to\infty}\frac{1}{\sqrt{1+1/n^2}} = 1 \neq 0.$

26.
$$s_n = \sum_{i=1}^{n}\frac{1}{4i^2-1} = \sum_{i=1}^{n}\left[\frac{1/2}{2i-1} - \frac{1/2}{2i+1}\right] \quad \text{(partial fractions)}$$
$$= \left(\frac{1}{2}\cdot 1 - \frac{1}{2}\cdot\frac{1}{3}\right) + \left(\frac{1}{2}\cdot\frac{1}{3} - \frac{1}{2}\cdot\frac{1}{5}\right) + \left(\frac{1}{2}\cdot\frac{1}{5} - \frac{1}{2}\cdot\frac{1}{7}\right) + \cdots + \left(\frac{1}{2}\cdot\frac{1}{2n-1} - \frac{1}{2}\cdot\frac{1}{2n+1}\right)$$
$$= \frac{1}{2} - \frac{1}{4n+2}, \text{ so } \sum_{n=1}^{\infty}\frac{1}{4n^2-1} = \lim_{n\to\infty} s_n = \frac{1}{2}.$$

27. Converges.
$$s_n = \sum_{i=1}^{n}\frac{1}{i(i+2)} = \sum_{i=1}^{n}\left[\frac{1/2}{i} - \frac{1/2}{i+2}\right] \quad \text{(partial fractions)}$$
$$= \left[\frac{1}{2} - \frac{1}{6}\right] + \left[\frac{1}{4} - \frac{1}{8}\right] + \left[\frac{1}{6} - \frac{1}{10}\right] + \cdots + \left[\frac{1}{2n-2} - \frac{1}{2n+2}\right] + \left[\frac{1}{2n} - \frac{1}{2n+4}\right]$$
$$= \frac{1}{2} + \frac{1}{4} - \frac{1}{2n+2} - \frac{1}{2n+4} \quad \text{(telescoping series).}$$
Thus $\sum_{n=1}^{\infty}\frac{1}{n(n+2)} = \lim_{n\to\infty}\left[\frac{1}{2} + \frac{1}{4} - \frac{1}{2n+2} - \frac{1}{2n+4}\right] = \frac{3}{4}.$

28. $\lim_{n\to\infty} a_n = \lim_{n\to\infty}\ln\left(\frac{n}{2n+5}\right) = \lim_{n\to\infty}\left(\frac{1}{2+5/n}\right) = \ln\frac{1}{2} \neq 0$ so the series diverges by the Test for Divergence.

29. Converges. $\sum_{n=1}^{\infty}\frac{3^n+2^n}{6^n} = \sum_{n=1}^{\infty}\left[\left(\frac{1}{2}\right)^n + \left(\frac{1}{3}\right)^n\right] = \frac{1/2}{1-1/2} + \frac{1/3}{1-1/3} = \frac{3}{2}$

30.
$$s_n = \sum_{i=1}^{n}\frac{2i+1}{i^2(i+1)^2} = \sum_{i=1}^{n}\left[\frac{1}{i^2} - \frac{1}{(i+1)^2}\right] \quad \text{(partial fractions)}$$
$$= \left(1 - \frac{1}{4}\right) + \left(\frac{1}{4} - \frac{1}{9}\right) + \cdots + \left(\frac{1}{n^2} - \frac{1}{(n+1)^2}\right) = 1 - \frac{1}{(n+1)^2}, \text{ so } \sum_{n=1}^{\infty}\frac{2n+1}{n^2(n+1)^2} = \lim_{n\to\infty} s_n = 1$$

31. Converges. $s_n = \left(\sin 1 - \sin\frac{1}{2}\right) + \left(\sin\frac{1}{2} - \sin\frac{1}{3}\right) + \cdots + \left[\sin\frac{1}{n} - \sin\frac{1}{n+1}\right] = \sin 1 - \sin\frac{1}{n+1}$, so
$$\sum_{n=1}^{\infty}\left(\sin\frac{1}{n} - \sin\frac{1}{n+1}\right) = \lim_{n\to\infty} s_n = \sin 1 - \sin 0 = \sin 1$$

32. $\lim_{n\to\infty}\frac{1}{5+2^{-n}} = \frac{1}{5} \neq 0$ so the series diverges by the Test for Divergence.

33. Diverges since $\lim_{n\to\infty}\arctan n = \frac{\pi}{2} \neq 0.$

34. $s_n = \sum_{i=1}^{n} \frac{1}{i(i+1)(i+2)} = \sum_{i=1}^{n} \left[\frac{1/2}{i} - \frac{1}{i+1} + \frac{1/2}{i+2}\right] = \sum_{i=1}^{n} \left[\frac{1/2}{i} - \frac{1/2}{i+1}\right] + \sum_{i=1}^{n} \left[-\frac{1/2}{i+1} + \frac{1/2}{i+2}\right]$,

both of which are clearly telescoping sums, so

$s_n = \left[\frac{1}{2} - \frac{1}{2(n+1)}\right] + \left[-\frac{1}{4} + \frac{1}{2(n+2)}\right] = \frac{1}{4} - \frac{1}{2(n+1)} + \frac{1}{2(n+2)}$. Thus

$\sum_{n=1}^{\infty} \frac{1}{n(n+1)(n+2)} = \lim_{n\to\infty} s_n = \frac{1}{4}$.

35. $s_n = (\ln 1 - \ln 2) + (\ln 2 - \ln 3) + (\ln 3 - \ln 4) + \cdots + [\ln n - \ln(n+1)] = \ln 1 - \ln(n+1) = -\ln(n+1)$ (telescoping series). Thus $\lim_{n\to\infty} s_n = -\infty$, so the series is divergent.

36. Write $\ln \frac{n^2-1}{n^2} = \ln \frac{(n-1)(n+1)}{n \cdot n}$. Then

$$s_n = \ln \frac{1\cdot 3}{2\cdot 2} + \ln \frac{2\cdot 4}{3\cdot 3} + \ln \frac{3\cdot 5}{4\cdot 4} + \cdots + \ln \frac{(n-2)n}{(n-1)(n-1)} + \ln \frac{(n-1)(n+1)}{n\cdot n}$$
$$= \ln\left(\frac{1\cdot 3}{2\cdot 2}\cdot\frac{2\cdot 4}{3\cdot 3}\cdot\frac{3\cdot 5}{4\cdot 4}\cdot\cdots\cdot\frac{(n-2)n}{(n-1)(n-1)}\cdot\frac{(n-1)(n+1)}{n\cdot n}\right) = \ln \frac{1}{2}\cdot\frac{n+1}{n}$$

Therefore $\sum_{n=1}^{\infty} \ln \frac{n^2-1}{n^2} = \lim_{n\to\infty} s_n = \ln \frac{1}{2}\left(1 + \frac{1}{n}\right) = \ln \frac{1}{2}$

37. $0.\overline{5} = 0.5 + 0.05 + 0.005 + \cdots = \frac{0.5}{1-0.1} = \frac{5}{9}$

38. $0.\overline{15} = 0.15 + 0.0015 + 0.000015 + \cdots = \frac{0.15}{1-0.01} = \frac{15}{99} = \frac{5}{33}$

39. $0.\overline{307} = 0.307 + 0.000307 + 0.000000307 + \cdots = \frac{0.307}{1-0.001} = \frac{307}{999}$

40. $1.1\overline{23} = 1.1 + 0.023 + 0.00023 + 0.0000023 + \cdots = 1.1 + \frac{0.023}{1-0.01} = \frac{11}{10} + \frac{23}{990} = \frac{556}{495}$

41. $0.123\overline{456} = \frac{123}{1000} + \frac{0.000456}{1-0.001} = \frac{123}{1000} + \frac{456}{999{,}000} = \frac{123{,}333}{999{,}000} = \frac{41{,}111}{333{,}000}$

42. $4.\overline{1570} = 4 + 0.1570 + 0.00001570 + \cdots = 4 + \frac{0.1570}{1-0.0001} = \frac{41{,}566}{9999}$

43. $\sum_{n=0}^{\infty}(x-3)^n$ is a geometric series with $r = x-3$, so it converges whenever $|x-3| < 1 \Rightarrow$ $-1 < x-3 < 1 \Leftrightarrow 2 < x < 4$. The sum is $\frac{1}{1-(x-3)} = \frac{1}{4-x}$.

44. $\sum_{n=0}^{\infty}(3x)^n$ is geometric with $r = 3x$, so converges for $|3x| < 1 \Leftrightarrow -\frac{1}{3} < x < \frac{1}{3}$ to $\frac{1}{1-3x}$.

45. $\sum_{n=2}^{\infty}\left(\frac{x}{5}\right)^n$ is a geometric series with $r = \frac{x}{5}$, so converges whenever $\left|\frac{x}{5}\right| < 1 \Leftrightarrow -5 < x < 5$. The sum is $\frac{(x/5)^2}{1-x/5} = \frac{x^2}{25-5x}$.

46. $\displaystyle\sum_{n=0}^{\infty}\left(\frac{1}{x}\right)^n$ is geometric with $r=\dfrac{1}{x}$ so converges when $\left|\dfrac{1}{x}\right|<1 \quad\Leftrightarrow\quad |x|>1 \quad\Leftrightarrow\quad x>1$ or $x<-1$, and the sum is $\dfrac{1}{1-1/x}=\dfrac{x}{x-1}$.

47. $\displaystyle\sum_{n=0}^{\infty}(2\sin x)^n$ is geometric so converges whenever $|2\sin x|<1 \quad\Leftrightarrow\quad -\frac{1}{2}<\sin x<\frac{1}{2} \quad\Leftrightarrow$ $n\pi-\frac{\pi}{6}<x<n\pi+\frac{\pi}{6}$, where the sum is $\dfrac{1}{1-2\sin x}$.

48. $\displaystyle\sum_{n=0}^{\infty}\tan^n x$ is geometric and converges when $|\tan x|<1 \quad\Leftrightarrow\quad -1<\tan x<1 \quad\Leftrightarrow\quad n\pi-\dfrac{\pi}{4}<x<n\pi+\dfrac{\pi}{4}$ (n any integer). On these intervals the sum is $\frac{1}{1-\tan x}$.

49. After defining f, We use `convert(f,parfrac);` in Maple or `Apart` in Mathematica to find that the general term is $\dfrac{1}{(4n+1)(4n-3)}=-\dfrac{1/4}{4n+1}+\dfrac{1/4}{4n-3}$. So the nth partial sum is

$$s_n=\sum_{k=1}^{n}\left(-\frac{1/4}{4k+1}+\frac{1/4}{4k-3}\right)$$

$$=\frac{1}{4}\left[\left(-\frac{1}{5}+1\right)+\left(-\frac{1}{9}+\frac{1}{5}\right)+\left(-\frac{1}{13}+\frac{1}{9}\right)+\cdots+\left(-\frac{1}{4n+1}+\frac{1}{4n-3}\right)\right]=\frac{1}{4}\left(1-\frac{1}{4n+1}\right).$$

The series converges to $\lim\limits_{n\to\infty}s_n=\frac{1}{4}$. This can be confirmed by directly computing the sum using `sum(f,1..infinity);` (in Maple) or `Sum[f,{n,0,Infinity}]` (in Mathematica).

50. $\dfrac{n^2+3n+1}{(n^2+n)^2}=\dfrac{1}{n^2}+\dfrac{1}{n}-\dfrac{1}{(n+1)^2}-\dfrac{1}{n+1}$, [using the `convert` command in Maple or `Apart` in Mathematica]. So the nth partial sum is

$$s_n=\sum_{k=1}^{n}\left(\frac{1}{k^2}+\frac{1}{k}-\frac{1}{(k+1)^2}-\frac{1}{k+1}\right)$$

$$=\left(1+1-\frac{1}{2^2}-\frac{1}{2}\right)+\left(\frac{1}{2^2}+\frac{1}{2}-\frac{1}{3^2}-\frac{1}{3}\right)+\cdots+\left(\frac{1}{n^2}+\frac{1}{n}-\frac{1}{(n+1)^2}-\frac{1}{n+1}\right)$$

$=2-\dfrac{1}{(n+1)^2}-\dfrac{1}{n+1}$. The series converges to $\lim\limits_{n\to\infty}s_n=2$.

This can be confirmed by directly computing the sum using `sum(f,1..infinity);` (in Maple) or `Sum[f,{n,0,Infinity}]` (in Mathematica).

51. Plainly $a_1=0$ since $s_1=0$. For $n\neq 1$, $a_n=s_n-s_{n-1}=\dfrac{n-1}{n+1}-\dfrac{(n-1)-1}{(n-1)+1}$ $=\dfrac{(n-1)n-(n+1)(n-2)}{(n+1)n}=\dfrac{2}{n(n+1)}$. Also $\displaystyle\sum_{n=1}^{\infty}a_n=\lim_{n\to\infty}s_n=\lim_{n\to\infty}\frac{1-1/n}{1+1/n}=1$.

52. $a_1=s_1=5/2$. For $n\neq 1$, $a_n=s_n-s_{n-1}=(3-n2^{-n})-\left[3-(n-1)2^{-(n-1)}\right]=\dfrac{2(n-1)}{2^n}-\dfrac{n}{2^n}=\dfrac{n-2}{2^n}$.

Also $\displaystyle\sum_{n=1}^{\infty}a_n=\lim_{n\to\infty}s_n=\lim_{n\to\infty}\left(3-\frac{n}{2^n}\right)=3$ because $\displaystyle\lim_{x\to\infty}\frac{x}{2^x}\overset{\text{H}}{=}\lim_{x\to\infty}\frac{1}{2^x\ln 2}=0$.

53. **(a)** The first step in the chain occurs when the local government spends D dollars. The people who receive it spend a fraction c of those D dollars, that is, Dc dollars. Those who receive the Dc dollars spend a fraction c of it, that is, Dc^2 dollars. Continuing in this way, we see that the total spending after n transactions is $S_n = D + Dc + Dc^2 + \cdots + Dc^{n-1} = \dfrac{D(1-c^n)}{1-c}$ by (3).

(b) $\lim\limits_{n\to\infty} S_n = \lim\limits_{n\to\infty} \dfrac{D(1-c^n)}{1-c} = \dfrac{D}{1-c}\lim\limits_{n\to\infty}(1-c^n) = \dfrac{D}{1-c} = \dfrac{D}{s} = kD$, since $0 < c < 1$ $\Rightarrow$ $\lim\limits_{n\to\infty} c^n = 0$. If $c = 0.8$, then $s = 1 - c = 0.2$ and the multiplier is $k = 1/s = 5$.

54. **(a)** Initially, the ball falls a distance H, then rebounds a distance rH, falls rH, rebounds r^2H, falls r^2H, etc. The total distance it travels is $H + 2rH + 2r^2H + 2r^3H + \cdots = H(1 + 2r + 2r^2 + 2r^3 + \cdots)$

$$= H\left[1 + 2r(1 + r + r^2 + \cdots)\right] = H\left[1 + 2r\left(\frac{1}{1-r}\right)\right] = H\left(\frac{1+r}{1-r}\right) \text{meters.}$$

(b) From Exercise 9.1.28, we know that a ball falls $\frac{1}{2}gt^2$ meters in t seconds, where g is the gravitational acceleration. Thus a ball falls h meters in $\sqrt{2h/g}$ seconds. The total travel time in seconds is

$$\sqrt{\frac{2H}{g}} + 2\sqrt{\frac{2H}{g}r} + 2\sqrt{\frac{2H}{g}r^2} + 2\sqrt{\frac{2H}{g}r^3} + \cdots = \sqrt{\frac{2H}{g}}\left[1 + 2\sqrt{r} + 2\left(\sqrt{r}\right)^2 + 2\left(\sqrt{r}\right)^3 + \cdots\right]$$

$$= \sqrt{\frac{2H}{g}}\left(1 + 2\sqrt{r}\left[1 + \sqrt{r} + \left(\sqrt{r}\right)^2 + \cdots\right]\right) = \sqrt{\frac{2H}{g}}\left[1 + 2\sqrt{r}\left(\frac{1}{1-\sqrt{r}}\right)\right] = \sqrt{\frac{2H}{g}}\frac{1+\sqrt{r}}{1-\sqrt{r}}$$

(c) It will help to make a chart of the time for each descent and each rebound of the ball, together with the velocity just before and just after each bounce. Recall that the time in seconds needed to fall h meters is $\sqrt{2h/g}$. The ball hits the ground with velocity $-g\sqrt{2h/g}$ (taking the upward direction to be positive) and rebounds with velocity $kg\sqrt{2h/g}$, taking time $k\sqrt{2h/g}$ to reach the top of its bounce, where its velocity is 0. At that point, its height is k^2h.

All these results follow from the formulas for vertical motion with gravitational acceleration $-g$:

$$\frac{d^2y}{dt^2} = -g \quad\Rightarrow\quad v = \frac{dy}{dt} = v_0 - gt \quad\Rightarrow\quad y = y_0 + v_0t - \tfrac{1}{2}gt^2.$$

number of descent	time of descent	speed before bounce	speed after bounce	time of ascent	peak height
1	$\sqrt{2H/g}$	$\sqrt{2Hg}$	$k\sqrt{2Hg}$	$k\sqrt{2H/g}$	k^2H
2	$\sqrt{2k^2H/g}$	$\sqrt{2k^2Hg}$	$k\sqrt{2k^2Hg}$	$k\sqrt{2k^2H/g}$	k^4H
3	$\sqrt{2k^4H/g}$	$\sqrt{2k^4Hg}$	$k\sqrt{2k^4Hg}$	$k\sqrt{2k^4H/g}$	k^6H
$\cdots$	$\cdots$	$\cdots$	$\cdots$	$\cdots$	$\cdots$

The total travel time in seconds is

$$\sqrt{\frac{2H}{g}} + k\sqrt{\frac{2H}{g}} + k\sqrt{\frac{2H}{g}} + k^2\sqrt{\frac{2H}{g}} + k^2\sqrt{\frac{2H}{g}} + \cdots = \sqrt{\frac{2H}{g}}\left(1 + 2k + 2k^2 + 2k^3 + \cdots\right)$$

$$= \sqrt{\frac{2H}{g}}\left[1 + 2k(1 + k + k^2 + \cdots)\right] = \sqrt{\frac{2H}{g}}\left[1 + 2k\left(\frac{1}{1-k}\right)\right] = \sqrt{\frac{2H}{g}}\frac{1+k}{1-k}.$$

Another Method: We could use part (b). At the top of the bounce, the height is $k^2h = rh$, so $\sqrt{r} = k$ and the result follows from part (b).

55. $\sum_{n=2}^{\infty}(1+c)^{-n}$ is a geometric series with $a=(1+c)^{-2}$ and $r=(1+c)^{-1}$, so the series converges when $\left|(1+c)^{-1}\right|<1 \quad\Rightarrow\quad |1+c|>1 \quad\Rightarrow\quad 1+c>1$ or $1+c<-1 \quad\Rightarrow\quad c>0$ or $c<-2$. We calculate the sum of the series and set it equal to 2: $\dfrac{(1+c)^{-2}}{1-(1+c)^{-1}}=2 \quad\Leftrightarrow\quad \left(\dfrac{1}{1+c}\right)^2=2-2\left(\dfrac{1}{1+c}\right) \quad\Leftrightarrow$ $1-2(1+c)^2+2(1+c)=0 \quad\Leftrightarrow\quad 2c^2+2c-1=0 \quad\Leftrightarrow\quad c=\frac{-2\pm\sqrt{12}}{4}=\frac{\pm\sqrt{3}-1}{2}$. However, the negative root is inadmissible because $-2<\frac{-\sqrt{3}-1}{2}<0$. So $c=\frac{\sqrt{3}-1}{2}$.

56. The area between $y=x^{n-1}$ and $y=x^n$ for $0\le x\le 1$ is $\displaystyle\int_0^1\left(x^{n-1}-x^n\right)dx=\left[\frac{x^n}{n}-\frac{x^{n+1}}{n+1}\right]_0^1$

$$=\frac{1}{n}-\frac{1}{n+1}=\frac{(n+1)-1}{n(n+1)}=\frac{1}{n(n+1)}.$$

As we can see from the diagram, as $n\to\infty$ the sum of the areas between the successive curves approaches the area of the unit square, that is, 1.

So $\displaystyle\sum_{n=1}^{\infty}\frac{1}{n(n+1)}=1.$

57. Let d_n be the diameter of C_n. We draw lines from the centers of the C_i to the center of D (or C), and using the Pythagorean Theorem, we can write

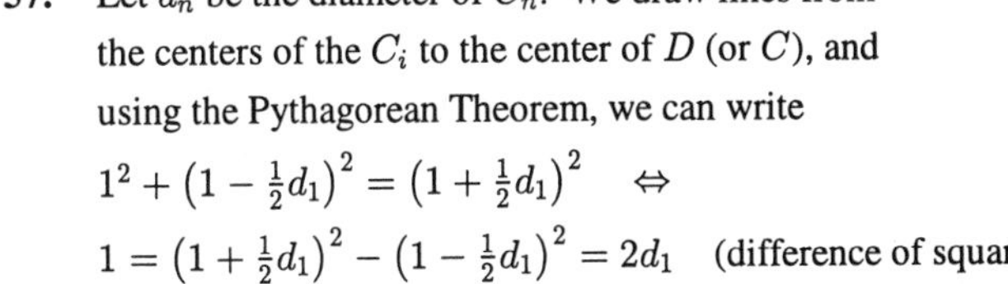

$1^2+\left(1-\frac{1}{2}d_1\right)^2=\left(1+\frac{1}{2}d_1\right)^2 \quad\Leftrightarrow$

$1=\left(1+\frac{1}{2}d_1\right)^2-\left(1-\frac{1}{2}d_1\right)^2=2d_1$ (difference of squares)

$\Rightarrow\quad d_1=\frac{1}{2}$. Similarly,

$1=\left(1+\frac{1}{2}d_2\right)^2-\left(1-d_1-\frac{1}{2}d_2\right)^2=(2-d_1)(d_1+d_2)$

$\Leftrightarrow\quad d_2=\dfrac{(1-d_1)^2}{2-d_1}$, $1=\left(1+\frac{1}{2}d_3\right)^2-\left(1-d_1-d_2-\frac{1}{2}d_3\right)^2 \quad\Leftrightarrow\quad d_3=\dfrac{[1-(d_1+d_2)]^2}{2-(d_1+d_2)}$, and in general,

$d_{n+1}=\dfrac{\left(1-\sum_{i=1}^{n}d_i\right)^2}{2-\sum_{i=1}^{n}d_i}$. If we actually calculate d_2 and d_3 from the formulas above, we find that they are $\dfrac{1}{6}=\dfrac{1}{2\cdot 3}$ and $\dfrac{1}{12}=\dfrac{1}{3\cdot 4}$ respectively, so we suspect that in general, $d_n=\dfrac{1}{n(n+1)}$. To prove this, we use induction: assume that for all $k\le n$, $d_k=\dfrac{1}{k(k+1)}=\dfrac{1}{k}-\dfrac{1}{k+1}$. Then $\displaystyle\sum_{i=1}^{n}d_i=1-\frac{1}{n+1}=\frac{n}{n+1}$ (telescoping sum). Substituting this into our formula for d_{n+1}, we get

$$d_{n+1}=\frac{\left[1-\dfrac{n}{n+1}\right]^2}{2-\left(\dfrac{n}{n+1}\right)}=\frac{\dfrac{1}{(n+1)^2}}{\dfrac{n+2}{n+1}}=\frac{1}{(n+1)(n+2)},$$ and the induction is complete.

Now, we observe that the partial sums $\sum_{i=1}^{n}d_i$ of the diameters of the circles approach 1 as $n\to\infty$; that is,

$\displaystyle\sum_{n=1}^{\infty}d_n=\sum_{n=1}^{\infty}\frac{1}{n(n+1)}=1$, which is what we wanted to prove.

58. $|CD| = b\sin\theta, |DE| = |CD|\sin\theta = b\sin^2\theta, |EF| = |DE|\sin\theta = b\sin^3\theta, \ldots$ Therefore $|CD| + |DE| + |EF| + |FG| + \cdots = b\sum_{n=1}^{\infty}\sin^n\theta = b\left(\frac{\sin\theta}{1-\sin\theta}\right)$ since this is a geometric series with $r = \sin\theta$ and $|\sin\theta| < 1$ since $0 < \theta < \frac{\pi}{2}$.

59. The series $1 - 1 + 1 - 1 + 1 - 1 + \cdots$ diverges (geometric series with $r = -1$) so we cannot say $0 = 1 - 1 + 1 - 1 + 1 - 1 + \cdots$.

60. If $\sum_{n=1}^{\infty} a_n$ is convergent then $\lim_{n\to\infty} a_n = 0$ by Theorem 6, so $\lim_{n\to\infty}\frac{1}{a_n} \neq 0$, and so $\sum_{n=1}^{\infty}\frac{1}{a_n}$ is divergent by (7).

61. $\sum_{n=1}^{\infty} ca_n = \lim_{n\to\infty}\sum_{i=1}^{n} ca_i = \lim_{n\to\infty} c\sum_{i=1}^{n} a_i = c\lim_{n\to\infty}\sum_{i=1}^{n} a_i = c\sum_{n=1}^{\infty} a_n$, which exists by hypothesis.

62. If $\sum ca_n$ were convergent, then $\sum(1/c)(ca_n) = \sum a_n$ would be also, by Theorem 8. But this is not the case, so $\sum ca_n$ must diverge.

63. Suppose on the contrary that $\sum(a_n + b_n)$ converges. Then by Theorem 8(c), so would $\sum[(a_n + b_n) - a_n] = \sum b_n$, a contradiction.

64. No. For example, take $\sum a_n = \sum n$ and $\sum b_n = \sum -n$, which both diverge, yet $\sum(a_n + b_n) = \sum 0$ converges with sum 0.

65. The partial sums $\{s_n\}$ form an increasing sequence, since $s_n - s_{n-1} = a_n > 0$ for all n. Also, the sequence $\{s_n\}$ is bounded since $s_n \leq 1000$ for all n. So by Theorem 10.1.10 , the sequence of partial sums converges, that is, the series $\sum a_n$ is convergent.

66. **(a)** RHS $= \dfrac{f_n f_{n+1} - f_n f_{n-1}}{f_n^2 f_{n-1} f_{n+1}} = \dfrac{f_{n+1} - f_{n-1}}{f_n f_{n-1} f_{n+1}} = \dfrac{(f_{n-1} + f_n) - f_{n-1}}{f_n f_{n-1} f_{n+1}} = \dfrac{1}{f_{n-1} f_{n+1}} =$ LHS

(b) $\sum_{n=2}^{\infty}\frac{1}{f_{n-1}f_{n+1}} = \sum_{n=2}^{\infty}\left(\frac{1}{f_{n-1}f_n} - \frac{1}{f_n f_{n+1}}\right)$ [from (a)]

$$= \lim_{n\to\infty}\left[\left(\frac{1}{f_1 f_2} - \frac{1}{f_2 f_3}\right) + \left(\frac{1}{f_2 f_3} - \frac{1}{f_3 f_4}\right) + \left(\frac{1}{f_3 f_4} - \frac{1}{f_4 f_5}\right) + \cdots + \left(\frac{1}{f_{n-1} f_n} - \frac{1}{f_n f_{n+1}}\right)\right]$$

$$= \lim_{n\to\infty}\left(\frac{1}{f_1 f_2} - \frac{1}{f_n f_{n+1}}\right) = \frac{1}{f_1 f_2} - 0 = 1 \text{ because } f_n \to \infty \text{ as } n \to \infty.$$

(c) $\sum_{n=2}^{\infty}\frac{f_n}{f_{n-1}f_{n+1}} = \sum_{n=2}^{\infty}\left(\frac{f_n}{f_{n-1}f_n} - \frac{f_n}{f_n f_{n+1}}\right)$ (as above) $= \sum_{n=2}^{\infty}\left(\frac{1}{f_{n-1}} - \frac{1}{f_{n+1}}\right)$

$$= \lim_{n\to\infty}\left[\left(\frac{1}{f_1} - \frac{1}{f_3}\right) + \left(\frac{1}{f_2} - \frac{1}{f_4}\right) + \left(\frac{1}{f_3} - \frac{1}{f_5}\right) + \left(\frac{1}{f_4} - \frac{1}{f_6}\right) + \cdots + \left(\frac{1}{f_{n-1}} - \frac{1}{f_{n+1}}\right)\right]$$

$$= \lim_{n\to\infty}\left(\frac{1}{f_1} + \frac{1}{f_2} - \frac{1}{f_n} - \frac{1}{f_{n+1}}\right) = 1 + 1 - 0 - 0 = 2 \text{ because } f_n \to \infty \text{ as } n \to \infty.$$

67. **(a)** At the first step, only the interval $\left(\frac{1}{3}, \frac{2}{3}\right)$ (length $\frac{1}{3}$) is removed. At the second step, we remove the intervals $\left(\frac{1}{9}, \frac{2}{9}\right)$ and $\left(\frac{7}{9}, \frac{8}{9}\right)$, which have a total length of $2 \cdot \left(\frac{1}{3}\right)^2$. At the third step, we remove 2^2 intervals, each of length $\left(\frac{1}{3}\right)^3$. In general, at the nth step we remove 2^{n-1} intervals, each of length $\left(\frac{1}{3}\right)^n$, for a length of $2^{n-1} \cdot \left(\frac{1}{3}\right)^n = \frac{1}{3}\left(\frac{2}{3}\right)^{n-1}$. Thus, the total length of all removed intervals is $\displaystyle\sum_{n=1}^{\infty} \frac{1}{3}\left(\frac{2}{3}\right)^{n-1} = \frac{1/3}{1 - 2/3} = 1$ (geometric series with $a = \frac{1}{3}$ and $r = \frac{2}{3}$).

Notice that at the nth step, the leftmost interval that is removed is $\left(\left(\frac{1}{3}\right)^n, \left(\frac{2}{3}\right)^n\right)$, so we never remove 0, so 0 is in the Cantor set. Also, the rightmost interval removed is $\left(1 - \left(\frac{2}{3}\right)^n, 1 - \left(\frac{1}{3}\right)^n\right)$, so 1 is never removed. Some other numbers in the Cantor set are $\frac{1}{3}$, $\frac{2}{3}$, $\frac{1}{9}$, $\frac{2}{9}$, $\frac{7}{9}$, and $\frac{8}{9}$.

(b) The area removed at the first step is $\frac{1}{9}$; at the second step, $2^3 \cdot \left(\frac{1}{9}\right)^2$; at the third step, $\left(2^3\right)^2 \cdot \left(\frac{1}{9}\right)^3$. In general, the area removed at the nth step is $\left(2^3\right)^{n-1}\left(\frac{1}{9}\right)^n = \frac{1}{9}\left(\frac{8}{9}\right)^{n-1}$, so the total area of all removed squares is $\displaystyle\sum_{n=1}^{\infty} \frac{1}{8}\left(\frac{8}{9}\right)^{n-1} = \frac{1/9}{1 - 8/9} = 1$.

68. **(a)**

a_1	1	2	4	1
a_2	2	3	1	4
a_3	1.5	2.5	2.5	2.5
a_4	1.75	2.75	1.75	3.25
a_5	1.625	2.625	2.125	2.875
a_6	1.6875	2.6875	1.9375	3.0625
a_7	1.65625	2.65625	2.03125	2.96875
a_8	1.67188	2.67188	1.98438	3.01563
a_9	1.66406	2.66406	2.00781	2.99219
a_{10}	1.66797	2.66797	1.99609	3.00391
a_{11}	1.66602	2.66602	2.00195	2.99805
a_{12}	1.66699	2.66699	1.99902	3.00098

The limits seem to be $\frac{5}{3}$, $\frac{8}{3}$, 2, and 3.

In general, we guess that the limit is $\dfrac{a_1 + 2a_2}{3}$.

(b) $a_{n+1} - a_n = \frac{1}{2}(a_n + a_{n-1}) - a_n = -\frac{1}{2}(a_n - a_{n-1}) = -\frac{1}{2}\left[-\frac{1}{2}(a_{n-1} - a_{n-2})\right] = \cdots = \left(-\frac{1}{2}\right)^{n-1}(a_2 - a_1)$.

Note that we have used the formula $a_k = \frac{1}{2}(a_{k-2} + a_{k-1})$ a total of $n - 1$ times in this calculation, once for each k between 3 and $n + 1$. Now we can write

$$a_n = a_1 + (a_2 - a_1) + (a_2 - a_1) + (a_3 - a_2) + \cdots + (a_{n-1} - a_{n-2}) + (a_n - a_{n-1})$$

$$= a_1 + \sum_{k=1}^{n-1}(a_{k+1} - a_k) = a_1 + \sum_{k=1}^{n-1}\left(-\tfrac{1}{2}\right)^{k-1}(a_2 - a_1), \text{ and so}$$

$$\lim_{n\to\infty} a_n = a_1 + (a_2 - a_1)\sum_{k=1}^{\infty}\left(-\frac{1}{2}\right)^{k-1} = a_1 + (a_2 - a_1)\left[\frac{1}{1 - (-1/2)}\right] = a_1 + \tfrac{2}{3}(a_2 - a_1) = \frac{a_1 + 2a_2}{3}.$$

69. **(a)** $s_1 = \frac{1}{1 \cdot 2} = \frac{1}{2}$, $s_2 = \frac{1}{2} + \frac{1}{1 \cdot 2 \cdot 3} = \frac{5}{6}$, $s_3 = \frac{5}{6} + \frac{3}{1 \cdot 2 \cdot 3 \cdot 4} = \frac{23}{24}$, $s_4 = \frac{23}{24} + \frac{4}{1 \cdot 2 \cdot 3 \cdot 4 \cdot 5} = \frac{119}{120}$.

The denominators are $(n+1)!$ so a guess would be $s_n = \frac{(n+1)! - 1}{(n+1)!}$.

(b) For $n = 1$, $s_1 = \frac{1}{2} = \frac{2! - 1}{2!}$, so the formula holds for $n = 1$. Assume $s_k = \frac{(k+1)! - 1}{(k+1)!}$. Then

$$\begin{aligned} s_{k+1} &= \frac{(k+1)! - 1}{(k+1)!} + \frac{k+1}{(k+2)!} \\ &= \frac{(k+1)! - 1}{(k+1)!} + \frac{k+1}{(k+1)!(k+2)} \\ &= \frac{(k+2)! - (k+2) + k + 1}{(k+2)!} \\ &= \frac{(k+2)! - 1}{(k+2)!} \end{aligned}$$

Thus the formula is true for $n = k + 1$. So by induction, the guess is correct.

(c) $\lim_{n\to\infty} s_n = \lim_{n\to\infty} \frac{(n+1)! - 1}{(n+1)!} = \lim_{n\to\infty}\left[1 - \frac{1}{(n+1)!}\right] = 1$ and so $\sum_{n=0}^{\infty} \frac{n}{(n+1)!} = 1$.

70.

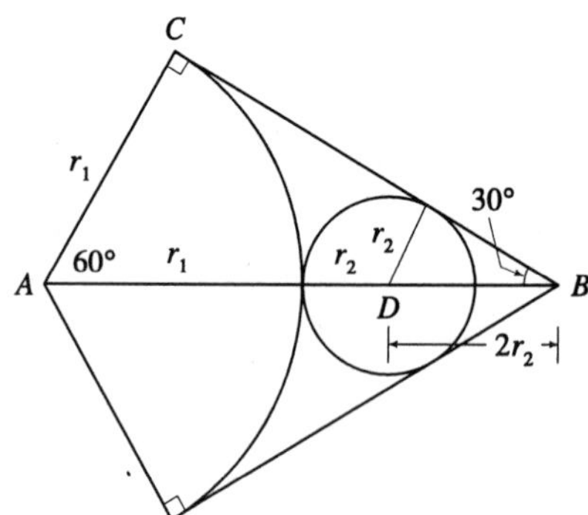

Let r_1 = radius of the large circle, r_2 = radius of next circle, and so on. From the figure we have $\angle BAC = 60°$, so $|AB| = 2r_1$ and $|DB| = 2r_2$. Therefore $2r_1 = r_1 + r_2 + 2r_2 = r_1 + 3r_2 \quad \Rightarrow \quad r_1 = 3r_2$.
In general, we have $r_{n+1} = \frac{1}{3}r_n$, so the total area is

$$\begin{aligned} A &= \pi r_1^2 + 3\pi r_2^2 + 3\pi r_3^2 + \cdots \\ &= \pi r_1^2 + 3\pi r_2^2\left(1 + \frac{1}{3^2} + \frac{1}{3^4} + \frac{1}{3^6} + \cdots\right) \\ &= \pi r_1^2 + 3\pi r_2^2 \cdot \frac{1}{1 - 1/9} \\ &= \pi r_1^2 + \tfrac{27}{8}\pi r_2^2 \end{aligned}$$

Now $r_1 = \frac{\tan 30°}{2} = \frac{1}{2\sqrt{3}} \quad \Rightarrow \quad r_2 = \frac{1}{6\sqrt{3}}$, so $A = \pi\left(\frac{1}{2\sqrt{3}}\right)^2 + \frac{27\pi}{8}\left(\frac{1}{6\sqrt{3}}\right)^2 = \frac{\pi}{12} + \frac{\pi}{32} = \frac{11\pi}{96}$.

EXERCISES 10.3

1. $\displaystyle\sum_{n=1}^{\infty}\frac{2}{\sqrt[3]{n}}=2\sum_{n=1}^{\infty}\frac{1}{n^{1/3}}$, which is a p-series, $p=\frac{1}{3}<1$, so it diverges.

2. $\displaystyle\sum_{n=1}^{\infty}\left(\frac{2}{n\sqrt{n}}+\frac{3}{n^3}\right)=2\sum_{n=1}^{\infty}\frac{1}{n^{3/2}}+3\sum_{n=1}^{\infty}\frac{1}{n^3}$, both of which are convergent p-series because $\frac{3}{2}>1$ and $3>1$, so $\displaystyle\sum_{n=1}^{\infty}\left(\frac{2}{n\sqrt{n}}+\frac{3}{n^3}\right)$ converges by (10.2.8).

3. $\displaystyle\sum_{n=5}^{\infty}\frac{1}{n^{1.0001}}$ is a p-series, $p=1.0001>1$, so it converges.

4. $\displaystyle\sum_{n=1}^{\infty}n^{-0.99}=\sum_{n=1}^{\infty}\frac{1}{n^{0.99}}$ which diverges since $p=0.99<1$.

5. $\displaystyle\sum_{n=5}^{\infty}\frac{1}{(n-4)^2}=\sum_{n=1}^{\infty}\frac{1}{n^2}$ is a p-series, $p=2>1$, so it converges.

6. $f(x)=\dfrac{1}{2x+3}$ is positive, continuous, and decreasing on $[1,\infty)$, so applying the Integral Test,

$$\int_1^{\infty}\frac{dx}{2x+3}=\lim_{t\to\infty}\left[\tfrac{1}{2}\ln(2x+3)\right]_1^t=\infty \quad\Rightarrow\quad \sum_{n=1}^{\infty}\frac{1}{2n+3}\text{ is divergent.}$$

7. Since $\dfrac{1}{\sqrt{x}+1}$ is continuous, positive, and decreasing on $[0,\infty)$ we can apply the Integral Test.

$$\int_1^{\infty}\frac{1}{\sqrt{x}+1}\,dx=\lim_{t\to\infty}\left[2\sqrt{x}-2\ln(\sqrt{x}+1)\right]_1^t \quad \left[\begin{array}{c}\text{using the substitution } u=\sqrt{x}+1,\text{ so}\\ x=(u-1)^2 \text{ and } dx=2(u-1)du\end{array}\right]$$

$$=\lim_{t\to\infty}\left(\left[2\sqrt{t}-2\ln\left(\sqrt{t}+1\right)\right]-(2-2\ln 2)\right).$$

Now $2\sqrt{t}-2\ln\left(\sqrt{t}+1\right)=2\ln\left(\dfrac{e^{\sqrt{t}}}{\sqrt{t}+1}\right)$ and so $\displaystyle\lim_{t\to\infty}\left[2\sqrt{t}-2\ln\left(\sqrt{t}+1\right)\right]=\infty$ (using l'Hospital's Rule) so both the integral and the original series diverge.

8. $f(x)=\dfrac{1}{x^2-1}$ is positive, continuous, and decreasing on $[2,\infty)$, so applying the Integral Test,

$$\int_2^{\infty}\frac{dx}{x^2-1}=\int_2^{\infty}\left(\frac{-1/2}{x+1}+\frac{1/2}{x-1}\right)dx=\lim_{t\to\infty}\left[\ln\left(\frac{x-1}{x+1}\right)^{1/2}\right]_2^t=\ln\sqrt{3}\quad\Rightarrow\quad\sum_{n=2}^{\infty}\frac{1}{n^2-1}\text{ converges.}$$

9. $f(x)=xe^{-x^2}$ is continuous and positive on $[1,\infty)$, and since $f'(x)=e^{-x^2}(1-2x^2)<0$ for $x>1$, f is decreasing as well. We can use the Integral Test. $\displaystyle\int_1^{\infty}xe^{-x^2}dx=\lim_{t\to\infty}\left[-\tfrac{1}{2}e^{-x^2}\right]_1^t=0-\left(-\frac{e^{-1}}{2}\right)=\frac{1}{2e}$ so the series converges.

10. $f(x) = \dfrac{x}{2^x}$ is positive and continuous on $[1, \infty)$, and since $f'(x) = \dfrac{1 - x \ln 2}{2^x} < 0$ when $x > \dfrac{1}{\ln 2} \approx 1.44$, f is eventually decreasing, so we can apply the Integral Test. Integrating by parts, we get

$$\int_1^\infty \frac{x}{2^x}\,dx = \lim_{t\to\infty}\left(-\frac{1}{\ln 2}\left[\frac{x}{2^x} + \frac{1}{2^x \ln 2}\right]_1^t\right) = \frac{1}{2\ln 2} + \frac{1}{2(\ln 2)^2}$$, since $\lim_{t\to\infty} \dfrac{t}{2^t} = 0$ by l'Hospital's Rule, and

so $\displaystyle\sum_{n=1}^{\infty} \frac{n}{2^n}$ converges.

11. $f(x) = \dfrac{x}{x^2+1}$ is continuous and positive on $[1, \infty)$, and since $f'(x) = \dfrac{1 - x^2}{(x^2+1)^2} < 0$ for $x > 1$, f is also decreasing. Using the Integral Test, $\displaystyle\int_1^\infty \frac{x}{x^2+1}\,dx = \lim_{t\to\infty}\left[\frac{\ln(x^2+1)}{2}\right]_1^t = \infty$, so the series diverges.

12. $f(x) = \dfrac{1}{2x^2 - x - 1} = \dfrac{1}{(2x+1)(x-1)}$ is continuous and positive on $[2, \infty)$, and since

$f'(x) = \dfrac{1 - 4x}{(2x^2 - x - 1)^2} < 0$ on $[2, \infty)$ the function is decreasing on $[2, \infty)$, so we can apply the Integral Test.

$$\int_2^\infty \frac{dx}{(2x+1)(x-1)} = \int_2^\infty \left(\frac{-2/3}{2x+1} + \frac{1/3}{x-1}\right)dx = \lim_{t\to\infty}\left[\tfrac{1}{3}\ln(x-1) - \tfrac{1}{3}\ln(2x+1)\right]_2^t$$

$$= \lim_{t\to\infty} \frac{1}{3}\left[\ln\left(\frac{t-1}{2t+1}\right) - \ln\frac{1}{5}\right] = \frac{\ln 5 - \ln 2}{3} < \infty$$ so the series converges.

13. $f(x) = \dfrac{1}{x \ln x}$ is continuous and positive on $[2, \infty)$, and also decreasing since $f'(x) = -\dfrac{1 + \ln x}{x^2 (\ln x)^2} < 0$ for $x > 2$, so we can use the Integral Test. $\displaystyle\int_2^\infty \frac{1}{x \ln x}\,dx = \lim_{t\to\infty}[\ln(\ln x)]_2^t = \lim_{t\to\infty}[\ln(\ln t) - \ln(\ln 2)] = \infty$, so the series diverges.

14. $f(x) = \dfrac{1}{4x^2+1}$ is continuous, positive and decreasing on $[1, \infty)$, so applying the Integral Test,

$$\int_1^\infty \frac{dx}{4x^2+1} = \lim_{t\to\infty}\left[\frac{\arctan 2x}{2}\right]_1^t = \frac{\pi}{4} - \frac{\arctan 2}{2} < \infty$$, so the series converges.

15. $f(x) = \dfrac{\arctan x}{1 + x^2}$ is continuous and positive on $[1, \infty)$. $f'(x) = \dfrac{1 - 2x \arctan x}{(1+x^2)^2} < 0$ for $x > 1$, since $2x \arctan x \geq \frac{\pi}{2} > 1$ for $x \geq 1$. So f is decreasing and we can use the Integral Test.

$$\int_1^\infty \frac{\arctan x}{1+x^2}\,dx = \lim_{t\to\infty}\left[\tfrac{1}{2}\arctan^2 x\right]_1^t = \frac{(\pi/2)^2}{2} - \frac{(\pi/4)^2}{2} = \frac{3\pi^2}{32}$$, so the series converges.

16. $f(x) = \dfrac{\ln x}{x^2}$ is continuous and positive for $x \geq 2$, and $f'(x) = \dfrac{1 - 2\ln x}{x^3} < 0$ for $x \geq 2$ so f is decreasing.

$\displaystyle\int_2^\infty \frac{\ln x}{x^2}\,dx = \lim_{t\to\infty}\left[-\frac{\ln x}{x} - \frac{1}{x}\right]_1^t$ (using integration by parts) $= 1$ (by L'Hospital's Rule). Thus

$\displaystyle\sum_{n=1}^{\infty} \frac{\ln n}{n^2} = \sum_{n=2}^{\infty} \frac{\ln n}{n^2}$ converges by the Integral Test.

17. $f(x) = \dfrac{1}{x^2 + 2x + 2}$ is continuous and positive on $[1, \infty)$, and $f'(x) = -\dfrac{2x+2}{(x^2+2x+2)^2} < 0$ for $x \geq 1$, so f is decreasing and we can use the Integral Test. $\displaystyle\int_1^\infty \frac{1}{x^2+2x+2}\,dx = \int_1^\infty \frac{1}{(x+1)^2+1}\,dx$ $= \lim\limits_{t\to\infty}[\arctan(x+1)]_1^t = \frac{\pi}{2} - \arctan 2$, so the series converges also.

18. $f(x) = \dfrac{1}{x \ln x \ln(\ln x)}$ is positive and continuous on $[3, \infty)$, and is decreasing since x, $\ln x$, and $\ln(\ln x)$ are all increasing; so we can apply the Integral Test. $\displaystyle\int_3^\infty \frac{dx}{x \ln x \ln(\ln x)} = \lim_{t\to\infty}[\ln(\ln(\ln x))]_3^t$ which diverges, and hence $\displaystyle\sum_{n=3}^\infty \frac{1}{n \ln n \ln(\ln n)}$ diverges.

19. We have already shown that when $p = 1$ the series diverges (in Exercise 13 above), so assume $p \neq 1$. $f(x) = \dfrac{1}{x(\ln x)^p}$ is continuous and positive on $[2, \infty)$, and $f'(x) = -\dfrac{p + \ln x}{x^2(\ln x)^{p+1}} < 0$ if $x > e^{-p}$, so that f is eventually decreasing and we can use the Integral Test.

$$\int_2^\infty \frac{1}{x(\ln x)^p}\,dx = \lim_{t\to\infty}\left[\frac{(\ln x)^{1-p}}{1-p}\right]_2^t \quad (\text{for } p \neq 1) \quad = \lim_{t\to\infty}\left[\frac{(\ln t)^{1-p}}{1-p}\right] - \frac{(\ln 2)^{1-p}}{1-p}.$$

This limit exists whenever $1 - p < 0 \quad \Leftrightarrow \quad p > 1$, so the series converges for $p > 1$.

20. As in Exercise 18 we can apply the Integral Test. $\displaystyle\int_3^\infty \frac{dx}{x \ln x (\ln\ln x)^p} = \lim_{t\to\infty}\left[\frac{(\ln\ln x)^{-p+1}}{-p+1}\right]_3^t$ (for $p \neq 1$; if $p = 1$ see Exercise 18) and $\displaystyle\lim_{t\to\infty}\frac{(\ln\ln t)^{-p+1}}{-p+1}$ exists whenever $-p + 1 < 0 \quad \Leftrightarrow \quad p > 1$, so the series converges for $p > 1$.

21. Clearly the series cannot converge if $p \geq -\frac{1}{2}$, because then $\lim\limits_{n\to\infty} n(1+n^2)^p \neq 0$. Also, if $p = -1$ the series diverges (see Exercise 11 above.) So assume $p < -\frac{1}{2}$, $p \neq -1$. Then $f(x) = x(1+x^2)^p$ is continuous, positive, and eventually decreasing on $[1, \infty)$, and we can use the Integral Test.

$$\int_1^\infty x(1+x^2)^p\,dx = \lim_{t\to\infty}\left[\frac{1}{2}\cdot\frac{(1+x^2)^{p+1}}{p+1}\right]_1^t = \lim_{t\to\infty}\frac{1}{2}\cdot\frac{(1+t^2)^{p+1}}{p+1} - \frac{2^p}{p+1}.$$ This limit exists and is finite $\Leftrightarrow \; p + 1 < 0 \; \Leftrightarrow \; p < -1$, so the series converges whenever $p < -1$.

22. If $p \leq 0$, $\lim\limits_{n\to\infty}\dfrac{\ln n}{n^p} = \infty$ and the series diverges, so assume $p > 0$. $f(x) = (\ln x)/x^p$ is positive and continuous and $f'(x) < 0$ for $x > e^{1/p}$, so f is eventually decreasing and we can use the Integral Test. Integration by parts gives $\displaystyle\int_1^\infty \frac{\ln x}{x^p}\,dx = \lim_{t\to\infty}\left[\frac{x^{1-p}[(1-p)\ln x - 1]}{(1-p)^2}\right]_1^t$ (for $p \neq 1$) $\displaystyle = \frac{1}{(1-p)^2}\left[\lim_{t\to\infty} t^{1-p}[(1-p)\ln t - 1] + 1\right]$ which exists whenever $1 - p < 0 \quad \Leftrightarrow \quad p > 1$. Since we have already done the case $p = 1$ in Exercise 19 (set $p = -1$ in that exercise), $\displaystyle\sum_{n=1}^\infty \frac{\ln n}{n^p}$ converges $\quad \Leftrightarrow \quad p > 1$.

23. Since this is a p-series with $p = x$, $\zeta(x)$ is defined when $x > 1$.

24. **(a)** $f(x) = 1/x^4$ is positive and continuous and $f'(x) = -4/x^5$ is negative for $x > 1$, and so the Integral Test applies. $\sum_{n=1}^{\infty} \frac{1}{n^4} \approx s_{10} = \frac{1}{1^4} + \frac{1}{2^4} + \frac{1}{3^4} + \cdots + \frac{1}{10^4} \approx 1.0820$. $R_{10} \le \int_{10}^{\infty} \frac{1}{x^4}\,dx$

$= \lim_{t\to\infty}\left[\frac{1}{-3x^3}\right]_{10}^{t} = \lim_{t\to\infty}\left(-\frac{1}{3t^3} + \frac{1}{3(10)^3}\right) = \frac{1}{3000}$, so the error is at most $0.000\overline{3}$.

(b) $s_{10} + \int_{11}^{\infty} \frac{1}{x^4}\,dx \le s \le s_{10} + \int_{10}^{\infty} \frac{1}{x^4}\,dx \quad\Rightarrow\quad s_{10} + \frac{1}{3(11)^3} \le s \le s_{10} + \frac{1}{3(10)^3} \quad\Rightarrow$

$1.082037 + 0.000250 = 1.082287 \le s \le 1.082037 + 0.000333 = 1.082370$, so we get $s \approx 1.08233$ with error ≤ 0.00005.

(c) $R_n \le \int_{n}^{\infty} \frac{1}{x^4}\,dx = \frac{1}{3n^3}$. So $R_n < 0.00001$ for $n > \sqrt[3]{(10)^5/3} \approx 32.2$, that is, for $n > 32$.

25. **(a)** $f(x) = \frac{1}{x^2}$ is positive and continuous and $f'(x) = \frac{-2}{x^3}$ is negative for $x > 1$, and so the Integral Test applies. $\sum_{n=1}^{\infty} \frac{1}{n^2} \approx s_{10} = \frac{1}{1^2} + \frac{1}{2^2} + \frac{1}{3^2} + \cdots + \frac{1}{10^2} \approx 1.54977$.

$R_{10} \le \int_{10}^{\infty} \frac{1}{x^2}\,dx = \lim_{t\to\infty}\left[\frac{-1}{x}\right]_{10}^{t} = \lim_{t\to\infty}\left(-\frac{1}{t} + \frac{1}{10}\right) = \frac{1}{10}$, so the error is at most 0.1.

(b) $s_{10} + \int_{11}^{\infty} \frac{1}{x^2}\,dx \le s \le s_{10} + \int_{10}^{\infty} \frac{1}{x^2}\,dx \quad\Rightarrow\quad s_{10} + \frac{1}{11} \le s \le s_{10} + \frac{1}{10} \quad\Rightarrow$

$1.549768 + 0.090909 = 1.640677 \le s \le 1.549768 + 0.1 = 1.649768$, so we get $s \approx 1.64522$ with error ≤ 0.005.

(c) $R_n \le \int_{n}^{\infty} \frac{1}{x^2}\,dx = \frac{1}{n}$. So $R_n < 0.001$ for $n > 1000$.

26. $f(x) = 1/x^5$ is positive and continuous and $f'(x) = -5/x^6$ is negative for $x > 1$, and so the Integral Test applies. Using (4), $R_n \le \int_{n}^{\infty} x^{-5}\,dx = \lim_{t\to\infty}\left[\frac{-1}{4x^4}\right]_{n}^{t} = \frac{1}{4n^4}$. If we take $n = 5$, then $s_5 = 1.03\overline{6}$ and $R_5 \le 0.0004$. So $s \approx s_5 \approx 1.037$.

27. $f(x) = x^{-3/2}$ is positive and continuous and $f'(x) = -\frac{3}{2}x^{-5/2}$ is negative for $x > 1$, so the Integral Test applies. Using (5), we need n such that

$$0.01 > \frac{1}{2}\left(\int_{n}^{\infty} x^{-3/2}\,dx - \int_{n+1}^{\infty} x^{-3/2}\,dx\right) = \frac{1}{2}\left(\lim_{t\to\infty}\left[\frac{-2}{\sqrt{x}}\right]_{n}^{t} - \lim_{t\to\infty}\left[\frac{-2}{\sqrt{x}}\right]_{n+1}^{t}\right) = \frac{1}{\sqrt{n}} - \frac{1}{\sqrt{n+1}} \quad\Leftrightarrow$$

$n > 13$. Then, again from (5),

$s \approx s_{14} + \frac{1}{2}\left(\int_{14}^{\infty} x^{-3/2}\,dx + \int_{15}^{\infty} x^{-3/2}\,dx\right) = 2.0872 + \frac{1}{\sqrt{14}} + \frac{1}{\sqrt{15}} \approx 2.6127$. Any larger value of n will also work. For instance, $s \approx s_{30} + \frac{1}{\sqrt{30}} + \frac{1}{\sqrt{31}} \approx 2.6124$.

28. $f(x) = \frac{1}{x \ln x}$ is positive and continuous and $f'(x) = \frac{-\ln x - 1}{x^2 \ln x}$ is negative for $x > 2$, so the Integral Test applies. Using (4), we need $0.01 > \int_{n}^{\infty} \frac{dx}{x \ln x} = \lim_{t\to\infty}\left[\frac{-1}{(\ln x)}\right]_{n}^{t} = \frac{1}{\ln n}$. This is true for $n > e^{100}$, so we would have to take this many terms, which would be problematic because $e^{100} \approx 2.7 \times 10^{45}$.

29. **(a)** From (1), with $f(x) = \frac{1}{x}$, $\frac{1}{2} + \frac{1}{3} + \frac{1}{4} + \cdots + \frac{1}{n} \le \int_1^n \frac{1}{x}\,dx = \ln n$, so

$s_n = 1 + \frac{1}{2} + \frac{1}{3} + \frac{1}{4} + \cdots + \frac{1}{n} \le 1 + \ln n$.

(b) By part (a), $s_{10^6} \le 1 + \ln 10^6 \approx 14.82 < 15$ and $s_{10^9} \le 1 + \ln 10^9 \approx 21.72 < 22$.

30. **(a)** $f(x) = \left(\frac{\ln x}{x}\right)^2$ is continuous and positive for $x > 1$, and since $f'(x) = \frac{2\ln x(1 - \ln x)}{x^3} < 0$ for $x > e$, we can apply the Integral Test. Using a CAS, we get $\int_1^\infty \left(\frac{\ln x}{x}\right)^2 dx = 2$.

(The Maple command is `int(f,x=1..infinity);` and the Mathematica command is `Integrate[f,{x,1,Infinity}]`). So the series also converges.

(b) Since the Integral Test applies, the error in $s \approx s_n$ is $R_n \le \int_n^\infty \left(\frac{\ln x}{x}\right)^2 dx = \frac{(\ln n)^2 + 2\ln n + 2}{n}$.

(c) By using a CAS to graph the functions $y = \frac{(\ln x)^2 + 2\ln x + 2}{x}$ and $y = 0.05$ on the same screen, we see that $\frac{(\ln n)^2 + 2\ln n + 2}{n} < 0.05$ for $n \ge 1373$.

$\left[\text{We can check our work by asking the CAS to solve the equation } \frac{(\ln x)^2 + 2\ln x + 2}{x} = 0.05.\right]$

(d) Using the CAS to sum the first 1373 terms, we get $s_{1373} \approx 1.94$.

31. $b^{\ln n} = e^{\ln b \ln n} = n^{\ln b} = \frac{1}{n^{-\ln b}}$. This is a p-series, which converges for all b such that $-\ln b > 1 \Leftrightarrow \ln b < -1$, so for $b < 1/e$.

32. **(a)** The sum of the areas of the n rectangles in the graph to the right is $1 + \frac{1}{2} + \frac{1}{3} + \cdots + \frac{1}{n}$. Now $\int_1^{n+1} \frac{dx}{x}$ is less than this sum because the rectangles extend above the curve $y = 1/x$, so

$\int_1^{n+1} \frac{1}{x}\,dx = \ln(n+1) < 1 + \frac{1}{2} + \frac{1}{3} + \cdots + \frac{1}{n}$, and since

$\ln n < \ln(n+1)$, $0 < 1 + \frac{1}{2} + \frac{1}{3} + \cdots + \frac{1}{n} - \ln n = t_n$.

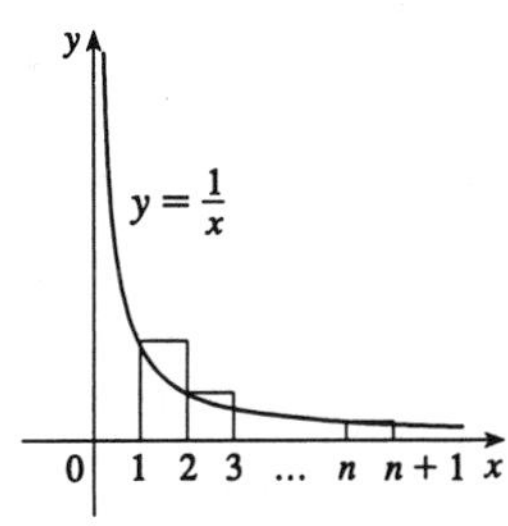

(b) The area under $f(x) = 1/x$ between $x = n$ and $x = n + 1$ is

$\int_n^{n+1} \frac{dx}{x} = \ln(n+1) - \ln n$, and this is clearly greater than the area of the inscribed rectangle in the figure to the right

$\left[\text{which is } \frac{1}{n+1}\right]$, so $t_n - t_{n+1} = [\ln(n+1) - \ln n] - \frac{1}{n+1} > 0$,

and so $t_n > t_{n+1}$, so $\{t_n\}$ is a decreasing sequence.

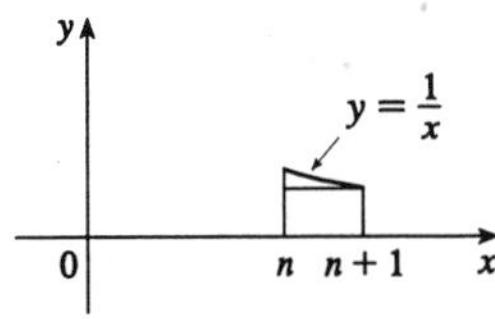

(c) We have shown that $\{t_n\}$ is decreasing and that $t_n > 0$ for all n. Thus $0 < t_n \le t_1 = 1$, so $\{t_n\}$ is a bounded monotonic sequence, and hence converges by Theorem 10.1.10.

EXERCISES 10.4

1. $\frac{1}{n^3+n^2} < \frac{1}{n^3}$ since $n^3+n^2 > n^3$ for all n, and since $\sum_{n=1}^{\infty} \frac{1}{n^3}$ is a convergent p-series ($p = 3 > 1$), $\sum_{n=1}^{\infty} \frac{1}{n^3+n^2}$ converges also by the Comparison Test [part (a).]

2. $\frac{3}{4^n+5} < \frac{3}{4^n}$ and $\sum_{n=1}^{\infty} \frac{3}{4^n}$ converges (geometric with $|r| = \frac{1}{4} < 1$) so by the Comparison Test, $\sum_{n=1}^{\infty} \frac{3}{4^n+5}$ converges also.

3. $\frac{3}{n2^n} \le \frac{3}{2^n}$. $\sum_{n=1}^{\infty} \frac{3}{2^n}$ is a geometric series with $|r| = \frac{1}{2} < 1$, and hence converges, so $\sum_{n=1}^{\infty} \frac{3}{n2^n}$ converges also, by the Comparison Test.

4. $\frac{1}{\sqrt{n}-1} > \frac{1}{\sqrt{n}}$ and $\sum_{n=2}^{\infty} \frac{1}{\sqrt{n}}$ diverges (p-series with $p = \frac{1}{2} < 1$) so $\sum_{n=2}^{\infty} \frac{1}{\sqrt{n}-1}$ diverges by the Comparison Test.

5. $\frac{1+5^n}{4^n} > \frac{5^n}{4^n} = \left(\frac{5}{4}\right)^n$. $\sum_{n=0}^{\infty} \left(\frac{5}{4}\right)^n$ is a divergent geometric series ($|r| = \frac{5}{4} > 1$) so $\sum_{n=0}^{\infty} \frac{1+5^n}{4^n}$ diverges by the Comparison Test.

6. $\frac{\sin^2 n}{n\sqrt{n}} \le \frac{1}{n\sqrt{n}} = \frac{1}{n^{3/2}}$ and $\sum_{n=1}^{\infty} \frac{1}{n^{3/2}}$ converges ($p = \frac{3}{2} > 1$) so $\sum_{n=1}^{\infty} \frac{\sin^2 n}{n\sqrt{n}}$ converges by the Comparison Test.

7. $\frac{3}{n(n+3)} < \frac{3}{n^2}$. $\sum_{n=1}^{\infty} \frac{3}{n^2} = 3\sum_{n=1}^{\infty} \frac{1}{n^2}$ is a convergent p-series ($p = 2 > 1$) so $\sum_{n=1}^{\infty} \frac{3}{n(n+3)}$ converges by the Comparison Test.

8. $\frac{1}{\sqrt{n(n+1)(n+2)}} < \frac{1}{\sqrt{n \cdot n \cdot n}} = \frac{1}{n^{3/2}}$ and since $\sum_{n=1}^{\infty} \frac{1}{n^{3/2}}$ converges ($p = \frac{3}{2} > 1$), so does $\sum_{n=1}^{\infty} \frac{1}{\sqrt{n(n+1)(n+2)}}$ by the Comparison Test.

9. $\frac{\sqrt{n}}{n-1} > \frac{\sqrt{n}}{n} = \frac{1}{n^{1/2}}$. $\sum_{n=2}^{\infty} \frac{1}{n^{1/2}}$ is a divergent p-series ($p = \frac{1}{2} < 1$) so $\sum_{n=2}^{\infty} \frac{\sqrt{n}}{n-1}$ diverges by the Comparison Test.

10. Use the Limit Comparison Test with $a_n = \frac{1}{\sqrt[3]{n(n+1)(n+2)}}$ and $b_n = \frac{1}{n}$.

$\lim_{n\to\infty} \frac{a_n}{b_n} = \lim_{n\to\infty} \frac{n}{\sqrt[3]{n(n+1)(n+2)}} = \lim_{n\to\infty} \frac{1}{\sqrt[3]{1(1+1/n)(1+2/n)}} = 1 > 0$, so since $\sum_{n=1}^{\infty} \frac{1}{n}$ diverges, so does $\sum_{n=1}^{\infty} \frac{1}{\sqrt[3]{n(n+1)(n+2)}}$.

11. $n^3+1>n^3 \quad\Rightarrow\quad \dfrac{1}{n^3+1}<\dfrac{1}{n^3} \quad\Rightarrow\quad \dfrac{n}{n^3+1}<\dfrac{n}{n^3} \quad\Rightarrow\quad \dfrac{n-1}{n^3+1}<\dfrac{n}{n^3}=\dfrac{1}{n^2}$. Now $\sum_{n=1}^{\infty}\dfrac{1}{n^2}$ is a convergent p-series ($p=2>1$) so $\sum_{n=1}^{\infty}\dfrac{n-1}{n^3+1}$ converges by the Comparison Test.

12. $\dfrac{n}{(n+1)2^n}<\dfrac{1}{2^n}$ and $\sum_{n=1}^{\infty}\dfrac{1}{2^n}$ is a convergent geometric series ($|r|=\frac{1}{2}<1$), so $\sum_{n=1}^{\infty}\dfrac{n}{(n+1)2^n}$ converges by the Comparison Test.

13. $\dfrac{3+\cos n}{3^n}\leq\dfrac{4}{3^n}$ since $\cos n\leq 1$. $\sum_{n=1}^{\infty}\dfrac{4}{3^n}$ is a geometric series with $|r|=\frac{1}{3}<1$ so it converges, and so $\sum_{n=1}^{\infty}\dfrac{3+\cos n}{3^n}$ converges by the Comparison Test.

14. $\dfrac{5n}{2n^2-5}>\dfrac{5n}{2n^2}=\dfrac{5}{2}\left(\dfrac{1}{n}\right)$ and since $\dfrac{5}{2}\sum_{n=1}^{\infty}\dfrac{1}{n}$ diverges (harmonic series) so does $\sum_{n=1}^{\infty}\dfrac{5n}{2n^2-5}$ by the Comparison Test.

15. $\dfrac{n}{\sqrt{n^5+4}}<\dfrac{n}{\sqrt{n^5}}=\dfrac{1}{n^{3/2}}$. $\sum_{n=1}^{\infty}\dfrac{1}{n^{3/2}}$ is a convergent p-series ($p=\frac{3}{2}>1$) so $\sum_{n=1}^{\infty}\dfrac{n}{\sqrt{n^5+4}}$ converges by the Comparison Test.

16. $\dfrac{\arctan n}{n^4}<\dfrac{\pi/2}{n^4}$ and $\dfrac{\pi}{2}\sum_{n=1}^{\infty}\dfrac{1}{n^4}$ converges ($p=4>1$) so $\sum_{n=1}^{\infty}\dfrac{\arctan n}{n^4}$ converges by the Comparison Test.

17. $\dfrac{2^n}{1+3^n}<\dfrac{2^n}{3^n}=\left(\dfrac{2}{3}\right)^n$. $\sum_{n=1}^{\infty}\left(\dfrac{2}{3}\right)^n$ is a convergent geometric series ($|r|=\frac{2}{3}<1$), so $\sum_{n=1}^{\infty}\dfrac{2^n}{1+3^n}$ converges by the Comparison Test.

18. Use the Limit Comparison Test with $a_n=\dfrac{1+2^n}{1+3^n}$ and $b_n=\dfrac{2^n}{3^n}$: $\lim_{n\to\infty}\dfrac{a_n}{b_n}=\lim_{n\to\infty}\dfrac{(1/2)^n+1}{(1/3)^n+1}=1>0$, so since $\sum_{n=1}^{\infty}b_n$ converges (geometric series with $|r|=\frac{2}{3}<1$), $\sum_{n=1}^{\infty}\dfrac{1+2^n}{1+3^n}$ converges also.

19. Let $a_n=\dfrac{1}{1+\sqrt{n}}$ and $b_n=\dfrac{1}{\sqrt{n}}$. Then $\lim_{n\to\infty}\dfrac{a_n}{b_n}=\lim_{n\to\infty}\dfrac{\sqrt{n}}{1+\sqrt{n}}=1>0$. Since $\sum_{n=1}^{\infty}\dfrac{1}{\sqrt{n}}$ is a divergent p-series ($p=\frac{1}{2}<1$), $\sum_{n=1}^{\infty}\dfrac{1}{1+\sqrt{n}}$ also diverges by the Limit Comparison Test.

20. Use the Limit Comparison Test with $a_n=\dfrac{1}{n^2-4}$ and $b_n=\dfrac{1}{n^2}$: $\lim_{n\to\infty}\dfrac{a_n}{b_n}=\lim_{n\to\infty}\dfrac{n^2}{n^2-4}=1>0$, and $\sum_{n=1}^{\infty}b_n$ converges ($p=2>1$) so $\sum_{n=1}^{\infty}\dfrac{1}{n^2-4}$ converges.

21. Let $a_n=\dfrac{n^2+1}{n^4+1}$ and $b_n=\dfrac{1}{n^2}$. Then $\lim_{n\to\infty}\dfrac{a_n}{b_n}=\lim_{n\to\infty}\dfrac{n^4+n^2}{n^4+1}=1$. Since $\sum_{n=1}^{\infty}\dfrac{1}{n^2}$ is a convergent p-series ($p=2>1$), so is $\sum_{n=1}^{\infty}\dfrac{n^2+1}{n^4+1}$ by the Limit Comparison Test.

22. Use the Limit Comparison Test with $a_n = \dfrac{3n^3 - 2n^2}{n^4 + n^2 + 1}$ and $b_n = \dfrac{1}{n}$. $\lim\limits_{n\to\infty} \dfrac{a_n}{b_n} = \lim\limits_{n\to\infty} \dfrac{3n^4 - 2n^3}{n^4 + n^2 + 1} = 3 > 0$, so since $\sum\limits_{n=1}^{\infty} b_n$ diverges, so does $\sum\limits_{n=1}^{\infty} \dfrac{3n^3 - 2n^2}{n^4 + n^2 + 1}$.

23. Let $a_n = \dfrac{n^2 - n + 2}{\sqrt[4]{n^{10} + n^5 + 3}}$ and $b_n = \dfrac{1}{\sqrt{n}}$. Then

$\lim\limits_{n\to\infty} \dfrac{a_n}{b_n} = \lim\limits_{n\to\infty} \dfrac{n^{5/2} - n^{3/2} + 2n^{1/2}}{\sqrt[4]{n^{10} + n^5 + 3}} = \lim\limits_{n\to\infty} \dfrac{1 - n^{-1} + 2n^{-2}}{\sqrt[4]{1 + n^{-5} + 3n^{-10}}} = 1$. Since $\sum\limits_{n=1}^{\infty} \dfrac{1}{\sqrt{n}}$ is a divergent p-series $(p = \frac{1}{2} < 1)$, $\sum\limits_{n=1}^{\infty} \dfrac{n^2 - n + 2}{\sqrt[4]{n^{10} + n^5 + 3}}$ diverges by the Limit Comparison Test.

24. Use the Limit Comparison Test with $a_n = \dfrac{n^2 - 3n}{\sqrt[3]{n^{10} - 4n^2}}$ and $b_n = \dfrac{1}{n^{4/3}}$.

$\lim\limits_{n\to\infty} \dfrac{a_n}{b_n} = \lim\limits_{n\to\infty} \dfrac{n^{10/3} - 3n^{7/3}}{\sqrt[3]{n^{10} - 4n^2}} = \lim\limits_{n\to\infty} \dfrac{1 - 3/n}{\sqrt[3]{1 - 4n^{-8}}} = 1 > 0$, so since $\sum\limits_{n=1}^{\infty} b_n$ converges $(p = \frac{4}{3} > 1)$, so does $\sum\limits_{n=1}^{\infty} \dfrac{n^2 - 3n}{\sqrt[3]{n^{10} - 4n^2}}$.

25. Let $a_n = \dfrac{n+1}{n2^n}$ and $b_n = \dfrac{1}{2^n}$. Then $\lim\limits_{n\to\infty} \dfrac{a_n}{b_n} = \lim\limits_{n\to\infty} \dfrac{n+1}{n} = 1$. Since $\sum\limits_{n=1}^{\infty} \dfrac{1}{2^n}$ is a convergent geometric series $(|r| = \frac{1}{2} < 1)$, $\sum\limits_{n=1}^{\infty} \dfrac{n+1}{n2^n}$ converges by the Limit Comparison Test.

26. Use the Limit Comparison Test with $a_n = \dfrac{2n^2 + 7n}{3^n(n^2 + 5n - 1)}$ and $b_n = \dfrac{1}{3^n}$.

$\lim\limits_{n\to\infty} \dfrac{a_n}{b_n} = \lim\limits_{n\to\infty} \dfrac{2n^2 + 7n}{n^2 + 5n - 1} = 2 > 0$, and since $\sum\limits_{n=1}^{\infty} b_n$ is a convergent geometric series $(|r| = \frac{1}{3} < 1)$, $\sum\limits_{n=1}^{\infty} \dfrac{2n^2 + 7n}{3^n(n^2 + 5n - 1)}$ converges also.

27. Let $a_n = \dfrac{\ln n}{n^3}$ and $b_n = \dfrac{1}{n^2}$. Then $\lim\limits_{n\to\infty} \dfrac{a_n}{b_n} = \lim\limits_{n\to\infty} \dfrac{\ln n}{n} = \lim\limits_{n\to\infty} \dfrac{1/n}{1} = 0$. So since $\sum\limits_{n=1}^{\infty} \dfrac{1}{n^2}$ converges (p-series, $p = 2 > 1$), so does $\sum\limits_{n=1}^{\infty} \dfrac{\ln n}{n^3}$ by part (b) of the Limit Comparison Test.

28. Use the Limit Comparison Test with $a_n = \dfrac{1}{\ln n}$ and $b_n = \dfrac{1}{n}$. $\lim\limits_{n\to\infty} \dfrac{a_n}{b_n} = \lim\limits_{n\to\infty} \dfrac{n}{\ln n} = \lim\limits_{n\to\infty} \dfrac{1}{1/n} = \infty$, so since $\sum\limits_{n=2}^{\infty} \dfrac{1}{n}$ diverges (harmonic series), $\sum\limits_{n=2}^{\infty} \dfrac{1}{\ln n}$ diverges.

29. Clearly $n! = n(n-1)(n-2)\cdots(3)(2) \geq 2 \cdot 2 \cdot 2 \cdot \cdots \cdot 2 \cdot 2 = 2^{n-1}$, so $\dfrac{1}{n!} \leq \dfrac{1}{2^{n-1}}$. $\sum\limits_{n=1}^{\infty} \dfrac{1}{2^{n-1}}$ is a convergent geometric series $(|r| = \frac{1}{2} < 1)$ so $\sum\limits_{n=1}^{\infty} \dfrac{1}{n!}$ converges by the Comparison Test.

30. $\dfrac{n!}{n^n} = \dfrac{1 \cdot 2 \cdot 3 \cdot \cdots \cdot (n-1)n}{n \cdot n \cdot n \cdot \cdots \cdot n \cdot n} \le \dfrac{1}{n} \cdot \dfrac{2}{n} \cdot 1 \cdot 1 \cdot \cdots \cdot 1$ for $n \ge 2$, so since $\sum_{n=1}^{\infty} \dfrac{2}{n^2}$ converges ($p = 2 > 1$), $\sum_{n=1}^{\infty} \dfrac{n!}{n^n}$ converges also by the Comparison Test.

31. Let $a_n = \sin\left(\dfrac{1}{n}\right)$ and $b_n = \dfrac{1}{n}$. Then $\lim\limits_{n\to\infty} \dfrac{a_n}{b_n} = \lim\limits_{n\to\infty} \dfrac{\sin(1/n)}{1/n} = \lim\limits_{\theta\to 0} \dfrac{\sin\theta}{\theta} = 1$, so since $\sum_{n=1}^{\infty} b_n$ is the harmonic series (which diverges), $\sum_{n=1}^{\infty} \sin\left(\dfrac{1}{n}\right)$ diverges as well by the Limit Comparison Test.

32. Use the Limit Comparison Test with $a_n = \dfrac{1}{n^{1+1/n}}$ and $b_n = \dfrac{1}{n}$. $\lim\limits_{n\to\infty} \dfrac{a_n}{b_n} = \lim\limits_{n\to\infty} \dfrac{n}{n^{1+1/n}} = \lim\limits_{n\to\infty} \dfrac{1}{n^{1/n}} = 1$ $\left(\text{since } \lim\limits_{x\to\infty} x^{1/x} = 1 \text{ by l'Hospital's Rule,}\right)$ so $\sum_{n=1}^{\infty} \dfrac{1}{n}$ diverges (harmonic series) $\Rightarrow$ $\sum_{n=1}^{\infty} \dfrac{1}{n^{1+1/n}}$ diverges.

33. $\sum_{n=1}^{10} \dfrac{1}{n^4 + n^2} = \dfrac{1}{2} + \dfrac{1}{20} + \dfrac{1}{90} + \cdots + \dfrac{1}{10{,}100} \approx 0.567975$. Now $\dfrac{1}{n^4 + n^2} < \dfrac{1}{n^4}$, so using the reasoning and notation of Example 7, the error is $R_{10} \le T_{10} = \sum_{n=11}^{\infty} \dfrac{1}{n^4} \le \int_{10}^{\infty} \dfrac{dx}{x^4} = \lim\limits_{t\to\infty} \left[-\dfrac{x^{-3}}{3}\right]_{10}^{t} = \dfrac{1}{3000} = 0.000\overline{3}$.

34. $\sum_{n=1}^{10} \dfrac{1 + \cos n}{n^5} = 1 + \cos 1 + \dfrac{1 + \cos 2}{32} + \dfrac{1 + \cos 3}{243} + \cdots + \dfrac{1 + \cos 10}{100{,}000} \approx 1.55972$. Now $\dfrac{1 + \cos n}{n^5} \le \dfrac{2}{n^5}$, so as in Example 7, $R_{10} \le T_{10} \le \int_{10}^{\infty} \frac{2}{x^5}\,dx = 2 \lim\limits_{t\to\infty} \left[-\frac{1}{4}x^{-4}\right]_{10}^{t} = 0.00005$.

35. $\sum_{n=1}^{10} \dfrac{1}{1 + 2^n} = \dfrac{1}{3} + \dfrac{1}{5} + \dfrac{1}{9} + \cdots + \dfrac{1}{1025} \approx 0.76352$. Now $\dfrac{1}{1 + 2^n} < \dfrac{1}{2^n}$, so the error is

$R_{10} \le T_{10} = \sum_{n=11}^{\infty} \dfrac{1}{2^n} = \dfrac{1/2^{11}}{1 - 1/2}$ (geometric series) ≈ 0.00098.

36. $\sum_{n=1}^{10} \dfrac{n}{(n+1)3^n} = \dfrac{1}{6} + \dfrac{2}{27} + \dfrac{3}{108} + \cdots + \dfrac{10}{649{,}539} \approx 0.283597$. Now $\dfrac{n}{(n+1)3^n} < \dfrac{n}{n \cdot 3^n} = \dfrac{1}{3^n}$, so the error is

$R_{10} \le T_{10} = \sum_{n=11}^{\infty} \dfrac{1}{3^n} = \dfrac{1/3^{11}}{1 - 1/3} \approx 0.0000085$.

37. Since $\dfrac{d_n}{10^n} \le \dfrac{9}{10^n}$ for each n, and since $\sum_{n=1}^{\infty} \dfrac{9}{10^n}$ is a convergent geometric series ($|r| = \frac{1}{10} < 1$),

$0.d_1 d_2 d_3 \ldots = \sum_{n=1}^{\infty} \dfrac{d_n}{10^n}$ will always converge by the Comparison Test.

38. Clearly if $p < 0$ the series diverges since $\lim\limits_{n\to\infty} \dfrac{1}{n^p \ln n} = \infty$. If $0 \le p \le 1$, then $n^p \ln n \le n \ln n$ $\Rightarrow$ $\dfrac{1}{n^p \ln n} \ge \dfrac{1}{n \ln n}$ and $\sum_{n=2}^{\infty} \dfrac{1}{n \ln n}$ diverges (Exercise 10.3.13), so $\sum_{n=2}^{\infty} \dfrac{1}{n^p \ln n}$ diverges. If $p > 1$, use the Limit Comparison Test with $a_n = \dfrac{1}{n^p \ln n}$ and $b_n = \dfrac{1}{n^p}$. $\sum_{n=2}^{\infty} b_n$ converges, and $\lim\limits_{n\to\infty} \dfrac{a_n}{b_n} = \lim\limits_{n\to\infty} \dfrac{1}{\ln n} = 0$, so $\sum_{n=2}^{\infty} \dfrac{1}{n^p \ln n}$ also converges. (Or use the Comparison Test since $n^p \ln n > n^p$ for $n > e$.) In summary, the series converges if and only if $p > 1$.

39. Since $\sum a_n$ converges, $\lim\limits_{n\to\infty} a_n = 0$, so there exists N such that $|a_n - 0| < 1$ for all $n > N \quad \Rightarrow \quad 0 \le a_n < 1$ for all $n > N \quad \Rightarrow \quad 0 \le a_n^2 \le a_n$. Since $\sum a_n$ converges, so does $\sum a_n^2$ by the Comparison Test.

40. Since $\lim\limits_{n\to\infty} \dfrac{a_n}{b_n} = 0$, there is a number $N > 0$ such that $\left|\dfrac{a_n}{b_n} - 0\right| < 1$ for all $n > N$, and so $a_n < b_n$ since a_n and b_n are positive. Thus, since $\sum b_n$ converges, so does $\sum a_n$ by the Comparison Test.

41. We wish to prove that if $\lim\limits_{n\to\infty} \dfrac{a_n}{b_n} = \infty$ and $\sum b_n$ diverges, then so does $\sum a_n$. So suppose on the contrary that $\sum a_n$ converges. Since $\lim\limits_{n\to\infty} \dfrac{a_n}{b_n} = \infty$, we have that $\lim\limits_{n\to\infty} \dfrac{b_n}{a_n} = 0$, so by part (b) of the Limit Comparison Test (proved in Exercise 40), if $\sum a_n$ converges, so must $\sum b_n$. But this contradicts our hypothesis, so $\sum a_n$ must diverge.

42. Let $a_n = \dfrac{1}{n^2}$ and $b_n = \dfrac{1}{n}$. Then $\lim\limits_{n\to\infty} \dfrac{a_n}{b_n} = \lim\limits_{n\to\infty} \dfrac{1}{n} = 0$, but $\sum b_n$ diverges while $\sum a_n$ converges.

43. $\lim\limits_{n\to\infty} na_n = \lim\limits_{n\to\infty} \dfrac{a_n}{1/n}$, so we apply the Limit Comparison Test with $b_n = \dfrac{1}{n}$. Since $\lim\limits_{n\to\infty} na_n > 0$ we know that either both series converge or both series diverge, and we also know that $\sum\limits_{n=0}^{\infty} \dfrac{1}{n}$ diverges (p-series with $p = 1$). Therefore $\sum a_n$ must be divergent.

44. First we observe that, by l'Hospital's Rule, $\lim\limits_{x\to 0} \dfrac{\ln(1+x)}{x} = \lim\limits_{x\to 0} \dfrac{1}{1+x} = 1$. Also, if $\sum a_n$ converges, then $\lim\limits_{n\to\infty} a_n = 0$ by Theorem 10.2.6. Therefore $\lim\limits_{n\to\infty} \dfrac{\ln(1+a_n)}{a_n} = 1$.

We are given that $\sum a_n$ is convergent and $a_n > 0$. Thus $\sum \ln(1 + a_n)$ is convergent by part (a) of the Limit Comparison Test.

45. Yes. Since $\sum a_n$ converges, its terms approach 0 as $n \to \infty$, so $\lim\limits_{n\to\infty} \dfrac{\sin a_n}{a_n} = 1$ by Theorem 2.4.4. Thus $\sum \sin a_n$ converges by the Limit Comparison Test.

46. Yes. Since $\sum a_n$ converges, its terms approach 0 as $n \to \infty$, so for some integer N, $a_n \le 1$ for all $n \ge N$. But then $\sum\limits_{n=1}^{\infty} a_n b_n = \sum\limits_{n=1}^{N-1} a_n b_n + \sum\limits_{n=N}^{\infty} a_n b_n \le \sum\limits_{n=1}^{N-1} a_n b_n + \sum\limits_{n=N}^{\infty} b_n$. The first term is a finite sum, and the second term converges since $\sum\limits_{n=1}^{\infty} b_n$ converges. So $\sum a_n b_n$ converges by the Comparison Test.

EXERCISES 10.5

1. $\sum_{n=1}^{\infty}(-1)^{n-1}\frac{3}{n+4}$. $b_n = \frac{3}{n+4} > 0$ and $b_{n+1} < b_n$ for all n; $\lim_{n\to\infty} b_n = 0$ so the series converges by the Alternating Series Test.

2. $-5+\sum_{n=0}^{\infty}(-1)^{n-1}\frac{5}{3n+2}$. $b_n = \frac{5}{3n+2}$ is decreasing and positive for all n, and $\lim_{n\to\infty}\frac{5}{3n+2} = 0$ so the series converges by the Alternating Series Test.

3. $\sum_{n=1}^{\infty}(-1)^{n}\frac{n}{n+1}$. $\lim_{n\to\infty}\frac{n}{n+1} = 1$ so $\lim_{n\to\infty}(-1)^n\frac{n}{n+1}$ does not exist and the series diverges by the Test for Divergence.

4. $\sum_{n=2}^{\infty}\frac{(-1)^n}{\ln n}$. $b_n = \frac{1}{\ln n}$ is positive and decreasing, and $\lim_{n\to\infty}\frac{1}{\ln n} = 0$, so $\sum_{n=2}^{\infty}\frac{(-1)^n}{\ln n}$ converges by the Alternating Series Test.

5. $\sum_{n=1}^{\infty}(-1)^{n-1}\frac{1}{n^2}$. $b_n = \frac{1}{n^2} > 0$ and $b_{n+1} < b_n$ for all n, and $\lim_{n\to\infty}\frac{1}{n^2} = 0$, so the series converges by the Alternating Series Test.

6. $\sum_{n=1}^{\infty}\frac{(-1)^n}{\sqrt{n+3}}$. $b_n = \frac{1}{\sqrt{n+3}}$ is positive and decreasing, and $\lim_{n\to\infty}\frac{1}{\sqrt{n+3}} = 0$, so the series converges by the Alternating Series Test.

7. $\sum_{n=1}^{\infty}(-1)^{n+1}\frac{n}{5n+1}$. $\lim_{n\to\infty}\frac{n}{5n+1} = \frac{1}{5}$ so $\lim_{n\to\infty}(-1)^{n+1}\frac{n}{5n+1}$ does not exist and the series diverges by the Test for Divergence.

8. $\sum_{n=2}^{\infty}\frac{(-1)^{n-1}}{n\ln n}$. $b_n = \frac{1}{n\ln n}$ is positive and decreasing for $n \geq 2$, and $\lim_{n\to\infty}\frac{1}{n\ln n} = 0$ so the series converges by the Alternating Series Test.

9. $\sum_{n=1}^{\infty}(-1)^{n}\frac{n}{n^2+1}$. $b_n = \frac{n}{n^2+1} > 0$ for all n. $b_{n+1} < b_n \Leftrightarrow \frac{n+1}{(n+1)^2+1} < \frac{n}{n^2+1} \Leftrightarrow$ $(n+1)(n^2+1) < \left[(n+1)^2+1\right]n \Leftrightarrow n^3+n^2+n+1 < n^3+2n^2+2n \Leftrightarrow 0 < n^2+n-1$, which is true for all $n \geq 1$. Also $\lim_{n\to\infty}\frac{n}{n^2+1} = \lim_{n\to\infty}\frac{1/n}{1+1/n^2} = 0$. Therefore the series converges by the Alternating Series Test.

10. $\sum_{n=1}^{\infty}(-1)^{n}\frac{n^2}{n^2+1}$. $\lim_{n\to\infty}\frac{n^2}{n^2+1} = 1$, so $\lim_{n\to\infty}(-1)^n\frac{n^2}{n^2+1}$ does not exist. Thus the series diverges by the Test for Divergence.

11. $\sum_{n=1}^{\infty}(-1)^{n-1}\frac{\sqrt{n}}{n+4}$. $b_n = \frac{\sqrt{n}}{n+4} > 0$ for all n. Let $f(x) = \frac{\sqrt{x}}{x+4}$. Then $f'(x) = \frac{4-x}{2\sqrt{x}(x+4)^2} < 0$ if $x > 4$, so $\{b_n\}$ is decreasing after $n = 4$. $\lim_{n\to\infty}\frac{\sqrt{n}}{n+4} = \lim_{n\to\infty}\frac{1}{\sqrt{n}+4/\sqrt{n}} = 0$. So the series converges by the Alternating Series Test.

12. $\sum_{n=1}^{\infty}(-1)^{n+1}\frac{n}{2^n}$. $b_n = \frac{n}{2^n} > 0$ and $b_n \geq b_{n+1} \quad\Leftrightarrow\quad \frac{n}{2^n} \geq \frac{n+1}{2^{n+1}} \quad\Leftrightarrow\quad 2n \geq n+1 \quad\Leftrightarrow\quad n \geq 1$ which is certainly true. $\lim_{n\to\infty}(n/2^n) = 0$ by l'Hospital's Rule, so the series converges by the Alternating Series Test.

13. $\sum_{n=2}^{\infty}(-1)^n\frac{n}{\ln n}$. $\lim_{n\to\infty}\frac{n}{\ln n} = \lim_{n\to\infty}\frac{1}{1/n} = \infty$ so the series diverges by the Test for Divergence.

14. $\sum_{n=1}^{\infty}(-1)^{n-1}\left(\frac{\ln n}{n}\right) = 0 + \sum_{n=2}^{\infty}(-1)^{n-1}\left(\frac{\ln n}{n}\right)$. $b_n = \frac{\ln n}{n} > 0$ for $n \geq 2$, and if $f(x) = \frac{\ln x}{x}$ then $f'(x) = \frac{1-\ln x}{x^2} < 0$ if $x > e$, so $\{b_n\}$ is eventually decreasing. $\lim_{n\to\infty}\frac{\ln n}{n} = \lim_{n\to\infty}\frac{1/n}{1} = 0$ so the series converges by the Alternating Series Test.

15. $\sum_{n=1}^{\infty}\frac{\cos n\pi}{n^{3/4}} = \sum_{n=1}^{\infty}\frac{(-1)^n}{n^{3/4}}$. $b_n = \frac{1}{n^{3/4}}$ is decreasing and positive, and $\lim_{n\to\infty}\frac{1}{n^{3/4}} = 0$ so the series converges by the Alternating Series Test.

16. $\sin\left(\frac{n\pi}{2}\right) = 0$ if n is even and $(-1)^k$ if $n = 2k+1$, so the series is $\sum_{n=0}^{\infty}\frac{(-1)^n}{(2n+1)!}$. $b_n = \frac{1}{(2n+1)!} > 0$, $\{b_n\}$ is decreasing, and $\lim_{n\to\infty}\frac{1}{(2n+1)!} = 0$, so the series converges by the Alternating Series Test.

17. $\sum_{n=1}^{\infty}(-1)^n\sin\left(\frac{\pi}{n}\right)$. $b_n = \sin\left(\frac{\pi}{n}\right) > 0$ for $n \geq 2$ and $\sin\left(\frac{\pi}{n}\right) \geq \sin\left(\frac{\pi}{n+1}\right)$, and $\lim_{n\to\infty}\sin(\pi/n) = \sin 0 = 0$, so the series converges by the Alternating Series Test.

18. $\sum_{n=1}^{\infty}(-1)^n\cos\left(\frac{\pi}{n}\right)$. $\lim_{n\to\infty}\cos\left(\frac{\pi}{n}\right) = \cos(0) = 1$, so $\lim_{n\to\infty}(-1)^n\cos\left(\frac{\pi}{n}\right)$ does not exist and the series diverges by the Test for Divergence.

19. $\frac{n^n}{n!} = \frac{n\cdot n\cdot\cdots\cdot n}{1\cdot 2\cdot\cdots\cdot n} \geq n \quad\Rightarrow\quad \lim_{n\to\infty}\frac{n^n}{n!} = \infty \quad\Rightarrow\quad \lim_{n\to\infty}\frac{(-1)^n n^n}{n!}$ does not exist. So the series diverges by the Test for Divergence.

20. $\frac{1}{\sqrt[3]{\ln n}}$ decreases and $\lim_{n\to\infty}\frac{1}{\sqrt[3]{\ln n}} = 0$, so by the Alternating Series Test the series converges.

21. Let $\sum b_n$ be the series for which $b_n = 0$ if n is odd and $b_n = 1/n^2$ if n is even. Then $\sum b_n = \sum 1/(2n)^2$ clearly converges (by comparison with the p-series for $p = 2$). So suppose that $\sum(-1)^{n-1}b_n$ converges. Then by Theorem 10.2.8(b), so does $\sum\left[(-1)^{n-1}b_n + b_n\right] = 1 + \frac{1}{3} + \frac{1}{5} + \cdots = \sum\frac{1}{2n-1}$. But this diverges by comparison with the harmonic series, a contradiction. Therefore $\sum(-1)^{n-1}b_n$ must diverge. The Alternating Series Test does not apply since $\{b_n\}$ is not decreasing.

22. If $p > 0$, $\dfrac{1}{(n+1)^p} \le \dfrac{1}{n^p}$ and $\lim_{n\to\infty} \dfrac{1}{n^p} = 0$, so the series converges by the Alternating Series Test. If $p \le 0$, $\lim_{n\to\infty} \dfrac{(-1)^{n-1}}{n^p}$ does not exist, so the series diverges by the Test for Divergence.

Thus $\sum_{n=1}^{\infty} \dfrac{(-1)^{n-1}}{n^p}$ converges $\Leftrightarrow$ $p > 0$.

23. Clearly $b_n = \dfrac{1}{n+p}$ is decreasing and eventually positive and $\lim_{n\to\infty} b_n = 0$ for any p. So the series will converge (by the Alternating Series Test) for any p for which every b_n is defined, that is, $n + p \ne 0$ for $n \ge 1$, or p is not a negative integer.

24. Let $f(x) = \dfrac{(\ln x)^p}{x}$. Then $f'(x) = \dfrac{(\ln x)^{p-1}(p - \ln x)}{x^2} < 0$ if $x > e^p$ so f is eventually decreasing for every p. Clearly $\lim_{n\to\infty} \dfrac{(\ln n)^p}{n} = 0$ if $p \le 0$, and if $p > 0$ we can apply l'Hospital's Rule $[\![p+1]\!]$ times to get a limit of 0 as well. So the series converges for all p (by the Alternating Series Test).

25. If $b_n = \dfrac{1}{n^2}$, then $b_{11} = \dfrac{1}{121} < 0.01$, so by Theorem 1, $\sum_{n=1}^{\infty} \dfrac{1}{n^2} \approx \sum_{n=1}^{10} \dfrac{1}{n^2} \approx 0.82$.

26. $b_6 = 1/6^4 \approx 0.00077 < 0.001$, so by Theorem 1, $\sum_{n=1}^{\infty} \dfrac{(-1)^{n+1}}{n^4} \approx s_5 = 1 - \dfrac{1}{16} + \dfrac{1}{81} - \dfrac{1}{256} + \dfrac{1}{625} \approx 0.948$.

27. $\sum_{n=0}^{\infty} (-1)^n \dfrac{2^n}{n!}$. Since $\dfrac{2}{n} < \dfrac{2}{3}$ for $n \ge 4$, $0 < \dfrac{2^n}{n!} < \dfrac{2}{1} \cdot \dfrac{2}{2} \cdot \dfrac{2}{3} \cdot \left(\dfrac{2}{3}\right)^{n-3} \to 0$ as $n \to \infty$, so by the Squeeze Theorem, $\lim_{n\to\infty} \dfrac{2^n}{n!} = 0$, and hence $\sum_{n=0}^{\infty} (-1)^n \dfrac{2^n}{n!}$ is a convergent alternating series. $\dfrac{2^8}{8!} = \dfrac{256}{40{,}320} < 0.01$, so $\sum_{n=0}^{\infty} (-1)^n \frac{2^n}{n!} \approx \sum_{n=0}^{7} (-1)^n \frac{2^n}{n!} \approx 0.13$.

28. $b_6 = \dfrac{6}{4^6} \approx 0.0015 < 0.002 \quad \Rightarrow \quad \sum_{n=0}^{\infty} \dfrac{(-1)^n n}{4^n} \approx s_5 = 0 - \dfrac{1}{4} + \dfrac{1}{8} - \dfrac{3}{64} + \dfrac{1}{64} - \dfrac{5}{1024} \approx -0.161$.

29. $\sum_{n=1}^{\infty} \dfrac{(-1)^{n-1}}{(2n-1)!}$. $b_5 = \dfrac{1}{(2 \cdot 5 - 1)!} = \dfrac{1}{362{,}880} < 0.00001$, so $\sum_{n=1}^{\infty} \dfrac{(-1)^{n-1}}{(2n-1)!} \approx \sum_{n=1}^{4} \dfrac{(-1)^{n-1}}{(2n-1)!} \approx 0.8415$.

30. $b_4 = \dfrac{1}{(2 \cdot 4)!} = \dfrac{1}{40{,}320} \approx 0.000025$ and $s_3 = 1 - \dfrac{1}{2} + \dfrac{1}{24} - \dfrac{1}{720} \approx 0.54028$, so, correct to four decimal places, $\sum_{n=0}^{\infty} \dfrac{(-1)^n}{(2n)!} \approx 0.5403$.

31. $\sum_{n=0}^{\infty} \dfrac{(-1)^n}{2^n n!}$. $b_6 = \dfrac{1}{2^6 6!} = \dfrac{1}{46{,}080} < 0.000022$, so $\sum_{n=0}^{\infty} \dfrac{(-1)^n}{2^n n!} \approx \sum_{n=0}^{5} \dfrac{(-1)^n}{2^n n!} \approx 0.6065$.

32. $b_8 = \dfrac{1}{8^6} < 0.0000038$ and $s_7 = 1 - \dfrac{1}{64} + \dfrac{1}{729} - \dfrac{1}{4096} + \dfrac{1}{15{,}625} - \dfrac{1}{46{,}656} + \dfrac{1}{117{,}649} \approx 0.9855537$, so correct to five decimal places, $\sum_{n=1}^{\infty} \dfrac{(-1)^{n-1}}{n^6} \approx 0.98555$.

33. $\sum_{n=1}^{\infty}\frac{(-1)^{n-1}}{n}=1-\frac{1}{2}+\frac{1}{3}-\frac{1}{4}+\cdots+\frac{1}{49}-\frac{1}{50}+\frac{1}{51}-\frac{1}{52}+\cdots$. The 50th partial sum of this series is an underestimate, since $\sum_{n=1}^{\infty}\frac{(-1)^{n-1}}{n}=s_{50}+\left(\frac{1}{51}-\frac{1}{52}\right)+\left(\frac{1}{53}-\frac{1}{54}\right)+\cdots$, and the terms in parentheses are all positive. The result can be seen geometrically in Figure 1.

34.

n	a_n	s_n
1	1	1
2	−0.125	0.875
3	0.03704	0.91204
4	−0.01563	0.89641
5	0.008	0.90441
6	−0.00463	0.89978
7	0.00292	0.90270
8	−0.00195	0.90074
9	0.00137	0.90212
10	−0.001	0.90112

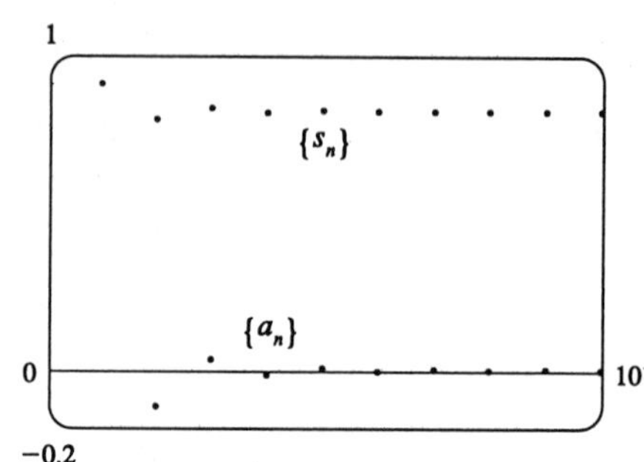

By Theorem 1, the error in the approximation $\sum_{n=1}^{\infty}\frac{(-1)^{n-1}}{n^3}\approx 0.90112$ is $|s-s_{10}|\le b_{11}=1/11^3\approx 0.0007513$.

35. **(a)** We will prove this by induction. Let $P(n)$ be the proposition that $s_{2n}=h_{2n}-h_n$. $P(1)$ is true by an easy calculation. So suppose that $P(n)$ is true. We will show that $P(n+1)$ must be true as a consequence.

$$h_{2n+2}-h_{n+1}=\left(h_{2n}+\frac{1}{2n+1}+\frac{1}{2n+2}\right)-\left(h_n+\frac{1}{n+1}\right)$$
$$=(h_{2n}-h_n)+\frac{1}{2n+1}-\frac{1}{2n+2}=s_{2n}+\frac{1}{2n+1}-\frac{1}{2n+2}=s_{2n+2},$$

which is $P(n+1)$, and proves that $s_{2n}=h_{2n}-h_n$ for all n.

(b) We know that $h_{2n}-\ln 2n\to\gamma$ and $h_n-\ln n\to\gamma$ as $n\to\infty$. So

$s_{2n}=h_{2n}-h_n=(h_{2n}-\ln 2n)-(h_n-\ln n)+(\ln 2n-\ln n)$, and

$\lim_{n\to\infty}s_{2n}=\gamma-\gamma+\lim_{n\to\infty}[\ln 2n-\ln n]=\lim_{n\to\infty}(\ln 2+\ln n-\ln n)=\ln 2$.

EXERCISES 10.6

1. $\displaystyle\sum_{n=1}^{\infty} \frac{1}{n\sqrt{n}} = \sum_{n=1}^{\infty} \frac{1}{n^{3/2}}$ is a convergent p-series ($p = \frac{3}{2} > 1$), so the given series is absolutely convergent.

2. $\displaystyle\sum_{n=1}^{\infty} \frac{(-1)^n}{n^{1/2}}$ converges by the Alternating Series Test, but $\displaystyle\sum_{n=1}^{\infty} \frac{1}{n^{1/2}}$ is a divergent p-series $\left(p = \frac{1}{2} < 1\right)$, so $\displaystyle\sum_{n=1}^{\infty} \frac{(-1)^n}{n^{1/2}}$ converges conditionally.

3. $\displaystyle\lim_{n\to\infty} \left|\frac{a_{n+1}}{a_n}\right| = \lim_{n\to\infty} \left|\frac{(-3)^{n+1}/(n+1)^3}{(-3)^n/n^3}\right| = 3 \lim_{n\to\infty} \left(\frac{n}{n+1}\right)^3 = 3 > 1$, so the series diverges by the Ratio Test.

4. $\displaystyle\lim_{n\to\infty} \left|\frac{a_{n+1}}{a_n}\right| = \lim_{n\to\infty} \left|\frac{(-3)^{n+1}/(n+1)!}{(-3)^n/n!}\right| = 3 \lim_{n\to\infty} \frac{1}{n+1} = 0 < 1$ so the series is absolutely convergent by the Ratio Test.

5. $\displaystyle\sum_{n=1}^{\infty} \frac{1}{2n+1}$ diverges (use the Integral Test or the Limit Comparison Test with $b_n = 1/n$), but since $\displaystyle\lim_{n\to\infty} \frac{1}{2n+1} = 0$, $\displaystyle\sum_{n=1}^{\infty} \frac{(-1)^{n+1}}{2n+1}$ converges by the Alternating Series Test, and so is conditionally convergent.

6. $\displaystyle\frac{1}{n^2+1} < \frac{1}{n^2}$ and $\displaystyle\sum_{n=1}^{\infty} \frac{1}{n^2}$ converges ($p = 2 > 1$), so $\displaystyle\sum_{n=1}^{\infty} \frac{1}{n^2+1}$ converges absolutely by the Comparison Test.

7. $\displaystyle\lim_{n\to\infty} \left|\frac{a_{n+1}}{a_n}\right| = \lim_{n\to\infty} \frac{1/(2n+1)!}{1/(2n-1)!} = \lim_{n\to\infty} \frac{1}{(2n+1)2n} = 0$, so by the Ratio Test the series is absolutely convergent.

8. $\displaystyle\lim_{n\to\infty} \left|\frac{a_{n+1}}{a_n}\right| = \lim_{n\to\infty} \left|\frac{(n+1)!/e^{n+1}}{n!/e^n}\right| = \frac{1}{e} \lim_{n\to\infty} (n+1) = \infty$, so the series diverges by the Ratio Test.

9. $\displaystyle\sum_{n=1}^{\infty} \frac{n}{n^2+4}$ diverges (use the Limit Comparison Test with $b_n = 1/n$). But since $\displaystyle 0 \le \frac{n+1}{(n+1)^2+4} < \frac{n}{n^2+4}$ $\Leftrightarrow$ $n^3 + n^2 + 4n + 4 < n^3 + 2n^2 + 5n$ $\Leftrightarrow$ $0 < n^2 + n - 4$ (which is true for $n \ge 2$), and since $\displaystyle\lim_{n\to\infty} \frac{n}{n^2+4} = 0$, $\displaystyle\sum_{n=1}^{\infty} (-1)^n \frac{n}{n^2+4}$ converges (conditionally) by the Alternating Series Test.

10. Let $\displaystyle a_n = \frac{\sqrt{n}}{n+1}$ and $\displaystyle b_n = \frac{1}{\sqrt{n}}$. Then $\displaystyle\lim_{n\to\infty} \frac{a_n}{b_n} = \lim_{n\to\infty} \frac{n}{n+1} = 1$, so since $\sum b_n$ diverges ($p = \frac{1}{2} < 1$), so does $\sum a_n$ by the Limit Comparison Test. But $\displaystyle\lim_{n\to\infty} \frac{\sqrt{n}}{n+1} = \lim_{n\to\infty} \frac{1}{\sqrt{n} + 1/\sqrt{n}} = 0$, so $\displaystyle\sum_{n=1}^{\infty} (-1)^{n-1} \frac{\sqrt{n}}{n+1}$ converges by the Alternating Series Test, and so is conditionally convergent.

11. $\displaystyle\lim_{n\to\infty} \frac{2n}{3n-4} = \frac{2}{3}$, so $\displaystyle\sum_{n=1}^{\infty} (-1)^n \frac{2n}{3n-4}$ diverges by the Test for Divergence.

12. $\lim_{n\to\infty}(-1)^n\dfrac{2^n}{n^2+1}$ does not exist, so $\sum_{n=1}^{\infty}(-1)^n\dfrac{2^n}{n^2+1}$ diverges by the Test for Divergence.

13. $\left|\dfrac{\sin 2n}{n^2}\right|\le\dfrac{1}{n^2}$ and $\sum_{n=1}^{\infty}\dfrac{1}{n^2}$ converges (p-series, $p=2>1$), so $\sum_{n=1}^{\infty}\dfrac{\sin 2n}{n^2}$ converges absolutely by the Comparison Test.

14. $\dfrac{\arctan n}{n^3}<\dfrac{\pi/2}{n^3}$ and $\sum_{n=1}^{\infty}\dfrac{\pi/2}{n^3}$ converges ($p=3>1$), so $\sum_{n=1}^{\infty}(-1)^n\dfrac{\arctan n}{n^3}$ converges absolutely by the Comparison Test.

15. $\lim_{n\to\infty}\left|\dfrac{a_{n+1}}{a_n}\right|=\lim_{n\to\infty}\left|\dfrac{2^{n+1}/[(n+1)3^{n+2}]}{2^n/(n3^{n+1})}\right|=\frac{2}{3}\lim_{n\to\infty}\dfrac{n}{n+1}=\frac{2}{3}<1$ so the series converges absolutely by the Ratio Test.

16. $\lim_{n\to\infty}\left|\dfrac{a_{n+1}}{a_n}\right|=\lim_{n\to\infty}\left|\dfrac{5^n/[(n+2)^2 4^{n+3}]}{5^{n-1}/[(n+1)^2 4^{n+2}]}\right|=\frac{5}{4}\lim_{n\to\infty}\left(\dfrac{n+1}{n+2}\right)^2=\frac{5}{4}>1$ so the series diverges by the Ratio Test.

17. $\lim_{n\to\infty}\left|\dfrac{a_{n+1}}{a_n}\right|=\lim_{n\to\infty}\dfrac{(n+2)5^{n+1}/[(n+1)3^{2(n+1)}]}{(n+1)5^n/(n3^{2n})}=\lim_{n\to\infty}\dfrac{5n(n+2)}{9(n+1)^2}=\frac{5}{9}<1$ so the series converges absolutely by the Ratio Test.

18. $\left|\cos\dfrac{n\pi}{6}\right|\le 1$, so since $\sum_{n=1}^{\infty}\dfrac{1}{n\sqrt{n}}$ converges ($p=\frac{3}{2}>1$), the given series converges absolutely by the Comparison Test.

19. $\lim_{n\to\infty}\left|\dfrac{a_{n+1}}{a_n}\right|=\lim_{n\to\infty}\dfrac{(n+1)!/10^{n+1}}{n!/10^n}=\lim_{n\to\infty}\dfrac{n+1}{10}=\infty$, so the series diverges by the Ratio Test.

20. $\lim_{n\to\infty}\left|\dfrac{a_{n+1}}{a_n}\right|=\lim_{n\to\infty}\dfrac{(n+1)!/(n+1)^{n+1}}{n!/n^n}=\lim_{n\to\infty}\dfrac{n^n}{(n+1)^n}=\lim_{n\to\infty}\dfrac{1}{(1+1/n)^n}=\dfrac{1}{e}<1$, so the series converges absolutely by the Ratio Test.

21. $\dfrac{|\cos(n\pi/3)|}{n!}\le\dfrac{1}{n!}$ and $\sum_{n=1}^{\infty}\dfrac{1}{n!}$ converges (Exercise 10.4.29), so the given series converges absolutely by the Comparison Test.

22. $\lim_{n\to\infty}\sqrt[n]{|a_n|}=\lim_{n\to\infty}\dfrac{1}{\ln n}=0<1$ so the series converges absolutely by the Root Test.

23. $\lim_{n\to\infty}\left|\dfrac{a_{n+1}}{a_n}\right|=\lim_{n\to\infty}\dfrac{(n+1)^{n+1}/5^{2n+5}}{n^n/5^{2n+3}}=\lim_{n\to\infty}\dfrac{1}{25}\left(\dfrac{n+1}{n}\right)^n(n+1)=\infty$, so the series diverges by the Ratio Test.

24. Since $\left\{\dfrac{1}{n\ln n}\right\}$ is decreasing and $\lim_{n\to\infty}\dfrac{1}{n\ln n}=0$, the series converges by the Alternating Series Test, but since $\sum_{n=2}^{\infty}\dfrac{1}{n\ln n}$ diverges by the Integral Test (Exercise 10.3.13), the given series converges only conditionally.

25. $\lim_{n\to\infty} \sqrt[n]{|a_n|} = \lim_{n\to\infty} \left|\frac{1-3n}{3+4n}\right| = \frac{3}{4} < 1$, so the series converges absolutely by the Root Test.

26. $\lim_{n\to\infty} \left|\frac{a_{n+1}}{a_n}\right| = \lim_{n\to\infty} \left|\frac{(-2)^{n+1}(n+1)^2/(n+3)!}{(-2)^n n^2/(n+2)!}\right| = \lim_{n\to\infty} \frac{2(n+1)^2}{n^2(n+3)} = 0 < 1$, so the series converges absolutely by the Ratio Test.

27. $\lim_{n\to\infty} \left|\frac{a_{n+1}}{a_n}\right| = \lim_{n\to\infty} \frac{(n+1)!/[1\cdot 3\cdot 5\cdot\cdots\cdot(2n+1)]}{n!/[1\cdot 3\cdot 5\cdot\cdots\cdot(2n-1)]} = \lim_{n\to\infty} \frac{n+1}{2n+1} = \frac{1}{2} < 1$, so the series converges absolutely by the Ratio Test.

28. $\lim_{n\to\infty} \left|\frac{a_{n+1}}{a_n}\right| = \lim_{n\to\infty} \left|\frac{\dfrac{1\cdot 4\cdot 7\cdot\cdots\cdot(3n+1)}{3\cdot 5\cdot 7\cdot\cdots\cdot(2n+3)}}{\dfrac{1\cdot 4\cdot 7\cdot\cdots\cdot(3n-2)}{3\cdot 5\cdot 7\cdot\cdots\cdot(2n+1)}}\right| = \lim_{n\to\infty} \frac{3n+1}{2n+3} = \frac{3}{2} > 1$ so the series diverges by the Ratio Test.

29. $\sum_{n=1}^{\infty} \frac{2\cdot 4\cdot 6\cdot\cdots\cdot(2n)}{n!} = \sum_{n=1}^{\infty} \frac{2^n n!}{n!} = \sum_{n=1}^{\infty} 2^n$ which diverges by the Test for Divergence since $\lim_{n\to\infty} 2^n = \infty$.

30. $\lim_{n\to\infty} \left|\frac{a_{n+1}}{a_n}\right| = \lim_{n\to\infty} \left|\frac{\dfrac{2^{n+1}(n+1)!}{5\cdot 8\cdot 11\cdot\cdots\cdot(3n+5)}}{\dfrac{2^n n!}{5\cdot 8\cdot 11\cdot\cdots\cdot(3n+2)}}\right| = \lim_{n\to\infty} \frac{2(n+1)}{3n+5} = \frac{2}{3} < 1$ so the series converges absolutely by the Ratio Test.

31. $\lim_{n\to\infty} \left|\frac{a_{n+1}}{a_n}\right| = \lim_{n\to\infty} \frac{(n+3)!/[(n+1)!\,10^{n+1}]}{(n+2)!/(n!\,10^n)} = \frac{1}{10}\lim_{n\to\infty} \frac{n+3}{n+1} = \frac{1}{10} < 1$ so the series converges absolutely by the Ratio Test.

32. $\lim_{n\to\infty} \sqrt[n]{|a_n|} = \lim_{n\to\infty} \frac{1}{\arctan n} = \frac{1}{\pi/2} = \frac{2}{\pi} < 1$ so the series converges absolutely by the Root Test.

33. By the recursive definition, $\lim_{n\to\infty} \left|\frac{a_{n+1}}{a_n}\right| = \lim_{n\to\infty} \left|\frac{5n+1}{4n+3}\right| = \frac{5}{4} > 1$, so the series diverges by the Ratio Test.

34. By the recursive definition, $\lim_{n\to\infty} \left|\frac{a_{n+1}}{a_n}\right| = \lim_{n\to\infty} \left|\frac{2+\cos n}{\sqrt{n}}\right| = 0 < 1$, so the series converges absolutely by the Ratio Test.

35. (a) $\lim_{n\to\infty} \left|\frac{1/(n+1)^3}{1/n^3}\right| = \lim_{n\to\infty} \frac{n^3}{(n+1)^3} = \lim_{n\to\infty} \frac{1}{(1+1/n)^3} = 1$. So inconclusive.

(b) $\lim_{n\to\infty} \left|\frac{(n+1)}{2^{n+1}}\cdot\frac{2^n}{n}\right| = \lim_{n\to\infty} \frac{n+1}{2n} = \lim_{n\to\infty} \left(\frac{1}{2}+\frac{1}{2n}\right) = \frac{1}{2}$. So conclusive (convergent).

(c) $\lim_{n\to\infty} \left|\frac{(-3)^n}{\sqrt{n+1}}\cdot\frac{\sqrt{n}}{(-3)^{n-1}}\right| = 3\lim_{n\to\infty} \sqrt{\frac{n}{n+1}} = 3\lim_{n\to\infty} \sqrt{\frac{1}{1+1/n}} = 3$. So conclusive (divergent).

(d) $\lim_{n\to\infty} \left|\frac{\sqrt{n+1}}{1+(n+1)^2}\cdot\frac{1+n^2}{\sqrt{n}}\right| = \lim_{n\to\infty} \left[\sqrt{1+\frac{1}{n}}\cdot\frac{1/n^2+1}{1/n^2+(1+1/n)^2}\right] = 1$. So inconclusive.

36. We use the Ratio Test:

$$\lim_{n\to\infty}\left|\frac{a_{n+1}}{a_n}\right| = \lim_{n\to\infty}\left|\frac{[(n+1)!]^2/[k(n+1)!]}{(n!)^2/(kn)!}\right| = \lim_{n\to\infty}\left|\frac{(n+1)^2}{[k(n+1)][k(n+1)-1]\cdots[kn+1]}\right|.$$ Now if $k = 1$, then this is equal to $\lim_{n\to\infty}\left|\frac{(n+1)^2}{(n+1)}\right| = \infty$, so the series diverges; if $k = 2$, the limit is $\lim_{n\to\infty}\left|\frac{(n+1)^2}{(2n+2)(2n+1)}\right| = \frac{1}{4} < 1$, so the series converges, and if $k > 2$, then the highest power of n in the denominator is larger than 2, and so the limit is 0, indicating convergence. So the series converges for $k \geq 2$.

37. **(a)** $\lim_{n\to\infty}\left|\frac{a_{n+1}}{a_n}\right| = \lim_{n\to\infty}\frac{|x|^{n+1}/(n+1)!}{|x|^n/n!} = |x|\lim_{n\to\infty}\frac{1}{n+1} = 0$, so by the Ratio Test the series converges for all x.

(b) Since the series of part (a) always converges, we must have $\lim_{n\to\infty}\frac{x^n}{n!} = 0$ by Theorem 10.2.6.

38. **(a)** $R_n = a_{n+1} + a_{n+2} + a_{n+3} + a_{n+4} + \cdots = a_{n+1}\left(1 + \frac{a_{n+2}}{a_{n+1}} + \frac{a_{n+3}}{a_{n+1}} + \frac{a_{n+4}}{a_{n+1}} + \cdots\right)$

$$= a_{n+1}\left(1 + \frac{a_{n+2}}{a_{n+1}} + \frac{a_{n+3}}{a_{n+2}}\frac{a_{n+2}}{a_{n+1}} + \frac{a_{n+4}}{a_{n+3}}\frac{a_{n+3}}{a_{n+2}}\frac{a_{n+2}}{a_{n+1}} + \cdots\right)$$

$$= a_{n+1}(1 + r_{n+1} + r_{n+2}r_{n+1} + r_{n+3}r_{n+2}r_{n+1} + \cdots) \quad (\bigstar)$$

$$\leq a_{n+1}(1 + r_{n+1} + r_{n+1}^2 + r_{n+1}^3 + \cdots) \quad \text{[since } \{r_n\} \text{ is decreasing]} \quad = \frac{a_{n+1}}{1 - r_{n+1}}$$

(b) Note that since $\{r_n\}$ is increasing and $r_n \to L$, we have $r_n < L$ for all n. So, starting with equation $\bigstar$, $R_n = a_{n+1}(1 + r_{n+1} + r_{n+2}r_{n+1} + r_{n+3}r_{n+2}r_{n+1} + \cdots) \leq a_{n+1}(1 + L + L^2 + L^3 + \cdots) = \frac{a_{n+1}}{1-L}$.

39. **(a)** $s_5 = \sum_{n=1}^{5}\frac{1}{n2^n} = \frac{1}{2} + \frac{1}{8} + \frac{1}{24} + \frac{1}{64} + \frac{1}{160} = \frac{661}{960} \approx 0.68854$. Now the ratios $r_n = \frac{a_{n+1}}{a_n} = \frac{n2^n}{(n+1)2^{n+1}} = \frac{n}{2(n+1)}$ form an increasing sequence, since

$$r_{n+1} - r_n = \frac{n+1}{2(n+2)} - \frac{n}{2(n+1)} = \frac{(n+1)^2 - n(n+2)}{2(n+1)(n+2)} = \frac{1}{2(n+1)(n+2)} > 0.$$ So by Exercise 38(b), the error is less than $\frac{a_6}{1 - \lim_{n\to\infty} r_n} = \frac{1/(6\cdot 2^6)}{1 - 1/2} = \frac{1}{192} \approx 0.00521$.

(b) The error in using s_n as an approximation to the sum is $R_n = \frac{a_{n+1}}{1/2} = \frac{2}{(n+1)2^{n+1}}$. We want $R_n < 0.00005 \Leftrightarrow \frac{1}{(n+1)2^n} < 0.00005 \Leftrightarrow (n+1)2^n > 20{,}000$. To find such an n we can use trial and error or a graph. We calculate $(11+1)2^{11} = 24{,}576$, so $s_{11} = \sum_{n=1}^{11}\frac{1}{n2^n} \approx 0.693109$ is within 0.00005 of the actual sum.

40. $\sum_{n=1}^{10} \frac{n}{2^n} = \frac{1}{2} + \frac{2}{4} + \frac{3}{8} + \cdots + \frac{10}{1024} \approx 1.988$. The ratios $r_n = \frac{a_{n+1}}{a_n} = \frac{n+1}{2n} = \frac{1}{2}\left(1 + \frac{1}{n}\right)$ form a decreasing sequence, so by Exercise 38 (a), using $a_{11} = \frac{11}{2048}$ and $r_{11} = \frac{12}{22} = \frac{6}{11}$, the error in the above approximation is less than $\frac{a_{11}}{1 - r_{11}} \approx 0.0118$.

41. By the Triangle Inequality (see Exercise 4.1.44) we have $\left|\sum_{i=1}^{n} a_i\right| \le \sum_{i=1}^{n} |a_i| \Rightarrow -\sum_{i=1}^{n} |a_i| \le \sum_{i=1}^{n} a_i \le \sum_{i=1}^{n} |a_i|$ $\Rightarrow -\lim_{n\to\infty} \sum_{i=1}^{n} |a_i| \le \lim_{n\to\infty} \sum_{i=1}^{n} a_i \le \lim_{n\to\infty} \sum_{i=1}^{n} |a_i| \Rightarrow -\sum_{n=1}^{\infty} |a_n| \le \sum_{n=1}^{\infty} a_n \le \sum_{n=1}^{\infty} |a_n| \Rightarrow \left|\sum_{n=1}^{\infty} a_n\right| \le \sum_{n=1}^{\infty} |a_n|$.

42. **(a)** Following the hint, we get that $|a_n| < r^n$ for $n \ge N$, and so since the geometric series $\sum_{n=1}^{\infty} r^n$ converges $(0 < r < 1)$, the series $\sum_{n=N}^{\infty} |a_n|$ will converge as well by the Comparison Test, and hence so does $\sum_{n=1}^{\infty} |a_n|$, so $\sum_{n=1}^{\infty} a_n$ is absolutely convergent.

(b) If $\lim_{n\to\infty} \sqrt[n]{|a_n|} = L > 1$, then there is an integer N such that $\sqrt[n]{|a_n|} > 1$ for all $n \ge N$, so $|a_n| > 1$ for $n \ge N$. Thus $\lim_{n\to\infty} a_n \ne 0$, so $\sum_{n=1}^{\infty} a_n$ diverges by the Test for Divergence.

43. **(a)** Since $\sum a_n$ is absolutely convergent, and since $|a_n^+| \le |a_n|$ and $|a_n^-| \le |a_n|$ (because a_n^+ and a_n^- each equal either a_n or 0), we conclude by the Comparison Test that both $\sum a_n^+$ and $\sum a_n^-$ must be absolutely convergent. (Or use Theorem 10.2.8.)

(b) We will show by contradiction that both $\sum a_n^+$ and $\sum a_n^-$ must diverge. For suppose that $\sum a_n^+$ converged. Then so would $\sum(a_n^+ - \frac{1}{2}a_n)$ by Theorem 10.2.8. But $\sum\left(a_n^+ - \frac{1}{2}a_n\right) = \sum\left[\frac{1}{2}(a_n + |a_n|) - \frac{1}{2}a_n\right] = \frac{1}{2}\sum|a_n|$, which diverges because $\sum a_n$ is only conditionally convergent. Hence $\sum a_n^+$ can't converge. Similarly, neither can $\sum a_n^-$.

44. Let $\sum b_n$ be the rearranged series constructed in the hint. [This series can be constructed by virtue of the result of Exercise 43(b).] This series will have partial sums s_n that oscillate in value back and forth across r. Since $\lim_{n\to\infty} a_n = 0$ (by Theorem 10.2.6), and since the size of the oscillations $|s_n - r|$ is always less than $|a_n|$ because of the way $\sum b_n$ was constructed, we have that $\sum b_n = \lim_{n\to\infty} s_n = r$.

EXERCISES 10.7

1. Use the Comparison Test, with $a_n = \dfrac{\sqrt{n}}{n^2+1}$ and $b_n = \dfrac{1}{n^{3/2}}$: $\dfrac{\sqrt{n}}{n^2+1} < \dfrac{\sqrt{n}}{n^2} = \dfrac{1}{n^{3/2}}$, and $\sum\limits_{n=1}^{\infty} \dfrac{1}{n^{3/2}}$ is a convergent p-series ($p = \frac{3}{2} > 1$), so $\sum\limits_{n=1}^{\infty} a_n = \sum\limits_{n=1}^{\infty} \dfrac{\sqrt{n}}{n^2+1}$ converges as well.

2. $\lim\limits_{n\to\infty} \cos n$ does not exist, so the series diverges by the Test for Divergence.

3. $\sum\limits_{n=1}^{\infty} \dfrac{4^n}{3^{2n-1}} = 3\sum\limits_{n=1}^{\infty} \left(\dfrac{4}{9}\right)^n$ which is a convergent geometric series ($|r| = \frac{4}{9} < 1$.)

4. $\lim\limits_{i\to\infty} \left|\dfrac{a_{i+1}}{a_i}\right| = \lim\limits_{i\to\infty} \dfrac{(i+1)^4/4^{i+1}}{i^4/4^i} = \dfrac{1}{4} \lim\limits_{i\to\infty} \left(\dfrac{i+1}{i}\right)^4 = \dfrac{1}{4} < 1$ so the series converges by the Ratio Test.

5. The series converges by the Alternating Series Test, since $a_n = \dfrac{1}{(\ln n)^2}$ is decreasing ($\ln x$ is an increasing function) and $\lim\limits_{n\to\infty} a_n = 0$.

6. Let $f(x) = x^2 e^{-x^3}$. Then f is continuous and positive on $[1, \infty)$, and $f'(x) = \dfrac{x(2-3x^3)}{e^{x^3}} < 0$ for $x \geq 1$, so f is decreasing on $[1, \infty)$ as well, and we can apply the Integral Test. $\int_1^{\infty} x^2 e^{-x^3}\,dx = \lim\limits_{t\to\infty} \left[-\frac{1}{3}e^{-x^3}\right]_1^t = \dfrac{1}{3e}$, so the series converges.

7. $\sum\limits_{k=1}^{\infty} \dfrac{1}{k^{1.7}}$ is a convergent p-series ($p = 1.7 > 1$).

8. $\lim\limits_{n\to\infty} \left|\dfrac{a_{n+1}}{a_n}\right| = \lim\limits_{n\to\infty} \dfrac{10^{n+1}/(n+1)!}{10^n/n!} = \lim\limits_{n\to\infty} \dfrac{10}{n+1} = 0 < 1$, so the series converges by the Ratio Test.

9. $\lim\limits_{n\to\infty} \left|\dfrac{a_{n+1}}{a_n}\right| = \lim\limits_{n\to\infty} \dfrac{(n+1)/e^{n+1}}{n/e^n} = \dfrac{1}{e} \lim\limits_{n\to\infty} \dfrac{n+1}{n} = \dfrac{1}{e} < 1$, so the series converges by the Ratio Test.

10. $\lim\limits_{m\to\infty} \dfrac{2m}{8m-5} = \dfrac{1}{4} \neq 0$ so the series diverges by the Test for Divergence.

11. Use the Limit Comparison Test with $a_n = \dfrac{n^3+1}{n^4-1}$ and $b_n = \dfrac{1}{n}$. $\lim\limits_{n\to\infty} \dfrac{a_n}{b_n} = \lim\limits_{n\to\infty} \dfrac{n^4+n}{n^4-1} = \lim\limits_{n\to\infty} \dfrac{1+1/n^3}{1-1/n^4} = 1$, and since $\sum\limits_{n=2}^{\infty} b_n$ diverges (harmonic series), so does $\sum\limits_{n=2}^{\infty} \dfrac{n^3+1}{n^4-1}$.

12. $\lim\limits_{n\to\infty} \sqrt[n]{|a_n|} = \lim\limits_{n\to\infty} \dfrac{n^2+1}{2n^2+1} = \dfrac{1}{2} < 1$, so the series converges (Root Test).

13. Let $f(x) = \dfrac{2}{x(\ln x)^3}$. $f(x)$ is clearly positive and decreasing for $x \geq 2$, so we apply the Integral Test.

$$\int_2^{\infty} \frac{2}{x(\ln x)^3}\,dx = \lim_{t\to\infty} \left[\frac{-1}{(\ln x)^2}\right]_2^t = 0 - \frac{-1}{(\ln 2)^2},$$ which is finite, so $\sum\limits_{n=2}^{\infty} \dfrac{2}{n(\ln n)^3}$ converges.

14. Let $f(x) = \dfrac{\sqrt{x}}{e^{\sqrt{x}}}$. Then $f(x)$ is continuous and positive, and $f'(x) = \dfrac{1-\sqrt{x}}{2\sqrt{x}e^{\sqrt{x}}} < 0$ on $[1,\infty)$, so $f(x)$ is decreasing and we can use the Integral Test. $\displaystyle\int_1^\infty \frac{\sqrt{x}}{e^{\sqrt{x}}}\,dx = \lim_{t\to\infty}\left[\frac{-2x-4\sqrt{x}-4}{e^{\sqrt{x}}}\right]_1^t = 0 - \left(-\frac{10}{e}\right) = \frac{10}{e}$ (using integration by parts and l'Hospital's Rule), so the series converges.

15. $\displaystyle\lim_{n\to\infty}\left|\frac{a_{n+1}}{a_n}\right| = \lim_{n\to\infty}\frac{3^{n+1}(n+1)^2/(n+1)!}{3^n n^2/n!} = 3\lim_{n\to\infty}\frac{n+1}{n^2} = 0$, so the series converges by the Ratio Test.

16. Let $a_n = \dfrac{3}{4n-5}$ and $b_n = \dfrac{1}{n}$. Then $\displaystyle\lim_{n\to\infty}\frac{a_n}{b_n} = \frac{3}{4}$, so since $\displaystyle\sum_{n=1}^\infty b_n$ diverges (harmonic series), so does $\displaystyle\sum_{n=1}^\infty \frac{3}{4n-5}$ by the Limit Comparison Test.

17. $\dfrac{3^n}{5^n+n} \le \dfrac{3^n}{5^n} = \left(\dfrac{3}{5}\right)^n$. Since $\displaystyle\sum_{n=1}^\infty\left(\frac{3}{5}\right)^n$ is a convergent geometric series ($|r| = \frac{3}{5} < 1$), $\displaystyle\sum_{n=1}^\infty \frac{3^n}{5^n+n}$ converges by the Comparison Test.

18. $\displaystyle\lim_{k\to\infty}\left|\frac{a_{k+1}}{a_k}\right| = \lim_{k\to\infty}\frac{(k+6)/5^{k+1}}{(k+5)/5^k} = \frac{1}{5}\lim_{k\to\infty}\frac{k+6}{k+5} = \frac{1}{5} < 1$, so the series converges by the Ratio Test.

19. $\displaystyle\lim_{n\to\infty}\left|\frac{a_{n+1}}{a_n}\right| = \lim_{n\to\infty}\frac{(n+1)!/[2\cdot5\cdot8\cdot\cdots\cdot(3n+5)]}{n!/[2\cdot5\cdot8\cdot\cdots\cdot(3n+2)]} = \lim_{n\to\infty}\frac{n+1}{3n+5} = \frac{1}{3} < 1$, so the series converges by the Ratio Test.

20. $\displaystyle\lim_{n\to\infty}\frac{n}{(n+1)(n+2)} = \lim_{n\to\infty}\frac{1}{n+3+2/n} = 0$, so since $\{b_n\}$ is a positive, decreasing sequence, $\displaystyle\sum_{n=1}^\infty\frac{(-1)^n n}{(n+1)(n+2)}$ converges by the Alternating Series Test.

21. Use the Limit Comparison Test with $a_i = \dfrac{1}{\sqrt{i(i+1)}}$ and $b_i = \dfrac{1}{i}$. $\displaystyle\lim_{i\to\infty}\frac{a_i}{b_i} = \lim_{i\to\infty}\frac{i}{\sqrt{i(i+1)}} = \lim_{i\to\infty}\frac{1}{\sqrt{1+1/i}} = 1$. Since $\displaystyle\sum_{i=1}^\infty b_i$ diverges (harmonic series) so does $\displaystyle\sum_{i=1}^\infty\frac{1}{\sqrt{i(i+1)}}$.

22. Let $a_n = \dfrac{n^2}{\sqrt{n^5+n^2+2}}$ and $b_n = \dfrac{1}{\sqrt{n}}$. Then $\displaystyle\lim_{n\to\infty}\frac{a_n}{b_n} = \lim_{n\to\infty}\frac{1}{\sqrt{1+n^{-3}+2n^{-5}}} = 1$, so since $\displaystyle\sum_{n=1}^\infty b_n$ diverges (p-series with $p = \frac{1}{2} < 1$), so does $\displaystyle\sum_{n=1}^\infty\frac{n^2}{\sqrt{n^5+n^2+2}}$ by the Limit Comparison Test.

23. $\displaystyle\lim_{n\to\infty}2^{1/n} = 2^0 = 1$, so $\displaystyle\lim_{n\to\infty}(-1)^n 2^{1/n}$ does not exist and the series diverges by the Test for Divergence.

24. $\dfrac{|\cos(n/2)|}{n^2+4n} < \dfrac{1}{n^2}$ and since $\displaystyle\sum_{n=1}^\infty\frac{1}{n^2}$ converges ($p = 2 > 1$), $\displaystyle\sum_{n=1}^\infty\frac{\cos(n/2)}{n^2+4n}$ converges absolutely by the Comparison Test.

25. Let $f(x) = \dfrac{\ln x}{\sqrt{x}}$. Then $f'(x) = \dfrac{2 - \ln x}{2x^{3/2}} < 0$ when $\ln x > 2$ or $x > e^2$, so $\dfrac{\ln n}{\sqrt{n}}$ is decreasing for $n > e^2$.

By l'Hospital's Rule, $\lim\limits_{n\to\infty} \dfrac{\ln n}{\sqrt{n}} = \lim\limits_{n\to\infty} \dfrac{1/n}{1/(2\sqrt{n})} = \lim\limits_{n\to\infty} \dfrac{2}{\sqrt{n}} = 0$, so the series converges by the Alternating Series Test.

26. Let $a_n = \dfrac{\tan(1/n)}{n}$ and $b_n = \dfrac{1}{n^2}$. Then $\lim\limits_{n\to\infty} \dfrac{a_n}{b_n} = \lim\limits_{n\to\infty} n \cdot \tan(1/n) = \lim\limits_{n\to\infty} \dfrac{\tan(1/n)}{1/n}$

$= \lim\limits_{n\to\infty} \dfrac{(-1/n^2)\sec^2(1/n)}{-1/n^2}$ (l'Hospital's Rule) $= \sec^2(0) = 1 > 0$, so since $\sum\limits_{n=1}^{\infty} b_n$ converges ($p = 2 > 1$),

$\sum\limits_{n=1}^{\infty} \dfrac{\tan(1/n)}{n}$ converges also, by the Limit Comparison Test.

27. The series diverges since it is a geometric series with $r = -\pi$ and $|r| = \pi > 1$. (Or use the Test for Divergence.)

28. Let $a_n = \dfrac{\sqrt[3]{n} + 1}{n(\sqrt{n} + 1)}$ and $b_n = n^{-7/6}$. Then $\lim\limits_{n\to\infty} \dfrac{a_n}{b_n} = \lim\limits_{n\to\infty} \dfrac{1 + n^{-1/3}}{1 + n^{-1/2}} = 1$, so since $\sum\limits_{n=1}^{\infty} b_n$ converges

($p = \frac{7}{6} > 1$), $\sum\limits_{n=1}^{\infty} \dfrac{\sqrt[3]{n} + 1}{n(\sqrt{n} + 1)}$ converges by the Limit Comparison Test.

29. $\sum\limits_{n=1}^{\infty} \dfrac{(-2)^{2n}}{n^n} = \sum\limits_{n=1}^{\infty} \left(\dfrac{4}{n}\right)^n$. $\lim\limits_{n\to\infty} \sqrt[n]{|a_n|} = \lim\limits_{n\to\infty} \dfrac{4}{n} = 0$, so the series converges by the Root Test.

30. $\lim\limits_{n\to\infty} \dfrac{2^{3n-1}}{n^2 + 1} = \infty$ (use l'Hospital's Rule twice) so the series diverges by the Test for Divergence. (Or use the Ratio Test.)

31. $\displaystyle\int_2^{\infty} \frac{\ln x}{x^2}\,dx = \lim_{t\to\infty} \left[-\frac{\ln x}{x} - \frac{1}{x}\right]_1^t$ (using integration by parts) $= 1$ (by L'Hospital's Rule). So $\sum\limits_{n=1}^{\infty} \dfrac{\ln n}{n^2}$

converges by the Integral Test, and since $\dfrac{k \ln k}{(k+1)^3} < \dfrac{k \ln k}{k^3} = \dfrac{\ln k}{k^2}$, the given series converges by the Comparison Test.

32. Since $\{1/n\}$ is a decreasing sequence, $e^{1/n} \le e^{1/1} = e$ for all $n \ge 1$, and $\sum\limits_{n=1}^{\infty} \dfrac{e}{n^2}$ converges ($p = 2 > 1$), so

$\sum\limits_{n=1}^{\infty} \dfrac{e^{1/n}}{n^2}$ converges by the Comparison Test. (Or use the Integral Test.)

33. $\lim\limits_{n\to\infty} \left|\dfrac{a_{n+1}}{a_n}\right| = \lim\limits_{n\to\infty} \dfrac{2^{n+1}/(2n+3)!}{2^n/(2n+1)!} = 2 \lim\limits_{n\to\infty} \dfrac{1}{(2n+3)(2n+2)} = 0$, so the series converges by the Ratio Test.

34. Let $f(x) = \dfrac{\sqrt{x}}{x+5}$. Then $f(x)$ is continuous and positive on $[1, \infty)$, and since $f'(x) = \dfrac{5 - x}{2\sqrt{x}(x+5)^2} < 0$ for

$x > 5$, $f(x)$ is eventually decreasing, so we can use the Alternating Series Test.

$\lim\limits_{n\to\infty} \dfrac{\sqrt{n}}{n+5} = \lim\limits_{n\to\infty} \dfrac{1}{n^{1/2} + 5n^{-1/2}} = 0$, so the series converges.

35. $0 < \dfrac{\tan^{-1} n}{n^{3/2}} < \dfrac{\pi/2}{n^{3/2}}$. $\displaystyle\sum_{n=1}^{\infty} \frac{\pi/2}{n^{3/2}} = \frac{\pi}{2}\sum_{n=1}^{\infty} \frac{1}{n^{3/2}}$ which is a convergent p-series ($p = \frac{3}{2} > 1$), so $\displaystyle\sum_{n=1}^{\infty} \frac{\tan^{-1} n}{n^{3/2}}$ converges by the Comparison Test.

36. $\displaystyle\lim_{n\to\infty} \sqrt[n]{|a_n|} = \lim_{n\to\infty} \frac{2n}{n^2} = \lim_{n\to\infty} \frac{2}{n} = 0$ so the series converges by the Root Test.

37. $\displaystyle\lim_{n\to\infty} \sqrt[n]{|a_n|} = \lim_{n\to\infty} \left(\frac{n}{n+1}\right)^{n^2/n} = \lim_{n\to\infty} \frac{1}{[(n+1)/n]^n} = \frac{1}{\lim\limits_{n\to\infty} (1+1/n)^n} = \frac{1}{e} < 1$ (see Equation 6.4.9 or 6.4*.8), so the series converges by the Root Test.

38. Note that $(\ln n)^{\ln n} = \left(e^{\ln \ln n}\right)^{\ln n} = \left(e^{\ln n}\right)^{\ln \ln n} = n^{\ln \ln n}$ and $\ln \ln n \to \infty$ as $n \to \infty$, so $\ln \ln n > 2$ for sufficiently large n. For these n we have $(\ln n)^{\ln n} > n^2$, so $\dfrac{1}{(\ln n)^{\ln n}} < \dfrac{1}{n^2}$. Since $\displaystyle\sum_{n=2}^{\infty} \frac{1}{n^2}$ converges ($p = 2 > 1$), so does $\displaystyle\sum_{n=2}^{\infty} \frac{1}{(\ln n)^{\ln n}}$ by the Comparison Test.

39. $\displaystyle\lim_{n\to\infty} \sqrt[n]{|a_n|} = \lim_{n\to\infty} \left(2^{1/n} - 1\right) = 1 - 1 = 0$, so the series converges by the Root Test.

40. Use the Limit Comparison Test with $a_n = \sqrt[n]{2} - 1$ and $b_n = 1/n$. Then $\displaystyle\lim_{n\to\infty} \frac{a_n}{b_n} = \lim_{n\to\infty} \frac{2^{1/n} - 1}{1/n} = \ln 2 > 0$ (by l'Hospital's Rule). So since $\sum b_n$ diverges (harmonic series), so does $\displaystyle\sum_{n=1}^{\infty} \sqrt[n]{2} - 1$.

Alternate Solution: $\sqrt[n]{2} - 1 = \dfrac{1}{2^{(n-1)/n} + 2^{(n-2)/n} + 2^{(n-3)/n} + \cdots + 2^{1/n} + 1} \begin{bmatrix}\text{rationalize} \\ \text{the numerator}\end{bmatrix} \geq \dfrac{1}{2n}$, and since $\displaystyle\sum_{n=1}^{\infty} \frac{1}{2n} = \frac{1}{2}\sum_{n=1}^{\infty} \frac{1}{n}$ diverges (harmonic series), so does $\displaystyle\sum_{n=1}^{\infty} \left(\sqrt[n]{2} - 1\right)$ by the Comparison Test.

EXERCISES 10.8

Note: "R" stands for "radius of convergence" and "I" stands for "interval of convergence" in this section.

1. (a) We are given that the power series $\sum_{n=0}^{\infty} c_n x^n$ is convergent for $x = 4$. So by Theorem 3 it must converge for at least $-4 < x \le 4$. In particular it converges when $x = -2$, that is, $\sum_{n=0}^{\infty} c_n(-2)^n$ is convergent.

(b) But it does not follow that $\sum_{n=0}^{\infty} c_n(-4)^n$ is necessarily convergent. [See the comments after Theorem 3. An example is $c_n = (-1)^n/(n4^n)$.]

2. We are given that the power series $\sum_{n=0}^{\infty} c_n x^n$ is convergent for $x = -4$ and divergent when $x = 6$. So by Theorem 3 it converges for at least $-4 \le x < 4$ and diverges for at least $x \ge 6$ and $x < -6$. Therefore:

(a) It converges when $x = 1$, that is, $\sum c_n$ is convergent.

(b) It diverges when $x = 8$, that is, $\sum c_n 8^n$ is divergent.

(c) It converges when $x = -3$, that is, $\sum c_n(-3)^n$ is convergent.

(d) It diverges when $x = -9$, that is, $\sum c_n(-9)^n = \sum(-1)^n c_n 9^n$ is divergent.

3. If $a_n = \dfrac{x^n}{n+2}$, then $\lim\limits_{n\to\infty}\left|\dfrac{a_{n+1}}{a_n}\right| = \lim\limits_{n\to\infty}\left|\dfrac{x^{n+1}}{n+3}\cdot\dfrac{n+2}{x^n}\right| = |x|\lim\limits_{n\to\infty}\dfrac{n+2}{n+3} = |x| < 1$ for convergence (by the Ratio Test). So $R = 1$. When $x = 1$, the series is $\sum_{n=0}^{\infty}\dfrac{1}{n+2}$ which diverges (Integral Test or Comparison Test), and when $x = -1$, it is $\sum_{n=0}^{\infty}\dfrac{(-1)^n}{n+2}$ which converges (Alternating Series Test), so $I = [-1, 1)$.

4. If $a_n = \dfrac{(-1)^n x^n}{\sqrt[3]{n}}$, then $\lim\limits_{n\to\infty}\left|\dfrac{a_{n+1}}{a_n}\right| = |x|\lim\limits_{n\to\infty}\left(\dfrac{n}{n+1}\right)^{1/3} = |x| < 1$ for convergence (by the Ratio Test), and $R = 1$. When $x = 1$, $\sum_{n=1}^{\infty} a_n = \sum_{n=1}^{\infty}\dfrac{(-1)^n}{\sqrt[3]{n}}$ which is a convergent alternating series, but when $x = -1$, $\sum_{n=1}^{\infty} a_n = \sum_{n=1}^{\infty}\dfrac{1}{n^{1/3}}$ which is a divergent p-series ($p = \frac{1}{3} < 1$), so $I = (-1, 1]$.

5. If $a_n = nx^n$, then $\lim\limits_{n\to\infty}\left|\dfrac{a_{n+1}}{a_n}\right| = \lim\limits_{n\to\infty}\left|\dfrac{(n+1)x^{n+1}}{nx^n}\right| = |x|\lim\limits_{n\to\infty}\dfrac{n+1}{n} = |x| < 1$ for convergence (by the Ratio Test). So $R = 1$. When $x = 1$ or -1, $\lim\limits_{n\to\infty} nx^n$ does not exist, so $\sum_{n=0}^{\infty} nx^n$ diverges for $x = \pm 1$. So $I = (-1, 1)$.

6. If $a_n = \dfrac{x^n}{n^2}$ then $\lim\limits_{n\to\infty}\left|\dfrac{a_{n+1}}{a_n}\right| = |x|\lim\limits_{n\to\infty}\left(\dfrac{n}{n+1}\right)^2 = |x| < 1$ for convergence (by the Ratio Test), so $R = 1$. If $x = \pm 1$, $\sum_{n=1}^{\infty}|a_n| = \sum_{n=1}^{\infty}\dfrac{1}{n^2}$ which converges ($p = 2 > 1$), so $I = [-1, 1]$.

7. If $a_n = \dfrac{x^n}{n!}$, then $\lim\limits_{n\to\infty}\left|\dfrac{a_{n+1}}{a_n}\right| = \lim\limits_{n\to\infty}\left|\dfrac{x^{n+1}/(n+1)!}{x^n/n!}\right| = |x| \lim\limits_{n\to\infty}\dfrac{1}{n+1} = 0 < 1$ for all x. So, by the Ratio Test, $R = \infty$, and $I = (-\infty, \infty)$.

8. Here the Root Test is easier. If $a_n = n^n x^n$ then $\lim\limits_{n\to\infty}\sqrt[n]{|a_n|} = \lim\limits_{n\to\infty} n|x| = \infty$ if $x \neq 0$ so $R = 0$ and $I = \{0\}$.

9. If $a_n = \dfrac{(-1)^n x^n}{n2^n}$, then $\lim\limits_{n\to\infty}\left|\dfrac{a_{n+1}}{a_n}\right| = \lim\limits_{n\to\infty}\left|\dfrac{x^{n+1}/[(n+1)2^{n+1}]}{x^n/(n2^n)}\right| = \left|\dfrac{x}{2}\right|\lim\limits_{n\to\infty}\dfrac{n}{n+1} = \left|\dfrac{x}{2}\right| < 1$ for convergence, so $|x| < 2$ and $R = 2$. When $x = 2$, $\displaystyle\sum_{n=1}^{\infty}\frac{(-1)^n x^n}{n2^n} = \sum_{n=1}^{\infty}\frac{(-1)^n}{n}$ which converges by the Alternating Series Test. When $x = -2$, $\displaystyle\sum_{n=1}^{\infty}\frac{(-1)^n x^n}{n2^n} = \sum_{n=1}^{\infty}\frac{1}{n}$ which diverges (harmonic series), so $I = (-2, 2]$.

10. If $a_n = n5^n x^n$ then $\lim\limits_{n\to\infty}\left|\dfrac{a_{n+1}}{a_n}\right| = 5|x|\lim\limits_{n\to\infty}\dfrac{n+1}{n} = 5|x| < 1$ for convergence (by the Ratio Test), so $R = \dfrac{1}{5}$. If $x = \pm\frac{1}{5}$, $|a_n| = n \to \infty$ as $n \to \infty$, so $\displaystyle\sum_{n=1}^{\infty} a_n$ diverges by the Test for Divergence and $I = \left(-\dfrac{1}{5}, \dfrac{1}{5}\right)$.

11. If $a_n = \dfrac{3^n x^n}{(n+1)^2}$, then $\lim\limits_{n\to\infty}\left|\dfrac{a_{n+1}}{a_n}\right| = \lim\limits_{n\to\infty}\left|\dfrac{3^{n+1}x^{n+1}}{(n+2)^2}\cdot\dfrac{(n+1)^2}{3^n x^n}\right| = 3|x|\lim\limits_{n\to\infty}\left(\dfrac{n+1}{n+2}\right)^2 = 3|x| < 1$ for convergence, so $|x| < \frac{1}{3}$ and $R = \frac{1}{3}$. When $x = \frac{1}{3}$, $\displaystyle\sum_{n=0}^{\infty}\frac{3^n x^n}{(n+1)^2} = \sum_{n=0}^{\infty}\frac{1}{(n+1)^2} = \sum_{n=1}^{\infty}\frac{1}{n^2}$ which is a convergent p-series ($p = 2 > 1$). When $x = -\frac{1}{3}$, $\displaystyle\sum_{n=0}^{\infty}\frac{3^n x^n}{(n+1)^2} = \sum_{n=0}^{\infty}\frac{(-1)^n}{(n+1)^2}$ which converges by the Alternating Series Test, so $I = \left[-\frac{1}{3}, \frac{1}{3}\right]$.

12. If $a_n = \dfrac{n^2 x^n}{10^n}$, then $\lim\limits_{n\to\infty}\left|\dfrac{a_{n+1}}{a_n}\right| = \dfrac{|x|}{10}\lim\limits_{n\to\infty}\left(\dfrac{n+1}{n}\right)^2 = \dfrac{|x|}{10} < 1$ for convergence (by the Ratio Test), so $R = 10$. If $x = \pm 10$, $|a_n| = n^2 \to \infty$ as $n \to \infty$, so $\displaystyle\sum_{n=0}^{\infty} a_n$ diverges (Test for Divergence) and $I = (-10, 10)$.

13. If $a_n = \dfrac{x^n}{\ln n}$, then $\lim\limits_{n\to\infty}\left|\dfrac{a_{n+1}}{a_n}\right| = \lim\limits_{n\to\infty}\left|\dfrac{x^{n+1}}{\ln(n+1)}\cdot\dfrac{\ln n}{x^n}\right| = |x|\lim\limits_{n\to\infty}\dfrac{\ln n}{\ln(n+1)} = |x|$ (using l'Hospital's Rule), so $R = 1$. When $x = 1$, $\displaystyle\sum_{n=2}^{\infty}\frac{x^n}{\ln n} = \sum_{n=2}^{\infty}\frac{1}{\ln n}$ which diverges because $\dfrac{1}{\ln n} > \dfrac{1}{n}$ and $\displaystyle\sum_{n=2}^{\infty}\frac{1}{n}$ is the divergent harmonic series. When $x = -1$, $\displaystyle\sum_{n=2}^{\infty}\frac{x^n}{\ln n} = \sum_{n=2}^{\infty}\frac{(-1)^n}{\ln n}$ which converges by the Alternating Series Test. So $I = [-1, 1)$.

14. If $a_n = \sqrt{n}\,(3x+2)^n$, then $\left|\dfrac{a_{n+1}}{a_n}\right| = \left|\dfrac{\sqrt{n+1}\,(3x+2)^{n+1}}{\sqrt{n}(3x+2)^n}\right| = \left|\sqrt{1+\dfrac{1}{n}}\cdot(3x+2)\right| \to |3x+2|$ as $n \to \infty$ so for convergence, $|3x+2| < 1 \quad\Rightarrow\quad \left|x + \frac{2}{3}\right| < \frac{1}{3}$ so $R = \frac{1}{3}$ and $-1 < x < -\frac{1}{3}$. If $x = -1$, the series becomes $\displaystyle\sum_{n=0}^{\infty}(-1)^n\sqrt{n}$ which is divergent by the Test for Divergence. If $x = -\dfrac{1}{3}$, the series is $\displaystyle\sum_{n=0}^{\infty}\sqrt{n}$ which is also divergent by the Test for Divergence. So $I = \left(-1, -\frac{1}{3}\right)$.

15. If $a_n = \dfrac{n}{4^n}(2x-1)^n$, then $\left|\dfrac{a_{n+1}}{a_n}\right| = \left|\dfrac{(n+1)(2x-1)^{n+1}}{4^{n+1}} \cdot \dfrac{4^n}{n(2x-1)^n}\right| = \left|\dfrac{2x-1}{4}\left(1+\dfrac{1}{n}\right)\right| \to \frac{1}{2}\left|x-\frac{1}{2}\right|$ as $n \to \infty$. For convergence, $\frac{1}{2}\left|x-\frac{1}{2}\right| < 1 \;\Rightarrow\; \left|x-\frac{1}{2}\right| < 2 \;\Rightarrow\; R = 2$ and $-2 < x - \frac{1}{2} < 2 \;\Rightarrow$ $-\frac{3}{2} < x < \frac{5}{2}$. If $x = -\frac{3}{2}$, the series becomes $\displaystyle\sum_{n=0}^{\infty} \frac{n}{4^n}(-4)^n = \sum_{n=0}^{\infty}(-1)^n n$ which is divergent by the Test for Divergence. If $x = \frac{5}{2}$, the series is $\displaystyle\sum_{n=0}^{\infty} \frac{n}{4^n}4^n = \sum_{n=0}^{\infty} n$, also divergent by the Test for Divergence. So $I = \left(-\frac{3}{2}, \frac{5}{2}\right)$.

16. If $a_n = \dfrac{(-1)^n x^{2n-1}}{(2n-1)!}$ then $\displaystyle\lim_{n\to\infty}\left|\frac{a_{n+1}}{a_n}\right| = \lim_{n\to\infty}\frac{x^2}{(2n+1)2n} = 0 < 1$ for all x. By the Ratio Test the series converges for all x, so $R = \infty$ and $I = (-\infty, \infty)$.

17. If $a_n = \dfrac{(-1)^n(x-1)^n}{\sqrt{n}}$, then $\displaystyle\lim_{n\to\infty}\left|\frac{a_{n+1}}{a_n}\right| = \lim_{n\to\infty}\left|\frac{(x-1)^{n+1}}{\sqrt{n+1}} \cdot \frac{\sqrt{n}}{(x-1)^n}\right| = |x-1|\lim_{n\to\infty}\sqrt{\frac{n}{n+1}} = |x-1| < 1$ for convergence, or $0 < x < 2$, and $R = 1$. When $x = 0$, $\displaystyle\sum_{n=1}^{\infty}\frac{(-1)^n(x-1)^n}{\sqrt{n}} = \sum_{n=1}^{\infty}\frac{1}{\sqrt{n}}$ which is a divergent p-series ($p = \frac{1}{2} < 1$). When $x = 2$, the series is $\displaystyle\sum_{n=1}^{\infty}\frac{(-1)^n}{\sqrt{n}}$ which converges by the Alternating Series Test. So $I = (0, 2]$.

18. If $a_n = \dfrac{(x-4)^n}{n5^n}$ then $\displaystyle\lim_{n\to\infty}\left|\frac{a_{n+1}}{a_n}\right| = \frac{|x-4|}{5}\lim_{n\to\infty}\frac{n}{n+1} = \frac{|x-4|}{5} < 1$ for convergence, or $-1 < x < 9$ and $R = 5$. When $x = 9$, $\displaystyle\sum_{n=1}^{\infty} a_n = \sum_{n=1}^{\infty}\frac{1}{n}$ which diverges (harmonic series), and when $x = -1$, $\displaystyle\sum_{n=1}^{\infty} a_n = \sum_{n=1}^{\infty}\frac{(-1)^n}{n}$ which converges by the Alternating Series Test, so $I = [-1, 9)$.

19. If $a_n = \dfrac{(x-2)^n}{n^n}$, then $\displaystyle\lim_{n\to\infty}\sqrt[n]{|a_n|} = \lim_{n\to\infty}\frac{x-2}{n} = 0$, so the series converges for all x (by the Root Test). $R = \infty$ and $I = (-\infty, \infty)$.

20. If $a_n = \dfrac{(-3)^n(x-1)^n}{\sqrt{n+1}}$ then $\displaystyle\lim_{n\to\infty}\left|\frac{a_{n+1}}{a_n}\right| = 3|x-1|\lim_{n\to\infty}\left(\frac{n+1}{n+2}\right)^{1/2} = 3|x-1| < 1$ for convergence, or $\frac{2}{3} < x < \frac{4}{3}$ and $R = \frac{1}{3}$. When $x = \frac{4}{3}$, $\displaystyle\sum_{n=0}^{\infty} a_n = \sum_{n=0}^{\infty}\frac{(-1)^n}{\sqrt{n+1}}$ which is a convergent alternating series, and when $x = \frac{2}{3}$, $\displaystyle\sum_{n=0}^{\infty} a_n = \sum_{n=0}^{\infty}\frac{1}{\sqrt{n+1}}$ which is a divergent p-series ($p = \frac{1}{2} < 1$), so $I = \left(\frac{2}{3}, \frac{4}{3}\right]$.

21. If $a_n = \dfrac{2^n(x-3)^n}{n+3}$, then $\displaystyle\lim_{n\to\infty}\left|\frac{a_{n+1}}{a_n}\right| = \lim_{n\to\infty}\left|\frac{2^{n+1}(x-3)^{n+1}}{n+4} \cdot \frac{n+3}{2^n(x-3)^n}\right| = 2|x-3|\lim_{n\to\infty}\frac{n+3}{n+4}$ $= 2|x-3| < 1$ for convergence, or $|x-3| < \frac{1}{2} \;\Leftrightarrow\; \frac{5}{2} < x < \frac{7}{2}$, and $R = \frac{1}{2}$. When $x = \frac{5}{2}$, $\displaystyle\sum_{n=0}^{\infty}\frac{2^n(x-3)^n}{n+3} = \sum_{n=0}^{\infty}\frac{(-1)^n}{n+3}$ which converges by the Alternating Series Test. When $x = \dfrac{7}{2}$, $\displaystyle\sum_{n=0}^{\infty}\frac{2^n(x-3)^n}{n+3} = \sum_{n=0}^{\infty}\frac{1}{n+3} = \sum_{n=3}^{\infty}\frac{1}{n}$, the harmonic series, which diverges. So $I = \left[\frac{5}{2}, \frac{7}{2}\right)$.

22. If $a_n = \dfrac{(x+1)^n}{n(n+1)}$ then $\lim\limits_{n\to\infty}\left|\dfrac{a_{n+1}}{a_n}\right| = |x+1|\lim\limits_{n\to\infty}\dfrac{n}{n+2} = |x+1| < 1$ for convergence, or $-2 < x < 0$ and $R = 1$. If $x = -2$ or 0, then $|a_n| = \dfrac{1}{n^2+n} < \dfrac{1}{n^2}$ so $\sum\limits_{n=1}^{\infty}|a_n|$ converges since $\sum\limits_{n=1}^{\infty}\dfrac{1}{n^2}$ does ($p = 2 > 1$), and $I = [-2, 0]$.

23. If $a_n = \left(\dfrac{n}{2}\right)^n (x+6)^n$, then $\lim\limits_{n\to\infty}\sqrt[n]{|a_n|} = \lim\limits_{n\to\infty}\dfrac{n(x+6)}{2} = \infty$ unless $x = -6$, in which case the limit is 0. So by the Root Test, the series converges only for $x = -6$. $R = 0$ and $I = \{-6\}$.

24. If $a_n = \dfrac{n\,x^n}{1\cdot 3\cdot 5\cdot\cdots\cdot(2n-1)}$ then $\lim\limits_{n\to\infty}\left|\dfrac{a_{n+1}}{a_n}\right| = |x|\lim\limits_{n\to\infty}\dfrac{n+1}{n(2n+1)} = 0$ for all x. So the series converges for all x $\Rightarrow$ $R = \infty$ and $I = (-\infty, \infty)$.

25. If $a_n = \dfrac{(2x-1)^n}{n^3}$, then $\lim\limits_{n\to\infty}\left|\dfrac{a_{n+1}}{a_n}\right| = |2x-1|\lim\limits_{n\to\infty}\left(\dfrac{n}{n+1}\right)^3 = |2x-1| < 1$ for convergence, so $\left|x - \frac{1}{2}\right| < \frac{1}{2}$ $\Leftrightarrow$ $0 < x < 1$, and $R = \frac{1}{2}$. The series $\sum\limits_{n=1}^{\infty}\dfrac{(2x-1)^n}{n^3}$ converges both for $x = 0$ and $x = 1$ (in the first case because of the Alternating Series Test and in the second case because we get a p-series with $p = 3 > 1$). So $I = [0, 1]$.

26. If $a_n = \dfrac{(-1)^n(2x+3)^n}{n\ln n}$ then $\lim\limits_{n\to\infty}\left|\dfrac{a_{n+1}}{a_n}\right| = |2x+3|\lim\limits_{n\to\infty}\dfrac{n\ln n}{(n+1)\ln(n+1)} = |2x+3| < 1$ for convergence, so $-2 < x < -1$ and $R = \frac{1}{2}$. When $x = -2$, $\sum\limits_{n=2}^{\infty}a_n = \sum\limits_{n=2}^{\infty}\dfrac{1}{n\ln n}$ which diverges (Integral Test), and when $x = -1$, $\sum\limits_{n=2}^{\infty}a_n = \sum\limits_{n=2}^{\infty}\dfrac{(-1)^n}{n\ln n}$ which converges (Alternating Series Test), so $I = (-2, -1]$.

27. If $a_n = \dfrac{x^n}{(\ln n)^n}$ then $\lim\limits_{n\to\infty}\sqrt[n]{|a_n|} = \lim\limits_{n\to\infty}\dfrac{|x|}{\ln n} = 0 < 1$ for all x, so $R = \infty$ and $I = (-\infty, \infty)$ by the Root Test.

28. If $a_n = \dfrac{2\cdot 4\cdot 6\cdot\cdots\cdot(2n)x^n}{1\cdot 3\cdot 5\cdot\cdots\cdot(2n-1)}$ then $\lim\limits_{n\to\infty}\left|\dfrac{a_{n+1}}{a_n}\right| = \lim\limits_{n\to\infty}|x|\left(\dfrac{2n+2}{2n+1}\right) = |x| < 1$ for convergence, so $R = 1$. If $x = \pm 1$, $|a_n| = \dfrac{2\cdot 4\cdot 6\cdot\cdots\cdot(2n)}{1\cdot 3\cdot 5\cdot\cdots\cdot(2n-1)} > 1$ for all n since each integer in the numerator is larger than the corresponding one in the denominator, so $\sum a_n$ diverges in both cases by the Test for Divergence, and $I = (-1, 1)$.

29. If $a_n = \dfrac{(n!)^k}{(kn)!}x^n$, then $\lim\limits_{n\to\infty}\left|\dfrac{a_{n+1}}{a_n}\right| = \lim\limits_{n\to\infty}\dfrac{[(n+1)!]^k(kn)!}{(n!)^k[k(n+1)]!}|x|$

$$= \lim_{n\to\infty}\frac{(n+1)^k}{(kn+k)(kn+k-1)\cdots(kn+2)(kn+1)}|x| = \lim_{n\to\infty}\left[\frac{(n+1)}{(kn+1)}\frac{(n+1)}{(kn+2)}\cdots\frac{(n+1)}{(kn+k)}\right]|x|$$

$$= \lim_{n\to\infty}\left[\frac{n+1}{kn+1}\right]\lim_{n\to\infty}\left[\frac{n+1}{kn+2}\right]\cdots\lim_{n\to\infty}\left[\frac{n+1}{kn+k}\right]|x| = \left(\frac{1}{k}\right)^k|x| < 1 \quad\Leftrightarrow\quad |x| < k^k$$ for convergence, and the radius of convergence is $R = k^k$.

30.

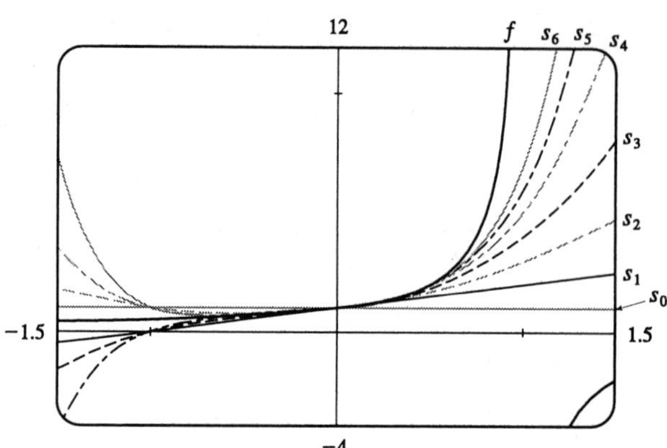

The partial sums definitely do not converge to $f(x)$ for $x \geq 1$, since f is undefined at $x = 1$ and negative on $(1, \infty)$, while all the partial sums are positive on this interval. The partial sums also fail to converge to f for $x \leq -1$, since $0 < f(x) < 1$ on this interval, while the partial sums are either larger than 1 or less than 0. The partial sums seem to converge to f on $(-1, 1)$. This graphical evidence is consistent with what we know about geometric series: convergence for $|x| < 1$, divergence for $|x| \geq 1$ (see Example 10.2.5).

31. **(a)** If $a_n = \dfrac{(-1)^n x^{2n+1}}{n!(n+1)!\,2^{2n+1}}$, then $\lim\limits_{n\to\infty}\left|\dfrac{a_{n+1}}{a_n}\right| = \left(\dfrac{x}{2}\right)^2 \lim\limits_{n\to\infty} \dfrac{1}{(n+1)(n+2)} = 0$ for all x. So $J_1(x)$ converges for all x; the domain is $(-\infty, \infty)$.

(b), (c) The initial terms of $J_1(x)$ up to $n = 5$ are

$$a_0 = \frac{x}{2},\ a_1 = -\frac{x^3}{16},\ a_2 = \frac{x^5}{384},\ a_3 = -\frac{x^7}{18{,}432},\ a_4 = \frac{x^9}{1{,}474{,}560},\ a_5 = -\frac{x^{11}}{176{,}947{,}200}.$$

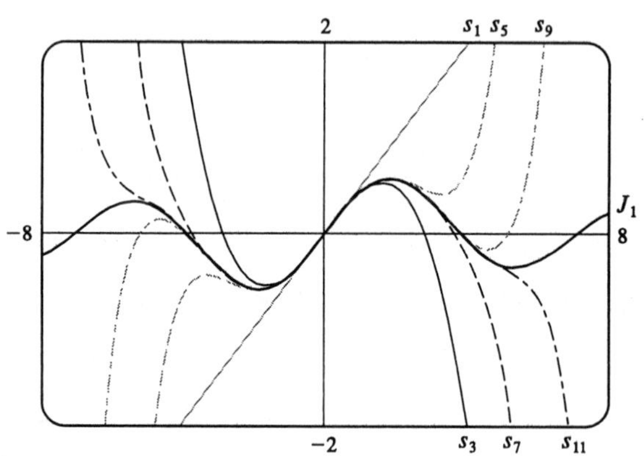

The partial sums seem to approximate $J_1(x)$ well near the origin, but as $|x|$ increases, we need to take a large number of terms to get a good approximation.

32. **(a)** $A(x) = 1 + \sum\limits_{n=1}^{\infty} a_n$ where $a_n = \dfrac{x^{3n}}{2 \cdot 3 \cdot 5 \cdot 6 \cdot \cdots \cdot (3n-1)(3n)}$, so

$$\lim_{n\to\infty}\left|\frac{a_{n+1}}{a_n}\right| = |x|^3 \lim_{n\to\infty} \frac{1}{(3n+2)(3n+3)} = 0 \text{ for all } x, \text{ so the domain is } \mathbb{R}.$$

(b), (c)

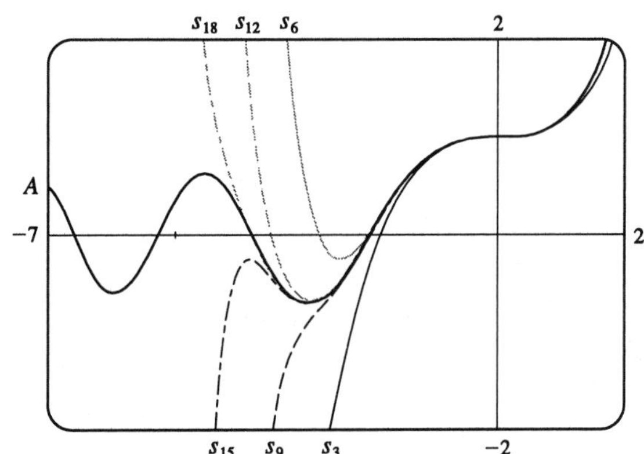

$s_0 = 1$ has been omitted from the graph. The partial sums seem to approximate $A(x)$ well near the origin, but as $|x|$ increases, we need to take a large number of terms to get a good approximation. Note that both Maple and Mathematica have two initially known Airy functions: `Ai(x)` and `Bi(x)` in Maple, `AiryAi(x)` and `AiryBi(x)` in Mathematica. The function we are concerned with here is given by $A(x) = \dfrac{\sqrt{3}\text{Ai}(x) + \text{Bi}(x)}{\sqrt{3}\text{Ai}(0) + \text{Bi}(0)}$.

33. $s_{2n-1} = 1 + 2x + x^2 + 2x^3 + \cdots + x^{2n-2} + 2x^{2n-1} = (1 + 2x)\left(1 + x^2 + x^4 + \cdots + x^{2n-2}\right)$

$= (1 + 2x)\dfrac{1 - x^{2n}}{1 - x^2} \to \dfrac{1 + 2x}{1 - x^2}$ as $n \to \infty$, when $|x| < 1$.

Also $s_{2n} = s_{2n-1} + x^{2n} \to \dfrac{1 + 2x}{1 - x^2}$ since $x^{2n} \to 0$ for $|x| < 1$. Therefore $s_n \to \dfrac{1 + 2x}{1 - x^2}$ by Exercise 10.1.70(a). Thus the interval of convergence is $(-1, 1)$ and $f(x) = \dfrac{1 + 2x}{1 - x^2}$.

34. $s_{4n-1} = c_0 + c_1x + c_2x^2 + c_3x^3 + c_0x^4 + c_1x^5 + c_2x^6 + c_3x^7 + \cdots + c_3x^{4n-1}$

$= \left(c_0 + c_1x + c_2x^2 + c_3x^3\right)\left(1 + x^4 + x^8 + \cdots + x^{4n-4}\right)$

$\to \dfrac{c_0 + c_1x + c_2x^2 + c_3x^3}{1 - x^4}$ as $n \to \infty$ for $|x| < 1$.

Also s_{4n}, s_{4n+1}, s_{4n+2} have the same limits $\left(\text{for example, } s_{4n} = s_{4n-1} + c_0x^{4n} \text{ and } x^{4n} \to 0 \text{ for } |x| < 1.\right)$ So the interval of convergence is $(-1, 1)$ and $f(x) = \dfrac{c_0 + c_1x + c_2x^2 + c_3x^3}{1 - x^4}$.

35. We use the Root Test on the series $\sum c_nx^n$. $\lim\limits_{n\to\infty} \sqrt[n]{|c_nx^n|} = |x| \lim\limits_{n\to\infty} \sqrt[n]{|c_n|} = c|x| < 1$ for convergence, or $|x| < 1/c$, so $R = 1/c$.

36. Since $\sum c_nx^n$ converges whenever $|x| < R$, $\sum c_nx^{2n} = \sum c_n(x^2)^n$ converges whenever $|x^2| < R \Leftrightarrow |x| < \sqrt{R}$, so the second series has radius of convergence $\sqrt{R}$.

37. $\sum(c_n + d_n)x^n = \sum c_nx^n + \sum d_nx^n$ on the interval $(-2, 2)$, since both series converge there. So the radius of convergence must be at least 2. Now since $\sum c_nx^n$ has $R = 2$, it must diverge either at $x = -2$ or at $x = 2$. So by Exercise 10.2.63, $\sum(c_n + d_n)x^n$ diverges either at $x = -2$ or at $x = 2$, and so its radius of convergence is 2.

EXERCISES 10.9

Note: "R" stands for "radius of convergence" and "I" stands for "interval of convergence" in this section.

1. $f(x) = \dfrac{1}{1+x} = \dfrac{1}{1-(-x)} = \sum_{n=0}^{\infty}(-1)^n x^n$ with $|-x| < 1 \Leftrightarrow |x| < 1$ so $R = 1$ and $I = (-1, 1)$.

2. $\dfrac{x}{1-x} = x\left(\dfrac{1}{1-x}\right) = x\sum_{n=0}^{\infty} x^n = \sum_{n=0}^{\infty} x^{n+1} = \sum_{n=1}^{\infty} x^n$ with $R = 1$ and $I = (-1, 1)$.

3. $f(x) = \dfrac{1}{1+4x^2} = \sum_{n=0}^{\infty}(-1)^n (4x^2)^n$ $\left(\begin{array}{c}\text{substituting } 4x^2 \text{ for } x \text{ in the}\\ \text{series from Exercise 1}\end{array}\right)$ $= \sum_{n=0}^{\infty}(-1)^n 4^n x^{2n}$, with $|4x^2| < 1$ so $x^2 < \frac{1}{4} \Leftrightarrow |x| < \frac{1}{2}$, and so $R = \frac{1}{2}$ and $I = \left(-\frac{1}{2}, \frac{1}{2}\right)$.

4. $\dfrac{1}{x^4+16} = \dfrac{1}{16}\left[\dfrac{1}{1+(x/2)^4}\right] = \dfrac{1}{16}\sum_{n=0}^{\infty}(-1)^n \left(\dfrac{x}{2}\right)^{4n} = \sum_{n=0}^{\infty} \dfrac{(-1)^n x^{4n}}{2^{4n+4}}$ for $\left|\dfrac{x}{2}\right| < 1 \quad \Leftrightarrow \quad |x| < 2$ so, $R = 2$ and $I = (-2, 2)$.

5. $f(x) = \dfrac{1}{4+x^2} = \dfrac{1}{4}\left(\dfrac{1}{1+x^2/4}\right) = \dfrac{1}{4}\sum_{n=0}^{\infty}(-1)^n \left(\dfrac{x^2}{4}\right)^n$ (using Exercise 1)

$= \sum_{n=0}^{\infty} \dfrac{(-1)^n x^{2n}}{4^{n+1}}$, with $\left|\dfrac{x^2}{4}\right| < 1 \Leftrightarrow x^2 < 4 \Leftrightarrow |x| < 2$, so $R = 2$ and $I = (-2, 2)$.

6. $\dfrac{1+x^2}{1-x^2} = 1 + \dfrac{2x^2}{1-x^2} = 1 + 2x^2 \sum_{n=0}^{\infty}(x^2)^n = 1 + \sum_{n=0}^{\infty} 2x^{2n+2} = 1 + \sum_{n=1}^{\infty} 2x^{2n}$, with $R = 1$ and $I = (-1, 1)$.

7. $\dfrac{x}{x-3} = 1 + \dfrac{3}{x-3} = 1 - \dfrac{1}{1-x/3} = 1 - \sum_{n=0}^{\infty}\left(\dfrac{x}{3}\right)^n = -\sum_{n=1}^{\infty}\left(\dfrac{x}{3}\right)^n$. For convergence, $\dfrac{|x|}{3} < 1 \quad \Leftrightarrow$ $|x| < 3$, so $R = 3$ and $I = (-3, 3)$.

Another Method: $\dfrac{x}{x-3} = -\dfrac{x}{3(1-x/3)} = -\dfrac{x}{3}\sum_{n=0}^{\infty}\left(\dfrac{x}{3}\right)^n = -\sum_{n=0}^{\infty}\dfrac{x^{n+1}}{3^{n+1}} = -\sum_{n=1}^{\infty}\dfrac{x^n}{3^n}$

8. $\dfrac{2}{3x+4} = \dfrac{1}{2}\left(\dfrac{1}{1+3x/4}\right) = \dfrac{1}{2}\sum_{n=0}^{\infty}(-1)^n\left(\dfrac{3x}{4}\right)^n = \sum_{n=0}^{\infty}\dfrac{(-1)^n 3^n x^n}{2^{2n+1}}$, $\left|\dfrac{3x}{4}\right| < 1$ so $R = \frac{4}{3}$ and $I = \left(-\frac{4}{3}, \frac{4}{3}\right)$.

9. $\dfrac{3x-2}{2x^2-3x+1} = \dfrac{3x-2}{(2x-1)(x-1)} = \dfrac{A}{2x-1} + \dfrac{B}{x-1} \quad \Leftrightarrow \quad A + 2B = 3$ and $-A - B = -2 \quad \Leftrightarrow$ $A = B = 1$, so $f(x) = \dfrac{3x-2}{2x^2-3x+1} = \dfrac{1}{2x-1} + \dfrac{1}{x-1} = -\sum_{n=0}^{\infty}(2x)^n - \sum_{n=0}^{\infty} x^n = -\sum_{n=0}^{\infty}(2^n+1)x^n$, with $R = \frac{1}{2}$. At $x = \pm\frac{1}{2}$, the series diverges by the Test for Divergence, so $I = \left(-\frac{1}{2}, \frac{1}{2}\right)$.

10. $\dfrac{x}{x^2-3x+2} = \dfrac{x}{(x-2)(x-1)} = \dfrac{A}{x-2} + \dfrac{B}{x-1} \quad \Leftrightarrow \quad A + B = 1$ and $-A - 2B = 0 \quad \Leftrightarrow \quad A = 2$, $B = -1$, so $f(x) = \dfrac{x}{(x-2)(x-1)} = \dfrac{2}{x-2} - \dfrac{1}{x-1} = -\dfrac{1}{1-x/2} + \dfrac{1}{1-x} = -\sum_{n=0}^{\infty}\left(\dfrac{x}{2}\right)^n + \sum_{n=0}^{\infty} x^n$ $= \sum_{n=0}^{\infty}(1-2^{-n})x^n$, with $R = 1$. At $x = \pm 1$, the series diverges by the Test for Divergence, so $I = (-1, 1)$.

11. $f(x) = \dfrac{1}{(1+x)^2} = -\dfrac{d}{dx}\left(\dfrac{1}{1+x}\right) = -\dfrac{d}{dx}\left(\sum_{n=0}^{\infty}(-1)^n x^n\right)$ (from Exercise 1)

$= \sum_{n=1}^{\infty}(-1)^{n+1}nx^{n-1} = \sum_{n=0}^{\infty}(-1)^n(n+1)x^n$ with $R = 1$.

12. $\dfrac{1}{1+x} = \sum_{n=0}^{\infty}(-1)^n x^n$ (geometric series with $R = 1$), so $f(x) = \ln(1+x) = \displaystyle\int \frac{dx}{1+x}$

$= \displaystyle\int\left[\sum_{n=0}^{\infty}(-1)^n x^n\right]dx = C + \sum_{n=0}^{\infty}(-1)^n\frac{x^{n+1}}{n+1} = \sum_{n=1}^{\infty}\frac{(-1)^{n-1}x^n}{n}$ [$C = 0$ since $f(0) = 0$], with $R = 1$.

13. $f(x) = \dfrac{1}{(1+x)^3} = -\dfrac{1}{2}\dfrac{d}{dx}\left[\dfrac{1}{(1+x)^2}\right] = -\dfrac{1}{2}\dfrac{d}{dx}\left(\sum_{n=0}^{\infty}(-1)^n(n+1)x^n\right)$ (from Exercise 11)

$= -\dfrac{1}{2}\sum_{n=1}^{\infty}(-1)^n(n+1)nx^{n-1} = \dfrac{1}{2}\sum_{n=0}^{\infty}(-1)^n(n+2)(n+1)x^n$ with $R = 1$.

14. $f(x) = x\ln(1+x) = x\left(\sum_{n=1}^{\infty}\dfrac{(-1)^{n-1}x^n}{n}\right)$ (by Exercise 12) $= \sum_{n=2}^{\infty}\dfrac{(-1)^n x^n}{n-1}$ with $R = 1$.

15. $f(x) = \ln(5-x) = -\displaystyle\int\frac{dx}{5-x} = -\frac{1}{5}\int\frac{dx}{1-x/5} = -\frac{1}{5}\int\left[\sum_{n=0}^{\infty}\left(\frac{x}{5}\right)^n\right]dx$

$= C - \dfrac{1}{5}\sum_{n=0}^{\infty}\dfrac{x^{n+1}}{5^n(n+1)} = C - \sum_{n=1}^{\infty}\dfrac{x^n}{n5^n}$

Putting $x = 0$, we get $C = \ln 5$. The series converges for $|x/5| < 1 \Leftrightarrow |x| < 5$. So $R = 5$.

16. $\tan^{-1}2x = 2\displaystyle\int\frac{dx}{1+4x^2} = 2\int\sum_{n=0}^{\infty}(-1)^n(4x^2)^n\,dx = 2\int\sum_{n=0}^{\infty}(-1)^n4^nx^{2n}dx$

$= C + 2\sum_{n=0}^{\infty}\dfrac{(-1)^n4^nx^{2n+1}}{2n+1} = \sum_{n=0}^{\infty}\dfrac{(-1)^n2^{2n+1}x^{2n+1}}{2n+1}$ for $|4x^2| < 1$ so $|x| < \frac{1}{2}$ and $R = \frac{1}{2}$.

17. $f(x) = \ln(1+x) - \ln(1-x) = \displaystyle\int\frac{dx}{1+x} + \int\frac{dx}{1-x} = \int\left[\sum_{n=0}^{\infty}(-1)^n x^n + \sum_{n=0}^{\infty}x^n\right]dx$

$= \displaystyle\int\sum_{n=0}^{\infty}2x^{2n}\,dx = \sum_{n=0}^{\infty}\frac{2x^{2n+1}}{2n+1} + C.$

But $f(0) = \ln 1 - \ln 1 = 0$, so $C = 0$ and we have $f(x) = \sum_{n=0}^{\infty}\dfrac{2x^{2n+1}}{2n+1}$ with $R = 1$.

18. We know that $\dfrac{1}{1-2x} = \sum_{n=0}^{\infty}(2x)^n$. Differentiating, we get $\dfrac{2}{(1-2x)^2} = \sum_{n=1}^{\infty}2^nnx^{n-1} = \sum_{n=0}^{\infty}2^{n+1}(n+1)x^n$, so

$f(x) = \dfrac{x^2}{(1-2x)^2} = \dfrac{x^2}{2}\cdot\dfrac{2}{(1-2x)^2} = \dfrac{x^2}{2}\sum_{n=0}^{\infty}2^{n+1}(n+1)x^n = \sum_{n=0}^{\infty}2^n(n+1)x^{n+2}$ or $\sum_{n=2}^{\infty}2^{n-2}(n-1)x^n$, with

$R = \frac{1}{2}$.

19. $f(x) = \ln(3+x) = \int \frac{dx}{3+x} = \frac{1}{3}\int \frac{dx}{1+x/3} = \frac{1}{3}\int \sum_{n=0}^{\infty} (-1)^n \left(\frac{x}{3}\right)^n dx$ (from Exercise 1)

$= C + \frac{1}{3}\sum_{n=0}^{\infty} \frac{(-1/3)^n}{n+1} x^{n+1} = \ln 3 + \frac{1}{3}\sum_{n=1}^{\infty} \frac{(-1/3)^{n-1}}{n} x^n = \ln 3 + \sum_{n=1}^{\infty} \frac{(-1)^{n-1}}{n3^n}$ $[C = f(0) = \ln 3]$

with $R = 3$. The terms of the series are $a_0 = \ln 3$, $a_1 = \frac{x}{3}$, $a_2 = -\frac{x^2}{18}$, $a_3 = \frac{x^3}{81}$, $a_4 = -\frac{x^4}{324}$, $a_5 = \frac{x^5}{1215}, \ldots$.

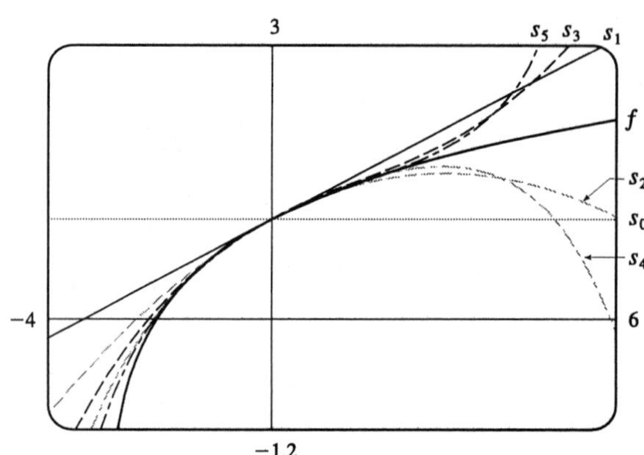

As n increases, $s_n(x)$ approximates f better on the interval of convergence, which is $(-3, 3)$.

20. $f(x) = \frac{1}{x^2+25} = \frac{1/25}{1+(x/5)^2} = \frac{1}{25}\sum_{n=0}^{\infty} (-1)^n \left(\frac{x}{5}\right)^{2n}$ (by Exercise 1) with $R = 5$. The terms of the series are

$a_0 = \frac{1}{25}$, $a_1 = -\frac{x^2}{625}$, $a_2 = \frac{x^4}{15{,}625}, \ldots$.

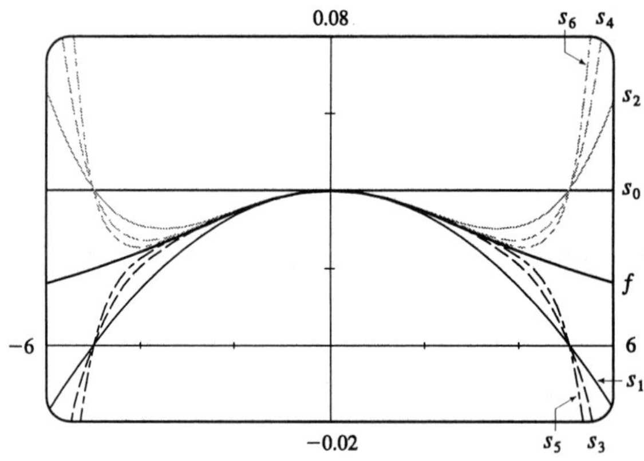

As n increases, the partial sums approximate f better on the interval of convergence, which is $(-5, 5)$.

21. $\int \frac{dx}{1+x^4} = \int \sum_{n=0}^{\infty} (-1)^n x^{4n}\, dx = C + \sum_{n=0}^{\infty} \frac{(-1)^n x^{4n+1}}{4n+1}$ with $R = 1$.

22. $\frac{1}{1+x^5} = \sum_{n=0}^{\infty} (-x^5)^n = \sum_{n=0}^{\infty} (-1)^n x^{5n} \quad\Rightarrow\quad \frac{x}{1+x^5} = \sum_{n=0}^{\infty} (-1)^n x^{5n+1} \quad\Rightarrow$

$\int \frac{x}{1+x^5}\, dx = C + \sum_{n=0}^{\infty} \frac{(-1)^n x^{5n+2}}{5n+2}$ with $R = 1$.

23. By Example 7, $\arctan x = \sum_{n=0}^{\infty}(-1)^n \frac{x^{2n+1}}{2n+1}$, so

$$\int \frac{\arctan x}{x}\,dx = \int \sum_{n=0}^{\infty}(-1)^n \frac{x^{2n}}{2n+1}\,dx = C + \sum_{n=0}^{\infty}(-1)^n \frac{x^{2n+1}}{(2n+1)^2} \text{ with } R = 1.$$

24. $\int \tan^{-1}(x^2)\,dx = \int \sum_{n=0}^{\infty}(-1)^n \frac{(x^2)^{2n+1}}{2n+1}\,dx$ (by Example 7) $= C + \sum_{n=0}^{\infty}(-1)^n \frac{x^{4n+3}}{(2n+1)(4n+3)}$ with $R = 1$.

25. We use the representation $\int \frac{dx}{1+x^4} = C + \sum_{n=0}^{\infty} \frac{(-1)^n x^{4n+1}}{4n+1}$ from Exercise 21, with $C = 0$. So

$\int_0^{0.2} \frac{dx}{1+x^4} = \left[x - \frac{x^5}{5} + \frac{x^9}{9} - \frac{x^{13}}{13} + \cdots\right]_0^{0.2} = 0.2 - \frac{0.2^5}{5} + \frac{0.2^9}{9} - \frac{0.2^{13}}{13} + \cdots$. Since the series is alternating, the error in the nth-order approximation is less than the first neglected term by Theorem 10.5.1. If we use only the first two terms of the series, then the error is at most $0.2^9/9 \approx 5.7 \times 10^{-8}$. So, to six decimal places, $\int_0^{0.2} \frac{dx}{1+x^4} \approx 0.2 - \frac{0.2^5}{5} \approx 0.199936$.

26. We use the representation $\int \tan^{-1}(x^2)\,dx = C + \sum_{n=0}^{\infty}(-1)^n \frac{x^{4n+3}}{(2n+1)(4n+3)}$ from Exercise 24, with $C = 0$:

$$\int_0^{1/2} \tan^{-1}(x^2)\,dx = \left[\frac{x^3}{3} - \frac{x^7}{21} + \frac{x^{11}}{55} - \frac{x^{15}}{105} + \frac{x^{19}}{171} - \cdots\right]_0^{1/2} = \frac{0.5^3}{3} - \frac{0.5^7}{21} + \frac{0.5^{11}}{55} - \frac{0.5^{15}}{105} + \frac{0.5^{19}}{171} - \cdots.$$

If we use only the first four terms of the series, then the error is at most $0.5^{19}/171 \approx 1.1 \times 10^{-8}$. So, to six decimal places, $\int_0^{1/2} \tan^{-1}(x^2)\,dx \approx \frac{1}{3}(0.5)^3 - \frac{1}{21}(0.5)^7 + \frac{1}{55}(0.5)^{11} - \frac{1}{105}(0.5)^{15} \approx 0.041303$.

27. We substitute x^4 for x in Example 7, and find that $\int x^2 \tan^{-1}(x^4)\,dx = \int x^2 \sum_{n=0}^{\infty}(-1)^n \frac{(x^4)^{2n+1}}{2n+1}\,dx$

$= \int \sum_{n=0}^{\infty}(-1)^n \frac{x^{8n+6}}{2n+1}\,dx = C + \sum_{n=0}^{\infty}(-1)^n \frac{x^{8n+7}}{(2n+1)(8n+7)}$. So

$\int_0^{1/3} x^2 \tan^{-1}(x^4)\,dx = \left[\frac{x^7}{7} - \frac{x^{15}}{45} + \cdots\right]_0^{1/3} = \frac{1}{7 \cdot 3^7} - \frac{1}{45 \cdot 3^{15}} + \cdots$. The series is alternating, so if we use only one term, the error is at most $1/(45 \cdot 3^{15}) \approx 1.5 \times 10^{-9}$. So $\int_0^{1/3} x^2 \tan^{-1}(x^4)\,dx \approx 1/(7 \cdot 3^7) \approx 0.000065$ to six decimal places.

28. $\int_0^{0.5} \frac{dx}{1+x^6} = \int_0^{0.5} \sum_{n=0}^{\infty}(-1)^n x^{6n}\,dx = \sum_{n=0}^{\infty}\left[\frac{(-1)^n x^{6n+1}}{6n+1}\right]_0^{1/2} = \sum_{n=0}^{\infty} \frac{(-1)^n}{(6n+1)2^{6n+1}}$

$= \frac{1}{2} - \frac{1}{7 \cdot 2^7} + \frac{1}{13 \cdot 2^{13}} - \frac{1}{19 \cdot 2^{19}} + \cdots$. If we use only three terms, the error is at most $\frac{1}{19 \cdot 2^{19}} \approx 1.0 \times 10^{-7}$. So, to six decimal places, $\int_0^{0.5} \frac{dx}{1+x^6} \approx \frac{1}{2} - \frac{1}{7 \cdot 2^7} + \frac{1}{13 \cdot 2^{13}} \approx 0.498893$.

29. Using the result of Example 6 with $x = -0.1$, we have

$\ln 1.1 = \ln[1 - (-0.1)] = 0.1 - \frac{0.01}{2} + \frac{0.001}{3} - \frac{0.0001}{4} + \frac{0.00001}{5} - \cdots$. If we use only the first four terms, the error is at most $\frac{0.00001}{5} = 0.000002$. So $\ln 1.1 \approx 0.1 - \frac{0.01}{2} + \frac{0.001}{3} - \frac{0.0001}{4} \approx 0.09531$.

30. We calculate $f'(x)=\sum_{n=1}^{\infty}\frac{(-1)^n 2nx^{2n-1}}{(2n)!}$ (the first term disappears), so

$$f''(x)=\sum_{n=1}^{\infty}\frac{(-1)^n(2n)(2n-1)x^{2n-2}}{(2n)!}=\sum_{n=1}^{\infty}\frac{(-1)^n x^{2(n-1)}}{[2(n-1)]!}=\sum_{n=0}^{\infty}\frac{(-1)^{n+1}x^{2n}}{(2n)!}\quad\text{(substituting } n+1 \text{ for } n\text{)}$$

$$=-\sum_{n=0}^{\infty}\frac{(-1)^n x^{2n}}{(2n)!}=-f(x)\quad\Rightarrow\quad f''(x)+f(x)=0.$$

31. **(a)** $J_0(x)=\sum_{n=0}^{\infty}\frac{(-1)^n x^{2n}}{2^{2n}(n!)^2}$, $J_0'(x)=\sum_{n=1}^{\infty}\frac{(-1)^n 2nx^{2n-1}}{2^{2n}(n!)^2}$, and $J_0''(x)=\sum_{n=1}^{\infty}\frac{(-1)^n 2n(2n-1)x^{2n-2}}{2^{2n}(n!)^2}$, so

$$x^2J_0''(x)+xJ_0'(x)+x^2J_0(x)=\sum_{n=1}^{\infty}\frac{(-1)^n 2n(2n-1)x^{2n}}{2^{2n}(n!)^2}+\sum_{n=1}^{\infty}\frac{(-1)^n 2nx^{2n}}{2^{2n}(n!)^2}+\sum_{n=0}^{\infty}\frac{(-1)^n x^{2n+2}}{2^{2n}(n!)^2}$$

$$=\sum_{n=1}^{\infty}\frac{(-1)^n 2n(2n-1)x^{2n}}{2^{2n}(n!)^2}+\sum_{n=1}^{\infty}\frac{(-1)^n 2nx^{2n}}{2^{2n}(n!)^2}+\sum_{n=1}^{\infty}\frac{(-1)^{n-1}x^{2n}}{2^{2n-2}[(n-1)!]^2}$$

$$=\sum_{n=1}^{\infty}(-1)^n\left[\frac{2n(2n-1)+2n-2^2n^2}{2^{2n}(n!)^2}\right]x^{2n}=\sum_{n=1}^{\infty}(-1)^n\left[\frac{4n^2-2n+2n-4n^2}{2^{2n}(n!)^2}\right]x^{2n}=0.$$

(b) $$\int_0^1 J_0(x)dx=\int_0^1\left[\sum_{n=0}^{\infty}\frac{(-1)^n x^{2n}}{2^{2n}(n!)^2}\right]dx=\int_0^1 dx-\int_0^1\frac{x^2}{4}\,dx+\int_0^1\frac{x^4}{64}\,dx-\int_0^1\frac{x^6}{2304}\,dx+\cdots$$

$$=\left[x-\frac{x^3}{3\cdot 4}+\frac{x^5}{5\cdot 64}-\frac{x^7}{7\cdot 2304}+\cdots\right]_0^1=1-\frac{1}{12}+\frac{1}{320}-\frac{1}{16{,}128}+\cdots.$$

Since $\frac{1}{16{,}128}\approx 0.000062$, it follows from Theorem 10.5.1 that, correct to three decimal places, $\int_0^1 J_0(x)dx\approx 1-\frac{1}{12}+\frac{1}{320}\approx 0.920$.

32. **(a)** $J_1(x)=\sum_{n=0}^{\infty}\frac{(-1)^n x^{2n+1}}{n!\,(n+1)!\,2^{2n+1}}$, $J_1'(x)=\sum_{n=0}^{\infty}\frac{(-1)^n(2n+1)x^{2n}}{n!\,(n+1)!\,2^{2n+1}}$, and

$$J_1''(x)=\sum_{n=1}^{\infty}\frac{(-1)^n(2n+1)(2n)x^{2n-1}}{n!\,(n+1)!\,2^{2n+1}}.$$

$x^2J_1''(x)+xJ_1'(x)+(x^2-1)J_1(x)$

$$=\sum_{n=1}^{\infty}\frac{(-1)^n(2n+1)(2n)x^{2n+1}}{n!\,(n+1)!\,2^{2n+1}}+\sum_{n=0}^{\infty}\frac{(-1)^n(2n+1)x^{2n+1}}{n!\,(n+1)!\,2^{2n+1}}+\sum_{n=0}^{\infty}\frac{(-1)^n x^{2n+3}}{n!\,(n+1)!\,2^{2n+1}}-\sum_{n=0}^{\infty}\frac{(-1)^n x^{2n+1}}{n!\,(n+1)!\,2^{2n+1}}$$

$$=\sum_{n=1}^{\infty}\frac{(-1)^n(2n+1)(2n)x^{2n+1}}{n!\,(n+1)!\,2^{2n+1}}+\sum_{n=0}^{\infty}\frac{(-1)^n(2n+1)x^{2n+1}}{n!\,(n+1)!\,2^{2n+1}}-\sum_{n=1}^{\infty}\frac{(-1)^n x^{2n+1}}{n!\,(n-1)!\,2^{2n-1}}-\sum_{n=0}^{\infty}\frac{(-1)^n x^{2n+1}}{n!\,(n+1)!\,2^{2n+1}}$$

$$=\frac{x}{2}-\frac{x}{2}+\sum_{n=1}^{\infty}(-1)^n\left[\frac{(2n+1)(2n)+(2n+1)-(n)(n+1)2^2-1}{n!\,(n+1)!\,2^{2n+1}}\right]x^{2n+1}=0.$$

(b) $$J_0'(x)=\sum_{n=1}^{\infty}\frac{(-1)^n(2n)x^{2n-1}}{2^{2n}(n!)^2}=\sum_{n=0}^{\infty}\frac{(-1)^{n+1}2(n+1)x^{2n+1}}{2^{2n+2}[(n+1)!]^2}=-\sum_{n=0}^{\infty}\frac{(-1)^n x^{2n+1}}{2^{2n+1}(n+1)!\,n!}=-J_1(x)$$

33. **(a)** We calculate $f'(x) = \sum_{n=1}^{\infty} \frac{nx^{n-1}}{n!} = \sum_{n=1}^{\infty} \frac{x^{n-1}}{(n-1)!} = \sum_{n=0}^{\infty} \frac{x^n}{n!} = f(x)$.

(b) By Theorem 6.5.2, the only solutions to the differential equation $\frac{df(x)}{dx} = f(x)$ are $f(x) = Ke^x$, but $f(0) = 1$, so $K = 1$ and $f(x) = e^x$.

Or: We could solve the equation $\frac{df(x)}{dx} = f(x)$ as a separable differential equation.

34. $\frac{|\sin nx|}{n^2} \le \frac{1}{n^2}$ so $\sum_{n=1}^{\infty} \frac{\sin nx}{n^2}$ converges by the Comparison Test. $\frac{d}{dx}\left(\frac{\sin nx}{n^2}\right) = \frac{\cos nx}{n}$, so when $x = 2k\pi$ (k an integer), $\sum_{n=1}^{\infty} f_n'(x) = \sum_{n=1}^{\infty} \frac{\cos(2kn\pi)}{n} = \sum_{n=1}^{\infty} \frac{1}{n}$ which diverges (harmonic series). $f_n''(x) = -\sin nx$, so $\sum_{n=1}^{\infty} f_n''(x) = -\sum_{n=1}^{\infty} \sin nx$ which converges only if $\sin nx = 0$, or $x = k\pi$ (k an integer).

35. If $a_n = \frac{x^n}{n^2}$, then $\lim_{n\to\infty}\left|\frac{a_{n+1}}{a_n}\right| = |x| \lim_{n\to\infty}\left(\frac{n}{n+1}\right)^2 = |x| < 1$ for convergence, so $R = 1$. When $x = \pm 1$, $\sum_{n=1}^{\infty}\left|\frac{x^n}{n^2}\right| = \sum_{n=1}^{\infty} \frac{1}{n^2}$ which is a convergent p-series ($p = 2 > 1$), so the interval of convergence for f is $[-1, 1]$.

By Theorem 10.9.2, the radii of convergence of f' and f'' are both 1, so we need only check the endpoints. $f'(x) = \sum_{n=1}^{\infty} \frac{nx^{n-1}}{n^2} = \sum_{n=0}^{\infty} \frac{x^n}{n+1}$, and this series diverges for $x = 1$ (harmonic series) and converges for $x = -1$ (Alternating Series Test), so the interval of convergence is $[-1, 1)$. $f''(x) = \sum_{n=1}^{\infty} \frac{nx^{n-1}}{n+1}$ diverges at both 1 and -1 (Test for Divergence) since $\lim_{n\to\infty} \frac{n}{n+1} = 1 \ne 0$, so its interval of convergence is $(-1, 1)$.

36. **(a)** $\sum_{n=1}^{\infty} nx^{n-1} = \sum_{n=0}^{\infty} \frac{d}{dx} x^n = \frac{d}{dx}\left[\sum_{n=0}^{\infty} x^n\right] = \frac{d}{dx}\left[\frac{1}{1-x}\right] = -\frac{1}{(1-x)^2}(-1) = \frac{1}{(1-x)^2}$, $|x| < 1$.

(b) **(i)** $\sum_{n=1}^{\infty} nx^n = x\sum_{n=1}^{\infty} nx^{n-1} = x\left[\frac{1}{(1-x)^2}\right]$ [from (a)] $= \frac{x}{(1-x)^2}$ for $|x| < 1$.

(ii) Put $x = \frac{1}{2}$ in (i): $\sum_{n=1}^{\infty} \frac{n}{2^n} = \sum_{n=1}^{\infty} n\left(\frac{1}{2}\right)^n = \frac{1/2}{(1-1/2)^2} = 2$.

(c) **(i)** $\sum_{n=2}^{\infty} n(n-1)x^n = x^2\sum_{n=2}^{\infty} n(n-1)x^{n-2} = x^2\frac{d}{dx}\left[\sum_{n=0}^{\infty} nx^{n-1}\right] = x^2\frac{d}{dx}\frac{1}{(1-x)^2}$

$= x^2\frac{2}{(1-x)^3} = \frac{2x^2}{(1-x)^3}$ for $|x| < 1$.

(ii) Put $x = \frac{1}{2}$ in (i): $\sum_{n=2}^{\infty} \frac{n^2-n}{2^n} = \sum_{n=2}^{\infty} n(n-1)\left(\frac{1}{2}\right)^n = \frac{2\left(\frac{1}{2}\right)^2}{\left(1-\frac{1}{2}\right)^3} = 4$

(iii) From (b)(ii) and (c)(ii) we have $\sum_{n=1}^{\infty} \frac{n^2}{2^n} = \sum_{n=1}^{\infty} \frac{n^2-n}{2^n} + \sum_{n=1}^{\infty} \frac{n}{2^n} = 4 + 2 = 6$.

EXERCISES 10.10

1.

n	$f^{(n)}(x)$	$f^{(n)}(0)$
0	$\cos x$	1
1	$-\sin x$	0
2	$-\cos x$	-1
3	$\sin x$	0
4	$\cos x$	1
...	...	...

$$\cos x = f(0) + f'(0)x + \frac{f''(0)}{2!}x^2 + \frac{f^{(3)}(0)}{3!}x^3 + \frac{f^{(4)}(0)}{4!}x^4 + \cdots$$
$$= 1 - \frac{x^2}{2!} + \frac{x^4}{4!} - \cdots = \sum_{n=0}^{\infty} \frac{(-1)^n x^{2n}}{(2n)!}$$

If $a_n = \dfrac{(-1)^n x^{2n}}{(2n)!}$, then

$$\lim_{n\to\infty}\left|\frac{a_{n+1}}{a_n}\right| = x^2 \lim_{n\to\infty} \frac{1}{(2n+2)(2n+1)} = 0 < 1 \text{ for all } x.$$

So $R = \infty$.

2.

n	$f^{(n)}(x)$	$f^{(n)}(0)$
0	$\sin 2x$	0
1	$2\cos 2x$	2
2	$-2^2 \sin 2x$	0
3	$-2^3 \cos 2x$	-2^3
4	$2^4 \sin 2x$	0
...	...	...

$f^{(n)}(0) = 0$ if n is even and $f^{(2n+1)}(0) = (-1)^n 2^{2n+1}$,

$$\text{so } \sin 2x = \sum_{n=0}^{\infty} \frac{f^{(n)}(0)}{n!} x^n = \sum_{n=0}^{\infty} \frac{f^{(2n+1)}(0)}{(2n+1)!} x^{2n+1}$$
$$= \sum_{n=0}^{\infty} \frac{(-1)^n 2^{2n+1} x^{2n+1}}{(2n+1)!}.$$

$$\lim_{n\to\infty}\left|\frac{a_{n+1}}{a_n}\right| = \lim_{n\to\infty} \frac{2^2|x|^2}{(2n+3)(2n+2)} = 0 < 1$$

for all x, so $R = \infty$ (Ratio Test).

3.

n	$f^{(n)}(x)$	$f^{(n)}(0)$
0	$(1+x)^{-2}$	1
1	$-2(1+x)^{-3}$	-2
2	$2 \cdot 3(1+x)^{-4}$	$2 \cdot 3$
3	$-2 \cdot 3 \cdot 4(1+x)^{-5}$	$-2 \cdot 3 \cdot 4$
4	$2 \cdot 3 \cdot 4 \cdot 5(1+x)^{-6}$	$2 \cdot 3 \cdot 4 \cdot 5$
...	...	...

So $f^{(n)}(0) = (-1)^n (n+1)!$ and

$$\frac{1}{(1+x)^2} = \sum_{n=0}^{\infty} \frac{(-1)^n (n+1)!}{n!} x^n$$
$$= \sum_{n=0}^{\infty} (-1)^n (n+1) x^n.$$

If $a_n = (-1)^n (n+1) x^n$, then

$$\lim_{n\to\infty}\left|\frac{a_{n+1}}{a_n}\right| = |x|, \text{ so } R = 1.$$

4.

n	$f^{(n)}(x)$	$f^{(n)}(0)$
0	$x/(1-x)$	0
1	$(1-x)^{-2}$	1
2	$2(1-x)^{-3}$	2
3	$3 \cdot 2(1-x)^{-4}$	$3 \cdot 2$
4	$4 \cdot 3 \cdot 2(1-x)^{-5}$	$4 \cdot 3 \cdot 2$
...	...	...

$f^{(n)}(0) = n!$ except when $n = 0$, so

$$\frac{x}{1-x} = \sum_{n=1}^{\infty} \frac{n!}{n!} x^n = \sum_{n=1}^{\infty} x^n.$$

$$\lim_{n\to\infty}\left|\frac{a_{n+1}}{a_n}\right| = |x| < 1 \text{ for}$$

convergence, so $R = 1$.

5.

n	$f^{(n)}(x)$	$f^{(n)}(0)$
0	$\sinh x$	0
1	$\cosh x$	1
2	$\sinh x$	0
3	$\cosh x$	1
4	$\sinh x$	0
...	...	...

So $f^{(n)}(0) = \begin{cases} 0 & \text{if } n \text{ is even} \\ 1 & \text{if } n \text{ is odd} \end{cases}$

and $\sinh x = \sum_{n=0}^{\infty} \frac{x^{2n+1}}{(2n+1)!}$. If $a_n = \frac{x^{2n+1}}{(2n+1)!}$ then

$$\lim_{n\to\infty}\left|\frac{a_{n+1}}{a_n}\right| = x^2 \lim_{n\to\infty} \frac{1}{(2n+3)(2n+2)} = 0 < 1$$

for all x, so $R = \infty$.

6.

n	$f^{(n)}(x)$	$f^{(n)}(0)$
0	$\cosh x$	1
1	$\sinh x$	0
2	$\cosh x$	1
3	$\sinh x$	0
...	...	...

$f^{(n)}(0) = \begin{cases} 1 & \text{if } n \text{ is even} \\ 0 & \text{if } n \text{ is odd} \end{cases}$

so $\cosh x = \sum_{n=0}^{\infty} \frac{x^{2n}}{(2n)!}$ with $R = \infty$,

by the Ratio Test.

7.

n	$f^{(n)}(x)$	$f^{(n)}(\pi/4)$
0	$\sin x$	$\sqrt{2}/2$
1	$\cos x$	$\sqrt{2}/2$
2	$-\sin x$	$-\sqrt{2}/2$
3	$-\cos x$	$-\sqrt{2}/2$
4	$\sin x$	$\sqrt{2}/2$
...	...	...

$$\begin{aligned} \sin x &= f\left(\tfrac{\pi}{4}\right) + f'\left(\tfrac{\pi}{4}\right)\left(x - \tfrac{\pi}{4}\right) + \frac{f''\left(\frac{\pi}{4}\right)}{2!}\left(x - \tfrac{\pi}{4}\right)^2 \\ &\quad + \frac{f^{(3)}\left(\frac{\pi}{4}\right)}{3!}\left(x - \tfrac{\pi}{4}\right)^3 + \frac{f^{(4)}\left(\frac{\pi}{4}\right)}{4!}\left(x - \tfrac{\pi}{4}\right)^4 + \cdots \\ &= \tfrac{\sqrt{2}}{2}\left[1 + \left(x - \tfrac{\pi}{4}\right) - \tfrac{1}{2!}\left(x - \tfrac{\pi}{4}\right)^2 - \tfrac{1}{3!}\left(x - \tfrac{\pi}{4}\right)^3 + \tfrac{1}{4!}\left(x - \tfrac{\pi}{4}\right)^4 + \cdots\right] \\ &= \frac{\sqrt{2}}{2}\sum_{n=0}^{\infty} \frac{(-1)^{n(n-1)/2}\left(x - \frac{\pi}{4}\right)^n}{n!} \end{aligned}$$

If $a_n = \frac{(-1)^{n(n-1)/2}\left(x - \frac{\pi}{4}\right)^n}{n!}$, then $\lim_{n\to\infty}\left|\frac{a_{n+1}}{a_n}\right| = \lim_{n\to\infty} \frac{\left|x - \frac{\pi}{4}\right|}{n+1} = 0 < 1$

for all x, so $R = \infty$.

8.

n	$f^{(n)}(x)$	$f^{(n)}\left(-\frac{\pi}{4}\right)$
0	$\cos x$	$\frac{\sqrt{2}}{2}$
1	$-\sin x$	$\frac{\sqrt{2}}{2}$
2	$-\cos x$	$-\frac{\sqrt{2}}{2}$
3	$\sin x$	$-\frac{\sqrt{2}}{2}$
4	$\cos x$	$\frac{\sqrt{2}}{2}$
...	...	...

$f^{(n)}\left(-\frac{\pi}{4}\right) = (-1)^{n(n-1)/2}\frac{\sqrt{2}}{2}$, so

$$\begin{aligned} \cos x &= \sum_{n=0}^{\infty} \frac{f^{(n)}\left(-\frac{\pi}{4}\right)}{n!}\left(x + \tfrac{\pi}{4}\right)^n \\ &= \sum_{n=0}^{\infty} \frac{(-1)^{n(n-1)/2}\sqrt{2}}{2 \cdot n!}\left(x + \tfrac{\pi}{4}\right)^n, \end{aligned}$$

with $R = \infty$ by the Ratio Test.

9.

n	$f^{(n)}(x)$	$f^{(n)}(1)$
0	x^{-1}	1
1	$-x^{-2}$	-1
2	$2x^{-3}$	2
3	$-3\cdot 2x^{-4}$	$-3\cdot 2$
4	$4\cdot 3\cdot 2x^{-5}$	$4\cdot 3\cdot 2$
$\cdots$	$\cdots$	$\cdots$

So $f^{(n)}(1)=(-1)^n n!$, and

$$\frac{1}{x}=\sum_{n=0}^{\infty}\frac{(-1)^n n!}{n!}(x-1)^n=\sum_{n=0}^{\infty}(-1)^n(x-1)^n.$$

If $a_n=(-1)^n(x-1)^n$ then

$$\lim_{n\to\infty}\left|\frac{a_{n+1}}{a_n}\right|=|x-1|<1 \text{ for convergence,}$$

so $0<x<2$ and $R=1$.

10.

n	$f^{(n)}(x)$	$f^{(n)}(4)$
0	$x^{1/2}$	2
1	$\frac{1}{2}x^{-1/2}$	2^{-2}
2	$-\frac{1}{4}x^{-3/2}$	-2^{-5}
3	$\frac{3}{8}x^{-5/2}$	$3\cdot 2^{-8}$
4	$-\frac{15}{16}x^{-7/2}$	$-15\cdot 2^{-11}$
$\cdots$	$\cdots$	$\cdots$

$$f^{(n)}(4)=\frac{(-1)^{n-1}1\cdot 3\cdot 5\cdots(2n-3)}{2^{3n-1}} \text{ for } n\geq 2, \text{ so}$$

$$\sqrt{x}=2+\frac{x-4}{4}+\sum_{n=2}^{\infty}\frac{(-1)^{n-1}1\cdot 3\cdot 5\cdot\cdots\cdot(2n-3)}{2^{3n-1}n!}(x-4)^n.$$

$$\lim_{n\to\infty}\left|\frac{a_{n+1}}{a_n}\right|=\frac{|x-4|}{8}\lim_{n\to\infty}\left(\frac{2n-1}{n+1}\right)=\frac{|x-4|}{4}<1$$

for convergence, so $|x-4|<4 \quad\Rightarrow\quad R=4$.

11. Clearly $f^{(n)}(x)=e^x$, so $f^{(n)}(3)=e^3$ and $e^x=\sum_{n=0}^{\infty}\frac{e^3}{n!}(x-3)^n$. If $a_n=\frac{e^3}{n!}(x-3)^n$ then

$$\lim_{n\to\infty}\left|\frac{a_{n+1}}{a_n}\right|=\lim_{n\to\infty}\frac{|x-3|}{n+1}=0 \text{ for all } x, \text{ so } R=\infty.$$

12.

n	$f^{(n)}(x)$	$f^{(n)}(2)$
0	$\ln x$	$\ln 2$
1	x^{-1}	$\frac{1}{2}$
2	$-x^{-2}$	$-\frac{1}{4}$
3	$2x^{-3}$	$\frac{2}{8}$
4	$-3\cdot 2x^{-4}$	$-\frac{3\cdot 2}{16}$
$\cdots$	$\cdots$	$\cdots$

$$f^{(n)}(2)=\frac{(-1)^{n-1}(n-1)!}{2^n} \text{ for } n\geq 1, \text{ so}$$

$$\ln x=\ln 2+\sum_{n=1}^{\infty}\frac{(-1)^{n-1}(x-2)^n}{n\cdot 2^n}.$$

$$\lim_{n\to\infty}\left|\frac{a_{n+1}}{a_n}\right|=\frac{|x-2|}{2}\lim_{n\to\infty}\frac{n}{n+1}=\frac{|x-2|}{2}<1 \text{ for}$$

convergence $\Leftrightarrow$ $|x-2|<2$, so $R=2$.

13. If $f(x)=\cos x$, then by Formula 9, $R_n(x)=\frac{f^{(n+1)}(z)}{(n+1)!}x^{n+1}$, where $0<|z|<|x|$. But $f^{(n+1)}(z)=\pm\sin z$ or $\pm\cos z$. In each case, $\left|f^{(n+1)}(z)\right|\leq 1$, so $|R_n(x)|\leq\frac{1}{(n+1)!}x^{n+1}\to 0$ as $n\to\infty$ by Equation 11. So $\lim_{n\to\infty}R_n(x)=0$ and, by Theorem 8, the series in Exercise 1 represents $\cos x$ for all x.

14. If $f(x)=\sin x$, then $R_n(x)=\frac{f^{(n+1)}(z)}{(n+1)!}\left(x-\frac{\pi}{4}\right)^n$, where $\left|z-\frac{\pi}{4}\right|<\left|x-\frac{\pi}{4}\right|$. But $f^{(n+1)}(z)=\pm\sin z$ or $\pm\cos z$. In each case, $\left|f^{(n+1)}(z)\right|\leq 1$, so $|R_n(x)|\leq\frac{1}{(n+1)!}\left(x-\frac{\pi}{4}\right)^n\to 0$ as $n\to\infty$ by Equation 11. So by Theorem 9, the series in Exercise 7 represents $\sin x$ for all x.

15. If $f(x) = \sinh x$, then $R_n(x) = \dfrac{f^{(n+1)}(z)}{(n+1)!}x^{n+1}$, where $0 < |z| < |x|$. But for all n, $\left|f^{(n+1)}(z)\right| \le \cosh z \le \cosh x$ (since all derivatives are either sinh or cosh, $|\sinh z| < |\cosh z|$ for all z, and $|z| < |x| \Rightarrow \cosh z < \cosh x$), so $|R_n(z)| \le \dfrac{\cosh x}{(n+1)!}x^{n+1} \to 0$ as $n \to \infty$ (by Equation 11). So by Theorem 9, the series represents $\sinh x$ for all x.

16. If $f(x) = \cosh x$, then $R_n(x) = \dfrac{f^{(n+1)}(z)}{(n+1)!}x^{n+1}$, where $0 < |z| < |x|$. But for all n, $\left|f^{(n+1)}(z)\right| \le \cosh z \le \cosh x$ (since all derivatives are either sinh or cosh, $|\sinh z| < |\cosh z|$ for all z, and $|z| < |x| \Rightarrow \cosh z < \cosh x$), so $|R_n(z)| \le \dfrac{\cosh x}{(n+1)!}x^{n+1} \to 0$ as $n \to \infty$. So by Theorem 9, the series represents $\cosh x$ for all x.

17. $e^{3x} = \sum_{n=0}^{\infty} \dfrac{(3x)^n}{n!} = \sum_{n=0}^{\infty} \dfrac{3^n x^n}{n!}$, with $R = \infty$.

18. $\sin 2x = \sum_{n=0}^{\infty} \dfrac{(-1)^n (2x)^{2n+1}}{(2n+1)!} = \sum_{n=0}^{\infty} \dfrac{(-1)^n 2^{2n+1} x^{2n+1}}{(2n+1)!}$, $R = \infty$

19. $x^2 \cos x = x^2 \sum_{n=0}^{\infty} \dfrac{(-1)^n x^{2n}}{(2n)!} = \sum_{n=0}^{\infty} \dfrac{(-1)^n x^{2n+2}}{(2n)!}$, $R = \infty$

20. $\cos(x^3) = \sum_{n=0}^{\infty} \dfrac{(-1)^n (x^3)^{2n}}{(2n)!} = \sum_{n=0}^{\infty} \dfrac{(-1)^n x^{6n}}{(2n)!}$, $R = \infty$.

21. $x \sin\left(\dfrac{x}{2}\right) = x \sum_{n=0}^{\infty} \dfrac{(-1)^n (x/2)^{2n+1}}{(2n+1)!} = \sum_{n=0}^{\infty} \dfrac{(-1)^n x^{2n+2}}{(2n+1)!\,2^{2n+1}}$, with $R = \infty$.

22. $xe^{-x} = x \sum_{n=0}^{\infty} \dfrac{(-x)^n}{n!} = \sum_{n=0}^{\infty} \dfrac{(-1)^n x^{n+1}}{n!} = \sum_{n=1}^{\infty} \dfrac{(-1)^{n-1} x^n}{(n-1)!}$, $R = \infty$.

23. $\sin^2 x = \frac{1}{2}[1 - \cos 2x] = \dfrac{1}{2}\left[1 - \sum_{n=0}^{\infty} \dfrac{(-1)^n (2x)^{2n}}{(2n)!}\right]$

$= \dfrac{1}{2}\left[1 - 1 - \sum_{n=1}^{\infty} \dfrac{(-1)^n (2x)^{2n}}{(2n)!}\right] = \sum_{n=1}^{\infty} \dfrac{(-1)^{n+1} 2^{2n-1} x^{2n}}{(2n)!}$, with $R = \infty$.

24. $\cos^2 x = \frac{1}{2}(1 + \cos 2x) = \dfrac{1}{2}\left[1 + \sum_{n=0}^{\infty} \dfrac{(-1)^n (2x)^{2n}}{(2n)!}\right] = \dfrac{1}{2}\left[1 + 1 + \sum_{n=1}^{\infty} \dfrac{(-1)^n 2^{2n} x^{2n}}{(2n)!}\right]$

$= 1 + \sum_{n=1}^{\infty} \dfrac{(-1)^n 2^{2n-1} x^{2n}}{(2n)!}$, $R = \infty$

Another method: Use $\cos^2 x = 1 - \sin^2 x$ and Exercise 23.

25. $\dfrac{\sin x}{x} = \dfrac{1}{x} \sum_{n=0}^{\infty} \dfrac{(-1)^n x^{2n+1}}{(2n+1)!} = \sum_{n=0}^{\infty} \dfrac{(-1)^n x^{2n}}{(2n+1)!}$ and this series also gives the required value at $x = 0$, so $R = \infty$.

26. $\dfrac{1 - \cos x}{x^2} = x^{-2}\left(1 - \sum_{n=0}^{\infty} \dfrac{(-1)^n x^{2n}}{(2n)!}\right) = x^{-2}\left(-\sum_{n=1}^{\infty} \dfrac{(-1)^n x^{2n}}{(2n)!}\right) = \sum_{n=1}^{\infty} \dfrac{(-1)^{n+1} x^{2n-2}}{(2n)!} = \sum_{n=0}^{\infty} \dfrac{(-1)^n x^{2n}}{(2n+2)!}$, since the series is equal to $\frac{1}{2}$ when $x = 0$; $R = \infty$.

27.

n	$f^{(n)}(x)$	$f^{(n)}(0)$
0	$(1+x)^{1/2}$	1
1	$\frac{1}{2}(1+x)^{-1/2}$	$\frac{1}{2}$
2	$-\frac{1}{4}(1+x)^{-3/2}$	$-\frac{1}{4}$
3	$\frac{3}{8}(1+x)^{-5/2}$	$\frac{3}{8}$
4	$-\frac{15}{16}(1+x)^{-7/2}$	$-\frac{15}{16}$
...	...	...

So $f^{(n)}(0) = \dfrac{(-1)^{n-1}1\cdot 3\cdot 5\cdots(2n-3)}{2^n}$ for $n \geq 2$, and $\sqrt{1+x} = 1 + \dfrac{x}{2} + \sum_{n=2}^{\infty}\dfrac{(-1)^{n-1}1\cdot 3\cdot 5\cdots(2n-3)}{2^n n!}x^n$.

If $a_n = \dfrac{(-1)^{n+1}1\cdot 3\cdot 5\cdot\cdots\cdot(2n-3)}{2^n n!}x^n$, then $\lim_{n\to\infty}\left|\dfrac{a_{n+1}}{a_n}\right| = \dfrac{|x|}{2}\lim_{n\to\infty}\dfrac{2n-1}{n+1} = |x| < 1$ for convergence, so $R = 1$.

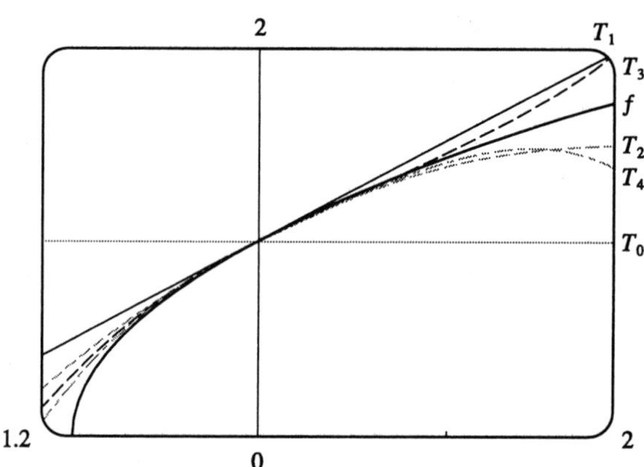

28.

n	$f^{(n)}(x)$	$f^{(n)}(0)$
0	$(1+2x)^{-1/2}$	1
1	$-\frac{1}{2}(1+2x)^{-3/2}(2)$	-1
2	$\frac{3}{2}(1+2x)^{-5/2}(2)$	3
3	$-3\cdot\frac{5}{2}(1+2x)^{-7/2}(2)$	$-3\cdot 5$
...	...	...

$f^{(n)}(0) = (-1)^n 1\cdot 3\cdot 5\cdot 7\cdots(2n-1)$, so

$$(1+2x)^{-1/2} = \sum_{n=0}^{\infty}\frac{f^{(n)}(0)}{n!}x^n = \sum_{n=0}^{\infty}\frac{(-1)^n 1\cdot 3\cdot 5\cdot\cdots\cdot(2n-1)}{n!}x^n.$$

$\lim_{n\to\infty}\left|\dfrac{a_{n+1}}{a_n}\right| = \lim_{n\to\infty}\dfrac{2n+1}{n+1}|x| = 2|x| < 1$ for convergence, so $R = \frac{1}{2}$.

Another method: Use Exercise 27 and differentiate.

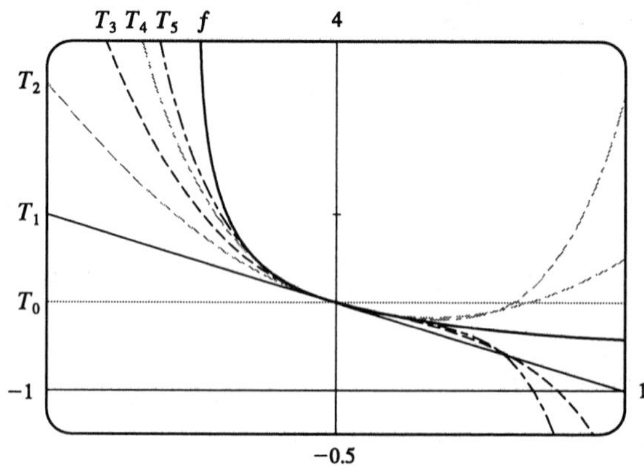

29. $f(x) = (1+x)^{-3} = -\frac{1}{2}\frac{d}{dx}\left[\frac{1}{(1+x)^2}\right] = -\frac{1}{2}\frac{d}{dx}\left[\sum_{n=0}^{\infty}(-1)^n(n+1)x^n\right]$ (from Exercise 5)

$= -\frac{1}{2}\sum_{n=1}^{\infty}(-1)^n n(n+1)x^{n-1} = \sum_{n=0}^{\infty}\frac{(-1)^n(n+1)(n+2)x^n}{2}$, with $R = 1$ since that is the R in Exercise 5.

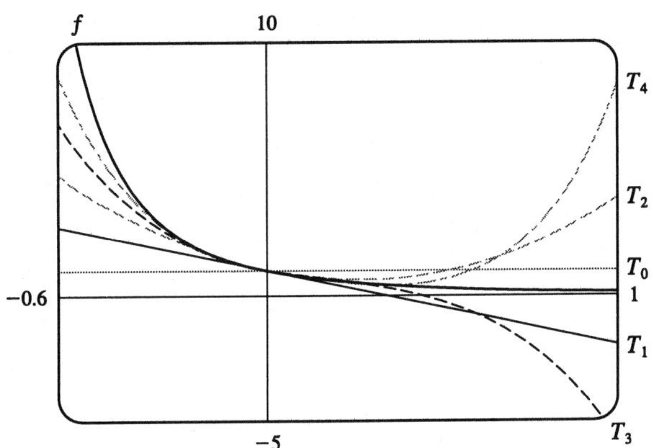

30. $2^x = \left(e^{\ln 2}\right)^x = e^{x\ln 2} = \sum_{n=0}^{\infty}\frac{(x\ln 2)^n}{n!} = \sum_{n=0}^{\infty}\frac{(\ln 2)^n x^n}{n!}$, $R = \infty$.

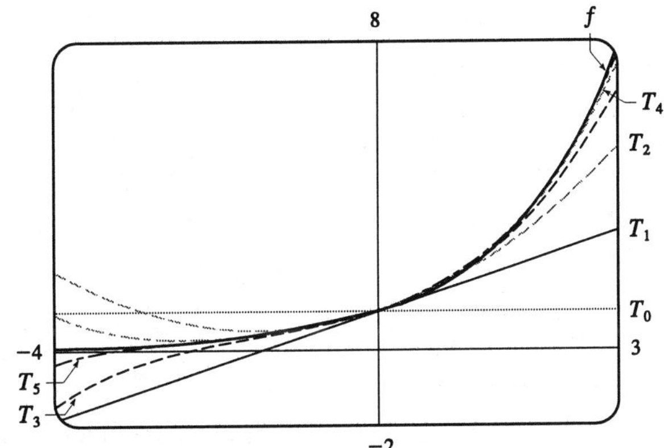

31. $\ln(1+x) = \int\frac{dx}{1+x} = \int\sum_{n=0}^{\infty}(-1)^n x^n\,dx = \sum_{n=1}^{\infty}\frac{(-1)^{n-1}x^n}{n}$ with $R = 1$, so $\ln(1.1) = \sum_{n=1}^{\infty}\frac{(-1)^{n-1}(0.1)^n}{n}$.

This is an alternating series with $b_5 = \frac{(0.1)^5}{5} = 0.000002$, so to five decimals,

$\ln(1.1) \approx \sum_{n=1}^{4}\frac{(-1)^{n-1}(0.1)^n}{n} \approx 0.09531$.

32. $\sin x = \sum_{n=0}^{\infty}\frac{(-1)^n x^{2n+1}}{(2n+1)!}$ so $\sin\frac{\pi}{60} = \frac{\pi}{60} - \frac{(\pi/60)^3}{3!} + \frac{(\pi/60)^5}{5!} - \cdots$

$= \frac{\pi}{60} - \frac{\pi^3}{1{,}296{,}000} + \frac{\pi^5}{93{,}312{,}000{,}000} - \cdots$. But $\frac{\pi^5}{93{,}312{,}000{,}000} < 10^{-8}$, so by Theorem 10.5.1, $\sin\frac{\pi}{60} \approx \frac{\pi}{60} - \frac{\pi^3}{1{,}296{,}000} \approx 0.05234$.

33. $\int \sin(x^2)\,dx = \int \sum_{n=0}^{\infty} (-1)^n \frac{(x^2)^{2n+1}}{(2n+1)!}\,dx = \int \sum_{n=0}^{\infty} \frac{(-1)^n x^{4n+2}}{(2n+1)!}\,dx = C + \sum_{n=0}^{\infty} \frac{(-1)^n x^{4n+3}}{(4n+3)(2n+1)!}$

34. $\frac{\sin x}{x} = \frac{1}{x}\sum_{n=0}^{\infty} \frac{(-1)^n x^{2n+1}}{(2n+1)!} = \sum_{n=0}^{\infty} \frac{(-1)^n x^{2n}}{(2n+1)!}$ so $\int \frac{\sin x}{x}\,dx = \int \sum_{n=0}^{\infty} \frac{(-1)^n x^{2n}}{(2n+1)!}\,dx$

$= C + \sum_{n=0}^{\infty} \frac{(-1)^n x^{2n+1}}{(2n+1)(2n+1)!}$, with $R = \infty$.

35. Using the series we obtained in Exercise 27, we get

$\sqrt{x^3+1} = 1 + \frac{x^3}{2} + \sum_{n=2}^{\infty} \frac{(-1)^{n-1} 1\cdot 3\cdot 5\cdot\dots\cdot(2n-3)}{2^n\, n!} x^{3n}$, so

$$\int \sqrt{x^3+1}\,dx = \int \left(1 + \frac{x^3}{2} + \sum_{n=2}^{\infty} \frac{(-1)^{n-1} 1\cdot 3\cdot 5\cdot\dots\cdot(2n-3)}{2^n\, n!} x^{3n}\right) dx$$

$$= C + x + \frac{x^4}{8} + \sum_{n=2}^{\infty} \frac{(-1)^{n-1} 1\cdot 3\cdot 5\cdot\dots\cdot(2n-3)}{2^n n!\,(3n+1)} x^{3n+1}.$$

36. $\int e^{x^3}\,dx = \int \sum_{n=0}^{\infty} \frac{x^{3n}}{n!}\,dx = C + \sum_{n=0}^{\infty} \frac{x^{3n+1}}{(3n+1)n!}$, with $R = \infty$.

37. Using our series from Exercise 33, we get $\int_0^1 \sin(x^2)\,dx = \sum_{n=0}^{\infty} \left[\frac{(-1)^n x^{4n+3}}{(4n+3)(2n+1)!}\right]_0^1 = \sum_{n=0}^{\infty} \frac{(-1)^n}{(4n+3)(2n+1)!}$

and $|c_3| = \frac{1}{75{,}600} < 0.000014$, so by Theorem 10.5.1, we have

$\sum_{n=0}^{2} \frac{(-1)^n}{(4n+3)(2n+1)!} \approx \frac{1}{3} - \frac{1}{42} + \frac{1}{1320} \approx 0.310.$

38. $\cos(x^2) = \sum_{n=0}^{\infty} \frac{(-1)^n (x^2)^{2n}}{(2n)!}$ so $\int_0^{0.5} \cos(x^2)\,dx = \int_0^{0.5} \sum_{n=0}^{\infty} \frac{(-1)^n x^{4n}}{(2n)!}\,dx = \sum_{n=0}^{\infty} \left[\frac{(-1)^n x^{4n+1}}{(4n+1)(2n)!}\right]_0^{0.5}$

$= 0.5 - \frac{(0.5)^5}{5\cdot 2!} + \frac{(0.5)^9}{9\cdot 4!} - \dots$, but $\frac{(0.5)^9}{9\cdot 4!} \approx 0.000009$, so by Theorem 10.5.1,

$\int_0^{0.5} \cos(x^2)\,dx \approx 0.5 - \frac{(0.5)^5}{5\cdot 2!} \approx 0.497.$

39. We first find a series representation for $f(x) = (1+x)^{-1/2}$, and then substitute.

n	$f^{(n)}(x)$	$f^{(n)}(0)$
0	$(1+x)^{-1/2}$	1
1	$-\frac{1}{2}(1+x)^{-3/2}$	$-\frac{1}{2}$
2	$\frac{3}{4}(1+x)^{-5/2}$	$3/4$
3	$-\frac{15}{8}(1+x)^{-7/2}$	$-15/8$
$\cdots$	$\cdots$	$\cdots$

$\frac{1}{\sqrt{1+x}} = 1 - \frac{x}{2} + \frac{3}{4}\left(\frac{x^2}{2!}\right) - \frac{15}{8}\left(\frac{x^3}{3!}\right) + \cdots \Rightarrow$

$\frac{1}{\sqrt{1+x^3}} = 1 - \frac{1}{2}x^3 + \frac{3}{8}x^6 - \frac{5}{16}x^9 + \cdots \Rightarrow$

$\int_0^{0.1} \frac{dx}{\sqrt{1+x^3}} = \left[x - \frac{1}{8}x^4 + \frac{3}{56}x^7 - \frac{1}{32}x^{10} + \cdots\right]_0^{0.1}$

$\approx (0.1) - \frac{1}{8}(0.1)^4$ by Theorem 10.28

since $\frac{3}{56}(0.1)^7 \approx 0.0000000054 < 10^{-8}$.

Therefore $\int_0^{0.1} \frac{dx}{\sqrt{1+x^3}} \approx 0.09998750.$

40. $\displaystyle\int_0^{0.5} x^2 e^{-x^2}\,dx = \int_0^{0.5}\sum_{n=0}^{\infty}\frac{(-1)^n x^{2n+2}}{n!}\,dx = \sum_{n=0}^{\infty}\left[\frac{(-1)^n x^{2n+3}}{n!(2n+3)}\right]_0^{1/2} = \sum_{n=0}^{\infty}\frac{(-1)^n}{n!(2n+3)2^{2n+3}}$ and since $c_2 = \frac{1}{1792} < 0.001$ we use $\displaystyle\sum_{n=0}^{1}\frac{(-1)^n}{n!(2n+3)2^{2n+3}} = \frac{1}{24} - \frac{1}{160} \approx 0.0354.$

41. As in Example 8(a), we have $e^{-x^2} = 1 - \dfrac{x^2}{1!} + \dfrac{x^4}{2!} + \dfrac{x^6}{3!} + \cdots$ and we know that $\cos x = 1 - \dfrac{x^2}{2!} + \dfrac{x^4}{4!} - \cdots$ from Equation 16. Therefore $e^{-x^2}\cos x = \left(1 - x^2 + \frac{1}{2}x^4 - \cdots\right)\left(1 - \frac{1}{2}x^2 + \frac{1}{24}x^4 - \cdots\right)$
$= 1 - \frac{1}{2}x^2 + \frac{1}{24}x^4 - x^2 + \frac{1}{2}x^4 + \frac{1}{2}x^4 + \cdots = 1 - \frac{3}{2}x^2 + \frac{25}{4}x^4 + \cdots$

42. $\sec x = \dfrac{1}{\cos x} = \dfrac{1}{1 - \frac{1}{2}x^2 + \frac{1}{24}x^4 - \cdots}.$

From the long division at right,

$\sec x = 1 + \frac{1}{2}x^2 + \frac{5}{24}x^4 + \cdots.$

$$\begin{array}{r|l}
 & 1 + \frac{1}{2}x^2 + \frac{5}{24}x^4 + \cdots \\
\hline
1 - \frac{1}{2}x^2 + \frac{1}{24}x^4 - \cdots & 1 \\
 & 1 - \frac{1}{2}x^2 + \frac{1}{24}x^4 - \cdots \\
\hline
 & \frac{1}{2}x^2 - \frac{1}{24}x^4 + \cdots \\
 & \frac{1}{2}x^2 - \frac{1}{4}x^4 + \cdots \\
\hline
 & \frac{5}{24}x^4 + \cdots \\
 & \frac{5}{24}x^4 + \cdots \\
\hline
 & \cdots
\end{array}$$

43. From Example 6 in Section 10.9, we have $\ln(1-x) = -x - \frac{1}{2}x^2 - \frac{1}{3}x^3 - \cdots,$ $|x| < 1$. Therefore

$y = \dfrac{\ln(1-x)}{e^x} = \dfrac{-x - \frac{1}{2}x^2 - \frac{1}{3}x^3 - \cdots}{1 + x + \frac{1}{2}x^2 + \frac{1}{6}x^3 + \cdots}.$

So by the long division at right,

$\dfrac{\ln(1-x)}{e^x} = -x + \dfrac{x^2}{2} - \dfrac{x^3}{3} + \cdots,\ |x| < 1.$

$$\begin{array}{r|l}
 & -x + \frac{1}{2}x^2 - \frac{1}{3}x^3 + \cdots \\
\hline
1 + x + \frac{1}{2}x^2 + \frac{1}{6}x^3 - \cdots & -x - \frac{1}{2}x^2 - \frac{1}{3}x^3 - \cdots \\
 & -x - x^2 - \frac{1}{2}x^3 - \cdots \\
\hline
 & \frac{1}{2}x^2 + \frac{1}{6}x^3 - \cdots \\
 & \frac{1}{2}x^2 + \frac{1}{2}x^3 + \cdots \\
\hline
 & -\frac{1}{3}x^3 + \cdots \\
 & -\frac{1}{3}x^3 + \cdots \\
\hline
 & \cdots
\end{array}$$

44. From Example 6 in Section 10.9 we have $\ln(1-x) = -x - \frac{1}{2}x^2 - \frac{1}{3}x^3 - \cdots,\ |x| < 1$. Therefore
$e^x \ln(1-x) = \left(1 + x + \frac{1}{2}x^2 + \cdots\right)\left(-x - \frac{1}{2}x^2 - \frac{1}{3}x^3 + \cdots\right)$
$= -x - \frac{1}{2}x^2 - \frac{1}{3}x^3 - x^2 - \frac{1}{2}x^3 - \frac{1}{2}x^3 - \cdots = -x - \frac{3}{2}x^2 - \frac{4}{3}x^3 - \cdots,\ |x| < 1.$

45. $\displaystyle\sum_{n=0}^{\infty}(-1)^n\frac{x^{4n}}{n!} = \sum_{n=0}^{\infty}\frac{\left(-x^4\right)^n}{n!} = e^{-x^4}$ by (12).

46. $\displaystyle\sum_{n=0}^{\infty}\frac{(-1)^n\pi^{2n}}{6^{2n}(2n)!} = \sum_{n=0}^{\infty}(-1)^n\frac{(\pi/6)^{2n}}{2n!} = \cos\frac{\pi}{6} = \frac{\sqrt{3}}{2}$ by (16).

47. $\displaystyle\sum_{n=0}^{\infty}\frac{(-1)^n\pi^{2n+1}}{4^{2n+1}(2n+1)!} = \sum_{n=0}^{\infty}\frac{(-1)^n(\pi/4)^{2n+1}}{(2n+1)!} = \sin\frac{\pi}{4} = \frac{1}{\sqrt{2}}$ by (15).

48. $\displaystyle\sum_{n=2}^{\infty}\frac{x^{3n+1}}{n!} = x\sum_{n=2}^{\infty}\frac{\left(x^3\right)^n}{n!} = x\left[\sum_{n=0}^{\infty}\frac{\left(x^3\right)^n}{n!} - 1 - x^3\right] = x\left(e^{x^3} - 1 - x^3\right)$ by (12).

49. $\displaystyle\sum_{n=0}^{\infty} \frac{x^{n+1}}{(n+1)!} = \frac{x}{1!} + \frac{x^2}{2!} + \frac{x^3}{3!} + \cdots = \left(1 + \frac{x}{1!} + \frac{x^2}{2!} + \frac{x^3}{3!} + \cdots\right) - 1 = e^x - 1$ by (12).

50. $\displaystyle\sum_{n=0}^{\infty} \frac{x^n}{2^n(n+1)!} = \sum_{n=0}^{\infty} \frac{(x/2)^n}{(n+1)!} = \frac{2}{x}\sum_{n=0}^{\infty} \frac{(x/2)^n}{(n+1)!} = \frac{2}{x}\left[(x/2) + \frac{(x/2)^2}{2!} + \frac{(x/2)^3}{3!} + \cdots\right] = \frac{2}{x}\left(e^{x/2} - 1\right)$

51. By (12), $e^x = 1 + x + \frac{x^2}{2!} + \frac{x^3}{3!} + \frac{x^4}{4!} + \cdots$, but for $x > 0$, all of the terms after the first two on the RHS are positive, so $e^x > 1 + x$ for $x > 0$.

52. From Exercises 6 and 16, $\cosh x = 1 + \frac{1}{2}x^2 + \frac{1}{24}x^6 + \cdots \geq 1 + \frac{1}{2}x^2$ for all x since there are only even powers of x on the RHS, so all of the remaining terms of the expansion are positive.

53. $$\lim_{x\to 0} \frac{\sin x - x + \frac{1}{6}x^3}{x^5} = \lim_{x\to 0} \frac{\left(x - \frac{1}{6}x^3 + \frac{1}{5!}x^5 - \frac{1}{7!}x^7 + \cdots\right) - x + \frac{1}{6}x^3}{x^5}$$
$$= \lim_{x\to 0} \frac{\frac{1}{5!}x^5 - \frac{1}{7!}x^7 + \cdots}{x^5} = \lim_{x\to 0}\left(\frac{1}{5!} - \frac{x^2}{7!} + \frac{x^4}{9!} - \cdots\right) = \frac{1}{5!} = \frac{1}{120},$$

since power series are continuous functions.

54. $$\lim_{x\to 0} \frac{1 - \cos x}{1 + x - e^x} = \lim_{x\to 0} \frac{1 - \left(1 - \frac{1}{2}x^2 + \frac{1}{4!}x^4 - \frac{1}{6!}x^6 + \cdots\right)}{1 + x - \left(1 + x + \frac{1}{2}x^2 + \frac{1}{3!}x^3 + \frac{1}{4!}x^4 + \frac{1}{5!}x^5 + \frac{1}{6!}x^6 + \cdots\right)}$$
$$= \lim_{x\to 0} \frac{\frac{1}{2}x^2 - \frac{1}{4!}x^4 + \frac{1}{6!}x^6 - \cdots}{-\frac{1}{2}x^2 - \frac{1}{3!}x^3 - \frac{1}{4!}x^4 - \frac{1}{5!}x^5 - \frac{1}{6!}x^6 - \cdots} = \lim_{x\to 0} \frac{\frac{1}{2} - \frac{1}{4!}x^2 + \frac{1}{6!}x^4 - \cdots}{-\frac{1}{2} - \frac{1}{3!}x - \frac{1}{4!}x^2 - \frac{1}{5!}x^3 - \frac{1}{6!}x^5 - \cdots}$$
$$= \frac{\frac{1}{2} - 0}{-\frac{1}{2} - 0} = -1, \text{ since power series are continuous functions.}$$

55. We must show that f equals its Taylor series expansion on I; that is, we must show that $\lim_{n\to\infty} |R_n(x)| = 0$. For $x \in I$, $|R_n(x)| = \left|\frac{f^{(n+1)}(z)}{(n+1)!}(x-a)^{n+1}\right| \leq \frac{M \cdot R^{n+1}}{(n+1)!} \to 0$ as $n \to \infty$ by (11).

56. (a) $f(x) = \begin{cases} e^{-1/x^2} & \text{if } x \neq 0 \\ 0 & \text{if } x = 0 \end{cases}$ so

$$f'(0) = \lim_{x\to 0} \frac{f(x) - f(0)}{x - 0} = \lim_{x\to 0} \frac{e^{-1/x^2}}{x} = \lim_{x\to 0} \frac{1/x}{e^{1/x^2}} = \lim_{x\to 0} \frac{x}{2e^{1/x^2}} \quad \left[\begin{array}{c}\text{using l'Hospital's Rule} \\ \text{and simplifying}\end{array}\right] = 0$$

Similarly, we can use the definition of the derivative and l'Hospital's Rule to show that $f''(0) = 0, f^{(3)}(0) = 0, \ldots, f^{(n)}(0) = 0$, so that the Maclaurin series for f consists entirely of zero terms. But since $f(x) \neq 0$ except for $x = 0$, we see that f cannot equal its Maclaurin series except at $x = 0$.

(b)

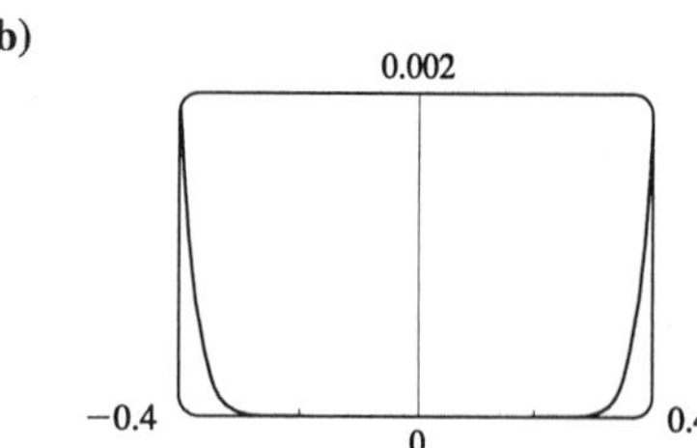

From the graph, it seems that the function is extremely flat at the origin. In fact, it could be said to be "infinitely flat" at $x = 0$, since all of its derivatives are 0 there.

EXERCISES 10.11

1. $$(1+x)^{1/2}=\sum_{n=0}^{\infty}\binom{1/2}{n}x^n=1+\left(\tfrac{1}{2}\right)x+\frac{\left(\frac{1}{2}\right)\left(-\frac{1}{2}\right)}{2!}x^2+\frac{\left(\frac{1}{2}\right)\left(-\frac{1}{2}\right)\left(-\frac{3}{2}\right)}{3!}x^3+\cdots$$
$$=1+\frac{x}{2}-\frac{x^2}{2^2\cdot 2!}+\frac{1\cdot 3\cdot x^3}{2^3\cdot 3!}-\frac{1\cdot 3\cdot 5\cdot x^4}{2^4\cdot 4!}+\cdots$$
$$=1+\frac{x}{2}+\sum_{n=2}^{\infty}\frac{(-1)^{n-1}1\cdot 3\cdot 5\cdot\cdots\cdot(2n-3)x^n}{2^n\cdot n!},\ R=1$$

2. $$(1+x)^{-3}=\sum_{n=0}^{\infty}\binom{-3}{n}x^n=1+\sum_{n=1}^{\infty}\frac{(-1)^n\cdot 3\cdot 4\cdot 5\cdot\cdots\cdot(n+2)x^n}{n!}\text{ with }R=1.$$

3. $$[1+(2x)]^{-4}=1+(-4)(2x)+\frac{(-4)(-5)}{2!}(2x)^2+\frac{(-4)(-5)(-6)}{3!}(2x)^3+\cdots$$
$$=1+\sum_{n=1}^{\infty}\frac{(-1)^n2^n4\cdot 5\cdot 6\cdots(n+3)}{n!}x^n$$
$$=\sum_{n=0}^{\infty}(-1)^n\frac{2^n(n+1)(n+2)(n+3)}{6}x^n,\ |2x|<1\quad\Leftrightarrow\quad|x|<\tfrac{1}{2}\text{ so }R=\tfrac{1}{2}.$$

4. $$\left(1+x^2\right)^{1/3}=\sum_{n=0}^{\infty}\binom{1/3}{n}x^{2n}=1+\frac{x^2}{3}+\frac{\left(\frac{1}{3}\right)\left(-\frac{2}{3}\right)}{2!}x^4+\frac{\left(\frac{1}{3}\right)\left(-\frac{2}{3}\right)\left(-\frac{5}{3}\right)}{3!}x^6+\cdots$$
$$=1+\frac{x^2}{3}+\sum_{n=2}^{\infty}\frac{(-1)^{n-1}\cdot 2\cdot 5\cdot 8\cdot\cdots\cdot(3n-4)x^{2n}}{3^n\,n!}\text{ with }R=1.$$

5. $$[1+(-x)]^{-1/2}=\sum_{n=0}^{\infty}\binom{-1/2}{n}(-x)^n=1+\left(-\tfrac{1}{2}\right)(-x)+\frac{\left(-\frac{1}{2}\right)\left(-\frac{3}{2}\right)}{2!}(-x)^2+\cdots$$
$$=1+\frac{x}{2}+\frac{1\cdot 3}{2^22!}x^2+\frac{1\cdot 3\cdot 5}{2^33!}x^3+\frac{1\cdot 3\cdot 5\cdot 7}{2^44!}x^4+\cdots=1+\sum_{n=1}^{\infty}\frac{1\cdot 3\cdot 5\cdot\cdots\cdot(2n-1)}{2^n\,n!}x^n,$$
so $$\frac{x}{\sqrt{1-x}}=x+\sum_{n=1}^{\infty}\frac{1\cdot 3\cdot 5\cdot\cdots\cdot(2n-1)}{2^n\,n!}x^{n+1}\text{ with }R=1.$$

6. $$(2+x)^{-1/2}=\frac{1}{\sqrt{2}}\left(1+\frac{x}{2}\right)^{-1/2}=\frac{\sqrt{2}}{2}\sum_{n=0}^{\infty}\binom{-1/2}{n}\left(\frac{x}{2}\right)^n$$
$$=\frac{\sqrt{2}}{2}\left[1+\left(-\frac{1}{2}\right)\left(\frac{x}{2}\right)+\frac{\left(-\frac{1}{2}\right)\left(-\frac{3}{2}\right)}{2!}\left(\frac{x}{2}\right)^2+\cdots\right]$$
$$=\frac{\sqrt{2}}{2}\left[1+\sum_{n=1}^{\infty}\frac{(-1)^n\cdot 1\cdot 3\cdot 5\cdot\cdots\cdot(2n-1)x^n}{2^{2n}\cdot n!}\right]\text{ with }|x/2|<1\text{ so }|x|<2\text{ and }R=2.$$

7. $$\left(1-x^4\right)^{1/4}=1+\left(\tfrac{1}{4}\right)\left(-x^4\right)+\frac{\left(\frac{1}{4}\right)\left(-\frac{3}{4}\right)}{2!}\left(-x^4\right)^2+\frac{\left(\frac{1}{4}\right)\left(-\frac{3}{4}\right)\left(-\frac{7}{4}\right)}{3!}\left(-x^4\right)^3+\cdots$$
$$=1-\frac{x^4}{4}-\sum_{n=2}^{\infty}\frac{3\cdot 7\cdot 11\cdot\cdots\cdot(4n-5)}{4^n\cdot n!}x^{4n}\text{ with }R=1.$$

8. $$[1+(-x^3)]^{-1/2} = \sum_{n=0}^{\infty}\binom{-1/2}{n}(-x^3)^n$$
$$= 1 + \left(-\tfrac{1}{2}\right)(-x^3) + \frac{\left(-\frac{1}{2}\right)\left(-\frac{3}{2}\right)}{2!}(-x^3)^2 + \cdots$$
$$= 1 + \sum_{n=1}^{\infty}\frac{1\cdot 3\cdot 5\cdot\cdots\cdot(2n-1)x^{3n}}{2^n\cdot n!}, \text{ so}$$
$$\frac{x^2}{\sqrt{1-x^3}} = x^2 + \sum_{n=1}^{\infty}\frac{1\cdot 3\cdot 5\cdot\cdots\cdot(2n-1)x^{3n+2}}{2^n\cdot n!} \text{ with } R=1.$$

9. $$(1-x)^{-5} = 1 + (-5)(-x) + \frac{(-5)(-6)}{2!}(-x)^2 + \frac{(-5)(-6)(-7)}{3!}(-x)^3 + \cdots$$
$$= 1 + \sum_{n=1}^{\infty}\frac{5\cdot 6\cdot 7\cdots(n+4)}{n!}x^n$$
$$= \sum_{n=0}^{\infty}\frac{(n+4)!}{4!\cdot n!}x^n \quad\Rightarrow$$
$$\frac{x^5}{(1-x)^5} = \sum_{n=0}^{\infty}\frac{(n+4)!}{4!\cdot n!}x^{n+5}\ \left(\text{or } \sum_{n=0}^{\infty}\frac{(n+1)(n+2)(n+3)(n+4)}{24}x^{n+5}\right) \text{ with } R=1.$$

10. $$\sqrt[5]{x-1} = -[1+(-x)]^{1/5} = -\sum_{n=0}^{\infty}\binom{1/5}{n}(-x)^n$$
$$= -\left[1 + \tfrac{1}{5}(-x) + \frac{\left(\frac{1}{5}\right)\left(-\frac{4}{5}\right)}{2!}(-x)^2 + \frac{\left(\frac{1}{5}\right)\left(-\frac{4}{5}\right)\left(-\frac{9}{5}\right)}{3!}(-x)^3 + \cdots\right]$$
$$= -1 + \frac{x}{5} + \sum_{n=2}^{\infty}\frac{4\cdot 9\cdots(5n-6)x^n}{5^n\cdot n!} \text{ with } R=1.$$

11. $$(8+x)^{-1/3} = \frac{1}{2}\left(1+\frac{x}{8}\right)^{-1/3} = \frac{1}{2}\left[1 + \left(-\frac{1}{3}\right)\left(\frac{x}{8}\right) + \frac{\left(-\frac{1}{3}\right)\left(-\frac{4}{3}\right)}{2!}\left(\frac{x}{8}\right)^2 + \cdots\right]$$
$$= \frac{1}{2}\left[1 + \sum_{n=1}^{\infty}\frac{(-1)^n 1\cdot 4\cdot 7\cdot\cdots\cdot(3n-2)}{3^n\cdot n!\,8^n}x^n\right] \text{ and } \left|\frac{x}{8}\right| < 1 \Leftrightarrow \quad |x| < 8, \text{ so } R = 8.$$

The first three Taylor polynomials are $T_1(x) = \frac{1}{2} - \frac{1}{48}x$, $T_2(x) = \frac{1}{2} - \frac{1}{48}x + \frac{1}{576}x^2$, and $T_3(x) = \frac{1}{2} - \frac{1}{48}x + \frac{1}{576}x^2 - \frac{4\cdot 7}{27\cdot 6\cdot 512}x^3 = \frac{1}{2} - \frac{1}{48}x + \frac{1}{576}x^2 - \frac{7}{41{,}472}x^3$.

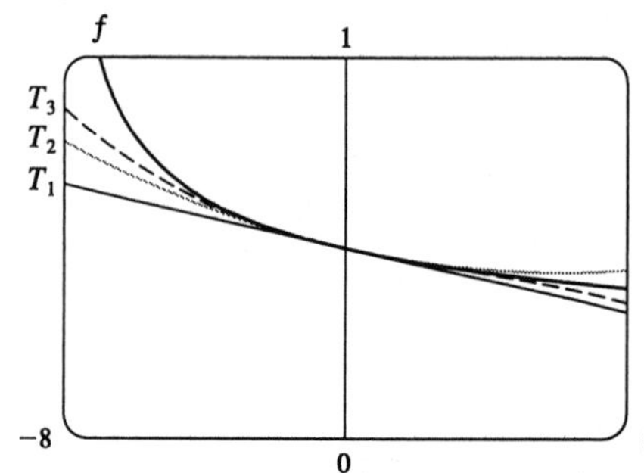

12. $(4+x)^{3/2} = 8\left(1+\frac{x}{4}\right)^{3/2} = 8\sum_{n=0}^{\infty}\binom{3/2}{n}\left(\frac{x}{4}\right)^n$

$$= 8\left[1+\frac{3}{2}\left(\frac{x}{4}\right)+\frac{\left(\frac{3}{2}\right)\left(\frac{1}{2}\right)}{2!}\left(\frac{x}{4}\right)^2+\frac{\left(\frac{3}{2}\right)\left(\frac{1}{2}\right)\left(-\frac{1}{2}\right)}{3!}\left(\frac{x}{4}\right)^3+\cdots\right]$$

$$= 8+3x+\sum_{n=2}^{\infty}\frac{(3)(1)(-1)\cdots(5-2n)x^n}{8^{n-1}\cdot n!} \text{ with } \left|\frac{x}{4}\right|<1, \text{ so } |x|<4 \text{ and } R=4.$$

The first three Taylor polynomials are $T_1(x) = 8+3x$, $T_2(x) = 8+3x+\frac{3}{16}x^2$, and $T_3(x) = 8+3x+\frac{3}{16}x^2-\frac{1}{128}x^3$.

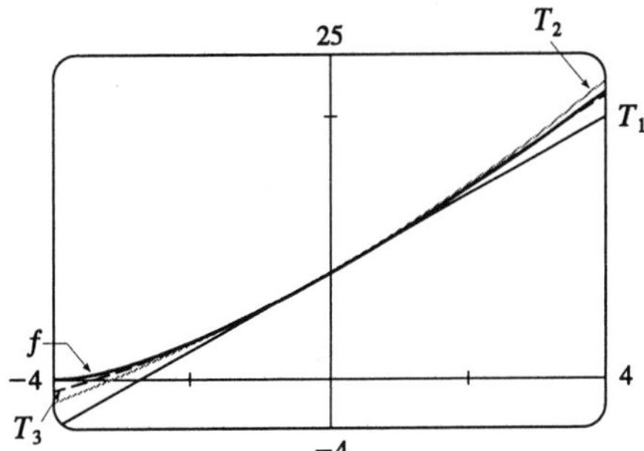

13. **(a)** $\left(1-x^2\right)^{-1/2} = 1+\left(-\frac{1}{2}\right)\left(-x^2\right)+\frac{\left(-\frac{1}{2}\right)\left(-\frac{3}{2}\right)}{2!}\left(-x^2\right)^2+\frac{\left(-\frac{1}{2}\right)\left(-\frac{3}{2}\right)\left(-\frac{5}{2}\right)}{3!}\left(-x^2\right)^3+\cdots$

$$= 1+\sum_{n=1}^{\infty}\frac{1\cdot3\cdot5\cdot\cdots\cdot(2n-1)}{2^n\cdot n!}x^{2n}$$

(b) $\sin^{-1}x = \int\frac{1}{\sqrt{1-x^2}}\,dx = C+x+\sum_{n=1}^{\infty}\frac{1\cdot3\cdot5\cdot\cdots\cdot(2n-1)}{(2n+1)2^n\cdot n!}x^{2n+1}$

$$= x+\sum_{n=1}^{\infty}\frac{1\cdot3\cdot5\cdots(2n-1)}{(2n+1)2^n\cdot n!}x^{2n+1} \text{ since } 0=\sin^{-1}0=C.$$

14. **(a)** $\left(1+x^2\right)^{-1/2} = \sum_{n=0}^{\infty}\binom{-1/2}{n}x^{2n} = 1+\sum_{n=1}^{\infty}\frac{(-1)^n\cdot1\cdot3\cdot5\cdot\cdots\cdot(2n-1)x^{2n}}{2^n\cdot n!}$

(b) $\sinh^{-1}x = \int\frac{dx}{\sqrt{1+x^2}} = C+x+\sum_{n=1}^{\infty}\frac{(-1)^n\cdot1\cdot3\cdot5\cdot\cdots\cdot(2n-1)x^{2n+1}}{2^n\cdot n!\,(2n+1)}$, but $C=0$ since $\sinh^{-1}0=0$, so $\sinh^{-1}x = x+\sum_{n=1}^{\infty}\frac{(-1)^n\cdot1\cdot3\cdot5\cdot\cdots\cdot(2n-1)x^{2n+1}}{2^n\cdot n!\,(2n+1)}$, $R=1$.

15. **(a)** $(1+x)^{-1/2} = 1+\left(-\frac{1}{2}\right)x+\frac{\left(-\frac{1}{2}\right)\left(-\frac{3}{2}\right)}{2!}x^2+\frac{\left(-\frac{1}{2}\right)\left(-\frac{3}{2}\right)\left(-\frac{5}{2}\right)}{3!}x^3+\cdots$

$$= 1+\sum_{n=1}^{\infty}\frac{(-1)^n1\cdot3\cdot5\cdot\cdots\cdot(2n-1)}{2^n\cdot n!}x^n$$

(b) Take $x=0.1$ in the above series. $\frac{1\cdot3\cdot5\cdot7}{2^4\,4!}(0.1)^4 < 0.00003$, so $\frac{1}{\sqrt{1.1}} \approx 1-\frac{0.1}{2}+\frac{1\cdot3}{2^2\cdot2!}(0.1)^2-\frac{1\cdot3\cdot5}{2^3\cdot3!}(0.1)^3 \approx 0.953.$

16. **(a)** $(8+x)^{1/3} = 2\left(1+\dfrac{x}{8}\right)^{1/3} = 2\displaystyle\sum_{n=0}^{\infty}\binom{1/3}{n}\left(\dfrac{x}{8}\right)^n$

$$= 2\left[1+\tfrac{1}{3}\left(\frac{x}{8}\right)+\frac{\left(\frac{1}{3}\right)\left(-\frac{2}{3}\right)}{2!}\left(\frac{x}{8}\right)^2+\frac{\left(\frac{1}{3}\right)\left(-\frac{2}{3}\right)\left(-\frac{5}{3}\right)}{3!}\left(\frac{x}{8}\right)^3+\cdots\right]$$

$$= 2\left[1+\frac{x}{24}+\sum_{n=2}^{\infty}\frac{(-1)^{n-1}\cdot 2\cdot 5\cdot\cdots\cdot(3n-4)x^n}{24^n\cdot n!}\right]$$

(b) $(8+0.2)^{1/3} = 2\left[1+\dfrac{0.2}{24}-\dfrac{(0.2)^2}{24^2}+\dfrac{2\cdot 5(0.2)^3}{24^3\cdot 3!}-\cdots\right] \approx 2\left[1+\dfrac{0.2}{24}-\dfrac{(0.2)^2}{24^2}\right]$ since

$\dfrac{2\cdot 5(0.2)^3}{24^3\cdot 3!} \approx 0.000001$, so $\sqrt[3]{8.2} \approx 2.0165$.

17. **(a)** $(1-x)^{-2} = 1+(-2)(-x)+\dfrac{(-2)(-3)}{2!}(-x)^2+\cdots = \displaystyle\sum_{n=0}^{\infty}(n+1)x^n$, so

$$\frac{x}{(1-x)^2} = \sum_{n=0}^{\infty}(n+1)x^{n+1} = \sum_{n=1}^{\infty}nx^n.$$

(b) With $x=\frac{1}{2}$ in part (a), we have $\displaystyle\sum_{n=1}^{\infty}\frac{n}{2^n} = \frac{1/2}{(1-1/2)^2} = 2.$

18. **(a)** $[1+(-x)]^{-3} = \displaystyle\sum_{n=0}^{\infty}\binom{-3}{n}(-x)^n = 1+(-3)(-x)+\frac{(-3)(-4)}{2!}x^2+\frac{(-3)(-4)(-5)}{3!}(-x)^3+\cdots$

$$= \sum_{n=0}^{\infty}\frac{3\cdot 4\cdot 5\cdot\cdots\cdot(n+2)}{n!}x^n = \sum_{n=0}^{\infty}\frac{(n+1)(n+2)}{2}x^n \quad\Rightarrow$$

$$(x+x^2)[1+(-x)]^{-3} = \sum_{n=0}^{\infty}\frac{(n+1)(n+2)}{2}x^{n+1}+\sum_{n=0}^{\infty}\frac{(n+1)(n+2)}{2}x^{n+2}$$

$$= x+\sum_{n=2}^{\infty}\left[\frac{n(n+1)}{2}+\frac{(n-1)n}{2}\right]x^n = x+\sum_{n=2}^{\infty}n^2x^n = \sum_{n=1}^{\infty}n^2x^n,\ -1<x<1.$$

(b) Setting $x=\frac{1}{2}$ in the last series above gives the required series, so $\displaystyle\sum_{n=1}^{\infty}\frac{n^2}{2^n} = \frac{1/2+(1/2)^2}{(1-1/2)^3} = 6.$

19. **(a)** $(1+x^2)^{1/2} = 1+\left(\frac{1}{2}\right)x^2+\dfrac{\left(\frac{1}{2}\right)\left(-\frac{1}{2}\right)}{2!}(x^2)^2+\dfrac{\left(\frac{1}{2}\right)\left(-\frac{1}{2}\right)\left(-\frac{3}{2}\right)}{3!}(x^2)^3+\cdots$

$$= 1+\frac{x^2}{2}+\sum_{n=2}^{\infty}\frac{(-1)^{n-1}1\cdot 3\cdot 5\cdot\cdots\cdot(2n-3)}{2^n\cdot n!}x^{2n}$$

(b) The coefficient of x^{10} in the above Maclaurin series is $\dfrac{f^{(10)}(0)}{10!}$, so $f^{(10)}(0) = 10!\left(\dfrac{1\cdot 3\cdot 5\cdot 7}{2^5\cdot 5!}\right) = 99{,}225.$

20. **(a)** $(1+x^3)^{-1/2} = \displaystyle\sum_{n=0}^{\infty}\binom{-1/2}{n}x^{3n} = 1+\sum_{n=1}^{\infty}\frac{(-1)^n 1\cdot 3\cdot 5\cdot\cdots\cdot(2n-1)x^{3n}}{2^n\cdot n!}$

(b) The coefficient of x^9 in the above series is $\dfrac{f^{(9)}(0)}{9!}$, so $\dfrac{f^{(9)}(0)}{9!} = \dfrac{(-1)^3 1\cdot 3\cdot 5}{2^3\cdot 3!}$ $\Rightarrow$

$f^{(9)}(0) = -\dfrac{9!\cdot 5}{8\cdot 2} = -113{,}400.$

21. **(a)** $g'(x) = \sum_{n=1}^{\infty} \binom{k}{n} n x^{n-1}$.

$$(1+x)g'(x) = (1+x)\sum_{n=1}^{\infty} \binom{k}{n} n x^{n-1} = \sum_{n=1}^{\infty} \binom{k}{n} n x^{n-1} + \sum_{n=1}^{\infty} \binom{k}{n} n x^{n}$$
$$= \sum_{n=0}^{\infty} \binom{k}{n+1} (n+1) x^{n} + \sum_{n=0}^{\infty} \binom{k}{n} n x^{n}$$
$$= \sum_{n=0}^{\infty} (n+1)\frac{k(k-1)(k-2)\cdots(k-n)}{(n+1)!} x^{n} + \sum_{n=0}^{\infty} \left((n)\frac{k(k-1)(k-2)\cdots(k-n+1)}{n!}\right) x^{n}$$
$$= \sum_{n=0}^{\infty} \frac{(n+1)k(k-1)(k-2)\cdots(k-n+1)}{(n+1)!}[(k-n)+n]x^{n}$$
$$= \sum_{n=0}^{\infty} \frac{k^2(k-1)(k-2)\cdots(k-n+1)}{n!}x^{n} = k\sum_{n=0}^{\infty}\binom{k}{n}x^n = kg(x). \text{ So } g'(x) = \frac{kg(x)}{1+x}.$$

(b) $h'(x) = -k(1+x)^{-k-1}g(x) + (1+x)^{-k}g'(x) = -k(1+x)^{-k-1}g(x) + (1+x)^{-k}\dfrac{kg(x)}{1+x}$

$$= -k(1+x)^{-k-1}g(x) + k(1+x)^{-k-1}g(x) = 0$$

(c) From part (b) we see that $h(x)$ must be constant for $x \in (-1,1)$, so $h(x) = h(0) = 1$ for $x \in (-1,1)$. Thus $h(x) = 1 = (1+x)^{-k}g(x) \quad \Leftrightarrow \quad g(x) = (1+x)^k$ for $x \in (-1,1)$.

22. **(a)** $4\sqrt{\dfrac{L}{g}}\displaystyle\int_0^{\pi/2} \frac{dx}{\sqrt{1-k^2\sin^2 x}} = 4\sqrt{\frac{L}{g}}\int_0^{\pi/2}\left[1+(-k^2\sin^2 x)\right]^{-1/2}dx$

$$= 4\sqrt{\frac{L}{g}}\int_0^{\pi/2}\left[1 - \frac{1}{2}(-k^2\sin^2 x) + \frac{\frac{1}{2}\cdot\frac{3}{2}}{2!}(-k^2\sin^2 x)^2 - \frac{\frac{1}{2}\cdot\frac{3}{2}\cdot\frac{5}{2}}{3!}(-k^2\sin^2 x)^3 + \cdots\right]dx$$
$$= 4\sqrt{\frac{L}{g}}\int_0^{\pi/2}\left[1 + \left(\frac{1}{2}\right)k^2\sin^2 x + \left(\frac{1\cdot 3}{2\cdot 4}\right)k^4\sin^4 x + \left(\frac{1\cdot 3\cdot 5}{2\cdot 4\cdot 6}\right)k^6\sin^6 x + \cdots\right]dx$$
$$= 4\sqrt{\frac{L}{g}}\left[\frac{\pi}{2} + \left(\frac{1}{2}\right)\left(\frac{1}{2}\cdot\frac{\pi}{2}\right)k^2 + \left(\frac{1\cdot 3}{2\cdot 4}\right)\left(\frac{1\cdot 3}{2\cdot 4}\cdot\frac{\pi}{2}\right)k^4 + \left(\frac{1\cdot 3\cdot 5}{2\cdot 4\cdot 6}\right)\left(\frac{1\cdot 3\cdot 5}{2\cdot 4\cdot 6}\cdot\frac{\pi}{2}\right)k^6 + \cdots\right]$$
$$= 2\pi\sqrt{\frac{L}{g}}\left[1 + \frac{1^2}{2^2}k^2 + \frac{1^2\cdot 3^2}{2^2\cdot 4^2}k^4 + \frac{1^2\cdot 3^2\cdot 5^2}{2^2\cdot 4^2\cdot 6^2}k^6 + \cdots\right]$$

At the fourth line, we have split up the integral and used the result from Exercise 7.1.40.

(b) The first of the two inequalities is true because all of the terms in the series are positive. For the second,

$$T = 2\pi\sqrt{\frac{L}{g}}\left[1 + \frac{1^2}{2^2}k^2 + \frac{1^2\cdot 3^2}{2^2\cdot 4^2}k^4 + \frac{1^2\cdot 3^2\cdot 5^2}{2^2\cdot 4^2\cdot 6^2}k^6 + \frac{1^2\cdot 3^2\cdot 5^2\cdot 7^2}{2^2\cdot 4^2\cdot 6^2\cdot 8^2}k^8 + \cdots\right]$$
$$\le 2\pi\sqrt{\frac{L}{g}}\left[1 + \tfrac{1}{4}k^2 + \tfrac{1}{4}k^4 + \tfrac{1}{4}k^6 + \tfrac{1}{4}k^8 + \cdots\right] = 2\pi\sqrt{\frac{L}{g}}\left[1 + \tfrac{1}{4}k^2\left(1 + k^2 + (k^2)^2 + (k^2)^3 + \cdots\right)\right]$$

The terms in brackets (after the first) form a geometric series with $a = \frac{1}{4}k^2$ and $r = k^2 = \sin^2\left(\frac{1}{2}\theta_0\right) < 1$.

So $T \le 2\pi\sqrt{\dfrac{L}{g}}\left[1 + \dfrac{k^2/4}{1-k^2}\right] = 2\pi\sqrt{\dfrac{L}{g}}\dfrac{4-3k^2}{4-4k^2}$.

(c) We substitute $L = 1$, $g = 9.8$, and $k = \sin(10°/2) \approx 0.08716$, and the inequality from part (b) becomes $2.01090 \le T \le 2.01093$, so $T \approx 2.0109$. The estimate $T \approx 2\pi\sqrt{L/g} \approx 2.0071$ differs by about 0.2%. If $\theta_0 = 42°$, then $k \approx 0.35837$ and the inequality becomes $2.0715 \le T \le 2.0810 \quad \Leftrightarrow \quad T \approx 2.076$. The one-term estimate is the same, and the discrepancy between the two estimates increases to about 3.3%.

EXERCISES 10.12

1.

n	$f^{(n)}(x)$	$f^{(n)}\left(\frac{\pi}{6}\right)$
0	$\sin x$	$\frac{1}{2}$
1	$\cos x$	$\frac{\sqrt{3}}{2}$
2	$-\sin x$	$-\frac{1}{2}$
3	$-\cos x$	$-\frac{\sqrt{3}}{2}$

$$T_3(x) = \sum_{n=0}^{3} \frac{f^{(n)}\left(\frac{\pi}{6}\right)}{n!}\left(x - \tfrac{\pi}{6}\right)^n$$
$$= \tfrac{1}{2} + \tfrac{\sqrt{3}}{2}\left(x - \tfrac{\pi}{6}\right) - \tfrac{1}{4}\left(x - \tfrac{\pi}{6}\right)^2 - \tfrac{\sqrt{3}}{12}\left(x - \tfrac{\pi}{6}\right)^3$$

2.

n	$f^{(n)}(x)$	$f^{(n)}\left(\frac{2\pi}{3}\right)$
0	$\cos x$	$-\frac{1}{2}$
1	$-\sin x$	$-\frac{\sqrt{3}}{2}$
2	$-\cos x$	$\frac{1}{2}$
3	$\sin x$	$\frac{\sqrt{3}}{2}$
4	$\cos x$	$-\frac{1}{2}$

$$T_4(x) = \sum_{n=0}^{4} \frac{f^{(n)}\left(\frac{2\pi}{3}\right)}{n!}\left(x - \tfrac{2\pi}{3}\right)^n$$
$$= -\tfrac{1}{2} - \tfrac{\sqrt{3}}{2}\left(x - \tfrac{2\pi}{3}\right) + \tfrac{1}{4}\left(x - \tfrac{2\pi}{3}\right)^2 + \tfrac{\sqrt{3}}{12}\left(x - \tfrac{2\pi}{3}\right)^3 - \tfrac{1}{48}\left(x - \tfrac{2\pi}{3}\right)^4$$

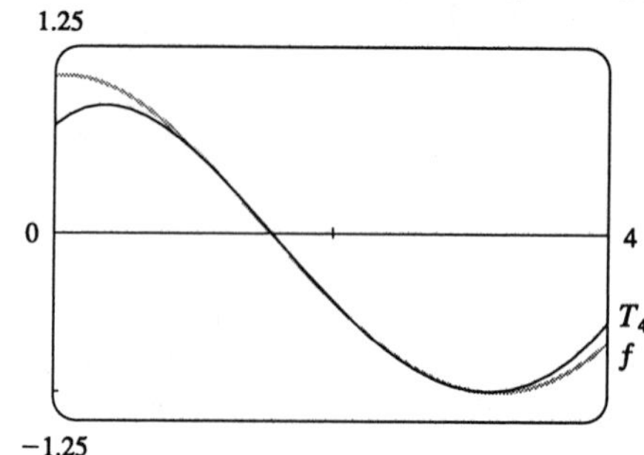

3.

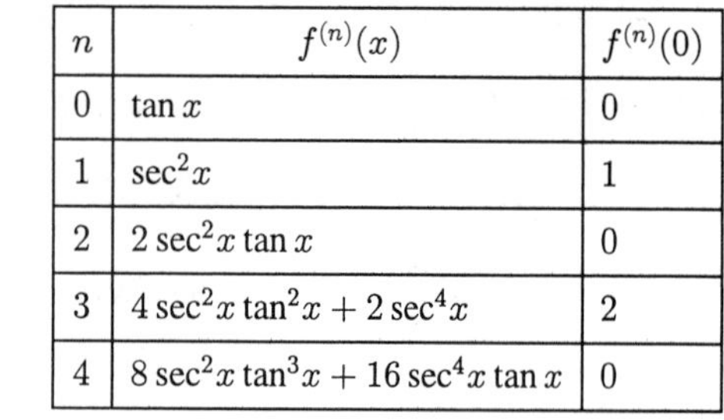

n	$f^{(n)}(x)$	$f^{(n)}(0)$
0	$\tan x$	0
1	$\sec^2 x$	1
2	$2\sec^2 x \tan x$	0
3	$4\sec^2 x \tan^2 x + 2\sec^4 x$	2
4	$8\sec^2 x \tan^3 x + 16\sec^4 x \tan x$	0

$$T_4(x) = \sum_{n=0}^{4} \frac{f^{(n)}(0)}{n!}x^n = x + \frac{2x^3}{3!} = x + \frac{x^3}{3}$$

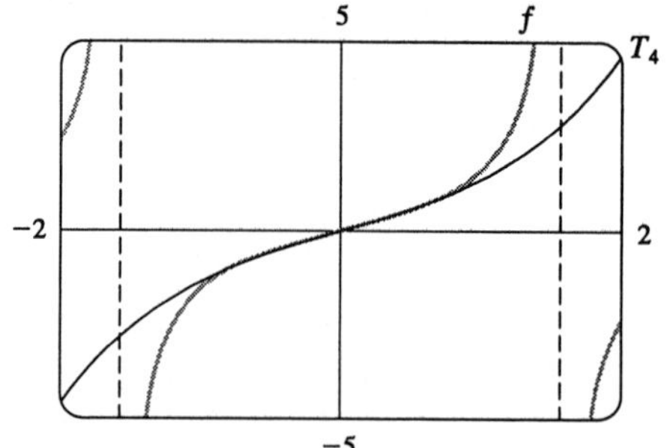

4.

n	$f^{(n)}(x)$	$f^{(n)}\left(\frac{\pi}{4}\right)$
0	$\tan x$	1
1	$\sec^2 x$	2
2	$2\sec^2 x \tan x$	4
3	$4\sec^2 x \tan^2 x + 2\sec^4 x$	16
4	$8\sec^2 x \tan^3 x + 16\sec^4 x \tan x$	80

$$T_4(x) = \sum_{n=0}^{4} \frac{f^{(n)}\left(\frac{\pi}{4}\right)}{n!}\left(x - \frac{\pi}{4}\right)^n$$
$$= 1 + 2\left(x - \frac{\pi}{4}\right) + 2\left(x - \frac{\pi}{4}\right)^2 + \frac{8}{3}\left(x - \frac{\pi}{4}\right)^3 + \frac{10}{3}\left(x - \frac{\pi}{4}\right)^4$$

5.

n	$f^{(n)}(x)$	$f^{(n)}(0)$
0	$e^x \sin x$	0
1	$e^x(\sin x + \cos x)$	1
2	$2e^x \cos x$	2
3	$2e^x(\cos x - \sin x)$	2

$$T_3(x) = \sum_{n=0}^{3} \frac{f^{(n)}(0)}{n!}x^n = x + x^2 + \tfrac{1}{3}x^3$$

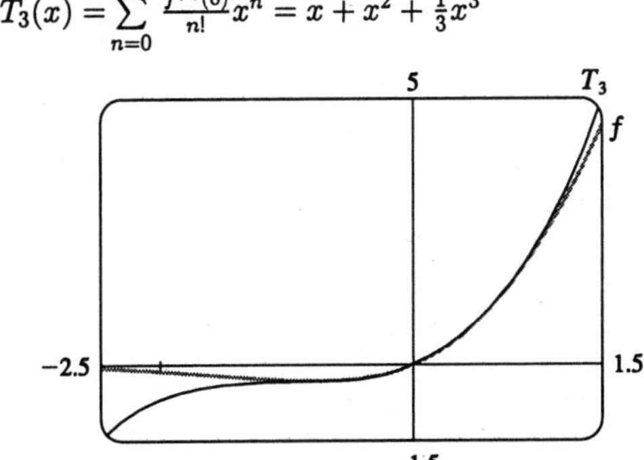

6.

n	$f^{(n)}(x)$	$f^{(n)}(9)$
0	$x^{1/2}$	3
1	$\frac{1}{2}x^{-1/2}$	$\frac{1}{6}$
2	$-\frac{1}{4}x^{-3/2}$	$-\frac{1}{108}$
3	$\frac{3}{8}x^{-5/2}$	$\frac{1}{648}$

$$T_3(x) = \sum_{n=0}^{3} \frac{f^{(n)}(9)}{n!}(x-9)^n$$
$$= 3 + \tfrac{1}{6}(x-9) - \tfrac{1}{216}(x-9)^2 + \tfrac{1}{3888}(x-9)^3$$

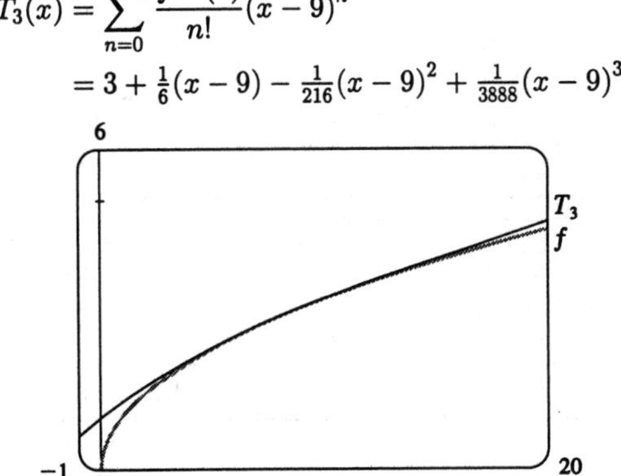

7.

n	$f^{(n)}(x)$	$f^{(n)}(8)$
0	$x^{-1/3}$	$\frac{1}{2}$
1	$-\frac{1}{3}x^{-4/3}$	$-\frac{1}{48}$
2	$\frac{4}{9}x^{-7/3}$	$\frac{1}{288}$
3	$-\frac{28}{27}x^{-10/3}$	$-\frac{7}{6912}$

$$T_3(x) = \sum_{n=0}^{3} \frac{f^{(n)}(8)}{n!}(x-8)^n$$
$$= \tfrac{1}{2} - \tfrac{1}{48}(x-8) + \tfrac{1}{576}(x-8)^2 - \tfrac{7}{41{,}472}(x-8)^3$$

8.

n	$f^{(n)}(x)$	$f^{(n)}\left(\frac{\pi}{3}\right)$
0	$\sec x$	2
1	$\sec x \tan x$	$2\sqrt{3}$
2	$\sec x \tan^2 x + \sec^3 x$	14
3	$\sec x \tan^3 x + 5\sec^3 x \tan x$	$46\sqrt{3}$

$$T_3(x) = \sum_{n=0}^{3} \frac{f^{(n)}\left(\frac{\pi}{3}\right)}{n!}\left(x-\tfrac{\pi}{3}\right)^n$$
$$= 2 + 2\sqrt{3}\left(x-\tfrac{\pi}{3}\right) + 7\left(x-\tfrac{\pi}{3}\right)^2 + \tfrac{23\sqrt{3}}{3}\left(x-\tfrac{\pi}{3}\right)^3$$

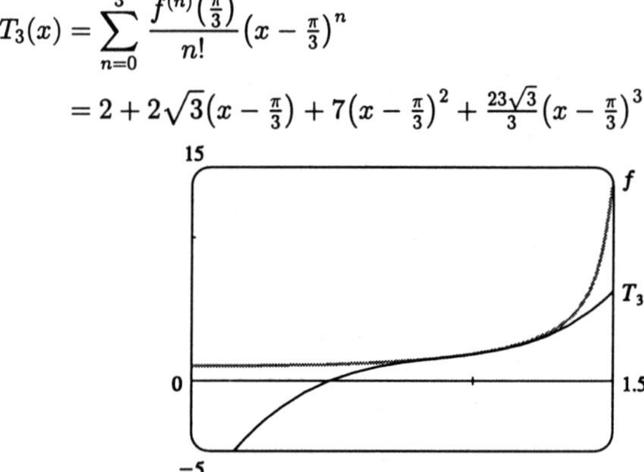

9.

n	$f^{(n)}(x)$	$f^{(n)}(0)$	$T_n(x)$
0	$\cos x$	1	1
1	$-\sin x$	0	1
2	$-\cos x$	-1	$1 - \frac{1}{2}x^2$
3	$\sin x$	0	$1 - \frac{1}{2}x^2$
4	$\cos x$	1	$1 - \frac{1}{2}x^2 + \frac{1}{24}x^4$

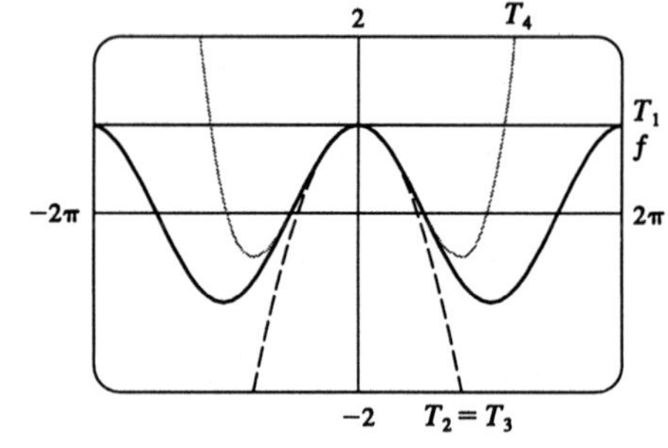

10.

n	$f^{(n)}(x)$	$f^{(n)}(1)$	$T_n(x)$
0	x^{-1}	1	1
1	$-x^{-2}$	-1	$1-(x-1) = 2-x$
2	$2x^{-3}$	2	$1-(x-1)+(x-1)^2$ $= x^2 - 3x + 3$
3	$-6x^{-4}$	-6	$1-(x-1)+(x-1)^2-(x-1)^3$ $= -x^3 + 4x^2 - 6x + 4$

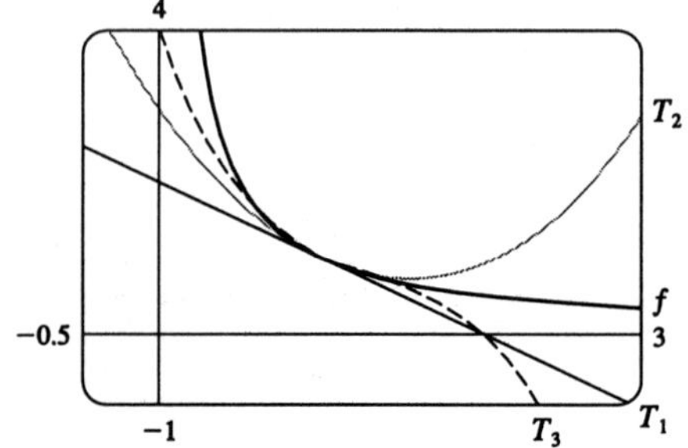

11. In Maple, we can find the Taylor polynomials by the following method: first define `f:=sec(x);` and then set `T2:=convert(taylor(f,x=0,3),polynom);`, `T4:=convert(taylor(f,x=0,5),polynom);`, etc. (The third argument in the `taylor` function is one more than the degree of the desired polynomial). We must `convert` to the type `polynom` because the output of the `taylor` function contains an error term which we do not want. In Mathematica, we use `Tn:=Normal[Series[f,{x,0,n}]]`, with `n=2, 4`, etc. Note that in Mathematica, the "degree" argument is the same as the degree of the desired polynomial. The eighth Taylor polynomial is $T_8(x) = 1 + \frac{1}{2}x^2 + \frac{5}{24}x^4 + \frac{61}{720}x^6 + \frac{277}{8064}x^8$.

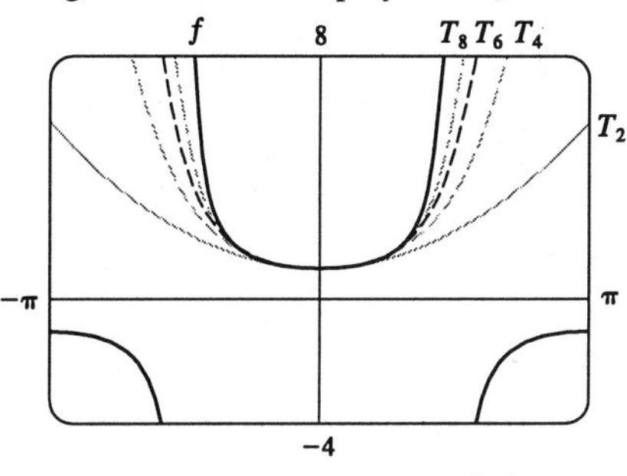

12. See Exercise 11 for the CAS commands used to generate the Taylor polynomials. The ninth Taylor polynomial for $\tan x$ is $T_9(x) = x + \frac{1}{3}x^3 + \frac{2}{15}x^5 + \frac{17}{315}x^7 + \frac{62}{2835}x^9$.

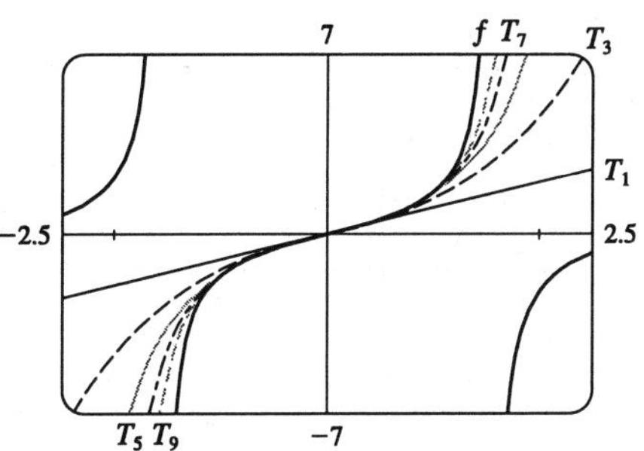

13. $f(x) = (1+x)^{1/2}$ $\quad f(0) = 1$

$f'(x) = \frac{1}{2}(1+x)^{-1/2}$ $\quad f'(0) = \frac{1}{2}$

$f''(x) = -\frac{1}{4}(1+x)^{-3/2}$

(a) $(1+x)^{1/2} \approx T_1(x) = 1 + \frac{1}{2}x$

(b) By Taylor's Formula, the remainder is

$$R_1(x) = \frac{f''(z)}{2!}x^2 = -\frac{1}{8(1+z)^{3/2}}x^2,$$ where z lies between 0 and x. Now $0 \le x \le 0.1 \Rightarrow 0 \le x^2 \le 0.01$, and $0 < z < 0.1 \Rightarrow 1 < 1 + z < 1.1$ so $|R_1(x)| < \frac{0.01}{8.1} = 0.00125$.

(c)

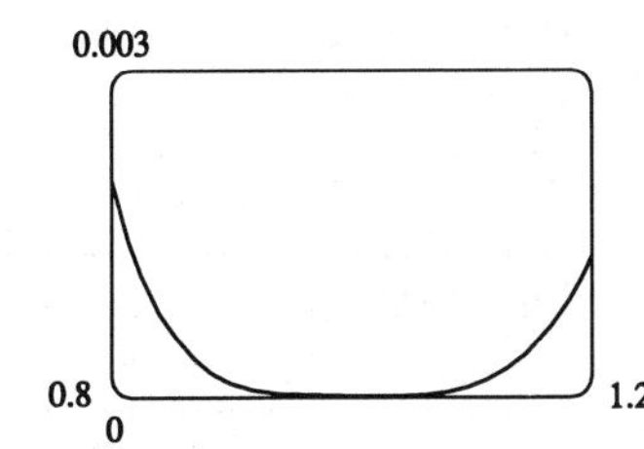

From the graph of $|R_1(x)| = \left|\sqrt{1+x} - \left(1 + \frac{1}{2}x\right)\right|$, it seems that the error is at most 0.0013 on $(0, 0.1)$.

14. **(a)** From Exercise 10, $1/x \approx T_3(x) = 1 - (x-1) + (x-1)^2 - (x-1)^3$.

(b) $f^{(4)}(x) = 24x^{-5}$, so

$$R_3(x) = \frac{24z^{-5}}{4!}(x-1)^4 = \frac{(x-1)^4}{z^5}$$ where z is between 1 and x. $0.8 \le x \le 1.2 \Rightarrow |x-1| \le 0.2$ and $z^5 > (0.8)^5 \Rightarrow |R_3(x)| < \frac{(0.2)^4}{(0.8)^5} < 0.005$.

(c)

0.003

0.8 0 1.2

From the graph, it seems that the error is less than 0.002 on $(0.8, 1.2)$.

15. $f(x) = \sin x \qquad f\left(\frac{\pi}{4}\right) = \frac{\sqrt{2}}{2}$

$f'(x) = \cos x \qquad f'\left(\frac{\pi}{4}\right) = \frac{\sqrt{2}}{2}$

$f''(x) = -\sin x \qquad f''\left(\frac{\pi}{4}\right) = -\frac{\sqrt{2}}{2}$

$f'''(x) = -\cos x \qquad f'''\left(\frac{\pi}{4}\right) = -\frac{\sqrt{2}}{2}$

$f^{(4)}(x) = \sin x \qquad f^{(4)}\left(\frac{\pi}{4}\right) = \frac{\sqrt{2}}{2}$

$f^{(5)}(x) = \cos x \qquad f^{(5)}\left(\frac{\pi}{4}\right) = \frac{\sqrt{2}}{2}$

$f^{(6)}(x) = -\sin x$

(c)

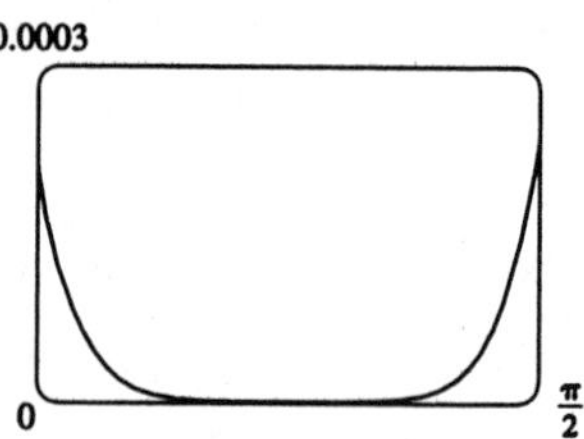

From the graph, it seems that the error is less than 0.00026 on $\left(0, \frac{\pi}{2}\right)$.

(a) $\sin x \approx T_5(x) = \frac{\sqrt{2}}{2} + \frac{\sqrt{2}}{2}\left(x - \frac{\pi}{4}\right) - \frac{\sqrt{2}}{4}\left(x - \frac{\pi}{4}\right)^2 - \frac{\sqrt{2}}{12}\left(x - \frac{\pi}{4}\right)^3 + \frac{\sqrt{2}}{48}\left(x - \frac{\pi}{4}\right)^4 + \frac{\sqrt{2}}{240}\left(x - \frac{\pi}{4}\right)^5$

(b) The remainder is $R_5(x) = \frac{1}{6!} f^{(6)}(z)\left(x - \frac{\pi}{4}\right)^6 = \frac{1}{720}(-\sin z)\left(x - \frac{\pi}{4}\right)^6$, where z lies between $\frac{\pi}{4}$ and x. Since $0 \le x \le \frac{\pi}{2}$, $-\frac{\pi}{4} \le x - \frac{\pi}{4} \le \frac{\pi}{4} \quad \Rightarrow \quad 0 \le \left(x - \frac{\pi}{4}\right)^6 \le \left(\frac{\pi}{4}\right)^6$, and since $0 < z < \frac{\pi}{2}$, $0 < \sin z < 1$, so $|R_5(x)| < \frac{1}{720}\left(\frac{\pi}{4}\right)^6 \approx 0.00033$.

16. $f(x) = \cos x \qquad f\left(\frac{\pi}{3}\right) = \frac{1}{2}$

$f'(x) = -\sin x \qquad f'\left(\frac{\pi}{3}\right) = -\frac{\sqrt{3}}{2}$

$f''(x) = -\cos x \qquad f''\left(\frac{\pi}{3}\right) = -\frac{1}{2}$

$f'''(x) = \sin x \qquad f'''\left(\frac{\pi}{3}\right) = \frac{\sqrt{3}}{2}$

$f^{(4)}(x) = \cos x \qquad f^{(4)}\left(\frac{\pi}{3}\right) = \frac{1}{2}$

$f^{(5)}(x) = -\sin x$

(c)

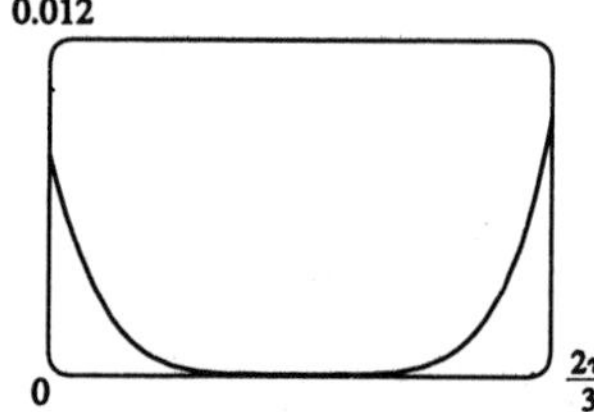

From the graph, it seems that the error is less than 0.01 on $\left(0, \frac{2\pi}{3}\right)$.

(a) $\cos x \approx T_4(x) = \frac{1}{2} - \frac{\sqrt{3}}{2}\left(x - \frac{\pi}{3}\right) - \frac{1}{4}\left(x - \frac{\pi}{3}\right)^2 + \frac{\sqrt{3}}{12}\left(x - \frac{\pi}{3}\right)^3 + \frac{1}{48}\left(x - \frac{\pi}{3}\right)^4$

(b) The remainder is $R_4(x) = \frac{1}{5!}(-\sin z)\left(x - \frac{\pi}{3}\right)^5$, where z is between x and $\frac{\pi}{3}$.

$0 \le x \le \frac{2\pi}{3} \quad \Rightarrow \quad \left|x - \frac{\pi}{3}\right| < \frac{\pi}{3} \quad \Rightarrow \quad |R_4(x)| < \frac{1}{5!}\left(\frac{\pi}{3}\right)^5 \approx 0.0105.$

17. $f(x) = \tan x \qquad f(0) = 0$

$f'(x) = \sec^2 x \qquad f'(0) = 1$

$f''(x) = 2\sec^2 x \tan x \qquad f''(0) = 0$

$f'''(x) = 4\sec^2 x \tan^2 x + 2\sec^4 x \qquad f'''(0) = 2$

$f^{(4)}(x) = 8\sec^2 x \tan^3 x + 16\sec^4 x \tan x$

(c)

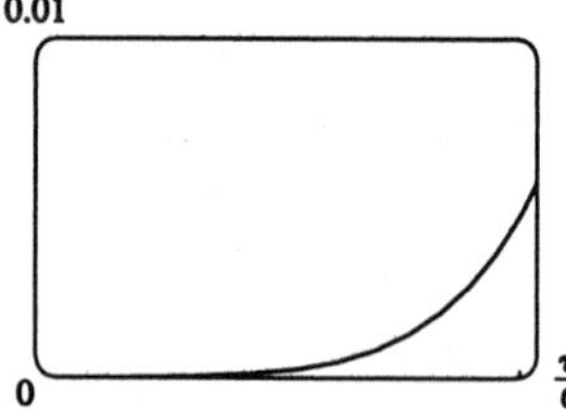

From the graph, it seems that the error is less than 0.006 on $(0, \pi)$

(a) $\tan x \approx T_3(x) = x + \frac{1}{3}x^3$

(b) $R_3(x) = \dfrac{f^{(4)}(z)}{4!}x^4 = \dfrac{8\sec^2 z \tan^3 z + 16\sec^4 z \tan z}{4!}x^4$

$= \dfrac{\sec^2 z \tan^3 z + 2\sec^4 z \tan z}{3}$ where z lies between 0 and x. Now $0 \le x^4 \le \left(\dfrac{\pi}{6}\right)^4$ and $0 < z < \dfrac{\pi}{6}$

$\Rightarrow \quad \sec^2 z < \frac{4}{3}$ and $\tan z < \frac{\sqrt{3}}{3}$, so $|R_3(x)| < \dfrac{\frac{4}{3}\cdot\frac{1}{3\sqrt{3}} + 2\cdot\frac{16}{9}\cdot\frac{1}{\sqrt{3}}}{3}\left(\dfrac{\pi}{6}\right)^4 = \dfrac{4\sqrt{3}}{9}\left(\dfrac{\pi}{6}\right)^4 < 0.06.$

18. $f(x) = (1+x^2)^{1/3}$ $\qquad f(0) = 1$

$f'(x) = \frac{2}{3}x(1+x^2)^{-2/3}$ $\qquad f'(0) = 0$

$f''(x) = \frac{2}{3}\left(1 - \frac{1}{3}x^2\right)(1+x^2)^{-5/3}$ $\qquad f''(0) = \frac{2}{3}$

$$f'''(x) = \frac{8x^3 - 72x}{27(1+x^2)^{8/3}}$$

(a) $\sqrt[3]{1+x^2} \approx T_2(x) = 1 + \frac{1}{3}x^2$

(b) $R_2(x) = \dfrac{8z^3 - 72z}{3!\,27(1+z^2)^{8/3}}x^3$, with z between 0 and x. $1 + z^2 > 1$ and $|z| < |x| \le 0.5$, so $|R_2(x)| < \dfrac{8(0.5)^3 + 72(0.5)}{3!\,27 \cdot 1}(0.5)^3 \approx 0.0285$ (using the Triangle Inequality). *Or:* $R_2(x) = \dfrac{|8z(z^2-9)|}{3!\,27(1+z^2)^{8/3}}|x|^3 \le \dfrac{8(0.5)(9)}{6 \cdot 27 \cdot 1} = \dfrac{1}{36} < 0.028$.

(c)

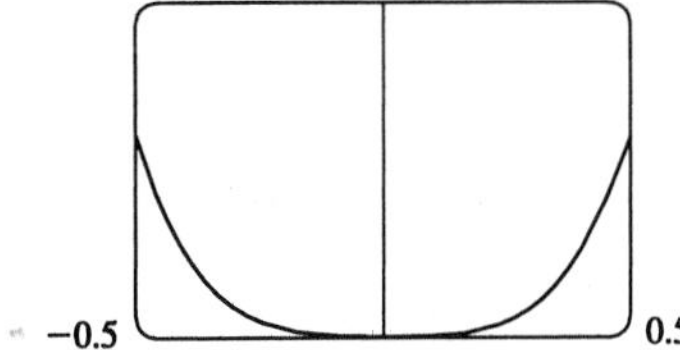

It seems that the error is less than 0.0061 on $(-0.5, 0.5)$

19. $f(x) = e^{x^2}$ $\qquad f(0) = 1$

$f'(x) = e^{x^2}(2x)$ $\qquad f'(0) = 0$

$f''(x) = e^{x^2}(2 + 4x^2)$ $\qquad f''(0) = 2$

$f'''(x) = e^{x^2}(12x + 8x^3)$ $\qquad f'''(0) = 0$

$f^{(4)}(x) = e^{x^2}\left(12 + 48x^2 + 16x^4\right)$

(a) $e^{x^2} \approx T_3(x) = 1 + x^2$

(b) $R_3(x) = \dfrac{f^{(4)}(z)}{4!}x^4 = \dfrac{e^{z^2}\left(3 + 12z^2 + 4z^4\right)}{6}x^4$, where z lies between 0 and x. $0 \le x \le 0.1 \quad \Rightarrow \quad |R_3(x)| < \dfrac{e^{0.01}(3 + 0.12 + 0.0004)}{6}(0.0001) < 0.00006$.

(c)

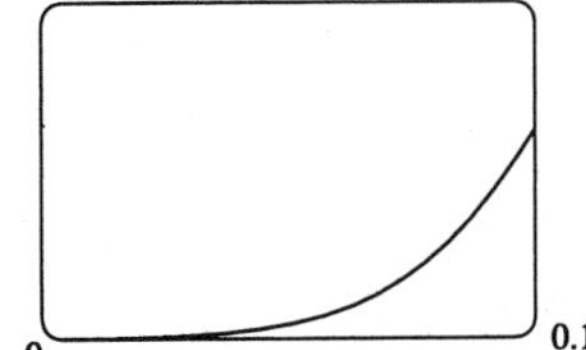

It appears that the error is less than 0.00005 on $(0, 0.1)$.

20. **(a)** Clearly $f^{(2n)}(0) = 1$ and $f^{(2n+1)}(0) = 0$, so

$$\cosh x \approx T_5(x) = 1 + \frac{x^2}{2} + \frac{x^4}{24}.$$

(b)

$$R_5(x) = \frac{f^{(6)}(z)}{6!}x^6 = \frac{\cosh z}{6!}x^6 \le \frac{\cosh(1)}{6!} \approx 0.00214$$

(c)

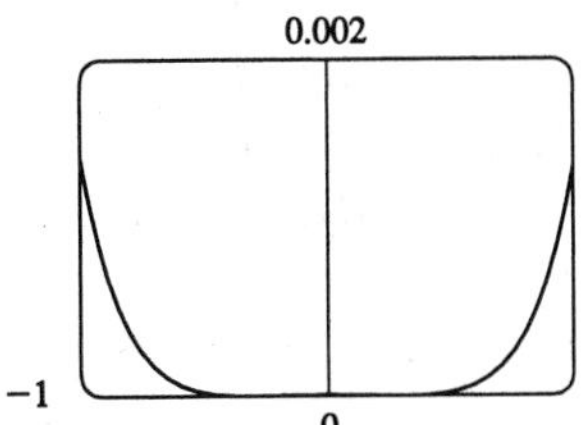

It appears that the error is less than 0.0015 on $(-1, 1)$.

21. $f(x) = x^{3/4}$ $\qquad$ $f(16) = 8$

$f'(x) = \frac{3}{4}x^{-1/4}$ $\qquad$ $f'(16) = \frac{3}{8}$

$f''(x) = -\frac{3}{16}x^{-5/4}$ $\qquad$ $f''(16) = -\frac{3}{512}$

$f'''(x) = \frac{15}{64}x^{-9/4}$ $\qquad$ $f'''(16) = \frac{15}{32{,}768}$

$f^{(4)}(x) = -\frac{135}{256}x^{-13/4}$

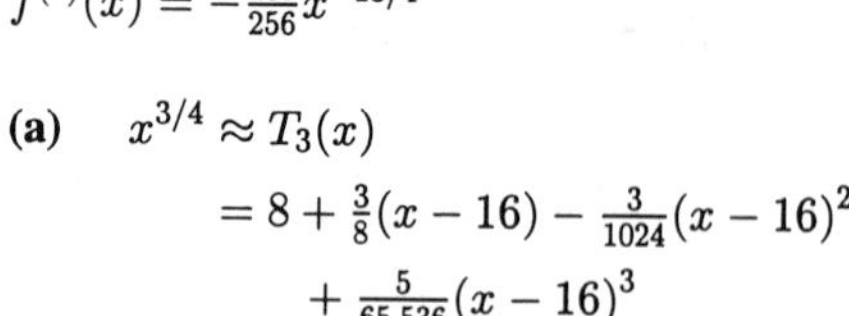

(a) $x^{3/4} \approx T_3(x)$
$= 8 + \frac{3}{8}(x-16) - \frac{3}{1024}(x-16)^2 + \frac{5}{65{,}536}(x-16)^3$

(c)

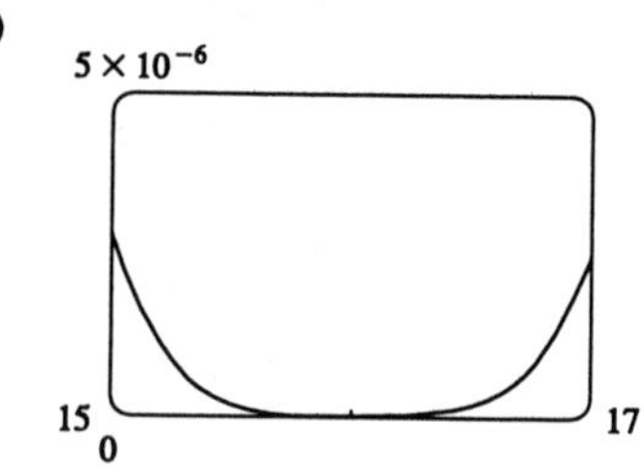

It appears that the error is less than 3×10^{-6} on $(15, 17)$.

(b) $R_3(x) = \dfrac{f^{(4)}(z)}{4!}(x-16)^4 = -\dfrac{135(x-16)^4}{256 \cdot 4!\, z^{13/4}}$, where z lies between 16 and x. $|x - 16| \leq 1$ and $z > 15$ $\Rightarrow$ $|R_3(x)| < \dfrac{135}{256 \cdot 24 \cdot 15^{13/4}} < 0.0000034.$

22. $f(x) = \ln x$ $\qquad$ $f(4) = \ln 4$

$f'(x) = x^{-1}$ $\qquad$ $f'(4) = \frac{1}{4}$

$f''(x) = -x^{-2}$ $\qquad$ $f''(4) = -\frac{1}{16}$

$f'''(x) = 2x^{-3}$ $\qquad$ $f'''(4) = \frac{1}{32}$

$f^{(4)}(x) = -6x^{-4}$

(a) $\ln x \approx T_3(x)$
$= \ln 4 + \frac{1}{4}(x-4) - \frac{1}{32}(x-4)^2 + \frac{1}{192}(x-4)^3$

(c)

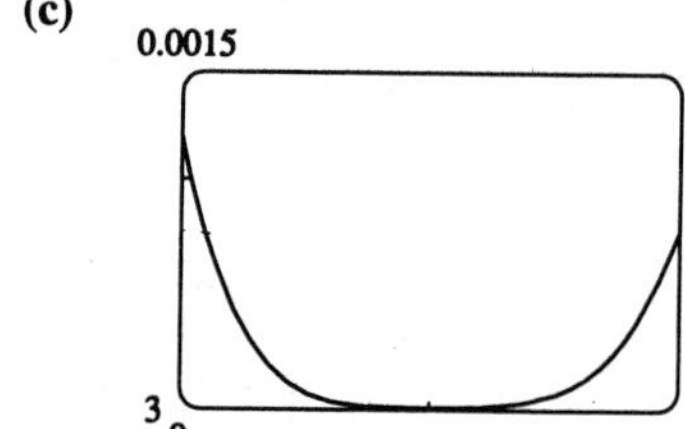

It appears that the error is less than 0.0013 on $(3, 5)$.

(b) $R_3(x) = \dfrac{-6z^{-4}}{4!}(x-4)^4$, where z lies between x and 4. $3 \leq x \leq 5$ $\Rightarrow$ $|x - 4| \leq 1$ and $z > 3$ $\Rightarrow$ $\dfrac{1}{z^4} < \dfrac{1}{3^4}$, so $|R_3(x)| < \dfrac{6}{4!\,3^4} = \dfrac{1}{324} \approx 0.0031.$

23. From Exercise 1, $\sin x = \frac{1}{2} + \frac{\sqrt{3}}{2}\left(x - \frac{\pi}{6}\right) - \frac{1}{4}\left(x - \frac{\pi}{6}\right)^2 - \frac{\sqrt{3}}{12}\left(x - \frac{\pi}{6}\right)^3 + R_3(x)$, where $R_3(x) = \dfrac{\sin z}{4!}\left(x - \frac{\pi}{6}\right)^4$ and z lies between $\frac{\pi}{6}$ and x. Now $35° = \left(\frac{\pi}{6} + \frac{\pi}{36}\right)$ radians, so the error is $\left|R_3\left(\frac{\pi}{36}\right)\right| < \frac{(\pi/36)^4}{4!} < 0.000003$. Therefore, to five decimal places, $\sin 35° \approx \frac{1}{2} + \frac{\sqrt{3}}{2}\left(\frac{\pi}{36}\right) - \frac{1}{4}\left(\frac{\pi}{36}\right)^2 + \frac{\sqrt{3}}{12}\left(\frac{\pi}{36}\right)^3 \approx 0.57358.$

24. From Exercise 16, $\cos x = \frac{1}{2} - \frac{\sqrt{3}}{2}\left(x - \frac{\pi}{3}\right) - \frac{1}{4}\left(x - \frac{\pi}{3}\right)^2 + \frac{\sqrt{3}}{12}\left(x - \frac{\pi}{3}\right)^3 + \frac{1}{48}\left(x - \frac{\pi}{3}\right)^4 + R_4(x)$. Now since $x = 69° = \left(\frac{\pi}{3} + \frac{\pi}{20}\right)$ radians, the error is $|R_4(x)| < \frac{(\pi/20)^5}{5!} < 8 \times 10^{-7}$. Therefore, to five decimal places, $\cos 69° \approx \frac{1}{2} - \frac{\sqrt{3}}{2}\left(\frac{\pi}{20}\right) - \frac{1}{4}\left(\frac{\pi}{20}\right)^2 + \frac{\sqrt{3}}{12}\left(\frac{\pi}{20}\right)^3 + \frac{1}{48}\left(\frac{\pi}{20}\right)^4 \approx 0.35837.$

25. All derivatives of e^x are e^x, so the remainder term is $R_n(x) = \dfrac{e^z}{(n+1)!}x^{n+1}$, where $0 < z < 0.1$. So we want $R_n(0.1) \leq \dfrac{e^{0.1}}{(n+1)!}(0.1)^{n+1} < 0.00001$, and we find that $n = 3$ satisfies this inequality. [In fact $R_3(0.1) < 0.0000046$.]

26. From Exercise 10.10.31, the Maclaurin series for $\ln(1+x)$ is $\sum_{n=1}^{\infty}\frac{(-1)^{n-1}}{n}x^n$. So $\ln 1.4=\sum_{n=1}^{\infty}\frac{(-1)^{n-1}}{n}(0.4)^n$.

Since this is an alternating series, the error is less than the first neglected term by Theorem 10.5.1, and we find that $|a_6|=0.4^6/6\approx 0.0007<0.001$. So we need the first five (non-zero) terms of the Maclaurin series for the desired accuracy.

27. $\sin x=x-\frac{1}{3!}x^3+\frac{1}{5!}x^5-\cdots$. By the Alternating Series Estimation Theorem, the error in the approximation $\sin x=x-\frac{1}{3!}x^3$ is less than $\left|\frac{1}{5!}x^5\right|<0.01 \quad\Leftrightarrow\quad |x^5|<1.2 \quad\Leftrightarrow\quad |x|<(1.2)^{1/5}\approx 1.037$.

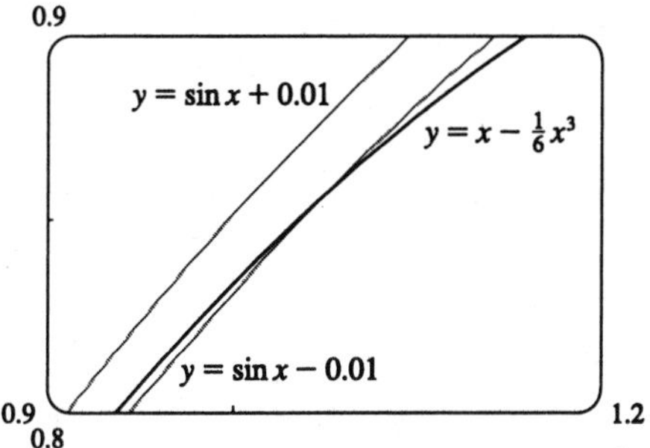

The graph confirms our estimate. Since both the sine function and the given approximation are odd functions, we only need to check the estimate for $x>0$.

28. By Theorem 10.5.1, the error is less than $\left|-\frac{1}{6!}x^6\right|<0.005 \;\Leftrightarrow\; x^6<3.6 \;\Leftrightarrow\; |x|<(3.6)^{1/6}\approx 1.238$.

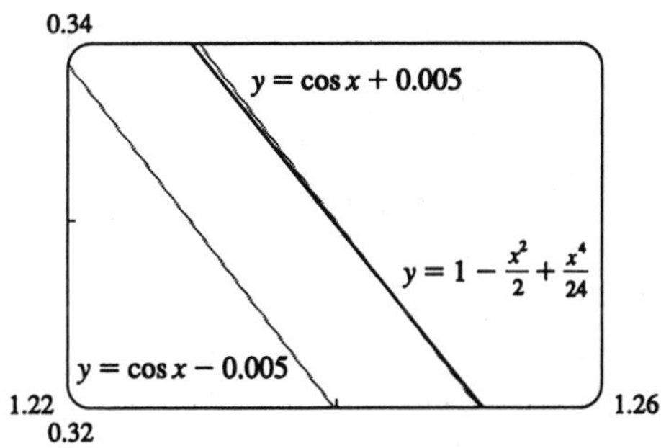

The graph confirms our estimate. Since both the cosine function and the given approximation are even functions, we only need to check the estimate for $x>0$.

29. Let $s(t)$ be the position function of the car, and for convenience set $s(0)=0$. The velocity of the car is $v(t)=s'(t)$ and the acceleration is $a(t)=s''(t)$, so the second degree Taylor polynomial is $T_2(t)=s(0)+v(0)t+\frac{a(0)}{2}t^2=20t+t^2$. We estimate the distance travelled during the next second to be $s(1)\approx T_2(1)=20+1=21$ m. The function $T_2(t)$ would not be accurate over a full minute, since the car could not possibly maintain an acceleration of 2 m/s^2 for that long (if it did, its final speed would be $140\text{ m/s}\approx 315\text{ mi/h}$!)

30. **(a)** We expand the denominator of Planck's Law using the Taylor series $e^x = 1 + x + \frac{x^2}{2!} + \frac{x^3}{3!} + \cdots$ with $x = \frac{hc}{\lambda kT}$, and use the fact that if λ is large, then all subsequent terms in the Taylor expansion are very small compared to the first one, so we can approximate using the Taylor polynomial T_1:

$$f(\lambda) = \frac{8\pi hc\lambda^{-5}}{e^{hc/(\lambda kT)} - 1} = \frac{8\pi hc\lambda^{-5}}{\left[1 + \frac{hc}{\lambda kT} + \frac{1}{2!}\left(\frac{hc}{\lambda kT}\right)^2 + \frac{1}{3!}\left(\frac{hc}{\lambda kT}\right)^3 + \cdots\right] - 1} \approx \frac{8\pi hc\lambda^{-5}}{\left(1 + \frac{hc}{\lambda kT}\right) - 1}$$

$$= \frac{8\pi kT}{\lambda^4}, \text{ which is the Rayleigh-Jeans Law.}$$

(b) To convert to μm, we substitute $\lambda/10^6$ for λ in both laws. We can see that the two laws are very different for short wavelengths (Planck's Law gives a maximum at $\lambda \approx 0.5$ μm; the Rayleigh-Jeans Law gives no extremum.) The two laws are similar for large λ.

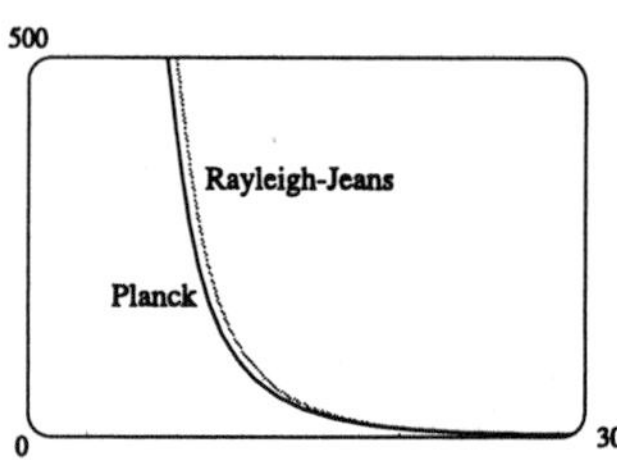

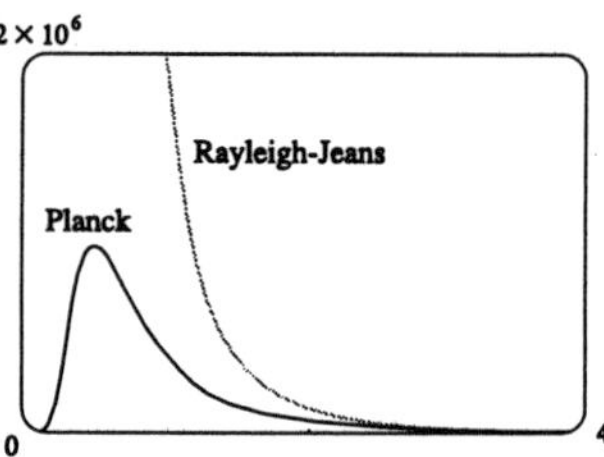

31. $E = \frac{q}{D^2} - \frac{q}{(D+d)^2} = \frac{q}{D^2} - \frac{q}{D^2(1 + d/D)^2} = \frac{q}{D^2}\left[1 - \left(1 + \frac{d}{D}\right)^{-2}\right].$

We use the Binomial Series to expand $(1 + d/D)^{-2}$:

$$E = \frac{q}{D^2}\left[1 - \left(1 - 2\left(\frac{d}{D}\right) + \frac{2\cdot 3}{2!}\left(\frac{d}{D}\right)^2 - \frac{2\cdot 3\cdot 4}{3!}\left(\frac{d}{D}\right)^3 + \cdots\right)\right]$$

$$= \frac{q}{D^2}\left[2\left(\frac{d}{D}\right) - 3\left(\frac{d}{D}\right)^2 + 4\left(\frac{d}{D}\right)^3 - \cdots\right] \approx 2qd \cdot \frac{1}{D^3} \text{ when } D \text{ is much larger than } d.$$

32. **(a)**
$\rho(t) = \rho_{20}e^{\alpha(t-20)}$ $\quad \rho(20) = \rho_{20}$
$\rho'(x) = \alpha\rho_{20}e^{\alpha(t-20)}$ $\quad \rho'(20) = \alpha\rho_{20}$
$\rho''(x) = \alpha^2\rho_{20}e^{\alpha(t-20)}$ $\quad \rho''(20) = \alpha^2\rho_{20}$

The linear approximation is
$T_1(t) = \rho(20) + \rho'(20)(t-20) = \rho_{20}[1 + \alpha(t-20)]$.
The quadratic approximation is

$$T_2(t) = \rho(20) + \rho'(20)(t-20) + \frac{\rho''(20)}{2}(t-20)^2$$
$$= \rho_{20}\left[1 + \alpha(t-20) + \tfrac{1}{2}\alpha^2(t-20)^2\right].$$

(b)

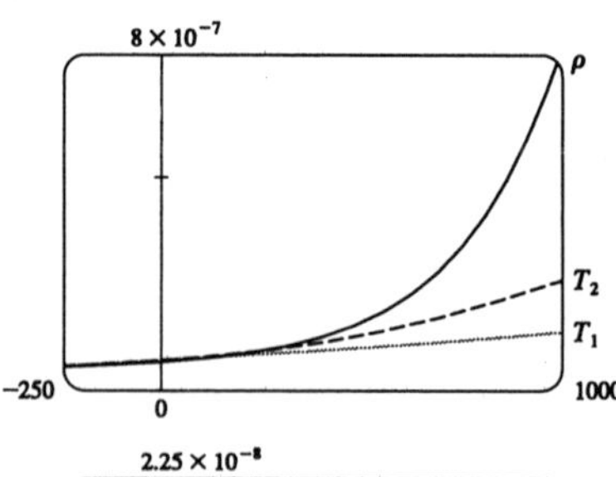

(c) From the graph, it seems that $T_1(t)$ is within 1% of $\rho(t)$, that is, $0.99\rho(t) \le T_1(t) \le 1.01\rho(t)$, for $-14°\text{C} \le t \le 58°\text{C}$.

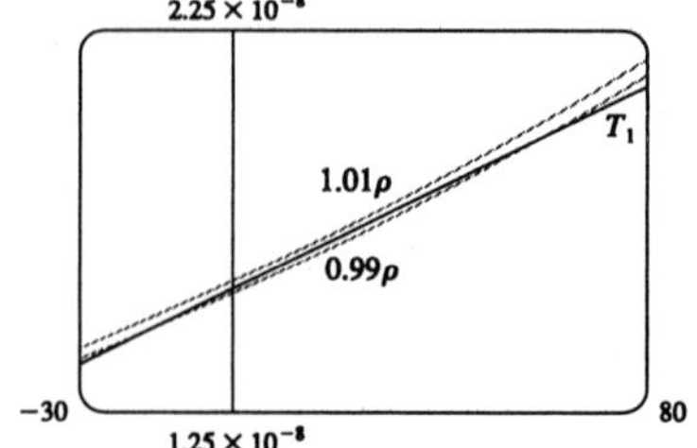

33. **(a)** If the water is deep, then $2\pi d/L$ is large, and we know that $\tanh x \to 1$ as $x \to \infty$. So we can approximate $\tanh(2\pi d/L) \approx 1$, and so $v^2 \approx gL/(2\pi) \quad \Leftrightarrow \quad v \approx \sqrt{gL/(2\pi)}$.

(b) From the calculations at right, the first term in the Maclaurin series of $\tanh x$ is x, so if the water is shallow, we can approximate $\tanh \frac{2\pi d}{L} \approx \frac{2\pi d}{L}$, and so $v^2 \approx \frac{gL}{2\pi} \cdot \frac{2\pi d}{L} \quad \Leftrightarrow \quad v \approx \sqrt{gd}$.

$$\begin{array}{ll} f(x) = \tanh x & f(0) = 0 \\ f'(x) = \operatorname{sech}^2 x & f'(0) = 1 \\ f''(x) = -2\operatorname{sech}^2 x \tanh x & f''(0) = 0 \\ f'''(x) = 2\operatorname{sech}^2 x(3\tanh^2 x - 1) & f'''(0) = -2 \end{array}$$

(c) Since $\tanh x$ is an odd function, its Maclaurin series is alternating, so the error in the approximation $\tanh \frac{2\pi d}{L} \approx \frac{2\pi d}{L}$ is less than the first neglected term, which is $\frac{|f'''(0)|}{3!}\left(\frac{2\pi d}{L}\right)^3 = \frac{1}{3}\left(\frac{2\pi d}{L}\right)^3$. If $L > 10d$, then $\frac{1}{3}\left(\frac{2\pi d}{L}\right)^3 < \frac{1}{3}\left(2\pi \cdot \frac{1}{10}\right)^3 = \frac{\pi^3}{375}$, so the error in the approximation $v^2 = gd$ is less than $\frac{gL}{2\pi} \cdot \frac{\pi^3}{375} \approx 0.0132gL$.

34. $T_n(x) = f(a) + \frac{f'(a)}{1!}(x-a) + \frac{f''(a)}{2!}(x-a)^2 + \cdots + \frac{f^{(n)}(a)}{n!}(x-a)^n$. Let $0 \le m \le n$. Then

$$T_n^{(m)}(x) = m!\,\frac{f^{(m)}(a)}{m!}(x-a)^0 + (m+1)(m)\cdots(2)\frac{f^{(m+1)}(a)}{(m+1)!}(x-a)^1$$
$$+ \cdots + n(n-1)\cdots(n-m+1)\frac{f^{(n)}(z)}{n!}(x-a)^{n-m}.$$

For $x = a$, all terms in this sum except the first one are 0, so $T_n^{(m)}(a) = \frac{m!\,f^{(m)}(a)}{m!} = f^{(m)}(a)$.

35. Using Taylor's Formula with $n = 1$, $a = x_n$, $x = r$, we get $f(r) = f(x_n) + f'(x_n)(r - x_n) + R_1(x)$, where $R_1(x) = \frac{1}{2}f''(z)(r-x_n)^2$ and z lies between x_n and r. But r is a root, so $f(r) = 0$ and Taylor's Formula becomes $0 = f(x_n) + f'(x_n)(r - x_n) + \frac{1}{2}f''(z)(r-x_n)^2$. Taking the first two terms to the left side and dividing by $f'(x_n)$, we have $x_n - r - \frac{f(x_n)}{f'(x_n)} = \frac{1}{2}\frac{f''(z)}{f'(x_n)}|x_n - r|^2$. By the formula for Newton's Method, we have $|x_{n+1} - r| = \left|x_n - \frac{f(x_n)}{f'(x_n)} - r\right| = \frac{1}{2}\frac{|f''(z)|}{|f'(x_n)|}|x_n - r|^2 \le \frac{M}{2K}|x_n - r|^2$ since $|f''(z)| \le M$ and $|f'(x_n)| \ge K$.

36. $q!(e - s_q) = q!\left(\frac{p}{q} - 1 - \frac{1}{1!} - \frac{1}{2!} - \cdots - \frac{1}{q!}\right) = p(q-1)! - q! - q! - \frac{q!}{2!} - \cdots - 1$, which is clearly an integer, and $q!(e - s_q) = q!\left[\frac{e^z}{(q+1)!}\right] = \frac{e^z}{q+1}$. We have $0 < \frac{e^z}{q+1} < \frac{e}{q+1} < \frac{e}{3} < 1$ since $0 < z < 1$ and $q > 2$, and so $0 < q!(e - s_q) < 1$, which is a contradiction since we have already shown $q!(e - s_q)$ must be an integer. So e cannot be rational.

REVIEW EXERCISES FOR CHAPTER 10

1. False. See the warning in Note 2 after Theorem 10.2.6.

2. True by Theorem 10.8.3. *Or:* Use the Comparison Test to show that $\sum a_n(-2)^n$ converges absolutely.

3. False. For example, take $a_n = (-1)^n/(n6^n)$.

4. True by Theorem 10.8.3.

5. False, since $\lim\limits_{n\to\infty}\left|\dfrac{a_{n+1}}{a_n}\right| = \lim\limits_{n\to\infty}\left|\dfrac{n^3}{(n+1)^3}\right| = \lim\limits_{n\to\infty}\dfrac{1}{(1+1/n)^3} = 1$.

6. True since $\lim\limits_{n\to\infty}\left|\dfrac{a_{n+1}}{a_n}\right| = \lim\limits_{n\to\infty}\left|\dfrac{n!}{(n+1)!}\right| = \lim\limits_{n\to\infty}\dfrac{1}{n+1} = 0$.

7. False. See the remarks after Example 3 in Section 10.4.

8. True since $\dfrac{1}{e} = e^{-1}$ and $e^x = \sum\limits_{n=0}^{\infty}\dfrac{x^n}{n!}$ and so $e^{-1} = \sum\limits_{n=0}^{\infty}\dfrac{(-1)^n}{n!}$.

9. False. A power series has the form $a_0 + a_1x + a_2x^2 + a_3x^3 + \cdots$.

10. False. See the discussion in Section 10.10 before Example 1.

11. True. See Example 8 in Section 10.1.

12. True, because if $\sum |a_n|$ is convergent, then so is $\sum a_n$ by Theorem 10.6.3.

13. True. By Theorem 10.10.5 the coefficient of x^3 is $\dfrac{f'''(0)}{3!} = \dfrac{1}{3} \quad\Rightarrow\quad f'''(0) = 2$.

Or: Use Theorem 10.9.2 to differentiate f three times.

14. False. Let $a_n = n$ and $b_n = -n$. Then $\{a_n\}$ and $\{b_n\}$ are divergent, but $a_n + b_n = 0$, so $\{a_n + b_n\}$ is convergent.

15. False. For example, let $a_n = b_n = (-1)^n$. Then $\{a_n\}$ and $\{b_n\}$ are divergent, but $a_nb_n = 1$, so $\{a_nb_n\}$ is convergent.

16. True by Theorem 10.1.10, since $\{a_n\}$ is decreasing and $0 < a_n \le a_1$ for all $n \quad\Rightarrow\quad \{a_n\}$ is bounded.

17. True by Theorem 10.6.3. $\left[\sum(-1)^n a_n\right.$ is absolutely convergent and hence convergent.$\left.\right]$

18. True. $\lim\limits_{n\to\infty}\dfrac{a_{n+1}}{a_n} < 1 \Rightarrow \sum a_n$ converges (Ratio Test) $\Rightarrow \lim\limits_{n\to\infty} a_n = 0$ (Theorem 10.2.6).

19. $\lim\limits_{n\to\infty}\dfrac{n}{2n+5} = \lim\limits_{n\to\infty}\dfrac{1}{2+5/n} = \dfrac{1}{2}$ and the sequence is convergent.

20. $\lim\limits_{n\to\infty}[5 - (0.9)^n] = 5 - 0 = 5$. Convergent

21. $\{2n+5\}$ is divergent since $2n+5 \to \infty$ as $n \to \infty$.

22. $\left\{\dfrac{n}{\ln n}\right\}$ diverges, since $\displaystyle\lim_{x\to\infty}\frac{x}{\ln x}\overset{H}{=}\lim_{x\to\infty}\frac{1}{1/x}=\infty$.

23. $\{\sin n\}$ is divergent since $\displaystyle\lim_{n\to\infty}\sin n$ does not exist.

24. $\left\{\dfrac{\sin n}{n}\right\}$ converges, since $-\dfrac{1}{n}\le\dfrac{\sin n}{n}\le\dfrac{1}{n}$ and $\pm\dfrac{1}{n}\to 0$ as $n\to\infty$, so $\displaystyle\lim_{n\to\infty}\frac{\sin n}{n}=0$ by the Squeeze Theorem.

25. $\left\{\left(1+\dfrac{3}{n}\right)^{4n}\right\}$ is convergent. Let $y=\left(1+\dfrac{3}{x}\right)^{4x}$. Then

$$\lim_{x\to\infty}\ln y=\lim_{x\to\infty}4x\ln(1+3/x)=\lim_{x\to\infty}\frac{\ln(1+3/x)}{1/(4x)}\overset{H}{=}\lim_{x\to\infty}\frac{\dfrac{1}{1+3/x}\left(-\dfrac{3}{x^2}\right)}{-1/(4x^2)}$$

$$=\lim_{x\to\infty}\frac{12}{1+3/x}=12,\text{ so }\lim_{x\to\infty}y=\lim_{n\to\infty}\left(1+\frac{3}{n}\right)^{4n}=e^{12}.$$

26. $\left\{\dfrac{(-10)^n}{n!}\right\}$ converges, since $\dfrac{10^n}{n!}=\dfrac{10\cdot 10\cdot 10\cdot\cdots\cdot 10}{1\cdot 2\cdot 3\cdot\cdots\cdot 10}\cdot\dfrac{10\cdot 10\cdot\cdots\cdot 10}{11\cdot 12\cdot\cdots\cdot n}\le 10^{10}\left(\dfrac{10}{11}\right)^{n-10}\to 0$ as $n\to\infty$, so $\displaystyle\lim_{n\to\infty}\frac{(-10)^n}{n!}=0$ (Squeeze Theorem). *Or:* Use (10.10.11).

27. We use induction, hypothesizing that $a_{n-1}<a_n<2$. Note first that $1<a_2=\frac{1}{3}(1+5)=\frac{5}{3}<2$, so the hypothesis holds for $n=2$. Now assume that $a_{k-1}<a_k<2$. Then $a_k=\frac{1}{3}(a_{k-1}+4)<\frac{1}{3}(a_k+4)<\frac{1}{3}(2+4)=2$. So $a_k<a_{k+1}<2$, and the induction is complete. To find the limit of the sequence, we note that $L=\displaystyle\lim_{n\to\infty}a_n=\lim_{n\to\infty}a_{n+1}\;\Rightarrow\;L=\frac{1}{3}(L+4)\;\Rightarrow\;L=2$.

28. $$\lim_{x\to\infty}\frac{x^4}{e^x}\overset{H}{=}\lim_{x\to\infty}\frac{4x^3}{e^x}\overset{H}{=}\lim_{x\to\infty}\frac{12x^2}{e^x}\overset{H}{=}\lim_{x\to\infty}\frac{24x}{e^x}\overset{H}{=}\lim_{x\to\infty}\frac{24}{e^x}=0.$$

Then we conclude from Theorem 10.1.2 that $\displaystyle\lim_{n\to\infty}n^4e^{-n}=0$. From the graph, it seems that $12^4e^{-12}>0.1$, but $n^4e^{-n}<0.1$ whenever $n>12$. So the smallest value of N corresponding to $\epsilon=0.1$ in the definition of the limit is $N=12$.

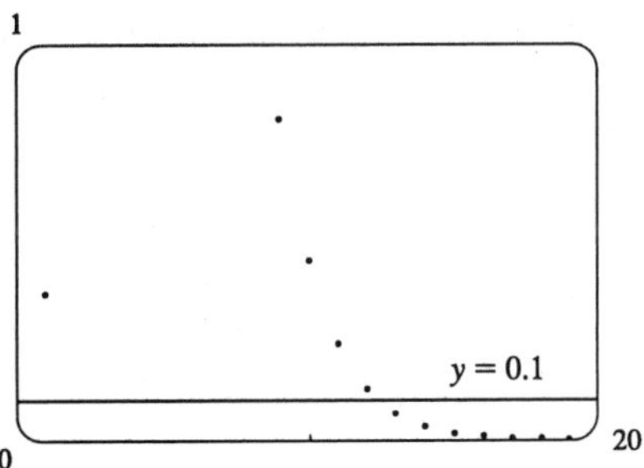

29. Use the Limit Comparison Test with $a_n=\dfrac{n^2}{n^3+1}$ and $b_n=\dfrac{1}{n}$. $\displaystyle\lim_{n\to\infty}\frac{a_n}{b_n}=\lim_{n\to\infty}\frac{n^2/(n^3+1)}{1/n}=\lim_{n\to\infty}\frac{1}{1+1/n^3}=1$. Since $\displaystyle\sum_{n=1}^{\infty}\frac{1}{n}$ (the harmonic series) diverges, $\displaystyle\sum_{n=1}^{\infty}\frac{n^2}{n^3+1}$ diverges also.

30. Let $a_n=\dfrac{n+n^2}{n+n^4}$ and $b_n=\dfrac{1}{n^2}$. Then $\displaystyle\lim_{n\to\infty}\frac{a_n}{b_n}=\lim_{n\to\infty}\frac{n^4+n^3}{n^4+n}=1>0$, so since $\displaystyle\sum_{n=1}^{\infty}b_n$ converges ($p=2>1$), $\displaystyle\sum_{n=1}^{\infty}\frac{n+n^2}{n+n^4}$ converges also by the Limit Comparison Test.

31. An alternating series with $a_n = \frac{1}{n^{1/4}}$, $a_n > 0$ for all n, and $a_n > a_{n+1}$. $\lim_{n\to\infty} a_n = \lim_{n\to\infty} \frac{1}{n^{1/4}} = 0$, so the series converges by the Alternating Series Test.

32. $\lim_{n\to\infty}\left|\frac{a_{n+1}}{a_n}\right| = \frac{1}{3}\lim_{n\to\infty}\left(\frac{n+1}{n}\right)^2 = \frac{1}{3} < 1$, so the series converges by the Ratio Test.

33. $\lim_{n\to\infty}\sqrt[n]{|a_n|} = \lim_{n\to\infty}\frac{n}{3n+1} = \frac{1}{3} < 1$, so series converges by the Root Test.

34. $\lim_{n\to\infty}\left(\frac{n-1}{n}\right)^{1/2} = 1$ so the series diverges by the Test for Divergence.

35. $\frac{|\sin n|}{1+n^2} \le \frac{1}{1+n^2} < \frac{1}{n^2}$ and since $\sum_{n=1}^{\infty}\frac{1}{n^2}$ converges (p-series with $p = 2 > 1$), so does $\sum_{n=1}^{\infty}\frac{|\sin n|}{1+n^2}$ by the Comparison Test.

36. $f(x) = \frac{1}{x(\ln x)^2}$ is continuous, positive, and decreasing on $[2,\infty)$, so we can use the Integral Test.

$\int_2^{\infty}\frac{dx}{x(\ln x)^2} = \lim_{t\to\infty}\left[\frac{-1}{\ln x}\right]_2^t = \frac{1}{\ln 2}$ so the series converges.

37. $\lim_{n\to\infty}\left|\frac{a_{n+1}}{a_n}\right| = \lim_{n\to\infty}\frac{1\cdot 3\cdot 5\cdot\cdots\cdot(2n-1)(2n+1)}{5^{n+1}(n+1)!}\cdot\frac{5^n\,n!}{1\cdot 3\cdot 5\cdot\cdots\cdot(2n-1)} = \lim_{n\to\infty}\frac{2n+1}{5(n+1)}$
$= \frac{2}{5} < 1$, so the series converges by the Ratio Test.

38. If $f(x) = \frac{\ln x}{\sqrt{x}}$ then $f'(x) = \frac{2-\ln x}{2x^{3/2}} < 0$ for $x > e^2$ so $\{a_n\}$ is eventually decreasing, and since $\lim_{n\to\infty}\frac{\ln n}{\sqrt{n}} = 0$ (use l'Hospital's Rule), the series converges by the Alternating Series Test.

39. $\lim_{n\to\infty}\left|\frac{a_{n+1}}{a_n}\right| = \lim_{n\to\infty}\frac{4^{n+1}}{(n+1)3^{n+1}}\cdot\frac{n3^n}{4^n} = \frac{4}{3}\lim_{n\to\infty}\frac{n}{n+1} = \frac{4}{3} > 1$ so the series diverges by the Ratio Test.

40. Use the Limit Comparison Test with $a_n = \frac{\sqrt{n+1}-\sqrt{n-1}}{n} = \frac{2}{n\left(\sqrt{n+1}+\sqrt{n-1}\right)}$ (rationalizing the numerator) and $b_n = \frac{1}{n^{3/2}}$. $\lim_{n\to\infty}\frac{a_n}{b_n} = \lim_{n\to\infty}\frac{2\sqrt{n}}{\sqrt{n+1}+\sqrt{n-1}} = 1$, so since $\sum_{n=1}^{\infty} b_n$ converges ($p = \frac{3}{2} > 1$), $\sum_{n=1}^{\infty} a_n$ converges also.

41. Consider the series of absolute values: $\sum_{n=1}^{\infty} n^{-1/3}$ is a p-series with $p = \frac{1}{3} < 1$ and is therefore divergent. But if we apply the Alternating Series Test we see that $a_{n+1} < a_n$ and $\lim_{n\to\infty} n^{-1/3} = 0$. Therefore $\sum_{n=1}^{\infty}(-1)^{n-1}n^{-1/3}$ is conditionally convergent.

42. $\sum_{n=1}^{\infty}\left|(-1)^{n-1}n^{-3}\right| = \sum_{n=1}^{\infty} n^{-3}$ is a convergent p-series ($p = 3 > 1$.) Therefore $\sum_{n=1}^{\infty}(-1)^{n-1}n^{-3}$ is absolutely convergent.

43. $\left|\dfrac{a_{n+1}}{a_n}\right| = \left|\dfrac{(-1)^{n+1}(n+2)3^{n+1}}{2^{2n+3}} \cdot \dfrac{2^{2n+1}}{(-1)^n(n+1)3^n}\right| = \dfrac{n+2}{n+1} \cdot \dfrac{3}{4} = \dfrac{1+(2/n)}{1+(1/n)} \cdot \dfrac{3}{4} \to \dfrac{3}{4} < 1$ as $n \to \infty$ so by the Ratio Test, $\displaystyle\sum_{n=1}^{\infty} \frac{(-1)^n(n+1)3^n}{2^{2n+1}}$ is absolutely convergent.

44. $\displaystyle\lim_{x\to\infty} \frac{\sqrt{x}}{\ln x} \overset{\text{H}}{=} \lim_{x\to\infty} \frac{1/(2\sqrt{x})}{1/x} = \lim_{x\to\infty} \frac{\sqrt{x}}{2} = \infty$. Therefore $(-1)^{n+1}\dfrac{\sqrt{n}}{\ln n}$ does not approach 0, so the given series is divergent by the Test for Divergence.

45. Convergent geometric series. $\displaystyle\sum_{n=1}^{\infty} \frac{2^{2n+1}}{5^n} = 2\sum_{n=1}^{\infty} \frac{4^n}{5^n} = 2\left(\frac{4/5}{1-4/5}\right) = 8.$

46. $\displaystyle\sum_{n=1}^{\infty} \frac{1}{n(n+3)} = \sum_{n=1}^{\infty} \left[\frac{1}{3n} - \frac{1}{3(n+3)}\right]$ (partial fractions).

$\displaystyle s_n = \sum_{i=1}^{n} \left[\frac{1}{3i} - \frac{1}{3(i+3)}\right] = \frac{1}{3} + \frac{1}{6} + \frac{1}{9} - \frac{1}{3(n+1)} - \frac{1}{3(n+2)} - \frac{1}{3(n+3)}$ (telescoping sum) so

$\displaystyle\sum_{n=1}^{\infty} \frac{1}{n(n+3)} = \lim_{n\to\infty} s_n = \frac{1}{3} + \frac{1}{6} + \frac{1}{9} = \frac{11}{18}.$

47. $\displaystyle\sum_{n=1}^{\infty} \left[\tan^{-1}(n+1) - \tan^{-1}n\right] = \lim_{n\to\infty} \left[\left(\tan^{-1}2 - \tan^{-1}1\right) + \left(\tan^{-1}3 - \tan^{-1}2\right) + \cdots + \left(\tan^{-1}(n+1) - \tan^{-1}n\right)\right]$

$\displaystyle = \lim_{n\to\infty} \left[\tan^{-1}(n+1) - \tan^{-1}1\right] = \tfrac{\pi}{2} - \tfrac{\pi}{4} = \tfrac{\pi}{4}$

48. $\displaystyle\sum_{n=0}^{\infty} \frac{(-1)^n x^n}{2^{2n} n!} = \sum_{n=0}^{\infty} \frac{(-x/4)^n}{n!} = e^{-x/4}$

49. $1.2 + 0.0\overline{345} = \dfrac{12}{10} + \dfrac{345/10{,}000}{1 - 1/1000} = \dfrac{12}{10} + \dfrac{345}{9990} = \dfrac{4111}{3330}$

50. This is a geometric series which converges whenever $|\ln x| < 1 \quad\Rightarrow\quad -1 < \ln x < 1 \quad\Rightarrow\quad e^{-1} < x < e$.

51. $\displaystyle\sum_{n=1}^{\infty} \frac{(-1)^{n+1}}{n^5} = 1 - \frac{1}{32} + \frac{1}{243} - \frac{1}{1024} + \frac{1}{3125} - \frac{1}{7776} + \frac{1}{16{,}807} - \frac{1}{32{,}768} + \cdots.$

Since $\dfrac{1}{32{,}768} < 0.000031$, $\displaystyle\sum_{n=1}^{\infty} \frac{(-1)^{n+1}}{n^5} \approx \sum_{n=1}^{7} \frac{(-1)^{n+1}}{n^5} \approx 0.9721.$

52. **(a)** $\displaystyle s_5 = \sum_{n=1}^{5} \frac{1}{n^6} = 1 + \frac{1}{2^6} + \cdots + \frac{1}{5^6} \approx 1.017305$. The series $\displaystyle\sum_{n=1}^{5} \frac{1}{n^6}$ converges by the Integral Test, so we estimate the remainder R_5 with (10.3.4): $\displaystyle R_5 \le \int_5^{\infty} \frac{dx}{x^6} = \left[-\frac{x^{-5}}{5}\right]_5^{\infty} = \frac{5^{-5}}{5} = 0.000064$. So the error is at most 0.000064.

(b) In general, $\displaystyle R_n \le \int_n^{\infty} \frac{dx}{x^6} = \frac{1}{5n^5}$. If we take $n = 9$, then $s_9 \approx 1.01734$ and $R_9 \le \dfrac{1}{5 \cdot 9^5} \approx 3.4 \times 10^{-6}$.

So to five decimal places, $\displaystyle\sum_{n=1}^{\infty} \frac{1}{n^5} \approx \sum_{n=1}^{9} \frac{1}{n^5} \approx 1.01734.$

Another Method: Use (10.3.5) instead of (10.3.4).

53. $\displaystyle\sum_{n=1}^{\infty}\frac{1}{2+5^n}\approx\sum_{n=1}^{8}\frac{1}{2+5^n}\approx 0.18976224$. To estimate the error, note that $\dfrac{1}{2+5^n}<\dfrac{1}{5^n}$, so the remainder term is $\displaystyle R_8=\sum_{n=9}^{\infty}\frac{1}{2+5^n}<\sum_{n=9}^{\infty}\frac{1}{5^n}=\frac{1/5^9}{1-1/5}\approx 6.4\times10^{-7}$ (geometric series with $a=1/5^9$ and $r=\frac{1}{5}$).

54. **(a)** $\displaystyle\lim_{n\to\infty}\left|\frac{a_{n+1}}{a_n}\right|=\lim_{n\to\infty}\frac{(n+1)^{n+1}(2n)!}{(2n+2)!\,n^n}=\lim_{n\to\infty}\left(1+\frac{1}{n}\right)^n\frac{1}{2(2n+1)}=e\cdot0=0<1$ so the series converges by the Ratio Test.

(b) The series in part (a) is convergent, so $\lim\limits_{n\to\infty}a_n=0$ by Theorem 10.2.6.

55. Use the Limit Comparison Test. $\displaystyle\lim_{n\to\infty}\left|\frac{\left(\frac{n+1}{n}\right)a_n}{a_n}\right|=\lim_{n\to\infty}\frac{n+1}{n}=\lim_{n\to\infty}\left(1+\frac{1}{n}\right)=1>0$. Since $\sum|a_n|$ is convergent, so is $\displaystyle\sum\left|\left(\frac{n+1}{n}\right)a_n\right|$ by the Limit Comparison Test.

56. $\displaystyle\lim_{n\to\infty}\left|\frac{a_{n+1}}{a_n}\right|=3x^2\lim_{n\to\infty}\frac{n+1}{n+2}=3x^2<1$ for convergence $\Rightarrow$ $|x|<\frac{1}{\sqrt3}$ so $R=\frac{1}{\sqrt3}$. If $x=\pm\frac{1}{\sqrt3}$ the series is $\displaystyle\sum_{n=1}^{\infty}\frac{(-1)^n}{n+1}$ which converges by the Alternating Series Test, so the interval of convergence is $\left[-\dfrac{1}{\sqrt3},\dfrac{1}{\sqrt3}\right]$.

57. $\displaystyle\lim_{n\to\infty}\left|\frac{a_{n+1}}{a_n}\right|=\lim_{n\to\infty}\left|\frac{x^{n+1}}{3^{n+1}(n+1)^3}\cdot\frac{3^nn^3}{x^n}\right|=\frac{|x|}{3}\lim_{n\to\infty}\left(\frac{n}{n+1}\right)^3=\frac{|x|}{3}<1$ for convergence (Ratio Test) $\Rightarrow$ $|x|<3$ and the radius of convergence is 3. When $x=\pm3$, $\displaystyle\sum_{n=1}^{\infty}|a_n|=\sum_{n=1}^{\infty}\frac{1}{n^3}$ which is a convergent p-series ($p=3>1$), so the interval of convergence is $[-3,3]$.

58. $\displaystyle\lim_{n\to\infty}\sqrt[n]{|a_n|}=\lim_{n\to\infty}\frac{x+1}{n}=0$ for all x so $R=\infty$ and $I=(-\infty,\infty)$.

59. $\displaystyle\lim_{n\to\infty}\left|\frac{a_{n+1}}{a_n}\right|=\lim_{n\to\infty}\left|\frac{2^{n+1}(x-3)^{n+1}}{\sqrt{n+4}}\cdot\frac{\sqrt{n+3}}{2^n(x-3)^n}\right|=2|x-3|\lim_{n\to\infty}\sqrt{\frac{n+3}{n+4}}=2|x-3|<1$ $\Leftrightarrow$ $|x-3|<\frac{1}{2}$ so the radius of convergence is $\frac{1}{2}$. For $x=\frac{7}{2}$ the series becomes $\displaystyle\sum_{n=0}^{\infty}\frac{1}{\sqrt{n+3}}=\sum_{n=3}^{\infty}\frac{1}{n^{1/2}}$ which diverges ($p=\frac{1}{2}<1$), but for $x=\frac{5}{2}$ we get $\displaystyle\sum_{n=0}^{\infty}\frac{(-1)^n}{\sqrt{n+3}}$ which is a convergent alternating series, so the interval of convergence is $\left[\frac{5}{2},\frac{7}{2}\right)$.

60. $\displaystyle\lim_{n\to\infty}\left|\frac{a_{n+1}}{a_n}\right|=\lim_{n\to\infty}\frac{(2n+2)(2n+1)}{(n+1)(n+1)}|x|=4|x|<1$ to converge, so $R=\frac{1}{4}$.

61. $f(x)=\sin x$ $\quad f\left(\frac{\pi}{6}\right)=\frac{1}{2}$

$f'(x)=\cos x$ $\quad f'\left(\frac{\pi}{6}\right)=\frac{\sqrt3}{2}$

$f''(x)=-\sin x$ $\quad f''\left(\frac{\pi}{6}\right)=-\frac{1}{2}$

$f'''(x)=-\cos x$ $\quad f'''\left(\frac{\pi}{6}\right)=-\frac{\sqrt3}{2}$

$f^{(4)}(x)=\sin x$ $\quad f^{(4)}\left(\frac{\pi}{6}\right)=\frac{1}{2}$

$\cdots \qquad \cdots$

$f^{(2n)}\left(\frac{\pi}{6}\right)=(-1)^n\cdot\frac{1}{2}$ and

$f^{(2n+1)}\left(\frac{\pi}{6}\right)=(-1)^n\cdot\frac{\sqrt3}{2}$.

$$\sin x=\sum_{n=0}^{\infty}\frac{f^{(n)}\left(\frac{\pi}{6}\right)}{n!}\left(x-\frac{\pi}{6}\right)^n$$

$$=\sum_{n=0}^{\infty}\frac{(-1)^n}{2(2n)!}\left(x-\frac{\pi}{6}\right)^{2n}+\sum_{n=0}^{\infty}\frac{(-1)^n\sqrt3}{2(2n+1)!}\left(x-\frac{\pi}{6}\right)^{2n+1}$$

62. $f(x) = \cos x \qquad f\left(\frac{\pi}{3}\right) = \frac{1}{2}$ $\qquad \cos x = \sum_{n=0}^{\infty} \frac{(-1)^n \left(x - \frac{\pi}{3}\right)^{2n}}{2(2n)!} + \sum_{n=0}^{\infty} \frac{(-1)^{n+1} \sqrt{3} \left(x - \frac{\pi}{3}\right)^{2n+1}}{2(2n+1)!}$

$f'(x) = -\sin x \qquad f'\left(\frac{\pi}{3}\right) = -\frac{\sqrt{3}}{2}$

$f''(x) = -\cos x \qquad f''\left(\frac{\pi}{3}\right) = -\frac{1}{2}$

$f'''(x) = \sin x \qquad f'''\left(\frac{\pi}{3}\right) = \frac{\sqrt{3}}{2}$

$f^{(4)}(x) = \cos x \qquad f^{(4)}\left(\frac{\pi}{3}\right) = \frac{1}{2}$

63. $\frac{1}{1+x} = \frac{1}{1-(-x)} = \sum_{n=0}^{\infty} (-1)^n x^n$ for $|x| < 1 \quad \Rightarrow \quad \frac{x^2}{1+x} = \sum_{n=0}^{\infty} (-1)^n x^{n+2}$ with $R = 1$.

64. $\left(1 - x^2\right)^{1/2} = \sum_{n=0}^{\infty} \binom{1/2}{n} \left(-x^2\right)^n = 1 - \frac{x^2}{2} - \sum_{n=2}^{\infty} \frac{1 \cdot 3 \cdot 5 \cdot \cdots \cdot (2n-3) x^{2n}}{2^n\, n!}$, $R = 1$.

65. $\frac{1}{1-x} = \sum_{n=0}^{\infty} x^n$ for $|x| < 1 \quad \Rightarrow \quad \ln(1-x) = -\int \frac{dx}{1-x} = -\int \sum_{n=0}^{\infty} x^n \, dx = C - \sum_{n=0}^{\infty} \frac{x^{n+1}}{n+1}$.

$\ln(1-0) = C - 0 \quad \Rightarrow \quad C = 0 \quad \Rightarrow \quad \ln(1-x) = -\sum_{n=0}^{\infty} \frac{x^{n+1}}{n+1} = \sum_{n=1}^{\infty} \frac{-x^n}{n}$ with $R = 1$.

66. $e^x = \sum_{n=0}^{\infty} \frac{x^n}{n!} \quad \Rightarrow \quad xe^{2x} = \sum_{n=0}^{\infty} \frac{x(2x)^n}{n!} = \sum_{n=0}^{\infty} \frac{2^n\, x^{n+1}}{n!}$, $R = \infty$

67. $\sin x = \sum_{n=0}^{\infty} \frac{(-1)^n x^{2n+1}}{(2n+1)!} \quad \Rightarrow \quad \sin\left(x^4\right) = \sum_{n=0}^{\infty} \frac{(-1)^n \left(x^4\right)^{2n+1}}{(2n+1)!} = \sum_{n=0}^{\infty} \frac{(-1)^n x^{8n+4}}{(2n+1)!}$ for all x, so the radius of convergence is ∞.

68. $10^x = e^{x \ln 10} = \sum_{n=0}^{\infty} \frac{(\ln 10)^n x^n}{n!}$, $R = \infty$

69. $(16 - x)^{-1/4} = \frac{1}{2}\left(1 - \frac{1}{16}x\right)^{-1/4} = \frac{1}{2}\left[1 + \left(-\frac{1}{4}\right)\left(-\frac{x}{16}\right) + \frac{\left(-\frac{1}{4}\right)\left(-\frac{5}{4}\right)}{2!}\left(-\frac{x}{16}\right)^2 + \cdots\right]$

$= \frac{1}{2} + \sum_{n=1}^{\infty} \frac{1 \cdot 5 \cdot 9 \cdot \cdots \cdot (4n-3)}{2 \cdot 4^n \cdot n! \cdot 16^n} x^n = \frac{1}{2} + \sum_{n=1}^{\infty} \frac{1 \cdot 5 \cdot 9 \cdot \cdots \cdot (4n-3)}{2^{6n+1}\, n!} x^n$ for $\left|-\frac{x}{16}\right| < 1 \quad \Rightarrow \quad R = 16$.

70. $(1 - 3x)^{-5} = \sum_{n=0}^{\infty} \binom{-5}{n} (-3x)^n = 1 + \sum_{n=1}^{\infty} \frac{5 \cdot 6 \cdot \cdots \cdot (n+4) \cdot 3^n\, x^n}{n!}$, $|3x| < 1$ so $R = \frac{1}{3}$.

71. $e^x = \sum_{n=0}^{\infty} \frac{x^n}{n!}$ so $\frac{e^x}{x} = \frac{1}{x} + \sum_{n=1}^{\infty} \frac{x^{n-1}}{n!}$ and $\int \frac{e^x}{x}\, dx = C + \ln|x| + \sum_{n=1}^{\infty} \frac{x^n}{n \cdot n!}$

72. $\left(1 + x^4\right)^{1/2} = \sum_{n=0}^{\infty} \binom{1/2}{n} \left(x^4\right)^n = 1 + \left(\frac{1}{2}\right)x^4 + \frac{\left(\frac{1}{2}\right)\left(-\frac{1}{2}\right)}{2!}\left(x^4\right)^2 + \frac{\left(\frac{1}{2}\right)\left(-\frac{1}{2}\right)\left(-\frac{3}{2}\right)}{3!}\left(x^4\right)^3 + \cdots$

$= 1 + \frac{1}{2}x^4 - \frac{1}{8}x^8 + \frac{1}{16}x^{12}$, so

$\int_0^1 \left(1 + x^4\right)^{1/2} dx = \left[x + \frac{1}{10}x^5 - \frac{1}{72}x^9 + \frac{1}{208}x^{13} - \cdots\right]_0^1 = 1 + \frac{1}{10} - \frac{1}{72} + \frac{1}{208} - \cdots$. This is an alternating series, so by Theorem 10.5.1, the error in the approximation $\int_0^1 \left(1 + x^4\right)^{1/2} dx \approx 1 + \frac{1}{10} - \frac{1}{72} \approx 1.086$ is less than $\frac{1}{208}$, sufficient for the desired accuracy. Thus, correct to two decimal places, $\int_0^1 \left(1 + x^4\right)^{1/2} dx \approx 1.09$.

73. **(a)** $f(x) = x^{1/2}$ $\quad f(1) = 1$

$f'(x) = \frac{1}{2}x^{-1/2}$ $\quad f'(1) = \frac{1}{2}$

$f''(x) = -\frac{1}{4}x^{-3/2}$ $\quad f''(1) = -\frac{1}{4}$

$f'''(x) = \frac{3}{8}x^{-5/2}$ $\quad f'''(1) = \frac{3}{8}$

$f^{(4)}(x) = -\frac{15}{16}x^{-7/2}$

$\sqrt{x} \approx T_3(x) = 1 + \frac{1}{2}(x-1)$
$\quad - \frac{1}{8}(x-1)^2 + \frac{1}{16}(x-1)^3$

(b)

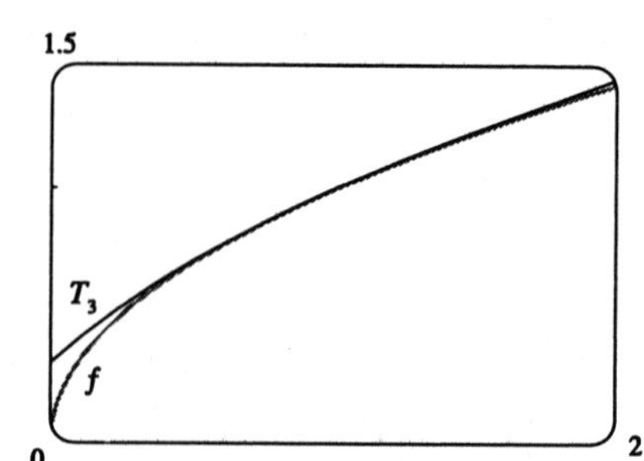

(c) By Taylor's Formula,

$$R_3(x) = \frac{f^{(4)}(z)}{4!}(x-1)^4 = -\frac{5(x-1)^4}{128z^{7/2}},$$

with z between x and 1. If $0.9 \le x \le 1.1$ then $0 \le |x-1| \le 0.1$ and $z^{7/2} > (0.9)^{7/2}$ so

$$|R_3(x)| < \frac{5(0.1)^4}{128(0.9)^{7/2}} < 0.000006.$$

(d)

5×10^{-6}

0.9 0 1.1

It appears that the error is less than 5×10^{-6} on $(0.9, 1.1)$.

74. **(a)** $f(x) = \sec x$ $\quad f(0) = 1$

$f'(x) = \sec x \tan x$ $\quad f'(0) = 0$

$f''(x) = \sec x \tan^2 x + \sec^3 x$ $\quad f''(0) = 1$

$f'''(x) = \sec x \tan^3 x + 5\sec^3 x \tan x$

$\sec x \approx T_2(x) = 1 + \dfrac{x^2}{2}$

(b)

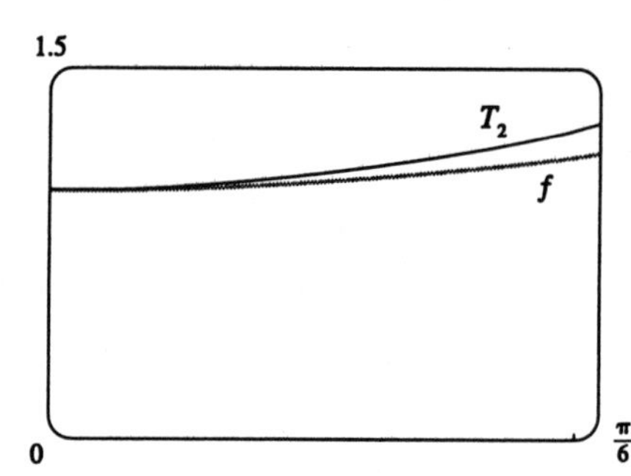

(c) $R_2(x) = \dfrac{\sec z \tan^3 z + 5\sec^3 z \tan z}{3!}x^3$

with z between 0 and x. If $0 < z < \frac{\pi}{6}$,

$1 < \sec z < \frac{2}{\sqrt{3}}$ and $0 < \tan z < \frac{1}{\sqrt{3}}$,

so $|R_2(x)| < \dfrac{2/9 + 40/9}{6}\left(\dfrac{\pi}{6}\right)^3 \approx 0.1117.$

(d)

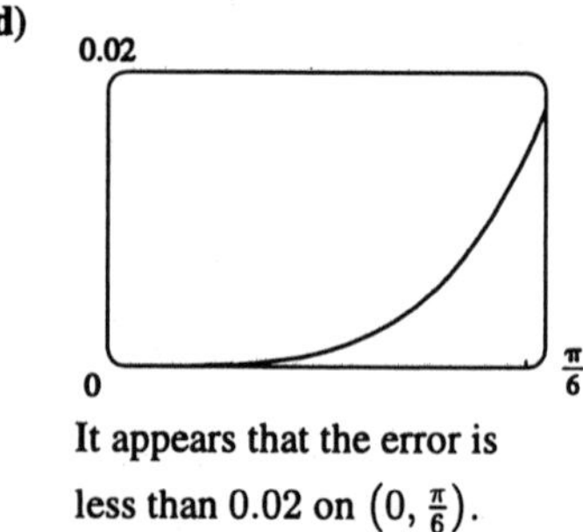

It appears that the error is less than 0.02 on $\left(0, \frac{\pi}{6}\right)$.

75. $e^x = \sum_{n=0}^{\infty} \frac{x^n}{n!} \Rightarrow e^{-1/x^2} = \sum_{n=0}^{\infty} \frac{(-1/x^2)^n}{n!} = 1 - \frac{1}{x^2} + \frac{1}{2x^4} - \cdots \Rightarrow$

$x^2\left(1 - e^{-1/x^2}\right) = x^2\left(\frac{1}{x^2} - \frac{1}{2x^4} + \cdots\right) = 1 - \frac{1}{2x^2} + \cdots \to 1$ as $x \to \infty$.

76. **(a)** $F = \frac{mgR^2}{(R+h)^2} = \frac{mg}{(1+h/R)^2} = mg\sum_{n=0}^{\infty}\binom{-2}{n}\left(\frac{h}{R}\right)^n$ (Binomial Series)

(b) We expand $F = mg\left[1 - 2(h/R) + 3(h/R)^2 - \cdots\right]$. This is an alternating series, so by (10.5.1), the error in the approximation $F = mg$ is less than $2mgh/R$, so for accuracy within 1% we want

$\left|\frac{2mgh/R}{mgR^2/(R+h)^2}\right| < 0.01 \Leftrightarrow \frac{2h(R+h)^2}{R^3} < 0.01$. This inequality would be difficult to solve

solve for h, so we substitute $R = 6{,}400$ km and plot both sides of the inequality. It appears that the approximation is accurate to within 1% for $h < 31$ km.

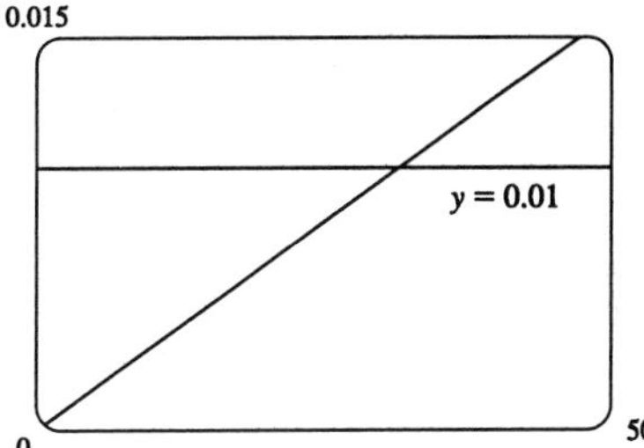

77. $f(x) = \sum_{n=0}^{\infty} c_n x^n \Rightarrow f(-x) = \sum_{n=0}^{\infty} c_n(-x)^n = \sum_{n=0}^{\infty}(-1)^n c_n x^n$

(a) If f is an odd function, then $f(-x) = -f(x) \Rightarrow \sum_{n=0}^{\infty}(-1)^n c_n x^n = \sum_{n=0}^{\infty} -c_n x^n$. The coefficients of any power series are uniquely determined (by Theorem 10.10.5), so $(-1)^n c_n = -c_n$. If n is even, then $(-1)^n = 1$, so $c_n = -c_n \Rightarrow 2c_n = 0 \Rightarrow c_n = 0$. Thus all even coefficients are 0.

(b) If f is even, then $f(-x) = f(x) \Rightarrow \sum_{n=0}^{\infty}(-1)^n c_n x^n = \sum_{n=0}^{\infty} c_n x^n \Rightarrow (-1)^n c_n = c_n$. If n is odd, then $(-1)^n = -1$, so $-c_n = c_n \Rightarrow 2c_n = 0 \Rightarrow c_n = 0$. Thus all odd coefficients are 0.

78. $e^x = \sum_{n=0}^{\infty} \frac{x^n}{n!} \Rightarrow e^{x^2} = \sum_{n=0}^{\infty} \frac{(x^2)^n}{n!} = \sum_{n=0}^{\infty} \frac{x^{2n}}{n!} = \sum_{k=0}^{\infty} \frac{f^{(k)}(0)}{k!}x^k \Rightarrow \frac{f^{(2n)}(0)}{(2n)!} = \frac{1}{n!} \Rightarrow f^{(2n)}(0) = \frac{(2n)!}{n!}$.

79. Let $f(x) = \sum_{m=0}^{\infty} c_m x^m$ and $g(x) = e^{f(x)} = \sum_{n=0}^{\infty} d_n x^n$. Then $g'(x) = \sum_{n=0}^{\infty} n d_n x^{n-1}$, so nd_n occurs as the coefficient of x^{n-1}. But also

$$g'(x) = e^{f(x)}f'(x) = \left(\sum_{n=0}^{\infty} d_n x^n\right)\left(\sum_{m=1}^{\infty} m c_m x^{m-1}\right)$$
$$= \left(d_0 + d_1 x + d_2 x^2 + \cdots + d_{n-1}x^{n-1} + \cdots\right)\left(c_1 + 2c_2 x + 3c_3 x^2 + \cdots + nc_n x^{n-1} + \cdots\right),$$ so the

coefficient of x^{n-1} is $c_1 d_{n-1} + 2c_2 d_{n-2} + 3c_3 d_{n-3} + \cdots + nc_n d_0 = \sum_{i=1}^{n} ic_i d_{n-i}$. Therefore $nd_n = \sum_{i=1}^{n} ic_i d_{n-i}$.

PROBLEMS PLUS (page 678)

1. It would be far too much work to compute 15 derivatives of f. The key idea is to remember that $f^{(n)}(0)$ occurs in the coefficient of x^n in the Maclaurin series of f. We start with the Maclaurin series for sin: $\sin x = x - \frac{x^3}{3!} + \frac{x^5}{5!} - \cdots$. Then $\sin(x^3) = x^3 - \frac{x^9}{3!} + \frac{x^{15}}{5!} - \cdots$ and so the coefficient of x^{15} is $\frac{f^{(15)}(0)}{15!} = \frac{1}{5!}$. Therefore, $f^{(15)}(0) = \frac{15!}{5!} = 6 \cdot 7 \cdot 8 \cdot 9 \cdot 10 \cdot 11 \cdot 12 \cdot 13 \cdot 14 \cdot 15 = 10{,}897{,}286{,}400$.

2. We use the problem-solving strategy of taking cases:

Case (i): If $|x| < 1$, then $0 \le x^2 < 1$, so $\lim_{n\to\infty} x^{2n} = 0$ (see Example 8 in Section 10.1) and

$$f(x) = \lim_{n\to\infty} \frac{x^{2n} - 1}{x^{2n} + 1} = \frac{0-1}{0+1} = -1.$$

Case (ii): If $|x| = 1$, that is, $x = \pm 1$, then $x^2 = 1$, so $f(x) = \lim_{n\to\infty} \frac{1-1}{1+1} = 0$.

Case (iii): If $|x| > 1$, then $x^2 > 1$, so $\lim_{n\to\infty} x^{2n} = \infty$ and

$$f(x) = \lim_{n\to\infty} \frac{x^{2n} - 1}{x^{2n} + 1} = \lim_{n\to\infty} \frac{1 - (1/x^{2n})}{1 + (1/x^{2n})} = \frac{1-0}{1+0} = 1.$$

Thus $f(x) = \begin{cases} 1 & \text{if } x < -1 \\ 0 & \text{if } x = -1 \\ -1 & \text{if } -1 < x < 1 \\ 0 & \text{if } x = 1 \\ 1 & \text{if } x > 1 \end{cases}$

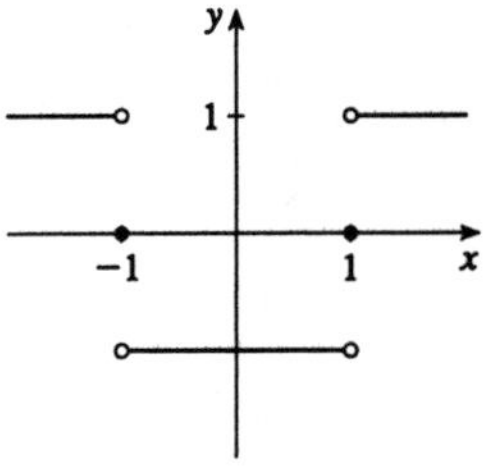

The graph shows that f is continuous everywhere except at $x = \pm 1$.

3. **(a)** From Formula 14a in Appendix D, with $x = y = \theta$, we get $\tan 2\theta = \frac{2\tan\theta}{1 - \tan^2\theta}$, so $\cot 2\theta = \frac{1 - \tan^2\theta}{2\tan\theta}$ $\Rightarrow$ $2\cot 2\theta = \frac{1 - \tan^2\theta}{\tan\theta} = \cot\theta - \tan\theta$. Replacing θ by $\frac{1}{2}x$, we get $2\cot x = \cot\frac{1}{2}x - \tan\frac{1}{2}x$, or $\tan\frac{1}{2}x = \cot\frac{1}{2}x - 2\cot x$.

(b) From part (a) we have $\tan\frac{x}{2^n} = \cot\frac{x}{2^n} - 2\cot\frac{x}{2^{n-1}}$, so the nth partial sum of the given series is

$$\begin{aligned} s_n &= \frac{\tan(x/2)}{2} + \frac{\tan(x/4)}{4} + \frac{\tan(x/8)}{8} + \cdots + \frac{\tan(x/2^n)}{2^n} \\ &= \left[\frac{\cot(x/2)}{2} - \cot x\right] + \left[\frac{\cot(x/4)}{4} - \frac{\cot(x/2)}{2}\right] + \left[\frac{\cot(x/8)}{8} - \frac{\cot(x/4)}{4}\right] \\ &\qquad + \cdots + \left[\frac{\cot(x/2^n)}{2^n} - \frac{\cot(x/2^{n-1})}{2^{n-1}}\right] \\ &= -\cot x + \frac{\cot(x/2^n)}{2^n} \quad \text{(telescoping sum).} \end{aligned}$$

Now $\dfrac{\cot(x/2^n)}{2^n} = \dfrac{\cos(x/2^n)}{2^n\sin(x/2^n)} = \dfrac{\cos(x/2^n)}{x} \cdot \dfrac{x/2^n}{\sin(x/2^n)} \to \dfrac{1}{x} \cdot 1 = \dfrac{1}{x}$ as $n \to \infty$ since $\dfrac{x}{2^n} \to 0$ for $x \neq 0$. Therefore, if $x \neq 0$ and $x \neq n\pi$, then $\displaystyle\sum_{n=1}^{\infty} \frac{1}{2^n} \tan \frac{x}{2^n} = \lim_{n\to\infty}\left(-\cot x + \frac{1}{2^n}\cot\frac{x}{2^n}\right) = -\cot x + \frac{1}{x}$.

If $x = 0$, then all terms in the series are 0, so the sum is 0.

4. $x = \displaystyle\int_1^t \frac{\cos u}{u}\,du$, $y = \displaystyle\int_1^t \frac{\sin u}{u}\,du$, so by Part 1 of the Fundamental Theorem of Calculus, we have $\dfrac{dx}{dt} = \dfrac{\cos t}{t}$ and $\dfrac{dy}{dt} = \dfrac{\sin t}{t}$. Vertical tangent lines occur when $\dfrac{dx}{dt} = 0 \iff \cos t = 0$. The parameter value corresponding to $(x, y) = (0, 0)$ is $t = 1$, so the nearest vertical tangent occurs when $t = \frac{\pi}{2}$. Therefore the arc length between these points is

$$L = \int_1^{\pi/2} \sqrt{\left(\frac{dx}{dt}\right)^2 + \left(\frac{dy}{dt}\right)^2}\,dt = \int_1^{\pi/2} \sqrt{\frac{\cos^2 t}{t^2} + \frac{\sin^2 t}{t^2}}\,dt = \int_1^{\pi/2} \frac{dt}{t} = [\ln t]_1^{\pi/2} = \ln \tfrac{\pi}{2}.$$

5. **(a)** At each stage, each side is replaced by four shorter sides, each of length $\frac{1}{3}$ of the side length at the preceding stage. Writing s_0 and ℓ_0 for the number of sides and the length of the side of the initial triangle, we generate the following table.

$s_0 = 3$	$\ell_0 = 1$
$s_1 = 3 \cdot 4$	$\ell_1 = \dfrac{1}{3}$
$s_2 = 3 \cdot 4^2$	$\ell_2 = \dfrac{1}{3^2}$
$s_3 = 3 \cdot 4^3$	$\ell_3 = \dfrac{1}{3^3}$
$\cdots$	$\cdots$

In general, we have $s_n = 3 \cdot 4^n$ and $\ell_n = \left(\frac{1}{3}\right)^n$, so the length of the perimeter at the nth stage of construction is $p_n = s_n\ell_n = 3 \cdot 4^n \cdot \left(\frac{1}{3}\right)^n = 3 \cdot \left(\frac{4}{3}\right)^n$.

(b) $p_n = \dfrac{4^n}{3^{n-1}} = 4\left(\frac{4}{3}\right)^{n-1}$. Since $\frac{4}{3} > 1$, $p_n \to \infty$ as $n \to \infty$.

(c) The area of each of the small triangles added at a given stage is one-ninth of the area of the triangle added at the preceding stage. Let a be the area of the original triangle. Then the area a_n of each of the small triangles added at stage n is $a_n = a \cdot \dfrac{1}{9^n} = \dfrac{a}{9^n}$. Since a small triangle is added to each side at every stage, it follows that the total area A_n added to the figure at the nth stage is

$A_n = s_{n-1} \cdot a_n = 3 \cdot 4^{n-1} \cdot \dfrac{a}{9^n} = a \cdot \dfrac{4^{n-1}}{3^{2n-1}}$. Then the total area enclosed by the snowflake curve is

$A = a + A_1 + A_2 + A_3 + \cdots = a + a \cdot \dfrac{1}{3} + a \cdot \dfrac{4}{3^3} + a \cdot \dfrac{4^2}{3^5} + a \cdot \dfrac{4^3}{3^7} + \cdots$. After the first term, this is a geometric series with common ratio $\frac{4}{9}$, so $A = a + \dfrac{a/3}{1 - \frac{4}{9}} = a + \dfrac{a}{3} \cdot \dfrac{9}{5} = \dfrac{8a}{5}$. But the area of the original equilateral triangle with side 1 is $a = \frac{1}{2} \cdot 1 \cdot \sin\frac{\pi}{3} = \frac{\sqrt{3}}{4}$.

So the area enclosed by the snowflake curve is $\frac{8}{5} \cdot \frac{\sqrt{3}}{4} = \frac{2\sqrt{3}}{5}$.

6. **(a)** The curve $x^4 + y^4 = x^2 + y^2$ is symmetric about both axes and about the line $y = x$ (since interchanging x and y does not change the equation) so we need only consider $y \geq x \geq 0$ to begin with. Implicit differentiation gives $4x^3 + 4y^3y' = 2x + 2yy' \quad \Rightarrow \quad y' = \dfrac{x(1-2x^2)}{y(2y^2-1)} \quad \Rightarrow \quad y' = 0$ when $x = 0$ and when $x = \pm\frac{1}{\sqrt{2}}$. If $x = 0$, then $y^4 = y^2 \quad \Rightarrow \quad y^2(y^2-1) = 0 \quad \Rightarrow \quad y = 0$ or ± 1. The point $(0,0)$ can't be a highest or lowest point because it is isolated. [If $-1 < x < 1$ and $-1 < y < 1$, then $x^4 < x^2$ and $y^4 < y^2 \quad \Rightarrow \quad x^4 + y^4 < x^2 + y^2$, except for $(0,0)$.] If $x = \frac{1}{\sqrt{2}}$, then $x^2 = \frac{1}{2}$, $x^4 = \frac{1}{4}$, so $\frac{1}{4} + y^4 = \frac{1}{2} + y^2 \quad \Rightarrow \quad 4y^4 - 4y^2 - 1 = 0 \quad \Rightarrow \quad y^2 = \frac{4 \pm \sqrt{16+16}}{8} = \frac{1 \pm \sqrt{2}}{2}$. But $y^2 > 0$, so $y^2 = \frac{1+\sqrt{2}}{2}$ $\Rightarrow \quad y = \pm\sqrt{\frac{1}{2}(1+\sqrt{2})}$. Near the point $(0,1)$, the denominator of y' is positive and the numerator changes from negative to positive as x increases through 0, so $(0,1)$ is a local minimum point. At $\left(\frac{1}{\sqrt{2}}, \sqrt{\frac{1+\sqrt{2}}{2}}\right)$, y' changes from positive to negative, so that point gives a maximum. By symmetry, the highest points on the curve are $\left(\pm\frac{1}{\sqrt{2}}, \sqrt{\frac{1+\sqrt{2}}{2}}\right)$ and the lowest points are $\left(\pm\frac{1}{\sqrt{2}}, -\sqrt{\frac{1+\sqrt{2}}{2}}\right)$.

(b) We use the information from part (a), together with symmetry with respect to the axes and the lines $y = \pm x$, to sketch the curve.

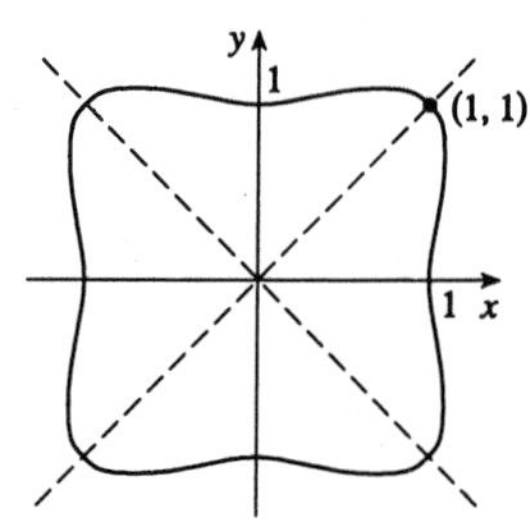

(c) In polar coordinates, $x^4 + y^4 = x^2 + y^2$ becomes $r^4\cos^4\theta + r^4\sin^4\theta = r^2$ or $r^2 = 1/(\cos^4\theta + \sin^4\theta)$. By the symmetry shown in part (b), the area enclosed by the curve is $A = 8 \cdot \dfrac{1}{2}\displaystyle\int_0^{\pi/4} r^2\,d\theta = 4\int_0^{\pi/4} \frac{d\theta}{\cos^4\theta + \sin^4\theta}$ (if we have a CAS, this can be evaluated to give $\sqrt{2}\pi$).

The usual Weierstrass substitution $t = \tan(\theta/2)$ leads to a complicated integrand, so we first simplify:

$$\cos^4\theta + \sin^4\theta = \left(1 - \sin^2\theta\right)^2 + \sin^4\theta = 1 - 2\sin^2\theta + 2\sin^4\theta = 1 - 2\sin^2\theta\left(1 - \sin^2\theta\right)$$
$$= 1 - 2\sin^2\theta\cos^2\theta = 1 - \tfrac{1}{2}\sin^2 2\theta.$$

Then we substitute $t = \tan 2\theta$, which gives $\theta = \frac{1}{2}\tan^{-1}t \quad \Rightarrow \quad d\theta = \dfrac{dt}{2(1+t^2)}$ and $\sin 2\theta = \dfrac{t}{\sqrt{1+t^2}}$. Also, $\theta \to \frac{\pi}{4} \quad \Rightarrow \quad t \to \infty$,

so we get the following improper integral:

$$A = 4\int_0^{\pi/4} \frac{d\theta}{1 - \frac{1}{2}\sin^2 2\theta} = 4\int_0^{\infty} \frac{1}{1 - \frac{1}{2}\left[t^2/(1+t^2)\right]}\,\frac{dt}{2(1+t^2)} = 4\int_0^{\infty}\frac{dt}{t^2+2}$$
$$= \lim_{x\to\infty} 4\left[\tfrac{1}{\sqrt{2}}\tan^{-1}\left(\tfrac{1}{\sqrt{2}}t\right)\right]_0^x = 2\sqrt{2}\lim_{x\to\infty}\tan^{-1}\left(\tfrac{1}{\sqrt{2}}x\right) = 2\sqrt{2}\cdot\tfrac{\pi}{2} = \sqrt{2}\pi.$$

7. $x^2 + y^2 \leq 4y \quad \Leftrightarrow \quad x^2 + (y-2)^2 \leq 4$, so S is part of a circle, as shown in the diagram. The area of S is

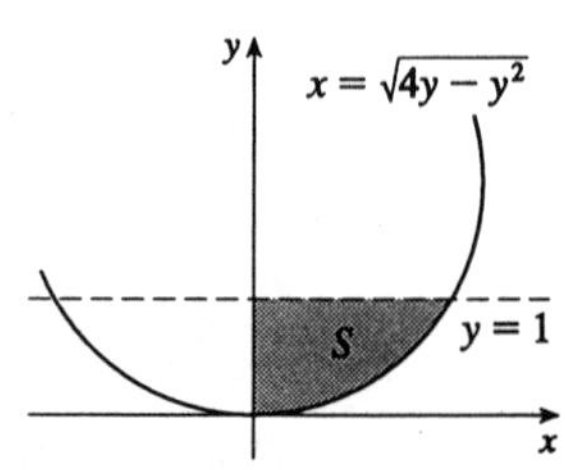

$\int_0^1 \sqrt{4y - y^2}\,dy = \int_{-2}^{-1}\sqrt{4 - v^2}\,dv \quad$ (put $v = y - 2$)

$= \left[\frac{1}{2}v\sqrt{4-v^2} + \frac{1}{2}(4)\sin^{-1}\left(\frac{1}{2}v\right)\right]_{-2}^{-1}$ (Formula 30) $= \frac{2\pi}{3} - \frac{\sqrt{3}}{2}$

Another Method (without calculus): Note that $\theta = \angle ABC = \frac{\pi}{3}$, so the area is (area of sector AOC) − (area of $\triangle ABC$) $= \frac{1}{2}(2^2)\frac{\pi}{3} - \frac{1}{2}(1)\sqrt{3} = \frac{2\pi}{3} - \frac{\sqrt{3}}{2}$.

8. **(a)** $\sin\theta = 2\sin\frac{\theta}{2}\cos\frac{\theta}{2} = 2\left(2\sin\frac{\theta}{4}\cos\frac{\theta}{4}\right)\cos\frac{\theta}{2} = 2\left(2\left(2\sin\frac{\theta}{8}\cos\frac{\theta}{8}\right)\cos\frac{\theta}{4}\right)\cos\frac{\theta}{2}$

$$= \cdots = 2\left(2\left(2\left(\cdots\left(2\left(2\sin\frac{\theta}{2^n}\cos\frac{\theta}{2^n}\right)\cos\frac{\theta}{2^{n-1}}\right)\cdots\right)\cos\frac{\theta}{8}\right)\cos\frac{\theta}{4}\right)\cos\frac{\theta}{2}$$

$$= 2^n\sin\frac{\theta}{2^n}\cos\frac{\theta}{2}\cos\frac{\theta}{4}\cos\frac{\theta}{8}\cdots\cos\frac{\theta}{2^n}$$

(b) $\sin\theta = 2^n\sin\frac{\theta}{2^n}\cos\frac{\theta}{2}\cos\frac{\theta}{4}\cos\frac{\theta}{8}\cdots\cos\frac{\theta}{2^n} \quad\Leftrightarrow\quad \frac{\sin\theta}{\theta}\cdot\frac{\theta/2^n}{\sin(\theta/2^n)} = \cos\frac{\theta}{2}\cos\frac{\theta}{4}\cos\frac{\theta}{8}\cdots\cos\frac{\theta}{2^n}.$

Now we let $n\to\infty$, using $\lim\limits_{x\to 0}\frac{\sin x}{x} = 1$ with $x = \frac{\theta}{2^n}$:

$$\lim_{n\to\infty}\left[\frac{\sin\theta}{\theta}\cdot\frac{\theta/2^n}{\sin(\theta/2^n)}\right] = \lim_{n\to\infty}\left[\cos\frac{\theta}{2}\cos\frac{\theta}{4}\cos\frac{\theta}{8}\cdots\cos\frac{\theta}{2^n}\right] \quad\Leftrightarrow\quad \frac{\sin\theta}{\theta} = \cos\frac{\theta}{2}\cos\frac{\theta}{4}\cos\frac{\theta}{8}\cdots.$$

(c) If we take $\theta = \frac{\pi}{2}$ in the result from part (b) and use the half-angle formula $\cos x = \sqrt{\frac{1}{2}(1+\cos 2x)}$ (see Formula 17a in Appendix D), we get

$$\frac{\sin\pi/2}{\pi/2} = \cos\tfrac{\pi}{4}\sqrt{\frac{\cos\frac{\pi}{4}+1}{2}}\sqrt{\frac{\sqrt{\frac{\cos\frac{\pi}{4}+1}{2}}+1}{2}}\sqrt{\frac{\sqrt{\frac{\sqrt{\frac{\cos\frac{\pi}{4}+1}{2}}+1}{2}}+1}{2}}\cdots \quad\Rightarrow$$

$$\frac{2}{\pi} = \frac{\sqrt{2}}{2}\sqrt{\frac{\frac{\sqrt{2}}{2}+1}{2}}\sqrt{\frac{\sqrt{\frac{\frac{\sqrt{2}}{2}+1}{2}}+1}{2}}\cdots = \frac{\sqrt{2}}{2}\,\frac{\sqrt{2+\sqrt{2}}}{2}\sqrt{\frac{\frac{\sqrt{2+\sqrt{2}}}{2}+1}{2}}\cdots$$

$$= \frac{\sqrt{2}}{2}\,\frac{\sqrt{2+\sqrt{2}}}{2}\,\frac{\sqrt{2+\sqrt{2+\sqrt{2}}}}{2}\cdots.$$

9. $a_{n+1} = \frac{a_n+b_n}{2}$, $b_{n+1} = \sqrt{b_n a_{n+1}}$. So $a_1 = \cos\theta$, $b_1 = 1 \;\Rightarrow\; a_2 = \frac{1+\cos\theta}{2} = \cos^2\frac{\theta}{2}$,

$b_2 = \sqrt{b_1 a_2} = \sqrt{\cos^2\frac{\theta}{2}} = \cos\frac{\theta}{2}$ since $-\frac{\pi}{2}\le\theta\le\frac{\pi}{2}$. Then

$$a_3 = \frac{1}{2}\left(\cos\frac{\theta}{2}+\cos^2\frac{\theta}{2}\right) = \cos\frac{\theta}{2}\cdot\frac{1}{2}\left(1+\cos\frac{\theta}{2}\right) = \cos\frac{\theta}{2}\cos^2\frac{\theta}{4} \quad\Rightarrow$$

$$b_3 = \sqrt{b_2 a_3} = \sqrt{\cos\frac{\theta}{2}\cos\frac{\theta}{2}\cos^2\frac{\theta}{4}} = \cos\frac{\theta}{2}\cos\frac{\theta}{4} \quad\Rightarrow$$

$$a_4 = \frac{1}{2}\left(\cos\frac{\theta}{2}\cos^2\frac{\theta}{4}+\cos\frac{\theta}{2}\cos\frac{\theta}{4}\right) = \cos\frac{\theta}{2}\cos\frac{\theta}{4}\cdot\frac{1}{2}\left(1+\cos\frac{\theta}{4}\right) = \cos\frac{\theta}{2}\cos\frac{\theta}{4}\cos^2\frac{\theta}{8} \quad\Rightarrow$$

$b_4 = \sqrt{\cos\frac{\theta}{2}\cos\frac{\theta}{4}\cos\frac{\theta}{2}\cos\frac{\theta}{4}\cos^2\frac{\theta}{8}} = \cos\frac{\theta}{2}\cos\frac{\theta}{4}\cos\frac{\theta}{8}$. By now we see the pattern:

$b_n = \cos\frac{\theta}{2}\cos\frac{\theta}{2^2}\cos\frac{\theta}{2^3}\cdot\cdots\cdot\cos\frac{\theta}{2^{n-1}}$ and $a_n = b_n\cos\frac{\theta}{2^{n-1}}$. (This could be proved by mathematical induction.) By Exercise 8(a), $\sin\theta = 2^{n-1}\sin\frac{\theta}{2^{n-1}}\cos\frac{\theta}{2}\cos\frac{\theta}{4}\cdots\cos\frac{\theta}{2^{n-1}}$.

So $b_n = \cos\frac{\theta}{2}\cos\frac{\theta}{2^2}\cos\frac{\theta}{2^3}\cdots\cos\frac{\theta}{2^{n-1}} \to \frac{\sin\theta}{\theta}$ as $n\to\infty$ by Exercise 8(b), and

$a_n = b_n\cos\frac{\theta}{2^{n-1}} \to \frac{\sin\theta}{\theta}\cdot 1 = \frac{\sin\theta}{\theta}$ as $n\to\infty$. So $\lim\limits_{n\to\infty} a_n = \lim\limits_{n\to\infty} b_n = \frac{\sin\theta}{\theta}$.

10. **(a)**

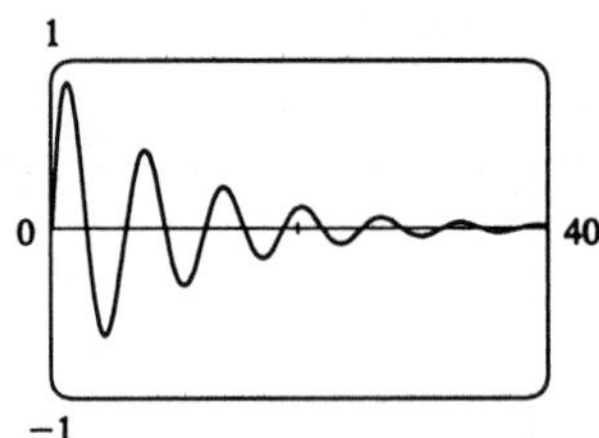

The x-intercepts of the curve occur where $\sin x = 0 \quad \Leftrightarrow \quad x = n\pi$, n an integer. So using the formula for disks (and either a CAS or $\sin^2 x = \frac{1}{2}(1 - \cos 2x)$ and Formula 99 to evaluate the integral), the volume of the nth bead is $V_n = \pi \int_{(n-1)\pi}^{n\pi} \left(e^{-x/10} \sin x\right)^2 dx = \pi \int_{(n-1)\pi}^{n\pi} e^{-x/5} \sin^2 x\, dx = \frac{250\pi}{101}\left(e^{-(n-1)\pi/5} - e^{-n\pi/5}\right)$.

(b) The total volume is

$$\pi \int_0^\infty e^{-x/5} \sin^2 x\, dx = \sum_{n=1}^\infty V_n = \frac{250\pi}{101} \sum_{n=1}^\infty \left[e^{-(n-1)\pi/5} - e^{-n\pi/5}\right] = \frac{250\pi}{101} \text{ (telescoping sum).}$$

Another Method: If the volume in part (a) has been written as $V_n = \frac{250\pi}{101} e^{-n\pi/5}\left(e^{\pi/5} - 1\right)$, then we recognize $\sum_{n=1}^\infty V_n$ as a geometric series with $a = \frac{250\pi}{101}\left(1 - e^{-\pi/5}\right)$ and $r = e^{-\pi/5}$.

11. We start with the geometric series $\sum_{n=0}^\infty x^n = \frac{1}{1-x}$, $|x| < 1$, and differentiate:

$$\sum_{n=1}^\infty n x^{n-1} = \frac{d}{dx}\left(\sum_{n=0}^\infty x^n\right) = \frac{d}{dx}\left(\frac{1}{1-x}\right) = \frac{1}{(1-x)^2} \text{ for } |x| < 1 \quad \Rightarrow \quad \sum_{n=1}^\infty n x^n = x \sum_{n=1}^\infty n x^{n-1} = \frac{x}{(1-x)^2}$$

for $|x| < 1$. Differentiate again: $\sum_{n=1}^\infty n^2 x^{n-1} = \frac{d}{dx} \frac{x}{(1-x)^2} = \frac{(1-x)^2 - x \cdot 2(1-x)(-1)}{(1-x)^4} = \frac{x+1}{(1-x)^3} \quad \Rightarrow$

$$\sum_{n=1}^\infty n^2 x^n = \frac{x^2+x}{(1-x)^3} \quad \Rightarrow \quad \sum_{n=1}^\infty n^3 x^{n-1} = \frac{d}{dx} \frac{x^2+x}{(1-x)^3}$$

$$= \frac{(1-x)^3(2x+1) - (x^2+x)3(1-x)^2(-1)}{(1-x)^6} = \frac{x^2+4x+1}{(1-x)^4} \quad \Rightarrow \quad \sum_{n=1}^\infty n^3 x^n = \frac{x^3+4x^2+x}{(1-x)^4}, \ |x| < 1.$$

The radius of convergence is 1 because that is the radius of convergence for the geometric series we started with. If $x = \pm 1$, the series is $\sum n^3(\pm 1)^n$, which diverges by the Test For Divergence, so the interval of convergence is $(-1, 1)$.

12. $|AP_2|^2 = 2$, $|AP_3|^2 = 2 + 2^2$, $|AP_4|^2 = 2 + 2^2 + \left(2^2\right)^2$, $|AP_5|^2 = 2 + 2^2 + \left(2^2\right)^2 + \left(2^3\right)^2, \ldots,$

$$|AP_n|^2 = 2 + 2^2 + \left(2^2\right)^2 + \cdots + \left(2^{n-2}\right)^2 = 2 + \left(4 + 4^2 + 4^3 + \cdots + 4^{n-2}\right) = 2 + \frac{4(4^{n-2} - 1)}{4 - 1} = \frac{2}{3} + \frac{4^{n-1}}{3}$$

(finite geometric sum with $a = 4$, $r = 4$). So

$$\tan \angle P_n A P_{n+1} = \frac{|P_n P_{n+1}|}{|AP_n|} = \frac{2^{n-1}}{\sqrt{\frac{2}{3} + \frac{4^{n-1}}{3}}} = \frac{\sqrt{4^{n-1}}}{\sqrt{\frac{2}{3} + \frac{4^{n-1}}{3}}} = \frac{1}{\sqrt{\frac{2}{3 \cdot 4^{n-1}} + \frac{1}{3}}} \to \sqrt{3} \text{ as } n \to \infty, \text{ so}$$

$\angle P_n A P_{n+1} \to \frac{\pi}{3}$ as $n \to \infty$.

13. **(a)** Let $a = \arctan x$ and $b = \arctan y$. Then, from the endpapers,

$$\tan(a-b) = \frac{\tan a - \tan b}{1 + \tan a \tan b} = \frac{\tan(\arctan x) - \tan(\arctan y)}{1 + \tan(\arctan x)\tan(\arctan y)} \quad\Rightarrow\quad \tan(a-b) = \frac{x-y}{1+xy} \quad\Rightarrow$$

$\arctan x - \arctan y = a - b = \arctan \dfrac{x-y}{1+xy}$ since $-\dfrac{\pi}{2} < \arctan x - \arctan y < \dfrac{\pi}{2}$.

(b) From part (a) we have $\arctan \frac{120}{119} - \arctan \frac{1}{239} = \arctan \dfrac{\frac{120}{119} - \frac{1}{239}}{1 + \frac{120}{119} \cdot \frac{1}{239}} = \arctan \dfrac{\frac{28{,}561}{28{,}441}}{\frac{28{,}561}{28{,}441}} = \arctan 1 = \frac{\pi}{4}$.

(c) Replacing y by $-y$ in the formula of part (a), we get $\arctan x + \arctan y = \arctan \dfrac{x+y}{1-xy}$. So

$$\begin{aligned} 4\arctan \tfrac{1}{5} &= 2\left(\arctan \tfrac{1}{5} + \arctan \tfrac{1}{5}\right) = 2\arctan \frac{\frac{1}{5} + \frac{1}{5}}{1 - \frac{1}{5}\cdot\frac{1}{5}} = 2\arctan \tfrac{5}{12} \\ &= \arctan \tfrac{5}{12} + \arctan \tfrac{5}{12} = \arctan \frac{\frac{5}{12} + \frac{5}{12}}{1 - \frac{5}{12}\cdot\frac{5}{12}} = \arctan \tfrac{120}{119}. \end{aligned}$$

Thus, from part (b), we have $4\arctan \frac{1}{5} - \arctan \frac{1}{239} = \arctan \frac{120}{119} - \arctan \frac{1}{239} = \frac{\pi}{4}$.

(d) From Example 7 in Section 10.9 we have $\arctan x = x - \dfrac{x^3}{3} + \dfrac{x^5}{5} - \dfrac{x^7}{7} + \dfrac{x^9}{9} - \dfrac{x^{11}}{11} + \cdots$, so

$\arctan \dfrac{1}{5} = \dfrac{1}{5} - \dfrac{1}{3\cdot 5^3} + \dfrac{1}{5\cdot 5^5} - \dfrac{1}{7\cdot 5^7} + \dfrac{1}{9\cdot 5^9} - \dfrac{1}{11\cdot 5^{11}} + \cdots$. This is an alternating series and the size of the terms decreases to 0, so by Theorem 10.5.1, the sum lies between s_5 and s_6, that is, $0.197395560 < \arctan \frac{1}{5} < 0.197395562$.

(e) From the series in part (d) we get $\arctan \dfrac{1}{239} = \dfrac{1}{239} - \dfrac{1}{3\cdot 239^3} + \dfrac{1}{5\cdot 239^5} - \cdots$. The third term is less than 2.6×10^{-13}, so by Theorem 10.5.1 we have, to nine decimal places, $\arctan \frac{1}{239} \approx s_2 \approx 0.004184076$. Thus $0.004184075 < \arctan \frac{1}{239} < 0.004184077$.

(f) From part (c) we have $\pi = 16\arctan \frac{1}{5} - 4\arctan \frac{1}{239}$, so from parts (d) and (e) we have

$16(0.197395560) - 4(0.004184077) < \pi < 16(0.197395562) - 4(0.004184075) \quad\Rightarrow$

$3.141592652 < \pi < 3.141592692$. So, to 7 decimal places, $\pi \approx 3.1415927$.

14. **(a)** Let $a = \operatorname{arccot} x$ and $b = \operatorname{arccot} y$. Then

$\cot(a-b) = \dfrac{1 + \cot a \cot b}{\cot b - \cot a} = \dfrac{1 + \cot(\operatorname{arccot} x)\cot(\operatorname{arccot} y)}{\cot(\operatorname{arccot} y) - \cot(\operatorname{arccot} x)} = \dfrac{1+xy}{y-x} \quad\Rightarrow\quad \operatorname{arccot} x - \operatorname{arccot} y = a - b = \operatorname{arccot} \dfrac{1+xy}{y-x}$.

(b) Applying the identity in part (a) with $x = n$ and $y = n+1$, we have

$$\operatorname{arccot}(n^2+n+1) = \operatorname{arccot}(1 + n(n+1)) = \operatorname{arccot} \frac{1+n(n+1)}{(n+1)-n} = \operatorname{arccot} n - \operatorname{arccot}(n+1).$$

Thus we have a telescoping series with nth partial sum

$$\begin{aligned} s_n &= [\operatorname{arccot} 0 - \operatorname{arccot} 1] + [\operatorname{arccot} 1 - \operatorname{arccot} 2] + \cdots + [\operatorname{arccot} n - \operatorname{arccot}(n+1)] \\ &= \operatorname{arccot} 0 - \operatorname{arccot}(n+1). \end{aligned}$$

Thus $\sum_{n=0}^{\infty} \operatorname{arccot}(n^2+n+1) = \lim\limits_{n\to\infty} [-\operatorname{arccot}(n+1)] = \frac{\pi}{2}$.

15. $u = 1 + \frac{x^3}{3!} + \frac{x^6}{6!} + \frac{x^9}{9!} + \cdots$, $v = x + \frac{x^4}{4!} + \frac{x^7}{7!} + \frac{x^{10}}{10!} + \cdots$, $w = \frac{x^2}{2!} + \frac{x^5}{5!} + \frac{x^8}{8!} + \cdots$.

The key idea is to differentiate: $\frac{du}{dx} = \frac{3x^2}{3!} + \frac{6x^5}{6!} + \frac{9x^8}{9!} + \cdots = \frac{x^2}{2!} + \frac{x^5}{5!} + \frac{x^8}{8!} + \cdots = w$.

Similarly, $\frac{dv}{dx} = 1 + \frac{x^3}{3!} + \frac{x^6}{6!} + \frac{x^9}{9!} + \cdots = u$, and $\frac{dw}{dx} = x + \frac{x^4}{4!} + \frac{x^7}{7!} + \frac{x^{10}}{10!} + \cdots = v$.

So $u' = w$, $v' = u$, and $w' = v$. Now differentiate the left hand side of the desired equation:

$$\frac{d}{dx}\left(u^3 + v^3 + w^3 - 3uvw\right) = 3u^2u' + 3v^3v' + 3w^3w' - 3(u'vw + uv'w + uvw')$$

$$= 3u^2w + 3v^2u + 3w^2v - 3(vw^2 + u^2w + uv^2) = 0 \quad \Rightarrow \quad u^3 + v^3 + w^3 - 3uvw = C.$$

To find the value of the constant C, we put $x = 1$ in the equation and get $1^3 + 0 + 0 - 3(1 \cdot 0 \cdot 0) = C \quad \Rightarrow$

$C = 1$, so $u^3 + v^3 + w^3 - 3uvw = 1$.

16. **(a)** If (a, b) lies on the curve, then there is some parameter value t_1 such that $\frac{3t_1}{1 + t_1^3} = a$ and $\frac{3t_1^2}{1 + t_1^3} = b$. If $t_1 = 0$, the point is $(0, 0)$, which lies on the line $y = x$. If $t_1 \neq 0$, then the point corresponding to $t = \frac{1}{t_1}$ is given by $x = \frac{3(1/t_1)}{1 + (1/t_1)^3} = \frac{3t_1^2}{t_1^3 + 1} = b$, $y = \frac{3(1/t_1)^2}{1 + (1/t_1)^3} = \frac{3t_1}{t_1^3 + 1} = a$. So (b, a) also lies on the curve.

[Another way to see this is to do part (e) first; the result is immediate.] The curve intersects the line $y = x$ when $\frac{3t}{1 + t^3} = \frac{3t^2}{1 + t^3} \Rightarrow t = t^2 \Rightarrow t = 0$ or 1, so the points are $(0, 0)$ and $\left(\frac{3}{2}, \frac{3}{2}\right)$.

(b) $\frac{dy}{dt} = \frac{(1 + t^3)(6t) - 3t^2(3t^2)}{(1 + t^3)^2} = \frac{6t - 3t^4}{(1 + t^3)^2} = 0$ when $6t - 3t^4 = 3t(2 - t^3) = 0 \quad \Rightarrow \quad t = 0$ or $t = \sqrt[3]{2}$, so there are horizontal tangents at $(0, 0)$ and $\left(\sqrt[3]{2}, \sqrt[3]{4}\right)$. Using the symmetry from part (a), we see that there are vertical tangents at $(0, 0)$ and $\left(\sqrt[3]{4}, \sqrt[3]{2}\right)$.

(c) Notice that as $t \to -1^+$, we have $x \to -\infty$ and $y \to \infty$. As $t \to -1^-$, we have $x \to \infty$ and $y \to -\infty$. Also $y - (-x - 1) = y + x + 1 = \frac{3t + 3t^2 + (1 + t^3)}{1 + t^3} = \frac{(t + 1)^3}{1 + t^3} = \frac{(t + 1)^2}{t^2 - t + 1} \to 0$ as $t \to -1$. So $y = -x - 1$ is a slant asymptote.

(d) $\frac{dx}{dt} = \frac{(1 + t^3)(3) - 3t(3t^2)}{(1 + t^3)^2} = \frac{3 - 6t^3}{(1 + t^3)^2}$ and from (b) we have $\frac{dy}{dt} = \frac{6t - 3t^4}{(1 + t^3)^2}$. So

$\frac{dy}{dx} = \frac{dy/dt}{dx/dt} = \frac{t(2 - t^3)}{1 - 2t^3}$. Also $\frac{d^2y}{dx^2} = \frac{\frac{d}{dt}\left(\frac{dy}{dx}\right)}{dx/dt} = \frac{2(1 + t^3)^4}{3(1 - 2t^3)^3} > 0 \quad \Leftrightarrow \quad t < \frac{1}{\sqrt[3]{2}}$. So the curve is concave upward there and has a minimum point at $(0, 0)$ and a maximum point at $\left(\sqrt[3]{2}, \sqrt[3]{4}\right)$.

Using this together with the information from parts (a), (b), and (c), we sketch the curve.

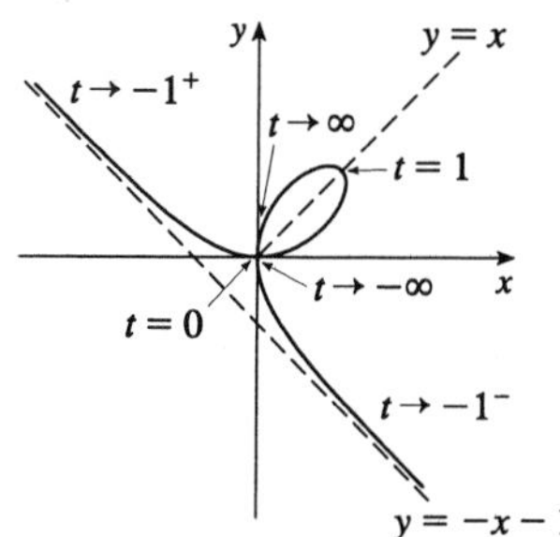

(e) $x^3+y^3=\left(\dfrac{3t}{1+t^3}\right)^3+\left(\dfrac{3t^2}{1+t^3}\right)^3=\dfrac{27t^3+27t^6}{(1+t^3)^3}=\dfrac{27t^3(1+t^3)}{(1+t^3)^3}=\dfrac{27t^3}{(1+t^3)^2}$ and

$3xy=3\left(\dfrac{3t}{1+t^3}\right)\left(\dfrac{3t^2}{1+t^3}\right)=\dfrac{27t^3}{(1+t^3)^2}$, so $x^3+y^3=3xy$.

(f) We start with the equation from part (e) and substitute $x=r\cos\theta$, $y=r\sin\theta$. Then $x^3+y^3=3xy \Rightarrow$ $r^3\cos^3\theta+r^3\sin^3\theta=3r^2\cos\theta\sin\theta$. For $r\neq 0$, this gives $r=\dfrac{3\cos\theta\sin\theta}{\cos^3\theta+\sin^3\theta}$. Dividing numerator and denominator by $\cos^3\theta$, we obtain $r=\dfrac{3\left(\dfrac{1}{\cos\theta}\right)\dfrac{\sin\theta}{\cos\theta}}{1+\dfrac{\sin^3\theta}{\cos^3\theta}}=\dfrac{3\sec\theta\tan\theta}{1+\tan^3\theta}$.

(g) The loop corresponds to $\theta\in\left(0,\frac{\pi}{2}\right)$, so its area is

$$A=\int_0^{\pi/2}\frac{r^2}{2}\,d\theta=\frac{1}{2}\int_0^{\pi/2}\left(\frac{3\sec\theta\tan\theta}{1+\tan^3\theta}\right)^2 d\theta=\frac{9}{2}\int_0^{\pi/2}\frac{\sec^2\theta\tan^2\theta}{(1+\tan^3\theta)^2}d\theta$$
$$=\frac{9}{2}\int_0^{\infty}\frac{u^2\,du}{(1+u^3)^2}\text{ (put } u=\tan\theta) = \lim_{b\to\infty}\tfrac{9}{2}\left[-\tfrac{1}{3}(1+u^3)^{-1}\right]_0^b=\tfrac{3}{2}.$$

(h) By symmetry, the area between the folium and the line $y=-x-1$ is equal to the enclosed area in the third quadrant, plus twice the enclosed area in the fourth quadrant. The area in the third quadrant is $\frac{1}{2}$, and since $y=-x-1 \Rightarrow r\sin\theta=-r\cos\theta-1 \Rightarrow r=-\dfrac{1}{\sin\theta+\cos\theta}$, the area in the fourth quadrant is $\dfrac{1}{2}\displaystyle\int_{-\pi/2}^{-\pi/4}\left[\left(-\frac{1}{\sin\theta+\cos\theta}\right)^2-\left(\frac{3\sec\theta\tan\theta}{1+\tan^3\theta}\right)^2\right]d\theta \overset{\text{(CAS)}}{=} \dfrac{1}{2}$. Therefore the total area is $\frac{1}{2}+2\left(\frac{1}{2}\right)=\frac{3}{2}$.

17. $(a^n+b^n+c^n)^{1/n}=\left(c^n\left[\left(\dfrac{a}{c}\right)^n+\left(\dfrac{b}{c}\right)^n+1\right]\right)^{1/n}=c\left[\left(\dfrac{a}{c}\right)^n+\left(\dfrac{b}{c}\right)^n+1\right]^{1/n}$. Since $0\le a\le c$, we have $0\le a/c\le 1$, so $(a/c)^n\to 0$ or 1 as $n\to\infty$. Similarly, $(b/c)^n\to 0$ or 1 as $n\to\infty$. Thus $\left(\dfrac{a}{c}\right)^n+\left(\dfrac{b}{c}\right)^n+1\to d$, where $d=1, 2$, or 3 and so $\left[\left(\dfrac{a}{c}\right)^n+\left(\dfrac{b}{c}\right)^n+1\right]^{1/n}\to 1$. Therefore $\lim\limits_{n\to\infty}(a^n+b^n+c^n)^{1/n}=c$.

18. **(a)** Let us find the polar equation of the path of the bug that starts in the upper right corner of the square. If the polar coordinates of this bug, at a particular moment, are (r,θ), then the polar coordinates of the bug that it is crawling toward must be $\left(r,\theta+\frac{\pi}{2}\right)$. (The next bug must be the same distance from the origin and the angle between the lines joining the bugs to the pole must be $\frac{\pi}{2}$.) The Cartesian coordinates of the first bug are $(r\cos\theta, r\sin\theta)$ and for the second bug we have

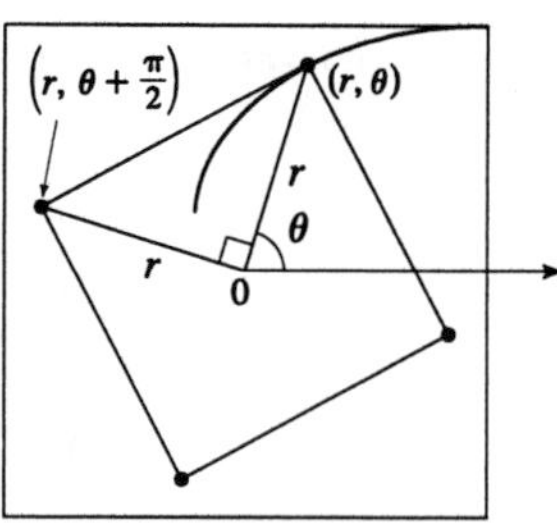

$x=r\cos\left(\theta+\frac{\pi}{2}\right)=-r\sin\theta$, $y=r\sin\left(\theta+\frac{\pi}{2}\right)=r\cos\theta$. So the slope of the line joining the bugs is $\dfrac{r\cos\theta-r\sin\theta}{-r\sin\theta-r\cos\theta}=\dfrac{\sin\theta-\cos\theta}{\sin\theta+\cos\theta}$. This must be equal to the slope of the tangent line at (r,θ), so by Equation 9.4.3 we have $\dfrac{(dr/d\theta)\sin\theta+r\cos\theta}{(dr/d\theta)\cos\theta-r\sin\theta}=\dfrac{\sin\theta-\cos\theta}{\sin\theta+\cos\theta}$. Solving for $\dfrac{dr}{d\theta}$, we get

$$\frac{dr}{d\theta}\sin^2\theta+\frac{dr}{d\theta}\sin\theta\cos\theta+r\sin\theta\cos\theta+r\cos^2\theta=\frac{dr}{d\theta}\sin\theta\cos\theta-\frac{dr}{d\theta}\cos^2\theta-r\sin^2\theta+r\sin\theta\cos\theta$$

$\Rightarrow\quad \dfrac{dr}{d\theta}\left(\sin^2\theta+\cos^2\theta\right)+r\left(\cos^2\theta+\sin^2\theta\right)=0 \quad\Rightarrow\quad \dfrac{dr}{d\theta}=-r$. Solving this differential equation as a separable equation (as in Section 8.1), or using Theorem 6.5.2 with $k=-1$, we get $r=Ce^{-\theta}$. To determine C we use the fact that, at its starting position, $\theta=\frac{\pi}{4}$ and $r=\frac{1}{\sqrt{2}}a$, so $\frac{1}{\sqrt{2}}a=Ce^{-\pi/4}\quad\Rightarrow$ $C=\frac{1}{\sqrt{2}}ae^{\pi/4}$. Therefore a polar equation of the bug's path is $r=\frac{1}{\sqrt{2}}ae^{\pi/4}e^{-\theta}$ or $r=\frac{1}{\sqrt{2}}ae^{(\pi/4)-\theta}$.

(b) The distance traveled by this bug is $L=\displaystyle\int_{\pi/4}^{\infty}\sqrt{r^2+(dr/d\theta)^2}\,d\theta$, where $\dfrac{dr}{d\theta}=\dfrac{a}{\sqrt{2}}e^{\pi/4}\left(-e^{-\theta}\right)$ and so $r^2+(dr/d\theta)^2=\frac{1}{2}a^2e^{\pi/2}e^{-2\theta}+\frac{1}{2}a^2e^{\pi/2}e^{-2\theta}=a^2e^{\pi/2}e^{-2\theta}$. Thus $L=\int_{\pi/4}^{\infty}ae^{\pi/4}e^{-\theta}\,d\theta$ $=ae^{\pi/4}\lim\limits_{t\to\infty}\int_{\pi/4}^{t}e^{-\theta}\,d\theta=ae^{\pi/4}\lim\limits_{t\to\infty}\left[-e^{-\theta}\right]_{\pi/4}^{t}=ae^{\pi/4}\lim\limits_{t\to\infty}\left[e^{-\pi/4}-e^{-t}\right]=ae^{\pi/4}e^{-\pi/4}=a.$

19. As in Section 10.9 we have to integrate the function x^x by integrating a series. Writing $x^x=\left(e^{\ln x}\right)^x=e^{x\ln x}$ and using the Maclaurin series for e^x, we have $x^x=e^{x\ln x}=\displaystyle\sum_{n=0}^{\infty}\frac{(x\ln x)^n}{n!}=\sum_{n=0}^{\infty}\frac{x^n(\ln x)^n}{n!}$. As with power series, we can integrate this series term-by-term: $\displaystyle\int_0^1 x^x\,dx=\sum_{n=0}^{\infty}\int_0^1\frac{x^n(\ln x)^n}{n!}\,dx=\sum_{n=0}^{\infty}\frac{1}{n!}\int_0^1 x^n(\ln x)^n\,dx$. We integrate by parts with $u=(\ln x)^n$, $dv=x^n\,dx$, so $du=\dfrac{n(\ln x)^{n-1}}{x}\,dx$ and $v=\dfrac{x^{n+1}}{n+1}$:

$$\int_0^1 x^n(\ln x)^n\,dx=\lim_{t\to0^+}\int_t^1 x^n(\ln x)^n\,dx=\lim_{t\to0^+}\left[\frac{x^{n+1}}{n+1}(\ln x)^n\right]_t^1-\lim_{t\to0^+}\int_t^1\frac{n}{n+1}x^n(\ln x)^{n-1}\,dx$$

$=0-\dfrac{n}{n+1}\displaystyle\int_0^1 x^n(\ln x)^{n-1}\,dx$ (where l'Hospital's Rule was used to help evaluate the first limit).

Further integration by parts gives $\displaystyle\int_0^1 x^n(\ln x)^k\,dx=-\frac{k}{n+1}\int_0^1 x^n(\ln x)^{k-1}\,dx$ and, combining these steps, we get $\displaystyle\int_0^1 x^n(\ln x)^n\,dx=\frac{(-1)^n n!}{(n+1)^n}\int_0^1 x^n\,dx=\frac{(-1)^n n!}{(n+1)^{n+1}}\quad\Rightarrow$

$$\int_0^1 x^x\,dx=\sum_{n=0}^{\infty}\frac{1}{n!}\int_0^1 x^n(\ln x)^n\,dx=\sum_{n=0}^{\infty}\frac{1}{n!}\frac{(-1)^n n!}{(n+1)^{n+1}}=\sum_{n=0}^{\infty}\frac{(-1)^n}{(n+1)^{n+1}}=\sum_{n=1}^{\infty}\frac{(-1)^{n-1}}{n^n}.$$

20. Let the series be S. Then every term in S is of the form $\dfrac{1}{2^m 3^n}$, $m, n \geq 0$, and furthermore each term occurs only once. So we can write $S = \sum_{m=0}^{\infty}\sum_{n=0}^{\infty} \dfrac{1}{2^m 3^n} = \sum_{m=0}^{\infty}\sum_{n=0}^{\infty} \dfrac{1}{2^m}\dfrac{1}{3^n} = \sum_{m=0}^{\infty}\dfrac{1}{2^m}\sum_{n=0}^{\infty}\dfrac{1}{3^n} = \dfrac{1}{1-\frac{1}{2}}\cdot\dfrac{1}{1-\frac{1}{3}} = 2\cdot\frac{3}{2} = 3.$

21. Call the series S. We group the terms according to the number of digits in their denominators:

$$S = \underbrace{\left(1+\frac{1}{2}+\cdots+\frac{1}{8}+\frac{1}{9}\right)}_{g_1} + \underbrace{\left(\frac{1}{11}+\cdots+\frac{1}{99}\right)}_{g_2} + \underbrace{\left(\frac{1}{111}+\cdots+\frac{1}{999}\right)}_{g_3} + \cdots$$

Now in the group g_n, there are 9^n terms, since we have 9 choices for each of the n digits in the denominator. Furthermore, each term in g_n is less than $\dfrac{1}{10^{n-1}}$. So $g_n < 9^n \cdot \dfrac{1}{10^{n-1}} = 9\left(\dfrac{9}{10}\right)^{n-1}$.

Now $\sum_{n=1}^{\infty} 9\left(\dfrac{9}{10}\right)^{n-1}$ is a geometric series with $a = 9$ and $r = \frac{9}{10} < 1$. Therefore, by the Comparison Test,

$$S = \sum_{n=1}^{\infty} g_n < \sum_{n=1}^{\infty} 9\left(\frac{9}{10}\right)^{n-1} = \frac{9}{1-\frac{9}{10}} = 90.$$

22. First notice that both series are absolutely convergent (p-series with $p > 1$.) Let the given expression be called x.

$$\text{Then } x = \frac{1+\frac{1}{2^p}+\frac{1}{3^p}+\frac{1}{4^p}+\cdots}{1-\frac{1}{2^p}+\frac{1}{3^p}-\frac{1}{4^p}+\cdots} = \frac{1+\left(2\cdot\frac{1}{2^p}-\frac{1}{2^p}\right)+\frac{1}{3^p}+\left(2\cdot\frac{1}{4^p}-\frac{1}{4^p}\right)+\cdots}{1-\frac{1}{2^p}+\frac{1}{3^p}-\frac{1}{4^p}+\cdots}$$

$$= 1+\frac{2\left(\frac{1}{2^p}+\frac{1}{4^p}+\frac{1}{6^p}+\frac{1}{8^p}+\cdots\right)}{1-\frac{1}{2^p}+\frac{1}{3^p}-\frac{1}{4^p}+\cdots} = 1+\frac{\frac{1}{2^{p-1}}\left(1+\frac{1}{2^p}+\frac{1}{3^p}+\frac{1}{4^p}+\cdots\right)}{1-\frac{1}{2^p}+\frac{1}{3^p}-\frac{1}{4^p}+\cdots}$$

$$= 1+2^{1-p}x. \text{ Therefore } x\left(1-2^{1-p}\right) = 1 \quad\Leftrightarrow\quad x = \frac{1}{1-2^{1-p}}.$$

23. If L is the length of a side of the equilateral triangle, then the area is $A = \frac{1}{2}L\cdot\frac{\sqrt{3}}{2}L = \frac{\sqrt{3}}{4}L^2$ and so $L^2 = \frac{4}{\sqrt{3}}A$. Let r be the radius of one of the circles when there are n rows of circles. The figure shows that

$$L = \sqrt{3}r + r + (n-2)(2r) + r + \sqrt{3}r$$
$$= r\left(2n-2+2\sqrt{3}\right), \text{ so } r = \frac{L}{2\left(n+\sqrt{3}-1\right)}.$$

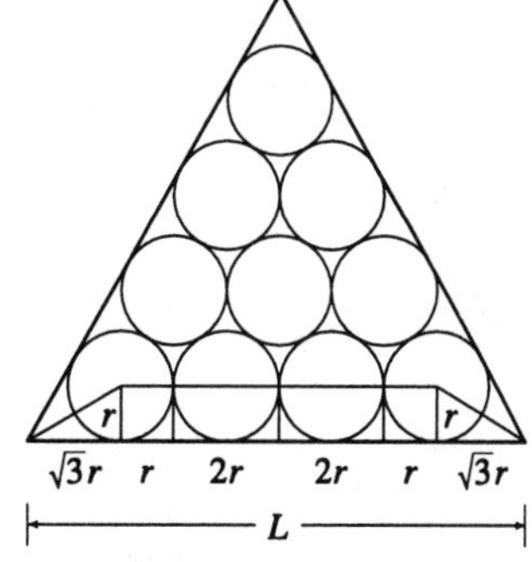

The number of circles is $1+2+\cdots+n = \dfrac{n(n+1)}{2}$ and so the total area of the circles is

$$A_n = \frac{n(n+1)}{2}\pi r^2 = \frac{n(n+1)}{2}\pi\frac{L^2}{4\left(n+\sqrt{3}-1\right)^2} = \frac{n(n+1)}{2}\pi\frac{4A/\sqrt{3}}{4\left(n+\sqrt{3}-1\right)^2} = \frac{n(n+1)}{\left(n+\sqrt{3}-1\right)^2}\frac{\pi A}{2\sqrt{3}}$$

$$\Rightarrow \frac{A_n}{A} = \frac{n(n+1)}{\left(n+\sqrt{3}-1\right)^2}\frac{\pi}{2\sqrt{3}} = \frac{1+1/n}{\left[1+\left(\sqrt{3}-1\right)/n\right]^2}\frac{\pi}{2\sqrt{3}} \to \frac{\pi}{2\sqrt{3}} \text{ as } n\to\infty.$$

24. Let's first try the case $k = 1$: $a_0 + a_1 = 0 \quad \Rightarrow$

$$\lim_{n\to\infty}\left(a_0\sqrt{n} + a_1\sqrt{n+1}\right) = \lim_{n\to\infty}\left(a_0\sqrt{n} - a_0\sqrt{n+1}\right)$$

$$= a_0 \lim_{n\to\infty}\left(\sqrt{n} - \sqrt{n+1}\right)\frac{\sqrt{n}+\sqrt{n+1}}{\sqrt{n}+\sqrt{n+1}} = a_0 \lim_{n\to\infty}\frac{-1}{\sqrt{n}+\sqrt{n+1}} = 0.$$

In general we have $a_0 + a_1 + \cdots + a_k = 0 \quad \Rightarrow \quad a_k = -a_0 - a_1 - \cdots - a_{k-1} \quad \Rightarrow$

$$\lim_{n\to\infty}\left(a_0\sqrt{n} + a_1\sqrt{n+1} + a_2\sqrt{n+2} + \cdots + a_k\sqrt{n+k}\right)$$

$$= \lim_{n\to\infty}\left(a_0\sqrt{n} + a_1\sqrt{n+1} + \cdots + a_{k-1}\sqrt{n+k-1} - a_0\sqrt{n+k} - a_1\sqrt{n+k} - \cdots - a_{k-1}\sqrt{n+k}\right)$$

$$= a_0\lim_{n\to\infty}\left(\sqrt{n} - \sqrt{n+k}\right) + a_1\lim_{n\to\infty}\left(\sqrt{n+1} - \sqrt{n+k}\right) + \cdots + a_{k-1}\lim_{n\to\infty}\left(\sqrt{n+k-1} - \sqrt{n+k}\right)$$

Each of these limits is 0 by the same sort of computation as in the case $k = 1$. So we have

$$\lim_{n\to\infty}\left(a_0\sqrt{n} + a_1\sqrt{n+1} + a_2\sqrt{n+2} + \cdots + a_k\sqrt{n+k}\right) = a_0(0) + a_1(0) + \cdots + a_{k-1}(0) = 0.$$

25. **(a)** f is continuous when $x \neq 0$ since x, $\sin x$, and π/x are continuous when $x \neq 0$. Also $|\sin(\pi/x)| \leq 1$ $\Rightarrow$ $|x\sin(\pi/x)| \leq |x|$ and $\lim_{x\to 0}|x| = 0$, so by the Squeeze Theorem we have $\lim_{x\to 0}|x\sin(\pi/x)| = 0$, so $\lim_{x\to 0} f(x) = \lim_{x\to 0} x\sin(\pi/x) = 0 = f(0)$. Therefore f is continuous at 0, and so f is continuous on $(-1, 1)$.

(b) Note that $f(x) = 0$ when $x = 0$ and when $\pi/x = n\pi \quad \Rightarrow \quad x = 1/n$, n an integer. Since $-1 \leq \sin(\pi/x) \leq 1$, the graph of f lies between the lines $y = x$ and $y = -x$ and touches these lines when

$$\frac{\pi}{x} = \frac{\pi}{2} + n\pi \quad \Rightarrow \quad x = \frac{1}{n+1/2}.$$

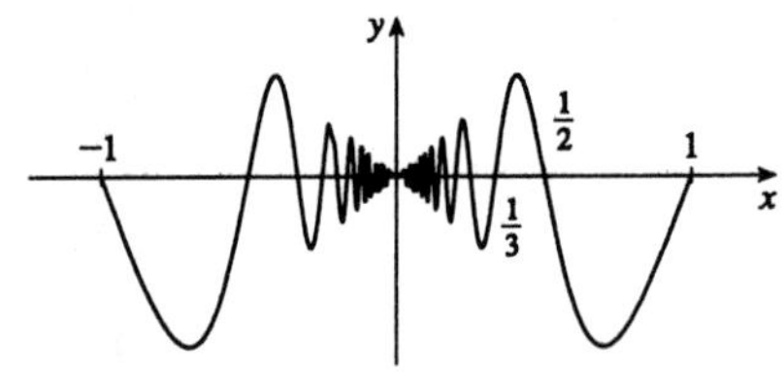

(c) The enlargement of the portion of the graph between $x = \dfrac{1}{n}$ and $x = \dfrac{1}{n-1}$ (the case where n is odd is illustrated) shows that the arc length from $x = \dfrac{1}{n}$ to $x = \dfrac{1}{n-1}$ is greater than $|PQ| = \dfrac{1}{n - 1/2} = \dfrac{2}{2n-1}$.

Thus the total length of the graph is greater than $2\displaystyle\sum_{n=1}^{\infty}\frac{2}{2n-1}$.

This is a divergent series (by comparison with the harmonic series), so the graph has infinite length.

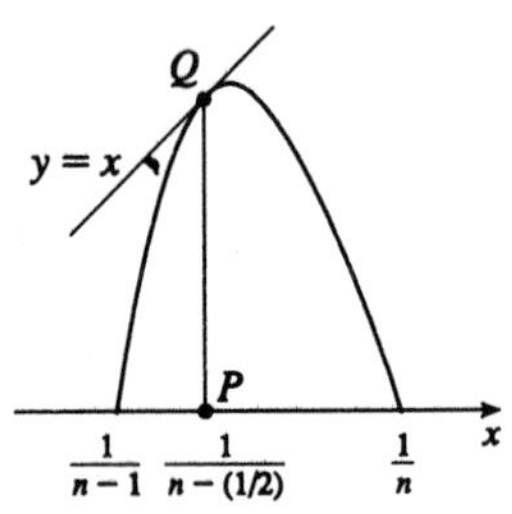

26. We calculate the next few terms of the sequence: $a_2 = \dfrac{1\cdot 0\cdot a_1 - (-1)a_0}{2\cdot 1} = \dfrac{1}{2}$, $a_3 = \dfrac{2\cdot 1\cdot a_2 - 0a_1}{3\cdot 2} = \dfrac{1}{6}$, $a_4 = \dfrac{3\cdot 2\cdot a_3 - 1a_2}{4\cdot 3} = \dfrac{1}{24}$. It seems that $a_n = \dfrac{1}{n!}$, so we try to prove this by induction. The first step is done, so assume $a_k = 1/k!$ and $a_{k-1} = 1/[(k-1)!]$. Then

$$a_{k+1} = \frac{k(k-1)a_k - (k-2)a_{k-1}}{(k+1)k} = \frac{\dfrac{k(k-1)}{k!} - \dfrac{k-2}{(k-1)!}}{(k+1)k} = \frac{(k-1)-(k-2)}{[(k+1)(k)](k-1)!} = \frac{1}{(k+1)!},$$ and the

induction is complete. Therefore $\sum_{n=0}^{\infty} a_n = \sum_{n=0}^{\infty} 1/n! = e$.

27. We use a method similar to that for Exercise 10.5.35.

Let s_n be the nth partial sum for the given series, and let h_n be the nth partial sum for the harmonic series. From Exercise 10.3.32 we know that $h_n - \ln n \to \gamma$ as $n \to \infty$. So

$$\begin{aligned} s_{3n} &= 1 + \frac{1}{2} - \frac{2}{3} + \frac{1}{4} + \frac{1}{5} - \frac{2}{6} + \cdots + \frac{1}{3n-2} + \frac{1}{3n-1} - \frac{2}{3n} \\ &= \left(1 + \frac{1}{2} + \frac{1}{3} + \frac{1}{4} + \cdots + \frac{1}{3n}\right) - \left(\frac{3}{3} + \frac{3}{6} + \frac{3}{9} + \cdots + \frac{3}{3n}\right) = h_{3n} - h_n \\ &= \ln 3 + (h_{3n} - \ln 3n) - (h_n - \ln n) \to \ln 3 + \gamma - \gamma = \ln 3 \end{aligned}$$

Note: The method suggested in the first printing of the text doesn't quite work. We can differentiate to get $f'(x) = 1 + x - 2x^2 + x^3 + x^4 - 2x^5 + \cdots = (1 + x - 2x^2) + x^3(1 + x - 2x^2) + x^6(1 + x - 2x^2) + \cdots$

$$= (1 + x - 2x^2)(1 + x^3 + x^6 + x^9 + \cdots) = \frac{1 + x - 2x^2}{1 - x^3} \quad (\text{if } |x| < 1) \quad = \frac{(1-x)(1+2x)}{(1-x)(1+x+x^2)}$$

$= \dfrac{1+2x}{1+x+x^2}$. Then $f(x) = \ln(1 + x + x^2) + C$. But $f(0) = 0$, so $C = 0$ and we have shown that

$x + \dfrac{x^2}{2} - \dfrac{2x^3}{3} + \dfrac{x^4}{4} + \dfrac{x^5}{5} - \dfrac{2x^6}{6} + \cdots = \ln(1 + x + x^2)$ for $-1 < x < 1$. As $x \to 1$, the limit of the right side is $\ln 3$. However, it is not so easy to show that the limit of the left side is $1 + \frac{1}{2} - \frac{2}{3} + \frac{1}{4} + \frac{1}{5} - \frac{2}{6} + \cdots$.

28. Place the y-axis as shown and let the length of each book be L. We want to show that the center of mass of the system of books lies above the table, that is, $\overline{x} < L$. The x-coordinates of the centers of mass of the books are $x_1 = \dfrac{L}{2}$, $x_2 = \dfrac{L}{2(n-1)} + \dfrac{L}{2}$,

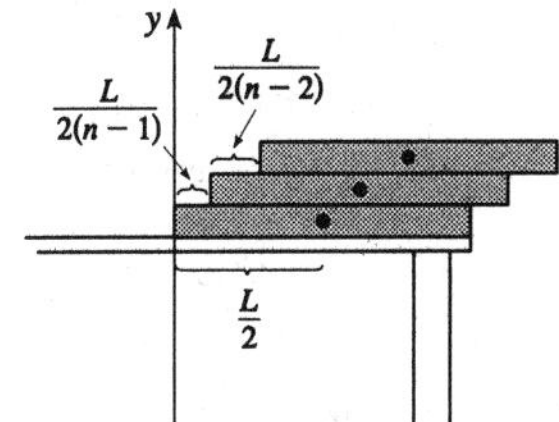

$x_3 = \dfrac{L}{2(n-1)} + \dfrac{L}{2(n-2)} + \dfrac{L}{2}$, and so on. Each book has the same mass m, so if there are n books, then

$$\begin{aligned} \overline{x} &= \frac{mx_1 + mx_2 + \cdots + mx_n}{mn} = \frac{x_1 + x_2 + \cdots + x_n}{n} \\ &= \frac{1}{n}\left[\frac{L}{2} + \left(\frac{L}{2(n-1)} + \frac{L}{2}\right) + \left(\frac{L}{2(n-1)} + \frac{L}{2(n-2)} + \frac{L}{2}\right)\right. \\ &\qquad \left. + \cdots + \left(\frac{L}{2(n-1)} + \frac{L}{2(n-2)} + \cdots + \frac{L}{4} + \frac{L}{2} + \frac{L}{2}\right)\right] \\ &= \frac{L}{n}\left[\frac{n-1}{2(n-1)} + \frac{n-2}{2(n-2)} + \cdots + \frac{2}{4} + \frac{1}{2} + \frac{n}{2}\right] = \frac{L}{n}\left[(n-1)\frac{1}{2} + \frac{n}{2}\right] \\ &= \frac{2n-1}{2n}L < L. \end{aligned}$$

This shows that, no matter how many books are added according to the given scheme, the center of mass lies above the table. It remains to observe that the series $\dfrac{1}{2} + \dfrac{1}{4} + \dfrac{1}{6} + \dfrac{1}{8} + \cdots = \dfrac{1}{2}\sum \dfrac{1}{n}$ is divergent (harmonic series), so we can make the top book extend as far as we like beyond the edge of the table if we add enough books.

29. $\displaystyle\int_0^1 \left(\frac{-x-1}{100}\right)\left(\frac{1}{x+1}+\frac{1}{x+2}+\frac{1}{x+3}\cdots+\frac{1}{x+100}\right)dx$

$$= \int_0^1 \frac{(-x-1)(-x-2)(-x-3)\cdots(-x-100)}{100!}\left(\frac{1}{x+1}+\cdots+\frac{1}{x+100}\right)dx$$

$$= \frac{1}{100!}\int_0^1 (x+1)(x+2)(x+3)\cdots(x+100)\left(\frac{1}{x+1}+\cdots+\frac{1}{x+100}\right)dx$$

$$= \frac{1}{100!}(x+1)(x+2)(x+3)\cdots(x+100)\Big|_0^1 \quad \text{(the Product Rule in reverse)}$$

$$= \frac{2\cdot 3\cdot\cdots\cdot 101}{100!} - \frac{100!}{100!} = 101 - 1 = 100.$$

30. Using the Maclaurin series for $\cosh x$ in Exercise 10.10.6 (proved in Exercise 10.10.16), the inequality $\cosh x \le e^{cx^2}$ becomes $\displaystyle\sum_{n=0}^{\infty}\frac{x^{2n}}{(2n)!} \le \sum_{n=0}^{\infty}\frac{(cx^2)^n}{n!} = \sum_{n=0}^{\infty}\frac{c^n x^{2n}}{n!}$. This will be true if $\dfrac{1}{(2n)!} \le \dfrac{c^n}{n!}$.

If $c \ge \frac{1}{2}$, then $\dfrac{c^n}{n!} \ge \dfrac{1}{2^n n!} \ge \dfrac{1}{(2n)!}$ because $(2n)! = 1\cdot 2\cdot 3\cdot\cdots\cdot(2n) \ge 2\cdot 4\cdot 6\cdot\cdots\cdot(2n) = 2^n n!$.

So if $c \ge \frac{1}{2}$, then $\cosh x \le e^{cx^2}$ for all x. Furthermore, the condition $c \ge \frac{1}{2}$ is necessary because $e^{cx^2} - \cosh x \ge 0 \;\Rightarrow\; (1 + cx^2 + \cdots) - (1 + \frac{1}{2}x^2 + \cdots) \ge 0 \;\Rightarrow$
$\dfrac{(1 + cx^2 + \cdots) - (1 + \frac{1}{2}x^2 + \cdots)}{x^2} \ge 0 \;\Rightarrow\; \displaystyle\lim_{x\to 0}\frac{(1 + cx^2 + \cdots) - (1 + \frac{1}{2}x^2 + \cdots)}{x^2} \ge 0 \;\Rightarrow\; c - \tfrac{1}{2} \ge 0$
$\Rightarrow\; c \ge \frac{1}{2}$.

31. We can write the sum as

$$\sum_{n=1}^{\infty}\frac{1}{n}\left[\sum_{m=1}^{\infty}\frac{1}{m}\left(\frac{1}{m+(n+2)}\right)\right] = \sum_{n=1}^{\infty}\frac{1}{n}\left[\frac{1}{n+2}\sum_{m=1}^{\infty}\left(\frac{1}{m} - \frac{1}{m+(n+2)}\right)\right] \quad \text{(partial fractions)}$$

$$= \frac{1}{2}\sum_{n=1}^{\infty}\left[\left(\frac{1}{n} - \frac{1}{n+2}\right)\sum_{m=1}^{n+2}\left(\frac{1}{m}\right)\right] \quad \left(\begin{array}{c}\text{partial fractions in the outer sum; all terms}\\ \text{beyond the } (n+2)\text{th cancel in the inner sum}\end{array}\right)$$

$$= \frac{1}{2}\left[\sum_{n=1}^{\infty}\left(\frac{1}{n}\sum_{m=1}^{n+2}\frac{1}{m}\right) - \sum_{n=3}^{\infty}\left(\frac{1}{n}\sum_{m=1}^{n}\frac{1}{m}\right)\right] \quad \text{(change the index)}$$

$$= \frac{1}{2}\left[\sum_{n=1}^{2}\left(\frac{1}{n}\sum_{m=1}^{n+2}\frac{1}{m}\right) + \sum_{n=3}^{\infty}\left(\frac{1}{n}\sum_{m=1}^{n+2}\frac{1}{m} - \frac{1}{n}\sum_{m=1}^{n}\frac{1}{m}\right)\right]$$

$$= \frac{1}{2}\left[1\left(1+\frac{1}{2}+\frac{1}{3}\right) + \left(\frac{1}{2}\right)\left(1+\frac{1}{2}+\frac{1}{3}+\frac{1}{4}\right) + \sum_{n=3}^{\infty}\frac{1}{n}\left(\frac{1}{n+1}+\frac{1}{n+2}\right)\right]$$

$$= \frac{1}{2}\left[\frac{11}{6} + \frac{25}{24} + \sum_{n=3}^{\infty}\left(\frac{1}{n} - \frac{1}{n-1}\right) + \sum_{n=3}^{\infty}\left(\frac{1/2}{n} - \frac{1/2}{n-2}\right)\right] \quad \text{(partial fractions)}$$

$$= \frac{1}{2}\left[\frac{11}{6} + \frac{25}{24} + \frac{1}{3} + \frac{1}{2}\left(\frac{1}{3}+\frac{1}{4}\right)\right] \quad \text{(both series telescope)} \quad = \frac{7}{4}.$$

32. **(a)** Since P_n is defined as the midpoint of $P_{n-4}P_{n-3}$, $x_n = \frac{1}{2}(x_{n-4} + x_{n-3})$ for $n \geq 5$. So we prove by induction that $\frac{1}{2}x_n + x_{n+1} + x_{n+2} + x_{n+3} = 2$. The case $n = 1$ is immediate, since $\frac{1}{2}0 + 1 + 1 + 0 = 2$. Assume that the result holds for $n = k - 1$. Then

$\frac{1}{2}x_k + x_{k+1} + x_{k+2} + x_{k+3} = \frac{1}{2}x_k + x_{k+1} + x_{k+2} + \frac{1}{2}(x_{k+3-4} + x_{k+3-3})$ (by above)

$= \frac{1}{2}x_{k-1} + x_k + x_{k+1} + x_{k+2} = 2$ (by the induction hypothesis). Similarly, for $n \geq 4$,

$y_n = \dfrac{y_{n-4} + y_{n-3}}{2}$, so the same argument as above holds for y, with 2 replaced by $\frac{1}{2}y_1 + y_2 + y_3 + y_4$ $= \frac{1}{2}1 + 1 + 0 + 0 = \frac{3}{2}$. So $\frac{1}{2}y_n + y_{n+1} + y_{n+2} + y_{n+3} = \frac{3}{2}$ for all n.

(b) $\lim\limits_{n\to\infty}\left(\frac{1}{2}x_n + x_{n+1} + x_{n+2} + x_{n+3}\right) = \frac{1}{2}\lim\limits_{n\to\infty} x_n + \lim\limits_{n\to\infty} x_{n+1} + \lim\limits_{n\to\infty} x_{n+2} + \lim\limits_{n\to\infty} x_{n+3} = 2$. Since all the limits on the left hand side are the same, we get $\frac{7}{2}\lim\limits_{n\to\infty} x_n = 2 \quad\Rightarrow\quad \lim\limits_{n\to\infty} x_n = \frac{4}{7}$. In the same way $\lim\limits_{n\to\infty} y_n = \frac{3}{7}$. So $P = \left(\frac{4}{7}, \frac{3}{7}\right)$.

33. **(a)** We prove by induction that $1 < a_{n+1} < a_n$. For $n = 1$,

$$a_{1+1} = \frac{3\left(\frac{3}{2}\right)^2 + 4\left(\frac{3}{2}\right) - 3}{4\left(\frac{3}{2}\right)^2} = \frac{3\left(\frac{9}{4}\right) + 6 - 3}{9} = \frac{13}{12},$$ so $1 < a_2 < a_1$. Assume for k. Then

$$a_{k+1} = \frac{3a_k^2 + 4a_k - 3}{4a_k^2} > 1 \quad\Leftrightarrow\quad 3a_k^2 + 4a_k - 3 > 4a_k^2 \quad\Leftrightarrow\quad 0 > a_k^2 - 4a_k + 3 = (a_k - 3)(a_k - 1)$$

$\Leftrightarrow \quad 1 < a_k < 3$, which is true by the induction hypothesis. So we have the first inequality. For the second, $a_{k+1} = \dfrac{3a_k^2 + 4a_k - 3}{4a_k^2} < a_k \quad\Leftrightarrow\quad 3a_k^2 + 4a_k - 3 < 4a_k^3 \quad\Leftrightarrow$

$0 < 4a_k^3 - 3a_k^2 - 4a_k + 3 = (4a_k - 3)\left(a_k^2 - 1\right) > 0$ since $a_k > 1$. So both results hold by induction.

(b) $\{a_n\}$ converges by part (a) and Theorem 10.1.10, so let $\lim\limits_{n\to\infty} a_n = L$. Then

$$L = \lim_{n\to\infty}\frac{3a_n^2 + 4a_n - 3}{4a_n^2} = \frac{3\left(\lim\limits_{n\to\infty} a_n\right)^2 + 4\left(\lim\limits_{n\to\infty} a_n\right) - 3}{4\left(\lim\limits_{n\to\infty} a_n\right)^2} = \frac{3L^2 + 4L - 3}{4L^2} \quad\Rightarrow$$

$4L^3 - 3L^2 - 4L + 3 = 0 \quad\Rightarrow\quad (4L - 3)\left(L^2 - 1\right) = (4L - 3)(L - 1)(L + 1) = 0$. Since $a_n > 1$ for all n, $L \geq 1$, but 1 is the only such root of this polynomial, so $L = 1$.

(c) Observe that $a_{n+1} = \dfrac{3a_n^2 + 4a_n - 3}{4a_n^2} \quad\Leftrightarrow\quad 4a_n^2 \cdot a_{n+1} - 3a_n^2 = 4a_n - 3 \quad\Leftrightarrow$

$a_n^2(4a_{n+1} - 3) = 4a_n - 3$. Substituting this for $n = 2$ into the same for $n = 1$ gives $a_1^2a_2^2(4a_3 - 3) = (4a_1 - 3)$. If we carry on these substitutions we get $a_1^2a_2^2\cdots a_n^2(4a_{n+1} - 3) = (4a_1 - 3)$

$\Leftrightarrow \quad a_1a_2\cdots a_n = \sqrt{\dfrac{4a_1 - 3}{4a_{n+1} - 3}}$. So $\lim\limits_{n\to\infty} a_1a_2\cdots a_n = \lim\limits_{n\to\infty}\sqrt{\dfrac{4a_1 - 3}{4a_{n+1} - 3}} = \sqrt{\dfrac{4\left(\frac{3}{2}\right) - 3}{4(1) - 3}} = \sqrt{3}$.

EXERCISES A

1. $|5-23| = |-18| = 18$

2. $|5| - |-23| = 5 - 23 = -18$

3. $|-\pi| = \pi$ because $\pi > 0$.

4. $|\pi - 2| = \pi - 2$ because $\pi > 2$.

5. $\left|\sqrt{5}-5\right| = -\left(\sqrt{5}-5\right) = 5 - \sqrt{5}$ because $\sqrt{5} - 5 < 0$.

6. $||-2| - |-3|| = |2-3| = |-1| = 1$

7. For $x < 2$, $x - 2 < 0$, so $|x-2| = -(x-2) = 2 - x$.

8. For $x > 2$, $x - 2 > 0$, so $|x-2| = x - 2$.

9. $|x+1| = \begin{cases} x+1 & \text{for } x+1 \geq 0 \quad \Leftrightarrow \quad x \geq -1 \\ -(x+1) & \text{for } x+1 < 0 \quad \Leftrightarrow \quad x < -1 \end{cases}$

10. $|2x-1| = \begin{cases} 2x-1 & \text{for } 2x-1 \geq 0 \quad \Leftrightarrow \quad x \geq \frac{1}{2} \\ 1-2x & \text{for } 2x-1 < 0 \quad \Leftrightarrow \quad x < \frac{1}{2} \end{cases}$

11. $|x^2+1| = x^2+1$ (since $x^2 + 1 \geq 0$ for all x).

12. Determine when $1 - 2x^2 < 0 \quad \Leftrightarrow \quad 1 < 2x^2 \quad \Leftrightarrow \quad x^2 > \frac{1}{2} \quad \Leftrightarrow \quad x < -\frac{1}{\sqrt{2}}$ or $x > \frac{1}{\sqrt{2}}$. Thus,

$$|1-2x^2| = \begin{cases} 2x^2 - 1 & \text{for } x < -\frac{1}{\sqrt{2}} \text{ or } x > \frac{1}{\sqrt{2}} \\ 1 - 2x^2 & \text{for } -\frac{1}{\sqrt{2}} \leq x \leq \frac{1}{\sqrt{2}} \end{cases}$$

13. $2x + 7 > 3 \quad \Leftrightarrow \quad 2x > -4$
$\Leftrightarrow \quad x > -2$, so $x \in (-2, \infty)$.

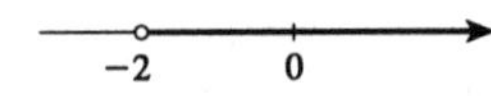

14. $3x - 11 < 4 \quad \Leftrightarrow \quad 3x < 15$
$\Leftrightarrow \quad x < 5$, so $x \in (-\infty, 5)$.

15. $1 - x \leq 2 \quad \Leftrightarrow \quad -x \leq 1$
$\Leftrightarrow \quad x \geq -1$, so $x \in [-1, \infty)$.

16. $4 - 3x \geq 6 \quad \Leftrightarrow \quad -3x \geq 2$
$\Leftrightarrow \quad x \leq -\frac{2}{3}$, so $x \in \left(-\infty, -\frac{2}{3}\right]$.

17. $2x + 1 < 5x - 8 \quad \Leftrightarrow \quad 9 < 3x$
$\Leftrightarrow \quad 3 < x$, so $x \in (3, \infty)$.

18. $1 + 5x > 5 - 3x \quad \Leftrightarrow \quad 8x > 4$
$\Leftrightarrow \quad x > \frac{1}{2}$, so $x \in \left(\frac{1}{2}, \infty\right)$.

19. $-1 < 2x - 5 < 7 \quad \Leftrightarrow \quad 4 < 2x < 12$
$\Leftrightarrow \quad 2 < x < 6$, so $x \in (2, 6)$.

20. $1 < 3x + 4 \leq 16 \quad \Leftrightarrow \quad -3 < 3x \leq 12$
$\Leftrightarrow \quad -1 < x \leq 4$, so $x \in (-1, 4]$.

21. $0 \leq 1 - x < 1 \quad \Leftrightarrow \quad -1 \leq -x < 0$
$\Leftrightarrow \quad 1 \geq x > 0$, so $x \in (0, 1]$.

22. $-5 \leq 3 - 2x \leq 9 \quad \Leftrightarrow \quad -8 \leq -2x \leq 6$
$\Leftrightarrow \quad 4 \geq x \geq -3$, so $x \in [-3, 4]$.

23. $4x < 2x + 1 \le 3x + 2$. So $4x < 2x + 1 \quad \Leftrightarrow \quad 2x < 1 \quad \Leftrightarrow \quad x < \frac{1}{2}$, and $2x + 1 \le 3x + 2 \quad \Leftrightarrow \quad -1 \le x$. Thus $x \in \left[-1, \frac{1}{2}\right)$.

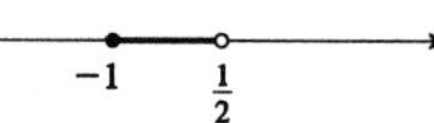

24. $2x - 3 < x + 4 < 3x - 2$. So $2x - 3 < x + 4 \quad \Leftrightarrow \quad x < 7$, and $x + 4 < 3x - 2 \quad \Leftrightarrow \quad 6 < 2x \quad \Leftrightarrow$ $3 < x$, so $x \in (3, 7)$.

25. $1 - x \ge 3 - 2x \ge x - 6$. So $1 - x \ge 3 - 2x \quad \Leftrightarrow \quad x \ge 2$, and $3 - 2x \ge x - 6 \quad \Leftrightarrow \quad 9 \ge 3x \quad \Leftrightarrow$ $3 \ge x$. Thus $x \in [2, 3]$.

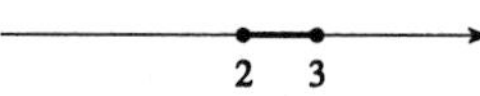

26. $x > 1 - x \ge 3 + 2x$. so $x > 1 - x \quad \Leftrightarrow \quad 2x > 1 \quad \Leftrightarrow \quad x > \frac{1}{2}$, and $1 - x \ge 3 + 2x \quad \Leftrightarrow \quad -2 \ge 3x \quad \Leftrightarrow$ $-\frac{2}{3} \ge x$. This contradicts $x > \frac{1}{2}$, so there is no solution.

27. $(x - 1)(x - 2) > 0$. *Case 1:* $x - 1 > 0 \quad \Leftrightarrow \quad x > 1$, and $x - 2 > 0 \quad \Leftrightarrow \quad x > 2$, so $x \in [1, \infty)$. *Case 2:* $x - 1 < 0 \quad \Leftrightarrow \quad x < 1$, and $x - 2 < 0 \quad \Leftrightarrow \quad x < 2$, so $x \in (-\infty, 1)$. Thus the solution set is $(-\infty, 1) \cup (2, \infty)$.

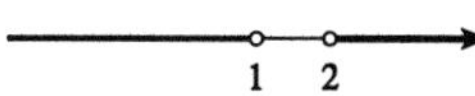

28. $(2x + 3)(x - 1) \ge 0$. *Case 1:* $2x + 3 \ge 0 \quad \Leftrightarrow \quad x \ge -\frac{3}{2}$, and $x - 1 \ge 0 \quad \Leftrightarrow \quad x \ge 1$, so $x \in [1, \infty)$. *Case 2:* $2x + 3 \le 0 \quad \Leftrightarrow \quad x \le -\frac{3}{2}$, and $x \le 1$, so $x \in \left(-\infty, -\frac{3}{2}\right]$. Thus the solution set is $\left(-\infty, -\frac{3}{2}\right] \cup [1, \infty)$.

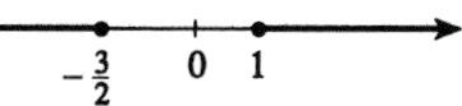

29. $2x^2 + x \le 1 \quad \Leftrightarrow \quad 2x^2 + x - 1 \le 0 \quad \Leftrightarrow \quad (2x - 1)(x + 1) \le 0$. *Case 1:* $2x - 1 \ge 0 \quad \Leftrightarrow \quad x \ge \frac{1}{2}$, and $x + 1 \le 0 \quad \Leftrightarrow \quad x \le -1$, which is impossible. *Case 2:* $2x - 1 \le 0 \quad \Leftrightarrow \quad x \le \frac{1}{2}$, and $x + 1 \ge 0 \quad \Leftrightarrow$ $x \ge -1$, so $x \in \left[-1, \frac{1}{2}\right]$. Thus the solution set is $\left[-1, \frac{1}{2}\right]$.

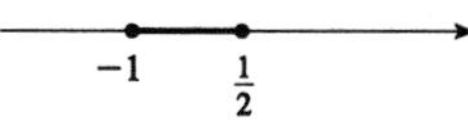

30. $x^2 < 2x + 8 \quad \Leftrightarrow \quad x^2 - 2x - 8 < 0 \quad \Leftrightarrow \quad (x - 4)(x + 2) < 0$. *Case 1:* $x > 4$ and $x < -2$, which is impossible. *Case 2:* $x < 4$ and $x > -2$. So the solution set is $(-2, 4)$.

31. $x^2 + x + 1 > 0 \quad \Leftrightarrow \quad x^2 + x + \frac{1}{4} + \frac{3}{4} > 0 \quad \Leftrightarrow \quad \left(x + \frac{1}{2}\right)^2 + \frac{3}{4} > 0$. But since $\left(x + \frac{1}{2}\right)^2 \ge 0$ for every real x, the original inequality will be true for all real x as well. Thus, the solution set is $(-\infty, \infty)$.

32. $x^2 + x > 1 \quad \Leftrightarrow \quad x^2 + x - 1 > 0$. Using the quadratic formula, we obtain $x^2 + x - 1 = \left(x - \frac{-1-\sqrt{5}}{2}\right)\left(x - \frac{-1+\sqrt{5}}{2}\right) > 0$. *Case 1:* $x - \frac{-1-\sqrt{5}}{2} > 0$ and $x - \frac{-1+\sqrt{5}}{2} > 0$, so that $x > \frac{-1+\sqrt{5}}{2}$. *Case 2:* $x - \frac{-1-\sqrt{5}}{2} < 0$ and $x - \frac{-1+\sqrt{5}}{2} < 0$, so that $x < \frac{-1-\sqrt{5}}{2}$. Thus the solution set is $\left(-\infty, \frac{-1-\sqrt{5}}{2}\right) \cup \left(\frac{-1+\sqrt{5}}{2}, \infty\right)$.

$(-1-\sqrt{5})/2 \quad 0 \quad (-1+\sqrt{5})/2$

33. $x^2 < 3 \quad \Leftrightarrow \quad x^2 - 3 < 0 \quad \Leftrightarrow \quad \left(x - \sqrt{3}\right)\left(x + \sqrt{3}\right) < 0$. *Case 1:* $x > \sqrt{3}$ and $x < -\sqrt{3}$, which is impossible. *Case 2:* $x < \sqrt{3}$ and $x > -\sqrt{3}$. Thus the solution set is $\left(-\sqrt{3}, \sqrt{3}\right)$.

Another Method: $x^2 < 3 \quad \Leftrightarrow \quad |x| < \sqrt{3} \quad \Leftrightarrow \quad -\sqrt{3} < x < \sqrt{3}$.

$-\sqrt{3} \quad 0 \quad \sqrt{3}$

34. $x^2 \geq 5 \quad \Leftrightarrow \quad x^2 - 5 \geq 0 \quad \Leftrightarrow \quad \left(x - \sqrt{5}\right)\left(x + \sqrt{5}\right) \geq 0$. *Case 1:* $x \geq \sqrt{5}$ and $x \geq -\sqrt{5}$, so $x \in \left[\sqrt{5}, \infty\right)$. *Case 2:* $x \leq \sqrt{5}$ and $x \leq -\sqrt{5}$, so $x \in \left(-\infty, -\sqrt{5}\right]$. Thus the solution set is $\left(-\infty, -\sqrt{5}\right] \cup \left[\sqrt{5}, \infty\right)$. *Another Method:* $x^2 \geq 5 \quad \Leftrightarrow \quad |x| \geq \sqrt{5} \quad \Leftrightarrow \quad x \geq \sqrt{5}$ or $x \leq -\sqrt{5}$.

$-\sqrt{5} \quad 0 \quad \sqrt{5}$

35. $x^3 - x^2 \leq 0 \quad \Leftrightarrow \quad x^2(x-1) \leq 0$. Since $x^2 \geq 0$ for all x, the inequality is satisfied when $x - 1 \leq 0 \quad \Leftrightarrow \quad x \leq 1$. Thus the solution set is $(-\infty, 1]$.

$0 \quad 1$

36. $(x+1)(x-2)(x+3) = 0 \quad \Leftrightarrow \quad x = -1, 2,$ or -3. Constructing a table:

Interval	$x+1$	$x-2$	$x+3$	$(x+1)(x-2)(x+3)$
$x < -3$	$-$	$-$	$-$	$-$
$-3 < x < -1$	$-$	$-$	$+$	$+$
$-1 < x < 2$	$+$	$-$	$+$	$-$
$x > 2$	$+$	$+$	$+$	$+$

Thus $(x+1)(x-2)(x+3) \geq 0$ on $[-3, -1]$ and $[2, \infty)$, and the solution set is $[-3, -1] \cup [2, \infty)$.

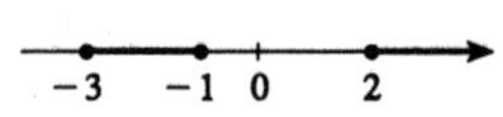

37. $x^3 > x \quad \Leftrightarrow \quad x^3 - x > 0 \quad \Leftrightarrow \quad x(x^2 - 1) > 0 \quad \Leftrightarrow \quad x(x-1)(x+1) > 0$. Constructing a table:

Interval	x	$x-1$	$x+1$	$x(x-1)(x+1)$
$x < -1$	$-$	$-$	$-$	$-$
$-1 < x < 0$	$-$	$-$	$+$	$+$
$0 < x < 1$	$+$	$-$	$+$	$-$
$x > 1$	$+$	$+$	$+$	$+$

Since $x^3 > x$ when the last column is positive, the solution set is $(-1, 0) \cup (1, \infty)$.

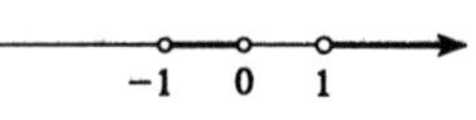

38. $x^3 + 3x < 4x^2 \quad \Leftrightarrow \quad x^3 - 4x^2 + 3x < 0 \quad \Leftrightarrow \quad x(x^2 - 4x + 3) < 0 \quad \Leftrightarrow \quad x(x-1)(x-3) < 0.$

Interval	x	$x-1$	$x-3$	$x(x-1)(x-3)$
$x < 0$	$-$	$-$	$-$	$-$
$0 < x < 1$	$+$	$-$	$-$	$+$
$1 < x < 3$	$+$	$+$	$-$	$-$
$x > 3$	$+$	$+$	$+$	$+$

Thus the solution set is $(-\infty, 0) \cup (1, 3)$.

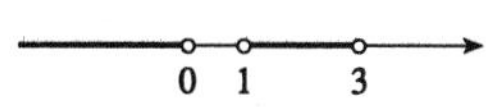

39. $1/x < 4$. This is clearly true for $x < 0$. So suppose $x > 0$. then $1/x < 4 \quad \Leftrightarrow \quad 1 < 4x \quad \Leftrightarrow \quad \frac{1}{4} < x$. Thus the solution set is $(-\infty, 0) \cup \left(\frac{1}{4}, \infty\right)$.

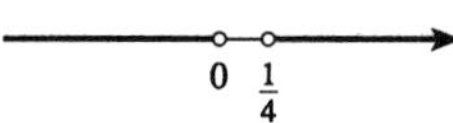

40. We solve the two inequalities separately and take the intersection of the solution sets. First, $-3 < 1/x$ is clearly true for $x > 0$. So suppose $x < 0$. Then $-3 < 1/x \quad \Leftrightarrow \quad -3x > 1 \quad \Leftrightarrow \quad x < -\frac{1}{3}$, so for this inequality, the solution set is $\left(-\infty, -\frac{1}{3}\right) \cup (0, \infty)$. Now $1/x \leq 1$ is clearly true if $x < 0$. So suppose $x > 0$. Then $1/x < 1 \quad \Leftrightarrow \quad 1 \leq x$, and the solution set here is $(-\infty, 0) \cup [1, \infty)$. Taking the intersection of the two solution sets gives the final solution set: $\left(-\infty, -\frac{1}{3}\right) \cup [1, \infty)$.

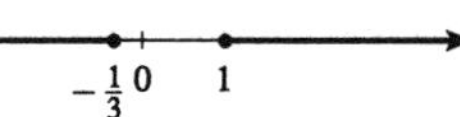

41. Multiply both sides by x. *Case 1:* If $x > 0$, then $4/x < x \quad \Leftrightarrow \quad 4 < x^2 \quad \Leftrightarrow \quad 2 < x$. *Case 2:* If $x < 0$, then $4/x < x \quad \Leftrightarrow \quad 4 > x^2 \quad \Leftrightarrow \quad -2 < x < 0$. Thus the solution set is $(-2, 0) \cup (2, \infty)$.

−2 0 2

42. $\dfrac{x}{x+1} > 3$. *Case 1:* If $x + 1 > 0$ (that is, $x > -1$), then $x > 3(x+1) \quad \Leftrightarrow \quad -3 > 2x \quad \Leftrightarrow \quad -\dfrac{3}{2} > x$, which is impossible in this case. *Case 2:* If $x + 1 < 0$ (that is, $x < -1$), then $x < 3(x+1) \quad \Leftrightarrow \quad -3 < 2x \quad \Leftrightarrow \quad -\frac{3}{2} < x$, so $-\frac{3}{2} < x < -1$. Thus the solution set is $\left(-\frac{3}{2}, -1\right)$.

−3/2 −1 0

43. $\dfrac{2x+1}{x-5} < 3$. *Case 1:* If $x - 5 > 0$ (that is, $x > 5$), then $2x + 1 < 3(x-5) \quad \Leftrightarrow \quad 16 < x$, so $x \in (16, \infty)$. *Case 2:* If $x - 5 < 0$ (that is, $x < 5$), then $2x + 1 > 3(x-5) \quad \Leftrightarrow \quad 16 > x$, so in this case $x \in (-\infty, 5)$. Combining the two cases, the solution set is $(-\infty, 5) \cup (16, \infty)$.

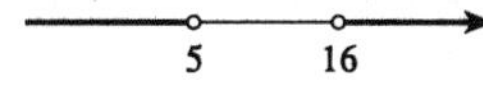

44. $\dfrac{2+x}{3-x} \leq 1$. *Case 1:* If $3 - x < 0$ (that is, $3 < x$), then $2 + x \geq 3 - x \quad \Leftrightarrow \quad 2x \geq 1 \quad \Leftrightarrow \quad x \geq \frac{1}{2}$, so $x \in (3, \infty)$. *Case 2:* If $3 - x > 0$ (that is, $3 > x$), then $2 + x \leq 3 - x \quad \Leftrightarrow \quad 2x \leq 1 \quad \Leftrightarrow \quad x \leq \frac{1}{2}$, so $x \in \left(-\infty, \frac{1}{2}\right] \cup (3, \infty)$.

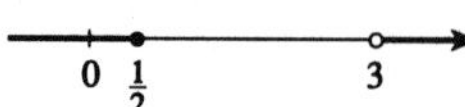

45. $\dfrac{x^2-1}{x^2+1} \geq 0$. Since $x^2 + 1 \geq 0$ for all real x, this inequality will hold whenever $x^2 - 1 \geq 0 \quad \Leftrightarrow$ $(x-1)(x+1) \geq 0$. *Case 1:* $x \geq 1$ and $x \geq -1$, so $x \in [1, \infty)$. *Case 2:* $x \leq 1$ and $x \leq -1$, so $x \in (-\infty, -1]$. Thus the solution set is $(-\infty, -1] \cup [1, \infty)$.

Another Method: $x^2 \geq 1 \quad \Leftrightarrow \quad |x| \geq 1 \quad \Leftrightarrow \quad x \geq 1$ or $x \leq -1$.

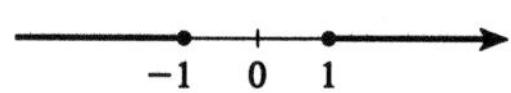

46. $\dfrac{x^2-2x}{x^2-2} > 0 \quad \Leftrightarrow \quad \dfrac{x(x-2)}{\left(x-\sqrt{2}\right)\left(x+\sqrt{2}\right)} > 0$. Call this quotient Q and construct a table:

Interval	x	$x-2$	$x-\sqrt{2}$	$x+\sqrt{2}$	Q
$x < -\sqrt{2}$	$-$	$-$	$-$	$-$	$+$
$-\sqrt{2} < x < 0$	$-$	$-$	$-$	$+$	$-$
$0 < x < \sqrt{2}$	$+$	$-$	$-$	$+$	$+$
$\sqrt{2} < x < 2$	$+$	$-$	$+$	$+$	$-$
$x > 2$	$+$	$+$	$+$	$+$	$+$

Thus the solution set is $\left(-\infty, -\sqrt{2}\right) \cup \left(0, \sqrt{2}\right) \cup (2, \infty)$.

47. $C = \frac{5}{9}(F - 32) \quad \Rightarrow \quad F = \frac{9}{5}C + 32$. So $50 \leq F \leq 95 \quad \Rightarrow \quad 50 \geq \frac{9}{5}C + 32 \leq 95 \quad \Rightarrow \quad 18 \leq \frac{9}{5}C \leq 63$ $\Rightarrow \quad 10 \leq C \leq 35$. So the interval is $[10, 35]$.

48. Since $20 \leq C \leq 30$ and $C = \frac{5}{9}(F - 32)$, then $20 \leq \frac{5}{9}(F - 32) \leq 30 \quad \Rightarrow \quad 36 \leq F - 32 \leq 54 \quad \Rightarrow$ $68 \leq F \leq 86$. So the interval is $[68, 86]$.

49. **(a)** Let T represent the temperature in degrees Celsius and h the height in km. $T = 20$ when $h = 0$ and T decreases by 10° C for every km. Thus $T = 20 - 10h$ when $0 \leq h \leq 12$.

(b) From (a), $T = 20 - 10h \quad \Rightarrow \quad h = 2 - T/10$. So $0 \leq h \leq 5 \quad \Rightarrow \quad 0 \leq 2 - T/10 \leq 5 \quad \Rightarrow$ $-2 \leq -T/10 \leq 3 \quad \Rightarrow \quad -30 \leq T \leq 20$. Thus the range of temperatures to be expected is $[-30, 20]$.

50. The ball will be at least 32 ft above the ground if $h \geq 32 \quad \Leftrightarrow \quad 128 + 16t - 16t^2 \geq 32 \quad \Leftrightarrow$ $16t^2 - 16t - 96 \leq 0 \quad \Leftrightarrow \quad 16(t-3)(t+2) \leq 0$. $t = 3$ and $t = -2$ are endpoints of the interval we're looking for, and constructing a table gives $-2 \leq t \leq 3$. But $t \geq 0$, so the ball will be at least 32 ft above the ground in the time interval $[0, 3]$.

51. $|2x| = 3 \quad \Leftrightarrow \quad$ either $2x = 3$ or $2x = -3 \quad \Leftrightarrow \quad x = \frac{3}{2}$ or $x = -\frac{3}{2}$.

52. $|3x+5| = 1 \quad \Leftrightarrow \quad$ either $3x + 5 = 1$ or -1. In the first case, $3x = -4$ so $x = -\frac{4}{3}$, and in the second case, $3x = -6 \quad \Leftrightarrow \quad x = -2$.

53. $|x+3| = |2x+1| \quad \Leftrightarrow \quad$ either $x + 3 = 2x + 1$ or $x + 3 = -(2x+1)$. In the first case, $x = 2$, and in the second case, $3x = -4 \quad \Leftrightarrow \quad x = -\frac{4}{3}$.

54. $\left|\dfrac{2x-1}{x+1}\right| = 3 \quad \Leftrightarrow \quad$ either $\dfrac{2x-1}{x+1} = 3$ or $\dfrac{2x-1}{x+1} = -3$. In the first case, $2x - 1 = 3x + 3 \quad \Leftrightarrow \quad x = -4$, and in the second case, $2x - 1 = -3x - 3 \quad \Leftrightarrow \quad x = -\frac{2}{5}$.

55. By (6), Property 5, $|x| < 3 \quad \Leftrightarrow \quad -3 < x < 3$, so $x \in (-3, 3)$.

56. By (6), Properties 4 and 6, $|x| \geq 3 \quad \Leftrightarrow \quad x \leq -3$ or $x \geq 3$, so $x \in (-\infty, -3] \cup [3, \infty)$.

57. $|x-4| < 1 \quad \Leftrightarrow \quad -1 < x - 4 < 1 \quad \Leftrightarrow \quad 3 < x < 5$, so $x \in (3, 5)$.

58. $|x-6| < 0.1 \quad \Leftrightarrow \quad -0.1 < x - 6 < 0.1 \quad \Leftrightarrow \quad 5.9 < x < 6.1$, so $x \in (5.9, 6.1)$.

59. $|x+5| \geq 2 \quad \Leftrightarrow \quad x + 5 \geq 2$ or $x + 5 \leq -2 \quad \Leftrightarrow \quad x \geq -3$ or $x \leq -7$, so $x \in (-\infty, -7] \cup [-3, \infty)$.

60. $|x+1| \geq 3 \quad \Leftrightarrow \quad x + 1 \geq 3$ or $x + 1 \leq -3 \quad \Leftrightarrow \quad x \geq 2$ or $x \leq -4$, so $x \in (-\infty, -4] \cup [2, \infty)$.

61. $|2x-3| \leq 0.4 \quad \Leftrightarrow \quad -0.4 \leq 2x - 3 \leq 0.4 \quad \Leftrightarrow \quad 2.6 \leq 2x \leq 3.4 \quad \Leftrightarrow \quad 1.3 \leq x \leq 1.7$, so $x \in [1.3, 1.7]$.

62. $|5x-2| < 6 \quad \Leftrightarrow \quad -6 < 5x - 2 < 6 \quad \Leftrightarrow \quad -4 < 5x < 8 \quad \Leftrightarrow \quad -\frac{4}{5} < x < \frac{8}{5}$, so $x \in \left(-\frac{4}{5}, \frac{8}{5}\right)$.

63. $1 \leq |x| \leq 4$. So either $1 \leq x \leq 4$ or $1 \leq -x \leq 4 \quad \Leftrightarrow \quad -1 \geq x \geq -4$. Thus $x \in [-4, -1] \cup [1, 4]$.

64. $0 < |x-5| < \frac{1}{2}$. Clearly $0 < |x-5|$ for $x \neq 5$. Now $|x-5| < \frac{1}{2} \quad \Leftrightarrow \quad -\frac{1}{2} < x - 5 < \frac{1}{2} \quad \Leftrightarrow$ $4.5 < x < 5.5$. So the solution set is $(4.5, 5) \cup (5, 5.5)$.

65. $|x| > |x-1|$. Since $|x|, |x-1| \geq 0$, $|x| > |x-1| \quad \Leftrightarrow \quad |x|^2 > |x-1|^2 \quad \Leftrightarrow$ $x^2 > (x-1)^2 = x^2 - 2x + 1 \quad \Leftrightarrow \quad 0 > -2x + 1 \quad \Leftrightarrow \quad x > \frac{1}{2}$, so $x \in \left(\frac{1}{2}, \infty\right)$.

66. $|2x-5| \leq |x+4| \quad \Leftrightarrow \quad |2x-5|^2 \leq |x+4|^2 \quad \Leftrightarrow \quad (2x-5)^2 \leq (x+4)^2 \quad \Leftrightarrow$ $4x^2 - 20x + 25 \leq x^2 + 8x + 16 \quad \Leftrightarrow \quad 3x^2 - 28x + 9 \leq 0 \quad \Leftrightarrow \quad (3x-1)(x-9) \leq 0$. *Case 1:* $x - 9 \geq 0$ (that is, $x \geq 9$), and $3x - 1 \leq 0 \quad \Leftrightarrow \quad x \leq \frac{1}{3}$, which is impossible in this case. *Case 2:* $x - 9 \leq 0$ (that is, $x \leq 9$), and $3x - 1 \geq 0 \quad \Leftrightarrow \quad x \geq \frac{1}{3}$, so $\frac{1}{3} \leq x \leq 9$. Thus the solution set is $\left[\frac{1}{3}, 9\right]$.

67. $\left|\dfrac{x}{2+x}\right| < 1 \quad \Leftrightarrow \quad \left(\dfrac{x}{2+x}\right)^2 < 1 \quad \Leftrightarrow \quad x^2 < (2+x)^2 \quad \Leftrightarrow \quad x^2 < 4 + 4x + x^2 \quad \Leftrightarrow \quad 0 < 4 + 4x$ $\Leftrightarrow \quad -1 < x$, so $x \in (-1, \infty)$.

68. $\left|\dfrac{2-3x}{1+2x}\right| \leq 4 \quad \Leftrightarrow \quad \left(\dfrac{2-3x}{1+2x}\right)^2 \leq 16 \quad \Leftrightarrow \quad (2-3x)^2 \leq 16(1-2x)^2$ (for $x \neq -\frac{1}{2}$) $\quad \Leftrightarrow$ $4 - 12x + 9x^2 \leq 16 + 64x + 64x^2 \quad \Leftrightarrow \quad 0 \leq 55x^2 + 76x + 12 = (5x+6)(11x+2)$. *Case 1:* $5x + 6 \geq 0$ $\Leftrightarrow \quad x \geq -\frac{6}{5}$ and $11x + 2 \geq 0 \quad \Leftrightarrow \quad x \geq -\frac{2}{11}$, so $x \in \left[-\frac{2}{11}, \infty\right)$. *Case 2:* $5x + 6 \leq 0 \quad \Leftrightarrow \quad x \leq -\frac{6}{5}$ and $11x + 2 \leq 0 \quad \Leftrightarrow \quad x \leq -\frac{2}{11}$, so $x \in \left(-\infty, -\frac{6}{5}\right]$. Thus the solution set is $\left(-\infty, -\frac{6}{5}\right] \cup \left[-\frac{2}{11}, \infty\right)$.

69. $a(bx-c) \geq bc \quad \Leftrightarrow \quad bx - c \geq \dfrac{bc}{a} \quad \Leftrightarrow \quad bx \geq \dfrac{bc}{a} + c = \dfrac{bc+ac}{a} \quad \Leftrightarrow \quad x \geq \dfrac{bc+ac}{ab}$

70. $a \leq bx + c < 2a \quad \Leftrightarrow \quad a - c \leq bx < 2a - c \quad \Leftrightarrow \quad \dfrac{a-c}{b} \leq x < \dfrac{2a-c}{b}$ (since $b > 0$)

71. $ax + b < c \quad \Leftrightarrow \quad ax < c - b \quad \Leftrightarrow \quad x > \dfrac{c-b}{a}$ (since $a < 0$)

72. $\dfrac{ax+b}{c} \leq b \quad \Leftrightarrow \quad ax + b \geq bc$ (since $c < 0$) $\quad \Leftrightarrow \quad ax \geq bc - b \quad \Leftrightarrow \quad x \leq \dfrac{b(c-1)}{a}$ (since $a < 0$)

73. $|(x+y)-5| = |(x-2)+(y-3)| \leq |x-2| + |y-3| < 0.01 + 0.04 = 0.05$

74. Use the Triangle Inequality: $|x+3| < \frac{1}{2} \quad \Rightarrow$
$|4x+13| = |4(x+3)+1| \leq |4(x+3)| + |1| = 4|x+3| + 1 < 4\left(\frac{1}{2}\right) + 1 = 3$
Alternate Solution: $|x+3| < \frac{1}{2} \quad \Rightarrow \quad -\frac{1}{2} < x + 3 < \frac{1}{2} \quad \Rightarrow \quad -2 < 4x + 12 < 2 \quad \Rightarrow$
$-1 < 4x + 13 < 3 \quad \Rightarrow \quad |4x+13| < 3$

75. If $a < b$ then $a + a < a + b$ and $a + b < b + b$. So $2a < a + b < 2b$. Dividing by 2, $a < \frac{1}{2}(a+b) < b$.

76. If $0 < a < b$, then $\dfrac{1}{ab} > 0$. So $a < b \quad \Rightarrow \quad \dfrac{1}{ab} \cdot a < \dfrac{1}{ab} \cdot b \quad \Leftrightarrow \quad \dfrac{1}{b} < \dfrac{1}{a}$.

77. $|ab| = \sqrt{(ab)^2} = \sqrt{a^2b^2} = \sqrt{a^2}\sqrt{b^2} = |a||b|$

78. $\left|\dfrac{a}{b}\right||b| = \left|\dfrac{a}{b} \cdot b\right| = |a|$ (using the result of Exercise 77). Dividing the equation through by $|b|$ gives $\left|\dfrac{a}{b}\right| = \dfrac{|a|}{|b|}$.

79. If $0 < a < b$, then $a \cdot a < a \cdot b$ and $a \cdot b < b \cdot b$ [using (2), Rule 3]. So $a^2 < ab < b^2$ and hence $a^2 < b^2$.

80. Following the hint, the Triangle Inequality becomes $|(x-y)+y| \leq |x-y| + |y| \quad \Leftrightarrow \quad |x| \leq |x-y| + |y|$
$\Leftrightarrow \quad |x-y| \geq |x| - |y|$.

81. Observe that the sum, difference and product of two integers is always an integer. Let the rational numbers be represented by $r = m/n$ and $s = p/q$ (where m, n, p and q are integers with $n \neq 0$, $q \neq 0$). Now $r + s = \dfrac{m}{n} + \dfrac{p}{q} = \dfrac{mq+pn}{nq}$, but $mq + pn$ and nq are both integers, so $\dfrac{mq+pn}{nq} = r + s$ is a rational number by definition. Similarly, $r - s = \dfrac{m}{n} - \dfrac{p}{q} = \dfrac{mq-pn}{nq}$ is a rational number. Finally, $r \cdot s = \dfrac{m}{n} \cdot \dfrac{p}{q} = \dfrac{mp}{nq}$ but mp and nq are both integers, so $\dfrac{mp}{nq} = r \cdot s$ is a rational number by definition.

82. **(a)** No. Consider the case of $\sqrt{2}$ and $\sqrt{3}$. Both are irrational numbers, yet $\sqrt{2} + \left(-\sqrt{2}\right) = 0$ and 0, being an integer, is not irrational.

(b) No. Consider the case of $\sqrt{2}$ and $\sqrt{2}$. Both are irrational numbers, yet $\sqrt{2} \cdot \sqrt{2} = 2$ is not irrational.

EXERCISES B

1. From the Distance Formula (1) with $x_1 = 1$, $x_2 = 4$, $y_1 = 1$, $y_2 = 5$, we find the distance to be $\sqrt{(4-1)^2 + (5-1)^2} = \sqrt{3^2 + 4^2} = \sqrt{25} = 5$.

2. $\sqrt{(5-1)^2 + [7-(-3)]^2} = \sqrt{4^2 + 10^2} = \sqrt{116} = 2\sqrt{29}$

3. $\sqrt{(-1-6)^2 + [3-(-2)]^2} = \sqrt{(-7)^2 + 5^2} = \sqrt{74}$

4. $\sqrt{(-1-1)^2 + [-3-(-6)]^2} = \sqrt{(-2)^2 + 3^2} = \sqrt{13}$

5. $\sqrt{(4-2)^2 + (-7-5)^2} = \sqrt{2^2 + (-12)^2} = \sqrt{148} = 2\sqrt{37}$

6. $\sqrt{(b-a)^2 + (a-b)^2} = \sqrt{(a-b)^2 + (a-b)^2} = \sqrt{2(a-b)^2} = \sqrt{2}|a-b|$

7. From (2), the slope is $\dfrac{11-5}{4-1} = \dfrac{6}{3} = 2$.

8. $m = \dfrac{-3-6}{4-(-1)} = -\dfrac{9}{5}$

9. $m = \dfrac{-6-3}{-1-(-3)} = -\dfrac{9}{2}$

10. $m = \dfrac{0-(-4)}{6-(-1)} = \dfrac{4}{7}$

11. Since $|AC| = \sqrt{(-4-0)^2 + (3-2)^2} = \sqrt{(-4)^2 + 1^2} = \sqrt{17}$ and $|BC| = \sqrt{[-4-(-3)]^2 + [3-(-1)]^2} = \sqrt{(-1)^2 + 4^2} = \sqrt{17}$, the triangle has two sides of equal length, and so is isosceles.

12. **(a)** $|AB| = \sqrt{(11-6)^2 + [-3-(-7)]^2} = \sqrt{5^2 + 4^2} = \sqrt{41}$,
$|AC| = \sqrt{(2-6)^2 + [-2-(-7)]^2} = \sqrt{(-4)^2 + 5^2} = \sqrt{41}$, and
$|BC| = \sqrt{(2-11)^2 + [-2-(-3)]^2} = \sqrt{(-9)^2 + 1^2} = \sqrt{82}$, so
$|AB|^2 + |AC|^2 = 41 + 41 = 82 = |BC|^2$, and so $\triangle ABC$ is a right triangle.

(b) $m_{AB} = \dfrac{-3-(-7)}{11-6} = \dfrac{4}{5}$ and $m_{AC} = \dfrac{-2-(-7)}{2-6} = -\dfrac{5}{4}$. Thus $m_{AB} \cdot m_{AC} = -1$ and so AB is perpendicular to AC and $\triangle ABC$ must be a right triangle.

(c) Taking lengths from (a), the base is $\sqrt{41}$ and the height is $\sqrt{41}$. Thus the area is $\frac{1}{2}bh = \frac{1}{2}\sqrt{41}\sqrt{41} = \frac{41}{2}$.

13. Label the points A, B, C, and D respectively. Then
$|AB| = \sqrt{[4-(-2)]^2 + (6-9)^2} = \sqrt{6^2 + (-3)^2} = 3\sqrt{5}$,
$|BC| = \sqrt{(1-4)^2 + (0-6)^2} = \sqrt{(-3)^2 + (-6)^2} = 3\sqrt{5}$,
$|CD| = \sqrt{(-5-1)^2 + (3-0)^2} = \sqrt{(-6)^2 + 3^2} = 3\sqrt{5}$, and
$|DA| = \sqrt{[-2-(-5)]^2 + (9-3)^2} = \sqrt{3^2 + 6^2} = 3\sqrt{5}$. So all sides are of equal length. Moreover, $m_{AB} = \dfrac{6-9}{4-(-2)} = -\dfrac{1}{2}$, $m_{BC} = \dfrac{0-6}{1-4} = 2$, $m_{CD} = \dfrac{3-0}{-5-1} = -\dfrac{1}{2}$, and $m_{DA} = \dfrac{9-3}{-2-(-5)} = 2$, so the sides are perpendicular. Thus, it is a square.

14. (a) $|AB| = \sqrt{[3-(-1)]^2 + (11-3)^2} = \sqrt{4^2+8^2} = \sqrt{80} = 4\sqrt{5}$,
$|BC| = \sqrt{(5-3)^2 + (15-11)^2} = \sqrt{2^2+4^2} = \sqrt{20} = 2\sqrt{5}$, and
$|AC| = \sqrt{[5-(-1)]^2 + (15-3)^2} = \sqrt{6^2+12^2} = \sqrt{180} = 6\sqrt{5} = |AB| + |BC|$.

(b) $m_{AB} = \dfrac{11-3}{3-(-1)} = \dfrac{8}{4} = 2$ and $m_{AC} = \dfrac{15-3}{5-(-1)} = \dfrac{12}{6} = 2$. Since the segments AB and AC have the same slope, A, B and C must be collinear.

15. The slope of the line segment AB is $\dfrac{4-1}{7-1} = \dfrac{1}{2}$, the slope of CD is $\dfrac{7-10}{-1-5} = \dfrac{1}{2}$, the slope of BC is $\dfrac{10-4}{5-7} = -3$, and the slope of DA is $\dfrac{1-7}{1-(-1)} = -3$. So AB is parallel to CD and BC is parallel to DA. Hence $ABCD$ is a parallelogram.

16. The slopes of the four sides are $m_{AB} = \dfrac{3-1}{11-1} = \dfrac{1}{5}$, $m_{BC} = \dfrac{8-3}{10-11} = -5$, $m_{CD} = \dfrac{6-8}{0-10} = \dfrac{1}{5}$, and $m_{DA} = \dfrac{1-6}{1-0} = -5$. Hence $AB \parallel CD$, $BC \parallel DA$, $AB \perp BC$, $BC \perp CD$, $CD \perp DA$, and $DA \perp AB$, and so $ABCD$ is a rectangle.

17. $x = 3$

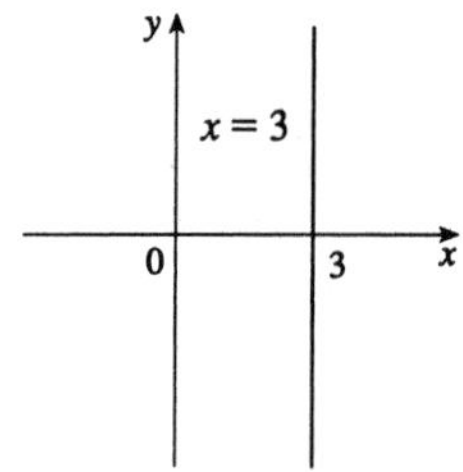

18. $y = -2$

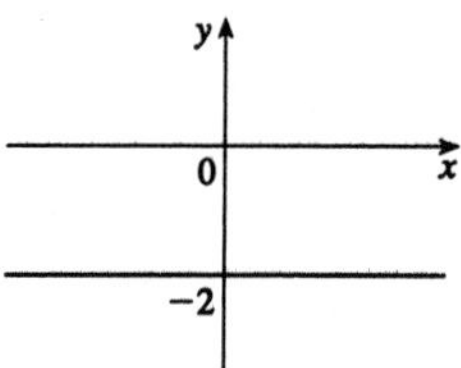

19. $xy = 0 \quad \Leftrightarrow \quad x = 0$ or $y = 0$

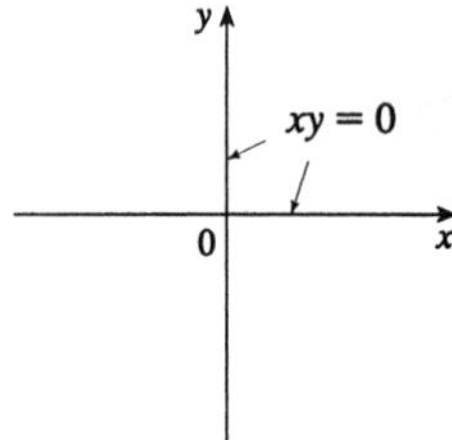

20. $|y| = 1 \quad \Leftrightarrow \quad y = 1$ or $y = -1$

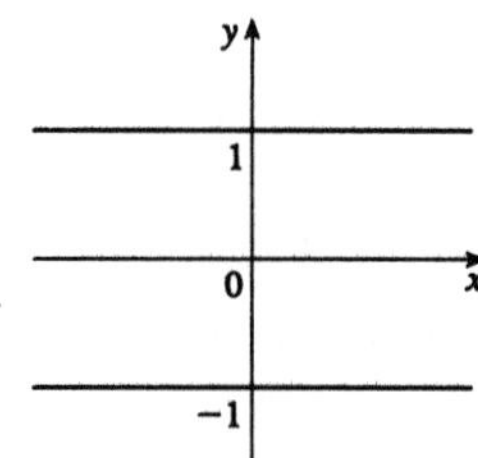

21. From (3), the equation of the line is $y - (-3) = 6(x-2)$ or $y = 6x - 15$.

22. $y - 4 = -3[x - (-1)]$ or $y = -3x + 1$

23. $y - 7 = \frac{2}{3}(x-1)$ or $2x - 3y + 19 = 0$

24. $y - (-5) = -\frac{7}{2}[x - (-3)]$ or $7x + 2y + 31 = 0$

25. The slope is $m = \dfrac{6-1}{1-2} = -5$, so the equation of the line is $y - 1 = -5(x-2)$ or $5x + y = 11$.

26. $m = \dfrac{3-(-2)}{4-(-1)} = 1$. So $y - 3 = 1(x-4)$ or $y = x - 1$.

27. From (4), the equation is $y = 3x - 2$.

28. The equation is $y = \frac{2}{5}x + 4$ or $2x - 5y + 20 = 0$.

29. Since the line passes through $(1, 0)$ and $(0, -3)$, its slope is $m = \dfrac{-3-0}{0-1} = 3$, so its equation is $y = 3x - 3$.

30. $m = \dfrac{6-0}{0-(-8)} = \dfrac{3}{4}$. So $y = \frac{3}{4}x + 6$ or $3x - 4y + 24 = 0$.

31. Since $m = 0$, $y - 5 = 0(x-4)$ or $y = 5$.

32. Vertical line $x = 4$

33. Putting the line $x + 2y = 6$ into its slope-intercept form $y = -\frac{1}{2}x + 3$, we see that this line has slope $-\frac{1}{2}$. So we want the line of slope $-\frac{1}{2}$ that passes through the point $(1, -6)$: $y - (-6) = -\frac{1}{2}(x-1) \quad \Leftrightarrow \quad y = -\frac{1}{2}x - \frac{11}{2}$ or $x + 2y + 11 = 0$.

34. $2x + 3y + 4 = 0 \;\Leftrightarrow\; y = -\frac{2}{3}x - \frac{4}{3}$, so $m = -\frac{2}{3}$. So the required line is $y = -\frac{2}{3}x + 6$ or $2x + 3y - 18 = 0$.

35. $2x + 5y + 8 = 0 \quad \Leftrightarrow \quad y = -\frac{2}{5}x - \frac{8}{5}$. Since this line has slope $-\frac{2}{5}$, a line perpendicular to it would have slope $\frac{5}{2}$, so the required line is $y - (-2) = \frac{5}{2}[x - (-1)] \quad \Leftrightarrow \quad y = \frac{5}{2}x + \frac{1}{2}$ or $5x - 2y + 1 = 0$.

36. $4x - 8y = 1 \quad \Leftrightarrow \quad y = \frac{1}{2}x - \frac{1}{8}$. Since this line has slope $\frac{1}{2}$, a line perpendicular to it would have slope -2, so the required line is $y - \left(-\frac{2}{3}\right) = -2\left(x - \frac{1}{2}\right) \quad \Leftrightarrow \quad y = -2x + \frac{1}{3}$ or $6x + 3y = 1$.

37. $x + 3y = 0 \quad \Leftrightarrow \quad y = -\frac{1}{3}x$, so the slope is $-\frac{1}{3}$ and the y-intercept is 0.

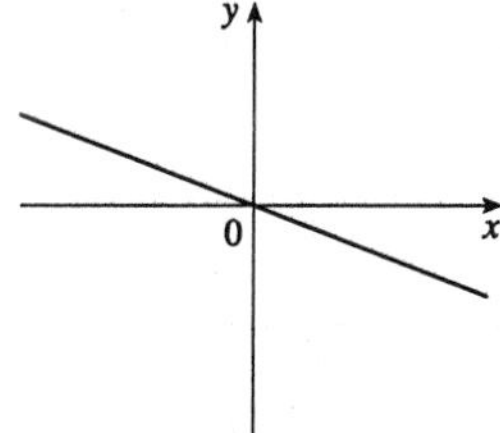

38. $2x - 5y = 0 \quad \Leftrightarrow \quad y = \frac{2}{5}x$, so the slope is $\frac{2}{5}$ and the y-intercept is 0.

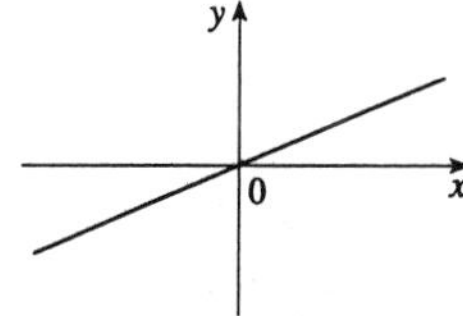

39. $y = -2$ is a horizontal line with slope 0 and y-intercept -2.

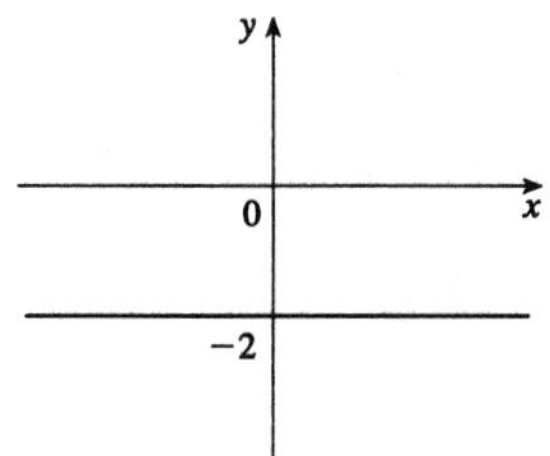

40. $2x - 3y + 6 = 0 \quad \Leftrightarrow \quad y = \frac{2}{3}x + 2$, so the slope is $\frac{2}{3}$ and the y-intercept is 2.

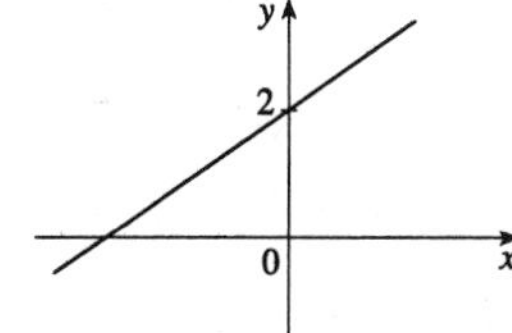

41. $3x - 4y = 12 \quad \Leftrightarrow \quad y = \frac{3}{4}x - 3$, so the slope is $\frac{3}{4}$ and the y-intercept is -3.

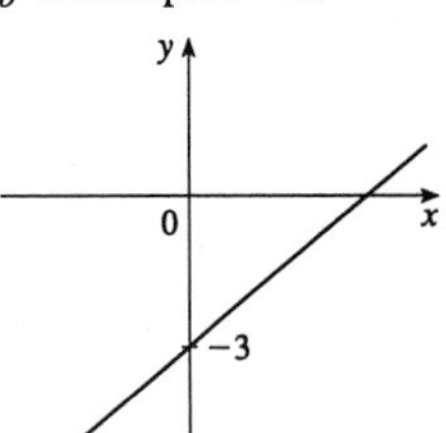

42. $4x + 5y = 10 \quad \Leftrightarrow \quad y = -\frac{4}{5}x + 2$, so the slope is $-\frac{4}{5}$ and the y-intercept is 2.

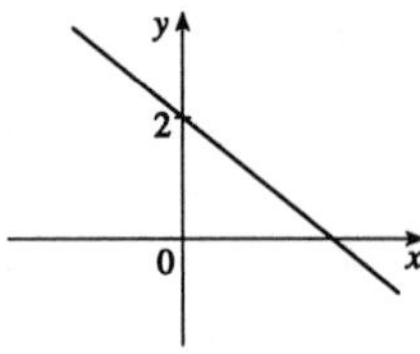

43. $\{(x, y) \mid x < 0\}$

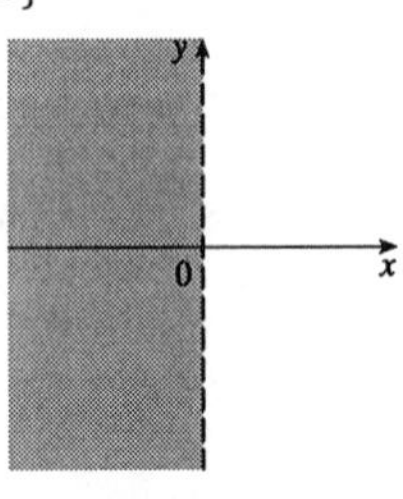

44. $\{(x, y) \mid y > 0\}$

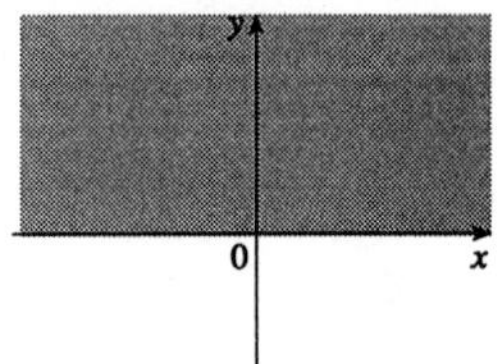

45. $\{(x, y) \mid xy < 0\} = \{(x, y) \mid x < 0 \text{ and } y > 0\}$
$\cup \{(x, y) \mid x > 0 \text{ and } y < 0\}$

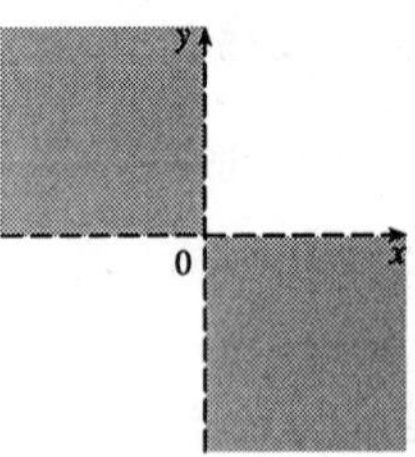

46. $\{(x, y) \mid x \geq 1 \text{ and } y < 3\}$

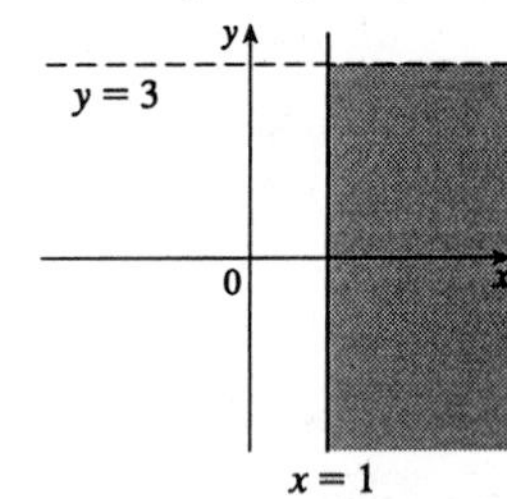

47. $\{(x, y) \mid |x| \leq 2\} = \{(x, y) \mid -2 \leq x \leq 2\}$

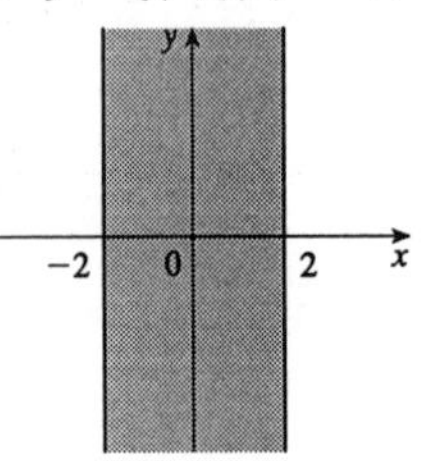

48. $\{(x, y) \mid |x| < 3 \text{ and } |y| < 2\}$

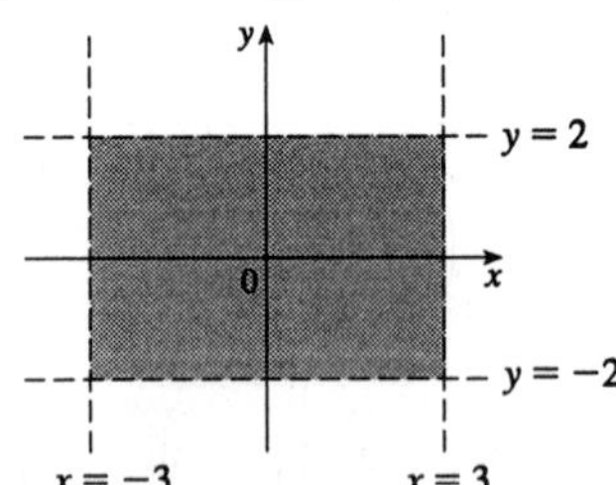

49. $\{(x, y) \mid 0 \leq y \leq 4, x \leq 2\}$

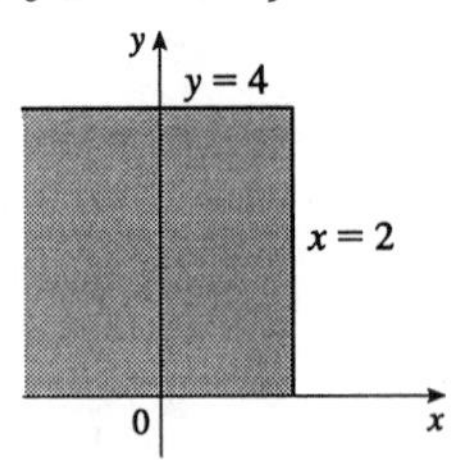

50. $\{(x, y) \mid y > 2x - 1\}$

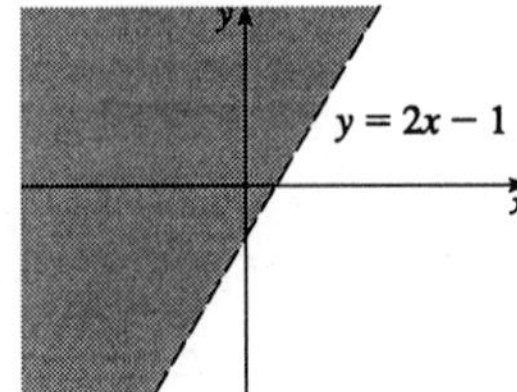

51. $\{(x,y) \mid 1+x \leq y \leq 1-2x\}$

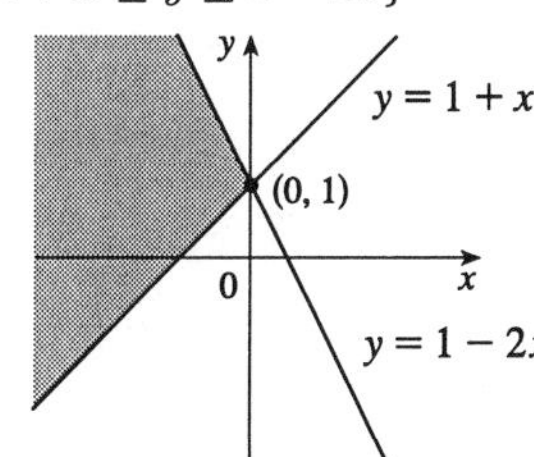

52. $\left\{(x,y) \mid -x \leq y < \dfrac{x+3}{2}\right\}$

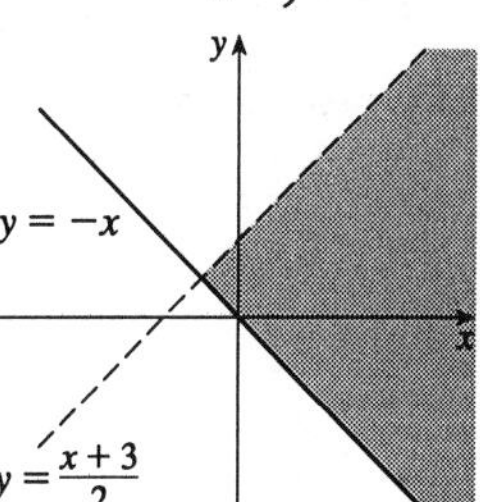

53. Let $P(0,y)$ be a point on the y-axis. The distance from P to $(5,-5)$ is $\sqrt{(5-0)^2+(-5-y)^2} = \sqrt{5^2+(y+5)^2}$. The distance from P to $(1,1)$ is $\sqrt{(1-0)^2+(1-y)^2} = \sqrt{1^2+(y-1)^2}$. We want these distances to be equal: $\sqrt{5^2+(y+5)^2} = \sqrt{1^2+(y-1)^2} \quad \Leftrightarrow \quad 5^2+(y+5)^2 = 1^2+(y-1)^2 \quad \Leftrightarrow$ $25+(y^2+10y+25) = 1+(y^2-2y+1) \;\Leftrightarrow\; 12y=-48 \;\Leftrightarrow\; y=-4$. So the desired point is $(0,-4)$.

54. Let M be the point $\left(\dfrac{x_1+x_2}{2}, \dfrac{y_1+y_2}{2}\right)$. Then

$|MP_1|^2 = \left(x_1 - \dfrac{x_1+x_2}{2}\right)^2 + \left(y_1 - \dfrac{y_1+y_2}{2}\right)^2 = \left(\dfrac{x_1-x_2}{2}\right)^2 + \left(\dfrac{y_1-y_2}{2}\right)^2$ and

$|MP_2|^2 = \left(x_2 - \dfrac{x_1+x_2}{2}\right)^2 + \left(y_2 - \dfrac{y_1+y_2}{2}\right)^2 = \left(\dfrac{x_2-x_1}{2}\right)^2 + \left(\dfrac{y_2-y_1}{2}\right)^2$. Hence $|MP_1| = |MP_2|$.

55. Using the midpoint formula of Exercise 54, we get

(a) $\left(\dfrac{1+7}{2}, \dfrac{3+15}{2}\right) = (4,9)$ **(b)** $\left(\dfrac{-1+8}{2}, \dfrac{6-12}{2}\right) = \left(\dfrac{7}{2}, -3\right)$

56. The midpoint M_1 of AB is $\left(\dfrac{1+3}{2}, \dfrac{0+6}{2}\right) = (2,3)$, the midpoint M_2 of BC is $\left(\dfrac{3+8}{2}, \dfrac{6+2}{2}\right) = \left(\dfrac{11}{2}, 4\right)$, and the midpoint M_3 of CA is $\left(\dfrac{8+1}{2}, \dfrac{2+0}{2}\right) = \left(\dfrac{9}{2}, 1\right)$. The lengths of the medians are

$|AM_2| = \sqrt{\left(\frac{11}{2}-1\right)+(4-0)^2} = \sqrt{\left(\frac{9}{2}\right)^2+4^2} = \sqrt{\frac{145}{4}} = \frac{\sqrt{145}}{2}$,

$|BM_3| = \sqrt{\left(\frac{9}{2}-3\right)^2+(1-6)^2} = \sqrt{\left(\frac{3}{2}\right)^2+(-5)^2} = \sqrt{\frac{109}{4}} = \frac{\sqrt{109}}{2}$, and

$|CM_1| = \sqrt{(2-8)^2+(3-2)^2} = \sqrt{(-6)^2+1^2} = \sqrt{37}$.

57. $2x-y=4 \quad \Leftrightarrow \quad y=2x-4 \quad \Rightarrow \quad m_1=2$ and $6x-2y=10 \quad \Leftrightarrow \quad 2y=6x-10 \quad \Leftrightarrow \quad y=3x-5$ $\Rightarrow \quad m_2=3$. Since $m_1 \neq m_2$, the two lines are not parallel [by 6(a)]. To find the point of intersection: $2x-4=3x-5 \quad \Leftrightarrow \quad x=1 \quad \Rightarrow \quad y=-2$. Thus, the point of intersection is $(1,-2)$.

58. $3x - 5y + 19 = 0 \Leftrightarrow 5y = 3x + 19 \Leftrightarrow y = \frac{3}{5}x + \frac{19}{5} \Rightarrow m_1 = \frac{3}{5}$ and $10x + 6y - 50 = 0 \Leftrightarrow$ $6y = -10x + 50 \Leftrightarrow y = -\frac{5}{3}x + \frac{25}{3} \Rightarrow m_2 = -\frac{5}{3}$. Since $m_1m_2 = \frac{3}{5}\left(-\frac{5}{3}\right) = -1$, the two lines are perpendicular. To find the point of intersection: $\frac{3}{5}m + \frac{19}{5} = -\frac{5}{3}x + \frac{25}{3} \Leftrightarrow 9x + 57 = -25x + 125 \Leftrightarrow$ $34x = 68 \Leftrightarrow x = 2 \Rightarrow y = \frac{3}{5} \cdot 2 + \frac{19}{5} = \frac{25}{5} = 5$. Thus the point of intersection is $(2, 5)$.

59. The slope of the segment AB is $\dfrac{-2-4}{7-1} = -1$, so its perpendicular bisector has slope 1. The midpoint of AB is $\left(\dfrac{1+7}{2}, \dfrac{4-2}{2}\right) = (4, 1)$, so the equation of the perpendicular bisector is $y - 1 = 1(x - 4)$ or $y = x - 3$.

60. **(a)** Side PQ has slope $\dfrac{4-0}{3-1} = 2$, so its equation is $y - 0 = 2(x - 1) \Leftrightarrow y = 2x - 2$. Side QR has slope $\dfrac{6-4}{-1-3} = -\dfrac{1}{2}$, so its equation is $y - 4 = -\frac{1}{2}(x - 3) \Leftrightarrow y = -\frac{1}{2}x + \frac{11}{2}$. Side RP has slope $\dfrac{0-6}{1-(-1)} = -3$, so its equation is $y - 0 = -3(x - 1) \Leftrightarrow y = -3x + 3$.

(b) M_1 (the midpoint of PQ) has coordinates $\left(\dfrac{1+3}{2}, \dfrac{0+4}{2}\right) = (2, 2)$. M_2 (the midpoint of QR) has coordinates $\left(\dfrac{3-1}{2}, \dfrac{4+6}{2}\right) = (1, 5)$. M_3 (the midpoint of RP) has coordinates $\left(\dfrac{1-1}{2}, \dfrac{0+6}{2}\right) = (0, 3)$. RM_1 has slope $\dfrac{2-6}{2-(-1)} = -\dfrac{4}{3}$ and hence equation $y - 2 = -\frac{4}{3}(x - 2)$ $\Leftrightarrow y = -\frac{4}{3}x + \frac{14}{3}$. PM_2 is a vertical line with equation $x = 1$. QM_3 has slope $\dfrac{3-4}{0-3} = \dfrac{1}{3}$ and hence equation $y - 3 = \frac{1}{3}(x - 0) \Leftrightarrow y = \frac{1}{3}x + 3$. PM_2 and RM_1 intersect where $x = 1$ and $y = -\frac{4}{3}(1) + \frac{14}{3} = \frac{10}{3}$, or at $\left(1, \frac{10}{3}\right)$. PM_2 and QM_3 intersect where $x - 1$ and $y = \frac{1}{3}(1) + 3 = \frac{10}{3}$, or at $\left(1, \frac{10}{3}\right)$, so this is the point where all three medians intersect.

61. **(a)** Since the x-intercept is a, the point $(a, 0)$ is on the line, and similarly since the y-intercept is b, $(0, b)$ is on the line. Hence the slope of the line is $m = \dfrac{b-0}{0-a} = -\dfrac{b}{a}$. Substituting into $y = mx + b$ gives $y = -\dfrac{b}{a}x + b \Leftrightarrow y + \dfrac{b}{a}x = b \Leftrightarrow \dfrac{y}{b} + \dfrac{x}{a} = 1$.

(b) Letting $a = 6$ and $b = -8$ gives $\dfrac{y}{-8} + \dfrac{x}{6} = 1 \Leftrightarrow 6y - 8x = -48 \Leftrightarrow 8x - 6y - 48 = 0 \Leftrightarrow$ $4x - 3y - 24 = 0$.

62. **(a)** Let d = distance traveled (in miles) and t = time elapsed (in hours). At $t = 0$, $d = 0$ and at $t = 50$ minutes $= 50 \cdot \frac{1}{60} = \frac{5}{6}$ h, $d = 40$. Thus we have two points: $(0, 0)$ and $\left(\frac{5}{6}, 40\right)$, so

$$m = \frac{40 - 0}{\frac{5}{6} - 0} = 48 \text{ and so } d = 48t.$$

(b)

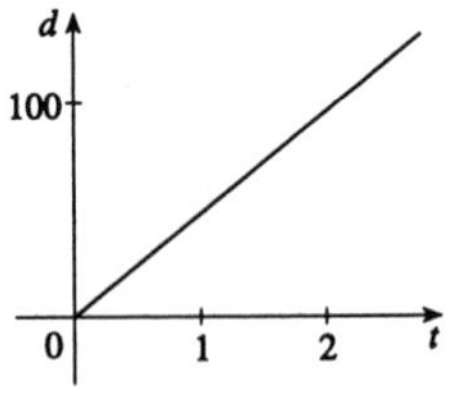

(c) The slope is 48 and represents the car's speed in mi/h.

EXERCISES C

1. From (1), the equation is $(x-3)^2+(y+1)^2=25$.

2. From (1), the equation is $(x+2)^2+(y+8)^2=100$.

3. The equation has the form $x^2+y^2=r^2$. Since $(4,7)$ lies on the circle, we have $4^2+7^2=r^2 \quad\Rightarrow\quad r^2=65$. So the required equation is $x^2+y^2=65$.

4. The equation has the form $(x+1)^2+(y-5)^2=r^2$. Since $(-4,-6)$ lies on the circle, we have $r^2=(-4+1)^2+(-6-5)^2=130$. So the required equation is $(x+1)^2+(y-5)^2=130$.

5. $x^2+y^2-4x+10y+13=0 \quad\Leftrightarrow\quad x^2-4x+y^2+10y=-13 \quad\Leftrightarrow$ $(x^2-4x+4)+(y^2+10y+25)=-13+4+25=16 \quad\Leftrightarrow\quad (x-2)^2+(y+5)^2=4^2$. Thus, we have a circle with center $(2,-5)$ and radius 4.

6. $x^2+y^2+6y+2=0 \quad\Leftrightarrow\quad x^2+(y^2+6y+9)=-2+9 \quad\Leftrightarrow\quad x^2+(y+3)^2=7$. Thus, we have a circle with center $(0,-3)$ and radius $\sqrt{7}$.

7. $x^2+y^2+x=0 \quad\Leftrightarrow\quad \left(x^2+x+\frac{1}{4}\right)+y^2=\frac{1}{4} \quad\Leftrightarrow\quad \left(x+\frac{1}{2}\right)^2+y^2=\left(\frac{1}{2}\right)^2$. Thus, we have a circle with center $\left(-\frac{1}{2},0\right)$ and radius $\frac{1}{2}$.

8. $16x^2+16y^2+8x+32y+1=0 \quad\Leftrightarrow\quad 16\left(x^2+\frac{1}{2}x+\frac{1}{16}\right)+16(y^2+2y+1)=-1+1+16 \quad\Leftrightarrow$ $16\left(x+\frac{1}{4}\right)^2+16(y+1)^2=16 \quad\Leftrightarrow\quad \left(x+\frac{1}{4}\right)^2+(y+1)^2=1$. Thus, we have a circle with center $\left(\frac{1}{4},-1\right)$ and radius 1.

9. $2x^2+2y^2-x+y=1 \quad\Leftrightarrow\quad 2\left(x^2-\frac{1}{2}x+\frac{1}{16}\right)+2\left(y^2+\frac{1}{2}y+\frac{1}{16}\right)=1+\frac{1}{8}+\frac{1}{8} \quad\Leftrightarrow$ $2\left(x-\frac{1}{4}\right)^2+2\left(y+\frac{1}{4}\right)^2=\frac{5}{4} \quad\Leftrightarrow\quad \left(x-\frac{1}{4}\right)^2+\left(y+\frac{1}{4}\right)^2=\frac{5}{8}$. Thus, we have a circle with center $\left(\frac{1}{4},-\frac{1}{4}\right)$ and radius $\frac{\sqrt{5}}{2\sqrt{2}}=\frac{\sqrt{10}}{4}$.

10. $x^2+y^2+ax+by+c=0 \quad\Leftrightarrow\quad \left(x^2+ax+\frac{1}{4}a^2\right)+\left(y^2+by+\frac{1}{4}b^2\right)=-c+\frac{1}{4}a^2+\frac{1}{4}b^2 \quad\Leftrightarrow$ $\left(x+\frac{1}{2}a\right)^2+\left(y+\frac{1}{2}b\right)^2=\frac{1}{4}(a^2+b^2-4c)$. For this to represent a nondegenerate circle, $\frac{1}{4}(a^2+b^2-4c)>0$ or $a^2+b^2>4c$. If this condition is satisfied, the circle has center $\left(-\frac{1}{2}a,-\frac{1}{2}b\right)$ and radius $\frac{1}{2}\sqrt{a^2+b^2-4c}$.

11. $y=-x^2$. Parabola

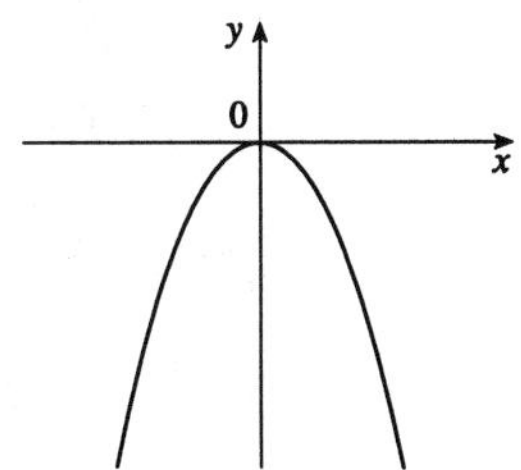

12. $y^2-x^2=1$. Hyperbola

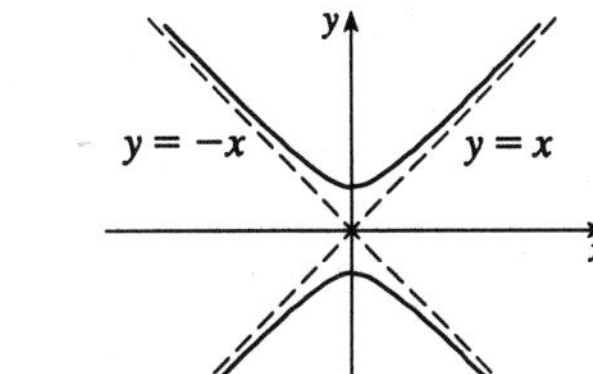

13. $x^2 + 4y^2 = 16 \quad \Leftrightarrow \quad \dfrac{x^2}{16} + \dfrac{y^2}{4} = 1$. Ellipse

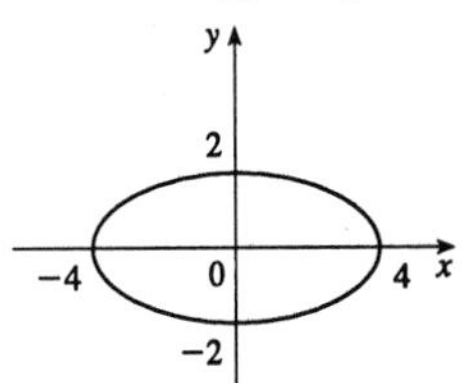

14. $x = -2y^2$. Parabola

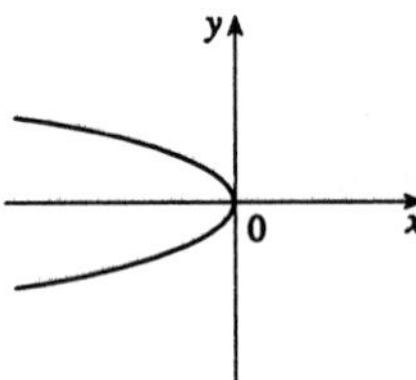

15. $16x^2 - 25y^2 = 400 \quad \Leftrightarrow \quad \dfrac{x^2}{25} - \dfrac{y^2}{16} = 1$.

Hyperbola

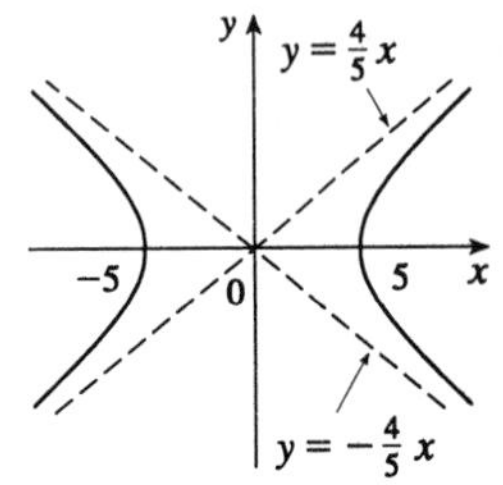

16. $25x^2 + 4y^2 = 100 \quad \Leftrightarrow \quad \dfrac{x^2}{4} + \dfrac{y^2}{25} = 1$.

Ellipse

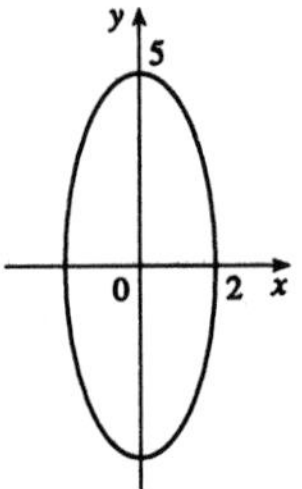

17. $4x^2 + y^2 = 1 \quad \Leftrightarrow \quad \dfrac{x^2}{1/4} + y^2 = 1$. Ellipse

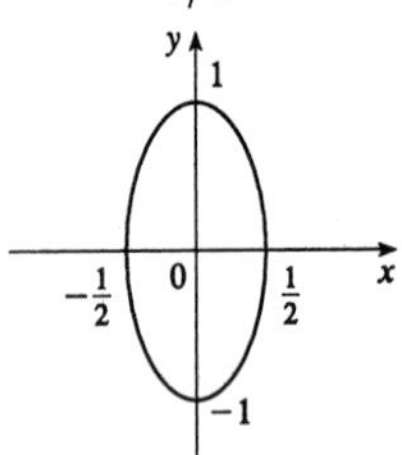

18. $y = x^2 + 2$. Parabola with vertex at $(0, 2)$

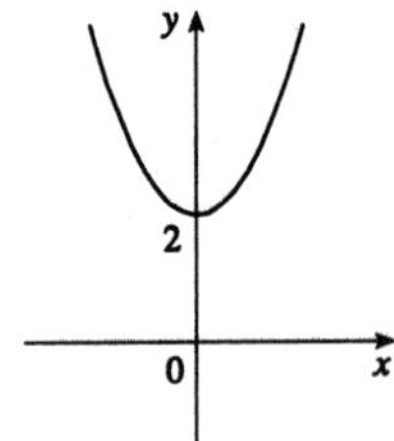

19. $x = y^2 - 1$. Parabola with vertex at $(-1, 0)$

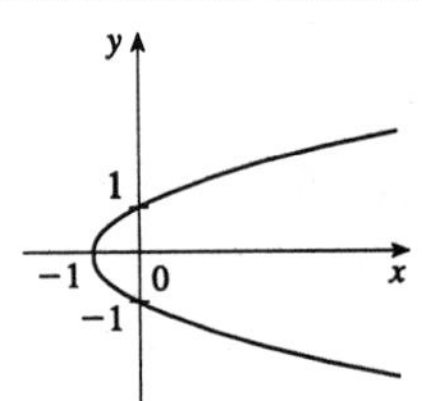

20. $9x^2 - 25y^2 = 225 \quad \Leftrightarrow \quad \dfrac{x^2}{25} - \dfrac{y^2}{9} = 1$. Hyperbola

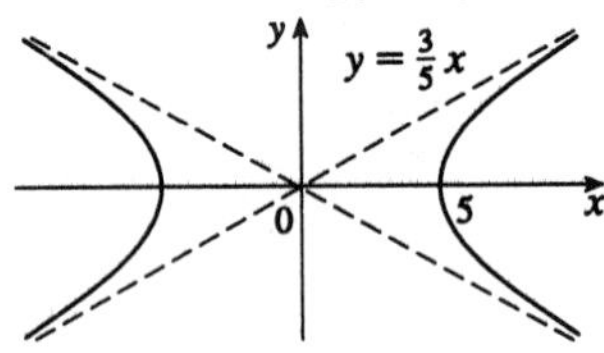

21. $9y^2 - x^2 = 9 \quad \Leftrightarrow \quad y^2 - \dfrac{x^2}{9} = 1$. Hyperbola

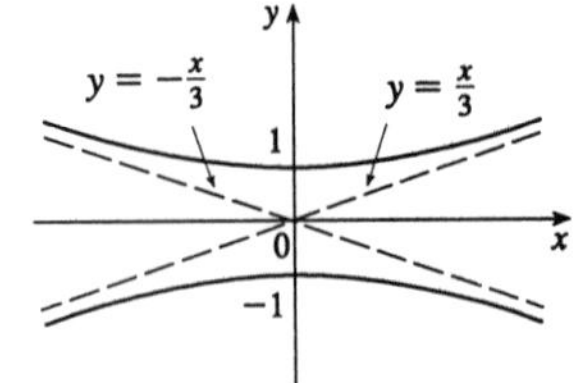

22. $2x^2 + 5y^2 = 10 \quad \Leftrightarrow \quad \dfrac{x^2}{5} + \dfrac{y^2}{2} = 1$. Ellipse

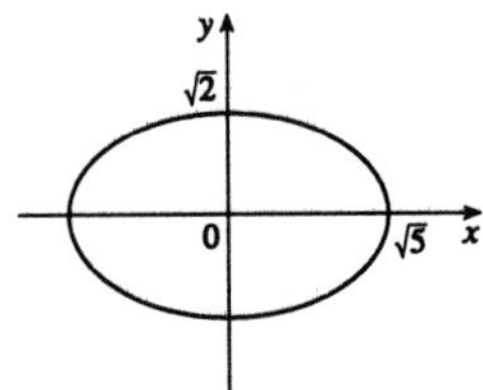

23. $xy = 4$. Hyperbola

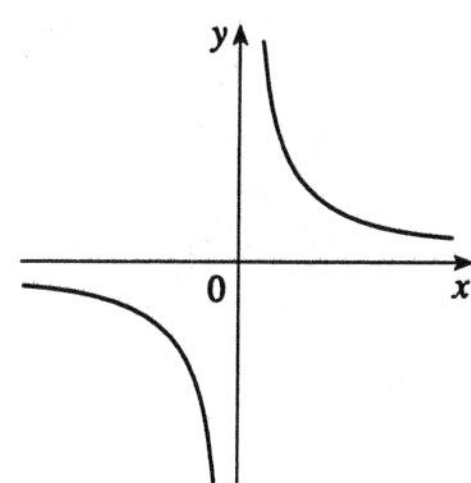

24. $y = x^2 + 2x = (x^2 + 2x + 1) - 1 = (x + 1)^2 - 1$. Parabola with vertex at $(-1, -1)$

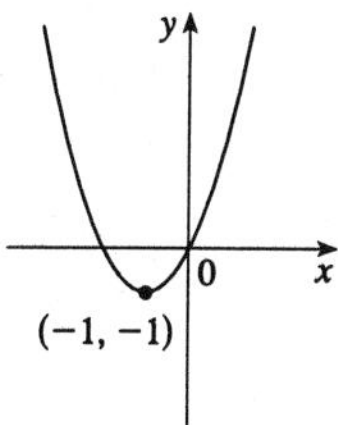

25. $9(x-1)^2 + 4(y-2)^2 = 36 \quad \Leftrightarrow$ $\dfrac{(x-1)^2}{4} + \dfrac{(y-2)^2}{9} = 1$. Ellipse centered at $(1, 2)$

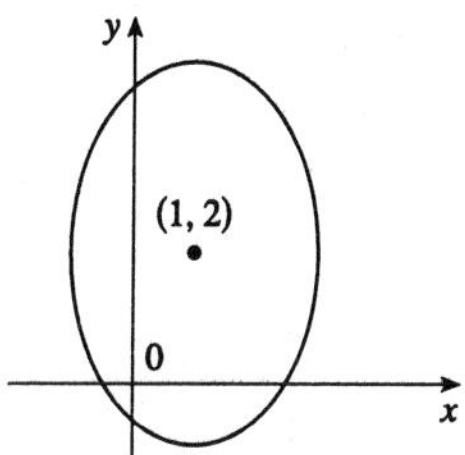

26. $16x^2 + 9y^2 = 108 \quad \Leftrightarrow$ $16x^2 + 9(y^2 - 4y + 4) = 108 + 36 = 144 \quad \Leftrightarrow$ $\dfrac{x^2}{9} + \dfrac{(y-2)^2}{16} = 1$. Ellipse centered at $(0, 2)$

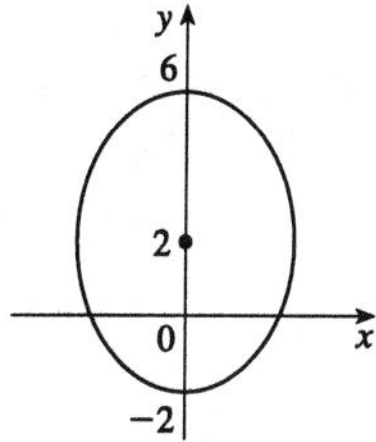

27. $y = x^2 - 6x + 13 = \left(x^2 - 6x + 9\right) + 4$ $= (x-3)^2 + 4$. Parabola with vertex at $(3, 4)$

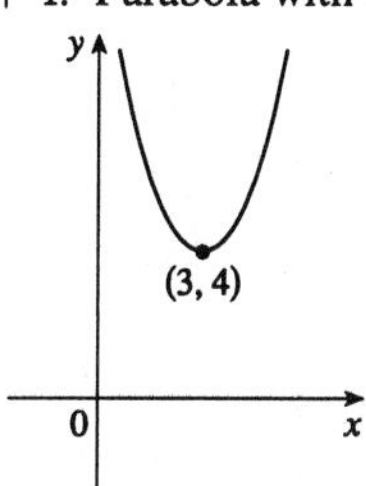

28. $x^2 - y^2 - 4x + 3 = 0 \quad \Leftrightarrow$ $(x^2 - 4x + 4) - y^2 = -3 + 4 = 1 \quad \Leftrightarrow$ $(x-2)^2 - y^2 = 1$. Hyperbola centered at $(2, 0)$

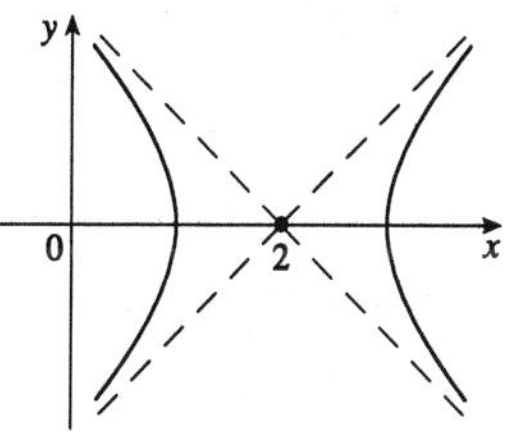

29. $x = -y^2 + 4$. Parabola with vertex at $(4, 0)$

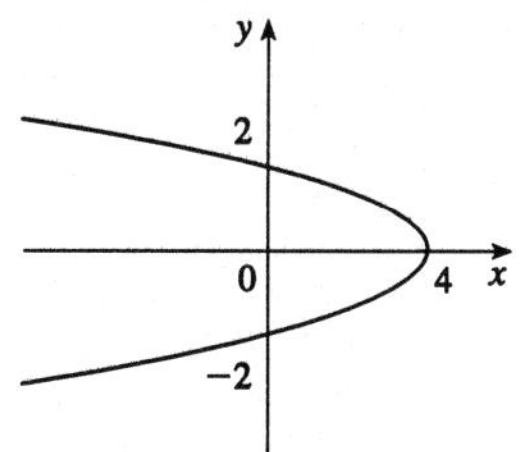

30. $y^2 - 2x + 6y + 5 = 0 \quad \Leftrightarrow$ $y^2 + 6y + 9 = 2x + 4 \quad \Leftrightarrow (y+3)^2 = 2(x+2)$. Parabola with vertex $(-2, -3)$

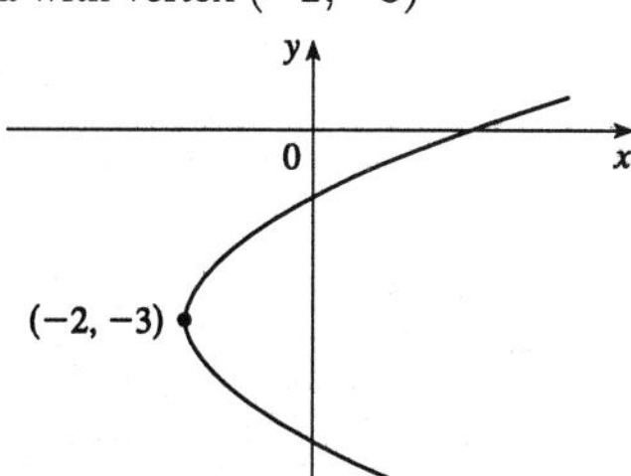

31. $x^2 + 4y^2 - 6x + 5 = 0 \quad \Leftrightarrow$

$(x^2 - 6x + 9) + 4y^2 = -5 + 9 = 4 \quad \Leftrightarrow$

$\dfrac{(x-3)^2}{4} + y^2 = 1$. Ellipse centered at $(3, 0)$

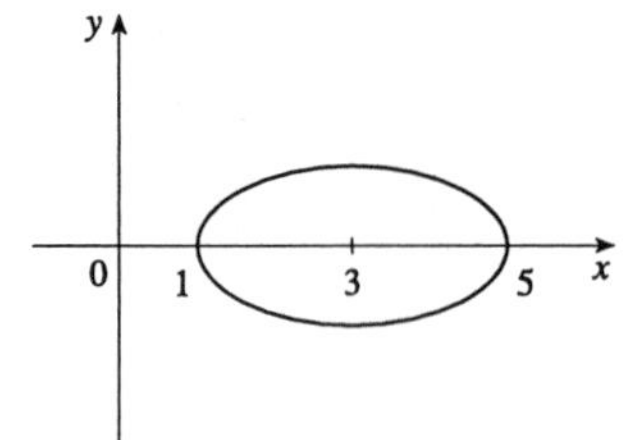

32. $4x^2 + 9y^2 - 16x + 54y + 61 = 0 \quad \Leftrightarrow$

$4(x^2 - 4x + 4) + 9(y^2 + 6y + 9) = -61 + 16 + 81$

$= 36 \quad \Leftrightarrow \quad \dfrac{(x-2)^2}{9} + \dfrac{(y+3)^2}{4} = 1.$

Ellipse centered at $(2, -3)$.

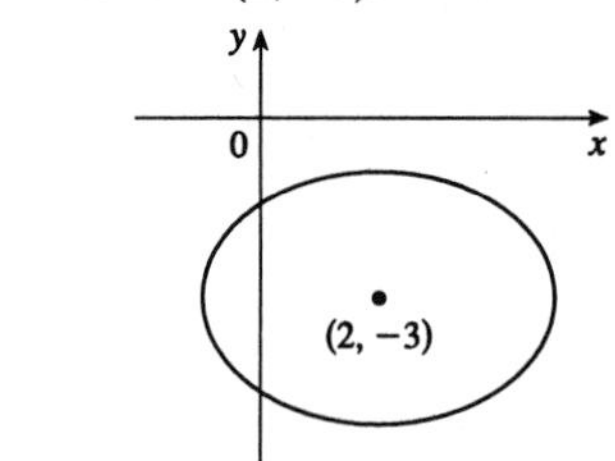

33. $y = 3x$ and $y = x^2$ intersect where $3x = x^2$

$\Leftrightarrow \quad 0 = x^2 - 3x = x(x-3)$,

that is, at $(0, 0)$ and $(3, 9)$.

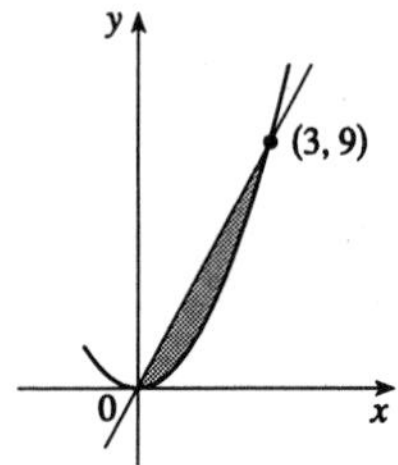

34. $y = 4 - x^2$, $x - 2y = 2$. Substitute y from the first equation into the second:

$x - 2(4 - x^2) = 2 \quad \Leftrightarrow \quad 2x^2 + x - 10 = 0$

$\Leftrightarrow \quad (2x+5)(x-2) = 0 \quad \Leftrightarrow$

$x = -\frac{5}{2}$ or 2. So the points of intersection are $\left(-\frac{5}{2}, -\frac{9}{2}\right)$ and $(2, 0)$.

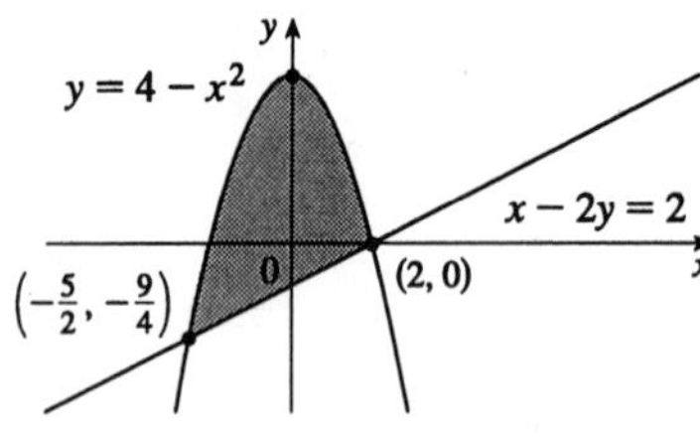

35. The parabola must have an equation of the form $y = a(x-1)^2 - 1$. Substituting $x = 3$ and $y = 3$ into the equation gives $3 = a(3-1)^2 - 1$, so $a = 1$, and the equation is $y = (x-1)^2 - 1 = x^2 - 2x$. Note that using the other point $(-1, 3)$ would have given the same value for a, and hence the same equation.

36. The ellipse has an equation of the form $\dfrac{x^2}{a^2} + \dfrac{y^2}{b^2} = 1$. Substituting $x = 1$ and $y = -\frac{10\sqrt{2}}{3}$ gives $\dfrac{1^2}{a^2} + \dfrac{\left(-10\sqrt{2}/3\right)^2}{b^2} = \dfrac{1}{a^2} + \dfrac{200}{9b^2} - 1$. Substituting $x = -2$ and $y = \frac{5\sqrt{5}}{3}$ gives $\dfrac{(-2)^2}{a^2} + \dfrac{\left(5\sqrt{5}/3\right)^2}{b^2} = \dfrac{4}{a^2} + \dfrac{125}{9b^2} = 1$. From the first equation, $\dfrac{1}{a^2} = 1 - \dfrac{200}{9b^2}$. Putting this into the second equation gives $4\left(1 - \dfrac{200}{9b^2}\right) + \dfrac{125}{9b^2} = 1 \quad \Leftrightarrow \quad 3 = \dfrac{675}{9b^2} \quad \Leftrightarrow \quad b^2 = \dfrac{625}{27} = 25$, so $b = 5$. Hence $\dfrac{1}{a^2} = 1 - \dfrac{200}{9(5)^2} = \dfrac{1}{9}$ and so $a = 3$. The equation of the ellipse is $\dfrac{x^2}{9} + \dfrac{y^2}{25} = 1$.

37. $\{(x,y) \mid x^2+y^2 \le 1\}$

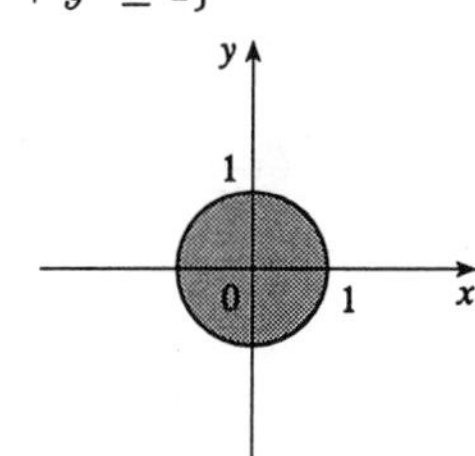

38. $\{(x,y) \mid x^2+y^2 > 4\}$

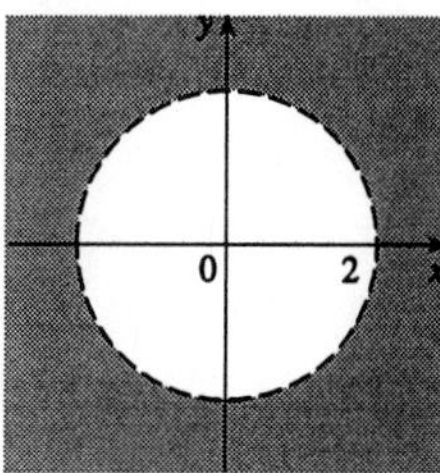

39. $\{(x,y) \mid y \ge x^2-1\}$

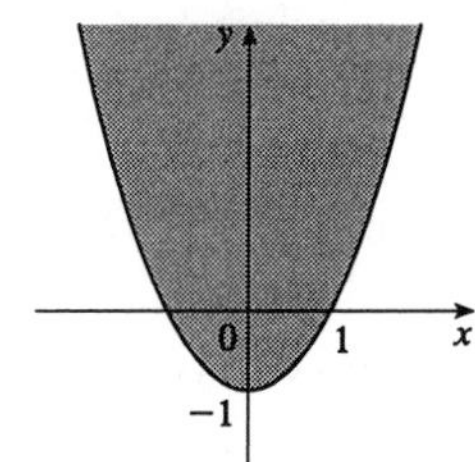

40. $\{(x,y) \mid x^2+4y^2 \le 4\}$

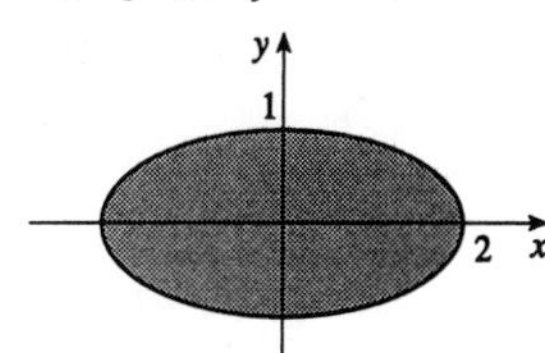

EXERCISES D

1. $210° = 210\left(\frac{\pi}{180}\right) = \frac{7\pi}{6}$ rad

2. $300° = 300\left(\frac{\pi}{180}\right) = \frac{5\pi}{3}$ rad

3. $9° = 9\left(\frac{\pi}{180}\right) = \frac{\pi}{20}$ rad

4. $-315° = -315\left(\frac{\pi}{180}\right) = -\frac{7\pi}{4}$ rad

5. $900° = 900\left(\frac{\pi}{180}\right) = 5\pi$ rad

6. $36° = 36\left(\frac{\pi}{180}\right) = \frac{\pi}{5}$ rad

7. 4π rad $= 4\pi\left(\frac{180}{\pi}\right) = 720°$

8. $-\frac{7\pi}{2}$ rad $= -\frac{7\pi}{2}\left(\frac{180}{\pi}\right) = -630°$

9. $\frac{5\pi}{12}$ rad $= \frac{5\pi}{12}\left(\frac{180}{\pi}\right) = 75°$

10. $\frac{8\pi}{3}$ rad $= \frac{8\pi}{3}\left(\frac{180}{\pi}\right) = 480°$

11. $-\frac{3\pi}{8}$ rad $= -\frac{3\pi}{8}\left(\frac{180}{\pi}\right) = -67.5°$

12. 5 rad $= 5\left(\frac{180}{\pi}\right) = \left(\frac{900}{\pi}\right)^\circ$

13. Using Formula 3, $a = r\theta = \frac{36\pi}{12} = 3\pi$ cm.

14. $a = r\theta = 10 \cdot 72\left(\frac{\pi}{180}\right) = 4\pi$ cm

15. Using Formula 3, $\theta = \frac{1}{1.5} = \frac{2}{3}$ rad $= \frac{2}{3}\left(\frac{180}{\pi}\right) = \left(\frac{120}{\pi}\right)^\circ$.

16. $a = r\theta \quad \Rightarrow \quad r = \dfrac{a}{\theta} = \dfrac{6 \cdot 4}{3\pi} = \dfrac{8}{\pi}$ cm

17.

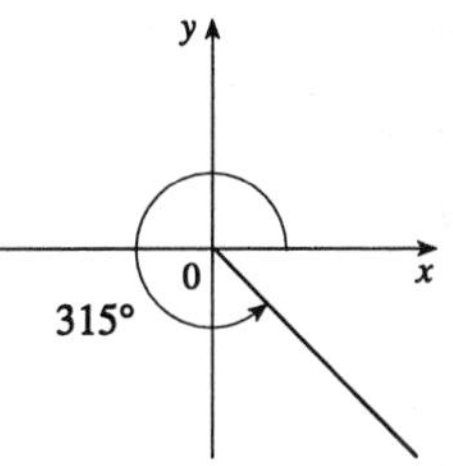

18.

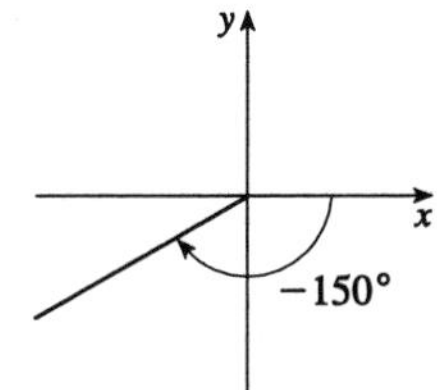

19.

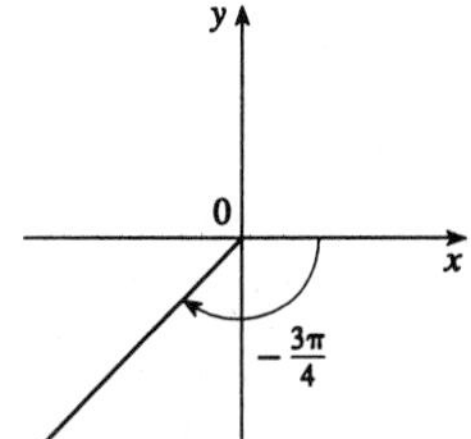

20.

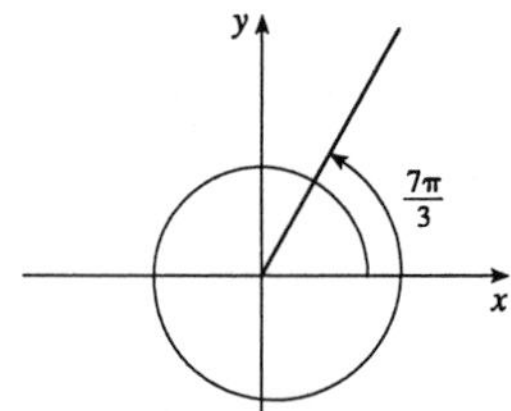

21.

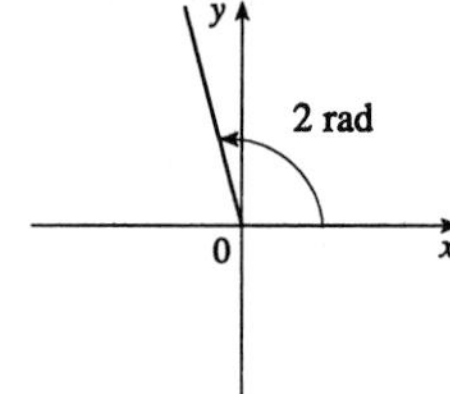

22.

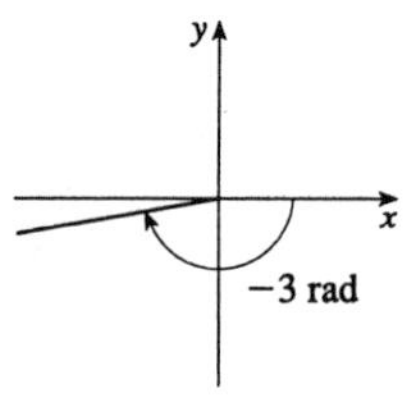

23.

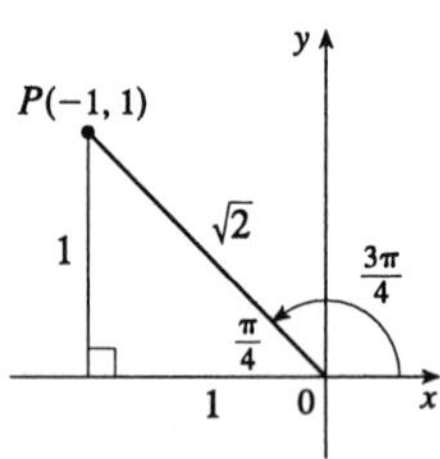

From the diagram we see that a point on the terminal line is $P(-1,1)$. Therefore taking $x=-1$, $y=1$, $r=\sqrt{2}$ in the definitions of the trigonometric ratios, we have $\sin\frac{3\pi}{4}=\frac{1}{\sqrt{2}}$, $\cos\frac{3\pi}{4}=-\frac{1}{\sqrt{2}}$, $\tan\frac{3\pi}{4}=-1$, $\csc\frac{3\pi}{4}=\sqrt{2}$, $\sec\frac{3\pi}{4}=-\sqrt{2}$, and $\cot\frac{3\pi}{4}=-1$.

24.

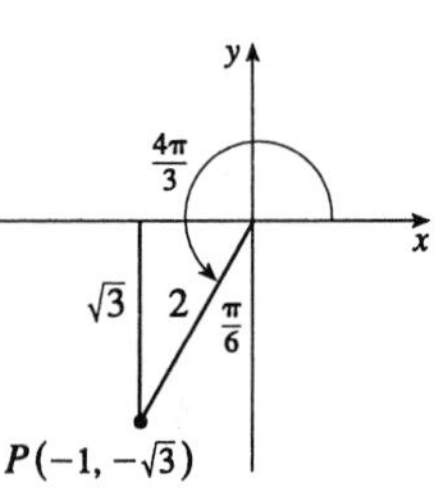

From the diagram and Figure 8, we see that a point on the terminal line is $P\left(-1,-\sqrt{3}\right)$. Therefore taking $x=-1$, $y=-\sqrt{3}$, $r=2$ in the definitions of the trigonometric ratios, we have $\sin\frac{4\pi}{3}=-\frac{\sqrt{3}}{2}$, $\cos\frac{4\pi}{3}=-\frac{1}{2}$, $\tan\frac{4\pi}{3}=\sqrt{3}$, $\csc\frac{4\pi}{3}=-\frac{2}{\sqrt{3}}$, $\sec\frac{4\pi}{3}=-2$, and $\cot\frac{4\pi}{3}=\frac{1}{\sqrt{3}}$.

25.

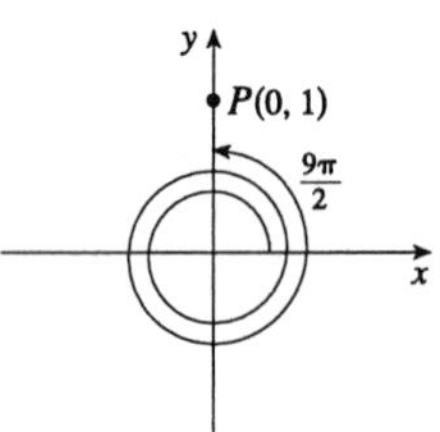

From the diagram we see that a point on the terminal line is $P(0,1)$. Therefore taking $x=0$, $y=1$, $r=1$ in the definitions of the trigonometric ratios, we have $\sin\frac{9\pi}{2}=1$, $\cos\frac{9\pi}{2}=0$, $\tan\frac{9\pi}{2}=y/x$ is undefined since $x=0$, $\csc\frac{9\pi}{2}=1$, $\sec\frac{9\pi}{2}=r/x$ is undefined since $x=0$, and $\cot\frac{9\pi}{2}=0$.

26.

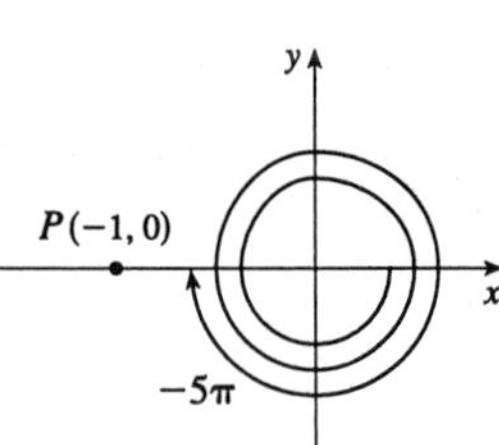

From the diagram, we see that a point on the terminal line is $P(-1,0)$. Therefore taking $x=-1$, $y=0$, $r=1$ in the definitions of the trigonometric ratios we have $\sin(-5\pi)=0$, $\cos(-5\pi)=-1$, $\tan(-5\pi)=0$, $\csc(-5\pi)$ is undefined, $\sec(-5\pi)=-1$, and $\cot(-5\pi)$ is undefined.

27.

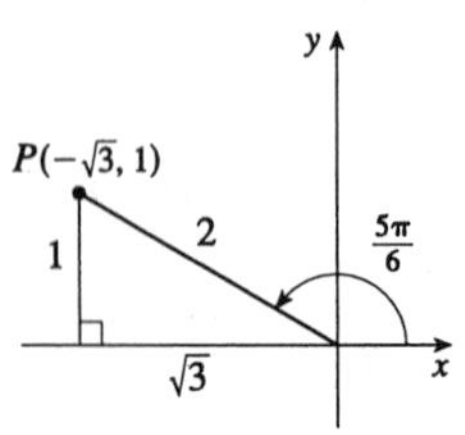

Using Figure 8 we see that a point on the terminal line is $P\left(-\sqrt{3},1\right)$. Therefore taking $x=-\sqrt{3}$, $y=1$, $r=2$ in the definitions of the trigonometric ratios, we have $\sin\frac{5\pi}{6}=\frac{1}{2}$, $\cos\frac{5\pi}{6}=-\frac{\sqrt{3}}{2}$, $\tan\frac{5\pi}{6}=-\frac{1}{\sqrt{3}}$, $\csc\frac{5\pi}{6}=2$, $\sec\frac{5\pi}{6}=-\frac{2}{\sqrt{3}}$, and $\cot\frac{5\pi}{6}=-\sqrt{3}$.

28. 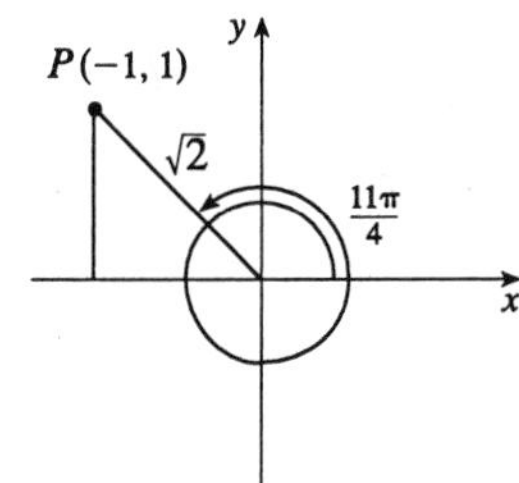

From the diagram, we see that a point on the terminal line is $P(-1, 1)$. Therefore taking $x = -1$, $y = 1$, $r = \sqrt{2}$ in the definitions of the trigonometric ratios we have $\sin\frac{11\pi}{4} = \frac{1}{\sqrt{2}}$, $\cos\frac{11\pi}{4} = -\frac{1}{\sqrt{2}}$, $\tan\frac{11\pi}{4} = -1$, $\csc\frac{11\pi}{4} = \sqrt{2}$, $\sec\frac{11\pi}{4} = -\sqrt{2}$, and $\cot\frac{11\pi}{4} = -1$.

29. $\sin\theta = y/r = \frac{3}{5} \quad\Rightarrow\quad y = 3$, $r = 5$, and $x = \sqrt{r^2 - y^2} = 4$ (since $0 < \theta < \frac{\pi}{2}$). Therefore taking $x = 4$, $y = 3$, $r = 5$ in the definitions of the trigonometric ratios, we have $\cos\theta = \frac{4}{5}$, $\tan\theta = \frac{3}{4}$, $\csc\theta = \frac{5}{3}$, $\sec\theta = \frac{5}{4}$, and $\cot\theta = \frac{4}{3}$.

30. Since $0 < \alpha < \frac{\pi}{2}$, α is in the first quadrant where x and y are both positive. Therefore $\tan\alpha = y/x = \frac{2}{1} \quad\Rightarrow\quad y = 2$, $x = 1$, and $r = \sqrt{x^2 + y^2} = \sqrt{5}$. Taking $x = 1$, $y = 2$, $r = \sqrt{5}$ in the definitions of the trigonometric ratios, we have $\sin\alpha = \frac{2}{\sqrt{5}}$, $\cos\alpha = \frac{1}{\sqrt{5}}$, $\csc\alpha = \frac{\sqrt{5}}{2}$, $\sec\alpha = \sqrt{5}$, and $\cot\alpha = \frac{1}{2}$.

31. $\frac{\pi}{2} < \phi < \pi \quad\Rightarrow\quad \phi$ is in the second quadrant, where x is negative and y is positive. Therefore $\sec\phi = r/x = -1.5 = -\frac{3}{2} \quad\Rightarrow\quad r = 3$, $x = -2$, and $y = \sqrt{r^2 - x^2} = \sqrt{5}$. Taking $x = -2$, $y = \sqrt{5}$, and $r = 3$ in the definitions of the trigonometric ratios, we have $\sin\phi = \frac{\sqrt{5}}{3}$, $\cos\phi = -\frac{2}{3}$, $\tan\phi = -\frac{\sqrt{5}}{2}$, $\csc\phi = \frac{3}{\sqrt{5}}$, and $\cot\theta = -\frac{2}{\sqrt{5}}$.

32. Since $\pi < x < \frac{3\pi}{2}$, x is in the third quadrant where x and y are both negative. Therefore $\cos x = x/r = -\frac{1}{3}$ $\Rightarrow\quad x = -1$, $r = 3$, and $y = -\sqrt{r^2 - x^2} = -\sqrt{8} = -2\sqrt{2}$. Taking $x = -1$, $r = 3$, $y = -2\sqrt{2}$ in the definitions of the trigonometric ratios, we have $\sin x = -\frac{2\sqrt{2}}{3}$, $\tan x = 2\sqrt{2}$, $\csc x = -\frac{3}{2\sqrt{2}}$, $\sec x = -3$, and $\cot x = \frac{1}{2\sqrt{2}}$.

33. $\pi < \beta < 2\pi$ means that β is in the third or fourth quadrant where y is negative. Also since $\cot\beta = x/y = 3$ which is positive, x must also be negative. Therefore $\cot\beta = x/y = \frac{3}{1} \quad\Rightarrow\quad x = -3$, $y = -1$, and $r = \sqrt{x^2 + y^2} = \sqrt{10}$. Taking $x = -3$, $y = -1$ and $r = \sqrt{10}$ in the definitions of the trigonometric ratios, we have $\sin\beta = -\frac{1}{\sqrt{10}}$, $\cos\beta = -\frac{3}{\sqrt{10}}$, $\tan\beta = \frac{1}{3}$, $\csc\beta = -\sqrt{10}$, and $\sec\beta = -\frac{\sqrt{10}}{3}$.

34. Since $\frac{3\pi}{2} < \theta < 2\pi$, θ is in the fourth quadrant where x is positive and y is negative. Therefore $\csc\theta = r/y = -\frac{4}{3} \quad\Rightarrow\quad r = 4$, $y = -3$, and $x = \sqrt{r^2 - y^2} = \sqrt{7}$. Taking $x = \sqrt{7}$, $y = -3$, and $r = 4$ in the definitions of the trigonometric ratios, we have $\sin\theta = -\frac{3}{4}$, $\cos\theta = \frac{\sqrt{7}}{4}$, $\tan\theta = -\frac{3}{\sqrt{7}}$, $\sec\theta = \frac{4}{\sqrt{7}}$, and $\cot\theta = -\frac{\sqrt{7}}{3}$.

35. $\sin 35^\circ = \dfrac{x}{10} \quad\Rightarrow\quad x = 10\sin 35^\circ \approx 5.73576$ cm

36. $\cos 40^\circ = \dfrac{x}{25} \quad\Rightarrow\quad x = 25\cos 40^\circ \approx 19.15111$ cm

37. $\tan\frac{2\pi}{5} = \dfrac{x}{8} \quad\Rightarrow\quad x = 8\tan\frac{2\pi}{5} \approx 24.62147$ cm

38. $\cos\frac{3\pi}{8} = \dfrac{22}{x} \quad\Rightarrow\quad x = \dfrac{22}{\cos\frac{3\pi}{8}} \approx 57.48877$ cm

39. **(a)** From the diagram we see that

$$\sin\theta = \frac{y}{r} = \frac{a}{c}, \text{ and } \sin(-\theta) = -\frac{a}{c} = -\sin\theta.$$

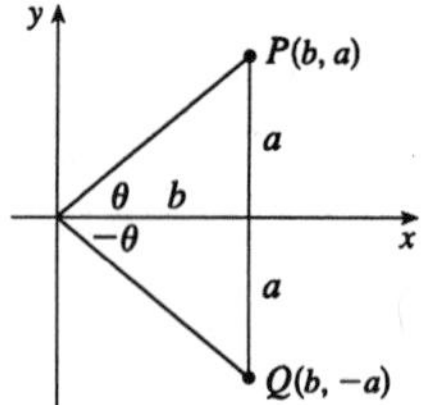

(b) Again from the diagram we see that

that $\cos\theta = \frac{x}{r} = \frac{b}{c} = \cos(-\theta)$.

40. **(a)** Using (12a) and (12b),

$$\tan(x+y) = \frac{\sin(x+y)}{\cos(x+y)} = \frac{\sin x\cos y + \cos x\sin y}{\cos x\cos y - \sin x\sin y} = \frac{\dfrac{\sin x\cos y}{\cos x\cos y} + \dfrac{\cos x\sin y}{\cos x\cos y}}{\dfrac{\cos x\cos y}{\cos x\cos y} - \dfrac{\sin x\sin y}{\cos x\cos y}} = \frac{\tan x + \tan y}{1 - \tan x\tan y}.$$

(b) From (10a) and (10b), we have $\tan(-\theta) = -\tan\theta$, so (14a) implies that

$$\tan(x-y) = \frac{\tan x + \tan(-y)}{1 - \tan x\tan(-y)} = \frac{\tan x - \tan y}{1 + \tan x\tan y}.$$

41. **(a)** Using (12a) and (13a), we have $\frac{1}{2}[\sin(x+y) + \sin(x-y)]$

$= \frac{1}{2}[\sin x\cos y + \cos x\sin y + \sin x\cos y - \cos x\sin y] = \frac{1}{2}(2\sin x\cos y) = \sin x\cos y.$

(b) This time, using (12b) and (13b), we have $\frac{1}{2}[\cos(x+y) + \cos(x-y)]$

$= \frac{1}{2}[\cos x\cos y - \sin x\sin y + \cos x\cos y + \sin x\sin y] = \frac{1}{2}(2\cos x\cos y) = \cos x\cos y.$

(c) Again using (12b) and (13b), we have $\frac{1}{2}[\cos(x-y) - \cos(x+y)]$

$= \frac{1}{2}[\cos x\cos y + \sin x\sin y - \cos x\cos y + \sin x\sin y] = \frac{1}{2}(2\sin x\sin y) = \sin x\sin y.$

42. Using (13b), $\cos\left(\frac{\pi}{2} - x\right) = \cos\frac{\pi}{2}\cos x + \sin\frac{\pi}{2}\sin x = 0\cdot\cos x + 1\cdot\sin x = \sin x.$

43. Using (12a), $\sin\left(\frac{\pi}{2} + x\right) = \sin\frac{\pi}{2}\cos x + \cos\frac{\pi}{2}\sin x = 1\cdot\cos x + 0\cdot\sin x = \cos x.$

44. Using (13a), $\sin(\pi - x) = \sin\pi\cos x - \cos\pi\sin x = 0\cdot\cos x - (-1)\sin x = \sin x.$

45. Using (6), $\sin\theta\cot\theta = \sin\theta\cdot\dfrac{\cos\theta}{\sin\theta} = \cos\theta.$

46. $(\sin x + \cos x)^2 = \sin^2 x + 2\sin x\cos x + \cos^2 x = (\sin^2 x + \cos^2 x) + \sin 2x$ [by (15a)] $= 1 + \sin 2x$ [by (7)]

47. $\sec y - \cos y = \dfrac{1}{\cos y} - \cos y$ [by (6)] $= \dfrac{1 - \cos^2 y}{\cos y} = \dfrac{\sin^2 y}{\cos y}$ [by (7)] $= \dfrac{\sin y}{\cos y}\sin y = \tan y\sin y$ [by (6)]

48. $\tan^2\alpha - \sin^2\alpha = \dfrac{\sin^2\alpha}{\cos^2\alpha} - \sin^2\alpha = \dfrac{\sin^2\alpha - \sin^2\alpha\cos^2\alpha}{\cos^2\alpha} = \dfrac{\sin^2\alpha(1 - \cos^2\alpha)}{\cos^2\alpha} = \tan^2\alpha\sin^2\alpha$ [by (7)]

49. $\cot^2\theta + \sec^2\theta = \dfrac{\cos^2\theta}{\sin^2\theta} + \dfrac{1}{\cos^2\theta}$ [by (6)] $= \dfrac{\cos^2\theta\cos^2\theta + \sin^2\theta}{\sin^2\theta\cos^2\theta} = \dfrac{(1 - \sin^2\theta)(1 - \sin^2\theta) + \sin^2\theta}{\sin^2\theta\cos^2\theta}$ [by (7)]

$= \dfrac{1 - \sin^2\theta + \sin^4\theta}{\sin^2\theta\cos^2\theta} = \dfrac{\cos^2\theta + \sin^4\theta}{\sin^2\theta\cos^2\theta}$ [by (7)] $= \dfrac{1}{\sin^2\theta} + \dfrac{\sin^2\theta}{\cos^2\theta} = \csc^2\theta + \tan^2\theta$ [by (6)]

50. $2\csc 2t = \dfrac{2}{\sin 2t} = \dfrac{2}{2\sin t\cos t}$ [by (15a)] $= \dfrac{1}{\sin t\cos t} = \sec t\csc t$

51. Using (14a), we have $\tan 2\theta = \tan(\theta + \theta) = \dfrac{\tan\theta + \tan\theta}{1 - \tan\theta\tan\theta} = \dfrac{2\tan\theta}{1 - \tan^2\theta}.$

52. $\dfrac{1}{1 - \sin\theta} + \dfrac{1}{1 + \sin\theta} = \dfrac{1 + \sin\theta + 1 - \sin\theta}{(1 - \sin\theta)(1 + \sin\theta)} = \dfrac{2}{1 - \sin^2\theta} = \dfrac{2}{\cos^2\theta}$ [by (7)] $= 2\sec^2\theta$

53. Using (15a) and (16a), $\sin x \sin 2x + \cos x \cos 2x = \sin x(2 \sin x \cos x) + \cos x(2\cos^2 x - 1)$
$= 2\sin^2 x \cos x + 2\cos^3 x - \cos x = 2(1 - \cos^2 x)\cos x + 2\cos^3 x - \cos x$ [by (7)]
$= 2\cos x - 2\cos^3 x + 2\cos^3 x - \cos x = \cos x$.

54. Working backward, we start with equations (12a) and (13a):
$\sin(x+y)\sin(x-y) = (\sin x \cos y + \cos x \sin y)(\sin x \cos y - \cos x \sin y)$
$= \sin^2 x \cos^2 y - \sin x \cos y \cos x \sin y + \cos x \sin y \sin x \cos y - \cos^2 x \sin^2 y$
$= \sin^2 x(1 - \sin^2 y) - (1 - \sin^2 x)\sin^2 y = \sin^2 x - \sin^2 x \sin^2 y - \sin^2 y + \sin^2 x \sin^2 y$ [by (7)]
$= \sin^2 x - \sin^2 y$.

55. $\dfrac{\sin\phi}{1-\cos\phi} = \dfrac{\sin\phi}{1-\cos\phi}\cdot\dfrac{1+\cos\phi}{1+\cos\phi} = \dfrac{\sin\phi(1+\cos\phi)}{1-\cos^2\phi} = \dfrac{\sin\phi(1+\cos\phi)}{\sin^2\phi}$ [by (7)] $= \dfrac{1+\cos\phi}{\sin\phi}$
$= \dfrac{1}{\sin\phi} + \dfrac{\cos\phi}{\sin\phi} = \csc\phi + \cot\phi$ [by (6)]

56. $\tan x + \tan y = \dfrac{\sin x}{\cos x} + \dfrac{\sin y}{\cos y} = \dfrac{\sin x \cos y + \cos x \sin y}{\cos x \cos y} = \dfrac{\sin(x+y)}{\cos x \cos y}$ [by (12a)]

57. Using (12a), $\sin 3\theta + \sin\theta = \sin(2\theta + \theta) + \sin\theta = \sin 2\theta\cos\theta + \cos 2\theta \sin\theta + \sin\theta$
$= \sin 2\theta \cos\theta + (2\cos^2\theta - 1)\sin\theta + \sin\theta$ [by (16a)] $= \sin 2\theta\cos\theta + 2\cos^2\theta\sin\theta - \sin\theta + \sin\theta$
$= \sin 2\theta\cos\theta + \sin 2\theta \cos\theta$ [by (15a)] $= 2\sin 2\theta\cos\theta$.

58. We use (12b) with $x = 2\theta$, $y = \theta$ to get $\cos 3\theta = \cos(2\theta + \theta) = \cos 2\theta \cos\theta - \sin 2\theta \sin\theta$
$= (2\cos^2\theta - 1)\cos\theta - 2\sin^2\theta\cos\theta$ [by (16a) and (15a)] $= (2\cos^2\theta - 1)\cos\theta - 2(1 - \cos^2\theta)\cos\theta$ [by (7)]
$= 2\cos^3\theta - \cos\theta - 2\cos\theta + 2\cos^3\theta = 4\cos^3\theta - 3\cos\theta$.

59. Since $\sin x = \frac{1}{3}$ we can label the opposite side as having length 1, the hypotenuse as having length 3, and use the Pythagorean Theorem to get that the adjacent side has length $\sqrt{8}$. Then, from the diagram, $\cos x = \frac{\sqrt{8}}{3}$. Similarly we have that $\sin y = \frac{3}{5}$.

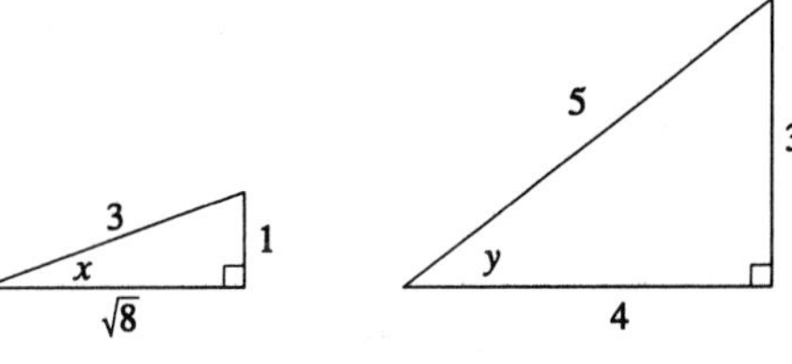

Now use (12a): $\sin(x+y) = \sin x \cos y + \cos x \sin y = \frac{1}{3}\cdot\frac{4}{5} + \frac{\sqrt{8}}{3}\cdot\frac{3}{5} = \frac{4}{15} + \frac{3\sqrt{8}}{15} = \frac{4+6\sqrt{2}}{15}$.

60. Use (12b) and the values for $\sin y$ and $\cos x$ obtained in Exercise 59 to get
$\cos(x+y) = \cos x\cos y - \sin x \sin y = \frac{\sqrt{8}}{3}\cdot\frac{4}{5} - \frac{1}{3}\cdot\frac{3}{5} = \frac{8\sqrt{2}-3}{15}$.

61. Using (13b) and the values for $\cos x$ and $\sin y$ obtained in Exercise 59, we have
$\cos(x-y) = \cos x \cos y + \sin x \sin y = \frac{\sqrt{8}}{3}\cdot\frac{4}{5} + \frac{1}{3}\cdot\frac{3}{5} = \frac{8\sqrt{2}+3}{15}$.

62. Using (13a) and the values for $\sin y$ and $\cos x$ obtained in Exercise 59, we get
$\sin(x-y) = \sin x\cos y - \cos x \sin y = \frac{1}{3}\cdot\frac{4}{5} - \frac{\sqrt{8}}{3}\cdot\frac{3}{5} = \frac{4-6\sqrt{2}}{15}$.

63. Using (15a) and the value for $\sin y$ obtained in Exercise 59, we have
$\sin 2y = 2\sin y \cos y = \dfrac{2\sin y}{\sec y} = 2\left(\frac{3}{5}\right)\left(\frac{4}{5}\right) = \frac{24}{25}$.

64. Using (16a), $\cos 2y = 2\cos^2 y - 1 = \dfrac{2}{\sec^2 y} - 1 = \dfrac{2}{(5/4)^2} - 1 = 2\left(\frac{4}{5}\right)^2 - 1 = \frac{32}{25} - 1 = \frac{7}{25}$.

65. $2\cos x - 1 = 0 \quad \Leftrightarrow \quad \cos x = \frac{1}{2} \quad \Rightarrow \quad x = \frac{\pi}{3}, \frac{5\pi}{3}$

66. $3\cot^2 x = 1 \quad \Leftrightarrow \quad 3 = 1/\cot^2 x \quad \Leftrightarrow \quad \tan^2 x = 3 \quad \Leftrightarrow \quad \tan x = \pm\sqrt{3} \quad \Rightarrow \quad x = \frac{\pi}{3}, \frac{2\pi}{3}, \frac{4\pi}{3}$, and $\frac{5\pi}{3}$.

67. $2\sin^2 x = 1 \quad \Leftrightarrow \quad \sin^2 x = \frac{1}{2} \quad \Leftrightarrow \quad \sin x = \pm\frac{1}{\sqrt{2}} \quad \Rightarrow \quad x = \frac{\pi}{4}, \frac{3\pi}{4}, \frac{5\pi}{4}, \frac{7\pi}{4}$.

68. $|\tan x| = 1 \quad \Leftrightarrow \quad \tan x = -1$ or $\tan x = 1 \quad \Leftrightarrow \quad x = \frac{3\pi}{4}, \frac{7\pi}{4}$ or $x = \frac{\pi}{4}, \frac{5\pi}{4}$.

69. Using (15a), $\sin 2x = \cos x \;\Rightarrow\; 2\sin x\cos x - \cos x = 0 \;\Leftrightarrow\; \cos x(2\sin x - 1) = 0 \;\Leftrightarrow\; \cos x = 0$ or $2\sin x - 1 = 0 \;\Rightarrow\; x = \frac{\pi}{2}, \frac{3\pi}{2}$ or $\sin x = \frac{1}{2} \;\Rightarrow\; x = \frac{\pi}{6}$ or $\frac{5\pi}{6}$. Therefore the solutions are $x = \frac{\pi}{6}, \frac{\pi}{2}, \frac{5\pi}{6}, \frac{3\pi}{2}$.

70. By (15a), $2\cos x + \sin 2x = 0 \quad \Leftrightarrow \quad 2\cos x + 2\sin x\cos x = 0 \quad \Leftrightarrow \quad 2\cos x(1 + \sin x) = 0 \quad \Leftrightarrow$ $\cos x = 0$ or $1 + \sin x = 0 \quad \Leftrightarrow \quad x = \frac{\pi}{2}, \frac{3\pi}{2}$ or $\sin x = -1 \quad \Rightarrow \quad x = \frac{3}{2}\pi$. So the solutions are $x = \frac{\pi}{2}, \frac{3\pi}{2}$.

71. $\sin x = \tan x \quad \Leftrightarrow \quad \sin x - \tan x = 0 \quad \Leftrightarrow \quad \sin x - \dfrac{\sin x}{\cos x} = 0 \quad \Leftrightarrow \quad \sin x\left(1 - \dfrac{1}{\cos x}\right) = 0 \quad \Leftrightarrow$ $\sin x = 0$ or $1 - \dfrac{1}{\cos x} = 0 \quad \Rightarrow \quad x = 0, \pi, 2\pi$ or $1 = \dfrac{1}{\cos x} \quad \Rightarrow \quad \cos x = 1 \quad \Rightarrow \quad x = 0, 2\pi$. Therefore the solutions are $x = 0, \pi, 2\pi$.

72. By (16a), $2 + \cos 2x = 3\cos x \quad \Leftrightarrow \quad 2 + 2\cos^2 x - 1 = 3\cos x \quad \Leftrightarrow \quad 2\cos^2 x - 3\cos x + 1 = 0 \quad \Leftrightarrow$ $(2\cos x - 1)(\cos x - 1) = 0 \quad \Leftrightarrow \quad \cos x = 1$ or $\cos x = \frac{1}{2} \quad \Rightarrow \quad x = 0,\ 2\pi$ or $x = \frac{\pi}{3}, \frac{5\pi}{3}$.

73. We know that $\sin x = \frac{1}{2}$ when $x = \frac{\pi}{6}$ or $\frac{5\pi}{6}$, and from Figure 13(a), we see that $\sin x \le \frac{1}{2} \quad \Rightarrow \quad 0 \le x \le \frac{\pi}{6}$ or $\frac{5\pi}{6} \le x \le 2\pi$.

74. $2\cos x + 1 > 0 \quad \Rightarrow \quad 2\cos x > -1 \quad \Rightarrow \quad \cos x > -\frac{1}{2}$. $\cos x = -\frac{1}{2}$ when $x = \frac{2\pi}{3}, \frac{4\pi}{3}$ and from Figure 13(b), we see that $\cos x > -\frac{1}{2}$ when $0 \le x < \frac{2\pi}{3}, \frac{4\pi}{3} < x \le 2\pi$.

75. $\tan x = -1$ when $x = \frac{3\pi}{4}, \frac{7\pi}{4}$, and $\tan x = 1$ when $x = \frac{\pi}{4}$ or $\frac{5\pi}{4}$. From Figure 14 we see that $-1 < \tan x < 1$ $\Rightarrow \quad 0 \le x < \frac{\pi}{4}, \frac{3\pi}{4} < x < \frac{5\pi}{4}$, and $\frac{7\pi}{4} < x \le 2\pi$.

76. We know that $\sin x = \cos x$ when $x = \frac{\pi}{4}, \frac{5\pi}{4}$, and from the diagram we see that $\sin x > \cos x$ when $\frac{\pi}{4} < x < \frac{5\pi}{4}$.

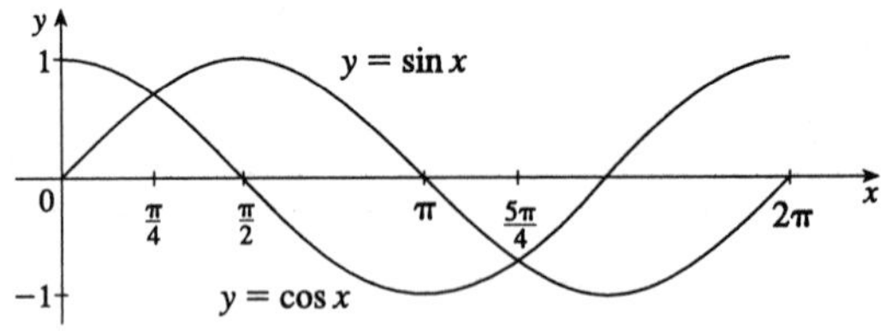

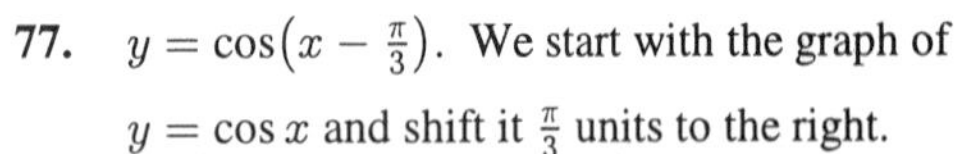
77. $y = \cos\left(x - \frac{\pi}{3}\right)$. We start with the graph of $y = \cos x$ and shift it $\frac{\pi}{3}$ units to the right.

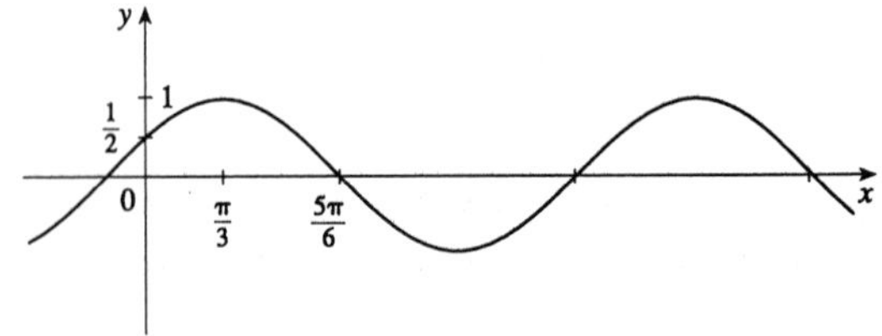

78. $y = \tan 2x$. Start with the graph of $y = \tan x$ with period π and compress it to a period of $\frac{\pi}{2}$.

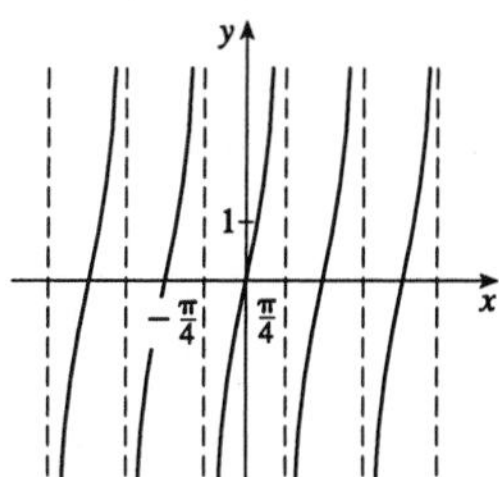

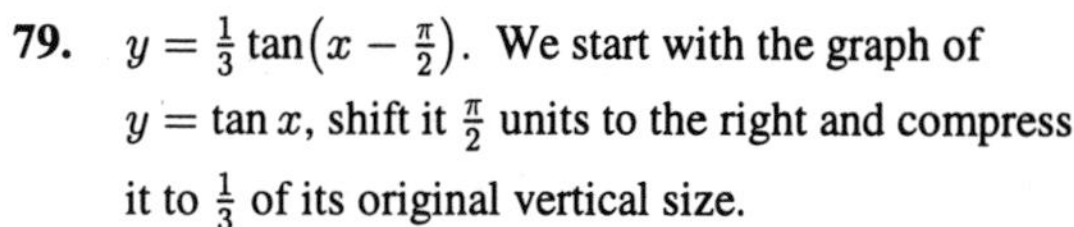

79. $y = \frac{1}{3}\tan\left(x - \frac{\pi}{2}\right)$. We start with the graph of $y = \tan x$, shift it $\frac{\pi}{2}$ units to the right and compress it to $\frac{1}{3}$ of its original vertical size.

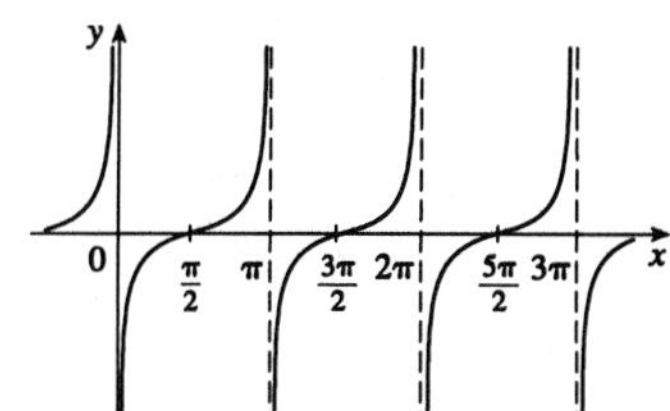

80. $y = 1 + \sec x$. Start with the graph of $y = \sec x$ and raise it by one unit.

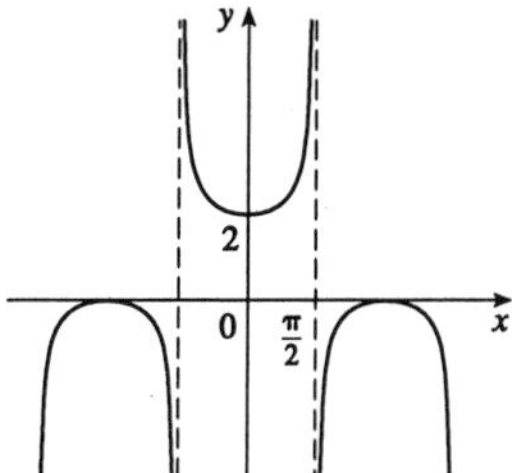

81. $y = |\sin x|$. We start with the graph of $y = \sin x$ and reflect the parts below the x-axis about the x-axis.

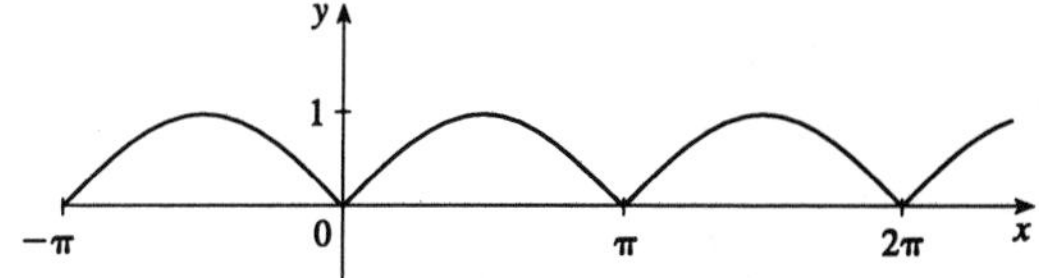

82. $y = 2 + \sin\left(x + \frac{\pi}{4}\right)$. Start with the graph of $y = \sin x$, and shift it $\frac{\pi}{4}$ units to the left and 2 units up.

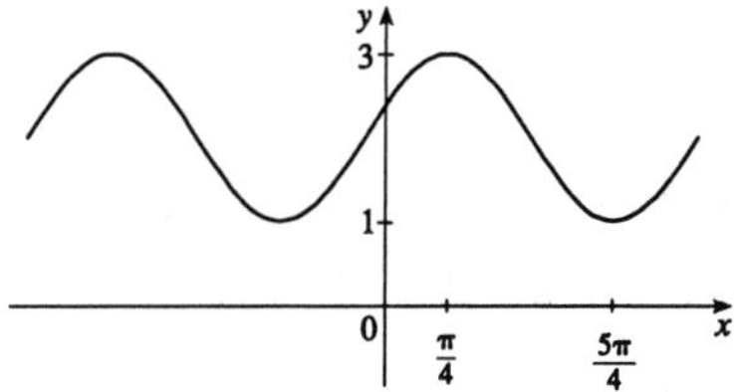

83. From the figure we see that $x = b\cos\theta$, $y = b\sin\theta$, and from the distance formula we have that the distance c from (x, y) to $(a, 0)$ is $c = \sqrt{(x-a)^2 + (y-0)^2} \quad\Rightarrow\quad c^2 = (b\cos\theta - a)^2 + (b\sin\theta)^2$
$= b^2\cos^2\theta - 2ab\cos\theta + a^2 + b^2\sin^2\theta = a^2 + b^2(\cos^2\theta + \sin^2\theta) - 2ab\cos\theta = a^2 + b^2 - 2ab\cos\theta$ [by (7)].

84. $|AB|^2 = |AC|^2 + |BC|^2 - 2|AC||BC|\cos\angle C = (820)^2 + (910)^2 - 2(820)(910)\cos 103° \approx 1{,}836{,}217 \quad\Rightarrow$
$|AB| \approx 1355$ m

85. Using the Law of Cosines, we have $c^2 = 1^2 + 1^2 - 2(1)(1)\cos(\alpha - \beta) = 2[1 - \cos(\alpha - \beta)]$. Now, using the distance formula, $c^2 = |AB|^2 = (\cos\alpha - \cos\beta)^2 + (\sin\alpha - \sin\beta)^2$. Equating these two expressions for c^2, we get $2[1 - \cos(\alpha - \beta)] = \cos^2\alpha + \sin^2\alpha + \cos^2\beta + \sin^2\beta - 2\cos\alpha\cos\beta - 2\sin\alpha\sin\beta \quad\Rightarrow$
$1 - \cos(\alpha - \beta) = 1 - \cos\alpha\cos\beta - \sin\alpha\sin\beta \quad\Rightarrow\quad \cos(\alpha - \beta) = \cos\alpha\cos\beta + \sin\alpha\sin\beta$.

86. $\cos(x + y) = \cos[x - (-y)]$, so take $\alpha = x$, $\beta = -y$ to get
$\cos[x - (-y)] = \cos x\cos(-y) + \sin x\sin(-y) = \cos x\cos y - \sin x\sin y$ (using Equations 10a and 10b).

87. In Exercise 86 we used the subtraction formula for cosine to prove the addition formula for cosine. Using that formula with $x = \frac{\pi}{2} - \alpha$, $y = \beta$, we get $\cos\left[\left(\frac{\pi}{2} - \alpha\right) + \beta\right] = \cos\left(\frac{\pi}{2} - \alpha\right)\cos\beta - \sin\left(\frac{\pi}{2} - \alpha\right)\sin\beta \quad\Rightarrow$
$\cos\left[\frac{\pi}{2} - (\alpha - \beta)\right] = \cos\left(\frac{\pi}{2} - \alpha\right)\cos\beta - \sin\left(\frac{\pi}{2} - \alpha\right)\sin\beta$. Now we use the identities given in the problem to get $\sin(\alpha - \beta) = \sin\alpha\cos\beta - \cos\alpha\sin\beta$.

88. If $0 < \theta < \frac{\pi}{2}$, we have the case depicted in the first diagram. In this case, we see that the height of the triangle is $h = a \sin \theta$. If $\frac{\pi}{2} \leq \theta < \pi$, we have the case depicted in the second diagram. In this case, the height of the triangle is $h = a \sin(\pi - \theta) = a \sin \theta$ (by the identity proved in Exercise 44). So in either case, the area of the triangle is $\frac{1}{2}bh = \frac{1}{2}ab \sin \theta$.

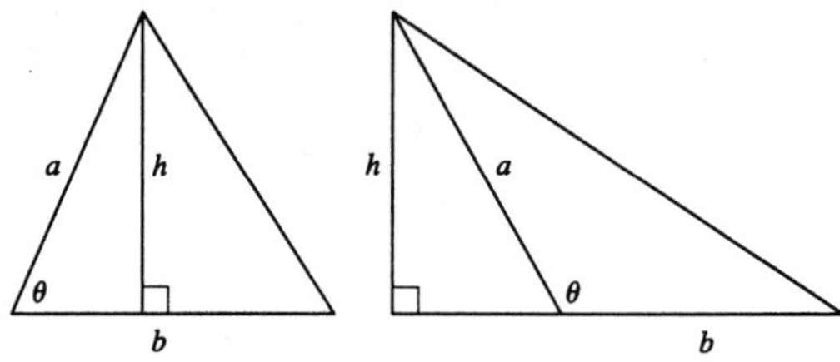

89. Using the formula derived in Exercise 88, the area of the triangle is $\frac{1}{2}(10)(3)\sin 107° \approx 14.34457$.

EXERCISES E

1. Let S_n be the statement that $2^n > n$.

1. S_1 is true because $2^1 = 2 > 1$.
2. Assume S_k is true, that is, $2^k > k$. Then $2^{k+1} = 2^k \cdot 2 > k \cdot 2 = 2k$ (since $2^k > k$). But $k > 1 \Rightarrow k + k > k + 1 \Rightarrow 2k > k + 1$, so that $2^{k+1} > 2k > k + 1$, which shows that S_{k+1} is true.
3. Therefore, by mathematical induction, $2^n > n$ for every positive integer n.

2. Let S_n be the statement that $3^n > 2n$.

1. S_1 is true because $3^1 = 3 > 2(1) = 2$.
2. Assume S_k is true, that is, $3^k > 2k$. Then $3^{k+1} = 3^k \cdot 3 > 2k \cdot 3 = 6k$. But for $k > 1$, $k > \frac{1}{2} \Rightarrow 4k > 2 \Rightarrow 6k > 2k + 2 \Rightarrow 6k > 2(k+1)$. So $3^{k+1} > 6k > 2(k+1) \Rightarrow S_{k+1}$ is true.
3. Therefore, by mathematical induction, $3^n > 2n$ for every positive integer n.

3. Let S_n be the statement that $(1+x)^n \geq 1 + nx$.

1. S_1 is true because $(1+x)^1 = 1 + (1)x$.
2. Assume S_k is true, that is, $(1+x)^k \geq 1 + kx$. Then $(1+x)^{k+1} = (1+x)^k(1+x) \geq (1+kx)(1+x)$ [since $(1+x)^k \geq 1 + kx$] $= 1 + x + kx + kx^2 \geq 1 + x + kx = 1 + (k+1)x$, which shows that S_{k+1} is true.
3. Therefore, by mathematical induction, $(1+x)^n \geq 1 + nx$ for every positive integer n.

4. Let S_n be the statement that for $0 \leq a < b$, $a^n < b^n$.

1. S_1 is true because $a < b \Rightarrow a^1 < b^1$.
2. Assume S_k is true, that is, for $0 \leq a < b$, $a^k < b^k$. Then $a^{k+1} = a^k a < b^k b = b^{k+1}$, which shows that S_{k+1} is true.
3. Therefore, by mathematical induction, for $0 \leq a < b$, $a^n < b^n$ for every positive integer n.

5. Let S_n be the statement that $7^n - 1$ is divisible by 6.

1. S_1 is true because $7^1 - 1 = 6$ is divisible by 6.
2. Assume S_k is true, that is, $7^k - 1$ is divisible by 6; in other words $7^k - 1 = 6m$ for some positive integer m. Then $7^{k+1} - 1 = 7^k \cdot 7 - 1 = (6m+1) \cdot 7 - 1 = 6(7m+1)$, which is divisible by 6, so S_{k+1} is true.
3. Therefore, by mathematical induction, $7^n - 1$ is divisible by 6 for every positive integer n.

6. Let S_n be the statement that $(a/b)^n = a^n/b^n$.

1. S_1 is true because $(a/b)^1 = a/b = a^1/b^1$.
2. Assume S_k is true, that is, $\left(\frac{a}{b}\right)^k = \frac{a^k}{b^k}$. Then $\left(\frac{a}{b}\right)^{k+1} = \left(\frac{a}{b}\right)^k \left(\frac{a}{b}\right) = \frac{a^k}{b^k}\frac{a}{b} = \frac{a^{k+1}}{b^{k+1}}$, which shows that S_{k+1} is true.
3. Therefore, by mathematical induction, $(a/b)^n = a^n/b^n$ for every positive integer n.

7. Let S_n be the statement that $1 + 3 + 5 + \cdots + (2n-1) = n^2$.

1. S_1 is true because $[2(1) - 1] = 1 = 1^2$.
2. Assume S_k is true, that is $1 + 3 + 5 + \cdots + (2k-1) = k^2$. Then $1 + 3 + 5 + \cdots + [2(k+1) - 1] = 1 + 3 + 5 + \cdots + (2k-1) + (2k+1) = k^2 + (2k+1) = (k+1)^2$, which shows that S_{k+1} is true.
3. Therefore, by mathematical induction, $1 + 3 + 5 + \cdots + (2n-1) = n^2$ for every positive integer n.

8. Let S_n be the statement that $2 + 6 + 12 + \cdots + n(n+1) = \frac{1}{3}[n(n+1)(n+2)]$.

1. S_1 is true because $1(1+1) = 2 = \frac{1}{3}[1(1+1)(1+2)]$.
2. Assume S_k is true, that is, $2 + 6 + 12 + \cdots + k(k+1) = \frac{1}{3}[k(k+1)(k+2)]$. Then
$$2 + 6 + 12 + \cdots + (k+1)[(k+1)+1] = 2 + 6 + 12 + k(k+1) + (k+1)(k+2)$$
$$= \frac{k(k+1)(k+2)}{3} + (k+1)(k+2) = \frac{k(k+1)(k+2) + 3(k+1)(k+2)}{3}$$
$$= \frac{(k+1)(k+2)[k+3]}{3} = \frac{(k+1)[(k+1)+1][(k+1)+2]}{3},$$ which shows that S_{k+1} is true.
3. Therefore, by mathematical induction, $2 + 6 + 12 + \cdots + n(n+1) = \frac{n(n+1)(n+2)}{3}$ for every positive integer n.

9. Let S_n be the statement that $\frac{1}{2} + \frac{1}{6} + \frac{1}{12} + \cdots + \frac{1}{n(n+1)} = \frac{n}{n+1}$.

1. S_1 is true because $\frac{1}{1(1+1)} = \frac{1}{1+1}$.
2. Assume S_k is true, that is, $\frac{1}{2} + \frac{1}{6} + \frac{1}{12} + \cdots + \frac{1}{k(k+1)} = \frac{k}{k+1}$. Then
$$\frac{1}{2} + \frac{1}{6} + \frac{1}{12} + \cdots + \frac{1}{(k+1)[(k+1)+1]} = \frac{1}{2} + \frac{1}{6} + \frac{1}{12} + \cdots + \frac{1}{(k+1)(k+2)}$$
$$= \frac{1}{2} + \frac{1}{6} + \frac{1}{12} + \cdots + \frac{1}{k(k+1)} + \frac{1}{(k+1)(k+2)} = \frac{k}{k+1} + \frac{1}{(k+1)(k+2)} = \frac{k(k+2)+1}{(k+1)(k+2)}$$
$$= \frac{k^2+2k+1}{(k+1)(k+2)} = \frac{(k+1)(k+1)}{(k+1)(k+2)} = \frac{(k+1)}{(k+1)+1}$$ which shows that S_{k+1} is true.
3. Therefore, by mathematical induction, $\frac{1}{2} + \frac{1}{6} + \frac{1}{12} + \cdots + \frac{1}{n(n+1)} = \frac{n}{n+1}$ for every positive integer n.

10. Let S_n be the statement that $a + ar + ar^2 + \cdots + ar^{n-1} = \dfrac{a(1-r^n)}{1-r}$.

1. S_1 is true because $ar^0 = a = \dfrac{a(1-r^1)}{1-r}$.
2. Assume S_k is true, that is, $a + ar + ar^2 + \cdots + ar^{k-1} = \dfrac{a(1-r^k)}{1-r}$. Then

$$a + ar + ar^2 + \cdots + ar^{(k+1)-1} = a + ar + ar^2 + \cdots + ar^k = a + ar + ar^2 + \cdots + ar^{k-1} + ar^k$$
$$= \frac{a(1-r^k)}{1-r} + ar^k = \frac{a(1-r^k) + (1-r)ar^k}{1-r} = \frac{a(1-r^k + r^k - r^{k+1})}{1-r} = \frac{a(1-r^{k+1})}{1-r},$$

which shows that S_{k+1} is true.

3. Therefore, by mathematical induction, $a + ar + ar^2 + \cdots + ar^{n-1} = \dfrac{a(1-r^n)}{1-r}$.

EXERCISES G

1. The computer results are from Maple, with `Digits:=16`. The last column shows the values of the sixth-degree Taylor polynomial for $f(x) = \csc^2 x - x^{-2}$ near $x = 0$. Note that the second arrangement of Taylor's polynomial is easier to use with a calculator:

$$T_6(x) = \tfrac{1}{3} + \tfrac{1}{15}x^2 + \tfrac{2}{189}x^4 + \tfrac{1}{675}x^6 = \left[\left(\tfrac{1}{675}x^2 + \tfrac{2}{189}\right)x^2 + \tfrac{1}{15}\right]x^2 + \tfrac{1}{3}$$

x	$f(x)_{\text{calculator}}$	$f(x)_{\text{computer}}$	$f(x)_{\text{Taylor}}$
0.1	0.33400107	0.3340010596845	0.33400106
0.01	0.333341	0.33334000010	0.33334000
0.001	0.3334	0.333333400	0.33333340
0.0001	0.34	0.3333334	0.33333333
0.00001	2.0	0.33334	0.33333333
0.000001	100 or 200	0.334	0.33333333
0.0000001	10000 or 20000	0.4	0.33333333
0.00000001	1000000	0	0.33333333

We see that the calculator results start to deteriorate seriously at $x = 0.0001$, and for smaller x, they are entirely meaningless. The different results "100 or 200" etc. depended on whether we calculated $\left[(\sin x)^2\right]^{-1}$ or $\left[(\sin x)^{-1}\right]^2$. With Maple, the result is off by more than 10% when $x = 0.0000001$ (compare with the calculator result!) A detailed analysis reveals that the values of the function are always greater than $\frac{1}{3}$, but the computer eventually gives results less than $\frac{1}{3}$.

The polynomial $T_6(x)$ was obtained by patient simplification of the expression for $f(x)$, starting with $\sin^2(x) = \frac{1}{2}(1 - \cos 2x)$, where $\cos 2x = 1 - \dfrac{(2x)^2}{2!} + \dfrac{(2x)^4}{4!} - \cdots - \dfrac{(2x)^{10}}{10!} + R_{12}(x)$.

Consequently, the exact value of the limit is $T_6(0) = \frac{1}{3}$. It can also be obtained by several applications of l'Hospital's Rule to the expression $f(x) = \dfrac{x^2 - \sin^2 x}{x^2 \sin^2 x}$ with intermediate simplifications.

2.

h	$\ln(1+h)/h$
10^{-2}	0.99503309
10^{-4}	0.99995000
10^{-6}	0.99999950
10^{-8}	$1-5\times10^{-9}$
7×10^{-11}	1.4285714
10^{-11}	0
-10^{-2}	1.0050336
-10^{-4}	1.00005
-10^{-6}	1.0000005
-10^{-8}	$1+5\times10^{-9}$
-7×10^{-11}	1.0
-7×10^{-12}	1.4285714

Taylor's expansion of $\ln x$ centered at $c=1$ is $\ln x=(x-1)-\frac{1}{2}(x-1)^2+\frac{1}{3}(x-1)^3-\frac{1}{4}(x-1)^4+\cdots$ and x is calculated as $1+h$ where h is small. We have the situation described in the text: the calculation of $(1+h)-1$, which should of course result in h, results instead in a rounded value of h, destroying some of its digits. Hence the strange behavior at $h=7\times10^{-11}$ and -7×10^{-12}. When $1+h$ is indistinguishable from 1, the calculated values are 0. The precise value is $\lim\limits_{h\to0}\dfrac{\ln(1+h)-\ln 1}{h}=\left[\dfrac{d}{dx}\ln x\right]_{x=1}=1.$ Instead of using this shortcut, we could use l'Hospital's Rule.

3. From $f(x)=\dfrac{x^{25}}{(1.0001)^x}$ (we may assume $x>0$; why?), we have $\ln f(x)=25\ln x-x\ln(1.0001)$ and $\dfrac{f'(x)}{f(x)}=\dfrac{25}{x}-\ln(1.0001)$. This derivative, as well as the derivative $f'(x)$ itself, is positive for $0<x<x_0=\dfrac{25}{\ln(1.0001)}\approx 249{,}971.015$, and negative for $x>x_0$. Hence the maximum value of $f(x)$ is $f(x_0)=\dfrac{x_0^{25}}{(1.0001)^{x_0}}$, a number too large to be calculated directly. Using decimal logarithms, $\log_{10}f(x_0)\approx124.08987757$, so that $f(x_0)\approx1.229922\times10^{124}$. The actual value of the limit is $\lim\limits_{x\to\infty}f(x)=0$; it would be wasteful and inelegant to use l'Hospital's Rule twenty-five times since we can transform $f(x)$ into $f(x)=\left(\dfrac{x}{(1.0001)^{x/25}}\right)^{25}$, and the inside expression needs just one application of l'Hospital's Rule to give 0.

4. $\sqrt{1-\cos x}=\sqrt{2\sin^2\left(\frac{1}{2}x\right)}=\sqrt{2}\cdot\left|\sin\left(\frac{1}{2}x\right)\right|$. Another form (not as generally valid as the first one): $\sqrt{1-\cos x}\cdot\sqrt{\dfrac{1+\cos x}{1+\cos x}}=\dfrac{|\sin x|}{\sqrt{1+\cos x}}$.

5. For $f(x)=\ln\ln x$ with $x\in[a,b]$, $a=10^9$, and $b=10^9+1$, we need $f'(x)=\dfrac{1}{x\ln x}$, $f''(x)=-\dfrac{\ln x+1}{x^2(\ln x)^2}$.

(a) $f'(b)<D<f'(a)$, where $f'(a)\approx4.8254942434\times10^{-11}$, $f'(b)\approx4.8254942383\times10^{-11}$.

(b) Let us estimate $f'(b)-f'(a)=(b-a)f''(c_1)=f''(c_1)$. Since f'' increases (its absolute value decreases), we have $|f'(b)-f'(a)|<|f''(a)|\approx5.0583\times10^{-20}$.

6. Applying the inequalities to $f(x)=x^{-1.001}$, we obtain $\dfrac{1000}{(N+1)^{0.001}}<r_{N+1}=\sum\limits_{N+1}^{\infty}\dfrac{1}{n^{1.001}}<\dfrac{1000}{N^{0.001}}$ and a sufficient condition for $r_{N+1}<5\times10^{-9}$, as required, is $\dfrac{1000}{N^{0.001}}\le5\times10^{-9}$, which happens for the first time when $N=\left(2\times10^{11}\right)^{1000}$. On the other hand, the remainders are decreasing and if $M=N-1$ then the left-hand part of the inequalities show that $r_{M+1}>5\times10^{-9}$. Compared with a power of 10 using logarithms with base 10, our value of $N>10^{11{,}301}$.

7. (a) The 11-digit calculator value of $192 \sin \frac{\pi}{96}$ is 6.2820639018, while the value (on the same device) of p before rationalization is 6.282063885, which is 1.68×10^{-8} less than the trigonometric result.

(b) $p = \dfrac{96}{\sqrt{2+\sqrt{3}} \cdot \sqrt{2+\sqrt{2+\sqrt{3}}} \cdot \sqrt{2+\sqrt{2+\sqrt{2+\sqrt{3}}}} \cdot \sqrt{2+\sqrt{2+\sqrt{2+\sqrt{2+\sqrt{3}}}}}}$, but of course we can avoid repetitious calculations by storing intermediate results in a memory: $p_1 = \sqrt{2+\sqrt{3}}$, $p_2 = \sqrt{2+p_1}$, $p_3 = \sqrt{2+p_2}$, $p_4 = \sqrt{2+p_3}$, and so $p = \dfrac{96}{p_1 p_2 p_3 p_4}$. According to this formula, a calculator gives $p \approx 6.2820639016$, which is within 2×10^{-10} of the trigonometric result. With `Digits:=16;` Maple gives $p \approx 6.282063901781030$ before rationalization (off the trig result by about 1.1×10^{-14}) and $p \approx 6.282063901781018$ after rationalization (error of about 1.7×10^{-15}), a gain of about 1 digit of accuracy for rationalizing. If we set `Digits:=100;`, the difference between Maple's calculation of $192 \sin \frac{\pi}{96}$ and the radical is only about 4×10^{-99}.

8. The identity $\ln a = b + \ln\left(1 + \dfrac{a - e^b}{e^b}\right)$ can help as follows: Suppose that b is the poor approximation of $\ln a$, and denote $h = \left(a - e^b\right)/e^b$; since some closeness of a to e^b is to be expected, h is a small number. This follows from the Mean Value Theorem: $\dfrac{a - e^b}{\ln a - b} = e^c$, where c lies between b and $\ln a$, hence $h = (\ln a - b) \cdot e^{c-b}$. The use of Taylor's expansion is straightforward. Going up to the third derivative, $\ln(1+h) = h - \frac{1}{2}h^2 + \dfrac{2}{3!(1+\theta h)^3} h^3$ $(0 < \theta < 1)$ means roughly that the number of valid digits after the correction $\ln a \approx b + \dfrac{a - e^b}{e^b} - \dfrac{1}{2}\left(\dfrac{a - e^b}{e^b}\right)^2$ triples.

9. (a) Let $A = \left[\frac{1}{2}\left(27q + \sqrt{729q^2 + 108p^3}\right)\right]^{1/3}$ and $B = \left[\frac{1}{2}\left(27q - \sqrt{729q^2 + 108p^3}\right)\right]^{1/3}$. Then $A^3 + B^3 = 27q$ and $AB = \frac{1}{4}[729q^2 - (729q^2 + 108p^3)]^{1/3} = -3p$. Substitute into the formula $A + B = \dfrac{A^3 + B^3}{A^2 - AB + B^2}$ where we replace B by $-\dfrac{3p}{A}$:

$$x = \tfrac{1}{3}(A+B) = \frac{27q/3}{\left[\frac{1}{2}\left(27q + \sqrt{729q^2 + 108p^3}\right)\right]^{2/3} + 3p + 9p^2\left[\frac{1}{2}\left(27q + \sqrt{729q^2 + 108p^3}\right)\right]^{-2/3}}$$

which almost yields the given formula; since replacing q by $-q$ results in replacing x by $-x$, a simple discussion of the cases $q > 0$ and $q < 0$ allows us to replace q by $|q|$ in the denominator, so that it involves only positive numbers. The problems mentioned in the introduction to this exercise have disappeared.

(b) A direct attack works best here. To save space, let $\alpha = 2 + \sqrt{5}$, so we can rationalize, using $\alpha^{-1} = -2 + \sqrt{5}$ and $\alpha - \alpha^{-1} = 4$ (check it!):

$$u = \frac{4}{\alpha^{2/3} + 1 + \alpha^{-2/3}} \cdot \frac{\alpha^{1/3} - \alpha^{-1/3}}{\alpha^{1/3} - \alpha^{-1/3}} = \frac{4\left(\alpha^{1/3} - \alpha^{-1/3}\right)}{\alpha - \alpha^{-1}} = \alpha^{1/3} - \alpha^{-1/3}$$

and we cube the expression for u: $u^3 = \alpha - 3\alpha^{1/3} + 3\alpha^{-1/3} - \alpha^{-1} = 4 - 3u$, $u^3 + 3u - 4 = (u-1)(u^2 + u + 4) = 0$, so that the only real root is $u = 1$. A check using the formula from part (a): $p = 3$, $q = -4$, so $729q^2 + 108p^3 = 14{,}580 = 54^2 \times 5$, and $x = \dfrac{36}{\left(54 + 27\sqrt{5}\right)^{2/3} + 9 + 81\left(54 + 27\sqrt{5}\right)^{-2/3}}$, which simplifies to the given form after reduction by 9.

10. **(a)** See (10.6.5) and the three lines that follow it.

(b) For $x = 99$, the remainder after n terms is $r_{n+1} = \sum_{j=n+1}^{\infty} \frac{99^j}{100^j+1} < \sum_{j=n+1}^{\infty} \frac{99^j}{100^j} = 100\left(\frac{99}{100}\right)^{n+1}$ and it is sufficient to make n so large that $100\left(\frac{99}{100}\right)^{n+1} < 5 \times 10^{-7}$; after some logarithmic manipulation, the answer is $n \geq 1901$. [Using the ideas of part (c), we can show that this is indeed the first time the error is so small.]

(c) Using the hint, $\sum_{n=1}^{\infty} \frac{x^n}{100^n} = \frac{x}{100} \cdot \frac{1}{1-\frac{x}{100}} = \frac{x}{100-x}$, so on subtracting,

$f(x) - \frac{x}{100-x} = \sum_{n=1}^{\infty}\left(\frac{x^n}{100^n+1} - \frac{x^n}{100^n}\right) = -\sum_{n=1}^{\infty} \frac{\left(\frac{1}{100}x\right)^n}{100^n+1} = -f\left(\frac{x}{100}\right)$, as claimed. Here the remainder at $x = 99$ is estimated as follows: $r_{n+1} = \sum_{j=n+1}^{\infty} \frac{\left(\frac{99}{100}\right)^j}{100^j+1} < \frac{(0.0099)^{n+1}}{0.9901}$, and to make the latter expression less than 5×10^{-7}, all it takes is $n \geq 3$. The calculation (not really necessary):

$$f(99) \approx \frac{99}{100-99} - \left[\frac{0.99}{100+1} + \frac{(0.99)^2}{100^2+1} + \frac{(0.99)^3}{100^3+1}\right] \approx 98.990099.$$

11. Proof that $\lim_{n\to\infty} a_n = 0$: From $1 \leq e^{1-x} \leq e$ it follows that $x^n \leq e^{1-x}x^n \leq x^n e$, and integration gives

$\frac{1}{n+1} = \int_0^1 x^n\,dx \leq \int_0^1 e^{1-x}x^n\,dx \leq \int_0^1 x^n e\,dx = \frac{e}{n+1}$, that is, $\frac{1}{n+1} \leq a_n \leq \frac{e}{n+1}$, and since $\lim_{n\to\infty} \frac{1}{n+1} = \lim_{n\to\infty} \frac{e}{n+1} = 0$, it follows from the Squeeze Theorem that $\lim_{n\to\infty} a_n = 0$. Of course, the expression $1/(n+1)$ on the left side could have been replaced by 0 and the proof would still be correct.

Calculations: Using the formula $a_n = \left[e - 1 - \left(\frac{1}{1!} + \frac{1}{2!} + \cdots + \frac{1}{n!}\right)\right] n!$ with an 11-digit pocket calculator:

n	a_n	n	a_n	n	a_n
0	1.7182818284	7	0.1404151360	14	−5.07636992
1	0.7182818284	8	0.1233210880	15	−77.1455488
2	0.4365636568	9	0.1098897920	16	−1235.3287808
3	0.3096909704	10	0.0988979200	17	−21001.589274
4	0.2387638816	11	0.0878771200	18	−378029.60693
5	0.1938194080	12	0.0545254400	19	−7182563.5317
6	0.1629164480	13	−0.2911692800	20	−143651271.63

It is clear that the values calculated from the direct reduction formula will diverge to $-\infty$. If we instead calculate a_n using the reduction formula in Maple (with `Digits:=16`), we get some odd results : $a_{20} = -1000$, $a_{28} = 10^{14}$, $a_{29} = 0$, and $a_{30} = 10^{17}$, for example. But for larger n, the results are at least small and positive (for example, $a_{1000} \approx 0.001$.) For $n >$ 32,175, we get the delightful `object too large` error message. If, instead of using the reduction formula, we integrate directly with Maple, the results are much better.

12. **(a)** The reversed formula was started at $n = 30$, knowing that $\frac{1}{31} \le a_{30} \le \frac{1}{31}e$, so that the gap between the bounds for a_{29} is reduced by a factor of 30; for a_{28} by an extra factor of 29, and so on, until the bounds for a_{20} are at a distance $\dfrac{e-1}{31 \cdot 30 \cdot 29 \cdot \cdots \cdot 21} < 5.1 \times 10^{-16}$. This is enough accuracy even when we use a computer. The result from an 11-digit calculator is $a_{20} \approx 0.049881742885$ to all 11 digits.

(b) Integration of $b_n = \int_0^1 x^{n-\theta} e^{1-x}\,dx$ by parts gives $b_n = (n-\theta)b_{n-1} - 1$ or, in reversed form, $b_{n-1} = \dfrac{1+b_n}{n-\theta} = \dfrac{1}{n-\theta} + \dfrac{b_n}{n-\theta}$ (★). Again, from $1 \le e^{1-x} \le e$ we conclude (multiplying by $x^{n-\theta}$ and integrating) $\dfrac{1}{n+1-\theta} \le b_n \le \dfrac{e}{n+1-\theta}$ hence $\lim\limits_{n\to\infty} b_n = 0$. Also, as we descend from a larger n to a smaller m using (★), the gap between the bounds for b_m becomes $\dfrac{e-1}{(m+1-\theta)(m+2-\theta)\cdots(n+1-\theta)}$ and this number (assuming that m is given) can be made as small as we please by choosing n sufficiently large. The results are shown at right. (For $m = 0, 1, \ldots, 5$ we can take $n = 11$ for 5 digits of accuracy.)

m	b_m
0	2.85335
1	0.90223
2	0.50372
3	0.34326
4	0.25861
5	0.20684

13. We can start by expressing e^x and e^{-x} in terms of $E(x) = (e^x - 1)/x$ ($x \ne 0$), where $E(0) = 1$ to make E continuous at 0 (by L'Hospital's Rule). Namely, $e^x = 1 + xE(x)$, $e^{-x} = 1 - xE(-x)$ and $\sinh x = \dfrac{1 + xE(x) - [1 - xE(-x)]}{2} = \frac{1}{2}x[E(x) + E(-x)]$, where the addition involves only positive numbers $E(x)$ and $E(-x)$, thus presenting no loss of accuracy due to subtraction.

Another form, which calls the function E only once: we write $\sinh x = \dfrac{(e^x)^2 - 1}{2e^x} = \dfrac{[1 + xE(x)]^2 - 1}{2[1 + xE(x)]} = \dfrac{x\left[1 + \frac{1}{2}|x|E(|x|)\right]E(|x|)}{1 + |x|E(|x|)}$, taking advantage of the fact that $\dfrac{\sinh x}{x}$ is an even function, so replacing x by $|x|$ does not change its value.

EXERCISES H

1. $(3 + 2i) + (7 - 3i) = (3 + 7) + (2 - 3)i = 10 - i$

2. $(1 + i) - (2 - 3i) = (1 - 2) + (1 + 3)i = -1 + 4i$

3. $(3 - i)(4 + i) = 12 + 3i - 4i - (-1) = 13 - i$

4. $(4 - 7i)(1 + 3i) = 4 + 12i - 7i - 21(-1) = 25 + 5i$

5. $\overline{12 + 7i} = 12 - 7i$

6. $2i(\frac{1}{2} - i) = i - 2(-1) = 2 + i \quad \Rightarrow \quad \overline{2i(\frac{1}{2} - i)} = \overline{2 + i} = 2 - i$

7. $\dfrac{2+3i}{1-5i} = \dfrac{2+3i}{1-5i} \cdot \dfrac{1+5i}{1+5i} = \dfrac{2+10i+3i+15(-1)}{1-25(-1)} = \dfrac{-13+13i}{26} = -\frac{1}{2} + \frac{1}{2}i$

8. $\dfrac{5-i}{3+4i} = \dfrac{5-i}{3+4i} \cdot \dfrac{3-4i}{3-4i} = \dfrac{15-20i-3i+4(-1)}{9-16(-1)} = \dfrac{11-23i}{25} = \frac{11}{25} - \frac{23}{25}i$

9. $\dfrac{1}{1+i} = \dfrac{1}{1+i} \cdot \dfrac{1-i}{1-i} = \dfrac{1-i}{1-(-1)} = \dfrac{1-i}{2} = \frac{1}{2} - \frac{1}{2}i$

10. $\dfrac{3}{4-3i} = \dfrac{3}{4-3i} \cdot \dfrac{4+3i}{4+3i} = \dfrac{12+9i}{16-9(-1)} = \frac{12}{25} + \frac{9}{25}i$

11. $i^3 = i^2 \cdot i = (-1)\,i = -i$

12. $i^{100} = (i^2)^{50} = (-1)^{50} = 1$

13. $\sqrt{-25} = \sqrt{25}\,i = 5i$

14. $\sqrt{-3}\sqrt{-12} = \sqrt{3}i\sqrt{12}i = \sqrt{3 \cdot 12}i^2 = -6$

15. $\overline{3+4i} = 3-4i$, $|3+4i| = \sqrt{3^2+4^2} = \sqrt{25} = 5$

16. $\overline{\sqrt{3}-i} = \sqrt{3}+i$, $\left|\sqrt{3}-i\right| = \sqrt{\left(\sqrt{3}\right)^2 + (-1)^2} = \sqrt{4} = 2$

17. $\overline{-4i} = \overline{0-4i} = 0+4i = 4i$, $|-4i| = \sqrt{0^2+(-4)^2} = 4$

18. Let $z = a+bi$, $w = c+di$.

(a) $\overline{z+w} = \overline{(a+bi)+(c+di)} = \overline{(a+c)+(b+d)i} = (a+c)-(b+d)i = (a-bi)+(c-di) = \overline{z}+\overline{w}$

(b) $\overline{zw} = \overline{(a+bi)(c+di)} = \overline{(ac-bd)+(ad+bc)i} = (ac-bd)-(ad+bc)i$.

On the other hand, $\overline{z}\,\overline{w} = (a-bi)(c-di) = (ac-bd)-(ad+bc)i = \overline{zw}$.

(c) Use mathematical induction and part (b): Let S_n be the statement that $\overline{z^n} = \overline{z}^n$. S_1 is true because $\overline{z^1} = \overline{z} = \overline{z}^1$. Assume S_k is true, that is $\overline{z^k} = \overline{z}^k$. Then $\overline{z^{k+1}} = \overline{z^{1+k}} = \overline{zz^k} = \overline{z}\,\overline{z^k}$ [part (b) with $w = z^k$] $= \overline{z}^1\overline{z}^k = \overline{z}^{1+k} = \overline{z}^{k+1}$ which shows that S_{k+1} is true. Therefore, by mathematical induction, $\overline{z^n} = \overline{z}^n$ for every positive integer n.

Another Proof: Use part (b) with $w = z$, and mathematical induction.

19. $4x^2+9=0 \;\Leftrightarrow\; 4x^2 = -9 \;\Leftrightarrow\; x^2 = -\frac{9}{4} \;\Leftrightarrow\; x = \pm\sqrt{-\frac{9}{4}} = \pm\sqrt{\frac{9}{4}}i = \pm\frac{3}{2}i.$

20. $x^4 = 1 \;\Leftrightarrow\; x^4-1=0 \;\Leftrightarrow\; (x^2-1)(x^2+1)=0 \;\Leftrightarrow\; x^2-1=0$ or $x^2+1=0 \;\Leftrightarrow\; x=\pm 1$ or $x = \pm i$.

21. By the quadratic formula, $x^2-8x+17=0 \;\Leftrightarrow\; x = \dfrac{8 \pm \sqrt{8^2-4(1)(17)}}{2(1)} = \dfrac{8\pm\sqrt{-4}}{2} = \dfrac{8\pm 2i}{2} = 4 \pm i.$

22. $x^2-4x+5=0 \;\Leftrightarrow\; x = \dfrac{4\pm\sqrt{16-4(1)(5)}}{2} = \dfrac{4\pm\sqrt{-4}}{2} = \dfrac{4\pm 2i}{2} = 2\pm i$

23. By the quadratic formula, $z^2+z+2=0 \;\Leftrightarrow\; z = \dfrac{-1\pm\sqrt{1-4(1)(2)}}{2(1)} = \dfrac{-1\pm\sqrt{-7}}{2} = -\frac{1}{2} \pm \frac{\sqrt{7}}{2}i.$

24. $z^2 + \frac{1}{2}z + \frac{1}{4} = 0 \quad \Leftrightarrow \quad 4z^2 + 2z + 1 = 0 \quad \Leftrightarrow$

$$z = \frac{-2 \pm \sqrt{4 - 4(4)(1)}}{2(4)} = \frac{-2 \pm \sqrt{-12}}{8} = \frac{-2 \pm 2\sqrt{3}i}{8} = -\frac{1}{4} \pm \frac{\sqrt{3}}{4}i$$

25. $r = \sqrt{(-3)^2 + 3^2} = 3\sqrt{2}$, $\tan\theta = \frac{3}{-3} = -1 \quad \Rightarrow \quad \theta = \frac{3}{4}\pi$ (since the given number is in the second quadrant). Therefore $-3 + 3i = 3\sqrt{2}\left(\cos\frac{3\pi}{4} + i\sin\frac{3\pi}{4}\right)$.

26. For $z = 1 - \sqrt{3}i$, $r = \sqrt{1^2 + \left(-\sqrt{3}\right)^2} = 2$ and $\tan\theta = \frac{-\sqrt{3}}{1} = -\sqrt{3} \quad \Rightarrow \quad \theta = \frac{5\pi}{3}$ (since z lies in the fourth quadrant). Therefore $1 - \sqrt{3}i = 2\left(\cos\frac{5\pi}{3} + i\sin\frac{5\pi}{3}\right)$.

27. $r = \sqrt{3^2 + 4^2} = 5$, $\tan\theta = \frac{4}{3} \quad \Rightarrow \quad \theta = \tan^{-1}\frac{4}{3}$ (since the given number is in the second quadrant). Therefore $3 + 4i = 5\left[\cos\left(\tan^{-1}\frac{4}{3}\right) + i\sin\left(\tan^{-1}\frac{4}{3}\right)\right]$.

28. For $z = 8i$, $r = \sqrt{0^2 + 8^2} = 8$ and $\tan\theta = \frac{8}{0}$ is undefined so that $\theta = \frac{\pi}{2}$. Therefore $8i = 8\left(\cos\frac{\pi}{2} + i\sin\frac{\pi}{2}\right)$.

29. For $z = \sqrt{3} + i$, $r = \sqrt{\left(\sqrt{3}\right)^2 + 1^2} = 2$, and $\tan\theta = \frac{1}{\sqrt{3}} \quad \Rightarrow \quad \theta = \frac{\pi}{6}$ so that $z = 2\left(\cos\frac{\pi}{6} + i\sin\frac{\pi}{6}\right)$. For $w = 1 + \sqrt{3}i$, $r = 2$, and $\tan\theta = \sqrt{3} \quad \Rightarrow \quad \theta = \frac{\pi}{3}$ so that $w = 2\left(\cos\frac{\pi}{3} + i\sin\frac{\pi}{3}\right)$. Therefore
$zw = 2 \cdot 2\left[\cos\left(\frac{\pi}{6} + \frac{\pi}{3}\right) + i\sin\left(\frac{\pi}{6} + \frac{\pi}{3}\right)\right] = 4\left(\cos\frac{\pi}{2} + i\sin\frac{\pi}{2}\right)$,
$z/w = \frac{2}{2}\left[\cos\left(\frac{\pi}{6} - \frac{\pi}{3}\right) + i\sin\left(\frac{\pi}{6} - \frac{\pi}{3}\right)\right] = \cos\left(-\frac{\pi}{6}\right) + i\sin\left(-\frac{\pi}{6}\right)$, and $1 = 1 + 0i = \cos 0 + i\sin 0 \quad \Rightarrow$
$1/z = \frac{1}{2}\left[\cos\left(0 - \frac{\pi}{6}\right) + i\sin\left(0 - \frac{\pi}{6}\right)\right] = \frac{1}{2}\left[\cos\left(-\frac{\pi}{6}\right) + i\sin\left(-\frac{\pi}{6}\right)\right]$.

30. For $z = 4\sqrt{3} - 4i$, $r = \sqrt{\left(4\sqrt{3}\right)^2 + (-4)^2} = \sqrt{64} = 8$ and $\tan\theta = \frac{-4}{4\sqrt{3}} = -\frac{1}{\sqrt{3}} \quad \Rightarrow \quad \theta = \frac{11\pi}{6} \quad \Rightarrow$
$z = 8\left(\cos\frac{11\pi}{6} + i\sin\frac{11\pi}{6}\right)$. For $w = 8i$, $r = \sqrt{0^2 + 8^2} = 8$ and $\tan\theta = \frac{8}{0}$ is undefined so that $\theta = \frac{\pi}{2} \quad \Rightarrow$
$w = 8\left(\cos\frac{\pi}{2} + i\sin\frac{\pi}{2}\right)$. Therefore $zw = 8 \cdot 8\left[\cos\left(\frac{11\pi}{6} + \frac{\pi}{2}\right) + i\sin\left(\frac{11\pi}{6} + \frac{\pi}{2}\right)\right] = 64\left(\cos\frac{\pi}{3} + i\sin\frac{\pi}{3}\right)$,
$z/w = \frac{8}{8}\left[\cos\left(\frac{11\pi}{6} - \frac{\pi}{2}\right) + i\sin\left(\frac{11\pi}{6} - \frac{\pi}{2}\right)\right] = \cos\frac{4\pi}{3} + i\sin\frac{4\pi}{3}$, and $1 = \cos 0 + i\sin 0 \quad \Rightarrow$
$1/z = \frac{1}{8}\left[\cos\left(0 - \frac{11\pi}{6}\right) + i\sin\left(0 - \frac{11\pi}{6}\right)\right] = \frac{1}{8}\left[\cos\left(\frac{\pi}{6}\right) + i\sin\left(\frac{\pi}{6}\right)\right]$.

31. For $z = 2\sqrt{3} - 2i$, $r = 4$, $\tan\theta = \frac{-2}{2\sqrt{3}} = -\frac{1}{\sqrt{3}} \quad \Rightarrow \quad \theta = -\frac{\pi}{6} \quad \Rightarrow \quad z = 4\left[\cos\left(-\frac{\pi}{6}\right) + i\sin\left(-\frac{\pi}{6}\right)\right]$. For $w = -1 + i$, $r = \sqrt{2}$, $\tan\theta = \frac{1}{-1} = -1 \quad \Rightarrow \quad \theta = \frac{3\pi}{4} \quad \Rightarrow \quad z = \sqrt{2}\left(\cos\frac{3\pi}{4} + i\sin\frac{3\pi}{4}\right)$. Therefore
$zw = 4\sqrt{2}\left[\cos\left(-\frac{\pi}{6} + \frac{3\pi}{4}\right) + i\sin\left(-\frac{\pi}{6} + \frac{3\pi}{4}\right)\right] = 4\sqrt{2}\left(\cos\frac{7\pi}{12} + i\sin\frac{7\pi}{12}\right)$,
$z/w = \frac{4}{\sqrt{2}}\left[\cos\left(-\frac{\pi}{6} - \frac{3\pi}{4}\right) + i\sin\left(-\frac{\pi}{6} - \frac{3\pi}{4}\right)\right] = \frac{4}{\sqrt{2}}\left[\cos\left(-\frac{11\pi}{12}\right) + i\sin\left(-\frac{11\pi}{12}\right)\right] = 2\sqrt{2}\left(\cos\frac{13\pi}{12} + i\sin\frac{13\pi}{12}\right)$,
and $1 = 1 + 0i = \cos 0 + i\sin 0 \quad \Rightarrow \quad 1/z = \frac{1}{4}\left[\cos\left(0 - \left(-\frac{\pi}{6}\right)\right) + i\sin\left(0 - \left(-\frac{\pi}{6}\right)\right)\right] = \frac{1}{4}\left(\cos\frac{\pi}{6} + i\sin\frac{\pi}{6}\right)$.

32. For $z = 4\left(\sqrt{3} + i\right) = 4\sqrt{3} + 4i$, $r = \sqrt{\left(4\sqrt{3}\right)^2 + 4^2} = \sqrt{64} = 8$ and $\tan\theta = \frac{4}{4\sqrt{3}} = \frac{1}{\sqrt{3}} \quad \Rightarrow \quad \theta = \frac{\pi}{6}$
$\Rightarrow \quad z = 8\left(\cos\frac{\pi}{6} + i\sin\frac{\pi}{6}\right)$. For $w = -3 - 3i$, $r = \sqrt{(-3)^2 + (-3)^2} = \sqrt{18} = 3\sqrt{2}$ and $\tan\theta = \frac{-3}{-3} = 1$
$\Rightarrow \quad \theta = \frac{5\pi}{4}$ (w is in the third quadrant) $\quad \Rightarrow \quad w = 3\sqrt{2}\left(\cos\frac{5\pi}{4} + i\sin\frac{5\pi}{4}\right)$. Therefore
$zw = 8 \cdot 3\sqrt{2}\left[\cos\left(\frac{\pi}{6} + \frac{5\pi}{4}\right) + i\sin\left(\frac{\pi}{6} + \frac{5\pi}{4}\right)\right] = 24\sqrt{2}\left(\cos\frac{17\pi}{12} + i\sin\frac{17\pi}{12}\right)$,
$z/w = \frac{8}{3\sqrt{2}}\left[\cos\left(\frac{\pi}{6} - \frac{5\pi}{4}\right) + i\sin\left(\frac{\pi}{6} - \frac{5\pi}{4}\right)\right] = \frac{4\sqrt{2}}{3}\left[\cos\left(-\frac{13\pi}{12}\right) + i\sin\left(-\frac{13\pi}{12}\right)\right]$, and $1 = \cos 0 + i\sin 0 \quad \Rightarrow$
$1/z = \frac{1}{8}\left[\cos\left(0 - \frac{\pi}{6}\right) + i\sin\left(0 - \frac{\pi}{6}\right)\right] = \frac{1}{8}\left[\cos\left(-\frac{\pi}{6}\right) + i\sin\left(-\frac{\pi}{6}\right)\right]$.

33. For $z = 1 + i$, $r = \sqrt{2}$, $\tan\theta = \frac{1}{1} = 1 \quad\Rightarrow\quad \theta = \frac{\pi}{4} \quad\Rightarrow\quad 1 + i = \sqrt{2}\left(\cos\frac{\pi}{4} + i\sin\frac{\pi}{4}\right)$. So by De Moivre's Theorem, $(1+i)^{20} = \left[\sqrt{2}\left(\cos\frac{\pi}{4} + i\sin\frac{\pi}{4}\right)\right]^{20} = \left(2^{1/2}\right)^{20}\left(\cos\frac{20\pi}{4} + i\sin\frac{20\pi}{4}\right) = 2^{10}(\cos 5\pi + i\sin 5\pi)$
$= 2^{10}[-1 + i(0)] = -2^{10} = -1024$.

34. For $z = 1 - \sqrt{3}i$, $r = \sqrt{1^2 + \left(-\sqrt{3}\right)^2} = 2$ and $\tan\theta = \frac{-\sqrt{3}}{1} = -\sqrt{3} \quad\Rightarrow\quad \theta = \frac{5\pi}{3}$ (since z lies in the fourth quadrant) $\quad\Rightarrow\quad z = 2\left(\cos\frac{5\pi}{3} + i\sin\frac{5\pi}{3}\right) \quad\Rightarrow\quad \left(1 - \sqrt{3}i\right)^5 = \left[2\left(\cos\frac{5\pi}{3} + i\sin\frac{5\pi}{3}\right)\right]^5$
$= 2^5\left(\cos\frac{5\cdot 5\pi}{3} + i\sin\frac{5\cdot 5\pi}{3}\right) = 2^5\left(\cos\frac{25\pi}{3} + i\sin\frac{25\pi}{3}\right) = 2^5\left(0.5 + \frac{\sqrt{3}}{2}i\right) = 16 + 16\sqrt{3}i$.

35. For $z = 2\sqrt{3} + 2i$, $r = 4$, $\tan\theta = \frac{2}{2\sqrt{3}} = \frac{1}{\sqrt{3}} \quad\Rightarrow\quad \theta = \frac{\pi}{6} \quad\Rightarrow\quad 2\sqrt{3} + 2i = 4\left(\cos\frac{\pi}{6} + i\sin\frac{\pi}{6}\right)$. So by De Moivre's Theorem,

$\left(2\sqrt{3} + 2i\right)^5 = \left[4\left(\cos\frac{\pi}{6} + i\sin\frac{\pi}{6}\right)\right]^5 = 4^5\left(\cos\frac{5\pi}{6} + i\sin\frac{5\pi}{6}\right) = 4^5\left[-\frac{\sqrt{3}}{2} + i(0.5)\right] = -512\sqrt{3} + 512i$.

36. For $z = 1 - i$, $r = \sqrt{2}$ and $\tan\theta = \frac{-1}{1} = -1 \quad\Rightarrow\quad \theta = \frac{7\pi}{4}$ (since z is in the fourth quadrant) $\quad\Rightarrow$
$1 - i = \sqrt{2}\left(\cos\frac{7\pi}{4} + i\sin\frac{7\pi}{4}\right) \quad\Rightarrow$
$(1-i)^8 = \left[\sqrt{2}\left(\cos\frac{7\pi}{4} + i\sin\frac{7\pi}{4}\right)\right]^8 = 2^4\left(\cos\frac{7\cdot 8\pi}{4} + i\sin\frac{7\cdot 8\pi}{4}\right) = 16(\cos 14\pi + i\sin 14\pi) = 16$.

37. $1 = 1 + 0i = \cos 0 + i\sin 0$. Using Equation 3 with $r = 1$, $n = 8$, and $\theta = 0$ we have

$$w_k = 1^{1/8}\left[\cos\left(\frac{0 + 2k\pi}{8}\right) + i\sin\left(\frac{0 + 2k\pi}{8}\right)\right] = \cos\frac{k\pi}{4} + i\sin\frac{k\pi}{4}, \text{ where } k = 0, 1, 2, \ldots, 7.$$

$w_0 = (\cos 0 + i\sin 0) = 1 \qquad w_4 = (\cos\pi + i\sin\pi) = -1$

$w_1 = \left(\cos\frac{\pi}{4} + i\sin\frac{\pi}{4}\right) = \frac{1}{\sqrt{2}} + \frac{1}{\sqrt{2}}i \qquad w_5 = \left(\cos\frac{5\pi}{4} + i\sin\frac{5\pi}{4}\right) = -\frac{1}{\sqrt{2}} - \frac{1}{\sqrt{2}}i$

$w_2 = \left(\cos\frac{\pi}{2} + i\sin\frac{\pi}{2}\right) = i \qquad w_6 = \left(\cos\frac{3\pi}{2} + i\sin\frac{3\pi}{2}\right) = -i$

$w_3 = \left(\cos\frac{3\pi}{4} + i\sin\frac{3\pi}{4}\right) = -\frac{1}{\sqrt{2}} + \frac{1}{\sqrt{2}}i \qquad w_7 = \left(\cos\frac{7\pi}{4} + i\sin\frac{7\pi}{4}\right) = \frac{1}{\sqrt{2}} - \frac{1}{\sqrt{2}}i$

38. $32 = 32(\cos 0 + i\sin 0)$, so

$w_k = 32^{1/5}\left[\cos\left(\frac{1}{5}(0 + 2k\pi)\right) + i\sin\left(\frac{1}{5}(0 + 2k\pi)\right)\right]$
$= 2\left(\cos\frac{2}{5}\pi k + i\sin\frac{2}{5}\pi k\right)$, $k = 0, 1, 2, 3, 4$.

$w_0 = 2(\cos 0 + i\sin 0) = 2$
$w_1 = 2\left(\cos\frac{2\pi}{5} + i\sin\frac{2\pi}{5}\right)$
$w_2 = 2\left(\cos\frac{4\pi}{5} + i\sin\frac{4\pi}{5}\right)$
$w_3 = 2\left(\cos\frac{6\pi}{5} + i\sin\frac{6\pi}{5}\right)$
$w_4 = 2\left(\cos\frac{8\pi}{5} + i\sin\frac{8\pi}{5}\right)$.

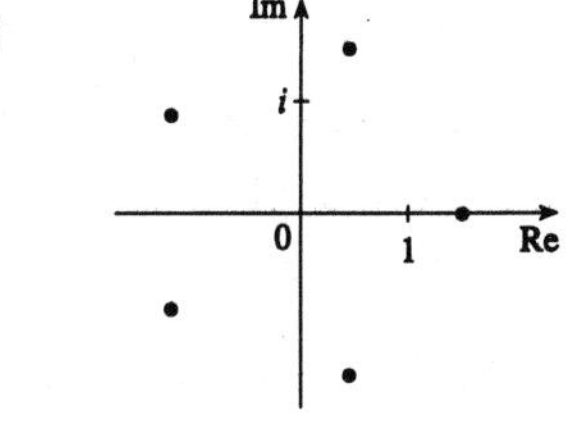

39. $0 = 0 + i = \cos\frac{\pi}{2} + i\sin\frac{\pi}{2}$. Using Equation 3 with $r = 1$, $n = 3$, and $\theta = \frac{\pi}{2}$, we have

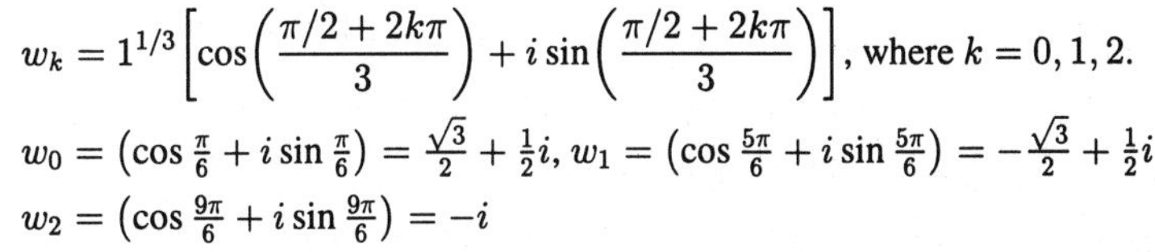

$$w_k = 1^{1/3}\left[\cos\left(\frac{\pi/2 + 2k\pi}{3}\right) + i\sin\left(\frac{\pi/2 + 2k\pi}{3}\right)\right], \text{ where } k = 0, 1, 2.$$

$w_0 = \left(\cos\frac{\pi}{6} + i\sin\frac{\pi}{6}\right) = \frac{\sqrt{3}}{2} + \frac{1}{2}i$, $w_1 = \left(\cos\frac{5\pi}{6} + i\sin\frac{5\pi}{6}\right) = -\frac{\sqrt{3}}{2} + \frac{1}{2}i$
$w_2 = \left(\cos\frac{9\pi}{6} + i\sin\frac{9\pi}{6}\right) = -i$

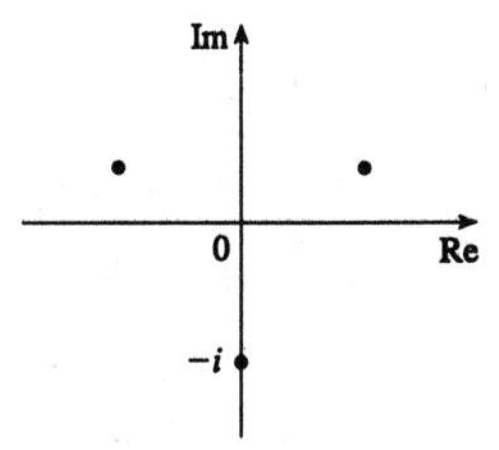

40. $1+i=\sqrt{2}\left(\cos\frac{\pi}{4}+i\sin\frac{\pi}{4}\right)$, so

$$w_k=\left(\sqrt{2}\right)^{1/3}\left[\cos\left(\frac{\pi/4+2k\pi}{3}\right)+i\sin\left(\frac{\pi/4+2k\pi}{3}\right)\right],\ k=0,1,2.$$

$$w_0=2^{1/6}\left(\cos\tfrac{\pi}{12}+i\sin\tfrac{\pi}{12}\right),$$

$$w_1=2^{1/6}\left(\cos\tfrac{3\pi}{4}+i\sin\tfrac{3\pi}{4}\right)=2^{1/6}\left(-\tfrac{1}{\sqrt{2}}+\tfrac{1}{\sqrt{2}}i\right)=-2^{-1/3}+2^{-1/3}i,$$

$$w_2=2^{1/6}\left(\cos\tfrac{17\pi}{12}+i\sin\tfrac{17\pi}{12}\right).$$

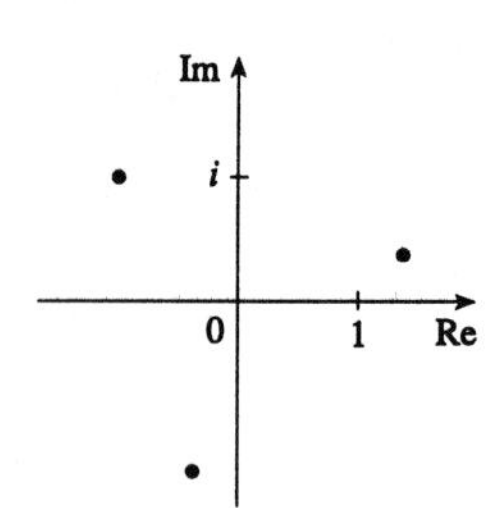

41. Using Euler's formula (6) with $y=\frac{\pi}{2}$, $e^{i\pi/2}=\cos\frac{\pi}{2}+i\sin\frac{\pi}{2}=i$.

42. Using Euler's equation (6) with $y=2\pi$ we have $e^{2\pi i}=\cos 2\pi+i\sin 2\pi=1$.

43. Using Euler's formula with $y=\frac{3\pi}{4}$, $e^{i3\pi/4}=\cos\frac{3\pi}{4}+i\sin\frac{3\pi}{4}=-\frac{1}{\sqrt{2}}+\frac{1}{\sqrt{2}}i$.

44. Using Euler's formula with $y=-\pi$ we have $e^{-i\pi}=\cos(-\pi)+i\sin(-\pi)=-1$.

45. Using Equation 7 with $x=2$ and $y=\pi$, $e^{2+i\pi}=e^2e^{i\pi}=e^2(\cos\pi+i\sin\pi)=e^2(-1+0)=-e^2$.

46. Using Equation 7 with $x=1$ and $y=2$ we have $e^{1+2i}=e^1e^{2i}=e(\cos 2+i\sin 2)=e\cos 2+(e\sin 2)\,i$.

47. Take $r=1$ and $n=3$ in De Moivre's Theorem to get

$[1(\cos\theta+i\sin\theta)]^3=1^3(\cos 3\theta+i\sin 3\theta)\quad\Rightarrow\quad(\cos\theta+i\sin\theta)^3=\cos 3\theta+i\sin 3\theta\quad\Rightarrow$

$\cos^3\theta+3(\cos^2\theta)(i\sin\theta)+3(\cos\theta)(i\sin\theta)^2+(i\sin\theta)^3=\cos 3\theta+i\sin 3\theta\quad\Rightarrow$

$(\cos^3\theta-3\sin^2\theta\cos\theta)+(3\sin\theta\cos^2\theta-\sin^3\theta)i=\cos 3\theta+i\sin 3\theta$. Equating real and imaginary parts gives

$\cos 3\theta=\cos^3\theta-3\sin^2\theta\cos\theta$ and $\sin 3\theta=3\sin\theta\cos^2\theta-\sin^3\theta$.

48. Using Equation 6,

$e^{ix}+e^{-ix}=(\cos x+i\sin x)+[\cos(-x)+i\sin(-x)]=\cos x+i\sin x+\cos x-i\sin x=2\cos x$.

Thus $\cos x=\dfrac{e^{ix}+e^{-ix}}{2}$. Similarly

$e^{ix}-e^{-ix}=(\cos x+i\sin x)-[\cos(-x)+i\sin(-x)]=\cos x+\sin x-\cos x-(-\sin x)=2i\sin x$.

Therefore $\sin x=\dfrac{e^{ix}-e^{-ix}}{2i}$.

49. $F(x)=e^{rx}=e^{(a+bi)x}=e^{ax+bxi}=e^{ax}(\cos bx+i\sin bx)=e^{ax}\cos bx+i(e^{ax}\sin bx)\quad\Rightarrow$

$$\begin{aligned}F'(x)&=(e^{ax}\cos bx)'+i(e^{ax}\sin bx)'=(ae^{ax}\cos bx-be^{ax}\sin bx)+i(ae^{ax}\sin bx+be^{ax}\cos bx)\\&=a[e^{ax}(\cos bx+i\sin bx)]+b[e^{ax}(-\sin bx+i\cos bx)]=ae^{rx}+b\left[e^{ax}\left(i^2\sin bx+i\cos bx\right)\right]\\&=ae^{rx}+bi[e^{ax}(\cos bx+i\sin bx)]=ae^{rx}+bie^{rx}=(a+bi)e^{rx}=re^{rx}.\end{aligned}$$

Now symbolic computation and mathematical typesetting is as accessible as your Windows™-based word processor!

Scientific WorkPlace™ 2.0 for Windows

Producing publication-quality course materials, journal articles, research reports, or full-length books has never been so easy!

"Scientific WorkPlace is a heavy-duty mathematical word processor and typesetting system that is able to expand, simplify, and evaluate conventional mathematical expressions and compose them as elegant printed mathematics. It gives working mathematicians LaTeX without pain."
—Roger Horn, University of Utah

"The thing I like most about Scientific WorkPlace is its basic simplicity and ease of use. With an absolute minimum of effort, one can begin to do things that LaTeX and Maple can help with."
—Barbara Osofsky, Rutgers University

Easy access to a powerful computer algebra system *inside* your word-processing documents!

Scientific WorkPlace is a revolutionary program that gives you a "work place" environment—a single place to do all your work. It combines the ease of use of a technical word processor with the typesetting power of TeX and the numerical, symbolic, and graphic computational facilities of the **Maple® V** computer algebra system. All capabilities are included in the program—you don't need to *own* or *learn* TeX, LaTeX, or Maple to use ***Scientific WorkPlace*—everything for super productivity is included in one powerful tool for just $495.00!**

With ***Scientific WorkPlace***, you can enter, solve, and graph mathematical problems right in your word-processing documents in seconds, with no clumsy cut-and-paste from equation editors or clipboards. Input mathematics as easily as you type in words. ***Scientific WorkPlace*** calculates answers quickly and accurately, then prints your work in impressive, professional-quality documents using TeX's internationally accepted mathematical typesetting standard. More than 200 document styles, AMS fonts, and a style editor are included.

Install *Scientific WorkPlace* and watch your productivity soar! You'll be creating professional-quality documents in a fraction of the time you would spend using any other program!

ORDER FORM

Yes! Please send me Scientific Workplace 2.0 for Windows

_____ copies (ISBN: 0-534-25596-5) @ $495.00 each _________

(Residents of AL, AZ, CA, CO, CT, FL, GA, IL, IN, KS, KY, LA, MA, MD, MI, MN, MO, NC, NJ, NY, OH, PA, RI, SC, TN, TX, UT, VA, WA, WI must add appropriate sales tax) Tax _________

Payment Options Handling _________

_____ Purchase Order enclosed. Please bill me.

_____ Check or Money Order enclosed. Total _________

_____ Charge my _____ VISA _____ MasterCard _____ American Express

Card Number ____________________________ Expiration Date __________

Signature____

Please ship to: (Billing and shipping address must be the same.)

Name____

Department __________________ School ___________________________

Street Address ____

City ____________________________ State______ Zip+4______________

Office phone number (_____) ____________________

You can fax your response to us at 408-375-6414 or e-mail your order to: info@brookscole.com or detach, fold, secure, and mail with payment.

SECURE WITH TAPE

CUT ALONG DOTTED LINE

FOLD HERE

NO POSTAGE
NECESSARY
IF MAILED
IN THE
UNITED STATES

BUSINESS REPLY MAIL

FIRST CLASS PERMIT NO. 358 PACIFIC GROVE, CA

POSTAGE WILL BE PAID BY ADDRESSEE

ATTN: MARKETING

Brooks/Cole Publishing Company
511 Forest Lodge Road
Pacific Grove, California 93950-9968

FOLD HERE

CUT ALONG DOTTED LINE

FOLD HERE

NO POSTAGE
NECESSARY
IF MAILED
IN THE
UNITED STATES

BUSINESS REPLY MAIL

FIRST CLASS PERMIT NO. 358 PACIFIC GROVE, CA

POSTAGE WILL BE PAID BY ADDRESSEE

ATTN: MARKETING

Brooks/Cole Publishing Company
511 Forest Lodge Road
Pacific Grove, California 93950-9968

FOLD HERE